2014 International Power Electronics Conference

(IPEC-Hiroshima 2014 ECCE-ASIA)

Hiroshima, Japan
18-21 May 2014

Pages 843-1633

IEEE Catalog Number:	CFP14CPB-POD
ISBN:	978-1-4799-2706-7

**Copyright © 2014 by the Institute of Electrical and Electronic Engineers, Inc
All Rights Reserved**

Copyright and Reprint Permissions: Abstracting is permitted with credit to the source. Libraries are permitted to photocopy beyond the limit of U.S. copyright law for private use of patrons those articles in this volume that carry a code at the bottom of the first page, provided the per-copy fee indicated in the code is paid through Copyright Clearance Center, 222 Rosewood Drive, Danvers, MA 01923.

For other copying, reprint or republication permission, write to IEEE Copyrights Manager, IEEE Service Center, 445 Hoes Lane, Piscataway, NJ 08854. All rights reserved.

***This publication is a representation of what appears in the IEEE Digital Libraries. Some format issues inherent in the e-media version may also appear in this print version.**

IEEE Catalog Number: CFP14CPB-POD
ISBN 13:   978-1-4799-2706-7

Additional Copies of This Publication Are Available From:

Curran Associates, Inc
57 Morehouse Lane
Red Hook, NY 12571 USA
Phone: (845) 758-0400
Fax: (845) 758-2633
E-mail: curran@proceedings.com
Web: www.proceedings.com

2014 International Power Electronics Conference (IPEC-Hiroshima 2014 ECCE-ASIA)

Hiroshima, Japan
18-21 May 2014

IEEE Catalog Number: CFP14CPB-POD
ISBN: 978-1-47992-706-7

TABLE OF CONTENTS

A NOVEL CONTROL SCHEME FOR THREE-LEVEL FULL-BRIDGE CONVERTER ACHIEVING LOW THD OUTPUT VOLTAGE...........66
Liu, Jilong ; Xiao, Fei ; Chen, Wei ; Yang, Guorun

PARALLEL CONNECTED THREE PHASE INVERTERS BASED ON MODULAR DESIGN AND DISTRIBUTED CONTROL...........72
Xiao, Fei ; Chen, Wei ; Liu, Jilong ; Wang, Hengli

EFFICIENCY INVESTIGATIONS OF A 3KW T-TYPE INVERTER FOR SWITCHING FREQUENCIES UP TO 100 KHZ...........78
Anthon, Alexander ; Zhang, Zhe ; Andersen, Michael A.E. ; Franke, Toke

MINIATURIZATION OF THE BOOST-UP TYPE ACTIVE BUFFER CIRCUIT IN A SINGLE-PHASE INVERTER...........84
Watanabe, Hiroki ; Koiwa, Kazuhiro ; Itoh, Jun-ichi ; Ohnuma, Yoshiya ; Miyawaki, Satoshi

TESTING FACILITY USING LARGE CAPACITY INVERTER...........92
Ishimaru, Yusuke ; Adachi, Mitsuo ; Tsukakoshi, Masahiko ; Nakamura, Ritaka ; Masuda, Hiroyuki ; Ogashi, Yoshihiro ; Tsuboi, Yuichi

PERFORMANCE EVALUATION UNDER THE ACTUAL OPERATING CONDITION OF A LARGE CAPACITY VSI INVERTER FOR STEEL MILL APPLICATIONS...........97
Mamun, Mostafa ; Yoshizawa, Daisuke ; Mukunoki, Makoto

A SOFT-SWITCHING SINGLE-PHASE UNIFIED POWER QUALITY CONDITIONER...........105
Jiang, Maoh-Chin ; Chang, Kai-Chi ; Lu, Kao-Yi ; Shih, Bing-Jyun ; Liu, Tai-Chun

NOVEL THREE-PHASE PWM AC-AC CONVERTERS SOLVING COMMUTATION PROBLEM...........110
Khan, Ashraf Ali ; Shin, Hyunhak ; Cha, Honnyong ; Kim, Heung-Geun

EXPERIMENTAL INVESTIGATION OF NORMALLY-ON TYPE BIDIRECTIONAL SWITCH FOR INDIRECT MATRIX CONVERTERS...........117
Sung, Kyungmin ; Iijima, Ryuji ; Nishizawa, Shinichi ; Norigoe, Isami ; Ohashi, Hiromichi

VISUALIZATION OF PWM WAVEFORMS OF OUTPUT VOLTAGE AND INPUT CURRENT FOR A DIRECT MATRIX CONVERTER...........123
Asai, Inami ; Takeshita, Takaharu

SPACE VECTOR MODULATION BASED ON VIRTUAL INDIRECT CONTROL FOR HIGH FREQUENCY AC-LINKED MATRIX CONVERTER...........130
Inoue, Keita ; Shioda, Masashi ; Katade, Motohumi ; Goto, Akira ; Morishita, Shin ; Itoh, Junichi ; Koiwa, Kazuhiro

A FUNDAMENTAL VERIFICATION OF A SINGLE-PHASE TO THREE-PHASE MATRIX CONVERTER WITH A PDM CONTROL BASED ON SPACE VECTOR MODULATION...........138
Nakata, Yuki ; Itoh, Jun-ichi

STEADY STATE CHARACTERISTICS OF THE BOOST-TYPE MATRIX CONVERTER FOR STAND-ALONE POWER SOURCE...........146
Nagano, Y. ; Yamamura, N. ; Ishida, M. ; Hirokado, K.

DESIGN PROCEDURE FOR OUTPUT CURRENT CONTROL AND DAMPING CONTROL OF MATRIX CONVERTER...........152
Takahashi, Hiroki ; Itoh, Jun-ichi

A NOVEL LCL FILTER PARAMETER DESIGN METHOD BASING ON RESONANT FREQUENCY OPTIMIZATION OF THREE-LEVEL NPC GRID CONNECTED INVERTER...........160
Li, Ning ; Wang, Yue ; Niu, Ruigen ; Guo, Wei ; Lei, Wanjun ; Wang, Zhao'An

DESIGN AND ANALYSIS OF ISOLATED BI-DIRECTIONAL DC/DC CONVERTER USING QUASI-RESONANT ZVS...........166
Noh, Yong-Su ; Won, Chung-Yuen ; Oh, Min-Seok ; Jeon, Jin-Yong ; Jung, Yong-Chae

AN ACTIVE-CLAMPING ZVS FLYBACK CONVERTER WITH INTEGRATED TRANSFORMER...........172
Lin, Jing-Yuan ; Lo, Yu-Kang ; Chiu, Huang-Jen ; Wang, Chao-Fu ; Lin, Chien-Yu

PFM AND PWM HYBRID CONTROLLED LLC CONVERTER...........177
Yamamoto, Junichi ; Zaitsu, Toshiyuki ; Abe, Seiya ; Ninomiya, Tamotsu

DISCUSSIONS ON VARIOUS VOLTAGE EQUALIZERS FOR EDLCS USING CW CIRCUIT...........183
Khant, Hlaing Kyi Pyar ; Matsui, Keiju ; Hasegawa, Masaru ; Yasubayashi, Mikio ; Umeno, Masayoshi ; Ooishi, Eiji

ISOLATION SYSTEM WITH WIRELESS POWER TRANSFER FOR MULTIPLE GATE DRIVER SUPPLIES OF A MEDIUM VOLTAGE INVERTER...........191
Kusaka, Keisuke ; Orikawa, Koji ; Itoh, Jun-ichi ; Morita, Kazunori ; Hirao, Kuniaki

STUDY AND IMPLEMENTATION OF A 15-W POWER AMPLIFIER FOR PIEZOELECTRIC ACTUATOR...........199
Lo, Yu-Kang ; Chiu, Huang-Jen ; Liu, Yu-Chen ; Lin, Chung-Yi ; Cheng, Shih-Jen ; Yang, CS

ISOLATED VOLTAGE-BOOSTING CONVERTER...........204
Hwu, K.I. ; Jiang, W.Z. ; Shieh, Jenn-Jong

HIGH VOLTAGE CONVERSION RATIO CASCADE BOOST CONVERTER WITH DC SNUBBER...........208
Lee, Yuang-Shung ; Yu, Ling-Chia ; Chou, Tzu-Han

DESIGN-ORIENTED ANALYSIS OF RESONANCE DAMPING AND HARMONIC COMPENSATION FOR LCL-FILTERED VOLTAGE SOURCE CONVERTERS...........216
Wang, Xiongfei ; Blaabjerg, Frede ; Loh, Poh Chiang

STATE-SPACE AVERAGE MODELING OF BIDIRECTIONAL DC-DC CONVERTER FOR BATTERY CHARGER USING LCLC FILTER......224
Moon, Sang-Ho ; Jou, Sung-Tak ; Lee, Kyo-Beum

A NEW SVPWM STRATEGY FOR INPUT SWITCHED MULTILEVEL CONVERTER......230
Xiong, Li ; Prasanna, U.R. ; Bilal, Akin ; Rajashekara, Kaushik

ESD RELIABILITY INFLUENCE OF A 60 V POWER LDMOS BY THE FOD-BASED (& DOTTED-OD) DRAIN......236
Chen, Shen-Li ; Lee, Min-Hua

ENHANCED TRANSVERSE-FLUX MOTOR WITH TORUS COILS......240
Tanaka, Junya ; Sakai, Kazuto

THE INFLUENCE OF MAGNETIC PROPERTIES OF PERMANENT MAGNET ON THE PERFORMANCE OF IPMSM FOR AUTOMOTIVE APPLICATION......246
Yoshioka, S. ; Morimoto, S. ; Sanada, M. ; Inoue, Y.

CHARACTERISTICS OF INTERIOR PERMANENT MAGNET SYNCHRONOUS MOTOR WITH IMPERFECT MAGNETS......252
Shinagawa, Syuhei ; Ishikawa, Takeo ; Kurita, Nobuyuki

STUDY OF STATOR STRUCTURE TO IMPROVE RELUCTANCE TORQUE FOR IPMSM WITH CONCENTRATED WINDING......258
Morikawa, R. ; Sanada, M. ; Morimoto, S. ; Inoue, Y.

DEVELOPMENT AND VERIFICATION OF ENERGY-ACCURATE SIMULATION MODELS FOR PERMANENT MAGNET SYNCHRONOUS MOTORS IN AUTOMATION SYSTEMS......264
Blank, Frederic ; Roth-Stielow, Jorg

COMPARISON OF THE RESISTANCE- AND INDUCTANCE-BASED SALIENCY OF A PMSM DUE TO A SHORT-CIRCUITED ROTOR WINDING......270
Graus, Johannes ; Rambetius, Alexander ; Hahn, Ingo

DESIGN AND OPTIMIZATION OF HIGH-SPEED SWITCHED RELUCTANCE MOTOR USING SOFT MAGNETIC COMPOSITE MATERIAL......278
Gaing, Zwe-Lee ; Kuo, Kuan-Yi ; Hu, Jia-Sheng ; Hsieh, Min-Fu ; Tsai, Ming-Hsiao

INFLUENCE OF PULSE WIDTH MODULATION (PWM) ON THE IRON LOSSES OF ELECTRICAL STEEL......283
Boehm, Andreas ; Hahn, Ingo

INVESTIGATION ON IRON LOSS CHARACTERISTICS IN STAR-CONNECTION AND DELTA-CONNECTION UNDER THREE PHASE PWM INVERTER EXCITATION......289
Odawara, Shunya ; Fujisaki, Keisuke ; Fukuhara, Shuhei

OPTIMIZATION ON ARRANGEMENT OF PERMANENT MAGNETS FOR MAGNETIC LEVITATION SYSTEM FOR THIN STEEL PLATE (FUNDAMENTAL CONSIDERATION ON LEVITATION PROBABILITY)......294
Ishii, Hirotaka ; Hasegawa, Shinya ; Narita, Takayoshi ; Oshinoya, Yasuo

EFFECT OF A MAGNETIC FIELD FROM THE HORIZONTAL DIRECTION ON A MAGNETICALLY LEVITATED STEEL PLATE (FUNDAMENTAL CONSIDERATIONS ON THE SHAPE ANALYSIS OF ULTRATHIN STEEL PLATE)......299
Kurihara, Takeshi ; Hasegawa, Shinya ; Narita, Takayoshi ; Oshinoya, Yasuo

NOVEL MAGNETIC STRUCTURE OF INTEGRATED DIFFERENTIAL-MODE AND COMMON-MODE INDUCTORS TO SUPPRESS DC SATURATION......304
Umetani, Kazuhiro ; Tera, Takahiro ; Shirakawa, Kazuhiro

A NOVEL CONTROL METHOD IN FLUX-WEAKENING REGION FOR EFFICIENT OPERATION OF INTERIOR PERMANENT MAGNET SYNCHRONOUS MOTOR......312
Ueda, K. ; Morimoto, S. ; Inoue, Y. ; Sanada, M.

IMPLEMENTATION OF THE MTPA AND MTPV CONTROL WITH ONLINE PARAMETER IDENTIFICATION FOR A HIGH SPEED IPMSM USED AS TRACTION DRIVE......318
Nguyen, Quoc Khanh ; Petrich, Matthias ; Roth-Stielow, Jorg

CORRECTION OF REFERENCE FLUX FOR MTPA CONTROL IN DIRECT TORQUE CONTROLLED INTERIOR PERMANENT MAGNET SYNCHRONOUS MOTOR DRIVES......324
Shinohara, Atsushi ; Inoue, Yukinori ; Morimoto, Shigeo ; Sanada, Masayuki

VOLTAGE REGULATION AND MAXIMUM OUTPUT POWER TRACKING OF A 4.5KW PERMANENT-MAGNET SYNCHRONOUS GENERATOR......330
Chang, Yuan-Chih ; Chang, Hsiu-Feng ; Dai, Wei-Fu ; Wu, Chun-Wei

A NOVEL FLUX-WEAKENING CONTROL METHOD BASED ON SINGLE CURRENT REGULATOR FOR PERMANENT MAGNET SYNCHRONOUS MOTOR......335
Fang, Xiaocun ; Hu, Taiyuan ; Lin, Fei ; Yang, Zhongping

PREDICTIVE CURRENT CONTROL METHOD IN INDUCTION MOTOR SPEED SENSORLESS DRIVE......341
Wei, Sun ; Yong, Yu ; Dianguo, Xu ; Jin, Xu ; Li, Ding

REAL-TIME IMPLEMENTATION OF AN ONLINE MODEL PREDICTIVE CONTROL FOR IPMSM USING PARALLEL COMPUTING ON FPGA......346
Leuer, Michael ; Bocker, Joachim

AN INTEGRAL SLIDING-MODE CONTROLLER FOR ENERGY EFFICIENCY IMPROVEMENT IN AC POWER SOURCE SUPPLIED AC MACHINE DRIVES......351
Shieh, Hsin-Jang ; Chen, Ying-Zuo

PERFORMANCE IMPROVEMENT OF ULTRA-HIGH-SPEED PMSM DRIVE SYSTEM BASED ON DTC BY USING SIC INVERTER......356
Togashi, Ryo ; Inoue, Yukinori ; Morimoto, Shigeo ; Sanada, Masayuki

MATHEMATICAL MODEL FOR HIGH-EFFICIENCY CONTROL OF PERMANENT-MAGNET SYNCHRONOUS MOTOR IN STATOR FLUX LINKAGE SYNCHRONOUS FRAME......363

Inoue, Tatsuki ; Inoue, Yukinori ; Morimoto, Shigeo ; Sanada, Masayuki

WIDE-SPEED-RANGE OPERATION OF DTC-BASED PMSM DRIVE SYSTEM USING MTPF CONTROL......370

Inoue, Yukinori ; Ichiya, Takahiro ; Morimoto, Shigeo ; Sanada, Masayuki

AN INDUSTRIAL LOW-VOLTAGE INVERTER FOR PRM CONTROL......376

Nakamura, M. ; Oka, T. ; Oishi, K.

OPTIMAL PULSE PATTERN DETERMINATION BASED ON PULSE HARMONIC MODULATION......383

Furukawa, Kimihisa ; Ajima, Toshiyuki ; Miyazaki, Hideki

METHOD FOR AUTO-TUNING OF CURRENT AND SPEED CONTROLLER IN IPMSM DRIVE SYSTEM BASED ON PARAMETER IDENTIFICATION......390

Tadokoro, D. ; Morimoto, S. ; Inoue, Y. ; Sanada, M.

COMPARATIVE STUDY OF PWM STRATEGIES FOR THREE-PHASE OPEN-END WINDING INDUCTION MOTOR DRIVES......395

Zhu, B. ; Prasanna, U.R. ; Rajashekara, K. ; Kubo, H.

10MW,3.3MWH ENERGY STORAGE SYSTEM CONSISTING OF 4000 FLYWHEELS CONTROLLED BY ICT NETWORK FOR SHORT CYCLE POWER FLUCTUATION COMPENSATION......403

Kato, Koji ; Ishigma, Satoru ; Nakajima, Yoichiro ; Arai, Haruki ; Ueda, Tetsuya ; Iwata, Tetsuki ; Ito, Yoichi ; Sugao, Kazumi

VERSATILE POWER TRANSFER STRATEGIES OF PV-BATTERY HYBRID SYSTEM FOR RESIDENTIAL USE WITH ENERGY MANAGEMENT SYSTEM......409

Choi, Seong-Chon ; Sin, Min-ho ; Kim, Dong-Rak ; Won, Chung-Yuen ; Jung, Yong-Chae

HIGH-EFFICIENCY AND COST-MINIMIZATION METHOD OF ENERGY STORAGE SYSTEM WITH MULTI STORAGE DEVICES FOR GRID CONNECTION......415

Haga, Hitoshi ; Shimao, Toshihiro ; Kondo, Seiji ; Kato, Koji ; Itoh, Youichi ; Arimatsu, Kenji ; Matsuda, Katsuhiro

BIDIRECTIONAL DC-DC CONVERTER WITH MULTIPLE SWITCHED-CAPACITOR CELLS......421

Lee, Yuang-Shung ; Huang, Hsin-Wei ; Chou, Tzu-Han

SWITCHED-CAPACITOR CHARGE EQUALIZATION CIRCUIT FOR SERIES-CONNECTED BATTERIES......429

Hsieh, Yao-Ching ; Cai, Zheng-Xiu ; Wu, Wen-Zhe

PERFORMANCE ANALYSIS OF UNITL-H6 INVERTER WITH SIC MOSFETS......433

Barater, Davide ; Buticchi, Giampaolo ; Concari, Carlo ; Franceschini, Giovanni ; Gurpinar, Emre ; De, Dipankar ; Castellazzi, Alberto

MAXIMUM POWER POINT TRACKING OF GRID-TIED PHOTOVOLTAIC POWER SYSTEMS......440

Lee, Ya-Ting ; Chiu, Chian-Song ; Chiu, Tse-Wei

A NEW VOLTAGE TYPE MAGNETICALLY COUPLED T-SOURCE INVERTER......446

Tran, Q.V. ; Low, K.S.

A HIGH EFFICIENCY HYBRID 7-LEVEL INVERTER WITH SINGLE DC SOURCE......452

Yanhong, Zhang ; Kazuya, Ogura ; Oi, Kazunobu

OPTIMAL IDLING CONTROL STRATEGY FOR THREE-PORT FULL-BRIDGE CONVERTER......458

Jiang, Yongjie ; Liu, Fuxin ; Ruan, Xinbo ; Wang, Lipeng

FILTER DESIGN FOR THREE-LEVEL GRID-CONNECTED INVERTER WITH LOW SWITCHING FREQUENCY......465

Ren, Kangle ; Zhang, Xing ; Wang, Fusheng ; Tu, Yunwu ; Wang, Lingxiang ; Deng, Lirong

A NOVEL EFFICIENT T TYPE THREE LEVEL NEUTRAL-POINT-CLAMPED INVERTER FOR RENEWABLE ENERGY SYSTEM......470

Wu, Wenlong ; Wang, Fei ; Wang, Yong

A NOVEL NEUTRAL POINT VOLTAGE AUTOMATIC BALANCING CARRIER-BASED MODULATION STRATEGY OF THREE-LEVEL NPC CONVERTER......475

Li, Ning ; Wang, Yue ; Niu, Ruigen ; Guo, Wei ; Lei, Wanjun ; Wang, Zhao'An

A HIGH VOLTAGE GAIN SWITCHED-COUPLED-INDUCTOR QUASI-Z-SOURCE INVERTER......480

Ahmed, Furqan ; Cha, Honnyong ; Kim, Su-Han ; Kim, Heung-Geun

A NOVEL CONTROL STRATEGY TO SUPPRESS DC CURRENT INJECTION TO THE GRID FOR THREE-PHASE PV INVERTER......485

Zhang, Tao ; He, Guofeng ; Chen, Min ; Xu, Dehong

CLC FILTER DESIGN OF A FLYBACK-INVERTER FOR PHOTOVOLTAIC SYSTEMS......493

Shin, Yesl ; Lee, June-Hee ; Lee, June-Seok ; Lee, Kyo-Beum

THREE-PHASE INVERTER TOPOLOGIES FOR GRID-CONNECTED PHOTOVOLTAIC SYSTEMS......498

Ozkan, Ziya ; Hava, Ahmet M.

A THREE-PORT TOPOLOGY COMPARISON FOR A LOW POWER STAND-ALONE PHOTOVOLTAIC SYSTEM......506

Mira, Maria C. ; Knott, Arnold ; Andersen, Michael A.E.

EFFECT OF CONVENTIONAL GRID-VOLTAGE FEEDFORWARD ON THE OUTPUT IMPEDANCE OF A THREE-PHASE PHOTOVOLTAIC INVERTER......514

Messo, T. ; Jokipii, J. ; Suntio, T.

POWER AMPLIFIER SUITABLE FOR PHOTOVOLTAIC CELL BOOSTER......522

Kohama, Teruhiko ; Sogawa, Yuki ; Tsuji, Satoshi

REALIZATION STUDY OF INTERLEAVED PV MICROINVERTER BY QUADRATURE-PHASE-SHIFT SPWM CONTROL......526

Hsieh, Hung-I ; Hsieh, Guan-Cyun ; Hou, Jiaxin

CURRENT SENSORLESS MPPT METHOD FOR A PV FLYBACK MICROINVERTERS USING A DUAL-MODE 532

Lee, June-Hee ; Lee, June-Seok ; Lee, Kyo-Beum

A NOVEL METHOD OF SUPPRESSING INRUSH CURRENTS OF SQUIRREL-CAGE INDUCTION MACHINE USING MATRIX CONVERTER IN WIND POWER GENERATION SYSTEMS 538

Yamada, Hiroaki ; Hanamoto, Tsuyoshi

NONLINEAR PITCH CONTROL DESIGN FOR LOAD REDUCTION ON WIND TURBINES 543

Xiao, Shuai ; Yang, Geng ; Geng, Hua

DEVICE LOADING OF MODULAR MULTILEVEL CONVERTER MMC IN WIND POWER APPLICATION 548

Popova, L. ; Pyrhonen, J. ; Ma, K. ; Blaabjerg, F.

A NOVEL OPTIMAL DESIGN OF DFIG CROWBAR RESISTOR DURING GRID FAULTS 555

Hu, Sheng ; Zou, XuDong ; Kang, Yong

DC-VOLTAGE REGULATION OF A FIVE LEVELS NEUTRAL POINT CLAMPED CASCADED CONVERTER FOR WIND ENERGY CONVERSION SYSTEM 560

Merahi, Farid ; Mekhilef, Saad ; Berkouk, El Madjid

A REACTIVE POWER SHARING METHOD BASED ON VIRTUAL CAPACITOR IN ISLANDING MICROGRID 567

Xu, Haizhen ; Zhang, Xing ; Liu, Fang ; Shi, Rongliang ; Yu, Changzhou ; Zhao, Wei ; Yu, Yong ; Cao, Wei

STORAGE CAPACITY PERFORMANCE FOR HYBRID PV/DIESEL SYSTEM IN SABAH MALAYSIA 573

Hidayat, Nabil M ; Kari, Mat Nasir ; Mohd Arif, Mohd Johari

NEW TECHNIQUES FOR MEASURING ISLANDED MICROGRID IMPEDANCE CHARACTERISTICS BASED ON CURRENT INJECTION 577

Hou, Lixiang ; Liu, Baoquan ; Shi, Hongtao ; Yi, Hao ; Zhuo, Fang

A GENERAL FRAMEWORK TO DESIGN OPERATION MODES OF DC MICROGRIDS WITHOUT COMMUNICATION LINKS 582

Pan, Miao ; Shen, Na ; Yang, Geng ; Morita, Kazunori ; Ogura, Kazuya ; Wu, Weiyang

IMPLEMENTATION DESIGN OF THE CONVERTER-BASED GALVANIC ISOLATION FOR LOW VOLTAGE DC DISTRIBUTION 587

Mattsson, A. ; Vaisanen, V. ; Nuutinen, P. ; Kaipia, T. ; Lana, A. ; Peltoniemi, P. ; Silventoinen, P. ; Partanen, J.

PEAK DETECTION METHOD USING TWO-DELTA OPERATION FOR SINGLE VOLTAGE SAG 595

Lee, Woo-Cheol ; Lee, Taeck-Kie

LINE LOSS MINIMIZATION IN RADIAL DISTRIBUTION SYSTEM USING MULTIPLE STATCOMS AND STATIC CAPACITORS 601

Miyazaki, Kensuke ; Takeshita, Takaharu

A NOVEL CONTROL METHOD FOR INDIVIDUAL DC VOLTAGE BALANCING IN H-BRIDGE CASCADED STATCOM 609

Xu, Rong ; Yu, Yong ; Yang, Rongfeng ; Qu, Lizhi ; Sun, Wei ; Xu, Dianguo

RESEARCH ON THE CONTROL STRATEGY OF STATCOM BASED ON MODULAR MULTILEVEL CONVERTER 614

Zhang, Wei ; Gao, Qiang ; Su, Bonan ; Jin, Miaoxin ; Xu, Dianguo ; Liu, Jianyu

FAULT DIAGNOSIS IN LARGE FORMAT LIFEPO4 ESS APPLICATION THROUGH DWT-BASED MRA 619

Kim, Jonghoon

COMPARISON OF DIFFERENT IGBT BASED DESIGNS OF POWER ELECTRONIC TRANSFORMER 624

Wang, Xinyu ; Ouyang, Shaodi ; Liu, Jinjun ; Meng, Fei ; Javed, Riffat

SEMI-ADAPTIVE HARMONIC CONTROL FOR POWER BALANCING DEVICE FOR AC TRACTION 629

Akagi, Masataka ; Tsuruta, Hironori ; Oso, Hiroshi

RESEARCH OF EFFICIENT MAIN POWER EQUIPMENT USING SIC POWER DEVICE 634

Shinbo, Mitsuo ; Sonoda, Hideki ; Ishida, Takahito ; Abiko, Hiroshi ; Shibanuma, Kenichi ; Chiba, Yoshinori

A HIGH PERFORMANCE CONTROL STRATEGY FOR THREE-LEVEL NPC EMU CONVERTERS 640

Song Kejian ; Wu Mingli ; Wang Hui ; Agelidis, Vassilios Georgios

A DESIGN OF INRUSH CURRENT IDENTIFICATION SYSTEM FOR HIGH-SPEED TRAIN'S TRACTION TRANSFORMER 647

Yu, Weikai ; Liu, Xiankai ; Zhang, Yuzhuo ; Cao, Yuan ; Ma, Weigang ; Hei, Xinhong ; Huang, Zhenhui ; Jiang, Dawang

CURRENT SOURCE INVERTER BASED CASCADED SOLID STATE TRANSFORMER FOR AC TO DC POWER CONVERSION 651

Roy, Sudhin ; De, Ankan ; Bhattacharya, Subhashish

EVALUATION OF HIGH VOLTAGE 15 KV SIC IGBT AND 10 KV SIC MOSFET FOR ZVS AND ZCS HIGH POWER DC-DC CONVERTERS 656

Moballegh, Shiva ; Madhusoodhanan, Sachin ; Bhattacharya, Subhashish

THE DIRECT YAW-MOMENT CONTROL TO FOLLOW THE NEUTRAL STEERING PATH REGARDLESS OF VELOCITY 664

Jang, Young-Jin ; Nam, Kwang-Hee

NEXT-GENERATION IGBT MODULE STRUCTURE FOR HYBRID VEHICLE WITH HIGH COOLING PERFORMANCE AND HIGH TEMPERATURE OPERATION 671

Morozumi, Akira ; Gohara, Hiromichi ; Momose, Fumihiko ; Saito, Takashi ; Nishimura, Yoshitaka ; Mochizuki, Eiji ; Takahashi, Yoshikazu

INTEGRATION OF PLUG-IN ELECTRIC VEHICLES IN POWER SYSTEMS USING CHARGING MODE SWITCHING 677

Wen-Tai Li ; Wen, Chao-Kai ; Chen, Jung-Chieh ; Teng, Jen-Hao ; Ting, Pangan

A NOVEL COMPENSATION METHOD FOR A MOTOR PHASE CURRENT SENSOR OFFSET ERROR VARIED DURING A VSI-MOTOR DRIVE ... 682

Tamura, Hiroshi ; Noto, Yasuo ; Ajima, Toshiyuki ; Itoh, Jun-ichi

INVESTIGATION OF CALCULATION METHOD OF LOSSES IN PWM INVERTER WITH VOLTAGE BOOSTER USING BOTH DC LINK VOLTAGE CONTROL AND FLUX WEAKENING CONTROL ... 689

Imakiire, Akihiro ; Hikita, Masayuki ; Yamamoto, Kichiro ; Yonemori, Ryo

DYNAMIC AND STEADY-STATE BEHAVIOR OF A PARALLELING THREE-PHASE AC-TO-DC CONVERTER WITH REDUCED DC BUS CAPACITOR ... 694

Kamnarn, Uthen ; Kanthaphayao, Yutthana ; Chunkag, Viboon

REACTIVE POWER LOSS OPTIMIZATION METHOD FOR BI-DIRECTIONAL ISOLATED DC-DC CONVERTERS ... 702

Wen, Huiqing

POWER SUPPLY FOR A WIRELESS SENSOR NETWORK: AIRLINER FLIGHT TEST CASE STUDY ... 707

Durand Estebe, P. ; Boitier, V. ; Bafleur, M. ; Dilhac, J-M. ; Berhouet, S.

A CONFIGURABLE THREE-PHASED INVERTER FOR TEACHING POWER ELECTRONICS ... 712

Kern, Ansgar

A BACHELOR-STUDENT PROJECT: BUCK-BOOST OPERATION OF AN INTEGRATED H-BRIDGE FOR VARIABLE-SPEED ENERGY STORAGE SYSTEMS USING MEASUREMENT COILS IN THE STATOR OF A DC-MACHINE ... 718

De Belie, Frederik ; Darba, Araz ; Melkebeek, Jan

DEVELOPMENT OF A WEB-BASED REMOTE EXPERIMENT SYSTEM FOR ELECTRICAL MACHINERY LEARNERS ... 724

Ishibashi, Makoto ; Fukumoto, Hisao ; Furukawa, Tatsuya ; Itoh, Hideaki ; Ohchi, Masashi

DEVELOPMENT OF POWER MEASUREMENT SYSTEM IN SIMULATED MICRO GRID SYSTEM FOR EDUCATION ... 730

Hira, Yuki ; Furukawa, Tatsuya ; Yakabe, Seichiro ; Fukumoto, Hisao ; Itoh, Hideaki ; Ohchi, Masashi

POWER ELECTRONIC TECHNOLOGIES FOR FLEXIBLE DC DISTRIBUTION GRIDS ... 736

De Doncker, Rik W.

2.5KV, 200KW BI-DIRECTIONAL ISOLATED DC/DC CONVERTER FOR MEDIUM-VOLTAGE APPLICATIONS ... 744

Matsuoka, Yuji ; Wada, Keiji ; Nakahara, Mizuki ; Takao, Kazuto ; Kyungmin Sung ; Ohashi, Hiromichi ; Nishizawa, Shinichi

POWER-LOSS BREAKDOWN OF A 750-V, 100-KW, 20-KHZ BIDIRECTIONAL ISOLATED DC-DC CONVERTER USING SIC-MOSFET/SBD DUAL MODULES ... 750

Akagi, Hirofumi ; Yamagishi, Tatsuya ; Tan, Nadia M.L. ; Kinouchi, Shin-ichi ; Miyazaki, Yuji ; Koyama, Masato

DESIGN CONSIDERATIONS OF A 15KV SIC IGBT ENABLED HIGH-FREQUENCY ISOLATED DC-DC CONVERTER ... 758

Tripathi, Awneesh ; Mainali, Krishna ; Patel, Dhaval ; Kadavelugu, Arun ; Hazra, Samir ; Bhattacharya, Subhashish ; Hatua, Kamalesh

COMMON-MODE CURRENTS IN MULTI-CELL SOLID-STATE TRANSFORMERS ... 766

Huber, Jonas E. ; Kolar, Johann W.

SINGLE-STAGE RECONFIGURABLE DC/DC CONVERTER FOR WIDE INPUT VOLTAGE RANGE OPERATION IN HEVS ... 774

Zeljkovic, Sandra ; Reiter, Tomas ; Gerling, Dieter

A TWO STAGE DC/DC CONVERTER WITH WIDE INPUT RANGE FOR EV ... 782

Peng Wen ; Changsheng Hu ; Haitao Yang ; Longlong Zhang ; Cheng Deng ; Yashun Li ; Dehong Xu

INTERMEDIATE AND LIGHT LOAD EFFICIENCY IMPROVEMENT OF A HIGH-POWER DENSITY BIDIRECTIONAL DC-DC CONVERTER IN HYBRID ELECTRIC VEHICLES WITH MR FLUID GAP INDUCTOR ... 790

Ahmed, Furqan ; Su-Han Kim ; Cha, Honnyong ; Kim, Dong-Hun ; Heung-Geun Kim

REGENERATIVE CONTROL OF BI-DIRECTIONAL DC-DC CONVERTER CONTROLLING VARIABLE DC-LINK FOR FCEV ... 796

Il-Kuen Won ; An-Yeol Ko ; Do-Yun Kim ; Chung-Yuen Won ; Young-Ryul Kim

LARGE DRIVING RANGE INCREASE OF SERIES CHOPPER BASED POWER TRAIN USING MOTOR TEST BENCH ... 801

Hosoyamada, Yu ; Takeda, Masashi ; Motoi, Naoki ; Kawamura, Atsuo

THE POWER ELECTRONICS PROGRAM AT BEIJING JIAOTONG UNIVERSITY ... 807

Fei Lin ; Zhongping Yang ; Zheng, T.Q.

EFFORTS FOR POWER ELECTRONICS EDUCATION IN A START-UP COMPANY ... 811

Hattori, Fumiya ; Imaoka, Jun ; Ishitobi, Manabu ; Nagai, Shinichiroh ; Yamamoto, Masayoshi

EDUCATION FOR THE ENGINEERS OF TRACTION POWER SUPPLY DIVISION IN EAST JAPAN RAILWAY COMPANY ... 817

Takino, Toshiaki ; Iwakami, Tetsuro

SUCCESSFUL ONLINE EDUCATION - GECKOCIRCUITS AS OPEN-SOURCE SIMULATION PLATFORM ... 821

Musing, Andreas ; Kolar, Johann W.

AN ELECTRIC VEHICLE PROJECT FOR ECO-RUN RACE ... 829

Yamagata, Shinichi ; Oda, Yoshinori ; Tanai, Masanobu ; Sung, Kyungmin

MULTI-LOOP CONTROLLER DESIGN FOR DIODE-ASSISTED BUCK-BOOST VOLTAGE SOURCE INVERTER ... 835

Yan Zhang ; Jinjun Liu ; Xiaolong Ma ; Junjie Feng

VOLUME 2

REAL-TIME SIMULATION OF WIND TURBINE CONVERTER-GRID SYSTEMS 843
Shah, Shahil ; Vieto, Ignacio ; Nian Heng ; Sun, Jian

TECHNOLOGIES FOR MITIGATING FLUCTUATION CAUSED BY RENEWABLE ENERGY SOURCES 850
Katoh, Shuji ; Ohara, Shinya ; Itoh, Tomomichi

RELIABILITY-ORIENTED ENERGY STORAGE SIZING IN WIND POWER SYSTEMS 857
Zian Qin ; Liserre, Marco ; Blaabjerg, Frede ; Poh Chiang Loh

A MULTI-LEVEL VIRTUAL CONDUCTOR AS A BACKBONE OF A DC POWER ROUTING SYSTEM 863
Ramadan, Husam A. ; Imamura, Yasutaka ; Kawachi, Konosuke ; Yang, Sihun ; Shoyama, Masahito

SEMI-NUMERICAL METHOD FOR LOSS-CALCULATION IN FOIL-WINDINGS EXPOSED TO AN AIR-GAP FIELD 868
Leuenberger, D. ; Biela, J.

LOSS REDUCTION OF LAMINATED CORE INDUCTOR USED IN ON-BOARD CHARGER FOR EVS 876
Tera, Takahiro ; Taki, Hiroshi ; Shimizu, Toshihisa

FEASIBLE EVALUATIONS OF COUPLED MULTILAYER CHIP INDUCTOR FOR POL CONVERTER 883
Imaoka, Jun ; Kimura, Shota ; Itoh, Yuki ; Yamamoto, Masayoshi ; Suzuki, Michiaki ; Kawano, Kenji

OPTIMAL INDUCTOR DESIGN FOR 3-PHASE VOLTAGE-SOURCE PWM CONVERTERS CONSIDERING DIFFERENT MAGNETIC MATERIALS AND A WIDE SWITCHING FREQUENCY RANGE 891
Burkart, Ralph M. ; Uemura, Hirofumi ; Kolar, Johann W.

COMPARATIVE ANALYSIS OF INDUCTOR CONCEPTS FOR HIGH PEAK LOAD LOW DUTY CYCLE OPERATION 899
Leibl, Michael ; Kolar, Johann W.

INITIAL POSITION ESTIMATION FOR IPMSMS USING COMB FILTERS AND EFFECTS ON VARIOUS INJECTED SIGNAL FREQUENCIES 907
Suzuki, Toshiki ; Tomita, Mutuwo ; Hasegawa, Masaru ; Doki, Shinji

ADAPTIVE SIGNAL INJECTION METHOD COMBINED WITH EEMF BASED POSITION SENSORLESS CONTROL OF IPMSM DRIVES 914
Ohnuma, Takumi ; Makaino, Yuki ; Saitoh, Ryoh

STUDY OF LOW SPEED SENSORLESS DRIVES FOR SPMSM BY CONTROLLING ELLIPTICAL INDUCTANCE 919
Maekawa, Sari ; Hinata, Toshifumi ; Suzuki, Nobuyuki ; Kubota, Hisao

SUPPRESSION OF INJECTION VOLTAGE DISTURBANCE FOR HIGH FREQUENCY SQUARE-WAVE INJECTION SENSORLESS DRIVE WITH REGULATION OF INDUCED HIGH FREQUENCY CURRENT RIPPLE 925
Dongouk Kim ; Yong-Cheol Kwon ; Seung-Ki Sul ; Jang-Hwan Kim ; Rae-Sung Yu

APPLICATION TREND OF SALIENCY-BASED SENSORLESS DRIVES 933
Yamazaki, Akira ; Ide, Kozo

SWITCHING-LEVEL SIMULATION MODEL OF MMC-BASED BACK-TO-BACK CONVERTER FOR HVDC APPLICATION 937
Byung Moon Han ; Jong kyou Jeong

POWER-CELL SWITCHING-CYCLE CAPACITOR VOLTAGE CONTROL FOR THE MODULAR MULTILEVEL CONVERTERS 944
Wang, Jun ; Burgos, Rolando ; Boroyevich, Dushan ; Bo Wen

A COMPARISON OF MODULAR MULTILEVEL ENERGY CONVERSION PROCESSES: DC/AC VERSUS DC/DC 951
Kish, Gregory J. ; Lehn, Peter W.

A NOVEL TOPOLOGY OF WIND POWER PLANT SUITABLE FOR DC POWER TRANSMISSION SYSTEMS 959
Nishikata, Shoji ; Tatsuta, Fujio ; Suzuki, Katsumi

AN IMPEDANCE-BASED APPROACH TO HVDC SYSTEM STABILITY ANALYSIS AND CONTROL DEVELOPMENT 967
Liu, Hanchao ; Shah, Shahil ; Sun, Jian

TOPOLOGY EVALUATION OF SLOTLESS BEARINGLESS MOTORS WITH TOROIDAL WINDINGS 975
Steinert, Daniel ; Nussbaumer, Thomas ; Kolar, Johann W.

WINDING ARRANGEMENT IN SINGLE-DRIVE BEARINGLESS MOTOR WITH RADIAL GAP 982
Sugimoto, Hiroya ; Tanaka, Seiyu ; Chiba, Akira ; Rahman, M.A.

DEVELOPMENT OF A ONE-AXIS ACTIVELY REGULATED BEARINGLESS MOTOR WITH A REPULSIVE TYPE PASSIVE MAGNETIC BEARING 988
Asama, Junichi ; Watanabe, Daisuke ; Oiwa, Takaaki ; Chiba, Akira

CONTROL CHARACTERISTICS OF 8/10 AND 12/14 BEARINGLESS SWITCHED RELUCTANCE MOTOR 994
Zhenyao Xu ; Dong-Hee Lee ; Jin-Woo Ahn

BASIC CHARACTERISTIC OF A TWO-UNIT OUTER ROTOR TYPE BEARINGLESS MOTOR WITH CONSEQUENT POLE PERMANENT MAGNET STRUCTURE 1000
Takemoto, Masatsugu

VOLTAGE RIPPLE ELIMINATION IN INDUCTOR-LESS AC-TO-AC CONVERTERS FOR MULTI-POLE PERMANENT MAGNET SYNCHRONOUS GENERATORS 1006
Tanaka, Koutaro ; Fujita, Hideaki

A NEW SVM METHOD TO REDUCE COMMON-MODE VOLTAGE IN DIRECT MATRIX CONVERTER 1013
Huu-Nhan Nguyen ; Hong-Hee Lee

EXPERIMENTAL VERIFICATION OF HIGH FREQUENCY LINK DC-AC CONVERTER USING PULSE DENSITY MODULATION AT SECONDARY MATRIX CONVERTER 1021
Itoh, Jun-ichi ; Oshima, Ryo ; Takahashi, Hiroki

LOSS ANALYSIS AND DESIGN METHOD FOR HIGH EFFICIENCY MATRIX CONVERTER 1028
Koiwa, Kazuhiro ; Goh Teck Chiang ; Itoh, Jun-ichi

CAPACITOR CLAMPED MULTI-LEVEL MATRIX CONVERTER 1036
Raju, Siddharth ; Mohan, Ned

EUROPEAN TRENDS AND TECHNOLOGIES IN TRACTION 1043
Drofenik, Uwe ; Canales, Francisco

CO-PHASE POWER SUPPLY SYSTEM FOR HSR 1050
Qunzhan Li ; Wei Liu ; Zeliang Shu ; Shaofeng Xie ; Fulin Zhou

THE APPLICATION OF ELECTRONIC FREQUENCY CONVERTER TO THE SHINKANSEN RAILYARD POWER SUPPLY 1054
Shimizu, Toshimasa ; Kunomura, Ken ; Kai, Masahiko ; Onishi, Mitsuru ; Masuzawa, Hiroshi ; Miyajima, Hiroki ; Otsuki, Midori ; Tsuruma, Yoshinori

APPLICATION EXAMPLES OF ENERGY SAVING MEASURES IN JAPANESE DC FEEDING SYSTEM 1062
Suzuki, Takashi ; Hayashiya, Hitoshi ; Yamanoi, Takashi ; Kawahara, Keiji

LITHIUM ION BATTERY APPLICATION IN TRACTION POWER SUPPLY SYSTEM 1068
Teshima, Masato ; Takahashi, Hirotaka

INTEGRATED ISOLATION AND VOLTAGE BALANCING LINK OF 3-PHASE 3-LEVEL PWM RECTIFIER AND INVERTER SYSTEMS 1073
Boillat, David O. ; Kolar, Johann W.

VOLTAGE STEP-UP CONVERTER BASED ON MULTISTAGE STACKED BOOST ARCHITECTURE (MSBA) 1081
Rufer, Alfred ; Barrade, Philippe ; Steinke, Gina

COMPARISON OF CASCADED MULTILEVEL CONVERTER TOPOLOGIES FOR AC/AC CONVERSION 1087
Ilves, Kalle ; Bessegato, Luca ; Norrga, Staffan

EVALUATION OF ISOLATED THREE-PHASE AC-DC CONVERTER USING MODULAR MULTILEVEL CONVERTER TOPOLOGY 1095
Nakanishi, Toshiki ; Itoh, Jun-ichi

SELF-DECOUPLED DUAL PICK-UP COILS WITH LARGE LATERAL TOLERANCE FOR ROADWAY POWERED ELECTRIC VEHICLES 1103
Choi, Su Y. ; Lee, Sung W. ; Lee, Eun S. ; Jeong, Seog Y. ; Gu, Beom W. ; Rim, Chun T.

CONTACTLESS POWER TRANSFER SYSTEM SUITABLE FOR LOW VOLTAGE AND LARGE CURRENT CHARGING FOR EDLCS 1109
Kudo, Takahiro ; Toi, Takahiro ; Kaneko, Yasuyoshi ; Abe, Shigeru

EXCITATION SYSTEM BY CONTACTLESS POWER TRANSFER SYSTEM WITH THE PRIMARY SERIES CAPACITOR METHOD 1115
Nozawa, Ryosuke ; Kobayashi, Ryota ; Tanifuji, Hikaru ; Kaneko, Yasuyoshi ; Abe, Shigeru

DESIGN OF FERRITE CORES OF INDUCTIVE POWER COLLECTION COILS FOR MOVING VEHICLES 1122
Shimode, Daisuke ; Murai, Toshiaki ; Sawada, Tadashi

TORQUE/CURRENT RATIO IMPROVEMENT AND VIBRATION REDUCTION OF SWITCHED RELUCTANCE MOTORS USING MULTI-STAGE STRUCTURE 1128
Matsui, Ryota ; Nakao, Noriya ; Akatsu, Kan

IMPROVEMENT OF EFFICIENCY BY STEPPED-SKEWING ROTOR FOR SWITCHED RELUCTANCE MOTORS 1135
Sugiura, Makoto ; Ishihara, Yuji ; Ishikawa, Hiroki ; Naitoh, Haruo

A SINGLE PHASE SRM DRIVEN BY COMMERCIAL AC POWER SUPPLY 1141
Aiso, Kohei ; Nakao, Noriya ; Akatsu, Kan

FAST ANALYTICAL MODEL OF SWITCHED RELUCTANCE MACHINE 1148
Smaka, Senad ; Masic, Semsudin ; Cosovic, Mirsad

DETAILED ANALYSIS AND A GENERAL DESIGN PROCEDURE OF DAMPED LCL FILTERS IN THREE PHASE VOLTAGE SOURCE CONVERTERS 1155
Baoquan Liu ; Shaohui Zhong ; Yixin Zhu ; Hao Yi ; Fang Zhuo

70 KHZ, 15 KW SILICON-CARBIDE MOSFET INVERTER FOR INDUSTRIAL INDUCTION HEATING SYSTEMS 1160
Komeda, Shohei ; Tsuboi, Yoshiki ; Fujita, Hideaki

A STUDY ON EFFICIENCY IMPROVEMENT OF HIGH-FREQUENCY CURRENT OUTPUT INVERTER BASED ON IMMITTANCE CONVERSION ELEMENT 1166
Suzuki, Shun ; Shimizu, Toshihisa

HIGH-SPEED SWITCHING METHOD OF MOSFET USING VOLTAGE BOOST AUXILIARY CIRCUIT FED BY GATE DRIVE POWER SUPPLY 1173
Noguchi, Toshihiko ; Murata, Munehiro

OPERATING STRATEGY FOR BI-DIRECTIONAL LLC RESONANT CONVERTER WITH SEAMLESS OPERATION 1179

Abe, Seiya ; Yamamoto, Junichi ; Zaitsu, Toshiyuki ; Ninomiya, Tamotsu

NEGATIVE SEQUENCE CURRENT INJECTION CONTROL ALGORITHM COMPENSATING FOR UNBALANCED PCC VOLTAGE IN MEDIUM VOLTAGE PMSG WIND TURBINES 1185

Jayoon Kang ; Daesu Han ; Suh, Yongsug ; Byoungchang Jung ; Jeongjoong Kim ; Jonghyung Park ; Youngjoon Choi

OPTIMIZATION OF AN OFF-GRID HYBRID SYSTEM FOR SUPPLYING OFFSHORE PLATFORMS IN ARCTIC CLIMATES 1193

Kalogera, Maria ; Bauer, Pavol

ACTIVE DAMPING CONTROL OF LLCL FILTERS FOR THREE-LEVEL T-TYPE GRID CONVERTERS 1201

Alemi, Payam ; Lee, Dong-Choon

DEVELOPING A NEW TOPOLOGY FOR THE DC-DC CONVERTER USED IN FUEL CELL-ELECTRIC DOUBLE LAYER CAPACITOR HYBRID POWER SOURCE SYSTEM FOR MOBILE DEVICES 1207

Tosaka, Shuhei ; Yamanaka, Tatsuya ; Katayama, Noboru ; Hayase, Masanori ; Dowaki, Kiyoshi ; Kogoshi, Sumio

MULTIPLE OUTPUT CHARGER BASED ON PHASE SHIFT FULL BRIDGE CONVERTER WITH NOVEL TIME DIVISION MULTIPLE CONTROL TECHNIQUE 1214

Van-Long Tran ; Woojin Choi

DC-BREAKER FOR A MULTI-MEGAWATT BATTERY ENERGY STORAGE SYSTEM 1220

Demetriades, Georgios D. ; Hermansson, Willy ; Svensson, Jan R ; Papastergiou, Konstantinos ; Larsson, Tomas

ENERGY MANAGEMENT METHOD USING THE IIR FILTER FOR PEMFC-SUPERCAPACITOR HYBRID POWER SOURCE 1227

Yamanaka, Tatsuya ; Katayama, Noboru ; Tosaka, Shuhei ; Kogoshi, Sumio

ADVANCED TORQUE AND CURRENT CONTROL TECHNIQUES FOR PMSMS WITH A REAL-TIME SIMULATOR INSTALLED BEHAVIOR MOTOR MODEL 1234

Tanabe, Ryo ; Akatsu, Kan

COMPENSATION OF THE CURRENT MEASUREMENT ERROR WITH PERIODIC DISTURBANCE OBSERVER FOR MOTOR DRIVE 1242

Yamaguchi, Takashi ; Tadano, Yugo ; Hoshi, Nobukazu

RAPID AND STABLE SPEED CONTROL OF SPMSM BASED ON CURRENT DIFFERENTIAL SIGNAL 1247

Kitajima, Jun ; Ohishi, Kiyoshi

PARALLEL CONNECTED MULTIPLE DRIVE SYSTEM USING SMALL AUXILIARY INVERTER FOR NUMBERS OF PMSM 1253

Nagano, Tsuyoshi ; Itoh, Jun-chi

A TRANSFORMER INRUSH REDUCTION TECHNIQUE FOR LOW-VOLTAGE RIDE-THROUGH OPERATION OF RENEWABLE CONVERTERS 1261

Hsin-Chih Chen ; Ping-Heng Wu ; Cheng, Po-Tai

A CELL CAPACITOR ENERGY BALANCING CONTROL OF MODULAR MULTILEVEL CONVERTER CONSIDERING THE UNBALANCED AC GRID CONDITIONS 1268

Jung, Jae-Jung ; Shenghui Cui ; Kim, Sungmin ; Sul, Seung-Ki

FAULT CURRENT LIMITATION USING THYRISTOR BASED DEVICES 1276

Komatsu, Wilson ; Giaretta, Antonio Ricardo ; de Miranda, Rubens Domingos ; Jardini, Jose Antonio ; Casolari, Ronaldo Pedro ; Vasquez-Arnez, Ricardo Leon ; Hojo, Toshiaki ; Carvalho, Eden Luiz ; Maezono, Paulo Koiti

DC-DC BOOST CONVERTER BASED MSHE-PWM CASCADED MULTILEVEL INVERTER CONTROL FOR STATCOM SYSTEMS 1283

Law, Kah Haw ; Dahidah, Mohamed S.A.

NOVEL PRINCIPLE FOR FLUX SENSING IN THE APPLICATION OF A DC + AC CURRENT SENSOR 1291

Schrittwieser, L. ; Mauerer, M. ; Bortis, D. ; Ortiz, G. ; Kolar, J.W.

UTILIZING VOLTAGE MEASUREMENT OF FET SWITCH FOR MPPT OF DC ENERGY SOURCE 1299

Kimura, Noriyuki ; Niijima, Koji ; Morizane, Toshimitsu ; Omori, Hideki

HIGH FREQUENCY TRANSFORMER BASED ON A COUPLED INDUCTOR TOPOLOGY WITH DIELECTRIC ISOLATION 1303

Amanci, Adrian Z. ; Dawson, Francis P. ; Ruda, Harry E.

CONCEPT AND EXPERIMENTAL EVALUATION OF A NOVEL DC- 100MHZ WIRELESS OSCILLOSCOPE 1309

Lobsiger, Yanick ; Ortiz, Gabriel ; Bortis, Dominik ; Kolar, Johann W.

INTRODUCTION AND EFFECTIVENESS OF STATCOM TO THE INDEPENDENT POWER SYSTEM OF JR EAST 1317

Omi, Masataro ; Kotegawa, Ryo ; Ando, Masato ; Masui, Takeshi ; Horita, Yasuhisa

THE ANALYSIS OF TIME-VARYING RESONANCES IN THE POWER SUPPLY LINE OF HIGH SPEED TRAINS 1322

Chu, Xi ; Lin, Fei ; Yang, Zhongping

FUZZY FEED-FORWARD CHARGE/DISCHARGE CONTROL OF STATIONARY ENERGY STORAGE SYSTEMS FOR DC ELECTRIC RAILWAYS 1328

Kikuchi, Takuya ; Taga, Hironori ; Takagi, Ryo

TRAIN GROUP CONTROL FOR ENERGY-SAVING DC-ELECTRIC RAILWAY OPERATION 1334

Watanabe, Shoichiro ; Koseki, Takafumi

TRANSFORMER-LESS UNIFIED POWER FLOW CONTROLLER USING THE CASCADE MULTILEVEL INVERTER 1342

Fang Zheng ; Shao Zhang ; Shuitao Yang ; Gunasekaran, Deepak ; Karki, Ujjwal

A NEW POWER FLOW CONTROLLER USING SIX MULTILEVEL CASCADED CONVERTERS FOR DISTRIBUTION SYSTEMS...............1350
Tsuruta, Ryoji ; Hosaka, Tatsuya ; Fujita, Hideaki

A PROPOSAL OF MODULAR MULTILEVEL CONVERTER APPLYING THREE WINDING TRANSFORMER...............1357
Tamada, Shunsuke ; Nakazawa, Yosuke ; Irokawa, Shoichi

BACK-TO-BACK SYSTEM FOR FIVE-LEVEL CONVERTER WITH COMMON FLYING CAPACITORS...............1365
Hasegawa, Isamu ; Urushibata, Shota ; Kondo, Takeshi ; Hirao, Kuniaki ; Kodama, Takashi ; Hui Zhang

HARMONIC MODELING OF A VEHICLE TRACTION CIRCUIT TOWARDS THE DC BUS...............1373
Haghbin, Saeid ; Karvonen, Andreas ; Thiringer, Torbjorn

AC/DC CONVERTER BASED ON INSTANTANEOUS POWER BALANCE CONTROL FOR REDUCING DC-LINK CAPACITANCE...............1379
Tokumasu, Akira ; Taki, Hiroshi ; Shirakawa, Kazuhiro ; Wada, Keiji

MODULAR CONVERTER ARCHITECTURE FOR MEDIUM VOLTAGE ULTRA FAST EV CHARGING STATIONS: DUAL HALF-BRIDGE-BASED ISOLATION STAGE...............1386
Vasiladiotis, Michail ; Bahrani, Behrooz ; Burger, Niklaus ; Rufer, Alfred

NEW INTERLEAVED CURRENT-FED RESONANT CONVERTER WITH SIGNIFICANTLY REDUCED HIGH CURRENT OUTPUT FILTER FOR EV AND HEV APPLICATION...............1394
Moon, Dongok ; Park, Junsung ; Choi, Sewan

15 PHASE INDUCTION MOTOR DRIVE WITH 1:3:5 SPEED RATIOS USING POLE PHASE MODULATION...............1400
Umesh B S ; Sivakumar K

MATHEMATICAL MODEL OF NOVEL WOUND-FIELD SYNCHRONOUS MOTOR SELF-EXCITED BY SPACE HARMONICS...............1405
Aoyama, Masahiro ; Noguchi, Toshihiko

DUAL PURPOSE NO VOLTAGE WINDING DESIGN FOR THE BEARINGLESS AC HOMOPOLAR AND CONSEQUENT POLE MOTORS...............1412
Severson, Eric ; Nilssen, Robert ; Undeland, Tore ; Mohan, Ned

HARVESTING ENERGY FROM SHIP ROLLING USING AN ECCENTRIC DISK REVOLVING IN A HULA-HOOP MOTION...............1420
Yu-Jen Wang

LOAD-INDEPENDENT CURRENT OUTPUT OF INDUCTIVE POWER TRANSFER CONVERTERS WITH OPTIMIZED EFFICIENCY...............1425
Zhang, Wei ; Wong, Siu-Chung ; Tse, Chi K. ; Chen, Qianhong

VOLTAGE CONTROL OF INDUCTIVE CONTACTLESS POWER TRANSFER SYSTEM WITH COAXIAL CORELESS TRANSFORMER FOR DC POWER DISTRIBUTION...............1430
Miiura, Yushi ; Ojika, Satoshi ; Ise, Tomofumi

CONTACTLESS HIGH POWER TRANSFORMER TECHNOLOGIES FOR RAILWAY VEHICLES...............1438
Kondo, Keiichiro ; Yamamoto, Kohei ; Kitazawa, Satochi

TWO-SWITCH VOLTAGE EQUALIZER BASED ON HALF-BRIDGE CONVERTER WITH MULTI-STACKED CURRENT DOUBLERS FOR SERIES-CONNECTED BATTERIES...............1444
Uno, Masatoshi ; Kukita, Akio

OPTIMAL ENERGY STORAGE SYSTEM PLANNING FOR MICROGRIDS WITH CONTRACT CAPACITY CONSTRAINT...............1452
Shu-Hung Liao ; Jen-Hao Teng ; Yung-Ching Huang ; Dong-Jing Lee

OPTIMAL ZERO SEQUENCE INJECTION IN MULTILEVEL CASCADED H-BRIDGE CONVERTER UNDER UNBALANCED PHOTOVOLTAIC POWER GENERATION...............1458
Yu, Yifan ; Konstantinou, Georgios ; Hredzak, Branislav ; Agelidis, Vassilios G.

SIMPLE METHOD FOR MEASURING OUTPUT IMPEDANCE OF A THREE-PHASE INVERTER IN DQ-DOMAIN...............1466
Jokipii, Juha ; Messo, Tuomas ; Suntio, Teuvo

ANALYSIS AND DESIGN OF POWER MANAGEMENT SCHEME FOR AN ON-BOARD SOLAR ENERGY STORAGE SYSTEM...............1471
Jiang, W. ; Yu, F.Y. ; Lin, Z.Y. ; Wu, G.F. ; Chen, H. ; Hashimoto, S

LVRT CONTROL STRATEGY OF CSC-DPMSG-WGS UNDER UNBALANCED GRID FAULTS...............1476
Meiqin Mao ; Yong Ding ; Shiting Weng ; Liuchen Chang

A NEW CURRENT CONTROL DROOP STRATEGY FOR VSI-BASED ISLANDED MICROGRIDS...............1482
Shoeiby, B. ; Davoodnezhad, R. ; Holmes, D.G. ; McGrath, B.P.

POWER EXCHANGE USING PFC FOR MICRO GRID...............1490
Sakai, Tomoyasu ; Takeda, Takashi ; Yukita, Kazuto ; Goto, Yasuyuki ; Ichiyanagi, Katsuhiro ; Morita, Hiroshi

DETERMINATION OF ROTOR TEMPERATURE FOR AN INTERIOR PERMANENT MAGNET SYNCHRONOUS MACHINE USING A PRECISE FLUX OBSERVER...............1501
Specht, Andreas ; Wallscheid, Oliver ; Bocker, Joachim

MONITORING CRITICAL TEMPERATURES IN PERMANENT MAGNET SYNCHRONOUS MOTORS USING LOW-ORDER THERMAL MODELS...............1508
Huber, Tobias ; Peters, Wilhelm ; Bocker, Joachim

ROBUST CURRENT CONTROL INSENSITIVE TO GAIN DEVIATION AND OFFSET OF INVERTER DC-LINK CURRENT SENSOR FOR SPMSM...............1516
Matsuura, Kei ; Ando, Itaru ; Ohishi, Kiyoshi ; Matsuhashi, Masataka

AUTO-TUNING METHOD OF INDUCTANCES FOR PERMANENT MAGNET SYNCHRONOUS MOTORS...............1522
Nomura, Naofumi ; Higuchi, Shinichi

AN IMPEDANCE-BASED STABILITY ANALYSIS METHOD FOR PARALLELED VOLTAGE SOURCE CONVERTERS 1529

Wang, Xiongfei ; Blaabjerg, Frede ; Loh, Poh Chiang

DYNAMIC CHARACTERISTICS AND STABILITY COMPARISONS BETWEEN VIRTUAL SYNCHRONOUS GENERATOR AND DROOP CONTROL IN INVERTER-BASED DISTRIBUTED GENERATORS 1536

Jia Liu ; Miura, Yushi ; Ise, Toshifumi

EMBEDDED LIMITATIONS AND PROTECTIONS FOR DROOP-BASED CONTROL SCHEMES WITH CASCADED LOOPS IN THE SYNCHRONOUS REFERENCE FRAME 1544

D'Arco, Salvatore ; Guidi, Giuseppe ; Suul, Jon Are

VIRTUAL SYNCHRONOUS GENERATOR CONTROL WITH DOUBLE DECOUPLED SYNCHRONOUS REFERENCE FRAME FOR SINGLE-PHASE INVERTER 1552

Hirase, Yuko ; Noro, Osamu ; Yoshimura, Eiji ; Nakagawa, Hidehiko ; Sakimoto, Kenichi ; Shindo, Yuji

CONTACTLESS DC CONNECTOR BASED ON GAN LLC CONVERTER FOR NEXT GENERATION DATA CENTERS 1560

Hayashi, Yusuke ; Toyoda, Hajime ; Ise, Toshifumi ; Matsumoto, Akira

ANALYSIS OF MIS-INTERRUPTION OF SEMICONDUCTOR BREAKER IN DC POWER FEEDING SYSTEM 1567

Murai, Kensuke ; Kanai, Yasuyuki ; Asakimori, Koki ; Babasaki, Tadatoshi

A RELIABLE ELECTRONIC CHOKE WITH NO NEED OF GAIN ADJUSTMENT FOR WIRE COMMUNICATION SYSTEM 1575

Katsuki, Akihiko ; Nakamura, Tatsuya ; Mizuki, Tatsuya ; Shibahara, Kohei ; Abe, Tomohiko ; Ikeda, Tomohiko ; Maeyama, Shigetaka

DESIGN OF NEW CONTROL STRATEGIES FOR A FOUR-LEG THREE-PHASE INVERTER TO ELIMINATE THE NEUTRAL CURRENT UNDER UNBALANCED LOADS 1580

Zhao-Qin Guo ; Panda, Sanjib Kumar ; Prasanna, I.V.

RESEARCH TRENDS OF MODULAR MULTILEVEL CASCADE INVERTER (MMCI-DSCC)-BASED MEDIUM-VOLTAGE MOTOR DRIVES IN A LOW-SPEED RANGE 1586

Okazaki, Yuhei ; Matsui, Hitoshi ; Hagiwara, Makoto ; Akagi, Hirofumi

AN INPUT SWITCHED MULTILEVEL INVERTER FOR OPEN-END WINDING INDUCTION MOTOR DRIVE 1594

Zhu, B. ; Jia, Y. ; Prasanna, U.R. ; Rajashekara, K. ; Kubo, H.

VARIABLE CARRIER FREQUENCY MIXED PWM TECHNIQUE BASED ON CURRENT RIPPLE PREDICTION FOR REDUCED SWITCHING LOSS 1601

Kubo, Hajime ; Yamamoto, Yasuhiro

SLIDING MODE PWM FOR EFFECTIVE CURRENT CONTROL IN SWITCHED RELUCTANCE MACHINE DRIVES 1606

Manolas, Iakovos ; Papafotiou, Georgios ; Manias, Stefanos N.

EXPERIMENTAL VERIFICATION OF AN EMC FILTER USED FOR PWM INVERTER WITH WIDE BAND-GAP DEVICES 1613

Itoh, Jun-ichi ; Araki, Takahiro ; Orikawa, Koji

PACKAGING FOR SIC POWER DEVICE 1621

Funaki, Tsuyoshi

SOLID STATE TRANSFORMER AND MV GRID TIE APPLICATIONS ENABLED BY 15 KV SIC IGBTS AND 10 KV SIC MOSFETS BASED MULTILEVEL CONVERTERS 1626

Madhusoodhanan, Sachin ; Tripathi, Awneesh ; Patel, Dhaval ; Mainali, Krishna ; Kadavelugu, Arun ; Hazra, Samir ; Bhattacharya, Subhashish ; Hatua, Kamalesh

VOLUME 3

GENERALIZED MODULAR MULTILEVEL CONVERTER AND MODULATION 1634

Hui Liu ; Loh, Poh Chiang ; Blaabjerg, Frede

AVERAGE POWER CONTROL OF DC BUS VOLTAGES OF CASCADED H-BRIDGE MULTILEVEL CONVERTERS 1639

Lee, Chia-Tse ; Chen, Hsin-Chih ; Ching-Wei Wang ; Ching-Hsiang Yang ; Cheng, Po-Tai

ANALYSIS AND COMPARISON OF HIGH POWER SEMICONDUCTOR DEVICE LOSSES IN 5MW PMSG MV WIND TURBINES 1646

Kihyun Lee ; Kyungsub Jung ; Seunghoo Song ; Suh, Yongsug ; Changwoo Kim ; Hyoyol Yoo ; Sunsoon Park

APPLICATION OF MODULAR MATRIX CONVERTER TO WIND TURBINE GENERATOR 1654

Inomata, Kentaro ; Hara, Hidenori ; Morimoto, Shinya ; Fujii, Junji ; Takeda, Kotaro ; Yamamoto, Eiji

FREE MOTION MECHANICAL POWER FACTOR; COMPARISON BETWEEN ROBOTS IN DIFFERENT STRUCTURE AND COORDINATE 1660

Mizoguchi, Takahiro ; Nozaki, Takahiro ; Ohnishi, Kouhei

ANALYSIS OF SETTLING BEHAVIOR AND DESIGN OF CASCADED PRECISE POSITIONING CONTROL IN PRESENCE OF NONLINEAR FRICTION 1665

Ruderman, Michael ; Iwasaki, Makoto

FIELD AND BENCH TEST EVALUATION OF RANGE EXTENSION CONTROL SYSTEM FOR ELECTRIC VEHICLES BASED ON FRONT AND REAR DRIVING-BRAKING FORCE DISTRIBUTIONS 1671

Fujimoto, Hiroshi ; Harada, Shingo ; Goto, Yuichi ; Kawano, Daisuke ; Sato, Koji ; Matsuo, Yusuke

VIBRATION SUPPRESSION OF INTEGRATED RESONANT AND TIME DELAY SYSTEM BY REFLECTED WAVE REJECTION 1679

Saito, Eiichi ; Oboe, Roberto ; Katsura, Seiichiro

THRUST CHARACTERISTICS IMPROVEMENT OF A CIRCULAR SHAFT MOTOR FOR DIRECT-DRIVE APPLICATIONS 1685

Omura, Mototsugu ; Shimono, Tomoyuki ; Fujimoto, Yasutaka

DESIGN OF A BEARINGLESS FLUX-SWITCHING SLICE MOTOR 1691

Gruber, Wolfgang ; Radman, Karlo ; Schob, Reto.T.

PROPOSAL OF A PERMANENT MAGNET HYBRID TYPE AXIAL MAGNETICALLY LEVITATED MOTOR 1697

Kurita, Nobuyuki ; Ishikawa, Takeo ; Takada, Hiromu ; Suzuki, Genri

COMPARISON OF HIGH SPEED BEARINGLESS DRIVE TOPOLOGIES WITH COMBINED WINDINGS 1701

Mitterhofer, Hubert ; Mrak, Branimir ; Gruber, Wolfgang

HIGH-SPEED MAGNETICALLY LEVITATED REACTION WHEEL DEMONSTRATOR 1707

Zwyssig, Christof ; Baumgartner, Thomas ; Kolar, Johann W.

STABILIZED SUSPENSION CONTROL CONSIDERING ARMATURE REACTION IN A D-Q AXIS CURRENT CONTROL BEARINGLESS MOTOR 1715

Ooshima, Masahide ; Kumakura, Yoshito

ANALYSIS AND DESIGN OF A HIGH-FREQUENCY ISOLATED DUAL-TANK LCL RESONANT AC-DC CONVERTER 1721

Du, Yimian ; Bhat, Ashoka K.S.

VERIFICATION OF LLC RESONANT CONVERTER APPLIED A CURRENT-BALANCING HIGH-FREQUENCY TRANSFORMER WITH MULTI-OUTPUT WINDINGS 1728

Araki, Jun ; Shinozaki, Ikki ; Funato, Hirohito ; Ogasawara, Satoshi ; Murakami, Daichi ; Hirota, Yukitsugu ; Mihara, Teruyoshi ; Mouri, Masayuki ; Okazaki, Fumihiro

LIGHT-LOAD EFFICIENCY IMPROVEMENT STRATEGY FOR LLC RESONANT CONVERTER UTILIZING A STEP-GAP TRANSFORMER 1734

Huang, Wen-Nan ; Lee, Shiu-Hui ; Chen, Ching-Guo

A NOVEL ACCURATE PRIMARY SIDE CONTROL (PSC) METHOD FOR HALF-BRIDGE (HB) LLC CONVERTER 1738

Jae-Bum Lee ; Kim, Chong-Eun ; Jae-Hyun Kim ; Cheol-O Yeon ; Young-Do Kim ; Moon, Gun-Woo

A SIMPLE CONTROL SCHEME FOR IMPROVING LIGHT-LOAD EFFICIENCY IN A FULL-BRIDGE LLC RESONANT CONVERTER 1743

Kim, Jae-Hyun ; Kim, Chong-Eun ; Lee, Jae-Bum ; Young-Do Kim ; Han-Shin Youn ; Moon, Gun-Woo

POWER CONDITIONER FOR STABILIZING POWER DISTURBANCE CAUSED OF WIND TURBINE GENERATOR SYSTEM 1748

Saga, Yasunao ; Fujii, Kansuke ; Yoda, Kazuyuki

A FRONT-TO-FRONT (FTF) SYSTEM CONSISTING OF MULTIPLE MODULAR MULTILEVEL CASCADE CONVERTERS FOR OFFSHORE WIND FARMS 1761

Sasongko, Firman ; Hagiwara, Makoto ; Akagi, Hirofumi

MODELLING, DESIGN AND CONTROL OF GRID CONNECTED CONVERTER FOR HIGH ALTITUDE WIND POWER APPLICATION 1775

Adhikari, Jeevan ; Rathore, Akshay K. ; Panda, S K

PRACTICAL STUDY OF A HIGH STEP-DOWN CONVERTER 1781

Jinno, Masahito ; Su, Hong-Wei ; Tsai, Jiung-Lin ; Matsuo, Hirofumi

GENERALIZED MODELING AND OPTIMIZATION OF A BIDIRECTIONAL DUAL ACTIVE BRIDGE DC-DC CONVERTER INCLUDING FREQUENCY VARIATION 1788

Jauch, Felix ; Biela, Jurgen

BALANCED DISCHARGING OF POWER BANK WITH BUCK-BOOST BATTERY POWER MODULES 1796

Moo, Chin-Sien ; Wu, Tsung-Hsi ; Hou, Chih-Hao ; Hsieh, Yao-Ching

Y-SOURCE IMPEDANCE-NETWORK-BASED ISOLATED BOOST DC/DC CONVERTER 1801

Siwakoti, Yam P. ; Town, Graham E. ; Loh, Poh Chiang ; Blaabjerg, Frede

MULTI-PHASE DC-DC CONVERTER WITH RIPPLE-LESS OPERATION FOR THERMO-ELECTRIC GENERATOR 1806

Kimura, Noriyuki ; Niijima, Koji ; Morizane, Toshimitsu ; Omori, Hideki

POSITION SENSORLESS START-UP METHOD OF SURFACE PERMANENT MAGNET SYNCHRONOUS MOTOR USING NONLINEAR ROTOR POSITION OBSERVER 1811

Hanamoto, Tsuyoshi ; Yamada, Hiroaki ; Okuyama, Yoshihiro

SENSORLESS CONTROL OF PMSM FOR THE WHOLE SPEED RANGE USING TWO-DEGREE-OF-FREEDOM CURRENT CONTROL AND HF TEST CURRENT INJECTION FOR LOW SPEED RANGE 1816

Seilmeier, Markus ; Piepenbreier, Bernhard

ELLIPSE-TRAJECTORY-ORIENTED VECTOR CONTROL FOR ENERGY EFFICIENT/WIDE-SPEED-RANGE DRIVES OF SENSORLESS PMSM 1824

Shinnaka, Shinji ; Amano, Yuki

DEVELOPMENT OF POSITION SENSORLESS CONTROL FOR PERMANENT-MAGNET SYNCHRONOUS GENERATOR DRIVE 1832

Chang, Yuan-Chih ; Lin, Chia-Yu ; Dai, Wei-Fu ; Wu, Chun-Wei

CONTROL OF A 750KW PERMANENT MAGNET SYNCHRONOUS MOTOR 1837

Liping Zheng ; Dong Le

REGIONAL SMART GRID OF ISLAND IN CHINA WITH MULTIFOLD RENEWABLE ENERGY 1842
Xu Cai ; Zheng Li

STABILIZING SMALL ISLAND POWER SYSTEM WITH RENEWABLES BY USE OF POWER CONDITIONING SYSTEMS - JAPANESE ISLAND SYSTEM CASE - 1849
Baba, Jumpei

POWER ELECTRONICS SOLUTIONS APPLIED TO A VARIETY OF DEMONSTRATIVE MICROGRID PROJECTS 1855
Ueda, Yoshinobu

MOVING TOWARDS THE SMART GRID: THE NORWEGIAN CASE 1861
Fosso, Olav B. ; Molinas, Marta ; Sand, Kjell ; Coldevin, Grete H.

POWER ELECTRONICS TECHNOLOGY IN SMART GRID PROJECTS -APPLICATIONS AND EXPERIENCES- 1868
Kobayashi, Takenori

EV AND HEV MOTOR DEVELOPMENT IN TOSHIBA 1874
Arata, Masanori ; Kurihara, Yoshihiro ; Misu, Daisuke ; Matsubara, Masakatsu

MOTOR STATOR WITH THICK RECTANGULAR WIRE LAP WINDING FOR HEVS 1880
Ishigami, Takashi ; Tanaka, Yuichiro ; Homma, Hiroshi

COMPARISON STUDY OF VARIOUS MOTORS FOR EVS AND THE POTENTIALITY OF A FERRITE MAGNET MOTOR 1886
Matsuhashi, Daiki ; Matsuo, Keisuke ; Okitsu, Takashi ; Ashikaga, Tadashi ; Mizuno, Takayuki

OPTIMAL FIELD EXCITATION CONTROL OF A CLAW POLE MOTOR FOR HYBRID ELECTRIC VEHICLE 1892
Azuma, M. ; Hazeyama, M. ; Morita, M. ; Kuroda, Y. ; Daikoku, A. ; Inoue, M.

A WIDE SPEED RANGE HIGH EFFICIENCY EV DRIVE SYSTEM USING WINDING CHANGEOVER TECHNIQUE AND SIC DEVICES 1898
Takatsuka, Yushi ; Hara, Hidenori ; Yamada, Kenji ; Maemura, Akihiko ; Kume, Tsuneo

PERFORMANCE COMPARISON OF A GAN GIT AND A SI IGBT FOR HIGH-SPEED DRIVE APPLICATIONS 1904
Tuysuz, Arda ; Bosshard, Roman ; Kolar, Johann W.

WIDE-BAND GAP DEVICES IN PV SYSTEMS - OPPORTUNITIES AND CHALLENGES 1912
Sintamarean, C. ; Eni, E. ; Blaabjerg, F. ; Teodorescu, R. ; Wang, H.

POWER ELECTRONICS EQUIPMENTS APPLYING NOVEL SIC POWER SEMICONDUCTOR MODULES 1920
Mino, Kazuaki ; Yamada, Ryuji ; Kimura, Hiroshi ; Matsumoto, Yasushi

EMI PREDICTION METHOD FOR SIC INVERTER BY THE MODELING OF STRUCTURE AND THE ACCURATE MODEL OF POWER DEVICE 1929
Maekawa, Sari ; Tsuda, Junichi ; Kuzumaki, Atsuhiko ; Matsumoto, Shuhei ; Mochikawa, Hiroshi ; Kubota, Hisao

SYSTEM INTEGRATION OF GAN TECHNOLOGY 1935
Ferreira, J.A. ; Popovic, J. ; van Wyk, J.D. ; Pansier, F.

POWER LOSSES OF MULTILEVEL CONVERTERS IN TERMS OF THE NUMBER OF THE OUTPUT VOLTAGE LEVELS 1943
Kashihara, Yugo ; Itoh, Jum-ichi

A LARGE CAPACITY 3-LEVEL IEGT INVERTER 1950
Yoshizawa, Daisuke ; Mukunoki, Makoto ; Omote, Kenichiro ; Hayashi, Makoto ; Isida, Takashi

VIBRATION SUPPRESSING CONTROL METHOD OF ANGULAR TRANSMISSION ERROR OF CYCLOID GEAR FOR INDUSTRIAL ROBOTS 1956
Yoshioka, Takashi ; Hirano, Yosei ; Ohishi, Kiyoshi ; Miyazaki, Toshimasa ; Yokokura, Yuki

AN ADVANCED POSITION CONTROL OF OVERHEAD CRANE BY SWAY SUPPRESSION METHOD EMULATING NATURAL DAMPING 1962
Kurabayashi, Toshiyuki ; Yang Chuan ; Murakami, Toshiyuki

A ROBOTIC CANE FOR WALKING ASSISTANCE 1968
Shimizu, Kyohei ; Smadi, Issam ; Fujimoto, Yasutaka

HAND POSITION ESTIMATION IN BINOCULAR VISUAL SPACE USING LINEAR APPROXIMATION OF KINEMATICS 1974
Komada, Satoshi ; Turpin, Santiago ; Hashimoto, Kento ; Yashiro, Daisuke ; Hirai, Junji

CONTACT STATE RECOGNITION BASED ON HAPTIC SIGNAL PROCESSING FOR ROBOTIC TOOL USE 1978
Matsuzaki, Ryohei ; Okuma, Jun ; Sakaino, Sho ; Tsuji, Toshiaki

RECENT TECHNICAL TRENDS IN MAGNETIC MATERIALS 1984
Wajima, Kiyoshi ; Toda, Hiroaki ; Kosaka, Takashi ; Marukawa, Yasuhiro ; Ishihara, Chio

MULTI-DOMAIN CO-SIMULATION WITH NUMERICALLY IDENTIFIED PMSM INTERWORKING AT HILS FOR ELECTRIC PROPULSION 1990
Park, Gyeong-Jae ; Jung, Hochang ; Kim, Yong-Jae ; Jung, Sang-Yong

RECENT TECHNICAL TRENDS IN PMSM 1997
Morimoto, Shigeo ; Asano, Yoshinari ; Kosaka, Takashi ; Enomoto, Yuji

RECENT TECHNICAL TRENDS IN SRM AND FSM 2004
Kano, Yoshiaki

RECENT TECHNICAL TRENDS IN VARIABLE FLUX MOTORS 2011
Toba, Akio ; Daikoku, Akihiro ; Nishiyama, Noriyoshi ; Yoshikawa, Yuichi ; Kawazoe, Yosuke

A GENERAL DISCRETE TIME MODEL TO EVALUATE ACTIVE DAMPING OF GRID CONVERTERS WITH LCL FILTERS 2019
Parker, S.G. ; McGrath, B.P. ; Holmes, D.G.

ANALYSIS AND REDUCTION OF POWER LOSSES IN PV CONVERTERS FOR GRID CONNECTION TO LOW-VOLTAGE THREE-PHASE THREE-WIRE SYSTEMS 2027
Amma, Ryosuke ; Fujita, Hideaki

DESIGN OF GRID CONNECTED PWM CONVERTERS CONSIDERING TOPOLOGY AND PWM METHODS FOR LOW-VOLTAGE RENEWABLE ENERGY APPLICATIONS 2034
Kantar, Emre ; Hava, Ahmet M.

PERFORMANCE OF DEAD TIME COMPENSATION METHODS IN THREE-PHASE GRID-CONNECTION CONVERTERS 2042
Mannen, Tomoyuki ; Fujita, Hideaki

D-S DIGITAL CONTROL FOR THREE-PHASE BI-DIRECTIONAL INVERTERS 2050
Wu, T.-F. ; Chang, C.-H. ; Lin, L.-C.

EXPECTATIONS OF NEXT-GENERATION POWER DEVICES FOR HOME AND CONSUMER APPLIANCES 2058
Kanouda, Akihiko ; Shoji, Hiroyuki ; Shimada, Takae ; Okubo, Toshikazu

APPLICATION TREND AND FORESIGHT OF SIC POWER DEVICES TO AIR CONDITIONERS 2064
Kamikura, Mamoru ; Murata, Yuichiro ; Kutsuki, Tomohiro ; Saito, Katsuhiko

RECENT TECHNICAL TRENDS AND FUTURE PROSPECTS OF IGBTS AND POWER MOSFETS 2068
Ogura, Tsuneo

RECENT DEVELOPMENT AND FUTURE PROSPECTS OF POWER SIC DEVICES 2074
Nakamura, T. ; Nakano, Y. ; Aketa, M. ; Hanada, T.

RECENT ADVANCES AND FUTURE PROSPECTS ON GAN-BASED POWER DEVICES 2075
Ueda, Tetsuzo

SCALING AND BALANCING OF MULTI-CELL CONVERTERS 2079
Kasper, Matthias ; Bortis, Dominik ; Kolar, Johann W.

HYBRID MODULATED UNIVERSAL SOFT-SWITCHING CURRENT-FED DC/DC CONVERTER FOR WIDE VOLTAGE REGULATION FOR PV/FUEL CELLS/BATTERY APPLICATIONS 2087
Moorthy, Radha Sree Krishna ; Rathore, Akshay Kumar

HIGH EFFICIENCY POWER CONVERTERS FOR BATTERY ENERGY STORAGE SYSTEMS 2095
Kawakami, Noriko ; Iijima, Yukihia ; Li, Haiqing ; Ota, Satoru

IMPLEMENTATION OF BRIDGELESS CUK POWER FACTOR CORRECTOR WITH POSITIVE OUTPUT VOLTAGE 2100
Yang, Hong-Tzer ; Chiang, Hsin-Wei

A NOVEL SYNCHRONOUS RECTIFIER METHOD FOR A LLC RESONANT CONVERTER WITH VOLTAGE-DOUBLER RECTIFIER 2108
Murata, Koji ; Kurokawa, Fujio

LATEST DEVELOPMENTS IN INCREASING THE POWER DENSITY OF TRACTION DRIVES 2113
Bakran, Mark-M. ; Marz, Andreas ; Laska, Bernd ; Krafft, Eberhard ; Korner, Olaf ; Nagel, Andreas

CATENARY AND STORAGE BATTERY HYBRID SYSTEM FOR ELECTRIC RAILCAR SERIES EV-E301 ...2120
Kono, Y. ; Shiraki, N. ; Yokoyama, H. ; Furuta, R.

TECHNOLOGY FOR ENERGY-SAVING RAILWAY OPERATION THROUGH POWER-LIMITING BRAKES—A CASE STUDY AT AN URBAN RAILWAY 2126
Koseki, Takafumi ; Watanabe, Shoichiro ; Hamazaki, Yasuhiro ; Kondo, Keiichiro ; Hasegawa, Tomonori ; Mizuma, Takeshi

AN OVERVIEW ON BRAKING ENERGY REGENERATION TECHNOLOGIES IN CHINESE URBAN RAILWAY TRANSPORTATION 2133
Yang, Zhongping ; Xia, Huan ; Wang, Bin ; Lin, Fei

TRACTION INVERTER THAT APPLIES COMPACT 3.3 KV / 1200 A SIC HYBRID MODULE 2140
Ishikawa, Katsumi ; Yukutake, Seigo ; Kono, Yasuhiko ; Ogawa, Kazutoshi ; Kameshiro, Norifumi

POWER ELECTRONIC-BASED PROTECTION FOR DIRECT-CURRENT POWER DISTRIBUTION IN MICRO-GRIDS 2145
Tseng, K.J. ; Luo, Guomin

A CONCEPT OF HIGH POWER DC/DC CONVERTER WITH DOUBLE LOW POWER OUTPUTS 2152
Hojo, Masahide ; Nishioka, Tomoya ; Yamanaka, Kenji

PERFORMANCE EVALUATION FOR GRID IMPEDANCE BASED ISLANDING DETECTION METHOD 2156
Liu, Ning ; Aljankawey, A.S. ; Diduch, C.P. ; Chang, L. ; Mao, Meiqin ; Yazdkhasti, Pegah ; Su, Jianhui

IDENTIFYING NATURAL DEGRADATION/AGING IN POWER MOSFETS IN A LIVE GRID-TIED PV INVERTER USING SPREAD SPECTRUM TIME DOMAIN REFLECTOMETRY 2161
Li, Qian ; Khan, Faisal H.

CONTROL METHOD FOR INDUCTIVE POWER TRANSFER WITH HIGH PARTIAL-LOAD EFFICIENCY AND RESONANCE TRACKING 2167
Bosshard, R. ; Kolar, J.W. ; Wunsch, B.

STANDARD MODELS FOR SMART GRID SIMULATIONS 2175
Noda, Taku ; Nagashima, Tomohiro ; Sekisue, Takayuki ; Kabasawa, Yuichiro ; Kato, Shinji ; Sekiba, Yoichi ; Tokuda, Hirokazu ; Kounoto, Masaaki

MODEL DEVELOPMENT FOR MOTOR DRIVE SYSTEM SIMULATIONS 2183
Ishikawa, Hiroki ; Abe, Takashi ; Kato, Toshiji ; Kubota, Yutaka ; Shimomura, Junichi ; Kohno, Yusuke ; Ikeda, Masahiro ; Umeda, Nobuhiro ; Kimura, Noriyuki ; Shigematsu, Koichi ; Inoue, Yukinori

PRACTICAL SIMULATION EXAMPLES OF AUTOMOTIVE AND POWER SUPPLY SYSTEMS 2189
Abe, Takashi ; Fukushima, Kentaro ; Sekisue, Takayuki ; Shigematsu, Koichi ; Ichihara, Junichi ; Kato, Toshiji ; Ishikawa, Hiroki ; Kouno, Yusuke ; Konoto, Masaaki ; Saito, Ryoji ; Nishida, Yasuyuki

ADMITTANCE MATRICES OF VOLTAGE SOURCE CONVERTERS FOR DISTRIBUTED GENERATORS 2195
Lian, K.L. ; Huang, T.D.

FPGA-BASED SIMULATION OF POWER ELECTRONICS USING ITERATIVE METHODS 2202
Zhang, Huiguo ; Sun, Jian

GALLIUM ARSENIDE IC TECHNOLOGY FOR POWER SUPPLIES ON CHIP 2208
Pala, Vipindas ; Peng, Han ; Hella, Mona ; Chow, T.Paul

SILICON ON NANOCRYSTALLINE AND MICROCRYSTALLINE DIAMOND STACKING STRUCTURE FOR POWER SUPPLY ON CHIP 2212
Yamada, Takatoshi ; Hasegawa, Masataka

A NOVEL LOAD REGULATION TECHNIQUE FOR POWER-SOC WITH PARALLEL CONNECTED POLS 2216
Abe, Seiya ; Matsumoto, Satoshi ; Hidaka, Akira ; Rikitake, Jungo ; Ninomiya, Tamotsu

MATRIX-POL ARCHITECTURE FOR INTEGRATED POWER SUPPLY 2222
Ishizuka, Yoichi ; Shibahara, Ryota ; Ninomiya, Tamotsu ; Tanaka, Kiminori ; Abe, Seiya

ON-CHIP BUCK CONVERTER WITH SPIRAL FERRITE INDUCTOR AND REDUCING IR DROP IN 3D STACKED INTEGRATION 2228
Fuketa, Hiroshi ; Shinozuka, Yasuhiro ; Ishida, Koichi ; Takamiya, Makoto ; Sakurai, Takayasu

DCM ANALYSIS OF A SINGLE SIC SWITCH BASED ZVZCS TAPPED BOOST CONVERTER 2232
Choi, Bo H. ; Lee, Eun S. ; Kim, Ji H. ; Rim, Chun T.

EFFECT OF INPUT AND OUTPUT TERMINAL SOURCES ON DYNAMIC BEHAVIOR OF SWITCHED-MODE CONVERTERS 2240
Suntio, T. ; Viinamaki, J. ; Jokipii, J. ; Messo, T. ; Sitbon, M. ; Kuperman, A.

A FULLY SOFT-SWITCHED MULTIPHASE DC-DC CONVERTER WITH REDUCED SWITCH COUNT FOR HIGH POWER APPLICATION 2247
Kim, Minjae ; Yang, Daeki ; Choi, Sewan

A STATIC CHARACTERISTIC ANALYSIS OF PROPOSED BI-DIRECTIONAL DUAL ACTIVE BRIDGE DC-DC CONVERTER 2252
Nagata, Shun ; Takasaki, Mika ; Furukawa, Yutaka ; Hirose, Toshiro ; Ishizuka, Yoichi

HYBRID BATTERY CHARGING SYSTEM COMBINING OBC WITH LDC FOR ELECTRIC VEHICLES 2260
Kim, Seonghye ; Kang, Feel-soon

TRANSIENT BEHAVIOR OF THE DUAL ACTIVE BRIDGE CONVERTER IN HIGH EFFICIENT ENERGY CONVERSION SYSTEM 2266
Aoyama, Kohei ; Motoi, Naoki ; Tsuruta, Yukinori ; Kawamura, Atsuo

STATE-OF-CHARGE ESTIMATION FOR LITHIUM-ION BATTERY PACK USING RECONSTRUCTED OPEN-CIRCUIT-VOLTAGE CURVE 2272
Chun, Chang Yoon ; Seo, Gab-Su ; Yoon, Sung Hyun ; Cho, Bo-Hyung

SYSTEM DESIGN OF ELECTRIC ASSISTED BICYCLE USING EDLCS AND WIRELESS CHARGER 2277
Itoh, Jun-ichi ; Noguchi, Kenji ; Orikawa, Koji

STUDY ON LOW-LOSS GATE DRIVE CIRCUIT FOR HIGH EFFICIENCY SERVER POWER SUPPLY USING NORMALLY-OFF SIC-JFET 2285
Katoh, Kaoru ; Ishikawa, Katsumi ; Hatanaka, Ayumu ; Ogawa, Kazutoshi ; Akiyama, Satoru ; Ogawa, Takashi ; Yokoyama, Natsuki ; Maru, Naoki ; Takahashi, Osamu ; Nishisu, Koji

A SHORT CIRCUIT PROTECTION METHOD BASED ON A GATE CHARGE CHARACTERISTIC 2290
Horiguchi, Takeshi ; Kinouchi, Shin-ichi ; Nakayama, Yasushi ; Oi, Takeshi ; Urushibata, Hiroaki ; Okamoto, Shoji ; Tominaga, Shinji ; Akagi, Hirofumi

HIGHLY RELIABLE 1200-V P-TYPE MOSFET FOR LEVEL-SHIFT CIRCUIT USED IN DRIVER IC 2297
Sakurai, Naoki ; Hakutou, Takuma ; Yura, Masashi

A NEW LEVEL UP SHIFTER FOR HVICS WITH HIGH NOISE TOLERANCE 2302
Akahane, Masashi ; Jonishi, Akihiro ; Yamaji, Masaharu ; Kanno, Hiroshi ; Tanaka, Takahide ; Nishio, Haruhiko ; Sumida, Hitoshi

OUTPUT RIPPLE MINIMIZATION OF SINGLE-STAGE POWER FACTOR CORRECTED BI-DIRECTIONAL BUCK AC/DC CONVERTER 2310
Veerasamy, Balaji ; Kitagawa, Wataru ; Takeshita, Takaharu

THREE-PHASE ISOLATED FULL-BRIDGE BOOST PFC WITH FLYBACK PASSIVE AUXILIARY CONVERTER 2318
Meng, Tao ; Yu, Shuai ; Ben, Hongqi ; Wei, Guo ; Sun, Shaohua

CONTROL AND EXPERIMENT OF A MODULAR PUSH-PULL PWM CONVERTER FOR A BATTERY ENERGY STORAGE SYSTEM 2323
Hagiwara, Makoto ; Akagi, Hirofumi

ACTIVE FRONT-END TOPOLOGY FOR 5 LEVEL MEDIUM VOLTAGE DRIVE SYSTEM WITH ISOLATED DC BUS 2330
Oka, Toshiaki ; Kusunoki, Hironobu ; Tsukakoshi, Masahiko ; Kleinecke, John ; Daskalos, Mike

A DUAL ACTIVE BRIDGE DC-DC CONVERTER WITH OPTIMAL DC-LINK VOLTAGE SCALING AND FLYBACK MODE FOR ENHANCED LOW-POWER OPERATION IN HYBRID PV/STORAGE SYSTEMS 2336
Poshtkouhi, Shahab ; Trescases, Olivier

NOVEL MODULAR MULTIPLE-INPUT BIDIRECTIONAL DC-DC POWER CONVERTER (MIPC) 2343
Hintz, Andrew ; Prasanna, Udupi.R. ; Rajashekara, Kaushik

SINGLE-SWITCH PWM CONVERTER INTEGRATING VOLTAGE EQUALIZER FOR PHOTOVOLTAIC MODULES UNDER PARTIAL SHADING..2351
Uno, Masatoshi ; Kukita, Akio

NEW DC RAIL SIDE SOFT-SWITCHING PWM DC-DC CONVERTER WITH VOLTAGE DOUBLER RECTIFIER FOR PV GENERATION INTERFACE..2359
Sayed, Khairy ; Kwon, Soon-Kurl ; Nishida, Katsumi ; Nakaoka, Mutsuo

MODELING METHOD OF STRAY MAGNETIC COUPLINGS IN AN EMC FILTER FOR A SIC SOLAR INVERTER..2366
Masuzawa, Takashi ; Hoene, Eckart ; Hoffmann, Stefan ; Lang, Klaus-Dieter

DC BUS VOLTAGE EMI MITIGATION IN THREE-PHASE ACTIVE RECTIFIERS USING A VIRTUAL NEUTRAL FILTER..2372
Parker, S.G. ; Segaran, D.S. ; Holmes, D.G. ; McGrath, B.P.

EFFECTS OF TRANSFORMER STRUCTURES ON THE NOISE BALANCING AND CANCELLATION MECHANISMS OF SWITCHING POWER CONVERTERS..2380
Hsieh, Hung-I ; Shih, Sheng-Fang

A NOVEL TECHNIQUE FOR REDUCING LEAKAGE CURRENT BY APPLICATION OF ZERO-SEQUENCE VOLTAGE..2385
Ayano, Hideki ; Murakami, Kouhei ; Matsui, Yoshihiro

AC-CHOPPERS USING INSTANTANEOUS VOLTAGE CONTROL TECHNIQUE TO SOLVE VOLTAGE SAG PROBLEMS..2392
Khomfoi, Surin

VOLTAGE REGULATION IN DISTRIBUTION SYSTEM USING THE COMBINED DVR..................2400
Nakamura, Sota ; Aoki, Mutsumi ; Ukai, Hiroyuki

NONLINEAR CONTROL OF THREE-PHASE FOUR-WIRE DYNAMIC VOLTAGE RESTORERS FOR DISTRIBUTION SYSTEM..2406
Jeong, Seon-Yeong ; Nguyen, Thanh Hai ; Lee, Dong-Choon ; Kim, Jang-Mok

VOLUME 4

DISTURBANCE CALCULATION BASED ON SPACE VECTOR DOT PRODUCT: APPLICATIONS TO COMPENSATORS..2413
de Carvalho, Kelly Caroline Mingorancia ; Ama, Naji Rajai Nasri ; Komatsu, Wilson ; Martinz, Fernando Ortiz ; Figueredo, Ricardo Souza ; Matakas, Lourenco

PROPOSAL OF 6TH RADIAL FORCE CONTROL BASED ON FLUX LINKAGE..................................2421
Kanematsu, Masato ; Miyajima, Takayuki ; Fujimoto, Hiroshi ; Hori, Yoichi ; Enomoto, Toshio ; Kondou, Masahiko ; Komiya, Hiroshi ; Yoshimoto, Kantaro ; Miyakawa, Takayuki

AIR GAP CONTROL OF MULTI-PHASE TRANSVERSE FLUX PERMANENT MAGNET LINEAR SYNCHRONOUS MOTOR BY USING INDEPENDENT VECTOR CONTROL......................................2427
Hwang, Seon-Hwan ; Bang, Deok-Je ; Kim, Ji-Won

MODIFIED DIRECT INSTANTANEOUS TORQUE CONTROL OF SWITCHED RELUCTANCE MOTOR WITH HIGH TORQUE PER AMPERE AND REDUCED SOURCE CURRENT RIPPLE..................2433
Suryadevara, Rohit ; Fernandes, B.G.

CONTROL OF WOUND FIELD SYNCHRONOUS MOTOR INTEGRATED WITH ZSI..................2438
Tajima, G. ; Kosaka, T. ; Matsui, N. ; Tonogi, K. ; Minoshima, N. ; Yoshida, T.

A NOVEL IPMSM MODEL FOR ROBUST POSITION SENSORLESS CONTROL TO MAGNETIC SATURATION..2445
Matsumoto, Atsushi ; Hasegawa, Masaru ; Doki, Shinji

MOTOR DRIVE SYSTEM USING NONLINEAR MATHEMATICAL MODEL FOR PERMANENT MAGNET SYNCHRONOUS MOTORS..2451
Iwaji, Yoshitaka ; Nakatsugawa, Junnosuke ; Sakai, Toshifumi ; Aoyagi, Shigehisa ; Nagura, Hirokazu

SENSORLESS-ORIENTED DESIGN OF IPMSM..2457
Kano, Yoshiaki

NOISE REDUCTION METHOD BY INJECTED FREQUENCY CONTROL FOR POSITION SENSORLESS CONTROL OF PERMANENT MAGNET SYNCHRONOUS MOTOR..2465
Taniguchi, Shun ; Yasui, Kazuya ; Yuki, Kazuaki

FORCE SENSORLESS BILATERAL CONTROL USING A DYNAMICAL ASYMMETRIC COMPENSATOR..................2470
Hama, Ryota ; Imai, Jun ; Takahashi, Akiko ; Funabiki, Shigeyuki

DESIGN OF M-IPD CONTROLLER OF MULTI-INERTIA SYSTEM USING DIFFERENTIAL EVOLUTION..................2476
Ikeda, Hidehiro ; Tsuyoshi, Hanamoto

A GUIDE TO DESIGN DISTURBANCE OBSERVER BASED MOTION CONTROL SYSTEMS..................2483
Sariyildiz, Emre ; Ohnishi, Kouhei

IDENTIFICATION OF TWO-MASS MECHANICAL SYSTEMS USING TORQUE EXCITATION: DESIGN AND EXPERIMENTAL EVALUATION..2489
Saarakkala, Seppo E. ; Hinkkanen, Marko

INDUCTOR LOSS CALCULATION OF COUPLED INDUCTORS FOR HIGH POWER DENSITY BOOST CONVERTER..2497
Itoh, Yuki ; Kimura, Shota ; Imaoka, Jun ; Yamamoto, Masayoshi

1.2KW DUAL-ACTIVE BRIDGE CONVERTER USING SIC POWER MOSFETS AND PLANAR MAGNETICS..................2503
De, D. ; Castellazzi, A. ; Lamantia, A.

ANALYSIS OF HYSTERESIS AND EDDY-CURRENT LOSSES FOR A MEDIUM-FREQUENCY TRANSFORMER IN AN ISOLATED DC-DC CONVERTER 2511
Nakahara, Mizuki ; Wada, Keiji

EXPERIMENTAL VERIFICATION OF CAPACITIVE POWER TRANSFER USING ONE PULSE SWITCHING ACTIVE CAPACITOR FOR PRACTICAL USE 2517
Kitabayashi, Tatsuaki ; Funato, Hirohito ; Kobayashi, Hiroya ; Yamaichi, Katsuya

A SINGLE-STAGE HIGH-PF DRIVER FOR SUPPLYING A T8-TYPE LED LAMP 2523
Cheng, Chun-An ; Chang, Chien-Hsuan ; Cheng, Hung-Liang ; Chung, Tsung-Yuan

ELIMINATION OF ELECTROLYTIC CAPACITOR IN AC-DC SYSTEM OF LED DRIVER 2529
Mustapa, Rijalul Fahmi ; Hidayat, Nabil M ; Tukiman, Rahayu

A NOVEL BRIDGELESS BOOST HALF-BRIDGE ZVS-PWM SINGLE-STAGE UTILITY FREQUENCY AC-HIGH FREQUENCY AC RESONANT CONVERTER FOR DOMESTIC INDUCTION HEATERS 2533
Mishima, Tomoakzu ; Nakagawa, Yuki ; Nakaoka, Mutsuo

APPLICATION OF VIRTUAL VALIDATION SYSTEM FOR INVERTER HEAT PUMP SYSTEM 2541
Kanamori, Masaki ; Noda, Koji ; Endo, Takahisa ; Suzuki, Nobuyuki

TEST SETUP FOR ACCELERATED TEST OF HIGH POWER IGBT MODULES WITH ONLINE MONITORING OF VCE AND VF VOLTAGE DURING CONVERTER OPERATION 2547
de Vega, Angel Ruiz ; Ghimire, Pramod ; Pedersen, Kristian Bonderup ; Trintis, Ionut ; Beczckowski, Szymon ; Munk-Nielsen, Stig ; Rannestad, Bjorn ; Thogersen, Paul

DESIGN OF HIGH-SPEED IGBT-BASED SWITCHING MODULES FOR PULSED POWER APPLICATIONS 2554
Kluge, Andreas ; Goehler, Lutz ; Gueldner, Henry ; Trompa, Thomas ; Mory, David ; Segsa, Karl-Heinz

COMPARATIVE SUITABILITY EVALUATION OF REVERSE-BLOCKING IGBTS FOR CURRENT-SOURCE BASED CONVERTER 2562
De, Ankan ; Roy, Sudhin ; Bhattacharya, Subhashish

NEW REVERSE-CONDUCTING IGBT (1200V) WITH REVOLUTIONARY COMPACT PACKAGE 2569
Takahashi, K. ; Yoshida, S. ; Noguchi, S. ; Kuribayashi, H. ; Nashida, N. ; Kobayashi, Y. ; Kobayashi, H. ; Mochizuki, K. ; Ikeda, Y. ; Ikawa, O.

AN IMPROVED MODULATED CARRIER CONTROL OF SINGLE-PHASE CCM BOOST PFC CONVERTER 2575
Kim, Hyejin ; Cho, Bo-Hyung ; Choi, Hangseok

MODIFIED INTERLEAVED CURRENT SENSORLESS CONTROL FOR THREE-LEVEL BOOST PFC CONVERTER WITH ASYMMETRIC LOADS 2580
Chen, Hung-Chi ; Liao, Jhen-Yu

A NOVEL CRITICAL-CONDUCTION-MODE BRIDGELESS INTERLEAVED BOOST PFC RECTIFIER 2587
Cao, Guoen ; Kim, Hee-Jun

ANALYSIS AND DESIGN OF A PUSH-PULL SINGLE-STAGE FLYBACK POWER FACTOR CORRECTOR 2593
Lo, Yu-Kang ; Chiu, Huang-Jen ; Liu, Yu-Chen ; Lin, Chung-Yi ; Cheng, Shih-Jen ; Yang, CS

LINEAR OVER-MODULATION STRATEGY FOR CURRENT CONTROL IN PHOTOVOLTAIC INVERTER 2598
Park, Yongsoon ; Sul, Seung-Ki ; Hong, Ki-Nam

DESIGN OF DECENTRALIZED VOLTAGE CONTROL FOR PV INVERTERS TO MITIGATE VOLTAGE RISE IN DISTRIBUTION POWER SYSTEM WITHOUT COMMUNICATION 2606
Lee, Tzung-Lin ; Yang, Shih-Sian ; Hu, Shang-Hung

STABILITY ANALYSIS AND ACTIVE DAMPING FOR LLCL-FILTER BASED GRID-CONNECTED INVERTERS 2610
Huang, Min ; Blaabjerg, Frede ; Loh, Poh Chiang ; Wu, Weimin

INTEGRATED COMMON AND DIFFERENTIAL MODE FILTER APPLIED TO A SINGLE-PHASE TRANSFORMERLESS PV MICROINVERTER WITH LOW LEAKAGE CURRENT 2618
Figueredo, Ricardo Souza ; de Carvalho, Kelly Caroline Mingorancia ; Matakas, Lourenco

DESIGN AND INTEGRATION OF INTERPHASE INDUCTORS FOR INTERLEAVED THREE PHASE VOLTAGE-SOURCE-INVERTERS IN DC-FED MOTOR DRIVE SYSTEMS 2626
Zhang, Xuning ; Boroyevich, Dushan ; Burgos, Rolando

A NOVEL TRANSFORMER MODEL USING MAGNETIC CIRCUIT 2632
Nakamurame, Fuminori ; Ise, Toshifumi

HARDWARE-IN-THE-LOOP SIMULATION OF A MACHINE MODEL WITH REAL-TIME ANIMATION 2638
Xiaojie Zhuang ; Hibino, Shinya ; Harakawa, Masaya ; Terabe, Ryosuke ; Ozaki, Takayuki ; Nagano, Tetsuaki

DEVELOPMENT OF REAL TIME DIGITAL SIMULATOR FOR SELF-COMMUTATED SVC TO SUPPRESS VOLTAGE FLICKER 2644
Terao, Yutaka ; Shishida, Yasuhiro ; Tsuruma, Yoshinori ; Ishizuka, Tomotsugu ; Aoyama, Fumio ; Yoshino, Teruo ; Kato, Yutaka ; Belanger, Jean

OPERATIONAL ASPECTS AND POWER ARCHITECTURE DESIGN FOR A MICROGRID TO INCREASE THE USE OF RENEWABLE ENERGY IN WIRELESS COMMUNICATION NETWORKS 2649
Kwasinski, Alexis ; Kwasinski, Andres

P+ MULTIPLE RESONANT CONTROL FOR OUTPUT VOLTAGE REGULATION OF MICROGRID WITH UNBALANCED AND NONLINEAR LOADS 2656
Kyungbae Lim ; Jaeho Choi ; Juyoung Jang ; Junghum Lee ; Jaesig Kim

130MVA-STATCOM FOR TRANSIENT STABILITY IMPROVEMENT 2663
Imanishi, Takao ; Nagatomo, Yoshinobu ; Iwasaki, Shinya ; Masaki, Kenji ; Fujii, Toshiyuki ; Ieda, Jun

IMPROVED DROOP CONTROLLER FOR MICROGRID INVERTER CONSIDERING THE LINE IMPEDANCE MISMATCHING 2668
Du Yan ; Liuchen Chang ; Meiqin Mao ; Jianhui Su ; Ning Liu

SUPPRESSION CONTROL METHOD FOR IRON LOSS OF MATRIX MOTOR UNDER FLUX WEAKENING UTILIZING INDIVIDUAL WINDING CURRENT CONTROL 2673

Hijikata, Hiroki ; Akatsu, Kan ; Miyama, Yoshihiro ; Arita, Hideaki ; Daikoku, Akihiro

PERFORMANCE ANALYSIS OF A NEW CONCENTRATEDWINDING INTERIOR PERMANENT MAGNET SYNCHRONOUS MACHINE UNDER FIELD ORIENTED CONTROL 2679

Nguyen, D. ; Dutta, R. ; Fletcher, J. ; Rahman, F. ; Lovatt, Howard

ONLINE PARTICLE SWARM OPTIMIZATION FOR SENSORLESS IPMSM DRIVES CONSIDERING PARAMETER VARIATION 2686

Song, Z.Q. ; Xiao, D. ; Rahman, M.F.

A DTC-PWM CONTROL SCHEME OF PMSM BASED ON 12-SECTORS DIVISION AND SPEED INFORMATION 2693

Yunchang Kwak ; Jin-Woo Ahn ; Dong-Hee Lee

CONTROL OF POWER FLOW BETWEEN THE WIND GENERATOR AND NETWORK 2700

Stumpf, Peter ; Nagy, Istvan ; Vajk, Istvan

ADVANCES IN NANOGRID TECHNOLOGY AND ITS INTEGRATION INTO RURAL ELECTRIFICATION IN INDIA 2707

Mishra, Santanu ; Ray, Olive

STUDY AND IMPLEMENTATION OF SEVEN-LEVEL INVERTER USING COUPLED INDUCTOR AND SWITCHED-CAPACITOR 2714

Yi-Chun Lin ; Jiann-Fuh Chen ; Wen-Chien Hsu ; Sheng-Kai Kao

CASCADED MULTILEVEL CONVERTER BASED BIDIRECTIONAL INDUCTIVE POWER TRANSFER (BIPT) SYSTEM 2722

Bac Xuan Nguyen ; Vilathgamuwa, D.M. ; Foo, Gilbert ; Ong, Andrew ; Sampath, Prasad K. ; Madawala, Udaya K.

UNDERSAMPLING CONTROL OF A BIDIRECTIONAL CASCADED BUCK+BOOST DC-DC CONVERTER 2729

Rosekeit, Martin ; Joebges, Philipp ; Lelie, Markus ; Sauer, Dirk Uwe ; De Doncker, Rik W.

SUB-MICROSECOND RESPONSE DIGITAL CONTROLLER FOR POL 2737

Nonaka, Hirotaka ; Ishizuka, Yoichi ; Mii, Kenji ; Takenami, Fumiaki ; Kanemoto, Daisuke

GAIN CONTROLLED HIGH EFFICIENCY POWER FACTOR CORRECTION CIRCUIT 2745

Yonezawa, Yu ; Nakao, Hiroshi ; Sasaki, Tomotake ; Matsui, Yoshinobu ; Nakashima, Yoshiyasu ; Kaneko, Junji ; Shimamori, Hiroshi ; Yoshino, Yukio ; Hisato, Hosoyama ; Atsushi, Manabe ; Motizuki, Shun ; Yamashita, Shigeharu

DESIGN OF QUASI-RESONANT FLYBACK CONVERTER CONTROL IC WITH DCM AND CCM OPERATION 2750

Kai-Hui Chen ; Tsorng-Juu Liang

LOAD TRANSIENT RESPONSE IMPROVEMENT BASED ON PID CONTROL 2754

Yau, Y.T. ; Hwu, K.I.

AN ACTIVE-CLAMPING FORWARD CONVERTER WITH NON-LINEAR STEP-DOWN CONVERSION 2758

Jing-Yuan Lin ; Yu-Kang Lo ; Huang-Jen Chiu ; Chao-Fu Wang ; Chien-Yu Lin

SWITCHING LOSS MINIMIZATION OF 3-PHASE INTERLEAVED BIDIRECTIONAL DC-DC CONVERTER 2763

Eui-Cheol Nho ; Jae-Hun Jung ; Hak-Soo Kim ; In-Dong Kim ; Heung-Geun Kim ; Tae-Won Chun

MODIFIED THREE-PHASE THREE-LEVEL DC-DC CONVERTER -ADOPTING ASYMMETRICAL DUTY CYCLE CONTROL 2768

Yue Chen ; Xuling Chen ; Liu, Fuxin ; Ruan, Xinbo

DEADBEAT CONTROL OF POWER LEVELING UNIT WITH BIDIRECTIONAL BUCK/BOOST DC/DC CONVERTER 2775

Hamasaki, Shin-ichi ; Mukai, Ryosuke ; Yano, Yoshihiro ; Tsuji, Mineo

DESIGN OF OPTIMIZED ON-OFF CONTROL TO IMPROVE EFFICIENCY OF PARALLELED CONVERTER SYSTEM 2781

Kohama, Teruhiko ; Sogawa, Yuki ; Tsuji, Satoshi

EFFICIENCY IMPROVEMENTS IN A SINGLE ACTIVE BRIDGE MODULAR DC-DC CONVERTER WITH SNUBBER CAPACITANCE OPTIMISATION 2787

Ting, Yeh ; de Haan, Sjoerd ; Ferreira, Jan A.

A WIRELESS POWER TRANSFER SYSTEM OPTIMIZED FOR HIGH EFFICIENCY AND HIGH POWER APPLICATIONS 2794

Bani Shamseh, Mohammad ; Kawamura, Atsuo ; Yuzurihara, Itsuo ; Takayanagi, Atsushi

NON-ITERATIVE LCL FILTER DESIGN FOR THREE-PHASE TWO-LEVEL VOLTAGE-SOURCE PWM CONVERTERS 2802

Byung-Geuk Cho ; Seung-Ki Sul

DSP-BASED INTERLEAVED BUCK POWER FACTOR CORRECTOR 2810

Yu-Chen Liu ; Tsan Chen ; Po-Jung Tseng ; Yu-Kang Lo ; Huang-Jen Chiu

THE AVERAGE MODEL OF A THREE-PHASE THREE-STAGE POWER ELECTRONIC TRANSFORMER 2815

Shaodi Ouyang ; Liu, Jinjun ; Wang, Xinyu ; Wang, Xiaojian ; Fei Meng ; Riffat, Javid

A MULTI-CARRIER PWM FOR AC-DC-AC CONVERTER WITHOUT DC LINK ELECTROLYTIC CAPACITOR 2821

Chung-Chuan Hou ; Hsin-Ping Su

A DECOUPLING OFFSET-BASED PWM CONTROL FOR A MULTILEVEL INVERTER UNDER DC VOLTAGE UNBALANCE 2826

Nho Van Nguyen ; Tam Khanh Tu Nguyen ; Lee, Hong-Hee

?-? PARETO OPTIMIZATION OF 3-PHASE 3-LEVEL T-TYPE AC-DC-AC CONVERTER COMPRISING SI AND SIC HYBRID POWER STAGE..2834

Uemura, Hirofumi ; Krismer, Florian ; Okuma, Yasuhiro ; Kolar, Johann W.

PRACTICAL INVESTIGATION OF THE GATE BIAS EFFECT ON THE REVERSE RECOVERY BEHAVIOR OF THE BODY DIODE IN POWER MOSFETS..2842

Lindberg-Poulsen, Kristian ; Petersen, Lars Press ; Ouyang, Ziwei ; Andersen, Michael A.E.

AN ONLINE VCE MEASUREMENT AND TEMPERATURE ESTIMATION METHOD FOR HIGH POWER IGBT MODULE IN NORMAL PWM OPERATION..2850

Ghimire, Pramod ; de Vega, Angel Ruiz ; Beczkowski, Szymon ; Munk-Nielsen, Stig ; Rannested, Bjorn ; Thogersen, Paul Bach

EVALUATION ON IRON LOSS CHARACTERISTICS IN SERIES CONNECTION AND PARALLEL CONNECTION OF LOADS WITH INVERTER EXCITATION..2856

Odawara, Shunya ; Fujisaki, Keisuke

LOSS AND THERMAL MODEL FOR POWER SEMICONDUCTORS INCLUDING DEVICE RATING INFORMATION..2862

Ma, K. ; Bahman, A.S. ; Beczkowski, S.M. ; Blaabjerg, F.

IMPROVING RELIABILITY OF IGBT SURFACE ELECTRODE FOR 200 C OPERATION................2870

Nishimura, Tomohiro ; Ikeda, Yoshinari ; Hokazono, Hiroaki ; Mochizuki, Eiji ; Takahashi, Yoshikazu

INFLUENCE OF CARRIER FREQUENCY ON IRON LOSS TAKING ACCOUNT OF DEAD TIME EFFECT................2874

Kogi, Ryosuke ; Odawara, Shunya ; Fujisaki, Keisuke

DECREASE OF SIC-BJT DRIVER LOSSES BY ONE-STEP COMMUTATION................................2881

Barth, Henry ; Hofmann, Wilfried

POWER PROFILE BASED SELECTION AND OPERATION OPTIMIZATION OF PARALLEL-CONNECTED POWER CONVERTER COMBINATIONS..2887

Vogt, T. ; Peters, A. ; Frohleke, N. ; Bocker, J. ; Kempen, S.

A NOVEL POWER LOSS CALCULATION METHOD FOR IGBTS IN POWER CONVERTERS VIA CHAOTIC SPWM CONTROL..2893

Boyu Wang ; Li, Hong ; Xiaojie You ; Trillion Zheng

LOSS ANALYSIS AND SOFT-SWITCHING CHARACTERISTICS OF FLYBACK-FORWARD HIGH GAIN DC/DC CONVERTER WITH GAN FET..2899

Zhang Yajing ; Zheng, Trillion Q. ; Li Yan

INSULATED METAL SUBSTRATE FOR POWER MODULES USING ANODIC OXIDE FILM OF ALUMINUM..2904

Tokuyama, Takeshi ; Kusukawa, Jumpei ; Nakatsu, Kinya

A FAST-TRANSIENT-RESPONSE BUCK CONVERTER WITH SPLIT-TYPE III COMPENSATION AND CHARGE-PUMP CIRCUIT TECHNIQUE..2910

Chen, Jiann-Jong ; Wei-Ting Hsu ; Jih-Hua Yu ; Hwang, Yuh-Shyan ; Cheng-Chieh Yu

ADVANTAGES OF LOW PARASITIC INDUCTANCE PACKAGES OF POWER MOSFET FOR SERVER POWER APPLICATIONS..2914

Wonsuk Choi ; Dongkook Son ; Dongwook Kim

MODULAR INTEGRATION OF A MATRIX CONVERTER..2920

Solomon, Adane Kassa ; Skuriat, Robert ; Castellazzi, Alberto ; Wheeler, Pat

A MODULAR NANOSECOND PULSE GENERATION SYSTEM FOR PLASMA-ASSISTED IGNITION................2926

Peng Gao ; Fletcher, John ; O'Byrne, Sean

DEVELOPMENT OF A SINGLE SWITCH CELL FOR MODULAR NANOSECOND PULSE GENERATION SYSTEMS..2932

Peng Gao ; Fletcher, John ; O'Byrne, Sean

ADVANTAGE OF SUPER JUNCTION MOSFET FOR POWER SUPPLY APPLICATION................2939

Tabira, K. ; Watanabe, S. ; Shimatou, T. ; Watashima, T. ; Takenoiri, S.

STUDY ON AN ACCURATE CALCULATION OF THE CONDUCTED EMI NOISE OF THE POWER CONVERTERS..2944

Omata, Shinpei ; Shimizu, Toshihisa

AN EXACT DISCRETE-TIME MODEL CONSIDERING DEAD-TIME NONLINEARITY FOR AN H-BRIDGE GRID-CONNECTED INVERTER..2950

Xie, Ruiliang ; Hao, Xiang ; Yang, Xu ; Chen, Wenjie ; Huang, Lang ; Chao Wang

THEORETICAL ANALYSIS OF THE DUALITY PRINCIPLE APPLIED TO INTERLEAVED TOPOLOGIES................2954

Caris, M.L.A. ; Huisman, H. ; Duarte, J.L.

A NEW IMPEDANCE MEASUREMENT METHOD BASED ON HIGH FREQUENCY COMPENSATION................2960

Yue, Xiaolong ; Zhuo, Fang ; Hao Yi

NUMERICAL AND EXPERIMENTAL INVESTIGATION OF PARASITIC EDGE CAPACITANCE FOR PHOTOVOLTAIC PANEL..2967

Wenjie Chen ; Xiaomei Song ; Hao Huang ; Xu Yang

VEHICLE INTERIOR NOISE CONTROL OF ULTRA-COMPACT ELECTRIC VEHICLE (FUNDAMENTAL CONSIDERATION USING RECTANGULAR ENCLOSURE)..2972

Kato, Taro ; Kato, Hideaki ; Oshinoya, Yasuo ; Suzuki, Ryosuke ; Hasegawa, Shinya

CONSIDERATION FOR THE PROPAGATION PATH OF CONDUCTIVE NOISE IN AIR CONDITIONERS................2977

Tokiwa, Tsuyoshi ; Kanamori, Masaki ; Endo, Takahisa ; Iida, Mikiya ; Ogasawara, Satoshi ; Yizhanyi Tang

IRON LOSS EVALUATION OF IRON POWDER CORE SUITABLE FOR INDUCTOR USED IN POWER CONVERTERS..2983

Mori, Tomohiro ; Igarashi, Kazunori ; Kanagawa, Kinji ; Yamashita, Nobuyuki ; Shimizu, Toshihisa ; Bizen, Yosio

OPTIMIZED TUNING METHOD OF STATIONARY FRAME PROPORTIONAL RESONANT CURRENT CONTROLLERS 2988

Martinz, Fernando Ortiz ; de Carvalho, Kelly Caroline Mingorancia ; Ama, Naji Rajai Nasri ; Komatsu, Wilson ; Matakas, Lourenco

INSTANTANEOUS POWER THEORY APPLIED TO POWER CONDITIONING UNDER DISTORTED MAINS VOLTAGES: A MATLAB/SIMULINK APPROACH 2996

Nicolae, Petre-Marian ; Popa, Lucian-Dinut ; Nicolae, Marian-Stefan ; Nicolae, Ileana-Diana

THE RESEARCH ON RELIABILITY AND REAL-TIME OF THE SCHEME OF PROCESS LAYER GOOSE NETWORK IN SMART SUBSTATION BASED ON ARTIFICIAL COBWEB TOPOLOGY STRUCTURE 3002

Liu, Xiaosheng ; Zhu, Honglin ; Xu, Dianguo ; Li, Yanxiang

EFFICIENCY IMPROVEMENT OF A SELF-START TYPE PERMANENT MAGNET SYNCHRONOUS MOTOR 3007

Saikusa, H. ; Arikawa, S. ; Higuchi, T. ; Yokoi, Y. ; Abe, T.

CONSIDERATION OF OPTIMAL NUMBER OF POLES AND FREQUENCY FOR HIGH-EFFICIENCY PERMANENT MAGNET MOTOR 3012

Misu, Daisuke ; Matsushita, Makoto ; Takeuchi, Katsutoku ; Oishi, Koji ; Kawamura, Mitsuhiro

BASIC STUDY ON THE SUITABLE STRUCTURE OF A PERMANENT MAGNET SYNCHRONOUS MOTOR WITH A POWDER MAGNETIC CORE 3018

Hashimoto, Shizuka ; Sanada, Masayuki ; Morimoto, Shigeo ; Inoue, Yukinori

CHARACTERISTICS OF A HALF-WAVE RECTIFIED BRUSHLESS SYNCHRONOUS GENERATOR 3024

Hirakawa, Yuki ; Higuchi, Tsuyoshi ; Yokoi, Yuichi ; Abe, Takashi

MODELING OF WOUND ROTOR SYNCHRONOUS MACHINES CONSIDERING HARMONICS, GEOMETRIC SALIENCIES AND SATURATION INDUCED SALIENCIES 3029

Rambetius, Alexander ; Luthardt, Sven ; Piepenbreier, Bernhard

DESIGN AND COMPARISON OF HIGH FREQUENCY TRANSFORMERS USING FOIL AND ROUND WINDINGS 3037

Iyer, Kartik V ; Robbins, William P ; Mohan, Ned

A METHOD TO CALCULATE THE PERFORMANCE OF LINEAR INDUCTION MOTORS USING SIMPLE TWO-PHASE MODEL 3044

Hirahara, Hideaki ; Yamamoto, Shu ; Ara, Takahiro ; Shimizu, Toshihisa

AN ESP DOWNHOLE PARAMETERS MONITORING SYSTEM BASED ON CURRENT LOOP TRANSMISSION METHOD 3050

Jin Miaoxin ; Zhang Wei ; Gao Qiang ; Xu Dianguo

BENDING MAGNETIC LEVITATION CONTROL FOR THIN STEEL PLATE (EXPERIMENTAL CONSIDERATION USING SLIDING MODE CONTROL) 3055

Yonezawa, Hikaru ; Narita, Takayoshi ; Oshinoya, Yasuo ; Marumori, Hiroki ; Hasegawa, Shinya

TRANSFORMER WINDING LOSSES WITH ROUND CONDUCTORS FOR DUTY-CYCLE REGULATED SQUARE WAVES 3061

Iyer, Kartik V ; Robbins, William P ; Basu, Kaushik ; Mohan, Ned

SIMULATION OF RESIN MOLDED TYPE SENSOR IN POLE SWITCH FOR POWER DELIBERY SYSTEMS 3067

Furukawa, Tatsuya ; Muta, Shoichiro ; Fukumoto, Hisao ; Itoh, Hideaki ; Ohchi, Masashi

ROBUST STARTUP CONTROL OF SENSORLESS PMSM DRIVES WITH SELF-COMMISSIONING 3072

Lin, Chiao-Chien ; Tzou, Ying-Yu

POSITION SENSORLESS CONTROL OF PMSM WITH A LOW-FREQUENCY SIGNAL INJECTION 3079

Nimura, Tomohiro ; Doki, Shinji ; Fujitsuna, Masami

A COMPARISON OF DIFFERENT SENSORLESS POSITION ACQUISITION METHODS AT LOW SPEEDS FOR A PERMANENT MAGNET SYNCHRONOUS MACHINE IN VEHICLE APPLICATIONS 3085

Lehmann, Oliver ; Zehelein, Matthias ; Schuster, Johannes ; Roth-Stielow, Jorg

STABILITY COMPARISON OF IPMSM SENSORLESS VECTOR CONTROL SYSTEMS USING EXTENDED EMF 3093

Tsuji, Mineo ; Mizusaki, Hiroshi ; Hamasaki, Sin-ichi

INDUCTION MACHINE BASED FLYWHEEL SPEED ESTIMATION AT STAND-BY MODE 3099

Liu, Rongqiang ; Xu, David

SYMMETRICAL SIGNALING SYSTEM FOR SENSOR-LESS SRM DRIVE 3106

Yamamoto, Kenji ; Takahashi, Hisashi ; Ushiro, Nobumasa ; Shirasawa, Koki

DIGITAL INTEGRATORS FOR CONDITION MONITORING: A DC AND MULTITONE SIGNAL ANALYSIS 3111

Peretti, L.

AUDIBLE NOISE REDUCTION METHOD IN IPMSM POSITION SENSORLESS CONTROL BASED ON HIGH-FREQUENCY CURRENT INJECTION 3119

Tauchi, Yuki ; Kubota, Hisao

A NOVEL DESIGN FOR INDUCTION MOTOR FLUX ESTIMATION USING IMPULSIVE OBSERVER 3124

Peng Wang ; Yan Li ; Jianwen Zhang ; Xu Cai ; Zhengzhi Han

LOAD TORQUE AND INERTIA SIMULATION BASED ON DOUBLE-STATOR PERMANENT-MAGNET SYNCHRONOUS MOTOR 3129

Zhe Wang ; Mingyan Wang ; Ben Guo ; Chai Feng

INDEPENDENT SPEED AND POSITION CONTROL OF TWO PERMANENT MAGNET SYNCHRONOUS MOTORS FED BY A FOUR-LEG INVERTER 3134

Kubo, Yuji ; Moroi, Takayuki ; Kouki, Matsuse ; Kubota, Hisao ; Rajashekara, Kaushik

MINIMIZATION OF STATOR CURRENTS FOR MONO INVERTER DUAL PARALLEL PMSM DRIVE SYSTEM 3140
Yongjae Lee ; Ha, Jung-Ik

PERFORMANCE COMPARISON OF INVERTER AND DRIVE CONFIGURATIONS WITH OPEN-END AND STAR-CONNECTED WINDINGS 3145
Neubert, Markus ; Koschik, Stefan ; De Doncker, Rik W.

INPUT CURRENT HARMONICS REDUCTION CONTROL FOR ELECTROLYTIC CAPACITOR LESS INVERTER BASED IPMSM DRIVE SYSTEM 3153
Abe, Kodai ; Ohishi, Kiyoshi ; Haga, Hitoshi

NONCONTACT GUIDE SYSTEM FOR TRAVELING ELASTIC STEEL PLATES (THEORETICAL STUDY ON THE SHAPE OF TRAVELING STEEL PLATE) 3159
Sakaba, Kouichi ; Hasegawa, Shinya ; Narita, Takayoshi ; Oshinoya, Yasuo

ACTIVE SEAT SUSPENSION FOR ULTRA-COMPACT VEHICLE (FUNDAMENTAL CONSIDERATION ON ELECTROMYOGRAM WHEN FALL FROM THE BUMP) 3162
Mashino, Masahiro ; Sunaga, Keita ; Hasegawa, Shinya ; Ishida, Masaki ; Kato, Hideaki ; Oshinoya, Yasuo

ADAPTIVE CURRENT TRACKING OF THREE-PHASE ACTIVE POWER FILTER USING BACKSTEPPING CONTROL 3168
Yunmei Fang ; Juntao Fei ; Shixi Hou ; Weili Dai

FAST IDENTIFICATION OF RESONANCE CHARACTERISTIC FOR 2-MASS SYSTEM WITH ELASTIC LOAD 3174
Ming Yang ; Liang Hao ; Dianguo Xu

AUTONOMOUS NAVIGATION SYSTEM BASED ON COLLISION DANGER-DEGREE FOR UNMANNED GROUND VEHICLE 3179
Yasuno, Takashi ; Tanaka, Daiki ; Kuwahara, Akinobu

A HIGH-PERFORMANCE BIDIRECTIONAL DC-DC CONVERTER FOR DC MICRO-GRID SYSTEM APPLICATION 3185
Shu-Wei Kuo ; Yu-Kang Lo ; Huang-Jen Chiu ; Shih-Jen Cheng ; Chung-Yi Lin ; Yang, CS

VOLUME 5

IMPROVEMENT IN EFFICIENCY OF LED LIGHTING SYSTEM 3190
Hwu, K.I. ; Jiang, W.Z. ; Jenn-Jong Shieh

COMPARISON AND EVALUATION OF VIBRATION-BASED PIEZOELECTRIC POWER GENERATORS 3194
Basari, Amat A. ; Awaji, Sosuke ; Hashimoto, Seiji ; Kasai, Makoto ; Suto, Kenji ; Kumagai, Shunji ; Kasai, Makoto ; Suto, Kenji ; Wei Jiang ; Shuren Wang

BATTERY SELECTION FOR HYBRID ENERGY SYSTEMS AND THERMAL MANAGEMENT IN ARCTIC CLIMATES 3200
Kalogera, Maria ; Bauer, Pavol

100KW PV PCS WITH NATURAL CONVECTION COOLING FOR OUTDOOR INSTALLATION 3207
Jin, Yasuhiro ; Matsuoka, Kazumasa ; Takahashi, Takehiro ; Takahashi, Nobuhiro

A NEW PLL BASED ON FAST POSITIVE AND NEGATIVE SEQUENCE DECOMPOSITION ALGORITHM WITH MATRIX OPERATION UNDER DISTORTED GRID CONDITIONS 3213
Shaohua Sun ; Hongqi Ben ; Tao Meng ; Jinyong Zhang

PERFORMANCE IMPROVEMENT OF PHOTOVOLTAIC POWER GENERATION SYSTEMS USING ON-OFF CONTROL METHODS 3218
Kenji, Matsumoto ; Nomura, Shinichi

LOW VOLTAGE PV POWER INTEGRATION INTO MEDIUM VOLTAGE GRID USING HIGH VOLTAGE SIC DEVICES 3225
Chattopadhyay, Ritwik ; Bhattacharya, Subhashish ; Foureaux, Nicole C. ; Silva, Sidelmo M. ; Braz Cardoso, F. ; de Paula, Helder ; Pires, Igor A. ; Cortizio, Porfirio C. ; Moraes, Lenin ; de S.Brito, Jose A.

A NOVEL GLOBAL MAXIMUM POWER POINT TRACKING METHOD FOR PHOTOVOLTAIC GENERATION SYSTEM OPERATING UNDER PARTIALLY SHADED CONDITION 3233
Jing-Hsiao Chen ; Yu-Shan Cheng ; Shun-Chung Wang ; Huang, Jia-Wei ; Liu, Yi-Hua

AN APPLICATION OF Z-SOURCE CONVERTER TO BATTERIES CHARGE WITH A PHOTOVOLTAIC SYSTEM 3239
Razik, H. ; Zitouni, Y. ; Maret, C.

PCS WITH SCANNING-TYPE MPPT CONTROL FOR INDUSTRIAL GRID-CONNECTED PV POWER GENERATION SYSTEM 3244
Itako, Kazutaka

FEASIBLE METHOD OF CALCULATING LEAKAGE REACTANCE OF 9-WINDING TRANSFORMER FOR HIGH-VOLTAGE INVERTER SYSTEM 3249
Fukumoto, Hisao ; Furukawa, Tatsuya ; Itoh, Hideaki ; Ohchi, Masashi

HIGH POWER HVDC-DC CONVERTERS FOR THE INTERCONNECTION OF HVDC LINES WITH DIFFERENT LINE TOPOLOGIES 3255
Schon, Andre ; Bakran, Mark-M.

CHARACTERIZATION OF A CURRENT SHUNT AND AN INDUCTIVE VOLTAGE DIVIDER FOR PMU CALIBRATION 3263
Kon, Saytaro ; Yamada, Tatsuji

DISTRIBUTED SERIES/HYBRID-SHUNT COMPENSATION FOR HARMONIC MITIGATION IN COMMERCIAL FACILITIES......3270

Diniz, Rogerio Azevedo ; Pires, Igor A. ; Franca, Gleisson J. ; Cardoso, Braz J.

ROBUST CONTROL DESIGN FOR THE VOLTAGE TRACKING LOOP OF A DVR......3278

Ferrari, Bruno Augusto ; Ama, Naji Rajai Nasri ; de Carvalho, Kelly Caroline Mingorancia ; Martinz, Fernando Ortiz ; Matakas, Lourenco

MULTI-PORT SOLID STATE TRANSFORMER FOR INTER-GRID POWER FLOW CONTROL......3286

Roy, Sudhin ; De, Ankan ; Bhattacharya, Subhashish

REACTIVE POWER CONTROL STRATEGY BASED ON DC CAPACITOR VOLTAGE CONTROL FOR ACTIVE LOAD BALANCER IN THREE-PHASE FOUR-WIRE DISTRIBUTION SYSTEMS......3292

Tint Soe Win ; Hisada, Yoshihiro ; Tanaka, Toshihiko ; Hiraki, Eiji ; Okamoto, Masayuki ; Lee, Seong Ryong

VOLTAGE SAG RIDE-THROUGH PERFORMANCE OF VIRTUAL SYNCHRONOUS GENERATOR......3298

Alipoor, Jaber ; Miura, Yushi ; Ise, Toshifumi

CONTROL OF DISTRIBUTED GENERATION SYSTEMS UNDER UNBALANCED VOLTAGE CONDITIONS......3306

Kabiri, R. ; Holmes, D.G. ; McGrath, B.P.

STABILITY ANALYSIS OF GRID-CONNECTED INVERTERS WITH LCL-FILTER BASED ON HARMONIC BALANCE AND FLOQUET THEORY......3314

Jing Bian ; Hong Li ; Zheng, Trillion Q.

COMPARATIVE EVALUATION OF PASSIVE DAMPING TOPOLOGIES FOR PARALLEL GRID-CONNECTED CONVERTERS WITH LCL FILTERS......3320

Beres, Remus ; Wang, Xiongfei ; Blaabjerg, Frede ; Bak, Claus Leth ; Liserre, Marco

STUDY AND IMPLEMENTATION OF A SEPIC LED DRIVER WITH ADJUSTABLE OUTPUT VOLTAGE......3328

Po-Jung Tseng ; Yu-Chen Liu ; Yu-Kang Lo ; Chiu, Huang-Jen ; Yun-Chu Chiu

AN INTERLEAVED SINGLE-STAGE LLC RESONANT CONVERTER USED FOR MULTI-CHANNEL LED DRIVING......3333

Chang, Chien-Hsuan ; Cheng, Chun-An ; Jinno, Masahito ; Cheng, Hung-Liang

A NOVEL TYPE OF WIRELESS V2H SYSTEM WITH BIDIRECTIONAL RESONANT SINGLE-ENDED INVERTER......3341

Fukuoka, Hiroki ; Iga, Yuichi ; Omori, Hideki ; Morizane, Tosimitsu ; Kimura, Noriyuki ; Nakaoka, Mutuo

DESIGN AND IMPLEMENTATION OF AN INTERLEAVED BCM BOOST PFC CONTROL IC......3346

Kuan-Hsien Chou ; Tsorng-Juu Liang ; Kai-Hui Chen ; Ji-Shiang Lee

LOW CAPACITIVE INDUCTORS FOR FAST SWITCHING DEVICES IN ACTIVE POWER FACTOR CORRECTION APPLICATIONS......3352

Hernandez, Juan C. ; Petersen, Lars P. ; Andersen, Michael A.E.

TEMPERATURE-ROBUST LC3 LED DRIVER WITH LOW THD, HIGH EFFICIENCY, AND LONG LIFE......3358

Lee, Eun S. ; Choi, Bo H. ; Cheon, Jun P. ; Kim, Bong C. ; Rim, Chun T.

OPTIMIZING REPULSIVE LORENTZ FORCES FOR A LEVITATING INDUCTION COOKER......3365

Zingerli, Claudius M. ; Nussbaumer, Thomas ; Kolar, Johann W.

DESIGN OF A MODULAR RESONANT CONVERTER FOR 25KV-8A DC POWER SUPPLY OF RF CAVITIES......3371

Siemaszko, Daniel ; Pittet, Serge ; Aguglia, Davide ; de Mallac, Louis

A NOVEL TRANSFORMER-LESS INTERLEAVED FOUR-PHASE HIGH STEP-DOWN DC CONVERTER WITH LOW SWITCH VOLTAGE STRESS......3379

Ching-Tasi Pan ; Chen-Feng Chuang ; Chia-Chi Chu ; Hao-Chien Cheng

EFFICIENCY IMPROVEMENT OF POWER SUPPLY WITH TRANSIENT CURRENT CIRCUIT USING DIGITAL CONTROL......3386

Takashita, Haruomi ; Shoyama, Masahito ; Yonezawa, Yu ; Nakashima, Yoshiyasu

ULTRA HIGH STEP-DOWN CONVERTER......3392

Yau, Y.T. ; Hwu, K.I.

DIGITAL CONTROL OF PWM INVERTER USING ULTRA HIGH SPEED NETWORK FOR FEEDBACK SIGNALS WITH COMMUNICATION DISTURBANCE OBSERVER BASED ON ROCKET I/O PROTOCOL......3397

Saito, Ryo ; Tsuchida, Kazuo ; Yokoyama, Tomoki

100 KHZ DC CHOPPER DIGITALLY GATE CONTROLLED WITH PARTIAL TURN- OFF SWITCHING USING SIC-MOSFET AND FPGA......3403

Tsuruta, Yukinori ; Kawamura, Atsuo

VARIABLE CARRIER DEADBEAT CONTROL WITH DIGITAL HYSTERESIS METHOD USING SOC-FPGA FOR UTILITY INTERACTIVE INVERTER......3410

Ohashi, Shunsuke ; Yoshida, Morito ; Yokoyama, Tomoki

A SPACE VECTOR MODULATION STRATEGY FOR THREE-LEVEL OPERATION BASED ON DUAL TWO-LEVEL VOLTAGE SOURCE INVERTERS......3417

Kumsuwan, Yuttana ; Srirattanawichaikul, Watcharin

INVESTIGATION ON THE PARALLEL OPERATION OF ALL-GAN POWER MODULE AND THERMAL PERFORMANCE EVALUATION......3425

Cheng, Stone ; Po-Chien Chou

FULL SILICON CARBIDE BOOST CHOPPER MODULE FOR HIGH FREQUENCY AND HIGH TEMPERATURE OPERATION......3432

Pettersson, Sami ; Kicin, Slavo ; Holm, Toni ; Bianda, Enea ; Canales, Francisco

DEVELOPMENT OF ULTRAHIGH VOLTAGE SIC POWER DEVICES...3440
Fukuda, Kenji ; Okamoto, Dai ; Harada, Shinsuke ; Tanaka, Yasunori ; Yonezawa, Yoshiyuki ; Deguchi, Tadayoshi ; Katakami,
Shuji ; Ishimori, Hitoshi ; Takasu, Shinji ; Arai, Manabu ; Takenaka, Kensuke ; Fujisawa, Hiroyuki ; Takei, Manabu ; Matsumoto,
Kazushi ; Ohse, Naoyuki ; Ryo, Mina ; Ota, Chiharu ; Takao, Kazuto ; Mizukami, Makoto ; Kato, Tomohisa ; Izumi, Toru ;
Hayashi, Toshihiko ; Nakayama, Koji ; Asano, Katsunori ; Okuyama, Hajime ; Kimoto, Tsunenobu

**HIGH SWITCHING PERFORMANCE OF 1.7KV, 50A SIC POWER MOSFET OVER SI IGBT FOR
ADVANCED POWER CONVERSION APPLICATIONS**..3447
Hazra, Samir ; De, Ankan ; Bhattacharya, Subhashish ; Lin Cheng ; Palmour, John ; Schupbach, Marcelo ; Hull, Brett ; Allen,
Scott

**CONTROL METHOD FOR FIVE LEVEL CONVERTER WITH COMMON FLYING CAPACITORS TO
AVOID VOLTAGE LEVEL SKIP**..3455
Wei Yan ; Hui Zhang ; Ogura, Kazuya ; Urushibata, Shota

**LOW-COMPLEXITY ANALYTICAL APPROXIMATIONS OF SWITCHING FREQUENCY HARMONICS OF
3-PHASE N-LEVEL VOLTAGE-SOURCE PWM CONVERTERS**...3460
Burkart, Ralph M. ; Kolar, Johann W.

**DYNAMIC VOLTAGE BALANCING ALGORITHM FOR MODULAR MULTILEVEL CONVERTER WITH
THREE-LEVEL FLYING CAPACITOR SUBMODULES**...3468
Dekka, Apparao ; Wu, Bin ; Zargari, Navid R.

MODULAR MEDIUM VOLTAGE DRIVE FOR DEMANDING APPLICATIONS..3476
Dujic, Drazen ; Wahlstroem, Jonas ; Marrero Sosa, Juan Alberto ; Fritz, Dominik

**ASYMMETRICAL FAULT RIDE-THROUGH OF THREE-PHASE PV SYSTEMS USING FOUR-WIRE DC-
AC CONVERTERS**...3482
Iyer, Shivkumar ; Bin Wu ; Yunwei Li ; Singh, B.N.

**OPERATION MODE ANALYSIS FOR SOLVING THE PARTIAL SHADOW IN A NOVEL PV POWER
GENERATION SYSTEM**..3489
Qi Zhang ; Xiangdong Sun ; Yanru Zhong ; Lie Guo ; Matsui, Mikihiko

**ANALYSIS OF PARTIAL POWER PROCESSING DISTRIBUTED MPPT FOR A PV POWERED ELECTRIC
AIRCRAFT**...3496
Marzouk, Ahmad Diab ; Fournier-Bidoz, Sebastien ; Yablecki, Jessica ; McLean, Kenneth ; Trescases, Olivier

**IMPACTS OF RECTIFIER CIRCUIT LOADS ON ISLANDING DETECTION OF PHOTOVOLTAIC
SYSTEMS**...3503
Yoshida, Yoshiaki ; Suzuki, Hirokazu

INDUCTION MOTOR MADE OF SMC...3509
Morimoto, Masayuki ; Inamori, Mamiko

**ESTIMATION AND COMPARISON OF THE WINDAGE LOSS OF A 60 KW SWITCHED RELUCTANCE
MOTOR FOR HYBRID ELECTRIC VEHICLES**..3513
Kiyota, Kyohei ; Kakishima, Takeo ; Chiba, Akira

**DEVELOPMENT OF HIGH-POWER PMASYNRM USING FERRITE MAGNETS FOR REDUCING RARE-
EARTH MATERIAL USE**..3519
Sanada, Masayuki ; Morimoto, Shigeo ; Inoue, Yukinori

**CONSIDERATION OF 10KW IN-WHEEL TYPE AXIAL-GAP MOTOR USING FERRITE PERMANENT
MAGNETS**...3525
Sone, Kodai ; Takemoto, Masatsugu ; Ogasawara, Satoshi ; Takezaki, Kenichi ; Hino, Wataru

**POWER CONTROL METHOD FOR MULTI-PARALLEL DC DISTRIBUTION SYSTEM THROUGH THE
EQUIVALENT CIRCUIT MODEL**..3532
Seok-Jin Hong ; Soo-Cheol Shin ; Hee-Jun Lee ; Chung-Yuen Won ; Taeck-Kie Lee

**A COMMUNICATION-LESS DISTRIBUTED VOLTAGE CONTROL STRATEGY FOR A MULTI-BUS AC
ISLANDED MICROGRID**...3538
Wang, Yanbo ; Yongdong Tan ; Chen, Zhe ; Wang, Xiongfei ; Tian, Yanjun

**AN ENHANCED LOAD POWER SHARING STRATEGY FOR LOW-VOLTAGE MICROGRIDS BASED ON
INVERSE-DROOP CONTROL METHOD**...3546
Yixin Zhu ; Fang Zhuo ; Baoquan Liu ; Hao Yi

**ADDING VIRTUAL RESISTANCE IN SOURCE SIDE CONVERTERS FOR STABILIZATION OF
CASCADED CONNECTED TWO STAGE CONVERTER SYSTEMS WITH CONSTANT POWER LOADS IN
DC MICROGRIDS**...3553
Mingfei Wu ; Lu, Dylan D.C.

**EXPANSION OF OPERATING RANGE AND IMPROVEMENT OF TORQUE RESPONSE OF PMSM DRIVE
BY USING MODEL PREDICTIVE CONTROL**...3557
N/A

**NONLINEAR MODEL PREDICTIVE TORQUE CONTROL OF A LOAD COMMUTATED INVERTER AND
SYNCHRONOUS MACHINE**..3563
Almer, Stefan ; Besselmann, Thomas ; Ferreau, Joachim

**MODEL PREDICTIVE CURRENT CONTROL FOR PMSM CONSIDERING NUMBER OF SWITCHING
OPERATIONS**...3568
Zanma, Tadanao ; Yasumura, Yuji ; Liu, KangZhi

**PREDICTIVE INDIRECT MATRIX CONVERTER FED TORQUE RIPPLE MINIMIZATION WITH
WEIGHTING FACTOR OPTIMIZATION**..3574
Uddin, Muslem ; Mekhilef, Saad ; Rivera, Marco ; Rodriguez, Jose

HIGH-POWER DENSITY HYBRID CONVERTER TOPOLOGIES FOR LOW-POWER DC-DC SMPS.............3582
Radic, Aleksandar ; Ahssanuzzaman, S.M. ; Mahdavikhah, Behzad ; Prodic, Aleksandar

COUPLED INDUCTOR BASED CURRENT-FED SWITCHED INVERTER FOR LOW VOLTAGE RENEWABLE INTERFACE 3587

Nag, Soumya Shubhra ; Mishra, Santanu Kumar

A SEMI-ISOLATED MULTI-INPUT CONVERTER FOR HYBRID PV/WIND POWER CHARGER SYSTEM 3592

Cheng-Wei Chen ; Kun-Hung Chen ; Chen, Yaow-Ming

HFL PV MICRO-INVERTER WITH FRONT-END CURRENT-FED CONVERTER AND HALF-WAVE CYCLOCONVERTER 3598

Nayanasiri, D.R. ; Vilathgamuwa, D.M. ; Maskell, D.L.

COMPREHENSIVE STUDY ABOUT STABILITY ISSUES OF MULTI-MODULE DISTRIBUTED SYSTEM 3604

Liu, Fangcheng ; Liu, Jinjun ; Zhang, Haodong ; Xue, Danhong ; Dou, Qinyun

CHARACTERISTICS STUDY OF NEURAL NETWORK AIDED DIGITAL CONTROL FOR DC-DC CONVERTER 3611

Maruta, Hidenori ; Motomura, Masashi ; Kurokawa, Fujio

ZERO CURRENT SWITCHING CURRENT-FED PARALLEL RESONANT PUSH-PULL (CFPRPP) CONVERTER 3616

Moorthy, Radha Sree Krishna ; Rathore, Akshay Kumar

CHARACTERISTICS OF TRANSMISSION CARRIER IN A NEW WIRE COMMUNICATION SYSTEM BY THE USE OF HIGH-RIPPLE DC-DC CONVERTER 3624

Katsuki, Akihiko ; Mizuki, Tatsuya ; Shibahara, Kohei ; Morita, Kosuke ; Masutomo, Kazufumi ; Maeyama, Shigetaka

5MHZ PWM-CONTROLLED CURRENT-MODE RESONANT DC-DC CONVERTER USING GAN-FETS 3630

Hariya, Akinori ; Yanagi, Hiroshige ; Ishizuka, Yoichi ; Matsuura, Ken ; Tomioka, Satoshi ; Ninomiya, Tamotsu

DESIGN AND PERFORMANCE EVALUATION OF DIGITAL CONTROL FOR LLC SERIES RESONANT DC-TO-DC CONVERTERS 3638

Pidaparthy, Syam Kumar ; Choi, Byungcho ; Jang, Jinhaeng

EXPERIMENTAL VERIFICATION OF NOISELESS SAMPLING FOR BUCK CHOPPER CIRCUIT WITH CURRENT CONTROL 3646

Takeuchi, Shun ; Wada, Keiji

CONTROL CHARACTERISTICS IMPROVEMENT OF FULL-BRIDGE DC-DC CONVERTER WITH SNUBBER CAPACITOR 3652

Domoto, Kazuhide ; Ishizuka, Yoichi ; Abe, Seiya ; Ninomiya, Tamotsu

DCM CONTROL METHOD OF BOOST CONVERTER BASED ON CONVENTIONAL CCM CONTROL 3659

Le Hoai Nam ; Orikawa, Koji ; Itoh, Jun-ichi

TECHNICAL ASSESSMENT OF LOAD COMMUTATION SWITCH IN HYBRID HVDC BREAKER 3667

Hassanpoor, Arman ; Hafner, Jurgen ; Jacobson, Bjorn

CONTROL OF HEXAGONAL MODULAR MULTILEVEL CONVERTER FOR 3-PHASE BTB SYSTEM 3674

Hamasaki, Shin-ichi ; Okamura, Kazuki ; Tsubakidani, Takashi ; Tsuji, Mineo

A SYNTHESIZED CAPACITORS VOLTAGE CONTROL FOR MODULAR MULTILEVEL CONVERTER IN HVDC APPLICATION 3680

Rongfeng Yang ; Shunke Sui ; Binbin Li ; Wei Wang ; Dianguo Xu

OPERATING PHASE AND FREQUENCY SELECTION OF LOW FREQUENCY AC TRANSMISSION SYSTEM USING CYCLOCONVERTERS 3687

Achara, Pichetjamroen ; Ise, Toshifumi

FAST ACTING DC CIRCUIT BREAKER FOR HVDC TRANSMISSION LINE BASED ON DC/DC CHOPPER 3695

Liangyi Tang ; Bin Wu ; Yaramasu, Venkata ; Weirong Chen ; Athab, Hussain S.

1700V SI-IGBT AND SIC-SBD HYBRID MODULE FOR AC690V INVERTER SYSTEM 3702

Haining Wang ; Ikawa, O. ; Miyashita, S. ; Nishimura, T. ; Igarashi, S.

SWITCHING SIMULATION OF SIC HIGH-POWER MODULE WITH LOW PARASITIC INDUCTANCE 3707

Yamamoto, Takashi ; Hasegawa, Kohei ; Ishida, Masaaki ; Takao, Kazuto

SWITCHING PERFORMANCE OF PARALLEL-CONNECTED POWER MODULES WITH SIC MOSFETS 3712

Colmenares, Juan ; Peftitsis, Dimosthenis ; Nee, Hans-Peter ; Rabkowski, Jacek

BUILT-IN RELIABILITY DESIGN OF A HIGH-FREQUENCY SIC MOSFET POWER MODULE 3718

Jianfeng Li ; Gurpinar, Emre ; Lopez-Arevalo, Saul ; Castellazzi, Alberto ; Mills, Liam

EXPERIMENTAL SWITCHING FREQUENCY LIMITS OF 15 KV SIC N-IGBT MODULE 3726

Kadavelugu, Arun ; Bhattacharya, Subhashish ; Ryu, Sei-Hyung ; Van Brunt, Edward ; Grider, Dave ; Leslie, Scott

SELECTION OF SUITABLE CARRIER-BASED PWM METHOD FOR MODULAR MULTILEVEL CONVERTER 3734

Ciftci, Baris ; Erturk, Feyzullah ; Hava, Ahmet M.

CONTROL AND EXPERIMENT OF A 380-V, 15-KW MOTOR DRIVE USING MODULAR MULTILEVEL CASCADE CONVERTER BASED ON TRIPLE-STAR BRIDGE CELLS (MMCC-TSBC) 3742

Kawamura, Wataru ; Hagiwara, Makoto ; Akagi, Hirofumi

A POWER ELECTRONIC TRANSFORMER WITH SINUSOIDAL VOLTAGES AND CURRENTS USING MODULAR MULTILEVEL CONVERTER 3750

Sahoo, Ashish Kumar ; Mohan, Ned

VARYING AND UNEQUAL CARRIER FREQUENCY PWM TECHNIQUES FOR MODULAR MULTILEVEL CONVERTERS 3758

Konstantinou, Georgios ; Darus, Rosheila ; Pou, Josep ; Ceballos, Salvador ; Agelidis, Vassilios G.

COMPARISON OF PHASE-SHIFTED AND LEVEL-SHIFTED PWM IN THE MODULAR MULTILEVEL CONVERTER 3764

Darus, Rosheila ; Konstantinou, Georgios ; Pou, Josep ; Ceballos, Salvador ; Agelidis, Vassilios G.

A SINGLE-PHASE POWER CONDITIONER WITH A BUCK-BOOST-TYPE POWER DECOUPLING CIRCUIT...3771
Yamaguchi, Shota ; Shimizu, Toshihisa

A NOVEL ASYMMETRICAL FLC-BASED MPPT TECHNIQUE FOR PHOTOVOLTAIC GENERATION SYSTEM...3778
Yi-Hsun Chiu ; Yu-Shan Cheng ; Yi-Hua Liu ; Shun-Chung Wang ; Zong-Zhen Yang

A NOVEL CURRENT LINK DISTRIBUTED MPPT PV SYSTEM - OVERALL SYSTEM PROTOTYPING AND EVALUATION..3784
Mikihiko ; Toru ; Akira ; Xiang-Dong Sun ; Byung-Gyu Yu

POWER FLOW CONTROL AND MPPT PARAMETER SELECTION FOR RESIDENTIAL GRID-CONNECTED PV SYSTEMS WITH BATTERY STORAGE...3789
Chokchai, Chuenwattanapraniti

A MAXIMUM POWER POINT TRACKING METHOD WITH RIPPLE CURRENT ORIENTATION.....................................3796
Moo, Chin-Sien ; Wu, Gwo-Bin

OUTPUT CHARACTERISTICS OF A SURFACE PERMANENT MAGNET-TYPE VERNIER MOTOR - COMPARISON OF TEST RESULTS AND CALCULATION...3801
Kataoka, Yasuhiro ; Takayama, Masakazu ; Anazawa, Yoshihisa ; Matsushima, Yoshitarou

TOPOLOGY OPTIMIZATION FOR SKEW OF SPMSM BY USING MULTI-STEP PARALLEL GA 3809
Kitagawa, Wataru ; Takeshita, Takaharu

LOSS MINIMIZATION DESIGN USING MAGNETIC EQUIVALENT CIRCUIT FOR A PERMANENT MAGNET SYNCHRONOUS MOTOR...3815
Sato, Daisuke ; Itoh, Jun-ichi

THE PROPOSAL OF A NEW MOTOR WHICH HAS A HIGH WINDING FACTOR AND A HIGH SLOT FILL FACTOR..3823
Makita, Shinji ; Ito, Yasuhide ; Aoyama, Tomohiro ; Doki, Shinji

VARIABLE LEAKAGE FLUX INTERIOR PERMANENT MAGNET SYNCHRONOUS MACHINE FOR IMPROVING EFFICIENCY ON DUTY CYCLE...3828
Minowa, Masanao ; Hijikata, Hiroki ; Akatsu, Kan ; Kato, Takashi

HISTORY AND TRENDS OF CONVERTER TECHNOLOGY FOR DC AND AC TRANSMISSION IN JAPAN 3834
Yoshino, Teruo

ACCURATE OUTPUT POWER CONTROL OF CONVERTERS FOR MICROGRIDS BASED ON LOCAL MEASUREMENT AND UNIFIED CONTROL...3842
Meiqin Mao ; Zheng Dong ; Yong Ding ; Liuchen Chang

IMPEDANCE-BASED ANALYSIS OF ACTIVE FREQUENCY DRIFT ISLANDING DETECTION METHOD FOR GRID-TIED INVERTER SYSTEM...3850
Wen, Bo ; Boroyevich, Dushan ; Burgos, Rolando ; Shen, Zhiyu ; Mattavelli, Paolo

DEVELOPMENT OF 200-MVAR CLASS THYRISTOR SWITCHED CAPACITOR SUPPORTING FAULT RIDE-THROUGH..3857
Ohtake, Asuka ; Fei Zhang ; Fujimoto, Takafumi ; Nakayama, Naoyuki

DETAILED ANALYSIS AND DESIGN OF A THREE-PHASE PHASE-MODULAR ISOLATED MATRIX-TYPE PFC RECTIFIER..3864
Cortes, Patricio ; Fassler, Lukas ; Bortis, Dominik ; Kolar, Johann W. ; Silva, Marcelo

AN ENERGY SAVING DRIVE METHOD OF AN INDUCTION MOTOR WITH THE SUPPRESSION OF SUDDEN ACCELERATION AND DECELERATION...3872
Asano, Yuji ; Inoue, Kaoru ; Kotera, Keito ; Kato, Toshiji

FIELD ORIENTED CONTROL OF SENSORLESS LINEAR INDUCTION MOTOR USING MATRIX CONVERTER...3877
Sayed, Mahmoud A. ; Mohamed, Essam Ebaid ; Mohamed, Tarek Hassan ; Takeshita, Takaharu

A STATOR-EQUATION-BASED REDUCED-ORDER OBSERVER FOR POSITION-SENSORLESS VECTOR CONTROL SYSTEM OF DOUBLY-FED INDUCTION MACHINES..3885
Smiththisomboon, Somrat ; Suwankawin, Surapong

INPUT CURRENT RIPPLE ANALYSIS OF INVERTER FED DUAL THREE-PHASE AC MOTORS 3893
Dahono, Pekik Argo ; Satria, Andri

OFFLINE EXTRACTION OF INDUCTION MACHINE PARAMETERS FOR CONTROL STRATEGY SYNTHESIS...3898
Koschik, Stefan ; Bauer, Florian ; De Doncker, R.W.

HIGH CURRENT PLANAR TRANSFORMER FOR VERY HIGH EFFICIENCY ISOLATED BOOST DC-DC CONVERTERS...3905
Pittini, Riccardo ; Zhe Zhang ; Andersen, Michael A.E.

HIGH VOLTAGE-GAIN INTERLEAVED BOOST DC-DC CONVERTER DISCARDED ELECTROLYTIC CAPACITOR...3913
Nha, Quang Trong ; Huang-Jen Chiu ; Yu-Kang Lo ; Pham Phu Hieu

PARALLEL BI-DIRECTIONAL DC-DC CONVERTER FOR ENERGY STORAGE SYSTEM....................................3920
Ouchi, Takayuki ; Kanoda, Akihiko ; Takahashi, Naoya

CHARGING SCENARIO OF SERIAL BATTERY POWER MODULES WITH BUCK-BOOST CONVERTERS 3928
Jhen-Yu Jian ; Chu-Shen Chang ; Moo, Chin-Sien ; Hau-Chen Yen

COMPARATIVE THERMAL PERFORMANCE EVALUATION OF SIC MOSFETS AND SI MOSFET FOR 1.2 KW 300 KHZ DC-DC BOOST CONVERTER AS A SOLAR PV PRE-REGULATOR 3933
Taekyun Kim ; Minsoo Jang ; Agelidis, Vassilios G.

TOLERANCE ANALYSIS OF A CONSTANT-ON TIME CURRENT-MODE VOLTAGE REGULATOR WITH ADAPTIVE VOLTAGE POSITION FEATURE 3938

Chih Wei Chen ; Dan Chen ; Shin Shiung Wang

FPGA-BASED DIGITAL-CONTROLLED POWER CONVERTER DESIGNED WITH UNIVERSAL INPUT MEETING 80 PLUS PLATINUM EFFICIENCY CODE AND STANDBY POWER CODE FOR SEVER POWER APPLICATIONS 3942

Lai, Yen-Shin ; Ho, Kung-Min

STATIC AND DYNAMIC ANALYSES OF DIGITAL PEAK CURRENT MODE DC-DC CONVERTER 3950

Kajiwara, Kazuhiro ; Kurokawa, Fujio ; Shibata, Yuichiro

EXTENDED DISCRETE CONTROL OF CLASS E AMPLIFIER IN ORDER TO ACHIEVE NOMINAL OPERATION 3955

Suetsugu, Tadashi ; Xiuqin Wei ; Kuga, Shotaro

ADAPTIVE POWER EFFICIENCY CONTROL BY COMPUTER POWER CONSUMPTION PREDICTION USING PERFORMANCE COUNTERS 3959

Kawaguchi, Shinichi ; Yachi, Toshiaki

Author Index

The 2014 International Power Electronics Conference

Real-Time Simulation of Wind Turbine Converter-Grid Systems

Shahil Shah, Ignacio Vieto, Nian Heng and Jian Sun

Department of Electrical, Computer and Systems Engineering
Rensselaer Polytechnic Institute, Troy, NY 12180-3590, USA
Telephone: (518) 276-8297; Fax: (518) 276-2387; E-mail: jsun@rpi.edu

Abstract– **This paper presents real-time simulation of both Type-III and Type-IV wind turbines connected to the power grid. Detailed switching models of the converter are used in conjunction with complete turbine and converter control models to evaluate the feasibility of real-time simulation using commercial real-time simulation platforms. In addition to time-domain responses, frequency-domain characteristics, such as inverter output impedance, and possible harmonic resonance as well as subsynchronous interactions with the grid are also determined and compared with offline simulation as well as experimental measurements using a scaled-down hardware to verify the ability of real-time simulation in predicting high-frequency dynamics and control instability of wind turbine converters. The developed real-time simulation models will be used to study wind turbine behavior under different and variable grid conditions, including control interactions with other power electronics devices connected to the power grid.**

I. INTRODUCTION

Real-time simulation provides a convenient and cost-effective tool for the development and testing of complex systems. It has found applications in power systems, motor drives, as well as automotive applications [1], and starts to be used in power electronics as well. One successful example is in the development of modulation and control methods for modular multilevel converters [2]. Another emerging application is the testing of control for grid-connected power electronics devices such as wind and solar inverters, energy storage units, high-voltage dc (HVDC) converters, etc. Since the operation and performance of each such device depends strongly on the characteristics of the power grid, it is essential to test the control functions under various grid parameters and conditions, which is very difficult to perform in the field. Simulating the power grid and the power electronics hardware in real-time enables their control and protection functions to be thoroughly tested in the lab before field commissioning, thereby significantly reducing the development cost.

One particular issue that has drawn the attention of the renewable energy industry in the last couple years is possible resonance between wind and commercial solar inverters and the grid [3]. Harmonic resonance, for example, may happen to Type-IV wind inverters connected to weak grids [4]. Analytical impedance models have been developed for grid-connected inverters and used to predict such resonance [5]. The analytical models indicate that the output impedance of a grid-connected

inverter is typically capacitive from the grid fundamental to the crossover frequency of the inverter current control loop. Furthermore, the phase angle of the output impedance in this frequency range can be less than -90° or even approach -180°, resembling the impedance characteristic of a capacitor with a negative resistor [6]. Based on the Nyquist stability criterion presented in [7] for grid-connected inverters, an under-damped or unstable resonance will be formed if the inverter output impedance intersects with the (inductive) grid impedance in this frequency range, resulting in either severe harmonic currents or tripping of the inverter.

The impedance-based small-signal analysis of harmonic resonance involving Type-IV wind turbines presented in [4-6] assumed a simple grid model that can be represented by an ideal voltage source behind an inductive impedance. A Type-IV wind turbine uses a permanent-magnet generator with full-power conversion (rectifier-inverter) interface to the grid [8]. Since the resonance behavior depends on the grid network impedance as well as the dynamics of other devices connected to the grid, there is a need to expand the grid model used in such resonance analysis to include nonlinearity and the time-varying nature of the grid which cannot be modeled by an impedance. Additionally, to test adaptive control and other advanced inverter control methods, that tune the control parameters in real-time in response to changing grid conditions, it is necessary to subject the controller and control algorithms to different and varying grid conditions in order to determine the stability of adaptation and possible interactions with the control of other devices on the grid. Real-time simulation provides a solution to both needs.

A Type-III wind turbine uses a doubly-fed induction generator (DFIG) and partial parallel power conversion to improve efficiency under variable wind speed. The back-to-back converters connected between the rotor windings and the grid are typically rated for 25-30% of the turbine power [9]. Subsynchronous resonance, also called subsynchronous interaction, has been reported for Type-III wind turbines and power grids that use series compensation capacitors [10, 11]. On the other hand, because of the low power rating of the converter, its impedance looking from the grid side is about 10 dB higher than the impedance of full-power converter used in Type-IV turbines. Therefore, although the rotor-connected back-to-back converters may still exhibit capacitive impedance on the grid side, it is less likely to intersect with the grid impedance within the control

978-1-4799-2706-7/14 $31.00 © 2014 IEEE

bandwidth of the converter. Nevertheless, the potential for harmonic resonance between the converter and the grid still exists, and it is of practical interest to identify the conditions for such resonance as well as to understand its effects on the generator and the grid. Real-time simulation again provides a convenient tool to study both subsynchronous and harmonic resonance involving Type-III wind turbines.

The objective of this paper is to report the development and initial test results of real-time simulation models of both Type-III and Type-IV wind turbines for the purposes identified before. The turbine models are developed using ideal switch models for the converters, and are implemented in Opal-RT OP5600 platform. In addition to the converter and grid, all control functions of the wind turbine converters are also implemented such that the complete system can be simulated. Our plan is to test and validate the full system models first, and then replace the control models by actual control hardware to perform hardware-in-the-loop simulation. An additional purpose of this work is to develop a platform for the study of resonance between Type-III wind turbines and the grid, including both subsynchronous interactions and harmonics resonance which have not been well understood.

The rest of the paper is organized as follows: Section II summarizes the features of the Opal-RT platform used in this study and general considerations in the modeling and real-time simulation of power converters. Section III and IV presents the implementation and initial test results of real-time simulation models for Type-III and Type-IV wind turbines, respectively. Section V concludes the work and discusses future work in using the real-time simulation platform.

II. SIMULATION PLATFORM AND MODELS

The OP5600 real-time simulator from Opal-RT [12] is used in this work. The platform allows the modeling of systems in Simulink on a Windows-based host PC before compilation and real-time simulation on the simulator. In addition to monitoring during simulation, the simulation conditions can also be changed online using the host PC. The OP5600 also provides multiple digital and analog I/O ports for interfacing with external measurement devices and hardware.

One difficulty in simulating power electronic circuits in real time is the high switching frequency and the need to capture the switching instants accurately in order to ensure overall simulation accuracy. Typically, with fixed time-step simulation, the integration time step needs to be 1% or smaller than a switching cycle. For a typical power converter operating at 20 kHz, this implies 1 μs or smaller time step, which is too small for the existing commercial real-time simulation products using CPU-based computing. Real-time simulator based on field-programmable gate array (FPGA) allows for steps well below 1 μs [13], but are difficult to program and are memory restrained.

OP5600 addresses this problem by generating PWM gate signals in an FPGA using RT-Events toolbox and applying the so-called time-step averaging method [14] to compensate for the switching events occurring within a time-step. Fig. 1 shows the

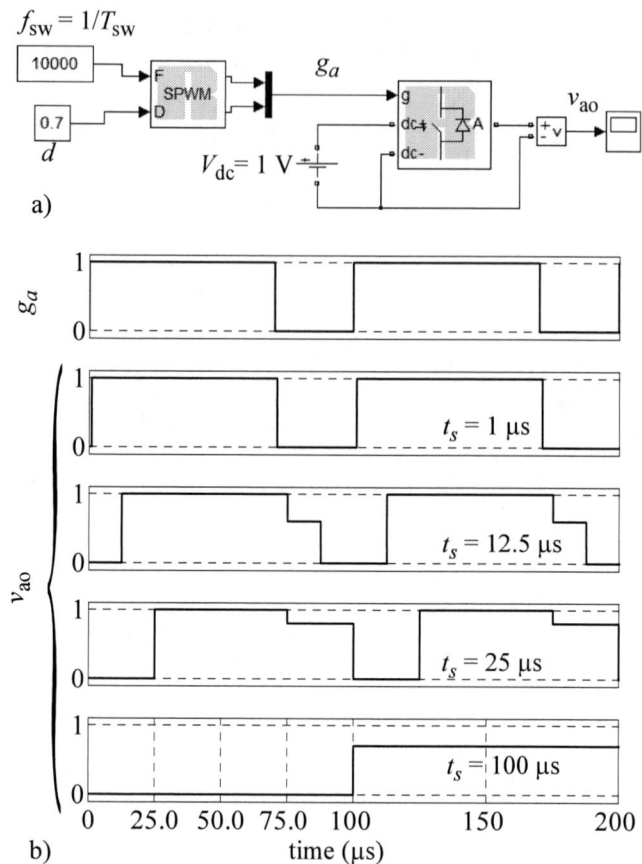

Fig. 1. Test of OP5600 for high-resolution PWM generation. a) test circuit, b) gate-signal and bridge output voltage for different simulation time-steps.

implementation of a single bridge converter and its PWM generator utilizing the RT-Events blocks and the corresponding simulation waveforms. The PWM generator produces accurate gating signals even when the simulation time-step is selected to be equal or larger than the switching period.

Another issue of working with large time step is the need to compensate the state variables for switching events that could have occurred inside the previous time-step. This can be accomplished by interpolating the state variables but at the expense of increased computational burden [15]. The method used in RT-Events block is to generate a so-called averaged state. For example, in Fig. 1, for the constant duty ratio of 0.7 and the input voltage of 1 V, the average output voltage over each switching period should be 0.7 V. This can be achieved by using ON and OFF switch states only when the time duration dT_{sw} is an integral multiple of the simulation time-step. For time-step of 1 μs, the average output voltage of 0.7 V is achieved without introducing an average state as shown in Fig. 1. For time-steps not satisfying the above condition, the RT-Events block introduces an average state where the output stays between 0 and V_{dc}, so that a desired total average voltage is produced. When the time-step is equal or higher than the switching period, the time-stamped bridge becomes equivalent to the averaged model [16], as seen by the output voltage for a time-step of 100 μs in Fig. 1.

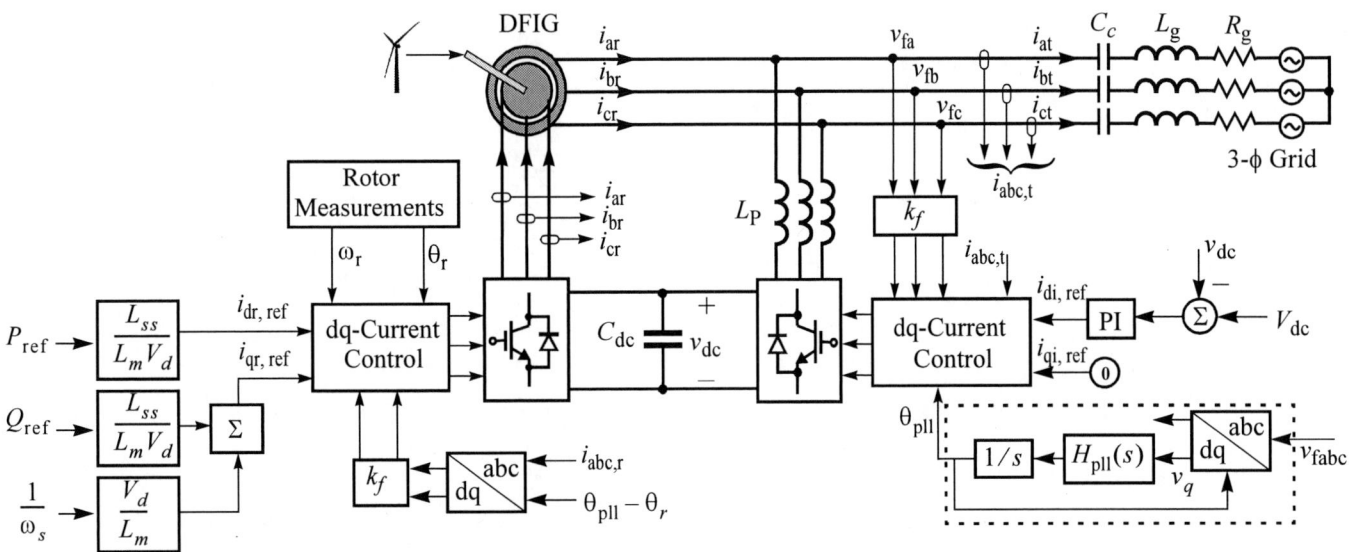

Fig. 2. Circuit and control diagram of the simulated Type-III turbine

III. TYPE-III WIND TURBINE

A. System Description

Fig. 2 shows the circuit and control functions of the simulated Type-III wind turbine. The induction generator is represented by a fourth order electrical circuit [17] and the grid is modeled by an ideal voltage source behind an impedance representing a series-compensated transmission line and step-down transformer. RT-Events PWM generator block discussed before is used to interface the control with the switches of back-to-back converters connected between the grid and the rotor windings. The speed control is achieved by controlling the generator output power using dq-domain current controllers of the rotor-side converter. The rotor-side converter control also includes active and reactive power control functions for the induction machine with the following compensation terms in the d- and q-axis

$$v_{\text{d-c}} = r_r i_{\text{dr}} - \frac{L_{\text{rr}} L_{\text{ss}} - L_{\text{m}}^2}{L_{\text{ss}}} i_{\text{qr}} s + \frac{L_{\text{m}}}{L_{\text{ss}}} v_d s \qquad (1)$$

$$v_{\text{q-c}} = r_r i_{\text{qr}} + \frac{L_{\text{rr}} L_{\text{ss}} - L_{\text{m}}^2}{L_{\text{ss}}} i_{\text{dr}} s \qquad (2)$$

where r_r is the rotor resistance, L_{ss} and L_{rr} the stator and rotor elf inductance, respectively, L_m the mutual inductance, i_{dr} the d-axis current, i_{qr} the q-axis current and v_d the d-axis voltage.

The grid-side converter regulates the intermediate dc-link voltage while keeping its reactive power at zero. The DFIG and grid parameters are taken from [18, 19] and are shown in Table I and Table II, respectively. The turns ratio in Table I refers to the stator to rotor windings ratio of the induction machine.

B. Subsynchronous Resonance

Subsynchronous resonance (or interaction) between Type-III wind turbines and grids with series compensation has been

TABLE I DFIG PARAMETERS

Parameter	Value	Parameter	Value
Rated Power	2 MVA	L_m	3.95 pu
Pole Pairs	3	L_{ls}	0.0923 pu
Turns Ratio	0.334	L_{lr}	0.09955 pu
R_r	0.00549 p.u.	R_s	0.004888 pu

TABLE II GRID AND SYSTEM PARAMETERS

Parameter	Value	Parameter	Value
Base Power	100 MVA	L_g	6.3 μH
Base Voltage	690 V	R_g	95 μΩ
frequency	60 Hz	V_{dc}	1200 V
L_p	0.24 mH	C_{dc}	14 mF

reported [11, 20]. To understand this behavior, we have simulated the implemented Type-III wind turbine model under different level of the series compensation provided in the grid. Fig. 3 shows the simulated stator and rotor currents when the transmission line uses 8% series compensation. The DFIG is operated at a constant speed of 1.2 p. u., resulting in the negative sequence currents at 12 Hz frequency in the rotor windings. For this case, the back-to-back converters carry 17% of the total output current. Both currents are as expected and no resonance is observed.

Fig. 4 shows the real-time simulation results when the amount of series compensation in the grid is increased to about 40%. A resonance below the grid fundamental frequency can be observed and the current amplitude reaches as high as 2 p. u. Interestingly, this subsynchronous resonance is not sustained and is damped out over time. On the other hand, it was observed that further

978-1-4799-2706-7/14 $31.00 © 2014 IEEE

The 2014 International Power Electronics Conference

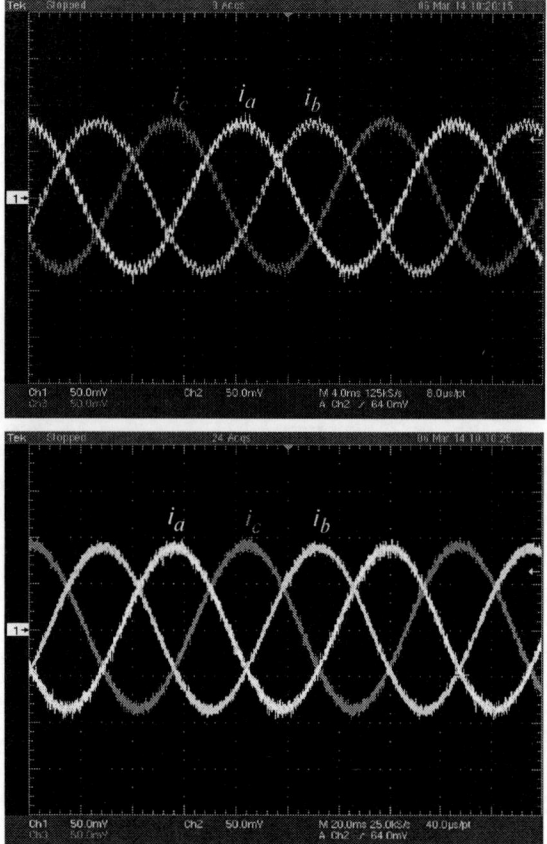

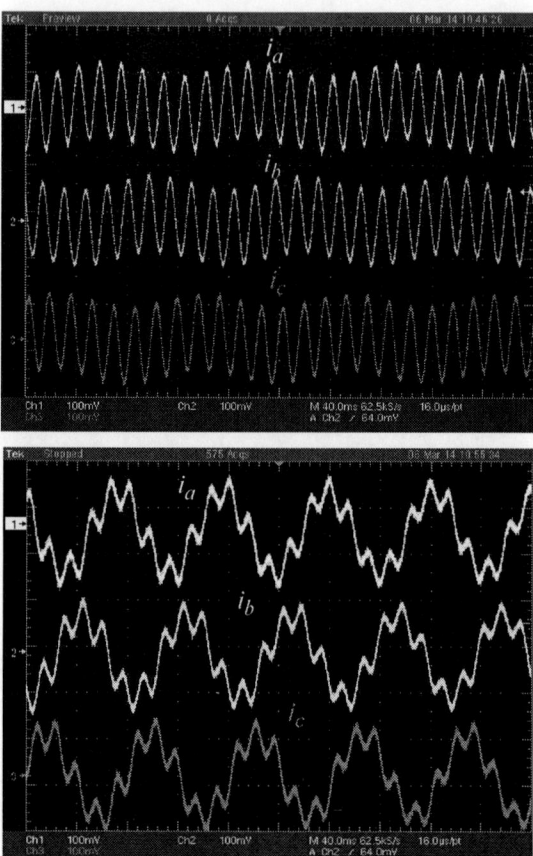

Fig. 3. Simulated Type-III turbine stator (upper) and rotor (lower) current waveforms with a lightly (8%) series-compensated transmission line.

Fig. 4. Simulated Type-III turbine stator (upper) and rotor (lower) current waveforms with a heavily (40%) series-compensated transmission line.

increasing the amount of series compensation will cause instability of the turbine control.

To understand the simulated current waveforms and the effects of subsynchronous interaction on DFIG as well as the converter operation, Fourier analysis is applied to determine the spectrum of both the stator and rotor currents. The results are shown in Fig. 5. As can be seen, the rotor current has a harmonic at 64 Hz with a magnitude of 8% of the rotor fundamental, while the stator has the subsynchronous component at 8 Hz. Recall that the DFIG is operating in the supersyncrhonous mode, with a speed of 1.2 p.u., such that the induced 12 Hz currents in the rotor are in the negative sequence. The 64 Hz harmonic in the rotor currents is also found to be of the negative sequence and can be explained as follows: With three pole pairs and operation at 60 Hz, the synchronous speed is 1200 rpm. The 8 Hz subharmonic in the stator current produces a harmonic magnetic field rotating at 160 rpm and in the same direction as that produced by the fundamental current. With the rotor rotating at 1.2 p.u. of the synchronous speed, that is, at 1440 rpm, the frequency of the currents induced in the rotor winding by the 160 rpm harmonic magnetic field is

$$\frac{1440 - 160}{1200} \times 60 = 64 \text{ Hz} \qquad (3)$$

which is the frequency of the measured rotor current harmonics.

As this example demonstrates, real-time simulation provides a

tool to conveniently study the resonance behavior of Type-III wind turbines under different grid conditions and to evaluate the effects of such resonance on the DFIG as well as the converter and its control. Additional simulation results, including possible harmonic resonance between the grid-side converter and the grid, will be reported in a future work along with the analytical impedance models to characterize the resonance behavior [7].

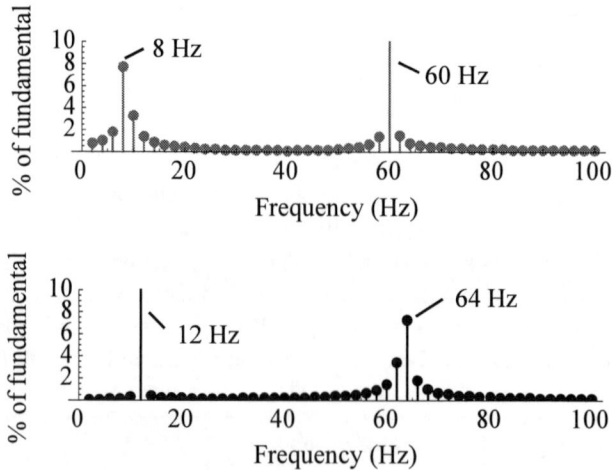

Fig. 5. Frequency spectrum of positive sequence stator currents (upper) and negative sequence rotor currents (lower).

978-1-4799-2706-7/14 $31.00 © 2014 IEEE

The 2014 International Power Electronics Conference

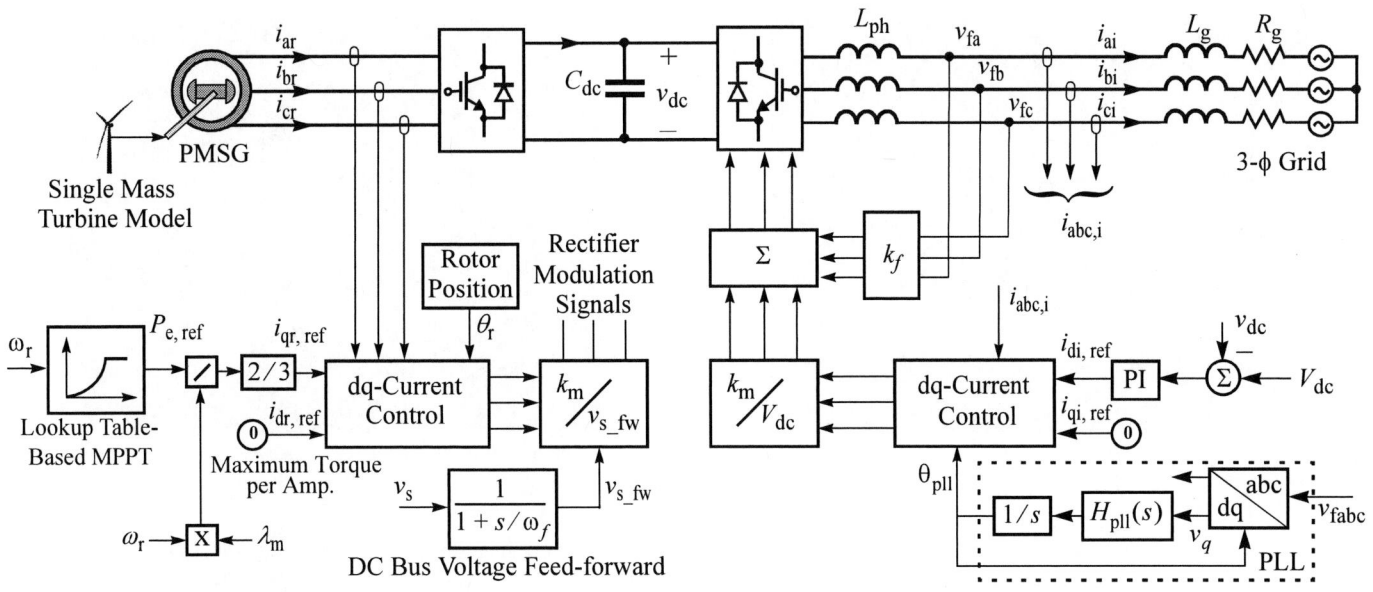

Fig. 6. Circuit and control diagram of the simulated Type-IV turbine.

IV. TYPE-IV WIND TURBINE

This section presents the implementation of a Type-IV wind-turbine shown in Fig. 6 for the real-time simulation. Mechanical dynamics and control architecture are the same as the GE 2.5 MW grid-connected wind-turbine [21]. The verification of real-time simulation in the frequency domain is also presented.

A. System Description

In Type-IV wind turbine of Fig. 6, a VSC-based turbine rectifier controls the power output from a permanent-magnet generator based upon the power reference supplied by a lookup table-based MPPT algorithm. Directly coupled multiple pole generator is used with the rated speed of 18 r.p.m. and the nominal frequency of generator output currents is 14.4 Hz. The control of active power is achieved by regulating the q-axis component of the generator stator currents, while the d-axis current component is kept at zero for the maximum torque per ampere control [22]. The bandwidths of the dq current control loops are designed to be 200 Hz (one-tenth of the VSC switching frequency). The grid-side converter regulates the dc-bus voltage and hence transfers the power supplied by the turbine rectifier to the utility grid. Commercial wind-turbines are controlled to supply either a constant reactive power to the grid, reactive power in proportion to the active power level, or reactive power to maintain the grid-voltage at the specified level based on the droop characteristics [21]. For simplicity, the reactive power supplied to the grid is maintained at zero level in the present study. Moreover, the dc-bus voltage control bandwidth is designed to be 20 Hz, PLL bandwidth to be 30 Hz, and the bandwidth of dq-current control loops is kept at 200 Hz, same as the rectifier. The filter-bus voltage feed-forward is included in the grid-side converter with a gain k_f to reject the disturbances in the grid voltage. The switching frequency of both converters is assumed to be 2 kHz and they are implemented by PWM generator and time-stamped bridge from

RT-Events toolbox discussed before, same as the converters in Type-III wind turbine. Simulation time-step is kept at 20 μs and one switch event is permitted within a time-step, which is sufficient for the selected switching frequency and the simulation time-step. The parameters of the simulated Type-IV wind turbine are given in TABLE III. To demonstrate the use of real-time simulation, Fig. 7 shows the simulated response of Type-IV turbine during a wind gust; the wind speed increases to 10 m/s from 5 m/s for a short duration. The output power and current amplitudes follow a first-order response depending on the inertia constant (H_{tg}) of the turbine-generator system.

TABLE III TYPE-IV WIND TURBINE PARAMETERS

Parameter	Value	Parameter	Value
Rated Power	2.5 MVA	L_m	1.2 mH
Pole Pairs	48	Rated Speed	18 r.p.m.
Flux per Pole	5.18 Wb.T	H_{tg}	4.18 s
L_{ph}	0.2 mH	R_{ph}	0.06 Ω
V_{dc}	2.5kV	C_{dc}	10 mF

B. Impedance Simulation and Experimental Validation

Verification of real-time simulation is usually done in the time domain. Frequency-domain accuracy is more critical for high-frequency stability analysis. To validate the implemented RT model, we have simulated a 10 kW inverter operating with a constant dc-bus voltage and extracted its output impedance responses. The inverter operates at 20 kHz and the simulation time step is 10 μs. These measurements are compared with that from an actual hardware and the analytical models presented in [6]. Fig. 8 shows the comparison of the positive-sequence (Z_p) and negative-

978-1-4799-2706-7/14 $31.00 © 2014 IEEE

The 2014 International Power Electronics Conference

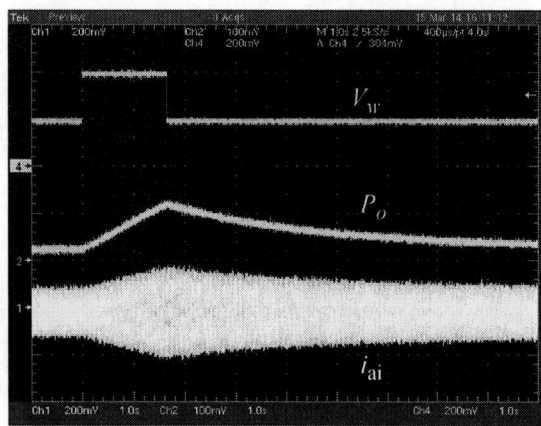

Fig. 7. Simulated behavior of Type-IV turbine during wind gust. Wind speed, V_w (5m/s per div.); output power, P_o (0.85 MW per div.); and phase-a output current, i_{ai} (1 kA per div.).

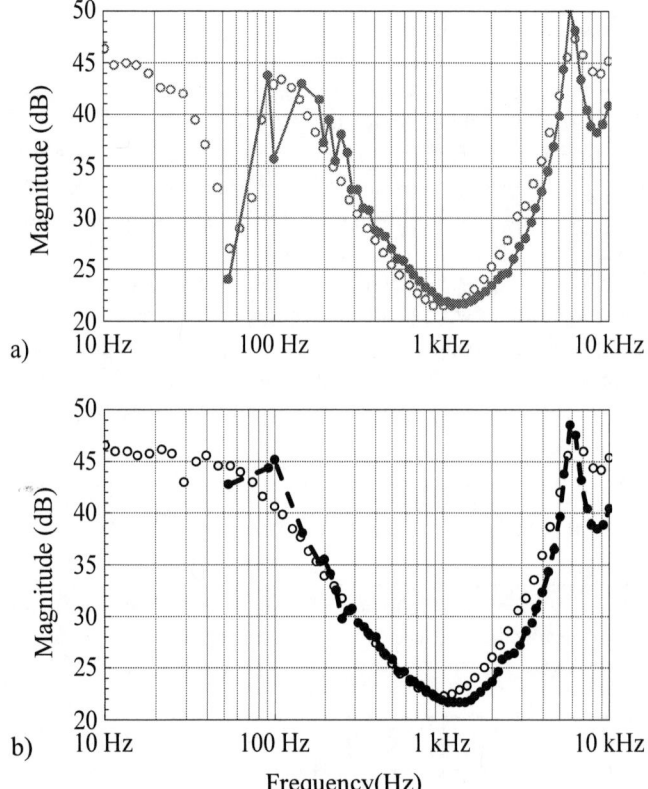

Fig. 8. Impedance response of the 10 kW inverter with constant dc bus voltage. a) Positive sequence impedance; b) negative-sequence impedance (dots: real-time simulation, dotted lines: actual hardware measurement).

sequence (Z_n) output impedances of the inverter. As can be seen, the responses obtained by real-time simulation agree with actual measurements at frequencies up to 6 kHz. This is about the limit of any time-invariant model of switching converters and is sufficient for the evaluation of different control stability and power quality problems in wind farms. For example, Fig. 9 shows the simulated grid current responses for this inverter when connected to a weak grid [6].

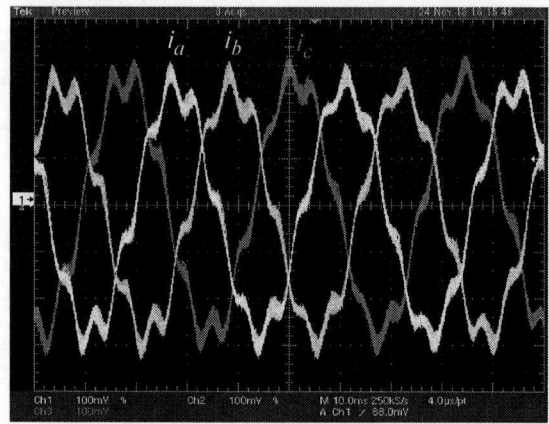

Fig. 9. Sustained harmonic resonance in the output currents of grid-connected inverter obtained from real-time simulation.

Time step has direct effects on the accuracy of real-time simulation and is the cause for the deviation of the simulated impedance responses above 6 kHz in Fig. 8. To further illustrate this effect, Fig. 10 shows the simulated positive-sequence impedance response of the above 10 kW inverter using a time step of 10 µs and 60 µs, respectively, and compares it against analytical model prediction [6].

V. CONCLUSIONS AND FUTURE WORK

This paper discussed real-time simulation of Type-III and Type-IV wind turbines. The utilities of real-time simulation in the study of turbine-grid interactions, particularly in the form of subsynchronous and harmonic resonances, are emphasized and demonstrated. Need of verification of real-time simulation in the frequency domain is discussed and the simulated models are verified by comparing the frequency domain impedance measurements for a scaled-down inverter with that from an actual hardware and analytical model predictions. Future work will include expansion of the simulation model to include more comprehensive grid models and multiple turbines to simulate a wind farm, as well as coupling with advanced control to perform hardware-in-the-loop simulation.

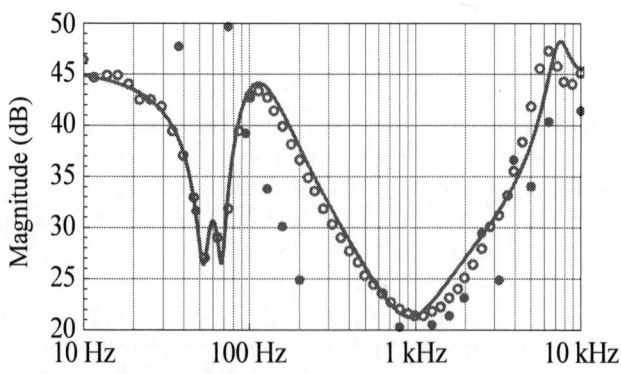

Fig. 10. Comparison of inverter positive-sequence output impedance response obtained by analytical model (solid line) and real-time simulation using 10 µs (circles) and 60 µs time-step (dots).

978-1-4799-2706-7/14 $31.00 © 2014 IEEE

ACKNOWLEDGMENT

This work is supported in part by the National Science Foundation under Award #ECCS-1002265.

REFERENCES

[1] C. Dufour, S. Abourida and J. Belanger, "Real-time simulation of permanent magnet motor drive on FPGA chip for high-bandwidth controller tests and validation", *2006 IEEE International Symposium on Industrial Electronics,* vol. 3, pp. 2591-2596.

[2] H. Saad, T. Ould-Bachir, J. Mahserdjian, C. Doufour, S. Dennetiere and S. Nguefeu, "Real-time simulation of MMCs using CPU and FPGA", *IEEE Trans. Power Electron.*, IEEE Xplore early access.

[3] X. Chen and J.Sun, "Characterization of inverter-grid interactions using a hardware-in-the-loop system test-bed," in *Proceedings of 2011 International Conference on Power Electronics (ECCE Asia 2011),* pp. 2180-2187, May 2011.

[4] M. Cespedes and J. Sun, "Modeling and mitigation of harmonic resonance between wind turbines and the grid," in *Proc. of IEEE Energy Conv. Cong. and Expo.*, 2011, pp. 2109-2116.

[5] M. Cespedes and J. Sun, "Renewable energy systems instability involving grid-parallel inverters," in *Proc. of Applied Power Electron. Conf. and Expo.*, 2009, pp. 1971-1977.

[6] M. Cespedes and J. Sun, "Impedance modeling and analysis of grid-connected voltage-source converters," *IEEE Trans. Power Electron.*, vol. 29, no. 3, pp. 1254-1261, Mar. 2014.

[7] J. Sun, "Impedance-based stability criterion for grid-connected inverters", *IEEE Trans. Power Electron.*, vol. 26, no. 11, pp. 3075-3078, Nov. 2011.

[8] A. Perdana, "A contribution towards the establishment of standardized models of wind turbines for power system stability studies", Ph. D. dissertation, Dept. of Energ. and Env, Chalmers Univ. of Tech., Goteborg, Sweden, 2008.

[9] J. Hu, Y. He, "Reinforced control and operation of DFIG-based wind-power-generation system under unbalanced grid voltage conditions", *IEEE Trans. on Ener. Conv.*, vol. 24, no. 4, pp. 905-915, Dec 2009.

[10] G. Irwin, A. Jindal, and A. Isaacs, "Sub-synchronous control interactions between type 3 wind turbines and series compensated ac transmission systems," in *Proc. IEEE Power Energy General Meet.*, Jul. 2011, pp. 1–6

[11] W. Wong and J. Daniel. (2010). Ercot CREZ reactive study. ABB [Online]. Available: http://www.ercot.com/content/meetings/rpg/keydocs/2010/0312/ABB_RPG_presentation.pdf

[12] Opal-RT Technologies, "OP5600 off-the-shelf Hardware-in-the-Loop (HIL) simulator," available at www.opal-rt.com/product/op5600.

[13] A. Myaing and V. Dinavahi, "FPGA-based real-time emulation of power electronic systems with detailed representation of device characteristics," *IEEE Trans. Ind. Electron.*, vol. 58, no. 1, Jan. 2011.

[14] K. L. Lian and P. W. Lehn, "Real-time simulation of voltage source converters based on time average method," *IEEE Trans. Power Sys.*, vol. 20, no. 1, pp. 110-118, Feb. 2005.

[15] B. D. Kelper, L. A. Dessaint, K. Al-Haddad, and H. Nakra, "A comprehensive approach to fixed-step simulation of switched circuits," *IEEE Trans. Power Electron.*, vol. 17, no. 2, pp. 216-224, March 2002.

[16] A. Yazdani and R. Iravani, *Voltage-Sourced Converter in Power Systems*, Hoboken, NJ: John Wiley & Sons, Inc., IEEE Press, 2010.

[17] P. Krause, O. Wasynczuk and S. Sudhoff, *Analysis of Electric Machinery and Drive Systems*, 2nd ed. Piscataway, NJ: IEEE Press, 2002.

[18] L. Fan, R. Kavasseri, Z. Lee, and C. Zhaa, "Modeling of DFIG-based wind farms for SSR analysis", *IEEE Trans. on Power Del.*, vol 25, no. 4, pp. 2073-2082, Oct. 2010.

[19] IEEE Committee Report, "First benchmark model for computer simulation of subsynchronous resonance," *IEEE Trans. on Power App. Systm.*, vol. 96, no.5, pp1565-1672, Sep./Oct. 1977.

[20] E. Larssen, "Wind Generators and series-compensated AC transmission lines", *IEEE PES T&D Conference*, May 2012.

[21] K. Clark, N. W. Miller, and J. Sanchez-Gazca, "Modeling of GE wind turbine generators for grid studies," GE Energy, Ver. 4.5, General Electric International, Inc., One River Road, Schenectady, NY, April 16, 2010.

[22] S. Li, T. A. Haskew, R. P. Swatloski, and W. Gathings, "Optimal and direct-current vector control of direct-driven PMSG wind turbines," *IEEE Trans. Power Electron*, vol.27, no.5, pp.2325-2337, May 2012.

Technologies for Mitigating Fluctuation Caused by Renewable Energy Sources

Shuji Katoh and Shinya Ohara
Hitachi Research Laboratory
Hitachi, Ltd, 7-2-1 Omika-cho, Hitachi, Ibaraki, Japan

Tomomichi Itoh
Infrastructure Systems Company
Hitachi, Ltd, 5-2-1 Omika-cho, Hitachi, Ibaraki, Japan

Abstract— In this paper, we would like to review and introduce technologies which mitigate problems in an electric power grid caused by installation of renewable energy sources (RES). We classified these technologies into three types of technologies in an analogy of "medical therapy" categorization, which are "causal therapy", "symptomatic therapy", and "infection prevention". As for the causal therapies, active-power-based control for reducing short-term output power fluctuations and a hybrid system of battery storage and wind turbines for long-term output power fluctuation are described. For the symptomatic therapy, a reactive power control is mentioned. In terms of the infection prevention, harmonic reduction technology is introduced.

Keywords— *Active-power-based cntrol, a hybrid system of battery storage and wind turbines, reactive power control, and harmonic reduction*

I. INTRODUCTION

Renewable energy sources (RES) such as photovoltaic (PV) generation systems and wind turbines are thought to be one of promising and attractive energy sources.

On the other hand, they are an intermittent power supply. Hence, they can bring some problems into the connecting grid; such as voltage fluctuation, frequency fluctuation, increase of harmonics, and instability of the power system. To introduce more RES and to achieve conservation of power quality at the same time, we have been developing new technologies. In this paper, we will introduce them in an analogy of "medical therapy".

They can be classified into three types of technologies. The first ones are technologies like "causal therapy". Underlying cause of these issues is output power fluctuation from RES which could lead to "the several diseases", such as voltage and frequency fluctuations. Thus, by mitigating the fluctuation of output power, "most of the diseases" are healed.

The second one is like "symptomatic therapy". It mitigates "a symptom" like voltage fluctuation.

The third one is like "infection prevention". It resolves problems caused by other reasons.

II. CAUSAL THERAPY" FOR OUTPUT POWER FLUCTUATIONS

A. Short-term fluctuations: Active-power-based Control or wind turbines

"Active-power-based control for wind turbine system" which can reduce the short-term output power fluctuations has been developed [1]. A wind turbine system works converting wind power into rotational power by its blades and at the same time it transforms the rotational power into electric power. Accordingly, changes in wind speed cause the rotational speed of the generator's rotor to vary. Because the generated electric power is proportional to the product of the torque and rotor speed, use of the conventional torque control tends to result in an output power fluctuation. On the other hand, the developed active-power-based control absorbs the changes of wind power as the rotational (mechanical) energy of the gears, the rotor, and blades, and supplies a smoothed active (electrical) power from the PCS (Power Conditioning Systems) to the grid according to the active power command.

Fig. 2.1 shows the example of operational data measured in a wind turbine that is equipped with the active-power-based control technique. The generated electrical power is maintained constant despite sudden changes in wind speed by allowing the rotor speed to vary. The result is a "grid-friendly wind turbine" that minimizes the disturbance to the power grid system.

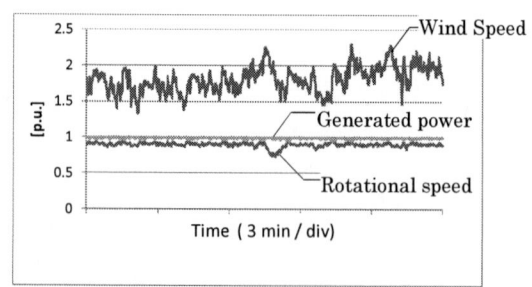

Fig. 2.1 Measured data of wind turbine equipped with the active-power-based control

B. Long term flucutations: Control for hybrid systems of battery storage and wind turbines

In order to mitigate the long-term output power fluctuation, a hybrid system of battery storage and wind turbines is suitable. In such a system, reduction of the life cycle cost of batteries is important.

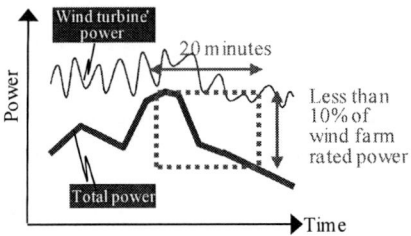

Fig. 2.2. Requirement for power fluctuation mitigation defined in one of Japanese grid codes

Hence, a control method for hybrid systems was developed, which has a stand-by mode for the battery storage's operation to reduce the life cycle cost of batteries [2].

Fig. 2.2 explains a grid code in one area in Japan, which obligates wind farms to confine their output power fluctuations in 20 minutes to be less than 10% of their rated output power [3].

Lifetime of batteries highly depends on the number of charge and discharge cycles [4]. Reduction of charge/discharge operations can extend battery lifetime.

Fig. 2.3 shows comparison between a conventional control method and our proposed control. In the conventional control, a first-order lag filter calculates a target value of total power from the wind turbine (WT) power, and the batteries charge or discharge to track the target value of the total power. As a result from the first-order lag filter, the target value of the total power and WT power differ almost all the time, so the batteries operate almost always. On the other hand, our proposed control keeps the battery storage on stand-by when the fluctuation of the wind farm output is less than 10%.

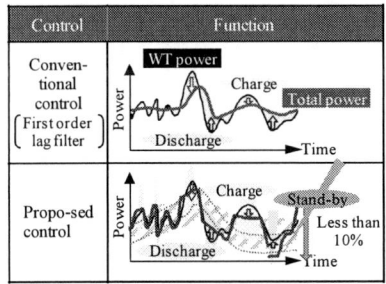

Fig. 2.3. Comparison of developed control and the conventional one

Performance of the developed control in the first commercial hybrid wind farm was evaluated [5]. The wind farm consists of eight 2-MW wind turbines, and has a nominal power of 15.44 MW. The battery storage has a maximum power of about 20% of the wind farm's nominal power. Valve-regulated lead-acid batteries are used in this wind farm.

Fig. 2.4 shows the measured data from the wind farm. The upper graph shows the output power of the wind farm. The lower graph shows the fluctuation of wind farm output, which, as we expected, is almost less than 10%. The graphs show also the operation of the battery storage. In the upper graph the battery storage did not operate from midnight to about five o'clock because fluctuation of the wind turbine output power was less than 10%.

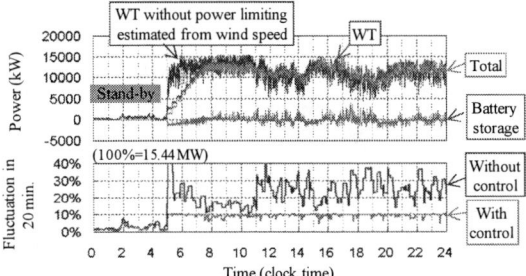

Fig. 2.4. Measured data at the wind farm equipped with developed control

III. "SYMPTOMATIC THERAPY" FOR VOLTAGE FLUCTUATIONS: REACTIVE POWER CONTROL OF WIND TURBINES TO MITIGATE THE VOLTAGE FLUCUTATOIN

One of well-known "symptoms" caused by the variation of RESs' output is voltage fluctuation. Reactive power control has possibility to mitigate the fluctuation. The appropriate amount of reactive power to mitigate the fluctuation depends on the grid impedance, especially the ratio between its resistance (R) and reactance (X). Our concept is to estimate the ratio of R and X of the grid online, and to provide appropriate reactive power based on the estimated ratio from the PCS (Power Conditioning System) in wind turbines [6].

Fig. 3.1 illustrates a model of a wind turbine. P_w and Q_w are the active and reactive power outputs from the wind turbine, respectively. The impedance (R + jX) contains also the impedance of the loads connected in the utility system. The impedance could vary depending on load condition. In Fig. 3.1, the relation between the voltage and power is given by the following equation.

$$\left(RP_w + XQ_w - V_r^2\right)^2 + (RQ_w - XP_w)^2 = V_r^2 V_s^2. \tag{1}$$

From Eq. (1), a simplified expression of the optimum reactive power Q_w is obtained as Eq. (2).

$$Q_w = -\frac{R}{X} P_w = -\alpha P_w \tag{2}$$

Eq. (2) indicates that when Q_w is in proportional to P_w with an adequate coefficient of α, the grid voltage fluctuation can be suppressed. Since R and X vary reflecting load changes in an order of hours, it is necessary to automatically tune the estimation of α to minimize the voltage fluctuation under any conditions.

978-1-4799-2706-7/14 $31.00 © 2014 IEEE   851

Table 3.1. The rule for automatic-tuning of α

Sign of $\Delta Pw \cdot \Delta Vr$	(Power factor)	Required Operation
Positive (+)	Smaller than Optimum Value	Increase α
Negative (-)	Larger than Optimum Value	Decrease α

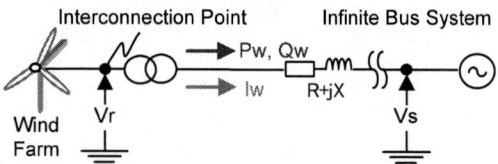

Fig.3.1. A wind turbine and power grid model

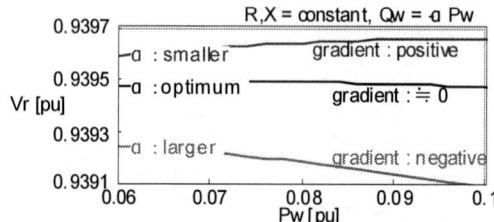

Fig. 3.2. A relationship between grid voltage and α

The control method estimates α by using active power fluctuation from the wind turbine itself. The grid voltage fluctuation ΔV_r caused by the active and reactive power of the wind turbine (P_w, Q_w) is defined by Eq. (3) , where variables are expressed in per-unit,

$$\Delta V_r = R P_w + X Q_w. \qquad (3)$$

In the case of $Q_w = -\alpha P_w$, Eq. (3) is expanded as follows,

$$\Delta V_r = R P_w + X(-\alpha P_w) = X\left(\frac{R}{X} - \alpha\right) P_w . \qquad (4)$$

Fig. 3.2 is a schematic figure of Eq. (4). When α is smaller than the optimum value, an increase of ΔP_w induces an increase of ΔV_r and a decrease of ΔP_w leads to a decrease of ΔV_r. On the other hand, if α is larger than the optimum value, an increase of ΔP_w induces a decrease of ΔV_r, and a decrease of ΔP_w leads to an increase of ΔV_r. Therefore, the optimum coefficient α can be tuned according to Table 3.1.

In this paper, as a tuning method of α, Eq. (5) is used

$$\alpha = \int_0^t K \cdot sgn(\Delta P_w) \cdot \Delta V_r dt \qquad (5)$$

where "sgn" denotes the signum function and the K is the control gain for automatic-tuning control.

Fig. 3.3 illustrates the block diagram for estimating the optimum coefficient α . LPFs (<u>L</u>ow <u>P</u>ass <u>F</u>ilter) attenuate noise and HPFs (<u>H</u>igh <u>P</u>ass <u>F</u>ilter) cut off the dc and low frequency components in the active power P_w and the grid voltage V_r at interconnection point.

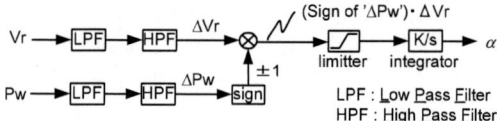

Fig. 3.3. The block diagram of the automatic tuning control

Simulation studies were carried out to confirm the validity of the control scheme as shown in Fig. 3.4. In order to simulate the behavior of the generated power from the wind turbine, the Kaimal turbulence model given in [7] was introduced to the simulation. The Kaimal model demonstrates typical wind speed fluctuations. Two load conditions of 1.5 MW and 5 MW were considered.

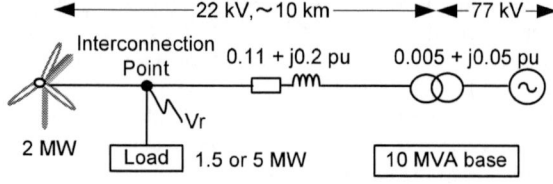

Fig. 3.4. Transmission system model for the simulation.

The graphs in Fig. 3.5 show, from top to bottom, the waveforms of wind speed, the active power P_w, α, the reactive power Q_w and the grid voltage V_r.
Figs. 3.5(a) and (b) show the simulation results with different load conditions. The optimum coefficient α depends on the load connected to the grid. The optimum value of α from Eq. (1) is 0.407 with a 1.5-MW load (Fig. 3.5(a)), and 0.452 with a 5-MW load (Fig. 3.5(b)).

The voltage control was activated at t = 100 s. One can observe that for both load conditions, the voltage fluctuations were suppressed and the parameter α converged to a constant value.

978-1-4799-2706-7/14 $31.00 © 2014 IEEE

The 2014 International Power Electronics Conference

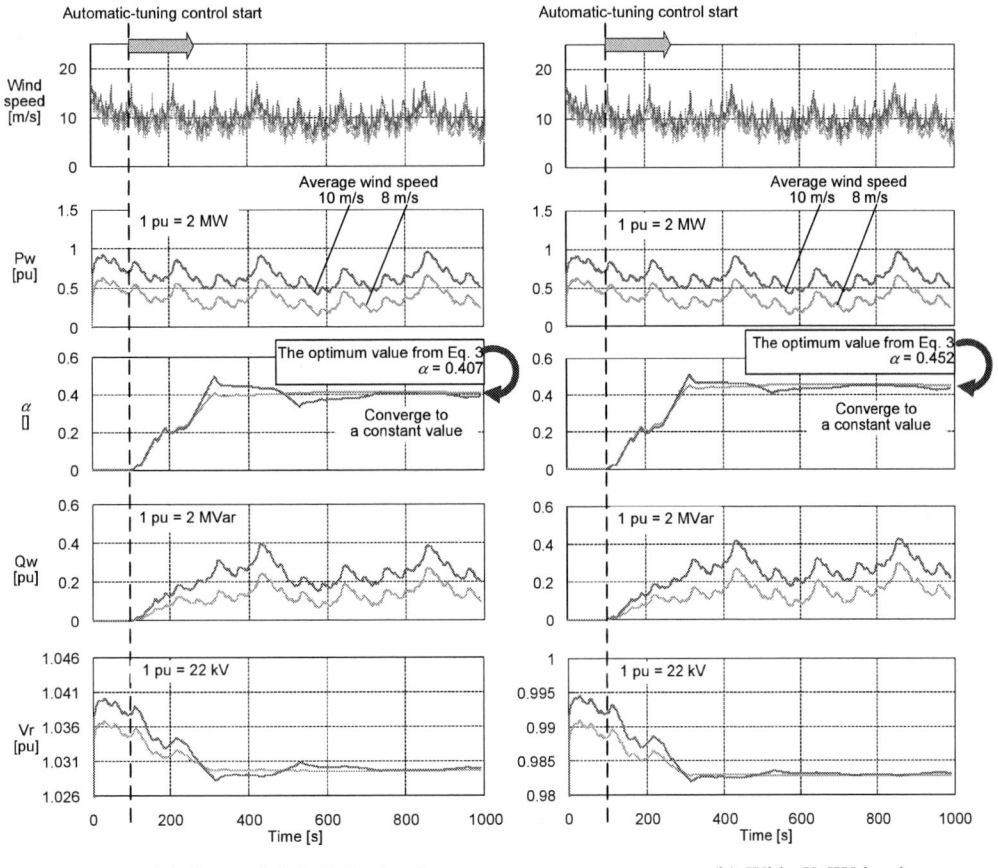

(a) changed.th 1.5MW-load

(b) With 5MW-load

Fig. 3.5. Waveforms of the automatic-tuning simulations

To confirm the validity of the proposed control scheme in an actual power system, experiments of scaled model of WTs, that is a 1/8000 model of 35-MW wind turbines (WTs), combined with a grid simulator were carried out with the Kansai Electric Power Co., Inc. (KEPCO). The grid simulator "APSA" (Advanced Power System Analyzer) shown in Fig. 3.6 is an analog simulator of KEPCO and it is possible to analyze the major behaviors of the whole power system of KEPCO. The scaled model consists of a motor which emulates input torque from wind, a doubly-fed generator (Fig. 3.7) and a PCS which drives the doubly-fed generator (Fig. 3.8).

Fig. 3.9 shows the single-line configuration of the power system APSA simulates. It is a part of KEPCO's 77-kV power system. Several hydroelectric power generators, loads, and the scaled model of WTs are connected to the grid. Fig. 3.10 shows experimental results. The graphs show the waveforms of output power from hydroelectric generators P_H, output power from WTs P_w and Q_w, the grid voltage V_r, and coefficient α from the top of Fig. 3.10. The generated power from the hydroelectric generators changes from 50 to 100% of their rated output power at $t = 4300$ s where the total rated output power of hydroelectric generators corresponds to 0.3 pu of the rated power of the power system under discussion.

When the power flow changes, the equivalent impedance

seen from the WTs also changes. The proposed control was activated at $t = 1200$ s.

Fig. 3.6. Advanced Power System Analyzer (APSA)

Fig. 3.7. Scaled model of WTs

978-1-4799-2706-7/14 $31.00 © 2014 IEEE   853

The 2014 International Power Electronics Conference

Fig. 3.8. PCS for controlling a scaled model of WTs

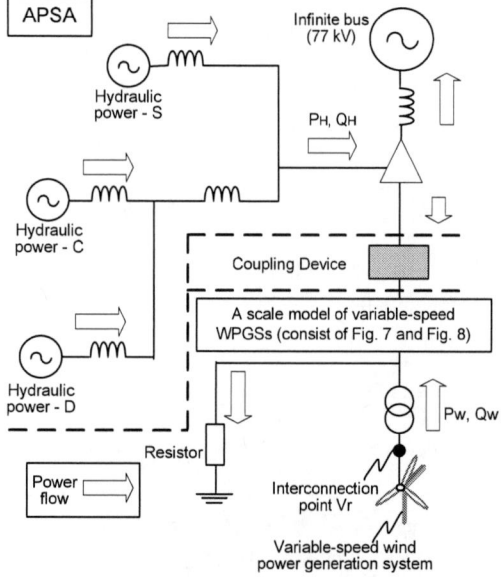

Fig. 3.9. APSA simulation model.

The waveforms show that the voltage fluctuations are suppressed from 3.5% to 0.6% by the reactive power control. The equivalent impedance of the grid changes depending on the increase from 0.15 to 0.3 pu of the active power P_H of hydroelectric generation plants. The voltage fluctuations are kept successfully below 0.4% even after a significant change of the active power P_H of the hydroelectric generation power plants. The validity of the proposed control is confirmed by experimental results as well as computer simulations.

KEPCO has its own rule that requires the generators connected to high-voltage grid to keep the voltage fluctuations caused by them less than 2%. The experimental results show that the wind farms with the proposed control scheme can comply with the requirement. Therefore, it is possible to say that more WTs can be connected to the grid with the proposed control scheme from the viewpoint of voltage fluctuations.

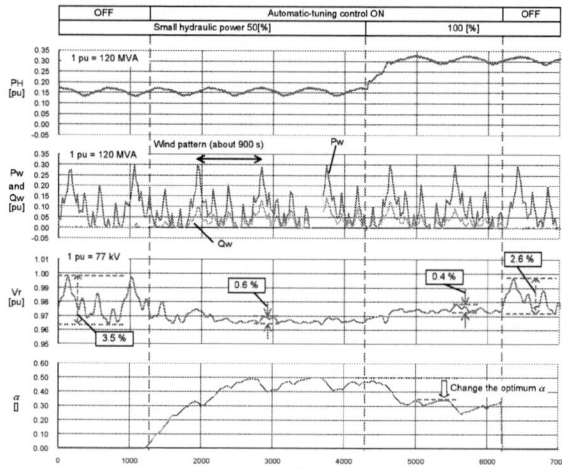

Fig. 3.10. Experimental results of optimum reactive power control

Therefore, from the viewpoint of voltage fluctuation, it is possible to say that the proposed control scheme enables the WTs to match the requirement from KEPCO at the same time.

IV. "INFECTION PREVENTION" FOR SUPPRESSION OF HARMONIC EMISSION

In most grid codes, harmonic current emission from a generating system is regulated [8]. Hence, filter technologies or multilevel converter technologies have been developed to make the output voltage waveform close to a pure sinusoidal wave. However, if the grid itself has a background voltage distortion, harmonic currents flow into the power conditioning system (PCS). To address this problem, our control scheme reduces harmonic currents even in distorted grid voltage.

Our concept, disturbance compensation control, is that the VSI (Voltage Source Inverter) produces harmonic voltages same as those contained in the grid voltage. One of the greatest benefits of this method is the low interference with the current control in the PCS. This control can be realized only using the measured grid voltage signals and that does not make any change in the characteristics of the existing current control.

From these reasons, the disturbance compensation control is adopted [9]. Fig. 4.1 shows the control block diagram of a 400-kW PCS using the proposed control.

The block diagrams surrounded by the thick dashed line are the new control blocks for the harmonic current reduction control. These control blocks extract the specific harmonic voltages by using Fourier coefficient

978-1-4799-2706-7/14 $31.00 © 2014 IEEE

expansions (FCEs). FCE has a great benefit that there is no interference with the other harmonic components in its output. This benefit is significant because the frequencies of the harmonic voltages are close to each other. FCE also allows us to extract both the positive- and negative-sequence components of the specific harmonic voltage easily.

By superimposing the compensation voltage references $\mathbf{v}_{gc}$ calculated with the outputs of the FCEs onto the reference signal of the output voltage of the VSI, the VSI can output the harmonic voltages, which are very similar to those contained in the grid voltage.

Fig. 4.2. is an experimental circuit diagram for the prototype of the PV generator system. The circuit consists of two PCSs. PCS1 represents a 3 x 400-kW PCS with the harmonic current reduction control. PCS2 represents the other PCSs. Each PCS is connected to PV simulators that simulate the characteristics of solar panels.

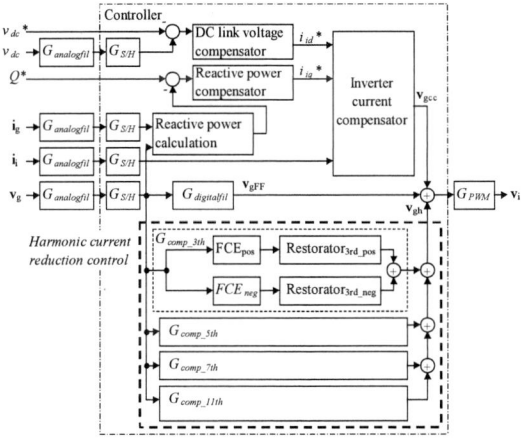

Fig. 4.1. Block diagram of the harmonic current reduction control

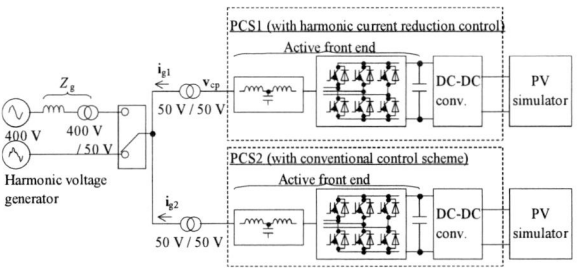

Fig. 4.2. Experimental circuit diagram.

At the grid side, two AC voltage sources can be connected. One is the commercial grid with grid impedance Z_g. The impedance Z_g simulates the impedance at the high voltage grid in Hokuto site. It is also possible to do grid fault experiments with the circuit. The other AC voltage source is the programmable harmonic voltage generator, which supplies desired harmonic voltages. It was used to evaluate the new control scheme.

To evaluate the performance of the new control, every parameter except the harmonic current reduction control in both PCSs is set the same.

Fig. 4.3. shows the results of Fourier analysis of the grid currents from PCS1 (i_{g1}) and PCS2 (i_{g2}). Here PCS1 denotes the power conditioning system equipped with the proposed control, and PCS2 denotes the one controlled in a conventional manner. It is clear from the figure that the new harmonic current reduction control suppressed the harmonic current successfully and the harmonics are lower than the target values.

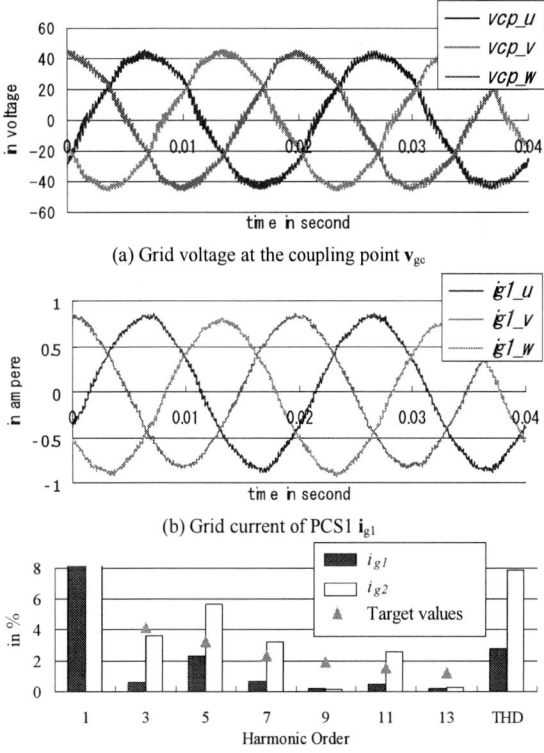

(a) Grid voltage at the coupling point $\mathbf{v}_{gc}$

(b) Grid current of PCS1 $\mathbf{i}_{g1}$

(c) Results of Fourier analysis $\mathbf{i}_{g1}$ and $\mathbf{i}_{g2}$

Fig. 4.3. Comparison of the proposing harmonic current reduction control and the conventional control

V. CONCLUSIONS

In this paper, technologies for mitigating issues caused by installation of RES were introduced.
Underlying cause of discussed issues is output power fluctuation from RES. For short-term fluctuation active-power-based control was proposed. For long-term fluctuation a battery storage system with stand-by mode is introduced.

Most possible issue caused by RES is voltage fluctuation. In order to mitigate it we proposed the optimum reactive power control which can estimate the optimal amount of reactive power even if the impedance of connecting power grid is changed.

Another issue caused by PCSs of RES is harmonic current emission. In order to reduce the harmonic current we proposed disturbance compensation control for disturbed grid voltage.

ACKNOWLEDGMENT

This paper describes the voltage fluctuation suppression technology studied with KEPCO. It was verified by APSA with KEPCO. The authors express great thanks to KEPCO.

This paper also describes a control technique developed in the Japanese national project "Verification of Grid Stabilization with Large-scale PV Power Generation Systems – Hokuto site". And the development of the control scheme is one of the results of the contract research offered by new energy and industrial technology development organization (NEDO). The authors express great thanks to all the people involved with this project, especially NTT Facilities.

REFERENCES

List only one reference per reference number according to the following samples:

[1] K. Sakamoto, T. Matsunobu, K. Sato, and S. Kondo, "Large Wind Turbine System and Smart Grid," *Hitachi Review*, vol. 60, no. 7, pp. 394-398, 2011.

[2] S. Oohara, N. Hoshino, K. Suzuki, and T. Hashimoto, "Battery Control Method for Hybrid Wind Power System," *Proceedings of PCIM 2011*.

[3] M. Young, "The PWM strategy on DC-DC converter," *IEEJ Journal of Industry Applications*, vol. 28, no. 15, pp. 123-129, 1989.

[4] G. Eason, B. Noble, and I. N. Sneddon, "On certain integrals of Lipschitz-Hankel type involving products of Bessel functions," *IEEE Trans. on Power Electronics*, vol. 247, no. 8, pp. 529-551, 1995.

[5] J. Clerk Maxwell, "A treatise on electricity and magnetism," *IEEE Trans. on Industry Applications*, vol. 589, no. 2, pp. 68-73, 2010.

[6] Y. Nakayama, S Ohara, M. Ichinose, M. Futami, T. Haswgawa, O. Iso, and M. Orui, "Optimum Reactive Power Control of Wind Power Generation," *Proceedings of PCIM 2011*.

[7] G. Eason, B. Noble, and I.N. Sneddon, "On certain integrals of Lipschitz-Hankel type involving products of Bessel functions," *Phil. Trans. Roy. Soc. London*, vol. A247, pp. 529-551, April 1955.

[8] J. Clerk Maxwell, *A Treatise on Electricity and Magnetism*, 3rd ed., vol. 2. Oxford: Clarendon, 1892, pp. 68-73.

[9] T. Ito, H. Miyata, M. Taniguchi, T. Aihara, N. Uchiyama, and H. Konishi, "Harmonic current reduction control for grid-connected PV generation systems," *Proceedings of IPEC*, pp. 1695-1700, 2010.

Reliability-oriented Energy Storage Sizing in Wind Power Systems

Zian Qin[1], Marco Liserre[2], Frede Blaabjerg[1], Poh Chiang Loh[1]

[1]Department of Energy Technology, Aalborg University
Pontoppidanstraede 101, DK-9220 Aalborg East, Denmark
zqi@et.aau.dk, fbl@et.aau.dk, pcl@et.aau.dk

[2]Christian-Albrechts-Universität zu Kiel
Kaiser Strasse 2, 24143 Kiel, Germany
ml@tf.uni-kiel.de

Abstract- **Energy storage can be used to suppress the power fluctuations in wind power systems, and thereby reduce the thermal excursion and improve the reliability. Since the cost of the energy storage in large power application is high, it is crucial to have a better understanding of the relationship between the size of the energy storage and the reliability benefit it can generate. Therefore, a reliability-oriented energy storage sizing approach is proposed for the wind power systems, where the power, energy, cost and the control strategy of the energy storage are all taken into account. With the proposed approach, the computational effort is reduced and the impact of the energy storage system on the reliability of the wind power converter can be quantified.**

Keywords- **wind turbine system, reliability evaluation, size of energy storage system.**

I. INTRODUCTION

The reliability of large scale wind turbine system (WTS) is critical in order to reduce cost of energy seen during life-time and it has attracted more and more attention in terms of the thermal performance investigation, comparison of the topologies from the thermal aspects, and also the control strategies [1]-[4]. Power fluctuations in wind power converters can lead to thermal variations of the power modules, and therefore need to be considered in order to save the cycle life. Energy storage technology, which is normally used to suppress the power fluctuations [5], thus can be employed to improve this problem. Actually, there are already commercial wind turbines with energy storage for ramp control, predictable power, and frequency regulation [6], which are operating with derivative functions of the power fluctuation smoothing.

As we know, the main barrier for the wind power application of energy storage technology is its high cost. Therefore, it is necessary to investigate the relationship between the size of the energy storage and the reliability benefit it can generate. In order to do that, an appropriate lifetime model is needed to evaluate the reliability of the wind power converter and the energy storage system. So far, the life-time models widely used are based on the number of the thermal cycles, like the Coffin-Manson model and its derivative versions [3], [7]. They can achieve good performance, when the power modules

work with a regular mission profile, for instance in railways, where only a few typical loading cycles are taken into account [8]. But in a wind power application, the mission profile of the power modules is more complicated due to the intermittent nature of the wind speed, the wind speed distribution, the issues like turbulences, and also controller in the power electronics. Therefore a time-domain stress-strain based life-time model is more appropriate to be used [9]. With this model, the shape instead of the number of the thermal cycles is considered; therefore make the life-time evaluation more accuracy.

In this paper, an improved reliability evaluation approach of the wind power converter based on the stress-strain lifetime model is illustrated when introducing energy storage. Then the impact of the size of the energy storage system on the reliability of the wind power converter is evaluated based on the proposed approach, where the cost, power, energy and control strategy of the energy storage system are all taken into account.

II. WIND TURBINE SYSTEM MODEL

The structure of a 3 MW PMSG based WTS is shown in Fig. 1. A conventional two-level Back-To-Back converter (BTB) is employed as a full-scale power interface from the generator to the grid, where the dc bus voltage is 1100 V and the ac distribution line-to-line voltage is 690 V. The switching frequency is fixed at 1.5 kHz for both generator side and grid side converters. The aerodynamic and generator parameters are obtained from [10] and they are also listed in Table I. The maximum power point tracking (MPPT) control is applied in the generator side converter, while the DC bus voltage is controlled to be constant by the grid side converter. The reactive power of the grid side converter is set to be zero in normal grid operation. It should be noted that the time constant of the MPPT control is much larger than the other controllers in the control system. The impact of the control system on the power fluctuation is therefore mainly due to the MPPT control, which has a 5 s time constant or above.

An ESS is connected to the DC link of the BTB, and

978-1-4799-2706-7/14 $31.00 © 2014 IEEE

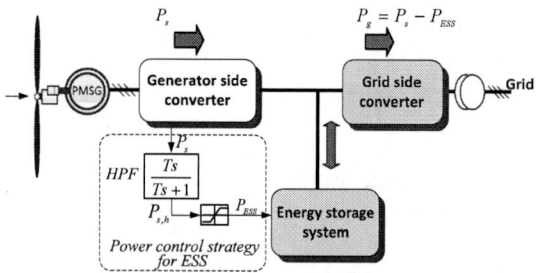

Fig.1. Wind turbine system with two-level back to back power converter and an energy storage system for power smoothing of grid side converter.

TABLE I.
WIND POWER SYSTEM PARAMETERS FOR 3 MW POWER LEVEL

Symbol	Name	Value
Aerodynamic parameters		
R	Blade radius	46.9 m
$V_{w,in}$	Cut-in wind speed	3 m/s
$V_{w,r}$	Rated wind speed	12.5 m/s
$V_{w,off}$	Cut-off wind speed	25 m/s
$C_{p\,max}$	Maximum power coefficient	0.48
λ_{opt}	Optimal tip speed ratio	8.1
PMSG parameters		
n_r	Rated rotor speed	15 rpm
N_g	Gear ratio	6.36
N_p	Number of pole pairs	20
ψ_m	Magnetic induced flux	2.8 Wb
L_s	Stator inductance	0.18 mH

its power control strategy is realized by a high pass filter, as shown in Fig. 1. The wind power P_s is first filtered by the high pass filter. Then the high frequency component $P_{s,h}$, which is actually the short term power fluctuation, is considered to be the charging/discharging power to the ESS. Since there is normally a power limitation to the ESS, the infinite power PESS is finally absorbed or generated by the ESS. The impact of the time constant T on the performance of the ESS will be discussed in Section V. It should be noted that the performance of the ESS can be influenced by its dynamic characteristic, which is normally several seconds to several minutes depending on the kind of the energy storage device used. However, in this study the fast energy storage devices like the supercapacitors and the lithium-ion batteries will be employed, and considering the relatively longer period of the short term power fluctuation, the time delay of the ESS is therefore ignored.

III. LIFE-TIME EVALUATION METHOD

The conventional thermal evaluation method is shown in Fig. 2. The wind speed V_W is firstly employed by the wind turbine and generator model to obtain the electrical parameters (V_G, I_G). Then the temperature of the hotspot is calculated by the electro-thermal model of the power devices. Finally, based on the profile of the junction temperature T_j, the life-time can be evaluated by the lifetime model.

The loss information is offered by the datasheet of the

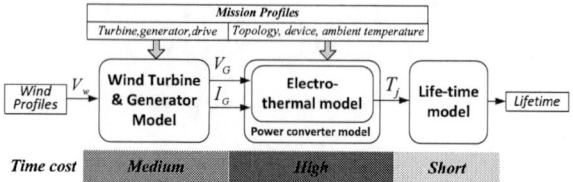

Fig. 2. Lifetime evaluation method based on the mission profile and the time cost.

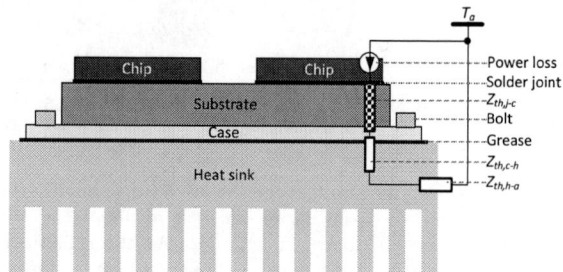

Note:
T_a: ambient temperture
$Z_{th,j-c}$: thermal impedence from junction to case
$Z_{th,c-h}$: thermal impedence from case to heat sink
$Z_{th,h-a}$: thermal impedence from heat sink to ambient

Fig. 3. Basic scheme of the power module and its thermal model.

power devices (HiPak^{TM} 5SNA 2400E170100), and its calculation approach can be found in [11], [12]. The thermal model of the power module is shown in Fig. 3 [13], where the power loss is considered as a current source, the relative temperature is like the voltage drop on the thermal impedance between the different areas in the power module. Normally, the basic thermal impedance network contains three parts, including the thermal impedances from junction to case $Z_{th,j-c}$, case to heat sink $Z_{th,c-h}$, and heat sink to ambient $Z_{th,h-a}$. The first part is provided by the datasheet of the power device. The second one depends on the thermal grease used and the compaction degree between the power module and the heat sink. Its value is therefore not able to have high accuracy, and the typical value is 0.003 K/W for IGBT and 0.006 K/W for the antiparallel diode. The third part of the thermal impedance from heat sink to the ambient $Z_{th,h-a}$ is much more complicated, which is related to the cooling approaches, the specification of the heat sink, and the environmental factors. Fortunately, in this study of energy storage evaluation the heat sink model with high precision is not necessary, because the temperature of the heat sink changes slowly compared with the junction temperature. It is therefore assumed to be constant at 50°C.

The conventional method works well for the thermal performance estimation in a short time scale. However, due to the random nature of the wind power, it is needed to utilize a relatively long wind profile in order to get some knowledge about the lifetime of the wind power converter in the field. But in this case the detailed electro-thermal model will cost much more computational effort. Fortunately, a good correlation between the power and temperature profile of the grid side converter is found when the conventional method mentioned above is used,

The 2014 International Power Electronics Conference

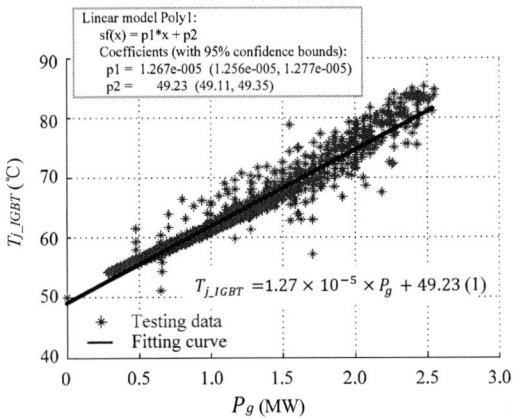

Fig. 4. Correlation between power and junction temperature in the grid side converter [14].

and it is shown in Fig. 4 [14]. Thus the detailed electro-thermal model is simplified by (1) for the temperature evaluation, where the computational effort can be reduced significantly.

IV. LIFE-TIME MODEL

When the temperature profile is obtained, life-time models can be used to evaluate the lifetime of the power converter, where the thermal cycling based life-time models are normally used and one of them, named Coffin-Manson-Arrhenius Model, is represented as the following,

$$N_f = \alpha \cdot (\Delta T_j)^{-n} \cdot e^{E_a/(k \cdot T_{jm})} \quad (2)$$

where N_f is the number of cycles to failure, ΔT_j is the amplitude of junction temperature cycling, T_{jm} is the maximum junction temperature, E_a is activation energy for degradation, and k is the Boltzmann constant. The other parameters like α and n are usually extracted based on experimental failure-accelerated test.

With this kind of lifetime model, the Rainflow Algorithm [15] is used firstly to draw out the characteristics of the thermal cycles, in terms of the number, the amplitude, the maximum value, and sometimes the cycle period. So the thermal cycles no matter in what shapes, if they have the same value of those parameters, will be considered as the same by the lifetime model. However, it may lead to big error in some cases. For example, the two thermal cycles in Fig. 5 have the same amplitude, maximum value, and cycle period. But for the second one, since the temperature changes very little in the beginning, the whole thermal cycle is thus equivalent to the narrow cycle on the right side. According to the test results in [8], where the thermal cycles with different periods will lead to different cycle life, the two thermal cycles #1 and #2 will therefore have different impact on the lifetime of the power modules. As a consequence, the lifetime model in the time domain, where the temperature in real time is used, can be more accurate for the lifetime evaluation.

The lifetime model in the time domain uses the deformation energy W_{tot} to evaluate the reliability, which

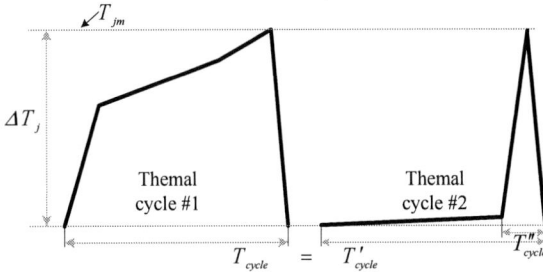

Fig. 5. Thermal cycles with the same characteristics but different shape.

TABLE II
MATERIAL PARAMETERS OF SEPARATED MODEL FOR SOLDER
60SN40PB [16], [17].

Elastic		Plastic	
G_0 (Pa)	G_1 (Pa)	C_p	m_p
1.310e10	0.559e8	2.3e13	5.6
Dislocation Controlled Creep						
C_l (K/s/Pa)	C_h (K/s/Pa)	α	n_l	n_h	Q_l (J/mol)	Q_h (J/mol)
2e-5	2.5e-1	1289	5.0	3.0	4.85e4	8.15e4
Diffusion Controlled Creep			
B_1 (K/Pa)	B_2 (K/Pa)	Q_b (J/mol)	Q_m (J/mol)
1.09e-17	2.06e-8	5.45e4	8.75e4

is defined as the area surrounded by the hysteresis stress-strain curve, accumulated in the solder layer leading to failures [9]. The strain contains the time-independent elastic and plastic strain, and time-dependent plasticity, called creep, which are defined as the following,

Elastic strain, $$\gamma_e = \frac{\tau}{G} \quad (3)$$

Plastic strain, $$\gamma_P = C_p \cdot \left(\frac{\tau}{G}\right)^{m_p} \quad (4)$$

Steady-state creep,

$$\dot{\gamma}_s = C_l \frac{G}{T}\left[\sinh\left(\alpha\frac{\tau}{G}\right)\right]^{n_l} \exp\left(-\frac{Q_l}{RT}\right)$$
$$+ C_h \frac{G}{T}\left[\sinh\left(\alpha\frac{\tau}{G}\right)\right]^{n_h} \exp\left(-\frac{Q_h}{RT}\right) \quad (5)$$

Transient creep,

$$\dot{\gamma}_t = B_1 \frac{G}{T}\left(\frac{\tau}{G}\right)\exp\left(-\frac{Q_b}{RT}\right) + B_2 \frac{G}{T}\left(\frac{\tau}{G}\right)\exp\left(-\frac{Q_m}{RT}\right) \quad (6)$$

where τ is the stress, γ is the strain, G is the shear modulus constant and $G=G_0-G_1(T-273K)$, C_p is the plastic strain coefficient, m_p is the stress sensitivity of the plastic strain, C_l and C_h are coefficients for the steady-state creep, α is the multiplier of stress, n is the stress sensitivity of creep rate, Q_b and Q_m are the activation energy, $R=8.31J/mol/K$ is the gas constant [18], γ_t is the transient creep strain, B_1 and B_2 are the creep coefficients. The parameters for the typical solder 60Sn40Pb in the power modules can be found in [16], [17] and they are also listed in Table II. The relationship between the shear

978-1-4799-2706-7/14 $31.00 © 2014 IEEE   859

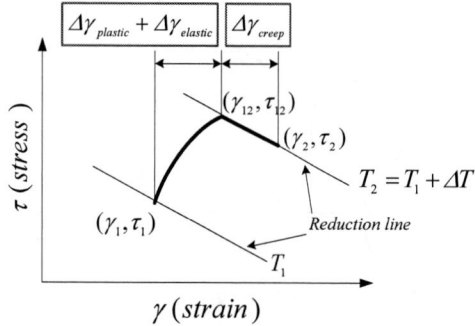

Fig. 6. Simulation step of the solder stress-strain curve.

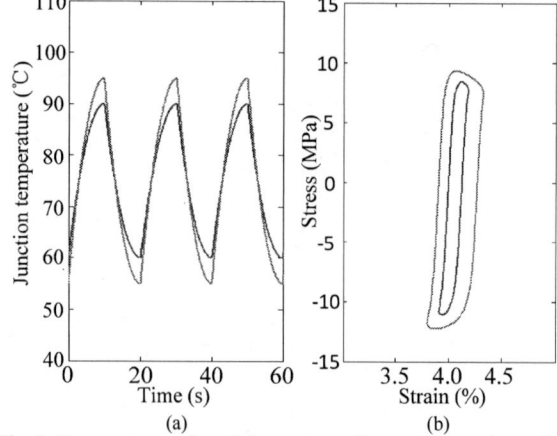

Fig. 7. Temperature cycle and the corresponding strain-stress curve. (a) temperature cycle (b) strain-stress curve.

$$\gamma + \frac{\tau}{K} = D_1 \cdot (T - T_0) \qquad (7)$$

stress and strain in the solder joints is also defined by the reduction lines, and it can be expressed by the following, where K is the reference assembly stiffness depending on the geometry and D_1 is a constant reflecting both the geometry and Coefficient of Thermal Expansion (CTE) mismatch between different layers connected by solder. These two parameters can be extracted based on the test data from the industry, where the most optimal K and D_1 will make the ratio of $W_{tot,k}$ between different cycling loads most close to 1. In this paper, the values are: K=1630, D_1=5.6e-4, which is based on the load-cycling capability test report of $Hipak^{TM}$ IGBT modules [8].

The stress-strain curve can be drawn according to the mission profile, and the detail for one step of the solder stress-strain curve is shown in Fig. 6. Due to the small temperature variation ΔT, the strain-stress point moves from (γ_1, τ_1) to (γ_2, τ_2). The process is divided into two parts, including the variation of the time-independent strain, $\Delta\gamma_{plastic}+\Delta\gamma_{elastic}$, and the variation of the time-dependent strain, $\Delta\gamma_{creep}$. Both the variation of the strain and stress can be calculated by (3)-(7). More details can be found in [19]. Fig. 7 shows an example of the temperature cycles and the corresponding strain-stress curves. It can be seen that since the amplitude of temperature cycles are unequal, the different deformation

energies are expressed by the area of the strain-stress curves.

V. SIZING OF THE ENERGY STORGAE SYSTEM

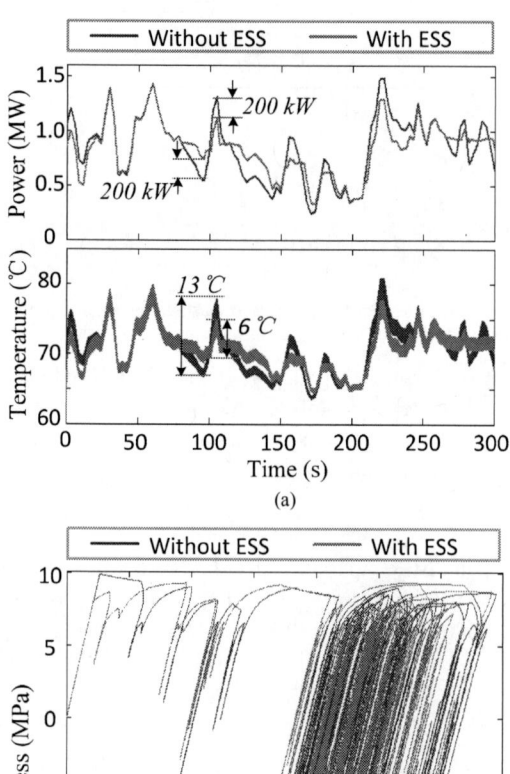

Fig. 8. Comparsion of the power and temperature with or without energy storage system. (a) temperature profile (b) strain-stress curve.

According to the approach mentioned above, the junction temperature of the power modules with the mission profile can be obtained first based on the simplified evaluation method. Afterwards, the strain-stress curve can be drawn and used for the calculation of the deformation energy W_{tot}. Since the impact of the loading on the lifetime of wind power converter can be quantified with this approach, the improvement of reliability caused by ESS can be calculated more accurately. One case is shown in Fig. 8 for detailed illustration. In Fig. 8(a), it can be seen that both the amplitudes of the power and temperature fluctuations are reduced when a 200 kW ESS is used, but there is not a quantified relationship to the reliability improvement. Fig. 8(b) is the corresponding strain-stress curves of the temperature profile. The strain-stress of the solder joint keeps increasing and decreasing as the temperature fluctuates. Moreover, by calculating the area of the strain-stress curves, the deformation energy is obtained and it is reduced to 0.62 p.u. using the ESS.

The 2014 International Power Electronics Conference

TABLE III. PARAMETERS OF THE ENERGY STORAGE TECHNOLOGIES [21-23].

	Supercap.	Li-ion
Energy density (Wh/kg)	5	150
Power density (W/kg)	10000	1500
Costs for energy (€/Wh)	15	0.8
Costs for power (€/kW)	600	900
Cycle life	>10⁶	>10³

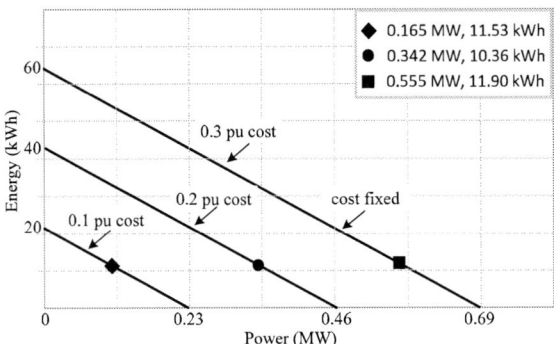

Fig. 9. Energy as a function of the power of the energy storage system when the cost is fixed.

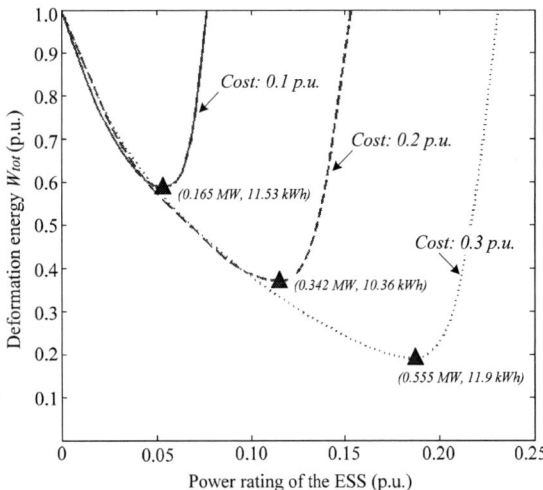

Fig. 10. Deformation energy as a function of the power rating of the energy storage system with various fixed cost.

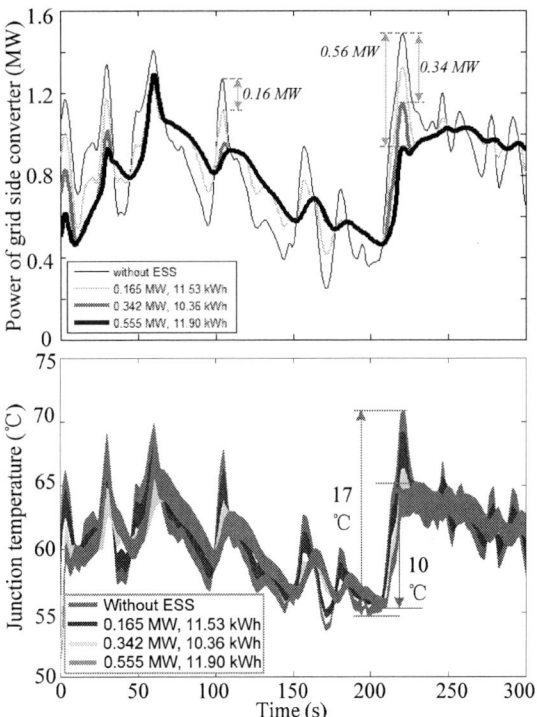

Fig. 11. Power and junction temperature of the grid side converter using energy storage.

Normally, with larger ESS the power will be smoother and thereby the higher reliability of the wind power converter can be gained. But at the same time the cost of the ESS will become higher, which is then a cost trade-off. Therefore, in order to locate the most optimal point in between, a set of cases are studied based on a low speed high turbulence wind profile, which belongs to turbulence factor IIB according to IEC 61400 [20]. The cost of the ESS is evaluated according to the following [21],

$$Cost_{total} = UnitCost_{power} * P + UnitCost_E * E \quad (8)$$

where the $UnitCost_E$ is the unit capacity cost and the $UnitCost_{power}$ is the unit power cost, including the unit power cost of the energy storage device and the bidirectional power converter, which usually costs 700 €/kW. Supercapacitors and Lithium-ion batteries are normally used for the short-term energy storage, and their characteristics are listed in Table III [21-23]. Compared with Li-ion, supercapacitors have much higher power density and longer cycle life. Therefore, it is chosen for this study.

According to (8), when the total cost of ESS is fixed, the power and energy of the ESS are still variable, and one of them increases will lead to the other one decrease, as shown in Fig. 9. Thus, among the different points with power and energy, there should be the most optimal one where the highest reliability of the wind power converter can be achieved. Moreover, if the cost of ESS is changed, the most optimal point will also change since the cost fixed line is shifted. A set of cases are studied, and as a consequence, the three curves in Fig. 10 are obtained. The curves represent the deformation energy as a function of the power of ESS with 0.1 pu, 0.2 pu and 0.3 pu cost, respectively, where the cost of the wind turbine system is 1 pu and 3 M€ (for a 3 MW WTS) [24]. As mentioned above, in each curve with fixed cost an

optimal point is found, which has the lowest deformation energy, and they are (0.165 MW, 11.53 kWh) for 0.1 pu cost, (0.342 MW, 10.36 kWh) for 0.2 pu cost and (0.555 MW, 11.9 kWh) for 0.3 pu cost, respectively. Further, among the optimal points, the energy of the ESS changes less than 10 %, but the power of the ESS is almost doubled every time the cost increases. It can therefore be concluded that the power is more critical for improved reliability than the energy of the ESS. Moreover, the deformation energy at the optimal point decreases significantly as the cost of the ESS increases, which are 0.57 pu for 0.1 pu cost, 0.35 pu for 0.2 pu cost and 0.16

978-1-4799-2706-7/14 $31.00 © 2014 IEEE

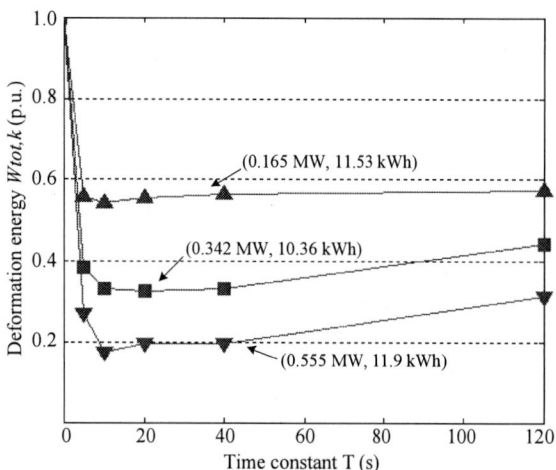

Fig. 12. Impact of the power control of the energy storage system on the deformation energy of the wind power converter.

pu for 0.3 pu cost, respectively.

The power and junction temperature behavior of the grid side converter using energy storage is indicated in Fig. 11. It can be seen that the power fluctuation is smoothed when the ESS is applied, and this smoothing ability is limited by the power rating of the ESS. Thus, ESS with higher power rating can compensate the power fluctuation more. Further, the junction temperature variation instead of the mean temperature is also improved by using ESS, and at some points it is reduced from 17 ℃ to 10 ℃.

It should be noted that all the results obtained in Fig. 10 and Fig. 11 are based on the control strategy in Fig. 1, and the time constant T of the high pass filter is fixed at 20 s. Thus, in order to prove that the results are globally valid, it is necessary to investigate the impact of T on the performance of the ESS. A set of cases with different time constant T are therefore studied, and they are set to be 5 s, 10 s, 20 s, 40 s, and 120 s, respectively, to cover all the time scale of short-term power fluctuations in the wind power converter. The corresponding deformation energies of the optimal points are shown in Fig. 12. It can be seen that, for all the cases the deformation energies decrease significantly when the time constant increases and is below 10 s. After that the deformation energies increase gradually but without large change even if the time constant increases to 120 s. Thus, it is concluded that the ESS shows the best performance when the time constant T of the high pass filter is between 10 s - 40 s. While during this interval the performance of the ESS is insensitive to the time constant.

VI. CONCLUSIONS

In the proposed approach, the reliability of wind power converter under a complicated wind profile can be quantified when introducing an ESS in a wind turbine system and the computational effort is reduced significantly. Moreover, it was found that the power level is much critical than energy of the energy storage system in order to improve the reliability of wind power converter. Finally, in a IIB (IEC 61400) wind condition, a supercapacitor bank (0.342 MW, 10.36 kWh), which costs 20% of the whole system, can save 65 % lifetime (deformation energy is reduced to 0.35 pu) of the wind power converter for a 3 MW wind turbine system.

REFERENCES

[1] B. Hahn, M. Durstewitz, K. Rohrig "Reliability of wind turbines – Experience of 15 years with 1500 WTs", *Wind Energy*, Spinger, Berlin, 2007.
[2] Y. Song, B. Wang, "Survey on Reliability of Power Electronic Systems," *IEEE Trans. Power Electron.*, vol.28, no.1, pp.591-604, Jan. 2013.
[3] H. Wang, M. Liserre, F. Blaabjerg, "Toward Reliable Power Electronics: Challenges, Design Tools, and Opportunities," *IEEE Ind. Electron. Mag.*, vol.7, no.2, pp.17-26, June 2013
[4] F. Blaabjerg, K. Ma, D. Zhou, "Power electronics and reliability in renewable energy systems," *in Proc. of ISIE' 2012*, pp.19-30, 2012.
[5] C. Abbey, G. Joos, "Supercapacitor Energy Storage for Wind Energy Applications," *IEEE Trans. Ind. Appl.*, vol.43, no.3, pp.769-776, 2007.
[6] GE 2.5 - 120 Wind turbine. Online available: http://www. ge-energy.com/products_and_services/products/ wind_turbine/ges_2.5_120_wind_turbine.jsp
[7] C. Busca, R. Teodorescu, F. Blaabjerg, S. Munk-Nielsen, L. Helle, T. Abeyasekera, P. Rodriguez, "An overview of the reliability prediction related aspects of high power IGBTs in wind power applications," *Microelectronics Reliability*, Vol. 51, no. 9-11, pp. 1903-1907, 2011.
[8] ABB Application Note, Load-cycling Capability of HiPakTM IGBT Modules, 2012.
[9] I. F. Kovačević, U. Drofenik, J. W. Kolar, "New physical model for lifetime estimation of power modules," *in Proc. of IPEC' 2010*, pp. 2106-2114, 2010.
[10] H. Li, Z. Chen, H. Polinder, "Optimization of Multibrid Permanent-Magnet Wind Generator Systems," *IEEE Trans. Energy Conversion*, vol.24, no.1, pp. 82-92, March 2009.
[11] F. Blaabjerg, U. Jaeger, S. Munk-Nielsen, "Power losses in PWM-VSI inverter using NPT or PT IGBT devices," *IEEE Trans. Power Electron.*, vol.10, no.3, pp. 358-367, May 1995.
[12] Z. Zhou, M. S. Khanniche, P. Igic, S. T. Kong, M. Towers, P. A. Mawby, "A fast power loss calculation method for long real time thermal simulation of IGBT modules for a three-phase inverter system," *in Proc. of EPE' 2005*, pp. P.1-P.10, 2005.
[13] ABB Application Note, Thermal Design of IGBT Modules, 2011.
[14] Z. Qin, M. Liserre, F. Blaabjerg, H. Wang, "Energy Storage System by Means of Improved Thermal Performance of a 3 MW Grid Side Wind Power Converter", *in Proc. of IECON' 2013*, pp.736-742, 2013.
[15] ASTM International, E1049-85 (2005) Standard practices for cycle counting in fatigue analysis, 2005.
[16] G. Wang, Z. Cheng, K. Becker and J. Wilde, "Applying ANAND model to represent the viscoplastic deformation behavior of solder alloys", *Journal of Electronic Packaging*, vol. 123, no. 3, pp. 247-254, Sept. 2001.
[17] X. Shi, Z. Wang, Q. Yang, and H. Pang, "Creep behaviour and deformation mechanism map of Sn-Pb eutectic solder alloy", *Journal of Engineering Materials and Technology*, vol. 125, pp. 81-87, Jan. 2003
[18] Wikipedia "Gas constant," May 2014.
[19] J. P. Clech, "Review and analysis of lead-free solder material properties," Online available at: http://www.metallurgy.nist.gov/ solder/clech/Introduction.htm.
[20] Wikipedia "IEC 61400," May 2014.
[21] S. M. Schoenung, "Energy storage systems cost update," Sandia National Laboratories, Albuquerque, 2011.
[22] H. Chen, T. N. Cong, W. Yang, C. Tan, Y. Li, Y. Ding, "Progress in electrical energy storage system: A critical review," *Progress in Natural Science*, vol. 19, Issue 3, pp. 291-312, March 2009.
[23] S. McCluer, J. Christin, "Comparing data center batteries, flywheels, and ultracapacitors," *White Paper 65*, Revision 2.
[24] M. Bolinger, R. Wiser, "Understanding trends in wind turbine prices over the past decade," Lawrence Berkeley National Laboratory, 2011.

The 2014 International Power Electronics Conference

A Multi-Level Virtual Conductor as a Backbone of a DC Power Routing System

Husam A. Ramadan, Yasutaka Imamura, Konosuke Kawachi, Sihun Yang, Masahito Shoyama

Kyushu University, Fukuoka, 819-0395 Japan

Email: husam_ramadan@ieee.org

Abstract − **A new control approach for bidirectional DC-DC (BDC) converter is proposed in this paper. This approach aims at controlling a BDC in such a way that makes it behaves like a multi-level virtual conductor. As a matter of fact, the voltage difference between the terminals of any conductor is zero volts. Conversely, the main target of the proposed control approach is to keep the voltage difference between the converter terminals constant at certain value. In other words, the proposed control approach permits the DC-DC converter to transfer the power between two nodes at different voltage levels. In this way, the converter performs like a conductor but unlike the normal conductor, it has voltage deference between its terminals. Thus, the authors call it virtual conductor. This virtual conductor is considered a base to a power routing in dc networks; since it has the ability of transferring the electric power between nodes at different voltage levels. Furthermore, it allows an easy plug-and-play feature. The proposed BDC system configuration has been investigated analytically, using simulation, and experimentally.**

I. INTRODUCTION

Bidirectional DC-DC converters (BDCs), recently, have gained a great attention due to the increasing need to systems with the capability of bidirectional energy transfer between two dc buses. BDCs have various applications that include energy storage in renewable energy systems, fuel cell energy systems, hybrid electric vehicles (HEVs) and uninterruptible power supplies (UPSs) [1]-[8].

The fluctuation nature of most renewable energy sources, like wind and solar, makes them unsuitable for standalone operation. A common solution to overcome this problem is to use an energy storage device besides the renewable energy resource to compensate these fluctuations and maintain a smooth and continuous power flow. As the most common and economical energy storage devices in medium-power range are batteries and super-capacitors, a DC-DC converter is usually required to allow energy exchange between storage device and the rest of the system.

Such converters must have bidirectional power flow capability with flexible control in all operating modes. Moreover, when integrating various renewable energy sources with different voltage levels into a dc grid, the main challenge is to have an easy plug and play system with a flexible dc power routing. This system should be capable of integrating such sources at different voltage levels. To face this challenge, a proposed approach based on a bidirectional DC-DC converter is introduced in this paper.

In this paper, a bidirectional DC-DC converter is investigated and controlled. It is considered that both input and output are independent current sources. The current source may represent a load, electric double layer capacitors (EDLCs), a battery, or even another bi-directional converter. Therefore, for such converter, it is required to control both V_1 and V_2 to keep the voltage difference between them at a certain value regardless of any variation that may be occurred to the currents I_1 and I_2.

II. BIDIRECTIONAL DC-DC CONVERTER

A. Circuit examples of a bi-directional DC-DC converter

Basic DC-DC converters such as buck and boost converters (and their derivatives) do not have bidirectional power flow capability. This limitation is due to the presence of diodes in their structure which prevents reverse current flow. In general, a unidirectional DC-DC converter can be turned into a bidirectional converter by replacing the diodes with a controllable switch in its structure [9]. Figure 1.a shows an example for a non-isolated bidirectional DC-DC converter. On the other hand, Fig. 1.b shows an example for an isolated bidirectional DC-DC converter.

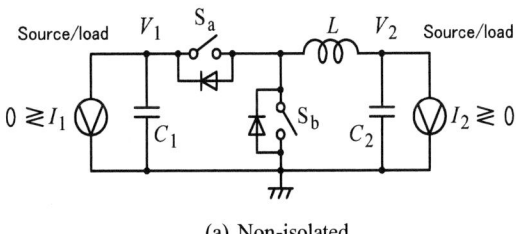

(a) Non-isolated.

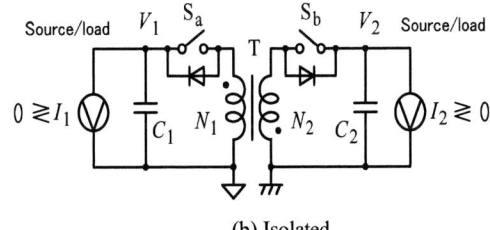

(b) Isolated.

Fig. 1. Circuit examples of a bi-directional DC-DC converter.

978-1-4799-2706-7/14 $31.00 © 2014 IEEE

B. Using a bi-directional DC-DC converter for charging and discharging of a battery

Figure 2 illustrate an example of a smart house integrates a renewable energy source (PV) and a storage battery. The PV is connected to the load via a maximum point power tracker (MPPT) and a unidirectional DC-DC converter. While, the battery is connected to the load via bidirectional DC-DC converter; since the power flow between the battery and the load is required to be bidirectional (charge/discharge). The coupling point Voltage V_N is controlled based on V_{Ref-N}.

III. DC POWER SYSTEM USING THE PROPOSED BI-DIRECTIONAL DC-DC CONVERTER (MULTI-LEVEL VIRTUAL CONDUCTOR)

Regarding the example of sec II.*B*, Fig. 2, for the sake of having a flexible dc power system; it should be easy to integrate different multi-level voltage sources and loads together. In other words, the coupling point is required to be a multi-level voltages point. The bidirectional dc-dc converter, with the proposed control approach, can play the role of a multi-level voltage coupling point. In this case, it is called a multi-level virtual conductor.

A. Implementation of multi-level virtual conductor using a bi-directional DC-DC converter

The proposed control approach for a bidirectional DC-DC is shown in Fig.3. The main target of this control approach is to keep the voltage difference between the converter terminals constant at certain value. As shown in Fig.3, V_1, V_2 are adjusted according to V_{Ref-1}, V_{Ref-2} respectively. Therefore, the duty ratios D_{sa}, D_{sb} of the converter are adjusted to the desired value when $(V_1-V_2) = (V_{Ref-1}-V_{Ref-2})$. Since V_{Ref-1} and V_{Ref-2} are constant values, then the difference between V_1 and V_2 is kept constant at steady state. For the former example of Fig.3 there was only one coupling point (V_N), but with this proposed approach; there are to coupling points V_1 and V_2. The application of this multi-level virtual conductor is shown in Fig. 4.

B. A multi-level dc power system using bi-directional DC-DC converter

The multi-level virtual conductor allows a flexible power transfer through an energy system having multiple energy sources with different voltage levels, energy storage equipment, and loads. Factually, it is impossible to use a conductor to connect such an energy system; however, a virtual conductor, having the voltage conversion function of a bidirectional DC-DC converter, can be used.

The configuration examples of the multi-level virtual conductor in dc power system are shown in Fig. 5. The series connection is presented in Fig. 5 (a), while the branch connection at a central node is revealed in Fig. 5

(b), and Fig. 5 (c) shows a loop connection. A further complex connection, grid connection, is shown in Fig 5 (d).

IV. SIMULATION AND EXPERIMENTAL RESULTS

The simulation circuit is shown in Fig. 6. The multi-level virtual conductor is connected at its one end to a bi-polar current source (load/ source), and at the other end to another bi-polar current source and a battery via bidirectional converter. The circuit parameters are shown in Table 1.

TABLE 1 THE SIMULATION CIRCUIT PARAMETERS

Symbol	Description	Value
V_B	Battery voltage	48 V
V_1	Voltage at node 1	24 V
V_2	Voltage at node 2	12 V
L_1	Inductance of the first converter	120µH
L_2	Inductance of the virtual conductor	80 µH
C_1	capacitance	100 µF
C_2	capacitance	200 µF
C_3	capacitance	330 µF
C_4	capacitance	100 µF
F	Switching frequency	100 kHz

The simulation results are shown in Fig. 7, and the experimental results are shown in Fig. 8. From these results it is obvious that the voltage deference between V_1 and V_2 is kept constant (12 V) regardless of the change in polarities of current s and powers

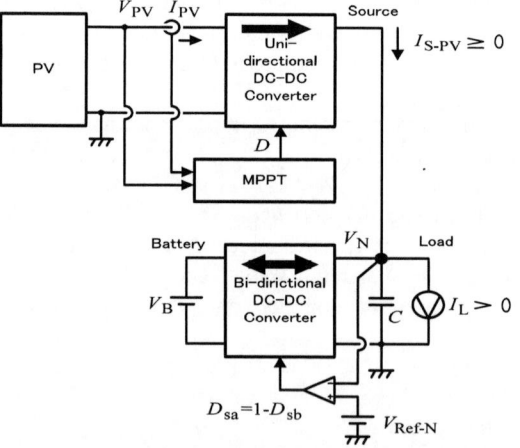

Fig. 2. Using a bi-directional DC-DC converter for charging and discharging of a battery.

978-1-4799-2706-7/14 $31.00 © 2014 IEEE

The 2014 International Power Electronics Conference

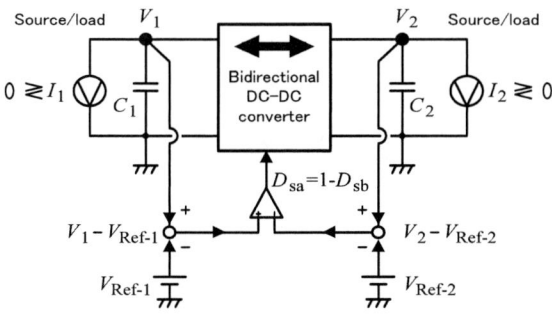

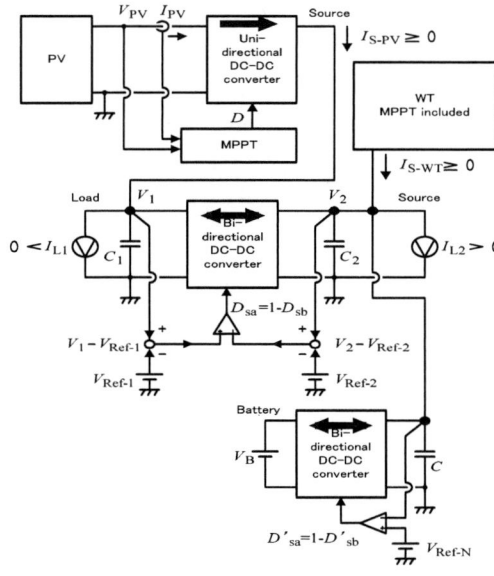

Fig. 3. Implementation of multi-level virtual conductor using a bi-directional DC-DC converter.

Fig. 4. A multi-level virtual conductor integrates different voltage sources (wind turbine, PV, battery) and loads.

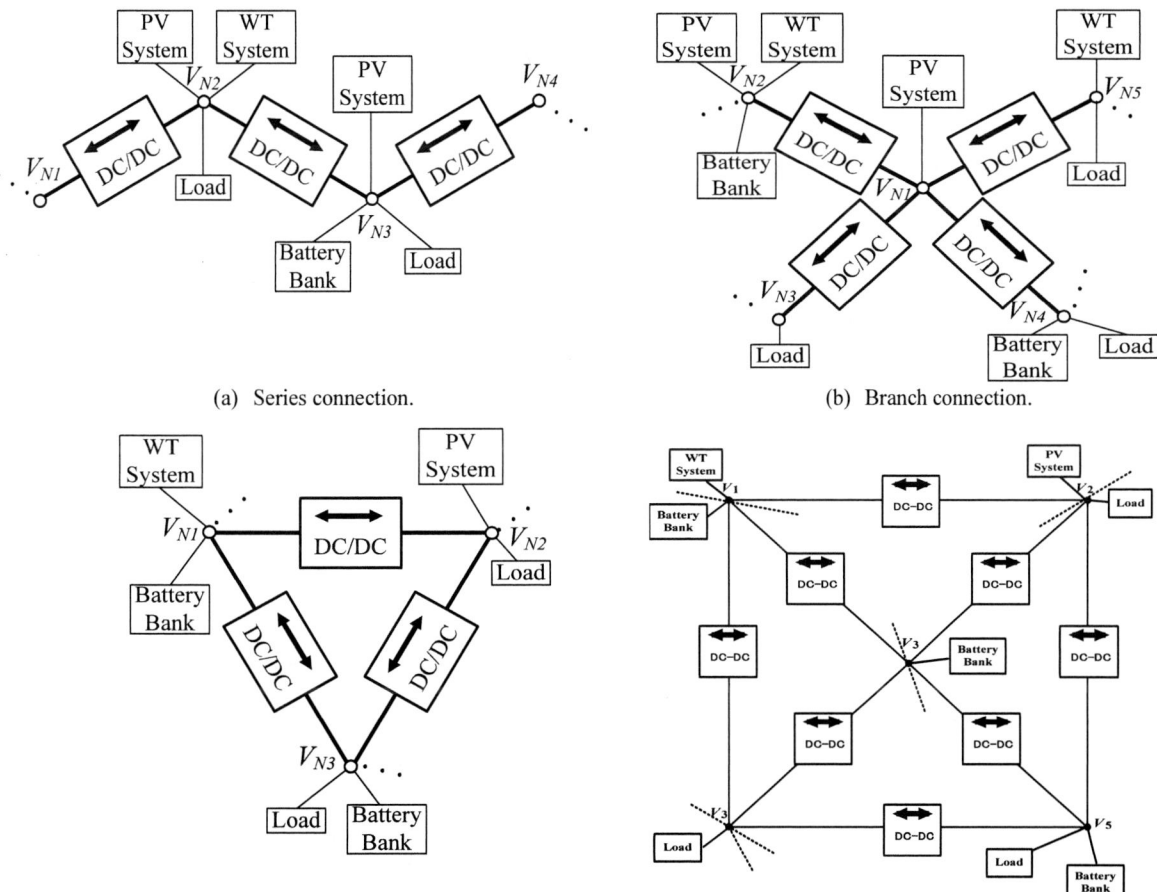

(a) Series connection.

(b) Branch connection.

(c) Loop connection.

(d) Grid connection

Fig. 5. The configuration examples of the multi-level virtual conductor in dc power system.

978-1-4799-2706-7/14 $31.00 © 2014 IEEE

865

The 2014 International Power Electronics Conference

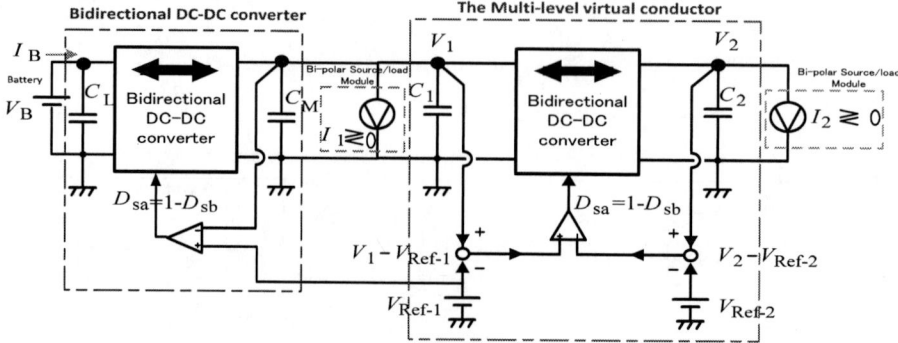

Fig. 6. The simulation circuit for a multi-level virtual conductor connected to two bipolar current sources and a battery

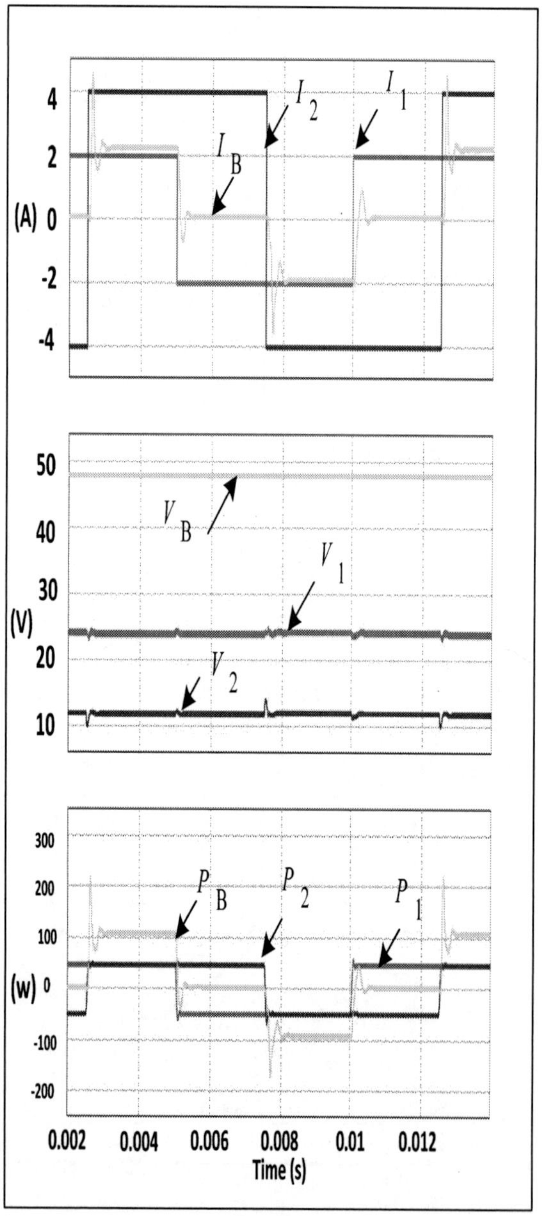

Fig. 7. Simulation results.

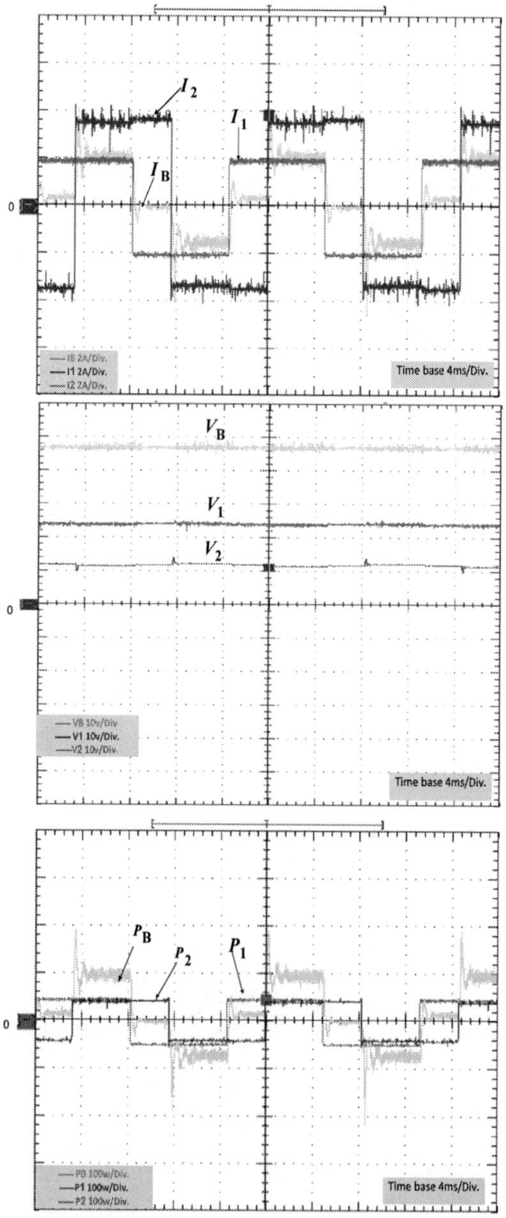

Fig. 8. Experimental results.

978-1-4799-2706-7/14 $31.00 © 2014 IEEE

V. Conclusion

This paper proposes a new bidirectional control approach leads to a performance of a multi-level virtual conductor. This virtual conductor is considered a base to power routing in dc networks. It allows an easy plug-and-play feature. This means that any terminal unit (load/source) can be safely and effectively connected / disconnected at its suitable voltage level. The basic idea is presented and a representative case study is addressed by simulation and experiment as well. Both of simulated and experimental results support the basic idea and prove its superiority.

References

[1] Yu Du, Lukic, et al, "Review of high power isolated bi-directional DC-DC converters for PHEV/EV DC charging infrastructure," (ECCE), 2011 IEEE, On pp. 553 – 560.

[2] Y. Imamura, H. A. Ramadan, S. Yang, G. M. Dousoky and M. Shoyama, "Seamless Dynamic Model for Bi-directional DC-DC Converter, "PEDS 2013, B4L-F , pp.1109-1113, Apr. 2013

[3] J. Moreno, M.E. Ortuzar, J.W. Dixon, "Energy management system for a hybrid electric vehicle, using ultracapacitors and neural networks," *IEEE Trans. on IES*, vol. 52, April 2006, pp. 614 – 623.

[4] M. Ortuzar, J. Moreno, J.W. Dixon, "Ultracapacitor-based auxiliary energy system for an electric vehicle: implementation and devaluation," *IEEE Trans. on IES*, vol. 52, Aug. 2007, pp. 2147 – 2156.

[5] D. Sable, F.C. Lee, B.H. Cho, "A Zero-Voltage-Switching bidirectional battery charger/discharger for the NASA EOS satellite," *in Proc. Of APEC*, 1992, pp.614-621.

[6] K. Asano, Y. Inaguma, H. Ohtani, E. Sato, M. Okamua, S. Sasaki,"High performance motor drive technologies for hybrid vehicles," *in Proc. of PCC, Nogoya*, Japan, April 2007, pp. 1606-1611.

[7] S. Aso, M. Kizaki, and Y. Nonobe, "Development of fuel cell hybrid vehicles in Toyota," *in Proc. of PCC*, Nogoya, Japan, April 2007, pp. 1606-1611.

[8] J. S. Lai, and D.J. Nelson, "Energy Management power converters in hybrid electric and fuel cell vehicles", *in Proc. of the IEEE*, Vol. 95, No. 4, April 2007, pp. 766 – 777.

[9] D. Dutta, S. Ganguli, "Design of A Bidirectional Dc-Dc Converter for Hybrid Electric Vehicles (Hev) Using Matlab,"*IJAREEIE*, Vol. 2, Issue 7, July 2013, pp. 2994-3002

Semi-Numerical Method for Loss-Calculation in Foil-Windings exposed to an Air-Gap Field

D. Leuenberger and J. Biela

Laboratory for High Power Electronic Systems (HPE)

ETH Zurich, Physikstrasse 3, CH-8092, Switzerland

Email: leuenberger@hpe.ee.ethz.ch

Abstract—**The calculation of Eddy current losses in foil windings exposed to a 2-D fringing field is a complex task, due to the current displacement along the height of the foil. For model based optimization of magnetic components, the loss calculation with a 2-D FEM simulation is not an option, due to the high computational effort. The existing alternative calculation methods, which allow for loss calculation with low computational effort, rely on approximations applicable only for a certain geometrical arrangement of the windings and the air gap. Therefore, a new semi-numerical method is developed to overcome these limitations. The method is based on the mirroring method and is applicable to arbitrary air gaps and winding arrangements. The accuracy of the new method is verified by measurements and the deviation of the model to the measured losses is below 15%.**

Keywords—*Air gap fringing field, Eddy current losses, Foil winding, Magnetic components*

I. INTRODUCTION

Foil windings feature better thermal properties and a higher copper filling factor than Litz or round wires. On the other hand, foil windings exposed to a 2-D magnetic fringing field are subject to current displacement in two directions, along the thickness *thck* as well as the width *wdth* of the foil, see fig. 1. To accurately predict the losses in these foil windings, the current displacement must be taken into account, which requires a 2-D field calculation in the winding window. A finite element simulation (FEM) is the most commonly applied approach to perform this 2-D field calculation. Though FEM suffers from long calculation times and difficult parametrization [1]. The work in [2], which applied a genetic algorithm to optimize a transformer with foil windings, reported calculation times of 20 hours even for a simplified FEM model considering only one harmonic component. However, for most applications it is inevitable to consider more than one harmonic component for accurate loss prediction. Therefore, FEM is not considered to be an ideal option for automatized model based optimization. Various alternative methods are proposed in literature to consider the effect of 2-D fringing fields. They can be categorized into two distinct approaches.

The first approach is to derive analytical formulas, which take into account the losses caused by the fringing field and allow for very high calculation speed. The derived formulas rely on an analytical solution of the Maxwell equations in the winding-window. However, to obtain analytically solvable differential-equations, approximations and restrictions to simple geometries are required. The solid-conductor-method proposed in [1], [3] and the method proposed by [4], [5] are the most known

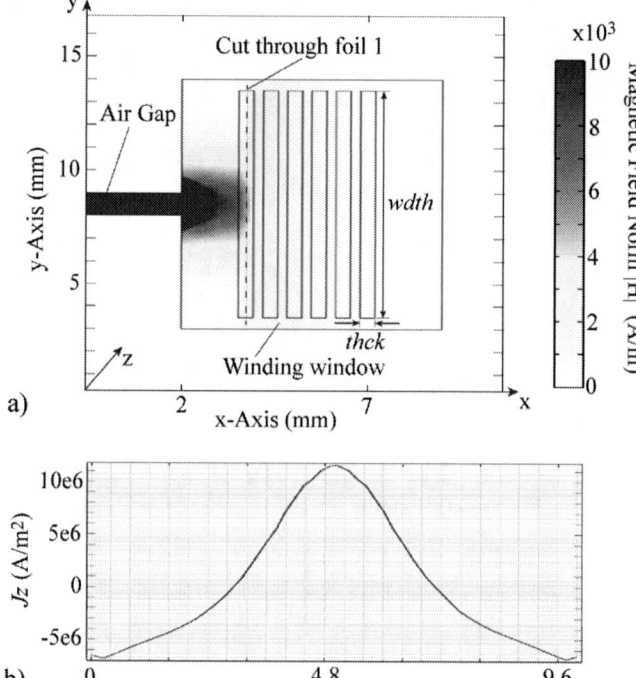

Figure 1. a) FEM Simulation: Foil windings in a winding window of a gapped magnetic core (I_{foil}=5A/30kHz) b) Non-homogeneous current density along foil 1 (see cut-line in fig.1a).The homogeneous current density would be 2.5e6A/m^2).

methods of this kind. The solid-conductor-method approximates the layers of a foil-winding as one unified solid conductor. The model is shown to be accurate for low frequencies, but at high frequencies the accuracy decreases, because the solid conductor exhibits different eddy-currents, than a foil winding would actually have. The method described in [4], [5] approximates the eddy-current as line-current-density located at the surface of the foil closest to the gap. This foil is assumed to absorb the whole fringing field. The air gap is also modelled as line-current-density (by Fourier-decomposition in space). For either of these two methods, the air gaps must be located at the inner core-leg and the area between the air gap and the foil-winding must be filled with air or a spacer only (no additional round-conductor winding).

The second approach is often referred to as semi-empirical or semi-numerical. A closed-form formula for the losses is derived from a set of prior FEM simulations. This approach tries to combine the advantages of the FEM-approach - high accuracy and no geometrical restrictions - and the advantages of the analytical-approach - high calculation speed. The squared-field-derivative method, proposed in [6] for round-wires is an example of such a method. The work in [7] derives a modified Dowells-formula [8] for losses in the foil-winding of a high-frequency transformer. The formula contains additional parameters, used to curve-fit the losses from 2-D FEM simulations and enables fast calculation of winding losses. Though, it is restricted to a certain geometry, analyzed prior by FEM simulations.

To be able to effectively perform model based optimization of magnetic components with foil windings, a method is needed, which features much lower calculation times than a FEM-simulation and on the same time is not subject to restrictions on air gap and winding arrangement as the existing analytical and semi-numerical approaches. To fulfil this need, in this work a novel, semi-numerical method is developed, which can be applied to arbitrary winding and air gap geometries. Regarding calculation times, the new method is in between the FEM- and the existing semi-empirical approaches. The developed method can be combined with the mirroring method and is a true 2-D field approximation for foil windings. The method is described in section II and its validation is given in section III.

II. FOIL- TO SQUARE-CONDUCTOR METHOD

In the following the principle of the calculation method is explained with the example of an inductor with foil windings. Figure 1a) shows a 2-D finite element simulation of an inductor with a sinusoidal winding current of $I_{z,foil}$=5A@30kHz. $|H|$ is the amplitude of the 2-D magnetic field introduced by the air gap. The x-component of the H-field, which is perpendicular to the foils, causes an Eddy current flowing in the y-z plane. The existence of the Eddy currents results in an inhomogeneous current distribution J_z, which is shown in fig. 1b) for the foil closest to the air gap. For accurate loss modelling, the investigated method must determine the non-homogeneous J_z of every foil of the winding.

The routine to perform this task, consists of two major parts. The first part is the calculation of the non-homogeneous current density $J_{z,foil}$ in a single foil by the following steps shown in fig. 2:

1) *Transformation to round-conductors:* The foil is transformed into area-equivalent round-conductors.

2) *Calculation of current $J_{z,round}$ in round-conductors:* The well known formulas for round conductors are used to calculate the Eddy current in each separate round conductor. The external magnetic field H_e is derived using the mirroring method, as will be explained in detail later in this section.

3) *Calculation of foil current $J_{z,foil}$:* The current density in the foil is derived from the current density of the round conductors, by postulating continuity of $J_{z,foil}$ at the boundary of adjacent round conductors.

4) *Discretize and average to square-conductors:* The foil is cut into area-equivalent square-conductors. To each square-conductor a current $I_{z,sqr,k}$ is attributed according to $J_{z,foil}$, whereas the current density is

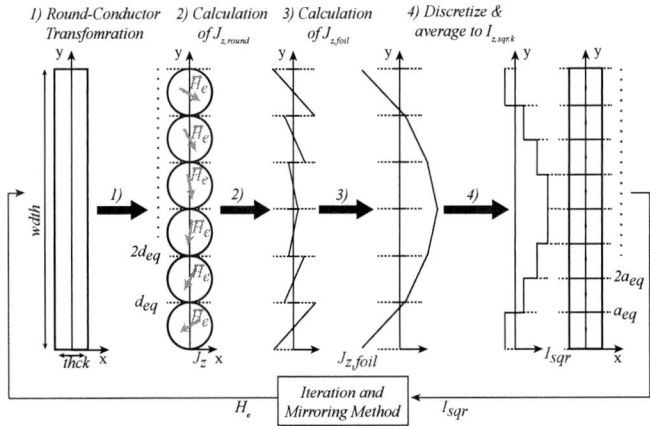

Figure 2. Overview calculation procedure for non-homogeneous current distribution in a foil, exposed to a 2-D transverse magnetic field: *1)* Round-Conductor Transformation, *2)* Calculation of round-conductor current density, *3)* Calculation of foil current density, *4)* Transformation to square-conductors and calculation of square-conductor currents.

approximated to be constant over the cross-section of each square-conductor. Unlike in step 1), a transformation to square-conductors is applied, as they represent the actual foil-winding more accurately. In this way, the current displacement in the foil is taken into account by the square-conductor currents and the mirroring method can be applied to calculate the H-field and the losses in the same way as for round- and Litz-wires [5], [9].

The second part of the routine considers the entire foil-winding and involves a numerical iteration to determine $J_{z,foil}$ in each foil of the winding, starting from the uniform distribution. The method is applicable to arbitrary air gaps and winding arrangements. As the mirroring method is based on a low-frequency approximation ([5] chapter 5.2.1), the model is applicable as long as the foil thickness $thck$ fulfills the following condition:

$$thck \leq 1.6 \cdot \delta, \tag{1}$$

where δ is the so called penetration- or skin-depth. The same condition can be alternatively expressed as a frequency limit at a given winding geometry:

$$f_{max} = \frac{2.56 \cdot \rho_c}{\mu \cdot \pi \cdot thck^2}, \tag{2}$$

where ρ_c is the electrical resistivity of the conductor material and μ the permeability of the material.

The following two sections describe both parts of the routine in detail.

A. Non-Homogeneous Current Density in a 2-D Transverse Field

The inhomogeneous foil current density $J_{z,foil}$ caused by the 2-D transverse field H_e is calculated with the procedure shown in fig. 2, which consists of four steps:

The 2014 International Power Electronics Conference

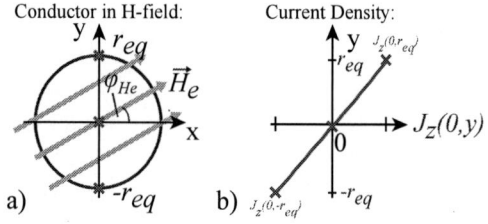

Figure 3. a) Round-conductor in a transverse H-field b) Linearized Eddy current density J_z in the round conductor evaluated on the y-axis: $(0,-r_{eq})$, $(0,0)$ and $(0,r_{eq})$.

1) Transformation to round-conductors: The foil-winding is transformed into a series of aligned equivalent round-conductors, using the equivalent DC-resistance transformation ([5], chapter 5.4.1 and [10]) and postulating equivalent width $wdth$ of the transformed winding, see fig. 2, *1)*. The constraints on surface and width can be expressed by $wdth \cdot thck = n_{cond} \cdot \pi \cdot (d_{eq}/2)^2$ and $n_{cond} \cdot d_{eq} = wdth$ from which follows:

$$d_{eq} = \frac{4 * thck}{\pi}. \tag{3}$$

2) Calculation of Current $J_{z,round}$ in round-conductors: With the foil decomposed into aligned round conductors, the Eddy-current caused by the external magnetic field $\vec{H}_e$ in a round conductor can be calculated using the formula derived in [11] formula (7-45):

$$\bar{J}_z(r,\varphi) = 4\mu_2 H_e j^{\frac{3}{2}} k \frac{J_1(j^{\frac{3}{2}}kr)}{F(j^{\frac{3}{2}}kr_{eq})} sin(\varphi - \varphi_{He}). \tag{4}$$

where

$$F(j^{\frac{3}{2}}kr_{eq}) = (\mu_1 + \mu_2) J_0(j^{\frac{3}{2}}kr_{eq}) + (\mu_1 - \mu_2) J_2(j^{\frac{3}{2}}kr_{eq}) \tag{5}$$

$$k = \sqrt{(2\pi f)\rho_1\mu_1} \tag{6}$$

and

μ_1	magnetic permeability of the conductor material
ρ_1	conductivity of the conductor material
μ_2	magnetic permeability of material around the conductor
$\vec{H}_e$	sinusoidal transverse magnetic field vector with amplitude H_e and φ_{He}
f	frequency of H_e.

Figure 3a) illustrates a single round-conductor exposed to a transverse H-field. The current density is derived on the y-axis by evaluating the current density at the three points $\left(J_z(r_{eq}, \frac{\pi}{2} - \varphi_{He}), J_z(r_{eq}, -\frac{\pi}{2} - \varphi_{He}), J_z(0,0)\right)$ and linear interpolation, as shown in fig. 3b). By assuming, that the aligned round-conductors are isolated from each other, the Eddy-current J_z along the y-axis is calculated separately for each round-conductor in the transformed foil winding, resulting in the current density shown in fig. 2, *2)*.

3) Calculation of foil current $J_{z,foil}$: The current density in the foil $\bar{J}_{z,foil}(y)$ ($y = [0, wdth]$) is derived from the Eddy currents of the separated round-conductors. Unlike before, the aligned round conductors are now assumed to be electrically connected. Under this condition, the current density must be continuous at the boundary between two conductors. This can be expressed as the following condition, which must hold true for all conductors

$$\bar{J}_{z,k}(d_{eq}, \frac{\pi}{2}) = \bar{J}_{z,k+1}(d_{eq}, -\frac{\pi}{2}); k = [1..n_{cond} - 1]. \tag{7}$$

As a consequence of (7), the derivative of the current density $\frac{dJ_{z,foil}(y)}{dy}$ is fully determined by the current density in the round conductors

$$\frac{dJ_{z,foil}(y)}{dy} = \frac{dJ_{z,k}(0, y_k)}{dy}; y_k = mod(y, d_{eq}), k = 1 + \frac{y}{d_{eq}}. \tag{8}$$

Note that $J_{z,k}$ is given in cartesian coordinates, for the sake of simplicity. A second condition for $J_{z,foil}(y)$ follows from the total current, flowing through the foil winding

$$\int \bar{J}_{z,foil}(y) dA = \bar{I}_{foil}. \tag{9}$$

The foil current density $\bar{J}_{z,foil}(y)$ can be calculated considering (8) and (9), which is shown schematically in fig. 2, *3)*.

4) Discretize and average to square-conductors: The foil is transformed into n_{sqr} square-conductors with the size $a_{eq} = thck$ as shown in fig. 2f). The dimension of the square-conductors is such, that they fulfill the low frequency approximation in (1). Hence for the mirroring method [12], the square-conductors can be treated in the same manner as windings of round conductors. It is assumed, that the current density in the foil is approximately linear across the crossection of the square conductors. With this assumption the current in each square conductor can be derived from the current distribution $\bar{J}_{z,foil}(y)$ by

$$\bar{I}_{z,sqr,k} = a_{eq}^2 \bar{J}_{z,foil}\left(\frac{a_{eq}}{2} + a_{eq}(k-1)\right) . k = [1..n_{sqr}]. \tag{10}$$

B. Numerical Iteration for entire Foil Winding

In section II-A a single foil exposed to a sinusoidal transverse field is modelled as n_{sqr} aligned square-conductors with a non-homogeneous current distribution $\bar{I}_{z,sqr,k}, k = [1..n_{sqr}]$. When a magnetic component with an entire foil winding is modelled, the correct determination of the square-conductor currents $\bar{I}_{z,sqr}$ becomes a non-trivial task. This is due to the fact, that a certain calculated $\bar{I}_{z,sqr}$ actually affects its root cause, being the external field $\vec{H}_e$. Therefore a numerical iteration is applied to determine $\bar{I}_{z,sqr}$.

The numerical iteration will again be explained on the example of the inductor with a foil winding, shown in fig. 1. Each foil is cut into n_{sqr} square conductors. Thus the whole foil winding is represented as $n_{sqr,tot} = n_{sqr} \cdot N_{foil}$ square conductors, where N_{foil} is the turns number of the foil winding. For the winding loss calculation, an arbitrary winding current waveform $i_{foil}(t)$ is decomposed into its complex spectrum by means of the Fourier transform. For each harmonic $\bar{I}_{foil}$ at frequency f_h,

978-1-4799-2706-7/14 $31.00 © 2014 IEEE

the iteration must be performed separately. The complex array $\bar{I}_{sqr}$ of size $(1 \times n_{sqr,tot})$ contains the current amplitudes of all square conductors. The starting point of the iteration is the uniform current-distribution:

$$\bar{I}_{sqr,0} = \left[\frac{\bar{I}_{foil}}{n_{sqr}} \cdots \frac{\bar{I}_{foil}}{n_{sqr}} \right]. \tag{11}$$

Figure 4 shows the overview of the numeric iteration. At the kth iteration, the latest current-distribution $\bar{I}_{sqr,k}$ is used as input to the mirroring method, to calculate the external H-field at the position of each square conductor $\bar{H}_{e,x}$ and $\bar{H}_{e,y}$. With the foil-to-square-conductor method described in section II-A, the current distribution $\bar{I}_{sqr,calc}$, that is caused by this external field, is calculated. The expression

$$\bar{I}_{sqr,k} = \bar{I}_{sqr,calc}. \tag{12}$$

is a sufficient condition for the correct current distribution. It describes the situation, where the physical root cause and its effect are in balance. Due to the miscellaneous approximations involved in this method, condition (12) can not be exactly fulfilled. The goal of the iteration is therefore to minimize the error

$$\bar{e}_{sqr} = |\bar{I}_{sqr,calc} - \bar{I}_{sqr,k}|. \tag{13}$$

This minimization could as well be treated as a purely mathematical problem and state of the art algorithms could be used to determine $\bar{I}_{sqr}$. Though the investigation of such algorithms and the comparison of their performance to the applied iteration-method is out of scope for this work. The applied iterative calculation method for $\bar{I}_{sqr}$ is based on a control-loop analogy. Further it is taken advantage of the fact, that the calculated current distribution $\bar{I}_{sqr,calc}$ exhibits, apart from a proportional scaling factor, approximately the same waveform as the correct current distribution $\bar{I}_{sqr,end}$. This makes it possible to adjust $\bar{I}_{sqr,k}$ by adding an increment $\bar{I}_{sqr,incr}$ derived from $\bar{e}_{sqr}$ at each iteration. Figure 5 illustrates this on the example of the inductor, shown in fig. 1. The norm of the current distributions in foil 1, $|\bar{I}_{sqr,k}|$ and $|\bar{I}_{sqr,calc}|$ are shown at the very beginning of the iteration and after 5 and 10 iteration steps. The current distribution, $|\bar{I}_{sqr,FEM}|$, derived from a 2-D FEM simulation, is shown as comparison. From the first iteration step on, $|\bar{I}_{sqr,calc}|$ and $|\bar{I}_{sqr,FEM}|$ exhibit similar waveforms and $|\bar{I}_{sqr,k}|$ approaches $|\bar{I}_{sqr,FEM}|$ with advancing iteration.

The detailed function to determine $\bar{I}_{incr}$ from $\bar{e}_{sqr}$ is split into three parts, shown in fig. 4, which are described in the following:

- *Variable Gain Controller*: To iteratively reduce the error $\bar{e}_{sqr}$, the current distribution is incremented by

$$\bar{\Delta}I_{sqr} = p_k \cdot \bar{e}_{sqr}. \tag{14}$$

- *Gain Adjustment*: The proportional gain p_k is adjusted in each iteration step, in order to limit the maximal current increment per iteration step to $I_{step,max,k}$, hence

$$p_k = \frac{I_{step,max,k}}{max(\bar{e}_{sqr})}. \tag{15}$$

The limit $I_{step,max,k}$ is initialized to the uniform current distribution

$$I_{step,max,0} = \frac{\bar{I}_{foil}}{n_{sqr}}. \tag{16}$$

During iteration $I_{step,max,k}$ is stepwise reduced to ensure, that $\bar{e}_{sqr}$ converges. If the averaged error over the whole winding $\bar{e}_{sqr,avg,k} = \sum \bar{e}_{sqr,k}/n_{sqr,tot}$ did not diminish compared to the error at the last iteration $\bar{e}_{sqr,avg,k-1}$, than $I_{step,max}$ is adjusted:

$$(e_{sqr,avg,k} > e_{sqr,avg,k-1}) \Rightarrow \\ I_{step,max,k+1} = \frac{I_{step,max,k}}{2}. \tag{17}$$

- *Phase Decoupling*: The feedback-loop introduces a phase-shift, see equation 4 of the foil-to-square-conductor method. To ensure, that the current increment $\bar{\Delta}I_{sqr}$ will actually compensate for the error $\bar{e}_{sqr}$, the phase-shift of the current increment must be compensated by:

$$\bar{I}_{incr} = \bar{\Delta}I_{sqr} \cdot e^{-j\varphi_{calc}}. \tag{18}$$

where φ_{calc} is the phase-shift of the foil-to-square-conductor method given by:

$$\varphi_{calc} = \angle \bar{I}_{sqr,calc} - \angle \bar{I}_{sqr,k}. \tag{19}$$

The iteration loop is executed and the current distribution is adjusted by

$$\bar{I}_{sqr,k+1} = \bar{I}_{sqr,k} + \bar{I}_{incr} \tag{20}$$

until $\bar{e}_{sqr}$ falls below a certain defined limit. In the case, where the foil-to-square-conductor method is applied for loss-calculation, the winding losses in the whole foil winding are taken as alternative convergence criteria.

$$P_{foil,tot,k} = \sum_{k=1}^{n_{sqr,tot}} P_{sqr,k}, \tag{21}$$

$$\Delta_{P,k} = 100 \frac{|P_{foil,tot,k} - P_{foil,tot,k-1}|}{|P_{foil,tot,k}|} \tag{22}$$

where $P_{sqr,k}$ are the Eddy-current losses in the kth square-conductor calculated as described in [5]. The iteration is stopped, if $\Delta_{P,k}$ stays below a certain threshold over 5 iterations:

$$max(\Delta_{P,k-5}, \dots, \Delta_{P,k}) \leq 0.1\% \Rightarrow \text{Stop-Iteration} \tag{23}$$

III. VALIDATION OF THE PROPOSED METHOD

The proposed method is validated on the example of a flyback-transformer for a PV-inverter. The whole calculation routine, including the foil-to-square-conductor method, is implemented as software program. The losses in the conductors are calculated according to [13], for Litz-wires, and [5], for square-conductors. The air-gap fringing field is modelled according to [12]. The Fourier decomposition of the flyback winding-currents is performed according to [10]. The validation is performed twofold, first with a FEM simulation and second with measured losses of a flyback-transformer.

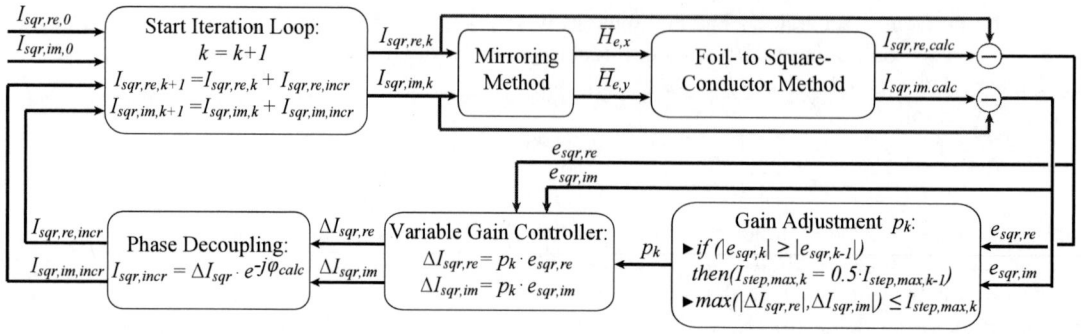

Figure 4. Overview numerical iteration for calculation of the non-homogeneous current distribution in foil windings.

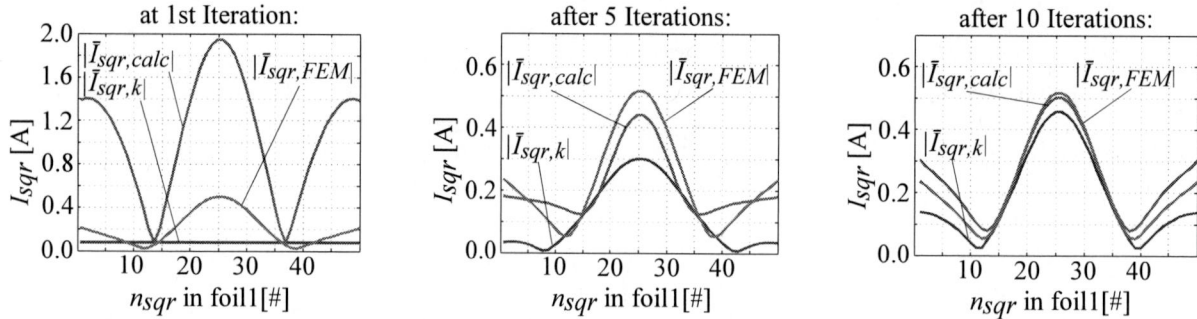

Figure 5. Numeric iteration for the current in foil 1 of the inductor with foil-winding (shown in fig. 1): $\bar{I}_{sqr,k}$ and $\bar{I}_{sqr,calc}$ at different iteration-steps in comparison to the current distribution $\bar{I}_{sqr,FEM}$ derived from the 2-D FEM simulation of the inductor.

Table I.   PARAMETERS 2-D FLYBACK TRANSFORMER WITH FOILS AND LITZ-WIRE.

Core	$a = 15\text{mm}, c = 7\text{mm}, d = 11\text{mm}$
Winding	Cu foil and Litz wires $N_1=5, thck_{w1}=0.2\text{mm}, wdth_{w1}=10mm$ $N_2 = 50, d_{s,w2} = 0.1\text{mm}, N_s = 7$
Air gap	$l_{airgap} = 1\text{mm}$

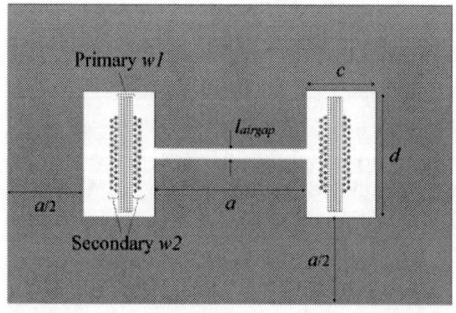

Figure 6. 2-D Flyback transformer with foils and Litz wires: E-core with air gap and 'sps' interleaved windings with foils on the primary and Litz wire for the secondary.

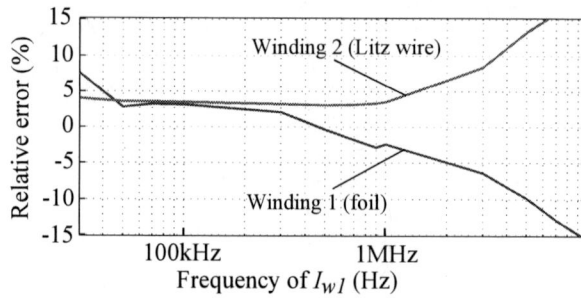

Figure 7. Flyback-Transformer as specified in tab. I, with sinusoidal excitation of the foil-winding ($I_{w1} = 5$A) and open circuit on the secondary: Relative difference between losses derived from the foil-to-square-conductor method and the 2D FEM simulation.

A. Validation with FEM Simulation

The model for foil-winding losses is compared to the conduction losses derived from a 2-D FEM simulation. The specification of the modelled transformer is given in table I and the 2-D winding arrangement is illustrated in fig. 6. The range of validity for the low frequency approximation of the mirroring method given by (2), is $f_{max,w1}$=270kHz for the foil

winding and $f_{max,w2}$=1MHz for the Litz-wire winding.

A first validation of the loss calculation is performed for the case of a sinusoidal current of 5A and various frequencies from 10kHz to 10MHz flowing through the foil winding. Whereas winding two is an open circuit. Figure 7 shows the deviation of the new loss model to the 2D FEM simulation (in percentage, normed to the FEM simulation values). The losses calculated with the new loss model exhibit good accordance to the 2D FEM simulation. The difference is below 7%, as long as the low frequency approximation is valid, and up to 15% in the whole considered frequency range.

The second validation is done by considering actual winding current waveforms of a DC-DC flyback converter operating in boundary conduction mode (BCM) at a switching frequency of 100kHz (8A peak, 0.75 duty cycle). To limit the FEM

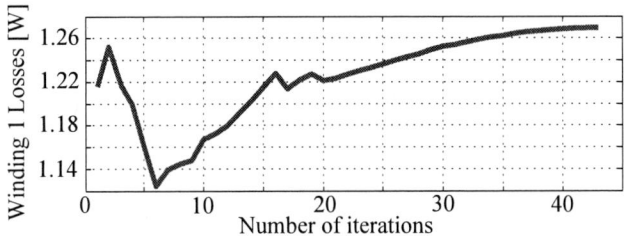

Figure 8. Foil-to-square-conductor Method Iteration, for flyback-transformer as specified in tab. I: Winding losses in the foil winding $P_{foil,totk,k}$ (see 21) with advancing iteration.

Table II. COMPARISON FOIL-TO-SQUARE-CONDUCTOR METHOD TO 2D FEM SIMULATION FOR FLYBACK TRANSFORMER WINDING LOSSES.

Calculation Method	2D FEM Simulation	Foil-to-Square-Cond. Method	
Computer	Server 8x paralleled Intel-Xeon-CPU E5-2660@2.2GHz 16GB RAM needed	Laptop Intel-Core i7-M620@2.67GHz	
Calculation time	~25min	19s	113s
Number of Harmonics	20	20	100
Number of Iterations	-	43	42
Winding Losses P_{w1}	1.32W	1.28W	1.33W
Winding Losses P_{w2}	0.576W	0.575W	0.67W

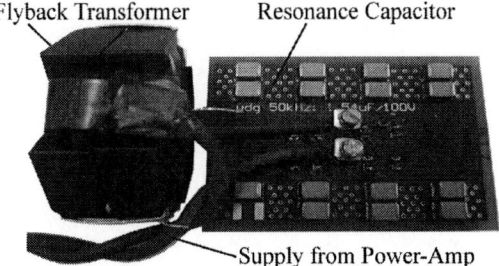

Figure 9. Experimental setup for winding loss measurement of the flyback transformer.

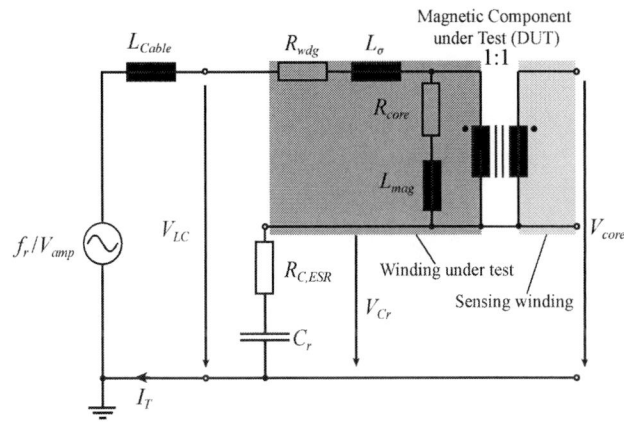

Figure 10. Overview measurement method for core- and winding-loss measurement of a magnetic component, according to [14].

Table III. WINDING LOSS MEASUREMENT SETUP: USED EQUIPMENT

Waveform Generator	Agilent 33522A
Power Amplifier	AE Techtron 7224
Oscilloscope	LeCroy WaveSurfer 24MXSB
Voltage Probes	LeCroy PP008
Current Probe	LeCroy AP015

calculation time, only harmonics from 100kHz to 2MHz are considered. Figure 8 shows the convergence of the total winding losses in the foil winding when the iteration is executed, as explained in section II-B. After 43 iterations the convergence criteria (23) is fulfilled and the iteration stops. The winding losses are listed in table II and compared to the results from the 2D FEM simulation. The difference in calculated winding losses is lower than 3.5%. To demonstrate the increase in calculation time, a second calculation-run of the foil-to-square-conductor method is performed and also listed in table II, taking into account a larger number of harmonics from 100kHz-10MHz. The winding losses in the foil increase marginally by 4% due to the higher order harmonic currents.

B. Validation with Measured Flyback Transformer

The model for foil-winding losses is verified by measurements on a flyback transformer. The transformer is built with a gapped RM low-profile core and foil windings on the primary and Litz wire on the secondary. First, the measurement setup is described in the following paragraph. The measurement results and the comparison are presented in III-B2.

1) Winding Loss Measurement Method and Setup: The measurement-methods proposed in [14] are applied for this verification, which allow to derive the losses in the foil-winding at a sinusoidal winding-current. Figure 9 shows the measurement setup consisting of the flyback transformer and a resonance capacitor. The schematic of the entire measurement setup is shown in fig. 10 and the applied measurement equipment is listed in table III. The primary winding is put in series with a capacitor C_{res} to form a resonant circuit together

with the transformers magnetizing inductance. A sinusoidal voltage source, realised with a signal generator and a power amplifier, drives the test current I_{test} through the primary winding. The secondary Litz-winding is left open circuit. The transformer has an additional sensing-winding (0.1mm² Cu round-wire) having the same turns-ratio as the primary winding, which is used for voltage measurements only. Hence no net current is flowing through the sensing-winding. The losses in the sensing winding are negligibly small, due to the low turns-number and the small conductor diameter.

The schematic in fig. 10 shows the T-equivalent circuit of the transformer, the parasitic cable inductance L_{Cable} and the resonance capacitor series resistance R_{CESR}. The equivalent components of the sensing winding are not relevant, because the winding does not carry any current. L_σ and L_{mag} are the magnetizing- and stray-inductance referred to the primary winding. The resistors R_{Core} and R_{wdg} model the losses in the core and the primary winding. The aim of the setup is to determine the resistive losses in R_{wdg}. This can be achieved by two distinct measurements.

- *The Resonant Method* as proposed in [15] and described in [14] section 1.2.4, allows to determine the total resistive losses of the resonant-circuit by operating the voltage-source at

$$f_r = \frac{1}{2\pi\sqrt{(L_{mag} + L_\sigma)C_r}}. \quad (24)$$

At this operating point the voltage over L_σ, L_{mag} and C_r cancel out and the voltage measured at the input of the resonant circuit V_{LC} only contains the resistive parts

$$V_{LC} = V_{R,wdg} + V_{R,core} + V_{R,C,ESR}. \quad (25)$$

Consequently, the losses in the resonance circuit can be split into three parts

$$P_{LC} = P_{R,wdg} + P_{R,core} + P_{R,C,ESR}. \quad (26)$$

The losses caused by the resonance current $P_{LC,fr}$ can be calculated from the measured voltage V_{LC} and current I_T by

$$P_{LC,fr} = \frac{1}{2}\hat{I}_{T,fr}\hat{V}_{LC,fr}\cos(\varphi_{I,T,fr} - \varphi_{V,LC,fr}), \quad (27)$$

where $\hat{I}_{T,fr}$, $\varphi_{I,T,fr}$ and $\hat{V}_{LC,fr}$, $\varphi_{V,LC,fr}$ are the amplitude and phase at the resonance frequency derived from the fourier transform.

- *The Capacitive Cancellation Core Loss Method* proposed in [14] section 2.1.2, can be applied to measure the core losses separately. Unlike the resonant method, where the winding losses are included in the measured losses. The voltage V_{core} is measured between the upper port of the sensing winding to ground, as shown in fig. 10. Note that the lower port of the sensing winding is connected to the resonance capacitor and hence V_{core} can be expressed as

$$V_{core} = V_{R,core} + V_{L,mag} + V_{R,C,ESR} + V_{C,r}. \quad (28)$$

The frequency of the voltage-source is chosen, such that $V_{L,mag} = -V_{C,r}$, which is the case at

$$f_r = \frac{1}{2\pi\sqrt{L_{mag}C_r}}. \quad (29)$$

The measured voltage only contains the resistive parts $V_{core} = V_{R,core} + V_{R,C,ESR}$ and the resistive losses caused by the resonance current can be calculated by:

$$P_{core,fr} = \frac{1}{2}\hat{I}_{T,fr}\hat{V}_{core,fr}\cos(\Delta\varphi_{fr}), \quad (30)$$

with

$$\Delta\varphi_{fr} = \varphi_{I,T,fr} - \varphi_{V,core,fr}. \quad (31)$$

The losses consists of the following two parts

$$P_{core,fr} = P_{R,core,fr} + P_{R,C,ESR,fr}. \quad (32)$$

To obtain the magnetic losses $P_{R,wdg,fm}$ at a certain frequency f_m the losses $P_{LC,fm}$ and $P_{core,fm}$ are measured as explained above. For the measured transformer the stray inductance L_σ is much smaller than L_{mag} and hence the same resonant capacitance C_r can be used for both measurements.

The magnetic losses can be obtained from the measurements with (27) and (32) by:

$$P_{R,wdg,fm} \simeq P_{LC,fm} - P_{core,fm}. \quad (33)$$

The losses in $R_{C,ESR}$ cancel out, though $R_{C,ESR}$ should not be much higher than R_{wdg} to obtain a good resolution of the measurement.

The accuracy of the performed loss measurements caused by the deviations in the current Δi, voltage- Δu and phase-angle measurements $\Delta\varphi$ can be deducted from (27),(30) using the second-order Taylor-series of the cosine $\left(\cos(x) = 1 - \frac{x^2}{2}\right)$ and neglecting deviation-coefficients of third order:

$$\Delta p_{meas} = \Delta v I + V\Delta i + \Delta v\Delta i - P\Delta\varphi^2. \quad (34)$$

The phase-deviation follows from the time-delay between the voltage- and current-probe by

$$\Delta\varphi = 2\pi \cdot 16ns \cdot f_{meas}, \quad (35)$$

whereas the time-delay is derived from a reference-measurement using a shunt-resistor. Voltage and current deviation are found to be

$$\Delta v \simeq 80\mu V, \Delta i \simeq 1mA. \quad (36)$$

This is above their theoretical resolution-limit of $63\mu V$ and $313\mu A$, due to the low signal-to-noise ratio at high scale-factors of the oscilloscope. An additional measurement error introduced by the parasitic inter-winding capacitance, described in [14] (2.11), is found to be negligible small, due to the relatively low measuring frequencies. The deviation in the measured winding losses follows from (33)

$$\Delta p_{wdg} \simeq \Delta p_{LC} + \Delta p_{core}, \quad (37)$$

where Δp_{core} and Δp_{LC} are calculated with (34).

2) Loss Measurements and Comparison: The flyback transformer loss model is parameterized to model the measured flyback transformer. The frequency limit, up to which the low-frequency 2-D field approximation used for winding loss calculation applies, is determined by (2) and equals

$$f_{max,low,freq} = 172kHz \quad (38)$$

for the investigated flyback transformer geometry. Above this frequency the modelled winding losses are subject to an increasing modelling error. In table IV the measured winding losses are compared to the losses calculated with the model at three different measuring points: 50kHz, 100kHz and 200kHz. Further the accuracy of the measured losses is determined with (37). The model exhibits a good accordance to the measured losses. The winding losses predicted by the model show a deviation below 15% for the measuring points at 50kHz and 100kHz. The measuring frequency of 200kHz is above $f_{max,low,freq}$ and accordingly the deviation increases to a value of 21%.

TABLE IV. FLYBACK TRANSFORMER WINDING LOSS MODEL VALIDATION.

Test Conditions:	Reference Measurements:		Loss Model:	Model Deviation:
f_{meas}	P_{wdg}	Δp_{wdg}	P_{wdg}	$\Delta_{model,to,meas}$
50kHz	0.0136W	$< \pm 3.9\%$	0.0132W	**-2.6%**
100kHz	0.0046W	$< \pm 6.1\%$	0.0052W	**14.9%**
200kHz	0.0022W	$< \pm 7.1\%$	0.0026W	**21.3%**

C. Calculation Time and Complexity

Calculation speed was a major motivation to develop the foil-to-square-conductor method. The achieved evaluation time of the whole loss-model of a flyback-transformer in DC-DC BCM operation (see III-A and table I for specifications), is 19s and 113s for a considered number of higher order harmonics of $n_h = 20$ and $n_h = 100$ on a laptop computer equiped with an Intel-Core-i7-620M@2.67GHz. Figure 11 shows the relative calculation time of the most dominant tasks of the loss-model. The calculation complexity of the foil-to-square-conductor method and the mirroring method scales linearly with n_h and the number of conductors, being $n_{sqr,tot}$ for the foil-to-square and $n_{sqr,tot} + n_{wdg,2}$ for the mirroring method ($n_{wdg,2}$ is the secondary turns number). Note, that all dominant tasks are in the iteration loop (see fig. 4). Thus the iteration itself is the most time-consuming part of the loss-model, whose calculation time depends linearly on the number of iterations $n_{num,it}$. Evaluations with different parameters showed, that the developed numerical iteration needs an average of $n_{num,it} \simeq 45$ to converge. To further reduce calculation time, an improved iteration-method would be most effective. While $n_{sqr,tot}$ and $n_{wdg,2}$ follow from the specifications, the number of harmonics n_h can be chosen as low as possible, depending on the considered current waveform. A further speed improvement can be achieved in the mirroring method by reducing the number of mirroring below the currently implemented 11x11 mirrored basic winding windows.

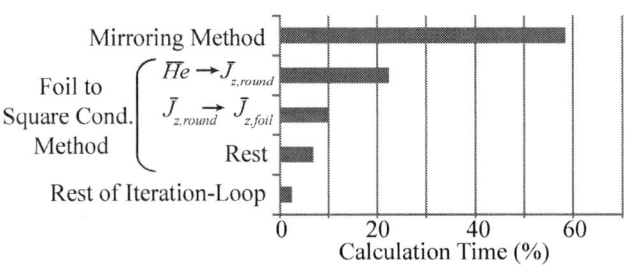

Figure 11. Flyback Transformer Model (specification as in table I at DC-DC BCM operation with harmonics from 100kHz to 2MHz): calculation complexity analysis.

IV. CONCLUSION

A new semi-numerical method is developed for loss calculation in foil windings exposed to a 2-D fringing field. Compared to existing calculation methods it features the advantage of much faster calculation speed compared to FEM simulations. For the considered example the calculation time is reduced by a factor of 75. At the same time the new method is not restricted to certain geometric arrangements as the existing analytical and semi-empirical methods. The analysis of the calculation complexity discloses the potential of further speed

improvement. The accuracy of the method is validated on the example of a flyback transformer by both FEM simulations and measurements on a test-setup. The method exhibits deviations below 15% in comparison to the measured losses.

REFERENCES

[1] P. Wallmeier, "Improved analytical modeling of conductive losses in gapped high-frequency inductors," *IEEE Transactions on Industry Applications*, vol. 37, no. 4, pp. 1045–1054, 2001.

[2] J. Zwysen, R. Gelagaev, J. Driesen, S. Gooossens, W. Vanvlasselaer, K and. Symens, and B. Schuyten, "Multiobjective design of a close-coupled inductor for a three-phase interleaved 140kw dc-dc converter," in *39th IECON, Vienna, 10-13 November*, 2013.

[3] P. Wallmeier, "Automatisierte Optimierung von induktiven Bauelementen für Stromrichteranlagen," Ph.D. dissertation, Universität Paderborn, 2001, Shaker-Verlag, ISBN 3-8265-8777-4.

[4] A. Van den Bossche and V. Valchev, "Eddy current losses and inductance of gapped foil inductors," in *IEEE, 28th IECON An. Conference*, vol. 2, 2002, pp. 1190–1195.

[5] ——, *Inductors and Transformers for Power Electronics*.

[6] C. Sullivan, "Computationally efficient winding loss calculation with multiple windings, arbitrary waveforms, and two-dimensional or three-dimensional field geometry," *IEEE Transactions on Power Electronics*, vol. 16, no. 1, pp. 142–150, 2001.

[7] F. Robert, P. Mathys, and J.-P. Schauwers, "A closed-form formula for 2D ohmic losses calculation in SMPS transformer foils," in *IEEE 14th APEC*, vol. 1, 1999, pp. 199–205.

[8] P. Dowell, "Effects of eddy currents in transformer windings," *Proceedings of the Institution of Electrical Engineers*, vol. 113, no. 8, pp. 1387–1394, 1966.

[9] J. Muhlethaler, J. W. Kolar, and A. Ecklebe, "Loss modeling of inductive components employed in power electronic systems," in *Proc. IEEE 8th IPEC (ECCE Asia)*, 2011, pp. 945–952.

[10] J. Zhang, W. Yuan, H. Zeng, and Z. Qian, "Simplified 2-d analytical model for winding loss analysis of flyback transformers," *Journal of Power Electronics*, vol. 12, no. 6, pp. 960 – 973, November 2012.

[11] J. Lammeraner and M. Stafl, *Eddy Currents*, I. B. LTD, Ed. SNTL Publisher of Technical Literature, 1966.

[12] J. Muehlethaler, M. Schweizer, B. Robert, J. Kolar, and A. Ecklebe, "Optimal design of LCL harmonic filters for three-phase PFC rectifiers," *IEEE Transactions on Power Electronics*, vol. 28, no. 7, pp. 3114–3125, 2013.

[13] F. Tourkhani and P. Viarouge, "Accurate analytical model of winding losses in round litz wire windings," *IEEE Transactions on Magnetics*, vol. 37, pp. 538–543, 2001.

[14] M. Mingkai, "High frequency magnetic core loss study," Ph.D. dissertation, Virginia Polytechnic Institute and State University, 2013.

[15] F. Dong Tan, J. Vollin, and S. Cuk, "A practical approach for magnetic core-loss characterization," *IEEE Transactions on Power Electronics*, vol. 10, pp. 124–130, 1995.

Loss Reduction of Laminated Core Inductor used in On-board Charger for EVs

Takahiro Tera, Hiroshi Taki
CORPORATE R&D DIV.2
DENSO CORPORATION
Aichi, Japan
takahiro_tera@denso.co.jp

Toshihisa Shimizu
Dept. of Electrical and Electronic Engineering
Tokyo Metropolitan University
Tokyo, Japan
shimizut@tmu.ac.jp

Abstract—**A laminated core has advantages such as low loss, high permeability, and high saturation flux density. For inductor applications, an air gap has to be inserted into the magnetic circuit to avoid saturation at lower magnetic field strengths. However, when an air gap is inserted, fringing flux leaks out of the air gap and concentrates on the surface of the core. As a result, the iron loss increases to 292% compared with an ungapped structure, and the inductance characteristic deteriorates. Although a chamfered structure known as a prior art reduces the iron loss to 147%, there is room for improvement. In this paper, we propose a method for optimizing the shape of the air gap. Consequently, the proposed structure reduces the iron loss to 113% and improves the inductance characteristic. The proposed method helps increase the efficiency and reduce the size of the on-board charger.**

Keywords—laminated core, fringing flux, iron loss, inductor

I. INTRODUCTION

With regard to the issue of global warming and fossil fuel depletion, the regulation of CO_2 emission has been tightened. In the automotive industry, vehicles with a combustion engine are converted into electric vehicles (EVs) or plug-in hybrid vehicles (PHVs) with a battery [1]. As for the battery charger for EVs/PHVs, there exist an on-board charger. The on-board charger has a power factor correction (PFC) converter and a DC-DC converter. The switched-mode power supply(SMPS) technology is used for these converters [2]. SMPS is becoming more important particularly in automotive applications such as the on-board charger for EVs. SMPS relies on the magnetic components for power conversion. SMPS works at frequencies much higher than the utility frequency, and increasing the frequency is crucial for reducing the size of the on-board SMPS. For reducing losses occurring in magnetic components, it is important to eliminate eddy currents. Therefore, powder cores including dust cores and laminated cores using amorphous and nanocrystalline materials have been proposed.

Powder cores have low permeability and are distributed-gap magnetic components in nature owing to their low density. There is a tendency to increase the core size because a reduction in the permeability of the core directly leads to a reduction in the inductance. They have a low loss at hundreds of kHz. The iron loss for the powder cores can be practically and accurately calculated, as described in [3]-[7], aiding design optimization.

On the other hand, laminated cores have a higher permeability than powder cores. The high permeability necessitates air gaps in the magnetic circuit when they are used for inductors. It is easy to downsize the core because the inductance relies on the air gap. If having no air gaps, these cores have a low loss at dozens of kHz. However, the magnetic flux crossing the air gap expands its cross sectional area, thereby causing fringing flux that induces in-plane eddy current and magnetic flux concentration on the surface of the core. As a result, the iron loss increases, and the inductance characteristic deteriorates.

As a well-known prior art, a structure with chamfered edges has been proposed, which reduces the iron loss by suppressing the fringing flux. However, it is insufficient to disperse the magnetic flux concentration for laminated cores owing to the high permeability of amorphous and nanocrystalline materials. A loss reduction method involving slitting of the core sheets has been proposed [8], which reduces the in-plane eddy current. However, it has no effect on dispersion of the magnetic flux.

The objective of this study is to clarify the behavior of the fringing flux, reduce the iron loss, and improve the inductance characteristics under DC-biased magnetization. In this paper, we propose a method for optimizing the shape of the air gap. In Section II, the specifications of the on-board charger and inductor are stated. In Section III, the loss calculation of the laminated core is explained, and the loss mechanism is determined using the model. In Section IV we propose an iron loss reduction method on the basis of the dispersion of magnetic flux by optimizing the shape of the air gap. In Section V, the proposed method is analyzed by simulation to verify the reduction of the iron loss and improvement of the inductance characteristic. In Section VI, the effect of the loss reduction is verified through an experiment, and it is confirmed that the proposed method is valuable for the loss reduction of the laminated core inductor used in the on-board charger for EVs. The conclusion of this study is stated in the final section.

II. CIRCUIT CONFIGURATION OF THE ON-BOARD CHARGER

Fig. 1 shows the charging system with an on-board charger. Fig. 2 shows the circuit configuration of the on-

978-1-4799-2706-7/14 $31.00 © 2014 IEEE

Fig. 1. Charging system with on-board charger.

TABLE I. SPECIFICATIONS OF CIRCUIT AND PFC INDUCTOR.

AC input voltage	100/200 V
AC input current	15 A
DC output power	3.0 kW
Inductance at 15 A	500 μH
Switching frequency	20 kHz

board charger. Although inductors are used in both the PFC converter and DC-DC converter, in this study, we have focused on the PFC inductor. The specifications of the circuit and the PFC inductor are listed in Table I. The input voltage and input current waveforms are shown in Fig. 3. The PFC converter controls the current to be in-phase with the voltage. The voltage frequency is the utility frequency of 50 Hz. The current waveform represents the sum of the sinusoidal current at 50 Hz and ripple current owing to converter switching at 20 kHz. It is sufficiently higher than 50 Hz, thus, the current at 50 Hz can be practically regarded as DC. The smaller the inductance, the larger is the ripple current, therefore, the switching loss and conduction loss of the switching devices increase. The target inductance value at 15 A is set to 500 μH. The structure of the designed inductor is shown in Fig. 4. The core characteristics are summarized in Table II. We use a FINEMET®️ core manufactured by Hitachi Metals, Ltd., composed of nanocrystalline soft magnetic material, because of the very small iron loss occurring within the core. It has high permeability and very thin sheet thickness. A general structure of the core has a right-angled edges near the air gap.

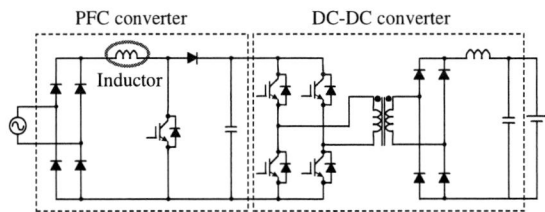

Fig. 2. Circuit configuration of the on-board charger.

TABLE II. CORE CHARACTERISTICS.

Relative permeability at 20 kHz	30000
Electric conductivity	8.3×10^5 S/m
Saturation flux density	1.23 T
Stacking factor	78 %
Thickness	18 μm

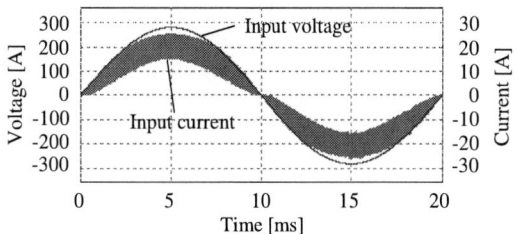

Fig. 3. Input voltage and current.

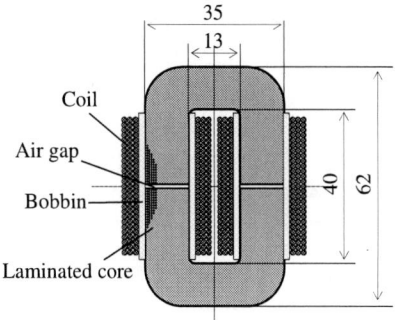

Fig. 4. Structure of the inductor.

III. LOSS CALCULATION OF THE LAMINATED CORE

A. Method for calculating the iron loss

We analyze the laminated core inductor by using the JMAG, which is finite element analysis software. Fig. 5 shows the one-eighth section model for the analysis of the iron loss. A coil is modeled as one block, thereby simplifying the actual strands of the magnet wire. The AC losses in the coil are not considered. Modeling within a single sheet of lamination is difficult owing to the limited computing power, therefore, several techniques are proposed [9]-[11]. We select a homogenization method [9]. By using the relative permeability μ and electric conductivity σ of the single laminated core, the relative permeability μ' and electric conductivity σ' of the homogenized model are expressed as,

$$1/\mu'_x = \alpha/\mu\mu_0 + (1-\alpha)/\mu_0$$

$$\mu'_{yz} = \alpha\mu\mu_0 + (1-\alpha)\mu_0$$

$$\sigma'_x = 0, \ \sigma'_{yz} = \alpha\sigma \quad (1)$$

where α is the stacking factor of the laminated core, μ_0 is the space permeability, subscript x indicates the direction orthogonal to the core sheets, and subscript yz indicates the direction in-plane with the core sheets. Here, the in-plane eddy current is calculated using (1). For accurately modeling the eddy current, we select a hexahedron mesh. After the magnetic analysis, the orthogonal eddy current loss and hysteresis loss are calculated by the Steinmetz's formula,

$$W_i = K_e B^2 f^2 + K_h B^\beta f \quad (2)$$

where W_i is the total iron loss per unit volume, K_e is the eddy current loss coefficient, K_h is the hysteresis loss coefficient, β is the fitting parameter, and B is the

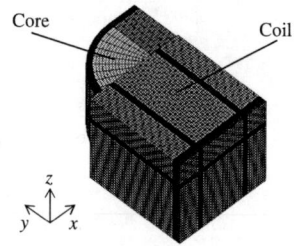

Fig. 5. Analysis model for iron loss.

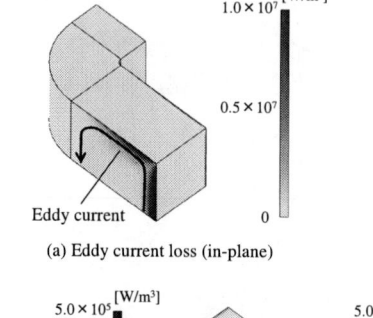

(a) Eddy current loss (in-plane)

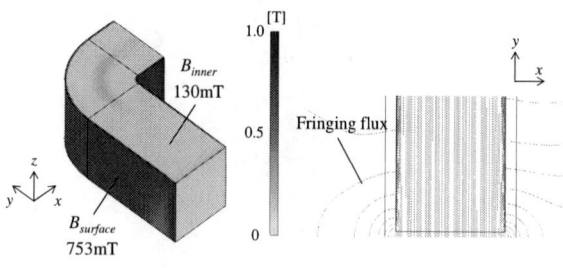

(a) Magnetic flux density distribution (b) Magnetic flux lines

Fig. 6. Magnetic flux of the general structure.

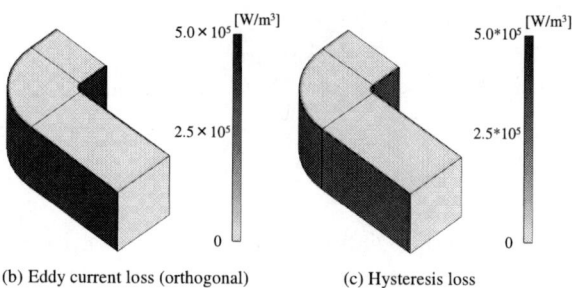

(b) Eddy current loss (orthogonal) (c) Hysteresis loss

Fig. 7. Iron loss distribution of the general structure.

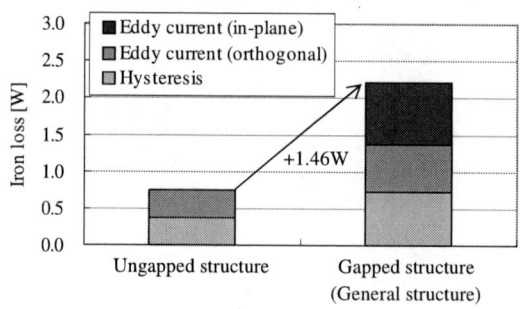

Fig. 8. Iron loss of the general structure.

peak value of the magnetic flux density for sinusoidal excitation at the switching frequency f. The minor loop under DC-biased magnetization is not considered here. The current for the analysis is set to AC 3.4 A, which approximates only the ripple current.

B. Loss mechanism caused by the fringing flux

Fig. 6(a) shows the magnetic flux density distribution of the model, where $B_{surface}$ is the maximum value of the magnetic flux density within the range of 1 mm from the surface, and B_{inner} is the maximum value in the inner region deeper than 1 mm from the surface. It shows that $B_{surface}$ is higher than B_{inner}. This implies that the fringing flux shown in Fig. 6(b) flows close to the surface because μ'_x is usually higher than μ'_{yz}. The result is particularly noticeable in the FINEMET® core because of the magnetic anisotropy with μ'_x 6,000 times higher than μ'_{yz}. As shown in Fig. 6(a), the concentrated magnetic flux at the surface leads to the increase in the iron loss.

Fig. 7 shows the the iron loss distribution. It shows that the in-plane eddy current flows on the surface of the core, and the orthogonal eddy current loss and hysteresis loss concentrate on the surface. Fig. 8 shows the result of the iron loss calculation in comparison with an the ungapped structure. Inserting an air gap increases the iron loss from 0.75 W to 2.21 W. The results indicate that it is necessary to minimize the in-plane eddy current on the surface of the core and disperse the magnetic flux evenly in the core.

IV. IRON LOSS REDUCTION BASED ON DISPERSION OF THE MAGNETIC FLUX

A. Conventional structure with chamfered edges

Fig. 9 shows the effect of the conventional structure with chamfered edges known as a prior art of the iron loss reduction. We designed the x-dimension as 1 mm, and the chamfer angle from the x-axis as 74 degrees. The magnetic flux is not completely dispersed, therefore, the iron loss reduces to 1.1 W.

B. Method for optimizing the shape of the air gap

In this paper, we have proposed the shape of an exponential function curve as shown in Fig. 11 and the following equation,

$$y_1 = y_g a b^{\frac{x + y_g x_{offset1} - x_{width}}{y_g}} + y_g(1 - a)$$

$$(x \geq x_{width} - y_g x_{offset1})$$

$$y_2 = y_g a b^{\frac{-x + y_g x_{offset2} - x_{width}}{y_g}} + y_g(1 - a)$$

$$(x \leq y_g x_{offset2} - x_{width}) \qquad (3)$$

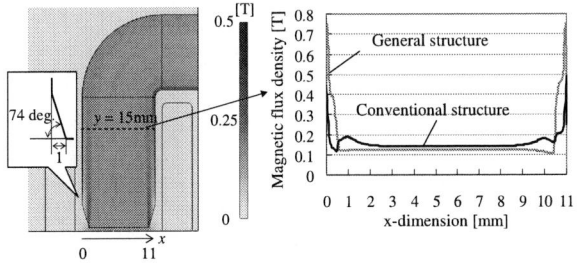

(a) Magnetic flux density distribution   (b) Comparison of magnetic flux density

Fig. 9. Effect of the conventional structure with chamfered edges.

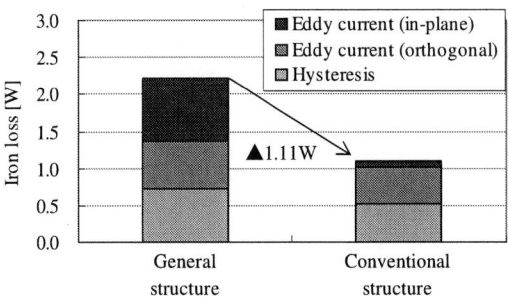

Fig. 10. Iron loss of the conventional structure.

where, a and b are the design parameters, $x_{offset1}$ and $x_{offset2}$ are offsets in the x-direction, y_g is half of the gap, x_{width} is half of the core width. The reference point in the x-direction is the center of the core width. The reference point in the y-direction is the center of the air gap. We select $a = 0.02$, $b = 5.21$, $x_{offset1} = 2.45$, and $x_{offset2} = 2.25$. The difference between $x_{offset1}$ and $x_{offset2}$ is because of the difference in the magnetic resistance between the inner perimeter and the outer perimeter, thus the fringing flux tends to leak out from the shorter magnetic path. The magnetic resistance of the proposed structure increases as a result of introducing a curved edge, therefore, the air gap of the proposed structure is adjusted from 1.1 mm to 0.9 mm to maintain the inductance at the same value. Fig. 12 shows the difference between the magnetic flux densities of the general structure and the proposed structure. It shows that the $B_{surface}$ in the y-direction at 15 mm reduces from 753 mT to 177 mT and B_{inner} increases owing to the reduced air gap.

We explain the effect of this shape by a simplified schematic shown in Fig. 13. Only six layers at the surface of the lamination are shown. In our simulation, approximately 480 sheets are modeled, where, i and B shown in the figure represent the in-plane eddy current and the magnetic flux density up to the sixth layer, respectively. Fig. 13(a) shows the general structure with a right-angled edge facing the air gap, i_{eddy1} and B_1 being the largest owing to the flux concentration in the first layer as discussed in Section III. As shown n Fig. 13(b), the core edge is chamfered to minimize i_{eddy1}. The dimensions d_x and d_y are selected to cover the area where the fringing flux accumulates. As a result, i_{eddy} is spread over the top five layers to form five small eddies, thereby reducing the i_{eddy1} loss considerably. Further, B_1 in the core is dispersed evenly, and B_1, B_2, B_3, and B_4 are substantially uniform. Therefore, the orthogonal eddy current is reduced. However, the still-remaining edge between the fifth and sixth layer of the core increases B_5 owing to the small magnetic resistance in this location, which results in an increase in the iron loss.

Further, the edge is altered to a curved form, as shown in Fig. 13(c). Here i and B up to the sixth layer are more evenly dispersed. For fully implementing the concept described above, we chose the shape of the exponential function curve, as shown in Fig. 11.

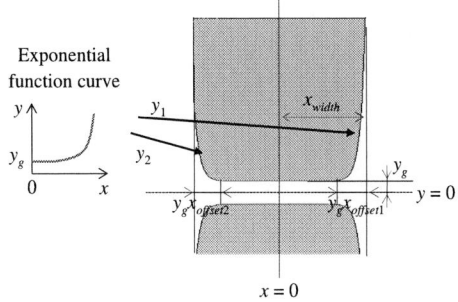

Fig. 11. Proposed structure with curved edges.

C. Effect of the parameters

The proposed equation (3) provides the optimal curve for the y_g, therefore, the change in y_g or x_{offset} lead to the concentration of the magnetic flux. Fig. 14 shows the calculation result obtained by changing the y_g to 0.25 mm, and 0.65 mm. The current value of each simulation is adjusted in order to maintain the magnetic flux density. When compared with the result of 0.45 mm, B_{inner} at 0.25 mm increases to 171 mT, and $B_{surface}$ at 0.65 mm

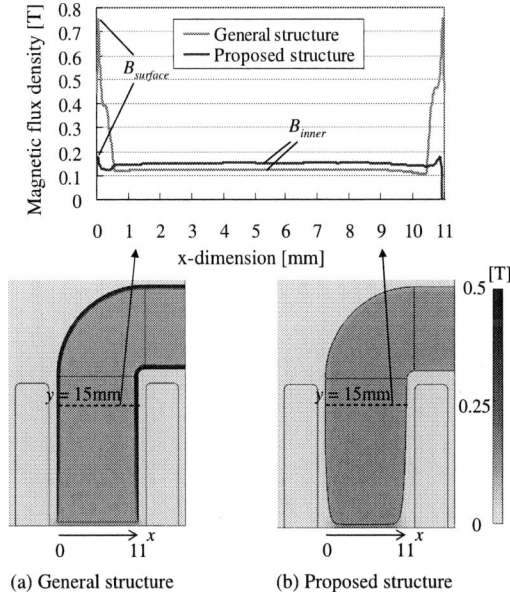

(a) General structure   (b) Proposed structure

Fig. 12. Effect of the proposed structure.

The 2014 International Power Electronics Conference

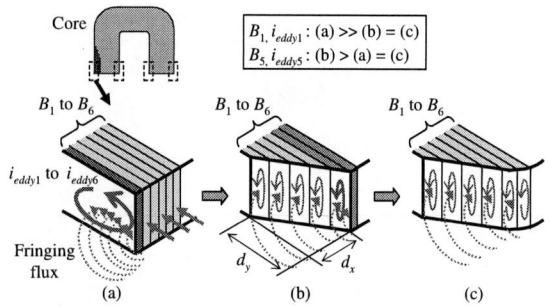

Fig. 13. Dispersion of the magnetic flux and minimization of the eddy current.

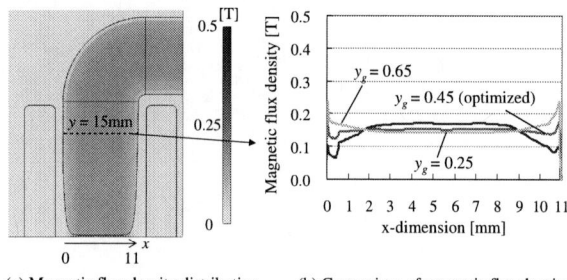

(a) Magnetic flux density distribution ($y_g = 0.25$) (b) Comparison of magnetic flux density

Fig. 14. Effect of the parameter y_g.

increases to 237 mT. In both the cases, the concentration of the magnetic flux leads to an increase in the iron loss. In particular, the increase in $B_{surface}$ causes the in-plane eddy current. Fig. 15 shows the simulation result obtained by changing $x_{offset1}$ to 1.75 mm. $B_{surface}$ increases to 440 mT, and causes an increase in the iron loss. On the other hand, for the parameter x_{width}, the proposed shape is applied as it is. Fig. 16 shows the simulation result obtained by changing the x_{width} twice. It is found that $B_{surface}$ and B_{inner} are the same. From the above, it is intended to provide an optimal curve for y_g. Further, it is applicable to the condition that x_{width} and y_g x_{offset} match.

V. SIMULATION RESULTS

A. Reduction of the iron loss

The loss reduction effect of the proposed structure is presented here. Fig. 17 shows the iron loss distribution

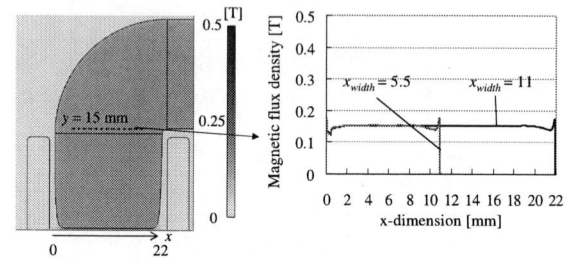

(a) Magnetic flux density distribution (b) Comparison of magnetic flux density

Fig. 16. Effect of the parameter x_{width}.

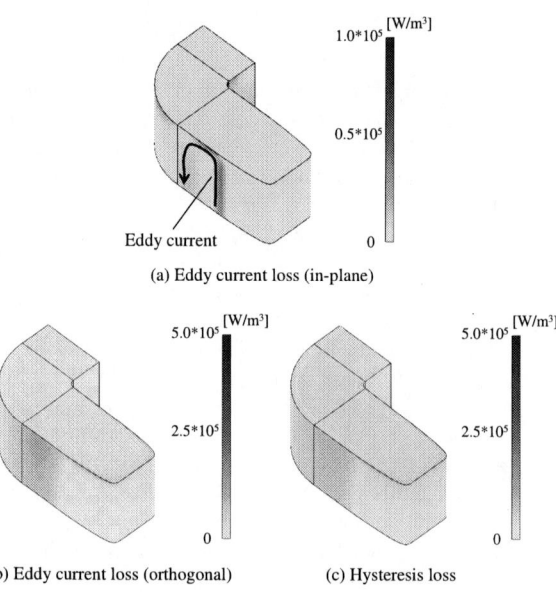

(a) Eddy current loss (in-plane)

(b) Eddy current loss (orthogonal) (c) Hysteresis loss

Fig. 17. Iron loss distribution of the proposed structure.

of the proposed structure. It shows that the in-plane eddy current loss is almost eliminated and the remaining 1% occurs at the unprocessed flat surface. The orthogonal eddy current loss and hysteresis loss are equalized in the distribution. The loss reduction effect of the proposed structure is presented in Fig. 18. The loss owing to the air gap reduces to 0.76 W. In this paper, the edges of the core are reshaped to equalize the magnetic flux density in the xy plane, thus, the magnetic flux concentration in the yz plane is still unabated. The remaining edges can be reshaped in the same way to reduce the loss further.

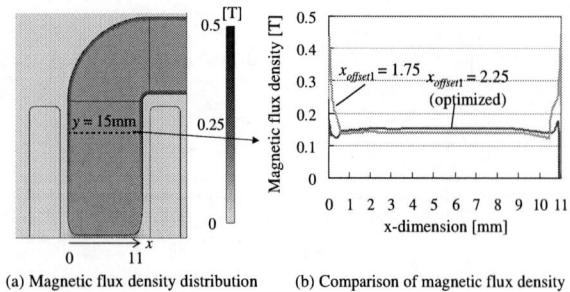

(a) Magnetic flux density distribution (b) Comparison of magnetic flux density

Fig. 15. Effect of the parameter x_{offset}.

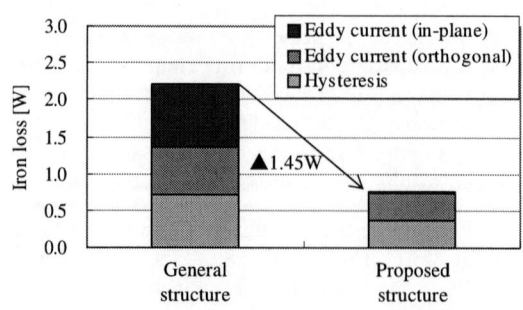

Fig. 18. Iron loss of the proposed structure.

978-1-4799-2706-7/14 $31.00 © 2014 IEEE

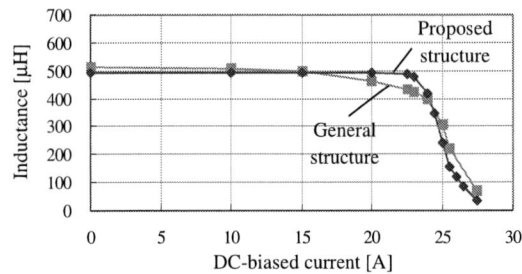

Fig. 19. Inductance characteristics under DC-biased magnetization.

B. Improvement of the inductance characteristic

We then proceed to the improvement of the inductance characteristic under DC-biased magnetization. The current for the analysis is derived as the sum of the AC current ripple ΔI and the specified DC current. The inductance L is calculated by the following equation,

$$L = \frac{\Delta \phi}{\Delta I} \qquad (4)$$

where, $\Delta \phi$ is the fluctuation in the flux linkage. Fig. 19 shows the inductance characteristics under DC-biased magnetization. It shows that the inductance of the proposed structure is improved under higher DC-biased currents when compared to the general structure. This implies that the general structure saturates gradually from where the magnetic flux is concentrated while the proposed structure saturates simultaneously owing to the effect of the dispersion of the magnetic flux. If an overcurrent protection is sufficient, the proposed structure using the FINEMET® core is suitable for inductor. This result contributes to the reduction of the switching loss and conduction loss, and downsizing.

VI. EXPERIMENTAL RESULTS

In this section, the loss reduction of the exponential function curved shape is confirmed experimentally. Fig. 20 shows a prototype of the proposed structure. The prototype is processed by grinding with a machining center. We measure the iron loss with the B-H analyzer SY-8219 manufactured by Iwatsu Test Instruments Corp. The number of turns of the exciting coil is 28, that of the detection coil is 12, and both the coils are wound in the same part. The measurement condition is similar to the simulation condition, and sinusoidal current is applied.

Fig. 21 shows the experimental result of the iron loss, and both the simulation result and experimental result are summarized in Table III. The percentage of the iron loss of each gapped structure to that of the ungapped structure is also shown. The general structure increases the iron loss to 2.19 W when the air gap is inserted, and it is consistent with the simulation result. The proposed structure increases the iron loss to 0.85 W, and there is a small error between the experimental result and the simulation result. The effect of the exponential function curved shape is confirmed.

Fig. 22 shows a comparison of the iron loss of the general structure and the proposed structure with the

Fig. 20. Prototype of the proposed structure.

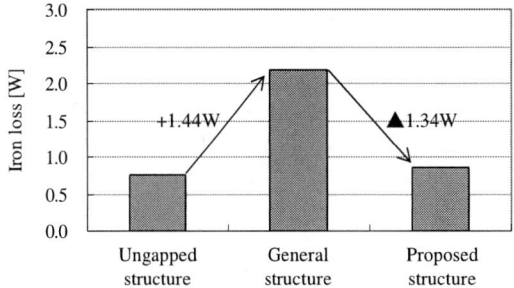

Fig. 21. Experimental result of iron loss.

magnetic flux density. As seen in Fig. 22, the effect of the proposed structure is slightly lower. Therefore, the shape error caused by the uneven surface of the laminated core and partial shorting of the processed surface can be considered. In the future work, we will improve these factors.

VII. CONCLUSION

In the present work, we proposed a method for optimizing the shape of the air gap in order to evenly disperse the magnetic flux in the core and minimize

TABLE III. SUMMARY OF RESULTS.

	Ungapped structure	General structure	Conventional structure	Proposed structure
Simulation	0.75 W (100%)	2.21 W (295%)	1.10 W (147%)	0.76 W (101%)
Experiment	0.75 W (100%)	2.19 W (292%)	Unmeasured	0.85 W (113%)

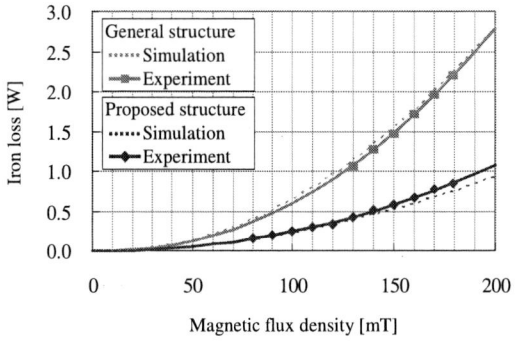

Fig. 22. Comparison of results.

the in-plane eddy current. We found that an exponential function curve is suitable. The effect of loss reduction is confirmed through an experiment. Inserting the air gap increases the iron loss to 292%. While the conventional structure with chamfered edges reduces the iron loss to 147%, the proposed structure reduces the iron loss to 113%. Furthermore, the proposed structure improves the inductance characteristic under DC-biased magnetization. The proposed method helps increase efficiency and reduce the size of the on-board charger.

In this study, the effect of the proposed structure was confirmed under the assumption that a sinusoidal current flows. In the actual PFC operation, the ripple current is almost a triangular wave, and the duty of the voltage changes. In addition, the iron loss changes under DC-biased magnetization. In the future work, we will verify the effect of loss reduction by considering the actual operation.

REFERENCES

[1] K. Yamamoto, "The background of electric vehicle spread", *International Electric Vehicle Technology Conference (EVTeC)*, 2011.

[2] D. Gautam, F. Musavi, M. Edington, W. Eberle, W. Dunford, "An automotive onboard 3.3-kW battery charger for PHEV Application", *IEEE Trans. Vehicular technology*, vol. 61, no.8, pp. 3466-3474, 2012.

[3] S. Iyasu, T. Shimizu, K. ishii, "A novel iron loss calculation method on power converters based on dynamic minor loop", *Proc. of European Conference on Power Electronics and Applications*, pp. 2016-2022, 2005.

[4] T. Shimizu, K. ishii, "An iron loss calculation method for AC filter inductors used on PWM inverters", *Proc. of Power Electronics Specialists Conference (PESC)*, pp. 1-7, 2006.

[5] T. Shimizu, S. Iyasu, "A practical iron loss calculation for AC filter inductors in PWM inverters", *IEEE Trans. Industial Electron.*, pp. 2000-2009, 2009.

[6] K. Venkatachalam, C. R. Sullivan, T. Abdallah, and H. Tacca, "Accurate prediction of ferrite core loss with nonsinusoidal waveforms using only Steinmetz parameters", *Proc. IEEE Workshop Comput. Power Electron.*, pp. 36-41, 2002.

[7] J. Muhlethaler, J. Biera, J. W. Kolar, A. Ecklebe, "Core loss under dc bias condition based on steinmetz parameters", *IEEE Trans. Power Electorn.*, vol. 27, no.2, pp. 953-963, 2012.

[8] S. Nogawa, M. Kuwata, D. Miyagi, T. Hayashi, H. Tounai, T. Nakau, N. Takahashi, "Study of eddy-current loss reduction by slit in reactor core", *IEEE Trans. Magn.*, vol. 41, no. 5, pp. 2024-2027, 2005.

[9] H. Kaimori, A. Kameari, K. Fujiwara, "FEM computation of magnetic field and iron loss in laminated iron core using homogenization method", *IEEE Trans. Magn.*, vol. 43, no. 4, pp. 1405-1408, 2007.

[10] K. Muramatsu, T. Okitsu, H. Fujitsu, F. Shimanoe, "Method of nonlinear magnetic field analysis taking into account eddy current in laminated core", *IEEE Trans. Magn.*, vol. 40, no. 2, pp. 896-899, 2004.

[11] S. Nogawa, M. Kuwata, T. Nakau, D. Miyagi, N. Takahashi, "Study of modeling method of laminaton of reactor core", *IEEE Trans. Magn.*, vol. 42, no. 4, pp. 1455-1458, 2006.

Feasible Evaluations of Coupled Multilayer Chip Inductor for POL Converter

Jun Imaoka, Shota Kimura, Yuki Itoh,
Masayoshi Yamamoto
Interdisciplinary Faculty of Science and Engineering
Shimane University
1060 Nishikawatsu, Matsue, Shimane 690-8504, Japan
yamamoto@ecs.shimane-u.ac.jp

Michiaki Suzuki, Kenji Kawano
TAIYO YUDEN CO., LTD
5607-2 Nakamurota, Takasaki, Gunma 370-3347, Japan
kkawano@jty.yuden.co.jp

Abstract— Point of load (POL) converters are required smaller size and high efficiency performance in IT industrial applications, etc. Interleaved technique, magnetic integration techniques and application of GaNFETs are well known as good approaches to satisfy these demands. Although coupled inductors for POL converters have been proposed in several studies, a coupled multilayer chip inductor has not examined because of the difficulty of its construction. In this paper, a novel coupled multilayer chip inductor for interleaved POL converters is proposed. This novel coupled inductor has a winding pair with inversely coupled in the magnetic core. Further, magnetic material of the coupled inductor is Fe-based metal composite powder (Fe-Si-Cr). In addition, Fe-based powder in the magnetic core has been processed electrical insulation by highly crystallized oxide nano-layer in order to reduce eddy-current losses. Its high efficiency performance is evaluated by interleaved step down chopper circuit prototype using normally off typed GaNFETs with 1MHz switching frequency.

Keywords— *Multilayer chip inductor, coupled inductor, POL converter, interleaved converter, GaN FET*

I. INTRODUCTION

The high efficiency and high power density POL (Point of load) converters have become more and more important in networking, telecommunications, and computing applications [1]~[5]. Usually, POL converters are directly built around digital equipments such as MCUs (Micro controller unit), FPGAs (Field-programmable gate array) and ASICs (Application specific integrated circuit) on mother boards, in order to reduce the influence parasitic components on the route between its output terminal and the digital equipment.

Mainly, POL converters are required the following three characteristics; 1. Stable low-voltages due to the necessity for the high speed response of these digital equipments. 2. High power density achievement in order to reduce the mounting area of this converter, and to enhance the application portability. 3. High efficiency

performance for improving the time duration of batteries. In general, POL converter is based on the single-phase step down non-isolated DC-DC converter, and the switching frequency of the converters which are constructed by discrete components is around 100~600 kHz to get decent efficiency [1]. For this reason, passive components such as inductors and capacitors are generally bulky, and they occupy a significant space in the converters. In order to achieve both tiny converter size and high efficiency, interleaved circuit topology, Gallium Nitride (GaN) devices and magnetic integration techniques have been applied to this converter. As attractive features of interleaved topology, this technique allows the use of small output filter capacitance while because of the small voltage ripple constraints [6]~[7]. Moreover, it contributes to the reduction of EMI contents and conduction losses [8]. On the other hand, Gallium Nitride (GaN) devices are frequently reported as low losses and high speed switching devices in comparison with conventional Silicon (Si) power devices [9]~[11]. Therefore, using this device is possible to operate at the high frequency switching and contribute to reduce dimensions of passive components without reducing power conversion efficiency.

Moreover, integrated magnetic components are applied to reduce the volume and the weight of magnetic components and to improve the performances of the interleaved converter [2], [12]~[17]. Integrated magnetic components have been applied owing some attractive features. First, higher frequency operation of the inductor currents can achieve by the effect of mutual induction. Second, DC fluxes which are generated in proportion to inductor average currents can be canceled by magnetic coupling.

Incidentally, there are two types of magnetic structures with respect to inductors in POL converters from the magnetic structural point of view. One is wire-wound typed, and the other is multilayer chip typed inductors. Although the first one has a large current-handling capacity by using wires, there are limitations to molding and processing of miniature drum cores. Therefore, this type may be unsuitable for tiny POL converters. In contrast, the latter is able to realize compact and low profile inductors by improving multilayer technology in recent years and achieving very high space factor for the windings. Though this type has not high inductance value as compared with the general wire-wound types, the multilayer chip inductor and high frequency driving by GaNFETs have an affinity for each other. However, "coupled multilayer chip inductor" has not examined yet because of the difficulty of its construction. Therefore, this paper proposes a novel coupled multilayer chip inductor and evaluates it by the interleaved POL converter using GaNFETs at 1MHz switching frequency.

The magnetic material of the proposed coupled inductor is Fe-based metal powder with high saturation flux density (B_{sat}). Generally, ferrite has been known as one of the favorite material of low loss under the high switching frequency condition. However, this material is unsuitable for high power density because B_{sat} is not enough high, and inductance value is rapidly decreased with an increasing current. Although Fe-based metal composite material has an excellent feature of high B_{sat}, it has the eddy current loss problem originated from the low -electrical resistivity. To solve this problem, the Fe-based magnetic material of proposed inductor has been processed electrical insulation by the highly crystallized oxide nano-layer without using resin [18]. The purpose of this study is to discuss the novel coupled multilayer chip inductor performances in case of the interleaved POL converter as a case study.

In section II, circuit analysis is carried out to show the effectiveness of the coupled inductor and its physical behavior. In section III, we will discuss the structure of the coupled multilayer chip inductor. Then, the introduction of the magnetic material and the result of its electrical insulation by the highly crystallized oxide nano-layer will be shown. In section IV, DC current superposition characteristic of the coupled inductor will be investigated. Finally, two coupled multilayer chip inductors with high and low coupling coefficient will prepare, and we evaluate these inductors to confirm the relationship between the coupling coefficient and its efficiency.

The effectiveness of the novel coupled multilayer chip inductor is discussed from the theoretical and experimental point of view.

II. INTERLEAVED POL CONVERTER WITH COUPLED MULTILAYER CHIP INDUCTOR

A. Steady state circuit analysis of the interleaved converter with coupled inductor

Fig. 1 shows the interleaved step down converter with a coupled inductor. This converter is composed of switches S_1, S_2, diodes D_1, D_2 and capacitors C_i, C_o for smoothing or decoupling in input and output side. Coupled inductor is mounted on the converter to reduce volume and eliminated dead space. The relationship

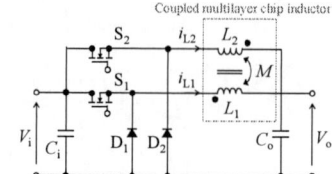

Fig.1. Interleaved step down non-isolated DC-DC converter with a coupled multilayer chip inductor.

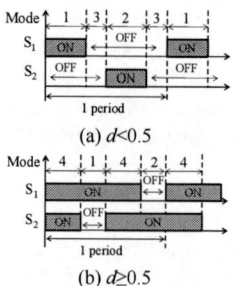

(a) $d<0.5$

(b) $d\geq0.5$

Fig. 2. Drive signals for interleaved step down converter.

between each winding is inversely coupled to educe good characteristic which is described in section I. Here, L_1 and L_2 are self-inductances in each phase, M is the mutual inductance between each winding.

Firstly, to show the relationship between M and inductor ripple current, circuit analysis is carried out. In the case of two phase interleaved converters, the switches are switched with 180 degrees and phase shift as shown in Fig. 2. Therefore, circuit analysis is performed separately for each mode. (It is notable that parasitic components such as capacitors and inductors on the PCB are not considered in order to show only the characteristics of coupled inductors.)

The relationships between the voltages across inductor windings v_{L1}, v_{L2} and inductor currents i_{L1}, i_{L2} are represented by the following;

$$\begin{cases} v_{L1} = L_1 \dfrac{di_{L1}}{dt} - M \dfrac{di_{L2}}{dt} \\ v_{L2} = L_2 \dfrac{di_{L2}}{dt} - M \dfrac{di_{L1}}{dt} \end{cases} \quad (1)$$

Generally, the inductances L_1, L_2 are designed and manufactured with the same self-inductance value. Thus, L_1 and L_2 are approximated as $L_1=L_2=L$.

<Mode1: S_1:on-state, S_2:off-state> In this mode, $v_{L1}=V_i-V_o$, $v_{L2}=-V_o$. Therefore, the slopes of inductor currents are given from (1).

$$\begin{cases} \dfrac{di_{L1\_mod1}}{dt} = \dfrac{1}{(L-M)} \cdot \left(\dfrac{1}{2}V_i - V_o\right) + \dfrac{1}{(L+M)} \cdot \dfrac{1}{2}V_i \\ \dfrac{di_{L2\_mod1}}{dt} = \dfrac{1}{(L-M)} \cdot \left(\dfrac{1}{2}V_i - V_o\right) - \dfrac{1}{(L+M)} \cdot \dfrac{1}{2}V_i \end{cases} \quad (2)$$

<Mode2: S_1:off-state, S_2:on-state> This mode is a symmetrical switching condition with mode1. Thus, relational equations for the inductor current is given by;

$$\begin{cases} \dfrac{di_{L1\_mod2}}{dt} = \dfrac{1}{(L-M)} \cdot \left(\dfrac{1}{2}V_i - V_o\right) - \dfrac{1}{(L+M)} \cdot \dfrac{1}{2}V_i \\ \dfrac{di_{L2\_mod2}}{dt} = \dfrac{1}{(L-M)} \cdot \left(\dfrac{1}{2}V_i - V_o\right) + \dfrac{1}{(L+M)} \cdot \dfrac{1}{2}V_i \end{cases} \quad (3)$$

<Mode3: S_1:off-state, S_2:off-state> In this mode, the switches in each phase are off-state together. Accordingly, the following equation is obtained.

$$\frac{di_{L1\_mod3}}{dt} = \frac{di_{L2\_mod3}}{dt} = \frac{1}{(L-M)} \cdot (-V_o) \qquad (4)$$

<Mode4: S_1:on-state, S_2:on-state> In the same way, two switches are on-state. Therefore, Slopes of inductor currents are as follows;

$$\frac{di_{L1\_mod4}}{dt} = \frac{di_{L2\_mod4}}{dt} = \frac{1}{(L-M)} \cdot (V_i - V_o) \qquad (5)$$

If these current behaviors summarize mathematically from (2)~(5), each current have the following relationships;

$$\begin{cases} \dfrac{di_{L1}}{dt} = \dfrac{di_{com}}{dt} + \dfrac{di_{wh}}{dt} \\[2mm] \dfrac{di_{L2}}{dt} = \dfrac{di_{com}}{dt} - \dfrac{di_{wh}}{dt} \\[2mm] \dfrac{di_{com}}{dt} = \dfrac{1}{(L-M)} \cdot \left((sl_1 + sl_2) \cdot \dfrac{1}{2} V_i - V_o \right) \\[2mm] \dfrac{di_{wh}}{dt} = \dfrac{1}{(L+M)} \cdot (sl_1 - sl_2) \cdot \dfrac{1}{2} V_i \end{cases} \qquad (6)$$

Here, sl_1 and sl_2 are logic functions which show switching conditions, and are given by following;

$$\begin{cases} sl_1 = 1\,(S_1 : \text{ON}), \quad sl_1 = 0\,(S_1 : \text{OFF}) \\ sl_2 = 1\,(S_2 : \text{ON}), \quad sl_2 = 0\,(S_2 : \text{OFF}) \end{cases} \qquad (7)$$

From (6), inductor current can separate the common current i_{com} in each phase, and the wheeling currents i_{wh} as current components. The common current is one of the current components which is related to the leakage inductance L_{lk} ($L_{lk}=L-M$), and can observe in all switching modes. The wheeling current is the current component which presents same absolute values and reverse direction in each phase. Mainly, i_{wh} relates to a transformer in the coupled inductor by the inductance value of $L+M$, and it generates in Mode 1 and 2.

Considering the above electrical behavior, the routes of each inductor current component in Mode 1 and 2 can be illustrated as shown in Fig. 3. As seen in this Fig. 3, the wheeling current i_{wh} flows between the windings. For these reasons, i_{wh} has no relation to the power supply to the output side. Therefore, only i_{com} has possibility to contribute power transfer to the output side. In other words, i_{com} contains DC current components, i_{wh} are AC current components.

B. Inductor ripple current of coupled inductor

Based on the above result of circuit analysis, the relationship between the inductor ripple current and each inductance are investigated theoretically in this subsection.

By using (6), the electrical behaviors of the inductor current including its components such as i_{com} or i_{wh} are illustrated in Fig. 4. As seen in this figure, i_{com} operates at a higher frequency in comparison with single-phase converters. In particular, change of the slope of i_{com} synchronizes with transition of switching modes. On the

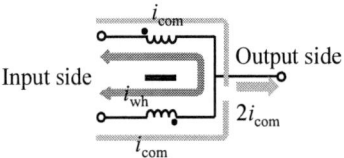

Fig. 3. Routes of each inductor current.

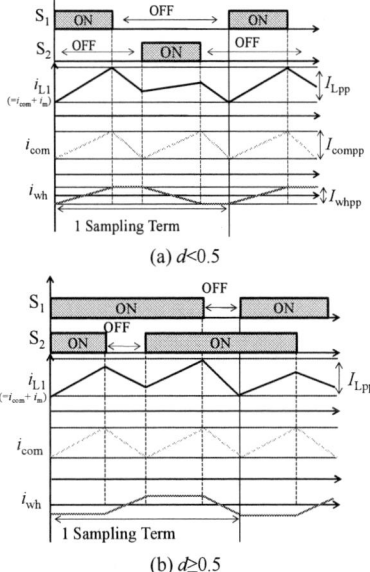

(a) $d<0.5$

(b) $d{\geq}0.5$

Fig.4. Inductor current waveforms. Upper waveform is inductor current, middle one is common current and lower one is wheeling current.

other hand, the change of the slope of i_{wh} occurs when the switching state is different in each phase.

Here, inductor ripple current I_{Lpp} is the sum of ripple currents I_{com} and I_{wh}. Therefore, I_{Lpp} is given by;

$$I_{Lpp} = \left| I_{compp} \right| + \left| I_{whpp} \right| \qquad (8)$$

When the ranges of $d<0.5$, the peak-to-peak amplitude of I_{Lpp} can calculate during Mode 1. Hence, ripple current values of I_{compp} and I_{whpp} can be obtained from (6);

$$\begin{cases} I_{compp\_d<0.5} = \dfrac{1}{L_{lk}} \cdot \left(\dfrac{1}{2} - d \right) \cdot V_i \cdot d \cdot T_s \\[3mm] I_{whpp\_d<0.5} = \dfrac{1}{(L_{lk}+2M)} \cdot \dfrac{1}{2} \cdot V_i \cdot d \cdot T_s \end{cases} \quad (d<0.5) \ (9)$$

Identically, I_{Lpp} is shown in Mode 4 in case of $d{\geq}0.5$. Therefore, the following relationships can be obtained;

$$\begin{cases} I_{compp\_d\geq0.5} = \dfrac{1}{L_{lk}} \cdot \left(d - \dfrac{1}{2} \right) \cdot V_i \cdot (1-d) \cdot T_s \\[3mm] I_{whpp\_d\geq0.5} = \dfrac{1}{(L_{lk}+2M)} \cdot \dfrac{1}{2} \cdot V_i \cdot (1-d) \cdot T_s \end{cases} \quad (d\geq0.5) \ (10)$$

By using (8)~(10), inductor ripple current can be calculated theoretically. To confirm the validity of (8)~(10), circuit simulation is carried out. Simulated circuit parameters are shown in TABLE I. The output voltage is fixed on 1V considering power supply to digital ICs, and the input voltage is varied greatly

TABLE I
CIRCUIT PARAMETERS FOR SIMULATION

Input voltage	V_i	10~1 V
Output voltage	V_o	1 V
Duty ratio	d	0.1~0.9
Switching frequency	f_s	1MHz
Self-inductance	L	2µH
Mutual inductance	M	1µH
Leak age inductance	L_{lk} (=L-M)	1µH

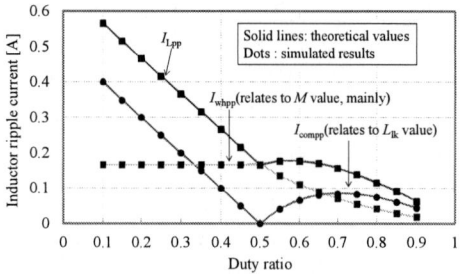

Fig. 5. Relationship between duty ratio and inductor ripple current.

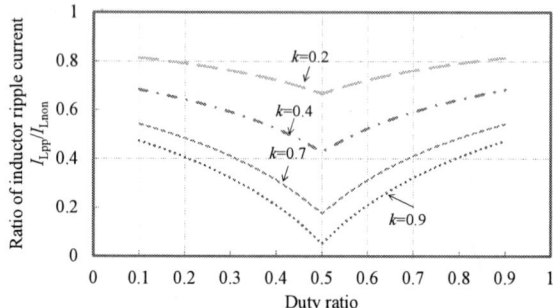

Fig. 6. Ratio of inductor ripple current (I_{Lpp}/I_{Lnon})

between 1V~10V to evaluate the relationship between duty ratio and inductor ripple current.

The simulated results and theoretical results in all ranges of duty ratio are shown in Fig. 5. In this figure, the solid line shows the theoretical value using (8)~(10), and the dots mean the simulated results. As seen in this figure, the theoretical values agree with and simulated values closely. Therefore, the validity of (8)~(10) is confirmed from simulated point of view. On the other hand, when focusing on the relationship between these currents and duty ratio, I_{compp} is the smallest at d=0.5, and effectively reduced around d=0.5. This is because the voltages across the leakage inductances in each phase are zero at d=0.5. On the other hand, I_{whpp} is a constant value in d<0.5, and reduces at d≥0.5. I_{Lpp} which is determined the sum of I_{compp} and I_{whpp} also has been confirmed from simulation results.

Then, to indicate the effectiveness of the coupled inductor, inductor ripple current is compared with non-coupled inductor in interleaved converters. The inductor ripple current I_{Lnon} of non-coupled inductor can be represented by the following equation;

$$I_{Lnon} = \frac{1}{L_{non}} \cdot V_i \cdot (1-d) \cdot d \cdot T_s \qquad (11)$$

Here, L_{non} is the self-inductance of non-coupled inductor. As comparison condition of ripple current, L_{lk} of the coupled inductor have equal values to L_{non}. In addition, the coupling coefficients of the coupled inductor are changed from 0.2 to 0.9 in order to investigate the effect of coupling.

Here, coupling coefficient k is given by:

$$k = \frac{M}{L} = \frac{M}{L_{lk} + M} \qquad (12)$$

Therefore, if (12) is substituted into (9) and (10) and then (8)/(11) is calculated, the ratio of inductor ripple current can be calculated. The result is shown in Fig. 6.

As seen in this figure, the high coupling coefficient is very effective for reducing ripple current as compared with non-coupled inductor. The reduction effect of inductor ripple current is effective with a focus on d=0.5, especially.

Therefore, coupled inductors have good characteristics from electrical points of view.

III. STRUCTURE OF THE COUPLED MULTILAYER CHIP INDUCTOR AND MAGNETIC MATERIAL

A. Magnetic structure

Structures of the coupled multilayer chip inductor are discussed in this section. Magnetic core structure is one of the important parts of the magnetic design. Generally, the windings of the multilayer chip inductors are covered by magnetic cores. The advantage of the multilayer chip inductors is to get higher winding factor. In addition, it is hard to influence to peripheral circuit such as control signals for power devices because this core structure covers magnetic core with high relative permeability (μ_r) in comparison with space permeability (μ_0).

And also, several studies have been discussed and proposed regarding magnetic coupling core structures for wire-wound type inductors. E-E or E-I core with typical wire-wound type are proposed in order to manufacture simply [13]. CCTT core split-winding structure is effective for reducing parasitic capacitance between the winding or the external leakage flux of windings [14]. E-E or E-I core structure with laminated or bifilar windings [15]~[16] and E-I-E core structure [17] for obtaining high coupling coefficients are proposed as well.

Some magnetic core structures for the coupled multilayer chip inductors can be considered based on the above references, and they are shown in Fig. 7 (a)~(c).

In the case of (a), although this structure is relatively easy to manufacture, high coupling coefficient cannot be obtained because magnetic flux has a straightness characteristic. In addition, making an air-gap for obtaining high coupling coefficient is difficult to keep constant quality or reliability of the product in case of multilayer chip inductors. (b) is effective for reducing the parasitic capacitances between the windings in the case of a relatively large core for high power applications. However, the internal structure of multilayer chip inductors is likely to be complicated.

978-1-4799-2706-7/14 $31.00 © 2014 IEEE

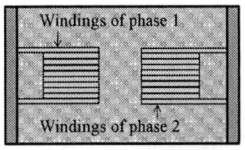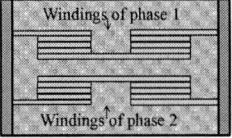

(a) E-E core structure   (b) Split-winding structure

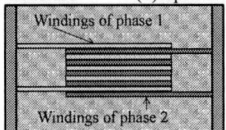

(c) Bifilar winding core structure

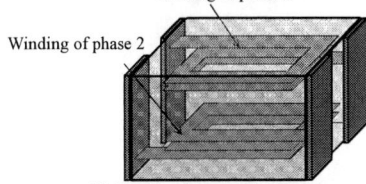

(d) Proposed magnetic structure

Fig. 7. Magnetic core structure for coupled multilayer chip inductor.

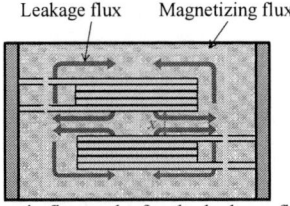

Fig. 8. Magnetic flux paths for the leakage flux and the magnetizing flux in the magnetic core.

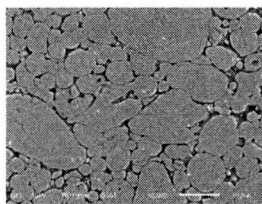

Fig. 9. Scanning Electron Microscope image.

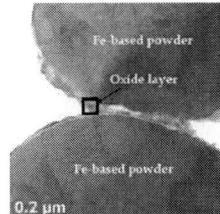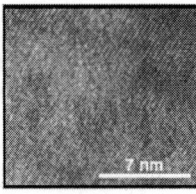

Fig. 10. the magnetic core and its electrical insulation.

Last is a structure (c). The advantage of the structure (c) is easy to obtain very high coupling coefficients. The effectiveness of this structure has been reported in the close-coupled inductor method which is separated into an energy storage inductor and a close-transformer [15]. However, the coupled inductor requires any leakage inductance for storing energy, and adjustment of the coupling coefficient is difficult in this structure. In addition, proximity effect loss of windings will increase at high switching frequency driving.

On the other hand, the proposed structure is shown in Fig. 7 (d). The proposed structure is based on E-I-E core structure which can achieve high coupling coefficient [17]. As advantages of this structure, there are two points. One is easy to manufacture. And another is relatively easy to obtain high coupling coefficient. Fig. 8 shows magnetic flux paths for the leakage flux and the magnetizing flux in the magnetic core. The leakage flux and the magnetizing flux are related to leakage inductance and mutual inductance value, respectively. The proposed coupled inductor can adjust the distance x between windings in each phase, and coupling coefficient can be designed and adjusted by winding distance x.

B. Magnetic core material and internal electrode material

Selection of magnetic materials is important to reduce iron losses and to downsize magnetic components. For high frequency converter such as POL and high power application, several magnetic materials such as amorphous, ferrite and Fe-based powder materials have been discussed and compared in [18]~[20]. Among the materials, ferrite materials have been known as one of the favorite materials that can achieve low-iron losses in high frequency and low-cost. However, the saturation magnetic flux density (B_{sat}) is not high enough, and inductance value is rapidly decreased with increasing current. Further, they cannot tolerate the high temperature because the curie temperature is around 200 °C.

On the other hand, a Fe-based composite material is an attractive material for the inductor, because of the higher B_{sat} than ferrite materials. Therefore, this material is effective for downsizing magnetic components on power converter. Based on the fact, references [15] have been achieved downsizing magnetic components by applying powder cores.

In this paper, Fe-Si-Cr chemical components are applied as magnetic material for powder core. The scanning electron microscope image of this Fe-based composite material is shown in Fig. 9. Fe-Si-Cr composite materials are better than iron losses characteristic in high frequency condition in comparison with other iron-materials.

And also, material of winding paste is silver (Ag). Silver (Ag) has a characteristic of lower electrical resistivity in comparison with other metals such as Cu, Fe. Therefore, reduction of the copper losses will achieve. For these reasons, the proposed coupled multilayer chip inductor has silver paste winding as an internal electrode material.

C. Electrical Insulation of iron powder

Although Fe-based composite materials have some good characteristics, they may not be suitable for high frequency condition because the electrical resistivity of the material is low in comparison with ferrite materials, etc. Hence, Fe-based powder cores have the possibility of increasing eddy current losses in high frequency condition.

Usually, Fe-based powder cores are conducted electric insulating treatment between powders by resin. However, if insulating layers by resin thicken, relative permeability (μ_r) of magnetic cores will be much smaller, and the result

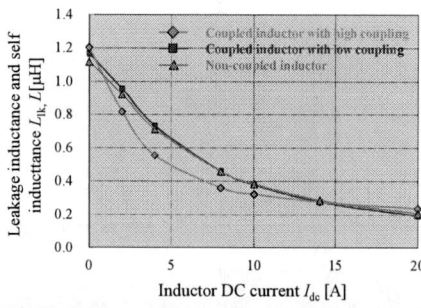

Fig. 11. DC superposition characteristics of the high coupling, low coupled inductors and non-coupled inductor.

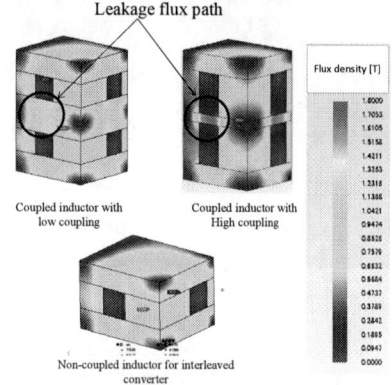

Fig. 12. Magnetic field simulation results.

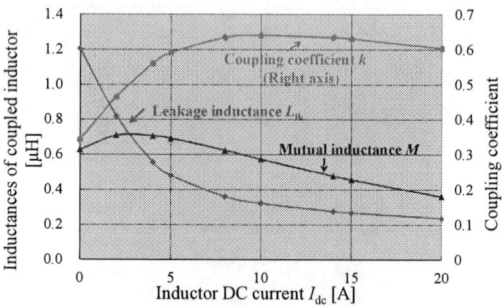

Fig. 13. DC superposition characteristics of leakage inductance and mutual inductance in case of coupled inductor with high coupling.

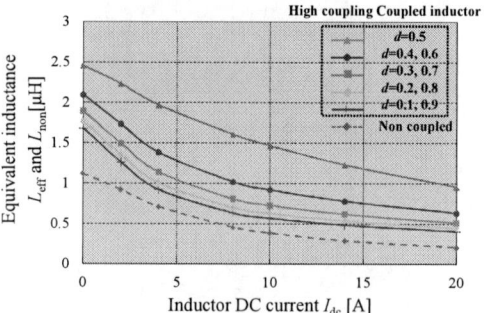

Fig. 14. Effective inductance comparison between the coupled inductor with high coupling and non-coupled inductor.

means an increasing number of winding turns to get requisite inductances. From the above, it is important to make to reduce thickness of the insulating layer in between powders.

For this problem, Fe-Si-Cr metal composite powder is processed electrical insulation by the highly crystallized oxide nano-layer. Fig. 10 shows Transmission Electron Microscope (TEM) images of the metal composite material for the proposed coupled multilayer chip inductor. This metal composite material has thin oxide layers with the thickness of a few hundred nanometers on the metal powder surface. Applying this insulating process by the highly crystallized oxide nano-layer, three advantages are obtained in comparison with a resin. One is that high μ_r can be obtained by thin insulating layer, and the result contribute to get high inductance or to reduce winding turns. Second is to get higher mechanical strength. While the powder with resin shows almost 8×10^4 Pa of mechanical strength, the powder with the highly crystallized oxide nano-layer show 1.5×10^5 Pa. Finally, it can obtain high breakdown voltage characteristic. The breakdown voltage of the highly crystallized oxide nano-layer is 3.6×10^4 V/m, and it is higher than the breakdown voltage of resin layer (2.5×10^4 V/m). The results suggest the highly crystallized oxide nano-layer has large electrical insulation. Considering the miniaturized chip inductor for POL converters, these attractive features are beneficial to improve size and product reliability from the magnetic point of view.

IV.   DC SUPERPOSITION CHARACTERISTICS OF COUPLED INDUCTOR

Proposed coupled inductor consists of powder metal composite. Therefore, there is need to investigate the DC

superposition characteristics for the coupled inductor with the proposed structure. In order to analyze, we conducted a simulation by JMAG (JSOL Corporation). In addition, we compared coupled inductor with high or low coupling coefficient and non-coupled inductor to confirm effectiveness of coupled inductor. Concerning to each inductor size, coupled inductor with high coupling and low coupling are almost the same size, and a non-coupled inductor per phase is half size of coupled inductors. In addition, inductor with high coupling has many winding turns as compared with low coupling inductor to get high inductance, and it has shortened distance x between windings to obtain high coupling coefficient. Further, the DC current condition at 0A, coupled inductor with high coupling has $L_{lk}=1.2\mu H$, $M=0.63\mu H$, and low coupling has $L_{lk}=1.17\mu H$, $M=0.17\mu H$. On the other hand, self-inductance L_{non} of non-coupled inductor is $1.12\mu H$. As seen in this condition, L_{lk} and L_{non} are almost the same value because these inductances are the inductances related to DC flux which generates according to DC current.

Fig. 11 shows DC superposition characteristics. In case of the coupled inductor with high coupling, the decreasing rate of the L_{lk} is relatively early as compared with other inductors. To confirm the reason, Fig.12 shows magnetic field simulation results of each inductor at 14A DC current condition. From this figure, it is understood that the leakage flux path of the coupled inductor with high coupling is high flux density. Therefore, L_{lk} decreases earlier, because the relative permeability μ_r of the magnetic material in this area decreases in an early stage. However, as understood from

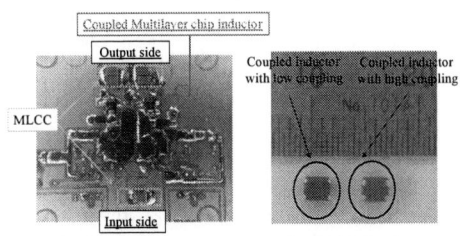

Fig. 15. View of prototyped interleaved POL converter and prototypes of coupled laminated chip inductor.

the analysis result in Section II, not only L_{lk} value but also M value contributes to inductor ripple current in case of the coupled inductor. Therefore, DC superposition characteristics of L_{lk} and M of the coupled inductor with high coupling are shown in Fig. 13. From this figure, it is understood that M value does not decrease easily because each winding are coupled inversely and DC magnetic fluxes are canceled effectively. In addition, to have high magnetic reactance in the leakage flux path contributes to keeping the high M at high current condition. As a result, coupling coefficient k increases at high DC current condition.

Then, DC superposition characteristics investigate including the effect of M. If to realize same inductor ripple current between coupled inductor and non-coupled inductor, the effective inductance $L_{eff\_coupled}$ of coupled inductor is defined by following equation, from (9), (10).

$$L_{non} = L_{eff\_coupled} = \begin{cases} \dfrac{2(1-d)\cdot L_{lk}\cdot(L_{lk}+2M)}{(1-2d)\cdot(L_{lk}+2M)+L_{lk}} & (d<0.5) \\[4mm] \dfrac{2d\cdot L_{lk}\cdot(L_{lk}+2M)}{(2d-1)\cdot(L_{lk}+2M)+L_{lk}} & (d\geq 0.5) \end{cases} \quad (13)$$

$L_{eff\_coupled}$ depends on duty ratio d, and M and L_{lk} values. By using (13), Fig. 14 shows an effective inductance comparison between the coupled inductor with high coupling and non-coupled inductor. As seen in this figure, $L_{eff\_coupled}$ increases wholly in comparison with self-inductance L_{non} of non-coupled inductor at all duty ranges. In addition, $L_{eff\_coupled}$ increases effectively at around duty 0.5 because the common ripple current I_{compp} does not occur in this duty range. Therefore, coupled inductor is effective to use in this duty range. Further, the result strongly suggests that L_{lk} is possible to reduce by high M, and contribute downsizing magnetic components. Or, high reduction effect of inductor ripple current can be obtained.

V. EXPERIMENTAL RESULTS

To confirm the effectiveness of the coupled multilayer chip inductor, experimental evaluation is conducted in this section. Circuit parameters for evaluation are shown in TABLE II.

First of all, two coupled multilayer chip inductors which have different coupling coefficient are prepared in order to investigate relationship the between power conversion efficiency and coupling coefficient. Magnetic parameters of the two inductors are shown in TABLE III.

Prototyped interleaved step down converter and two coupled inductors which have different coupling

TABLE II
CIRCUIT PARAMETER

Input voltage	V_i	6 V
Output voltage	V_o	4 V
Switching frequency	f_s	1MHz
Output power	P_o	4.5W

TABLE III
MAGNETIC PARAMETERS FOR EVALUATION

High coupling typed coupled inductor		
Self-inductance	L	1.75µH
Mutual inductance	M	0.61µH
Leakage inductance	L_{lk} (L-M)	1.14µH
Coupling coefficient	k	0.35
Size	$L \times W \times T$	2.47×2.01×1.83
Low coupling typed coupled inductor		
Self-inductnace	L	0.89µH
Mutual inductance	M	0.12µH
Leakage inductance	L_{lk} (L-M)	0.77µH
Coupling coefficient	k	0.13
Size	$L \times W \times T$	2.46×2.00×1.58

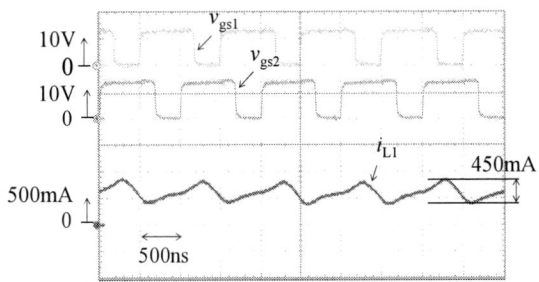

Fig. 16. Experimental waveforms using coupled inductor with high coupling coefficient.

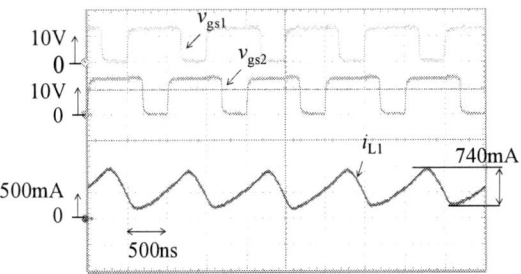

Fig. 17. Experimental waveforms using coupled inductor with low coupling coefficient.

coefficient are shown in Fig. 15. In addition, C_i and C_o have been applied to Multi-layered ceramic capacitors (MLCC) for smoothing or decoupling considering low ESR and ESL characteristic [5], [21]~[22].

Experimental waveforms when using high coupling inductor and low one are shown in Fig.16 and Fig. 17, respectively. In these figures, v_{g1} and v_{g2} show the gate signals in each phase respectively, and i_{L1} shows inductor current of the coupled inductors. This current is observed by shunt resistor, and we use 1MHz of bandwidth limiting function of an oscilloscope to reduce noise. From these figures, it is clear that the ripple of inductor current is controlled in high coupling inductor because of high mutual inductance. The efficiency of interleaved POL converter with coupled inductor with high coupling is 90% at 4.5W output. On the other hand, the efficiency of low coupling one is 89% at the same condition. When using the high coupling coefficient coupled inductor, low inductor ripple current performance are achieved, and conduction losses can be reduced. Therefore, efficiency of the coupled inductor with high coupling is increased.

VI. CONCLUSIONS

In this paper, the coupled multilayer chip inductor for POL converter was proposed and evaluated.

In the section II, inductor ripple current analysis conducted to confirm the effect of magnetic coupling. As a result, to divide inductor ripple current into common current ripple and wheeling ripple current were succeeded. Especially, high mutual inductance is effective for reducing wheeling ripple current. In addition, it is understood that common ripple current does not almost occurred in duty 0.5 or around. The validity of this analysis result is confirmed by circuit simulation. In section III, magnetic core structure and magnetic material discussed. The proposed magnetic core structure has simple structure, it is contributing to easy manufacture and design. In addition, proposed coupled inductor has Fe-based metal composite, and this metal composite have treated, processed by the highly crystallized oxide nano-layer in order to reduce eddy-current losses.

In section IV, DC superposition characteristics of proposed coupled inductors are investigated. Although the leakage inductance is decreased easily, mutual inductance does not decrease easily because each winding is coupled inversely and DC magnetic fluxes are canceled effectively. As a result, effective inductance of coupled inductor which has same ripple current with non-coupled inductor is increased by high mutual inductance. And also, its effect increases around duty ratio 0.5. Finally, two coupled multilayer chip inductors which have different coupling coefficients are prepared and evaluated at 1MHz switching frequency using GaNFETs. Coupled inductor with high coupling coefficient achieved power conversion efficiency 90%. This result is higher than the coupled inductor with low coupling coefficient.

From the above, coupled multilayer chip inductor with high coupling is effective for POL converter applications.

REFERENCES

[1] Fred C. Lee, Quang Li, "High-Frequency Integrated Point-of-Load Converters: Overview," *IEEE Trans. on Power Electron.*, vol. 28, no.9, pp. 4127-4136, 2013.

[2] Qiang Li, Yan Dong, Fred C. Lee, Gilham, D., "High-Density Low-Profile Coupled Inductor Design for Integrated Point-of-Load Converters," *IEEE Trans. on Power Electron.*, vol. 28, no. 1, pp. 547-554, 2013.

[3] Yipeng Su, Wenli Zhang, Qiang Li, Fred C. Lee and Mingkai Mu, "High Frequency Integrated Point of Load (POL) Module with PCB Embedded Inductor Substrate," In *Proc. IEEE Energy Conversion Cong. and Expo.*, *(ECCE2013)*, pp. 1243- 1250, 2013.

[4] Jong-Jae Lee, Bong-Hwan Kwon, "DC–DC Converter Using a Multiple-Coupled Inductor for Low Output Voltages," *IEEE Trans. on Ind. Electron.*, vol. 54, no.1, pp. 467-478, 2007.

[5] Antoniettea De Nardo, MicolaFemia, Giovanni Petrone, Giovanni Spagnuolo, "Optimal Buck Converter Output Filter Design for Point of Load Applications," *IEEE Trans. on Ind. Electron.*. vol. 57, no.4, pp. 1330-1341, 2013.

[6] J. Imaoka, M. Yamamoto, Y.Nakamura, and T.kawashima, "Analysis of Output Capacitor Voltage Ripple in Multi-Phase Transformer-Linked Boost Chopper Circuit," *IEEJ Journal of Ind. Applications*, Vol.2, pp.252-260, 2013.

[7] M. Pavlovsk´y, G. Guidi, and A. Kawamura, "Assessment of Coupled and Independent Phase Designs of Interleaved Multiphase Buck/Boost DC-DC Converter for EV Power Train," *IEEE Trans. on Power Electron.*, vol. 29, no. 6, pp. 2693-2704, 2013.

[8] P. Zumel, O. García, J. A. Cobos and J. Uceda, "EMI Reduction by Interleaving of Power Converters," in *Proc. IEEE Appl. Power Electron. Conf.and Expo.,(APEC)*, vol. 2. pp.688-694, 2004.

[9] O. Machida, M. Yanagihara, E. Chino, S. Iwakami, N. Kaneko, H. Goto, and K. Ohtsuka, "Evaluation of reverse conduction GaN FETs," *Inst. Electron. Inf. Commun. Eng.*, vol. 108, no. 377, pp. 29–34, 2009.

[10] J. Delaine, P. Jeannin, D. Frey, K. Guepratte: "High Frequency DC-DC Converter Using GaN Device," In *Proc. IEEE Appl. Power Electron. Conf.and Expo. (APEC)*, pp. 1754-1761. (2012)

[11] J. Everts, J. Das, J. Van den Keybus, J. Genoe, M. Germain, and J. Driesen: "A high-efficiency, high-frequency boost converter using enhancement mode GaN DHFETs on silicon," In *Proc. IEEE Energy Conver. Cong. and Expo. (ECCE2010)*, pp. 3296–3302 2010.

[12] W. Wen, and Yim-Shu Lee: "A Two-Cannel Interleaved Boost Converter with Reduced Core Loss and Copper Loss", In *Proc. IEEE Annual Power Electronics Spec. Conf.*, pp. 1003-1009, 2004.

[13] P. Wong, P. Xu, B. Yang and F. C. Lee, "Performance Improvements of Interleaving VRMs with Coupling Inductors", *IEEE Trans. on Power Electron.*, Vol.16, no.4, pp.499-507, 2001.

[14] K. Hartnett, J. Hayes, and M. Egan: "CCTT-Core Split-Winding Integrated Magnetic for High-power DC-DC Converters", *IEEE Trans. on Power Electron.*, pp. 4970-4984, 2013.

[15] M. Hirakawa, M. Nagano, Y. Watanabe, K. Andoh, S. Nakatomi and S. Hashino: "High Power Density DC/DC Converter using the Close-Coupled Inductors", In *Proc. IEEE Energy Conver. Cong. and Expo. (ECCE2009)*, pp.1760-1767, 2009.

[16] M. Hirakawa, Y. Watanabe, M. Nagano, K. Andoh, S. Nakatomi and S. Hashino, and T. Shimizu: "High Power Density DC/DC Converter using Extreme Close-Coupled Inductors for Electric Vehicles", In *Proc. International Power Electron. Conf. (IPEC)*, pp.2941-2948, 2010.

[17] J. Imaoka, M. Yamamoto: "Novel integrated Magnetic core structure suitable for Transformer-Linked Interleaved Boost chopper circuit" In *Proc. IEEE Energy Conver. Cong. and Expo. (ECCE2012)*, pp. 3279- 3284, 2012.

[18] K. Shiroki, K. Kawano, H. Matsuura, H. Kishi: "New Type Metal Composite Material for SMD power inductor", In *Proc. The 11th International conference on ferrites (ICF11)*, section number: 17pP-12, 2013.

[19] Y. Han, G. Cheung, A. Li, C. R. Sullivan, D. J. Perreault: "Evaluation of Magnetic Materials for Very High Frequency Power Applications", *IEEE Trans. on Power Electron.*, Vol. 27, no.1, pp.425-434, 2012.

[20] M. S. Rylko, K.J. Hartnett, J. G. Hayes, M.G. Egan: "Magnetic Material Selection for High Power High Frequency Inductors in DC-DC Converters" In *Proc. IEEE Applied Power Electronics Conference and Exposition (APEC)*, pp. 2034-2049. 2009.

[21] R. Miftakhutdinov, "Optimal Design of Interleaved Synchronous Buck Converterat High Slew-Rate Load Current Transients", In *Proc. IEEE Power Electron. Spec. Conf.*, Vol.3. pp1741-1718. 2001.

[22] A. V. Peterchev, S. R. Sanders, " Design of Ceramic-Capacitor VRM'S with Estimated Load Current Feedforward", In *Proc. IEEE Power Electron. Spec. Conf.*, Vol.6. pp4325-4332. 2004.

The 2014 International Power Electronics Conference

Optimal Inductor Design for 3-Phase Voltage-Source PWM Converters Considering Different Magnetic Materials and a Wide Switching Frequency Range

Ralph M. Burkart, Hirofumi Uemura and Johann W. Kolar
Power Electronic Systems Laboratory
ETH Zurich, Physikstrasse 3
Zurich, 8092, Switzerland
burkart@lem.ee.ethz.ch

Abstract—**In this paper, an optimization regarding volume, efficiency and costs of AC boost inductors in 3-phase PWM converters based on detailed multi-domain models is presented. The optimization is performed for a wide switching frequency range of 5-80 kHz and a wide current ripple range of 5-100 %, considering ferrite, amorphous and powder core materials in combination with round, litz, foil and flat wire windings. The shown analysis and optimization identifies the best core material/winding type combinations for both thermally and efficiency-constrained inductor designs. Furthermore, the investigations reveal that simplified scaling assumptions, e.g. a proportional relationship between the inductor volume and the inverse of the frequency or the stored energy, are only accurate in special cases.**

I. INTRODUCTION

Magnetic components represent an integral part of almost any modern power electronic converter and are essential for their proper operation. Commonly known examples are transformers for galvanic isolation and voltage adaption, AC and DC chokes in switched-mode power converters or differential-(DM) and common-mode (CM) inductors for EMI filtering. Typically, the magnetic components do not only use a considerable share of the total converter volume but also cause high relative costs with respect to other components [1]–[5]. Consequently, performance optimizations (e.g. regarding volume, cost, efficiency) of magnetic components is a commonly seen topic in literature. Beyond the optimization of application-specific magnetic components with comparably few and narrow design parameter ranges, there is also an interest for general scaling laws and knowledge about the trade-offs of magnetics as illustrated by the following examples.

In [6]–[8], inductor scaling laws are considered to minimize the volume of DC chokes in DC/DC converters or the inductive components of EMI filters. All contributions assume a simplified linear relationship between volume and stored magnetic energy. In [9] and [10], multi-level multi-cell PFC AC/DC converters are discussed. There, the number of employed converter cells in series and/or parallel has a significant impact on the AC choke(s) requirements. As a consequence, a detailed understanding of the inductor performance trade-offs within wide frequency and current-ripple ranges is essential for the proper optimization of such systems. Finally, with the introduction of SiC power transistors the frequency-volume-dependency of inductors has recently also gained importance, e.g. for photovoltaic power converters. In this area, it has

not yet quantitatively been clarified whether the volume and cost savings on inductors due to higher switching frequencies enabled by SiC can compensate the increased semiconductor costs [2], [11]–[15].

The above discussion and examples motivate for a systematic investigation of the trade-offs, scaling laws and limitations of magnetic components based on detailed multi-domain physical models rather than simplified assumptions. An early attempt to achieve this can be found in [16], where an optimization tool for litz wire ferrite core transformers is employed for performance investigations for frequencies from 100 kHz to 1 MHz. [17] presents a systematic optimization algorithm for inductors in different application areas with foil and round wire. Advanced loss and reluctance models are described but results are not shown. In [18], the volume scaling of DC/DC converter chokes is presented. Both frequency (5-80 kHz) and current ripple ratio (10-50 %) variations and different core materials are considered while no information is given on the modeling. Finally, [19]–[21] present DC choke optimizations for wide frequency (20-150 kHz) and current ripple ratio (10-220 %) ranges, forced and natural convection cooling and a wide variety of core materials. However, only foil windings are considered and the results are given as only indicative core-window area-products from which the underlying geometries and the final volumes cannot be uniquely inferred.

In this paper, a systematic and comprehensive performance analysis is presented. It includes the consideration of

- a wide frequency range of 5-80 kHz,
- a wide current ripple ratio range of 5-100 %,
- ferrite, amorphous iron and powder core materials and
- round, litz, foil and flat wires.

In contrast to the aforementioned studies, AC boost inductors in 20 kW 3-phase 3-level voltage-source PWM converters are analyzed. The objective is the investigation of scaling laws regarding

- the volume of thermally constrained designs,
- the volume of efficiency-constrained designs and
- total inductor costs,

employing advanced and experimentally verified multi-domain models and loss data from measurements. Natural convection cooling and a fixed inductor geometry are considered.

In **Sec. II**, the general system specifications are described. **Sec. III** presents a detailed discussion of the employed models.

978-1-4799-2706-7/14 $31.00 © 2014 IEEE

891

The 2014 International Power Electronics Conference

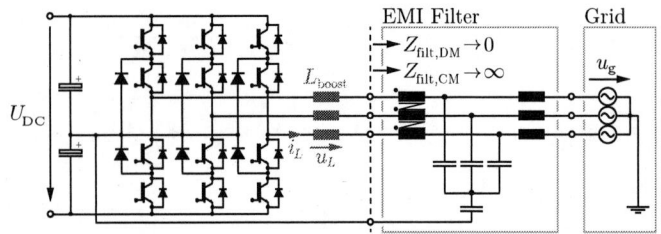

Fig. 1: Typical 3-phase 3-level voltage-source PWM inverter in a grid-side application. For the investigation of the performance trade-offs of the boost inductors L_{boost} regarding volume, efficiency and costs, the specifications listed in **Tab. I** are considered. The EMI filter is assumed to be ideal, i.e. zero DM and infinite CM impedance $Z_{filt,DM} \rightarrow 0$ and $Z_{filt,CM} \rightarrow \infty$, respectively, is seen from the converter side.

Tab. I: Specifications of the converter system depicted in **Fig. 1**.

General	Rated power	P_r	20 kW
	Fundamental phase current	$\hat{I}_{L(1)}$	41 A
	DC-link voltage	U_{DC}	650 V
	Grid line-voltage	$\hat{U}_g$	$(230 \cdot \sqrt{2})$ V
	Grid frequency	$f_g = f_1$	50 Hz
Control	Modulation signal [22]		Sinusoidal with 3rd harmonic injection
	Modulation depth	M	1.0
	Switching frequency	f_{sw}	5-80 kHz

Finally, in **Sec. IV**, the results of the comparative evaluation and the investigated scaling laws are presented.

II. SYSTEM SPECIFICATIONS

Fig. 1 shows the analyzed boost inductors in a typical grid-side application (e.g. solar inverters, input stage of uninterruptible power suplies). A 3-phase 3-level voltage-source PWM converter employing a sinusoidal PWM control scheme with third harmonic injection was selected [22]. The EMI filter, which is normally required to meet the applicable EMC directives, is assumed to feature ideal characteristics, i.e. the DM impedance seen from the converter side approaches zero while the CM impedance is very high and suppresses any CM currents. Accordingly, the inductor voltage u_L equals the whole DM voltage generated by the converter, while the generated CM voltage fully applies to the EMI filter. Note that this assumption is usually met with good accuracy for optimized filters [23].

III. INDUCTOR MODELING

In this section, the employed multi-domain models are discussed in detail. **Sec. III-A** presents the used reluctance model. **Sec. III-B** and **Sec. III-C** describe the core and winding loss and cost models. Finally, in **Sec. III-D** the considered thermal model is presented.

It is expected, that the fundamental trade-offs (e.g. losses versus volume) and physics of inductors are largely invariant from the selected core geometry (e.g. E-cores, U-cores). Therefore, for simplicity reasons, the symmetric EE-core inductor geometry as depicted in **Fig. 2** is assumed throughout this work. The available effective window area is equal for all analyzed winding types.

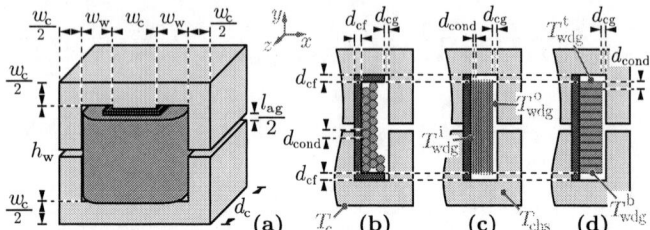

Fig. 2: Assumed inductor geometry. (a) Symmetric EE-core geometry with equal air gap lengths for all legs, center leg width w_c, core depth d_c, total air gap length l_{ag}, window width w_w and window height h_w. (b) Round and litz wire winding geometry with conductor diameter (copper and insulation) d_{cond}, coil former with constant thickness d_{cf} and gap between core and winding d_{cg}. (c) Foil and (d) flat wire winding geometry with equal gaps to the core as for round/litz wire. The figure furthermore depicts the location of the modeled temperatures of the thermal network model (**Fig. 5**).

A. Reluctance Model

The reluctance model of the shown inductor in **Fig. 2** can be described by the equation

$$ N_{wdg} \cdot i_L = \Phi \cdot \left(R_{mag,c} \left(\frac{\Phi}{k_{fill,c} A_c} \right) + R_{mag,ag} \right), \quad (1) $$

where N_{wdg} is the number of winding turns. The above model is non-linear due to the non-linear dependency of the core reluctance $R_{mag,c}$ from the the flux density $B = \frac{\Phi}{k_{fill,c} A_c}$,

$$ R_{mag,c}(B) = \frac{l_{mag,c}}{\mu_0 \, \mu_r(B) \, k_{fill,c} A_c}. \quad (2) $$

The product $k_{fill,c} A_c$ is the effective core cross section taking into account the core filling factor. The relative core permeability μ_r can equivalently be expressed as a function of the magnetic field H or the flux density B. An accurate calculation of the magnetic path length $l_{mag,c}$ including the treatment of the core corner sections can be found in [24]. The reluctance $R_{mag,ag}$ of the air gaps is calculated based on a 3D application of the Schwartz-Christoffel-transformation, which takes into account the 3D geometry and gives accurate results also for large relative air gap lengths. Formulas and experimental verifications can be found in [25].

B. Core Models

1) Core Materials: **Tab. II** lists the considered core materials along with the physical parameters as used in this work. The selected materials are commonly used in industry and literature and are thus representative candidates for each of the considered fundamental material types: ferrites, tape-wound amorphous iron cores and powder cores. High-performance nanocrystalline materials, such as the Vitroperm500F from Vacuumschmelze, have not been analyzed due their high relative costs [5] and the generally poor availability of cut cores (such as E-cores) as a result of the material's brittleness. Iron powder cores and laminated steel have been neglected for this first analysis due to the comparably high core losses. A detailed discussion of the selected material's properties along with a wide range of other materials can be found in [26].

2) Core Loss Models: For the given application, the material is excited by a LF 50 Hz (ideally) sinusoidal voltage and superimposed HF square-wave voltage pulses resulting from the switched-mode operation of the converter. Therefore, on

978-1-4799-2706-7/14 $31.00 © 2014 IEEE

the one hand, the core loss model comprises a LF component where the generalized Steinmetz equation (GSE) is used,

$$P_{c,LF} = k_{fill,c} V_c \cdot k \, f_g^\alpha \, \hat{B}^\beta \,, \tag{3}$$

where k, α and β are the Steinmetz parameters. On the other hand, the a HF component is included in the core loss model which employs the improved improved generalized Steinmetz equation (i^{2}GSE) for square-wave excitations [24],

$$P_{c,HF} = k_{fill,c} V_c \cdot f_g \sum_i \overline{k}_i \left| \frac{1}{\Delta T_i} \right|^{\alpha_i} |\Delta B_i|^{\beta_i} \,, \tag{4}$$

with

$$\overline{k} = \frac{k_i}{(2\pi)^{\alpha_i - 1} \int_0^{2\pi} |\cos\theta|_i^\alpha \, 2^{\beta_i - \alpha_i} \mathrm{d}\theta} \,. \tag{5}$$

The parameters k_i, α_i and β_i are operating-point-dependent Steinmetz parameters which take into account the influence of the premagnetization H_{DC} and the tempereture T_c. ΔB_i denotes the peak-to-peak flux density swings of the piecewise linear HF flux segments and T_i the corresponding time intervals ($\sum T_i = 1/f_g$).

3) Calculation of Core Losses: The approach taken for the core loss calculation is illustrated in **Fig. 3**.

The sinusoidal LF loss contribution is normally small. Therefore, for reasons of simplicity, the Steinmetz parameters for sinusoidal excitations provided in the respective data sheets are used. In contrast, the HF losses are mostly dominant and are caused by square-wave voltage pulses. Therefore, more care is required to obtain accurate results. For this work, the hybrid loss map approach presented in [24] is employed. The approach includes experimentally determined losses for different operating points characterized by the quadruple ($\Delta B, f, T_c, H_{DC}$). For any operating point of interest ($\Delta B_i, f_i, T_{c,i}, H_{DC,i}$), the closest measurements in the loss map are used to extract (local) Steinmetz parameters ($\overline{k}_i, \alpha_i, \beta_i$) which can then be used for a loss interpolation by means of the i^{2}GSE (4). A detailed analysis of the performed loss measurements and the achievable accuracy is presented in [24].

The LF flux amplitude $\hat{B}$ can be calculated by means of the reluctance model (1), while the linear HF flux density swings ΔB_i can be calculated by

$$\Delta B_i = \frac{\Delta \Psi_i}{N_{wdg} \, k_{fill,c} \, w_c \, d_c} = \frac{1}{N_{wdg} \, k_{fill,c} \, A_c} \cdot \int_{t_{i-1}}^{t_i} u_L(t) \, \mathrm{d}t \,. \tag{6}$$

The volt-second pulses $\int_{t_{i-1}}^{t_i} u_L(t) \, \mathrm{d}t$ can either be obtained by means of a circuit simulator or an equivalent model-based algorithm as described in [27].

4) Cost Model: A linear cost model for the core material costs is assumed as proposed in [5],

$$\Sigma_c = \sigma_{mat,c} W_c \,, \tag{7}$$

where $\sigma_{mat,c}$ are relative costs per core weight W_c (cf. **Tab. II**).

5) Model Limitations: The employed core models exhibit the following limitations:

- **Loss maps:** loss maps are so far only available for the materials N87 and 2605SA1. Therefore, for the KoolMµ materials, the premagnetization- and

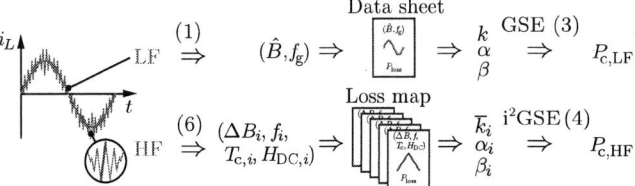

Fig. 3: Core loss computation approach. Using the reluctance model (1), the fundamental LF flux density amplitude $\hat{B}$ can be calculated from the current waveform $i_L(t)$. The LF core losses $P_{c,LF}$ are then calculated based on the data sheet Steinmetz parameters for sinusoidal excitations and the GSE (3). For each linear HF flux density segment ($\Delta B_i, f_i, T_{c,i}, H_{DC,i}$), operating-point-dependent Steinmetz parameters ($\overline{k}_i, \alpha_i, \beta_i$) are extracted from a loss map consisting of core loss measurements for different operating points. Finally, the HF core losses $P_{c,HF}$ can be estimated with the i^{2}GSE (4) for square-wave excitations.

Tab. II: Properties of the considered core materials, taken from [5], [28]–[30]. B_{sat} denotes the saturation flux density. $\mu_r(H=0)$ is the initial permeability, $T_{chs,max}$ the maximum permissible hot spot operating temperature, $\lambda_{th,c}$ the thermal conductivity, $k_{fill,c}$ the effective core fill factor (mainly relevant for laminated materials), ρ_c the material volumetric density and $\sigma_{mat,c}$ the relative costs. Soft saturation is given if the material's *B-H* curve has a soft roll-off, which makes the materials suitable for non-linear inductance designs.

Core material:	N87	2605SA1	KoolMµ 40µ/ 60µ
Type	Ferrite bulk	Amorphous laminated	Powder bulk
Manufacturer	EPCOS	Metglas	Magnetics
Composition	MnZn	Fe-B-Si	Fe-Al-Si
Air gap	Yes	Yes	No
Loss map	Yes	Yes	No
Soft saturation	No	Yes	Yes
B_{sat} @ $T_{chs,max}$ (T)	0.39	1.42	1.00
$\mu_r(H=0)$ @ $T_{chs,max}$	1590	55650	39/58
$T_{chs,max}$ (°C)	100	150	200
$\lambda_{th,c}$ (W/m K)	4.18	9/5 [1]	8
$k_{fill,c}$	1.00	0.83	1.00
ρ_c (kg/m^3)	4850	7180	6000/6800
$\sigma_{mat,c}$ (€/kg)	5.5	16	20

[1] Along/perpendicular to lamination

temperature-independent Steinmetz parameters for sinusoidal excitation from the data sheet were used for both LF and HF loss calculations. This is supported by measurements on comparable powder core materials which suggest, that the premagnetization-dependency of the KoolMµ materials is negligible [24].

- **Gap losses:** several contributions have reported increased losses of tape-wound materials (such as 2605SA1) in the presence of air gaps and fringing fields [26], [31], [32]. However, detailed systematic data or models, in particular for the 2505SA1 material, could not be found in literature and was neither available through own measurements. Therefore, in order to prevent significant underestimations of the respective losses, the zero-air-gap loss-map-based calculated 2505SA1 core losses were multiplied by a factor $k_{2605SA1} = 1.5$ throughout this work.
- **Non-linearity:** for simplicity reasons, the HF excitations (inductor volt-second pulses $\int_{t_{i-1}}^{t_i} u_L(t) \, \mathrm{d}t$) are calculated assuming a linear, current-independent inductance L_0. In

the case of non-linear inductances $L \equiv L(i_L)$, the 3-phase system shown in **Fig. 1** becomes asymmetric. As a result, involved analysis and/or simulations would be required to determine the correct waveform of u_L.

C. Winding Models

1) Winding types: The considered winding types and properties are listed in **Tab. III**. For the winding geometries refer to **Fig. 2**.

2) Winding Loss Models: Using the Maxwell equations, it can be shown that the general solution for the losses of an arbitrarily shaped single conductor x (solid or litz wire) in the presence of a sinusoidal current with frequency f are always of the form

$$P_{\text{wdg}}^x(T_{\text{wdg}}, f) = R_{\text{DC}}(T_{\text{wdg}}) \cdot X^*\big(\xi_x(T_{\text{wdg}}, f)\big) \cdot \hat{I}_{L(f)}^2 , \quad (8)$$

where R_{DC} is the conductor DC resistance. ξ_x is a function of the conductor copper diameter $d_{\text{cond,cop}}$ and the skin depth δ_{cop},

$$\xi_x = \frac{d_{\text{cond,cop}}}{\sqrt{2}\,\delta_{\text{cop}}} = d_{\text{cond,cop}} \cdot \sqrt{\frac{\pi f \mu_0 \, \sigma_{\text{cop}}(T_{\text{wdg}})}{2}} . \quad (9)$$

For the different winding types, different approaches were chosen to determine the unknown geometry-dependent function X^*.

- **Round and litz:** for round and litz wire conductors, well-known and tractable analytical solutions exist based on 1D field approximations. The solutions for X^* distinguish between skin and proximity effect losses and are of the form,

$$X^{\{ro,li\}}(\xi_{\{ro,li\}}) =$$
$$\big[F_{\text{skin}}^{\{ro,li\}}(\xi_{\{ro,li\}}) + G_{\text{prox}}^{\{ro,li\}}(\xi_{\{ro,li\}})\big] \cdot \overline{H}_{\text{ext}(f)}^2 . \quad (10)$$

The exact expressions for $F_{\text{skin}}^{\{ro,li\}}$ and $G_{\text{prox}}^{\{ro,li\}}$ can e.g. be found in [17] or [24]. The approach to calculate the required normalized external field $\overline{H}_{\text{ext}(f)}$ is presented below.

- **Foil and flat:** Investigations, e.g. found in [17] or [24], show that an accurate analytical determination of X^* is highly complex for foil and flat wire geometries. This is mainly due to the fact that 2D fields must be described as 1D approximation result in large errors. As a consequence, a pragmatic and more accurate approach based on finite-element (FEM) simulations was pursued to model X^* by means of interpolated FEM results. Note that the FEM simulations must only sweep over a few *relative* geometric parameters and a sufficiently wide range of frequencies, as the relationship between f, T_{wdg} and $d_{\text{cond,cop}}$ is known (9). This reduces the computational effort by $\mathcal{O}(n^2)$.

3) Calculation of Winding Losses: The winding losses must be calculated for each winding turn and frequency separately, using the above derived loss formula (8). Furthermore, it must be distinguished between the part of the winding inside the core window and the part outside the window.

The 50 Hz fundamental current amplitude $\hat{I}_{L(1)}$ is given in **Tab. I**, while the HF switching frequency harmonics $\hat{I}_{L(f)}$ of

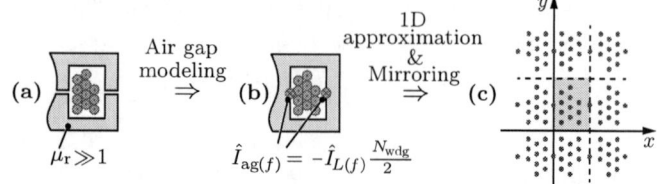

Fig. 4: Method of images (mirroring) used for the calculation of the normalized external fields $\overline{H}_{\text{ext}(f)}$ in round and litz wire windings. **(a)** Round wire winding inside the winding window. **(b)** The fringing field of the air gaps can be modeled with equivalent currents generating a similar field distribution. **(c)** The impact of the core material ($\mu_r \gg 1$, e.g. $\mu_r > 20$) is imitated by mirrored conductors. The individual conductors are modeled as point-currents (1D approximation) and the external field $\overline{H}_{\text{ext}(f)}$ in the center of a conductor of interest can be calculated with (11). In this work, 3 mirroring steps are performed as an analysis has shown that the accuracy cannot significantly be improved with a higher number of steps.

Tab. III: Properties of the considered winding types. The geometric data is taken from standards and data sheets. Note that d_{cond} includes both copper thickness and insulation increase, i.e. $d_{\text{cond}} = d_{\text{cond,cop}} + d_{\text{cond,ins}}$. Litz wires from [33] with strand diameters $d_{\text{cond,s}}$ of $\{40,50,71,100,200,355\}$ µm (AWG $\{46,44,41,38,32,27\}$) were considered . The number of strands N_{strand} for a given diameter d_{cond} was calculated based on fitted data sheet information [33]. The thermal conductivities are taken from data sheets, simulations and measurements. Finally, the cost data is from [5]. IACS copper with an electrical conductivity of $\sigma_{\text{cop}}(T = 20\,^\circ\text{C}) = 5.8 \cdot 10^8$ S and a density of $\rho_{\text{cop}} = 8890\,\text{kg/m}^3$ was assumed.

Winding type:	Round	Litz	Foil	Flat
$d_{\text{cond,min}}$ **(mm)**	0.50	0.50	0.05	1.25
$d_{\text{cond,max}}$ **(mm)**	5.00	5.00	1.25	$(1{:}2)^{1)}$
Insulation type	Enamel [34]	Silk [33]	Kapton [35]	Enamel/air
Ins. increase (mm)	0.025-0.049	0.0375	0.025	0.25
$T_{\text{wdghs,max}}$ **(°C)**	$150^{2)}$	$150^{2)}$	$150^{2)}$	$150^{2)}$
$\lambda_{\text{th,wdg}}$ **(W/m K)**	$1.0^{3)}$	$0.3^{3)}$	$401/0.37^{4)}$	$401/0.03^{4)}$
$\sigma_{\text{mat,wdg}}$ **(€/kg)**	10.0	$\frac{15.0}{\frac{A_{\text{strand}}}{\text{mm}^2}+0.45}$	20.0	10.0
$\Sigma_{\text{mat,wdg}}^{\text{fc}}$ **(€/unit)**	1.0	1.0	2.0	2.0
$\sigma_{\text{lab,wdg}}$ **(€/kg)**	7.0	7.0	14.0	21.0
$\Sigma_{\text{lab,wdg}}^{\text{fc}}$ **(€/unit)**	2.0	2.0	2.5	4.0

[1)] Ratio between d_{cond} and conductor width
[2)] Limited due to the considered coil formers [28] ($\lambda_{\text{th,cf}} = 0.3$ W/mK)
[3)] For a hexagonal winding arrangement including air (cf. **Fig. 2**)
[4)] IACS copper / insulation material

the given system can be computed using the analytical methods presented in [22] or by means of simulations.

The required normalized field amplitude $\overline{H}_{\text{ext}(f)}$ in (10) at the position $\vec{r}$ of a conductor of interest can be approximated by means of a superposition of all occurring fields at this position,

$$\overline{H}_{\text{ext}(f)}(\vec{r}) = \frac{1}{\hat{I}_{L(f)}} \cdot \left| \sum_i^{N^*} \vec{H}_{i(f)}(\vec{r}) \right| . \quad (11)$$

The N^* field sources partly result from the $(N_{\text{wdg}}-1)$ currents in the neighboring winding turns as shown in **Fig. 4(a)**. The remaining sources result from the employed method of images (mirroring) as described in [24], which can accurately take into account the impact of air gap fringing fields **Fig. 4(b)** and the core material **Fig. 4(c)** on $\overline{H}_{\text{ext}(f)}$.

An accuracy analysis based on 2D FEM simulations showed very good agreement with the litz and round wire loss models described above. The observed error was always below 5 % for high-permeability materials (N87 and 2605SA1) and below 10 % for the low-permeability materials (KoolMμ 40μ and 60μ). A similar analysis and results can be found in [24]. The results of the accuracy analysis imply that despite the underlying assumption of the mirroring method of ideal high-permeability core materials with $\mu_r \to \infty$, $\mu_r > 20$ is sufficient in practice to achieve high accuracy.

The statistical mean error introduced by the interpolation approach for foil and flat wire conductors is below 3 % with maximum errors of 20 % in special cases. Note that no extrapolation was necessary for all results shown in this work.

4) Cost Model: The winding costs can be estimated using the model presented in [5],

$$\Sigma_{\text{wdg}} = (\sigma_{\text{mat,wdg}} + \sigma_{\text{lab,wdg}}) \cdot W_{\text{wdg}} + \Sigma_{\text{mat,wdg}}^{\text{fc}} + \Sigma_{\text{lab,wdg}}^{\text{fc}} . \quad (12)$$

The parameters $\sigma_{\text{mat,wdg}}$ and $\sigma_{\text{lab,wdg}}$ are specific material and labor costs per weight W_{wdg} and $\Sigma_{\text{mat,wdg}}^{\text{fc}}$ and $\Sigma_{\text{lab,wdg}}^{\text{fc}}$ are fixed material and labor costs (e.g. coil former, connectors).

5) Model Limitations: The main limitation of the presented winding models is the calculation of the HF current harmonics $\hat{I}_{L(f)}$ for which the analytical method presented in [22] is employed. The method is accurate for linear inductors (error < 5 %) but cannot take into account non-linear inductor behavior.

D. Thermal Model

The employed thermal model in this work is based on the empirical models presented in [36]. Modifications were introduced in order to account for anisotropic, direction-dependent thermal properties of the used materials.

1) Thermal Resistance Network: The thermal resistance network of the modified thermal model is shown in **Fig. 5**. The underlying inductor geometry is depicted in **Fig. 2**. The winding hot spot temperature is, based on measurements, assumed to be on the outer surface of the winding, while the core hot spot location is assumed to be in the center of the core. Different winding surface temperatures (inner, outer, top, bottom) are modeled, whereas, for simplicity reasons, a uniform core surface temperature is assumed.

2) Heat Transfer Mechanisms: 3 different types of heat transfer mechanisms are considered for the resistances shown in **Fig. 5**. For the thermal constants and parameters used in the below equations, please refer to **Tab. V**.

- **Conduction:** the thermal resistance which models conducted heat transfer is given by

$$R_{\text{th,cond}} = \frac{l_*}{\lambda_{\text{th},*} \cdot A_*} , \quad (13)$$

where $\lambda_{\text{th},*}$ is the material's thermal conductivity (cf. **Tab. II** and **Tab. III**) and l_* and A_* the (direction-dependent) length and cross section of the heat conductor to be modeled.

- **Radiation:** radiated heat from an object 1 to an object 2 can be modeled with the thermal resistance,

$$R_{\text{th,rad}} = \frac{1}{h_{\text{rad},*} \cdot A_*} = \frac{T_{1,*} - T_{2,*}}{\epsilon_{1,*} \cdot \sigma \left(T_{1,*}^4 - T_{2,*}^4 \right)} \cdot \frac{1}{A_*} . \quad (14)$$

where $h_{\text{rad},*}$ represents the heat transfer coefficient, A_* is the overlapping area of the two objects and $T_{1,*} > T_{2,*}$ are the respective temperatures.

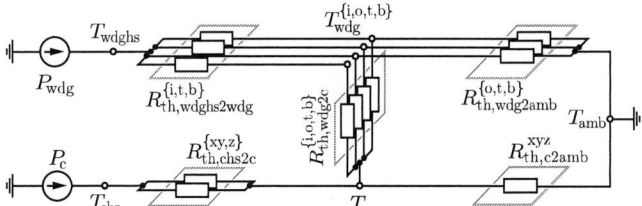

Fig. 5: Anisotropic modified thermal model based on the empirical models in [36]. P_{wdg} and P_{c} are the total winding and total core losses, respectively. $R_{\text{th,wdghs2wdg}}^{\{i,t,b\}}$ are the thermal resistances from the winding hot spot, which is assumed to be on the outer winding surface, to the respective inner, top and bottom winding surface. $R_{\text{th,wdg2c}}^{\{i,o,t,b\}}$ and $R_{\text{th,wdg2amb}}^{\{o,t,b\}}$ are the resistances from the winding surfaces to the core surface and to the ambient, respectively. The inner winding surface is not exposed to the ambient. $R_{\text{th,chs2c}}^{\{xy,z\}}$ and $R_{\text{th,c2amb}}^{xyz}$ are the resistances from the core hot spot to the core surface with uniform temperature T_{c} and from the surface to the ambient. The corresponding geometry of this network is depicted in **Fig. 2**.

Tab. IV: Modeled heat transfer mechanisms in the thermal network depicted in **Fig. 5**. Conducted heat transfer between the winding and the core is only applicable where the winding touches the core. This depends on the specific winding geometry and type (cf. **Fig. 2**).

Resistances:	$R_{\text{th,wdghs2wdg}}^{\{i,t,b\}}$	$R_{\text{th,chs2c}}^{\{xy,z\}}$	$R_{\text{th,wdg2c}}^{\{i,o,t,b\}}$	$R_{\text{th,wdg2amb}}^{\{o,t,b\}}$	$R_{\text{th,c2amb}}^{xyz}$
Conduction	✓	✓	Partly		
Radiation		✓		✓	✓
Convection				✓	✓

Tab. V: Thermal parameters and constants employed in the empirical thermal model of this work (taken from [36]).

Emissivity of enameled copper	ϵ_{wdg}	0.8
Emissivity of the core materials	ϵ_{wdg}	0.9
Stephan-Boltzmann constant	σ	$5.67 \cdot 10^{-8}$ W/m² K⁴
Vertical-positioning coefficient	C_{v}	1.58
Reference pressure (sea level)	p_{ref}	101.32 kPa
Reference ambient temperature	$T_{\text{amb,ref}}$	$(25 + 273.15)$ K

- **Natural convection:** natural convection is considered where the transfer coefficient $h_{\text{rad},*}$ in (14) changes to

$$h_{\text{conv},*} = C_{\text{v}} \left(\frac{p}{p_{\text{ref}}} \right)^{0.477} \left(\frac{T_{\text{amb}}}{T_{\text{amb,ref}}} \right)^{-0.218} \frac{(T_* - T_{\text{amb}})^{0.225}}{L_{\text{ch}}^{0.285}} . \quad (15)$$

In the above equation (15), L_{ch} is the characteristic length which for the given inductor geometry was approximated with

$$L_{\text{ch}} \approx d_{\text{c}} + (h_{\text{w}} - 2d_{\text{cf}}) + 2\sqrt{w_{\text{w}}^2 + (\frac{w_{\text{c}}}{2} + d_{\text{cf}})^2} . \quad (16)$$

Most of the thermal resistances depicted in the thermal network model in **Fig. 5** combine more than one of the above mechanisms. The total resistances shown in **Fig. 5** can be calculated by the parallel connection of the individual type-specific resistances,

$$\frac{1}{R_{\text{th,tot}}} = \frac{1}{R_{\text{th,cond}}} + \frac{1}{R_{\text{th,rad}}} + \frac{1}{R_{\text{th,conv}}} . \quad (17)$$

The applicable types of heat transfer of each of the thermal resistances depend on the specific geometry and/or winding type and are listed in **Tab. IV**.

The 2014 International Power Electronics Conference

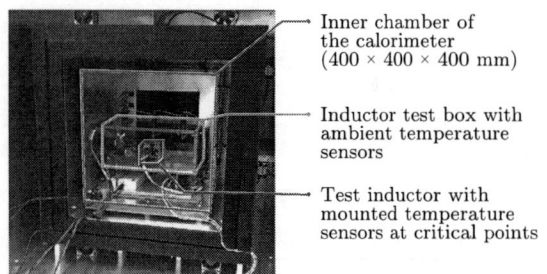

Inner chamber of
the calorimeter
(400 × 400 × 400 mm)

Inductor test box with
ambient temperature
sensors

Test inductor with
mounted temperature
sensors at critical points

Fig. 6: Experimental inductor loss and temperature measurement setup.

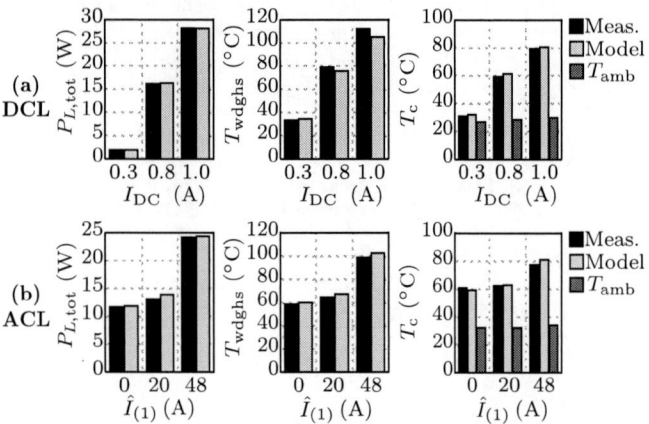

Fig. 7: Results of the experimental verification of the proposed thermal and loss models based on the inductor designs of **Tab. VI**. **(a)** Verification of the thermal model only, using different DC currents I_{DC} in conjunction with a known DC winding resistance $R_{DC}(T_{wdg})$. **(b)** Combined verification of the thermal and loss models, using different 50 Hz LF AC currents $\hat{I}_{(1)}$ superimposed by a 9.5 A/16 kHz peak-to-peak HF current. Deviations of less than 10 % between actual and calculated temperatures/losses were observed.

Tab. VI: Test inductor design parameters for the experimental verification of the thermal and loss models shown in **Fig. 7**. The high DC resistance of the DCL inductor winding results from the large number of winding turns ($N_{wdg} > 300$) which was used to experimentally determine the thermal conductivity $\lambda_{th,wdg}$ of hexagonal round wires (cf. **Tab. III**).

Inductor name:	DCL	ACL
Core material	N87	2605SA1
Core size	E80/38/20	AMCC06R3
	(1 set × 3 stacked)	(2 sets × 2 stacked)
Air gap length l_{ag}	None	1.0 mm
Winding type	Round	Round
$R_{DC}(T_{wdg} = 25\,°C)$	21.2 Ω	7.8 mΩ
LF current	0 Hz (DC)	50 Hz
HF current	None	16 kHz / 9.5 A

3) Experimental verification: In order to verify the accuracy of the thermal model, the temperature and loss measurement setup depicted in **Fig. 6** has been implemented. Measurements have been performed on several inductors, such as the two test inductors with parameters as listed in **Tab. VI**. In a first experiment (DCL), only DC currents without superimposed HF currents were used **Fig. 7(a)**. This allows to evaluate the performance of the thermal model only due to the negligibly uncertainty regarding the driving losses. In a second experiment (ACL), LF AC currents with a superimposed HF current component were used to verify the performance

Tab. VII: Considered optimization constraints in addition to the core, winding and system constraints/specifications (**Tab. I-III**). The geometric constraints reflect typical properties of available EE-cores [28]–[30] with a stacking factor of 3 cores.

General	Ambient Temperature	T_{amb}	40 °C
	Ambient air pressure	T_{amb}	97.7 kPa
	Max. inductance decrease	$L_{min}/L_0 = L(I_{L,max})/L(0)$	50 %
	Max. flux density	B_{max}/B_{sat}	90 %
Geometry	Window ratio	h_w/w_w	3
	Core width ratio	w_c/w_w	$\sqrt{2}$
	Core depth	d_c/w_c	3
	Coil former ratio	d_{cf}/w_w	1/5
	Core gap ratio	d_{gc}/w_w	1/20
	Max. air gap ratio	l_{ag}/w_c	1/2

of the combined loss and thermal models **Fig. 7(b)**. In all analyzed operating points deviations of less than 10 % between measurement and calculation were observed.

IV. COMPARATIVE EVALUATION

In this section, the results of a comparative evaluation of the selected materials and winding types regarding achievable performance (volume, efficiency, costs) is presented. Furthermore, general scaling laws are investigated. For simplicity reasons, a fixed inductor geometry as described in **Tab. VII** is assumed where only the boxed inductor volume V_L is continuously varied. It is expected that the volume most significantly affects the inductor performance, while other geometric parameters have only a subordinate influence. The parameters in **Tab. VII** were chosen based on typical available E-cores. The chosen stacking factor/core depth d_c yields a close to maximum area product for a given volume V_L.

A. Performance Analysis

In **Fig. 8**, the minimum achievable volumes are compared where the required efficiency is unconstrained. As a result, most designs operate at the thermal limit. An exception are the KoolMµ designs for low-frequency/low-ripple conditions due to the missing degree of freedom of the air gap and the maximum permissible inductance drop constraint of 50 % (cf. **Tab. VII**). Amongst all possible options, the combinations of flat and foil windings with 2605SA1 cores achieve the lowest volumes. The main limitation of N87 is the low saturation flux density B_{sat} and low maximum hot spot temperature $T_{chs,max}$ (cf. **Tab. II**) which cannot be compensated by the almost negligible core losses. Despite similar core losses, 2605SA1 outperforms the KoolMµ materials due to a higher B_{sat}, better temperature exploitation and the fact. Moreover, like the KoolMµ materials, 2605SA1 also features a soft roll-off of the B-H curve (cf. **Tab. II**, [29]). This feature can be utilized to decrease the inductor volume by means of allowing the inductance to drop a certain percentage (here 50 %) at the inductor peak current [30], [37].

Similar performance results are obtained for efficiency-constrained designs as depicted in **Fig. 9**. However, 2605SA1 partly looses its dominance in high-ripple conditions where N87 and the KoolMµ materials feature a higher efficiency and hence lower volumes.

In **Fig. 10**, a cost analysis of the thermally-constrained designs (**Fig. 8**) is shown. Despite slightly higher resulting

978-1-4799-2706-7/14 $31.00 © 2014 IEEE

The 2014 International Power Electronics Conference

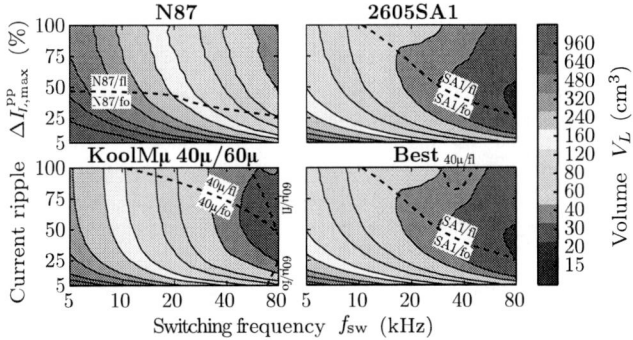

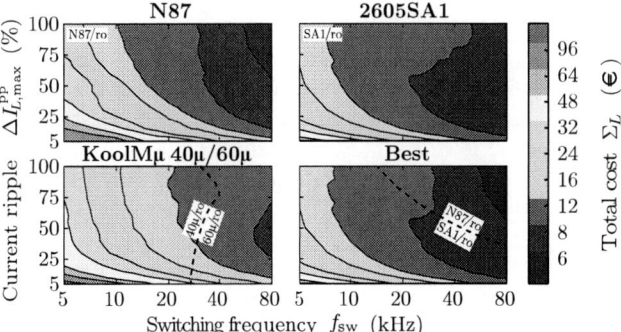

Fig. 8: Achievable minimum boxed volumes V_L as a function of the maximum peak-to-peak current ripple ratio $\Delta I_{L,\max}^{\mathrm{pp}}$ and the converter switching frequency f_{sw} (log$_2$-scaling for f_{sw} and V_L). All designs except the KoolMµ inductors are thermally constrained. For each material, only the best performing winding options are shown. These are without exception foil ("fo") and flat ("fl") wires mainly due to the superior filling factors and thermal advantages. Flat wire generally shows lower HF losses and is therefore better suited in high frequency/high ripple conditions. As recommended in the data sheet [30], the lower permeability powder core material KoolMµ 40µ is better suited for inductors with large energies $E_L = \int_0^{I_{L,\max}} L(i_L) i_L \, \mathrm{d}i_L$. The best overall option for all operating conditions is 2605SA1 with a foil/flat wire winding mainly due to the highest saturation flux density, a high maximum core temperature and comparably low core losses. The inductor losses here (best case) range from 7.5 W (smallest design) to 51 W (largest design).

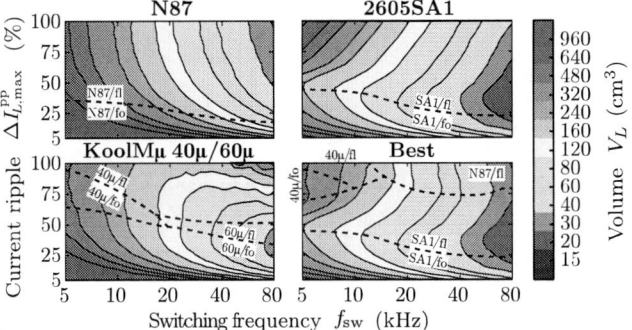

Fig. 9: Achievable minimum volumes V_L for efficiency-constrained designs with a maximum permissible loss power of $P_L \leq 7\,\mathrm{W}$ ($\eta \geq 99.9\,\%$), considering all 3 inductors L_{boost}). By inspection of **Fig. 8**, it can be seen that partially significantly larger volumes results than for thermally constrained designs, while flat and foil windings still offer the best performance for all materials. Regarding the overall optimum, powder (KoolMµ) and the low-loss ferrite (N87) cores partly supersede the 2605SA1-based designs for the high-ripple conditions. This can mainly be explained by the fact, that 2605SA1 cannot exploit the temperature advantage as for thermally constrained designs. For the best inductor designs, the winding hot spot temperatures T_{wdghs} are between 53 °C and 139 °C.

Fig. 10: Achievable minimum total inductor costs Σ_L for the *volume* optimized designs depicted in **Fig. 8**. It can be seen that round wires ("ro") generally outperform other winding types, which implies that the lower volume achieved by e.g. flat or foil (**Fig. 8**) windings is not sufficient to compensate for the higher losses. For high frequency/high ripple conditions, the volume difference between N87-based designs (larger) with respect to 2605SA1-based designs (smaller) becomes small enough to yield lower costs for N87.

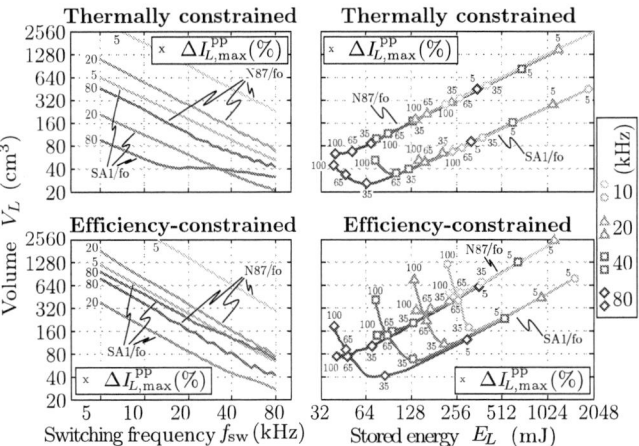

Fig. 11: Scaling analysis of the inductor volume V_L as a function of the converter switching frequency f_{sw} and the maximum stored inductor energy $E_L = \int_0^{I_{L,\max}} L(i_L) i_L \, \mathrm{d}i_L$ for efficiency- and thermally constrained designs. None of the depicted combinations of foil winding with N87 and 2605SA1 have a constrained air gap length or winding geometry. The volumes scale approximately with f^{-1} in a limited range, where the HF losses (in particular the core losses) are small when compared to the total losses. The same applies to the volume-energy scaling. The low-loss N87 shows a largely proportional dependency between stored energy and volume in a wide range of the realized inductance L and current ripples ΔI_L. On the other hand, the lossier 2605SA1 clearly deviates from this proportionality in high ripple/high frequency conditions.

volumes than for flat/foil wire solutions, round wires in combination with 2605SA1 or N87 are generally the least expensive option.

B. Scaling Laws

Fig. 11 shows the scaling of the volume as a function of the frequency and the maximum stored inductor energy. The examples of foil windings in combination with N87 and 2605SA1 demonstrate, that the often used simplified scalings laws – volume proportional to the inverse of the frequency $V_L \propto f^{-1}$ or proportional to the stored energy $V_L \propto E_L$ – are only good approximations in special cases where the HF core and winding losses are small when compared to the total losses. This is mostly the case for N87 and 2605SA1 in

low frequency/low ripple conditions. For N87 it is expected, that the simplified scaling laws also loose accuracy for higher frequencies or AC only applications. The scaling of the KoolMµ materials is generally highly non-linear (**Fig. 8** and **Fig. 9**).

V. CONCLUSION

In this paper, a systematic multi-objective analysis and comparison regarding volume, efficiency and costs of AC boost chokes in 3-phase 3-level switched-mode PWM converters is presented. Ferrite, amorphous and powder cores in combination with round, litz, foil and flat wire windings were considered. Detailed multi-domain models, which are verified by means of experimental measurements and FEM

978-1-4799-2706-7/14 $31.00 © 2014 IEEE

simulations, were used.

The comparative analysis showed a generally superior performance of the amorphous material 2605SA1 for this kind of application. The combination with flat wire and foil windings allows for the lowest achievable volumes in a wide frequency and current ripple range, while a combination with round wire results in the lowest costs. An investigation of the volume scaling as a function of the frequency and the stored inductor energy revealed, that the often used assumptions of a volume proportional to the inverse of the frequency or to the stored energy are only accurate in special cases, i.e. under the condition of low relative HF losses.

Future research continuing this work can be divided into 4 categories.

- **Applications:** the derived models can be used to investigate the trade-offs and Pareto-fronts of both standard and novel applications, e.g. multi-level multi-cell converters or high-switching-frequency SiC PV inverters.
- **Investigations:** more detailed investigations can be performed with the optimization tool as described in this work, e.g. the analysis of the influence of different core geometries (other core shapes, varying core proportions) or the consideration of other core materials.
- **Model extensions:** in order to overcome the limitations of the employed models, loss maps with experimentally determined losses for the powder core materials should be established. Furthermore, the influence of air gaps and fringing fields on tape-wound core losses must be investigated in detail. The present linear approach for the calculation of the HF currents and flux density swings should be replaced by an approach which enhances the accuracy for non-linear, current-dependent inductors. Moreover, the thermal model can be modified to include forced convection cooling and heat sinking of the core and winding.
- **Model verifications:** Finally, implementations of inductors made of different materials and winding types should be used for a combined verification of the derived models, which have so far mostly been separately verified.

REFERENCES

[1] G. Gong, M. L. Heldwein, U. Drofenik, J. Miniböck, K. Mino, and J. W. Kolar, "Comparative Evaluation of Three-Phase High-Power-Factor AC-DC Converter Concepts for Application in Future More Electric Aircraft," *IEEE Trans. IE*, vol. 52, no. 3, pp. 727–737, June 2005.

[2] B. Burger, D. Kranzer, and O. Stalter, "Cost Reduction of PV-Inverters with SiC-DMOSFETs," in *Proc. CIPS*, pp. 1–6, June 2008.

[3] J. Biela, U. Badstübner, and J. W. Kolar, "Impact of Power Density Maximization on Efficiency of DC-DC Converter Systems," *IEEE Trans. PE*, vol. 24, no. 1, pp. 288–300, Jan. 2009.

[4] Y. Wang, S. W. H. De Haan, and J. Ferreira, "Potential of Improving PWM Converter Power Density with Advanced Components," in *Proc. EPE*, pp. 1–10, Sept. 2009.

[5] R. M. Burkart and J. W. Kolar, "Component Cost Models for Multi-Objective Optimizations of Switched-Mode Power Converters," in *Proc. ECCE*, Sept. 2013.

[6] J. Kassakian, M. F. Schlecht, and G. C. Verghese, *Principles of Power Electronics*. Prentice Hall, 1991.

[7] D. L. Loree and J. P. O'Loughlin, "Design Optimization of L-C Filters," in *Proc. PMS*, pp. 137 –140, June 2000.

[8] K. Jalili and S. Bernet, "Design of LCL Filters of Active-Front-End Two-Level Voltage-Source Converters," *IEEE Trans. IE*, vol. 56, no. 5, pp. 1674–1689, May 2009.

[9] H. Ertl, J. W. Kolar, and F. C. Zach, "Analysis of a Multilevel Multicell Switch-Mode Power Amplifier Employing the "Flying-Battery" Concept," *IEEE Trans. IE*, vol. 49, no. 4, pp. 816–823, Aug. 2002.

[10] M. Kasper, D. Bortis, and J. Kolar, "Scaling and Control of Multi-Cell Converters," in *Proc. IPEC*, May 2014.

[11] D. Kranzer, C. Wilhelm, F. Reiners, and B. Burger, "Application of Normally-off SiC-JFETs in Photovoltaic Inverters," in *Proc. EPE*, pp. 1–6, Sept. 2009.

[12] C. Wilhelm, D. Kranzer, and B. Burger, "Development of a Highly Compact and Efficient Solar Inverter with Silicon Carbide Transistors," in *Proc. CIPS*, March 2010.

[13] S. Araújo, C. Nöding, B. Sahan, and P. Zacharias, "Exploiting the Benefits of SiC by Using 1700 V Switches in Single-Stage Inverter Topologies Applied to Photovoltaic Systems," in *Proc. PCIM Europe*, May 2011.

[14] R. Mallwitz, C. Althof, S. Buchhold, and E. Kiel, "First 99% PV Inverter with SiC JFETs on the Market – Future Role of SiC," in *Proc. PCIM Europe*, May 2012.

[15] R. M. Burkart and J. W. Kolar, "Comparative Evaluation of SiC and Si PV Inverter Systems Based on Power Density and Efficiency as Indicators of Initial Cost and Operating Revenue," in *Proc. COMPEL*, June 2013.

[16] S. A. Mulder, "On the Design of Low-Profile High Frequency Transformers," in *Proc. HFPC*, pp. 141–159, 1990.

[17] P. Wallmeier, N. Fröhleke, and H. Grotstollen, "Automated Optimization of High Frequency Inductors," in *Proc. IECON*, vol. 1, pp. 342–347 vol.1, Aug. 1998.

[18] H. Sartori, H. Hey, and J. Pinheiro, "An Optimum Design of PFC Boost Converters," in *Proc. EPE*, pp. 1–10, Sept. 2009.

[19] B. Lyons, J. Hayes, and M. Egan, "Magnetic Material Comparisons for High-Current Inductors in Low-Medium Frequency DC-DC Converters," in *Proc. APEC*, pp. 71–77, Feb. 2007.

[20] M. Ryłko, B. Lyons, K. Hartnett, J. Hayes, and M. Egan, "Magnetic Material Comparisons for High-Current Gapped and Gapless Foil Wound Inductors in High Frequency DC-DC Converters," in *Proc. EPE-PEMC*, pp. 1249–1256, Sept. 2008.

[21] M. Ryłko, K. Hartnett, J. Hayes, and M. Egan, "Magnetic Material Selection for High Power High Frequency Inductors in DC-DC Converters," in *Proc. APEC*, pp. 2043–2049, Feb. 2009.

[22] R. M. Burkart and J. W. Kolar, "Low-Complexity Analytical Approximations of Switching Frequency Harmonics of 3-Phase N-Level Voltage-Source PWM Converters," in *Proc. IPEC*, May 2014.

[23] D. O. Boillat, J. W. Kolar, and J. Mühlethaler, "Volume Minimization of the Main DM/CM EMI Filter Stage of a Bidirectional Three-Phase Three-Level PWM Rectifier System," in *Proc. ECCE*, pp. 2008–2019, Sept. 2013.

[24] J. Mühlethaler, "Modeling and Multi-Objective Optimization of Inductive Power Components," Ph.D. dissertation, ETH Zurich, 2012.

[25] J. Mühlethaler, J. W. Kolar, and A. Ecklebe, "A Novel Approach for 3D Air Gap Reluctance Calculations," in *Proc. ECCE Asia*, pp. 446–452, May 2011.

[26] M. Ryłko, B. Lyons, J. Hayes, and M. Egan, "Revised Magnetics Performance Factors and Experimental Comparison of High-Flux Materials for High-Current DCDC Inductors," *IEEE Trans. PE*, vol. 26, no. 8, pp. 2112–2126, Aug. 2011.

[27] M. Schweizer, T. Friedli, and J. W. Kolar, "Comparison and Implementation of a 3-Level NPC Voltage Link Back-to-Back Converter with SiC and Si Diodes," in *Proc. APEC*, pp. 1527–1533, Feb. 2010.

[28] EPCOS. SIFERRIT Material N87. [Online]. Available: http://www.epcos.com, Feb. 2014.

[29] Metglas. POWERLITE Inductor Cores. [Online]. Available: http://www.metglas.com, Feb. 2014.

[30] Magnetics. Powder Core Catalog 2013. [Online]. Available: http://www.magnetics.com, Feb. 2014.

[31] H. Fukunaga, T. Eguchi, Y. Ohta, and H. Kakehashi, "Core Loss in Amorphous Cut Cores with Air Gaps," *IEEE Trans. Magn.*, vol. 25, no. 3, pp. 2694–2698, 1989.

[32] B. Cougo, A. Tüysüz, J. Mühlethaler, and J. W. Kolar, "Increase of Tape Wound Core Losses Due to Interlamination Short Circuits and Orthogonal Flux Components," in *Proc. IECON*, vol. 1, pp. 123–128 vol.1, 2011.

[33] Pack Feindrähte. RUPALIT Classic Litz Wires. [Online]. Available: http://www.pack-feindraehte.de, Dec. 2013.

[34] *Specifications for Particular Types of Winding Wires – Part 0-1: General Requirements – Enamelled Round Copper Wire*, IEC 60317-0-1, 2008.

[35] Bridgeport Magnetics Group. LAMINAX Winding Conductors. [Online]. Available: http://www.alphacoredirect.com, Feb. 2014.

[36] A. van den Bossche and V. C. Valchev, *Inductors and Transformers for Power Electronics*. Taylor & Francis, 2005.

[37] A. Stadler and C. Gulden, "Efficient Nonlinear Inductors for PV Inverters and Active PFC," in *Proc. CIPS*, March 2012.

Comparative Analysis of Inductor Concepts for High Peak Load Low Duty Cycle Operation

Michael Leibl and Johann W. Kolar
Power Electronic Systems Laboratory
ETH Zurich
8092 Zurich, Switzerland
leibl@lem.ee.ethz.ch

Abstract—In this work a compact-model-based inductor design procedure is presented. The losses, temperatures, and the total cost of ownership (TCO) of an inductor are expressed analytically. All geometric dependencies are summarized in a set of parameters which are calculated using finite element analysis (FEA). Therefore the model does not depend on the actual inductor geometry, instead it only relies on a generalized set of parameters which contain all geometric information. Different inductor geometries only result in different values of parameters. The dynamic thermal model is verified using time dependent FEA, the high frequency winding loss model is verified by measurements. The inductor model is then used to study the effect of changes in inductor geometry on the performance of the device. It is shown that for applications requiring high peak but low average load substantial cost reductions are achieved if the inductor's geometry is optimized and the right inductor topology is selected.

Keywords—inductor, geometry, optimization

I. INTRODUCTION

The intended application of this work requires high peak power for several seconds which is supplied by a three-level PFC rectifier system. However, the average power demand is only about 1 % of the peak power. This load profile allows to make use of the thermal time constants of the passive components of the system. Savings in terms of inductor volume can be achieved since the inductor thermal time constants are in the range of several 10 s. The thermal time constants of semiconductor devices are in the range of 100 ms, so no savings are expected here. However, since the semiconductor losses of the rectifier are usually dominated by the switching losses one is able to reduce the chip area by reducing the switching frequency. Of course this comes at the price of high values of inductance.

It becomes obvious that the inductor is a key component in such a design and should be subject to optimization. However, also in continuous load scenarios an optimized inductor design is essential since it has been shown that inductors have a major impact on overall system cost [1]. Core element of such an optimization is a model that renders thermal and electromagnetic effects accurately but fast enough to

be called iteratively. Resolving geometric dependencies precisely requires complex thermal and magnetic equivalent networks as in [2] and [3]. Using finite element analysis (FEA) based models, one achieves highest accuracy but the computational effort increases dramatically. To reach high accuracy and at the same time low computational effort, a compact model-based approach is proposed. The structure of the model is shown in Fig. 1. All geometry dependencies are described using a few parameters which are calculated from FEA simulations. The actual thermal and loss models are expressed as analytic equations. This way the actual inductor model and the optimization routine only depend on a generalized set of parameters, which makes it possible to investigate a wide range of different inductor topologies and to compare their performance in a fair way.

In Section II the compared inductor topologies are described and the range of proportions which are considered in this investigation are given. In Section III the applied loss models are described and in Section IV the dynamic thermal model is presented. In order to be able to compare different core and winding materials and to find the right relation between efficiency and material cost a cost model is needed which is presented in Section V. Finally in Section VI the optimization results are shown and discussed.

II. INDUCTOR TOPOLOGIES

The inductor topologies which have been compared in this work are shown in Fig. 2a-d. The scope of the analysis is limited to axisymmetric geometries to allow fast FEA simulation. However, the investigation could be easily extended by additional 3D simulations of inductors having arbitrary shape.

A. Geometric Proportions

The key idea is to describe the dimensions of the inductors using a set of proportions and one additional value as scaling quantity. In this case the volume V of the cylinder that encloses the inductor assembly is specified together with 2 or 3 proportions. All dimensions are derived from such a set of proportions and the cylinder volume V as scaling quantity. Describing the geometry using proportions allows to run the FEM simulation only once for each set of proportions and to

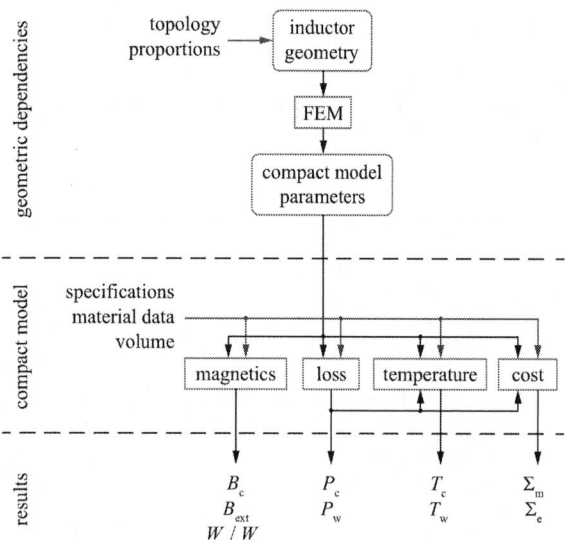

Figure 1. The proposed compact modeling approach uses analytic expressions to calculate flux densities, losses, temperatures and inductor cost. All geometric dependencies, except the volume of the inductor are concentrated in a set of parameters which are calculated using a FEM simulation.

scale the obtained parameters for the required volume. The geometric proportions p_i are defined in Fig. 2.

B. Permeability and Ratio of Thermal Conductivity

For each inductor geometry two different FEA studies have to be calculated. First the magneto-static field distribution is solved. The result will be used to calculate the necessary number of turns and the proximity losses in the winding. The field distribution depends on the core's permeability μ_r, therefore the magneto-static simulation is solved for a parametric sweep of the relative permeability in the range $1 \leq \mu_r \leq 10^6$. The solutions are be interpolated for permeabilities not on the sweep grid.

Second a static thermal study has to be solved in order to determine the average thermal resistances of winding to surface and core to surface. The thermal resistances depend on the ratio of the thermal conductivity of the winding to thermal conductivity of the core. This ratio $p_{wc} = \frac{\lambda_w}{\lambda_c}$ is swept through the range $10^{-3} \leq p_{wc} \leq 10$.

C. Scaling Laws

For a given set of proportions one can derive two scaling laws for all magnetic fields and thermal resistances obtained from a FEA simulation. Assuming the simulation was performed on a geometry with volume V_0, current density J_0, and core thermal conductivity λ_{c0}. All magnetic field quantities H_{i0} and thermal resistances $R_{th,i0}$ obtained at volume V_0 from the simulation can be scaled for a volume V using the following expressions:

$$H_i = k_{Hi} J V^{\frac{1}{3}} \qquad k_{Hi} = \frac{H_{i0}}{J_0 V_0^{\frac{1}{3}}}$$

$$R_{th,i} = k_{Rth,i} V^{-\frac{1}{3}} \lambda_c^{-1} \qquad k_{Rth,i} = R_{th,i0} V_0^{\frac{1}{3}} \lambda_{c0}$$

The scaling law for magnetic field quantities relies on the fact that if a current density J is specified the magnetic field is always expressed by an expression such as $H = \frac{JA}{d}$ with specific area A and distance d. If no proportion of the geometry is changed, the magnetic field therefore scales with $\frac{V^{\frac{2}{3}}}{V^{\frac{1}{3}}} = V^{\frac{1}{3}}$.

The scaling law for thermal resistances within a particular volume can be derived from $R_{th} = \frac{l}{\lambda A}$. For a particular value of λ, the thermal resistance scales with $\frac{V^{\frac{1}{3}}}{V^{\frac{2}{3}}} = V^{-\frac{1}{3}}$ if no geometric proportions are changed.

III. Loss Models

A. Winding Loss Model

In order to simplify the calculation of the winding losses the scope of the investigation is limited to windings made of round conductors or high frequency litz wire. In the case of a winding made from solid conductors it is further assumed that the conductors are evenly spaced within the winding region. In case of litz wire it is assumed that the strands are evenly spaced throughout the winding region. (see Fig. 2e) These assumptions are true for closely packed windings of round conductors or for a winding made from rectangular profile litz wire. The inner proximity effect which is observed on solitary litz wires, vanishes with this assumption, since no unique partitioning of the array of strands into bundles of strands is possible. The loss per unit length of a single strand or round conductor could be expressed using Ferreira's solution for a cylindrical conductor exposed to an external magnetic field ([4], [5]). However, if Ferreira's solution is applied the proximity effect could be overestimated by over 80 % if the conductors of the winding are closely stacked. For windings having high filling factor Dowell's solution gives better results, however it underestimates the losses if the filling factor is low [6]. A modified version of Dowell's proximity loss factor is proposed in [7] which also takes into account horizontal and vertical distances between the conductors. For the investigation at hand it is assumed that horizontal and vertical distance are equal and therefore determined by the filling factor k_{cu}. Using the improved proximity loss factor and Ferreira's factor for the skin effect, the losses per unit length in a single strand with diameter d_s are expressed as

$$P_s' = R_s'(F\hat{I}_s^2 + k_{cu} A_s \hat{G} \hat{H}^2). \qquad (1)$$

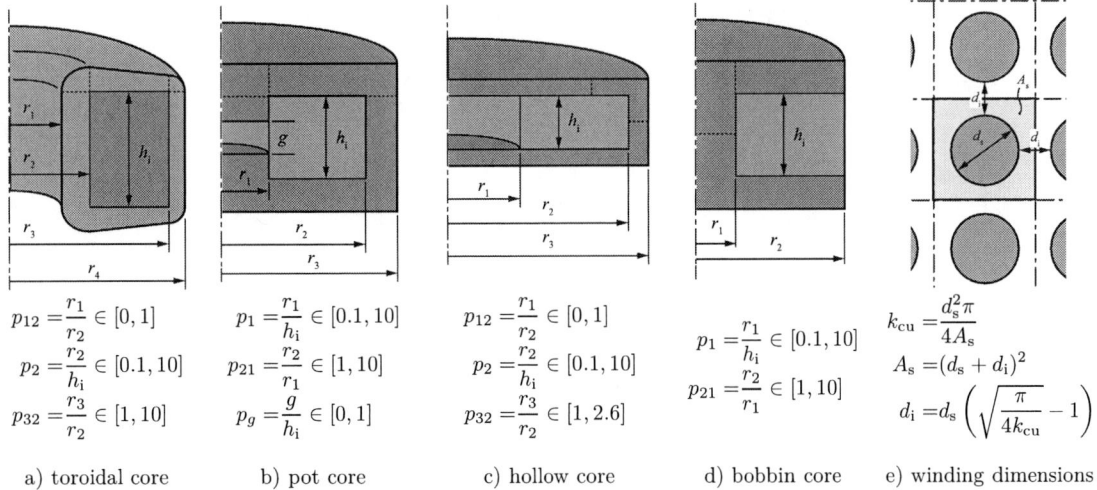

$$p_{12} = \frac{r_1}{r_2} \in [0,1]$$
$$p_2 = \frac{r_2}{h_i} \in [0.1, 10]$$
$$p_{32} = \frac{r_3}{r_2} \in [1, 10]$$

$$p_1 = \frac{r_1}{h_i} \in [0.1, 10]$$
$$p_{21} = \frac{r_2}{r_1} \in [1, 10]$$
$$p_g = \frac{g}{h_i} \in [0, 1]$$

$$p_{12} = \frac{r_1}{r_2} \in [0,1]$$
$$p_2 = \frac{r_2}{h_i} \in [0.1, 10]$$
$$p_{32} = \frac{r_3}{r_2} \in [1, 2.6]$$

$$p_1 = \frac{r_1}{h_i} \in [0.1, 10]$$
$$p_{21} = \frac{r_2}{r_1} \in [1, 10]$$

$$k_{cu} = \frac{d_s^2 \pi}{4 A_s}$$
$$A_s = (d_s + d_i)^2$$
$$d_i = d_s \left(\sqrt{\frac{\pi}{4 k_{cu}}} - 1 \right)$$

a) toroidal core   b) pot core   c) hollow core   d) bobbin core   e) winding dimensions

Figure 2. Inductor topologies that are compared in this work (a-d). The geometries are simulated within the given ranges of proportions. The dashed lines indicate cross sectional areas that are equal. The winding is assumed to be a matrix of evenly spaced conductors, defined by filling factor k_{cu} and strand diameter d_s (e).

With skin and proximity factors $F(\xi)$ and $\hat{G}(X, d_i)$

$$F = \frac{\xi}{4\sqrt{2}} \left(\frac{\text{ber}_0(\xi)\text{bei}_1(\xi) - \text{ber}_0(\xi)\text{ber}_1(\xi)}{\text{ber}_1(\xi)^2 + \text{bei}_1(\xi)^2} \right.$$
$$\left. + \frac{-\text{bei}_0(\xi)\text{ber}_1(\xi) - \text{bei}_0(\xi)\text{bei}_1(\xi)}{\text{ber}_1(\xi)^2 + \text{bei}_1(\xi)^2} \right)$$

$$\hat{G} = \frac{\pi}{16} \left(3(1-w)k^{-3} X \frac{\sinh(kX) - \sin(kX)}{\cosh(kX) + \cos(kX)} \right.$$
$$\left. + \frac{w}{2} \frac{X}{X^{-3} + b^3} \right)$$

$$\xi = \frac{d_s}{\sqrt{2}\delta}$$

$$X = \frac{d_s}{\delta}$$

$$\delta = \frac{1}{\sqrt{\pi f \sigma \mu}}$$

the constants k, b and w are given in [7]. One can now derive the average loss density in the winding. Since the magnetic field usually varies throughout the winding volume V_w, the spatial root mean square (SRMS) of the magnetic field H_w in the winding

$$H_{wsrms} = \sqrt{\frac{1}{V_w} \int_{V_w} H_w^2 dV},$$ (2)

has to be used to obtain the average loss density in the winding region. The filling factor k_{cu} of the winding is defined as ratio of total copper cross section per winding cross section. Also the current density J is defined as total winding current (NI) per winding cross section. Then the occupied cross section of a single strand is given as

$$A_s = \frac{d_s^2 \pi}{4 k_{cu}},$$ (3)

and the current per strand is

$$I_s = J A_s.$$ (4)

Using the strand resistance per unit length

$$R_s' = \frac{1}{\sigma k_{cu} A_s},$$ (5)

the average winding loss density is obtained as

$$p_w = \frac{P_s'}{A_s} = \frac{1}{\sigma k_{cu}} \left(F \hat{J}^2 + \frac{k_{cu} \hat{G}}{A_s} \hat{H}_{wsrms}^2 \right).$$ (6)

Neglecting external fields, the magnetic field in all regions of the inductor is proportional to the current density, with a factor k_{Hwsrms} which is calculated using FEA,

$$H_{wsrms}(t) = k_{Hwsrms} V^{\frac{1}{3}} J(t).$$ (7)

The time dependent shape function $k_J(t)$ of the current density is defined by unifying the current density to its peak value J_{max},

$$k_J(t) = \frac{J(t)}{J_{max}}.$$ (8)

The Fourier coefficients of the current density shape function are expressed as

$$\hat{k}_{J(n)} = \frac{2}{T} \int_0^T k_J(t) e^{-j \frac{2\pi n}{T} t} dt.$$ (9)

By inserting (9), (8) and (7) in (6) one obtains

$$p_w = \frac{1}{\sigma k_{cu}} \left(\left(\frac{\hat{k}_{J(0)}}{2} \right)^2 + \sum_{n=1}^{\infty} F \hat{k}_{J(n)}^2 \right.$$
$$\left. + k_{Hwsrms}^2 V^{\frac{2}{3}} \sum_{n=1}^{\infty} \frac{k_{cu} \hat{G}}{A_s} \hat{k}_{J(n)}^2 \right) J_{max}^2.$$

This result can be expressed as

$$p_w = \left(c_{skin} + k_{Hwsrms}^2 V^{\frac{2}{3}} c_{prox} \right) J_{max}^2$$ (10)

with the constants c_{skin} and c_{prox}, which are not depending on any core dimension, but only on current

The 2014 International Power Electronics Conference

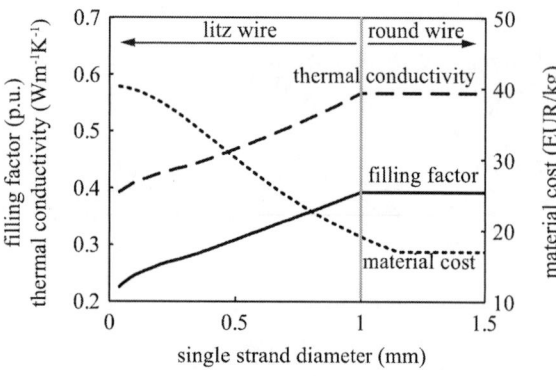

Figure 3. Filling factor, thermal conductivity and mass specific material cost of a winding as function of the strand diameter

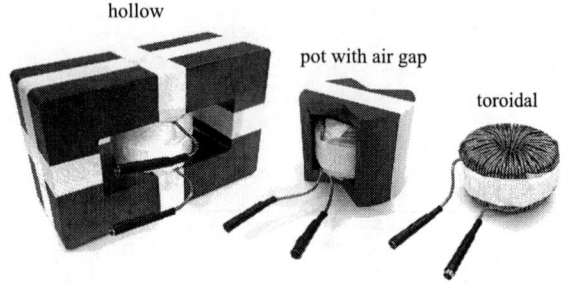

Figure 4. Picture of the inductors used to verify the winding loss model. Instead of a cylindrical hollow core, four C-shaped ferrite cores are used to resemble the same magnetic field distribution in the winding as it would occur in a real cylindrical core.

shape and winding properties such as filling factor and strand diameter,

$$c_{\text{skin}} = \frac{1}{\sigma k_{\text{cu}}} \left(\left(\frac{\hat{k}_{J(0)}}{2} \right)^2 + \sum_{n=1}^{\infty} F \hat{k}_{J(n)}^2 \right) \quad (11)$$

$$c_{\text{prox}} = \frac{1}{\sigma A_{\text{s}}} \sum_{n=1}^{\infty} \hat{G} \hat{k}_{J(n)}^2. \quad (12)$$

1) Winding Properties: In case of litz wire windings the filling factor itself is a function of the strand diameter. Moreover the thermal conductivity of the winding is a function of the filling factor and therefore depends on the strand diameter. Also the weight specific material cost depends on the strand diameter. All these dependencies are illustrated in Fig. 3. The filling factor k_{cu} for strand diameters smaller than 1 mm has been calculated by using averaged manufacturer's data [8], further a factor of $\frac{\pi}{4}$ for stacking of the wires is assumed and an additional factor of 50 % accounts for space taken up by tolerances, bobbin and layer isolation. Therefore, the maximum filling factor is $0.5 \cdot \frac{\pi}{4}$. The thermal conductivity of litz wire has been determined using a 2D FEM simulation for different values of filling factor. The weight specific cost is calculated using the function given in [1].

2) Experimental Verification: The accuracy of the described method that uses the spatial root mean square of the magnetic field together with the assumption of evenly distributed conductors is studied on a set of test inductors. The test devices resemble the topologies *toroidal, hollow* and, *pot with air gap*. A picture of the inductors is shown in Fig. 4. All inductors use a winding made of round magnet wire with a diameter of 1.7 mm. The resistance of the winding has been measured for frequencies starting at 100 Hz up to $\frac{f_0}{10}$, with the resonance frequency of the inductor f_0. The measurements are compared to the AC resistance calculated with the winding loss model (Fig. 5). It is shown that the relative error of the model is less than 15 % for all three topologies.

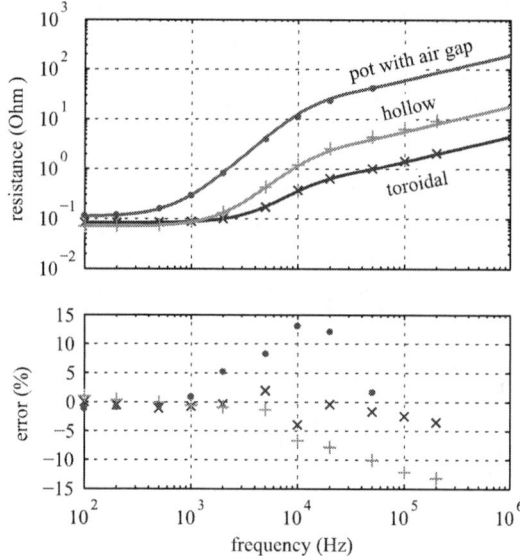

Figure 5. Measured and calculated winding resistance of the test inductors as function of the frequency as well as relative error between measured and calculated values.

B. Core Loss Model

Manufacturers of magnetic materials usually provide loss density data for a specific frequency and flux density range. The loss measurements are carried out using sinusoidal waveforms and a common fitting equation, referred to as the original Steinmetz equation (OSE), with the parameters k, α, β, the frequency f and the flux density amplitude $\hat{B}$, which provides an expression for the core loss density

$$p_{\text{c}} = k f^{\alpha} \hat{B}^{\beta}. \quad (13)$$

Additionally effective core area A_{e} and effective core volume V_{e} are often specified. To relate effective and geometric dimensions, we introduce the core area filling factor $k_{\text{fea}} = \frac{A_{\text{e}}}{A_{\text{c}}}$ and the core volume filling factor $k_{\text{fev}} = \frac{V_{\text{e}}}{V}$. Both are determined for each core material by averaging their values determined from multiple core data sheets as provided by the manufacturer [9, 10, 11, 12, 13, 14]. Especially for laminated

978-1-4799-2706-7/14 $31.00 © 2014 IEEE

Material Name	k ($\frac{\mathrm{Ws}^\alpha}{\mathrm{m}^3\mathrm{T}^\beta}$)	α	β	ρ ($\frac{\mathrm{kg}}{\mathrm{m}^3}$)	k_{fea} %	k_{fev} %	σ_{c} ($\frac{\mathrm{EUR}}{\mathrm{kg}}$)
Epcos N27	11.70	1.32	2.32	4800	97	98	5.50
Epcos N87	0.08	1.78	2.84	4850	97	98	5.50
Vitroperm 500F	0.01	1.78	2.08	7350	65	75	23.00
Metglas 2605SA1	1.38	1.51	1.74	7180	82	89	16.00
Kaschke K2004	11.25	1.29	2.11	4800	97	99	5.50
Kaschke K2006	1.24	1.50	2.64	4800	97	99	5.50
Kaschke K2008	0.30	1.68	2.98	4800	97	99	5.50
Micrometals -02	2360	1.05	2.37	5000	95	95	32.19
Micrometals -08	6156	1.02	2.38	6500	95	95	36.27
Micrometals -14	1537	1.10	2.36	5200	95	95	37.89
Micrometals -18	3359	1.08	2.30	6600	95	95	31.47
Micrometals -19	2399	1.12	2.30	6800	95	95	13.48
Micrometals -26	401	1.29	2.22	7000	95	95	10.16
Micrometals -30	2212	1.08	1.99	6000	95	95	13.41
Micrometals -34	2036	1.11	2.15	6200	95	95	14.49
Micrometals -35	1894	1.09	1.99	6300	95	95	13.07
Micrometals -40	329	1.31	2.14	6900	95	95	10.31
Micrometals -45	507	1.24	2.21	7200	95	95	10.40
Micrometals -52	1463	1.14	2.21	7000	95	95	10.96
Magnetics MPP 14u	0.28	1.87	2.50	8000	95	95	229.32
Magnetics MPP 26u	0.79	1.65	2.34	8000	95	95	165.31
Magnetics MPP 60u	156	1.12	2.05	8000	95	95	227.08
Magnetics MPP 125u	1.11	1.56	2.06	8000	95	95	224.18
Magnetics MPP 147u	1.02	1.57	2.00	8000	95	95	156.70
Magnetics MPP 160u	1.02	1.57	2.00	8000	95	95	143.42
Magnetics MPP 173u	1.02	1.57	2.00	8000	95	95	143.42
Magnetics MPP 200u	0.35	1.68	2.09	8000	95	95	143.42
Magnetics MPP 300u	0.35	1.68	2.09	8000	95	95	321.94
Magnetics MPP 550u	7.04	1.47	2.13	8000	95	95	515.92
Magnetics High Flux 14u	9.33	1.54	2.31	7600	95	95	120.14
Magnetics High Flux 26u	12.70	1.49	2.21	7600	95	95	109.65
Magnetics High Flux 60u	53.95	1.32	2.22	7600	95	95	88.43
Magnetics High Flux 125u	9.57	1.47	2.23	7600	95	95	122.04
Magnetics High Flux 147u	26.36	1.41	2.30	7600	95	95	110.09
Magnetics High Flux 160u	26.36	1.41	2.30	7600	95	95	110.17
Magnetics Kool Mu 26	5.00	1.46	2.09	5500	95	95	36.89
Magnetics Kool Mu 40	5.00	1.46	2.09	5500	95	95	36.89
Magnetics Kool Mu 60	26.03	1.29	2.01	5500	95	95	36.89
Magnetics Kool Mu 75	26.03	1.29	2.01	5500	95	95	22.88
Magnetics Kool Mu 90	26.03	1.29	2.01	5500	95	95	22.88
Magnetics Kool Mu 125	1.18	1.63	2.20	5500	95	95	27.45
Magnetics XFlux	9.52	1.50	1.80	7500	95	95	33.57
GOES M130-27S (50Hz)	9.44	1.50	2.20	7600	95	95	2.50
GOES M165-35S (20kHz)	11.21	1.58	1.90	7600	95	95	2.50

Table I.   CORE MATERIAL PROPERTIES

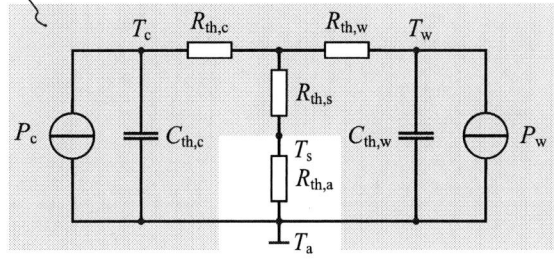

Figure 6.   General thermal network for average core and winding termperatures T_{c} and T_{w}.

cores the effective quantities deviate significantly from their geometric counterparts. For ferrite and powder cores however, the difference is small. The Steinmetz parameters, filling factors and weight specific prices of the core materials used in the optimization are listed in Tab. I.

All field quantities are usually given as geometric quantities, i.e. the flux density is defined as magnetic flux per cross sectional area. However the quantities that have to be used in the OSE are effective quantities. Therefore, the filling factors have to be considered in the OSE if geometric quantities are used and (13) is rewritten as

$$p_{\mathrm{c}} = \frac{k_{\mathrm{fev}}}{k_{\mathrm{fea}}^{\beta}} k f^{\alpha} \hat{B}^{\beta}. \tag{14}$$

The improved generalized Steinmetz equation (iGSE) [15] extends the OSE to allow loss calculation for non-sinusoidal flux density waveforms. If the waveform is purely sinusoidal it gives the same result as the OSE. Using the core filling factors the iGSE is given as

$$p_{\mathrm{c}} = \frac{k_{\mathrm{fev}} k_{\mathrm{i}}}{k_{\mathrm{fea}}^{\beta} T} \int_{0}^{T} \left| \frac{\mathrm{d}B}{\mathrm{d}t} \right|^{\alpha} (\Delta B)^{\beta-\alpha} \mathrm{d}t. \tag{15}$$

With the approximation for the constant

$$k_{\mathrm{i}} = \frac{k}{2^{\beta+1}\pi^{\alpha-1}\left(0.2761 + \frac{1.7061}{\alpha+1.354}\right)}. \tag{16}$$

If we assume constant permeability, the magnetic flux density is proportional to the current density and

according to the scaling law for magnetic fields it is given as

$$B(t) = \mu_{\mathrm{r}}\mu_{0}k_{H\mathrm{cavg}}V^{\frac{1}{3}}k_{J}(t)J_{\max}. \tag{17}$$

By inserting (17) into (15) and by introducing

$$c_{\mathrm{core}} = \frac{k_{\mathrm{fev}} k_{\mathrm{i}}}{k_{\mathrm{fea}}^{\beta} T} \int_{0}^{T} \left| \frac{\mathrm{d}k_{J}}{\mathrm{d}t} \right|^{\alpha} (\Delta k_{J})^{\beta-\alpha} \mathrm{d}t, \tag{18}$$

the loss density in the core is expressed as

$$p_{\mathrm{c}} = (\mu_{\mathrm{r}}\mu_{0}k_{H\mathrm{cavg}})^{\beta} V^{\frac{\beta}{3}} c_{\mathrm{core}} J_{\max}^{\beta}. \tag{19}$$

Similar to the expression for the winding loss density, all material dependency together with the shape function of the current density is summarized in the constant c_{core}. All geometry dependency is summarized in the parameter $k_{H\mathrm{cavg}}$ which expresses the average magnetic field in the core by using the scaling law from II-C.

IV.   THERMAL MODEL

The aim of a compact dynamic thermal model for the conductor is to calculate the time dependent temperatures in core and winding for any load scenario. This way one is able to determine the exact volume of the inductor which is necessary in order to not exceed the material specific maximum temperatures. The problem is split into two parts, one that models the conductive heat transfer from core and winding to the surface, the conductive heat transfer between core and winding, and the thermal capacitances of core and winding. The surface of the inductor is assumed to be isothermal and the second part of the model therefore describes the surface temperature rise, considering radiation as well as natural and forced convection.

A. Dynamic Thermal Network

The equivalent thermal network of a core-winding arrangement is shown in Fig. 6. It consists of two heat sources P_{c}, P_{w} representing core and winding losses. Since the thermal capacitances $C_{\mathrm{th,c}}$ and $C_{\mathrm{th,w}}$ of core and winding are spread within the same volume as the according heat source, each of them is represented by a single capacitance in parallel to the according heat source. The temperature rise that occurs from the inside of core and winding to the surface is

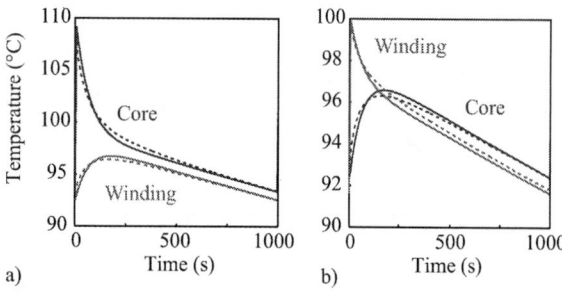

a)   b)

Figure 7. Average temperatures of core and winding of a toroidal inductor with $r_1 = 10\,\mathrm{mm}$, $r_2 = 30\,\mathrm{mm}$, $r_3 = 60\,\mathrm{mm}$ and $h_\mathrm{i} = 20\,\mathrm{mm}$ subject to a periodic 5 s loss pulse of 200 W applied to the core (a) and applied to the winding (b). Solid: temperature response calculated using equivalent circuit. Dashed: direct FEM simulation.

modelled by the thermal resistances $R_\mathrm{th,c}$, $R_\mathrm{th,w}$ and $R_\mathrm{th,s}$. The actual temperatures in core and winding could be calculated using a circuit simulation tool or simply by decomposing the loss waveforms $P_\mathrm{c}(t)$ and $P_\mathrm{w}(t)$ into their Fourier series and evaluating the transfer functions that result from the circuit. In order to verify the use of the equivalent circuit, the temperature response to a loss power pulse of 5 s duration and 1000 s period time is calculated using the equivalent circuit and compared to the output of a time dependent FEM simulation. The results are shown in Fig. 7. In particular for scenarios with high peak and low average load the results obtained by using the equivalent circuit show good agreement with the temperature waveforms simulated directly using FEA.

The thermal resistances $R_\mathrm{th,c}$, $R_\mathrm{th,w}$ and $R_\mathrm{th,s}$ are calculated from two steady state FEM simulations. The first simulates the average temperatures T_cc0 and T_wc0 of core and winding with $P_\mathrm{c} = P_\mathrm{c0}$ and $P_\mathrm{w} = 0\,\mathrm{W}$, the second the temperatures T_cw0 and T_ww0 with $P_\mathrm{c} = 0\,\mathrm{W}$ and $P_\mathrm{w} = P_\mathrm{w0}$, assuming isothermal surface temperature of $T_\mathrm{s} = 0\,\mathrm{K}$. Using these results one can calculate the average core and winding temperature rise to surface temperature for any values of core and winding loss P_c and P_w.

$$T_\mathrm{c} = \frac{T_\mathrm{cc0}}{P_\mathrm{c0}}P_\mathrm{c} + \frac{T_\mathrm{cw0}}{P_\mathrm{w0}}P_\mathrm{w} = R_\mathrm{th,cc0}P_\mathrm{c} + R_\mathrm{th,cw0}P_\mathrm{w} \tag{20}$$

$$T_\mathrm{w} = \frac{T_\mathrm{wc0}}{P_\mathrm{c0}}P_\mathrm{c} + \frac{T_\mathrm{ww0}}{P_\mathrm{w0}}P_\mathrm{w} = R_\mathrm{th,wc0}P_\mathrm{c} + R_\mathrm{th,ww0}P_\mathrm{w} \tag{21}$$

The thermal resistances of the equivalent network at volume V_0 of the simulated inductor geometry are therefore given as

$$R_\mathrm{th,c0} = R_\mathrm{th,cc0} - R_\mathrm{th,cw0} \tag{22}$$

$$R_\mathrm{th,w0} = R_\mathrm{th,ww0} - R_\mathrm{th,wc0} \tag{23}$$

$$R_\mathrm{th,s0} = R_\mathrm{th,cw0} = R_\mathrm{th,wc0} \tag{24}$$

and subject to the scaling law for thermal resistances as stated in II-C. E.g. the thermal resistance $R_\mathrm{th,c}$ for

an inductor of volume V and thermal conductivity of the core λ_c is calculated as $R_\mathrm{th,c} = \frac{k_{R\mathrm{th,c}}}{V^{\frac{1}{3}}\lambda_\mathrm{c}}$ with $k_{R\mathrm{th,c}} = R_\mathrm{th,c0}V_0^{\frac{1}{3}}\lambda_\mathrm{c0}$. Of course with all proportions unchanged and the ratio between the thermal conductivities of core and winding of the inductor being equal to the ratio that was used in the FEM simulation ($\frac{\lambda_\mathrm{w}}{\lambda_\mathrm{c}} = \frac{\lambda_\mathrm{w0}}{\lambda_\mathrm{c0}}$).

The values of the thermal capacitances in the equivalent circuit equal the actual heat capacitances of core and winding and are simply calculated using the volume specific heat capacities.

The resistance $R_\mathrm{th,a}$ represents the thermal resistance from the surface A_sur of the inductor to ambient air. It is given as

$$R_\mathrm{th,a} = \frac{1}{A_\mathrm{sur}(h_\mathrm{rad} + h_\mathrm{nat} + h_\mathrm{for})} \tag{25}$$

with the heat transfer coefficients h_rad, h_nat and h_for for radiation, natural and forced convection. If laminar flow is assumed and the inductor surface is simplified to a cylinder of height h and radius r they are calculated as [16]

$$h_\mathrm{rad} = 5 \cdot 10^{-8}\,\frac{\mathrm{W}}{\mathrm{K^4 m^2}}(T_\mathrm{s}^2 + T_\mathrm{a}^2)(T_\mathrm{s} + T_\mathrm{a}) \tag{26}$$

$$h_\mathrm{nat} = \frac{0.82\,\frac{\mathrm{W}}{\mathrm{K^{\frac{5}{4}} m^{\frac{7}{4}}}}h^{\frac{3}{4}} + 0.68\,\frac{\mathrm{W}}{\mathrm{K^{\frac{5}{4}} m^{\frac{7}{4}}}}r^{\frac{3}{4}}}{h + r}(T_\mathrm{s} - T_\mathrm{a})^{\frac{1}{4}} \tag{27}$$

$$h_\mathrm{for} = 2.72\,\frac{\mathrm{W}\sqrt{\mathrm{s}}}{\mathrm{Km^2}}\sqrt{\frac{v}{r}}. \tag{28}$$

V. Cost Model

When designing a cost optimized inductor the trade-off between power density and efficiency is important. A compact design offers low material cost, but throughout the operational life of the device high additional costs for the cumulated loss energy occur. A large device on the other side offers high efficiency and therefore low energy cost at the price of higher material cost. The optimum volume of a device is found if the total cost of ownership $\Sigma(V) = \Sigma_\mathrm{e}(V) + \Sigma_\mathrm{m}(V)$, i.e. the sum of material cost Σ_m and energy cost Σ_e, is minimized.

A model to estimate the material cost of inductive components can be found in [1]. It includes weight depending parts with the core and winding weights m_c, m_w, the weight specific material costs σ_c and σ_w, the weight specific labour cost σ_lab and a fix cost share σ_f,

$$\Sigma_\mathrm{m} = \sigma_\mathrm{c}m_\mathrm{c} + (\sigma_\mathrm{w} + \sigma_\mathrm{lab})m_\mathrm{w} + \sigma_\mathrm{f}. \tag{29}$$

The cumulated loss energy is calculated by defining a certain load scenario $P(t)$, assuming a lifetime T_l and the price of energy σ_e, which is assumed to be $0.15\,\mathrm{EUR/kWh}$. With the load dependent efficiency $\eta(P)$ the energy cost is expressed as

$$\Sigma_\mathrm{e} = \sigma_\mathrm{e}\int_0^{T_\mathrm{l}}(1 - \eta(P(t)) \cdot P(t)\mathrm{d}t. \tag{30}$$

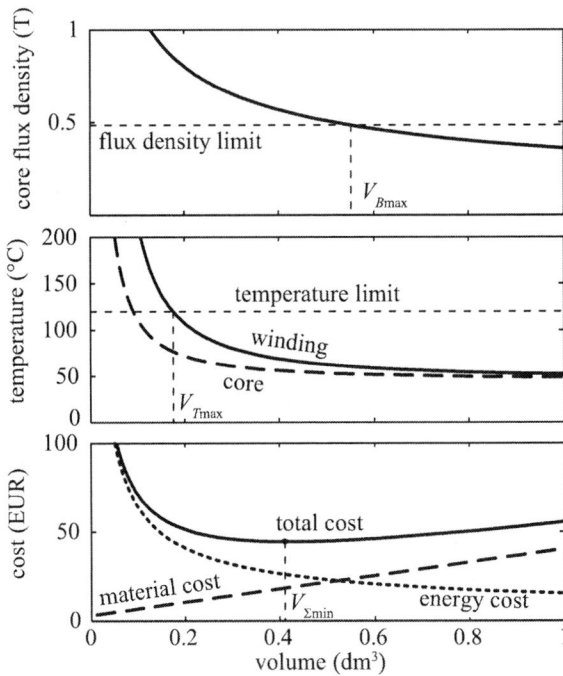

Figure 8. Core flux density, maximum temperatures and TCO of an inductor as function of volume. The volume should be optimized to reach a minimum of TCO while matching the constraints $V > V_{T\max}$ and $V > V_{B\max}$.

VI. RESULT OF GEOMETRY OPTIMIZATION

The benefit of geometry optimization of inductors is demonstrated on a three phase Vienna rectifier as shown in Fig. 9. The rectifier is assumed to operate in one of two load scenarios, also defined in Fig. 9. The boost inductors of the system are optimized for minimum TCO. The proportions of each inductor topology are swept within the ranges given in Fig. 2, resulting in 900 different geometry samples for each topology. For each sample the volume is adjusted such that all temperatures and flux densities are in valid range and that the TCO is minimized (Fig. 8). The volume optimization is performed for all core materials given in Tab. I and for strand diameters in the range of $50\,\mu$m to 3 mm. The material combination with minimum cost is finally selected for the geometry sample. The TCO of all optimized inductor samples is illustrated in Fig. 10 and Fig. 11. It is observed that the TCO of a boost inductor which is optimized for the pulse load scenario is less than 15 % of the one optimized for continuous load. Further if one would build an inductor with volume optimized for the continuous load profile but with the same proportions as the best sample in the pulse load scenario the TCO would rise by more than 100 % compared to the geometry with the optimum proportions for the continuous load profile. Also, vice versa, if the best geometry for continuous load is used in the pulse load scenario TCO increase of more than 100 % occurs. The differences are substantial for the topology *pot with air gap* but the effect is also observed with all

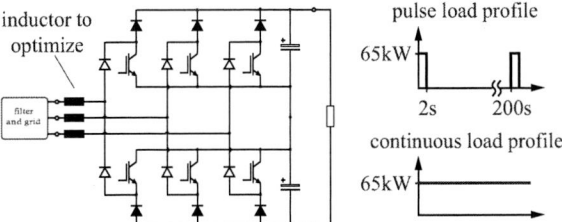

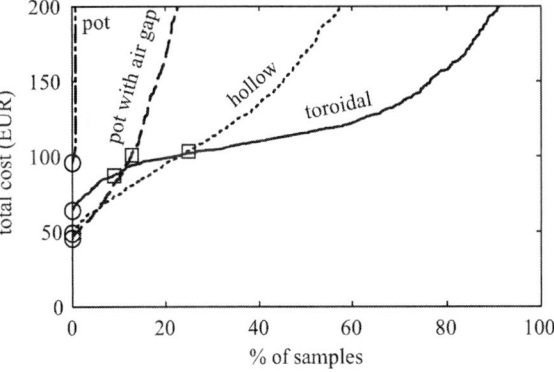

Figure 9. The boost inductors of a three-phase, three-level rectifier are optimized for a pulse and a continuous load scenario. The allowed range of grid voltages is 290 V to 530 V, the DC-link voltage 800 V. Using $50\,\mu$H boost inductance value at 28 kHz switching frequency results in a worst case peak-peak current ripple of less than 20 %. The peak current, the inductor has to be designed for is 200 A. For the TCO optimization a life time of 7 years is assumed with energy price of 0.15 EUR/kWh.

Figure 10. TCO of all optimized inductor samples for the pulse load scenario. The best sample is a *pot with air gap* type inductor, closely followed by *hollow* inductors. The best sample of each topology is marked with circle. Squares mark the performance of the inductor samples which are best for the continuous load scenario. For all topologies a considerable difference in the performance of the continuous load optimized and the pulse load optimized inductors is observed.

other topologies. Details of the best samples for each load scenario, as well as a scaled drawing of the cross section are shown in Fig. 12.

VII. CONCLUSIONS

A generalized compact model for different inductor topologies, featuring a dynamic thermal model and high frequency loss models for core and winding has been demonstrated. Summarizing all geometric dependencies and analytically expressing all relevant quantities as function of the inductor volume allows fast calculation of the cost function and therefore fast optimization of a high number of samples. The winding loss model has been verified by measurements for three different inductor topologies. The dynamic thermal model fits well with time discrete FEM simulation. The model has been used to optimize the TCO of the boost inductors of a Vienna rectifier for different load scenarios. It is shown that substantial cost savings are possible if all geometric degrees of freedom are included in the optimization.

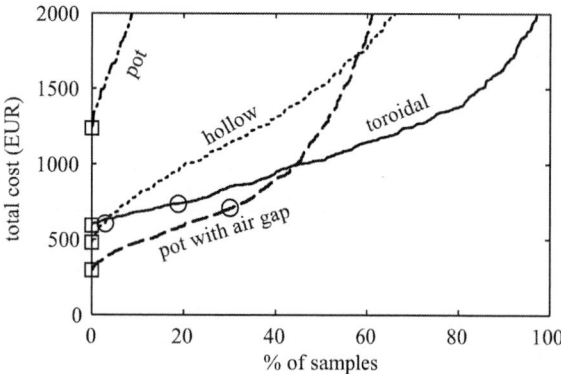

Figure 11. TCO of all optimized inductor samples for the continuous load scenario. The TCO is less sensitive to changes in the proportions, however the samples that provide best performance in pulse load scenarios (circles) show considerably higher TCO than the samples with optimum proportions.

	pulse scenario	continuous scenario
core material	EPCOS N27	EPCOS N27
strand diameter	$50\,\mu m$	$50\,\mu m$
efficiency	98.8 %	99.9 %
material cost	23 EUR	200 EUR
energy cost	22 EUR	96 EUR

Figure 12. Details of the optimum inductors for pulse- and continuous load scenarios

For applications that require high peak load, but only low average load inductor geometries having big air gaps provide minimum TCO. Contrary to continuous load scenarios in which inductors with rather small air gap perform best. TCO optimization allows to find the optimum volume of an inductor, i.e. to find the best compromise between high power density and high efficiency.

REFERENCES

[1] R. Burkart and J. Kolar, "Component cost models for multi-objective optimizations of switched-mode power converters," in *Proceedings of the IEEE Energy Conversion Congress and Exposition (ECCE USA)*, 2013.

[2] P. Wallmeier, "Automatisierte Optimierung von induktiven Bauelementen für Stromrichteranwendungen," Ph.D. dissertation, University of Paderborn, 2001.

[3] A. Brockmeyer, "Dimensionierungswerkzeug für magnetische Bauelemente in Stromrichteranwendungen," Ph.D. dissertation, RWTH Aachen, 1997.

[4] C. Sullivan, "Optimal choice for number of strands in a litz-wire transformer winding," *IEEE Transactions on Power Electronics*, vol. 14, no. 2, pp. 283–291, 1999.

[5] J. Muhlethaler, J. Kolar, and A. Ecklebe, "Loss

modeling of inductive components employed in power electronic systems," in *Proceedings of the 8th IEEE International Conference on Power Electronics and Exhibition (ICPE ECCE Asia)*, 2011, pp. 945–952.

[6] X. Nan and C. Sullivan, "An improved calculation of proximity-effect loss in high-frequency windings of round conductors," in *Proceedings of the 34th annual Power Electronics Specialist Conference (PESC)*, vol. 2, June 2003, pp. 853–860 vol.2.

[7] ——, "Simplified high-accuracy calculation of eddy-current loss in round-wire windings," in *Proceedings of the 35th annual Power Electronics Specialists Conference (PESC)*, vol. 2, June 2004, pp. 873–879 Vol.2.

[8] Pack Feindrähte, "Rupalit Classic Specifications," http://www.pack-feindraehte.de/de/index.html, 2014, [Online; accessed 24-March-2014].

[9] EPCOS, "EPCOS N27 and N87 properties," http://www.epcos.com/ferrites, 2014, [Online; accessed 24-March-2014].

[10] Kaschke, "Produktkatalog," http://www.kaschke.de/, 2010, [Online; accessed 24-March-2014].

[11] Vacuumschmelze, "VAC Vitroperm 500F properties," http://www.vacuumschmelze.de/, 2014, [Online; accessed 24-March-2014].

[12] Metglas, "Metglas 2605SA1 properties," http://www.metglas.com/, 2014, [Online; accessed 24-March-2014].

[13] Micrometals, "Micrometals Power Conversion Catalog," http://www.micrometals.com/, 2014, [Online; accessed 24-March-2014].

[14] Magnetics, "Magnetics Powder Core Catalog," http://www.mag-inc.com/, 2012, [Online; accessed 24-March-2014].

[15] K. Venkatachalam, C. Sullivan, T. Abdallah, and H. Tacca, "Accurate prediction of ferrite core loss with nonsinusoidal waveforms using only steinmetz parameters," in *Proceedings of the IEEE Workshop on Computers in Power Electronics*, 2002, pp. 36–41.

[16] *Heat Transfer – Thermal Management of Electronics*. CRC Press, 2009.

978-1-4799-2706-7/14 $31.00 © 2014 IEEE

Initial Position Estimation for IPMSMs Using Comb Filters and Effects on Various Injected Signal Frequencies

Toshiki Suzuki
Department of Electrical Engineering
Chubu University
1200 Matsumoto-cho Kasugai Aichi 487-8501, Japan

Masaru Hasegawa
Department of Electrical Engineering
Chubu University
1200 Matsumoto-cho Kasugai Aichi 487-8501, Japan
mhasega@isc.chubu.ac.jp

Mutuwo Tomita
Department of Electrical and Computer Engineering
Gifu National College of Technology
2236-2 Kamimakuwa Motosu Gifu 501-0495, Japan

Shinji Doki
Department of Electrical Engineering and Computer Science
Nagoya University
Furo-cho, Chikusa-ku, Nagoya, 464-8601, Japan

Abstract—This paper proposes fast initial position estimation of IPMSMs (Interior Permanent Magnet Synchronous Motors), which is based on intentional pulse voltage injection and comb filters. Most initial position estimation methods utilize the amplitude or differential values of high-frequency currents and the performance of sensorless control depends on the algorithm to reconstruct these values. Although Fourier series expansion which requires some Low-Pass Filters(LPFs) is utilized in general, the filters degrade the response of position estimation. This paper focuses on the algorithm for initial position estimation of IPMSM at standstill, and proposes a new algorithm using comb filters which can rapidly calculate amplitude values of high-frequency currents and can improve the initial position estimation performance. In addition, this paper experimentally shows the performances of the initial position estimation, starting characteristics of sensorless control and characteristics of sensorless control including standstill under the load.

Keywords—IPMSM, sensorless control, initial position estimation, pulse voltage injection, comb filter

I. INTRODUCTION

The position sensorless control of Interior Permanent Magnet Synchronous Motors (IPMSMs) has already been utilized for many applications from the view points of the reliability, the increases of torque-volume density. Generally, position estimation in the middle and high-speed regions is mostly based on the extended electromotive force (EEMF)[1] and the various flux definitions[2][3][4]. On the other hand, high-frequency signal injection is a common strategy that can be utilized for drives in low-speed regions including standstill. Various position estimation methods in the low-speed regions including standstill have been studied so far. Most of these methods utilize high-frequency signal injection to estimate rotor position of the motor with magnetic-saliency. These signals for the saliency excitation are roughly classified into two types: sinusoidal voltage injection[5][6][7], and

pulse voltage injection[8][9][10][11]. In general, the frequency of the sinusoidal voltage is usually much lower than the frequency of the pulse voltage. In the case of the sinusoidal voltage injection, these injected frequency components are close to the bandwidth of the auto current regulator (ACR). Hence, the lower-response frequency of the ACR is required to be designed at the sacrifice of current control performances. Band stop filters (BSFs) and band pass filters (BPFs) are often used to eliminate and extract the injected frequency components, respectively, which deteriorates the transient performance of current controls[12]. On the other hand, in the case of the pulse voltage injection, the higher frequency can be selected than the bandwidth of ACR and therefore the above problems can be avoided. From this viewpoint, the pulse voltage injection approach would be desirable.

In the position estimation based on signal injection, the current responses for the injected voltage needs to be demodulated from the detected stator currents. The demodulation algorithm is classified into two approaches from the viewpoint of signal processing: the differential value[10] and the amplitude value of high-frequency components[5]. In the case using the differential value, noise in the sampled data would cause inaccurate position estimation, and the case using the amplitude value requires some low-pass filters (LPFs) for the amplitude calculation, which degrade the response of position estimation. Hence, fast initial position estimation would be impossible[11]. Although position estimation based on the amplitude value has the fatal problem in the position estimation response, these can avoid to fail position estimation due to noise. Therefore, it is an attractive effort to realize faster position estimation based on this approach.

From the above discussion, the combination of the high-frequency pulse voltage injection and the demodulation algorithm based on amplitude value is superior to other approaches in terms of the operation reliability.

For this problem, the authors have already proposed the initial position estimation method of IPMSM using comb filters which can solve the fatal problem of the demodulation. The relationship between the response of initial position estimation and the injected frequency, however, has not been revealed[13]. This paper experimentally shows the performances of the initial position estimation with various injected signal frequencies based on the proposed algorithm, starting characteristics of sensorless control and characteristics of sensorless control including standstill under the load.

II. CONVENTIONAL METHODS FOR INITIAL POSITION ESTIMATION AND THEIR PROBLEMS

A. Position Estimation Method with High Frequency Voltage Injection[7]

The methods for the initial rotor position estimation are mostly based on high-frequency signal injection to induce magnetic-saliency. Therefore, the current trajectory for high-frequency voltage is elliptical and includes rotor position information in the amplitude values of high-frequency currents[7]. This section briefly reviews the algorithm based on the amplitude value of high-frequency currents[7] and points out its problems.

In [7], the initial position estimation is realized as the follows:

1) First, high-frequency pulse voltage is injected in the γ−axis shown in Fig.1, which is aligned with the estimated rotor position, and the stator currents are detected in given $\gamma - \delta$ axes.

2) As shown in Fig.1, the detected currents are transformed into two components on the $\gamma_m - \delta_m$ axes which are at 45° from the $\gamma-$ axis[7].

3) Next, I_{γ_m} and I_{δ_m}, the amplitude values of the high-frequency components in currents i_{γ_m} and i_{δ_m}, are obtained. In [7], the Fourier series expansion is carried out as the demodulation strategy, which requires some LPFs to detect the averaged amplitudes of current response for the injected frequency component.

4) According to Fig.1(a), the difference between I_{γ_m} and I_{δ_m} obviously indicates incomplete rotor position estimation. On the other hand, Fig.1(b) shows that the vector diagram under no rotor position estimation error. Consequently, converging $I_{\delta_m} - I_{\gamma_m}$ into 0 by the phase-locked-loop (PLL), the initial position estimation is finalized.

B. Problems of signal amplitude based approach

It should be noted that the response of position estimation depends on the algorithm used to reconstruct the amplitude values. Figure 2 shows a typical block diagram of amplitude value calculation algorithm based on Fourier series and several LPFs.

As pointed out in [7], the cut-off frequencies of LPFs should be lower enough than f_h. In general, the

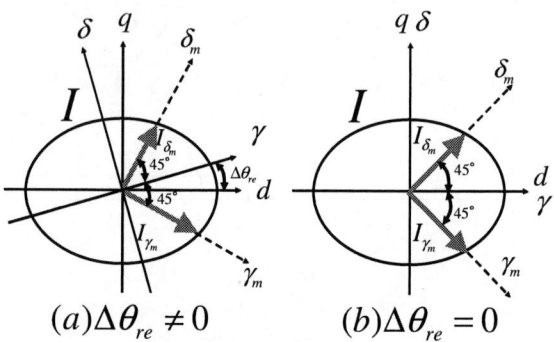

Fig. 1. High frequency current vector diagram

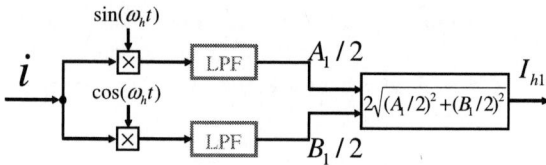

Fig. 2. Block diagram of amplitude value calculation[7]

low cut-off frequencies of LPFs cause large phase delay and restrict the control response. Namely, LPFs for the amplitude value calculation seriously degrades position estimation response. Therefore, an algorithm to obtain the amplitude values in high-frequency components without any LPFs is necessary to improve position estimation response.

III. PROPOSED ALGORITHM OF FAST AMPLITUDE CALCULATION USING COMB FILTERS

This section proposes a new algorithm using comb filters which can rapidly calculate amplitude value of high-frequency current, yielding to improve the performance of initial position estimation.

A. Frequency Characteristics of the Comb Filter

To estimate initial rotor position quickly, the high-frequency component in the detected stator current needs to be rapidly extracted. Namely, all one has to do is that the rapid DC component detection and rejection from the stator current. From the above, this subsection demonstrates the frequency characteristics of a comb filter, which is the basis of the proposed high performance in the initial position estimation.

A comb filter adds the input signal to its delayed signal. The transfer function of the comb filter is given by

$$H(z) = 1 + z^{-N_{delay}}. \qquad (1)$$

Hence, the gain characteristics of $H(e^{j\omega\tau})$ are expressed by

$$|H(e^{j\omega\tau})| = \sqrt{2 + 2\cos(\omega N_{delay}\tau)} \qquad (2)$$

The 2014 International Power Electronics Conference

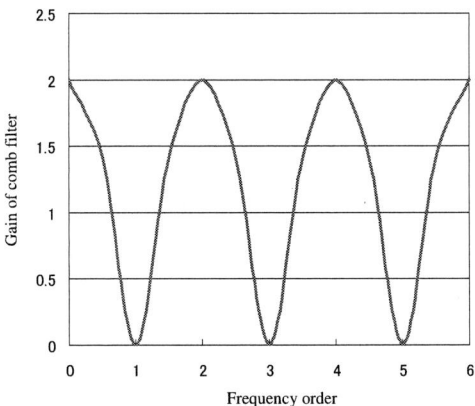

Fig. 3. Frequency characteristics of comb filter

Figure 3 illustrates the frequency characteristics of the comb filter from (2), of which the gain is repeatedly increasing and decreasing. It can be seen from the gain characteristics that DC component has perfectly passed and the odd-order harmonic components are completely rejected. In order to use these gain characteristics for DC component detection and calculate the amplitude value of high-frequency currents, it is desirable that the inputs of comb filters be in triangular waveform consisting of both the DC component and the odd-order frequency components. Consequently, it reveals that the comb filter is well-suitable for input signals with triangular waveform. As a result, the comb filter realizes that

1) The DC component can be easily extracted by this filter with less computational efforts, and
2) this filter has fewer phase delay characteristics, which would contribute faster initial position estimation.

B. Proposed Algorithm for Detection and Rejection of the DC Component Using Comb Filters

In this subsection, the proposed algorithm to detect and reject DC component is discussed in the time domain. In general, the current for the injected high-frequency pulse voltage flows as triangular waveform with DC component, $i = i_h + I_0$, as shown in Fig.4(a). Hence, the following DC component detection algorithm is feasible under assumption of this current waveform.

In order to extract I_0 rapidly, the following procedures are constructed:

1) The measured current $i = i_h + I_0$ is intentionally delayed for $N/2$ samples as shown in Fig. 4(a), where N is the sample number for the period of the high-frequency pulse voltage.
2) Using the comb filter in Fig.5, I_0 can be obtained by the average of both $z^{-\frac{N}{2}}(i_h + I_0)$ and $i_h + I_0$,
3) Finally, the AC component i_h can be extracted by subtracting the pre-obtained I_0 from the measured current i.

It should be noted that the proposed algorithm can detect the DC component only by the delay and the summation operations in comb filters. Therefore, the proposed algorithm can rapidly reject the DC component without any LPFs.

C. Proposed Algorithm for Amplitude Value Calculation Using Additional Comb Filters

Next, the amplitude value calculation of high-frequency current i_h is proposed. From Fig.4(b), the amplitude value of i_h obviously equals to the doubled DC component of $|i_h|$, so that the amplitude value of i_h can be obtained as the following three procedures:

1) the absolute value calculation of i_h,
2) the delay operation of $|i_h|$ for $\frac{N}{4}$ samples, and
3) the summation of $|i_h|$ and $z^{-\frac{N}{4}}|i_h|$ using the comb filter.

Figure 5 shows the block diagram of the proposed DC component rejection and amplitude calculation. As shown in this figure, the proposed demodulation algorithm is constructed without any LPFs which cause large phase delay and restrict the position estimation response. In addition, the proposed demodulation algorithm using comb filters can reconstruct the amplitude value with less computational efforts. As a result, faster position estimation response can be expected.

IV. EXPERIMENTS

This section demonstrates the experimental results to evaluate the performance of the position estimation based on the proposed algorithm with various injected frequencies and shows that the proposed algorithm can improve the performance of sensorless control including standstill.

A. Results of Initial Position Estimation with Various Injected Frequencies

Figure 6 shows the sensorless control system with the proposed position estimator based on the comb filters. In this figure, auto speed regulator (ASR) was stopped during a first initial position estimation period. The 1.5kW IPMSM with the concentrated winding was used as the test motor, of which parameters are shown in table I. The stator currents were detected by 12 bit A/D converter. The switching frequency of the PWM inverter was set at 10kHz, and the current control period and position estimation period were set $100\mu s$. PI gains of the position estimator were set K_p=0.5 and K_I=10000. The high-frequency pulse voltage was injected into the voltage reference in γ−axis. The amplitudes of the injected voltage were set at the values to flow high-frequency currents with same amplitude at each frequencies. Namely, the amplitude of high-frequency currents in the estimated synchronously rotating frame was 255mA, which is 2.94% and 1.33% of the motor rating and full scale of A/D converter, respectively, in all high-frequency condition at 417Hz, 500Hz, 625Hz, 833Hz, and 1250Hz. The auto

978-1-4799-2706-7/14 $31.00 © 2014 IEEE   909

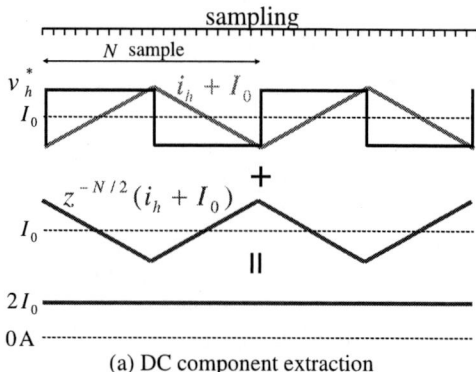

(a) DC component extraction

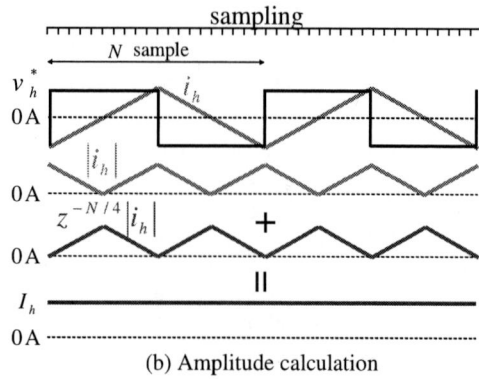

(b) Amplitude calculation

Fig. 4.　Algorithm to calculate amplitude value

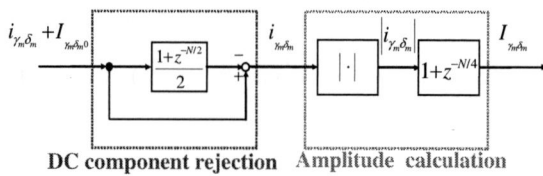

Fig. 5.　Block diagram of DC component rejection and amplitude calculation using comb filters

TABLE I.　PARAMETERS OF THE IPMSM

Parameters	Symbol	Value
Rated Power	P_n	1.5kW
Rated Speed	ω_{rm}	3600min^{-1}
Rated Line Voltage	V_n	200V
Rated Phase Current	I_n	5.0A
Winding Resistance	R	0.550Ω
d-axis inductance	L_d	8.31mH
q-axis inductance	L_q	14.8mH
Back-EMF Constant	K_E	0.230V.s/rad
Pole Pairs	P	2

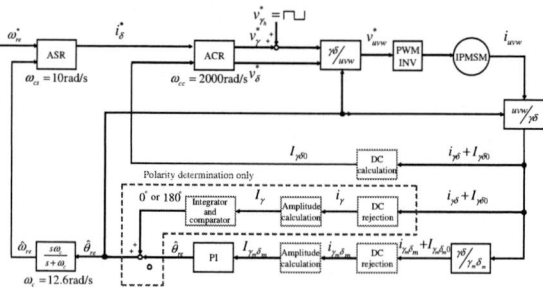

Fig. 6.　Block diagram of salient-pole position estimation and polarity determination

current regulator (ACR) was constantly designed for the response frequency of 2000rad/s in all conditions. The following experiments were for evaluation of the response of initial position estimation using comb filters at each injected frequencies. Figures 7 and 8 show the results of the amplitude calculation with the injected voltage of 19V-417Hz and 25V-1250Hz. It should be noted that the ACR was designed as the response frequency is 2000rad/s. Hence, the component of 417Hz is suppressed by ACR, on the other hand, the component of 1250Hz in the current remains. As shown in Fig.7, the result of amplitude calculation I_{γ_m} have large AC component, compared with the results of 1250Hz pulse voltage injection in Fig.8. This is because the high frequency currents i_{γ_m}, i_{δ_m} are far from triangular waveform by ACR. We have confirmed that, however, the average of the amplitude calculation is approximately equal to the amplitude of the high frequency current component.

Figures 9 and 10 show the initial position estimation performances with the injected frequencies of 417Hz and 1250Hz. In this figures, I_{γ_m} and I_{δ_m} illustrate the results of the amplitude calculation of i_{γ_m} and i_{δ_m}, respectively. θ_{re} is the rotor position, and was intentionally kept at -90°, $\hat{\theta}_{re}$ and $\hat{\theta}_{re} + 180$ mean the estimated positions with polarity ambiguity. Polarity determination based on the magnetic saturation in the $d-$axis begins in 20ms after the salient-pole position estimation sequence starts. It can be seen from Figs.9 and 10, the salient-pole position estimation is completed within 12.1ms and 3.9ms at $\theta_{re} = -90°$, respectively. which means that the response of salient-pole position estimation depends on the delay length N_{delay}, which is uniquely decided by the frequency of the injected voltage, f_h. The response of position estimation and acoustic noise caused by the voltage injection are in the relationship of trade-off.

Figure 11 shows the results of position estimation and the total time for the salient-pole position estimation time and the polarity determination time. As shown in Fig.11, the position estimation performance depends on the initial rotor position and the rotor position of ±90° requires the longest time for position estimation. The estimated position with the injected frequencies of 417Hz and 1250Hz almost agrees with the real rotor position and the error of position estimation was 7° at most. We confirmed, however, the AC components of estimated position with the injected frequencies of 417Hz is larger than that with 1250Hz.

Figure 12 shows the frequency dependence of the salient-pole position estimation time and the polarity determination time at the rotor position of ±90°, which requires the longest time for position estimation. It can

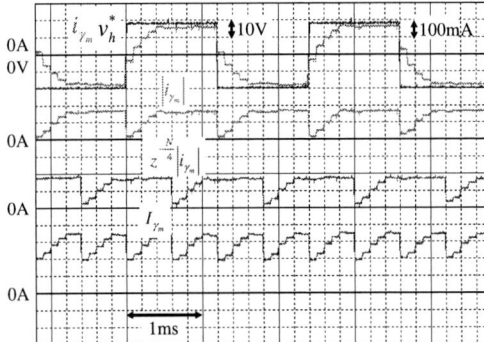

Fig. 7. Results of the amplitude calculation(f_h=417Hz)

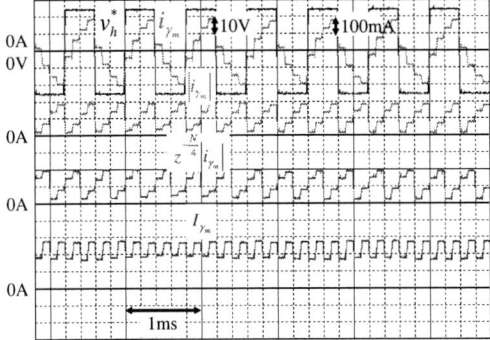

Fig. 8. Results of the amplitude calculation(f_h=1250Hz)

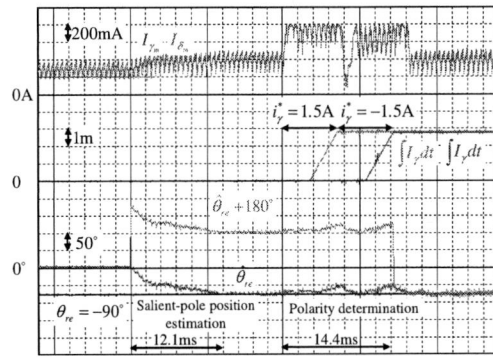

Fig. 9. Performance of initial salient-pole position estimation (f_h=417Hz)

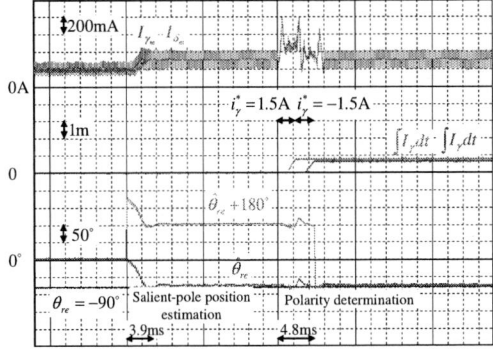

Fig. 10. Performance of initial salient-pole position estimation (f_h=1250Hz)

be seen from Fig.12 that the response of salient-pole position estimation depends on the delay length N_{delay}, which is uniquely decided by the frequency of the injected voltage, f_h. The time for the polarity determination is consistent with 6 periods of the injected high-frequency pulse voltage. Therefore, Fig.12 concludes that higher f_h is desirable in order to improve the response of initial position estimation.

B. *Starting Characteristics of Sensorless Control under No-load*

This subsection evaluates starting characteristics of sensorless control under the no-load in Fig.6. The amplitude and the frequency of the injected high frequency pulse voltage f_h were 19V-417Hz or 10V-1250Hz, respectively. PI gains of the position estimator were set K_p=0.5, K_I=2000 at 417Hz and K_p=0.5, K_I=10000 at 1250Hz, respectively, because the AC components of the estimated position with the injected frequencies of 417Hz was larger than that with 1250Hz. Estimated speed $\hat{\omega}_{re}$ used for speed control was obtained by pseudo differential operation of estimated position $\hat{\theta}_{re}$. The cut-off frequency used for the pseudo differential operation was set 5Hz(31.4rad/s) by trial and error. The speed commands were stepwise given as 0min^{-1} to 60min^{-1}(1.67% speed), 180min^{-1}(5.00% speed), 360min^{-1}(10.0% speed). The ASR was constantly designed for the response frequency of 10rad/s in all conditions. The current command in δ axis i_δ^* was

determined by ASR and the current command in γ axis i_γ^* was set 0A. Figures 13 to 17 show the results of speed step responses with various injected voltages. The results of the 0 to 360min^{-1}(10.0% speed) sensorless control with injected frequency 417Hz was not shown because it was impossible. In these figures, ω_{re}^*, ω_{re} and $\hat{\omega}_{re}$ are speed command, real speed and estimated speed, respectively, These results with injected frequency of 1250Hz lead us to the conclusion that position estimation is possible in a range of speed from standstill to the region which can switch to the estimation method based on EEMF and the various flux definitions.

C. *Characteristics of Sensorless Control Including Standstill under load*

In this subsection, characteristics of sensorless control with injected voltage 10V-1250Hz including standstill under the load was evaluated. IPMSM was rotated at 0min^{-1} or 100min^{-1}(2.78% speed) by sensorless control and then the vector controlled induction motor coupled with IPMSM applied 50% step loads to the IPMSM. Figures 18 and 19 show the characteristics of sensorless control including standstill under the load step. In these figures, τ is detected torque by torque sensor, ω_{re}^*, ω_{re} and $\hat{\omega}_{re}$ are speed command, real speed and estimated speed, θ_{re} and $\hat{\theta}_{re}$ are rotor position and estimated position, respectively. From Fig.18, the speed returns to

The 2014 International Power Electronics Conference

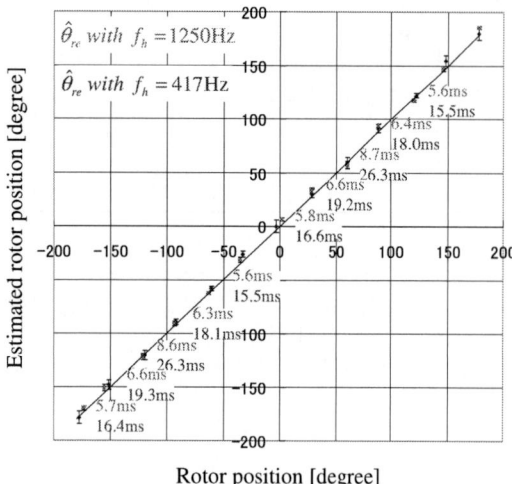

Fig. 11. Evaluation of position estimation for various initial rotor position

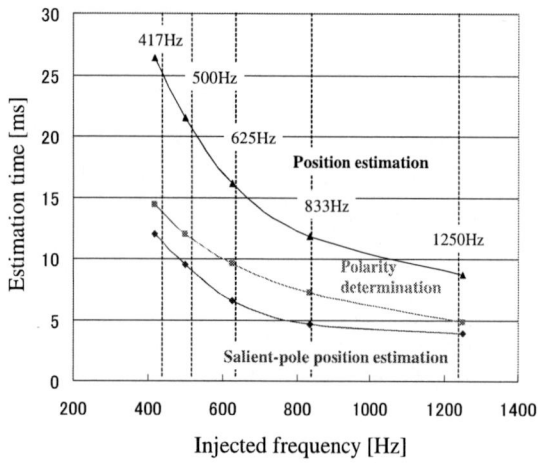

Fig. 12. Frequency dependence of initial position estimation

the speed command 0min^{-1} within 1.3 seconds although the IPMSM inversely rotates just after the load is applied to IPMSM. From Fig.19, the speed returns to the speed command 100min^{-1} within 1.8 seconds.

As a result of these experiments, it is found out that sensorless control including standstill under the load is possible. In this section, performance of initial position estimation with various injected frequencies, starting characteristics of sensorless control under no-load and characteristics of sensorless control including standstill under the load were evaluated. These results demonstrate that the proposed algorithm can improve the performance of sensorless control including standstill.

V. Conclusions

This paper has noted the problem of initial position estimation methods and has proposed the new algorithm using comb filters that can rapidly reconstruct amplitude

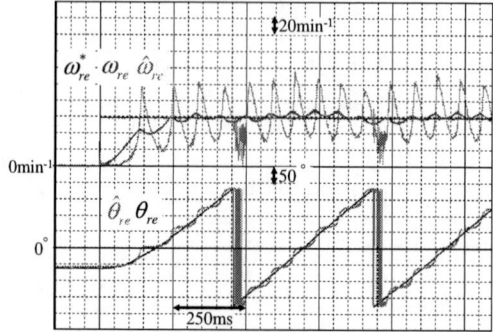

Fig. 13. Starting characteristics of sensorless control under no-load (f_h = 417Hz, 0 to 60min^{-1}(1.67% speed))

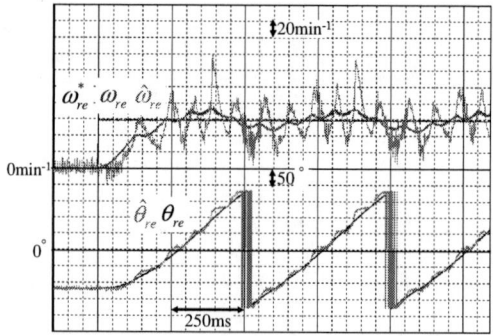

Fig. 14. Starting characteristics of sensorless control under no-load (f_h = 1250Hz, 0 to 60min^{-1}(1.67% speed))

values of high-frequency currents and can improve the response of initial position estimation. This paper is concluded as follows:

1) It is proposed that DC component rejection and amplitude calculation algorithm using comb filters have less phase delay and provide faster initial position estimation.

2) The response of salient-pole position estimation depends on the delay length N_{delay}, which is uniquely decided by the frequency of the injected voltage, f_h.

3) Higher f_h is desirable in order to improve the response of initial position estimation in the proposed method.

4) Position estimation is possible in a range of speed from standstill to the region which can switch the estimation method based on EEMF and the various flux definitions.

5) Sensorless speed control is realized superior in disturbance torque characteristic.

References

[1] Shinji Ichikawa, Zhiqian Chen, Mutuwo Tomita, Shinji Doki, and Shigeru Okuma: "Sensorless Controls of Salient-Pole Permanent Magnet Synchronous Motors Using Extended Electromotive Force Models", *IEEJ Trans. on Industry Applications*, Vol.122-D, No.12, pp.1088-1096 (2002) (in Japanese)

[2] Masaru Hasegawa and Keiju Matsui: "IPMSM Position Sensorless Drives Using Robust Adaptive Observer on Stationary Reference

978-1-4799-2706-7/14 $31.00 © 2014 IEEE   912

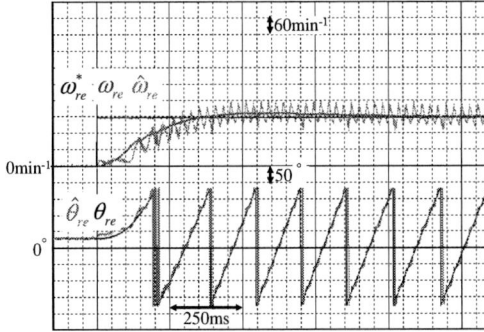

Fig. 15. Starting characteristics of sensorless control under no-load (f_h = 417Hz, 0 to 180min^{-1}(5.00% speed))

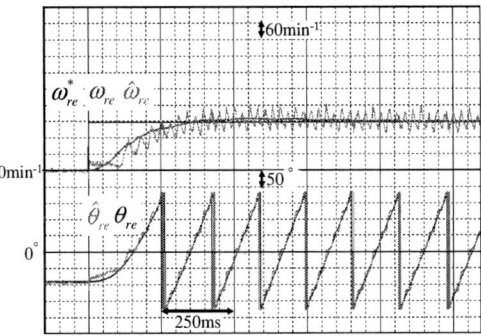

Fig. 16. Starting characteristics of sensorless control under no-load (f_h = 1250Hz, 0 to 180min^{-1}(5.00% speed))

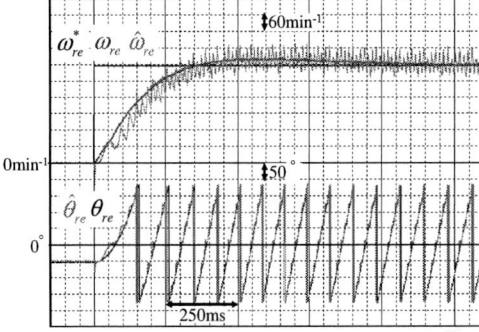

Fig. 17. Starting characteristics of sensorless control under no-load (f_h = 1250Hz, 0 to 360min^{-1}(10.0% speed))

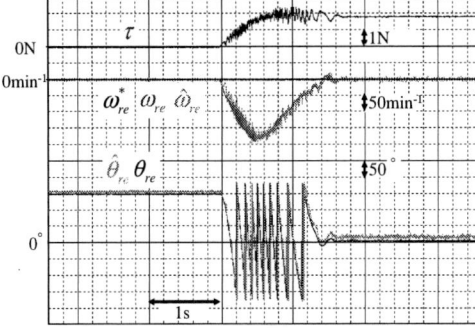

Fig. 18. Results of sensorless control under load impact (0min^{-1}, 50% load)

Frame", *IEEJ Trans. on Electrical and Electronic Engineering*, Vol.3, No.1 pp.120-127 (2008)

[3] Yasuhiro Yamamoto, Yasuhiro Yoshida, and Tadashi Ashikaga: "Sensor-less Control of PM Motor using Full Order Flux Observer", *IEEJ Trans. on Industry Applications*, Vol.124, No.8, pp.743-749 (2004) (in Japanese)

[4] Atsushi Matsumoto, Masaru Hasegawa, and Keiju Matsui: "Position Sensorless Control of IPMSMs Based on a Novel Flux Model Suitable for Maximum Torque Control", *IEEJ Trans. on Industry Applications*, Vol.132-D, No.1, pp.67-77 (2012) (in Japanese)

[5] Masato Ito and Yoshihiko Kinpara: "Direct Rotor-Position Estimation Method for Salient Pole PM Motor by Using High-Frequency Voltage", *IEEJ Trans. on Industry Applications*, Vol.131, No.6, pp.785-792 (2011) (in Japanese)

[6] Jung-IK Ha and Seung-Ki Sul: "Sensorless Field-Orientation Control of an Induction Machine by High-Frequency Signal Injection", *IEEE Trans. on Industry Applications*, Vol.35, No.1, pp.45-51 (1999)

[7] Sadayuki Sato, Hideaki Iura, and Kozo Ide: "Three Years of Industrial Experience with Sensorless IPMSM Drive based on High Frequency Injection Method", *Proc. of IEEE Int'l Sympo. on Sensorless Control for Electrical Drives*, P17 (2011)

[8] Takaharu Takeshita, Makoto Ichikawa, Nobuyuki Matsui, Eiji Yamada, and Ryoji Mizutani: "Initial Rotor Position Estimation of Sensorless Salient-pole Brushless DC Motor", *IEEJ Trans. on Industry Applications*, Vol.116, No.7, pp.736-742 (1996) (in Japanese)

[9] Sohji Murakami, Takayuki Shiota, Motomichi ohto, Kozo Ide, and Masaki Hisatsune: "Encoderless Servo Drive With Adequately Designed IPMSM for Pulse-Voltage-Injection-Based Position Detection", *IEEE Trans. on Industry Applications*, Vol.48, No.6, pp.1922-1930 (2012)

[10] Daigo Kaneko, Yoshitaka Iwaji, Kiyoshi Sakamoto, and Tsunehiro Endoh: "Initial Rotor Position Estimation of Interior Per-

manent Magnet Synchronous Motor", *IEEJ Trans. on Industry Applications*, Vol.123, No.2, pp.140-148 (2003) (in Japanese)

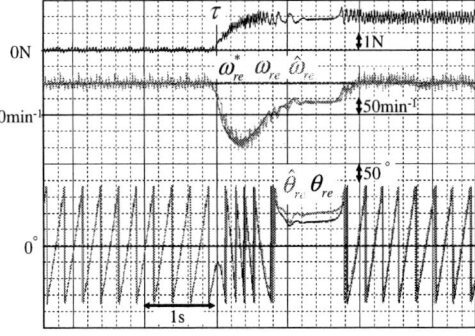

Fig. 19. Results of sensorless control under load impact (100min^{-1}, 50% load)

[11] Young-doo Yoon, Seung-ki Sul, Shinya Morimoto, and Kozo Ide: "High Bandwidth Sensorless Algorithm for AC Machines Based on Square-wave Type Voltage Injection", *IEEE Trans. on Industry Applications*, Vol.47, No.3, pp.2123-2130 (2011)

[12] Suk-Hwa Jung, Shinji Doki, Shigeru Okuma and Masami Fujitsuna: "Sensorless Control Method with Relatively Low Frequency Signal Injection for Low Acoustic Noise", in Proc. of EPE'11, 2011

[13] Toshiki Suzuki, Masaru Hasegawa and Mutuwo Tomita: "Fast Initial Position Estimation of IPMSMs Using Comb Filters" , *IEEJ Trans. of IA*, Vol.3, No2, pp.104-111 (2014)

Adaptive Signal Injection Method combined with EEMF based Position Sensorless Control of IPMSM Drives

Takumi Ohnuma, Yuki Makaino, and Ryoh Saitoh
Department of Digital Engineering
Numazu National College of Technology
Numazu city Shizuoka, Japan
ohnuma@numazu-ct.ac.jp, a12112@ccst.numazu-ct.ac.jp, d09119@ccst.numazu-ct.ac.jp

Abstract—This paper deals with an amplitude adjusting method for a signal injection in position sensorless controls of IPMSM (Interior Permanent Magnet Synchronous Motor) drives. Signal injection is necessary for position estimation at standstill and low speeds. Response signals to the injected signals have position information caused by a saliency of the IPMSMs. The position information appears in inductances depending on the rotor positions, therefore, time derivative of motor currents has to be measured directly or indirectly. The position estimation method discussing in this paper is based on an EEMF (Extended Electromotive Force) model. Although the EEMF model is mainly used in middle and high speed regions of IPMSM drives, it can be applied all speed regions by combining with signal injection methods. We apply a signal injection method to stabilize the estimation and reduce the effects of disturbances. Then the amplitude of the signal currents is adjusted so as to keep an enough amplitude of EEMF against disturbances. Signal settings become easier because the lower limit of EEMF can be adjusted against a degree of disturbances.

I. INTRODUCTION

Interior permanent magnet synchronous motors (IPMSMs) are popular in various applications such as industrial machines, home appliances and hybrid electric vehicles. Their high efficiency and compact designs become more and more important against the background of the world wide energy issues. Because rotor position informations are necessary to control IPMSMs properly, position sensors are required in high performance drives. However, position sensors require delicate wirings and are easy to break compared to the machines. Therefore, position sensorless controls of IPMSMs have been studied [1]–[7] to enhance reliability and also to reduce product cost. Position sensorless control techniques have been pull the widespread of IPMSM drive systems.

Existing position sensorless control techniques for IPMSMs can mainly be divided into two categories. In the middle and high speeds, electromotive force is available in order to estimate rotor positions [1], [2]. The electromotive force vector is always orthogonal to the main flux of the permanent magnet. The electromotive force can be estimated from the motor currents and voltages. On the other hand, in low speed regions, signal injections are necessary for position

estimation because the electromotive force decreases [3]–[5]. Response signals to the injected signals have position information caused by a saliency of the IPMSMs. The position information appears in inductances depending on the rotor positions, therefore, time derivative of motor currents have to be measured directly or indirectly. In generally, since an additional signal current causes a torque ripple, the signal frequency has to be set to higher than mechanical response of motor. If the signal frequency is much higher than mechanical response, it requires high response controllers and sensors. The measurement accuracy depends on sample timing and resolution of the current sensors. Moreover, losses, noise and torque ripple due to the signal currents need to be taken into account. For the reasons above, it has been difficult to give any specific guidelines for signal amplitude and frequencies. The settings are empirically adjusted for respective systems. These limitations disturb a spread of the sensorless drives of IPMSMs.

The position estimation method discussing in this paper is based on an EEMF (Extended Electromotive Force) model [6]. Although the EEMF model is mainly used in middle and high speed regions of IPMSM drives, it can be applied all speed regions by combining with signal injection methods [7], [8]. We apply a signal injection method to stabilize the estimation and reduce the effects of disturbances. In this paper, we design the signal injection method with two policies. First one is a low frequency signal which within a bandwidth of observer and current controller. As mentioned above, high frequency signal requires high response controllers and sensors. In addition, since the EEMF observer has a low-pass filter characteristic, high frequency signal injection is not suitable. Therefore, we use the signal frequency within several tens of Hz to several hundreds of Hz. Second one is a signal current adjusting method. Since a speed electromotive force is unavailable at low speed, position information of EEMF is easily affected by harmonic disturbances due to dead times, on-state voltages in switching devices, parameter errors of EEMF model, cogging torque and so on. In our designed method, the amplitude of the signal currents is adjusted so as to keep an enough amplitude of EEMF signals against disturbances. Signal settings become easier because the lower limit of EEMF can be adjusted against a degree of disturbances.

II. EEMF BASED POSITION SENSORLESS CONTROL

A. EEMF model

EEMF model [8] on the stationary reference frame (α-β axes) is written as (1),(2).

$$\mathbf{v} = (R + pL_q)\mathbf{i} + \mathbf{e} \qquad (1)$$

$$\mathbf{e} = \omega_{re}\mathbf{J}\boldsymbol{\lambda} - (L_q - L_d)\dot{i}_d \begin{bmatrix} -\sin\theta_{re} \\ \cos\theta_{re} \end{bmatrix} \qquad (2)$$

$$\boldsymbol{\lambda} = \{K_E - (L_q - L_d)i_d\} \begin{bmatrix} \cos\theta_{re} \\ \sin\theta_{re} \end{bmatrix} \qquad (3)$$

$$\mathbf{i} = [i_\alpha i_\beta]^T$$
$$\mathbf{v} = [v_\alpha v_\beta]^T$$
$$\mathbf{e} = [e_\alpha e_\beta]^T$$

Here, R is the stator resistance, L_d and L_q are d- and q-axis inductances, K_E is the EMF constant, p and "˙" represent time derivative, ω_{re} is the rotor angular frequency, and θ_{re} is the rotor position, respectively. In (3), $\boldsymbol{\lambda}$ is the extended flux and (2) is determined by differentiating $\boldsymbol{\lambda}$ with respect to time. In (2), $\mathbf{e}$ is the EEMF. Rotor position can be estimated from the phase information of EEMF with the following disturbance observer.

From the EEMF model, a linear state equation is derived as follows:

$$\begin{bmatrix} \dot{\mathbf{i}} \\ \dot{\mathbf{e}} \end{bmatrix} = \begin{bmatrix} -\frac{R}{L_q}\mathbf{I} & -\frac{1}{L_q}\mathbf{I} \\ \mathbf{0} & \omega_{re}\mathbf{J} \end{bmatrix} \begin{bmatrix} \mathbf{i} \\ \mathbf{e} \end{bmatrix} + \begin{bmatrix} \frac{1}{L_q}\mathbf{I} \\ \mathbf{0} \end{bmatrix}\mathbf{v} + \begin{bmatrix} \mathbf{0} \\ \mathbf{W} \end{bmatrix} \qquad (4)$$

where

$$\mathbf{W} = -(L_q - L_d)(\omega_{re}\ddot{i}_d\mathbf{J} - \ddot{i}_q) \begin{bmatrix} \cos\theta_{re} \\ \sin\theta_{re} \end{bmatrix} \qquad (5)$$

In (5), $\mathbf{W}$ is ignored as a modeling error ($\mathbf{W} \approx \mathbf{0}$) because $\mathbf{W}$ cannot be represented in the linear state equation.

From (4), a minimal order observer of EEMF model is given by the following equations:

$$\dot{\hat{\mathbf{i}}} = -\frac{R}{L_q}\mathbf{i} - \frac{1}{L_q}\hat{\mathbf{e}} + \frac{1}{L_q}\mathbf{v}$$
$$\dot{\hat{\mathbf{e}}} = \omega_{re}\mathbf{J}\hat{\mathbf{e}} + \mathbf{G}(\dot{\hat{\mathbf{i}}} - \dot{\mathbf{i}}) \qquad (6)$$

Here, "ˆ" represents estimated values in the observer, and G is a gain of the observer. The rotor position θ_{re} can be calculated by (7) as the phase angle of EEMF vector from the β-axis.

$$\hat{\theta}_{re} = \tan^{-1}\left(\frac{-\hat{e}_\alpha}{\hat{e}_\beta}\right) \qquad (7)$$

B. Position Information in the EEMF

The first term of EEMF in (2) is directly proportional to the rotor speed. On the other hand, the second term is generated by the differential of the d-axis current. In a steady state, the differential of the d-axis current does not occur. Therefore, we can generate this term by signal injecting to current. Even if the rotor speed is near zero, it gives the position information effectively for the position sensorless control. This property is useful for standstill and low speeds drives.

III. SIGNAL INJECTION METHOD

Since a speed electromotive force is unavailable at low speed, position information of EEMF is easily affected by harmonic disturbances due to dead times, on-state voltages in switching devices, parameter errors of EEMF model, cogging torque and so on. Therefore, it is necessary to keep the EEMF amplitude by signal injecting [9], [10]. However, the signal injection causes a torque ripple. To avoid an effect of the torque ripple, there are two approaches for separating the injection signal current from the fundamental excitation. First one is a high frequency injection method (separation of signal frequency), and second one is a signal injection method using an insensitive current component to the torque ripple (separation of current vector phase). The high frequency method is effective against the torque ripple. However, the observer to extract position information has low-pass filter characteristics. Therefore, it is not easy to detect the signal of small amplitude and a very high frequency through the observer. In other words, this method limits an effective operation range of sensorless control. In addition, high frequency signal injection causes a electromagnetic noise. In contrast, the signal injection method using an insensitive current component to the torque ripple makes it possible to use any frequency of signals. Lower frequency signals which within a bandwidth of the current controller are easier to generate, detect and avoid the carrier band. This method is based on a maximum torque control frame. However, the signal current adjusting method is not studied enough. In the following sections, we discuss the signal injection method based on maximum torque control frame, a relation between the EEMF and the signal current and the signal current adjusting method.

A. signal injection method based on maximum torque control frame

A maximum torque control frame, which is denoted as f-t axes, is defined a rotation angle ϕ from the d-q axes. The slopes of tangential lines along the constant torque curves. Figure 8 shows the definition of f-t axes. The t- and f-axes show a torque component and flux components, respectively. Therefore, we can easily make a insensitive current component to the torque ripple by using this control frame. The reference [9] reported that the signal injection to the f-axis current.

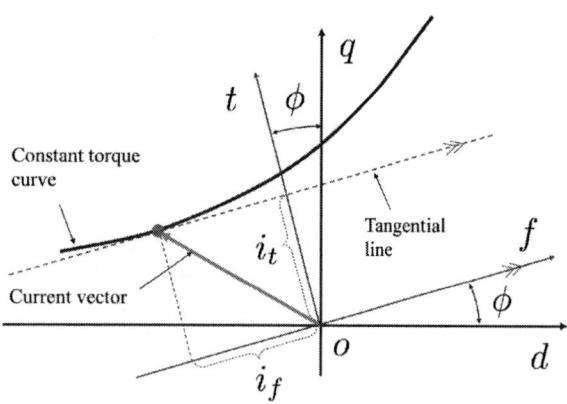

Fig. 1.   Definition of the Maximum Torque Control Frame.

B. The relation between EEMF and signal current

The current references based on maximum torque control frame are written as (8).

$$\begin{cases} i_f = I_i \sin \omega_i t \\ i_t = I_t. \end{cases} \tag{8}$$

To transform f-t axes currents to d-q axes currents, we obtain

$$\begin{cases} i_d = I_i \sin \omega_i t \cos \phi - I_t \sin \phi \\ i_q = I_i \sin \omega_i t \sin \phi - I_t \cos \phi. \end{cases} \tag{9}$$

Here, I_i is the signal amplitude, ω_i is the signal angular frequency and I_t is a current reference which is calculated from a torque reference. Assuming the rotation angle ϕ and the t-axis current is constant, we obtain

$$\dot{i}_d = \omega_i I_i \cos \omega_i t \cos \phi \tag{10}$$

To substitute (10) for (2), EEMF is represented as (11).

$$\mathbf{e} = \left\{ -E_{di}\cos\omega_i t\mathbf{I} + (E_0 - E_{qi}\sin\omega_i t)\mathbf{J} \right\} \begin{bmatrix} \cos\theta_{re} \\ \sin\theta_{re} \end{bmatrix} \quad (11)$$

where

$$\begin{cases} E_0 = \omega_{re}(K_E + (L_q - L_d)I_t \sin\phi) \\ E_{di} = \omega_i (L_q - L_d)I_i \cos\phi \\ E_{qi} = \omega_{re}(L_q - L_d)I_i \cos\phi. \end{cases} \tag{12}$$

Here, E_0 is the fundamental amplitude, the E_{di} and E_{qi} are d- and q-axes modulated signal amplitudes, respectively. The E_0 is direct proportional to the rotor speed. In contrast, the E_{di} and E_{qi} are generated by signal injection. Therefore, we can generate EEMF in any speed. In other words, we can obtain position information at standstill and low speed.

C. Signal current adjusting method

In the previous section, we showed the relation between EEMF and signal current. One of the simple adjusting method is to increase the signal current with rotor speed decrease. However, the EEMF amplitude changes with not only motor speed but also motor load. If the signal current does not appropriate, it causes some bad effects. In this section, we show the signal current adjusting method based on EEMF amplitude.

We consider a situation of EEMF amplitude becomes lower than a lower limit e_{min}. Then, we define a auxiliary EEMF e^*_{aux} as follows:

$$e^*_{aux} = e_{min} - \omega_{re}\{K_E + (L_q - L_d)I_t \sin\phi\} \tag{13}$$

This e^*_{aux} is the EEMF amplitude increment which is generated by signal injection. However, signal amplitude reference I^*_i has a upper limit. Therefore, it is necessary to set the upper limit of signal amplitude I_{max}. Considering these conditions, we obtain following equations.

$$I^*_i = \begin{cases} 0 & ; e^*_{aux} < 0 \\ \sqrt{e^*_{aux} \frac{2\sqrt{2}}{(L_q - L_d)\omega_i \cos\phi}} & ; 0 \leq e^*_{aux} < e_{amax} \\ I_{imax} & ; e_{amax} \leq e^*_{aux} \end{cases} \tag{14}$$

where

$$e_{amax} = I^2_{imax} \frac{(L_q - L_d)\omega_i \cos\phi}{2\sqrt{2}} \tag{15}$$

IV. EXPERIMENTS

A. System setup

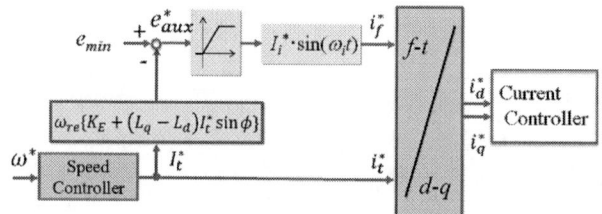

Fig. 2.   Block diagram of signal current adjusting method.

Figure 2 shows the block diagram of signal current adjusting method. The signal frequency ω_i of the f-axis current change is set to 100 Hz which is selected within a bandwidth of the current controller. These current references in the f-t axes are transformed into the d-q axes with the rotation angle ϕ. Controllers were executed with a digital signal processor (Texas Instruments: TMS320VC33). Bandwidth of PI current controllers are 2000 rad/s. Carrier frequency of the PWM inverter is set to 10 kHz, and the control period is set to 100 μs. Table I shows nominal parameters of the test IPMSM and which is used in the observer. The IPMSM is controlled with a sensorless vector control.

978-1-4799-2706-7/14 $31.00 © 2014 IEEE

TABLE I. MOTOR PARAMETERS

Number of pole pairs	3
Stator resistance	0.774 Ω
d-Axis inductance	8.90 mH
q-Axis inductance	11.96 mH
EMF constant	0.2284 V/rad/s
Rated current	6.3 A
Rated voltage	170 V
Rated speed	1750 r/min
Rated torque	8.2 N·m
Rated capacity	1.5 kw

B. Experimental results

Figure 3 shows the waveforms of the reference speed ω_{re}, amplitude of the estimated EEMF, amplitude of the signal current, position estimation error and W-phase current. The speed reference changes from 500 r/min to 0 r/min at a slope of 100 r/min/s. Both deceleration and acceleration are stable in a wide speed range including zero speed. The lower limit e_{min} is set to 10.0 V, the motor load is about 20 % of the rated torque and the upper limit of signal amplitude I_{max} is 4.0 A.

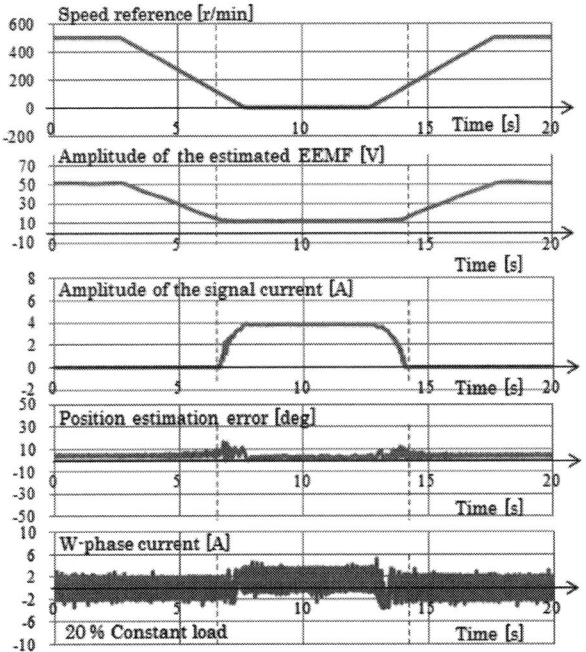

Fig. 3. The result of deceleration and acceleration test.

We verified that the signal amplitude automatically increases with decreasing EEMF amplitude. In addition, EEMF amplitude was kept higher than e_{min} even if motor speed becomes zero. The position estimation error increased temporarily when signal injection began. It is mainly caused by the disturbances such as dead time effect and zero crossing of the phase current. However, position sensorless control could

continue stably.

Moreover, we verified that the signal injection to the f-axis does not generate the torque ripple. We gave the current signal of 2.0 A and 100 Hz into the f-axis. The motor condition is 0 r/min and constant load. Figure 4 shows the waveforms of d- and q-axis currents and motor torque.

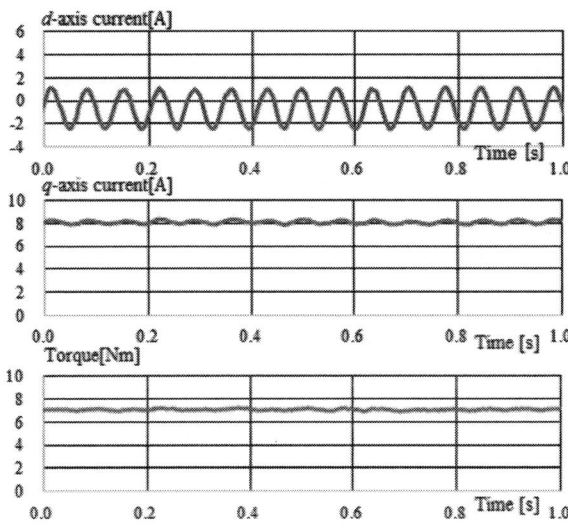

Fig. 4. The waveforms of torque and d- and q-axis currents with signal injection on f-t axes.

This result shows that the signal injection to the f-axis does not generate the torque ripple.

Then, we have carried out the sensorless drive test on all speed-torque(S-T) map. The upper limit of signal amplitude I_{max} is 4.0 A, the signal frequency is set to 300 Hz. Figure 5 shows the result of EEMF amplitude on S-T map.

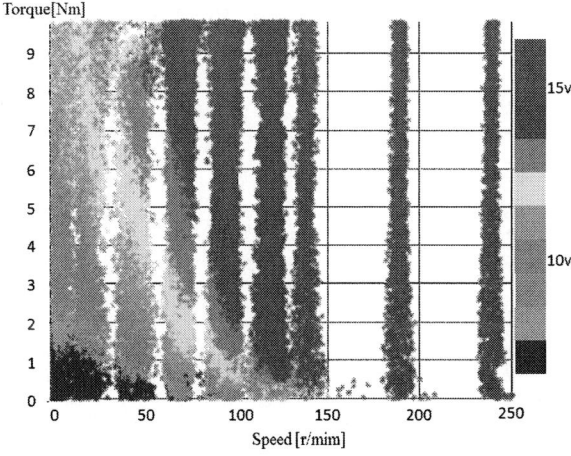

Fig. 5. The result of EEMF amplitude with the proposed method.

Our proposed method makes the sensorless drive possible

in all speed and 0% to 120% of rated load. However, EEMF amplitude becomes lower than EEMF lower limit at the light load and low speed. This is mainly caused by signal phase delay of current controller and parameter errors. The details of the errors will discuss in the future paper.

C. Conclusion

This paper presented the amplitude adjusting method for the signal injection in position sensorless controls of IPMSM drives. We described the EEMF models which is suitable for position sensorless controls combined with low frequency signal injection and the signal injection method based on maximum torque control frame. Then, we proposed signal current adjusting method. The validity of our proposed method for keeping EEMF amplitude was verified from experimental results.

REFERENCES

[1] L. Jones and J.Lang, "A state observer for the permanent-magnet synchronous motor", *IEEE Transaction on Industrial Engineering*, vol.IE-36, No.3, pp.374-382, 1989.

[2] R. Wu and G.Slemon, "A permanent magnet motor drive without a shaft sensor", *IEEE Transaction on Industrial Applications*, vol.IA-27, No.5, pp.1005-1011, 1991.

[3] P. L. Jansen and R. D. Lorenz, "Transducerless Position and Velocity Estimation in Induction and Salient AC Machines", *IEEE Trans. Industry Applications*, vol.31, No.2, pp.240-247, 1995.

[4] M. J. Corley, R. D. Lorenz, "Rotor Position and Velocity Estimation for a Salient-Pole Permanent Magnet Synchronous Machine at Standstill and high Speeds", *IEEE Trans. Industry Applications*, vol.34, No.4, pp.784-789, 1998.

[5] T. Noguchi, K. Takehana, and S. Kondo, "Mechanical-Sensorless Robust Control of Permanent-Magnet Synchronous Motor Using Phase Information of Harmonic Reactive Power", *IEEE Trans. Industry Applications*, vol.37, No.6, pp.1786-1792, 2001.

[6] Z. Chen, T. Mutuwo, D. Shinji and O. Shigeru, "An extended electromotive force model for sensorless control of interior permanent-magnet synchronous motors", *IEEE Transaction on Industrial Electronics*, vol.50, No.2, pp.288-295, 2003.

[7] T. Ohnuma, D. Shinji, O. Shigeru, "Extended EMF Observer for Wide Speed Range Sensorless Control of Salient-pole Synchronous Motor Drives", *in Proc. IEEE ICEM'10*, pp.1-6, 2010.

[8] T. Ohnuma, "Extended EMF model for Signal Injection Method", *in Proc. IAS Japan*, vol. 1-194, 2011 (in Japanese).

[9] T. Ohnuma, S. Doki and S. Okuma, "Signal injection method without torque ripple based on maximum torque control frame", *IEEJ Transaction on Electrical and Engineering*, vol. 8, no. 1, pp. 87-93, 2013.

[10] T. Ohnuma, S. Doki, S. Okuma, "Signal Injection Method without Torque Variation for Salient-pole Synchronous Motors", *in Proc. EPE'09*, paper 0465, 2009.

Study of Low Speed Sensorless Drives for SPMSM by Controlling Elliptical Inductance

Sari Maekawa
TOSHIBA CORPORATION FUCHU OPERATIONS
1, Toshiba-Cho, Fuchu-Shi, Tokyo, Japan 183-8511
sari1.maekawa@toshiba.co.jp

Toshifumi Hinata, Nobuyuki Suzuki
TOSHIBA CORPORATION YOKOHAMA COMPLEX
33, Shin-Isogo-Cho,Isogo-Ku,Yokohama-Shi, Kanagawa,
Japan 235-0017
nobuyuki17.suzuki@toshiba.co.jp

Hisao Kubota
Graduate School of Science and Technology,
Meiji University
1-1-1 Higashimita, Tamaku, Kawasaki, 214-8571, Japan
kubota@isc.meiji.ac.jp

Abstract— The rotor position sensor is required to control Permanent Magnet Synchronous Motor (PMSM). However, the rotor position sensor, for example, Resolvers or Pulse Generator (PG) have some disadvantages which are an increase of total volume of the motor, fault of power supply lines and signal lines of these sensors, and maintenance. To overcome these difficulties, sensorless control methods have been developed [1-10]. From standstill, Sensorless drive method for saliency PMSMs has been already proposed. However, applying these conventional techniques is not adequate to Surface Permanent Magnet Synchronous Motor (SPMSM) because of a few saliency of SPMSM. This paper proposes a low speed sensorless drive method for SPMSM which has small salient magnets. The low speed range is zero to 2-3% of rated speed. The study proposes the principle of anisotropy of the inductance of SPMSM generated by the d-axis current and magnetic saturation by the magnet flux.

Keywords— Sensorless, SPMSM, Saliency, Standstill.

I. INTRODUCTION

Recently, the demand of the position sensorless control is increasing in motor drive. In sensorless drive of the low-speed range which is zero to 2-3% of rated speed, various methods using the saliency of the rotator of PMSM are proposed[1]-[10]. However, SPMSM or Interior Permanent Magnet Synchronous Motor (IPMSM) with low salient pole ratio has a difficulty to detect a pole position by the saliency.

This paper proposes a new phase detection at low saliency of SPMSM. To achieve a control without position sensors, the paper focuses on a magnetic saliency which is increased by the magnetic saturation. Increasing $+I_d$ allows to cause a magnetic saturation. Also, this paper investigates following items and mention to detect phase information of driving motor.

(I)   Phase detection at low S/N ratio by AC signal conversion of $\sin 2\Delta\theta$

(II)   Elliptical shape of inductance by magnetic saturation

II. POSITION SENSORLESS CONTROL METHOD

A. Basic principle

Let assume to $d_c q_c$-axis $\Delta\theta$ is shifted from dq-axis shown in Fig. 1. A voltage amplitude V_h from three hundred Hz to1kHz is shown by Eq. (2).A frequency ω_h is on a $d_c q_c$-axis. θ is actual magnetic pole position and θ_c is the estimation position.

$$\theta = \theta_c + \Delta\theta \quad\quad\quad\quad\quad\quad\quad\quad\quad\quad\quad (1)$$

$$V_{dch} = V_h \cos \omega_h t$$
$$\quad\quad\quad\quad\quad\quad\quad\quad\quad\quad\quad (2)$$
$$V_{qch} = V_h \sin \omega_h t$$

When a rotation speed of SPMSM is low and ω_h is high, the high frequency current I_{qc} and I_{dc} is shown in Eq. (3). I_{qc} and I_{dc} draw the ellipse rotated by ω_h and its direction of a long-axis corresponds with d-axis. Although this ellipse is the form of current, the characteristic of dq-axis inductance causes it.

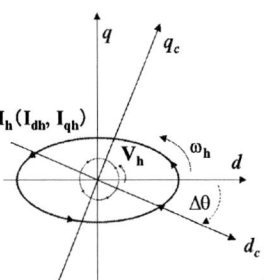

Fig. 1.   The high-frequency voltage and current in the $d_c q_c$-axis and dq coordinate.

$$I_{dch} = \frac{V_h}{2\omega_h L_d L_q} \{ (L_q + L_d)\sin\omega_h t$$
$$+ (L_d - L_q)\sin 2\Delta\theta \cos\omega_h t$$
$$+ (L_q - L_d)\cos 2\Delta\theta \sin\omega_h t \} \quad \cdots\cdots\cdots (3)$$
$$I_{qch} = \frac{V_h}{2\omega_h L_d L_q} \{ -(L_d + L_q)\cos\omega_h t$$
$$-(L_d - L_q)\sin 2\Delta\theta \sin\omega_h t$$
$$-(L_d - L_q)\cos 2\Delta\theta \cos\omega_h t \}$$

$\sin 2\Delta\theta$ indicates a magnetic pole position in Eq.(3). Eq.(3) is multiplied by $\cos\omega_h t$ and $\sin\omega_h t$. Eq.(3) shows a low-pass filter which has a cutoff-frequency of $2\omega_h$. These subtractions obtain the error signal Err_{dcqc} which is changed according to estimated error of 2θ.

$$E_{rr\,dcqc} = I_{qch\_LPF} - I_{dch\_LPF}$$
$$\quad\quad\quad\quad\quad\quad\quad \cdots\cdots\cdots\cdots\cdots (4)$$
$$= \frac{V_h}{\omega_h L_d L_q}(L_d - L_q)\sin 2\Delta\theta$$

Err may be controlled to zero and the position θ_c which is estimated [3].

B. Noise factor and influence in a low salient pole ratio

Eq.(4) is not assumed to have switching noise, quantization error of AD converter, mechanical eccentric error and unbalance error of tri-phase currents. If Err is high compared to detect errors, the estimated algorithm may be operated. However, in case of saliency ratio < 1.1, Err is too low and even if the amplitude V_h of high-frequency voltage is high, sufficient S/N ratio is not obtained. The error of frequency higher than the rotation speed, such as a switching noise, is controllable by low gain of position estimate PLL. However, at the low frequency error and the DC error are high, zero crossing does not occur according to 2θ. The estimation by PLL is difficult. Fig. 2 shows the waveform of Err of IPMSM at a salient pole ratio 1.1, and the angle θ is obtained by position sensor. θ_c of Eq. (1) is zero in Fig.2. Err_{xy} is a signal which changes synchronizing with the position θ as described in a Eq.(5).

$$E_{rr\,\alpha\beta} = I_{qch\_LPF} - I_{dch\_LPF}$$
$$\quad\quad\quad\quad\quad\quad\quad \cdots\cdots\cdots\cdots\cdots (5)$$
$$= \frac{V_h}{\omega_h L_d L_q}(L_d - L_q)\sin 2\theta$$

The paper proposes a method to detect a phase from Eq.(9) modeled as a single-phased signal. A signal of Eq.(9) is not connected to PLL.

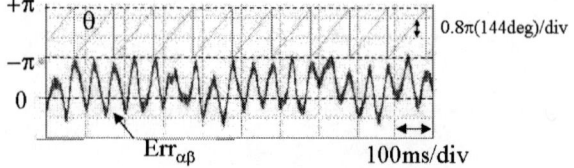

Fig. 2. Characteristics of Err when SN ratio of saliency is low.

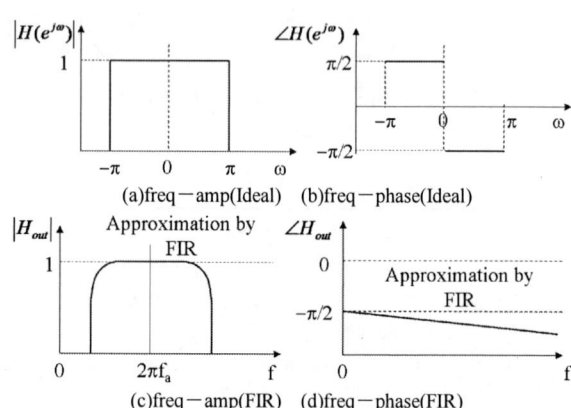

Fig. 3. Characteristics of the Hilbert transform.

C. Improvement in the S/N ratio by Hilbert transform

This study applies Hilbert transform for a phase detecting method of a single-phase AC signal in Eq.(6). Let $-j = e^{-j\pi 2}$ and $j = e^{j\pi 2}$ in Eq.(6) changes to the following equation.

$$H(e^{j\omega}) = \begin{cases} -j & : 0 < \omega < \pi \\ j & : -\pi < \omega < 0 \end{cases} \quad\cdots\cdots\cdots\cdots (6)$$

$$|H(e^{j\omega})| = 1 \quad\cdots\cdots\cdots\cdots\cdots\cdots\cdots (7)$$

$$\angle H(e^{j\omega}) = \begin{cases} -\dfrac{\pi}{2} & : 0 < \omega < \pi \\ \dfrac{\pi}{2} & : -\pi < \omega < 0 \end{cases} \quad\cdots\cdots\cdots (8)$$

Eq.(7) describes an amplitude characteristics. Eq.(8) describes a phase characteristics. Hilbert transform shifts a phase to $\pi/2$ of an input signal as a positive frequency component. The output amplitude is normalized to 1. Fig.3 (a) shows the amplitude characteristics of Hilbert transform filter, and (b) shows the phase characteristic. Fig.3 (c) and (d) are the filter characteristics expressed in discrete time (FIR filter). A horizontal axis is a linear scale and a phase characteristic has delay.

As H which is an output of Hilbert transform, low and high frequency component are removed. The 2nd outputs H_{out2} and H_{out1} are calculated by Eq.(9).

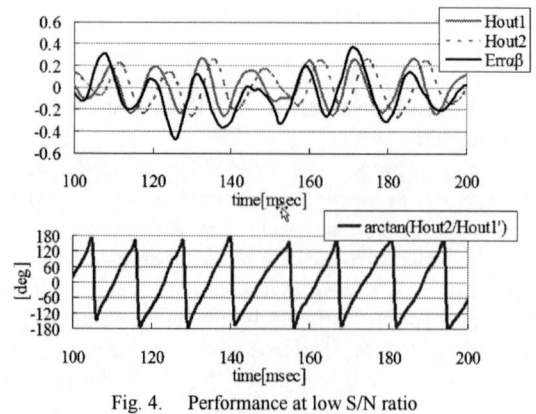

Fig. 4. Performance at low S/N ratio by Hilbert transform

$$2\theta = \arctan\left(\frac{H_{out2}}{H_{out1}}\right) \quad \dots\dots\dots\dots\dots\dots\dots\dots(9)$$

The phase 2θ of AC signal may be detected. Fig. 4 shows Err_{xy}, H_{out1}, H_{out2}, and a result of arctangent calculation. The S/N ratio of the input signal is low. Since the error of DC offset and low frequency component are removed by Hilbert transform, a phase of Err may be detected.

D. AC conversion for Err Component by Constant rotation axis

Eq.(5) is transformed by constant-axis($\alpha\beta$-axis). However, in case of motors at a standstill, since a frequency component of Err is out of frequency band of our proposed Hilbert transform. Therefore Hilbert transform is not able to apply to motors at a standstill. Hilbert transform of Eq.(5) cannot detect an AC signal in all frequency range. The proposed method applies Err_{xy} of Eq.(12) changed at θ_H defined by Eq.(10),(11). A position is estimated using Err_{xy} and the Hilbert transform filter.

$$\omega_H = \omega_c + 2\pi f_a \quad \dots\dots\dots\dots\dots\dots\dots\dots(10)$$

$$\theta_H = \int \omega_H dt \quad \dots\dots\dots\dots\dots\dots\dots\dots(11)$$

$$E_{rrxy} = I_{yh\_LPF} - I_{xh\_LPF} \quad \dots\dots\dots\dots\dots(12)$$
$$= \frac{V_h}{\omega_h L_d L_q}(L_d - L_q)\sin 2(\theta - \theta_H)$$

Fig. 5 shows dq-axis, $d_c q_c$-axis, and xy-axis transformed by θ_H. Although θ_H rotates at the speed ω_H, if it sees from the estimated axis-$d_c q_c$, it will rotate at the fixed speed $2\pi f_a$(in Eq.(10)). ω_c of Eq.(10) is rotation speed calculated from the position to estimate.

Fig. 6 shows the position θ, the estimated position θ_H, the position err $\Delta\theta$ and Err_{xy}. Fig. 5 (a) shows θ_H rotation information. Fig. 6 (b) shows Err where $\theta_H = 0$, or fixed-axis of $\alpha\beta$-axis. Since $\theta_c = \theta_H$ changed by θ_H, Err_{dcqc} of Eq.(4) is described by Err_{xy} of Eq.(12). Therefore, Eq.(12) is AC signal in the total frequency range including a standstill, and the frequency of $4\pi f_a$.

An amplitude of Eq.(12) is replaced by A in Eq.(15). As a result, H_{out1} and H_{out2} are described by Eq.(13) and Eq.(14) with a delay of Hilbert transform filter θ_d. The signal which processed delay θ_d of H_{out1} is H_{out1}. H_{out1} is generated with the filter, and the phase is only delayed

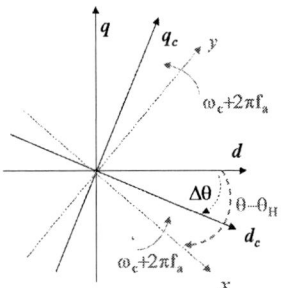

Fig. 5. Relationship dq-axis , $d_c q_c$ axis and xy-axis by θ_H.

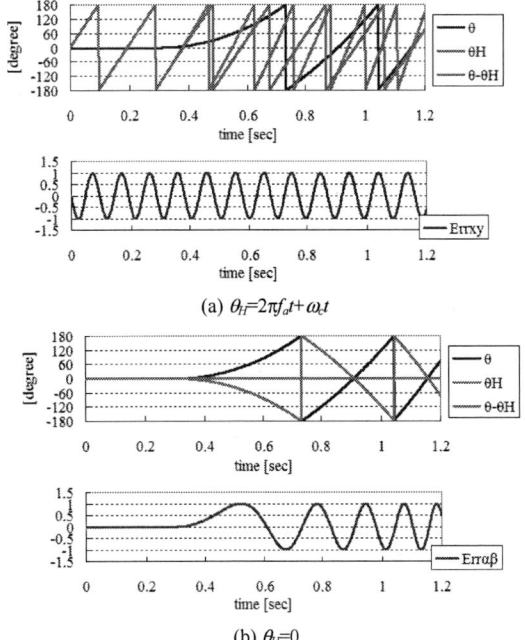

(a) $\theta_H = 2\pi f_a t + \omega_c t$

(b) $\theta_H = 0$

Fig. 6. Characteristic of Err as seen from θ_H.

without changing any amplitudes.

$$H_{out1} = A\cos\{2(\theta - \theta_H) + \theta_d\} \quad \dots\dots\dots\dots\dots(13)$$

$$H_{out2} = A\sin\{2(\theta - \theta_H) + 2\theta_d\} \quad \dots\dots\dots\dots\dots(14)$$

$$A = \frac{V_h}{\omega_h L_d L_q}(L_d - L_q) \quad \dots\dots\dots\dots\dots(15)$$

$$H_{out1}' = A\cos\{2(\theta - \theta_H) + 2\theta_d\} \quad \dots\dots\dots\dots\dots(16)$$

$$\theta_1 = \arctan\left(\frac{H_{out2}}{H_{out1}'}\right) \quad \dots\dots\dots\dots\dots(17)$$

As a result, the delay of H_{out1}' is the same $2\theta_d$ as H_{out2} by Eq.(16). And an angle θ_1 which includes a position is

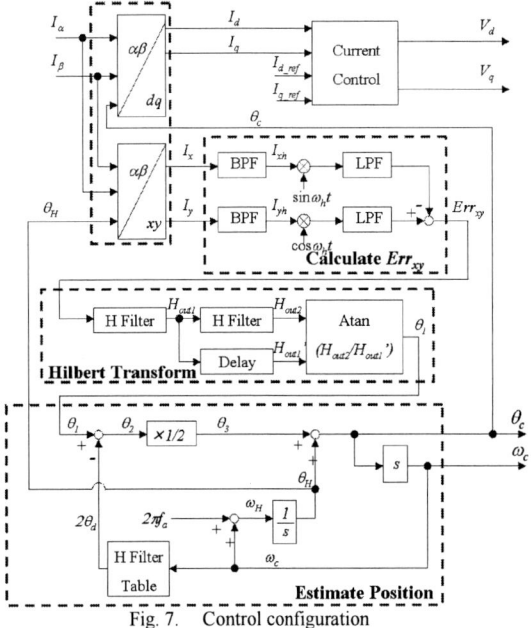

Fig. 7. Control configuration

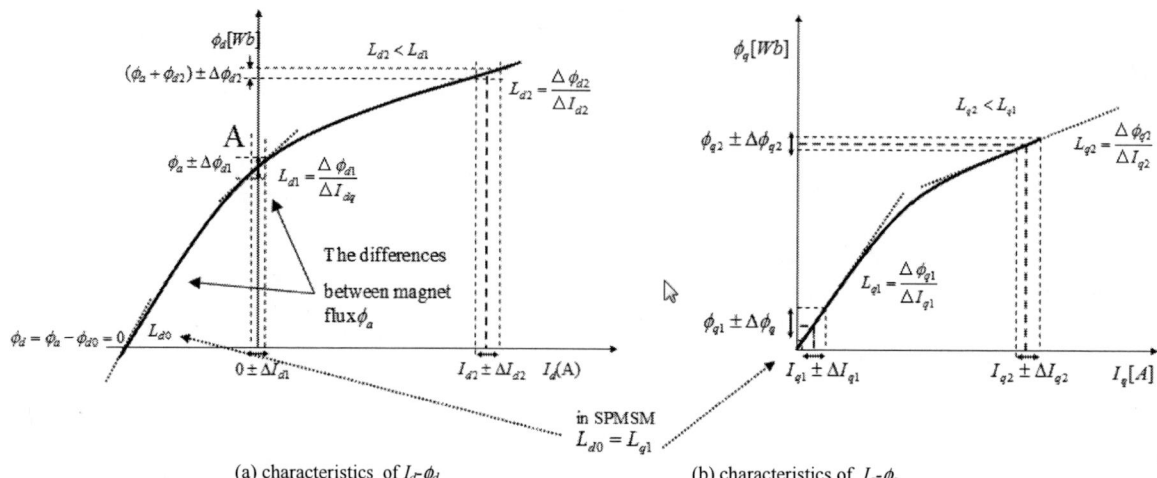

(a) characteristics of I_d-ϕ_d (b) characteristics of I_q-ϕ_q

Fig. 8.   Flux characteristics and current of SPMSM.

calculated by Eq.(17). Furthermore, the frequency is changed into 1/2 and the estimated position θ_c is calculated by adding θ_H. Fig.7 shows coordinate conversion and the composition of the Hilbert transform and the estimated part of pole position.

E.   Magnetic saliency of SPMSM

Geometrically, there is a few saliency in SPMSM, since arc shaped magnets are fixed on the surface of SPMSM. Theoretically, in case of SPMSM, L_d should be equal to L_q. However, when a magnetic saturation is occurred at d-axis magnetic circuit, L_d becomes increasingly likely to be smaller than L_q [3].

Fig.8 (a)(b) show characteristics of dq-axis magnetic flux ϕ_d and ϕ_q vs dq-axis-current I_d, I_q in SPMSM. A horizontal axis is dq-axis current I_d and I_q. The vertical axis is dq-magnetic flux ϕ_d and ϕ_q. The slope of the horizontal axis to a vertical axis is inductance.

When I_d does not flow, d-axis magnetic flux ϕ_d is generated by the magnetic force F of permanent magnets. By this magnetic flux, if a few saturation of an iron core occurs, the slope may be decreased at a point of A Fig.8 (a). For this reason, the slope of point A is smaller than the slope by the origin of Fig.8 (b) showing the q-axis characteristic. This causes the difference between L_{d1} and L_{q1} in a figure.

In the outer rotor type concentrated winding SPMSM shown in Fig. 9, the salient pole ratio at no-load is 1.03. In the SPMSM, Fig. 10 shows the high frequency current I_{dch}, I_{qch} of the Eq.(3) at the condition of impressing the

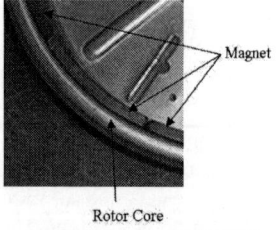

Fig. 9.   Rotor of SPMSM.

Fig. 10.   Elliptic shape at 300Hz drawn by current at no-load of test SPMSM.

high frequency voltage V_{dch}, V_{qch} of Eq.(2)

F.   Adjustments of magnetic saliency with magnetic saturation

A position is enable to be estimated with Err_{xy} of Eq.(12). However, testing SPMSM has a condition which a saliency is disappeared by the saturation of q-axis magnetic circuit at a full load. In this paper, magnetic saturation is intentionally generated by flowing $+I_d$, and the shape of an inductance ellipse is controlled. As a result, position sensorless control is realized by maintaining a suitable saliency.

The d-axis magnetic flux characteristics is shown in Fig. 8 (a). If d-axis current $+ I_d$ flow, magnetic saturation may progress and slope of d-axis magnetic flux ϕ_d may decrease. In case of IPMSM, if $+ I_d$ flows, torque/current ratio may decrease. However, in case of SPMSM, the torque/current ratio does not decrease rather than IPMSM.

Fig. 11 shows an elliptic shape and long-axis/short-axis ratio of high-frequency current when I_d and I_q are controlled with the position sensor of SPMSM in Fig. 11. At the origin, a salient pole ratio is 1.03. In contrast, when $+I_d$ is flowed, it will increase. Furthermore, the saliency has disappeared in the point of I_d= 0p.u, I_q=0.8p.u , and the point of I_d=0.2p.u ,I_q=0.2p.u. Thus, it is necessary to avoid the range to which a salient pole ratio including these two points decreases.

By the proposed method using Hilbert transform, when there are 1.03 or more salient pole ratios, a position can be estimated. This value was verified in an

The 2014 International Power Electronics Conference

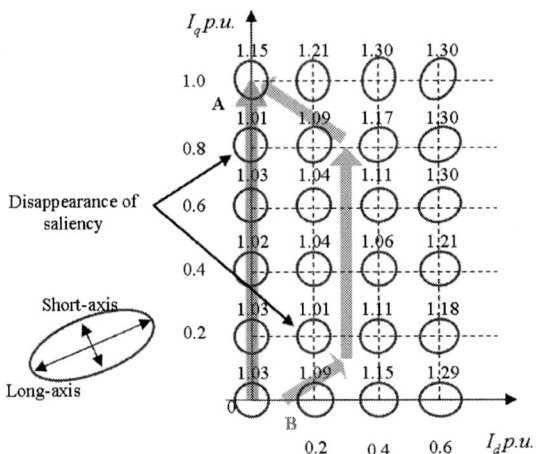

Fig. 11. Experimental results for saliency ratio of elliptic shape in $I_d>0$ area.

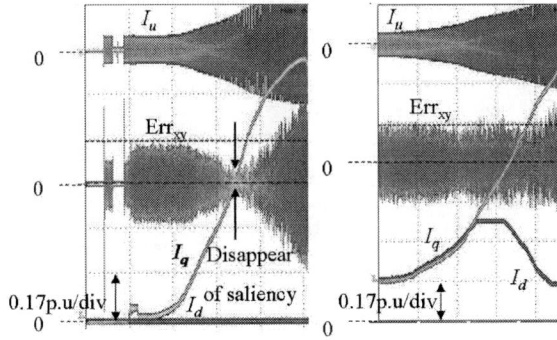

(a)$I_d = 0$ (b)+I_d (maintain a salient pole ratio 1.05)

Fig. 12. Amplitude characteristic of Err_{xy} to increase I_q.

experiment by SPMSM of Fig. 9.

The current command I_{d_ref}, I_{q_ref} shown by B of Fig. 11 is selected from the above verification. Fig. 12 (a) shows Err_{xy} when I_q is increased from 0p.u to 1.0p.u in the direction of A. Fig. 12.(b) shows the case of the direction of B. In Fig.12(a), the amplitude of Err_{xy} decreases to zero. It causes a loss of synchronism at a sensorless control. In contrast, in Fig. 12(b), the amplitude of Err_{xy} is maintained at a steady state.

G. Adjustments of inductance taking into consideration magnetic saturation and cross-coupling effects.

Fig.10 shows the inductance ellipse at no-load, and its direction of a long-axis corresponds with d-axis. However, the elliptical direction of a long-axis changes according to I_d and I_q. It has been reported that this cause is influence of magnetic saturation and cross-coupling[11]-[14].

The inductance of dq-axis is classified into eight sorts as shown in Eq. (18)-(21). As a result, dq-axis voltage equation can be expressed Eq.(22). ΔI_d and ΔI_q are a derivative value of current. $\Delta\phi_d$ and $\Delta\phi_q$ are a derivative value of magnetic flux.

I_{dh} and I_{qh} used for position estimate are greatly

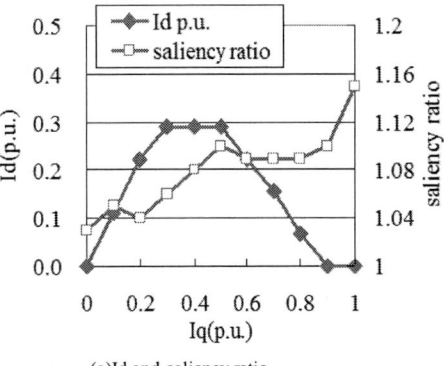

(a)Id and saliency ratio

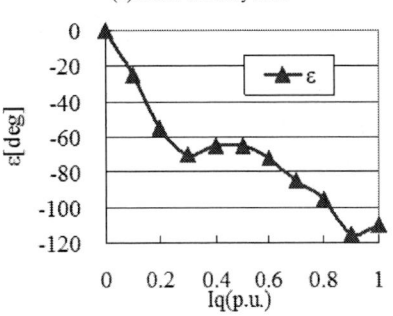

(b) Correction angle

Fig. 13. Correction I_d, saliency ratio and angle with respect to I_q.

influenced by derivative inductance L_{dd}', L_{qq}', L_{dq}', L_{qd}'.

<<dq-axis self inductances>>

$$L_{dd} = \frac{\phi_d}{I_d} \ , \ L_{qq} = \frac{\phi_q}{I_q} \quad\quad\quad\quad\quad\quad\quad\quad(18)$$

$$L_{dd}'= \frac{\Delta\phi_d}{\Delta I_d} \ , \ L_{qq}'= \frac{\Delta\phi_q}{\Delta I_q} \quad\quad\quad\quad\quad\quad(19)$$

<<dq-axis cross-coupling inductances>>

$$L_{dq} = \frac{\phi_d}{I_q} \ , \ L_{qd} = \frac{\phi_q}{I_d} \quad\quad\quad\quad\quad\quad\quad(20)$$

$$L_{dq}'= \frac{\Delta\phi_d}{\Delta I_q} \ , \ L_{qd}'= \frac{\Delta\phi_q}{\Delta I_d} \quad\quad\quad\quad(21)$$

$$V_d = RI_d + pL_{dd}'I_d + pL_{dq}'I_q - \omega\left(L_{qq}I_q + L_{qd}I_d\right)$$
$$V_q = RI_q + pL_{qq}'I_q + pL_{qd}'I_d + \omega\left(L_{dd}I_d + L_{dq}I_q\right) + \omega\phi_a$$
$$\quad\quad\quad\quad\quad\quad\quad\quad\quad\quad\quad\quad\quad\quad\quad(22)$$

Therefore, the slope of inductance ellipse can be described by the following equation using the derivative inductances.

$$\varepsilon = \arctan(\frac{L_{dq}'}{L_{dd}'-L_{qq}'}) \quad\quad\quad\quad\quad\quad(23)$$

The estimated position is adjusted using slope of ellipse. Fig.13 shows I_d to I_q selected by route-B of Fig.11, and the adjusted angle ε.

III. EXPERIMENTAL RESULTS

A. Direct start from a standstill

An experimental condition is shown in Table.1. Fig. 14

978-1-4799-2706-7/14 $31.00 © 2014 IEEE   923

Table 1. Experimental parameters

Item	Value
Rating output [W]	300
Inverter DC link voltage [V]	200
Switching frequency [kHz]	5.0
Resolution of the current detection	4p.u./12bit (0.001p.u/LSB)
Amplitude of the high-frequency current	0.1p.u.
Frequency of the high-frequency current[Hz]	300
Pole pair number of motor	24

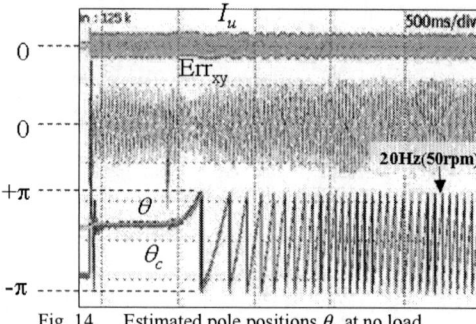

Fig. 14. Estimated pole positions θ_c at no load.

shows the test result of proposed method at a driving start of SPMSM. The position θ, estimate position θ_c, I_u, and Err_{xy} are shown in Fig.14. High frequency voltage V_h is expressed so that high frequency current may become 0.1p.u. The starting characteristic of Fig. 14 shows SPMSM is under controlled with our proposd methods . The salient pole ratio is 1.03, and its Err_{xy} is proportional to L_q-L_d

B. Characteristics of position estimation at full load

Fig.15 shows Err_{xy}, an angle error $\Delta\theta$, and I_d, I_q at full load. Although the elliptical angle is adjusted by I_d, An estimated error $\Delta\theta$ is 15 degrees. In SPMSM, since only magnet torque is outputted, torque is 96.6%.

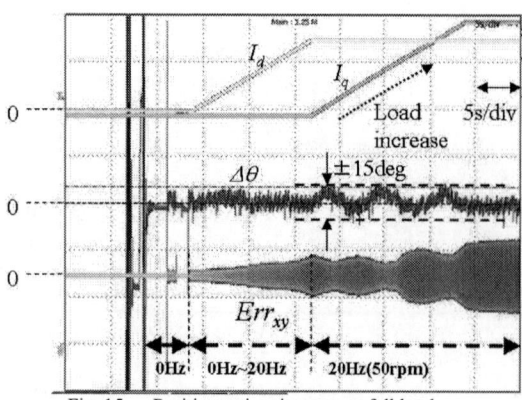

Fig. 15. Position estimation error at full load.

IV. CONCLUSION

This study achieves to detect a pole position of SPMSM with a low saliency at low speed ranges.

(I) SPMSM pole position is able to be estimated at a minimum salient pole ratio 1.03 by transforming AC of sin2$\Delta\theta$.

(II) A salient pole ratio of SPMSM is able to be adjusted in accordance with a magnetic saturation by +I_d.

As a result, SPMSM is enable to be driven at a low speed without any position sensors at full load. By adjusting +I_d , this propose method may be applicable to different salient pole ratio.

REFERENCES

[1] T. Yokoyama, H. Kubota: "Pole Position Estimation at Low Speed for High-Speed and Low Saliency Ratio PMSM Using One Third Carrier Frequency Component", *T. IEE Japan, Industry Applications*(12-01) , SPC11-177 (2011) (in Japanese)

[2] S. Maekawa, K. Nagai, Y. Hasegawa: "Positional detection method of low salient ratio IPMSM", *Annual Conference of IEE Japan*, Vol. 4, No. 98, p. 165 (2009) (in Japanese)

[3] Y. Nakazawa: "A Position Sensorless Control for Permanent Magnet Reluctance Motor", *The Papers of Technical Meeting IEE Japan*, Vol. VT-02, No.1-13, pp.67-72(2002)

[4] M. Schroedl: "Sensorless Control of AC Machines at Low Speed and Standstill Based on the "INFORM" Method", *1996 IEEE Industry Applications Society Annual Meeting*, pp.270-277(1996)

[5] S. Ostlund, M. Brokemper: "Sensorless rotor-position detection from zero to rated speed for an integrated PM synchronous motor drive", *IEEE Trans. Industry Applications*, Vol. 32, No. 5, pp.1158-1165(1996)

[6] T. Aihara, A. Toba, T. Yanase, A. Mashimo, K. Endo: "Sensorless torque control of salient-pole synchronous motor at zero-speed operation", *IEEE Trans. Power Electronics*, Vol. 14, No. 1, pp.202-208(1999)

[7] N. Bianchi, S. Bolongnani, J. Ji-Hoon, S. Seung-Ki: "Comparison of PM Motor Structures and Sensorless Control Techniques for Zero-Speed Rotor Position Detection", *IEEE Trans. Power Electronics*, Vol. 22, No. 6, pp.2466-2475(2007)

[8] W. Chuanyang, X. Longya: "A novel approach for sensorless control of PM machines down to zero speed without signal injection or special PWM technique", *IEEE Trans. Power Electronics*, Vol. 19, No. 6, pp.1601-1607(2004)

[9] S. Ogasawara, T. Matsuzawa, H. Akagi: "A Position-Sensorless IPM Motor Drive System Using a Position Estimation Based on Magnetic Saliency", *T. IEE Japan*, Vol.118-D, No.5, pp.652-660 (1998) (in Japanese)

[10] D. Kaneko, Y. Iwaji, K. Sakamoto, T. Endo: "Initial Rotor Position Estimation of Interior Permanent Magnet Synchronous Motor", *T. IEE Japan*, Vol.123-D, No.2, pp.140-148 (2003) (in Japanese)

[11] E. Armando, P. Guglielmi: "Accurate Modeling and Performance Analysis of IPM-PMASR Motors", *IEEE Trans. Industry Applications*, Vol. 45, No. 1, pp.123-130(2009)

[12] B. Stumberger, G. Stumberger, D. Dolinar: "Evaluation of Saturation and Cross-Magnetization Effects in Interior Permanent-Magnet Synchronous Motor", *IEEE Trans. Industry Applications*, Vol. 39, No. 5, pp.1264-1271(2003)

[13] G. Armandoz, J. Poza, M. Rodriguez: "Modeling of Cross-Magnetization Effect in Interior Permanent Magnet Machines", *Electrical Machines, 2008. ICEM 2008. 18th International Conference on*, pp.1-6(2008)

[14] Y. Kano, T. Kosaka, N. Matsui, T. Nakanishi: "Sensorless-Oriented Design of Concentrated-Winding IPM Motor for General Industrial Applications", *T. IEE Japan*, Vol.130-D, No.2, pp.119-128 (2010) (in Japanese)

Suppression of Injection Voltage Disturbance for High Frequency Square-Wave Injection Sensorless Drive with Regulation of Induced High Frequency Current Ripple

Dongouk Kim, Yong-Cheol Kwon, and Seung-Ki Sul
School of Electrical & Computer Engineering
Seoul National University
Seoul, Korea
kdw@eepel.snu.ac.kr, dydcjfe@eepel.snu.ac.kr, sulsk@plaza.snu.ac.kr

Jang-Hwan Kim and Rae-Sung Yu
Manufacturing Technology Center
SAMSUNG ELECTRONICS Company, Ltd
Suwon, Korea
janghwane.kim@samsung.com, rs.yu@samsung.com

Abstract—**In square wave voltage signal injection sensorless drive, the injection voltage can be distorted by the inverter nonlinearity effects especially when the injection voltage is low. If that happens, High Frequency (HF) current signal which contains the rotor position information could be also distorted, which directly leads to an error in the position estimation. This paper analyzes the effects of the inverter nonlinearity to injection voltage, to induced current ripple, and to the position estimation performance in sequence and proposes a voltage injection method to minimize the impact of the inverter nonlinearity by the regulation of HF current ripple. By simulations and experiments, performance of the proposed method has been verified. The experimental results show 34.9% reduction of noise input in the position estimation and 19.7% improvement of the position estimation performance under 15% of rated voltage signal injection**

NOMENCLATURE

Superscript "s"	stationary reference frame.
Superscript "r"	rotor reference frame.
Superscript "*"	reference value.
θ_r	electrical rotor position.
$\hat{\theta}_r$	estimated value of θ_r.
$\tilde{\theta}_r$	rotor position estimation error, $\theta_r - \hat{\theta}_r$.
T_{sw}, f_{sw}	switching period and frequency.
T_{samp}, f_{samp}	sampling period and frequency.
T_h, f_h	signal injection period and frequency.
L_{ds}, L_{qs}	Synchronous inductances.

I. INTRODUCTION

Sensorless control techniques made it possible to realize accurate estimation of rotor position without additional position sensors. Especially at standstill and low speed, saliency based methods [1]-[5] are normally used. Since the rotor position determines spatial distribution of inductance, it can be estimated by injecting signal and analyzing its response. Through several researches and developments, the saliency based methods have been evolved to be compatible to low end sensored servo drives. The control bandwidth of recently reported square-wave signal injection sensorless drive could extend to 40Hz in the case of speed control loop and 10Hz in the case of position control loop [5].

But additional signal injection causes undesired effects such as acoustic noise, reduced voltage margin of inverter, and losses coming from induced current ripple. These problems can be relieved by reducing injection voltage magnitude, resulting in less induced current ripple. However, the inverter nonlinearity effects would be getting larger as the magnitude of the square wave is getting smaller. Since the inverter nonlinearity effects induce considerable error in the rotor position estimation [6]-[7], it is difficult to reduce the injection voltage under a certain limiting value.

In principle, the inverter nonlinearity effects are mainly determined by the switching sequence (on and off sequences) and phase current at switching instant. In Ref. [8]-[10], several compensation methods had been proposed, where compensation factors were calculated from the sampled current. But due to the time delay from the current sampling to the PWM switching of inverter which is about $1.5 \cdot T_{samp}$ [11], it is inappropriate to use the sampled current for the compensation under the square-wave voltage injection sensorless drive where the current varies rapidly. Ref. [12] proposed a compensation method that was done from the prediction of the current at the switching instant. But this algorithm is vulnerable to the variation of drive system parameters since the prediction depends on the parameters. Accurate prediction of the current at the switching instant during the square-wave injection is almost impossible. In Ref. [13], High Frequency (HF) current ripple coming from signal injection was processed to calculate and feed-forward HF component of voltage distortion. But, this method mainly focused on sinusoidal voltage injection whose frequency was in several hundreds of Hz range where T_h was at least 20 times of T_{samp}. In square-wave voltage injection sensorless drive where T_h is only 2 or 4 times of T_{samp}, feed-forwarding based on sensed and filtered current signal can be inaccurate due to the $1.5 \cdot T_{samp}$ time delay.

This paper proposes a voltage signal injection technique to cope with the degradation of position estimation performance caused by the inverter nonlinearity effects in square-wave signal injection sensorless drive. Contrary to the conventional voltage injection methods [10], the proposed method does not inject voltage with fixed magnitude, but varies the magnitude of the injection voltage to regulate the magnitude of the induced current ripple.. By the regulation, the noise component included in position information can conspicuously decrease under the constraint of the same rms value of the current ripple. In other words, the injection voltage can be decreased without degrading the sensorless control performance. The effectiveness of the proposed method is verified by series of experiments.

The 2014 International Power Electronics Conference

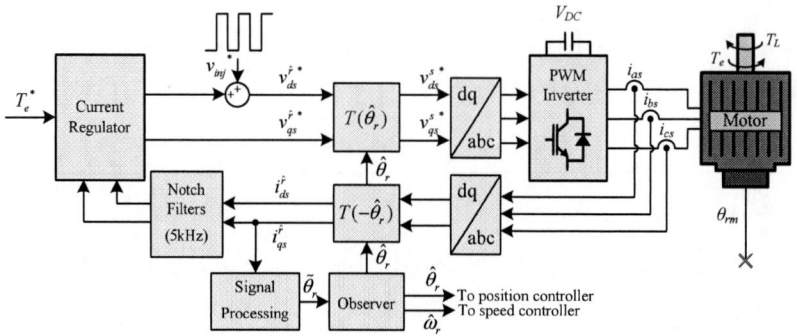

Fig. 1. Block diagram of conventional square-wave voltage injection sensorless control.

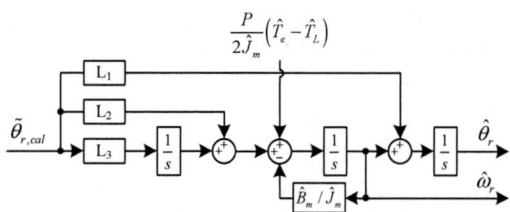

Fig. 2. Position and speed observer based on mechanical model.

II. POSITION ESTIMATION BY SQUARE-WAVE INJECTION

In square-wave injection sensorless control, square-wave voltage expressed as (1) is injected to estimated d-axis as shown in Fig. 1.

$$\begin{bmatrix} v^{\hat{r}}_{dsh} \\ v^{\hat{r}}_{qsh} \end{bmatrix} = \begin{bmatrix} v_{inj}^* \\ 0 \end{bmatrix}, \text{where } v_{inj}^* = \begin{cases} V_h, & \text{half duty} \\ -V_h, & \text{otherwise} \end{cases}. \quad (1)$$

Then the injection voltage in real rotor reference frame can be deduced as (2).

$$\begin{bmatrix} v^r_{dsh} \\ v^r_{qsh} \end{bmatrix} = T_{\tilde{\theta}_r} \begin{bmatrix} v^{\hat{r}}_{dsh} \\ v^{\hat{r}}_{qsh} \end{bmatrix} = \begin{bmatrix} \cos\tilde{\theta}_r & \sin\tilde{\theta}_r \\ -\sin\tilde{\theta}_r & \cos\tilde{\theta}_r \end{bmatrix} \begin{bmatrix} v_{inj}^* \\ 0 \end{bmatrix} = \begin{bmatrix} v_{inj}\cos\tilde{\theta}_r \\ -v_{inj}\sin\tilde{\theta}_r \end{bmatrix}. \quad (2)$$

And current ripple induced by injection voltage in estimated rotor reference frame can be calculated by (3)-(4).

$$\begin{bmatrix} \Delta i^r_{dsh} \\ \Delta i^r_{qsh} \end{bmatrix} = \int_t^{t+T_{samp}} \begin{bmatrix} L_{ds} & 0 \\ 0 & L_{qs} \end{bmatrix}^{-1} \begin{bmatrix} v^r_{dsh} \\ v^r_{qsh} \end{bmatrix} d\tau = T_{samp} \begin{bmatrix} \dfrac{1}{L_{ds}}\cos\tilde{\theta}_r \cdot v_{inj}^* \\ -\dfrac{1}{L_{qs}}\sin\tilde{\theta}_r \cdot v_{inj}^* \end{bmatrix}. \quad (3)$$

$$\begin{bmatrix} \Delta i^{\hat{r}}_{dsh} \\ \Delta i^{\hat{r}}_{qsh} \end{bmatrix} = T_{\tilde{\theta}_r}^{-1} \begin{bmatrix} \Delta i^r_{dsh} \\ \Delta i^r_{qsh} \end{bmatrix} = T_{samp} \begin{bmatrix} \left(\dfrac{1}{L_{ds}}\cos^2\tilde{\theta}_r + \dfrac{1}{L_{qs}}\sin^2\tilde{\theta}_r\right) \cdot v_{inj}^* \\ \left(\dfrac{1}{2}\dfrac{(L_{qs}-L_{ds})}{L_{ds}L_{qs}}\sin 2\tilde{\theta}_r\right) \cdot v_{inj}^* \end{bmatrix}. \quad (4)$$

From q-axis current ripple in (4), $\tilde{\theta}_r$ can be estimated by (5).

$$\tilde{\theta}_{r,cal} = \Delta i^{\hat{r}}_{qsh} \cdot \frac{L_{ds}L_{qs}}{T_{samp}v_{inj}^*(L_{qs}-L_{ds})} = \frac{1}{2}\sin 2\tilde{\theta}_r \approx \tilde{\theta}_r. \quad (5)$$

Then, rotor position and speed can be estimated by a PID observer shown in Fig. 2 based on mechanical model using $\tilde{\theta}_{r,cal}$ as an input to the observer. The observer works by

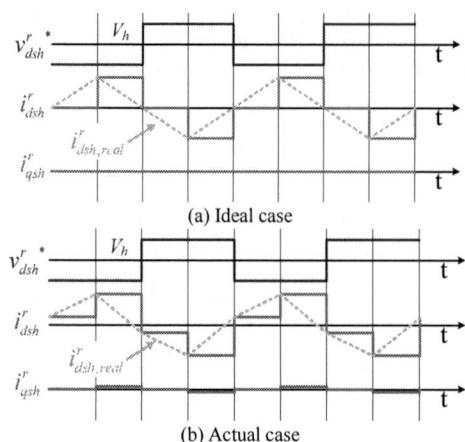

(a) Ideal case

(b) Actual case

Fig. 3. Distortion of d-axis current ripple during square-wave injection.

eliminating q-axis current ripple which corresponds to the position estimation error.

III. DISTORTION OF VOLTAGE AND CURRENT BY INVERTER NONLINEARITY EFFECTS

The rotor position estimation through (1)-(5) can be deteriorated by the inverter nonlinearity effects. Assume that square-wave voltage with uniform magnitude is injected to actual d-axis. Ideally, d-axis current ripple induced by square-wave voltage must be regular like Fig. 3(a) and there should be no current ripple in q-axis. Note that i^r_{dsh} and i^r_{qsh} in Fig. 3 are sample and hold waveforms. However, actual experimental current waveforms are different from the ideal ones. In Fig. 3(b), d-axis current ripple waveform is quite distorted and unexpected current ripple appears in q-axis due to the inverter nonlinearity effects. This current ripple in q-axis causes position estimation error, which will be discussed in the next section. Since the position information is extracted from the sensed current ripple, the irregular current ripple can cause position estimation error. This section investigates how the current waveform in Fig. 3(b) is induced by the inverter nonlinearity effects during square-wave voltage injection sensorless drive.

Inverter itself is not completely linear and its output pole voltage may be different from its reference. Pole voltage error from the inverter nonlinearity can be defined as (6).

$$v_{err} = v_{an}^* - v_{an\_real} \quad (6)$$

978-1-4799-2706-7/14 $31.00 © 2014 IEEE

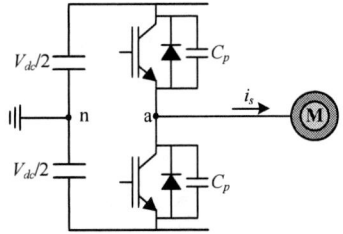

Fig. 4. One phase arm of an inverter feeding a motor.

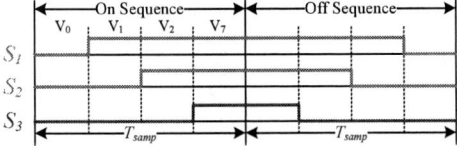

Fig. 5. Gating signals at on and off sequences.

Three main sources of the inverter nonlinearity that cause the voltage error are to be explained one by one.

A. Diode conduction characteristic during dead time [9]

Fig. 4 is a conceptual diagram describing one phase arm of an inverter feeding a motor. In the figure, the phase current denoted as i_s doesn't flow through active switching device during the dead time but only flows through diode. If the current flows to the load during the dead time, the lower diode is turned on. However, if the current flows to inverter, the upper diode is turned on. The voltage distortion by diode conduction characteristic during the dead time is mainly determined by polarity of phase current and switching sequence as expressed in (7), where the dead time is denoted as T_{dead}.

$$v_{err\_DT} = \begin{cases} V_{dc}\dfrac{T_{dead}}{T_{samp}} & (i_s>0,\ \text{on sequence}) \\[2mm] 0 & (i_s<0,\ \text{on sequence}) \\[2mm] 0 & (i_s>0,\ \text{off sequence}) \\[2mm] -V_{dc}\dfrac{T_{dead}}{T_{samp}} & (i_s<0,\ \text{off sequence}) \end{cases} \qquad (7)$$

Fig. 5 describes gating signals of each arm during on and off sequences.

B. Parasitic capacitance effect during dead time [14]

If the magnitude of current is near zero during the dead time, the pole voltage is affected by charging and discharging phenomena of the parasitic capacitor denoted as C_p in Fig. 3. Let $i_c(=V_{dc}/T_{dead}\cdot C_{st})$ the critical value of i_s when the dead time is equal to the charging or discharging time of C_p. The pole voltage error induced by parasitic capacitance effect can be expressed as follows.

$$v_{err\_PC} = \begin{cases} -(\dfrac{C_{st}V_{dc}^2}{2T_{samp}})\dfrac{1}{i_s} & (i_s<0,\ |i_s|>|i_c|,\ \text{on sequence}) \\[2mm] (V_{dc}+\dfrac{T_{dead}}{2C_{st}}i_s)\dfrac{T_{dead}}{T_{samp}} & (i_s<0,\ |i_s|\le|i_c|,\ \text{on sequence}) \\[2mm] -(\dfrac{C_{st}V_{dc}^2}{2T_{samp}})\dfrac{1}{i_s} & (i_s>0,\ |i_s|>|i_c|,\ \text{off sequence}) \\[2mm] -(V_{dc}-\dfrac{T_{dead}}{2C_{st}}i_s)\dfrac{T_{dead}}{T_{samp}} & (i_s>0,\ |i_s|\le|i_c|,\ \text{off sequence}) \end{cases} \qquad (8)$$

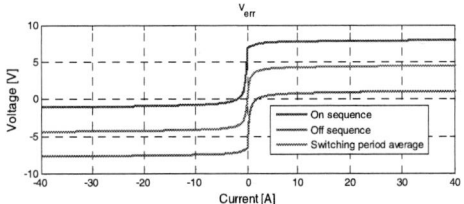

Fig. 6. Pole voltage error by inverter nonlinearity.

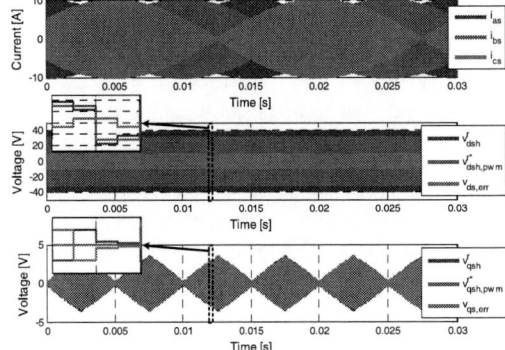

Fig. 7. Voltage distortion in rotor reference frame. (zero DC current, V_h=30V).

C. Voltage drop of switching device

Every semiconductor devices have intrinsic voltage drop when conducting current. The pole voltage error of switching device by the voltage drop can be represented as a function of current flowing through it as expressed in (9).

$$\begin{aligned} v_{err\_Switch} &= g(i_{Switch}) \\ v_{err\_Diode} &= g(i_{Diode}) \end{aligned} \qquad (9)$$

Inverter nonlinearity is affected by three main sources aforementioned. The total pole voltage error is the sum of the three error terms, that is

$$v_{err} = v_{err\_DT} + v_{err\_PC} + \frac{v_{err\_Switch}+v_{err\_Diode}}{2}. \qquad (10)$$

In (10), duty ratio of the switch is assumed as a half. The total pole voltage error with V_{dc}=144V, T_{dead}=2μs, C_p=5nF, and V-I characteristics from switching device, PM300CL1A060, is shown in Fig. 6.

The voltage error in rotor reference frame, i.e., $v_{ds,err}$ and $v_{qs,err}$, can be calculated through coordinate transformation of pole voltage error in (10). Fig. 7 shows the tendency of d-q voltage error with zero DC current, V_h=30V, f_h=0.5·f_{sw}, f_{samp}=2·f_{sw}. In the figure, $v_{dsh}^r = v_{dsh,pwm}^{r*} - v_{ds,err}$. As shown in Fig. 6, the pole voltage error rapidly varies near zero current. Thus multiple zero crossing of phase current in Fig. 7 induces rapid variation of the voltage error. The waveform of $v_{ds,err}$ is periodic and is synchronized to v_{inj}^* since the current pulsation is in d-axis. And the absolute value of v_{dsh}^r has the frequency of 2·f_h. On the other hand, $v_{qs,err}$ is not synchronized to v_{inj}^* and thus it is difficult to expect the overall waveform of v_{qsh}^r. Although $v_{qs,err}$ is smaller than $v_{ds,err}$ in its magnitude, q-axis current ripple induced by $v_{qs,err}$

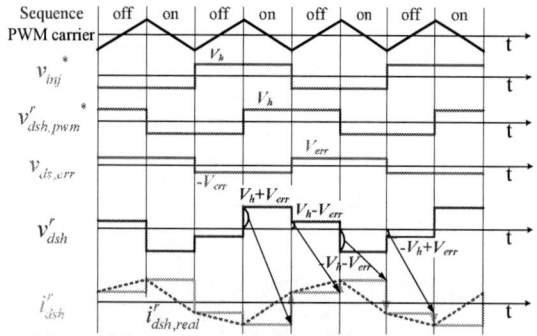

Fig. 8. Step-by-step procedure of current distortion.

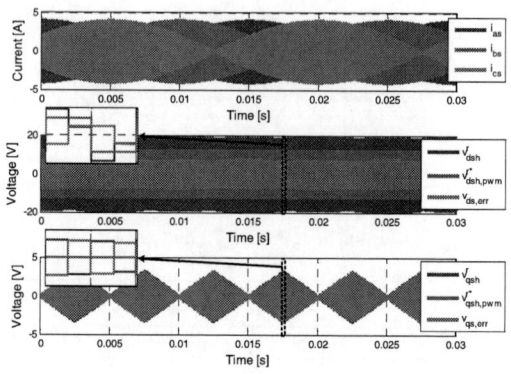

Fig. 9. Voltage distortion in rotor reference frame. (zero DC current, V_h=12V)

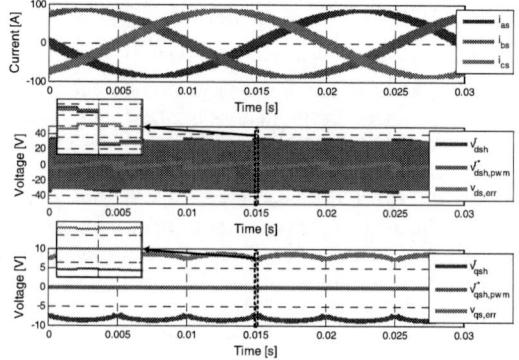

Fig. 10. Voltage distortion in rotor reference frame. (rated current at q-axis, V_h=30V)

significantly causes rotor position estimation error, which will be analyzed in the next section.

From these observations, the current ripple distortion in Fig. 3(b) can be explained by Fig. 8. By the voltage error induced by the inverter nonlinearity effects, the actual injection voltage is distorted and the sampled current waveform like Fig. 3(b) could be generated.

Fig. 9 shows the voltage error with zero DC current, V_h=12V, f_h=0.5·f_{sw}, f_{samp}=2·f_{sw}. In this case, current ripple magnitude decreased from that in Fig. 7 due to reduced injection voltage. However, $v_{ds,err}$ and $v_{qs,err}$ in Fig. 9 are almost the same as that in Fig. 7. This is because the voltage error in Fig. 6 is saturated as the current increases. If the current ripple magnitude is larger than a certain value, $v_{ds,err}$ and $v_{qs,err}$ do not change much. Focusing on the actual injection voltage, $v_{dsh,real}^r$ is more distorted in Fig. 9 than in Fig. 7 since $v_{dsh}^{r\,*}$ is much larger than $v_{ds,err}$ in Fig. 9. $v_{qsh,real}^r$ in Fig. 7 and 9 are almost the same since $v_{qsh}^{r\,*}$ is zero in both cases.

Fig. 10 shows the voltage error with rated current at q-axis, V_h=30V, f_h=0.5·f_{sw}, f_{samp}=2·f_{sw}. In this condition, pulsation occurs in $v_{ds,err}$ only six times per fundamental cycle when a phase current multiply crosses zero. And $v_{qs,err}$ is static without pulsation since there is large DC current without HF ripples in q-axis.

IV. EFFECTS OF CURRENT RIPPLE DISTORTION ON ROTOR POSITION ESTIMATION

Considering the inverter nonlinearity effects, injection voltage terms (1)-(2) can be modified to (11)-(12) where voltage error terms are added.

$$\begin{bmatrix} v_{dsh}^{\hat{r}} \\ v_{qsh}^{\hat{r}} \end{bmatrix} = \begin{bmatrix} v_{inj}^{*} - v_{ds,err} \\ -v_{qs,err} \end{bmatrix}. \tag{11}$$

$$\begin{bmatrix} v_{dsh}^{r} \\ v_{qsh}^{r} \end{bmatrix} = \begin{bmatrix} \left(v_{inj}^{*} - v_{ds,err}\right)\cos\tilde{\theta}_r - v_{qs,err}\sin\tilde{\theta}_r \\ -\left(v_{inj}^{*} - v_{ds,err}\right)\sin\tilde{\theta}_r - v_{qs,err}\cos\tilde{\theta}_r \end{bmatrix}. \tag{12}$$

The current ripple in estimated rotor reference frame can be deduced through (13)-(14).

$$\begin{bmatrix} \Delta i_{dsh}^{r} \\ \Delta i_{qsh}^{r} \end{bmatrix} = T_{samp} \begin{bmatrix} \dfrac{1}{L_{ds}}\left\{\left(v_{inj}^{*} - v_{ds,err}\right)\cos\tilde{\theta}_r - v_{qs,err}\sin\tilde{\theta}_r\right\} \\ \dfrac{1}{L_{qs}}\left\{-\left(v_{inj}^{*} - v_{ds,err}\right)\sin\tilde{\theta}_r - v_{qs,err}\cos\tilde{\theta}_r\right\} \end{bmatrix}. \tag{13}$$

$$\begin{bmatrix} \Delta i_{dsh}^{\hat{r}} \\ \Delta i_{qsh}^{\hat{r}} \end{bmatrix} = T_{samp} \begin{bmatrix} \left(\dfrac{\cos^2\tilde{\theta}_r}{L_{ds}} + \dfrac{\sin^2\tilde{\theta}_r}{L_{qs}}\right)\cdot\left(v_{inj}^{*} - v_{ds,err}\right) \\ -\dfrac{\sin 2\tilde{\theta}_r}{2}\cdot\dfrac{L_{qs} - L_{ds}}{L_{ds}L_{qs}}\cdot v_{qs,err} \\ \dfrac{\sin 2\tilde{\theta}_r}{2}\cdot\dfrac{L_{qs} - L_{ds}}{L_{ds}L_{qs}}\cdot\left(v_{inj}^{*} - v_{ds,err}\right) \\ -\left(\dfrac{\sin^2\tilde{\theta}_r}{L_{ds}} + \dfrac{\cos^2\tilde{\theta}_r}{L_{qs}}\right)\cdot v_{qs,err} \end{bmatrix}$$

$$\approx T_{samp} \begin{bmatrix} -\dfrac{L_{qs} - L_{ds}}{L_{ds}L_{qs}}\cdot v_{qs,err}\cdot\tilde{\theta}_r + \dfrac{v_{inj}^{*} - v_{ds,err}}{L_{ds}} \\ \dfrac{L_{qs} - L_{ds}}{L_{ds}L_{qs}}\cdot\left(v_{inj}^{*} - v_{ds,err}\right)\cdot\tilde{\theta}_r - \dfrac{v_{qs,err}}{L_{qs}} \end{bmatrix}. \tag{14}$$

In (14), $v_{ds,err}$ and $v_{qs,err}$ appear in q-axis current ripple that is necessary signal to estimate rotor position. Applying the computation in (5), the extracted signal becomes as follows.

$$\tilde{\theta}_{r,cal} \approx k_{scale}\cdot\tilde{\theta}_r + N,$$

$$\text{where} \begin{cases} k_{scale} = 1 - \dfrac{v_{ds,err}}{v_{inj}^{*}} \\ N = -\dfrac{L_{ds}v_{qs,err}}{v_{inj}^{*}\left(L_{qs} - L_{ds}\right)} \end{cases}. \tag{15}$$

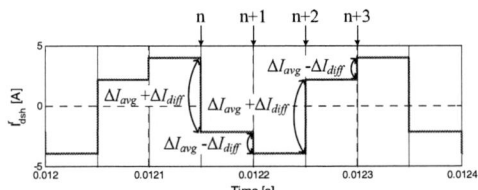

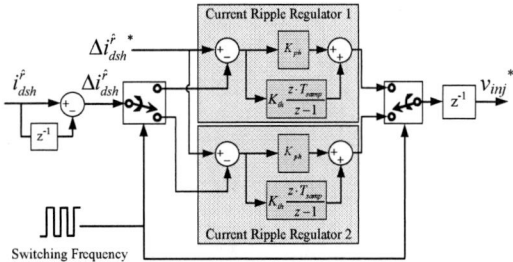

Fig. 13. Block diagram of proposed scheme.

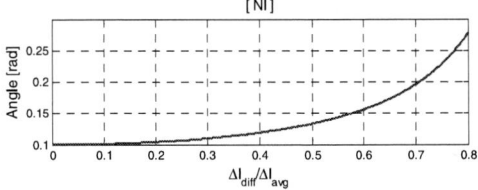

Fig. 12. Noise index according to $\Delta I_{diff} / \Delta I_{avg}$.

Fig. 11. d-axis current ripple waveform. (zero DC current, V_h =12V)

In (15), the first term related with $\tilde{\theta}_r$ is an effective signal that helps the position estimation and the second term works as a noise input denoted as n in Fig. 2. Due to the inverter nonlinearity effects, k_{scale} is not the unity but fluctuates as $v_{ds,err}$ pulsates. This is equivalent to the fluctuation of observer gains, L_1-L_3, which significantly degrades the position estimation performance. And the noise term denoted as n generated by $v_{qs,err}$ directly leads to position estimation error. Since v_{inj}^* is a square-wave signal with frequency f_h, only AC signal of $v_{qs,err}$ with frequency f_h works as a noise input. In the case of Fig. 9 when the injection voltage is low, both k_{scale} and N increase, which means more fluctuation observer gains and more noise input to the observer. This is why the position estimation performance is degraded when the injection voltage is low. In the case of Fig. 10, large DC value in $v_{qs,err}$ is filtered out and does not contribute to the noise input. For this reason, the position estimation performance is less influenced by the inverter nonlinearity effects in load condition.

V. PROPOSED VOLTAGE INJECTION METHOD

Rearranging the d-axis equation in (14), v_{inj}^* can be expressed as (16).

$$ v_{inj}^* \approx \frac{L_{ds}}{T_{samp}} \Delta i_{dsh}^r + \frac{L_{qs} - L_{ds}}{L_{qs}} \cdot v_{qs,err} \cdot \tilde{\theta}_r + v_{ds,err}. \qquad (16) $$

Substituting (16) into the q-axis equation in (14), q-axis current ripple can be expressed as a function of Δi_{dsh}^r and $v_{qs,err}$.

$$ \Delta i_{qsh}^r \approx \frac{L_{qs} - L_{ds}}{L_{qs}} \cdot \Delta i_{dsh}^r \cdot \tilde{\theta}_r - \Delta i_{qsh,err}. \qquad (17) $$

$\Delta i_{qsh,err}$ in (17) indicates $T_{samp} / L_{qs} \cdot v_{qs,err}^r$ which is the ripple component induced by q-axis voltage distortion. In (17), the effective term is proportional to Δi_{dsh}^r. Then the computation of $\tilde{\theta}_{r,cal}$ can be improved by (18).

$$ \tilde{\theta}_{r,cal} = \Delta i_{qsh}^r \cdot \frac{L_{qs}}{\left(L_{qs} - L_{ds}\right) \Delta i_{dsh}^r} \approx \tilde{\theta}_r - \frac{L_{qs}}{\left(L_{qs} - L_{ds}\right)} \cdot \frac{\Delta i_{qsh,err}}{\Delta i_{dsh}^r}. \qquad (18) $$

Dividing Δi_{qsh}^r by Δi_{dsh}^r and inductance terms, the fluctuation of observer gains can be suppressed since k_{scale} becomes the

unity in (18). From the last term in (18) a Noise Index (NI) can be defined as (19).

$$ \mathrm{NI} = \frac{\Delta i_{qsh,err}}{\Delta i_{dsh}^r}. \qquad (19) $$

NI is the source of the noise input to the observer in Fig. 2.

The key idea of the proposed algorithm is to reduce NI by regulating $\left|\Delta i_{dsh}^r\right|$. By regulating the magnitude of Δi_{dsh}^r, NI can be reduced in the moving average sense. For example, the waveform of Δi_{dsh}^r in the same condition as Fig. 9 is shown in Fig. 11. The magnitude of Δi_{dsh}^r can be expressed as (20).

$$ \left|\Delta i_{dsh}^r\right| = \Delta I_{avg} + (-1)^n \Delta I_{diff}, \qquad (20) $$

where n means sampling count. For constant magnitude of Δi_{dsh}^r like the case in Fig. 3(a), ΔI_{diff} would be zero. ΔI_{diff} becomes larger with low injection voltage like the case of Fig. 11 because the magnitude of $v_{ds,err}$ is comparable to V_h. Assume that

$$ \left|\Delta i_{qsh}^r\right| = k \cdot \Delta I_{avg}. \qquad (21) $$

Then the moving average of NI for two sampling periods can be calculated as

$$ [\mathrm{NI}] = \frac{k}{1 - \left(\Delta I_{diff} / \Delta I_{avg}\right)^2}. \qquad (22) $$

With k=0.1, the moving average of NI is shown in Fig. 12. In Fig. 11 $\Delta I_{diff} / \Delta I_{avg}$=0.541 and [NI]=0.141. If Δi_{dsh}^r is regulated, ΔI_{diff} becomes zero and [NI]=0.1. By the regulation of Δi_{dsh}^r, the noise input can be reduced by 29.1% under the constraint of the same peak-to-peak current ripple. The effect of constant Δi_{dsh}^r becomes larger with lower injection voltage, i.e., smaller magnitude of V_h.

Fig. 13 shows the implementation of the current ripple regulation. Instead of injecting pulsating voltage with fixed magnitude in Fig. 1, the injection voltage is adjusted by PI regulator in order to fix the magnitude of Δi_{dsh}^r. There are two current ripple regulators work together alternately. Because Δi_{dsh}^r has the same magnitude for every other sampling instant as shown in Fig. 11, one regulator can be used for even sampling instants and the other can be used for odd sampling instants.

$$ \begin{aligned} \left|\Delta i_{dsh}^r\right|[n] &= \left|\Delta i_{dsh}^r\right|[n+2] \\ &\neq \left|\Delta i_{dsh}^r\right|[n+1]. \end{aligned} \qquad (23) $$

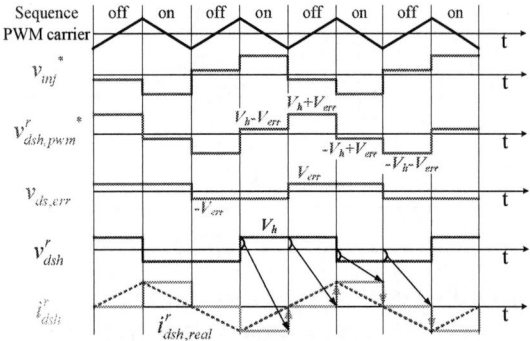

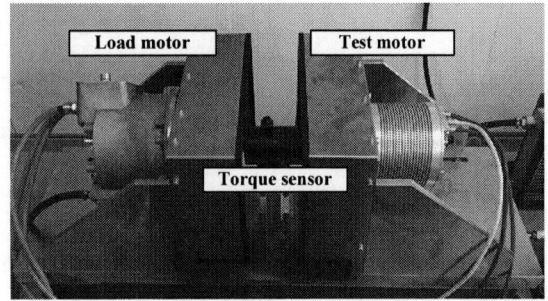

Fig. 14. Procedure of current ripple regulation.

Fig. 15. Experimental setup.

Fig. 14 shows how Δi_{dsh}^r is regulated by the proposed scheme. In the steady state, the current ripple is regulated, which means that the actual injection voltage has the waveform as shown in Fig. 14. In other words, $v_{ds,err}$ can be completely compensated by the proposed scheme.

VI. EXPERIMENTAL RESULTS

The hardware platform to evaluate the proposed scheme is constructed based on a TMS28335 DSP. The power devices used in prototype inverter is MITSUBISHI PM300CAL1A06 IPM. f_{sw}=10kHz, f_{samp}=20kHz, f_h=5kHz, V_{dc}=144V, and T_{dead} is set as 2μs. Fig. 15 shows experimental setup. The target motor on the right side of the test bed is IPMSM whose parameters are specified in Table I.

TABLE I.　PARAMETERS OF IPMSM

IPMSM Parameter	Value
Rated power	8 kW
Rated current	60 A_rms
Pole number	10
Back-EMF constant	0.0281 V·s
Winding resistance	0.01 Ω
Synchronous inductances	L_{ds}: 143 μH, L_{qs}: 216 μH

A. Regulation of HF Current Ripple

Fig. 16 shows the regulation of HF current ripple by the proposed scheme. Using conventional voltage injection method, the waveform of i_{dsh}^r is irregular due to the inverter nonlinearity effects. Applying the proposed voltage injection method, i_{dsh}^r is kept regular thanks to adjusted v_{inj}^* as shown in Fig. 16(b).

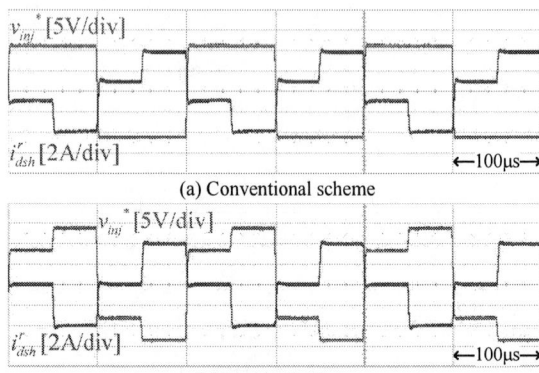

(a) Conventional scheme

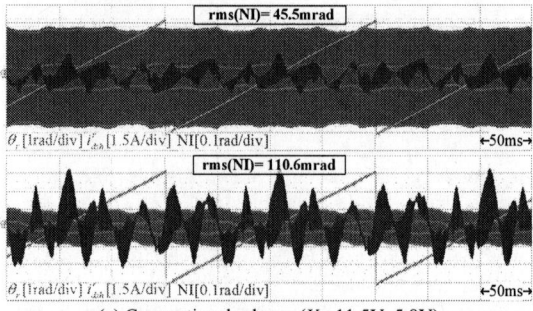

(b) Proposed scheme

Fig. 16. Experiment A: Injection voltage and current ripple.

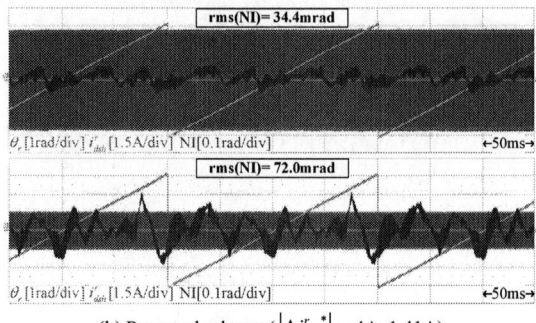

(a) Conventional scheme (V_h=11.5V, 5.8V)

(b) Proposed scheme ($\left| \Delta i_{dsh}^{r\,*} \right|$ = 4A, 1.41A)

Fig. 17. Experiment B: Measurement of NI. (60r/min, zero current)

B. Reduction of Noise Index

The noise index defined in (19) can be experimentally measured. Under sensored operation, $\tilde{\theta}_r$ is equal to zero. In this condition, q-axis current ripple is induced only by the inverter nonlinearity effects, that is $\Delta i_{qsh}^r = \Delta i_{qsh,err}$. Dividing Δi_{qsh}^r by Δi_{dsh}^r, NI can be calculated.

Fig. 17 shows experimental waveforms of noise indices in no load condition. In the figure, rms values of the current ripples are matched for fair comparison considering the losses and acoustic noises. V_h=11.5V and 5.8V correspond to $\left| \Delta i_{dsh}^{r\,*} \right|$ = 4A and 1.41A, respectively. It is evident in the figure that noise indices are conspicuously attenuated by the proposed voltage injection technique. In the three cases, the percentages of NI reduction are 24.4% and 34.9%.

Fig. 18 shows the same noise index waveforms in full load condition. Thanks to reduced number of zero crossings of phase currents in this condition, the inverter nonlinearity effects are smaller. Comparing Fig. 17(a) and Fig. 18(a), it can be known that although the injection voltage is not

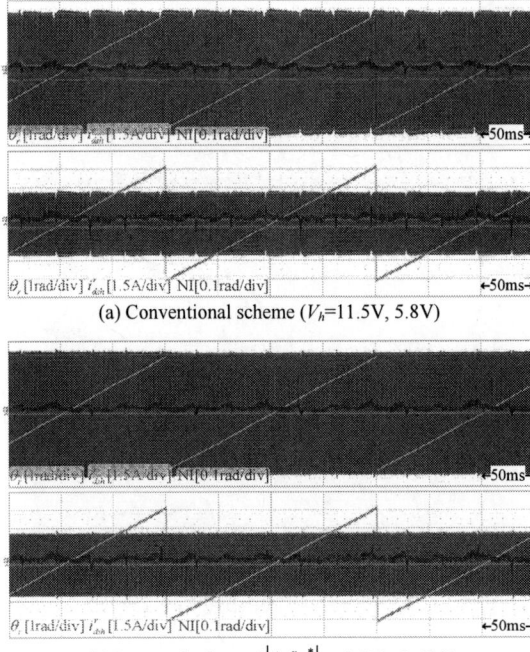

(a) Conventional scheme (V_h=11.5V, 5.8V)

(b) Proposed scheme ($\left|\Delta i_{dsh}^{r\,*}\right|$ = 4.89A, 2.47A)

Fig. 18. Experiment B: Measurement of NI. (60r/min, rated current)

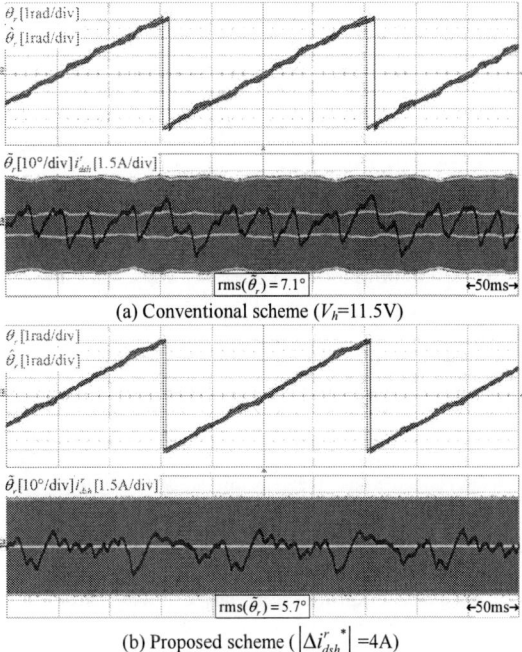

(a) Conventional scheme (V_h=11.5V)

(b) Proposed scheme ($\left|\Delta i_{dsh}^{r\,*}\right|$ =4A)

Fig. 19. Experiment C: Steady-state position estimation under sensorless drive. (60r/min, zero current)

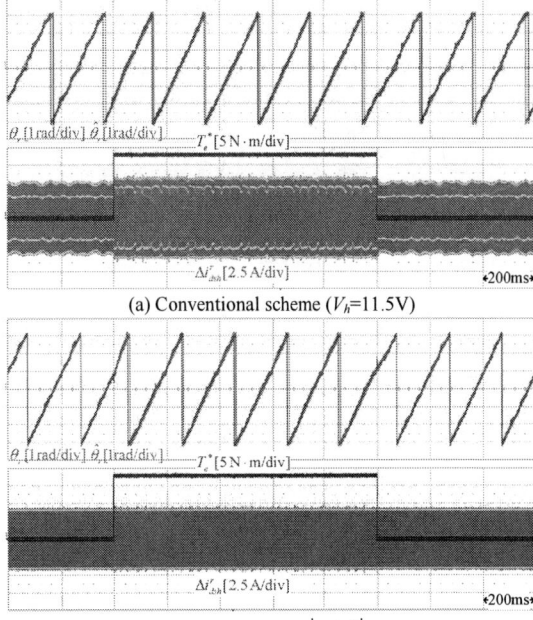

(a) Conventional scheme (V_h=11.5V)

(b) Proposed scheme ($\left|\Delta i_{dsh}^{r\,*}\right|$ =4A)

Fig. 20. Experiment C: Transient-state position estimation under sensorless drive. (60r/min)

increased, current ripple is increased in load condition. And as discussed in section IV, $v_{qs,err}$ has DC component only in load condition. For this reason, generation of noise input is very small in load condition. This fact is reflected in Fig. 18. The waveforms of NI in both Fig. 18(a) and (b) are almost the same. The proposed scheme is more effective in no load and light load condition.

C. Improvement of Position Estimation Performance under Sensorless Drive

Fig. 19 shows the position estimation in the steady state. In the figure, conventional scheme indicates fixed voltage injection method explained in section II. And proposed scheme indicates the current ripple regulation and the computation of $\tilde{\theta}_{r,cal}$ in (18). In both cases, rms values of the current ripple are matched. Thanks to reduced NI, the position estimation under closed-loop sensorless operation has been improved. In case of the proposed scheme, rms value of position estimation error is decreased by 19.7% without increasing losses.

Fig. 20 shows the other experimental results. Under sensorless operation at 60r/min, torque command is jumped up to the rated value and return back to zero. The magnitude of the current ripple is kept uniform by the proposed scheme, while that varies depending on load condition by the conventional scheme. Due to the non-uniform magnitude of the current ripple, position estimation performance varies depending on load condition. In Fig. 20(a), 6th harmonic component in the estimated rotor position is quite relieved during the rated torque period. However, the position estimation performance is kept relatively consistent by the proposed scheme.

VII. CONCLUSIONS

In square wave voltage signal injection sensorless control, the injection voltage is distorted by voltage error induced by the inverter nonlinearity effects. And the HF current ripple is also distorted. Since the position information is extracted from the HF current signal, the position estimation performance can be seriously degraded.

To improve this phenomenon, a voltage injection method which regulates the magnitude of the induced current ripple has been proposed. By the proposed method, the noise input in the position estimation has been reduced by 24.4% and 34.9% in no load condition depending on injection voltage. And the position estimation performance under sensorless operation was improved by 19.7% in no load condition.

REFERENCES

[1] T. Ohtani, N. Takada, and K. Tanaka, "Vector Control of Induction Motor without Shaft Encoder," *IEEE Trans. Ind. Appl.,* vol. 28, no. 1, pp. 157-164, Jan./Feb. 1992.

[2] J. Holtz, "Sensorless position control of induction motors-an emerging technology," *IEEE Trans. Ind. Electron,* vol. 45 no. 6, pp. 840 -851, Dec. 1998.

[3] P. L. Jansen and R. D. Lorenz, "Transducerless position and velocity estimation in induction and salient AC machines," *IEEE Trans. Ind. Appl.,* vol. 31, no. 2, pp. 240-247, Mar./ Apr. 1995.

[4] J. I. Ha, K. Ide, T. Sawa, and S. K. Sul, "Sensorless rotor position estimation of an interior permanent-magnet motor from initial states," *IEEE Trans. Ind. Appl.,* vol. 39, no. 3, pp. 761–767, May./Jun. 2003.

[5] Y. D. Yoon, S. K. Sul, Morimoto, K. Ide, "High-Bandwidth Sensorless Algorithm for AC Machines Based on Square-Wave-Type Voltage Injection," *IEEE Trans. Ind. Appl.,* vol. 47, no. 3, pp. 1361-1370, May./Jun. 2011.

[6] J. M. Guerrero, M. Leetmaa, F. Briz, A. Zamarron, and R. D. Lorenz, "Inverter nonlinearity effects in high-frequency signal-injection-based sensorless control methods," *IEEE Trans. Ind. Appl.,* vol. 41, no. 2, pp. 618–626, Mar./Apr. 2005.

[7] C. Silva, G. M. Asher, and M. Sumner, "Influence of dead-time compensation on rotor position estimation in surface mounted PM machines using HF voltage injection," in *Proc. IEEE Power Conversion Conf.,* Osaka, Japan. pp. 1279–1284, 2002.

[8] J.-W. Choi and S.-K. Sul, "A new compensation strategy reducing voltage/current distortion in PWM VSI systems operating with low output voltages," *IEEE Trans. Ind. Appl.,* vol. 31, no. 5, pp. 1001–1008, Sept./Oct. 1995.

[9] J. W. Choi and S. K. Sul, "Inverter output voltage synthesis using novel dead time compensation," *IEEE Trans. Power Electron.,* vol. 11, no. 2, pp. 221–227, Mar. 1996

[10] N. Urasaki, T. Senjyu, T. Kinjo, T. Funabashi, and H. Sekine, "Dead time compensation for permanent magnet synchronous motor drive taking zero-current clamp and parasitic capacitance effects into account," *IEE Proc. Electric Power Appl.,* vol. 152, no. 4, pp. 845–853, Jul. 2005.

[11] Bon-Ho Bae, Seung-Ki Sul, "A compensation method for time delay of full-digital synchronous frame current regulator of PWM AC drive," *IEEE Trans. Ind. Appl.,* vol. 39, no. 3, pp. 802–810, May 2003.

[12] Hyung-Min Ryu, Jung-Ik Ha, Seung-Ki Sul, Kozo Ide, Yoichi Yamamoto, Eiji Watanabe, "Compensation of Voltage Distortion in PWM-VSI by Prediction of Stator Currents at Switching Point," Conf. Rec. *of IEEJ, Annual Meeting,* vol.3, pp. 87-90. Apr. 1999.

[13] C. H. Choi and J. K. Seok, "Compensation of zero-current clamping effects in high-frequency-signal-injection-based sensorless PM motor drives," *IEEE Trans. Ind. Appl.,* vol. 43, no. 5, pp. 1258–1265, Sept./Oct. 2007.

[14] K. Wiedmann, F. Wallrapp, and A. Mertens, "Analysis of inverter nonlinearity effects on sensorless control for permanent magnet machine drives based on high-frequency signal injection," in *Proc. 13th Eur. Conf. Power Electron. Appl. (EPE),* 2009, pp. 1–10.

978-1-4799-2706-7/14 $31.00 © 2014 IEEE

Application Trend of Saliency-Based Sensorless Drives

Akira Yamazaki
YASKAWA ELECTRIC CORPORATION
Yukuhashi, Japan

Kozo Ide
YASKAWA ELECTRIC CORPORATION
Yukuhashi, Japan

Abstract— **This paper introduces a hybrid sensorless control of interior permanent magnet synchronous machines (IPMSM) combining high frequency injection and back EMF based method. As successful applications, a compressor, an injection molding machine and a textile machine are introduced in this paper.**

Keywords— *Hybrid Sensorless Control, Synergetic Effect, Direct Drive Application*

I. INTRODUCTION

Permanent magnet synchronous motor （IPMSM） miniaturization, drive and energy saving is possible has attracted attention, spread to the market is progressing. Yaskawa have rolled out the drive for IPMSM first in the world in 1997. Later, to accommodate a wide variety of applications, have been efforts to improve control performance. Especially, sensorless control is possible to realize a variable speed and constant torque without speed sensor, and contribute to the expansion of application. Therefore it's becoming more important.

Sensorless control method is classified into two types from the principles of the position estimation. The high frequency injection method (HFIM) is effective for position and speed sensorless AC motor drives operated in ultra-low speed region including zero speed, but has still shortcoming in practical use for the industrial application due to torque ripple and the acoustic noise. Meanwhile, the back EMF based method has been widely used in commercial products, but the controllable speed range is limited to more than about 5% of the rated speed. To operate the sensorless AC motor drives in the entire speed range, hybrid sensorless control combining both methods is one of the solutions, and several algorithms have been proposed [1]-[4]. The synergy effect to suppress the torque ripple produced by HFIM was first explained with an induction machine in [1]. Yasukawa has adopted and evolved a hybrid sensorless control [5].

In this paper, Yaskawa's sensorless motor control techniques are described and several application examples for PM motor drives are introduced later chapter.

II. HYBRID SENSORLESS CONTROL

The HFIM is effective for position and speed estimation in ultra-low speed region including zero stator frequency,

due to torque ripple and the acoustic noise the method has shortcoming in practical use in the industrial application. Meanwhile, the back EMF based method is able to estimate position and speed without acoustic and additional torque ripple, but the controllable speed range is more than about 5% of the rated speed. In order to realize position and speed estimation in entire speed range, a hybrid method is proposed. Fig. 1 shows the proposed hybrid method combining HFIM and back EMF based method. The key technique is the changeover which is done only in the internal speed . The internal speed consists of the estimated speed of both the HFIM and the back EMF based method. The internal estimated speed is set as follows:

$$\hat{\omega}_{cmb} = \hat{\omega}_0 + G_1 \cdot \boldsymbol{K} \cdot \frac{\hat{\varepsilon}_\gamma \hat{\varepsilon}_\delta}{\hat{\varepsilon}_\gamma^2 + \hat{\varepsilon}_\delta^2} + G_2 \cdot \hat{\omega}_{HFIM} \quad (1)$$

where $\boldsymbol{K}$ is the controller involving an integration ,G1 and G2 are weighting factors. The weighting factors are regulated along with the command and/or the estimated speed.

The internal estimated speed is only employed in the machine model of the back EMF based method, and not

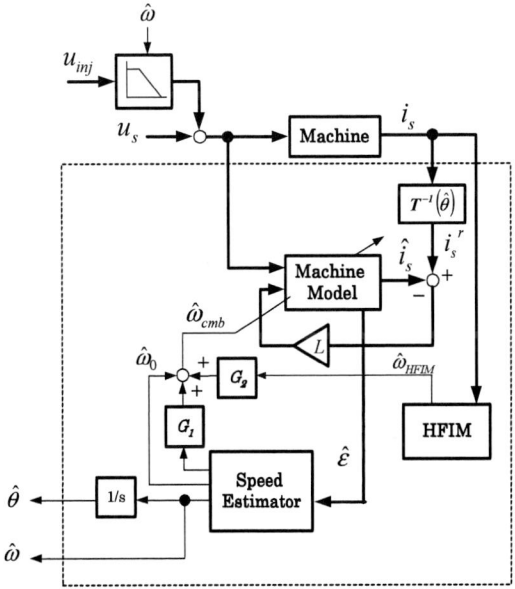

Fig. 1. Position and speed estimation with a hybrid method

employed to estimate the position of the magnet flux. The position is estimated by integration of the estimated speed , which is calculated by (2).

$$\hat{\omega} = \hat{\omega}_0 + K \cdot \frac{\hat{\varepsilon}_\gamma \hat{\varepsilon}_\delta}{\hat{\varepsilon}_\gamma^2 + \hat{\varepsilon}_\delta^2} \qquad (2)$$

The same changeover concept is described in [1], and the advantage of this method is followings:

· The response of the position and the speed estimation is able to keep constant in entire speed range
· The seamless changeover is achieved because the feedback position and speed to outer controller have no changeover

The additional torque ripple and the acoustic noise due to the injected high-frequency signal of HFIM are reduced because of the filtering effect of the back EMF based method

In 2008 the first commercial general purpose inverter with sensorless control based on hybrid method was launched. Now, five years have passed, and the application field has been expanded. Next chapter introduces four years of industrial experience with sensorless IPMSM drive based on hybrid method. As successful special applications, a compressor, an injection molding machine, and a textile machine are introduced there.

III. APPLICATION EXAMPLE

3.1 Compressor Fig. 2 shows an air compressor system. The conventional system employed an induction machine driven with commercial power supply. Getting the importance to save energy, the PM machine driven by inverter system is replaced with the conventional one. The feature of saving-space is one of the attractive features to employ PM machine. The sensorless drive is required in case of direct drive with the reason that the position sensor may fail in the environment of high temperature, humidity, and mechanical vibration.

The air pressure of the compressor is controlled by un-loader and check valves. In such case, the machine is required to start under heavy load condition. Opening un-loader valve, the machine connected to the compressor acts at 3% of the rated speed sustaining load about 200 to 300% of the rated torque.

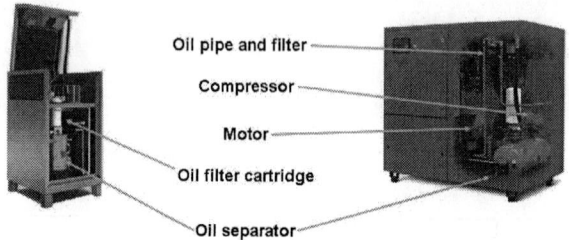

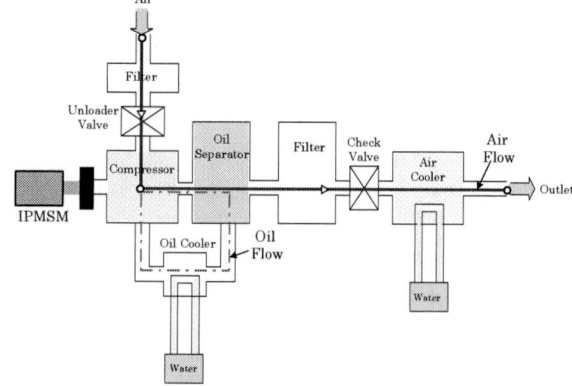

Fig.2   Air compressor configuration

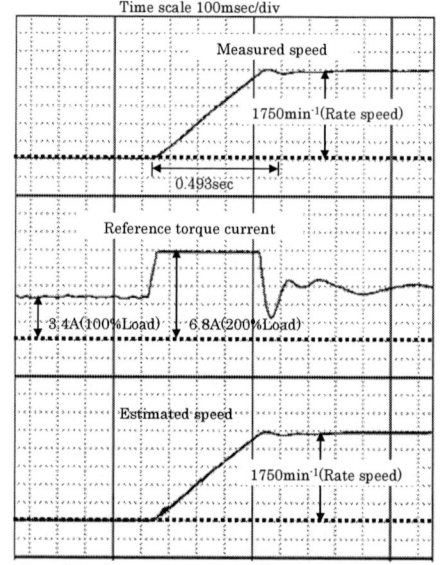

Fig.3   Acceleration test results with 100% of rated torque on the compressor: top: Real speed wave-form, middle: Reference torque current wave-form, bottom: Estimated speed waveform

The specification of 0.4kW IPMSM is shown in Table 1. Fig. 3 shows the waveform of the speed and torque. Before starting the machine, the load torque was applied to the rating. In this condition, the machine was controlled to zero speed, and the torque needed 200% of the rated value in the acceleration of the speed. Then, the speed reached to the rated speed within 0.5sec. The hybrid sensorless control performed well under tested severe load conditions.

The air pressure of the compressor is controlled by un-

Table 1
Specification of the IPMSM for applied the compressor

Item	Specification
Stator Structure	Concentrated Winding
Slot combinations	10poles 12slots
Rated power	0.4kW
Rated speed	1750 min^{-1}
Rated frequency	145.8Hz
Rated torque	2.18 Nm
Rated current	2.0A
EMF factor	40.0 mV/min^{-1}(per phase, 20deg)
Staor resistance	2.07Ω
q-axis inductance	38.0mH
d-axis inductance	27.5mH

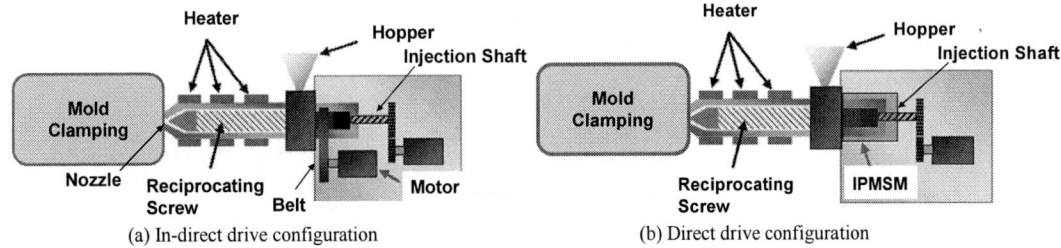

(a) In-direct drive configuration   (b) Direct drive configuration

Fig. 4 Injection molding machine

loader and check valves. In such case, the machine is required to start under heavy load condition. Opening un-loader valve, the machine connected to the compressor acts at 3% of the rated speed sustaining load about 200 to 300% of the rated torque.

3.2 Injection Molding Machine Fig.4 shows the standard structure of the injection molding machine. In this application, there are controlled processes such as molding, plasticization, injection, and extrusion, where servo drives through timing belts and gears is usually employed shown in Fig. 4(a). As the sensorless-drive, the plasticization is focused attention on. In the process, the reciprocating screw kneads melted plastic uniformly, and the back-pressure is controlled to make value of plastic and the pressure in the heat cylinder constant. The drive through belts and gears causes the following problem:

· Torque ripple causes due to mechanical factor
· Precise back-pressure control is difficult because of the indirect connection between the machine and the load cell
· Centering of the screw is difficult due to radial force in case of timing belts connected

In order to solve the solution, direct drive without belts and gears are required. The hybrid sensorless control system is employed to realize the direct drive. Fig .4(b) shows the system configuration. The injection molding machine has the hollowed rotor connected to the screw driving shaft in the plasticizing unit.

For the sensorless control, the following conditions are required in this application:

· Initial pole position is detected at stand-still
· Starting under heavy load condition and the response to reach the rated speed in 0.1 to 0.2s

Table 2
Specification of the IPMSM for the injection molding machine

Item	Specification
Stator Structure	Distributed Winding
Slot combinations	24poles 54slots
Rated power	2.8 / 4.2 / 6.0kW
Rated speed	360 min⁻¹
Rated frequency	72Hz
Rated torque	74 / 111 /159 Nm
Rated current	15.0 / 20.0 / 27.0A
EMF factor	198 / 220 / 222 mV/min⁻¹(per phase, 20deg)
Stator resistance	0.758 / 0.482 / 0.271Ω
q-axis inductance	4.16 / 3.51 / 2.55mH
d-axis inductance	2.77 / 2.34 / 1.69mH

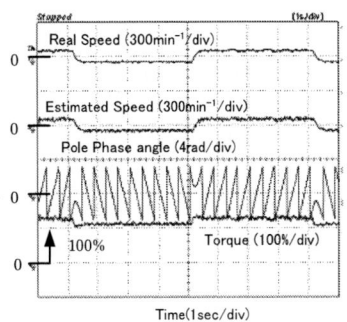

Fig. 5 Waveform in speed control mode

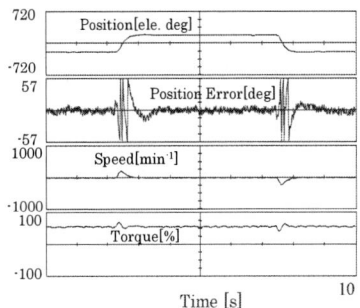

Fig. 6 Waveform in position control mode

· Servo-locked position control keeping zero speed under rebound torque caused in injection mode

The sensorless control system met all conditions. The initial pole position was detected within the error angle of 10deg (elec.) and/or 1deg (mech.). Fig. 5 shows the waveform in speed control mode, where the constant heavy load about 100% of the rated torque is kept in constant. The speed control mode needs the controllable speed range from zero to the rated speed. Starting of the machine under heavy load was possible. The response of the speed acceleration and deceleration were achieved within 0.2s. Fig. 6 shows the waveform in position control mode, where the position varies along with the command under the heavy load of about 80% of the rated torque. The performance of the servo-locked position control was also satisfied to suppress the movement against the rebound torque caused in injection mode.

The specification of the tested IPMSM is shown in Table 2.

3.3 Textile Machine Fig.7 shows the standard structure of the textile machine. Textile machinery

978-1-4799-2706-7/14 $31.00 © 2014 IEEE

market is an important market, which accounts for about 10% of the inverter annual sales in Japan. In recent years, the domestic market has become a difficult situation due to reduced production of synthetic fiber, including acrylic staple fiber. However, 80% of the production volume is for export, 70% of them, is for China. Future, the rise of India and Pakistan market is also attracting attention, expansion of the global market is expected.

Textile machinery market is promoting the introduction of early induction motor. It's the pioneer of the variable speed drive. Recently, the motor drive of the PMM has been increasing in order to respond to the request such as miniaturization, high accuracy and energy-saving machine

Textile machine weave fabric by repeating the following operations which is referred to as one pitch a series of operations.

1. Split up and down the warp. Open so that weft can passes through a stretch between

2. The transport to the other side the weft between the warp which is open

3. The hit to the front in the reed weft passed through, to incorporate the weft and warp.

Textile machine is continuously operated basically except stopped by supplementation of the yarn. To reduce the woven irregularity by stop and restart, following performance is required.

· Restarting the motor idling,

· Application specifications such as momentary power failure measures

· Environmental measures by finless structure and braking resistor less

In addition, the motor drive performance whitch accelerate and decelerate high inertia load and high torque load up to the target rotation speed in short time is required. Therefore, the following specifications are required for motor control , when replaced the IPMSM from induction machine

· 300 [%] torque limit acceleration at Start-up

· 10 [%] improvement of efficiency at steady operation @ IM rate

The specification of the tested IPMSM is shown in Table 3.

Table 3
Specification of the IPMSM for the textile machine

Item	Specification
Stator Structure	Distributed Winding
Pole Pairs	3
Rated power	22.0kW
Rated speed	2020min^{-1}
Rated frequency	101Hz
Rated torque	105.0Nm
Rated current	52.5A
EMF factor	73.4mVrms/min^{-1}
Stator resistance	0.073Ω
q-axis inductance	3.71mH
d-axis inductance	2.57mH

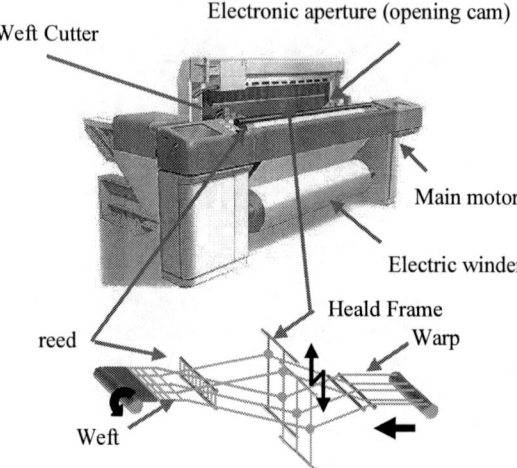

Fig. 7   Textile machine configuration

By acceleration at Torque limit state 300[%], the motor can be accelerated to the target speed without affecting the feeding of the weft. During steady operation, by diminishing the speed response and suppressing unwanted torque fluctuations, thereby energy saving of more than 10% IM ratio is achieved.

IV. CONCLUSIONS

This paper presents a hybrid sensorless control of IPMSM for several applications. The control combines both high frequency injection and back EMF based method, and realizes position, speed, and torque control in the entire speed range. As successful applications, a compressor, an injection molding machine and a textile machine were introduced. Experimental results showed the effectiveness to the required specification of the control modes. Expansion into other applications are expected by using the sensorless control according to the results.

REFERENCES

[1] K. Ide, J-I. Ha, M. Sawamura, H. Iura, and Y. Yamamoto, "A novel hybrid speed estimator of flux observer for induction motor drive" in Proc. IEEE ISIE'02, vol. 3, L'Aquila, Italy, pp. 822-827. July 2002.

[2] E. Robeischl, M. Schroedl, and K. Krammer, "Position-sensorless biaxial position control with industrial PM motor drive based on INFORM- and back EMF model," in Proc. IEEE IECON'02, vol. 1, Sevillia, Spain, pp. 668-673, Nov. 2002.

[3] C. Silvia, G. M. Asher, and M. Sumner, "An hf signal-injection based observer for wide speed range sensorless PM motor drives including zero speed," in Proc. EPE'03, vol. 1, Toulouse, France, pp. 1-9, Sep. 2003.

[4] M. Tursini, P. Petrella, and F. Prasiliti, "Sensorless control of an IPM synchronous motor for city-scooter applications," in Conf. Rec. IEEE IAS Ann. Meeting, vol. 3, Salt Lake City, UT, pp. 1472-1479, Oct. 2003.

[5] Sadayuki Sato, Hideaki Iura, Kozo Ide and Seung-ki Sul: "Three Years of Industrial Experience withsensorless IPMSM Drive based on High Frequency Injection Method", Sensorless Control for Electrical Drives(SLED) IEEE, pp.74-79 (2011)

The 2014 International Power Electronics Conference

Switching-Level Simulation Model of MMC-based Back-to-Back Converter for HVDC Application

Byung Moon Han, Senior Member IEEE
Department of Electrical Engineering
Myongji University
Yongin, Korea
erichan@mju.ac.kr

Jong kyou Jeong, Student Member KIPE
Department of Electrical Engineering
Myongji University
Yongin, Korea
jjuk486@naver.com

Abstract— **Switching-level simulation model of MMC-based BTB (back-to-back) converter for HVDC (high voltage DC transmission) application is very important to analyze its operation and transient. However, switching-level simulation model for the actual MMC-based BTB converter is very difficult because it consists of more than 150 sub-modules for each arm. In this paper a switching-level simulation model for the 11-level MMC-based BTB converter was developed with PSCAD/EMTDC software. It has 12 sub-modules for the positive arm and another 12 sub-modules for the negative arm. The developed simulation model includes the DC-voltage balance algorithm, the circulating-current reduction algorithm, the harmonic reduction algorithm, and the redundancy operation algorithm. So, the developed simulation model can be utilized to design the MMC-based BTB converter and to develop its protection scheme.**

Keywords— *HVDC(high voltage DC transmission), VSC(voltage source converter), MMC(modular multi-level converter), SM(sub-module), DC voltage balance, circulating current, redundancy module, harmonic reduction.*

I. INTRODUCTION

In the early stage the voltage source converter for HVDC consists of many switching devices connected in series and operated in pulse width modulation (PWM) switching, in which system efficiency is rather low due to the switching losses [1,2]. In order to improve the system efficiency, MMC-based HVDC converter was proposed and operated [3].

MMC is comprised of many sub-modules (SM) connected in series, in which each sub-module has a configuration of half-bridge converter [4]. MMC generates the line-frequency voltage waveform by switching each sub-module with low switching frequency and low harmonic level, which results in high efficiency [5,6,7]. 400MW/200kV MMC-based HVDC system was installed near San Francisco in the United States for the first time in the world [8].

Switching-level simulation model is required for analyzing the operation of MMC-based HVDC converter to be newly installed, and deriving the protection schemes against related system faults. However, the actual MMC-based HVDC converter has 150 SMs per

each arm, which is a big burden to analyze its operation in switching level.

Instead of individually considering switching and passive elements comprising each SM, a simulation model for the MMC-based HVDC converter with equivalent source was proposed, in which the simulation model consists of 150 equivalent sources connected in series to form one arm [9-11]. In this case, the complexity of simulation model may be simplified but it is impossible to analyze actual operation of MMC-based HVDC converter in switching-level. Therefore, this model is not effective to utilize in analyzing and designing the MMC-based HVDC converter.

This paper describes the development of a simulation model of an MMC-based BTB HVDC system with 12 SMs per each arm in order to analyze an MMC-based HVDC system operation at switching-level and design an MMC converter. The developed simulation model adopted each SM's DC voltage equilibrium maintenance algorithm, an algorithm for the inhibition of cyclic current, an algorithm to reduce harmonic wave of alternating current (AC) output voltage, and an algorithm to bypass a SM which became out of order.

II. MMC(MODULAR MULTI-LEVEL CONVERTER)

A. Basic Structure

Fig. 1. shows the circuit diagram of a three-phase MMC, in which each arm is composed of SMs and arm reactors connected in series. An SM consists of two semiconductor switches and one capacitor. It forms the output terminal voltage with positive, negative, and zero voltage according to the ON/Off operation of the upper and lower end switches. When current flows into or flows out from the capacitor, the capacitor voltage goes up or down. That means the capacitor voltage changes according to the operation state of both switches in SM.

Without appropriate switching motion, the capacitor voltage at each SM in MMC is different, as a result failing to form SM's output terminal voltage appropriately. This leads to increasing total harmonic distortions (THDs) of MMC's output voltage. Therefore,

978-1-4799-2706-7/14 $31.00 © 2014 IEEE   937

maintaining all SM capacitor's voltages at a certain value is very important.

When the output voltage of MMC is formed using the staircase modulation method, the level of output voltage is determined by the number of SMs comprising an arm. If the number of SMs comprising the arm is n, the voltage level of output voltage is formed at n+1. Each level of output voltage is determined by the number of SMs under ON state in the upper and lower arms. In order for output voltage to form positive n/2-level including the neutral point, all SMs of the upper arm should be in ON state and all SMs of the lower arm should be in OFF state. On the contrary, in order for output voltage to form negative n/2-level, all SMs of the upper arm should be in OFF state and all SMs of the lower arm should be in ON state.

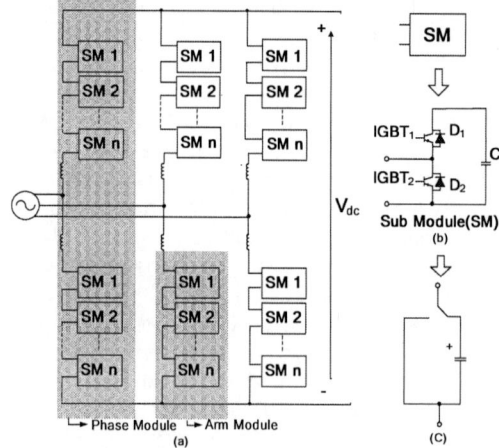

Fig. 1. MMC Structure, (a)3-phase MMC System, (b)SM(sub-module), (c)equivalent circuit

B. Output Voltage Modulation

The method to form MMC's output voltage is largely divided into a PWM method and a staircase modulation method [13]. PWM method is inappropriate to apply for a high-power system due to switching loss. Therefore, this paper formed the output voltage using a staircase modulation method. In MMC's output voltage, each SM comprising the upper and lower arms should perform ON/OFF operation at an appropriate moment so that the wave form is generated close to sine waves. This paper applied equal area method (EAM) shown in Fig. 2. to minimize harmonics of the output voltage [14].

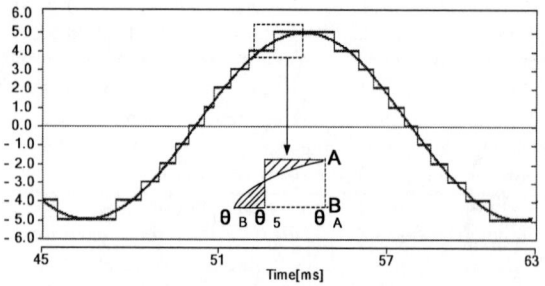

Fig. 2. EAM for Harmonic Reduction

If the two slashed areas have the equal area, the wave form of output voltage which may be formed with the number of SMs comprising an arm contains a minimum level of harmonic. Equation (1) is a formula to calculate which makes their areas the same.

$$\int_{\theta_B}^{\theta_S} (m_a \sin\theta - B)d\theta - \int_{\theta_S}^{\theta_A} (A - m_a \sin\theta)d\theta = 0 \quad (1)$$

Equation (1) is simplified into equation (2).

$$\theta_S = \frac{A\theta_A - B\theta_B + m_a\cos\theta_A - m_a\cos\theta_B}{A - B} \quad (2)$$

θA and θB in equation (2) can be derived as equation (3).

$$\theta_A = \sin^{-1}\frac{A}{m_a}, \theta_B = \sin^{-1}\frac{B}{m_a} \quad (3)$$

Generalizing equation (2) so that it may be applied even when the number of SMs changes leads to equation (4)

$$\theta_n = (n\theta_{n/m} - (n-1)\theta_{(n-1)/M})$$
$$+ Mm_a(\cos\theta_{n/M} - \cos\theta_{(n-1)/M}) \quad (4)$$

Where, n=1,2, ···,n, A= -M/n, B= -M/n-1, and M is equal to n/2 (n: the number of SMs in the arm)

C. SM Capacitor Voltage Balancing

When SM is in ON condition, the voltage of capacitor is formed at the output terminal. If the capacitor voltages of all SMs of three phase upper and lower arms fail to maintain a 1/n of DC-link voltage, the output voltage of MMC is not formed close to the sine wave and the THD of output increases. And another adverse effect is uneven voltage stress given to the semiconductor switch.

To maintain the capacitor voltage constant, each SM capacitor voltage in each arm is measured at every control cycle. It is aligned in ascending or descending order through a sorting algorithm. If the MMC output voltage is determined, the numbers of SMs to be in ON or OFF state should be decided. When SM is in ON condition, the capacitor is charged or discharged according to the direction of current. If current flows in the direction of charging, the SMs with lower capacitor voltage aligned by the sorting algorithm become in ON state to maintain a certain voltage. If current flows in the direction of discharging, the SMs with high capacitor voltage aligned by the sorting algorithm become in OFF state to maintain a certain voltage. Fig. 3. displays an algorithm for the DC voltage balancing aimed at maintaining the capacitor voltage of all SMs at a certain level.

The 2014 International Power Electronics Conference

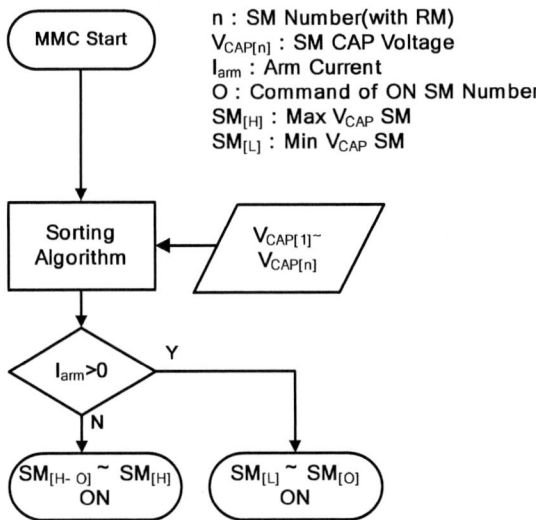

n : SM Number(with RM)
$V_{CAP[n]}$: SM CAP Voltage
I_{arm} : Arm Current
O : Command of ON SM Number
$SM_{[H]}$: Max V_{CAP} SM
$SM_{[L]}$: Min V_{CAP} SM

Fig. 3. SM Capacitor Voltage Balancing Algorithm

D. Circulating Current Suppression

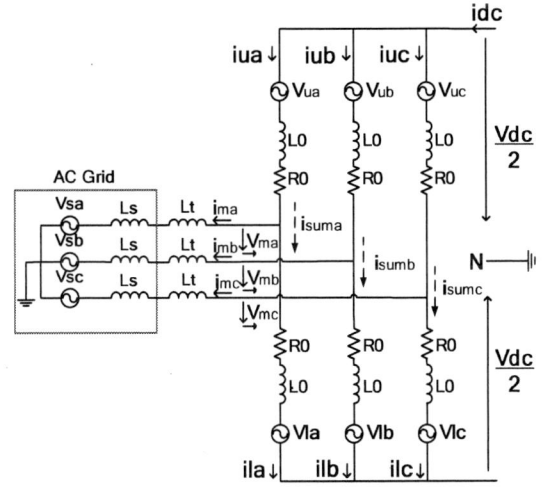

Fig. 4. Circulating current of MMC

The voltage difference between the upper and lower arms occurs cyclically and accordingly internal circulating current is created. Fig. 4. shows internal circulating current through the MMC's equivalent circuit.

Circulating current distorts the arm current and increases voltage ripple of SM capacitor. In order to suppress them, there is a method to increase the rating of arm reactor. However, large arm reactor decreases the rapidness of control response and increases system cost. Circulating current suppression algorithm was proposed and analyzed in [15, 16]. The circulating current flows through each phase of an MMC with a frequency characteristic twice as high as the grid frequency. The circulating currents are measured in the order of phase A, C, and B and they are transformed into d and q components through the d-q transform process. This transformed current is compared with the reference value of zero. The reference voltage to suppress the circulating current is generated through PI control and inverse d-q

transform as shown in Fig. 5(a).

The generated CCSC reference voltage is combined with the reference voltage generated in the current controller, as shown in Fig. 5(b). After then it generates a reference phase voltage with new phase A, B, C. By applying the reference phase voltage with new A, B, and C phases to the EAM, the output voltage to suppress the circulating current and minimize the harmonics is generated.

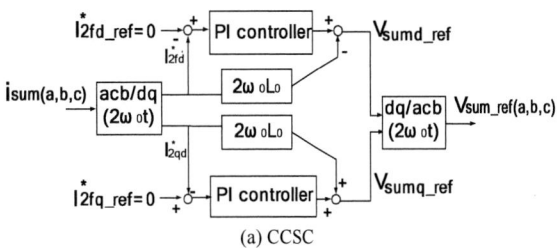

(a) CCSC

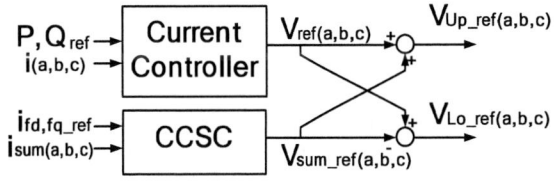

(b) Current controller with CCSC
Fig. 5. Current controller

E. Redundancy Module Operation Algorithm

If the number of SMs for an arm is increased to n+1 or n+2, only n SMs among n+1 or n+2 SMs are operated during the normal operation. When one SM becomes out of order, the SM is bypassed with a mechanical switch, and one or two redundancy modules is involved in operation to form the output voltage for maintaining normal operation. Redundancy module does not remain without any operation when all SMs are under normal operation. After all voltages of n+2 SM capacitors are measured and aligned by a sorting algorithm, the number of SMs to be ON state is determined among n+2 SMs. Through this activity, all capacitor voltages of n+2 SMs form a certain voltage. Therefore, when one SM is excluded because of malfunction, n SMs among a total of n+1 SMs are turned on to stably form output voltage.

However, when the arm is composed of n+2 SMs and two redundancy modules are set as certain SMs, the output voltage with n normal SMs excluding the redundancy modules are formed. When one SM is out of order, redundancy module should be inserted but the capacitor of the redundancy module is not charged. So, a certain amount of time is needed to charge it to higher than a certain voltage, and the voltage formation cannot be made rapidly by inserting the redundancy module. Fig. 6. shows a process of inserting the redundancy module.

978-1-4799-2706-7/14 $31.00 © 2014 IEEE   939

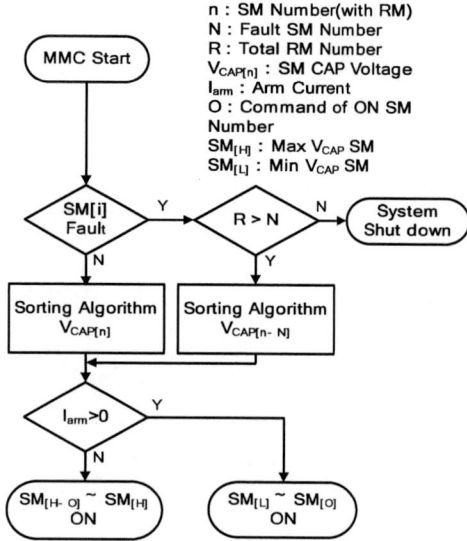

n : SM Number(with RM)
N : Fault SM Number
R : Total RM Number
V_CAP[n] : SM CAP Voltage
I_arm : Arm Current
O : Command of ON SM Number
SM[H] : Max V_CAP SM
SM[L] : Min V_CAP SM

Fig. 6. Redundancy module operation algorithm

III. MMC CONTROLLER

The MMC controller is composed of a master controller, a current controller, an arm controller, and an SM controller. The master controller generates the reference value of active power and reactive power using energy management algorithm. The generated reference value is delivered to the current controller. The current controller consists of algorithm to reduce the circulating current and algorithm to control the output current. The reference voltage for phase A, B, and C are generated and delivered to the arm controller. The arm controller has algorithm to reduce the harmonics of the output voltage, algorithm to balance the capacitor voltage in SM, and the redundancy algorithm of SM. The SM controller provides the ON/OFF switching command to the SM to be in ON state.

Fig. 7. shows a current controller which controls the magnitude and phase of the output voltage to control active and reactive power of the back-to-back MMC. The current controller receives the command for active and reactive power from the master controller and generates three phase reference voltages. To look into the operation of current controller in more detail, the reference value of active current is compared with the measured active current and generates the active power component of the converter through the PI control. This active power component is combined with the measured active power component and is compensated to the extent of cross component supplied from reactive current component, and then reference active voltage to be output by the converter is generated. Meanwhile, the reference value of reactive current is compared with the measured reactive current and passes through the PI control and generates reactive voltage component of the converter. This reactive voltage component is combined with the measured reactive voltage component and then compensated to the extent of cross component supplied from reactive current component and generates the reference reactive power v_q^* of the converter. When the

active and reactive reference voltages formed in this way are d-q reverse transformed, the reference phase voltage to be output v_a^*, v_b^*, v_c^* are generated. In the staircase modulation, the reference phase voltage generated in the current controller is input into the algorithm for the reduction of harmonic waves of the output voltage (EAM) and generates gate pulse which enables ON and OFF of SMs at the point where harmonic waves of MMC's output voltage become minimal.

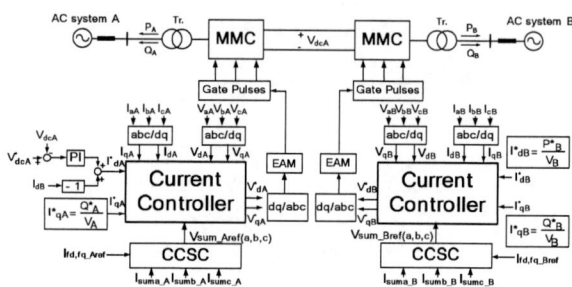

Fig. 7. Current controller

IV. 11-LEVEL HVDC SIMULATION MODEL

A. HVDC Simulation Model

Fig. 8. shows system configuration of 25MVA class 11-level MMC HVDC analyzed in this paper. Two MMCs are connected with Back-to-Back (BTB) structure and the DC link voltage is controlled at ±10kV. In the MMC HDVC system, one MMC conducts the control operation to maintain the DC link voltage at a certain level through active power control and selectively performs the power factor control or the grid voltage stabilizing control through reactive power control. The other MMC controls power flow in bidirectional mode through the active power control and performs the power factor control or the grid voltage stabilizing control through reactive power control.

In order to form the output voltage of 11 levels, the number of SMs comprising an arm should be 10. In the simulation model, two redundancy modules are added to each arm and the total number of SMs comprising the arm is 12. The arm reactor suppresses the circulating current during normal operation and suppresses the fault current at DC side. The transformer raises the MMC's output voltage to the grid voltage level with electrical isolation.

A simulation model for 25MVA 11-level MMC HVDC System was developed with PSCAD/EMTDC. The system parameters used in simulation are presented in Table 1. The developed simulation model includes the DC-voltage balance algorithm, the circulating-current suppression algorithm, the harmonic reduction algorithm, and the redundancy operation algorithm. All control algorithms were implemented using user defined models comprised of C codes.

978-1-4799-2706-7/14 $31.00 © 2014 IEEE

The 2014 International Power Electronics Conference

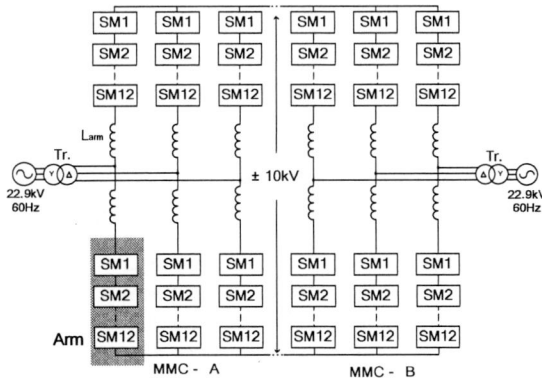

Fig. 8. 25MVA 11-level MMC HVDC System

TABLE I
SYSTEM PARAMETER

Item	Values
Rated Power	25 MVA
DC Link Voltage	20 kV
SM Capacitor	9100 uF
Arm reactor	2.5 mH
Line-Line Output Voltage	11.0 kV
Tr.(Y, 22.9kV//□, 11.0kV) Leakage Reactance	5.5 mH
Number of SM per Arm	12 ea

B. Active and Reactive Power Control

In the developed simulation model, active and reactive power control was verified according to appropriate scenarios. The voltage and current at major points were checked to confirm the reasonable operation. Also the voltage ripples appearing in the designed SM capacitor were verified.

Table 2 shows the reference value change of active and reactive power according to time. Fig. 9. shows the graph of the reference values and the measured values of active and reactive voltage according to time. The total simulation time is divided into 10 modes and analysis of each mode's operation condition was made through analyses of extended waveform and THDs.

TABLE II
SIMULATION TIME TABLE

Item	Values									
Mode	1	2	3	4	5	6	7	8	9	10
Paref	20M		0M		-20M		0M			
Qaref	0M	15M		0M		-15M		0M		
Qbref	0M								15M	-15M

The MMC at receiving end controls DC link voltage at a constant value of 20kV and it was verified that it maintained constant value even under a transient condition where active power is controlled in both ways. Moreover, the reactive voltage of the grid connected to each MMC well followed the reference value of reactive power control.

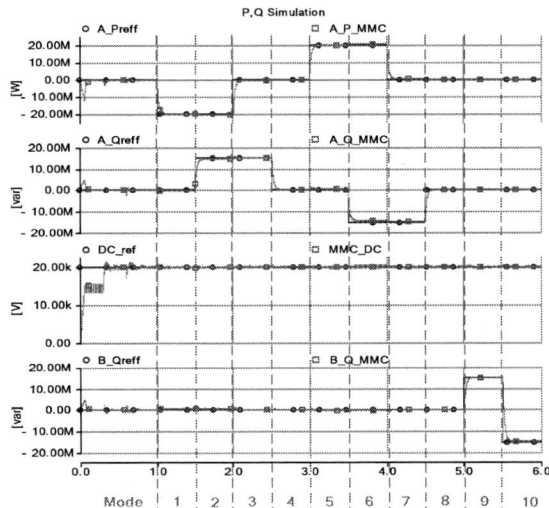

Fig. 9. P & Q Control of 25MVA 11-level MMC HVDC System

Fig. 10(a) shows the magnified view for the mode-1 section where active power operated at ration among all sections of Fig. 9. The voltage and current waveforms were observed, and the THDs were measured, and the performance of EAM, the output voltage formation algorithm, was verified. Fig. 10(b) shows the magnified view of the mode-2 section where both active and reactive power operated at ration among all sections of Fig. 9. The voltage and current waveforms were observed and the THDs were measured.

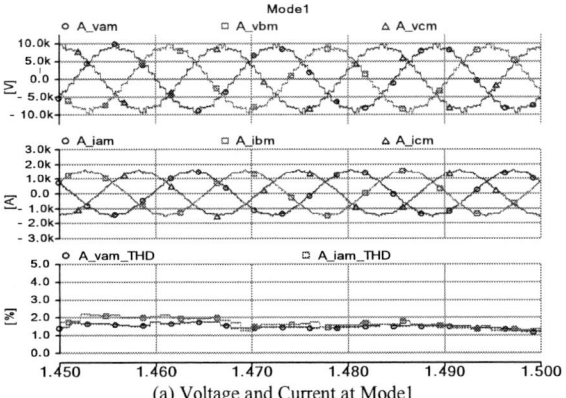

(a) Voltage and Current at Mode1

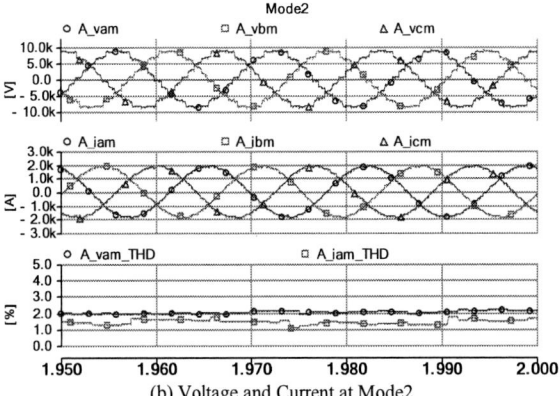

(b) Voltage and Current at Mode2

Fig. 10. Extended Waveform of Output Voltage and Current

978-1-4799-2706-7/14 $31.00 © 2014 IEEE   941

Fig. 11. is the resulting waveform of applying the capacity of SM capacitor designed above to the simulation. The capacitor was designed with allowable voltage ripple of SM set at 10% and the validity of capacitor design formula was verified through simulation result analysis. It is shown that a large voltage ripple of SM capacitor appears when the load is greatest.

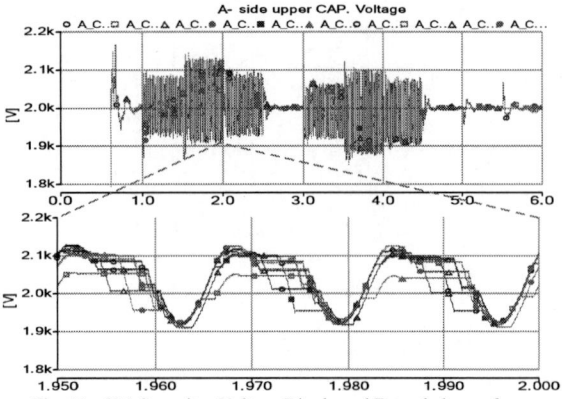

Fig. 11. SM Capacitor Voltage Ripple and Extended waveform

C. The Simulation Applying CCSC

Fig. 12. shows waveforms to verify the CCSC performance by applying CCSC to the simulation for the performance improvement of MMC by comparing the waveforms before and after operation. The first waveform shows circulating current waveform flowing in phase A and the circulating current's frequency is at 120Hz twice as high as the grid frequency. The size of circulating current before and after CCSC operation differs markedly. The second waveform shows the arm current flowing into the upper arm. It shows a distorted waveform containing harmonics before the operation of CCSC but improves into sine waves after the operation of CCSC. The third waveform shows capacitor voltage ripples of SMs in upper arm. When the circulating current is suppressed, the voltage ripple at the capacitor has become smaller by about 40V. The fourth and fifth waveforms are the 3-phase output voltage and output current, which did not directly affect the MMC's active and reactive power control before and after the operation of CCSC.

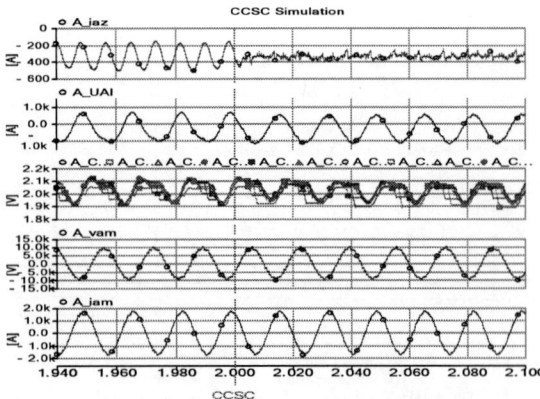

Fig. 12. Simulated waveform of MMC using CCSC

D. Redundancy Module Operation Simulation

The designed 11-level MMC includes two additional redundancy modules and therefore it has 12SMs in one arm. The 10 SMs are necessary to form the 11-level output voltage. Even though problems with a maximum of 2 SMs occur, it may continue to operate normally. Fig. 13. shows simulation results of the output voltage, output current, and the voltage ripples at SM capacitor to verify whether MMC operates normally when up to two SMs become out of order consecutively. When an SM becomes out of order, the ripple voltage at the SM capacitor rises slightly from the transient condition. However, the output voltage and current maintain the condition before a fault occurs in the SM.

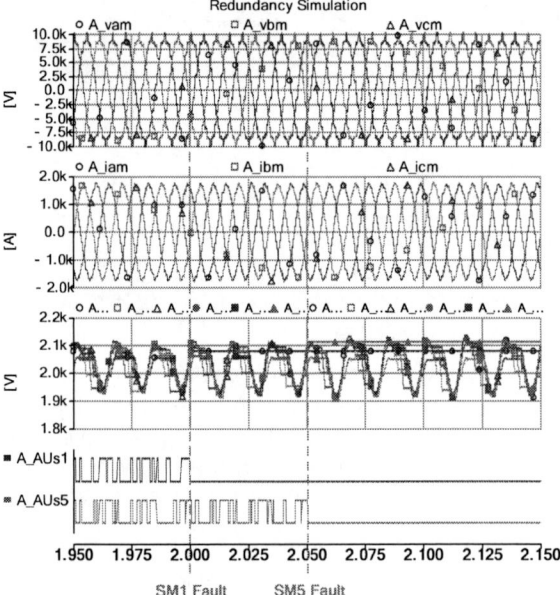

Fig. 13. Redundancy Module Operation

V. CONCLUSION

A switching-level simulation model of MMC-based BTB (back-to-back) converter for HVDC application is very important to analyze its normal operations and transient phenomena. However, the switching-level simulation model for the commercial MMC-based BTB converter is very difficult because it consists of more than 150 sub-modules for each arm.

In this paper a switching-level simulation model for the 11-level MMC-based BTB converter was developed with PSCAD/EMTDC software. It has 12 sub-modules for the positive arm and another 12 sub-modules for the negative arm.

The developed simulation model adopts the DC-voltage balance algorithm, the circulating-current reduction algorithm, the harmonic reduction algorithm, and the redundancy operation algorithm. So, the developed simulation model can be utilized to design the MMC-based BTB converter and to develop its protection algorithm.

REFERENCES

[1] U. Axelsson, A. Holm, C. Liljegren, M. Aberg, K. Eriksson, O. Tollerz "The Gotland HVDC Light project-experiences from trial and commercial operation", Eletricity Distribution, 2001. Part1:Contributions. CIRED. 16th International Conference and Exhibition on (IEE Conf. Publ No. 482)

[2] B. Jacobson , P. Kalsson , G. Asplund , L. Harnefors and T. Jonsson "VSC-HVDC transmission with cascaded two-level converters", CIGRE Session, pp.B4 -B110 2010

[3] C D Barker, N M Kirby, "Reactive power loading of components within a modular multi-level VSC HVDC converter", IEEE Electric power and energy conference, Winnipeg, Canada, 2011

[4] R. Marquardt "Modular multilevel convertertopologies with DC-short circuit current limitation", Proc.8th Int. Power Electron. ECCE Asia Conf., pp.1425 -1431 2011

[5] R. Marquardt "Modular multilevel converter: An universal concept for HVDC-Networks and extended DC-Bus-applications", Proc. Int. Power Electron. Conf., 2010, Jun. 2010

[6] R. Marquardt, A. Lesnicar, "New Concept for High Voltage - Modular Multilevel Converter," PESC 2004, Aachen.

[7] R. Marquardt, A.Lesnicar, "A new modular voltage source inverter topology," EPE 2003, Toulouse.

[8] H.J. Knaak, " Modular Multilevel Converters and HVDC/FACTS: a success story, " in Conf EPE 2011, Birmingham.

[9] U. N. Gnanarathna , A. M. Gole and R. P. Jayasinghe "Efficient modeling of modular multilevel HVDC converters (MMC) on electromagnetic transient simulation programs", IEEE Trans. Power Del., vol. 26, no. 1, pp.316 –324 2011

[10] J. Peralta, H. Saad, S. Dennetiee, J. Mahseredjian, and S. Nguefeu, "Detailed and averaged models for a 401-level MMC-HVDC system," IEEE Trans. Power Del., vol. 27, no. 3, pp. 1501–1508, Jul. 2012.

[11] B. Han, H. Kim, and S. Baek, "Performance analysis of SSSC based on three-level multi-bridge PWM inverter", Elsevier Sci. Elect. Power Syst. Rese., vol. 61, no. 3, pp.195 -202 2002

[12] G. Ding, M. Ding and G.Tang "An Innovative Modular Multilevel Converter Topology and Modulation Control Scheme for the first VSC-HVDC project in China", in Proc. 16th Power Systems Computation Conference (PSCC), 14-18 July 2008, Glasgow, United Kingdom.

[13] B. Han , B. Bae , S. Baek and G. Jang "New configuration of UPQC for medium-voltage application", IEEE Trans. Power Del., vol. 21, no. 3, pp.1438 –1444 2006

[14] B. Han , S. Baek , H. Kim and G. Karady "Dynamic characteristic analysis of SSSC based on multibridge inverter", IEEE Trans. Power Del., vol. 17, no. 2, pp.623 -629 2002

[15] Q. Tu, Z. Xu and L. Xu "Reduced switching-frequency modulation and circulating current suppression for modular multilevel converters", IEEE Trans. Power Del., vol. 26, no. 3, pp.2009 -2017 2011

[16] Q. Tu, Z. Xu, and J. Zhang, "Circulating current suppressing controller in modular multilevel converter," in Proc. IECON 2010 - 36th Annual Conf. IEEE Industrial Electronics Society, 2010, pp. 3198-3202.

Power-Cell Switching-Cycle Capacitor Voltage Control for the Modular Multilevel Converters

Jun Wang, Rolando Burgos, Dushan Boroyevich, Bo Wen

Center for Power Electronics Systems
Virginia Polytechnic Institute and State University
Blacksburg, VA 24061, USA
junwang@vt.edu

Abstract-This paper presents a power-cell Switching-Cycle Capacitor Voltage Control (SCCVC) approach to control the capacitor voltage ripple for the Modular Multilevel Converters (MMC). The capacitor voltage will be following the reference in one switching cycle by taking advantage of the degree of freedom in the control, the circulating current. The capacitor voltages resonate with the inductors in the converter circuit when the capacitors are inserted into the converter arm, and can reach the reference voltage at the end of the insertion time interval by proper control of the initial arm current. The SCCVC decouples the capacitor voltage ripple magnitude apart from the line frequency to certain extent, which not only provides a significant reduction of the capacitance value in designing the MMC parameters, but also enable the MMC to deliver very low frequency or even DC line currents. The control principle of the SCCVC and the resonance behavior between the capacitors and inductors are shown in this paper and finally the simulation results validate the control feasibility.

I. INTRODUCTION

The Modular Multilevel Converter (MMC), shown in Fig.1, has been increasingly considered for medium voltage and high voltage variable frequency applications due to its favorable features of high modularity and scalability [1]-[5]. The conventional operation mode of the MMC requires the capacitors in power modules to buffer the power fluctuations at line frequency and second order harmonic frequency, inherently resulting in a very large capacitance as well as the fact that the capacitor voltage ripple magnitude is inversely proportional to the phase current frequency [1] and will become infinite at zero frequency (DC), which is a big issue at the start of the motor where the phase current frequency growing from zero is required. A. J. Korn proposed a method to shift the arm current towards a higher frequency and therefore to reduce the capacitor energy ripple magnitude, by injecting high frequency sinusoidal circulating currents and adding high frequency common mode voltages on the three phases during the low frequency operation of the MMC [2]. This approach is applicable in the motor drive applications with quadratic torque load, although the torque and current has to be derated due to increased semiconductor loss. But the impact on the bearings of the motor made by the common mode voltages used in this method remains to be an issue, and the capacitance value and size remain as

large as the conventional designs. In this paper the proposed SCCVC shifts the frequency component of the arm current towards the switching frequency to reduce the required capacitance value and to enable the converter operation at very low frequency and DC.

II. PRINCIPLE OF THE SCCVC

The concept of SCCVC can be illustrated with the simplest MMC circuit of single phase and single cell per arm, as shown in Fig.2. The two semiconductor switches in each module are represented by a Single-Throw-Double-Pole (STDP) switch. The goal of the converter is to deliver power from DC sources V_{dc} to the AC source v_S with a sinusoidal current i_{ph} flowing through it, controlled by proper operations of the two STDP switches whose switching functions are defined as (1). S_{Ux} and S_{Lx} are the switching functions of the single-throw-double-pole switch that can be '1' or '0', where x is a phase identifier, given by (1). For simplicity, "x=a, b, c" will be omitted in all the equations that contains x. The phase identifier will not be shown in the single phase case.

$$S_{U(L)x} = \begin{cases} 1 & \text{when switched to the upper pole} \\ 0 & \text{when switched to the lower pole} \end{cases} \quad (1)$$

The addition of the upper and lower arm current is labeled as i_{sum} given by (2), while the subtraction between the upper and lower arm current is the current flowing to the phase branch, given by (3). Since the sum current is equivalently flowing in the loop between DC-link and two arms, the half of its values is usually defined as the circulating current i_{cir}.

$$i_{sum} = i_U + i_L = 2i_{cir} \quad (2)$$

$$i_{ph} = i_U - i_L \quad (3)$$

Based on circuit analysis and previously defined sum and phase current, the current state equations can be derived as (4) and (5).

$$\frac{d}{dt}i_{sum} = \frac{V_{dc} - (S_U v_U + S_L v_L)}{L} \quad (4)$$

$$\frac{d}{dt}i_{ph} = \frac{(S_L v_L - S_U v_U)/2 - v_S}{L_O + L/2} \quad (5)$$

978-1-4799-2706-7/14 $31.00 © 2014 IEEE   944

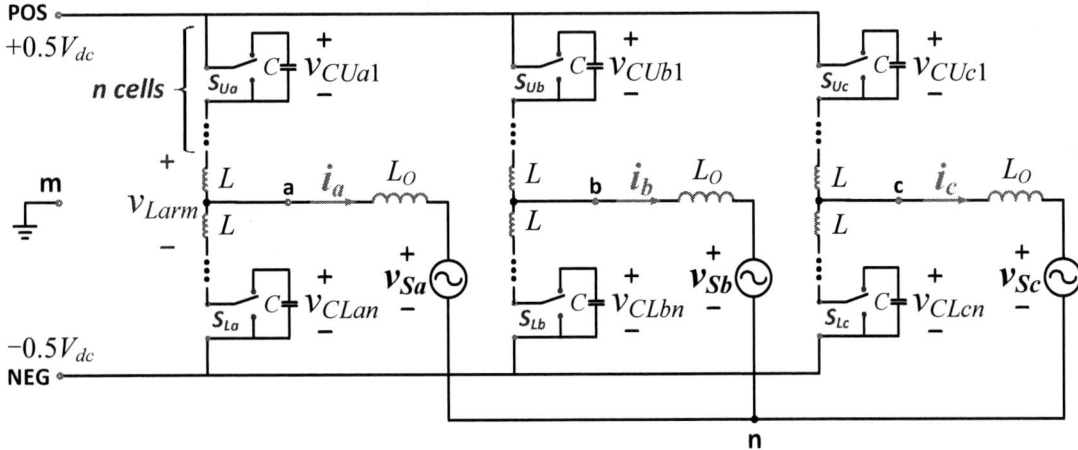

Fig. 1 Schematic of the Modular Multilevel Converter

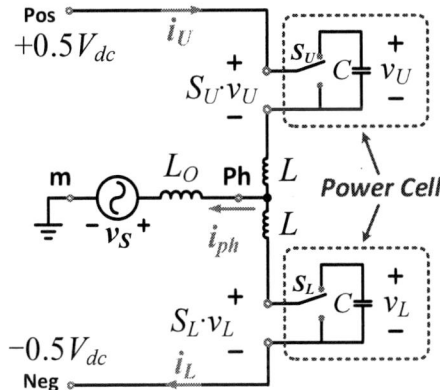

Fig. 2 The single-phase MMC with single module per arm

The capacitor voltage equations are derived as shown in (6) and (7).

$$\frac{d}{dt}v_U = \frac{1}{C}S_U i_U = S_U \frac{i_{sum}+i_{ph}}{2C} \qquad (6)$$

$$\frac{d}{dt}v_L = \frac{1}{C}S_L i_L = S_L \frac{i_{sum}-i_{ph}}{2C} \qquad (7)$$

It is observed that the critical state variables of phase current and capacitor voltages are controlled by the equations of (5), (6) and (7), which determine whether the converter can fulfill its function of power delivery with sinusoidal phase current and balanced capacitor voltages. However, the remaining equation (4) tells that on basis of the function fulfillment, the sum (circulating) current can be regulated as desired, which offers a degree of freedom to influence the arm currents that charge and discharge the capacitor. Therefore, the SCCVC is proposed motivated by this observation.

$$S_U + S_L \neq 1 \qquad (8)$$

As illustrated in Fig.3, there are totally four switching states of the switching function combinations for the converter shown in Fig.2. The state ① and ②, with only one capacitor connected in the circuit, control the energy transfer to the load. The state ③ and ④ control the sum (circulating) current flowing between the DC sources and the capacitors. In conventional control approaches of the MMC, the states that satisfy (8), the states ③ and ④ in Fig.15, are usually intentionally to be avoided such that there is not high peaking currents in the arms, while the states ① and ② are used to control the phase current. However from time to time, there do exist the states of (8) caused by the circulating current control loop and generated by the comparison between carrier and modulation signals. In the proposed SCCVC control approach, the specific sequence arrangements of all the four states in a switching cycle in terms of the capacitor voltages and arm currents control have been taken into considerations, and then the states ③ and ④ become critical and useful.

The concept of this new control method is to taking advantage of the resonance between the arm inductors and cell capacitor to control the capacitor voltage back to the reference by giving proper initial arm current. Fig.4 shows the comparison between the conventional control method and the SCCVC to tell how the capacitor voltage balancing in one switching cycle is achieved. Assuming that the phase current is in its positive half cycle, in the figure of left-hand side the alternating use of states ① and ② gradually controls the average value of the phase current to be sinusoidal, while the sum current is not really impact by the two states. The unbalanced charges from one switching cycle to another make the capacitor voltages deviate from their original values, and finally back to their initial states after one line cycle. In the figure of right-hand side the state ③ is inserted right before the state ① to control the lower arm current to an offset value such that in the time interval of the state ① the average current flowing through the lower capacitor is controlled toward zero. Similarly, the SCCVC arranges the state ④ right before the state ② to control the upper arm current to an offset value such that in the time duration of the state ② the average current flowing through the upper capacitor is controlled toward zero.

The 2014 International Power Electronics Conference

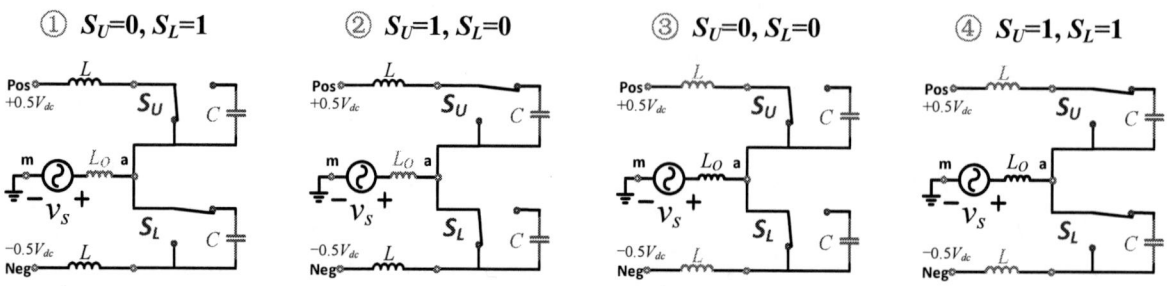

Fig. 3 The switching states of the single-phase MMC with single module per arm

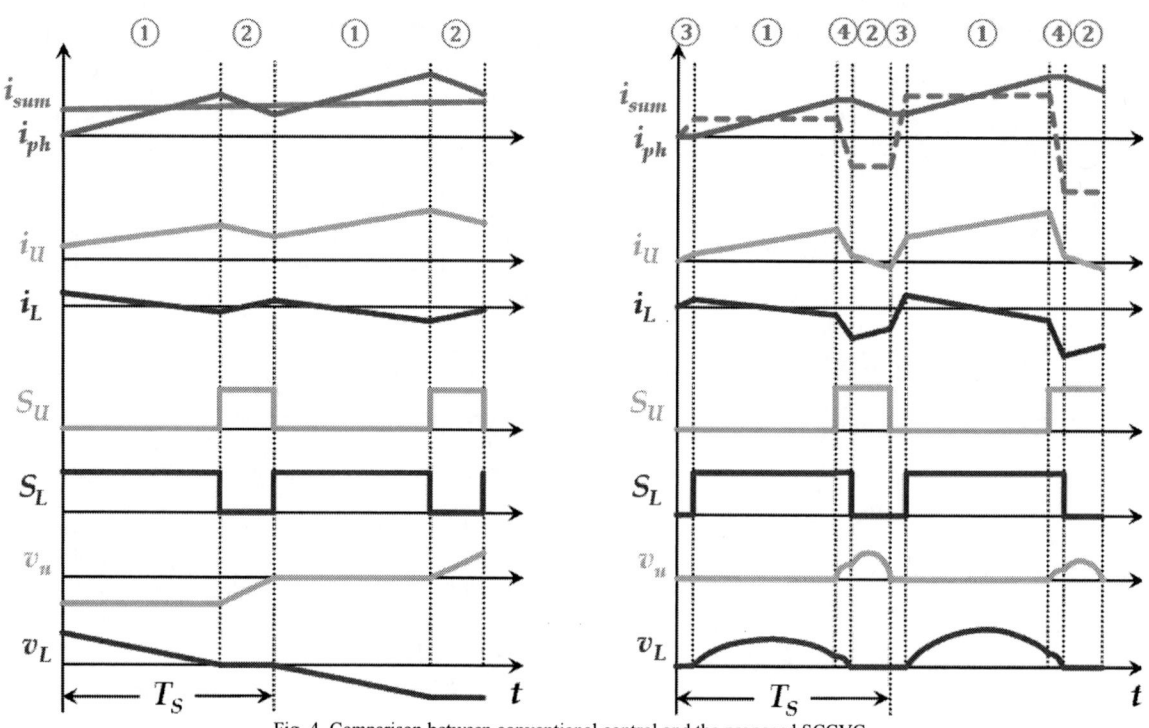

Fig. 4 Comparison between conventional control and the proposed SCCVC

III. GENERALIZATION OF THE SCCVC TO THE MULTI-CELL CASES

This method can be easily extended to the situations when the phase current is negative by inserting state ④ instead of ③ before state ①, and similarly putting state ③ before state ②. The SCCVC can also be directly extended to the three-phase case without any efforts since the arm inductors and cell capacitors are totally independent from one phase to another.

When the situation comes to multi-cell per arms, a simply extended method of the SCCVC is proposed in Fig.5, taking a 2-cell-per-arm case as demonstration. The concept is still to introduce the states that can quickly change the arm current such that the capacitor voltage can be brought to the reference at the next time interval. Assuming again that the phase current is in its positive half cycle, short delays are introduced at every edge of the two lower cell switches. If a proper duration of delay is placed at "a" on S_{L1}, then the capacitor voltage of v_{CL2}

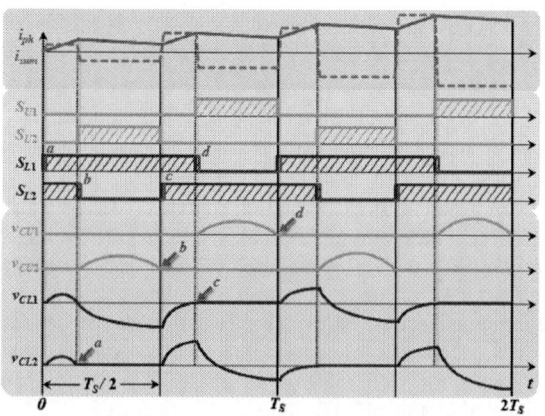

Fig. 5 The SCCVC in 2-cell-per-arm case

can resonate back to the reference voltage where an arrow of "a" is point toward. Afterward, the capacitor voltage v_{CL2} can stay at the reference voltage when the capacitor is bypassed. Similar mechanisms work well with the delays where "b", "c" and "d" are labeled.

978-1-4799-2706-7/14 $31.00 © 2014 IEEE   946

$$
i_U = -\left(
\begin{aligned}
&\left(\frac{\omega_0 V_s \cos(2\pi \cdot k/CR)}{(\omega_{r1}^2-\omega_0^2)(2L_o+L)}\left(1+\frac{(C_L-C_U)}{2(\omega_{r1}^2-\omega_{r2}^2)C_U C_L L_1}-\frac{1}{(\omega_{r1}^2-\omega_{r2}^2)C_L L_2}\right)\right.\\
&\left.+\frac{I_{U0}}{2}+\frac{(C_L-C_U)I_{U0}}{2(\omega_{r1}^2-\omega_{r2}^2)C_U C_L L_1}+\frac{I_{L0}}{(\omega_{r1}^2-\omega_{r2}^2)C_L L_2}\right)\cdot\cos(\omega_{r1}t)+\left(\frac{\omega_0 V_s \cos(2\pi \cdot k/CR)}{(\omega_{r2}^2-\omega_0^2)(2L_o+L)}\left(\frac{1}{2}-\frac{(C_L-C_U)}{2(\omega_{r1}^2-\omega_{r2}^2)C_U C_L L_1}+\frac{1}{(\omega_{r1}^2-\omega_{r2}^2)C_L L_2}\right)\right.\\
&\hspace{10cm}\left.+\frac{I_{U0}}{2}-\frac{(C_L-C_U)I_{U0}}{2(\omega_{r1}^2-\omega_{r2}^2)C_U C_L L_1}-\frac{I_{L0}}{(\omega_{r1}^2-\omega_{r2}^2)C_L L_2}\right)\cdot\cos(\omega_{r2}t)\\
&-\left(\frac{0.5V_{dc}}{\omega_{r1}L}\left(1+\frac{(C_L-C_U)}{2(\omega_{r1}^2-\omega_{r2}^2)C_U C_L L_1}+\frac{1}{(\omega_{r1}^2-\omega_{r2}^2)C_L L_2}\right)\right.\\
&\left.-\frac{\omega_{r1}V_s \sin(2\pi \cdot k/CR)}{(\omega_{r1}^2-\omega_0^2)(2L_o+L)}\left(1-\frac{(C_L-C_U)}{2(\omega_{r1}^2-\omega_{r2}^2)C_U C_L L_1}+\frac{1}{(\omega_{r1}^2-\omega_{r2}^2)C_L L_2}\right)\right)\cdot\sin(\omega_{r1}t)-\left(\frac{0.5V_{dc}}{\omega_{r2}L}\left(\frac{1}{2}-\frac{(C_L-C_U)}{2(\omega_{r1}^2-\omega_{r2}^2)C_U C_L L_1}-\frac{1}{(\omega_{r1}^2-\omega_{r2}^2)C_L L_2}\right)\right.\\
&\hspace{4cm}\left.-\frac{\omega_{r1}C_U V_{U0}}{(\omega_{r1}^2-\omega_{r2}^2)C_L L_1}+\frac{\omega_{r1}V_{L0}}{(\omega_{r1}^2-\omega_{r2}^2)L_2}+\frac{\omega_{r1}^3 C_U V_{U0}}{(\omega_{r1}^2-\omega_{r2}^2)}\right)\cdot\left(-\frac{\omega_{r2}V_s \sin(2\pi \cdot k/CR)}{(\omega_{r2}^2-\omega_0^2)(2L_o+L)}\left(\frac{1}{2}+\frac{(C_L-C_U)}{2(\omega_{r1}^2-\omega_{r2}^2)C_U C_L L_1}-\frac{1}{(\omega_{r1}^2-\omega_{r2}^2)C_L L_2}\right)\right)\cdot\sin(\omega_{r2}t)\\
&\hspace{8cm}+\frac{\omega_{r2}C_U V_{U0}}{(\omega_{r1}^2-\omega_{r2}^2)C_L L_1}-\frac{\omega_{r2}V_{L0}}{(\omega_{r1}^2-\omega_{r2}^2)L_2}-\frac{\omega_{r2}^3 C_U V_{U0}}{(\omega_{r1}^2-\omega_{r2}^2)}\\[6pt]
&+\frac{\omega_0 V_s}{(2L_o+L)}\left(\left(\frac{1}{\omega_{r1}^2-\omega_0^2}\right)\left(-\frac{1}{2}-\frac{(C_L-C_U)}{2(\omega_{r1}^2-\omega_{r2}^2)C_U C_L L_1}+\frac{1}{(\omega_{r1}^2-\omega_{r2}^2)C_L L_2}\right)\right.\\
&\left.\hspace{3cm}+\left(\frac{1}{\omega_{r2}^2-\omega_0^2}\right)\left(-\frac{1}{2}+\frac{(C_L-C_U)}{2(\omega_{r1}^2-\omega_{r2}^2)C_U C_L L_1}-\frac{1}{(\omega_{r1}^2-\omega_{r2}^2)C_L L_2}\right)\right)\cdot\cos(\omega_0 t+2\pi \cdot k/CR)
\end{aligned}
\right)
\tag{14}
$$

As observed from the Fig.5, the initial values of capacitor voltages when they begin to resonate toward the reference are no longer at the reference because the capacitor voltages will be impacted by the arm current when the other capacitors are in their courses of balancing. The resonant interactions among the multiple capacitors and the two arm inductors in one phase-leg require mathematical solutions to analyze.

IV. RESONANT BEHAVIOR ANALYSIS AND PARAMETER DESIGN GUIDELINE

In Fig.6, a 2-cell-per-arm MMC converter ($N_{Cell}=2$) at certain switching state is shown. From the circuit diagram, the state-space differential equation is derived as (9) using i_U, i_L, v_U and v_L as state variables that are of clear physical definitions.

$$
\frac{d}{dt}\begin{bmatrix} i_U \\ i_L \\ v_U \\ v_L \end{bmatrix}=
\begin{bmatrix}
0 & 0 & -\frac{1}{L_1} & \frac{1}{L_2} \\
0 & 0 & -\frac{1}{L_2} & -\frac{1}{L_1} \\
\frac{1}{C_U} & 0 & 0 & 0 \\
0 & \frac{1}{C_L} & 0 & 0
\end{bmatrix}
\begin{bmatrix} i_U \\ i_L \\ v_U \\ v_L \end{bmatrix}+
\begin{bmatrix}
\frac{0.5V_{dc}}{L}-\frac{V_s \sin(\omega_0 t+2\pi \cdot k/CR)}{2L_o+L} \\
\frac{0.5V_{dc}}{L}+\frac{V_s \sin(\omega_0 t+2\pi \cdot k/CR)}{2L_o+L} \\
0 \\
0
\end{bmatrix}
\tag{9}
$$

where the parameters are defined as (10) - (13).

$$
N_U=\sum_{i=1}^{N_{Cell}} S_{Ui}, \quad N_L=\sum_{i=1}^{N_{Cell}} S_{Li}
\tag{10}
$$

$$
v_U=\sum_{i=1}^{N_{Cell}} S_{Ui}\cdot v_{Ui}, \quad v_L=\sum_{i=1}^{N_{Cell}} S_{Li}\cdot v_{Li}
\tag{11}
$$

$$
\frac{1}{C_U}=\frac{N_U}{C}, \quad \frac{1}{C_L}=\frac{N_L}{C}
\tag{12}
$$

$$
L_1=\frac{L(2L_O+L)}{L_O+L}, \quad L_2=\frac{L(2L_O+L)}{L_O}
\tag{13}
$$

The complete solution of the i_U is shown as (14) where two resonant frequencies are found in (15) and (16).

$$
\omega_{r1}=\sqrt{\frac{C_U+C_L}{2C_U C_L L_1}+\frac{\sqrt{C_L^2 L_2^2+4C_U C_L L_1^2-2C_U C_L L_2^2+C_U^2 L_2^2}}{2C_U C_L L_1 L_2}}
\tag{15}
$$

$$
\omega_{r2}=\sqrt{\frac{C_U+C_L}{2C_U C_L L_1}-\frac{\sqrt{C_L^2 L_2^2+4C_U C_L L_1^2-2C_U C_L L_2^2+C_U^2 L_2^2}}{2C_U C_L L_1 L_2}}
\tag{16}
$$

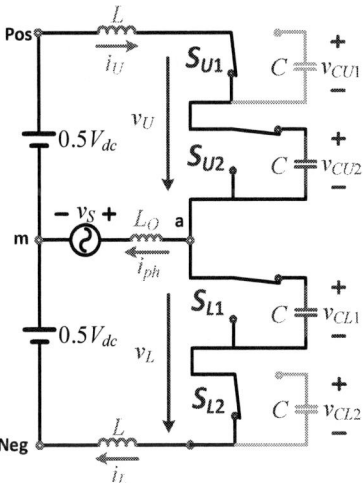

Fig. 6 Circuit diagram of a MMC with three cells per arm

Simplification can be made when $N_U=0$ or $N_L=0$ and one of the two terms in (12) equals to zero that further simplify the coefficient matrix in (9). With similar derivation procedure, the lower arm current i_L can also be derived as in (17). The approximation of the two resonant frequencies are shown in (18) and (19) indicating the physical meaning of the resonant behavior of capacitors and inductors. The faster resonance between the series of the two equivalent arm capacitors and the series of the two arm inductors determines the basic behaviors of the circuit, where the lower frequency resonance and the fundamental component in (14) have minor but not negligible impact on the time-domain response.

$$
\omega_{r1}\approx\sqrt{\frac{1}{2L}\left(\frac{1}{C_U}+\frac{1}{C_L}\right)}
\tag{18}
$$

$$
\omega_{r2}\approx\sqrt{\frac{1}{4L_O}\left(\frac{1}{C_U}+\frac{1}{C_L}\right)}
\tag{19}
$$

With the derived two arm current expressions, one can easily calculate the any capacitor voltage response during the time interval which can be used to calculate the initial arm current value that are needed to balance the capacitor voltages. Meanwhile, the impact to the other capacitors when the control is trying to balance one capacitor can also be simply derived.

$$
i_L = -\left(
\begin{aligned}
&\left(\frac{\omega_0 V_s \cos(2\pi \cdot k/CR)}{(\omega_{r1}^2-\omega_0^2)(2L_o+L)}\left(-\frac{1}{2}-\frac{(C_U-C_L)}{2(\omega_{r1}^2-\omega_{r2}^2)C_U C_L L_1}+\frac{1}{(\omega_{r1}^2-\omega_{r2}^2)C_U L_2}\right)\right) \\
&\left.+\frac{I_{L0}}{2}+\frac{(C_U-C_L)I_{L0}}{2(\omega_{r1}^2-\omega_{r2}^2)C_U C_L L_1}+\frac{I_{U0}}{(\omega_{r1}^2-\omega_{r2}^2)C_U L_2}\right)\cdot\cos(\omega_{r1}t) \\[4pt]
&-\left(\frac{-0.5V_{dc}}{\omega_{r1}L}\left(\frac{1}{2}+\frac{(C_U-C_L)}{2(\omega_{r1}^2-\omega_{r2}^2)C_U C_L L_1}+\frac{1}{(\omega_{r1}^2-\omega_{r2}^2)C_U L_2}\right)\right. \\
&-\frac{\omega_{r1}V_s \sin(2\pi \cdot k/CR)}{(\omega_{r1}^2-\omega_0^2)(2L_o+L)}\left(\frac{1}{2}+\frac{(C_U-C_L)}{2(\omega_{r1}^2-\omega_{r2}^2)C_U C_L L_1}-\frac{1}{(\omega_{r1}^2-\omega_{r2}^2)C_U L_2}\right) \\
&\left.+\frac{\omega_{r1}V_{U0}}{(\omega_{r1}^2-\omega_{r2}^2)L_2}+\frac{\omega_{r1}C_L V_{L0}}{2}-\frac{\omega_{r1}(C_L-C_U)V_{L0}}{2(\omega_{r1}^2-\omega_{r2}^2)C_U L_1}\right)\cdot\sin(\omega_{r1}t) \\[4pt]
&+\frac{\omega_0 V_s}{(2L_o+L)}\left(\left(\frac{1}{\omega_{r1}^2-\omega_0^2}\right)\left(\frac{1}{2}+\frac{(C_U-C_L)}{2(\omega_{r1}^2-\omega_{r2}^2)C_U C_L L_1}-\frac{1}{(\omega_{r1}^2-\omega_{r2}^2)C_U L_2}\right)\right. \\
&\left.+\left(\frac{1}{\omega_{r2}^2-\omega_0^2}\right)\left(\frac{1}{2}-\frac{(C_U-C_L)}{2(\omega_{r1}^2-\omega_{r2}^2)C_U C_L L_1}+\frac{1}{(\omega_{r1}^2-\omega_{r2}^2)C_U L_2}\right)\right)\cdot\cos(\omega_0 t+2\pi\cdot k/CR)
\end{aligned}
\right.
$$

$$
\begin{aligned}
&+\left(\frac{\omega_0 V_s \cos(2\pi \cdot k/CR)}{(\omega_{r2}^2-\omega_0^2)(2L_o+L)}\left(-\frac{1}{2}+\frac{(C_U-C_L)}{2(\omega_{r1}^2-\omega_{r2}^2)C_U C_L L_1}-\frac{1}{(\omega_{r1}^2-\omega_{r2}^2)C_U L_2}\right)\right. \\
&\left.+\frac{I_{L0}}{2}-\frac{(C_U-C_L)I_{L0}}{2(\omega_{r1}^2-\omega_{r2}^2)C_U C_L L_1}-\frac{I_{U0}}{(\omega_{r1}^2-\omega_{r2}^2)C_U L_2}\right)\cdot\cos(\omega_{r2}t) \\[4pt]
&-\left(\frac{-0.5V_{dc}}{\omega_{r2}L}\left(\frac{1}{2}-\frac{(C_U-C_L)}{2(\omega_{r1}^2-\omega_{r2}^2)C_U C_L L_1}-\frac{1}{(\omega_{r1}^2-\omega_{r2}^2)C_U L_2}\right)\right. \\
&-\frac{\omega_{r2}V_s \sin(2\pi \cdot k/CR)}{(\omega_{r2}^2-\omega_0^2)(2L_o+L)}\left(\frac{1}{2}-\frac{(C_U-C_L)}{2(\omega_{r1}^2-\omega_{r2}^2)C_U C_L L_1}+\frac{1}{(\omega_{r1}^2-\omega_{r2}^2)C_U L_2}\right) \\
&\left.-\frac{\omega_{r2}V_{U0}}{(\omega_{r1}^2-\omega_{r2}^2)L_2}+\frac{\omega_{r2}C_L V_{L0}}{2}+\frac{\omega_{r2}(C_L-C_U)V_{L0}}{2(\omega_{r1}^2-\omega_{r2}^2)C_U L_1}\right)\cdot\sin(\omega_{r2}t)
\end{aligned}
\tag{17}
$$

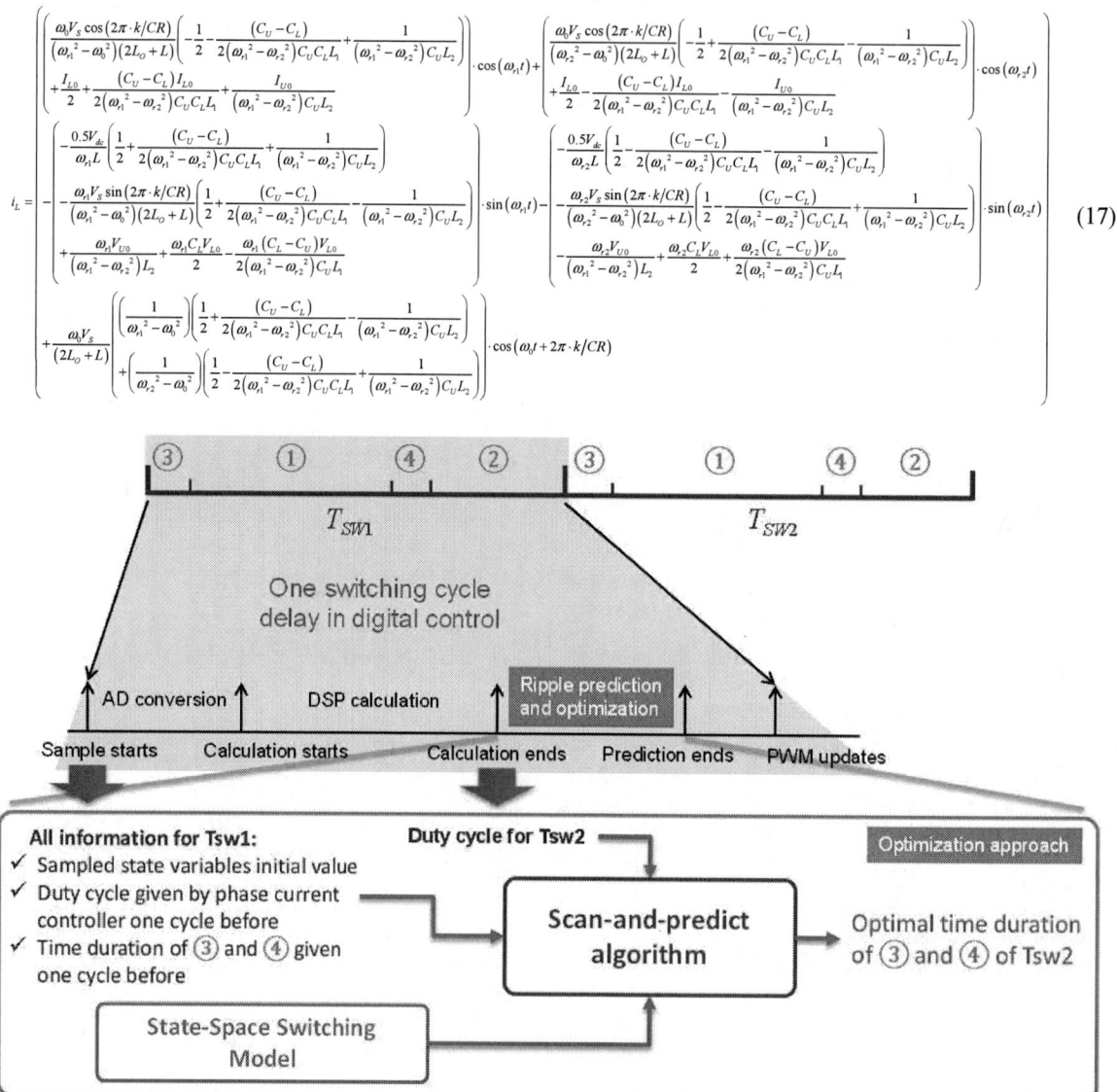

V. APPLICATION OF PREDICTIVE CONTROL IN DIGITAL CONTROLLER

It is well known that there always exists one switching cycle delay in the digital controller, which fundamentally restrains the application of the SCCVC since the variables are calculated based on the converter information in the previous switching cycle and has to be used in the next cycle. One solution is that to use the expressions in (14) and (17) to predict what the system response will be in the next cycle and use the data sampled in the previous cycle to conduct the calculations.

As per the complicated expression (14) and (17) for even a single module MMC circuit, another alternative Model-Predictive-Control (MPC) approach is proposed for the realization of the SCCVC, whose fundamental scheme is show in Fig. 7. At the beginning of the first switching cycle T_{SW1}, all the state variable values have been sampled, and the switching function pattern has been determined one cycle before. Therefore, the state variable values at the end of T_{SW1} can be calculated by iterations by using the state-space switching model derived in [6]. Afterward, the averaged line current controller will calculate the duty cycle needed for the next switching cycle, T_{SW2}. By scanning different time durations for state ③ and ④, the same iteration approach as used in T_{SW1} can be used to predict the optimal time durations that yield the values of state variables satisfying that $(v_L-V_{Lref})^2$ reaches its minimum value. The same process can be repeated when it comes to the next switching cycle.

Need to notice that due to the insertion of state ③ and ④, the originally calculated duty cycles could be slightly changed, and thus proper compensation need to be made after the time durations of state ③ and ④ are determined. As they are very short, the duty cycle only need to be slightly changed before it is given to the modulator, which has very little change in the state variable response

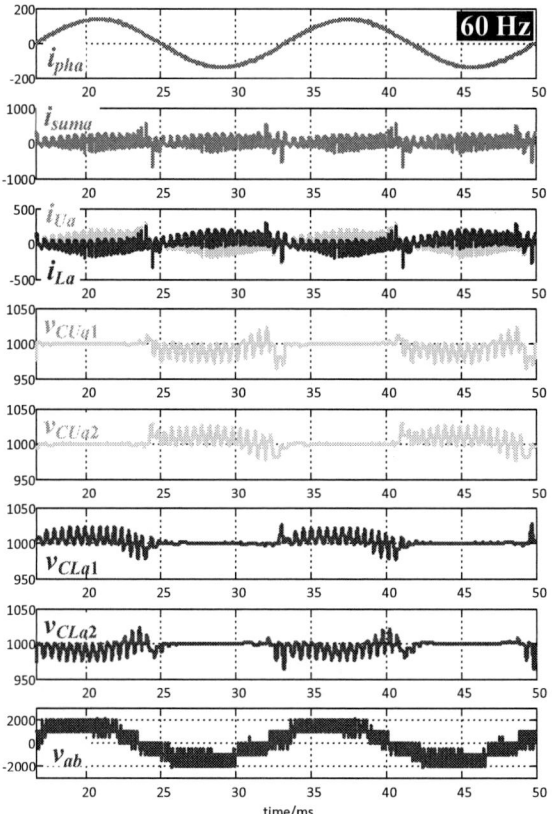

Fig. 8. The three-phase MMC with SCCVC, f_0=60 Hz

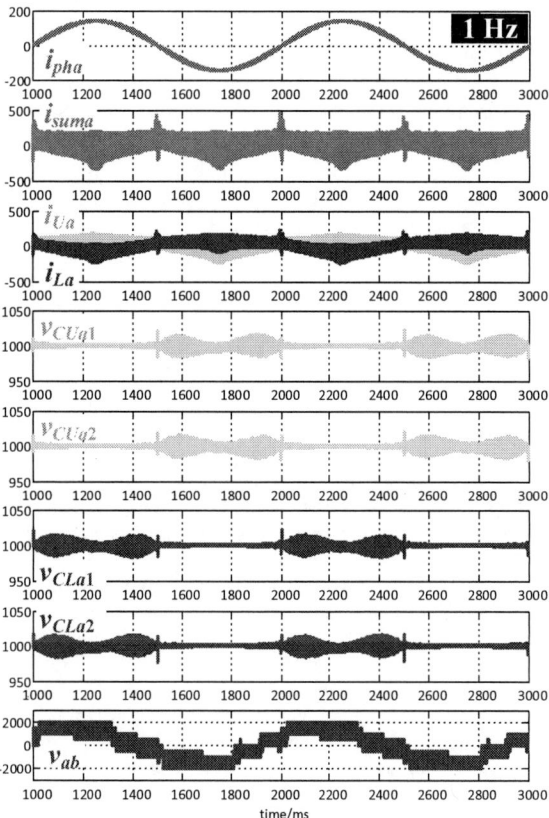

Fig. 9. The three-phase MMC with SCCVC, f_0=1 Hz

since their resonant frequencies are much longer than the time length of state ③ and ④.

VI. SIMULATION RESULT AND SEMICONDUCTOR LOSS EVALUATION

The specifications of the example simulation are shown in TABLE. I. The cell DC-link voltage is selected as 1 kV in order to enable the use of low cost 1.7 kV IGBT module. The cell count in each arm is selected to be two to simply demonstrate the operation waveforms, but not too complicated to be observed and understood. Following the similar control approach, the cell number can be scaled to larger to fit higher grid voltages.

The simulation results are shown in Fig. 8. The line current are controlled to follow the current requirement as specified. There are quite a number of intermediate voltages levels (±500V, ±1500V) caused by the newly used states, shown on the line-to-line voltage vab waveform. In the SCCVC control the duty cycle are still maintained as it was originally required for the averaged line current control, and thus there is actually no low order harmonic distortions on the line-to-line voltage. Therefore, there is no distortion occurring on the phase current. The sum current are controlled to indirectly regulate the cell capacitor voltages to make sure that at the end of each switching cycle the capacitor voltage will go back to the reference voltage, 1 kV in this case. Fig. 9, where the line frequency is 1 Hz, demonstrates that the

TABLE. I Converter Specifications in Simulations

Apparent power	200 kVA
Power factor	Unity
Line-to-line grid voltage	1140 V
Line current	100 A
DC-bus voltage	2000 V
Cell count per arm	2
Cell DC-link voltage	1000 V
Cell DC-link voltage ripple	-50 V ~ +50 V
Capacitance in each cell	400 μF
Arm inductance	20 μH
Line inductance	2000 μH
Line frequency	60 Hz
Switching frequency	33*60 Hz

SCCVC nearly eliminates the dependency on the line frequency to balance the cell capacitor voltages. The capacitor voltage ripples are almost the same as the 60 Hz case, and there are actually no low order harmonic distortions on the line-to-line voltage either. As observed in the specification, the capacitance value is 400 μF and the arm inductance value is 20 μH, which is significantly reduced compared to the conventional case where 4 mF capacitance and 1 mH arm inductance are needed to achieve the same capacitor voltage ripples.

VII. Conclusions

This paper presented a brand new approach to control the capacitor voltage of the MMC within a switching cycle, demonstrating benefits in greatly reducing the capacitance and arm inductance of the MMC. It also provides the MMC with the capability to operate at lower-frequency-high-torque load conditions if the motor drive applications are considered, which can be a promising method that expands the application domains of the MMC.

References

[1] M. Hagiwara, K. Nishimura, and H. Akagi, "A Medium-Voltage Motor Drive With a Modular Multilevel PWM Inverter," Power Electronics, IEEE Transactions on, vol. 25, pp. 1786-1799, 2010.

[2] A. J. Korn, M. Winkelnkemper, and P. Steimer, "Low output frequency operation of the Modular Multi-Level Converter," in Energy Conversion Congress and Exposition (ECCE), 2010 IEEE, 2010, pp. 3993-3997.

[3] A. J. Korn, M. Winkelnkemper, P. Steimer, and J. W. Kolar, "Direct modular multi-level converter for gearless low-speed drives," in Power Electronics and Applications (EPE 2011), Proceedings of the 2011-14th European Conference on, 2011, pp. 1-7.

[4] A. Antonopoulos, L. Angquist, S. Norrga, K. Ilves, and H. P. Nee, "Modular multilevel converter ac motor drives with constant torque form zero to nominal speed," in Energy Conversion Congress and Exposition (ECCE), 2012 IEEE, 2012, pp. 739-746.

[5] K. Ilves, A. Antonopoulos, L. Harnefors, S. Norrga, L. Angquist, and H. P. Nee, "Capacitor voltage ripple shaping in modular multilevel converters allowing for operating region extension," in IECON 2011 - 37th Annual Conference on IEEE Industrial Electronics Society, 2011, pp. 4403-4408.

[6] J. Wang, E. Farr, R. Burgos, D. Boroyevich, R. Feldman, A. Watson, et al., "State-space switching model of modular multilevel converters," in Control and Modeling for Power Electronics (COMPEL), 2013 IEEE 14th Workshop on, 2013, pp. 1-10.

The 2014 International Power Electronics Conference

A Comparison of Modular Multilevel Energy Conversion Processes: DC/AC versus DC/DC

Gregory J. Kish and Peter W. Lehn
Department of Electrical and Computer Engineering
University of Toronto
Toronto, Canada

Abstract—Utilizing strings of cascaded submodules for multilevel dc/ac conversion has become well known due to the widespread popularity of the modular multilevel converter (MMC). Recently, it has been shown that these submodule strings can be adapted to achieve single-stage dc/dc conversion. Although the modular string configurations for dc/ac versus dc/dc conversion are similar, the underlying power transfer mechanisms employed by each are significantly different. This paper compares the dc/ac and dc/dc energy conversion processes for series-cascaded submodules and highlights their key similarities and differences. Converter operation and control requirements for dc/ac versus dc/dc conversion are discussed and contrasted. Case study simulations performed in PLECS illustrate both energy conversion processes for a single submodule string.

Keywords—*DC/AC Conversion, DC/DC Conversion, HVDC, Modular Multilevel Converter.*

I. INTRODUCTION

The well known modular multilevel converter (MMC) [1]–[3] uses strings of cascaded submodules (SMs) to perform dc/ac conversion. These SMs constitute modular voltage cells (i.e. each SM contains a local energy storage capacitor) that can be individually switched "in" or "out" along the string to create a multilevel ac voltage waveform. By stacking in series within each string the requisite number and type of SMs, any desired multilevel ac voltage can be synthesized. This modular string architecture has become the basic building block for many converter topologies [2], [4], [5] and is particularly appealing for use in HVDC transmission [6], [7].

Although attractive for dc/ac conversion, using strings of cascaded SMs (or SM based modular strings [8], [9]) to perform dc/dc conversion has, to date, required a front-to-front (i.e. dc/ac-ac/dc) converter configuration [8]–[11]. This implies the use of two dc/ac conversion stages and an intermediate ac transformer, all of which must be rated for the full input power. As each stage processes the same input power this results in: 1) poor utilization of total installed SM rating and 2) reduced overall conversion efficiency. Consequently, the cascading of two complete dc/ac stages to form a dc/dc converter structure is a relatively costly option.

Recently, it has been shown that conventional strings of series-cascaded SMs can be adapted for single-stage bidirectional dc/dc conversion [12], [13]. Elimination of the traditional intermediate ac link is achieved by exploiting circulating ac currents to maintain power balance of the SM capacitors.

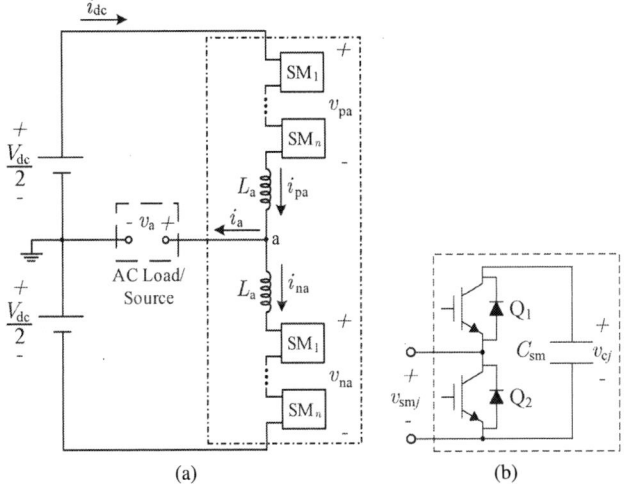

Fig. 1. String of series-cascaded SMs comprising two arms for dc/ac conversion: (a) modular structure, (b) switching cell for j th half-bridge SM.

This new power balancing mechanism enables replacement of two cascaded dc/ac converters with a single dc/dc architecture (termed the DC-MMC [13]), thus opening the door to a new class of energy conversion well suited for HVDC applications.[1]

This paper compares and contrasts the new form of single-stage dc/dc conversion introduced in [12], [13] with the well known dc/ac conversion process. Key similarities and differences between each are identified and discussed. In particular it is shown that three physical modifications enable a conventional dc/ac string of series-cascaded SMs to be adapted for single-stage dc/dc conversion. However, although sharing similar modular structures, performing dc/dc versus dc/ac conversion requires significantly different power transfer mechanisms. Operation and control requirements for each form of energy conversion are thus discussed and contrasted. A comparison of total SM cell voltage requirements for dc/ac versus dc/dc conversion is also provided.

This work focuses on cascaded SMs in a single string configuration, as this is the simplest modular structure enabling both dc/ac and dc/dc multilevel energy conversion. Additional converter structures can be realized by using multiple paralleled SM strings as illustrated in section II-C.

---

[1]Other recent works exploiting a similar power balancing mechanism for dc/dc conversion have also emerged [14]–[16].

978-1-4799-2706-7/14 $31.00 © 2014 IEEE

II. MODULAR MULTILEVEL ARCHITECTURES

A. Single-String: DC/AC Conversion

The conventional single-string architecture utilizing series-cascaded half-bridge SMs for dc/ac conversion is shown in Fig. 1. This structure is well recognized as one phase of the three-phase dc/ac MMC. The string (commonly referred to as a leg [17]) is comprised of an upper arm and lower arm, where an arm is defined as a cascaded set of n half-bridge SMs. A choke L_a is included in each arm to accommodate the switching action of the SMs.

Either a passive ac load or an ac source can be connected at midpoint node "a" as shown. Connecting an ac load restricts power transfer from dc to ac terminals, while interfacing to an ac source enables bidirectional power transfer capability.

B. Single-String: DC/DC Conversion

Fig. 2 [12], [13] shows the structure of Fig. 1(a) adapted for single-stage dc/dc conversion. The string now comprises two pairs of arms, where each pair consists of an inner arm and an adjacent outer arm. The arms are series-stacked in symmetric relation about the string midpoint, node "o1", with the inner arms flanked by the outer arms. Thus, each inner arm and outer arm employs m and k half-bridge SMs, respectively. Output ac filtering is provided by a coupled inductor, where L_f denotes its magnetizing inductance. Although non-coupled inductors can be used, the implementation in Fig. 2 permits a large ac impedance (via the magnetizing branch) to be realized while also eliminating dc flux in the core. Filter capacitors C_f[2] are a practical consideration to sink high-frequency switching harmonics. Midpoint inductor L_r plays an integral role in establishing circulating ac currents needed to maintain power balance of the SM capacitors.

Either a bipolar dc load or dc source can be coupled across the inner arms as shown in Fig. 2. Imposing a load restricts dc power transfer from the input to output terminals. Connecting a dc source at the output terminals enables bidirectional dc power transfer capability. The string's ideal *output-to-input voltage conversion ratio*, D, and its complement, D', are defined as

$$D \triangleq \frac{V_{dc,o}}{V_{dc,i}} \qquad (1)$$

$$D' \triangleq 1 - D. \qquad (2)$$

In [12], [13] it is shown that step-up voltage conversion (i.e. $D > 1$) can be achieved by employing full-bridge SMs in the outer arms of Fig. 2. However, to be consistent with conventional dc/ac conversion requiring only half-bridge SMs, this work focuses on step-down operation (i.e. $D < 1$).

Observe the single-string configurations in Fig. 1(a) and Fig. 2 share a similar modular structure. This similarity stems from the fact the dc/ac structure can be adapted for single-stage dc/dc conversion by imposing three key modifications:

---

[2]Capacitors C_f can alternatively be placed across inner arms if so desired.

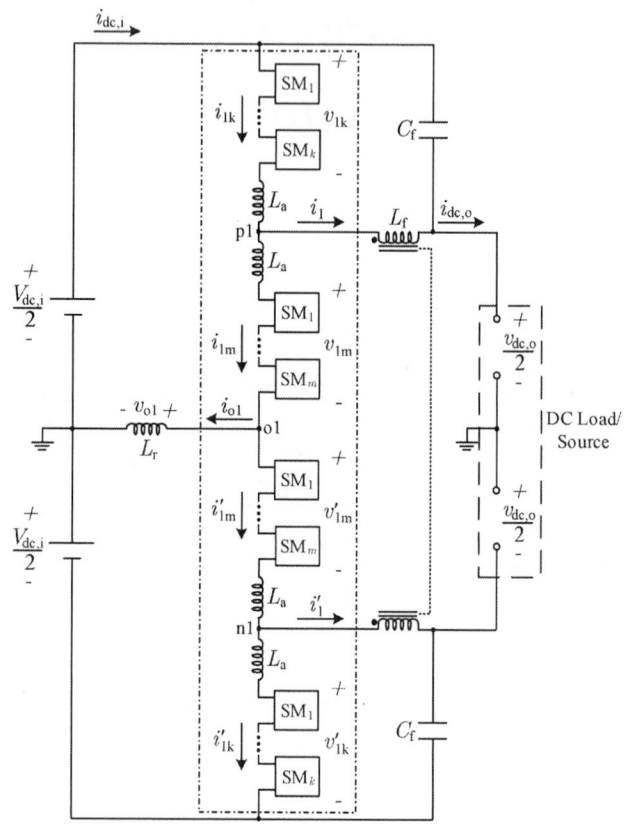

Fig. 2. String of series-cascaded SMs comprising four arms for single-stage dc/dc conversion.

1) Each arm of n SMs is partitioned into a pair of arms: an inner arm of m SMs and an outer arm of k SMs;
2) A path, enabled by L_r in this case, is created to allow the flow of circulating ac currents.
3) A filter network, comprised by passive elements L_f and C_f in this case, is added to prevent ac currents from propagating to the output.

C. Multi-String Converter Architectures

The single-string architectures in Fig. 1(a) and Fig. 2 are the simplest modular structures enabling dc/ac and single-stage dc/dc multilevel energy conversion. Moreover they simplify analysis as only one string need be studied. However, although useful from an analytical perspective, these architectures impose fundamental frequency ac currents to flow into the dc line, e.g. i_{dc} in Fig. 1(a) has an ac component. In practice dc/ac converters employ multiple strings [1], [17] to prevent ac currents from having to enter the dc network.

Fig. 3 illustrates example dc/ac and single-stage dc/dc [12], [13] converter architectures. The multi-string architectures can eliminate ac current flow in the dc network(s) by imposing appropriate phase shifts between ac quantities of the different strings, e.g. the 3-phase MMC uses a phase shift between strings of 120°. For consistency with established literature the dc/ac converters are referred to as being multi-phase. As previously highlighted, MMC based dc/dc conversion has

The 2014 International Power Electronics Conference

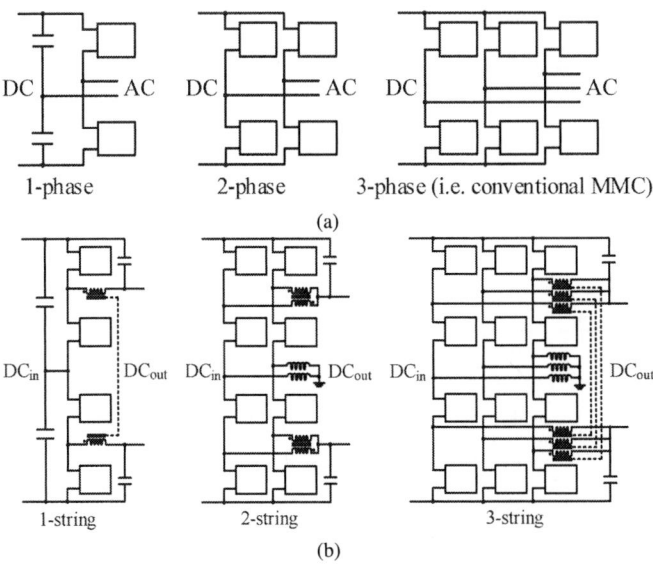

Fig. 3. Example modular multilevel converter architectures: (a) dc/ac, (b) single-stage dc/dc.

traditionally required two dc/ac stages in a front-to-front arrangement. For example, two 2-phase dc/ac converters from Fig. 3(a), along with an intermediate ac transformer, can be arranged in a dc/ac-ac/dc configuration [10].

As shown in Fig. 3(b) the 2-string and 3-string dc/dc converters are formed by interleaving multiple SM strings. In general, an arbitrary number of strings can be interleaved. Note the 2-string exploits a coupled inductor at each output pole (due to the even number of strings) while the 3-string[3] retains the filter arrangement in Fig 2. Of notable interest in Fig. 3(b) is the 1-string dc/dc converter. Here only one SM string is needed, as ac currents are allowed to circulate within the converter structure by installing two series-stacked capacitors as shown [13]. This architecture, a minor modification of Fig. 2, is particularly interesting as neither input nor output dc terminals are adversely impacted by significant ac ripple content at the midpoint node. This is not the case for the 1-phase dc/ac structure in Fig. 3(a) as the midpoint node is directly coupled to the ac output terminal.

III. IDEAL ENERGY CONVERSION PROCESSES

This section compares the ideal dc/ac and single-stage dc/dc energy conversion processes. A large number of cascaded SMs are assumed such that individual arms can be modeled as ideal voltage sources [18]–[20]. AC voltages and currents are represented by their fundamental frequency components. Resistance terms are neglected (i.e. lossless energy conversion). Ideal ac output filtering is assumed for the dc/dc architecture in Fig. 2.

With reference to Figs. 1(a) and 2 the notation $\tilde{i}$ represents the fundamental frequency component of i, i.e. $i(t) = I_{dc} + \tilde{i}(t)$ where $\tilde{i}(t) = \hat{I}\cos(w_m t + \theta_i)$. $\mathbf{I}$ is the rms phasor for $\tilde{i}$, i.e.

---

[3]Different passive (or even active) output filtering solutions are possible; for example, zig-zag grounding transformers can be exploited [16].

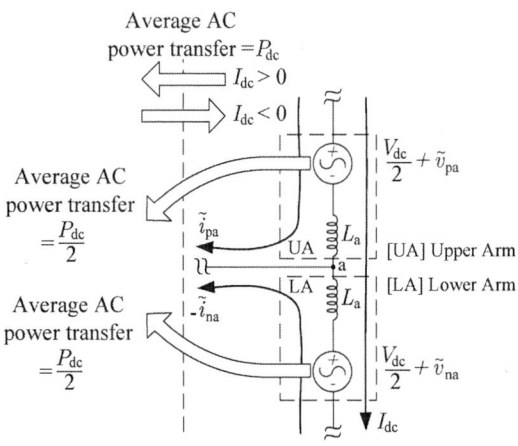

Fig. 4. Ideal dc/ac conversion process for Fig. 1(a).

$\mathbf{I} = (\hat{I}/\sqrt{2})\angle\theta_i$. For simplicity of notation, in Figs. 1(a) and 2: 1) average input current is denoted by I_{dc}, and 2) average input voltage is denoted by V_{dc}. Thus, both the total average power transfer between dc and ac terminals in Fig. 1(a) and total average power transfer between dc terminals in Fig. 2 are denoted by $P_{dc} = V_{dc}I_{dc}$.

A. DC/AC Conversion

Fig. 4 shows the ideal dc/ac conversion process for Fig. 1(a). Voltages and currents are separated into their dc and ac components, and are summarized in Table I. In Fig. 4 each arm supplies one-half of the total ac output current. When coupled to a passive ac load with both real and reactive parts, each arm delivers one-half of the average ac power transfer and either supplies (inductive load) or absorbs (capacitive load) one-half of the load vars. In the case of an ac source, the arms can also absorb equal amounts of average ac power if required. Based on the above discussion, the upper and lower arms operate identically in that they equally share the real and reactive power requirements of the ac load/source. Thus, as a result of this symmetry each arm inherently operates at the same ac power factor. Note, due to ac load/source requirements, the fundamental frequency of the ac arm voltages and currents is constrained to conventional 50/60 Hz.

To summarize, the key power flow related characteristics of Fig. 4 are: 1) fundamental frequency of ac voltages and currents is fixed to conventional 50/60 Hz, 2) upper and lower arms each supply 0.5 times the total average ac power transfer, and 3) vars demand of the ac load/source is split equally between upper and lower arms.

B. Single-Stage DC/DC Conversion

The ideal single-stage dc/dc conversion process, which is described in detail in [13], is shown in Fig. 5. Table I summarizes the dc and ac voltages and currents for each arm. By chosen convention each inner arm ac voltage is scaled by $Me^{j\Phi}$ relative to the adjacent outer arm. The composite reactive load formed by L_r (not shown in Fig. 5) and each

978-1-4799-2706-7/14 $31.00 © 2014 IEEE   953

TABLE I. ARMS VOLTAGES AND CURRENTS FOR SINGLE-STRING DC/AC AND DC/DC ARCHITECTURES

Quantities	DC/AC Structure in Fig. 4	DC/DC Structure in Fig. 5	
Arms Voltages	$v_{\mathrm{pa}} = \frac{V_{\mathrm{dc}}}{2} + \hat{V}\cos(w_{\mathrm{m}}t)$	$v_{1k} = D'\frac{V_{\mathrm{dc}}}{2} + \hat{V}\cos(w_{\mathrm{m}}t);$	$v_{1m} = D\frac{V_{\mathrm{dc}}}{2} + M\hat{V}\cos(w_{\mathrm{m}}t + \Phi)$
	$v_{\mathrm{na}} = \frac{V_{\mathrm{dc}}}{2} - \hat{V}\cos(w_{\mathrm{m}}t)$	$v'_{1k} = D'\frac{V_{\mathrm{dc}}}{2} - \hat{V}\cos(w_{\mathrm{m}}t);$	$v'_{1m} = D\frac{V_{\mathrm{dc}}}{2} - M\hat{V}\cos(w_{\mathrm{m}}t + \Phi)$
Arms Currents	$i_{\mathrm{pa}} = I_{\mathrm{dc}} + \hat{I}\cos(w_{\mathrm{m}}t + \theta_{\mathrm{i}})$	$i_{1k} = I_{\mathrm{dc}} + \hat{I}\cos(w_{\mathrm{m}}t + \theta_{\mathrm{i}});$	$i_{1m} = -(D'/D)I_{\mathrm{dc}} + \hat{I}\cos(w_{\mathrm{m}}t + \theta_{\mathrm{i}})$
	$i_{\mathrm{na}} = I_{\mathrm{dc}} - \hat{I}\cos(w_{\mathrm{m}}t + \theta_{\mathrm{i}})$	$i'_{1k} = I_{\mathrm{dc}} - \hat{I}\cos(w_{\mathrm{m}}t + \theta_{\mathrm{i}});$	$i'_{1m} = -(D'/D)I_{\mathrm{dc}} - \hat{I}\cos(w_{\mathrm{m}}t + \theta_{\mathrm{i}})$

TABLE II. AVERAGE AND RIPPLE POWERS FOR INDIVIDUAL ARMS IN SINGLE-STRING DC/AC AND DC/DC ARCHITECTURES

Arms Power *	DC/AC Structure in Fig. 4 **	DC/DC Structure in Fig. 5 **
Average Value	$P_{\mathrm{pa}}^0 = \frac{V_{\mathrm{dc}}}{2}I_{\mathrm{dc}} + \frac{\hat{V}\hat{I}}{2}\cos(\theta_{\mathrm{i}})$	$P_{1k}^0 = D'\frac{V_{\mathrm{dc}}}{2}I_{\mathrm{dc}} + \frac{\hat{V}\hat{I}}{2}\cos(\theta_{\mathrm{i}})$ $P_{1m}^0 = -D'\frac{V_{\mathrm{dc}}}{2}I_{\mathrm{dc}} + M\frac{\hat{V}\hat{I}}{2}\cos(\Phi\text{-}\theta_{\mathrm{i}})$
1st harmonic (i.e. w_{m})	$P_{\mathrm{pa}}^1 = \frac{V_{\mathrm{dc}}}{2}\hat{I}\cos(w_{\mathrm{m}}t + \theta_{\mathrm{i}}) + I_{\mathrm{dc}}\hat{V}\cos(w_{\mathrm{m}}t)$	$P_{1k}^1 = D'\frac{V_{\mathrm{dc}}}{2}\hat{I}\cos(w_{\mathrm{m}}t + \theta_{\mathrm{i}}) + I_{\mathrm{dc}}\hat{V}\cos(w_{\mathrm{m}}t)$ $P_{1m}^1 = D\frac{V_{\mathrm{dc}}}{2}\hat{I}\cos(w_{\mathrm{m}}t + \theta_{\mathrm{i}}) - (D'/D)MI_{\mathrm{dc}}\hat{V}\cos(w_{\mathrm{m}}t + \Phi)$
2nd harmonic (i.e. $2w_{\mathrm{m}}$)	$P_{\mathrm{pa}}^2 = \frac{\hat{V}\hat{I}}{2}\big[\cos(\theta_{\mathrm{i}})\cos(2w_{\mathrm{m}}t) + \sin(\text{-}\theta_{\mathrm{i}})\sin(2w_{\mathrm{m}}t)\big]$	$P_{1k}^2 = \frac{\hat{V}\hat{I}}{2}\big[\cos(\theta_{\mathrm{i}})\cos(2w_{\mathrm{m}}t) + \sin(\text{-}\theta_{\mathrm{i}})\sin(2w_{\mathrm{m}}t)\big]$ $P_{1m}^2 = M\frac{\hat{V}\hat{I}}{2}\big[\cos(\Phi\text{-}\theta_{\mathrm{i}})\cos(2(w_{\mathrm{m}}t + \Phi)) + \sin(\Phi\text{-}\theta_{\mathrm{i}})\sin(2(w_{\mathrm{m}}t + \Phi))\big]$

* Powers are defined as being into (i.e. absorbed by) each arm, e.g. $P_{\mathrm{pa}} = v_{\mathrm{pa}}i_{\mathrm{pa}}$.

** Due to symmetry only the upper arm in Fig. 4 and positive pole inner and outer arms in Fig. 5 are analyzed.

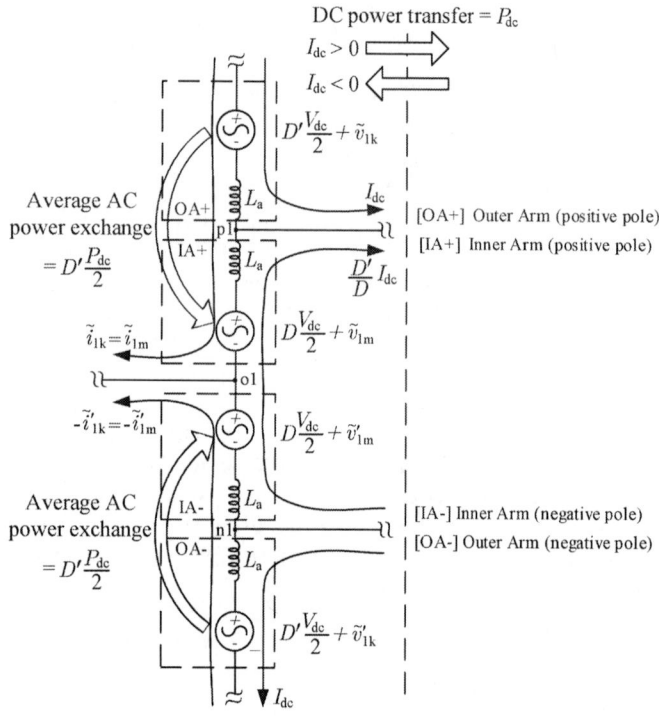

Fig. 5. Ideal single-stage dc/dc conversion process for Fig. 2.

L_{a} establishes circulating ac currents that allow the exchange of average ac power between each inner arm and the adjacent outer arm, in a near lossless manner. To achieve the desired ac current M, $\hat{V}$ and Φ are appropriately selected. This power transfer mechanism is exploited to maintain power balance of the SM capacitors, thereby eliminating the need for a traditional intermediate ac link. Interestingly, while the presence of circulating ac currents is a largely undesirable

phenomenon for cascaded SM based dc/ac conversion [20], it becomes the key enabling power balancing mechanism by which single-stage dc/dc conversion can be realized.

Each inner arm and the adjacent outer arm in Fig. 5 exchange an average ac power of $0.5D'P_{\mathrm{dc}}$, where P_{dc} is the total dc power transfer between input and output terminals. This can be verified by evaluating power balance criteria $P_{1k}^0 = 0$ and $P_{1m}^0 = 0$ in Table II, which tabulates both the average and ripple powers for individual arms in Fig. 4 and Fig. 5. Similar to the dc/ac conversion process, all arms in Fig. 5 see one-half of the total ac current leaving the midpoint node. However, as each arm in Fig. 4 is now partitioned into an inner arm and an adjacent outer arm, ac voltage components of inner and outer do not necessarily have to be equal. This implies each inner arm and the adjacent outer arm do not have to operate at the same ac power factor. Moreover, as the circulating ac currents serve only to maintain power balance of the SM capacitors, the fundamental frequency is not constrained to conventional 50/60 Hz (unlike in Fig. 4). In general any type of modulation can be used (i.e. not merely sinusoidal) as the ac quantities remain internal to the converter.

In Fig. 5 the dc components of the arm voltages and currents are functions of D. This is fundamentally different from the dc/ac conversion process in Fig. 4, where each of the arms carry the same amount of dc current and have the same dc voltage component. Consequently, this impacts the distribution of ac ripple power between arms, i.e. all arms are not necessarily subjected to the same ripple power. Moreover, the unequal splitting of vars support between inner and outer arms further impacts the distribution of ac ripple power. This asymmetry between inner and outer arms is easily visualized by examining the appropriate rows of Table II. Specifically, parameters D, M, $\hat{V}$ and Φ all influence the active and reactive power injections of each arm.

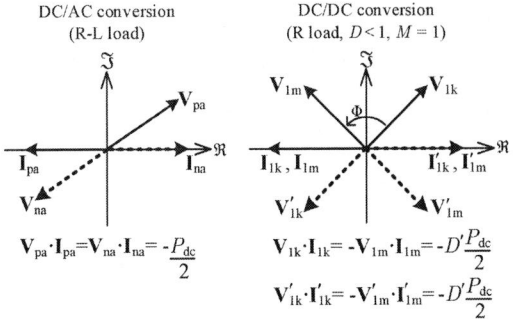

Fig. 6. Example fundamental frequency rms phasor diagrams illustrating power transfer mechanisms employed to achieve dc/ac (ref. Fig. 4) versus dc/dc (ref. Fig. 5) conversion for a single string of series-cascaded SMs.

To summarize, the key power flow related characteristics of Fig. 5 are: 1) fundamental frequency of ac voltages and currents is a free design parameter, 2) allocation of vars support between inner and outer arms can be assigned arbitrarily, 3) each pair of inner and outer arms exchange an average ac power equal to $0.5D'$ times the total dc power transfer, and 4) ac ripple power can be unequally distributed between arms.

C. Example Power Flows Visualization

Fig. 6 provides example phasor diagrams that illustrate the power transfer mechanisms employed by the ideal conversion processes in Figs. 4 and 5. In this example the inner and outer arm voltages in Fig. 5 have equal magnitudes (i.e. $M = 1$ and thus arms equally share vars requirements); however, as mentioned this does not have to be the case. By using phasor dot products, the ac power flows for each energy conversion process are easily visualized. Bidirectional power flow (i.e. $I_{dc} < 0$) can be accommodated by appropriately adjusting the ac arm voltages, e.g. for dc/dc conversion simply reflect voltage phasors about the imaginary axis.

It is important to stress here that, as shown by Fig. 6, both the dc/ac and single-stage dc/dc conversion processes benefit from natural cancellation of ac voltages along the SM string. That is, $\mathbf{V}_{pa} + \mathbf{V}_{na} = 0$ and $\mathbf{V}_{1k} + \mathbf{V}_{1m} + \mathbf{V}'_{1m} + \mathbf{V}'_{1k} = 0$.

IV. OPERATION AND CONTROL REQUIREMENTS

In Fig. 7 the arm current stresses for Fig. 4 and Fig. 5, normalized to I_{dc}, are compared using the parameters in Table I and considering $P_{dc} = 1$ [pu]. For ease of comparison the converter fundamental frequency vars consumption is neglected (i.e. assume small L_a in Fig. 1(a) and small L_a, L_r in Fig. 2), and $\hat{V}$ is maximized for each dc/dc operating point. Observe in Fig. 7(a) the peak ac current, $\hat{I}$, has a minimum value of 2 [pu] for a resistive load and increases with worsening load power factor, PF_{load}. The resulting peak arm current stresses are shown. The "nominal operating point" corresponding to $\hat{V}/(V_{dc}/2)$ slightly less than 1 [pu] accounts for prescribed headroom to avoid converter over-modulation.

Fig. 7(b) reveals peak arm current stresses for single-stage dc/dc conversion are comparable to those in Fig. 7(a)

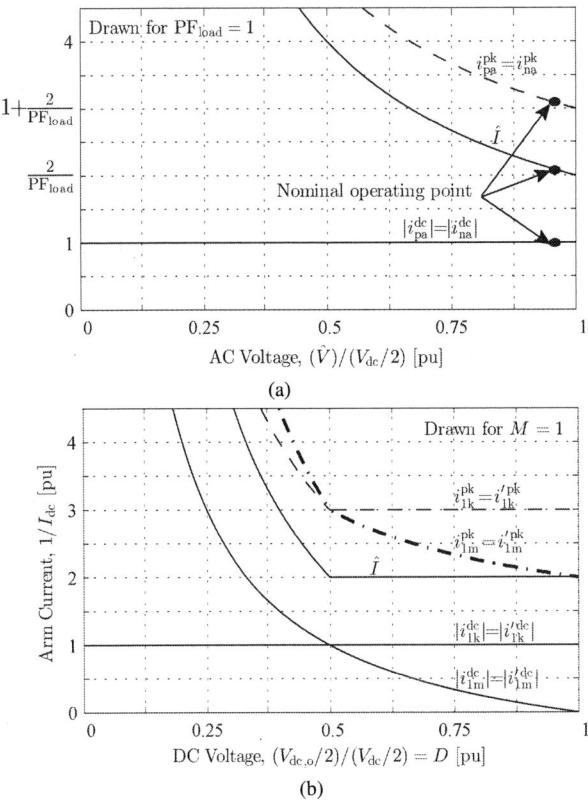

Fig. 7. Arm current stresses for $P_{dc} = 1$ [pu] (peak stresses shown by dashed lines), neglecting converter fundamental frequency vars consumption: (a) dc/ac conversion, (b) single-stage dc/dc conversion.

(at nominal operating point) when operating in the vicinity of $D = 0.5$. Stresses increase substantially for smaller D. However, inner arm peak current stresses are reduced as D increases from 0.5 to 1. This plot clearly illustrates the converter architecture in Fig. 2 is best suited for dc voltage conversion ratios ranging from unity to somewhat less than 0.5.

In Fig. 7 the plots are drawn assuming negligible converter fundamental frequency vars consumption. Of course, accounting for non-zero L_a and L_r will cause $\hat{I}$ in both diagrams to increase slightly. In Fig. 7(b), which is drawn for $M = 1$ (i.e. arms have same ac voltage magnitude), an increase in $\hat{I}$ implies a larger circulating ac current is needed for a given P_{dc}. To counter this increase in $\hat{I}$, thereby reducing losses and increasing power transfer capability, the ac voltage ratio M can be adjusted to improve the arms power factor as needed.

Based on the above discussion, Table III shows the ac modulation strategy proposed for Fig. 2 to minimize circulating ac currents. For values of D around 0.5 (quantified by $D = 0.5 \pm \alpha$), all arms should operate at the same ac power factor as $D \approx D' \approx 0.5$. However, for larger (or smaller) values of D the outer arms (or inner arms) should operate at unity power factor. The premise behind this strategy is to maximize the available ac arms voltage, such that circulating ac currents needed for a given average ac power exchange (ref.

TABLE III. PROPOSED AC MODULATION STRATEGY FOR DC/DC STRUCTURE IN FIG. 2 TO MINIMIZE REQUIRED CIRCULATING AC CURRENTS

DC Voltage Ratio	Vars Allocation Between Arms	AC Arms Voltage Magnitudes *	Total SM Voltage Requirements **
$(0.5+\alpha) \le D < 1$ (larger values of D)	Outer arms: unity power factor Inner arms: generate all vars	$M > 1$, where $M\cos(\Phi) = -1$ (inner arms have larger ac voltages)	$1 + D'(1+M)$ [pu]
$0.5 \le D < (0.5+\alpha)$ (values of D near 0.5)	Equally split vars generation between outer and inner arms	$M = 1$ (all arms have equal magnitude ac voltages)	$1 + 2D'$ [pu]
$(0.5-\alpha) < D < 0.5$ (values of D near 0.5)	Equally split vars generation between outer and inner arms	$M = 1$ (all arms have equal magnitude ac voltages)	$1 + 2D$ [pu]
$0 < D \le (0.5-\alpha)$ (smaller values of D)	Outer arms: generate all vars Inner arms: unity power factor	$M < 1$, where $\cos(\Phi) = -M$ (outer arms have larger ac voltages)	$1 + D(1 + \dfrac{1}{M})$ [pu]

* M and Φ are defined in Table I for inner arms voltages v_{1m} and v'_{1m}. ** Refer to Fig. 8 for illustration of SM voltage requirements.

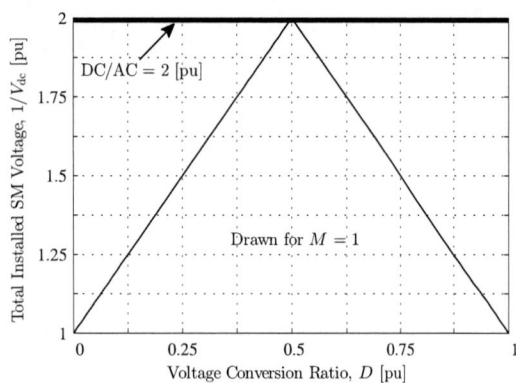

Fig. 8. Total installed SM voltage requirements for single-stage dc/dc conversion, using half-bridge SMs. Depicted curves are defined in Table III.

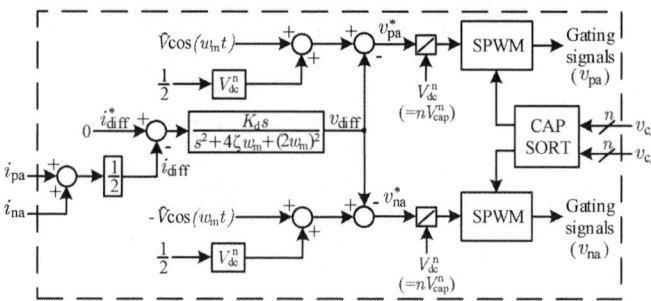

Fig. 9. Open loop control with second harmonic circulating ac currents suppression for dc/ac architecture in Fig. 1(a).

Fig. 5) are minimized. For example, for large D the maximum outer arms ac voltage is limited to $\hat{V}/(V_{dc}/2) = D'$ (where $D' << D$). In this case, the outer arms should operate at unity power factor to minimize the required $\hat{I}$. To achieve this the inner arms ac voltage must be larger than the outer arms, requiring $M > 1$.

Total installed SM voltage requirements corresponding to Table III and Fig. 7(b) are given in Fig. 8. Note that 2 [pu] installed SM voltage is inherently needed for dc/ac conversion. In comparison, fewer SMs are employed for single-stage dc/dc conversion. Although only half-bridge SMs are considered in this work, full-bridge SMs can also be exploited in the outer arms to increase available ac voltage headroom [13]. Such a strategy is particularly beneficial at values of D approaching unity.

The preceding discussions have outlined key differences and similarities in operational requirements for dc/ac versus single-stage dc/dc modular multilevel energy conversion. Control requirements are now contrasted. Fig. 9 and Fig. 10 show the basic control loops necessary to enable open loop control of Fig. 1(a) and Fig. 2, respectively. For dc/ac conversion, direct modulation is employed for fundamental frequency quantities while a supplemental current control loop suppresses undesirable second harmonic circulating ac currents. Balancing of SM capacitor voltages in Fig. 9 is achieved using the ubiquitous SM capacitor voltage sort and selection algorithm [19].

The control algorithm for single-stage dc/dc conversion,

although sharing a similar control structure to Fig. 9, renders use of significantly different power transfer mechanisms. In Fig. 10 separate control blocks are implemented for positive and negative dc poles. To balance SM capacitor voltages the sort and selection algorithm is alone insufficient; some form of regulation of average ac power exchange between inner and outer arms is required. This is achieved via closed loop-control of fundamental frequency circulating ac currents, which acts on the difference in SM capacitor voltages between arms. Weighting parameter $k_q \in [0,1]$ allows arbitrary splitting of vars generation between inner and outer arms, by appropriately distributing voltages v_{diff+} and v_{diff-} as shown. Key values include $k_q = 0$ (inner arms supply all vars), $k_q = 0.5$ (split vars equally between arms) and $k_q = 1$ (outer arms supply all vars). Note, for control implementation the ac parameter M discussed earlier has effectively been replaced by k_q.

It should be highlighted that Fig. 10 is a more general formulation of the open loop control first proposed in [13]. Specifically, the open loop control in [13] corresponds to Fig. 10 where: 1) $k_q = 0$, 2) reference signals v^*_{1m}, v'^*_{1m} are generated solely by the proportional-resonant compensators, 3) control feedback of circulating ac currents utilizes only the outer arms current measurements and 4) balancing of SM capacitor voltages assumes $k = m$. It is also interesting to note that for dc/ac conversion, suppression of second harmonic circulating ac currents is an optional feature to reduce the rms value of the arms currents and thus avoid additional losses [19]. However, in Fig. 10 control of fundamental frequency circulating ac currents is necessary to ensure SM capacitor voltages are regulated to their nominal values [13].

The 2014 International Power Electronics Conference

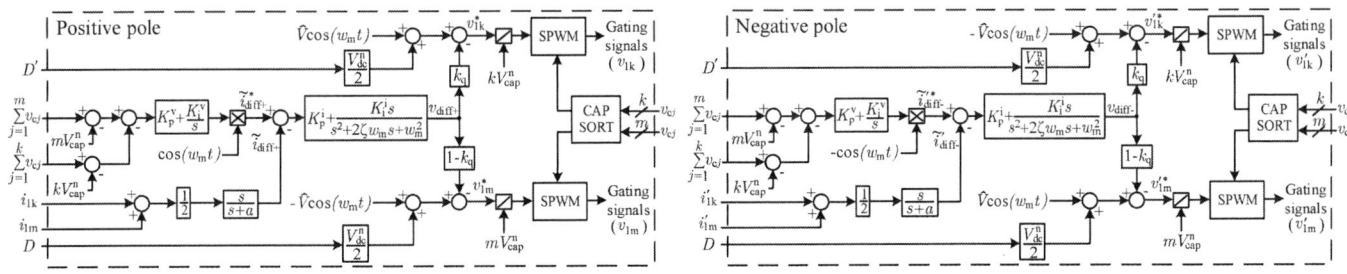

Fig. 10. Open loop control with fundamental frequency circulating ac current regulation for dc/dc architecture in Fig. 2, where weighting parameter $k_q \in [0,1]$ allows arbitrary splitting of vars generation between inner and outer arms.

TABLE IV.   PLECS SIMULATION DATA

General Parameters	Value	
DC input voltage, V_{dc}	± 8.8 kV	
Average power transfer, P_{dc}	7 MW	
SM capacitor, C_{sm}	14 mF	
SM voltage, V_{cap}^n	2.2 kV	
Fundamental frequency, w_m	377 rad/s	
Carrier frequency, f_c	4 kHz	
DC/AC Parameters	**Value**	
Arm choke, L_a	0.5 mH	
SMs per arm, n	8	
AC voltage, $\hat{V}$	8.58 kV$_{pk}$	
DC/DC Parameters	**Value**	
	$D = 0.5$	$D = 0.75$
SMs per arm, k/m	4/4	2/5
AC voltage, $\hat{V}$	4.3 kV$_{pk}$	2.15 kV$_{pk}$
Vars allocation, k_q	0.5	0
Arm choke, L_a	0.5 mH	
Midpoint string inductor, L_r	0.5 mH	
Filter magnetizing inductance, L_f	990 mH	
Filter capacitor, C_f	15 μF	

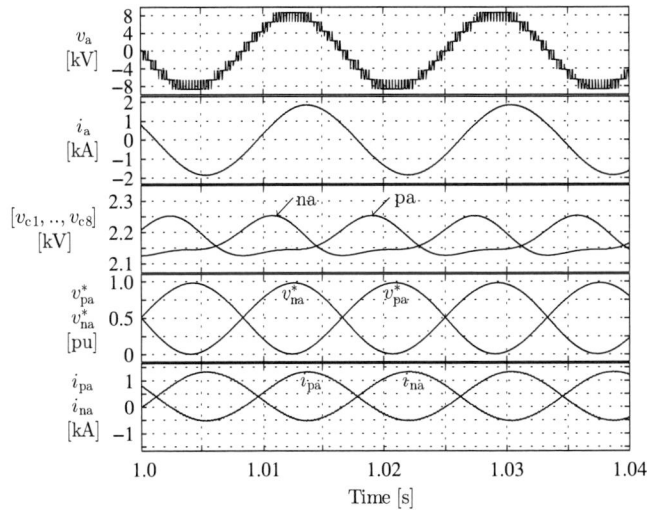

Fig. 11.   DC/AC conversion process for Fig. 1(a) with $n = 8$; P_{dc}=7 MW for ac load at 0.9 lagging power factor.

V.   CASE STUDY SIMULATIONS

Simulation results for switched models of Fig. 1(a) and Fig. 2 implemented in PLECS are presented. Open loop controls in Fig. 9 and Fig. 10 are adopted. For each scenario $V_{dc} = \pm 8.8$ kV and $P_{dc} = 7$ MW. For dc/ac conversion $2n = 16$ half-bridge SMs are used. For single-stage dc/dc conversion $2(k+m) = 16$ and $2(k+m) = 14$ half-bridge SMs are employed to achieve $D = 0.5$ and $D = 0.75$, respectively. Simulation parameters are given in Table IV.

The dc/ac conversion process is demonstrated in Fig. 11. Here a passive R-L load consumes 7 MW at a power factor of 0.9 lagging. Observe the symmetry in upper and lower arm quantities as the arms equally share the active and reactive power requirements of the load.

Fig. 12 shows the single-stage dc/dc conversion process for $D = 0.5$ (8.8 kV / 17.6 kV ratio). For $P_{dc} = 7$ MW, each pair of inner and outer arms must exchange 1.75 MW of average ac power. k_q is set to 0.5 to split the vars requirements of L_a and L_r equally between arms. Note the similarity in waveforms between Fig. 12 and Fig. 11. This similarity is a result of the chosen dc/dc operating point $D = 0.5$, which coincides with the inherent dc/ac operation where all arms have equal dc voltage components.

The single-stage dc/dc conversion process for $D = 0.75$ (13.2 kV / 17.6 kV ratio) is shown in Fig. 13. Each pair of inner and outer arms exchange only 0.875 MW of average ac power for the same $P_{dc} = 7$ MW. k_q is set to 0 to operate the outer arms at unity power factor. Observe fewer total SMs are employed for $D = 0.75$ ($k + m = 7$) as compared to $D = 0.5$ ($k + M = 8$). Observe also the SM capacitor voltage ripple between inner and outer arms is now unequal.

VI.   CONCLUSION

A comparison is carried out between the established dc/ac and recently introduced single-stage dc/dc MMC based on series-cascaded half-bridge SMs. It is shown that three physical modifications allow the well known single-string dc/ac structure to be adapted for single-stage dc/dc conversion. However, the power transfer mechanisms employed for each energy conversion process are significantly different.

Peak arm current stresses for single-stage dc/dc conversion are comparable to those for dc/ac conversion when operating in the vicinity of $D = 0.5$. However, peak current stresses are reduced as D increases, clearly demonstrating the dc/dc architecture is best suited for voltage conversion ratios ranging

978-1-4799-2706-7/14 $31.00 © 2014 IEEE   957

The 2014 International Power Electronics Conference

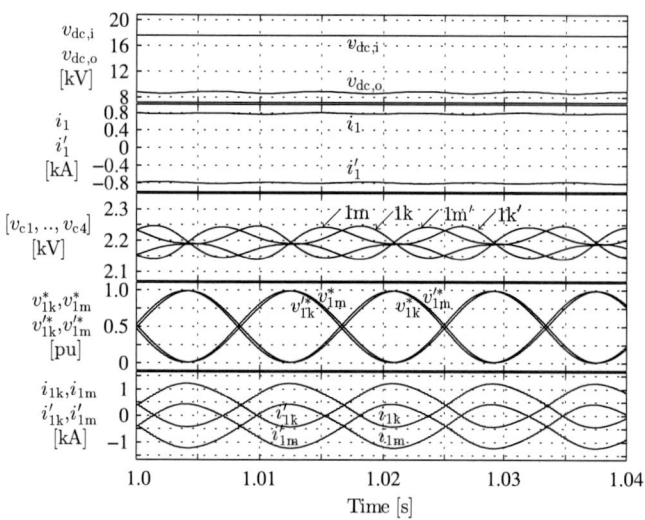

Fig. 12. Single-stage dc/dc conversion process for Fig. 2 with $k = m = 4$ and $D = 0.5$ (vars allocation: $k_q = 0.5$); P_{dc}=7 MW for resistive load.

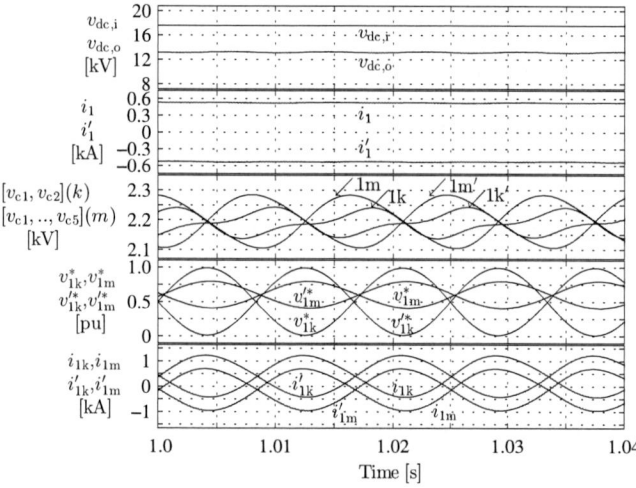

Fig. 13. Single-stage dc/dc conversion process for Fig. 2 with $k = 2$, $m = 5$ and $D = 0.75$ (vars allocation: $k_q = 0$); P_{dc}=7 MW for resistive load.

from 1 to 0.5. For this range of conversion ratios, it is also shown that fewer total number of SMs can be used for dc/dc conversion in comparison to dc/ac conversion.

Control requirements for dc/ac versus single-stage dc/dc conversion are also compared and contrasted. An open loop control scheme for dc/dc conversion is proposed whereby vars allocation between converter arms can be assigned arbitrarily in order to maximize conversion efficiency.

REFERENCES

[1] A. Lesnicar and R. Marquardt, "An innovative modular multilevel converter topology suitable for a wide power range," in *IEEE Bologna Power Tech Conference Proceedings*, vol. 3, Jun. 2003, pp. 1–6.

[2] M. Hagiwara and H. Akagi, "Control and experiment of pulsewidth-modulated modular multilevel converters," *IEEE Trans. Power Electron.*, vol. 24, no. 7, pp. 1737–1746, Jul. 2009.

[3] S. Rohner, S. Bernet, M. Hiller, and R. Sommer, "Modulation, losses, and semiconductor requirements of modular multilevel converters," *IEEE Trans. Ind. Electron.*, vol. 57, no. 8, pp. 2633–2642, Aug. 2010.

[4] C. Oates, "A methodology for developing 'Chainlink' converters," in *13th European Conference on Power Electronics and Applications*, Sep. 2009, pp. 1–10.

[5] M. Glinka and R. Marquardt, "A new AC/AC multilevel converter family," *IEEE Trans. Ind. Electron.*, vol. 52, no. 3, pp. 662–669, Jun. 2005.

[6] N. Ahmed, S. Norrga, H.-P Nee, A. Haider, D. Van Hertem, L. Zhang, and L. Harnefors, "HVDC supergrids with modular multilevel converters - The power transmission backbone of the future," in *9th International Multi-Conference on Systems, Signals and Devices*, Mar. 2012, pp. 1–7.

[7] R. Marquardt, "Modular multilevel converter: An universal concept for HVDC-networks and extended DC-bus-applications," in *International Power Electronics Conference*, Jun. 2010, pp. 502–507.

[8] C.D. Barker, C.C. Davidson, D.R. Trainer and R.S. Whitehouse, "Requirements of DC-DC Converters to facilitate large DC grids," in *CIGRE Symp., Paris, France*, Aug. 2012, pp. 1–10.

[9] T. Luth, M. Merlin, T. Green, F. Hassan, and C. Barker, "High Frequency Operation of a DC/AC/DC System for HVDC Applications," *IEEE Trans. Power Electron.*, vol. PP, no. 99, pp. 1–9, Nov. 2013.

[10] S. Kenzelmann, A. Rufer, D. Dujic, F. Canales, and Y.R. de Novaes, "Isolated DC/DC structure based on modular multilevel converter," *IEEE Trans. Power Electron.*, vol. PP, no. 99, pp. 1–10, Feb. 2014.

[11] I.A. Gowaid, G.P. Adam, A.M. Massoud, S. Ahmed, D. Holliday, and B.W. Williams, "Quasi two-level operation of modular multilevel converter for use in a high-power DC transformer with DC fault isolation capability," *IEEE Trans. Power Electron.*, vol. PP, no. 99, pp. 1–16, Feb. 2014.

[12] G.J. Kish and P.W. Lehn, "A modular bidirectional DC power flow controller with fault blocking capability for DC networks," in *14th IEEE Workshop on Control and Modeling for Power Electronics*, Jun. 2013, pp. 1–7.

[13] G.J. Kish, M. Ranjram and P.W. Lehn, "A modular multilevel DC/DC converter with fault blocking capability for HVDC interconnects," *IEEE Trans. Power Electron.*, vol. PP, no. 99, pp. 1–15, Dec. 2013.

[14] J.A. Ferreira, "The multilevel modular DC converter," *IEEE Trans. Power Electron.*, vol. 28, no. 10, pp. 4460–4465, Oct. 2013.

[15] A. Schön and M-M. Bakran, "A new hvdc-dc converter for the efficient connection of hvdc networks," in *PCIM Europe Conference*, May 2013, pp. 525–532.

[16] S. Norrga, L. Angquist, and A. Antonopoulos, "The polyphase cascaded-cell DC/DC converter," in *IEEE Energy Conversion Congress and Exposition*, Sep. 2013, pp. 4082–4088.

[17] M. Hagiwara, R. Maeda, and H. Akagi, "Control and analysis of the modular multilevel cascade converter based on double-star chopper-cells (MMCC-DSCC)," *IEEE Trans. Power Electron.*, vol. 26, no. 6, pp. 1649–1658, Jun. 2011.

[18] H.-J. Knaak, "Modular multilevel converters and HVDC/FACTS: A success story," in *Proceedings of the 14th European Conference on Power Electronics and Applications*, Aug. 2011, pp. 1–6.

[19] A. Antonopoulos, L. Angquist, and H.-P. Nee, "On dynamics and voltage control of the modular multilevel converter," in *13th European Conference on Power Electronics and Applications*, Sep. 2009, pp. 1–10.

[20] Q. Tu, Z. Xu, and L. Xu, "Reduced switching-frequency modulation and circulating current suppression for modular multilevel converters," *IEEE Trans. Power Del.*, vol. 26, no. 3, pp. 2009–2017, Jul. 2011.

978-1-4799-2706-7/14 $31.00 © 2014 IEEE

A Novel Topology of Wind Power Plant suitable for DC Power Transmission Systems

Shoji Nishikata, Fujio Tatsuta, Katsumi Suzuki
Tokyo Denki University
5 Senju-Asahi-cho, Adachi-ku,Tokyo l0l-8457, Japan
e-mail: west@cck.dendai.ac.jp

Abstract—**A promising topology to interconnect the wind power plants is proposed. Advantages of series-connected system over ordinary parallel system are first discussed, and control strategies for the series-connected system are studied to realize an optimum use of the wind energy. The dynamic performances of the series-connected system are investigated through simulation and experiments. An effective method to realize high quality of output power is also shown. Finally, a novel topology of the wind turbine/generators connected in series suitable for the DC power transmission to enlarge the scale of the wind power plant without an increase in system voltage level is introduced.**

Keywords—*energy resources, thyristor, wind power, wind turbine/generator*

I. INTRODUCTION

Wind energy is one of the most expected renewable energies, and its effective usage is important because the amount of energy that we can consume is limited.

Since the output power obtained from one wind turbine/generator is at most the order of several MW, there is a need to interconnect plural wind turbine/generators to generate a large amount of power as produced in the ordinary thermal or nuclear power plants.

In the future offshore wind power plants of 1 GW size are imaged [1], and hence, interconnecting method for the wind turbine/generators is an important issue when we use wind energy as a reliable energy.

Nowadays typical wind power plants are of the parallel-connected type, in which the output terminals of individual wind turbine/generators are connected in parallel [2]. Since the wind power plant is usually provided with a substation when the wind generators are operated in parallel, it follows that an increase in cost and losses as the plants move farther offshore.

On the other hand, a wind power plant consisting of wind generators connected in series is proposed [3], [4]. Because the series-connected system does not need a substation, the cost and losses can be reduced when compared with the parallel-connected ones.

This paper discusses control strategies for the series-connected wind power plants and a topology suitable for DC transmission.

Advantages of series-connected system are first discussed, and control strategies for the system are

studied for an optimum use of the wind energy. The dynamic performances of the system are then explored, and a method to realize high quality of output power is given. A topology suitable for the DC power transmission is finally discussed to enlarge the scale of the wind power plants.

II. BASIC CONFIGURATIONS OF WIND POWER PLANTS CONNECTED IN PARALLEL OR SERIES

When we construct wind power plants, the converted outputs from individual wind generators are usually connected in parallel, as shown in Fig. 1. In this system the synchronous generators with permanent magnets (or field windings) are used as wind turbine generators. It is noted that DFIG (Doubly Fed Induction Generator) systems can be used instead of synchronous generator systems. Since a parallel-connected system requires installing a grid-side inverter and an output filter for every wind generator, and the frequency and phase of the voltage of each output terminal have to be controlled to coincide with each other, the system configuration becomes complicated. As in the figure, voltage-source PWM inverter (and/or converter), in which controlled turn-off semiconductor devices such as IGBTs are used as switching devices, is mainly employed, and an output filter for eliminating higher-order harmonics and a large smoothing capacitor, that has to be replaced periodically,

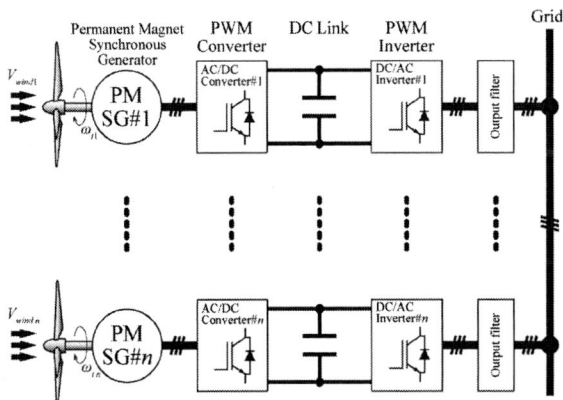

Fig. 1. Configuration of parallel-connected system.

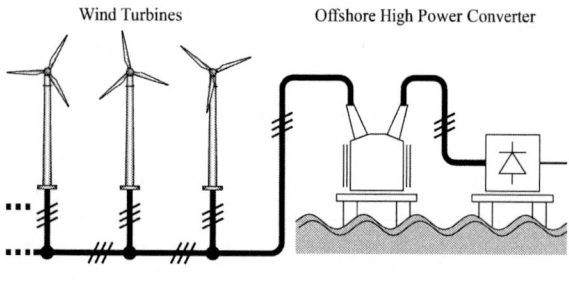

Fig. 2. Parallel-connected system used in offshore application.

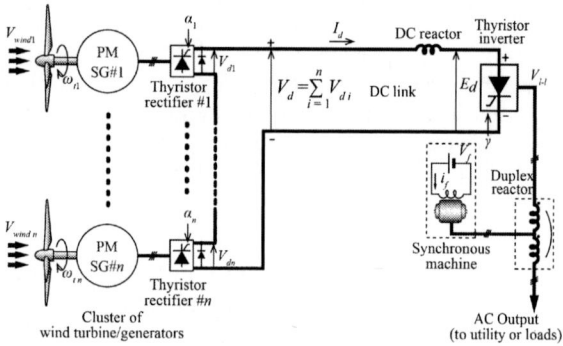

Fig. 3. Configuration of series-connected system.

are needed. Hence, there exist some issues to be solved from the viewpoints of reliability, costs, and so on.

Moreover, since the site suitable for wind power plants should be far distant from urban areas or offshore, long distance power transmissions are required for wind power utilization. In such a case, DC transmission system is more advantageous than AC system. Because wind resources are abundant in the offshore areas, it is predicted that a large amount of electricity will be supplied from wind power in the near future [5].

In Fig. 2 an offshore wind power plant is shown, in which several wind generators are connected in parallel. As in the figure the output terminals of the wind generators are connected to the low voltage side of the offshore grid (typically, 690V[2]), and a high-power transformer is required in the substation to boost the voltage level suitable for transmission. To reduce losses in the low voltage offshore grid the location of substation should be close to the wind turbines, resulting in increases in the construction and maintenance costs of the whole system.

Fig. 3 shows a wind power plant consisting of plural wind turbine/generators connected in series [3]. In this system the rectified output of individual wind generator is connected in series, and the combined DC power is converted to AC through a current-source type thyristor inverter. In this configuration, only one inverter is enough, making the system very simple compared to the conventional systems of Fig. 1, and the configuration carries the advantage that DC transmission line can be used because of the DC link without a substation. Also, it is noted that thyristors can be used as switching devices

in the inverter and rectifiers, improving the system reliability in comparing to the case of parallel system.

It should be recognized that synchronous generators with wound rotors can be used as wind generators instead of PMSGs, and in this case through adjusting the field currents as well as control angles, the DC output voltages of the rectifiers can be controlled, extending the freedom of control.

One concern about series-connected system is that the system may come to a halt when one of the turbine/generators is lost for some reason, including the lack of wind. However, such a system shutdown can be prevented through the bypass diodes connected to each rectifier in parallel. That is, once one of the output voltages of the rectifiers is lost for some reason, the corresponding diode turns into ON-state immediately, continuing DC-link current to flow and enabling the whole system to operate without interruption.

III. Control Strategy for Series-Connected System

It is essential to utilize wind power captured by the wind turbines as much as possible.

In order to realize an optimum use of the wind energy, the tip speed ratio λ of each wind turbine should be kept constant to obtain a large performance coefficient C_p, conversion ratio of the turbine output to captured wind power [6]. Here, λ is defined as blade tip speed divided by the wind velocity. In Fig. 4 an example of C_p-λ characteristics is given, which was obtained from the experimental results for the wind turbine installed in our campus [7].

In wind power plants since the wind velocities flowing to the individual wind turbines are basically different, proper control strategies to obtain constant tip speed ratio have to be adopted. In addition, in the series-connected system controlling the DC link current is crucial because the same current flows through the DC link.

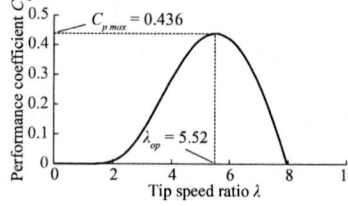

Fig. 4. Performance coefficient C_p versus tip speed ratio λ.

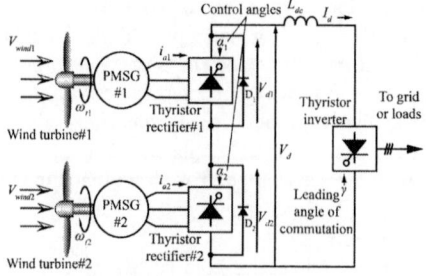

Fig. 5. Wind power plant consisting of two turbine/generators.

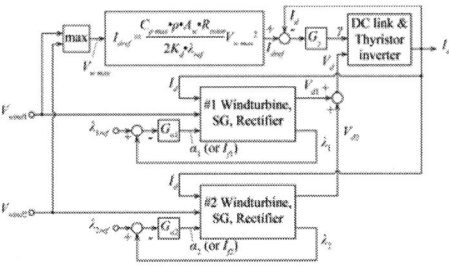

V_{wind1}, V_{wind2} : wind velocities flowing to wind turbines #1 and #2, λ_1, λ_2 : tip speed ratios for turbine #1 and #2, λ_{1ref}, λ_{2ref} : reference values for λ_1 and λ_2, α_1, α_2: control angles for thyristor rectifiers #1 and #2, γ: leading angle of commutation for thyristor inverter, V_d, I_d : DC link voltage and DC link current, V_{d1}, V_{d2} : DC output voltages of thyristor rectifiers #1 and #2, G_γ, G_{a1}, G_{a2}, : PI controllers.

Fig. 6. Closed-loop control system for constant tip speed ratio.
[System A]

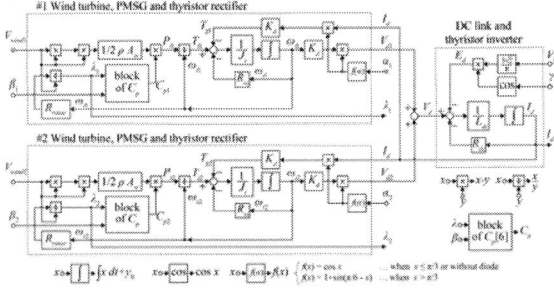

β_1, β_2, β : pitch angles, R_{rotor}: blade radius, ρ: air density, A_w : rotor swept area, C_{p1}, C_{p2}: performance coefficients, P_{t1}, P_{t2} : output powers of wind generators, T_{t1}, T_{t2} : torques of wind turbines, T_{g1}, T_{g2} : torques of wind generators, ω_{t1}, ω_{t2} : angular velocities of wind generators, J_1, J_2 : moment inertias of mechanical systems of turbine/generators #1 and #2, $R_{\omega 1}$, $R_{\omega 2}$: damping coefficients of turbine/generators #1 and #2, R_{dc}, L_{dc} : resistance and inductance in the DC link, V_{L-L}: inverter output voltage (line-to-line), E_d: inverter DC side voltage, K_d : coefficient of proportion between output voltage of the thyristor rectifier when $\alpha=0°$ and angular velocity of PMSG.

Fig. 7. Simulation block diagram for two wind turbine/generators and inverter.

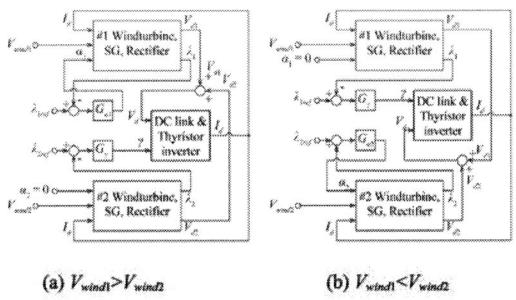

(a) V_{wind1}>V_{wind2} (b) V_{wind1}<V_{wind2}

Fig. 8. Closed-loop control system for constant tip speed ratio.
[System B]

As basic discussion the case of a wind power plant consisting of two wind turbine/PMSGs is considered here as given in Fig. 5.

One of the closed-loop control systems for constant tip speed ratio is shown in Fig. 6, we call it system A, in which the block diagrams for wind turbines, PMSG, thyristor rectifiers, and thyristor inverter are given in Fig.

7. [8]

In the system shown in Fig. 6, there are three closed-control loops; one for DC link current, and two for tip speed ratios.

That is, DC link current I_d is controlled with the leading angle of the inverter γ based on the following equation:

$$\gamma = G_\gamma \left(I_{dref} - I_d \right) \quad\text{.. (1)}$$

where, I_{dref}:reference value for I_d.

Here, I_{dref} is determined depending on the maximum wind velocity between the two wind turbines [3], and is given by:

$$I_{dref} = \frac{C_{pmax} \cdot \rho \cdot A_w \cdot R_{rotor}}{2 \cdot K_d \cdot \lambda_{ref}} V_{wmax}{}^2 \quad\text{.................. (2)}$$

where, C_{pmax}, λ_{ref}, V_{wmax}: maximum performance coefficient of the wind turbines (see Fig. 4), reference value of tip speed ratios($=\lambda_{1ref}=\lambda_{2ref}=\lambda_{op}$), and the maximum wind velocity among the wind turbines [m/s], respectively.

Meanwhile, the tip speed ratios of each turbine are separately controlled with closed-loop controllers as in the figure, namely, tip speed ratio λ_1 (λ_2) for #1(#2) turbine is controlled with the control angle α_1 (α_2) of the rectifier #1(#2). For the case of synchronous generators with wound rotors, the field currents I_{f1} and I_{f2} can be used as a substitute for α_1 and α_2, as shown in the figure.

As another control system for realizing constant tip speed ratios of the wind turbines one given in Fig. 8, called system B, is also available. [9]

In Fig. 8 there are two control systems. These are applied depending on the wind condition.

That is, Fig. 8 (a) is used for the case of V_{wind1}>V_{wind2}, while (b) for the case of V_{wind1}<V_{wind2}.

When V_{wind1}>V_{wind2}, for example, the control system (a) is applied. In this case, tip speed ratio λ_1 for #1 wind turbine is controlled to be constant with DC link current I_d, which is controlled by leading angle of commutation γ with the control angle $\alpha_1=0$ as in the figure, and simultaneously, tip speed ratio λ_2 for #2 turbine is controlled with the control angle α_2 of the rectifier #2. On the contrary, for the case when V_{wind1}<V_{wind2}, the control variables are replaced to obtain constant tip speed ratios as in (b).

IV. DYNAMIC PERFORMANCES OF SERIES-CONNECTED SYSTEM

Here, dynamic performances of the closed-control systems for series-connected wind turbine generators shown in Figs. 6 (system A) and 8 (system B) are investigated.

The same wind turbine/PMSGs (SUBARU 15/40) as in [7] are used for the simulation. The wind velocities flowing to each turbine are assumed to be those shown in Fig. 9, which were acquired based on the natural wind observed at the wind turbine of our university [10]. Table I gives general specifications of the turbine.

The 2014 International Power Electronics Conference

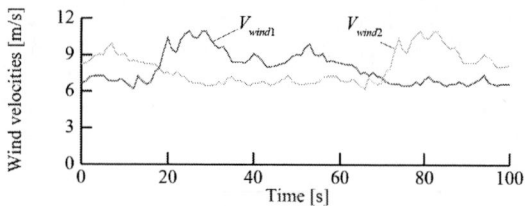

Fig. 9. Wind velocities.

TABLE I
GENERAL SPECIFICATIONS OF SUBARU 15/40

Rotor Position	Up-wind	Number of Blades	3	Rotational Speed (min⁻¹)	18~68
Hub Height (m)	21	Rated Wind Velocity (m/s)	11	Type of Generator	PM SG
Rotor Diameter (m)	15	Rated Power (kW)	40	Speedup Gear Ratio	5.5

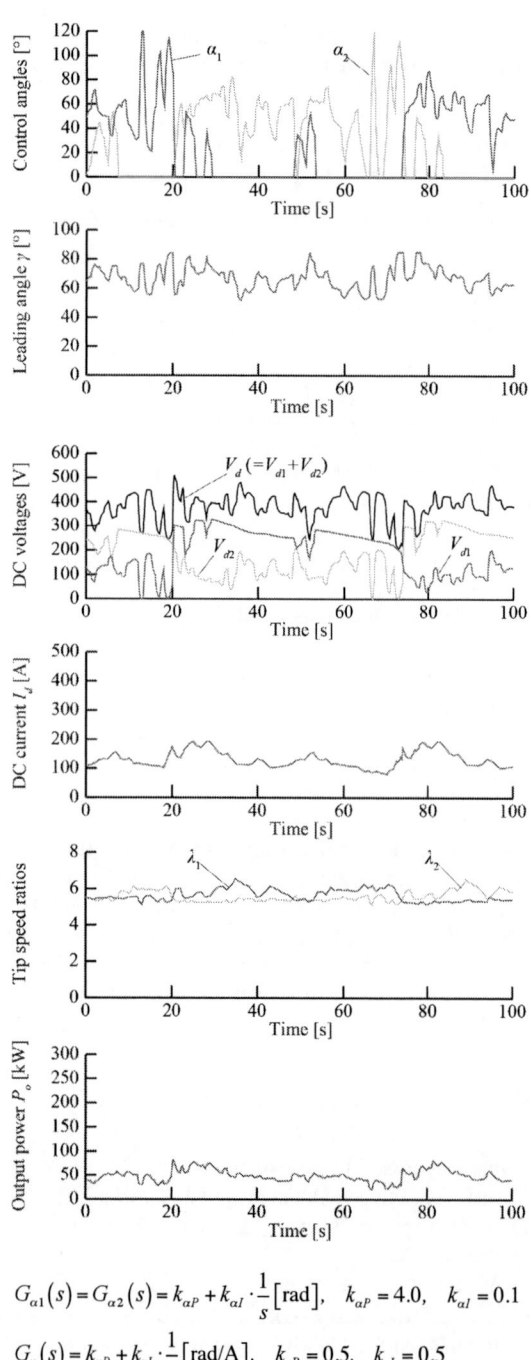

$$G_{\alpha1}(s) = G_{\alpha2}(s) = k_{\alpha P} + k_{\alpha I} \cdot \frac{1}{s} \,[\text{rad}], \quad k_{\alpha P} = 4.0, \quad k_{\alpha I} = 0.1$$

$$G_{\gamma}(s) = k_{\gamma P} + k_{\gamma I} \cdot \frac{1}{s} \,[\text{rad/A}], \quad k_{\gamma P} = 0.5, \quad k_{\gamma I} = 0.5$$

Fig. 10. Dynamic responses for system A.

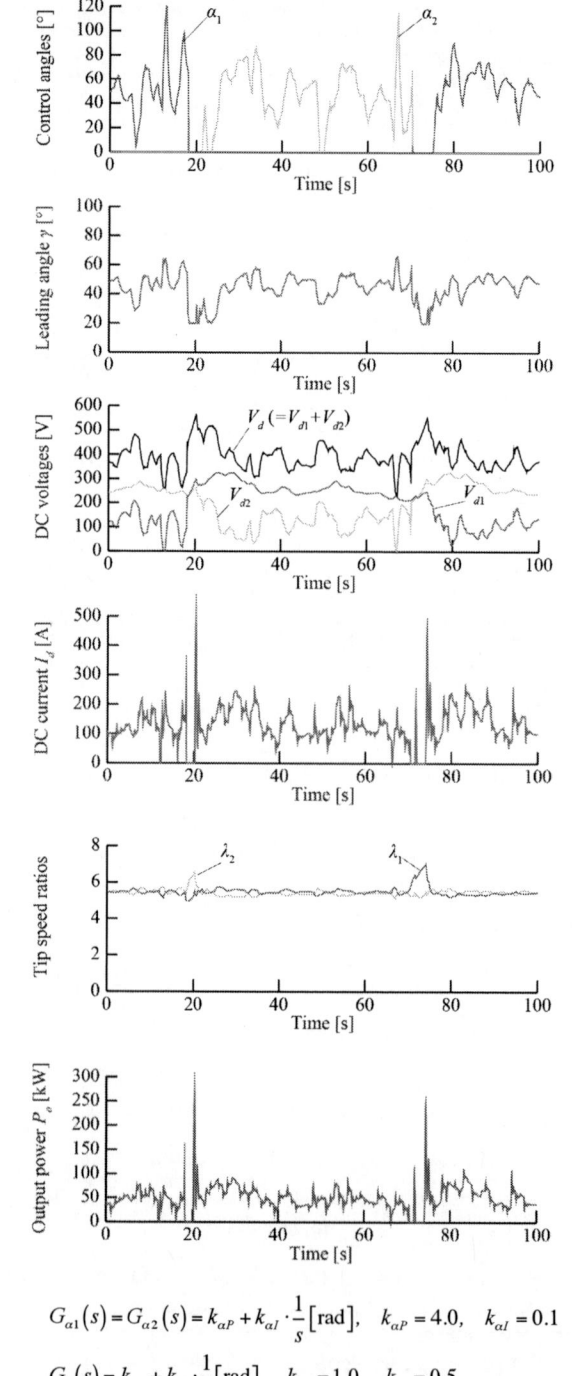

$$G_{\alpha1}(s) = G_{\alpha2}(s) = k_{\alpha P} + k_{\alpha I} \cdot \frac{1}{s} \,[\text{rad}], \quad k_{\alpha P} = 4.0, \quad k_{\alpha I} = 0.1$$

$$G_{\gamma}(s) = k_{\gamma P} + k_{\gamma I} \cdot \frac{1}{s} \,[\text{rad}], \quad k_{\gamma P} = 1.0, \quad k_{\gamma I} = 0.5$$

Fig. 11. Dynamic responses for system B.

978-1-4799-2706-7/14 $31.00 © 2014 IEEE

Fig. 10 shows the simulated dynamic performances for system A. In this figure the parameters of the PI controllers in Fig. 6 are given. The simulation was executed using MATLAB/Simulink (The MathWorks, Inc.).

The dynamic responses of control angles α_1 and α_2, leading angle for the inverter γ, resultant DC output voltage V_d, rectified DC output voltages V_{d1}, V_{d2}, DC-link current I_d, tip speed ratios λ_1, λ_2, and instantaneous output power $(=V_d \times I_d)$ are shown here. It is seen in Fig. 10 that λ_1, λ_2 are kept almost constant because the control system shown in Fig. 6 is worked properly. It is recognized that the responses of V_{d1}, V_{d2} and DC-link current I_d change in accordance with the change in the wind velocity.

In Fig. 11 the simulated dynamic performances for system B (Fig. 8) are shown. The dynamic responses for the same variables as Fig. 10 are shown in the figure. It is seen that as a whole similar responses to those for system A are obtained with considerable change in DC-link current I_d. The large change in I_d is due to the changes in the leading angle for the inverter γ and the resultant DC output voltage V_d as can be seen in the figure. Also, it is shown that there are some periods of time in which no DC-link current flows, and this fact indicates the presence of periods with no generator torque, resulting in unexpected increases in the tip speed ratios as in the figure.

Therefore, it is seen that although both control systems A (Fig. 6) and B (Fig. 8) are useful for keeping the tip speed ratios λ_1, λ_2 almost constant, realizing an optimum use of the wind energy, when system B is used a deterioration in control quality is inevitable due to the interruption of DC-link current. Also, it is clear that the same control method as Fig. 6 can be applied to the wind power plants consisting of more than two wind turbine/generators.

V. EXPERIMENTAL INVESTIGATIONS

To confirm the validity of the proposed wind turbine generator system, we constructed an experimental set consisting of two wind turbines, for which DC motors are used to simulate the behavior of the wind turbine [11], and two synchronous generators.

That is, in the series-connected system of two wind turbine/SGs shown in Fig. 5, the DC motors were substituted for wind turbines, and the synchronous generators with wound rotors (with field windings) were used as wind turbine generators in the experimental system, and control system A (Fig. 6) was applied.

The ratings of DMs and SGs are shown in Table II. The parameters of the wind turbines assumed as simulator are shown in Table III. These parameters were selected based on the ratings of the DC motors. The relationships between C_p and λ are assumed to be the same as Fig. 4.

Fig. 12 shows an example of experimental results of the system (gray solid lines) when wind velocities are fluctuated as in (a) [12]. These wind data were acquired

TABLE II
RATINGS OF DMs AND SGs USED FOR EXPERIMENTAL INVESTIGATIONS

DC Motor (DM)	
Output power	2.2kW
Voltage	100V
Current	29A
Rotational speed	1500min^{-1}
Number poles	4
Synchronous Generator (SG)	
Apparent power	2.0kVA
Voltage	200V
Current	5.77A
Rotational speed	1500min^{-1}
Number poles	4

Note - The ratings of two sets of DM and SG are the same.

TABLE III
PARAMETERS OF THE WIND TURBINES
(COMMON IN BOTH #1 AND #2 DM/SGs)

Blade radius R_{rotor}	1.05 m
Air density ρ	1.225 kg/m^3
Rated wind velocity	10 m/s
Rated output power	925 W

based on the natural wind observed at the wind turbine of our university. In this case, the reference values of the tip speed ratios λ_{1ref}, λ_{2ref} were set to be 5.52. It is shown in (b) and (c) that the tip speed ratios were almost kept constant by controlling the field currents as in (d) and (e), and as well DC-link current I_d in (f). From (g) it is clarified that the output power of each wind turbine generator was combined effectively.

In Fig. 12 the black solid lines indicate simulated results obtained based on the block diagram of Fig. 6, and it is shown that a good agreement between experimental and simulated results is obtained, showing the effectiveness of the proposed control strategy as well as the validity of the simulation model.

VI. IMPROVEMENT OF THE QUALITY OF OUTPUT POWER

In the proposed series-connected system, the output current is of trapezoidal-like because the current-source type power converter is used. An example of instantaneous waveforms of inverter output voltage and current is shown in Fig. 13(a) and (b) for the experimental system, and a lot of harmonic components were included in the waveforms, deteriorating the quality of the output power.

However, when a set of synchronous machine and duplex reactor is incorporated to the output side of the inverter as in Fig. 3, considerable reduction of the harmonics is obtained as shown in Fig. 13(c) and (d) [13]. In this experiment, we used a currently available set of synchronous machine and duplex reactor.

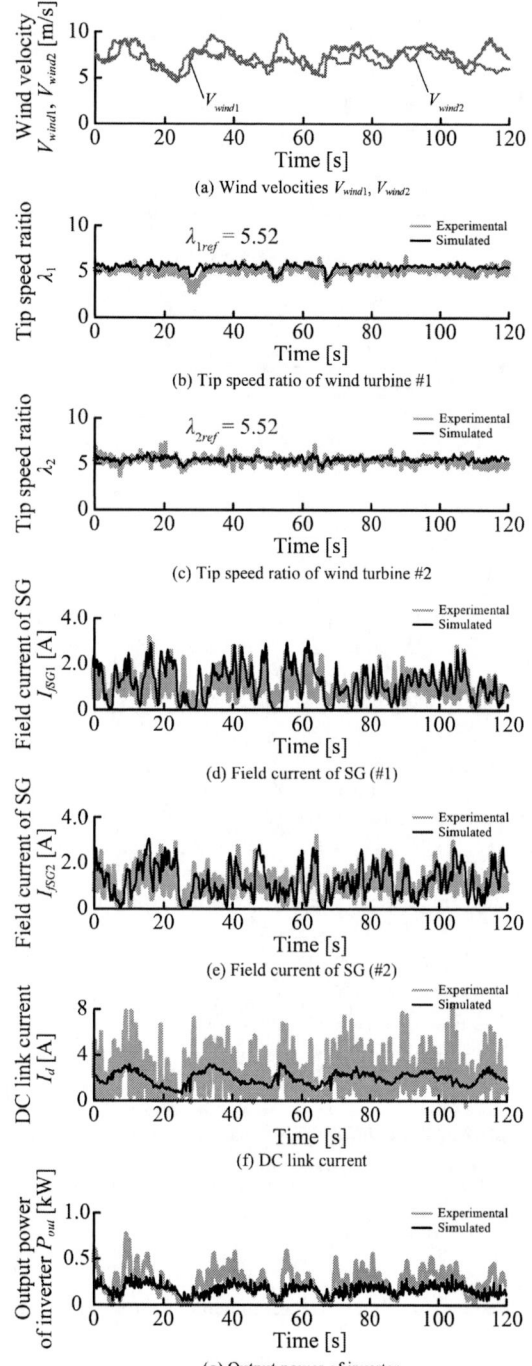

(a) Wind velocities V_{wind1}, V_{wind2}

(b) Tip speed ratio of wind turbine #1

(c) Tip speed ratio of wind turbine #2

(d) Field current of SG (#1)

(e) Field current of SG (#2)

(f) DC link current

(g) Output power of inverter

Fig. 12. Comparisons between experimental and simulated dynamic responses.

The realization of such high-quality output current is because that currents flow in the damper windings of the synchronous machine automatically so as to cancel the harmonic components of the flux linkages of the damper windings due to the inverter output current. As a result the armature current waveform eventually became as shown in Fig. 13(e). Detailed discussion on the harmonics reduction is given in [13].

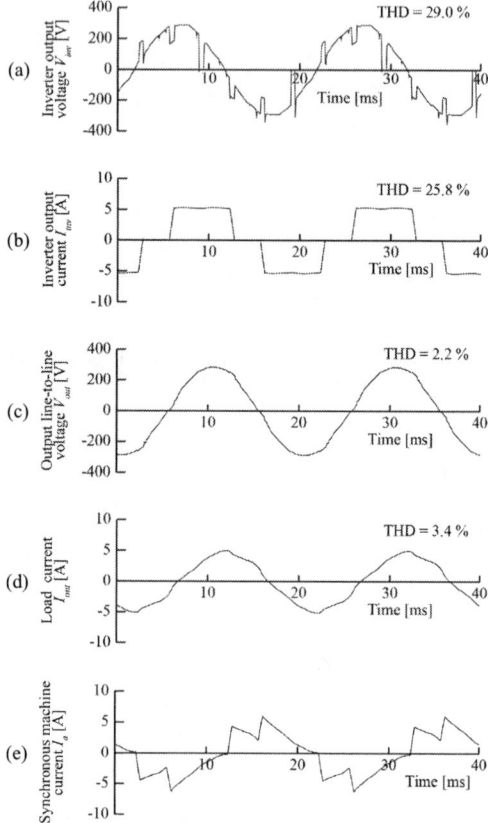

Fig. 13. An example of instantaneous voltage and current waveforms.

It should be noticed here that, as a synchronous machine connected to the output-side of the inverter through duplex reactor, a high-speed machine with small number of poles, for which the machine size is small, is enough to get output current of high quality.

VII. TOPOLOGY TO INCREASE PLANT OUTPUT WITHOUT INCREASING VOLTAGE LEVEL OF SYSTEM COMPONENTS

In the series topology shown in Fig. 3 the more the number of series connected wind turbines, the larger the total plant output. However, since the dielectric strength increases for the cases when the number of wind turbines is increased, careful considerations should be made for the dielectric strength of the individual components. A novel topology to double the plant output without increasing the dielectric strength is proposed here.

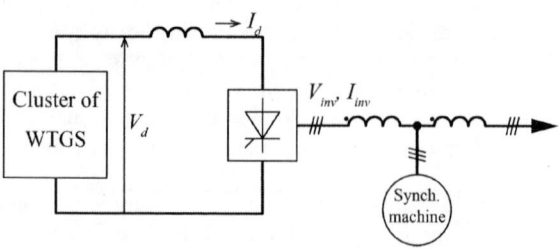

Fig. 14. Simplified representation for series-connected wind turbine/generators shown in Fig. 3.

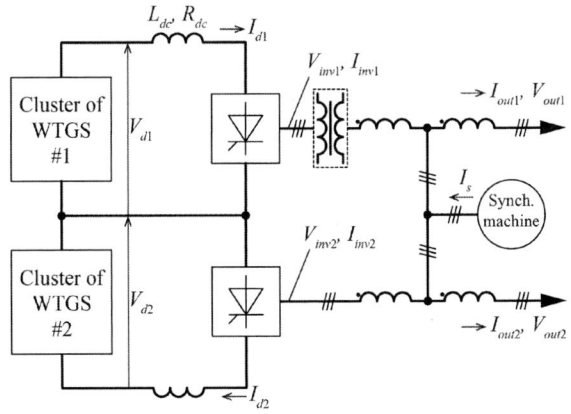

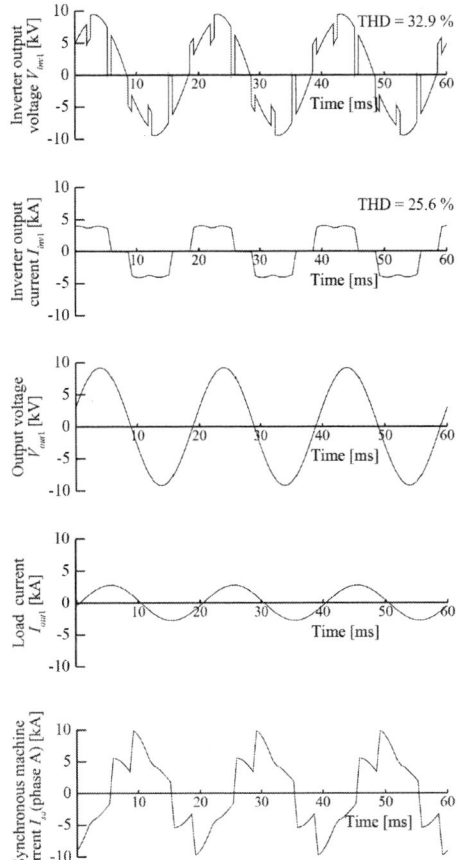

Fig. 15. Configuration of two clusters of series-connected wind turbine/generators with synchronous machine and duplex reactor.

Fig. 14 shows a simplified representation for series-connected wind turbine/generators and a thyristor inverter, in which the block of 'Cluster of WTGS' represents a cluster of wind turbine/generators connected in series.

Fig. 15 shows the configuration of wind power plant to realize the increase in the output power without increasing the voltage level of the system components. This system consists of two sets of the wind power plant shown in Fig. 3. In addition, an isolation transformer is inserted in one of the inverter output terminals to avoid interference in the two clusters of series-connected systems.

In Fig. 15, if the DC link voltages in each cluster of WTGS are the same level as that in Fig. 3, the voltage level of the wind power plant of Fig. 15 is the same as Fig. 3, doubling in the output power of the whole plant without increasing the system voltage level. Also, the configuration of the proposed topology is very simple because only one synchronous machine is used.

Fig. 16 shows an example of simulation results for the proposed wind power plant. PSIM (Powersim Inc.) is used to obtain these waveforms.

System parameters and the operating points for the simulation are chosen as follows:

L_{dc}=5mH, R_{dc}=0.6Ω
V_{d1}=8kV=V_{d2}
(estimated number of turbines = 10 for each cluster),
I_{d1} = 3.7kA = I_{d2}
Leading angle of commutation of inverter γ = 50deg
(for both inverters)
Output of each cluster:
Output voltage (l-l) = 6.5kV, Output current = 1.9kA
Synchronous machine voltage (l-l) = 6.6kV

In this simulation the waveforms for inverter output voltage (in which voltage jumps and sinks appear due to the commutations of the thyristors in the inverter) and current, system output voltage and current, and armature

Fig. 16. Simulated voltage and current waveforms of the proposed wind power plant.

current of the synchronous machine are shown from top to bottom, and these results are similar to the experimental ones in Fig. 13.

From these results it is clarified that a series-connected wind power plant to increase the output power without an increase in the voltage level of the system components can be realized, and that the output voltage of sinusoidal waveform is obtained by means of the function of the duplex reactors connected to the synchronous machine.

VIII. CONCLUSIONS

Control strategies and a topology of series-connected wind power plants using current-source type thyristor inverter have been discussed.

As basic discussion control strategies for wind power plant consisting of two turbine/generators connected in series have been studied, and it has been shown that the control system having a loop for DC-link current control by means of leading angle of inverter is effective for an optimum use of the wind energy, and that the usefulness of the control system was verified through experimental investigations.

It has also been shown that the quality of output power of the wind power plant is improved greatly when a set of synchronous machine with small number of poles and duplex reactor is incorporated to the output side of the thyristor inverter.

Finally, a topology to double the output power of the wind power plant without increasing the voltage level of the system has been proposed, and the simulation results have shown the validity of the topology.

In the proposed series-connected wind power plant, the synchronous machine connected to the inverter output has the ability to control the active and reactive power for the grid or loads, and these issues should be investigated to bring out more advanced applications and/or topologies of the proposed system. These are left for future study.

ACKNOWLEDGMENT

This work was partially supported by Research Institute for Science and Technology of Tokyo Denki University Q13E-02/Japan. We would like to thank all members of Nishikata laboratory at Tokyo Denki University for their cooperation of this work.

REFERENCES

[1] H. J. Bahirat, B. A. Mork, and H. Kr. Høidalen, "Comparison of wind farm topologies for offshore applications," *IEEE Power and Energy Society General Meeting*, pp. 1–8, July 2012.

[2] M. Liserre, R. Cárdenas, M. Molinas, and J. Rodríguez, "Overview of multi-MW wind turbines and wind parks," *IEEE Trans. on Ind. Electron.*, vol. 58, no. 4, pp. 1081–1095, Apr. 2011.

[3] S. Nishikata and F. Tatsuta, "A new interconnecting method for wind turbine/generators in a wind farm and basic performances of the integrated system," *IEEE Trans. on Ind. Electron.*, vol. 57, no. 2, pp. 468–475, Feb. 2010.

[4] M. Popat, BinWu, and N. R. Zargari, "A novel decoupled interconnecting method for current-source converter-based offshore wind farms," *IEEE Trans. on Power Electron.*, vol. 27, no. 10, pp. 4224–4233, Oct. 2012.

[5] NAVIGANT Research, *A BTM WIND REPORT: World Market Update 2012*, March 2013.

[6] Siegfried Heier: Grid Integration of Wind Energy Conversion Systems, second edition, John Wiley & Sons, Ltd, 2006, p.44.

[7] T. Teshirogi, and S. Nishikata, "Effects of system parameters on the performance characteristics of a wind turbine generating system using a current-source thyristor inverter," *IEEE Trans. on Industry Applications*, vol.47, no.1, pp.252-257, JAN/FEB 2011.

[8] F. Tatsuta and S. Nishikata, "Improvement of dynamic performances of tip speed ratios of series connected wind turbine generators," *Proc. of 15th International Conference on Electrical Machines and Systems* (ICEMS2012), Sapporo, Japan, pp. 1–4, October 2012.

[9] F. Tatsuta, and S. Nishikata, "Dynamic performance analysis of a wind turbine generating system with series connected wind generators and bypass diodes using a current source thyristor inverter," *The 2010 International Power Electronics Conference* (IPEC2010), Sapporo, Japan, pp. 1830-1836, June 2010.

[10] F. Tatsuta, and S. Nishikata, "Dynamic performance analysis of a wind turbine generating system with series connected wind generators using a current source thyristor inverter," *13th European Conference on Power Electronics and Applications* (EPE2009), Barcelona, Spain, pp. 1–9, September 2009.

[11] I. Tsoumas, E. Tsimplostephanakis, E. Tatakis, and A. Safacas, "Control of a WECS using a simulated wind turbine and an asynchronous generator – Maximum power tracking," *Proc. of 15th International Conference on Electrical Machines* (ICEM2002), Brugge, Belgium, pp. 1–6, August 2002.

[12] A. Takemura, F. Tatsuta, H. Yokoyama, and S. Nishikata, "Studies on field current control method for constant tip speed ratios of series connected wind turbine generators in a wind farm," *Proc. of 15th International Conference on Electrical Machines and Systems* (ICEMS2012), Sapporo, Japan, pp. 1–5, October 2012.

[13] F. Tatsuta, and S. Nishikata, "Studies on a hybrid wind turbine generating system using a current-source thyristor inverter", *IEEJ Transactions on Industry Applications*, Vol.132, No.9 pp.875-883, September 2012.

An Impedance-Based Approach to HVDC System Stability Analysis and Control Development

Hanchao Liu, Shahil Shah and Jian Sun

Department of Electrical, Computer and Systems Engineering
Rensselaer Polytechnic Institute, Troy, NY 12180-3590, USA
Telephone: (518) 276-8297; Fax: (518) 276-2387; E-mail: jsun@rpi.edu

Abstract– **High-voltage dc (HVDC) transmission system stability has traditionally been studied using state-space models and eigenvalue analysis. The state-space approach requires design details of each component of the HVDC grid, which is usually not available, and does not support local control development at individual terminals. This paper presents an impedance-based approach to HVDC system stability analysis that is easier to apply and supports local and adaptive control development. Representation of a HVDC system by an impedance equivalent circuit is presented first. Analytical impedance models are then developed for different HVDC converters to support system impedance analysis. The ability of impedance-based analysis to predict and mitigate different control and system instability problems are demonstrated and verified by detailed circuit simulation.**

I. Introduction

Control development for traditional point-to-point HVDC connections based on line-commutated converters (LCC) has emphasized steady-state power flow and operation under different grid conditions [1]. Stability of the dc link has not been a major concern because of the simple system topology and slow control functions. Control instability of individual terminals is a potential problem when connected to a weak grid, which has been analyzed using state-space models and eigenvalue analysis [2, 3]. This approach has influenced the more recent development in multi-terminal HVDC systems based on voltage-source converters (VSC) [4, 5]. Although stability is considered as a much more significant problem due to the fast dynamics of VSC control, the basic approach is still eigenvalue analysis based on state-space models of the entire HVDC grid. Establishing such state-space models requires design details of each terminal but such design details are usually not available to system owners, particularly when multiple suppliers are involved. The complex system model also makes it difficult to associate any unstable (or marginally stable) mode with physical design parameters, which is essential for solving an instability problem. Additionally, state-space analysis can only be performed centrally and in advance. It does not support local control development at individual terminals, especially when online adaptation based on real-time grid topology and operation conditions is required.

This paper presents an impedance-based approach to HVDC system stability analysis which solves the aforementioned problems. The impedance-based approach has been used in power electronics to study converter-filter interactions [6] and stability of

distributed power systems where multiple loads are powered from a central voltage source [7]. The method has not been used in HVDC system, partially because of the lack of proper impedance models. Impedance models of line-commutated converters have been developed by either neglecting the effects of phase control (constant firing angle assumption) [8] or with significant simplification of control functions [9, 10]. However, such simplistic models are accurate only in a limited frequency range and are not suitable for stability analysis. Recently, the impedance-based stability criterion has been extended to current-source converters [11]. Meantime, new analytical models have been developed for VSC and LCC converters dc-terminal impedance [12] using the principle of harmonic linearization [13], making it possible now to study HVDC system stability based on impedance.

The objective of this paper is to demonstrate the ability of impedance-based approach to detect various control and system instability problems in LCC and VSC based point-to-point HVDC systems. The rest of the paper is organized as follows: Section II reviews the impedance-based approach for the stability analysis and discusses its applicability to the HVDC link stability analysis. Section III presents the small-signal impedance models for LCC based terminals and their application to evaluate the dc-link stability in LCC based systems. Section IV does the same for VSC based systems. Section V summarizes the work.

II. HVDC System Instability

This section uses an example to illustrates possible instability in HVDC systems and introduces the impedance-based system stability analysis method. Fig. 1 depicts a LCC-based HVDC system that will be studied. Both the rectifier and the inverter are rated for 500 kV and 1000 MW, and each uses a 12-pulse LCC converter. The ac grid at each terminal is modeled by an *RL* circuit having the same short circuit ratio (SCR) of 2.5. The transformer is rated for 345/211 kV and 230/211 kV, respectively. *G*(*s*) is the transfer function of the current/voltage sensing circuit. It is assumed that the inverter controls the dc link voltage, while the rectifier regulates its output current. A dq-domain PLL is used to track the ac bus voltage reference angle for phase control. The system parameters are presented in [13].

For the purpose of illustration, the rectifier current compensator is designed such that it operates stably when feeding current into an ideal voltage source. Similarly, the inverter voltage compensator is designed such that it operates stably when fed by an ideal

The 2014 International Power Electronics Conference

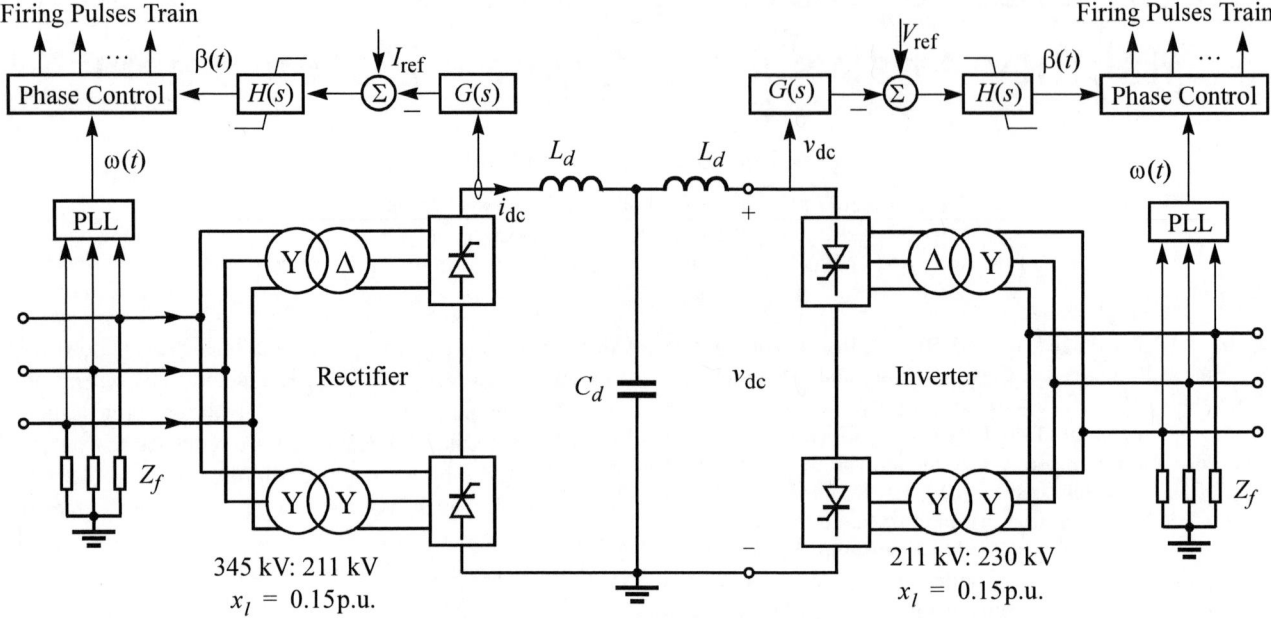

Fig. 1. Simplified diagram of a LCC-based point-to-point HVDC system.

current source. Simulation of each converter confirmed this design objective. However, it is found that when they are connected as shown in Fig. 1, the dc link voltage and current become unstable.

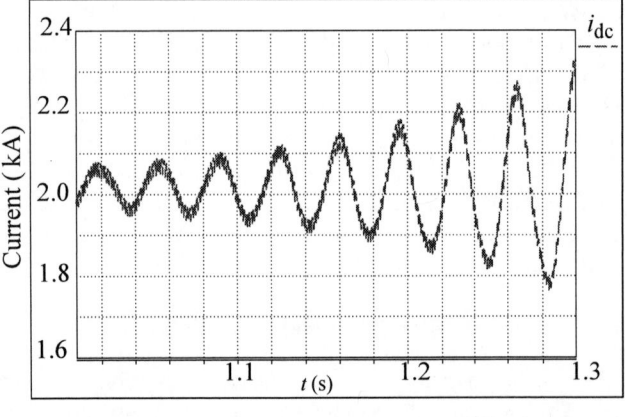

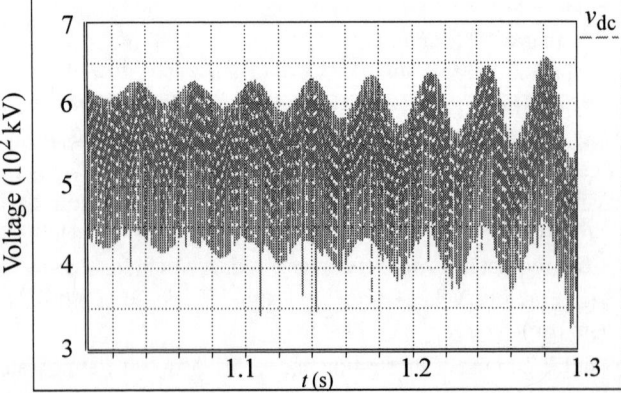

Fig. 2. DC link current and voltage responses of the example LCC point-to-point HVDC system depicted in Fig. 1.

Fig. 2 shows the simulated unstable behavior when the dc current ramps up to its rated value during start-up.

To identify the cause of system instability, Fig. 3 shows an equivalent circuit model of the system where the rectifier is modeled as current source in parallel with an output impedance Z_{rec}, while the inverter is modeled as voltage source in series with an input impedance Z_{inv}. Note that this equivalent circuit resembles that of an inverter connected to a non-ideal grid [11]. Therefore, the impedance-based stability criterion developed in [11] for grid-connected inverters can also be applied to determine point-to-point HVDC system stability. Specifically, an HVDC system consisting of a current-controlled rectifier and voltage-controlled inverter is stable if

1) The rectifier control is stable when feeding current into an ideal voltage source;

2) The inverter control is stable when being fed by an ideal current source; and

3) The ratio of the inverter input impedance to the rectifier output impedance satisfies the Nyquist stability criterion.

In the following two sections, we will use this impedance-based criterion to study LCC and VSC HVDC system stability. Since the method requires an impedance model for each converter

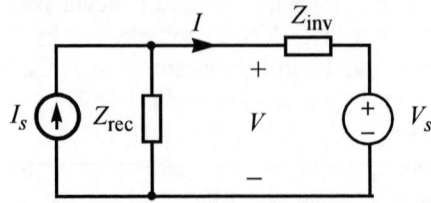

Fig. 3. Small-signal impedance model of a rectifier-inverter system.

terminal, each section will start from the development of the rectifier output impedance and the inverter input impedance, followed by numerical simulation of possible instability and its characterization using the impedance models.

III. LCC HVDC SYSTEMS

A. Impedance Modeling

Fig. 4 depicts detailed ac side equivalent impedance of the LCC HVDC rectifier terminal. The ac grid is modeled by an RL circuit with short circuit ratio (SCR) set to 2.5 and 75 degrees at the fundamental frequency (60 Hz). The phase control design is presented in Fig.1. The ac filters are designed to absorb 11th and 13th characteristics harmonics current and provide low impedance path at 3rd harmonics to avoid low-oder harmonics resonance problems.

The modeling process and circuit parameters have been presented in [13] and shown in (1), where $L(s) = H(s)G(s)$, k_T is the voltage ratio of the converter transformer; Z_{ac} represents the ac side impedance seen by the converter; and $\alpha = \alpha_0 + 5\pi/6$, α_0 is steady-stage firing angle. The dynamic of the PLL is characterized by the loop gain

$$T_{pll}(s) = H_{pll}(s)/[s + V_1 H_{pll}(s)] \qquad (2)$$

where V_1 is the amplitude of the grid voltage and $H_{pll}(s)$ is the PLL compensator transfer function.

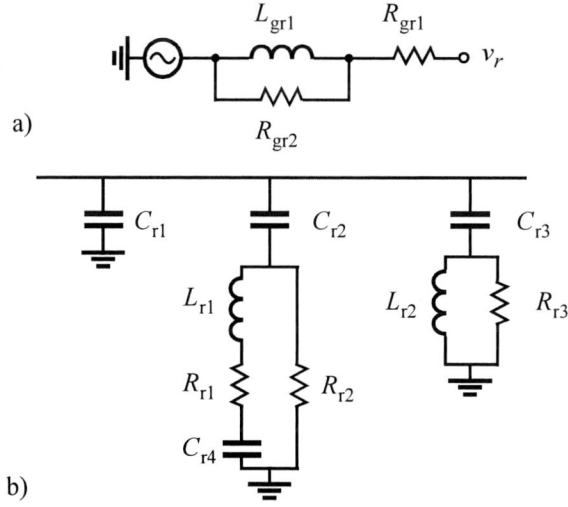

Fig. 4. LCC rectifier ac side impedance: a) ac grid equivalent circuit; b) rectifier filters circuit.

Fig. 5 shows the validation of the impedance model of the rectifier. The impedance model of the inverter and its validation are presented in (5) and Fig. 6 in the following page.

B. Stability Analysis

As explained in Section II, the dc link stability of the HVDC system depends on the ratio between the inverter impedance Z_{inv} and the equivalent impedance Z_{eqv} which is the dc network impedance seen by the inverter including the dc link impedance (L_d and C_d) and the rectifier output impedance. In the unstable case (Case I) presented in Section II, the dc current compensator transfer function is designed to be

$$H(s) = 1.6 + 19.3/s, \qquad (3)$$

and the dc voltage compensator transfer function is

$$H(s) = 2 + 0.2/s. \qquad (4)$$

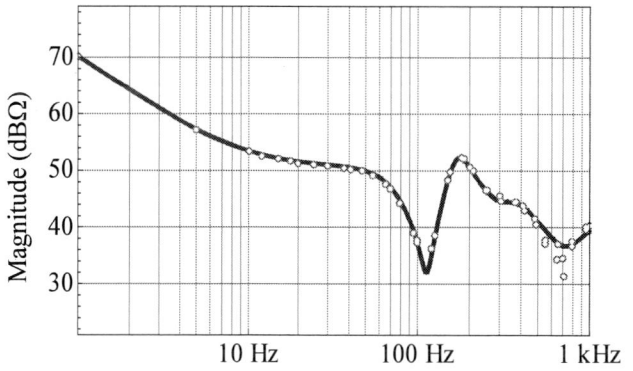

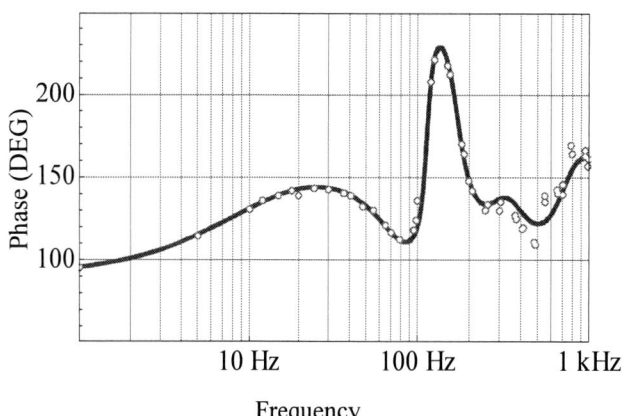

Fig. 5. LCC rectifier output impedance responses. Solid lines: model prediction; Dots: point by point simulation results.

$$Z_{rec}(s) = \left\{ \frac{6\sqrt{3}\cos\alpha V_1}{k_T\pi} + \frac{36jI_0}{k_T^2\pi^2}[Z_{ac}(s+j2\pi f_1) - Z_{ac}(s-j2\pi f_1)] \right\} \frac{L(s) + T_{pll}(s)[Z_{ac}(s+j2\pi f_1)e^{-j\alpha} + Z_{ac}(s-j2\pi f_1)e^{j\alpha}]}{\frac{\pi k_T}{2\sqrt{3}} + jT_{pll}(s)I_0[Z_{ac}(s+j2\pi f_1)e^{-j\alpha} - Z_{ac}(s-j2\pi f_1)e^{j\alpha}]}$$

$$- \frac{36}{k_T^2\pi^2}[Z_{ac}(s-j2\pi f_1) + Z_{ac}(s+j2\pi f_1)] \qquad (1)$$

The 2014 International Power Electronics Conference

$$Z_{\text{inv}}(s) = \cfrac{2T_{\text{pll}}(s)[Z_{\text{ac}}(s+j2\pi f_1)e^{-j\alpha}+Z_{\text{ac}}(s-j2\pi f_1)e^{j\alpha}]/(k_T\pi)}{L(s)/\sqrt{3}-\cfrac{1+jI_0T_{\text{pll}}(s)[Z_{\text{ac}}(s+j2\pi f_1)e^{-j\alpha}-Z_{\text{ac}}(s-j2\pi f_1)e^{j\alpha}]}{-3\sqrt{3}\cos\alpha\, V_1+\{18jI_0[Z_{\text{ac}}(s+j2\pi f_1)-Z_{\text{ac}}(s-j2\pi f_1)]\}/(k_T\pi)}}$$

$$+\cfrac{6jI_0[Z_{\text{ac}}(s+j2\pi f_1)-Z_{\text{ac}}(s-j2\pi f_1)]/(k_T\pi)-\sqrt{3}\cos\alpha\, V_1}{\cfrac{6[Z_{\text{ac}}(s-j2\pi f_1)+Z_{\text{ac}}(s+j2\pi f_1)]/[L(s)k_T^2\pi^2]}{\pi k_T+j2\sqrt{3}I_0T_{\text{pll}}[Z_{\text{ac}}(s+j2\pi f_1)e^{-j\alpha}-Z_{\text{ac}}(s-j2\pi f_1)e^{j\alpha}]}} \tag{5}$$

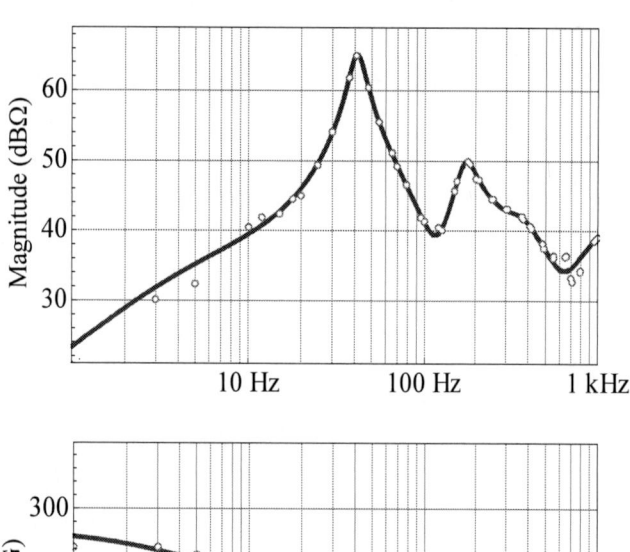

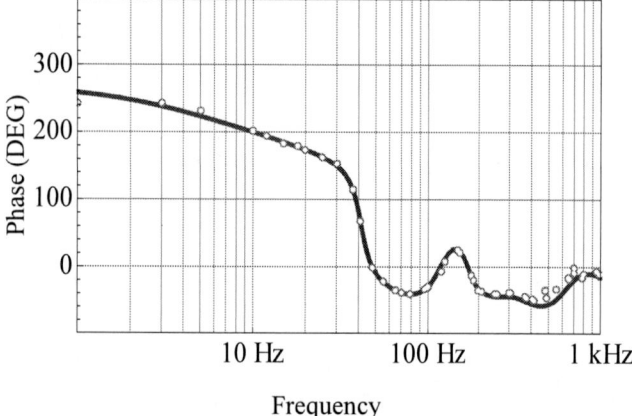

Fig. 6. LCC inverter input impedance response. Solid lines: model prediction; Dots: point by point simulation results.

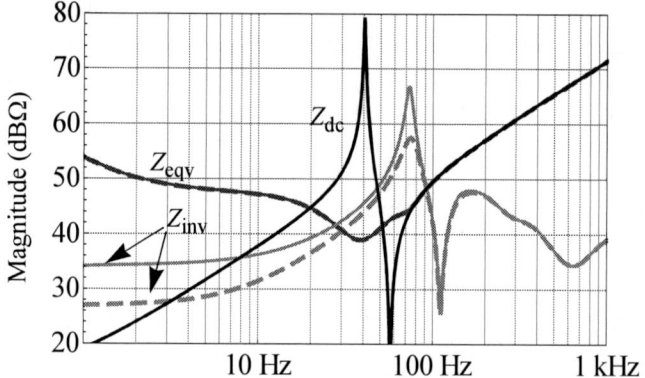

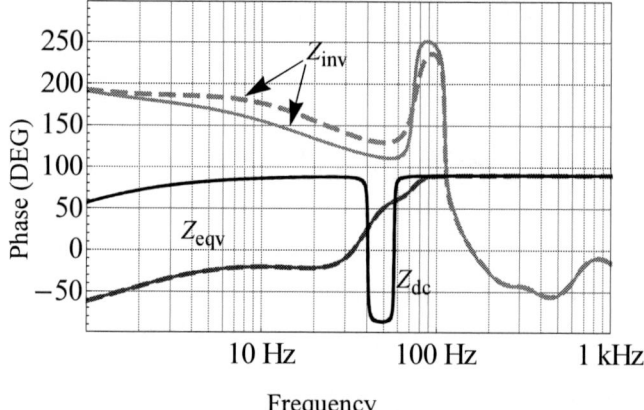

Fig. 7. Impedance analysis of the LCC system. Dashed line: Case I. Solid lines: Case II.

In addition to the unstable case, a stable system design case is also presented as Case II in which the dc voltage control compensator transfer function is designed to be

$$H(s) = 0.2 + 0.2/s . \tag{6}$$

In both of two cases, the inverter PLL compensator is

$$H_{\text{pll}}(s) = 4.25 \times 10^{-3} + 4/s . \tag{7}$$

In the Bode plots of Fig. 7, Z_{eqv} includes the large dc smoothing reactor but still shows capacitive at the low frequency. This capacitive behavior is due to integral gain of the current controller. From the Fig. 7, it can be observed that in the unstable case I the impedance Z_{inv} and Z_{eqv} intersect at 30 Hz, with a phase

margin of 25 degrees which is corresponding to the resonance frequency of 30 Hz in Fig.2. In Case II, the two impedances Z_{inv} and Z_{eqv} intersect at 28 Hz, with a phase margin of 45 degrees which shows a stable system design.

In order to show the effect of the rectifier control, the passive dc link impedance $Z_{\text{dc}}(s)$ is also shown in Fig. 7, where

$$Z_{\text{dc}} = sL_d + \frac{sL_d}{1+s^2L_dC_d} . \tag{8}$$

It can be found that in the low frequency range where the resonance happens, the impedance of the rectifier greatly changed the dc impedance seen by the inverter.

IV. VSC HVDC Systems

A 300 MW, 300 kV point-to-point VSC HVDC system is shown in Fig. 8. Similar to the LCC system, the inverter terminal regulates the dc link voltage and the rectifier terminal controls the power flow on the dc link. Moreover, the rectifier terminal is designed to operate stably when supplying power to an ideal voltage source and the inverter terminal is designed to operate stably when being fed by an ideal current source. Following subsections develop terminal impedance models and apply them to the dc link stability analysis.

A. VSC Rectifier Output Impedance Modeling

The circuit diagram of VSC based rectifier along with control implementation is shown in Fig. 9. The VSC rectifier controls the active power flow on HVDC link by outer power control loop with 1 Hz bandwidth. The inner dq-current control is designed with 200 Hz bandwidth (10% of the switching frequency). To model the rectifier output impedance, a small perturbation voltage signal, $\hat{v}_s$, is injected in the dc-link voltage. Based on power balance between the input and output of the VSC, we have

$$v_s i_s = \frac{3}{2}(v_{id} i_d + v_{iq} i_q). \tag{9}$$

The modeling process is simplified by assuming the ideal grid and neglecting the perturbations in the grid voltage. The assumption makes ac-side current perturbations to depend only on the rectifier bridge input voltages. It also ignores the effect of PLL dynamics. The ac-side current perturbations are mapped back to the dc-side current using (9). The ratio of injected voltage perturbation to the resulting dc current perturbation gives the output impedance. The resulting rectifier output impedance is found to be

$$Z_r(s) = \cfrac{V_s^2}{P_0 + \cfrac{s}{s + \omega_c[1 + 3V_p H_p(s)/2]}\left(\cfrac{3V_p^2}{2Z_{ph}} - P_o\right)} \tag{10}$$

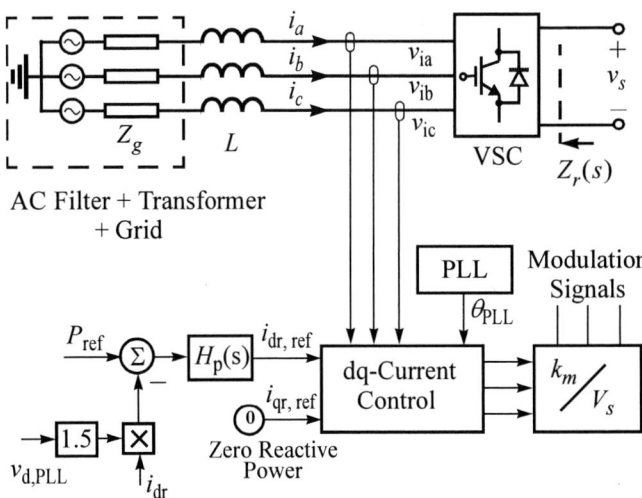

Fig. 9. VSC rectifier circuit with active power control.

where
- V_s is the nominal dc-link voltage;
- P_0 is the nominal output power;
- ω_c is the dq-current control bandwidth;
- V_p is the nominal phase voltage amplitude;
- $H_p(s)$ is the active power loop compensator and
- Z_{ph} is the phase reactor impedance.

The simplified expression of impedance is obtained by assuming that the PI current control compensator has zero placed to cancel the pole introduced by the phase inductor impedance and its gain is designed to give bandwidth of ω_c. Fig. 10 shows the comparison of the model prediction and the impedance obtained by point-by-point simulations for different grid conditions. As can be seen that the model is accurate for strong grid, but weak grid introduces resonant peaks not captured by the model. For large dc-link capacitor at the front of VSC rectifier, these resonant peaks will not be seen by the cable and the HVDC inverter; hence model can be used for dc link stability analysis.

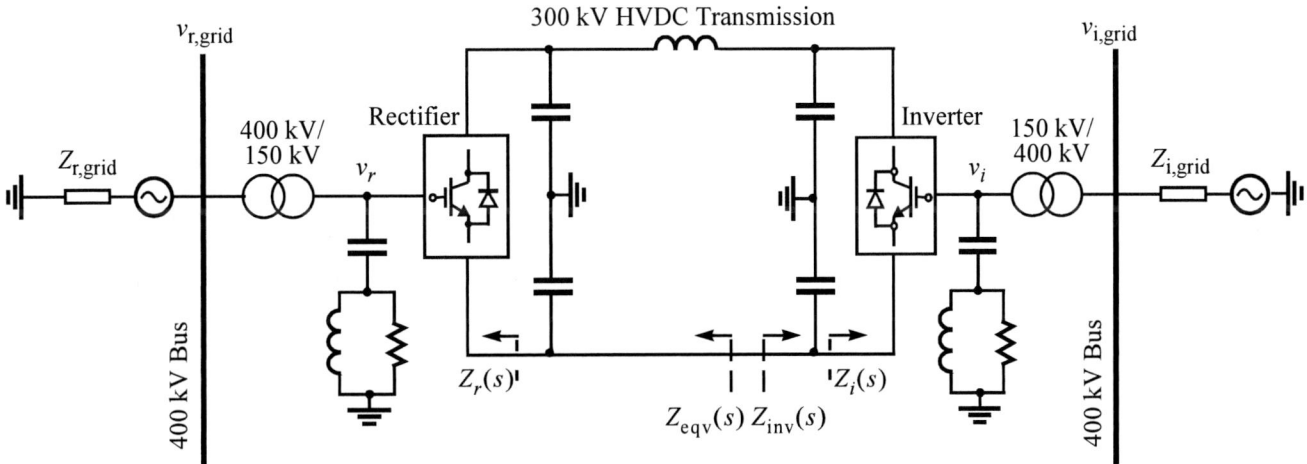

Fig. 8. Single-line diagram of a point-to-point VSC-HVDC system.

The 2014 International Power Electronics Conference

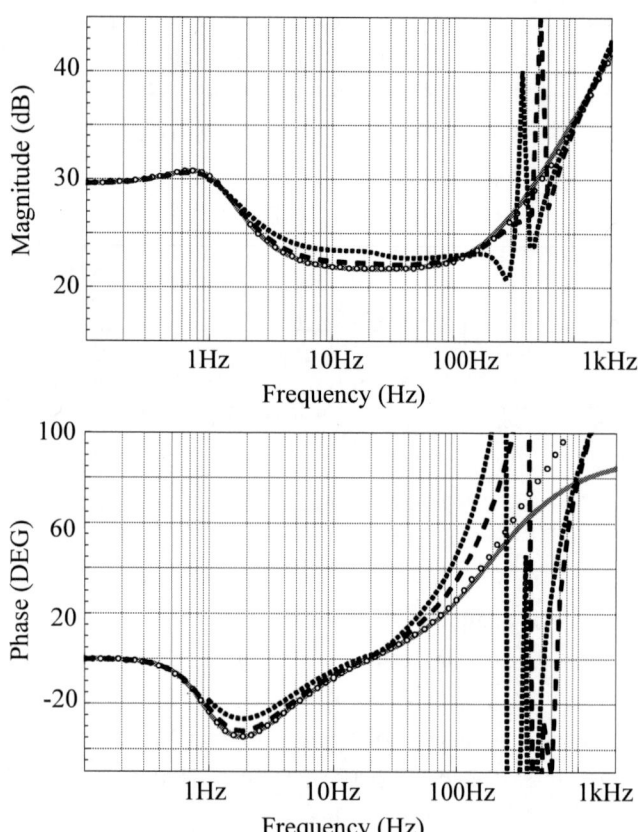

Fig. 10. Comparison of VSC rectifier input impedance responses: predicted by analytical model (solid lines) and point-by-point simulation for SCR of 15.0 (circles), 5.0 (dashed lines), and 2.5 (dotted lines).

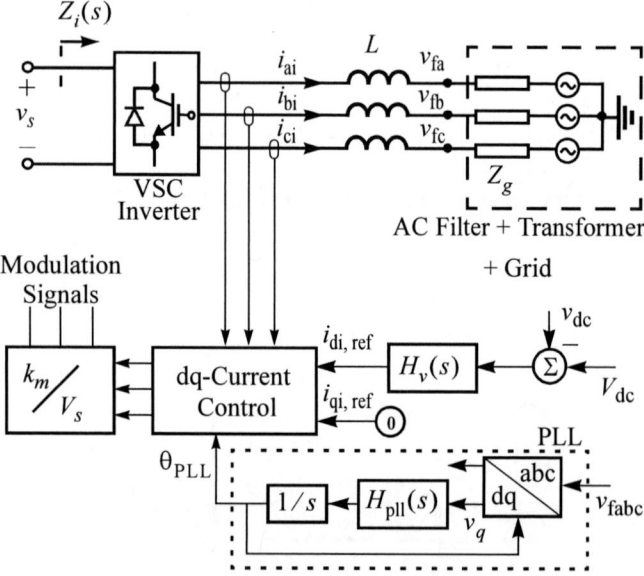

Fig. 11. VSC inverter circuit with dc-bus voltage control.

B. VSC Inverter Input Impedance Modeling

Same as VSC rectifier, input impedance of VSC inverter shown in Fig. 11, is developed and is given by (11),

$$Z_i(s) = \frac{-V_s^2}{P_i - \frac{s}{s + \omega_c}\left[1 + \frac{V_s}{V_p}H_i(s)H_v(s)\right]\left(\frac{3V_p^2}{2Z_{\text{ph}}} + P_i\right)} \quad (11)$$

where different parameters are same as discussed for rectifier, except P_i is the inverter terminal input power and $H_v(s)$ is the dc link voltage compensator.

As the design of voltage compensator, $H_v(s)$, depends on the size of dc link capacitor located at front of the VSC, the dc link capacitor is included for the verification of the impedance model. Fig. 12 compares model predictions with the impedance obtained by point-by-point simulations for different grid conditions. It shows the effect of weak grid on the accuracy of the reduced-order model. For grid with SCR of 2.5, the impedance model exhibit errors in the mid-frequency region, but it is reasonably accurate for moderately weak and strong grids. Moreover, with practical size of terminal capacitor, the model is accurate even for weak grid (SCR = 2.5) as shown in Fig. 13. It is important to note that below the dc bus voltage control bandwidth (20 Hz in present design), the input impedance is inductive with the inductance depending on the integral gain of $H_v(s)$; beyond the voltage control bandwidth the input impedance is dominated by the dc-link capacitor.

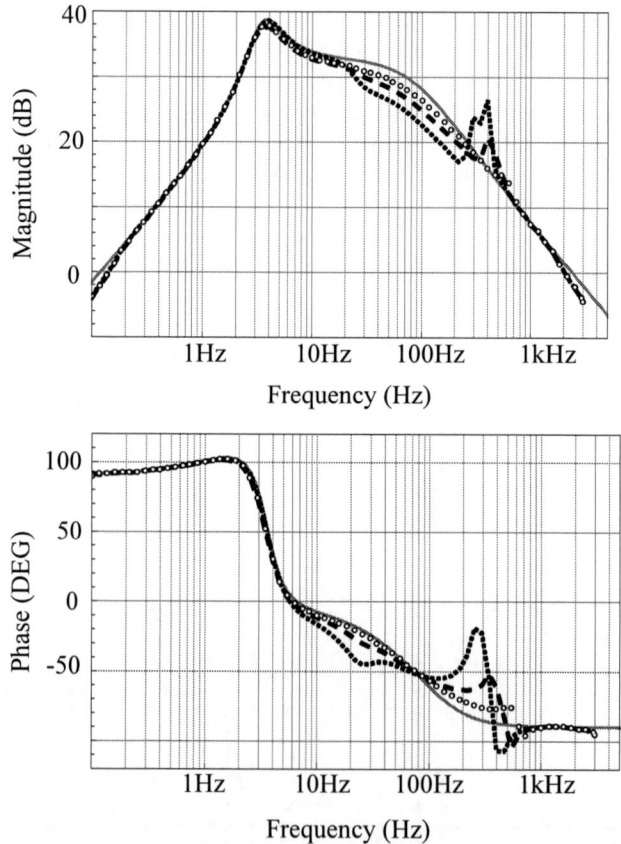

Fig. 12. Comparison of VSC inverter input impedance responses (including terminal capacitor of 67 µF): predicted by analytical model (solid lines) and point-by-point simulation results for SCR of 15.0 (circles), 5.0 (dashed lines), and 2.5 (dotted lines).

978-1-4799-2706-7/14 $31.00 © 2014 IEEE

The 2014 International Power Electronics Conference

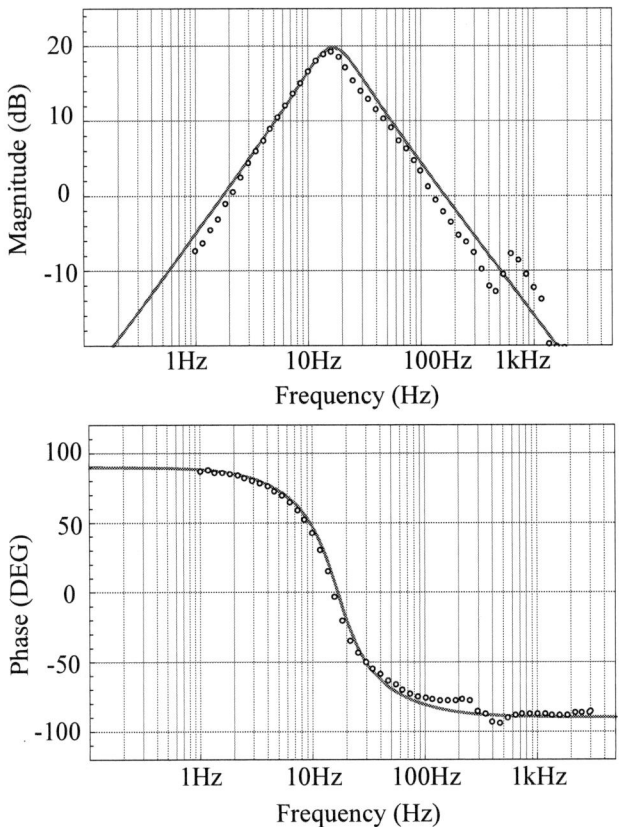

Fig. 13. Comparison of VSC inverter input impedance responses (including terminal capacitor of 1 mF): predicted by analytical model (solid lines) and point-by-point simulation results for SCR of 2.5 (circles).

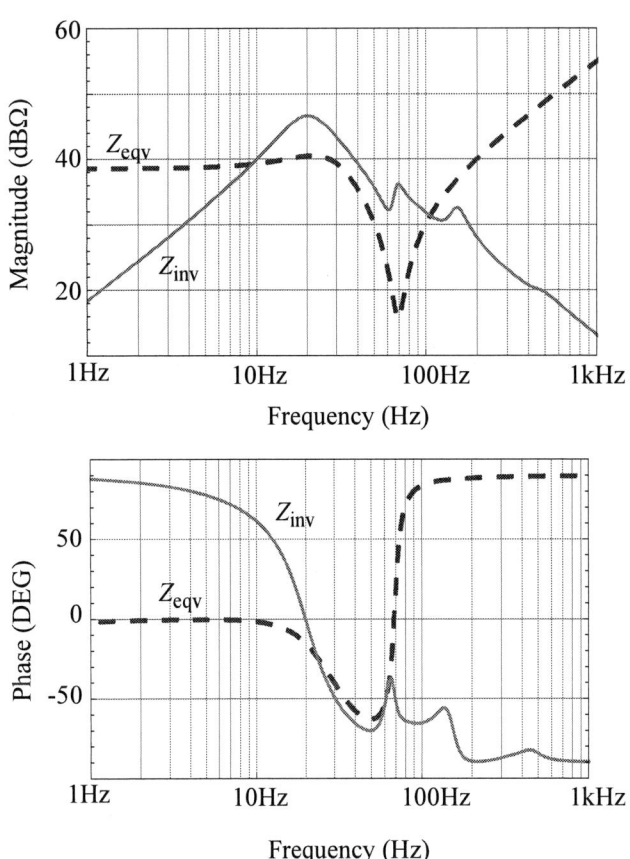

Fig. 14. Impedance-based analysis of VSC system; inverter input impedance (solid lines), equivalent impedance seen by inverter terminal (dashed lines).

C. VSC HVDC System Stability

The dc-link stability in VSC systems is analyzed using impedance models from (10) and (11). Same as LCC based systems, the dc-link stability depends on the ratio of inverter input impedance (Z_{inv}) to the impedance of the remaining network seen by inverter (Z_{eqv}) as shown in Fig. 8. The Bode plots in Fig. 14 compare Z_{inv} and Z_{eqv} and the impedance-based analysis predicts a damped resonance at the natural frequency of 110 Hz with 30 degrees of phase margin. Fig. 15 shows the simulated response of the dc-link voltage when at t = 1 s, the rectifier output power is increased from 20 MW to 300 MW, confirming the predicted resonance.

V. SUMMARY

This paper showed the applicability of impedance-based approach to the stability analysis of point-to-point HVDC systems. Both LCC and VSC are considered and impedance models required for stability analysis are developed. Impedance-based approach to stability can provide an analytical tool for the control design of HVDC terminal to potential unstable modes and ensure dc link stability for different operation conditions.

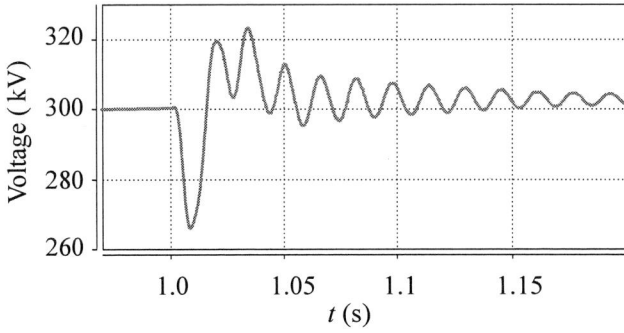

Fig. 15. Time domain simulation of the damped resonance.

REFERENCES

[1] S. Shah, R. Hassan and J. Sun, "HVDC transmission system architectures and control - a review," in *Proc. of 2013 IEEE COMPEL Workshop*, June 2013.

[2] D. Jovcic, N. Pahalawaththa, and M. Zvahir, "Stability analysis of HVDC control loops," in *IEE Proceedings - Generation, Transmission and Distribution*, vol. 146, no. 2, pp. 143-148, March 1999.

[3] A. E. Hammad, "Analysis of second harmonic instability for the Chateauguay HVDC/SVC scheme," *IEEE Trans. on Power Delivery*, vol. 7, pp. 410-415, Jan. 1992.

[4] R. S. Whitehouse, "Technical challenges of realizing multi-terminal networking with VSC," in *Proceedings of EPE* 2011, Sep. 2011.

[5] G. O. Kalcon, G. P. Adam, O. Anaya-Lara, S. Lo and K. Uhlen, "Small-signal stability analysis of multi-terminal VSC-based dc transmission system," *IEEE Trans. Power Sys.*, vol. 27, no. 4, pp. 1818-1830, Nov. 2012.

[6] R. D. Middlebrook, "Input filter considerations in design and application of switching regulators," in Proc. *IEEE Ind. Appl. Soc. Annu. Meeting*, 1976, pp. 366-382.

[7] X. Feng, J. Liu, and F. C. Lee, "Impedance specification for stable dc distributed power systems," *IEEE Trans. Power Electron.*, vol. 17, no. 2, pp. 157-162, Mar. 2002.

[8] F. Luo, J. Li, Z. Xu, Y. Li, J. Zhang and S. Liu, "Study on impedance frequency characteristics of HVDC filter commutate converter," in *Proc. of 2008 Electric Utility Deregulation and Restructuring and Power Technologies Conference*, pp. 1652-1656, April 2008.

[9] A. R. Wood and J. Arrillaga, "Composite resonance: a circuit approach to the waveform distortion dynamics of an HVDC converter," *IEEE Trans. Power Delivery*, vol. 10, pp. 1882-1888, Oct. 1995.

[10] A. R. Wood and J. Arrillaga, "The frequency dependent impedance of an HVDC converter," IEEE Trans. Power Delivry, vol. 10, pp. 1635-1641, July 1995.

[11] J. Sun, "Impedance-based stability criterion for grid-connected inverters," *IEEE Trans. Power Electronics*, vol. 27, no. 11, pp. 3075-3078, November 2011.

[12] H. Liu and J. Sun, "Modeling and analysis of dc-link harmonic instability in LCC HVDC systems," in *Proc. of* 2013 *IEEE COMPEL Workshop*, June 2013.

[13] H. Liu and J. Sun, "DC terminal impedance modeling of LCC HVDC converters," in *Proc. of 2013 IEEE COMPEL Workshop*, June 2013.

978-1-4799-2706-7/14 $31.00 © 2014 IEEE

Topology Evaluation of Slotless Bearingless Motors with Toroidal Windings

Daniel Steinert
Power Electronic Systems Laboratory
ETH Zurich, Switzerland
steinert@lem.ee.ethz.ch

Thomas Nussbaumer
Levitronix GmbH
Zurich, Switzerland
nussbaumer@levitronix.com

Johann W. Kolar
Power Electronic Systems Laboratory
ETH Zurich, Switzerland
kolar@lem.ee.ethz.ch

Abstract— In this paper, different winding and magnet topologies are analyzed and compared for a slotless bearingless disk drive with toroidal windings. Basis of the studies is a six phase motor with a diametrically magnetized one–pole-pair rotor. Due to the absence of mechanical bearings, the motor is suitable for applications with high purity and special chemical demands. Its slotless design results in low losses even at high rotational speeds. To improve the operational behavior of the rotor in different applications, the influence of higher pole pair numbers on the passive bearing stiffness is examined. Possible winding configurations for these rotors are presented and evaluated for their bearing and motor performance. Based on the results, a further prototype was built and is presented in this paper.

Keywords— Active magnetic bearing, bearingless motor, high speed drive, slotless motor

I. INTRODUCTION

In a bearingless slice motor, developed by Schoeb and Barletta [1], a disk- or ring-shaped rotor is spatially suspended and rotated without any mechanical contact. Due to the lack of lubrication and abrasion as well as the possibility to hermetically isolate rotor and stator from the environment, bearingless motors are used for example in semiconductor manufacturing [2], chemical or biological industries [3], [4], medicine, or in high speed applications [5]. Furthermore, the disk topology results in an enhanced compactness due to its low axial length.

In the presented slotless bearingless motor concept a disk shaped rotor is levitated magnetically without mechanical contact in the middle of an annular stator (cf. Fig. 1). The coils, which are wound toroidally on the stator iron, can generate both bearing forces and drive torque. [6]–[8]

Due to the absence of mechanical contact between rotor and stator, mechanical bearings and shaft feed-throughs can be omitted. This allows for a hermetic encapsulation of rotor and stator, which results in high resistance against chemically aggressive fluids and gases and in increased lifetime compared to conventional motors. The absence of abrasion and lubrication enables operation in high purity applications.

However, magnetic bearings have a significantly lower bearing stiffness than mechanical ones, which limits the possible applications. In pumps and blowers for instance, axial and radial forces resulting from the differential pressure of the fluid act on the impeller. This means

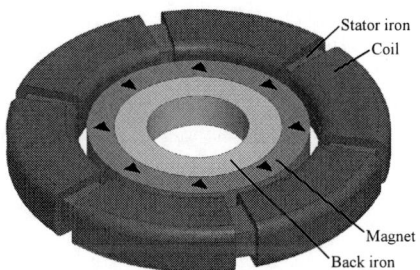

Fig. 1. Principle setup of a slotless bearingless disk drive with six coils and a diametrically magnetized permanent magnet rotor. The coils generate both, force and torque, simultaneously. A six phase inverter is necessary for operation.

possibly a considerable movement of the magnetically levitated rotor and limits the achievable pump, blower or turbine performance [9].

To improve both the passive bearing stiffness and the active bearing and motor performance of the bearingless motor, different rotor magnetizations, rotor pole pair numbers and winding configurations will be examined in this paper.

At first, the magnetic air gap field distributions for diametrical, radial and Halbach magnetization of the rotor are evaluated. The influence of the pole pair number on the passive bearing stiffnesses and on the drive and bearing performance is shown.

As only specific winding configurations are capable of generating torque and force independently for a given rotor magnetization, we derive the criteria for developing suitable configurations. The winding concepts differ with regard to the coil number and the interconnection of the coils. Additionally, it has to be distinguished between combined coils, which generate force and torque with one set of coils, and separate coils for motor and bearing. By 2D and 3D finite element simulations, selected winding configurations are compared regarding their bearing and motor capabilities, as well as passive stiffnesses and losses. Additionally, the influence of the pole pair number on the iron losses is estimated. At the end, a prototype with two pole pair rotor will be presented and compared to the existing one pole pair motor.

II. WORKING PRINCIPLE OF THE SLOTLESS BEARINGLESS DISK DRIVE

In [6] the authors present the detailed working principles as well as simulation and test results of the bearingless slotless disk drive with six coils and a diametrically magnetized one-pole-pair rotor. The presented prototype is the basis for the topology evaluation in this paper.

The disk type bearingless motor is passively stable in tilting and axial direction and has to be actively controlled in radial direction. The magnetic field in the magnetic gap, as is shown in Fig. 2 for a one- and a two-pole-pair rotor, generates reluctance forces between the stator iron and the rotor. Therefore, a deflection in the axial z-direction will lead to a counteracting force

$$dF_z = -c_z \cdot dz \qquad (1)$$

due to the axial stiffness constant c_z similarly to a spring constant. With this definition, a positive value of the stiffness leads to a stabilizing force.

Similarly, the tilting of the rotor results in a counteracting torque. At a rotor with pole pair number $p = 1$ it has to be distinguished between rotation around the axis of magnetization (angle α) or perpendicular to it (angle β). The tilting stiffnesses c_α and c_β are then defined by

$$c_\alpha = -\frac{dT_\alpha}{d\alpha} \text{ and } c_\beta = -\frac{dT_\beta}{d\beta}. \qquad (2)$$

When the rotor is displaced from its center position in radial direction, a radial force will act in the same direction. Therefore, the radial stiffnesses

$$c_d = -\frac{dF_x}{dx_d} \text{ and } c_q = -\frac{dF_y}{dx_q} \qquad (3)$$

are destabilizing. Here, x_d means a displacement in the direction of the magnetization and x_q perpendicular to the magnetization.

A one-pole-pair rotor results in anisotropic radial and tilting stiffness. Deflection of the rotor in the direction of magnetization (d-axis) results in a higher attractive force than deflection perpendicular to the magnetization (q-axis). The same holds for the tilting stiffness. A rotation around the d-axis results in a lower torque than rotation around q-axis. This results in a broad resonance frequency range, as is shown in [10]. Rotors with higher pole pair numbers, however, show almost isotropic stiffnesses in all directions.

$$\begin{aligned} p = 1 &: c_\alpha < c_\beta \text{ and } c_q < c_d \\ p \geq 2 &: c_\alpha \approx c_\beta \text{ and } c_q \approx c_d \end{aligned} \qquad (4)$$

To stabilize the passively instable radial position, an actively controlled force has to be applied on the rotor. It

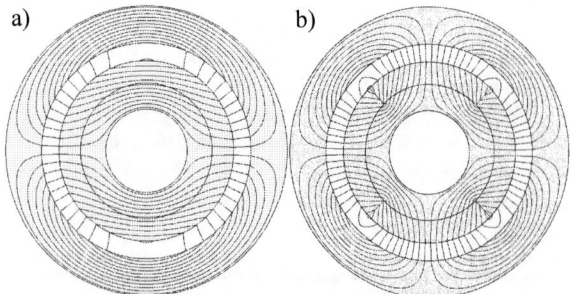

Fig. 2. Simulated magnetic field lines with a diametrically magnetized one-pole-pair rotor (a) and a two-pole-pair rotor with radial magnetization (b). The outer ring is the slotless stator, the rotor consists of the magnet ring and a backiron on its inner side. The coils of the motor are not shown here.

can be shown that the bearing force F_{bng} as well as the drive torque T

$$\vec{F}_{bng} = k_F \cdot \hat{I}_B \begin{pmatrix} \cos(\varphi_F) \, \vec{e}_x \\ \sin(\varphi_F) \, \vec{e}_y \end{pmatrix} \qquad (5)$$

$$T = k_T \cdot \hat{I}_D \qquad (6)$$

are proportional to the peak bearing current $\hat{I}_B$ or drive current $\hat{I}_D$ with the force coefficient k_F and the drive coefficient k_T. The force direction can be controlled by the phase shift φ_F between electrical rotor angle and bearing current.

III. EVALUATION OF PASSIVE BEARING PROPERTIES FOR DIFFERENT MAGNET CONFIGURATIONS

Depending on the application, considerably large magnetic gaps are necessary, for instance if a chemically resistant wall has to be inserted between rotor and stator. This drastically reduces the passive stiffnesses of the bearing, as shown in [7]. Therefore, a strong magnetic field in the air gap between magnet and stator iron is important. Moreover, the active force and torque generation depends especially on the fundamental wave of the flux density distribution in the air gap.

With 2D FE simulations, the magnetic field in the magnetic gap between rotor and stator is simulated. Fig. 3 shows the evaluated magnet configurations with one and two pole pairs, as well as the harmonic analysis of the magnetic field distribution in the middle of the magnetic gap.

For a rotor with one pole pair ($p = 1$), only the diametrical magnetization is of interest, as it yields the best results with a purely sinusoidal magnetic air gap field. However, as shown before, the stiffnesses are anisotropic leading to an unfavorable resonance behavior.

Going to higher pole pair numbers, different magnetization schemes are possible. The radial magnetization yields the highest flux density, however also higher harmonics in the field distribution. A back iron carries the flux at the inner side of the rotor and therefore enhances the flux density in the magnetic gap.

The Halbach magnetization has lower harmonics and does not need a back iron, but the fundamental wave is reduced by 18 %. The tangential distribution yields an even lower fundamental wave and higher harmonics and therefore promises the lowest performance.

With a fixed geometry (cf. Table I) 3D simulations have been conducted with different pole pair numbers and magnetization schemes. It shows that the highest passive stiffness values can be achieved with $p = 2$. The axial stiffness (cf. Fig. 4) increases by 39% and the mean tilting stiffness (cf. Fig. 5) increases by 37% with a radial magnetization compared to the diametrical magnetization. Furthermore, the tilting stiffnesses are isotropic for the $p \geq 2$ magnetizations. This means an increase of the minimal tilting stiffness by 300%, when changing the magnetization from one to two pole pairs.

When the space inside the hollow shaft rotor has to be used in the application, e.g. in axial blowers, the back iron can be omitted. This results in a weaker air gap field and in reduced stiffnesses. A radial magnetized two-pole-pair

The 2014 International Power Electronics Conference

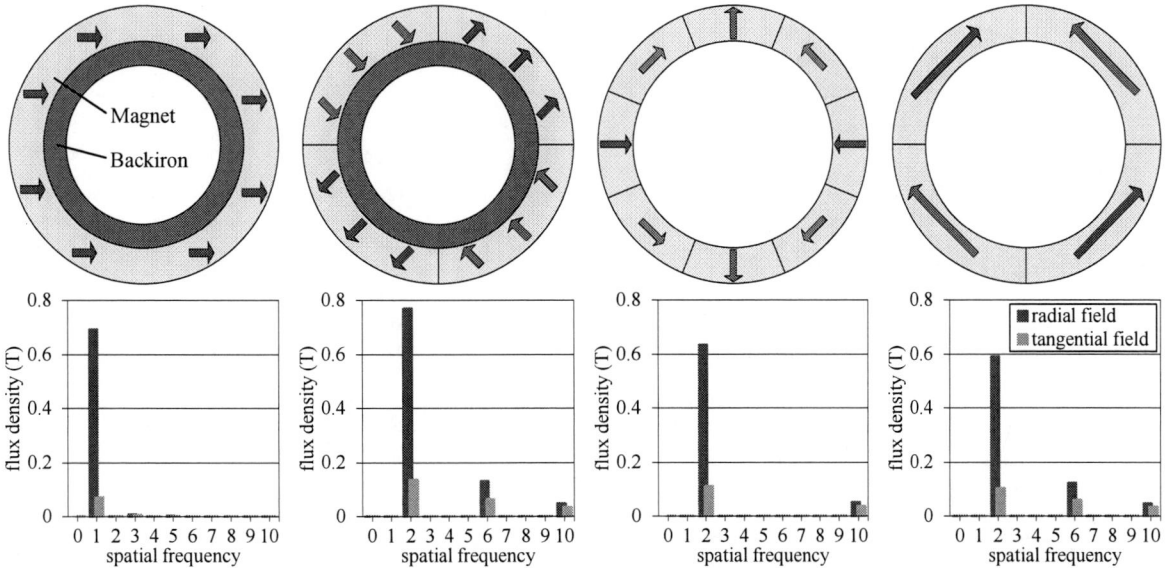

Fig. 3. Schematic drawing of the magnet configurations suitable for application in the bearingless disk motor (top) and FFT of the flux density distribution in the middle of the air gap between rotor magnet and stator iron (bottom). F.l.t.r: Diametrical magnetization (one pole pair), radial magnetization, Halbach magnetization and tangential magnetization (each two pole pairs).

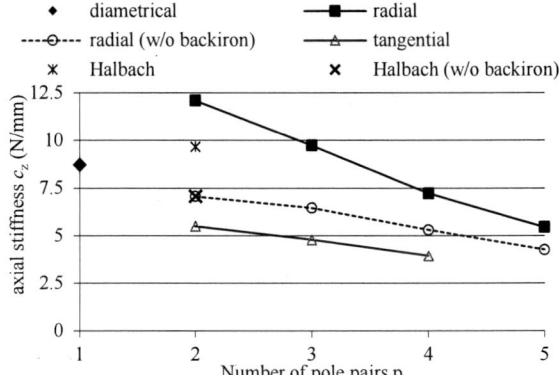

Fig. 4. Axial stiffness simulated with 3D FEM for different magnet configurations.

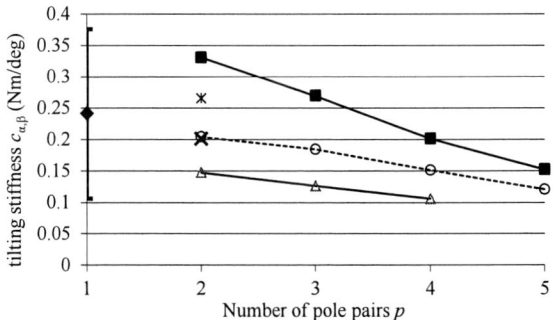

Fig. 5. Tilting stiffness simulated with 3d FEM for different magnet configurations. For the diametrical magnetization the mean value is given and the highest and lowest tilting stiffness is indicated.

TABLE I
GEOMETRY PARAMETERS OF THE EXAMINED TOPOLOGIES

Parameter	Value
outer magnet diameter	97 mm
outer stator diameter	156 mm
inner stator diameter	116 mm
magnetic gap	9.5 mm
magnet height	15 mm
stator height	12.5 mm

rotor without back iron shows about 40% lower stiffness. A Halbach magnetized rotor without back iron has nearly the same stiffnesses as a rotor with radial magnetization. Tangential magnetization always shows the weakest stiffnesses.

Summarizing, the highest passive bearing stiffnesses can be achieved with a radial magnetized rotor with two pole pairs. A further increase of the pole number reduces the stiffnesses. If a rotor without backiron shall be used, radial and Halbach magnetization are equal.

IV. WINDING CONCEPTS

To generate torque acting on the rotor, the stator coils have to generate a magnetic armature reaction field in the magnetic gap with a pole pair number

$$p_{\mathrm{drv}} = p \tag{7}$$

that is equal to the pole pair number p of the rotor magnetic field. A force will be generated by a stator field with a pole pair number

$$p_{\mathrm{bng}} = p \pm 1 . \tag{8}$$

The bearing forces consist of reluctance and Lorentz forces that act in the same direction when the stator pole pair number is greater than the rotor field, otherwise both forces partially cancel each other out. [2]

Therefore, to achieve high bearing forces, a magnetic air gap field with $p_{\mathrm{bng}} = p + 1$ has to be generated by the set of coils.

To derive a winding configuration that is capable of generating both torque and force independently, it has to be distinguished between combined coils and separated coils.

A. Winding criteria for combined coils

With combined coils, both torque and force are generated by the same set of coils. Because of the different pole numbers for bearing and drive, the winding scheme

978-1-4799-2706-7/14 $31.00 © 2014 IEEE   977

The 2014 International Power Electronics Conference

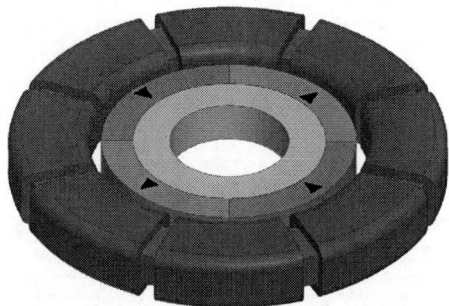

Fig. 6. Bearingless motor with a two-pole-pair rotor and eight combined coils for force and torque. An eight-phase inverter is necessary for operation.

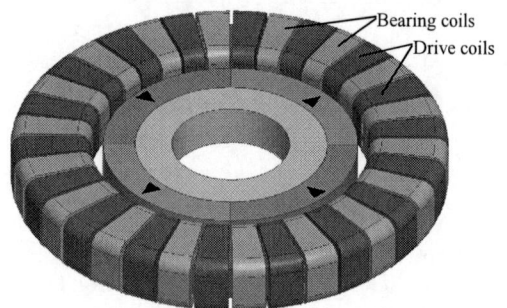

Fig. 7. Bearingless motor with a two-pole-pair rotor and 36 coils in total. Each 18 coils are either for bearing or motor operation. Motor and bearing operation can be done by two independent three-phase inverters.

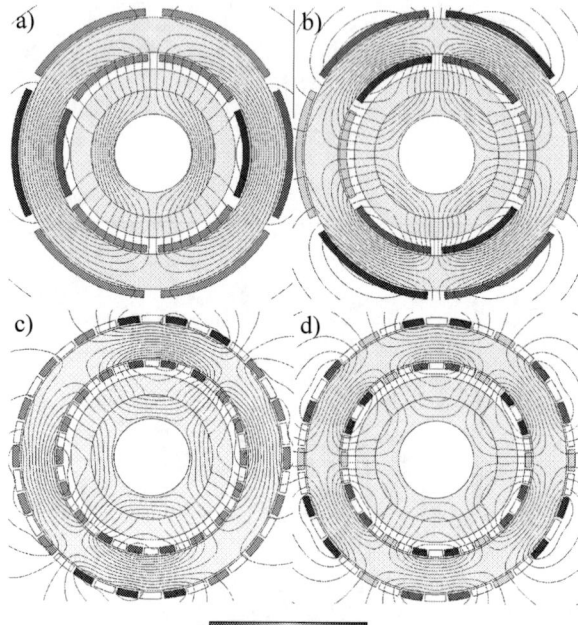

500 A turns ▬▬▬ -500 A turns

Fig. 8. Simulated armature reaction field due to the coil currents indicated by different shades of blue and red. The magnetic field of the rotor is not shown here. The upper figures show the six coil topology with combined coils with a drive current (a) that generates a one pole pair field, and a bearing current (b) with a two pole pair field. In the 2x18 coil topology, the drive currents generate a two-pole-pair field (c) and the bearing currents generate a three-pole pair field (d). Bearing and drive currents are combined in the six coil topology, and separated on different coil sets in the 2x18 coil topology.

contains no repetitive elements. This means that the number of phases m equals the number of coils N. However, one coil can be separated in two coils with reversed winding direction, which are connected in series, as it is proposed in [10]. This will have no direct effect on the feasibility of the winding configuration and is not examined in this paper.

To generate a field with a given pole pair number p, at least

$$N \geq 2 \cdot p \qquad (9)$$

coils are necessary to avoid aliasing. Additionally, the number of phases divided by the pole pair number for the bearing field

$$\frac{m}{p_{\mathrm{bng}}} \neq 1, 2, 4 \qquad (10)$$

must not equal 1, 2 or 4. Otherwise, the phase shift between two phases would be a multiple of 90° and at times when one phase current is zero it wouldn't be possible to control both radial degrees of freedom. Therefore, there is one direction, where no force can be generated at a certain time and no stable bearing operation would be possible.

With the criteria in (9), (10) at least seven coils are necessary for a rotor with pole pair number $p = 2$. As eight coils can be connected in two star points, this configuration can be realized by a controller with eight half bridges and six current sensors (cf. Fig. 6).

B. Winding criteria for separated coils

With separated coils, the magnetic field for each bearing and motor operation is generated by an independent set of coils. Therefore, two independent

inverters can be used. Then, the winding configuration consists of repetitive elements corresponding to the respective pole pair number of bearing and motor field.

For the bearing field, at least three phases per pole pair are necessary, as two degrees of freedom have to be controlled and two coils per pole pair would result in a phase shift of 90° as already explained.

To maintain manufacturability, the coil numbers for bearing and motor have to be the same. Therefore, winding configurations with nine coils or eighteen coils (cf. Fig. 7) each for bearing and motor are possible.

C. Simulation results of proposed topologies

For the four mentioned topologies, the armature reaction field caused by drive and bearing currents is simulated and analyzed by FFT analysis. Fig. 8 shows the field lines of the armature reaction field for the 6-coil-topologie and the 2x18-coil topology. Fig. 9 shows the FFT of the armature reaction field in the magnetic gap for all four topologies of interest.

At the 6-coil-topology, the drive current shows a field with solely one first harmonic, which corresponds to the first harmonic of the diametrical rotor and generates a torque according to (7). The bearing current generates a force on the rotor due to its second harmonic armature reaction field. The higher harmonics are of low amplitude and can be neglected.

For the 8-coil-topology, the second and sixth harmonic of the drive field fits very well to the diametrical rotor, which shows a similar spectrum. The force on the rotor is generated by the third and fifth harmonic of the bearing

978-1-4799-2706-7/14 $31.00 © 2014 IEEE

978

field interacting with the second and sixth harmonic of the rotor field. The results are similar to the 2x18-coil topology, where bearing and drive coils are separated.

For the 2x9-coil topology, however, it can be seen that the drive field shows a fifth and seventh harmonic, which generate a force in combination with the sixth harmonic of the rotor field. Additionally, the bearing field has a sixth harmonic, which generates a torque. Therefore, force and torque are interconnected, which most likely will make the control of this topology more difficult.

Subsequently, the 8-coil-topology with combined windings and the 2x9- and 2x18-coil-topologies with separated windings have been simulated in 3D FE simulations and compared to the 6-coil topology with diametrical rotor. The results are shown in Table II. It shows that the three topologies with two-pole-pair rotor gain in force and torque compared to the 6-coil topology. The 2x18-coil topology is better in force generation than the 8-coil topology, but has lower torque output.

The simulation of the 2x9-coil topology confirmed the interconnection of bearing and drive. It shows, that a drive current generates a force corresponding to a bearing current with an amplitude of 25% of the drive current. Vice versa, the bearing current generates a torque ripple of 15.6% corresponding drive current. This interconnection might lead to instable behavior of the motor, if the control does not compensate this interconnection.

Summarizing, with the 8- and 2x18-coil topology, promising alternatives to the diametrical magnetized 6-coil topology are developed. The passive stiffnesses as well as the active bearing and drive performance exceed the values of the 6-coil topology. As the 2x18-coil topology shows the best bearing force performance, it was decided to build up a prototype of this new topology. This prototype is presented and compared to the 6-coil prototype in section VI.

TABLE II
COMPARISON OF PROPOSED TOPOLOGIES WITH DIAMETRICAL AND RADIAL MAGNETIZED ROTOR

p	N	m	k_{T} $\left(\frac{\text{Nm}}{\text{A/mm}^2}\right)$	k_{F} $\left(\frac{\text{N}}{\text{A/mm}^2}\right)$	c_z $\left(\frac{\text{N}}{\text{mm}}\right)$	c_α $\left(\frac{\text{Nm}}{\text{deg}}\right)$	c_β $\left(\frac{\text{Nm}}{\text{deg}}\right)$
1	6	6	0.27	4.93	8.7	0.11	0.38
2	8	8	0.31	5.59	12.1	0.33	0.33
2	2x9	2x3	0.31	5.94	12.1	0.33	0.33
2	2x18	2x3	0.29	6.63	12.1	0.33	0.33

k_{T} ... Torque per peak current density in the coils
k_{F} ... Force per peak current density in the coils
c_z ... Axial stiffness
c_α... Stiffness against tilting around d-axis
c_β... Stiffness against tilting around q-axis

TABLE III
COMPARISON OF 36-COIL TOPOLOGY WITH DIFFERENT ROTOR MAGNETIZATIONS

Magnetization	Back iron	k_{T} $\left(\frac{\text{Nm}}{\text{A/mm}^2}\right)$	k_{F} $\left(\frac{\text{N}}{\text{A/mm}^2}\right)$	c_z $\left(\frac{\text{N}}{\text{mm}}\right)$	c_α $\left(\frac{\text{Nm}}{\text{deg}}\right)$
Radial	yes	0.29	6.63	12.1	0.33
Halbach	yes	0.26	6.08	9.7	0.27
Tangential	yes	0.29	6.63	12.1	0.33
Radial	no	0.21	4.85	7.1	0.20
Halbach	no	0.22	5.05	7.1	0.20
Tangential	no	0.22	4.32	5.5	0.15

V. LOSSES

In [6] it is shown, that the main losses of the slotless bearingless disk drive, especially at the proposed prototype, consist of iron losses in the stator. According to [11], these iron losses

$$P_{\text{Fe}} = P_{\text{Hy}} + P_{\text{Ed}} \qquad (11)$$

can be separated into hysteresis losses P_{Hy} and eddy current losses P_{Ed}. The hysteresis loses

$$P_{\text{Hy}} = c_{\text{Hy}} \cdot f_{\text{el}} \cdot \hat{B}^{1.6} \cdot m_{\text{Fe}} \qquad (12)$$

depend linearly on the electric frequency f_{el}, whereas the

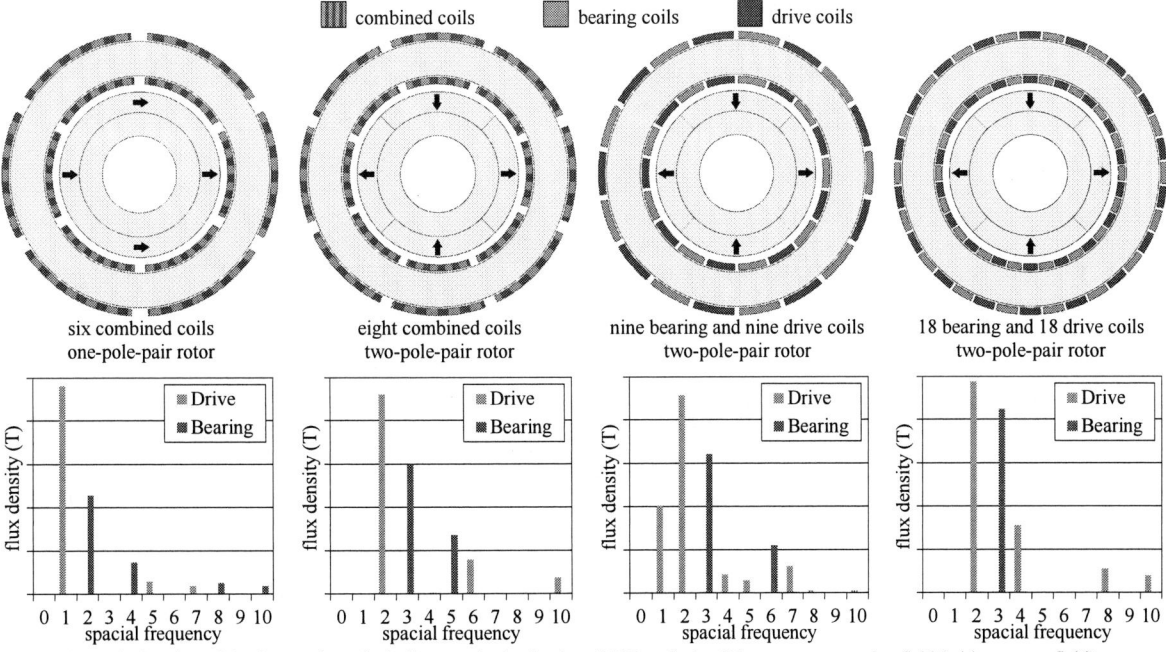

Fig. 9: Schematic drawing of the four evaluated winding topologies (top) and FFT analysis of the armature reaction field (without rotor field) generated by drive or bearing currents in the middle of the magnetic gap.

eddy current losses

$$P_{\mathrm{Ed}} = c_{\mathrm{Ed}} \cdot f_{\mathrm{el}}^2 \cdot \hat{B}^2 \cdot d_{\mathrm{Fe}}^2 \cdot m_{\mathrm{Fe}} \qquad (13)$$

increase quadratically with the frequency. Both loss components are supposed to scale linearly with the mass of the stator iron m_{Fe} and also depend on the peak flux density $\hat{B}$ in the stator. The coefficients c_{Ed} and c_{Hy} describe material properties.

Therefore, increasing the pole pair number of the rotor from $p = 1$ to $p = 2$ will double the electric frequency. Thus, the hysteresis losses will be twice as high whereas the eddy current losses will be even four times higher. Therefore, higher pole pair numbers are supposed to be less suitable for high speeds.

To better compare the losses of the two prototypes, which will be presented in the next section, we introduce two new coefficients $c_{\mathrm{n,1}}$ and $c_{\mathrm{n,2}}$ being the coefficients of a quadratic approximation of the motor losses according to the rotational speed and normalized to the stator iron mass. Therefore, the motor losses can be written as

$$P = c_{\mathrm{n,1}} \cdot m_{\mathrm{Fe}} \cdot n + c_{\mathrm{n,2}} \cdot m_{\mathrm{Fe}} \cdot n^2 . \qquad (14)$$

VI. TEST RESULTS

The first prototype ('prototype 1') with a diametrical magnetized one-pole-pair rotor and six combined coils has already been built and successfully tested (Fig. 10) [6]. Stable operation is possible for up to 20 000 rpm, which is the mechanical limit of the rotor.

Additionally, a second prototype ('prototype 2') with a two-pole-pair rotor and 2x18 separated coils for bearing and drive is built (cf. Fig. 11) and will now be compared to prototype 1. As higher iron losses are expected at the same rotational speed, prototype 2 was built with a significantly bigger rotor outer diameter, so that higher circumferential speeds can be achieved with lower rotational speeds.

While prototype 1 has a rotor diameter of 102 mm, the new prototype has a rotor size of 164.8 mm. The bigger size also enhances the manufacturability especially with the high number of coils that have to be placed on the stator of prototype 2. The geometrical details of both prototypes are compared in Table IV.

With prototype 2, rotational speeds of up to 12 000 r/min are reached [12]. This is the limit of mechanically safe operation. The prototype showed stable bearing behavior in standstill and during rotation. Rigid body resonances were observed between 1000-2000 r/min. Outside of this range, no resonances occurred. The

Fig. 10. Test setup of the slotless bearingless disk drive with six coils and diametrically magnetized rotor.

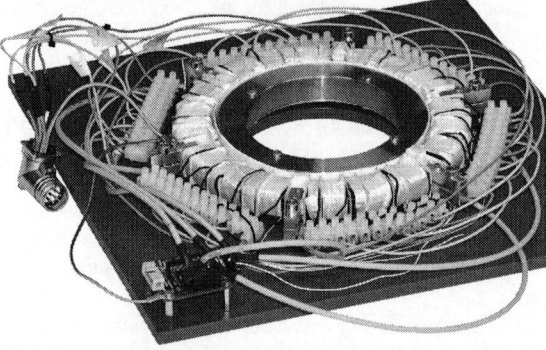

Fig. 11. Test setup of the slotless bearingless disk drive with 36 coils and a two pole pair, radial magnetized rotor.

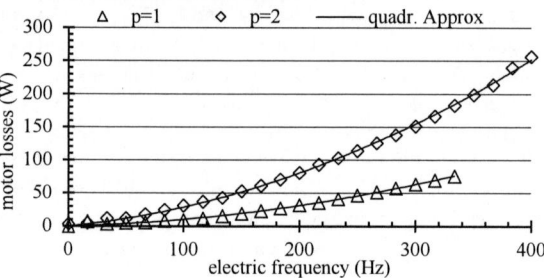

Fig. 12 Measured losses of the two prototypes shown for different electrical frequencies. The losses of prototype 2 are more than twice as high as protoyte 1 at the same electrical frequency, although it rotates half as fast due to the two pole pairs.

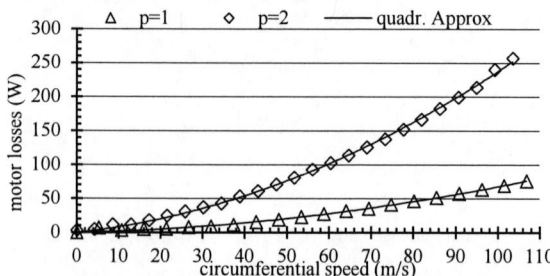

Fig. 13 Measured losses of the two prototypes shown for different circumferential speeds. As prototype 2 is bigger than prototype 1, it has a higher circumferential speed as the smaller prototype at the same rotational speed. The circumferential speed is important for the application like blowers or pumps.

TABLE IV
GEOMETRIC DETAILS OF BOTH PROTOTYPES

	Prototype 1	Prototype 2	
pole pairs	1	2	
maximum speed	20 000	12 000	(r/min)
rotor diameter	102	164.8	(mm)
magnet thickness	12	10.2	(mm)
magnet height	15	20	(mm)
backiron thickness	12	8	(mm)
stator inner diameter	116	181.8	(mm)
stator outer diameter	156	221.8	(mm)
stator height	12.5	14	(mm)
stator mass	0.81	1.35	(kg)
magnetic air gap	9.5	11	(mm)

TABLE V
COEFFICIENTS OF QUADRATIC APPROXIMATION
OF THE MEASURED NO LOAD LOSSES

	Prototype 1	Prototype 2	
$c_{\mathrm{Hy,}}$	0.0231	0.0553	(W Hz^{-1} kg^{-1} B$^{-1.6}$)
c_{Ed}	1501	2017	(W Hz^{-2} kg^{-1} B$^{-2.3}$ m^{-2})
$c_{\mathrm{n,1}}$	$0.95 \cdot 10^{-3}$	$4.20 \cdot 10^{-3}$	(W rpm^{-1} kg^{-1})
$c_{\mathrm{n,2}}$	$1.88 \cdot 10^{-7}$	$9.17 \cdot 10^{-7}$	(W rpm^{-1} kg^{-1})

separation of bearing and drive coils proved to be advantageous during implementation of drive and bearing, as the control of both three-phase current systems is independent. Additionally, the coils for bearing and drive could be adapted in size and number of turns to their respective function. It also can be demonstrated, that the stiffnesses of the two pole pair rotor are isotropic and significantly stronger than in prototype 1. Measurements of the axial stiffness fit nearly exactly to the simulated values.

The losses, however, proved to be even higher in prototype 2 than expected. In Fig. 12 the measured losses of both prototypes are plotted against the electric frequency. This means, that prototype 1 rotates twice as fast as prototype 2 at the same electric frequency. It can be seen that the losses in prototype 2 are more than two times higher at the same frequency. When plotting the losses against the circumferential speed (cf. Fig. 13), the losses are even worse compared to prototype 1. This is important for applications as blowers or pumps, where high circumferential speeds are necessary.

The measured losses can be approximated very well with the quadratic equations (11) and (14). The parameters are shown in Table V. The peak flux density in the stator of both prototypes was simulated and with the different mass of both stators, the coefficients are calculated. It shows, that the coefficients c_{Ed} and c_{Hy} are not constant for both prototypes as would be expected for material constants, which shows that other loss mechanisms are present. The quadratic coefficient is 34% higher in prototype 2, the linear loss coefficient is even 140% higher. However, the biggest loss component is the quadratic loss term. It is supposed, that other loss terms in addition to the iron losses (such as copper and air friction losses) cannot be completely neglected in the analysis. Additionally, the radially magnetized rotor in prototype 2 produces higher flux harmonics in the air gap field, which might also cause higher losses.

VII. Conclusion

Summarizing, it can be said that at disk type motors with sufficiently large outer diameter, a two-pole-pair rotor will lead to significantly higher passive bearing stiffnesses. Different coil topologies were proposed, which are possible with a two-pole-pair rotor. All topologies feature a higher number of coils, which leads to a bigger effort during manufacturing. A motor with 36 coils was built to demonstrate the feasibility of a two-pole-pair machine. It showed a very stable operational behavior and high isotropic stiffnesses. However, it showed also that the losses of a two-pole-pair rotor are much bigger than with a one-pole-pair rotor. Especially when high speeds are required, the one-pole-pair rotor might be chosen, whereas the two-pole-pair rotor is for applications with bigger demands on the passive stiffness.

References

[1] R. Schoeb and N. Barletta, "Principle and Application of a Bearingless Slice Motor," *JSME Int. J. Ser. C Mech. Syst. Mach. Elem. Manuf.*, vol. 40, no. 4, pp. 593–598, Dezember 1997.

[2] F. Zürcher, T. Nussbaumer, and J. W. Kolar, "Motor Torque and Magnetic Levitation Force Generation in Bearingless Brushless Multipole Motors," *IEEEASME Trans. Mechatron.*, vol. 17, no. 6, pp. 1088–1097, Dec. 2012.

[3] B. Warberger, R. Kaelin, T. Nussbaumer, and J. W. Kolar, "50-Nm/2500-W Bearingless Motor for High-Purity Pharmaceutical Mixing," *IEEE Trans. Ind. Electron.*, vol. 59, no. 5, pp. 2236–2247, May 2012.

[4] T. Reichert, T. Nussbaumer, and J. W. Kolar, "Bearingless 300-W PMSM for Bioreactor Mixing," *IEEE Trans. Ind. Electron.*, vol. 59, no. 3, pp. 1376–1388, Mar. 2012.

[5] S. Silber, J. Sloupensky, P. Dirnberger, M. Moravec, M. Reisinger, and W. Amrhein, "High Speed Drive for Textile Rotor Spinning Applications," *IEEE Trans. Ind. Electron.*, vol. Early Access Online, 2013.

[6] D. Steinert, T. Nussbaumer, and J. Kolar, "Slotless Bearingless Disk Drive for High-Speed and High-Purity Applications," *IEEE Trans. Ind. Electron.*, vol. Early Access Online, 2014.

[7] D. Steinert, T. Nussbaumer, and J. W. Kolar, "Concept of a 150 krpm Bearingless Slotless Disc Drive with Combined Windings," in *Proceedings of the IEEE International Electric Machines and Drives Conference (IEMDC 2013)*, Chicago, USA, 2013.

[8] H. Mitterhofer, W. Gruber, and W. Amrhein, "On the High Speed Capacity of Bearingless Drives," *IEEE Trans. Ind. Electron.*, vol. 61, no. 6, pp. 3119–3126, Jun. 2014.

[9] H. Mitterhofer, B. Mrak, and W. Amrhein, "Suitability investigation of a bearingless disk drive for micro turbine applications," in *2013 IEEE Energy Conversion Congress and Exposition (ECCE)*, 2013, pp. 2480–2485.

[10] H. Mitterhofer, W. Amrhein, and H. Grabner, "Comparison of two- and four-pole rotors for a high speed bearingless drive," presented at the ISMB 2012, The 13th International Symposium on Magnetic Bearings, Arlington, USA, 2012.

[11] M. T. Bartholet, T. Nussbaumer, S. Silber, and J. W. Kolar, "Comparative Evaluation of Polyphase Bearingless Slice Motors for Fluid-Handling Applications," *IEEE Trans. Ind. Appl.*, vol. 45, no. 5, pp. 1821–1830, Sep. 2009.

[12] P. Peralta Fierro, "High Speed Slotless Bearingless Axial Blower," Master thesis, ETH Zurich, Zurich, 2014.

Winding Arrangement in Single-Drive Bearingless Motor with Radial Gap

Hiroya Sugimoto, *Member, IEEE*, Seiyu Tanaka, and
Akira Chiba, *Fellow, IEEE*
Department of Electrical and Electric Engineering
Tokyo Institute of Technology
Tokyo, Japan
Sugimoto@belm.ee.titech.ac.jp

M. A. Rahman, *life Fellow, IEEE*
Faculty of Engineering and Applied Science
Memorial University of Newfoundland
St. John's, Canada

Abstract— A novel single-drive bearingless motor with high passive stiffness and wide gap factor has been proposed. The single-drive bearingless motor has only one set of three-phase windings. It generates both torque and axial suspension force independently with only one three-phase inverter and one displacement sensor. Therefore, the single-drive bearingless motors have the advantages of low cost and small size. Only axial direction z-axis is actively positioned. The other axes, radial movements x and y, and tilting movements θ_x and θ_y, are passively stable by repulsive passive magnetic bearings. The stator consists of six C-shaped cores and one set of three-phase windings. A novel V-shaped winding structure is proposed to generate the torque and active axial force simultaneously. In this paper, the winding arrangement is presented. In addition, the mathematical calculation is carried out.

Keywords— *bearingless motor, magnetic bearing, single-drive, one-axis actively positioning*

I. INTRODUCTION

Bearingless motors have advantages of no wear, no lubricant, no-pollution and maintenance-free because a magnetic bearing function is magnetically integrated in a single motor. Therefore, the bearingless motors have been applied in centrifugal pumps, contamination-free ventricular assist devices, high purity pharmaceutical mixing devices, rotating stages, flywheels and high speed motors [1]-[7].

In a bearingless pump [8], the rotor and the stator surfaces must be covered by a metallic or plastic can. This can must be as thick as possible to avoid the damage from chemical fluid. Therefore, a wide magnetic gap between the stator and the rotor is required. A cylindrical radial gap structure is necessary to assemble the reliable can. The five degrees of freedom (5DOF) actively positioned bearingless motors with wide cylindrical radial gap have been proposed [9]. However, in the case of 5DOF bearingless motor, high cost and large motor size are unavoidable, because five displacement sensors and amplifiers, three units of a three-phase inverter and one single-phase inverter are needed. For reducing cost a motor with active positioning of two degree-of-freedom has been studied and applied in the semiconductor industry [1]-[4], [8] and [10]. In addition, active

positioning of only one degree-of-freedom (1DOF) with magnetic bearing has been studied [11]-[17].

In the one-axis active magnetic suspension, the radial and tilting directions are passively positioned. The stiffness in the passive suspension is an important performance index. Passive magnetic bearings (PMBs) with attractive or repulsive forces have been proposed. In case of the attractive passive radial and tilting magnetic bearings in [11]-[13], the rotor axial length is long with respect to the rotor radius to stabilize the radial and tilting directions. In case of the repulsive passive radial and tilting magnetic bearings (RPMBs) in [14]-[17], the radial gap are utilized effectively because the passive radial and tilting directions are positioned by additional two PMBs installed in tandem to a motor. The passive stiffness is high compared with the attractive PMBs. In the references [14]-[17], novel radial double air-gap structures have been proposed, however the mechanical structure of the double air-gaps is rather complicated.

Fig. 1 shows the radial stiffness k_r versus gap factor. The horizontal axis is magnetic gap factor, i.e., the ratio g / R, where g is effective magnetic gap length and R is the rotor radius. In [16], the gap factor is the largest. The gap factor and the measured radial stiffness are approximately 0.18 and 2.1 N/mm, respectively. These values are in [17] are approximately 0.02 and 29 N/mm. The radial stiffness is the largest. In [11], these values are approximately 0.1 and 12.5 N/mm, positioned between [16] and [17]. From [11], [16] and [17], it is seen that the radial stiffness is decreased in high (g / R) machines because flux density is decreased when the magnetic gap is wide. It is seen that improvement of both the gap factor and the radial

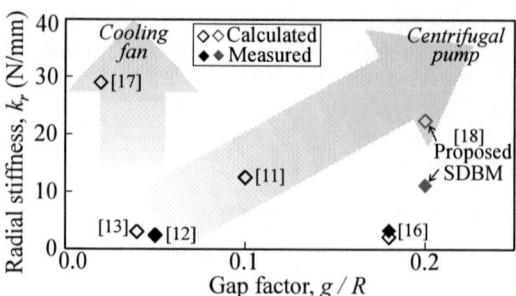

Fig. 1. Radial stiffness map with respect to gap factor.

stiffness is a difficult problem. Typically, the radial stiffness is reduced when the gap factor is increased because flux density in the gap is reduced. Moreover, in case of the RPMBs, unstable axial force is generated. This unstable axial force is increased when the radial stiffness is increased. Thus, considerable axial suspension force is required to overcome the unstable force. In case of cooling fan application, large gap factor is not required. However, large radial stiffness is required for the vibration reduction. On the other hand, in case of the centrifugal pump application, both large gap factor and high radial stiffness are required.

In [18], a novel bearingless motor with $g / R = 0.2$ and $k_r = 21.2$ N/mm is presented, as indicated by "Proposed SDBM" in Fig. 1. The radial stiffness has been improved with enhanced axial force generation. The SDBM is short for a Single-Drive Bearingless Motor. The single-drive bearingless motor is similar to the 1DOF active positioning bearingless motor. However, the advantage of SDBM is that both the axial suspension force and torque are generated in a single unit of a three-phase inverter [13]. The winding is only one set of three-phase winding. The axial force and torque are regulated by the d- and q-axis currents, respectively. Therefore, the winding must be designed to generate the axial force and torque simultaneously.

This paper presents a new winding arrangement. The winding is wound around the stator so as to generate the flux in the axial direction. In the proposed machine, a part of the winding is bent as V-shape so that the Lorentz force torque is generated because the vertical current component is generated in the winding. In this paper, the mathematical calculations are carried out.

II. STRUCTURE OF SINGLE-DRIVE BEARINGLESS MOTOR

Fig. 2 illustrates xz cross-sectional view of a proposed bearingless motor. The single-drive bearingless motor is installed in the center, and the repulsive passive magnetic bearings (RPMBs) are installed at both ends of the SDBM. Each RPMB is constructed with the two permanent magnets. The S-pole of two permanent magnets is faced so that the radial stiffness is improved. The displacement sensor and the hall sensors are installed in the negative and positive z-axis directions, respectively. In the design, the radial air-gap length is 1 mm. The axial and radial touch down length is 0.2 mm.

Table I shows the performance parameters. The inertia in the tilting direction is calculated. The other parameters are measured in the experiments. The radial and tilting stiffness are one of the best performances in the 1DOF bearingless motors and the thrust magnetic bearing motors [18]. The maximum radial restoring force is approximately eight times as high as the rotor weight when the rotor is displaced by the touch down length of 0.2 mm. Therefore, the proposed SDBM has high passive stiffness.

Fig. 3 shows a principle of the axial active suspension force generation in xz cross-sectional view. The stator consists of six divided C-shaped cores, and the cores are

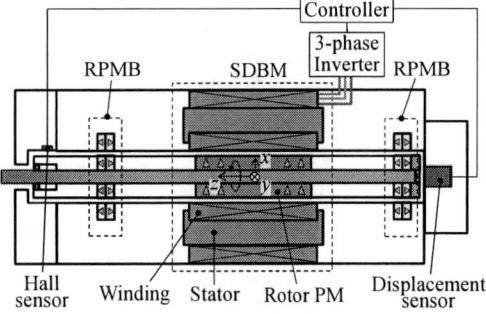

Fig. 2. Cross-sectional view of the prototype machine.

TABLE I
PERFORMANCE PARAMETERS

Parameter		Value
Rotor mass	m	0.039 kg
Inertia in the tilting direction	J	2.37×10^{-5} kgm^2
Winding resistance per phase	R_w	4.7 Ω
Winding inductance per phase	L_w	15.7 mH
Active force per d-axis current	K_{zi}	2.39 N/A
Passive radial stiffness	k_r	11.2 N/mm
Passive tilting stiffness	k_θ	15.4 Nm/rad
Torque per q-axis current	k_T	0.6 mNm/A

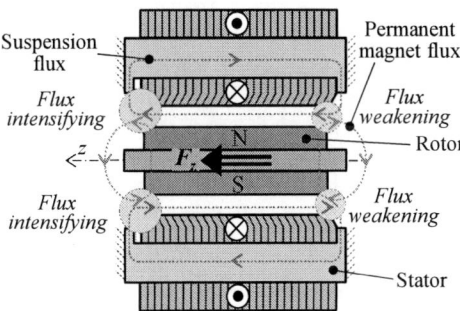

Fig. 3. Principle of axial suspension force generation.

installed in the circumference. The one set of three-phase windings are wound around the stator cores. The rotor is two-pole cylindrical permanent magnet. The permanent magnet fluxes at both rotor ends are circulated in the air-gap as shown by green arrows in Fig. 3. The suspension flux mainly flows in the air-gap in the z-direction as shown by red arrows. When the suspension flux is superimposed on the permanent magnet flux, the flux intensifying and weakening occur in the air-gaps at both axial ends. Therefore, the axial suspension force is generated in the positive z-direction. In the proposed structure, the Lorentz force is also generated. Thus, the axial active suspension force is sum of the Maxwell and the Lorentz forces.

Fig. 4 shows the axial position and the rotational speed regulation system. The axial suspension force and the rotating torque are regulated by the d- and q-axis currents, respectively. Therefore, these current references are generated from the feedback signal of the axial displacement and the rotational speed. The rotor axial position is detected by the displacement sensor. The error signal between the reference z^* and the measured feedback signal z is input in the Proportional-Differential-Integral (PID) controller, and then, the d-axis current reference i_d^* is produced. On the other hand, the rotor rotational angular position is detected by two hall sensors installed in the x- and y-axis directions. The rotational angle θ and the rotational speed ω are calculated from the feedback signals. The q-axis current reference i_q^* is generated by the PI controller. To compensate the current lag caused by the coil inductance, additional PI current feedback loop is adopted. The d- and q-axis voltage commands v_d^* and v_q^* are generated by the PI controller. These voltage commands are transformed into the three-phase voltage references v_u^*, v_v^* and v_w^* with the rotor rotational angular position θ. The three-phase voltage source inverter regulated by the voltage commands provides the three-phase currents i_u, i_v and i_w. The u- and w-phase currents i_u and i_w are detected by two current sensors, and then, the currents are transformed into d- and q-axis currents i_d and i_q as the current feedback signals. Therefore, the axial position and the rotating torque can be regulated by one three-phase voltage source inverter, one displacement sensor and two hall sensors.

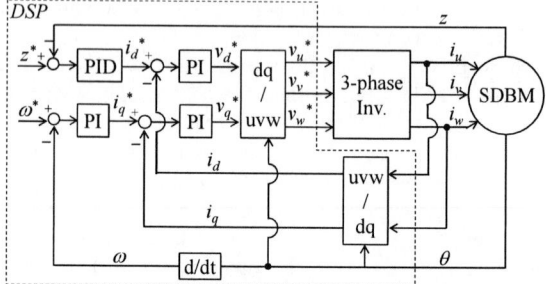

Fig. 4. Regulation system of the rotor axial position and the rotational speed.

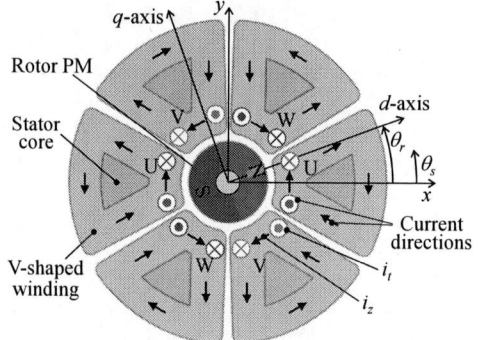

Fig. 5. Winding arrangement with V-shaped structure.

III. Winding Arrangement

Fig. 5 shows the detail of the winding arrangement of the proposed SDBM. Only one set of three-phase two-pole winding is installed. The windings are wound around the six stator cores. The black arrows in the windings indicate the positive current direction. Only the axial suspension force is generated by the x-y windings. To generate the rotating torque around the z-axis the current component in the z-axis is necessary. The torque is generated by the current component perpendicular to the current direction by the Lorentz force.

Fig. 6 shows the proposed V-shaped winding structure. One stator core with the winding seen from the air-gap side is shown. The coils facing to the air-gap is bent in the V-shape as shown in Fig. 6, so that the current component i_t in the z-axis direction is generated for the torque generation. The torque is increased with an increase of the fold angle ϕ. However, the axial active suspension force is reduced because the radial magnetomotive force (MMF) is reduced with an increase of the fold angle.

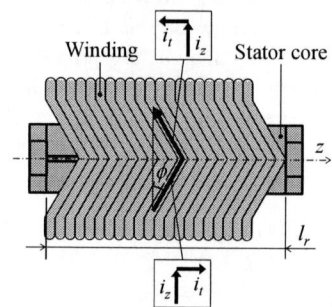

Fig. 6. V-shaped winding structure seen from air-gap side.

IV. Mathematical Torque Calculation

Fig. 7 shows the imaginary winding of the z-axis current component in the V-shaped winding. Let us define θ_r and θ_s as the rotor rotational angular position and stator coordinate angular position, respectively. The coils are wound around the stator cores. When the wire radius is reduced, the number of layers is increased. As a result, the number of winding is increased. The coil sides facing to the air-gap are bent on the z-axis so that the current directions of the imaginary windings are generated as shown in Fig. 6. The Lorentz force torque per one layer of U-phase T_{Lu} in the imaginary winding is given by a radius of each winding layer r_n, an axial length of the imaginary winding l_{zn}, U-phase current i_u and flux density distribution at each coil in the radial direction B_{rn}

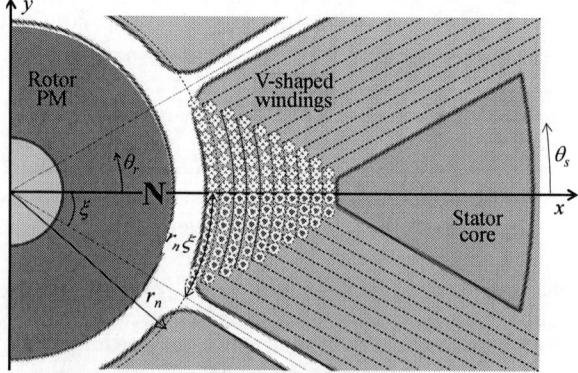

Fig. 7. Considerations of the imaginary winding in the circumferential direction.

as,

$$T_{Lu} = -r_n l_{zn} \int i_u B_{rn} d\theta_s \ . \tag{1}$$

The three-phase currents i_u, i_v, i_w and flux density distribution at each coil B_{rn} are written as

$$i_u = i_0 \sin\theta_r,$$
$$i_v = i_0 \sin(\theta_r - 2\pi/3),$$
$$i_w = i_0 \sin(\theta_r - 4\pi/3),$$
(2)
$$B_{rn} = B_n \cos(\theta_s - \theta_r),$$
(3)

where i_0 and A_n are the amplitudes of the current and the flux density. In the calculation, the current and the flux density distribution are assumed to be sinusoidal, respectively. Fig. 8 shows the flux density distribution in the radial direction with respect to the stator coordinate angular position. Note that the flux density at the center in the axial direction is shown. The flux density distributions in case of $r = 7$ mm and $r = 8$ mm are slightly distorted by influence of the stator C-shaped cores. However, in this paper, the flux density distribution with respect to the stator coordinate angular position is assumed to be a sinusoidal function. In addition, it is supposed that the amplitude B_n of the flux density in the radial direction varies with respect to the radius of the winding layer.

As seen from Fig. 7, the number of imaginary coils in the winding layers is reduced with an increase of the radius of the winding layer because the arc length facing to the air-gap is reduced. Therefore, the number of turns $N_{\theta n}$ of imaginary coils in the circumference direction is calculated from the arc length and the wire diameter $2r_c$ as follows:

$$N_{\theta n} = \frac{r_n \xi - 2r_c(n-1)}{2r_c},$$
(4)

where n is the layer number. $r_1 \xi$ is the arc length facing to the air-gap in the first winding layer. In case of the numerator in (4) of second layer ($n = 2$), a wire diameter $2r_c$ is subtracted from the arc length $r_2 \xi$.

Table II shows the radius of the winding layers, the number of imaginary coils in the circumference direction and the amplitude of the flux density. The decimal point is included in $N_{\theta n}$ because of imaginary coil.

Fig. 9 shows the axial length of the imaginary coils. Black arrows indicate current directions in the V-shaped windings seen from air-gap side. White and orange arrows indicate i_z and i_t components, respectively. For example, in case of the third layer, the black arrows are divided into 7 components as illustrated in Fig. 9 because $N_{\theta 3}$ is equal to 7. The length l_{it} of one of i_t component is given as

$$l_{it} = \frac{r_n \xi \tan\phi}{N_{\theta n}}.$$
(5)

Therefore, the axial length l_{zn} of the imaginary winding is expressed by l_{it} and the number of turns N_z in the axial direction is given as follows:

$$l_{zn} = N_z l_{it} = \frac{N_z r_n \xi \tan\phi}{N_{\theta n}}.$$
(6)

Note that the N_z is limited by the axial length of the stator core l_r. Actual total axial length of the V-shaped winding

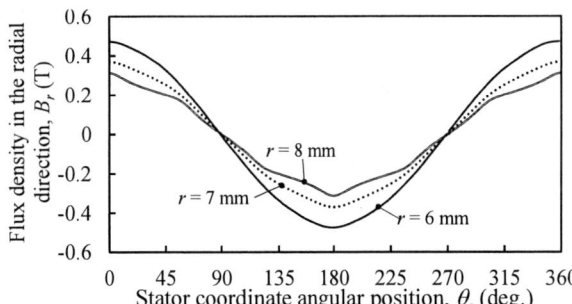

Fig. 8. Flux density distribution in the radial direction with respect to the stator coordinate angular position.

TABLE II
THE NUMBER OF IMAGINARY WINDING

Layer	r_n (mm)	$N_{\theta n}$	B_n (T)
1	6.2	6.49	0.46
2	6.6	5.91	0.41
3	7.0	5.33	0.37
4	7.4	4.75	0.34
5	7.8	4.17	0.32
6	8.2	3.59	0.31
7	8.6	3.01	0.30
8	9.0	2.42	0.31
9	9.4	1.84	0.33

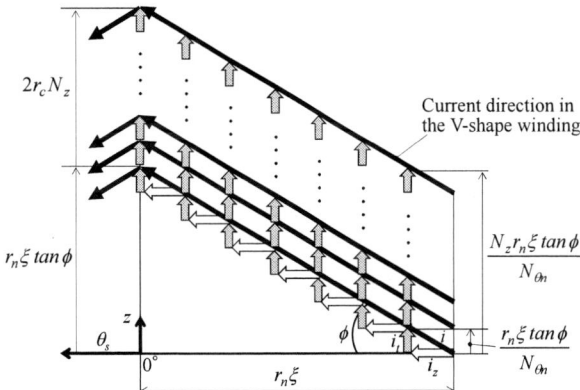

Fig. 9. Axial length of i_t component in the imaginary winding.

is limited by the core length as,

$$l_r \geq r_n \xi \tan\phi + 2r_c N_z > 0,$$

This can be solved for N_z as,

$$0 < N_z \leq \frac{l_r - r_n \xi \tan\phi}{2r_c}.$$
(7)

The equation (7) indicates that the number of coils N_z in the axial direction is reduced with an increase of the fold angle ϕ. The Lorentz force torque is function of N_z and ϕ, thus, there is optimal point with respect to the fold angle ϕ. The Lorentz force torques in the U-, V- and W-phase T_{Lu}, T_{Lv}, T_{Lw} and the total torque T_L are given by

$$T_{Lu} = -2\sum_{n=1}^{k} r_n l_{zn}\left(\int_0^{\pi/6} i_u B_{rn} d\theta_s + \int_{-\pi/6}^0 -i_u B_{rn} d\theta_s\right),$$

$$T_{Lv} = -2\sum_{n=1}^{k} r_n l_{zn}\left(\int_{2\pi/3}^{5\pi/6} i_v B_{rn} d\theta_s + \int_{\pi/2}^{2\pi/3} -i_v B_{rn} d\theta_s\right), \quad (8)$$

$$T_{Lw} = -2\sum_{n=1}^{k} r_n l_{zn}\left(\int_{4\pi/3}^{3\pi/2} i_w B_{rn} d\theta_s + \int_{7\pi/6}^{4\pi/3} -i_w B_{rn} d\theta_s\right),$$

$$T_L = T_{Lu} + T_{Lv} + T_{Lw}. \quad (9)$$

Fig. 10 shows the calculated Lorentz force torque with respect to the fold angle in the V-shaped winding at the rated current of $i_q = 1.22$ A. The black and blue curves indicate the calculated torque and the number of coils in the axial direction, respectively. In the calculation, the torque is increased with an increase of the fold angle. Hence, the torque can be improved by an increase of the fold angle. However, the torque is reduced after $\phi = 48$ degree because the number of turns N_z in the axial direction is reduced. The reduced N_z results in the reduction of the axial active force, therefore, the fold angle must be designed with consideration of both torque and force. In the prototype machine, the torque is 0.7 mNm at rated current when the fold angle is approximately 20 degree. In case of $\phi = 48$ degree, the torque can be twice as high as that of the prototype machine. On the other hand, N_z is reduced by 66 %, thus, maximum current at the start-up may be approximately 1.7 times as large as the prototype machine with $\phi = 20$ degree. Thus, compromising of torque and axial force is necessary.

V. COMPARISON OF CALCULATION AND MEASUREMENT

Fig. 11 shows the prototype machine. The machine parameters are previously shown in Table I. The wire radius r_c is 0.2 mm. The number of turns N_z in the axial direction is 56. The number of layers in the radial direction is 9. The fold angle is approximately 20 degree. At the right shaft end, an eddy current type displacement sensor is facing to detect the shaft axial position. On the other hand, two sensors are also seen. These sensors are installed to monitor the radial and tilting shaft movements only for monitoring purpose.

Fig. 12 shows the static torque measurement method. The static torque is measured by a pulley-balance method. A string is set on the pulley installed at the end of the shaft. A weight is hung at one end of the string, and a spring balance device is hung at another end of the string. When the rotor is magnetically suspended, the reading value of the spring balance device is a nominal value. When the q-axis current is supplied, the torque is generated. The product of the reading value of the spring balance device and the pulley radius is the static torque. In Fig. 12, M is a load mass. Let us assume g is gravity acceleration, then, Mg is the force caused by the weight. F_T is tangential force by generated the torque. Sum of the forces is equal to the spring force F_l, as,

$$F_l = Mg + F_T. \quad (10)$$

Let us assume the pulley radius and the string radius are

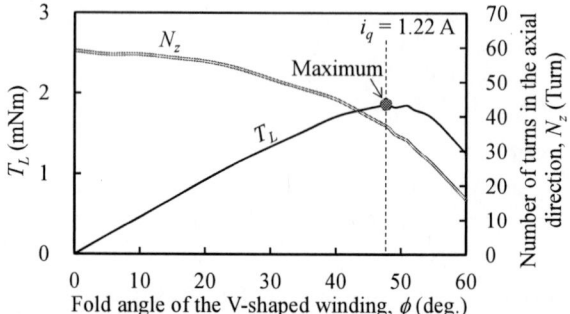

Fig. 10. Lorentz torque with respect to the winding fold angle.

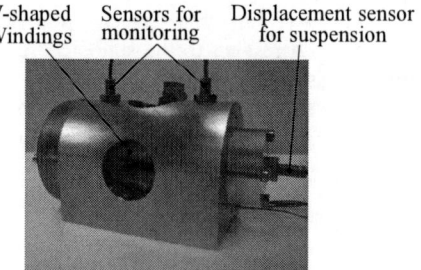

Fig. 11. Prototype machine.

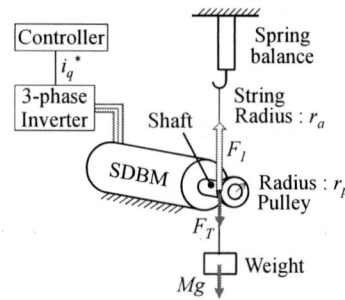

Fig. 12. Measurement method of the torque.

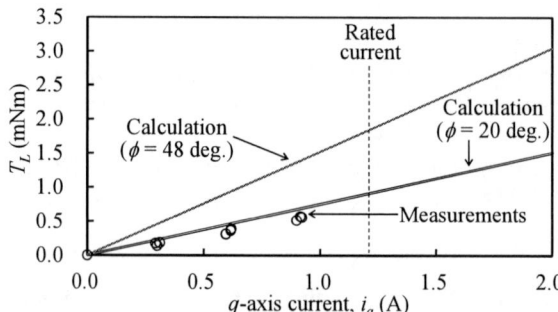

Fig. 13. Lorentz torque with respect to the winding fold angle.

r_p and r_a, respectively, thus, the tangential force is given by

$$F_T = \frac{T_L}{r_p + r_a}. \quad (11)$$

By substituting (10) into (9), the static torque is given as follows:

$$T_L = (F_l - Mg)(r_p + r_a). \quad (12)$$

Fig. 13 shows the calculated and measured torques

with respect to the q-axis current. The plots are measured values by the aforementioned method. The black and red curves indicate the calculated torque at $\phi = 20$ degree and $\phi = 48$ degree, respectively. The calculated torque at $\phi = 20$ degree is close to the measured torque as expected. Therefore, it is found that the proposed mathematical expression is valid for the torque estimation of the proposed bearingless motor with V-shaped winding. Hence, it is possible that the torque improvement can be realized by adjustment of the fold angle ϕ.

VI. CONCLUSION

This paper has presented a novel winding arrangement in the proposed single-drive bearingless motor with cylindrical radial air-gap. The axial suspension force and the rotating torque are generated simultaneously with only one set of three-phase winding driven by one unit of a three-phase inverter. The winding has a V-shaped structure to generate the Lorentz force torque. The mathematical calculation is carried out. The relationship between the fold angle and the torque is shown.

In the experiment, it is found that the derived mathematical expression is effective for the torque estimation because the measured static torque is within 10 % error. The torque improvement may be realized by an adjustment of the fold angle in the V-shaped winding.

ACKNOWLEDGMENT

This work was supported by JSPS KAKENHI Grant Number 24246046.

REFERENCES

[1] M. T. Bartholet, T. Nussbaumer, S. Silber, and J. W. Kolar, "Comparative Evaluation of Polyphase Bearingless Slice Motor for Fluid-Handling Applications", *IEEE Trans. Ind. Appl.*, vol.45, no.5, pp. 1821-1830, 2009.

[2] Toru Masuzawa, Toshiyuki Kita, and Yohji Okada, "An Ultradurable and Compact Rotary Blood Pump with a Magnetically Suspended Impeller in the Radial Direction", *Artificial Organs*, Vol.25, No.5, pp.395-399, 2001.

[3] T. Reichert, T. Nussbaumer, and J. W. Kolar, "Bearingless 300-W PMSM for bioreactor mixing," *IEEE Trans. Ind. Electron.*, vol. 59, no. 3, pp. 1376–1388, Mar. 2012.

[4] W. Gruber and W. Amrhein, "Bearingless Segment Motor With Five Stator Elements—Design and Optimization", *IEEE Trans. Ind. Appl.*, vol.45, no.4, pp.1301-1308, 2009.

[5] M. Ooshima, S. Kobayashi, and H. Tanaka, "Magnetic suspension perfonnance of a bearingless motor/generator for flywheel energy storage systems", *in Proc. Power and Energy Society General Meeting*, pp.1-4, July 2010.

[6] H. Mitterhofer, B. Mrak, and W. Amrhein, "Sutability Investigation of a Bearingless Disk Drive for Micro Turbine Applications", *in Proc. Energy Convers. Congr. and Expo. (ECCE)*, pp.2362-2367, 2013.

[7] A. Chiba, T. Fukao, O. Ichikawa, M. Oshima, M. Takemoto, and D.G. Dorrell, " Magnetic Bearings and Bearingless Drives", Newnes Elsevier ISBN 07506 5727 8, 2005.

[8] M. Neff, N. Barletta, and R. Schob, "Bearingless Centrifugal Pump for Highly Pure Chemicals", *in Proc. 8th Int. Symp. Magnetic Bearings*, pp.283-287, 2002.

[9] M. Takemoto, S. Iwasaki, H. Miyazaki, A. Chiba, and T. Fukao, "Experimental Evaluation of Magnetic suspension Characteristics in 5-axis Active Control Type Bearingless Motor without a Thrust Disk for Wide-gap Condition", *in Proc. Energy Convers. Congr. and Expo. (ECCE)*, pp. 2362-2367, 2009.

[10] T. Nussbaumer, P.Karutz, F. Zurcher, and J. W. Kolar, "Magnetically levitated slice motors—An overview," *IEEE Trans. Ind. Appl.*, vol. 47, no. 2, pp. 754–766, Mar./Apr. 2011.

[11] I. D. Silva, J. R. Cardoso, and O. Horikawa, "Design Considerations for Achieving High Radial Stiffness in an Attraction-Type Magnetic Bearing With Control in a Single Direction," *IEEE Trans. Magn.*, vol. 47, no. 10, pp. 4112–4115, Oct. 2011.

[12] J. Kuroki, T. Shinshi, L. Li, and A. Shimokohbe, "Miniaturization of a one-axis-controlled magnetic bearing," *Precis. Eng.*, vol. 29, no. 2, pp. 208–218, Apr. 2005.

[13] J. Asama, Y. Hamasaki, T. Oiwa, and A. Chiba, "A Novel Concept of a Single-Drive Bearingless Motor", *IEEE Trans. Industrial Electronics*, vol. 60, no. 1, pp. 129-138, Jan. 2013.

[14] S. Yang and M. Huang, "Design and Implementation of a Magnetically Levitated Single-Axis Controlled Axial Blood Pump", *IEEE Trans. Industrial Electronics*, vol. 56, no. 6, pp. 2213-2219, Jun. 2009.

[15] Q. D. Nguen and S. Ueno, "Modeling and control of salient-pole permanent magnet axial-gap self-bearing motor," *IEEE/ASME Trans. Mechatronics*, vol. 16, no. 3, pp. 518–526, Jun. 2011.

[16] T. Ohji, Y. Katsuda, K. Amei, and M. Sakui, "Structure of One-Axis Controlled Repulsive Type Magnetic Bearing System With Surface Permanent Magnets Installed and Its Levitation and Rotation Tests", *IEEE Trans. Magn.*, vol. 47, no. 12, pp. 4734-4739, Dec. 2011.

[17] W.. Bauer and W. Amrhein, "Electrical design and winding selection for a bearingless Axial-Force/Torque Motor", *in Proc. International Symposium on Power Electronics, Electrical Drives, Automation and Motion (SPEEDAM)*, pp. 1224-1229, 2012.

[18] H. Sugimoto, S. Tanaka, A. Chiba, and J. Asama, "Design and Test Result of Novel Single-Drive Bearingless Motor with Cylindrical Radial Gap", *in Proc. Energy Convers. Congr. and Expo. (ECCE)*, pp. 2362-2367, 2013.

Development of a One-Axis Actively Regulated Bearingless Motor with a Repulsive Type Passive Magnetic Bearing

Junichi Asama, *Member, IEEE*, Daisuke Watanabe,
Takaaki Oiwa
Department of Mechanical Engineering
Shizuoka University
Hamamatsu, Japan
tjasama@ipc.shizuoka.ac.jp

Akira Chiba, *Fellow, IEEE*
Department of Electrical & Electronic Engineering
Tokyo Institute of Technology
Tokyo, Japan
chiba@ee.titech.ac.jp

Abstract— This paper describes a novel one-axis actively regulated bearingless motor with a spinning top-shaped rotor, which is intended for a use as a cooling fan. This motor requires only one set of three-phase winding and thus one three-phase inverter to realize both rotation and magnetic suspension, and thus it is dubbed a single-drive bearingless motor. An axial position of the rotor is actively regulated and radial and tilting motions are passively stabilized by the passive magnetic bearing. Torque and axial suspension force are regulated by *q*-axis and *d*-axis currents, respectively. In this paper, the structure and principle, design, and test results of the proposed bearingless motor are presented.

Keywords— *Bearingless motor, magnetic bearing, magnetic suspension, single-drive, lorentz force, axial gap.*

I. INTRODUCTION

In magnetic bearings and bearingless motors, three translational and two tilting motions of a rotor must be magnetically stabilized for magnetic suspension. Conventional 5-axis actively regulated magnetic bearings and bearingless motors require a number of electromagnets, windings, displacement sensors, and power transistors. Reduction in the number of actively regulated axes is one possible solution to reduce the size, power consumption, and cost of these machines.

One-axis actively regulated magnetic bearings have been developed [1-3]. Only the axial position of the rotor is actively regulated in these machines. On the other hand, the two-radial and two-tilting motions are passively stabilized. Although successful magnetic suspension of the rotor was achieved, an additional motor unit must be needed for rotation. Therefore, one single-phase inverter or one linear servo-amplifier for magnetic suspension control and one three-phase inverter for speed regulation are generally required in these machines.

A bearingless motor is defined as a motor which magnetically integrates the bearing function in one unit. Nguyen, *et al.* developed an axial-gap type one-axis actively regulated bearingless motor [4]. The disk-shaped rotor is sandwiched with two stators. A repulsive type

passive magnetic bearing which consists of permanent magnets (PMs) stabilizes the 4-axis motion of the rotor. In this machine, two three-phase inverters are used for active regulation in both axial motion and rotation. Gruber, *et al.* adopted another unique approach to the one-axis active regulation of the bearingless motor [5]. This motor has four independently regulated coreless coils to generate both axial suspension force and torque.

For further simplification, cost reduction, downsizing, and energy saving, the authors have proposed a one-axis actively regulated bearingless motor which employs only one three-phase inverter and one set of three-phase winding, dubbed a single-drive bearingless motor (SDBelM) [6]. This motor has a long shaft-shaped rotor which is sandwiched between the axial gap stators. The suspension force in the axial direction and torque are actively regulated by the *d*-axis and *q*-axis currents, respectively. The remaining two-radial and two-tilting motions are passively stabilized by the magnetic couplings between the rotor and the stator.

Based on this single-drive principle, this paper focuses on a novel one-axis actively regulated SDBelM with a spinning top-shaped rotor, which is intended for a use as a cooling fan. This motor combines an axial gap surface-mounted permanent magnet (SPM) motor with coreless coils and a repulsive type passive magnetic bearing. The two-radial and two-tilting motions are passively stabilized by the repulsive type passive magnetic bearing. Details will be explained in the next section. In [7] and [8], an axial gap SPM motor was used for magnetic suspension and rotation using six linear power amplifiers. However, these papers focused on the radial active suspension. Moreover, further reduction in power devices is needed for energy saving.

To our knowledge, the proposed simple structure and principle of the SDBelM and is new to the field of electric machines. In this paper, the structure and principle of the proposed SDBelM, design consideration based on the theoretical calculation and three-dimensional finite element method (3D-FEM), and experimental results are presented.

978-1-4799-2706-7/14 $31.00 © 2014 IEEE

II. Structure and Principle

A. Structure of Proposed SDBelM

Fig. 1 shows a structure of the proposed SDBelM with a spinning top-shaped rotor. This motor combines an axial gap SPM motor with coreless coils and a repulsive type passive magnetic bearing. The rotor has axially magnetized PM pieces with a back yoke. These PM pieces contribute to generation of both torque and axial suspension force. Details will be described in the next section. An axially magnetized PM bar is attached at the center of the rotor for the passive magnetic bearing. The stator consists of a PM ring and three-phase coreless coils which are circumferentially located under the rotor PM pieces and fixed to the stator.

These PM ring and PM rod are magnetized in the same axial direction to compose a repulsive type passive magnetic bearing. As shown in the left of Fig. 2, when the rotor moves to the radial direction, resulting magnetic repulsive force, which works as restoring force, is acted on the rotor. As a result, the rotor is automatically aligned at the center position. In a similar manner, the restoring torque is generated when the rotor is tilted, as shown in the right of Fig. 2. Therefore, the two-radial and two-tilting motions of the rotor are passively stabilized by this repulsive type magnetic bearing.

B. Generation of Suspension Force and Torque

The axial motion of the rotor is magnetically unstable because of the repulsive type passive magnetic bearing, and thus must be actively regulated. In this section, the suspension force and torque of the proposed SDBelM are theoretically derived. Due to the Fleming's left-hand rule, the Lorentz force acting on the coil, $\mathbf{F}$, are generally described as,

$$\mathbf{F} = (\mathbf{i} \times \mathbf{B}) l \tag{1}$$

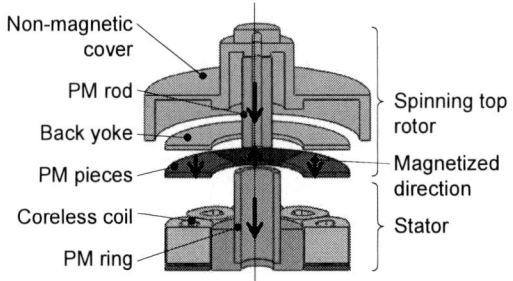

Fig. 1. Structure of the proposed single-drive bearingless motor with a spinning top-shaped rotor.

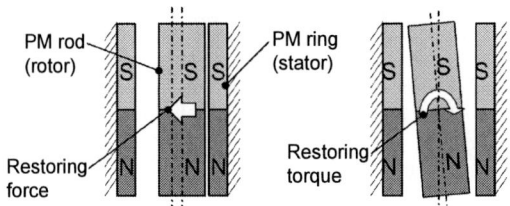

Fig. 2. Passive stabilization of the repulsive type magnetic bearing.

where $\mathbf{i}$, $\mathbf{B}$, and l are coil current, magnetic flux density, and length of the copper wire, respectively. In the axial gap SPM motor, these matrices can be written in the cylindrical coordinates as follows:

$$\mathbf{F} = \begin{pmatrix} F_r \\ F_\theta \\ F_z \end{pmatrix} \quad \mathbf{i} = \begin{pmatrix} i_r \\ i_\theta \\ i_z \end{pmatrix} \quad \mathbf{B} = \begin{pmatrix} B_r \\ B_\theta \\ B_z \end{pmatrix} \tag{2}$$

where the subscripts, r, θ, and z indicate the components in the radial, circumferential, and axial directions, respectively. In this stator structure, the axial component of the coil current, i_z, is zero. Then, (1) can be simplified as,

$$\begin{pmatrix} F_r \\ F_\theta \\ F_z \end{pmatrix} = \begin{pmatrix} i_\theta B_z l \\ -i_r B_z l \\ i_r B_\theta l - i_\theta B_r l \end{pmatrix} \tag{3}$$

The circumferential component of the Lorentz force, F_θ, contributes to torque generation, and the axial component of the force, F_z, corresponds to the axial suspension force. Based on (3), suspension force and torque, which are regulated by d-axis and q-axis currents, respectively, are theoretically derived.

Fig. 3 shows a simplified theoretical model with 2-pole/3-coil configuration. One coreless coil is modeled with four segmented wires (#1~#4). The lengths of these segmented wires are l_1, l_2, l_3, and l_4, where l_1 is equal to l_3. The center of the U-phase is aligned with X-axis, and a circumferential angle from the X-axis is defined as θ, which is fixed to the stator coordinates. The wires #1 and #3 of the U-phase are located at $\theta = \pi/3$ and $\theta = -\pi/3$, respectively. V-phase and W-phase are modeled in the similar manner to the U-phase. On the top of the three-phase winding, an axially magnetized two-pole PM is located. A rotational angle, ωt, is defined as an angle from X-axis to the center of the rotor S-pole. The symbol of S-pole indicates that the magnetic flux flows into the S-pole PM in the positive Z-axis. For example, when $\omega t = 0$, U-phase winding are exposed to the positive axial magnetic field. For simplification, it is assumed that the

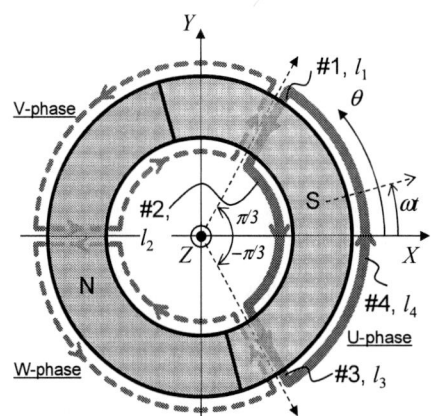

Fig. 3. Simplified theoretical model of the axial gap SPM motor with 2-pole/3-coil configuration.

distribution of the magnetic flux density is sinusoidal in this theoretical calculation as,

$$B_{ro} = -B_{rom}\cos(\theta - \omega t)$$
$$B_{ri} = B_{rim}\cos(\theta - \omega t)$$
$$B_{\theta} = -B_{\theta m}\sin(\theta - \omega t)$$
$$B_z = B_{zm}\cos(\theta - \omega t)$$

(4)

where the subscript m indicates the magnitude of each component. The symbols B_{ro} and B_{ri} are the radial flux components in the outward and inward directions, respectively. The radial outward direction is positive in this cylindrical coordinates.

Based on (3), the circumferential component of the force, F_{θ}, contributes to torque and is generated by an interaction between the radial component of the coil current, i_r, and the axial component of the magnetic flux density, B_z, as shown in Fig. 4 (a). Then, we can obtain the following equations:

$$F_{\theta} = -i_r B_z l$$
$$= -\{(-i_u + i_v)B_{z(\theta=\pi/3)}l_r + (-i_v + i_w)B_{z(\theta=\pi)}l_r$$
$$+ (-i_w + i_u)B_{z(\theta=5\pi/3)}l_r\}$$

(5)

where $l_r = l_1 = l_3$, and i_u, i_v, and i_w are three-phase currents fed into the windings. The three-phase currents are related to d-axis and q-axis currents by

$$\begin{pmatrix} i_u \\ i_v \\ i_w \end{pmatrix} = \sqrt{\frac{2}{3}} \begin{pmatrix} \cos\omega t & -\sin\omega t \\ \cos(\omega t - \frac{2\pi}{3}) & -\sin(\omega t - \frac{2\pi}{3}) \\ \cos(\omega t + \frac{2\pi}{3}) & -\sin(\omega t + \frac{2\pi}{3}) \end{pmatrix} \begin{pmatrix} i_d \\ i_q \end{pmatrix}$$

(6)

Substituting (4) and (6) into (5) yields

$$F_{\theta} = -\frac{3B_{zm}}{\sqrt{2}}i_q$$

(7)

It is noted that the circumferential force, F_{θ}, which contributes to torque generation, can be controlled independently by the q-axis current, i_q. This Lorentz force, F_{θ}, is generated on the coil. Since the coreless coils are fixed to the stator, resulting reactive forces are acted on the PMs of the rotor.

In a similar manner, the suspension force, F_z, is theoretically calculated. Each component of F_z in (3) can be expressed as,

$$i_r B_{\theta} l = (-i_u + i_v)B_{\theta(\theta=\pi/3)}l_r + (-i_v + i_w)B_{\theta(\theta=\pi)}l_r$$
$$+ (-i_w + i_u)B_{\theta(\theta=5\pi/3)}l_r$$

(8)

$$i_{\theta}B_r l = i_u \int_{\frac{\pi}{3}}^{\frac{\pi}{3}} (B_{ro}l_4 - B_{ri}l_2)d\theta$$
$$+ i_v \int_{\frac{\pi}{3}}^{\pi} (B_{ro}l_4 - B_{ri}l_2)d\theta + i_w \int_{\pi}^{\frac{5\pi}{3}} (B_{ro}l_4 - B_{ri}l_2)d\theta$$

(9)

Fig. 4 (b) shows axial suspension force generation where the circumferential component of the current, i_{θ}, and the radial component of the PM flux density, B_r, are interacted. Substituting (4) and (6) into (8) and (9) yields

$$F_z = \frac{3}{\sqrt{2}}(B_{\theta m}l_r + B_{rim}l_2 + B_{rom}l_4)i_d$$

(10)

It is noted that the axial suspension force, F_z, can be controlled independently by the d-axis current, i_d. In this SDBelM, torque and suspension force can be independently regulated by q-axis and d-axis currents. Therefore, only one three-phase inverter and one set of three-phase winding are needed for magnetic suspension and rotation in this SDBelM.

III. DESIGN CONSIDERATION

A. Model and Condition of 3D-FEM Analysis

We designed the geometry and dimensions of the proposed SDBelM using a commercially available 3D-FEM software (JMAG, JSOL Corp., Japan). Fig. 5 shows a cross-sectional view of the analysis model for calculation of torque and suspension force. Outer diameters of the rotor and stator coil are both 40 mm. A magnetic gap between the PM and coil is 0.5 mm. A height and a radial length of the coil are 8 mm and 10 mm, respectively. The symbol h_i and h_p indicate a height of the iron yoke and a height of the PM pieces, respectively. The symbol, a, indicates an inner radial

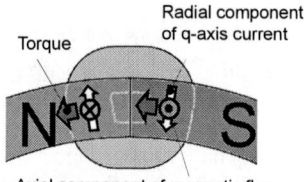

(a) Torque generation.

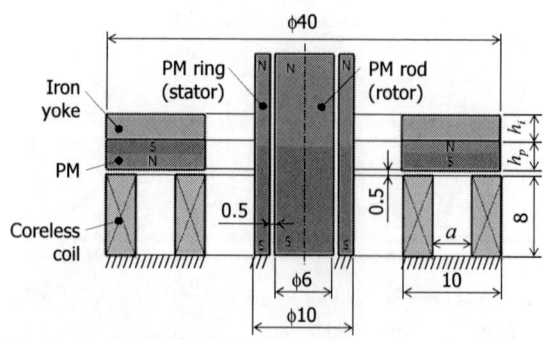

(b) Suspension force generation in the axial direction.

Fig. 4. Principle of suspension force and toque generation.

Fig. 5. Cross-sectional view of the analysis model for calculation of torque and suspension force.

distance of the coreless coil. In the analysis, neodymium-iron-boron magnet (N35 grade) and pure iron (less than 0.030 % carbon) are selected as materials of the PM and iron yoke, respectively.

B. Calculation of Suspension Force and Torque

Fig. 6 shows calculated suspension force and torque when the pole/coil combination of the motor and an inner radial distance of the coreless coil, a, are changed. As shown in Fig. 6 (a), the calculated torque increases as an increase in the number of poles. The calculated suspension force tends to be decreased as an increase in the number of poles, as shown in Fig. 6 (b). However, the peak value can be observed when the inner radial distance is changed. The targeted torque is 10 mNm for an application to a cooling fan. Therefore, one possible selection for the first prototype is the 6-pole/9-coil combination with a=4mm.

C. Calculation of Radial and Tilting Stiffnesses

The repulsive type passive magnetic bearing consist of a PM rod and a hollow PM ring. As shown in Fig. 4, outer diameters of the rod and ring are determined to be 6 mm and 10 mm, respectively. A radial air-gap is 0.5 mm. In the 3D-FEM analysis, neodymium-iron-boron magnet (N35 grade) is selected as material of these PMs. Fig. 7 shows the calculated radial and tilting stiffnesses of the repulsive type passive magnetic bearing. As can be seen, the gradients are both positive and these values are 3.3 N/mm and 4.7 mNm/deg., respectively.

IV. EXPERIMENTS

A. Fabrication

Fig. 8 shows a fabricated test machine. The iron yoke, PM pieces, and PM rod are attached with an aluminum

cover. The iron height, h_i, and PM height, h_p, are both 1 mm. An outer diameter of the aluminum cover is 49 mm. Total weight of the rotor is 39 g. The number of turns in one coreless coil is 468 with a wire diameter of 0.2 mm. Measured resistance of one coreless coil is 7.2 Ω. The coreless coils are molded with epoxy resin. Hall elements are embedded at the bottom of the coils to detect the rotational angle of the rotor. The hollow PM ring is attached at the center of the stator.

Eddy current type displacement sensors (PU-03A, AEC Corp., Japan) are used to measure axial (z), radial (r), and tilting (θ) motions of the rotor. A custom-made motor driver (Myway Plus Corp., Japan), which composes a three-phase voltage source pulse-width-modulation inverter, is used. This motor driver includes a controller board using a microprocessor. A minor proportional-integral loop regulates the three-phase current. A proportional-integral-derivative controller is adopted for active magnetic suspension control. Detailed control scheme can be seen in [6].

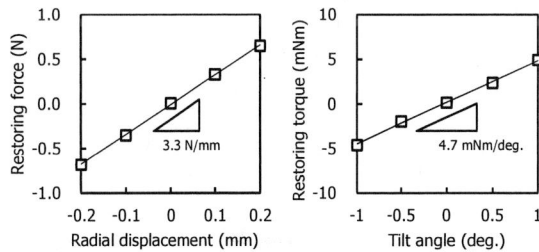

Fig. 7. Calculated radial and tilting stiffnesses of the repulsive type passive magnetic bearing.

Fig. 8. Fabricated spinning top-shaped rotor (left) and coreless coil stator (right).

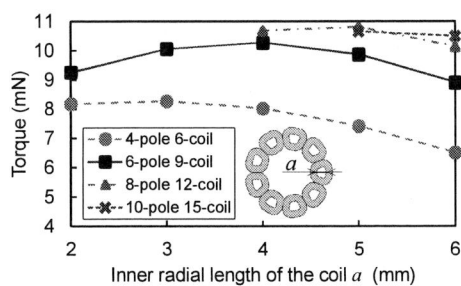

(a) Calculated torque.

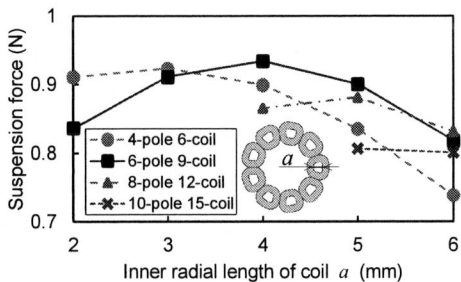

(b) Calculated suspension force.

Fig. 6. Calculated results with 3D-FEM analysis.

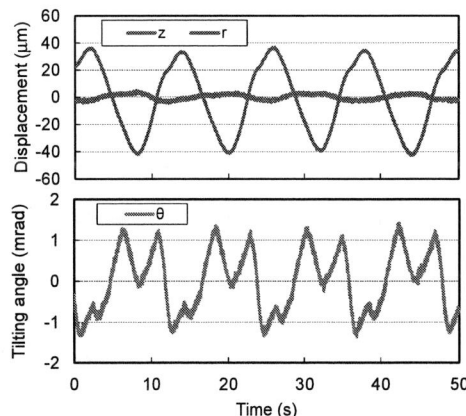

Fig. 9. Measured rotor vibration in axial (z), radial (r), and tilting (θ) directions at 5,000 rpm.

978-1-4799-2706-7/14 $31.00 © 2014 IEEE

B. Rtating Tests

By tuning control parameters, the rotor was successfully levitated. The vibration amplitudes of the rotor were evaluated with three times the standard deviation, 3σ, and these values in the axial, radial, and tilting directions at 0 rpm are 1.3 μm, 0.5 μm, and 0.04 mrad, respectively. The rotor was driven up to 5,000 rpm under no load condition. Fig. 9 shows the rotor vibration in the axial (z), radial (r), and tilting (θ) directions at 5,000 rpm. The vibration amplitude in the actively regulated axial direction was 6.2 μm. On the other hand, the amplitude in the passively stabilized radial and tilting directions were 77.6 μm and =2.3 mrad, respectively.

Fig. 10 shows sum of the vibration amplitudes in the actively regulated axial (z) direction and passively stabilized radial (r) and tilting (θ) directions. The vibrations were measured at every 250 rpm under steady state condition. In the radial direction, the rotor was mechanically contacted with the stator in the speed range from 1500 rpm to 2250 rpm because of the radial resonance. The rotor was also contacted in the tilting direction in the ranges from 250 rpm to 1000 rpm and from 2000 rpm to 2250 rpm. However, at the rotational speed of more than 2500 rpm, the rotor was stably levitated and rotated without any mechanical contact. Further vibration suppression and improvement of damping will be addressed in the future.

Fig. 11 shows measured power consumption of the fabricated SDBelM for both magnetic suspension and rotation under no load condition. When the rotational speeds are 2000 rpm and 2250 rpm, the power consumption are relatively large and are approximately 1.0 W and 1.7 W, respectively. This may be caused by an interference of the radial and/or tilting motion to the axial direction. This motion interference works as disturbance in the axial direction, and thus, the power consumption for the axial vibration suppression is increased. However, the power consumption at other rotational speeds is less than 0.5 W.

V. CONCLUSIONS

We proposed a novel one-axis actively regulated bearingless motor, dubbed a single-drive bearingless motor, with a spinning top-shaped rotor. The proposed SDBelM combines an axial gap SPM motor with coreless coils and a repulsive type passive magnetic bearing. It requires only one three-phase inverter and one set of three-phase winding for non-contact rotation. The rotor axial position and rotation are controlled by the d-axis and q-axis currents, respectively. The radial and tilting motions are passively stabilized by the passive magnetic bearing. The control method for magnetic suspension and rotation with d-axis and q-axis currents, respectively, was theoretically derived. The test machine was fabricated and tested. The rotor was successfully levitated and driven up to 5000 rpm. The vibration amplitudes in the axial (z), radial (r), and tilting (θ) directions were 6.2 μm, 77.6 μm, and 2.3 mrad, respectively.

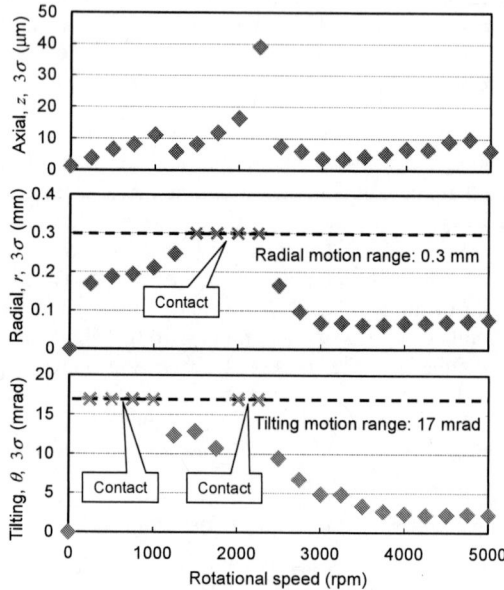

Fig. 10. Vibration amplitudes of the rotor in the axial (z), radial (r), and tilting (θ) directions.

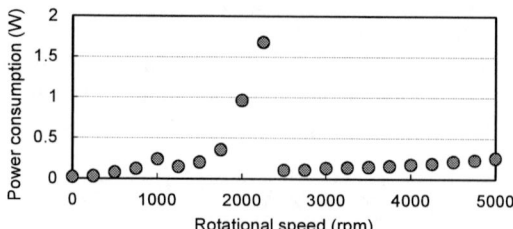

Fig. 11. Measured power consumption of the fabricated SDBelM for both magnetic suspension and rotation.

ACKNOWLEDGMENT

The authors would like to acknowledge the contribution of Mr. Syo Yamazaki, a former graduate student in the Department of Mechanical Engineering, Graduate School of Engineering, Shizuoka University, Japan, to this research. This work was supported by Grants-in-Aid for Scientific Research of the Japan Society for the Promotion of Science (23686041).

REFERENCES

[1] T. Ohji, *et al.*, "Performance of Repulsive Type Magnetic Bearing System Under Nonuniform Magnetization of Permanent Magnet," *IEEE Transactions on Magnetics*, vol. 36, no. 5, pp. 3696-3698, 2000.

[2] A. Yumoto, *et al.*, "A One-DOF Controlled Magnetic Bearing for Compact Centrifugal Blood Pumps," *Motion and Vibration Control*, pp. 357-366, 2009.

[3] I. Silva and O. Horikawa, "An Attraction-Type Magnetic Bearing with Control in a Single Direction," *IEEE Transactions on Industry Applications*, vol. 36, no. 4, pp. 1138-1142, 2000.

[4] Q. D. Nguyen and S. Ueno, "Analysis and Control of Nonsalient Permanent Magnet Axial Gap Self-Bearing Motor," *IEEE Transactions on Industrial Electronics*, vol. 58, no. 7, pp. 2644-2652, 2011.

[5] W. Bauer and W. Amrhein, "Electrical design and winding selection for a bearingless Axial-Force/Torque Motor,"

Proceedings of the 2012 International Symposium on Power Electronics, Electrical Drives, Automation and Motion, pp. 1224-1229, 2012.

[6] J. Asama, et al., "Proposal and Analysis of a Novel Single-Drive Bearingless Motor," *IEEE Transactions on Industrial Electronics*, vol. 60, no. 1, pp. 129-138, 2013.

[7] Y. Okada, H. Konishi, H. Kanebako, and C. W. Lee, "Lorentz Force Type Self-Bearing Motor", *Proceedings of the Seventh International Symposium on Magnetic Bearings*, pp. 353-358, August 23-25, 2000, ETH Zurich

[8] S. H. Park and C. W. Lee, "Lorentz Force-Type Integrated Motor-Bearing System in Dual Rotor Disk Configuration", *IEEE/ASME Transactions on Mechatronics*, Vol. 10, No. 6, pp. 618-625, 2005.

Control Characteristics of 8/10 and 12/14 Bearingless Switched Reluctance Motor

Zhenyao Xu, Dong-Hee Lee, Jin-Woo Ahn
Dept. of Mechatronics Engineering
Kyungsung University
Busan, Korea
zhenyao87@163.com

Abstract— **In this paper, two novel bearingless switched reluctance motors (BLSRMs), 8/10 hybrid type and 12/14 hybrid type, are presented. The two BLSRMs have separate torque and suspending force poles. Due to the independent characteristics between the torque and suspending force, the torque control can be decoupled from the suspending force control. In this paper, suspending force control characteristics of the two hybrid BLSRMs are compared. Although the two hybrid BLSRMs have excellent decoupling characteristics, the suspending force control performance is dependent on the torque characteristics. In the suspending force control comparison, a simple PID and hysteresis current control methods are employed. To verify the validity of the control strategy, tests are executed to the two types, and the control characteristics of the two types are presented and compared.**

Keywords— *Bearingless motors, force control, magnetic suspension, switched reluctance motorrs.*

I. INTRODUCTION

Many modern industrial applications, such as high speed machine tools, turbomolecular pumps, centrifugal pumps, compressors, flywheel energy storage and aerospace need high or ultra-high speed machines. However, many problems may arise, when traditional mechanical bearings are used to bear the shaft of high or ultra-high speed machine. For instance, the mechanical bearing can cause increased frictional drag, thermal problems and heavy wear in high-speed motors, which may affect the efficiency, bearing life and maintenance of the machine. Further, the lubrication oil that is required by the mechanical bearings cannot be used in high vacuum, ultra high temperature and low temperature environments [1]-[2]. In order to solve these problems caused by the mechanical bearings, magnetic bearing motors are researched [3]-[4]. Magnetic bearing motors have the advantages such as friction-free, abrasion-free, seal-free, lubrication-free, high speed, high precision, long life, easy-to-implement active control, and so on. However, high power density is not easy to be implemented due to the complex structure. The separated controllers are also needed because the torque and suspension part are separated in the magnetic bearing motor, which will lead to high cost.

Bearingless switched reluctance motor (BLSRM) is developed on the basis of magnetic bearing motor and switched reluctance motor (SRM). Hence, BLSRM not only extends theory and application of the bearingless motor, but also inherits high-speed performance and adaptability to harsh environment of SRM. Therefore, BLSRM can achieve the operation of high speed or ultra-high speed [5]-[6].

Recently, several structures of BLSRMs, such as 12/8 double winding type, 12/8 single winding type, 8/6 single winding type and Morrison type, are proposed [7]-[10]. But all the structures proposed in [7]-[10] are based on general SRM structure, so the available suspending force region is limited, and the torque control is coupled with the suspending force control. To realize stably suspension in these structures, complex mathematics equations have to be derived [11]-[14], and the mathematics equations are built up without considering the magnetic saturation, which increases the difficulty in controlling.

In order to expand the available suspending force region and reduce the control difficulty of conventional BLSRMs, an 8/10 hybrid stator poles BLSRM is proposed [6]. There are two types of stator poles in this motor: torque and suspending force poles. Compared with conventional BLSRM, the suspending force performance is improved and the air-gap is easier to control, but in this structure only half of the stator poles are used for the torque, output power density is very low. Moreover, the effect of torque current on the suspending force is somewhat large when the torque and suspending force windings are excited simultaneously.

In order to further improve the performances of BLSRM, a novel 12/14 hybrid stator pole type BLSRM is proposed [15]. The proposed structure also has separated torque and suspending force poles. Further, in the 12/14 type, short flux paths are taken and no flux reversal exists in the stator core. Compared to the 8/10 hybrid stator pole type BLSRM, the output torque is significantly improved and the air-gap is easier to control.

In this paper, the suspending force characteristics of the two BLSRMs are compared. In the suspending force control comparison, the BLSRMs mathematical model is not used in the control scheme, only a simple PI controller is used to regulate the speed of the proposed BLSRM, and two PID controllers are used to generate the desired suspending force commands to keep the rotor in the centre position. Based on the control scheme, the experimental system is constructed and experiments are performed to the two BLSRMs. Finally, the control characteristics of the two BLSRM are presented.

II. THE NOVEL BLSRMs WITH HYBRID STATOR POLE

A. 8/10 Hybrid Stator Pole Type BLSRM

Fig. 1 shows the structure of 8/10 hybrid stator pole type BLSRM. Different from the conventional structures, two types of stator poles are included in this structure. P_{A1} and P_{A2} are the torque poles of the phase A, and P_{B1} and P_{B2} are the torque poles of the phase B. The x-direction suspending force is generated by currents which flow in the windings of the suspending force poles P_{xp} and P_{xn}. That is, the current i_{xp} in the P_{xp} stator pole generates positive x-direction suspending force, and the current i_{xn} in the P_{xn} stator pole generates the negative x-direction suspending force. Similarly, the suspending forces for the y-direction are generated by the currents i_{yp} and i_{yn}, which flow in suspending force poles P_{yp} and P_{yn}, respectively. Meanwhile, to get a continuous suspending force, the suspending force pole arc needs to be wider than one rotor pole pitch. In that way, the aligned area between the suspending force and the rotor pole is always the same which may decrease the effect of suspending force current to the torque.

B. 12/14 Hybrid Stator Pole Type BLSRM

A novel 12/14 hybrid stator pole type BLSRM is proposed in Fig. 2. The proposed structure is similar to the structure of the 8/10 type. There are also torque and suspending force poles in the 12/14 type. But in the 12/14 type short flux paths are taken and no flux reversal exists in the stator core. Windings on the torque poles P_{A1}, P_{A2}, P_{A3} and P_{A4} are connected in series to construct phase A, and windings on the torque poles P_{B1}, P_{B2}, P_{B3} and P_{B4} are connected in series to construct phase B. Similar to the 8/10 type, the windings on the suspending force poles P_{xp}, P_{xn}, P_{yp} and P_{yn} are independently controlled to construct four suspending forces in the x- and y-directions. In order to get the continuous suspending force, the suspending force pole arc is also selected not to be less than one rotor pole pitch.

C. Torque and Suspending Force Characteristics of the two BLSRMs

As is well-known, the torque and suspending force windings are always excited simultaneously in BLSRMs. Hence, to obtain the characteristics of the proposed structure, finite element method (FEM) is employed to get the characteristics of the proposed structures.

Fig. 3 shows the torque profiles with fixed torque current (i_A=2A) and various suspending force currents. In Fig. 3, PU is per unit. 1 PU stands for one rotor pole pitch. That is, 1 PU stands for 36 and 26 mechanical degrees in the 8/10 and 12/14 types, respectively. From Fig. 3, it can be seen that the effect of suspending force current on the torque is very small in the two structures. Therefore, the suspending force current has almost no effect on torque.

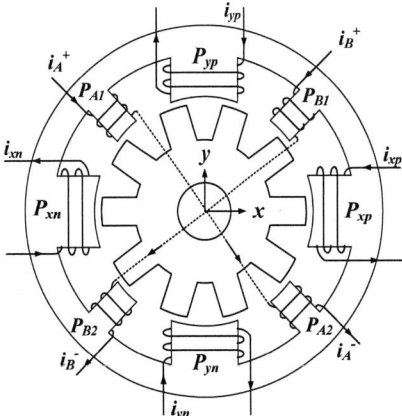

Fig. 1. Structure of 8/10 hybrid stator pole type BLSRM.

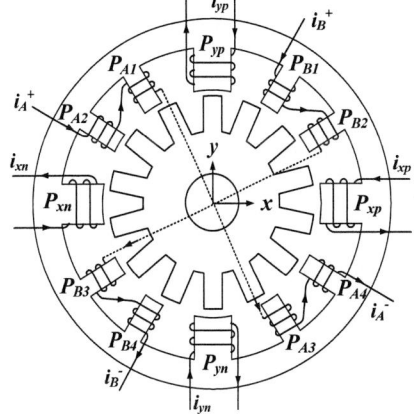

Fig. 2. Structure of 12/14 hybrid stator pole type BLSRM.

Fig. 4 shows the suspending force profiles with fixed suspending force current (i_{yp}=2A) and various torque currents. As seen in Fig. 4, when the torque winding current is zero, the suspending force has excellent linearity with respect to the rotor position in the two motors. Moreover, the available suspending force region in the proposed two structures is the whole rotor pole pitch, which is wider than the suspending force in conventional BLSRMs. It also can be seen from Fig. 4 that the effect of torque current on the suspending force is somewhat large in the 8/10 type, while in the 12/14 type, it is small enough to be ignored when compared to the suspending force generated by the suspending force winding.

From the above analysis, it can be seen that the proposed two structures cannot perfectly decouple the torque and suspending force control, but compared to conventional BLSRMs, the coupling effect is extremely reduced, especially in the 12/14 type, the torque control is almost decoupled from the suspending force control.

III. CONTROL SCHEME OF THE PROPOSED BLSRMs

According to the analysis in Section II, a control scheme for the two BLSRMs is proposed in Fig. 5. As shown in Fig. 5, mathematics model is not used in the control scheme, only a PI type speed controller is adopted

The 2014 International Power Electronics Conference

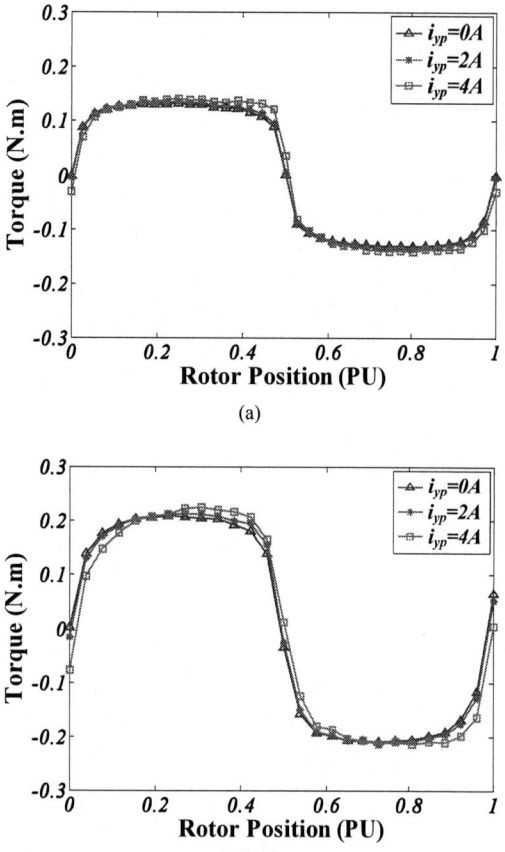

(a)

(b)

Fig. 3. Torque profiles with fixed torque current ($i_A = 2A$) and various suspending force currents ($i_{yp} = 0$, 2, and 4A). (a) 8/10 type. (b) 12/14 type.

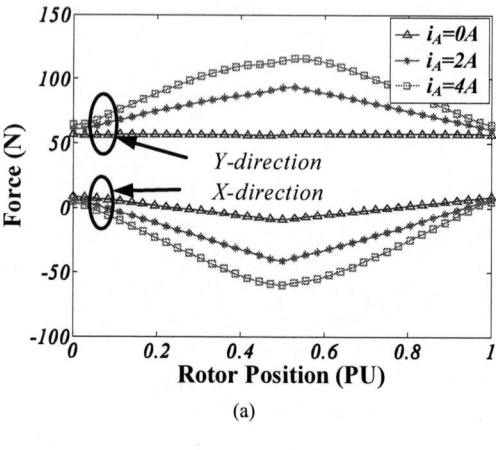

(a)

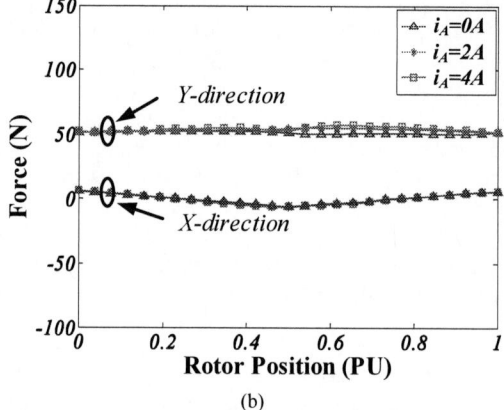

(b)

Fig. 4. Suspending force profiles with fixed suspending force current ($i_{yp}=2A$) and various torque currents ($i_A=0$, 2, and 4A). (a) 8/10 type. (b) 12/14 type.

to regulate the motor speed, and two independent close-loop PID air-gap displacement controllers, one for x-direction and the other for y-direction, are used to generate the desired suspending force commands F_x^* and F_y^* to keep the rotor at the center position. Further, in the control scheme, the actual current values of the exciting phase of the suspending force winding can be controlled through the hysteresis method according to the command current signals.

A. Command Current of the Suspending Force Winding

Fig. 6 shows the suspending force generation principle of the proposed BLSRMs. As shown in Fig. 6, if the winding on the suspending force pole P_{xp} is excited, the suspending force F_x in the positive x-direction can be generated. If the winding on the suspending force pole P_{yp} is excited, the suspending force F_y in the positive y-direction can be generated. In the same way, if the windings on the suspending force pole P_{xp} and P_{yp} are excited simultaneously, the force F will be generated, which is the synthesis force of F_x and F_y. Furthermore, the value and direction of the force F can be regulated by changing the values of the currents in the two windings. Therefore, when controlling values of the currents in the four suspending force windings, the desired resultant

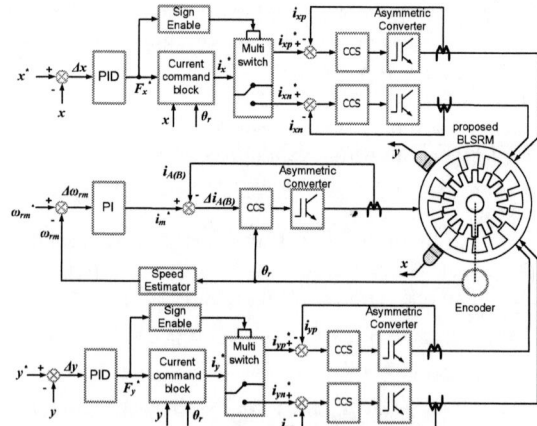

Fig. 5. Control Scheme for the two BLSRMs.

force in any arbitrary direction and magnitude can be obtained to compensate for the unbalanced pull force caused by the non-uniform air-gap.

According to the suspending force generation principle in the proposed BLSRMs, the force F, in any direction and magnitude, can be generated by the windings on the two poles, i.e. one each in the x- and y-directions. For instance, assume that F is the desired force, as shown in Fig. 6. In this case, because F is in the first quadrant, pole

978-1-4799-2706-7/14 $31.00 © 2014 IEEE

996

P_{xp} and P_{yp} are selected to produce the suspending force. Table I shows the excitation poles for various forces. In the table, F_x and F_y are used to determine the suspending force control poles. After choosing the force control poles, the command current i_x^* and i_y^* for these two poles can be calculated by a 3-dimensional look-up table.

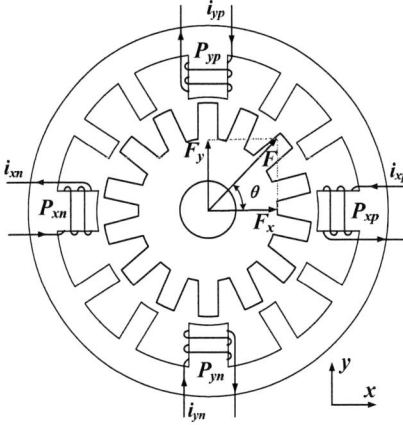

Fig. 6. Suspending force generation principle of proposed BLSRMs.

TABLE I
SELECTION OF SUSPENDING FORCE CONTROL POLES

Desired force	Suspending force control pole
$F_x > 0$ & $F_y > 0$	P_{xp} & P_{yp}
$F_x > 0$ & $F_y < 0$	P_{xp} & P_{yn}
$F_x < 0$ & $F_y > 0$	P_{xn} & P_{yp}
$F_x < 0$ & $F_y < 0$	P_{xn} & P_{yn}

B. Switching Control Strategy

According to the above suspending force current calculations, the asymmetric converter based on hysteresis control method is applied to control the winding current. Fig. 7 shows the switching rules of the hysteresis control method. As shown in Fig. 7, for any rotor position, currents in the x-and y-directions are selected as inputs to the current controller. Meanwhile, which suspending force winding is excited is determined by Table I. For instance, when $F_x^* > 0$, the multi-switch will connect with the winding on suspending force pole P_{xp}, and then the winding on suspending force pole P_{xp} is excited. Accordingly, according to the current error, the switching state S of this winding can be selected as 1, 0, or -1, in which the switching state corresponds to the operating modes of the asymmetric converter. Switching state 1 corresponds to magnetization mode, in which two power switches are excited simultaneously; switching state 0 corresponds to freewheeling mode, in which the winding is short-circuit through an power switch and a diode; and switching state -1 corresponds to demagnetization mode, in which two power switches are turned off and the current flows through two diodes.

IV. EXPERIMENTAL RESULTS

To verify the proposed control scheme, the tests are executed to the two types. Fig. 8 shows the structures of the test prototypes. And the main parameters of the prototypes are shown in Table II.

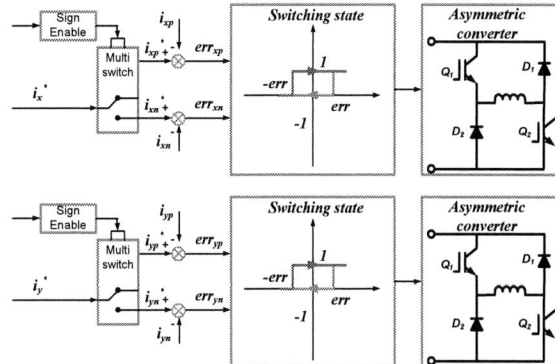

Fig. 7. Switching rules of the hysteresis control method.

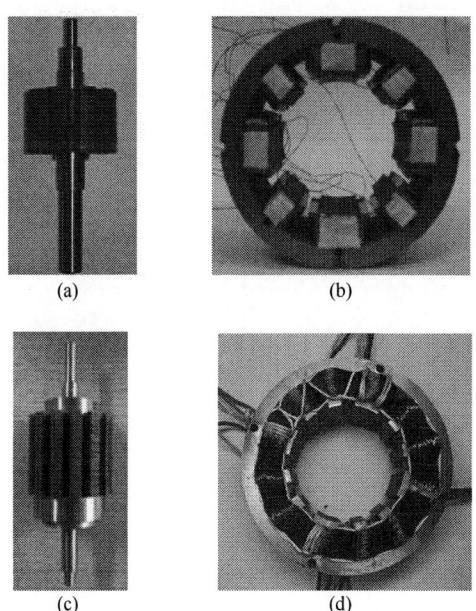

Fig. 8. Prototype of proposed BLSRMs. (a) Rotor of 8/10 type. (b) Stator of 8/10 type. (c) Rotor of 12/14 type. (d) Stator of 12/14 type.

TABLE II
DIMENSIONS OF THE TWO BLSRMS

Parameter	8/10	12/14
Number of phases	2	2
Number of torque stator poles	4	8
Number of suspending force stator poles	4	4
Number of rotor poles	10	14
Torque pole arc (degree)	18	12.85
Suspending force pole arc (degree)	36	25.7
Rotor pole arc (degree)	18	12.85
Length of axial stack (mm)	40	40
Outer diameter of outer stator (mm)	112	112
Yoke thickness of outer stator (mm)	10	7.7
Inner diameter of outer stator (mm)	64	60.2
Length of air gap (mm)	0.3	0.3
Yoke thickness of rotor (mm)	9.7	9.7
Shaft diameter (mm)	18	18

Based on the control scheme, the experimental system is constructed, as shown in Fig. 9. In the experimental system, the eddy current displacement sensors are used to detect the position of the rotor. The linear range of the sensor is from 0.5mm to 1.5mm, the sensitivity is 5V/mm. Meanwhile, to apply the suspending force load on the shaft, a line is used to connect the shaft and suspending force load. So when the rotor rotates, there will be a friction between the line and shaft. The experiments are realized by a Texas Instruments (TI) TMS320F2812 digital signal processor (DSP).

Fig. 10 shows the experimental results of the suspending force control in the two types with the rotor stationary. In the experiment, 2N suspending force load is applied to the shaft in both the negative x- and y-directions, respectively. From Fig. 10 it can be seen that as the suspending force control is applied, the rotor moves to its balanced position immediately.

Figs. 11 and 12 show the experimental results of the two types of BLSRMs with speed variation. Although the rotor is suspended at the center position, there are still frictions in the experiment system. One is the aerodynamic drag; another is the siding friction, which is caused by the suspending force load. Because the rotation friction is proportional to the speed, the rotation friction increases with increased speed and decreases with decreased speed. As shown in the figures, with the speed increase, to maintain the desired speed, the torque current has to increase to overcome the corresponding rotation friction, vice versa. From the air-gap displacement in the x- and y-directions, it is found that the rotor can still be steadily suspended when the motor speed is changed. But at the instant of the torque current increasing, the suspending force current in 8/10 type changes a little, while the suspending force current in the 12/14 type is almost same as before, which has a good match with the FEM analysis results in Fig. 4.

V. CONCLUSIONS

In this paper, the two hybrid BLSRMs are presented. The static characteristics are analyzed, such as torque and

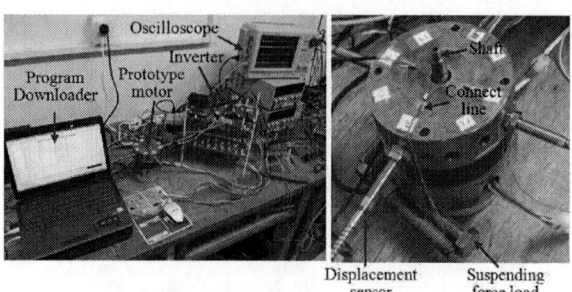

Fig. 9. Experimental set-up.

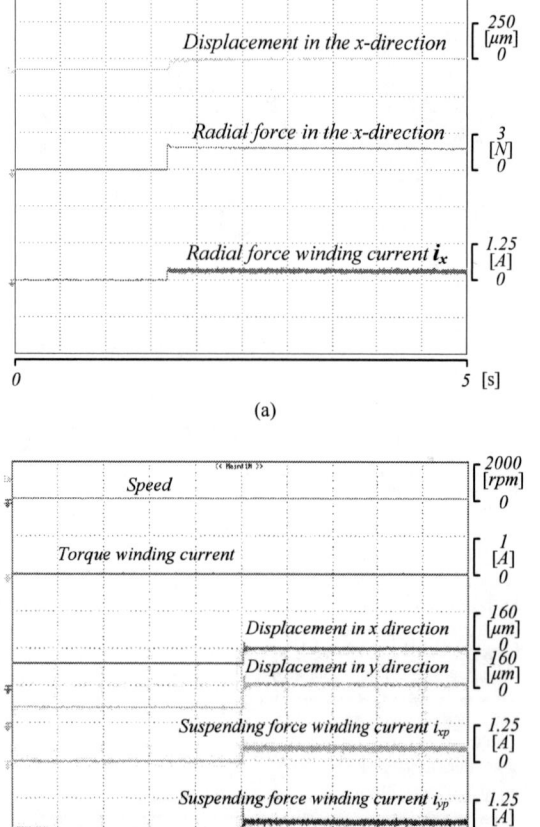

(a)

(b)

Fig. 10. Experimental results of suspending force control with rotor stationary. (a) 8/10 type. (b) 12/14 type.

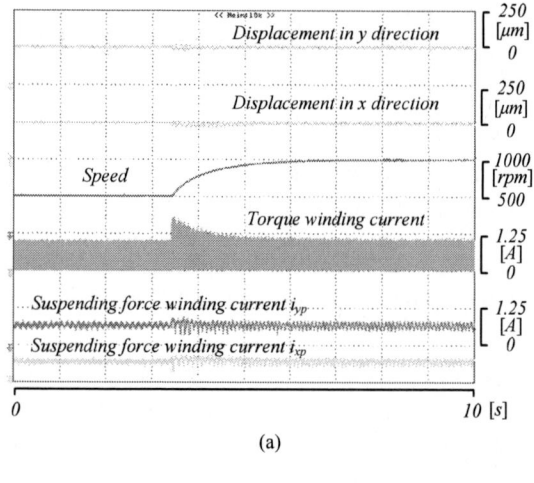

(a)

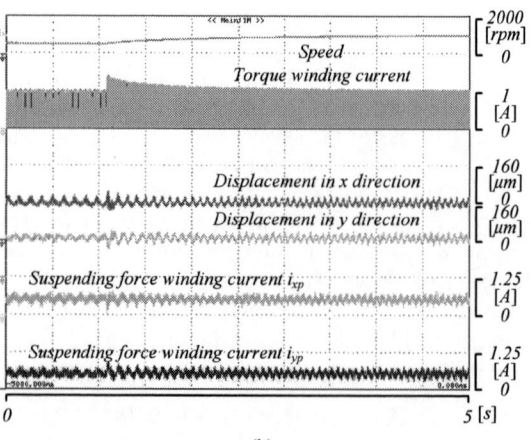

(b)

Fig. 11. Experimental results of suspending force control with speed increasing. (a) 8/10 type. (b) 12/14 type.

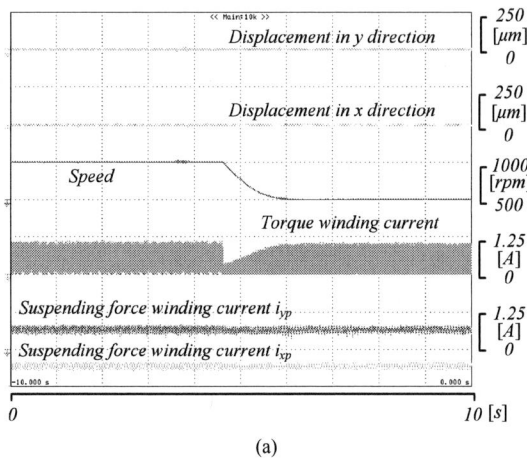

(a)

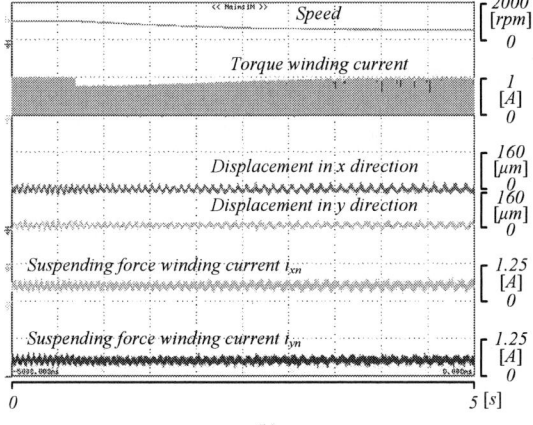

(b)

Fig. 12. Experimental results of suspending force control with speed decreasing. (a) 8/10 type. (b) 12/14 type.

suspending force. The suspending force control characteristics are also presented and compared. Although, the two hybrid BLSRMs have excellent decoupling characteristics between the torque and suspending force, the suspending force control performance is dependent on the torque characteristics. The experimental results show that the control scheme is valid for the two types, and the decoupling effect is better in the 12/14 type.

ACKNOWLEDGMENT

This work was supported by BK21 plus.

REFERENCES

[1] R. Bosch, "Development of a bearingless electric motor," *in Proc. Int. Conf. Electric Machines* (*ICEM'88*), Pisa, Italy, 1988, pp.373-375.

[2] J. Bichsel, "The bearingless electrical machine," *in Proc. Int. Symp. Magn. Suspension. Technol*, NASA Langley Res. Center, Hampton, 1991, pp. 561-573.

[3] J. X. Shen, K. J. Tseng, D. M. Vilathgamuwa, and W. K. Chan, "A novel compact PMSM with magnetic bearing for artificial heart application," *IEEE Transactions on Industry Applications*, vol. 36, no. 4, pp. 1061-1068, July/August 2000.

[4] C. C. Wang, and Y. D. Yao, "Bias-magnetic force for vibration reduction of magnetic bearing motors," *IEEE Transactions on Magnetics*, vol. 43, no. 6, pp. 2486-2488, June 2007.

[5] S. Ayari, M. besbes, M. Lecrivain, and M. Gabsi, "Effectes of the air gap eccentricity on the SRM vibrations," *in Proc. Int. Conf. Electr. Mach. and Drives*, 1999, pp.138-140.

[6] Dong-Hee Lee and Jin-Woo Ahn, "Design and analysis of hybrid stator bearingless SRM", *Journal of Electrical Engineering & Technology*, vol. 6, no. 1, pp.94-103, 2011.

[7] M. Takemoto, H. Suzuki, A. Chiba, et al., "Improved analysis of a bearingless switched reluctance motor," *IEEE Transaction on Industrial Application*, vol. 37, no. 1, pp. 26-34, Feb. 2001.

[8] Liu W, Yang S M, "Modeling and control of a self-bearing wwitched reluctance motor," *Fourtieth IAS Annual Meeting*, Hong Kong, 2005, pp. 2720-2725.

[9] L. Chen and W. Hofmann, "Analytically computing winding currents to generate torque and levitation force of a new bearingless switched reluctance motor", *in Proc.12th EPE-PEMC*, Aug. 2006, pp.1058-1063.

[10] Carlos R. Morrison, Mark W. Siebert and Eric J Ho, "Electromagnetic forces in a hybrid magnetic-bearing switched-reluctance motor", *IEEE Transaction on Magnetics*, vol.44, no.12, December, 2008.

[11] M. Takemoto, A. Chiba, and T. Fukao, "A method of determining the advanced angle of square-wave currens in a bearingless switched reluctance motor," *IEEE Transaction on Industrial Application*, vol. 37, no. 6, pp. 1702 -1709, Nov./Dec., 2001.

[12] Lin F C, Yang S M, "Self-bearing control of a switched reluctance motor using sinusoidal currents," *IEEE Transaction on Power Electronics*, vol. 22, no. 6, pp. 2518-2526, 2007.

[13] Feng-Chieh Lin and Sheng-Ming Yang, "An approach to producing controlled radial force in a switched reluctance motor", *IEEE Transaction on Industrial Electronics*, vol. 54, no. 4, pp.2137-2146, 2007.

[14] L. Chen, W. Hofman, "Speed regulation technique of one bearingless 8/6 switched reluctance motor with simpler single winding structure," *IEEE Transactions on Industrial Electronics*, vol. 59, no. 6, pp. 2592-2600, June 2012.

[15] Zhenyao Xu, Fengge Zhang, Dong-Hee Lee, and Jin-Woo Ahn, "Design and analysis of novel 12/14 hybrid pole type bearingless switched reluctance motor with short flux path", *Journal of Electrical Engineering & Technology*, vol. 7, no. 5, pp.705-713, 2012.

Basic Characteristic of A Two-Unit Outer Rotor Type Bearingless Motor with Consequent Pole Permanent Magnet Structure

Masatsugu Takemoto
Graduate School of Information Science and Technology
Hokkaido University

Abstract— **Bearingless motors (BelMs), that can realize shaft suspending without mechanical contacts and lubrication by electromagnetic force, have been proposed. However, in outer rotor type BelMs, one-unit structure with small shaft output alone have been developed. Therefore, this paper introduces a two-unit outer rotor type BelM with consequent-pole permanent magnet structure (CPPM) for high shaft output and the magnetic suspension characteristics of the proposed motor with experimental results. It can be verified that the proposed motor is equipped with the sufficient performance in order to make for practical use.**

Keywords— *Bearingless Motors, magnetic bearings, magnetic levitation, outer rotor motors.*

I. INTRODUCTION

From environmental and economical standpoints, more advantageous features such as small size, high efficiency, high rotational speed and maintenance-free are important issues for motors. As one of various approaches, bearingless motors (BelMs), that can realize shaft suspending without mechanical contacts and lubrication by electromagnetic force, have been proposed [1-12]. Many BelMs are equipped with two kinds of windings in same stator. Motor windings and suspension windings can generate suspension force and torque in same stator simultaneously.

In outer rotor type BelMs, one-unit structure alone have been developed [9-12]. The structure discussed in [9] has one BelM unit of 2-axis active position regulation and two thrust magnetic bearings of 3-axis passive position regulation. In addition, in [10-12], thrust magnetic bearing function of 3-axis passive position regulation is included in one BelM unit. In these structures, the rotor position in the conical direction is regulated by passive control. Accordingly, these structures are limited to flat structure and small shaft output. More stable suspension control and flexible structure have been required with high shaft output in industry applications.

The principal authors have proposed a two-unit outer rotor type BelM with consequent pole permanent magnet structure (CPPM) [13]. Two BelM units in the proposed motor can actualize 4-axis active position regulation. The

proposed motor is designed by 3D-FEM analysis and prototyped. In this paper, magnetic suspension characteristics of the prototype motor are measured and investigated by load experiments in the radial and thrust direction. In addition, the radial and thrust displacement profiles of the rotor is verified under rotational condition. It is shown that the prototype motor can be equipped with the very stable magnetic suspension characteristics.

II. STRUCTURE OF THE PROPOSED MOTOR

A. Proposed Motor Structure

Fig. 1 shows the structure of CPPM type BelM. The proposed motor has outer rotor structure with a two-unit BelM and two thrust magnetic bearings. The BelM units that generate radial suspension force and torque are center in the proposed motor. The thrust magnetic bearings are arranged at the upper and lower sides of the BelM units. Four-axis active position regulation can be realized by the BelM units. The thrust magnetic bearings can control the thrust direction position passively. Accordingly, the proposed motor can realize completely non-contact operation.

Fig. 2 shows the sectional view of unit 1 in the BelM and U-phase stator winding configuration. The rotor in which eight radial permanent magnets are equally spaced is CPPM type structure. The eight radial permanent magnets are magnetized in the same radial direction. The excitation flux ψ_{m16} generated by the radial permanent magnets flows via salient iron poles between the permanent magnets. Accordingly, the BelM is a sixteen-pole motor. The U-phase stator winding is composed of the motor winding N_{m16U} and the suspension winding N_{s2U}. As shown in Fig. 2(b), the motor winding is a sixteen-pole concentrated winding configuration. On the other hand, the suspension winding is a two-pole distributed winding configuration. Similarly, V and W-phase stator windings are wound respectively at equal angular interval of 120°.

Fig. 3 shows the sectional view of the lengthwise direction of the BelM rotor structure, and Fig. 4 shows the flux path of the bias permanent magnets. The proposed motor has low ripple characteristics of

978-1-4799-2706-7/14 $31.00 © 2014 IEEE

The 2014 International Power Electronics Conference

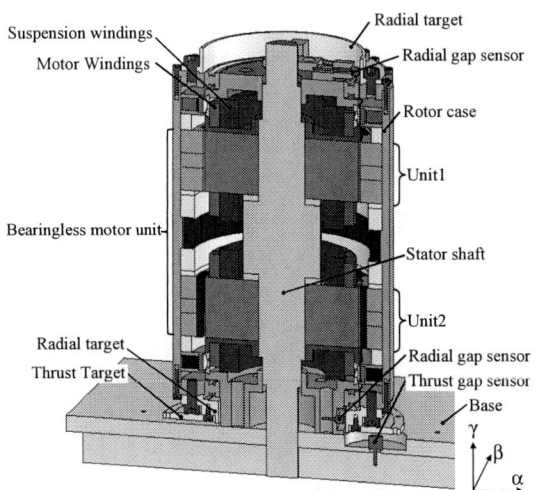

Fig. 1 Structure of two-unit outer rotor type BelM with CPPM structure.

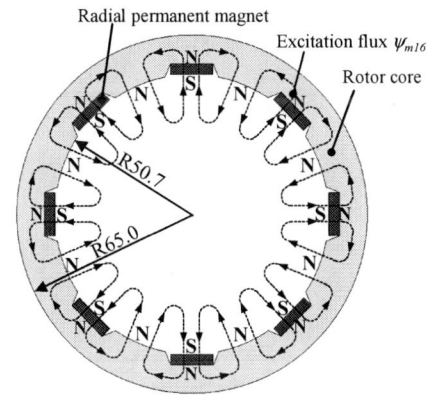

(a) Rotor structure (Unit1)

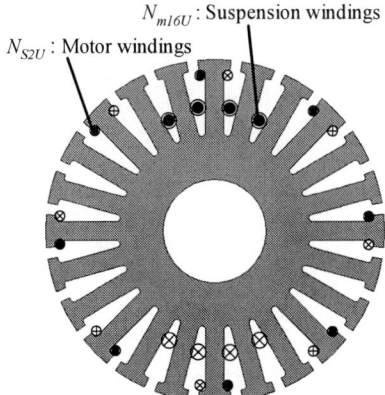

(b) Stator windings configuration

Fig. 2 Sectional view of BelM unit.

suspension force and torque, because the rotor structure is equipped with the trilaminar skew shifted at intervals of 5°. Moreover, the excitation flux ψ_{m16} is increased with the bias flux ψ_{bi} generated by bias permanent magnets between two rotors of BelM units as shown in Fig. 4, since the bias permanent magnets are magnetized in the thrust direction.

Fig. 5 shows the rotor and stator of a fabricated

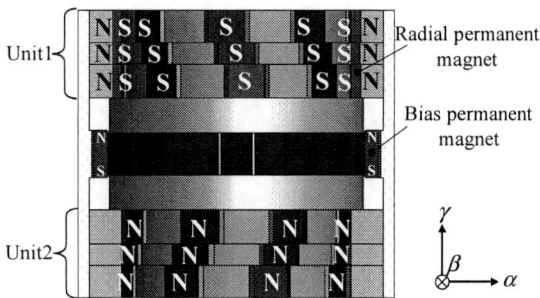

Fig. 3 BelM rotor structure of trilaminar skew.

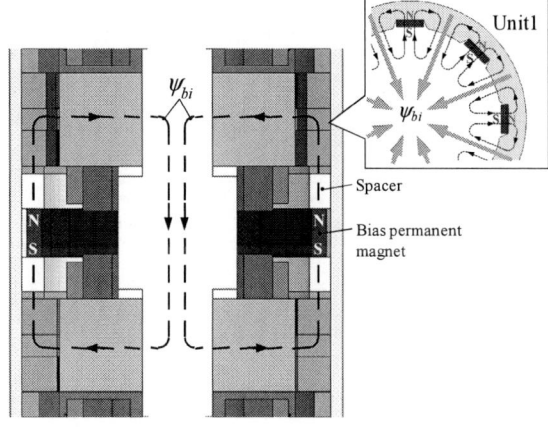

Fig. 4 Flux path of bias permanent magnet.

prototype motor. Table I shows the specification of the prototype motor.

B. Principle of Radial Suspension Force Generation

Fig. 6 shows the principle of radial suspension force generation. The α and β axes are defined as the perpendicular coordinates which are fixed to the stator. The suspension windings $N_{\alpha l}$ and $N_{\beta l}$ are the equivalent two-phase windings converted from the suspension windings N_{s2U}, N_{s2V}, and N_{s2W} by three-phase to two-phase transformation. If the suspension current flows in the suspension windings $N_{\alpha l}$, the 2-pole suspension flux $\psi_{\alpha l}$ is generated as shown in Fig. 6. As a result, the flux density of the left-side air gap is dense, and the right side is sparse. Accordingly, this superimposed magnetic field of the ψ_{m16}, ψ_{bi}, and $\psi_{\alpha l}$ results in the suspension force F_α acting on the rotor toward the positive direction in the α axis.

Even if the rotor rotates, there is no need to change the direction of the suspension flux $\psi_{\alpha l}$ in order to produce the suspension force F_α because the magnetic pole of the rotor salient iron magnetized by ψ_{m16} and ψ_{bi} is not changed. Furthermore, no suspension force is produced with a motor torque flux. Therefore, the CPPM type structure can suspend the rotor shaft with dc currents in the suspension windings independently from the motor operation such as the rotor angular position and torque.

978-1-4799-2706-7/14 $31.00 © 2014 IEEE   1001

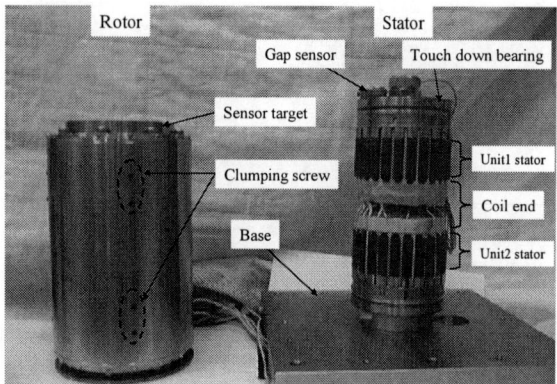

Fig. 5 Rotor and stator of fabricated prototype motor.

TABLE I
SPECIFICATION OF PROTOYPE MOTOR

Rotor weight	103N
Touch down	0.15mm
Gap length	1.0mm
Rotor length	235.8mm
Stator diameter	99.4mm
Rotor diameter	140mm
Motor unit length	38mm
number of stator slots	24
number of turns of motor winding (3phases, 0.7mmf, 2 pararells)	9turns
number of turns of suspension winding (3phases, 0.6mmf, 2 pararells)	14turns

C. Principle of Thrust Force Generation

Fig. 7 shows the structure of the thrust passive magnetic bearing and the principle of thrust force generation. The stator is a C-form core and the rotor of torus structure is constructed from thrust permanent magnets sandwiched by two rotor ring discs. The thrust flux ψ_γ generated from the thrust permanent magnets flows into a stator and returns to the thrust permanent magnets. If the rotor is displaced toward negative direction in the γ axis, thrust force toward positive direction is generated by magnetic attractive force. The torus structure can prevent the radial force acting on the rotor. Accordingly, the rotor can be suspended passively on the balanced position between rotor weight and thrust force.

III. EXPERIMENTAL SYSTEM

Fig. 8 shows the experimental system of the prototype motor. The suspension characteristics of the prototype motor were measured by radial load and thrust load under levitation condition. In order to realize the stable levitation, the rotor radial position is controlled in the two-unit BelM, and the rotor thrust position is suspended by the thrust passive magnetic bearings. In this situation, the disturbance force was applied to generate the suspension force as shown in Fig. 8. The suspension force can be derived from the disturbance force.

Relationship between suspension current and radial

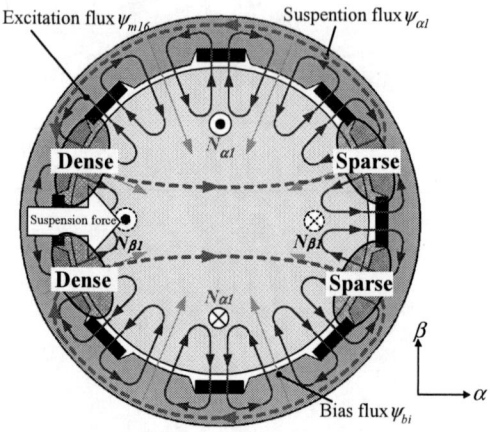

Fig. 6 Principle of radial suspension force generation.

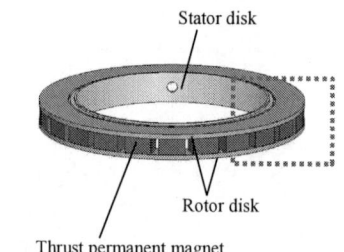

(a) Structure of thrust passive magnetic bearing.

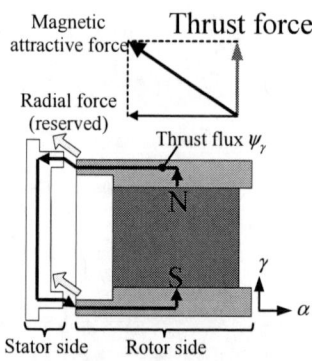

(b) Principal of thrust force generation.

Fig. 7 Structure and operating principle of thrust magnetic bearing.

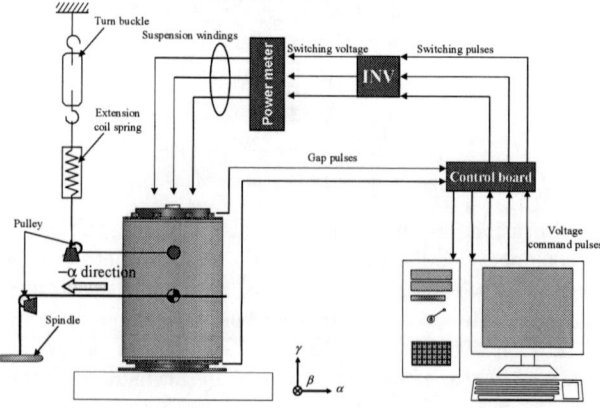

Fig. 8 Measurement system of prototype motor.

The 2014 International Power Electronics Conference

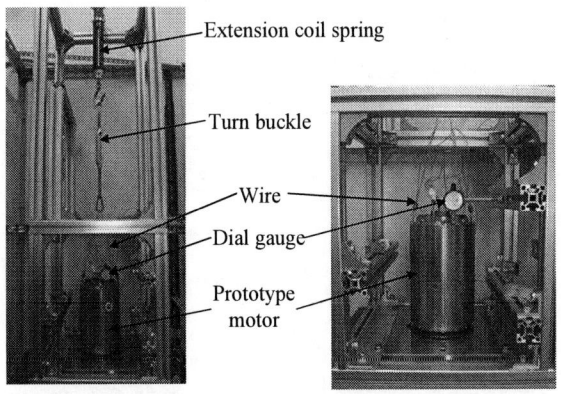

Fig. 9 Photographs of measurement system of passive thrust force.

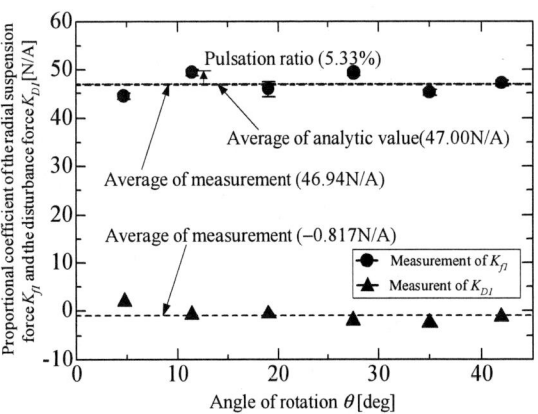

Fig. 12 Relationship of between K_{fl}, K_{DI} and θ under no load condition.

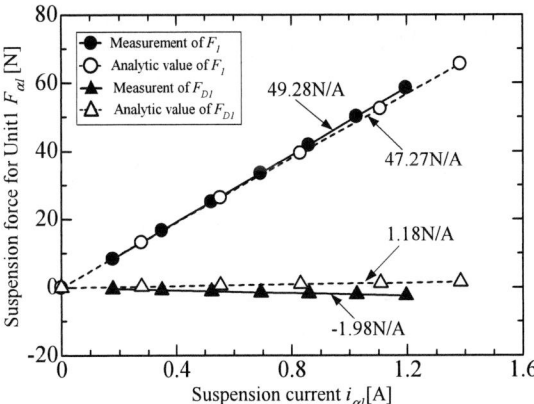

Fig. 10 Relationship of between $F_{\alpha l}$ and $i_{\alpha l}$ under no load condition.

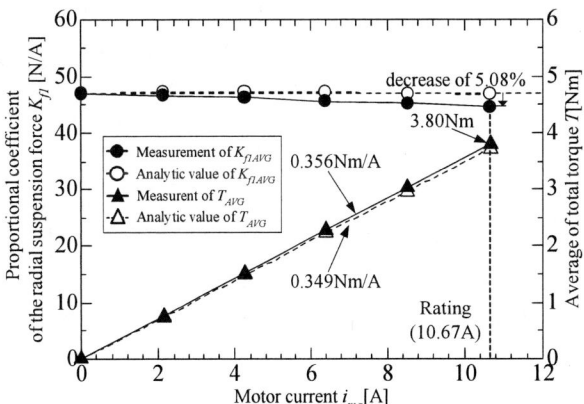

Fig. 13 Relationship of between $K_{fl.AVG}$, T_{AVG} and i_{mq}.

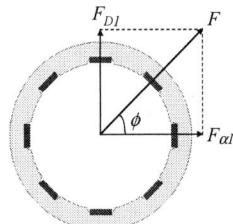

Fig. 11 Generating direction of suspension force.

suspension force is measured by varying the weight load, and the radial load generated by the weight is 1.3 times (137N) larger than rotor weight. The experiment was carried out at each rotor angular position. In addition, torque was also measured by varying motor current under the radial load. The detailed experiment conditions of the radial load are as follows:

- Radial load : 0~14 kg (2 kg increments)
- Motor torque current i_{mq} = 0~10.67 A (0~rated value)
- Rotor angular position θ = 4.75, 11.50, 19.00, 27.50, 35.00, 42.00°

Similarly, the thrust force characteristic was measured with extension coil spring load toward the thrust direction. Fig. 9 shows the photographs of the measurement system of the passive thrust force. Rotor thrust displacement is detected by a dial gauge.

IV. EXPERIMENTAL RESULTS OF MAGNETIC SUSPENSION CHARACTERISTICS

A. Radial Suspension Characteristics under No Load Condition of Electric motor

Fig. 10 shows the relationship the suspension force $F_{\alpha l}$ and the suspension current $i_{\alpha l}$ for Unit 1. As shown in Fig. 11, in spite of generating the suspension force toward the α direction, the generating direction of the actual suspension force generally inclines from the α direction because of the space harmonic component of the magnetic flux in the air gap and the mechanical error in the prototype. The force which is cause of inclining is the disturbance force F_{DI}. The proposed motor is the good characteristic, so that the disturbance force F_{DI} is small. The rotor angular position θ is fixed at 27.50°. From Fig. 10, the relationship of the suspension force $F_{\alpha l}$ and the suspension current $i_{\alpha l}$ is linearly increasing, and the disturbance force F_{DI} is little generated. Accordingly, the relationship between the suspension force $F_{\alpha l}$ and the suspension current $i_{\alpha l}$ can be represented as a slope coefficient of K_{fl}. Similarly, the characteristic of the disturbance force F_{DI} can be expressed as a slope coefficient of K_{DI}.

978-1-4799-2706-7/14 $31.00 © 2014 IEEE

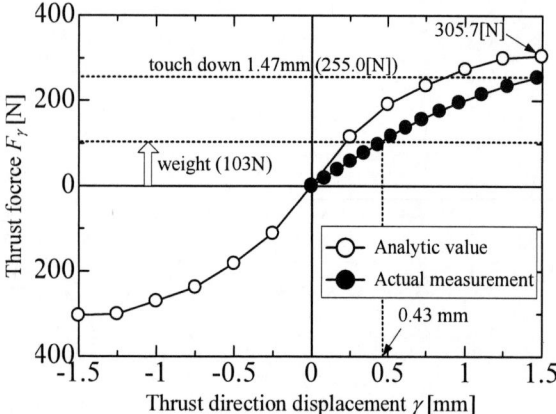

Fig. 14 Relationship between γ and F_γ

Fig. 12 shows the relationship between the slope coefficients K_{fl}, K_{D1} versus the rotor angular position θ measured with the prototype motor. It is known that the radial suspension force can be stably generated in all of the rotor angular position, because the pulsation ratio of the slope coefficient K_{fl} is by only 5.33%. Besides, the disturbance force F_{D1} can be negligible, because the average value of the slope coefficient K_{D1} is −0.819N/A slightly and the pulsation of the slope coefficient K_{D1} is very small. Therefore, it can be verified that the proposed motor can realize very stable suspension with dc currents. The average value of the slope coefficient $K_{fl.AVG}$ is 46.94N/A. Accordingly, the maximum suspension force of a single BelM unit is 260.09N, because the rated suspension current is 5.54A. Thus, a single BelM unit can generate the sufficient suspension force which is 2.53 times of the whole rotor weight. The proposed motor is equipped with the powerful radial suspension performance.

B. Radial Suspension Characteristics under Load Condition of Electric motor

Fig 13 shows the relationship between the average slope coefficient $K_{fl.AVG}$ and the average torque T_{AVG} versus the motor torque current i_{mq}. At the rated value of the motor torque current i_{mq}, the maximum average torque T_{AVG} is 3.80 Nm and the decrease ratio of the average slope coefficient $K_{fl.AVG}$ from no torque load condition is 5.08% slightly. The suspension characteristic of the proposed motor has the very small influence from the motor operation such as torque. It is found that the proposed motor is equipped with the very stable radial suspension characteristics.

In addition, from Fig. 10 and Fig. 13, these measurements of the proposed motor are very well in agreement with analytic values with 3D-FEM. In designing the proposed motor, 3D-FEM is effective.

C. Thrust Suspension Characteristics

Fig. 14 shows the relationship between the thrust direction displacement γ and thrust force F_γ. Although an

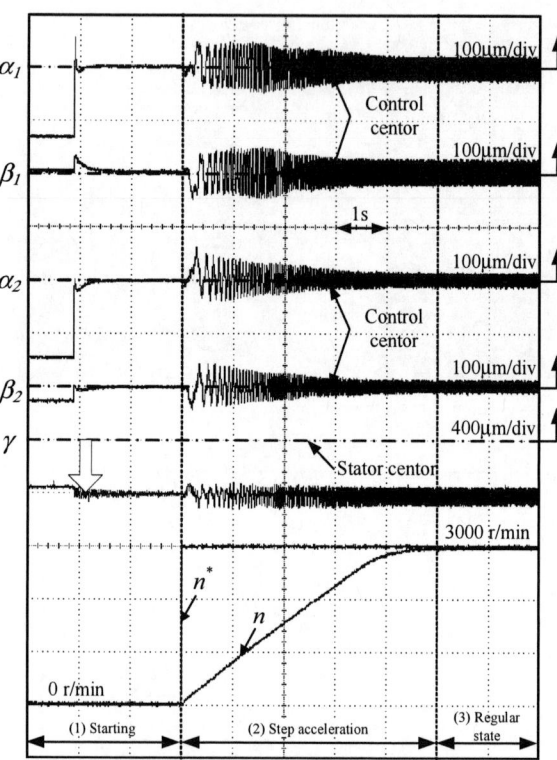

Fig. 15 Rotor displacement profiles on starting control from touchdown and acceleration.

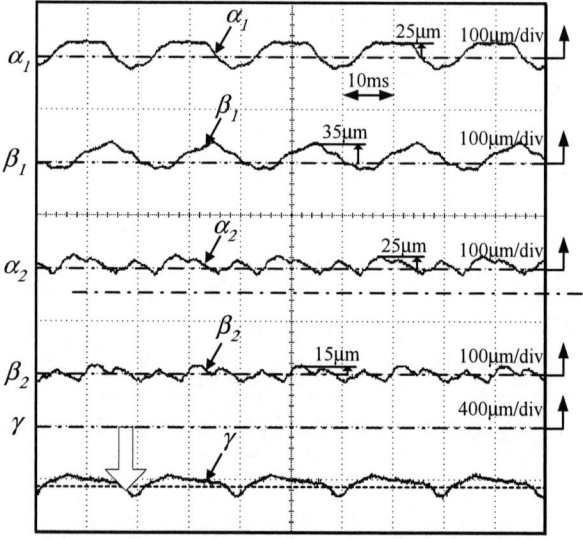

Fig. 16 α-, β-, and γ-gap sensor output under 3000 r/min.

error is between experimental results and analysis results, the tendency is well in agreement. It seems that the cause with this error is the fabrication error. However, the maximum thrust force of 255.0N is generated at the touch-down displacement $\gamma = 1.47$ mm. The prototype motor has sufficient thrust force characteristics, because the maximum thrust force is 2.48 times larger than the whole rotor weight (103N). The rotor can be passively suspended at $\gamma = 0.43$ mm of the balanced position between the whole rotor weight and the thrust force. This displacement magnitude is 1/3 or less of the touch-down displacement. Accordingly, the proposed motor is equipped with the stable thrust suspension characteristic.

D. Suspension Control under rotational Condition

Fig. 15 shows the experimental result of the rotor displacement profiles on the occasion of the starting suspension control from touchdown and the step acceleration. In this experiment, the rotor thrust displacement is detected without mechanical contact by the eddy-current type gap sensor shown in Fig. 1.

Firstly, the starting suspension control from the touchdown in the period of (1) in Fig. 15 is explained. A margin between the touchdown bearing and the rotor in the radial direction is about 150 μm on one side. Before beginning the suspension control, the rotor is touched down in the radial direction. After beginning the suspension control, the rotor radial positions can be promptly controlled at the control center by the proposed BelM units. In the touchdown state, the proposed BelM units can generate the radial suspension force which fully exceeds unbalanced magnetic attraction force caused by the permanent magnets. On the other hand, the rotor position in the γ direction does not change at about 400 μm.

Next, the suspension control under the step acceleration in the period of (2) in Fig. 15 is examined. The rotational speed command n^* is accelerated from 0 r/min to 3000 r/min. During acceleration, the rotor radial displacements $\alpha_1, \beta_1, \alpha_2, \beta_2$ are increased to about 60 μm at the maximum, and the center of the rotor thrust displacement γ is constant about 400 μm on all speed range. Therefore, the operation of the prototype is stable in spite of the sudden step acceleration.

Finally, the rotor displacement waveforms which are the regular state of 3000 r/min in the period of (3) in Fig. 15 are checked in detail. Fig. 16 shows the rotor displacement waveforms to which the period of (3) in Fig. 15 was enlarged. Although the rotor displacement waveforms are changing periodically, the proposed motor can realize completely non-contact operation. The maximum of the rotor radial displacements is 35 μm, this maximum displacement is 1/4 or less of the touch-down displacement. The rotor thrust displacement in the γ direction, which is regulated by the thrust passive magnetic bearings, is descending by about 400 μm by the rotor weight, and the amplitude of the rotor thrust displacement is about 80 μm. This rotor thrust displacement is small enough compared with about 1.5 mm which is touchdown width.

Consequently, it is found that the proposed motor can operate the stable suspension control under all the operating condition.

V. CONCLUSIONS

In this paper, the two-unit outer rotor type BelM with the consequent pole permanent magnet structure was introduced. The two BelM units in the proposed motor can actualize 4-axis active position regulation in the radial direction, the rotor position in the thrust direction can be regulated passively by the thrust passive magnetic bearings. The magnetic suspension characteristics of the prototype motor was measured and investigated by load experiments in the radial and thrust direction. In addition, the radial and thrust displacement profiles of the rotor was verified under rotational condition. It is shown that the prototype motor can be equipped with the very stable magnetic suspension characteristics.

ACKNOWLEDGMENT

I would like to thank Mr. Takashi Kono who carried out the big contribution to this research result.

REFERENCES

[1] M. T. Bartholet, T. Nussbaumer, D. Krähenbühl, Franz Zürcher, and J. W. Kolar, "Modulation Concepts for the Control of a Two-Phase Bearingless Slice Motor Utilizing", *IEEE Tran. on Industry Applications*, Vol. 46, No. 2, pp. 831-840, Mar./Apr. 2010.

[2] H. Grabner, W. Amrhein, S. Silber, and W. Gruber, "Nonlinear Feedback Control of a Bearingless Brushless DC Motor", *IEEE Tran. on Mechatronics*, Vol. 15, No. 1, pp. 40-47, Feb. 2010.

[3] Q. D. Nguyen and S. Ueno, "Analysis and Control of Nonsalient Permanent Magnet Axial Gap Self-Bearing Motor", *IEEE Tran. on Industrial Electronics*, Vol. 58, No. 7, pp. 2644-2652, Jul. 2011.

[4] V. F. Victor, F. O. Quintaes, J. S. B. Lopes, L. d. S. Junior, A. S. Lock, and A. O. Salazar, "Analysis and Study of a Bearingless AC Motor Type Divided Winding, Based on a Conventional Squirrel Cage Induction Motor", *IEEE Tran. on Magnetics*, Vol. 48, No. 11, pp. 3571-3574, Nov. 2012.

[5] J. Asama, D. Kanehara, T. Oiwa,and A. Chiba, "Suspension Performance of a Two-Axis Actively Regulated Consequent-Pole Bearingless Motor", *IEEE Tran. on Energy Conversion*, Vol. 28, No. 4, pp. 894-901, Dec. 2013.

[6] H. Sugimoto, Y. Uemura, A. Chiba, and M. A. Rahman, "Design of Homopolar Consequent-Pole Bearingless Motor With Wide Magnetic Gap", *IEEE Tran. on Magnetics*, Vol. 49, No. 5, pp. 2315-2318, May 2013.

[7] M. Nakagawa, Y. Asano, A. Mizuguchi, A. Chiba, C. X. Xuan, M. Ooshima, M. Takemoto, T. Fukao, O. Ichikawa, and D. G.. Dorrell, "Optimization of Stator Design in a Consequent-Pole Type Bearingless Motor Considering Magnetic Suspension Characteristics", *IEEE Tran. on Magnetics*, Vol. 42, No. 10, pp. 3422-3424, Oct. 2006.

[8] J. Amemiya, A. Chiba, D. G. Dorrell, and T. Fukao, "Basic Characteristics of a Consequent-Pole-Type Bearingless Motor", *IEEE Tran. on Magnetics*, Vol. 41, No. 1, Jan. 2005.

[9] T. Yamada, Y. Nakano, J. Asama, A. Chiba, T. Fukao, T. Hoshino, and A. Nakajima, "Outer Rotor Consequent-Pole Bearingless Motor With Improved Start-Up Characteristics", *IEEE Tran. on Magnetics*, Vol. 44, No. 11, May. 2008.

[10] T. Reichert, T. Nussbaumer, and J. W. Kolar, "Bearingless 300-W PMSM for Bioreactor Mixing", *IEEE Tran. on Industrial Electronics*, Vol. 59, No. 3, pp. 1376-1388, Mar. 2012.

[11] T. Reichert, T. Nussbaumer, and J. W. Kolar, "Investigation of Exterior Rotor Bearingless Motor Topologies for High-Quality Mixing Applications", *IEEE Tran. on Industry Applications*, Vol. 48, No. 6, pp. 2206-2216, Nov./Dec. 2012.

[12] T. Reichert, J. W. Kolar, and T. Nussbaumer, "Stator Tooth Design Study for Bearingless Exterior Rotor PMSM", *IEEE Tran. on Industry Applications*, Vol. 49, No. 4, pp. 1515-1522, July/Aug. 2013.

[13] S. Kobayashi, M. Takemoto, Y. Tanaka, A. Chiba, and T. Fukao, "A Proposal of a Consequent-pole Permanent Magnet Type Bearingless Motor Equipped with Outer Rotor Structure for 4-axes Active Suspension Control," *IEEJ Technical Meeting on Rotating Machinery*, RM-07-64, pp. 87-92, 2007 (in Japanese).

The 2014 International Power Electronics Conference

Voltage Ripple Elimination in Inductor-Less AC-to-AC Converters for Multi-Pole Permanent Magnet Synchronous Generators

Koutaro Tanaka and Hideaki Fujita
Department of Electrical and Electronic Engineering
Tokyo Institute of Technology
Tokyo, Japan
hf@ieee.org

Abstract—This paper discusses reduction of the output voltage ripples in an inductor-less three-phase to single-phase converter for a multi-pole three-phase permanent-magnet synchronous generator having open-end stator windings. The converter consists of only two three-phase bridge converters, and no output filter inductor is installed. The analysis in this paper is conducted to reveal the output voltage ripples based on the equivalent circuit of the generator. Thus, the output voltage ripples are caused by the stability problems in the current feedback control due to a low zero-sequence impedance and the distorted EMF. To solve these problems, this paper proposes a new current control method based on $\alpha\beta0$ transformation and feed forward harmonic compensation. The proposed new control method has been examined in a 2-kW prototype of the inductor-less converter. As a result, the high-frequency ringings and 1.25 kHz harmonic voltage are suppressed, and the voltage THD has been also reduced from 8.4% to 3.7%.

Keywords—Filter inductor, open-end windings, permanent-magnet synchronous generator, single-phase loads, voltage ripples, zero-sequence voltage.

I. INTRODUCTION

Recently, low-power generation systems using renewable energy resources are also expected as an emergency power supply, such as micro wind turbines, micro hydro, and so on. For these applications, efficient and low-cost power converters are required. The power converter is required to obtain an ac output voltage with a constant amplitude and a constant frequency. However, a PWM inverter generates a discontinuous voltage pulse train, and thus, requires an output ac inductor. The ac inductor is usually large and heavy, and relatively expensive, because it is made of iron. Moreover, boost up capability is requested for the power converters, to reduce the rotating speed of the combustion engine for energy saving.

Application of matrix converters has been discussed [1]. A single-phase to three-phase matrix converter has been proposed for driving a three-phase induction motor in [2]. This circuit can be employed as a generator interface circuit for a single-phase load or single-phase power grid, and it is possible to keep the rms value of the output load voltage even under a low speed operation. However, the circuit uses nine bidirectional ac switches composed by 18 IGBTs. These switches also has

over voltage problem when all switching devices are turned off, and it is impossible to use low-cost intelligent power modules (IPMs), three-phase bridge module and/or gate driving ICs.

The author have proposed a new inductor-less three-phase to single-phase boost power converter [3]. The power converter consists of two three-phase bridge converters and a three-phase permanent-magnet synchronous generator having open-end stator windings. Although the proposed circuit is very similar to the dual inverter topology [4][5], the single-phase load is connected between the positive dc terminals of the two three-phase bridge converters. The converter requires no ac inductor to suppress the switching ripples, and has the capability of boosting up the output load voltage higher than the input generator voltage.

This paper proposes a new control method to suppress the harmonic component in the output load voltage of the inductor-less power converter. The new control method introduces the $\alpha\beta0$ transformation to achieve a stable current feedback and feedforward compensation for the zero-sequence harmonic component in the EMF. The viability of the proposed method has been verified by theoretical analysis as well as experiments using a 2-kW prototype. As a result, a high-frequency ringings and 1.25 kHz harmonic voltage are suppressed, and the voltage THD has been also reduced from 8.4% to 3.7%.

II. INDUCTOR-LESS THREE-PHASE TO SINGLE-PHASE BOOST CONVERTER

A. Main circuit

Fig. 1 shows the circuit configuration of the proposed inductor-less three-phase to single-phase power converter. The proposed circuit consists of two three-phase bridge converters and a multi-pole permanent magnet synchronous generator with open-end stator windings. Each winding ends are connected to the ac terminals of the three-phase bridge converters, and two filter capacitors are connected to the dc side of the three-phase bridge converters. The negative dc terminals of the two three-phase bridge converters are commonly connected, and a single-phase load or the utility grid is connected between the positive dc terminals. This circuit topology makes it possible to reduce the size, weight, and initial cost of the

978-1-4799-2706-7/14 $31.00 © 2014 IEEE   1006

The 2014 International Power Electronics Conference

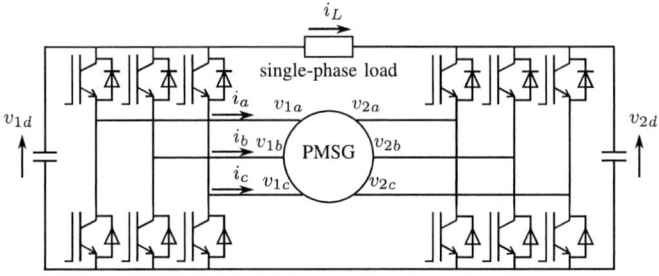

Fig. 1. System configuration of the proposed inductor-less three-phase to single-phase boost converter.

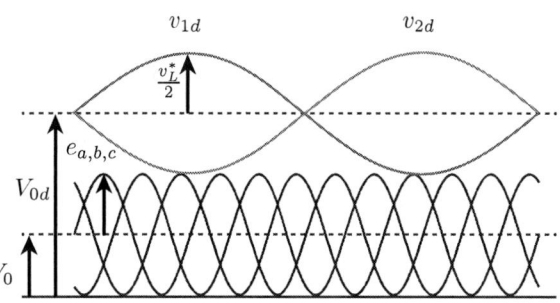

Fig. 2. Relationship of the generator EMFs and the dc capacitor voltages.

power converter because neither inductor nor transformer is employed.

B. Operating principle

Fig. 2 shows the relationship between the three-phase electromotive force induced by the synchronous generator and the dc capacitor voltages across the two filter capacitors. The filter capacitor voltages v_{1d} and v_{2d} have to include a dc offset voltage V_{0d} and a 50-Hz sinusoidal voltage. The sinusoidal component in v_{1d} should have an opposite phase angle to that in v_{2d}, because the voltage difference between the two filter capacitor is applied across the single-phase load. Then, the output voltage for the single-phase load is given by

$$v_L = v_{1d} - v_{2d}. \tag{1}$$

Moreover, the three-phase ac terminal voltages of each three-phase bridge converters should be positive and lower than the corresponding filter capacitor voltage to control the ac current flowing through the synchronous generator. Therefore, requirements of the offset voltages V_{0d} and V_0 are given by

$$V_0 > \frac{\sqrt{2}}{2} E, \tag{2}$$

$$V_{0d} = \frac{v_{1d} + v_{2d}}{2} > V_0 + \frac{\sqrt{2}}{2}(E + V_L). \tag{3}$$

In case of $E = 100$ V and $V_L = 100$ V, the minimum offset voltage is about $V_{0d} = 212$ V, and then the maximum filter capacitor voltage is about 283 V. Thus, it is possible to use low-cost 600-V IGBT modules or intelligent power modules (IPMs) as the three-phase bridge converters.

C. Control method

Fig. 3 shows the control block diagram for the inductor-less three-phase to single-phase power converter. The converter controls the two filter capacitor voltages by manipulating the positive- and zero-sequence current flowing through the generator. The generator power is given by

$$p_G = 3EI, \tag{4}$$

where I is the positive sequence current. The generator power is stored in the two filter capacitors and consumed in the single-phase load. Thus, to maintain the capacitor voltages, the

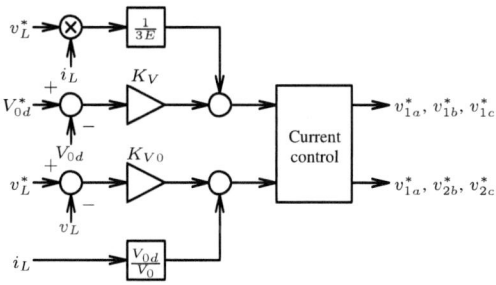

Fig. 3. Control block diagram.

reference of the positive sequence current should be calculated as

$$I^* = \frac{p_G}{3E} = \frac{v_L^* i_L}{3E} + K_V(V_{0d}^* - V_{0d}), \tag{5}$$

where v_L^* is the load voltage reference, i_L is the load current, K_V is a feedback gain for the capacitor voltage control, and V_{0d}^* is the reference of V_{0d}.

On the other hand, the zero-sequence current i_0 delivers an amount of power from one filter capacitor to the other. The delivered power is given by

$$p_0 = 3V_0 i_0. \tag{6}$$

The delivered power have to cancel circulating power caused by the load current and to control the load voltage. Thus, the reference of i_0^* is calculated by

$$i_0^* = \frac{p_0}{3V_0} = \frac{V_{0d} i_L}{3V_0} + K_{V0}(v_L^* - v_L), \tag{7}$$

where K_{V0} is the control gain of the load voltage feedback. The generator current feedback control manipulates the six voltage reference for the two three-phase bridge converter, v_{1a}^* through v_{2c}^* to regulate the three-phase generator currents i_a, i_b, and i_c.

Fig. 4 shows the block diagram of the current controller for the inductor-less three-phase to single-phase power converter. The three-phase current references are calculated from the

978-1-4799-2706-7/14 $31.00 © 2014 IEEE

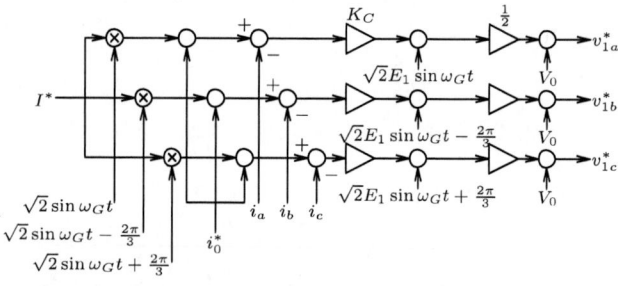

Fig. 4. Block diagram of the current controller for the inductor-less three-phase to single-phase power converter.

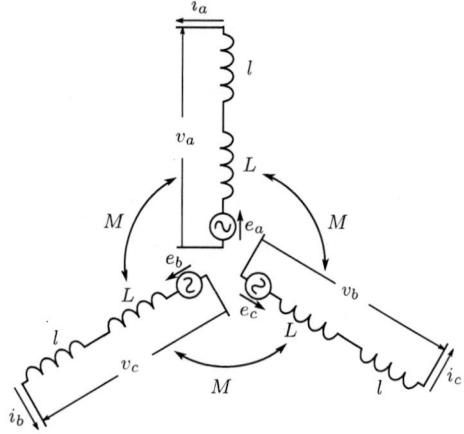

Fig. 5. Equivalent circuit of generator.

positive- and zero-sequence current reference, I^* and i_0^* as

$$i_a^* = i_0^* + \sqrt{2}I^* \sin(\omega t) \qquad (8)$$

$$i_b^* = i_0^* + \sqrt{2}I^* \sin\left(\omega t - \frac{2\pi}{3}\right) \qquad (9)$$

$$i_c^* = i_0^* + \sqrt{2}I^* \sin\left(\omega t + \frac{2\pi}{3}\right) \qquad (10)$$

In the current feedback control, the offset voltage V_0 and generator EMF are added to the voltage reference as a feed forward compensation. Assuming the two three-phase bridge converter share the generator EMF, voltage references for the two three-phase bridge converters are given by

$$\begin{cases} v_{1a}^* = V_0 - \frac{e_a}{2} + \frac{K_C}{2}(i_a^* - i_a) \\ v_{1b}^* = V_0 - \frac{e_b}{2} + \frac{K_C}{2}(i_b^* - i_b) \\ v_{1c}^* = V_0 - \frac{e_c}{2} + \frac{K_C}{2}(i_c^* - i_c) \end{cases} \qquad (11)$$

$$\begin{cases} v_{2a}^* = V_0 + \frac{e_a}{2} - \frac{K_C}{2}(i_a^* - i_a) \\ v_{2b}^* = V_0 + \frac{e_b}{2} - \frac{K_C}{2}(i_b^* - i_b) \\ v_{2c}^* = V_0 + \frac{e_c}{2} - \frac{K_C}{2}(i_c^* - i_c) \end{cases} \qquad (12)$$

where K_C is the feedback gain for the generator current.

III. ANALYSIS OF CONTROL CHARACTERISTIC

Fig. 5 shows an equivalent circuit of a surface-mounted permanent-magnet synchronous generator (PMSG). The volt-

age sources e_a, e_b and e_c represent the electromotive force (EMF) induced by the rotating magnetic flux of the permanent magnets mounted on the rotor surface. The inductance L is the self inductance of the stator windings, and ℓ is the leakage inductance. And then, mutual inductance exists between the three-phase stator windings, because they are placed with an electrical angle of $2\pi/3$ rad. Since the coupling coefficient is $k = \cos\frac{2}{3}$, the mutual inductance is given by

$$M = kL = -\frac{L}{2}. \qquad (13)$$

Therefore, the terminal voltage of the PMSG is expressed by using the inductance matrix as

$$\begin{bmatrix} v_a \\ v_b \\ v_c \end{bmatrix} = \begin{bmatrix} v_{1a} \\ v_{1b} \\ v_{1c} \end{bmatrix} - \begin{bmatrix} v_{2a} \\ v_{2b} \\ v_{2c} \end{bmatrix}$$

$$= \begin{bmatrix} e_a \\ e_b \\ e_c \end{bmatrix} - \frac{d}{dt}[L]\begin{bmatrix} i_a \\ i_b \\ i_c \end{bmatrix}, \qquad (14)$$

where $[L]$ is the inductance matrix represented by

$$[L] = \begin{bmatrix} L+\ell & -\frac{L}{2} & -\frac{L}{2} \\ -\frac{L}{2} & L+\ell & -\frac{L}{2} \\ -\frac{L}{2} & -\frac{L}{2} & L+\ell \end{bmatrix}. \qquad (15)$$

A. Harmonic components in the generator EMF

Fig. 6 shows the measured generator EMF at the rated speed of 3600 rpm. The measured generator EMF is not a sinusoidal waveform but a trapezoidal wave shape because it includes harmonic component induced by the magnetic flux density profile. The harmonic components included in the EMF are listed in Table I. Odd order harmonics do not appear in the EMF because the magnet arrangement is symmetrical. The third and ninth order harmonics are included and appear at the terminal voltage unlike general three-phase generators, because this generator has the open-end windings.

The three-phase EMF can be expressed by using the harmonic components E_n in Table I, as given by

$$e_a(t) = \sum \sqrt{2}E_n \sin n\omega t \qquad (16)$$

$$e_b(t) = \sum \sqrt{2}E_n \sin\left(n\omega t - \frac{2\pi n}{3}\right) \qquad (17)$$

$$e_c(t) = \sum \sqrt{2}E_n \sin\left(n\omega t + \frac{2\pi n}{3}\right), \qquad (18)$$

where n is the harmonic order.

The line-to-line EMF shown in Fig. 6 is represented by

$$e_{ab}(t) = e_a(t) - e_b(t). \qquad (19)$$

Then, the a-phase harmonics cancel the b-phase harmonics out at the third and ninth harmonic frequency. Thus, the line-to-line EMF does not contain third and ninth order harmonics, and the dominant harmonic EMFs are fifth and seventh components.

The 2014 International Power Electronics Conference

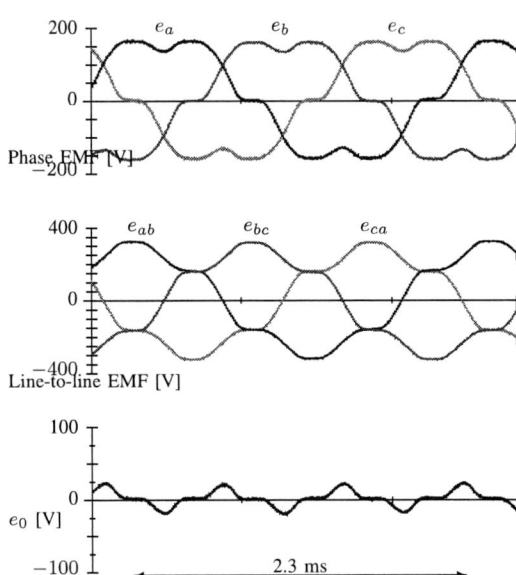

Fig. 6. Generator EMF.

TABLE I. GENERATOR EMF

symbol	harmonic order	frequency	rms value
E_1	first	420 Hz	119 V
E_3	third	1260 Hz	9.30 V
E_5	fifth	2100 Hz	-16.2 V
E_7	seventh	2940 Hz	-5.00 V
E_9	ninth	3780 Hz	-4.03 V

On the other hand, the zero-sequence EMF is given by

$$e_0(t) = \frac{e_a(t) + e_b(t) + e_c(t)}{3}. \tag{20}$$

The fundamental and the fifth and seventh harmonic components are eliminated, and the third and ninth harmonic components remains in the zero-sequence EMF. The zero-sequence components would affect to the current feedback control as a disturbance.

B. stability problem

Application of the three-phase to two-phase coordinate, or $\alpha\beta0$ coordinate, to the generator terminal voltage yields

$$\begin{bmatrix} v_0 \\ v_\alpha \\ v_\beta \end{bmatrix} = [C] \begin{bmatrix} v_a \\ v_b \\ v_c \end{bmatrix}, \tag{21}$$

where v_α and v_β are the transformed two-phase components and v_0 is the zero-sequence component in the terminal voltage, and $[C]$ is the transform matrix defined by

$$[C] = \sqrt{\frac{2}{3}} \begin{bmatrix} \frac{1}{\sqrt{2}} & \frac{1}{\sqrt{2}} & \frac{1}{\sqrt{2}} \\ 1 & -\frac{1}{2} & -\frac{1}{2} \\ 0 & \frac{\sqrt{3}}{2} & -\frac{\sqrt{3}}{2} \end{bmatrix}. \tag{22}$$

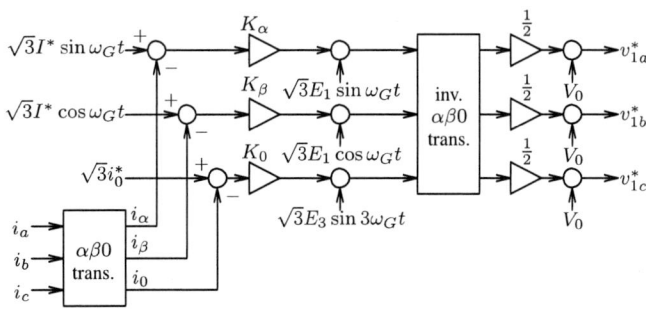

Fig. 7. Current control block of one of the two three-phase bridge converters.

Multiplying the matrix in (22) to (14) derives the terminal voltage/current equation on the $\alpha\beta0$ coordinate as

$$\begin{bmatrix} v_0 \\ v_\alpha \\ v_\beta \end{bmatrix} = \begin{bmatrix} e_0 \\ e_\alpha \\ e_\beta \end{bmatrix}$$
$$- \frac{d}{dt} \begin{bmatrix} \ell & 0 & 0 \\ 0 & L_S & 0 \\ 0 & 0 & L_S \end{bmatrix} \begin{bmatrix} i_0 \\ i_\alpha \\ i_\beta \end{bmatrix}, \tag{23}$$

where L_S is the synchronous inductance, that is $L_S = \frac{3L}{2} + \ell$. The above equation implies that the inductance for zero-sequence current is just the leakage inductance ℓ, which is smaller than the synchronous inductance L_S. A conventional three-phase current control method uses the same gain for three-phase current i_a, i_b and i_c, and the feedback gain is usually designed based on the synchronous inductance L_S. In this case, the conventional current control would cause a stability problem because the feedback gain for i_0 may exceed the critical gain.

IV. A NEW CURRENT CONTROL METHOD

Fig. 7 shows the block diagram of the proposed current control method. In the propose method, the $\alpha\beta0$ transformation is applied to decouple the zero-sequence current i_0 from the positive sequence currents included in i_α and i_β. As a result, this method allows to set a different feedback gain for i_0, and enables to improve the stability in the current feedback not only for i_0 but also for i_α and i_β.

The estimated EMF is added to the voltage reference as a feedforward compensation. The third-order zero-sequence EMF E_3 is injected as well as the fundamental EMF E_1 in Fig. 7. This compensation makes it possible to reduce the harmonic current caused by third-order zero-sequence EMF, resulting in eliminating the harmonic components in the output load voltage. The final manuscript will include the details of the stability of the current control and effect of the feedforward compensation.

V. EXPERIMENTAL RESULTS

Figs. 8-13 show experimental results of the 2-kW prototype of the inductor-less three-phase to single-phase power

TABLE II. CIRCUIT PARAMETERS FOR EXPERIMENTAL PROTOTYPE

rated power		3 kW
generator voltage	E	210 V (L-L)
generator frequency	f_G	420 Hz
synchronous inductance	L	1 mH(20%)
leakage inductance	l	0.55 mH(20%)
internal resistance	r	0.44 Ω
dc capacitor	C	100 μ F
switching frequency	f_{SW}	15 kHz
load voltage	v_{out}	100 V, 50Hz

converter. The circuit parameters of the prototype is shown in Table II. A 14-poles three-phase permanent-magnet synchronous generator was used in the following experiments. The generator was driven by an inverter-fed induction motor at the rated speed of 3600 rpm, and then, the generator frequency was 420 Hz. Two six-in-one IGBT power module (CM75TU-12F: Mitsubishi) was used as switching devices for the three-phase bridge converters. The switching frequency of the converter was set to be $f_{sw} = 15$ kHz.

Fig. 8 shows experimental waveforms of the three-phase current feedback control under a 2-kW resistive load condition. The harmonic spectrum of the load voltage is shown in Fig. 9. The filter capacitor voltages v_{1d} and v_{2d} included an ac component and a dc component. The ac components in v_{2d} was controlled to have an opposite phase angle against v_{1d}. Thus, the voltage difference between v_{1d} and v_{2d}, that is, the output load voltage v_L, had a sinusoidal wave shape. The rms value of v_L was 100 V, because the ac components in both v_{1d} and v_{2d} were regulated at 50 V. However, the output load voltage v_L included a high-frequency voltage ripples, whose frequency was about 3.5 kHz. The generator current i_a, i_b and i_c included a large amount of high-frequency current ripples which was caused by the instability of the three-phase current feedback control. In this case, the current feedback control gain $K_C = 10$ V/A was too high for the zero-phase current, which exhibits the same impact with a gain of $K_0 = 10$ V/A in the $\alpha\beta0$ current control, because the zero-sequence inductance was smaller than the synchronous inductance. A lot of high-frequency harmonic components appear around 3.5 kHz. These high-frequency components were caused by a stability problem in the current feedback control. Therefore, the voltage THD was 8.38%.

Fig. 10 is experimental waveforms when the $\alpha\beta0$ current control method was applied. The harmonic spectrum of the load voltage is shown in Fig. 11. In this case, the feedback gain for the zero-sequence component was reduced to be $K_0 = 6$ V/A, which was almost equal to the critical damping gain. Therefore, high-frequency current ripples were eliminated from the generator currents i_a, i_b and i_c, and then, high-frequency components were suppressed in the capacitor voltages v_{1d} and v_{2d} as well as the output load voltage v_L. However, a harmonic component still remains in the load voltage v_L, whose frequency was about 1260 Hz. This harmonic component was caused by the third-order zero-sequence component included in the generator EMF. Thus, the voltage THD was reduced to 5.9%. It would be difficult for voltage

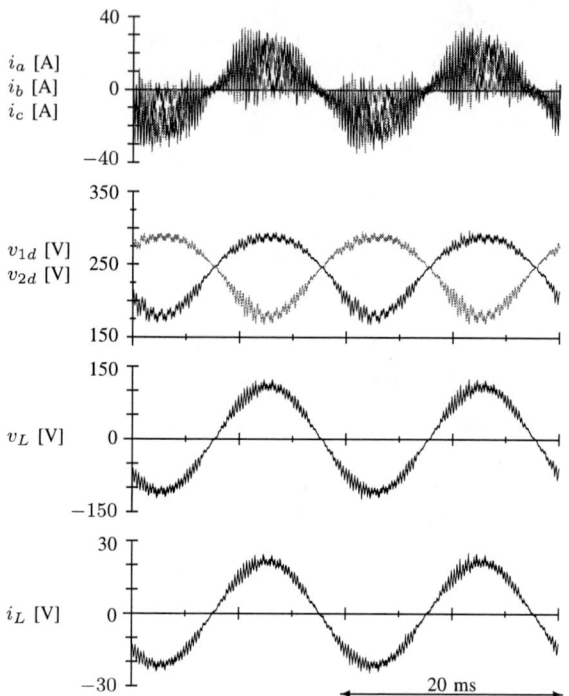

Fig. 8. Voltage and Current waveforms under a 2 kW load condition when a conventional three-phase current control method is applied where the feedback gain was $K_C = 10$ V/A.

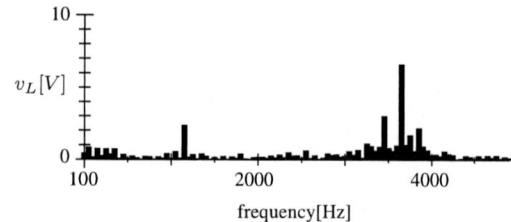

Fig. 9. Voltage spectrum under a 2 kW load condition when a conventional three-phase current control method is applied where the feedback gain was $K_C = 10$ V/A.

and/or current feedback to suppress the harmonic component, because the harmonic frequency is too high and the high feedback gain would cause a instability problem.

Figs. 12 and 13 show experimental waveform and harmonic spectrum when both both $\alpha\beta0$ current control. The 1260-Hz harmonic component was eliminated from the generator currents i_a, i_b and i_c, and thus, the harmonic voltages in the capacitor voltages v_{1d} and v_{2d} were reduced. Then the output load voltage v_L included almost no 1260-Hz harmonic component, although a small amount of harmonic component was remaining in the capacitor voltages v_{1d} and v_{2d}. The remaining harmonic voltages were almost in the same phase angle which can be considered as a common mode voltage, and thus, they did not appear in the output load voltage v_L. As a results, the feedforward compensation makes it possible to eliminate the harmonic component not only from the generator

978-1-4799-2706-7/14 $31.00 © 2014 IEEE

The 2014 International Power Electronics Conference

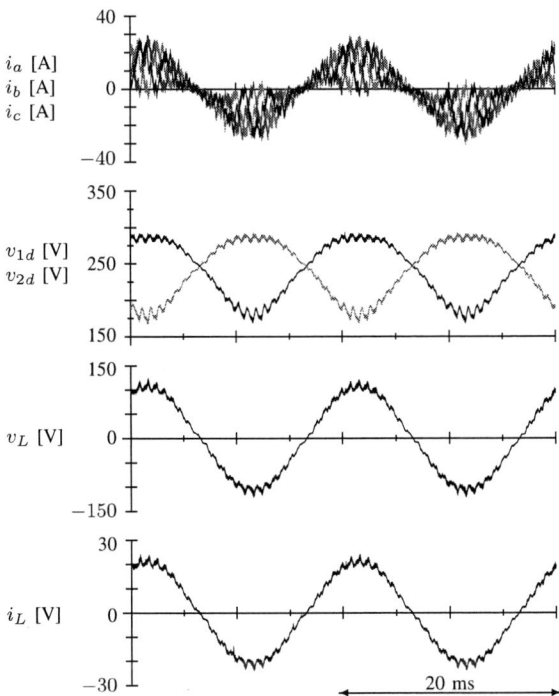

Fig. 10. Voltage and current waveforms under a 2 kW load condition when the $\alpha\beta0$ current control method is applied where $K_\alpha = 10$ V/A, $K_\beta = 10$ V/A, and $K_0 = 6$ V/A.

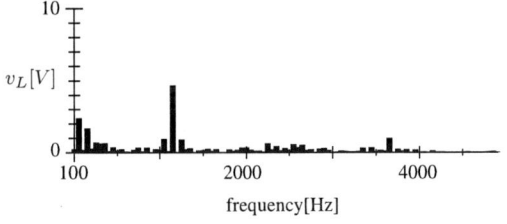

Fig. 11. Voltage spectrum under a 2 kW load condition when the $\alpha\beta0$ current control method is applied where $K_\alpha = 10$ V/A, $K_\beta = 10$ V/A, and $K_0 = 6$ V/A.

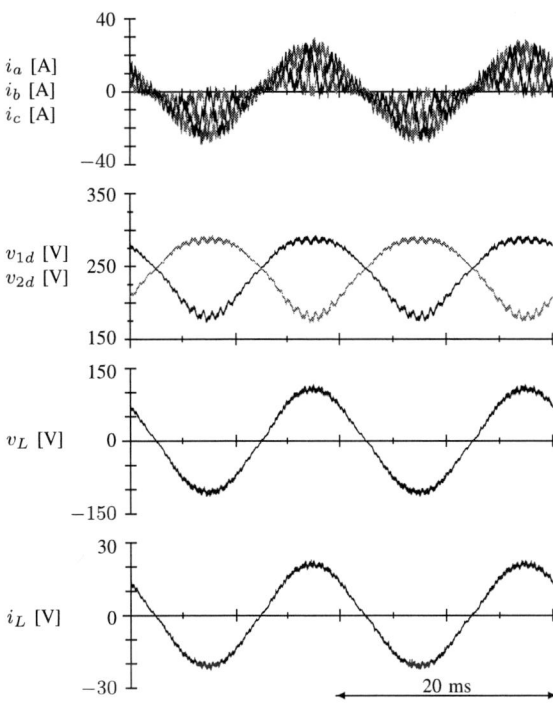

Fig. 12. Current and voltage waveforms under a 2 kW load condition when both $\alpha\beta0$ current control and feedfoward compensation methods are applied where $K_\alpha = 10$ V/A, $K_\beta = 10$ V/A, and $K_0 = 6$ V/A.

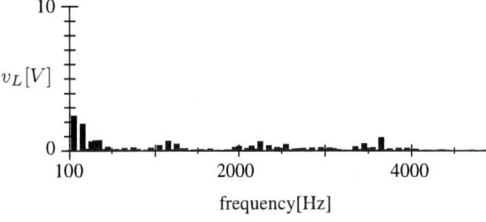

Fig. 13. Voltage spectrum under a 2 kW load condition when both $\alpha\beta0$ current control and feedfoward compensation methods are applied where $K_\alpha = 10$ V/A, $K_\beta = 10$ V/A, and $K_0 = 6$ V/A.

currents i_a, i_b and i_c but also from the output load voltage v_L. The voltage THD is 3.66%.

VI. CONCLUSION

This paper has discussed the output load voltage ripples in an inductor-less three-phase to single-phase converter. Analysis in this paper has clarified that output voltage ripples are caused by the stability problems in the current feedback control due to a low zero-sequence impedance. This paper proposes a new current control method based on $\alpha\beta0$ transformation to solve the problems. In addition, a feed forward harmonic compensation is also introduced to eliminates the harmonic distortion caused by the harmonic voltage in the generator EMF. A 2-kW prototype of the inductor-less converter demonstrates the viability of the proposed control method. As a result,it is confirmed that the new control method makes it possible to

reduce both high-frequency ringings around 3.5 kHz caused by the instability and 1260 kHz harmonic voltage induced by the harmonics in the EMF.

REFERENCES

[1] Y. Miura, S. Horie, S. Kokubo, T. Amano, T. Ise, T. Momose, Y. Sato, "Application of three-phase to single-phase matrix converter to gas engine co-generation system," *in Proc. of IEEE Energy Conversion Congress and Exposition,* pp. 3290-3297, 2009.

[2] M. Saito, T. Takeshita, N. Matsui, "A single to three phase matrix converter with a power decoupling capability," *in Proc. of IEEE Power Electronics Specialists Conference,* vol. 3, pp. 2400-2405, 2004.

[3] H. Fujita, "An inductor-less three-phase to single-phase boost converter for multi-pole permanent magnet synchronous generators," *in Proc. of International Future Energy Electronics Conference,* pp. 598-604, 2013.

[4] H. Stemmler, P. Guggenbach, "Configurations of high-power voltage source inverter drives," *in Proc. of European Conference on Power Electronics and Applications,* vol. 5, pp. 7–14, 1993.

[5] V. T. Somasekhar, K. Gopakumar, M. R. Baiju, "Dual two-level inverter scheme for an open-end winding induction motor drive with a single dc power supply and improved dc bus utilisation," *IEE Proceedings – Electric Power Applications,* vol. 151, no. 2, pp. 230-238, 2004.

[6] Capaldi, P. ; Ist. Motori, Naples, Italy ; Del Pizzo, A. ; Rizzo, R. ; Spina, I, "Torque-oscillations damping in micro-cogeneration units with high brake mean pressure engines and PM-brushless generators" *Energy Conference and Exhibition (ENERGYCON), 2012 IEEE International ,* pp. 121 - 126 , 9-12 Sept. 2012 .

[7] Joon-Hwan Lee ; Hyundai Mobis Co., Ltd., Seoul, South Korea ; Seung-Hwan Lee ; Seung-Ki Sul, "Variable-Speed Engine Generator With Supercapacitor: Isolated Power Generation System and Fuel Efficiency" *Industry Applications, IEEE Transactions on* volume. 45 pp. 2130 - 2135 , Nov.-dec. 2009 .

[8] Lohdefink, P. ; Inst. ELSYS, Georg-Simon-Ohm Hochschule Nurnberg, Nu?rnberg, Germany ; Grillenberger, M. ; Dietz, A. ; Groger, A. more authors "Sensorless vector control of a permanent magnet synchronous generator for micro hydro power " *Education and Research Conference (EDERC), 2012 5th European DSP* pp. 252 - 256 , 13-14 Sept. 2012 .

[9] Goto, H. ; Tohoku Univ., Sendai, Japan ; Hai-Jiao Guo ; Ichinokura, O, "A micro wind power generation system using permanent magnet reluctance generator " *Power Electronics and Applications, 2009. EPE '09. 13th European Conference on* pp. 1 - 8 ,8-10 Sept. 2009 .

[10] Leuchter, J. ; Univ. of Defence Kounicova, Brno ; Refucha, V. ; Krupka, Z. ; Bauer, P, "Dynamic Behavior of Mobile Generator Set with Variable Speed and Diesel Engine" *Power Electronics Specialists Conference, 2007. PESC 2007. IEEE* pp. 2287 - 2293 ,17-21 June 2007 .

The 2014 International Power Electronics Conference

A New SVM Method to Reduce Common-mode Voltage in Direct Matrix Converter

Huu-Nhan Nguyen[*], Hong-Hee Lee[**]

School of Electrical Engineering, University of Ulsan, South Korea
[*]huunhan.vn@gmail.com, [**]hhlee@ulsan.ac.kr

Abstract— **This paper focuses on the common-mode voltage (CMV) reduction in direct matrix converter (MC) using direct space vector modulation (DSVM) (SVM) method. Even though a previous DSVM method can reduce the CMV peak value to 42% compared to the traditional DSVM, it has some drawbacks such as high switching loss, high RMS value of CMV, etc. This paper proposes a new DSVM that can mitigate the peak value of the CMV to 42% compared to the traditional DSVM method with a better performance, lower RMS value of CMV, leakage current and reduced switching losses. The proposed method is easy to carry out via software without any additional hardware. Simulation results are shown to verify the correctness and effectiveness of the proposed method.**

Keywords— **Matrix converter, AC-AC power converter, direct space vector modulation method, common-mode voltage reduction.**

I. INTRODUCTION

In recent years, the MC has received considerable interest because it provides several advantages such as the lack of dc-link capacitor, sinusoidal input and output current waveforms, the bi-directional power flow capability, four quadrants operation and controllable input power factor [1]–[3]. The MCs are categorized into direct MC (DMC) and indirect MC (IMC). A DMC consists of nine bi-directional switches that allow each output phase to be connected to any input phase. One passive LC input filter is required to eliminate the harmonic components of input currents. In order to damp the oscillations in input currents, it is necessary to incorporate a damping resistor connected in parallel with filter inductor. A common three-phase to three-phase DMC configuration is shown in Fig. 1.

CMV and its sharp edges are responsible for a main source of the shaft voltage, bearing current and leakage current which are main cause of the motor winding failures and the motor bearing damages. In addition, leakage current creates noise problems that effect on other equipments nearby the power converter [4]. Therefore, the techniques to reduce CMV in power converter are very important for lifetime of motor bearing as well as motor winding.

Although several methods to reduce CMV have been proposed, most of these methods investigated the IMC [5], [6] or analyzed the DMC based on the indirect SVM [7], [8]. A method to reduce the CMV in DMC based on DSVM method was proposed by using the suitable couple of active space vectors instead of zero space vectors in [9]. The maximum voltage transfer ratio in this method is unaffected and remains at 0.866. However, this method has some drawbacks

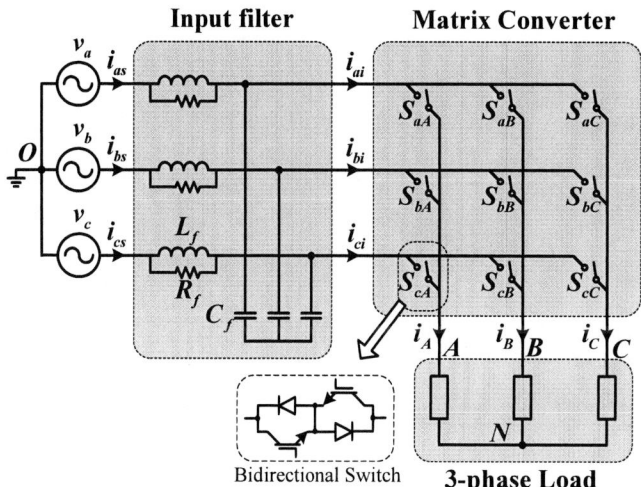

Fig. 1. A common three-phase to three-phase DMC configuration.

such as the high total harmonic distortion in input currents, high leakage current and switching losses. In order to overcome the drawbacks of this method, a new DSVM method to reduce common-mode voltage for DMC is proposed by using three couples of nearest active space vectors to generate the reference output voltage vector. Simulation results are given to compare the performance of the conventional CMV reduction and the proposed DSVM method in terms of peak and RMS value of CMV, input current performance and switching losses.

II. TRADITIONAL DSVM METHOD WITHOUT CMV REDUCTION

In balanced operation, the instantaneous input voltages are as follows:

$$v_i = \begin{bmatrix} v_a \\ v_b \\ v_c \end{bmatrix} = V_i \begin{bmatrix} \cos(\omega_i t) \\ \cos(\omega_i t - 2\pi/3) \\ \cos(\omega_i t + 2\pi/3) \end{bmatrix} . \qquad (1)$$

where V_i and ω_i are the amplitude and the angular frequency of the input voltage, respectively.

With the relevant symbols shown in Fig. 1, the space vector of input and output voltages, input and output currents can be defined respectively by the following equations:

$$\vec{v}_i = 2(v_a + v_b e^{j2\pi/3} + v_c e^{j4\pi/3})/3 = V_i e^{j\alpha_i} \qquad (2)$$

978-1-4799-2706-7/14 $31.00 © 2014 IEEE

The 2014 International Power Electronics Conference

TABLE I.
POSSIBLE SWITCHING CONFIGURATIONS IN DMC

	Switching Configurations			Output Voltage		Input Current	
	No	A	B	C	V_o	α_o	I_i	β_i
Group I	+1	a	b	b	$2v_{ab}/3$	0	$2i_A/\sqrt{3}$	$-\pi/6$
	-1	b	a	a	$-2v_{ab}/3$	0	$-2i_A/\sqrt{3}$	$-\pi/6$
	+2	b	c	c	$2v_{bc}/3$	0	$2i_A/\sqrt{3}$	$\pi/2$
	-2	c	b	b	$-2v_{bc}/3$	0	$-2i_A/\sqrt{3}$	$\pi/2$
	+3	c	a	a	$2v_{ca}/3$	0	$2i_A/\sqrt{3}$	$7\pi/6$
	-3	a	c	c	$-2v_{ca}/3$	0	$-2i_A/\sqrt{3}$	$7\pi/6$
	+4	b	a	b	$2v_{ab}/3$	$2\pi/3$	$2i_B/\sqrt{3}$	$-\pi/6$
	-4	a	b	a	$-2v_{ab}/3$	$2\pi/3$	$-2i_B/\sqrt{3}$	$-\pi/6$
	+5	c	b	c	$2v_{bc}/3$	$2\pi/3$	$2i_B/\sqrt{3}$	$\pi/2$
	-5	b	c	b	$-2v_{bc}/3$	$2\pi/3$	$-2i_B/\sqrt{3}$	$\pi/2$
	+6	a	c	a	$2v_{ca}/3$	$2\pi/3$	$2i_B/\sqrt{3}$	$7\pi/6$
	-6	c	a	c	$-2v_{ca}/3$	$2\pi/3$	$-2i_B/\sqrt{3}$	$7\pi/6$
	+7	b	b	a	$2v_{ab}/3$	$4\pi/3$	$2i_C/\sqrt{3}$	$-\pi/6$
	-7	a	a	b	$-2v_{ab}/3$	$4\pi/3$	$-2i_C/\sqrt{3}$	$-\pi/6$
	+8	c	c	b	$2v_{bc}/3$	$4\pi/3$	$2i_C/\sqrt{3}$	$\pi/2$
	-8	b	b	c	$-2v_{bc}/3$	$4\pi/3$	$-2i_C/\sqrt{3}$	$\pi/2$
	+9	a	a	c	$2v_{ca}/3$	$4\pi/3$	$2i_C/\sqrt{3}$	$7\pi/6$
	-9	c	c	a	$-2v_{ca}/3$	$4\pi/3$	$-2i_C/\sqrt{3}$	$7\pi/6$
Group II	0_a	a	a	a	0	x	0	x
	0_b	b	b	b	0	x	0	x
	0_c	c	c	c	0	x	0	x
Group III	r_1	a	b	c	x	x	x	x
	r_2	a	c	b	x	x	x	x
	r_3	c	a	b	x	x	x	x
	r_4	b	a	c	x	x	x	x
	r_5	b	c	a	x	x	x	x
	r_6	c	b	a	x	x	x	x

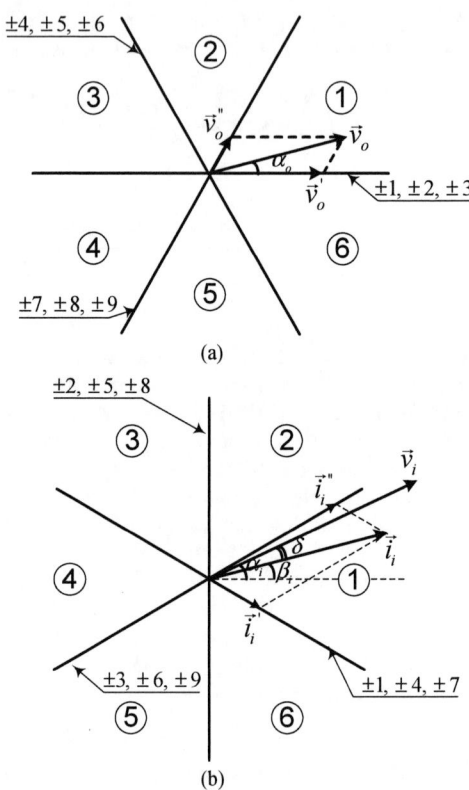

Fig. 2. (a) The location of output voltage vector.

(b) The location of input current vector.

$$\vec{v}_o = 2(v_A + v_B e^{j2\pi/3} + v_C e^{j4\pi/3})/3 = V_o e^{j\alpha_o} \quad (3)$$

$$\vec{i}_i = 2(i_{ai} + i_{bi} e^{j2\pi/3} + i_{ci} e^{j4\pi/3})/3 = I_i e^{j\beta_i} \quad (4)$$

$$\vec{i}_o = 2(i_A + i_B e^{j2\pi/3} + i_C e^{j4\pi/3})/3 = I_o e^{j\beta_o} . \quad (5)$$

Hereinafter any space vector is simply referred as vector.

There are only 27 configurations to satisfy two main rules in MC: input phases should never be short-circuited and the output currents should not be interrupted. The 27 possible switching configurations (SCs) are listed in Table I. These SCs are categorized into three groups:

1) Group I consists of active vectors.

2) Group II consists of zero vectors.

3) Group III consists of rotating vectors.

The traditional DSVM method for MC was presented in [2], [3]. In principle, the DSVM method is based on the selection of four suitable active vectors to synthesize the reference output voltage vector. The zero SCs are applied to complete the sampling period. For instance, both $\vec{v}_o$ and $\vec{i}_i$ are lying in

sector 1 as shown in Fig. 2(a) and (b)) ($k_v = 1$, $k_i = 1$, where k_v and k_i are the output voltage sector and the input current sector, respectively). The desired output voltage vector $\vec{v}_o$ can be synthesized from two vectors $\vec{v}_o'$ and $\vec{v}_o''$. The $\vec{v}_o'$ and $\vec{v}_o''$ components can be synthesized by using two of six possible SCs $\pm 1, \pm 2, \pm 3$ and $\pm 7, \pm 8, \pm 9$, respectively. Furthermore, SCs $\pm 1, \pm 4, \pm 7$ and $\pm 3, \pm 6, \pm 9$ are used to generate input current vector located in sector 1. Among these SCs, four common SCs with two higher voltage magnitudes are chosen. According to Fig. 3, when the input current vector $\vec{i}_i$ is in sector 1, two higher input line-to-line voltages are v_{ab} and v_{ac}.

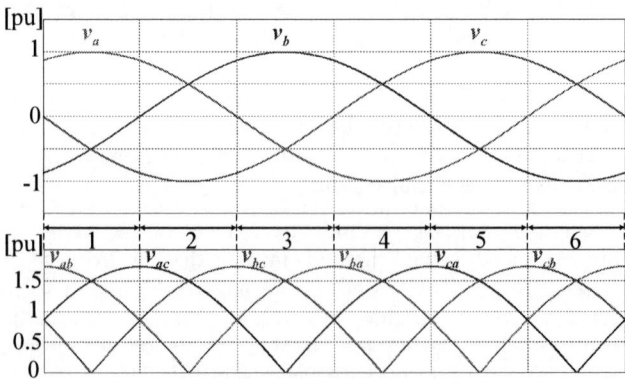

Fig. 3. Input phase and line-to-line voltage at main power supply.

978-1-4799-2706-7/14 $31.00 © 2014 IEEE

TABLE II.
SELECTED SCS FOR EACH COMBINATION OF OUTPUT VOLTAGE AND INPUT CURRENT SECTORS IN TRADITIONAL DSVM METHOD

		Input Current Vector Sector																						
		1				2				3				4				5				6			
Output Voltage Vector Sector	1	−7	+9	+1	−3	+9	−8	−3	+2	−8	+7	+2	−1	+7	−9	−1	+3	−9	+8	+3	−2	+8	−7	−2	+1
	2	+4	−6	−7	+9	−6	+5	+9	−8	+5	−4	−8	+7	−4	+6	+7	−9	+6	−5	−9	+8	−5	+4	+8	−7
	3	−1	+3	+4	−6	+3	−2	−6	+5	−2	+1	+5	−4	+1	−3	−4	+6	−3	+2	+6	−5	+2	−1	−5	+4
	4	+7	−9	−1	+3	−9	+8	+3	−2	+8	−7	−2	+1	−7	+9	+1	−3	+9	−8	−3	+2	−8	+7	+2	−1
	5	−4	+6	+7	−9	+6	−5	−9	+8	−5	+4	+8	−7	+4	−6	−7	+9	−6	+5	+9	−8	+5	−4	−8	+7
	6	+1	−3	−4	+6	−3	+2	+6	−5	+2	−1	−5	+4	−1	+3	+4	−6	+3	−2	−6	+5	−2	+1	+5	−4
Duty cycles		d_1	d_2	d_3	d_4	d_1	d_2	d_3	d_4	d_1	d_2	d_3	d_4	d_1	d_2	d_3	d_4	d_1	d_2	d_3	d_4	d_1	d_2	d_3	d_4

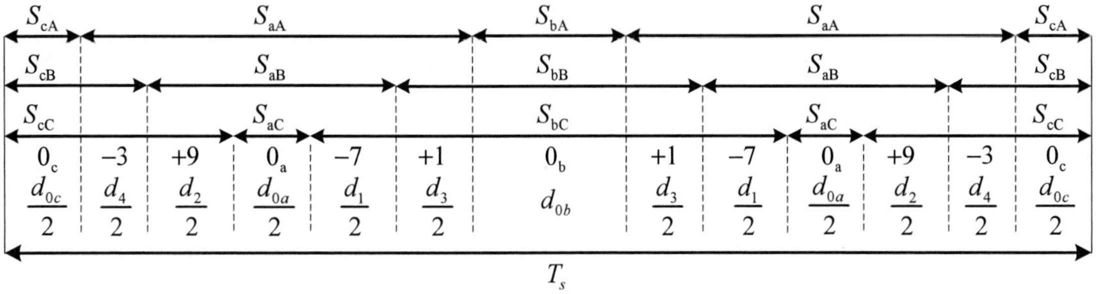

Fig. 4. Switching pattern of the traditional DSVM method.

From these analyses, four SCs $+1, -3, -7, +9$ are selected for the case of $k_v = 1$, $k_i = 1$.

Following the same above stated principle, the selected SCs for all combinations of output voltage and input current sectors are listed in Table II.

The duty cycles of SCs -7 (d_1) and $+9$ (d_2) in case of $k_v = 1$, $k_i = 1$ should satisfy the following relationship:

$$\frac{2}{3} v_{ab} d_1 + \frac{2}{3} v_{ac} d_2 = v_o^{''} = \frac{2}{\sqrt{3}} V_o \sin \alpha_o . \tag{6}$$

Since magnitude of input current is independent of the phase angle of output current, so:

$$\frac{d_1}{\sin(\pi/6 - \beta_i)} = \frac{d_2}{\sin(\pi/6 + \beta_i)} . \tag{7}$$

Solving equations (6) and (7), the duty cycles for SCs -7 and $+9$ are respectively as follows:

$$d_1 = \frac{2q}{\sqrt{3}} \frac{\sin \tilde{\alpha}_o \sin(\pi/6 - \tilde{\beta}_i)}{\cos \delta} \tag{8}$$

$$d_2 = \frac{2q}{\sqrt{3}} \frac{\sin \tilde{\alpha}_o \sin(\pi/6 + \tilde{\beta}_i)}{\cos \delta} \tag{9}$$

where $q = V_o / V_i$ is the voltage transfer ratio, δ is the delay angle of the input current vector compared to the input voltage vector and $\tilde{\alpha}_o$, $\tilde{\beta}_i$ are defined as follows:

$$\tilde{\alpha}_o = \alpha_o - (k_v - 1)\pi/3 \tag{10}$$

$$\tilde{\beta}_i = \beta_i - (k_i - 1)\pi/3 . \tag{11}$$

By similar analysis, the duty cycles for SCs $+1$ (d_3) and -3 (d_4) are as follows:

$$d_3 = \frac{2q}{\sqrt{3}} \frac{\sin(\pi/3 - \tilde{\alpha}_o) \sin(\pi/6 - \tilde{\beta}_i)}{\cos \delta} \tag{12}$$

$$d_4 = \frac{2q}{\sqrt{3}} \frac{\sin(\pi/3 - \tilde{\alpha}_o) \sin(\pi/6 + \tilde{\beta}_i)}{\cos \delta} . \tag{13}$$

Finally, the zero SCs are applied to complete the sampling period with the duty cycle:

$$d_0 = 1 - (d_1 + d_2 + d_3 + d_4)$$
$$= 1 - \frac{2q}{\sqrt{3}} \frac{\cos(\tilde{\alpha}_o - \pi/6) \cos \tilde{\beta}_i}{\cos \delta} . \tag{14}$$

The SCs in the traditional DSVM can be arranged symmetrically as shown in Fig. 4.

All duty cycles must satisfy the following constraints:

$$0 \le d_i \le 1; \quad i = 0, .., 4 . \tag{15}$$

This leads to the well-known restriction:

$$q \le \frac{\sqrt{3}}{2} \cos \delta . \tag{16}$$

III. DSVM METHODS WITH CMV REDUCTION

A. CMV Analysis

In a three-phase AC system, the CMV is defined as the voltage between the motor neutral point (N) and the power source ground (O), as shown in Fig. 1. The CMV v_{cm} can be expressed as:

978-1-4799-2706-7/14 $31.00 © 2014 IEEE

$$v_{cm} = v_{NO} . \qquad (17)$$

The CMV v_{cm} is derived as follows:

$$v_A - v_{cm} = Ri_A + L(di_A / dt) \qquad (18)$$
$$v_B - v_{cm} = Ri_B + L(di_B / dt) \qquad (19)$$
$$v_C - v_{cm} = Ri_C + L(di_C / dt) \qquad (20)$$

where v_A, v_B and v_C are the MC output phase voltage with respect to ground.

Assuming $i_A + i_B + i_C = 0$ and adding (18)–(20), we can obtain (21).

$$v_{cm} = \frac{v_A + v_B + v_C}{3} . \qquad (21)$$

Since v_A, v_B and v_C are obtained by 18 SCs available output voltage vectors, the CMV depends only on the switching states of DMC. From Table I, the possible CMVs are calculated as follows:

$$v_{cm} = \begin{cases} V_i / \sqrt{3}, & \text{group I} \\ V_i , & \text{group II} \\ 0 , & \text{group III (not used)} \end{cases} \qquad (22)$$

According to (22), group I always provides the lower CMV compared to group II. Therefore, CMV can be reduced by replacing the zero vectors in group II with the active vectors in group I.

B. Conventional DSVM Method to Reduce CMV for DMC

A DSVM method to reduce CMV for DMC was proposed in [9]. Instead of zero vectors, the authors used two active vectors which have the same magnitude but opposite direction as shown in Fig. 5. They do not change the synthesized reference output voltage and input current vectors. The

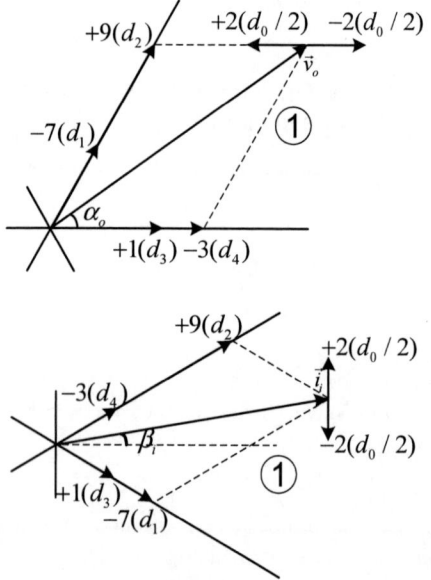

Fig. 5. The conventional DSVM method with CMV reduction using a couple of active vectors.

selections and duty cycles of four main active SCs in this method are totally same as those in the traditional DSVM method. Two extra active SCs to replace zero SC have the half of the duty cycle of zero SC, i.e., $d_0 / 2$. Therefore, the range of voltage transfer ratio is unaffected by the CMV reduction method in [9]. However, the use of auxiliary active vectors to replace zero vector can be a cause of high input current distortion, high switching loss in this method. In order to overcome these problems, a new SVM method is suggested in this paper.

C. Proposed DSVM Method to Reduce CMV for DMC

The proposed DSVM method is also based on the use of suitable active vectors instead of zero vectors. In this method, a group of three couples of active vectors is utilized to synthesize the reference output voltage vector as shown in Fig. 6. Six selected vectors are the closest to the reference output voltage vector. The definition of input current sectors is same as in the traditional DSVM. Output voltage sectors are redefined as shown in Fig. 6(a). Resuming the example related to the case of $k_v = 1$, $k_i = 1$, as shown in Fig. 3, the amplitudes of the selected vectors should be based on v_{ab} and v_{ac}. Combining with Table I, six selected SCs in this case are $-4, +6, +1, -3, -7, +9$. Table III summaries the selected SCs for all cases when output voltage and input current vectors locate

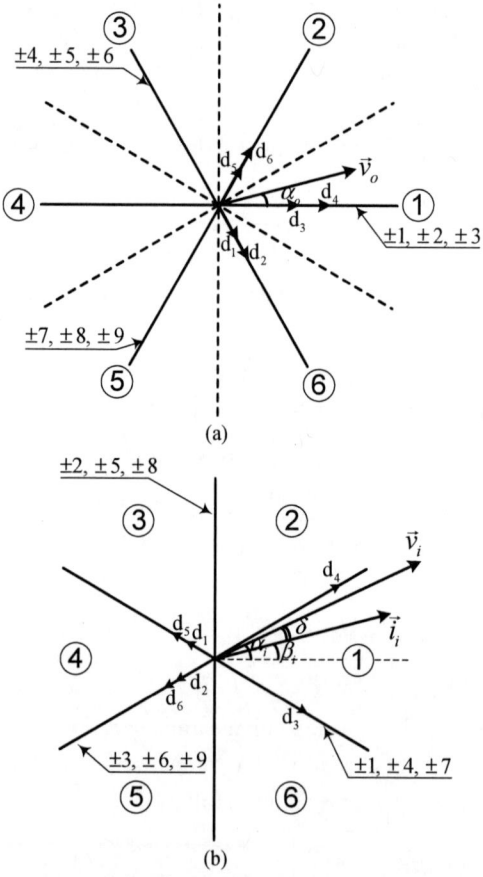

Fig. 6. The proposed DSVM method using three couples of active vectors
(a) Output voltage vector modulation principle.
(b) Input current vector modulation principle.

TABLE III. SWITCHING TABLE FOR THE PROPOSED DSVM METHOD TO REDUCE CMV

		Input Current Vector Sector																																	
		1						2						3						4						5						6					
Output Voltage Vector Sector	1	−4	+6	+1	−3	−7	+9	+6	−5	−3	+2	+9	−8	−5	+4	+2	−1	−8	+7	+4	−6	−1	+3	+7	−9	−6	+5	+3	−2	−9	+8	+5	−4	−2	+1	+8	−7
	2	+1	−3	−7	+9	+4	−6	−3	+2	+9	−8	−6	+5	+2	−1	−8	+7	+5	−4	−1	+3	+7	−9	−4	+6	+3	−2	−9	+8	+6	−5	−2	−1	+8	−7	−5	+4
	3	−7	+9	+4	−6	−1	+3	+9	−8	−6	+5	+3	−2	−8	+7	+5	−4	−2	+1	+7	−9	−4	+6	+1	−3	−9	+8	+6	−5	−3	+2	+8	−7	−5	+4	+2	−1
	4	+4	−6	−1	+3	+7	−9	−6	+5	+3	−2	−9	+8	+5	−4	−2	+1	+8	−7	−4	+6	+1	−3	−7	+9	+6	−5	−3	+2	+9	−8	−5	+4	+2	−1	−8	+7
	5	−1	+3	+7	−9	−4	+6	+3	−2	−9	+8	+6	−5	−2	−1	+8	−7	−5	+4	+1	−3	−7	+9	+4	−6	−3	+2	+9	−8	−6	+5	+2	−1	+8	−7	+5	−4
	6	+7	−9	−4	+6	+1	−3	−9	+8	+6	−5	−3	+2	+8	−7	−5	+4	+2	−1	−7	+9	+4	−6	−1	+3	+9	−8	−6	+5	+3	−2	−8	+7	+5	−4	−2	+1
Duty cycles		d_1	d_2	d_3	d_4	d_5	d_6	d_1	d_2	d_3	d_4	d_5	d_6	d_1	d_2	d_3	d_4	d_5	d_6	d_1	d_2	d_3	d_4	d_5	d_6	d_1	d_2	d_3	d_4	d_5	d_6	d_1	d_2	d_3	d_4	d_5	d_6

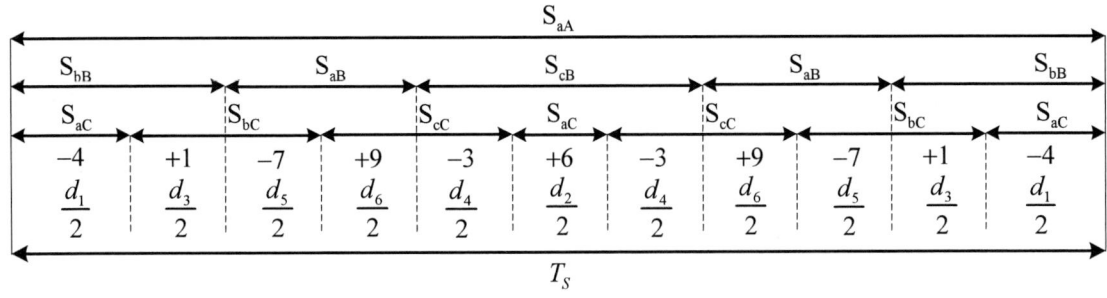

Fig. 7. Switching pattern of the proposed DSVM method to reduce CMV.

in any sector with the proposed DSVM method. For example, in case of $k_v = 1$, $k_i = 1$ (k_v and k_i are the output voltage sector and the input current sector, respectively), the duty cycles of SCs become −4 (d_1), +6 (d_2), +1 (d_3), −3 (d_4), −7 (d_5) and +9 (d_6) and the following six constraint equations should be satisfied:

$$d_1 + d_2 + d_3 + d_4 + d_5 + d_6 = 1 \tag{23}$$

$$\left(\frac{2}{3}v_{ab}d_1 + \frac{2}{3}v_{ac}d_2 \right)\frac{1}{2} + \left(\frac{2}{3}v_{ab}d_3 + \frac{2}{3}v_{ac}d_4 \right) +$$
$$+ \left(\frac{2}{3}v_{ab}d_5 + \frac{2}{3}v_{ac}d_6 \right)\frac{1}{2} = V_o \cos \alpha_o \tag{24}$$

$$\left(\frac{2}{3}v_{ab}d_1 + \frac{2}{3}v_{ac}d_2 \right)\left(-\frac{\sqrt{3}}{2} \right) + \left(\frac{2}{3}v_{ab}d_5 + \frac{2}{3}v_{ac}d_6 \right)\frac{\sqrt{3}}{2} = V_o \sin \alpha_o \tag{25}$$

$$d_1 \sin \left(\pi / 6 + \beta_i \right) = d_2 \sin \left(\pi / 6 - \beta_i \right) \tag{26}$$

$$d_3 \sin \left(\pi / 6 + \beta_i \right) = d_4 \sin \left(\pi / 6 - \beta_i \right) \tag{27}$$

$$d_5 \sin \left(\pi / 6 + \beta_i \right) = d_6 \sin \left(\pi / 6 - \beta_i \right) \tag{28}$$

The first equation is clearly related to the completion of the sampling period. The equations (24) and (25) are required to generate the reference output voltage vector, whereas (26), (27) and (28) are written so as to control the input power factor.

Using $\tilde{\alpha}_o$ and $\tilde{\beta}_i$ instead of α_o and β_i defined in (10) and (11), the general solution of the system of equations (23)–(28) is:

$$d_1 = \frac{\sin(\pi/6 - \tilde{\beta}_i)}{\cos \tilde{\beta}_i}\left(1 - q\frac{\cos \tilde{\alpha}_o \cos \tilde{\beta}_i}{\cos \delta_i} - \frac{q}{\sqrt{3}}\frac{\sin \tilde{\alpha}_o \cos \tilde{\beta}_i}{\cos \delta_i} \right) \tag{29}$$

$$d_2 = \frac{\sin(\pi/6 + \tilde{\beta}_i)}{\cos \tilde{\beta}_i}\left(1 - q\frac{\cos \tilde{\alpha}_o \cos \tilde{\beta}_i}{\cos \delta_i} - \frac{q}{\sqrt{3}}\frac{\sin \tilde{\alpha}_o \cos \tilde{\beta}_i}{\cos \delta_i} \right) \tag{30}$$

$$d_3 = \frac{\sin(\pi/6 - \tilde{\beta}_i)}{\cos \tilde{\beta}_i}\left(-1 + 2q\frac{\cos \tilde{\alpha}_o \cos \tilde{\beta}_i}{\cos \delta_i} \right) \tag{31}$$

$$d_4 = \frac{\sin(\pi/6 + \tilde{\beta}_i)}{\cos \tilde{\beta}_i}\left(-1 + 2q\frac{\cos \tilde{\alpha}_o \cos \tilde{\beta}_i}{\cos \delta_i} \right) \tag{32}$$

$$d_5 = \frac{\sin(\pi/6 - \tilde{\beta}_i)}{\cos \tilde{\beta}_i}\left(1 - q\frac{\cos \tilde{\alpha}_o \cos \tilde{\beta}_i}{\cos \delta_i} + \frac{q}{\sqrt{3}}\frac{\sin \tilde{\alpha}_o \cos \tilde{\beta}_i}{\cos \delta_i} \right) \tag{33}$$

$$d_6 = \frac{\sin(\pi/6 + \tilde{\beta}_i)}{\cos \tilde{\beta}_i}\left(1 - q\frac{\cos \tilde{\alpha}_o \cos \tilde{\beta}_i}{\cos \delta_i} + \frac{q}{\sqrt{3}}\frac{\sin \tilde{\alpha}_o \cos \tilde{\beta}_i}{\cos \delta_i} \right) \tag{34}$$

The corresponding double-side switching pattern of the proposed DSVM method is shown in Fig. 7.

All duty cycles must be positive and lower than unity:

$$0 \le d_i \le 1; \quad i = 1,..,6 . \tag{35}$$

(36) can be obtained when all duty cycles in (29)-(34) satisfy (35).

$$\frac{2}{3}\cos \delta \le q \le \frac{\sqrt{3}}{2}\cos \delta . \tag{36}$$

IV. SIMULATION RESULTS

Numerical simulations for the system shown in Fig. 1 are carried out by using PSIM 9.0 software to verify the effectiveness of the proposed DSVM method. The system parameters used for simulation are: power supply 100 Vrms/60 Hz (line-to-line), balanced three-phase load $R = 25\Omega$, $L = 20$mH output frequency $f_o = 50$Hz , voltage transfer ratio with $q = 0.7$.

Fig. 8, 10 and 12 show the input/output waveforms with the traditional, previous CMV reduction and proposed methods, respectively. Fig. 9, 11 and 13 show the CMV wave forms and

The 2014 International Power Electronics Conference

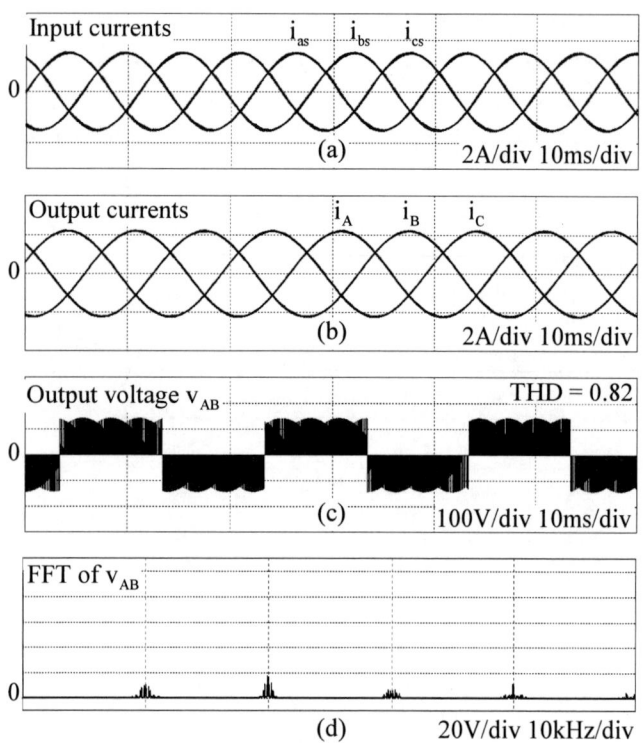

Fig. 8. (a) Three-phase input currents. (b) Three-phase output currents. (c) Output line-to-line voltage. (d) FFT for output line-to-line voltage using the traditional method with $q = 0.7$ and $f_o = 50Hz$.

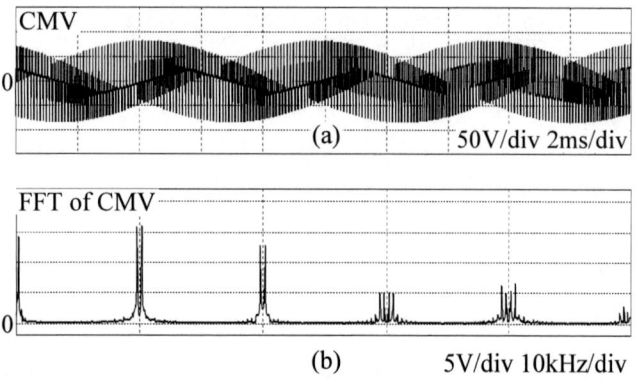

Fig. 9. (a) CMV waveform and (b) its FFT using the traditional method with $q = 0.7$ and $f_o = 50Hz$.

their corresponding frequency spectrums with the traditional, previous CMV reduction and proposed methods, respectively. Due to the zero SCs, the traditional method generates a CMV with peak value to be V_i. After eliminating zero SCs, the previous CMV reduction method in [9] and the proposed method provide CMV with a peak value $V_i/\sqrt{3}$. The CMV peak values of three DSVM methods are shown in Table IV. Both the CMV reduction method in [9] and the proposed method reduce the peak value of CMV by 42% compared to the traditional DSVM method.

According to the FFT analyses of CMV shown in Fig. 11(b) and Fig. 13(b), the proposed method shows lesser harmonic components at high frequencies compared to the CMV reduction method in [9]. The leakage current is directly related

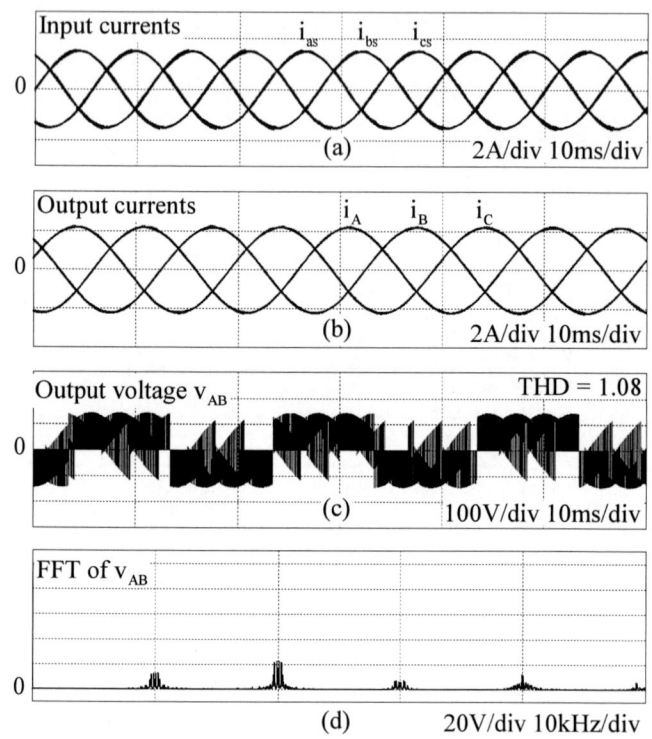

Fig. 10. (a) Three-phase input currents. (b) Three-phase output currents. (c) Output line-to-line voltage. (d) FFT for output line-to-line voltage using the previous CMV reduction method in [9] with $q = 0.7$ and $f_o = 50Hz$.

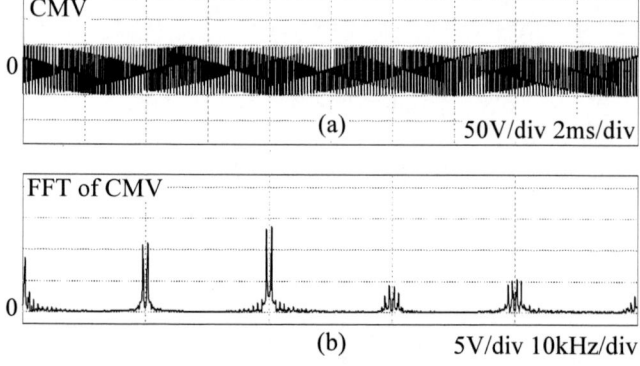

Fig. 11. (a) CMV waveform and (b) its FFT using the previous CMV reduction method in [9] with $q = 0.7$ and $f_o = 50Hz$.

TABLE IV. THE COMPARISON OF PEAK VALUE OF CMV FOR THREE DSVM METHODS

Method	Peak value of CMV [pu]
Traditional	1
CMV reduction in [9]	$1/\sqrt{3}$
Proposed	$1/\sqrt{3}$

with FFT analysis and RMS value of CMV. Fig. 14(a) presents a comparison of the RMS value versus the voltage transfer ratio for three DSVM methods. As can be seen, the proposed DSVM method provides the lowest RMS value of CMV. At $q = 0.67$, the RMS value of CMV is reduced by 20% and 29%

The 2014 International Power Electronics Conference

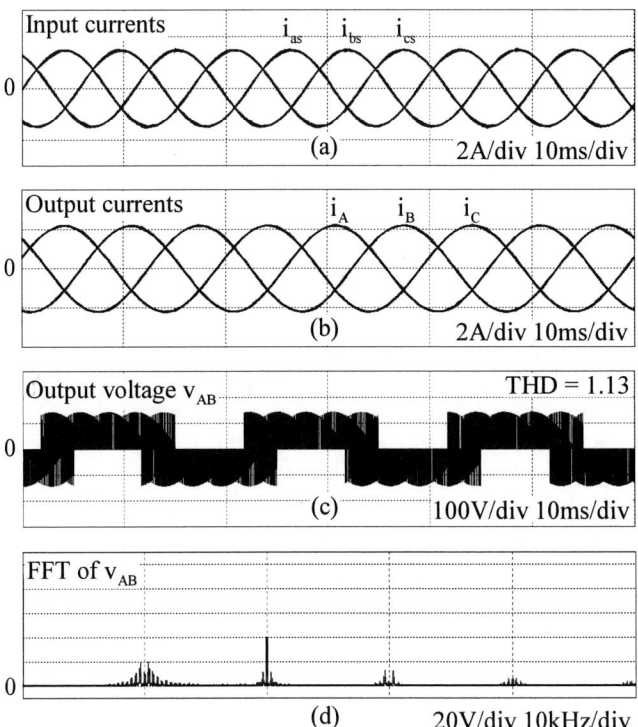

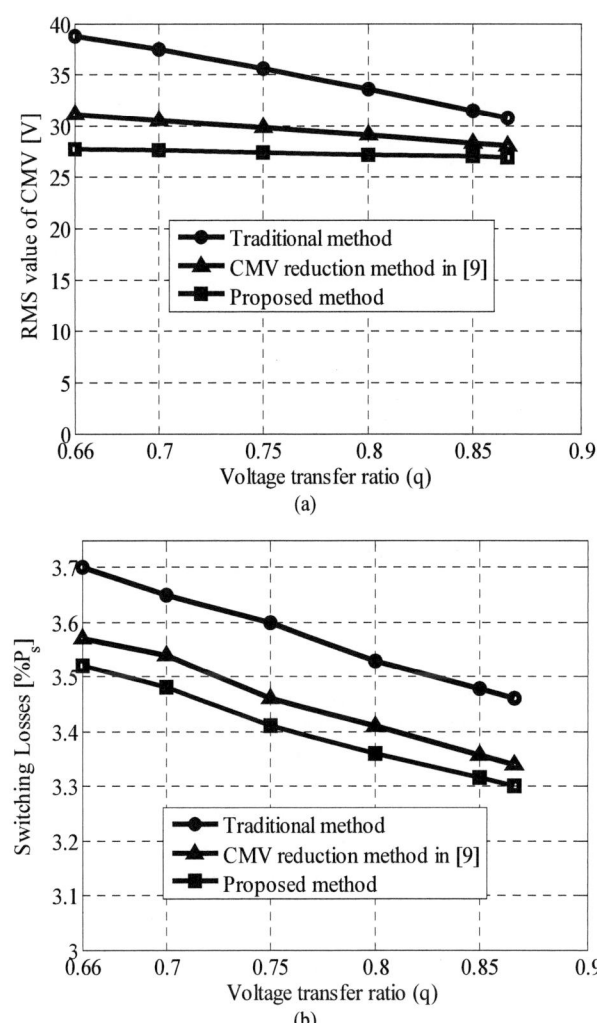

Fig. 12. (a) Three-phase input currents. (b) Three-phase output currents. (c) Output line-to-line voltage. (d) FFT for output line-to-line voltage using the proposed CMV reduction method with $q = 0.7$ and $f_o = 50Hz$.

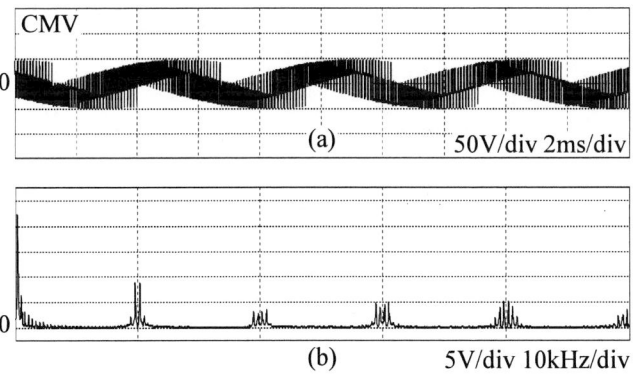

Fig. 14. Comparison between the traditional, conventional CMV reduction and proposed methods in terms of (a) RMS value of CMV; (b) Switching losses.

Fig. 13. (a) CMV waveform and (b) its FFT using the proposed CMV reduction method with $q = 0.7$ and $f_o = 50Hz$.

when the previous and proposed CMV reduction DSVM methods are applied, respectively. Therefore, the new DSVM method generates lowest leakage current among the CMV reduction methods.

Switching losses evaluate the efficiency of a power converter. The lower ratio of switching losses and source power is, the more efficient power converter achieves and vice versa. In this paper, the switching losses are calculated by using the thermal module in Psim 9.0 software with IGBT type G4PF50W. The comparison of switching losses in three DSVM methods is shown in Fig. 14b), which shows that the proposed DSVM method provides lowest switching losses.

Fig 15 gives a comparison of input performance in three DSVM methods for THD of input current versus the voltage transfer ratio. From this comparison, the proposed DSVM

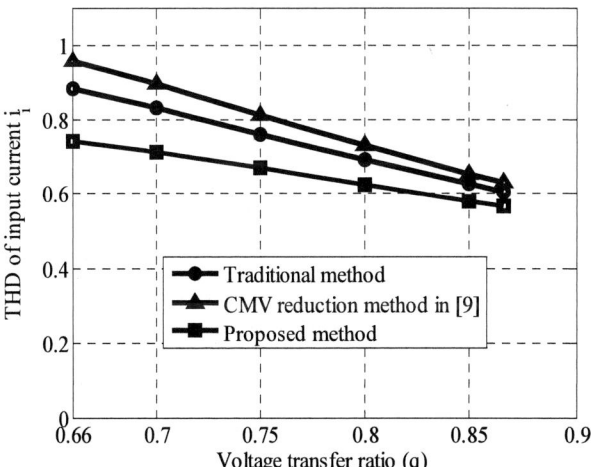

Fig. 15. Comparison of performance between the traditional, conventional CMV reduction and proposed methods in terms of THD of input current.

method generates a better quality of input current waveform with a lower THD compared to traditional and conventional CMV reduction DSVM method.

V. CONCLUSION

This paper has been presented a new DSVM method that can reduce the peak value of CMV by 42% and RMS value of CMV to 29% compared to those of the traditional DSVM in DMC. This method overcomes the limitations of the previous CMV reduction DSVM method regarding RMS value of CMV, switching losses and input current quality. From the simulation results and comparisons, the proposed DSVM method can reduce CMV better than the traditional method and previous CMV reduction method in [9] by generating lower RMS value of CMV, switching losses and higher input current quality. However, the voltage transfer ratio in the proposed DSVM method is limited from 0.667 to 0.866. Therefore, this method is suitable for the case of high voltage transfer ratio. Finally, simulation results are given to verify the advantages of the proposed DSVM method.

ACKNOWLEDGMENT

This work was partly supported by the National Research Foundation of Korea Grant funded by the Korean Government (No. 2010-0025483) and the Network-based Automation Research Center (NARC) funded by the Ministry of Trade, Industry & Energy.

REFERENCES

[1] P. W. Wheeler, J. Rodriguez, J. C. Clare, L.Empringham, and A. Weinstein, "Matrix converters: A technology review," *IEEE Trans. on Industrial Electronics, vol. 49, no. 2, pp. 276–288, April 2002.*

[2] Domenico Casadei, Giovanni Serra, Angelo Tani, and Luca Zarri "Matrix Converter Modulation Strategies: A New General Approach Based on Space Vector Representation of Switch State," *IEEE Trans. on Industrial Electronics, vol. 49, no. 2, pp. 370–381, April 2002.*

[3] Hoang M. Nguyen, Hong-Hee Lee, and Tae-Won Chun, "Input Power Factor Compensation Algorithms Using a New Direct-SVM Method for Matrix Converter," *IEEE Trans. on Industrial Electronics, vol. 58, no. 1, pp. 232–243, January 2011.*

[4] Jun-Koo Kang, Tsuneo Kume, Hidenori Hara and Eiji Yamamoto, "Common-Mode Voltage Characteristics of Matrix Converter-Driven AC Machines," *Industry Applications Conference, 2382 - 2387 vol. 4, 2005.*

[5] Tuyen D. Nguyen and Hong-Hee Lee, "Modulation Strategies to Reduce Common-Mode Voltage for Indirect Matrix Converters," *IEEE Trans. on Industrial Electronics, vol. 59, no. 1, pp. 129–140, January 2012.*

[6] Tuyen D. Nguyen and Hong-Hee Lee, "A New SVM Method for an Indirect Matrix Converter with Common-mode Voltage Reduction," *IEEE Trans. on Industrial Informatics, 27 March, 2013.*

[7] Laszo Huber, and Dusan Borojevic, "Space Vector Modulated Three-phase to Three-phase Matrix Converter with Input Power Factor Correction," *IEEE Trans. on Industry Applications, vol. 31, no 6, pp. 1234–1246, November/December 1995.*

[8] Han Ju Cha and Prasad N. Enjeti, "An Approach to Reduce Common-Mode Voltage in Matrix Converter," *IEEE Trans. on Industry Applications, vol. 39, no. 4, July/August 2003.*

[9] Hoang M. Nguyen, Hong-Hee Lee and Tae-Won Chun, "A novel method of common-mode voltage reduction in Matrix Converters," *International Journal of Electronics, vol. 99, no. 1, pp. 1–14, January 2012.*

Experimental Verification of High Frequency Link DC-AC Converter using Pulse Density Modulation at Secondary Matrix Converter.

Jun-ichi Itoh, Ryo Oshima and Hiroki Takahashi
Dept. of Electrical, Electronics and Information Engineering
Nagaoka University of Technology
Nagaoka, Niigata, Japan
ryo_oshima@stn.nagaokaut.ac.jp

Abstract— This paper verifies an isolated DC-AC power converter using a single phase to three phase matrix converter in experiment. A matrix converter does not require the large reactors and the large smoothing capacitors in a DC-link part. Furthermore, the proposed control method enables zero voltage switching of the matrix converter by implementing a phase shift control on the primary inverter and a pulse density modulation on the secondary matrix converter. In this paper, the fundamental operation of the converter is demonstrated by the experiment. From the experimental results, the total harmonic distortion in the output voltage is less than 5% in the entire range. In addition, a maximum efficiency of 90.9 % is achieved at an output power of 1.5kW.

Keywords— *dc-ac converter, high frequency link converter, pulse density modulation, zero voltage switching.*

I. INTRODUCTION

Recently, from the view point of global warming and environmental problems, the renewable energy systems are focused on. However, in a renewable energy system, especially a wind turbine and a photovoltaic cell, a power fluctuation occurs owing to the meteorological conditions. Therefore, an energy storage system using a battery is necessary in order to suppress the power fluctuation.

The battery energy storage system requires a DC-AC converter to connect the grid and the battery. In addition, the DC-AC converter requires the isolation by a transformer in order to protect the system from failure and noise. However, an isolation transformer designed for the commercial frequency is bulky and heavy. Hence, in order to reduce the volume of the isolation transformer, a high frequency AC link converter has been researched.

In past works, two typical high frequency link DC-AC converter topologies have been proposed. First one consists of a high frequency isolated DC-DC converter and a three-phase inverter for the grid connection [1-4]. This circuit configuration reduces the volume of the isolation transformer because the full bridge inverter operates in the DC-DC converter at the high frequency. However, the DC link capacitors between the rectifier

and three-phase inverter are bulky in order to smooth the DC link voltage. Thus, it is difficult to reduce the volume of the system and to realize long lifetime. Moreover, the total efficiency decreases because of three times power conversion.

Second one is the high frequency link converter with a matrix converter for the secondary power conversion in order to reduce the volume of the capacitor [5-7]. This circuit topology has advantages as follows: (i) the number of power conversion is reduced to two times. Thus, the efficiency is higher than the former type. (ii) The system achieves smaller size and long lifetime because the DC-link capacitor is not required. However, when PWM (pulse width modulation) control is used to the matrix converter in the secondary side, the switching loss of the matrix converter is increased because of a hard switching. Moreover, the conduction loss of the inverter is increased because the freewheeling current flows at the inverter while the matrix converter outputs zero vectors.

This paper proposes an isolated DC-AC converter which adopts a matrix converter with pulse density modulation (PDM) and the suppression control of freewheeling current. By using the PDM, zero voltage switching (ZVS) is achieved. Then, the fundamental operation of the proposed circuit is demonstrated in the simulations and the experiments. The remainder of this paper is organized as follows. First, the circuit configuration of the high frequency link converter is described. Second, the control scheme is explained in detail. Third, a problem of freewheeling current occurs at the primary inverter and the suppression control of freewheeling current are explained. Fourth, a fundamental operation of the proposed method is shown in simulation. Finally, the operation of the proposed circuit is demonstrated under the condition of an input DC voltage of 200 V and an AC link frequency of 50 kHz. In addition, the efficiency and the total harmonic distortion (THD) of the output voltage are evaluated in order to clarify the validity of the proposed system.

978-1-4799-2706-7/14 $31.00 © 2014 IEEE

II. CIRCUIT CONFIGURATION

Fig. 1 shows the configuration of the conventional circuit. The conventional circuit comprises a bidirectional DC-DC converter with an isolation transformer such as dual active bridge converter [8-9]. However, the two-stage topology requires the bulky DC capacitors in order to smooth the DC link voltage.

Fig. 2 shows the main circuit configuration of the proposed DC-AC converter with the matrix converter. The proposed circuit comprises a full bridge inverter with phase shift control at the primary side of the transformer and the single-phase to three phase matrix converter at the secondary side. It is noted that the PDM is applied to the matrix converter only. The proposed system achieves high efficiency because the number of the power conversions is reduced owing to the matrix converter. In addition, the advantages of the matrix converter are long lifetime and the reduced volume owing to the absence of the large DC-link capacitor and an initial charge circuit. However, it is difficult to reduce the switching loss due to the hard switching when the conventional PWM method is applied to the matrix converter.

In order to reduce the switching loss of the matrix converter by implementation of ZVS, the PDM is applied to the matrix converter. In addition, for simplicity of the PDM, the primary inverter provides three-level voltage including zero voltage with the phase shift control. As a result, the matrix converter achieves ZVS when the switches turn in the zero voltage periods and a commutation scheme of the matrix converter is simplified.

III. CONTROL STRATEGIES

A. Phase shift control

Fig. 3 shows a control block diagram of the phase shift control for the primary inverter. This control is composed of a carrier generator, a phase delay circuit, and two comparators. Duty ratio D of the primary inverter to adjust the input voltage of the matrix converter is controlled by the phase delay.

Fig. 4 shows an operation principle of the phase shift control for the primary inverter. The inverter outputs three-level voltage including zero voltage by the phase shift control. Therefore, the matrix converter achieves ZVS if the matrix converter turns in the zero voltage period generated by the primary inverter. Actually, the phase delay is achieved by adjusting carrier delay time T_{PD} given by (1).

$$T_{PD} = \frac{D}{2 \times f_{c\_inv}} \quad \cdots \cdots \cdots \cdots \cdots \cdots \cdots \cdots \cdots \cdots (1)$$

where, f_{c_inv} is carrier frequency of the primary inverter.

B. Pulse density modulation method based on space vector modulation

Fig. 5 shows a control block diagram of a PDM for the matrix converter. This control is composed of space vector modulation (SVM), a clock (CLK) generator for

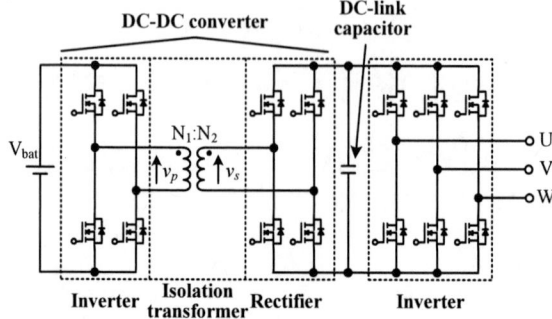

Fig.1. High frequency link DC-AC converter with rectifier and inverter

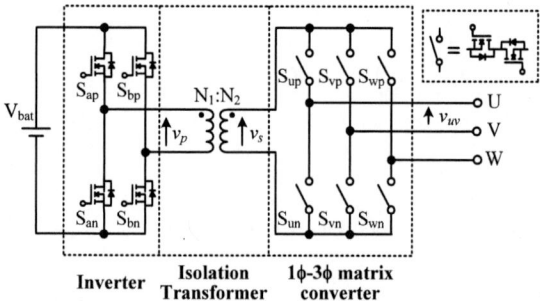

Fig.2. High frequency link DC-AC converter with matrix converter

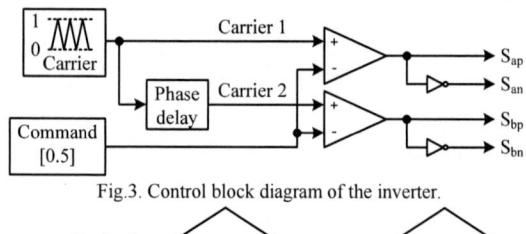

Fig.3. Control block diagram of the inverter.

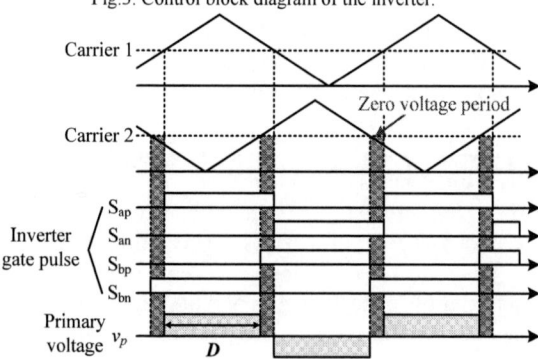

Fig.4. Operation principle of the phase shift control for the primary inverter.

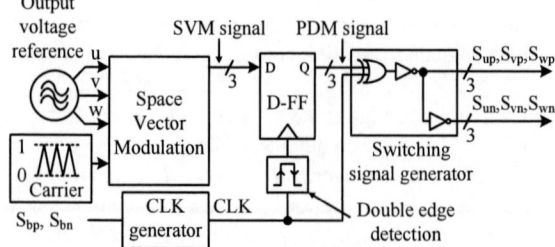

Fig.5. Control block diagram of the matrix converter.

the PDM, a delay (D-FF) and a switching signal generator. In order to generate the gate signal by the PDM, the D-FF is used to quantize the duty references generated by the SVM. Moreover, the CLK to drive the D-FF is synchronized with the zero voltage periods owing to the phase shift control in order to achieve the PDM and ZVS easily.

Fig. 6 shows operation waveforms of the PDM which enables ZVS of the matrix converter. The PDM controls the density and the pole of the constant width pulse. In addition, these pulse signals are used as the minimum unit of the output voltage waveform [10]. It should be noted that the input voltage of the matrix converter (secondary voltage of the transformer) is high frequency square waveform. Therefore, a half cycle of the input voltage is used as one pulse of the PDM. Then, the half cycle period of the input voltage is detected easily by the gate pulses S_{bp} and S_{bn} of the primary inverter. As a result, the PDM and the ZVS of the matrix converter are implemented without switching loss by the control scheme in Fig. 5.

IV. SUPPRESSION CONTROL OF FREEWHEELING CURRENT

This chapter describes a problem and the principle of a freewheeling current mode at the primary side. In addition, a suppression control of freewheeling current is presented in order to improve system efficiency.

Fig. 7 shows operation modes of the DC-AC converter. The red line (state 1) in Fig. 7(a) shows a current path when the primary inverter outputs positive voltage and the matrix converter selects (100) vector. It should be noted that this case is not the freewheeling current mode. In the state 1, the transformer current contributes to provide the DC battery power to the load because the primary inverter outputs positive voltage.

In Fig. 7(b), the red line shows a current pathway when the matrix converter chooses a vector except the zero vector though the primary inverter outputs zero voltage. This paper defines this mode as the freewheeling current mode. The problem of the freewheeling current mode means increasing conduction loss of the primary inverter and copper loss of the transformer. The principle of the increasing losses is following. In this state, the DC battery power is not transmitted to the load because the primary inverter outputs zero voltage. However, the primary and the secondary current of the transformer flow since the matrix converter selects a vector not the zero vector and the load is inductive. As a result, a current pathway in the primary side is constructed and the conduction loss of the primary inverter and the copper loss of the transformer are generated although the transmitted power from the DC battery to the load is zero. Therefore, the freewheeling current mode needs to be suppressed in order to improve the system efficiency.

Fig. 7(c) shows the principle of the suppression method of the freewheeling current mode. The

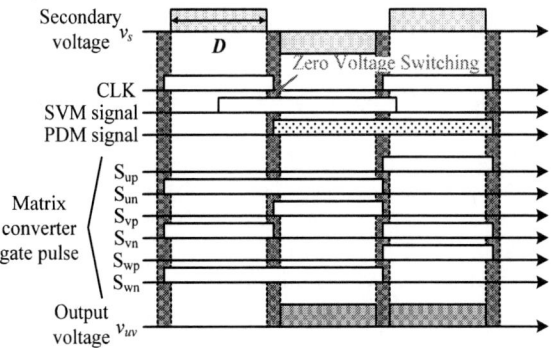

Fig.6. Operation principle of the pulse density modulation for the matrix converter.

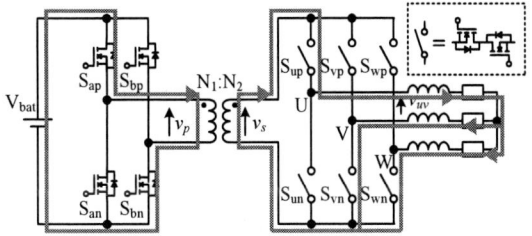

(a) The output current pathway when the DC-AC converter supplies electric power from the source to the RL load. (State 1)

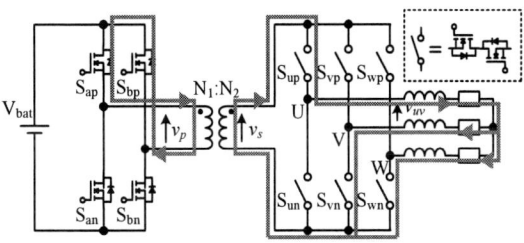

(b) The current pathway when the matrix converter supplies electric power while the inverter outputs zero vector. (State 2)

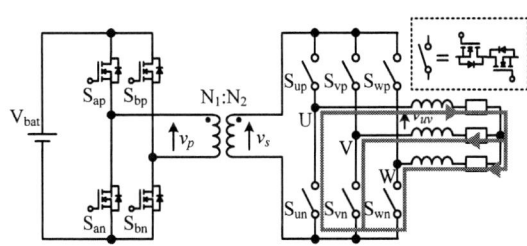

(c) The current pathway after changed the output current pathway. (State 3)

Fig.7 Switching transition of the DC-AC converter

TABLE I
SIMULATION AND EXPERIMENTAL CONDITIONS

Element	Symbol	Value
Battery voltage	V_{bat}	200 V
Carrier frequency of inverter	f_{c_inv}	50 kHz
Carrier frequency of matrix converter	f_{c_mc}	5 kHz
Modulation frequency of matrix converter	f_{m_mc}	50 Hz
Turn ratio	$N_1:N_2$	1:2
Duty of primary voltage	D	0.9 p.u.
Load power	P_{load}	3 kW

suppression method focuses on the output voltage in the freewheeling current mode. In Fig. 7 (b), the output voltage is zero because the primary inverter outputs zero voltage though the matrix converter chooses a vector except the zero vector. Therefore, the freewheeling current mode is equivalent to the zero vector of the matrix converter. In addition, if the matrix converter selects the zero vector, the secondary current of the transformer does not flow. It results in reducing the conduction loss of the primary inverter and the copper loss of the transformer. Hence, in order to prevent the freewheeling current mode, the matrix converter should select the zero vector while the primary inverter outputs zero voltage such as Fig. 7 (c).

V. SIMULATION RESULTS

Table I lists the simulation parameters. The DC-AC converter outputs three-phase voltage of 200V and 50 Hz for the grid connection. However, the simulation is implemented with a RL load for simplicity.

Fig. 8 shows the input and output waveforms of the matrix converter with the proposed method in simulations. It should be noted that the cut-off frequency of the low pass filter to observe the output voltage waveform is 1 kHz. As shown in figure, it is confirmed that the output current of the matrix converter is sinusoidal waveform. Then, the output current THD is 2.79%.

Fig. 9 shows the enlarged waveform of Fig. 8. The secondary voltage of the transformer is three-level voltage of 50 kHz. In addition, the output voltage waveform consists of the secondary voltage pulses of the transformer and the density of the pulses is controlled by the PDM. Thus, the operation of the PDM in the matrix converter is confirmed.

Fig. 10 shows the efficiency characteristic for the output power in simulation. It is noted that this loss simulation calculates only semiconductor losses on the proposed circuit. The efficiency is 97.8% at the maximum point. Moreover, the efficiency is greater than 95% in the entire range. Therefore, it is confirmed that the high efficiency is obtained by ZVS.

VI. EXPERIMENTAL RESULTS

A. Fundamental operation

In order to confirm the fundamental operation of the proposed method, the proposed circuit is demonstrated with the 3-kW prototype circuit in experiment. The experimental conditions are same as the simulation.

Fig. 11 shows the input and output waveforms of the matrix converter in experiment. As the result, the output current is sinusoidal and the output current THD is obtained by 4.28%.

Fig. 12 shows the enlarged waveforms of Fig. 11ff. From the result, it is confirmed that the secondary voltage of the transformer is 50kHz and three-level, which has

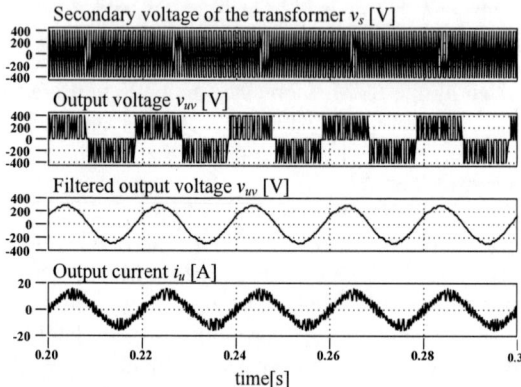

Fig.8. The input and output waveforms of the secondary matrix converter in simulation.

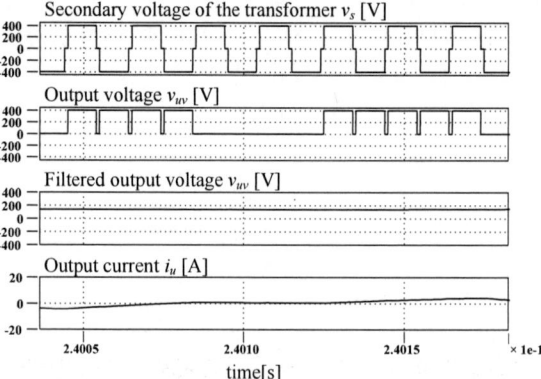

Fig.9. Enlarged waveforms of Fig.8.

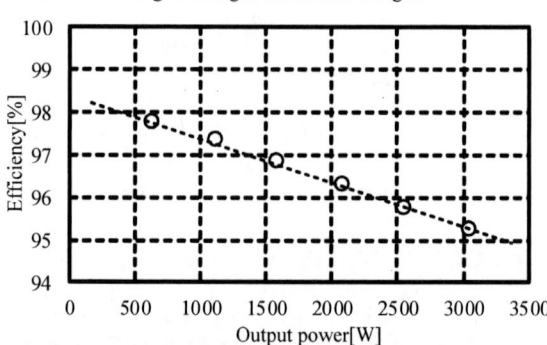

Fig.10. Efficiency characteristics except the transformer obtained from R-L load experiment that is subjected to the output power in simulation.

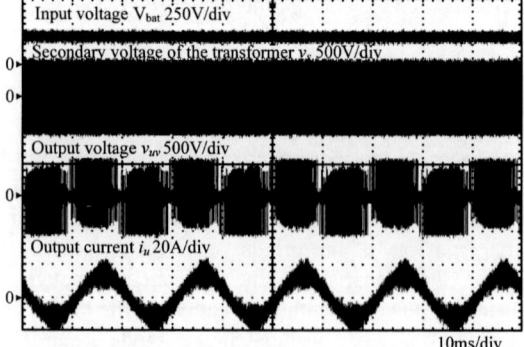

Fig.11. Experimental waveforms of input and output of the DC-AC converter.

the zero voltage period. In addition, the output voltage waveform consists of the secondary voltage pulses of the transformer. Moreover, the density of the pulse is controlled by the PDM. Thus, the PDM of the matrix converter is confirmed.

Fig. 13 shows the ZVS operation of the matrix converter in the proposed system. From the results, when the secondary voltage of the transformer is zero voltage, the switching of the matrix converter is implemented and the ZV is achieved.

Fig. 14 shows output voltage THD characteristic for the output power. As shown in Fig. 13, the output voltage THD is 3.76% at the minimum point and less than 5% all of the output power range. Therefore, the proposed system can be connected to the grid if the interconnection inductors are set to several percent of the rated power capacity.

B. Effectiveness of the suppression control of freewheeling current

Fig. 15 shows the primary voltage and current of the transformer regarding the suppression control of the freewheeling current mode in experiment. Fig. 15 (a) shows a result without the suppression control and Fig. 15 (b) shows a result with the control. It should be noted that the duty ratio D of the primary inverter is set to 0.5 p.u. to observe its difference easily. From Fig. 15 (a), the primary current flows while the primary inverter outputs zero voltage because of the freewheeling current mode. However, as shown Fig. 15 (b), the primary current of the transformer does not flow during the zero voltage period because the suppression control forces the matrix converter to select the zero vector while the primary inverter provides zero voltage. The voltage and current fluctuation in Fig. 15 is caused by a LC resonance due to the capacitance between primary side and secondary side of the transformer and wiring inductance.

Fig.16 shows the efficiency characteristic with respect to the output power with RL loads. A dotted line shows the efficiency characteristic with the suppression control of freewheeling current and the solid line shows efficiency characteristic with the suppression method. From the results, the efficiency without the suppression control of freewheeling current mode is 89.8% at maximum point. However, the suppression control of freewheeling current improves the efficiency. The maximum efficiency with the suppression control is obtained by 90.9 %. Especially, the suppression control is more effective in light load region because the exciting current of transformer is reduced. Therefore, the suppression control of the freewheeling current improves the efficiency of 4.4% at a maximum. Thus, the validity of the suppression control of the freewheeling current mode is confirmed.

Fig.17 shows loss analysis results at the maximum efficiency point and the rated power in experiment with the suppression control of freewheeling current. Then, the

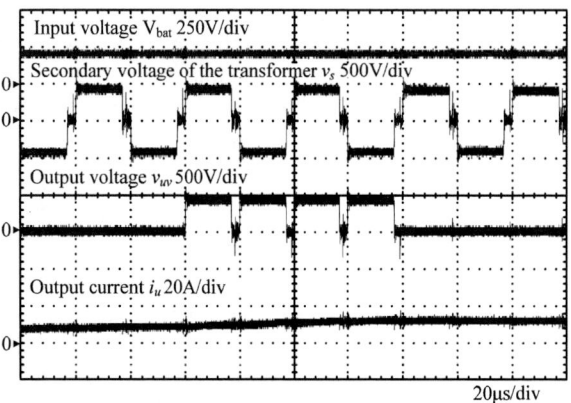

Fig.12. Enlarged waveforms of Fig.10.

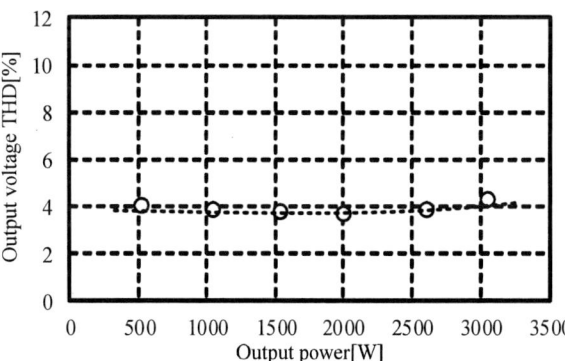

Fig.13. Experimental waveforms when the gate signals are changed in the zero voltage period.

Fig.14. THD characteristics of output voltage obtained from R-L load experiment that is subjected to the output power.

rated power is 3kW and the maximum efficiency is obtained at 1.5kW.

It should be noted that matrix converter loss includes the snubber loss. Then, snubber voltage is 880V at the rated power because the snubber absorbs the leakage inductance energy of the transformer and commutation failure of the matrix converter. As a result, the snubber loss is obtained by 12.9W and accounts for 35% of the matrix converter loss. However, if the commutation sequence is modified, it is possible to reduce the snubber loss because the snubber voltage increase is suppressed. Consequently, the efficiency improvement of DC-AC converter will be expected.

On the other hand, the main loss of the DC-AC

converter is the inverter loss at both maximum efficiency point and the rated power. Therefore, it is necessary to reduce the inverter loss in order to improve the efficiency of the DC-AC converter. As a method to increase the efficiency, soft switching technique will be introduced in the future.

VII. CONCLUSION

This paper verified the isolated DC-AC converter using a matrix converter in simulations and experiments. The matrix converter in the proposed system employs the PDM and achieves ZVS by combination with the high frequency inverter with the phase shift control. From the simulation result, it is confirmed the fundamental operation and the maximum efficiency of proposed DC-AC converter of 97.8% is obtained. In addition, a 3-kW prototype of the proposed circuit was tested. From the experimental results, the ZVS operation and the PDM of the matrix converter were confirmed. Moreover, the output voltage THD is less than 5% in the entire range.

Besides, the efficiency characteristics with respect to the output power is evaluated. In addition, a loss analysis of the experimental result is implemented. A maximum efficiency of 90.9 % is achieved at 1.5kW output power with the suppression control of freewheeling current. It is confirmed that the efficiency of 1 % is improved by applying the suppression control of freewheeling current. And, from the loss separation results, the inverter loss is the main loss of the DC-AC converter.

In the future work, in order to achieve the high efficiency of DC-AC converter, it will be performed the soft switching at the inverter. Finally, the prototype circuit will be connected to the grid and evaluated in experiment.

REFERENCES

[1] H. Ertl, J. W. Kolar, and F. C. Zach, "A novel multicell DC-AC converter for applications in renewable energy systems," IEEE Trans. Ind. Electron., vol. 49, no. 5, pp. 1048–1057, 2002.

[2] R. P. Torrico-Bascope, D. S. Oliveira, Jr., C. G. C. Branco, and F. L.M. Antunes, "A UPS with 110-V/220-V input voltage and high-frequency transformer isolation," IEEE Trans. Ind. Electron., vol. 55, no. 8, pp.2984–2996, 2008.

[3] C. Rodriguez and G. Amaratunga, "Long-lifetime power inverter for photovoltaic AC modules," IEEE Trans. Ind. Electron., vol. 55, no. 7, pp. 2593–2601, 2008.

[4] R. P. T. Bascope, D. S. Oliveira, C. G. C. Branco, and F. L. M. Antunes, "A UPS with 110 V/220 V input voltage and high-frequency transformer isolation," IEEE Trans. Ind. Electron., vol. 55, no. 8, pp. 2984–2996, Aug. 2008.

[5] I. Yamato, N. Tokunaga, Y. Matsuda, Y. Suzuki, and H. Amaro, "High frequency link dc–ac converter for UPS with a new voltage clamper," in Proc. IEEE Power Electron. Spec. Conf., pp. 749–756, 1990.

[6] K. Inagaki, T. Furuhashi, A. Ishiguro, M. Ishida, S. Okuma, Y. Uchikawa, "A Waveform Control Mehod of AC to DC converters with High-Frequency Links," IEEJ Trans. D,Vol.110, No.5, pp.525-533, 1990.

[7] K. Inagaki and S. Okuma, "High frequency link DC/AC converters using three phase PWM cycloconverters for

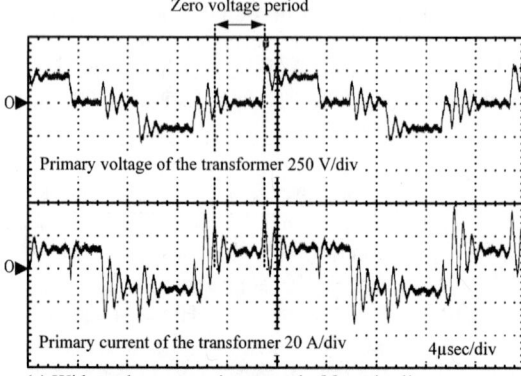
(a) Without the suppression control of freewheeling current.

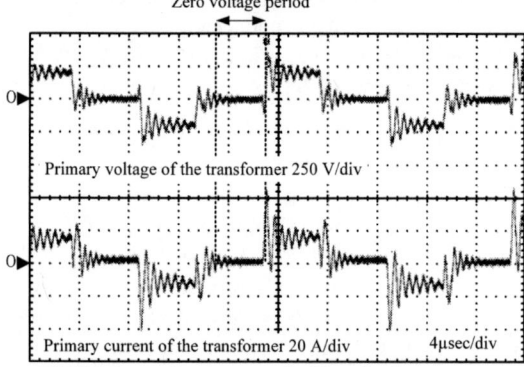
(b) With the suppression control of freewheeling current.
Fig.15 Experimental waveforms of primary voltage and current of the transformer with the suppression control of freewheeling current.

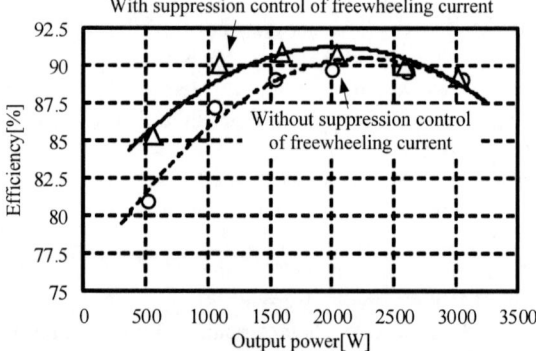
Fig.16 efficiency characteristics of the DC-AC converter obtained from R-L load experiment that is subjected to the output power.

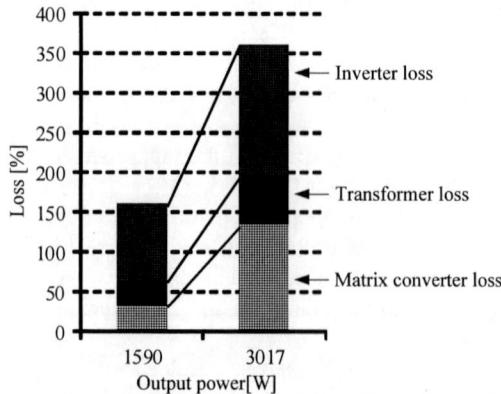
Fig.17 Loss analysis results of the DC-AC converter with the suppression control of freewheeling current.

uninterruptable power supplies," in Proc. Telecommun. Energy Conf., pp. 580–586, 1991.

[8] H. Bai and C. Mi, "Eliminate reactive power and increase system efficiency of isolated bidirectional dual-active-bridge dc–dc converters using novel dual-phase-shift control," IEEE Trans. Power Electron., vol. 23, no. 6, pp. 2905–2914, 2008.

[9] A. K. Jain and R. Ayyanar, "PWM control of dual active bridge: Comprehensive analysis and experimental verification," in Proc. 34th IEEE IECON, Orlando, FL, Nov. 10–13, pp. 909–915, 2008.

[10] Y. Nakata, J. Itoh, "Pulse Density Modulation Control using Space Vector Modulation for a Single-phase to Three-phase Indirect Matrix Converter," IEEE ECCE 2012, Raleigh, P3905, pp. 1753-1759, 2012.

Loss Analysis and Design Method for High Efficiency Matrix Converter

Kazuhiro Koiwa, Goh Teck Chiang, and Jun-ichi Itoh

Department of Electrical engineering
Nagaoka University of Technology, NUT
Niigata, Japan
newkoiwa@stn.nagaokaut.ac.jp

Abstract— This paper discusses loss analysis formulas to achieve high efficiency and high power density for matrix converter. In this paper, the conduction loss and the switching loss of the matrix converter are derived theoretically based on virtual AC-DC-AC control method. Then, the validity of the equations is confirmed in simulation and experiment. From the experimental results, the maximum efficiency is 97.9% with 2-phase modulation at rated power (Devices: MOSFET R6046FNZ). In addition, it is confirmed that the total loss error between the calculation and experimental result is 8.65%. Finally, the relationship between the efficiency and power density is discussed by a pareto front curve. The power density is calculated from the volume of the switching device, heat-sink, input inductor and filter capacitor. As the result, the efficiency and maximum power density in the matrix converter are 97.0% and 9.7 kW/dm³ when the switching frequency is set to 85 kHz.

Keywords— *loss analysis, pareto-front curve, matrix converter, maximum power density design*

I. INTRODUCTION

Recently, the high efficiency and high power density of the power converter have been discussed. In order to achieve the maximum power density of the power converter, the prototype is repeatedly manufactured after the design and decision of the specification of the power converter. However, the number of manufacturing the prototype causes high development cost and long development period. In order to solve this problem, the front loading design which estimates the loss, volume and Ematrix converter before manufacturing the prototype, is introduced to the power converter design [1][2]. According to the front loading design, the prototype can be manufactured by only one time owing to the evaluation of the performance. As a result, the developing cost and process can be reduced.

The matrix converter which can convert an AC power supply voltage directly into an AC output voltage of variable amplitude and frequency without any large energy storages, such as electrolytic capacitors, has been actively studied recently [3]-[11]. The following advantages are found as compared with the back-to-back (BTB) converter, which consist of a PWM rectifier and a PWM inverter; (i) reduced size, light-weight and long life-time owing to the absence of the large electrolytic capacitor in the main circuit; (ii) high efficiency because

of less switching devices in the turn-on current path; (iii) The matrix converter can avoid rising of temperature of the switching devices in the low output frequency operation. The matrix converter is expected to be applied in the future to renewable energy field such as hybrid electric vehicle systems, the wind power generator systems and others.

Further, high efficiency and high power density of the matrix converter have been discussed [7]-[9] based on front loading design. In general, it is necessary to optimize the switching device and the switching frequency by analyzing the loss. There are two methods for loss analysis as follows: (i) the loss is calculated from the device current and the drain-source voltage by using a circuit simulator [10]. (ii) the loss is theoretically obtained from the derived formulas by using the device parameters [11]. However, in the method (i), it is difficult to optimize the converter loss when the loss is analyzed in a lot of conditions. This is because a lot of simulation time is needed in order to find the minimum loss condition by trial-and-error. On the other hand, in the method (ii), it is easy to optimize the circuit because analysis time and the trial time can be reduced. For this reason, loss analysis using formulas for the multi-level converter has been discussed [12].

Similarly, the loss analysis method for the AC-AC direct converter has been discussed [13][14]. In the method of Ref. [14], the conduction loss and switching loss of the indirect matrix converter have been derived. However, the loss analysis method for the matrix converter has not been mentioned in experiment.

In this paper, in order to design high efficiency power converter easily, the conduction loss and the switching loss of the matrix converter are derived by formulas. In addition, the matrix converter is optimized by defining the loss in terms of high efficiency and high power density. Concretely, in order to obtain the power density, the volume of the input filter and the heat-sink are calculated. Further, a pareto-front curve is mentioned in order to find the optimization point between the efficiency and power density. The remainder of this paper is organized as follows. First, the conduction loss and the switching loss are introduced theoretically. Then, the operation of the matrix converter will be demonstrated by the experiment with a 2-kW RL-load. In addition, the validity of the conduction and switching loss equations is

978-1-4799-2706-7/14 $31.00 © 2014 IEEE

confirmed. Finally, the maximum power density design method of the matrix converter is explained.

II. CIRCUIT CONFIGURATION AND CONTROL STRATEGY

Fig. 1-(a) shows the circuit configuration of the matrix converter. Note that the bi-directional switches in the matrix converter can be used with two IGBTs or MOSFETs. In addition, in order to suppress the input filter resonance, the damping resistor is connected to the input inductor in parallel. In comparison to a BTB system, the matrix converter owing to the absence of the large electrolytic capacitor in the main circuit has the advantages as follows: (i) reduced size, light-weight and long life-time; (ii) high efficiency because of less switching devices in the turn-on current path; (iii) The matrix converter can avoid rising of temperature of the switching devices in the low output frequency operation. Thus, it is easy for the matrix converter to be higher efficiency and higher power density.

Fig. 1-(b) shows the control diagram of the matrix converter. In this paper, the virtual indirect control [5][6] is adopted to the matrix converter. Accordingly, the switching pattern, in which the switching changes directly from the maximum phase to the minimum phase is not taken into the consideration. This reason is because the switching pattern is that the middle phase is certainly gone through. Note that the maximum, middle and minimum phase depend on the relationship among each input phase voltage. In order to simplify the loss analysis, the three phase modulation is adopted. Furthermore, the dead-time and the commutation are not taken into the consideration. The input current command in the virtual rectifier is also set to unity power factor.

In order to design high efficiency and high power density of the matrix converter, it is necessary to optimize the switching device and the switching frequency by analyzing the loss. There are two method to analyze the loss. (i) the loss is calculated from the device current and the drain-source voltage by using a circuit simulator. (ii) the loss is theoretically obtained from the derived formulas by using the device parameters.

Fig. 2 shows the loss analysis method by simulator. The device voltage v_{ce} and current i_c are detected. Next, the conduction loss and switching loss are calculated from the measurement value and loss data table. In this method, it is easy to analyze the loss of the matrix converter by using the loss model of the switching device. However, it is difficult to optimize the converter loss because the loss is analyzed in a lot of conditions. This is because a lot of simulation time is needed in order to find the minimum loss condition by trial-and-error.

III. DERIVATION OF MATRIX CONVERTER LOSS

On the other hand, in the method by defining the loss of the matrix converter, it is easy to optimize the circuit because analysis time and the trial time can be reduced. In this paper, the loss analysis is mentioned by formula to

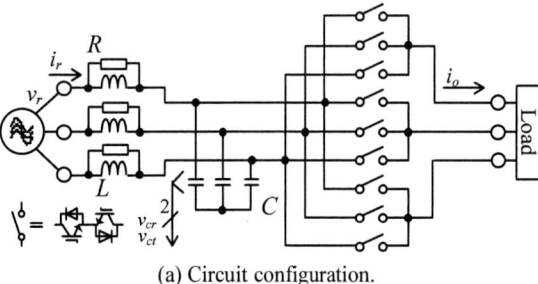

(a) Circuit configuration.

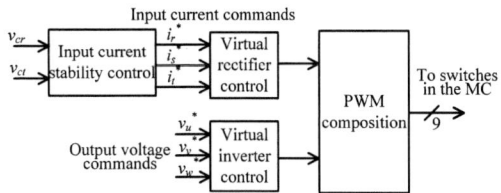

(b) Control diagram in the matrix converter.

Fig. 1. Circuit configuration and control diagram of the matrix converter. The virtual indirect control is adopted to the matrix converter. Accordingly, the switching pattern, in which the switching changes directly from the maximum phase to the minimum phase is not taken into the consideration.

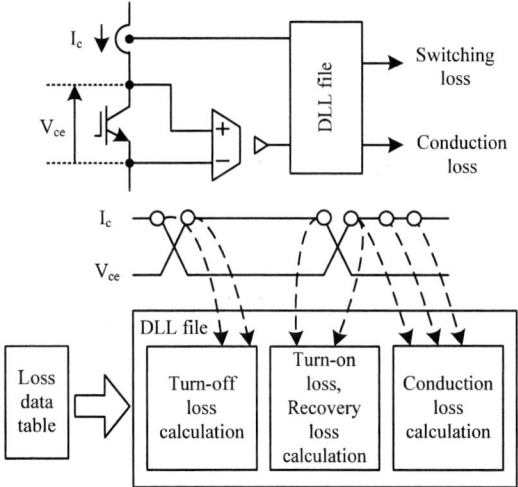

Fig. 2. Loss analysis method by simulator. The conduction loss and switching loss are calculated from the measurement value (v_{ce}, i_c) and loss data table. In this method, it is easy to analyze the loss of the converter. However, the simulation time is long. In addition, the loss data table is obtained from the datasheet of the switching device or the switching test.

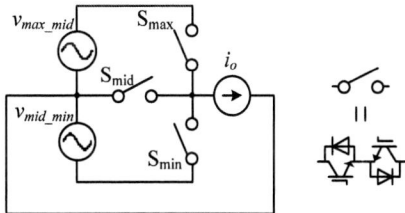

Fig. 3. Equivalent single phase model of the matrix converter. The matrix converter losses are discussed as the maximum phase, middle phase and minimum phase switches. As the result, it is easy to introduce the conduction loss and the switching loss of the matrix converter.

978-1-4799-2706-7/14 $31.00 © 2014 IEEE

design the high efficiency and high power density.

Fig. 3 shows the equivalent single-phase model of the matrix converter for derivation of loss. When the relationship among the input phase voltages is R>S>T, the maximum phase switch S_{max} is the R-phase switch, the middle phase switch is the S-phase switch and the minimum phase switch S_{min} is the T-phase switch. In other words, the matrix converter loss is derived by each S_{max}, S_{mid} and S_{min}.

A. Conduction loss

The conduction loss of the maximum phase switch in the matrix converter P_{con_max} is expressed by (1).

$$P_{con\_max} = \frac{1}{\pi} \int_{\theta_o}^{\pi+\theta_o} D_{max} \cdot v_{on} \cdot i_o \, d\omega_o t \quad \ldots\ldots(1)$$

Note that D_{max} is on-duty of the maximum phase switch and ω_o is the output angular frequency. In addition, v_{on}, which is the on-state voltage of the device, is calculated by (2).

$$v_{on} = k_{con1} i_o + k_{con2} \quad \ldots\ldots\ldots\ldots\ldots\ldots\ldots(2)$$

k_{con1} and k_{con2} are obtained from the on-state voltage characteristic in the datasheet. The instantaneous value of the load current i_o is expressed by (3) from the maximum load current I_o and the load angle θ_o.

$$i_o = I_o \sin(\omega_o t - \theta_o) \quad \ldots\ldots\ldots\ldots\ldots\ldots(3)$$

P_{con_max} can be calculated from (1). However, it is difficult to derive the loss because D_{max} depends on the input current command and the output voltage command. In this paper, the total conduction loss per the output phase P_{con} is derived. Consequently, the on-duty command per the output single-phase is expressed as

$$D_{max} + D_{mid} + D_{min} = 1 \quad \ldots\ldots\ldots\ldots\ldots\ldots(4)$$

where D_{mid} and D_{min} are the on-duties of S_{mid} and S_{min}. From (4), it can be revealed that only one switch is turned on and two switches are not on-state at same time. Assume that the switching devices with same characteristics are used. Thus, P_{con} is expressed by (5).

$$P_{con} = \frac{1}{\pi} \int_{\theta_o}^{\pi+\theta_o} v_{on} \cdot i_o \, d\omega_o t$$

$$= \frac{1}{2} k_{con1} I_o^2 + \frac{2}{\pi} k_{con2} I_o \quad \ldots\ldots\ldots\ldots(5)$$

B. Switching loss

In this section, the turn-on loss of the maximum phase P_{ton_smax} is derived. First, the general turn-on loss is expressed as

$$P_{sw\_loss} = \frac{1}{T} \int_0^{\frac{T}{2}} \frac{f_s}{V_s} e_{on} v_{sw} \, d\omega_o t \quad \ldots\ldots\ldots\ldots(6)$$

where T is the cycle of i_o, f_s is the switching frequency, V_s is the tested voltage when the switching loss was measured, and v_{sw} is the drain-source voltage of the device. In addition, the instantaneous turn-on loss e_{on} is expressed by (7).

$$e_{on} = k_{ton1} i_o + k_{ton2} \quad \ldots\ldots\ldots\ldots\ldots\ldots\ldots(7)$$

Similarly, k_{ton1} and k_{ton2} are obtained from the turn-on

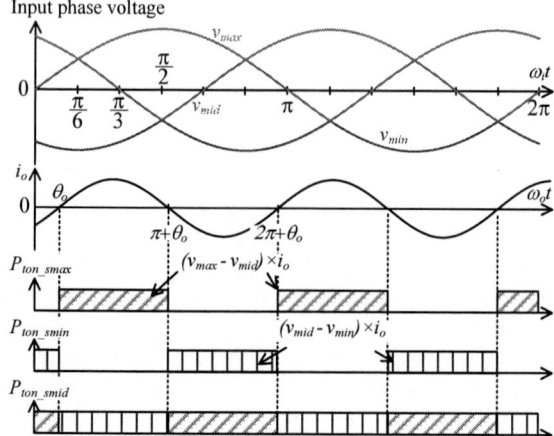

Fig. 4. Integral period of each switching loss of the matrix converter. The integral period is different among each switching devices owing to the polar of the output current. The turn-on loss of the maximum phase switch P_{ton_smax} occurs when the polar of the output current i_o is positive. Additionally, P_{ton_smax} is calculated from the output current and the voltage between the maximum phase and medium phase (v_{max} - v_{mid}).

TABLE 1. DEVICE PARAMETERS AND SIMULATION CONDITION TO CALCULATE THE MATRIX CONVERTER LOSSES.

Parameters of switching device (SK80GM063)		
On-state voltage characteristic	k_{con1} (V/A)	0.018
	k_{con2} (V)	0.977
Switching loss characteristic	k_{ton1} (mJ/A)	0.05
	k_{ton2} (J)	0.0
Input power factor		1.0
Maximum input line voltage V_{in}		283 V
Switching frequency f_s		10 kHz
Output frequency f_o		90 Hz
Input frequency f_i		50 Hz

loss characteristic in the datasheet.

Fig. 4 shows the integral period to calculate the switching loss. The integral period is different among each switching devices due to the polar of i_o. In other words, P_{ton_smax} occurs when the polar of i_o is positive. Furthermore, the drain-source voltage v_{sw} is different among each switching devices. Assume that the switching pattern is that the middle phase is certainly gone through. Accordingly, the drain-source voltage of S_{max} is differential voltage (v_{max} - v_{mid}) between the maximum phase voltage v_{max} and the middle phase voltage v_{mid}. Thus, P_{ton_smax} is expressed by (8).

$$P_{ton\_smax} = \frac{f_s}{4\pi^2 V_s} \cdot \int_{\theta_o}^{\pi+\theta_o} \left\{ \int_0^{\frac{\pi}{6}} e_{on}(\omega_o t) \cdot (v_{max}(\omega_i t) - v_{mid}(\omega_i t)) d\omega_i t \right.$$

$$\left. + \cdots + \int_{\frac{\pi}{6}}^{2\pi} e_{on}(\omega_o t) \cdot (v_{max}(\omega_i t) - v_{mid}(\omega_i t)) \, d\omega_i t \right\} d\omega_o t$$

$$\ldots\ldots\ldots\ldots\ldots\ldots\ldots\ldots\ldots\ldots\ldots(8)$$

Note that ω_i is the input angular frequency. By substituting (7) into (8), P_{ton_smax} is calculated by (9).

$$P_{ton\_smax} = \frac{3 f_s V_{in}}{2\pi^2 V_s} (2 k_{ton1} I_o + k_{ton2} \pi) \quad \ldots\ldots\ldots(9)$$

Similarly, the middle phase and minimum phase turn-on loss are expressed.

$$P_{ton\_smid} = \frac{3f_s V_{in}}{\pi^2 V_s}(2k_{ton1}I_o + k_{ton2}\pi) \quad \ldots\ldots\ldots(10)$$

$$P_{ton\_s\min} = \frac{3f_s V_{in}}{2\pi^2 V_s}(2k_{ton1}I_o + k_{ton2}\pi) \quad \ldots\ldots(11)$$

Based on these formulas such as (5) and (9), it is easy to design the matrix converter in terms of efficiency and reduced size. It is noted that the design method depends on the priority of the high efficiency or high power density. When the matrix converter is designed in terms of high power density, the switching frequency is increased. As the result, the input LC filter becomes smaller. However, the size of the heat-sink is large because the switching loss increases. Further, the switching loss is more dominant than conduction loss. Thus, the switching device, which has small parameters k_{ton1}, k_{ton2}, is selected in reference to the datasheets. Accordingly, high power density is obtained even if the switching frequency is increased. This is because the volume of heat-sink can be reduced owing to lower switching loss. On the other hand, when the matrix converter is designed in terms of high efficiency, the conduction loss is dominated owing to lower switching frequency. For this reason, the switching device, which has small parameters k_{con1}, k_{con2}, is selected from the datasheet. As the result, high efficiency can be obtained because conduction loss is low. Besides, the switching device in the matrix converter can be selected from the efficiency, which is required in an application. First, the circuit specification such as the rated power, rated voltage, rated current, and switching frequency is arbitrarily determined. Second, the on-state voltage parameters k_{con1}, k_{con2} and switching loss parameters k_{ton1}, k_{ton2}, that the demanded efficiency is obtained, are calculated by using (5) and (9). Next, based on the calculated parameters, the switching device is selected from the datasheet. Finally, the efficiency in experiment agrees with that of the design value when the matrix converter is manufactured by using the selected device.

C. *Validity of Matrix Converter Loss Equations*

In order to validate the conduction and switching loss equations, the calculation results are compared with the loss simulation results which are obtained by PLECS. Table 1 lists the calculation parameters. In the loss simulation, the stray inductance and the stray capacitance are not taken into consideration.

Fig. 5 shows the conduction and switching losses of the matrix converter by the theoretical calculation and the simulation. As a result, the calculation result of the conduction loss agrees well with the simulation result. The error between the calculation and simulation is within 0.02%. Thus, the validity of the derived equations of the conduction loss is confirmed. On the other hand, the error of the switching loss between the calculation and simulation is 2.4% as the load current is 20 A. This is caused by the switching ripple. It is confirmed that the error becomes smaller by increasing the switching

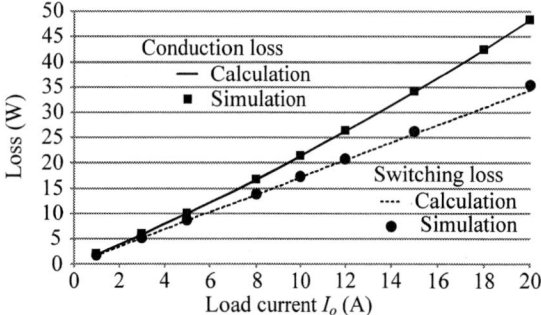

Fig. 5. Conduction loss and switching loss - load current characteristics between calculation and simulation. The error of the conduction loss between the calculation and simulation is within 0.02%. The error of the switching loss between the calculation and simulation is 2.4% as the load current is 20 A. This is caused by the switching ripple. It is confirmed that the error becomes smaller by increasing the switching frequency.

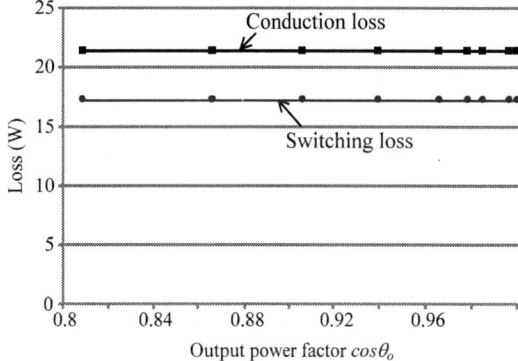

(a) Conduction loss and switching loss - output power factor characteristics between calculation and simulation.

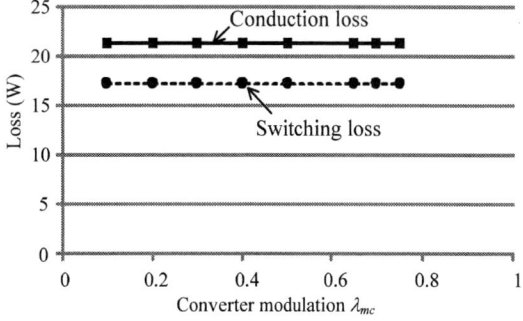

(b) Conduction loss and switching loss - modulation ratio characteristics between calculation and simulation.

Fig. 6. The matrix converter loss characteristics based on the output power factor and the modulation index in order to confirm the validity of the loss equation of the matrix converter. The conduction loss and switching loss do not depend on the output power factor and the modulation index.

frequency.

Fig. 6-(a) shows the matrix converter losses based on the output power factor. In addition, Fig. 6-(b) shows the matrix converter losses based on the modulation ratio. As the results, the conduction loss and switching loss do not depend on the output power factor and the modulation index. Thus, the converter loss is not changed when the

TABLE 2. DEVICE PARAMETERS AND EXPERIMENTAL CONDITIONS. THE NO-LOAD LOSS AND SNUBBER LOSS WERE MEASURED IN EXPERIMENT.

Parameters of R6046FNZ			Input line voltage V_{in}	200 V
On-state voltage characteristic	k_{con1} (V/A)	0.08	Output power P_{out}	2 kW
	k_{con2} (V)	0.0	Output frequency f_o	40 Hz
Turn-on loss characteristic	k_{ton1} (J/A)	9×10^{-6}	Commutation method	4-step voltage
	k_{ton2} (J)	0.0		
Turn-off loss characteristic	k_{toff1} (J/A)	3×10^{-5}	Load inductance	5 mH
	k_{toff2} (J)	3×10^{-6}	Input inductor	2 mH
No load loss and snubber loss		10 W	Filter capacitor	18.9 μF

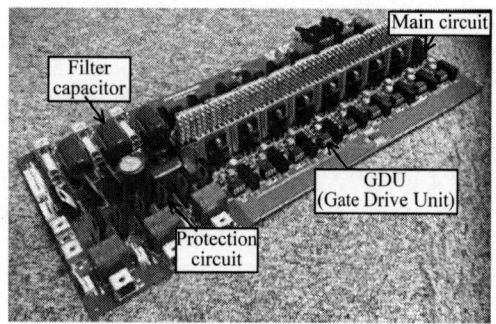

Fig. 7. Board of a 2-kW matrix converter. This board consists of the main circuit, the filter capacitor, the protection circuit for the surge voltage and the drive circuit. In addition, the size of this board is 358 mm×155 mm×40 mm. R6046FNZ is used as the switching device.

output power factor becomes low. In other words, the efficiency is degraded because the output power becomes lower. In order to improve the efficiency, the output power factor and the modulation index are increased.

IV.   EXPERIMENTAL RESULTS

High efficiency matrix converter is made by using the derived equations of the conduction loss and the switching loss. It is noted that the MOSFET(R6046FNZ) is used in experiment. Table 2 lists the parameters of the MOSFET and the experimental conditions. The switching loss of a MOSFET is twice as high as that of an IGBT. This reason is because the switching loss of a MOSFET does not depend on the pole of i_o.

Fig. 7 shows the matrix converter board by using the MOSFET. This board consists of the main circuit, the filter capacitor, the protection circuit for the surge voltage and the drive circuit. In addition, the size of this board is 358 mm×155 mm×40 mm. IC-0526B (RYOSAN) is used as the heat-sink.

Fig. 8 shows the operation waveforms by using 2-kW RL-load. Note that the output line voltage was observed through a low pass filter (LPF) that the cutoff frequency is 1.5 kHz. In addition, two phase modulation was adopted in the virtual inverter control. In addition, the four-step voltage commutation is adopted as the commutation method [15]. As a result, it is accomplished that unity power factor is obtained and the input current THD (Total Harmonic Distortion) is 7.4%. Thus, fundamental operation of the matrix converter can be confirmed.

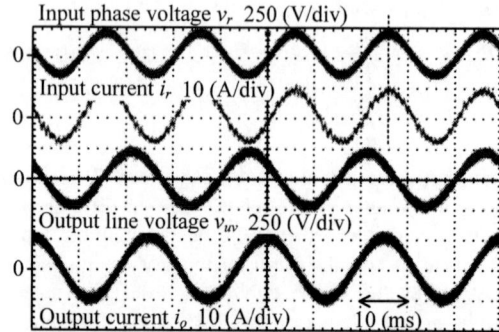

Fig. 8. Steady operation at rated power by the experiment. As a result, it is accomplished that unity power factor is obtained and the input current THD is 7.4%.

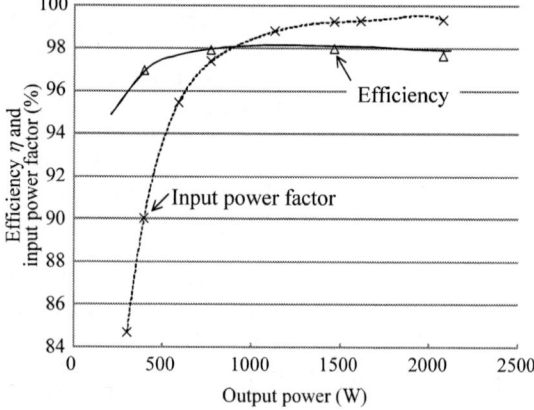

Fig. 9. Efficiency and input power factor characteristics. It is confirmed that the maximum efficiency is 97.9%. In addition, unity power factor is obtained when the output power is over 1 kW.

Fig. 9 shows the efficiency and input power factor characteristics. As the result, it is confirmed that the maximum efficiency is 97.9% at 1.5-kW output power. In addition, unity power factor is obtained when the output power is over 1 kW. However, in the light-load, the input power factor is degraded. This is because the input

Fig. 10 shows the loss characteristics for the output power. It can be confirmed that the maximum error of the total loss is 8.65% at 600-W output power. Thus, the calculation results almost agree to the simulation and experimental results. In this paper, the recovery loss of FWD is not taken into consideration. Thus, the error between the calculation result and the experimental result is caused by the recovery loss of the diode.

V.   DESIGN OF HIGH EFFICIENCY AND POWER DENSITY

Based on these equations, the maximum power density design method for the matrix converter is introduced in this section.

Fig. 11 shows the flow chart to design the matrix converter for the maximum power density. First, the volume of the input LC filter is introduced from the loss analysis results. Note that Table 3 lists the device

The 2014 International Power Electronics Conference

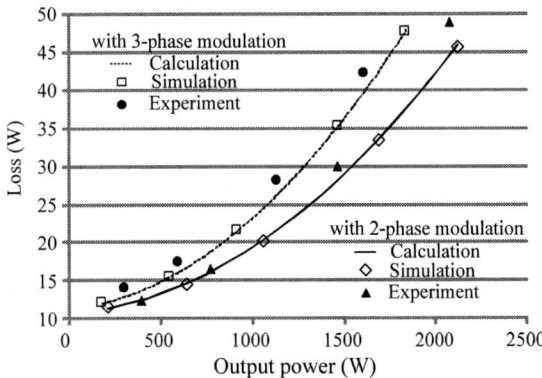

Fig. 10. Total loss characteristics comparison among calculation, simulation and experimental result. It can be confirmed that the maximum error of the total loss is 8.65% at 600-W output power. The error between the calculation result and the experimental result is caused by the recovery loss of the diode.

parameters to calculate the loss. Concretely, the capacitance value C_f is calculated from the voltage ripple V_{rip} on the filter capacitor. As the result, the volume of the filter capacitor Vol_C can be calculated. On the other hand, the inductance value of the input inductor L is decided from C_f and the cut-off frequency f_c. Moreover, the volume of the input inductor is calculated based on an Area Product [16]. Second, the volume of the heat-sink Vol_{heat} is calculated. It is noted that the thermal resistance R_{th} and the cooling system performance index (CSPI), which is the cooling performance of the heat-sink, are needed in order to calculate Vol_{heat}. After that, the Pareto-front curve is mentioned from the volume of the system and the efficiency. From the Pareto-front curve, f_s at the maximum power density is obtained. In other words, the specification of the matrix converter is decided. Finally, the inductor, the capacitor and the heat-sink are selected. It is noted that Vol_{heat} is recalculated when the required heat-sink is not found.

A.   Input LC filter

First, the volume of the input LC filter in the matrix converter is calculated. Assume that the cutoff frequency is one-tenth of the switching frequency and the filter capacitor voltage ripple V_{rip} is 3% or less. The capacitance of the filter capacitor C_f is expressed as

$$C_f = \frac{I_o}{\pi \omega_s V_{rip}} \sin \pi D \quad \text{................................(12)}$$

where ω_s is the switching angular frequency and D is on-duty of the converter. In the matrix converter, D cannot be obtained because D varies in the time. In this paper, C_f is designed in the worst case. Thus, D is set to 0.5. Accordingly, the volume of the filter capacitor Vol_C is obtained from electrostatic energy E as

$$E = \frac{1}{2}\varepsilon_0\varepsilon_s k_z^2 Vol_C = \frac{1}{2}C_f V_{in}^2 \quad \text{................(13)}$$

where ε_s and k_z which are the dielectric constant of the material and the breakdown voltage coefficient depend

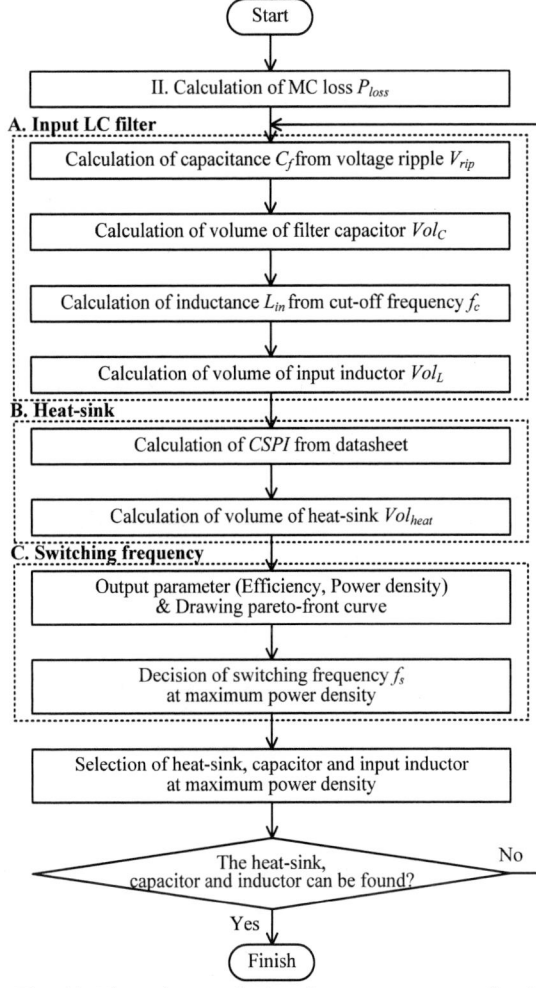

Fig. 11. Flow chart to design the matrix converter for the maximum power density based on front loading design. First, the volume of the input filter and the heat-sink is calculated. Second, the pareto-front curve is mentioned from the volume of the input LC filter and the heat-sink. Next, the switching frequency is selected at maximum power density point. Thus, the specification of the matrix converter can be decided.

TABLE 3. DESIGN CONDITIONS TO DESIGN MAXIMUM POWER DENSITY FOR THE MATRIX CONVERTER.

Voltage ripple V_{rip}	8.49V (3%)	Input line voltage V_{in}	200 V
Input current I_{in}	28.9 A	Output power P_{out}	2 kW
Space factor K_u	0.5	Dielectric constant of the material ε_s	2.2
Current density J	4 A/mm²	Breakdown voltage coefficient k_z	2e7 V/m
Flux density B_m	1.23Wb/m²	Switching frequency f_s / cut-off frequency f_c	5
Core coefficient K_v	17.3	Cooling system	Natural Air
Junction temperature T_j	65 deg	Ambient temperature T_a	25 deg

on the material of the capacitor. In addition, ε_0 is the dielectric constant of the vacuum. On the other hand, the volume of the inductor is calculated from area product [16] by (14).

$$Vol_L = K_v \left[\frac{2W}{K_u B_m J} \right]^{\frac{3}{4}} \quad\ldots\ldots\ldots\ldots\ldots\ldots(14)$$

Note that the core coefficient K_v depends on the type of the core.

Next, the volumes of the boost-up inductor and the DC link capacitor in the BTB system are calculated. The DC link capacitance C_{dc} is designed when the DC link voltage variation ΔU is 3% or less [17].

$$C_{dc} = \frac{P_n}{\left(E_{dc}\Delta U \pm \frac{1}{2}\Delta U^2 \right) f_s} \quad\ldots\ldots\ldots\ldots(15)$$

In addition, the inductance of the boost-up inductor is calculated when the input current ripple is 5% or less.

$$L_B = \frac{E_{dc}}{\pi \omega_s I_{rip}} \sin \pi D = \frac{E_{dc}}{\pi \omega_s I_{rip}} \quad\ldots\ldots\ldots(16)$$

Similarly, the volumes of the boost-up inductor and the DC link capacitor are obtained by using (13) and Ref. [16].

B. Heat-sink

Finally, the volume of the heat-sink Vol_{heat} is calculated from cooling system per index ($CSPI$) [18]. Note that the $CSPI$ is set to 14.2 from the datasheet of the heat-sink. As a result, Vol_{heat} is expressed as

$$Vol_{heat} = \frac{1}{CSPI \times R_{th}} \quad\ldots\ldots\ldots\ldots\ldots(17)$$

where thermal resistance R_{th} is expressed from the loss, maximum junction temperature and ambient temperature.

$$R_{th} = \frac{T_j - T_a}{P_{loss}} \quad\ldots\ldots\ldots\ldots\ldots\ldots\ldots(18)$$

Fig. 12 shows the volume characteristic. As the result, the volume of the input inductor and filter capacitor is reduced when the switching frequency increases because the cut-off frequency becomes high. In other words, the input inductance and filter capacitance are reduced. On the other hand, the volume of the heat-sink is increased owing to the switching loss.

C. Switching frequency

Fig. 13 shows the pareto-front curve of the matrix converter and the BTB system. Table 3 lists the calculation condition to decision the specification of the matrix converter at maximum power density point. It is noted that the conduction loss (P_{conS}, P_{conD}) and switching loss P_{tonS} of a BTB system are calculated as

$$P_{conS} = k_{con1}I_o^2 \left(\frac{1}{8} + \frac{\lambda}{3\pi}\cos\theta \right) + k_{con2}I_o \left(\frac{1}{2\pi} + \frac{\lambda}{8}\cos\theta \right)$$
$$\ldots\ldots\ldots\ldots\ldots\ldots\ldots\ldots\ldots\ldots\ldots\ldots(19)$$

$$P_{conD} = k_{con1}I_o^2 \left(\frac{1}{8} - \frac{\lambda}{3\pi}\cos\theta \right) + k_{con2}I_o \left(\frac{1}{2\pi} - \frac{\lambda}{8}\cos\theta \right)$$
$$\ldots\ldots\ldots\ldots\ldots\ldots\ldots\ldots\ldots\ldots\ldots\ldots(20)$$

$$P_{tonS} = \frac{E_{dc}f_s}{2\pi V_s} \left(2k_{ton1}I_o + k_{ton2}\pi \right) \ldots\ldots\ldots(21)$$

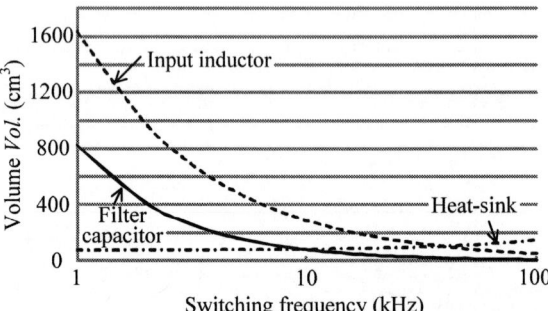

Fig. 12. Volume characteristics for switching frequency. The volume of the input inductor and filter capacitor is reduced when the switching frequency increases because the cut-off frequency becomes high. On the other hand, the volume of the heat-sink is increased owing to the switching loss.

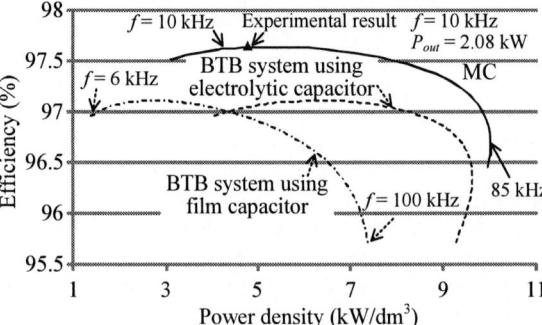

Fig. 13. Pareto front curves of the matrix converter and the BTB system (Rated power = 2 kW). It can be confirmed that the maximum power density in the matrix converter and the BTB system can be accomplished at 85 kHz and 100 kHz, respectively. The efficiency and the maximum power density in the matrix converter are 97.0% and 9.7 kW/dm³, respectively.

where λ is the modulation index, $\cos\theta$ is the output power factor, and E_{dc} is the DC link voltage [17]. Note that the efficiency characteristic for the power density was obtained when the switching frequency was varied from 6 kHz to 100 kHz. In addition, the output power is 2 kW. Moreover, the volume takes the input LC filter, the switching device and the heat-sink into consideration. Furthermore, the film capacitor is used as the filter capacitor in the matrix converter and the DC link capacitor in the BTB system.

As shown in Fig. 13, it can be confirmed that the maximum power density in the matrix converter and the BTB system can be accomplished at 85 kHz and 100 kHz, respectively. The efficiency and the maximum power density in the matrix converter are 97.0% and 9.7 kW/dm³, respectively. On the other hand, the efficiency and the power density in the BTB system are 95.7% and 7.4 kW/dm³, respectively. Thus, it is confirmed that the efficiency and the power density in the matrix converter are higher than that of the BTB system, and the optimized frequency depends on the converter type. In other words, it is necessary to determine the switching frequency of each converters by using the pareto-front curve. Furthermore, the experimental result is plotted in Fig. 13 as the switching frequency and the output power are 10 kHz and 2.08 kW. It is noted that the efficiency in the experiment is almost agree to the calculation result. However, the power density error between the calculation and the experiment is 10.0%. This is because the input LC filter in the experiment is not optimized.

Based on these results, it is easy to optimize the high efficiency and high power density of the matrix converter by using the pareto-front curve which takes many trial times.

VI. CONCLUSIONS

In this paper, the derivation methods of the conduction loss and the switching loss in the matrix converter were proposed. Additionally, the maximum power density for the matrix converter was discussed.

As the result, it was confirmed that the total loss error between the calculation and experimental result is 8.65% at maximum point. Moreover, it was confirmed that the maximum efficiency is 97.9%. Thus, the validity of the calculation formulas of the matrix converter loss could be confirmed. Furthermore, the efficiency and the maximum power density in the matrix converter are 97.0% and 9.7 kW/dm^3, respectively. On the other hand, the efficiency and the maximum power density in the BTB system are 95.7% and 7.4 kW/dm^3. Thus, it is confirmed that the efficiency and the power density in the matrix converter are higher than that of the BTB system, and the optimized frequency depends on the converter type. Based on these results, it is easy to optimize the high efficiency and high power density of the matrix converter by using the pareto-front curve which takes many trial times.

In future work, the total loss and power density will be compared between the matrix converter and the BTB system in experiment.

ACKNOWLEDGMENT

A part of this study was supported by industrial technology grant program in 2011 from new energy and industrial technology development organization (NEDO) of Japan.

REFERENCES

[1] U. Badstuebner, J. Miniboeck, J. W. Kolar: "Experimental verification of the efficiency/power-density (η-ρ) Pareto Front of single-phase double-boost and TCM PFC rectifier systems", APEC, pp. 1050-1057 (2013)

[2] F. Zare, G. F. Ledwich: "Reduced Layer Planar Busbar for Voltage Source Inverters", IEEE Transactions, Vol. 17, No. 4, pp. 508-516 (2002)

[3] P. W. Wheeler, J. Rodriguez, J. C. Clare, L. Empringham: "Matrix Converters: A Technology Review" IEEE Transactions, Vol. 49, No. 2, pp274-288, 2002.

[4] M. Mengoni, L. Zarri, A. Tani, G. Rini, G. Serra, D. Casadei: "A Modulation Strategy for Matrix Converter with Extended Control Range and Reduced Switching Power Losses", IEEE Energy Conversion Congress and Exposition, Vol. , No. , pp. 2721-2728 (2013)

[5] J. Itoh, I. Sato, H. Ohguchi, K. Sato, A. Odaka, N. Eguchi: "A Control Method for the Matrix Converter Based on

Virtual AC/DC/AC Conversion Using Carrier Comparison Method", IEEJ Trans. D, Vol. 124, No. 5, pp. 457-463 (2004)

[6] J. Itoh, H. Kodachi, A. Odaka, I. Sato, H. Ohguchi, H. Umida: "A High Performance Control Method for the Matrix Converter Based on PWM generation of Virtual AC/DC/AC Conversion", JIASC IEEJ, Vol. , No. , pp. I-303-I-308 (2004)

[7] Thomas Friedli, Johann W. Kolar, Jose Rodriguez, Patrick W. Wheeler: "Comparative Evaluation of Three-Phase AC–AC Matrix Converter and Voltage DC-Link Back-to-Back Converter Systems", IEEE Trans., Vol. 59, No. 12, pp. 4487-4510 (2012)

[8] Bo Wen, X. Zhang, Q. Wang, R. Burgos, P. Mattavelli, D. Boroyevich: "Comparison of Three-Phase AC-AC Matrix Converter and Voltage DC-Link Back-to-Back Converter Topologies Based on EMI Filter", ECCE US, Vol. , No. , pp. 2698-2706 (2013)

[9] R. Moghe, R. P. Kandula, A. Iyer, D. Divan: "Loss comparison between SiC, hybrid Si/SiC, and Si devices in direct AC/AC converters", ECCE, pp. 3848-3855 (2012)

[10] J. Itoh, T. Iida, A. Odaka:" Realization of High Efficiency AC link Converter System based on AC/AC Direct Conversion Techniques with RB-IGBT" Industrial Electronics Conference, Paris, PF-012149,2006

[11] K. Koiwa, J. Itoh: "Efficiency Evaluation of a Matrix Converter with a Boost-Up AC Chopper in an Adjustable Drive System", IEEJ Journal of Industry Applications, Vol. 3, No. 1, pp. 26-34 (2014)

[12] Y. Kashihara, J. Itoh: "Parformance Evaluation among Four types of Five-level Topologies using Pareto Front Curves ", ECCE US, pp. 1296-1303 (2013)

[13] F. Schafmeister, C. Rytz, Johann W. Kolar: "Analytical calculation of the conduction and switching losses of the conventional matrix converter and the (very) sparse matrix converter", APEC2005, Vol. 2, No. , pp. 875-881 (2005)

[14] R. Lai, F. Wang, R. Burgos, Y. Pei, D. Boroyevich, B. Wang, T. A. Lipo, V. D. Immanuel, K. J. Karimi: "A Systematic Topology Evaluation Methodology for High-Density Three-Phase PWM AC-AC Converters", IEEE Trans., Vol. 23, No. 6, pp. 2665-2680 (2008)

[15] K. Kato, J. Itoh: "Development of a Novel Commutation Method which Drastically Suppresses Commutation Failure of a Matrix Converter", IEEJ Trans. D, Vol. 127, No. 8, pp. 829-836 (2007)

[16] Wm. T. mclyman: "Transformer and inductor design handbook", Marcel Dekker Inc., Vol. , No. , pp. (2004)

[17] J. Xu, Y. Sato: "A Method to Determine Minimum DC-Link Capacitance in PWM Rectifier-Inverter Systems", IEEJ Transactions, Vol. 133, No. 8, pp. 804-811 (2013)

[18] U. DROFENIK, G. LAIMER, J. W. KOLAR: "Theoretical Converter Power Density Limits for Forced Convection Cooling", Proceedings of the International PCIM Europe Conference, Vol. , No. , pp. 608-619 (2005)

Capacitor Clamped Multi-level Matrix Converter

Siddharth Raju and Ned Mohan

Department of Electrical and Computer Engineering
University of Minnesota
Minneapolis MN 55455

Abstract— Multi-level matrix converter (MMC) is a three phase ac-ac converter that can synthesize three or more levels at the output voltage while maintaining desired input power factor. In this paper a new converter, Capacitor clamped Multi-level Matrix converter (CMMC) has been proposed as an addition to the MMC family. In comparison to conventional matrix converter, the CMMC offers an additional level at the output voltage. This leads to lower THD and voltage stress. This additional voltage is obtained by means of clamp capacitors which are maintained at half of input line to line voltages. An indirect space vector based modulation strategy is presented for generating desired output voltage at desired input power factor while balancing the clamp capacitor voltage at desired voltage level. The effectiveness of the proposed converter has been verified by means of numerical simulation.

Keywords— *capacitor clamped, matrix converter, multi-level, space vector PWM*

I. INTRODUCTION

Matrix converter (MC) is a three phase ac-ac converter, an alternative to the commonly used back to back converter. Similar to the latter, the MC can synthesize required three phase output voltage while maintaining desirable input power factor. Unlike the back-back converter, MCs do not require limited lifetime and bulky DC bus capacitor. The growing need for reliable and compact three phase motor drives has led to intense research into considering MC's as an alternative to the traditional three phase ac-ac converter namely the back to back converter. In spite of these massive advantages the large scale industrial implementation of MC's remains limited due to its disadvantages such as, low immunity to input disturbances, complex commutation, zero ride-through capability, higher semiconductor devices and limited output voltage magnitude.

Various control strategies and modifications have been proposed to conventional MC in [2-5] to overcome the challenges mentioned above. These converters either lie in one of the two categories, indirect matrix converter (IMC) and direct MC based on the presence or absence of a DC link having their own set of merits and demerits. In [6] a sparse indirect matrix converter has been discussed and has same functionality as a direct MC but lesser switches. A topology with much lower switch count has been discussed in [7] called the ultra-sparse matrix converter but the input current and voltage displacement is limited

to 30°. Both of these converter, though have lesser switches, they have higher conduction losses due to more series connected devices along any conduction path in comparison to conventional MC. The IMC topologies also offer the added advantage of zero current commutation as explained in [8-9]. In the two stage IMC, the converter side switches are commuted during zero current state of the inverter which leads to safer and easier commutation and lowered switching losses. As opposed to the IMN, the direct matrix converters offers features such as bi-directional power flow, and low conduction losses due to lesser series connected switches in a conduction path. High frequency component in the common mode voltage has been associated with early bearing failure in motors. MCs have been promising in reducing/eliminating common mode voltage as in [10]. Apart from the broad MC classification, other topologies have been proposed with specific advantages such as zero common mode voltage and greater than unity voltage transfer ratio as illustrated in [11-12].

The application of MC has been limited to low power application due to high switching losses and voltage stresses associated with high power medium voltage applications. In dc-ac conversion, multi-level converters have proven to be an effective solution to this limitation. The same concept has been extended to MC called the multi-level matrix converter (MLMC) as briefly presented in [13]. These are mainly of three types [14-16]: flying capacitor MC, modular matrix converter and diode clamped MC. Each of the converters has their roots in multi-level inverter topologies [17]. In flying capacitor MC the converter consists of 18 bidirectional switches and 9 line connected capacitors connected through the bi-directional switches and is derived from flying capacitor multi-level inverter. The modular matrix converter is a hybrid of cascaded H bridge inverter and MC. Each bi-directional switch of a direct MC is replaced by a capacitor connected H bridge cell. Diode clamped multi-level inverter topology has been modified to form the diode clamped MLMC.

The flying capacitor MC discussed earlier falls under the category of direct matrix converter. It has 9 capacitors, the voltages of which are maintained at half input line-line voltage. In this paper an indirect matrix converter topology using the same principle of flying capacitor MC has been presented. The significant advantage of the indirect topology as opposed to direct topology has been discussed in detail in the following sections. In a conventional MC,

978-1-4799-2706-7/14 $31.00 © 2014 IEEE   1036

the output voltage is switched between input line voltage and zero. In the flying capacitor MC, the capacitors are maintained at half line voltage thus offering half line voltage at the output in addition to line and zero voltage levels. This leads to 3 distinct voltage levels at the DC link. The proposed converter offers much lesser THD than conventional MC. In addition it also offers the possibility of limited ride through capability without any modifications. ITMC topology has been discussed in first section. The control of the above mentioned converter carried out using indirect space vector PWM for the converter and inverter has been explained in Sections 3. Following this, simulation results for the proposed converter have been presented to verify the effectiveness of the converter.

II. INDIRECT CAPACITOR CLAMPED MULTI-LEVEL MATRIX CONVERTER TOPOLOGY

The CMMC topology is an indirect matrix converter as opposed to the direct topology of flying capacitor MC. The conversion from direct to indirect topology leads to a significant reduction in number of capacitors and switches needed. The proposed capacitor clamped Matrix Converter (ITMC) requires 24 switches and 3 capacitors as opposed to 36 switches and 9 capacitors required by flying capacitor MC. This reduction in number of capacitor not only leads to more reliability and lesser size but also easier control and greater limit on the input current harmonics due to capacitor charging and discharging due to voltage balancing. The CMMC topology consists of 9 bidirectional switches and 3 clamp capacitors forming the front end converter and a conventional 2 level inverter as shown in Fig. 1. These capacitors are of the same order of magnitude as that of filter capacitors and are in no way bulky as those needed for back to back converter. The voltage across the capacitor is maintained at half of input line to line voltage across which the capacitor is connected. In conventional IMC, the DC bus voltage can be one of the input line-line voltage or zero. There is a limitation on which of the three possible line to line voltage can be switched, which would be the two of the voltages with highest magnitude for unity power factor. Thus the DC bus oscillates between peak of line-line voltage and zero which in turn is reflected on the output.

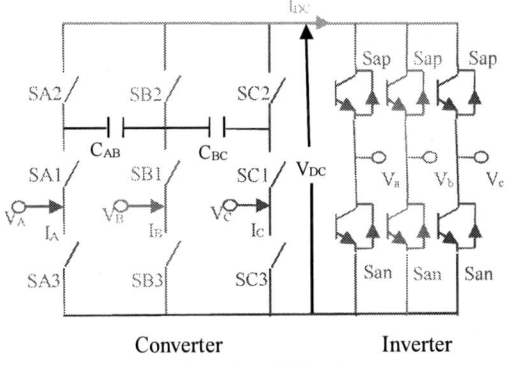

Fig. 1: CMMC Topology

In CMMC, the availability of half line voltage leads to greater number of possibility of DC bus voltage. When the load requires lower voltage, merely switching on half line voltage and zero across the DC bus would suffice and this leads to an extreme reduction THD and voltage stresses on the inverter side as opposed to IMC. The converter has 9 bi-direction switches which leads to 2^9 possible switch configurations and the inverter has six switches leading to 2^6 possible configurations. Input of the converter is connected to a voltage source and hence should not be shorted while the output which is typically connected to a motor is an inductive load and must never be left open. In addition to this the DC bus must never be shorted. This constraint as summarized in (1-2) limits the number of valid switching configuration to 21 on converter side and 8 on the inverter.

$$S_{Aj} + S_{Bj} + S_{Cj} + S_{Nj} = 1 \qquad j \in \{1,2,3,4\} \quad (1)$$
$$S_{ip} + S_{in} = 1 \qquad j \in \{a,b,c\} \quad (2)$$

Some of the converter switching configurations are redundant i.e. they generate the same voltage on the DC link but either charge or discharge the clamp capacitor. This principle is used to maintain the capacitor voltage at half line voltage and is explained in detail in the following section.

III. CMMC MODULATION TECHNIQUE

In conventional space vector PWM for IMC, the DC bus is maintained at 1.5 times that of input phase voltage peak. The converter side sole function is to regulate the input current angle. The inverter side controls the output voltage magnitude. This kind of role assignment fails to hold good in case of CMMC especially during higher modulation index for the following reason. During lower modulation index, the CMMC uses solely the half line voltage and zero. This zero vector can be either applied from the converter side or the inverter side. In most cases this zero state is applied on the inverter side and the converter zero is eliminated and the modulation index is proportionally reduced. This has an advantage of zero current commutation. During higher modulation index, the CMMC solely uses a combination of half line voltages and full line voltage to generate desired voltage at the output. The absence of a zero vector is the key to reduction of THD. Thus the zero vector state can't be used on the inverter side as it would defeat the whole purpose. For generalization, the zero vector state of the inverter is completely ignored and only the active vectors are used to follow the space vector hexagon. The effective increase in the output voltage due to this absence of zero vector is compensated for by reducing the modulation index appropriately. Thus in CMMC, the DC voltage magnitude and the input current phase angle is controlled by the converter while the inverter merely follows the outer space vector hexagon, the detailed analysis of which is discussed in the following subsections.

a) Converter Space Vector PWM:

As mentioned earlier there are 21 valid switching configurations for the converter. In this section, a space vector based modulation strategy is presented to synthesize required DC bus voltage and sinusoidal input current by use of the valid switch configurations.

The main principle of operation is the availability of half line to line voltage on the DC bus. Since these voltages are not readily available, they are synthesized by charging and discharging of the clamp capacitors. The clamp capacitors C_{AB} and C_{BC} are maintained at half of line voltage V_{AB} and V_{BC} respectively. If the capacitor voltage goes above the reference voltage it is discharged by connecting to the load and if it is less than the reference it is charged by connecting in series with the input source terminal and the load. Pictorial representation of this process is presented in Fig. 2. Fig. 2a and Fig. 2b show the charging and discharging process for maintaining average DC bus voltage be $1/2V_{AB}$. As observed, this can be achieved either by turning on switches SA1, SB2 and SB3 or SA2, SB1 and SB, the former charging the capacitor while the latter discharging it. A similar process for maintaining $1/2V_{AC}$ on the DC bus is shown in Fig. 2c and Fig. 2d, the first being charging and the other being discharging state.

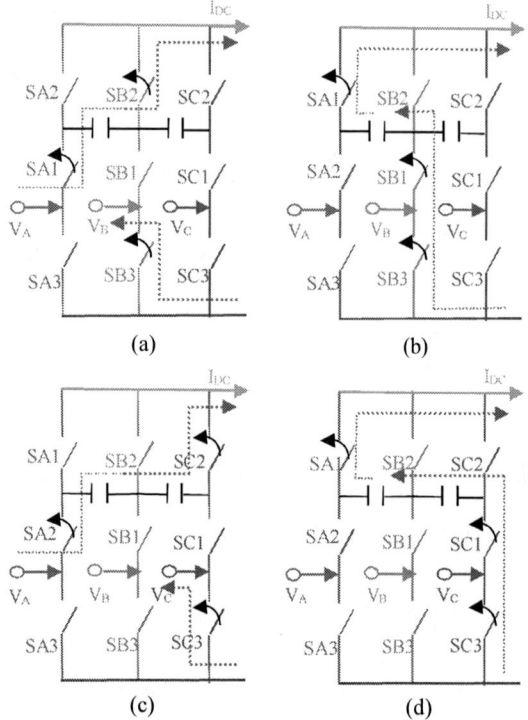

Fig. 2: Switching Configuration DC bus voltage and Capacitor state: (a) $1/2V_{AB}$ and charging, (b) $1/2V_{AB}$ and discharging, (c) $1/2V_{AC}$ and charging, and (d) $1/2V_{AC}$ and discharging

Under balanced load and source conditions, generating sinusoidal output voltage from a constant average DC bus voltage, leads to the input current to be inevitably sinusoidal in a MC. This same condition fails to hold for the CMMC due to the presence of clamp capacitors. The capacitor charging and discharging leads to harmonics in the input current. The size of these capacitor determine the effect on input current harmonics as well as the required switching frequency. A large value of capacitance would

predominantly charge during positive dv/dt leading to higher input current and discharge during negative dv/dt leading to close to zero input current component which causes excessive harmonics in the input. If the capacitance were small, it would charge and discharge multiple time in every switching cycle. The average current through the capacitor would be a small fraction of the input current required to support the load which would be almost zero and does not affect the input current waveform. But there is a limitation on how fast the capacitor can be charged and discharged due to the practical limitations on how fast the devices can be switched. A favorable situation would be to charge and discharge every alternate switching cycle. This ensures there aren't more switching vectors changes over a time period as compared to IMC. The clamp capacitor value can be determined from (3) where ΔV is clamp capacitor voltage fluctuation tolerance, Δt is the switching time period and I_{DC} is the DC bus current. A larger tolerance means smaller capacitor but higher output voltage distortion.

$$C = \frac{\Delta t}{\Delta V} I_{DC} \qquad (3)$$

There are six switch configurations for the IMC which generates non zero voltage on the DC bus and are referred to as the active vectors. As opposed to IMC the CMMC has 18 such non zero active vectors. Of these, there are six configurations similar to IMC which leads to line-line voltage on the DC bus. These states leave the clamp capacitors "hanging" and do not contribute to balancing of these capacitors. The absence of neutral line allows for the possibility of transforming the three dependent input current variables into two independent variables namely the input current space vector as per (4). The switching configuration leading to line voltage on the DC bus and the corresponding input current space vector is given in Table I. The last column indicates the clamp capacitor state (C: Charging, D: Discharging and NC: No Change).

$$I_i = I_A + I_B e^{i\frac{2\pi}{3}} + I_C e^{i\frac{4\pi}{3}} \qquad (4)$$

TABLE I
Vectors Contributing to Outer Hexagon

Vector	Switches ON	V_{DC}	Input SV		Cap St
			$|I_i|$	α_i	
1	SA1,SA2,SB3	V_{AB}	$\sqrt{3}I_{dc}$	$-\pi/6$	NC
2	SA1,SA2,SC3	V_{AC}	$\sqrt{3}I_{dc}$	$\pi/6$	NC
3	SB1,SB2,SC3	V_{BC}	$\sqrt{3}I_{dc}$	$\pi/2$	NC
4	SB1,SB2,SA3	V_{BA}	$\sqrt{3}I_{dc}$	$5\pi/6$	NC
5	SC1,SC2,SA3	V_{CA}	$\sqrt{3}I_{dc}$	$7\pi/6$	NC
6	SC1,SC2,SB3	V_{CB}	$\sqrt{3}I_{dc}$	$-\pi/2$	NC

Switching configurations involving the clamp capacitor produce half line voltage on the DC bus. As discussed earlier, there are redundant states which produce same DC bus voltage but either charger or discharge the capacitors. When the capacitors are charging, the input current space vector magnitude is $\sqrt{3}I_{dc}$ same as those of vectors in Table I and while discharging it is zero. In both these cases the DC bus voltage is half of input line to line voltage. Half line voltage vectors must produce half line current for sinusoidal input currents which implies that the charging and discharging vectors must be used more or less equally.

978-1-4799-2706-7/14 $31.00 © 2014 IEEE

A proper choice of clamp capacitor as explained above ensures this condition. The half line voltage switching vectors which form the inner hexagon and the corresponding capacitor state is given is Table II. The inner hexagon vectors produce half the voltage as that of outer hexagon vector. Under ideal condition, that is almost equal use of charging and discharging vectors, the average input current for half hexagon vectors is half that of outer hexagon but this does not lead to any reduction in the input current harmonics as it is merely an average of $\sqrt{3}I_{dc}$ and zero.

TABLE II
Vectors Contributing to Inner Hexagon

| Vector | | Switches ON | V_{DC} | Input SV | | Cap St |
				$|I_i|$	α_i	
7	7a	SA1,SB2,SB3	$V_{AB}/2$	$\sqrt{3}I_{dc}$	$-\pi/6$	C
	7b	SA2,SB1,SB3	$V_{AB}/2$	0	$-\pi/6$	D
8	8a	SA1,SC2,SC3	$V_{AC}/2$	$\sqrt{3}I_{dc}$	$\pi/6$	C
	8b	SA2,SC1,SC3	$V_{AC}/2$	0	$\pi/6$	D
9	9a	SB1,SC2,SC3	$V_{BC}/2$	$\sqrt{3}I_{dc}$	$\pi/2$	C
	9b	SB2,SC1,SC3	$V_{BC}/2$	0	$\pi/2$	D
10	10a	SB1,SA2,SA3	$V_{BA}/2$	$\sqrt{3}I_{dc}$	$5\pi/6$	C
	10b	SB2,SA1,SA3	$V_{BA}/2$	0	$5\pi/6$	D
11	11a	SC1,SA2,SA3	$V_{CA}/2$	$\sqrt{3}I_{dc}$	$7\pi/6$	C
	11b	SC2,SA1,SA3	$V_{CA}/2$	0	$7\pi/6$	D
12	12a	SC1,SB2,SB3	$V_{CB}/2$	$\sqrt{3}I_{dc}$	$-\pi/2$	C
	12b	SC2,SB1,SB3	$V_{CB}/2$	0	$-\pi/2$	D

The input current space vector for all valid switching configuration is shown in Fig. 3. The inner hexagon is formed by half line voltage vectors while the outer hexagon is formed by full line voltage vectors. There are three zero vectors obtained by turning on all the switches in any one of the vertical arm.

The various regions and the switching vector used within each sector is shown in Fig. 4. Depending on the position of the reference, a minimum of three vectors are required to enclose it. The closest three vectors lead to minimum possible THD. The region is chosen as such with the exception of region 4. In region 4, the best choice for minimum THD in output voltage would be to choose two outer hexagon vectors and one inner hexagon vector.

Under this case, only one of the clamp capacitor voltage would be under control while the other is left floating for every 60°. This would lead to excessive surge currents when transfer of region occurs which would lead to input current spikes every 60° and in turn higher input current harmonics. To avoid this, at any sector, both inner hexagon vectors must be employed for complete capacitor voltage control. The ratio by which the inner hexagon vectors are shared determines the point at which the virtual vector V_1, resultant of these two vectors lies. The computationally easiest solution would be to share these equally and hence the resultant virtual vector V_1 lies in-between the line connecting them as shown. The duty cycle of each vector is given in Table III.

b)   Inverter Space Vector PWM:

The inverter is a conventional two level inverter consisting of 6 switches with freewheeling diodes. There are six active space vector configuration and three zero states. The inverter output voltage space vector for various possible switching patterns is shown in Fig. 5. The space vector PWM for a two level inverter has been discussed in numerous literature and only a brief account of its employment with regards to CMMC is discussed.

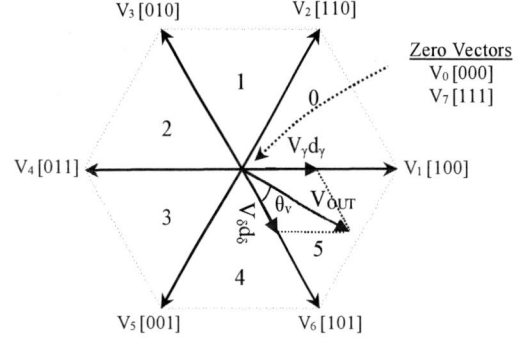

Fig. 5: Inverter Space Vector

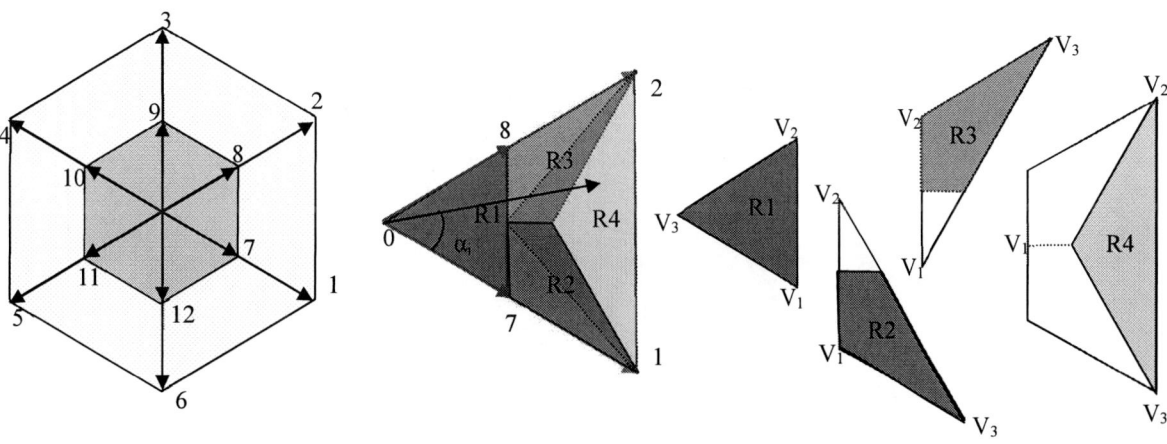

Figure 3: Converter Space Vector   Figure 4: Regions within a Sector

978-1-4799-2706-7/14 $31.00 © 2014 IEEE

TABLE III
Duty Cycle for Converter Switching Configuration

R	Duty cycle		
	δ_1	δ_2	δ_3
R_1	$\frac{4}{\sqrt{3}} m'_c \cos(\alpha_i + \pi/6)$	$\frac{4}{\sqrt{3}} m'_c \sin(\alpha_i)$	$1 - \frac{4}{\sqrt{3}} m'_c \cos(\alpha_i - \pi/6)$
R_2	$\frac{4}{\sqrt{3}} m'_c \cos(\alpha_i + \pi/6)$	$2 - 4m'_c \cos(\alpha_i)$	$\frac{4}{\sqrt{3}} m'_c \cos(\alpha_i - \pi/6) - 1$
R_3	$2 - 4m'_c \cos(\alpha_i - \pi/3)$	$\frac{4}{\sqrt{3}} m'_c \sin(\alpha_i)$	$\frac{4}{\sqrt{3}} m'_c \cos(\alpha_i - \pi/6) - 1$
R_4	$2 - \frac{4}{\sqrt{3}} m'_c \cos(\alpha_i - \pi/6)$	$\sqrt{\frac{7}{3}} m'_c \cos(\alpha_i - \pi/2.53) - 0.5$	$\sqrt{\frac{7}{3}} m'_c \cos(\alpha_i + \pi/16.5) - 0.5$

As briefly mentioned in the beginning of this section, the inverter merely control the output voltage space vector angle. The magnitude is controlled by the converter side. The inverter cannot have a zero component as it would increase THD at higher modulation index. This limitation leads to the inverter reference space vector being limited to the hexagon formed by sharing of active vectors which effectively control the space vector angle. The time period for which zero state should have been exerted is shared proportionally with two of the active states as given by (4-5). This leads to pulsation in the output voltage space vector. This is compensated for by proportionally reducing the convert modulation index m_c as given by (6). The proportional sharing of the vectors leads to the advantage of not having to measure or calculate the actual voltage distortion that needs to be compensated.

$$d_\delta = \frac{\sin(\frac{\pi}{3} - \theta_v)}{\sin\left(\frac{\pi}{3} - \theta_v\right) + \sin(\theta_v)} \tag{4}$$

$$d_\gamma = \frac{\sin(\theta_v)}{\sin\left(\frac{\pi}{3} - \theta_v\right) + \sin(\theta_v)} \tag{5}$$

$$m'_c = m_c * (\sin\left(\frac{\pi}{3} - \theta_v\right) + \sin(\theta_v)) \tag{6}$$

IV. SIMULATION RESULTS

The working of the proposed converter using the proposed modulation strategy has been carried out using MATLAB based simulation. The input voltage, output load and the converter switching frequency is as mentioned in Table 4.

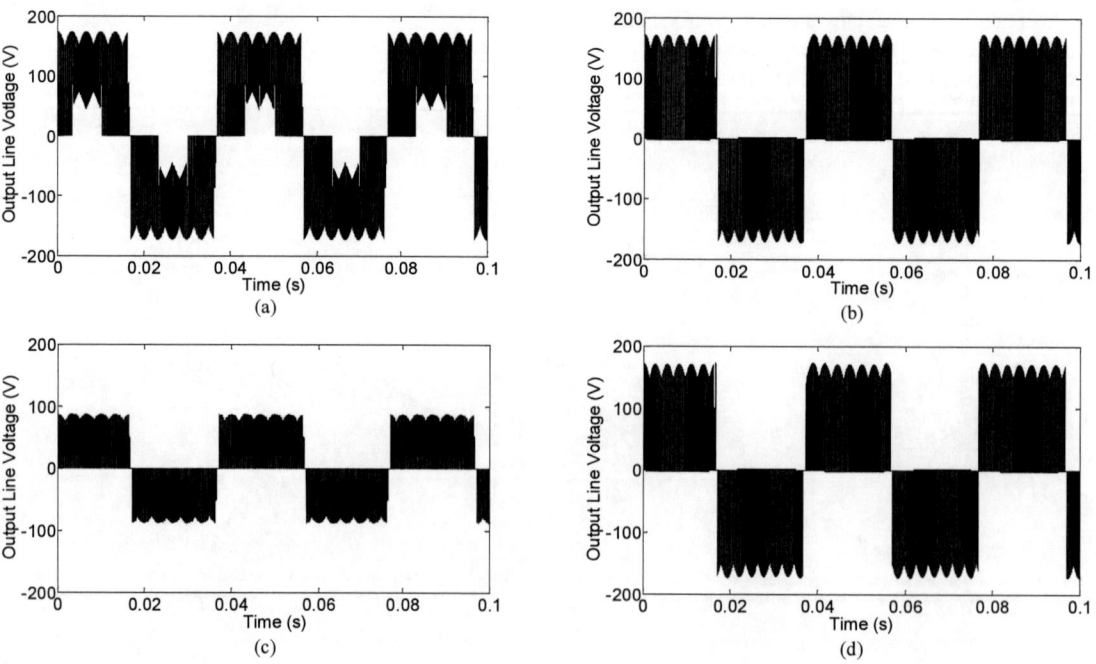

Fig.5: Output line to line voltage of CMMC (on left) and MC (on right) for modulation index of 0.75: (a and b) and 0.25: (c and d)

978-1-4799-2706-7/14 $31.00 © 2014 IEEE

The 2014 International Power Electronics Conference

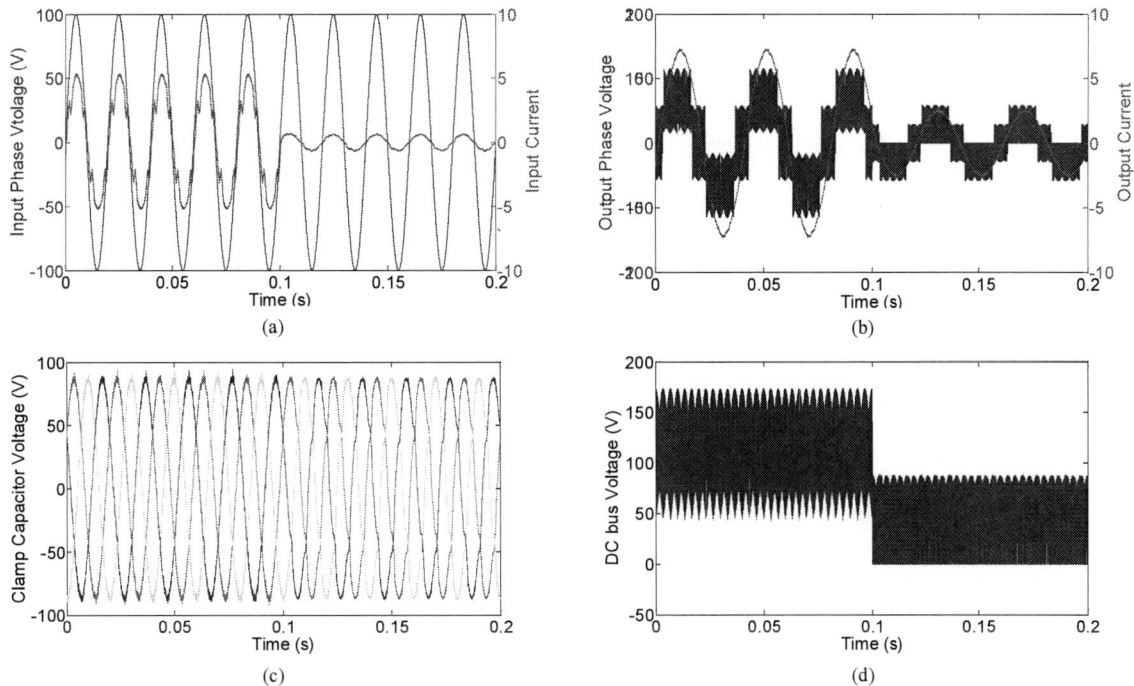

Fig.6 (a): Input phase voltage and current, (b): Output phase voltage and current, (c): Clamp capacitor voltage and (d): DC bus voltage

TABLE IV
Values for Simulation Parameters

Parameter	Value
Input Voltage	100 V, 60 Hz
RL Load	10 Ω, 15 mH
Output Frequency	25 Hz
Switching Frequency	6 kHz

In Fig. 5a and Fig. 5c, the output line to line voltage of CMMC for modulation index of 0.75 and 0.25 respectively is shown and similarly for MC in Fig. 5b and Fig. 5d. As it can be observed, the voltage level of MC is limited and hence irrespective of the modulation index the output voltage always switches between input line to line voltage and zero. In case of CMMC, at lower modulation index, the output switches between half line to line voltage and zero and at higher modulation index it switches between half line to line voltage and line to line voltage. This leads to significant reduction in THD out voltage stresses in CMMC at the output terminal in comparison to MC.

The input current, output phase voltage and current, clamp capacitor voltage and DC bus voltage for modulation step of 0.75 to 0.25 at 0.1s is shown in Fig. 6. As it can be observed, the capacitor voltage is maintained at half of input line to line voltage. As it can be seen, the peak of DC bus voltage steps down by half at lower modulation index and in turn so does the output phase voltage.

V. CONCLUSIONS

A novel capacitor clamped MC has been proposed in this paper as an alternative to the flying capacitor MC. It

is a much compact and reliable solution due to 33% reduction in number of switches needed and 66% reduction in the capacitor needs. The proposed converter is three level converter as opposed to two levels of conventional MC. This additional level leads to drastic reduction in the output voltage stresses and THD. An indirect space vector based modulation scheme has been presented to generate desired output voltage and input power factor while balancing the clamp capacitor voltage. This topology finds major use in medium voltage high power drives where conventional MC use is limited due to higher stresses and low quality output waveform especially at lower modulation index. The feasibility of the proposed converter has been verified using MATLAB based simulation. Though the proposed converter requires more discrete components compared to conventional MC, this disadvantage is far outweighed by its advantage.

REFERENCES

[1] M. Venturini, "A new sine wave in sine wave out, conversion technique which eliminates reactive elements," in *Proc. POWERCON 7*, 1980, pp. E3_1–E3_15.

[2] T. Satish, K. K. Mohapatra, N. Mohan, "Modulation methods based on a novel carrier-based PWM scheme for matrix converter operation under unbalanced input voltages," in *Proc. APEC'06*, 2006, pp. 127-132.

[3] J. Mahlein, J. Igney, J. Weigold, M. Braun, D. Simon, "Matrix converter commutation strategies with and without explicit input voltage sign measurement", *IEEE Trans. Ind. Electron.*, vol. 49, pp. 407-414, Apr. 2002.

[4] J.W. Kolar, F. Schafmeister, S. D. Round, H. Ertl, "Novel three-phase ac-ac sparse matrix converter," *IEEE Trans. Power Electron.*, vol. 22. pp. 1649-1661, Sept. 2007.

978-1-4799-2706-7/14 $31.00 © 2014 IEEE

[5] S. Thuta, K. K. Mohapatra, N. Mohan, "Matrix converter over-modulation using carrier-based control: maximizing the voltage transfer ratio," in *Proc. PESC'08*, 2008, pp. 1727-1733.

[6] M. Y. Lee, C. Klumpner, P. Wheeler, "Experimental evaluvation of the indirect three-level sparse matrix converter," in *Proc. PEMD'08*, 2008, pp. 50-54.

[7] J. Schonberger, T. Friedli, S. D. Round, J. W. Kolar, "An ultra sparse matrix converter with a novel active clamp circuit," in *Proc. PCC'07*, 2007, pp. 784-791.

[8] L. Wei, T. A. Lipo, "A novel matrix converter topology with simple commutation," in Conf. *Rec. IEEE IAS Annual Meeting*, 2001, pp. 1749-1754.

[9] B. Wang, G. Venkataramanan, "A carrier based PWM algorithm for indirect matrix converter," in *Proc. PESC'06*, 2006, pp. 1-8

[10] H. J. Cha, P. N. Enjeti, "An approach to reduce common mode voltage in matrix converter," *IEEE Trans. Ind. Application*, vol. 39, pp. 1151-1159, Aug. 2003.

[11] S. Mahadevan, S. Raju, R. Muthu, "Elimination of common mode voltage using phase shifted dual source matrix converter with improved modulation index," *COMPEL: The Intl. Journ. for Comp. and Math. in Elect. Eng.*, vol. 32 Iss: 6, pp.2006 – 2026, 2013

[12] R. K. Gupta, K. K. Mohapatra, A. Somani, N. Mohan, "Direct-matrix-converter-based drive for a three-phase open-end-winding AC machine with advanced features," *IEEE trans. Ind. Electron.*, vol. 57, pp. 4031-4042, Dec. 2010.

[13] P. Wheeler, X. Lie, M. Y. Lee, L. Empringham, C. Klumpner, and J. Clare, "A review of multi-level matrix converter topologies," in *Proc. Power Electronics Machines & Drives Conf.*, 2008, pp. 286-290.

[14] S. Angkititrakul and R. W. Erickson, "Control and Implementation of a New Modular Matrix Converter," in *Proc. IEEE Applied Power Electronic Conf.*, 2004, pp. 813–819.

[15] Y. Shi, X. Yang, Q. He, and Z. Wang, "Research on a novel capacitor clamped multilevel matrix converter," *IEEE Trans. Power Electronics, vol.20 pp. 1055-1065*, Sept. 2005.

[16] M. Y. Lee, P. Wheeler, C. Klumpner, "A new modulation method for the three-level-output-stage matrix converter," in *Proc. Power Conversion Conf.*, 2007, pp. 776-783.

[17] C. Newton and M. Summer, "Multi-level convertors a real solution to medium/high-voltage drives?," *Power Eng. Journal*, vol. 12, pp. 21-26, Feb. 1998.

European Trends and Technologies in Traction

Uwe Drofenik, Francisco Canales
ABB Corporate Research,
CH-5405 Dättwil, Switzerland
uwe.drofenik@ch.abb.com

Abstract-**Advanced semiconductors allow realization of traction systems with minimized weight and size, increased modularity and higher efficiency. These trends are driven by a growth in transportation volume, increasing passenger numbers, environmental concerns, and the desire for higher passenger comfort. After discussing trends and general traction systems, we go into the technical details of the power electronics transformer which has the potential to replace low-frequency transformers on board which are among the most bulky components in today' traction chains.**

I. INTRODUCTION

With advanced semiconductors like IGBTs designed for high voltage and high power, it became possible to replace components like the GTO and introduce new concepts in traction allowing further system optimization concerning weight and size. Also, system efficiency can be improved which is strongly desired due to environmental aspects and high energy cost.

Since rail transportation is competing with individual transport (cars) on short distances and airplanes on long distances, increasing passenger comfort is a strong trend requiring increased levels of auxiliary power on trains. Power consumers on board of trains are e.g. air conditioning, automatic doors, lights and laptops and other electronic equipment of passengers. From increased power requirements of auxiliary supplies, a trend of employing modular converter systems based on medium frequency transformers and soft-switched highly efficient converters can be seen (e.g. Bordline M [1]). This results in light-weight and efficient supplies with reduced volume. More space for passengers and/or freight is an important issue on trains which translates directly into economic benefits.

After modular medium frequency transformers combined with power electronics, also known as "power electronics transformer (PET)", have been employed successfully in on-board auxiliary supplies, the idea is being tested to employ such modular PET-based converter systems for the propulsion of the motors of the train. The main idea is to increase the transformer frequency significantly by applying power electronics so that the transformer size, weight and cost will shrink accordingly. This is especially an issue in some European countries which employ a 16.7Hz AC-grid resulting in very large transformers on board of trains. Other benefits of PET in traction are flexible design due to modularity and higher efficiency. Cost should not exceed established solutions based on large low-frequency transformers. Reliability issues due to higher component numbers, especially of the power electronics converters, might require redundant converter modules and/or increased service and maintenance.

Another idea being tested is transformer-less topologies, which are modular and require increased insulation in the propulsion motors. The insulation, which is provided by the transformer in traditional designs has now to be considered in the motors which requires new motor designs.

A future trend will be the application of wide band gap power semiconductors in traction. With next generation SiC power semiconductors, it will be possible to switch high frequencies with very low losses at blocking voltages of several kV. This would allow very compact and highly efficient PET-based converters. Another degree of optimization for low-frequency operation of unipolar SiC devices is to further reduce the conduction losses compared to Si-devices by increasing the chip area.

TABLE I
RAIL VOLTAGES IN EUROPE

Year of introduction	Type	Voltage [kV], frequency [Hz]
1890	DC	0.75 / 1.5 / 3.0
1910	1-phase AC	15
1940	1-phase AC	25

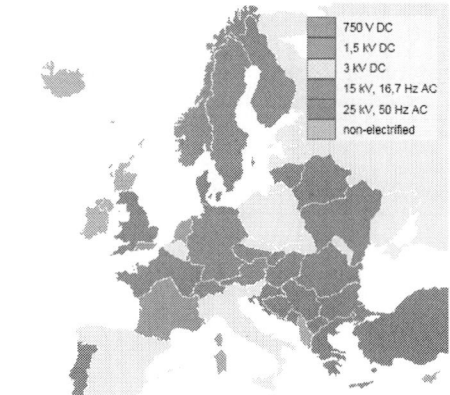

Fig. 1. Rail electrification in Europe [2]. 25kV line for high-speed trains in France, Spain, Italy, etc. are not shown.

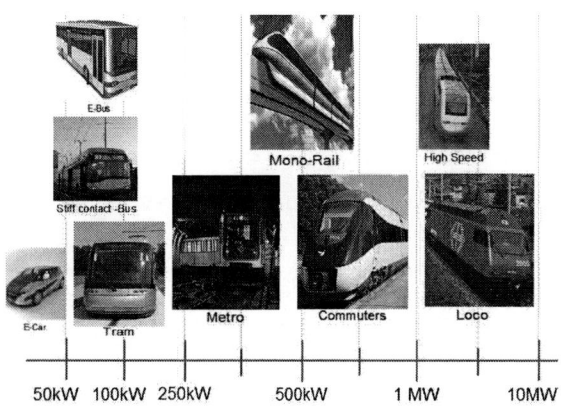

Fig. 2. Power levels in traction.

978-1-4799-2706-7/14 $31.00 © 2014 IEEE

Trolleybus systems compete with light trains on short distance passenger transportation in cities. A trolleybus gets electrical power from an overhead line like a tram. Recently, a new kind of electrical bus was introduced in Switzerland which charges its batteries at stops [3]. Here, no large batteries are required, and there are no electrical overhead lines which are considered as visually not attractive by many people and require certain maintenance.

The railway system in Europe is composed from a wide range of different systems, from DC electrification of different voltage levels, and from AC electrification of different voltages and different frequencies, see Fig. 1. The systems do not only differ from country to country but also inside countries.

For trains crossing country borders, it is important to be able to handle different electrical grids. For vendors of train equipment, the variety of grids in Europe (AC or DC, voltage level, frequency) provides a challenge for the design of optimized traction systems. Due to the high cost of redesigning electrical railway infrastructure, the European railway electrification system will remain fragmented, and the traction system on board has to handle this.

Figure 2 shows different power levels in traction. In the following section, different basic concepts of traction power supply systems will be discussed.

II. TRACTION ARCHITECTURE

A traction system for light trains is shown in Fig. 3 with properties given in Tab. 2. The overhead line provides low voltage DC (600..750V), and no transformer is needed. The DC/AC inverters for the propulsion motors are directly connected to the train-internal DC bus which is directly connected to the DC supply grid via line filters. A DC/DC converter with a medium frequency transformer provides galvanic separation for auxiliary systems at relatively low power (35kVA) and low voltage (750V).

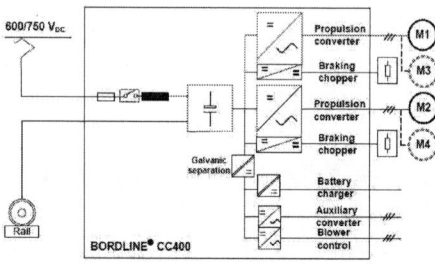

Fig. 3. Traction system for light trains (Bordline CC 400, see [1]).

TABLE II
TRACTION SYSTEM (BORDLINE CC 400) FOR LIGHT TRAINS (FIG. 3)

Propulsion output	2x 150kW (0..500Vac)
Braking chopper	2x 250kW at wheel
Auxiliary converter	3x 400V, 50Hz, 35kVA
Dimensions	1.6 x 1.8 x 0.43 m³
Weight	550kg

An example of a modular traction converter for high power applications like high speed trains or freight trains is shown in Fig. 4 with key properties given in Tab. 3. Such a multi-system can handle 15kV AC or 25kV AC, or 1.5kV DC or 3kV DC. The different grids are handled by a large low frequency transformer and/or switches. The power converters which provide speed-variable three-phase AC power for the propulsion motors and auxiliary power are designed in a modular way and

can be combined in various ways to serve applications that differ in required power.

The power electronics converters on the motor side are compact and modular. The transformer at the grid side is the remaining large and bulky component which is especially large and heavy if it operates in the 15kV rail grid at 16.7Hz AC which is employed in many European countries, see Fig. 1.

Replacing this large low-frequency 16.7Hz-transformer, which is typically of low efficiency (slightly above 90%) due to cost and shape requirements, by a potentially light-weight and efficient power electronics transformer (PET) has the potential to improve the whole traction system significantly. Scaling laws, optimization and design flexibility of the PET technology from a medium frequency transformer's point of view are discussed in sections IV and V in detail.

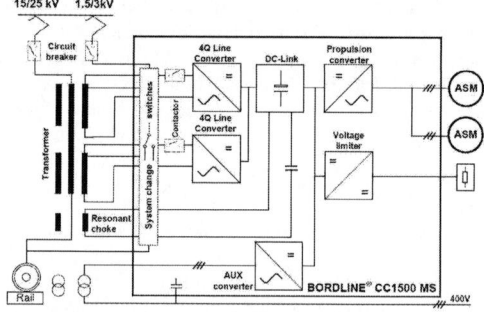

Fig. 4. Multi-system for (high-speed) locomotive (Bordline CC 1500, see [1] for details).

TABLE III
TRACTION SYSTEM (BORDLINE CC 1500) FOR MULTISYSTEM LOCOMOTIVE (FIG. 4)

Propulsion output	2200kW (0..2700Vac)
Auxiliary converter	360kVA
Dimensions	3.2 x 1 x 2 m³
Weight	2400kg

Fig. 5. Dual-mode diesel-electric locomotive EURODual (photo: Vossloh, see [3]).

Another interesting development are traction systems with combined diesel-electric dual-mode technology. An example is the locomotive EURODual with an electric and a supplementary auxiliary diesel drive, shown in Fig. 5. This enables to serve sidings or sections without a catenary wire by diesel and use the electric drive on electrified railway lines, which increases operational efficiency since locomotives do not have to be exchanged depending on the supply system. In addition, CO_2 emissions can be reduced compared to a pure diesel-electric vehicles.

III. CATENARY FREE ELECTRIC BUS FOR MASS TRANSIT

Trolleybus systems compete with light trains on short distance passenger transportation in cities. A new kind of electrical bus was introduced in Geneva, Switzerland, which charges its batteries at stops. Therefore, no large batteries are required, and there are no overhead lines for power supply.

The advantages are
- reduced "visual pollution" by overhead lines,
- providing greater route flexibility,
- and higher operating reliability and less maintenance due to missing overhead lines.

For the bus shown in Fig. 6, the onboard lithium batteries have to be charged at stops within 15 seconds. The power to be handled by the converters is approximately 400kW. At the end of the bus line there is a longer charging period between 3 and 4 minutes. For details see [4].

Fig. 6. TOSA (French abbreviation for: Trolleybus Optimisation Systeme Alimentation): Catenary free electrical bus for mass transit in Geneva, Switzerland [4].

IV. POWER ELECTRONICS TRANSFORMER

Based on the trends discussed in the previous section, one of the components with a lot of optimization potential in a traction system is the low frequency transformer which connects the converter to the grid. This transformer is especially bulky in the 15kV/16.7Hz grid. By increasing the frequency by several orders of magnitude, the transformer can be reduces in size and cost accordingly. Such so-called "power electronics transformers (PET)" have been introduced successfully in traction auxiliary power supply systems for power levels up to several hundreds of kW, and look promising for replacing line frequency transformer based traction converters feeding large power to the propulsion motors. The power levels are up to a few MW, and operating voltages up to 35kV (and higher) have to be guaranteed.

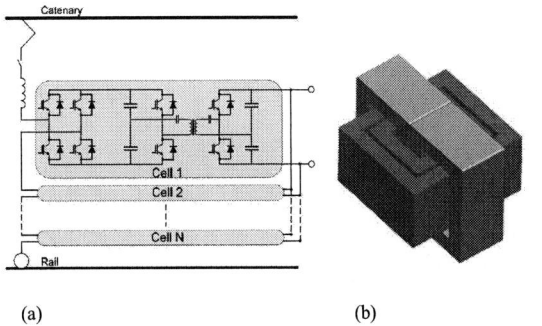

Fig. 7. (a) Resonant multi-cell topology for PET [5], [6], (b) general geometry of a shell-type medium frequency transformer (MFT).

One of the most promising converter topologies in terms of efficiency, reliability and simplicity is a modular approach employing zero current switching in resonant operation. The individual cells are series connected at the input side and parallel connected at the output side, see Fig. 7(a) and/or [5], [6] for details of the converter operation.

If there are e.g. 7 cells (N=7) and each cell's transformer has a 1:1 turns ratio, the converter system shown in Fig. 7(a) steps down the catenary voltage by a factor 7. The stationary insulation requirement for the medium frequency transformer in the cells is up to the catenary voltage level (plus a margin). Obviously, the number of cells is a significant design decision, and there are technology-dependent optimum designs.

A basic transformer design for the medium frequency transformer of each individual cell is shown in Fig. 7(b). In the following, we will perform all calculations for discussing scaling trends, optimization and design challenges based on this shell-type transformer model.

Generally, the shell-type transformer has a higher magnetic coupling than the core-type transformer, which has no center leg, and where the two windings are located on the outer core legs. Therefore, the stray inductance of the shell-type transformer is generally lower than the stray inductance of the core-type transformer. It is easier to increase the stray inductance (e.g. by adding a stray path) than to reduce the stray inductance, and since the stray inductance is an important parameter of the resonant topology (Fig.7), it is beneficial to select a transformer type which allows easy adaption of the stray inductance.

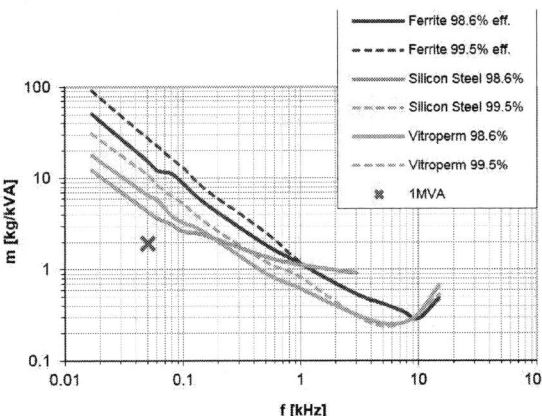

Fig. 8. Weight-optimized MFT (180kVA, turns ratio 7:1, 27kV insulation) scaling of the power density for different core materials. Insulation and cooling: air-insulation, forced convection.

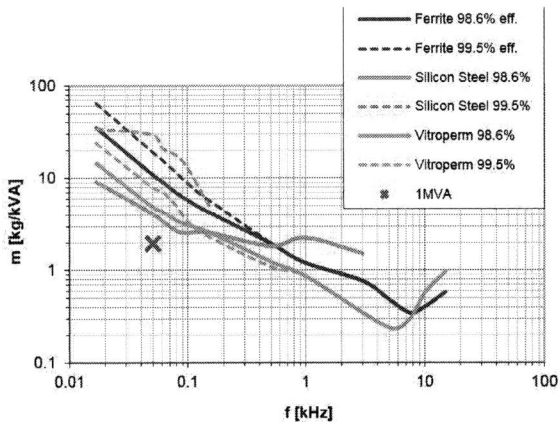

Fig. 9. Weight-optimized MFT (300kVA, turns ratio 4:1, 27kV insulation) scaling of the power density for different core materials. Insulation and cooling: air-insulation, forced convection.

For optimizing the system design, one has to optimize each single system component. Therefore, a systematic transformer optimization is required. All of the following curves and characteristics are based on design procedures as outlined in detail in [7] with minimum weight as optimization target.

The electrical-thermal transformer model includes saturation, inhomogeneous flux density, skin- and proximity-effect, temperature-dependency of winding resistance, non-sinusoidal and high-frequency core- and winding losses, inhomogeneous thermal conductivity in the core, convective cooling of transformer surfaces, and electro-static stress between wires and winding layers. There are several free optimization parameter (e.g. core geometry and turns number) which allow optimization. Boundary conditions like maximum core- and windings-temperatures, maximum losses and outer shape of the design have to be considered including mechanical and production constraints. Scaling factors for material properties are introduced to consider 30 years lifetime. The optimum parameter set defines an optimum transformer design.

Fig.8 and Fig.9 show normalized weight of examples of weight-optimized transformer designs dependent on operating frequency. In Fig.8, the converter consists of 7 cells, each at 3.3kV with 180kVA, realizing a converter of 1.2MVA with an insulation voltage of 27kV. In Fig.9, there are just 4 cells, each at 5.4kV with 300kVA, realizing a converter with equal properties (1.2MVA with an insulation voltage of 27kV). The 7-cell converter could be realized with 6.5kV Si-semiconductors allowing maximum switching frequencies of around 2.5kHz. Employing SiC devices of this voltage class, one could increase the switching frequency into the range of 20kHz and higher. The 4-cell converter would require 10kV SiC-devices, and these could operate at 20kHz and higher, allowing to take full advantage of the reduced size of the transformer. The curves in Fig.8 and Fig.9 are not perfectly straight and/or smooth because some design criterions and conditions (e.g. turns ratio) are discrete.

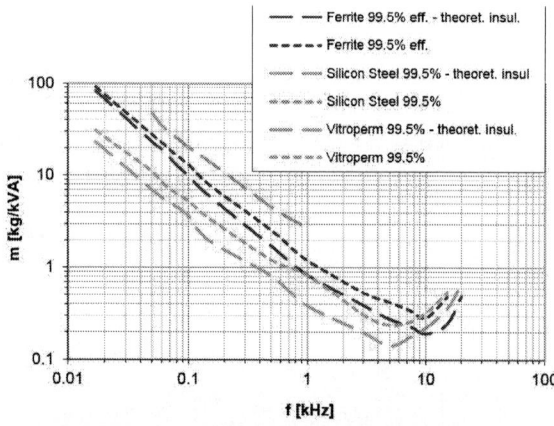

Fig. 10.   Weight-optimized MFT (1MW, turns ratio 1:45, 10kV insulation) scaling of the power density for different core materials. Cooling and insulation: oil-immersed, natural convection.

One interesting issue is the reliability of the converter when taking into account the redundancy of cells, which allows ongoing operation even in case of failure of a cell. In case of a large number of cells, it is easier to allow redundancy of at least one cell without significantly over-dimensioning the converter. On the other hand, increasing the number of cells also increases

the total number of components which reduces reliability. Another issue is that designing a large number of cells allows to employ low-voltage semiconductors which might be available at lower cost. All these issues are important design criterions, and are dependent on the application.

The insulation in our design example consist of air in gaps around the high-voltage winding. This makes cooling via forced convection relatively easy. A fan blows air through the gaps between the two windings of the MFT, the gaps between windings and core, and around the core surface, see Fig.7(b) for the general transformer design. In order to guarantee insulation between high-voltage and low-voltage winding e.g. in case of environments containing dust, the required insulation distances have to be increased by a certain factor. Theoretically, air shows an insulation capability of 3kV/mm, but in our design example we calculate with a reduced insulation capability of just 1kV/mm.

The required insulation distances grow accordingly, resulting in a larger and more expensive transformer design than compared to a theoretical minimum. A comparison of the size-increase due to applying a factor three on the insulation distances is shown in Fig.10. In case of mainly conductive cooling (e.g. solid insulation or indirect oil cooling) the insulation distance has a significant impact of the internal thermal resistances of the transformer, and the effect will be much more significant.

Each point on a curves in Fig.8 and Fig.9 represents a different design which is optimized for this certain frequency. There are two sets of curves in the figures, a solid one for designs with a minimum required transformer efficiency of 98.5%, and a dotted one for designs with a minimum required transformer efficiency of 99.5%. Each curve has been calculated for a certain minimum required efficiency, but the actually achieved efficiency might be higher. At high frequencies, the transformer can only be further reduced in size if one accepts a falling efficiency. If this is not accepted, one has to increase the transformer size at higher frequencies again which does not make sense, but explains the rising curves after a minimum at high frequency.

The red cross symbol in the figures shows an existing 1MW distribution transformer (scaled down to single-phase for direct comparison from the existing three-phase transformer). The design associated with the red cross symbol is oil-cooled and oil-insulated which is, at low frequencies, more effective than air-cooling. Therefore, the red cross symbol is not located on (or above) the green line (Silicon steel at 50Hz) which represents air-cooling, but below. Since cooling- and insulation-systems are different, a direct comparison does not make sense, but it allows to quickly evaluate the performance of air-cooled PETs over the frequency range.

It is very important to note that all these curves are not general curves. They are strongly dependent on cooling, electrical insulation requirements, voltage- and power level.

Generally, there are different core materials to be considered. The most important ones are listed in Table IV. In different frequency ranges different core materials are of advantage as can be directly seen in Fig.8 and Fig.9. In case of weight-optimization one would switch from Silicon steel to nanocristalline core materials which typically perform best in the range 1kHz – 10kHz. Above 10kHz Ferrite cores might be the best option but might be difficult to use, especially in case of larger power and size, due to their brittleness. If one considers the comparably high cost of nanocrystalline cores, a cost-optimization might change the picture significantly in favor of Silicon steel at lower frequencies and Ferrite at high frequencies. Amorphous cores (e.g. METGLAS) are not listed in the table, because they typically cannot provide better performance in any frequency range compared to one of the three core materials listed in Table IV. This is only true for medium frequency transformers. In case of e.g. filter inductors, amorphous cores show excellent performance at comparably low cost.

TABLE IV
MAGNETIC CORE MATERIALS FOR PET

Core material	Silicon Steel	Nanocristalline (e.g. Vitroperm)	Ferrite
Saturation	1.7T	1.1T	0.3T
Permeability	1,500	40,000	2,300
Losses	Very high above a few 100Hz.	Relatively low up to 5kHz, very high above 10kHz.	Extremely low up to (several) 100kHz.
Thermal	Very good thermal conductivity, 30K/Wm.	Rather poor thermal conductivity, below 1K/Wm.	Good thermal conductivity, about 5K/Wm.
Mechanical	Easy to handle, high design flexibility, very robust, allows large cores.	Robust, allows large cores, pressure-sensitive, cuts are very expensive and/or critical for performance.	Very brittle, tolerances in mm-range, does not allow single-piece large cores.
Cost	Low cost.	Very high cost.	Moderate.

TABLE V
INSULATION SYSTEM FOR PET

Insulation material	Air	Oil	Solid (e.g. Epoxy)	SF6, Vacuum, Two-Phase Liquid
Required distances	Very large (3kV/mm), strict standards.	Small distances (10kV/mm).	Small distances (10kV/mm).	Theoretically very small.
Cooling	Simple and very efficient, can be easily increased.	Easy cooling if convection is possible, otherwise as typical for PET difficult (indirect water).	Very difficult cooling (conduction 0.2K/Wm).	Gas/vacuum: Very difficult Two-Phase: Excellent
Mechanical and other	Large distances result in large cores. Works better at high power levels, limits for brittle core materials (Ferrite).	Self-healing, environmental issues are a concern in certain applications.	Robust, but danger of cracking in insulation resulting in partial discharge and insulation failure.	Containment might be a challenge: special standards, size limitations.
Reliability	Very high, Fan required.	Highly reliable, but might be critical in case of indirect water.	Serious reliability issue due to thermal cycling and cracking.	Partial discharge in bubbles (two-phase cooling) might destroy the liquid chemically.
Cost	Low.	Moderate.	Moderate.	Very high.

Table V lists different possible insulation systems. They all have to be evaluated especially from the viewpoint of cooling, furthermore, considering the high voltages between primary and secondary winding. Fig.11 shows the typical voltage stress on the insulation of a MFT of one cell in the converter topology shown in Fig.7. All cells experience the same AC voltage plus a DC voltage component which is at maximum for the "highest" cell in the topology. This kind of voltage stress is a challenge for insulation systems and another justification of a safety factor in the required insulation distances.

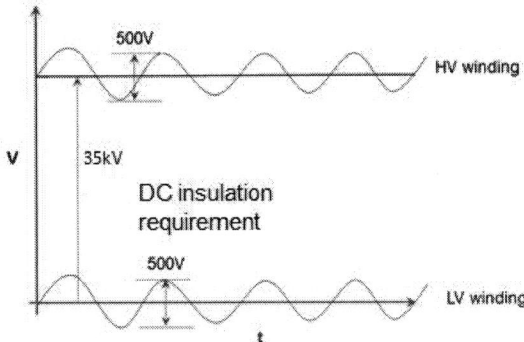

Fig. 11. Insulation stress on the MFT due to the topology consisting of series-connected cells on the high-voltage side, see Fig.7 for the converter topology.

Combinations of core materials (listed in Table IV) and insulations systems (listed in Table V), considering adequate cooling systems, result in different transformers which have advantages and disadvantages dependent on optimization target and application. Systematic design analysis as presented in Fig.8 and Fig.9 allows finding the best medium frequency transformer and its best operating frequency for a given application. In section V, the Pareto front analysis is discussed which provides another useful tool for comparing different MFT technologies and finding the best design.

V. SCALING AND PARETO FRONT ANALYSIS OF THE POWER ELECTRONICS TRANSFORMER

For finding the optimum system design, it is necessary to optimize transformers for different cooling methods, core materials, cell numbers (power- and voltage levels) and frequencies. Furthermore, in the PET-topology of Fig.7(a), it is important to achieve a certain mains- and stray inductance of the transformer for low-loss operation of the power semiconductors. In order to compare a large number of design options Pareto-front diagrams of optimized transformer designs are plotted in terms of efficiency versus power density.

Typical system operating frequencies are given in Table VI. The transformers in such systems have to be designed accordingly. A typical Pareto front plot for a 1MVA oil-immersed (oil-cooling, oil-insulation) transformer optimization is show in Fig.12. For a given frequency, e.g. 50Hz, transformer design optimizations are performed with different required minimum efficiencies. This results in curves like the dark green solid curve in Fig.12 labelled "50Hz Silicon Steel". Performing such optimizations at different frequencies gives a group of such curves, one per selected efficiency. The upper and lower boundary point of each curve of the group give, if connected, a new curve, the so-called Pareto front. Transformer designs inside the Pareto front represent optimized designs which can be realized, and points directly on the Pareto front curve represent theoretical optimum designs which allow to evaluate how much potential for design improvement is theoretically available.

TABLE VI
TYPICAL TRANSFORMER FREQUENCIES

Railway	16.7Hz
Distribution grid	50/60Hz
Ship, airplane	400Hz – 800Hz
Medium voltage PET	2kHz – 20kHz
AC-bus power supply for computer, control boards, satellites	up to 100kHz and higher

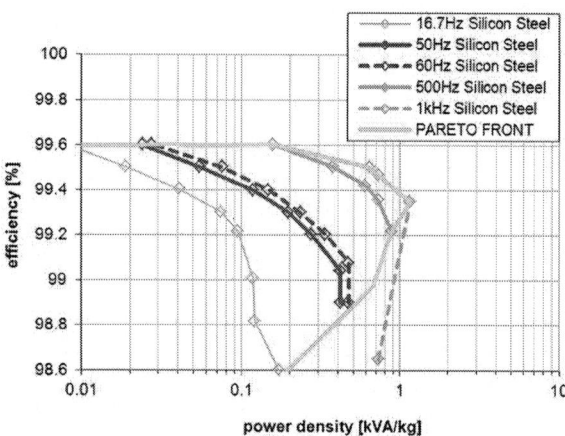

Fig. 12. Example of a Pareto front of weight-optimized MFT (1MW, turns ratio 1:45, 10kV insulation) with a silicon steel core for various operating frequencies. Insulation and cooling: oil-immersed, natural convection.

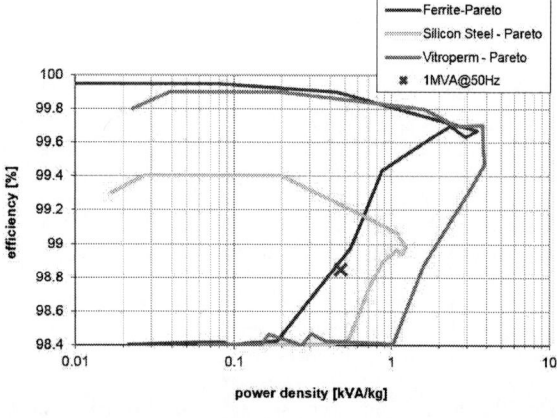

Fig. 13. Pareto front of weight-optimized MFT (1.2MW, turns ratio 4:1, 27kV insulation) as discussed in Fig.8 for different core materials. Insulation and cooling: air-insulation, forced convection.

Pareto front plots showing transformer efficiency over power density are shown in Fig.13 and Fig.14 for the design examples shown in Fig.8 and Fig.9. The design points in the upper right are of maximum efficiency and maximum power density (minimum weight at given power), but one has to make sure

that the associated frequency makes sense for the power semiconductors of the converter.

The designs in the examples (Fig.13, Fig.14) are all based on the same insulation (air) and cooling (forced convection, air), but employ different core materials: Ferrite, Silicon Steel and Nanocristalline cores ("Vitroperm"). This allows a direct and convenient comparison of different core materials. Employing different cooling and/or insulation systems will give additional Pareto front plots, which could all be overlaid, and allow a direct performance comparison of a large number of designs of different technologies.

The frequency information in the Pareto plots of Fig.13 and Fig.14 is "lost" but could be easily added by labelling points on the Pareto curves with the according frequencies.

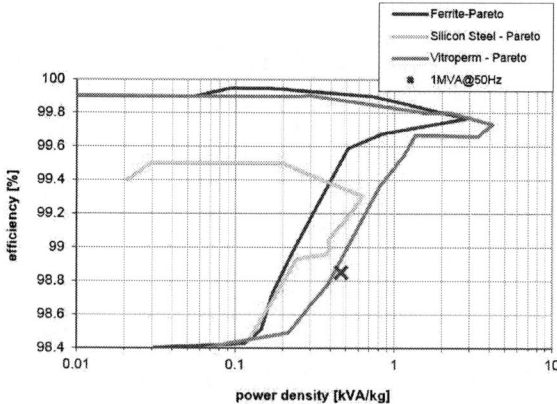

Fig. 14. Weight-optimized MFT (1.2MW, turns ratio 7:1, 27kV insulation) as discussed in Fig.9 for different core materials. Insulation and cooling: air-insulation, forced convection.

VI. SUMMARY

For historical reasons, the European rail system is based on different grids. Trains crossing country borders have to take this into account by power conversion systems which are capable of handling this efficiently. A major component with optimization potential is the line-frequency transformer which is typically optimized for shape and low cost, but with the drawback of low efficiency typically slightly above 90%. This could be improved by highly-efficient medium frequency transformers combined with power electronics converter cells. In order to get the maximum advantage from the available technologies, systematic design optimization has to be performed as discussed in detail in the paper on the example of a 1.2MVA transformer unit with a voltage insulation requirement of 27kV.

REFERENCES

[1] http://www.abb.com/railway
[2] http://en.wikipedia.org/wiki/Rail_transport_in_Europe
[3] http://www.vossloh-innotrans.com/de/unsere_themen/innovative_antriebstechnologien/dual_mode/dual_mode_1.html
[4] http://www.tosa2013.com
[5] D. Drazen, F. Kieferndorf, F. Canales, "Power Electronics Transformer Technology for Traction Applications – An Overview," Electronics, Vol. 16, No.1, pp. 50-56, June 2012
[6] D. Dujic, F. Kieferndorf, F. Canales, U. Drofenik, "Power electronic traction transformer technology," 7th International Power Electronics and Motion Control Conference (IPEMC), vol.1, pp. 636-642, June 2-5, 2012

[7] U. Drofenik, "A 150kW Medium Frequency Transformer Optimized for Maximum Power Density", Conference on Integrated Power Electronics Systems (CIPS), March 6, 2012

[8] T.M. Jahns, V. Blasko, "Recent advances in power electronics technology for industrial and traction machine drives," Proceedings of the IEEE , vol.89, no.6, pp.963-975, June 2001

[9] S. Bernet, "Recent developments of high power converters for industry and traction applications," IEEE Trans. on Power Electronics, vol.15, no.6, pp.1102-1117, Nov 2000

[10] W. Gunselmann, "Technologies for increased energy efficiency in railway systems," European Conference on Power Electronics and Applications 2005

[11] J. Taufiq, "Power electronics technologies for railway vehicles", International Power Conversion Conference (PCC), Nagoya, Japan, 2007

[12] M.-M. Bakran, "A power electronics view on rail transportation applications," 13th European Conference on Power Electronics and Applications (EPE) 2009, Sept. 2009

[13] T. Koseki, "Technical trends of railway traction in the world," International Power Electronics Conference (IPEC) 2010, June 21-24, 2010

[14] L. Eun-Kyu, "Traction technologies for railways in Korea," International Power Electronics Conference (IPEC) 2010, June 21-24, 2010

[15] E. Masada, "Railway technologies in the next decade and power electronics," International Power Electronics Conference (IPEC) 2010, June 21-24, 2010

[16] A. Rufer, N. Schibli, C. Chabert, C. Zimmermann, "Configurable front-end converters for multicurrent locomotives operated on 16 2/3 Hz AC and 3kV DC systems", IEEE Trans. on Power Electronics, vol. 18, no. 5, pp. 1186-1193, 2003

[17] M. Steiner, "Seriegeschaltete Gleichspannungszwischenkreis-umrichter in Traktionsanwendungen am Wechselspannungs-fahrdraht", Phd thesis, ETH Zurich, Switzerland, 2000

[18] M. Steiner, H. Reinold, "Medium frequency topology in railway applications", 12th European Conf. on Power Electronics and Applications (EPE) 2007, Aalborg, Denmark

[19] B. Engel, M. Victor, G. Bachmann, A. Falk, "15kV/16.7Hz energy supply system with medium frequency transformer and 6.5kV IGBTs in resonant operation", 10th European Conf. on Power Electronics and Applications (EPE) 2003, Toulouse, France

[20] M. Mermet-Guyennet, "New power technologies for traction drives", International Symposium on Power Electronics, Electrical Drives, Automation and Motion (SPEEDAM) 2010, Pisa, Italy

[21] M. Claessens, D. Dujic, F. Canales, J.K. Steinke, P. Stefanutti, C. Vetterli: "Traction transformation – A power electronic traction transformer (PETT)", ABB Review, 1/12, pp. 11-17, 2012

[22] M. Mermet-Guyennet, A. Castellazzi, J. Fabre, P. Ladoux, "Electrical Analysis and Packaging Solutions for High-Current Fast-Switching SiC Components," 7th International Conference on Integrated Power Electronics Systems (CIPS), March 6-8, 2012

Co-phase power supply system for HSR

Qunzhan Li, Wei Liu, Zeliang Shu, Shaofeng Xie, Fulin Zhou

School of Electrical Engineering
Southwest Jiaotong University
Chengdu, China
Liuwei_8208@swjtu.cn

Abstract—**Poor power quality and phase splitting are the main issues in electric railway. Co-phase traction power supply system is adopted in electrified railway for active power balance, reactive power compensation and harmonic filter. The research development and application of co-phase traction power supply system in China are depicted detailed in this paper. Experiments and pilot project proved that good power quality, feasible, reliable and economical traction power system can be implemented in co-phase traction power supply system.**

Keywords—Co-phase system, active power balance, reactive power compensation, traction power supply.

I. INTRODUCTION

There are several disadvantages for traditional traction power supply system. Reactive power, harmonic current and negative sequence current injections will cause poor quality problems. Phase split will also create speed loss and influence the reliability of the overhead catenary system [1, 2].

As shown in Fig 1, a co-phase traction power supply system is constituted by traction transformer and active power balance conditioner (APC). Compared with railway static power conditioner (RPC)[3, 4] applied in Japan, the phase split in front of traction substation is eliminated. Only single phase current feeds to the traction network. APC connects between feeding phase and another phase of traction transformer. The function of APC includes active power balancing, reactive power compensation and harmonic filtering.

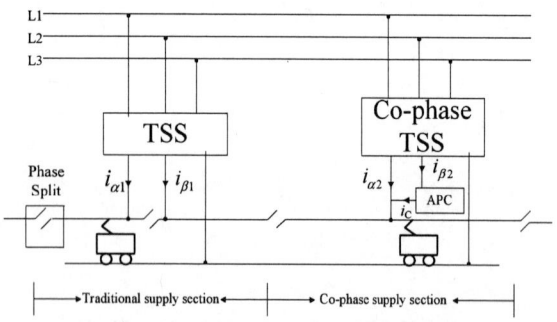

Fig 1 Connection scheme for traditional traction power supply system and co-phase traction power supply system

Compared with traditional traction power supply system, half of the phase splits are cut down. The investment is reduced because of neutral sections in front

of the substation are canceled. The traction performance is improved by reducing power and speed loss in high-speed and heavy load railway. The capacity utilization ratio of traction transformer is increased because of the active power balanced between secondary windings. The power quality issues to utility grid can be solved comprehensively [5, 6, 7].

II. CO-PHASE TRACTION POWER SUPPLY SYSTEM

As shown in Fig.2, A co-phase traction power supply system transmits power from power grid to traction network. A balanced transformer (YNvd) transfers power from three phase to two phase. The secondary windings of YNvd have two phases of 90° difference. Phase A supplies the traction load, and phase B is connected with an active power balance conditioner. A single phase back to back converter is used for active power balancing, reactive power compensation and harmonic filtering. In the connection, only phase A directly connected to the traction network between catenary and track. Therefore, it is a kind of co-phase traction power supply connection scheme.

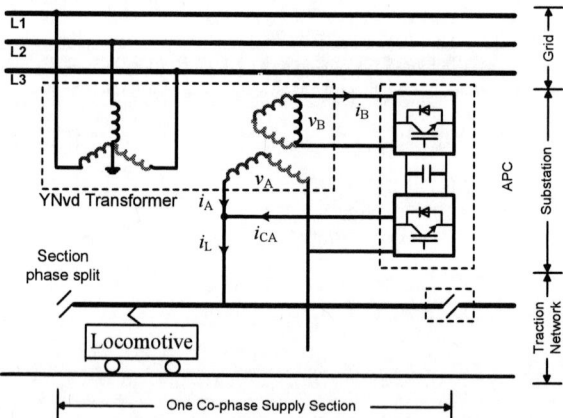

Fig 2 Co-phase traction power supply system including a balance transformer and APC.

A 10MVA co-phase traction power supply device has been built at Meishan substation which is 70km at southwest of Chengdu, China. The project was supported by the National Science and Technology Pillar Program during the Eleventh Five-year Plan Period of China. The

The 2014 International Power Electronics Conference

project is undertaken by Southwest Jiaotong University which is started at 2007 and completed at 2010. Xuji Power Co., CSR Zhuzhou Institute Co., Tsinghua University and Chengdu Railway Bureau have cooperated well together to research, design, manufactured the co-phase traction power device and retrofited the traction substation.

Connection diagram of Meishan co-phase traction substation is depicts in Fig 3. Single phase back to back converter is proposed for active power balancing, reactive power compensation and harmonic filtering. Layout of Meishan substation is described in Fig 4. A impedance matching balance traction transformer is used in substation. 1# APC is manufactured by CSR Zhuzhou Institute Co. 2# APC is manufactured by Xuji Power Co. Different single phase converter topologies and control scheme are verified through two APCs.

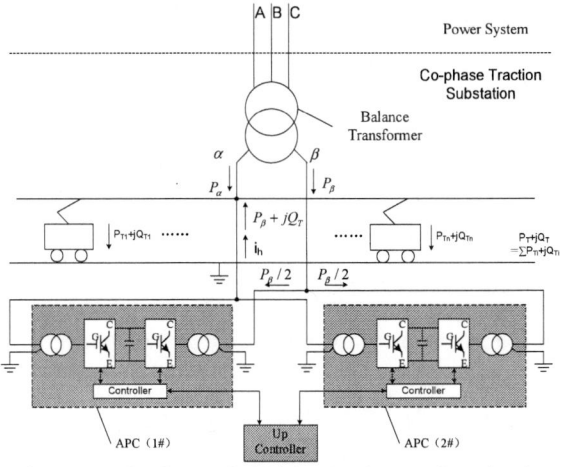

Fig 3 Connection diagram for Meishan Co-phase traction substation

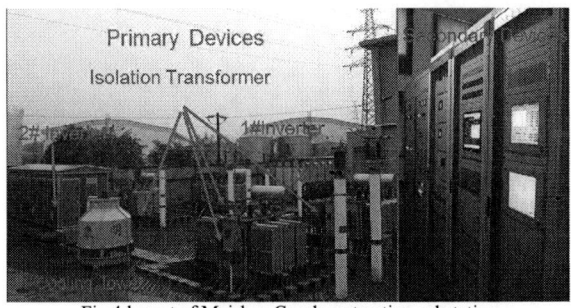

Fig 4 layout of Meishan Co-phase traction substation

In Oct 28, 2010, the world first co-phase traction power supply device has been put into trial operation at Meishan traction substation. So far, co-phase traction power supply system has been operated for 27 months, 13680 hours. AC-DC locomotives are applied in this railway. Before the co-phase traction power supply operated, power factor, harmonic distortion and voltage unbalance are the predominant power quality issues.

Fig 5 shows the power factor monitoring results at Meishan substation. The average power factor of Meishan substation before co-phase traction device

operation is about 0.77. After co-phase device operation, the power factor is compensated to 0.97.

Fig 6 gives the THD measured by FLUKE 43B power quality analyzer. The load current is seriously distorted with THD up to 27.8%. Because of the load influence, the THD of phase α compensating current is up to 7.4%. Meanwhile, the THD of phase β compensating current is only 2.7%.

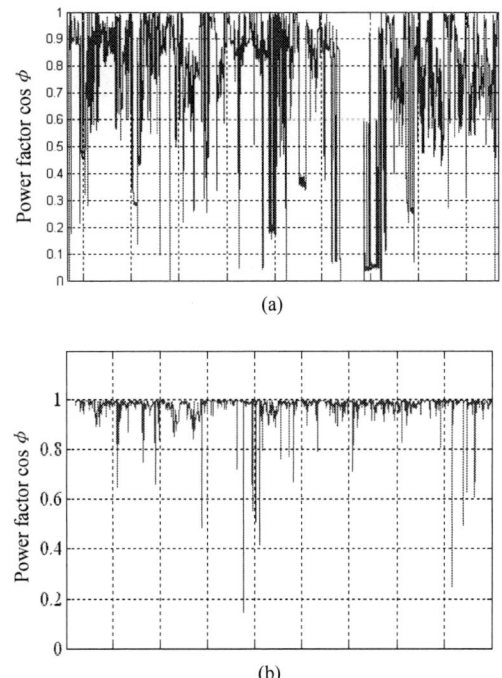

Fig5 Power factor result. (a) power factor at Meishan substation before co-phase device operation. (b) power factor at Meishan-substation after co-phase device operation.

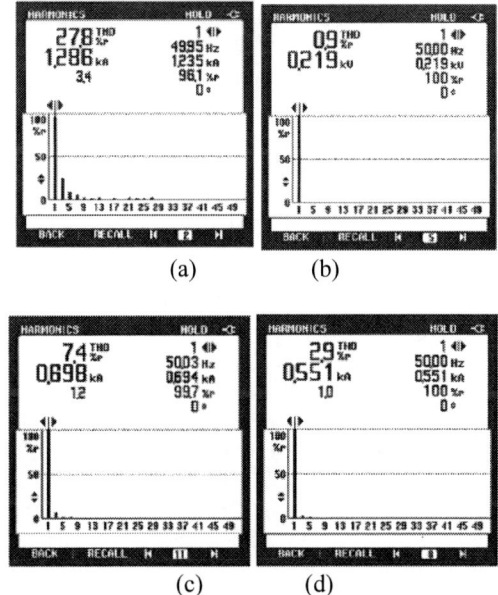

Fig 6 (a) current THD of load of phase A. (b) voltage THD of phase A. (c) APC compensation current of phase α (d) APC compensation current of phase β.

The voltage unbalance index result is shown in Table 1. The APCs transmit one half load required active power from phase β to phase α. The voltage unbalance index has been improved.

Table 1 Voltage unbalance index result

Voltage unbalance index	95% probability unbalance index (%)	Max unbalance index (%)
Before co-phase device operation	0.63	1.06
After co-phase device operation	0.28	0.50

III. COMBINED CO-PHASE TRACTION POWER SUPPLY SYSTEM

A. Principle of combined co-phase traction power supply system

But we still need to be aware of the shortages of Meishan co-phase traction power supply system:

(1) The structure of traction transformer and co-phase traction power supply device are bind together.

(2) Normal traction power supply and train operation will be influenced when co-phase traction power device is out of service.

(3) The capacity of co-phase power supply device needs in accordance with the traction transformer's capacity.

(4) Because of these shortages, high cost of co-phase traction power supply system is the bottle neck in the application.

Currently, the key to the implementation of co-phase power supply is the phase of traction network voltage, which is decided by the traction port of the traction transformer of a certain connection mode. The simplest and most economical connection mode is single-phase traction transformer among all kinds of traction transformer in traction substation.

As shown in Fig 7, a combined co-phase traction power supply device for substation with a single-phase and three phases modular includes a traction transformer (TT), a compensation transformer (CPT) and its co-phase compensation device (CPD) and a spare traction transformer (BTT). The primary sides of the traction transformer, the compensation transformer and the spare traction transformer are connected to high voltage bus; the traction transformer and the secondary side of the backup traction transformer as well as one side of the co-phase compensation device are connected to the traction bus. Under normal circumstances, the traction transformer, the compensation transformer and its co-phase compensation device work, and the spare traction transformer work at standby mode. When the traction transformer quits working, the spare traction transformer begins to work; when the compensation transformer or the co-phase compensation device quits running, the traction transformer can work independently for a while. The spare traction transformer can also replace the traction transformer to work and can run on the different line voltage from the traction transformer.

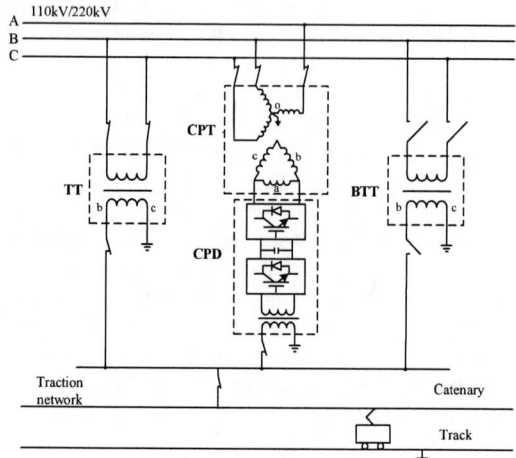

Fig 7 Combined co-phase traction power supply system

The characteristics of a combined co-phase traction power supply device with a single-phase and three-phase modular are as follows: The traction transformer and the spare traction transformer apply single-phase connection which sends the line voltage of power system to the traction bus. The compensation transformer uses YNd11 three-phase standard wiring. The capacity of the three-phase winding is set to be one heavy two light as demanding.

The capacity of co-phase compensation device is determined by the traction load power which causes the exceeding of the national standard limits on the imbalance degree of the three-phase voltage.

The operation principle of combined co-phase traction power supply system is:

When power requirement of traction load is less than two times of co-phase compensation device capacity, half of the traction load active power is supported by co-phase compensation device, so the negative sequence current can be fully compensated.

When the power requirement of traction load is large than two times of co-phase compensation device capacity, co-phase compensation device operates at its rated capacity, remaining negative sequence current still injects into power grid. But three phase voltage unbalance factor caused by injected negative sequence current achieves the nation power quality standard.

Compared with the existing co-phase power supply technology, the benefits of combined co-phase traction power supply system are described as follow:

The capacity of AC-DC-AC converter is minimized and effectively reduces investment in co-phase substation equipment.

The new combination of the traction transformer and the co-phase compensation device is proposed, which has improved the flexibility of the traction substation operation, especially in high speed railway.

Combined co-phase traction power supply system can improve the power utilization ratio of the equipment. The capacity utilization ratio of the single-phase traction transformer is 1. According to the design of one heavy winding load two light winding load, the utilization rate of the three-phase winding YNd11 connection compensation transformer is 75%.

The compensation transformer uses YNd11 standard wiring that can be more easily manufactured. At the same time, the primary side can carry out large current grounding and the secondary side three-phase winding can also provide balanced three-phase power electricity.

B. Application of combined co-phase traction power supply system

The combined co-phase traction power supply device will be installed at Shayu traction substation in Shanxi south central heavy load railway corridor in China, which is a 91.8km length railway with two traction substation. Two types of locomotives are operated for passenger and freight transportation as shown in Table 2. Combined co-phase traction power supply device can helps to solve the power quality issues, such as power quality, harmonic current and negative sequence current injection.

Table 2 Traction load characteristic

Locomotive Type	Rated Power	Power Factor	Current HR(%)	
Passenger Locomotive SS9	4800kW	≥0.81	3rd	20
			5th	10
			7th	7
Freight Locomotive HXD1	9600kW	≥0.97	3rd	3.5
			5th	0.2
			7th	0.2

The main circuit of combined co-phase compensation device is shown in Fig 8. HMT represents high voltage matching transformer connected between 220kV power grid and 10kV multiple connected back to back inverter units. TMT represents traction matching transformer connected between 55kV AT traction power supply system and back to back inverters. Multiple inverter units are parallel connected at 660V TMT low voltage winding side, and series connected to HMT's 10kV winding through inductors L1 and L2. To ensure high reliability, inverter units bypass technology is implemented at the software design. When an inverter units faults, the remaining inverter units keep normal operation to support continuous traction power supply.

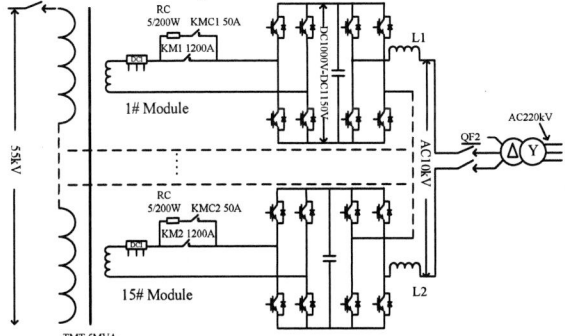

Fig 8 Main circuit of multiple connected combined co-phase compensation device

IV. CONCLUSION

Co-phase traction power supply system is a revolutionary in electrified railway. The system can transfer active power between secondary winding, compensate reactive power and filter harmonic current. Theory and practice show that the co-phase power supply technology does well in removing electricity phase split in the exit of traction substation, eliminating power supply bottlenecks, increasing power supply performance, improving power quality, as well as strengthening traction power supply performance.

REFERENCES

[1] C. Tsai-Hsiang,Y. Wen-Chih, and H. Yen-Feng, A systematic approach to evaluate the overall impact of the electric traction demands of a high-speed railroad on a power system. Vehicular Technology. IEEE transactions on , vol. 57. No 4, pp. 1378-1384, 1998

[2] C.Shi-Lin, R.J, and H. Pao-Hsiang, Traction system unbalance problem analysis methodologies. Power Delivery, IEEE transaction on, vol.19, no.4, pp. 1877-1883, 2004

[3] T.Uzuka and S.Ikedo, Railway static power conditioner field test. Quarterly Report of RTRI, vol. 45, no.2, pp. 64-67, 2004

[4] S.Zhou, J. Xinjian, A novel active power quality compensator topology for electrified railway. Power Electronics, IEEE Transactions on, vol. 19, no.4, pp. 1036-1042, 2006

[5] Zeliang Shu, Shaofeng Xie and Qunzhan Li, Single-phase back-to-back converter for active power balancing, reactive power compensation and harmonic filtering in traction power system. IEEE transaction on power electronics, vol. 26, No.2. pp. 334-343 2011

[6] Zeliang Shu, Shaofeng Xie and Qunzhan Li, Development and implementation of a prototype for co-phase traction power supply system.

[7] Zeliang Shu, Shaofeng Xie, Ke Lu, etc., Digital detection, control and distribution system for co-phase traction power supply application. IEEE transactions on industrial electronics, vol 60. no.5, pp 1831-1839, 2013

The application of electronic frequency converter to the Shinkansen railyard power supply

Toshimasa Shimizu*, Ken Kunomura*, Masahiko Kai**, Mitsuru Onishi*, Hiroshi Masuzawa**
*Construction Department, ** Shinkansen Operations Division Electrical Engineering Department
*, **Central Japan Railway Company
Kanagawa, Japan

Hiroki Miyajima***, Midori Otsuki***, Yoshinori Tsuruma****
***Toshiba Corporation
Kanagawa, Japan
****Toshiba-Mitsubishi Electric Industrial Systems Corporation
Tokyo, Japan

Abstract— In the Tokaido Shinkansen, we built new Electronic Frequency Converter (EFC) to replace Rotary Frequency Changer. The new EFC is individually operated in basic feeding pattern, and there appears the necessity for over-current tolerance in the converters. We describe the application of EFC to the Shinkansen railyard in this paper, showing the development of the control scheme which can control the over-current.

At the same time, the new EFC has unique feature in connecting existing power supplies apart from the EFC, for covering various feeding situations. We also describe the method of interconnection.

Keywords— *over-current suppression control, interlink equipments, electronic frequency converter, Tokaido Shinkansen*

I. INTRODUCTION

Although the Tokaido Shinkansen railway line extends over both 50 Hz and 60 Hz electric power system regions in Japan, the trains are driven by 60 Hz single phase electric power supply. Therefore, in the 50 Hz region, the electric power frequency has to be converted from 50 Hz to 60 Hz. For this purpose, the Tokaido Shinkansen has equipped frequency conversion systems, mainly using Rotary Frequency Changer (the following are referred to as "RFC"). And these ten years, technical development enabled us to introduce static frequency converters (Electronic Frequency Converter and the following are referred to as "EFC"), as Tsunashima and Numazu EFCs.

Recently, we built the third static frequency converter. The new one is called the Oi EFC and placed in the Oi railyard, as the overage replacement of the Hamamatsucho RFC.

The Oi EFC functions as a power supply for the trains' load of the Oi railyard, and performs as the following electric power supply.

(1) Individual power supply to the Oi railyard
(2) Interconnected power supply, connecting to the existing power supplies
(3) Individual power supply to the Tokyo Station serving rest trains

(4) Individual power supply between Tokyo station and Osaki Substation (SS), besides Oi railyard

II. SYSTEM OUTLINE

The configuration of the Oi frequency conversion substation is shown in Fig. 1.

Considering redundancy, the EFC consists of two units. The rated capacity of each unit is 30 MVA, and the total rated capacity is 60 MVA. Since the main circuit and control system are identical and independent, operation with either one unit is possible. Moreover, the two units are linked together with flexibility. This duplicated structure improves the reliability of the whole system.

Each unit contains a rectifier in three-phase 50 Hz side and inverter in single-phase 60 Hz side. Input and output voltages are 22 kV, and EFC units are connected to other systems using rectifier/inverter transformers.

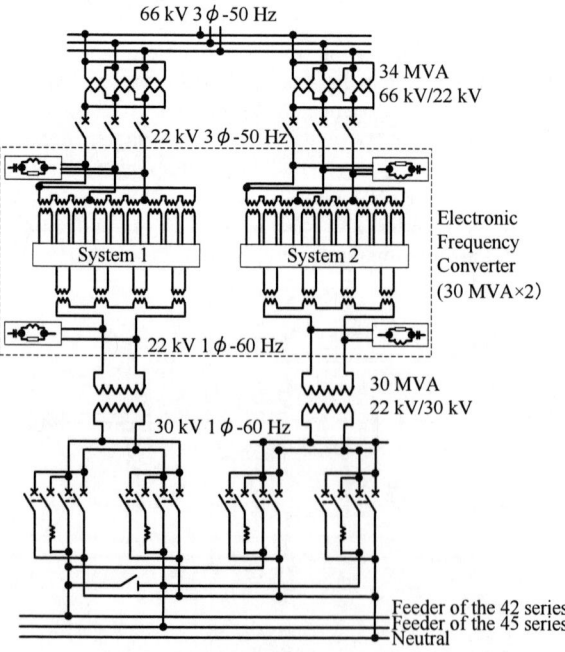

Fig. 1. Circuit diagram of the Oi EFC

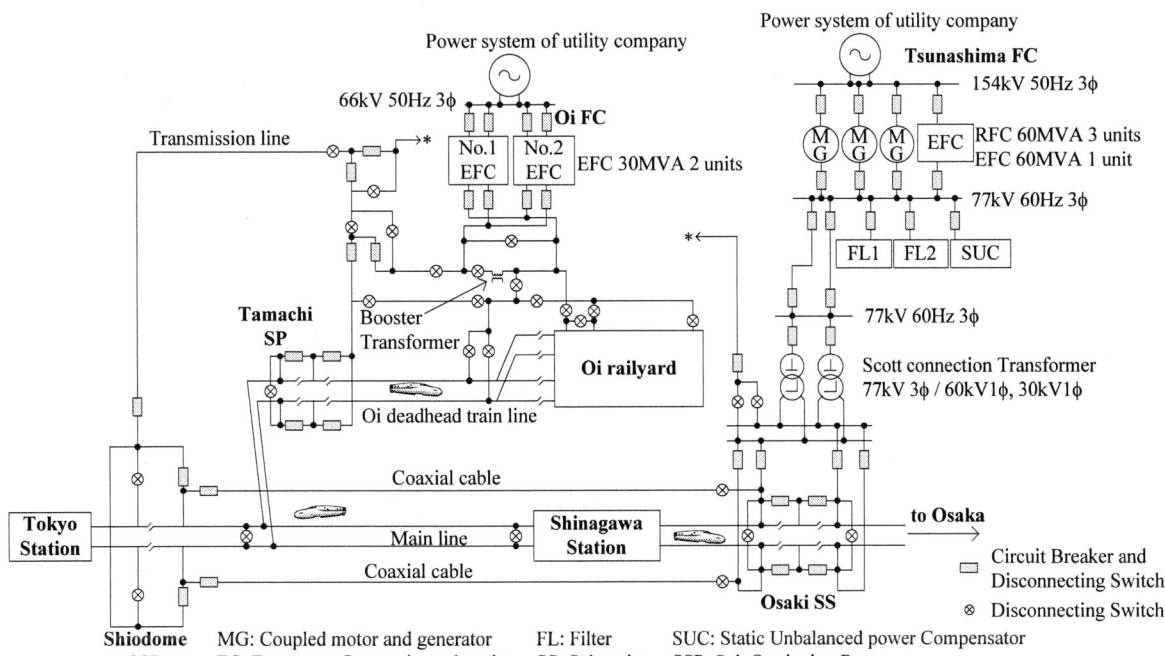

Fig. 2. The power system of the Tokaido Shinkansen in Tokyo area

III. COMPOSITION OF A STATIC FREQUENCY CONVERTER

EFC's ratings are shown in TABLE 1. The ratings are the same as the Numazu EFC, previously constructed. The rectifier and the inverter consist of single-phase bridges using injection-enhanced gate transistors (IEGTs) rated at 4500 V-2100 A with three IEGTs connected in series in each arm.

A triangular carrier signal with a frequency that is five times as the line voltage frequency is used for pulse width modulation (PWM). The carrier signals are shifted from each other to reduce harmonic components in the output voltage.

All control functions are executed in parallel on the processors. In particular, the main control processing is executed by Digital Signal Processors (DSPs). The parallel processing enables high-speed control with accurate and stable power feeding, such as train load which can change radically.

IV. BASIC EMPLOYMENT

Fig. 2 shows the power system of the Tokaido Shinkansen in Tokyo area. The Tsunashima Frequency conversion substation (Tsunashima FC; having three RFCs and one EFC) supplies train load power from Tokyo station to Hiratsuka SS (about fifty kilometers from Tokyo station). The new Oi EFC, replacing Hamamatsucho RFC existed near the Shiodome Sub Sectioning Post (SSP), has to be constructed for several purposes of feeding patterns. We describe the purposes below.

TABLE I
ELECTRONIC FREQUENCY CONVERTER RATINGS

Item	Ratings and Specifications
Rated capacity	30 MVA×2 series
Rectifier transformer	Rated capacity: 34 MVA×2 Δ/3 levels, 3 phase open connection Primary voltage: 22 kV Secondary voltage: 4 kV-3 levels
Rectifier AC rated voltage	4000V×3 levels-50 Hz(3 phase)
Rectifier configuration	Single-phase bridges×3 phase×3 converters multiple structure
Switching method of rectifier unit	5 pulses PWM
Valve devices of converter	IEGT（4500 V-2100 A）
DC rated voltage	6900 V
Inverter AC rated voltage	4100 V×4 levels-60 Hz(Single-phase)
Inverter configuration	Single-phase bridges×2 parallels×4 converters multiple structure
Inverter transformer	Rated capacity: 30 MVA×2 Single-phase/ 4 levels, single-phase connection Primary voltage: 22 kV Secondary voltage 4.1 kV×4 levels
Switching method of inverter unit	5 pulses PWM
Valve devices of inverter	IEGT（4500 V-2100 A）

A. Individual power supply to the Oi railyard

In basic feeding pattern, the Oi EFC is operated individually, and supplies power to the Oi railyard with both two units. In this operation mode (called "individual-mode": basic operation mode and shown in Fig. 3), the EFC is driven by "individual voltage control mode," controlling to keep feeder voltage to 26 kV with using AVR. And, in individual-mode, the EFC controls the outputs of each unit to equal with using ACR.

Moreover in individual mode, the EFC has typical equipment designed for the characteristic of high inrush current by the Oi railyard's load. The high inrush occurs on closing the feeder circuit breakers to start feeding, as tens of train transformers operate at one time. This occurs only in large railyard such as the Oi which has over 50 trains hold. The typical equipment, resistance equipped circuit breakers are applied to the feeder circuit in parallel, and reduce the inrush current below the over current relay's operation level.

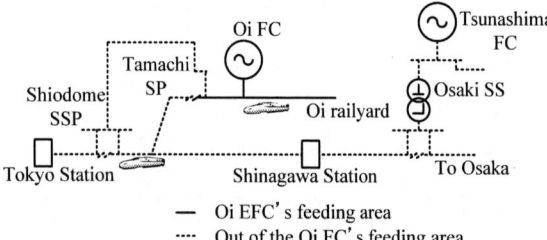

Fig. 3. Configuration of feeding in individual power supply to the Oi railyard

B. Interconnected power supply, connecting to existing power supplies

In maintenance, the Oi EFC drives as interconnected power supply, connecting to existing power supplies.

When the one unit of the Oi EFC is out of operation, the Oi EFC interconnects the Tsunashima FC's power supply system, and maintains the Oi railyard power supply (shown in Fig. 4). Another occasion, when the Tsunashima FC is troubled, the Oi EFC interconnects the Tsunashima FC or further existing power sources, the Nishisagami FC (having three 60MVA RFCs).

For these operations, the Oi EFC supplies active/reactive powers with the control of fixed power factor, and 28kV-controlling AVR operation.

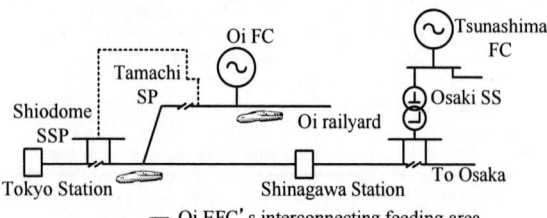

Fig. 4. Configuration of feeding in interconnected power supply, connecting to Tsunashima FC

C. Individual power supply to the Tokyo Station serving rest train

In cases of the trains' delay or similar occasions occurred, Tokaido Shinkansen sometimes offers trains for travelers to rest, typically in Tokyo Station. On the other hand, equipments' repairs should be done continuously in non-operating time (set from 0:00 am to 6:00 am). Therefore, the power system in Tokyo area has the ability to limit feeding areas from Tokyo station to Shiodome SSP.

Additionally, because the Oi railyard has to be feed for long time compared to main line, to maintenance trains, the Oi EFC has the typical ability to enable flexible feeding patterns overcoming these situations. For this, we developed changing sequence of power systems to limit feeding area, without stopping the power supply with using interlink equipments.

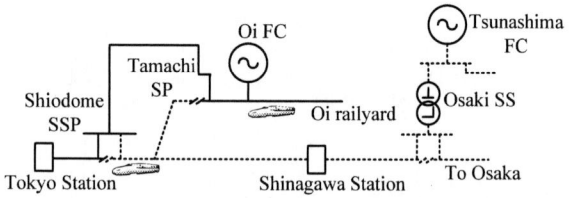

— Oi EFC's Feeding area in serving rest train in Tokyo Station
···· Out of the Oi EFC's feeding area, where can do maintenances

Fig. 5. Configuration of feeding in individual power supply to the Tokyo Station serving rest train

D. Individual power supply between Tokyo station and Osaki Substation, besides Oi railyard

For the feeding when Osaki SS's feeder bus has some accident, the Oi EFC has the ability for the individual power supply covering main line from Tokyo Station to Osaki SS, Oi deadhead train line, and the Oi railyard, though the restriction of the train load needed for stable operation (shown in Fig. 6). This ability reinforces the flexibility of the Tokaido Shinkansen operation.

In this operation, the Oi EFC is driven by the same way as individual-mode.

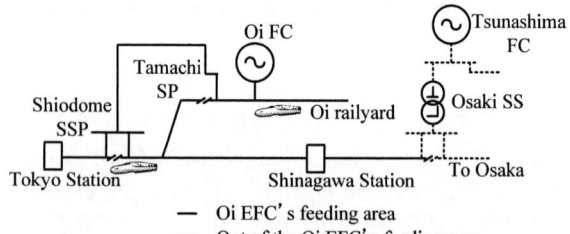

— Oi EFC's feeding area
···· Out of the Oi EFC's feeding area

Fig. 6. Configuration of feeding in individual power supply for Tokyo Area

V. CONTROLS

Fig. 7 shows the control block diagram. This section focuses on the control functions in interconnected mode operation and individual mode operation.

A. Interconnected mode operation

In the interconnected mode operation, controls are followed the policy of the former Numazu EFC. AC voltage control is executed on a rotating frame based on the phase signal detected in the feeder voltage. Basically, a PI controller determines the output power to keep the feeder voltage within a valid voltage range. In addition, following features are adopted:

(1) Voltage reference compensation: To save output power capacity, the controller reduces the output voltage reference when the output power of the EFC exceeds a fixed value.

(2) Slope reactance: The controller responds with a 10% (connected to Tsunashima FC) and 2.5% (connected to Nishisagami FC) slope reactance to prevent unnecessary output.

(3) Over-voltage prevention: To prevent an over-voltage caused by sudden change in load power, the controller reduces the power reference in a short time.

The AC current controller adjusts the output voltage of the inverter, so that the output power of the EFC follows the active power and reactive power.

B. Individual mode operation

In the individual operation mode, the EFC is operated as a voltage source. The feeder voltage of the railyard is controlled and kept equal to the voltage fixed beforehand as reference. This operation is similar to the Numazu FC's single mode operation, but the Oi EFC has to feed a large amount of train load as referred. Therefore, newly controls are needed, typically in a short fault.

When a short fault occurs, the output current of the EFC exceeds the rating current. Therefore, the circuit breaker should be opened immediately to protect the power systems. On the other hand, for easier identification of the failure point, over-current indicators

such as the fault situation recorder to limit the fault area, located around the feeding circuits, should operate. Since the over-current indicators operate after a set time, it is desirable to keep over-current condition for a moment until the over-current indicators operate. To fulfill these conflicting needs, we developed an over-current suppression control in the EFC's individual operation. When a short fault occurs, the controller radically steps down the output voltage of the EFC and keeps the output current below the over-current level of the EFC until the over-current indicators operate. After that, the protecting sequences are triggered by a protection relays such as over-current relays or distance relays, and open the feeding circuit breaker to protect the EFC.

Fig. 8 shows the control block diagram of output voltage control in individual operation (one of a series), and we explain the outline of the individual operation controls as follows.

(1) Automatic Voltage Regulator

The output voltage of the EFC is mainly controlled by the Automatic Voltage Regulator (AVR), so as to be kept equal to the voltage fixed beforehand as reference.

(2) Over-Current Suppression Control

The basic principle of this control is radical step down of the output voltage according to excess amount of the feeding current above the set value. To obtain quick response, we applied the squared value of the excess amount of current. The excess amount's value is calculated by a dead band. The width of the dead band is correspondingly set to EFC's ratings. The control value is calculated by multiplied value of the excess amount and the control gain K together. In addition, the control value is smoothened by the first order lag element. The lower limit of the first order lag element is set to input value itself. This configuration brings a radical step down and moderate recovery of the output voltage.

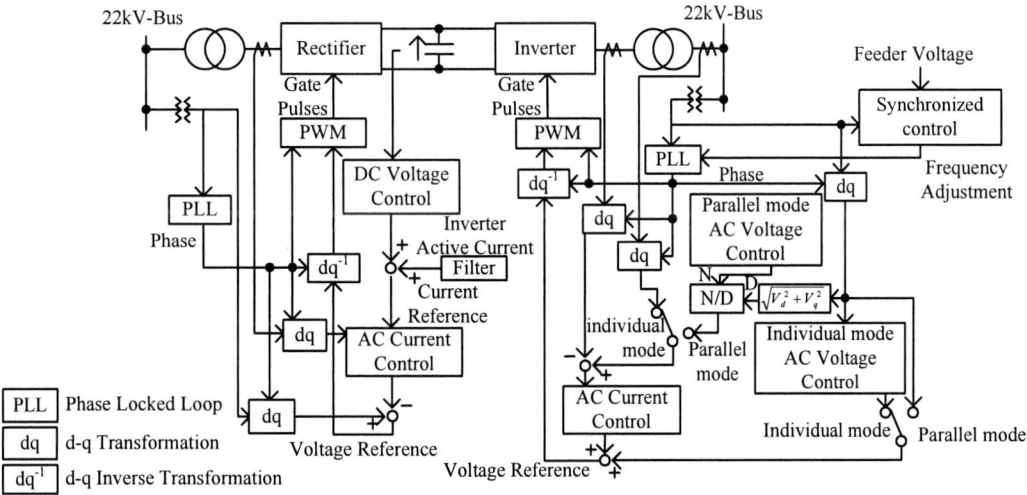

Fig. 7. The control block diagram of output voltage control in individual operation

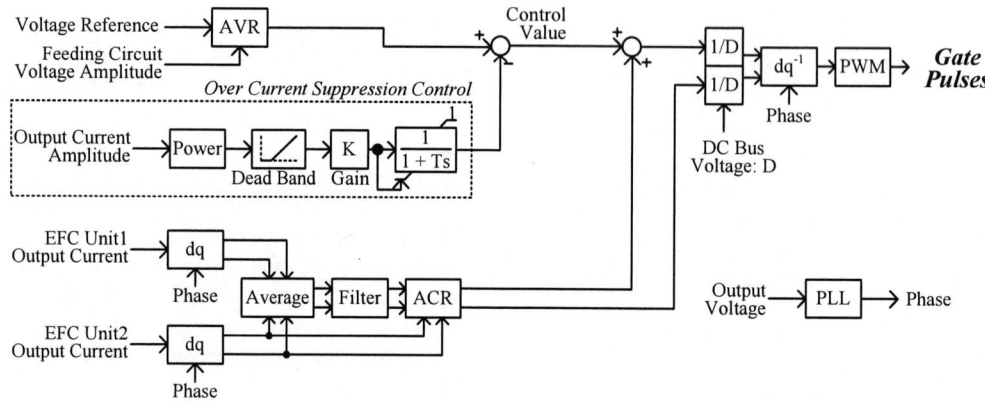

Fig. 8. The control block diagram of output voltage control in individual operation

(3) Automatic Current Regulator

Automatic Current Regulator (ACR) is set for the purpose of equalization of output currents of two EFC units and prevention of sudden change of output current of the EFC. On the rotating frame, the mean value of the output currents of both EFC units is smoothened by the low pass filter and set as the current command value. The ACR controls the output current of the EFC to track the command value. By setting the smoothed value as the command value, a sudden change of the output current caused by fluctuation of the load power is effectively suppressed.

(4) Output voltage calculation

The resultant values are converted from the rotating frame into the rest frame. The output voltage of the EFC is proportional to the DC bus voltage, so the output voltage command value is linearized by dividing by the DC bus voltage detected value. The linearized value is modulated by pulse width modulation for generating gate pulses.

For the purpose of easier troubleshooting, a waveform recording system is implemented in the EFC. The recording system automatically stores waveforms such as feeding voltage, output current, when fault occurs. We can obtain waveforms at the time of a short fault and analyze the waveforms. And, this function utilized already, when ground fault occurred.

This function utilized already. In June 2013, the feeder-to-ground fault occurred. The output current of the EFC exceeded the rating current value of the EFC. The over-current suppression control was activated and the output voltage of the EFC decreased immediately. The output current of the EFC was kept as much as the over-current level while the over-current suppression control was activated. The over-current suppression control was activated until the feeding circuit breaker was opened by the over-current relay and the distance relay, it took about 5 cycles. After the feeding circuit breaker was opened, the output voltage of the EFC rose moderately and the EFC kept operation. In addition, the over-current indicator properly operated. As a result, the interruption time of electric supply in the railyard was minimized by speedy identifying the failure point and removal of the failure factor.

VI. OPERATION CONTROL WITH INTERCONNECTED POSTS USING OPTICAL INTERLINK EQIPMENTS

As referred in IV section, the Oi EFC has several operation patterns, and each operation has different control factors. Table II shows the control factors in each operation patterns. Each factor was optimized by simulation, and changing control factors for fit operation pattern is inevitable for stable operation.

But, the operation patterns can't be cleared only by feeder circuit of the Oi EFC. Even in operation mode, parallel mode has two interconnected feeding patterns: interconnected by one EFC unit or by both EFC units. To decide the EFC's control factors, to know interconnected posts' feeding statements (circuit breakers, disconnecting switches, and related equipments) is required.

At the same time, when the fault occurs or radical inrush occurs with circuit breakers' operation, the Oi EFC should be protected from the feeder power system, because the control of the inverter could be diffused when the interconnecting power source lost. For the way to protect EFCs, gate block (GB) should be operated firstly. This protection sequence should be achieved in 100ms from a fault, but former transfer trip equipments with electrical signals can't do this.

TABLE II
CONTROL FACTORS OF THE OI EFC

Operation mode	Interconnected power supply	Referenced Voltage	Slope reactance
Individual	-----	26	-----
Parallel (interconnected)	Tsunashima FC	28	0.1
	Nishisagami FC	28	0.025

TABLE III
SPECIFICATIONS OF THE OPTICAL INTERLINK EQUIPMENT

Item	Ratings and Specifications
Transfer method	Optical fiber cable method Wavelength Division Multiplex: WDM
Communication method	Full duplex transmission, cyclic transmission
Communication speed	1.578 Mbps
Transfer format	76 bit / 1 frame
Maximum transferring distance	40 km
Multiplexed signals	21
Time for transfer	1 section transfer: about 9.0 ms 2 section transfer: about 14.0 ms 3 section transfer: about 18.0 ms 4 section transfer: about 22.0 ms (calculated time)
Interface for Switchboard	DC 100 V (non-voltage contact output)

TABLE IV
SIGNALS OF INTERLINK

No	Direction	Signal name
1	Outbound	Transfer trip signal (trip: open only the feeding circuit breakers)
2		Transfer trip signal (trip and re-feed)
3		Transfer trip signal (trouble: open both the feeding circuit breakers and disconnecting switches)
4	Inbound	Transfer trip signal (trip)
5		Transfer trip signal (trip and re-feed)
6		Transfer trip signal (trouble)
7	Direction (Inbound & Outbound)	Transfer trip signal (tie-open: open the disconnecting switches to tie inbound and outbound feeders)
8		Transfer trip signal (GB: gate block power devices)
9		Transfer trip signal (Stop: stop the EFCs)
10	For specific post	Transfer state signal 1
11		Transfer state signal 2
12		Transfer state signal 3
13		Transfer state signal 4
14		Over-post transfer trip signal 1
15		Over-post transfer trip signal 2
16		Over-post transfer trip signal 3
17		Over-post transfer trip signal 4
18		Linked close signal
19		Transfer-trip signal (EK : earthquake)
20		Operation signal for locator
21	Test	Test signal

Then, we built the new interlink equipments with optical signals. The new equipments were composed for "one step" interlink between each posts, however they were partitioned by other post. The "one step" means that the signals are converted only in sending and receiving post. And, this realized the cut-off of the converting time lost, which is referred often in the weakness of the optical networks. Table III shows the specification of the optical transfer trip equipment. For GB signal usage, the optical transfer trip equipment fulfills the demand of transfer time.

And, within the new equipments, we introduced the interlink signals besides transfer trip signals, typically in transfer state signals (shown in Table IV). These signals enable the Oi EFC to judge the operation factors, and to prepare faults fit for supplying power system. For

example, in the Oi EFC's interconnected operation to the Tsunashima FC's power system, the EFC judges the operating mode by basically two states, as shown in Fig. 9. One is the Osaki SS's feeding circuits closing state, and set for transfer state signal 1 transferred from the Osaki SS to the Oi EFC. Second is the Tamachi SP's connecting state of both main line side and the Oi railyard side, and set for transfer state signal 1 transferred from the Tamachi SP to the Oi EFC.

These interlink signal equipments finally achieve the feeding posts as one electrical system. In addition to the previous paragraph, these enabled the Oi EFC to have typical ability for changing sequence of power systems to limit feeding area, without stopping power supply, referred in IV-(3). These equipments tell the statement of supplying power system, and the Oi EFC can control feeding ways delicately.

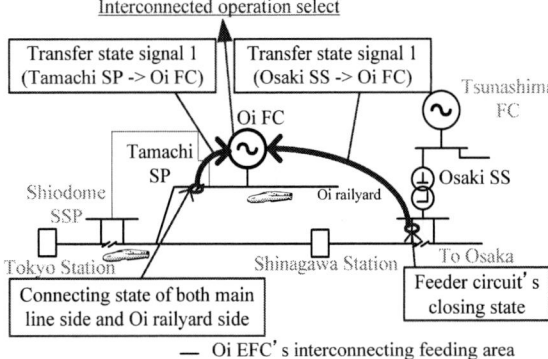

Fig. 9. Image of the transfer state signal using

VII. COMBINATION EXAMINATION

A. Electric Power System Analysis

For the inspection of the EFC's performance, we took electric power system analysis before construction. The analysis covered the individual operation's stability with radical changes of train load, interconnected operation's matching with interconnecting power system, and behaviors in faults, which is hard to test in field. The over-current suppression control is also analyzed, and the result is shown in Fig. 10. For ensuring over-current suppression control, we choose heavy load feeding patterns as referred in IV-D section in this analysis.

The result shows below. When the fault occurs, over current suppression control operate, and the reference voltage of the Oi EFC is declined. Then, though the feeder voltage of the Oi EFC drops abruptly, the output current of the Oi EFC is kept under 2 p.u., which is stable operating level of the Oi EFC. And, at the same time, failure current is kept. This means the EFC's protection from faults and identification of failure points are both achieved.

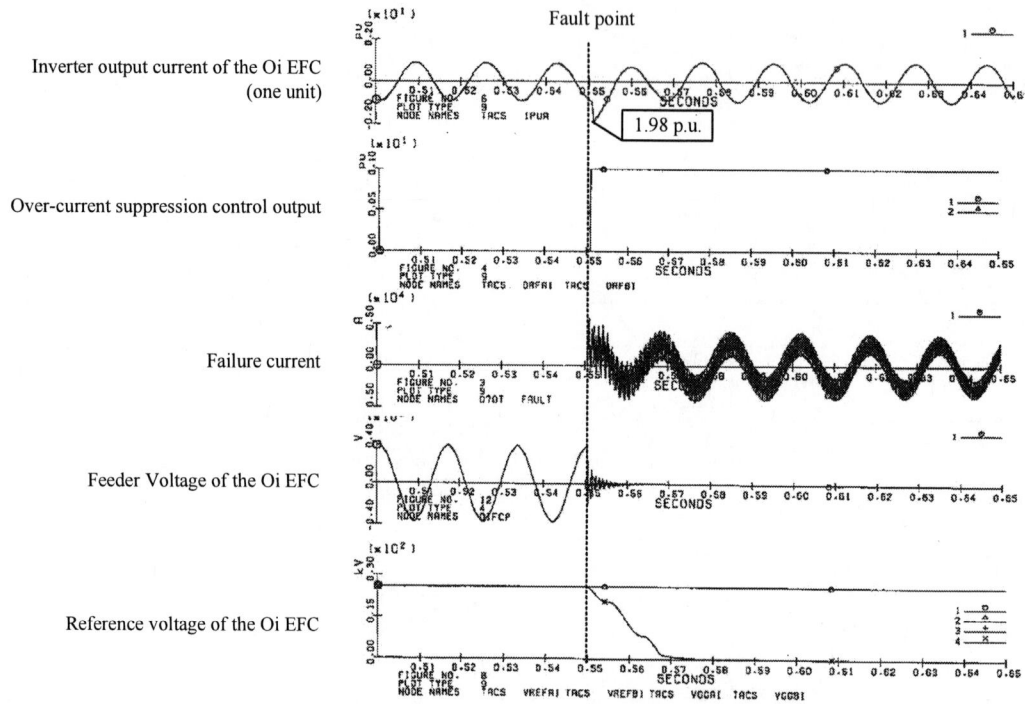

Fig. 10. Analysis result of the over-current suppression control

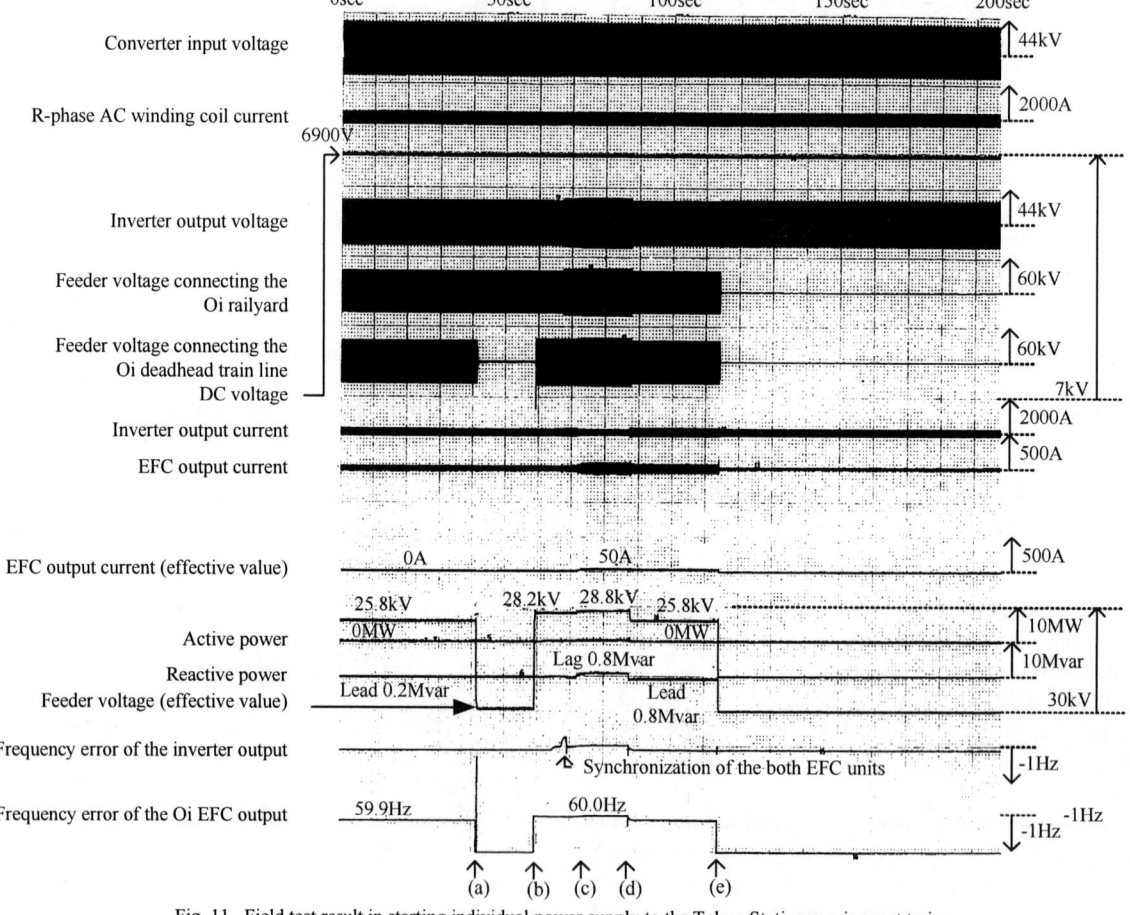

Fig. 11. Field test result in starting individual power supply to the Tokyo Station serving rest train

B. Field Test with Interlink Equipments

Fig. 11 shows the result of the field test in starting individual power supply to the Tokyo Station serving rest train. In this test, changing sequence with continuous feeding in the Tokyo station and the Oi railyard are achieved by interconnection of related posts, using interlink equipments.

In the start of the sequence, the Oi EFC is operated individually as shown in Fig. 3 (controlled for 26 kV). On the other side, the Tokyo station is fed by Tsunashima EFC's power system (controlled for 28 kV). The first step, the Tsunashima EFC's power system is led near to the Oi EFC, where the EFC can directly connect with circuit breaker closing ((a): cut off the connecting line for the Oi deadhead train line, (b): Tsunashima FC's power supply drawn). Then, the EFC synchronizes to the Tsunashima FC's power system, with holding individual operation and 26 kV-reference voltage, and is interconnected by closing circuit breaker ((c)). In this situation, the power system of Tsunashima FC is kept for 28 kV, but the Oi EFC is operated for keeping 26 kV. So, recovery operation obstruction control was introduced to the EFC. Finally, Tsunashima FC's power system is cut off, and the limitation of feeding area is executed ((d)). In the (e) section of Fig. 11, the Oi EFC's feeding was cut for finish field test.

With this sequence, the Oi EFC's operating area can change without inrush, and stable operation achieved. But, the composition of feeding circuits are so complicated to know only in the Oi EFC, and the interlink equipments enabled this sequence with transmitting circuit's conditions.

VIII. CONCLUSIONS

In this paper, we describe the introduction of the Oi EFC to Tokaido Shinkansen. The EFC enabled individual operation to feed the Oi railyard, with introducing new control, over-current suppression control. With this development of reducing faults-current, we obtained toughness against short faults in individual operation of power devices, which can't stand over loads. So, this development can lead up to the downsizing of the EFCs. And, we describe the interconnect operation with the Oi EFC, using interlink signals. The components enabled the feeding posts placed apart to one electrical system, and make the Oi EFC as flexible power supply.

REFERENCES

[1] M. Kitagawa, et al., "A 4500V injection enhanced Insulated Gate Bipolar Transistor (IEGT) in a mode similar to a thyristor", IEDM Tech. Dig., pp.679 -682(1993)

[2] M. Ohki, et al., "Electric Frequency Converter", *Toshiba Review*, Vol. 59, No. 11, pp. 35-38 (2004-11) (in Japanese)

[3] K. Ito, et al., "Electronic Frequency Converter", Proc. of the 2004 JIAS Conf., III, pp.347-352(2004)(in Japanese)

[4] K. Kunomura, et al., "Electronic Frequency Converter", *The 2005 International Power Electronics Conference (IPEC-Niigata 2005)*, pp. 2187-2191 (2005)

[5] K. Kunomura, et al., "Electronic Frequency Converter that Controls A.C. Voltage using Fixed Power Factor Method", *Proc. of Technical Meeting on Power Engineering, IEE Japan*, PE-08-171, pp. 37-42 (2008) (in Japanese)

[6] K. Kunomura, et al., "Electric Frequency Converter Feeding Single-Phase Circuit for Shinkansen Trains", *Toshiba Review*, Vol. 64, No. 9, pp. 45-48 (2009-9) (in Japanese)

[7] K. Kunomura, et al.: "Electronic Frequency Converter Controls A.C. Voltage Using Fixed Power Factor Method", *IEE Japan Trans. IA*, Vol. 129, No. 7, 2009, pp. 768-774 (2009) (in Japanese)

[8] K. Kunomura, et al., "Electronic Frequency Converter Feeding Single Phase Circuit and Controlling Feeder Voltage with Fixed Power Factor Method for Shinkansen," *IEEE Trans. on Power Electronics*, vol.27, No.9, September 2012

Application Examples of Energy Saving Measures in Japanese DC Feeding System

Takashi Suzuki, Hitoshi Hayashiya
Electrical & Signal Network System Department
East Japan Railway Company
Tokyo, Japan

Takashi Yamanoi, Keiji Kawahara
Electrical Engineering Department
West Japan Railway Company
Osaka, Japan

Abstract— **In this paper, some application examples of energy saving measures in two representative Japanese railway companies, mainly in d.c. electric traction power supply system, are introduced. The effect of electric energy storage systems installed to substations is mentioned especially. In addition, it is also described about the potential of regenerative braking energy, which will be able to be utilized by installation of energy storage systems or other measures.**

Keywords— *energy storage system, lithium ion battery, regenerative brake, traction power supply*

I. INTRODUCTION

It is important for railway company to make an effort continuously to reduce the energy for operating trains in Japan, especially after the Great East Japan Earthquake in 2011, although the electric train system is regarded as an efficient means of transportation. In this paper, some application examples of energy saving measures conducted or examined in East Japan Railway Company (JR East) and West Japan Railway Company (JR West) are introduced.

II. REDUCTION OF RESISTANCE IN FEEDING CIRCUIT

In Japan, d.c. 1.5kV feeding system is applied to about two-thirds of electrified conventional railway lines. The majority of lines in, around and connecting big cities, such as Tokyo and Osaka, are operated in d.c. 1.5kV system except for high speed line (Shinkansen). Reduction of resistance in feeding circuit is effective in saving traction energy, because such a low voltage is used while the power is MW class. The effectiveness is explained in following two aspects. First, it can reduce feeding loss, and second, it contributes to avoid limitation of regenerative brake power caused by high voltage at the pantograph of train. The tie feeding system is one measure for this concept that can be conducted with comparatively less cost. Fig. 1 shows the circuit of tie feeding system.

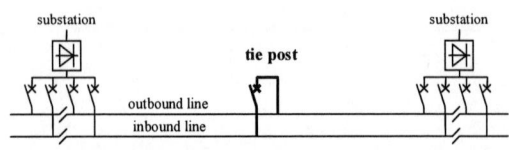

Fig. 1 Circuit of tie feeding system

In JR West, tie feeding system have been installed to some lines. For example, two tie posts were built in Yamatoji line in 2010 [1]. The location is shown in Fig. 2. The 'package type tie post' was contrived for these two posts. All facilities, such as circuit breaker, disconnecting switch, control unit, power unit and so on, were put into a metal cubicle. Peculiar features of the post are as follows:

· It is so compact that it is placed in only about 10m² and it can be carried by small truck, as shown in Fig. 3.
· The work of connecting control wires is completed in factories.
· Time for construction can be shortened.

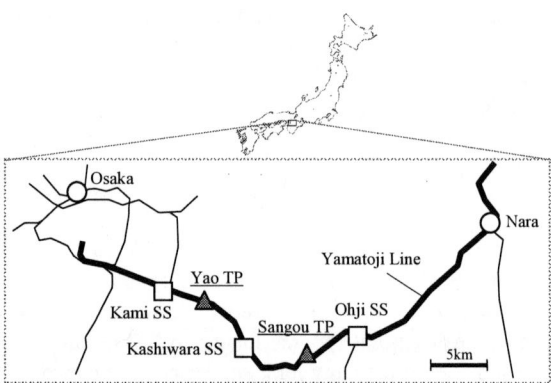

Fig.2 Location of tie posts and Yamatoji line

Fig.3 Photo of tie post installed in Yamatoji line

Fig. 4 shows comparison of traction energy fed from Kami substation before and after installation of the tie feeding system. Kami substation is located next to the installed tie post as shown in Fig. 2. Fig. 4 indicates that the feeding energy decreased about 9% by installation of tie feeding system. It has been reported that the amount of traction energy fed from substations located in the section the tie posts installed (including Kami SS) decreased about 3%.

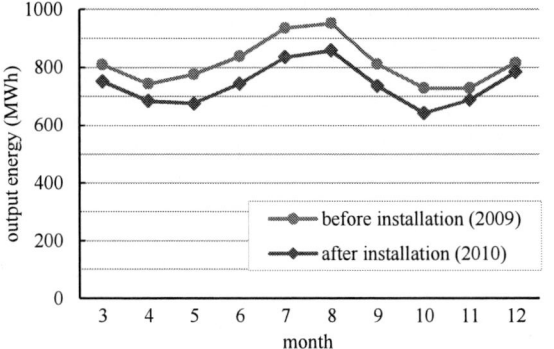

Fig.4 Output Energy from Kami Substation before and after Installation of Tie Feeding System

III. CONTRIVANCE IN TRAIN CONTROL BY DRIVERS

After the great earthquake in 2011, the condition of electric power supply in Japan has become severe because of the suspension of nuclear power plants. In such circumstances, some train drivers in JR West implemented 'energy saving driving'. In this way of driving, accelerating time was shortened by several seconds while the train could run on prescribed schedule by making use of margin time. Fig. 5 shows the image of this driving method. As a result, energy saving ratio reached up to about 30% in the best case.

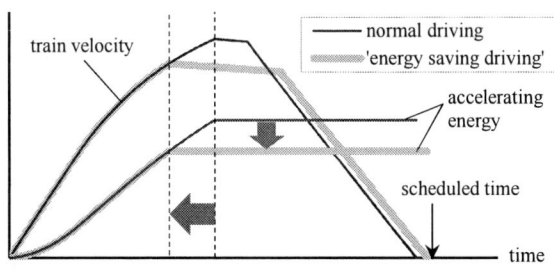

Fig.5 Image of the 'energy saving driving'

IV. ENERGY STORAGE BY BATTERIES

A. Energy Storage Systems in Japanese Railway

Recently, it is becoming economically reasonable in some conditions to introduce batteries on wayside for the sake of storing the regenerative brake energy and assisting in the traction power. It has been brought by development of technology and reduction of price for batteries depending on the spread of hybrid and electric vehicles. Fig. 6 shows electric energy storage systems installed to d.c. traction power supply system in Japan. Most of them consist of Ni-MH or Li-ion batteries.

There are some purposes for installing energy storage system.
1. To compensate for catenary voltage drop (as the alternative to substation)
2. To stabilize regenerative braking force
3. As a power source for emergency running in case of power failure
4. Energy saving by absorbing regenerative energy

Although more than 10 energy storage systems have been installed as shown in Fig. 6, only a few systems have been installed just for the 4th purpose yet.

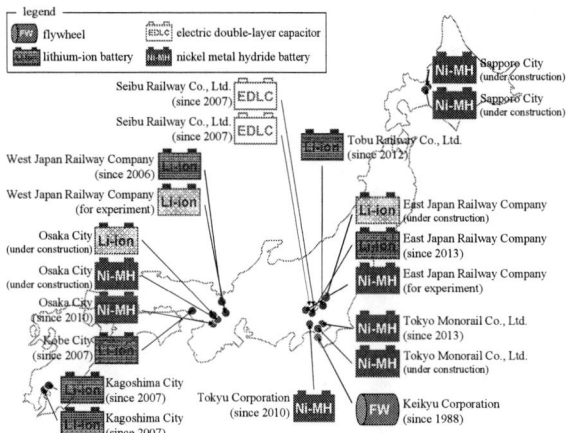

Fig. 6 Energy storage systems installed to d.c. traction power supply system in Japan [2](revised)

B. Examples of JR East

In JR East, two types of energy storage systems were installed just for energy saving, one consists of Ni-MH battery and another consists of Li-ion battery. Fig. 7 shows frequency of train and altitude on Chuo line and Ohme line, where two energy storage systems were installed.

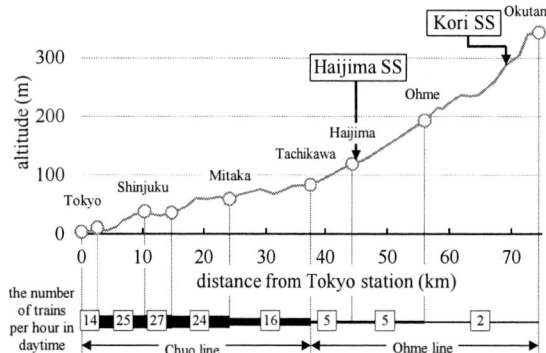

Fig. 7 The location of energy storage systems and features of lines

The first system with Ni-MH battery was installed to Kori substation in 2012 [3]. The purpose of the installation was an experiment (so it has been removed in 2013). So the location of Kori was not suitable as the place where the energy storage system could work effectively as indicated in Fig. 7. It is one of the features of Ni-MH battery system that it can be directly connected to d.c. 1.5kV feeding line without DC/DC converters.

The battery is charged or discharged automatically depending on the voltage of feeding line. The purpose of the experiment was to grasp the behavior of such a 'control-less' system. As a result of experimental operation, the following have been observed.

· About 30-70% of the charged energy was not from braking trains, i.e. regenerative braking energy, but directly from the rectifier.

· The volume of that 'direct charge' did not proportional to charged energy as shown in Fig. 8. In other words, the ratio of direct charge changed depending on the volume of charged energy.

The reason for high ratio of direct charge seems to be the low frequency of train. After discharging, the batteries are charged automatically from rectifier before next braking train approaches under such condition. It has been reported that the ratio of direct charge was around 10% on heavy load line in other railway company where similar Ni-MH battery system was installed [4].

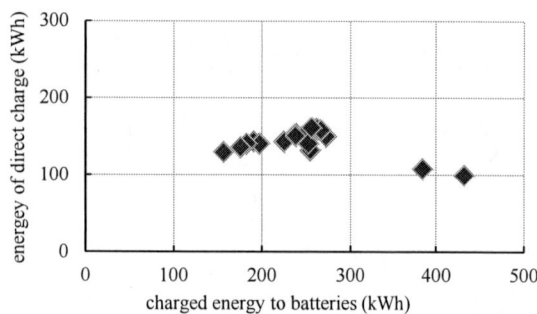

Fig. 8 Relation between energy charged directly from rectifier and charged energy to batteries per day (10:00 a.m. - 4:00 p.m.)

The second system was installed to Haijima substation in 2013 as equipment for practical use. It consists of Li-ion battery. TABLE 1 shows the specification and Fig. 9 is the photo of the system. The location of Haijima is suitable for the system in the aspect of train frequency and the geography.

In contrast with Ni-MH system at Kori, this Li-ion system has DC/DC converters as controller of charge and discharge. Therefore, tuning the threshold voltage of charge and discharge is substantial.

Fig. 10 shows progress of the tuning. After the first change of threshold voltage in the late May, the volume of discharged energy drastically increased.

TABLE 1
SPECIFICATION OF ENERGY STORAGE SYSTEM
AT HAIJIMA SUBSTATION

rated power	2000kW
rated voltage	1650V
rated current	1200A
rated capacity	76.12kWh (4 modules in series and 20 in parallel)
rated voltage*	173V (48 cells in series)
rated capacity*	5.5Ah

*:value for one battery module

Fig. 9 Energy storage system with Li-ion batteries at Haijima substation

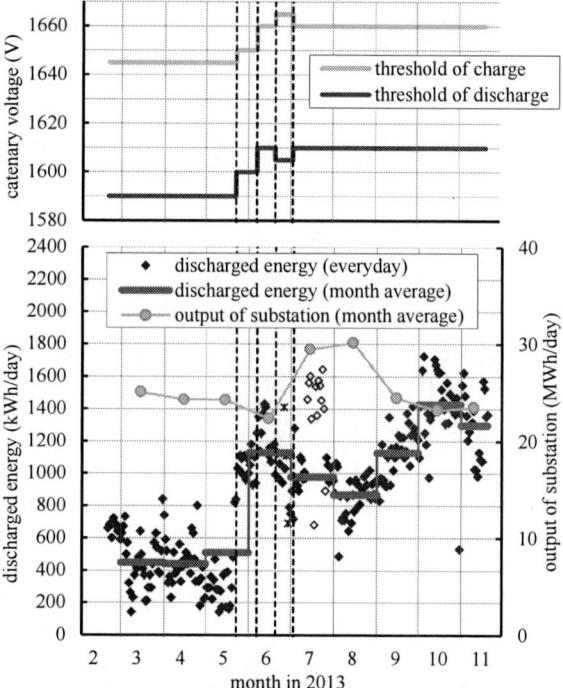

Fig. 10 Progress of tuning threshold voltage and change of discharged energy from batteries and output of substation.

Fig. 11 is the comparison between before and after the tuning in charged / discharged energy and state of charge (SOC) of batteries for every hour on a day. As shown in Fig. 11 (a), SOC hardly changed within high level and the volume of charged and discharged energy was small. Especially in daytime, the batteries seldom worked. It was expected that there were few opportunity for discharge because the threshold voltage of discharge was too low for the condition of Haijima substation. So the threshold voltage of discharge was raised step by step in the tuning, considering the threshold voltage of charge, which had to change related to that of discharge, did not become too high. Finally, the threshold voltage of discharge and charge was raised by 20V and 15V respectively from the original value. Such a small change of threshold voltage brought drastic change on behavior of the energy storage system.

Fig. 12 shows the relation between the discharged energy from the battery and output of the substation. It indicates that the lighter the load of substation is, the bigger the discharged energy is. It is supposed that regenerative energy tends to be consumed by other trains in the condition where load of the substation is heavy. The change of output of subtation depends on the seasons as shown in Fig. 10. The change seems to be brought by difference of the load of air conditioner on trains.

The discharged energy from storage system, which can be regarded as equivalent to saved energy, has reached around 1.6MWh/day as a result of tuning while that was about 0.4MWh/day at the beginning of operation. Considering this result, it has been estimated that the ratio of discharged energy to output energy from the substation to be about 4% on average, and total amount of saved energy to be around 500MWh for a year.

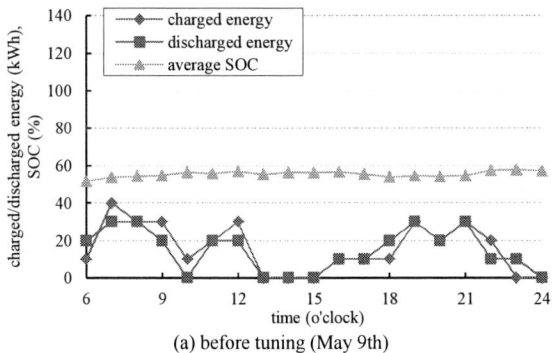

(a) before tuning (May 9th)

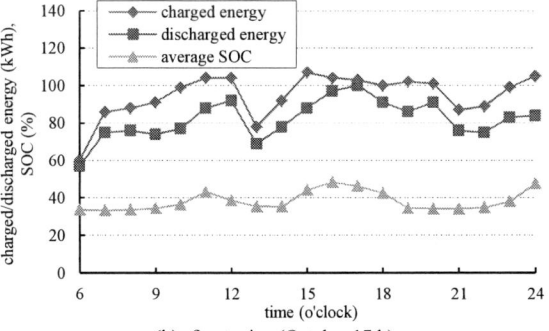

(b) after tuning (October 17th)

Fig. 11 Comparison between before and after the tuning in charged / discharged energy from batteries and SOC.

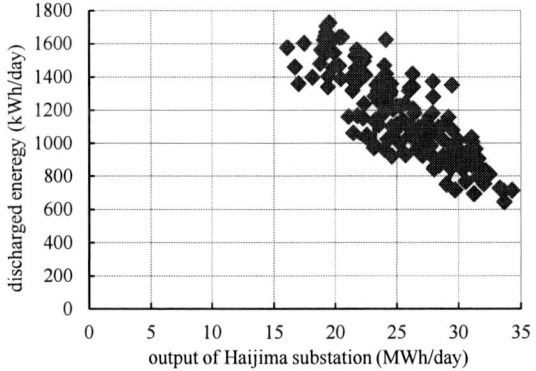

Fig.12 Relation between the discharged energy from battery and output energy from Haijima substation

C. Examples of JR West

In JR West, Li-ion battery system has been operated at Shin-Hikida substation since 2006 [5]. TABLE 2 shows the specification and Fig. 13 is the photo of the system. The primary purpose of this system is to compensate for feeding voltage drop in case of the receiving electric power fail at the substation caused by accidents or maintenance in power trans-mission line. Although it was not the main purpose, the effect of saving energy has also been brought. The amount of saved energy has been about 300kWh a day. That is equivalent to about 3% of output from this substation.

TABLE 2
SPECIFICATION OF ENERGY COMPENSATION SYSTEM
AT SHIN-HIKIDA SUBSTATION

Energy storage part		
medium of energy storage		Li-ion battery
rated capacity*		60AH (178 cells in series)
maximum voltage		780V
minimum voltage		500V
maximum current*		570A
Power conversion part		
type of main circuit		bi-directional DC/DC converter
type of control		current control by PWM
bus side	nominal voltage	DC1500V
	variation range of voltage	DC900-1850V
	maximum input current*	DC119A (at DC1700V)
	maximum output current*	DC347A (at DC1000V)
battery side	nominal voltage	DC641V
	variation range of voltage	DC500-780V
	maximum charge current*	DC300A
	maximum discharge current*	DC570A

*:value for one unit (the system consists of 3 units)

Fig.13 Energy storage system with Li-ion batteries at Shin-Hikida substation

Study and experiment have been conducted continuously, aiming to get more effect of energy saving by tuning the control parameter such as the threshold voltage of charge and discharge [6].

Fig. 14 is the control pattern of charge and discharge. This system equips 'adjusting charge/discharge mode' for the sake of keeping the state of charge of batteries within appropriate range for voltage drop compensation. Fig. 15 shows the result of experiments on varying threshold voltage of charge and discharge. As a result of both lowering threshold voltage of charge (Vc_0 and Vc_1) and raising threshold voltage of discharge (Vd_0 and Vd_1), the system worked more actively and the effect of energy saving improved.

Furthermore, another study is being carried out in order that the system will be able to be installed to other sites with more reasonable cost.

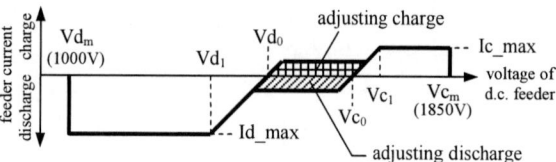

Vc_0 / Vd_0 : start voltage of charge / discharge
Vc_1 / Vd_1 : voltage for maximum charge / discharge
Vc_m / Vd_m : limit voltage of charge / discharge
Ic_max / Id_max : maximum charge / discharge current

Fig.14 Control pattern of charge and discharge

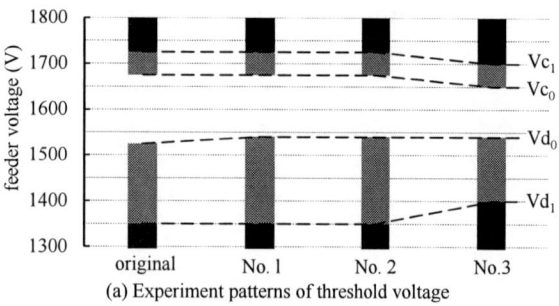

(a) Experiment patterns of threshold voltage

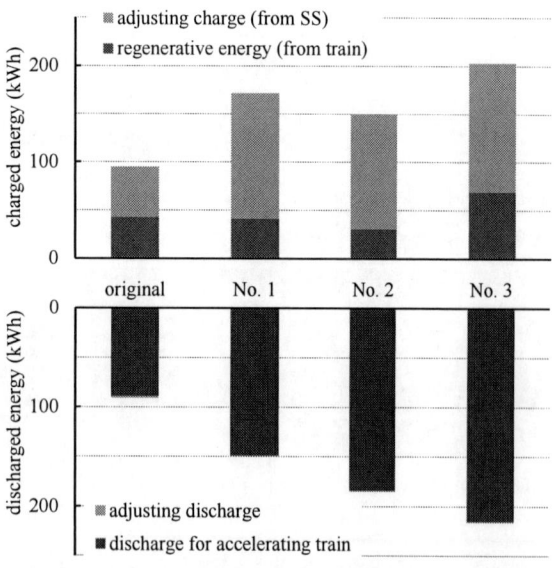

(b) Charged and discharged energy per day (9:00 a.m. - 6:00 p.m.)
Fig. 15 Result of experiments on varying threshold voltage of charge and discharge

V. POTENTIAL OF REGENERATIVE BRAKING ENERGY

In many railway lines, except for high density lines such as commuting lines in the metropolitan area, limitation of regenerative braking power often occurs because the voltage of contact line is elevated when there are not any accelerating trains as load near around the braking train. It is effective to introduce some devices for the sake of utilizing regenerative braking energy in such railway line. Energy storage system is one of the solutions for this purpose.

A suitable line condition exists for introduction of such devices because of the following reasons.
· The regenerative braking power will be consumed effectively by the nearby trains where the density of train is high enough.
· The regenerative braking energy to be utilized will be small in the line where the train density is low and the length of train set is short.

A study about the line condition has been conducted by JR West [7]. The scope of the study is the lines of JR West around Osaka area. Energy consumption by train per km and per hour (kWh/km/h), and unutilized rate of regenerative energy were indicated with considering the train operation interval as a parameter. And then, the product of the energy consumption and the unutilized rate was calculated as expected value of usable energy. Fig. 16 shows the result. It has been indicated that the suitable headway of train is around 7.5 minutes in the condition of this study.

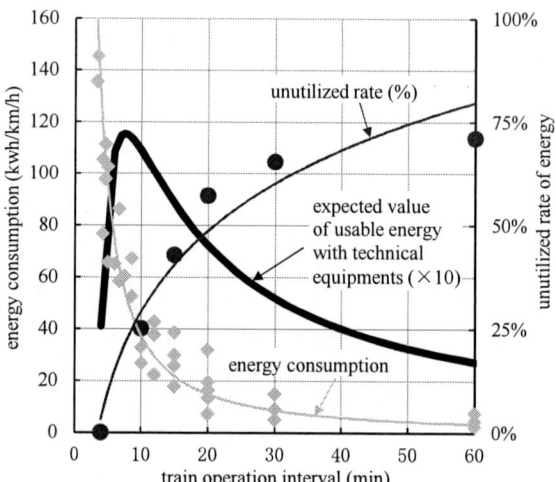

Fig. 16 Expected value of usable regenerated energy

In another study [8], which was also conducted by JR West, the potential volume of regenerative braking energy to be utilized has been indicated for each running section between stations. The major result of this study is shown in Fig. 17. The result is based on the measurement on the train.

The long bars in Fig. 17 indicate that the locations will be suitable points for the installation of some devices for regenerative energy utilizing such as energy storage system. It is supposed that the following reasons make the bars long.

· Distance from the previous stopping station is long, including the case that rapid service trains stop at the station.
· There are not enough trains as load of regenerative energy because the frequency of train is low. But if the frequency is too low, the bar becomes short.

Unutilized regenerative energy from Kyoto to around Kobe is small in average because the frequency of train is very high. In other area, except for near the both edge, there is comparatively large amount of unutilized energy. It will be worth installing energy utilizing devices in such areas.

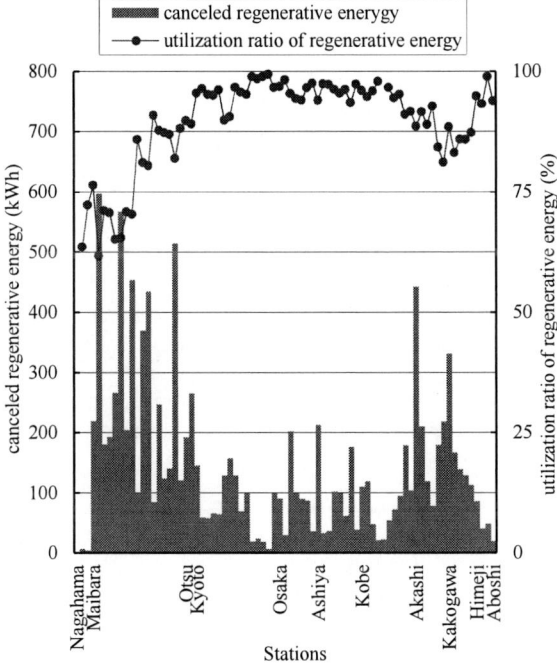

Fig. 17 The annual unutilized regenerative energy and utilization ratio of regenerative energy for each running section (from a station to the next station)

VI. CONCLUSION

In this paper, the energy saving measures which was conducted by the two Japanese railway companies have been introduced. It has been reported that several percent of the traction energy was saved by each measures. Especially, the energy storage system has recently become popular. As mentioned in detail, tuning the parameter of battery system and studies on the location or the structure of the systems are current topics. The interesting result of measurement about the potential volume of regenerative braking energy has been also referred.

REFERENCES

[1] M. Matsui, "The effect of introduction of tie post", *Trans. of Japan Railway Electrical Engineering Association*, Vol. 22, No. 4, pp. 21-25, 2011. (in Japanese)

[2] H. Hayashiya, "Possibility of energy saving by introducing energy conversion and energy storage technologies in traction power supply system", *15th European conference on power electronics and applications*, No. LS5d-73, 2013.

[3] H. Ogura, M. Nakahira, S. Kikuchi, H. Hayashiya, A. Terashima, T. Miyazaki, E. Yoshiyama, "Analysis of charging and discharging as power storage system of Nickel metal hydride battery", *J-RAIL2013*, S3-1-6, 2013. (in Japanese)

[4] M. Torizuka, S. Kodama, K. Hosoya, E. Yoshiyama, "Stationary Energy Storage System for Railways with High-Capacity Nickel- Metal Hydride Battery GIGACELL®", *Trans. of Japan Railway Electrical Engineering Association*, Vol. 22, No. 4, pp. 14-20, 2011. (in Japanese)

[5] T. Aihara, "Introduction of the power compensation system with Li-ion batteries", *Trans. of Japan Railway Electrical Engineering Association*, Vol. 18, No. 4, pp. 9-12, 2007. (in Japanese)

[6] Y. Ota, I. Tokuda, Y. Tsuji, K. Kyuuhei, T. Aihara and Y. Nakamura, "The effective use of compensator for DC railway", *2011 IEEJ annual conference*, 5-088, 2011. (in Japanese)

[7] T. Yamanoi, K. Ito, Y. Nakamura and K. Kawahara, "Study of introductory indices such as electric-power-storage equipment in DC feeding system," *2013 IEEJ Annual Conference*, 5-089, 2013. (in Japanese)

[8] T. Yamanoi, K. Kawahara, Y. Kani and Y. Kodama, "Measurement of utilized regenerative power in DC feeding system," *2013 IEEJ Industry Applications Society Conference*, 5-23, 2013. (in Japanese)

Lithium ion battery application in traction power supply system

Masato Teshima
Hitachi, Ltd,.
Omika works Power electronics design Dep
5-2-1,Omika,Hitachi city, Japan
masato.teshima.yo@hitachi.com

Hirotaka Takahashi
Hitachi, Ltd,.
Rail systems company Power supply system Dep
1-18-13,Soto-kanda,Chiyoda-ku,Tokyo, Japan
Hirotaka.takahashi.nf@hitachi.com

Abstract-
Technology development has been promoted aiming for an environment-friendly traction power supply system for railways. Among other things, a stationary energy storage system using lithium ion batteries for hybrid cars has been put into practical use since 2007. This device temporarily stores regenerative power generated from an train when it is stopped and the power is able to be applied for energy saving, measures against regeneration canceled and compensation for a voltage drop. The implementation planning for a stationary energy storage system (SESS) has been popular in recent years. The implementation of the SESS at railway companies can contribute to the reduction in the global warming gas emissions.

I. INTRODUCTION

In recent years, many railway companies have implemented the stationary energy storage system (SESS) in and outside Japan.

Trial to apply the SESS to the power supply equipment of d.c electric railways started earlier. Battery posts were introduced as parallel operation with such devices as rotary convertors and synchronous generators for the purpose of load leveling of power from 1912 to 1927 in Japan. Electricity was supplied by a battery power supply for the brief electric load for trains whereas the parallel operation of the battery and the rotary convertor was carried out during the morning and evening rush hours. Therefore, it was necessary for the batteries to have large capacity and to have larger installation area than the power supply equipment room. These battery posts were used for about 15 years, but abolished in 1927 due to the enhancement of substations and improvement in power supply situation.

Meanwhile, since the increase in the transportation capacity along with the start of the high economic growth period brought about the increase in demand of railways as well as caused the voltage drop and regeneration canceled on the railway truck, a power storage device which combined a storage medium and a power convertor was developed as a countermeasure against the voltage drop and regeneration canceled. The battery posts using lead batteries were installed in the Nakashima Station yard of Kabe Line in 1980 and the effectiveness of the battery posts as the measures against the voltage drop was confirmed through the tests carried out for a 3-year period.

Hitachi, Ltd. developed SESS using the lithium ion batteries for hybrid cars in 2005 and implemented a field test at Seishin-Yamate Line of Kobe city subway. Since the test results were excellent, the SESS was installed officially at Itayado substation of the Kobe city subway in 2007. It was the world's first implementation of the SESS which applied the lithium ion batteries in subways.

Here we describe the outline of the applied lithium ion batteries, control technology of the stationary energy storage system and the verification of energy-saving effectiveness on the locations where the system has already been introduced as a practical matter regarding a stationary energy storage system which applied the lithium ion batteries.

II. PRODUCT OVERVIEW

A. Outline of Energy Storage Device

The following characteristics are required as the energy storage device which is necessary for the SESS for the direct-current railways.

　　1) High power
　　2) High capacity
　　3) Long lifetime (Charge and discharge cycle lifetime)

The regenerative power from the trains has the steep rise characteristics. Therefore, the energy storage device with a high-rate characteristics where an output density is high is required in order to charge the regenerative power efficiently, whereas an energy density as the energy storage device is required in order to achieve the energy-saving effect.

The energy storage device is still a costly part despite of the acceleration of cost reduction from the start of the SESS development. In the case of the SESS, the number of charge and discharge times per hour is about 20 times at the most. Therefore, the energy storage device with the long charge and discharge cycle lifetime is necessary as the SESS when the life cycle cost is taken into consideration. The storage density of commonly-used industrial lithium ion batteries is relatively high. However, they also have the disadvantage of low output density and short cycle lifetime. In general, the energy density and output density of the lithium ion batteries are in the

978-1-4799-2706-7/14 $31.00 © 2014 IEEE

opposite relations, and electrode materials, etc. differ depending on their use.

TABLE 1 and Fig. 1 show the specifications of the lithium ion batteries for hybrid cars applied as the batteries for the SESS. Since the batteries for hybrid cars have the high output density, they are the most suitable for the charge of even the steep regenerative load.

The characteristic of these lithium ion batteries is the realization of the reduction in size and weight with the output density heightened up to 3,000W/kg.

Fig. 1. Lithium-ion battery module

TABLE 1
Specifications of storage battery module

	Value
Rating	173V 5.5Ah
Number of cell	48 (3.6V/cell)
Nominal energy	950Wh
Maximum output current	150A (27.5C)
Cooling	Natural cooling/Forced air cooling
Dimensions (W x D x H in mm)	318 x 618 x 103
Weight (kg)	24.5

For the purpose of the output enhancement in the aspect of materials, the metal ratio with the lithium of manganese anode electrode is optimized and reaction resistance of the surface of active materials is reduced by crystallinity control.

The manganese anode electrode material has the higher exothermic onset temperature compared with the cobalt anode electrode material. In addition, it has a smaller degree of thermal runaway and has high safety even when it is exposed to a high-temperature state as in over charge, etc. since a calorific value is low. For the case when the over charge control should become uncontrollable with a controller, fuses for excessive current protection are provided in a module unit inside the battery module to enhance safety.

B. Outline of Stationary Energy Storage System

TABLE 2 describes the outline of the product specifications of the SESS being marketed.

The SESS is composed of the blocks of a converter panel and battery panel. The major characteristic of this device is that it can be installed at any place without limiting the arrangement place.

TABLE 2
Specifications of Stationary Energy Storage System

	Value
Reference standards	IEC / EN / JEC
Rated System Capacity (kW)	500 / 1000 / 2000 / 3000
Nominal voltage (V)	DC 600 / 750 / 1500
Type of control	Automatic voltage regulator system with current limit function
Type of cooling	Natural air cooling or Forced air cooling

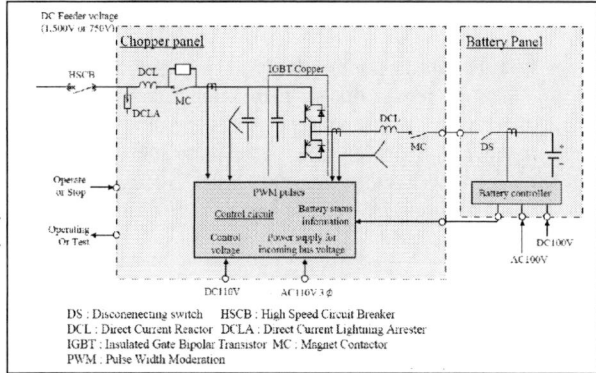

DS : Disconenecting switch   HSCB : High Speed Circuit Breaker
DCL : Direct Current Reactor DCLA : Direct Current Lightning Arrester
IGBT : Insulated Gate Bipolar Transistor MC : Magnet Contactor
PWM : Pulse Width Moderation

Fig. 2. Circuit diagram of SESS

The convertor uses IGBT (Insulated Gate Bipolar Transistor) element of 3,300V, 1,200A for 1500Vdc system. The convertor has multiple configuration with bidirectional chopper to attempt to suppress the ripple current to a feeder line side and the battery side.

Based on RAMS standards, it has the configuration which early recovery is considered if a fault should occur and trace-back function enabling an analysis of the main cause of the fault occurrence.

The lithium ion battery module is made 4 series with a parallel form conforming to a required absorption current aiming for the standardization of batteries in the case of 1500Vdc system. A battery on the individual lithium ion battery monitors its operation status including SOC (State of Charge), SOH (State of Hysteresis) and battery temperature, etc. and protects the battery. This information is sent to the convertor side and to realize the optimum operation control for the lithium ion batteries.

The control mode of the SESS is separate from the control whose purpose is the countermeasure against the regeneration canceled as well as energy saving.

The control mode map described in Fig. 3 is for the countermeasure against the regeneration canceled.

This control mode map consists of 4 control modes. The following is the operation of each mode.

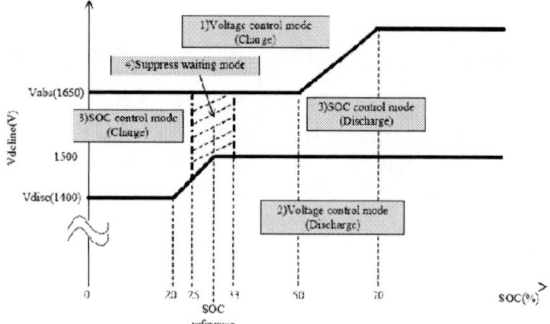

Fig. 3. Control mode map of SESS (at SOC Control)

1) Charge voltage control mode
   When the feeder line voltage Vdcline is increased to the charge onset voltage Vabs or above by the regenerative power from trains, the increase in the feeder line voltage is suppressed by charging the battery to make the feeder line voltage close to Vabs.

2) Discharge voltage control mode
   When the feeder line voltage Vdcline is decreased to the discharge onset voltage Vdisc or below by the powering of trains, the increase in the feeder line voltage is suppressed by discharging the battery to make the feeder line voltage close to Vdisc.

3) State of Charge (SOC) control mode
   When the feeder line voltage Vdcline is Vdisc or more and Vabs or less, the charge and discharging are carried out toward the feeder line to make SOC close to SOC command value (SOC reference). Utilization rate of the batteries is enhanced and the batteries' longevity is extended by preventing them from over charge and over discharge.

4) Suppress waiting mode
   This mode stops the switching of the chopper when both the feeder line voltage Vdcline and the State of Charge SOC are within the specified range. No-load loss (switching loss and loss at reactor or battery by the ripple current) is controlled by stopping the switching in the state where no operation is necessary.

As the measures to prevent batteries from over charge and over discharge, when the SOC becomes the specified value or more, the charge onset voltage Vabs is made higher and when the state of charge becomes lower than the specified value, the discharge onset voltage Vdisc is made lower.

The SOC is able to be controlled in the range of 20% to 50% by this characteristic.

In addition, the charge and discharge onset voltage should be varied with slopes. This makes the variation of the charge and discharge current small; therefore, the impact on the feeder line voltage applied during the protection operation against over charge and over discharge can be lowered.

The standard of the charge and discharge load pattern of the SESS is as shown in Fig. 4. The characteristic of this load pattern is to charge for 10 seconds at 1PU and then charge for 10 seconds at 0.5PU considering the rise at the initial stage of regeneration, which is the characteristic of the regenerative power of the trains. It is designed to be able to repeat the charge and discharge with a cycle of 180 seconds. This characteristic makes the regenerative power to be absorbed fully and prevents the regeneration canceled.

It is also available the voltage correction function where voltage elements of the power incoming side are applied to the device to prevent over charge and over discharge caused due to the voltage variation at the power incoming side.

In addition, the schedule control function is available in order to execute the optimum charge and discharge control since the regenerative volume varies substantially depending on the train operation diagram. The emergency running mode function of the trains is also manageable when a power failure occurs during power incoming.

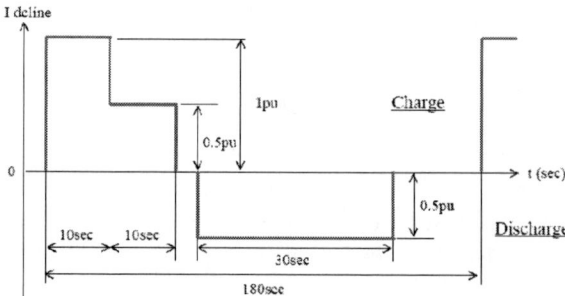

Fig. 4. Generic load pattern of SESS

III. INSTALLATION AT KOREA SEOUL METRO LINE9

A. Outline of the Installation at Seoul Metro Line 9

In recent years, the South Korean government and local government have positioned the railway transportation as an effective transportation means for the global warming prevention including the reduction in CO_2, etc. and have been promoting further energy saving, showing a great interest in the implementation of the technology regarding an effective utilization of the regenerative power generated when the trains are stopped or the speed is reduced.

A regenerative power storage device using electric double layer capacitors is introduced as the storage media

978-1-4799-2706-7/14 $31.00 © 2014 IEEE   1070

domestically in South Korea and some railway companies are verifying the effect.

Seoul Metro Line 9 is the urban railway by the South Korean social infrastructure improvement business utilizing private capital for the first time.

Seoul Metro Line 9 Corporation manages trains, electricity, machine and equipment, etc. by receiving a commission from Seoul city, farming out the operation and maintenance to the Seoul Line 9 Operation Corporation.

The SESS has been installed at 2 places: 909 and 921 substations. (Fig. 5)

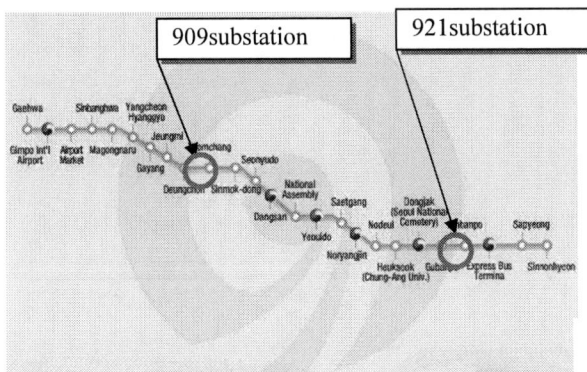

Fig. 5. Line map in SEOUL METRO LINE9

TABLE 3 describes the specifications of delivered B-CHOP.

The characteristic of the traction power supply equipment of Seoul Metro Line 9 is that most substations have 3 units of 3000kW rectifier set and 2 units are in operation at all times.

The current running train has 4 cars (2M2T) per train. Therefore, the equipment capacity has a wide margin from the viewpoint of the operation equipment capacity.

TABLE 3
Specifications of SESS

	Value
System	
Rated capacity	1000kW
Nominal voltage	1500Vdc
Rated current	606A
Battery	
Type	Lithium ion
Module capacity	5.5Ah / 173V
Configuration	4series, 10parallel
Maximum current	1500A
Storage capacity	38kWh
DC/DC Converter	
Switching device	IGBT
Control	PWM
Configuration	4parallel
Switching frequency	720Hz (2440Hz)

There was rarely the situation where the feeding bus voltage at the substations dipped below 1500V throughout the day.

For the setup of the charge and discharge onset voltage of the SESS, it is necessary to set up taking into consideration the no-load voltage of the rectifier set and the regenerative narrowing voltage at the train side. In the case of Seoul Metro Line 9, the setup for the regenerative narrowing onset voltage at the train side was as low as 1720V. Also, since the no-load voltage measured at local substations was in the vicinity of 1630V, it was necessary for the margin of the setup value of the charge and discharge onset voltage to be 100V and under.

When the margin of charge and discharge onset voltage was set for 100V and under, it was observed by the actual measurement that when the SOC control, which has the standard control characteristic, was implemented, loss became large and the energy-saving effect was reduced. Therefore, a highly-efficient charge and discharge control method (ECO operation mode: Fig. 6) which was newly improved from the SOC control method was added, which made it possible to reduce the loss when the setup was 100V and under after this control method was applied. The load loss was reduced by 6.5kW compared to the conventional product.

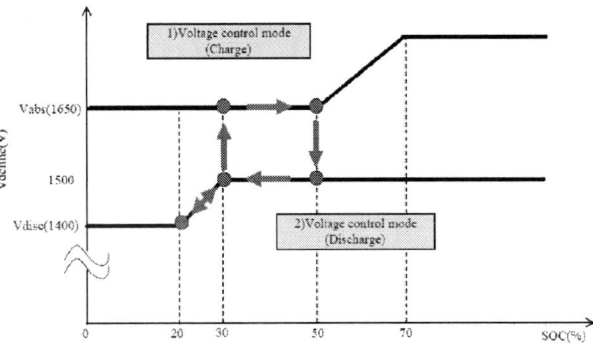

Fig. 6. Control mode map of SESS (at ECO Mode)

In order to verify the highly accurate energy saving effect, the electric energy when the SESS was in operation and out of operation was measured and compared on a weekly basis in May, 2013. The energy saving effect is heavily influenced by the number of trains in service and the number of passengers. Therefore, the number of passengers was checked and the comparison was carried out along with the number of trains in service in the verification. As a result, the average of 1500kWh per day and the reduction effectiveness of 16.7% were achieved by the SESS operation on weekdays. Meanwhile, the result of the average of 667kWh per day and the reduction effectiveness of 10% were achieved on weekends.

During the measurements this time, the numbers of passengers on the trains when the SESS was in operation and out of operation were recorded and compared. The number of passengers when the SESS was out of operation was fewer by 8%. However, the consumption electric energy at the time of SESS operation with a large number of passengers was less than that at the time when the SESS was out of operation. Therefore, by the installation of the SESS, the energy saving effectiveness was confirmed since the storage power was able to be utilized efficiently as the powering power.

At a maximum reduction effectiveness of 25.2% was achieved for the peak demand as shown in Fig. 7.

Moreover, the impact of the electric energy at neighboring substations was checked, but for the measurement this time, almost no variation was confirmed as a result when the SESS was in operation and out of operation.

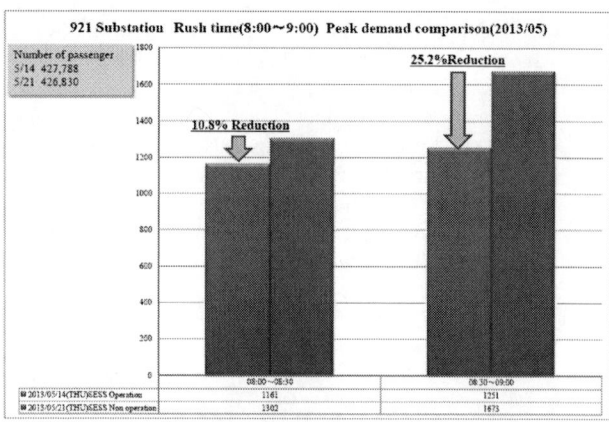

Fig. 7. Peak demand comparison

IV. CONCLUSIONS

Today, the SESS utilizing the lithium ion batteries has stared operation as the countermeasures against regeneration canceled and for energy saving. By considering essentially from the characteristics of the batteries, it can also be fully used as the power supply for rescuing trains to the nearest stations when a power failure occurs at electric power companies since the storage energy can be secured. Hitachi, Ltd. implemented a emergency train running test at the time of the power failure of power incoming by using this SESS in January, 2014, and the target result was achieved by running 10-car train for 2.7km.

We are willing to continually carry out the technology innovation in this field and contribute to solving the environment and energy issues.

REFERENCES

[1] H.Takahashi, "Effective Use of regenerative power." Railway & Electrical Engineering, Japan Railway Electrical Engineering Association (Jun. 2005) in Japanese.

[2] T. Ito et al., "Development of system for Absorption of Regenerative Power Using Lithium-ion Batteries." Transportation and Electric Railway Committee of the Institute of Electrical Engineers of Japan (Sep 2005) in Japanese.

[3] H.Takahashi et al., "Establishment of Methods for Effective Use of Regenerative power," Rolling stock & Technology, (Feb. 2007) in Japanese.

[4] H.Takahashi et al "Current and Future Applications for Regenerative Energy Storage System" Hitachi Review 61,(2012)
H.Takahashi et al., "Establishment of Methods for Effective Use of Regenerative power," Rolling stock & Technology, (Feb. 2012) in Japanese.

Integrated Isolation and Voltage Balancing Link of 3-Phase 3-Level PWM Rectifier and Inverter Systems

David O. Boillat and Johann W. Kolar

Power Electronic Systems Laboratory (PES), ETH Zurich, Switzerland
boillat@lem.ee.ethz.ch, www.pes.ee.ethz.ch

Abstract—For a 3-phase pulse width modulated high-bandwidth AC voltage source, this paper presents a series resonant DC–DC converter (SRC) with a high-frequency transformer operated by only two half-bridges interconnecting a 3-phase 3-level rectifier and 3-phase plus neutral conductor 3-level inverter stage. On the primary side, one terminal of the transformer is connected through the resonant capacitor and inductor to one bridge-leg output and the other terminal is connected to the DC-link midpoint. On the secondary side, both terminals directly connect to the second bridge-leg output and the DC-link midpoint. With the proposed SRC, the galvanic isolation and the balancing of the capacitor voltages of the inverter-side split DC-link can be achieved also for an unequal loading of the DC-link capacitors. The AC source needs to supply not only passive symmetrical 3-phase loads, but also passive or active single-phase, 2-phase, DC and asymmetrical 3-phase loads. Hence, the unequal loading of the DC-link capacitors can be generated, for example, in case a DC load is connected to the 4-wire inverter stage.

The operation principle of the SRC is described in detail for an unequal loading of the DC-link capacitors. Moreover, design guidelines for the suggested SRC are derived and finally the theoretical analysis is successfully verified by measurements conducted on a 1 kW proof-of-concept SRC prototype.

Keywords: *Galvanic Isolation, DC-Link Voltage Balancing, Series Resonant Converter, Dual Active Bridge, PWM Converter.*

I. INTRODUCTION

For bidirectional mains connected AC voltage sources consisting of a mains-side rectifier and a load-side inverter stage, a galvanically isolated DC–DC converter is inserted between the two stages to avoid a ground current in the case of grounded loads. International standards, e.g. IEC 60335-1, IEC 60950-1 and IEC 61140, limit the ground current for safety reasons; the maximum allowed ground current (e.g. 3.5 mA$_{rms}$) depends on the rated nominal current and on the specific application, in which the power converter is utilized. Commonly, the DC–DC converter is realized with a Dual Active Bridge Converter (DABC) [1]–[4] employing two active full-bridges as shown in **Fig. 1**, which depicts the simplified equivalent circuit of the considered 3-phase Pulse Width Modulated (PWM) high-bandwidth AC voltage source [5]–[7]. In **Table I** the electrical specifications of the AC source are summarized. A 3-phase plus neutral conductor inverter stage is utilized to not only power passive symmetrical 3-phase loads, but also passive as well as active single-phase, 2-phase, DC and asymmetrical 3-phase loads. The choice of the inverter stage switching frequency of

$f_{s,os} = 48$ kHz is motivated in [8] and the rectifier stage switching frequency of $f_{s,is} = 20$ kHz is chosen to be higher than the highest audible frequency. For such switching frequencies, it is shown in [9] that a 3-level Neutral Point Clamped (NPC) bridge-leg realized with 600 V IGBTs has lower losses than a standard 2-level bridge-leg built with 1200 V IGBTs. Thus, 3-level NPC bridge-legs are employed for the rectifier and the inverter stage, requiring also split DC-links.

The alternative isolation option, to insert a 50 Hz or 60 Hz line-frequency transformer between the mains and the rectifier stage [10], is typically discarded as it would significantly increase the size and the weight of the system [11].

As mentioned above, the AC source is also used to power DC loads. For example, a grounded DC load connected between phase A and the (grounded) neutral conductor N (cf. **Fig. 1**), which requires a positive voltage, takes its energy only from the upper DC-link capacitor $C_{dc,3}$ because the neutral conductor N is directly connected to the inverter-side DC-link midpoint. This leads to an unequal average loading of $C_{dc,3}$ and $C_{dc,4}$, and accordingly to an increasing positive voltage difference between $U_{dc,4}$ and $U_{dc,3}$ over time. Thus, to avoid this voltage difference and to ensure balanced DC-link voltages $U_{dc,3}$ and $U_{dc,4}$ in all operating points and for all loads, a balancer circuit with controlled current i_{bal}, as depicted in **Fig. 1**, is typically provided [12]. Balanced rectifier-side DC-link voltages $U_{dc,1}$ and $U_{dc,2}$ can be achieved with proper control of the rectifier stage, consisting of a 3-level NPC converter which allows to control the neutral-point potential [13], [14].

TABLE I Electrical specifications of the high-bandwidth AC source depicted in **Fig. 1**.

Nominal power P_{nom}	10 kW
Nominal mains voltage U_{mains}	400 V$_{ll,rms}$
Nominal rectifier stage current I_{nom}	14.5 A$_{rms}$
Nominal DC-link voltage $U_{dc,nom}$	700 V
Mains frequency f_{mains}	50 Hz
Rectifier stage switching (carrier) frequency $f_{s,is}$	20 kHz
DC–DC converter switching frequency $f_{s,dc}$	20 kHz
Inverter stage switching (carrier) frequency $f_{s,os}$	48 kHz

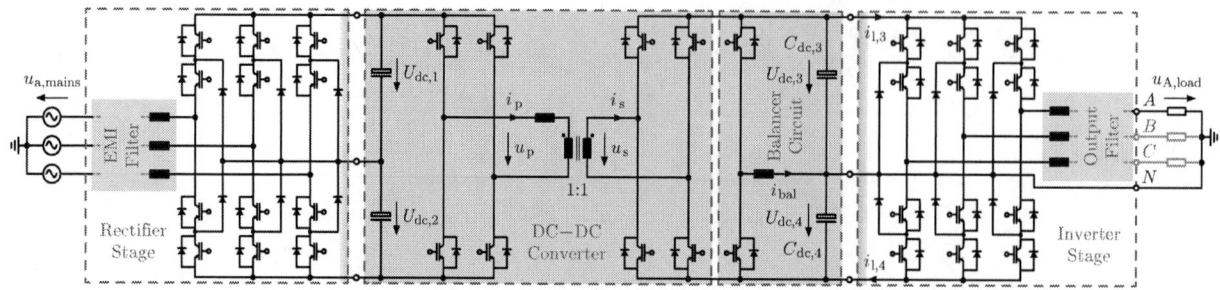

Fig. 1 Simplified equivalent circuit of the considered high-bandwidth AC source [5]–[7] with an integrated isolated DC–DC converter for galvanic isolation of the rectifier and inverter stage, and a balancer circuit to ensure equal DC-link voltages $U_{dc,3}$ and $U_{dc,4}$ also in case asymmetrical loads are supplied by the inverter stage.

As shown in **Fig. 1**, to achieve the galvanic isolation and the DC-link voltage balancing, typically two full-bridges and a half-bridge for the balancer circuit are employed. To reduce the number of power semiconductors and gate drive units from ten to four, the topology presented in **Fig. 2** is proposed [15]. Instead of a DABC, a <u>S</u>eries <u>R</u>esonant <u>C</u>onverter (SRC) [3], [16]–[19], with the resonant elements C_{res} and L_{res} is employed. In this topology, on the primary side, one terminal of the transformer is connected through the resonant capacitor and inductor to primary-side bridge-leg output and the other terminal is connected to the DC-link midpoint. On the secondary side, both terminals directly connect to the secondary-side bridge-leg output and the DC-link midpoint. The system is operated in a half-cycle discontinuous-conduction mode as further explained in **Section II**. Even though two active bridge-legs are used to allow a bidirectional power flow in all operating points, only one bridge-leg is switched and the other one operates as a diode rectifier. This reduces the complexity to run the converter compared to a DABC and motivates the selection of the SRC concept.

The load and the inverter stage are in **Fig. 2** represented by R_{34} and R_4 such that the same power flows through the DC–DC converter occur as if the inverter stage would be connected. The possible unequal average loading of the DC-link capacitors $C_{dc,3}$ and $C_{dc,4}$, as explained previously, is represented by the additional resistor R_4. The resistor R_{34} loads both capacitors $C_{dc,3}$ and $C_{dc,4}$ equally.

For an unequal average loading of the DC-link capacitors $C_{dc,3}$ and $C_{dc,4}$, the average values of the load currents $i_{l,3}$ and $i_{l,4}$ are different and the high-frequency transformer establishes a magnetizing current which is in average equal to the average of $i_{l,4} - i_{l,3}$ as further elaborated in **Section II**. Thus, the transformer is realized with an air-gap and integrates the galvanic isolation and DC-link voltage balancing.

For the DC–DC converter depicted in **Fig. 2**, 2-level half-bridges are employed instead of 3-level NPC bridge-legs as used for the rectifier and the inverter stages. A 2-level

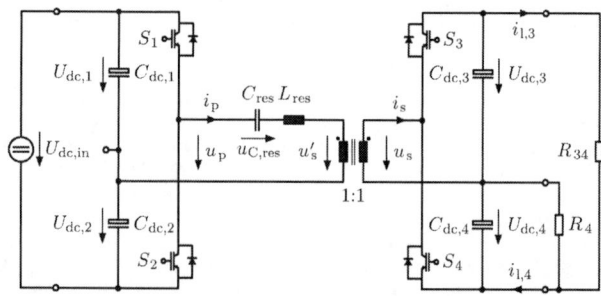

Fig. 2 Proposed DC–DC converter to achieve a galvanic isolation and a DC-link voltage balancing for split DC-links [15]. A series resonant converter (SRC) topology with only two active half-bridges is employed to reduce the number of power semiconductors and gate drive units compared to the standard solutions (cf. **Fig. 1**). Illustratively, the lower DC-link capacitor $C_{dc,4}$ is additionally loaded by R_4. It is assumed that the voltages $U_{dc,1}$ and $U_{dc,2}$ are balanced by the rectifier stage [13], [14].

bridge-leg with two 1200 V IGBTs has lower conduction losses but higher switching losses than a 3-level bridge-leg with four 600 V IGBTs [20], [21]. However, to reduce the number of power semiconductors and gate drive units, 2-level bridge-legs are selected. Furthermore, thw switching losses can be kept low due to the SR operation. The principle of operation for the proposed SRC with 3-level NPC bridge-legs can be found in [15].

Section II analyzes in detail the operation of the proposed SRC for an unequal loading of $C_{dc,3}$ and $C_{dc,4}$. The design guidelines for a 1 kW proof-of-concept SRC prototype are presented in **Section III**. The theoretical analysis of the converter is supported by measurements in **Section IV** and **Section V** concludes the paper.

A high-bandwidth AC source is the targeted application of the analyzed DC–DC converter shown in **Fig. 2**. However, the isolated DC–DC converter could also be used for mobile systems such as trains [22], electric cars [22], [23] and aerospace applications [2].

II. Operation Principle of the Half-Cycle Discontinuous-Conduction-Mode Series Resonant DC–DC Converter with Half-Bridges and Unequal Loading

The operation principle of a SRC run in Half-Cycle Discontinuous-Conduction Mode (HCDCM) with a half-bridge at the input and a full-bridge at the output, i.e. without the proposed DC-link balacing feature, is investigated in [19]. For analyzing the proposed system, in a first step, a symmetrical loading of the inverter-side DC-link capacitors is assumed, i.e. $R_4 \rightarrow \infty$ in **Fig. 2**. Moreover, the magnetizing current and the transformer stray inductance are neglected ($L_\mu \rightarrow \infty$). With these assumptions, the primary and the secondary currents are identical for a transformer turns ratio of 1:1. The voltage source $U_{\mathrm{dc,in}}$ depicted in **Fig. 2** represents the DC-link voltage controlled rectifier stage; the partial rectifier-side DC-link voltages $U_{\mathrm{dc,1}}$ and $U_{\mathrm{dc,2}}$ are balanced by the 3-level NPC rectifier stage.

For a power flow from the rectifier to the inverter stage, the primary-side bridge-leg with switches S_1 and S_2 (cf. **Fig. 2**) is actively switched with a duty-cycle d of 50% (including the interlocking time). The gate signals for S_1 and S_2 are phase-shifted by 180°. This leads to the primary voltage u_{p} depicted in **Fig. 3(a)** over two switching periods $T_{\mathrm{s}} = 1/f_{\mathrm{s,dc}}$. Because of the switching, a resonant current pulse is generated. The positive resonant pulse is denoted by $i_{\mathrm{s,pp}}$ and charges the upper DC-link capacitor $C_{\mathrm{dc,3}}$; the negative pulse is denoted by $i_{\mathrm{s,np}}$ and charges the lower DC-link capacitor $C_{\mathrm{dc,4}}$. These pulses are shown in **Fig. 3(a)** and defined as

$$i_{\mathrm{s,pp}} = \begin{cases} i_{\mathrm{s}} & \text{if } i_{\mathrm{s}} \geq 0 \\ 0 & \text{else} \end{cases}, \quad i_{\mathrm{s,np}} = \begin{cases} i_{\mathrm{s}} & \text{if } i_{\mathrm{s}} \leq 0 \\ 0 & \text{else} \end{cases}. \quad (1)$$

On the secondary side, the switches S_3 and S_4 are not operated or operated with synchronous rectification, i.e. the secondary-side bridge-leg acts as a diode rectifier. In HCDCM, the resonant current pulses reach zero before a new current pulse is excited, i.e. $t_{\mathrm{psz}} \leq T_{\mathrm{s}}/2$ [cf. **Fig. 3(a)**], i.e. the SRC is operated below the resonant frequency $f_{\mathrm{res}} = 1/\left(2 \cdot \pi \cdot \sqrt{L_{\mathrm{res}} \cdot C_{\mathrm{res}}}\right)$. Because of the diode rectification, once the current pulses reach zero at t_{psz}, the diodes avoid that the currents reverse direction. Accordingly, the secondary current is zero until a new current pulse is excited.

For a finite magnetizing inductance, with the equivalent circuit of the transformer referred to the primary side as depicted in **Fig. 3(e)** and for an unequal loading of $C_{\mathrm{dc,3}}$ and $C_{\mathrm{dc,4}}$, the primary voltage u_{p}, the secondary voltage u_{s}, and the secondary current i_{s} are depicted in **Fig. 3(b)**. In steady-state, the average values of $i_{\mathrm{s,pp}}$ and $i_{\mathrm{s,np}}$ over T_{s} are equal to the positive average load current $i_{\mathrm{l3,avg}}$ of $C_{\mathrm{dc,3}}$ and the negative average load current $i_{\mathrm{l4,avg}}$ of $C_{\mathrm{dc,4}}$ over the same period respectively. Thus,

$$i_{\mathrm{s,pp,avg}} = i_{\mathrm{l3,avg}}, \quad i_{\mathrm{s,np,avg}} = -i_{\mathrm{l4,avg}} \quad (2)$$

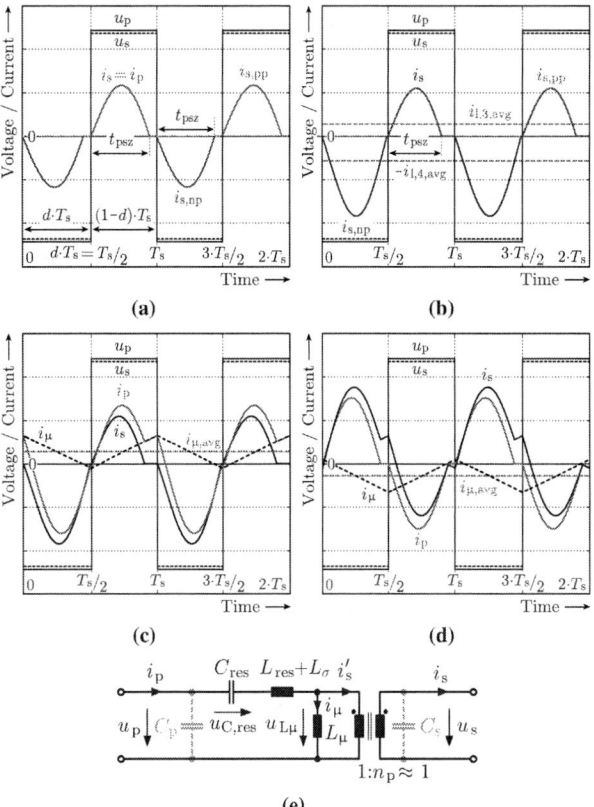

Fig. 3 Exemplary primary voltage u_{p}, secondary voltage u_{s}, primary current i_{p} and secondary current i_{s} of the SRC given in **Fig. 2** for $L_\mu \rightarrow \infty$ and equal average load currents $i_{\mathrm{l3,avg}} = i_{\mathrm{l4,avg}}$ **(a)**; u_{p}, u_{s} and i_{s} **(b)** and additionally i_{p} and magnetizing current i_μ **(c)** for a finite magnetizing inductance and an unequal *loading* of the inverter-side DC-link capacitors. Exemplary u_{p}, u_{s}, i_{p}, i_{s} and i_μ for an unequal power *fed* into the inverter-side DC-link capacitors **(d)**. T_{s} denotes the switching period and the equivalent circuit of the resonant components and the high-frequency transformer referred to the primary side is given in **(e)**.

and accordingly the average of the secondary current $i_{\mathrm{s,avg}}$ over T_{s} is given by (discontinuous-conduction-mode)

$$i_{\mathrm{s,avg}} = i_{\mathrm{s,pp,avg}} + i_{\mathrm{s,np,avg}} = i_{\mathrm{l3,avg}} - i_{\mathrm{l4,avg}}. \quad (3)$$

In the following, the averages of currents and voltages of the SRC are always taken over one switching period T_{s} what is not explicitly mentioned any more.

For the case shown in **Fig. 3(b)**, $i_{\mathrm{l4,avg}}$ is larger than $i_{\mathrm{l3,avg}}$ and accordingly $i_{\mathrm{s,avg}}$ is negative [cf. Eq. (3)]. On the primary side, the average of the primary current i_{p} is zero in steady-state; otherwise, C_{res} would be continuously charged or discharged:

$$i_{\mathrm{p,avg}} = i_{\mu,\mathrm{avg}} + i'_{\mathrm{s,avg}} = 0. \quad (4)$$

This means that an average magnetizing current $i_{\mu,\mathrm{avg}}$ flows which compensates the average value of the secondary current

i_s, and hence

$$i_{\mu,\text{avg}} = -i'_{s,\text{avg}} = i'_{l,4,\text{avg}} - i'_{l,3,\text{avg}}, \qquad (5)$$

as depicted in **Fig. 3(c)**. In the above equation the secondary-side load currents are transformed to the transformer primary side and accordingly assigned by a " ′ ". This notation is used throughout the rest of the paper.

In steady-state, the voltage

$$u_{L\mu} = \frac{L_\mu}{L_\mu + L_\sigma + L_{\text{res}}} \cdot \begin{cases} U_{\text{dc},1} - u_{C,\text{res}} & 0 \le t < \frac{T_s}{2} \\ -U_{\text{dc},2} - u_{C,\text{res}} & \frac{T_s}{2} \le t < T_s \end{cases}, \qquad (6)$$

applied to the transformer magnetizing inductance L_μ [cf. **Fig. 3(e)**] is because of (4) also zero.

Thus, DC-link voltage balancing is achieved by the average magnetizing current $i_{\mu,\text{avg}}$ and accordingly the magnetizing inductance of the high-frequency transformer takes over the functionality of the balancer inductance shown in **Fig. 1**. Therefore, the SRC integrates the DC-link voltage balancing and the galvanic isolation. In the case at hand, no voltage step-down or step-up is required, thus, a transformer turns ratio of 1:1 is employed.

Because of the average magnetizing current $i_{\mu,\text{avg}} \neq 0$, the transformer needs to store energy and accordingly is preferably built with a low-permeability material or with a discrete air-gap. Alternatively, an inductor could be placed in parallel to the transformer primary or secondary winding which would take over a large part of the magnetizing current [cf. (5)] flowing through the transformer without the additional inductor.

For an ideal switching, the average rectifier-side DC-link voltages $U_{\text{dc},1,\text{avg}}$ and $U_{\text{dc},2,\text{avg}}$ are with the help of the magnetizing current equal in average. Based on the more detailed analysis carried out in **Section III**, it is observed that the higher the load currents the lower the average inverter-side DC-link voltages for fixed rectifier-side DC-link voltages. The reason is that the average values of the resonant current pulses [cf. Eq. (1)] increase with the higher load currents. Accordingly, to generate resonant pulses with higher average values, larger voltage excitations are required [19] meaning lower average inverter-side DC-link voltages. Thus, for unequal load currents $i_{l,3}$ and $i_{l,4}$, different average DC-link voltages $U_{\text{dc},3,\text{avg}}$ and $U_{\text{dc},4,\text{avg}}$ result.

Ideally, i.e. without including the parasitic capacitances C_p and C_s [cf. **Fig. 3(e)**] between the bridge-leg outputs and the corresponding midpoints, the largest average inverter-side DC-link voltage difference is reached for $i_{l,3,\text{avg}} \neq 0$ and $i_{l,4,\text{avg}} = 0$ or vice versa. Including the mentioned capacitors, an oscillatory charge reversal of these capacitances occurs during a switching transient. If during the transient the voltage across C_s is in absolute value larger than the DC-link voltage $U_{\text{dc},3}$ or $U_{\text{dc},4}$ plus the diode forward voltage drop u_{df}, the DC-link capacitor $C_{\text{dc},3}$ or $C_{\text{dc},4}$ is slightly charged. Thus, $C_{\text{dc},3}$ or $C_{\text{dc},4}$ charges to the value of the transient peak voltage across

C_s minus u_{df}. This problem can be diminished by placing symmetrizing resistors across all DC-link capacitors which are loading the capacitors just enough to compensate the slight charging. As demonstrated by measurements (cf. **Section IV**), symmetrizing resistors of 22 kΩ could be employed for the 1 kW prototype which are increasing the losses by only about 1 W.

Another way to go around the issue is to actively switch the secondary-side bridge-leg in-phase with the primary-side bridge-leg, which would increase the total losses only by a few percent, as could be shown by a more detailed analysis. In case only a unidirectional power flow from the rectifier to the inverter stage is required, the secondary-side bridge-leg could be realized with only diodes and therefore symmetrizing resistors would be necessary[1].

Concluding, ideally a perfect DC-link voltage balancing is given on the rectifier side but not on the inverter side. However, because large resonant current pulses can be generated with a small excitation voltage, the voltage difference between $U_{\text{dc},3,\text{avg}}$ and $U_{\text{dc},4,\text{avg}}$ is for example at maximum 10% of $(U_{\text{dc},3,\text{avg}} + U_{\text{dc},4,\text{avg}})/2$ for the built prove-of-concept converter (cf. **Section IV**).

For feeding power back from the inverter to the rectifier side, the secondary-side bridge-leg is switched and the primary-side bridge-leg is operated as diode rectifier. In this case, again the magnetizing inductance of the transformer directly acts as a balancer inductor. Illustratively, the primary and secondary voltages as well as the primary and secondary currents are depicted in **Fig. 3(d)** for this case.

It is important to note that the resonant capacitor C_{res} needs to be placed on the side of the transformer where the DC-link capacitors are equally loaded (i.e. the rectifier side for the case at hand).

III. Design Guidelines for the Series Resonant DC–DC Converter with Unequal Loading

The resonant current pulses [cf. **Fig. 3(b)**] are computed analytically in this section to provide physical insight into the operating behavior of the SRC depicted in **Fig. 2**. For the sake of brevity, the subsequent equations are shown for the positive current pulse $i_{s,pp}$ of the secondary current [cf. **Fig. 3(a)**] and for a power flow from the rectifier to the inverter stage. The negative current pulse can be computed analogously.

By superimposing the magnetizing current i_μ and the secondary current i'_s, the primary current i_p is obtained. The average of i_μ is given by the average difference between the load current $i_{l,3,\text{avg}}$ and $i_{l,4,\text{avg}}$; the peak-to-peak

---

[1]A further option would be to control the duty-cycle d of the switches of the primary-side bridge-leg [cf. **Fig. 3(a)**] as presented in [15], which would allow to equalize the inverter-side DC-link voltages $U_{\text{dc},3}$ and $U_{\text{dc},4}$ in average.

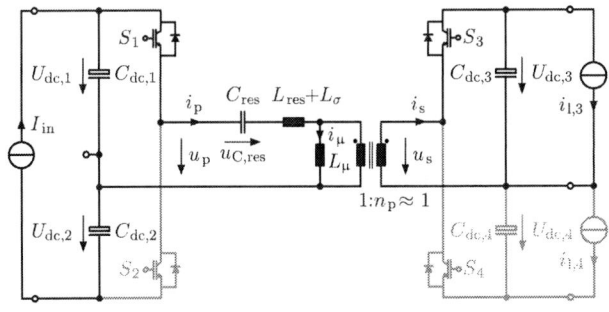

(a)

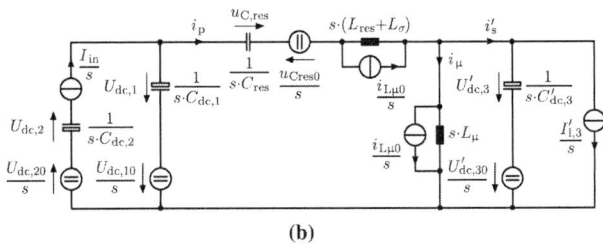

(b)

Fig. 4 Equivalent circuit employed for the theoretical analysis of the SRC of **Fig. 2** for a positive resonant current pulse [cf. **Fig. 3(b)**] **(a)**, and resulting equivalent circuit in the frequency domain based on the approach described in [19] **(b)**. The equivalent circuits are only valid for $0 \leq t \leq t_{psz}$ [cf. **Fig. 3(b)**]. The constant voltage and current sources in **(b)** result from the initial conditions (especially $i_{L\mu0} \neq 0$), which are marked with an additional "0". The input current $I_{in} = P_{in}/U_{dc,in}$ and the load current $i_{l,3} = I_{l,3}$ are assumed to be constant.

current ripple of i_μ can be assessed using (6) while setting $U_{dc,1}(t) = U_{dc,2}(t) = U_{dc,in}/2$ and neglecting $u_{C,res}(t)$ for $U_{dc,in}/2 \gg u_{C,res}(t) \; \forall \; t$.

As described in the last section, the SRC depicted in **Fig. 2** differs from standard SRC (cf. [19] for example) by its much lower magnetizing inductance. Thus, the magnetizing inductance needs to be included into the modeling of the converter. Based on the approach explained in [19], the equivalent circuit in the frequency domain depicted in **Fig. 4** can be employed to compute the secondary current $i_s'(s)$. Adding the magnetizing inductance in the modeling enforces to set an initial condition on the magnetizing current, i.e. $i_{L\mu0}$, and thus also on the primary current.

In the frequency domain, the current $i_s'(s)$ can be computed using nodal and mesh equations. Transforming $i_s'(s)$ into the time domain and assuming a constant load current, $i_{l,3} = I_{l,3} =$ const., leads to

$$
\begin{aligned}
i_s'(t) = \; & I_{l,3}' + \\
& I_{s,1} \cdot \sin\left(\omega_1 \cdot t\right) + I_{c,1} \cdot \cos\left(\omega_1 \cdot t\right) + \\
& I_{s,2} \cdot \sin\left(\omega_2 \cdot t\right) + I_{c,2} \cdot \cos\left(\omega_2 \cdot t\right).
\end{aligned} \tag{7}
$$

In the above equation the two angular frequencies depend only

on the circuit parameters and are with the simplification of $C_{dc,1} = C_{dc,2}$ given by

$$
\begin{aligned}
\omega_1 &= \sqrt{\frac{1}{2 \cdot C_{1res} \cdot L_{\sigma res}} + \frac{L_{\sigma res} + L_\mu - \tilde{L}}{2 \cdot C_{dc,3}' \cdot L_{\sigma res} \cdot L_\mu}}, \\
\omega_2 &= \sqrt{\frac{1}{2 \cdot C_{1res} \cdot L_{\sigma res}} + \frac{L_{\sigma res} + L_\mu + \tilde{L}}{2 \cdot C_{dc,3}' \cdot L_{\sigma res} \cdot L_\mu}},
\end{aligned} \tag{8}
$$

where $L_{\sigma res} = L_\sigma + L_{res}$, $C_{1res} = C_{dc,1} \cdot C_{res}/(C_{dc,1} + C_{res})$ and

$$
\tilde{L} = \sqrt{L_{\sigma res}^2 - 2 \cdot \tilde{C}_{f1} \cdot L_{\sigma res} \cdot L_\mu + \tilde{C}_{f2}^2 \cdot L_\mu^2} \tag{9}
$$

with the scaling factors

$$
\begin{aligned}
\tilde{C}_{f1} &= \frac{C_{dc,1} \cdot (C_{dc,3}' - C_{res}) + C_{dc,3}' \cdot C_{res}}{C_{dc,1} \cdot C_{res}} \; \text{and} \\
\tilde{C}_{f2} &= \frac{C_{dc,1} \cdot (C_{dc,3}' + C_{res}) + C_{dc,3}' \cdot C_{res}}{C_1 \cdot C_{res}}.
\end{aligned} \tag{10}
$$

The coefficients $I_{s,1}$, $I_{c,1}$, $I_{s,2}$ and $I_{c,2}$ are a function of the initial conditions:

$$
\begin{aligned}
I_{s,k} &= \frac{C_{dc,3}' \cdot \left(\frac{U_{dc,30}'}{C_{1res}} + \left(L_\mu \cdot u_{dc,p} - (L_{\sigma res} + L_\mu) \cdot U_{dc,30}' \right) \cdot \omega_k^2 \right)}{2 \cdot L_{\sigma res} \cdot L_\mu \cdot C_{dc,3}' \cdot \omega_k^3 - \left(L_{\sigma res} + L_\mu \cdot \tilde{C}_{f2} \right) \cdot \omega_k}, \\
I_{c,k} &= \frac{L_\mu \cdot \left(\frac{C_{dc,1}}{C_{res}} \cdot (i_{l,3}' + i_{L\mu0}) + \left(i_{l,3}' \cdot (1 - C_{dc,1} \cdot L_{\sigma res} \cdot \omega_k^2) - \Delta i_{in} \right) \right)}{\frac{C_{dc,1}}{C_{dc,3}'} \cdot \left(2 \cdot L_{\sigma res} \cdot L_\mu \cdot C_{dc,3}' \cdot \omega_k^2 - (L_{\sigma res} + L_\mu \cdot \tilde{C}_{f2}) \right)},
\end{aligned} \tag{11}
$$

where $k = 1, 2$, $u_{dc,p} = u_{Cres0} + U_{dc,10}$ and $\Delta i_{in} = I_{in} - i_{L\mu0}$. $I_{s,k}$ depends only on the initial conditions of the voltages and $I_{c,k}$ only on the initial conditions of the currents.

Moreover, it is noted that the resonant current pulses contain two frequency components. For the built hardware (cf. **Section IV**), the frequencies are $f_1 = \omega_1/(2 \cdot \pi) \approx 1$ kHz and $f_2 = \omega_2/(2 \cdot \pi) \approx 24$ kHz. Furthermore, $i_{l,3}' + I_{c,1} + I_{c,2} = 0$ follows directly from the fact that the current on the secondary side cannot reverse direction after a current pulse because of the secondary-side diodes. Moreover, (7) can be simplified to

$$
i_s'(t) = I_{l,3}' + I_1 \cdot \sin\left(\omega_1 \cdot t + \phi_1\right) + I_2 \cdot \sin\left(\omega_2 \cdot t + \phi_2\right) \tag{12}
$$

with $I_k = \sqrt{I_{s,k}^2 + I_{c,k}^2}$ and $\phi_k = \text{atan2}\left(I_{c,k}, I_{s,k}\right)$.

The time t_{psz} [cf. **Fig. 3(b)**, exemplary shown for the positive resonant current pulse $i_{s,pp}$] when the secondary current $i_s(t)$ reaches zero can be computed numerically by equating (12) to zero. t_{psz} increases with the load current $i_{l,3}$ and thus the resonant elements C_{res} and $L_{\sigma res}$ can be selected such that for the chosen switching frequency and for the maximum load current $i_{l,3,max}$ or $i_{l,4,max}$ (i.e. for loading only $C_{dc,3}$ or $C_{dc,4}$), the resonant current pulses reach zero before a new current pulse is generated, i.e. $t_{psz} \leq T_s/2$[2].

In conclusion, a first design of the SRC, which is supported by circuit simulations, can be obtained as follows. To reduce

---

[2]It is noted that the shape of the resonant current pulses and the time t_{psz} depend also on the parasitic resistive parts between one and the other split DC-links and the symmetrizing resistors connected across the DC-link capacitors.

the switching losses, the switching frequency $f_{s,dc}$ can be chosen such that it is just higher than the highest audible frequency. For the prototype (cf. **Section IV**), $f_{s,dc} = 20$ kHz is selected. In a next step, the maximum difference in the load current [cf. (5)] needs to be determined, which is given by the application in which the converter is employed. For the built 1 kW hardware, this maximum difference is set to 333 W. The capacitances of the DC-link capacitors can be determined such that the DC-link voltage ripple across one capacitor is limited, e.g. to 2.5% of $U_{dc,in}/2$.

On the one hand, the larger the value of the transformer's magnetizing inductance the smaller the peak-to-peak ripple of i_μ and accordingly the lower the high-frequency copper losses. On the other hand, there is also an advantage in having a reduced magnetizing inductance, especially in case the switches are implemented with MOSFETs. For a power flow from the rectifier to the inverter stage [cf. **Fig. 3(c)**], a good option is, as it is done for the 1 kW prototype, to choose the magnetizing inductance L_μ just large enough that soft-switching (zero-voltage switching) can be achieved in all operating points. Therefore, the magnetizing current has to become negative before the positive voltage is applied at the output of the primary-side bride-leg as exemplary shown in **Fig. 3(c)** and vice versa. Thus, the peak-to-peak magnetizing current ripple $\Delta i_{\mu,pp}$ needs to be slightly larger than two times the maximum average magnetizing current $i_{\mu,avg,max}$, i.e.

$$\Delta i_{\mu,pp} > 2 \cdot \left| i_{\mu,avg,max} \right| = 2 \cdot \left| i'_{l,3,avg,max} - i'_{l,4,avg,min} \right|. \quad (13)$$

With the above condition, also soft-switching is obtained in all operating points for an inverse power flow from the inverter to the rectifier stage as exemplary given in **Fig. 3(d)**.

To transfer the same amount of energy, the shorter the time t_{psz} the larger the peaks of the resonant current pulses for the same average DC-link voltages and switching frequency $f_{s,dc}$. This increases the primary and secondary rms currents and therefore also the copper losses. Accordingly, $t_{psz} = T_s/2$ is selected for the longest duration of the resonant current pulses. In a next step, the resonant elements can be determined. To achieve a compact converter, L_{res} should be as small as possible. Therefore, $L_{\sigma res}$ can be realized by only the stray inductance L_σ of the high-frequency transformer, i.e. $L_{\sigma res} = L_\sigma$, as it is done for the prototype. L_σ is fixed depending on the transformer design and hence C_{res} can be calculated. For a compact realization of the converter, the values of L_σ and C_{res} should be determined simultaneously to avoid high C_{res} values resulting for low L_σ.

IV. EXPERIMENTAL VERIFICATION

To verify the theoretically analyzed operation principle of the proposed SRC with integrated DC-link voltage balancing and galvanic isolation, a 1 kW proof-of-concept prototype is built. The electrical specifications and the circuit parameters of the system are summarized in **Table II**. The switched

TABLE II Electrical specifications and circuit parameters of the 1 kW SRC prototype with integrated DC-link voltage balancing and galvanic isolation.
*): To assess $L_{\sigma res}$ (cf. **Section III**), the inductances of the connecting wires, connectors and PCB tracks need to be accounted as well. For the built hardware these inductances were calculated to 0.31 µH.

Electrical Specifications	
Nominal power	1 kW
Max. asym. loading	333 W
Total DC-link voltage	150 V
Switching freq. $f_{s,dc}$	20 kHz
Circuit Parameters	
DC-link capacitors $C_{dc,x}$, $x = 1, 2, 3, 4$	
Measured capacitance	133 µF (2×68 µF)
Type	EPCOS MKT 100 V_{DC}
Resonant capacitor C_{res}	
Measured capacitance	30.2 µF (7×4.7 µF)
Type	WIMA MKS 2 30 V_{AC}
High-frequency transformer	
Core	$3 \times 2 \times$ E 42/21/15, EPCOS N27
Number of turns	$N_1 = N_2 = 17$ (bifilar windings)
Air-gap / coupling	1.7 mm / $k = 0.994 \rightarrow n_p = 0.994$
Magnetizing / stray inductance	$L_\mu = 153.1$ µH / $L_\sigma = 1.8$ µH$^{*)}$

bridge-leg is realized with 200 V MOSFETs (OptiMOS^{TM}3 IPP320N20N3 G from Infineon Technologies AG) and the secondary-side diode rectifier stage is implemented with Schottky Diodes (MBR40250TG from On Semiconductor®). MOSFETs are selected because the relatively large forward voltage drops of IGBTs in comparison to the DC-link voltage levels. Additionally, due to the zero-voltage switching, less switching losses are expected for MOSFETs than for IGBTs.

For the measurements, the equivalent circuits depicted in **Fig. 5** are employed. Five cases are investigated: for the cases $I-IV$, the power flow is from the rectifier to the inverter stage [cf. **Fig. 5(a)**]; for case V, the power flow is in the opposite direction [cf. **Fig. 5(b)**].

- *Case I* - Operation at "no load" [cf. **Fig. 6(a)**]: as mentioned in **Section II**, a minimum loading of $C_{dc,3}$ and $C_{dc,4}$ is necessary, which is achieved by setting $R_3 = R_4 \rightarrow \infty$ and $R_{34} = 22$ kΩ. This case represents the idle case, when no load is connected to the output of the SRC.
- *Case II* - Symmetrical operation at 1 kW [cf. **Fig. 6(b)**]: the inverter-side DC-link capacitors are loaded with $R_3 = R_4 = 22.5$ Ω ($R_{34} \rightarrow \infty$).
- *Case III* - Asymmetrical operation at 1 kW [cf. **Fig. 6(c)**]: $C_{dc,3}$ and $C_{dc,4}$ are loaded with $R_3 = 16.8$ Ω ($\rightarrow 330$ W) and $R_4 = 8.4$ Ω ($\rightarrow 660$ W) respectively ($R_{34} \rightarrow \infty$).
- *Case IV* - Asymmetrical operation at 330 W [cf. **Fig. 6(d)**]: the inverter-side DC-link capacitors

The 2014 International Power Electronics Conference

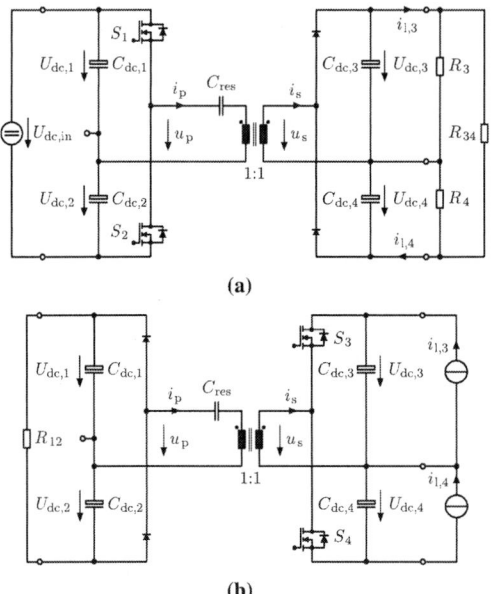

(a)

(b)

Fig. 5 Equivalent circuits of the measurement setups: for a power flow from the inverter stage to the rectifier stage **(a)**, and for a power flow from the inverter stage to the rectifier stage **(b)**.

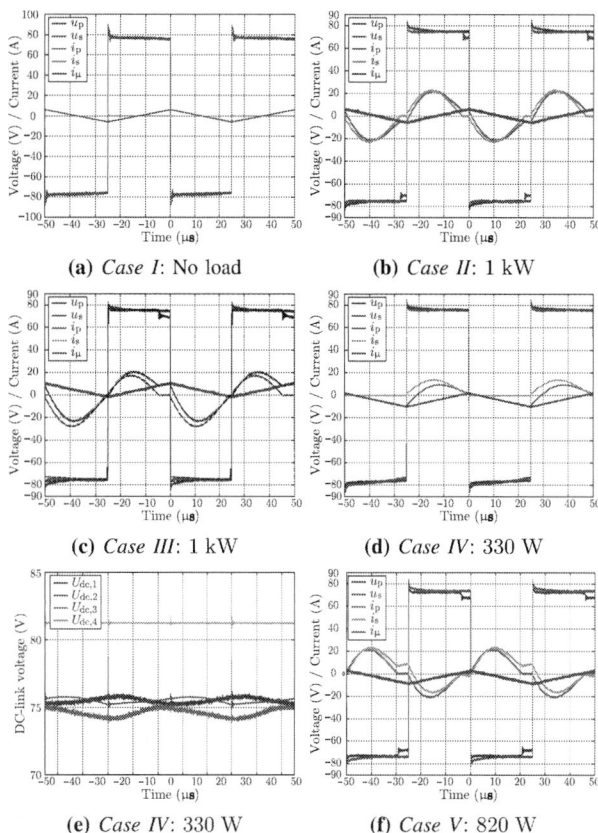

(a) *Case I*: No load **(b)** *Case II*: 1 kW

(c) *Case III*: 1 kW **(d)** *Case IV*: 330 W

(e) *Case IV*: 330 W **(f)** *Case V*: 820 W

Fig. 6 On the 1 kW prototype (cf. **Table II**) measured primary voltage u_p, primary current i_p, secondary voltage u_s and secondary current i_s (cf. **Fig. 5**) for the five cases mentioned in the text: case I **(a)**, case II **(b)**, case III **(c)**, case IV **(d)** and case V **(f)**. For case IV, the DC-link voltages ($U_{dc,1,avg} = 75.5$ V, $U_{dc,2,avg} = 75.3$ V, $U_{dc,3,avg} = 74.6$ V, $U_{dc,4,avg} = 81.2$ V) are given in **(e)**. The transformer magnetizing current was calculated as $i_\mu = i_p - i'_s$.

are loaded with $R_3 = 16.8\ \Omega\ (\to 330$ W) and $R_4 = 22\ k\Omega\ (R_{34} \to \infty)$. Because of the issue mentioned in **Section II**, $C_{dc,4}$ was slightly loaded.

- *Case V* - Asymmetrical operation at 820 W [cf. **Fig. 6(f)**]: $i_{l,3} = 4$ A ($\to 300$ W) and $i_{l,4} = 7.1$ A ($\to 520$ W) is set. On the rectifier side, $C_{dc,1}$ and $C_{dc,2}$ are equally loaded by $R_{12} = 25\ \Omega$.

For the above mentioned cases, the measurements are shown in **Fig. 6**. Comparing the experimental results to **Fig. 3**, the measured voltages and currents are in good agreement with the theoretical analysis conducted throughout this paper. Furthermore, as exemplary depicted in **Fig. 6(c)**, the measurements are matching with the simulations (dashed black curves) carried out in GeckoCIRCUITS.

For case IV, the DC-link voltages $U_{dc,1}$, $U_{dc,2}$, $U_{dc,3}$ and $U_{dc,4}$ on the rectifier side and on the inverter side are given in **Fig. 6(e)**. As explained in **Section II**, the maximum DC-link voltage unbalance on the inverter side occurs when only one DC-link capacitor is loaded. The average voltage difference between $U_{dc,4}$ and $U_{dc,3}$ is 6.6 V and therefore 8.5% referred to $(U_{dc,3,avg} + U_{dc,4,avg})/2 = 77.9$ V.

V. Conclusion

In this paper, a series resonant DC–DC converter (SRC) integrating the split DC-link voltage balancing for unequal loadings of the DC-link capacitors and the galvanic isolation is presented for a high-bandwidth AC voltage source. The

SRC interconnects a 3-phase 3-level rectifier stage with a 3-phase plus neutral conductor 3-level inverter stage which needs also to supply asymmetrical loads. The galvanic isolation is required to avoid a ground current and is achieved with a high-frequency transformer.

The proposed SRC is operated in half-cycle discontinuous-conduction-mode and consists of two half-bridges, from which only one is switched depending on the power flow direction. The bridge-leg output and the DC-link midpoint on the primary side are connected through the resonant capacitor, the resonant inductor and the primary winding of the transformer; on the secondary side, they are directly connected through the transformer secondary winding.

Supplying asymmetrical loads, the inverter-side DC-link capacitors are not equally loaded simultaneously. In this case, the voltage balancing across the DC-link capacitors is achieved by establishing an average (over one switching cycle)

magnetizing current which is equal to the difference of the load currents. Accordingly, the high-frequency transformer integrates the DC-link voltage balancing and the galvanic isolation. It needs to store energy and hence should be realized with an air-gap or a low-permeability material for a compact realization.

For standard solutions, the DC-link voltage balancing is achieved with an explicit balancer circuit and the galvanic isolation with a dual active bridge converter employing two full-bridges. Thus, the proposed SRC reduces the number of power semiconductors from 10 to 4 compared to the conventional topology.

Ideally, the DC-link voltage balancing on the rectifier side can be achieved perfectly. On the inverter side, the difference of the DC-link voltages depends on the average load currents. However, this voltage difference can be restricted to less than 10%, referred to the voltage across one DC-link capacitor.

Design guidelines are presented showing that the SRC can achieve soft-switching (zero-voltage switching) in all operating conditions if the peak-to-peak magnetizing current ripple is slightly larger than two times the maximum average magnetizing current (which corresponds to twice the maximum load current difference).

The theoretical analysis is supported by measurements conducted on a 1 kW proof-of-concept prototype matching very well with the theory and the simulations for no load, symmetrical loading and asymmetrical loading for a bidirectional power flow.

REFERENCES

[1] R. De Doncker, D. Divan, and M. Kheraluwala, "A Three-Phase Soft-Switched High-Power-Density DC/DC Converter for High-Power Applications," *IEEE Transactions on Industry Applications*, vol. 27, no. 1, pp. 63–73, 1991.

[2] M. Kheraluwala, R. Gascoigne, D. Divan, and E. Baumann, "Performance Characterization of a High-Power Dual Active Bridge DC-to-DC Converter," *IEEE Transactions on Industry Applications*, vol. 28, no. 6, pp. 1294–1301, 1992.

[3] R. Steigerwald, R. De Doncker, and M. Kheraluwala, "A Comparison of High-Power DC–DC Soft-Switched Converter Topologies," *IEEE Transactions on Industry Applications*, vol. 32, no. 5, pp. 1139–1145, 1996.

[4] F. Krismer and J. Kolar, "Efficiency-Optimized High-Current Dual Active Bridge Converter for Automotive Applications," *IEEE Transactions on Industrial Electronics*, vol. 59, no. 7, pp. 2745–2760, 2012.

[5] D. O. Boillat and J. W. Kolar, "Modeling and Experimental Analysis of a Coupling Inductor Employed in a High Performance AC Power Source," in *Proc. 1st Intern. Conf. on Renewable Energy Research and Applications (ICRERA)*, 2012.

[6] D. O. Boillat, J. W. Kolar, and J. Mühlethaler, "Volume Minimization of the Main DM/CM EMI Filter Stage of a Bidirectional Three-Phase Three-Level PWM Rectifier System," in *Proc. of the IEEE Energy Conversion Congress and Exposition (ECCE USA)*, Denver, Colorado, USA, Sept. 2013.

[7] P. Cortés, D. O. Boillat, and J. W. Kolar, "Phase-Oriented Control of a Modular 3-Phase 3-Level 4-Leg Inverter AC Power Source Supplying Floating or Grounded Loads," in *Proc. of the IEEE Energy Conversion Congress and Exposition (ECCE USA)*, Denver, Colorado, USA, Sep. 2013.

[8] D. O. Boillat, T. Friedli, J. Mühlethaler, J. W. Kolar, and W. Hribernik, "Analysis of the Design Space of Single-Stage and Two-Stage *LC* Output Filters of Switched-Mode AC Power Sources," in *Proc. IEEE Power and Energy Conf. at Illinois (PECI)*, 2012, pp. 1–8.

[9] M. Schweizer and J. Kolar, "Design and Implementation of a Highly Efficient Three-Level T-Type Converter for Low-Voltage Applications," *IEEE Transactions on Power Electronics*, vol. 28, no. 2, pp. 899–907, 2013.

[10] P. Hammond, "A New Approach to Enhance Power Quality for Medium Voltage AC Drives," *IEEE Transactions on Industry Applications*, vol. 33, no. 1, pp. 202–208, 1997.

[11] S. Inoue and H. Akagi, "A Bidirectional DC–DC Converter for an Energy Storage System With Galvanic Isolation," *IEEE Transactions on Power Electronics*, vol. 22, no. 6, pp. 2299–2306, 2007.

[12] H. Ertl, J. W. Kolar, and F. C. Zach, "Analysis of an Extended DC-DC Flyback Converter with Inherent Input Voltage Symmetrization," in *Proc. 38th Int. Conf. Power Conversion (PCIM)*, Nurenberg, Germany, May 2008, pp. 501–506.

[13] N. Celanovic and D. Boroyevich, "A Comprehensive Study of Neutral-Point Voltage Balancing Problem in Three-Level Neutral-Point-Clamped Voltage Source PWM Inverters," *IEEE Transactions on Power Electronics*, vol. 15, no. 2, pp. 242–249, 2000.

[14] J. Pou, R. Pindado, D. Boroyevich, and P. Rodriguez, "Evaluation of the Low-Frequency Neutral-Point Voltage Oscillations in the Three-Level Inverter," *IEEE Transactions on Industrial Electronics*, vol. 52, no. 6, pp. 1582–1588, 2005.

[15] D. O. Boillat and J. W. Kolar, "Vorrichtung und Verfahren zur galvanischen Trennung und Spannungssymmetrierung in einem bidirektionalen DC-DC Wandler mit hochfrequentem AC-AC Übertrager," CH Patent, 2013.

[16] R. King and T. Stuart, "Modeling the Full-Bridge Series-Resonant Power Converter," *IEEE Transactions on Aerospace and Electronic Systems*, vol. AES-18, no. 4, pp. 449–459, 1982.

[17] R. Steigerwald, "High-Frequency Resonant Transistor DC–DC Converters," *IEEE Transactions on Industrial Electronics*, vol. IE-31, no. 2, pp. 181–191, 1984.

[18] A. Esser and H.-C. Skudelny, "A New Approach to Power Supplies for Robots," *IEEE Transactions on Industry Applications*, vol. 27, no. 5, pp. 872–875, 1991.

[19] D. Rothmund, J. Huber, and J. W. Kolar, "Operating Behavior and Design of the Half-Cycle Discontinuous-Conduction-Mode Series-Resonant-Converter with Small DC Link Capacitors," in *Proc. of the 14th IEEE Workshop on Control and Modeling for Power Electonics (COMPEL)*, Salt Lake City, USA, June 2013.

[20] Infineon Technologies, "IKW40N60H3 - 600 V High Speed DuoPack IGBT in Trench and Fieldstop Technology with Soft, Fast Recovery Anti-Parallel Diode," Data Sheet - Infineon, 2010.

[21] ——, "IKW40N120H3 - 1200 V High Speed DuoPack: IGBT in Trench and Fieldstop Technology with Soft, Fast Recovery Anti-Parallel Diode," Data Sheet - Infineon, 2010.

[22] M. Pavlovsky, S. de Haan, and J. Ferreira, "Reaching High Power Density in Multikilowatt DC–DC Converters With Galvanic Isolation," *IEEE Transactions on Power Electronics*, vol. 24, no. 3, pp. 603–612, 2009.

[23] J. Walter and R. De Doncker, "High-Power Galvanically Isolated DC/DC Converter Topology for Future Automobiles," in *Proc. of the 34th Annual IEEE Power Electronics Specialist Conference (PESC)*, vol. 1, 2003, pp. 27–32 vol.1.

Voltage Step-Up Converter based on Multistage Stacked Boost Architecture (MSBA)

Alfred Rufer, Philippe Barrade, Gina Steinke

Ecole Polytechnique Fédérale de Lausanne, EPFL
CH1015 Lausanne, Switzerland
alfred.rufer@epfl.ch, philippe.barrade@epfl.ch, gina.steinke@epfl.ch

Abstract- **A non-isolated DC-DC converter is presented, based on the current diverter principle. The converter is foreseen for larger photovoltaic plants or other DC collection networks for renewable energy sources. The topology allows the DC-DC conversion as step-up converter typically from LV to MV levels. The Multistage Stacked Boost Architecture allows a conversion to voltage levels higher than the blocking capability of the used semiconductor devices.**

Keywords- DC-DC Converter, Non-Isolated Boost, PV application, Energetic Macroscopic Representation, EMR, Control, Real-Time simulation

I. INTRODUCTION

Renewable energy sources like photovoltaic plants have grown significantly in importance and size during the last decades. As a consequence, the problem of electrical energy transportation over a distance of hundreds or thousands of meters under good energy efficiency is addressed. When the AC systems allow a simple elevation of the voltage level through electromagnetic transformation, the cable wiring of complex arrangements with a high number of low frequency transformers can lead to problems of stability or resonances. DC systems often use power electronic step-up topologies, as proposed in [1], [2], [3], [4] and [5]. Several step-up topologies use medium frequency transformer based voltage elevation as the examples of [1] and [2], using power electronic conversion at the primary and secondary side of the transformer, leading to reduced efficiency. Centralized DC to AC conversion for the final injection into the grid seems to be an appropriated solution, with an alternative to bridge longer distances with HVDC links. In such a context, the collecting should be based on MV DC technique, as a solution to avoid too much ohmic losses.

In the category of non-isolated step-up conversion circuits, this paper presents an original topology, based on the principle of balancing the voltage of series connected capacitors using a so-called current diverter. Such a topology has been proposed for the charge balance of series connected batteries [6] or for active balancing of series connected supercapacitors [7] [8]. Figure 1 gives the scheme of a one-to-plus-and-minus-two step-up converter, where the input stage is a classical MPPT

boost converter. The step-up ratio can be extended by adding simply more stages. But global considerations on the resulting energy efficiency must be done and compared with MV transformer based solutions.

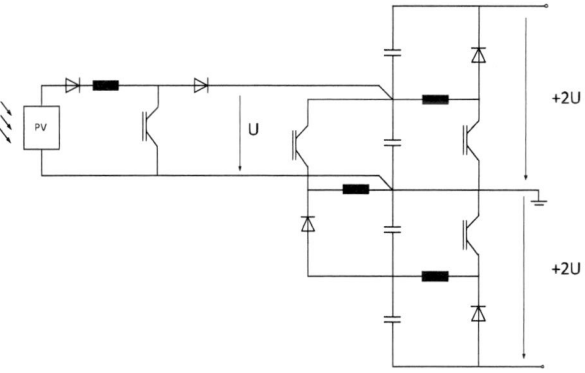

Figure 1: The 1 to +2/-2 step-up converter with input MPPT stage

A. The principle of the active balancing for series connected capacitors

In figure 2, the elementary principle of an active balancing circuit for series connected capacitors is represented.

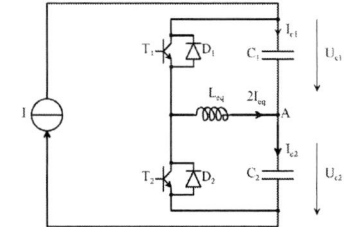

a) Elementary cell

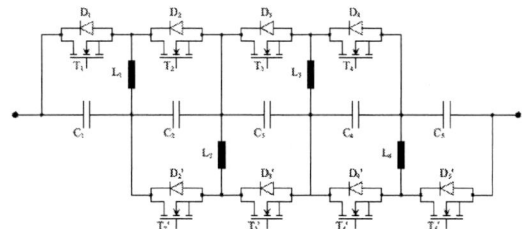

b) Multistage system

Figure 2: Active balancing circuits,

In the simple case of two capacitors, the balance of the two individual voltages can be imposed through the duty cycle of a pair of complementary switching transistors. The represented scheme corresponds to the case of two supercapacitors charged through a common current source. In such a case, the balancing circuit is used in order to compensate dispersion of the values of the capacitors, which would lead to voltage unbalance when the elements are charged through a common current source. In that case, the applied duty cycle is chosen equal to 50%, and imposes in the middle point of the transistor series-connection a voltage mean value equal to the half of the sum of the two voltages U_{c1} and U_{c2}. The current in the inductor automatically flows in the direction that balances the capacitor voltages, and comes automatically to zero when balance is reached. The basic topology of figure 2 is known as « current diverter », [6], allowing the diverted current path in the case of an unbalanced series connection.

The elementary scheme of figure 2a can be extended to a high number of series connected cells, as represented in figure 2b. In this case, the elementary cells with their current diverters as represented in figure 2a cannot be simply connected in further series, but the principle of balancing the capacitor voltages « two-by-two » can only be obtained if the current diverting circuits are connected in an alternating way, allowing to balance the capacitors by imbricated pairs.

In figures 2a) and 2b), the current in the inductors can circulate in two directions, when switching transistors and their anti-paralleled diodes are used for all stages. This bidirectional flow will allow the transfer of energy from one capacitor to the other, from the highest charged one to the lower one.

II. VOLTAGE STEP-UP CONVERTER

A. Application of the diverter-principle to voltage step-up

In figure 3a, a simple topology is represented where the output voltage of the circuit, U_s , can be set to the double value of the input circuit voltage U_i. In this scheme, the input voltage of that voltage-doubler is imposed through a boost converter, from which the input voltage is given by a PV panel. The control of the voltage U_i is achieved via the duty cycle of the boost converter, through the use of a current limitation controller R_i, cascaded with a superimposed voltage controller R_u. Regarding the operation of the output stage that elevates the voltage to the double value, the ratio is imposed through the fixed duty cycle of 50% of that stage. Based on the principle of the current diverter applied to multiple stages (Figure 2b), this converter can be extended to a step-up structure for higher ratios (Figure 3b). In the topology of figure 3b, the power injection has been done at the lowest level of the capacitor chain, but can be done at any other level as it is chosen in the scheme of Figure 1 in order to get a symmetric output voltage system. One must note in Figure 3b that the reduced number of active switches do

not allow the current reversibility compared to the topologies proposed in Figure 2.

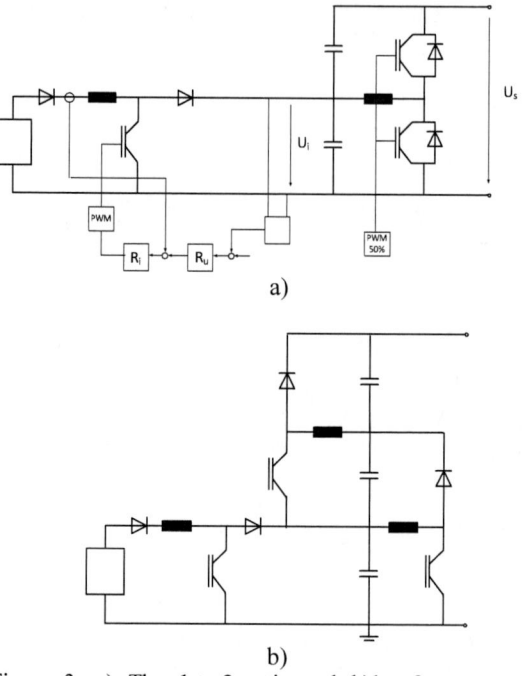

Figure 3: a) The 1-to-2 ratio and b)1-to-3converter with associated control circuits

B. Energetic Macroscopic Representation of the Multistage Stacked Boost Architecture (MSBA)

The converter represented in Figure 3 can be operated in stable conditions under the use of the current and voltage controllers as represented there. However, the operation of the output stage with constant duty cycle can lead to oscillating conditions due to energy exchange between capacitors and inductors. Such resonant properties have already been described for the so-called "series-combined connected boost and buck-boost dc-dc converter" [9].

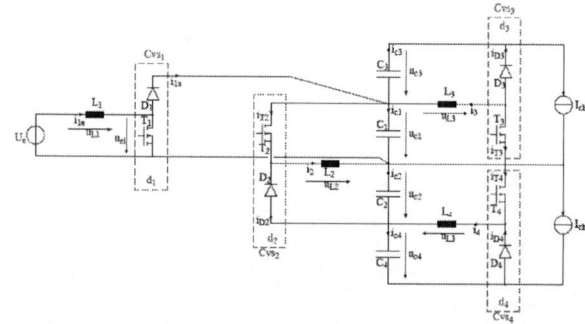

Figure 4: Representation of the complete topology with sub-converters

In order to better understand the complex structure with plural internal state-variables and complex couplings, a systematic step-by-step approach is chosen, based on the Energetic Macroscopic Representation (EMR). The further step will then be dedicated to the Inversion Based Control of the structure [10].

978-1-4799-2706-7/14 $31.00 © 2014 IEEE

First, the global scheme of Figure 1 is divided in sub-converters according the representation of Figure 4. The sub-converters are designated by C_{vs1}, C_{vs2}, C_{vs3} and C_{vs4}.

C. The elementary converter cell

The second step of the investigation is related to the elementary cell of the converter C_{vs1}. As represented in Figure 5a, the cell is composed by two capacitors, with an active balancing circuit using an inductor and two switches. The represented circuit is using a single transistor and a single diode, and as a result the current in the inductor can only be positive. That circuit can be realized also for a bidirectional inductor current, but the represented topology is chosen according the unidirectional power flow of the step-up converter shown in Figure 1 dedicated to the photovoltaic application. The voltage of capacitor C_1 of the elementary cell is given by

$$u_{c1} = \frac{1}{C_1}\int -i_{T2}dt$$

The transistor T_2 has the following conduction conditions, where the duty cycle is defined as d_2

$$\begin{cases} u_{2T} = d_2 u_{c1} \\ i_{T2} = d_2 i_2 \end{cases} \quad \text{with} \quad 0 \le d_2 \le 1$$

Then the current in the inductor L_2 is also described by the elementary differential equation:

$$i_{L2} = \frac{1}{L_2}\int (u_{2T} - u_{2D})dt$$

The diode D_2 has the following conduction conditions, where the duty cycle is defined as d_2

$$\begin{cases} u_{2D} = (1 - d_2)u_{c2} \\ i_{D2} = (1 - d_2)i_2 \end{cases} \quad \text{with} \quad 0 \le d_2 \le 1$$

And finally, the lower capacitor has a similar description as capacitor C_1:

$$u_{c2} = \frac{1}{C_2}\int i_{D2}dt$$

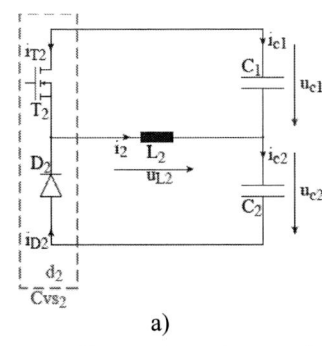

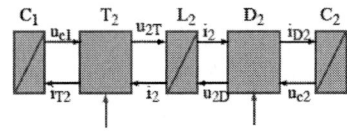

b)

Figure 5 a) Elementary cell Cvs2   b) Energetic Macroscopic Representation of the elementary cell

The Energetic Macroscopic Representation translates the behaviour of the components into a graph, using the elementary symbols of this method [10]. The elementary converter-cell together with the corresponding EMR diagram are given in Figure 5a and 5b.

D. Model of the whole converter

Then, the EMR translation is applied with similar rules to the whole converter, including the output load and the input boosting stage. All components are described by their corresponding macroscopic block-diagram. In recall to the basic elements of EMR, the interface between the blocks is defined through action and reaction quantities. System components with behaviour of integral causality are represented by square blocks with a diagonal bar. Their output quantities can be assimilated to state variables.

The EMR representation of the whole converter with load is given in Figure 6. In this representation all individual converter components are represented, together with the internal couplings. The input source is represented by the green block in the middle of the figure. The upper and lower loads are represented through the left and right green blocks.

As already explained, the association within the elementary cells of inductors and capacitors interfaced by converter-legs with constant duty cycles leads to an oscillation sensitive structure. Such typical oscillatory conditions are illustrated by simulation results where the input voltage and the output load have been changed. The capacitors voltage transients are given in figure 7.

The damping by passive parasitic elements (resistances of inductances) would have as consequence a reduced energy efficiency of the converter. Therefore, the full control of the converter internal state variables in preferred.

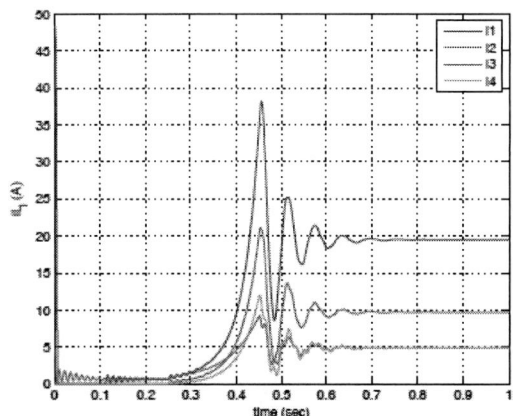

Figure 7: Voltage and current oscillations inside the converter

E. Inversion based control

The control of the converter and especially the control of the internal quantities can be realized according numerous different methods. In this investigation, a very pragmatic method is chosen. This method is based on the principle of inversion of the EMR graph [10] following some identified tuning paths indicated in yellow in Figure 8.

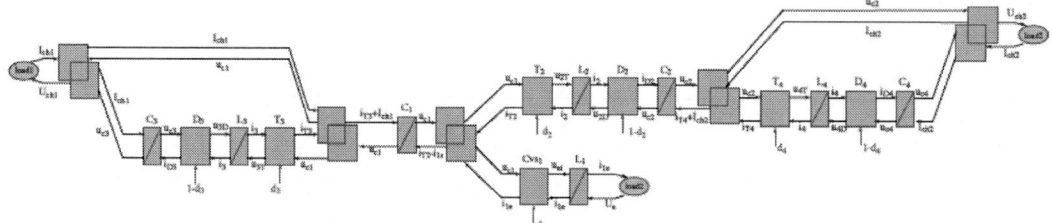

Figure 6: EMR representation of the complete converter with input source and load

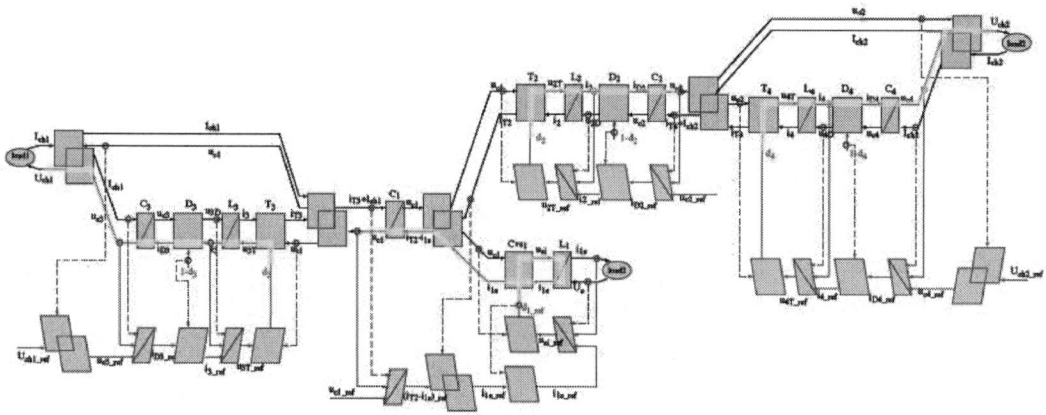

Figure 8 Graphic representation of the converter's EMR with associated Inversion Based Control

The control includes proportional-integral controllers according cascaded structure for each elementary converter cell. The inductor currents and capacitors voltages are individually and locally controlled. It leads to a so-called Maximal Control Structure of the system, where a non-negligible number of sensors must be implemented.

III. RESULTS FROM NUMERICAL SIMULATION AND REAL-TIME HIL

The complete converter with its control has first been simulated with a numeric circuit simulation tool (Portunus). After these first simulations, the real time behaviour of the controlled system has been tested. For this purpose, the converter was implemented in a Typhoon HIL600 real-time simulator [12]. The control functions have been implemented using a dedicated digital controller [11]. The interface between control electronics and real-time converter simulation has used a dedicated module where the control signals are transmitted by fiber optic, and the analogic values are provided by LEM-sensor emulation [11]. The HIL experimentation environment of LEI is illustrated in Figure 9.

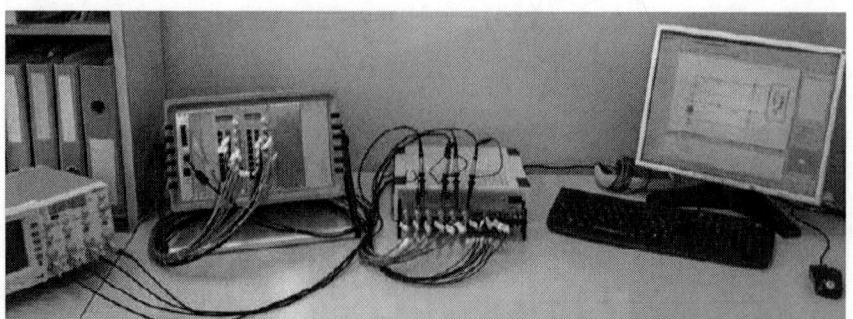

Figure 9: Real-time experimentation environment with digital controller and HIL simulator.

978-1-4799-2706-7/14 $31.00 © 2014 IEEE

The nominal data of the realized converter are:

U_{in} :	100 V
Uout+:	200 V
Uout-:	200 V
Iload:	5A

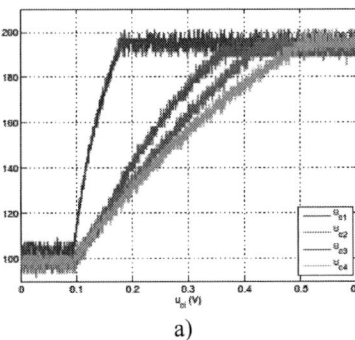

a)

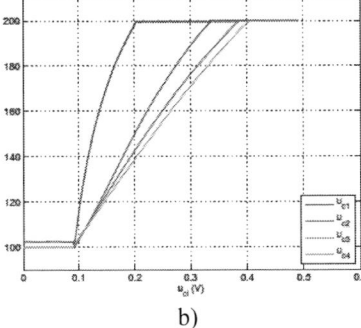

b)

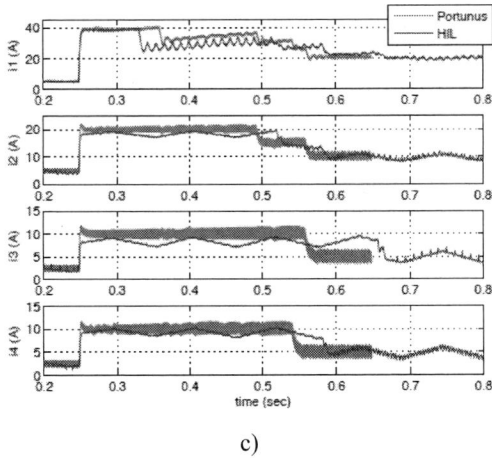

c)

Figure 10: Simulation and real-time experimentation
Capacitor voltage rise recorded from the HIL
simulation
a) Capacitor voltage rise recorded from the numeric
simulation (Portunus)
b) Inductor currents recorded from Portunus simulation
and real-time HIL simulation

IV. EXPERIMENTAL RESULTS

A small scale converter has been realized in order to make experimental verification and comparison with the numeric and real-time simulations. The realized converter is represented in Figure 11, and its electric scheme is identical to the scheme of Figure 1. The converter is completed with optical receivers and LEM transducers in order to have a fully compatible interface concept for the numeric controller.

Figure 11 Small power converter experimental set-up

Figure 12 shows the different currents recorded on the small scale converter while connecting the load resistor. The inductor labels correspond to the labels indicated in figure 4.

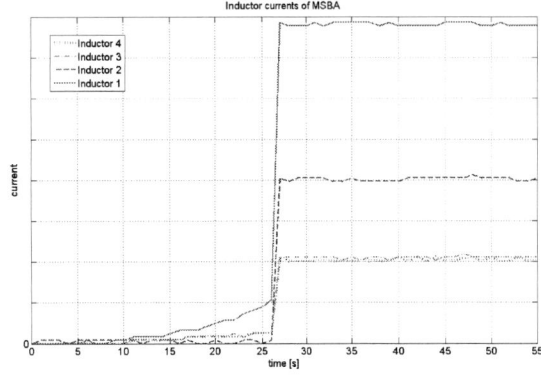

Figure 12 Experimental results

V. CONCLUSIONS

A dedicated DC-DC step-up converter has been analysed, where a Multistage Stacked Boost Architecture allows to generate output voltages in the MV range, even if the used switching devices don't have the corresponding blocking capability. The converter is analysed using the Energetic Macroscopic Representation. The control of the system is realized according the Inversion Based control method. The converter and its control have been verified by numeric simulation and also experimented with a DSP based controller interfaced to a real-time HIL system.

REFERENCES

[1] C. Arrioja, S. Kenzelmann and, Reversible DC/DC Converter as interface between Low and Medium Voltage DC Networks, PCIM 2011 : International Conference on Power

Electronics, Intelligent Motion and Power Quality, Nuremberg, Germany, 17-19 May 2011.

[2] Kenzelmann, S, Rufer A., Vasiladiotis, M., Dujic, D, Canales, F.,A Versatile DC-DC Converter for Energy Collection and Distribution using the Modular Multilevel Converter, EPE 2011 : 14th European Conference on Power Electronics and Applications, Birmingham, UK, 30 August - 1 September 2011.

[3] Wuhua Li ; Xiangning He, Review of Nonisolated High-Step-Up DC/DC Converters in Photovoltaic Grid-Connected Applications, IEEE Transactions on Industrial Electronics, Volume: 58 , Issue: 4, 2011 , Page(s): 1239 – 1250.

[4] Hwu, K.I. ; Chuang, C.F. ; Tu, W.C., High Voltage-Boosting Converters Based on Bootstrap Capacitors and Boost Inductors, IEEE Transactions on Industrial Electronics, Volume: 60 , Issue: 6 2013 , Page(s): 2178 – 2193.

[5] Meneses, D. ; Blaabjerg, F. ; García, O. ; Cobos, J.A., Review and Comparison of Step-Up Transformerless Topologies for Photovoltaic AC-Module Application IEEE Transactions on Power Electronics , Volume: 28 , Issue: 6, 2013, Page(s): 2649 – 2663.

[6] Kutkut, N.H, A modular non-dissipative current diverter for EV battery charge equalization, Applied Thirteenth Annual Power Electronics Conference and Exposition, 1998. APEC '98. 1998., Volume: 2, Page(s): 686 - 690 vol.2.

[7] P. Barrade, S. Pittet and A. Rufer, Energy storage system using a series connection of supercapacitors, with an active device for equalizing the voltages, IPEC 2000 : International Power Electronics Conference, Tokyo, Japan, 3-7 April 2000.

[8] P. Barrade, S. Pittet and A. Rufer, Series connection of supercapacitors, with an active device for equalizing the voltages. PCIM 2000 : International Conference on Power Electronics, Intelligent Motion and Power Quality, NÃ¼rnberg, Germany, 6-8 June 2000.

[9] Duran-Gomez, J.L. ; Garcia-Cervantes, E. ; Lopez-Flores, D.R. ; Enjeti, P.N. ; Palma, L., Analysis and evaluation of a series-combined connected boost and buck-boost dc-dc converter for photovoltaic application, Applied Power Electronics Conference and Exposition, 2006. APEC '06. Twenty-First Annual IEEE.

[10] Bouscayrol, A. ; Guillaud, X. ; Delarue, P. ; Lemaire-Semail, B., Energetic Macroscopic Representation and Inversion-Based Control Illustrated on a Wind-Energy-Conversion System Using Hardware-in-the-Loop Simulation, Industrial Electronics, IEEE Transactions on, Volume: 56 , Issue: 12, 2009 , Page(s): 4826 – 4835

[11] N.Cherix, S. Delalay, P.Barrade, A. Rufer, Fail-safe Control Platform for Power Electronic Applications in R&D Environments, EPE 2013, European Conference on Power Electronics and Applications, Lille, France, Sept. 2013.

[12] Majstorovic, D. ; Celanovic, Ivan ; Teslic, N.D. ; Celanovic, N. ; Katic, V.A., Ultralow-Latency Hardware-in-the-Loop Platform for Rapid Validation of Power Electronics Designs, Industrial Electronics, IEEE Transactions on, Volume: 58 , Issue: 10, 2011 , Page(s): 4708 - 4716

Comparison of Cascaded Multilevel Converter Topologies for AC/AC Conversion

Kalle Ilves, Luca Bessegato, and Staffan Norrga
School of Electrical Engineering
KTH Royal Institute of Technology
Stockholm, Sweden
Email: ilves@kth.se, lucabe@kth.se, norrga@kth.se

Abstract—**This paper presents a simplified qualitative comparison of previously presented cascaded multilevel converter topologies for ac-ac conversion with particular emphasis on motor drive applications. Performance criteria such as the pulsation of the stored energy in the cell capacitors and the total required semiconductor rating are derived by analytical methods. The main conclusion is that the back-to-back connected modular multilevel converter operates best at synchronous speed, whereas the modular matrix converter and Hexverter are better suited for low-frequency output. However, by injecting circulating currents in the phase arms the operating region can be extended for all of the studied topologies.**

I. INTRODUCTION

The market for medium voltage (MV) motor drives is expanding rapidly, propelled by demand from industry for improved process control. The energy savings that can be archived by variable-frequency operation also contributes significantly to this growth. Also within renewable energy applications such as grid integration of large wind turbines, the interest for MV drives is rapidly increasing. The conversion function that should be fulfilled is ac-to-ac conversion where both ac voltages lie in the range 2.3 kV through 6.9 kV. On the motor side variable frequency and variable voltage operation is normally required. Bidirectional power flow is sometimes desired such that regenerative operation is possible. Reliability is generally of high importance since a converter failure may halt an entire industrial process rapidly incurring costs of the same magnitude as the cost of the drive System. In terms of topologies the field is currently dominated by three-level neutral-point-clamped (NPC) converters [1] and topologies derived from the NPC, such as the five-level active NPC (ANPC). These are all characterized by the fact that there is a single common dc link taking up the full dc voltage. Generally, this dc link is split by a number of series-connected dc capacitors whereby the different voltage levels are created.

Within high-voltage applications, such as VSC-based high-voltage direct current-transmission (HVDC) and flexible ac-transmission Systems (FACTS), a dramatic shift has taken place over the last ten years towards cell-based topologies in which cascaded strings of converter cells act as controllable voltage sources. A crucial question at this point is whether cell-based topologies will also gain a more prominent role for motor drive applications.

The first cell-based converter to be proposed for motor drives was based on having chains of cells where each of them is fed power from a dedicated transformer winding [2]. Depending on the design of the cells, the converter can have capability for bidirectional power flow. Medium voltage drives operating according to this principle have been available in the market for more than fifteen years, however with a limited market share. The fact that a specialized transformer, with many secondary windings, is needed, is likely to increase the cost of this converter.

In the year 2001 a cell-based converter was presented which in terms of structure resembles a matrix converter, but where the bidirectional switches where replaced by strings of full-bridge cells [3]. This is a bidirectional converter that performs ac/ac conversion without an intermediate dc link. However, the need for nine strings of full-bridges adds cost to the System, which has so far limited the commercial feasibility. In a three-phase to single-phase version, with six cell strings this converter has successfully been employed in interfaces between the three-phase power grid and low-frequency railway power supply networks.

Shortly afterwards, a converter which instead resembles the conventional two-level VSC, but where each valve is replaced by a string of half-bridges [4] was introduced. This topology, which has met with great success as part of HVDC transmission stations, performs bidirectional ac/dc conversion and thus needs to be connected in a back-to-back configuration in a motor drive System fed from the ac mains. Low-frequency operation of this converter, needed in a motor drive, presents significant challenges since each phase leg operates essentially as a single-phase converter. The slow power pulsations of each phase arm could lead to excessive requirements for the amount capacitive energy storage in the cells. This issue has been addressed extensively in literature [5], [6]. These references describe ways of redistributing power between the phase arms using currents circulating between the phase legs such that the cell capacitors do not have to handle the full power pulsation of each phase.

A later addition to the family of cell-based ac/ac converters is the so called *Hexverter* [7]. It resembles the multilevel matrix converter described in [3] in that it is a direct ac/ac topology employing strings of full bridges. However, in this case only six strings are required and these are connected in

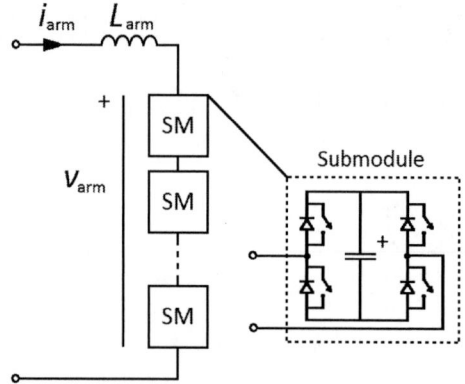

Fig. 1. Illustration and definition of arm quantities.

a hexagonal configuration.

II. Dimensioning Criteria and Terminology

Cascaded converter structures can be considered to be consisting of two or more arms, where an arm is a string of series-connected converter cells, often referred to as submodules. One arm with full-bridge submodules is illustrated in Fig. 1. The current flowing through each arm is referred to as the arm current, and the sum of the voltages inserted by the cascaded converters in each arm is referred to as the arm voltage. Furthermore, in order to limit parasitic currents, each arm also has an arm inductor with the inductance L_{arm}.

The cascaded submodules can generate multilevel voltage waveforms with excellent harmonic performance at very low switching frequencies. As a consequence, the current rating of the semiconductors is generally not limited by thermal aspects and the rms current will not be a dimensioning factor. Instead, the dimensioning factor for the current rating will be the peak value of the arm current, $\hat{I}_{\text{arm}}$.

The voltage rating of the semiconductors must be chosen such that they can handle the voltage fluctuations in the submodule capacitors, switching transients, and various fault cases. However, in order to perform a qualitative comparison of the different converter topologies, it is assumed that the combined voltage rating of the semiconductors will be directly proportional to the peak-value of the arm voltage, $\hat{V}_{\text{arm}}$.

In order to compare the semiconductor requirements in the different topologies, the combined power rating of the semiconductors, S_{sm}, will be normalized with respect to the rated power of the converter. Accordingly,

$$\frac{S_{\text{sm}}}{S_{\text{conv}}} = \frac{n_{\text{arm}} n_{\text{sm}} \hat{V}_{\text{arm}} \hat{I}_{\text{arm}}}{S_{\text{conv}}}, \tag{1}$$

where n_{arm} is the number of arms, n_{sm} is the number of semiconductors per submodule, and S_{conv} is the rated power of the converter. For simplicity, it will be assumed that the rated power is equal to S_{conv} on both sides. That is,

$$S_{\text{conv}} = 3 \frac{\hat{V}_1 \hat{I}_1}{2} = 3 \frac{\hat{V}_2 \hat{I}_2}{2}. \tag{2}$$

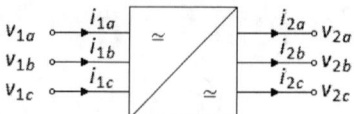

Fig. 2. Phase currents and phase voltages of System 1 (left) and System 2 (right).

There are various harmonic injection methods that have been proposed for the different topologies that may influence the dimesnioning of the capacitors and the rated power of the semiconductors. Some of these methods are injecting harmonics in the arm currents circulating inside the converter to reduce the energy ripple in the arms [8], [9]. There are also other harmonic injection methods that inject both voltages and currents to redistribute the energy in the converter [5]–[7]. These methods can be used when it is necessary to operate the converter at a frequency where the energy ripple normally becomes very large or unstable. The focus of this paper is, however, nominal operation at rated power. Therefore, the latter harmonic injection methods are not considered in this paper.

A. Voltages and Currents

The phase voltages in the two interconnected ac-systems are referred to as v_{1a}, v_{1b}, and v_{1c} for the first system and v_{2a}, v_{2b}, and v_{2c} for the second system. The directions of the currents are defined such that a positive power transfer indicates that the direction of the power flow is from System 1 to System 2, as shown in Fig. 2.

The amplitude and angular frequency of the alternating voltages in System 1 are $\hat{V}_1$ and ω_1. Similarly, the amplitude and angular frequency of the voltages in System 2 are $\hat{V}_2$ and ω_2. Accordingly,

$$v_{1a} = \hat{V}_1 \cos(\omega_1 t + \theta_1) \tag{3}$$
$$v_{2a} = \hat{V}_2 \cos(\omega_2 t + \theta_2), \tag{4}$$

where θ_1 is the inital phase of System 1 and θ_2 is the initial phase of System 2. It is assumed that both Systems are symmetric. Thus, v_{1b} and v_{1c} are equal to v_{1a} in frequency and magnitude but phase-shifted by $\pm \frac{2\pi}{3}$, respectively. Similarly, v_{2b} and v_{2c} are equal to v_{2a} in frequency and magnitude but phase shifted by $\pm \frac{2\pi}{3}$, respectively. The relation between the amplitudes of the phase voltages $\hat{V}_1$ and $\hat{V}_2$ is defined by the modulation index m as

$$m = \frac{\hat{V}_2}{\hat{V}_1}. \tag{5}$$

The angular displacements between the phase current and the phase voltages are denoted as φ_1 and φ_2 for System 1 and System 2, respectively. Accordingly,

$$i_{1a} = \hat{I}_1 \cos(\omega_1 t + \theta_1 - \varphi_1) \tag{6}$$
$$i_{2a} = \hat{I}_2 \cos(\omega_2 t + \theta_2 - \varphi_2) \tag{7}$$

where $\hat{I}_1$ is the amplitude of the phase current in System 1 and $\hat{I}_2$ is the amplitude of the phase current in System 2.

978-1-4799-2706-7/14 $31.00 © 2014 IEEE

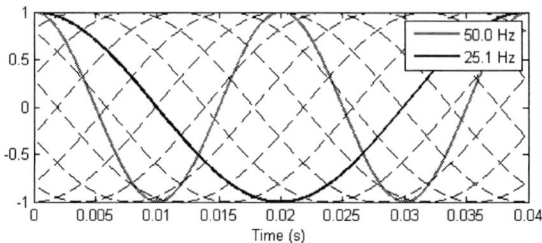

Fig. 3. Two signals, 50 Hz (red), and 25.1 Hz (black). The dashed lines indicate how the position of the 25.1-Hz signal changes after 50 periods of the 50-Hz signal.

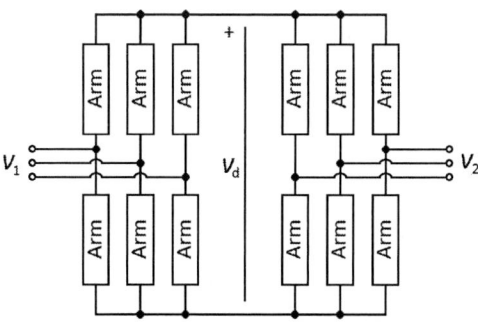

Fig. 4. A back-to-back configuration of the M2C

B. Dimensioning Case

Since the harmonic components are the same in all arms, with the exception that the angular displacements are different, the analysis is focusing on one arm. The dimensioning case is then the worst-case scenario when different possible phase shifts between System 1 and System 2 are considered. In order to simplify this process, instead of considering different values of θ_2, the frequency f_2 is chosen to be a non-integer number. For example, when observing a 50 Hz and a 25.1 Hz waveform it will appear as if the phase of the 25.1 Hz signal is slowly changing over time. This is illustrated in Fig. 3 where the 25.1 Hz signal is plotted in relation to the 50 Hz signal every 50th period (that is, between each dashed 25.1 Hz signal there is actually 48 additional waveforms that are not shown).

In order to simplify the nomenclature, the index a is hereafter omitted in the phase quantities. Thus, the analysis is based on four representative phase quantities v_1, v_2, i_1, and i_2. And the dimensioning case is identified by using a non-integer frequency in System 2 rather than trying different possible phase shifts θ_1 and θ_2.

C. Energy Variations

The energy variations in the arms play an important role in the dimensioning of the converter and affects the size, weight, and cost of the submodule capacitors. Assuming that a proper controller is used which can obtain a zero tracking-error at steady state, the arm voltage and arm current are known since they are equal to their reference values. This means that the energy variations in an arm can be directly calculated by integrating the product of the arm voltage and arm current over time. Accordingly,

$$w_{\text{arm}} = \int v_{\text{arm}} i_{\text{arm}} \, dt. \tag{8}$$

It is obvious that the energy variations will increase if the power transfer is increased. Therefore, all calculated values will be normalized with the apparent power transfer S_{conv}.

III. MODULAR MULTILEVEL CONVERTER

A back-to-back configuration of the modular multilevel converter (MMC) [4] is shown in Fig. 4. The arm voltages are composed of a direct voltage and an alternating component.

Due to the direct component in the arm voltages, there is no need to insert negative voltages in the arms. Thus, the submodules in the MMC can be implemented using half-bridges.

Since the back-to-back MMC consists of two bi-directional ac-dc converters, the definition of the modulation index in (5) is not sufficient. Instead, in the back-to-back MMC, the modulation index is defined as the peak-to-peak variation of the alternating component divided by the pole-to-pole dc-link voltage. In the converter connected to System 1, the modulation index is denoted m_1, and the modulation index of the converter connected to System 2 is denoted m_2. The semiconductor expenditure and energy ripple is increasing when the modulation index is reduced [10], [11]. Thus, in order to minimize the losses and semiconductor expenditure the dc-link voltage should be chosen such that the largest value of m_1 and m_2 is equal to unity.

As with the modulation indices, the arm quantities in the two converters must also be treated separately. This is done by using the indices 1 and 2 for arms that are connected to System 1 and System 2, respectively. Accordingly, the arm voltages can be expressed as

$$v_{\text{arm1,2}}^{(MMC)} = \frac{\hat{V}_{1,2}}{m_{1,2}} \mp \hat{V}_{1,2} \cos(\omega_{1,2}t + \theta_{1,2}). \tag{9}$$

where the minus and plus signs are for the upper and lower arms, respectively. The corresponding arm currents are given by

$$\begin{aligned}
i_{\text{arm1,2}}^{(MMC)} = &\frac{\hat{I}_{1,2}}{4} m_{1,2} \cos(\varphi_{1,2}) \\
&\pm \frac{\hat{I}_{1,2}}{2} \cos(\omega_{1,2}t + \theta_{1,2} - \varphi_{1,2}),
\end{aligned} \tag{10}$$

where the plus sign is used for the lower arms connected to System 1 and the upper arms connected to System 2, and the minus sign is used for the remaining arms.

The instantaneous arm power is given by the product of the arm voltage and arm current. In order for the System to be stable, the time average of this product must be zero. This criterion can always be satisfied in the back-to-back MMC, provided that the actove power input of System 1 equals to the active power output to System 2.

A. Energy Variations

The energy variations are given by

$$w_{\text{arm1,2}}^{(MMC)} = \int v_{\text{arm1,2}}^{(MMC)} i_{\text{arm1,2}}^{(MMC)} \, dt. \tag{11}$$

In order to reflect the total energy storage requirements in the converter, the energy variations per arm must be scaled by the number of arms in the converter. Accordingly, the total energy variations that the back-to-back MMC must be dimensioned to handle are given by

$$w_{tot}^{(MMC)} = 6w_{\text{arm1}}^{(MMC)} + 6w_{\text{arm2}}^{(MMC)}. \tag{12}$$

The energy variations in the converter arms can be affected by injecting harmonic components in the arm currents. In this way, the operating range of the converter can be extended by either shaping the capacitor voltage ripple such that overmodulation is avoided [12], or by reducing the peak-to-peak energy variations [8], [9], [13], [14]. In this paper, the measure of the energy storage requirements is the peak-to-peak energy ripple. Therefore, the harmonic injection method considered will be the one described in [14] due to its simplicity. By this method, the current flowing between the dc terminals of each phase leg is adjusted to match the energy pulsation in the phase power. Accordingly, when the considered harmonic injection method is used, the arm currents are given by

$$i_{\text{arm1,2}}^{(SHI)} = \frac{\hat{I}_{1,2}}{2} \Bigg[\pm \cos(\omega_{1,2}t + \theta_{1,2} - \varphi_{1,2})$$
$$+ m_{1,2} \cos(\omega_{1,2}t + \theta_{1,2}) \cos(\omega_{1,2}t + \theta_{1,2} - \varphi_{1,2}) \Bigg], \tag{13}$$

where the superscript (SHI) indicate that the considered second-order harmonic injection method is used. The total energy variations that must be handled by the converter can then be expressed as

$$w_{tot}^{(SHI)} = 6 \int v_{\text{arm1}}^{(MMC)} i_{\text{inj1}}^{(MMC)} \, dt$$
$$+ 6 \int v_{\text{arm2}}^{(MMC)} i_{\text{inj2}}^{(MMC)} \, dt. \tag{14}$$

B. Semiconductor Requirements

According to (10), the peak current that must be handled by the arms is given by

$$\hat{I}_{\text{arm1,2}}^{(MMC)} = \frac{\hat{I}_{1,2}}{4} m_{1,2} \cos(\varphi_{1,2}) + \frac{\hat{I}_{1,2}}{2}. \tag{15}$$

Similarly, the peak arm-voltage is given by

$$v_{\text{max1,2}}^{(MMC)} = \frac{\hat{V}_{1,2}}{m_{1,2}} + \hat{V}_{1,2}. \tag{16}$$

Accordingly, the product of the maximum voltage and current that must be handled by the arm is

$$S_{\text{arm1,2}}^{(MMC)} = \frac{\hat{V}_{1,2}\hat{I}_{1,2}}{2} \left[1 + \frac{1}{m_{1,2}} + \frac{1}{2}(1 + m_{1,2}) \cos(\varphi_{1,2}) \right]. \tag{17}$$

Thus, it can be concluded that the required power rating of the semiconductors is minimized if the maximum modulation index is used on both sides. That is, $\hat{V}_1 = \hat{V}_2$.

Assuming that active power transfer is the dimensioning case, i.e. $\cos(\varphi_{1,2}) = 1$,

$$S_{\text{arm1,2}}^{(MMC)} = 3\frac{\hat{V}_{1,2}\hat{I}_{1,2}}{2} = S_{\text{conv}}. \tag{18}$$

With 12 arms, and two semiconductors per submodule, the combined power rating of all semiconductor devices in the converter is

$$S_{\text{sm}}^{(MMC)} = 24 S_{\text{conv}}. \tag{19}$$

If the harmonic injection method in (13) is used, the peak arm-current at $\cos(\varphi) = 1$ is increased to

$$\hat{I}_{\text{arm1,2}}^{(SHI)} = \frac{\hat{I}_{1,2}}{2}(1 + m_{1,2}) \tag{20}$$

Accorddingly, if active power transfer is the dimensioning case, the peak arm-current is increased by 33% if the considered harmonic injection method is used. Accordingly, the current rating of the semiconductors must be increased as well.

If no harmonic injection method is used, the rms-value of the arm currents at active power transfer are given by

$$I_{\text{RMS1,2}}^{(MMC)} = \sqrt{\frac{\hat{I}_{1,2}^2 m_{1,2}^2}{16} + \frac{\hat{I}_{1,2}^2}{8}}. \tag{21}$$

However, if the harmonic inejction method is used, the rms value of the arm currents are increased to

$$I_{\text{RMS1,2}}^{(SHI)} = \sqrt{\frac{\hat{I}_{1,2}^2 m_{1,2}^2}{16} + \frac{\hat{I}_{1,2}^2}{8} + \frac{\hat{I}_{1,2}^2 m_{1,2}^2}{32}}. \tag{22}$$

Accordingly, at active power transfer and the modulation indices 1.0, the rms value of the arm currents is increased by 8%.

IV. MODULAR MATRIX CONVERTER

The modular matrix converter can be considered as three groups of Y-connected arms where the midpoint of each group is connected to one of the phases in System 2 as shown in Fig. 5. The arm voltages in the modular matrix converter can be expressed as

$$v_{\text{arm}}^{(MAT)} = \hat{V}_1 \cos(\omega_1 t + \theta_1) - \hat{V}_2 \cos(\omega_2 t + \theta_2). \tag{23}$$

Similarly, the currents can be expressed as

$$i_{\text{arm}}^{(MAT)} = \frac{1}{3}[\hat{I}_1 \cos(\omega_1 t + \theta_1 - \varphi_1) + \hat{I}_2 \cos(\omega_2 t + \theta_2 - \varphi_2)]. \tag{24}$$

Assuming that neither ω_1 nor ω_2 is zero, and that $\omega_1 \neq \omega_2$, the time-average of the arm power can be expressed as

$$P_{\text{arm}}^{(MAT)} = \frac{\hat{V}_1\hat{I}_1}{6} \cos(\varphi_1) - \frac{\hat{V}_2 I_2}{6} \cos(\varphi_2), \tag{25}$$

meaning that the arm energies can be kept balanced if the active power input in System 1 equals the active power output

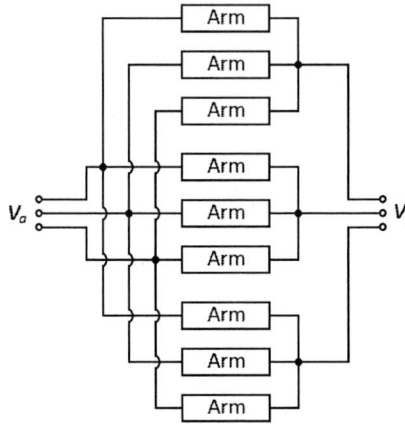

Fig. 5. A matrix converter based on cascaded converters

to System 2. However, in the case when $\omega_1 = \omega_2$, there is an additional term in the time average of the instantaneous arm power. Namely,

$$P_{f_1=f_2}^{(MAT)} = P_{\text{arm}} + \frac{\hat{V}_1 \hat{I}_2}{6} \cos(\theta_2 - \theta_1 - \varphi_2) \\ - \frac{\hat{V}_2 \hat{I}_1}{6} \cos(\theta_2 - \theta_1 + \varphi_1). \tag{26}$$

Accordingly, except for the specific case when the two terms in (26) cancel out, the arm energies will be unstable when $\omega_1 = \omega_2$. This problem can, however, be solved to some extent by injecting additional harmonic components in the arm currents [15].

A. Energy Variations

The energy variations in the arms of the matrix converter are given by

$$w_{\text{arm}}^{(MAT)} = \int v_{\text{arm}}^{(MAT)} i_{\text{arm}}^{(MAT)} \, dt. \tag{27}$$

Since there are 9 arms in the modular matrix converter, the total energy fluctuations that must be handled by the submodule capacitors are

$$w_{\text{tot}}^{(MAT)} = 9 \int v_{\text{arm}}^{(MAT)} i_{\text{arm}}^{(MAT)} \, dt. \tag{28}$$

The energy ripple is increasing at lower frequencies and can even become unstable if the arm currents are chosen as (24). In [16], it was proposed that this problem can be alleviated by operating in a so called *instantaneous power mode* (IPM). Although IPM can operate successfully at low frequencies, it becomes unstable when $\omega_2 = \frac{1}{3}\omega_1$. In order to solve this, an alternative method was proposed in [17]. In this case the arm currents can be expressed as

$$I_{\text{arm}}^{(LOW)} = \frac{1}{3} \{ \hat{I}_1 \cos(\omega_1 t + \theta_1 - \varphi_1) \\ + \hat{I}_2 \cos(\omega_2 t + \theta_2 - \varphi_2) \\ + m\hat{I}_2 \cos[(\omega_1 + 2\omega_2)t + \theta_1 + 2\theta_2 - \varphi_2] \}. \tag{29}$$

This harmonic injection method can reduce the capacitor voltages at low frequencies without the instability problem at $\omega_2 = \frac{1}{3}\omega_1$. Accordingly, when the harmonic injection in [17] is used, the total energy variations are given by

$$w_{\text{tot}}^{(LOW)} = 9 \int v_{\text{arm}}^{(MAT)} i_{\text{arm}}^{(LOW)} \, dt. \tag{30}$$

1) Near Synchronous Frequency: In [15], a harmonic injection method was proposed to alleviate the problems that occur when ω_1 and ω_2 are close or equal. It was proposed that the arm currents are controlled as

$$I_{\text{arm}}^{(SYN)} = \frac{1}{3} \{ \hat{I}_1 \cos(\omega_1 t + \theta_1 - \varphi_1) \\ + \hat{I}_2 \cos(\omega_2 t + \theta_2 - \varphi_2) \\ - \hat{I}_1 \cos[(2\omega_2 - \omega_1)t + 2\theta_2 - \theta_1 + \varphi_1] \\ - \hat{I}_2 \cos[(2\omega_1 - \omega_2)t - \theta_2 + 2\theta_1 + \varphi_2] \}. \tag{31}$$

Accordingly, the total energy variations are given by

$$w_{\text{arm}}^{(SYN)} = \int v_{\text{arm}}^{(MAT)} i_{\text{arm}}^{(SYN)} \, dt. \tag{32}$$

Although it is possible to eliminate the low-frequency components in (32), it will contain a component

$$w_{2(f_1-f_2)}^{(SYN)} = K\{ \hat{V}_2 \hat{I}_2 \sin[2(\omega_1 - \omega_2) + 2(\theta_1 - \theta_2) + \varphi_2] \\ - \hat{V}_1 \hat{I}_1 \sin[2(\omega_1 - \omega_2) + 2(\theta_1 - \theta_2) - \varphi_1] \} \tag{33}$$

where

$$K = \frac{1}{12(\omega_1 - \omega_2)}. \tag{34}$$

That is, in order to eliminate the low-frequency components φ_1 must be equal to $-\varphi_2$. The total energy variations are then given by

$$w_{\text{tot}}^{(SYN)} = 9 \int v_{\text{arm}}^{(MAT)} i_{\text{arm}}^{(SYN)} \, dt. \tag{35}$$

B. Semiconductor Ratings

From (23) it can be concluded that the peak arm voltage is

$$\hat{V}_{\text{arm}}^{(MAT)} = \hat{V}_1 + \hat{V}_2 = (1 + m)\hat{V}_1. \tag{36}$$

As each arm conducts one third of the phase current from both sides the peak arm-current is

$$\hat{I}_{\text{arm}}^{(MAT)} = \frac{1}{3} \left(\hat{I}_1 + \hat{I}_2 \right). \tag{37}$$

If active power transfer is the dimensioning case, the product $\hat{V}_1 \hat{I}_1$ must be equal to $\hat{V}_2 \hat{I}_2$. Consequently, the peak arm-current can be expressed as

$$\hat{I}_{\text{arm}}^{(MAT)} = \frac{1}{3} \left(1 + \frac{1}{m} \right) \hat{I}_1. \tag{38}$$

The rated power of each arm is then given by

$$S_{\text{arm}} = \frac{1}{3} \left(2 + m + \frac{1}{m} \right) \hat{V}_1 \hat{I}_1 \tag{39}$$

The 2014 International Power Electronics Conference

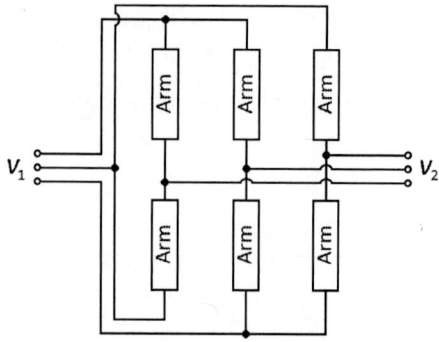

Fig. 6. A hexverter with three delta-connected phase legs.

It is concluded that the semiconductor requirements are minimized for $m = 1$, i.e. $\hat{V}_1 = \hat{V}_2$. Accordingly, since there are 4 devices per submodule and 9 arms, the minimum combined power rating of the semiconductors is

$$S_{\text{sm}}^{(MAT)} = 32 S_{\text{conv}}. \tag{40}$$

If the harmonic injectioin method in (29) is used, the peak arm-current at active power transfer is increased. Accordingly,

$$\hat{I}_{\text{arm}}^{(LOW)} = \frac{1}{3} \left[\hat{I}_1 + \hat{I}_2 (1 + m) \right]. \tag{41}$$

That is, at unity modulation index the peak arm-current is increased by 50%. Similarly, by using the method in Section II-B to identify the dimensioning case it can be calculated that when the harmonic injection method in (31) is used, the peak value of the arm currents are increased by 54% at unity modulation index and active power transfer.

The rms-value of the arm currents without any harmonic injection is given by

$$I_{\text{RMS}}^{(MAT)} = \frac{1}{3} \sqrt{\frac{\hat{I}_1^2}{2} + \frac{\hat{I}_2^2}{2}}. \tag{42}$$

However, with the two harmonic injection methods, the rms value is increased to

$$I_{\text{RMS}}^{(LOW)} = \frac{1}{3} \sqrt{\frac{\hat{I}_1^2}{2} + \frac{\hat{I}_2^2}{2} + \frac{\hat{I}_2^2 m^2}{2}}, \tag{43}$$

and

$$I_{\text{RMS}}^{(SYN)} = \frac{\sqrt{2}}{3} \sqrt{\frac{\hat{I}_1^2}{2} + \frac{\hat{I}_2^2}{2}}. \tag{44}$$

Accordingly, at unity modulation index, the two harmonic injection methods increase the rms-value by 22% and 41%, respectively.

V. HEXVERTER

The Hexverter can be considered as three delta-connected phase legs as shown in Fig. 6. The arm voltages in the Hexverter can be expressed as

$$v_{\text{arm}}^{(HEX)} = \hat{V}_1 \cos(\omega_1 t + \theta_1) - \hat{V}_2 \cos(\omega_2 t + \theta_2). \tag{45}$$

That is, the amplitude of the alternating components is equal to the amplitude of the line-to-neutral voltage in each system. The voltages in adjoining arms are, however, phase shifted by 60 degrees so that the sum is equal to the line-to-line voltage. The arm current can be expressed as [7]

$$\begin{aligned} i_{\text{arm}}^{(HEX)} =& \frac{1}{\sqrt{3}} \hat{I}_1 \cos\left(\omega_1 t + \theta_1 + \frac{\pi}{6} - \varphi_1 \right) \\ &+ \frac{1}{\sqrt{3}} \hat{I}_2 \cos\left(\omega_2 t + \theta_2 - \frac{\pi}{6} - \varphi_2 \right) \end{aligned} \tag{46}$$

It is observed that there is a phase-shift between the arm voltages and the arm currents also when $\cos(\varphi) = 1$.

The time average of the instantaneous arm power is given by

$$P_{\text{arm}}^{(HEX)} = \frac{\hat{V}_1 \hat{I}_1}{2\sqrt{3}} \cos\left(\varphi_1 - \frac{\pi}{6} \right) - \frac{\hat{V}_2 \hat{I}_2}{2\sqrt{3}} \cos\left(\varphi_2 + \frac{\pi}{6} \right) \tag{47}$$

That is, in order for the system to be stable φ_1 must be equal to $-\varphi_2$. This problem can, however, be circumvented by adding direct voltage and current components to the arm quantities as described in [7].

A. Energy Variations

Since there is a total of six arms in the hexverter the total energy variations that must be handled by the submodule capacitors is given by

$$w_{\text{tot}}^{(HEX)} = 6 \int v_{\text{arm}}^{(HEX)} i_{\text{arm}}^{(HEX)} dt. \tag{48}$$

The direct components in the arm quantities are not considered in this paper since the focus of the comparison is rated conditions with mainly active power transfer. Therefore, it is instead assumed that $\varphi_1 = -\varphi_2$.

B. Semiconductor Requirements

The amplitude of the arm voltages in the hexverter are the same as for the matrix converter. That is,

$$\hat{V}_{\text{arm}}^{(HEX)} = \left(\hat{V}_1 + \hat{V}_2 \right) = (1 + m) V_1. \tag{49}$$

From (46) it can be concluded that the peak arm-current is

$$\hat{I}_{\text{arm}}^{(HEX)} = \frac{1}{\sqrt{3}} \left(\hat{I}_1 + \hat{I}_2 \right) = \frac{1}{\sqrt{3}} \left(1 + \frac{1}{m} \right) \hat{I}_1. \tag{50}$$

Thus, the rated power of each arm can be expressed as

$$S_{\text{arm}}^{(HEX)} = \frac{1}{\sqrt{3}} \left(2 + m + \frac{1}{m} \right) \hat{V}_1 \hat{I}_1. \tag{51}$$

As with the two other converter types, the semiconductor requirements are minimized when $\hat{V}_1 = \hat{V}_2$. Accordingly, with 4 devices per submodule and 6 arms, the combined power rating of all semiconductors is

$$S_{\text{sm}}^{(HEX)} = 37 S_{\text{conv}}. \tag{52}$$

It was proposed in [7] that dc-components can be added to the arm voltages and arm currents in order to balance

978-1-4799-2706-7/14 $31.00 © 2014 IEEE

TABLE I. NORMALIZED CURRENTS AND SEMICONDUCTOR RATINGS

	$S_{\text{sm}}/S_{\text{conv}}$	Rms-current	Peak-current
Back-to-back MMC	24	0.4330	0.7500
Back-to-back MMC (SHI)	32	0.4677	1.0000
Modular matrix conv.	32	0.3333	0.6667
Modular matrix conv. (LOW)	48	0.4082	1.0000
Modular matrix conv. (SYN)	49	0.4714	1.0264
Hexverter	37	0.5774	1.1547

TABLE II. RELATIVE INCREASE IN ARM CURRENTS

	Rms-current	Peak-current
Back-to-back MMC with (SHI)	1.08	1.33
Modular Matrix Converter with (LOW)	1.22	1.50
Modular Matrix Converter with (SYN)	1.41	1.54

TABLE III. CONSIDERED OPERATING CONDITIONS

	$V_{LL}(rms)$	$I_{ph}(rms)$	S	φ	Frequency
System 1	6.6 kV	1.2 kA	10 MVA	$-15°$	50 Hz
System 2	6.0 kV	1.4 kA	10 MVA	$+15°$	1–80 Hz

the arm energies. This method is, however, not considered here. Instead, the arm energies are balanced by controlling the reactive power as suggested in [7]. The dimensioning example in [7] indicate that this yields better results in both power rating of the semiconductors as well as energy ripple in the arm capacitors. Accordingly, the rms-value of the only arm currents that are considered here is

$$I_{\text{RMS}}^{(HEX)} = \frac{1}{\sqrt{3}} \sqrt{\frac{\hat{I}_1^2}{2} + \frac{\hat{I}_2^2}{2}}. \tag{53}$$

VI. DISCUSSION

A. Arm Currents and Semiconductor Ratings

It is evident that there are clear differences in the minimum requirements of semiconductor ratings for the considered topologies. In terms of semiconductor ratings, the back-to-back MMC stands out with a minimum combined power rating of 24 times S_{conv}. This is significantly, lower than the 32 and 37 times S_{conv} that are are required in the modular matrix converter and Hexverter, respectively. It was also observed that the harmonic injection methods for the back-to-back MMC and the modular matrix converter will increase not only the rms-value of the arm currents but also the peak value. Thus, applying the considered harmonic injection methods at rated operating conditions will increase the minimum requirements of the semiconductor ratings. The normalized values of the arm currents and semiconductor ratings are listed in Table I. The values are calculated for active power-transfer at unity modulation index and the currents are normalized with respect to the amplitude of the phase currents.

In order to obtain a clearer view of how the harmonic injection methods affect the arm currents, the peak-values and rms-values can be compared to the nominal currents when no harmonic injection method is used. These values are listed in Table II for active power transfer at unity modulation index. It is observed that the impact of the harmonic injection methods are more severe for the modular matrix converter, thus limiting the benefits of using these methods at rated conditions. In the back-to-back MMC it appears to be feasible to utilize the considered harmonic injection method, even at rated conditions. Considering the increase in semiconductor expenditure, it can be observed in Table I that the required power rating of the semiconductors in the back-to-back MMC with harmonic injection is the same as for the modular matrix converter without any harmonic injection.

B. Energy Storage

In order to evaluate the differences in the energy-storage requirements between the considered topologies, the sum of the peak-to-peak energy variations in the converter arms is considered. The energy variations are compared at $15°$ power angle which corresponds to $\cos(\varphi) = 0.97$. The operating conditions at which the comparison is made are listed in Table III. The power angle is capacitive in System 1 and inductive in System 2. In this way, the hexverter can operate without injecting additional voltages and currents in the arm quantities, and the harmonic injection method for the matrix converter can maintain stability at synchronous frequency. This is an aspect where there are clear differences between the three converter topologies. The possibility of controlling the reactive power is limited in both the hexverter and the modular matrix converter. In the modular matrix converter, however, this limitation mainly applies when the frequencies of System 1 and 2 are close or equal. In the back-to-back MMC, however, the two converters connected to System 1 and System 2 can operate independently in terms of reactive power flow.

The sum of the peak-to-peak variations in the arm energies are shown in Fig. 7 as functions of the frequency in System 2. These variations were calculated using the method described in Section II-B. The solid lines indicate the energy variations at nominal operation whereas the dashed and dotted lines show the total peak-to-peak energy variations when the considered harmonic injection methods are used. It is observed that if no harmonic injection is used, both the hexverter and the modular matrix converter become unstable at synchronous frequency operation. For the hexverter it is, however, possible to use additional components in the arm-voltages and -currents, or by altering the reactive power flow. This is not considered here since it has already been concluded that synchronous frequency operation with the hexverter is not advisable [7]. Hence it can be concluded that the back-to-back MMC is most suitable for synchronous speed, not only from an energy-ripple point of view, but also when considering the power rating of the semiconductors.

At lower frequencies both the Hexverter and the modular matrix converter appear to be more suitable than the back-to-back MMC. The modular matrix converter, however, have both lower semiconductor ratings and energy ripple compared to the Hexverter. Considering the dramatic increase in the energy-ripple at lower frequencies, the modular matrix converter with harmonic injection can be a competitive alternative for low-speed drives.

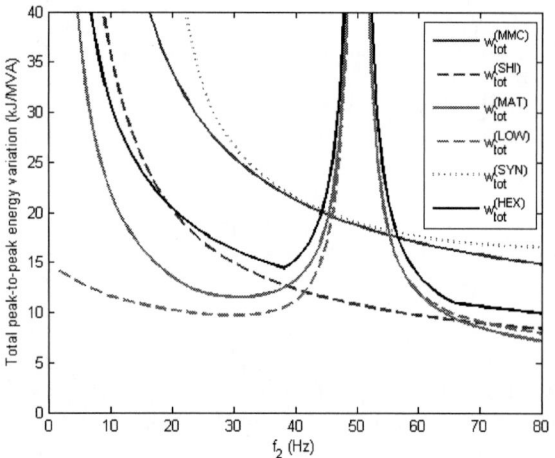

Fig. 7. Total peak-peak energy variation kJ/MVA in a back-to-back MMC (blue), modular matrix converter (red), and Hexverter (black) at the operating conditions specified in Table III.

It is noteworthy that the enrgy-ripple in the back-to-back MMC with harmonic injection is comparable to the modular matrix converter in the range 60–80 Hz. According to Table I, the combined power rating of the semiconductors is also the same for these two cases. Thus, in order to evaluate the most suitable topology for the higher frequency-range, a more detailed study is required in order to determine how the losses are are affected.

VII. CONCLUSIONS

This paper presents a comparison of three cascaded converter topologies for ac/ac conversion. It was found that the back-to-back MMC allows the lowest semiconductor ratings among the considered topologies. Thus, considering the increased energy ripple in the modular matrix converter and Hexverter, the back-to-back MMC is found to be the most suitable solution for synchronous frequency operation. The Hexverter and the modular matrix converter are, however, both better suited at low-frequency operation than the back-to-back MMC. The qualitative study in this paper indicate that the modular matrix converter allows both lower semiconductor ratings as well as lower energy ripple than the Hexverter. The energy-ripple in the modular matrix converter is lower compared to the back-to-back MMC also in the higher frequency-range. It was, however, observed that if harmonic injection is used in the back-to-back MMC, both the semiconductor ratings and the energy ripple become comparable to the modular matrix converter in the higher frequency-range. Accordingly, in order to obtain a clearer view of the most suitable topology for the higher frequency-range, a more detailed study that pays closer attention to the conduction losses is required.

REFERENCES

[1] A. Nabae, I. Takahashi, and H. Akagi, "A new neutral-point-clamped PWM inverter," *IEEE Trans. Ind. Appl.*, vol. IA-17, no. 5, pp. 518–523, 1981.

[2] P. W. Hammond, "A new approach to enhance power quality for medium voltage drives," in *Petroleum and Chemical Industry Conference, 1995. Record of Conference Papers., Industry Applications Society 42nd Annual*, 1995, pp. 231–235.

[3] R. Erickson and O. Al-Naseem, "A new family of matrix converters," in *Industrial Electronics Society, 2001. IECON '01. The 27th Annual Conference of the IEEE*, vol. 2, 2001, pp. 1515–1520.

[4] R. Marquardt, A. Lesnicar, and J. Hildinger, "Modulares Stromrichterkonzept für Netzkupplungsanwendung bei hohen Spannungen," in *ETG-Fachtagung, Bad Nauheim, Germany*, 2002.

[5] A. J. Korn, M. Winkelnkemper, and P. Steimer, "Low output frequency operation of the modular multi-level converter," in *Energy Conversion Congress and Exposition (ECCE), 2010 IEEE*, 2010, pp. 3993–3997.

[6] A. Antonopoulos, L. Ängquist, S. Norrga, K. Ilves, and H.-P. Nee, "Modular multilevel converter ac motor drives with constant torque from zero to nominal speed," in *IEEE Energy Conversion Congress and Exposition Conf Proc.*, Raleigh, 2012.

[7] L. Baruschka, "A new three-phase ac/ac modular multilevel converter with six branches in hexagonal configuration," *IEEE Trans. Ind. Appl.*, vol. 49, no. 3, pp. 1400–1410, 2013.

[8] R. Picas, J. Pou, S. Ceballos, J. Zaragoza, G. Konstantinou, and V. Agelidis, "Optimal injection of harmonics in circulating currents of modular multilevel converters for capacitor voltage ripple minimization," in *ECCE Asia, 2013 IEEE*, June 2013, pp. 318–324.

[9] S. Engel and R. De Doncker, "Control of the modular multi-level converter for minimized cell capacitance," in *Power Electronics and Applications (EPE 2011), Proceedings of the 2011-14th European Conference on*, 2011.

[10] K. Ilves, S. Norrga, and H.-P. Nee, "On energy variations in modular multilevel converters with full-bridge submodules for ac-dc and ac-ac applications," in *Power Electronics and Applications (EPE), 2013 15th European Conference on*, Sept 2013.

[11] L. Baruschka and A. Mertens, "Comparison of cascaded h-bridge and modular multilevel converters for bess application," in *Energy Conversion Congress and Exposition (ECCE), 2011 IEEE*, Sept 2011, pp. 909–916.

[12] K. Ilves, S. Norrga, L. Harnefors, and H.-P. Nee, "On energy storage requirements in modular multilevel converters," *IEEE Trans. Power Electron.*, vol. 29, no. 1, pp. 77–88, Jan 2014.

[13] K. Ilves, A. Antonopoulos, L. Harnefors, S. Norrga, L. Angquist, and H.-P. Nee, "Capacitor voltage ripple shaping in modular multilevel converters allowing for operating region extension," in *IECON 2011 - 37th Annual Conference oft the IEEE Industrial Electronics Society*, Nov 2011, pp. 4403–4408.

[14] M. Winkelnkemper, A. Korn, and P. Steimer, "A modular direct converter for transformerless rail interties," in *Industrial Electronics (ISIE), 2010 IEEE International Symposium on*, July 2010, pp. 562–567.

[15] W. Kawamura, M. Hagiwara, and H. Akagi, "A broad range of frequency control for the modular multilevel cascade converter based on triple-star bridge-cells (mmcc-tsbc)," in *Energy Conversion Congress and Exposition (ECCE), 2013 IEEE*, Sept 2013, pp. 4014–4021.

[16] A. J. Korn, M. Winkelnkemper, P. Steimer, and J. W. Kolar, "Direct modular multi-level converter for gearless low-speed drives," in *Power Electronics and Applications (EPE 2011), Proceedings of the 2011-14th European Conference on*, 2011.

[17] W. Kawamura and H. Akagi, "Control of the modular multilevel cascade converter based on triple-star bridge-cells (mmcc-tsbc) for motor drives," in *Energy Conversion Congress and Exposition (ECCE), 2012 IEEE*, 2012, pp. 3506–3513.

Evaluation of Isolated Three-phase AC-DC Converter using Modular Multilevel Converter Topology

Toshiki Nakanishi

Dept. of Electrical, Electronics and Information Engineering
Nagaoka University of Technology
Nagaoka, Niigata, Japan
nakanishi@stn.nagoakaut.ac.jp

Jun-ichi Itoh

Dept. of Electrical, Electronics and Information Engineering
Nagaoka University of Technology
Nagaoka, Niigata, Japan
itoh@vos.nagaokaut.ac.jp

Abstract— This paper discusses an isolated three-phase AC-DC converter using Modular Multilevel Converter (MMC) topology for a medium voltage application. The MMC at a primary stage of an isolated AC-DC converter directly obtains the high frequency AC voltage from the three-phase AC voltage of the commercial frequency. In addition, the proposed system obtains the low total harmonic distortion (THD) in the input current and maintains the cell capacitor voltage at constant. The comparative evaluations of the system design based on the rated voltage of switching devices are considered. In particular, the numbers of cells and switching devices, the electrostatic energy of the DC capacitors, the THD of the input current and device losses are evaluated when high rated voltage devices: 2.5-kV IGBTs, 3.3-kV IGBTs and 3.3-kV SiC-MOSFETs are applied to the proposed system.

Keywords— *Modular Multilevel Converter; Capacitor Voltage Control; High Frequency Transformer; H-bridge Cell; Isolated Three-phase AC-DC converter;*

I. INTRODUCTION

Recently, a DC distribution network system in datacenters has been actively researched in order to achieve size reduction of the network [1-3]. Moreover, in order to obtain the high efficiency operation, a medium voltage such as 6.6 kV is employed as the power source. In addition, DC voltage of 1 kV or less is supplied to a load such as a server supply by an isolation transformer and a rectifier. However, the conventional system becomes large due to passive components. A diode bridge or a PWM rectifier which is constructed from high rated voltage devices is applied in the conventional system. In general, it is difficult to operate in high switching frequency for a high rated voltage device. Thus, large passive components are required due to low switching frequency in order to suppress harmonic distortion of input current and voltage ripple of output DC voltage.

On the other hand, the multilevel converter is one of effective solution for the problem of above. Recently, the multilevel converter topologies have been actively discussed for the medium voltage application such as at

3.3 kV or 6.6 kV [4-10]. Especially, Modular Multilevel Converter (MMC) has been actively studied as one of the multilevel converter topologies [5]. Advantages of the MMC are as follows; (i) the circuit configuration is simpler than that of other multilevel converters due to cascade connection of cells. Each of the cells is constructed from switching devices and a DC capacitor, (ii) the MMC can reduce the harmonic distortions because the MMC output the multilevel waveform as with other multilevel converters. (iii) low voltage devices can be applied to each cells because the output voltage becomes low by increasing a number of cells. In addition, the MMC achieves the size reduction of passive components due to high frequency operation by low voltage devices. From the advantages of the above, the MMC topology can be applied in high capacity power applications such as Static Synchronous Compensator (STATCOM) and medium range of adjustable speed drive systems [11-15].

However, the high power converters which use MMC result that the volume of the converters becomes large due to the isolation transformer which is operated with line-frequency. The isolation transformer is applied in order to separate the grid side and the load side in order to protect the server supply. One of the effective ways to reduce the transformer size is applying high frequency operation. The size reduction of the transformer by using the multi-module converter has been reported [16-17]. In back-to-back (BTB) systems, multiple bidirectional isolated DC-DC converters are considered and consequently the size reduction of the transformer is achieved due to the usage of the high frequency transformer to each isolated DC-DC converter [9]. However, in principle, it is difficult to achieve high efficiency because the number of switching devices drastically increases with increasing the number of stages. Moreover, a DC-DC converter which has high frequency AC link using MMC has been proposed [18]. In the BTB system, the size reduction of the transformer can be achieved because the transformer is used in high frequency AC link [10]. However, the number of cells is large because MMC has to be applied to both the input and the output sides.

This paper discusses an isolated three-phase AC-DC converter using MMC and a control strategy. Three-phase to single-phase MMC topology is used at a primary stage in the isolated AC-DC converter. The proposed system directly converts three-phase voltage into high frequency single-phase voltage. Furthermore, the volume of the proposed system becomes small and simple due to a high frequency single-phase transformer.. It is confirmed by the simulation that the proposed system achieves low THD and the cell capacitor voltage remains at constant. In addition, the comparative evaluations of the system design based on the rated voltage of switching devices are considered. In particular, the numbers of cells and switching devices, the electrostatic energy of the DC capacitors, the THD of the input current and the device losses are evaluated when high rated voltage devices: 2.5-kV IGBTs, 3.3-kV IGBTs and 3.3-kV SiC-MOSFETs are applied into the proposed system.

II.   MAIN CIRCUIT CONFIGURATION

Fig. 1(a) shows the circuit configuration of the isolated three-phase AC-DC converter using MMC topology. Due to the cascade connection of cells, the proposed converter can achieve a multilevel voltage waveform and the rated voltage of each cell is reduced as well. Thus, increasing the number of cascaded cells are effective because the harmonic distortions of an input current can be reduced, and low voltage rating switching devices can be used.

Furthermore, the proposed system can directly convert an input three-phase voltage into a high frequency single-phase voltage. Therefore, the volume of the proposed system becomes small due to the use of the high frequency transformer. In addition, the topology is symmetry from input and output side. Thus, the proposed system can control each of group independently. In this paper, the group in the upper side is defined as an arm group A, and the group in the lower side is defined as an arm group B.

Fig. 1(b) shows the configuration of an H-bridge cell.

The output voltage of each cell is controlled by pulse width modulation (PWM). Moreover, it is necessary to ensure the capacitor voltage at constant for a stable operation of the main circuit.

III.   CONTROL STRATEGY

Fig. 2 shows the control block diagram of the proposed converter. The proposed control is applied to each group A and B, respectively as shown in Fig. 1. The proposed control block diagram is separated into two blocks, the capacitor voltage control block and the input current control block. Moreover, two types of control are used in the capacitor voltage control block, an average control and a balance control [19].

A. Average Voltage Control

The average voltage control corrects the error between the average value of the capacitors voltage in each arm group and the voltage command is generated from a PI regulator. The average value of the capacitor voltage v_{c_ave} is given by (1)

$$v_{c\_ave} = \frac{2}{3n} \sum_{m=1}^{n/2} \left(v_{crm} + v_{csm} + v_{ctm} \right) \tag{1}$$

where n is the number of cells in each leg and v_{crm}, v_{csm} and v_{ctm} are the capacitor voltage in the each arm. The number of cells is set $n/2$ in order to calculate the average value of the capacitors voltage in each arm group because each arm group is independently controlled.

B. Balance Voltage Control

The average voltage control is used to keep the voltage of all capacitors at constant level. However, an unbalance voltage which occurs among the capacitors cannot be suppressed by the average voltage control because the average control only corrects the error between the average value of the capacitors voltage in each group and subjects to the voltage command. Therefore, the balance voltage control is used in order to correct the error between each of the capacitors voltage and the voltage

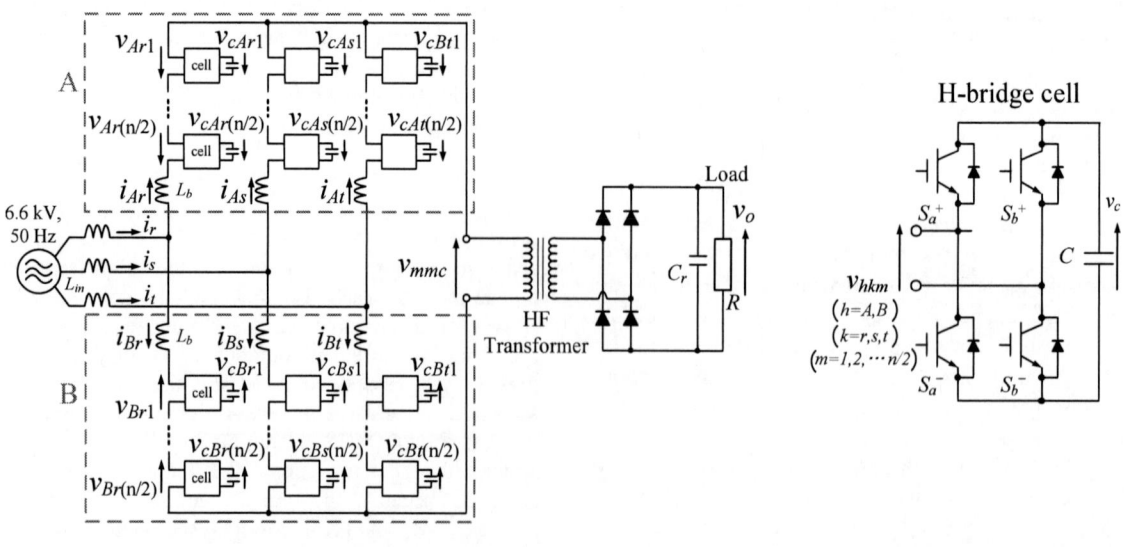

(a) Main circuit configuration.   (b) H-bridge cell.

command. The command of the balance control is given by (2)

$$v_{ce\_Akm}{}^* = K_C \left(v_c{}^* - v_{c\_km}\right) i_k \tag{2}$$

$$k = r, s, t \quad m = 1, 2, \ldots n/2$$

where v_{crm}, v_{csm} and v_{ctm} are the capacitor voltage in each arm respectively, K_C is the gain of the balance voltage control, k is the index of each phase and m is the index of number of cells.

Moreover, it is not necessary to eliminate the error completely because the balance control is an auxiliary control for the average voltage control. Therefore, the balance control can be constructed only by using proportional control.

C. Input Current Control

The input current flows separately to the upper side arm and the lower side arm in each phase. Therefore, the input current can be determined by controlling the current which flow into the buffer reactors. In addition, in the input current control, the linkage operation is not needed between the arm group A and B. Furthermore, the proposed system is able to control the arm groups A and B respectively.

The input current control is implemented by PI regulators and the compensators for the cross terms of i_d and i_q which are provided from the transformation of rotating frame. The input current control is separated into the active current control (i_d) and the reactive current control (i_q). The reactive current command $i_q{}^*$ is set to zero in order to control the input power factor to 1 and to reduce the absolute value of the input current. Therefore, the same input current command is given to each cell because the cascaded cells control the common arm current.

D. Output Voltage Control

The same command of the MMC output voltage control is given to each cell in order to convert a three-phase input voltage to a high frequency voltage. The output voltage of a cell is $v_{mmc}{}^*/n$ when the command of the MMC output voltage control is $v_{mmc}{}^*$ and the number of cells in a leg is n.

Finally, the output voltage of the cell is given by (3). Note that the sign of the MMC output voltage command $v_{mmc}{}^*$ has to be changed because the direction of cell output voltage is different in the group A and the group B.

$$v_{Akm}{}^* = \frac{1}{n}\left(2v_{Ak}{}^{**} + v_{mmc}{}^*\right) - v_{ce\_Akm}{}^*$$

$$v_{Bkm}{}^* = \frac{1}{n}\left(2v_{Bk}{}^{**} - v_{mmc}{}^*\right) - v_{ce\_Bkm}{}^* \tag{3}$$

$$k = r, s, t \quad m = 1, 2, \ldots n/2$$

E. Capacitor Voltage Command

The MMC output voltage command $v_{mmc}{}^*$ is added in order to obtain the high frequency output voltage, which is added to the output block of the input current control. The change of voltage values in each cell depends on the number of cells at each leg since the cells are connected to the load in parallel. In addition, each capacitor voltage also depends on the input and output voltage. The capacitor voltages command $v_c{}^*$ is given by (4).

$$v_c{}^* \geq \frac{1}{n}\left(2\sqrt{\frac{2}{3}}E + v_{mmc}{}^*\right) \tag{4}$$

where E is the effective value of the input line to the line voltage, $v_{mmc}{}^*$ is the maximum value of the MMC output voltage command and n is the number of cells at each leg. From (4), the capacitor voltage of each cell becomes smaller by increasing the number of cells. Therefore, the capacitor voltage command becomes lower.

IV. EVALUATION OF THE PROPOSED SYSTEM

A. Evaluation Condition

Table 1 shows the parameters of the switching devices with the high rated voltage. In the proposed system, from (4), the capacitor voltage depends on the input current control and the MMC output voltage control. Therefore, the rated voltage of the switching devices depends on the capacitor voltage of the cell. In the case where the high rated voltage devices are applied, it is possible to reduce the number of the circuit components in the proposed

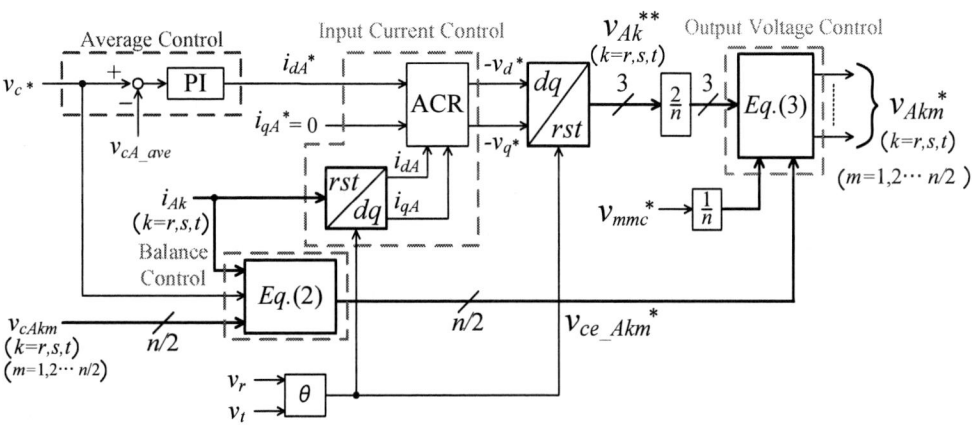

Fig. 2. Control block diagrams of the proposed circuit for the arm group A.

system. On the other hand, in the case where the low rated voltage devices are applied and the number of cells is increased, it is possible to suppress the harmonic distortion of the input current due to the high speed switching and multiple voltage levels. Thus, the design factors : the number of cells, the capacitor voltage and the THD of the input current are different. Therefore, comparing of the circuit performance based on the rated voltage of switching devices is required. In order to compare the circuit performance with different devices, the number of high rated voltage IGBTs and SiC MOSFETs which the rated voltage is 3.3-kV are employed.

The conditions to compare the circuit performance are as follow.

a) The rated voltages of IGBTs are 2.5-kV, 3.3-kV and 6.5-kV, respectively.

b) The capacitor voltage is set 20% more than the value which is calculated by (4). Moreover, the rated voltage is set 30% more than the capacitor voltage.

c) The output power is set at 400 kW and the output voltage is set at 450 V.

d) MMC output voltage command $v_{mmc}{}^{*}$ is set 2.6 kV, 1 kHz.

e) Buffer reactor L_b is 0.5 mH (the percentage of impedance $\%Z = 0.31\%$). Moreover, the interconnection reactor is 4 mH (the percentage of impedance $\%Z = 0.125$). Both reactors are common value in all system.

f) Switching frequency is chosen so that the proposed system can achieve the efficiency of 95% with considering of the devices loss: the conduction loss, the switching loss and the conduction loss of the free wheeling diode (FWD).

B. Comparative Evaluation of the proposed system

The number of cells, switching devices and DC capacitors are estimated by referring (4) and condition (b) is listed in previous part.

Table 2 shows the comparison result of the cell numbers and switching devices that depends on the rated voltage of the switching device.

In the case where IGBTs with the rated voltage of 2.5-kV are applied, 10 cells are required per leg. Moreover, 120 units of IGBT are required. On the other hand, 2.5-kV IGBT can be set as a high response angular frequency due to high speed switching. Therefore, it is possible to reduce the values of buffer reactors and DC capacitors, respectively.

On the other hand, in the case where IGBTs with the rated voltage of 3.3-kV are applied, 8 cells per leg and total of 96 units of IGBT are required. The number of cells and switching devices can be reduced compared with the case of 2.5-kV IGBTs. However, the switching frequency is limited because the switching speed is lower than the switching frequency of 2.5-kV IGBTs. Therefore, it is necessary to increase the value of DC capacitors since the response angular frequency of the control system is low.

In the case where IGBTs with the rated voltage of 6.5-kV are applied, 4 cells per leg and 48 units of IGBT are

Table 1. Parameters of switching devices and switching frequency.

	Device Parameters		f_s
2.5-kV IGBT IXBK64N250	V_{CE}=2.5 kV, I_c = 156 A E_{on} = 30.2 mJ @ Ic = 140 A, V_{CE} = 1.25kV, R_{Gon} = 1 Ω, V_{GE} = ±15 V E_{off} = 15.31 mJ @ Ic = 140 A, V_{CE} = 1.25kV, R_{Goff} = 1 Ω, V_{GE} = ±15 V V_{CEsat} = 2.5 V @ Ic = 64 A, V_{GE} = 15 V		5 kHz
2.5-kV_FWD SD203N/R25S20	V_{RRM} = 2.5 kV, I_c = 200 A V_{FM} = 1.65 V @ Ipk = 628 A		
3.3-kV_IGBT FF200R33KF2C	V_{CE}=3.3kV, I_c= 200A E_{on} = 235 mJ @ Ic = 200 A, V_{CE} = 1.8kV, R_{Gon} = 7.5 Ω, V_{GE} = ±15 V E_{off} = 215 mJ @ Ic = 200 A, V_{CE} = 1.8kV, R_{Goff} = 5.6 Ω, V_{GE} = ±15 V V_{CEsat} = 3.4 V @ Ic = 200 A, V_{GE} = 15 V		1.38 KHz
3.3-kV_FWD FF200R33KF2C	V_{RRM}= 3.3 kV, I_c= 200 A V_F = 2.8 V @ I_F = 200 A		
6.5-kV IGBT FZ400R65KE3	V_{CE}=6.5 kV, I_c=400 A E_{on} = 3450 mJ @ Ic = 400 A, V_{CE} = 3.6kV, R_{Gon} = 1.9 Ω, V_{GE} = ±15 V E_{off} = 2250 mJ @ Ic = 400 A, V_{CE} = 3.6kV, R_{Goff} = 13 Ω, V_{GE} = ±15 V		Under 1 kHz

Table 2. Comparison for numbers of cells and switching devices depending on the rated voltage of the switching device.

Rated Voltage	Numbers of Cells @leg	Numbers of Switching devices
2.5-kV	10	120
3.3-kV	8	96
6.5-kV	4	48

required. The number of cells and switching devices can be significantly reduced compared to 2.5-kV IGBTs and 3.3-kV IGBTs. The switching speed of 6.5-kV IGBT is lower than that of 2.5-kV IGBT and 3.3-kV IGBT. Furthermore, the large values of DC capacitor and buffer reactors are needed because angular frequency responses of the average control and the input current control become low, respectively. In addition, the large inductance of buffer reactors are also applied since the response angular frequency of the input current control becomes low. In the proposed system, MMC output is a high frequency voltage. However, the voltage drop becomes large since a voltage-dividing condition occurs between the high frequency transformer and buffer reactors. In the worst case, it is difficult to transmit a high frequency power to secondary side of the transformer when the buffer reactor is large. Therefore, it is necessary to reduce the inductance of buffer reactors to transmit the high frequency power.

C. Evaluation of proposed system by appling SiC MOSFETs

SiC-MOSFET with the rated voltage of 10 kV has been developed [20]. SiC devices have following advantages: high rated voltage, low loss, high switching speed and high operating temperature. Therefore, Employing SiC devices to high power applications are considered nowadays, especially, in the high voltage and high power conversion system. However, the *dv/dt* at winding wires of a high frequency transformer becomes greatly large. Thus, the insulation degradation among winding wires may occur when the proposed system is operated by using high rated voltage devices. From the previous topic, it is needed to set the switching frequency in order to modulate the operation frequency of the high frequency transformer at 1 kHz. Therefore, it is necessary to consider the operation of the proposed system when 3.3-kV SiC-MOSFET is applied into the proposed system. Note that the device data of 3.3-kV SiC-MOSFET was reported [21].

Table 3 shows the simulation parameters. The switching frequency of the SiC-MOSFET is set to 15 kHz. Note that the ideal elements are used in the switching device. In addition, the output voltage command is a square-wave of 2.3 kV, 1 kHz.

Fig. 3 shows the input phase voltage and current waveforms. From Fig. 3, it is confirmed that the unity power factor can be obtained in the input stage. Moreover, the total harmonic distortion (THD) of the input current is approximately 1.74%. The input current waveform includes few pulse beats. However, the pulse beats does not affect the THD of the input current since the frequency of the pulse beats is very high compared to the fundamental frequency.

Fig. 4 shows the output voltage waveform. From the result, the output voltage is constantly controlled approximately at 450 V. Moreover, the ripple voltage is obtained 0.4%. Thus, the power conversion from three-phase AC voltage into a DC voltage is confirmed.

Fig. 5 shows the cell capacitor voltage waveforms at the r-phase. The cell capacitor voltage is controlled according to the capacitor voltage command v_c^* by the average control and the balance control. In the capacitor voltage, the ripples include two frequency components: 100 Hz and 1 kHz. Then, the ripple of 1 kHz is less than that of 100 Hz. Furthermore, the ripple of 100 Hz comprises a majority of the capacitor voltage ripple. However, the maximum ripple of each capacitor voltage is obtained by 2.0%. Thus, it is confirmed that the voltage ripple does not affect the circuit operation in the proposed system.

Fig. 6 shows the output voltage waveforms of the MMC. It is confirmed that the high frequency AC voltage of 1 kHz is obtained. However, an error occurs between the output voltage of the MMC and the voltage command. The error depends on the voltage drop in each buffer reactor. Moreover, the output voltage waveform is dropped when the voltage polarity changes because the high frequency current flows in all diode in the rectifier circuit due to the leakage inductance condition of the high frequency transformer. The output voltage depends on the voltage drop of the leakage inductance.

D. Evaluation of total electrostatic energy

In general electrical products, the size of an electrolytic capacitor affects the system size. On the other hand, the size of a DC capacitor depends on the electrostatic energy because the size of the electrolytic

Table 3 Simulation condition

Output power P_O	400 kW	Input voltage rms	6.6 kV
Input voltage frequency	50 Hz	Output voltage	450 V
Number of cell per leg *n*	8	Load *R*	0.5 Ω
Turn ration pri.: sec.	1 : 1.5	Carrier frequency	15 kHz
Interconnection reactor L_{in}		4 mH (%Z = 1.15%)	
Buffer reactor L_b		0.5 mH (%Z = 0.31%)	
DC capacitor C		1,600 μF	
Output capacitor C_r		3,000 μF	
Switching device		3.3-kV SiC-MOSFET	

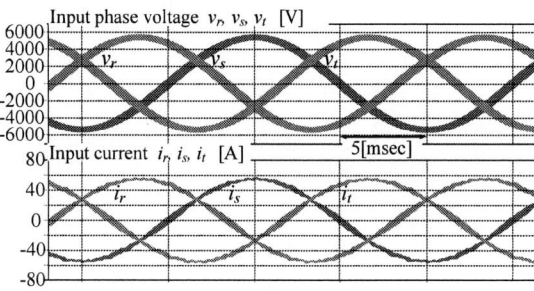

Fig. 3. Waveforms of the input phase voltage and current.

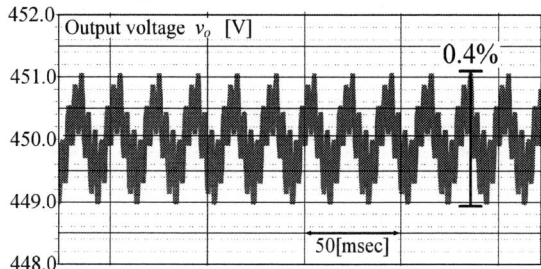

Fig. 4. Waveform of the output voltage at the diode bridge rectifier

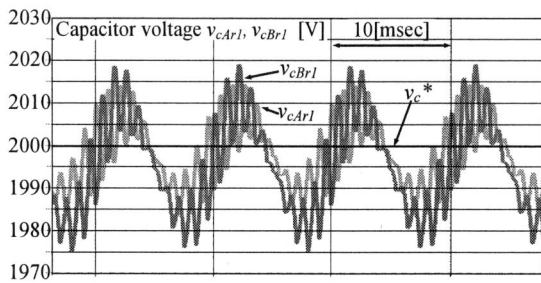

Fig. 5. Waveform of the capacitor voltage.

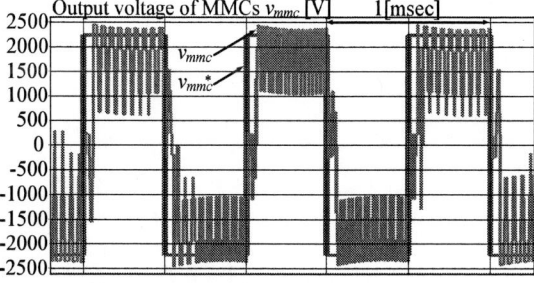

Fig. 6. Waveform of the MMC output voltage.

capacitor depends on the charge voltage of the DC capacitor and its capacitance. Thus, the system size can be evaluated by total electrostatic energy of the DC capacitor. The total electrostatic energy W_C is given by (5).

$$Wc = \frac{1}{2}CV_c^2 \times N \qquad (5)$$

where C is a capacitance of each capacitor, N is the number of capacitors in MMC and V_C is the average value of the capacitor voltage. V_C is equal to the capacitor voltage command v_c^* when all capacitor voltage is kept constant by the average control and the balance controls.

On the other hand, the capacitance of the DC capacitor is designed in order to suppress the voltage ripple. At the DC capacitor, the ripple frequency of the voltage ripple becomes 100 Hz occurs because each cell rectifies the input current of the commercial frequency. In this paper, the capacitance of the DC capacitor is designed in order to suppress the voltage ripple of 100 Hz. Moreover, the ripple is suppressed within 1% against the average value of the capacitor voltage. Furthermore, it is necessary to set the capacitance of the DC capacitor to satisfy the requirement. In order to control the input current, the buffer reactor current i_a and the cell output voltage v_{sc} is given by (6).

$$i_a = \frac{\sqrt{2}}{2} I \sin \omega t$$
$$v_{sc} = 2\sqrt{\tfrac{2}{3}} E \sin \omega t \qquad (6)$$

where I is the effective value of the input current, E is the effective value of the input line voltage and ω is an angular frequency of the input voltage. Moreover, the current which flows to the DC capacitor is i_{dc}. The DC capacitor voltage v_c is kept at constant when $v_c i_{dc} = v_s i_a$ holds because input power and the output power are equal. From (6), the DC capacitor current i_{dc} is given by (7).

$$i_{dc} = \frac{EI}{\sqrt{3}v_c n}(1 - \cos 2\omega t) \qquad (7)$$

The second term of (7) is the current ripple Δi_{dc}. In addition, the voltage ripple Δv_c can be calculated by the impedance of DC capacitor and the current ripple Δi_{dc}. From the relationship between the voltage ripple Δv_c and the capacitance of DC capacitor for the voltage ripple suppression is given by (8). Note that v_c is equal to the voltage command v_c^* in each system. From (8), the capacitance of each system can be calculated.

$$C = \frac{EI}{\sqrt{3}\Delta v_c n v_c^* \omega} \qquad (8)$$

Fig. 7 shows the comparison result of the total energy which all capacitor charges. From the comparison result, the total electrostatic energy is the largest when 3.3-kV IGBTs are applied. In the case of applying 3.3-kV IGBTs, the capacitance should be larger than that is calculated by (8) since the response angular frequency of the average

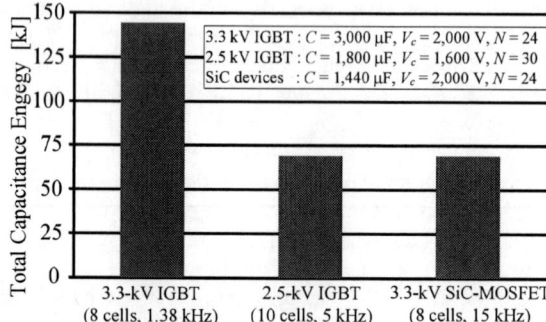

Fig. 7. Comparison of total electrostatic energy in DC capacitors

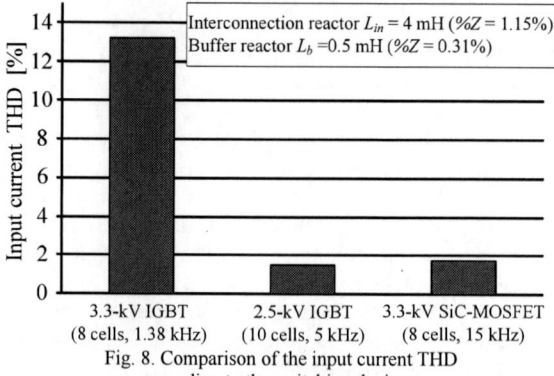

Fig. 8. Comparison of the input current THD according to the switching devices.

control becomes low.

In the case of applying 3.3-kV SiC-MOSFETs into the proposed system, the total electrostatic energy is the lowest since the number of cells and the ripple voltage against the average value of the capacitor voltage are reduced. Furthermore, by applying 3.3-kV SiC-MOSFETs to the system, the size reduction of the proposed system can be achieved since the number of circuit components can be reduced and the power loss in switching devices can be suppressed.

In the case of applying 2.5-kV IGBTs into the proposed system, the total electrostatic energy is nearly identical to that of 3.3-kV SiC MOSFETs because the capacitance C and the capacitor voltage V_C become low. In conclusion, the size of DC capacitors can be reduced when 2.5-kV IGBTs are applied into the proposed system as well as 3.3-kV SiC MOSFETs.

E. Comparison with respect to THD of Input Current

Fig. 8 shows the comparison result of the input current THD. From Fig. 8, the THD is the lowest in the case where IGBTs with the rated voltage of 2.5 kV. When 2.5-kV IGBTs are applied into the proposed system, the number of cells is the highest. Furthermore, the harmonic distortion is suppressed because the input voltage becomes multilevel waveform due to multi-cells. On the other hand, when 3.3-kV SiC-MOSFETs are applied into the proposes system, the low THD of the input current is achieved as well as 2.5-kV IGBTs because it is possible to set a higher response angular frequency of the input current control due to the high switching speed. Finally, the THD is the highest in cases where IGBTs with the rated voltage of 3.3-kV. When 3.3-kV IGBTs are applied into the proposed system, it is not possible to set a high response angular frequency because the switching speed of the IGBT is very low. Thus, it is difficult to improve the harmonic distortion for the system which is applied 2.5-kV IGBTs when the same values of the interconnection reactor and the buffer reactor are applied as well as the system which 2.5-kV IGBTs or 3.3-kV SiC-MOSFETs are applied.

F. Comparison Evaluation of the device loss

Fig. 9 shows the comparison result of the device losses: the conduction loss, the switching loss and the conduction loss of the FWD against the output power. Note that a SiC Junction Barrier Schottky Diode is applied as the FWD of 3.3-kV SiC-MOSFETs. Note that the device data of a SiC Junction Barrier Schottky Diode was reported [22].

In the case of applying 3.3-kV IGBTs to the proposed system, the switching loss is larger than the other system because turn-on loss and turn-off loss are large compared with the system which 2.5-kV IGBTs and 3.3-kV SiC-MOSFETs from Table 1. Moreover, the system using 3.3-kV IGBTs cannot achieve the efficiency of 95%. It is possible to achieve the efficiency of 95% when the low switching speed is set. However, the system cannot produce the high frequency output voltage when the low switching speed is set. In addition, in the system that is applied with low switching frequency devices, the value of the DC capacitor is needed to be increase in order to maintain the voltage of DC capacitors constantly.

In the case of applying 2.5-kV IGBTs, the efficiency of 95% can be achieved when the switching frequency is 5 kHz. Furthermore, it is possible to obtain the high frequency output voltage and keep the voltage of DC capacitors at constant. Moreover, when 2.5-kV IGBTs are applied into the proposed system, more cells are required rather than other systems. However, the system is able to achieve the efficiency of 95%. Thus, in the system which is applied 2.5-kV IGBTs, the number of cells is not significantly affected the efficiency.

In the case of applying 3.3-kV SiC-MOSFETs to the proposed system, the loss at switching devices is the lowest because both the turn-on loss and turn-off loss are lower than that of both IGBTs. Furthermore, the efficiency is the highest. Moreover, when the losses of the devices become half comparing to the system which is applied to 2.5-kV IGBTs at switching frequency of 5 kHz, the volume of the heat-sink can be dramatically reduced.

In conclusion, by applying 2.5-kV IGBTs or 3.3-kV SiC-MOSFETs, the high efficiency can be achieved in the proposed system. Moreover, SiC-MOSFET is the most suitable for the proposed system because SiC-MOSFET has the high performance features. Thus, the proposed system using 3.3-kV SiC-MOSFET reduces the harmonic distortion of the input current, can achieve the high efficiency and the size reduction of the system. On the other hand, 2.5-kV IGBTs also has good performance in terms of the efficiency and the suppression of the harmonic distortion.

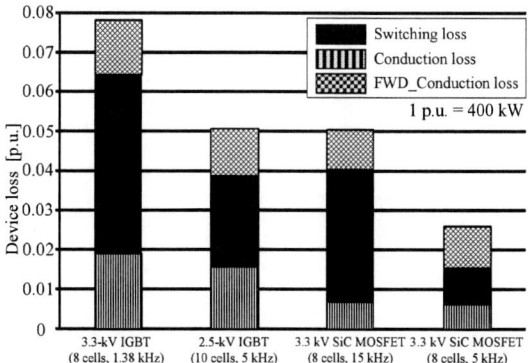

Fig. 9. Comparison of the switching loss and the conduction loss.

V.   CONCLUSION

In this paper, an isolated three-phase AC-DC converter using MMC and the control strategy were proposed for a medium voltage application. The proposed system could directly convert an input three-phase voltage into a high frequency single-phase voltage. Therefore, the proposed system could reduce the volume of the isolated transformer since the single-phase high frequency transformer could be applied to the high frequency link.

The proposed system is evaluated with numbers of different rated voltage devices: a 2.5-kV IGBT, a 3.3-kV IGBT and a 3.3-kV SiC-MOSFETs. As a result, when 3.3-kV SiC-MOSFETs is applied into the proposed system, the total electrostatic energy in DC capacitors is the lowest compared to the other devices. Furthermore, 3.3-kV SiC-MOSFETs could achieve the size reduction of the proposed system when the common value of energy density in all capacitor is considered. Moreover, the system using 3.3-kV SiC-MOSFETs could achieve low THD input current waveform which is approximately 2.36% and the maximum ripple factor of each capacitor voltage is approximately 2.0%. Finally, from comparison evaluation of the efficiency when device losses are considered in the proposed system, it is confirmed that the efficiency of 95% can be achieved when 2.5-kV IGBTs and 3.3-kV SiC-MOSFET are applied into the proposed system. In conclusion, in the cases of 2.5-kV IGBTs and 3.3-kV SiC-MOSFETs, the proposed system can achieve the size reduction, the low THD of input current and high efficiency. Moreover, the system which is applied SiC-MOSFETs can achieve better efficiency compared with 2.5-kV IGBT even the same switching frequency is applied.

In future works, the optimized design method of the high frequency transformer for the high power application will be considered. In addition, the volume evaluation results of the proposed system compared to the conventional system will be shown.

REFERENCES

[1]   E. Taylor , M. Korytowski and G. Reed ; "Voltage Transient Propagation in AC and DC Datacenter Distribution Architectures",

Energy Conversion Congress and Exposition (ECCE), pp. 1998 - 2004, (2012)

[2] J. Lago and M. L. Heldwein ; "Operation and Control-Oriented Modeling of a Power Converter for Current Balancing and Stability Improvement of DC Active Distribution Networks", IEEE Trans. on Power Electronics, vol. 26, No. 3, pp. 877–885, (2011)

[3] Y. Morishita, T. Ishikawa, I. Yamaguchi, S. Okabe, G. Ueta and S. Yanabu ; "Applications of DC Breakers and Concepts for Superconducting Fault-Current Limiter for a DC Distribution Network", IEEE Trans. on Applied Superconductivity, vol. 19, No. 4, pp. 3658–3664, (2009)

[4] N. Hatti, K. Hasegawa and H. Akagi ; "A 6.6-kV Transformerless Motor Drive Using a Five-Level Diode-Clamped PWM Inverter for Energy Savings of Pumps and Blowers", IEEE Trans. on Power Electronics, vol. 24, No. 3, pp. 796–803, (2009)

[5] N. Hatti, Y. Kondo and H. Akagi ; "Five-Level Diode-Clamped PWM Converters", IEEE Trans. on Industry Applications, vol. 44, No. 4, pp. 1268–1276, (2008)

[6] P. K. Steimer and M. Winkelnkemper, "Transformerless Multi-Level Converter based Medium Voltage Drives", ECCE2011 pp. 3435 -3441, (2011)

[7] R.l Crosier, S. Wang and M. Jamshidi, "A 4800-V Grid-Connected Electric Vehicle Charging Station that Provides STACOM-APF Functions with A Bi-directional, Multi-level,Cascaded Converter", APEC2012, pp. 1508-1515R. Nicole, "Title of paper with only first word capitalized," J. Name Stand. Abbrev., in press.

[8] H. M. Pirouz and M. T. Bina, "New Transformerless STATCOM Topology for Compensating Unbalanced Medium-Voltage Loads", Power Electronics and Applications, 2009. EPE '09. 13th European Conference, (2009)

[9] H. Li, Thomas L. Baldwin, C. A. Luongo, and D. Zhang , "A Multilevel Power Conditioning System for Superconductive Magnetic Energy Storage", IEEE Trans. on Applied Superconductivity, Vol. 15, No. 2, pp. 1943-1946 (2005)

[10] F. Z. Peng, W. Qian and D. Cao ; "Recent Advances in Multilevel Converter/Inverter Topologies and Applications", Power Electronics Conference (IPEC), 2010 International, pp. 492-501 (2010)

[11] M. Glinka and R. Marquardt, "A new ac/ac multilevel converter family", IEEE Trans. Industrial Electronics, vol. 52, No. 3, pp. 662–669, (2005)

[12] M. Hagiwara and H. Akagi, "Control and Experiment of Pulsewidth-Modulated Modular Multilevel Converters", IEEE Trans. on Power Electronics, Vol. 24, No. 7, pp. 1737-1746 (2009)

[13] M. Hagiwara, R. Maeda and H. Akagi, "Negative-Sequence Reactive-Power Control by a PWM STATCOM Based on a Modular Multilevel Cascade Converter (MMCC-SDBC)", IEEE Trans. on Industry Applications, Vol. 48, No. 2, pp .720-729 (2012)

[14] M. Vasiladiotis, S Kenzelmann, N. Cherix and A. Rufer, "Power and DC Link Voltage Control Considerations for Indirect AC/AC Modular Multilevel Converters", Power Electronics and Applications (EPE 2011), Proceedings of the 2011-14th European Conference, (2011)

[15] Y. Hayashi, T. Takeshita, M. Muneshima and Y. Tadano ; "Independent control of input current and output voltage for Modular Matrix Converter", Industrial Electronics Society, IECON 2013 - 39th Annual Conference of the IEEE, pp. 888-893 (2013)

[16] H. Akagi and R. Kitada, "Control and Design of a Modular Multilevel Cascade BTB System Using Bidirectional Isolated DC/DC Converters", IEEE Trans. on Power Electronics, Vol. 26, No. 9, pp. 2457-2464 (2009)

[17] J. Shi, W. Gou, H. Yuan T. Zhao and A. Q. Huang ; "Research on Voltage and Power Balance Control for Cascaded Modular Solid-State Transformer", IEEE Trans. on Power Electronics, vol. 26, No. 4, pp. 1154–1166, (2011)

[18] S Kenzelmann, A. Rufer, M. Vasiladiotis, D. Dujic, F. Cnales and Y. R. de Novaes, "A versatile DC-DC converter for energy collection and distribution using the Modular Multilevel Converter", Power Electronics and Applications (EPE 2011), Proceedings of the 2011-14th European Conference, (2011)

[19] T. Nakanishi and J. Itoh, "Evaluation of control methods for isolated three-phase AC-DC converter using modular multilevel converter topology", ECCE-Asia Downunder pp. 52-58 (2009)

[20] J. Wang, T. Zhao, J. Li, A. Q. Huang, R. Callanan, F. Husna and A. Agarwal ; "Characterization, Modeling, and Application of 10-kV SiC MOSFET", IEEE Trans. on Electron Devices, Vol. 55, No. 8, pp. 1798-1806 (2008)

[21] R. Lai, L. Wang, J. Sabate, A, Elasser and L. Stevanovic, "High-Voltage High-Frequency Inverter using 3.3 kV SiC MOSFETs", Power Electronics and Motion Control Conference (EPE/PEMC), 2012, DS2b.6-1 - DS2b.6-5 (2012)

[22] T. Duong, A. Hefner, K. Hobart, S.H. Ryu, D. Grider, D. Berning, J. M. Ortiz-Rodriguez, E. Imhoff, J. Sherbondy, "Comparison of 4.5 kV SiC JBS and Si PiN Diodes for 4.5 kV Si IGBT Anti-parallel Diode Applications", Applied Power Electronics Conference and Exposition (APEC), 1057 - 1063 (2011)

978-1-4799-2706-7/14 $31.00 © 2014 IEEE

The 2014 International Power Electronics Conference

Self-decoupled Dual Pick-up Coils with Large Lateral Tolerance for Roadway Powered Electric Vehicles

Su Y. Choi, Sung W. Lee*, and Eun S. Lee,
KAIST, Daejeon, Korea
Samsung Electronics*, Suwon, Korea
(suchoi@kaist.ac.kr, leesungwoo@gmail.com,
eunsoo86@kaist.ac.kr)

Seog Y. Jeong, Beom W. Gu, and Chun T. Rim
KAIST, Daejeon, Korea
(seogyong@kaist.ac.kr, kbw755@kaist.ac.kr,
ctrim@kaist.ac.kr)

Abstract- **Self-decoupled dual pick-up coils for roadway powered electric vehicle (RPEV) with large lateral tolerance and low EMF for pedestrians are proposed. The accurate equivalent circuits and mathematical models for the dual pick-up coils are completely developed to decouple the two adjacent pick-up coils for the both cases with and without core plates. The proposed models are verified by simulations and experiments. It is found that the adjacent pick-up coils can be decoupled regardless of the existence of core plates by overlapping the pick-up coils and the overlapping is nearly same for the both cases with and without core plates.**

I. INTRODUCTION

Environmental pollutions and the exhaustion of oil resources are becoming serious issue where conventional vehicles account for one of the largest portion of the pollutions. As a solution to the problems, electric vehicles (EV) such as hybrid electric vehicle (HEV), plug-in HEV (PHEV), and pure EV (PEV) have been developed, but they have not been widely used due to the battery problems, which come from its high price, heaviness, large installation space, and frequent charging due to low energy capacity.

In order to mitigate the battery problems, roadway powered electric vehicle (RPEV) using inductive power transfer system (IPTS) has been proposed [1]–[15]. With a virtue of the IPTS, the RPEV can be powered during its operation and it can be free from the battery problems. In general, the IPTS adopts the resonant circuits to maximize the power transfer, which is a challenging issue due to the static and dynamic characteristics of the IPTS in practice; however, the dynamic and static behaviors of the IPTS can be characterized by using the phasor transformation [16]-[19]. In the IPTS for RPEV, the sufficient air-gap between a power supply rail and pick-up coils, the large lateral tolerance between the centers of the power supply rail and pick-up coils, and the low electromagnetic field (EMF) are important issues. In order to meet the three requirements, the narrow width I-type power supply rail with alternative magnetic poles, was proposed [8]-[9]. However, the width of the pick-up coils should be enlarged to improve the large lateral tolerance and it results in the higher EMF level from the

pick-up coils than the ICNIRP guideline for pedestrians [20]. Moreover, the inductance of the pick-up coils is exactly proportional to its width, which leads to the high voltage stress on the compensation capacitors in resonant condition and the narrow -3dB bandwidth BW_{3dB} due to high quality factor Q. As a remedy for the problem, the dual pick-up coils with active EMF cancel coils for the I-type power supply rail were proposed [10], [13], [15]. Each of the wide pick-up coils is separated to reduce the inductance of the pick-up coils and the additional active EMF cancel coils, which were wound the reverse direction with the pick-up coils for decreasing the EMF for pedestrians, were adopted. Although the pick-up coils with active EMF cancel coils have successfully reduced the EMF level and the inductance, the lateral tolerance was deteriorated due to the low magnetic coupling between the two pick-up coils.

In this paper, the self-decoupled dual pick-up coils, which were introduced in a few researches of the magnetic resonance imaging (MRI) [21] and the EV stationary charging systems [22]-[23], are proposed for RPEVs, as shown in Fig. 1. With a virtue of the self-decoupled dual pick-up coils, not only large air gap but also large lateral tolerance can be achieved. In addition, the low voltage stress of compensation capacitors and low EMF without the active EMF cancel coils can be obtained, as shown in Fig. 1(b). The equivalent circuits and mathematical models of pick-up coils are proposed for the both cases with and without core plates which adopted the finite-width magnetic mirror models to analyze the effect on the existence of core plates [24]. The proposed models are verified by simulations and experiments.

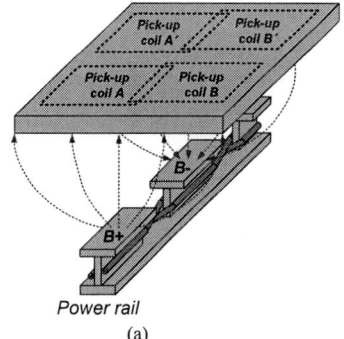

(a)

978-1-4799-2706-7/14 $31.00 © 2014 IEEE

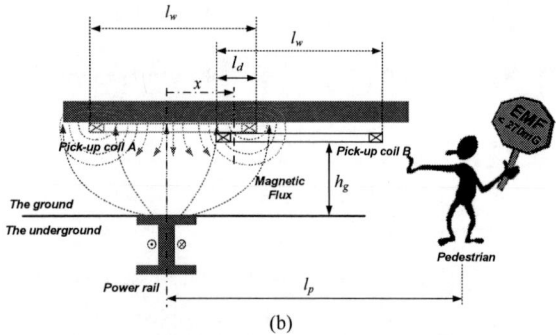

(b)

Fig. 1. The proposed dual pick-up coils for the large lateral tolerance and low EMF level with the I-type power supply rail. (a) Bird's view. (b) Front view: pick-up coils A are coupled with the power supply rail but pick-up coils B are not.

II. SELF-DECOUPLED DUAL PICK-UP COILS WITH I-TYPE POWER SUPPLY RAIL

A. Problem of the Dual Pick-up Coils

As mentioned in the previous section, the air-gap and lateral tolerance can be enlarged with the proposed dual pick-up coils, as shown in Fig. 1. As the power supply rail and the pick-up coils are well aligned, the pick-up coils A and B have almost the same power transfer. When the pick-up coils are misaligned with the power supply rail, one of the pick-up coils, which located on the power supply rail, can only transfer power, as shown in Fig. 1(b). However, the magnetic coupling between the two pick-up coils should be considered, which induces voltage to the other pick-up coil and leads to the high EMF for pedestrians.

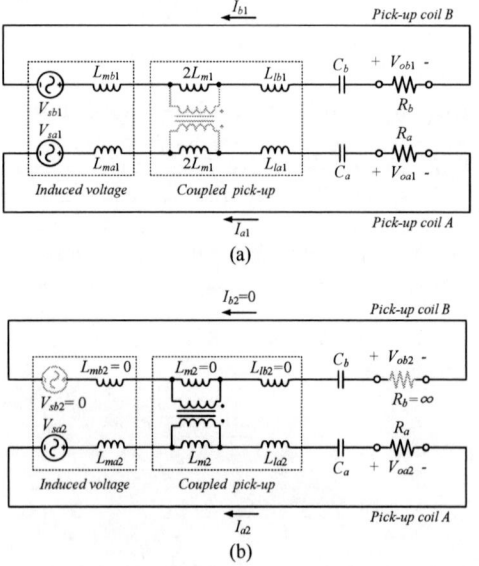

(a)

(b)

Fig. 2. The equivalent circuit models for the coupled dual pick-up coils when the power supply rail and the pick-up coils are (a) Well-aligned ($0 < x < 0.5l_d$) and (b) Misaligned ($0.5l_d < x < l_w - 0.5l_d$).

When the series-series resonant condition for the IPTS and current controlled power supply rail are used, the

equivalent circuit models for both the aligned and misaligned cases of the coupled dual pick-up coils can be obtained, as shown in Fig. 2. Here, the induced voltages V_{sa} and V_{sb} are defined as phasor form in the steady state as follows:

$$V_{sa} = j\omega_s L_{ma} I_p n , \qquad (1a)$$

$$V_{sb} = j\omega_s L_{mb} I_p n . \qquad (1b)$$

V_{sa} and V_{sb} are the induced voltages for pick-up coils A and B, where L_{ma} and L_{mb} are the magnetizing inductances between the power supply rail and two pick-up coils A and B, respectively. I_p and n are the power supply rail current and turn ratio. L_m, L_{la}, and L_{lb} are the magnetizing inductance between the two pick-up coils, and the leakage inductances, respectively. C_a and C_b are the compensation capacitor for pick-up side series resonant condition, and R_a and R_b are the effective load resistance of the two pick-up coils, respectively.

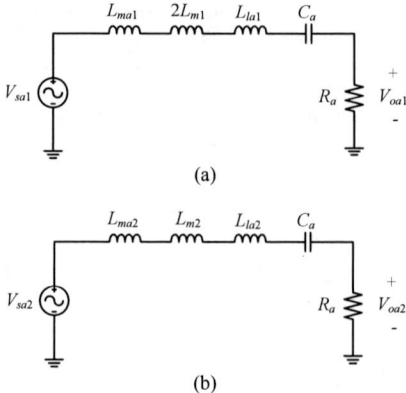

(a)

(b)

Fig. 3. The equivalent simplified circuits for the coupled dual pick-up coils A when (a) Well-aligned and (b) Misaligned.

Depending on the alignments of the pick-up coils A, the equivalent circuit of the pick-up coils A changes, as shown in Fig. 3 and it results in the two different resonant conditions in the pick-up coils A. In case when the pick-up coils A are well aligned, the resonant frequency ω_{r1} is obtained as

$$\omega_{r1} = \frac{1}{\sqrt{(L_{ma1} + 2L_{m1} + L_{la1})C_a}} , \qquad (2)$$

and when the pick-up coils A are misaligned, the resonant frequency ω_{r2} is obtained as

$$\omega_{r2} = \frac{1}{\sqrt{(L_{ma2} + L_{m2} + L_{la2})C_a}} . \qquad (3)$$

In order to obtain the sufficient power regardless of the pick-up alignments, there are two possible solutions; the one is to use the adaptive frequency control system on the power supply rail to make pick-up coils in the resonant condition for both the aligned and misaligned cases. The another one is to minimize the gap between two different resonant frequencies ω_{r1} and ω_{r2}. However, in practice, the first solution is hard to be used because many vehicles have different resonant conditions on the single power supply rail. On the other hand, the second solution is easy to be implemented and it needs no active component to satisfy the resonant condition, so the second one is adopted in this paper. Although the L_{m1} and L_{m2} can be easily nullified by increasing the distance between two pick-up coils, it is not good solution because of the limited power transferring from a power supply rail when the pick-up coils are well aligned ($x = 0$). Therefore, as the solution for the problem, the self-decoupled method, which can obtain the L_{m1} and L_{m2} as '0', is proposed.

B. Self-decoupled Dual Pick-up Coils without Core

The self-decoupled dual pick-up coils are composed of coplanar rectangular power cables with core plates, as shown in Fig. 1. To develop the preliminary analytic model, the pick-up coils A and B are simplified as only two pick-up coils without core plates, as shown in Fig. 4.

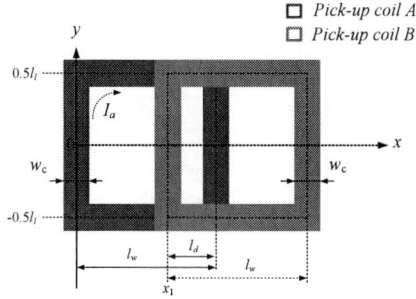

Fig. 4. The top view of the pick-up coil model without core plates.

The mutual magnetic flux ϕ_m is the sum of the ϕ_1, ϕ_2, ϕ_3, and ϕ_4; the ϕ_1, ϕ_2, ϕ_3, and ϕ_4 are induced from the left, right, bottom and top of the pick-up coil A, respectively, as shown in Fig. 5.

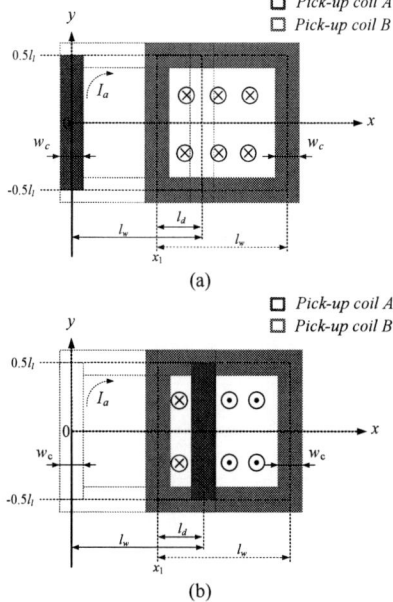

(a)

(b)

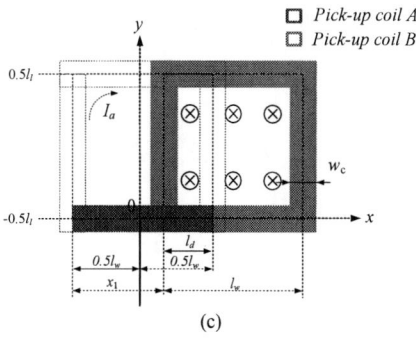

(c)

Fig. 5. The two pick-up coil models for the (a) ϕ_1, (b) ϕ_2 and (c) ϕ_3.

The magnetic flux from the bottom side of the pick-up coil A is equal to the top side, so the top side is not shown in the Fig. 5. In order to make L_{m1} and L_{m2} '0', the ϕ_m should be '0' by changing the left side of the pick-up coil B x_1 to find the magnetic decoupling position where the ϕ_m is '0'.

$$L_m = \frac{\phi_m}{I_a} . \tag{4}$$

The ϕ_1 is obtained as

$$\phi_1 = \int_{x_1}^{x_1+l_w} \int_{-0.5l_l}^{0.5l_l} B_1(x,y)\,dy\,dx , \tag{5}$$

where the magnetic flux densities $B_1(x,y)$ which can be found using the Biot-Savart Law is defined as

$$B_1(x,y) = \frac{\mu_o I_a}{4\pi x}\left(\frac{y - l_l/2}{\sqrt{(y-l_l/2)^2 + x^2}} - \frac{y + l_l/2}{\sqrt{(y+l_l/2)^2 + x^2}} \right) . \tag{6}$$

Applying (6) into (5), the ϕ_1 is obtained for $0 < x_1 < l_w$ as follow:

$$\phi_1 = -\frac{\mu_o I_a}{2\pi}\left[\sqrt{l_l^2 + x_1^2} + l_w - \sqrt{l_l^2 + (x_1+l_w)^2} - l_l \cdot \ln\frac{l_w + x_1}{x_1} \right] . \tag{7}$$

With the same method as calculated above, all the ϕ_1, ϕ_2, ϕ_3, and ϕ_4 can be obtained as $\phi_m = (\phi_1 + \phi_2 + 2\phi_3)$ and the ϕ_m is determined for $0 < x_1 < l_w$ as follow:

$$\phi_m = -\frac{2\mu_o I_a}{\pi}\left[\frac{2\sqrt{l_l^2 + x_1^2} - \sqrt{l_l^2 + (x_1 + l_w)^2} - \sqrt{l_l^2 + (l_w - x_1)^2}}{2} \right.$$
$$\left. + \frac{l_l}{4}\cdot\ln\frac{x_1^2}{l_w^2 - x_1^2} - \frac{(l_w - x_1)}{2}\cdot(\ln\frac{l_l - 0.5w_c}{0.5w_c} - 2) \right] . \tag{8}$$

The calculated L_m and the simulated L_m using the 3D finite-element analysis (FEA) simulation are shown in Fig. 6; the I_a is set to 1 A, the w_c is set to 0.5 mm, the l_w is set to 50 cm and the l_l is varied from 10 cm to 90 cm for the calculations and the simulations. As indicated in Fig. 6 with the red dotted circles, the L_m can be zero, and it means that the two pick-up coils can be decoupled by only changing the x_1. When the two pick-up coils are decoupled, the equivalent circuit can be obtained, as shown in Fig. 7(a) and the simplified circuit for the pick-up coil A can be obtained, as shown in Fig. 7 (b).

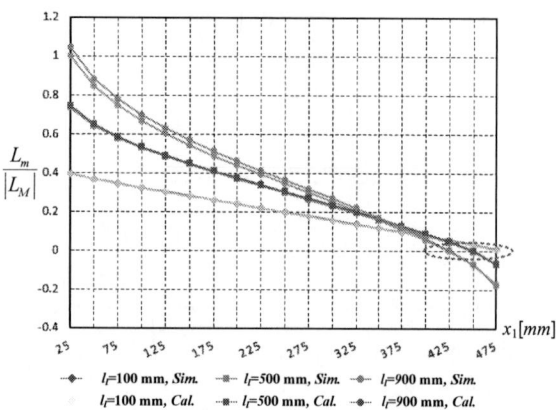

Fig. 6. The 3D FEA simulated L_m without core plates and the calculated L_m for various length l_l of the pick-up coil A and B. The L_M is the maximum mutual inductance between the two pick-up coils for $l_l = 900$ mm, $l_w = 500$ mm and $x_1 = 25$ mm without core plate.

Therefore, the resonant frequency of the pick-up coils for both the aligned and misaligned cases can be derived as

$$\omega_r = \frac{1}{\sqrt{(L_{ma} + L_{la})C_a}}. \qquad (9)$$

The resonant condition does not change for the lateral displacement $0 < x < l_w - 0.5l_d$. Therefore, the lateral tolerance can be increased from $0.5l_d$ to $l_w - 0.5l_d$ by using the self-decoupled dual pick-up coils. In addition, the EMF for pedestrians is negligible when the pick-up coil A is on the power supply rail, because the pick-up coil B which is close to pedestrians is completely decoupled from not only the power supply rail but also the pick-up coil A.

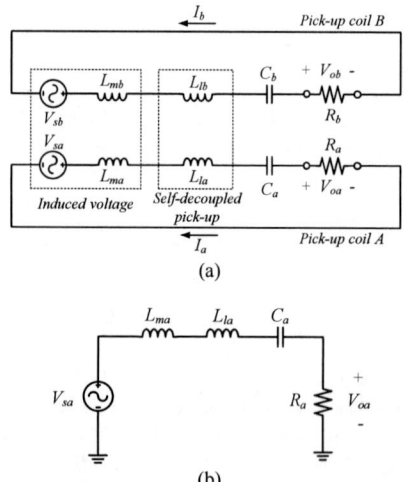

Fig. 7. The equivalent circuit model for the self-decoupled dual pick-up coils.

C. Self-decoupled Dual Pick-up Coils with Core

In practice, core plates on the pick-up coils are used to increase the magnetic coupling between the power supply

rail and the pick-up coils. The core plates have an influence on not only the coupling between the power supply rail and the pick-up coils but also the coupling between the two pick-up coils A and B. Due to its complex structure with the core plates, the magnetic flux density within the pick-up coil B induced by the pick-up coil A is difficult to be directly calculated. To simplify the problem, the magnetic mirror model, which provides the equation of the magnetic flux density when the pick-up coils are on the core plates, is adopted. According to the mirror model, the magnetic flux with core plates is ideally two times larger than the flux without the core plates [24]. The ideal condition of the magnetic mirror model is that the pick-up coils should be attached to core plates and the width of the core plates should be infinitely large. In the proposed dual pick-up coil model, the pick-up coils are just on core plates, so it is reasonable that the magnetic flux closely follows the ideal model of the magnetic mirror model and the ideal model, which have two times large magnetic flux, is adopted. In other words, the mutual magnetic flux between the two pick-up coils with core plates can be obtained by multiplying '2' and the (8), which is the mutual magnetic flux without core plates. And the small error correction factor 'α' is additionally considered as

$$\phi_m = -\frac{4\alpha\mu_o I_a}{\pi}\left[\frac{2\sqrt{l_l^2 + x_1^2} - \sqrt{l_l^2 + (x_1 + l_w)^2} - \sqrt{l_l^2 + (l_w - x_1)^2}}{2}\right.$$
$$\left. + \frac{l_l}{4}\cdot\ln\frac{x_1^2}{l_w^2 - x_1^2} - \frac{(l_w - x_1)}{2}\cdot(\ln\frac{l_l - 0.5w_c}{0.5w_c} - 2)\right]. \quad (10)$$

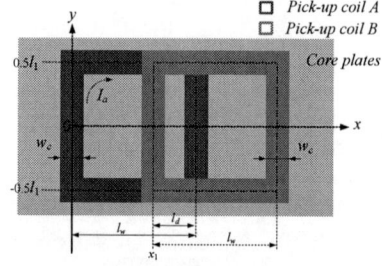

Fig. 8. The top view of the pick-up coil model with core plates.

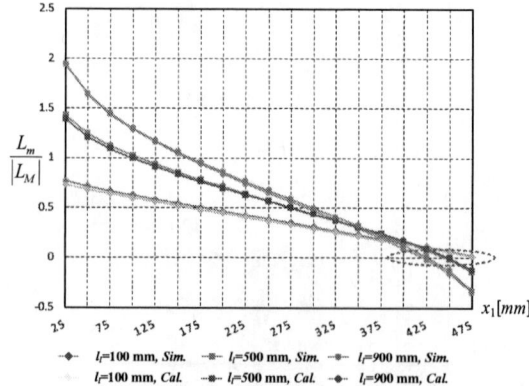

Fig. 9. The 3D FEA simulated L_m with core plates and the calculated L_m for various length l_l of the pick-up coil A and B. The L_M is the maximum mutual inductance between the two pick-up coils for $l_l = 900$ mm, $l_w = 500$ mm and $x_1 = 25$ mm with core plates.

To verify the model (10), the 3D FEM simulations were done and its results were compared with the (10), as shown in Fig. 9, and the 'α' is obtained as 0.925 from the calculation results. As shown in Fig. 9, although the L_m with the core plate is about two times larger than the without core plate case, the mutual decoupling point is nearly same in both cases.

III. EXPERIMENTAL VERIFICATIONS

The proposed self-decoupled pick-up coils with and without core plates were fabricated for experiments, as shown in Fig. 10. Each of the pick-up coil sets includes two pick-up coils A and B, and the pick-up width l_w and length l_l are determined as 62 cm and 36 cm, respectively, and the 20 turns are used for the each pick-up coil set.

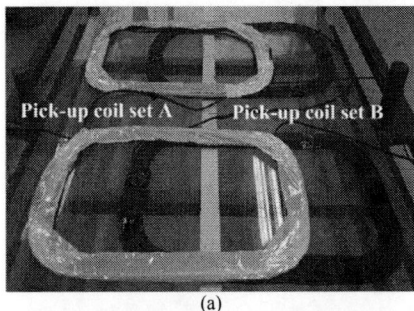

(a)

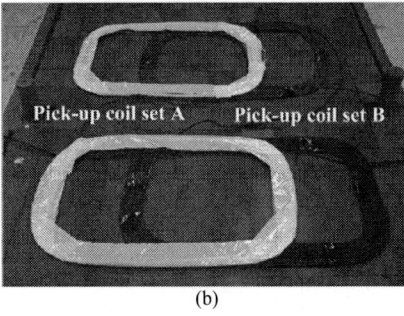

(b)

Fig. 10. Proposed self-decoupled pick-up coil sets for experiments. (a) Pick-up coil sets without core plates. (b) Pick-up coil sets with core plates.

The mutual inductances between the pick-up coil set A and the pick-up coil set B along with the x_1 were measured with and without core plates, as shown in Fig. 11. As expected from the magnetic mirror model and FEA simulation results, the mutual inductance with core plates is nearly two times larger than that without core plates and the x_{1_d} to meet the condition $L_m=0$ is almost same for the both cases with and without core plates. Here, the x_{1_d} shows a little discrepancy between the experiment results and simulation results due to the longer l_w and the additional effects on the magnetic fluxes from the adjacent pick-up coils A' and B', which were not considered in the simulations for simple analysis.

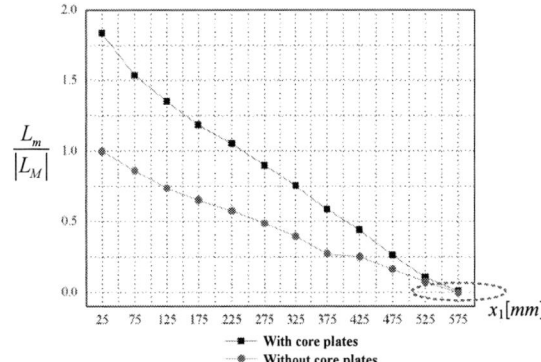

Fig. 11. Measured mutual inductances for various position of the pick-up B coils with/without core plates. The L_M is the maximum mutual inductance between the two pick-up coils for $l_l = 360$ mm, $l_w = 620$ mm and $x_1 = 25$ mm without core plates.

IV. CONCLUSIONS

The proposed self-decoupled dual pick-up coils for the I-type power supply rail of roadway powered electric vehicle (RPEV), which achieved the large air-gap as well as the lateral tolerance, and the low EMF for pedestrians, have been verified by calculations, FEA simulations and experiments in this paper. Throughout this paper, it is found that the adjacent pick-up coils can be decoupled regardless of the existence of core plates by overlapping the pick-up coils and the overlapping is nearly same for the both cases with and without core plates.

REFERENCES

[1] J. G. Bolger, "Urban electric transportation systems: the role of magnetic power transfer," *IEEE WESCON94 Conference*, 1994, pp. 41-45.

[2] G. A. Covic and J. T. Boys, "Modern trends in inductive power transfer for transportation applications," *IEEE Journal of Emerging and Selected Topics in Power Electronics*, vol. 1, no. 1, pp. 28-41, Mar. 2013.

[3] J. Meins, "German activities on contactless inductive power transfer," *IEEE Energy Conversion Congress and Exposition (ECCE), 2013.*

[4] Omer C. Onar, John M. Miller, Steven L. Campbell, Chester Coomer, Cliff. P. White, and Larry E. Seiber, "A novel wireless power transfer for in-motion EV/PHEV charging," *IEEE Applied Power Electronics Conference & Exposition (APEC)*, 2013, pp. 3073-3080.

[5] John M. Miller, Omer C. Onar, and P. T. Jones, "ORNL developments in stationary and dynamic wireless charging," *IEEE Energy Conversion Congress and Exposition (ECCE), 2013.*

[6] N. P. Suh, D. H. Cho, and Chun T. Rim, "Design of on-line electric vehicle (OLEV)," *Plenary lecture at the 2010 CIRP Design Conference*, 2010, pp. 3-8.

[7] S. W. Lee, J. Huh, C. B. Park, N. S. Choi, G. H. Cho, and Chun T. Rim, "On-line electric vehicle (OLEV) using inductive power transfer system," *IEEE Energy Conversion Congress and Exposition (ECCE)*, 2010, pp. 1598-1601.

[8] J. Huh, S. W. Lee, C. B. Park, G. H. Cho, and Chun T. Rim, "High performance inductive power transfer system with narrow rail width for on-line electric vehicles," *IEEE Energy Conversion Congress and Exposition (ECCE)*, 2010, pp. 647-651.

[9] J. Huh, S. W. Lee, W. Y. Lee, G. H. Cho, and Chun T. Rim, "Narrow-width inductive power transfer system for on-line electrical vehicles (OLEV)," *IEEE Trans. on Power Electron.*, Vol. 26, No. 12, pp. 3666-3679, Dec. 2011.

[10] S. W. Lee, W. Y. Lee, J. Huh, H. J. Kim, C. B. Park, G. H. Cho, and Chun T. Rim, "Active EMF cancellation method for I-type pick-up of On-Line Electric Vehicles (OLEV)," *IEEE Applied*

Power Electronics Conference & Exposition (APEC), 2011, pp. 1980-1983.

[11] Su Y. Choi, J. Huh, W. Y. Lee, S. W. Lee, and Chun T. Rim, "New cross-segmented power supply rails for roadway powered electric vehicles," *IEEE Trans. on Power Electron*, Vol. 28, no. 12, pp. 5832-5841, Dec. 2013.

[12] Su Y. Choi, J. Huh, Woo Y. Lee, Jung G. Cho, and Chun T. Rim, "Asymmetric coil sets for wireless stationary EV chargers with large lateral tolerance by dominant field analysis," *IEEE Trans. on Power Electron.*, Accepted for publication.

[13] Su Y. Choi, Beom W. Gu, Sung W. Lee, Woo Y. Lee, J. Huh, and Chun T. Rim, "Generalized active EMF cancel methods for wireless electric vehicles," *IEEE Trans. on Power Electron.*, Accepted for publication.

[14] Su Y. Choi, Beom W. Gu, Seog Y. Jeong, and Chun T. Rim "The history of wireless power transfer systems for roadway powered electric vehicles," *IEEE Journal of Emerging and Selected Topics in Power Electronics*, Submitted.

[15] N. P. Suh, D. H. Cho, Chun T. Rim, J. W. Kim, G. H. Jung, J. Huh, K. H. Lee, Y. D. Son, J. Y. Choi, E. H. Park, Y. J. Cho, J. C. Jang, Y. H. Kim, and H. G. Kim, "Collector device for electric vehicle with active cancellation of EMF," KR Patent 10-1038759, May 27, 2011.

[16] Chun T. Rim and G. H. Cho, "New approach to analysis of quantum rectifier-inverters," *IEEE Electronic Letters*, Vol. 25, No. 25, pp. 1744-1745, Dec. 1989.

[17] Chun T. Rim, "Unified general phasor transformation for AC converters," *IEEE Trans. on Power Electronics*, Vol. 26, No. 9, pp. 2465-2475, Sept. 2011.

[18] Jin Huh, W. Y. Lee, Su Y. Choi, G, H, Cho, and Chun T. Rim, "Frequency-domain circuit model and analysis of coupled magnetic resonance systems," *Journal of Power Electronics*, Vol. 13, No. 2, March, 2013.

[19] Sung W. Lee, B. Choi, and Chun T. Rim, "Dynamics characterization of the inductive power transfer system for on-line electric vehicles by Laplace phasor transform," *IEEE Trans. on Power Electronics*, Vol. 28, No. 12, pp. 5902-5909, Dec. 2013.

[20] International Commission on Non-Ionizing Radiation Protection, "Guidelines for limiting exposure to time-varying electric and magnetic fields (1 Hz to 100 kHz)," Health physics 99(6), pp. 818-836, 2010.

[21] D. Kwiat, S. Saoub, and S. Einav, "Calculation of the mutual induction between coplanar circular surface coils in magnetic resonance imaging," *IEEE Trans. Biomedical Engineering*, vol. 39, pp. 433-436, 1992.

[22] G. A. Covic, M. L. G. Kissin, D. Kacprzak, N. Clausen, and H. Hao, "A bipolar primary pad topology for EV stationary charging and highway power by inductive coupling," *IEEE Energy Convers. Congr. Expo.*, 2011, pp. 1832-1838.

[23] A. Zaheer, D. Kacprzak, and G. A. Covic, "A bipolar receiver pad in a lumped IPT system for electric vehicle charging applications," *IEEE Energy Convers. Congr. Expo.*, 2012, pp. 283-290.

[24] W. Y. Lee, J. Huh, S. Y. Choi, X. V. Thai, J. H. Kim, E. A. Al-Ammar, M. A. El-Kady, and Chun T. Rim, "Finite-width magnetic mirror models of mono and dual coils for wireless electric vehicles," *IEEE Trans. Power Electron.*, vol. 28, no. 3, pp. 1413-1428, Mar. 2013.

Contactless Power Transfer System Suitable for Low Voltage and Large Current Charging for EDLCs

Takahiro Kudo, Takahiro Toi, Yasuyoshi Kaneko, Shigeru Abe

Department of Electrical and Electronic Systems
Saitama University
Saitama, Japan

Abstract— The efficiency of a contactless power transformer is determined by the output load. We propose a high-efficiency contactless power transfer system for a low-voltage and large-current load. The proposed system is composed of a contactless power transfer system with series and series resonant capacitors and a current-doubler rectifier circuit. This paper presents the characteristic of the proposed system that charged electric double layer capacitors as a low-voltage and large-current load, as well as the experimental results.

Keywords— *Contactless power transfer system, current-doubler rectifier, EDLCs, efficiency*

I.   INTRODUCTION

A contactless power transfer system has many advantages, including the convenience of being cordless and its safety during high-power charging [1, 2]. Because no wear, spark, or contact failure occurs in a contactless power transfer system, it has been applied to practical use in home electric appliances [3, 4]. Charging of electronic devices has become easy using the contactless power transfer system. Therefore, the capacity of a power storage device can possibly be reduced by increasing the number of charging times.

The energy capacity of an electric double layer capacitor (EDLC) is smaller than that of a secondary battery, but the power density of EDLCs is higher than that of the secondary battery [5]. In addition, the lifetime of EDLCs is longer than that of the secondary battery because no chemical change occurs during charging and discharging.

Conventionally, the battery used in an automated guided vehicle (AGV) is a secondary battery. However, the method of using a contactless power transfer system with EDLCs has recently attracted attention because it can solve the issues of short battery lifetime and of the charging time [6]. If a battery is charged quite often by a contactless power transfer system during luggage loading and unloading, the EDLCs can be applied in the AGV. To increase the charging efficiency at large current, EDLCs with low internal resistance must be used.

In this study, we consider a circuit for a contactless power transfer system with EDLCs. We also investigate the effectiveness of this system by experiments.

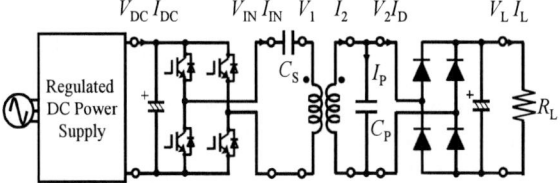

Fig. 1. Contactless power transfer system with SP topology.

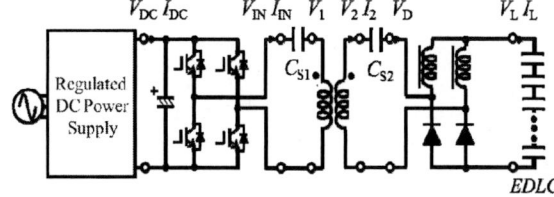

Fig. 2. Contactless power transfer system with SS topology.

Two issues must be addressed in the contactless power transfer system with EDLCs. First, the efficiency of a contactless power transformer reduces during the EDLC charging. A contactless power transfer system has a maximum-efficiency load resistance R_{Lmax}. Because the EDLC charging for the AGV has a low voltage and large current, the equivalent load resistance R_L during charging is significantly smaller than R_{Lmax}. Therefore, the efficiency of a contactless power transformer significantly decreases during the EDLC charging. Second, in charging the EDLCs, a constant current is desired. In a constant-voltage charging of EDLCs, the theoretical charging efficiency is a very low at 50%. Therefore, a constant-current charging is desirable for the output of a contactless power transfer system.

To solve these issues, development of a system is required, which combines a current-doubler rectifier and a contactless power transfer system with primary and secondary series resonant capacitors (SS topology) [7, 8]. R_{Lmax} of the SS topology is k^2 (k is a coupling factor) times smaller than that of the conventional topology of the primary series and secondary parallel resonant capacitors (SP topology) [9].

Furthermore, the equivalent load resistance R_L, which includes that of the current-doubler rectifier circuit, is 1/16 times smaller than that of a full-wave rectifier circuit. By adopting a current-doubler rectifier circuit and the SS

978-1-4799-2706-7/14 $31.00 © 2014 IEEE   1109

topology, the contactless power transfer system can achieve a high-efficiency EDLC charging.

In addition, the SS topology possesses the characteristic of an immittance converter, and constant-current charging is easy. Therefore, the SS topology is suitable for constant-current charging of the EDLCs.

In this paper, we clarify the efficiency issues of a contactless power transfer system for a low-voltage and large-current load. Then, we propose a new method, which uses the current-doubler rectifier circuit and SS topology in lieu of the full-wave rectifier circuit and the SP topology. Further, we show that the immittance converter characteristic of the SS topology is suitable for constant-current charging of the EDLCs. We investigate the characteristic of the proposed method by experiments and demonstrate its usefulness and novelty.

Section II describes the comparison between the SP and SS topologies in a contactless power transfer system. Section III describes the contactless power transfer system for the EDLC charging. Section IV presents the comparison between the theoretical and experimental values.

II.   COMPARISON OF THE SP AND SS TOPOLOGIES

A. Series and Parallel Resonant Capacitor Topology

Figure 1 shows the schematic diagram of a contactless power transfer system with an SP topology. A full-bridge inverter is used as a high-frequency power supply. A full-wave rectifier is used as a rectifier circuit at the secondary side to increase the efficiency. The cores are made of ferrite, and litz wires are used for the windings. Figure 2 shows the schematic diagram of a contactless power transfer system with an SS topology. A current-doubler rectifier is used as a rectifier circuit on the secondary side as a low-voltage and large-current load to increase the EDLC efficiency.

Figure 3 shows the detailed equivalent circuit of the SP topology. It consists of a T-shaped equivalent circuit in which primary series capacitor C_S, secondary parallel capacitor C_P, and load resistance R_L are added. The primary values are converted into secondary equivalent values using the turn ratio $a = N_1/N_2$. Because winding resistances r'_1 and r_2 and ferrite core loss r'_0 are considerably lower than leakage reactances x'_1 and x_2 and mutual reactance x'_0 at the resonant frequency, the ferrite core loss is ignored. Here, M is the mutual inductance ($x'_0 = \omega_0 M/a$). The rectifier is also omitted, and the secondary circuit for the analysis consists of C_P and R_L.

To achieve resonance at the input frequency f_0 ($= \omega_0/2\pi$) with the self-inductance of the secondary winding L_2, which is equivalent to the addition of mutual reactance x'_0 and leakage reactance x_2, the secondary parallel capacitor C_P is given by

$$\frac{1}{\omega_0 C_P} = \omega_0 L_2 = x_P = x'_0 + x_2 . \tag{1}$$

The primary series capacitor C_S (C'_S denotes its secondary equivalent) is determined by

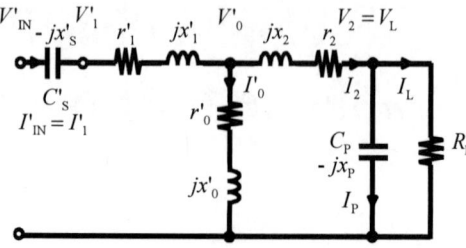

Fig. 3. Detailed equivalent circuit for the SP topology.

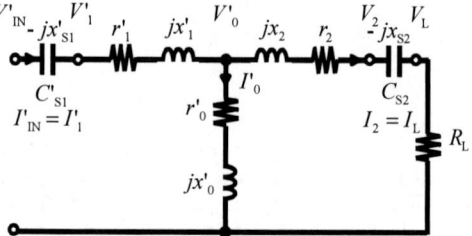

Fig. 4. Detailed equivalent circuit for the SS topology.

$$\frac{1}{\omega_0 C'_S} = x'_S = x'_1 + \frac{x'_0 x_2}{x'_0 + x_2} . \tag{2}$$

V_{IN} and I_{IN} can be expressed as

$$V'_{IN} = bV_2 = bV_L, \quad I'_{IN} = I_L/b, \quad b = \frac{x'_0}{x'_0 + x_2}. \tag{3}$$

These equations suggest that the equivalent circuit of a transformer with these capacitors is the same as that of an ideal transformer at the resonant frequency.

By ignoring the ferrite core loss ($r_0 = 0$), the efficiency can be approximated by

$$\eta_{SP} = \frac{R_L I_L^2}{R_L I_L^2 + r'_1 I_{IN}^2 + r_2 I_2^2} = \frac{R_L}{R_L + \frac{r'_1}{b^2} + r_2 \left\{ 1 + \left(\frac{R_L}{x_P} \right)^2 \right\}}. \tag{4}$$

The maximum efficiency η_{maxSP} is obtained when $R_L = R_{LmaxSP}$.

$$R_{LmaxSP} = x_P \sqrt{\frac{1}{b^2} \frac{r'_1}{r_2} + 1}, \eta_{maxSP} = \frac{1}{1 + \frac{2r_2}{x_P} \sqrt{\frac{1}{b^2} \frac{r'_1}{r_2} + 1}} \tag{5}$$

The coupling factor k, primary winding Q_1, and secondary winding Q_2 are represented by

$$k = \frac{M}{\sqrt{L_1 L_2}}, \quad Q_1 = \frac{\omega L_1}{r_1}, \quad Q_2 = \frac{\omega L_2}{r_2}. \tag{6}$$

Here, L_1 is the self-inductance of the primary winding ($L_1 = a^2(x'_1 + x'_0)/\omega_0$).

Then, these equations can be expressed using k and Q.

$$R_{Lmax SP} = \frac{r_2 Q_2}{k} \sqrt{\frac{Q_2}{Q_1}} \quad \eta_{maxSP} = \frac{1}{1 + \frac{2}{k\sqrt{Q_1 Q_2}} \sqrt{1 + \frac{Q_1}{Q_2} k^2}}. \tag{7}$$

If k is smaller than 0.3 and $Q_1 \cong Q_2$, then

$$\frac{1}{k^2} \frac{Q_2}{Q_1} \gg 1 . \tag{8}$$

978-1-4799-2706-7/14 $31.00 © 2014 IEEE   1110

Thus, these equations can be expressed using k and Q, i.e.,

$$R_{\text{Lmax SP}} = \frac{r_2 Q_2}{k} \sqrt{\frac{Q_2}{Q_1}} \quad \eta_{\text{max SP}} \cong \frac{1}{1 + \dfrac{2}{k\sqrt{Q_1 Q_2}}} . \quad (9)$$

To increase the efficiency, Q and k must be increased, and we can clearly see that a coil with a high Q value has a small loss.

B. Series and Series Resonant Capacitor Topology

Figure 4 shows the typical detailed equivalent circuit for the SS topology. It consists of a T-shaped equivalent circuit in which primary series capacitor C_{S1}, secondary series capacitor C_{S2}, and R_L are added.

To achieve resonance with the self-inductance of the primary winding L_1 and the secondary winding L_2, C_{S1} and C_{S2} are given by

$$\frac{1}{\omega_0 C'_{\text{S1}}} = x'_{\text{S1}} = x'_0 + x'_1 \quad \frac{1}{\omega_0 C_{\text{S2}}} = x_{\text{S2}} = x'_0 + x_2 \quad (10)$$

and V'_{IN} and I'_{IN} can be expressed as

$$V'_{\text{IN}} = -jx'_0 I_L \quad I'_{\text{IN}} = -j\frac{1}{x'_0} V_L . \quad (11)$$

These equations suggest that the equivalent circuit of a transformer with capacitors has the same characteristics as the immittance converter at the resonant frequency.

Ignoring the ferrite core loss ($r_0 = 0$), the efficiency can be approximated by

$$\eta_{\text{ss}} = \frac{R_L I_L^2}{R_L I_L^2 + r'_1 I'^2_1 + r_2 I_2^2} = \frac{R_L}{R_L + r_2 + r'_1 \left(\dfrac{R_L}{x'_0}\right)^2} . \quad (12)$$

The maximum efficiency η_{maxSS} is obtained when $R_L = R_{\text{LmaxSS}}$.

$$R_{\text{Lmax SS}} = x'_0 \sqrt{\frac{r_2}{r'_1}} , \eta_{\text{max SS}} = \frac{1}{1 + \dfrac{2r_2}{x'_0}\sqrt{\dfrac{r'_1}{r_2}}} \quad (13)$$

Then, these equations can be expressed using k and Q.

$$R_{\text{Lmax SS}} = kr_2 \sqrt{Q_1 Q_2} , \eta_{\text{max SS}} = \frac{1}{1 + \dfrac{2}{k\sqrt{Q_1 Q_2}}} . \quad (14)$$

η_{maxSS} in (14) is almost equal to η_{maxSP} in (9). In addition, R_L of the SS topology is k^2 ($k < 1$) times smaller than that of the SP topology when Q_1 and Q_2 are approximately equal. For example, in the system with $k = 0.3$, R_{Lmax} of the SS topology is approximately 1/10 times smaller than that of the SP topology.

III. CONTACTLESS POWER TRANSFER SYSTEM FOR EDLC CHARGING

A. Summary of the Proposed System using EDLCs

Figure 5 shows a contactless power transfer system for EDLC charging. The contactless power transformer with the SS topology is connected to a full-bridge inverter power supply of a commercial power supply input as well as to a current-doubler rectifier circuit with two inductors and two diodes. The average voltage of the EDLC is V_L, and its charging power is P_L. Here, the EDLC is represented by the equivalent load resistance R_L (= $V_L{}^2 / P_L$). When the inverter is

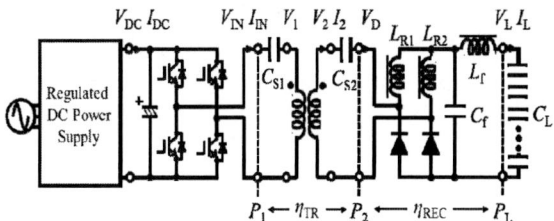

Fig. 5. Experimental circuit for the EDLC charging.

TABLE I
COMPARISON OF THE CHARACTERISTICS

System	R_{Lmax}	Output of rectifier circuit		R_L
		Voltage V_L	Current I_L	
SS and current-doubler rectifier circuit	$kr_2 Q$	$\dfrac{1}{2\sqrt{2}}V_D$	$2\sqrt{2}I_2$	$\dfrac{kr_2 Q}{8}$
SP and full-wave rectifier circuit	$\dfrac{r_2 Q}{k}$	$\sqrt{2}V_2$	$\dfrac{1}{\sqrt{2}}I_D$	$\dfrac{2r_2 Q}{k}$

Fig. 6. Transformer.

TABLE II
TRANSFORMER SPECIFICATION

Transformer	
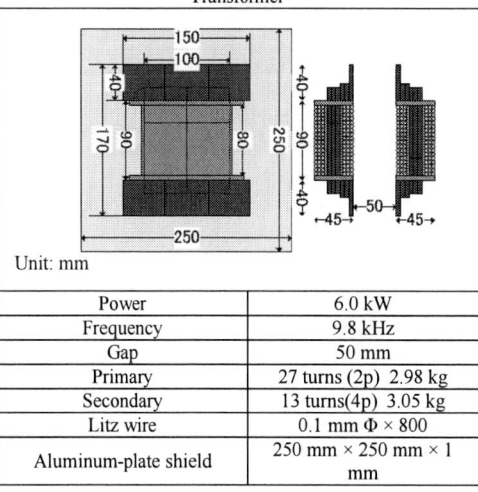 Unit: mm	
Power	6.0 kW
Frequency	9.8 kHz
Gap	50 mm
Primary	27 turns (2p) 2.98 kg
Secondary	13 turns(4p) 3.05 kg
Litz wire	0.1 mm $\Phi \times 800$
Aluminum-plate shield	250 mm × 250 mm × 1 mm

driven at constant voltage, the contactless transformer with the SS topology exhibits immittance converter characteristics. Therefore, the current in the current-doubler rectifier circuit is constant, and the EDLC is charged by a constant current.

B. Comparison with Conventional Contactless Power Transfer System (SP topology + Full bridge)

Figure 1 shows the configuration of a conventional contactless power transfer system for electric vehicles. The contactless power transformer with the SP topology is connected to the full-bridge inverter power supply with a commercial power supply input as well as to the full-wave circuit.

The EDLC voltage V_L used in the AGV is low, and the charging power P_L is large. Therefore, the equivalent load resistance R_L $(= V_L^2/P_L)$ is very small. To drive a contactless power transformer at high efficiency, decreasing the load resistance R_{Lmax} at maximum transformer efficiency is necessary. In the conventional system (Figure 1), the number of turns in the secondary windings can be changed to adjust R_{Lmax}. However, in the case of small R_L of the EDLC, adjusting R_{Lmax} is difficult. Table I lists the comparison of R_{Lmax} of the proposed system (Figure 2) with that of the conventional system (Figure 1). Table I also shows R_L at maximum efficiency, which includes that of the rectifier circuit.

First, we make a comparison of the capacitor topologies. If Q_1 and Q_2 are represented as Q ($Q_1 = Q_2$), R_{Lmax} of the SS topology is k^2 ($k < 1$) times smaller than that of the SP topology.

Second, we make a comparison of the rectifier circuits. To drive a contactless power transformer at high efficiency, the load resistance, which includes that of the rectifier circuit, must be equal to R_{Lmax}. Therefore, we consider the load resistance, including that of the rectifier circuit. Table I lists the output voltage and output current of each rectifier circuit using the fundamental-wave effective value of the input voltage and input current. In the full-wave rectifier circuit, the peak value of the input voltage is considered as an output voltage. Therefore, the load resistance, which includes that of the full-wave rectifier circuit, is 0.5 times smaller than the equivalent load resistance R_L. When $R_L = 2R_{Lmax}$, the transformer efficiency is maximum.

On the other hand, in the current-doubler rectifier, the output current is twice that of the peak value of the input current. Therefore, the load resistance, including that of the current-doubler rectifier circuit, is eight times smaller than R_L. When $R_L = 1/8R_{Lmax}$, the transformer efficiency is maximum.

In comparison with the combination of the capacitor topology and the rectifier circuit, R_L of the proposed system (SS topology + current-doubler rectifier circuit) is $k^2/16$ times smaller than that of the conventional system (SP topology + full-wave rectifier circuit). Therefore, driving a contactless power transformer at high efficiency is possible when the EDLC is charged.

IV. EXPERIMENTAL RESULT

A. Specification of the Contactless Power Transformer

Figure 6 shows the contactless power transformer for EDLC charging. Table II lists the transformer specification. The cores are made of ferrite, and litz wires are used for the

TABLE III
TRANSFORMER PARAMETERS

f_0 [kHz]	9.8	k	0.32
r_1 [mΩ]	88.7	b	0.32
r_2 [mΩ]	20.6	R_{LmaxSS} [Ω]	1.12
l_0 [μH]	73.9	R_{LmaxSP} [Ω]	11.2
l_1 [μH]	157.2	R'_{LmaxSS} [Ω]	0.14
l_2 [μH]	37.1	η_{max} [%]	96.2
C_{S1} [μF]	1.14	Q_1	160
C_{S2} [μF]	4.87	Q_2	162

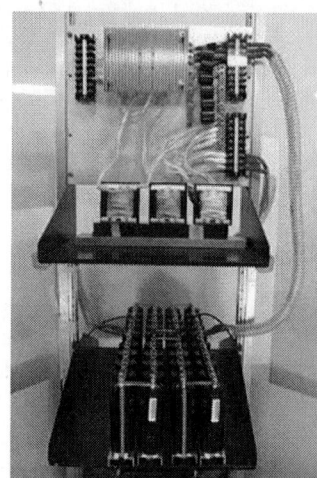

Fig. 7. Current-doubler rectifier with LC filter and EDLCs.

TABLE IV
CURRENT-DOUBLER RECTIFIER AND EDLC SPECIFICATIONS

Current-doubler rectifier circuit		EDLCs*	
L_{R1} [μH]	100	C_L [F]	175
L_{R2} [μH]	100	DCR [mΩ]	16
C_f [μF]	108	R_{leak} [kΩ]**	1.34
L_f [μH]	16	$R_{divider}$ [Ω]	100

*Low-resistance EDLCs made by NICHICON CORPORATION

**Average value of one cell

windings. In addition, the leakage flux is shielded using an aluminum plate attached to the back of the transformer. Table III lists the transformer parameters. The maximum efficiency η_{max} and load resistance R_{Lmax} of the two topologies are calculated from (9) and (14). We found that R_{LmaxSS} is much smaller than R_{LmaxSP}. The equivalent load resistance R'_{LmaxSS} is very small at 0.14 Ω by combining the current-doubler rectifier circuit and the SS topology. Driving a contactless power transformer at high efficiency is possible when the EDLC is charged.

B. Specification of the Current-doubler Rectifier and EDLC

Figure 5 shows the experimental circuit for the EDLC charging. Figure 7 shows the current-doubler rectifier circuit and the EDLC. Table IV lists their specifications.

The 2014 International Power Electronics Conference

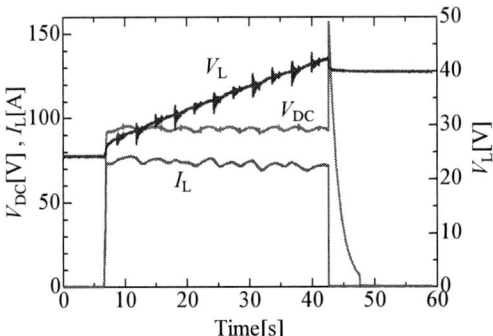

Fig. 8. Charging voltage and current for EDLCs.

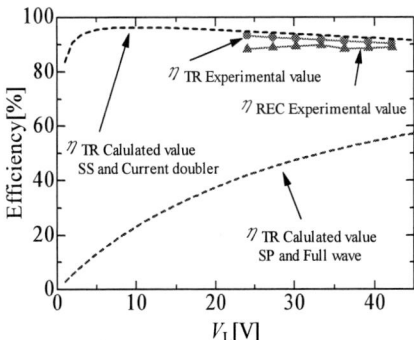

Fig. 10. Efficiency of each topology.

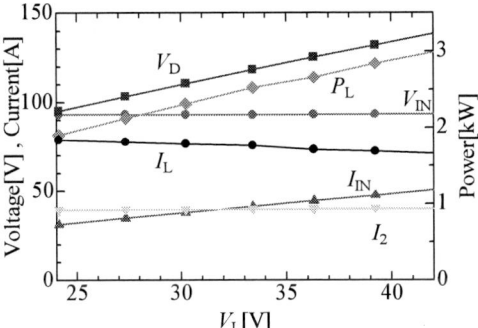

Fig. 9. Experimental results for transformer.

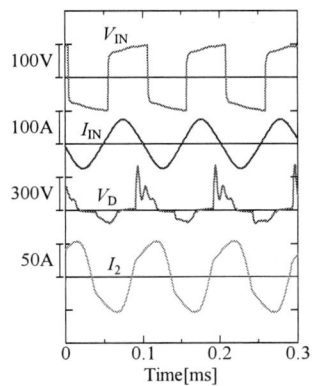

Fig. 11. Waveforms of transformer ($C_{S2} = 4.87\ \mu$ F).

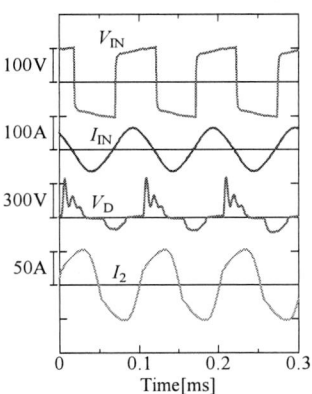

Fig. 12. Waveforms of the transformer ($C_{S2} = 3.7\ \mu$ F).

The current-doubler rectifier circuit and the EDLC are designed by considering that the inductor current is one-half of the output current, and the diode current is equal to the output current. In addition, to protect the EDLC from heating by the current ripple, an LC filter is connected to the rectifier circuit output.

The EDLC rated voltage is 2.5 V, and the rated capacity of one EDLC is 1750 F. The EDLCs have a low resistance and are connected in a 20-series, 2-parallel configuration, for a total of 40 EDLCs. Therefore, the capacity of the EDLC C_L is 175 F, and the internal resistance (direct-current resistance) is 16 mΩ. To prevent voltage unbalance due to variations in the leakage resistance and capacitance, a dividing resistor (R_{divider} = 100 Ω) is connected parallel to each cell.

A full-bridge inverter is used as a high-frequency power supply at 9.8 kHz. The capacitance of the connected capacitor is calculated from (9). The charging voltage V_L of the EDLC is 24–42 V, and the charging current I_L is 75 A.

C. Experimental Result for EDLC Charging

The experimental results are shown below. Figure 8 shows the time course of the charging voltage and current of the EDLCs. Figure 9 shows the various voltages and currents of the transformer and charging power P_L. Figure 10 shows the efficiency of each part during the charging. Figure 11 shows the waveforms of the transformer at V_L = 42 V (C_S = 4.87 μF)

Figures 8 and 9 show that charging current I_L of the EDLC is 75 A, and rapid charging is completed in approximately 35 s. When the input voltage of the inverter V_{DC} is constant, I_L is

TABLE V
FUNDAMENTAL POWER FACTOR OF THE TRANSFORMER
INPUT AND OUTPUT

C_{S2} [μF]	Transformer input power factor	Transformer output power factor	η_{TR} [%]
4.87	0.863	0.746	90.3
3.7	0.944	0.75	92.2

constant. This contactless power transfer system exhibits immittance converter characteristic, which achieves constant-current output using constant-voltage driving.

Figure 10 shows the transformer efficiency and the rectifier circuit efficiency during charging. The dashed lines are

978-1-4799-2706-7/14 $31.00 © 2014 IEEE

obtained from (4) and (11). R_L is the load resistance, including that of the rectifier circuit. Figure 10 shows that the proposed system has high efficiency for a low-voltage and large-current load EDLC.

From the experimental results, the average transformer efficiency is 91.7%, and the average rectifier circuit efficiency is 88.9%. When the EDLCs are charged by the conventional system (SP topology + full-wave rectifier circuit), significant reduction in the transformer efficiency is apparent.

D. Influence of Inductor in the Current-doubler Rectifier

The current-doubler rectifier circuit has two inductors. Therefore, we investigate the effects of the inductor on the entire circuit. Because the inductor component is included in the load of the proposed system, it becomes an inductive load. Therefore, the phase of the current is advanced relative to that of the voltage at the transformer input by approximately 28.2° (Figure 11). The advanced phase of the current causes failure of the inverter power supply. Changing the value of the secondary series capacitor C_{S2} will prevent excessive advance of the current phase. C_{S2} is then changed from 4.87 to 3.7 μF.

Figure 12 shows the experimental results of the transformer waveform with V_L = 42 V (C_{S2} = 3.7 μF). These waveforms are subjected to a discrete Fourier transform processing, and the extracted fundamental-wave power factor is 9.8 kHz. Table V lists the results when C_{S2} is changed. Figure 12 shows that the advanced phase of the current decreases to 14.1° from 28.2°. The fundamental wave power factor also improves to 0.944 from 0.863. The average transformer efficiency improves to 93.3%, and the average rectifier circuit efficiency improves to 91.2% with the improvement in the fundamental-wave power factor.

From the above results, the effects of the inductor must be considered when a current-doubler rectifier circuit is used as a load of the contactless power transfer system. The power factor at the transformer input needs to be monitored. However, this problem is solved by changing the value of the secondary series capacitor C_{S2}.

V. CONCLUSION

When EDLCs are to be charged at low voltage and large current, the current-doubler rectifier circuit and contactless power transfer system with an SS topology are suitable. In the proposed method, we have demonstrated that the contactless power transfer system has high efficiency. The constant-current charging of EDLCs is easier than that of the conventional system (SP topology + full-wave rectifier circuit).

When a current-doubler rectifier circuit is used, a problem arises in that the phase of the current is advanced compared with that of the voltage at the inverter output. However, by adjusting the value of the secondary series capacitor C_{S2} of the contactless power transfer system, the problem can be solved.

We have validated the performance of the proposed system by rapid-charging experiments of EDLCs. A 175-F EDLC was charged in 35 s at a constant current of 75 A by a contactless power transformer with a gap length of 50 mm and a 6-kW rating. As a result, the achieved average transformer efficiency was as high as 93.3%.

REFERENCES

[1] M. Chigira, Y. Nagatsuka, Y. Kaneko, S. Abe, T. Yasuda, and A. Suzuki, "Small-size light-weight transformer with new core structure for contactless electric vehicle power transfer system," *in Proc. IEEE ECCE 2011*, pp. 260-266, 2011.

[2] K. Kusaka and J. Itoh, "Input impedance matched AC-DC converter in contactless power transfer for EV charger," ICEMS2012, LS2A-2, 2012.

[3] G. A. J. Elliott, S. Raabe, G. A. Covic, and J. T. Boys, "Multi-phase pick-ups for large lateral tolerance contactless power transfer systems," *IEEE Trans. Ind. Electron.*, vol. 57, no. 5, pp. 1590-1598, 2010.

[4] A. Zaheer, M. Budhia, K. Kacprzak, and G. A. Covic, "Magnetic design of a 300W under-floor contactless power transfer system," *in Proc. 37th Annu. Conf. IEEE Ind. Electron. Soc.*, Melbourne, Australia, Nov. 7–10, pp. 1343-1348, 2011.

[5] M. Okamura, "Battery system using electric double-layer capacitor," *J. IEE Jpn.*, vol. 120, no. 10, pp. 610-613, 2000.

[6] T. Tabata, T. Yamamoto, S. Hori, and N. Inagaki, "Trial production and efficiency evaluation of wireless charging system for AGV," *in Proc. IEICE Technical Committee on Wireless Power Transfer*, WPT2012-15, 2012.

[7] H. Nakano, Y. Higuchi, and K. Hirachi, "Current doubler rectifier," *IEEJ Trans. IA*, vol. 116, no. 10, pp. 1081-1082, 1996.

[8] T. Imura, T. Uchida, and Y. Hori, "Basic experimental study on helical antennas of contactless power transfer for electric vehicles by using magnetic resonant couplings," *in Proc. IEEE Vehicle Power and Propulsion Conference*, pp. 936-940, 2009.

[9] T. Tohi, Y. Kaneko, and S. Abe, "Maximum efficiency of contactless power transfer systems using k and Q," *in Proc. IEEJ Technical Meeting on Semiconductor Power Converter*, SPC11-179, 2011.

Excitation System by Contactless Power Transfer System with the Primary Series Capacitor Method

Ryosuke Nozawa, Ryota Kobayashi, Hikaru Tanifuji, Yasuyoshi Kaneko, Shigeru Abe
Department of Electrical and Electronic Systems
Saitama University
Saitama, Japan

Abstract—**For power transfer to the rotor circuit of excitation-type synchronous motors, we propose a contactless power transfer system with a primary series capacitor (S topology) to compensate for the leakage reactance. This system does not contain secondary resonant and smoothing capacitors and can reduce the number of components in the rotor circuit. The value of the primary resonant capacitor is determined such that the input power factor is set to one for the operating frequency. The rotating shaft of the rotor is covered with an aluminum sheet to reduce the loss due to the leakage flux. Power transfer tests were performed. A high efficiency of 91.8% was achieved when power was transferred to the rotor circuit of the synchronous motor.**

Keywords—*contactless power transfer system, electromagnetic shielding, rotary transformer, synchronous motor*

I. INTRODUCTION

We study an excitation-type synchronous motor using a contactless power transfer system for the partial development of a variable magnetic flux motor. Conventionally, slip rings and a brushless excitation system have been used as the excitation method in excitation-type synchronous motors and generators. The slip ring is limited by the rotational speed and creates dust and wear during sliding because it has contact points. On the other hand, a brushless excitation system does not have contact points, but it is not able to excite the rotor at the start of operation. To overcome these problems, excitation using a contactless power transfer system has been proposed [1]–[5]. A contactless power transfer system has no limit to the rotational speed and does not create dust, wear, sparks, or contact failure. Further, the contactless power transfer system has the advantages of being clean and maintenance free and is able to excite the rotor of motor at the start of operation.

A contactless power transfer system using a rotary transformer has been used in a video deck, resolver, etc.; however, these systems are for dealing with small power such as signal transmission. When developing a contactless power transfer system for an excitation-type synchronous motor, it is necessary to consider miniaturizing and reducing the weight of the secondary circuit and loss due to the shaft of magnetic material. It is necessary for the secondary circuit to be small, lightweight, and maintenance free for mechanical strength and maintainability if it is placed in the interior of the rotor. A

high-frequency magnetic field generated by a contactless power transfer system causes iron loss of the magnetic materials surrounding the transformers and decreases the transmission efficiency.

In this paper, we investigate the resonant capacitor topology for leakage reactance compensation and the omission of the smoothing capacitor, considering miniaturization and weight reduction of the secondary side. In addition, we also investigate the magnetic shielding of the shaft.

For the resonant capacitor topology, we examine the characteristics of the contactless power transfer system with a primary series resonant capacitor (S topology) and compare with the series and parallel capacitor topology (SP topology). In a contactless power transfer system with a large air gap for an electric vehicle, the SP topology is often used for leakage reactance compensation [6], [7]. On the other hand, the S topology is able to omit the secondary capacitor and achieve miniaturization and weight reduction of the secondary circuit. Using the S topology, we derive equations to calculate the value of the primary series capacitor C_{OS}, the maximum transformer efficiency η_{TRmax}, and its load resistance R_{Lmax}.

For omission of the smoothing capacitor, the inductance of the excitation windings around the rotor of the synchronous motor is used. The effectiveness of an aluminum sheet has been described to magnetically shield the motor shaft [8]. In order to determine the most effective shape for the aluminum shield for the S topology, we performed simulations and power transfer experiments.

II. CONTACTLESS POWER TRANSFER SYSTEM

A. Series and Parallel Resonant Capacitor Topology (SP Topology)

Fig. 1(a) shows a schematic diagram of the contactless power transfer system for the SP topology. A full-bridge inverter is used as a high-frequency ($f_0 = 50$ kHz) power supply. Fig. 1(b) shows a detailed equivalent circuit. The primary values are converted into secondary equivalent values using the turns ratio $a = N_1/N_2$. Because the winding resistances and ferrite-core loss are considerably lower than the mutual and leakage reactances at the resonant frequency, the winding resistances (r'_1 and r_2) and the ferrite-core loss r'_0

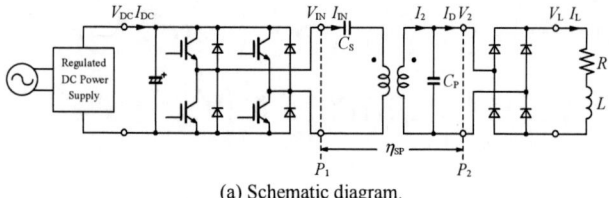

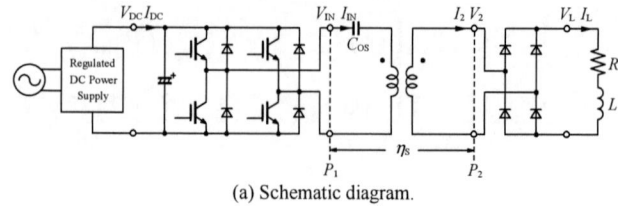

(a) Schematic diagram.

(a) Schematic diagram.

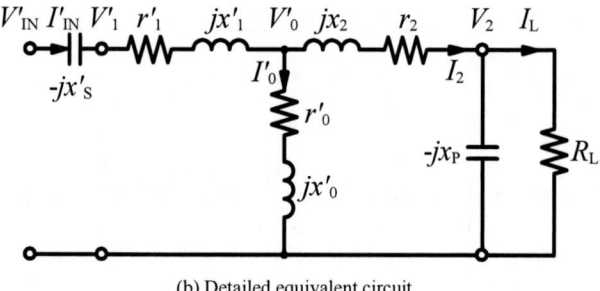

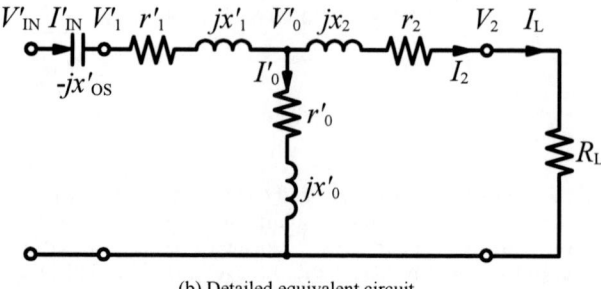

(b) Detailed equivalent circuit.

(b) Detailed equivalent circuit.

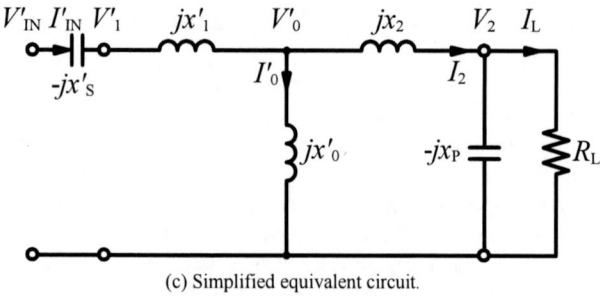

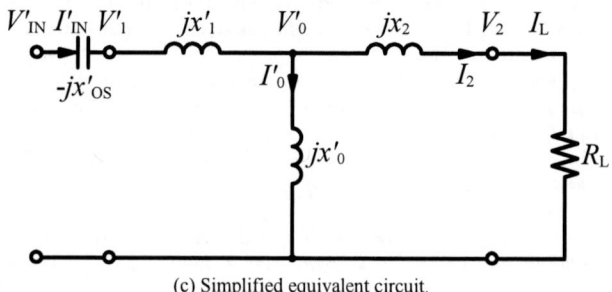

(c) Simplified equivalent circuit.

(c) Simplified equivalent circuit.

Fig. 1. Contactless power transfer system for the SP topology.

Fig. 2. Contactless power transfer system for the S topology.

can be ignored. Fig. 1(c) shows the simplified equivalent circuit that ignores r'_1, r'_2, and r'_0 from the detailed equivalent circuit.

To achieve resonance of the input frequency f_0 ($=\omega_0/2\pi$) with the self-inductance of the secondary winding L_2, which is equivalent to adding a mutual reactance x'_0 and a leakage reactance x_2, the secondary parallel capacitor C_P is given by

$$\frac{1}{\omega_0 C_P} = x_P = x'_0 + x_2 \tag{1}$$

The value of the primary series capacitor C_S (C'_S denotes its secondary equivalent) is determined when the imaginary part of the impedance is zero, and the inverter output power factor of the fundamental wave is to be one. C_S is given by

$$\frac{1}{\omega_0 C'_S} = x'_S = \frac{x'_0 x_2}{x'_0 + x_2} + x'_1 \tag{2}$$

The input voltage V'_{IN} and the input current I'_{IN} can be expressed as

$$V'_{IN} = bV_2, \quad I'_{IN} = I_L / b, \quad b = \frac{x'_0}{x'_0 + x_2} \tag{3}$$

These equations show that the equivalent circuit of a transformer with these capacitors is the same as an ideal transformer at the resonant frequency.

Ignoring r'_0, the efficiency of the transformer for the SP topology in Fig. 1(b) is defined by

$$\eta_{SP} = \frac{R_L I_L^2}{R_L I_L^2 + r'_1 I'^2_{IN} + r_2 I_L^2} = \frac{1}{R_L + \frac{r'_1}{b^2} + r_2 \left\{ 1 + \left(\frac{R_L}{x_P} \right)^2 \right\}} \tag{4}$$

The maximum transformer efficiency η_{maxSP} is obtained when $R_L = R_{LmaxSP}$.

$$R_{LmaxSP} = x_P \sqrt{\frac{1}{b^2} \frac{r'_1}{r_2} + 1} \tag{5}$$

$$\eta_{maxSP} = \frac{R_L}{1 + \frac{2r_2}{x_P} \sqrt{\frac{1}{b^2} \frac{r'_1}{r_2} + 1}} \tag{6}$$

The coupling factor k, the quality factor of the primary winding Q_1, and the quality factor of the secondary winding Q_2 are expressed as

$$k = \frac{M}{\sqrt{L_1 L_2}}, \quad Q_1 = \frac{\omega L_1}{r_1}, \quad Q_2 = \frac{\omega L_2}{r_2} \tag{7}$$

Then, the equations for R_{LmaxSP} and η_{maxSP} can be expressed in terms of k and Q.

$$R_{Lmax} = \frac{r_2 Q_2}{k} \sqrt{\frac{Q_2}{Q_1} + k^2} \tag{8}$$

$$\eta_{maxSP} = \frac{1}{1 + \frac{2}{k} \sqrt{\frac{1}{Q_1 Q_2} + \frac{k^2}{Q_1^2}}} \tag{9}$$

The 2014 International Power Electronics Conference

(a) External form.

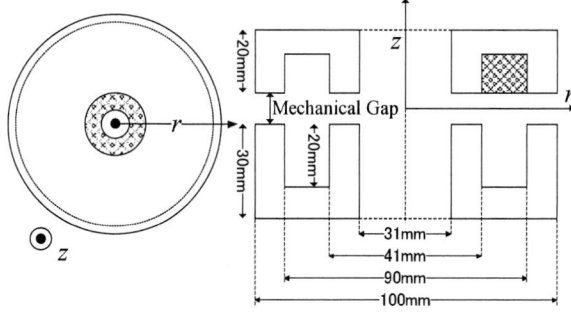

(b) Transformer dimensions.

Fig. 3. Rotary contactless power transformer.

TABLE I. TRANSFORMER SPECIFICATIONS

Rated power		1.0 kW
Mechanical gap		2 mm
Windings wires	Primary	24T1p
	Secondary	10T1p
Weight	Primary	890 g
	Secondary	610 g
Litz wire		0.1 mmφ × 800
Ferrite core		TDK PE90

TABLE II. TRANSFORMER PARAMETERS

Situation		With shaft and aluminum sheet (1.5 mm)			Without shaft
f_0 [kHz]		50			
gap [mm]		1	2	3	2
r_1 [mΩ]		188.1	103.3	85.6	88.6
r_2 [mΩ]		6.87	23.3	28.6	19.7
l_0 [µH]		123.8	110.1	101.1	185.8
l_1 [µH]		60.8	63.2	68.7	48.8
l_2 [µH]		8.46	8.80	8.80	8.97
k		0.70	0.67	0.64	0.79
Q_1		308	527	623	832
Q_2		1370	376	289	657
SP	C_S [µF]	0.106	0.103	0.099	0.114
	C_P [µF]	0.338	0.363	0.384	0.246
	R_{LmaxSP} [Ω]	30.1	14.26	12.2	19.53
	η_{maxSP} [%]	99.54	99.14	98.99	99.54
	η_{SP} [%]*	97.16 (99.42)	96.95 (99.14)	97.2 (98.97)	—
	pf_1*	0.88	0.91	0.93	—
	pf_2*	0.63	0.62	0.62	—
S	C_{OS} [µF]	0.071	0.072	0.071	0.071
	R_{LmaxS} [Ω]	9.91	11.13	11.28	17.3
	η_{maxS} [%]	98.60	98.91	98.91	99.48
	η_S [%]*	97.67 (98.60)	97.16 (98.90)	96.35 (98.90)	—
	pf_1*	0.82	0.81	0.77	—
	pf_2*	0.87	0.86	0.86	—

* experimental value
( ) Calculated value of efficiency
pf: power factor

B. Primary Series Capacitor Topology (S Topology)

Fig. 2(a) shows a schematic diagram of the contactless power transfer system for the S topology. A full-bridge inverter is used as a high-frequency (f_0 = 50 kHz) power supply. Fig. 2(b) shows a detailed equivalent circuit, and Fig. 2(c) shows the simplified equivalent circuit ignoring r'_1, r'_2, and r'_0 from the detailed equivalent circuit.

For the simplified equivalent circuit in Fig. 2(c), the impedance seen from the primary input is given by

$$Z = \frac{R_L x_0'^2}{R_L + (x_0' + x_2)^2} + j\left\{\frac{x_0'\left[R_L^2 + x_2(x_0' + x_2)\right]}{R_L^2 + (x_0' + x_2)^2} + x_1'\right\} \quad (10)$$

The value of the primary series capacitor C_{OS} is determined when the imaginary part of the impedance is zero, and the inverter output power factor of the fundamental wave is to be one. C_{OS} is given by

$$\frac{1}{\omega_0 C_{OS}} = x_{OS}' = \frac{x_0'\left[R_L^2 + x_2(x_0' + x_2)\right]}{R_L^2 + (x_0' + x_2)^2} + x_1' \quad (11)$$

Therefore, the relationship between the input and the output is expressed as

$$I_{IN}' = -j\frac{1}{x_0'}V_L + \frac{x_0' + x_2}{x_0'}I_L \quad (12)$$

Ignoring r'_0, the efficiency of the transformer for the S topology in Fig. 2(b) is defined as

$$\eta_s = \frac{R_L I_L^2}{R_L I_L^2 + r_1' I_{IN}'^2 + r_2 I_L^2} = \frac{R_L}{\frac{r_1'}{x_0'^2}\left[R_L^2 + (x_0' + x_2)^2\right] + R_L + r_2} \quad (13)$$

The maximum transformer efficiency η_{maxS} is obtained when $R_L = R_{LmaxS}$.

$$R_{LmaxS} = \sqrt{(x_0' + x_2)^2 + \frac{x_0'^2 r_2}{r_1'}} \quad (14)$$

$$\eta_{maxS} = \frac{1}{1 + \frac{2r_1'}{x_0'^2}\sqrt{(x_0' + x_2)^2 + \frac{x_0'^2 r_2}{r_1'}}} \quad (15)$$

The equations for R_{LmaxS} and η_{maxS} expressed in terms of k and Q are as follows:

978-1-4799-2706-7/14 $31.00 © 2014 IEEE

The 2014 International Power Electronics Conference

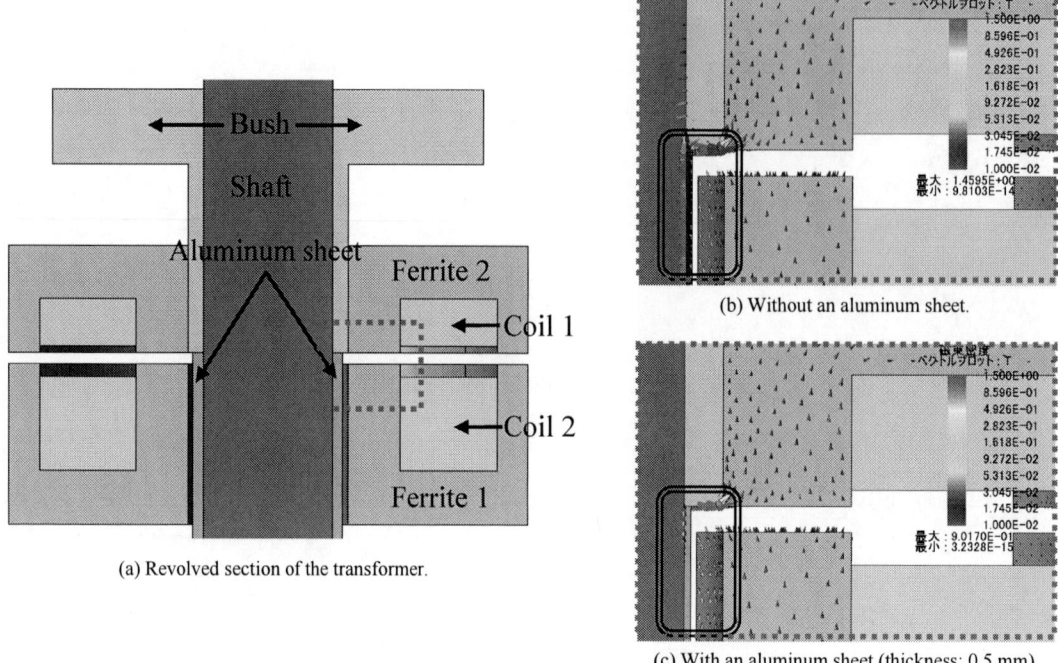

(a) Revolved section of the transformer.

(b) Without an aluminum sheet.

(c) With an aluminum sheet (thickness: 0.5 mm).

Fig. 4. Calculated results for the magnetic field analysis.

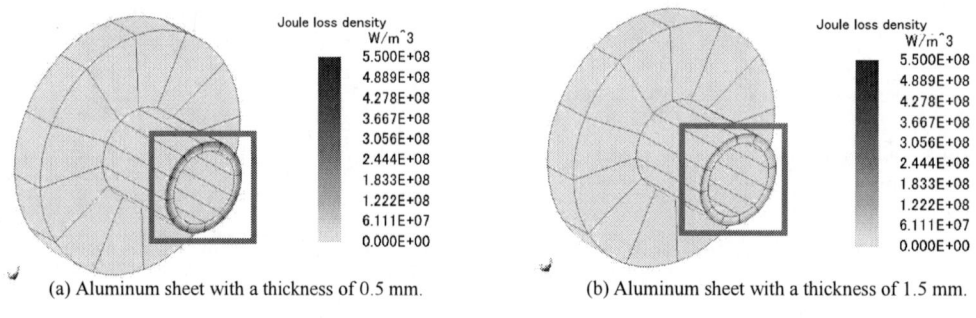

(a) Aluminum sheet with a thickness of 0.5 mm.

(b) Aluminum sheet with a thickness of 1.5 mm.

Fig. 5. Calculated results for the Joule loss density.

$$R_{\mathrm{LmaxS}} = kr_2\sqrt{Q_1 Q_2 + \frac{Q_2^2}{k^2}} \qquad (16)$$

$$\eta_{\mathrm{maxS}} = \frac{1}{1 + \dfrac{2}{k}\sqrt{\dfrac{1}{Q_1 Q_2} + \dfrac{1}{k^2 Q_1^2}}} \qquad (17)$$

III. MAGNETIC SHIELDING OF THE SHAFT

A. Rotary Contactless Power Transformer

Fig. 3(a) shows the exterior of the rotary contactless power transformer, and Fig. 3(b) shows the dimensions of the transformer. Table I summarizes the specifications of the transformer. The cores are made of ferrite, and Litz wires (0.1 mmφ × 800) are used for the windings. The gap of between the primary side and the secondary side of the transformer is 2

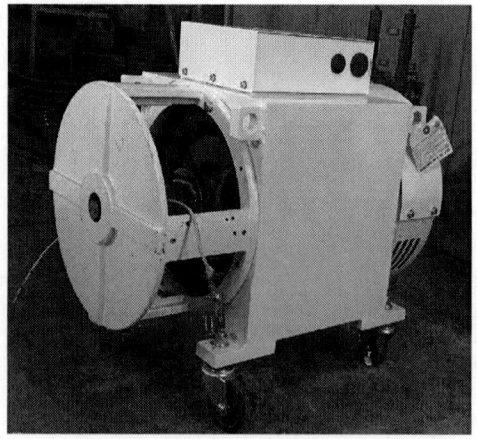

Fig. 6. 8 kW synchronous motor.

978-1-4799-2706-7/14 $31.00 © 2014 IEEE

The 2014 International Power Electronics Conference

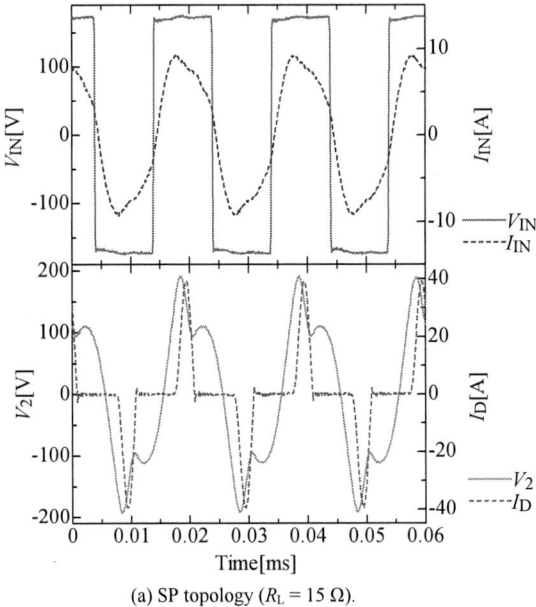

(a) SP topology (R_L = 15 Ω).

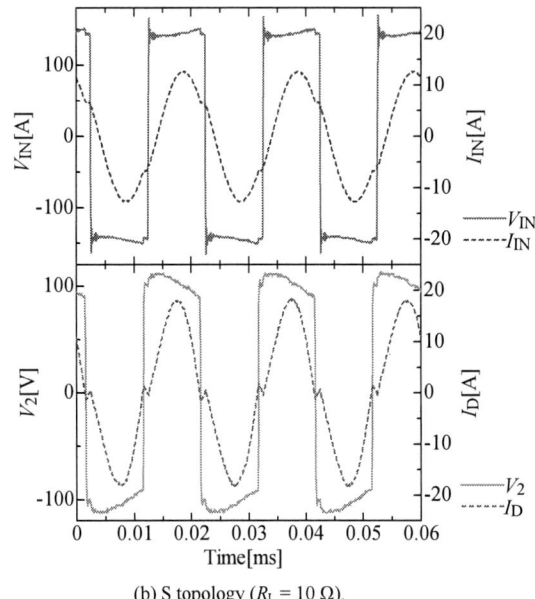

(b) S topology (R_L = 10 Ω).

Fig. 7. Input and output waveforms.

mm. Table II lists the parameters of the transformers measured by an LCR meter. The value of resonant capacitors C_P, C_S, and C_{OS} are calculated from (1), (2), and (11), respectively. Because the efficiency of the contactless power transfer system is altered by changing the gap, we also measured the transformer parameters of for gaps of 1 mm and 3 mm.

B. Magnetic Shielding of the Shaft Using an Aluminum Sheet

For a large synchronous motor, it is necessary to support both ends of the shaft of the rotor. Therefore, by placing the transformer inside of the synchronous motor, the shaft is necessary to penetrate the center of transformer. The 8 kW synchronous motor in Fig. 6 uses a carbon steel (S45C) shaft with a diameter of 25 mm. The primary transformer (transmitter) has a gap of 2.5 mm between the outer diameters of the shaft. The secondary transformer (receiver) is fastened to the shaft by bushing made of carbon steel (S45C).

A high-frequency magnetic field generated by the contactless power transfer system causes iron loss of the magnetic materials surrounding the transformers, and the transformer efficiency decreases [5], [8]. Therefore, we investigated the effectiveness of an aluminum sheet for magnetically shielding the shaft in this system. Fig. 5 shows the results of magnetic field analyses of the shaft with and without an aluminum sheet, as analyzed by the magnetic field analysis software "JMAG." The thickness of the aluminum sheet is 0.5 mm, which is thicker than the skin depth at 50 kHz. The leakage flux into the shaft from the transformers is reduced by the aluminum sheet. Thus an improvement in transformer efficiency can be realized when reducing the loss due to the leakage flux.

Furthermore, we carried out a magnetic field analysis of the

shaft with a 1.5-mm-thick aluminum sheet to reduce the leakage flux into the pointed end of the bushing. Fig. 6 shows the calculated results for the Joule loss density of the bushing. We confirmed that the loss of the pointed end of the bushing is reduced by increasing the thickness of the aluminum sheet.

We performed 1 kW power transfer experiments with the S topology to confirm the effectiveness of the aluminum sheet. The inverter output frequency f_0 was 50 kHz, and the thicknesses of the aluminum sheets are 1.5 mm and 0.5 mm. The smoothing capacitor and load resistance (10 Ω) are connected to the rectifier. From the results of the experiments, the transformer efficiencies are 97.0%, 95.3%, and 92.3% with a 1.5-mm-thick aluminum sheet, with a 0.5-mm-thick aluminum sheet, and without an aluminum sheet, respectively. These results mean that the transformer efficiency is improved by the presence of the aluminum sheet for magnetically shielding the shaft, and the loss of the pointed end of the bushing is reduced by using a thicker aluminum sheet. This shielding method is effective for reducing the loss and leakage flux into the shaft in this system.

IV. EXPERIMENTAL RESULTS

A. S Topology Versus SP Topology

In the S topology, the value of C_{os} is altered by load fluctuations, and the power factor of the primary side (inverter output) is lower. However, the load of the contactless power transfer system is constant for power transfer to the excitation windings of the rotor. Therefore, the S topology is useful in this case. In addition, the S topology has advantages such as compactness and no maintenance on the secondary side.

We performed 1 kW power transfer experiments for the S

978-1-4799-2706-7/14 $31.00 © 2014 IEEE

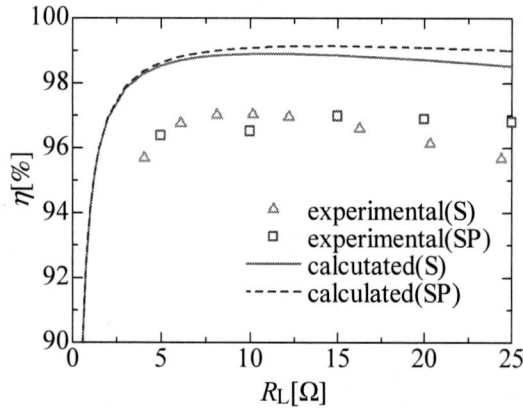

Fig. 8. Efficiency as a function of the load resistance.

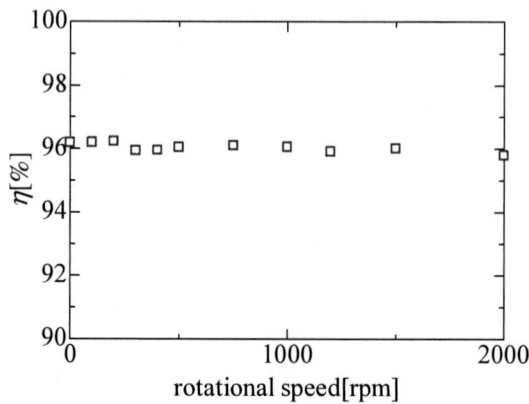

Fig. 9. Characteristics for changes in the rotational speed.

and SP topologies. The inverter output frequency f_0 was 50 kHz, and the thicknesses of the aluminum sheets were 1.5 mm. The smoothing capacitor and load resistance were connected to rectifier. Fig. 7 shows the results of experiments. In the SP topology, V_2 is a sine wave, and I_D flows into the rectifier when V_L is larger than the voltage of the smoothing capacitor. Therefore, the power factor of the secondary side is low because the time that I_D flows into the rectifier becomes short. On the other hand, I_2 flows into the rectifier at all times because V_2 of the S topology is a rectangular wave. When the system is connected to the smoothing capacitor and load resistance, the power factor of the secondary side of the S topology is higher than that of the SP topology.

B. Characteristics with a Change in the Gap Length (S Topology Versus SP Topology)

The gap between the transformers might be changed by vibration during rotation. In order to confirm the characteristics with a change in the gap length, we performed 1 kW power transfer experiments for a change in the gap length with the S and SP topologies. The inverter output frequency f_0 was 50 kHz, and the thickness of aluminum sheet was 1.5 mm. The smoothing capacitor and load resistance (10 Ω) were connected to rectifier. The gap is varied from 1 mm to 3 mm.

The results of experiments are listed in Table II. The experimental values of the transformer efficiency are lower than the calculated values because the iron loss was ignored. However, the transformer efficiencies of the two topologies are approximately constant when the gap changes from 2 mm.

C. Characteristics with a Load-Resistance Change (S Topology Versus SP Topology)

In order to confirm the characteristics with a load-resistance change, we performed 1 kW power transfer experiments with a load-resistance change for the S and SP topologies. The thickness of the aluminum sheet was 1.5 mm, and the smoothing capacitor and load resistance (10 Ω) were connected to rectifier.

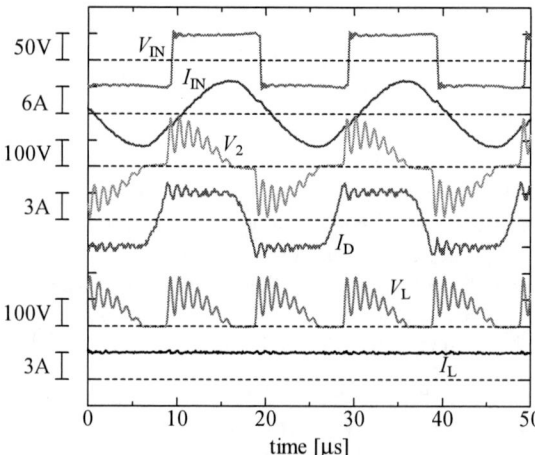

Fig. 10. Waveforms connected to the inductive load.

Figure 8 shows the transformer efficiency as a function of the load resistance. The experimental values for the transformer efficiency of the S and SP topologies are similar to the calculated curves. The experimental values of the transformer efficiency for the two topologies are at an approximately equal level. R_{Lmax} of the S topology is smaller than that of the SP topology.

D. Characteristics with a Change in the Rotational Speed (S Topology)

In order to confirm the characteristics for a change in the rotational speed, we performed 1 kW power transfer experiments for rotating secondary transformer and circuit conditions. The rotational speed was varied from 0 rpm to 2000 rpm, and the thickness of aluminum sheet was 1.5 mm. The smoothing capacitor and load resistance (10 Ω) were connected to rectifier. Fig. 9 shows the relationship between the rotational speed and the transformer efficiency. The results show that the transformer efficiencies are greater than 96% and are about same under rotating conditions. Therefore, the transformer efficiency is not affected by the rotational speed.

E. Power Transfer to the Inductive Load (S Topology)

We performed experiments for excitation windings around the rotor of the 8 kW synchronous motor in Fig. 6. The parameters of excitation windings of the rotor are $R = 20\ \Omega$, $L = 600$ mH, and an excitation current of 3 A. In order to adjust the value of C_{OS} to the load resistance, C_{OS} is set to 0.063 μF. When placing a contactless power transfer system into the interior of the synchronous motor, a secondary transformer and circuit need to be constructed on the rotor side of the motor. On the secondary side, the smoothing capacitor of the rectifier is omitted, and the secondary current is smoothed by the inductance of the excitation windings. Fig. 10 shows the experimental results under excitation-current conditions (3 A, 184 W) that excite the windings of the rotor. The voltage at the excitation winding V_L is not constant owing to the omission of the smoothing capacitor. Further, the field current I_L is smoothed by the inductance of excitation windings and is constant. The input current is in phase with the input voltage, and the input power factor is 0.83. The transformer efficiency η_{TR} is 91.8%, which is lower than the case of power transfer to the load resistance (10 Ω) with a smoothing capacitor, because the resistance of the excitation windings is 20 Ω, which is shifted from R_{LmaxS}. These results reveal that the S topology is effective for power transfer to an inductive load.

V.   CONCLUSION

In this paper, we proposed a resonant circuit with a primary series capacitor (S topology) for miniaturization and weight reduction of the secondary rotating part. The S topology uses only the primary series capacitor; thus the secondary circuit is compact and maintenance free. We derived η_{TRmaxS} and R_{LmaxS}, and we performed power-transfer experiments to investigate the validity of the proposed system.

The experimental results demonstrated that the transformer efficiency of the S topology is equal to that of the SP topology, and the S topology has high power factor performance at the normal gap length (2 mm). When the gap changes from 1 mm to 3 mm, the transformer efficiency remains approximately constant.

The results also showed that the transformer efficiency and power factor of the inverter output are not affected by the rotational speed or inductive load. Therefore, the S topology is effective for power transfer to an inductive load. A high efficiency of 91.8% is achieved when power is transferred to the rotor circuit of the synchronous motor.

In proposed system, an aluminum sheet for magnetically shielding the shaft is effective for reducing the loss and leakage flux into shaft.

REFERENCES

[1]   J. P. C. Smeets, L. Encica, and E. A. Lomonova, "Comparison of winding topologies in a pot core rotating transformer," in *12th Int. Conf. Optim. Electr. Electron. Equip. (OPTIM)*, 2010, pp. 103-110.

[2]   J. P. C. Smeets, D. C. J. Krop, J. W. Jansen, and E. A. Lomonova, "Contactless power transfer to a rotating disk," in *IEEE Int. Symp. Ind. Electron. (ISIE)*, 2010, pp. 748-753.

[3]   D. Hirschmann, C. P. Dick, S. Richter, and R. W. De Doncker, "Design of contactless rotary energy transmission for an industrial application," in *IEEE Power Electron. Specialists Conf.*, 2008, pp. 4314-4319.

[4]   A. Abdolkhani and A. P. Hu, "A contactless slipring system by means of axially travelling magnetic field," in *IEEE Energy Convers. Congress Expo. (ECCE)*, 2012, pp. 1796-1803.

[5]   A. Abdolkhani, A. P. Hu, G. A. Covic, and M. Moridnejad, "Contactless slipring system based on rotating magnetic field principle for rotary applications," in *IEEE Energy Convers. Congress Expo. (ECCE)*, 2013, pp. 2566-2573.

[6]   T. Yamanaka, Y. Kaneko, S. Abe, and T. Yasuda, "10 kW contactless power transfer system for rapid charger of electric vehicle," presented at EVS26 Int. Battery, Hybrid Fuel Cell Elect. Veh. Symp., Los Angeles, CA, May 6-9, 2012.

[7]   H. Takanashi, Y. Sato, Y. Kaneko, S. Abe, and T. Yasuda "A large air gap 3 kW wireless power transfer system for electric vehicles," in *IEEE Energy Convers. Congress Expo. (ECCE)*, 2012, pp. 269-274.

[8]   H. Tanifuji, R. Nozawa, Y. Kaneko, and S. Abe, "Characteristic analysis and improvement efficiency on contactless rotary transformer," in *Annu. Conf. IEEJ Ind. Appl. Soc.*, 2012, pp. I-415-I-418 (in Japanese).

Design of Ferrite Cores of Inductive Power Collection Coils for Moving Vehicles

Daisuke Shimode, Toshiaki Murai, and Tadashi Sawada
Central Japan Railway Company
Tokyo, Japan

Abstract— **This paper describes the design of ferrite cores on the secondary coils for the movable IPC system. The characteristics of secondary coils with various thicknesses of ferrite cores are examined analytically and the test coils are manufactured to verify the analytical results.**

Keywords— *inductive power collection coils, ferrite cores, bonding cores.*

I. INTRODUCTION

Recently, high frequency Inductive Power Collection (IPC) systems are studied in order to realize contactless electric power supply to automobiles, railways, household electrical goods and artificial hearts [1]-[4]. We have examined a movable IPC system with infinitely long cable-type primary coils which are laid parallel to the moving direction of the vehicle and confirmed that installing the ferrite cores only in the secondary coils is effective in improving the performance. Furthermore, we have proposed a secondary coil with unique structure, which has conductor with thinner cross-sections and uneven shaped ferrite cores that has over-hanged edges and removed central part [5].

In this paper, the performances of the secondary coils with various thicknesses of uneven shaped ferrite cores are examined analytically and experimentally. Since large cores are generally made by bonding small cores, the gaps generated by bonding the cores reduce the performances of secondary coils. We have also manufactured the secondary coils with the output power of 12 kW and evaluated their performances in terms of coupling degree, magnetic saturation, and temperature rise.

II. CONFIGURATION OF MOVABLE IPC SYSTEM

Fig.1 shows the basic structure of movable IPC system. The movable IPC system has infinitely long primary coils, which are laid parallel to the moving direction. The IPC system utilizes the principle of electromagnetic induction. Without contacting, the primary coil feeds power to the moving secondary coils by utilizing high frequency current. The magnetic gap between the two coils should be kept large to avoid hazardous contact caused by vertical vibration at high speed and since the secondary coils are installed on the vehicle, their weight should be light. Moreover, self-inductance of secondary coils should be reduced because high frequency IPC coil

generates large reactance voltage compared from common electrical machines. Therefore, we have proposed a secondary coil with unique structure as shown in Fig. 2[5], which has conductor with thinner cross-sections to reduce self-inductance, and uneven shaped ferrite cores to increase induced voltage effectively. The trial secondary coils verified to obtain the output power of 6 kW.

However, the discussions on the thickness of ferrite cores are not sufficient in the previous paper. The area of ferrite cores on the flat conductor of secondary coils is large, and their thickness should be small to reduce their weight. As these large ferrite cores cannot be molded at low cost, they are generally manufactured by bonding small cores. The gaps generated by bonding the cores increase their magnetic resistance, so the influence of the gap on the performances of the secondary coils also should be studied.

In addition, the temperature rise of secondary coils has not been also examined enough. The stationary IPC system can easily improve the temperature rise of secondary coils because they can increase the number of turns of primary coil and reduce the magnetic gap between primary and secondary coils. On the other hand, the movable IPC system cannot reduce the magnetic gap, which causes hazardous touch at traveling and cannot increase the number of turns of primary coil by economical problem. However the wind at travelling can cool the secondary coils in the movable IPC system, so the effect of the wind on their temperature rise also should be studied.

Thus we discuss the above-mentioned problems by using the finite-element method of the JMAG STUDIO 10.0. As a numerical example, the primary coil current is assumed to be 250A at 20kHz of frequency, the magnetic gap between primary and secondary coils 150-250mm, the magneto-motive force (m.m.f.) of secondary coils 800-1000A. The width of primary and secondary coils is selected to 500mm to obtain the highest induced voltages per weight at the maximum magnetic gap of 250 mm [5]. In order to verify the analytical results, we manufacture the test coils, which can obtain 12 kW of output power.

978-1-4799-2706-7/14 $31.00 © 2014 IEEE

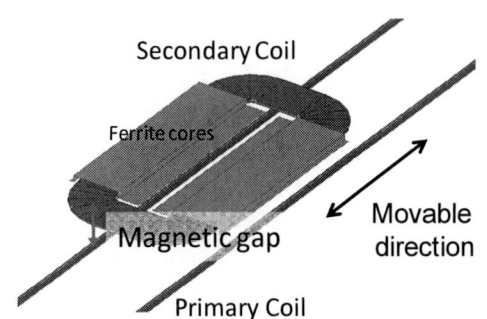

Fig.1. Basic structure of movable IPC system

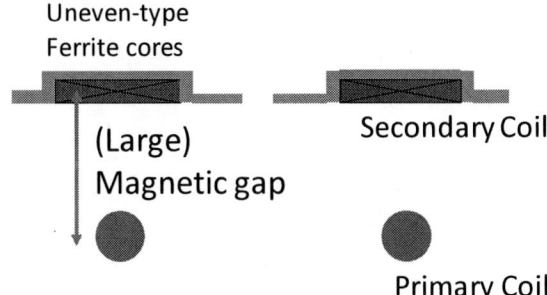

Fig. 2. Cross section of proposed ferrite cores (uneven cores)

III. STUDY OF THICKNESS OF FERRITE CORES

A. Requirements and estimations

The IPC system can be analyzed by the equivalent circuit for transformer as shown in Fig. 3. Since both the primary and the secondary circuits improve their power factor and efficiency by connecting resonant capacitors in series, it is important that their self-inductances do not fluctuate to keep their resonance.

When the IPC circuit does not keep the resonance perfectly, the maximum output power (Pmax) is expressed by

$$P_{max} = \frac{(\omega M I_1)^2}{2 \times (\omega L_2 - 1/\omega C_2)^2} \qquad (1)$$

where

$$R_L = \sqrt{R_2^2 + (\omega L_2 - 1/\omega C_2)^2} \qquad (2)$$

on condition that

$$R_2 \ll \left(\omega L_2 - 1/\omega C_2\right). \qquad (3)$$

The current and impedances on the primary circuit are assumed to be constant because we would like to concentrate on the performances of secondary coils. From the expression (1), the maximum power can

increase by increasing the mutual inductance or reducing the reactance of secondary circuit. Although, the reactance of secondary circuits could be approximately adjusted to be 0 by selecting the proper capacity of resonant capacitor (C_2), the self-inductance of secondary coil is fluctuated by the following two factors. The first factor is that the magnetic permeability of ferrite cores depends on their temperature and second factor is the magnetic saturation of ferrite cores generated by secondary coil's current.

The fluctuation of the self-inductance by changing their temperature is approximately evaluated to be about 2 % of the whole self-inductance as described later. Thus a part of self-inductance cannot be avoided to remain on the secondary circuit, so the self-inductance should be reduced to obtain larger power. Furthermore, if the magnetic saturation of ferrite cores also fluctuate the self-inductance which has already been shifted, the output power is considerably reduced, for example, when the shift of self-inductance of 2 % by magnetic saturation add to the original shift of 2 %, the total shift of 4 % reduce the maximum output power to 25 % of that at the original shift from the equation (1). This means that the self–inductance of secondary coils is strongly required not to be fluctuated by magnetic saturation in their ferrite cores.

Since the larger output power is obtained, as the mutual inductance becomes larger and the self-inductance becomes smaller on the secondary coils, the coupling degree k is defined as the expression (4) to evaluate the performances of secondary coils exclusively.

$$k = \frac{M}{\sqrt{L}} \qquad (4)$$

The coupling degree k is coupling factor dividing self-inductance of the primary coils, which is utilized to evaluate the performance of transformer, because the length of the primary coil for the movable IPC system is decided by economical factor instead of using coupling factor. This paper does not consider the self-inductance of the primary coils.

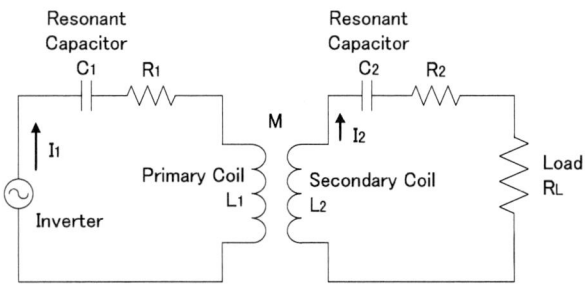

Fig. 3. Equivalent circuits of the IPC system

B. Coupling degree

The secondary coils with ferrite cores whose thickness are 2, 5, and 10 mm are examined analytically. Fig. 4 shows the analytical model of the movable IPC system. Fig. 5 shows the analytical results of the mutual inductance M, self-inductance L, and the coupling degree k at the magnetic gap of 160 mm for the secondary coils without bonding gap expressed as ratio from the thickness of 10 mm and the other results are also similarly expressed in the following chapters. We found that the differences in the M, L, and k by the thickness of ferrite cores without bonding gap are very small, for example, the differences in the M between 2 mm and 10 mm is 2 %, the L 3%, and the k 2 %.

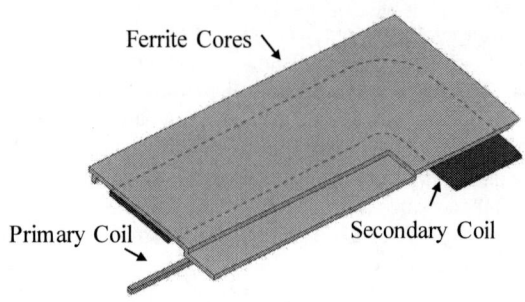

Fig. 4. Analytical model of the movable IPC system (1/4 partial model)

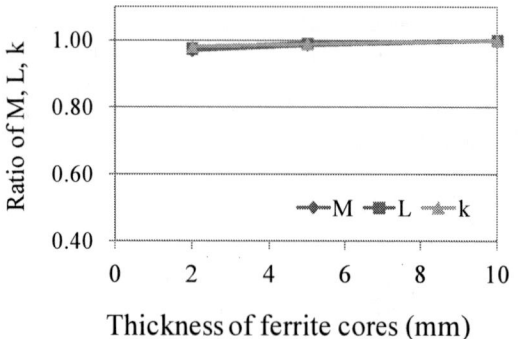

Fig. 5. Analytical results of the mutual inductance, the self-inductance, and the coupling degree of the secondary coils with various thicknesses of the ferrite cores without bonding gaps

C. Self inductance

As we have described before, their output power is greatly reduced if the self-inductances is fluctuated in the secondary coils. Thus this chapter examines the characteristics of self-inductance against m.m.f. of secondary coil. Fig. 6 shows the analytical results of the self-inductance L against the m.m.f. in the secondary coils with 2, 5, and 10mm of ferrite cores. These results indicate that the secondary coils with thinner cores reduce the self-inductance at the lower m.m.f., for example, the changing point of self-inductance on the 2mm-core-coil is the m.m.f. of 950A. As the value of 950A is close to the required the m.m.f. of 800A at the gap of 160mm for the output power of 12kW, the 2mm-coil has small margin for the fluctuation of their self-inductance.

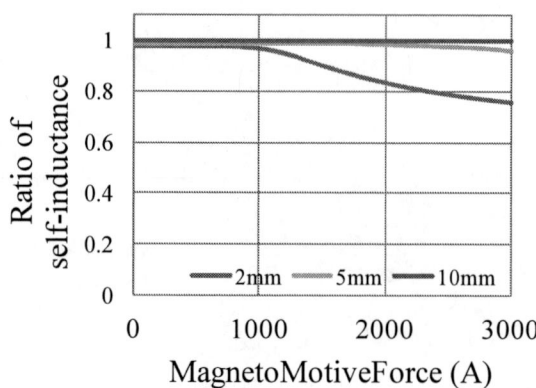

Fig. 6. Analytical results of the self-inductance of the secondary coil with various thicknesses of ferrite cores

D. Influence of bonding gap

This chapter discuss about the influence of bonding gap on the performances of secondary coils. As mentioned above, large flat ferrite cores are manufactured by bonding the small flat cores because the ferrite cores are molded by press machines, whose sizes are limited by economical problems. The bonding gaps between bonded ferrite cores are minute but the difference of the magnetic permeability between the gaps and ferrite cores is extremely large. Practically, the small rectangular ferrite cores are bonded on the flat conductor, so the combined large ferrite core has parallel and perpendicular bonding gaps to the moving direction. The only parallel gaps affect the performances because the main magnetic fluxes by primary and secondary coils are perpendicular to the moving direction. Fig. 7 shows the parallel bonding gaps to the moving direction in the uneven ferrite core. The bonding gaps exist not only on the conductor but also at the bending points.

Fig. 8, Fig. 9, and Fig. 10 show the analytical results of the mutual inductance M, the self inductance L, and the coupling degree k against the bonding gaps The 2mm-core-coil has the decreasing ratios of 13% and 9 % for the mutual inductance and the coupling degree, respectively, at the bonding gap of 0.2 mm, while the 10mm-core-coil has those of 3 % and 3 % respectively. These results indicate that the thinner ferrite cores have larger influence of performances by the bonding gaps than the thicker cores.

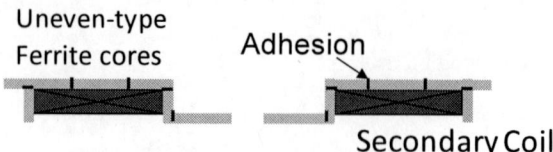

Fig. 7. Parallel bonding gaps in the ferrite cores

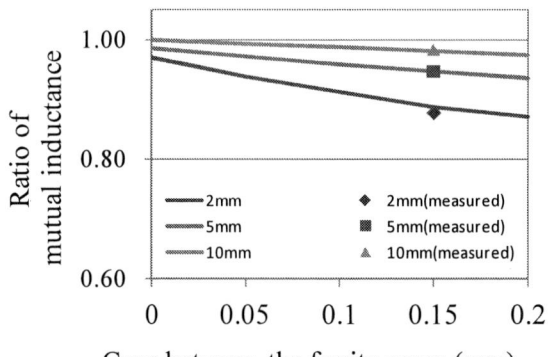

Fig. 8. Analytical results of the mutual inductance against bonding gap between ferrite cores

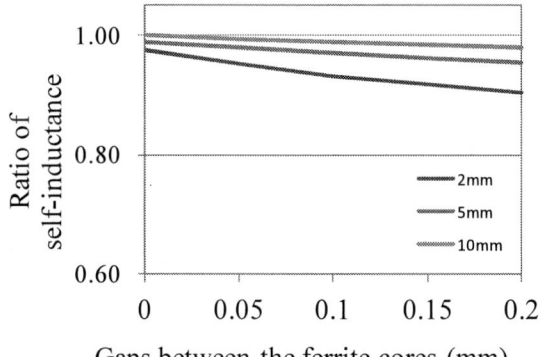

Fig. 9. Analytical results of the self-inductance against bonding gaps between ferrite cores

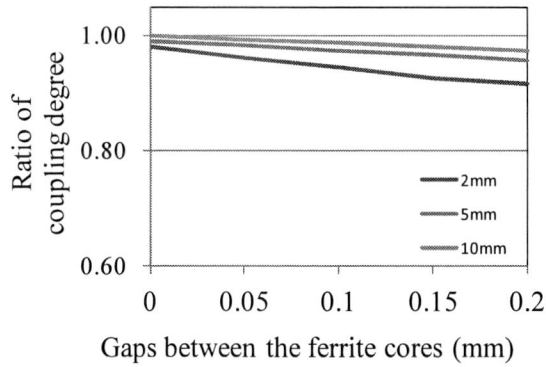

Fig. 10. Analytical results of the coupling degree against bonding gap between ferrite cores

E. Temperature rise of coil

This chapter examines the temperature rise of secondary coils, which are assumed to be installed under the bottom of vehicles. When the vehicles move at the speed of 100-250 km/h, the winds under the coils are assumed to blow at the velocity of 30-75 km/h, which are 30 % of the vehicle running speed. The maximum temperature rise of coils is assumed to be limited 40 degree, which is subtracted the outside temperature of 50 degree from the maximum coil temperature of 90 degree.

The core losses are estimated by the flux densities calculated by the FEM analysis. On the temperature analysis, the wind assumes to cool the only bottom side of secondary coil, whose heat transfer coefficient (α) is defined as the equation (5) [6].

$$\alpha = 7.14 \times v^{0.78} \qquad (5)$$

where v is the velocity of wind.

Fig. 11 shows the flux distributions in the uneven ferrite core. Since the centre of ferrite core, where the fluxes are concentrated, becomes the maximum temperature rise point, we estimate the temperature rise at this point in the following examination. Fig. 13 shows the analytical results of the temperature rises of 2mm-core and 10mm-core at the vehicle speed of 100 and 250km/h. The 2mm-core has larger temperature rise than the 10mm-core-coil at no wind, because the 2mm-core has about 5 times as much core losses as 10-mm-core. However, the temperature rise of 2mm-core at 100 km/h decrease to 35 % of that at no wind, and at the 250 km/h to 25 %. As a result, the temperature rise of 2mm-core is less than the temperature rise limit.

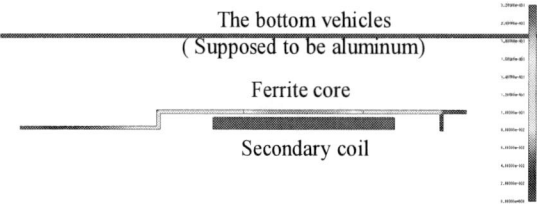

Fig. 11. Flux distribution in ferrite cores

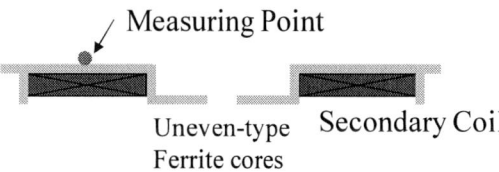

Fig. 12. Measuring point at the temperature rise test

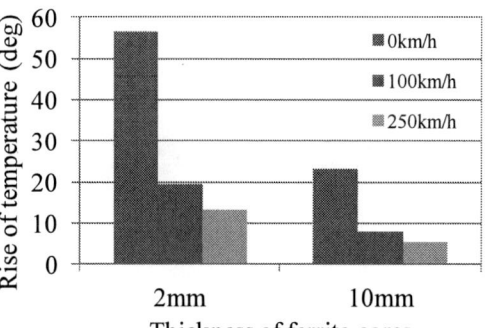

Fig. 13. Analytical results of temperature rise at 100km/h and 250km/h

IV. EXPERIMENTAL RESULTS

A. Coupling degree

In order to verify the above-mentioned analytical results, we manufactured 3 types of test secondary coils, which could obtain the output power of 12kW. Fig. 14 shows the overview of the test secondary coils, whose ferrite core's thicknesses are 2, 5, and 10 mm, respectively.

First, the bonding gaps are predicted by compared the experimental mutual inductances with analytical ones. The analytical mutual inductances were shown in Fig. 8, and the experimental one of 2mm-core-coil is 0.87, that of 5mm-core-coil 0.94, that of 10mm-core-coil 0.98. As these experimental values agree well with the analytical values at the bonding gap of 0.15 mm, the bonding gap of test coils might be 0.15 mm and these experimental results are also plotted at 0.15 mm in Fig. 8.

Secondary, we estimate the coupling degree of test coils. Fig. 15 shows the analytical and experimental results of the coupling degree of test coils whose bonding gaps are assumed to be 0.15 mm. This figure shows that the experimental values agree well with the analytical values. The difference in the coupling degree between the 2mm-core-coil and the 10mm-core-coil with bonding gaps is 5 %. These results verify that the bonding gap might reduce the coupling degree in the secondary coils with thinner ferrite cores.

B. Output power and temperature rise of coil

Thirdly, we confirm the output power and the temperature rise of test coils. The power collecting test was conducted by using the electric circuit as shown in Fig. 16. Fig. 17 shows the waveform examples of m.m.f. of the primary and the secondary coils, the output current and the output voltage at the 2mm-core-coil. These experimental outputs show that the secondary coil can obtain the output power of 12 kW and the efficiency of 0.9. In addition, the secondary coil's m.m.f. of 900A flows in stable, which does not exceed the changing point for the self-inductance as shown in Fig. 6.

Fig. 18 shows analytical and experimental results of temperature rise with no wind. This figure shows that the experimental values agree well with the analytical values, so our analytical methods for the temperature rise are confirmed and the temperature rise with wind-blow can be predicted accurately.

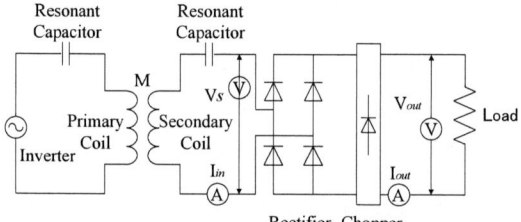
Fig. 16. Configuration of a secondary circuit

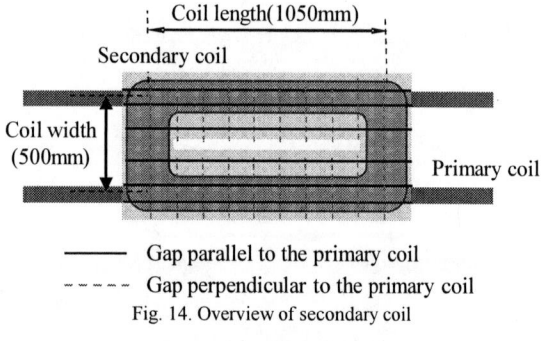

— Gap parallel to the primary coil

---- Gap perpendicular to the primary coil

Fig. 14. Overview of secondary coil

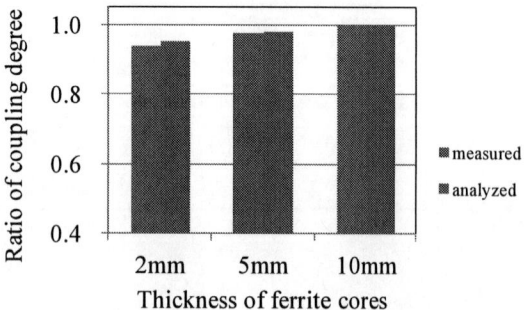

Fig. 15. Analytical and experimental results of coupling degree k

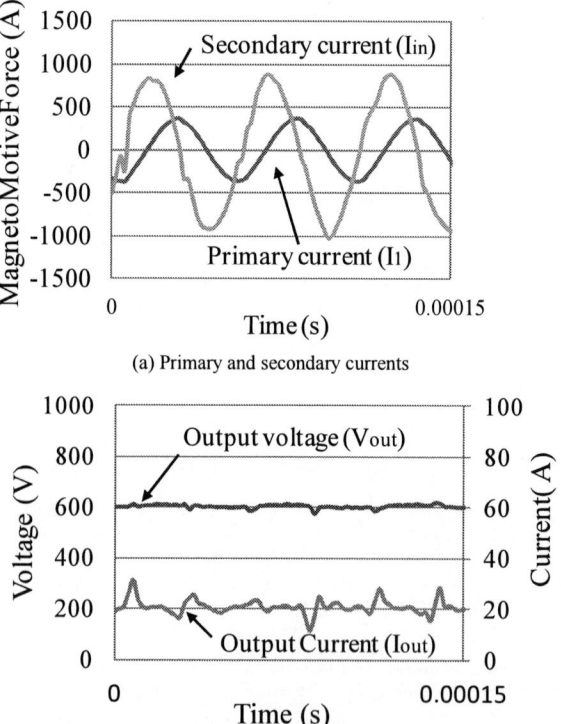

(a) Primary and secondary currents

(b) Output voltage and currents

Fig. 17. Waveform examples at the output power of 12kW

978-1-4799-2706-7/14 $31.00 © 2014 IEEE

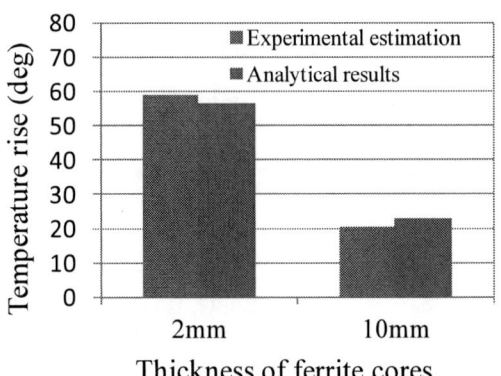

Fig. 18. Analytical and experimental results of temperature rise test with no wind

Finally, the fluctuations of self-inductance by changing the coil temperature are experimentally examined. Fig. 19 shows the experimental results of the self-inductances of secondary coils at changing the temperature from 50 to 20 Celsius in the 2mm-core-coil and the 10mm-core-coil, which are measured on the natural cooling after the temperature rise tests. The self-inductance of 2mm-core-coil changes 2 % of the whole self-inductance, though that of 10mm-core-coil changes 0.2 %. This result indicates that thinner ferrite cores have larger fluctuation of self-inductance by the fluctuation of temperature than the thicker cores.

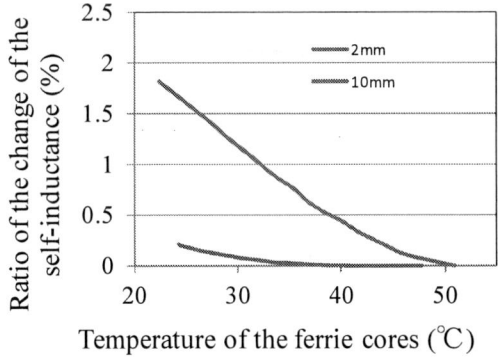

Fig. 19. Experimental results of the fluctuation of the self-inductance against the temperature of the ferrite cores

V. CONCLUSIONS

In the movable IPC system, the secondary coils should design for their self-inductance not to be fluctuated, because the fluctuation of self-inductance drastically reduces their collection power. The thickness of ferrite cores should be thinner as possible not to saturate their fluxes and to reduce their weight, but thinner ferrite cores are easily affected by the fluctuation of temperature and the bonding gaps. Although, the thinner ferrite cores have larger core losses, the wind-blow can remarkably reduce their temperature rise. The trial manufactured secondary coils with 12 kW of output power verify these analytical results.

REFERENCES

[1] K.Kobayashi, N.Yoshida, Y.Kamiya, Y.Daisho and S.Takahashi, "Development of a Non-contact Rapid Charging Inductive Power Supply System for Electric-driven Vehicles," Vehicle Power and Propulsion Conference, Sep. 2010

[2] M.Bauer, P.Becker and Q.Zheng, "Inductive Power Supply(IPS)for the Transrapid ," Maglev 2006, pp471-476, Sep. 2006

[3] F.Sato, J.Murakami, H.Matsuki, S.Kikuchi, K.Harakawa and T.Satoh, "Stable Energy Transmission on Moving Loads Utilizing New CLPS," *IEEE Trans. on Magn.*, vol.32, No5, vol.32, no.5, pp5034-5036, Sep.1996

[4] H.Miura, S.Arai, Y.Kakubari, F.Sato, H.Matsuki and T.Sato, "Improvement of the Transcutaneous Energy Transmission System Utilizing Ferrite Cored Coils for Artificial Hearts," *IEEE Trans. on Magn.*, vol.42, no.10, pp3578-3580, Oct.2006.

[5] D.Shimode, T.Murai, S.Fujiwara, "A Study of Coil Structure of Inductive Power Collection System for Moving Vehicle", Proceedings of the International Future Energy Electronics Conference 2013

[6] Electrical Engineering Handbook, 1st ed., The Institute of Electrical Engineers of Japan, 1988, pp.567.

Torque/Current Ratio Improvement and Vibration Reduction of Switched Reluctance Motors Using Multi-stage Structure

Ryota Matsui
Shibaura Institute of Technology,
3-7-5, Toyosu, Koto-ku, Tokyo,
Japan
e10106@shibaura-it.ac.jp

Noriya Nakao
Shibaura Institute of Technology,
3-7-5, Toyosu, Koto-ku, Tokyo,
Japan
nb12105@shibaura-it.ac.jp

Kan Akatsu
Shibaura Institute of Technology,
3-7-5, Toyosu, Koto-ku, Tokyo,
Japan
akatsu@shibaura-it.ac.jp

Abstract- **This paper presents a multi–stage Switched Reluctance (SR) motor. In this proposed machine, the numbers of stator and rotor poles are equal in each stage. Because of the structure the proposed SR motor can achieve high torque/current ratio in operation. Also, the low-vibration operation is possible by shifting the excitation firing angle. Effectiveness of the proposed SR motor is verified by performing some simulations and experiments.**

Keywords— Switched reluctance motor, torque/current raito, vibration.

I. INTRODUCTION

Switched Reluctance (SR) motors have low-cost, simple, and rugged structure without magnet and rotor field winding. Because of their structure they are applicable in harsh operating conditions, especially high speed and high temperature (high torque) drives. For the past two decades, techniques related to these machines have been an active area of research. However, in general, a torque density of SR motors is smaller than of permanent magnet synchronous motors which have been widely used in various industrial applications. Furthermore acoustic noise and vibration levels are high in SR motors due to magnetic radial forces [1]-[2]. To expand the industrial acceptance of them the high torque performance and radial force reduction are essential requirements in SR motors.

In previous researches, various strategies for improving torque density of SR motors have been proposed. In [3] and [4], high torque SR machines have been achieved based on the novel machine design techniques. In [5] and [6], Grain-Oriented (GO) magnetic steel sheets have been applied to the structure. SR motors with GO magnetic steel sheets have the high torque density because they generate not only reluctance but also magnetic anisotropy torques. However, the above researches include no verification of vibration and acoustic noise levels.

As previously noted, the vibration and acoustic noise reduction is also an essential requirement in SR motors. According to previous researches, vibration levels significantly depend on electromagnetic radial forces. The radial forces can be reduced by modifying the rotor-stator structures [7]-[8]. Also, the double stator [9] and skewed rotor and stator structure [10] are effective in distributing the forces which lead to the vibration generation. However, the above researches include no verification of torque density.

This paper proposes a novel SR motor which can achieve not only high torque/current ratio but also low vibration levels. The proposed machine consists of skewed multi-stage structure and the numbers of stator and rotor poles are equal in each stage. In conventional multi-phase SR motors, stator windings are sequentially excited (not simultaneously). Therefore instantaneous torque is generated in some pairs of the stator windings. On the other hand, the proposed SR motor can generate larger instantaneous torques than the conventional machines because all stator windings are simultaneously excited in each stage. However, in case that the numbers of stator and rotor poles are equal, the continuous torque production is impossible. To generate continuous torques, the multi-stage structure is skewed in the proposed machine. In previous research, the torque is improved by the two-phase two-stack SR motor [11].

In this research, the proposed SR motor uses GO magnetic steel sheets at rotor and stator poles because torque/current ratio is more improved. Also, the vibration due to the electromagnetic radial force can be suppressed by shifting the excitation firing angle in the proposed SR machine.

This paper presents a double-stage SR motor which has 12 stator and 12 rotor poles in each stage. A part of the stator and rotor poles are made by laminating GO magnetic steel sheets. The performances of this machine are compared with the conventional three-phase machine and the proposed SR motor with the GO magnetic steel sheets [6]. Effectiveness of the proposed SR motor is verified by performing some simulations and experiments.

II. THREE-PHASE SR MOTOR

In this section, the structure and drive principle of the conventional machine is introduced. Also, a cause of the vibration of the SR motors and the vibration reduction method of the previous researches is introduced.

978-1-4799-2706-7/14 $31.00 © 2014 IEEE

The 2014 International Power Electronics Conference

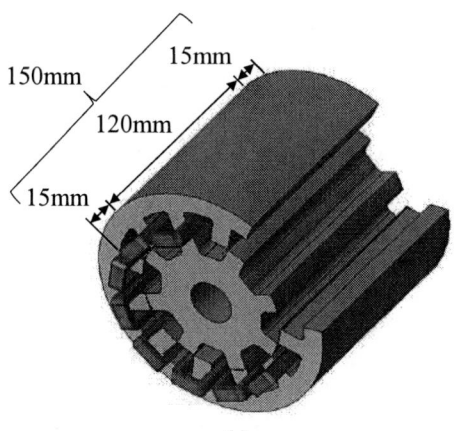

(a)

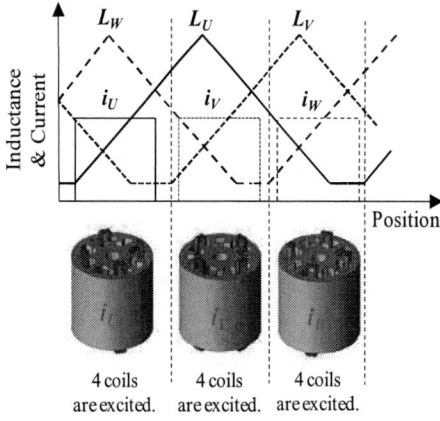

(b)

Fig. 1 Conventional three-phase SR motor (a) Structure of the SR motor (Stator and rotor diameters are 136mm and 83mm, respectively. Stator windings are concentrated around the teeth (i.e. non-overlapping).) (b)Drive principle of the conventional SR motor.

A. Structure and drive principle

Fig. 1 shows the structure and drive principle of the conventional multi-phase SR motors. In conventional multi-phase SR motors, the numbers of the rotor and stator poles are generally different each other and the SR motors have insulated circuits which consist of some coils in their stator. In case of the conventional three-phase SR motor which have 8 rotor and 12 stator poles, each phase winding consists of 4 coils connected in series and/or parallel (i.e. phase U, V, and W).

When the stator windings are excited, the rotor poles are pulled from the positions where the rotor poles are equidistant from the two adjacent stator poles (unaligned position) to reduce the reluctance. Therefore SR motors generate positive torques when the inductance increases.

In case of conventional three-phase SR motors, the inductance of them varies with position. In this conventional SR motor, 4 coils of one phase winding are simultaneously excited when the inductance increases in the one phase and the electromagnetic torque is generated by the field of the 4 stator poles. In order to obtain continuous torque, each phase winding is sequentially excited as shown in Fig. 1 (b).

B. The vibration production mechanism

As shown in Fig. 1 (b), each phase winding is sequentially excited in the conventional three-phase SR motor. In this time, the electromagnetic force of the radial direction (radial force) is generated apart from the force of the rotation direction (torque). This radial force is deformed the stator and generates the vibration at the stator when the radial force is rapidly changed at stopping excitation. Also, the vibration is mathematically modeled as follows [13].

$$\frac{1}{m}\frac{d^2 F_r}{dt^2} = \frac{d^2 a}{dt^2} + 2\zeta\omega_n\frac{da}{dt} + \omega_n^2 a \,, \qquad (1)$$

where, m is a mass of the stator, F_r is the radial force, a is an acceleration which is generated at stator, ζ is an attenuation coefficient, ω_n is a natural frequency. The

second derivative of the radial force is dominant for the generation of the vibration. When applied voltage changes from the positive value to the negative or zero, a time derivative of the radial force becomes large and the vibration (natural vibration) is excited.

Some research have reported that the vibration can be reduced by using a two-stage commutation method [12] − [14]. In this method, the two stage voltage is applied in order to cancel the generated vibration which is caused by the applied voltage. In the second stage, the voltage which shifts half cycle of the natural frequency from the first stage is applied. Therefore, the vibration is cancelled because the vibration phase of the first and second stages is opposite each other [12] - [14]. However, this method can be used only for non-modulated voltage apply such as one-pulse voltage control. For example, since the hysteresis current control cannot determine the voltage, it is very hard to generate the voltage for vibration reduction.

III. DOUBLE-STAGE SR MOTOR

In this section, the structure and drive principle of the proposed machine are introduced.

Fig. 2 shows the structure and drive principle of the proposed double-stage SR motor. The proposed SR motor has the double-stage structure skewed by 15 degrees and the numbers of the rotor and stator poles are equal in each stage. This machine has an insulated circuit which consists of some coils in each stage (i.e. phase A and B). The stator windings are simultaneously excited in each stage. When the rotor and stator are at alignment position in the upper stage, the lower stage rotor poles are at the unaligned position because of the skewed structure. Therefore, to obtain continuous torque, each stage is alternately energized in the proposed SR motor.

In order to increase a torque, the proposed SR motor has rotor and stator poles which are made by laminating GO magnetic steel sheets. GO magnetic steel sheets has high permeability under the strong magnetic saturation [6]. Therefore, SR motors with GO magnetic steel sheets

978-1-4799-2706-7/14 $31.00 © 2014 IEEE   1129

The 2014 International Power Electronics Conference

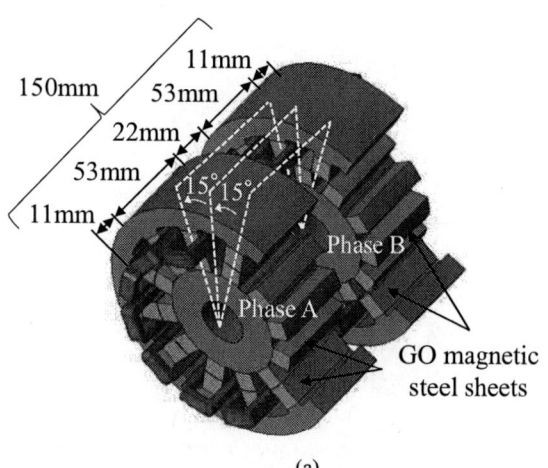

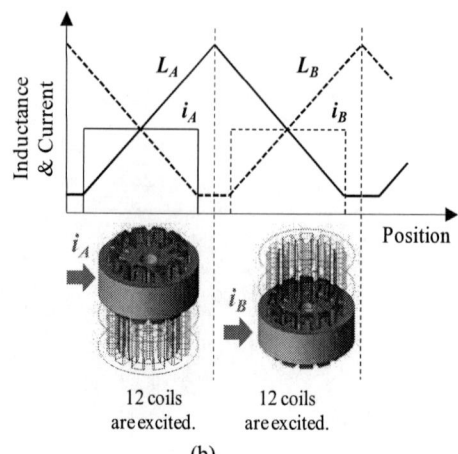

(a)   (b)

Fig. 2 Proposed double-stage SR motor (a) Structure of the proposed SR motor with GO magnetic steel sheets (Stator and rotor diameters are 136mm and 83mm, respectively.) (b)Drive principle of the proposed SR motor.

can generate higher torque than it with non-GO magnetic steel sheets.

The drive principle of the proposed SR motor is basically same as the conventional three-phase machine. The proposed SR motor is a two-phase machine which has the skewed double-stage stator. Each stage has a phase winding which consists of 12 coils connected in series and/or parallel (i.e. phase A and B). In this machine, 12 coils of one phase is simultaneously excited when the inductance increases in the one phase and the electromagnetic torque is produced by the 12 coils. Considering the number of simultaneously-exciting coils, the proposed SR motor has larger effective air-gap area for electromagnetic torques than the conventional machine (conventional: 4 coils, proposed: 12 coils). The proposed SR motor can improve the torque by not only changing materials of the motor but also changing the structure.

IV. THEORETICAL VERIFICATION OF TORQUES

In this section, effectiveness of the proposed SR motor is verified by comparing with the conventional SR motor. The validation is implemented by performing some theoretical calculations, simulations with the FEA and experiments.

A. Theoretical calculation

As previously noted, the proposed SR motor can achieve higher torque production than the conventional one. The advantage is verified based on theoretical calculations of both instantaneous and average torques. Instantaneous and average torques are defined as shown in Fig. 3. Table I shows the specification of SR motors for theoretical verification. In this validation, the proposed and conventional SR motors have same volume and stack and coil-end length. Also, the stator and rotor pole width and material are same in both machines.

Assuming a same excitation current, the instantaneous torque generated by the stator field due to simultaneously-excited coils is given as follow:

TABLE I
SPECIFICATION OF SR MOTORS FOR THEORETICAL VERIFICATION.

		Conventional SR motor	Proposed SR motor
Number of simultaneously-excited coils	N_c	4	12
Motor stack length including coil-end [mm]	h_m	150	150
Coil-end length [mm]	h_c	30	60
Number of stages	N_s	1	2
Number of phases	N_p	3	2
Number of rotor poles	N_r	8	12

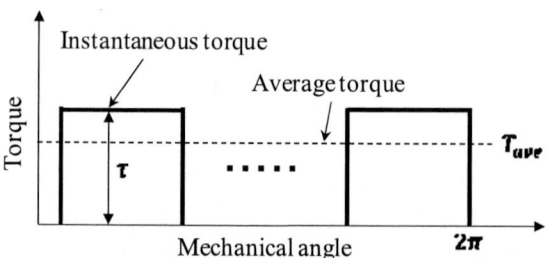

Fig. 3 Instantaneous torque and Average torque.

$$\tau = \frac{N_c\left(h_m - h_c\right)}{N_p^2 N_s}\tau_1 , \qquad (2)$$

where τ_1 is the instantaneous torque generated by the field of a 1-meter-length stator tooth. h_m and h_c are the motor stack length including coil-end length and total coil-end length, respectively. N_c, N_p, and N_s are the number of simultaneously-excited coils, phases, and stages, respectively. In case of the conventional SR motor, the instantaneous torque is $53.3\tau_1$. On the other hand, in case of the proposed machine, it is $135\,\tau_1$. Therefore, the proposed SR motor can achieve 2.53 times larger instantaneous torque than the conventional one at same excitation current.

The average torque is calculated as follow:

978-1-4799-2706-7/14 $31.00 © 2014 IEEE   1130

The 2014 International Power Electronics Conference

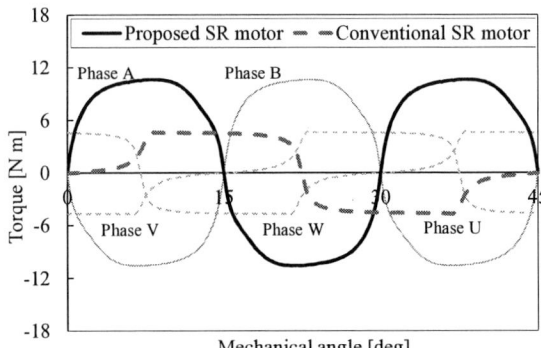

Fig. 4 Static torque waveforms.

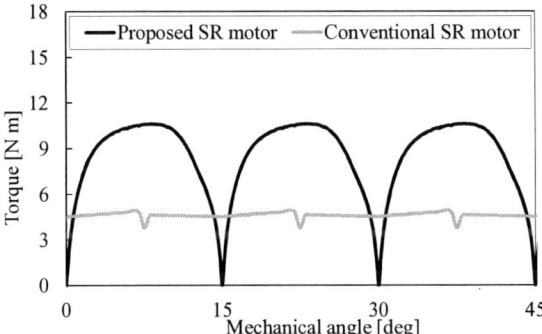

Fig. 5 Instantaneous torque waveforms.
(Average torque of the proposed SR motor is 8.45 Nm,
conventional SR motor is 4.62 Nm.)

$$T_{ave} = \frac{1}{2\pi}\int_0^{2\pi}\frac{N_r N_p}{N_r}\tau d\theta = N_p \tau , \qquad (3)$$

where θ and N_r are the mechanical angle and number of rotor poles, respectively. In case of the conventional SR motor, the average torque is $160\tau_1$. On the other hand, in case of the proposed machine, it is $270\tau_1$. Therefore, the proposed SR motor can achieve 1.69 times higher average torque than the conventional one at same excitation current.

B. Simulation with the FEA

In this validation of this section, the proposed and conventional SR motors have same volume and stack and coil-end lengths. Also, the stator and rotor pole width and material are same in both machines in the same terms of preceding paragraph.

Fig. 4 shows static torque waveforms. In this result, 54A DC current is distributed into each phase. Therefore, 18A DC current is applied to the conventional SR motor (=54A/3phase) and 27A DC current is applied proposed SR motor (=54A/2phase), respectively. As shown in Fig. 4, the proposed SR motor generates about 2.2 times higher instantaneous torque than the conventional SR motor.

Fig. 5 shows instantaneous torque waveforms when 38A RMS current is applied to the conventional SR and proposed motors. As shown in Fig. 5, the proposed SR motor generate about 1.83 times higher average torque than the conventional SR motor.

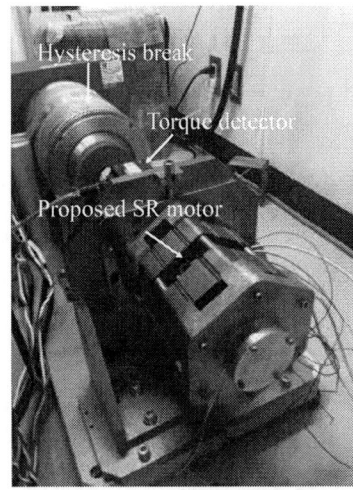

Fig. 6 The proposed SR motor with non-GO sheets on the test bed.

(a)   (b)

Fig. 7 Manufactured proposed SR motor with GO sheets (a) rotor (b) stator.

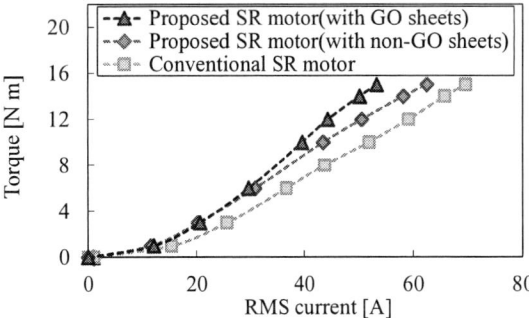

Fig. 8 Current-torque characteristic.

In these validations, the characteristics are calculated by using non-oriented steel. Thus, the proposed SR motor with GO magnetic steel sheets can be expected to generate further high torques.

C. Experiment

Fig. 6 shows the proposed SR motor with non-GO magnetic steel sheets (proposed SR motor with non-GO sheets) on the test bed. The shaft is connected to a hysteresis break via a 50 Nm torque detector. Fig. 7 shows the rotor and stator of the proposed SR motor with GO magnetic steel sheets (the proposed SR motor with GO sheets).

Fig. 8 shows a current-torque characteristic of the conventional SR motor, proposed SR motor with non-GO sheets and proposed SR motor with GO sheets. The

978-1-4799-2706-7/14 $31.00 © 2014 IEEE

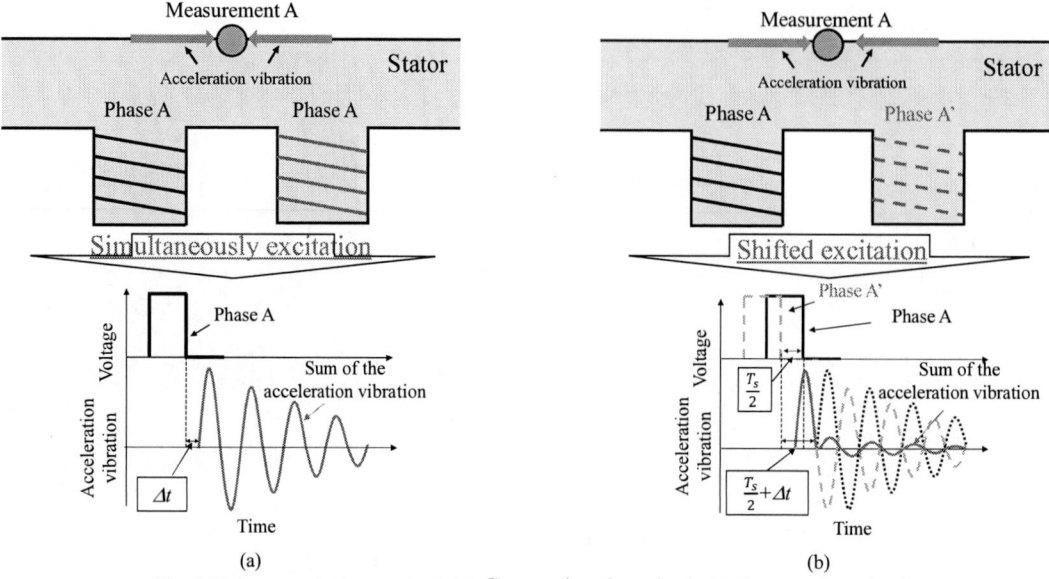

Fig. 9 Voltage excitation method. (a) Conventional method. (b) Proposed method.

driving method is a hysteresis control. Motor speed is 1000r/min and DC link voltage is 250V due to reduce the influence of back-EMF. Also, the excited angle is decided by changing turn-on angle = 0 ∼ 30 [deg], turn-off angle = 50 ∼ 90 [deg] to achieve the most largest torque/current ratio.

As shown in Fig. 8, the proposed SR motor with non-GO sheets has higher torque-current characteristic than the conventional motor. Also, the proposed SR motor with GO sheets can achieve the highest torque-current characteristic in Fig. 8.

V. THE METHOD OF VIBRATION REDUCTION

In this section, a method of vibration reduction (proposed method) is introduced. And the validation is implemented by performing an experiment.

A. Method of vibration reduction

The proposed SR motor has double-stage (the upper and lower stages) and single-phase electrification each stage, as shown in Fig. 2. Therefore, the proposed SR motor can be regarded as two of single-phase SR motor independently excited side by side.

Fig. 9 shows a method of excitation of conventional and proposed method and Fig. 10 shows windings of the proposed method. In this case, Δt is a transmission delay from the time when the vibration is observed at "measurement A" to it the vibration is generated as shown in Fig. 9 and T_s is a cycle of the natural frequency. In the proposed SR motor, the winding is simultaneously excited in each phase as shown in Fig. 1. However, the vibration level is high because the acceleration which is generated from each teeth is summed, as shown in Fig. 9 (a). However, in this proposed method, the winding is connected with 6 coils and makes double-phase in each stage (Fig. 10 shows only upper stage). When the proposed SR motor is driven by the proposed method, the winding of phase A is excited as with the conventional

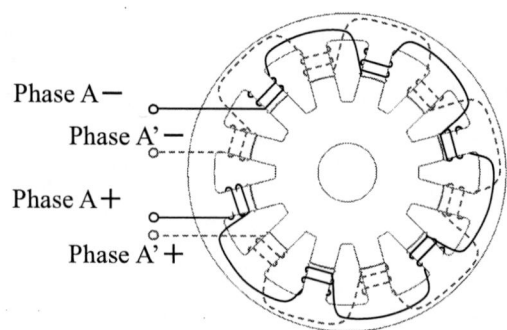

Fig. 10 Windings of the proposed method (upper stage).

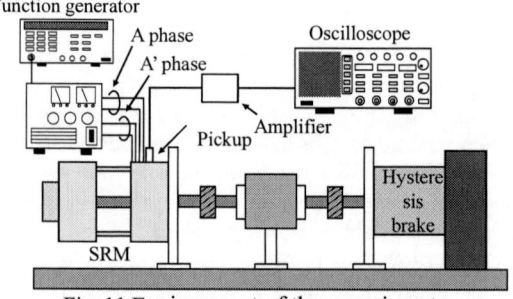

Fig. 11 Environment of the experiment.

method and phase A' is excited to be shifted half a cycle of the natural frequency (=T_s/2) in each stage as shown in Fig. 9 (b). Therefore, the vibration can be cancelled because a phase of acceleration is opposite each other. Although, two controllers are required to realize the proposed method, the vibration can be reduced. Also this method is possibility when the proposed SR motor is driven by using the single pulse voltage drive or hysteresis drive because this method only shifts the phase of the voltage in each stage. Therefore, the method can use in widely driving area. Oppositely, the conventional

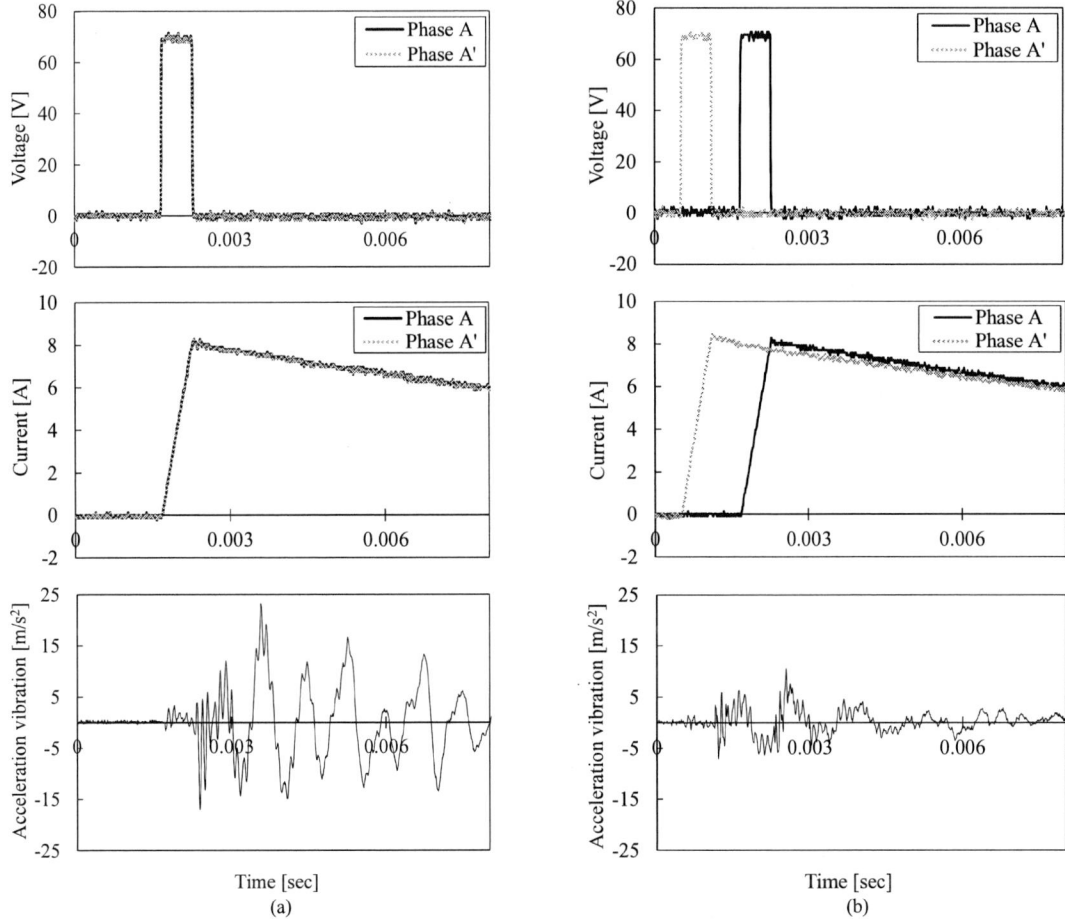

Fig. 12 Waveforms of the acceleration, voltage and current of phase A and phase A'. (a) Conventional method. (b) Proposed method.

three-phase SR motor cannot use this method due to the interference in phases.

B. Experimental evaluation

Fig. 11 shows an environment of the experiment. The voltage pulse waveform from a pulse generator is applied to SR motor on the test bed to theoretically validate the proposed method. The acceleration of the motor is detected by an acceleration pickup and measured with an oscilloscope. In this experiment, a firing angle of the voltage is manually shifted and measured the acceleration because the exact natural frequency cannot be measured. Also, the measurement point of the acceleration is" measurement A" which is a middle of two teeth in Fig. 9.

Fig. 12 shows the waveform of the acceleration, current and voltage of phase A and phase A'. As shown in this result, the proposed method has smaller acceleration than it of the conventional method.

VI. CONCLUSION

This paper proposed the double-stage SR motor with GO magnetic steel sheets. This machine can achieve higher torque production and indicates the possibility that the acceleration vibration is reduced. The effectiveness has been verified by performing some simulations and experiments. These results show that the proposed machine have high torque/current ratio and small vibration level in operation.

The proposed method to increase the torque/current ratio and reduction of the vibration can be used for many applications which require high torque/current ratio and low vibration. The proposed SR motor will be widely used in many different fields.

VII. REFERENCES

[1] D. E. Cameron, J. H. Lang and S. D. Umans, "The Origin and Reduction of Acoustic Noise in Doubly Salient Variable-Reluctance Motors", *IEEE Trans. on Industry Application,* vol. 28, no. 6, pp. 1250-1255, 1992.

[2] R. S. Colby, F. M. Mottier and T. J. E. Miller, "Vibration Modes and Acoustic Noise in a Four-Phase Switched Reluctance Motor," *IEEE Trans. on Industry Application*, vol. 32, no. 6, pp. 1357-1364, 1996.

[3] A. Chiba, M. Takeno, N. Hoshi, M. Takemoto, S. Ogasawara and M. A. Rahman, "Consideration of Number of Series Turns in Switched-Reluctance Traction Motor Competitive to HEV IPMSM", *IEEE Trans. on Industry Applications*, vol. 48, no. 6, pp. 2333-2340, 2012.

[4] P. A. Watterson, W. Wu, B. A. Kalan, H. C. Lovatt, G. Prout, J. B. Dunlop and S. J. Collocott, "A Switched-

Reluctance Motor/Generator for Mild Hybrid Vehicles", in Proc. *ICES.* pp. 2808-2813, 2008.

[5] Y. Matsuo, T. Higuchi, T. Abe, Y. Miyamoto and M. Ohto, "Characteristics of a Novel Segment Type Switched Reluctance Motor using Grain-Oriented Electric Steel", Int. Conf. *Electrical Machines and Systems (ICEMS)*, 2011.

[6] Y. Sugawara and K. Akatsu, "Characteristics of a Switched Reluctance Motor using Grain-Oriented Electric Steel Sheet", *ECCE Asia Downunder (ECCE Asia),* pp. 1105-1110, 2013.

[7] J. Li, X. G. Song, D. Choi and Y.H. Cho, "Research on Reduction of Vibration and Acoustic Noise in Switched Reluctance Motors", *ELECTROMOTION*, 2009.

[8] M. Sanada, S. Morimoto, Y. Takeda and N. Matsui, "Novel Rotor Pole Design of Switched Reluctance Motors to Reduce the Acoustic Noise", *IEEE. Industry Applications Conference*, vol. 1, pp. 107-113, 2000.

[9] A. H. Isfahani and B. Fahimi, "Multi-Physics Analysis of Double Stator Switched Reluctance Machines", *IEEE Energy Conversion Congress & Exposition (ECCE),* pp. 2827-2833, 2013.

[10] H. Y. Yang, Y. C. Lim and H. C. Kim, "Acoustic Noise/Vibration Reduction of a Single-Phase SRM Using Skewed Stator and Rotor", *IEEE Trans. on Industry Electronics*, vol. 60, no. 10, pp. 4292-4300, 2013.

[11] M. Crivii and M. Jufer, "Two-phase Two-Stack SR motor", Proc. of ICEM '98, Instanbul, vol. 3, pp. 1670-1673

[12] C. Pollock and C. Yao, "Acoustic Noise Cancellation Techniques for Switched Reluctance Drives", *IEEE Trans. on Industry Electronics*, vol. 33, no. 2, pp. 477-484, 1997.

[13] A. Michaelides and C. Pollock, "Reduction of Noise and Vibration in Switched Reluctance Motors : New Aspects," Industry Application Conference, 1996

[14] H. Makino, T. Kosaka and N, Matsui, "Some considerations on Noise and Vibration Reduction Control of SR motor", *Conference record of the technical meeting of IEE Japan,* RM-13-133, pp. 45-50, 2013 (in Japanese).

The 2014 International Power Electronics Conference

Improvement of Efficiency by Stepped-Skewing Rotor for Switched Reluctance Motors

Makoto Sugiura, Yuji Ishihara*, Hiroki Ishikawa, and Haruo Naitoh

Dept. of Electrical, Electronic and Computer Engineering
Gifu University
Gifu, JAPAN
ishikawa@gifu-u.ac.jp

Abstract—**This paper presents efficiency improvement of switched reluctance motors (SRMs) by stepped-skewing rotor (SSR). The cross sectional configuration of the SRM tested including size and shape of salient poles is almost same as those of a conventional SRM. The tested SRM is divided into 3 stacks, and only rotor is skewed. The skew angle has been designed to reduce both of torque ripple and radial force. The experiments have been carried out to confirm the effectiveness of efficiency improvement of the SRM with SSR. As a result, the highest efficiency of the SRM with SSR is achieved compared with that of the conventional SRM.**

Keywords— *Switched reluctance motors, stepped-skewing rotor.*

I. INTRODUCTION

The advantages of switched reluctance motors (SRMs) are permanent magnet free, simple structure, robust, inexpensive, reliable and suitable for higher speed applications, such as electric vehicle (EV) and hybrid EV applications. Some of SRMs has been developed for HEV and EV [1][2]. Requirements of motors for HEV and EV are higher efficiency, higher power density, lower cost, and so fourth. In reference [1], the developed SRM has realized as almost same level of efficiency as interior permanent magnet synchronous motors for HEV and EV.

SRMs with larger pole number can realize higher efficiency in larger output power range [3]-[5]. Most of SRMs for HEV and EV application have more poles. SRMs with smaller poles number, however, have an advantage in higher speed applications because such SRMs demands lower output frequency of drive circuit than SRMs with larger pole number under same rotational speed.

It is significant to improve efficiency of SRMs with smaller pole number. In case of such SRMs, configuration and core material are often changed to improve efficiency [6][7].

This paper deals with effects of stepped-skewing rotor (SSR) in SRMs. The tested SRM has 6 stator poles and 4 rotor poles. The cross sectional configuration of the rotor including motor size and shape of salient poles of the SRM tested is almost same as those of a conventional SRM. The tested SRM is divided into 3 stacks, and only

rotor is skewed. The skew angle is designed to reduce torque ripple and to distribute radial force.

In this paper, improvement in efficiency of the SRM with SSR is reported. Some experiments have been demonstrated to confirm the effectiveness of efficiency improvement, and the experimental results of the SRM with SSR are compared with those of the conventional SRM.

II. CONCEPT OF SWITCHED RELUCTANCE MOTORS WITH STEPPED-SKEW ROTOR

SRMs consist of a salient stator with excitation windings and a salient rotor with neither winding nor magnet. Absence of windings and magnets in rotor poles brings merits of simple structure, robust, inexpensive, reliable, and suitable for higher speed applications.

A. Problems of the conventional SRM

Cross sectional view of a conventional 6/4 SRM in laboratory is shown in Fig. 1. Rotor position θ is defined that aligned position is zero and the position is positive in clockwise as shown in Fig. 1.

Assuming that rotor position is shown in Fig. 1, current flows through winding on stator poles A and A' to rotate the rotor in clockwise direction. Just before aligned position, the current is turned off, and next winding on stator poles B and B' is energized to pull rotor poles Y and Y'.

Attractive force by energized winding can be split into torque and radial force. The magnitude of attractive force depends on rotor position, the magnitude of motor current, and magnetic saturation level of cores. Torque and radial

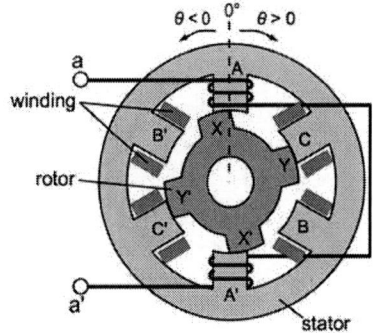

Fig. 1. Cross sectional view of an SRM.

978-1-4799-2706-7/14 $31.00 © 2014 IEEE   1135

force are not constant even if attractive force is constant. The radial force, therefore, causes stator vibration, and torque has ripple component.

Figure 2 shows an example of current and torque waveforms of conventional SRM with one-voltage-pulse control. Motor current is not constant even though winding voltage is constant from θ =-40° to θ =-10°. As a result, torque has large ripple component. It can be seen that torque decreases to a lower level around aligned position, especially, in Fig. 2(b). At the position, the SRM is driven by two-phase conduction mode. A rotor pole is aligned position. The attractive force is nearly equal to radial force, and less torque is generated. The next rotor pole is unaligned position, and smaller attractive force is generated. This is a reason of lower torque generation around aligned position. The proposed SRM with SSR is effective in order to the problem because next stack can compensate for lack of torque.

B. Design of skew angle for the SRM with SSR

The proposed SRM with SSR has 6 stator poles, and the mechanical angle θ_s between stator poles is 60° to be compared with characteristics of the conventional SRM. The skew angle of the rotor is designed to reduce both of torque ripple and vibration as follows.

The number of stacks is chosen as three to avoid larger scale drive system because the number of drive circuit is proportional to the number of stacks.

The rotor skew angle θ_{skew} is designed to reduce torque ripple and distribute radial force. In case of θ_{skew}=10° in Fig. 3, torque ripple can be reduced because rotor position of each stack is not aligned simultaneously. Motor current, however, often flows through same phase winding of each stack concurrently as shown in Fig. 4,

and radial force is generated in same direction.

Figure 5 illustrates an example of rotor position of each stack at θ_{skew}=20°. In this case, rotor poles of each stack arrive in aligned position at different timing, and torque ripple reduction can be achieved as shown in Fig. 6. Motor current flows through different phase winding of each stack, and radial force can be distributed.

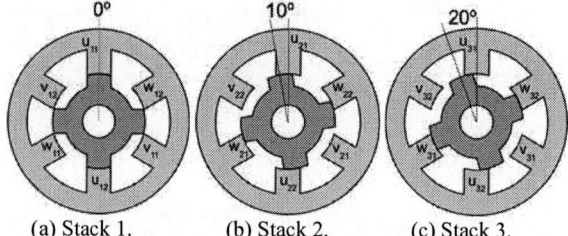

(a) Stack 1.   (b) Stack 2.   (c) Stack 3.

Fig. 3. Rotor position of each stack at θ_{skew}=10°.

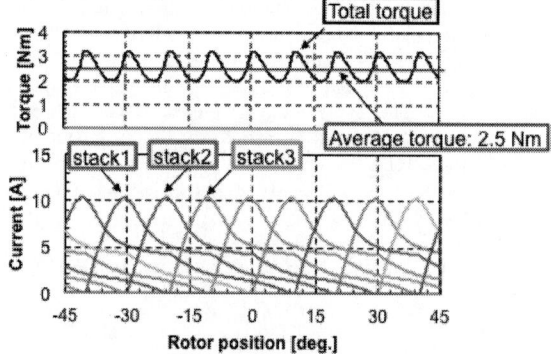

Fig. 4. Waveforms of torque and phase-u current of each stack at θ_{skew}=10°.

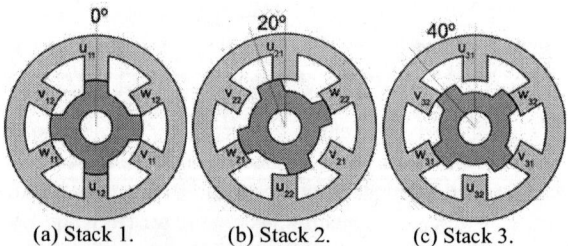

(a) Stack 1.   (b) Stack 2.   (c) Stack 3.

Fig. 5. Rotor position of each stack at θ_{skew}=20°.

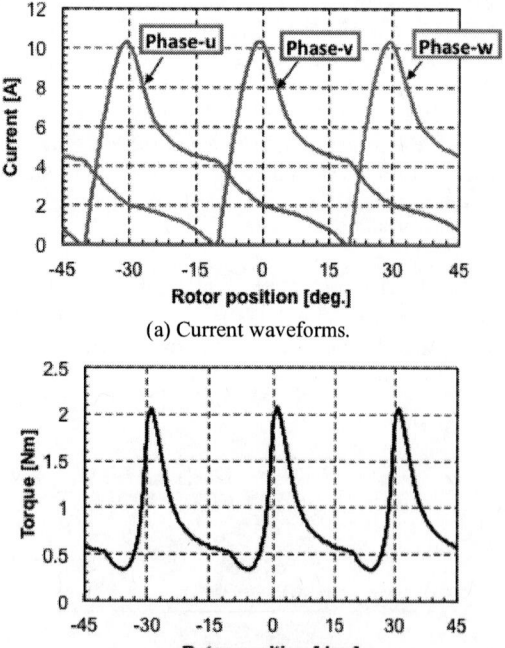

(a) Current waveforms.

(b) Torque wavwform.

Fig. 2. An example of simulation results of the conventional SRM with one-voltage-pulse control.

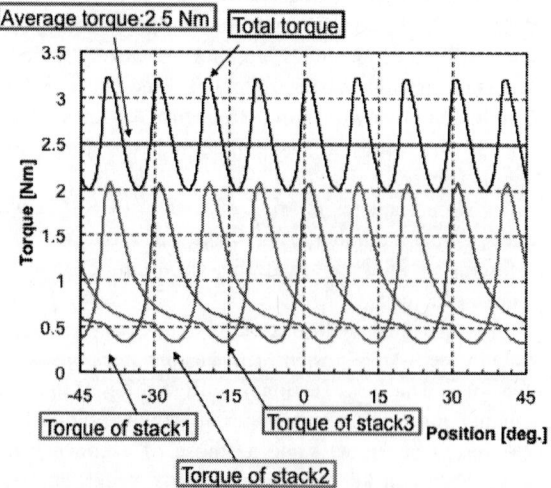

Fig. 6. Torque waveform at θ_{skew}=20°.

978-1-4799-2706-7/14 $31.00 © 2014 IEEE

The θ_{skew} is set to 30º in Fig. 7. Motor current always flows through different phase winding of each stack, and radial force distribution can be realized. Torque ripple is, however, larger than that of the conventional SRM as shown in Fig. 8 because motor current in each stack is synchronized completely.

From above consideration, θ_{skew} is designed as 20º.

C. Configuration of the SRM with SSR

Figure 9 shows outline of the tested 6/4 SRM with SSR, and the specifications of the SRM with SSR and the conventional SRM are listed in Table I. Stator and rotor size of the tested SRM with SSR is designed to be almost same as that of the conventional SRM in Fig. 1. The magnetization characteristic of iron core of the tested SRM with SSR is also almost same as that of the conventional SRM.

The rotor is divided into 3 stacks with stepped-skewing as shown in Fig. 9. The skew angle is designed as every 20 degrees for each stack, as shown in Fig. 5 to reduce torque ripple and to distribute radial force. The stator is also divided into 3 stacks without skewing. Each separated stator pole has a winging as shown in Fig. 10.

Currents through windings in each stack are controlled depending on rotor position of each stack. Assuming that the rotor position of each stack is shown in Fig. 5, currents flow through windings of v_{11} and v_{12} in stack 1, u_{21} and u_{22} in stack 2, and w_{31} and w_{32} in stack 3. As a result, the rotor rotates in clockwise. In this case, radial force is distributed in 3 directions, and vibration of the motor can be reduced.

D. Characteristics of the SRM with SSR

The comparison of characteristics is shown in Figs. 11 and 12. The aligned rotor position is defined as zero degree. The characteristics can be obtained by 2-D finite elements analysis (FEA). The reasons are follows.

The 3-D FEA have been carried out as first step to confirm flux linkage from a stack to the next stack. As a result, the flux linkage between stacks is very small, and the effect of mutual induction can be ignored. The 3-D FEA requires much calculation time to obtain useful characteristics of SRM with SSR, and the characteristics can be obtained from superposition of the results by 2-D FEA for each stack with shorter calculation time.

Figure 11(a) shows the characteristics of a phase winding inductance of the conventional SRM with respect to motor current, and (b) shows that of a stack in the SRM with SSR. The motor current range is set to 0-10A. The characteristics have similarity shape, but each value of the SRM with SSR is about one-third smaller than that of the conventional SRM because stack length of the SRM with SSR is also about one-third shorter than that of the conventional SRM as listed in Table I.

Figure 12(a) shows the torque characteristics of the conventional SRM, and (b) shows that of a stack in the SRM with SSR. The shape of torque characteristics is also similar, and the magnitude of a stack in the SRM with SSR is about one-third smaller than that of the conventional SRM. Total torque of the SRM with SSR is sum of generated torque in each stack.

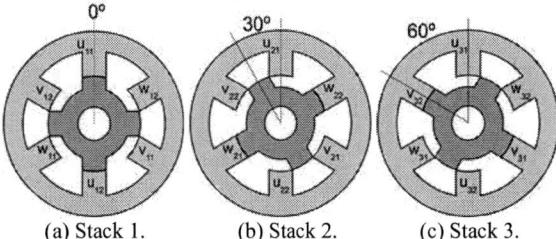

(a) Stack 1.   (b) Stack 2.   (c) Stack 3.

Fig. 7. Rotor position of each stack at θ_{skew}=30º.

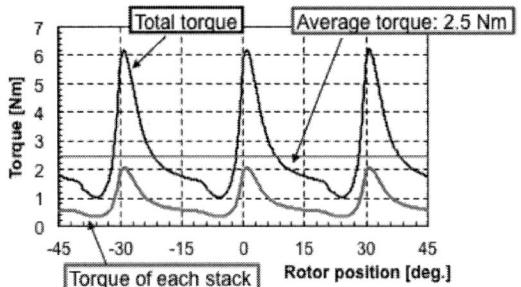

Fig. 8. Torque waveform at θ_{skew}=30º.

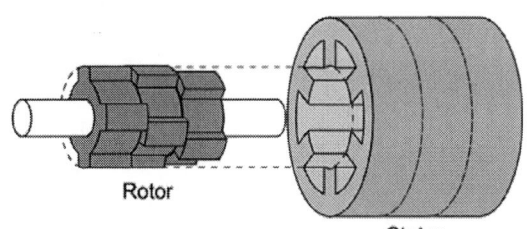

Fig. 9. Tested SRM with stepped-skewing rotor.

TABLE I
SPECIFICATIONS OF THE TESTED SRM.

	Conventional	With SSR
Rated power	1.5kW	
Stator poles	6	
Rotor poles	4	
Outer diameter of stator	155mm	
Outer diameter of rotor	79.4mm	
Turn number	80turns/pole	
Stack length	80mm	3x30mm
Winding resistor	0.524Ω/phase	0.375Ω/phase
Iron core	50RM400	35H360

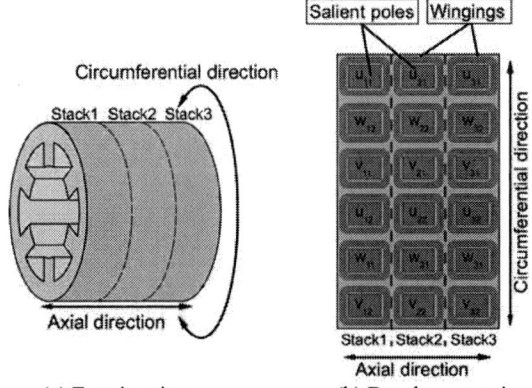

(a) Exterior view.   (b) Development view.

Fig. 10. Salient poles and windings in stator of the SRM with SSR.

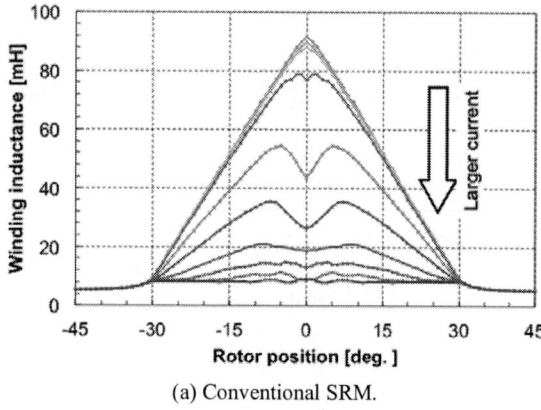

(a) Conventional SRM.

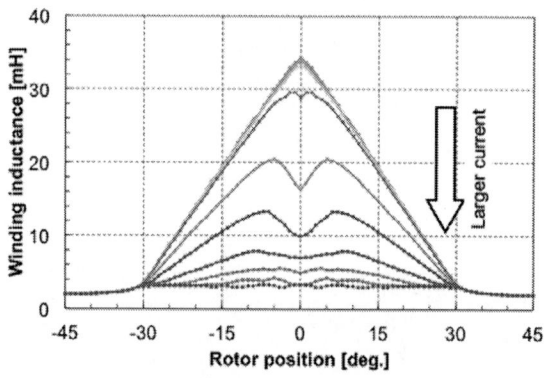

(b) a stack in SRM with SSR.

Fig. 11. Comparison of a phase winding inductance characteristics by FEA.

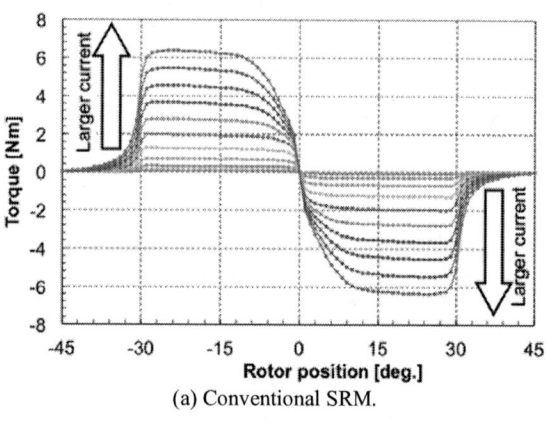

(a) Conventional SRM.

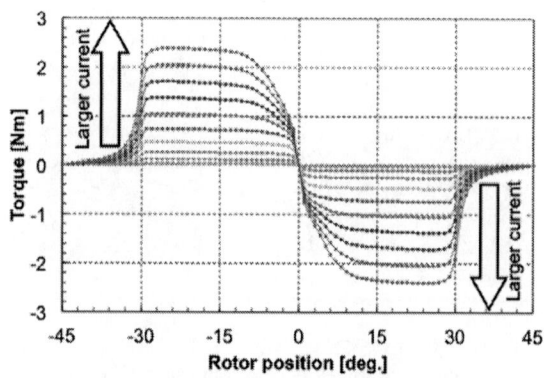

(b) a stack in SRM with SSR.

Fig. 12. Comparison of torque characteristics by FEA.

III. EXPERIMENTAL RESULTS

The experiments have been carried out in order to confirm the effectiveness of efficiency improvement effect of the SRM with SSR.

Figure 13 illustrates the typical asymmetric H-bridge SRM drive circuit. For the SRM with SSR drives, a drive circuit is connected to a stack as shown in Fig. 14.

Motor current is controlled by one-voltage-pulse control and flat-topped current control. The one-voltage-pulse control is the simplest current control for SRMs with only turn-on angle θ_{on} and turn-off angle θ_{off}. The θ_{on} and θ_{off} are fixed in the experiments as listed in Table II, and source voltage is adjusted to match load conditions. In the flat-topped current control, current is controlled to be constant from θ_{on} to θ_{off} with hysteresis current regulator.

Figure 15 represents the phase voltage and phase current waveforms of stack 1 in the SRM with SSR at 500rpm and torque reference T^*=2[Nm] with one-voltage-pulse control. The source voltage E is adjusted as 16 V to match the load condition. Figure 16 shows current waveforms of phase-u in each stack. From Figs. 15 and 16, it is confirmed that the SRM with SSR is operated properly. Figures 17 and 18 show the phase voltage and current waveform of the conventional SRM with one-voltage-pulse control at E=60[V] and with flat-topped current control at E=50[V], respectively. The winding inductance of the conventional SRM is larger than that of the SRM with SSR as shown in Fig. 11. The higher source voltage E of the conventional SRM is, therefore, needed than that of the SRM with SSR to operate under same load condition.

The efficiency of SRM with SSR for rotor speed with respect to reference torque T^* are shown in Fig. 19. The maximum efficiency 90.2% is achieved at 1500rpm and T^*=1 [Nm].

Figure 20 shows comparison results of the efficiency of SRM with SSR with that of the conventional SRM at T^*=2[Nm]. In case of the conventional SRM, the efficiency with flat-topped current control is higher than that with one-voltage-pulse control. The efficiency of SRM with SSR under one-voltage-pulse control is improved substantially compared with that of the conventional SRM with both of the controls.

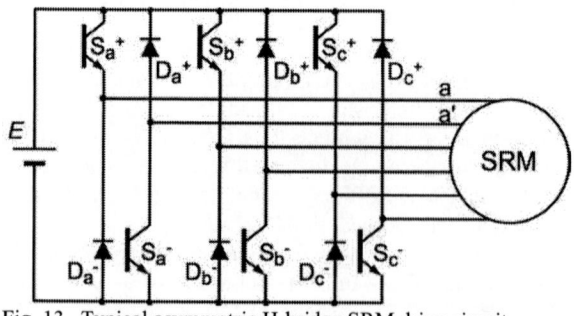

Fig. 13. Typical asymmetric H-bridge SRM drive circuit.

The 2014 International Power Electronics Conference

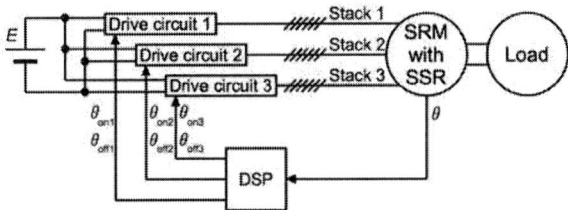

Fig. 14. Experimental setup.

TABLE II
EXPERIMENTAL CONDITIONS

Turn-on position	-40°
Turn-off position	-10°

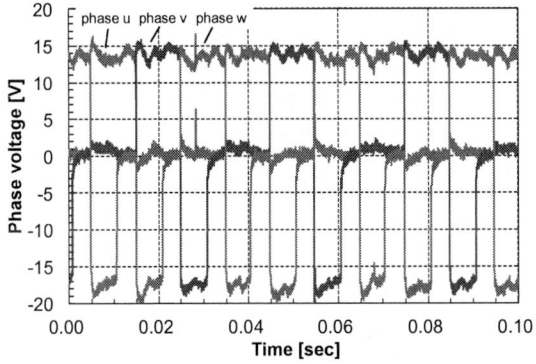

(a) Voltages across phase windings.

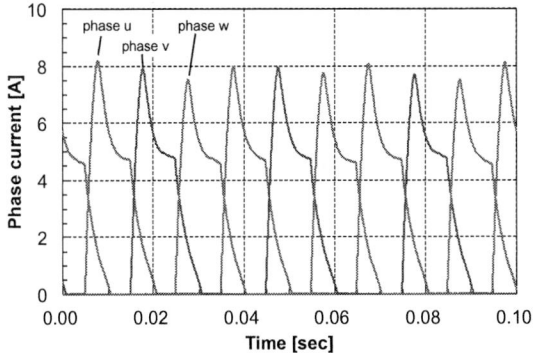

(b) Phase currents.

Fig. 15. Waveforms of phase current in stack 1 of the SRM with SSR at 500rpm and T^*=2Nm.

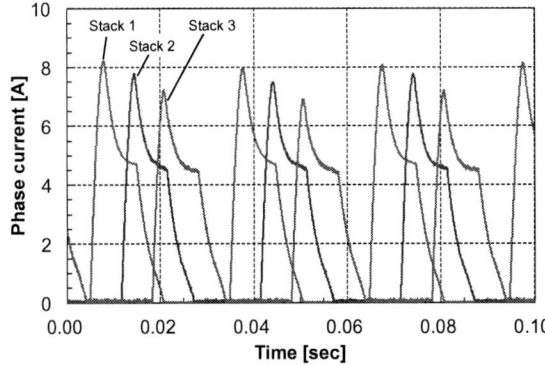

Fig. 16. Waveforms of phase-u current in each stack at 500rpm and T^*=2Nm.

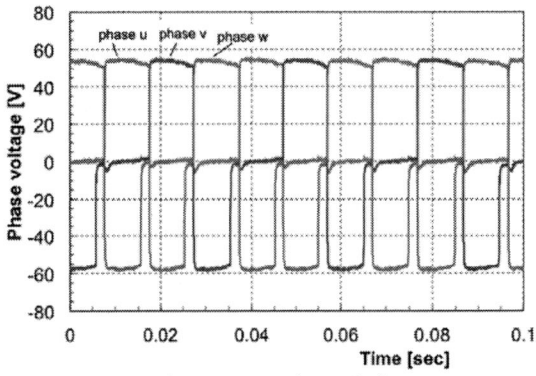

(a) Voltages across phase windings.

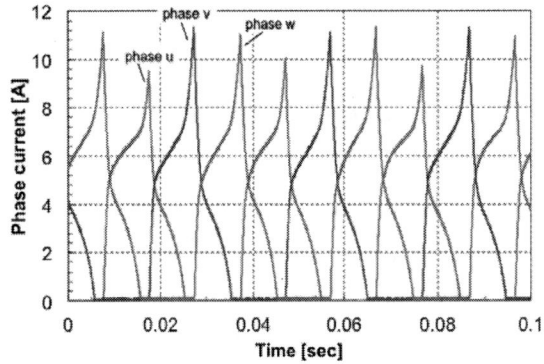

(b) Phase currents.

Fig. 17. Experimental waveforms of the conventional SRM with one-voltage pulse control at 500rpm and T^*=2Nm.

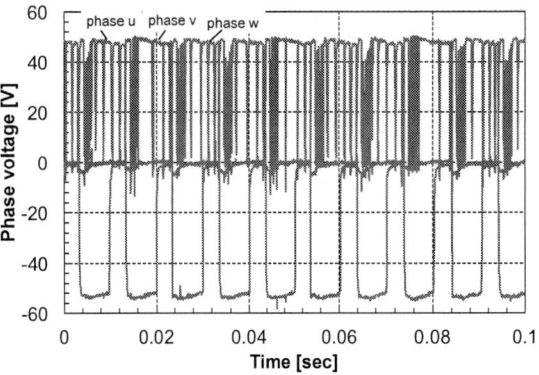

(a) Voltages across phase windings.

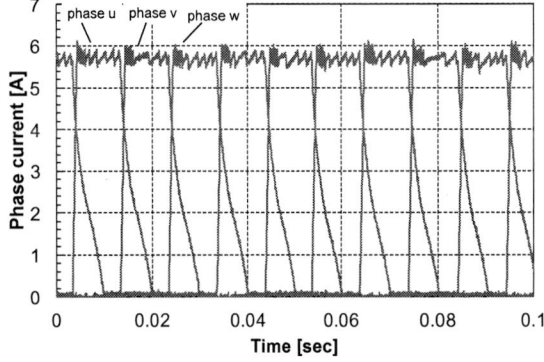

(b) Phase currents.

Fig. 18. Experimental waveforms of the conventional SRM with flat-topped current control at 500rpm and T^*=2Nm.

978-1-4799-2706-7/14 $31.00 © 2014 IEEE   1139

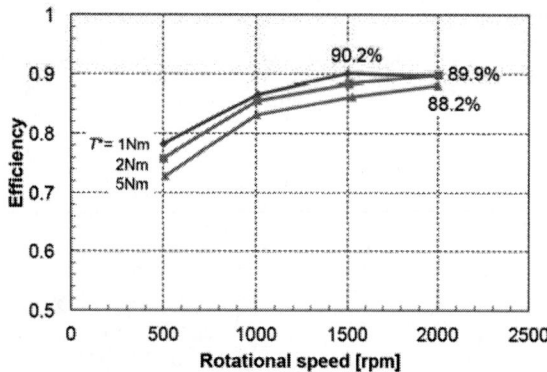

Fig. 19. Efficiency characteristics of the SRM with SSR under one-voltage-pulse control.

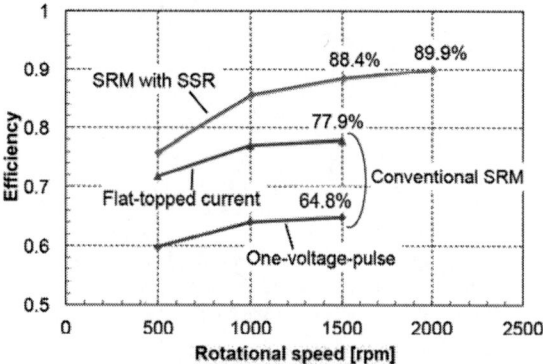

Fig. 20. Comparison of efficiency between the conventional SRM and the SRM with SSR at T^*=2Nm.

As a reason of higher efficiency realization in the proposed SRM, iron loss is reduced. Compared with the conventional SRM with one-voltage-pulse control, the SRM with SSR can generate same level of torque by smaller current. Current peak of the SRM with SSR can be also smaller than that of conventional SRM as shown in Figs. 15 and 17. The conventional SRM with flat-topped current control has larger iron loss by current ripple depending on switching operation. The efficiency of the proposed SRM with SSR, therefore, can realize the highest efficiency among them.

IV. CONCLUSIONS

This paper has reported the improvement of efficiency of the SRM with SSR. The cross sectional configuration of the rotor including motor size and shape of salient poles of the SRM tested is almost same as those of a conventional SRM. The designed SRM is divided into 3 stacks, and only rotor is skewed. The skew angle is designed as 20 degrees to reduce both of torque ripple and radial force.

The experiments have been carried out in order to confirm the effectiveness of efficiency improvement effect of the SRM with SSR. As a result, the maximum efficiency of more than 90% has been achieved. Compared with the conventional SRM, the efficiency of the proposed SRM with SSR is improved more than 10 points.

Winding inductance of a stack in the proposed SRM with SSR is about one-third smaller than that of the conventional SRM. Back EMF coefficient is also one-third smaller. The proposed SRM with smaller parameter can realize to expand drive condition because of smaller back EMF at higher speed. Especially, application of instantaneous current control, such as reduction of torque ripple and eddy current loss[8], [9], to the proposed SRM can be realize less current distortion at higher speed range. It is, therefore, possible to construct more sensitive speed and position control systems for the proposed SRM with SSR.

ACKNOWLEDGMENT

This research was supported by a Research and Development Program for Innovative Energy Efficiency Technology from New Energy and Industrial Technology Development Organization (NEDO) of Japan.

REFERENCES

[1] Kyohei Kiyota, and Akira Chiba, "Design of Switched Reluctance Motor Competitive to 60-kW IPMSM in Third-Generation Hybrid Electric Vehicle," *IEEE Transactions o Industry Applications*, VOL. 48, NO. 6, pp. 2303-2309, 2012

[2] Saurabh P. Nikam, Vandana Rallabandi, and B. G. Fernandes, "A High-Torque-Density Permanent-Magnet Free Motor for in-Wheel Electric Vehicle Application," *IEEE Transactions on Industry Applications*, VOL. 48, NO. 6, pp. 2287-2295, 2012.

[3] Hassan Moghbelli, Gayle E. Adams, and Richard G. Hoft, IEEE "Performance of a 10-Hp Switched Reluctance Motor and Comparison with Induction Motors," *IEEE Transactions on Industry Applications*, VOL. 21, NO. 3, pp. 531-538, 1991.

[4] Pavol Rafajdus, Valeria Hrabovcova and Peter Hudakt, "Investigation of Losses and Efficiency in Switched Reluctance Motor," *EPE-PEMC 2006*, pp. 296-301, 2006.

[5] Vladan P. Vuji˘ci´c, "Minimization of Torque Ripple and Copper Losses in Switched Reluctance Drive," *IEEE Transactions on Power Electronics*, VOL. 27, NO. 1, pp. 388-399, 2012.

[6] Keunsoo Ha, Cheewoo Lee, Jaehyuck Kim, R. Krishnan, and Seok-Gyu Oh, "Design and Development of Low-Cost and High-Efficiency Variable-Speed Drive System With Switched Reluctance Motor," *IEEE Transactions on Industry Applications*, VOL. 43, NO. 3, pp. 703-713, 2007.

[7] Akira Chiba, Hiroaki Hayashi, Kensaku Nakamura, Shinya Ito, Kanokvate Tungpimolrut, Tadashi Fukao, M. Azizur Rahman, and Mitsunobu Yoshida, "Test Results of an SRM Made From a Layered Block of Heat-Treated Amorphous Alloys," *IEEE Transactions on Industry Applications*, VOL. 44, NO. 3, pp. 699-706, 2008.

[8] Hiroki Ishikawa, Yuki Kobayashi, and Haruo Naitoh, "A Novel Position Control System with Torque Ripple Reduction for SRMs," *Proceedings of the 37th Annual Conference of the IEEE Industrial Electronics Society*, pp. 1651-1656, November 7-10, 2011.

[9] Hiroki Ishikawa, Masahide Uraji, and Haruo Naitoh, "A New Current Control for Eddy Current Loss Reduction of Switched Reluctance Motors," *IEEE-COMPEL 2013*, 2013.

A Single Phase SRM Driven by Commercial AC Power Supply

Kohei Aiso
Shibaura Institute of Technology
Department of Electrical
Engineering and Computer Science
3-7-5, Toyosu, Koto-ku, Tokyo,
Japan
ma13001@shibaura-it.ac.jp

Noriya Nakao
Shibaura Institute of Technology
Department of Electrical
Engineering and Computer Science
3-7-5, Toyosu, Koto-ku, Tokyo,
Japan
nb12105@shibaura-it.ac.jp

Kan Akatsu
Shibaura Institute of Technology
Department of Electrical
Engineering
3-7-5, Toyosu, Koto-ku, Tokyo,
Japan
akatsu@shibaura-it.ac.jp

Abstract-**This paper presents a novel single phase switched reluctance motor (SRM) which can be directly driven by commercial AC power supply without any switching circuit and sensor. This proposed SRM is optimized based on a simple design technique. The optimized machine has aimed to be higher motor efficiency than conventional induction motors (IMs). In addition, the novel SRM can achieve the start-up with directly connecting commercial AC power supply. In this paper, the performance of the prototype single phase SRM is verified by performing simulations and experiments.**

Keywords— commercial power AC supply, single phase SRM , SRM designe method , start up charactersitics,

I. INTRODUCTION

Single phase induction motors (IMs) have been widely used in home appliances such as fans, refrigerators, washing machines, hoods cooker, pumps and so on because direct drive by connecting commercial AC power supply is possible [1]. The manufacturing cost of their drive system is low because no circuit and sensor is required. However the motor efficiency is much lower than other types of electric motors. To reduce the electric power consumption by electric machines developing high efficiency single phase motors without any drive circuits and sensors is an essential requirement in the industrial field.

From the view point of the efficiency, Permanent magnet (PM) machines are expected as a candidate of IMs because they have high motor efficiency. However, PM motors are more expensive than IMs because of using rare earth magnets. Furthermore, PM motor drives generally require the rectifier and switching circuits and some sensors [2].

In the previous researches, some single phase switched reluctance motors (SRMs) for industrial applications have been presented. They have advantages such as low manufacturing cost and simple structure. In addition, these machines can achieve higher motor efficiency than IMs [3]. In conventional single phase SRMs, the numbers of stator and rotor poles are equal (For example, 2/2, 4/4, 6/6, and 8/8). The driving methodology of the single phase SRMs is that all stator coils are simultaneously excited to generate positive a torque when the stator and the rotor poles are in unaligned position, the excitation current is immediately turned off when the rotor poles are aligned with the stator poles, and the rotor keeps moving due to stored kinetic energy [4]. Consequently, the torque is discontinuous in the conventional SR motor drive. Because of this torque production mechanism, the start-up of single phase SRMs is difficult when rotor and stator poles are completely aligned or the rotor is in a position where negative torques are generated by armature excitation currents. Furthermore, their drives also require rectifier and switching circuits and sensors to implement the current control.

To overcome the start-up problem of conventional single phase SRMs, some researches which improve the motor structure have been reported [5]-[7]. In [5], an advanced design method of single phase SRMs has been introduced. This technique gives the geometry parameter calculation process for the optimal machine design to achieve the smooth torque production. In [6], a novel skewed motor structure overcomes the self-starting problem. The arc of rotor pole is changed to produce starting torque. In [7], a saturation-based starting method is used to start up. In the method, the saturable areas such as a simple hole are added on the rotor reduce the flux, it is possible to produce continuous torque at every rotor position and consequently to start by shifting the aligned position with varied current. Meanwhile it has reported that the smooth start-up is possible with a permanent magnet on the stator in [8]. The stator magnet pulls the rotor away from the alignment at stand-still (i.e. the magnet position is adjusted to achieve the maximum torque production at starting). In [9], magnetically insulated attraction elements which are called a shorting ring form a shade pole on the rotor to achieve start-up. If the ring is aligned with the stator pole, a change in stator coil excitation will produce a current in the ring which makes a magnetic flux of producing torque.

However, although the above techniques have been discussed on SRM drive system including switching

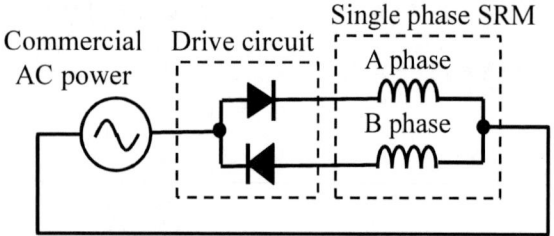

Fig. 1 Drive system of a novel single phase SRM

circuits and sensors, they are unsuited for low-price industry field since the system is complex.

In this research, a novel single phase SRM has been proposed [10]. Fig. 1 shows a drive system of a novel single phase SRM. The proposed machine can be driven by directly connecting to commercial AC power supply. To generate the continuous torque and to achieve the smooth start-up, stator poles are sequentially energized by exciting two insulated coils in the proposed SRM drive. Its drive circuit consists of only two diodes and it generates the two-phase excitation current from commercial AC power supply. Therefore, no switching device is needed in the circuit and the manufacturing-cost is low. In this paper, the optimal design parameter calculation of the proposed single phase SRM is also presented. The performance is verified by performing simulations and experiments. The results show the proposed SRM can be a candidate against the commonly-used IMs.

II. PROPOSED SINGLE PHASE SRM

As described in the above, typical single phase SRMs deliver the discontinuous shaft torque. Due to the torque production mechanism, the start-up is difficult when rotor and stator poles are completely aligned or the rotor is at a position where the negative torque is generated by excitation currents. To overcome the above problem, a novel motor structure and drive circuit are proposed in this research.

A. Machine Structure

Fig. 2 shows the proposed motor structure and the design parameters are shown in Table I. The proposed machine has 6 and 12 rotor and stator poles respectively. The armature concentrate windings are wound on each stator pole and the two-phase circuit is formed. The stator poles are sequentially energized by exciting two-phase windings. Therefore, the proposed SRM requires separated two-phase unipolar current.

B. Drive Circuit

As shown in Fig.1, the circuit consists of only two diodes and it rectifies the commercial AC power source into separated two-phase unipolar voltage. The SRM deliver a continuous shaft torque by applying the sequential current excitation and the smooth start-up is possible. In addition, the manufacturing cost of the drive

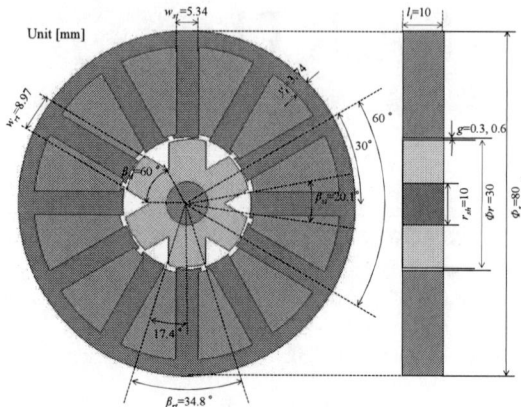

Fig. 2 Proposed motor structure

TABLE I

DIMENSION OF THE DESIGNED MOTOR OF SINGLE PHASE SRM

Number of rotor poles	6
Number of stator poles	12
Stator outer diameter : Φ_s [mm]	80
Rotor radius : r [mm]	15
Stack length : l_l [mm]	10
Air gap length : g [mm]	0.3,0.6
Shaft diameter : r_{sh} [mm]	10
Stator pole arc angle : β_{st} [deg]	20.1
Slot area : S_l [mm^2]	84.2
Number of turn : N [turn/slot]	1442

system is low because the circuit includes no switching device.

C. Principles of torque generation

The output torque of an SRM is expressed as follow:

$$T = \frac{1}{2} \frac{\partial L(i,\theta)}{\partial \theta} i^2 \qquad (1)$$

where T, L, θ, and i are the output torque, inductance, rotor position, and phase current, respectively. Then, rotor position θ is expressed as follow based on electric angle.

$$\theta = P\theta_m \qquad (2)$$

where P is pairs of poles, θ_m is mechanical angle. As described in (1), the output torque depends on the self-inductance variation and the phase current. Therefore, the SRM generates positive torque when the inductance increases. Fig. 3 shows the excitation current and inductance waveform for the proposed single phase SRM. In this paper, the unaligned and aligned positions are defined as shown in this figure (unaligned: 0 degrees, aligned: 90 degrees in the electric angle). An SRM delivers the positive torque in each phase when the rotor position is in the range from 0 to 90 degrees. The proposed SRM can achieve the smooth torque production by applying separated two-phase currents which are shifted 90 degrees each other in the electric angle.

978-1-4799-2706-7/14 $31.00 © 2014 IEEE   1142

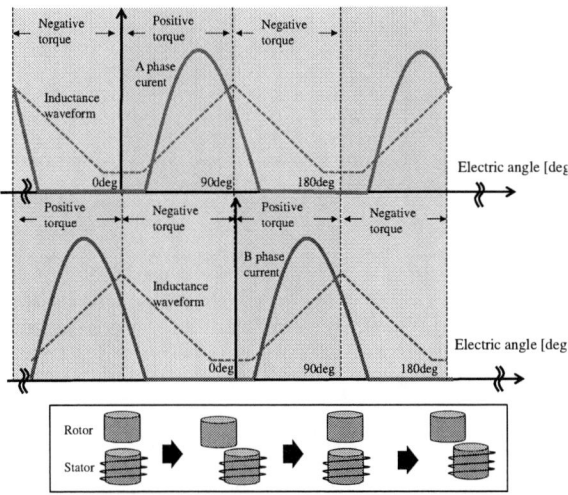

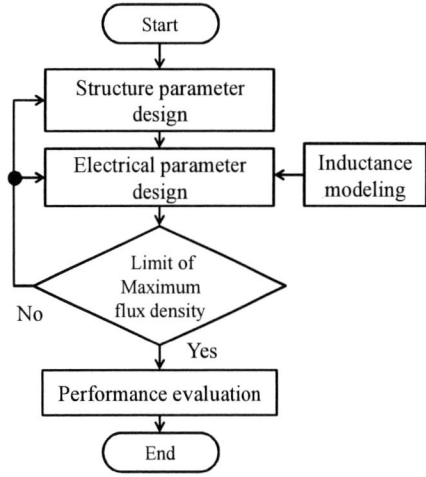

Fig. 4 Design flow chart

Fig. 3 Output current waveform and inductance waveform of the proposed single phase SRM

III. DESIGN METHOD OF SINGLE PHASE SRM

In this section, the design method for the proposed single phase SRM is presented. Fig. 4 shows the design flow chart. As shown in Fig. 4, the design flow chart is categorized into three main parts which are structure parameter design, electrical parameter design and performance evaluation. In the sturucture parameter design part, the detail of the motor structure is determined. In the electrical prameter design part, the magnetomotive force which includes number of turns and current is determined. In the performance evaluation part, the mechanical output, motor efficiency, copper loss of the design motor are calculated and evaluated.

In the single phase SRM driven by the voltage source, the mechanical output and motor efficiency depend on number of turns because output current depends on the inductance distribution and coils resistance which are determined by number of turns. Therefore, the electric parameter design is mainly presented in this section. In the electric parameter design, the keys to establish the efficient machine design are the inductance modeling, and performance calculation.

A. Structure parameter design

In the design of the single phase SRM, there are more than ten design parameters of motor structure such as stator teeth width: w_{st}, stator yoke length: y_{sh}, rotor teeth width: w_{rt}, and so on as shown in Fig. 1. However, their simultaneous design is difficult. To simplify the complexity, the design parameters are considered as a function of two variable parameters: rotor radius: r, stator pole arc angle: β_{st}. In the proposed SRM design, these two parameters are simultaneously determined to maximize the motor efficiency. The detail design of structure parameter such as rotor radius, stator pole arc angle is presented in [10].

B. Electrical parameter design

In the electrical parameter design, the number of turns which achieves maximum motor efficiency is determined. The motor efficiency is calculated by the effective power and mechanical output. Then, inductance modeling is important because the inductance is used to calculate the output current. Therefore, their performance significantly depends on the magnetic permeability distribution in the motor. Generally, the self-inductance depends on number of turns, the position and current (i.e. the inductance is effected by the spatial harmonics and magnetic saturation) and its spatial distribution includes the components of AC and DC as follow:

$$L = L_{DC} + L_{AC} \qquad (3)$$

where L_{DC} and L_{AC} are DC component and AC component. The accurate inductance distribution is obtained by the finite element analysis (FEA). However, the inductance must be calculated in each machine design. Therefore, it takes a long time to implement the design process as shown in Fig. 4 if the inductance calculation is implemented by the FEA. In this paper, since to calculate the current from the inputted voltage is the first objective, it is assumed that the inductance spatial distribution is negligible only to obtain the current. Adding that, the proposed machine has a lot of coils wound on the stator teeth to reduce the current, the inductance variation including spatial harmonics is negligible because the DC component of inductance is much larger than AC component of that ($L_{DC} \gg L_{AC}$). Therefore, inductance is expressed as follow:

$$L \cong L_{DC} \qquad (4)$$

Also, as shown in Fig. 4, the maximum flux density is limited to a value to ignore the magnetic saturation effect. Therefore, the inductance is expressed as a function of only number of turns.

$$L = k_{Lrms}(S_t / l_m)N^2 \qquad (5)$$

where S_t is cross-section area of stator teeth, l_m is flux

The 2014 International Power Electronics Conference

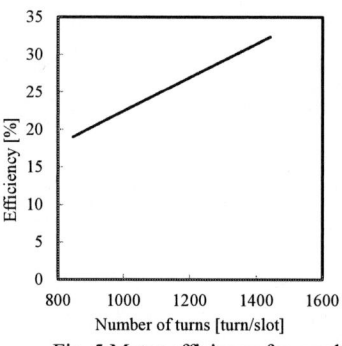

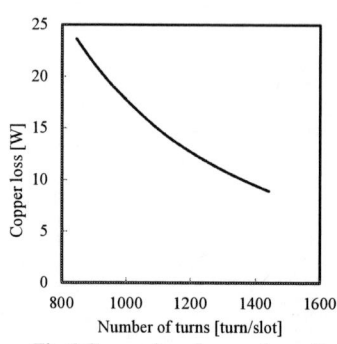

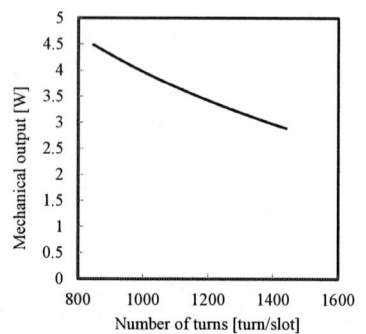

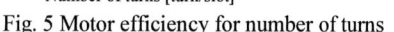

Fig. 5 Motor efficiency for number of turns Fig.6 Copper loss for number of turns Fig.7 Mechanical output for number of turns

path length, N is number of turns, k_{Lrms} is a constant of proportion which are based on the FEA result [10].

C. Performance evaluation

In the performance evaluation part, the copper loss, mechanical output and motor efficiency are calculated and evaluated to determine the optimum number of turns which achieves high motor efficiency. First, the output current and torque are calculated based on inductance expressed as (5) to calculate the mechanical output, copper loss and motor efficiency.

Here, the output current is derived from the voltage equation. The voltage equation is expressed as follow when commercial AC power supply is input.

$$V = Ri + L\frac{di}{dt} \qquad (6)$$

where V and R are the voltage amplitude and winding resistance. Generally, it is difficult to solve the equation because the inductance has non-linear characteristic such as spatial harmonics and magnetic saturation. However, (6) can be simply solved in the design method because the inductance can be assumed as a constant in (5). Therefore, the output current is expressed as follow when (5) is substituted into (6).

$$i(t) = \frac{V_0}{6\sqrt{R^2 + (\omega L)^2}}\left[\frac{\omega L e^{-\frac{R}{L}t}}{\sqrt{R^2 + (\omega L)^2}} + \sin\left\{\omega t - \tan^{-1}\left(\frac{\omega L}{R}\right)\right\}\right] \qquad (7)$$

where ω is electric angular velocity. Therefore the effective current is calculated by (7). The torque of SRM is expressed as (1). Generally, although the average torque is obtained by integrating instantaneous values calculated from (1), it is difficult to accurately calculate the instantaneous values of the output torque by (1) because inductance distributions have non-linear characteristic. However, in the design method, the inductance can be assumed constant (5). Therefore, the average torque is expressed as follow by the production of the inductance (5) and the squared effective current.

$$T_e = k_{Te}LI_{rms}^2 \qquad (8)$$

where T_e is average torque, I_{rms} is effective current, k_{Te} is a coefficient which is based on the FEA result [10]. Therefore, the copper loss is expressed as follow:

$$P_c = RI_e^2 \qquad (9)$$

The mechanical power is expressed as follow:

$$P_m = T_e \times \omega_m \qquad (10)$$

where P_m is the mechanical output, ω_m is mechanical angular velocity. The effective power is expressed as follow:

$$P_e = \frac{1}{T}\int_0^T v(t)i(t)dt \qquad (11)$$

where P_e is the effective power, v is single phase voltage. Therefore, the motor efficiency is calculated by (10) and (11).

D. Design result

Fig. 5 shows the calculated motor efficiencies for the each number of turns. As shown in Fig. 5, the highest motor efficiency 32% is achieved when the number of turn is 1442 turn/slot with the 60% space factor. Fig. 6 and Fig. 7 show the calculated copper loss and mechanical output for each number of turns, respectively. As shown in Figs. 5 and 6, the motor efficiency is increased as increasing the number of turns because the copper loss is decreased. Oppositely, as shown in Fig. 7, the mechanical output increases as the number of turn decreases. Therefore, the motor efficiency and mechanical output are relation of a trade-off. In the paper, as shown in Fig. 1, the number of turns is determined to 1442 turn/slot to achieve high motor efficiency. Fig. 8 shows a comparison result of the designed motor with a general commercial IM with shading coil as same motor volume. As shown in Fig. 8, the calculated efficiency is much higher than it of IMs with shading coil.

978-1-4799-2706-7/14 $31.00 © 2014 IEEE

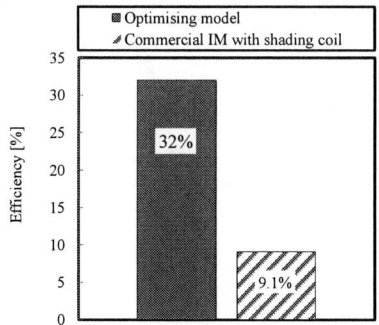

Fig. 8 Efficeincy comparison designed motor and
commercial IM with shading coil

Fig. 9 Prototype single phase SRM

IV.   EXPERIMENTAL RESULTS

In this section, the designed motor performances including starting characteristics, output current, static torque and motor efficiency are evaluated by experiment. Fig. 9 shows a picture of a prototype motor. The motor is manufactured based on the design method as described above section. However, the number of turns of the motor is different from that of the calculated one. The number of turns of the prototype motor is 700 turn/slot because the calculated number of turns cannot be wound as 60% winding space factor because of the manufacturing limitation. The winding space factor of the prototype motor is approximately 26%.

A. Starting characteristics

Starting characteristics which include starting rotation speed and output torque are verified. Fig. 10 shows an experiment setup to measure starting characteristics. As shown in Fig. 10, the rotation speed and output torque are measured by torque meter (TYPE: TMB304, Rated torque: 1Nm) and these measured values are monitored by PC-based measuring instrument (WE7000 produced by Yokogawa electric corporation) which synchronizes with the time when commercial AC power supply is inputted. Also, the position sensor is used only for measuring the rotation speed, the speed is not applied to the SRM drive. A fan is used as a load of the motor. Fig. 11 shows the starting rotation speed and Fig. 12 shows the starting output torque. Start timing is the time when the commercial AC power is supplied. As shown in Figs. 11 and 12, the prototype motor can start up by supplying commercial AC power supply. Also, the rotation speed is converged to synchronous speed; 500rpm.

B. Current and voltage of each phase

The current and voltage are measured by PC-based measuring instrument (WE7000). Fig. 13 and 14 show the current waveforms and the voltage waveforms of each phase, respectively. It is confirmed that the experimental output current is rectified in each phase by the additional diode.

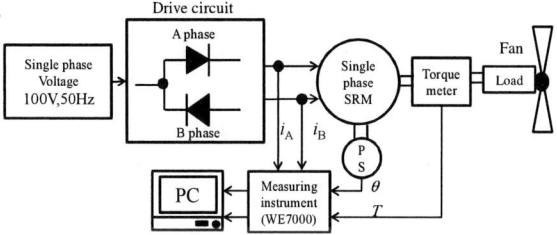

Fig. 10 Experiment setup for measuring
starting characteristics

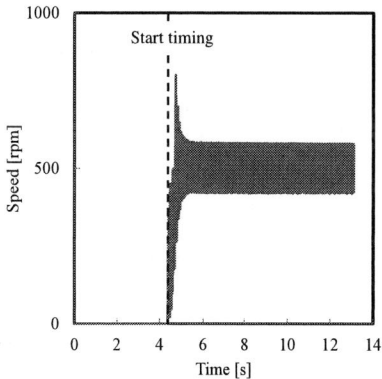

Fig. 11 Starting rotation speed

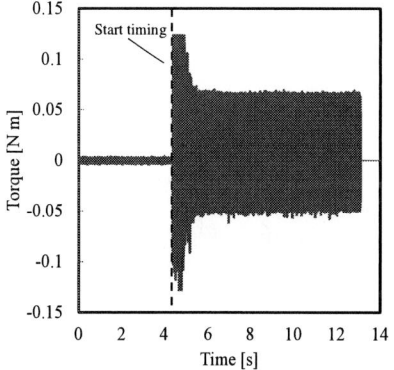

Fig. 12 Starting torque

The 2014 International Power Electronics Conference

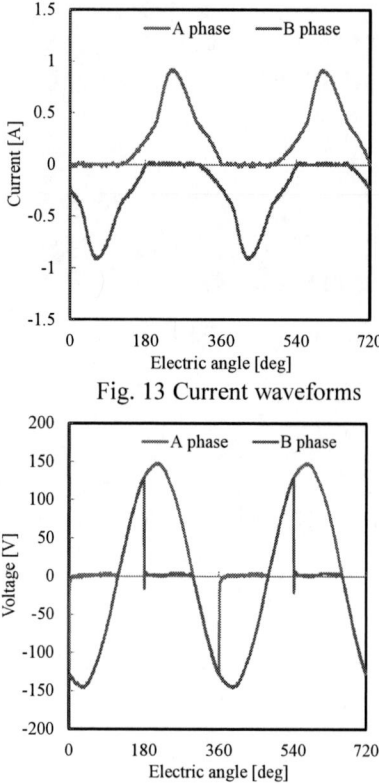

Fig. 13 Current waveforms

Fig.14 Voltage waveforms

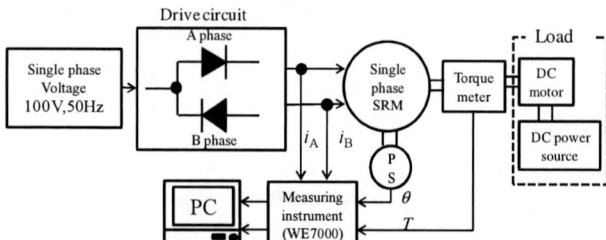

Fig.15 Experiment setup for measuring static torque.

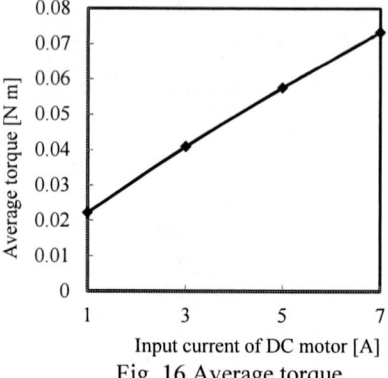

Fig. 16 Average torque

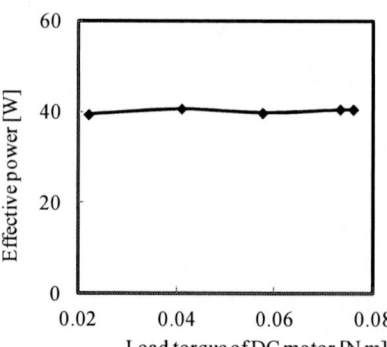

Fig. 17 Maximum torque waveform

Fig. 18 Effective power for the load torque of DC motor

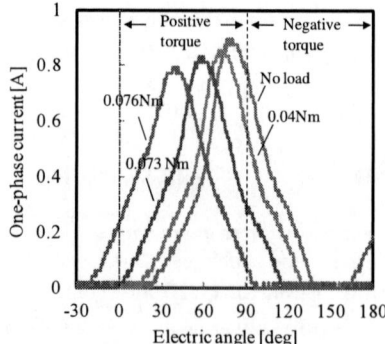

Fig. 19 One-phase current waveforms for the load torque of DC motor

C. Static torque

Fig. 15 shows experiment setup to measuring the static torque. As shown in Fig. 15, the torque is measured by the torque meter at the steady state; the rotation speed is synchronous speed 500rpm. The DC motor is used as a load. Fig. 16 shows the measured average torque when the load torque of DC motor is increased until losing the synchronies. As shown in Fig. 16, the average torque increases when the input current of DC motor is increased, the maximum torque is 0.076Nm. Fig. 17 shows the maximum torque waveform. As shown in this figure, continuous positive torque is confirmed. Fig. 18 shows the effective input power for the each load torque of DC motor. As shown in this figure, the effective input power is almost constant regardless the output torque increases. This reason is verified from the phase current

978-1-4799-2706-7/14 $31.00 © 2014 IEEE 1146

TABLE II
PERFORMANCE OF DESIGNED MOTOR AND PROTOTYPE MOTOR

	Number of turns [turn/slot]	Average torque [N m]	Copper loss [W]	Motor efficiency [%]
Designed motor	1442	0.055	8.9	32
Prototype motor	700	0.076	28.8	10

waveforms in Fig. 19. As described in section II-C, the positive torque is generated when current is excited in 0 to 90 degrees, oppositely the negative torque is generated when current is excited in 90 to 180 degrees. As shown in Figs. 18 and 19, the phase of current automatically shifts to the position which generates the positive torque according to a load of the motor. Therefore, the output torque increases almost without changing the effective input power.

D. Motor efficiency

The motor efficiency is calculated by the obtained average torque and the effective input power. Fig. 20 shows the motor efficiency. As shown in this figure, the motor efficiency increases as increasing the load torque because the average torque increases without changing the effective power. The maximum motor efficiency is approximately 10%. Table II shows the performance comparison of the prototype motor and calculated design motor. As shown in Table II, the motor efficiency is lower than the designed motor which efficiency is 32%. The copper loss of the prototype motor is larger than that of designed motor because the output current of prototype is larger. Therefore, it is important to design large coil area which can wind many number of turns to achieve high motor efficiency in the design of single phase SRMs. In the further improved design of the motor decreasing pole number will be proposed to get large coil area.

V. CONCLUSION

As an alternative to single phase IMs, this paper has presented a novel single phase SRM which can be directly driven by commercial AC power supply by simple circuit which has two diodes without any switching circuit and sensor. To achieve high motor efficiency, the simple design method of single phase SRM has presented. Also, it has presented that the motor efficiency increases when number of turns is increased. In the experiment, it has verified that the designed motor can start-up and generate continuous positive torque. Adding that, the phenomenon that the phase of the current is automatically adjusted to generate positive torque is verified. To achieve higher motor efficiency, large coil area to get number of turns is required. The further improvement will be tested in nearly future.

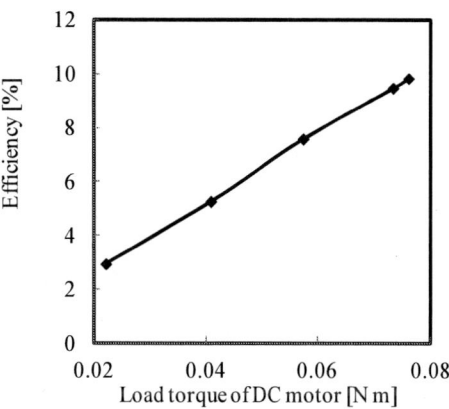

Fig. 20 Motor efficiency

REFERENCES

[1] Tian-Hua Liu, Ming-Tsan Lin, and Hann-Chung Wu, "A Single Phase Induction Motor Drive with Efficiency and Torque Improvement", *Industrial Electronics, 1997. ISIE '97*, PP 637-642, July/1997

[2] Shin-Myung Jung, Jin-Sik Park, Hag-Wone Kim, Associate Member, Kwan-Yuhl Cho, Myung-Joong Youn, "An MRAS-Based Diagnosis of Open-Circuit Fault in PWM Voltage-Source Inverters for PM Synchronous Motor Drive Systems", *IEEE Trans on. Power Electronics*, Vol. 8, pp.2514-2526, May 2013.

[3] M.R..Harris, T.J.E Miller, "Comparison of Design and Performance Parameters in Switched Reluctance and Induction Motors", *Fourth International Conference on Electrical Machines and Drives*, pp.303-307, sep. 1989.

[4] R.Krishnan, "Switched Reluctance Motor drive", CRC PRESS pp.16-17,2001.

[5] Jong-Han Lee, Eun-Woong Lee, Jun-Ho Kim, Dong-Ju Lee "Design of the single phase SRM considering the torque ripple", *PEDS 2005. International Conference on Power Electronics and Drives Systems*, Vol.2, pp.1171-1176, Nov.2005.

[6] M.Asgar, A.Siadatan, and E.Afjei, "A swappable single phase switched reluctance motor with bifilar drive converter", Power Electronics and Drive Systems Technology (PEDSTC), pp.261-265, Feb. 2012.

[7] M.D.Hennen, R.W. De Doncker, and Nisai H. Fuengwarodsakul,"Single-Phase Switched Reluctance Drive With Saturation-Based Starting Method", *IEEE Trans on Power Electronics* ,Vol.26, No.5, pp.1337-1343, May 2011.

[8] Jun-Ho Kim, Eun-Woong Lee, and Jong-Han Lee,"Design of the Starting Device Installed in the Single-Phase Switched Reluctance Motor", *IEEE Transactions on Magnetics*,Vol.43, No.1, pp.203-213, March 2013.

[9] J.E.Fletcher, A.Helal, and B.W.Williams,"Starting the Single-Phase Switched Reluctance Motor Using Rotor Shorting Rings", *IEEE Trans on Energy conversion*,Vol.21, No.4, pp.848-854, Dec. 2006.

[10] Yasuei Yoneoka, Kan Akatsu, "An Optimized Design of High-Efficiency Switched Reluctance Motor with Single-Phase Input Operation", *2011 International Conference on Electrical Machines and Systems (ICEMS)*, pp.1-6, Aug.2011.

978-1-4799-2706-7/14 $31.00 © 2014 IEEE

Fast Analytical Model of Switched Reluctance Machine

Senad Smaka, Semsudin Masic, Mirsad Cosovic
Department of Power Electrical Engineering
Faculty of Electrical Engineering
University of Sarajevo
Sarajevo, Bosnia and Herzegovina
ssmaka@etf.unsa.ba; smasic@etf.unsa.ba; mcosovic@etf.unsa.ba

Abstract— This paper describes fast analytical model for computation of the switched reluctance machine's (SRM) nonlinear magnetization characteristic and torque lookup table. The flux-tube and the gage-curve methods are used to develop this fast analytical model. Presented model is used for computation of the magnetization characteristic and torque lookup table of three and four-phase SRMs. The simulation results obtained using proposed analytical model are compared to the results of magnetostatic finite-element analysis (FEA) for a three-phase 12/8 SRM. Experimental verification of the analytical model is also presented for the same 12/8 SRM prototype.

Keywords— *analytical model; flux-tube; gage-curve; switched reluctance machine.*

I. INTRODUCTION

Switched reluctance machine (SRM) has highly saturated doubly salient pole structure. The requirement for predicting the steady state and dynamic performance of SRM is to generate its magnetization characteristic $\psi(I, \theta)$ and torque lookup table $T(I, \theta)$, where ψ is the flux linkage, I is the excitation current, θ is the rotor mechanical angular position, and T is the torque.

The magnetization characteristic can be obtained by magnetostatic FEA [1]–[3] or by measurements done on the existing machine. However, both methods are not particularly suitable to implement during initial stages of the machine's design process.

An alternative approach is to develop analytical model of SRM to compute its magnetization characteristic. This model has to be faster than FEA but still accurate enough.

The SRM modeling using analytical methods have been reported in [4]–[18]. The most common models are based on the use of: magnetic equivalent circuit (MEC) [4]–[6]; implementation of basic laws of physics on simplified motor geometry [7]–[9]; analytical equations that approximate magnetization characteristic [10]–[15]; several predefined flux-tubes [16]–[18].

This paper presents fast analytical model of SRM, which is intended to be a part of a sizing-design estimation process of the machine. The analytical model will be discussed in Section II. Proposed model combines two already known techniques, flux-tube and gage-curve. Flux-tube method is used to compute $\psi(I)$ characteristics at the aligned and the unaligned rotor position. Gage-

curve method is used to compute $\psi(I)$ characteristics at other rotor positions.

The comparison of the analytical results to FEA is given in Section III for the 12/8 SRM prototype. In Section IV, analytical results are compared with measurements for the 12/8 SRM. Finally, the conclusions are presented in Section V.

II. ANALYTICAL MODEL

A. Implementation of flux-tube model

The flux-tube based models that enable computation of flux linkage characteristics in arbitrary rotor positions are reported in literature, but the achieved level of accuracy in intermediate rotor positions can be a quite low [19].

In this paper flux-tube model is used only for computation of $\psi(I)$ characteristics at the aligned and the unaligned rotor position, where substantial accuracy can be obtained.

As in [16], two flux-tubes are assumed to be sufficient to represent the actual flux paths at the aligned position and seven flux-tubes are used at the unaligned position. The considered flux-tubes are shown in Fig. 1.

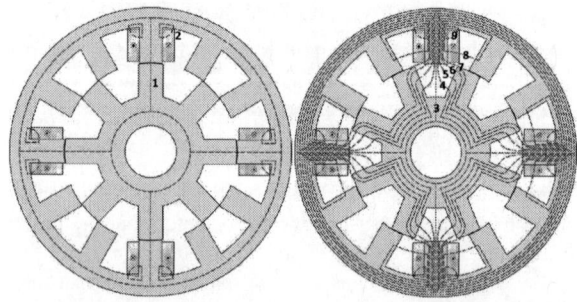

a. aligned position   b. unaligned position

Fig. 1. Assumed flux tubes – aligned and unaligned position.

Assumptions common to all flux-tube models are also applied here: the air gap flux lines consists of concentric arcs and straight-line parts; the flux lines enter and leave iron normally; the flux lines in the poles are parallel to the pole axes; the flux lines in the stator and rotor yokes are concentric.

978-1-4799-2706-7/14 $31.00 © 2014 IEEE

Flowchart for the computation of $\psi(I)$ curves at the aligned and the unaligned position is shown in Fig. 2.

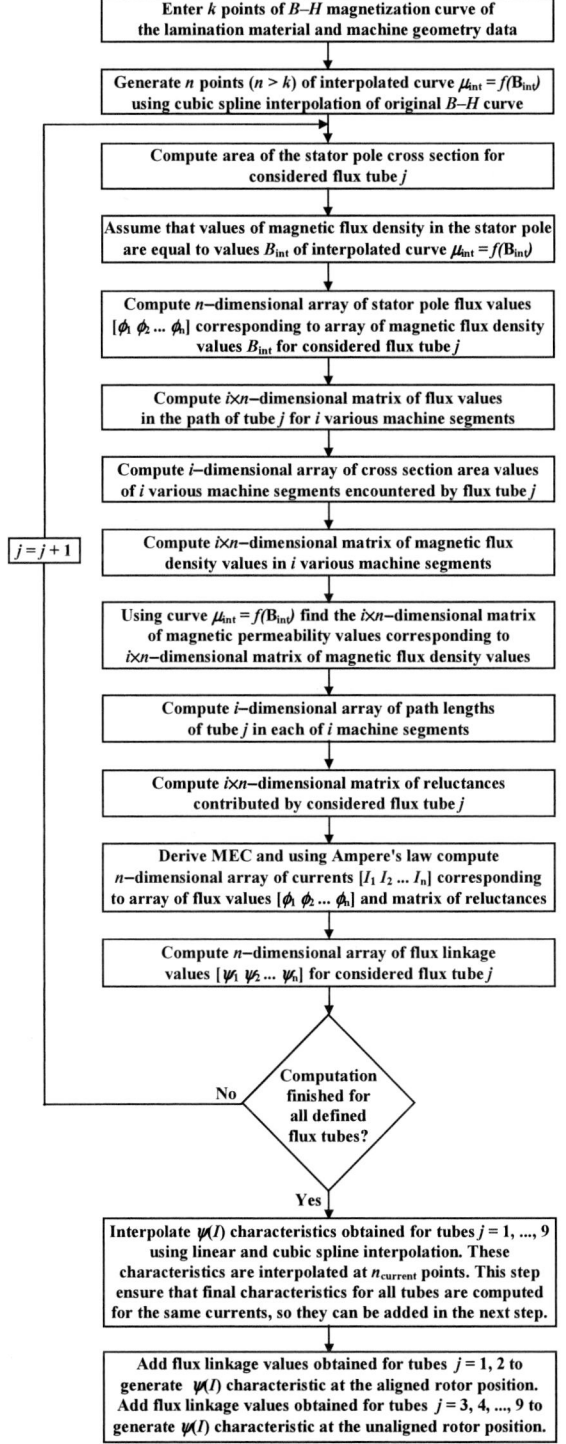

Fig. 2. Flowchart for the computation of flux linkage at the aligned and unaligned rotor position.

Model input data are: B–H curve of the lamination material, numbers of stator and rotor poles N_s and N_r, stator and rotor poles arc angles β_s and β_r, stator and rotor outer radii R_s and R_r, air gap length δ, stator and rotor yokes radii R_{sy} and R_{ry}, shaft radius R_{sh}, stack length l_{stk}, number of turns per phase winding N_{ph}, and number of phases m.

The SRM geometry parameters for analytical computation of flux linkage characteristics using flux-tube model are emphasized in Fig. 3.

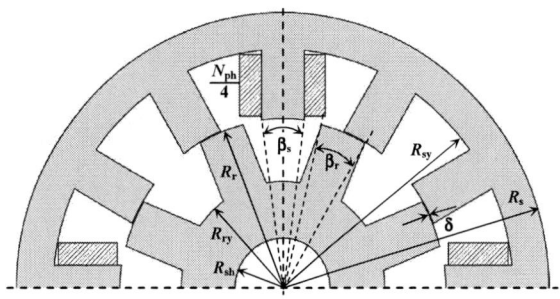

Fig. 3. SRM geometry parameters for flux-tube model.

The results of FEA or measurements in several characteristic points are not needed, which is the advantage in comparison to classical gage-curve models described in [10] and [12]. Also, implementation of flux-tube method is not based on the use of iterative computation like in [16]–[18]. Instead of using numerical iteration with a prescribed value of error to compute magnetomotive force, the flux linkage contributed by each flux-tube in computed for the assumed values of stator pole magnetic flux density.

Usually, only a small number of points k define original B–H curve of the lamination material. In order to achieve an acceptable level of computational accuracy, it was necessary to use cubic spline interpolation to generate interpolated B–H curve with more defined points n ($n > k$). Interpolated magnetization curve is recalculated to generate $\mu_{int} = f(B_{int})$ curve. Flux linkage characteristics are computed for n values of magnetic flux density. Since the computational time is rising with n, this number should be chosen carefully.

Equations derived for flux-tubes No. 1 and No. 6 are given in order to explain the computation procedure. The list of equations used for all flux-tubes is given in [20].

Flux-tube No. 1

The areas of cross-section penetrated by the flux-tube No. 1 are given in Table I. The lengths of the individual machine segments encountered by flux-tube No. 1 are given in Table II.

TABLE I
AREAS OF CROSS-SECTION FOR FLUX-TUBE NO. 1

Machine segment	Equation
Stator pole	$A_{sp1} = (R_r + \delta) \cdot \beta_s \cdot l_{stk}$
Stator yoke	$A_{sy1} = (R_s - R_{sy}) \cdot l_{stk}$
Rotor pole	$A_{rp1} = R_r \cdot \beta_r \cdot l_{stk}$
Rotor yoke	$A_{ry1} = (R_{ry} - R_{sh}) \cdot l_{stk}$
Air gap	$A_{\delta 1} = \dfrac{A_{sp1} + A_{rp1}}{2}$

TABLE II
LENGTHS OF MACHINE SEGMENTS FOR FLUX-TUBE NO. 1

Machine segment	Equation
One stator pole	$l_{sp1} = \dfrac{R_s + R_{sy}}{2} - (R_r + \delta)$
One quarter of stator yoke	$l_{sy1} = \dfrac{R_s + R_{sy}}{2} \cdot \dfrac{2\pi \cdot m}{N_s}$
One rotor pole	$l_{rp1} = R_r - \dfrac{R_{ry} + R_{sh}}{2}$
One quarter of rotor yoke	$l_{ry1} = \dfrac{R_{ry} + R_{sh}}{2} \cdot \dfrac{2\pi \cdot m}{N_s}$
One air gap	$l_{\delta 1} = \delta$

The array of magnetic flux density values in the stator pole $[B_{sp1}]_{1\times n}$ is assumed to be equal to the magnetic flux density array $[B_{int}]_{1\times n}$ of interpolated curve $\mu_{int} = f(B_{int})$

$$\begin{bmatrix} B_{sp1} \end{bmatrix} = \begin{bmatrix} B_{sp11} & B_{sp12} & ... & B_{sp1n} \end{bmatrix} = \begin{bmatrix} B_{int} \end{bmatrix} \quad (1)$$

The array of magnetic flux values in the stator pole $[\phi_1]_{1\times n}$ is calculated as

$$\begin{bmatrix} \phi_1 \end{bmatrix} = \begin{bmatrix} B_{sp1} \end{bmatrix} \cdot A_{sp1} \quad (2)$$

The arrays of magnetic flux density values in other parts of SRM are determined as

$$\begin{bmatrix} B_{sy1} \end{bmatrix} = \frac{[\phi_1]}{2 \cdot A_{sy1}} \; ; \; \begin{bmatrix} B_{rp1} \end{bmatrix} = \frac{[\phi_1]}{A_{rp1}} \; ;$$
$$\begin{bmatrix} B_{ry1} \end{bmatrix} = \frac{[\phi_1]}{2 \cdot A_{ry1}} \; ; \; \begin{bmatrix} B_{\delta 1} \end{bmatrix} = \frac{[\phi_1]}{A_{\delta 1}} \; . \quad (3)$$

Then on the basis of $\mu_{int} = f(B_{int})$ curve, the arrays of magnetic permeability $[\mu_{sp1}]_{1\times n}$, $[\mu_{sy1}]_{1\times n}$, $[\mu_{rp1}]_{1\times n}$, and $[\mu_{ry1}]_{1\times n}$, corresponding to the arrays of magnetic flux density values computed by (1) and (3), are obtained.

The magnetic equivalent circuit created for one quarter of flux-tube No. 1 is shown in Fig. 4.

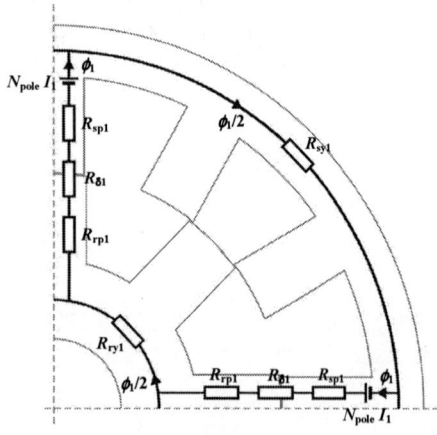

Fig. 4. MEC for one part of flux-tube No. 1.

Now it is possible to compute the arrays of reluctances related to various laminated parts of the magnetic circuit $[R_{sp1}]_{1\times n}$, $[R_{sy1}]_{1\times n}$, $[R_{rp1}]_{1\times n}$, and $[R_{ry1}]_{1\times n}$. These arrays are computed using arraywise right division in Matlab® as

$$\begin{bmatrix} R_{sp1} \end{bmatrix} = \frac{L_{sp1}}{[\mu_{sp1}] \cdot A_{sp1}} \; ; \; \begin{bmatrix} R_{sy1} \end{bmatrix} = \frac{L_{sy1}}{[\mu_{sy1}] \cdot A_{sy1}} \; ;$$
$$\begin{bmatrix} R_{rp1} \end{bmatrix} = \frac{L_{rp1}}{[\mu_{rp1}] \cdot A_{rp1}} \; ; \; \begin{bmatrix} R_{ry1} \end{bmatrix} = \frac{L_{ry1}}{[\mu_{ry1}] \cdot A_{ry1}} \; . \quad (4)$$

The reluctance in the air gap is computed as follows

$$R_{\delta 1} = \frac{L_{\delta 1}}{\mu_0 \cdot A_{\delta 1}} \quad (5)$$

The array of currents $[I_1]_{1\times n}$ corresponding to the array of magnetic flux density values $[B_{sp1}]_{1\times n}$ is computed using binary addition and arraywise multiplication in Matlab® by means of expression

$$[I_1] = \frac{[\phi_1]}{2N_{pole}} \left\{ 2 \left(\begin{bmatrix} R_{sp1} \end{bmatrix} + \begin{bmatrix} R_{rp1} \end{bmatrix} + R_{\delta 1} \right) + \frac{\begin{bmatrix} R_{sy1} \end{bmatrix}}{2} + \frac{\begin{bmatrix} R_{ry1} \end{bmatrix}}{2} \right\} \quad (6)$$

where N_{pole} is number of turns per pole ($N_{pole} = N_{ph} \cdot m / N_s$).

The array of flux linkage values of flux-tube No. 1 corresponding to the array of currents $[I_1]$ is

$$\begin{bmatrix} \psi_1 \end{bmatrix} = \begin{bmatrix} \phi_1 \end{bmatrix} \cdot N_{ph} \quad (7)$$

Flux-tube No. 6

In Fig. 5, the detailed sketch of flux-tube No. 6 and magnetic equivalent circuit created for this flux-tube are shown.

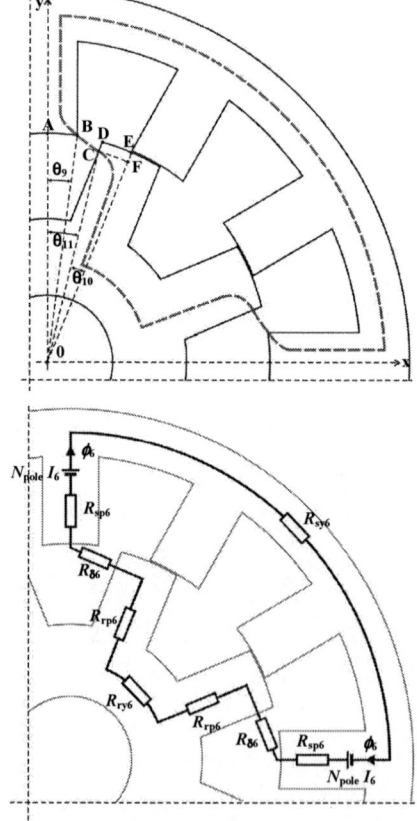

Fig. 5. Detailed sketch and MEC of flux-tube No. 6.

The areas of cross-section penetrated by the flux-tube No. 6 are given in Table III.

TABLE III
AREAS OF CROSS-SECTION FOR FLUX-TUBE NO. 6

Machine segment	Equation
Stator pole	$A_{sp6} = \left[(R_r + \delta) \cdot \dfrac{\beta_s}{32} + \dfrac{R_{sy} - (R_r + \delta)}{16} \right] \cdot l_{stk}$
Stator yoke	$A_{sy6} = (R_s - R_{sy}) \cdot l_{stk}$
Rotor pole	$A_{rp6} = \dfrac{R_r - R_{ry}}{4} \cdot l_{stk}$
Rotor yoke	$A_{ry6} = (R_{ry} - R_{sh}) \cdot l_{stk}$
Air gap	$A_{\delta6} = \dfrac{A_{sp6} + A_{rp6}}{2}$

The lengths of flux-tube No. 6 in one stator pole l_{sp6} and in the stator yoke l_{sy6} are: $l_{sp6} = l_{sp1}$ and $l_{sy6} = l_{sy1}$.

The lengths of flux-tube No. 6 in one rotor pole l_{rp6} and in the rotor yoke l_{ry6} are computed by means of expressions

$$l_{rp6} = \frac{7}{8} \cdot (R_r - R_{ry}) \tag{8}$$

$$l_{ry6} = 2 \cdot \left(R_{sh} + \frac{R_r - R_{ry}}{2} \right) \cdot \frac{2\pi}{N_r} \tag{9}$$

The length in the air gap $l_{\delta6}$ is given as an arch between points B and C, as shown in Fig. 5. The coordinates of point B are

$$x_B = (R_r + \delta) \cdot \sin\theta_9 = (R_r + \delta) \cdot \sin\left(\frac{\beta_s}{2} \right)$$
$$y_B = (R_r + \delta) \cdot \cos\theta_9 = (R_r + \delta) \cdot \cos\left(\frac{\beta_s}{2} \right) \tag{10}$$

The lengths of arcs CF and DE shown in Fig. 5 are

$$l_{CF} = \left(R_r - \frac{R_r - R_{ry}}{8} \right) \cdot \theta_{10} \; ; \quad l_{DE} = R_r \cdot \frac{\beta_r}{2} \; . \tag{11}$$

Angles θ_{10} and θ_{11} are calculated as follows

$$\theta_{10} = \frac{R_r \cdot \dfrac{\beta_r}{2}}{R_r - \dfrac{R_r - R_{ry}}{8}} \; ; \quad \theta_{11} = \frac{\pi}{N_r} - \theta_{10} \; . \tag{12}$$

The coordinates of point C are

$$x_C = \left(R_r - \frac{R_r - R_{ry}}{8} \right) \cdot \sin\left[\frac{\pi}{N_r} - \frac{4 \cdot \beta_r \cdot R_r}{8R_r - (R_r - R_{ry})} \right]$$
$$y_C = \left(R_r - \frac{R_r - R_{ry}}{8} \right) \cdot \cos\left[\frac{\pi}{N_r} - \frac{4 \cdot \beta_r \cdot R_r}{8R_r - (R_r - R_{ry})} \right] \tag{13}$$

The length of flux-tube No. 6 in the air gap is

$$l_{\delta6} = \sqrt{(x_C - x_B)^2 + (y_C - y_B)^2} \tag{14}$$

where the coordinates x_B, y_B, x_C, and y_C are given by (10) and (13).

The reluctances in various segments of the machine encountered by flux-tube No. 6 are calculated in a similar way as in (4) and (5).

The array of currents $[I_6]_{1\times n}$ corresponding to the array of magnetic flux density values in the stator pole $[B_{sp6}]_{1\times n}$ is computed as follows

$$[I_6] = \frac{[\phi_6]}{N_{ph}} \left\{ 2 \left([R_{sp6}] + [R_{rp6}] + R_{\delta6} \right) + [R_{sy6}] + [R_{ry6}] \right\}$$

The array of flux linkage values of flux-tube No. 6 corresponding to the array of currents $[I_6]$ is

$$[\psi_6] = n_{tube6} \cdot [\phi_6] \cdot N_{ph} \tag{15}$$

where n_{tube6} is the number of flux-tubes No. 6 at the unaligned position. For example, $n_{tube6} = 4$ for SRM 12/8 (see Fig. 1.b).

B. Implementation of gage-curve method

In this paper, gage-curve method is used for analytical computation of flux linkage characteristics in intermediate rotor positions.

Gage-curve method is introduced in [10]. The $\psi(I)$ characteristics at the unaligned and the aligned position are approximated using three precalculated points, first-order, and second-order functions. These characteristics are used to determine flux linkage characteristics at other rotor positions.

In this paper, $\psi(I)$ characteristics at the unaligned and the aligned position are calculated by flux-tube method instead of using their approximation. Then, these characteristics are exploited to compute $\psi(I,\theta)$ curves at intermediate rotor positions using equations given in [10]. These equations are based on the analysis of $\psi(\theta)$ curve corresponding to a fixed value of current I (Fig. 6.).

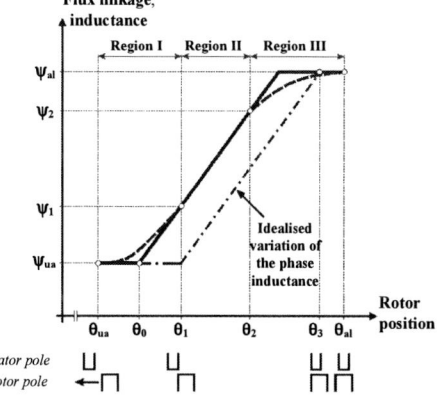

Fig. 6. $\psi(\theta)$ curve corresponding to a fixed value of current I.

The rotor positions are divided into three regions. The equations for modeling $\psi(I,\theta)$ characteristics for any chosen value of excitation current in these regions are:
- region I ($\theta_{ua} < \theta < \theta_1$)

$$\psi(I_i, \theta) = k_{ai} \cdot (\theta_1 - \theta_{0i}) + L_u \cdot I_i + \frac{X_i \cdot (\theta - \theta_1)}{Z_i - (\theta - \theta_1)} \quad (16)$$

- region II $(\theta_1 \leq \theta \leq \theta_2)$

$$\psi(I_i, \theta) = \psi_{uai} + k_i \cdot [\psi_{ali} - \psi_{uai}] \quad (17)$$

- region III $(\theta_2 < \theta \leq \theta_{al})$

$$\psi(I_i, \theta) = k_{ai} \cdot (\theta_2 - \theta_{0i}) + L_u \cdot I_i + \frac{X_i' \cdot (\theta - \theta_2)}{Z_i' - (\theta - \theta_2)} \quad (18)$$

where I_i ($i = 1, 2, ..., n_{current}$) represent actual value of excitation current selected for computation, while ψ_{uai} and ψ_{ali} are the flux linkages in the unaligned and the aligned position, respectively, corresponding to I_i.

The boundaries between regions are defined as

$$\theta_{ua} = 0°; \quad \theta_1 = \frac{\pi}{N_r} - \frac{\beta_r + \beta_s}{2}; \quad \theta_3 = \frac{\pi}{N_r} - \frac{|\beta_r - \beta_s|}{2};$$

$$\theta_2 = \frac{\theta_1 + \theta_3}{2}; \quad \theta_{al} = \frac{\pi}{N_r}.$$

The coefficients in (16)–(18) are determined as

$$Z_i = \frac{\left[k_{ai} \cdot (\theta_1 - \theta_{0i}) + L_u \cdot I_i - \psi_{uai}\right] \cdot (\theta_1 - \theta_{ua})}{k_{ai} \cdot (\theta_1 - \theta_{ua}) - \left[k_{ai} \cdot (\theta_1 - \theta_{0i}) + L_u \cdot I_i - \psi_{uai}\right]}$$

$$X_i = k_{ai} \cdot Z_i$$

$$k_i = \frac{\theta - \theta_{0i}}{\theta_3 - \theta_1}$$

$$Z_i' = \frac{\left[\psi_{ali} - k_{ai} \cdot (\theta_2 - \theta_{0i}) - L_u \cdot I_i\right] \cdot (\theta_{al} - \theta_2)}{k_{ai} \cdot (\theta_{al} - \theta_2) - \left[\psi_{ali} - k_{ai} \cdot (\theta_2 - \theta_{0i}) - L_u \cdot I_i\right]}$$

$$X_i' = k_{ai} \cdot Z_i'$$

The unaligned inductance L_u can be found from the $\psi(I)$ characteristic at the unaligned position.

The offset angle θ_{0i} is given as

$$\theta_{0i} = \theta_1 - \frac{\psi(I_{i-1}, \theta) \cdot \theta_{al}}{\psi_m \cdot \xi}$$

where ψ_m is maximum flux linkage at the aligned position corresponding to maximum value of excitation current, ξ is an empirical coefficient with the value in the range from 8 to 12, and $\psi(I_{i-1}, \theta)$ is the flux linkage in the position θ obtained for excitation current I_{i-1} selected in the previous step of computation.

Parameter k_{ai} is given as

$$k_{ai} = \frac{\psi_{ali} - \psi_{uai}}{\theta_3 - \theta_1}$$

C. Computation of torque lookup table $T(I, \theta)$

The torque lookup table can be calculated on the basis of magnetic coenergy W_m', using first central difference approximation method of numerical differentiation, as:

$$T = \left. \frac{\partial W_m'(I, \theta)}{\partial \theta} \right|_{I=const.} \quad (19)$$

where the magnetic coenergy can be obtained using numerical integration method based on Simpson's rule as

$$W_m' = \int_0^I \psi(I, \theta) \, dI \quad (20)$$

III. COMPARISON OF ANALYTICAL MODEL AND FEA

To investigate whether the proposed analytical model is suitable to predict SRM magnetization characteristic and torque lookup table, an experimental 12/8 SRM is used for analytical and numerical computation. The basic parameters of 12/8 SRM are given in [21].

Fig. 7 shows the flux linkage and torque lookup table computed by both analytical model and 2-D FEA. Ansoft Maxwell® software is used for numerical computation.

The magnetization characteristics obtained using analytical model is fairly matching with those obtained by FEA. However, there is a discrepancy between analytical and FEA results for higher torques. Actual torque curves tend not to be flat in the region II. The flat part of torque curves corresponding to the linear part of $\psi(I, \theta)$ characteristics modeled by (17). It is necessary to improve modeling of $\psi(I, \theta)$ curve in the region II in order to boost the accuracy of proposed model. The analytical model is proved as computationally very fast in comparison to FEA.

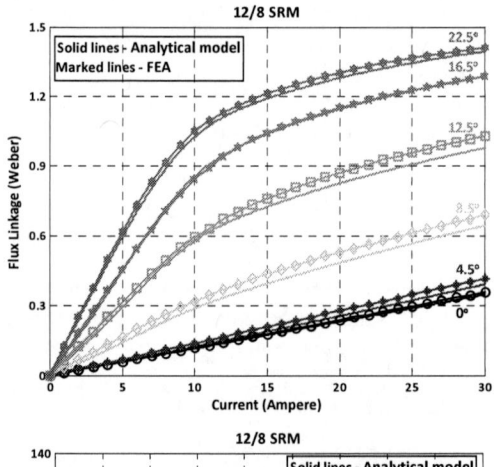

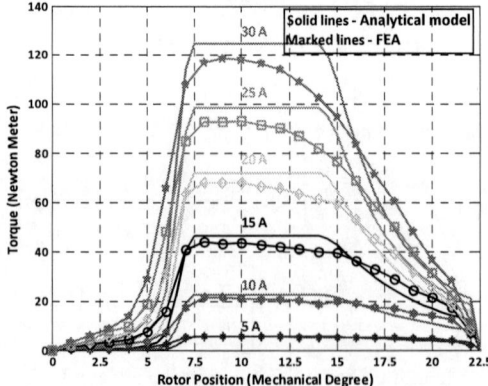

Fig. 7. Flux linkage and torque lookup table computed by analytical model and FEA for 12/8 SRM.

IV. EXPERIMENTAL VERIFICATION OF ANALYTICAL MODEL

The photo of the test setup used for experimental verification of analytically computed magnetization characteristic and torque lookup table is shown in Fig. 8.

Fig. 8. Experimental setup photo.

The three-phase full-wave rectifier is used as power supply during standstill test, while a classic three-phase asymmetric bridge converter feeds the 12/8 SRM during normal operation. The 12/8 SRM was loaded with eddy current brake WEKA Power LPS 800 LK, 4500 rpm, 0–800 Nm. An opto-interrupter with slotted disk is employed to determine the rotor position. The torque is obtained from a strain gauge force sensor that is a part of the test bench. Power analyzer LEM NORMA D6133M is used to measure the electrical quantities. Phase current is measured using current channel with triaxial shunt 61I3. Voltage channel 61U1 is used to measure applied voltage.

Standstill test is realized as follows. The rotor of the 12/8 SRM is blocked on a chosen position. Step voltage is applied to one of the three phases of the motor thus resulting in phase winding direct current flow. Voltage and current waveforms were visualized and recorded on a power analyzer. The produced torque is transmitted to a force sensor using a lever arm and obtained results are also recorded. Then, all measuring quantities are transferred and stored on a PC. Direct method for magnetization characteristic measurement described in [22] is used. It consists of voltage and phase current waveform storing and off-line flux linkage calculation from the measured voltage and current using a numerical integration

$$\psi(t) = \int \left[v(t) - R \cdot i(t) \right] dt \qquad (21)$$

where R is the phase resistance, $v(t)$ and $i(t)$ are stored waveforms of voltage and phase current.

Stored voltage and phase current waveforms are postprocessed in Matlab® according to (21). Computed and measured magnetization characteristics are shown in Fig. 9 where only a few measuring points are presented.

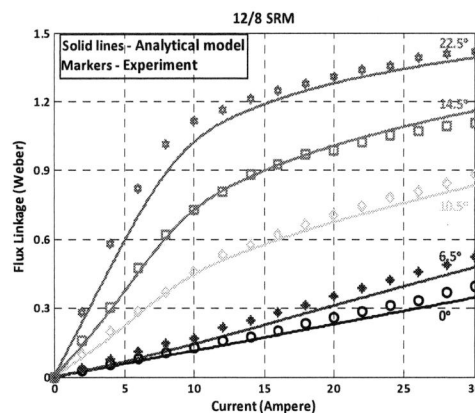

Fig. 9. Computed and measured magnetization characteristics $\psi(I,\theta)$.

Table IV gives a comparison of the flux linkages at unaligned, partially overlapped, and aligned position obtained analytically and experimentally for 20 A excitation.

TABLE IV
FLUX LINKAGE AT VARIOUS ROTOR POSITIONS FOR 20 A EXCITATION

Rotor position	Analytical (Wb)	Measurement (Wb)
Unaligned (0°)	0.23	0.26
Partially overlapped (10.5°)	0.68	0.71
Aligned (22.5°)	1.28	1.31

Fig. 10 shows comparison between computed and measured torque lookup table obtained for five excitation currents. Torque measurements are performed for 24 rotor positions.

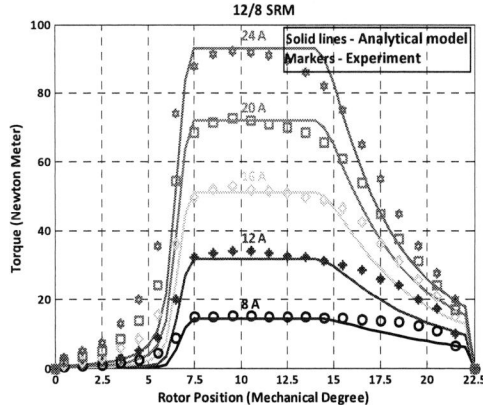

Fig. 10. Computed and measured torque lookup table $T(I,\theta)$.

Table V gives a comparison of standstill torque at various rotor positions obtained analytically and experimentally for 20 A excitation.

TABLE V
TORQUE AT VARIOUS ROTOR POSITIONS FOR 20 A EXCITATION

Rotor position	Analytical (Nm)	Measurement (Nm)
No overlapping (6.5°)	37.57	54.50
Half aligned (≈14.5°)	70.13	65.50
Significant overlapping (20.5°)	20.20	24.10

Good agreements in Fig. 9 and in Table IV are evident. However, the discrepancy between analytical and experimental results is higher for torque lookup table, especially when the rotor is near the unaligned position (0°).

V. CONCLUSIONS

Fast analytical model for computation of nonlinear magnetization characteristics and torque lookup table of SRM is presented in this paper.

The model is proved as computationally fast. For example, computation of the 12/8 SRM magnetization characteristics $\psi(I,\theta)$ and torque lookup table $T(I,\theta)$ for 31 values of current I and 46 values of angle θ lasted approximately nine seconds when the proposed analytical method is used while it lasted more than three hours when FEA software is used on a same PC.

Comparison of magnetization characteristics computed using analytical model with both FEA and experiment shows that the proposed analytical model is accurate enough. However, there is a discrepancy between the analytical and FEA and experimental results, especially for a higher torques.

The improvement of presented analytical model's accuracy especially regarding torque lookup table should be future work. The idea is to model variation of flux-linkage with rotor position in the central region II using second-order function instead of using linear variation as in this paper.

Since the high speed PC's and sophisticated FEA software are available, coupling of FEA and gage-curve method is also one of the possibilities. In this case FEA software will be used to compute flux linkage characteristics at the extreme rotor positions while gage-curve method will be used to compute magnetization characteristics at the arbitrary rotor positions like in this paper. Modeling time will increase but it should be acceptable. Computational time also will increase but model's accuracy can be improved.

The rotation interval between the unaligned and aligned position can be divided into higher number of regions instead of using only three regions as in this paper. Investigation of the influence of this parameter on the model accuracy is recommended.

REFERENCES

[1] R. Arumugam, D. A. Lowther, R. Krishnan, J. F. Lindsay, "Magnetic Field Analysis of a Switched Reluctance Motor Using a Two Dimensional Finite Element Model," *IEEE Transactions on Magnetics*, Vol. Mag-21, No. 5, pp. 1883–1885, September 1985.

[2] M. Moallem, C. M. Ong, "Predicting the Steady-State Performance of a Switched Reluctance Machine," *IEEE Transactions on Industry Applications*, Vol. 27, No. 6, pp. 1087–1097, November/December 1991.

[3] K. N. Srinivas, R. Arumugam, "Analysis and Characterization of Switched Reluctance Motors: Part I – Dynamic, Static, and Frequency Spectrum Analyses," *IEEE Transactions on Magnetics*, Vol. 41, No. 4, pp. 1306–1320, April 2005.

[4] A. Deihimi, S. Farhangi, G. Henneberger, "A general nonlinear model of switched reluctance motor with mutual coupling and multiphase excitation," *Electrical Engineering 84 (2002)*, Springer Verlag, pp. 143–158.

[5] S. H. Mao, D. Dorrell, M. C. Tsai, "Fast Analytical Determination of Aligned and Unaligned Flux Linkage in Switched Reluctance Motors Based on a Magnetic Circuit Model," *IEEE Transactions on Magnetics*, Vol. 45, No. 7, pp. 2935–2942, July 2009.

[6] D. Lin, P. Zhou, S. Stanton, Z. J. Cendes, "An Analytical Circuit Model of Switched Reluctance Motors," *IEEE Transactions on Magnetics*, Vol. 45, No. 12, pp. 5368–5375, December 2009.

[7] A. Radun, "Analytical Calculation of the Switched Reluctance Motor's Unaligned Inductance," *IEEE Transactions on Magnetics*, Vol. 35, No. 6, pp. 4473–4481, November 1999.

[8] A. Radun, "Analytically Computing the Flux Linked by a Switched Reluctance Motor Phase When the Stator and Rotor Poles Overlap," *IEEE Transactions on Magnetics*, Vol. 36, No. 4, pp. 1996–2003, July 2000.

[9] H. C. Lovatt, "Analytical model of a classical switched-reluctance motor," *IEE Proc.-Electr. Power Appl., IEE Proceedings online no. 20040545*, pp. 1–7, 2004.

[10] T. J. E. Miller, M. McGilp, "Nonlinear theory of the switched reluctance motor for rapid computer-aided design," *IEE Proceedings*, Vol. 137, Pt. B, No. 6, pp. 337–347, November 1990.

[11] D. A. Torrey, X.-M. Niu, E. J. Unkauf, "Analytical modelling of variable-reluctance machine magnetization characteristics," *IEE Proc.-Elect. Power Appl.*, Vol. 42, No.1, pp. 14–22, January 1995.

[12] C. Roux, M. M. Morcos, "A Simple Model for Switched Reluctance Motors," *IEEE Power Engineering Review*, pp. 49–52, October 2000.

[13] H. Le-Huy, P. Brunelle, "A Versatile Nonlinear Switched-Reluctance Motor Model in Simulink using Realistic and Analytical Magnetization Characteristics," *Proceedings of 31st Annual Conference of IEEE Industrial Electronics Society IECON 2005*, paper number RF-001757, November 2005.

[14] B. Schinnerl, D. Gerling, "Novel Analytical Calculation Method for the Non-Linear Ψ-i Characteristics of Switched-Reluctance-Machines in the Aligned Rotor Position," *Proceedings of the IEEE International Electric Machines and Drives Conference IEMDC 2006*, pp. 793–796, 2006.

[15] B. Schinnerl, D. Gerling, "Novel Analytical Calculation Method for the Non-Linear Ψ-i Characteristic of Switched-Reluctance-Machines in Arbitrary Rotor Positions," *Proceedings of the European Conference on Power Electronics and Applications EPE 2007*, ISBN 9789075815108, 2007.

[16] R. Krishnan, "Switched Reluctance Motor Drives: Modeling, Simulation, Analysis, Design, and Applications," *CRC Press LLC*, Chapter 2.3, 2001.

[17] P. Rafajdus, I. Zrak, V. Hrabovcová, "Analysis of the Switched Reluctance Motor (SRM) Parameters," *Journal of Electrical Engineering*, Vol. 55, No. 7–8, pp. 195–200, July 2004.

[18] N. K Sheth, K. R. Rajagopal, "Calculation of the Flux-Linkage Characteristics of a Switched Reluctance Motor by Flux Tube Method," *IEEE Transactions on Magnetics*, Vol. 41, No. 10, pp. 4069–4071, October 2005.

[19] E. Pădurariu, L. Someşan, I.-A. Viorel, "Switched Reluctance Motor Analytical Models, Comparative Analysis," *Proceedings of the 12th International Conference on Optimization of Electrical and Electronic Equipment OPTIM 2010*, pp. 285–290, 2010.

[20] S. Smaka, "'Optimizacija dizajna električnog motora za pogon hibridnog vozila," *Ph.D. Disertation*, Univerzitet u Sarajevu, Elektrotehnički Fakultet, Sarajevo, 2012.

[21] H. J. Wehner, "Betriebseigenschaften, Ausnutzung und Schwingungsverhalten bei geschalteten Reluktanzmotoren," *Ph.D. Dissertation*, Der Technischen Fakultät der Universität Erlangen-Nürnberg, Erlangen, Germany, 1997.

[22] C. S. Dragu, R. Belmans, "Measurement of Magnetic Characteristics of Switched Reluctance Motor," *Proceedings of the EPE 2003*, pp. 1–10, 2003.

The 2014 International Power Electronics Conference

Detailed Analysis and a General Design Procedure of Damped *LCL* Filters in Three Phase Voltage Source Converters

Baoquan Liu, Shaohui Zhong, Yixin Zhu, Hao Yi* and Fang Zhuo

School of Electrical Engineering and Automation
Xi'an Jiaotong University
Xi'an, Shaanxi, China
*Corresponding author E-mail: comeliu299@163.com

Abstract— **Detailed analysis and the parameter design of damped *LCL* filters applied in three phase voltage source converters is carried out in this work. Relations between the *LCL* filter parameters (L_1, L_2, C and R) and the corresponding transfer function parameters, including ω_s, ω_c, ω_z and ω_R, are investigated mathematically and demonstrated explicitly. Based on this methodology, a reasonable and optimal design procedure is proposed. Simulation results are provided to validate the performance of the *LCL* filter and the design procedure.**

Keywords—**Damped *LCL* Filters, Voltage Source Converter, Design Procedure**

I. INTRODUCTION

Three phase voltage source converter (VSC) is widely used in diverse application scenarios as an important topology (SVG, UPS and APF etc.). Typically, a simple series inductor is applied as the interface between the VSC and the utility grid. To attenuate switching frequency harmonics, especially in large power level, higher switching frequency or larger inductance is always required. As an alternative, *LCL* filter can perform satisfactorily, benefiting the harmonic suppression [1-5].

A *LCL* filter consists of two inductors and a capacitor and then systems incorporating *LCL* filters are third-order. To damp the inherent resonance, a resistor is always added in series with the capacitor. A good criterion to design *LCL* filter parameters can reduce the size of the installed reactive elements and can also reduce the power dissipations.

Literatures have explored this issue using different approaches and the calculated viewpoints are also not the same. [6] investigates the *LCL* filter on basis of the THD reduction; [7] design the *LCL* filter elements in a experiential way ; [8] analyzes the filter mathematically in a systematical way. Also their conclusions do not coincide with each other. This paper proposes a novel methodology for the analysis of a damped *LCL* filter and based on which, a general design procedure of each element is provided. Simulation results have verified the

good performance of the designed *LCL* filter.

II. MODEL ANALYSIS

A. Model of a VSC with a LCL Filter

Fig.1 a) shows a three phase VSC with a *LCL* filter and it consists of a standard VSC driven by an ideal DC voltage source. The system is assumed to be balanced during operation. A single phase PWM converter is considered for the analysis and design of the *LCL* filter, as showed in Fig.1 .b). The system has a fixed switching frequency denoted as ω_s and a constant DC voltage V_{DC} without any disturbance. The main purpose of this work is to determine the optimal values of L_1, L_2, C and R.

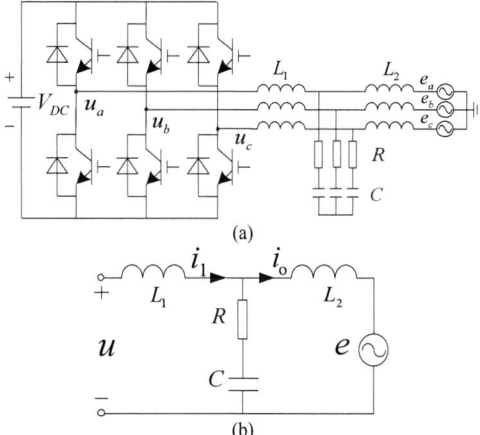

(a)

(b)

Fig. 1 System model, (a) three phase VSC with a *LCL* filter;
(b) Single phase *LCL* filter.

B. Basic Analysis of a LCL Filter

The transfer function of the *LCL* filter in Fig.1 b) can be expressed as (1), which has one zero point and three poles. Its zero point is expressed as (2).

$$G(s) = \frac{i_o}{u-e} = \frac{sRC+1}{s \times \dfrac{1}{L_1+L_2} \times (\dfrac{L_1 L_2 C}{L_1+L_2} s^2 + RCs^2 + 1)} \quad (1)$$

978-1-4799-2706-7/14 $31.00 © 2014 IEEE

$$\omega_z = 1/RC \tag{2}$$

If we define $L=L_1+L_2$, $k= L_1/L_2$, its resonant point can be written as (3):

$$\omega_R = \sqrt{\frac{L_1+L_2}{L_1 L_2 C}} = (k+1)\sqrt{\frac{1}{kLC}} \tag{3}$$

This system has got a damping character due to the resistor and the damping ratio can be derived as (4):

$$\xi = \frac{RC}{2}\sqrt{\frac{L_1+L_2}{L_1 L_2 C}} = \frac{RC}{2}(k+1)\sqrt{\frac{1}{kLC}} \tag{4}$$

Then considering expression (2), (3) and (4) we can conclude equation (5):

$$\frac{\omega_R}{\omega_z} = 2\xi \tag{5}$$

For the third order system of the *LCL* filter, it has three kinds of bode plots according to the value of ξ, as showed in Fig.2 a). For a damped *LCL* filter system, to get rid of the resonance hazard, ξ should be larger than 0.5. Otherwise, the output current harmonics introduced by the input voltage variations at the resonant point will be enlarged. And $\xi > 0.5$ means that the bode plot of a damped *LCL* filter is as line 1 and 2 showed in Fig.1 a) and expression (6) always satisfied.

$$\omega_R \geq \omega_z \tag{6}$$

For the third order system, the cross frequency in Bode plot of the *LCL* filter with the same total inductance can be derived as (7):

$$\omega_c = 1/L \tag{7}$$

With consideration of (7), the *LCL* filter has three types as displayed in in Fig.1 b).

- $\omega_z < \omega_R < \omega_c$, as line 1;

- $\omega_z < \omega_c < \omega_R$, as line 2;

- $\omega_c < \omega_z < \omega_R$, as line 3.

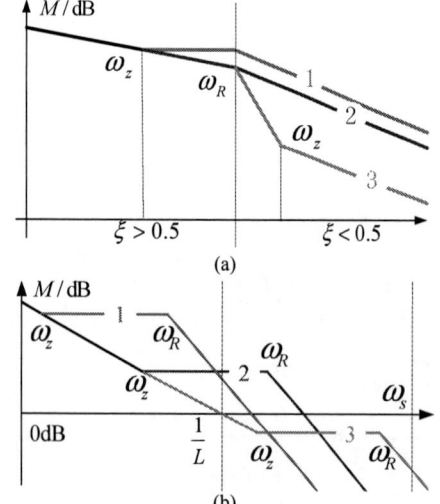

Fig. 2 Simplified bode plots of the *LCL* filter, (a) Bode plot types;
(b) Designed types.

To determine which curve of the three is the preference for the designer, we consider the capacitor of the LCL filter under the three conditions. Two scenarios at the demarcation points is mainly considered, which means when ω_z and ω_R are just coincide with the cross frequency ω_c, respectively. Then we can get the total capacitance expression (8) and their differential (9):

$$\begin{cases} C_1 = \dfrac{(k+1)^2}{4k\xi^2}L, & (\omega_c = \omega_z) \\[3mm] C_2 = \dfrac{(k+1)^2}{k}L, & (\omega_c = \omega_R) \end{cases} \tag{8}$$

$$\begin{cases} \dfrac{dC_1}{dk} = \dfrac{(k+1)(k-1)}{4k^2\xi^2}L, & (\omega_c = \omega_z) \\[3mm] \dfrac{dC_2}{dk} = \dfrac{(k+1)(k-1)}{k^2}L, & (\omega_c = \omega_R) \end{cases} \tag{9}$$

Through (9), we know that the total capacitance gets its minimum value under $k=1$. And at the demarcation points, the capacitance is:

$$\begin{cases} C_1 = \dfrac{1}{\xi^2}L, & (\omega_c = \omega_z) \\[3mm] C_2 = 4L, & (\omega_c = \omega_R) \end{cases} \tag{10}$$

As to the application of a *LCL* filter, the total capacitance is always small than $4L$. This indicates that $\omega_c < \omega_z < \omega_R$ is always fulfilled and its Bode plot is as line 3 in Fig.2 (b) shows.

From the analysis above, we can conclude that the designed damped *LCL* filter is -20dB/dec in low frequency segment from zero to ω_z, and is 0 dB/dec from ω_z to ω_R. In the high frequency segment, it becomes -40dB/dec because of the introduced zero point.

III. DESIGN OF THE PROPORTIONAL FACTOR

With regard to expression (5), the system transfer function of the *LCL* filter (1) can be rewrite as (11):

$$G(s) = \frac{\dfrac{2\xi}{\omega_R}s+1}{\dfrac{s}{L}\times(\dfrac{1}{\omega_R^2}s^2 + \dfrac{2\xi}{\omega_R}s+1)} \tag{11}$$

Expression (11) indicates that the performance of the *LCL* filter is determined by three variables: the total inductance L, the damping ration ξ and the location of the resonant point ω_R. The total inductance L is always determined by the designer according to the system switching frequency and the harmonic attenuation requirement. This means the location of ω_c is just designer oriented. As to the damping ratio ξ of the system, it is also determined by the designer on basis of the limitation of the resonance peak. Then if the location of ω_R is fixed, the whole system is determined.

Regarding expression (3), (12) can be derived:

$$\begin{cases} C = \dfrac{(k+1)^2}{kL\omega_R^2} \\[3mm] \dfrac{dC}{dk} = \dfrac{1}{L\omega_R^2}\dfrac{(k+1)(k-1)}{k^2} \end{cases} \quad (12)$$

The first expression in (12) indicates that to determine the location of ω_R, proper values of k and C should be calculated first. Then the second expression in (12) indicates that when k is equal to 1, the capacitance can get its minimum value. To get a more clear understanding in a intuitional way, the trends of C and dC/dk in a large scale is needed.

Fig.3 shows an example with the total inductance of the filter $L=0.2$mH and the location of ω_R is denoted by $\omega_R = a \times \omega_c$. Under different resonant frequencies, the relation between k and C is showed as Fig.3 a) and the relation between k and dC/dk is showed as Fig.3 b).

Fig.3 (a) and (b) evince that when $k<1$, C and dC/dk varies sharply. It seems that it is not applicable to choose k in this segment. This is also concluded in [8] and $k=3\sim7$ is recommended in it. However, it is obvious that if $k=L_1/L_2 <1$, it means $L_1<L_2$ and $1/k=L_2/L_1 >1$. Note expression (12), if the variable k is replaced by $k^* = 1/k = L_2/L_1$, we can get (13).

$$\begin{cases} C = \dfrac{(k^*+1)^2}{k^*L\omega_R^2} \\[3mm] \dfrac{dC}{dk^*} = \dfrac{1}{L\omega_R^2}\dfrac{(k^*+1)(k^*-1)}{(k^*)^2} \end{cases} \quad (13)$$

Expression (13) indicates that it is just the same form with that of (12). This means that the performance in segment $k<1$ is just the same with that in segment $k>1$. And it is also confirmed that $k=1$ is the optimal choice for the LCL filter design.

According to the above analysis, for the proportional factor k is determined, the location of ω_R is on related to the total capacitance as the first expression in (12) shows. The designer can chose different values of C to calculate different locations of ω_R according to design specifications, and different LCL filter plots are generated.

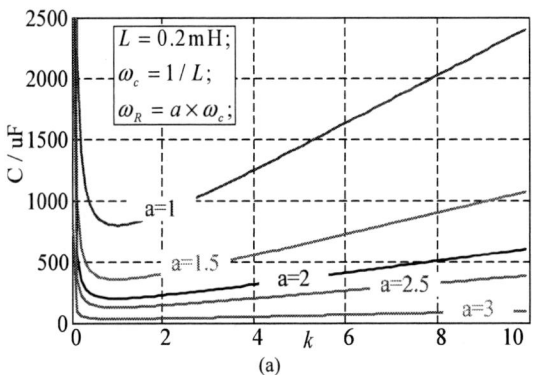

(a)

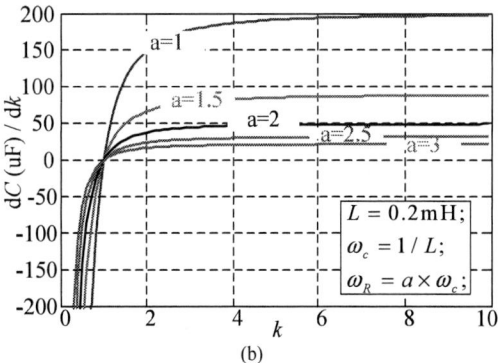

(b)

Fig. 3 Relations between k and C, k and dC/dk; (a) k and C; (b) k and dC/dk.

IV. DESIGN PROCEDURE OF A LCL FILTER

A general design procedure of damped LCL filters is provided here on basis of the above analysis.

A. Step 1: Total Inductance of the LCL Filter

As illustrated above, the cross frequency of the LCL filter is $\omega_c = 1/L$ and the total inductance L can be determined according to the system bandwidth, low frequency magnitude, commercial cost etc. And in a common design, equation (14) should be satisfied, where ω_s is the switching frequency of the VSC.

$$\omega_c = \frac{1}{5}\omega_s \sim \frac{1}{10}\omega_s \quad (14)$$

B. Step 2: Location of the Resonant Point

Then we come to determine the location of the resonant frequency ω_R. To get a satisfactory attenuation performance of the high frequency harmonics, especially the switching frequency ripple, equation (14) is always satisfied or even lower. For a -40dB attenuation of the switching ripple, $\omega_R < \dfrac{1}{10}\omega_s$ should be satisfied.

$$\omega_R < \frac{1}{5}\omega_s \quad (15)$$

C. Step 3: Total Capacitance of the LCL Filter

For $\omega_R = (k+1)\sqrt{\dfrac{1}{kLC}}$, the proportional factor k is designed as 1 and it is rewrite as $\omega_R = 2\sqrt{\dfrac{1}{LC}}$. When ω_R and L has been figured out, the capacitance can be derived as expression (16).

$$C = \frac{4}{L\omega_R^2} \quad (16)$$

In practical application, when L is determined, C can be used to regulate the position of ω_R to get a good performance of the system.

D. Step 4: Damping Resistor of the LCL Filter

As to the damping resistance, the system zero point is $\omega_z = \dfrac{1}{RC} = \dfrac{1}{2\xi}\omega_R = \dfrac{1}{\xi}\sqrt{\dfrac{1}{LC}}$ (k=1) we can get expression (17).

$$R = \xi\sqrt{\frac{L}{C}} \qquad (17)$$

Through the acquired L, C and the necessary damping ratio ξ, the damping resistance can be calculated then.

V. SIMULAITON RESULTS

An application scenario is demonstrated in this work and the specifications are displayed in Table. I.

TABLE I
SPECIFICATIONS OF THE SCENARIO

Element	Value
DC voltage	750VDC
AC voltage	380VAC
Switching Frequency	f_s=10kHz
Cross Frequency	$f_c \approx$900Hz
Resonant Frequency	$f_R = f_s/5$=2kHz
Total Inductance	L=0.2mH
Total Capacitance	C=128uF
Damping ratio	ξ=0.707
Damping Resistance	R=0.88Ω

The Bode plot of the designed LCL filter and the initial L filter are showed as Fig.4. The total inductance of the LCL filter and the L filter is 0.2mH and the cross frequency is about 900Hz. In Fig.4, we can see that the damped LCL has a much smaller magnitude at high frequency segments and this indicates a better saturation effect of the harmonics than the L filter. The two filters are almost the same at low frequency segments and then the LCL filter does not degrade the performance of the designed control loop. It can be concluded that the damped LCL filter has a significant improvement and can perform better than the L filter with the same inductance in total.

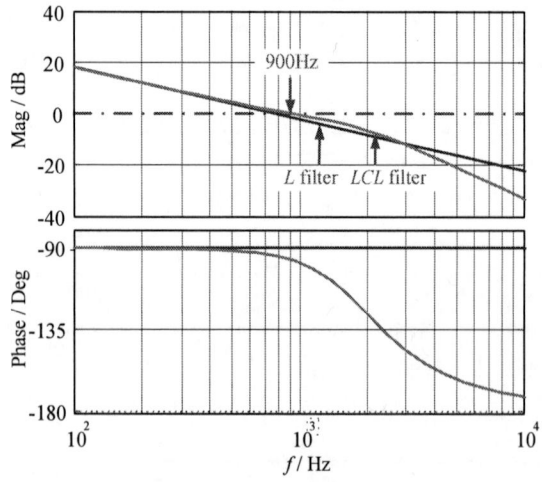

Fig. 4 Bode plot of he designed LCL filter.

Different proportional factor k is chosen for a comparison and the results are showed in Fig.5. k=1:19 (L_1=0.01mH, L_2=0.19mH) and k=19:1(L_1=0.19mH,

L_2=0.01mH) have the same bode plot and it is the same for k=3:1 (L_1=0.15mH, L_2=0.05mH) and k=1:3 (L_1=0.05mH, L_2=0.15mH). Though all the filters have almost the same low frequency character, the LCL filter with k=1:1 has the best performance in harmonic suppression because it has the lowest magnitude at high frequency segments.

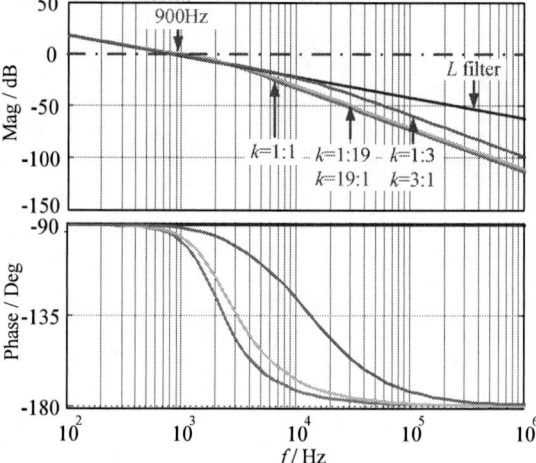

Fig. 5 Bode plot of different LCL filters with the same total inductance.

And an open loop simulation is carried out to verify the design procedure and the performance of the LCL filter, compared with L filter, as displayed in Fig.5.The comparison is carried out with single L filter and LCL filter under open loop, and the THD of the output current is reduced from 13.3% to 3.1%. There is a significant attenuation of the high frequency harmonics especially the switching frequency ripples.

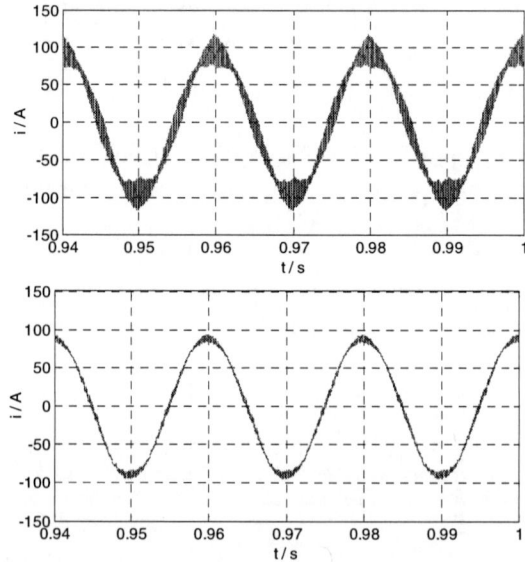

Fig. 5 Phase A current of the VSC with L and LCL filter, (a) 0.2mH L filter (b) LCL filter.

VI. CONCULSION

This work proposed a novel methodology and investigated the LCL filter mathematically and proposed a general and reasonable design procedure to determine the parameters of the filter system. $L_1/L_2=1$ is the optimal choice for the inductance allocation and L, C and R is determined according to the bandwidth, resonant frequency and the attenuation performance of the system. Simulation result of the performance of the LCL filter, compared with a simple L filter, has validated the design procedure.

REFERENCES

[1] Ning He, Dehong Xu, Ye Zhu, Jun Zhang, Guoqiao Shen, Yangfan Zhang, Jie Ma, and Changjin Liu, "Weighted Average Current Control in a Three-Phase Grid Inverter With an LCL Filter," *IEEE Transactions on Power Electronics*, vol. 26, no. 6, pp. 2785-2797, June. 2013.

[2] Guoqiao Shen, Dehong Xu, , Luping Cao, and Xuancai Zhu, "An Improved Control Strategy for Grid-Connected Voltage Source Inverters With an LCL Filter," *IEEE Transactions on Power Electronics*, vol. 23, no. 4, pp. 1899-1906, July. 2008.

[3] Ivan Jorge Gabe, Vin'ıcius Foletto Montagner, and Humberto Pinheiro, "Design and Implementation of a Robust Current Controller for VSI Connected to the Grid Through an LCL Filter," *IEEE Transactions on Power Electronics*, vol. 24, no. 6, pp. 1444-1452, June. 2009.

[4] Joerg Dannehl, Marco Liserre, and Friedrich Wilhelm Fuchs, "Filter-Based Active Damping of Voltage Source Converters With LCL Filter," *IEEE Transactions on Industrial Electronics*, vol. 58, no. 8, pp. 3623-3633, August. 2011.

[5] Yi Tang , Poh Chiang Loh, Peng Wang , Fook Hoong Choo, and Feng Gao, "Exploring Inherent Damping Characteristic of LCL-Filters for Three-Phase Grid-Connected Voltage Source Inverters," *IEEE Transactions on Power Electronics*, vol. 27, no. 3, pp. 1433-1443, March. 2012.

[6] Min-Young Park *, Min-Hun Chi**, Jong-Hyoung Park*, Heung-Geun Kim*, Tae-Won Chun***and Eui-Cheol Nho, "LCL-filter Design for Grid-Connected PCS Using Total Harmonic Distortion and Ripple Attenuation Factor," *The 2010 International Power Electronics Conference*, pp. 1688-1694, 2010.

[7] Liserre, M., Blaabjerg, F. and Hansen, S. "Design and control of an LCL-filter based three-phase active rectifier," *Industry Applications Conference*, pp. 299-307, 2001

[8] Kamran Jalili and Steffen Bernet,, "Design of LCL Filters of Active-Front-End Two-Level Voltage-Source Converters," *IEEE Transactions on Industrial Electronics*, vol. 56, no. 5, pp. 1674-1689, May. 2009.

The 2014 International Power Electronics Conference

70 kHz, 15 kW Silicon-Carbide MOSFET Inverter for Industrial Induction Heating Systems

Shohei Komeda, Yoshiki Tsuboi, and Hideaki Fujita
Department of Electrical and Electronic Engineering
Tokyo Institute of Technology
Tokyo, Japan
hf@ieee.org

Abstract—This paper discusses the applicability of SiC MOSFET to high-frequency induction heating applications. In this paper, a 15-kW, 70-kHz SiC MOSFET inverter prototype is constructed to evaluate the performance of the SiC MOSFETs. Power loss breakdown is also conducted from the experimental results. As a results, the inverter prototype has exhibited a power conversion efficiency as high as 98.5% at the rated power operation, and demonstrated the possibility of applications of SiC MOSFETs to high-frequency induction heating systems.

Keywords—*High-frequency inverters, Induction heating, Power conversion efficiency, SiC-MOSFETs.*

I. INTRODUCTION

Recently, it is strongly promoted to utilize renewable energy resources, such as wind power, solar power, geothermal, wave, and so on. Theses renewable energies are widely converted to a form of electrical power because the resources usually exist in a place far from resident areas. For this reason, smart grid and the next-generation HVDC are researched to improve the flexibility of the electricity and to achieve effective electric power transmissions. In the consumption, electricity is already used effectively for motor drives, lightings, heat pumps, and induction heatings. However, most of industrial heaters still use a gas, oil, and coal to obtain a high temperature and a large heat. Replacing the industrial heaters with heat pumps and/or induction heaters is the most important mission as the next step of the effective utilization of the renewable energy resource.

Induction heaters [1]-[18] commonly consists of resonant circuits and high-frequency power sources. It is requested to improve the efficiency of the high-frequency power source and to increase its operating frequency, in order to expand the applicable power range and heating performance.

Recently, wide band-gap semiconductors are expected as the material for the next generation switching power devices. Especially, silicon carbide (SiC) and gallium nitride (GaN) have superior electrical properties for power devices. SiC MOSFETs are already available and put into the market. SiC MOSFETs have a very low on-state resistance as well as a very fast switching capability. SiC MOSFETs are discussed mainly on applications to converters for motor drives and PV conversion systems. It is also required to discuss the applicability of SiC

MOSFETs to high-frequency induction heating systems [19][20]. In case of the induction heating applications, PWM operation is not required for switching devices. However, operating frequencies of the induction heaters are much higher than those in motor drive and PV conversion applications.

This paper presents the experimental confirmation of the applicability of SiC MOSFETs to high-frequency induction heating systems. In this paper, a 15-kW, 70-kHz SiC MOSFET inverter prototype is constructed and evaluated for experimental verifications. Especially, attention has been paid to the "self turn-on" phenomena, which is caused by the resonance between the gate stray inductance and the input capacitance of the SiC MOSFET. As a results, it is clarified that a trade-off relation exists between the rise time and the self turn-on duration. Thus, it is possible to minimize the total switching power loss by adjusting the gate resistor in the gate drive circuit. The inverter prototype has exhibited a power conversion efficiency as high as 98.5% at the rated power operation, and demonstrated the possibility of applications of SiC MOSFETs to high-frequency induction heating.

II. CIRCUIT CONFIGURATION

Fig. 1 show circuit configuration of a SiC MOSFET inverter prototype developed in this paper. The circuit parameters and specifications are listed in Table I. The inverter consists of an H-bridge inverter, a matching transformer, and a series resonant circuit. The H-bridge inverter uses two SiC-MOSFET modules (BSM100D12: 1200 V, 100 A, ROHM). In the module, two SiC MOSFETs and two SiC Schottky barrier diodes (SBDs) are assembled in the two-in-one package. The major electrical characteristics of the SiC MOSFET module are presented in Table II. The MOSFET has a very low on-state resistance of 20 mΩ, and is expected to be significant reduction in on-state loss. The dc link of the inverter is connected to the dc power supply whose voltage is set to $V_{dc} = 600$ V. A dc capacitor $C_{dc} = 20$ μF is also connected to the dc link.

Fig. 2 is the photograph of the H-bridge inverters. Two modules are placed in parallel on an aluminum heat sink. Four gate drive circuits are put on the modules and connected to the gate and control source terminals. Fig. 3 shows the circuit schematic of the gate drive circuit. Each gate drive circuit uses a MOSFET drivers

978-1-4799-2706-7/14 $31.00 © 2014 IEEE   1160

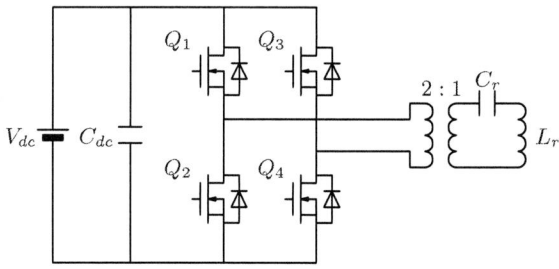

Fig. 1. Circuit configuration of the SiC MOSFET inverter prototype.

TABLE I.   CIRCUIT PARAMETERS OF THE EXPERIMENTAL SETUP

DC-link voltage	V_{dc}	600 V
DC-link current	I_{dc}	25 A
DC capacitor	C_{dc}	20 μs
Inverter fundamental voltage	V_{out1}	540 V
Inverter current	I_{out}	30 A
Turn ratio of transformer	$n_1 : n_2$	2 : 1
Coil voltage	V_L	530 V
Coil current	I_r	60 A
Resonant capacitor	C_r	0.3 μF
Working coil inductance	L_r	20 μH
Equivalent resistance	R_r	3 Ω (hot)
	R_r	2 Ω (cold)
Resonant frequency	f_r	65 kHz
Quality factor	Q_r	2.9 (hot)
		4.4 (cold)
Switching frequency	f_{sw}	70-100 kHz
Dead time	T_{DT}	800 ns

Fig. 2. SiC MOSFET modules and gate drive circuits.

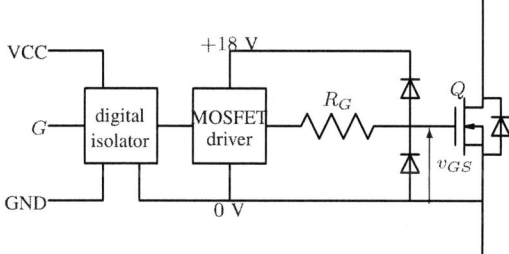

Fig. 3. Gate drive circuit.

(IXDD614: IXYS) and a digital isolator (ADuM6402: Analog Devices, 25 Mbps). Two clamping diodes are attached to the gate terminal of the MOSFET to protect the gate terminal from overvoltage. In the following experiments, four dc power supplies were used to provide a dc voltage of +18 V to date drivers, and thus, the gate-to-source voltage is +18 V for turn-on and 0 V for turn-off. The gate resistor was set to $R_G = 6.7\ \Omega$.

Fig. 4 shows the photograph of the working coil. The output ac terminals of the inverter is connected to the series resonant circuit L_r and C_r through the matching transformer. The turn ratio of the matching transformer is 2 : 1, and the magnetic core of the matching transformer is made of Ferrite. The working coil acts as a resonant inductor L_r which is made of a 5-mm diameter copper pipe for water cooling, and an iron workpiece with a diameter of 100 mm is inserted in the working coil. Film capacitors are used as the resonant capacitor C_r. Thus, the resonant frequency of the series resonant circuit was 65

kHz. the operating frequency of the inverter was adjusted from 68 to 85 kHz for controlling the heating power.

III.   EXPERIMENTAL RESULTS

Figs. 5 and 6 show experimental waveforms of the SiC MOSFET inverter prototype, which was measured under a 16-kW operating condition. Fig. 5 is waveforms of the output voltage v_{out} and the output current i_{out}. The output voltage v_{out} had a trapezoidal voltage wave shape, where any overshoots or voltage surges does not appear. The rise and fall times of the output voltage were about 200 ns and dv/dt was only 6kV/μs under this operating condition. In this experiment, the drain current was 15 A at the turn-off transition, the output capacitance $C_{oss} = 1000$ pF at $V_{dc} = 300$ V. From these parameters, dv/dt of the output

TABLE II.   ELECTRICAL CHARACTERISTICS OF THE SiC
MOSFET MODULE BSM100D12

Drain-to-source voltage	V_{DS}	1200 V
Drain current	I_D	100 A
		(pulse) 200 A
Gate-to-source voltage	V_{GS}	22 V
		-6 V
Threshold voltage	V_{th}	2.7 V
On-state resistance	R_{DSon}	20 mΩ
Rise time	t_r	50 ns
Fall time	t_f	160 ns

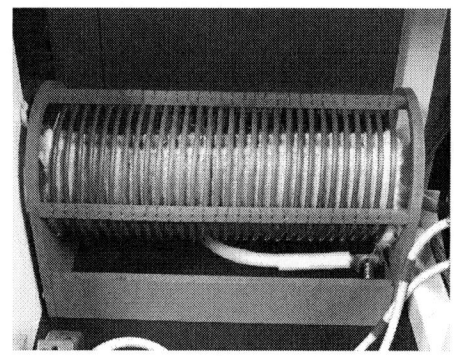

Fig. 4. Working coil.

978-1-4799-2706-7/14 $31.00 © 2014 IEEE

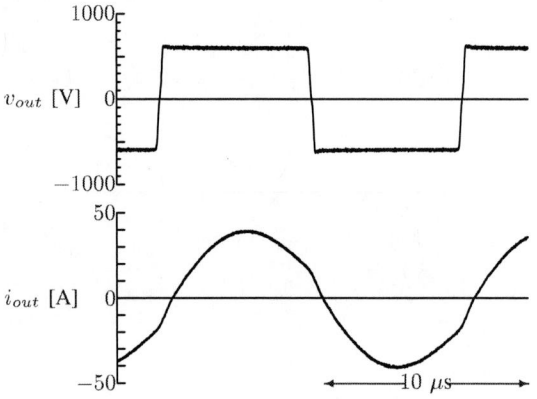

Fig. 5. Experimental waveforms of the output voltage and current at a 16-kW operation.

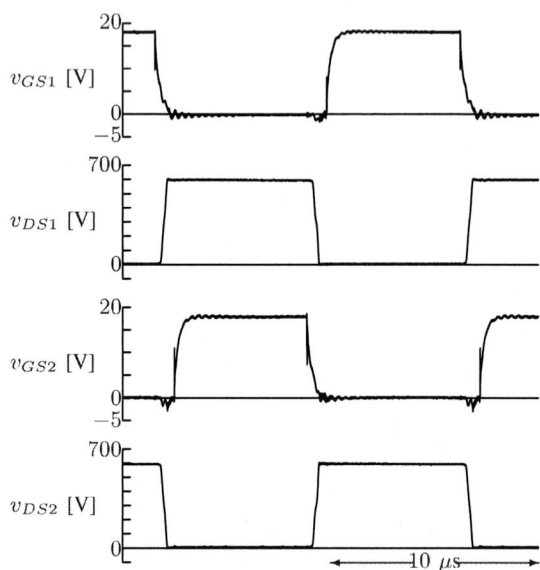

Fig. 6. Experimental waveforms of the gate-to-source and drain-to-source voltages at a 16-kW operation.

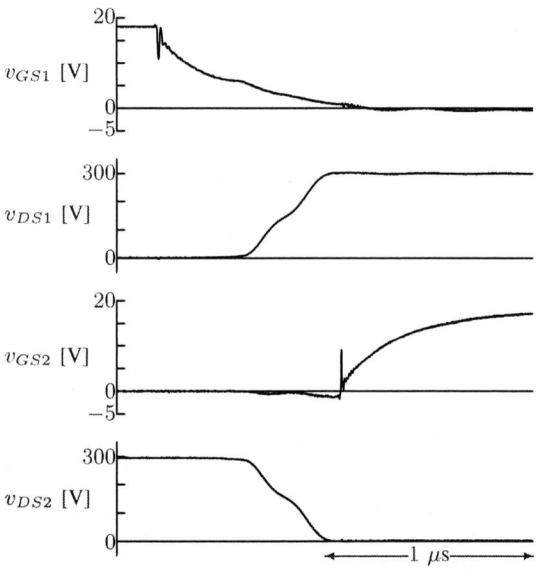

Fig. 7. Closed-up waveforms of the gate-to-source and drain-to-source voltages in case of $R_G = 10\ \Omega$.

voltage can be calculated as

$$\frac{dv_{DS}}{dt} = \frac{I_{Doff}}{2C_{oss}} = \frac{15\ \text{A}}{2000\ \text{pF}} = 7.5\ \text{kV}/\mu\text{s},$$

where I_{Doff} is the turn-off current, and C_{oss} is the effective output stray capacitance of the upper and lower SiC MOSFETs. and agrees well with the experimentally measured rise and fall times. This implies that The SiC MOSFETs can tuned off in a very short time, and almost all the load current just charges the output capacitance up. And then, the electric charge stored in the output capacitance is discharged or recovered by the resonant current in the next switching transition, and is not consumed in the MOSFETs, but effectively recovered to the dc link.

The output current i_{out} was a almost sinusoidal waveform. A small distortion was observed around the turn-off transition, which was caused by the relatively low quality factor of $Q = 3$ in the series resonant circuit. However, no high-frequency ringing appeared in the output current. In this case, the fundamental component of the output current i_{out} was 30 A, and the output power was about 16 kW. The measured power conversion efficiency reached 98.5% in this experimental situation.

Moreover, the power factor was 90% in this case, even when the SiC MOSFET inverter is operated at a high switching frequency of 70 kHz. This high power factor can be achieved by the shorten dead time of 800 ns. In general, the dead time is set to be 3 μs or longer if IGBT modules were used. Then the power factor would be lower than 22% because of such the long dead time.

Fig. 6 shows experimental waveforms of gate-to-source voltage v_{GS} and drain-to-source voltage v_{DS} across both upper and lower SiC MOSFETs Q_1 and Q_2. These four voltage waveforms are measured at the same time by using an optically-isolated acquisition system (DM-8000: IWATSU, 500 MHz, 1 GS/s, 2 ch/units). In these experiments, the dc-link voltage was adjusted to 600 V, and the operating frequency was set to 68 kHz. The gate-to-source voltage v_{GS1} was similar to a waveform of a first-order response. The upper and lower MOSFETs Q_1

and Q_2 are symmetrically operated, and their waveforms are almost the same.

Figs. 7-9 are the closed-up waveforms around the turn-off of the upper MOSFET Q_1. In these experiments, the dc-link voltage was reduced to $V_{dc} = 300$ V to observe the effect of the gate resistor R_G in the gate drive circuit. Therefore, the turn-off current also decreased to about 10 A.

Figs. 7 shows the waveforms in case of the gate resistor $R_G = 10\ \Omega$. The gate-to-source voltage of the upper MOSFET, v_{GS1} decreased with a time constant

The 2014 International Power Electronics Conference

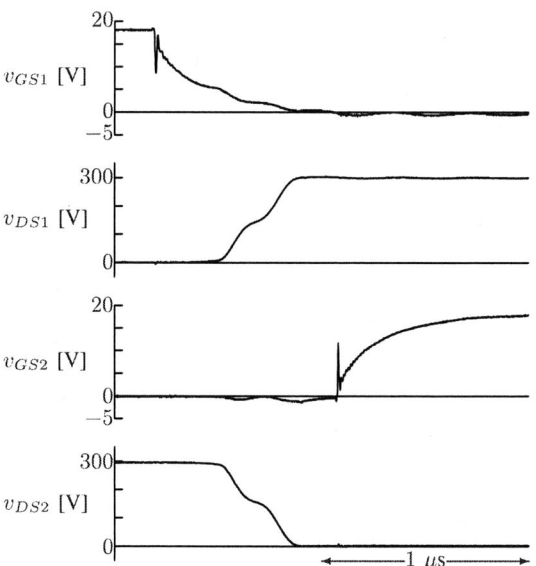

Fig. 8. Closed-up waveforms of the gate-to-source and drain-to-source voltages in case of $R_G = 6.7\ \Omega$.

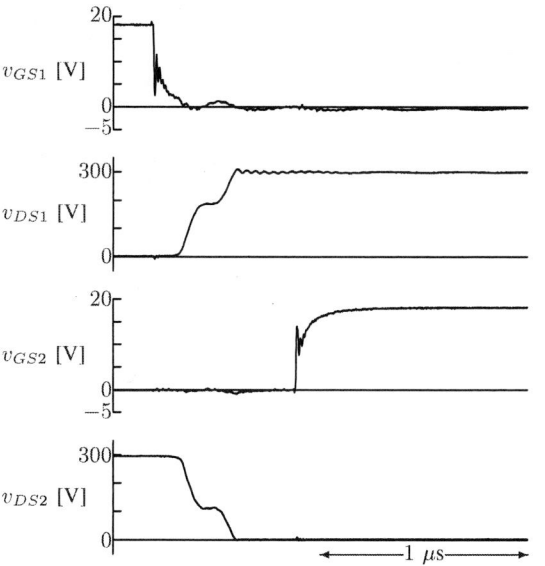

Fig. 9. Closed-up waveforms of the gate-to-source and drain-to-source voltages in case of $R_G = 1.1\ \Omega$.

of 1 μs, because the input stray capacitance is about $C_{iss} = 10,000$ pF. The gate-to-source voltage v_{GS1} has a flat part around $v_{GS1} = 7$ V which is caused by the Miller effect. Miller effect usually occurs in Si MOSFETs where dv_{DS}/dt is high. In case of the SiC MOSFET, the reverse stray capacitance is large where the drain-to-source voltage is low, and thus, a significant Miller effect occurred at such the low drain-to-source voltage.

The drain-to-source voltage v_{DS1} has almost no surge voltage when v_{DS1} reaches the dc-link voltage of $V_{dc} = 300$ V. The drain-to source voltage v_{DS1} had a strange voltage waveform, where the dv_{DS1}/dt became very low

TABLE III. SWITCHING POWER LOSS AGAINST THE GATE RESISTOR

Gate resistor R_G [Ω]	Switching loss P_{sw} [W]	Rise time t_r [ns]	Self turn-on t_{st} [ns]
1.1	77.1	110	120
6.7	70.7	230	80
10	75.8	300	70

around $v_{DS1} = 150$ V. This is referred to as "self turn-on" phenomena, which is induced by the resonance between the input stray capacitance C_{iss} and a stray inductance inside the MOSFET module. This phenomenon would slightly increase the switching power loss in the MOS-FETs. The gate resistor was reduced to $R_G = 6.7\ \Omega$ in Fig. 8, and $R_G = 1.1\ \Omega$ in Fig. 9. In Fig. 9, dv_{DS1}/dt was almost zero where $v_{DS1} = 200$ V. The gate-to-source voltage v_{GS1} had an oscillation even when the gate resistor is smaller than that in Figs. 7 and 8. The oscillation caused a serious self turn-on and an additional switching power loss.

Table III shows the relationship between the gate resistor and switching power loss. In Table III, the rise time t_r is calculated from the maximum value of dv_{DS1}/dt and the dc link voltage V_{dc}. Thus, the duration of the self tune-on t_{st} is excluded from the rise time t_r.

Reduction of the gate resistor R_G makes it possible to shorten the rise time t_r. This behavior is similar to that in a conventional Si MOSFETs, and usually short rise time results in a low switching power loss. On the other hand, the self turn-off duration t_{st} is lengthened according to reducing the gate resistor R_G. As a result, the switching power loss at $R_G = 6.7\ \Omega$ is the smallest among the three cases in Table III. It is important to adjust the resistance value of the gate resistor R_G for minimizing the switching power loss, paying attention not only to the rise time t_r but also to the duration of the self tune-on t_{st}.

In the case of $R_G = 6.7\ \Omega$, the additional energy loss caused the self turn-on phenomena can be calculate as

$$150\ \text{V} \times 10\ \text{A} \times 80\ \text{ns} = 0.12\ \text{mJ},$$

and it may cause a switching power loss as

$$0.12\ \text{mJ} \times 70\ \text{kHz} \times 4 = 34\ \text{W},$$

which is almost half of the switching power loss. However, the additional switching power loss caused by the self turn-off is only 0.85% of the output power of 4 kW at a half of the rated dc voltage.

Figs. 10-12 show the results of the power measurements when frequency control is adopted to the experimental SiC MOSFET inverter prototype. The dc link voltage was $V_{dc} = 600$ V, and the gate resistor is selected to be $R_G = 6.7\ \Omega$ to minimized the switching power loss. As shown in Fig. 10, the output power is adjustable by the switching frequency. The maximum output power can be obtained around 70-kHz switching frequency, and it reaches 15 kW. The output power can be reduced to 7.5 kW at a switching frequency of 84 kHz, which is almost a half of the maximum power.

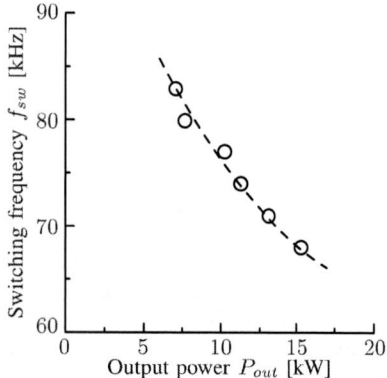

Fig. 10. Relationship between the switching frequency and the output power.

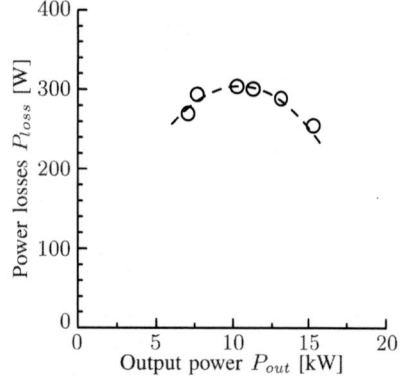

Fig. 11. Relationship between the switching frequency and the power loss.

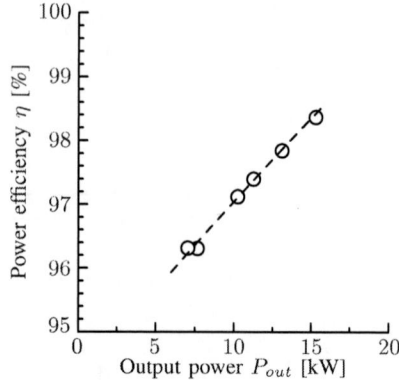

Fig. 12. Relationship between the switching frequency and the power loss.

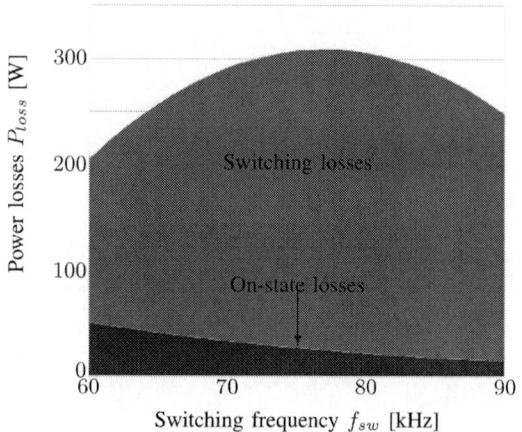

Fig. 13. Relationship between the switching frequency and the power loss.

Fig. 11 shows the power losses in the SiC MOSFET inverter. The power losses is only 250 W under the maximum power operating condition at 70 kHz. On the other hand, power losses reach its peak value of 300 W under a 75-kHz operating condition, in which the output power is about 12 kW. Under this operating condition, the turn-off current is relatively large even in a high switching frequency. The cooling devices should be designed based on the maximum power loss of 300 W. However, the maximum power loss is only 2% of the maximum output power of 15 kW.

Fig. 12 is the measurement results of the power conversion efficiency in the SiC MOSFET inverter. The highest efficiency of 98.5% is obtained under the maximum power operating condition of 15 kW. As presented in Fig. 11, the total power losses change in a range from 250 W to 300 W, even when the output power are adjusted from the full load to a half load. Thus, the power conversion efficiency is almost proportional to the output power in the measured power range.

Fig. 13 is the measured power losses consumed in the SiC MOSFETs and the analytical results of the loss breakdown. The on-state power losses are calculated from the rms values of the output current i_{out}, as

$$P_{on} = 2R_{on}I_{out}, \qquad (1)$$

where I_{out} is the rms value of the output current i_{out}.

while the switching power losses P_{sw} are obtained by subtraction of the on-state losses from the measured total losses P_{loss}, as given by

$$P_{sw} = P_{loss} - P_{on}. \qquad (2)$$

The SiC MOSFETs used in the experimental prototype have an on-state resistance as low as 20 mΩ, resulting in quite small on-state power losses less than 50 W. On the other hand, the switching power loss is much greater than the on-state power losses in the measured power range. The switching power losses are non-negligible because the inverter is operated at a high switching frequency of 70 kHz.

IV. CONCLUSION

This paper has discussed the experimental confirmation of the applicability of SiC MOSFETs to high-frequency induction heating systems. A 15-kW, 70-kHz SiC MOSFET inverter prototype has been constructed and evaluated for experimental verifications. The SiC MOSFET requires a very short dead time of 800 ns, and it achieves a stable operation in a switching frequency range from 70 to 85 kHz. A self turn-off phenomena have been observed in the drain-to-source voltage. It has been clarified that adjustment of the gate resistor makes it possible to minimize the total switching power loss caused by both rise time and self turn-off duration. As a results, the inverter prototype has exhibited a power conversion efficiency as high as 98.5% at the rated power operation, and demonstrated the possibility of applications of SiC MOSFETs to high-frequency induction heating.

REFERENCES

[1] W. G. Hurley and J. G. Kassakian, "Induction heating of circular ferromagnetic plates," *IEEE Trans. on Mag.*, vol.15, no.3, pp.1174–1181, 1979.

[2] F. W. Curtis, *High Frequency Induction Heating*, McGraw-Hill, New York, 1950.

[3] L. C. Meng, K. W. E. Cheng, and K. W. Chan, "Systematic Approach to High-power and Energy-efficient Industrial Induction Cooker System: Circuit Design, Control Strategy and Prototype Evaluation," in *IEEE Trans. on Power Electron.*, vol. 26, no. 12, pp. 3754–3765, Dec. 2011.

[4] J. C. Lewis, "Control Device for parallel induction heating coils," *US patent*, no. 4307278, Dec. 22, 1981.

[5] J. H. Simcock, "Multiple zone induction heating," *US patent*, no. 5059762, Oct. 22, 1991.

[6] Y. C. Jung, "Dual half bridge series resonant inverter for induction heating appliance with two loads," *Electronics Letters*, vol. 35, no. 16, pp. 1345-1346, 1999.

[7] T. J. Bowers, C. F. Der, and J. D. Parker, "Multi-zone induction heating system with bidirectional switching network," *US patent*, no. 6078033, Jun. 20, 2000.

[8] O. S. Fishman, R. K. Lampi, J. H. Mortimer, V. A. Peysakhovich, "Induction heating device and process for the controlled heating of a non-electrically conductive material," *US patent*, no. 6121592, Sep. 19, 2000.

[9] O. Lucia, J. M. Burdio, L. A. Barragan, J. Acero, I. Millan, "Series resonant multi-inverter for multiple induction heaters," *IEEE Trans. on Power Electron.*, vol. 25, no. 11, pp. 2860–2868, 2010.

[10] I. Millan, J. M. Burdo, J. Acero, O. Luca, and D. Palacios, "Resonant inverter topologies for three concentric planar windings applied to domestic induction heating," in *Electron. Lett.*, vol. 46, no. 17, pp. 1225–1226, 2010.

[11] F. Forest, E. Laboure, F. Costa, and J. Y. Gaspard, "Principle of a multi-load/single converter system for low power induction heating," *IEEE Trans. on Power Electron.*, vol. 15, no. 2, pp. 223–230, 2000.

[12] J. I. Rodriguez and S. B. Leeb, "A Multilevel Inverter Topology for Inductively Coupled Power Transfer," *IEEE Trans. on Power Electron.*, vol. 21, no. 6, pp. 1607–1617, 2006.

[13] J. I. Rodriguez and S. B. Leeb, "Nonresonant and resonant frequency-selectable induction-heating targets," *IEEE Trans. on Power Electron.*, vol. 57, no. 9, pp. 3095–3108, 2010.

[14] D. S. Schatz and J.M. Dorrenbacher, "Frequency selected, variable output inductor heater system," *U.S. Patent*, 6 316 754 B1, Nov. 13, 2001.

[15] H. Fujita, N. Uchida, and K. Ozaki, "A new zone-control induction heating system using multiple inverter units applicable under mutual magnetic coupling conditions," in *IEEE Trans. on Power Electron.*, vol. 26, no. 7, pp. 2009–2017, 2011.

[16] H. N. Pham, H. Fujita, K. Ozaki, and N. Uchida, "Phase angle control of high-frequency resonant currents in a multiple inverter system for zone-control induction heating," in *IEEE Trans. on Power Electron.*, vol. 26, no. 11, pp. 3357–3366, 2011.

[17] H. N. Pham, H. Fujita, K. Ozaki, and N. Uchida, "Estimating method of heat distribution using three-dimensional resistance matrix for zone-control induction heating systems," in *IEEE Trans. on Power Electron.*, vol. 27, no. 7, pp. 3374–3382, 2012.

[18] H. N. Pham, H. Fujita, N. Uchida, and K. Ozaki, "Dynamic analysis and control of a zone-control induction heating system," in *IEEE Trans. on Power Electron.*, vol. 28, no. 3, pp. 1297–1307, 2013.

[19] J. Avellaned, C. Bernal,A. Otin, P. Molina, J. M. Burdio, "Half-bridge resonant inverter with SiC cascode applied to domestic induction heating," in *Proc. of IEEE Applied Power Electronics Conference and Exposition*, pp.122-127, 2013.

[20] J. Jordán, V. Esteve, E. Sanchis-Kilders, E. J. Dede, E. Maset, J. B. Ejea, and Agustin Ferreres, "A comparative performance study of a 1200 V Si and SiC MOSFET intrinsic diode on an induction heating inverter," *IEEE Transactions on Power Electron.*, vol.29, no.5, pp.2550-2562, 2014.

A Study on Efficiency Improvement of High-frequency Current Output Inverter based on Immittance Conversion Element

Shun SUZUKI, Toshihisa SHIMIZU
Dept. Tokyo Metropolitan University
TMU
Tokyo, Japan
Email: shimizut@tmu.ac.jp

Abstract- **Wireless power transmission utilizing magnetic field resonance has been given increasing attention by power electronics engineers and industry. Actually, many technical papers that relate to the transmission coils of wireless power transmission based on magnetic field resonance have been published and verified the high-efficiency power transmission characteristics. However, the 10 MHz class high-frequency power converter suitable for high-power transmission systems has not been studied sufficiently, although small power rating converters, such as the class-E converter with a 74.9 W output power, have been proposed. Hence, the authors propose a new circuit topology for high-power high-frequency inverter systems suitable for high-power wireless power transmission.**

In this paper, the authors propose a novel high-frequency current output type inverter circuit topology utilizing an immittance conversion element and wireless transmission coils, which utilize an LC anti-resonant circuit suitable for the proposed current output type inverter.

As a result of the prototype test, the proposed inverter provides a high conversion efficiency of more than 77.6 % at 11.33 MHz operation with a 123 W output power rating. Also, the proposed transmission coils suitable for current output type inverters provide a high transmission efficiency of more than 94.2 % at 11.33 MHz operation with 20 mm transmission gaps and a 46.1 W output power.

Keywords- **current output inverter, immittance conversion, wireless power transmission, magnetic field resonance**

I. INTRODUCTION

In recent years, wireless power transmission based on magnetic field resonance technology has been studied extensively. Fig. 1 shows the system configuration. In this system, 10 MHz class high-frequency power is used for the intermediate-range wireless power transmission because the transmission efficiency of the transmission coils based on the magnetic field resonance phenomena exceeds 90 % [1]-[3]. In order to prevent the EMI, Electro-Magnetic Interference, problem, the transmission frequency is usually set to the ISM bandwidth, Industry-Science-Medical bandwidth, such as 13.56 MHz, in practical systems.

So far, the previous studies on wireless power transmission systems utilizing magnetic field resonance focused extensively on the transmission coils and the rectifier design [1]-[4]. However, only a few studies on increasing the conversion efficiency and the power rating of high-frequency inverters suitable for transmission systems utilizing magnetic field resonance have been reported.

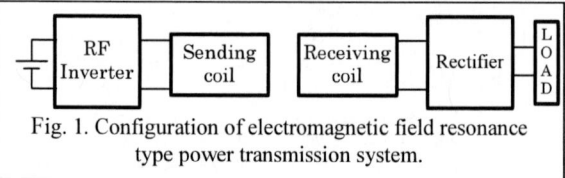

Fig. 1. Configuration of electromagnetic field resonance type power transmission system.

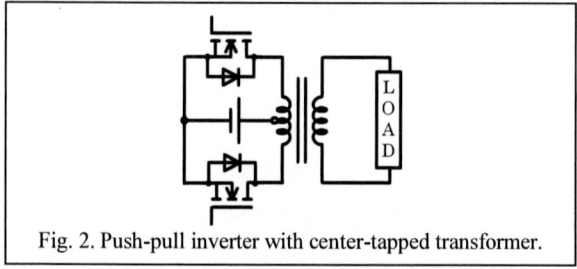

Fig. 2. Push-pull inverter with center-tapped transformer.

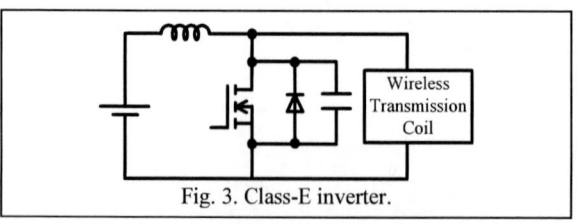

Fig. 3. Class-E inverter.

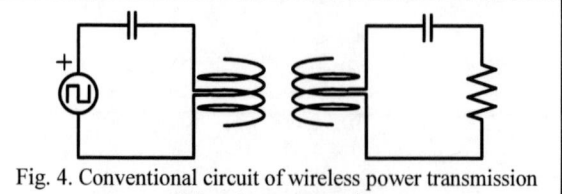

Fig. 4. Conventional circuit of wireless power transmission system utilizing electromagnetic resonance.

978-1-4799-2706-7/14 $31.00 © 2014 IEEE

In the previous studies, either linear amplifiers or conventional inverters have been applied in the transmission system, and the conversion efficiency of those power sources is potentially low at the very high frequency condition. Hence, the wireless power transmission system cannot provide a high efficiency for the entire system, even though the efficiency of the transmission coils is very high.

A conventional inverter that utilizes a push-pull inverter with a center-tapped transformer, as shown in Fig. 2, has been used. The inverter has a sufficient power rating for this application, but the conversion efficiency at 10 MHz operation cannot be high due to the large iron loss of the center-tapped transformer. Another example is a class-E multi-resonant inverter with ZVS, Zero Voltage Switching, as shown in Fig. 3. In this case, the transmission system achieves more than 74 % efficiency with a 74.9 W output power rating. However, it is difficult to increase the power rating because of the restriction of the circuit topology [5].

The authors propose an inverter topology suitable for high-power high-efficiency operation at an output frequency of around 10 MHz in order to apply the wireless power transmission technology, especially for high-power home appliances and electric vehicles. The basic structure of the inverter has been proposed for the application of electric ballast for HID Lamps and plasma generators [6]-[10].

The current output type inverter, which utilizes an immittance conversion element, is proposed by taking the above situation into account. One of the distinctive characteristics of the inverter is that a constant output current is supplied from the output terminal of the inverter by the effect of an immittance conversion element, which converts the voltage source to a current source. Also, the output current can easily be increased by connecting the output terminals of the multiple inverters in parallel, thus increasing the power rating of the inverter.

The second characteristic of this inverter is the resonant gate-drive circuit, which provides low driving power at high-frequency switching. In addition, transmission coils suitable for current output type inverters are studied because the conventional transmission coils can be driven by the voltage source, as shown in Fig. 4.

In this paper, LC, L designates inductor and C designates capacitor, parallel resonance type transmission coils with a high quality factor are proposed, and the power transmission characteristics with the high-frequency AC current source condition are studied. First, the circuit configuration of the proposed inverter and the magnetic field anti-resonance transmission coils are explained. Then, the operation principle of the immittance conversion element and the characteristics of converting the input voltage to the output current are explained. The operation principle of magnetic field anti-resonance coils is also explained. Finally, the usefulness of the proposed inverter and the anti-resonance transmission coils are confirmed through experiments utilizing an 11.33 MHz, 100 W prototype setup.

II. CIRCUIT CONFIGURATION

Fig. 5 and Table 1 show the circuit configuration and parameters of the proposed inverter, respectively. The circuit consists of a resonant gate-drive circuit, which is surrounded by a red dotted line, and a push-pull inverter circuit based on the immittance conversion element, which is surrounded by a blue dotted line.

The push-pull inverter circuit consists of two MOSFET switches Q_1 and Q_2, an immittance conversion element, which is composed of transmission lines T_1 and T_2, and a load Z_L, which represents the anti-resonance transmission coil. The length of the two distributed constant lines T_1 and T_2 is adjusted to 1/4 of the propagation wavelength of the inverter operation frequency. The sending ends of the two lines are connected in series so as to produce the center-tapped configuration, but the receiving ends are connected in inverse parallel.

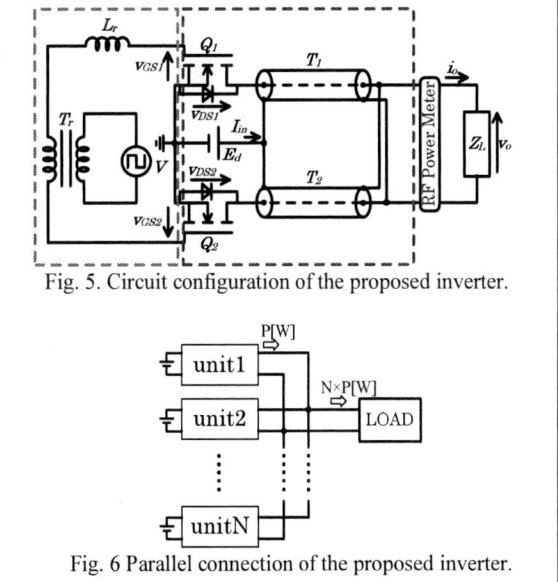

Fig. 5. Circuit configuration of the proposed inverter.

Fig. 6 Parallel connection of the proposed inverter.

Table 1. Circuit parameters of the wireless power transmission system

Name	Symbol	Specification
Parameters of the inverter		
Si-MOSFET	Q_1, Q_2	IPP50CN10N (100 V/20 A)
Immittance Conversion Element	T_1, T_2	10 D-2 V ($Z_\omega = 50\ \Omega, x = 4.37\ \mathrm{m}$)
DC Voltage Source	E_d	40 V
Operation Frequency	f_s	11.33 MHz
Parameters of the coils and capacitors for the anti-resonant circuit		
Resonant Inductor	L	200 nH
Resonant Capacitor	C	100 pF-1.5 nF
Load Resistor	R_L	25-200 Ω

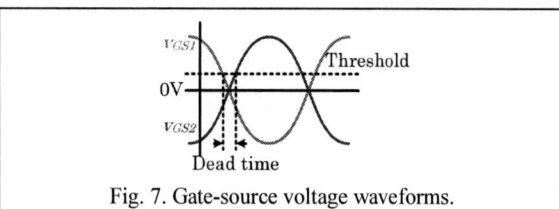

Fig. 7. Gate-source voltage waveforms.

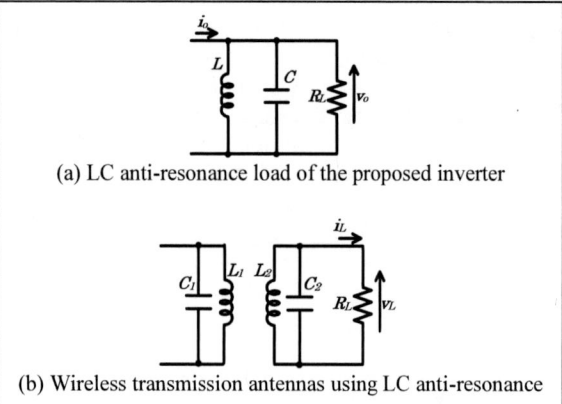

(a) LC anti-resonance load of the proposed inverter

(b) Wireless transmission antennas using LC anti-resonance

Fig. 8. Load circuit configuration of the proposed inverter.

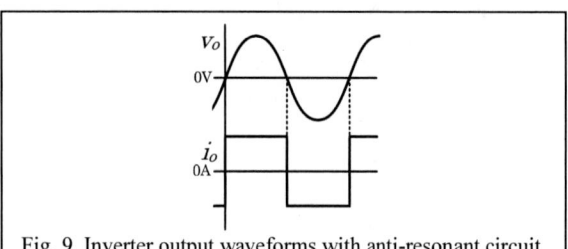

Fig. 9. Inverter output waveforms with anti-resonant circuit.

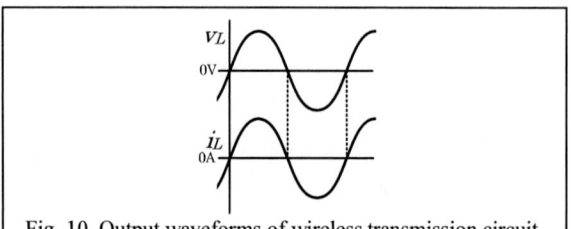

Fig. 10. Output waveforms of wireless transmission circuit.

In addition, the output power ratings of the inverter can be increased by connecting the output terminals with the same inverter output terminal in parallel, as shown in Fig. 6.

The waveforms of the gate-source voltage on both MOSFETs are shown in Fig. 7. The gate-source voltages of both MOSFETs are sinusoidal waveforms with a phase difference of 180 degrees, and thus, each MOSFET turns on alternately. In addition, the energy stored in the gate capacitance of one MOSFET is transferred to the gate capacitance of another MOSFET; hence, the energy supplied from the gate-drive circuit can be reduced significantly.

Fig. 8(a) shows the circuit configuration of the anti-resonant circuit, which is a typical load suitable for the current-source inverter. The circuit is composed of an inductor L, a capacitor C, and a load resister R_L. When the anti-resonant frequency is adjusted to the inverter output frequency, a sinusoidal voltage waveform appears at the output terminal, as shown in Fig. 9, and the amplitude of the output voltage is increased depending on the quality factor of the LC resonant circuit.

Fig. 8(b) shows the circuit configuration of the wireless power transmission coils, which consists of a sending circuit and a receiving circuit. The sending circuit is composed of a parallel connection of an inductor L_1 and a capacitor C_1, and the receiving circuit is composed of a parallel connection of an inductor L_2, a capacitor C_2, and a load resister R_L. When the anti-resonant frequency of the wireless transmission coil is adjusted to the inverter frequency, the transmission efficiency of the transmission coil is very high owing to the high quality factor of the parallel resonant circuit. The output waveforms of the receiving circuit are shown in Fig. 10. The waveforms of both the voltage and the current are sinusoidal because the transmission coils resonate.

III. OPERATION PRINCIPLE

Fig. 11 shows the equivalent circuit configuration of the proposed inverter. The DC voltage source E_d is replaced with two voltage sources E_1 and E_2, and those are relocated to each side. The MOSFETs Q_1 and Q_2 are replaced with ideal switches S_1 and S_2, which are turned on and off alternately with a duty ratio of 0.5. In order to make the understanding of the operation principle easier, the parallel LC anti-resonant circuit is removed, only the resistor R_L is connected to the load terminal, and one voltage source E_2 on the S_2 side is assumed to be zero.

At the beginning of the inverter operation, it is assumed that no transient voltage and current exist in the transmission lines T_1 and T_2 or the resistor R_L. At the first switching cycle T_{SW}, the operation of the inverter is classified into four modes. The detailed operations for each mode are as follows:

<Mode 1: $0 \leq t \leq T_{SW}/4$> When S_1 is turned on at $t = 0$, a DC voltage $E_1 = E$ is induced to the sending terminal of T_1, and the voltage wave propagates toward the receiving terminal of T_1. The propagation wave arrives at the receiving terminal at $t = T_{SW}/4$, and a reflection/transmission phenomena occurs at the resistor R_L based on the reflection coefficient Γ. Here, Γ is given by Eq. (1).

$$\Gamma = -\frac{Z_\omega}{Z_\omega + 2R_L} \tag{1}$$

Hence, the input voltage v_{1i} from T_1, the reflection voltage v_{1r} toward T_1, the transmission voltage v_{1t} toward T_2, and the resultant output voltage v_o at R_L are expressed as follows:

$$v_{1i}(T_{SW}/4) = E \tag{2}$$

978-1-4799-2706-7/14 $31.00 © 2014 IEEE

$$v_{1r}(T_{SW}/4) = \Gamma E \tag{3}$$

$$v_{1t}(T_{SW}/4) = (1 + \Gamma)E \tag{4}$$

$$\begin{aligned} v_o(T_{SW}/4) &= v_{1i}(T_{SW}/4) + v_{1r}(T_{SW}/4) \\ &= (1 + \Gamma)E \\ &= (1 + \Gamma)v_{1i}(T_{SW}/4) \end{aligned} \tag{5}$$

The output voltage v_o has a positive value because the absolute value of the reflection coefficient is smaller than 1.

<Mode 2: $T_{SW}/4 \leq t \leq T_{SW}/2$> The reflection voltage wave and the transmission voltage wave travel on transmission lines T_1 and T_2, respectively, and arrive at the sending terminals of T_1 and T_2 at $t = T_{SW}/2$, respectively. At this moment, S_1 is turned off, and S_2 is turned on.

At the sending terminal of T_1, the arrived voltage wave, which is the same as $v_{1r}(T_{SW}/4) = \Gamma E$, reflects with a reflection coefficient of $\Gamma = 1$, and the reflected voltage v_{S1r} propagates again toward T_1.

$$v_{S1r}(T_{SW}/2) = \Gamma E \tag{6}$$

Hence, the voltage v_{S1} at the sending terminal of T_1 and the voltage v_{DS1} at switch S_1 are

$$\begin{aligned} v_{S1}(T_{SW}/2) &= v_{1r}(T_{SW}/4) + v_{S1r}(T_{SW}/2) \\ &= 2\Gamma E \end{aligned} \tag{7}$$

$$v_{DS1}(T_{SW}/2) = E - v_{S1}(T_{SW}/2) = (1 - 2\Gamma)E \tag{8}$$

At the sending terminal of T_2, the arrived voltage wave, which is same as $v_{1t}(T_{SW}/4) = (1 + \Gamma)E$, reflects with a reflection coefficient of $\Gamma = -1$, and the reflected voltage v_{S2r} propagates again toward T_2.

$$v_{S2r}(T_{SW}/2) = -(1 + \Gamma)E \tag{9}$$

Hence, the voltage v_{S2} at the sending terminal of T_2 and the voltage v_{DS2} at switch S_2 are

$$v_{S2}(T_{SW}/2) = v_{1t}(T_{SW}/4) + v_{S2r}(T_{SW}/2) = 0 \tag{10}$$

$$v_{DS2}(T_{SW}/2) = 0 \tag{11}$$

<Mode 3: $T_{SW}/2 \leq t \leq 3\,T_{SW}/4$> The reflection voltage waves v_{S1r} and v_{S2r} from the sending ends of T_1 and T_2, respectively, arrive at the receiving ends of T_1 and T_2 at $t = 3\,T_{SW}/4$. The input voltage v_{1i} from T_1 and the input voltage v_{2i} from T_2 are expressed as

$$v_{1i}(3\,T_{SW}/4) = \Gamma E \tag{12}$$

$$v_{2i}(3\,T_{SW}/4) = -(1 + \Gamma)E \tag{13}$$

The transmission voltage $v_{1t}(3\,T_{SW}/4)$, which transmits from T_1 and T_2, and the reflection voltage $v_{2r}(3\,T_{SW}/4)$, which reflects from T_2 and T_2, are expressed as

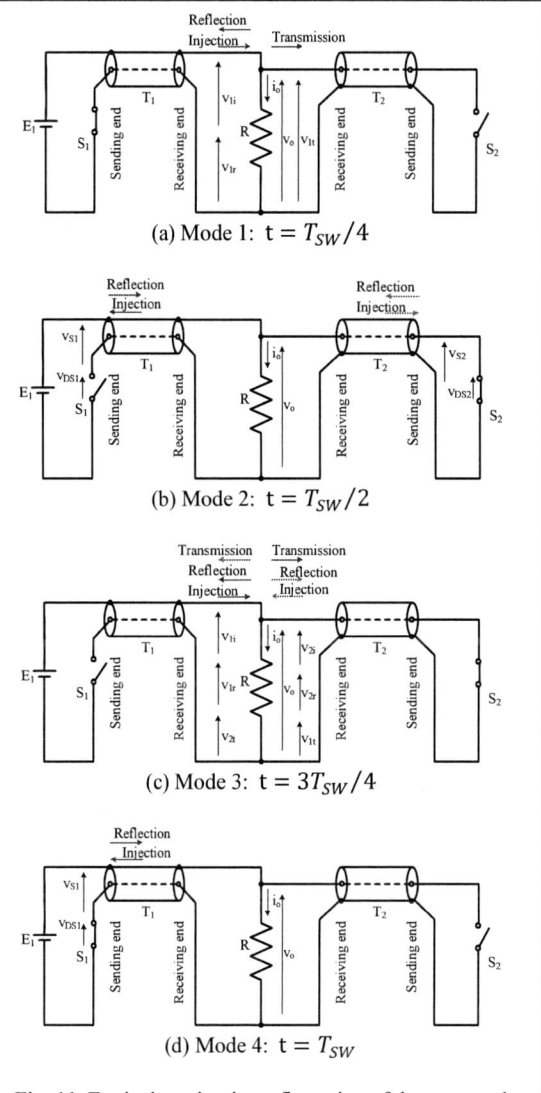

(a) Mode 1: $t = T_{SW}/4$

(b) Mode 2: $t = T_{SW}/2$

(c) Mode 3: $t = 3T_{SW}/4$

(d) Mode 4: $t = T_{SW}$

Fig. 11. Equivalent circuit configuration of the proposed inverter circuit.

$$v_{1t}(3\,T_{SW}/4) = (1 + \Gamma)\Gamma E \tag{14}$$

$$v_{2r}(3\,T_{SW}/4) = -(1 + \Gamma)\Gamma E \tag{15}$$

As a result, the propagation voltage toward T_2 is diminished. On the other hand, the transmission voltage $v_{2t}(3\,T_{SW}/4)$, which transmits from T_2 and T_1, and the reflection voltage $v_{1r}(3\,T_{SW}/4)$, which reflects from T_1 and T_1, are expressed as

$$v_{2t}(3\,T_{SW}/4) = -(1 + \Gamma)^2 E \tag{16}$$

$$v_{1r}(3\,T_{SW}/4) = \Gamma^2 E \tag{17}$$

Hence, the propagation wave toward T_1 is expressed as

$$v_{2t}(3\,T_{SW}/4) + v_{1r}(3\,T_{SW}/4) = -(1 + 2\Gamma)E \tag{18}$$

The resultant output voltage v_o at R_L is expressed as follows:

$$v_o(3\,T_{SW}/4) = v_{1i}(3\,T_{SW}/4) + v_{1r}(3\,T_{SW}/4)$$
$$+ v_{2t}(3\,T_{SW}/4) = -(1+\Gamma)E \quad (19)$$
$$= -(1+\Gamma)v_{1i}(T_{SW}/4)$$

It should be noted that the output voltage v_o expressed by Eq. (19) has an inverse polarity from the one expressed by Eq. (5).

<Mode 4: $3\,T_{SW}/4 \leq t \leq T_{SW}$> The propagation wave toward T_1, which is expressed by Eq. (18), travels on the transmission line T_1 and arrives at the sending terminal of T_1 at $t = T_{SW}/2$. At this moment, S_1 is turned on. Since the impedance of the voltage source E_1 is zero, the arrived voltage wave is reflected with a reflection coefficient of $\Gamma = -1$, and the reflection voltage v_{S1r} is expressed as

$$v_{S1r}(T_{SW}) = -(1+2\Gamma)E \quad (20)$$

Hence, the voltage v_{S1} at the sending terminal of T_1 and the voltage v_{DS1} at switch S_1 are expressed as follows:

$$v_{S1}(T_{SW}) = E + \{v_{2t}(3\,T_{SW}/4)$$
$$+ v_{1r}(3\,T_{SW}/4)\} + v_{S1r}(T_{SW})$$
$$= E - (1+2\Gamma)E \quad (21)$$
$$+ (1+2\Gamma)E = E$$

$$v_{DS1}(T_{SW}) = 0 \quad (22)$$

At the second interval, both the supply voltage E_1 and the reflection voltage $v_{DS1}(T_{SW}) = (1+\Gamma)E$ for the previous operation mode (Mode 4) propagate toward the receiving terminal of T_1, and those operate as the input voltage v_{1i}.

$$v_{1i}(T_{SW}/4 + T_{SW}) = E + (1+2\Gamma)E \quad (23)$$

In the same manner, the input voltage for the third interval is expressed as follows:

$$v_{1i}(T_{SW}/4 + 2T_{SW})$$
$$= E$$
$$+ (1+2\Gamma)\{E + (1+2\Gamma)E\} \quad (24)$$
$$= E + (1+2\Gamma)E$$
$$+ (1+2\Gamma)^2 E$$

Hence, the input voltage of the n-th interval is

$$v_{1i}(n) = \sum_{k=1}^{n} E(1+2\Gamma)^2 \quad (25)$$

The output voltages for Modes 1 and 3 for the second interval are respectively

$$v_o(T_{SW}/4 + T_{SW}) = (1+\Gamma)v_{1i}(T_{SW}/4 + T_{SW}) \quad (26)$$

$$v_o(3\,T_{SW}/4 + T_{SW})$$
$$= -(1+\Gamma)v_{1i}(T_{SW}/4 + T_{SW}) \quad (27)$$

Hence, the output voltage of the n-th interval is

$$v_o(n) = \sum_{k=1}^{n} E(1+\Gamma)(1+2\Gamma)^{k-1} \quad (28)$$

From the above description, the amplitudes of the input voltage V_i, the output voltage V_o, and the output current I_o for the steady-state condition are expressed as follows:

$$V_{1i} = \lim_{n \to \infty} v_{1i}(n) = -\frac{E}{2\Gamma} = \frac{Z_\omega - 2R_L}{2Z_\omega}E \quad (29)$$

$$V_o = \lim_{n \to \infty} v_o(n) = -\frac{E(1+\Gamma)}{2\Gamma} = \frac{R_L}{Z_\omega}E \quad (30)$$

$$I_o = \frac{V_o}{R_L} = \frac{E}{Z_\omega} \quad (31)$$

In the case where $E_1 = 0$ and $E_2 = E$, the same results are obtained, and the resultant output voltage and output current have twice the values of Eqs. (30) and (31), respectively.

It should be noted that the output current depends only on the characteristic impedance of the transmission line and the DC bus voltage of the inverter. In other words, the output voltage of the proposed inverter is converted to the current source at the output terminal of the transmission lines. This is because the transmission line is called the immittance conversion element.

In the case where the anti-resonant circuit is connected to the output terminal of the inverter, as shown in Fig. 5 and Fig. 7(a), the output voltage is formed as sinusoidal, and the anti-resonant voltage is boosted to a high voltage. Accordingly, the output current on the receiving end of the immittance conversion element has a square waveform.

B. Operation principle of wireless power transmission circuit utilizing magnetic field anti-resonance

In a conventional magnetic field resonance circuit such as that shown in Fig. 4, the transmission efficiency of the wireless power transmission circuit depends on both the coupling coefficient of the two coils and the quality factor of the sending and receiving coils. The quality factor represents the amplification ratio of the voltage or current in each coil. In LC parallel resonance type transmission coils, as shown in Fig. 8(b), the quality factors of the sending circuit and receiving circuit at the anti-resonant frequency are given by Eq. (32).

$$Q = \omega_0 CR = R\sqrt{\frac{C}{L}} \quad (32)$$

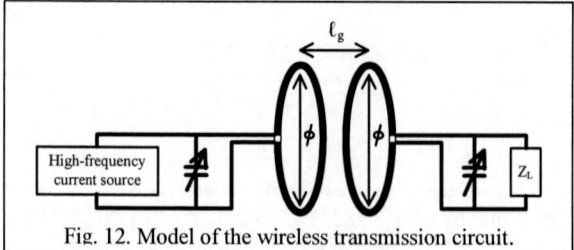

Fig. 12. Model of the wireless transmission circuit.

Fig. 12 shows a model of the transmission circuit utilizing magnetic field resonance. In this model, the coupling coefficient depends on the transmission gaps ℓ_g and the areas of interlinkage magnetic flux in the transmission coils, which are determined by the diameter ϕ. Hence, one-turn coils are used in order to low inductance and High capacitance, then, the transmission circuit gives highly transfer.

IV. EXPERIMENTAL RESULTS

A. Experiment results for the proposed inverter

Fig. 13 shows the output waveforms of the inverter for the condition of $F_{out} = 11.33$ MHz, $Z_L = 100\,\Omega$, and $E_d = 40$ V. In this case, the squared current waveform with an amplitude of $I_o = 1.3$ A and the sinusoidal voltage waveform with an amplitude of $V_o = 150$ V appear at the output terminal, and the sinusoidal current flows into the load resistor. Fig. 14 shows that the output current remains at a constant value, even when the load resistance is changed. This means that the proposed inverter operates as a high-frequency AC current source. Fig. 15 shows the operation waveforms of the input voltage v_{D1-2}, the input current waveforms i_{in}, the output voltage v_o, and the output current i_o for the immittance conversion element. We can see that there is a 90 degree phase shift between the input voltage and the output current, and this proves that the immittance conversion is performed properly.

Fig. 16 shows the measured result for the conversion efficiency. In this case, the input voltage is changed, and the output resistor is kept constant at $Z_L = 100\,\Omega$. More than 74 % of the conversion efficiency is achieved in the entire output power range, and a maximum efficiency of 77.6 % is achieved at a maximum power of 123 W.

Fig. 17 shows a loss analysis for the proposed inverter at maximum conversion efficiency. It can be seen that the switching loss is relatively small owing to the ZVS switching operation of the MOSFETs. On the other hand, the loss of the coaxial cables accounts for most of the inverter loss. A detailed examination of the cause of the cable loss will be performed in the future.

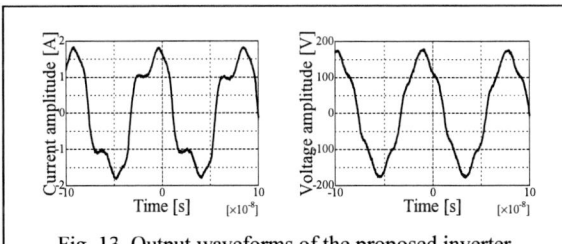

Fig. 13. Output waveforms of the proposed inverter.

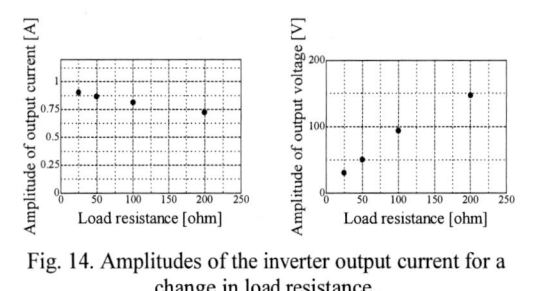

Fig. 14. Amplitudes of the inverter output current for a change in load resistance.

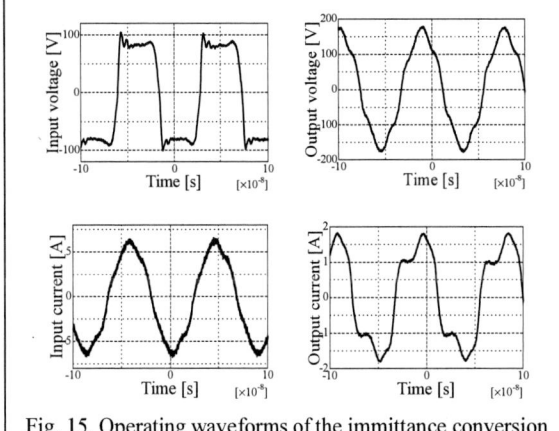

Fig. 15. Operating waveforms of the immittance conversion element.

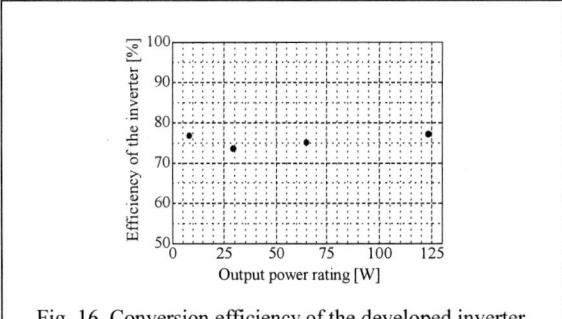

Fig. 16. Conversion efficiency of the developed inverter.

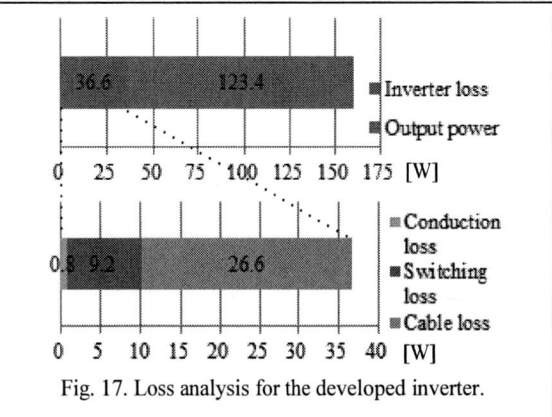

Fig. 17. Loss analysis for the developed inverter.

B. Experiment results for the wireless power transmission circuit utilizing magnetic field anti-resonance

The authors performed the transmission test using magnetic field resonance coils suitable for the proposed high-frequency current source. Here, the operating frequency is $F_{out} = 11.33\,\text{MHz}$, the load resistor is $Z_L = 50\,\Omega$, the input power is $P_{in} = 49.5\,\text{W}$, the diameter of the transmission coils is $\phi = 10\,\text{cm}$, and the air gap of the coil is $\ell_g = 2\,\text{cm}$. A maximum transmitting efficiency of 94.2 % is achieved at an output power of 48 W. The efficiency evaluation and loss analysis of the entire system will be topics for future research.

V. CONCLUSIONS

A current output type high-frequency inverter utilizing an immittance conversion element and a wireless power transmission circuit suitable for the inverter was proposed. By utilizing the immittance conversion element, both a center-tapped configuration and high-frequency current output operation of the inverter circuit were realized. The center-tapped configuration enables the use of a resonant-type gate-drive circuit, which provides the minimum gate-drive power. The effectiveness of the proposed inverter was verified through the experimental setup with a 123 W output power. A conversion efficiency of more than 74 % was achieved in the entire output power range, and a maximum efficiency of 77.6 % was achieved at a maximum power of 123 W.

The operation principle of the newly proposed wireless power transmission circuit suitable for the current output inverter was presented. The power transmission efficiency of the transmission coil was measured and reached a maximum value of 94.2 %.

Through the above discussion, it is confirmed that the proposed system has sufficient potential for high-frequency and high-power wireless power transmission. In order to provide a much higher efficiency of the inverter, a study on the loss reduction of the coaxial cable will be conducted. Also, verification of the parallel connection operation of the proposed inverter in order to increase the power rating of the system will be addressed in the future.

REFERENCES

[1] A. Kurs, A. Karalis, R. Moffatt, J. D. Joannopoulos, P. Fisher, M. Soljačić: "Wireless Power Transfer via Strongly Coupled Magnetic Resonances", Science, Vol.317, pp.83-86, Jan.2007.

[2] A. Karalis, J. D. Joannopoulos, M. Soljačić: "Efficiency wireless non-radiative mid-range energy transfer", Annals of Physics, vol. 323, no. 1, pp.34-48, Jan 2008.

[3] T. Imura, T. Uchida, Y. Hori: "Flexibility of Contactless Power Transfer using Magnetic Resonance Coupling to Air Gap and Misalignment for EV", The papers of Technical Meeting on Vehicle Technology, World Electric Vehicle Association Journal, Vol. 3, 2010.

[4] K. Kusaka, J. Itoh: "Reduction of Reflected Power Loss in an AC-DC Converter for Wireless Power Transfer System", IEEJ Journal of Industry Applications, Vol. 2, No. 4, pp. 195-203, 2013.

[5] T. Hosotani: "A Novel Design Theory Using Coupling Coefficient for the ZVS Resonant Wireless Power Transfer with High-Frequency Power Electronics", IEICE Technical Report, WPT2012-23, pp.17-22, (in Japanese).

[6] T. Shimizu, M.Shioya: "Characteristics of Electric Power Transmission on High-Frequency Inverter Having Distributed Constant Line", IEEE Trans. on Industrial Electronics, Vol. 38, No. 2, pp. 119-131, 1992.

[7] H. Ohguchi, T. Shimizu, et.al.: "A High Frequency Electric Ballast for HID Lamp based on a 1/4 Long Distributed Constant Line", Trans. on Power Electronics, Vol. 13, No. 6, pp. 1023-1029,1998.

[8] H. Ohguchi, R. Shimotaya, T. Shimizu, H. Takagi, and M. Ito, "13.56MHz Current Source Inverter based on Immittance Conversion Topology", Trans. on IEE Japan, Vol. 121-D, No. 7, pp. 805-8013, 2001.

[9] H. Kinjyo, T. Shimizu: "A novel high-frequency current output inverter based on an immittance conversion element and a hybrid MOSFET-SiC diode switch", Power Electronics Specialist Conference, 2003, PESC '03, 2003 IEEE 34th Annual.

[10] Y. Sakamoto, K. Wada, T. Shimizu: "A 13.56MHz Current-output-type Inverter Utilizing an Immittance Conversion Element", Power Electronics and Motion Control Conference, 2008, EPE-PEMC 2008, 13th.

High-Speed Switching Method of MOSFET Using Voltage Boost Auxiliary Circuit Fed by Gate Drive Power Supply

-Applications to Chopper and Half-Bridge Inverter and Their Operation Characteristics -

Toshihiko Noguchi

Department of Electrical and Electronic Engineering,
Graduate School of Engineering,
Shizuoka University
3-5-1 Johoku, Naka-Ku, Hamamatsu,
Shizuoka 432-8561, Japan
ttnogut@ipc.shizuoka.ac.jp

Munehiro Murata

Department of Electrical and Electronic Engineering,
Graduate School of Engineering,
Shizuoka University
3-5-1 Johoku, Naka-Ku, Hamamatsu,
Shizuoka 432-8561, Japan
f0330137@ipc.shizuoka.ac.jp

Abstract— **This paper describes a high-speed switching method of MOSFETs applied to a chopper and a half bridge inverter. By using a voltage boost auxiliary circuit fed by the gate drive power supply, the turn-off time of the MOSFET can be effectively reduced, which enables a high-frequency drive and reduction of the switching loss. It was confirmed through experimental tests that the turn-off dv/dt of the MOSFET was drastically improved by 9 times with the proposed method, especially in the low-load range.**

Keywords—MOSFET, turn-off time, high-speed switching, auxiliary circuit, chopper, inverter

I. INTRODUCTION

In recent years great attention has been paid on next-generation semiconductor devices such as SiC (Silicon Carbide) based MOSFETs and diodes because they are very promising and attractive as power switching devices for varieties of near-future power converters. The SiC based devices have the following significant features, which completely surpasses conventional Si (Silicon) based semiconductor devices:

(1) approximately ten times higher voltage ratings than Si based devices;

(2) 10^5 V/μs-class switching speed;

(3) as low on-resistance as 1/200 to 1/500 of conventional MOSFETs; and

(4) over 300 °C junction temparature.

These are very attractive features to improve the power density as well as the efficiency of the power converters dramatically. However, there are still many problems to be solved from the viewpoint of practical implementation of such SiC based MOSFETs. One of the problems is an extremely high-speed operation without sacrificing the power converter efficiency. In general, the parasitic capacitors of the MOSFET disturb the high-speed switching operation because the turn-on

and the turn-off times become longer due to the parasitic capacitors. To make matters worse, as the on-resistance becomes lower and the drain current becomes higher, the MOSFET tends to increase its parasitic capacitance. Similarly in the case of the SiC power devices, the parasitic capacitance is one of the great concerns, which may fade away the above excellent features.

It is necessary to reduce both of the turn-on time and the turn-off time to achieve high-speed switching of the MOSFET. Particuraly the turn-on time can be adjusted by controlling electrical charge to the parasitic input capacitance. Conventional approaches have been taking some simple techniques to speed up the turn-on time, e.g., reducing the gate resistance of the MOSFET, and connecting a speed-up capacitor in parallel with the gate resistor. Recently another new gate drive circuit is proposed, which introduces inductive impulse superposition technique to drive the high-speed MOSFET. The superposed impulse gate current allows exteremely fast charge to the parasitic input capacitor of the MOSFET. On the other hand, it is difficult to speed up the turn-off behaviour by tuning the gate drive circuit because the turn-off time is mainly determined by the charging time of the output capacitor across the source and the drain.

The percentage of the switching loss to the whole power loss becomes large in the high-frequency operation of the power converters. The parasitic output capacitor of the MOSFET used in the power converter such as the inverter cannot be discharged perfectly in the low-load range during the dead time period. As a result, the short circuit current flows through the switching device to charge up the output capacitor when the switching device turns on. The short circuit current is the major part of the switching loss it the light-load operations. In general, there are two manners to operate the power switching devices, i.e., one is a soft switching technique and the other is a hard switching technique. The former technique is basically employed to

978-1-4799-2706-7/14 $31.00 © 2014 IEEE

The 2014 International Power Electronics Conference

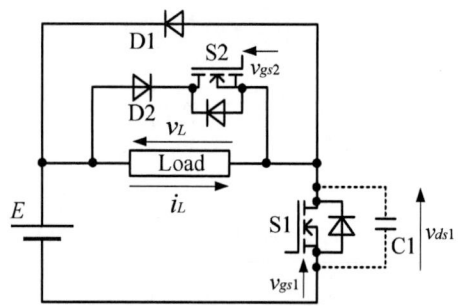

Fig. 1. Existing proposed buck chopper with auxiliary circuit.

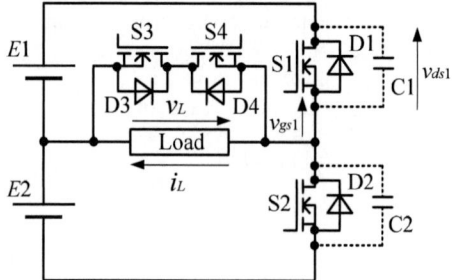

Fig. 2. Existing proposed half-bridge inverter with auxiliary circuit.

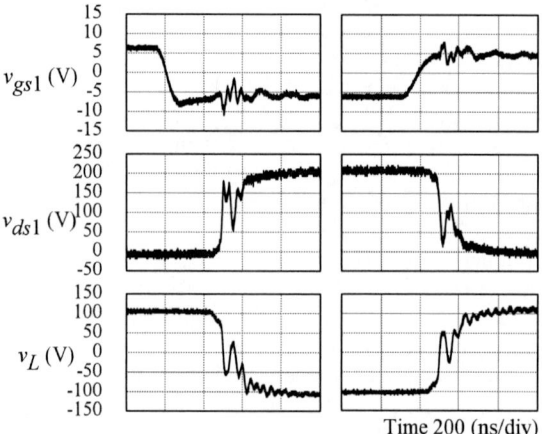

Fig. 3. Experimental waveforms of existing proposed inverter.

reduce the switching loss as well as the radiation and the conduction noises. However, the technique can reduce the switching loss by lowering dv/dt and di/dt, so it is rather difficult to achieve a high-frequency operation of MHz-class because the lowered dv/dt and di/dt increase the switching transient time.

The authors have proposed a new switching method of the MOSFET using a switching assist auxiliary circuit. The proposed method can reduce the turn-off time by connecting the auxiliary circuit in parallel with the load. However, the method has a drawback because it requires modification of the main circuit. Therefore, this paper proposes another new approach to achieve the high-speed switching operation of the MOSFET that has a large parasitic output capacitor. By using a voltage boost auxiliary circuit fed by the gate drive power supply, the high-frequency drive can be effectively carried out without any modification of the main circuit. The

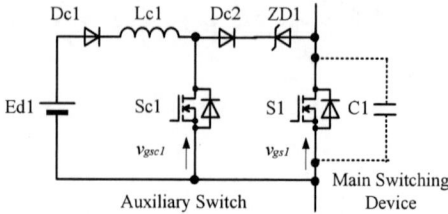

Fig. 4. Voltage boost auxiliary circuit fed by gate drive power supply.

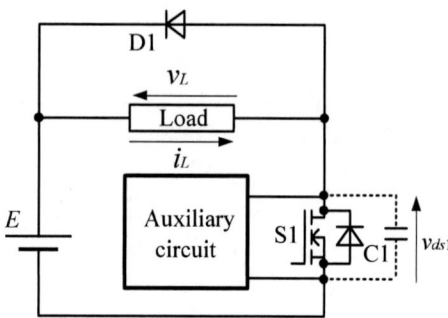

Fig. 5. Proposed buck chopper with auxiliary circuit.

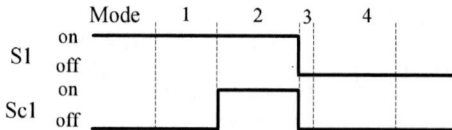

Fig. 6. Switching sequence of proposed chopper with auxiliary circuit.

purpose of this auxiliary circuit is not soft-switching but a hard-switching operation, and it makes not only the high-speed switching but also the high-efficiency operation possible by enhancing dv/dt. The proposed method is applied to a chopper and a half-bridge inverter as well in this paper.

II. DRAWBACKS OF EXISTING AUXILIARY CIRCUIT

Figure 1 shows a chopper with the existing auxiliary circuit proposed by the authors, which consists of an additional auxiliary diode D2 and an auxiliary switch S2. Figure 2 shows another application of the existing auxiliary circuit to a half-bridge inverter, which consists of auxiliary switches S3 and S4. C1 and C2 are parasitic output capacitors of the main devices. The existing proposed method can reduce the turn-off time by shorting the load as soon as the main devices are turned off. However, they have some drawbacks as described below. First, modification of the main circuit is indispensable. Secondly, since the output capacitor of the main device is charged quickly by a high-di/dt short circuit current, the resonance is caused, depending on the inductance of the current path and the output capacitor as shown in Fig. 3. The oscillation may cause malfunction when turning on the main devices. Thirdly, the on-timing control of the auxiliary devices is critically severe. If the auxiliary devices should be turned on before the main device is turned off, not only the main device but also the auxiliary devices would be destroyed by the short circuit current.

978-1-4799-2706-7/14 $31.00 © 2014 IEEE

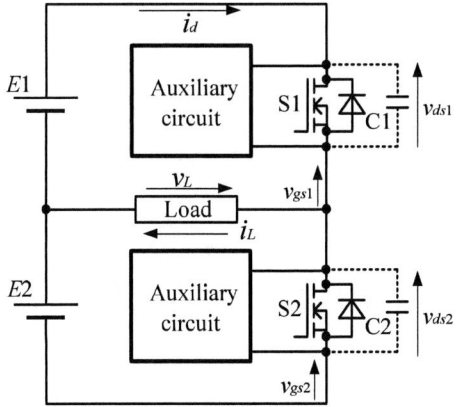

Fig. 7. Proposed half-bridge inverter with auxiliary circuit.

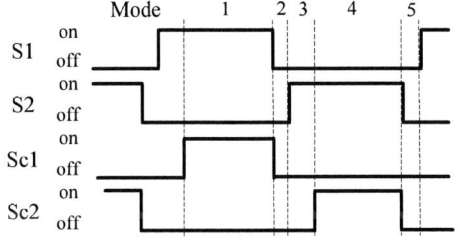

Fig. 8. Switching sequence of proposed inverter with auxiliary circuit.

III. CIRCUIT CONFIGURATIONS AND OPERATIONS

A. Chopper with Proposed Auxiliary Circuit

Figure 4 shows the voltage boost auxiliary circuit fed by the gate drive power supply. The proposed circuit is merely added to the gate drive circuit, so the modification of the main circuit is not required. The auxiliary circuit is composed of a gate drive power supply Ed1, auxiliary diodes Dc1 and Dc2, an inductor Lc1, a zener diode ZD1, and an auxiliary transistor Sc1. The gate drive power supply and the power source of the auxiliary transistor are on a common potential. Therefore, the auxiliary transistor and the main switching device can be driven by only a single gate drive power supply. A low-current rating device can be used as the auxiliary transistor to make the high-speed drive possible. The zener diode ZD1 is required to prevent the auxiliary circuit from circulating the current through S1 while S1 is turned on. Three zener diodes are actually connected in parallel because the individual current rating is low.

Figure 5 shows a buck chopper with the proposed auxiliary circuit. C1 is a parasitic output capacitor of the main switching device S1. The turn-off time of the conventional circuit is governed by a time constant determined by the load and the parasitic output capacitor of the main switching device. The turn-off time directly corresponds to a charging time of the parasitic output capacitor; thus, if the load resistance is high, the turn-off time becomes longer. Particularly it is difficult to speed up the switching time in the low-load range. In the proposed circuit, Sc1 is turned on prior to turning off S1 to store the energy in the inductor Lc1. Sc1 is turned off as soon as S1 is turned off to transfer the energy from Lc1 to C1.

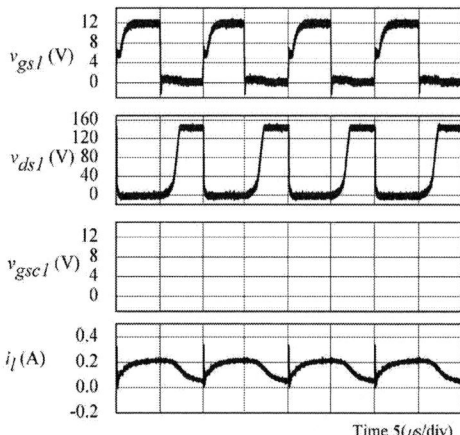

Fig. 9. Experimental waveforms of conventional buck chopper.

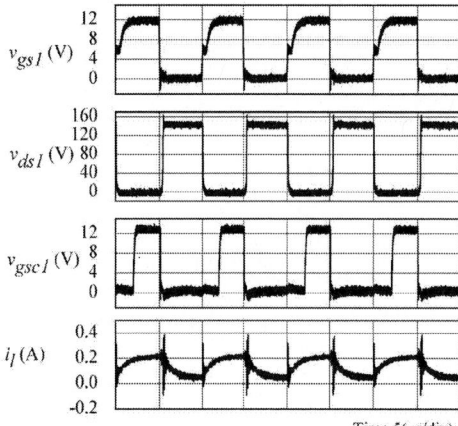

Fig. 10. Experimental waveforms of proposed buck chopper.

It makes the high-speed switching possible regardless of the load condition. Figure 6 shows the switching sequence of the proposed chopper. In the mode 1, the current flows through the path $E{\rightarrow}$Load${\rightarrow}$S1${\rightarrow}E$. In the mode 2, the energy is charged in Lc1 through the path Ed1${\rightarrow}$Dc1${\rightarrow}$Lc1${\rightarrow}$Sc1${\rightarrow}$Ed1 by turning on Sc1. In the mode 3, the current flow commutates to the path Ed1${\rightarrow}$Dc1${\rightarrow}$Lc1${\rightarrow}$Dc2${\rightarrow}$ZD1${\rightarrow}$C1${\rightarrow}$Ed1 by turning off Sc1, so C1 is charged rapidly. The charging time of C1 is determined by the resonance frequency of Lc1 and C1, having nothing to do with the load current. The load current in the mode 4 circulates in the loop of Load${\rightarrow}$D1${\rightarrow}$Load.

B. Half-Bridge Inverter with Proposed Auxiliary Circuit

Figure 7 shows another application of the proposed auxiliary circuit to the half-bridge inverter. C1 and C2 are parasitic output capacitors of the main switching devices S1 and S2. During the dead time period, the charging time and the discharging time of C1 and C2 are governed by a time constant determined by the load and each of the parasitic output capacitors. If the dead time period is too short, turning on S1 dissipates the energy stored in C1, and the short circuit current flows through S1 by the path $E1{\rightarrow}$S1${\rightarrow}$C2${\rightarrow}$E2${\rightarrow}$E1; hence the total efficiency decreases. In the proposed circuit,

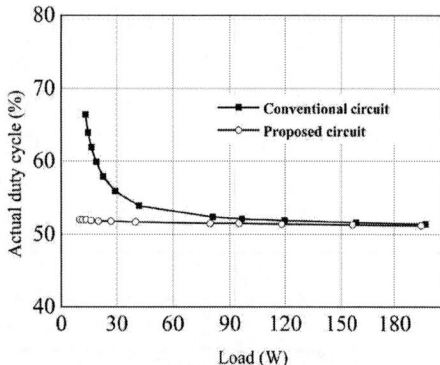

Fig. 11. Load-actual duty cycle characteristic.

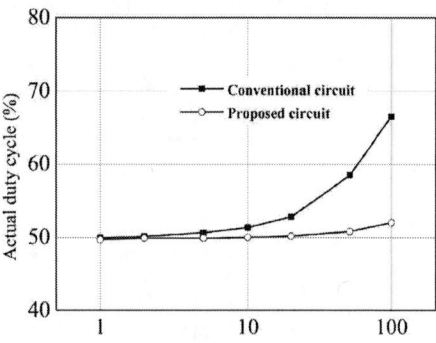

Fig. 12. Frequency-actual duty cycle characteristic.

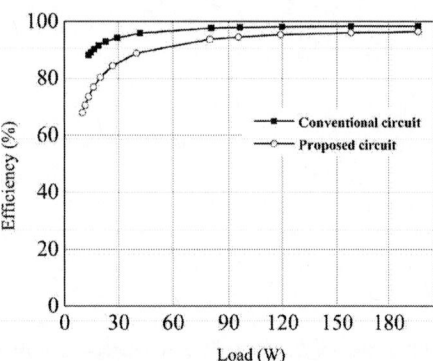

Fig. 13. Load-efficiency characteristic.

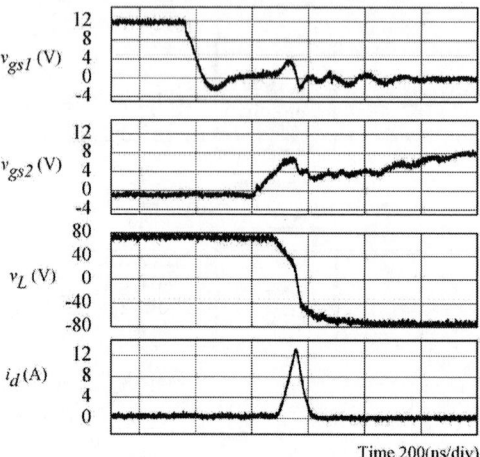

Fig. 14. Experimental waveforms of conventional inverter.

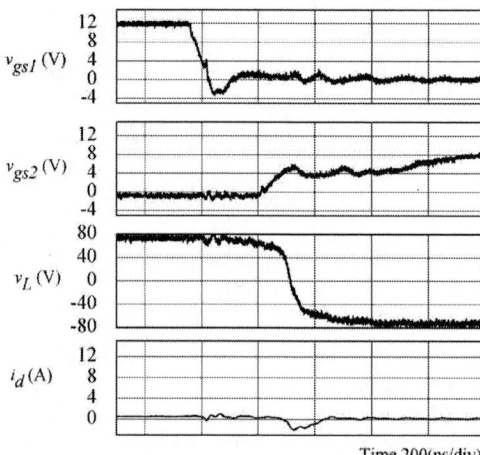

Fig. 15. Experimental waveforms of proposed inverter.

IV. EXPERIMENTAL TEST RESULTS

A. Operation Characteristics of Chopper

The experimental setup was built to confirm the operation characteristics of the proposed method. The experimental circuit has the same configuration as shown in Fig. 4, and is connected across the source and the drain of the main switching device. The main power supply voltage E is 140 V, S1 is STY60NM60 (ST microelectronics, Coss = 2000 pF), Ed1 is 12 V, Sc1 is STP12NM60 (ST microelectronics), Dc1 and Dc2 are D06S60 (ST microelectronics), the free-wheeling diode D1 is D12S60 (ST microelectronics), ZD1 is three parallel connected 1N5349BG (ON Semiconductor), Lc1 is 12 μH, and the load is inductive of which power factor is kept at 0.85.

Figures 9 and 10 show the output waveforms of the conventional and the proposed circuits under 13-W output condition. In order to achieve this operating condition, a 800-Ω and 0.8-mH RL load, a 100-kHz switching frequency, a 50 % duty cycle, and a 3-μs on-time of Sc1 are introduced to the system. In these figures, it can be found that the turn-off

the auxiliary switch Sc1 is turned on prior to turning on S1 to store the energy in Lc1 like the proposed chopper does. Figure 8 shows the switching sequence of the proposed inverter. Sc1 is turned off as soon as S1 is turned on to transfer the energy from Lc1 to C1. By turning off Sc1, C1 is rapidly charged. Therefore, the short circuit current through the two of the main switching devices is effectively reduced, compared with the conventional circuit. This operation is performed in the same manner when S2 is turned off. The charging time is determined by the resonance frequency of Lc1 and C1, so dv/dt of the drain-source voltage becomes slightly lower than that of the conventional circuit. Therefore, the proposed circuit operates in a more stable fashion than the conventional circuit.

The 2014 International Power Electronics Conference

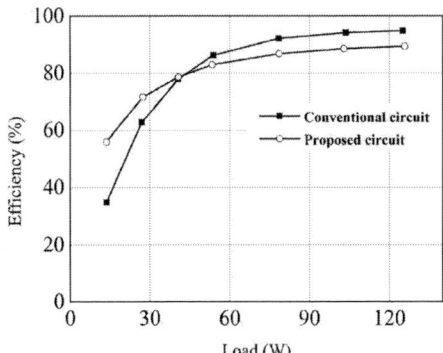

Fig. 16. Load-efficiency characteristic.

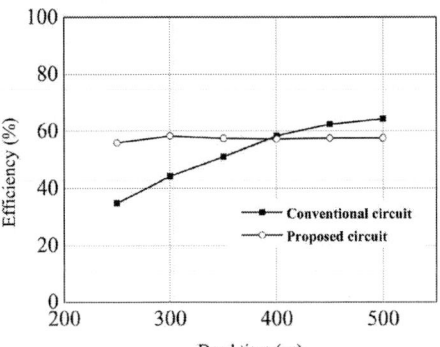

Fig. 17. Dead time-efficiency characteristic.

time of the proposed circuit is much shorter than that of the conventional circuit. Since C1 is charged quickly by using the proposed auxiliary circuit, the actual duty cycle of $vds1$ is improved from 66.4 % to 52.0 %. In addition, the turn-off dv/dts of $vds1$ are 0.16 kV/μs and 1.49 kV/μs, respectively, so the turn-off dv/dt of the proposed circuit is 9 times higher than that of the conventional circuit. Figure 11 shows a characteristic between the load power and the actual duty cycle of the main switching device. It can be seen in the figure that the proposed circuit always operates around 50 % even in the low-load range, while the conventional circuit can not achieve 50 % duty cycle operation in the range. This is because the output capacitor cannot be charged due to the low-load current in the low-load range. The actual duty cycle of the proposed circuit, however, is slightly deviated from 50 %, which corresponds to the charge time determined by the resonance frequency of Lc1 and C1. Figure 12 shows a frequency to the actual duty cycle characteristic. The actual duty cycle of the conventional circuit is far from 50 % in the high-frequency drive. On the other hand, the actual duty cycle of the proposed circuit can maintain around 50 % for any load range. Figure 13 shows a load to total efficiency characteristic including the auxiliary circuit power consumption. The efficiency of the proposed circuit is lower than that of the conventional circuit because the power loss in the auxiliary circuit is added to the total loss of the conventional circuit. It is confirmed that the efficiency of the proposed circuit is 15 pt lower than the conventional circuit.

It has been confirmed through the above experimental tests that the high-speed switching operation is possible by using the proposed auxiliary circuit technique.

B. Operation Characteristics of Half-Bridge Inverter

The experimental setup was built in the similar way as the chopper. The main power supply voltage $E1$ and $E2$ are 70 V, Dc1 is IDH12S60C (ST microelectronics), the inductors Lc1 and Lc2 are 4.5 μH, and other devices are same as the components used in the previously described chopper.

Figures 14 and 15 show the output waveforms of the conventional and the proposed circuits for 14-W output power. A 200-Ω and 0.2-mH RL load, a 100-kHz switching frequency, a 50 % duty cycle, a 250-ns dead time, and a 2-μs on-time of Sc1 and Sc2 are applied to the circuits. It is confirmed from Fig. 14 that the short circuit current i_d flows through the path $E1{\rightarrow}C1{\rightarrow}S2{\rightarrow}E2{\rightarrow}E1$. The short circuit is caused because one of the output capacitors C1 is not discharged during the dead time period. The gate voltage $vgs1$ oscillates just after S1 is turned off due to the high-di/dt of the short circuit current flowing through the parasitic inductance along the current path. On the other hand, the short circuit current is remarkably reduced by the proposed circuit as shown in Fig.15. The output capacitor is charged quickly during the dead time period by using the proposed auxiliary circuit regardless of the load current. Therefore, the resonance of $vgs1$ is reduced owing to the low-di/dt because the short circuit current does not flow. The turn-off times of S1 with and without the proposed auxiliary circuit are 450 ns and 390 ns, respectively. Figure 16 shows a load to total efficiency characteristic including the auxiliary circuit power consumption. It is confirmed from the figure that the total efficiency of the proposed circuit is higher than that of the conventional circuit in the low-load range. In the conventional circuit, the charging time and the discharging time of the output capacitor becomes longer in the low-load range, and short circuit current occurs across the DC bus. To make matters worse, the short circuit current flows though the main switching device when the device is turned on. However, the proposed circuit can charge and discharge the output capacitor quickly during the dead time period even in the low-load range, so the short circuit current can be dramatically reduced. The total efficiency is also improved by 9 points from 62.7 % to 71.7 % for 26-W output. Figure 17 shows a characteristic between the dead time and the total efficiency. The total efficiency of the conventional circuit is higher than that of the proposed circuit when the dead time is long because the output capacitor can be sufficiently charged and discharged even in the conventional circuit. In other words, the proposed circuit is significantly effective when the dead time should be shortened.

Generally, the dead-time period should be less than 5 % of the switching operation period. Therefore, it is required to determine the dead-time period less than 500-ns, when the operation frequency is 100-kHz, for example. In the case of the high-frequency drive such as 100-kHz, the output capacitors cannot be fully charged and discharged during the dead time period, so the proposed method is effective to make

978-1-4799-2706-7/14 $31.00 © 2014 IEEE

the high-speed and high-frequency operation possible. Since the conduction loss and the switching loss are dominant factors among the power conversion losses and the conduction loss is almost same level between the conventional and the proposed circuits, the efficiency improvement of the proposed method mainly owes reduction of the switching loss. The proposed technique is considered to be effective and efficient for the MHz-class high-frequency power converters.

V. CONCLUSION

This paper has described a new high-speed switching operation technique of the MOSFET using the voltage boost auxiliary circuit fed by the gate drive power supply. The proposed circuit requires no modification of the main circuit. The technique has been applied to a chopper and a half-bridge inverter and their operation characteristics have been examined through several experimental tests.

As a result, the proposed circuit surpasses the conventional one particularly in the low-load range from various viewpoints. In the case of the chopper application, the actual duty cycle was improved by 14.4 pt from 66.4% to 52.0% when the duty cycle command of the gate signal was 50 %. Furthermore, it was confirmed that the turn-off dv/dt was enhanced from 0.16 kV/μs to 1.49 kV/μs; hence it meant that the proposed technique realized 9 times high-speed switching of the main switching device in comparison with the conventional circuit.

On the other hand, the half-bridge inverter with the proposed auxiliary circuit is capable to shorten the turn-off time from 450 ns to 390 ns even though as low-output-power as 14 W is demanded. It was confirmed that the total efficiency including the proposed auxiliary circuit power consumption was improved by 9 points from 62.7 % to 71.7 % for 26-W output condition.

The proposed method is very effective to drive the MOSFET with a large parasitic capacitance. Particularly, if multiple MOSFETs are connected in parallel to enhance the current rating, the method is more suitable to such high-power applications.

ACKNOWLEDGMENT

The authors wish to express their gratitude to Mr. T. Sasaya, Mr. T. Kimura, and Mr. S. Katayama, DENSO Research Laboratories, DENSO Corporation for valuable financial and technical support to their power electronics laboratory, Department of Electrical and Electronic Engineering, Faculty of Engineering, Shizuoka University.

REFERENCES

[1] T. Noguchi, S. Yajima and H. Komatsu : "Development of Gate Drive Circuit for Next-Generation Ultra High-Speed Switching Devices", IEEJ Trans., Vol. 129-D, No. 1, pp. 46-52 (2009) (in Japanese)

[2] T. Noguchi and T. Mizuno: "High-Speed Switching Method of MOSFET Using Auxiliary Circuit Shorting Load: Application and Operation Characteristics of Chopper", IEEJ Trans., Vol. 132-D, No. 5, pp. 598-559 (2012) (in Japanese)

[3] T. Noguchi and T. Mizuno: "High-Speed Switching Method of MOSFET Using Auxiliary Circuit Shorting Load: Application and Operation Characteristics of High-Frequency Half-Bridge Inverter", IEEJ Trans., Vol. 132-D, No. 11, pp. 1080-1081 (2012) (in Japanese)

[4] M. Ishigaki and H. Fujita : "A Resonant Gate-Drive Circuit Capable of High-Frequency and High-Efficiency Operation", IEEJ Trans, Vol. 127-D, No. 10, pp. 1090-1096 (2007) (in Japanese)

[5] R.W. De Doncker and J.P. Lyons : "The Auxiliary Resonant Commutated Pole Converter", Conf. Rec. of the IEEE-IAS Annual Meeting, Vol. 2, pp. 1228-1235 (1990)

[6] M. Murata, T. Mizuno and T. Noguchi: "High-Speed Switching Method of MOSFET Using Voltage Boost Auxiliary Circuit Fed by Gate Drive Power Supply", IEEJ Proc. IAS Annual Conference, Vol.1 , pp. 369-370 (2013) (in Japanese)

[7] N. Hoshi and A. Matsui: "Improvement of Power Conversion Efficiency of Soft-Switching Inverter in Range of Low Output Power by Adjustable Dead Time Control", IEEJ Trans., Vol.131-D, No.5, pp.679-684 (2011) (in Japanese)

[8] K. Shirakawa, K. Wada and T. Shimizu: "An Issue on 200 kHz Class High Frequency Switching of a PWM Inverter", IEEJ Proc. IAS Annual Conference, Vol. 1, pp. 277-280 (2006) (in Japanese)

[9] H. Mochikawa and T. Koyama: "Innovation Circuit Technology for More Compact and Effective Inverters", Toshiba Review 2006 vol. 61, No. 11, pp. 32-35 (2006) (in Japanese)

The 2014 International Power Electronics Conference

Operating Strategy for Bi-directional LLC Resonant Converter with Seamless Operation

Seiya Abe
International Centre for the Study of East Asian
Development (ICSEAD)
Kitakyushu, Japan
abe@icsead.or.jp

Junichi Yamamoto
Texas Instruments Japan Ltd.
Tokyo, Japan
J_yamamoto1@ti.com

Toshiyuki Zaitsu
Texas Instruments Japan Ltd.
Tokyo, Japan
t-zaitsu@ti.com

Tamotsu Ninomiya
International Centre for the Study of East Asian
Development (ICSEAD)
Kitakyushu, Japan
t_ninomiya@icsead.or.jp

Abstract— **This paper proposes the simple and seamless operation technique for bi-directional LLC resonant converter. Generally, bi-directional operation of LLC resonant converter is complicated and difficult to control. Because, the switching frequency has to be changed across fsr (series resonant frequency) and the switching sequences (the switching timing between high voltage side and low voltage side) has to be changed due to resonant characteristics as well. In order to simplify the switching scheme between forward and reverse direction, the switching frequency has been moved to higher frequency than series resonant frequency where ZVS condition is maintained. As a result, the simple switching sequence has been achieved for both forward and reverse direction. The experimental results for 400Vin, 48Vo, 300W board achieves 93% efficiency and validated this method is useful practically.**

Keywords— *LLC resonant converter, Bi-directional, Seamless,*

I. INTRODUCTION

Recently, the effective utilization of the renewable energy is one of the most important topics in order to improve the environmental issue. Many power management systems such as PV system have included batteries to ensure a stable supply of electrical energy [1-2]. Figure 1 (a) shows the example of home PV system. This system consists of boost converter with MPPT, DC-AC inverter connected to commercial line and battery system. In this battery system, the bi-directional DC-DC converter has a key role in renewable power management system. The requirements for bi-directional DC-DC converter are isolated, high efficiency, low noise, small size and so on. Moreover, the CC-CV (Constant Current-Constant Voltage) control is also required in high specification bi-directional DC-DC converters which lead to higher cost. However, most of the existing battery

charge system includes the charger in home PV system as shown in Fig. 1 (b). Therefore, the only function required for existing PV system is DC transformer.

The LLC resonant converter is one of topology to achieve isolated, high efficiency and small size [3-6].

However, the LLC resonant converter has two difficulties as follows;

(1) Complicated synchronous rectification
(2) Large switching frequency change when it works as bi-directional converter.

Regarding (1), the synchronous switch drive of the conventional PWM converter such as buck converter can be realized easily due to the complementary switching timing. On the other hand, switching timing of the LLC resonant converter has dependency on the relation between magnetizing current and resonant current.

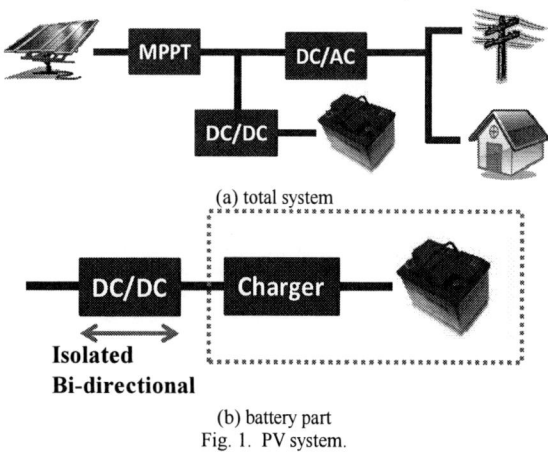

(a) total system

Isolated
Bi-directional

(b) battery part
Fig. 1. PV system.

978-1-4799-2706-7/14 $31.00 © 2014 IEEE

Hence, the switch drive of LLC resonant converter is very difficult compared with conventional PWM converter. Generally, the switches are driven by using same signals such as transformer voltage, switch current, switch voltage. In this case, the circuit configuration and its control system become complicated which lead to higher-cost.

Regarding (2), there is a big difference of the voltage characteristics between forward and reverse direction in LLC resonant converter. Then, the switching frequency has to be changed largely among forward direction and reverse direction for soft switching operation. This makes the control system complicated as well.

In order to reduce the complexity of synchronous driving and control system, then achieve simple seamless operation, this paper propose unique switching scheme of the bi-directional LLC converter.

II. Basic Operation of Conventional LLC Resonant Converter

A. Power Stage Operation

Figure 2 shows the bi-directional LLC resonant converter, and Fig. 3 shows the ac equivalent model to analyze converter characteristics [14-17]. Figure 4 show the impedance and voltage characteristics of forward and reverse direction. As shown in Fig. 4 (a), the two resonance peaks appear at fsr (series resonant frequency) and fpr (parallel resonant frequency) in the output impedance, and one resonance peak appears in the input impedance.

For ZCS (Zero-Current-Switching) operation of low voltage side switches (Q3, Q4), the switching frequency fs should be fs<fsr. Meanwhile, for ZVS (Zero-Voltage-Switching) operation of high voltage side switches (Q1, Q2), the switching frequency fs should be fs>fzinp (fzinp : peak frequency of Zin).

Hence, the switching frequency range for soft-switching operation (ZVS & ZCS) is fzinp<fs<fsr. As shown in Fig. 4(b) of voltage conversion ratio, the peak of conversion ratio is dumped, and the peak frequency is sifted to high frequency side at heavy load. This frequency characteristic has large dependency on load resistance.

Therefore, switching frequency fs has to change largely, which is one of the drawbacks of LLC resonant converter.

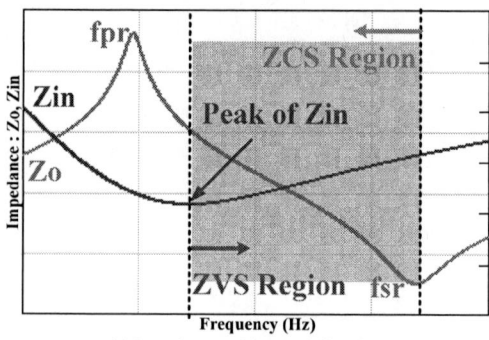

(a) Impedance of forward direction

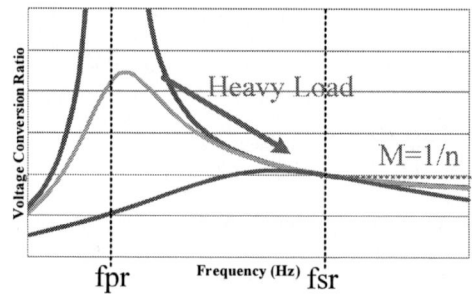

(b) voltage ratio of forward direction

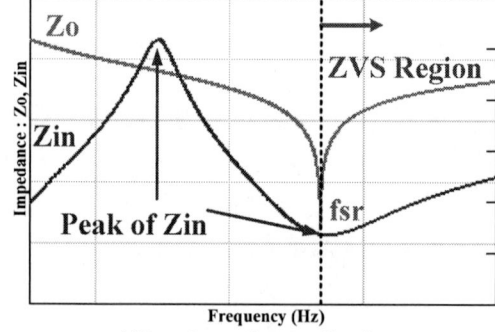

(c) Impedance of reverse direction

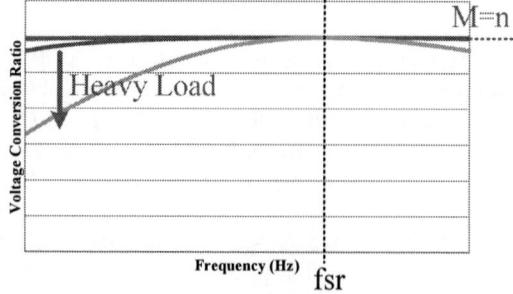

(d) voltage ratio of reverse direction
Fig. 4. Impedance and voltage conversion ratio characteristics.

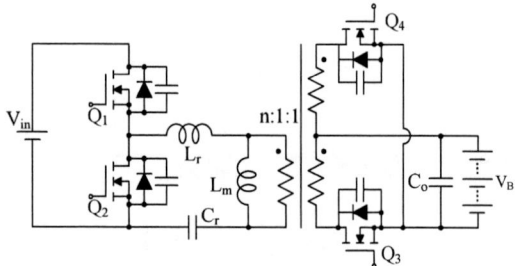

Fig. 2. LLC resonant converter.

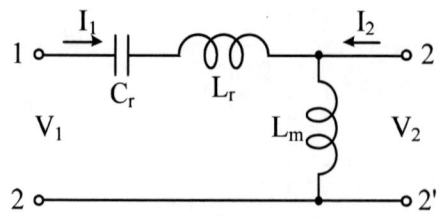

Fig. 3. AC equivalent model.

Figure 4 (c) shows the impedance characteristics of reverse direction. One resonance peak appears at fpr in the output impedance, and two resonance peaks appear at fsr and fpr in the input impedance. The frequency of all resonance peaks of both impedance do not depend on load resistance. In reverse direction, the frequency range of fs>fsr is good for soft-switching operation due to inductive behavior of resonant circuit. Figure 4 (d) shows the voltage conversion ratio of reverse direction. The peak frequency does not change from fsr. Also, the voltage conversion ratio around fsr does not change, and does not show much dependency on load resistance as well.

This means that the regulation capability is not expected but good enough as DC transformer function in reverse direction.

B. Switching Sequence

Figure 5 show the switching sequences of both side switches [18]. In forward direction, the duty ratio of low voltage side switches has to be changed for appropriative drain current conduction by the input and output voltage, and the both side switches operate in phase.

In soft switching region (fzin<fs<fsr), the duty ratio is increasing with increasing fs, and the duty ratio reaches to maximum of 0.5 at fsr. Over fsr, the duty ratio is saturated at D=0.5.

In reverse direction, the duty ratio of both side switches are set to be full (D=0.5), and do not need to be changed, but need to operate in phase-shift due to push-pull resonant operation. The phase of high voltage side switches are lead when fs is lower than fsr meanwhile the phase are lag when fs is higher than fsr.

From above discussions, the signal detection and power flow detection are needed in order to optimize the switching sequence which makes control circuit complicated and noise problem. Therefore, it is preferable that switching frequency is set to fs<fsr or fs>fsr to simplify the switching sequence.

III. PROPOSAL FOR SIMPLE AND SEAMLESS OPERATION

In order to simplify the control scheme suitable for bi-directional operation, the switching frequency fs is set to the range of fs>fsr as shown in Fig. 6 (only red area), where the fs does not need to change across the fsr. In this case, ZVS operation of high voltage side switches is still maintained on forward direction. Meanwhile, ZCS operation of battery side switches is not achieved. However, it is not so big problem practically because it keeps high efficiency. The great benefit of simplification and seamless operation is obtained by this technique.

In forward direction, the duty ratio of battery side is simply fixed as D=0.5. ZCS operation of battery side switches is not achieved. However, ZVS operation of high voltage side can be achieved.

In reverse direction, the switching frequency can stay with the same as forward direction case because the battery voltage is almost constant. Furthermore, the duty ratio can be set simply as D=0.5. These are the great benefit of proposed method.

IV. EXPERIMENTAL VERIFICATIONS

The prototype board is implemented to verify the proposed control method. The circuit parameters and specifications are shown in Table. 1.

A. Forward direction

Figure 7 shows the experimental key waveforms. Figure 7 (a) shows the conventional control which the fs is lower than fsr. Q3 is set to be turned off exactly same timing when id3 (the resonant current flowing through Q3) reaches zero (ZCS operation).

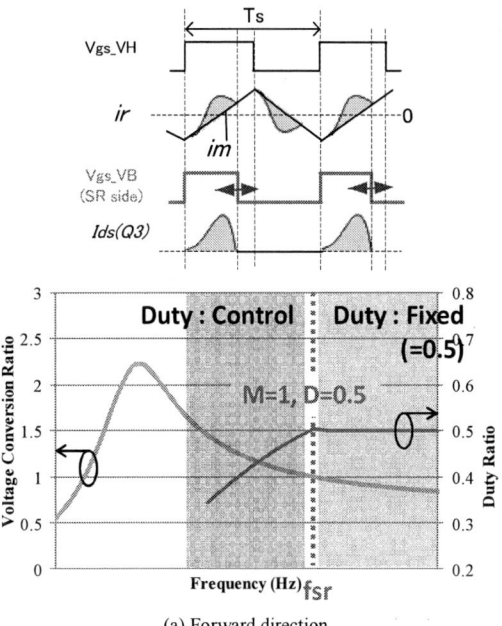

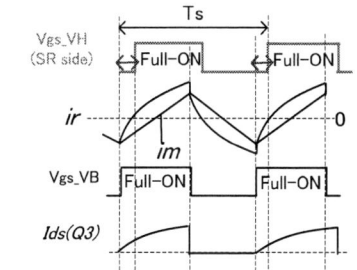

(a) Forward direction

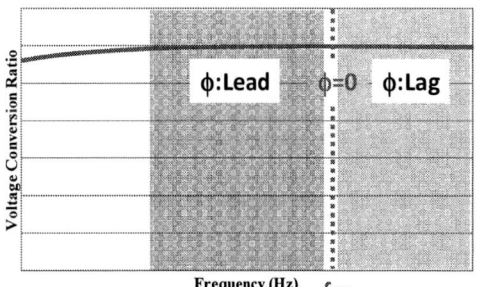

(b) Reverse direction
Fig. 5. Switching sequences

978-1-4799-2706-7/14 $31.00 © 2014 IEEE   1181

Table 1 The circuit parameters and specifications

Symbol	Parameters	Value
VH	High voltage	380V
VB	Battery voltage	48V
Pr	Rated power	300W
n	Turns ratio of transformer	4
Lm	Magnetizing inductance	200uH
Lr	Resonant inductance	20uH
Cr	Resonant capacitor	100nH
fsr	Series resonant frequency	112kHz

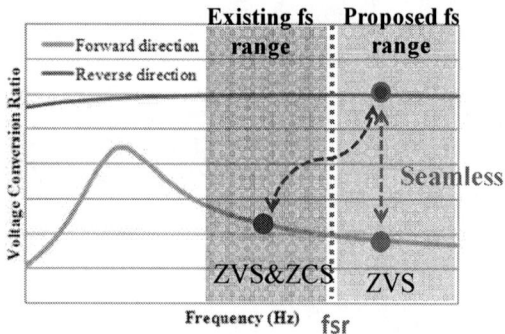

Fig. 6. Operating frequency range

On the other hand, this is the proposed operation in forward direction, it is observed that switch Q3 is forced to turn off before id3 (the resonant current flowing through Q3) reaches zero (non-ZCS operation). Regarding the influence of non-ZCS operation, a little reverse recovery current flowing through Q3, Q4 at battery side is observed.

However, this is not a big problem in a practical application because the efficiency keeps high such 90% as shown in Fig. 8 which is same level with ZCS operation.

B. Reverse direction

Figure 9 shows the experimental key waveforms of reverse direction case when fs>fsr. The surge voltage occurs during turn-off time across switch Q3. This surge voltage is caused by resonant phenomenon between leakage inductance and parasitic capacitor. In reverse direction, the resonant tank is arranged by high voltage side, and the battery side has push-pull operation. Therefore, the surge voltage is generated essentially regardless of resonant type or PWM type.

This is the drawback of this method, it needs some snubber circuit to reduce this surge voltage. However, the ZVS operation of battery side switch can be achieved by resonant phenomenon in resonant type circuit. This is chief advantage of this circuit.

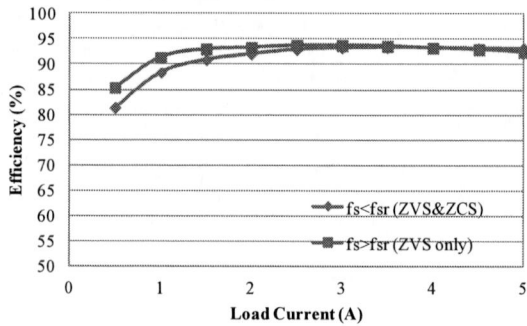

Fig. 8. Efficiency characteristics (Forward direction)

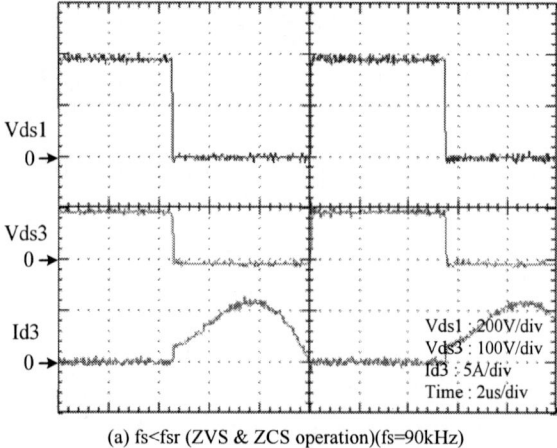

(a) fs<fsr (ZVS & ZCS operation)(fs=90kHz)

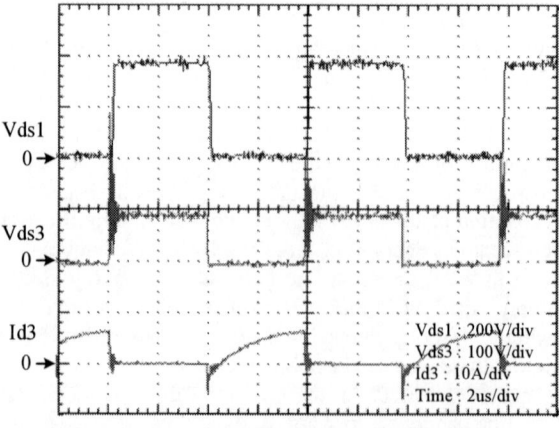

Fig. 9. Key waveforms (Reverse direction)(fs=130kHz)

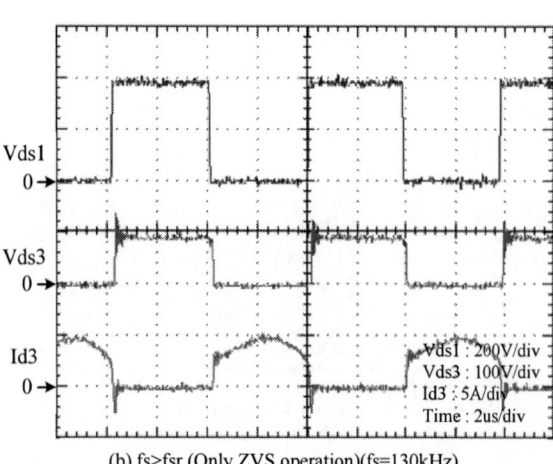

(b) fs>fsr (Only ZVS operation)(fs=130kHz)

Fig. 7. Key waveforms (Forward direction)

Figure 10 shows the high side voltage characteristic. The voltage is slightly decreased by the influence of in-phase operation. Moreover, the efficiency is over 91% at rated.

C. Improving of surge problem

In forward direction, regarding the influence of non-ZCS operation at fs>fsr, a little reverse recovery current flowing through Q3, Q4 at battery side is observed. The surge voltage of Q3 ,Q4 occurs for this reverse recovery current as shown in Fig. 7 (b). This is drawback of this operating strategy.

On the other hand, the large surge voltage of Q3, Q4 occurs in reverse direction as shown in Fig. 9. This surge voltage is caused by essential circuit operation which is push-pull operation. In this case, the snubber circuit is one of effective measure in order to suppress surge voltage of Q3, Q4.

When the snubber circuit is installed, the drawback in forward direction mode is also suppressed.

Figure 11 shows the experimental key waveforms installing snubber circuit. The surge voltage can be reduced dramatically in both direction cases.

V. CONCLUSIONS

This paper presents the simple and seamless operation of bi-directional LLC resonant converter. The simplified control and seamless operation can be achieved by setting the switching frequency region to fs>fsr while conventional LLC resonant converter is operating in fs<fsr. The power conversion efficiency is around 93% on forward direction and is around 91% on reverse direction. Moreover, the ZVS operation can be achieved in both modes. It is confirmed that the proposed method works well practically for seamless operation of LLC bi-directional operation.

REFERENCES

[1] R. Li, et al, "Analysis and design of improved full-bridge bidirectional DC-DC converter," PESC'04, pp. 521-526, 2004
[2] S. B. Kjaer, J. K. Pedersen, and F. Blaabjerg, "A review of single-phase grid-connected inverters for photovoltaic modules," IEEE Trans. Ind. Appl., vol. 41, no. 5, pp. 1292-1306, 2005.
[3] M. Jain, M. Daniele, and P. K. Jaine, "A bidirectional DC-DC converter topology for low power application", IEEE Trans. Power electron., vol. 15, no. 4, pp. 595-606, jul. 2000.

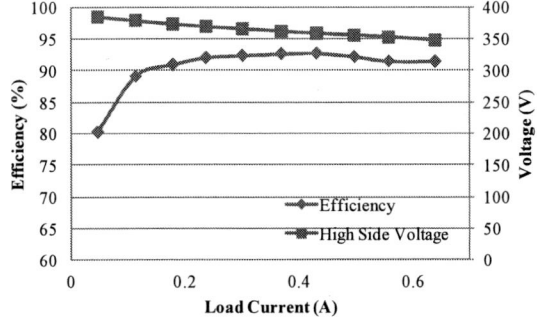

Fig. 10. Voltage and efficiency characteristics (Reverse direction)

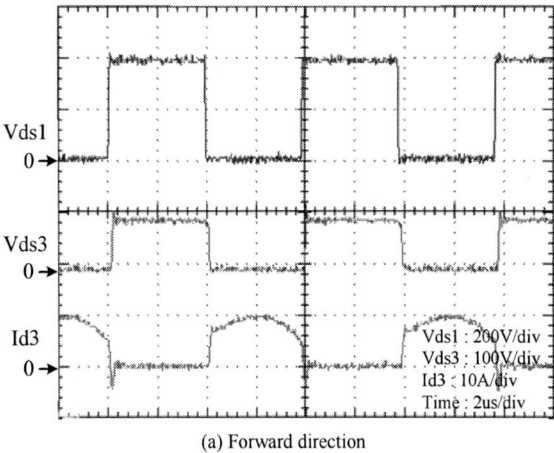

(a) Forward direction

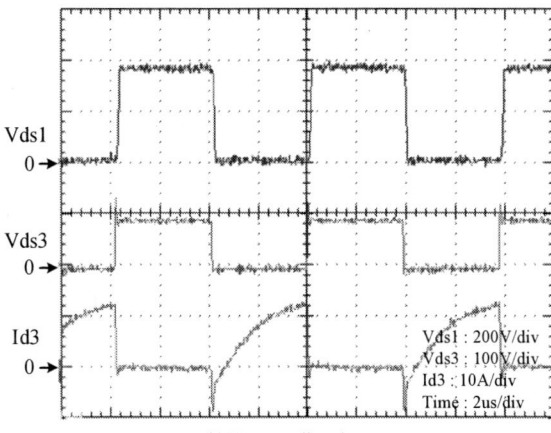

(b) Reverse direction

Fig. 11. Key waveform (with snubber circuit)(fs=130kHz)

[4] B. Yang, F. C. Lee, A. J. Zhang, G. Huang, "LLC resonant converter for front end DC/DC conversion," APEC '02, pp. 1108-1112, 2002.
[5] Z. Pavlovic, J. A. Oliver, P. Alou, O. Garcia, J. A. Cobos,"Bidirectional Dual Active Bridge Series Resonant Converter with Pulse Modulation," IEEE APEC'12, pp. 503-508, 2012
[6] J. H. Jung, H. S. Kim, J. H. Kim, M. H. Ryu, J. W. Baek,"High Efficiency Bidirectional LLC Resonant Converter for 380V DC Power Distribution System Using Digital Control Scheme," APEC'12, pp. 532-538, 2012
[7] R. L. Steigerwald, "A Comparison of Half-Bridge Resonant Converter Topology," IEEE Trans. On PE, Vol. 3, No. 2, pp. 174-182, 1988.
[8] C. Zhao, L. Bao-hong, J. Cao, Y. Chen, X. Wu, Z. Qian, "A Novel Primary Current Detecting Concept for Synchronous Rectified LLC Resonant Converter," ECCE'09, pp. 766-770, 2009
[9] X. Wu, B. Li, Z. Qian, R. Zhao, "Current Driven Synchronous Rectifier with Primary Current Sensing for LLC Converter,"ECCE'09, pp. 738-743
[10] D. Fu, Y. Liu, F. C. Lee, M.g Xu, "An Improved Novel Driving Scheme of Synchronous Rectifiers for LLC Resonant Converters,"APEC'08, pp. 510-516, 2008
[11] W. Feng, D. Huang, P. Mattavelli, D. Fu, F. C. Lee, "Digital Implementation of Driving Scheme for Synchronous Rectification LLC Resonant Converter," ECCE'10, pp. 256-263, 2010

[12] W. Feng, P. Mattavelli, F. C. Lee, D. Fu, "LLC Converters with Automatic Resonant Frequency Tracking Based on Synchronous Rectifier (SR) Gate Driving Signals,"APEC'11, pp. 1-5, 2011

[13] G. Zhang, J. Zhang, C. Zhao, X. Wu, Z. Qian, "LLC Resonant DC/DC Converter with Current-Driven Synchronized Voltage-Doubler Rectifier,"ECCE'09, pp. 744-749, 2009

[14] V. Vorperian, S. Cuk, "A Complete DC Analysis of The Series Resonant Converter, " PESC'82, pp. 85-100, 1982

[15] Ashoka K. S. Bhat,"A Generalized Steady-State Analysis of Resonant Converters Using Two-Port Model and Fourier-Series Approach," IEEE Trans. on P. E., vol. 13, No. 1, pp.142-151, 1998

[16] V. Vorperian, "High-Q Approximation in The Small-Sgnal Analysis of Resonant Converters," PESC'85, pp. 707-715, 1985

[17] R. L. Steigerwald, "A Comparison of Half-Bridge Resonant Converter Topology," IEEE Trans. On PE, Vol. 3, No. 2, pp. 174-182, 1988.

[18] S. Abe, T. Zaitsu, J. Yamamoto, S. Ueda, S. Yang, M. Shoyama, T. Ninomiya, "Adaptive Driving of Synchronous Rectifier for LLC Converter without Signal Sensing," APEC'13 , pp. 1370-1375, 2013.

978-1-4799-2706-7/14 $31.00 © 2014 IEEE

The 2014 International Power Electronics Conference

Negative Sequence Current Injection Control Algorithm Compensating for Unbalanced PCC Voltage in Medium Voltage PMSG Wind Turbines

Jayoon Kang, Daesu Han, and Yongsug Suh

Chonbuk Nat'l Univ., Dept. of Elec. Eng.
664-14 Duckjin-dong, Jeonju, Korea
e-mail: {kangja, rimonia, ysuh}@jbnu.ac.kr

Byoungchang Jung, Jeongjoong Kim
Jonghyung Park, and Youngjoon Choi

Power & Industrial Systems R&D Center, Hyosung Co.
e-mail: {changwin, j2kim, bell, swot87}@hyosung.com

Abstract— This paper proposes a control algorithm for permanent magnet synchronous generator with a back-to-back three-level neutral-point clamped voltage source converter in a medium-voltage offshore wind power system under unbalanced grid conditions. The proposed control algorithm particularly compensates for the unbalanced grid voltage at the point of common coupling in a collector bus of offshore wind power system. This control algorithm has been formulated based on the symmetrical components in positive and negative rotating synchronous reference frames under generalized unbalanced operating conditions. Negative sequential component of ac input current is injected to the point of common coupling in the proposed control strategy. The amplitude of negative sequential component is calculated to minimize the negative sequential component of grid voltage at PCC under the limitation of current capability in a voltage source converter. The proposed control algorithm makes it possible to provide a balanced voltage at the point of common coupling resulting in the generated power of high quality from offshore wind power system under unbalanced network conditions.

Keywords— *unbalanced grid, PCC, PMSG, converter, wind power system*

I. INTRODUCTION

Wind power system is one of the fastest growing renewable energy systems. Wind power installation has been increasing both in number and size of individual wind turbine unit. In large scaled MW-range wind turbines, PMSG (Permanent Magnet Synchronous Generator) type wind turbine involving a full-scaled PCS (Power Conditioning System) becomes a dominant choice due to its superb performance in active and reactive power generation.

Recently, grid codes about LVRT and operation under grid unbalances become more strict. In general, unbalanced current is caused by unbalanced grid conditions, and it leads to unbalanced voltage at PCC (Point of Common Coupling). These unbalanced voltage conditions generate ripple and distortion of dc-link and grid current [1]. Most of previous studies regarding unbalance grid input focused on stabilizing the operation of wind turbine PCS itself, i.e. reducing the harmonics in ac input current and dc link voltage, or compensating unbalanced grid current [1 and 2]. There have been several studies trying to solve PCC unbalanced problem instead of PCS itself. Some papers proposed a

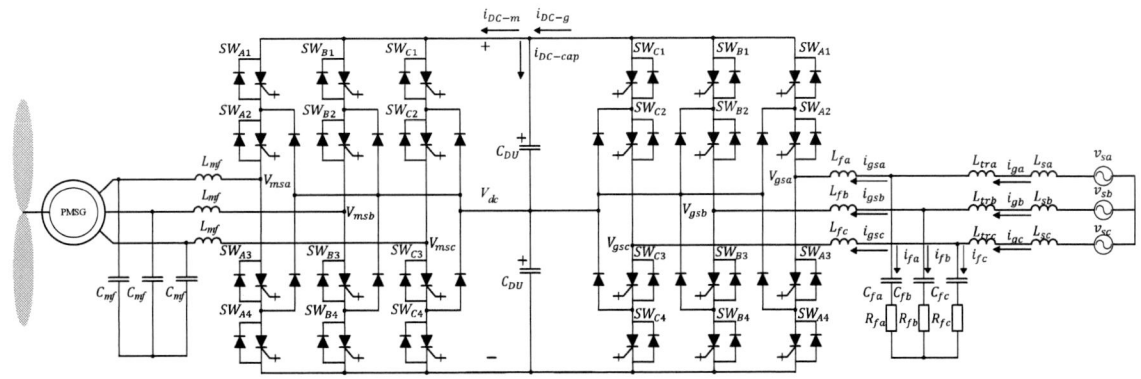

Fig. 1. PMSG wind turbine with a back-to-back 3-Level NPC VSC.

978-1-4799-2706-7/14 $31.00 © 2014 IEEE

compensating solution employing an additional active filter at PCC [3]. Other papers proposed compensation for unbalanced voltage at PCC using STATCOM [4]. There has been a little work dealing with PCC unbalance problem solely by PCS itself without employing additional active filter or STATCOM.

This paper proposes a control algorithm to actively compensate for the unbalanced grid voltage at PCC. Negative sequence current is injected to the grid to cancel the negative sequential component of grid voltage at PCC. The amplitude of injected negative sequence current is computed under the limit of maximum input current of PCS. Negative sequential component of input current is regulated separately in the clockwise-rotating synchronous reference frame irrespective of positive sequential component of input current. Proposed control algorithm performs under the current capability limit of PCS and also improves the quality of output power from wind farm. Detailed calculation result as well as simulation result of 2.7MW wind turbine of PMSG type are provided to validate the proposed control algorithm in this paper.

This paper is structured in five main sections. Section II describes the modeling of PMSG under grid unbalance. Section III discusses various grid unbalance types and its general unbalance factor. Section IV presents negative sequence current injection control algorithm. Finally, Section V and VI provide the simulation and experimental result to validate the proposed control algorithm, respectively.

II. MODELING OF PMSG UNDER GRID UNBALANCE

Under unbalanced grid voltage, PMSG can be effectively modeled by using both positive and negative sequence components of voltages and currents in grid-side converter and input filter as shown in Fig. 1. The positive and negative sequence components in a CCW-rotating and CW-rotating synchronous reference frame, respectively, are expressed as followings;

$$V_{gsd}^{p}-\omega L_f I_{gsq}^{p}+L_f\frac{d}{dt}I_{gsd}^{p}=V_{sd}^{p}+\omega(L_{tr}+L_s)I_{gq}^{p}-(L_{tr}+L_s)\frac{d}{dt}I_{gd}^{p} \quad (1)$$

$$-\omega C_f V_{fq}^{p}+C_f\frac{d}{dt}V_{fd}^{p}=I_{gd}^{p}-I_{gsd}^{p} \quad (2)$$

$$V_{gsq}^{p}+\omega L_f I_{gsd}^{p}+L_f\frac{d}{dt}I_{gsq}^{p}=V_{sq}^{p}-\omega(L_{tr}+L_s)I_{gd}^{p}-(L_{tr}+L_s)\frac{d}{dt}I_{gq}^{p} \quad (3)$$

$$\omega C_f V_{fd}^{p}+C_f\frac{d}{dt}V_{fq}^{p}=I_{gq}^{p}-I_{gsq}^{p} \quad (4)$$

$$V_{gsd}^{n}+\omega L_f I_{gsq}^{n}+L_f\frac{d}{dt}I_{gsd}^{n}=V_{sd}^{n}-\omega(L_{tr}+L_s)I_{gq}^{n}-(L_{tr}+L_s)\frac{d}{dt}I_{gd}^{n} \quad (5)$$

$$\omega C_f V_{fq}^{n}+C_f\frac{d}{dt}V_{fd}^{n}=I_{gd}^{n}-I_{gsd}^{n} \quad (6)$$

$$V_{gsq}^{n}-\omega L_f I_{gsd}^{n}+L_f\frac{d}{dt}I_{gsq}^{n}=V_{sq}^{n}+\omega(L_{tr}+L_s)I_{gd}^{n}-(L_{tr}+L_s)\frac{d}{dt}I_{gq}^{n} \quad (7)$$

$$-\omega C_f V_{fd}^{n}+C_f\frac{d}{dt}V_{fq}^{n}=I_{gq}^{n}-I_{gsq}^{n} \quad (8)$$

Based on the model given in (1)-(8), the equivalent circuit of grid input side of PMSG corresponding to positive and negative sequential components can be generated as shown in Fig. 2 and 3.

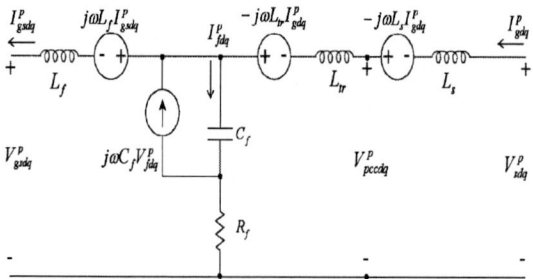

Fig. 2. Equivalent circuit of positive sequence dq components in CCW-rotating synchronous reference frame.

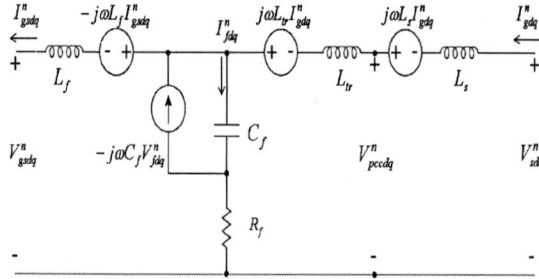

Fig. 3. Equivalent circuit of negative sequence dq components in CW-rotating synchronous reference frame.

III. UNBALANCE FACTOR

Unbalance Factor (*UF*) is a metric factor describing the depth of unbalance. In this paper, *UF* is newly defined to be the ratio of magnitude of negative sequence to that of positive sequence component as shown in (9). Unbalance depth of grid and PCC voltage are quantitatively expressed employing a newly defined *UF* in this paper. In general, unbalance type is classified into four different cases as shown in Table I. The newly defined *UF* can effectively describe the depth of unbalance in all four cases. As an example, the correlation of newly defined *UF* and the per unit value of single-phase sag voltage of Type B are given in Table II.

$$UF=\sqrt{\frac{(V_{sd}^{n})^2+(V_{sq}^{n})^2}{(V_{sd}^{p})^2+(V_{sq}^{p})^2}} \quad (9)$$

978-1-4799-2706-7/14 $31.00 © 2014 IEEE

TABLE I
UNBALANCE TYPE

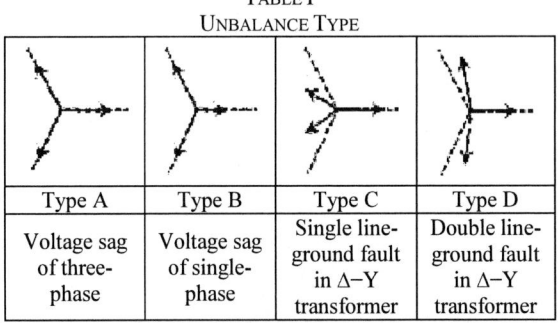

Type A	Type B	Type C	Type D
Voltage sag of three-phase	Voltage sag of single-phase	Single line-ground fault in Δ−Y transformer	Double line-ground fault in Δ−Y transformer

TABLE II
UF ACCORDING TO UNBALANCE DEGREE OF TYPE B

Voltage sag (pu)	1	0.9	0.8	0.7	0.6	0.5	0.3	0.1
UF (%)	0	3	7	11	15	25	36	50

IV. NEGATIVE – SEQUENCE CURRENT INJECTION (NCI) CONTROL ALGORITHM

A. Calculation of Reference Current

This paper proposes *Negative sequence Current Injection* (NCI) algorithm to actively compensate for the unbalance of PCC voltage by controlling negative sequence current. Balanced PCC voltage implies that the negative sequential components of corresponding voltage must be zero. Negative sequential components of ac input current (I_{gd}^n and I_{gq}^n) are regulated, in other words, injected to the grid in such a way that the negative sequential components of PCC voltage (V_{pccd}^n and V_{pccq}^n) are cancelled.

In general, ac input current regulator requires four current references (I_{gd}^p, I_{gq}^p, I_{gd}^n and I_{gq}^n). These four current reference values are calculated to meet certain control laws. In this paper, four control laws are formulated to satisfy the balanced PCC voltage and active/reactive power generation conditions.

First and second control laws are to keep PCC voltage balanced. In Fig. 2 and 3, the negative sequential components of ac input current (I_{gd}^n and I_{gq}^n) are related with the negative sequential components of PCC voltage (V_{pccd}^n and V_{pccq}^n) as the following;

$$I_{gq}^n = \frac{V_{sd}^n - V_{pccd}^n}{\omega L_s} \tag{10}$$

$$I_{gd}^n = -\frac{V_{sq}^n - V_{pccq}^n}{\omega L_s} \tag{11}$$

If V_{pccd}^n and V_{pccq}^n are to be zero, then I_{gd}^n and I_{gq}^n are simplified as in (12) and (13).

$$I_{gq}^n = \frac{V_{sd}^n}{\omega L_s} \tag{12}$$

$$I_{gd}^n = -\frac{V_{sq}^n}{\omega L_s} \tag{13}$$

The third and fourth control laws are to meet the demand of active and reactive power generation. The average values of instantaneous input active and reactive power can be formulated in terms of sequential components of ac input voltage and current as followings.

$$\frac{2}{3}P_{so}^{in} = V_{sd}^p I_{gd}^p + V_{sq}^p I_{gq}^p + V_{sd}^n I_{gd}^n + V_{sq}^n I_{gq}^n \tag{14}$$

$$\frac{2}{3}Q_{so}^{in} = V_{sq}^p I_{gd}^p - V_{sd}^p I_{gq}^p - V_{sd}^n I_{gd}^n + V_{sq}^n I_{gq}^n \tag{15}$$

After applying (12) and (13) into (14) and (15) and rearranging terms lead to (16).

$$-\frac{2}{3}\begin{bmatrix} P_{so}^{in} \\ Q_{so}^{in} \end{bmatrix} = \begin{bmatrix} V_{sd}^p & V_{sq}^p \\ V_{sq}^p & -V_{sd}^p \end{bmatrix}\begin{bmatrix} I_{gd}^p \\ I_{gq}^p \end{bmatrix} + \frac{1}{\omega L_s}\begin{bmatrix} V_{sd}^n & V_{sq}^n \\ -V_{sd}^n & V_{sq}^n \end{bmatrix}\begin{bmatrix} -V_{sq}^n \\ V_{sd}^n \end{bmatrix} \tag{16}$$

Let's introduce the parameter of k_{pf} to correlate Q_{so}^{in} with P_{so}^{in} as in (17). For example, under the condition of unity power factor, i.e. zero value of Q_{so}^{in}, the parameter of k_{pf} is equal to zero.

$$Q_{so}^{in} = k_{pf} P_{so}^{in} \tag{17}$$

Finally I_{gd}^p and I_{gq}^p can be obtained from (16).

$$I_{gd}^p = \frac{1}{(V_{sd}^p)^2 + (V_{sq}^p)^2}\{\frac{2}{3}P_{so}^{in}(-V_{sd}^p - V_{sq}^p k_{pf}) \\ + \frac{(-V_{dq}^p + V_{sq}^p)V_{sd}^n(-V_{sq}^n) - (V_{dq}^p + V_{sq}^p)V_{sd}^n V_{sq}^n}{\omega L_s}\} \tag{18}$$

$$I_{gq}^p = \frac{1}{(V_{sd}^p)^2 + (V_{sq}^p)^2}\{\frac{2}{3}P_{so}^{in}(-V_{sq}^p + V_{sd}^p k_{pf}) \\ + \frac{(-V_{dq}^p - V_{sq}^p)V_{sd}^n(-V_{sq}^n) + (V_{dq}^p - V_{sq}^p)V_{sd}^n V_{sq}^n}{\omega L_s}\} \tag{19}$$

As a result, the positive and negative sequential components of ac input grid currents (I_{gd}^p, I_{gq}^p, I_{gd}^n and I_{gq}^n) are calculated from four equations of (12), (13), (18), and (19). They are used as current reference values in a dual-frame current regulator [2].

B. Control Block Diagram

The overall control structure consists of two nested regulating loops; the outer dc link voltage regulating loop and the inner ac current regulating loop. The outer voltage regulating loop determines the reference signals for the inner current loop according to the four control laws described in Section IV-A. The inner current loop is made up of two parallel *dq* synchronous frame current regulators; one for the positive sequence and the other for the negative sequence. Fig. 4 shows the entire control scheme block diagram.

A detailed control block diagram of the cascade dual current regulator with a voltage regulator is shown in Fig.

The 2014 International Power Electronics Conference

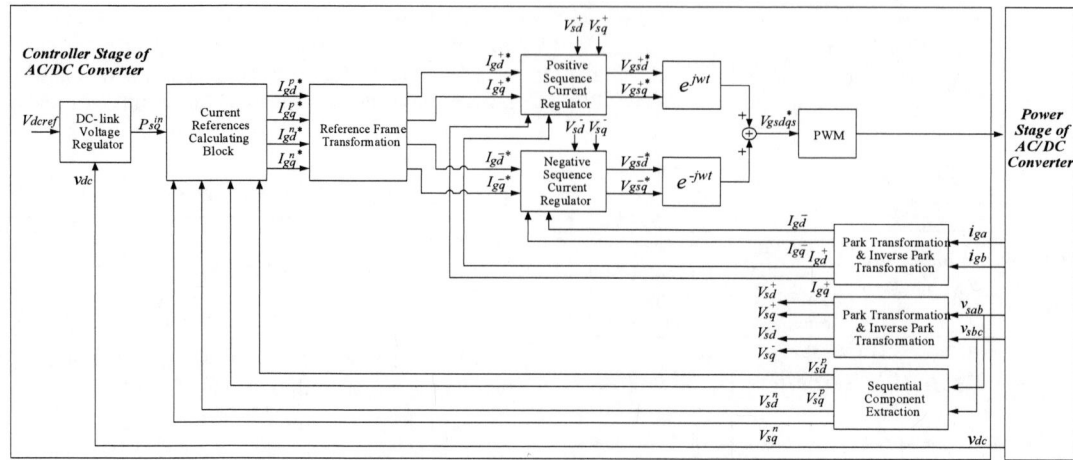

Fig. 4. Overall control block diagram.

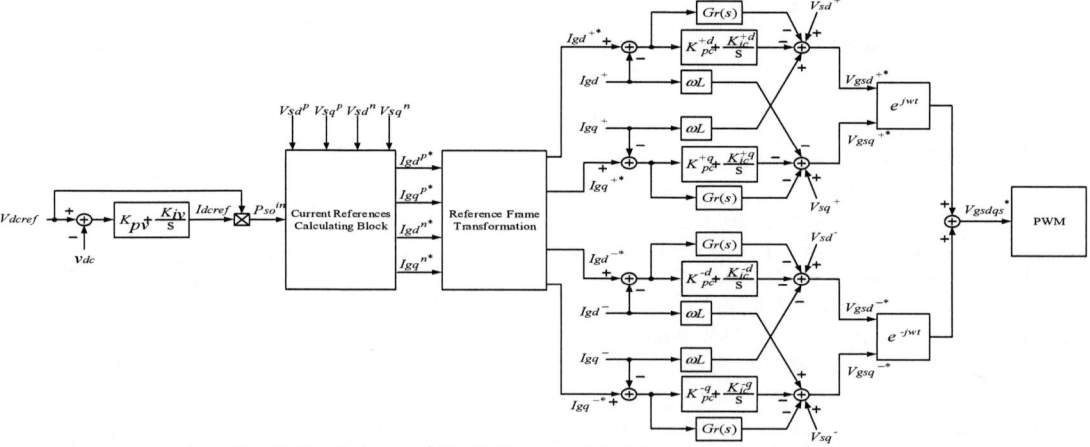

Fig. 5. Detailed control block diagram of dual frame current regulator.

5. In general, it is required that the bandwidth of current regulation loop is high enough to maintain the fast dynamics of complete system. Sequence separation method of input current typically involves the low pass filter or notch filter to filter out the negative (positive) sequential component of twice input frequency in positive (negative) synchronous reference frame. These low pass filter and notch filter in the current feedback path usually undermine the bandwidth of current regulation loop or phase margin of the system [2]. In this paper, instead of separating the positive and negative sequence component from the total signal, these two components are rotated together in a counterclockwise direction to be regulated in a positive synchronous reference frame and in a clockwise direction to be regulated in a negative synchronous reference frame, respectively [2]. The measured three-phase ac input current quantities are directly transformed to either positive or negative rotating synchronous reference frame resulting in the ac signal of twice line frequency imposed on the dc signal without being decomposed into separate positive and negative sequential components. However, the four current references that are calculated based on four control laws are dc signals separated in positive/negative sequence synchronous reference frames. Therefore, in order to match the calculated current

reference values (I_{gd}^p, I_{gq}^p, I_{gd}^n and I_{gq}^n) to measured current quantities (I_{gd}^+, I_{gq}^+, I_{gd}^- and I_{gq}^-) in the same transformation frames, the reference values should be transformed accordingly.

V. SIMULATION

A. Case1: UF = 3%, Power = 60% of rated power

The NCI control algorithm which is proposed in this paper is verified through the simulation. Parameters of PMSG employed in the simulation are summarized in Table III. Simulation result is presented under the two different cases; Case 1 and Case 2. The first case (Case 1) is about the operating condition under which the grid unbalance of type B (voltage sag of single-phase) with the unbalance factor (UF) of 3% occurs and the wind turbine generates 60% of rated power. The second case (Case 2) is about the operating condition under which grid unbalance of type B with the UF of 7% occurs and the wind turbine generates 30% of rated power. These two representative operating conditions are chosen to effectively verify the performance of proposed NCI control algorithm under the most frequently encountered

978-1-4799-2706-7/14 $31.00 © 2014 IEEE

grid unbalance case and the range of output power generation from wind turbines.

TABLE III
PARAMETERS OF PMSG WIND TURBINE SYSTEM

Parameter	Value
Rated power (P_{rated})	2.7 MW
Rated line voltage ($V_{llrated}$)	3300 V
Rated ac input current (I_{rated})	520 A
Frequency (f_{in})	60 Hz
DC link voltage (V_{DC})	5200 V
DC link capacitance (C_{DC})	6 mF
Converter switching frequency (f_{sw})	1020 Hz
Grid side line inductance (L_s)	1.07 mH (0.1 pu)
Transformer leakage inductance (L_{tr})	0.54 mH (0.05 pu)
Filter inductance (L_f)	0.08 mH (0.007 pu)
Filter capacitance (C_f)	0.18 mF (0.27 pu)
Filter resistance (R_f)	0.614 Ω (0.15 pu)

Fig. 6 and 7 show the waveforms of current reference signals (I_{gd}^{+}, I_{gq}^{+}, I_{gd}^{-}, and I_{gq}^{-}) under the balanced and unbalanced grid condition of Case 1, respectively. The four control laws described in Section IV-A determine the constant values of four current reference signals (I_{gd}^{p}, I_{gq}^{p}, I_{gd}^{n}, and I_{gq}^{n}). These constant values are then transformed into the counterclockwise (clockwise) rotating synchronous reference frame having the ac signal of twice line frequency due to negative (positive) sequential components imposed on the dc signal due to positive (negative) sequential components.

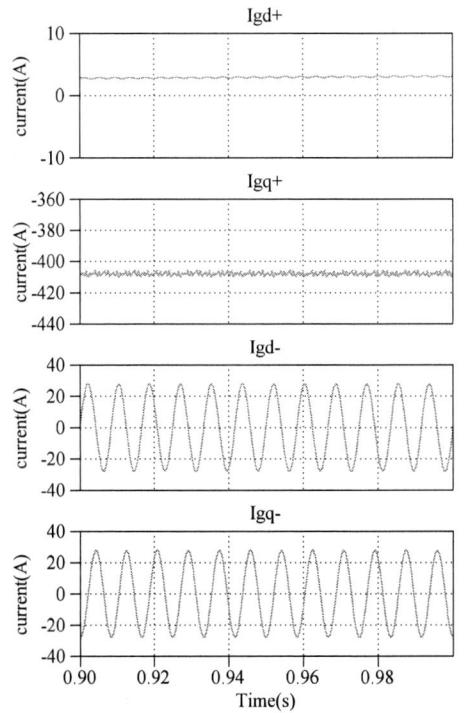

Fig. 6. Four current references under balanced grid condition.
($I_{gd(avg)}^{+}$=2A, $I_{gq(avg)}^{+}$=-407A, $I_{gd(avg)}^{-}$=0A, $I_{gq(avg)}^{-}$=0A)

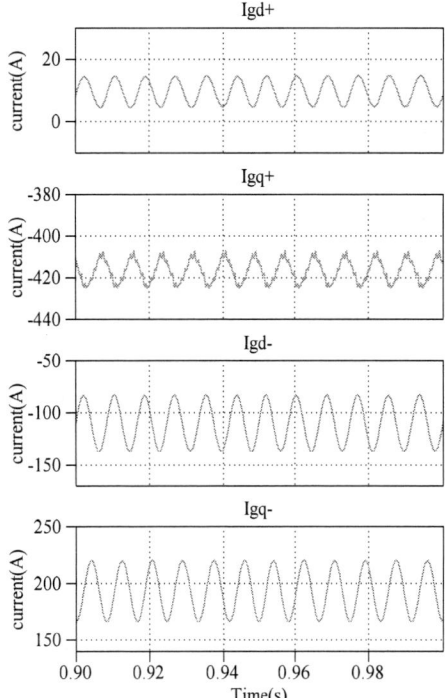

Fig. 7. Four current references under case1.
($I_{gd(avg)}^{+}$=9A, $I_{gq(avg)}^{+}$=-416A, $I_{gd(avg)}^{-}$=-110A, $I_{gq(avg)}^{-}$=193A)

Under the grid unbalance and output power generation condition of case 1, the three-phase line voltages at PCC are illustrated in Fig. 8 and 9. Waveforms in Fig. 8 are obtained under the condition of case 1 without NCI compensating algorithm being employed, i.e. the conventional single-frame current regulator of positive sequential component of ac input current only. Waveforms in Fig. 9 are obtained under the condition of case 1 with NCI compensating algorithm being employed. It is noted from the frequency spectrum in Fig. 8 that the amplitudes of three-phase line voltages are unbalanced. The corresponding numerical data are summarized in Table IV for the sake of readers' convenience. In Fig. 8, the unbalance factor of line voltages at PCC is close to that of grid input voltages, i.e. UF=3%. As shown in Fig. 9 and Table IV, the proposed NCI control algorithm actively compensates for the unbalanced grid voltage. As a result NCI control algorithm makes the three-phase line voltages at PCC balanced having the UF of almost zero.

The compensation of grid unbalanced voltage at PCC is made possible by the injection of negative sequence current to the grid. Figure 10 provides the waveforms of three-phase ac input currents under the condition of case 1 without NCI control algorithm being considered. Figure 11 shows the waveforms of three-phase ac input currents under the condition of case 1 with NCI control algorithm considered. It is clearly noted from Fig. 11 that the three-phase ac input currents become more unbalanced as compared to those of Fig. 10 because of injected negative sequential component of input current. The peak amplitude of input current at maximum is increased from 433A in Fig. 10 to 618A in Fig. 11. This increase of amplitude of ac input current gives a rise to a larger power loss in switching devices of converter. Therefore, proposed NCI control algorithm may suffer from this increasing loss factor due to additionally injected negative sequential component of input currents. However, as long as the peak current of ac input current having a injected negative sequential component is maintained below the value of rated condition (735A$_{pk}$), the disadvantage of NCI control algorithm doesn't necessarily undermine the system efficiency and requires enlarged cooling system. This operating condition can be satisfied when the wind turbine generates the output power less than the rated power of 2.7MW. For example, operating condition of case 1 in this paper has the output power generation at 60% of rated value. In general, the mean value of wind speed is below the rated wind speed of 12m/s in a typical offshore wind turbine.

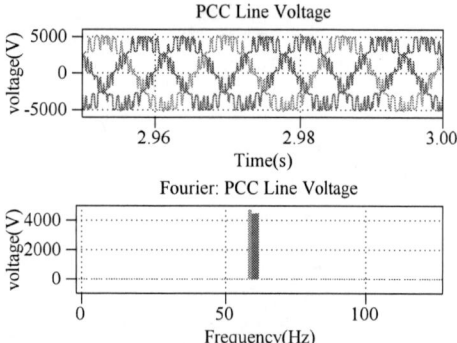

Fig. 8. PCC line voltages without compensation algorithm under case1.

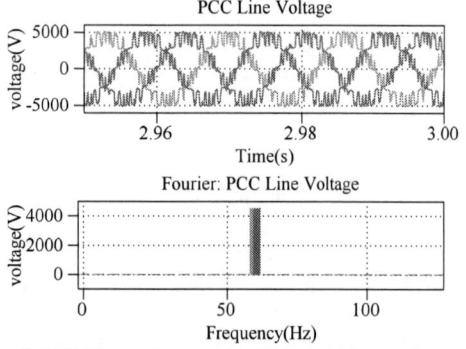

Fig. 9. PCC line voltages with NCI algorithm under case1.

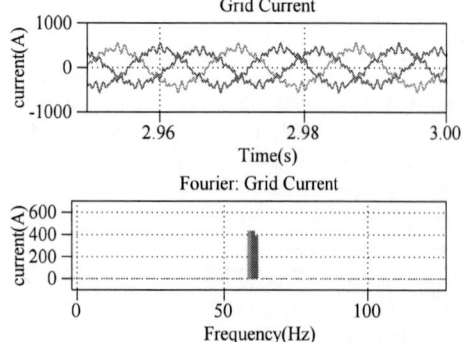

Fig. 10. Grid currents without compensation algorithm under case1.

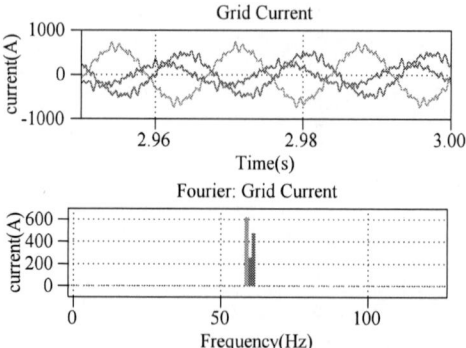

Fig. 11. Grid currents with NCI algorithm under case1.

TABLE IV
COMPARISON OF LINE VOLTAGES AT PCC AND GRID
CURRENTS (CASE1: UF=3%, POWER=60%)

Variables	Conventional algorithm	NCI algorithm
PCC line voltage v_{ab}	4674 V$_{pk}$	4519 V$_{pk}$
PCC line voltage v_{bc}	4433 V$_{pk}$	4519 V$_{pk}$
PCC line voltage v_{ca}	4453 V$_{pk}$	4518 V$_{pk}$
UF of PCC voltage	3%	0%
Grid current i_a	433 A$_{pk}$	618 A$_{pk}$
Grid current i_b	430 A$_{pk}$	249 A$_{pk}$
Grid current i_c	386 A$_{pk}$	471 A$_{pk}$

B. Case2: UF = 7%, Power = 30% of rated power

In this section, the simulation result of proposed NCI control algorithm is described for the operating condition under which the grid unbalance of type B with the UF of 7% occurs and the wind turbine generates 30% of rated power. As compared to case 1, the depth of grid unbalance in case 2 is deeper. In addition, the level of output power is decreased from 60% to 30% of rated power. This is to ensure that the maximum peak current of ac input current for NCI control algorithm is maintained to be below the rated current value.

Figure 12 presents the waveforms of current reference signals ($I_{gd}^+, I_{gq}^+, I_{gd}^-,$ and I_{gq}^-) under case 2. The amplitudes of negative sequential component are larger than those of case 1, in other words, larger injected negative sequential components of input current.

Waveforms of line voltages at PCC and input currents without NCI control algorithm being employed are given in Fig. 13 and 15, respectively. Similarly, waveforms of line voltages at PCC and input currents with NCI control algorithm employed are illustrated in Fig. 14 and 16, respectively. Table V summarizes the numerical data corresponding to waveforms in Fig. 13-16. Simulation results provided in Fig. 13-16 exhibit the similar operational characteristic of NCI control algorithm to that of case 1; active compensation for balancing line voltage at PCC at the cost of larger amplitude of ac input current due to injected negative sequential component of current. Because UF of case 2 is larger than that of case 1, the required amount of negative sequential component of input current for the case 2 tends to be larger than case 1.

Therefore, in order to stay below the current capability of grid side converter, the level of output power for the case 2 is decreased from 60% to 30% of rated value.

Simulation results for two different cases successfully verify the operational characteristics of proposed NCI control algorithm. NCI control algorithm can actively compensate for the line unbalance at PCC in a collector bus of wind turbines, therefore improve the power quality of wind farm. Although NCI control algorithm requires larger amplitude of ac input current leading to a higher power loss factor in switching devices, the proposed control algorithm can be an effective solution complying with the given current capability of grid side converter under the derated output power conditions.

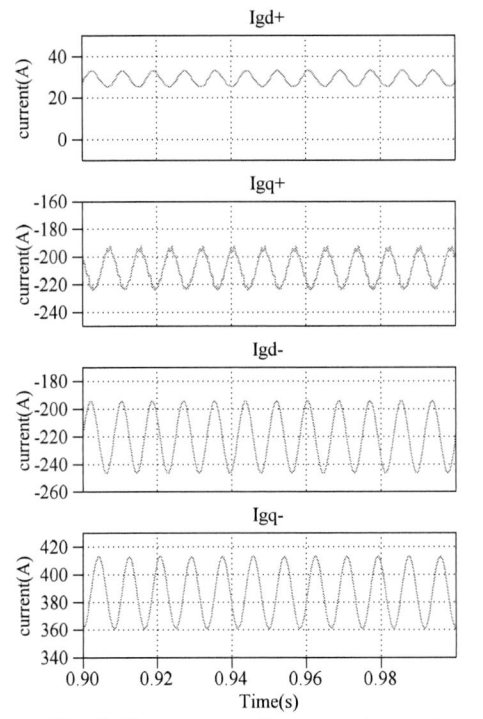

Fig. 12. Four current references under case2.
($I_{gd(avg)}^{+}=29A$, $I_{gq(avg)}^{+}=-208A$, $I_{gd(avg)}^{-}=-220A$, $I_{gq(avg)}^{-}=386A$)

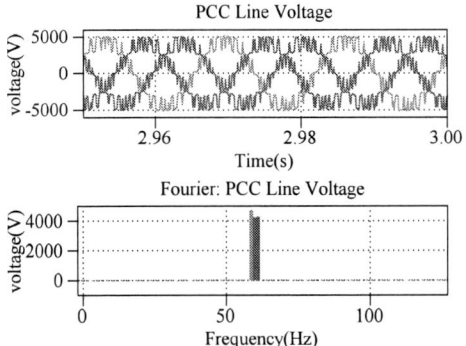

Fig. 13. PCC line voltages without compensation algorithm under case2.

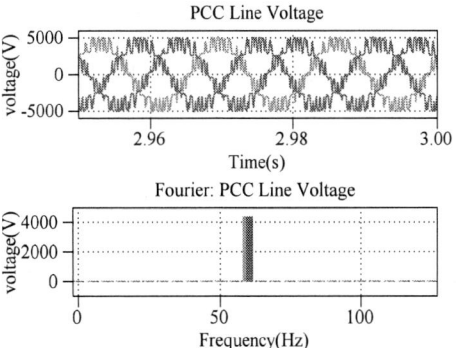

Fig. 14. PCC line voltages with NCI algorithm under case2.

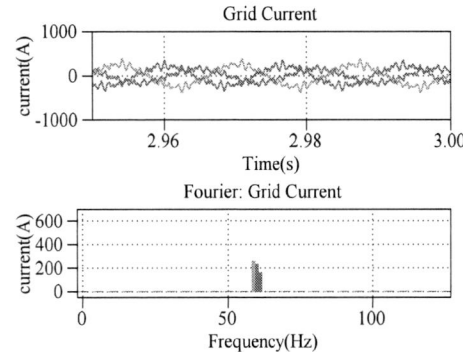

Fig. 15. Grid currents without compensation algorithm under case2.

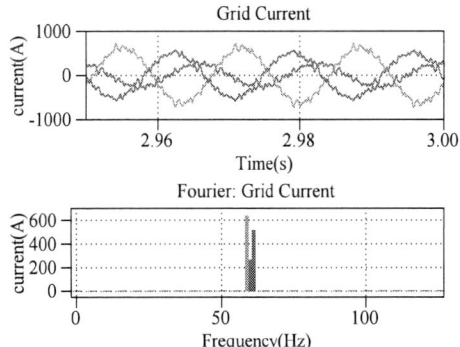

Fig. 16. Grid currents with NCI algorithm under case2.

TABLE V
COMPARISON OF LINE VOLTAGES AT PCC AND GRID
CURRENTS (CASE2: UF=7%, POWER=30%)

Variables	Conventional algorithm	NCI algorithm
PCC line voltage v_{ab}	4677 V_{pk}	4374 V_{pk}
PCC line voltage v_{bc}	4204 V_{pk}	4374 V_{pk}
PCC line voltage v_{ca}	4256 V_{pk}	4373 V_{pk}
UF of PCC voltage	7%	0%
Grid current i_{a}	259 A_{pk}	633 A_{pk}
Grid current i_{b}	231 A_{pk}	262 A_{pk}
Grid current i_{c}	158 A_{pk}	515 A_{pk}

978-1-4799-2706-7/14 $31.00 © 2014 IEEE

VI. EXPERIMENT

Experimental setup for a test-bed of 36kW PMSG to validate the proposed NCI control algorithm is illustrated in Fig. 17 and 18. The parameters of test-bed are summarized in Table VI. The experimental verification is under progress and its result will be reported in future publications.

TABLE VI
PARAMETERS OF TEST-BED

Parameter	Value
Rated power (P_{rated})	36 kW
Rated line voltage ($V_{llrated}$)	460 V
Rated ac input current (I_{rated})	46 A
Frequency (f_{in})	60 Hz
DC link voltage (V_{DC})	858 V
DC link capacitance (C_{DC})	1.23 mF
Converter switching frequency (f_{sw})	10 kHz
Filter inductance (L_f)	2 mH (0.128pu)

Fig. 17. Experimental set-up of 36kW test-bed.

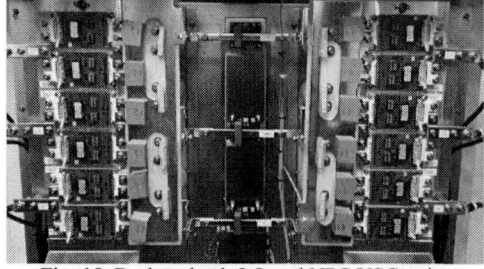

Fig. 18. Back-to-back 3-Level NPC VSC unit.

VII. CONCLUSION

This paper proposes a negative sequence current injection algorithm to actively compensate for voltage unbalance at PCC. The algorithm is to cancel the negative sequential component of voltage by injecting the appropriate negative sequential component of input current. As the depth of unbalance becomes severe, the necessary magnitude of negative sequence current also increases over the current capability of PCS. When the output power of wind turbine gets smaller at low wind speed, the injection of negative sequence current becomes more effective under the current limit of PCS. In a wind farm consisting of multiple wind turbines connected to a collector bus, the contribution of negative

sequence current generated by each wind turbine can be summed to compensate for the unbalance at PCC. The proposed control algorithm makes it possible for wind farms to generate a high quality output power under unbalance grid disturbance.

ACKNOWLEDGMENT

This work was supported by the National Research Foundation of Korea (NRF) grant funded by the Korea government (MSIP) (No. 2010-0028509)

REFERENCES

[1] S. Yang, S. Kim, J. Choi, I. Choy, S. Song, S. Lee, and D. Lee, "A compensation of the grid current unbalance and distortion caused by the grid voltage unbalance and distortion in 3-phase bi-directional DC to AC inverter," *The Transactions of the Korean Institute of Power Electronics*, vol. 18, no. 2, pp. 161-168, April 2013.

[2] Y. Suh and T. A. Lipo, "Control scheme in hybrid synchronous stationary frame for PWM AC/DC converter under generalized unbalanced operating conditions," *IEEE Transactions on Industry Applications*, vol. 42, no. 3, pp. 825-835, May/June 2006.

[3] W. Lee, T. Lee, and D. Hyun, "A three-phase parallel active power filter operating with PCC voltage compensation with consideration for an unbalanced load," *IEEE Transactions on Power Electronics*, vol. 17, no. 5, pp. 807-814, September 2002.

[4] K. Li, J. Liu, Z. Wang, and B. Wei, "Strategies and operating point optimization of STATCOM control for voltage unbalance mitigation in three-phase three-wire systems," *IEEE Transaction on Power Delivery*, vol. 22, no. 1, pp. 413-422, January 2007.

[5] Y. Suh, Y. Go, and D. Rho "A comparative study on control algorithm for active front-end rectifier of large motor drives under unbalanced input," *IEEE Transactions on Industry Applications*, vol. 47, no. 3, pp.1419-1431, May/June 2011.

[6] J. Hu and Y. He, "Modeling and control of grid-connected voltage-sourced converters under generalized unbalanced operation conditions." *IEEE Transactions on energy conversion*, vol. 23, no. 3, pp. 903-913, September 2008.

[7] H. Song and K. Nam, "Dual current control scheme for PWM converter under unbalanced input voltage conditions," *IEEE Transactions on Industrial Electronics*, vol. 46, no. 5, pp. 953-959, October 1999.

[8] Y. Suh, Y. Go, and D. Rho "A comparative study on control algorithm for active front-end rectifier of large motor drives under unbalanced input," *IEEE Transactions on Industry Applications*, vol. 47, no. 3, pp. 1419-1431, May/June 2011.

[9] A. V. Stankovic and T. A. Lipo, "A novel control method for input output harmonic elimination of the PWM boost type rectifier under unbalanced operating conditions," *IEEE Transactions on Power Electronics*, vol. 16, no. 5, pp. 603-611, September 2001.

[10] Mohamed Y.A.-R.I. and El-Saadany, E.F, "A control scheme for PWM voltage-source distributed-generation inverters for fast load-voltage regulation and effective mitigation of unbalanced voltage disturbances," *IEEE Transactions on Industrial Electronics*, vol. 55, no. 5, pp. 2072-2084, May 2008.

[11] M. Savaghebi, A. Jalilian, J.C. Vasquez, and J.M. Guerrero, "Secondary control scheme for voltage unbalance compensation in an islanded droop-controlled microgrid," *IEEE Transactions on Smart Grid*, vol. 3, no. 2, pp. 797-807, June 2012.

[12] J. Im, S. Song and S. Kang, "Analysis and compensation of PCC voltage variations caused by wind turbines power fluctuations," Journal of Power Electronics, vol.13, No.5, pp. 854-860, September 2013.

Optimization of an Off-grid Hybrid System for Supplying Offshore Platforms in Arctic Climates

Maria Kalogera*, Pavol Bauer, Senior Member IEEE **

Delft University of Technology
*e-mail: Kalogera.maria@gmail.com
**e-mail: p.bauer@tudelft.nl

Abstract— **A hybrid power system placed in Beaufort Sea in order to power supply the offshore oil and gas platforms is investigated in this paper. The proposed micro-grid includes various energy sources such as photovoltaic panels, wind turbines, diesel generator as a secondary power source and stationary storage system. Simulation of each component and the proposed topology is discussed with the energy management strategy and the implementation of different scenarios that show relevant characteristics of the strategies. Finally, a multi-objective optimization of the system is carried out and the results obtained are used to construct a Pareto front, for the best trade-off between emissions and cost.**

Keywords— *Arctic, microgrid, multi-objective optimization, solar, storage, sustainable energy, thermal management, wind.*

I. INTRODUCTION

New opportunities and challenges are emerging across the Arctic and it is considered that the vast Arctic region is probably the last remaining unexplored source of hydrocarbons in the planet [1]. Energy companies first started activities in the Arctic around a century ago, but the majority of the Arctic's resources remain untapped [2]. Extended operational activities of oil companies in Arctic have necessitated the need for power supplying the offshore platforms. As such, a hybrid power system for power supplying the offshore platforms in the Arctic will be examined in the framework of this paper. It focuses on the combination of wind, solar, diesel generator and energy storage systems for sustainable power generation of offshore platforms in harsh Arctic conditions.

First, a prefeasibility study of the system will be discussed, presenting the meteorological data of the specific site at Beaufort Sea, as well as the power requirements of the system. Further, the connecting topology of the hybrid system will be presented and the simulation techniques used for the various components of the system will be shown. Following up, the type of batteries used for storage will be discussed. Later on, the power control strategy will be analyzed and finally a multi-objective optimization will be attempted. In the end, conclusions are drawn upon the specification of the proposed hybrid system.

II. FEASIBILITY STUDY OF THE HYBRID SYSTEM

Climatic conditions verify the availability and magnitude of wind and solar energy at particular site. Pre-feasibility studies are based on weather data [3] and load requirements for specific site. After specifying that a hybrid system is viable, the planning, modeling and designing of the system is set forward.

A. Location

With current oil reserves struggling with surging global energy demand, the Arctic and its seas are seen as being of utmost importance. The paper will focus on the Beaufort Sea in line with Shell's Beaufort Sea exploration Program. The geographic region is seen in the following picture and is bounded by the following coordinates:

69° 57' 0" N – 71° 30' 0" N latitude	141° 48' 0" W – 156° 0' 0" W longitude.

B. Ambient Temperature Profile

The monthly temperature profile is generated based on the data from NASA Surface Meteorology and Solar Energy [4] and is shown in Fig. 1.

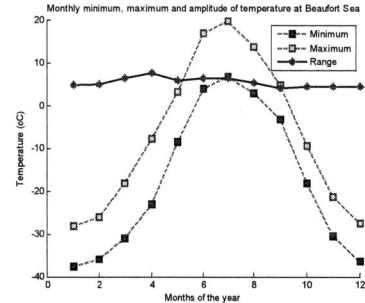

Fig. 1 Temperature Profile in the Beaufort Sea

C. Hours of daylight at Beaufort Sea

A primary section heading is enumerated by a Roman numeral followed by a period and is centered above the text. A primary heading should be in small caps. The first letter of each important word is capitalized.

The hourly solar irradiation profile in Beaufort Sea is

shown in Fig. 2. This profile stemmed from the available data extracted by the USA National Renewable Energy Laboratory for various years [5].

Fig. 2 also shows the hours of daylight at the Beaufort Sea. As illustrated, for more than eighty-three days the sun does not set in Beaufort Sea and provides more than 20 hours of daylight.

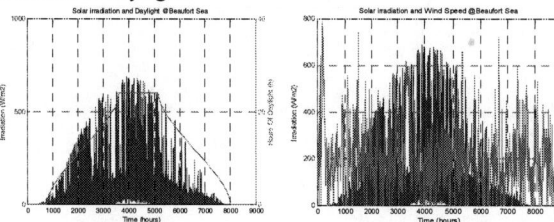

Fig. 2 Hours of daylight-Solar irradiation at the Beaufort Sea

Fig. 2 Solar Irradiation and Wind speed profile at Beaufort Sea

D. Wind Profile at Beaufort Sea

Analyzing the wind profile of the selected region is vital, as the power produced from the wind turbines depends on the wind speed. Contrary to the solar irradiation profile, the wind speed does not follow a particular pattern. To visualize the solar irradiation and wind speed profile at the Beaufort Sea a graph representing both of them is also shown in Fig. 3.

From Fig. 3, it is clear that is hardly possible to get high irradiation and wind speed at the same time. Thus, both wind turbines and solar panels are considered for the hybrid system so as to capture the renewable energy potential.

III. POWER DEMAND OF THE HYBRID SYSTEM

Remote monitoring is a key issue in the operation of unmanned offshore platforms.[6] Incessant monitoring implies that the power system should be trustworthy, well-engineered and capable of operating in an immense range of conditions. Some of the applications that the power system should be able to cover include; RTU/Scada, Flow monitoring, Telemetry, Drilling Meters, Cathodic Protection, Natural Gas Automation, Gas Flow Measurement, Data Recording, Control Valves, Process control equipment and Lightning [7]. For the design of the offshore platform the load profile is of supreme importance. The system size must be such that the power demand is covered. A histogram with the load profile that needs to be covered is shown in the Fig. 4.

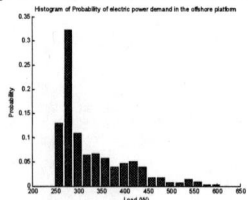

Fig. 3 Histogram of probability of occurrence of Electric Power Demand

Fig. 5 Mixed-coupled Hybrid Power Systems

It is based on a typical Direct Current (DC) load profile of the existing North Sea Cutter platform. As shown in Fig. 4, the load varies between $250W$ and $400W$ while the probability of occurrence of being higher is very low. The daily average current demand of the instrumentation of the offshore platform is $13.85\ A$ in a voltage of $24V$ as specified by Shell. Thus, the load profile accounts for $332.4W$ on average which results in $\frac{7978Wh}{day}$.

IV. SUSTAINABLE MICROGRID

A. Topology

The hybrid power systems can be classified depending on their coupling technologies as Direct Current (DC) or Alternating Current (AC) coupled systems. [8] However, no configuration scores best for all the cases.[9] Based on different scenarios and different energy sources connections, it has been proved that the Mixed coupling Hybrid Power System gives the highest power consumption efficiency and is shown in Fig. 5 [10].

B. Sources

1) Solar Power

For the solar panels, a model was developed in MATLAB/Simulink based on the mathematic equations suggested by D, Sorensen, Hansen, & Bindner [11].

a) Effect of irradiation levels

As expected and validated from Fig. 6, as the irradiance level incident on the PV changes, the short circuit current (Isc) changes proportional to the irradiance.

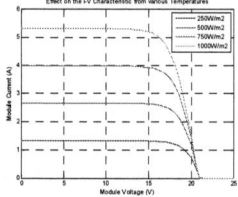 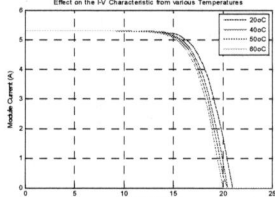

Fig. 4 Dependence of current and voltage on incident sunlight levels

Fig. 7 Dependence of current and voltage on temperature for sunlight level of $1000W/m^2$

However, the open circuit voltage (Voc) remains almost the same.

b) Effect of Temperature

S Solar cells are sensitive to temperature and increase in temperature reduces the band gap of the semiconductor. This reduction in the band gap can be realized as the energy change of the electrons in the material. Thus, less energy is required to break the bond. Consequently, rise in temperature reduces the band gap. As seen in the Fig. 7, as the temperature of the PV module changes, the open circuit voltage (Voc) changes inversely proportional to the temperature. The linear decrease of the open circuit voltage, is the dominant effect with increasing cell's temperature, leading to less efficient solar cells. However, the short circuit (Isc) remains almost the same. Summarizing, higher cell operating temperatures reduce cell output, efficiency and lifetime, while operating conditions result in higher operating voltages, which is advantageous for the case studied.

c) Maximum Power Point Tracking Algorithms

Various methods for MPPT Algorithms exist in literature, each of them being advantageous in specific aspects. In this paper the fractional open-circuit voltage technique is used. This technique is based on the linear relationship between VMPP and VOC. For every irradiation level as well as working temperature of the cell there is a specific voltage value (Vmpp) at which the solar module produces the maximum amount of power. [14] In other words: $\frac{V_{MPP}}{V_{OC}} \approx K < 1.$. This ratio was calculated to be 0.80 for the selected photovoltaic module TSM85-EX 85Wp Solar Module. Although this described method comprises an easy and cheap operation technique, the optimal value of the constant ration K is difficult to choose. This difficulty can be overcome by dynamically adjusting the value of K.

2) Wind Power

a) Background

The rapid global development of wind power capacity influenced the cost of wind power over the last 20 years. Improved cost-effectiveness and larger turbines were the main drivers for the development of economics of wind-power systems [15].

b) Wind Turbines Installation in Arctic Climates

When it comes to the critical point of installing wind turbines in Arctic Climates, added costs and performance reliability are coupled with the economic risks and a special framework should be considered to assess it. These risks may include the increased capital costs of the project, due to the necessity for higher equipment, installation costs and augmented maintenance costs because of potential more frequent repairs caused by the cold temperatures. Undoubtedly, increased long-term exposure to icing may also result in fatigue loading, early failures and power reduction. Clearly, it is of utmost importance to assess all the risks and mitigate risk strategies so as to ensure enhanced operation of the wind turbines [17].

c) Wind Turbine Simulation Concept

A complete (offshore) wind turbine system can be thought of as being constructed of a number of coupled mass spring-damper systems [18]. Such simulation concept is considered sufficient for this study. It is based upon computation of the acceleration of the angular velocity of the rotor, via a spring damper system. From Newton's law or using Lagrange equations of motions, the second-order differential equations of torsional-dynamics is found as:

$$\ddot{a}J = ka - c\dot{a} + T$$

Where:

$\ddot{\alpha}$= angular acceleration $[\frac{rad}{s^2}]$ | $\dot{a}$= angular velocity $[\frac{rad}{s}]$

α= angle [rad] | J= rotational moment of inertia [kg m^2]

C= damping $[\frac{Nm\,s}{rad}]$ or $[\frac{kg\,m}{s\,rad}]$ | k= rotational spring $[\frac{Nm\,s}{rad}]$

T= Torque [Nm]

The rotational stiffness of the spring is set to zero. The damping factor represents the energy that is taken out of the system by the generator. The torque represents the fact that inserts energy into the system. A simple expression for the angular acceleration can be defined as follows: $\ddot{a} = -\frac{c}{J}\dot{a} + \frac{T}{J}$

In order to obtain values for c and T the power curve of the wind turbine is used. In four steps the damping coefficient is included.

1. Calculation of the tip speed ratio (λ). $\lambda = \frac{\dot{a}_{rated}\,R}{v}$
2. Calculation for every wind speed an equilibrium rotational velocity $\dot{\alpha}(v)| = \frac{\lambda\,v}{R}$
3. Calculation of the aerodynamic torque at each wind speed $T(v) = \frac{P(v)}{\dot{a}(v)}$ for each wind speed, where P(v) comes from power curve
4. Calculation of the equilibrium damping factor
5. $c(v) = \frac{T(v)}{\dot{a}(v)}$ for each wind speed

To solve the equation of motion, the dynamic model is programmed in Simulink to solve the mathematical expressions. The result of this dynamic simulation is saved in a structure which is then directly put into Simulink. So, the input from workspace has already taken into consideration the wind speed profiles, the power curve of the wind turbine and a dynamic simulation of its response. The output active power is fed into the controller and is monitored whereas the reactive output power is fed into the dumping system. A lossless converter is used to connect the wind turbine to the ac bus.

3) Diesel Generator

a) Type Selection

The selected generator is a low-speed 4-pole diesel generator. Diesel generators are preferred as they have higher fuel efficiency than gasoline engines, they are more robust and diesel fuel is more easily stored for extended periods of time. For Antarctic applications, DFA (Diesel Fuel Arctic) is used. This is a jet-fuel with additives to adjust centane rating and lubricity. The diesel generator model is comprised of a model of the diesel engine and a model of an electrical induction generator rotating on one shaft. The diesel engine offers the active power by converting the fuel into mechanical energy, which is represented by the mechanical torque on the rotating shaft [19]. The choice of the induction generator is based upon the fact that they are widely and commercially available, inexpensive and easy to operate with. There is an extensive use of them especially in small hybrid power plants. Typically, the induction generators are mostly appropriate for hydro and wind power systems. However, they can also be efficiently used with prime movers driven by biogas, diesel, natural gas, alcohol motors and gasoline. Induction generators operate exceptional in motor or generator mode and are the most inexpensive compared to the other generators [20], [21]. It is also very easy to operate them in parallel with large power systems, because the utility grid controls voltage and frequency while static and reactive compensating capacitors can be used for correction of the power factor and harmonic reduction [21], [22], [23]. Induction generators comprise robust construction features and offer natural protection for short-circuits. Their solid motor facilitates the absorption of abrupt speed changes due to load or primary source changes. What is more important, demagnetization path of their iron core enables the damping of any current surge without fear of demagnetization; thus making induction machines favorable compared to magnet-based generators [21], [24]. The self-excitation phenomenon of the induction machine is also a very essential feature, which can be achieved when a capacitor bank with an appropriate value is properly connected across the machine terminals [25].

b) Simulation of the diesel generator

The diesel engine model is depicted in Fig. 8 and its concept is relatively simple. The wref (pu) is the input speed setting value which is the reference speed and ω(pu) is the actual speed of the generator in per unit values. These two speeds are taken as inputs to the system and by calculating their difference the correspondent error is computed. This is transferred to the control system, to the two transfer functions and ultimately the output speed of the diesel engine is converted to torque through an integral unit. The whole system implements an additional time delay to be more realistic as the diesel engine is a large time delay system. The signal of the mechanical torque is an input to the electrical generator model, which then changes the rotational speed of the rotor. The rotational speed of the rotor is then fed to the diesel engine model to compute the error with the reference speed and provide the accordingly adjusted signal of the torque. The procedure is followed by the multiplication of the speed signal in order to reach the machinery power signal. The Pmech(pu) is the per unit of diesel engine output power used to drive the generator [26].

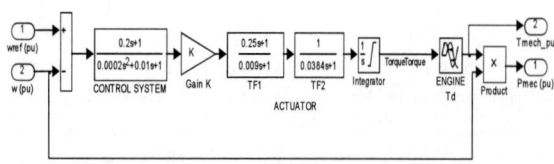

Fig. 8 Simple diesel engine model

c) Simulation of the induction machine

The induction generator model is a 3rd order reduced model developed in the q-d axis arbitrary reference frame. The transients of the stator are neglected, since a fully transient model not only would lead to a

significant increase of the simulation time but also such a model is not considered necessary in our case of study. In this paper, the main interest lies on the power flow between different components and their appropriate control. Consequently, a 3rd order model is considered to be adequate for the purpose of the current project. The mathematical equations of the machine are expressed in the Stator flux reference frame, *SFO*. Therefore the following are valid:

1. The machine mathematical model equations are expressed in the direct (d) and quadrature (q) reference frame which is oriented to the stator flux and rotates with the synchronous speed.
2. The q axis is leading for 90° from the d axis.
3. The voltage of the stator in the q axis V_{qs}, is equal to the voltage of the stator whereas V_{qs} is 0 since the position of the d axis is the same as the maximum stator flux. The rotor of the induction machine is assumed to be short circuited which means that the rotor voltage in the dq reference frame is 0.
4. A motor convention is used which means that the power is positive when the machine consumes power.

On the whole, taking into consideration the theory of the induction machines, the machine equation in the SFO reference frame are the following:

$$\frac{\omega_e}{\omega_b} \cdot \psi_{ds} = v_{qs} - R_S \cdot \frac{X'_{rr}}{D} \cdot \psi_{qs} + \frac{R_S \cdot X_m}{D} \psi'_{qr}$$

$$\left[\frac{\omega_e^2 D^2 + \omega_b^2 R_S^2 (X'_{rr})^2}{\omega_e \omega_b D^2}\right] \psi_{qs} == $$
$$-v_{ds} + \frac{R_S (X'_{rr})}{D} \frac{\omega_b}{\omega_e} v_{qs} + \frac{R_S^2 (X'_{rr})}{D} \frac{\omega_b}{\omega_e} \psi'_{qr} - \frac{R_S X_m}{D} \psi'_{dr},$$

$$\frac{p}{\omega_b} \psi'_{qr} = v'_{qr} - \frac{R'_r X_{ss}}{D} \psi'_{qr} - \left(\frac{\omega_e - \Omega_{gen}}{\omega_b}\right) \psi'_{dr} + \frac{R'_r X_m}{D} \psi_{qs}$$

Where: $p = \frac{d}{dt}$, $\omega_e = \omega_s = 2\pi f_e$, the synchronous speed of the induction machine,

$$X_{ss} = X_{s\sigma} + X_m, \quad \bigg| \quad X'_{rr} = X'_{r\sigma} + X_m, \quad \bigg| \quad D = X_{ss} X'_{rr} - X_m^2$$

The reactances are given by the following formulas:

$$X_{s\sigma} = \omega_s L_{S\sigma}, \quad \bigg| \quad X'_{r\sigma} = \omega_s L'_{r\sigma}, \quad \bigg| \quad X_m = \omega_s L_M$$

Where $L_{S\sigma}$ is the stator inductance and $L'_{r\sigma}$ the rotor inductance expressed in the stator circuit.

The mechanical motion equation expressed in the per unit system as well, is included below: $p\Omega_{gen} = \frac{T_e - T_m}{2H}$. The stator currents in terms of q-d axis fluxes can be written as:

$$i_{qs} = \frac{1}{D}(X'_{rr}\psi_{qs} - X_m\psi'_{qr}), \quad \bigg| \quad i_{ds} = \frac{1}{D}(X'_{rr}\psi_{ds} - X_m\psi'_{dr})$$

While the rotor currents (with respect to the stator) are determined by the equations below:

$$i'_{qr} = \frac{1}{D}(-X_m\psi_{qs} + X_{ss}\psi'_{qr})$$

$$i'_{dr} = \frac{1}{D}(-X_m\psi_{ds} + X_{ss}\psi'_{dr})$$

The electric torque can be computed by two equivalent equations:

$$T_e = \psi_{qs}i_{qs} - \psi_{qs}i_{ds} \text{ OR} \quad\mid\quad T_e = \psi'_{qr}i'_{dr} - \psi'_{dr}i'_{qr}$$

The stator and rotor active power as well as the power produced from the generator are also easily determined:

$$P_S = v_{qs}i_{qs} + v_{ds}i_{ds} \quad\mid\quad P_r = v'_{qr}i'_{qr} + v'_{dr}i'_{dr}$$

$$P = P_S + P_r = v_{qs}i_{qs} + v_{ds}i_{ds} + v'_{qr}i'_{qr} + v'_{dr}i'_{dr}$$

Where $v, i, r, L, \omega, \varphi, t, H, T, K, \psi$ stand for voltage, current, resistance, inductance, speed, flux linkage, time, inertia, torque, damping factor and stator flux respectively. While the suffixes q, d, s, r, k, m and e stand for q and d axis quantity, stator and rotor quantity, damper winding, mechanical, and electrical quantity respectively.

4) Energy Storage
a) Background

The simulation of the operation of an autonomous hybrid system requires among others a mathematic model to describe the function of the battery system. There are various models approaching the properties of the batteries. The most popular are the electrochemical model of Doyle, the Peukert model, the Rakhmatov Model, the KiBaM model, the stochastic version of KiBaM, the stochastic model of Chiaserrini and Rao and the generic battery model developed by Mathworks [27] Each of these models is distinguished by its accuracy and its possibility to determine the lifetime of a particular battery type. The battery model used in the framework of this paper is the generic battery model developed by Mathworks. The main advantage of this model is that the parameters used can be easily obtained from manufacturer data books or from simple testing. What is more, the model is scalable and can be implemented even if curves or other data are not available. In addition to this, the validity of the model has been tested [28].

C. Hybrid System set-up

1) Power Management System

Considering the inherent intermittency of renewable energy techniques which are reliant on irradiance levels and wind speed, it is unlikely that the load will be covered by the Solar and/or Wind Power. Therefore, the use of a controller is of utmost importance so as to determine which strategy should be followed each time. At any time the controller should define whether the back-up generator should kick in or not, and whether the battery will be charging or not.

2) Controller Input and Output data

The controller has only three parameters as inputs, namely the; On/off signal to the backup generator, Power demand of the battery, Amount of power to dump.
Initially, the load/ wind and solar power should be known. After defining these input data, the controller determines whether there is surplus or deficit of power. If there is excess of power then the system checks the state of charge of the battery. If this is higher than 90%, which means that the battery is fully charged, then the excess energy is dumped, otherwise it can be stored. The SOC is an important parameter as it also determines whether the backup generator should be on or off. If the SOC of the battery is less than 25%, which means that the battery becomes depleted then the backup generator should be turned on. To prevent power shortages, the diesel generator should be turned on before the battery becomes fully depleted. It could also be the case, that when the SOC may still decrease even if the diesel generator is running. This happens when the backup generator rated power is below peak load power.

3) Diesel Generator Controller

The purpose of the diesel generator controller is that it will send the signal to the generator to kick in or not, depending on the SOC of the battery, which constitutes its input. When the SOC of the battery drops below a predefined level, the diesel generator is turned on. At the same time, a timer is set, counting back the time that is set as the minimum operation time of the generator.
If this time runs out, then the output signal will be 0 (engine turns off). When the SOC drops below the set limit, the timer will reset. Thus, the controller keeps the output signal at 1 as long as the SOC remains above the SOC "diesel on and the "min time engine" on.

4) Battery Controller

The main control concept of the battery controller is that when the battery reaches the maximum allowed SOC is restricted from charging for 60 seconds. If the power extracted from the renewable energy sources is higher than the load for these times and the SOC is higher than 90% then the time span that the battery is forbidden to discharge becomes even larger. Charging of the battery is done only in case the state of charge drops below 90% for the last 60seconds.

5) Dump of Energy Controller

The flowchart indicates the dump of energy controller as well. It is based on the simple equation:

$$P_{dump} = signal_{dump} * P_{pv} + P_{engine\_produced} + P_{battery\_produced} - P_{load}$$

When the production from renewables exceeds the load and the battery is fully charged, the signal of the dump controller is set to 1 which means that the excess power is dumped. If not, the signal is 0 which means that no power is dumped.

V. OPTIMIZATION OF THE SYSTEM

A. Introduction

In this section, the optimal design for the hybrid PV-Wind-Diesel System will be presented. Such optimization includes specifying optimum dimensions for the system components such as number of solar panels, battery banks and wind turbines. Determining these parameters is challenging due to the complexity of the decision space variables, the objective functions and the conflicting objectives such as cost-Green House Gas (GHG) emissions, cost-unmet load fraction etc. Several methods for determining optimum characteristics have been studied based on numerical [29], probabilistic [30] [31] and heuristic techniques [32] while many software programs have also been introduced. Among them the most popular is the HOMER software introduced by NREL. Due to the conflicting objectives, the Pareto Optimality is usually introduced to show the obtained set of possible solutions. This Pareto frontier is shown after a searching process in which the objectives are evaluated separately. From the solutions obtained, the decision maker can choose the most appropriate for his/her case.

B. Optimization technique proposal

The optimization selected is a parametric based technique. Its operational technique is as follows. The model takes as inputs different types and sizing of the photovoltaic panels and wind turbines. More specifically, for each type of photovoltaic module the inputs of the model are:

The module maximum power under standard conditions	W
The open short circuit current under standard conditions	A
The module open circuit voltage under standard conditions	V
The number of cells connected in series	–
The number of cells connected in parallel	–
The cell working temperature	°C
The open circuit voltage temperature coefficient for a single cell	$V/°C$

While for the wind turbines different sizings and models were selected for the optimization, each of which came with revolutions per minute (RPM), rated wind speed, diameter of the blade, and power curve. For each of those combinations of wind/solar power, the simulation model in Matlab/Simulink is running and the results of it are saved in a struct. After obtaining the results of the model for all the possible combinations, two different functions are running. These are the cost objective function and the fuel emissions function the concept of which is explained detailed below. Finally, the cost and the fuel emissions are presented together in a Pareto frontier graph for each feasible combination of the wind turbines and solar panels.

C. Cost objective function

The objective function studied here is to minimize the total cost of the system during its lifetime. The latter is commonly considered to be the same as the life of the solar panels, as they have longest life time. The typically utilized cost objective functions comprise the net present cost of the investment (NPC) which includes the cost of installing added to the discounted present values of all future costs throughout the life span of the installation. A general equation of the objective function includes the Capital Costs (CC), the annualized Operation and Maintenance (O&M) cost, the Installation Cost, the replacement cost of the wind turbines, the PV modules and the battery banks and the operation cost of the diesel generators throughout the system lifetime. Costs of

inverters, rectifiers, cable costs and control system costs comprise only a small share of the system costs. For that reason in the framework of this thesis, they are not taken into consideration.

$$OF_{cost} = CC + OMC + RC_{batteries} + FC + EC_{CO2}$$

Where:

OF_{cost}	Objective function cost
CC	Capital Cost
OMC	Annual operation and maintenance cost of the hybrid system
$RC_{batteries}$	Cost from the replacement of the batteries during the investment period.
FC	Fuel Costs
EC_{CO2}	CO_2 emissions cost

D. Capital Cost

The investment cost includes the cost of plant, machinery and equipment, and the cost of studies and labor work. In this study, an estimation of specific investment costs per kW for each component of the hybrid plant separately and the values are presented in Table I.

The investment cost in present value, which is the time of the starting of the operation of the system, is formulated by the following equation:

$$CC = P_{wind} \cdot WC_{\frac{\epsilon}{kW}} + P_{PV} \cdot PVC_{\frac{\epsilon}{kW}} + Q_{battery} \cdot QC_{\frac{\epsilon}{kWh}}$$
$$+ P_{conv} \cdot CONC_{\frac{\epsilon}{kW}} + VC$$

E. Operation and Maintenance Costs

The annual operating and maintenance costs (O&M) include expenses for the labor salaries, maintenance and repair expenses of batteries, central inverter of the wind turbines and photovoltaic panels, and the annual premium for the facilities.

In the calculations the annual O & M costs are generally considered to be stable and equal to a percentage of the investment cost. These costs are considered to increase only by following the rate of inflation.

$$OMC = OMC_{\%CC} \cdot CC \cdot (\frac{1+fr}{1+ir})^t$$

F. Replacement cost of the batteries

Replacement cost of the batteries, include the cost for their replacement, during the lifetime of the investment, since the typical life of battery is less than twenty years. For the future purchase cost of the batteries it is considered, given the technological maturity of the technology, that the annual cost reduction will not be significant and that will lie within the inflation levels. As such, the assumption that the value of the battery in current values remains constant throughout the economic life of the investment is made. Under these assumptions, the present value of replacement cost in one year t* is calculated from the equation below, where it should be pointed that the calculation is done only for the years t* during which the replacement of the batteries should be done.

$$RC_{battery} = (Q_{battery} \cdot QC_{\frac{\epsilon}{kWh}}) \cdot \frac{1}{(1+ir)^{t^*}}$$

TABLE I
ESTIMATION OF SPECIFIC INVESTMENT COSTS PER KW FOR EACH COMPONENT OF THE HYBRID SYSTEM

WC	Unit Investment Cost of Wind Turbines	0.06	€/W
PVC	Unit Investment Cost of the Photovoltaics	0.52	€/W
QC	Unit Investment Cost of the Batteries	0.15	€/Wh
CONC	Unit Investment Cost of the Converter	0.05	€/W
OMC	O&M Cost as percentage of CC	2	%
Ir	Discount Rate	0.06	%
N	Lifetime of the investment	20	Years
Fr	General inflation rate	2	%
fr_{fuel_cost}	Inflation rate of fuel	3	%
LIF	Annual load growth rate	3	%
$fuel_{cost\,20}$	Fuel cost in year 2013	1.45	€/W
emission:	Cost of emissions right in 2013	25	€/tn
EF	CO2 emissions rate/liter diesel coefficient	2.77 2	Kg/lt
d_{diesel}	Diesel Density	0.85	Kg/lt
VC	Various costs (% of the investment cost)	10	%
$OMC_{gener\iota}$	O%M cost of the Diesel Generator	0.01 6	€/kWh

As far as the maintenance cost of the system in concerned, there is a wide range reported and it was also observed that O&M costs do not necessarily decrease when increasing the project size. Older installations tend to have higher yearly O&M costs. Newer installations are better designed and have lower installation and lifetime O&M costs. A comparison between solar and wind shows that solar systems require less yearly maintenance and so the O&M costs are lower in that case.

G. Fuel Costs

To calculate the total cost of fuel the simulation should be made for all the lifespan of the investment (N =20 years), in which an annual growth rate of the load should be taken into consideration. However, to reduce the time complexity of the simulation the two decades of the investment were divided into 4 five years and each is representative for 5 years. For more reliable results it is considered that both the load and the fuel consumption increase by the constant annual growth rate of the load. During the optimization process, in any step the parametric optimization runs the simulation 4 times, once for every five years and the total cost of the fuel is the sum of the cost of each five years. Fuel costs each five years is the sum of the costs during each year and this depends on the annual consumption and fuel cost, and the

rate of inflation of fuel prices.

By combining these assumptions the fuel cost equation is formulated as following:

$$FC_i = \sum_{j=sy}^{5i-1} [fuel_{consumption_{sy}}(1+LIF)^{j-sy} \cdot fuel_{cost\,2013}(\frac{1+fr_{fuel\_cost}}{1+ir})$$

A liter of diesel fuel generates $2.772kg$ of CO_2, so the CO_2 direct emissions for the diesel generator are taken as $2.772 \cdot fuel_{consumption}(kg - CO_2)$. To calculate the cost of CO2 emission rights the same 4decades-concept is applied. Moreover, it is assumed that the cost of rights emitted per tonne of CO2 is kept constant throughout the decades.

$$Cost\_CO_2 rights = \sum_{j=sy}^{5i-1} [fuel_{consumption_{sy}}(1+LIF)^{j-sy} \cdot \frac{EF}{d_{diesel}} emissions\_cost_{2013}(\frac{1}{(1+ir)^j})$$

H. Results

Upon applying the multi-objective optimization approach suggested in section V.B, the configuration of the system elements that optimize both the total levelized cost and the CO_2 emissions are obtained. In Fig. 9, the set of possible trade-offs are shown.

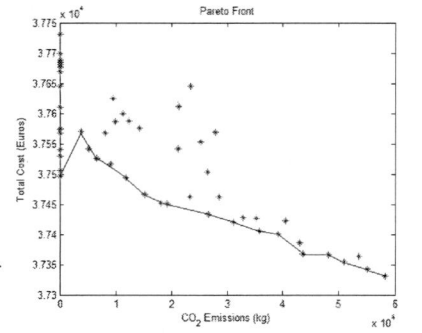

Fig. 9 Pareto front of the case study

The points, which lie on the Pareto front represent the best feasible trade-offs. It is worth mentioning that the solution with the least cost is close to the solution with the least emissions level.

In the Table II, the system configurations that achieve minimum cost and CO_2 emissions are shown.

TABLE II
SYSTEM CONFIGURATION FOR COST AND EMISSIONS MINIMIZATION

Objective	Number of PV Panels	Number of Wind Turbines	Power of each Wind Turbine
Cost Minimization	0	1	400
Emissions Minimization	0	3	250

The wind-battery-diesel combination achieves the best results for emissions and cost while the photovoltaic panels were not selected. This is logical outcome as the cost per energy unit delivered for the PV modules is higher than for the wind turbines, as shown in the Table

IV.

TABLE IV
COST AND CO_2 EMISSIONS PER KWH DELIVERED BY EACH TYPE OF
POWER GENERATION/ STORAGE UNITS

Power generation/ storage unit type	Cost/ energy unit delivered ($€/kWh$)	Emissions/ energy unit delivered ($kg - CO_2/kWh$)
Diesel Generator	0.21	0.6
PV Module	0.11	0.13
Wind Turbine	0.03	0.02
Battery	0.12	0.02

VI. CONCLUSIONS

Power supplying of offshore platforms in Arctic climates can be made feasible with hybrid energy systems. The combination of solar and wind power together with conventional diesel energy supply as a back-up source was considered. The configuration used for the setup of the system was discussed and the simulation method for each of the sources was presented. Chemical storage was considered for energy storage as it presents the best trade-off for efficiency, reliability, maturity and cost. From the various chemical storage types, NiCd batteries were selected based on various criteria. Finally the optimization of the system was presented for specifying optimum sizing for the system components such as number of solar panels, battery banks and wind turbines. The configuration of the system elements that optimize both the total levelized cost and the CO_2 were presented in a graph.

REFERENCES

[1] USGS, "U.S Geological Survey World Petroleum Assessment 2000- Description and Results," 2000.

[2] Shell, "Arctic exploration and production," available online at: http://www.shell.com/home/content/future_energy/meeting_demand/arctic/exploration_production/, 2012.

[3] Lu L, Burnett J. Yang HX,., 2003, pp. 1813-24.

[4] NASA, "Nasa surface Meterology and Solar Energy," Atmosperic Science Data Center , Online available at: https://eosweb.larc.nasa.gov/cgi-bin/sse/grid.cgi?&=&num=333160&lat=69.4&hgt=100&submit=Submit&veg=17&email=kalogera.maria@gmail.com&sitelev=-999&p=grid_id&step=2&lon=152.1 2013.

[5] NREL, "USA National Renewable Energy Laboratory ," Available online at: http://rredc.nrel.gov/solar/old_data/nsrdb/1991-2005/tmy3/by_state_and_city.html 2013.

[6] V. Idichandy, C. Ganapathy Lalu Mangal, "Structural monitoring of offshore platforms using impulse and relaxation response," in Applied Ocean Research.: ISSN 0141-1187, 10.1016/0141-1187(85)90014-8. (http://www.sciencedirect.com/science/article/pii/0141118785900148), 1985, pp. 14-23.

[7] PTT Explor and Prod PLC Roongwit Rongsopa, "Hybrid Power Generation for Offshore Wellhead Platform: A Starting Point for Offshore Green Energy," in International Petroleum Technology Conference , Bangkok, 2012.

[8] L. Anton, G. Bopp, K. V. Dohlen K. Preiser, "The hybrid energy system in Kaysersberg," in Photovoltaics, Hydropower and Gas Generator, Barcelona, Spain, 1997, pp. 1106-1109.

[9] Leake E. Weldemariam, Evert Raijen, Praveen Kumar Pavol Bauer, "Connecting Topologies of Stand Alone Micro Hybrid Power Systems," in 2011 VDE VERLAG GMBH. Berlin, Germany: ISBN 978-3-8007-3344-6, 2010, ch. ISSN 2191 3358, pp. 304-310.

[10] Leake E. Weldemariam, Evert Raijen P. Bauer, "Stand Alone Micro Grids," in 33rd International Telecommunications Energy Conference (INTELEC), Amsterdam, The Netherlands, 9-13 October 2011.

[11] Poul Sørensen, Lars H. Hansen, Henrik Bindner Anca D. Hansen, "Models for a Stand-Alone PV System," Risø National Laboratory, Roskilde, 2000.

[12] International Electrotechnical Commission, "Photovoltaic Devices Part 1: Measurement of Photovoltaic Current-Voltage Characteristics," Geneva, Switzerland , 2006.

[13] International Electrotechnical Commision, "Photovoltaic Devices Part 3: Measurement Principles for Terrestrial Photovoltaic (PV) solar Devices with Regerence Spectral Irradiance Data ," Geneva, Switzerland , 2008.

[14] J. J. Schoeman and J. D. van Wyk, "A simplified maximal power controller for terrestial photovoltaic panel arrays," in IEEE Power Electron. Specialists Conf., 1982, pp. 361-367.

[15] Poul-Erik Morthorst, Shimon Awerbuch Soren Krohn, "The economics of Wind Energy," 2009.

[16] Tractebel Energy Engineering, Riso National Laboratory, Kvaerner Oil and Gas, Energi & Miljoe Undersoegelser Garrad Hassan, "Offshore Wind Energy: Ready to Power a Sustainable Europe, Final Report," 2001.

[17] Timo Laakso, "Wind Energy Projects in Cold Climates," FInland , 2005.

[18] David-Pieter Molenaar Jan van der Tempel, "Wind Turbine Structural Dynamics – A Review of the Principles for Modern Power Generation, Onshore and Offshore," Wind Engineering, vol. 4, pp. 211-220, 2002.

[19] A. Jalilvand, R. Noroozian B. Sedaghat, "Design of a multilevel control strategy for integration of stand-alone wind/diesel system," International Journal of Electrical Power & Energy Systems, vol. 35, no. 1, pp. 123-137, February 2012.

[20] FELIX A AUTOR FARRET M. GODOY AUTOR SIMOES, Renewable Energy Systems: Design and Analysis with Induction Generators/ Volume 4 of Power Electronics and Applications Series, illustrated ed.: CRC PressINC, 2004, 2004.

[21] R.L Lawrence, Principles of Alternating Current Machinery , 2nd ed. The University of Michigan, Michigan : McGraw-Hill Book Company, Inc., 1920, 2006.

[22] S.J. Chapman, Electric Machinery Fundamentals, 5th ed. New York, USA: McGraw-Hill Companies,Incorporated, 2011, 2011.

[23] Alexander Suss Langsdorf, Theory of alternating-current machinery. the University of Wisconsin - Madison: McGraw-Hill book company, inc., 1937, 2007.

[24] M.G Simões F.A.Farret, Integration of Alternative Sources of Energy.: John Wiley & Sons, 2006, 2006.

[25] R. IBTIOUEN, S. MEKHTOUB, O. TOUHAMI, N. TAKORABET A. NESBA, "Autonomous Induction Generator/Rectifier as Regulated DC Power Supply for Hybrid Renewable Energy Systems," in 7th WSEAS Int. Conference on Mathematical Methods and Computational Techniques in Electrical Engineering, Sofia , 27-29/10/05, pp. 162-167.

[26] Le Luo, "Control and Modeling of Diesel Generator Set in Electric Propulsion Ship," Information Technology and Computer Science, vol. 2, pp. 31-37, March 2011.

[27] B.R. Havenkort M.R. Jongerden, "Battery Modeling," University of Twente, Twente, The Netherlands , Technical Report 2012.

[28] J.A.Oliver, I. Reglero, J.A.Cobos R.Prieto, "Generic Battery Model based on a Paramaetric Implementation," Universidad Potecnica de Madrid / Centro de Electronica Industrial, Madrid, 2010.

[29] Nehrir M. H., Venkataramanan G., Gerez V. Kellogg W., "Optimal unit sizing for a hybrid wind/photovoltaic generating system," Electical Power Systems research, pp. 35-38, 2010.

[30] and Salameh Z. M. Borowy B. S., "Optimum Photovoltaic Array Size for a Hybrid Wind/PV System ," IEEE Transactions on Energy Conversion, vol. 3, pp. 484-488, Sept 1994.

[31] Salameh Z. M., Borowy B. S., "Methodology for Optimally Sizing the Combination of a battery bank and pv array in a wind/pv hybrid system," IEEE Transactions on Energy Conversion, vol. 2, no. 11, pp. 367-375, 2010.

[32] S. M. Hakimi S. M. Moghaddas, "Optimal sizing of a stand-alone hybrid power system via particle swarm," Renewable Energy, vol. 34, pp. 1855-1862, 2010.

978-1-4799-2706-7/14 $31.00 © 2014 IEEE

The 2014 International Power Electronics Conference

Active Damping Control of LLCL filters for Three-level T-type Grid Converters

Payam Alemi and Dong-Choon Lee

Dept. of Electrical Eng. Yeungnam University,
280, Daehak-Ro, Gyeongsan, Gyeongbuk, Korea
E-mail: payamalemi@ynu.ac.kr and dclee@yu.ac.kr

Abstract—In this paper, an active damping control scheme using the PR (proportional and resonant) regulator is proposed to eliminate the resonance for the grid-connected T-type three-level PWM converter systems with LLCL filter by which the total inductance can be reduced compared with the conventional LCL filters. The simulation results have verified the validity of the proposed control algorithm for a 25-kW three-phase T-type converter.

Keywords—grid-connected converter, LLCL filter, PR control.

I. INTRODUCTION

Multilevel PWM converters are being increasingly used for high power applications [1]. It is known that the T-type three-level converter is more efficient than other three-level converter topologies for the medium switching frequency range (8kHz $< f_{sw} <$ 20kHz) [2], [3]. To reduce the current harmonics around the switching frequency (2-15 kHz), a high value of input inductance is needed. However, a large inductance deteriorates the system dynamics. Due to the better attenuation performance of switching frequency harmonics, the LCL filter is preferred to the L filter for grid-connected PWM converters [3]-[4]. However, the LCL filter has a resonance problem. The LCL filter requires more complex current control strategies to maintain the system stability. More attention has been paid to the possible instability of the system caused by the zero impedance at the resonance frequency. To suppress the resonance in the LCL filter, active damping and passive damping methods have been proposed [4]-[10]. Although the passive damping method has advantages like reliability and simplicity, it has also a disadvantage of additional power loss [3], [5]. On the other hand, many different active damping methods have been discussed in the literature [6]-[8]. The active damping methods are more preferable, where the control algorithm is modified to stabilize the system without any additional power loss.

To decrease the total inductance of the LCL filter, a new LLCL filter has been proposed in [9], which is used for a lower grid-side inductor. There exists also a resonance problem in the LLCL filter. A passive damping method to suppress the resonance in the LLCL filter has been proposed for the single-phase power converter [10], where the power loss can be an issue in the case of high power range.

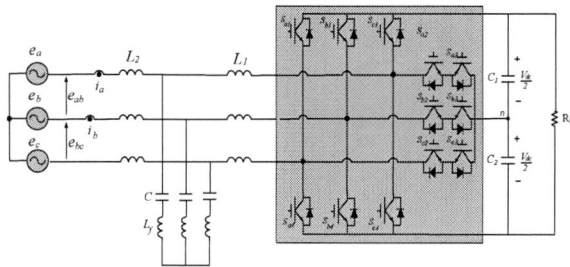

Fig. 1. Grid-connected three-level T-type converter with LLCL filter.

In this paper, an active damping scheme using the PR controllers is proposed to damp out the resonance in the LLCL filter. The PR controllers gives an infinite gain at a specified resonant frequency, at which the steady-state error is eliminated. So, the resonance phenomenon can be damped out.

First, the parameters of the LLCL filter are designed and then the performance of the PR controllers is investigated. The system modeling and control structures are presented in the subsequent sections. Finally, the simulation and experimental results are presented to verify the effectiveness of the proposed control algorithm.

II. SYSTEM DESCRIPTION

The three-level T-type AC/DC PWM converter is shown in Fig.1, which is connected to the grid through the LLCL filter. The grid is modeled as an ideal sinusoidal voltage source without line impedances. A resistive load is connected to the DC output terminal of the converter. The system parameters are listed in Table I.

A. Converter Modeling

For the PWM converter, the voltage equations are expressed in a synchronous reference frame as

$$e_{de} = Ri_{de} + L\frac{di_{de}}{dt} - \omega Li_{qe} + v_{de} \tag{1}$$

$$e_{qe} = Ri_{qe} + L\frac{di_{qe}}{dt} + \omega Li_{de} + v_{qe} \tag{2}$$

where,

e_{de}, e_{qe} : *dq*-axis grid voltages,

i_{de}, i_{qe} : *dq*-axis converter currents,

978-1-4799-2706-7/14 $31.00 © 2014 IEEE   1201

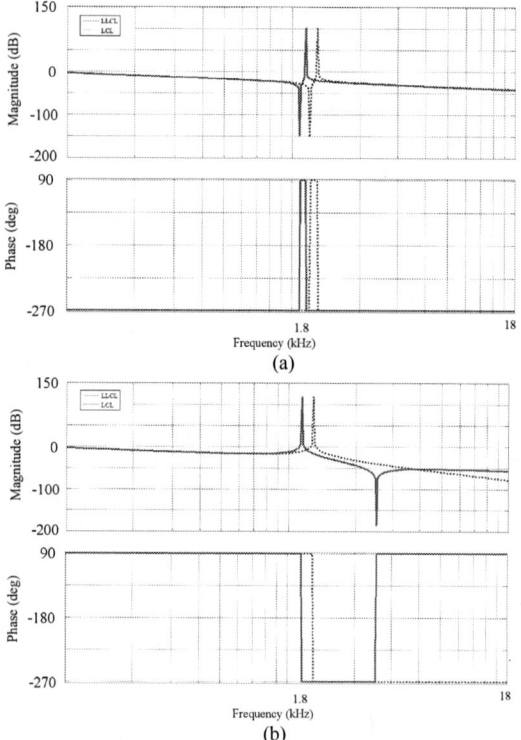

(a)

(b)

Fig. 2. Bode plots of the transfer functions in LCL and LLCL filter. (a) $G_1(s) = I_c(s)/V_i(s)$, (b) $G_2(s) = I_g(s)/V_i(s)$

v_{de}, v_{qe} : dq-axis converter voltages,

ω : angular frequency of the grid voltage,

R : resistance of inductors, ($R_1 + R_2$),

L : sum of grid- and converter-side inductors ($L_1 + L_2$).

By aligning the q-axis of the synchronous reference frame to the grid voltage, $e_{de} = 0$. The DC-link voltage dynamics can be expressed by

$$C \frac{dv_{dc}}{dt} = i_{DC} - i_L \qquad (3)$$

where,

v_{dc} : DC-link voltage,

i_{DC} : converter side DC current,

i_{DC} : load current,

in which the losses in the filter and the converter are neglected.

B. Modeling of LLCL Filter

In the LLCL filter, the inductor (L_f) in a series with the capacitor (C) is designed to result in a zero-impedance in the capacitor branch at the switching frequency [5]. So, the switching-frequency related harmonic currents do not flow into the grid, by which the grid-side inductor size, L_2, can be reduced. The value of L_f is much lower than L_1 and L_2. Since the grid is an ideal sinusoidal voltage source, the transfer functions of the grid current and the converter current to the inverter voltage are expressed, respectively, as

$$G_2(s) = \frac{I_c(s)}{V_i(s)} = \frac{(L_2 + L_f)Cs^2 + 1}{(L_f C(L_1 + L_2) + L_1 L_2 C)s^3 + (L_1 + L_2)s} \qquad (4)$$

$$G_1(s) = \frac{I_g(s)}{V_i(s)} = \frac{L_f Cs^2 + 1}{(L_f C(L_1 + L_2) + L_1 L_2 C)s^3 + (L_1 + L_2)s} \qquad (5)$$

From (4) and (5), the resonance frequency is given by

$$f_{res} = \frac{1}{2\pi} \sqrt{\frac{L_1 + L_2}{(L_1 \cdot L_2 \cdot C) + (L_1 + L_2)L_f C}} \qquad (6)$$

The frequency behaviors of the converter current and the line current to the converter voltage are illustrated in Fig. 2(a) and (b), respectively. The LLCL filter has very low impedance around the switching frequency, which is due to the appropriate design of L_f in series with the capacitor.

C. Parameter Design of LLCL Filter

In the design of the filters, it is needed to consider some limitations on the parameter values as follows [6], [8]:

1) The required current ripple in the converter side decides the converter-side inductor L_1.

2) The grid-side inductor L_2 is limited based on the IEEE 519-1992 standard recommendation, in which the harmonic currents higher than the 35th – order component should be less than 0.3% of the fundamental component.

3) The lower limit of the total inductance should be in the following range:

$$\frac{V_{dc}}{8i_{rp_{Max}} \cdot f_{sw}} \leq L \qquad (7)$$

where,

$i_{rp_{Max}}$: maximum ripple magnitude in the rated current

f_{sw} : switching frequency.

4) The capacitor value C is determined based on the reactive power which is absorbed under the rated condition, in which the upper limit is expressed as

$$C = \frac{5\% P_{Rated}}{V_g^2 \omega} \qquad (8)$$

where

V_g : grid fundamental RMS voltage,

ω : fundamental frequency in radian per second.

5) The series inductor L_f is designed based on the LLCL filter capacitor. For the zero impedance at the switching frequency,

$$L_f = \frac{1}{\omega_{sw}^2 \cdot C} \qquad (9)$$

The 2014 International Power Electronics Conference

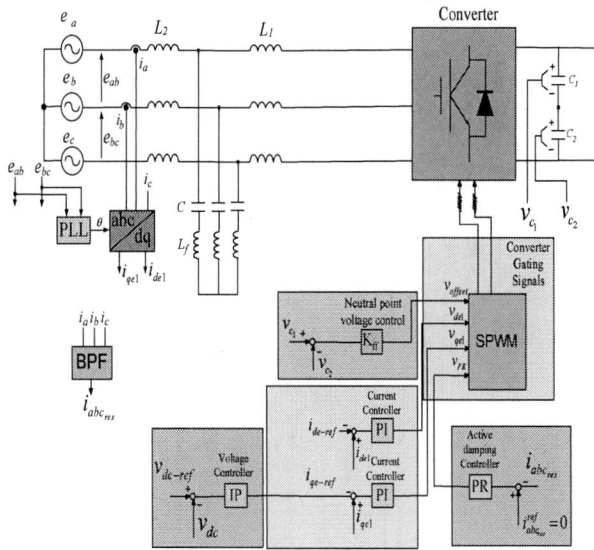

Fig. 3. Control block diagram of the grid-connected PWM converter with LLCL filter.

6) The resonance frequency should be in the range between ten times the fundamental frequency and one half of the switching frequency.

III. CONTROL STRUCTURE

Fig. 3 shows the control block diagram of the grid-connected T-type converter with the LLCL filter.

A. DC Voltage and Input Current Control

The integral-proportional (IP) controllers are used for an outer DC-link voltage control. The proportional-integral (PI) controllers are used for the inner dq–axis current control. The q-axis current reference is generated from the DC-link voltage controller. The d-axis current is controlled to zero for a unity power factor. The PI current controllers are equipped with the anti-wind-up function. The outputs of the dq-axis current controllers can be expressed in (10) and (11), respectively,

$$v_{de1-ref} = k_{p1}(-i_{de1-ref} + i_{de1}) + k_{i1}\int((-i_{de1-ref} + i_{de1})$$
$$+\frac{1}{k_{p1}}(-v_{de1-ref} + v_{de1}))dt + E_{d1} - L\omega i_{qe1} \qquad (10)$$

and

$$v_{qe1-ref} = k_{p2}(-i_{qe1-ref} + i_{qe1}) + k_{i2}\int((-i_{qe1-ref} + i_{qe1})$$
$$+\frac{1}{k_{p2}}(-v_{qe1-ref} + v_{qe1}))dt + E_{q1} + L\omega i_{de1} \qquad (11)$$

where,

K_{p1}, K_{p2} : proportional gains,

K_{i1}, K_{i2} : integral gains,

i_{de-ref}, i_{qe-ref} : dq-axis current references.

The PI gains are given, respectively, by [11]

$$K_{p1}, K_{p2} = L\omega_c \qquad (12)$$

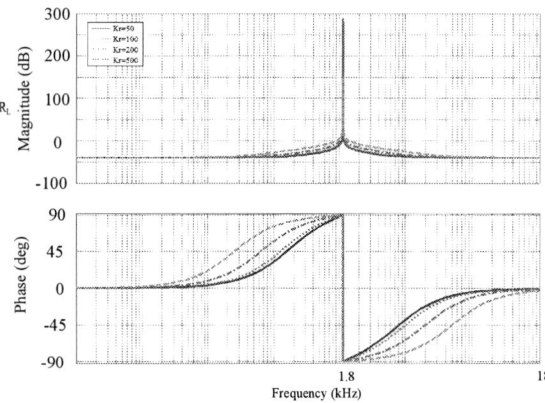

Fig. 4. Bode plots of the resonant controller for different gains (K_r).

$$K_{i1}, K_{i2} = R\omega_c \qquad (13)$$

where ω_c is the cut-off frequency of the current control loop (400 rad/s).

The neutral-point voltage control of the three-level converter is added, which utilizes the offset voltage that is found from the difference between the upper and lower capacitor voltages in the DC-link [12].

B. Active Damping Control

The function of the PR controllers is controlling the AC reference signals, and therefore the steady-state errors can be eliminated at the specified resonant frequency. The transfer function of the PR controllers is given by

$$G_{PR}(s) = K_p + \frac{K_r s}{s^2 + \omega_{res}^2} \qquad (14)$$

where K_p and K_r are the proportional and resonant gains, respectively, and ω_{res} is the resonance frequency. The magnitude and phase characteristics of the open-loop transfer function for the PR controllers with respect to the different resonant gains K_r are shown in Fig. 4, where the resonance frequency is set to 1.8 kHz. As illustrated, the higher resonant gain K_r leads to the wider bandwidth.

In this work, the resonance frequency of the LLCL filter is lower than a half of the switching frequency (4 kHz). The band-pass-filer (BPF) with the cut-off frequency of 1.8 kHz and the bandwidth of 200Hz are applied to extract the resonant component of the grid currents.

IV. SIMULATION RESULTS

The simulation has been performed under the condition as listed in Table I and II for the T-type three-level PWM converter and filter parameters, respectively. The converter-side inductor is 1.2 mH. The filter capacitor is 35 μF. The series inductor is 45 μH.

Fig. 5(a) shows the resonance phenomenon without any active damping control. However, it is damped out by the active damping method as shown in Fig. 5(b).

978-1-4799-2706-7/14 $31.00 © 2014 IEEE

TABLE I
PWM CONVERTER PARAMETERS

Parameters	Value
Power rating	25kW
Input AC voltage	$380\,V_{rms}/60\,Hz$
DC-link voltage	600 V
Switching frequency	4 kHz
DC-link capacitance	2,500 μF

TABLE II
FILTER PARAMETERS

Parameters	LLCL		LCL	
	L_1	L_2	L_1	L_2
Converter-side inductor-L_1(mH)	1.2	0.2	1.2	0.8
Grid-side inductor-L_2 (mH)	0.8			0.2
Capacitor (μF)	35		35	
Series inductor (μH)	45		0	

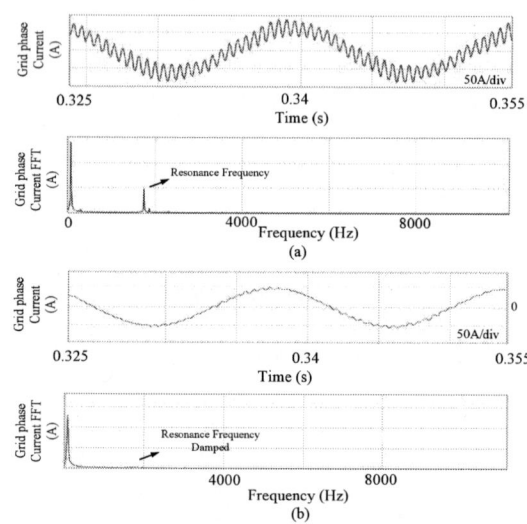

Fig. 5. Performance of T-type PWM converter with LLCL filter. (a) without active damping, (b) with active damping.

The grid and converter currents and the FFT spectrums are shown in Fig. 6 for the different cases of the LCL and LLCL filter. The grid current harmonics should meet the IEEE standard 519-1992. For a weak grid the, $I_{SC}/I_L < 20$ and the THD should be less than 5%, whereas for a stiff grid, $I_{SC}/I_L = 100 \sim 1000$ and the THD should be less than 15% [13]. In the case I of the LCL filter (L_1=1.2 mH, L_2= 0.8 mH), the magnitude of the dominant harmonic component in the grid current is lower than 0.11 A (0.26% of the fundamental component). The total harmonic distortion (THD) factors are listed in Table III, which is 3.21% (Fig. 6(a)). Fig. 6(b) illustrates the case II of the LCL filter (L_1=1.2 mH, L_2= 0.2 mH), where the dominant harmonic magnitude in the grid current is lower than 0.83 A (1.43 % of the fundamental current) with THD (7.02%). It is obvious that the case II of the LCL filter doesn't meet the requirement of IEEE standard. However, in the LLCL filter, the switching frequency ripples are suppressed by adding the inductor in series with the capacitor, which makes the zero impedance in the capacitor branch at the switching frequency. So the grid-side inductor will be sized based on the double switching frequency. In the case III of the LLCL filter (L_1 =1.2 mH, L_2= 0.2 mH, and L_f= 0.045 mH) shown in Fig. 6(c), the magnitude of the dominant harmonic in the grid current is lower than 0.1 A (0.21 % of the fundamental component) in which the THD is 2.64%. The LLCL filter with a lower converter-side inductor is also investigated in the case IV (L_1 =0.8 mH, L_2= 0.2 mH, and L_f= 0.045 mH) in which the dominant harmonic component in the grid current is lower than 0.17 A (0.29 % of the fundamental component) and the THD is 4.2% (Fig. 6(d)). It is shown that the LLCL filter can meet the IEEE standard even in the low converter-side inductor (Table. III).

Fig. 7 shows the filter performance in the case of load

application. The 25-kW resistive load is applied, where the grid current is increased up to 58-A. The current control performance is shown in Fig. 7(b). The DC-link voltage is fluctuated at the points where the converter q-axis current changes abruptly. The undershoot in the DC-link voltage is lower than 10% as shown in Fig .7(c). The converter current is illustrated in the Fig. 7(d), which is more distorted than the grid current.

Fig. 8 shows the performance of the converter in the case of unbalanced grid voltages, where one of the grid voltages is decreased by 20% as shown in Fig. 8(a). The grid current is shown in Fig. 8(b), where the switching ripples are met from the IEEE standard of 0.3%. The fluctuations in the DC-link voltage do not exceed the 10%, which is acceptable even at this condition (Fig. 8(c)).

V. EXPERIMENTAL RESULTS

A brief experiment has been performed for a 3-kW converter system. Under this condition, the LLCL filter is redesigned by the same design procedure described in the previous section (Table. IV).

Fig. 9 shows the performance of PWM converter with LLCL filter. The waveforms of the three-phase grid currents are shown in Fig. 9(a), where the filter resonance phenomenon is damped by the PR controllers. The THD of the currents is 2.72%, which is lower than the standard limit. The converter input currents are illustrated in Fig. 9(b) which are more distorted than the grid currents.

Fig. 10 illustrates the FFT spectra of currents. The FFT spectrum of the grid current is illustrated in Fig. 10(a), in which the magnitude of the switching-frequency-related harmonics is lower than 0.025 A (0.20 % of the fundamental component). Fig. 10(b) illustrates the FFT

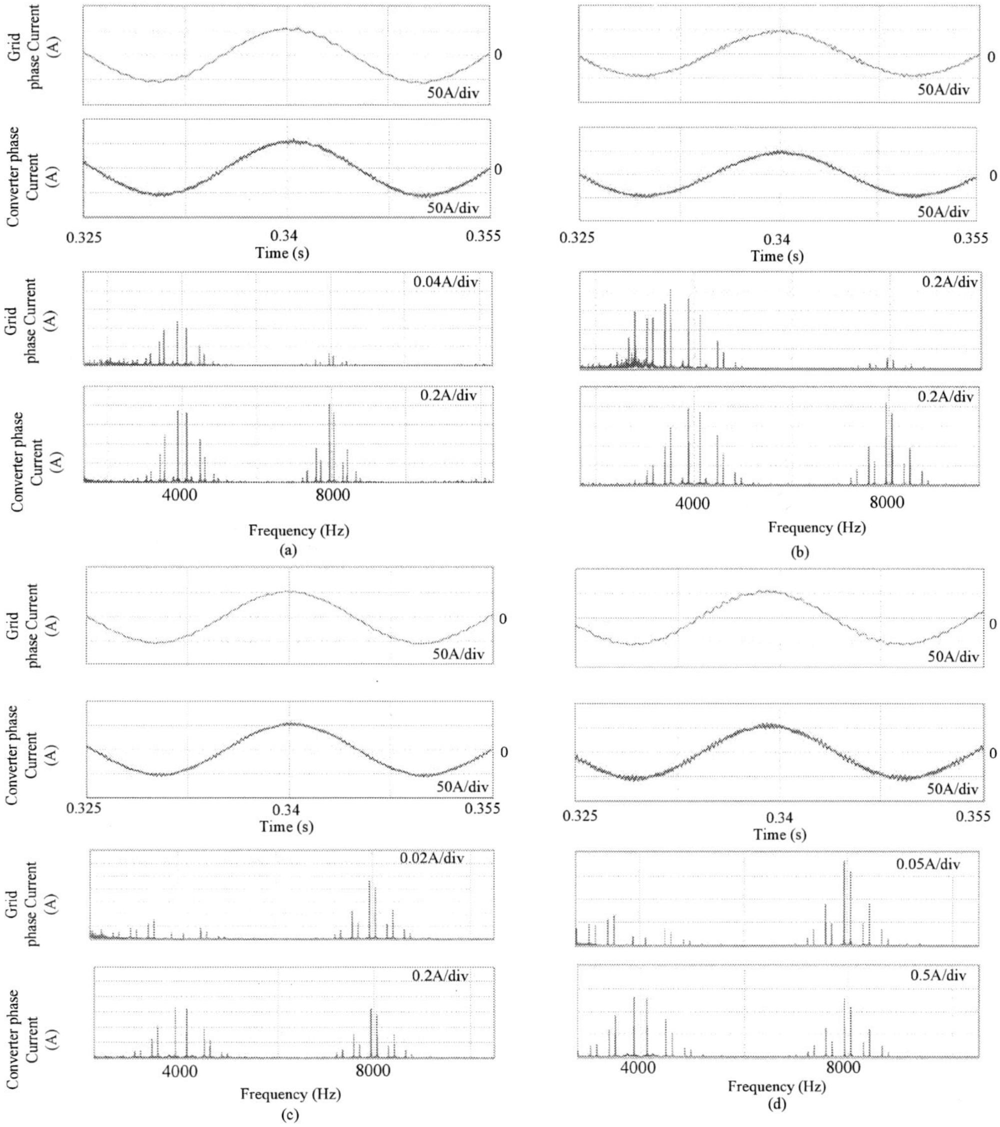

The 2014 International Power Electronics Conference

Fig. 6. The grid and converter currents with FFT spectra in different cases. (a) LCL filter: L₁=1.2 mH, L₂=0.8 mH. (b) LCL filter: L₁=1.2 mH, L₂=0.2 mH. (c) LLCL filter: L₁=1.2 mH, L₂=0.2 mH, Lf=0.045 mH. (d) LLCL filter: L₁=0.8 mH, L₂=0.2 mH, Lf=0.045 mH.

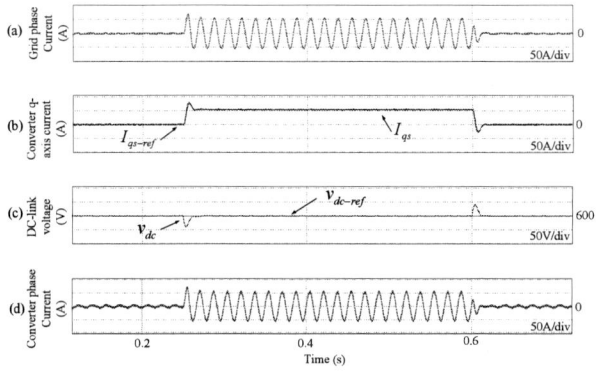

Fig. 7. Performance of PWM converter with LLCL filter. (L1 =1.2 mH, L2 = 0.2 mH, and Lf = 0.045 mH).

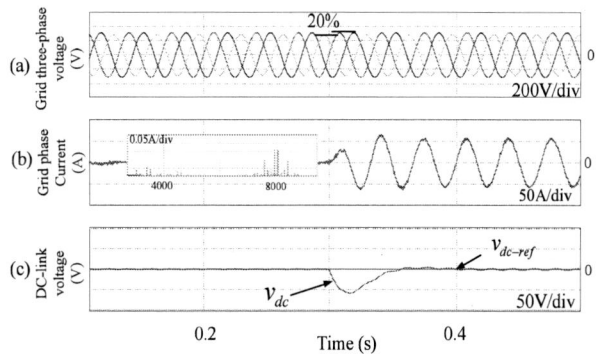

Fig. 8. T-type PWM converter performance with LLCL filter in the case of 20% drop in grid voltage. (L1 =1.2 mH, L2 = 0.2 mH, and Lf = 0.045 mH)

978-1-4799-2706-7/14 $31.00 © 2014 IEEE

TABLE III
THD AND SIDEBAND HARMONICS MAGNITUDE ON THE GRID CURRENT IN DIFFERENT CASES

Inductor size (mH)	Case I: LCL		Case II: LCL		Case III: LLCL			Case IV: LLCL		
	L_1	L_2	L_1	L_1	L_1	L_2	L_f	L_1	L_2	L_f
	1.2	0.8	1.2	0.2	1.2	0.2	0.045	0.8	0.2	0.045
THD (%)	3.21		7.02		2.64			4.2		
Dominant harmonic magnitude per fundamental (%)	0.26		1.43		0.21			0.29		

TABLE IV
DESIGN OF PWM CONVERTER CONNECTED LLCL FILTER (3 KW)

PWM converter		LLCL filter	
Parameters	Value	Parameter	value
Power rating	3kW	L_1	3 mH
Input AC voltage	220 V_{rms} / 60 Hz	L_2	0.4 mH
DC-link voltage	340 V	C	20 μF
Switching frequency	5 kHz	L_f	50 μH

spectrum of the converter current, which contains higher harmonic components.

VI. CONCLUSION

In this paper, the active damping method using the PR controllers for the LLCL filter connected to the T-type three-level converter has been proposed. The LLCL filter parameters are decided by the design procedure to have a resonance frequency of filter in the acceptable range. In the LLCL filter, the grid-side inductor has been reduced by 75% (from 0.8-mH to 0.2-mH), compared with the LCL filter, with which the THD and dominant harmonic component of the grid current have met the IEEE standard. The FFT spectra of the grid current have shown that the switching ripples of the grid currents are acceptable with the LLCL filter even in the case of 33% (from 1.2-mH to 0.8-mH) reduction of the converter-side inductor. The undershoot of the DC-link voltage is lower than 10% at the load transient in the grid unbalance condition. The damping effect of the PR controllers on the LLCL filter has been verified by simulation and experimental results.

ACKNOWLEDGMENT

This research was supported by Basic Science Research Program through the National Research Foundation of Korea (NRF) funded by the Ministry of Education, Science and Technology (2012R1A1A4A01).

REFERENCES

[1] J. Rodriguez, J.-S. Lai, and F. Z. Peng, "Multilevel inverters: A survey of topologies, controls, and applications," *IEEE Trans. on Ind. Applicat.*, vol. 49, no. 3, pp. 724-738, Aug. 2002.

[2] M. Schweizer, and J. W. Kolar, "Design and implementation of a highly efficient three-level T-type converter for low-voltage applications," *IEEE Trans. on Power Electron.*, vol. 28, no. 2, pp. 899-907, Feb. 2013.

[3] M. Liserre, F. Blaabjerg, and S. Hansen, "Design and control of an LCL-filter-based three-phase active rectifier," *IEEE Trans. on Ind. Applicat.*, vol. 41, no. 5, pp. 1281-1291, Sep/Oct. 2005.

[4] E. Twining and D. G. Holmes, "Grid current regulation of a three-phase voltage source inverter with an LCL input filter," *IEEE Trans. on Power Electron.*, vol. 18, no. 3, pp. 888-895, Feb. 2003.

[5] R. P-Alzola, M. Liserre, F. Blaabjerg, R. Sebastian, J. Dannehl, and F. Wilhelm Fuchs, "Analysis of the passive damping losses in

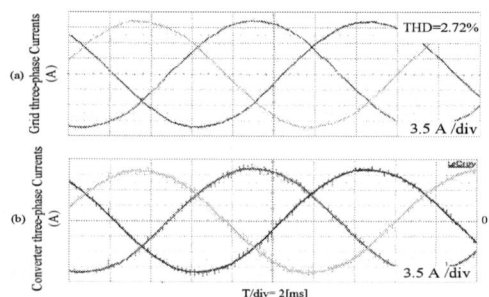

Fig. 9. Experimental results of PWM converter with LLCL filter. (a) Grid currents. (b) Converter input currents.

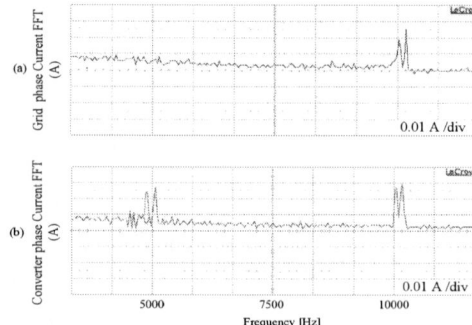

Fig. 10. FFT spectra of (a) Grid current. (b) Converter input current.

LCL-filter-based grid converters," *IEEE Trans. on Power Electron.*, vol. 28, no. 6, pp. 2642-2646, June. 2013.

[6] J. Dannehl, C. Wessels, and F. Fuchs , "Limitations of voltage-oriented PI current control of grid-connected PWM rectifiers with *LCL* filters," *IEEE Trans. Ind. Electron.*, vol. 56, no. 2, pp. 380–388, Feb. 2009.

[7] I. J. Gabe, V. F.Montagner, and H. Pinheiro, "Design and implementation of a robust current controller for VSI connected to the grid through an LCL filter," *IEEE Trans. on Power Electron.*, vol. 24, no. 6, pp. 1444-1452, Jun. 2009.

[8] G. Hu, C. Chen, and D. Shanxu, "New active damping strategy for LCL-filter-based grid-connected inverters with harmonics compensation," *Journal of Power Electronics*, vol. 13, no. 2, pp. 287-295, Mar. 2013.

[9] W. Wu and F. Blaabjerg, "An LLCL power filter for single-phase grid-tied inverter," *IEEE Trans. on Power Electron.*, vol. 27, no. 2, pp. 782-789, Feb. 2012.

[10] W. Wu, Y. He, T. Tang, and F. Blaabjerg, "A new design method for the passive damped LCL and LLCL filter based single phase grid tied inverter," *IEEE Trans. on Industry Applicat.*, vol. 60, no. 10, pp. 4339-4349, Oct. 2013.

[11] J.-I. Jang and D.-C. Lee, "High performance control of three-phase PWM converters under nonideal source voltage," *Proc. in.IEEE ICIT*, May/Jun. 2011, pp.2791-2796.

[12] U.-M. Choi and H.-B. Lee, "Neutral-point voltage balancing method for three-level inverter system with a time-offset estimation scheme," *Journal of Power Electronics*, vol. 13, no. 2, pp. 243-249, Mar. 2013.

[13] M. Z. Lowenstein, and J. Hibbard, "Modeling and application of passive-harmonic trap filters for harmonic reduction and power factor improvement," *Proc. of IEEE IAS Annual Meeting*, pp. 1570-1578, Oct.1993.

Developing a new topology for the DC-DC converter used in Fuel Cell-Electric Double Layer Capacitor Hybrid Power Source System for Mobile Devices

SHUHEI TOSAKA, TATSUYA YAMANAKA, NOBORU KATAYAMA, MASANORI HAYASE, KIYOSHI DOWAKI, SUMIO KOGOSHI

Tokyo University of Science, 2641 Yamazaki, Noda, Chiba, 278-8510 Japan

Abstract— The polymer electrolyte membrane fuel cell (PEMFC) has a potential to be an alternative power source for smartphones and other mobile electronic applications. However, PEMFCs cannot quickly response against fast load change of electronic devices. To resolve this issue, we have proposed a hybrid power source system, in which super capacitors and a PEMFC are combined using DC-DC converter. The converter is controlled by a micro-controller so that the power flows are managed. However, in low load condition, voltage became slightly higher than the target value and the super capacitor was not able to be charged. To challenge these drawbacks, in this study, we develop a new type topology for the converter, which use three switching devices, and control method. We also demonstrated the behavior of the system. The results showed that the drawbacks mentioned above is resolved and power conversion efficiency of the converter is improved under various load profile.

I. Introduction

Nowadays, smartphones are gradually spreading all around the world, replacing conventional cellphones. However, due to higher power consumption of multi-functional smartphones, limited energy capacity of conventional secondary battery such as a Li-ion battery is one of main problem in both developing countries and developed ones. The polymer electrode membrane fuel cell (PEMFC) has a potential to be an alternative power source for smartphones and other mobile electronic applications. Recently, a miniature PEMFC that achieved an active area of 3 mm x 3 mm and a peak power density of 145 mW/cm^2 was produced onto silicone substrate with micro-fabrication technology [1]. Using a PEMFC for mobile devices, there are some advantages. PEMFCs enable us to save time for recharging Li-ion batteries. Using Bio-H$_2$O as a fuel for a PEMFC may contribute to CO$_2$ emission reduction [2]. However, PEMFCs cannot quickly response against load change of electronic devices, and under this condition, the catalysts and polymer electrode membrane, which are essential components for PEMFCs, may suffer critical damage in short time [3]. To resolve this issue, we have proposed a hybrid power source system, in which the super

capacitors (SCs) and a PEMFC are combined using DC-DC converter [4]. Compared to other secondary batteries, super capacitors are characterized by extremely higher output response and long lifetime against charge/discharge cycles. However, in low load condition, load voltage became slightly higher than the target value and the super capacitor was not able to be charged. To challenge these drawbacks, in this study, we develop a new type topology for the converter, which use three switching device, and control method. We also demonstrated the behavior of the system.

II. Proposed system

Proposed DC-DC converter and its control method are described in this section.

2.1. Circuit Structure

The circuit diagram of the converter proposed in previous study is illustrated in Figure 1. The circuit has two switching devices and a diode. They are connected in series, and between them, a FC, SC and load are

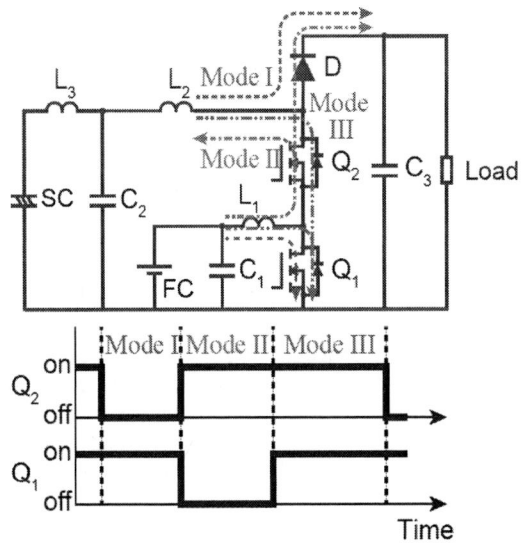

Figure 1: Circuit diagram and switching timing chart of previous converter.

The 2014 International Power Electronics Conference

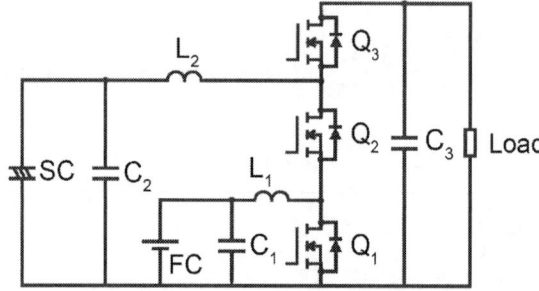

Figure 2: Circuit diagram of proposed converter.

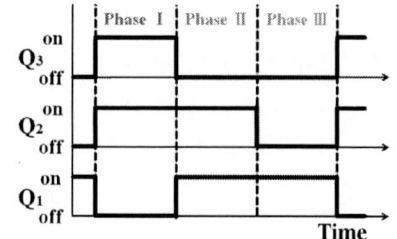

Figure 3: Switching timing chart of proposed converter in mode I.

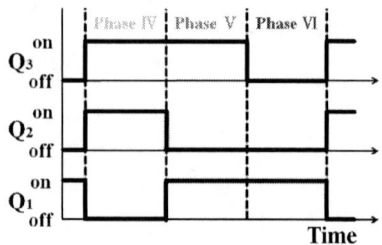

Figure 4: Switching timing chart of proposed converter in mode II.

Table 1: Circuit parameters

Elements	Specification	Quantity
EDLC	5.6F/2.5V	5
Inductance L1	47μH	1
Inductance L2	47μH	1
Capacitor C1/C2/C3	100μF	1
MOSFET	TOSHIBA TPCA8023-H	3

connected via switching noise reduction filters consisting of inductors and capacitors. By controlling the duty ratios of switching devices, the charge/discharge state of the SC is changed, which compensates the difference between the load demand and output power from the FC. However, in low load condition, load voltage became slightly higher than the target value and the super capacitor was not able to be charged. To challenge these drawbacks, we modify the circuit structure of the converter of the previous study. The circuit diagram of the converter proposed in this study is illustrated in Figure 2 and important parameters are shown in Table 1. In this circuit, diode used in the previous converter is replaced with a MOSFET. This replacement allows bidirectional current between the load and the super capacitor and resolve the drawbacks mentioned above.

2.2. Operation Mode

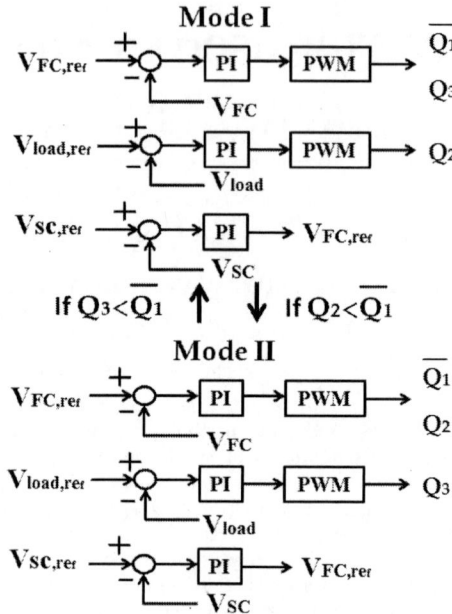

Figure 5: Control block diagram.

By controlling voltage of the FC, SC and load by micro-controller, voltage of the load is highest, and that of the FC is lowest. Detailed control method is described in the next section. In this paper, current flowing to the load, flowing from the FC, flowing from the SC is defined as positive current. Switching timing chart of proposed converter is shown in figure 3, 4. The control algorithm for the hybrid power source system has two control modes, mode I and mode II. Mode I has three operation phases. At the beginning of phase I, if the SC is charged, current from the FC flows to the SC and load. If the SC discharges electricity, the currents from the FC and SC flow to the load. After that, the current values of the SC and load decrease as time goes on. Either may be negative value. If the current of the load becomes negative value, currents from the FC and load flow to the SC. At the beginning of phase II, if the SC is charged, current from the FC flows to the SC and ground. If the SC discharges electricity, currents from the FC and SC flow to the ground. After that, the current of the SC and the current flowing to the ground increases as time goes on. At the beginning of phase III, if the SC is charged, current from the FC flows to the SC though body diode of Q_2 and ground. If the SC discharges electricity, current from the FC flows to the ground and current from the SC flows to the load though body diode of Q_3. After that, the current flowing to the ground increases and the current of the SC gets closer to zero as time goes on.

The Mode II has also three operation phases. In this mode, the SC cannot discharge electricity. In phase IV, the current directions are the same as the phase I. At the beginning of phase V, current from the FC flows only to the ground and current from the load flows to the SC. After that, the current of the SC decreases and the current flowing to the ground increases as time goes on. At the beginning of phase VI, current from FC flows to the ground and the SC though body diode of Q_2. After that, the current of the SC gets closer to zero current and the current flowing to the ground increases as time goes on.

2.3. Control Method

Figure 5 shows control block diagram. This study employs a micro-controller (RL78G13, Renesas Electronics) to control switching devices digitally. As mentioned in section 2.2, there are two control modes, mode I and mode II. In mode I, the load voltage V_{load} and the FC voltage V_{FC} are periodically captured and quantized at 10-bit resolution by the analogdigital converter (ADC) on the micro-controller. Each voltage value is compared with each reference value $V_{load,ref}$, and $V_{FC,ref}$, and the difference between the captured value, and the reference value is minimized by the proportional integral (PI) control method, which varies duty ratio of MOSFETs. V_{load} is determined by duty ratio of Q_2. V_{FC} is determined by that of Q_1. To avoid diode loss of body diode of Q_2, Q_3 due to current from the FC, off time of Q_1 synchronizes with on time of Q_3. The load is getting small, on time of Q_2 is decreased and may be exceed the off time of Q_1. If the exceeding occurs, the system enters mode II from mode I. In mode II, on time of Q_3 is changed while on time of Q_2 is fixed. Without that, the loss of the body diode of Q_2 and Q_3 would lead to the decreasing of system efficiency. Consequently, V_{load} is stabilized due to this mode changing and controlling of Q_3. V_{FC} is determined by duty ratio of Q_1. Under the same purpose of mode I, to avoid diode loss, off time of Q_1 synchronizes with on time of Q_2. However, in both mode I and mode II, if the V_{FC} is maintained at the constant value, the super capacitor voltage V_{SC} may get out of the rated voltage range because of over charging or discharging. To avoid this, by applying PI control to the V_{SC} and varying reference voltage of FC, V_{SC} is confined around its target value. Note that the time constant for the PI controller should be much smaller than that of other PI controllers to avoid frequently changing of the FC voltage, which may degrade the FC. Based on this strategy, the sampling rate for the super capacitor voltage is decided to be 0.1 seconds, same as previous study. More detailed relationship between frequency and degradation of FCs is presented in [3]. The operation waveforms that are actually measured by oscilloscope are showed in figure 6-11. They includes switching waveforms, the load voltage V_{load}, the FC voltage V_{FC}, the SC voltage V_{SC}, the inductance voltages V_{L1} and V_{L2}, the voltage of Q_3 source V_2, ,the voltage of Q_2 source V_1. We also measure SCs current I_{SC} that is measured by a bipolar super supply. If it is negative value, SCs are charged. In figure 6-11, V_{FC} and V_{load} are constant values due to controlling DC-DC converter. V_{SC} is constant value because a bipolar power supply is used as the SCs. V_{FC} is 2.4V. V_{load} is 4V. V_{SC} are around 2.5V, 3.0V, 3.5V because counter electromotive force of wire between the bipolar power supply and the circuit board occurs. I_{load}

are 0A or 0.8A. Noise on V_{load}, V_{FC}, V_{SC} are observed in these results. They may be high frequency harmonics of MOSFET's switching.

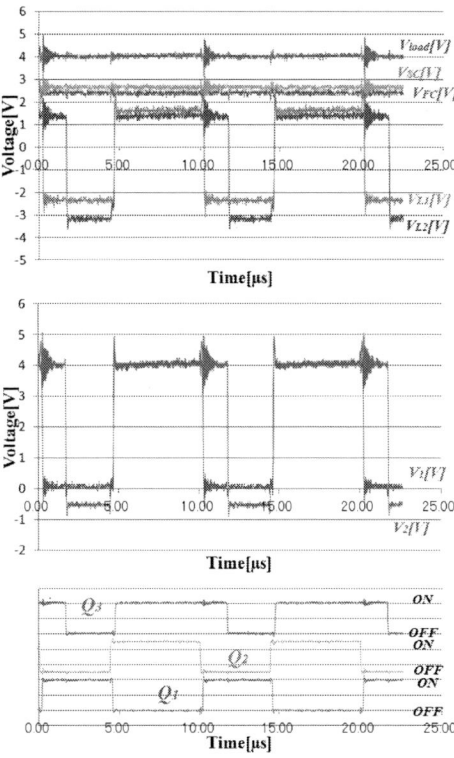

Figure 6: Operation waveforms (V_{SC}= 2.647V, I_{load}=0A, I_{SC} = -0.864A, MODE II)

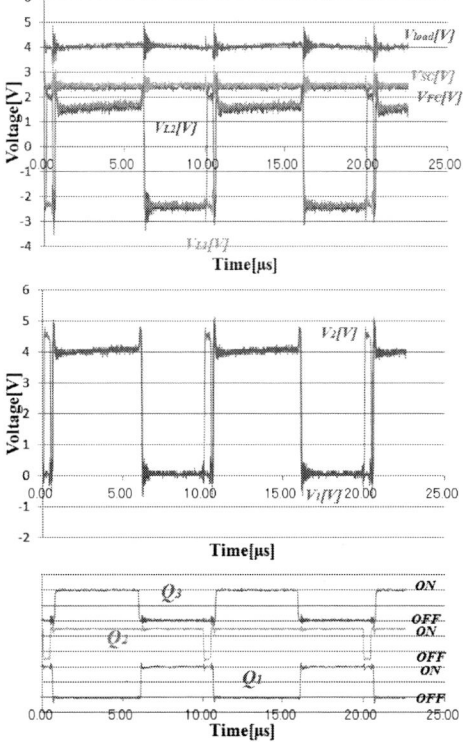

Figure 7: Operation waveforms (V_{SC}=2.499V, I_{load}=0.8A, I_{SC} = 0.325A, MODE I)

The 2014 International Power Electronics Conference

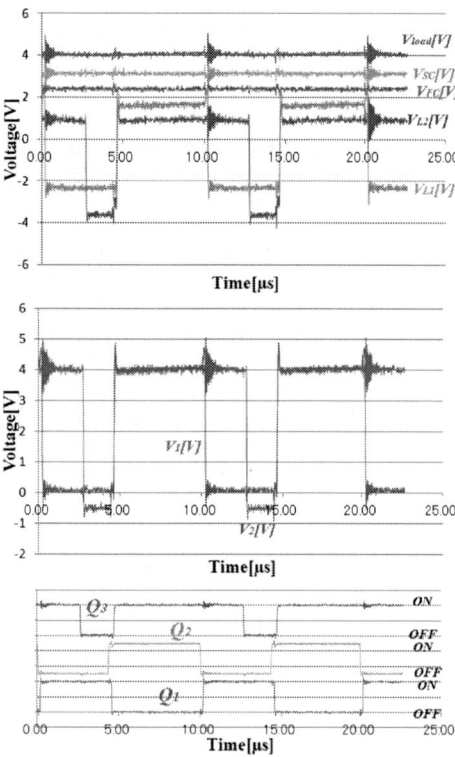

Figure 8: Operation waveforms (V_{SC}=3.120V, I_{load}=0A, I_{SC}= -0.755A, MODE II)

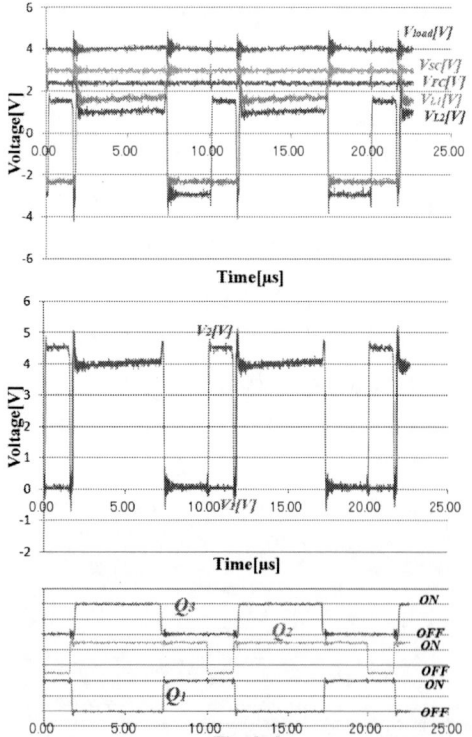

Figure 9: Operation waveforms (V_{SC}=2.995V, I_{load}=0.8A, I_{SC}= 0.269A, MODE I)

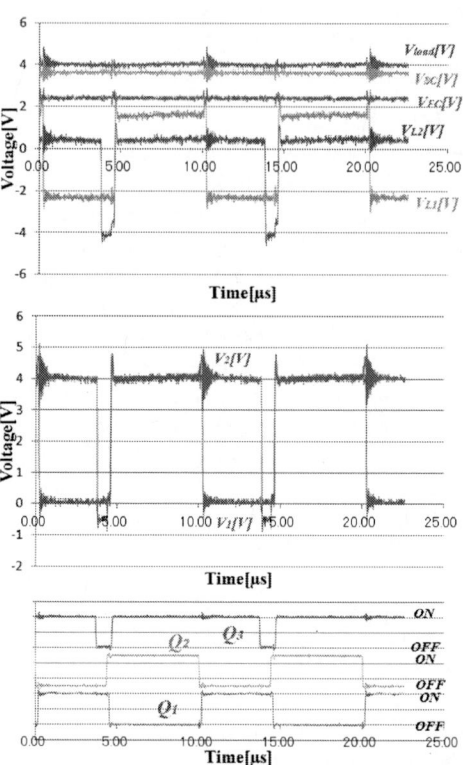

Figure 10: Operation waveforms (V_{SC}=3.615V, I_{load}=0A, I_{SC}= -0.668A, MODE II)

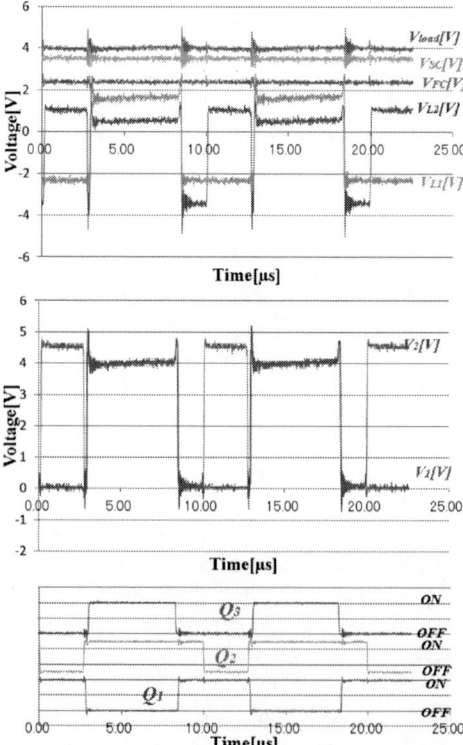

Figure 11: Operation waveforms (V_{SC}=3.507V, I_{load}=0.8A, I_{SC}= 0.230A, MODE I)

978-1-4799-2706-7/14 $31.00 © 2014 IEEE

III. Experimental

On the assumption that the proposed power source system is applied for smartphones, the experiments in this study were carried out. A DC-power supply connected in series with a resistance of 0.2 Ω was used as a pseudo FC. Note that the actual resistance value is slightly larger than this value. Open circuit voltage (OCV) of the pseudo FC is 2.8V. The OCV is determined on the assumption that four FCs, each of the FC has the OCV of 0.7V, are connected in series. An electric load that is programed to reproduce the load patterns of a smartphone was employed. The schematic dialog of the measuring system for the power consumption (the load patterns) of a smartphone [5] is showed in fig. 12. The load patterns were pre-measured from an actual smartphone that operate under six conditions including internet, calling, SNS, game, music, and standby modes, which are frequently used as typical applications. The duration time of each load pattern is 130 seconds. Electric double layer capacitors (EDLCs) were used as super capacitors because the voltage management of EDLCs is easy. Five EDLCs that has a capacitance of 56 F are connected in series. Consequently the total capacitance is 11.2 F. The switching frequency is maintained at 100kHz by microcontroller. Microcontroller is connected with the load through jumper pins to compare the characteristics when the power of microcontroller is supplied with the load voltage to that of when supplied with an external power supply.

IV. Results and Discussion

The experimental results in which a microcontroller is not connected with the load are shown in Figure 13-18. The results are respectively obtained when the smartphone operates at the Internet, SNS, game, calling, music, and standby modes. The results include the waveforms of load voltage V_{load}, FC voltage V_{FC}, EDLC voltage V_{EDLC}, load current I_{load}, EDLC current I_{EDLC}, and calculated efficiency that excludes power consumption of microcontroller. The average power conversion efficiency is 94.5% when the internet load profile was used, 95.7% for the SNS load profile, 95.0% for the game load profile, 94.9% for the calling load profile, 93.0% for the music load profile, 78.5% for the standby load profile. In music load profile, conversion efficiency is lower than previous one. Compared to the previous converter [4], FC voltage change is smaller. Thus, we think that using this converter, degradation of FC will be suppressed compared to previous one. As can be seen from the waveforms of standby in figure 18, V_{load} is maintained at the target value. Figure 19 shows the waveforms in the standby load profile, in which the initial EDLC voltage is lower. This result illustrate that the EDLC can be charged in low load condition. It is due to bidirectional current through Q_3. The experimental results in which a microcontroller is connected with the load are shown in Figure 20-25. The average power conversion efficiency that includes power consumption

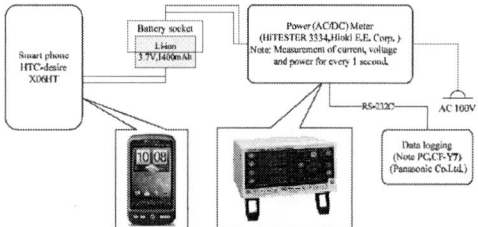

Figure 12: The schematic dialog of the measuring system for the measuring system for the power consumption of a smartphone [5].

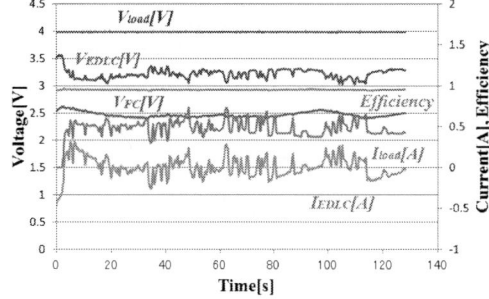

Figure 13: Waveforms when the smartphone is operated in the Internet mode. (Microcontroller is not connected with the load)

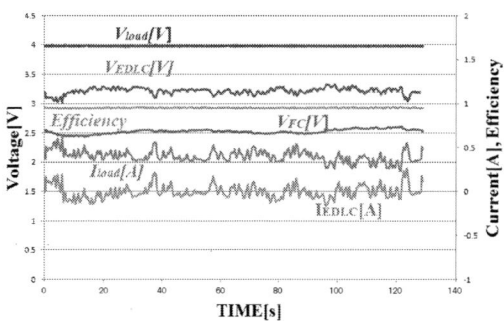

Figure 14: Waveforms when the smartphone is operated in the SNS mode. (Microcontroller is not connected with the load)

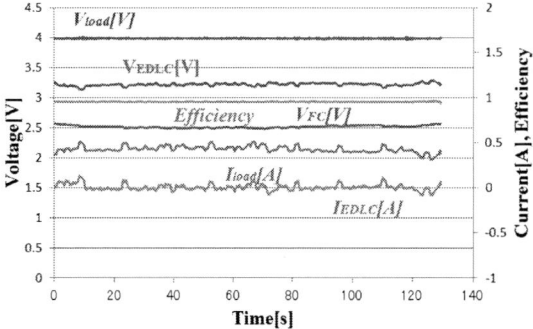

Figure 15: Waveforms when the smartphone is operated in the game mode. (Microcontroller is not connected with the load)

of the microcontroller is 91.4% when the internet load profile was used, 92.4% for the SNS load profile, 92.0% for the game load profile, 89.5% for the calling load profile, 81.0% for the music load profile, 39.4% for the standby load profile.

The 2014 International Power Electronics Conference

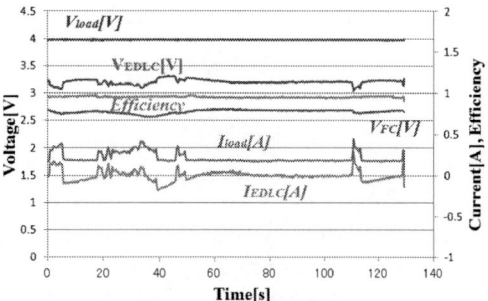

Figure 16: Waveforms when the smartphone is operated in the calling mode. (Microcontroller is not connected with the load)

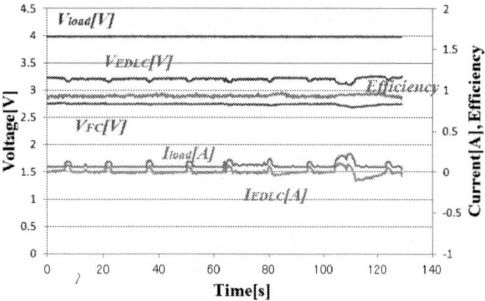

Figure 17: Waveforms when the smartphone is operated in the music mode. (Microcontroller is not connected with the load)

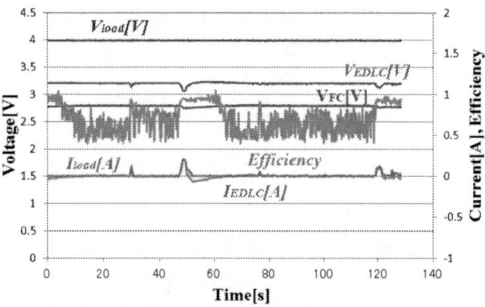

Figure 18: Waveforms when the smartphone is operated in the standby mode. (Microcontroller is not connected with the load)

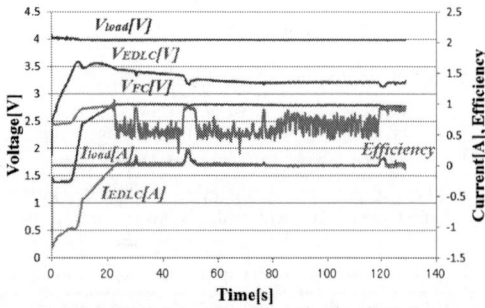

Figure 19: Waveforms when the smartphone is operated in the standby mode in which initial EDLC voltage is lower. (Microcontroller is not connected with the load)

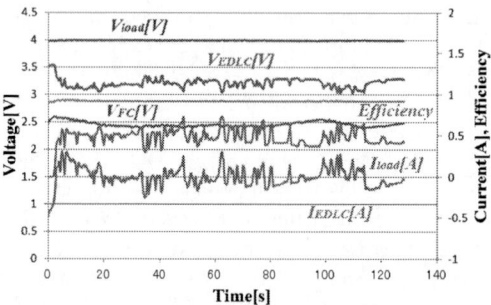

Figure 20: Waveforms when the smartphone is operated in the Internet mode. (Microcontroller is connected with the load)

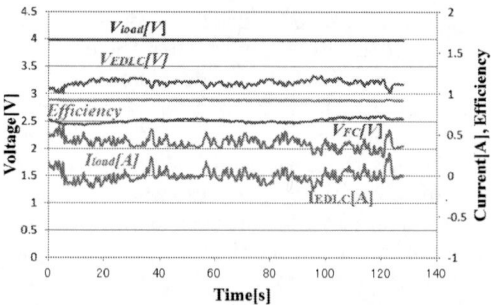

Figure 21: Waveforms when the smartphone is operated in the SNS mode. (Microcontroller is connected with the load)

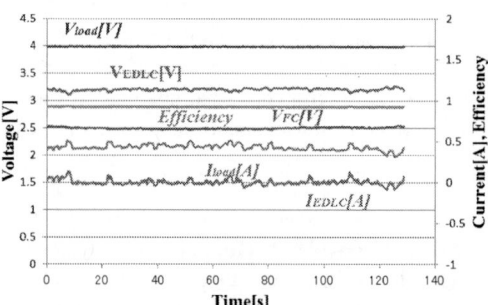

Figure 22: Waveforms when the smartphone is operated in the game mode. (Microcontroller is connected with the load)

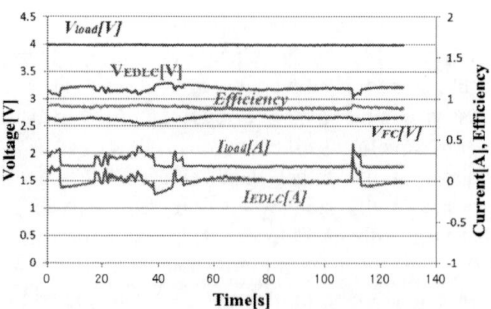

Figure 23: Waveforms when the smartphone is operated in the calling mode. (Microcontroller is connected with the load)

978-1-4799-2706-7/14 $31.00 © 2014 IEEE   1212

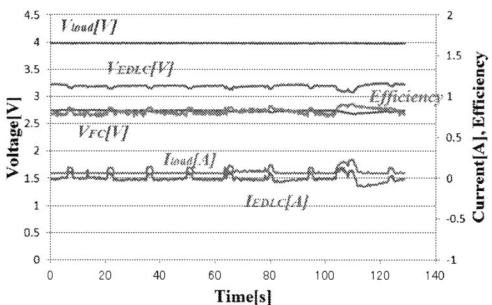

Figure 24: Waveforms when the smartphone is operated in the music mode. (Microcontroller is connected with the load)

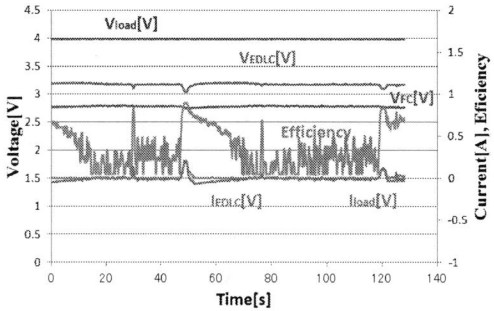

Figure 25: Waveforms when the smartphone is operated in the standby mode. (Microcontroller is connected with the load)

V. Conclusion

In this study, we improved the power conversion efficiency of the DC-DC converter of the hybrid power source system and proposed a new control method for the converter. The system behavior was also demonstrated. The experimental results showed that load voltage is maintained at the target value and the super capacitors can be charged under low load condition, in which, previous converter was not able to charge EDLC sufficiency. The power conversion efficiency that excludes power consumption of the microcontroller is 94.5% when the internet load profile was used, 95.7% for the SNS load profile, 95.0% for the game load profile, 94.9% for the calling load profile, 93.0% for the music load profile, 78.5% for the standby load profile. For all load profiles except the music profile, conversion efficiency is higher than that of the converter proposed in previous study. It is due to bidirectional current through Q_3. The average power conversion efficiency that includes power consumption of the microcontroller is 91.4% when the internet load profile was used, 92.4% for the SNS load profile, 92.0% for the game load profile, 89.5% for the calling load profile, 81.0% for the music load profile, 39.4% for the standby load profile. Our future work includes to improve the conversion efficiency especially in the standby mode and to reduce the power consumption of the microcontroller.

References

[1] M. Hayase, T. Fujii, and J. G. Alves Brito-Neto, A Miniature Fuel Cell with Porous Pt Layer Formed on a Si Substrate,Journal of The Electrochemical Society,Vol. 158, No. 4, 2011, pp. B355-B359.

[2] K. Dowaki, K. Hogen, Tanimukti, Ivan Prasetya, A Feasibility Study on the PEFC (Polymer Electrolyte Fuel Cell) Smartphone Operated by Bio-H2 Fuel using LCA Methodology, Journal of Life Cycle Assessment Japan, Vol. 8, Nob. 1, 2012, pp. 14-25. (in Japaneses)

[3] S. Zhang, X. Z. Yuan, J. N. C. Hin, H. Wang, K. A. Friedrich, and M. Schulze, A review of platinum-based catalyst layer degradation in proton exchange membrane fuel cells, Journal of Power Sources, Vol. 194, No. 2, 2009, pp. 588–600.

[4] S. Tosaka, N. Katayama, K,Dowaki, S. Kogoshi, "Energy Management of Fuel Cell-Elctric Double Layer Capacitor Hybrid Power Source System for Mobile Devices," Proceedings of the International Conference on Electrical Engineering 2014, pp. 890- 893, 2013.

[5] Kiyoshi Dowaki, Masanori Hayase, Noboru Katayama, "A Life Cycle Inventory Analysis on FC-Iphone fueled by Bio-H2" in proceedings of the 9th Meeting of the Institute of Life Cycling Assessment, Japan, May 2013.

The 2014 International Power Electronics Conference

Multiple Output Charger based on Phase Shift Full Bridge Converter with Novel Time Division Multiple Control Technique

Van-Long Tran
Department of Electrical Engineering
Soongsil University
Seoul, Korea
tranvanlong988@gmail.com

Woojin Choi
Department of Electrical Engineering
Soongsil University
Seoul, Korea
cwj777@ssu.ac.kr

Abstract— **Multiple output converters (MOCs) are widely used for applications which require various levels of the output voltages due to their benefits in cost, volume, and efficiency. However, most of the MOCs developed so far can regulate only one output tightly and require as many secondary windings in the transformer as the number of the outputs. In this paper, a novel Time Division Multiple Control (TDMC) method to regulate all the outputs in high precision is proposed and applied for the multiple output battery charger based on the phase shift full bridge topology to charge a multiple number of batteries at one time. The proposed converter can charge three different kinds of batteries or same kind of batteries in different state of charges (SOCs) by using constant current/constant voltage (CC/CV) charge mode independently. At the same time it can provide an even degree of tight regulation for each output to satisfy the strict ripple requirement of the battery. The validity and feasibility of the proposed method are verified through the experiments.**

Keywords— *TDMC, Multiple output battery charger, Phase shift full bridge, and ZVS-ZCS*

I. INTRODUCTION

Multiple output converters (MOCs) are widely used and applied for many applications such as: voltage regulated module (VRM) of the portable devices, household equipment, multiple output power supply, and telecommunication system. One of the most promising applications of the MOCs would be the multiple battery charger due to its advantages in terms of cost, volume, efficiency and space for the installation.

However, most of the MOCs developed so far are unable to regulate the outputs in high precision and thus hardly satisfy the strict ripple requirements of the charge applications [1]. Researches on this topic have been conducted and several methods have been proposed in order to overcome the disadvantages mentioned above [2-5]. One method is to use the pre and post regulator to regulate the multiple output but it is quite complex in control [2]. Another method is to use the controlled current source and to connect it to each output on a time shared basis [3-4]. However, those methods require a large inductor which is bulky and expensive and the cross regulation problem exists. The other method is to regulate

the two outputs by controlling the duty cycle and frequency, respectively [5]. However, the number of converter outputs is limited only by two and the regulation performance is not good enough. Since all the above methods tried to regulate all the outputs in one switching period, the cross regulation problem exists. Also they require as many number of the secondary windings in the high frequency transformer as the number of the outputs if applied to the isolated converter topologies.

In this paper, a novel Time Division Multiple Control (TDMC) method which can precisely and independently control the multiple outputs with only one secondary winding is proposed and applied for the multiple output charger based on the phase shift full bridge topology. The proposed multiple battery charger is able to charge a multiple number of batteries at different state of charges (SOCs) by using constant current (CC) and constant voltage (CV) charge modes. The CC/CV charge method offers the best performance to fully charge the batteries [6]. Since the TDMC method can control each output independently, while one battery is being charged by CC mode, the others can be charged by CV mode so that three batteries in different state-of-charge can be charged at one time. Also the strict ripple requirement of the battery can be satisfied since the cross regulation problem between the outputs does not exist. The operation principle of the proposed TDMC method and the proposed topology will be detailed in the following sections. The effectiveness of the proposed method will be verified by the simulation and experimental results.

II. PROPOSED TDMC METHOD AND OPERATION PRINCIPLE OF THE MULTIPLE CHARGER

Fig.1 shows the proposed multiple charger based on the phase shift full bridge converter including R-C model of the Li-Po batteries with time division multiple control (TDMC) method. As shown in Fig. 1, the proposed converter was developed by modifying the conventional phase shift full bridge converter. Three different output circuits shares the secondary winding of the transformer and an active switch is added to each output in order to perform the TDMC method.

978-1-4799-2706-7/14 $31.00 © 2014 IEEE

The 2014 International Power Electronics Conference

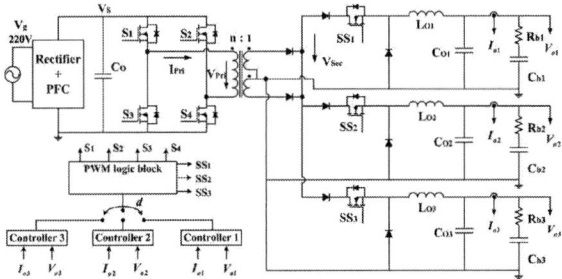

Fig. 1. Proposed multiple charger based on Time Division Multiple Control method for three Li-Po batteries

Fig. 2 shows the overview of the PWM scheme of the proposed converter with TDMC method. As shown in Fig. 2, the PWM is performed at the frequency of f_s for the primary switch and at $f_s/3$ for each of the secondary switches. The proposed method can regulate all of the outputs in one sampling time ($3T_S$). Each output is controlled independently and precisely during one switching period T_S (interrupt cycle) by way of the secondary switches. In one switching period, just one secondary switch is turned on to allow the primary side to control each of the outputs by shifting the phase between the leading and lagging switches. Similarly, another output is regulated in the next switching period. Thus, all of the outputs are controlled in one sampling time ($3T_S$). In the proposed method, since each output is regulated independently within one switching cycle (T_S), no cross regulation problem. Thus it is possible to provide an even degree of the tight regulation for all the outputs.

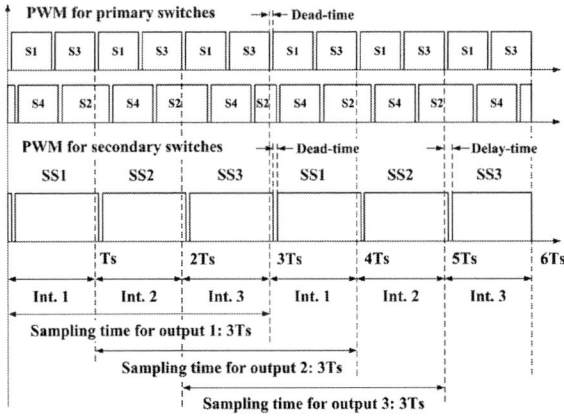

Fig. 2. Switching waveforms for the primary and secondary switches of the proposed converter with TDMC method

The primary side operates as a conventional phase shift full bridge converter and the TDMC method is implemented at the secondary side through the active switches. The conventional phase shift PWM scheme is used to get the zero voltage switching (ZVS) at all the primary side switches by shifting the phase of the gate signals between leading-leg switches (S_1 and S_3) and lagging-leg switches (S_2 and S_4). However, in the proposed converter the primary side operates at the switching frequency of fs and the each secondary side operates at the switching frequency of $fs/3$ to achieve the TDMC method. Each secondary output circuit operates one by one sequentially with the help of the additional

switches (SS_1, SS_2 and SS_3) during one switching cycle (T_S) of the primary side. The operation of the proposed converter at each output consists of eight modes and is detailed as follows:

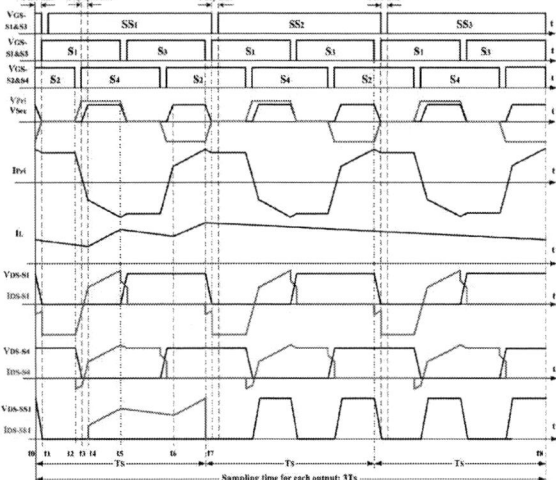

Fig. 3. Voltage and current waveforms at the primary and secondary sides of the proposed converter in the steady-state

1) Mode 1 ($t_0 \sim t_1$, Fig. 4(a))

At $t=t_0$, since the switch 3 is turned off, the primary current I_{Pri} starts to charge the output capacitance of the switch 3 and discharge the output capacitance of switch 1. The voltage of the switch 1 is decreased to zero and then its body diode conducts thereby achieving zero voltage switching condition. The transformer secondary voltage V_{Sec} becomes zero at $t=t_1$. The output inductor current ramps down with a slope, V_o/L, and reaches zero at $t=t_4$.

2) Mode 2 ($t_1 \sim t_2$, Fig. 4(b))

The switch 1 achieves ZVS turn-on at $t=t_1$, the primary current I_{Pri} freewheels through the switch 1 and switch 2 during this mode. As a result, the transformer secondary voltage V_{Sec} remains at zero.

3) Mode 3 ($t_2 \sim t_3$, Fig. 4(c))

The switch 2 is turned off at $t=t_2$, then the primary current I_{Pri} starts to charge the output capacitance of the switch 2 and discharge the output capacitance of switch 4. The voltage on the switch 4 is discharged to zero during this mode, and then its body diode conducts to achieve the zero voltage switching condition. The transformer secondary voltage V_{Sec} remains at zero in this mode.

4) Mode 4 ($t_3 \sim t_4$, Fig. 4(d))

The switch 4 achieves ZVS turn-on at $t=t_3$. The transformer secondary voltage V_{sec} remains at zero until the primary current I_{Pri} reverses its direction and rises to reach the reflected output inductor current I_L/n at $t=t_4$. The switch SS1 is turned on with ZVS at $t=t_4$ and no power is delivered to the output in this mode.

5) Mode 5 ($t_4 \sim t_5$, Fig. 4(e))

At $t=t_4$, the transformer secondary voltage V_{sec} is equal to V_S/n, the output inductor current starts to ramp up with the slope $(V_{Sec} - V_o)/L$ until $t=t_5$. The primary winding current I_{Pri} is equal to the reflected output inductor current I_L/n during this mode. In this mode, the power is delivered to the secondary.

978-1-4799-2706-7/14 $31.00 © 2014 IEEE   1215

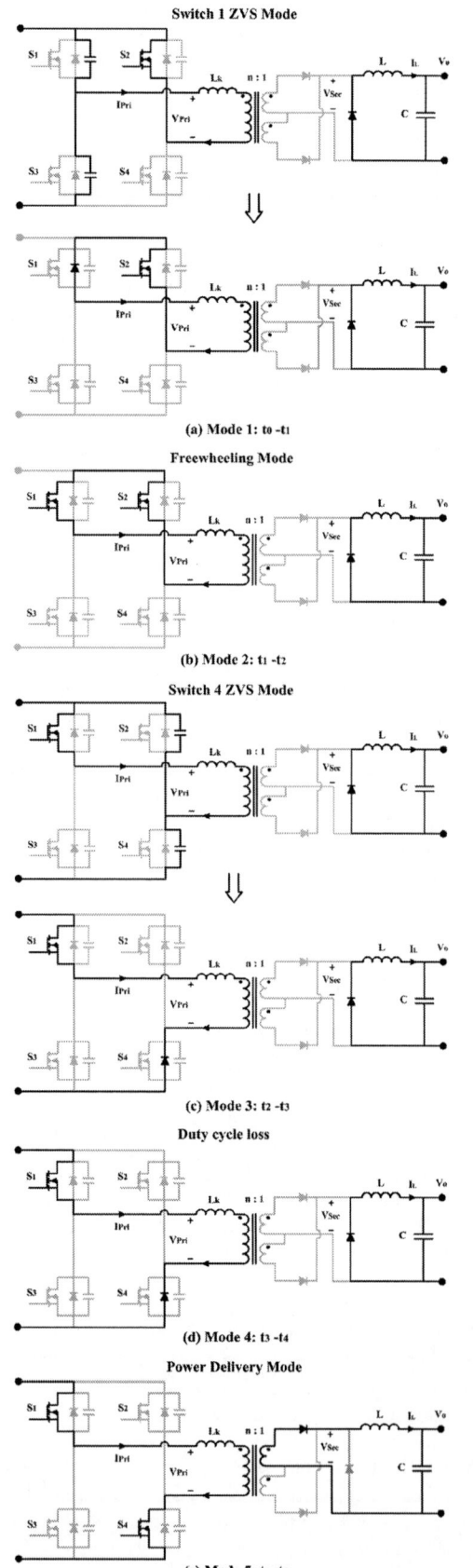

(a) Mode 1: t₀ -t₁

(b) Mode 2: t₁ -t₂

(c) Mode 3: t₂ -t₃

(d) Mode 4: t₃ -t₄

(e) Mode 5: t₄ -t₅

Fig. 4. Five operation modes of the proposed multiple charger for a half of switching cycle at the primary side

Similarly, the switches 2 and 3 can achieve the zero voltage switching in the other half of the switching cycle, during the mode 6 (t_5~t_6) and mode 7(t_6~t_7).

6) Mode 6 (t_5~t_6)

At t=t_5, the inductor current starts to ramp down with a slope V_o/L until t=t_6. The switches 2 and 3 can achieve ZVS turn-on during this mode.

7) Mode 7 (t_6~t_7)

The inductor current starts to ramp up again with the slope $(V_{Sec} - V_o)/L$ and the power is transferred from the primary to the secondary side during this mode. At t=t_7, the inductor current starts to ramp down with the slope, V_o/L, and the secondary winding voltage becomes zero. As a result, the switch SS_1 can be turned off with zero current switching (ZCS) condition.

8) Mode 8 (t_7~t_8)

The inductor current starts to ramp down and it keeps ramping down until the sampling time ends at t=t_8.

In order to achieve the ZVS at all the switches during overall charge process, the sufficient energy for the soft switching and the suitable dead-time are essential even in the light load condition at the end of charge process. To guarantee the soft switching condition the total inductive energy (E_L) stored in the leakage inductor should be larger than the total capacitive energy (E_C) of the switches during mode 1 and 3. The total capacitive energy in a switching leg can be expressed as (1). The energy stored in the inductive components for the ZVS of the leading leg and lagging leg switches can be expressed as (2) and (3), respectively.

$$E_C = C_{OSS} \times V_S^2 \tag{1}$$

$$E_{L\_Lead} = \frac{1}{2} L_M I_{M,pk}^2 + \frac{1}{2} L I_{L,\max}^2 + \frac{1}{2} L_k (I_{M,pk} + \frac{I_{L,\max}}{n})^2 \tag{2}$$

$$E_{L\_Lag} = \frac{1}{2} L_k (I_{M,pk} + \frac{I_{L,\min}}{n})^2 \tag{3}$$

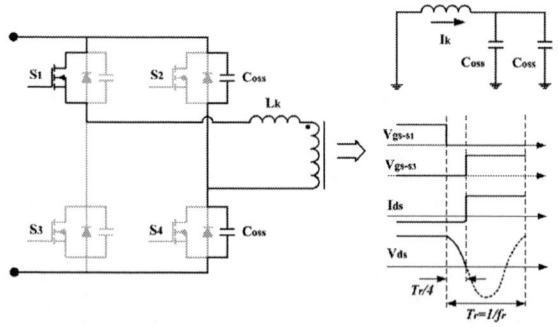

Fig. 5. Equivalent resonant circuit and ZVS transition during dead-time at each primary switch

As shown in (2) and (3), the inductive energy available for ZVS of the leading switches 1 and 3 is higher than that of the lagging switches 2 and 4, in which L_M and I_M are the magnetizing inductance and the magnetizing current of the transformer, and C_{OSS} is the effective output capacitance of each switch. As a result, the leading switches 1 and 3 have a wider ZVS range compared to the lagging switches 2 and 4. Thus, the magnetizing

978-1-4799-2706-7/14 $31.00 © 2014 IEEE

inductance of the transformer should be designed to guarantee the ZVS for the lagging leg switches even with the light load at the end of CV charge.

One more thing essential for the ZVS of the switches is that the dead-time should be enough to make the voltage transition of the switches completely based on the resonant frequency, T_r, as (4), as shown in Fig. 5.

$$T_{Dead-time} \geq \frac{T_r}{4} = \pi \sqrt{\frac{L_k C_{OSS}}{2}} \quad (4)$$

As shown in (4), when one of the switches can achieve ZVS condition, the leakage inductance of the transformer (L_k) and the effective output capacitance of the switches (C_{OSS}) cause the resonant frequency (f_r). As a result, the dead-time should be larger than $T_r/4$ to guarantee the voltage transition of the switch completely before it is turned-on.

III. STATE SPACE AVERAGE MODELING OF THE PROPOSED CHARGER FOR CC/CV MODE CHARGE

As shown in Fig. 1, three secondary circuits have the same configuration. Since the same control method can be applied to each secondary circuit in a time-shared basis, the control-to-output transfer function with only one secondary circuit need to be derived. The battery load is modeled by a resistor R_b in series with a capacitor C_b having an initial voltage, V_{Cb}. The state space average model of the proposed multiple charger at each output including R-C equivalent circuit model of the Li-Po battery is shown in Fig. 6, where D is the phase shift period between the leading-leg and lagging-leg.

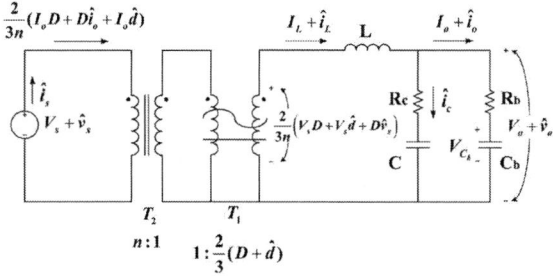

Fig. 6. State space average modeling of the proposed multiple charger for each secondary output

By perturbing the state space average model to produce large (DC) and small (AC) signal terms, the small signal model can be obtained as (5) to (9) by separating the AC terms in the state space average model.

$$3L\frac{d\hat{i}_L}{dt} = 2\frac{V_s}{n}\hat{d} + 2\frac{D}{n}\hat{v}_s - 3\hat{v}_o \quad (5)$$

$$\hat{i}_L = \hat{i}_c + \hat{i}_o \quad (6)$$

$$\hat{i}_c = C\frac{d(\hat{v}_o - \hat{i}_c R_c)}{dt} \quad (7)$$

$$\hat{v}_o = \frac{1}{C_b}\int \hat{i}_o dt + R_b \hat{i}_o \quad (8)$$

$$\hat{i}_s = \frac{2}{3n}(D\hat{i}_o + I_o\hat{d}) \quad (9)$$

The control-to-output voltage and control-to-output current transfer functions can be obtained as (10) and (11) by using (5) to (9).

$$G_{vd} = \frac{\hat{v}_o}{\hat{d}}\bigg|_{\hat{v}_s=0} = \frac{V_o}{D} \times \frac{(1+a_1 s)(1+a_2 s)}{s^3 b_3 + s^2 b_2 + s b_1 + 1} \quad (10)$$

$$G_{id} = \frac{\hat{i}_o}{\hat{d}}\bigg|_{\hat{v}_s=0} = \frac{V_{C_b} + I_o R_b}{DR_b} \times \frac{(1+a_1 s)(1+a_2 s)}{s^3 b_3 + s^2 b_2 + s b_1 + 1} \quad (11)$$

where,

$$a_1 = R_b C_b; \quad a_2 = R_c C; \quad b_1 = R_c C + R_b C_b$$

$$b_2 = LC + LC_b + R_b R_c C C_b; \quad b_3 = R_b LC C_b + R_c LC C_b$$

As shown in (10) and (11), the control-to-output voltage and control-to-output current transfer functions of the proposed multiple charger are obtained as the third order system with two zeros and three poles at the left half plane. In the numerator, one zero is caused by R-C model of the battery and the other is caused by the ESR of the output capacitor. The R-C model of the battery also causes a pole and the L-C output filter causes a double pole in the denominator.

IV. DESIGN OF THE DUAL LOOP CONTROLLER FOR THE CHARGE FUNCTION OF THE PROPOSED MULTIPLE CHARGER

In order to verify the derived transfer functions for the design of the controller, Bode plot of the control-to-output transfer functions are drawn by using both of AC sweep and transfer functions in PSIM as shown in Fig. 7. As shown in the figure, the results are well matched, thereby proving the accuracy of the derived transfer functions.

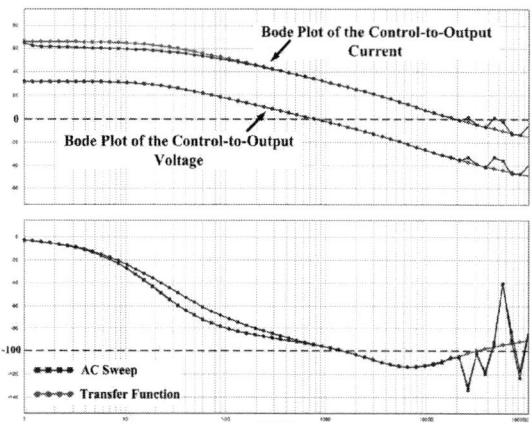

Fig. 7. Bode plot of control-to-output transfer functions drawn by AC sweep and transfer function in PSIM for each output

As shown in Fig. 7, both control-to-output transfer functions show enough phase margin all over the frequency range. Thus, it is just required to put one pole at the origin to improve the gain at the low frequency range and one zero at the desired crossover frequency to make the closed-loop stable. The PI controller can be used for both of the control-to-output transfer functions of the proposed multiple charger.

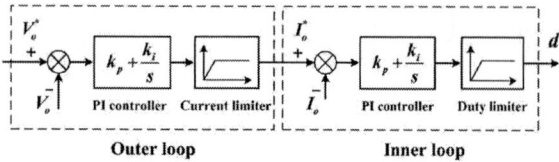

Outer loop   **Inner loop**

Fig. 8. Block diagram of the dual control loop for each output

In this application, CC/CV mode charge is performed by using a dual loop control, where the inner control loop serves for the output current control and the outer loop serves for the output voltage control as shown in Fig. 8. In the dual control loop design, the inner control loop requires a higher bandwidth, typically 5~10 times higher than that of the outer loop to guarantee that the inner loop will not affect to the outer loop performance [7].

In this case, the closed output voltage loop is judged stable with 67.7° phase margin and 1.04kHz bandwidth, and the closed output current loop is judged stable with 62.3° phase margin and 10.2kHz bandwidth. The bandwidth of the inner loop is 10 times higher than that of the outer loop to guarantee the stability of the dual control loop.

V. EXPERIMENTAL RESULTS AND DISCUSSIONS

TABLE I
SPECIFICATIONS OF THE PROPOSED MULTIPLE CHARGER

P_o	Power rating	250 [W]
V_G	Input voltage	220 [VAC]
V_{o1}, V_{o2}, and V_{o3}	Charge voltage (CV)	12.6 [V]
I_{o1}, I_{o2}, and I_{o3}	Charge current (CC)	6 [A](0.5C)
f_S	Switching frequency	100 [kHz]
L_{o1}, L_{o2}, and L_{o3}	Output inductor	280 [μH]
C_{o1}, C_{o2}, and C_{o3}	Output capacitor	1000 [μF]
R_b	Battery ESR	0.116 [Ω]
C_b	Battery capacitance	21500 [F]
ΔV_{o1}, ΔV_{o2}, and ΔV_{o3}	Output voltage ripple	2[%](252mV)
ΔI_{o1}, ΔI_{o2}, and ΔI_{o3}	Output current ripple	5[%](0.6A)

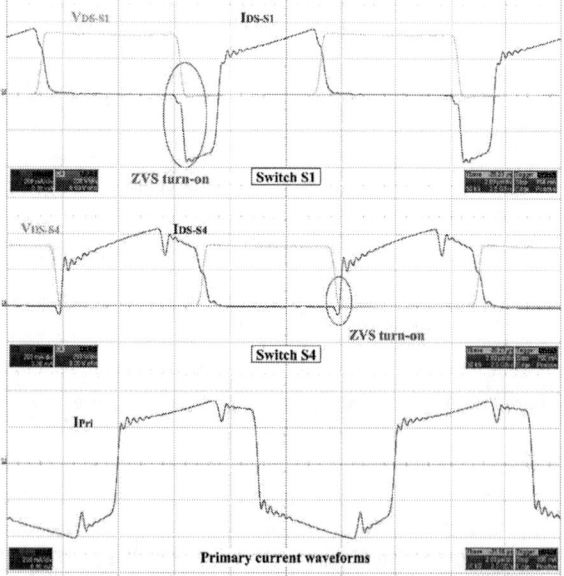

Fig. 9. Voltage and current waveforms of the switches S1 and S4 and the primary current waveform of the transformer

The specification of the proposed charger is summarized in the Table I. The operation and control

method of the proposed multiple charger based on the phase shift full bridge topology are verified by charging the three Li-Po battery packs in different SOCs simultaneously. The Li-Po battery pack configuration is 3S3P meaning 3 cells are in series and 3 strings of these in parallel. The nominal voltage of the battery is 11.1[V] and its charge current is 12[A](1C). For the control of the proposed charger, the digital signal processor (DSP) TMS320F28335 from TI was used to achieve the high speed switching and calculation.

Fig. 9 shows voltage and the current waveforms of the switch S1 and S4, and the primary current waveform of the transformer captured at the end of CV charge mode. As shown in Fig. 9, ZVS turn-on is achieved at the switches S1 and S4 under the light load condition. The ZVS turn-on is also achieved at the switches S2 and S3 similarly though it is not shown here. As a result, ZVS turn-on of the primary switches can be achieved all over the charge process.

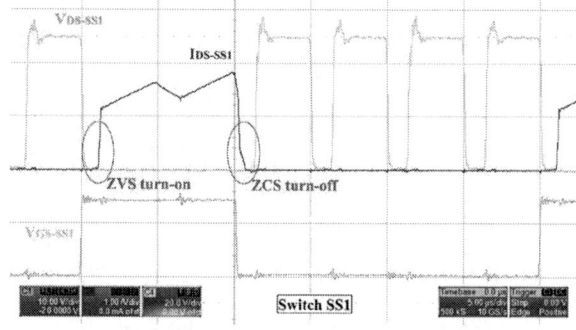

Fig. 10. ZVS turn-on and ZCS turn-off waveforms of the secondary side switch SS1 and its gate signal

Fig. 10 shows the ZVS turn-on and ZCS turn-off waveforms of the secondary switches at the outputs. As shown in the figure, since ZVS turn-on and ZCS turn-off can be easily achieved at each secondary switch all over the charge process, there are no switching losses at the additional secondary switches.

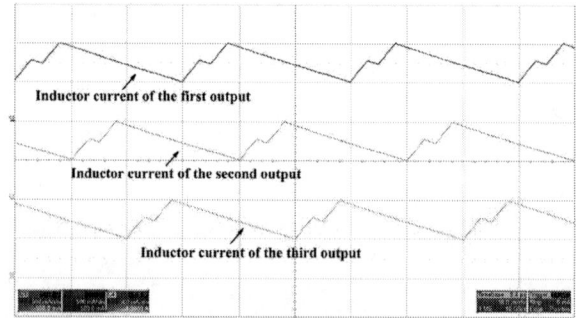

Fig. 11. The inductor current waveforms at the outputs of the proposed multiple converter

Fig. 11 shows the inductor current waveforms at the outputs of the proposed converter. As shown in the figure, only one of the secondary switches is turned on to regulate each output independently and tightly.

It is observed that the proposed converter with TDMC method satisfies the ripple requirements of the Li-Po batteries during overall period of the charge process. As shown in Fig. 12 and Fig. 13, the voltage ripple is less than 2% (0.252V) and the current ripple is less than 5% (0.6A). Both are lower than the maximum allowable ripple values suggested by the battery manufacturer [1].

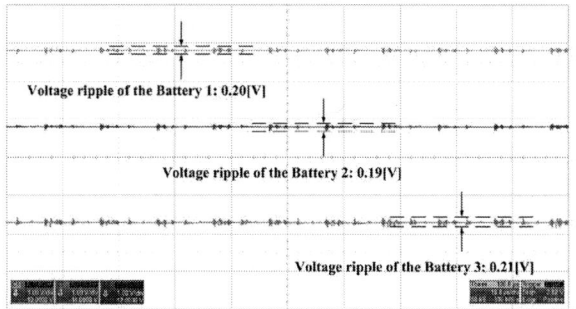

Fig. 12. The output voltage waveforms of the proposed charger

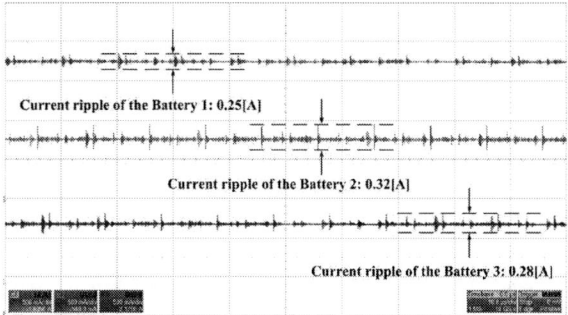

Fig. 13. The output current waveforms of the proposed charger

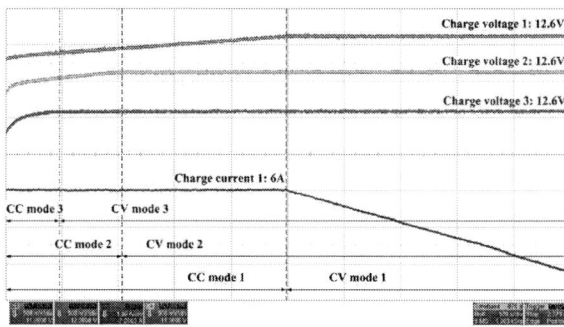

Fig. 14. Charge voltage and current profiles of the battery loads by using the proposed converter

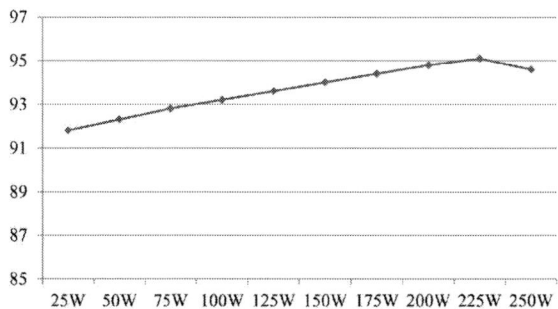

Fig. 15. Efficiency profile of the proposed multiple converter

Fig. 14 shows the charge current and voltage profiles of the three Li-Po batteries in different SOCs during the charge process by the proposed multiple converter

respectively. The figure shows that the proposed converter and the TDMC method work properly to charge three batteries independently and precisely by CC/CV charge modes. It is also shown that the all three batteries can be successfully charged with no cross regulation problem simultaneously by the proposed charger and the TDMC method.

Fig. 15 shows the efficiency profile of the proposed charger. It varies from 91.8% to 95% according to the load and the maximum efficiency is 95% at 90% load.

VI. CONCLUSION

In this paper, a multiple output battery charger based on the novel Time Division Multiple Control (TDMC) technique is proposed and applied to the phase shift full bridge topology to charge for three Li-Po batteries. The proposed charger can regulate all three outputs precisely and independently with only one secondary winding of the transformer. The ZVS turn-on at the primary switches and ZVS turn-on and ZCS turn-off of the secondary switches are achieved all over charge process thereby reducing the switching losses. With the help of a digital signal processor capable of high speed operation, the TDMC method can be simply implemented. If the proposed method is applied for the EV charger, it can provide several advantages in terms of volume, cost, efficiency and the installation area.

ACKNOWLEDGMENT

This work was supported by the Human Resources Development program (No.20124030200070) of the Korea Institute of Energy Technology Evaluation and Planning (KETEP) grant funded by the Korea government Ministry of Knowledge Economy.

REFERENCES

[1] Quest Battery, "Harding Battery Handbook for Lithium Polymer," Section 6 [Online].
Available: http://www.hardingenergy.com/handbook/

[2] Youhao Xi, and Praveen K. Jain, "A Forward Converter Topology With Independently and Precisely Regulated Multiple Output," *IEEE Transactions On Power Electronics*, vol. 18, pp.648–658, 2003.

[3] Sombuddha Chakraborty, Amit Kumar Jain, and Ned Mohan, "A Novel Converter Topology for Multiple Individually Regulated Outputs," *IEEE Transactions On Power Electronics*, vol. 21, pp.361–369, 2006.

[4] Daniele Trevisan, Paolo Mattavelli, and Paolo Tenti, "Digital Control of Single-Inductor Multiple-Output Step-Down DC-DC Converters in CCM," *IEEE Transactions On Power Electronics*, vol. 55, pp.3476–3483, 2008.

[5] J. Sebastian, J. Uceda, M.A. Perez, M. Rico, and F. Aldana, "A Very Simple Method to Obtain One Additional Fully Regulated Output in Zero-Current-Switched Quasiresonant Converter," *Power Electronics Specialists Conference*, San Antonio, vol. 21, pp.536–542, 1990.

[6] Dearborn. S, "Charging Li-Ion Battery for Maximum run times," *Power Electronics Technology Magazine*, 2005.

[7] M.E. Hervas, S. Vazquez, M. Reyes, J.M. Carrasco, and E. Dominguez, "A Dual-Loop PI controller for a DC/DC Full-Bridge Power Converter with ZVS modulation," *IEEE 35th Annual Conference of the Industrial Electronics*, pp. 37-41, 2009.

DC-breaker for a Multi-Megawatt Battery Energy Storage System

Georgios D. Demetriades, Willy Hermansson,
Jan R Svensson, Konstantinos Papastergiou
ABB AB
Corporate Research
721 78 Västerås, SWEDEN

Tomas Larsson
ABB AB
Power Systems, HVDC
721 64 Västerås, SWEDEN

Abstract— Energy storage will gain importance in the future power system due to increased amount of renewables. In this paper, a utility scale Battery Energy Storage System (BESS) for megawatt utilizing a high voltage and a high power Voltage Source Converter (VSC) is proposed to be able to inject/consume the required power and energy to the grid. The batteries are both series and parallel connected on the DC-side of the VSC. In order to obtain high reliability, high availability and safety of the BESS, this paper proposes and analyzes a distributed Solid-State DC Breaker protecting the BESS and its batteries.

Keywords— Battery Energy Storage System, Distributed Solid-State DC-Breaker, Multi-Megawatt, VSC, protection

I. INTRODUCTION

The integration of energy storage systems into transmission and distribution networks has the potential to provide significant benefits at all points in the supply chain. Increasing penetration of distributed generation (DG), particularly the ones based on renewable energy resources, is driving the need for distributed energy storage to provide services that will allow existing network assets to continue to deliver reliable, high quality electricity [1, 2]. Energy storage systems are deployed across the world using hydro-pumped storage, lead-acid batteries and compressed-air energy storage. Existing systems tend to provide utility level assistance to solve specific problems. The installed base of storage solutions remains low due to high capital costs, lack of proven methods for providing revenue streams and the relatively unproven technology of distributed storage systems.

There are several technologies able to provide energy storage and their characteristics make them more or less suitable to operate at particular power and energy levels [3] and [4].

The Li-Ion battery technology was chosen for our Battery Energy Storage System (BESS) due to benefits on: high-power density, many cycles before end-of-life, mature technology, high round-trip efficiency, and high-charge retention. The batteries are both series and parallel connected on the DC-side of the high-voltage voltage source converter (VSC) [5,6] to build up the required voltage and current level as shown in Fig. 1.

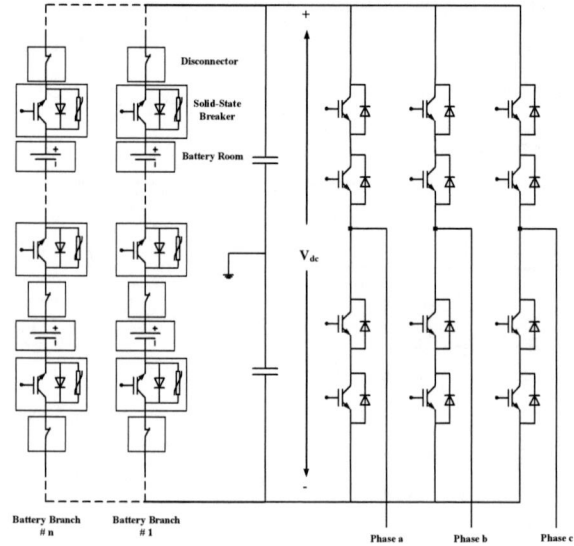

Fig. 1. Scheme of multi-megawatt battery Energy Storage System using solid-state DC breaker.

The maximum number of parallel battery branches is limited by the rated current of the VSC. Each branch is composed of series-connected batteries. For safety, they are divided into sections and each section has its own enclosed room. The battery rooms are series connected via a DC Breaker, which is distributed into each battery room to prevent short-circuits around and in the branches. The system requirements and specifications corresponding to a DC-link voltage of 80 kV are summarized in Table 1.

This paper proposes and analyzes a distributed Solid-State DC Breaker protecting a multi-mega BESS and its batteries.

TABLE 1
SYSTEM REQUIREMENTS AND SPECIFICATIONS OF BATTERY PART OF MULTI-MEGAWATT BESS

Current [kA]	Power [MW]	# of Battery Rooms/Branch	# of Branches	Prospective short-circuit current [kA]
0,6	42	27	4	12
1.2	84	27	8	12
1.8	125	27	12	12

II. DC BREAKERS

Avoiding and handling short-circuits in power system networks are important to provide safety and achieve high reliability, availability and power quality. Normally, Mechanical Circuit Breakers (MCB), [7, 8, 19], are used to interrupt fault currents in AC grids. After detecting short-circuit or an over-load, the opening of the electrical contacts in MCB takes several line cycles. An arc appears across the contact gap due to current flowing through it. The arc impact of the fault current is limited and the arc can only be extinguished at current zero-crossings, assuming that the plasma is significantly cooled down to avoid arc re-ignitions.

In DC, zero crossing of fault currents will not occur, resulting in a continuously increased fault current and traditional MCBs will not work. In order to interrupt DC-fault currents, dedicated DC Breakers are needed. A number of configurations [30] on different DC Breaker for specific applications exists. The three major configurations are: (a) Mechanical breakers employing a resonance circuit; (b) Solid-state breakers and (c) Hybrid breakers [10-20]. Fig. 2 illustrates the different configurations [20-29] and the major features are briefly summarized in Table 2.

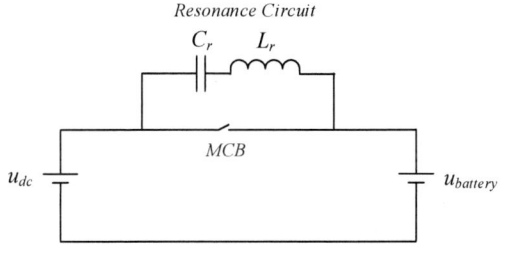

(a) Mechanical Circuit Breakers employing a resonance circuit

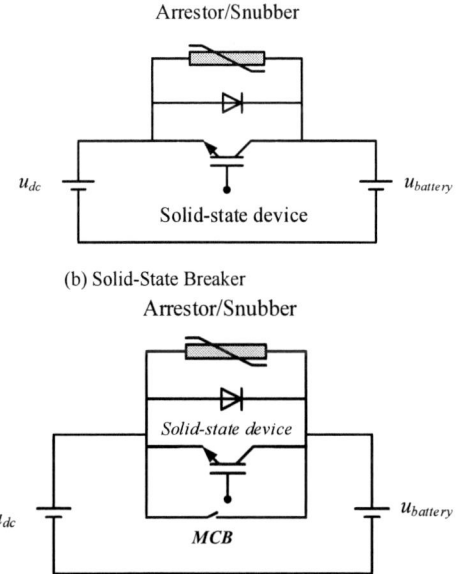

(b) Solid-State Breaker

(c) Hybrid-circuit Breaker

Fig. 2. DC Breaker configurations.

TABLE 2
MAJOR FEATURES OF DC BREAKER CONFIGURATIONS

Type	Complexity	Speed	Current	Losses	Cost
a	High	Low	High	Low	Moderate
b	Moderate	High	Moderate	High	Moderate
c	High	Moderate	Moderate	Low	High

Notice that the configurations employing mechanical mechanisms, e.g. mechanical breaker, are suffering in terms of limited number of interruptions.

One major requirement concerning the BESS application is the interruption speed in case of a fault to avoid malfunctions and high fault currents. Fault currents can be interrupted before reaching unmanageable levels by using fast enough dc-circuit breakers. Therefore, the solid-state breaker configuration has been selected.

To reduce costs, total losses and handle all possible short-circuit faults in the DC Breaker, the solid-state circuit breakers have been distributed to one breaker in each battery room. As a result, and by synchronizing switching of the circuit breakers, the component costs have been comparably reduced.

A. Solid-State DC Breaker

In Fig. 3 the basic operation of the solid-state DC Breaker with an arrester at short-circuit fault is shown.

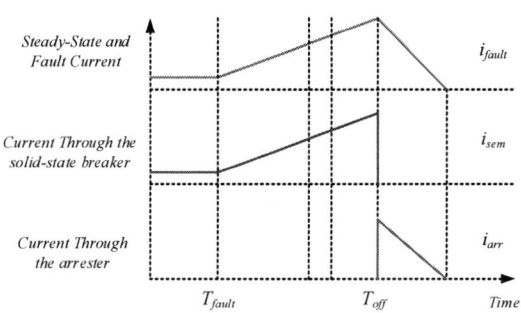

Fig.3. Basic operation of the solid-state DC Breaker.

During normal operation of the system, i.e. during charging or discharging of the batteries, the current is flowing through the solid-state dc-breaker. When a fault has occurred at T_{fault}, the current increases with the slope determined by the circuit parameters (inductance and resistance) in the current path. At the time instant T_{off} the controllable switch is turned off. Since inductance is present in the current path a considerable amount of inductive energy is stored. At turn-off the stored energy has to be removed by the overvoltage limiting elements, e.g. arrester or snubber circuit, which is placed in parallel with the solid-state semiconductor switch/device. This implies a current flow through the overvoltage element, shown in Fig. 3. The stored inductive energy is given by

$$E_{stored} = \tfrac{1}{2} L \, i_{fault}^{2} \qquad (1)$$

As shown, the stored energy is proportional to the inductance enclosed in the current path and to the square of the fault current peak. To reduce the energy, and as a result relax the requirements of the energy absorbing arrangement, either the loop inductance is reduced or the peak value of the fault current is limited. The peak value of the fault current can be reduced by increasing the fault detection speed and the interruption speed of the solid-state DC breaker.

In Fig. 4 the configuration that has been chosen for a multi-megawatt BESS is shown. Due to the high DC-link voltages (up to 80 kV) and a battery room voltage of 3 kV, 4.5 kV, 1.8 kA StakPak [31] IGBTs have been chosen as semiconductor switch. Moreover, water cooling is used. A Valve Control Unit (VCU) controls all IGBTs via the Gate Drive Units (GDU) and the measured battery branch currents are fed to the VCU. This current is used for short-circuit protection. As shown in Fig. 4, an RCD snubber has been used to absorb the excess amount of energy stored in the current path at IGBT turn-off.

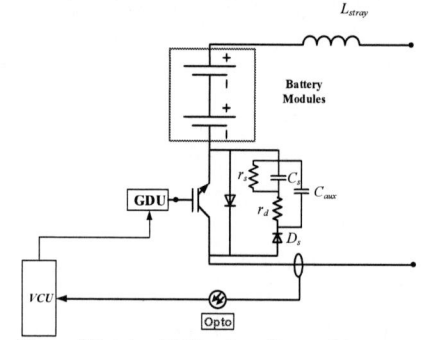

Fig.4. The solid-state DC Breaker for multi-megawatt BESS applications.

In Fig. 5, the geometrical arrangement of the battery modules in the battery room is shown. The total loop stray inductance and the stray inductance of the battery module arrangement are illustrated in Fig. 5(a). In Fig. 5(b) the stray components, i.e., stray inductance and capacitance, and the resistance, are shown and are distributed in the overall geometry of the battery room.

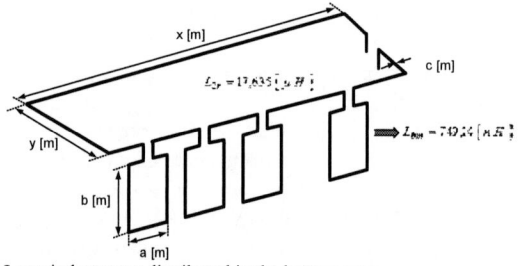

(a) Loop inductances distributed in the battery room

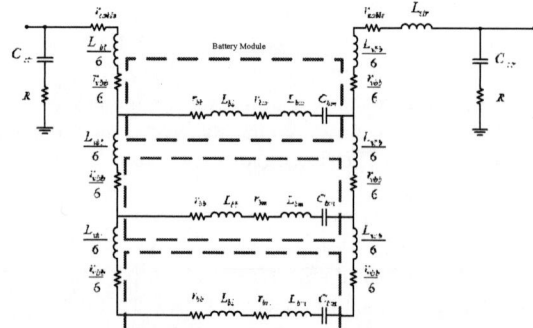

(b) The stray components are distributed in the battery room

Fig. 5. The stray components distributed in the battery room.

In order to calculate the stray inductance corresponding to different parts of the battery room Eq.(2) has been used.

$$
L_{stray} = N^2 \, \frac{\mu_0 \, \mu_r}{\pi}
\begin{bmatrix}
-2\left(w+h\right)+2\sqrt{h^2+w^2} \\[4pt]
-h\ln\left(\dfrac{h+\sqrt{h^2+w^2}}{w}\right) \\[8pt]
-w\ln\left(\dfrac{w+\sqrt{h^2+w^2}}{h}\right)+ \\[8pt]
h\ln\left(\dfrac{2\,w}{\alpha}\right)+w\ln\left(\dfrac{2\,h}{\alpha}\right)
\end{bmatrix}
\tag{2}
$$

where w is the length of the loop, h is the width of the loop, α is the diameter of the cable, N is the number of turns, μ_r is the magnetic permeability of the material and μ_0 is the magnetic permeability of the air. In Fig. 6(a) the different values of stray inductances corresponding to different loop geometries are shown. Similarly in Fig. 6(b) the stored energy corresponding to different values of stray inductances and currents are illustrated. Thus, the snubber circuit has to be designed in order to handle the stored energy corresponding to the stray components of the battery room and for a certain maximum current.

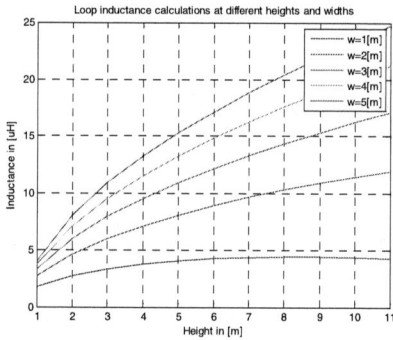

(a) Stray inductance calculations for different loops

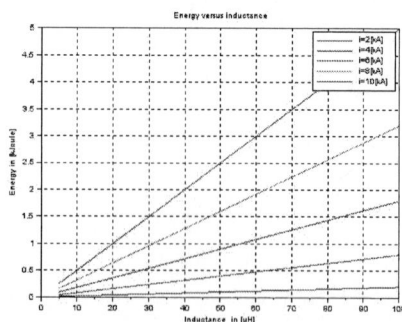

(b) Energy calculations

Fig. 6.(a) Stray inductance calculations corresponding to different heights and widths of the loop. (b) Energy calculations corresponding to different stray inductances and currents.

B. *RCD Snubber Design and considerations*

The snubber absorbs the stored energy in the loop that otherwise could cause an overvoltage across the semiconductor of the solid-state DC Breaker and destroy it when it is turned off during high fault currents. The main components of the protection network are the resistor marked r_s and the capacitor marked C_s. The diode D_s will only allow current to flow in one direction hence blocking the resonant current from flowing back and forth to the battery room and the parasitic inductances of the system. The capacitor C_{aux} is used to alleviate the effect of parasitic inductance within the snubber network. The resistor r_d is used to discharge the snubber capacitors after the IGBT turn-off.

The RCD snubber circuit is shown in Fig. 7. Note that the RCD snubber is not a lossless snubber arrangement. If choosing a lossless snubber arrangement, a considerably larger capacitor across the semiconductor is needed. This implies that the footprint of the solid-state DC Breaker will considerably increase.

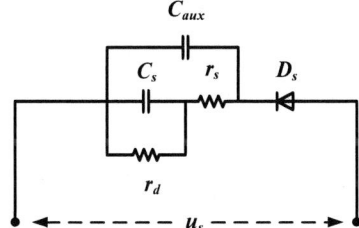

Fig. 7. The proposed RCD snubber.

Once the solid-state DC Breaker is turned-off, the short-circuit current is redirected through the snubber. The snubber voltage u_s is the resultant voltage across C_s and the dumping resistor r_s.

The voltage rise across C_s can be approximated by using the energy balance between the stored inductive energy and the snubber capacitance, as shown in Eq.(3), assuming that resistive losses are negligible.

$$\frac{1}{2} L_{stray}\, i_{fault}^{\,2} = \frac{1}{2} C_s\, \Delta u^2 \tag{3}$$

The voltage across the dumping resistor is given by

$$u_{r_s} = i_{fault}\, r_s \tag{4}$$

By combining Eq.(3) and Eq.(4), the minimum snubber capacitance C_s is given by

$$C_s = \frac{L_{stray}\, i_{fault}^{\,2}}{\left(u_s - u_{r_s}\right)^2} \tag{5}$$

where u_s is the maximum allowed voltage across the solid-state dc-circuit breaker and L_{stray} of 23,1 µH is a typical stray inductance value corresponding to a battery room.

The value of the capacitor C_{aux} can be defined by

$$C_{aux} = \frac{L_{stray\,snubber}\, i_{res\,fault}^{\,2}}{u_s^{\,2}} \tag{6}$$

where, $L_{stray\,snubber}$ is the stray inductance of the snubber circuit, and $i_{res\,fault}$ is the peak resonant fault current through the snubber.

C. Thermal Design and Considerations

Since the charge and discharge current continuously flows through the solid-state dc-circuit breakers it is of importance to calculate the losses and perform a thermal analysis. The losses will entirely originate from the on-state losses of the IGBT since the dc-circuit breaker is not switched on and off regularly, besides turning on at the start of the discharge period and turning off at the end. However, the turn-on and turn-off are normally done at zero current and zero voltage across the solid-state DC Breaker resulting in zero switching losses. The conduction losses have been calculated by using

$$P_{IGBT} = i_{DC}\, V_{ce0} + i_{DC}^2\, r_{CE} \tag{7}$$

where i_{DC} is the current, V_{ce0} is the emitter threshold voltage of the IGBT and r_{CE} is its slope resistance.

The conduction losses during discharging of the batteries are presented in Fig. 8 for different current levels.

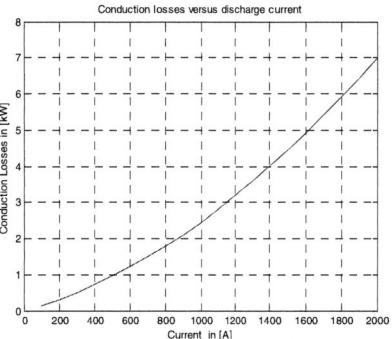

Fig.8 Conduction losses during discharging of the batteries.

The analysis of thermal behavior of the solid-state DC Breaker uses the dynamic thermal network, presented in Fig. 9, have been calculated by using Eq.(8).

$$Z_{th\_x(t)} = Z_{th\_x} \left(1 - e^{-\frac{t}{\tau_{th\_x(t)}}} \right) \tag{8}$$

where $Z_{th\_x(t)}$ is the dynamic thermal impedance at a time instant $\tau_{th\_x(t)}$. The $Z_{th\_x(t)}$ is given in K/kW and $\tau_{th\_x(t)}$ is given in seconds.

Knowing the conduction losses during discharge and the value of the dynamic thermal impedances, the maximum allowed cooling water temperature can be estimated. The results are shown Table 3.

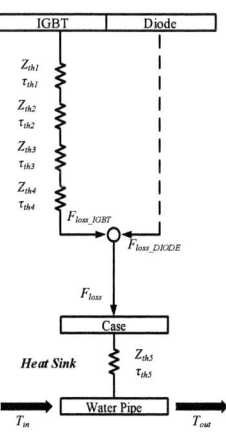

Fig.9. Dynamic thermal network of semiconductor switch of solid-state DC Breaker.

TABLE 3
ALLOWED WATER TEMPERATURE OF SEMICONDUCTOR SWITCH OF
SOLID-STATE DC BREAKER AT 1500 A.

Chip Temperature in Celsius degrees	Allowed Water Temperature in Celsius degrees
110	52
125	67

III. SYSTEM SIMULATIONS

Simulations have been performed in order to study the behavior of the breaker and confirm the theoretical results. The cases that have been studied are summarized in Table 4.

TABLE 4
DIFFERENT SIMULATION CASES FOR SOLID-STATE DC BREAKER

	Stray Inductance [μH]	Turn-off Current in [kA]	Snubber Capacitance C_s in [μF]
Case 1	23.1	3.4	0
Case 2	23.1	3.5	90
Case 3	23.1	6.8	90
Case 4	23.1	6.8	200

In Fig. 10 the simulation results corresponding to Case 1 shown in Table 4 are illustrated. Note that the voltage of the battery room is assumed to be 3 kV.

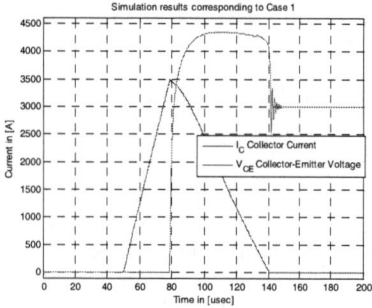

Fig.10. Simulation results corresponding to Case 1.

The waveforms shown are collector current (blue) and collector-emitter voltage (red). As shown, the peak collector-emitter voltage V_{CE} is 4.3 kV and the turn-off energy, E_{off}, is estimated at around 430 J. The voltage stress across the component is high and might result into a component failure.

In Fig. 11 the simulation results corresponding to Case 2, as shown in Table 4, are displayed.

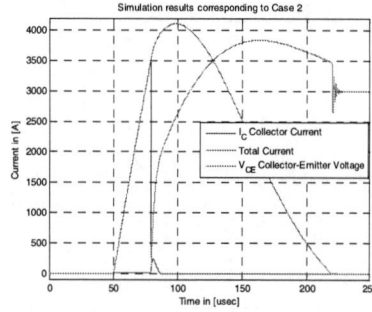

Fig. 11. Simulation results corresponding to Case 2.

The waveforms shown are collector current (blue), collector-emitter voltage (red) and total current (green). Thus the collector current and the snubber current. As shown in Fig.11 the peak collector-emitter voltage, V_{CE},

is 3.8 kV and the turn-off energy, E_{off}, is 1.1 J. On the other hand the energy dissipation in the snubber resistor, E_r, is estimated at around 455 J.

Case 3, as shown in Table 4, corresponds to a situation, where the detection system is malfunctioning. Thus, delays in the measuring system or failure of the measuring system to detect a fault condition. In this case the fault current will be comparably increased. In Fig. 12 the simulation results are illustrated.

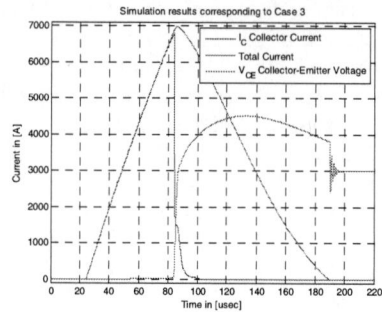

Fig.12. Simulation results corresponding in Case 3.

The waveform shown are collector current (blue) and collector-emitter voltage (red). As shown, the peak collector-emitter voltage, V_{CE}, is 4.55 kV and the turn-off energy, E_{off}, is estimated at around 3.8 J. Additionally, the energy dissipation in the snubber resistor, E_r, is 770 J. In order to mitigate the risk of measuring system malfunction the snubber capacitance must be increased. This has been considered in Case 4 as shown in Table 4.

In Fig.13 the results corresponding in Case 4 are illustrated. As shown, the peak collector-emitter voltage, V_{CE}, has been limited to 3.7 kV and the turn-off energy, E_{off}, is 3.8 J. The energy dissipation in the snubber resistor, E_r, has been increased to 1210 J. The voltage stress across the component is manageable but the snubber capacitor has been considerably increased. This implies in a comparably increase of the footprint of the solid-state DC Breaker.

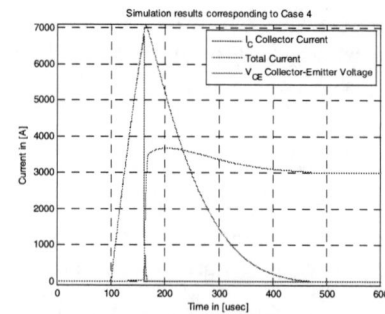

Fig.13. Simulation results corresponding in Case 4.

The simulation results for the different cases are summarized in Table 5.

TABLE 5
SUMMARY OF SIMULATION RESULTS FOR SOLID-STATE DC BREAKER
WITH 3KV BATTERY ROOM VOLTAGE

	Peak V_{CE} in [kV]	IGBT E_{off} in [J]	E_r in [J]
Case 1	4.3	430	0
Case 2	3.8	1.1	455
Case 3	4.55	3.8	770
Case 4	3.7	3.8	1210

As already mentioned in the article, the stray inductance of the system has an impact on the dimensioning and the performance of the breaker. By reducing the stray inductance of the system the performance of the breaker can be considerably improved.

IV. EXPERIMENTAL RESULTS

An experimental set-up has been assembled in order to prove the feasibility of the solid-state DC Breaker as well as to study the behavior of the breaker during fault and transient conditions. 4.5 kV, 1.8 kA StakPak IGBTs have been used as semiconductor switches. The experimental set-up is shown in Fig.14.

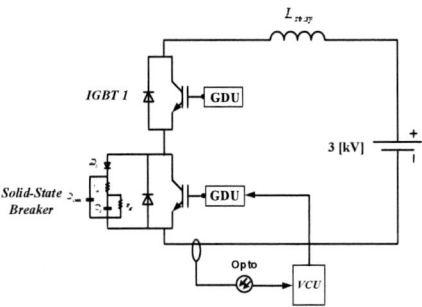

Fig.14. Experimental set-up to prove feasibility of the solid-state DC Breaker.

The sequence starts by turning on the solid-state DC Breaker. The current through the breaker is zero since it is not connected to the battery with IGBT1 in off-state. In order to emulate a fault, the upper switch, IGBT 1, is turned on. The battery is short-circuited by the solid-state DC Breaker and IGBT 1 via an inductor representing the stray capacitance of a battery room. The short-circuit detection unit is turning-off the breaker when the rate of increase of the current is exceeding a defined threshold value.

In Fig. 15 the results corresponding to a stray inductance of 25 µH are illustrated. In the upper graph, the collector current (blue) and the snubber current (red) are shown. Similarly, in the lower graph the collector-emitter voltage is shown. As displayed, the turn-off current is 3.5 kA and the peak collector-emitter voltage is 3.5 kV. Notice that the snubber capacitance is 90 µF, the auxiliary capacitance is 6 µF and the snubber resistor is 0.5 Ω.

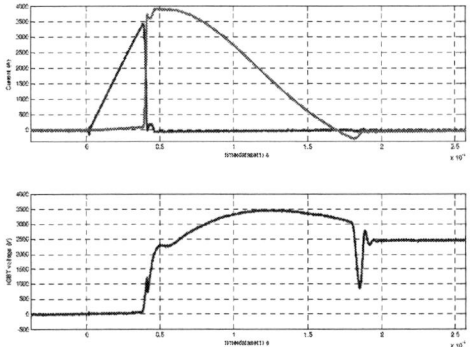

Fig.15. Measured results showing collector and snubber current (top). The collector-emitter voltage (bottom). X-axis 50 µs per division.

The impact of the auxiliary capacitor has been studied experimentally as well. Thus, the auxiliary capacitor has been decreased to 3 µF. The obtained results for the case with a battery room voltage of 1500 V and turn-off at 1.5 kA are shown in Fig.16. The results corresponding to the lower auxiliary capacitance are plotted in blue.

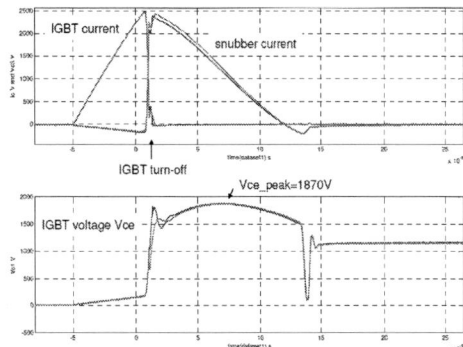

Fig.16. Experimental results obtained by reducing the auxiliary capacitor from 6 µF (red) to 3 µF (blue). X-axis 50 µs per division.

As shown in Fig.16, the results are almost identical. The major deviation are the high-frequency oscillations appearing on the collector-emitter voltage during the turn-off of the solid-state DC Breaker. Those oscillations can reached unacceptable levels and resulting in a breaker failure. The high-frequency oscillations are depending on the stray inductance of the snubber circuit.

Fig.17 illustrates the operation of the short circuit detection unit for a case with a battery voltage of 1500V and turn-off at 1.8 kA. The detection unit is designed by a Rogowski coil measuring the di/dt of the fault current. When the di/dt is exceeding the threshold level, the solid-state DC Breaker is turned off. In Fig. 17 the signal from the Rogowski coil is plotted in blue in the lower graph. In the upper graph, with fault collector current (in blue) and snubber current (in red), the di/dt of the fault current is estimated around 46 A/µs which is above the threshold value. The high di/dt has been successfully detected and the breaker has been turned-off as shown in the figure after approximately 40 µs, which in this case was the "delay" time set in order to avoid spurious detections. The middle graph shows the collector-emitter voltage.

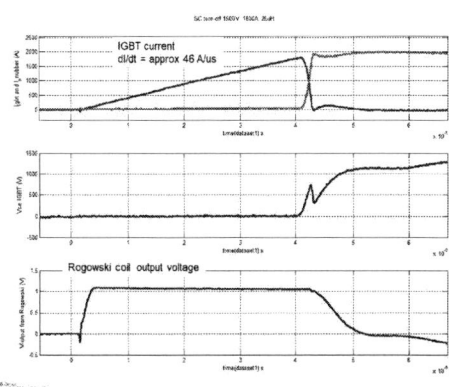

Fig.17. Measurements illustrating the short-circuit detection unit of solid-state DC Breaker. X-axis 10 µs per division.

V. CONCLUSIONS

In the paper DC Breakers have been briefly discussed. The system requirements and the specifications of the multi-megawatt BESS have been summarized and the major challenges in terms of breaker stresses and critical design parameters have been presented. The solid-state DC Breaker has been chosen since the response time is considerably lower and exhibit unlimited operations. One of the most important parameters affecting the operation of the solid-state DC Breaker is the stray inductance of the system. High stray inductance implies in current limitation but on the other hand the energy stored in the stray inductance has to be absorbed by the absorbing circuit in order to avoid over voltages across the breaker. Over voltages will lead to breaker failures. The solid-state DC Breaker has been validated both theoretically and experimentally.

Finally, it is worth noticing that the presented solid-state DC Breaker is not a universal solution. In the specific application it has been proven to be the optimum solution.

REFERENCES

[1] J. P. Barton and D. G. Infield, 2004, "Energy storage and its use with intermittent renewable energy", *IEEE Transaction on Energy Conversion. vol. 19, 441-448.*

[2] J.O.G. Tande, 2000, "Exploitation of wind-energy resources in proximity to weak electric grids", Appl. Energ. vol. 65, 395-401.

[3] P.J. Hall and E.J. Bain, 2008, "Energy-storage technologies and electricity generation", Energ Policy.vol. 36(12), 4352-4355.

[4] W. Jewell, et al, 2004, Evaluation of Distributed Electric Energy Storage and Generation, Final Report. PSERC Publication 04-25, Wichita State University, USA.

[5] R. Grünbaum, et al, 2001, "SVC Light: Evaluation of first installation at Hagfors, Sweden," CIGRÉ conference, paper 13/14/36-03, Paris, France.

[6] T. Larsson, B. Ratering-Schnitzler, 2000, "SVC Light: A utility's aid to restructuring its grid," IEEE PES winter meeting, Singapore.

[7] W. Rieder, "Circuit breakers Physical and engineering problems I - Fundamentals," *IEEE spectr.* , vol. 7, no. 7, pp. 35-43, July 1970.

[8] R.G. Hoerauf, D.D. Shipp, "Characteristics and applications of various arc interrupting methods ,"in *Proc. Rec. IEEE Paper and Pulp Ind. Technol. Conf.*, Jun 1991, pp.151- 163.

[9] W. Rieder, "ARC-Circuit Interaction Near Current Zero and Circuit Breaker Testing," *IEEE Trans. Power App. Syst.*, no.3, pp.705-713, May 1972.

[10] A.M.S. Atmadji, J.G.J. Sloot, "Hybrid switching: a review of current literature,"in *Proc. Rec. IEEE EMPD Conf.*, vol.2, pp.683-688, 3-5 Mar 1998.

[11] N.Y.A. Shammas, "The role of semiconductor devices in high power circuit breaker applications," in *Proc. Rec.Pow. Syst.and Electromag. Compatibility of WSEAS*, pp. 257-262, 23-25 Aug. 2005

[12] N.Y.A. Shammas, "Combined conventional and solid-state device breakers," in *Proc. Rec. IET Power Semicond. Devices conf.*, pp.1-5, 14 Dec 1994.

[13] S.B. Tennakoon, P.M. McEwan,"Performance evaluation of an internally energized thyristor circuit breaker with rapid auto reclose," in *Proc. Rec. IEEE APEC* , pp.686-690, 7-11 Mar 1993.

[14] P.M. McEwan, S.B. Tennakoon, "A two-stage DC thyristor circuit breaker," *IEEE Tarns. Power Electron.*, vol.12, no.4, pp.597-607, Jul 1997.

[15] C. Meyer, R.W. De Doncker, "Solid-state circuit breaker based on active thyristor topologies," *IEEE Trans. Power Electron.*, vol.21, no.2, pp. 450- 458, March 2006.

[16] J. Zyborski, J. Czucha, M. Sajnacki, "Thyristor circuit breaker for overcurrent protection of industrial d.c. power installations," *IET Trans. Elec. Eng.*, vol.123, no.7, pp.685-688, July 1976.

[17] Mu Jian-guo, Wang Li, Hu Jie, "Research on main circuit topology for a novel DC solid -state circuit breaker," in *Proc. Rec. IEEE ICIEA Conf.*, pp.926-930, 15-17 June 2010.

[18] T. Ueda, M. Morita, H. Arita, Y. Kida, Y. Kurosawa, T. Yamagiwa, "Solid-state current limiter for power distribution system," *IEEE Trans. Power Del.*, vol.8, no.4, pp.1796-1801, Oct 1993.

[19] L. Klingbeil, W. Kalkner, Ch. Heinrich, "Fast acting solid state circuit breaker using state of art power electronics devices,"

[20] T.F. Podlesak, J.A. McMurray, J.L. Carter, "An all solid state fault interrupter using gto's," in *Proc. Rec. IEEE Pulsed Power Conf.*, pp.344-347, 1989.

[21] R.K. Smith, P.G. Slade, M. Sarkozi, E.J. Stacey, J.J. Bonk, H. Mehta, "Solid- state distribution current limiter and circuit breaker: application requirements and control strategies," *IEEE Trans. Power Del.*, vol.8, no.3, pp.1155-1164, July 1993.

[22] C. Meyer, S. Schroder, R.W. De Doncker, "Solid-state circuit breakers and current limiters for medium-voltage systems having distributed power systems," *IEEE Trans. Power Electron.*, vol.19, no.5, pp. 1333-1340, Sept. 2004.

[23] S.B. Tennakoon, P.M. McEwan, "Short-circuit interruption performance of thyristor circuit breakers," in *Proc. Rec. IEEE APEC*, pp.832-838 vol.2, 13-17 Feb 1994.

[24] L. Palav, A.M. Gole, "On using the solid state breaker in distribution systems," in *Proc. Rec. IEEE Electr and Comput. Eng. Conf.*, vol.2, pp.693-696, 24-28 May 1998.

[25] T. F. Podlesak, H. Singh, K. Fonda, J. Creedon, G.L. Schofield, F. O. Johnson, "Megawatt high speed solid state circuit breaker for pulse power applications," in *Proc. Rec. IEEE Pulsed Power Conf,* vol.2, pp.984, 21-23 Jun 1993.

[26] C. Meyer, M. Kowal, R.W. De Doncker, "Circuit breaker concepts for future high-power DC-applications," in *Proc. Rec. IEEE IAS Conf.*, vol.2, pp. 860- 866, 2-6 Oct. 2005.

[27] M. Takeda, H. Yamamoto, Y. Hosokawa, I. Kamiyama, "A low loss solid-state transfer switch using hybrid switch devices ," in *Proc. Rec. IEEE IPEMC Conf.*, vol.1, pp.235-240, 2000

[28] J. Czucha, T. Lipski, J. Zyborski, "Hybrid current limiting interrupting device for 3-phase 400 V AC applications, "in *Proc. Rec. IET Conf.*, pp.161-166, 10-12 Nov 1998.

[29] T. Genji, O. Nakamura, M. Isozaki, M. Yamada, T. Morita, M. Kaneda, "400 V class high-speed current limiting circuit breaker for electric power system", *IEEE Trans. Power Del.*, vol.9, no.3, pp.1428-1435, Jul 1994.

[30] R.Kapoor, A. Shukla, G.D. Demetriades, " State of art of power electronics in circuit breaker technology", *IEEE Energy Conversion Congress and Exposition 2012, pp. 615-622, Sept. 2012*

[31] ABB Semiconductors, " StakPak for industrial Applications", *http://new.abb.com/products/semiconductors/insulated-gate-bipolar-transistor-(igbt)-and-diode-modules/stakpak*

978-1-4799-2706-7/14 $31.00 © 2014 IEEE

Energy Management Method Using the IIR Filter for PEMFC-supercapacitor Hybrid Power Source

Tatsuya Yamanaka
Dept. of Electrical Engineering
Tokyo University of Science
Chiba, Japan
j7310154@ed.tus.ac.jp

Shuhei Tosaka
Dept. of Electrical Engineering
Tokyo University of Science
Chiba, Japan
j7313649@ed.tus.ac.jp

Noboru Katayama
Dept. of Electrical Engineering
Tokyo University of Science
Chiba, Japan
katayama@rs.tus.ac.jp

Sumio Kogoshi
Dept. of Electrical Engineering
Tokyo University of Science
Chiba, Japan
kogoshi@ee.noda.tus.ac.jp

Abstract— **A miniature proton exchange membrane fuel cell (PEMFC) is expected to be applied to small electronic devices such as smartphones. However, PEMFC cannot response quickly against fast load changes of electronic devices. Hybrid power source system using fuel cell (FC) and supercapacitor (SC) with multi-port bidirectional DC-DC converter has been proposed. In this work, the energy management method has been proposed. In this method, the SC voltage in the range of lower frequencies was extracted by using finite impulse response (FIR) low-pass filter (LPF) and determined the reference value for the FC through proportional integral (PI) control method. In this study, infinite impulse response (IIR) filter, which would reduce the number of calculations compared to a FIR filter, was used as LPF. A simplified numerical model was constructed to ensure the system stability. It was confirmed that the system with the IIR filter worked properly.**

Keywords— *IIR filter, Multi-port bidirectional DC-DC converter, Fuel cell, Supercapacitor.*

I. Introduction

As smartphones are spreading around the world, the battery durability is one of the main problems because the multi-functional smartphones require more power consumption. The proton exchange membrane fuel cell (PEMFC), which is environmentally friendly and has higher energy density than the conventional secondary batteries such as Li-ion batteries, is expected to be an alternative power source for many applications.

However, PEMFCs cannot response quickly against fast load changing of electronic devices, and under this condition, the catalyst and polymer electrolyte membrane, which are essential components for PEMFCs, may suffer critical damage in short time [1]. Besides, operating PEMFCs under particular frequencies and power accelerate the degradations of the polymer electrolyte membrane [2], [3], [4]. Consequently, FC voltage

requires to be changed as slowly as possible. On the other hand, supercapacitors (SCs) are characterized by extremely higher output response and long life time against charge/discharge cycles. Hybrid power source systems using FCs with SCs for many applications such as vehicles, forklifts et al. have been proposed [5], [6].

Recently, a miniature PEMFC that achieved an active area of 3 mm x 3 mm and a peak power density of 145 mW/cm^2 was produced [7] and expected to be applied to small electronic devices such as smartphones, laptop et al. [8]. To realize a PEMFC-SC hybrid power source system, multi-port bidirectional DC-DC converter has been developed to manage the energy flow among the FC, the SC and the load, and to regulate the voltages of their devices [9]., Many energy management strategies have been proposed [10], [11]. The FC surplus power against the required load power was charged to the SC and the insufficiency of the FC power was supplied from the SC. However, state of charge (SoC) of the SC has to be managed properly to avoid overcharging or over discharging. The average power consumption of smartphones changes from 0.1 W to 2 W during data transmission [12]. Thus, the FC output power needs to be changed. Besides, the reference value of the FC is necessary to change slowly because the FC cannot follow fast load changes. To realize this, a low pass filter (LPF) was applied to extract the SC voltage in the range of lower frequencies and proportional-integral (PI) controller was used in order to realize a feedback loop for the reference value of the FC voltage in our previous work [13]. A finite impulse response (FIR) filter or a first order filter, which are stable and easy to apply, was used as the LPF. However, the FIR filter requires higher orders and long calculation time. A first order filter has poor cut-off characteristics. For mobile use, the micro controller for the digital control would be limited. Long time occupation of CPU time for the filter calculation

978-1-4799-2706-7/14 $31.00 © 2014 IEEE

may cause interruptions of calculations for other control loops such as the load voltage regulation.

In this study, a new energy management method was proposed for smartphones. The infinite impulse response (IIR) filter, which would reduce the number of calculations compared with that of the FIR filter and has excellent cut-off characteristic, was employed as the LPF in the control loop for the SC voltage. To ensure the stability of the system with the IIR filter, a simplified numerical model was constructed with MATLAB/Simulink. After that, waveforms that were measured in advance from an actual smartphone under typical usages were used as the load power consumptions.

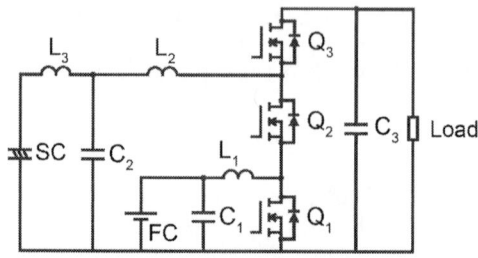

Fig. 1 Multi-port bidirectional DC-DC converter in our work.
The number of switching devices are reduced compared with the conventional topology.

II. Proposed Method

A. Multi-port Bidirectional DC-DC Converter

The multi-port bidirectional DC-DC converter, to which we apply the proposed method, and the operation mode are shown in Fig. 1 and Fig.2, respectively. It has three switching devices that are connected in series. The FC, the SCs and the load are connected via switching noise reduction filters consisting of inductors and capacitors are connected between the switching devices. The device parameters used at the converter are listed in TABLE I. The SCs are charged and discharged to compensate the difference between the load demand and output power from the FC by controlling the duty ratio of switching devices. The switching devices are controlled by a micro-controller (RL78G, Renesas Electronics). The operation mode has been explained in our literature [14].

B. New Energy Management Method

New energy management method we propose for the converter is illustrated in Fig. 3. The FC voltage v_{FC} and the load voltage v_{load} are periodically captured and quantized at 10-bit resolution by analogue-digital converter (ADC). Each voltage value is compared with each reference value v_{FC_ref} and v_{load_ref} and the difference between the captured value and the reference value is minimized by PI control method. The SC voltage v_{SC} is also periodically captured. The captured value for v_{SC} passes through LPF to extract v_{SC} in the range of lower frequencies and the filtered value is compared with the reference SC voltage value v_{SC_ref}. v_{FC_ref} is determined after the compared value is processed with a PI controller. Thus, the FC voltage is changed slowly, which would avoid the particular frequencies that could accelerate the FC degradation. Besides, Combining of the FC with the SCs would compensate the FC poor characteristics of load following. The switching frequency is fixed at 100 kHz by the micro-controller. It is important to reduce calculation time for the LPF because the micro-controller cannot do controlling the switching devices and calculation for the LPF simultaneously.

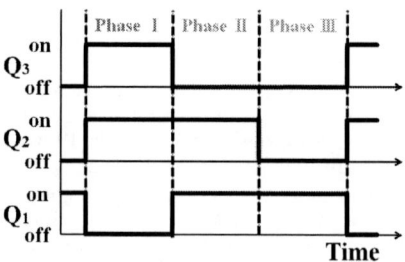

(a) Mode I (charging mode)

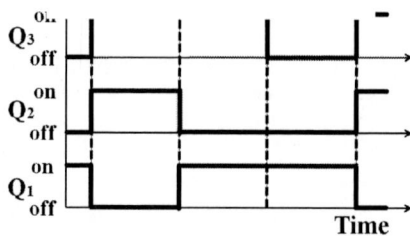

(b) Mode II (discharging mode)

Fig. 2. Operating mode

TABLE I
CURCUIT PARAMETER LISTS

Elements	Value
Inductance L_1	47 μH
Inductance L_2	10 μH
Capacitors $C_1/C_2/C_3$	100 μF
MOSFET $Q_1/Q_2/Q_3$	TOSHIBA TPCA8203-H

C. Low Pass Filter

The IIR filter was used as a LPF in this energy management method. There exist kinds of IIR filters such as the Butterworth filter, the Chebyshev filter, the Bessel filter et al. and these IIR filters have different cut-off and phase characteristics. In this study, the Chebyshev type II filter, which has a excellent cut-off characteristic with relatively lower orders compared with other IIR filters in spite of having ripple in the cut-off frequency band, was used as the LPF. The IIR filter would reduce the number of calculations but is prone to be unstable because IIR filter has feedback loops. The following equation describes the transfer function of IIR filter.

$$G_{IIRfilter} = \frac{\sum\limits_{i=0}^{n} b_i z^{-i}}{1 + \sum\limits_{i=1}^{n} a_i z^{-i}} \tag{1}$$

where a_i are the feedback coefficients and b_i are the feedfoward coefficients.

The transfer function of FIR filter is described as

$$G_{FIRfilter} = \sum\limits_{i=0}^{n} a_i z^{-i} \tag{2}$$

where a_i and b_i are coefficients which determine the characteristic of the filter.

III. Modeling

A numerical model was constructed with MATLAB/Simulink to ensure the system with the IIR filter stability because IIR filter is prone to be unstable. From here, the constructed model is detailed. The energy of the SC v_{SC} is described as

$$w_{SC} = \int_0^t \eta_{SCport}^s \left(\eta_{FCport} p_{FC} - p_{load} \right) d\tau \tag{3}$$

where p_{FC} and p_{load} are output power from the FC and the consumption power of the load, respectively. η_{FCport} and η_{SCport} are the power conversion efficiencies of the multi-port bidirectional DC-DC converter connected with the FC and the SC, respectively and are supposed as 0.9. The value s defined as 1 when the SC is being charged, or -1 when discharged. The output power p_{FC} is described as

$$p_{FC} = v_{FC} i_{FC} \tag{4}$$

where relation between the FC voltage v_{FC} and the FC current i_{FC} is generally known as

$$v_{FC} = E - r i_{FC} - a \ln b i_{FC} - c \exp d i_{FC} \tag{5}$$

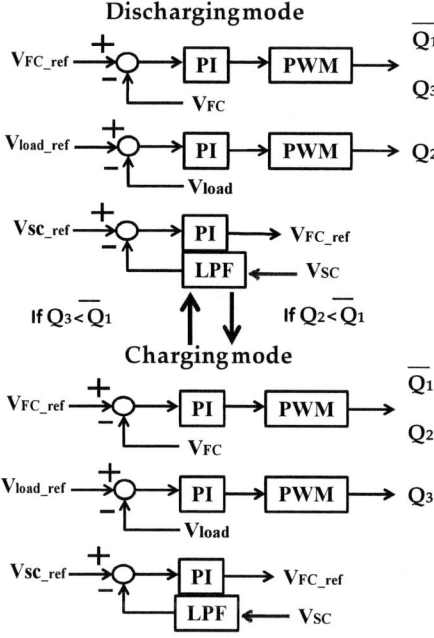

Fig. 3. New energy management method

where E is open circuit voltage, r is the ohmic resistance and a, b, c and d are the coefficients that determine the activation loss in the third term and the diffusion loss in the forth term. The activation loss and the diffusion loss are considered as constant values when the FC operates in the middle range of the power. Hence, (5) is approximated as

$$v_{FC} = E' - r' i_{FC} \tag{6}$$

where E' and r' are determined by fitting to the measured I-V characteristic. Using (5), the p_{FC} can be described as

$$p_{FC} = \frac{E' - v_{FC}}{r'} v_{FC} \tag{7}$$

The terminal voltage of the SC is described as

$$v_{SC} = \sqrt{\frac{2 w_{SC}}{c_{SC}}} - r_{ESR} i_{SC} \tag{8}$$

where r_{ESR} and i_{SC} are the equivalent series resistance and the SC current, respectively.

The PI controller block H is described as

$$H = K_p + K_i \frac{z}{z-1} \tag{9}$$

TABLE II
PARAMETERS FOR THE CONTROL LOOP, THE IIR FILTER AND THE SIMULATION

Parameter	Value	Description
K_p	0.3	Proportional coefficient of the PI control loop.
K_i	0.003	Integral coefficient of the PI control loop
a_1	-1.8849	Coefficients of the denominator of the transfer function for
a_2	0.8918	the IIR filter
b_0	0.0961	Coefficients of the numerator of the transfer function for the
b_1	-0.1854	IIR filter
b_2	0.0961	
V_{FCmax}	2.8 [V]	Limited voltage of the FC
V_{FCmin}	1.4 [V]	
SC	11.2 [F]	SC capacitance
$V_{SC\_reference}$	3.2 [V]	Reference voltage of the SC
f_s	3.3 [Hz]	Sampling frequency
R_0	0.45 [Ω]	Internal ohmic resistance of the SC

where K_p and K_i are proportional and integral coefficients, respectively.

IV. Simulation

To investigate that the system with the IIR filter works properly, waveforms when a smartphone is operating under the typical usages including load waveforms at the standby mode, music mode, calling mode, game mode, Internet mode and when using social networking service (SNS). As shown in Fig. 4, the waveforms were measured from the actual smartphone (HTC-desire, X06HT) in advance [12]. The coefficients for the IIR filter, used and the PI controller are listed in TABLE II. a_i and b_i are the coefficients designed for the IIR filter that behaves as a 2nd order Chebyshev type II filter, which has a excellent cut-off characteristic in spite of the low orders. The upper cut-off frequency is set to 0.1 Hz to avoid the particular frequencies that could accelerate FC degradation and fast FC output power changes. The FIR filter has 30th orders to realize the equivalent characteristic with the 2nd Chebyshev type II filter. Fig. 5 shows the frequency and characteristic of the IIR filter used in this study. The IIR filter has a 20 dB attenuation ratio at 0.1 Hz. The sampling frequency and the reference voltage for the SCs are 3.3 Hz and 3.2 V, respectively.

V. Results and Discussion

The voltage and the current waveforms of the FC, the SCs and the load when the load extracted from the smartphone's actual load waveform is applied are shown in Fig. 6-11. The results show that the system behavior with the IIR filter works properly and the energy flow among the FC, the SC and the load are managed as well as that with the FIR filter. Due to LPF, the operating point of the FC voltage moves slower smoothly in spite of the intense change of the load current. Accordingly, the SC supplies power that contains high frequency components while the FC supplies lower. No divergence and oscillation has been observed. The IIR
filter is able to realize the function with fewer

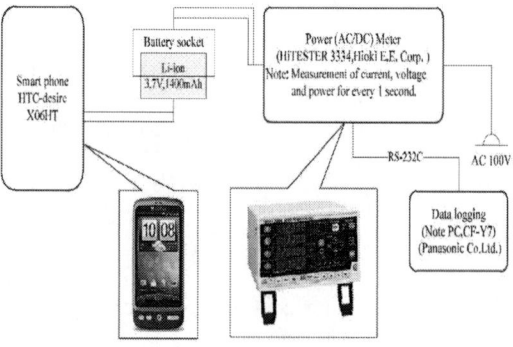

Fig. 4. Schematic diagram of the system for the smartphone's power consumption measurement [12].

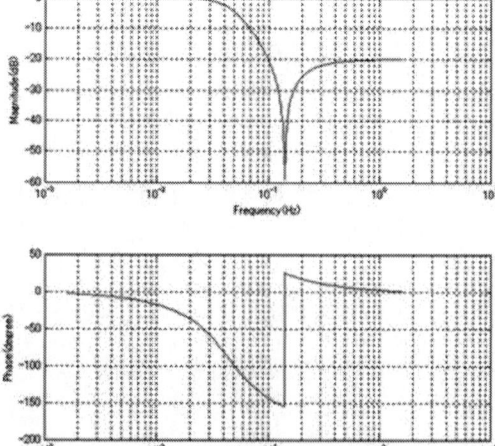

Fig. 5. Frequency characteristic of the IIR filter used in this study

calculations than the FIR filter. In this study, the IIR filter contains 4 additions and 5 multiplications while the FIR

The 2014 International Power Electronics Conference

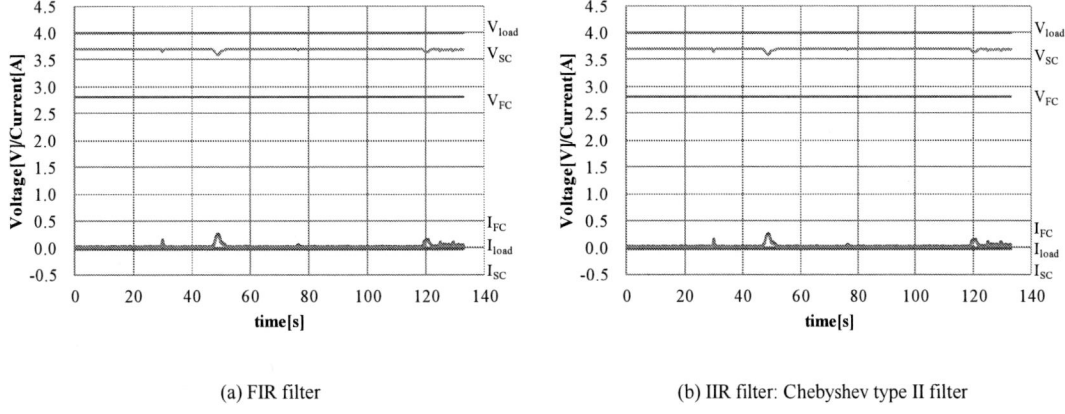

(a) FIR filter

(b) IIR filter: Chebyshev type II filter

Fig. 6. The voltage and current of the FC, the SC and the load when a waveform at the standby mode

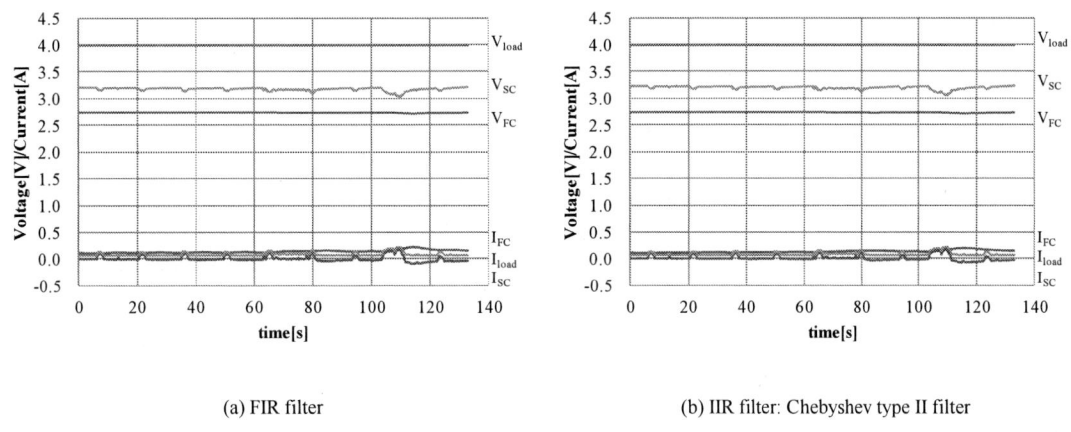

(a) FIR filter

(b) IIR filter: Chebyshev type II filter

Fig. 7. The voltage and current of the FC, the SC and the load when a waveform at the music mode

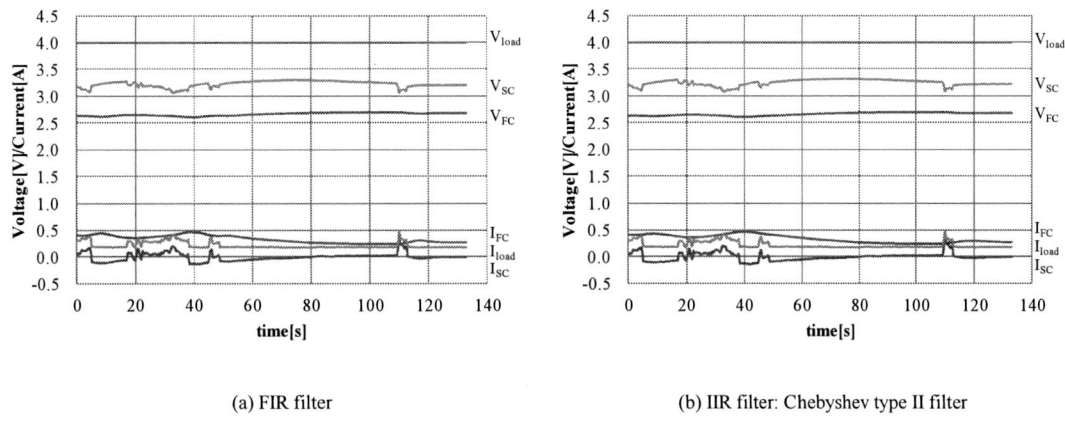

(a) FIR filter

(b) IIR filter: Chebyshev type II filter

Fig. 8. The voltage and current of the FC, the SC and the load when a waveform at the calling mode

978-1-4799-2706-7/14 $31.00 © 2014 IEEE   1231

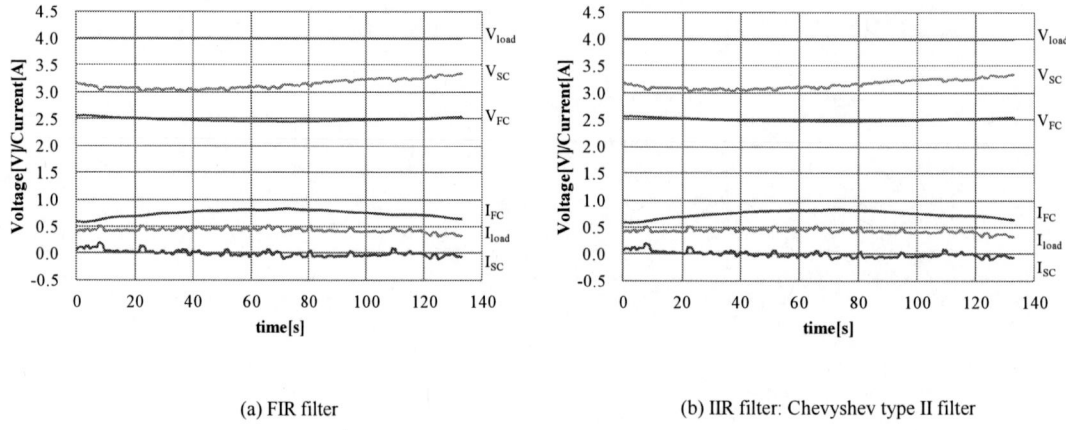

(a) FIR filter   (b) IIR filter: Chevyshev type II filter

Fig. 9. The voltage and current of the FC, the SC and the load when a waveform at the game mode

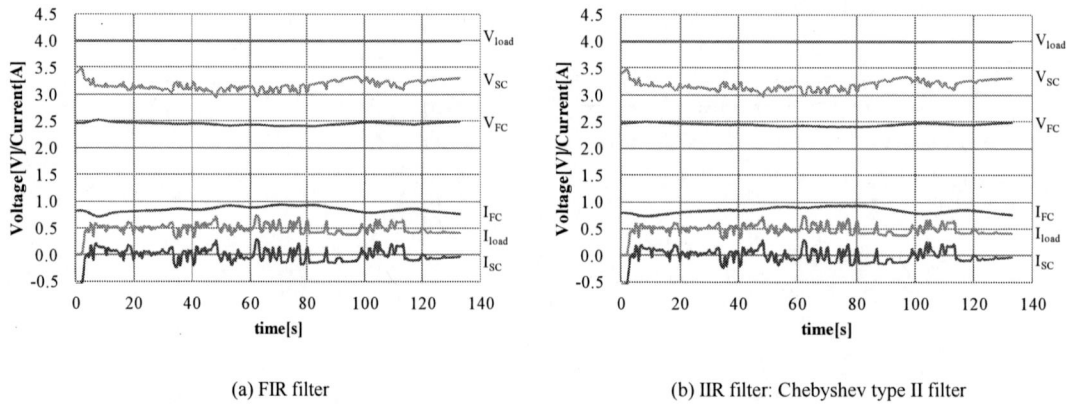

(a) FIR filter   (b) IIR filter: Chebyshev type II filter

Fig. 10. The voltage and current of the FC, the SC and the load when a waveform at the internet mode

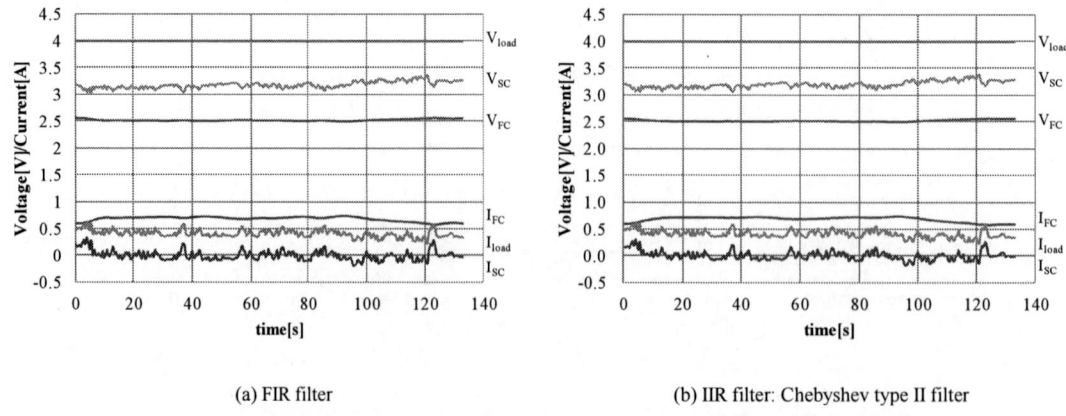

(a) FIR filter   (b) IIR filter: Chebyshev type II filter

Fig.11 . The voltage and current of the FC, the SC and the load when a waveform at the SNS mode

filter contains 30 additions and 31 multiplications. For mobile use, the micro-controller for the digital controls would be limited. Long time occupations of CPU time for the filter calculation may cause interruptions of calculations for other control loops. Therefore, this reduction is beneficial.

VI. Conclusions

Energy management method using the IIR filter for PEMFC-supercapacitor hybrid power source was proposed. It was confirmed that the system with the IIR filter worked properly and the energy flow among the FC, the SC and the load managed well. Due to the IIR filter, the number of calculations was reduced dramatically compared with the FIR filter and that would not interrupt other control loops by occupying CPU time. The FC voltage is changed so slowly that it could avoid the particular frequencies that could accelerate the FC degradation. Our future work is to implement the proposed system to the actual multi-port bidirectional DC-DC converter and ensure if the system works as properly as the simulation results.

Acknowledgment

This work was supported by Grant-in-Aid for Young Scientists (B) 25820109 from the Japan Society for the Promotion of Science.

References

[1] S. Zhang, X. Z. Yuan, J. N. C. Hin, H. Wang, L. A. Friedrich and M. Schulze, A Review of Platinum-based Catalyst Layer Degradation in Proton Exchange Membrane Fuel Cells," *Journal of Power Sources*, vol. 194, no. 2, pp. 588-600, 2009.

[2] R. M. Darling and J. P. Meyers, "Mathematical Model of Platinum Movement in PEM Fuel Cells," *Journal of Electrochemical Society*, vol. 152, no. 1, pp. A242-A247, 2005.

[3] M. Uno and K. Tanaka, "Pt/C catalyst degradation in proton exchange membrane fuel cells due to high-frequency potential cycling induced by switching power converters," *Journal of Power Sources*, vol. 196, no. 23 , pp. 9884-9889, 2011.

[4] H. Tao, A. Kotsopoulos, J. Duarte and M. Hendrix, "Family of multiport bidirectional DC-DC converters," *IEE Electric Power Applications*, vol. 153, no. 3, pp. 451-458, 2006.

[5] L. Solero, A. Lidozzi and J. Pomilio, "Design of multiple-input power converter for hybrid vehicles,", *IEEE Transactions on Power Electronics*, vol. 20, no. 5, pp. 1007-1016, 2005.

[6] T. M. Keränen, H. Karimäki, J. Viitakangas, J. Vallet, J, Ihone, P. Hyötylä, H, Uusalo and T. Tingelöf, "Development of integrated fuel cell hybrid power source for forklifts," *Journal of Power Sources*, vol. 196, no. 21 , pp. 9058-9068, 2011.

[7] M. Hayase, T. Fujii and J. G. A. Brito-Neto, "A Miniature Fuel Cell with Porous Pt Layer Formed on a Si Substrate," *Journal of Electrochemical Society*, vol. 158, no. 4, pp. B355-358, 2011.

[8] M. Weiland, S. Wagner, R. Hahn and H. Reichl, "Design and evaluation of a passive self-breathing micro fuel cell for autonomous portable applications," *International Journal of Hydrogen Energy*, vol. 38, no. 11, pp. 440-446, 2011.

[9] N. Katayama and S. Kogoshi, "Fuel Cell/Electric Double Layer Capacitor Hybrid Power Source Using a Multi-port Bidirectional DC-DC converter," *IEEJ Transactions IA*, vol. 130, no. 11, pp. 1279-1280, 2010.

[10] H. Tao, A. Kotsopoulos, J. Duarte and M. Hendrix, "Family of multiport bidirectional DC-DC converter," *IEE Electric Power Applications*, vol. 153, no. 3, pp. 451-458, 2006.

[11] Z. Zhang, Z. Ouyang, OC Thomsen, M. A. E. Anderson, "Analysis and Design of a Bidirectional Isolated DC-DC Converter for Fuel Cells and Supercapacitors Hybrid System," *Power Elecronix, IEEE Transactions on*, vol. 27, no. 2, pp. 848-859, 2012.

[12] K. Dowaki, M. Hayase, N. Katayama, "A Life Cycle Inventory Analysis on FC-Iphone fueled by Bio-H_2," *in proceedings of the 8th Meeting of the Institute of Life Cycling Assessment*, 2013.

[13] N. Katayama and S. Kogoshi, "Energy Control Method for Fuel Cell-Electric Double Layer Capacitor Hybrid Power Source System," *Power Electronics and Drive System (PEDS), IEEE 10th International Conference on*, pp. 72-77, 2013.

[14] S. Tosaka, N. Katayama and S. Kogoshi, "Developing a new topology for the DC-DC converter used in Fuel Cell-Electric Double Layer Capacitor Hybrid Power Source System for Mobile Devices, *IPEC-Hiroshima 2014*, in press.

Advanced Torque and Current Control Techniques for PMSMs with a Real-time Simulator Installed Behavior Motor Model

Ryo Tanabe
Shibaura Institute of Technology
Department of Electrical Engineering and Computer Science
Tokyo, Japan
E10085@shibaura-it.ac.jp

Kan Akatsu
Shibaura Institute of Technology
Department of Electric Engineering
Tokyo, Japan
akatsu@sic.shibaura-it.ac.jp

Abstract—This paper presents advanced torque and current control techniques for Permanent Magnet Synchronous Motors (PMSMs) with a real-time simulator which has a nonlinear motor model. This model is called a "behavior motor model" and is developed based on finite element analysis results to take into account nonlinear characteristics including spatial harmonics and magnetic saturation. The real-time simulator is implemented within a circuit simulator coupling the behavior motor model with switching circuits in real time and it can be applied to torque and current control as an advanced controller. In this paper, the torque and current control of a PMSM with the proposed system is presented. Effectiveness of this technique is verified by performing some simulations and experiments.

Keywords— Behavior motor model, PMSMs drive system, Real time simulator,

I. INTRODUCTION

Permanent Magnet Synchronous Motors (PMSMs) have been widely used in various industrial applications due to their high efficiency and high power density characteristics. In general, PMSMs drive systems are developed based on simplified mathematical model. In this model, the distributions of the inductance and linkage flux density due to rotor magnets are assumed as sinusoidal. Consequently, the *dq* transformation method gives simple voltage and torque equation expressed with constant motor parameters. Because of this modeling, PMSMs drive system can be easily constructed by controller design techniques for DC machines. However, PMSMs typically have nonlinear characteristics such as the spatial harmonics and magnetic saturation [1][2]. The nonlinear characteristics degrade the performance of torque and current control based on simplified mathematical model because this drive system is controlled as a linear system.

In PMSMs which have the nonlinear characteristics, the pulsating torque is caused by interaction between excitation current and spatial harmonics in the flux linkage. Torque ripples may degrade speed and position control and cause large acoustic noises and vibrations. In previous researches, the optimal current command for the torque smoothness are achieved by developing torque controllers with mathematical model which has spatially dependent machine parameters has been proposed [3]-[5]. However, this technique requires amount of off line data for look up tables. If the magnetic saturation is also taken into account the machine parameters depend on not only the position but also current amplitude. Therefore, the model construction becomes much complicated.

Furthermore, the realization of the optimal excitation current to achieve torque smoothness requires high performance current controller. The current feed-forward control gives optimal voltage commands to achieve desired current based on inverse PMSMs model. The current feed-forward controller has already been developed in the linear system [6]-[8]. However, in the case of nonlinear machine, the inverse model cannot give optimal voltage commands because of the nonlinearity of the machine. As described in [9], precise voltage equation model can be obtained by considering the nonlinear characteristics. However, as previously noted, the model construction is much complicated.

This paper presents advanced torque and current control techniques by using a real time simulator (high speed calculator) which has a precise PMSMs model. Precise PMSMs model is created by Finite Element Analysis (FEA) which can take into account the spatial harmonics and magnetic saturation. In the previous research, this model is called as a "behavior motor model" and this model is applied to a circuit simulator such as MATLAB/Simulink. In this paper, the coupling analysis between the behavior motor model and the circuit simulator is installed in the real time simulator as a controller of PMSM drive system. Therefore, the controller enables to achieve the advanced control method by using the analysis results. This paper presents torque ripple control and high performance current control based on a feed forward control of PMSM by using the proposed technique. Effectiveness of them is verified by performing some experiments.

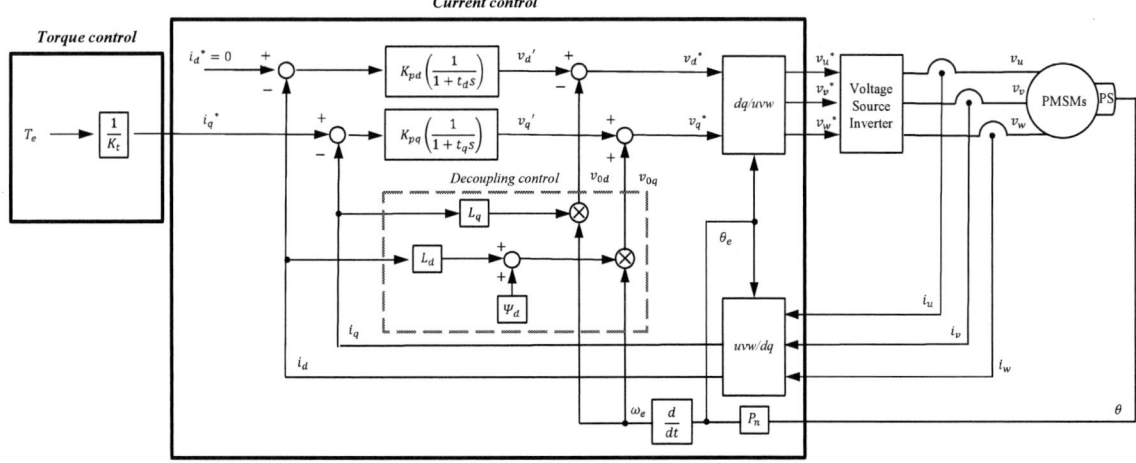

Fig. 1 Linear PMSMs control system

II. CONVENTIONAL LINEAR CONTROL SYSTEM

In this section, the linear model of PMSM is firstly introduced. In general, the control system is developed by simplified mathematical model which presences linear characteristics of PMSM in dq reference frame. The torque equation is described as follow:

$$T_e = P_n \left\{ \Psi_d + \left(L_d - L_q \right) i_d \right\} i_q \tag{1}$$

where P_n, Ψ_d, L_d, L_q, i_d and i_q, are the number of pole pairs, d-axis flux linkage due to the rotor magnets, d-axis inductance, q-axis inductance, d-axis armature current and q-axis armature current, respectively. In case the d-axis current is a constant value, the torque has a linear characteristic with q-axis current.

The voltage differential equation is described as follow:

$$\begin{bmatrix} v_d \\ v_q \end{bmatrix} = R \begin{bmatrix} i_d \\ i_q \end{bmatrix} + \begin{bmatrix} L_d & 0 \\ 0 & L_q \end{bmatrix} \frac{d}{dt} \begin{bmatrix} i_d \\ i_q \end{bmatrix} + \omega_e \left(\begin{bmatrix} 0 & -L_q \\ L_d & 0 \end{bmatrix} \begin{bmatrix} i_d \\ i_q \end{bmatrix} + \begin{bmatrix} 0 \\ \Psi_d \end{bmatrix} \right) \tag{2}$$

where R, p and ω_e are the armature resistance, and angular mechanical velocity, respectively. The current dynamics is described by differential equation based on (2) as follow:

$$\frac{d}{dt} \begin{bmatrix} i_d \\ i_q \end{bmatrix} = \begin{bmatrix} -\dfrac{R}{L_d} & \dfrac{\omega L_q}{L_d} \\ -\dfrac{\omega L_d}{L_q} & -\dfrac{R}{L_q} \end{bmatrix} \begin{bmatrix} i_d \\ i_q \end{bmatrix} + \begin{bmatrix} \dfrac{1}{L_d} & 0 \\ 0 & \dfrac{1}{L_q} \end{bmatrix} \begin{bmatrix} v_d \\ v_q \end{bmatrix} - \begin{bmatrix} 0 \\ \dfrac{\omega_e \Psi_d}{L_q} \end{bmatrix} \tag{3}$$

Fig. 1 shows the block diagram of PMSMs control system which has torque and current controller based on (1) and (3).

A. Conventional torque control

The torque control generates current command due to torque constant under an assumption that the relationship between torque and current is a linear characteristic.

Therefore, the current command is generated as constant value according to torque command. However, in this control, the pulsating torque is caused because the relationship between current and torque is not linear by nonlinear characteristics of PMSMs.

B. Conventional current control

The current control works to achieve the current command. In this control, the PI controller and decoupling controller are designed based on (2). However, this controller may exhibit bandwidth limit and is not suited to cover the full critical range of harmonics frequencies. Therefore, the desired current is not completely excited. Adding that, this control cannot achieve a fast response because of its bandwidth limitation.

III. PROPOSED NONLINEAR CONTROL SYSTEM

In this section, a proposed nonlinear control system based on FEA is described. The proposed control system is developed based on a behavior motor model which presences nonlinear characteristics of PMSMs. Therefore, this control system can take into account nonlinear characteristics. The behavior motor model has 3-D table data of inductance, magnet flux and torque as the function depend on current amplitude, current phase and rotor position [10]. In general, the construction of 3D-table data requires some complex processes such as the data interpolation. However, in this research, acquisition, interpolation, and construction of the machine parameter data are automatically implemented by computer software with the FEA (JMAG-RT produced by JSOL Corporation is used in this research). In addition, the behavior motor model is easily applied to the circuit simulator such as MATLAB/Simulink. This coupling method is implemented as a controller of PMSM drive system by using a real time simulator. This controller can take into account nonlinearities of not only tested machines but also external drive circuits.

978-1-4799-2706-7/14 $31.00 © 2014 IEEE   1235

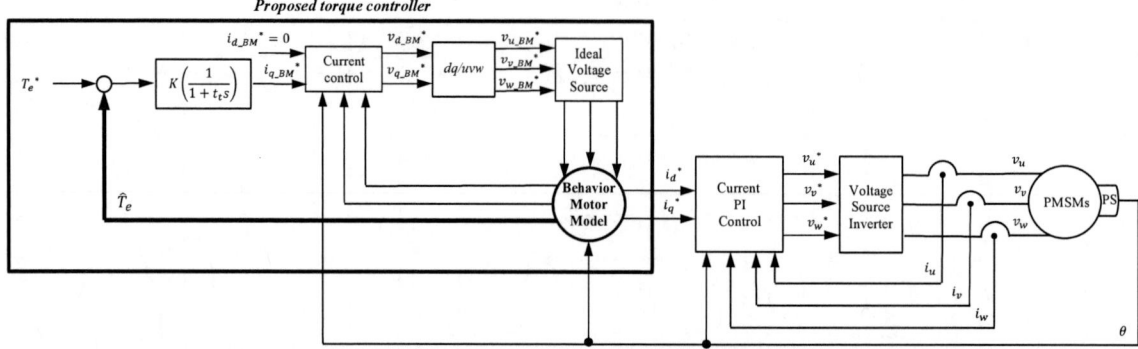

Fig. 2. Proposed torque ripple control system

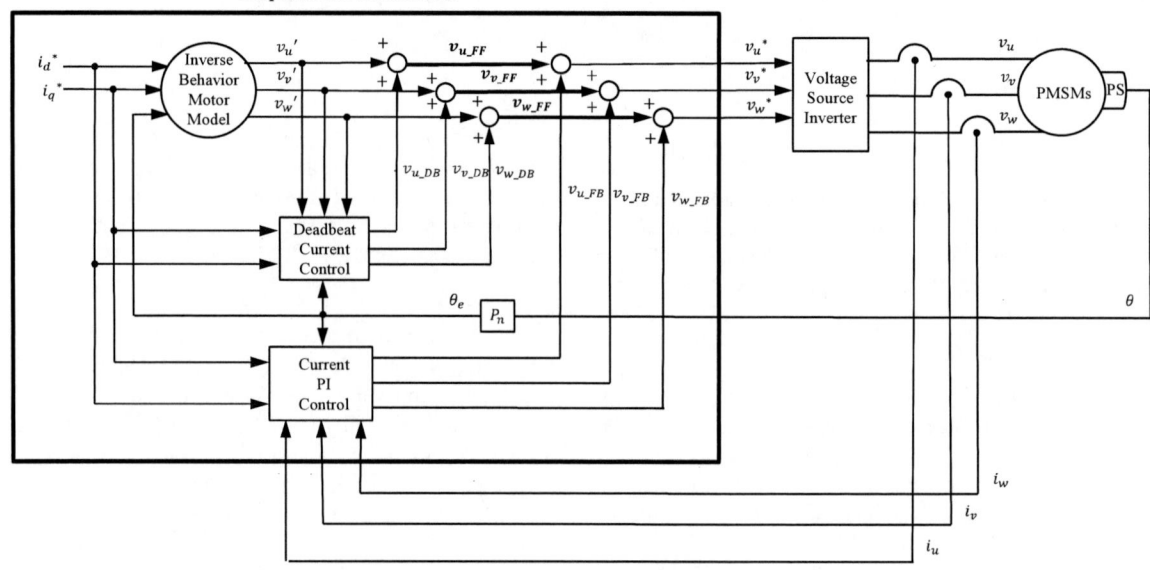

(a) Perspective of proposed current control system

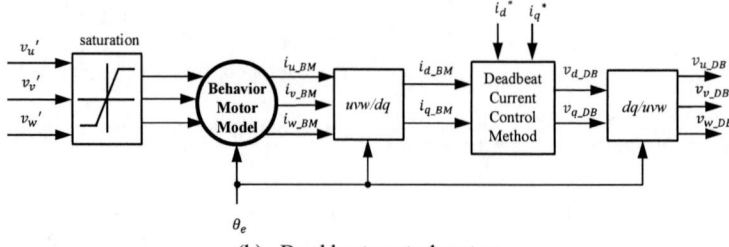

(b) Dead beat control system

Fig. 3. Proposed current control system

A. Proposed torque control

Fig. 2 shows the system of a proposed torque ripple control. In the proposed torque ripple control, the instantaneous torque is estimated by the behavior motor model from the simulated excitation current. This estimated torque includes characteristics of torque ripple. The current command generated by the torque controller is the optimal current to achieve torque smoothness. Therefore, the torque ripple is reduced compared with the

conventional one. In addition, the excitation current which is applied the behavior motor model is estimated in the real time simulator. Hence, the gain of torque PI controller can be high because noises are not included in the estimated current.

B. Proposed current control

Fig. 3 shows the system of proposed current control including a dead beat control which is implemented in the proposed technique. The proposed current control is

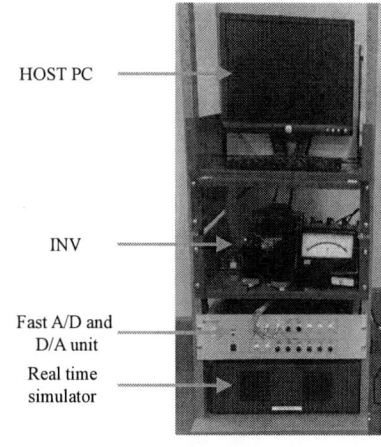

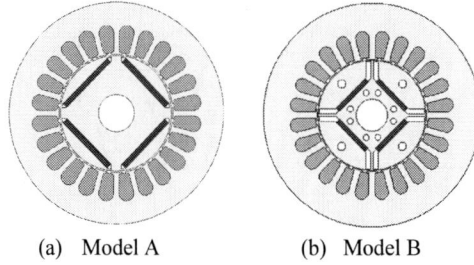

(a) Model A (b) Model B

Fig. 5. Tested Motor models

(a) PMSMs drive system

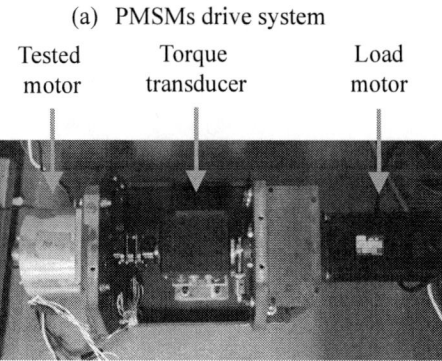

Tested motor Torque transducer Load motor

(b) Test bed

Fig. 4 Experimental setup

TABLE I. SPECIFICATION OF CONTROLLED MOTORS

	Model A	Model B
Rated voltage [V]	100	165
Rated current [Arms]	7	3
Rated speed [rpm]	1200	1800
Magnetic linkage flux Ψ_d [mWb]	250.0	158.0
d-axis inductance L_d [mH]	5.0	10.7
q-axis inductance L_q [mH]	10.0	26.3
Winding resistance R [ohm]	0.46	0.814
Pole pairs P	2	2
Stator slots	24	24

developed based on the feed-forward control with dead beat control. The optimal voltage command to achieve desired current is given by inverse PMSM model as shown in Fig. 3(a). In this research, this system is called as an "inverse behavior motor model". This model enables to generate optimal voltage command with precise nonlinear motor parameters. By using this voltage command the behavior motor model estimates instantaneous current to give for the dead beat current controller as shown in Fig. 3(b). Please note that the estimated current can take into account the voltage saturation by using the calculated voltage command from the inverse behavior motor model. In general, the dead beat controller generates much high voltage command to achieve the high speed current response. However, due to the voltage limitation the response is not achieved and the controller does not correctly work. In the proposed controller, the voltage limit is considered from the voltage reference, the controller does not need to search the possible output voltage with the limitation. Adding that designing the control system including the dead beat control method is usually difficult because of noises and differential operation. However, in the proposed current control, since the instantaneous current is estimated by the behavior motor model, the actual current detected by the current sensor is not used for the dead beat control algorithm. Therefore, the dead beat control works for

stability by using the signal generated in the real time simulator. Hence, the desired current can be excited and this control achieves fast response. In addition, the current PI controller will work to compensate disturbances which are not considered in the real time simulator such as the error of the modeling by the temperature variation.

IV. EXPERIMENTAL VALIDATION

In this section, to verify an effectiveness of the drive system, advanced control techniques including torque ripple control and current control are performed on 0.8 kW IPMSMs by using the real time simulator installed behavior motor model.

Fig. 4 shows the experimental set up. Developed drive system by MATLAB/Simulink with the behavior motor model is installed in the real time simulator (LT-RTSimII produced by DSP Technology Co.Ltd is used in this research). The real time simulator communicates with fast A/D converter to take the analog signal of actual system into the real time simulator. Also, the real time simulator outputs the PWM pulses to the voltage source inverter. The tested motor is rotated at a constant speed by a servo motor system. In addition, the behavior motor models and tested motor are synchronized each other by detecting the rotor position.

A. Verification of the behavior motor model

Fig. 5 shows tested motor models. Specifications of the tested motors are shown in TABLE I. Model A and Model B are used for torque ripple control and current control tests.

Fig. 6 shows the characteristics comparison between the behavior motor model and tested motor of model A. As shown in Fig. 6(a), the cogging torque waveform calculated by the behavior motor model corresponds with

The 2014 International Power Electronics Conference

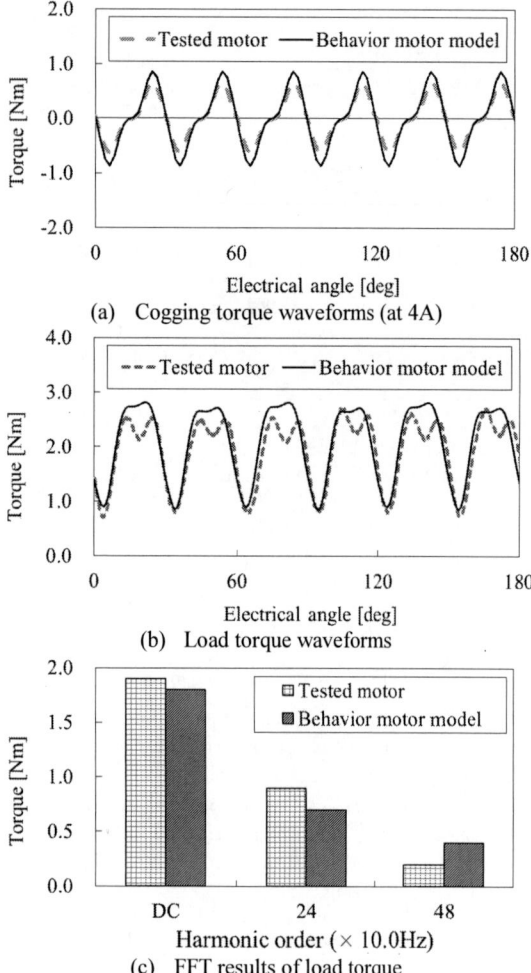

(a) Cogging torque waveforms (at 4A)

(b) Load torque waveforms

(c) FFT results of load torque

Fig. 6. Characteristics comparison of Model A

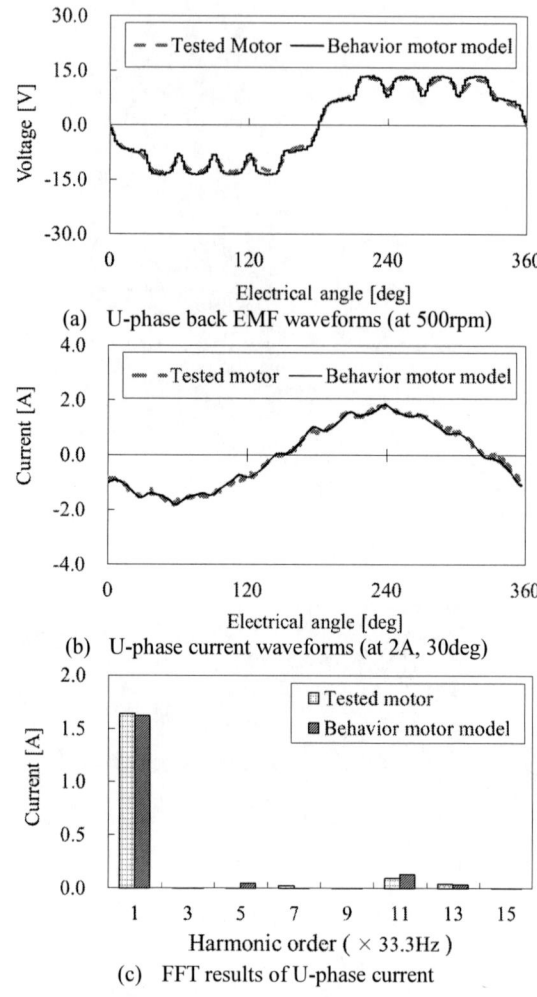

(a) U-phase back EMF waveforms (at 500rpm)

(b) U-phase current waveforms (at 2A, 30deg)

(c) FFT results of U-phase current

Fig. 7. Characteristics comparison of Model B

the tested motor output. As shown in Fig. 6(b) and (c), the behavior motor model simulates 24th and 48th harmonics component which are main component of torque ripples. From these results, the behavior motor model can calculate precise output torque includes ripples. Therefore, this model gives accurately estimated instantaneous torque.

Fig. 7 shows the characteristics comparison between the behavior motor model and tested motor of model B. As shown in Fig. 7(a), the U-phase back EMF waveform calculated by the behavior motor model corresponds with output of tested motor. As shown in Fig. 7(b) and (c), The U-phase current waveform calculated by the behavior motor model is in agreement with excitation current of the tested motor. From these results, the behavior motor model can calculate precise instantaneous motor behavior including nonlinear characteristics. Therefore, this model enables to design the inverse model and to estimate the instantaneous current of the tested motor in the current control.

B. Experimental results of the proposed torque ripple control

Fig. 8 shows the experimental results of the conventional torque control and the proposed torque ripple control. The condition of this verification is shown in TABLE II. As shown in Fig. 8(a) and (b), it is confirmed that the torque ripple is compensated by the proposed torque ripple control. Especially as shown in Fig. 8(c), the 24th and 48th harmonics components are adequately reduced. As shown in Fig. 8(d), the pulsating current as the optimal current to achieve torque smoothness is confirmed. Therefore, using the estimated instantaneous torque by the behavior motor model enables to reduce the torque ripple. From these results, the proposed torque ripple control which uses the real time simulator installed behavior motor model is effective to reduce the torque ripple of PMSMs.

978-1-4799-2706-7/14 $31.00 © 2014 IEEE

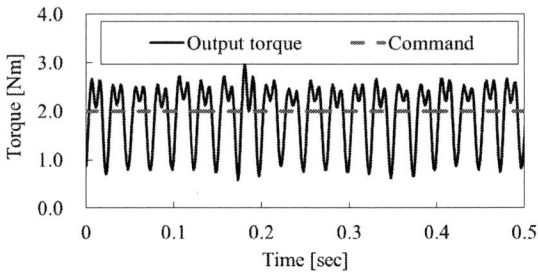

(a) Output torque waveform of conventional torque control(without torque ripple control)

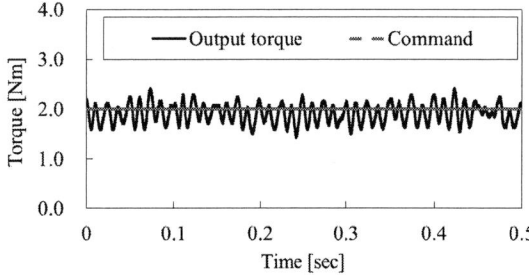

(b) Output torque waveform of proposed torque ripple control

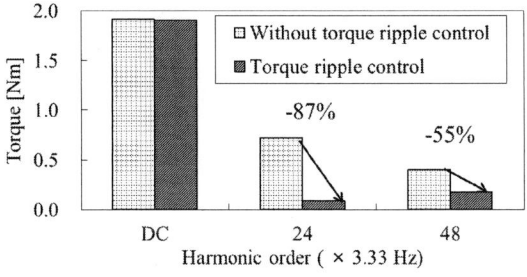

(c) FFT results of output torque

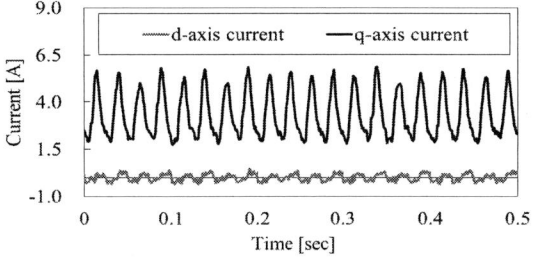

(d) *dq*-axis current waveforms of proposed torque ripple control

Fig. 8. Experimental results of torque ripple control

C. Experimental results of the proposed current control

Fig. 9 shows the experimental results of the conventional current control and the proposed current control in the steady state. The condition of this verification is shown in TABLE III. As shown in Fig. 9(a), (b) and (c), in the conventional current control, U-phase current is distorted by spatial harmonics. On the other hands, in the proposed current control, the harmonics in U-phase current are reduced in comparison with conventional one. Also as shown in Fig. 9(d) and (e), in the proposed current control, the voltage command is pulsating because the voltage commands are generated to achieve the desired current.

Fig. 10 shows the experimental results of the conventional current control and proposed current control in the transient state. The condition of this verification is shown in TABLE IV. As shown in Fig. 10(a), the proposed current control can achieve the fast response in comparison with the conventional one. Also as shown in Fig. 10(b), the proposed current control utilizes the maximum level of the limited voltage until current agrees with command to achieve the fast response because the dead beat controller gives optimal voltage command to correspond with command on next sampling step.

From these results, the proposed current control which uses the real time simulator installed the inverse behavior motor model can achieve good tracking performance in steady state, and can achieve fast response in transient state.

TABLE II
EXPERIMENTAL CONDITION OF TORQUE RIPPLE CONTROL

Torque Command [Nm]	2.0
Motor Speed [rpm]	100
Current Phase [degree]	0
DC Link Voltage [V]	100
Carrier Frequency [kHz]	10
Simulator Sample Time [μsec]	50

TABLE III
EXPERIMENTAL CONDITION OF CURRENT CONTROL IN STEADY STATE

d-axis Current Command [A]	0
q-axis Current Command[A]	2
Motor Speed [rpm]	1500
Dead time [μsec]	4
DC Link Voltage [V]	200
Carrier Frequency [kHz]	10
Simulator Sample Time [μsec]	50

TABLE IV
EXPERIMENTAL CONDITION OF CURRENT CONTROL IN TRANSIENT STATE

d-axis Current Command [A]	0
q-axis Current Command[A]	0.5→2
Motor Speed [rpm]	300
Dead time [μsec]	4
DC Link Voltage [V]	200
Carrier Frequency [kHz]	10
Simulator Sample Time [μsec]	50

The 2014 International Power Electronics Conference

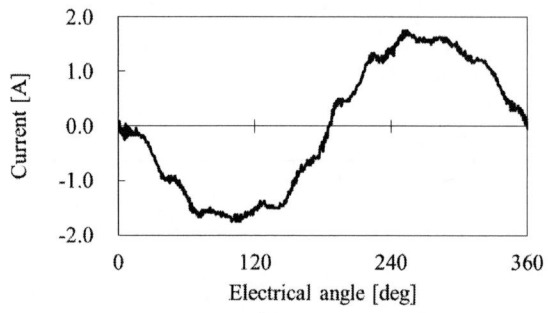

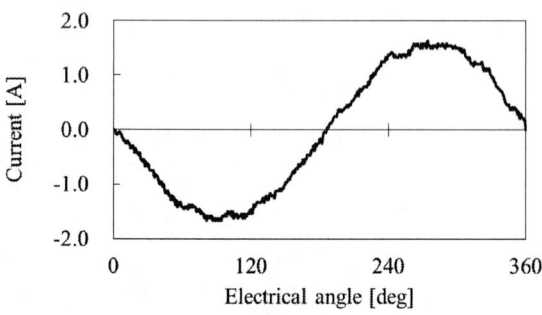

(a) U-phase current waveform of conventional current control

(b) U-phase current waveform of proposed current control

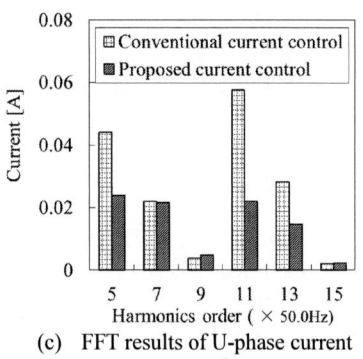

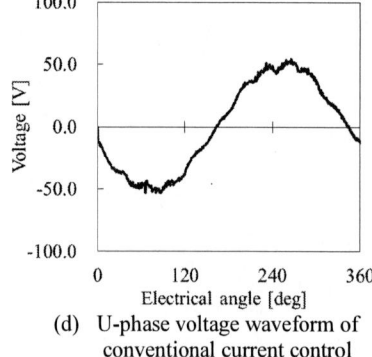

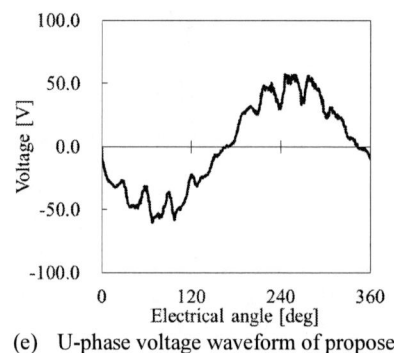

(c) FFT results of U-phase current waveform

(d) U-phase voltage waveform of conventional current control

(e) U-phase voltage waveform of proposed current control

Fig. 9. Experimental results of current control in steady state

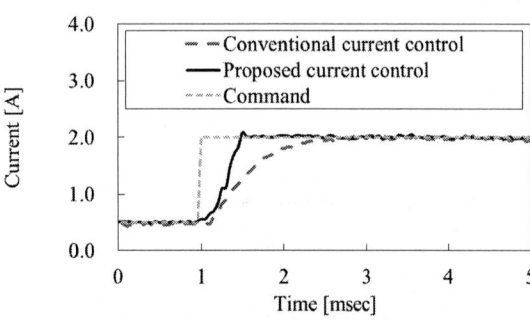

(a) q-axis current waveforms

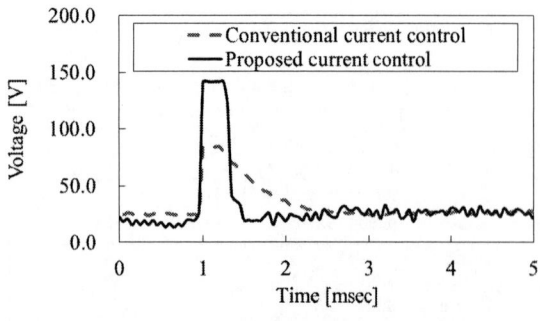

(b) q-axis voltage waveforms

Fig. 10. Experimental results of current control in transient state

V. CONCLUSION

This paper presented the advanced torque ripple control and current control based on precise motor model called as "behavior motor model". The circuit simulator installed the behavior model is applied in real time simulator to develop the advanced PMSM drive system, the system can realize these advanced control owe to the nonlinear characteristics of tested motor are taken into account. The proposed torque ripple control based on the instantaneous torque estimation by the behavior motor model can give the optimal current command to achieve smooth torque because the behavior motor model can estimate the instantaneous torque includes torque ripples according to spatial harmonics. The proposed current control based on the inverse behavior motor model can achieve the desired current because the inverse model developed precise motor parameters can generate the optimal voltage commands. Adding that, this control achieved fast current response by the dead beat controller with consideration of the voltage limitation. The effectiveness of them is verified by performing some experiments.

In conclusion, the proposed advanced torque ripple control and current control implemented by the real time simulator proved higher performance than the conventional ones.

978-1-4799-2706-7/14 $31.00 © 2014 IEEE

REFERENCES

[1] B. Stumberger, B. Kreˇca, and B. Hribernik, "Determination of parameters of synchronous motor with permanent magnets from measurement of load conditions," *IEEE Trans. Energy Convers.*, vol. 14, no. 4, pp. 1413–1416,Dec. 1999.

[2] B. Stumberger, G. Stumberger, D. Dolinar, A. Hamler, and M. Trlep, "Evaluation of saturation and cross-magnetization effects in interior permanent-magnet synchronous motor," *IEEE Trans. Ind. Appl.*, vol. 39, no. 5, pp. 1264–1271, Sep./Oct. 2003

[3] S. Clenet, Y. Lefèvre, N. Sadowski, S. Astier, and M. Lajoie, "Compensation of permanent magnet motors torque ripple by means of current supply wave shapes control determined by finite element method," *IEEE Trans. Magn.*, vol. 29, no. 2, pp. 2019–2023, Mar. 2003.

[4] N. Nakao, and K. Akatsu, "Torque ripple control for synchronous motors using instantaneous torque estimation," *IEEE* Energy Conversion Congress and Exposition, Phoenix, AZ, pp.2452-2459, Sept. 2011.

[5] Bo Guan, Yifan Zhao and Yi Ruan, "Torque Ripple Minimization in Interior PM Machines Using FEM and Multiple Reference Frames," *IEEE Conference on the 1st Industrial Electronics and Applications 2006 (ICIEA 2006)*, Singapore, May 24-26, 2006, pp. 1-6.

[6] P. Blaha and P. VacIavek, "Adaptive control of pm synchronous motor using dead-beat current controllers," in Electrical Machines and Systems, 2009. ICEMS 2009. International Conference on, nov. 2009, pp. I -5.

[7] D.G Holmes and D.A. Martin,"Implementation of a direct digital predictive current controller for single and three phase voltage source inverters," *IEEE* IAS Annual Meeting, pp.906-913, 1996.

[8] L. Springob and J. Holtz, "High-bandwidth current control for torque ripple compensation in PM synchronous machines," *IEEE Trans. Ind. Electron.*, vol. 45, no. 5, pp. 713–721, Oct. 1998.

[9] Y. Kano, K. Watanabe, T. Kosaka, and N. Matsui, "A Novel Approach for Circuit-Field-Coupled Time-Stepping Electromagnetic Analysis of Saturated Interior PM Motors," *IEEE Trans. Ind. Appl.* vol. 45, no. 4, pp. 1325–1333, July/Aug. 2009.

[10] K. Narita, Y. Sakashita, T. Yamada and K. Akatsu, "Iron Loss Calculation of PM Motor by Coupling Analysis between Magnetic Field Simulator and Control Simulator," proc. Of 11th Int.Conf, on Electrical Machines ans d systems (ICEMS 2008), Wuhan(Chine),Oct, 2008, pp828

[11] M. Usui, N. Nakao, and K. akatsu "Motor control methods with behavior model based on FEA results," Power Electronics and drive System (PEDS), 2013 IEEE 10th International Conference on 22-55 April 2013, pp.445-450

Compensation of the Current Measurement Error with Periodic Disturbance Observer for Motor Drive

Takashi Yamaguchi*, Yugo Tadano*, and Nobukazu Hoshi**

* MEIDENSHA CORPORATION, 2-8-1 Ohsaki Shinagawa-ku, Tokyo 141-8565, Japan

** Tokyo University of Science, 2641, Yamazaki, Noda-shi, Chiba 278-8510, Japan

Abstract- **Output current of general inverter is controlled based on the information of current sensors which are placed on the output. However, these sensor outputs have gain error and offset which are depending on the peripheral components and the accuracy characteristics of current sensor. Current control performance can be deteriorated a control method which uses a periodic disturbance observer to suppress vibration. We have developed a periodic disturbance observer as a control method for vibration suppression. Focusing on the current sensor error, it may cause the vibration of a specific frequency. Then, this paper proposes a current sensor error compensation control method by applying a periodic disturbance observer. To suppress the harmonic current online by this approach, the current control performance can be improved.**

Keywords - current sensor error, periodic disturbance observer, torque ripple

I. INTRODUCTION

AC motors such as permanent magnet synchronous motor (PMSM) and so on can be driven by three-phase inverter with vector control. In general, vector control is established by feeding back output current value from the current sensors.

Two or three current sensors are used in order to obtain three-phase current values. In motor applications, two of three-phase current are measured by two sensors; and the other current is calculated by considering three-phase equilibrium condition. This current sensor system is configured with analog circuit and a sensor body in general. Thus, gain error and offset error in the detection value are caused by the accuracy of analog circuits and the characteristics of sensor. The offset error causes harmonic current with synchronous frequency. On the other hand, gain error causes harmonic current with double of synchronous frequency. Current control performance is degraded by the harmonic current; and, it also has an adverse effect on the higher control system, such as position control and torque control which require the control accuracy of current control

subsystem. When the inverter output is connected to a motor, the harmonic currents cause the torque ripple with one, two, and higher times electrical synchronous frequency[1][2].

Sensor error is dependent on variation of the voltage and current value and temperature, and a precision of components and a sensor body characteristic. Offset error can be adjusted when the running stopped. On the other hand, it is difficult to compensate the gain error and the error caused by the operating conditions in advance. Thus, it is necessary to perform online compensation for high accuracy. The method for current error compensation have been proposed by reference such as [2]-[4]. In case of three-phase AC motor current sensing applications, it may detect only two phases. Thus, change of the compensation method is required according to the number of phase detection.

We have proposed periodic disturbance observer compensation method[5][6] as a vibration suppression control method. Basic configuration of this method is same as the disturbance observer. Feature of this method is to suppress only a specific frequency component. Specifically, this method performs system identification for each frequency component, and uses the inverse system model to estimate vibration of the frequency and compensate vibration. Despite the relatively simple control structure, this method can obtain good suppression effect at the target frequency.

This paper proposes a compensation scheme of current sensor errors. In consideration of a current sensor error generating specific order harmonic current, this method suppresses the vibration by applying a periodic disturbance observer. It enables the simultaneous compensation of the offset and gain errors, and the corresponding dynamic variations of the error online. Simulation result shows the effectiveness of the proposed method.

II. *dq*-AXIS CURRENT WITH CURRENT SENSOR ERROR

In this chapter, we analyze the adverse effect on *dq*-axis current by gain error and offset error of the current sensor at the case of three-phase-detection system.

978-1-4799-2706-7/14 $31.00 © 2014 IEEE

Equation (1) shows output current of the inverter i_u, i_v, and i_w. Here, θ : rotation angle, φ : phase angle, I : current peak value.

$$\begin{pmatrix} i_u \\ i_v \\ i_w \end{pmatrix} = \begin{pmatrix} I\sin(\theta+\varphi) \\ I\sin\left(\theta+\varphi-\dfrac{2}{3}\pi\right) \\ I\sin\left(\theta+\varphi+\dfrac{2}{3}\pi\right) \end{pmatrix} \quad \text{... (1)}$$

Equation (2) shows three-phase current i_u^{sense}, i_v^{sense}, and i_w^{sense} which have gain error α, β, and γ and offset error Δi_u, Δi_v, and Δi_w.

$$\begin{pmatrix} i_u^{sense} \\ i_v^{sense} \\ i_w^{sense} \end{pmatrix} = \begin{pmatrix} i_u \\ i_v \\ i_w \end{pmatrix} + \begin{pmatrix} \Delta i_u \\ \Delta i_v \\ \Delta i_w \end{pmatrix} + \begin{pmatrix} (\alpha-1)\cdot i_u \\ (\beta-1)\cdot i_v \\ (\gamma-1)\cdot i_w \end{pmatrix} \quad \text{................................ (2)}$$

Equation (3) shows a common dq-transformation which transform three-phase alternating current $\boldsymbol{i_{uvw}}$ to dq-axis current $\boldsymbol{i_{dq}}$.

$$\begin{pmatrix} i_d \\ i_q \end{pmatrix} = \begin{pmatrix} \cos\theta & -\sin\theta \\ \sin\theta & \cos\theta \end{pmatrix} \cdot \sqrt{\frac{2}{3}} \cdot \begin{pmatrix} 1 & -\dfrac{1}{2} & -\dfrac{1}{2} \\ 0 & \dfrac{\sqrt{3}}{2} & -\dfrac{\sqrt{3}}{2} \end{pmatrix} \cdot \begin{pmatrix} i_u \\ i_v \\ i_w \end{pmatrix} \quad \text{..... (3)}$$

Equation (4) shows dq-axis current i_d^{sense}, i_q^{sense} with gain error and offset error which resulting from equations (1), (2) and (3).

$$\begin{aligned} \begin{pmatrix} i_d^{sense} \\ i_q^{sense} \end{pmatrix} = & \begin{pmatrix} i_d \\ i_q \end{pmatrix} \\ & + \frac{1}{\sqrt{6}} \cdot \begin{pmatrix} (2\Delta i_u - \Delta i_v - \Delta i_w) & -\left(\sqrt{3}\Delta i_v - \sqrt{3}\Delta i_w\right) \\ \left(\sqrt{3}\Delta i_v - \sqrt{3}\Delta i_w\right) & (2\Delta i_u - \Delta i_v - \Delta i_w) \end{pmatrix} \cdot \begin{pmatrix} \cos\theta \\ \sin\theta \end{pmatrix} \\ & + \frac{I}{\sqrt{6}} \cdot (\alpha+\beta+\gamma-3) \cdot \begin{pmatrix} \sin\varphi & \cos\varphi \\ -\cos\varphi & \sin\varphi \end{pmatrix} \cdot \begin{pmatrix} \cos2\theta \\ \sin2\theta \end{pmatrix} \\ & + \frac{I}{6\sqrt{2}} \cdot \begin{pmatrix} 3(\beta-\gamma) & (2\alpha-\beta-\gamma) \\ -(2\alpha-\beta-\gamma) & 3(\beta-\gamma) \end{pmatrix} \cdot \begin{pmatrix} \cos\varphi \\ \sin\varphi \end{pmatrix} \end{aligned} \quad \text{..... (4)}$$

From equation (4), the offset error generates vibration with the synchronous frequency $1f$ to $\boldsymbol{i_{dq}}$, gain error generates $2f$ vibration and DC component.

III. PERIODIC DISTURBANCE OBSERVER[6]

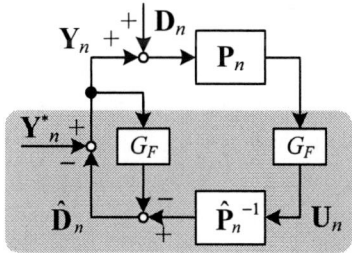

Fig. 1. Periodic Disturbance Observer

Fig. 1 shows a control block diagram for the n-th order

frequency component of the periodic disturbance observer (PDO). Here, $\mathbf{Y}_n^*$: n-th order disturbance compensation command vector, $\mathbf{Y}_n$: n-th order disturbance compensation value vector, $\mathbf{D}_n$: n-th order disturbance vector, $\hat{\mathbf{D}}_n$: n-th order disturbance estimation vector , $\mathbf{U}_n$: n-th order disturbance detection vector, $\mathbf{P}_n$: system of control target, $\hat{\mathbf{P}}_n$: system model. In addition, $\mathbf{X} = X_A + jX_B$ is complex notation of each vector.

Frequency component of the n-th order component that is suppression target extracts the cosine/sine coefficient by passing it through a Low-Pass Filter (LPF) G_F to the product of the cosine/sine value synchronizing with the target frequency of the detected value. In order to obtain exact value, use of Fourier transform is required in general. However, we use LPF instead of Fourier transform because of easy calculation. Parameters determining the cutoff frequency and filter order is determined by experimentally investigated from the residual noise appearing on the detection value passed through the filter. The filter frequency is corresponding to the frequency of general disturbance observer, and then cutoff frequency has trade-off between stability and response.

The inverse system model $\left(\hat{P}_{An} + j\hat{P}_{Bn}\right)^{-1}$ multiplies by the extracted cosine/sine components and the target frequency components of the periodic disturbance are estimated in the form of complex vector. The system model $\hat{\mathbf{P}}_n = \hat{P}_{An} + j\hat{P}_{Bn}$ is determined by system identification from the compensation command to detection value. The target frequency components can be expressed by simple one-dimensional complex vector. The system model construction method and identification method do not matter in any form, if it can be expressed form of one-dimensional complex vector for target frequency finally. The periodic disturbance compensation value is generated by multiplying the cosine/sine value by difference between command value (usually 0 value) and the estimated value. The details of the control configuration are described in [6].

IV. dq-AXIS COMPENSATION CURRENT GENERATION

If an error has occurred in the current sensor disposed in the three-phase alternating current, the Automatic Current Regulator (ACR) recognizes control deviation equality with error of the detection value by the expression (4). Therefore, the vibration current i_{dq}^{sense} which is occurred by the sensor error is suppressed by the ACR. As a result, the vibration is absorbed by the voltage command v_{dq}^{ref}, and vibrations of the torque T and the actual current occur. Therefore, it cannot be directly observed the influence of sensor error from i_{dq}^{sense}.

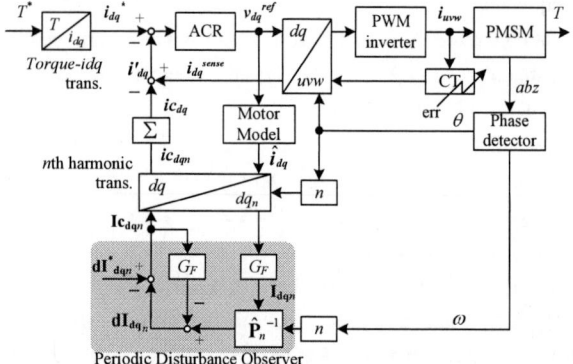

Fig. 2. Current Measurement Error Correction with PDO

In this method, the virtual current value $\hat{i}_{dq}$ is calculated from $v_{dq}{}^{ref}$ by the circuit equation. Harmonic current is observed by this virtual current, and set to vibration suppression target. Fig. 2 shows the overall control configuration of the current sensor error compensation based on PDO. As the circuit equation, common equation of PMSM expressed as (5) is applied. Here, R : motor resistance, L_d : d-axis inductance, L_q : q-axis inductance, Φ : magnetic flux.

$$\begin{pmatrix} \hat{i}_{dn} \\ \hat{i}_{qn} \end{pmatrix} = \begin{pmatrix} \dfrac{1}{R + sL_d} \cdot \left(v_d + \omega \hat{i}_q L_q \right) \\ \dfrac{1}{R + sL_q} \cdot \left(v_q - \omega \hat{i}_d L_d - \omega \Phi \right) \end{pmatrix} \quad \text{...............} \quad (5)$$

Then perform harmonic coordinate transformation of $\hat{i}_{dq}$, and extracts a frequency component of the vibration suppression target through G_F. G_F is second-order LPF and the cut-off frequency was set to $2\pi[rad/s]$. Filter order and the parameter of the cut-off frequency are determined by experimentally based on the residual noise appearing on the detection value after filtering. This cut-off frequency corresponds to the filter frequency of the common disturbance observer. The controller configuration has trade-off of stability and response. Further, the G_F is used for the detection of the vibration components, then it does not affect the fundamental response of the current.

Frequency component extraction and coordinate conversion to $\mathbf{I_{dqn}}$ from $\hat{i}_{dq}$ is expressed in equation (6). Here, $\mathcal{L}$: Laplace operator.

$$\begin{pmatrix} I_{dn} \\ I_{qn} \end{pmatrix} = G_F \cdot \mathcal{L}\left[\begin{pmatrix} \cos n\theta & \sin n\theta \\ -\sin n\theta & \cos n\theta \end{pmatrix} \cdot \begin{pmatrix} \hat{i}_{dn} \\ \hat{i}_{qn} \end{pmatrix} \right] \quad \text{..................} \quad (6)$$

Compensation value $\mathbf{Ic_{dqn}}$ is calculated by applying the PDO as n-th order disturbance detection vector $\mathbf{I_{dqn}}$. Equation (7) shows the coordinate system conversion to the compensation value ic_{dqn} in dq coordinate system from the frequency components of the compensation value $\mathbf{Ic_{dqn}}$.

$$\begin{pmatrix} ic_{dn} \\ ic_{qn} \end{pmatrix} = \begin{pmatrix} \cos n\theta & -\sin n\theta \\ \sin n\theta & \cos n\theta \end{pmatrix} \cdot \begin{pmatrix} Ic_{dn} \\ Ic_{qn} \end{pmatrix} \quad \text{...............} \quad (7)$$

From equation (4), harmonic current of 1,2f is generated by offset and gain errors. Control system is configured in simultaneous parallel by applying n=1,2 as the suppression order. Finally, the current error compensation ic_{dq} is generated by summing the ic_{dqn} obtained for each order, and compensate for the current detection value by subtracting from $i_{dq}{}^{sense}$.

Also, for each of the parameters used in equation (5), the exact value is not necessary. Absolute amount of $\hat{i}_{dq}$ is not important; parameter precision does not give a direct effect on sensor error compensation precision, because PDO computes the compensation value automatically. It is possible to calculate the system model of PDO which influences the control stability of the PDO from motor model equation (5) and the known values including the current controller. Then, it is possible to maintain enough control stability of PDO.

V. SIMULATION CONSIDERATION

This chapter verifies the validity of the proposed method by simulation. Error of the current sensor was set to the offset error and gain error as shown in TABLE 1. These errors were set as a condition that the sum of each phase error is zero. This method compensates the sensor error which appears in the harmonic current. Therefore, it is not possible to compensate the sensor errors appearing in the DC component by the equilibrium error. For equilibrium error, we assume to be adjusted by observing the DC value of the torque.

TABLE 1   CURRENT ERROR SETTING

Error Target	U phase	V phase	W phase
Offset Error	+1A	+2A	-3A
Gain Error	+5%	+10%	-15%

For simulation results, Fig. 3 shows time response of the shaft torque, Fig. 4 shows the FFT analysis result of the torque before and after error compensation, Figs. 5 and 6 show the three phase current waveform of before and after error compensation, respectively. The current waveform is not current sensor value, it is actual value of motor current. From Figs. 3 and 4, it can be confirmed that the torque ripple is suppressed that occurred 1,2f. From the three-phase current waveform at Figs. 5 and 6, the unbalance from gain error and offset error has been eliminated by the compensation.

978-1-4799-2706-7/14 $31.00 © 2014 IEEE

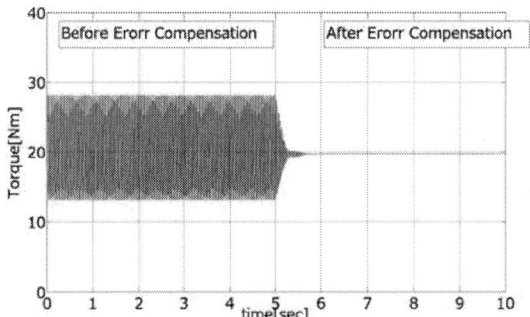

Fig. 3. Torque Waveform

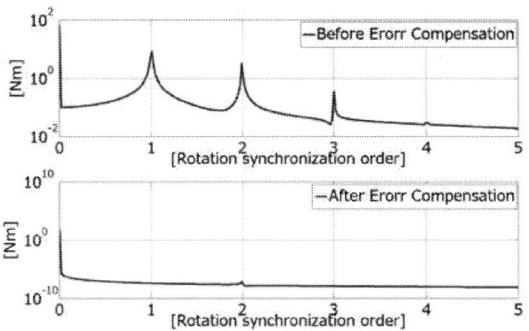

Fig. 4. FFT Analysis of the Torque

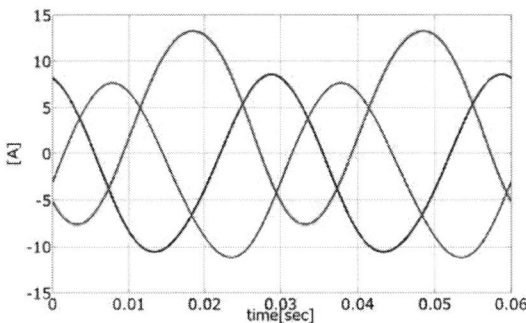

Fig. 5. Current Waveform Before Compensation

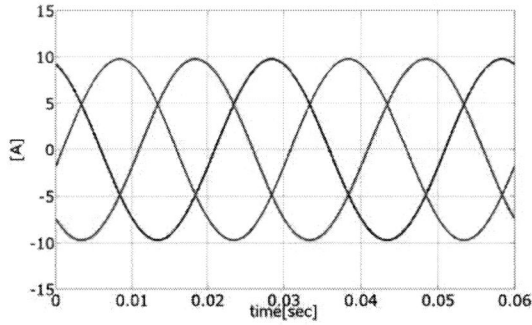

Fig. 6. Current Waveform After Compensation

VI. EXPERIMENTAL RESULT

We verified the effectiveness of the proposed method using the actual machine. TABLE 2 shows the actual machine parameters. Because of in the actual machine, detection circuit and sensor body have some errors. However, since

the adjustment is applied manually in order to reduce the error of the detection circuit to zero, it is not enough to produce a large torque ripple. For easy verification of the effectiveness by the proposed method, we set further current sensor error on the gain and offset as shown in TABLE 3 .The error can be set arbitrarily. From the above, the sensor error correction by the proposed method performs for the adjustment error and manual error shown in the TABLE 3.

TABLE 2   SYSTEM CONDITION

rated power	2.20[kW]
rated torque	42.0[Nm]
rated speed	500[min^{-1}]
poles	8
DC voltage	310 [V]
rated current	115[A]
motor inertia	0.16[kg m^2]
PDO control period	200[us]

TABLE 3   CURRENT ERROR SETTING

Error Target	U phase	V phase	W phase
Offset Error	+1.36A	-1.36A	-0.54A
Gain Error	+10%	-20%	+20%

For experimental results, Fig. 7 shows time response of the shaft torque, Fig. 8 shows the FFT analysis result of the torque before and after error compensation, Figs. 9 and 10 show the three phase current waveform of before and after error compensation, respectively. The current waveform is not current sensor value, but actual value of motor current. From Figs. 7 and 8, it can be confirmed that the torque ripple is suppressed that occurred 1,2*f*. From the three-phase current waveform shown in Fig. 9 and 10, the unbalance from gain error and offset error has been eliminated by the compensation.

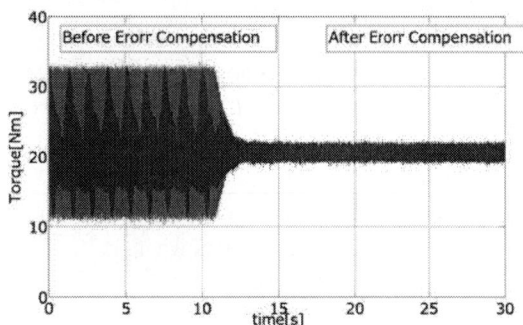

Fig. 7. Torque Waveform

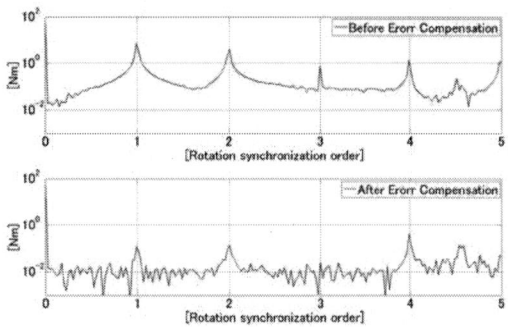

Fig. 8. FFT Analysis of the Torque

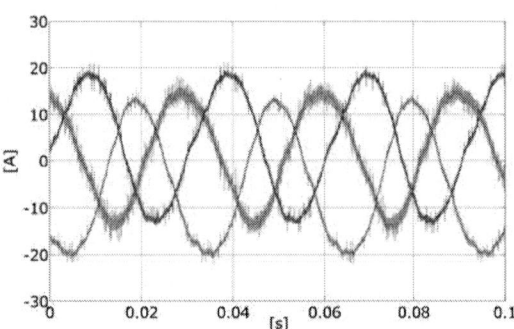

Fig. 9. Current Waveform Before Compensation

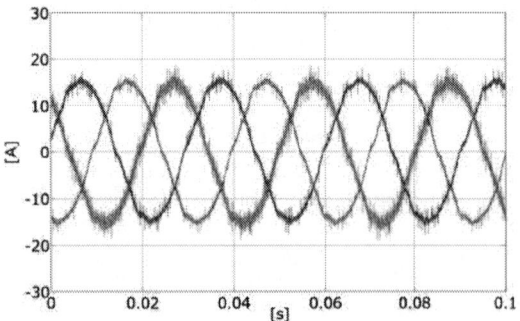

Fig. 10. Current Waveform After Compensation

VII. CONCLUSION

This paper proposed current sensor error compensation method based on the periodic disturbance observer compensation method. By compensating current sensor errors, the vibration of dq-axis current is suppressed, and then current control accuracy is possible to improve more.

The experimental result showed the effectiveness of the proposed method. In experiment and simulation, we investigated on the assumption of inverter motor applications in this paper. Influence of the current sensor error is not limited to motor applications, it will appear as current harmonics as well, such as the power converter, and it becomes many problems. The proposed method is also applicable to these cases, we considered can be corrected a current sensor. We plan to verify its effectiveness in the future work.

REFERENCES

[1] Arinori. S, Zanma. T, Doki. S, Ishida. M: "Current Compensation Signal in Suppression Control for frame vibration of PMSM by Sensorless Control", *Power Conversion Conference (PCC Nagoya 2007)*, pp874-878(2007)

[2] Naoki Miyamoto, Kiyoshi Ohishi, "Correction Method of Current Measurement Error caused by Offset and Gain Deviation of Current Sensor for SPMSM", IEEJ Trans. IA, Vol. 133 (2013) No. 6 P 627-628 (in Japanese)

[3] Yutaro Uenaka, Masaki Sazawa, Kiyoshi Ohishi, Takahashi Kenji, "Method for On-line Estimation of Electrical Motor Parameter Variation and Current Sensor Offset for SPM Motor", IEEJ Trans. IA, Vol. 131 (2011) No. 10 P 1193-1202 (in Japanese)

[4] Jeong-seong Kim, Doki. S, Ishida. M: "Improvement of IPMSM sensorless control performance by suppression of harmonics on the vector control using Fourier transform and repetitive control", *Industrial Electronics Conference (IECON 2002)*, pp597-602(2002)

[5] Y. Tadano, T. Akiyama, M. Nomura and M. Ishida: "Periodic Learning Suppression Control of Torque Ripple Utilizing System Identification for Permanent Magnet Synchronous Motors", *IEEE International Power Electronics Conference (IPEC-Sapporo 2010)*, pp.1363-1370 (2010)

[6] Yugo Tadano, Takao Akiyama, Masakatsu Nomura, Muneaki Ishida, "Torque Ripple Suppression Control Based on the Periodic Disturbance Observer with a Complex Vector Representation for Permanent Magnet Synchronous Motors", IEEJ Trans. IA, Vol. 132 (2012) No. 1 P 84-93 (in Japanese)

Rapid and Stable Speed Control of SPMSM Based on Current Differential Signal

Jun Kitajima
Nagaoka University of Technology
1603-1 Kamitomioka, Nagaoka,
Niigata 940-2188, JAPAN
Email: s125019@stn.nagaokaut.ac.jp

Kiyoshi Ohishi
Nagaoka University of Technology
1603-1 Kamitomioka, Nagaoka,
Niigata 940-2188, JAPAN
Email: ohishi@vos.nagaokaut.ac.jp

Abstract—In AC servo system, the output voltage is restricted by inverter output limitations. Voltage saturation occurs as a result of this limitation. The settling time of current and speed increases because of voltage saturation. This paper proposes a new control system for improving the transient response of surface permanent magnet synchronous motor (SPMSM) by using the proposed flux-weakening control based on current differential signal. As a result of considering the current differential signal, the suppression of voltage saturation is realized with optimal d-axis current in a transient state. The proposed speed control system realizes the quick and stable speed and current response. The effectiveness of proposed control method is confirmed by the numerical simulation results and the experimental results.

Index Terms—Flux-Weakening Control, Voltrage Saturation, Space Vector Modulation, SPMSM

I. INTRODUCTION

AC servo motors, which are used in industrial applications, are required to have a quick and stable response of speed and torque. Generally, a speed servo system based on vector control of AC servo motor consists of several PI controllers, and these outputs are restricted by a current limiter and an inverter voltage output limitation. Voltage saturation occurs becuse of existence of the inverter voltage output limitation. In votage saturation region, control performance is often degraded for a quick speed reference, which is caused by PWM inverter voltage limitation.

In order to overcome these problems caused by voltage saturation, we need to increase the inverter voltage utilization. Inverter modulation schemes are used to improve the voltage utilization of a three-phase inverter. For example, two phase modulation, third harmonics injection and space vector pulse width modulation (SVPWM). Inverters using SVPWM output voltage as the linear region inside of the inverter voltage limitation hexagon [1], [2]. A stable speed and current responses are obtained as a result of the inverter operating in the linear region. The SVPWM based inverter has the large voltage region of the precise sinusoidal phase voltage in the linear region. However, SVPWM based inverter also has a voltage saturation for large quick current reference. It often has some problems because of voltage saturation. For example, the settling time of the speed response becomes longer, and the current response doesn't coincide with its reference.

A flux-weakening control has often been used to supress voltage saturation [3]-[7]. However, conventional flux-weakening control outputs excessive d-axis current reference in transient state. Because conventional flux-weakening control is based on steady state motor voltage equation.

This paper proposes a new flux-weakening control method with optimal d-axis current that improves the transient response in which operates only under a saturated-voltage condition, while the inverter operates in the linear region. The effectiveness of proposed method is confirmed by the numerical simulated results and the experimental results.

II. SUPPRESSION METHOD FOR VOLTAGE SATURATION BY CONVENTIOANL FLUX-WEAKENING CONTROL

A. Control scheme

When voltage saturation occurs, flux-weakening control is operated by using the d-axis current and suppresses the voltage saturation. If the voltage vector exceeds the inscribed circle of the inverter output limitation hexagon, voltage saturation occurs as shown in Fig. 1(a). In flux-weakening control, in order to keep the inverter voltage vector on the inscribed circle as shown in Fig. 1(b), the inverter control system determines the desired d-axis current reference. The d-axis current reference is determined by an equation that is derived from the motor voltage equation. On d-q axis, SPMSM voltage equations are expressed as eq.(1).

$$\begin{bmatrix} v_d \\ v_q \end{bmatrix} = \begin{bmatrix} R_a + PL_a & -\omega_e L_a \\ \omega_e L_a & R_a + PL_a \end{bmatrix} \begin{bmatrix} i_d \\ i_q \end{bmatrix} + \begin{bmatrix} 0 \\ \omega_e \Phi_a \end{bmatrix} \quad (1)$$

where, i_d : d-axis current, i_q : q-axis current, v_d : d-axis voltage, v_q : q-axis voltage, Φ_a : linkage flux of permanent magnet, R_a : winding resistance, L_a : winding inductance, ω_e : electrical angler speed, $P = d/dt$. The supplied motor voltage is restricted by the inverter output limitation. Eq.(2) shows the limitation of motor voltage.

$$\left(\frac{V_{DC}}{\sqrt{2}} \right)^2 \geq v_d^2 + v_q^2 \quad (2)$$

V_{DC} is DC link voltage. The left side of eq.2 describes an inscribed circle of the inverter output limitation hexagon. Eq.(2) is transformed into eq.(3) by substituting each axis

The 2014 International Power Electronics Conference

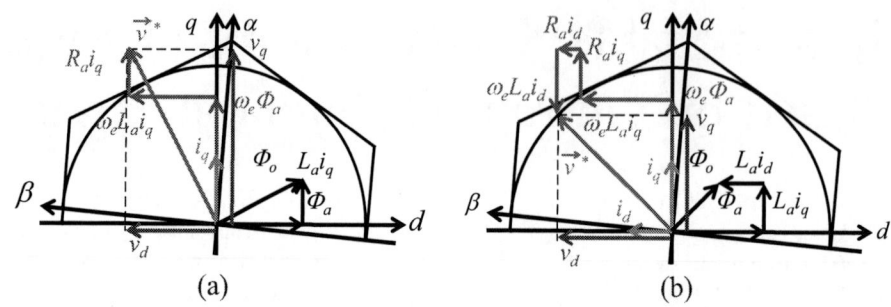

Fig. 1. Voltage vector diagram of SPMSM

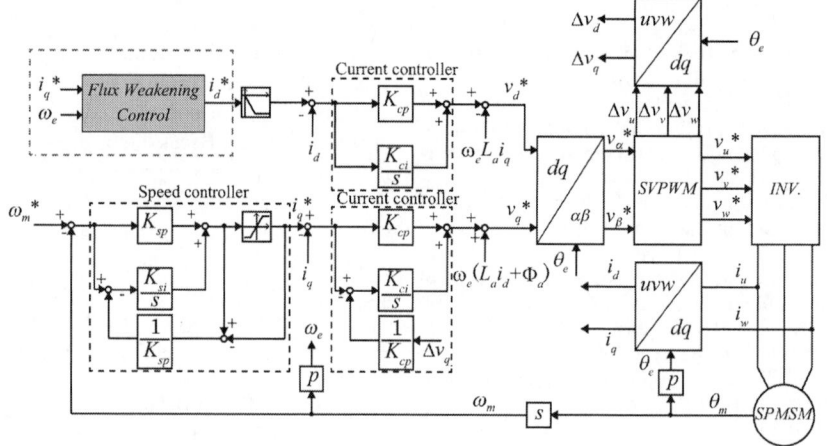

Fig. 2. Total system of speed servo system

$$i_d^* = \frac{-\omega_e^2 L_a \Phi_a + \sqrt{(\omega_e^2 L_a \Phi_a)^2 - (R_a^2 + \omega_e^2 L_a^2)\left[2R_a i_q^* \omega_e \Phi_a + \omega_e^2 \Phi_a^2 + (R_a^2 + \omega_e^2 L_a^2) i_q^{*2} - \left(\frac{V_{DC}}{\sqrt{2}}\right)^2\right]}}{R_a^2 + \omega_e^2 L_a^2} \tag{5}$$

voltage from eq.(1). When the current differential signals in eq.(3) are omitted, the simple voltage limitation equation is induced in eq.(4)

$$\left(\frac{V_{DC}}{\sqrt{2}}\right)^2 = \left(R_a i_d^* + \frac{di_d^*}{dt} - \omega_e L_a i_q^*\right)^2$$
$$+ \left(R_a i_q^* + \frac{di_q^*}{dt} + \omega_e\left(L_a i_d^* + \Phi_a\right)\right)^2 \tag{3}$$

$$\left(\frac{V_{DC}}{\sqrt{2}}\right)^2 \geq \left(R_a i_d^* - \omega_e L_a i_q^*\right)^2 + \left[R_a i_q^* + \omega_e\left(L_a i_d^* + \Phi_a\right)\right]^2 \tag{4}$$

The d-axis current reference equation, that outputs d-axis current reference during voltage saturation period, is derived by solving eq.(4) for i_d^*. The equation determining the d-axis current reference is expressed as eq.(5).

B. System structure

Fig. 2 shows a structure of speed servo system considering the windup phenomenon. In the speed servo system, the speed controller makes a speed response ω_m track the speed reference ω_m^* and determines the q-axis current reference i_q^*. If i_q^* is restricted by the current limitation, the integral calculation PI regulator is corrected by feeding back the saturated value of the q-axis current Δi_q within the current sampling time. When the voltage reference exceeds the inverter output voltage limitation, the inverter cannot output the original voltage reference. Its output is limited by the voltage limitation value. The integral calculation of the current controller should be also corrected by considering the voltage saturation to prevent an unstable response [8], [9]. Fig.3 shows the block diagram of a PI controller considering output variable saturation. When the PI controller output variable y is saturated by the limiter, the difference Δy between the controller output variable y

978-1-4799-2706-7/14 $31.00 © 2014 IEEE   1248

and the plant input variable $\tilde{y}$ is calculated. The calculated value is multiplied by the conditioning gain $F(s)$. Then, its calculated value is fed back to the integral state, and the state variable of the PI controller is regulated. Accordingly, the PI controller output variable prevents integrator windup from occurring. Here, a ideal PI controller without limiter is defined as shown in Fig.4. The PI controller output variable $\tilde{y}$ is not saturated by the limiter, and the equation for $\tilde{y}$ is defined in eq.(6). Moreover, the equation for the control variable u is defined in eq.(7), and the equation for the control variable y in Fig.3 is defined in eq.(8).

$$\tilde{y} = \frac{sK_P + K_I}{s}(\tilde{u} - u) \tag{6}$$

$$u = \tilde{u} - \frac{s}{sK_P + K_I} \cdot \tilde{y} \tag{7}$$

$$y = \frac{sK_P + K_I}{s}(u^* - u) - \frac{K_I \cdot F(s)}{s}(y - \tilde{y}) \tag{8}$$

The command $\tilde{u}$ is expressed by eq.(7) and eq.(8).

$$\tilde{u} = u^* - \frac{1}{K_P}\frac{s + K_I \cdot F(s)}{s + \frac{K_I}{K_P}}(y - \tilde{y}) \tag{9}$$

When the conditioning gain $F(s) = \frac{1}{K_P}$, $\tilde{u}$ is expressed as eq.(10). In eq.(10), $\tilde{u}$ is achieved without influence of integral calculation when the PI controller output variable is saturated by the limiter. And, when $\tilde{y}$ is equal to y, $\tilde{u}$ coincide with u immediately.

$$\tilde{u} = u^* - \frac{1}{K_P}(y - \tilde{y}) \tag{10}$$

The anti-windup PI controller is applied to the speed controller and the q-axis current controller. The integral state and the state variable of the PI controller are regulated as

$$\tilde{e} = e - \frac{\Delta y}{K_P} \tag{11}$$

$$y = K_P \cdot e + \frac{K_I}{s} \cdot \tilde{e} \tag{12}$$

Thus, a PI controller can avoid the influences of windup phenomenon [10]. The anti-windup control is not applied to the d-axis current controller because, when the voltage saturation value is fed back to the d-axis current controller,

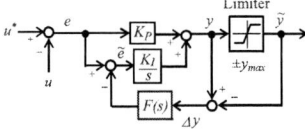

Fig. 3. Block diagram of PI controller considering output variable saturation.

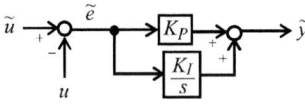

Fig. 4. Block diagram of PI controller.

decoupling control is disturbed by the feedback at the control with $i_d = 0$[11]. However, d-axis current controller is compensated from voltage saturation because flux-weakening control works as one kind of anti-windup control. The d-axis current reference is calculated by eq.(5) using ω_e and i_q^*. On condition that the calculated value of i_d^* is negative, it is used as the d-axis current reference. Its upper limit of i_d^* is set as operating control with $i_d = 0$ when the calculated value is positive. Additionally, the d-axis current is almost transformed into the power loss because the d-axis current does not contribute to the torque and the output in the SPMSM. Permanent magnets have a risk of permanent demagnetization due to heating. Thus, its lower limit is set to -2.42 A to avoid permanent demagnetization.

C. Numerical simulation results

Fig. 5 shows the numerical simulation results on condition of $i_d = 0$. Fig. 6 shows the numerical simulation results of the conventional method for the system shown in Fig. 2. In Fig. 5, voltage saturation occurs in the transient state. In this case, the q-axis current cannot track its reference, and the settling time of the speed response becomes longer. In contrast, in Fig.6, voltage saturation is suppressed by outputting the d-axis current and operating the flux-weakening control in the transient state. As a result of the suppressed voltage saturation, the q-axis current can track its reference, and a fast settling time of the speed response is obtained.

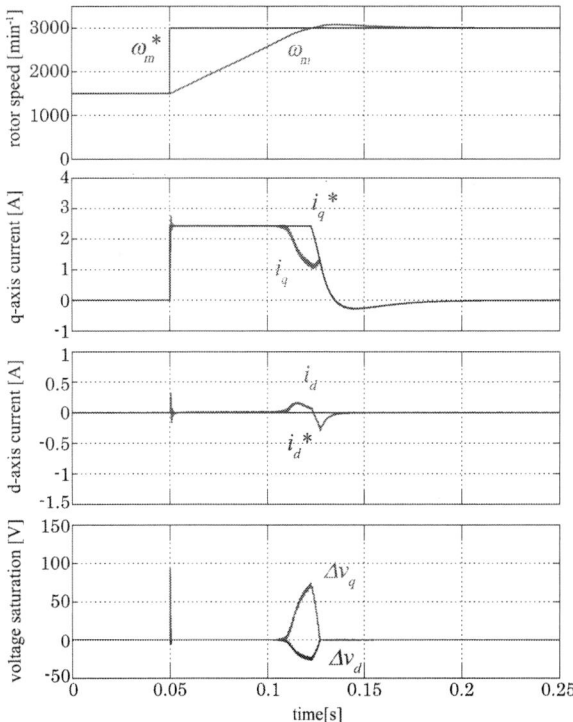

Fig. 5. Simulation results of the control with $i_d = 0$.

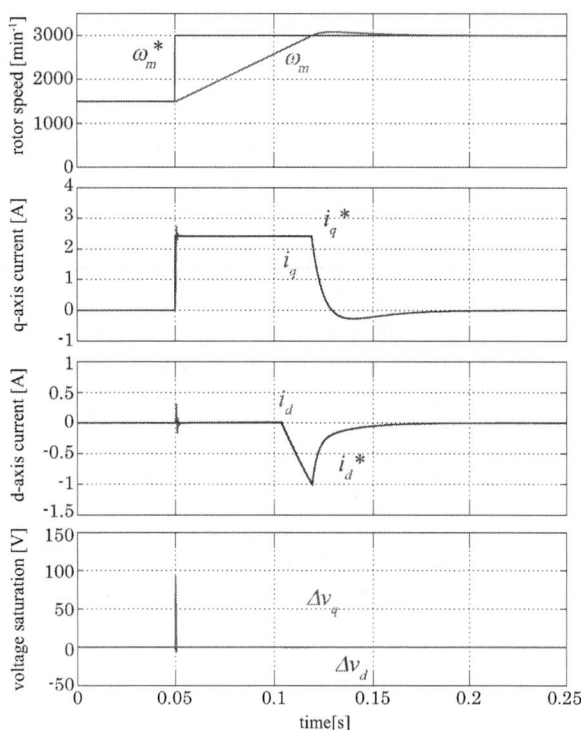

Fig. 6. Simulation results of the conventional method.

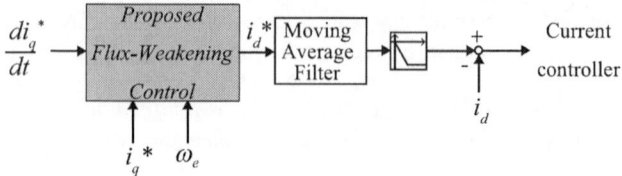

Fig. 7. Block diagram of proposed flux-weakening control with current differential signal.

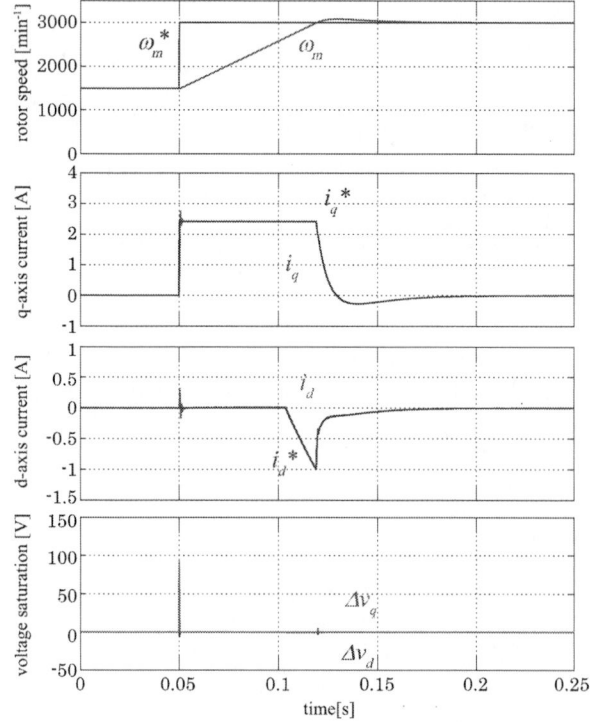

Fig. 8. Simulation results of the proposed flux-weakening control considering current differential signal.

III. PROPOSED FLUX-WEAKENING CONTROL

A. Control scheme

On condition of conventional flux-weakening control, excessive d-axis current reference is calculated. Although the equation is simple. For the proposed flux-weakening control considering a current differential signal, it is possible to calculate a more optimal d-axis current reference in a transient state. Eq.(13) is derived by considering q-axis current differential term in eq.(3).

$$\left(\frac{V_{DC}}{\sqrt{2}}\right)^2 = \left(R_a i_d^* - \omega_e L_a i_q^*\right)^2$$
$$+ \left[R_a i_q^* + L_a \frac{di_q^*}{dt} + \omega_e \left(L_a i_d^* + \Phi_a\right)\right]^2 (13)$$

In eq.(13), d-axis current differential term is omitted to prevent oscillation of calculated d-axis current reference. If d-axis current differential term is included in eq.(13), huge d-axis current reference is caluculated in next sampling when d-axis current reference change largely. And, the amount of change become zero when the d-axis current reference is restricted by d-axis current limitter. Obviously, the d-axis current differential signal becomes zero and the calculated d-axis current reference becomes small. As a result of such calculation is done repeatedly, the d-axis current reference becomes oscillated. Eventually, flux-weakening control doesn't not work properly. Therefore, d-axis current differential term is not included in eq.(13). The d-axis current reference as shown

in eq.(14) is obtained by solving eq.(13) for i_d^*.

$$i_d^* = -\alpha + \sqrt{\alpha^2 - \beta} \qquad (14)$$

$$\alpha = \frac{B}{A}$$
$$\beta = \frac{C}{A}$$
$$A = R_a^2 + \omega_e^2 L_a^2$$
$$B = \omega_e L_a \left(L_a \frac{di_q^*}{dt} + \omega_e \Phi_{fa}\right)$$
$$C = A i_q^{*2} + \frac{2B R_a}{\omega_e L_a} i_q^* + \frac{B^2}{\omega_e^2 L_a^2} - \left(\frac{V_{DC}}{\sqrt{2}}\right)^2$$
$$\frac{di_q}{dt} = \frac{i_{q(n-1)}^* - i_{q(n-2)}^*}{T}$$

Where, T is sampling period.

B. System structure

Fig. 7 shows the proposed block diagram having the d-axis current reference calculation considering the current differential signal. In the calculation of proposed method, i_d^* is determined by using i_q^*, ω_e and q-axis current differential signal di_q^*/dt. However, the i_d^* often has the noisy value because of the current differential signal. Hence, i_d often oscillates because of the noisy value of i_d^*. Moreover, the voltage saturation caused by the oscillated current response occurs. Therefore, the response of the q-axis current does not coincide with its reference and the settling time of the rotor speed becomes longer in voltage saturation region. In the proposed method, a moving average filter is used for i_d^* to suppress the oscillation phenomenon. The system structure is the same as in Fig. 2, except the part of the d-axis current reference calculation.

C. Numerical simulation results

Fig. 8 shows the numerical simulation results of proposed method considering a current differential signal. As a result of operating flux-weakening control, voltage saturation is suppressed, the q-axis current response tracks its reference and a fast settling time of the speed response is obtained. Moreover, the proposed method considering the differential signal suppresses voltage saturation with a smaller d-axis current in comparison with the conventional method omitting a current differential signal.

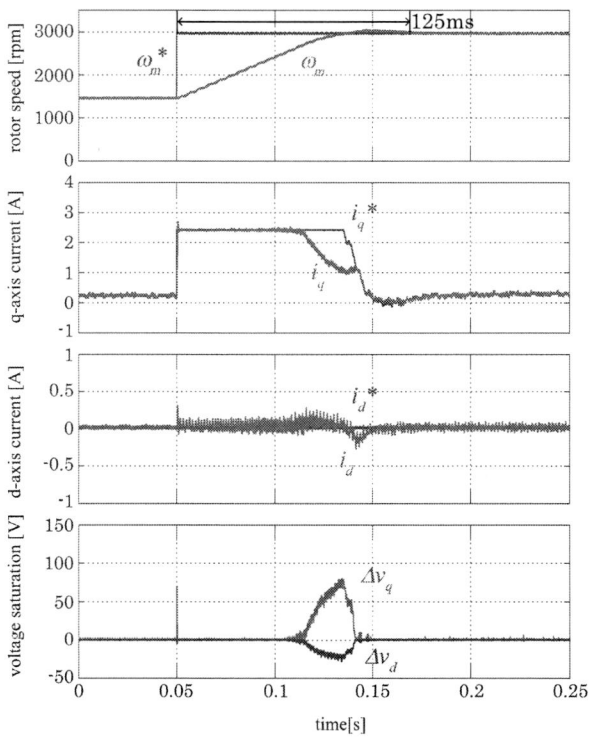

Fig. 9. Experimental results on condition of $i_d = 0$.

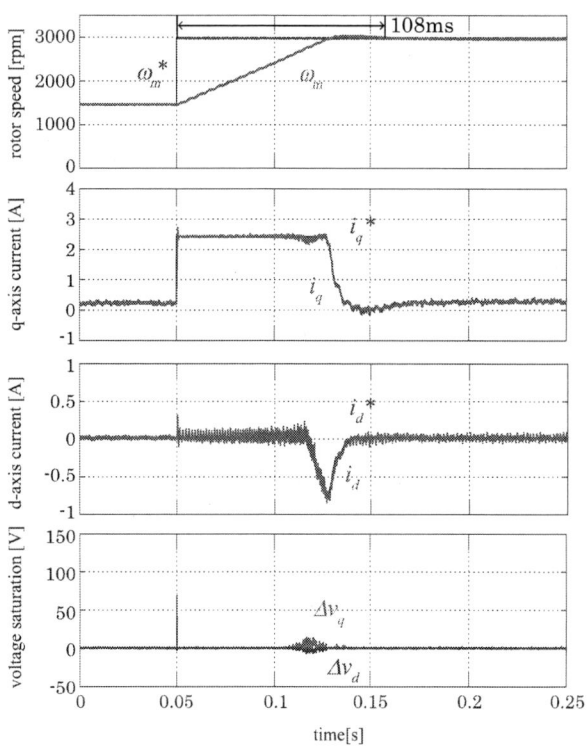

Fig. 10. Experimental results of conventional method.

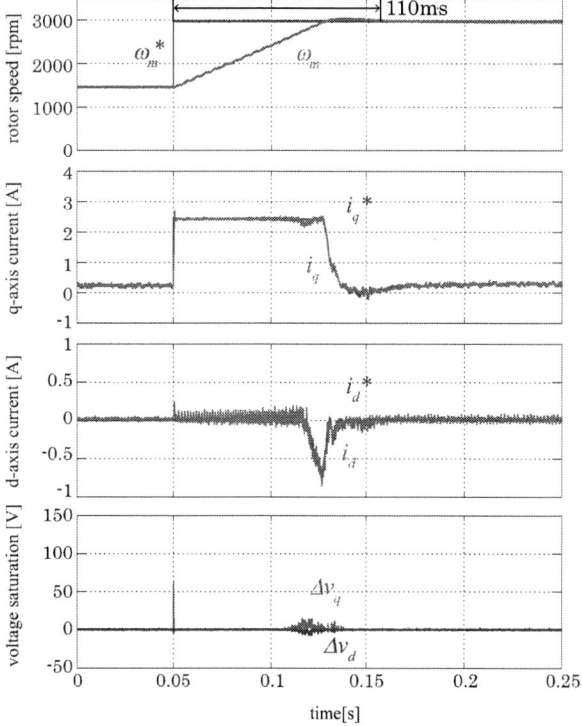

Fig. 11. Experimental results of proposed method with current differential signal.

IV. Experimental Results

A. Conventional flux-weakening control

Table I summarizes the specification of the tested motor used in the experiment. The experiment is applied a speed step reference of 1500 min^{-1}, which is not in the voltage saturation region, to 3000 min^{-1}, which is in the voltage saturation region. Fig. 9 shows the experimental results of the control with $i_d = 0$ and Fig.10 shows the experimental results of the conventional flux-weakening control, respectively. In Fig. 9, the large voltage saturation occurs in a transient state. i_q cannot track its reference, and the settling time of the speed response becomes long because of the voltage saturation. This settling time is 125 ms. In Fig. 10, the voltage saturation is suppressed by outputting i_d. The q-axis current response tracks its reference and a fast settling time whose response of 108 ms is obtained. Compared with the simulation results, voltage saturation slightly occurs around 0.12 s in Fig. 10. The reason is thought to be a reduction in the DC link voltage due to flowing large current and the DC link capacitor discharging when the motor is accelerated.

B. Proposed method with the current differential signal

Fig. 11 shows the experimental results of the proposed flux-weakening control considering a current differential signal. In Fig. 11, voltage saturation is suppressed with smaller d-axis current in comparison with Fig. 10. The proposed method makes 27% reduction of d-axis power loss because of d-axis current reduction. The settling time of the speed response is 110 ms, it is similar to the conventional method. TableII shows a comparison of each experimental results.

TABLE I
RATED VALUES OF TESTED PMSM

Rated power	200[W]
Rated speed	3000[min^{-1}]
Rated torque	0.64[Nm]
Pole pair	4[-]
Rated q-axis current	2.42[A]

TABLE II
COMPARISON OF THE RESPONSES FOR EACH CONTROL METHOD

Control method	Settling time[ms]	D-axis power loss[%]
Conventional method	108	100
Proposed method	110	73

V. Conclusion

In this paper, a new flux-weakening control method based on current differential signal is proposed. Two flux weakening controls are compared. The conventional flux-weakening contorl outputs excessive d-axis current reference in transient state due to omitting current differential terms. In the proposed flux-weakening control, optimal d-axis current reference is calculated by considering q-axis current differential signal. It is confirmed that both methods achieve good responses in voltage saturation region according to the numerical simulation and experiment. Furthermore, the proposed method suppress voltage saturation with smaller i_d in comparison with the conventional method.

References

[1] K.Takahashi, K.Ohishi and T.Kanmachi:"Space Vector Modulation Inverter Considering Acceleration Torque and Voltage Saturation for Speed Servo System", T.IEE Japan, Vol.129-D, No.10, pp.1013-1014 (2009) (in Japanese)

[2] K. Takahashi, K. Ohishi and T. Kanmachi:"Driving Method with Priority of d-axis Voltage for Interior Permanent Magnet Synchronous Motor", T.IEE Japan, Vol.131-D, No.9, pp.1103-1111 (2011) (in Japanese)

[3] S. Morimoto, M. Sanada, Y. Tanaka :"Wide-Speed Operation of Interior Permanent Magnet Synchronous Motors with High-Performance Current Regulator"IEEE TRANSACTION ON INDUSTRY APPLICATION, VOL 30, NO. 4, JULY/AUGUST 1994

[4] T. Kwon, S. Sul :"Novel Antiwindup of Current Regulator of a Surface-Mounted Permanent-Magnet Motor for Flux-Weakening Control"IEEE TRANSACTION ON INDUSTRY APPLICATION, VOL 42, NO. 5, SEPTEMBER/OCTOBER 2006

[5] Jiunn-Jiang Chen and Kan-Ping Chin :"Minimum Copper Loss Flux-Weakening Control of Surface Mounted Permanent Magnet Synchronous Motors"IEEE TRANSACTIONS ON POWER ELECTRONICS, VOL. 18, NO. 4, JULY 2003

[6] Hesong Liu, Z. Q. Zhu, Essam Mohamed, Yongling Fu, and Xiaoye Qi :"Flux-Weakening Control of Nonsalient Pole PMSM Having Large Winding Inductance, Accounting forResistive Voltage Drop and Inverter Nonlinearities"IEEE TRANSACTIONS ON POWER ELECTRONICS, VOL. 27, NO. 2, FEBRUARY 2012

[7] Ping-Yi Lin and Yen-Shin Lai, Senior :"Voltage Control Technique for the Extension of DC-Link Voltage Utilization of Finite-Speed SPMSM Drives"IEEE TRANSACTIONS ON INDUSTRIAL ELECTRONICS, VOL. 59, NO. 9, SEPTEMBER 2012

[8] K.Ohishi, E.Hayasaka, T.Nagano and M.Hrakawa:"Speed Servo System Considering Voltage Saturation of Indirect Vector Contorl System",T.IEE Japan, Vol.122-D, No.2, pp.120-127 (2002) (in Japanese)

[9] K. Kneko and K. Ohishi:"Design Method of Limitation Compensator Keeping Response Performance for PI Speed Control System of SPMSM", T.IEE Japan, Vol.133-D, No.5, pp.526-535 (2013-5) (in Japanese)

[10] K. Ohishi and T. Mashimo:"Design Method of Digital Robust Speed Servo System Considering Output Saturations", T.IEE Japan, Vol.119-D, No.1, pp.88-96 (2002-1) (in Japanese)

[11] Y. Egashira and K. Ohishi:"Speed Control Method for PM Motor Considering Voltage Saturation", T.IEE Japan, Vol.126-D, No.2, pp.206-207 (2006-2) (in Japanese)

Parallel Connected Multiple Drive System Using Small Auxiliary Inverter for Numbers of PMSM

Tsuyoshi Nagano and Jun-chi Itoh
Dept. of Energy and Environmental
Nagaoka University of Technology
Nagaoka, Niigata, Japan
ngn244@stn.nagaokaut.ac.jp
itoh@vos.nagaokaut.ac.jp

Abstract— **This paper discusses a multi-parallel drive system for permanent magnet synchronous motors (PMSMs). This system proposes to use two different functions of inverter, a main inverter to drive parallel connected PMSMs and an auxiliary inverter to suppress the vibration of the motor speed. However, the power capacity of the auxiliary inverter depends on the speed response. In this paper, the performance of the proposed system is discussed. From the frequency characteristics, it is clarified that the power capacity of the auxiliary inverter is lower than 10% of the main inverter when the frequency components included in the speed command are less than 12 rad/s. As a result, the experimental results demonstrated that the proposed system can reduce the speed vibration from 400 r/min to nearly 0 r/min with damping control in the prototype of 1500-W PMSM drive system. The experimental results well agree with the theoretical analysis results.**

Keywords— *Permanent magnet synchronous motor, Parallel operation, Damping control.*

I. INTRODUCTION

Recently, the adjustable speed drive techniques for the permanent magnet synchronous motor (PMSM) have attracted a lot of attentions among the researchers in terms of the energy saving [1]-[3]. PMSM has high efficiency and high power density in comparison with the induction motor (IM) because PMSM does not have the excitation circuit. On the other hand, IMs can be driven in parallel by using only one inverter. Parallel connected multiple induction motor drive systems are applied to industry in terms of cost reduction and simplify of the system, because this system only uses large power capacity inverter to multiple induction motors [4]-[7].

The efficiency of the motor drive system is improved by replacing IM with PMSM. However, it is difficult to drive the parallel connected PMSMs by using only one inverter because the torque vibration occurs due to the resonance between a synchronous reactance and the inertia moment [8]. When the parallel connected PMSMs are driven by one inverter, the current cannot be controlled corresponding to each pole of PMSM because the pole position is different from the pole position of other PMSMs. Therefore, the torque vibration will occur in this multiple motor drive system.

The damping control that uses the current feedback has been proposed [9]. This method suppresses the torque vibration with adding the vibrational component in the effective current to the output phase command of the voltage command. However, this method cannot be applied to the parallel connected PMSMs because one inverter cannot control each current corresponding to each pole of the PMSM.

There are some literatures which two PMSMs are driven by one power converters [10]-[15]. These converters can drive two PMSMs independently. However, these converters have some problems. For instance, the five-leg inverter [10]-[12] and the nine-switch inverter [13]-[14] is driven at low voltage utilization ratio in comparison with a case when one PMSM is driven by these converters because DC-link voltage is distributed to two PMSMs. Moreover, the current capacity of the switching device increases because there is a common leg. The four-leg inverter using the neutral point of the DC link capacitor is less number of the switches than the other topologies [15]. However, this converter requires the capacitor voltage balancing control because the capacitor neutral-point voltage is unbalanced by the motor current. Therefore, it is difficult to apply these converters to several motors because the output voltage of the inverter is limited and increasing the number of switching devices. Therefore, for this reason, these converters are not suitable for driving multi-parallel connected PMSMs. So, the authors have proposed a drive system for multi-parallel connected PMSMs by the auxiliary inverter and motor windings [16]-[17].

In this paper, the acceleration performance of the proposed system is evaluated from the frequency characteristic for the speed command, the acceleration test and the simulation results that two sets of parallel connected PMSMs are driven. From the frequency characteristics of the transfer function from the speed command to the rotational speed, it is clarified that the resonance is suppressed by the damping control applying to the auxiliary inverter. Moreover, the minimum power capacity of the auxiliary inverter is clarified from the frequency characteristics of the transfer function from the

978-1-4799-2706-7/14 $31.00 © 2014 IEEE

speed command to the output power of the auxiliary inverter.

This paper is organized as follows; first, the configuration and the control strategy of the proposed parallel drive system are introduced. Next, the suppression effect of the damping control from the frequency characteristics from the speed command to rotational speed in the proposed system. In addition, the stability of the proposed system with the damping control is clarified in the Routh-Hurwitz stability criterion. Next, the minimum power capacity of the auxiliary inverter clarified from the frequency characteristics from the speed command to the output power of each inverter in the proposed system. Finally, the simulation and experimental results demonstrate that the proposed system can drive multiple of parallel connected PMSMs effectively by suppressing the torque vibration.

II. PROPOSED SYSTEM AND CONTROL STRATEGY

Fig. 1 shows the configuration of the proposed system. In PMSMs, the auxiliary windings which are used in the damping control (with the auxiliary inverter) are placed in the slots together with the main windings. The proposed system uses two different power rating inverters. The first one is the large power capacity inverter for the main windings to control the speed of the parallel connected PMSMs. The second one is a small power capacity inverter for the auxiliary windings to suppress the torque vibration. In term of the effectiveness of the proposed system, it is very important that the power capacity of the auxiliary inverter is enough small.

Fig. 2 shows the control block diagram of the proposed system. In the proposed system, the V/f control is applied to the main inverter and the field-oriented control and the damping control are applied to the auxiliary inverter. Each of the auxiliary inverter controls the current in auxiliary windings of the PMSM in order to suppress the torque and speed vibrations. Since these vibrations are caused by the phase difference between the rotational coordinates of the inverter and of the PMSM, it can be suppressed by the damping control in the auxiliary inverter.

Fig. 3 shows the relationship between the d-q rotating frame and the γ-δ rotating frame for a V/f control. In the control of the auxiliary invereter, the direction of the flux vector with permanent magnet is defined as d-axis as same as the conventional FOC. In the V/f control, the output voltage vector is defined as δ-axis, the axis which lags by π/2 rad from δ-axis can be defined as γ-axis. The lag of the load angle φ occurs between the d-q rotating frame and γ-δ rotating frame as shonwn in Fig. 3. Therefore, the load angle φ, rotational speed ω_e and the speed command ω^* can be expressed as

$$p\varphi = \omega_{re} - \omega^* \quad ....(1)$$

where p is differential operator.

When the vibration of the speed is caused by the resonance between the inertia moment and the synchronous reactance, the load angle φ also is vibrating as shown (1). Then, the changes of the load angle $p\varphi$ is

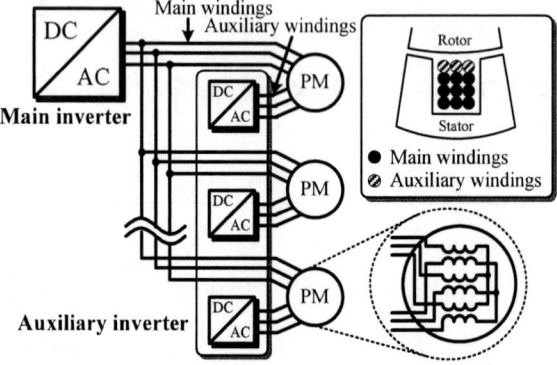

Fig. 1.Configuration of the proposed system. The proposed system uses two different power rating inverters, the large power capacity inverter to drive the motors and the small power capacity inverter to suppress the vibration.

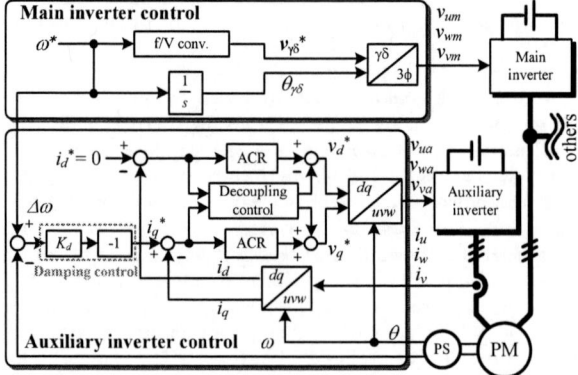

Fig. 2. Control block diagram of the proposed system. The V/f control is applied to the main inverter. The auxiliary inverters uses the field-oriented control for the damping control.

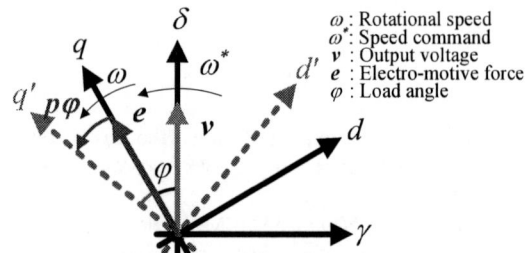

Fig. 3. Relationship between the d-q rotating frame and the γ-δ rotating frame. The γ-δ rotating frame lags by the load angle φ from the d-q rotating frame.

the difference between the rotational speed and the speed command. In order to compensate the changes of the load angle $p\varphi$, the q-axis current command i_q^* is calculated from the damping controller as shown in Fig. 2. As a result, the vibration of the speed and torque caused by the resonance are suppressed by compensating the changes of the load angle $p\varphi$ with the current controlled by FOC and the damping control.

III. EVALUATION OF SUPPRESSION PERFORMANCE

The damping control has to suppress the speed and torque vibration caused by the resonance between the inertia moment and the synchronous reactance. In this chapter, the suppression effect of the damping control is clarified from the frequency characteristics from the speed command to rotational speed in the proposed system.

Fig. 4 shows the simplified system in order to confirm the operation of the damping control. The proposed system uses the PMSM placed the auxiliary windings in the slots together with the main (conventional) windings, and so the mutual magnetic interference occurs between the main and the auxiliary windings. Due to the above reason, the control for the auxiliary inverter becomes complicated. Therefore, the proposed system is validated using a model where two PMSMs are connected in series via single shaft. Then, the rear end of the main PMSM is connected to the load machine. It means that the magnetic coupling was neglected in the simulation and the experiment.

The voltage equation of the motor that is connected to the main inverter is represented as the voltage equation in the γ-δ rotating frame because the main inverter is driven by the V/f control. Note that the γ-δ rotating frame lags by the load angle φ from the d-q rotating frame as shown in Fig. 3. On the other hand, the voltage equation of the motor that is connected to the auxiliary inverter is represented as the voltage equation in the d-q reference frame because the auxiliary inverter control is FOC. Therefore, the voltage equation of the motor that is connected to the main inverter in the estimated rotating γ-δ frame is given by

$$\begin{bmatrix} v_{M\gamma} \\ v_{M\delta} \end{bmatrix} = \begin{bmatrix} R_M + pL_{dM} & -\omega^* L_{qM} \\ \omega^* L_{dM} & R_M + pL_{qM} \end{bmatrix} \begin{bmatrix} i_{M\gamma} \\ i_{M\delta} \end{bmatrix} + \omega_{re} \psi_{mM} \begin{bmatrix} \sin\varphi \\ \cos\varphi \end{bmatrix} \dots (2)$$

On the other hand, the voltage equation of the motor that is connected to the auxiliary inverter in d-q reference frame can be given by

$$\begin{bmatrix} v_{Ad} \\ v_{Aq} \end{bmatrix} = \begin{bmatrix} R_A + pL_{dA} & -\omega_{re} L_{qA} \\ \omega_{re} L_{dA} & R_A + pL_{qA} \end{bmatrix} \begin{bmatrix} i_{Ad} \\ i_{Aq} \end{bmatrix} + \begin{bmatrix} 0 \\ \omega_{re} \psi_{mA} \end{bmatrix} \dots (3)$$

Moreover, torque and speed equations can be given by

$$T = T_M - T_A = \frac{3}{2} P_{fM} \psi_{mM} \left(i_{M\gamma} \sin\varphi + i_{M\delta} \cos\varphi \right) - \frac{3}{2} P_{fA} \psi_{mA} i_{Aq} \dots (4)$$

$$p\omega_{re} = \frac{1}{J} \left(P_{fM} T_M - P_{fA} T_A \right) \dots (5)$$

where p is differential operator, R is armature resistance, L is synchronous reactance, P_f is pairs of the pole, Ψ_m is magnet flux linkage, φ is the load angle, and J is the inertia moment of motors, Suffix 'A' represents the parameter of the motor that is connected to the auxiliary inverter, 'M' represents the parameter of the motor that is connected to the main inverter.

Equations (2) and (3), (4) are non-linear so that these

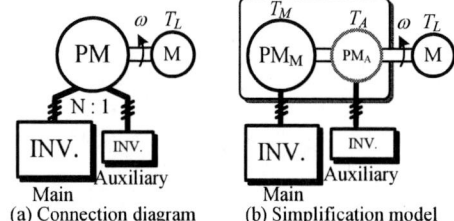

(a) Connection diagram   (b) Simplification model

Figs.4 Analysis and verification model of the PMSM in addition the auxiliary windings for damping control.

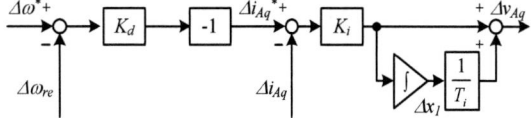

Fig. 5. State variable diagrams of the auxiliary inverter control. If the current control response is faster than the damping control response, the loop gain in the current control is 1. Therefore, the q-axis voltage command Δv_{Aq}^* can be given by (10).

equations are linearized about the steady state. The state equation after the linearization can be given by (6).

Equation (6) shows the 6th order of the state equation that is complicated to evaluate the stability. In order to simplify this equation, it is assumed that the mechanical time constant is larger than electrical time constant, then (6) can be approximated as the 2nd order state equation, as (7).

$$\begin{bmatrix} p\Delta\omega_{re} \\ p\Delta\varphi \end{bmatrix} = \begin{bmatrix} -\dfrac{3}{2} \dfrac{P_{fM}^{\,2} \psi_{mM}}{J} \dfrac{\psi_{mM} R_M}{\omega_0^2 L_{dM} L_{qM}} & \dfrac{3}{2} \dfrac{P_{fM}^{\,2} \psi_{mM}^{\,2}}{JL_{qM}} \\ -1 & 0 \end{bmatrix} \begin{bmatrix} \Delta\omega_{re} \\ \Delta\varphi \end{bmatrix}$$

$$+ \begin{bmatrix} \dfrac{3}{2} \dfrac{P_f^{\,2} \psi_{mM} R_M}{J\omega_0^2 L_d L_q} & -\dfrac{3}{2} \dfrac{P_f^{\,2} \psi_{mA}}{JR_A} & 0 \\ 0 & 0 & 1 \end{bmatrix} \begin{bmatrix} \Delta v_{M\delta} \\ \Delta v_{Aq} \\ \Delta\omega_1 \end{bmatrix} \dots (7)$$

where suffix '0' is the steady state value.

Fig. 5 shows the state variable diagrams of the auxiliary inverter control that is applying the damping control. In Fig. 5, the q-axis voltage command Δv_{Aq}^*, it is assumed that the loop gain in the current control is 1 (assuming that the current control response is faster than the damping control response), can be given by

$$\Delta v_{Aq} = -K_d \left(\Delta\omega^* - \Delta\omega_{re} \right) \dots (8)$$

$$p\Delta x_1 = 0 \dots (9)$$

where K_d is the damping gain.

Moreover, in Fig. 3, the δ-axis voltage command $\Delta v_{M\delta}^*$ and the γ-axis voltage command $\Delta v_{M\gamma}^*$ of the main

$$\begin{bmatrix} p\Delta i_{M\gamma} \\ p\Delta i_{M\delta} \\ p\Delta i_{Ad} \\ p\Delta i_{Aq} \\ p\Delta\omega_{re} \\ p\Delta\varphi \end{bmatrix} = \begin{bmatrix} -\dfrac{R_M}{L_M} & \omega_0 & 0 & 0 & -\dfrac{\psi_{mM}}{L_M}\sin\varphi_0 & -\dfrac{\omega_0\psi_{mM}}{L_M}\cos\varphi_0 \\ -\omega_0 & -\dfrac{R_M}{L_M} & 0 & 0 & -\dfrac{\psi_{mM}}{L_M}\cos\varphi_0 & \dfrac{\omega_0\psi_{mM}}{L_M}\sin\varphi_0 \\ 0 & 0 & -\dfrac{R}{L_A} & 0 & 0 & 0 \\ 0 & 0 & 0 & -\dfrac{R}{L_A} & 0 & 0 \\ \dfrac{3}{2}\dfrac{P_{fM}^{\,2}\psi_{mM}}{J}\sin\varphi_0 & \dfrac{3}{2}\dfrac{P_{fM}^{\,2}\psi_{mM}}{J}\cos\varphi_0 & 0 & -\dfrac{3}{2}\dfrac{P_{fA}^{\,2}\psi_{mA}}{J} & 0 & \dfrac{3}{2}\dfrac{P_{fM}^{\,2}\psi_{mM}}{J}\left(i_{M\gamma 0}\cos\varphi_0 - i_{M\delta 0}\sin\varphi_0\right) \\ 0 & 0 & 0 & 0 & -1 & 0 \end{bmatrix} \begin{bmatrix} \Delta i_{M\gamma} \\ \Delta i_{M\delta} \\ \Delta i_{Ad} \\ \Delta i_{Aq} \\ \Delta\omega_{re} \\ \Delta\varphi \end{bmatrix} + \begin{bmatrix} \dfrac{1}{L_M} & 0 & 0 & 0 & i_{M\delta 0} \\ 0 & \dfrac{1}{L_M} & 0 & 0 & -i_{M\gamma 0} \\ 0 & 0 & \dfrac{1}{L_A} & 0 & 0 \\ 0 & 0 & 0 & \dfrac{1}{L_A} & 0 \\ 0 & 0 & 0 & 0 & 0 \\ 0 & 0 & 0 & 0 & 1 \end{bmatrix} \begin{bmatrix} \Delta v_{M\gamma} \\ \Delta v_{M\delta} \\ \Delta v_{Ad} \\ \Delta v_{Aq} \\ \Delta\omega^* \end{bmatrix}$$

$$\dots (6)$$

inverter are given by

$$\Delta v_{M\gamma} = 0 \quad\text{...(10)}$$

$$\Delta v_{M\delta} = \psi_{mM}\Delta\omega^* \quad\text{...............................(11)}$$

Assuming as $\delta_0 = 0$, $i_{M\gamma0} = 0$, $i_{M\delta0} = 0$, ωL is larger than R, substituting (8), (10) and (11) into (7) yields the state equation of

$$p\mathbf{x} = \mathbf{A}\mathbf{x} + \mathbf{B}\mathbf{u} \quad\text{...............................(12)}$$

where $\mathbf{x} = \begin{bmatrix}\Delta\omega_{re} & \Delta\delta\end{bmatrix}$, $\mathbf{u} = \Delta\omega^*$

$$\mathbf{A} = \begin{bmatrix} -\dfrac{3}{2}\dfrac{P_{fM}^{2}\psi_{mM}}{J}K_d & \dfrac{3}{2}\dfrac{P_{fM}^{2}\psi_{mM}^{2}}{JL_{qM}} \\ -1 & 0 \end{bmatrix}, \quad \mathbf{B} = \begin{bmatrix} \dfrac{3}{2}\dfrac{P_{fM}^{2}\psi_{mM}}{J}K_d \\ 1 \end{bmatrix}$$

Fig. 6 shows the frequency characteristic of the transfer function from the speed command to rotational speed of the V/f control from the main motor. Table 1 shows the analysis and experimental conditions. In Fig. 6, the magnitude becomes lower at the resonance frequency ω_h by applying the damping control in comparison to the proposed system without the damping control. It means that the suppression effect becomes higher.

The transfer function is derived from the 2nd state equation of (12). The speed command $\Delta\omega^*$ to the rotational speed $\Delta\omega_e$ transfer function is given by

$$\frac{\Delta\omega_{re}}{\Delta\omega^*} = \frac{\dfrac{3}{2}\dfrac{P_f^{2}\psi_{mM}}{J}}{s^2 + \dfrac{3}{2}\dfrac{P_f^{2}\psi_{mM}}{J}K_d s + \dfrac{3}{2}\dfrac{P_f^{2}\psi_{mM}^{2}}{JL_{qM}}} \times$$

$$\left\{\frac{\psi_{mM}}{L_{qM}} - s\left(K_d + \frac{(i_{M\gamma0}\cos\varphi_0 + i_{M\delta0}\sin\varphi_0)}{\omega_0} - \frac{\psi_{mM}}{L_{qM}\omega_0}\sin\varphi_0\right)\right\}$$

$$\text{...(13)}$$

The damping factor ζ and the natural angular frequency ω_h are given by

$$\zeta = \sqrt{\frac{3}{2}\frac{P_{fM}K_d}{2}}\sqrt{\frac{L_{qM}}{J}} \quad\text{...........................(14)}$$

$$\omega_n = \sqrt{\frac{3}{2}\frac{P_{fM}\psi_{mM}}{\sqrt{JL_{qM}}}} \quad\text{...............................(15)}$$

Without the damping control ($\zeta = 0$), the magnitude at the resonance frequency rises. By contrast, by applying the damping control ($\zeta = 0.3$), the magnitude at the resonance frequency decreases to 5 dB. The high suppression effect is obtained by increasing the damping gain as shown (14) because the damping factor increases.

Table 1. Analysis and experimental conditions

	PM$_M$	PM$_A$
Rated power [W]	1500	750
Rated speed [min^{-1}]	1800	
Rated current [A]	8.2	4
Number of pole pairs	3	3
Armature resistance [Ω]	1.55	1.98
d-axis inductance [mH]	11.5	15.2
q-axis inductance [mH]	23	33.2
Electro-motive force constant [Vs/rad]	0.368	0.338
Inertia moment [kgm^2]	0.0051	0.0026
Rotational speed in stationary state [rad/s]	900	

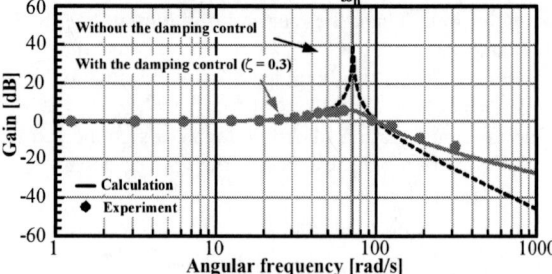

Fig. 6. Frequency characteristics of the speed command to the rotational speed in the proposed system. Table I shows the evaluated condition. By applying the damping control, the magnitude at the resonance frequency decreases to 5 dB.

Therefore, the suppression effect of the damping control is clarified from the frequency characteristics from the speed command to rotational speed in the proposed system. The proposed damping control can suppress the resonance between the moment inertia and synchronous reactance as show in Fig. 6. In the experiment, when the constant speed command adds sinusoidal wave as the speed command, the amplitude in the frequency of the sinusoidal wave is drawn by observing the speed response. The gain between the speed commands and rotational speed in the experimental results has been plotted in Fig. 6 as dot marks. The experimental results well agree with the analysis results.

IV.   STABILITY OF PROPOSED SYSTEM

In the previous section, it is confirmed that the PMSM can suppress the resonance by applying the damping control in the proposed system. In this chapter, the stability of the proposed system with the damping control

$$\begin{bmatrix} p\Delta i_{Main\gamma} \\ p\Delta i_{Main\delta} \\ p\Delta i_{Auxd} \\ p\Delta i_{Auxq} \\ p\Delta\omega_{re} \\ p\Delta\delta \\ p\Delta x_{acr\_d} \\ p\Delta x_{acr\_q} \end{bmatrix} = \begin{bmatrix} -\dfrac{R}{L} & \omega_0 & 0 & 0 & -\dfrac{\psi_m}{L}\sin\delta_0 & -\dfrac{\omega_0\psi_m}{L}\cos\delta_0 & 0 & 0 \\ -\omega_0 & -\dfrac{R}{L} & 0 & 0 & -\dfrac{\psi_m}{L}\cos\delta_0 & \dfrac{\omega_0\psi_m}{L}\sin\delta_0 & 0 & 0 \\ 0 & 0 & -\dfrac{R}{L}-\dfrac{K_i}{L} & 0 & 0 & 0 & \dfrac{1}{LT_i} & 0 \\ 0 & 0 & 0 & -\dfrac{R}{L}-\dfrac{K_i}{L} & -\dfrac{K_d K_i}{L} & 0 & 0 & \dfrac{1}{LT_i} \\ \dfrac{3}{2}\dfrac{P_f^{2}\psi_m}{J}\sin\delta & \dfrac{3}{2}\dfrac{P_f^{2}\psi_m}{J}\cos\delta & 0 & \dfrac{3}{2}\dfrac{P_f^{2}\psi_m}{J} & 0 & \dfrac{3}{2}\dfrac{P_f^{2}\psi_m}{J}(i_\gamma\cos\delta - i_\delta\sin\delta) & 0 & 0 \\ 0 & 0 & 0 & 0 & -1 & 0 & 0 & 0 \\ 0 & 0 & -K_i & 0 & 0 & 0 & 0 & 0 \\ 0 & 0 & 0 & -K_i & -K_d K_i & 0 & 0 & 0 \end{bmatrix} \begin{bmatrix} \Delta i_{Main\gamma} \\ \Delta i_{Main\delta} \\ \Delta i_{Auxd} \\ \Delta i_{Auxq} \\ \Delta\omega_{re} \\ \Delta\delta \\ \Delta x_{acr\_d} \\ \Delta x_{acr\_q} \end{bmatrix} + \begin{bmatrix} i_{\delta0} \\ \dfrac{\psi_m}{L}-i_{\gamma0} \\ 0 \\ \dfrac{K_d K_i}{L} \\ 0 \\ 1 \\ 0 \\ K_d K_i \end{bmatrix}\Delta\omega^*$$

$$\text{...(16)}$$

is clarified in the Routh-Hurwitz stability criterion.

The state equation of the proposed system which is applied with the damping control is given by (16).

There are two integrators in the auxiliary inverter control because the field-oriented control has d-axis and q-axis current control as shown in Fig. 3. Thus, the state variables becomes from six to eight state variables when the auxiliary inverter is applied with the damping control.

The Routh-Hurwitz stability criterion is based on ordering the coefficients of the characteristic polynomial which have the real coefficients of (17).

$$a_0 s^n + a_1 s^{n-1} + \cdots + a_{n-1} z + a_n = 0 \quad \dots\dots\dots\dots (17)$$

The proposed system is stable when all roots the characteristic polynomial of (17) is negative. In Hurwitz stability criterion, the necessary and sufficient conditions for all roots of the characteristic polynomial of (17) are as follows;

1) There are All coefficients (a_0, a_1, a_2,...., a_n).
2) All coefficients have same sign.
3) All Hurwitz determinants (Δ_1, Δ_2,.., Δ_{n-1}) are positive.

$$\Delta_1 = a_1$$

$$\Delta_2 = \begin{vmatrix} a_1 & a_3 \\ a_0 & a_2 \end{vmatrix}$$

$$\Delta_{n-1} = \begin{vmatrix} a_1 & a_3 & a_5 & \cdots & a_{2n-3} \\ a_0 & a_2 & a_4 & \cdots & a_{2n-4} \\ 0 & a_1 & a_3 & \cdots & a_{2n-5} \\ 0 & a_0 & a_2 & \cdots & a_{2n-6} \\ 0 & 0 & a_1 & \cdots & a_{2n-7} \\ \vdots & \vdots & \vdots & \ddots & \vdots \\ 0 & 0 & 0 & \cdots & a_{n-1} \end{vmatrix}$$

Table 2 shows calculation of Hurwitz determinants and Table 3 shows the coefficient of the characteristic polynomial in the state equation of the proposed system which is applied with the damping control of (16). The analysis condition is shown in Table 1. All Hurwitz determinants are positive as shown Table 2. Moreover, there are All coefficient which have same sign. Therefore, because of the necessary and sufficient conditions, the proposed system is stable.

V. MINIMUM POWER CAPACITY OF AUXILIARY INVERTER

In the previous section, it is confirmed that the PMSM can be stabilized by applying the damping control in the proposed system. However, if the damping gain K_d increases to obtain the larger suppression effect, the output power of the auxiliary inverter becomes larger. Moreover, the power capacity of the auxiliary inverter depends on the speed command response. Thus, in this section, the minimum power capacity of the auxiliary inverter is clarified from frequency characteristics.

Fig. 7 shows the frequency characteristic of the transfer function from speed command to output power of the main inverter, the auxiliary inverter, and the sum of the inverters in the proposed system. In the proposed system, the main torque cannot be controlled directly due

Table 2 Calculation of Hurwitz determinants

Hurwitz determinants	Results	
Δ_1	8458	> 0
Δ_2	2.138×10^{11}	> 0
Δ_3	9.616×10^{21}	> 0
Δ_4	5.479×10^{35}	> 0
Δ_5	9.657×10^{51}	> 0
Δ_6	5.691×10^{70}	> 0
Δ_7	3.486×10^{91}	> 0

Table 3 Coefficient of the characteristic polynomial

Coefficients		Coefficients	
a_0	1	a_4	7.404×10^{13}
a_1	8458	a_5	2.765×10^{16}
a_2	3.339×10^7	a_6	9.413×10^{18}
a_3	6.866×10^{10}	a_7	9.888×10^{20}
		a_8	9.143×10^{22}

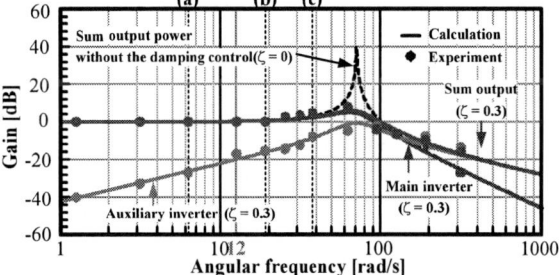

Fig. 7. Frequency characteristics of the transfer function from speed command to output power of the main inverter, the auxiliary inverter, and the sum of the inverters in the proposed system.

to the V/f control. However, by calculating the torque transfer function from the speed one, the frequency characteristic of each output power is derived. If the electrical and mechanical losses are neglected, the output power of the auxiliary inverter depends on the mechanical output of the motor. In other words, the transfer function of the output power for the auxiliary inverter is derived from the mechanical output transfer function of the motor. However, the mechanical output is nonlinear because the mechanical output is the product of two variables, the rotational speed and the torque. Due to this reason, the output power transfer function of the auxiliary inverter is derived after the mechanical output is linearized. The speed command to the output power of the auxiliary inverter transfer function is given by

$$\frac{\Delta P_A}{\Delta \omega^*} = \frac{s^2 \omega_0 K_d \psi_{mA}}{s^2 + \dfrac{P_{fM}{}^2 \psi_{mM}}{J} K_d s + \dfrac{P_{fM}{}^2 \psi_{mM}{}^2}{J L_{qM}}} \quad \dots\dots\dots (18)$$

In Fig. 7, the output power of the auxiliary inverter rises near the resonance frequency. On the other hand, when the frequency components into the speed command are lower and lower, the output power of the auxiliary inverter is suppressed more. In other words, the auxiliary inverter does not contribute steady state i.e. constant speed. Nevertheless, although the output power of the main inverter is constant because the main inverter drives the motor. Note that the damping factor is designed by the equations which are derived by the authors.

Figs. 8 show the experimental results that each output power is measured at each operating point in Fig. 5 when 0.5 p.u. of the constant speed command adds sinusoidal wave as the speed command. Assuming that the inverter output voltage is equal to the voltage command, each output power is calculated from the voltage command and the measured current. The waveforms in Fig. 8(a), (b) and (c) mention the operation at the point (a), (b) and (c) in Fig. 7, respectively. Comparing with each result, the amplitude ratio between the output power of the auxiliary inverter and the main inverter is changed corresponding to the frequency components included in the speed command. When the frequency component included in the speed command is lower than 32 rad/s, the output power of the auxiliary inverter can become lower than 10% of the main inverter as shown Fig. 7. This boundary condition is decided by inertia moment J, q-axis inductance L_{qM}, magnet flux linkage Ψ_m and the damping gain K_d as shown (14). When the damping gain K_d increases, the output power of the auxiliary inverter increase from 10% of the main inverter at 32 rad/s because the damping factor increases. Therefore, the results show that the power capacity of the auxiliary inverter can be designed to nearly several percent of the main inverter when the proposed system is applied to the application which requires a slow speed response such as fan applications. In this session, in order to confirm the validity of proposed system in terms of acceleration which means the rate of the speed command, the relationship between the output power of the auxiliary inverter and the overshoot of the rotational speed for speed command is demonstrated in the experiment.

The output power of the auxiliary inverter was discussed by frequency characteristics. However, the acceleration or deceleration time is set as the speed command in practically. In this session, the output power of the auxiliary inverter is discussed by acceleration which means the rate of the speed command.

The time response of the rotational speed is derived from the inverse Laplace transform of the ramp response of (13). In this section, in order to simplify (13), it is assumed as $\varphi_0 = 0$, $i_{M\gamma0} = 0$ and $i_{M\delta0} = 0$ in (13). Moreover, the overshoot of the rotational speed can be derived from this time response, as given by

$$\Delta\omega_{re\_Overshoot} = \frac{\alpha \sin\left(\omega_n t_{peak}\sqrt{1-\zeta^2}\right)}{\omega_n\sqrt{1-\zeta^2}\exp\left(\zeta\omega_n t_{peak}\right)} \quad\text{......(19)}$$

$$t_{peak} = \frac{1}{2}\frac{\log\left(2\zeta^2-1+2\zeta\sqrt{\zeta^2-1}\right)}{\omega_n\sqrt{\zeta^2-1}} \quad\text{......(20)}$$

where α [rad/s^2] is acceleration.

The overshoot of the rotational speed can be calculated by using (19).

Similarly, the time response of the output power of the auxiliary inverter during the acceleration is also derived from the inverse Laplace transform of the ramp response

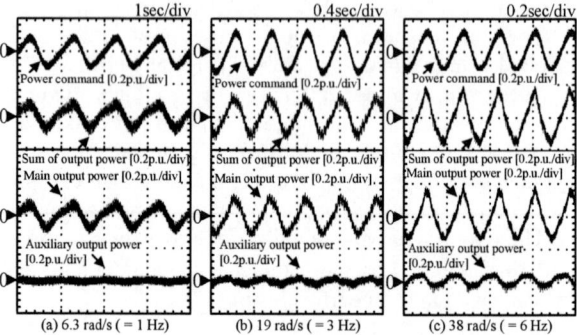

(a) 6.3 rad/s (= 1 Hz)   (b) 19 rad/s (= 3 Hz)   (c) 38 rad/s (= 6 Hz)

Figs. 8. Experimental results that each output power is measured at each operating point ((a) is at 1 Hz, (b) is at 3 Hz, (c) is at 6 Hz) in Fig. 7 when 0.5 p.u. of the constant speed command adds sinusoidal wave as the speed command.

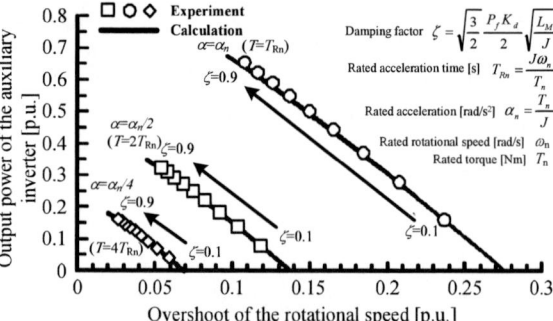

Fig. 9. Relationship between the output power of the auxiliary inverter and the overshoot of the rotational speed the against speed command. Table 1 shows also the evaluated condition. The higher suppression effect is obtained in exchange for the increase of the output power of the auxiliary inverter by increasing the damping gain.

of (18). In this section, in order to simplify (18), it is assumed as $\delta_0 = 0$, $i_{M\gamma0} = 0$, $i_{M\delta0} = 0$ in (18). Moreover, the maximum output power of the auxiliary inverter can be derived from this time response, as given by

$$\Delta P_{aPeak} = \frac{P_f\Psi_{mA}K_d\omega_0\alpha\sin\left(\omega_n\sqrt{1-\zeta^2}\,t_{peak}\right)}{\omega_n\sqrt{1-\zeta^2}\exp\left(\zeta\omega_n t_{peak}\right)} \quad\text{..(21)}$$

The maximum output power of the auxiliary inverter can be calculated by (21).

Fig. 9 shows the relationship among the damping gain K_d, the maximum output power of the auxiliary inverter during the acceleration and the overshoot of the rotational speed when the PMSM is accelerated during one, half, quarter of the rated acceleration α_n on the condition as shown in Table 1. The damping gain is standardized by the damping factor ζ as given by (14). The output power of the auxiliary inverter rises with the increase of the damping factor. On the other hand, the overshoot of the rotational speed decreases with the increase of the damping factor, as shown in Fig. 9. The maximum output power of the auxiliary inverter (i.e. the power capacity of the auxiliary inverter) depends on the acceleration factor α in the speed command as shown in Fig. 9. The experimental results well agree with the calculation.

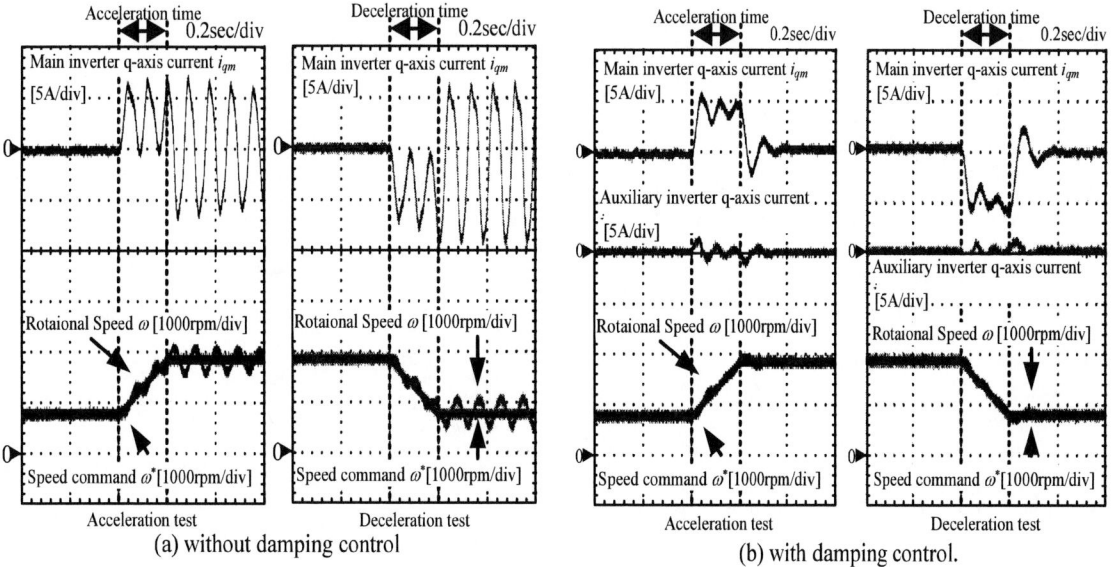

(a) without damping control   (b) with damping control.

Fig. 10. Acceleration and deceleration test without/with damping control in motor-generator set. (a) After the acceleration, the 400 r/min - speed vibration is maintained. (b) The speed vibration is reduced from 400 r/min to nearly 0 r/min in compared with (a)

VI.   MOTOR DRIVE PERFORMANCE EVALUATION

In this chapter, in order to confirm the effectiveness of the damping control, the experiments are conducted with a motor –generator set.

Fig. 10 shows the experimental results that illustrate the motor speed vibration when the proposed system is applied (a) without the damping control and (b) with the damping control ($\zeta = 0.3$) in an acceleration and deceleration test. In this experiment, it is difficult to measure the torque response directly. Therefore, the speed vibration is evaluated instead of the torque response. In Fig. 10(a), the proposed system is implemented without the damping control. The speed vibration occurs during the acceleration. After the acceleration, a 400 r/min of speed vibration and the 10 A_{p-p} of current vibration in q-axis of the main inverter are maintained. On the other hand, Fig. 10(b) demonstrates the experimental results, where the proposed system is implemented with the damping control ($\zeta = 0.3$). The effectiveness of the auxiliary inverter from the results confirms that the speed vibration is reduced from 400 r/min to nearly 0 r/min in compared with the acceleration test of Fig. 10(a). The 10 A_{p-p} of the current vibration in the q-axis of the main inverter is suppressed as well. Then, the output power of the auxiliary inverter is suppressed to 13% of the main inverter when the damping factor is designed at 0.3. In the same way as the acceleration test of Fig. 10(a), after the deceleration, a 500 r/min of speed vibration and the 15 A_{p-p} of current vibration in q-axis of the main inverter are maintained as shown in the deceleration test of Fig. 10(a). In contrast, it is confirmed that, as shown the deceleration test of Fig. 10(b), the speed vibration is reduced from 500 r/min to nearly 0 r/min and the current vibration is reduced from 15 A_{p-p} to 0 A_{p-p} in compared to Fig. 10(a) in the same

way as acceleration test. Nevertheless, it is confirmed that the q-axis current of the auxiliary inverter flows only during acceleration and deceleration. Moreover, the maximum q-axis current of the auxiliary inverter is 20% of the q-axis current of the main inverter. Therefore, it is confirmed that the auxiliary inverter can suppress the speed vibration via auxiliary windings with a small q-axis current of the auxiliary inverter even if in the acceleration and deceleration test.

VII.   EVALUATION OF PARALLEL DRIVE SYSTEM

In previous chapter, the effectiveness of the damping control in the proposed system is discussed. This chapter discusses the effectiveness of the proposed system when two PMSM are driven.

Fig. 9 shows the simulation models that are used to verify the operation of two sets of parallel connected PMSM. Since it is difficult to construct the prototype of the drive system using two PMSMs, the simulation is used to evaluate the validity of the proposed controller as first step.

Fig. 10 shows the simulation results when two sets of parallel connected PMSMs are driven by the proposed system with the damping control as shown in Fig. 8. Smooth acceleration progresses are confirmed in the two set of parallel connected PMSMs. Besides, when the rated motor speed, load step applies to PMSM1 at 0.38s, and later also applies to PMSM2 at 0.42s, the operation of the two auxiliary inverters can be observed from the output power. The maximum output power of the auxiliary inverter is approximately 0.25 p.u. of the rated power of the main inverter. Although the power capacity of main inverter increases with the increase of parallel units, the power capacity of auxiliary inverter does not change according to the numbers of parallel units.

978-1-4799-2706-7/14 $31.00 © 2014 IEEE

VIII. CONCLUSION

This paper discusses the acceleration performance of the proposed parallel connected multiple PMSM drive system from the frequency characteristics and the acceleration test. From the frequency characteristics of the speed response, it was clarified that the resonance is suppressed by applying the damping control. In addition, the output power of the auxiliary inverter is suppressed to 10% of the main motor in the speed response of lower than 32 rad/s in the prototype of 1500-W PMSM drive system. The results show that the power capacity of the auxiliary inverter is designed to nearly several percent of the main inverter when the proposed system is applied to the application which requires a slow speed response such as fan applications.

In the future work, the driving several units of the motor-generator in parallel with the proposed system drives will be evaluated in the experimental system.

ACKNOWLEDGMENT

A part of this study was supported by Industrial Technology Grant Program in 2011 from New Energy and Industrial Technology Development Organization (NEDO) of Japan.

REFERENCES

[1] M. J. Corley, and R. D. Lorenz, "Rotor Position and Velocity Estimation for a Salient-Pole Permanent Magnet Synchronous Machine at Standstill and High Speeds", IEEE Trans. Industry Applications, Vol. 34, No. 4, pp. 784-789 (1998)

[2] Rahman, M.F., Zhong, L.,Khiang Wee Lim : "A direct torque-controlled interior permanent magnet synchronous motor drive incorporating field weakening", IEEE Trans. Industry Applications, Vol.34, No.6, pp.1246-1253 (1998)

[3] M. Nasir Uddin, Tawfik S. Radwan, and M. Azizur Rahman,: "Performance of interior permanent magnet motor drive over wide speed range", IEEE Trans. Energy Conversion, Vol.17, No.1, pp.79-84 (2002)

[4] Chan-Hee Choi, Jul-Ki Seok, and Lorenz, R.D. : "Wide-Speed Direct Torque and Flux Control for Interior PM Synchronous Motors Operating at Voltage and Current Limits", IEEE Trans. Industry Applications, Vol. 49, No.1, pp.119-117 (2013)

[5] P. M. Kelecy, R. D. Lorenz, "Control methodology for single inverter, parallel connected dual induction motor drives for electric vehicles," in Proc. IEEE PESC'94, pp.987–991 (1994)

[6] Matsuse, K. ; Kouno, Y. ; Kawai, H. ; Yokomizo, S.: "A speed-sensorless vector control method of parallel-connected dual induction motor fed by a single inverter", IEEE Trans. Industry Applications., Vol.38, pp.1566-1571 (2002)

[7] J. D. Ma, B. Wu, N. R. Zargari, Steven C. Rizzo, "A Space Vector Modulated CSI-Based AC Drive for Multimotor Applications", IEEE Trans. POWER ELECTRONICS, VOL.16, NO.4, pp.535-544 (2001)

[8] A. Bouscayrol, M. Pietrzak-David, P. Delarue, R. Peña-Eguiluz, P. V. Kestelyn, "Weighted Control of Traction Drives With Parallel-Connected AC Machines", IEEE Trans. Industrial Electronics, Vol.53, No.6, pp.1799-1806 (2006)

[9] J, Itoh, N, Nomura, H, Ohsawa: "A Comparison between V/f Control and Position-Sensorless Vector Control for the Permanent Magnet Synchronous Motor", Proc. of the Power Conversion Conference PCC Osaka 2002, Vol. 3, pp.1310 - 1315 (2002)

[10] Nozawa, Y, et al : "Performance for Position Control of Two Permanent Magnet Synchronous Motors with the Five-Leg Inverter.", IECON'06, pp.1182-1187 (2006)

[11] Ibrahim, Z.; Lazi, J.M. ; Sulaiman, M.: "Independent speed sensorless control of dual parallel PMSM based on Five-Leg Inverter", Proc. of the 9th International Multi-Conference on Systems, Signals and Devices, pp.1-6 (2012)

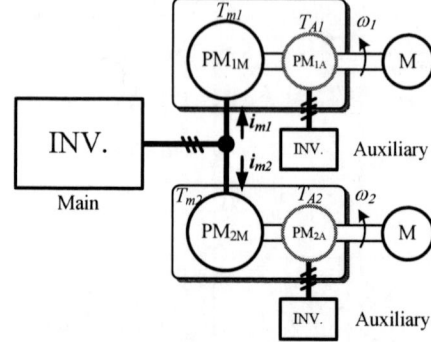

Fig. 9. Simulation model for parallel connected dual motor drive. In order to neglect the magnetic coupling, two parallel drive of the proposed system is validated using Fig. 8.

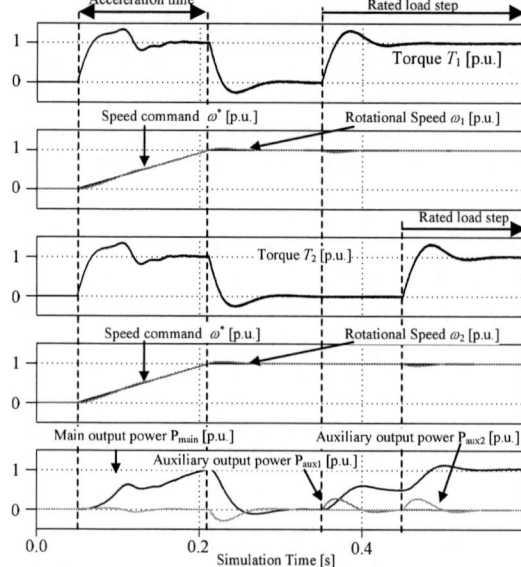

Fig. 10. Simulation results for parallel motor drive with damping control. Smooth acceleration progresses are confirmed in the two parallel connected PMSMs. After the rated motor speed, load step applies to each PMSM, stable operation of each PMSM can be confirmed.

[12] S. Mohammad et al.: "Space Vectors Modulation for Nine-Switch Converters", IEEE Trans. Power Electronics., Vol.25, pp.1488-1496 (2010)

[13] F. Gao, L. Zhang, D. Li, P. C. Loh, Y. Tang, H. Gao, "Optimal Pulsewidth Modulation of Nine-Switch Converter", IEEE Trans. Power Electronics, Vol. 25, No. 9, pp.2331-2342 (2010)

[14] Shibata, M., Hoshi, N.: "Novel inverter topologies for two-wheel drive electric vehicles with two permanent magnet synchronous motors", 12th European Conference on Power Electronics and Applications (2007)

[15] Mori, T., Tanaka, H., Kubo, Y., Matsuse, K., "Independent vector control of two permanent magnet synchronous motors fed by a four-leg inverter", Proc. of the 2012 IEEE International Conference on Power Electronics, Drives and Energy Systems (PEDES), pp.1-6 (2012)

[16] T. Nagano, et al. "Design of Multi-Parallel Drive Technique for System with Numbers of Permanent Magnet Synchronous Motors", The 10th IEEE International Conference on Power Electronics and Drive Systems, pp. 193-198 (2013)

[17] T. Nagano, et al. "Verification of Parallel Connected Multiple Motor Drive System with Numbers of Permanent Magnet Synchronous Motors", 15th European Conference on Power Electronics and Applications (2013)

The 2014 International Power Electronics Conference

A Transformer Inrush Reduction Technique for Low-Voltage Ride-Through Operation of Renewable Converters

Hsin-Chih Chen   Ping-Heng Wu   Po-Tai Cheng

Center for Advanced Power Technologies
Department of Electrical Engineering
National Tsing Hua University
Hsinchu 30013, TAIWAN.

Abstract—**The low-voltage ride-through (LVRT) capability has become one of the most important issues since more renewable and distributed energy resource are installed in the grid. The LVRT capability means the converter should remain grid-connected and inject required current to support gird voltage. On the other hand, the sudden reduction of grid voltage may result in magnetic flux deviation in the step-up transformer as grid voltage sag occurs. The deviation of magnetic decays very slowly due to the high efficiency transformer, and it would lead to inrush current as grid voltage restores to normal level. This paper proposes a method to mitigate the magnetic flux deviation and limit the peak current without excelling the current capability of semiconductor devices by using a close-loop current control during LVRT operation. The proposed method can manage the converter's output peak current and reduce the risk of inrush current as grid voltage restores.**

Index Terms—**Low-voltage ride-through, flux compensation, inrush current, distributed energy resources**

I. Introduction

Renewable and distributed energy resources (DERs) systems becomes popular in these years since the petroleum energy expensive. More DERs systems are installed in the utility grid, their effect on the power system's stability become an important issue. Thus, the grid operators start laying down the specific requirements for the grid-connected DERs systems.

The Low-voltage ride-through (LVRT) capability is one of the most important issues in the grid codes [1], [2], [3], [4], [5]. In wind power system, the power converters should remain connected to the grid and inject the required current to support grid voltages. Several converter control strategies have been reported [6], [7], [8] for the LVRT functionalities.

In wind power generation system, the grid-side converter is connected to the utility through a step-up transformer [9], [10]. In LVRT operation, the sudden reduction of grid voltages often lead to the magnetic flux deviation in the step-up transformer. The magnetic flux deviation decays very slowly due to high efficiency nature of the transformer, and it easily leads to inrush current in grid side of the step-up transformer as grid voltages restore to the normal level [11], [12]. This inrush current results in unbalanced magnetomotive force (MMF), which can exert significant axial forces on transformer windings and wreck its insulation [13]. In order to reduce the occurrence of this phenomenon, this paper proposes an inrush

current mitigation method for the grid-connected converters during the LVRT operation.

The grid voltage becomes unbalanced in LVER operation, Hsu et al. [8] presented a peak current limit control (PCLC) to meet the grid requirement and manege the peak current. Based on PCLC method, this paper presents a flux compensation technique with peak current regulator to limit the peak current without excelling the current capacity of semiconductor devices. Thus, the stress of both the transformer and the converter can be managed appropriately.

II. Fundamental Active and Reactive Current Strategy

The overall system configuration of the converter and overall control block diagram are shown in Fig. 1. A three-level neutral-point-clamped (NPC) converter, which is typical for MW class wind power system, is connected to the grid through a delta-wye transformer.

A. Converter voltages and currents

The voltages become unbalanced as voltage faults, the phase voltages and currents can be expressed as Equation (1) and Equation (2), where V^p and V^n are the positive- and negative-sequence components of the voltage. I^p and I^n represent the positive- and negative-sequence currents. θ_1 and θ_2 are the phase angles of the positive- and negative-sequence voltages with respect to the reference axis. θ_p and θ_n are the phase angles of the positive- and negative-sequence currents in reference to their voltage components.

$$
\begin{aligned}
v_a &= V^p \cos(\omega t + \theta_1) + V^n \cos(-\omega t + \theta_2) \\
v_b &= V^p \cos(\omega t - \frac{2\pi}{3} + \theta_1) + V^n \cos(-\omega t - \frac{2\pi}{3} + \theta_2) \\
v_c &= V^p \cos(\omega t + \frac{2\pi}{3} + \theta_1) + V^n \cos(-\omega t + \frac{2\pi}{3} + \theta_2)
\end{aligned}
\tag{1}
$$

$$
\begin{aligned}
i_a &= I^p \cos(\omega t + \theta_p) + I^n \cos(-\omega t + \theta_n) \\
i_b &= I^p \cos(\omega t - \frac{2\pi}{3} + \theta_p) + I^n \cos(-\omega t - \frac{2\pi}{3} + \theta_n) \\
i_c &= I^p \cos(\omega t + \frac{2\pi}{3} + \theta_p) + I^n \cos(-\omega t + \frac{2\pi}{3} + \theta_n)
\end{aligned}
\tag{2}
$$

978-1-4799-2706-7/14 $31.00 © 2014 IEEE

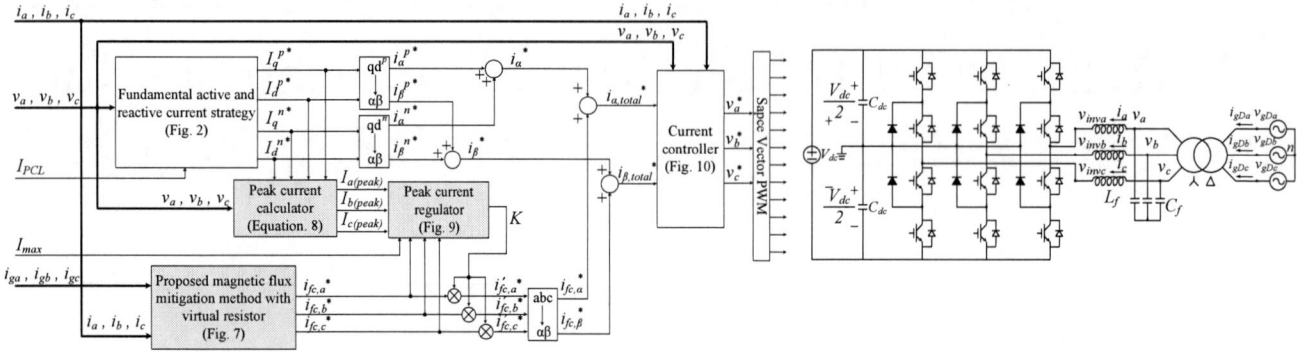

Fig. 1. Overall system configuration control block diagram.

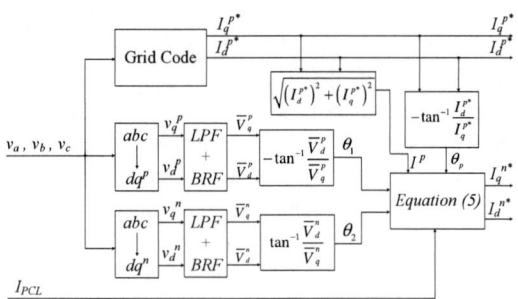

Fig. 2. Block diagram of PCLC.

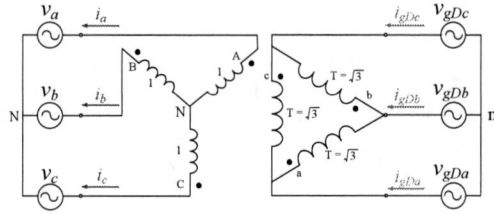

Fig. 3. Three-phase circuit diagram.

The transformation for the stationary reference frame, the positive- and negative-sequence synchronous reference frames are given as:

$$\begin{bmatrix} v_\alpha \\ v_\beta \end{bmatrix} = \begin{bmatrix} \frac{2}{3} & -\frac{1}{3} & -\frac{1}{3} \\ 0 & -\frac{1}{\sqrt{3}} & \frac{1}{\sqrt{3}} \end{bmatrix} \begin{bmatrix} v_a \\ v_b \\ v_c \end{bmatrix}$$

$$\begin{bmatrix} v_q^p \\ v_d^p \end{bmatrix} = \begin{bmatrix} \cos(\omega t) & -\sin(\omega t) \\ \sin(\omega t) & \cos(\omega t) \end{bmatrix} \begin{bmatrix} v_\alpha \\ v_\beta \end{bmatrix} \quad (3)$$

$$\begin{bmatrix} v_q^n \\ v_d^n \end{bmatrix} = \begin{bmatrix} \cos(\omega t) & \sin(\omega t) \\ -\sin(\omega t) & \cos(\omega t) \end{bmatrix} \begin{bmatrix} v_\alpha \\ v_\beta \end{bmatrix}$$

These synchronous frame values ($v_q^p, v_d^p, v_q^n, v_q^n$) include 2ω ripples due to the unbalanced voltages, thus low-pass filter (LPF) and band-reject filter (BRF) are employed to extract their DC components $\bar{V}_q^p, \bar{V}_d^p, \bar{V}_q^n, \bar{V}_d^n$ respectively. The phase angles can be calculated as:

$$\theta_1 = -\tan^{-1}(\frac{\bar{V}_d^p}{\bar{V}_q^p}) \qquad \theta_2 = -\tan^{-1}(\frac{\bar{V}_d^n}{\bar{V}_q^n})$$

$$\theta_p = -\tan^{-1}(\frac{I_d^{p*}}{I_q^{p*}}) \qquad \theta_n = -\tan^{-1}(\frac{I_d^{n*}}{I_q^{n*}}) \quad (4)$$

B. LVRT with peak current limit

Fig. 2 shows the control block diagram of the peak current limit control (PCLC) method. This method has been presented in [8], which can use to meet grid code and fully utilize the ampere capacity of grid-connected converter. The PCLC method can pre-define ampere capacity I_{PCLC} which is allocated for fulfilling the LVRT requirement. The grid code

requires the positive-sequence active current I_q^{p*} and reactive current I_d^{p*} injection during sags, the PCLC method calculates the fundamental negative-sequence reactive current (I_d^{n*}) to ensure the peak current of each phase are not higher than the budget of I_{PCLC},

$$I_q^{n*} = 0$$

$$I_d^{n*} = -I^p \cos(\alpha + k\frac{4\pi}{3}) + \sqrt{(I^p)^2[\cos^2(\alpha + k\frac{4\pi}{3})] + I_{PCLC}^2}$$

$$\text{where } k = \begin{cases} 0, -\pi/3 \le \alpha < \pi/3 \\ 1, \pi/3 \le \alpha < \pi \\ -1, \pi \le \alpha < 5\pi/3 \end{cases}, \alpha = -\theta_p - \theta_n$$

(5)

Note that the negative-sequence reactive current is inductive (ie: $\theta_n = -\frac{\pi}{2}$) which can reduce the negative-sequence component of grid voltage.

III. PROPOSED TRANSFORMER FLUX MITIGATION METHOD

Based upon the PCLC method, this paper proposed a transformer magnetic flux compensation technique. The magnetic flux deviation is caused by sudden change of grid voltage as fault occurs, and the flux deviation decays very slowly. This paper proposed a state-feedback current control technique to mitigate the deviation quickly, and hence reduce the risk of inrush current when grid voltage recovers to normal level.

A. Transformer circuit model

Fig. 3 shows the circuit diagram of delta-wye step-up transformer, where v_a, v_b, v_c are output voltages of the converter, i_a, i_b, i_c are output currents of the converter, $v_{gDa}, v_{gDb}, v_{gDc}$

The 2014 International Power Electronics Conference

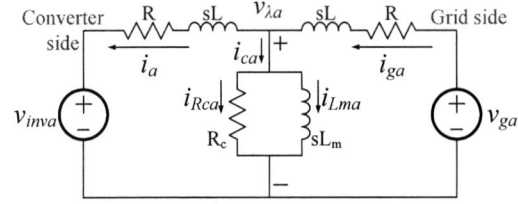

Fig. 4. The single-phase equivalent circuit diagram of a-phase.

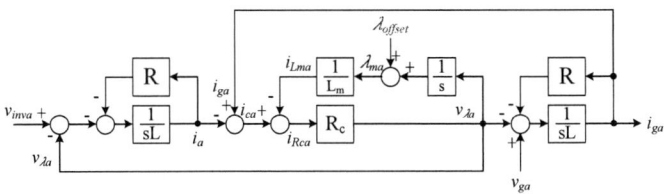

Fig. 5. Block diagram of transformer model in a-phase as voltage sags occur.

are grid voltages, and $i_{gDa}, i_{gDb}, i_{gDc}$ are grid currents. The turns ratio of the transformer is $\sqrt{3}$ for simplicity. The voltages and currents of the delta side and wye side are related as:

$$
\begin{bmatrix} v_{ga} \\ v_{gb} \\ v_{gc} \end{bmatrix} = \frac{1}{\sqrt{3}} \begin{bmatrix} v_{gDa} - v_{gDb} \\ v_{gDb} - v_{gDc} \\ v_{gDc} - v_{gDa} \end{bmatrix}
$$
$$
\begin{bmatrix} i_{gDa} \\ i_{gDb} \\ i_{gDc} \end{bmatrix} = \frac{1}{\sqrt{3}} \begin{bmatrix} i_{ga} - i_{gb} \\ i_{gb} - i_{gc} \\ i_{gc} - i_{ga} \end{bmatrix}
$$

(6)

Fig. 4 shows the a-phase transformer equivalent circuit of Fig. 3, where L is the leakage inductance, R is the winding resistor, L_m means the magnetizing inductance, and R_c represents the core loss. The voltages and currents can also be expressed as block diagram in Fig. 5. As voltage sags occur, the sudden reduction grid voltage results in the magnetic flux deviation (λ_{offset}). TABLE I shows the parameters of transformer equivalent circuit, and Fig. 6 shows the response of the transformer flux linking λ_{mk}, assumes the step input disturbance $\lambda_{offset} = 0.5Wb$ occurs to the transformer. The response waveform shows that it takes more than 90 seconds for λ_{mk} to decay to $0.01Wb$. Thus, if the significant flux offset still remains in the transformer, it can easily lead to inrush current as grid voltage restores to normal level.

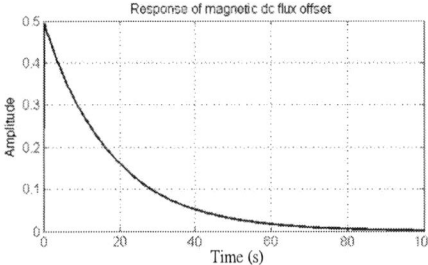

Fig. 6. The natural response of the transformer DC flux linkage λ_{mk}.

TABLE I
TRANSFORMER PARAMETERS

Transformer parameters	
Winding resistance (R)	0.3035Ω
Core loss of the magnetic core material (R_c)	3083.37Ω
Winding leakage inductance (L)	$294.43\mu H$
Magnetizing inductance (L_m)	$2.6691H$

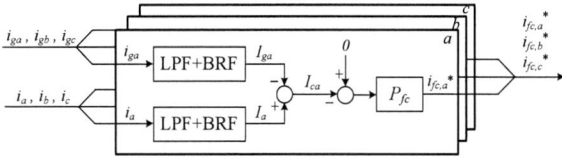

Fig. 7. Block diagram of proposed magnetic flux mitigation method.

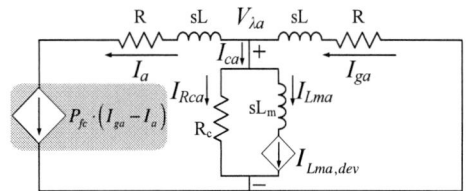

Fig. 8. The equivalent circuit of a-phase with proposed flux mitigation method as voltage sags occur.

B. The proposed method to mitigate the magnetic DC flux deviation

This paper proposes a state-feedback current control technique to accelerate the decay of transformer magnetic flux deviation. Fig. 7 shows the control block diagram of the proposed method. Fig. 8 shows the proposed method can be implemented as a current-controlled current-sources of DC component at converter side of the transformer to mitigate the flux deviation.

In Fig. 5, the magnetic flux deviation (λ_{offset}) results in current offset ($I_{Lma,dev}$, $I_{Lmb,dev}$, $I_{Lmc,dev}$) in the transformer's equivalent circuit. Thus, the converter side currents (i_a, i_b, i_c) and grid side currents (i_{ga}, i_{gb}, i_{gc}) are taken for the offset current calculation. A low-pass filter (LPF) and a band-reject filter (BRF) are employed to filter out the fundamental components and smooth the output.

After the core current deviation (I_{ca}, I_{cb}, I_{cc}) has been known, a proportional controller is used to generate the compensating current command ($i_{fc,a}{}^*$, $i_{fc,b}{}^*$, $i_{fc,c}{}^*$) of each phase to reduce the flux deviation.

$$
i_{fc,a}{}^* = P_{fc} \cdot (0 - (I_a - I_{ga}))
$$
$$
i_{fc,b}{}^* = P_{fc} \cdot (0 - (I_b - I_{gb}))
$$
$$
i_{fc,c}{}^* = P_{fc} \cdot (0 - (I_c - I_{gc}))
$$

(7)

C. Peak current control

The proposed method generates compensating current to mitigate the magnetic flux offset of the step-up transformer. However, the compensating current may result in high peak current in converter side of the transformer.

The 2014 International Power Electronics Conference

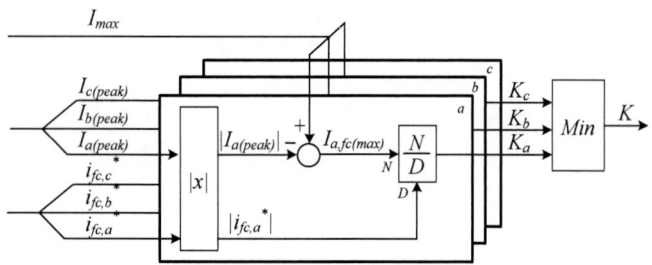

Fig. 9. Block diagram of peak current regulator.

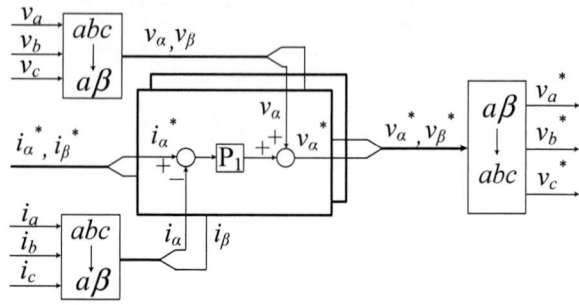

Fig. 10. Block diagram of current controller.

Fig. 9 shows control block diagram the peak current regulator. The proposed method employs the peak current regulator to scale the peak current without higher than I_{max}, which is designed based on the current capacity of semiconductor devices.

Because the fundamental injection current has negative-sequence component, the remaining current capacity of each phase are different. The peak currents of each phase ($I_{a(peak)}$, $I_{b(peak)}$, $I_{c(peak)}$) can be calculated as Equation (8) [8] and the maximum current capacity of proposed flux compensation can be calculated by Equation (9).

$$I_{a(peak)} = \sqrt{(I^p)^2 + (I^n)^2 + 2I^p I^n \cos(\alpha)}$$

$$I_{b(peak)} = \sqrt{(I^p)^2 + (I^n)^2 + 2I^p I^n \cos\left(\alpha + \frac{4\pi}{3}\right)} \quad (8)$$

$$I_{c(peak)} = \sqrt{(I^p)^2 + (I^n)^2 + 2I^p I^n \cos\left(\alpha - \frac{4\pi}{3}\right)}$$

After the maximum current capacities of each phase have been calculated, the K_a, K_b, and K_c gains is used to scale the compensating current within the current limit (I_{max}).

$$K_a = \frac{I_{a,fc(max)}}{|i_{fc,a}^*|}$$

$$K_b = \frac{I_{b,fc(max)}}{|i_{fc,b}^*|} \quad (10)$$

$$K_c = \frac{I_{c,fc(max)}}{|i_{fc,c}^*|}$$

In order to make sure all the peak currents without excelling than pre-defined value (I_{max}), the scaled gain (K) should be selected as the minimum. Moreover, the scaled gain (K) will be set at 1 as the highest current without higher than I_{max}, which means keep the original value.

$$K = \min(K_a, K_b, K_c)$$
$$(if\ K \geq 1,\ K = 1) \quad (11)$$

Finally, the output compensating current commands will be

$$I_{a,fc(max)} = I_{max} - I_{a(peak)}$$
$$I_{b,fc(max)} = I_{max} - I_{b(peak)} \quad (9)$$
$$I_{c,fc(max)} = I_{max} - I_{c(peak)}$$

regulated as

$$i'_{fc,a}{}^* = i_{fc,a}{}^* \cdot K$$
$$i'_{fc,b}{}^* = i_{fc,b}{}^* \cdot K \quad (12)$$
$$i'_{fc,c}{}^* = i_{fc,c}{}^* \cdot K$$

After all the current commands have been decided, the current controller is employed to regulate the converter's output current and it's shown as Fig. 10.

IV. LABORATORY TEST RESULTS

The system configuration and overall control block diagram are shown in Fig. 1, the parameters of the step-up transformer has been shown in TABLE I, and TABLE II shows the parameters of filters and controllers. The system parameters are shown as follows.

- The AC voltage is $220\ V_{rms}$ (line-to-line), and the frequency is 60 Hz. The DC bus voltage is 400V.
- The rated capacity is 1kVA, the rated current is 3.7 A.
- The switching frequency of converter is 10kHz, sampling frequency is 20kHz.
- The output filter $L_f = 4mH$, $C_f = 6.8\mu F$.
- The pre-defined peak value of PCLC method is set at $I_{PCLC} = 1\,p.u. = 3.7\,A$, the pre-defined maximum peak current is set at $I_{max} = 1.4\,p.u. = 5.2\,A$
- In LVRT operation, the reactive positive-sequence current commands are set at $I_q^{p*} = 0$ and $I_d^{p*} = 0.7\,p.u. = 2.6\,A$

The voltage sag has been defined in IEEE P1668 [14]. The type B (single phase 50% voltage reduction) and type E (two phases 50% voltage reduction) are tested in this paper, where the sags duration are 27.5 times fundamental cycle.

Fig. 11(a) shows the converter's phase voltages waveform as the voltage sag occurs under type B voltage sag test. The 50% voltage sag happens at the grid side of transformer, the response is like two phase voltage sag at the converter

TABLE II
TRANSFORMER PARAMETERS

Filter and controller parameters	
Cut-off frequency of LPF (ω_{LPF})	40Hz
Cut-off frequency of BRF (ω_{BRF})	60Hz
Quality factor of BRF (Q)	0.5
Proportional gain of proposed flux mitigation method (P_{fc})	10.7
Proportional gain of current controller (P_1)	70

978-1-4799-2706-7/14 $31.00 © 2014 IEEE

side. Fig. 11(b) shows the current commands of PCLC, the method confines the peak current to pre-defined value (I_{PCLC}) and the current waveform is unbalanced. Fig. 12(a) shows the converter's phase currents waveform without peak current regulator, and Fig. 12(b) shows the converter's phase current with proposed peak current regulator. Fig. 13(a) and Fig. 13(b) show the comparison between with proposed flux compensation and without proposed flux compensation. In Fig. 13(a), the proposed flux compensation method reduces the inrush current as grid-voltage recovers to normal level.

Fig. 14(a) shows the converter's phase voltage waveform, Fig. 14(b) shows the converter's PCLC current command, and Fig. 14(c) shows the converter's current waveforms during LVRT operation under type E voltage sag. Fig. 14(c) shows the peak current regulator limits the peak current at I_{max}.

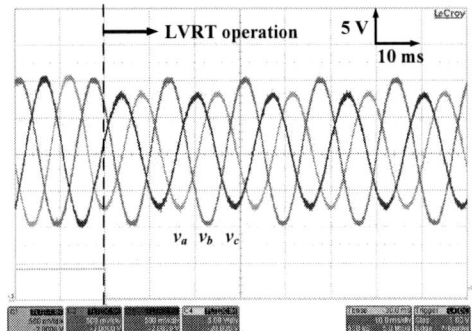

(a) Converter's phase voltage waveform.

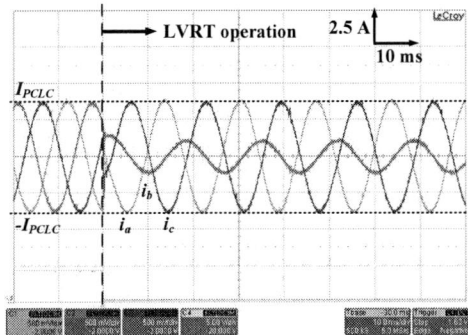

(b) Converter's PCLC current command.

Fig. 11. Circuit diagram of grid-connected step-up transformer.

V. DISCUSSION

The laboratory test result shows the proposed method injects the compensating current to reduce the inrush current. In the proposed method, a low-pass filter (LPF) and a band-reject filter (BRF) are employed to extract the core current deviation then using a proportional controller to generate the compensating current command. The cut-off frequency of low-pass filter (ω_{LPF}) and the proportional gain (P_{fc}) are the parameters that can be regulator easily.

Fig. 15(a) and Fig. 15(b) show the root locus of the dominate pole which are variable P_{fc} and ω_{LPF}. The parameters of this paper are shown as TABLE I and TABLE II, where the operation point of the laboratory test result is marked as red point in the Fig. 15.

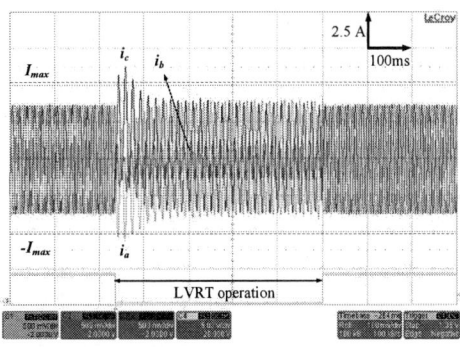

(a) The converter's current waveform without peak current regulator.

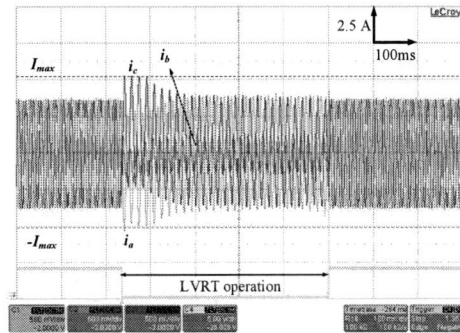

(b) The converter's current waveform with peak current regulator.

Fig. 12. The converter's current waveforms during LVRT operation.

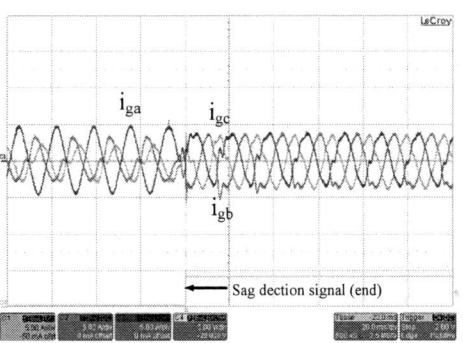

(a) Grid's phase currents waveform with flux compensation.

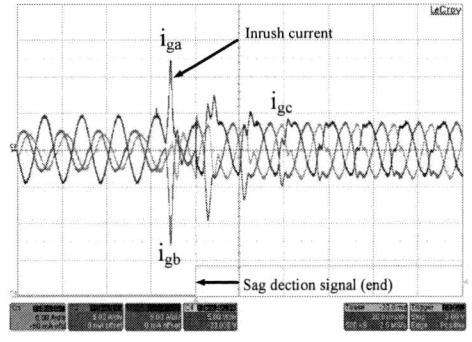

(b) Grid's phase currents waveform without flux compensation.

Fig. 13. Grid side current waveform as the voltage recovers to normal level.

The 2014 International Power Electronics Conference

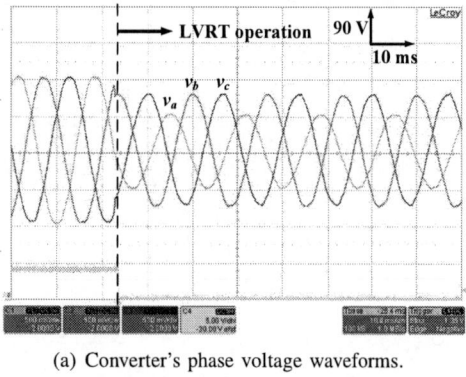

(a) Converter's phase voltage waveforms.

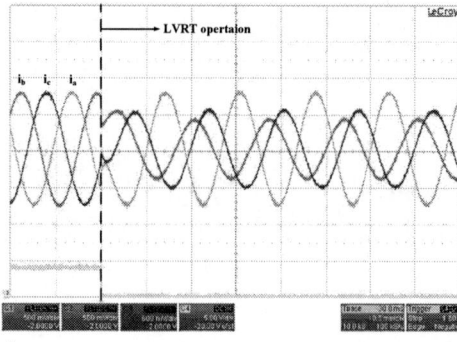

(b) Converter's PCLC current command.

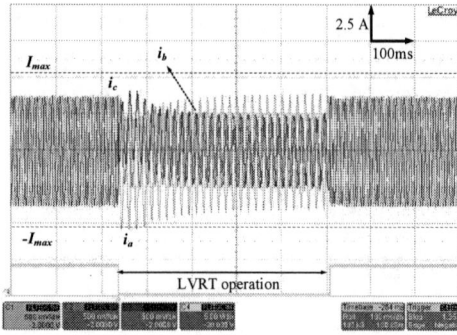

(c) Converter's phase current waveforms.

Fig. 14. The waveforms during type E voltage sag.

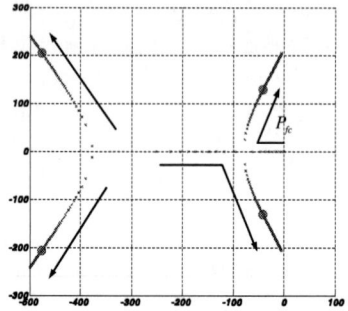

(a) The diagram of dominate pole (variable of P_{fc}).

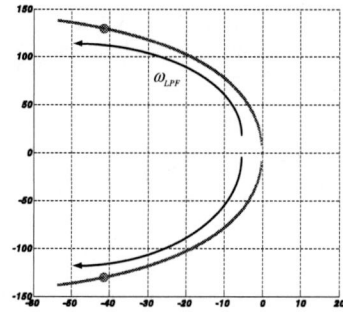

(b) The diagram of dominate pole (variable of ω_{LPF}).

Fig. 15. The diagram of root locus.

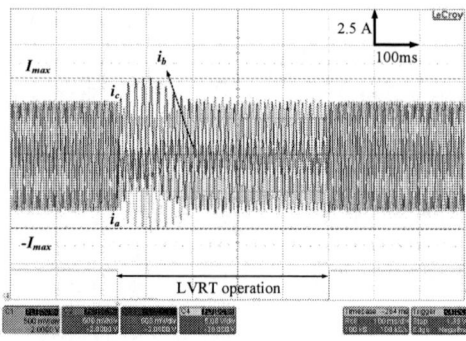

Fig. 16. The converter's phase currents as the ω_{LPF} is set at 5Hz.

In order to extract the magnetic flux mitigation exactly, the cut-off frequency of LPF should be selected as low as possible. However, Fig. 15(b) shows the compensating response become slow as the cut-off frequency of LPF is decreased. Fig. 16 shows the converter's current waveforms as the ω_{LPF} is set at 5Hz and the proportional gain P_{fc} is set at 10.7 under type B 50% voltage reduction. Fig. 16 and Fig. 14(c) verify the compensating response becomes slow as the cut-off frequency of LPF is decreased.

VI. CONCLUSION

As grid side voltage sag occurs, the sudden reduction of grid voltage usually results in magnetic flux deviation in the step-up transformer. The flux deviation decays very slowly since the high efficiency transformer and it would lead inrush current in grid side of the transformer as grid voltage restores to normal level. The inrush current would generate unbalanced magnetomotive force to damage the insulation of the transformer windings.

This paper presents an inrush current mitigation method based on state-feedback current control. The proposed method injects compensating current to accelerate the decay of the flux deviation during LVRT operation. Moreover, the proposed method includes a peak current regulator to limit the peak current without higher than pre-defined value which is design by the current capacity of the semiconductor devices. The laboratory test result shows the inrush current is reduced by proposed method and the converter's current is limited.

REFERENCES

[1] "Grid code high and extra high voltage," E.ON Netz GmbH, Bayreuth, April 2006. [Online]. Available: http://www.eon-netz.com

[2] "The grid code, issue 4 revision 2," National Grid Electricity Transmission plc, Great Britain, March 2010. [Online]. Available: http://www.nationalgrid.com/uk

[3] Energinet, "Wind turbines connected to grids with voltages below 100 kv," Regulation TF 3.2.6, Denmark, 19 May 2010.

[4] C. Schauder, "Impact of ferc 661-a and ieee 1547 on photovoltaic inverter design," in *Proc. IEEE Power and Energy Society General Meeting*, 2011, pp. 1–6.

[5] "Test procedures for protection measures of grid-connected photovoltaic inverters," Japan Electrical Safety and Environment Technology Laboratories.

[6] S. Alepuz, S. Busquets-Monge, J. Bordonau, J. A. Martinez-Velasco, C. A. Silva, J. Pontt, and J. Rodriguez, "Control strategies based on symmetrical components for grid-connected converters under voltage dips," *IEEE Trans. Ind. Electron*, vol. 56, pp. 2162–2173, June 2009.

[7] A. S. Magueed, Fainan A. and J. Svensson, "Transient performance of voltage source converter under unbalanced voltage dips," *Proc. IEEE PESC*, vol. 2, pp. 1163–1168, 2004.

[8] C.-W. Hsu, C.-T. Lee, and P.-T. Cheng, "A low-voltage ride-through techniquefor grid-connected converters of distributed energy resources," *IEEE Trans. Ind. Appl.*, vol. 47, pp. 1821–1832, July-Aug 2011.

[9] Z. Chen, J. M. Guerrero, and F. Blaabjerg, "A review of the state of the art of power electronics for wind turbines," *IEEE Trans. Power Electronics.*, vol. 24, no. 8, pp. 1859–1875, Aug 2009.

[10] F. Blaabjerg, M. Liserre, and K. Ma, "Power electronics converters for wind turbine systems," *IEEE Trans. Ind. Appl*, vol. 48, no. 2, pp. 708–719, Mar.-Apr. 2012.

[11] J. Pedra, L. Sainz, F. Corcoles, and L. Guasch, "Symmetrical and unsymmetrical voltage sag effects on three-phase transformers," *IEEE Transactions on Power Delivery*, vol. 20, no. 2, pp. 1683–1691, Apr. 2005.

[12] M. Hagiwara, P. V. Pham, and H. Akagi, "Calculation of dc magnetic flux deviation in the converter-transformer of a self-commutated btb system during single-line-to-ground faults," *IEEE Transactions on Power Electronics*, vol. 23, no. 2, pp. 698–706, Mar. 2008.

[13] M. Steurer and K. Frohlich, "The impact of inrush currents on the mechanical stress of high voltage power transformer coils," *IEEE Trans.Power Del*, vol. 17, pp. 155–160, Jan. 2002.

[14] *IEEE IAS P1668 Voltage Sag Ride-through Working Group, Recommended Practice for Voltage Sag and Interruption Ride-Through Testing for End-Use Electrical Equipment Less Than 1,000 Volts*. Wiley-IEEE Press.

The 2014 International Power Electronics Conference

A Cell Capacitor Energy Balancing Control of Modular Multilevel Converter Considering the Unbalanced AC Grid Conditions

Jae-Jung Jung, Shenghui Cui, Sungmin Kim, *Student Member, IEEE*, and Seung-Ki Sul, *Fellow, IEEE*

School of Electrical & Computer Engineering
Seoul National University
Seoul, Republic of Korea
Email: jaejung.jung@eepel.snu.ac.kr

Abstract— This paper presents a control scheme for the regulation of cell capacitor energy balancing of a Modular Multilevel Converter (MMC) for HVDC transmission systems, considering the unbalanced AC grid conditions. It is essential that the MMC balancing control should be valid not only for the balanced normal operations but also for the asymmetrical grid fault conditions. This paper proposes the control scheme that has the ability of seamless mode change between balanced and unbalanced grid condition. Applying the proposed method, the capacitor energy balancing operation is successfully realized with improved dynamic responses. Finally, the simulation results verify the validity of the proposed method.

Keywords— *HVDC transmission system, modular multilevel converter (MMC), unbalanced grid fault, single-line-to-ground fault, capacitor energy balancing.*

I. INTRODUCTION

Modular Multilevel Converter (MMC) is a promising and appropriate topology for voltage sourced converter (VSC)-based HVDC transmission. Compared to the two- or three-level standard VSCs whose valves include series connected switches, the MMC for VSC-HVDC allows a reduction of the power losses with very low switching frequency, as well as low dv/dt, low harmonics, and modularity, etc. Due to these advantages of the MMC, many researches about it have been conducted in many aspects [1]-[4]. As one of many concerns about the high power transmission system, grid faults occurrence has to be addressed in the case of MMC based HVDC transmission system. Many researches have been mainly focused on MMC modeling, control strategy, and modulation with balanced grid conditions. And some literatures have discussed the dynamic performance of MMC under grid faults [5]-[9]. This paper deals with the balancing of cell capacitor voltage under asymmetrical grid voltage (e.g., single line-to-ground (SLG) fault, which is the dominant fault mode in the practical transmission lines).

Vector current control derived in the *dq* synchronous reference frame (SRF) has been widely used in three-phase AC/DC converter. To deal with unbalanced grid

conditions, a dual current control scheme including positive- and negative-sequence current controller based on vector control concept was introduced in [10], and applied to two-level and three-level grid-connected VSCs. This control scheme had been also applied to the control of MMC-based HVDC system in [8]-[9]. When asymmetrical faults occur at the AC system, there would be negative-sequence components in grid voltages. Therefore, using the dual current controller, negative-sequence current components in grid can be eliminated and the three-phase currents could be kept balanced during the asymmetrical faults [10].

This paper is proposing a control strategy of circulating current for balancing the cell capacitor energy in an MMC under balanced and unbalanced grid condition. The proposed balancing controller exploits the negative sequence components in output phase voltages which should be generated inevitably to eliminate the negative-sequence grid currents in the faulty conditions. Therefore, the proposed method can improve the dynamic performances of MMC under the grid fault condition. Besides, the devised method is universally valid under both balanced and unbalanced grid conditions without any mode transition. So, the fault ride-through capability of the MMC-HVDC system can be enhanced by the proposed control schemes. Finally, to investigate the performance of the proposed controllers, the results of time-domain simulation studies are presented.

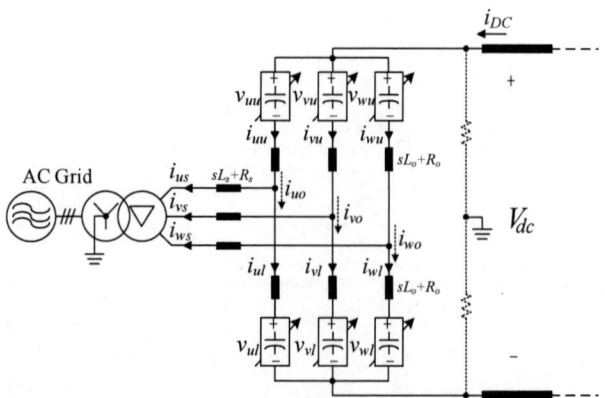

Fig. 1. Simplified schematic of an MMC in HVDC application.

978-1-4799-2706-7/14 $31.00 © 2014 IEEE   1268

II. CONFIGURATION AND BASIC PRINCIPLE OF THE MMC

The circuit configuration of an MMC in HVDC application is shown as Fig. 1. The three-phase MMC is composed of three legs, and each leg has two arms and two arm inductors. An arm usually consists of numerous half-bridge based sub-module called as cell and the whole arm can be modeled as a high bandwidth controlled voltage source with a capacitor tank. The AC side of MMC is connected to AC grid through a Y/Δ transformer for preventing the zero-sequence phase currents from flowing into the converter. For detailed mathematical description of the model of MMC, the equivalent circuit diagram can be depicted as Fig. 2. In Fig. 2, at first, i_{xu} and i_{xl} are upper and lower arm currents, respectively, and i_{xs} is the grid current, where the notation 'x' denotes a phase among u, v, and w.

In accordance with the conventional definitions [11]-[12], the upper and lower arm, the leg and the circulating current can be defined and deduced as (1)-(4), respectively.

$$i_{xu} = \frac{1}{2}i_{xs} + i_{xo}. \tag{1}$$

$$i_{xl} = -\frac{1}{2}i_{xs} + i_{xo}. \tag{2}$$

$$i_{xo} = \frac{i_{xu} + i_{xl}}{2} = i_{xo,cir} + \frac{1}{3}i_{DC}. \tag{3}$$

$$i_{xo,cir} = i_{xo,cirDC} + i_{xo,cirAC}. \tag{4}$$

The phase voltage can be used to regulate the grid current and the leg current i_{xo} can be regulated by the leg internal voltage. The phase v_{xs} and leg internal voltage v_{xo} are defined as (5) and (6), respectively.

$$v_{xs} = -\frac{1}{2}(v_{xu} - v_{xl}). \tag{5}$$

$$v_{xo} = \frac{1}{2}\{V_{dc} - (v_{xu} + v_{xl})\} = (R_o + L_o\frac{d}{dt})i_{xo}. \tag{6}$$

Therefore, the upper and lower arm voltage reference should be (7) and (8) for the desired performances of MMC. Namely, the arm voltage references are sum of independent components for regulating DC bus current, AC output current, and leg current.

$$v_{xu}^* = \frac{V_{dc}^*}{2} - v_{xs}^* - v_{xo}^*. \tag{7}$$

$$v_{xl}^* = \frac{V_{dc}^*}{2} + v_{xs}^* - v_{xo}^*. \tag{8}$$

From the above descriptions, the mechanisms for dealing with the capacitor energy in MMC can be derived. Both DC bus voltage and AC power reference should be determined to keep the total capacitor energy in MMC. And, the controllable component of circulating current has to be used to balance the capacitor energy among six arms. To balance the arm capacitor energy, the power flow into a leg should be considered. Power that flows into upper and lower arms in the same leg are:

$$P_{xu} = v_{xu}^* i_{xu} = (\frac{V_{dc}^*}{2} - v_{xs}^* - v_{xo}^*)(\frac{1}{2}i_{xs} + i_{xo}). \tag{9}$$

$$P_{xl} = v_{xl}^* i_{xl} = (\frac{V_{dc}^*}{2} + v_{xs}^* - v_{xo}^*)(-\frac{1}{2}i_{xs} + i_{xo}). \tag{10}$$

Then, the sum and difference of the power between upper and lower arms are:

$$P_x^\Sigma = P_{xu} + P_{xl} = V_{dc}^* i_{xo} - v_{xs}^* i_{xs} - 2v_{xo}^* i_{xo}. \tag{11}$$

$$P_x^\Delta = P_{xu} - P_{xl} = \frac{1}{2}V_{dc}^* i_{xs} - 2v_{xs}^* i_{xo} - v_{xo}^* i_{xs}. \tag{12}$$

And these two equations have to be considered for understanding of the subsequent proposed balancing control. The sum of upper and lower arm energy, E_x^Σ, should be controlled by the first power term in right side of (11). Thus, a DC component of leg current, i_{xo}, is injected to regulate E_x^Σ. The difference of upper and lower arm energy, E_x^Δ, should be counterbalanced by the second term in right side of (12). So, a fundamental frequency component of leg current i_{xo} is injected to control E_x^Δ.

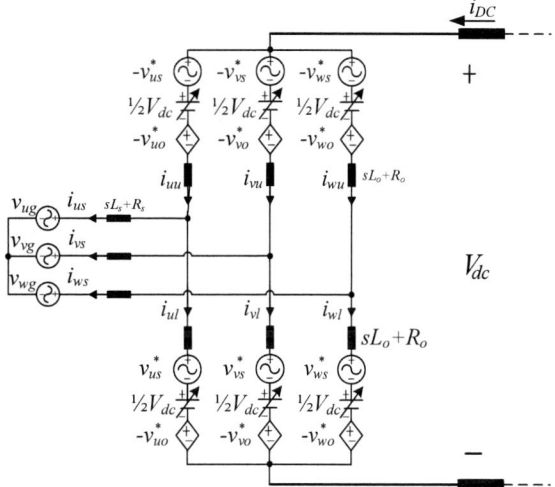

Fig. 2. Equivalent circuit diagram of the MMC.

III. MMC MODELING IN VSC-HVDC TRANSMISSION

In [11], a new modified model of MMC for energy control of cell capacitor has been presented, which divides the MMC model like Fig. 2 into AC grid current model, DC bus current model, and circulating current model. According to the modified model of MMC, the AC grid power control and DC bus voltage control can be decoupled.

The equivalent circuit diagram in Fig. 2 can be divided into three aforementioned simplified equivalent models like Fig. 3. First, for phases x and y, according to KVL, transients of upper and lower arm currents can be derived as (13) and (14).

$$-v_{xs}^* + \frac{1}{2}V_{dc}^* - v_{xo}^* + (L_o\frac{d}{dt} + R_o)i_{xu} + (L_s\frac{d}{dt} + R_s)i_{xs} + v_{xg}$$
$$= -v_{ys}^* + \frac{1}{2}V_{dc}^* - v_{yo}^* + (L_o\frac{d}{dt} + R_o)i_{yu} + (L_s\frac{d}{dt} + R_s)i_{ys} + v_{yg}. \tag{13}$$

The 2014 International Power Electronics Conference

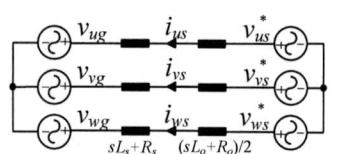

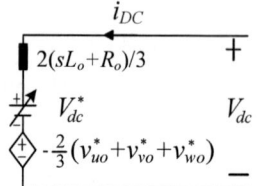

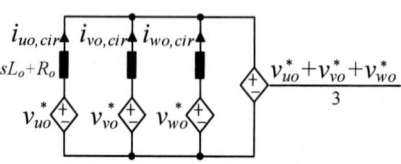

(a) Equivalent model to describe AC grid current.   (b) Equivalent model to describe DC bus current.   (c) Equivalent model to describe circulating current.

Fig. 3. Simplified equivalent models of AC grid current, DC bus current, and circulating current.

$$v_{xs}^* + \frac{1}{2}V_{dc}^* - v_{xo}^* + (L_o\frac{d}{dt} + R_o)i_{xl} - (L_s\frac{d}{dt} + R_s)i_{xs} - v_{xg}$$
$$= v_{ys}^* + \frac{1}{2}V_{dc}^* - v_{yo}^* + (L_o\frac{d}{dt} + R_o)i_{yl} - (L_s\frac{d}{dt} + R_s)i_{ys} - v_{yg}. \quad (14)$$

Both of (13) and (14) become same one equation of (15) which is basis of AC grid current model, using the definition of leg current in (3) and leg internal voltage in (6).

$$v_{xs}^* - (L_o\frac{d}{dt} + R_o)i_{xs}/2 - (L_s\frac{d}{dt} + R_s)i_{xs} - v_{xg}$$
$$= v_{ys}^* - (L_o\frac{d}{dt} + R_o)i_{ys}/2 - (L_s\frac{d}{dt} + R_s)i_{ys} - v_{yg}. \quad (15)$$

According to (15), a simplified equivalent model to describe the AC grid current can be extracted from Fig. 2, as shown in Fig. 3(a).

Second, the instantaneous DC bus voltage equations of *u*-, *v*-, and *w*-phase by KVL for Fig. 2 can be described as (16).

$$\begin{cases} (-v_{us}^* + \frac{V_{dc}^*}{2} - v_{uo}^*) + (L_o\frac{d}{dt} + R_o)(i_{uu} + i_{ul}) + (v_{us}^* + \frac{V_{dc}^*}{2} - v_{uo}^*) = V_{dc} \\ (-v_{vs}^* + \frac{V_{dc}^*}{2} - v_{vo}^*) + (L_o\frac{d}{dt} + R_o)(i_{vu} + i_{vl}) + (v_{vs}^* + \frac{V_{dc}^*}{2} - v_{vo}^*) = V_{dc} \\ (-v_{ws}^* + \frac{V_{dc}^*}{2} - v_{wo}^*) + (L_o\frac{d}{dt} + R_o)(i_{wu} + i_{wl}) + (v_{ws}^* + \frac{V_{dc}^*}{2} - v_{wo}^*) = V_{dc} \end{cases} . (16)$$

From summation of three equations in (16), DC bus voltage can be deduced as (17).

$$V_{dc}^* + \frac{2}{3}(L_o\frac{d}{dt} + R_o)i_{dc} - \frac{2}{3}(v_{uo}^* + v_{vo}^* + v_{wo}^*) = V_{dc}. \quad (17)$$

According to (17), a simplified equivalent model to describe the DC bus current can be extracted from Fig. 2, as shown in Fig. 3(b).

Meanwhile, (18) can be derived after canceling an *x*-phase equation of (16) and (17) by the DC bus voltage, V_{dc}.

$$(L_o\frac{d}{dt} + R_o)(i_{xo} - \frac{i_{dc}}{3}) = v_{xo}^* - \frac{v_{uo}^* + v_{vo}^* + v_{wo}^*}{3}. \quad (18)$$

A circulating current $i_{xo,cir}$ is defined as the difference between the leg current i_{xo} and a third of the DC bus current that identically flows into each phase:

$$i_{xo,cir} = i_{xo} - \frac{i_{dc}}{3}. \quad (19)$$

Therefore, (20) can be derived from (18) and (19).

$$v_{xo}^* - (L_o\frac{d}{dt} + R_o)i_{xo,cir} = \frac{v_{uo}^* + v_{vo}^* + v_{wo}^*}{3}. \quad (20)$$

And finally, according to (20), a simplified equivalent model to describe the circulating current can be extracted from Fig. 2, as shown in Fig. 3(c).

The average of three phase leg internal voltages (v_{uo}^*, v_{vo}^*, v_{wo}^*) is defined as (21) which is the common mode leg internal voltage $v_{o,com}^*$.

$$v_{o,com}^* = \frac{v_{uo}^* + v_{vo}^* + v_{wo}^*}{3}. \quad (21)$$

As shown in Fig. 3(b), the common mode leg internal voltage $v_{o,com}^*$ affects the real DC bus voltage V_{dc}. Furthermore, $v_{o,com}^*$ of the circulating current model also affects circulating currents as shown in Fig. 3(c). So, in the case of weak DC bus condition in HVDC transmission system, the control dynamics and system performance would significantly depend on the common mode leg internal voltage $v_{o,com}^*$. And, $v_{o,com}^*$ should be controlled as null. Based on above MMC modeling which can completely decouple AC current regulation, DC current regulation, and cell capacitor energy control, Section IV proposes the capacitor energy balancing strategy under balanced and unbalanced grid conditions.

IV. CAPACITOR ENERGY BALANCING STRATEGY UNDER BALANCED AND UNBALANCED GRID CONDITIONS

A. Overall Control Scheme in MMC System Considering Unbalanced AC Grid Conditions

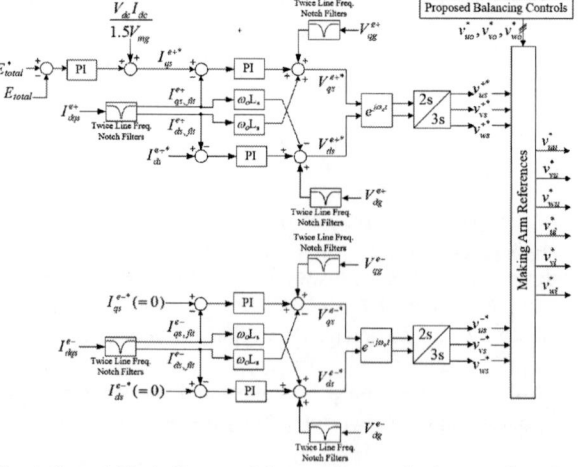

Fig. 4. Control block diagram of the overall control scheme under AC grid unbalanced conditions.

978-1-4799-2706-7/14 $31.00 © 2014 IEEE   1270

Under the unbalanced grid faults, one of countermeasures to the faults is balancing the grid currents. This can minimize the fault current that may make detrimental effect on the converter. To prevent this fault current effect, the negative-sequence current should be eliminated by setting the references of negative-sequence current controller as zero ($I_{dqs}^{e-*} = 0$). The negative-sequence current controller in VSC generates the corresponding negative-sequence output voltage references to suppress the unbalanced currents [8]. Meanwhile, as shown like Fig. 1, Y/Δ transformer connected to AC grid excludes the argument about the zero-sequence component from the converter. Based on above observation, the overall control scheme under unbalanced grid condition can be depicted as shown in in Fig. 4.

This paper proposes a capacitor energy balancing control method that can be applied under unbalanced grid condition as well as balanced grid condition. Especially, this method achieves improved dynamic behavior of the MMC by using the negative-sequence component in output phase voltage of MMC system that is made from negative sequence current controller. For example, if SLG fault occurs at the line side of the MMC, the positive-sequence voltage reduces. In conventional balancing method which only considers positive-sequence component in output phase voltage [8]-[9], the balancing controller should increase the magnitude of circulating current for eliminating the considerable unbalanced capacitor energy. Moreover, a considerable twice fundamental frequency current flows through DC transmission line under SLG fault. However, the proposed method can improve the dynamics by exploiting the negative-sequence output phase voltages from the negative sequence current controller.

B. Balancing of Three Phase Leg Energy

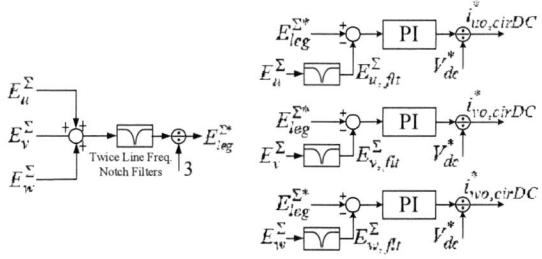

(a) Calculation of leg energy reference. (b) Inter-leg energy balancing controller.
Fig. 5. Control block diagram of the proposed leg capacitor energy balancing controller.

The proposed strategy for the comprehensive cell capacitor energy control in [11] makes the leg capacitor energy balanced and eliminates upper and lower arm capacitor energy difference with excellent performance in view point of both dynamics and stability even under the weak DC grid. The three phase leg energy balancing control is identical to that of proposed method in [11]. The leg capacitor energy balancing controller is shown in Fig. 5. The leg capacitor energy reference is updated as the average capacitor energy of three phase legs at every

sampling instant as shown in Fig. 5(a). The philosophy of using the average energy of three legs as the reference is balancing inner-converter leg capacitor energy by only controlling the circulating currents, without affecting the energy flows from AC grid or DC bus into the whole three phase converter. And, the capacitor energy among three legs can be balanced by DC component of circulating currents as shown by the outputs of controllers in Fig. 5(b). Because the leg capacitor energy reference is calculated and updated at every sampling period, it can be valid that the common mode leg internal voltage $v_{o,com}^*$ is inherently null, as (22).

$$\sum_{x=u,v,w}(E_{leg}^{\Sigma*} - E_{x,flt}^{\Sigma}) = 0. \Rightarrow \sum_{x=u,v,w} i_{xo,cirDC}^* = 0. \Rightarrow \sum_{x=u,v,w} v_{xo,DC}^* = 0. \quad (22)$$

C. Balancing of Upper and Lower Arm Capacitor Energy under Balanced and Unbalanced Grid Conditions

As with many conventional researches about energy balancing control [11]-[12], the second term in right side of (12) is employed for balancing of upper and lower arm capacitor energy. Thus, the energy difference between upper and lower arms can be eliminated by injecting a fundamental frequency on leg currents or circulating currents. Like the preceding about nullifying the common mode leg internal voltage $v_{o,com}^*$, the common mode AC circulating current reference at fundamental frequency should be null as (23).

$$\sum_{x=u,v,w} v_{xo,AC}^* = 0 . \Rightarrow \sum_{x=u,v,w} i_{xo,cirAC}^* = 0. \quad (23)$$

Considering the grid fault condition, the output phase voltage reference has the negative sequence components which should be generated inevitably to eliminate the negative grid currents as referred at Fig 4.

First, assuming unbalanced grid condition, the positive- and negative-sequence component of the output phase voltages are defined by (24) and (25), respectively.

$$\begin{cases} v_{us}^{+*} = V_{ms}^+ \sin(\omega_o t + \phi_v^+) \\ v_{vs}^{+*} = V_{ms}^+ \sin(\omega_o t + \phi_v^+ - 2\pi/3) \\ v_{ws}^{+*} = V_{ms}^+ \sin(\omega_o t + \phi_v^+ + 2\pi/3) \end{cases} \quad (24)$$

$$\begin{cases} v_{us}^{-*} = V_{ms}^- \sin(\omega_o t + \phi_v^-) \\ v_{vs}^{-*} = V_{ms}^- \sin(\omega_o t + \phi_v^- + 2\pi/3) \\ v_{ws}^{-*} = V_{ms}^- \sin(\omega_o t + \phi_v^- - 2\pi/3) \end{cases} \quad (25)$$

Because the capacitor energy should be balanced for three phases, three Degree of Freedoms (DOFs) are needed to regulate arm energy difference, P_u^Δ, P_v^Δ, and P_w^Δ independently. And, the positive- and negative-sequence circulating currents provide three DOFs for arm energy balancing. Therefore, controllable positive- and negative-sequence circulating current references for balancing are defined as (26) and (27), respectively.

$$\begin{cases} i_{uo,cirAC}^{+*} = I_{cirAC}^+ \sin(\omega_o t + \phi_i^+) \\ i_{vs,cirAC}^{+*} = I_{cirAC}^+ \sin(\omega_o t + \phi_i^+ - 2\pi/3) \\ i_{ws,cirAC}^{+*} = I_{cirAC}^+ \sin(\omega_o t + \phi_i^+ + 2\pi/3) \end{cases} \quad (26)$$

$$\begin{cases} i_{uo,cirAC}^{-*} = I_{cirAC}^{-} \sin(\omega_o t + \phi_i^{-}) \\ i_{vs,cirAC}^{-*} = I_{cirAC}^{-} \sin(\omega_o t + \phi_i^{-} + 2\pi/3) \\ i_{ws,cirAC}^{-*} = I_{cirAC}^{-} \sin(\omega_o t + \phi_i^{-} - 2\pi/3) \end{cases} \tag{27}$$

The counterbalancing terms, $v_{xs}^* i_{xo,cir}$, to eliminate the difference between upper and lower arm capacitor energy are derived for three phases as (28)-(31). Eq. (28)-(31) are produced by multiplication between (24)-(25) and (26)-(27). Only DC component which plays a significant role in balancing control is presented.

$$\begin{cases} P_u^{++\Delta} = v_{us}^{+*} i_{uo,cirAC}^{+*} \Big|_{DC} = \dfrac{V_{ms}^+ I_{cirAC}^+}{2} \cos(\phi_v^+ - \phi_i^+) \\ P_v^{++\Delta} = v_{vs}^{+*} i_{vo,cirAC}^{+*} \Big|_{DC} = \dfrac{V_{ms}^+ I_{cirAC}^+}{2} \cos(\phi_v^+ - \phi_i^+) \\ P_w^{++\Delta} = v_{ws}^{+*} i_{wo,cirAC}^{+*} \Big|_{DC} = \dfrac{V_{ms}^+ I_{cirAC}^+}{2} \cos(\phi_v^+ - \phi_i^+) \end{cases} \tag{28}$$

$$\begin{cases} P_u^{+-\Delta} = v_{us}^{+*} i_{uo,cirAC}^{-*} \Big|_{DC} = \dfrac{V_{ms}^+ I_{cirAC}^-}{2} \cos(\phi_v^+ - \phi_i^-) \\ P_v^{+-\Delta} = v_{vs}^{+*} i_{vo,cirAC}^{-*} \Big|_{DC} = \dfrac{V_{ms}^+ I_{cirAC}^-}{2} \cos(\phi_v^+ - \phi_i^- + 2\pi/3) \\ P_w^{+-\Delta} = v_{ws}^{+*} i_{wo,cirAC}^{-*} \Big|_{DC} = \dfrac{V_{ms}^+ I_{cirAC}^-}{2} \cos(\phi_v^+ - \phi_i^- - 2\pi/3) \end{cases} \tag{29}$$

$$\begin{cases} P_u^{-+\Delta} = v_{us}^{-*} i_{uo,cirAC}^{+*} \Big|_{DC} = \dfrac{V_{ms}^- I_{cirAC}^+}{2} \cos(\phi_v^- - \phi_i^+) \\ P_v^{-+\Delta} = v_{vs}^{-*} i_{vo,cirAC}^{+*} \Big|_{DC} = \dfrac{V_{ms}^- I_{cirAC}^+}{2} \cos(\phi_v^- - \phi_i^+ - 2\pi/3) \\ P_w^{-+\Delta} = v_{ws}^{-*} i_{wo,cirAC}^{+*} \Big|_{DC} = \dfrac{V_{ms}^- I_{cirAC}^+}{2} \cos(\phi_v^- - \phi_i^+ + 2\pi/3) \end{cases} \tag{30}$$

$$\begin{cases} P_u^{--\Delta} = v_{us}^{-*} i_{uo,cirAC}^{-*} \Big|_{DC} = \dfrac{V_{ms}^- I_{cirAC}^-}{2} \cos(\phi_v^- - \phi_i^-) \\ P_v^{--\Delta} = v_{vs}^{-*} i_{vo,cirAC}^{-*} \Big|_{DC} = \dfrac{V_{ms}^- I_{cirAC}^-}{2} \cos(\phi_v^- - \phi_i^-) \\ P_w^{--\Delta} = v_{ws}^{-*} i_{wo,cirAC}^{-*} \Big|_{DC} = \dfrac{V_{ms}^- I_{cirAC}^-}{2} \cos(\phi_v^- - \phi_i^-) \end{cases} \tag{31}$$

In (28)-(31), if the positive-sequence circulating current is in phase with the output voltage, namely $\phi_v^+ = \phi_i^+$, then I_{cirAC}^+, I_{cirAC}^-, ϕ_i^- provides three DOFs for arm energy balancing under the restriction of nullification of (21). In case of (28) and (31), because the power are equally distributed among three phases, these terms contribute to counterbalancing the upper and lower arm energy of three phases, E_{com}^Δ. On the other hand, in the case of (29) and (30), the differential components of the three phase arm energy difference can be eliminated, which are E_d^Δ and E_q^Δ of three phase upper and lower arm capacitor energy differences. The three power terms (P_{com}^Δ, P_d^Δ, and P_q^Δ) are deduced as (32)-(34).

$$P_{com}^\Delta = \frac{dE_{com}^\Delta}{dt} = \frac{d}{dt}\left\{\frac{1}{3}(E_u^\Delta + E_v^\Delta + E_w^\Delta)\right\}$$
$$= \frac{1}{3}(P_u^{++\Delta} + P_u^{--\Delta} + P_v^{++\Delta} + P_v^{--\Delta} + P_w^{++\Delta} + P_w^{--\Delta}) \tag{32}$$
$$= \frac{V_{ms}^+ I_{cirAC}^+}{2}\cos(\phi_v^+ - \phi_i^+) + \frac{V_{ms}^- I_{cirAC}^-}{2}\cos(\phi_v^- - \phi_i^-)$$
$$= \frac{1}{2}(V_{ms,d}^{e+} I_{cirAC,d}^{e+} + V_{ms,q}^{e+} I_{cirAC,q}^{e+} + V_{ms,d}^{e-} I_{cirAC,d}^{e-} + V_{ms,q}^{e-} I_{cirAC,q}^{e-}).$$

$$P_d^\Delta = \frac{dE_d^\Delta}{dt} = \frac{d}{dt}\left(\frac{2}{3}E_u^\Delta - \frac{1}{3}E_v^\Delta - \frac{1}{3}E_w^\Delta\right)$$
$$= \frac{2}{3}(P_u^{+-\Delta} + P_u^{-+\Delta}) - \frac{1}{3}(P_v^{+-\Delta} + P_v^{-+\Delta}) - \frac{1}{3}(P_w^{+-\Delta} + P_w^{-+\Delta}) \tag{33}$$
$$= \frac{V_{ms}^+ I_{cirAC}^-}{2}\cos(\phi_v^- - \phi_i^+) + \frac{V_{ms}^- I_{cirAC}^+}{2}\cos(\phi_v^- - \phi_i^+)$$
$$= \frac{1}{2}(V_{ms,d}^{e+} I_{cirAC,d}^{e-} + V_{ms,q}^{e+} I_{cirAC,q}^{e-} + V_{ms,d}^{e-} I_{cirAC,d}^{e+} + V_{ms,q}^{e-} I_{cirAC,q}^{e+}).$$

$$P_q^\Delta = \frac{dE_q^\Delta}{dt} = \frac{d}{dt}\left(\frac{\sqrt{3}}{3}E_v^\Delta - \frac{\sqrt{3}}{3}E_w^\Delta\right)$$
$$= \frac{\sqrt{3}}{3}(P_v^{+-\Delta} + P_v^{-+\Delta}) - \frac{\sqrt{3}}{3}(P_w^{+-\Delta} + P_w^{-+\Delta}) \tag{34}$$
$$= \frac{V_{ms}^+ I_{cirAC}^-}{2}\sin(\phi_i^- - \phi_v^+) + \frac{V_{ms}^- I_{cirAC}^+}{2}\sin(\phi_v^- - \phi_i^+)$$
$$= \frac{1}{2}(-V_{ms,q}^{e+} I_{cirAC,d}^{e-} + V_{ms,d}^{e+} I_{cirAC,q}^{e-} + V_{ms,q}^{e-} I_{cirAC,d}^{e+} - V_{ms,d}^{e-} I_{cirAC,q}^{e+}).$$

For simplifying the equations, the d axis is oriented as the axis where all positive-sequence phase voltage lies ($V_{ms,q}^{e+} = 0$) and the q axis component of positive-sequence circulating current is controlled as null ($I_{cirAC,q}^{e+} = 0$) in accordance with the aforementioned synchronization of $\phi_v^+ = \phi_i^+$, as (35)-(37).

$$P_{com}^\Delta = \frac{1}{2}(V_{ms,d}^{e+} I_{cirAC,d}^{e+} + V_{ms,d}^{e-} I_{cirAC,d}^{e-} + V_{ms,q}^{e-} I_{cirAC,q}^{e-}). \tag{35}$$

$$P_d^\Delta = \frac{1}{2}(V_{ms,d}^{e+} I_{cirAC,d}^{e-} + V_{ms,d}^{e-} I_{cirAC,d}^{e+}). \tag{36}$$

$$P_q^\Delta = \frac{1}{2}(V_{ms,d}^{e+} I_{cirAC,q}^{e-} + V_{ms,q}^{e-} I_{cirAC,d}^{e+}). \tag{37}$$

Eq. (35)-(37) can be expressed in matrix form as (38).

$$\begin{bmatrix} P_{com}^\Delta \\ P_d^\Delta \\ P_q^\Delta \end{bmatrix} = \frac{1}{2}\begin{bmatrix} V_{ms,d}^{e+} & V_{ms,d}^{e-} & V_{ms,q}^{e-} \\ V_{ms,d}^{e-} & V_{ms,d}^{e+} & 0 \\ V_{ms,q}^{e-} & 0 & V_{ms,d}^{e+} \end{bmatrix}\begin{bmatrix} I_{cirAC,d}^{e+} \\ I_{cirAC,d}^{e-} \\ I_{cirAC,q}^{e-} \end{bmatrix}. \tag{38}$$

Therefore, the references of the circulating current can be calculated like (39), where the determinant D is calculated as (40).

$$\begin{bmatrix} I_{cirAC,d}^{e+*} \\ I_{cirAC,d}^{e-*} \\ I_{cirAC,q}^{e-*} \end{bmatrix} = \frac{2}{D}\begin{bmatrix} (V_{ms,d}^{e+})^2 & -V_{ms,d}^{e+} V_{ms,d}^{e-} & -V_{ms,d}^{e+} V_{ms,q}^{e-} \\ -V_{ms,d}^{e+} V_{ms,d}^{e-} & (V_{ms,d}^{e+})^2 - (V_{ms,q}^{e-})^2 & V_{ms,d}^{e-} V_{ms,q}^{e-} \\ -V_{ms,d}^{e+} V_{ms,q}^{e-} & V_{ms,d}^{e-} V_{ms,q}^{e-} & (V_{ms,d}^{e+})^2 - (V_{ms,d}^{e-})^2 \end{bmatrix}\begin{bmatrix} P_{com}^{\Delta *} \\ P_d^{\Delta *} \\ P_q^{\Delta *} \end{bmatrix}. \tag{39}$$

$$D = 3V_{ms,d}^{e+}\left\{(V_{ms,d}^{e+})^2 - (V_{ms,d}^{e-})^2 - (V_{ms,q}^{e-})^2\right\}. \tag{40}$$

And then, based on (39), the proposed balancing controller can be implemented as Fig. 6.

To regulate the circulating current and produce leg voltage references, Proportional, Integral, and Resonant (PIR) controllers have been used.

The 2014 International Power Electronics Conference

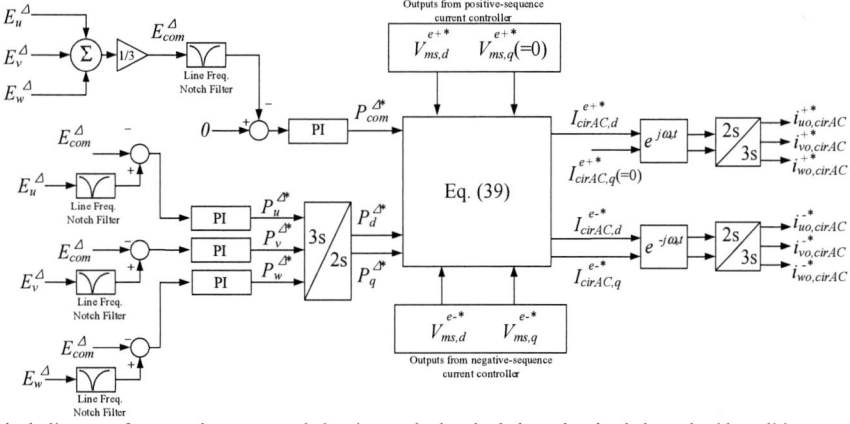

Fig. 6. Control block diagram of proposed arm energy balancing method under balanced and unbalanced grid conditions.

V. SIMULATION RESULTS

TABLE I   SIMULATION PARAMETERS

Quantity	Values
Number of cells per arm	3
Rated DC bus voltage (V_{dc}^{*})	1.55kV
Rated cell capacitor voltage (V_{cell}^{*})	517V
Cell capacitor (C_{cell})	4.4mF
Grid line-to-line voltage (RMS)	550Vrms
Arm inductor inductance (L_o)	2.0 mH
Arm inductor resistance (R_o)	5.0 mΩ
DC bus R-L load inductance (L_{load})	10.0 mH
DC bus R-L load resistance (R_{load})	155 Ω
Sampling frequency (f_{samp})	8 kHz
Switching frequency (f_{sw})	4 kHz

A 4-level MMC model has been established by using time-domain simulation program, PSIM. The simulation parameters are given in Table I, referred to Fig. 1. The DC side of the MMC is connected to RL load with $L_L = 10mH$, $R_L = 150Ω$, emulating the practical long DC transmission line.

A single line-to-ground fault is applied on AC grid at t=1.5s and lasts for 0.3s as shown in Fig 7(a). The simulations have been carried out in two cases, with and without the proposed method.

First, Fig. 7 shows dynamic performance of an MMC system with the controller described in Section IV in response to SLG fault. Fig. 7(b) shows the DC bus voltage and DC bus current. The DC bus voltage and current are sustained at the set level, even when the fault occurs. It is shown that the power flows constantly to DC line. Therefore, the magnitude of grid currents under fault condition is increased for constant power transmission, as shown in Fig. 7(c). Furthermore, the second harmonic components in DC bus voltage and current caused by negative sequence voltage are eliminated because the decoupling balancing method in [11] was applied to the

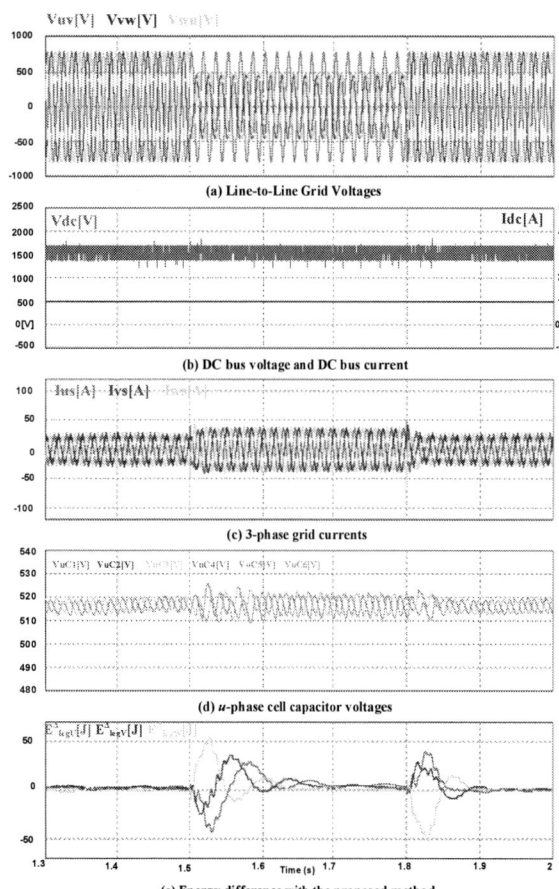

Fig. 7. Dynamic responses with the proposed control scheme under a single line-to-ground fault at grid.

proposed method basically. Fig. 7(d) shows *u*-phase cell capacitor voltages, and it can be seen that the maximum pulsation of cell capacitor voltages is around 17V (3%), which is much lower than allowable boundary. At last, energy differences between upper and lower arm capacitors at *u*, *v*, *w* leg with the proposed control scheme are shown in Fig. 7(e). From the figure, it can be seen that the energy difference are well regulated within 0.1s after SLG fault.

978-1-4799-2706-7/14 $31.00 © 2014 IEEE

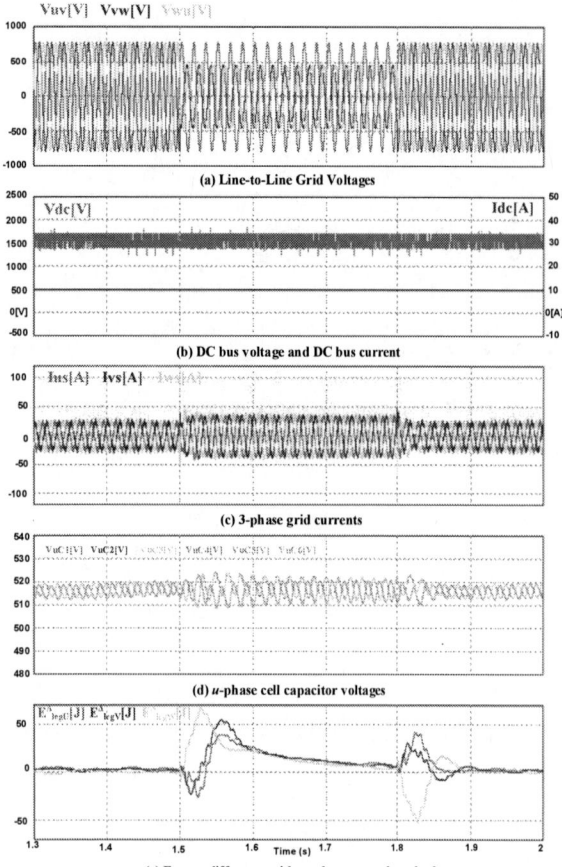

(a) Line-to-Line Grid Voltages

(b) DC bus voltage and DC bus current

(c) 3-phase grid currents

(d) u-phase cell capacitor voltages

(e) Energy difference without the proposed method

Fig. 8. Dynamic responses without the proposed control scheme under the same operating condition in Fig. 7.

In contrast, when the proposed method was not incorporated to the existing balancing controller under the same operating conditions, the energy differences between upper and lower arm capacitor are eliminated too slowly compared to that with proposed method, as shown in Fig. 8. And, the energy difference are regulated after 0.3s after SLG fault as shown in Fig. 8(e). Therefore, the proposed method could reduce the transient time by 60%.

Apart from the SLG fault conditions, to highlight the balancing performance of the proposed method, it is assumed that the negative-sequence components in grid flow into the system, although it is extreme fault situation. The magnitude of negative sequence voltage which is twice as large as positive-sequence voltage is applied on AC grid at t=1.5s, as shown in Fig. 9(a). The DC bus voltage and current, grid currents, and cell capacitor voltages fluctuate under allowable boundary because the large unbalance component in AC grid adversely affects the MMC system. However, the proposed method utilizes the negative component of output voltage for balancing, and then the upper and lower arm energy balancing is accomplished as shown in Fig. 9(e). On the other hand, the control scheme without the proposed method is also performed in simulation under the same conditions. In this case, the energy difference between upper and lower

arm is considerably large and unbalanced. So, this system could not endure the same unbalance in Fig. 9(a), as shown in Fig. 10.

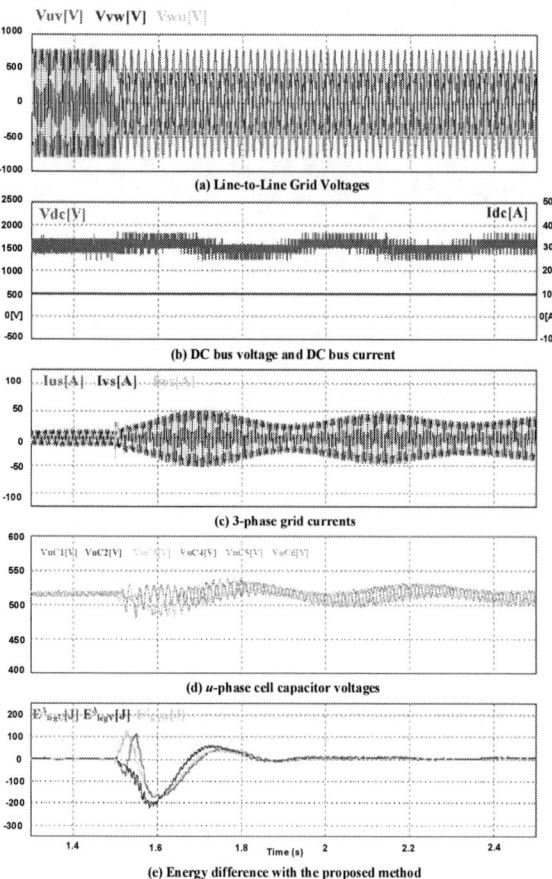

(a) Line-to-Line Grid Voltages

(b) DC bus voltage and DC bus current

(c) 3-phase grid currents

(d) u-phase cell capacitor voltages

(e) Energy difference with the proposed method

Fig. 9. Dynamic responses with the proposed control scheme under the condition that negative component is as twice as the positive component in AC grid voltage.

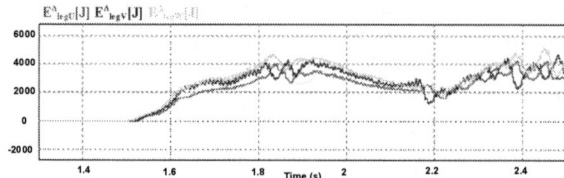

Fig. 10. Energy difference between upper and lower arm without the proposed control scheme under the same AC grid condition in Fig. 9.

VI. CONCLUSIONS

A control strategy of the MMC for capacitor energy balancing has been presented, which is universally valid under both balanced and unbalanced grid condition without any transition in control modes. By exploiting the inevitable negative-sequence component in output phase voltage for capacitor balancing, the dynamics of cell voltage balancing have been improved under asymmetric fault conditions.

The simulation study in this paper mainly focused on the dynamic response of the proposed method under the single line-to-ground terminal bus fault. The validity of the proposed method has been supported by computer simulation results. The fault ride-through capability of the MMC could be much enhanced if the proposed control method is incorporated in the existing cell voltage balancing controller.

REFERENCES

[1] S. Allebrod, R. Hamerski, and R. Marquardt, "New transformerless, scalable modular multilevel converters for HVDC-transmission," in *Proc. IEEE PESC*, 2008, pp. 174-179.

[2] A. Lesnicar and R. Marquardt, "An innovative modular multilevel converter topology suitable for a wide power range," in *Proc. IEEE Boogna Power Tech*, 2008, vol. 3, pp. 1-6.

[3] D. Peftitsis, G. Tolstoy, A. Antonopoulos, J. Rabkowski, J. K. Lim, M. Bakowski, L. Angquist, and H. P. Nee, "High-power modular multilevel converters with SiC JFETs," in *Proc. IEEE ECCE*, 2010, pp. 2148-2155.

[4] AmirnaserYazdani, and Reza Iravani, "Voltage-Sourced Converters in Power Systems –Modeling, Control, and Applications," *IEEE Press*, 2011.

[5] Soto-Sanchez, D.; Green, T.C., "Control of a modular multilevel converter-based HVDC transmission system," *Power Electronics and Applications (EPE 2011), Proceedings of the 2011-14th European Conference on*, vol., no., pp.1,10, Aug. 30 2011-Sept. 1 2011.

[6] X. Chen, C. Zhao, C. Cao, "Research on the fault characteristics of HVDC based on modular multilevel converter," Electrical Power and Energy Conference (EPEC), 2011 IEEE, pp.91,96, 3-5 Oct. 2011.

[7] Bordignon, P.; Marchesoni, M.; Parodi, G.; Vaccaro, L. "Modular multilevel converter in HVDC systems under fault conditions", Power Electronics and Applications (EPE), 2013 15th European Conference on, On page(s): 1 – 10

[8] Minyuan Guan; Zheng Xu, "Modeling and Control of a Modular Multilevel Converter-Based HVDC System Under Unbalanced Grid Conditions," *Power Electronics, IEEE Transactions on*, vol.27, no.12, pp.4858,4867, Dec. 2012.

[9] Qingrui Tu; Zheng Xu; Yong Chang; Li Guan, "Suppressing DC Voltage Ripples of MMC-HVDC Under Unbalanced Grid Conditions," *Power Delivery, IEEE Transactions on*, vol.27, no.3, pp.1332,1338, July 2012.

[10] Yazdani, A.; Iravani, R., "A unified dynamic model and control for the voltage-sourced converter under unbalanced grid conditions," *Power Delivery, IEEE Transactions on*, vol.21, no.3, pp.1620,1629, July 2006.

[11] S.H. Choi, S. Kim, J. J. Jung, and S. K. Sul, "A Comprehensive Cell Capacitor Energy Control Strategy of a Modular Multilevel Converter (MMC) without a Stiff DC Bus Voltage Source," *2014 Applied Power Electronics Conference and Exposition*, 2014-Mar., 2014.

[12] Jae-Jung Jung, Hak-Jun Lee, and Seung-Ki Sul, "Control strategy for improved dynamic performance of variable-speed drives with the Modular Multilevel Converter," in *IEEE Energy Conversion Congress and Exposition*, pp. 1481-1488, 2013.

Fault Current Limitation Using Thyristor Based Devices

Wilson Komatsu, Antonio Ricardo Giaretta,
Rubens Domingos de Miranda,
José Antonio Jardini
Polytechnic School of the University of Sao Paulo
São Paulo, Brazil
wilsonk@usp.br

Ronaldo Pedro Casolari, Ricardo Leon Vasquez-Arnez
Foundation for the Technological Development of the
Engineering Sciences
Toshiaki Hojo, Eden Luiz Carvalho Jr.,
Paulo Koiti Maezono
Transmissoras Brasileiras de Energia/Empresa Amazonense
de Transmissão de Energia S.A.

Abstract— **This paper presents analysis, simulation and experimental results of two short circuit current limiters, the Thyristor Inserted Reactor and the Thyristor Bypassed Capacitor. The purpose of the fault current limiters is to reduce the overstress of circuit breakers provoked by the network expansion and the consequent increase in fault currents. The simulation and experimental prototype results demonstrate the effectiveness of these devices to limit fault currents, deferring significantly the replacement of circuit breakers with surpassed fault current interruption capacity.**

Keywords— Fault current limiter, thyristor bypassed capacitor, thyristor inserted reactor.

I. INTRODUCTION

The continuous growth of power generation and network interconnection in today's power systems result in higher fault currents. As a consequence, various network components can reach and even exceed their rated short circuit capacity. To deal with this effect, two main types of solutions are commonly considered [1]:
- Temporary solutions, which are normally applied for short periods of time or during contingency situations;
- Long lasting solutions, which require detailed studies and longer execution periods, and which may end up in the complete substitution of the component. Fault current limiting devices, characterized by having low impedance during normal operation and high impedance during short circuit periods, can be considered within this group.

With the continuous progress and cost reduction of the solid-state devices, the limitation of fault currents through Fault Current Limiters (FCL) appear as one of the most economical solutions compared to the complete upgrade of exceeded capacity installations, with the additional advantage of easier implementation.

According to a study released by [1] various high-voltage circuit-breakers in many substations of the Brazilian southeast region have their short-circuit current interruption capacity exceeded and many others will soon

This work was financially supported by TBE/EATE (Transmissoras Brasileiras de Energia/Empresa Amazonense de Transmissão de Energia S.A) under the R&D program established by ANEEL (Brazilian Energy Regulatory Agency).

be exceeded. Table 1 shows the situation of three of these substations. Substation C, for example, shows a high percentage of overstressed circuit breakers. This is one of the main reasons that led to the development of the research presented in this work.

On the short-circuit limiting subject, [2] and [3] present a comprehensive study on the FCL technology for medium and high voltage systems. [4] presents one of the earliest references on this issue, where the combined concept of a FCL and a dynamic series compensator are shown. [5] present a comparative survey of the research activities and emerging technologies of solid-state FCLs for power distribution systems. [6] also presents a description of some fault current limiters with emphasis on the Interphase Power Controller and its application to limit short circuit currents.

TABLE 1
CIRCUIT-BREAKER CONDITION AT THREE SUBSTATIONS CHOSEN FROM [1] (DEC. 2011).

Substation	Voltage level	No. of Circuit-breakers	(%) of Circuit breakers and their condition
A	138 kV	21 (100%)	57 (%) OK [1]
			43 (%) Alert [2]
			0 (%) Exceeded [3]
B	138 kV	13 (100%)	77 (%) OK
			15 (%) Alert
			8 (%) Exceeded
C	345 kV	13 (100%)	7.5 (%) OK
			7.5 (%) Alert
			85 (%) Exceeded

[1] Circuit breakers with more than 20% of total capacity available.
[2] Circuit breakers with less than 20% of total capacity available.
[3] Circuit breakers with exceeded short circuit withstanding capacity.

A solid-state FCL based on a single-phase bridge that might be applied at medium and high voltage levels is proposed by [7]. The use of a series connected FCL reactor jointly with a series line compensating capacitor is presented by [8]. [9] proposes two resonant-circuit based topologies (series and parallel topologies) as FCLs, applying thyristors as power switches. [10] discusses the insertion of inductive impedance to the line, through a thyristor based device, but does not present simulation or experimental results. [11] uses two Flexible AC Transmission Systems (FACTS) devices, namely the

Static Synchronous Series Compensator (SSSC) and the Unified Power Flow Controller (UPFC), to limit fault currents in a compensated line. At present, despite the practical installation and operation of FACTS devices by some utilities, none of these devices perform this additional fault current limitation task.

This work presents a proposal of an experimental setup for FCL topology and control studies and also presents the simulated and experimental results.

II. ANALYSIS OF CHOSEN FAULT CURRENT LIMITER CIRCUITS

The capacity for handling large currents makes the power thyristor, or Silicon Controlled Rectifier (SCR), a device particularly suitable for FCL application. Another advantage lies in the possibility of series and parallel connection of SCRs to reach the desired voltage and current level. Although IGBT switches are proposed for distribution level FCL circuits [5], for transmission level power requirements the choice of SCRs is more adequate, in the authors' vision.

Based on the study on FCL circuits, two FCL topologies were chosen and are investigated here: (a) the Thyristor Inserted Reactor (TIR), shown in Fig. 1(a), and (b) Thyristor Bypassed Capacitor (TBC), shown in Fig. 1(b).

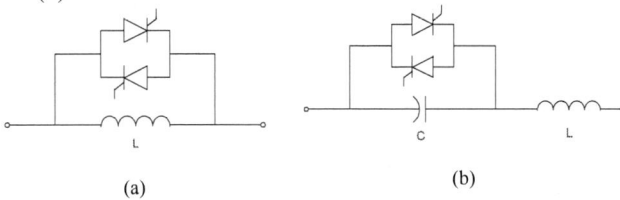

(a)   (b)

Fig. 1. Topologies of: (a) Thyristor Inserted Reactor (TIR), (b) Thyristor Bypassed Capacitor (TBC).

The difference between the TIR and the Thyristor Controlled Reactor (TCR), often found in the literature, is that in the TCR the reactance is variable with the changing thyristor firing angle, and in the TIR, the reactor is either fully inserted or removed, as its purpose is simply to limit the short-circuit current and not to be a line power flow control.

The operating principles of the two FCL devices (i.e. TIR and TBC schemes) are briefly described here.

A. Thyristor Inserted Reactor (TIR)

The reactor L, shown in Fig. 1(a), is bypassed by the thyristors during normal operation and only inserted when the line current (or its increase rate) exceeds a certain level. The value of the reactance depends on the system short-circuit impedance, the line impedance and the maximum fault current allowed. This device can be considered the simplest of both FCLs as it only involves a reactor in parallel with the thyristors. During normal operation the power thyristors are continuously conducting, thus, the FCL presents very low losses. A disadvantage is the need to keep thyristors operating constantly during normal operation.

B. Thyristor Bypassed Capacitor (TBC)

The TBC, shown in Fig. 1(b), has its thyristors blocked under normal operation conditions. The capacitor C and the reactor L are tuned to the mains frequency; therefore equivalent impedance is very low, mainly due to the reactor's ohmic resistance. In the event of a short-circuit event, the capacitor is shunted by the thyristors and reactor L impedance limits the short-circuit current. A disadvantage of this configuration is the need to synchronize thyristor's firing instant with zero voltage over capacitor C, in order to avoid high current capacitor discharge on the thyristors. Another disadvantage are the ohmic losses of the reactor.

It can be noticed that the FCL topologies of Fig. 1 are single-phase ones. Therefore separated measurement and action on each of the three phases of the AC system is required. Most faults in AC systems are single-phase ones, but usually FCL circuits act almost simultaneously on all phases in these single-phase events, in order to avoid line impedance unbalances. For the proposed experimental circuits, only one single-phase limitation was implemented, as the extension to three phases is immediate.

III. SIMULATION RESULTS FOR TIR AND TBC TOPOLOGIES

The modeling and simulations of both FCL devices were done using the PSCAD/EMTDC® program.

A. Thyristor Inserted Reactor (TIR)

The basic simulation circuit for the TIR topology is shown in Fig. 2, with simulation results in Fig. 3.

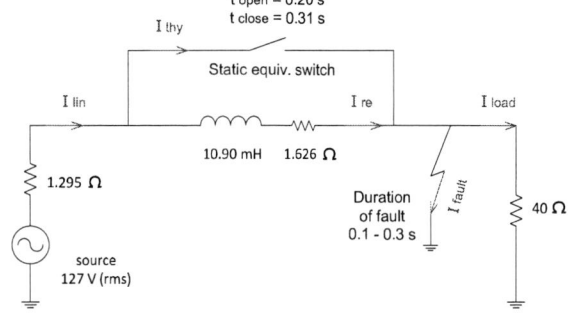

Fig. 2. Simplified single-phase system for simulation of the TIR topology.

The values applied in the simulation follows the ones chosen for the experimental prototype[1]. The fault is applied at $t = 0.1$s and removed at $t = 0.3$s, shunting the load during the fault (Fig. 3a) and reaching a maximum non-limited value of 130A (Fig. 3b). At $t = 0.2$s the thyristors are turned off (Fig. 3c) forcing the fault current to pass through the reactor, which limits peak fault current to approximately 35A (Fig. 3d). The system returns to its normal operative condition after the

_____

[1] Transmission level voltages, as those found in [1], were also investigated with simulation, but in this work low voltage systems were studied to validate both simulated and experimental setup.

extinction of the fault. In this simulation, the fault current was reduced to about 27%.

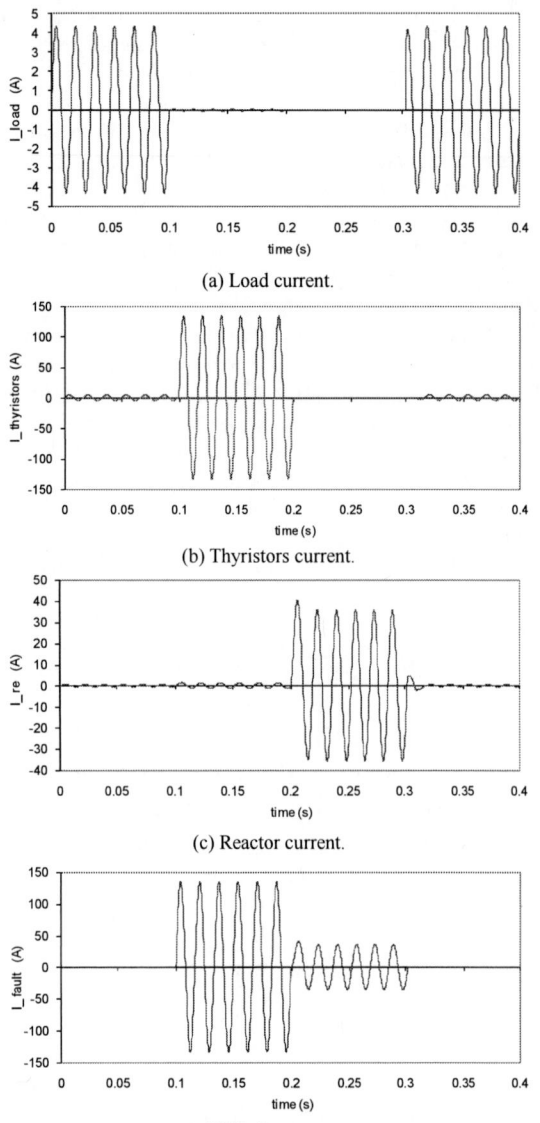

(a) Load current.

(b) Thyristors current.

(c) Reactor current.

(d) Fault current.

Fig. 3. Simulation results for the TIR topology. Fault applied at t=0.1s, reactor inserted at t=0.2s, fault removed at t=0.3s.

B. Thyristor Bypassed Capacitor (TBC)

The basic simulation circuit for the TBC topology is shown in Fig. 4, with simulation results in Fig. 5.

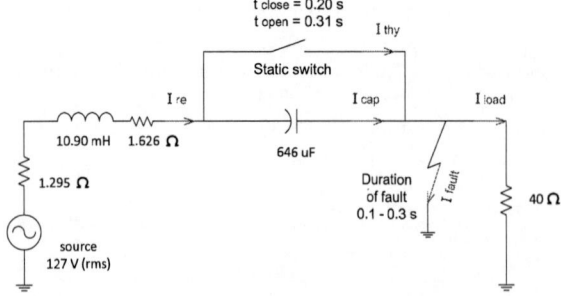

Fig. 4. Simplified single-phase system for simulation of the TBC topology.

The fault is applied at t = 0.1s and removed at t = 0.3s, reaching a maximum non-limited value of 60A (Fig. 5b). At t = 0.2s the thyristors are turned on (Fig. 5b and 5c), shunting the capacitor (Fig. 5e) and limiting peak fault current to approximately 35A (Fig. 5d). The system returns to its normal operative condition after the extinction of the fault. In this simulation, the fault current was reduced to about 58%.

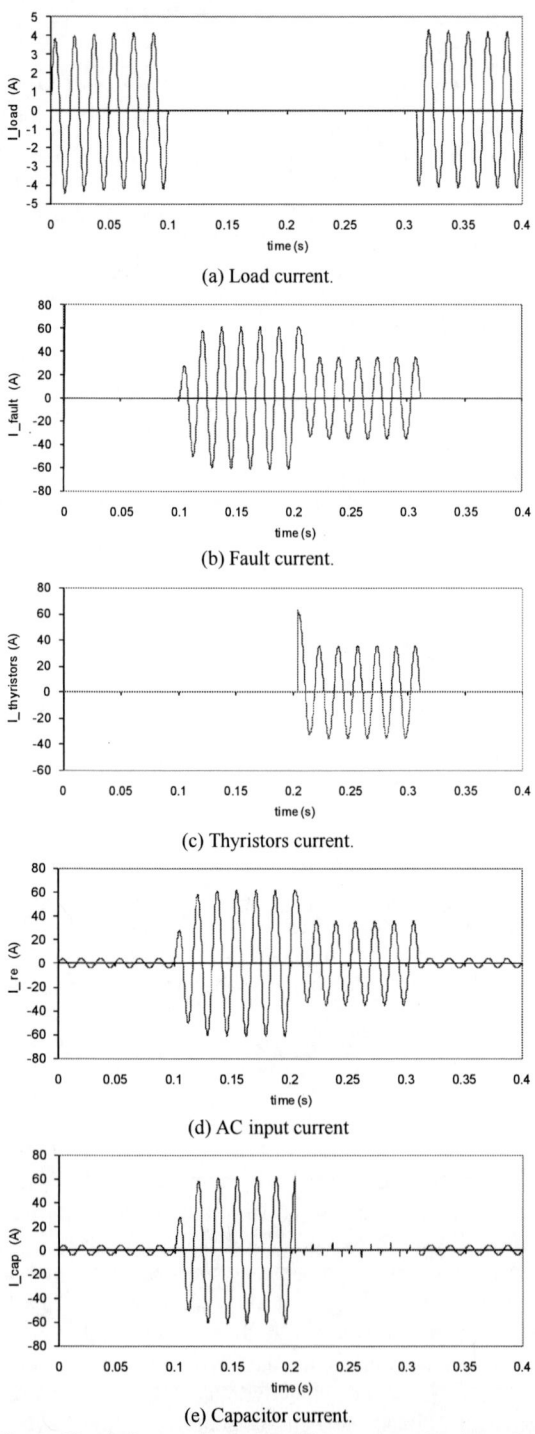

(a) Load current.

(b) Fault current.

(c) Thyristors current.

(d) AC input current

(e) Capacitor current.

Fig. 5. Simulation results for the TBC topology. Fault applied at t=0.1s, capacitor removed at t=0.2s, fault removed at t=0.3s.

IV. EXPERIMENTAL PROTOTYPE FOR FAULT CURRENT LIMITER CIRCUITS

The experimental prototype is presented in Fig. 6, and provides both FCL topologies (TIR and TBC) at the same power panel with adequate power and control connections, as shown in block diagrams of Figs. 7 and 8.

Fig. 6. Experimental prototype for current limiter circuits.

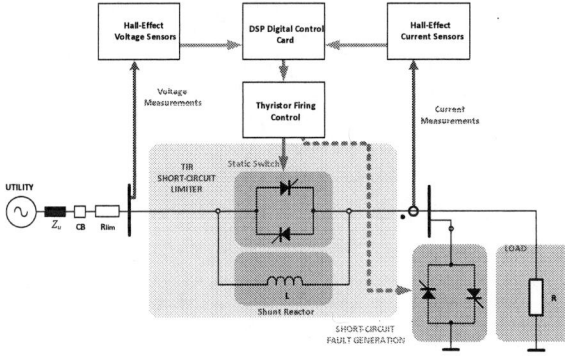

Fig. 7. Block diagram of the experimental topology of the Thyristor Inserted Reactor (TIR) fault current limiter.

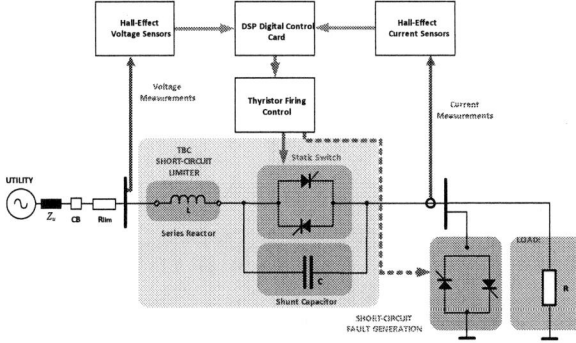

Fig. 8. Block diagram of the experimental topology of the Thyristor Bypassed Capacitor (TBC) fault current limiter.

The experimental setup is fed by three-phase (220V, 60Hz) mains, including neutral wire. It allows implementation of a required three-phase PLL (phase locked loop) algorithm [12] and prevents feeding the controller from the same FCL loaded circuit phase.

The short-circuit event is single-phase (line to neutral), generated through a static switch composed by two antiparallel thyristors (Short-Circuit Fault Generation on Figs. 7 and 8). It can be preset from 0° to 180° of the AC phase voltage and is limited by the controller to six AC cycles. The current value is limited by the series resistance R_{lim} (Figs. 7 and 8), to avoid undesirable tripping of the installation[2].

Air-core reactor L (to avoid saturation with the fault current) is used for both TIR and TBC topologies[3]. Capacitance C, only for the TBC topology, is provided by a bank of dry polypropylene capacitors[4]. Hall Effect voltage and current sensors provide galvanic isolation between power and control. The control system uses a Digital Signal Processor (DSP) Texas TMS28335, which acquires AC voltages and currents, runs fault limitation algorithms and synchronization with AC mains, and generates firing signals for the thyristors.

The short-circuit event is manually triggered and automatically detected by FCL limiting algorithm, which monitors AC load current[5] and acts according to the specific FCL circuit.

V. EXPERIMENTAL RESULTS FOR TIR AND TBC TOPOLOGIES

A. Results without fault current limiting devices

In order to establish short-current reduction effect of FCL devices, a short circuit, not limited by FCLs, is shown in Figs. 9 and 10.

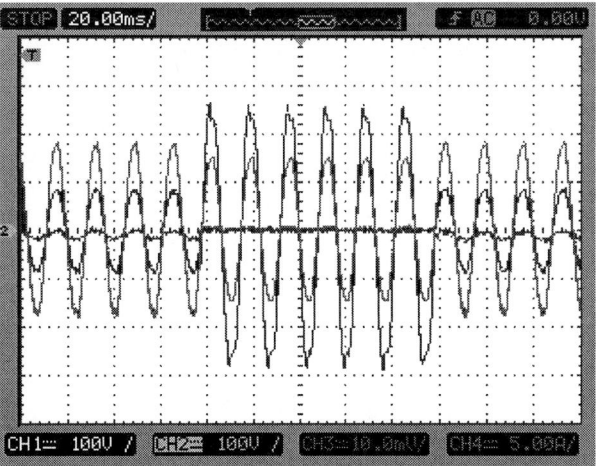

Fig. 9. Experimental results without FCL current limiting. Short-circuit instant at 0° of the AC phase voltage. Input voltage (green, 100V/div), load voltage (yellow, 100V/div), input current (blue, 50A/div), load current (magenta, 5A/div).

---

[2] The value of the current limiting resistor R_{lim}=954mΩ, was experimentally obtained, as it depends on the short-circuit current level of the AC mains, Notice that simulated value of R=1.295 Ω (Figs. 2 and 4) includes total AC mains resistance.

[3] Two series connected air-core inductors provide L=10.90mH with resistance R_L=1.626Ω. L is the "Shunt Reactor" in the TIR topology (Fig. 4) and "Series Reactor" in the TBC topology (Fig. 5).

[4] Total value of the capacitor bank is C=646µF. Capacitor bank is called "Shunt Capacitor" in the TBC topology (Fig. 5).

[5] Nominal load is R=40Ω, generating about I_{nom}=4.5A peak current. During a non-limited fault, peak current value can surpass 100A.

Figs. 9 and 10 clearly show the software determined short circuit duration of six AC cycles. Fig. 9 also shows the AC line regulation effect, as short-circuit provokes a line to neutral voltage value reduction (voltage sag).

Input peak current on Figs. 9 and 10 current exceeds 100A and is distorted. This distortion is due to current probe (Tektronix A6303) saturation, as this fault current is beyond the probe specification (100A).

The antiparallel thyristors of the Short-Circuit Fault Generation (Figs. 7 and 8) shunts load R, as shown in Fig. 10, were the load R current goes to zero.

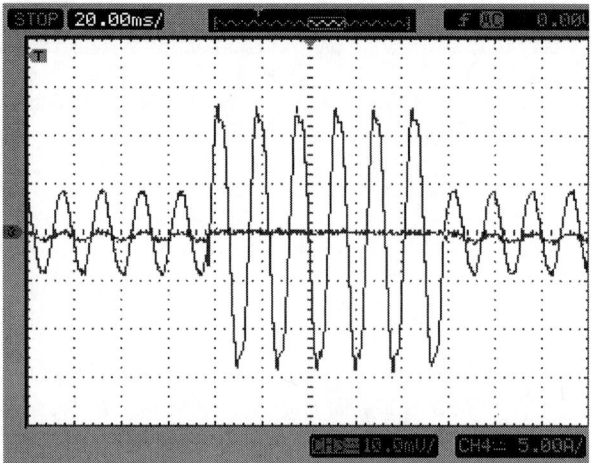

Fig. 10. Experimental results without FLC current limiting. Short-circuit instant at 0° of the AC phase voltage. Input current (blue, 50A/div), load current (magenta, 5A/div).

B. Thyristor Inserted Reactor (TIR)

In normal conditions, the load R of Fig. 7 is fed by the phase to neutral voltage, with nominal current, and the limiting reactor L is kept shunted by the Static Switch (Fig. 7). A program subroutine continuously calculates the squared value of the current entering the FCL circuit[6]. If this value surpasses a predetermined value, the Static Switch is turned off, and the reactor L limits short-circuit current. When measured input current decreases below a certain value[7], the control system recognizes the end of fault and the Static Switch is again turned on, shunting reactor L. The software determined short circuit duration of six AC cycles is independent from the current measuring and Static Switch triggering algorithms.

The current values for nominal load, non-limited fault and limited fault are predetermined by computer simulation and experimental measurements, taking into account sustainable currents[8] in the laboratory electrical circuit and per-unit (pu) scaling from real world high-voltage (HV) and medium-voltage (MV) power systems

---

[6] The analog to digital (AD) sampling rate is 24kHz (400 samples per cycle, at 60Hz). In each sampling, the newest instantaneous value of the input current is squared and added to a sampling window of 400 samples, the oldest sample is discarded, and the average value of the window is calculated. Therefore, a moving average measurement and calculation of the squared instantaneous current is performed.

[7] The non-fault current must be lower than the FLC limited current.

[8] Currents which do not generate laboratory circuit CB tripping.

studies[9]. Fig. 11 shows the fault current limitation performed by the TIR scheme.

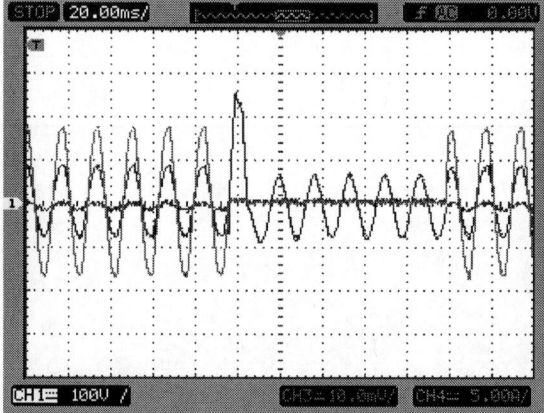

Fig. 11. Experimental results of the TIR topology. Short-circuit instant is at 0° of the phase to neutral AC voltage input. Load voltage (yellow, 100V/div), input current (blue, 50A/div), load current (magenta, 5A/div).

In Fig. 11, load R voltage (yellow) and current (magenta) are null during the fault duration. From the short-circuit instant, TIR entrance current (blue) increases until the control system detects fault current, and turns off the Static Switch, inserting reactor L into the circuit, and reducing short-circuit current to a level above nominal non-fault condition. After six AC cycles, short-circuit is automatically finished, current reduction is detected and Static Switch is turned on, shunting reactor L. Load voltage and current return to non-fault levels.

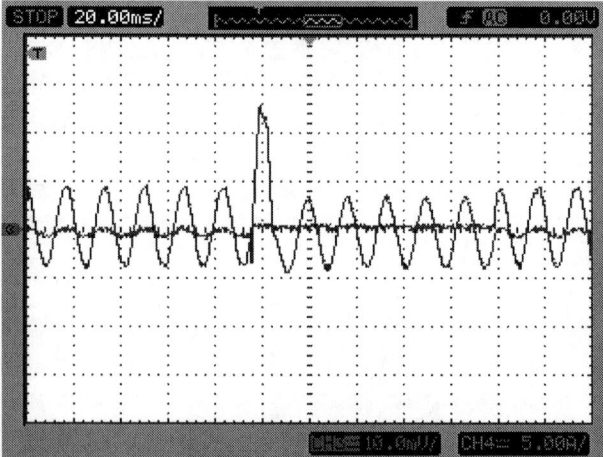

Fig. 12. Experimental results of the TIR topology. Short-circuit instant at 0° of the AC phase voltage. Input current (blue, 50A/div), load current (magenta, 5A/div).

Fig. 12 shows load and input currents, and the short-circuit limitation effect can be clearly observed, comparing it with Figs. 9 and 10.

---

[9] Briefly, L value for TIR topology can be derived from desired short-circuit current level reduction and C value for TBC topology is chosen to allow series resonance with L, keeping the same L value of the TIR topology.

C. Thyristor Bypassed Capacitor (TBC)

Operation in TBC mode is slightly different from TIR mode. Fault current measurement routine is the same as in TIR, but once it is detected, another routine waits for the next zero voltage value on the series capacitor C (Fig. 8) to turn-on the Static Switch, shunting capacitor C. Therefore overcurrent through TBC thyristors from capacitor discharge is minimized.

In order to provide thyristors turn-on at zero voltage on the series capacitor C, its voltage is sampled by the system (at a 24kHz rate). A software flag is raised when capacitor voltage is near zero, which is the condition where it is safer to turn-on thyristors to shunt the capacitor C. Notice that due to analog measurement, sampling rate and AD process, it is very difficult to detect the real zero-crossing of the capacitor voltage, so a voltage band around zero must be set for approximate detection of the zero-crossing region. Therefore, a near zero crossing switching is achieved. Consequently, transient capacitor C discharge current could appear through TBC's Static Switch thyristors. A different approach, using a PLL to track the capacitor C voltage in order to establish a more precise zero voltage crossing, is hard to implement because capacitor C instantaneous voltage is highly dependent on its current, and it is difficult to establish PLL tracking in this condition, within authors' previous experiences.

Short-circuit event is the same from the TIR case, six cycles duration of a phase to neutral short-circuit.

Some results corresponding to the TBC scheme are presented in Figs. 13 and 14, and again repeating some waveforms in the figures for clarification.

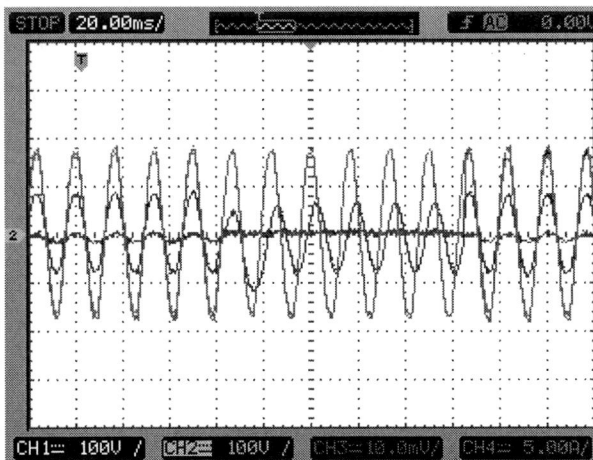

Fig. 13. Experimental results of the TBC topology. Short-circuit instant at 0° of the AC phase voltage. Input voltage (green, 100V/div), load voltage (yellow, 100V/div), input current (blue, 50A/div), load current (magenta, 5A/div).

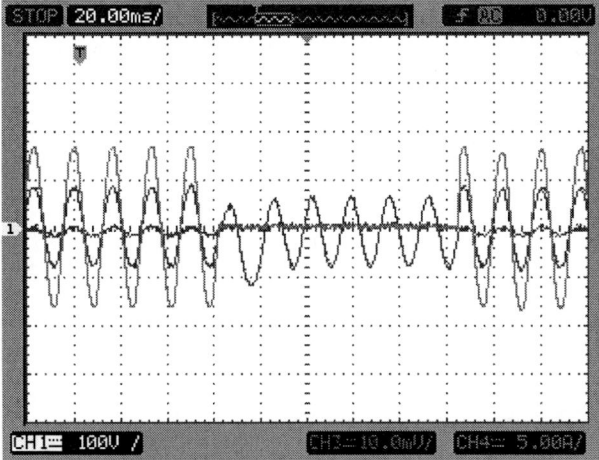

Fig. 14. Experimental results of the TBC topology. Short-circuit instant at 0° of the AC phase voltage. Load voltage (yellow, 100V/div), input current (blue, 50A/div), load current (magenta, 5A/div).

Figs. 13 and 14 shows the circuit behavior for the (phase to neutral) short-circuit at load R. Load R voltage (yellow) and current (magenta) are zero during the six cycles fault duration, as expected. From the short-circuit instant, FCL entrance current (blue) increases until the control system detects overcurrent, and turns on the TBC Static Switch, bypassing capacitor C and putting only the reactor L in series with the short-circuit, reducing short-circuit current to a level above nominal non-fault condition. After six cycles the short-circuit is automatically finished, the system detects current reduction below fault level and turns off TBC's Static Switch, putting back shunt capacitor C. Load voltage and current return to nominal levels.

More results corresponding to the TBC scheme are presented in Figs. 15 and 16.

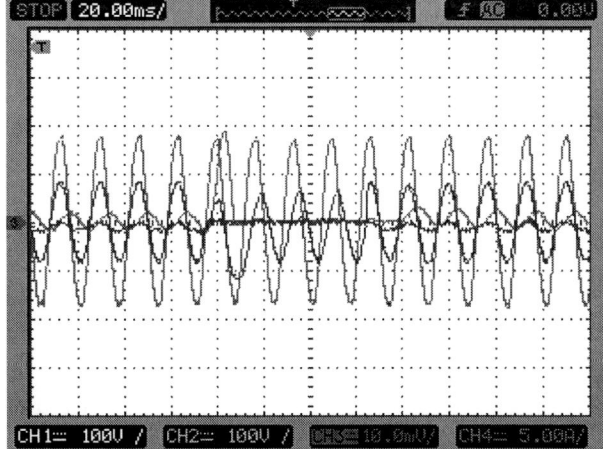

Fig. 15. Experimental results of the TBC topology. Short-circuit instant at 0° of the AC phase voltage. TBC input voltage (green, 100V/div), TBC capacitor voltage (yellow, 100V/div), input current (blue, 50A/div), load current (magenta, 5A/div).

Fig. 15 shows voltage on TBC capacitor C (yellow). Usually, voltage drop on TBC capacitor is relatively low with nominal current, but current increase due to short-circuit causes immediate capacitor voltage increase, as shown. The TBC's control algorithm must wait until next zero voltage on capacitor C to fire the Static Switch thyristors and short-circuit it. After shunting capacitor C, only the series reactor L remains, effectively reducing short-circuit current value.

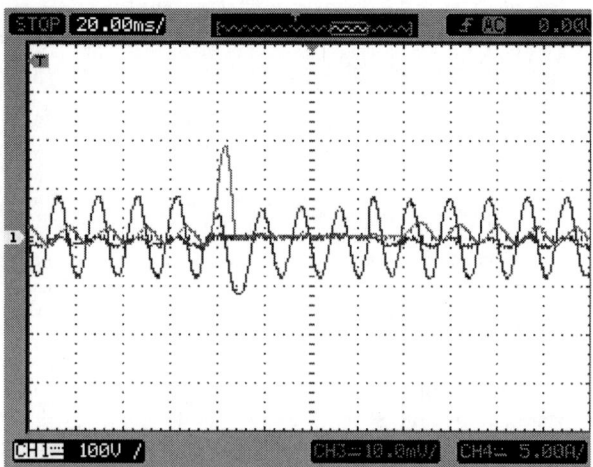

Fig. 16. Experimental results of the TBC topology. Short-circuit instant at 0° of the AC phase voltage. TBC capacitor voltage (yellow, 100V/div), input current (blue, 50A/div), load current (magenta, 5A/div).

Fig. 16 shows capacitor voltage behavior, as well as FCL input current (blue) and load current (magenta) during the short-circuit event. As in the TIR topology case, the provoked short-circuit event shunts the load, and its current is reduced to zero during the event.

One can notice that TBC topology results, in this particular setup, in lower short-circuit peak currents before FCL action, compared with TIR topology. The presence of the series inductor in the TBC in non-fault conditions can explain this behavior, as series inductor L has non-negligible ohmic resistance (R_L=1.626Ω). Also, perfect series resonance between L and C in the TBC topology is not achieved due to components values variation. Notice also that in the TIR topology for the non-fault condition there is only the Static Switch, with very low impedance, in series with the load.

VI. CONCLUSIONS

The use of power electronics based FCLs can be efficient and cost-effective ways to reduce short-circuit current levels. The use of FCLs can delay significantly, or even avoid, the complete substitution of CBs when short-circuit currents increase due to power system growth.

Authors' studies on FCL topologies lead to the need for the implementation of a low power experimental setup for FCL topologies, which allows modifications on control software for system behavior verification, as well as helping to validate computer simulation results.

The simulated and experimental setups results show the effective reduction of short-circuit current levels. The

analysis of these experimental results allows enlightening on the understanding of the FCL's inner behavior.

ACKNOWLEDGMENT

The authors are grateful to FAPESP (Fundação de Amparo à Pesquisa do Estado de São Paulo) for financial support.

REFERENCES

[1] ONS Report, "Short Circuit Level Evaluation (Period 2008-2011)," Vol. 1, Brazilian National System Operator (ONS), 2009 (in Portuguese). Available: http://www.ons.org.br.

[2] CIGRE Working Group 10. Study Committee A3 (High Voltage Equipment), "Fault Current Limiters in Electrical Medium and High Voltage Systems", 2003.

[3] V. Gor, D. Povh, Lu Yichuan, E. Lerch, D. Retzmann, K. Sadek and G. Thumm. SCCL – A New Type of FACTS Based Short-Circuit Current Limiter for Application in High-Voltage Systems, CIGRE, 2004.

[4] G.G. Karady. Concept of a Combined Short Circuit Limiter and Series Compensator, *IEEE Transactions on Power Delivery*, vol. 6, no. 3, pp. 1031-1037, 1991.

[5] A. Abramovitz and K. M. Smedley. Survey of Solid-State Fault Current Limiters, *IEEE Transactions on Power Electronics*, vol. 27, no. 6, pp. 2770-2782, 2012.

[6] A.M. Monteiro. A Study on the Short-circuit Current Limiting Devices with Focus on the IPC (Interphase Power Controller), M.Sc. Diss., Federal Univ. of Rio de Janeiro (in Portuguese), 2005.

[7] G. Chen, D. Jiang, Z. Lu, and Z. Wu. A New Proposal for Solid State Fault Current Limiter and Its Control Strategies, in Proc. IEEE Power Engineering Society General Meeting, vol. 2, pp. 1468–1473, 2004.

[8] S. Sugimoto, J. Kida, H. Arita, C. Fukui, C. and T. Yamagiwa. Principle and Characteristics of a Fault Current Limiter with Series Compensation, *IEEE Transactions on Power Delivery*, vol. 11, no. 2, pp. 842-847, 1996.

[9] M. M. Lanes, H. A. C. Braga and P. G. Barbosa. Resonant Circuit-Based Short Circuit Current Limiter Controlled by Power Electronics Devices. *IEEE Latin America Transactions*, vol. 5, no. 5, pp. 311-320, 2007.

[10] K. Renz, G. Thumm, and St. WeiB. Thyristor Control for Fault Current Limitation, in IEE Colloquium on Fault Current Limiters, London, 1995.

[11] R.L. Vasquez-Arnez, L.C. Zanetta Jr., and F.A. Moreira Effective Limitation of Line Fault Currents by Means of the Series-Connected VSC-Based FACTS Devices. *Control & Automation Magazine*, vol. 17, no. 4. pp. 459-466. 2006. Available: http://www.scielo.br/pdf/ca/v17n4/a07v17n4.pdf.

[12] T.M. Terrazas, F.P. Marafão, T.C. Monteiro, A.R. Giaretta, L. Matakas Jr. and W. Komatsu. "Reference Generator for Voltage Controlled Power Conditioners," in Proc. of COBEP 2011, vol. 01, pp. 513-519, 2011.

DC-DC Boost Converter Based MSHE-PWM Cascaded Multilevel Inverter Control for STATCOM Systems

Kah Haw Law
Dept. of Electrical and Electronic Engineering
The University of Nottingham Malaysia Campus (UNMC)
Semenyih, Selangor, MALAYSIA
keyx1lkw@nottingham.edu.my

Mohamed S. A. Dahidah
School of Electrical and Electronic Engineering
Newcastle University, UK
mohamed.dahidah@newcastle.ac.uk

Abstract— **This paper investigates the performance of Static Synchrounous Compensator (STATCOM) with the new variation of multilevel selective harmonic elimination pulse width modulation (MSHE-PWM). The proposed MSHE-PWM optimizes both the dc-voltage levels and the switching angles, enabling more harmonics to be eliminated without affecting the structure of the inverter circuit. The method also provides constant switching angles and linear pattern of dc voltage levels over the modulation index range. This in turns eliminates the required tedious steps for manipulating the off-line calculated switching angles and therefore, easing the implementation of the MSHE-PWM for dynamic systems. The required variable dc voltage levels are obtained from simple dc-dc boost converter. Current and voltage closed loop controllers are implemented for both the STATCOM and the boost converter to meet the reactive power (VAR) demand for different loading conditions. The proposed approach is validated through both simulation and experimental results.**

Keywords— Cascaded multilevel inverter , boost dc-dc converter, MSHE-PWM, STATCOM; reactive power compensation

I. INTRODUCTION

The large penetration of renewable resources and large growth of modern nonlinear loads have introduced new challenges to existing power networks [1]. Nevertheless, advanced power semiconductor devices has opened opportunities for improving the operation and management of power system networks via flexible AC transmission system (FACTS) controllers such as static synchronous compensator (STATCOM) to enhance neighboring utilities and regions for more economical and reliable exchange of power [2]-[9].

STATCOM has been an effective solution for reactive power (VAR) compensation using static converters and the multilevel cascaded H-bridge inverters (MCHIs) have been an attractive topology for STATCOM systems due to their modularity, extensibility, and control simplicity [7]-[10]. Various modulation techniques have been proposed in the literature including carrier-based pulse width modulation (CB-PWM) [11] and [12], multilevel space vector modulation (SVM) [13] and [14], and multilevel selective harmonic elimination PWM (MSHE-PWM) [15]-[24]. Despite its simplicity, the CB-PWM technique does not offer any direct manipulation over the

harmonic contents and also exhibits high switching losses; making its employment in high-power applications is a big concern where high losses are intolerable.

MSHE-PWM offer of a drastic low equivalent switching frequency compared to CB-PWM technique, leads to low switching power loss and good harmonic performance has made it a competitive solution for medium- and high-power conversion systems such as STATCOM [7]-[9]. The method was initially studied for two-level voltage source converters and then extended to various multilevel converters. For instance, a generalized formulation for SHE-PWM control suitable for high-power CHI converters with both equal and non-equal dc sources is discussed in [15]. Another generalized formulation of quarter-wave symmetrical SHE problems is presented in [16]. Other approaches have also been reported where the SHE-PWM waveforms defined based on the well-known carrier-based PWM were proposed in [17] and [18]. Particle swarm optimization (PSO) technique was also employed to provide solutions to the multilevel SHE problem where a minimization of the total harmonic distortion (THD) was the criteria for both equal [19] and non-equal dc sources [20]. The symmetry requirements of the waveform were later alleviated to produce a larger solution pool [22]-[24].

On the other hand, the choice of STATCOM control strategies has a crucial influence on its compensation objective as well as performance requirements. Various control methods have been developed and reported in the literature which can be broadly classified into frequency or time domain approaches. However, most recent algorithms development adopts time-domain approach due to its ability to compute the reference currents and precise trace of the load changes. The well-known *abc*-to-*dq* transformation or else known as synchronous rotating reference frame (SRF) which incorporates decoupling feed-forward/feedback systems with appropriate control algorithms is commonly used to achieve lower steady state errors and acceptable transient characteristics [10]. An integrated control which incorporates both the voltage and current controls to compensate the unbalance problem caused by the load and the voltage source was proposed in [5]. Another based on Modified Selective Swapping (MSS) algorithm was presented in [7] and [8] to provide reactive power compensation. The method was able to reduce the dc-link voltage ripple but at the expenses of higher switching frequency. Reference [6] reports an interesting comparison between three types of balance control strategy, namely, the proposed active voltage vector superposition, modulation index regulation, and phase shift angle regulation. It was concluded that the proposed method offers good control performance with strong regulation capability followed by the phase shift angle and modulation index regulation methods.

This work exploits the benefits of the newly developed MSHE-PWM control technique proposed by the authors in [21] to STATCOM system based on MCHI. The paper also extends the previous work [3] by employing dc-dc boost converter to provide variable dc voltage levels to each H-bridge cell of the MCHI. Although only five-level inverter is considred in this paper, however the technique is applicable with any number of levels. The STATCOM system is studied with different loading conditions to show the effectiveness of the proposed control method.

The paper is organized as follows: Section II presents the analysis and the formulation of the proposed MSHE-PWM method applied to five-level CHI. The proposed STATCOM system along with its control is presented in Section III. Simulation and experimentally validated results are reported in Section IV. Finally, conclusions are summarized in Section V.

II. FIVE-LEVEL SHE-PWM WITH VARIABLE DC VOLTAGE LEVELS

Fig. 1 shows the proposed three-phase five-level voltage source converter and its associated MSHE-PWM output voltage waveform. Let N_1 be the number of switching angles between zero and v_{DC1}, which must be an odd number and N_2 be the number of switching angles between v_{DC1} and v_{DC2}, which can be odd or even number. Furthermore, the dc voltage levels are also variables and need to be optimized increasing the number of harmonics that can be controlled and/or eliminated [21]. The generalized set of equations for any number of switching angles that describe the fundamental component and the higher order harmonics for a symmetrically defined five-level SHE-PWM waveform is given as follows.

$$\left(v_{DC_1} \sum_{i=1}^{N_1} (-1)^{i+1} \cos(\alpha_i) + v_{DC_2} \sum_{i=N_1+1}^{N_1+N_2} (-1)^i \cos(\alpha_i) \right) = \frac{m_i * \pi}{4}$$

$$\left(v_{DC_1} \sum_{i=1}^{N_1} (-1)^{i+1} \cos(5\alpha_i) + v_{DC_2} \sum_{i=N_1+1}^{N_1+N_2} (-1)^i \cos(5\alpha_i) \right) = 0$$

$$\vdots$$

$$\left(v_{DC_1} \sum_{i=1}^{N_1} (-1)^{i+1} \cos(n\alpha_i) + v_{DC_2} \sum_{i=N_1+1}^{N_1+N_2} (-1)^i \cos(n\alpha_i) \right) = 0 \quad (1)$$

where m_i is the modulation index.

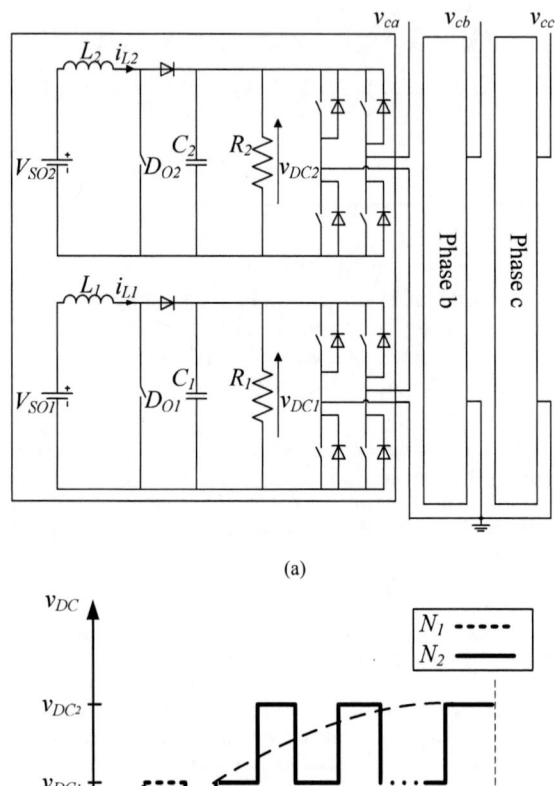

(a)

(b)

Fig. 1. (a) Three-phase five-level cascaded inverter with boost converter. (b) Five-level SHE-PWM waveform (quarter-cycle shown).

The optimal switching angles and the dc voltage levels are obtained by solving (1) when it is subject to the constraints of equations (2) and (3).

$$\left(0 < \alpha_1 < \alpha_2 < ... < \alpha_{N_1+N_2} < \frac{\pi}{2} \right) \quad (2)$$

$$v_{min} \leq v_{DC_M} \leq v_{max} \quad (3)$$

Multiple sets of solutions for various waveforms were acquired, however, only two cases (i.e. $N_1/N_2 = 3/5$ and $N_1/N_2 = 3/8$) are reported in this paper and a solution set for each is presented in Fig. 2(a) and Fig. 2(b), respectively, while more solutions for the same multilevel waveforms can be found in [21].

The 2014 International Power Electronics Conference

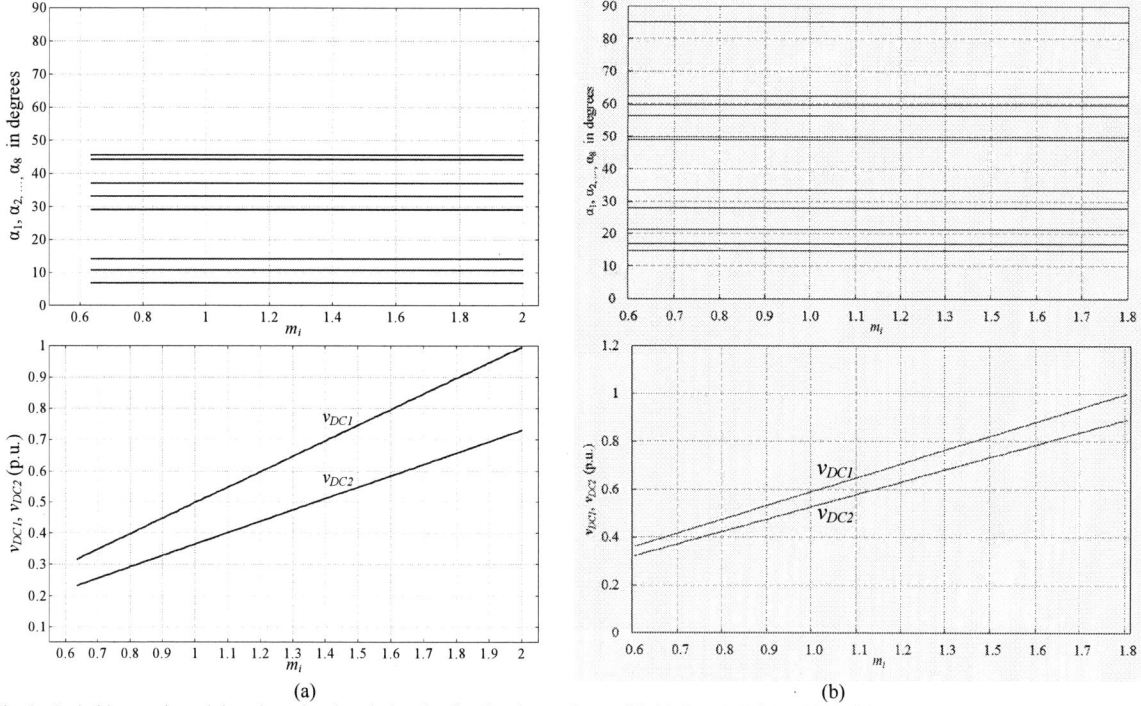

(a)   (b)

Fig. 2. Switching angle and dc voltage levels solution for five-level waveform with (a) Case I: $N_1/N_2 = 3/5$ and $0.64 < m_i < 2.0$), (b) Case II: $N_1/N_2 = 3/8$ and $0.63 < m_i < 1.75$.

III. THE PROPOSED STATCOM: KEY OPERATION

Fig. 3 shows the proposed STATCOM system connected to a three-phase 415 V_{rms} power system feeding an RL load at different loading conditions (i.e. power factors). The control algorithm and current controllers of the STATCOM developed in [3] have been employed to this work, where the resultant dq-voltage reference values are illustrated as follows:

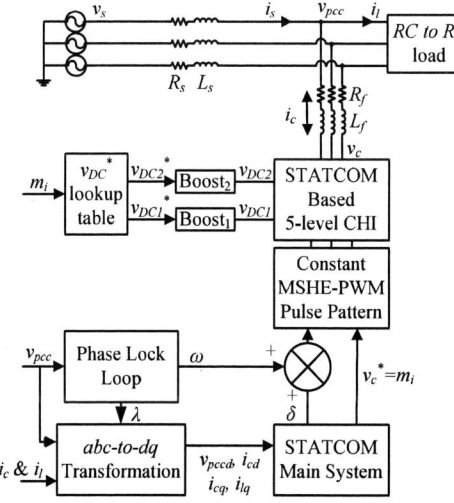

Fig. 3. Block diagram of the STATCOM system with associated proposed control scheme.

$$v_{cd}^*(k) = v_{pccd}(k) - R_f i_{cd}(k) + \omega L_f \left[\frac{1}{2}i_{cq}(k) + \frac{1}{2}i_{lq}(k)\right]$$
$$-K_{p\_id}\left[-i_{cd}(k)\right] \tag{4}$$

$$v_{cq}^*(k) = -R_f i_{cq}(\text{k}) - \omega L_f \left[\frac{1}{2}i_{cd}(k)\right]$$
$$-K_{p\_iq}\left[i_{lq}(k) - i_{cq}(k)\right] \tag{5}$$

where v_{pccd} defines the d-axis grid voltage, R_f and L_f, are the coupling resistance and inductance, respectively, ω defines the fundamental grid voltage frequency, i_{cd} and i_{cq}, are the STATCOM active and reactive current, respectively, i_{cd} is the STATCOM current, and i_{lq} is the reference reactive current extracted from the load side.

From (4) and (5), the respective proportional gain $K_{p\_i(d,q)}$ of the P-controller is given by:

$$K_{p\_i(d,q)} = \frac{L_f}{T_{i(d,q)}} + \frac{R_f}{2} \tag{6}$$

where T_i defines the selected controller operating rate.

The desired STATCOM output voltage magnitude v_c^* and its phase angle δ with respect to v_{pcc} is given as follows:

$$v_c^* = \sqrt{(v_{cd}^*)^2 + (v_{cq}^*)^2} \tag{7}$$

$$\delta = tan^{-1}\left(\frac{v_{cq}^*}{v_{cd}^*}\right) \tag{8}$$

The proposed method relies on the availability of the variable dc voltage levels which can be easily obtained by advanced dc-dc converters. With the rapid growth in semiconductor devices industry and advanced materials such as nanocrystalline soft magnetic core that offers high saturation flux density (more than 1.2 T) and high relative permeability (over 10 000 at 100 kHz), leading to an extremely low core loss. The combination of the magnetic core with the latest trench-gate IGBTs and super-junction MOSFETs has made it possible to improve the system efficiency of the dc-dc converters up to 97% or higher [27]. In the near future, the emergence of silicon carbide (SiC) switching devices and a new magnetic core material will allow the system efficiency to reach higher than 99% [28]. This revolution leads to produce relatively medium- to high-voltage dc-dc converters with switching frequencies as high as 50 kHz [29] which have been proposed for different grid-connected converter applications [27]-[29].

Compared with dc-dc buck converters [3], the boost type possess high efficient operation with reduced current ratings and increased the bandwidth of the converter resulting in smaller size of LC output filter and simpler control design; therefore they have been chosen to provide variable dc voltage levels for each H-bridge inverters in accordance to the resultant modulation index m_i. The voltage loop controller of the dc-dc converter is designed in continuous conduction mode (CCM) using the state space averaging modeling technique with Venable approach [31]. From Fig. 1, by considering the small deviation and the steady state operating point of the boost converter, the linear, time-invariant state space equation over the switching period t_{sw} based on two modes of operation (i.e. when D_O on and off) is defined by:

$$\frac{di_L}{dt} = -v_{DC}\left(\frac{1-D_o}{L}\right) + V_{DC}\left(\frac{d}{L}\right) + v_{SO}\left(\frac{1}{L}\right) \quad (9)$$

$$\frac{dv_{DC}}{dt} = i_L\left(\frac{1-D_o}{C}\right) - v_{DC}\left(\frac{1}{RC}\right) - I_L\left(\frac{d}{C}\right) \quad (10)$$

where variables with the uppercase (I_L, D_o, V_{DC}, and V_{SO}) define the steady state operating point while those with the lowercase (i_L, d, and v_{DC}) define the small perturbations or deviation from the steady state operating point.

By taking Laplace transform, a linear DC output voltage $v_{DC}(s)$ in (11) is formed by equating (9) and (10) as follows:

$$v_{DC}(s) = \left(\frac{v_{SO}(s)}{V_{SO}(s)^2(1-D_O)} + \frac{d(s)}{(1-D_O)^2}\left(1 - \frac{sL}{(1-D_O)^2 R}\right)\right)$$

$$* \left(\frac{V_{SO}(s)}{\left(\frac{s^2 LC}{(1-D_O)^2} + \frac{sL}{(1-D_O)^2 R} + 1\right)}\right) \quad (11)$$

where the first bracket defines the boost converter's input voltage deviation at the operating point and the second bracket defines the boost converter's transfer function $G_{conv}(s)$ (i.e., Fig. 4) for designing the voltage closed loop controller $G_v(s)$.

From (11), the right-half plane zero occurs is related to the effective value of the filter inductance L and the load resistance R. To mitigate the instability effect caused by the right-half plane zero, higher operating switching frequency f_{sw} must be selected to reduce the size of filter inductance L and hence, optimize the voltage loop performance with a proper selection of the crossover frequency for better transient response.

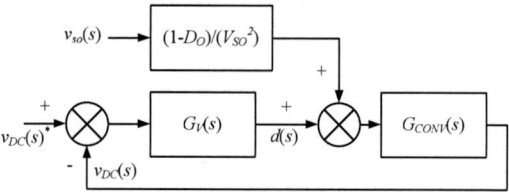

Fig. 4. Block diagram of the boost converter with voltage loop feedback control.

As for P, PI, and PID controllers, the $G_v(s)$ shown in Fig. 4 is basically an error amplifier for the voltage mode closed-loop controller to generate the duty cycles demand d for the switching device of the boost converter. From [31], the transfer function of $G_v(s)$ is defined as:

$$G_v(s) = \frac{A}{s}\left(\frac{1 + \dfrac{s}{\omega_z}}{1 + \dfrac{s}{\omega_p}}\right) \quad (12)$$

where A is the gain of the error amplifier, ω_z (i.e. ω_{cross}/K) and ω_p (i.e. $\omega_{cross}*K$), respectively, are the zero frequency and pole frequency based on the selected crossover frequency and K factor value [31].

IV. RESULTS AND DISCUSSION

The proposed STATCOM is first investigated through various simulation examples using Matlab/Simulink software package [32] to investigate the dynamic, transient, and harmonics performances of the proposed MSHE-PWM method. A small scale low-voltage laboratory prototype based on five-level CHI is also developed and experiments are conducted to validate the simulation and theoretical findings.

A. Simulation Results

The simulation study of the proposed STATCOM system is carried out with the system parameters listed in Table I. The MSHE-PWM waveforms for both cases (i.e., 3/5 and 3/8) and the associated spectra are presented in Fig. 5(a) and Fig. 5(b), respectively, where it is clear that all the intended low-order non-triplen harmonics are eliminated and the next harmonic appears in the spectrum would be the 31st and 41st, respectively. Furthermore, triplen harmonics will be absent between line-to-line voltage.

It is worth nothing that with the distribution ratios of 3/5 and 3/8, the effective switching frequency of the inverter will be 1.6 kHz and 2.2 kHz, respectively. Therefore, when compared to the CB-PWM based STATCOM with a typical switching frequency ranges

978-1-4799-2706-7/14 $31.00 © 2014 IEEE

between 5 kHz and 10 kHz [25] and [26], the proposed method offers less switching losses as well as smaller cooling and power dissipation requirement, which considerably reduce the volume and weight of the overall system and hence increase its reliability and performance.

TABLE I
SYSTEM PARAMETERS USED IN COMPUTER SIMULATION

Boost converter steady state parameters:
dc input voltage source: $V_{so} = 50$ V Switching frequency: $f_{sw} = 25$ kHz Filter inductance: $L = 6.5$ µH Filter capacitance: $C = 0.3$ mF Load resistance: $R = 10$ Ω Crossover frequency: $\omega_{cross} = 114000$ rad/s Gain of the error amplifier: $A = 3.32$ Zero frequency: $\omega_z = 2221$ rad/s Pole frequency: $\omega_p = 58516$ rad/s
Single-phase power system parameters: $S_{base} = 1.44$ kVA, $V_{base} = 240$ V$_{rms}$, $I_{base} = 6$ A$_{rms}$, $Z_{base} = 40$ Ω
Single phase power rating = 1 per unit (p.u) Single phase source voltage: $v_s = 1$ p.u Source resistance: $R_s = 0.398$ Ω $= 0.00995$ p.u Source inductance: $L_s = 12.669$ mH $= 0.00025$ p.u Coupling resistance: $R_f = 3.98$ Ω $= 0.0995$ p.u Coupling inductance: $L_f = 126.69$ mH $= 0.0025$ p.u RC load A: $P = 1008$ W $= 0.7$ p.u, $Q = 1008$ VAR $= 0.7$ p.u RL load B: $P = 1008$ W $= 0.7$ p.u, $Q = 1008$ VAR $= 0.7$ p.u d-axis Proportional gain: $K_{pid} = 0.21$ q-axis Proportional gain: $K_{pid} = 1.63$

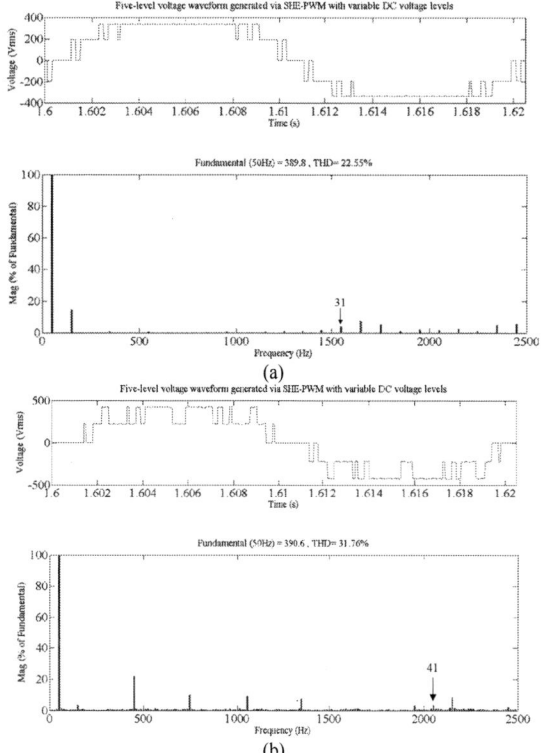

Fig. 5. Simulation results of (a) Case I and (b) Case II of single-phase (i.e. line-to-neutral) five-level MSHE-PWM output voltage waveform and its spectrum for load B.

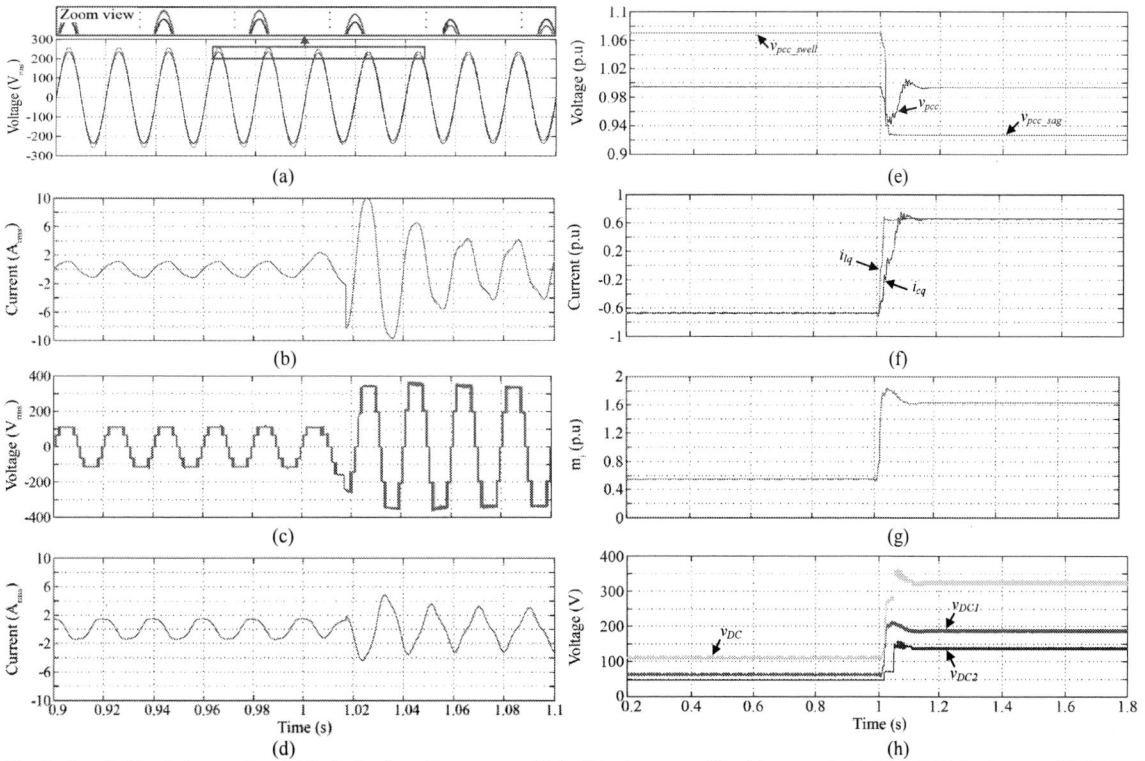

Fig. 6. Case I: Simulation results of AC single-phase line-to-neutral (a) grid voltage v_{pcc}, (b) grid current i_s, (c) STATCOM voltage v_c, (d) STATCOM current i_c, and dc quantities of (e) grid voltage magnitude v_{pcc}, (f) ramp change of measured i_{cq} and reactive current reference i_{lq}, (g) modulation indexes m_i, and (h) boost converters output voltages v_{DC1} and v_{DC2} at different loading conditions.

The dynamic and transient stability responses of the proposed STATCOM system (i.e., for Case I) along with the associated dc-dc boost converters are presented in Fig. 6 under different loading conditions. Specifically, Figs. 6(a)-6(d) show the line-to-neutral voltages and currents responses for the changes from capacitive (i.e., load A) to inductive (i.e., load B) characteristics. Fig. 6(e) illustrates the grid voltage swell and sag by about 7% from the nominal voltage value) when the load changed from load A to load B, respectively at the time of 1 second. These voltage swell and sag are then restored back to a unity by the proposed STATCOM system via reactive current compensation. Fig. 6(f) shows the dynamic and transient responses of the STATCOM reactive current i_{cq} which is consistently tracking the ramp-change of the load reactive current i_{lq}. Similar observation can be also seen from Fig. 6(h) where the dynamic and steady state characteristics of each dc-dc boost converter's dc output voltage (i.e., v_{DC1} and v_{DC2}) in respond to the resultant modulation index m_i of Fig. 6(g) provided by the STATCOM's dq vector current controller is demonstrated.

The STATCOM's performance under Case II is also investigated and as expected a better dynamic response is observed as shown in Fig. 7. However this was at the cost of higher switching frequency (i.e., 2.2 kHz compared to 1.6 kHz), which increases the switching losses. Nevertheless, this is comparably still less than the typical frequency of CB-PWM.

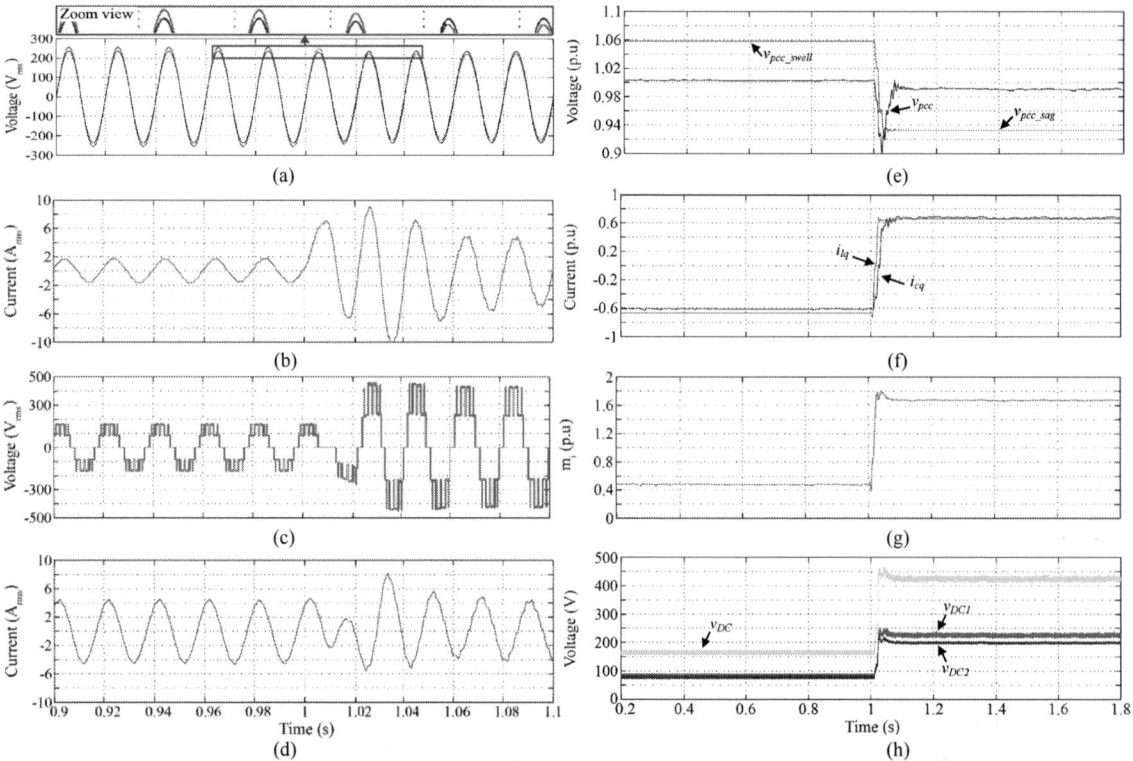

Fig. 7. Case II: Simulation results of AC single-phase line-to-neutral (a) grid voltage v_{pcc}, (b) grid current i_s, (c) STATCOM voltage v_c, (d) STATCOM current i_c, and dc quantities of (e) grid voltage magnitude v_{pcc}, (f) ramp change of measured i_{cq} and reactive current reference i_{lq}, (g) modulation indexes m_i, and (h) boost converters output voltages v_{DC1} and v_{DC2} at different loading conditions.

B. Experiment Results

A small scale laboratory prototype based on a single-phase (one-leg) STATCOM system is developed as shown in Fig. 8 and tested to validate the theoretical and the simulation findings. A five-level CHI forms the heart of the proposed STATCOM system which is studied with the system parameters of Table II.

TABLE II
SYSTEM PARAMETERS USED IN THE EXPERIMENTAL WORK

Single-phase power system parameters:
S_{base} = 90 VA, V_{base} = 60 V$_{rms}$, I_{base} = 1.5 A$_{rms}$, Z_{base} = 40 Ω
Single phase power rating = 1 p.u
Single phase source voltage: v_s = 1 p.u
Coupling inductance: L_f = 126.69 mH = 0.0025 p.u
R load A: P = 63 W = 0.7 p.u
RL load B: P = 63 W = 0.7 p.u, Q = 64 VAR = 0.7 p.u

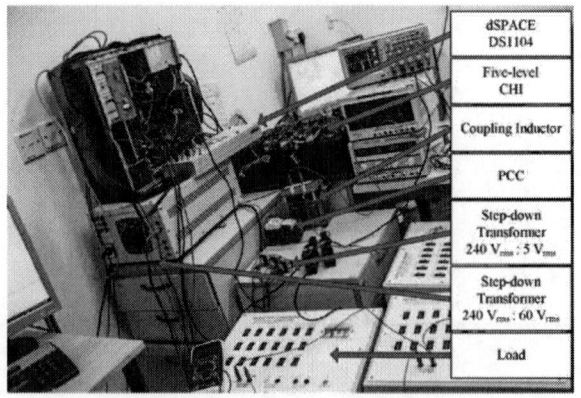

Fig. 8. Experimental set up of the single-phase (one-leg) STATCOM system based on the five-level CHI.

978-1-4799-2706-7/14 $31.00 © 2014 IEEE

A dSPACE DS1104 processor board with CP1104 and CLP1104 connector panels is used as the main control platform. It enables the link between analogue and digital interface by introducing the dSPACE DS1104 input-output (I/O) interface blocks into the Matlab/Simulink model. With the Matlab/Simulink real-time workshop function, the Matlab/Simulink model with the dSPACE interface blocks are automatically converted into C-code, compiled, and then linked to the real-time dSPACE DS1104 processor board for digital signal processing. The application of the dSPACE graphical user interface (GUI) software allows the user to alter and monitor the model behavior and the control parameters in real time.

From Fig. 8, a step-down transformer from 240 V_{rms} to 5 V_{rms} is used to scale down the grid voltage v_{pcc} to a safe and compatible voltage level before feeding it to the I/O connectors of the panel. Another step-down transformer 240 V_{rms} to 60 V_{rms} is utilized to step-down the overall experiment voltage to meet the security and safety codes imposed by the laboratory. Two current clamp-on probes (Tektronix A622) are used to measure both the load current i_l and the STATCOM output current i_c. A digital real-time oscilloscope (Tektronix TDS2004C) with four channels is used to capture and display the selected output waveforms simultaneously. Multiple plotter instruments from the dSPACE controldesk's element library under "Data Acquisition" group are utilized to monitor the behavior of the whole STATCOM control variables.

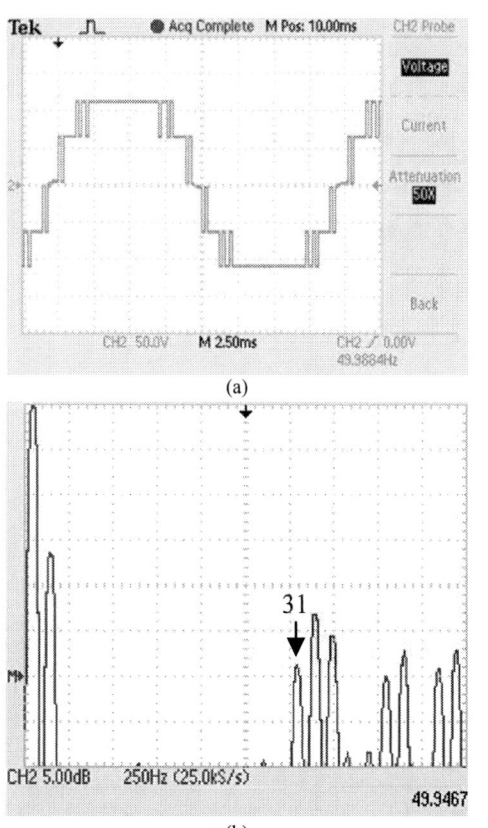

(a)

(b)

Fig. 9. Experimental implementation of line-to-neutral MSHE-PWM (a) five-level output voltage waveform and (b) the associated spectrum.

The real-time implementation of line-to-neutral five-level SHE-PWM output voltage waveform (i.e., for Case I) and the associated spectra are shown in Fig. 9, which is in a good match with the simulation results presented in Fig. 5(a).

Once again, it is confirmed that the proposed MSHE-PWM method offers a tight control of the low-order non-triplen harmonics for the full range of the inverter operation.

Fig. 10 presents the experimental waveforms of the grid voltage v_{pcc} (i.e., channel 1), the five-level STATCOM output voltage v_c (i.e., channel 2), the STATCOM output current i_c (i.e., channel 3), and the reactive current references i_{lq} (i.e., channel 4) for both capacitive and inductive modes.

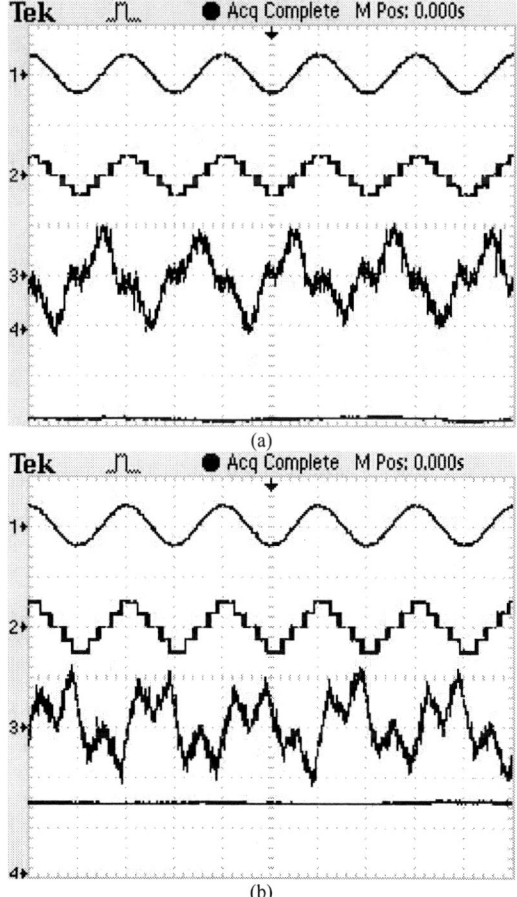

(a)

(b)

Fig. 10. Experimental results of a STATCOM in (a) inductive mode and (b) capacitive mode operation.

From channel 3 of Fig. 10(a) and Fig. 10(b), the phase distortion in STATCOM current is caused by the 3rd harmonic component which can be observed from Fig. 5(a) and Fig. 9 [33].

V. CONCLUSION

This paper present the performance of STATCOM system based on a new variation of multilevel SHE-PWM control technique by optimizing the dc source levels. The proposed formulation of the MSHE-PWM provides an output with a wider range of modulation index by making the dc voltage levels variant without affecting the number of harmonics being eliminated. It further provides constant switching angles and linear pattern of dc voltage levels over the full range of the modulation index, which in turns eliminates the tedious steps of the off-line calculations of switching angles and eases the implementation of the MSHE-PWM for dynamic systems such as STATCOM. Variable dc voltage sources were obtained through dc-dc boost converters to supply the required adjustable dc voltage levels for the H-bridge cells of the cascaded inverter. The performance of the system verified with different loading conditions where a good response was acheived and the prilimary experimental results is presented in this digest. Although only boost converter was considered in this paper, however, other topologies can be equally applied and this will be investigated in future works for potential maximization of the system efficiency and performance.

REFERENCES

[1] J. Dixon, L. Moran, J. Rodriguez, and R. Domke, "Reactive power compensation technologies: state-of-the-art review," in *Proc. IEEE*, vol. 93, no. 12, pp. 2144-2164, Dec. 2005.

[2] L.K. Haw, M.S.A. Dahidah, and N. Marium, "Cascaded multilevel inverter based STATCOM with power factor correction feature," IEEE Conf. on Sustainable Utilization and Development in Engineering and Technology, pp. 1-7, 2011.

[3] L. K. Haw, M. S. A. Dahidah, and H. A. F. Almurib, "SHE-PWM cascaded multilevel inverter with adjustable DC voltage levels control for STATCOM applications," *IEEE Trans. Power Electron.*, vol. 29, **Early Access.**

[4] Q. Song, W. Liu, and Z. Yuan, "Multilevel optimal modulation and dynamic control strategies for STATCOMs using cascaded multilevel inverters", IEEE Transl. on Power Delivery, vol. 22, No. 3, pp. 1937-1946, July 2007.

[5] Y. Xu, L. M Tolbert, J. D. Kueck, and D. T. Rizy, "Voltage and current unbalance compensation using a static var compensator." *IET Power Electron.*, vol. 3, no. 6, pp. 977-988, Nov. 2010.

[6] Z. Liu, B. Liu, S. Duan, and Y. Kang, "A novel DC capacitor voltage balance control method for cascade multilevel STATCOM," *IEEE Trans. Power Electron.*, vol. 27, no. 1, pp. 14-27, Jan. 2012.

[7] B. Gultekin, C. O. Gercek, T. Atalik, M. Deniz, N. Bicer, M. Ermis, K. N. Kose, E. Koc, I. Cadirci, A. Acik, Y. Akkaya, H. Toygar, and S. Bideci, "Design and implementation of a 154-kV ±50-Mvar transmission STATCOM based on 21-level cascaded multilevel converter," *IEEE Trans. Ind. Applicat.*, vol. 48, no. 3, pp. 1030-1045, May/Jun. 2012.

[8] B. Gultekin and M. Ermis, "Cascaded multilevel converter-based transmission STATCOM: system design methodology and development of a 12kV ±12MVAr power stage," *IEEE Trans. Power Electron.*, vol. 28, no. 11, pp. 4930-4950, Nov. 2013.

[9] K. H. Law, M. S. A Dahidah, G. Konstantinou, and V. G. Agelidis, "SHE-PWM cascaded multilevel converter with adjustable DC sources control for STATCOM applications," in *IEEE 7th Int. Power Electron. and Motion Control*, 2012, pp. 330-334.

[10] S. Kouro, M. Malinowski, and K. Gopakumar, J. Pou, L.G. Franquelo, B. Wu, J. Rodriguez, M.A. Pérez, and J.I. Leon, "Recent advances and industrial applications of multilevel converters," IEEE Transl. on Industrial Electronics, vol. 57, No. 8, pp.2553-2580, August 2010.

[11] R. Naderi and A. Rahmati, "Phase-shifted carrier PWM technique for general cascaded inverters," *IEEE Trans. Power Electron.*, vol. 23, no. 3, pp. 1257-1269, May 2008.

[12] L. M. Tolbert, F. Z. Peng, and T. G. Habetler, "Multilevel PWM methods at low modulation indices," *IEEE Trans. Power Electron.*, vol. 15, no. 4, pp. 719-725, Jul. 2000.

[13] W. Yao, H. Hu, and Z. Lu, "Comparisons of space-vector modulation and carrier-based modulation of multilevel inverter," *IEEE Trans. Power Electron.*, vol. 23, no. 1, pp. 45-51, Jan. 2008.

[14] A. K. Gupta and A. M. Khambadkone, "A general space vector PWM algorithm for multilevel inverters, including operation in overmodulation range," *IEEE Trans. Power Electron.*, vol. 22, no. 2, pp. 517-526, Mar. 2007.

[15] M. S. A. Dahidah and V. G. Agelidis, "Selective harmonic elimination PWM control for cascaded multilevel voltage source converters: a generalized formula," *IEEE Trans. Power Electron.*, vol. 23, no. 4, pp. 1620-1630, Jul. 2008.

[16] W. Fei, X. Ruan, and B. Wu, "A generalized formulation of quarter-wave symmetry SHE-PWM problems for multilevel inverters," *IEEE Trans. Power Electron.*, vol. 24, no. 7, pp. 1758-1766, Jul. 2009.

[17] M. S. A. Dahidah and V. G. Agelidis, "Single-carrier sinusoidal PWM-equivalent selective harmonic elimination five-level inverter control," *Electric Power Syst. Research*, vol. 78, no. 11, pp. 1826- 1836, Jun. 2008.

[18] V. G. Agelidis, A. I. Balouktsis, and M. S. A. Dahidah, "A five-level symmetrically defined selective harmonic elimination PWM strategy: analysis and experimental validation," *IEEE Trans. Power Electron.*, vol. 23, no. 1, pp. 19-26, Jan. 2008.

[19] R. N. Ray, D. Chatterjee, and S. K. Goswami, "Harmonics elimination in a multilevel inverter using the particle swarm optimisation technique," *IET Power Electron.*, vol. 2, no. 6, pp. 646–652, Nov. 2009.

[20] H. Taghizadeh and M. Tarafdar Hagh, "Harmonic elimination of cascade multilevel inverters with nonequal DC sources using particle swarm optimization," *IEEE Trans. Ind. Electron.*, vol. 57, no. 11, pp. 3678-3684, Nov. 2010.

[21] M. S. A. Dahidah, G. Konstantinou, and V. G. Agelidis, "SHE-PWM and optimized DC voltage levels for cascaded multilevel inverters control," in *IEEE Symp. Ind. Electron. and Applicat.*, 2010, pp. 143-148.

[22] W. Fei, X. Du, and B. Wu, "A generalized half-wave symmetry SHE-PWM formulation for multilevel voltage inverters," *IEEE Trans. Ind. Electron.*, vol. 57, no. 9, pp. 3030-3038, Sep. 2010.

[23] M. S. A. Dahidah, G. Konstantinou, N. Flourentzou, and V. G. Agelidis, "On comparing the symmetrical and non-symmetrical selective harmonic elimination pulse-width modulation technique for two-level three-phase voltage source converters," *IET Power Electrons*, vol. 3, pp. 829-842, 2010.

[24] M. S. A. Dahidah, V. G. Agelidis, and M. V. Rao, "On abolishing symmetry requirements in the formulation of a five-Level selective harmonic elimination pulse-width modulation technique," *IEEE Trans. Power Electron.*, vol. 21, no. 6, pp. 1833-1837, Nov. 2006.

[25] K. Sano, and M. Takasaki, "A transformerless D-STATCOM based on a multivoltage cascade converter requiring no dc sources," *IEEE Trans. on Power Electron.*, vol. 27, no. 6, pp. 2783 – 2795, Jun. 2012.

[26] H. Akagi, H. Fujita, S. Yonetani, and Y. Kondo, "A 6.6-kV transformerless STATCOM based on a five-level diode-clamped PWM converter: system design and experimentation of a 200-V 10-kVA laboratory model," *IEEE Trans. Ind. Applicat.*, vol. 44, no. 2, pp. 672 – 680, Mar./Apr. 2008.

[27] M. N. Slepchenkov, K. M. Smedley, and J. Wen, "Hexagram-converter-based STATCOM for voltage support in fixed-speed wind turbine generation systems," *IEEE Trans. Ind. Electron.*, vol. 58, no. 4, pp. 1120-1131, Apr. 2011.

[28] T. Jimichi, H. Fujita, and H. Akagi, "A dynamic voltage restorer equipped with a high-frequency isolated DC–DC converter," *IEEE Trans. Ind. Applicat.*, vol. 47, no. 1, pp. 169 – 175, Jan./Feb. 2011.

[29] L. Zhang, K. Sun, Y. Xing, L. Feng, and H. Ge, "A Modular Grid-Connected Photovoltaic Generation System Based on DC Bus," *IEEE Trans. Power Electron.*, vol. 26, no. 2, pp. 523 – 531, Feb. 2011.

[30] F. Deng, and Z. Chen, "Control of Improved Full-Bridge Three-Level DC/DC Converter for Wind Turbines in a DC Grid," *IEEE Trans. Power Electron.*, vol. 28, no. 1, pp. 314 – 324, Jan. 2013.

[31] H. D. Venable, "The K factor: A new mathematical tool for stability analysis and synthesis" [Online]. Available: http://www.icwic.com/icwic/data/pdf/cd/cd057/Switching.%20DC-DC%20Regulator,%20Controller/1184.pdf.

[32] MathWorks. [Online]. Available: http://www.mathworks.com/.

[33] Eric Coates. (2013). Negative feedback & distortion [Online]. Available: http://www.learnabout-electronics.org/Amplifiers/amplifiers34.php

978-1-4799-2706-7/14 $31.00 © 2014 IEEE

Novel Principle for Flux Sensing in the Application of a DC + AC Current Sensor

L. Schrittwieser*, M. Mauerer*, D. Bortis*[†], G. Ortiz*[†] and J. W. Kolar*
*Power Electronic Systems Laboratory, ETH Zurich, Switzerland
[†]Enertronics GmbH, Switzerland
Email: bortis@lem.ee.ethz.ch

Abstract—**Magnetostriction describes the geometrical change in length of a ferromagnetic material in dependence of its internal magnetic flux density value. By detecting the vibrations caused by these dimensional changes with a piezo-electric transducer, the instantaneous value of the magnetic flux inside a magnetic core can be sensed from DC up to a few kilohertz. This principle, together with a high bandwidth current transformer, was utilized in order to construct a current sensor capable of measuring currents ranging from DC to several MHz. As will be shown in this paper, an additional sinusoidal AC-excitation of the core material provides higher sensitivity of the length measurement and overcomes the high-pass characteristic of the piezo sensor. In order to prove the principle and to demonstrate the capabilities of this new sensor, a series of experimental measurements and implementation results are presented.**

I. INTRODUCTION

Precise current measurement is a mandatory requirement of modern power electronic systems as it enables the implementation of high performance current control loops, monitoring and safe shut-down in case the maximum allowed current value has been exceeded, among others.

Depending on the specific application and the required performance of the measurement system, the existing current measurement concepts can be classified according to their key operating principles. **Fig. 1** gives an overview of the most common current measurement methods that are applicable in power electronics whereby their key features are presented in the following.

A. Isolated Current Measurement Concepts

If a galvanic isolation between the current to be measured and the sensor is required, the measurement principle is typically based on *Ampère's Law* where effects caused by the magnetic field of the current are exploited. There, Rogowski coils or AC current transformers are commonly used when only AC currents need to be measured, e.g. in power transmission systems [1][2].

On the other hand, if AC as well as DC current components need to be captured, several techniques are applicable. Magneto-resistive sensors make use of the fact that some materials change their resistance in the presence of a magnetic field [3]. Magneto-optical current sensors exploit the *Faraday Effect* and are usually applied in high-current applications [4]. There are many methods involving a saturable magnetic material in order to measure current. Sensors of this kind can be operated in open or closed loop systems and usually require more than one magnetic core [5]. Current can also be measured by introducing a semiconductor Hall-effect

sensor in the magnetic path in order to directly measure the magnetic field caused by the current. These current sensors can be operated in open or closed loop configurations as well and they can be combined with a current transformer to achieve a higher bandwidth [6].

B. Non-isolated Current Measurement Concepts

If no galvanic isolation is required, the current can be measured by using *Ohm's Law* and a shunt resistor [6]. In order to achieve a high bandwidth of several tens of megahertz or more, special construction techniques for the shunt are required in order to reduce parasitic effects [7].

C. Proposed Concept

The new current measurement technique is based on the magnetostriction effect. Other sensors based on inverse magnetostriction have been presented [8]. However, the new sensor directly measures the change in length of a magnetic core to gain information about its magnetic flux density. This information is then used together with a current transformer in order to accurately measure both static and time-varying current components.

Compared to a current probe that utilizes a hall sensor, this system does not require any interruption of the magnetic path in order to introduce the flux sensor. The operation principle and the theoretical background of this new technology are presented in the following.

II. THEORY OF OPERATION

The proposed current sensor is based on a current transformer which provides a galvanic isolation as described in the previous section. Since current transformers have a high-pass characteristic given by their magnetizing inductance and the burden resistance, no DC currents can be measured.

As described in [9] and [10], in order to be able to measure also the low frequency components of the current, often a closed-loop flux compensation circuit consisting of a flux sensor (e.g. a Hall effect sensor inserted into the air gap of the magnetic core), an op-amp circuit and a compensation winding is added. The current through the compensation winding is controlled in such a way that an operation of the magnetic core at zero flux density can be ensured (cf. **Fig. 2 a)**). This way, the current in the compensation winding is proportional (depending on the measurement to compensation winding's turns ratio) to the input current i_{in}.

With this configuration, frequency components above the current transformer's lower cut-off frequency ($i_{in,HF}(t)$

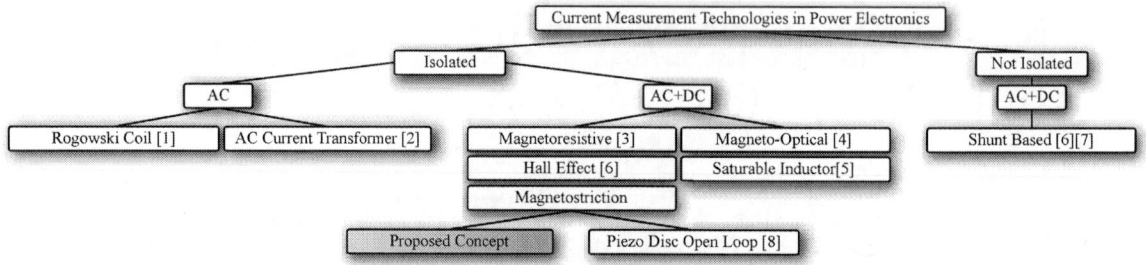

Fig. 1: Overview of previously presented current measurement technologies utilized in power electronic systems.

would cause a magnetic flux $B_{\mathrm{in,HF}}(t)$ in the core, which is compensated by the high frequency current $i_{\mathrm{m,HF}}(t)$ flowing in the measuring winding and through the burden resistor R_{B}. Due to the core material's high permeability, the resulting magnetizing flux is very small at these frequencies, thus no signal will be measured by the flux sensor. On the other hand, any frequency components which are below the current transformer's lower cut-off frequency ($i_{\mathrm{in,LF}}(t)$) would not induce a voltage in the burden winding. However, they are captured by the flux sensor and therefore will be compensated by the current $i_{\mathrm{m,LF}}(t)$ in the compensation winding, allowing a measurement with bandwith ranging from DC to several MHz.

It has to be emphasized that for this configuration, no frequency response matching of the flux sensor and the current transformer is necessary. However, in order to be able to compensate this low frequency flux, the current transformer's lower cut-off frequency has to be well below the flux sensor's upper cut-off frequency. Then, the measurement error caused by the magnetizing inductance of the transformer is negligible.

Instead of using an additional compensation winding, the compensation current can also be fed directly into the measurement winding as shown in **Fig. 2 b)**. Both, the low frequency compensation current and the current transformed to the measurement winding, are then forced through the same winding, thus also through the burden resistor R_{B}. Consequently, both current components $i_{\mathrm{m,LF}}(t)$ and $i_{\mathrm{m,HF}}(t)$ are inherently added and the voltage across the burden resistor R_{B} is directly proportional to the current

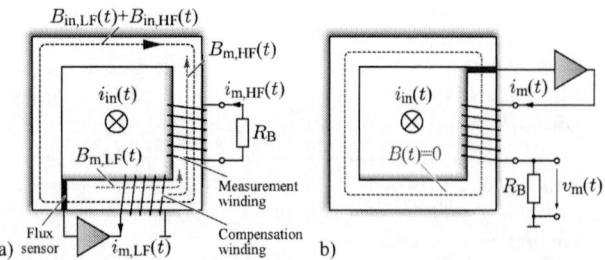

Fig. 2: Flux compensated current transducer concept: **a)** A flux sensor is inserted in the magnetic path and its signal is used to compensate the low frequency components of the main current. The high frequency components are measured by a current transformer. **b)** Integrated flux measurement and current transformer.

$i_{\mathrm{in}}(t)$ [9].

A. Magnetostriction-based Flux Sensing

The flux sensor, typically a hall-element, is the crucial part of an actively compensated current transformer design since its signal allows the controller to close the loop. Another possible way to sense the magnetic flux is to use the magnetostriction phenomenon, as proposed in this paper.

Basically, magnetostriction is the relative change in length $\Delta l/l$ of a magnetic material when a magnetic flux density B is applied. As described in [11] and [12], the change of length is proportional to the square of the magnetic flux density for flux densities much smaller than the saturation flux density B_{S}. The common magnitude for this effect is typically given as saturation magnetostriction λ_{S} which describes the relative change in length reached at saturation flux density (cf. (1)).

$$\frac{\Delta l}{l}(B) \approx B^2 \cdot \frac{\lambda_{\mathrm{S}}}{B_{\mathrm{S}}^2} \qquad \text{for} \quad B \ll B_{\mathrm{S}}. \qquad (1)$$

The magnetostriction can be measured e.g. with an electromechanical transducer which converts the change in length into an electrical output voltage. Possible electromechanical transducers are strain gauges or piezo elements. A strain gauge changes its electric resistance depending on the applied elongation, thus constant elongations can be captured. Piezoelectric transducers, on the other hand, feature an inherent high pass characteristic. Consequently, only elongations above its lower cut-off frequency can be measured. Piezoelectric sensors are available with a higher sensitivity than strain gauges. As a consequence, a piezo sensor was used for the implementation of the prototype.

Due to the dependency on B^2 in (1), a direct measurement of the transformer core's change in length $\Delta l/l$ results in a non-linear sensor output signal $v_{\mathrm{s}}(t)$. In addition, the sign of the flux and therefore the sign of the magnetization current, is not preserved. Thus, even if the quadratic relationship between the flux and the change in length could be linearised, the requirement of the flux's sign would make it impossible to control the magnetic flux to zero without additional precautions taken.

The aforementioned drawbacks can be overcome by injecting an AC voltage signal through an additional winding in the current sensor's magnetic core, as shown in **Fig. 4)** and as will be discussed in the following.

B. AC Excitation

An additional winding N_{ex}, connected to a sinusoidal voltage source v_{AC}, is added to the core in order to create a new sinusoidal component $B_{ex}(t) = \widehat{B}_{ex} \sin(\omega_{ex}t)$ in the core flux (cf. **Fig. 4**).

Since the frequency components above the current transformer's lower cut-off frequency $B_{HF}(t)$ are already compensated by the measuring winding, the flux measured by the flux sensor is $B(t) = B_{LF}(t) + \widehat{B}_{ex} \sin(\omega_{ex}t)$. In **Fig. 3 a)** the resulting spectrum of the flux $B(t)$, with an upper cutoff frequency of ω_{LF}, is shown. Based on (1) and the given spectrum of the flux $B(t)$, the spectrum of the flux sensor's output signal $v_s(t)$ can be calculated as

$$
\begin{aligned}
v_s(t) \propto B^2(t) &= (B_{LF}(t) + \widehat{B}_{ex} \sin(\omega_{ex}t))^2 \\
&= B_{LF}^2(t) + \frac{\widehat{B}_{ex}^2}{2} + \underbrace{2B_{LF}(t)\widehat{B}_{ex}\sin(\omega_{ex}t)}_{\text{AM Modulated Excitation}} \\
&\quad + \frac{\widehat{B}_{ex}^2}{2}\sin(2\omega_{ex}t)
\end{aligned}
\tag{2}
$$

and is illustrated in **Fig. 3 b)**. As can be noticed from (2) and **Fig. 3 b)**, with the superposition of the excitation flux $B_{ex}(t)$ and the low frequency flux $B_{LF}(t)$, new frequency components are introduced. The spectrum of the sensor voltage differs from the spectrum of the core flux density in several aspects; a new component occurs at twice the excitation frequency $2\omega_{ex}$ due to the squaring of the excitation signal $B_{ex}(t)$. Furthermore, the low frequency signal's amplitude is squared and the bandwidth is doubled to $2\omega_{LF}$. In addition, the DC-value is increased by $\widehat{B}_{ex}^2/2$. The most important difference, however, is the fact that the excitation frequency component gets modulated by the original low frequency spectrum of $B_{LF}(t)$. This means that the spectrum of $B_{LF}(t)$ is now centered around the excitation frequency ω_{ex} (cf. middle term in (2) and **Fig. 3 b)**) which is in accordance with measurements published in [13].

For this reason, the modulated signal which is now purely AC, can easily be measured with a piezoelectric transducer, whereby the transducer's lower cut-off frequency has to be below $\omega_{ex} - \omega_{LF}$. In addition, based on **Fig. 3 b)**, it can be noticed that the excitation frequency ω_{ex} has to be at least three times higher than the lower cut-off frequency of the transformer ω_{LF}. Otherwise, the spectrum centered around the excitation frequency ω_{ex} would overlap with the spectrum of $B_{LF}^2(t)$ which would result in a distortion of the original signal $B_{LF}(t)$. In this case, the modulated signal can be used to measure the low frequency signal $B_{LF}(t)$. Hence, the desired spectrum centered around the excitation frequency ω_{ex} has to be isolated by the use of a suitable bandpass filter (cf. **Fig. 4**). The spectrum of the filtered output signal $v_{BP}(t)$ is shown in **Fig. 3 c)**.

In a further step, the bandpass filtered signal $v_{BP}(t) \propto \widehat{B}_{ex}\sin(\omega_{ex}t)$, whose amplitude depends linearly on $B_{LF}(t)$, can be demodulated by the multiplication with $\sin(\omega_{ex}t)/\widehat{B}_{ex}$. This yields the demodulator output $v_D(t)$ containing the intended component $B_{LF}(t)$ as well as a copy

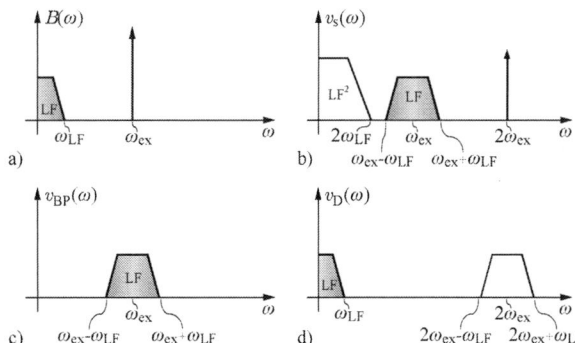

Fig. 3: Resulting spectra with AC-excitation, spectrum of **a)** the magnetic flux, **b)** the flux sensor's output signal, **c)** the bandpass filtered signal and **d)** the demodulated signal.

of this spectrum centered around $2\omega_{ex}$ as shown in **Fig. 3 d)**. With a subsequent low pass filter, this copy can be discarded.

This principle allows the sensing of positive as well as negative magnetization currents without applying an additional offset current. Furthermore, the amplitude $\widehat{B}_{ex}$ of the excitation signal gives a degree of freedom which allows to scale the magnetostriction signal amplitude independently of the magnitude of $B_{LF}(t)$. Also, no DC signal has to be measured by the magnetostriction sensor. This allows the usage of piezo transducers which have an inherent high pass characteristic and it eliminates problems with DC drifts and offsets of amplifiers in the signal path. However, the AC excitation flux in the core will induce a voltage in the measurement winding and will therefore disturb the measured signal. This has to be compensated by using a second transformer as it will be explained in the following.

C. Compensation Transformer

Since the AC-excitation signal is also transformed to the other windings, i.e. the burden winding as well as the conductor whose current has to be measured, a second transformer T_2 with the same winding arrangement as for transformer T_1 is connected in series to T_1, with the only difference that the orientation of the excitation winding is reversed (cf. **Fig. 5**). Consequently, in both cores of T_1 and T_2, the same AC-excitation signal with opposite sign is impressed, which - assuming identical properties of the two transformers - is cancelled out in the other windings. However, due to production and assembly tolerances prevalent in a real system, the two transformers will not be

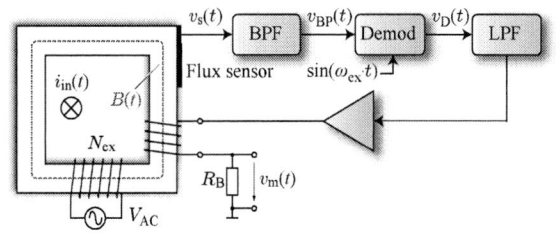

Fig. 4: Block diagram of AC-excitation and signal processing of the magnetostriction-based flux sensor.

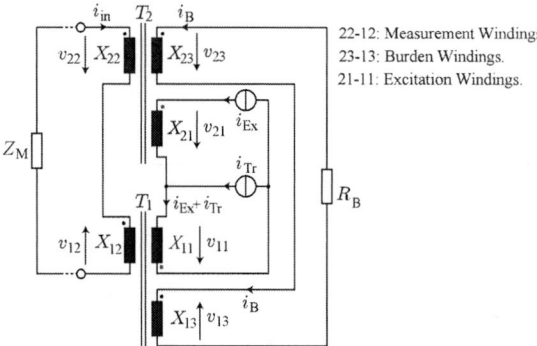

22-12: Measurement Windings.
23-13: Burden Windings.
21-11: Excitation Windings.

Fig. 5: Schematic representation of the current transducer comprising measurement windings, burden windings and excitation windings. These last ones are used in order to inject the AC high frequency signal and to perform the required trimming in the magnetic core.

exactly identical. This leads to an incomplete cancellation of the excitation signal in both the burden winding and in the measured conductor. This problem can be minimized by introducing an additional current source $i_{\mathrm{Tr}}(t)$ at the connection point of the two excitation windings (cf. **Fig. 5**). The following section will reveal that with this additional current source, the undesired effects introduced by the non-identical transformers can be significantly reduced.

D. Trimming

As mentioned above, the current source $i_{\mathrm{Tr}}(t)$ in **Fig. 5** can be used to trim the system in such a way that undesired signals, introduced by the slightly different transformers, can be eliminated. The aim of this investigation is to analyze the effects introduced by the excitation current on the current in the burden winding i_{B} as well as the current in the measurement winding i_{in}. In order to investigate the possibilities and limitations of this approach, some simplifications are made. The excitation voltage source is replaced with a current source $i_{\mathrm{Ex}}(t)$. Furthermore, the compensation voltage source in the burden winding path is omitted since it operates independently from the excitation and trimming system. It is also assumed that an arbitrary impedance Z_{m} is connected to the measurement winding. This is the impedance of the circuit providing the current i_{in} which is measured. Voltage sources in series with Z_{m} and current sources in parallel with it do not affect the trimming system and hence are omitted. Additionally, the transformers are assumed to be linear.

Both transformers have three windings, denoted as follows: Winding 1 is the excitation winding, winding 2 is the measurement winding and winding 3 is the burden winding. With reference to **Fig. 5**, T_2, a three-winding transformer can be described, for AC steady state analysis, in the following way:

$$
\begin{bmatrix} \underline{v}_{21} \\ \underline{v}_{22} \\ \underline{v}_{23} \end{bmatrix} = \begin{bmatrix} \underline{X}_{21} & \underline{X}_{21\_2} & \underline{X}_{21\_3} \\ \underline{X}_{22\_1} & \underline{X}_{22} & \underline{X}_{22\_3} \\ \underline{X}_{23\_1} & \underline{X}_{23\_2} & \underline{X}_{23} \end{bmatrix} \cdot \begin{bmatrix} \underline{i}_{\mathrm{Ex}} \\ \underline{i}_{\mathrm{M}} \\ \underline{i}_{\mathrm{B}} \end{bmatrix},
$$

whereby:

$$
\underline{X}_{\mathrm{xyz}} = j\omega \cdot L_{\mathrm{xyz}}.
$$

L_{21}, L_{22} and L_{23} are the self inductances of winding 1, 2 and 3 respectively. L_{21_2} is the mutual inductance between winding 1 and 2. Note that the inductance matrix is symmetric, e.g. $L_{21\_2} = L_{22\_1}$.

Using these equations, the circuit in **Fig. 5** can be analyzed to determine the currents in the burden and measurement windings caused by the excitation and trimming currents. The result is shown in equations (3) and (4).

$$
\underline{i}_{\mathrm{B}} = \frac{i_{\mathrm{Ex}}(\underline{X}_{13\_1} - \underline{X}_{23\_1}) - i_{\mathrm{M}}(\underline{X}_{13\_2} + \underline{X}_{23\_2}) + i_{\mathrm{Tr}} \cdot \underline{X}_{13\_1}}{R_{\mathrm{B}} + \underline{X}_{13} + \underline{X}_{23}} \tag{3}
$$

$$
\underline{i}_{\mathrm{in}} = \frac{i_{\mathrm{Ex}}(\underline{X}_{12\_1} - \underline{X}_{22\_1}) - i_{\mathrm{B}}(\underline{X}_{12\_3} + \underline{X}_{22\_3}) + i_{\mathrm{Tr}} \cdot \underline{X}_{12\_1}}{\underline{Z}_{\mathrm{M}} + \underline{X}_{12} + \underline{X}_{22}} \tag{4}
$$

Both equations reveal that the disturbance, caused by the excitation current, depends on the difference of the transformer's mutual inductances. This holds for both currents, in the burden winding as well as in the measurement winding. It can be seen that, if the transformers have identical mutual inductances, no disturbance will be caused by the excitation current.

Assuming non-identical transformers, the trimming current i_{Tr} allows the elimination of the coupled excitation signal in either the measurement or the burden winding, but not in both. This is due to the fact that the measurement and burden windings in the two transformers do not necessarily show the same difference in mutual inductances with respect to the excitation windings and there is only one degree of freedom, the trimming current i_{Tr}. Hence the excitation signal can only be cancelled in one winding. The goal is to eliminate the excitation signal in the burden winding since the current in the burden winding represents the output signal of the system and is directly measured. Hence it can be used as an input for the trimming system. The prototype system has shown that the remaining disturbance in the measurement winding, due to the asymmetry of the transformers, can be neglected even for small $\underline{Z}_{\mathrm{M}}$ since it is possible to produce two very similar transformers.

Combining equations (3) and (4) yields a rather complex expression for $\underline{i}_{\mathrm{B}}$. However, as $\underline{i}_{\mathrm{B}}$ depends linearly on $\underline{i}_{\mathrm{Tr}}$, there exists a unique solution to the equation $\underline{i}_{\mathrm{B}} = 0$. Therefore, the trimming system allows to drive the current in the burden winding to zero, eliminating any coupled excitation currents.

Since there are many parameters involved in this system, a calculation of the necessary trimming current is complex. The prototype of the current probe therefore uses a search algorithm to determine the required phase and amplitude of the trimming current in order to reduce the disturbing burden current to a minimum. The algorithm works by applying several equally spaced current phase angles with constant amplitude. In the first sweep, the tested phase angles span over the full parameter range, i.e. $[0, 2\pi]$. For each point, the amplitude of the excitation frequency component is measured using a matched filter. Once all points have been measured, the phase angle resulting in the lowest distortion, is selected. The next sweep will cover only a part of the parameter space in order to narrow down the search. As the parameter space is two dimensional, the same sweeping technique is applied to the amplitude of the trimming signal.

III. IMPLEMENTATION RESULTS

In order to verify the new current measurement principle, a compact prototype has been built. It is capable of measuring currents up to $\pm 20\,\mathrm{A}$ with a bandwidth of DC to $20\,\mathrm{MHz}$. As described above, two magnetic cores, showing a relatively high magnetostriction, are the key components.

Fig. 6 depicts the core setup of the prototype. In the picture, the second transformer core is behind the first core. The two cores are connected to the circuitry depicted in **Fig. 7** which performs all the required signal processing.

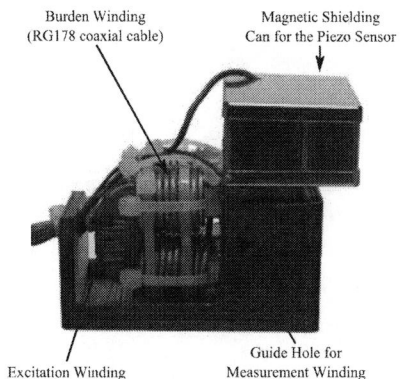

Fig. 6: Current transducer components' arrangement.

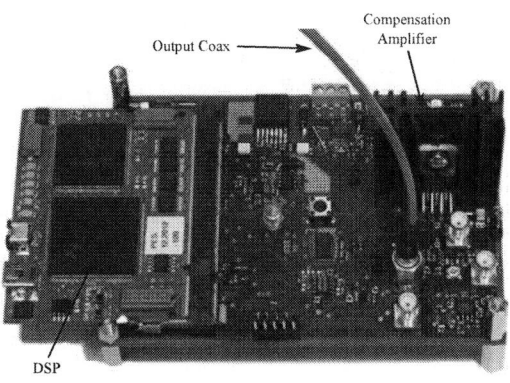

Fig. 7: Analog and digital circuits utilized for signal processing.

A. Transformer Core

For the two transformers, *AMCC-4* C-shaped cores made from the amorphous material *2605SA1* from *Hitachi Metals/Metglas* were used. This material was selected as it shows the highest magnetostriction of the tested materials. Furthermore, cores of this material suitable to build a current transformer with are available. **Fig. 8** illustrates the arrangement of such a transformer. For the burden winding, an *RG178* coaxial cable was used in order to shield the burden winding from external electric fields. The current transformer itself, without flux compensation, has lower cut-off frequency of $\approx 530\,\mathrm{Hz}$. **Table I** lists the measured properties of the

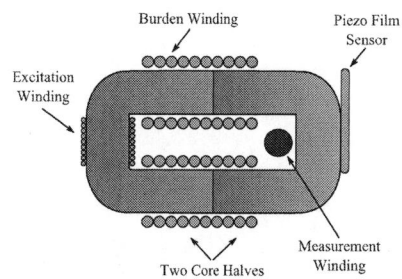

Fig. 8: Winding arrangement of one of the transformers and the respective piezo transducer.

two transformer cores. Note that k is the coupling factor between the windings. The well known relationship between the mutual inductance and the coupling factor is as follows:

$$k = \frac{L_{21\_2}}{\sqrt{L_{21}L_{22}}} \tag{5}$$

An ideal transformer has coupling factors of 1. The coupling factors involving the measurement winding depend on the position of the measurement winding within the transformer as well as its length. As long as the winding arrangement is similar for both transformers, the coupling factors are almost identical as well and as a consequence, so are the mutual inductances.

B. Piezo Film Sensor

In order to measure the core's magnetostriction, an electrically shielded *SDT1-028k* piezo film sensor from *Measurement Specialities* is adhered to one of the two magnetic cores. The sensor, as depicted in **Fig. 9**, is approximately $30\,\mathrm{mm}$ long. Its terminals and cable are electrically shielded, as is the sensor element itself. This reduces electrical interference to the sensor output. However, the sensor is not shielded against magnetic fields. As a consequence, care must be taken in order to avoid the generation of faulty sensor output singals due to stray magnetic fields. In the prototype, the sensor was rigidly placed inside a magnetic shielding can.

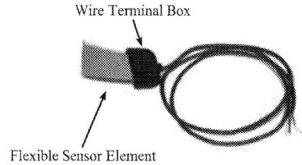

Fig. 9: *SDT1-028k* piezo film sensor.

C. Excitation and Trimming

An excitation frequency of $16\,\mathrm{kHz}$ was selected as the piezo sensor shows a good response around this frequency. This is likely due to a mechanical resonance of the two core halves. The bandwidth of this resonance proved to be high enough to enable successful operation of the system. A peak excitation flux density of $\approx 250\,\mathrm{mT}$ was selected which is well below the material's saturation flux density of $1.56\,\mathrm{T}$.

978-1-4799-2706-7/14 $31.00 © 2014 IEEE

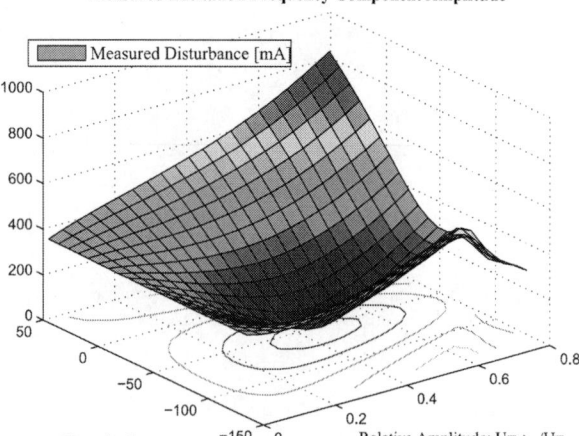

Measured Excitation Frequency Component Amplitude

Fig. 10: Excitation frequency disturbance in the burden resistor as function of trimming voltage.

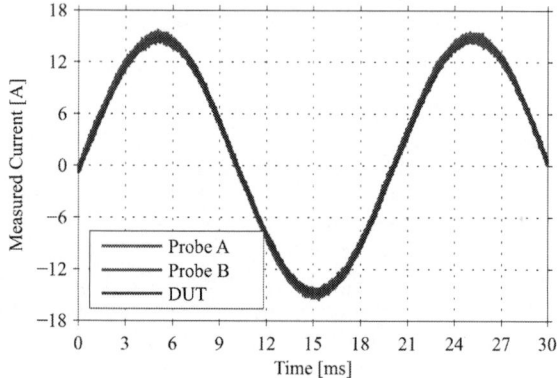

Fig. 11: Experimental measurement of an 8A, 50Hz sine signal utilizing the proposed magnetostriction-based current sensor.

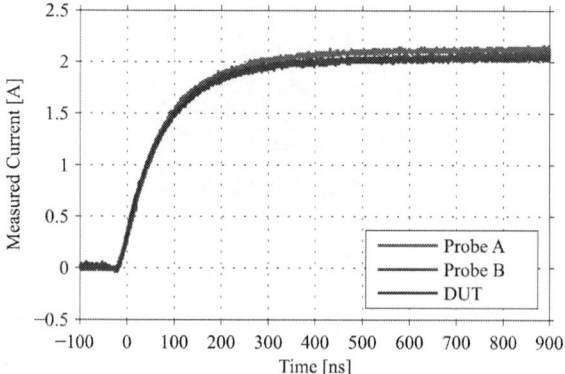

Fig. 12: Response to a 2 A current step. This test shows the fast dynamic performance of the sensor (note the timescale).

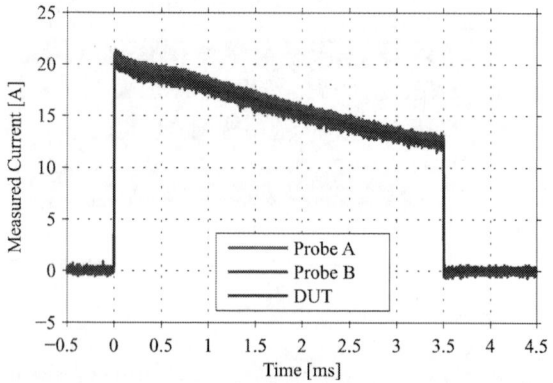

Fig. 13: Response to a 20 A current step. During this test, both the flux measurement concept and the current transformer are utilized. During the initial step, the current transformer is active whereas during the decaying current ramp the flux sensor is utilized to measure the signal.

Using a higher excitation flux density would yield a higher output signal from the piezo sensor at the cost of increased core losses and power consumption.

A measurement from the prototype showing the measured excitation frequency component in the output signal as a function of trimming current phase and amplitude is given in **Fig. 10**. In the prototype, the trimming current source is implemented as a voltage source with a defined series impedance. Note that there is, as expected, a single point where the excitation signal vanishes from the burden winding. This point is found and selected by the algorithm described above. The bend in the plot results from an unstable amplifier which was not designed for operation with high relative trimming voltage.

D. Signal Processing

A digital signal processor handles the signal generation, demodulation, flux control, measurement routines and trimming. The two filters shown in **Fig. 4** are implemented as digital filters. The band-pass is a 10th order Chebyshev Type 2 filter with a passband of $11\,\mathrm{kHz}\ldots21\,\mathrm{kHz}$ providing $50\,\mathrm{dB}$ attenuation for frequencies below $7\,\mathrm{kHz}$ and above $33\,\mathrm{kHz}$. The low-pass is a 4th order Chebyshev Type 2 filter with $1\,\mathrm{dB}$ attenuation at $10\,\mathrm{kHz}$ and $50\,\mathrm{dB}$ attenuation above $32\,\mathrm{kHz}$. A standard PI controller is used as flux controller.

TABLE I: Measured transformer parameters.

Parameter	T1	T2
Turns Measurement Winding	1	1
Turns Burden Winding	20	20
Turns Excitation Winding	9	9
$L_{\mathrm{Measurement}}[\mu\,\mathrm{H}]$	0.77	0.78
$L_{\mathrm{Burden}}[\mu\,\mathrm{H}]$	297	302.8
$L_{\mathrm{Excitation}}[\mu\,\mathrm{H}]$	64.6	64.3
$k_{\mathrm{Measurement-Burden}}$	0.855 - 0.92	0.858 - 0.914
$k_{\mathrm{Measurement-Excitation}}$	0.83 - 0.92	0.849 - 0.927
$k_{\mathrm{Burden-Excitation}}$	0.924	0.927

E. Performance

The prototype has been compared to commercially available current probes. **Fig. 11** shows the measurement of a

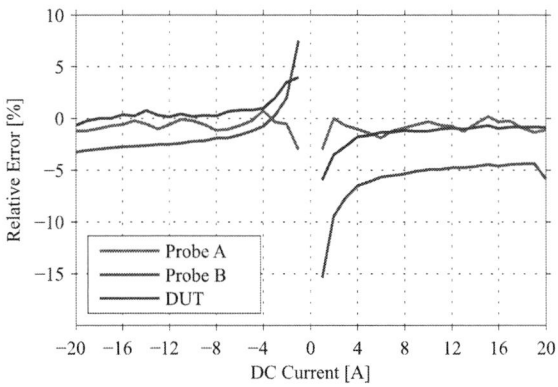

Fig. 14: Comparison of relative measurement errors with other common current probes.

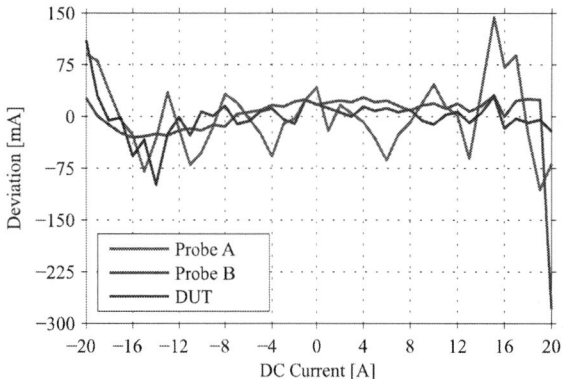

Fig. 15: Deviation of the measurement from its linear best first order fit for different current values and different current sensor manufacturers.

50 Hz, 8 A sine current. *DUT* denotes the magnetostriction based prototype. Note that 50 Hz is well below the lower cut-off frequency of the current transformer, thus, this signal is measured using the magnetostriction based compensation system only.

Fig. 12 depicts the response to a quickly rising 2 A current step. This plot solely shows the high-frequency response of the current probe, the dynamics are too fast for the magnetostriction system to take action. Thus, this response was generated by the current transformer only.

In **Fig. 13**, the step response to a much longer 20 A current step is shown. In this plot, the timescale is long enough that both, the current transformer and the flux compensation circuit, are involved in the current measurement.

Additionally, **Fig. 14** shows the relative measurement errors of the three current probes for DC-currents from −20 A to 20 A. A *Yokogawa WT3000* precision power analyzer, with an maximum error of ±2 mA over the full range, has been used as reference. The same data has been used to derive the linearity properties of all three probes. Least squares regression has been used to determine the best-fit line for each probe. **Fig. 15** shows the difference between the measured current and the best-fit lines of the probes. The best-fit integral nonlinearity (INL), defined as the maximum deviation between measured value and best-fit line, is given in **Table II**.

TABLE II: Linearity measurement results.

Probe	Best-Fit INL	Gain Error	Offset
Current Probe A	145 mA	−0.74 %	−2 mA
Current Probe B	279 mA	−3.77 %	−138 mA
DUT	126 mA	−0.41 %	−68 mA

F. Implementation Challenges

The successful implementation of a prototype proves the feasibility of this new flux sensing technique. However, several aspects require careful attention in order to ensure reliable and stable operation. First of all, the principle is based on magnetostriction, and hence a measurement of mechanical strain. It is therefore important to isolate the strain sensor from other mechanical influences, such as

vibrations, that might generate erroneous sensor signals. Additionally, it is important that the sensor is immune to electromagnetic interference from the measuring signal and the excitation winding. This requires careful electric as well as magnetic shielding of the strain sensor. Furthermore, a solid magnetic core having no gap is preferable since it eliminates mechanical vibrations that might arise from the contacting core halves.

IV. CONCEPT EXTENSIONS

Since there is no necessity for an air gap in the magnetic core, this technique of magnetic flux measurement can be utilized in other applications such as DC flux measurement in power distribution transformers or isolated DC-DC converters. As described in [14], isolated DC-DC converters may require some form of transformer core flux measurement in order to prevent core saturation. The technique described in this paper might be used to accomplish this. As the transformer is already excited by the DC-DC converter, no additional excitation winding is needed. This means that only a piezo sensor, measuring the core's magnetostriction, and analog amplification is required. As presented in the previous sections, the main part of the signal processing can be done digitally and thus requires no additional components in a digitally controlled converter.

V. SUMMARY

A new and magnetic flux sensing principle has been proven in the application of a DC + AC current probe. This principle is based on measuring the changes in length of a magnetic component during its operation due the magnetostriction phenomenon. In order to detect these length changes, a piezo-electric transducer was adhered to the surface of a magnetic core. In addition, an AC signal was injected through an external winding in order to shift the input current's spectrum to higher levels, thus enabling the utilization of the piezo sensor.

By implementing a feedback loop comprising the piezo-electric sensor and the appropriate analog and digital circuitry, an accurate measurement of the input current in the low frequency spectrum (including DC) was achieved. The combination of this magnetic flux measurement and the typically high bandwidth of a current transformer allowed to realize a current sensor able to measure currents up to 20 A and with a bandwidth ranging from DC to 20 MHz.

REFERENCES

[1] W. F. Ray and C. R. Hewson, "High performance Rogowski current transducers," in *Proc. of the Industry Applications Conference*, Oct. 8-12, 2000, pp. 3083–3090 vol.5.

[2] F. Costa, E. Laboure, F. Forest and C. Gautier, "Wide bandwidth, large AC current probe for power electronics and EMI measurements," *IEEE Transactions on Industrial Electronics*, vol. 44, no. 4, pp. 502–511, August 1997.

[3] G. Laimer and J. W. Kolar, "Design and experimental analysis of a DC to 1 MHz closed loop magnetoresistive current sensor," in *Proc. of the Applied Power Electronics Conference and Exposition (APEC), Austin, Texas*, vol. 2, Mar. 6-10, 2005, pp. 1288–1292 Vol. 2.

[4] A. Papp and H. Harms, "Magnetooptical current transformer. 1: Principles," *Applied Optics*, vol. 19, no. 22, pp. 3729–3734, 1980.

[5] P. Ripka, "Review of Fluxgate Sensors," *Sensors and Actuators A: Physical*, vol. 33, no. 4, pp. 129–141, 1992.

[6] S. Ziegler, R. C. Woodward, H. H.-C. Iu, and L. J. Borle, "Current Sensing Techniques: A Review," *Sensors Journal, IEEE*, vol. 9, no. 4, pp. 354–376, Apr. 2009.

[7] C. M. Johnson and P. R. Palmer, "Current Measurement Using Compensated Coaxial Shunts," *Science, Measurement and Technology, IEE Proceedings*, vol. 141, no. 6, pp. 471–480, 1994.

[8] F. Koga, T. Tadatsu, J. Inoue and I. Sasada, "A New Type of Current Sensor Based on Inverse Magnetostriction for Large Current Detection," *IEEE Transactions on Magnetics*, vol. 45, no. 10, pp. 4506–4509, Oct. 2009.

[9] LEM, "Isolated Current and Voltage Transducers (3rd edition)," LEM Components, 8 Chemin des Aulx, CH-1228 Plan-les-Ouates, 2004, cH24101.

[10] J. R. Leehey, L. Kushner and W. S. Brown, "DC Current Transformer," in *Proc. 13th Annu. Power Electron. Spec. Conf., Cambridge, MA, USA*, Jun. 1982, pp. 438–444.

[11] T. Zhang, C. Jiang, H. Zhang and H. Xu, "Giant Magnetostrictive Actuators for Active Vibration Control," *Smart Materials and Structures*, vol. 13, no. 3, pp. 473–477, Jun. 2004.

[12] T. Hilgert, L. Vandevelde and J. Melkebeek, "Comparison of Magnetostriction Models for Use in Calculations of Vibrations in Magnetic Cores," *IEEE Transactions on Magnetics*, vol. 44, no. 6, pp. 874–877, 2008.

[13] Y. Ishihara, H. Maeda, K. Harada and T. Todaka, "Performance of the Magnetostriction of a Silicon Steel Sheet with a Bias Field," *Journal of Magnetism and Magnetic Materials*, vol. 160, pp. 149–150, Jul. 1996.

[14] G. Ortiz, J. Mühlethaler and J. W. Kolar, "Magnetic Ear - Based Balancing of Magnetic Flux in High-Power Medium-Frequency Dual-Active-Bridge converter Transformer-Cores," in *Proc. of the 8th International Conference on Power Electronics (ECCE), Jeju, Korea*, May 30 - June 3 2011, pp. 1307–1314.

The 2014 International Power Electronics Conference

Utilizing Voltage Measurement of FET Switch for MPPT of DC Energy Source

Noriyuki Kimura*, Koji Niijima*, Toshimitsu Morizane* and Hideki Omori*

(Osaka Institute of Technology)

Abstract- **This paper proposes the new simple current measurement method. It utilizes the voltage measurement of the MOS-FET switching device at the dc-dc converter to measure the current instead of conventional measuring methods, such as the DC-CT or the shunt resistance. The precise measurement is not necessary to perform the Maximum Power Point Tracking (MPPT) control of dc energy source, such as the thermoelectric generation (TEG). The simulation and experimental results show the good feasibility of proposed measurement method.**

I. INTRODUCTION

To overcome the energy problem, energy harvesting is one of the solutions. Especially the electric generation from the wasted heat energy is long term object. To gather a certain amount of energy, parallel connection of many generators as shown in Fig. 1is necessary. And each generator must have the maximum power point tracking ability.

One of the power generation methods from the waste heat energy is the thermoelectric generation by Seebeck effect of thermoelectric generator (TEG) [1]. The TEG can generate the electric power at low-temperature difference. However the thermoelectric generation has some problems, such as the energy conversion efficiency, thermal leakage, and internal resistance. It is necessary to improve the energy conversion efficiency from thermal energy to electric energy by tracking the maximum power point of TEG [2]. The TEG has low output voltage, though it is constructed from the module with series connected Seebek devices as shown in Fig. 2(a). Then it usually has to have a boost chopper circuit as shown in Fig.2(b). This boost chopper can step up the output voltage to the high voltage enough to connect to user load or utility system and may have the ability of maximum power point tracking (MPPT). TEG has the linear voltage-current characteristics as shown in Fig. 2(c) and the equivalent circuit is depicted with the constant dc voltage source Edc and the inner resistance Rin in series as shown in Fig. 2(b). To realize MPPT, measurement of the output voltage and the output current is required. The voltage measurement at the dc source terminal is easy. The measurement of current is usually realized by using the current transformer or the low resistance shunt resistor. The former is expensive and the latter cause the energy loss. Therefore, the authors have proposed to utilize the resistance of MOS-FET in the conduction mode to sense the current[3]. For MPPT control, precise measurement of the current is not necessary. Increment of the current is enough information for MPPT control.

II. PROPOSED SYSTEM

To realize MPPT, measurement of the output voltage and the output current is required. The voltage measurement at the dc source terminal is easy. The measurement of current is realized by two methods conventionally.

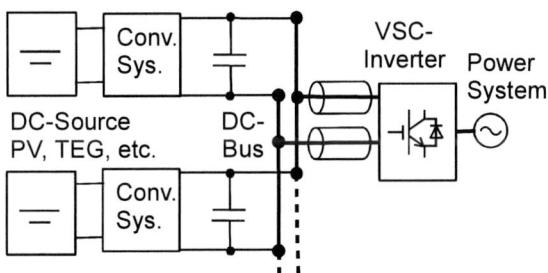

Fig. 1. Schematic diagram of parallec connected dc source

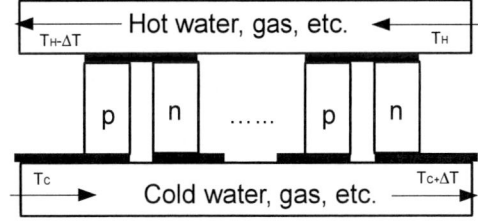

(a) Themo-electric generator (TEG) module

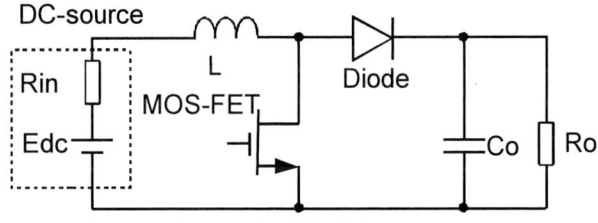

(b)Typical output circuit for DC energy source

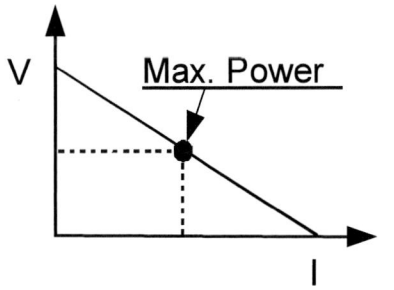

(c) Example of output characteristics

Fig. 2.   Example of DC energy source

One of them is using the dc-current transformer. This is expensive, since the electronic circuit is necessary to measure the dc magnetic flux intensity. Another is using the low resistance shunt resistor. This causes the energy loss, even though it is small. And both, of course, need the cost increase of assembly.

The authors propose to utilize the resistance of MOS-FET in the conduction mode (on-state) to sense the current, as shown in Fig. 3. The principle is as same as the shunt resistor method.

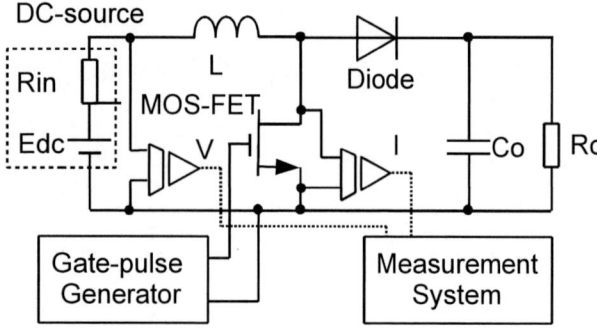

Fig. 3.   Proposed current measurement with on-state FET voltage

III. EXPERIMENTAL SYSTEM

However, some questions may occur about using MOS-FET, e.g. the linearity of resistance, the temperature change, noise suppression and timing control.

Experimental setup is shown in Fig. 4. It is required to know the average current and average voltage to measure the output power. Obviously, the center point of the on-state FET voltage represents the average current.

In our experimental equipment, triangular wave is used to generate the gate pulse. And the triangular wave is generated by integrating square wave. Then the falling edge of this square wave indicates the center point of the on-state FET voltage as shown in Fig. 6. Here, the switching period is about 1 mS, that means the switching frequency is 1 kHz.

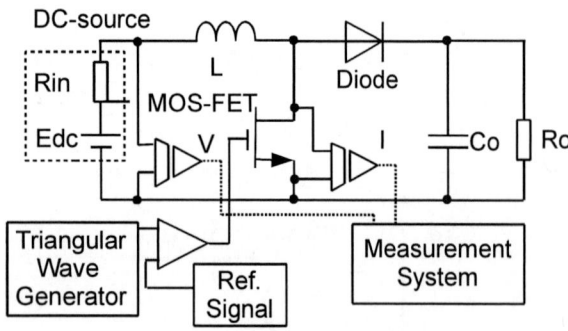

Fig. 4. Experimental setup of proposed measurement

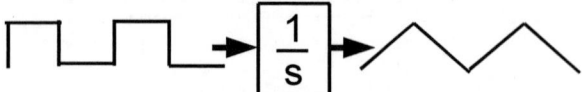

Fig. 5. Generation of triangular wave

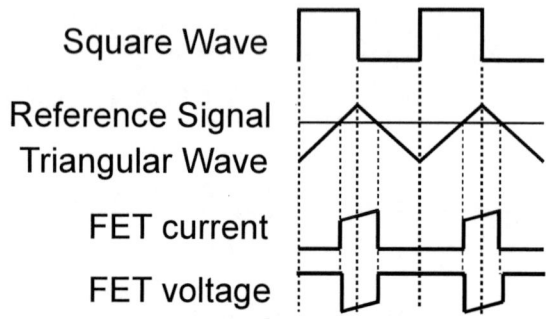

Fig. 6. Timing of square wave and FET current waveform

IV. EXPERIMENTS OF CURRENT MEASUREMENT

Table I shows the parameters of experimental circuit of dc current measurement. Fig. 7 shows the experimental results. Fig. 7(a) shows the voltage between drain and source of MOS-FET (2SK1520). On-state voltage changes from 0.2V to 0.3V. Fig. 7(b) shows the drain current of MOS-FET (2SK1520). The drain current changes from 0.7A to 1.05A. On-state resistance is approximately 0.3ohm.

Fig. 8 shows the voltage current characteristics of the MOS- FET (2SK1520) at on-state. The measured results show the good linearity and certify the constant resistance. Experimental results shown in Fig. 9 verify the coincidence between the timing of the square wave falling (down) edge and the center point of the on-state FET voltage. A/D conversion timing can be determined by the falling (down) edge of the square (rectangular) wave used for generating the triangular wave.

V. EXPERIMENTS OF MPPT CONTROL

Table II shows the parameters of experimental circuit of MPPT. Experimental results of Maximum Power Point Tracking (MPPT) control is shown in Fig. 10-13.

Purple line in Fig. 10 is D-A conversion output from the DSP controller. The output voltage changes from +5V to -5V, or vice versa, at the timing of the interrupt. It is confirmed that the interrupt occurs at the center point during the gate-on signal.

Light-cyan line is the output terminal voltage of the TEG. The shape is the reversed phase of the current, since when the current increases, the voltage increases.

TABLE I.   PARAMETERS OF CURRENT MEASUREMENT EXPERIMENT

Item	Value
Ro	98Ω
Co	3300uF
L	2mH
Edc	6V
FET	2SK1520

The 2014 International Power Electronics Conference

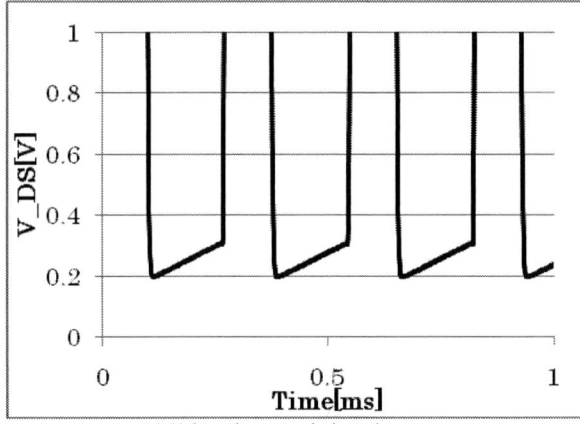

(a) Voltage between drain and source

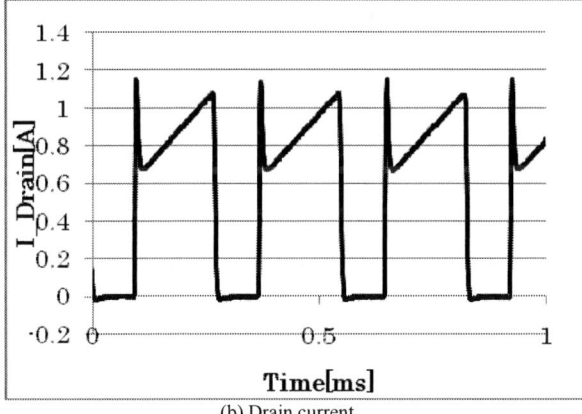

(b) Drain current

Fig. 7. Experimental results of proposed measurement (FET: 2SK1520)

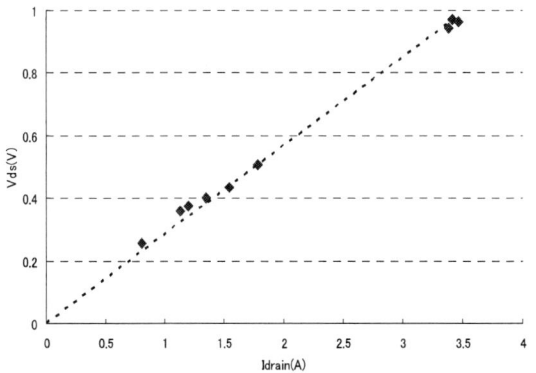

Fig. 8. Experimental results of proposed measurement (FET: 2SK1520)

Yellow line is the voltage between the drain and the source of the FET. It is very near to zero at the on-state.

Noise at the switching is very large because the ground level is common.

Fig. 11 shows the voltage of power supply and input terminal at steady state. Fig. 11(a) and (b) are in case of the voltages of the power supply are 5V and 6V, respectively. The voltages at the output terminal are 2.2V and 2.8V, respectively. They are slightly smaller than

theoretically expected values, which are the half of the voltage of the power supply.

Fig. 12 shows the voltage of power supply and input terminal at sudden change of the voltage of power supply from 7V to 6V at 1.2s. Then the MPPT program changes the duty factor gradually with the time interval of 50mS. The voltage at output terminal changes gradually from 3.2V to 2.8V. The response time is about 0.7s.

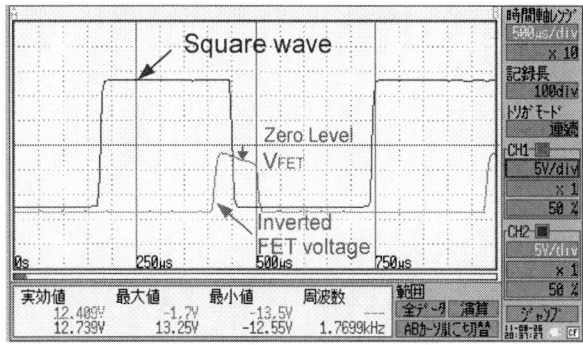

Fig. 9. Experimental results of proposed measurement (FET: 2SK1520) : Square wave and inverted FET voltage

TABLE II.   PARAMETERS OF MPPT EXPERIMENT

Item	Value
Rin	2.5Ω
Ro	$10\Omega/30\Omega$
Co	3300uF
L	1.535mH
Edc	5~7V
FET	2SK1520

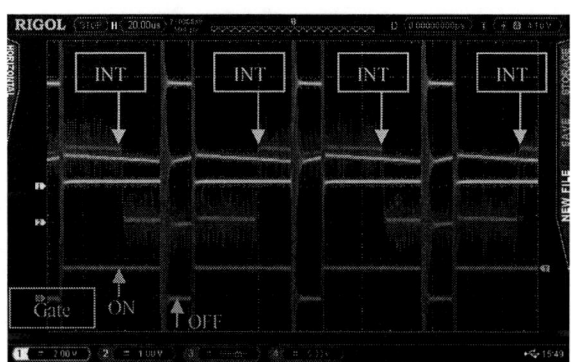

Fig. 10.   Experimental results of MPPT control at load change

Ch.1: FET voltage, ch.2: DC source output voltage, ch.3: D/A output, ch.4: Gate input

Fig. 13 shows the voltage of power supply and input terminal at sudden change of the load resistance between 10ohm and 30ohm at 1.2s.

Fig 13(a) shows the voltage drop soon after the load change because of increasing of the current. The output

978-1-4799-2706-7/14 $31.00 © 2014 IEEE   1301

voltage converges to 2.2V at 2s. Therefore the convergence time of the MPPT control is around 0.7s. It is slightly smaller than the duty before the change. The cause may be the precision of the current measurement, since the voltage value is very small. Proper amplifier may be applied here.

Fig 13(b) shows the voltage rise soon after the load change because of decreasing of the current. The output voltage converges to 2.3V at 2s.

Therefore the convergence time of the MPPT control is around 0.8S for load change. It may be fast enough to consider the change of the output characteristics of the thermoelectric power generator (TEG), which may not be changed abruptly in seconds.

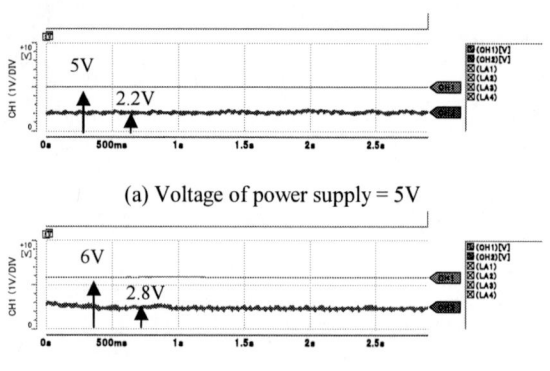

(a) Voltage of power supply = 5V

(b) Voltage of power supply = 6V

Fig. 11. Experimental results of MPPT control (Voltage of power supply and input terminal at steady state)

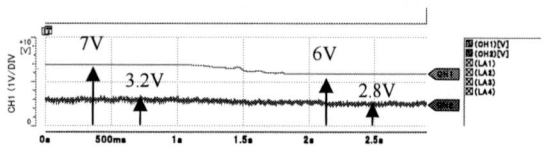

Fig. 12. Experimental results of MPPT control at inner voltage change

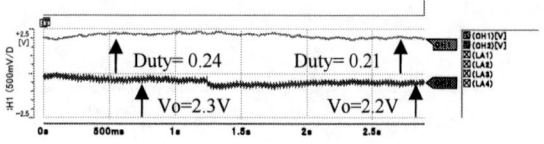

(a) Load resistance decreased from 30ohm to 10ohm

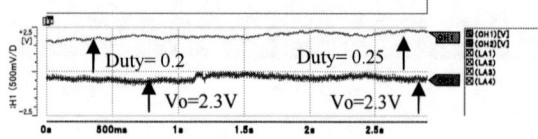

(b) Load resistance increased from 10ohm to 30ohm

Fig. 13. Experimental results of MPPT control at load change

VI. CONCLUSIONS

In this paper, new current measurement method, which uses the on-state FET voltage, has been investigated.

Experimental results are shown and they verify the feasibility of proposed measurement method.

A/D conversion and the timing of measurement has been considered and the possibility to use the falling (down) edge of the square (rectangular) wave used for generating the triangular wave is certified.

Experimental results certify that the little noise is observed in measured signal and the A/D conversion time is short enough to perform MPPT (Maximum Power Point Tracking) control with the proposed measurement method.

Experimental results of MPPT show good performance of the proposed system. Hill-climb (Perturbation) method is set at arbitrary time interval and duty factor change. It still can have the response of 0.7 S which may be fast enough for most of the applications.

This technology is expected to be effective to reduce the cost of current measurement system and to increase the reliability of the total control system. Authors also expect that it contributes making single control chip for DC energy source.

In near future, dc distribution system in office building or at home will be developed widely. In that system, many dc-dc converters would be installed. This method will be expected to reduce the cost of the dc-dc converter controller. Optimization in realistic system is the next step.

REFERENCES

[1] Singh, B.; Tan, L.; Date, A.; Akbarzadeh, A. Power Generation from Salinity Gradient Solar Pond Using Thermoelectric Generators for Renewable Energy Application. In Proceeding of the 2012 IEEE International Conference on Power and Energy (PECon), Kota Kinabalu, Malaysia, 2-5 December 2012; pp. 89-92, 2012.

[2] Hiroaki Yamada, Koji Kimura, Tsuyoshi Hanamoto, Toshihiko Ishiyama, Tadashi Sakaguchi and Tsuyoshi Takahashi, "A Novel MPPT Control Method of Thermoelectric generation with Single Sensor", Appl. Sci. 2013, vol.3(No.2), pp. 545-558; doi:10.3390/app3020545, 2013.

[3] Noriyuki Kimura, Yoshinori Sakoda, and Toshimitsu Morizane, "DC-DC Converter Current Measurement for MPPT Control", Proceedings of the 2011 IEEE 8th International Conference on Power Electronics - ECCE Asia (ICPE 2011-ECCE Asia), Jeju Korea, ThH1-4, May 30 - June 03, 2011.

[4] Myway Plus Corporation: - Digital Control System PE-Expert3:
http://www.myway.co.jp/en/products/pe_expert3.html

High Frequency Transformer Based on a Coupled Inductor Topology with Dielectric Isolation

Adrian Z. Amanci and Francis P. Dawson
Electrical and Computer Engineering
University of Toronto
Toronto, Canada
Email: zsa.amanci@mail.utoronto.ca

Harry E. Ruda
Materials Science and Engineering
University of Toronto
Toronto, Canada
Email: harry.ruda@utoronto.ca

Abstract—**This paper presents a new high-frequency coupled inductor topology in conjunction with a dielectric isolation device which provides improved efficiency and reduced input currents over conventional magnetic transformers. The main focus of the paper will be the design and operation of this new passive hybrid system, along with a comparison with a regular magnetic high frequency transformer based on simulations and experimental results.**

Keywords: dielectric isolation, mutually coupled inductors, high frequency transformers.

I. INTRODUCTION

The increased use of high frequency transformers is associated with the increase in operating frequency for power converters which provides two important advantages: a reduction in the converter footprint and mass, and an increase in the power density of the device [1], [2].

An unwanted consequence of this move towards higher switching frequencies is a reduction in the efficiency of the transformers used. The benefit of transformer size reduction is also more difficult to achieve because of issues such as winding loss, in the form of skin and proximity effects, and core loss in the form of Eddy currents and hysteresis [3]-[5]. These limitations have driven research in the area of capacitive coupling power transfer [6]-[7], but the transformer remains the main tool used for galvanic isolation and source-load matching.

The transformer winding loss can be reduced with the use of Litz wire or copper foil, but results in an increased manufacturing cost. The increased core loss experienced at MHz frequencies by most ferromagnetic materials decreases transformer efficiency substantially. Because of this, only a few select materials with low relative permittivity (3-20 range) can be considered as suitable magnetic material candidates [8].

So far, there have been two main paths taken to improve the efficiency of the transformer at high frequencies: changing the transformer layout, or changing the magnetic materials used. The standard layouts of high-frequency transformers are the flat core or racetrack topologies which have the advantage of a lower footprint, while providing

similar loss to a regularly wound transformer. Unfortunately, the disadvantages outlined above (winding loss, core loss) remain a limiting issue and are only exacerbated by the higher frequencies of operation (MHz range) [9], [10]. The other avenue taken for improving transformer efficiency has been the search for better engineered magnetic materials.

While several research groups have reported improvements on different manufactured materials [11], [12], their relative permittivity has not been significantly altered. In fact, the operation of many high frequency magnetic transformers (power loss density and quality factor) does not offer a significant performance increase over air-core transformers [13]. The issues of high losses in the MHz frequency range remains an ongoing issue for power transformers.

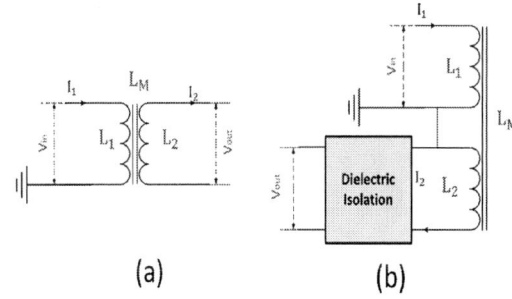

Fig. 1. Topology of (a) classic transformer and (b) proposed topology

The design alternative proposed in this paper is different from most other attempts at improving transformer performance. A different winding arrangement has been implemented in conjunction with an additional component rather than the conventional approach of changes to the geometry or material of a classical transformer. The proposed design alternative, denoted as the Mutual Branch Topology (MBT), is presented in Fig.1(b), while the classic design topology is illustrated in Fig.1(a). The new passive hybrid system uses two coupled inductors (in a series-parallel arrangement) together with a dielectric coupling isolation device to obtain the same transfer characteristics as a classical magnetic transformer at high frequencies (100kHz-1MHz), but with

978-1-4799-2706-7/14 $31.00 © 2014 IEEE

improved efficiency.

The advantages offered by the MBT, are improved efficiencies and lower input current than for an equivalent classical transformer. The main disadvantage of the MBT is that it requires an additional component, in the form of a galvanic isolation device such as the one presented in [14].

This paper is organized as follows: section II presents the principles of operation behind this new passive hybrid system and an equivalent circuit model. Section III presents a simulation case study in which the new model is compared to the classic transformer model. Section IV provides experimental results for a given configuration, and section V concludes the paper.

II. OPERATING PRINCIPLES

The simplest representation of a conventional transformer is having two windings on the same magnetic core, as shown in Fig. 2(a). Under ideal conditions (no losses and perfect coupling) the relationship between the voltage/currents associated with the terminals of the device (presented in circuit form representation in Fig. 2(b)) is described by the following two phasor form equations:

$$V_1 = j\omega L_1 I_1 \pm j\omega L_M I_2 \qquad (1)$$

$$V_2 = \pm j\omega L_M I_1 + j\omega L_2 I_2 \qquad (2)$$

where L_1, L_2 are the self inductances of the two windings and L_M represents their mutual inductance. Non-idealities

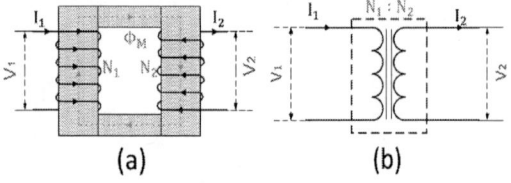

Fig. 2. Model of a two winding transformer: (a) physical implementation and (b) electrical equivalent circuit

can be modeled into the system with, for example, the help of leakage inductances (to account for leakage fields), series resistance (to account for winding loss) and parallel branch conductance (for the magnetic core loss). The analysis presented will focus on the winding loss since air core inductors were used in the experimental implementation.

The electrical circuit analysis of a classical transformer can be made easier when the two coupled windings are replaced by an equivalent circuit of uncoupled inductors. One of these equivalent circuit implementations is called the T-equivalent circuit, as shown in figure 3(b). In order to obtain the same current-voltage characteristics at the input and output terminals of a T-model, the following discrete inductance values need to be used:

$$L_A = L_1 \mp L_M \qquad (3)$$

$$L_B = L_2 \mp L_M \qquad (4)$$

$$L_C = \pm L_M \qquad (5)$$

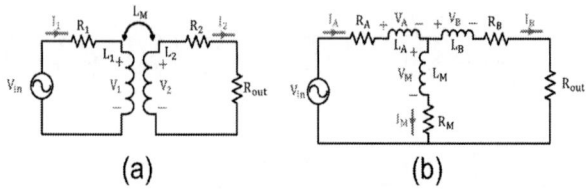

Fig. 3. Electrical circuit diagram of the (a) traditional two winding transformer and the (b) T-model implementation

The T-model is used as an analysis tool of the actual transformer, but there are no practical transformers implemented by using three discrete reactive components (two inductors and a capacitor), for the following reasons:

- Matching between the discrete inductors will be harder to achieve due to tolerance levels.

- The overall power losses of the T-implementation are larger than for the two-winding transformer.

The Mutual Branch Topology (MBT) has a similar architecture to the T-model. However, unlike the T-model, the MBT has mutual coupling between its two inductors, as shown in figure 4. Inductor L_z shown in Fig. 4 has negative reactance, and was implemented with a Multi-Layer Ceramic Isolation Device (MLCID) [14], which provides the required negative reactance and galvanic isolation for the MBT. The equations used to describe the behavior of the

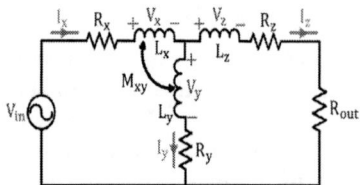

Fig. 4. Electrical Representation of the Mutual Branch Topology

system are provided in equation 6 on the following page. By selecting an appropriate value of the inductor magnitudes (L_x, L_y and L_z), and the mutual coupling term (M_{xy}), we are able to obtain the appropriate voltage and current step-up or step-down ratio. The 'best case scenario', that is, highest efficiency configuration was obtained with the help of a MATLAB developed optimization script, which is applied on a case by case basis. The input variables for the MATLAB script are the turns ratio desired for the MBT and the value of the series inductance L_x; the values for L_y, L_z and M_{xy} are then computed to provide optimal efficiency for a given load spectrum.

III. SIMULATIONS RESULTS

This section presents a case study for the operation of a 4:1 step-down transformer for three separate implementations: (1) the traditional model, (2) the T-model and (3)

978-1-4799-2706-7/14 $31.00 © 2014 IEEE   1304

The 2014 International Power Electronics Conference

$$
\begin{bmatrix} V_{in} \\ 0 \\ 0 \\ 0 \end{bmatrix} = \begin{bmatrix} R_x & R_y & 1 & 1 \\ j\omega L_x & j\omega M_{xy} & -1 & 0 \\ j\omega M_{xy} & j\omega L_y & 0 & -1 \\ j\omega L_z + R_z + R_{out} & -(j\omega L_z + R_z + R_y + R_{out}) & 0 & -1 \end{bmatrix} \cdot \begin{bmatrix} I_x \\ I_y \\ V_x \\ V_y \end{bmatrix} \tag{6}
$$

the MBT model. This case study was undertaken in order to illustrate the advantages provided by the MBT model. The parameters of the simulation are provided in Table I. V_{in} and f_0 are the input voltage and operating frequency, N_1 and N_2 are the turns ratios, and K_c and R_{MLT} are the inductance and resistance associated with one winding turn; L_1, L_2, R_1, R_2 are the parameters of the classical transformer, L_A, R_A, L_B, R_B, L_M, R_M are the parameters of the T-model, and L_x, R_x, L_y, R_y, L_z, R_z, k_{xy}, M_{xy} are the parameters of the MBT. The winding resistance was

TABLE I.   SIMULATION PARAMETERS FOR A 4:1 TRANSFORMER
AT 5MHZ

Simulation Parameters			
V_{in}	40V RMS	f_0	5MHz
N_1	40	N_2	10
K_c	0.25nH/turn	R_{MLT}	5mΩ/turn
L_1	400nH	R_1	200mΩ
L_2	25nH	R_2	50mΩ
L_A	300nH	R_A	173mΩ
L_B	-75nH	R_B	87mΩ
L_M	100nH	R_M	100mΩ
L_x	300nH	R_x	173mΩ
L_y	40nH	R_y	63mΩ
k_{xy}	0.9	M_{xy}	98.6nH
L_z	-	R_z	-

computed based on an identical resistance per turn for all three models, thus ensuring that the comparison performed is fair. Note that L_z has no values listed in Table 1, since the physically measured transfer function and losses of the integrated dielectric structures were used in this simulation, rather than the ideal values.

The formulae used to compute the power loss of the three different models are provided in equations (8) - (10):

$$
P_{loss}^{(1)} = R_1 \cdot |I_1|^2 + R_2 \cdot |I_2|^2 \tag{7}
$$

$$
P_{loss}^{(2)} = R_A \cdot |I_A|^2 + R_B \cdot |I_B|^2 + R_M \cdot |I_M|^2 \tag{8}
$$

$$
P_{loss}^{(3)} = R_x \cdot |I_x|^2 + R_y \cdot |I_y|^2 + R_z \cdot |I_z|^2 \tag{9}
$$

The MBT model provides better efficiency over the two-winding transformer and T-model, as shown in figure 5. The simulations were performed for the same parameters displayed in Table I.

The reason for the increased efficiency and decreased total losses of the MBT model can be seen from the lower current magnitudes, when compared to the two-winding or the T-model implementation, as exemplified in Fig. 6. Fig. 6 highlights an important advantage of the MBT implementation: the magnitude of the currents flowing through the inductor network is lower when compared to the classic transformer topology. This advantage is observed in a plot of

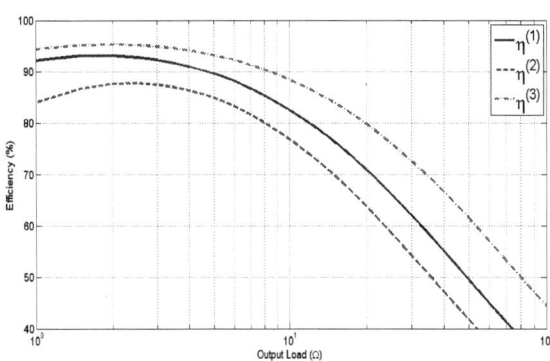

Fig. 5.   Efficiency of 5MHz implementations as a function of output load

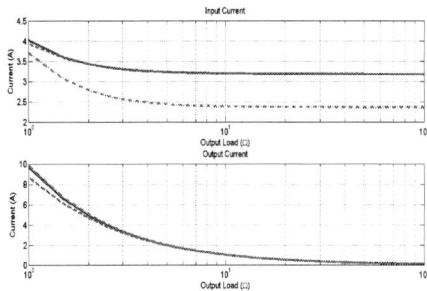

Fig. 6.   Magnitude of currents as a function of output load for the three models

the efficiencies of the three models: the two-winding model ($\eta^{(1)}$), the T-model ($\eta^{(2)}$) and the MBT model ($\eta^{(3)}$). While the efficiencies for high loads are very similar for the given parameters, there is a clear advantage of the MBT in terms of losses for light loads.

Fig. 7 displays the plot of the currents as a function of time of the three discussed models as a function of time for a 2.5Ω output load. Note that as previously mentioned, the magnitude of the input current is reduced for the MBT model, as opposed to the classic transformer topology.

Table II displays the parameters used for a second case study. This second case study shows that the advantages offered by the MBT model are not only valid for high frequency transformers since efficiency can also be achieved at lower (100kHz range) frequencies. The equivalent capacitance of the dielectric isolation device required for the MBT is approximately $0.33\mu F$. Figures 8 and 9 display the current magnitudes and the efficiency as a function of load for the three models tested at 100kHz. Note that in this lower operating frequency case, the efficiency difference between

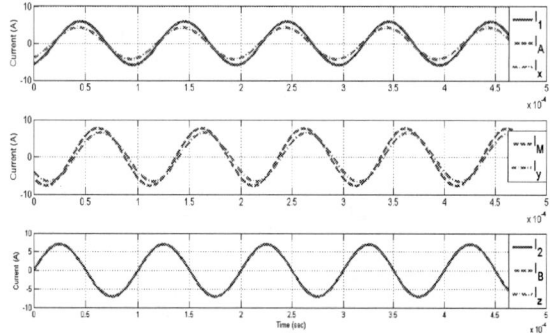

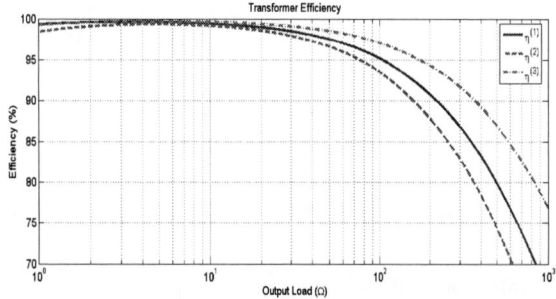

Fig. 7. Time domain simulation of the model currents for a 2.5Ω load

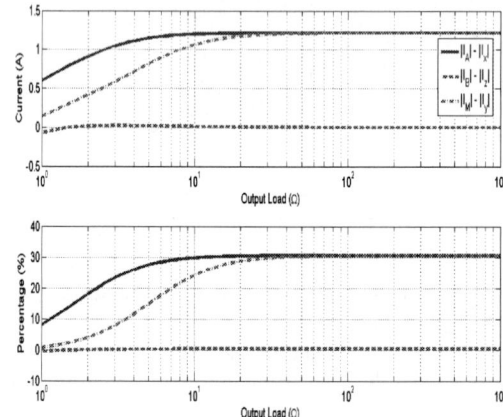

Fig. 9. Difference between the current magnitudes in the three compared models: (a) magnitude and (b) percentage difference

the MBT and 2-winding transformer is not as pronounced. This is because the self and mutual inductance associated with the windings is larger. Theoretically, the MBT model would have no advantage when dealing with infinitely large mutual and self inductances (the ideal case). Nevertheless, an improvement in the overall efficiency and reduction in magnetizing currents is possible even for lower operating frequencies.

TABLE II. SIMULATION PARAMETERS OF STEP-DOWN CASE STUDY AT 100KHZ

V_{in}	100V RMS	f_0	100kHz
N_1	400	N_2	100
K_c	2.5nH/turn	R_{MLT}	0.05mΩ/turn
L_1	400μH	R_1	20mΩ
L_2	25μH	R_2	5mΩ
L_A	300μH	R_A	17.3mΩ
L_B	-75μH	R_B	8.7mΩ
L_M	100μH	R_B	10mΩ
L_x	330μH	R_x	18.2mΩ
L_y	50μH	R_y	6.3mΩ
k_{xy}	0.9	M_{xy}	103.4 μH
L_z	-300μH	R_z	17.3mΩ

implemented on the same PCB as part of two identical step-down DC-AC resonant converters which were operated at 1MHz. The DC-AC converters can be separated into three main power components: (I) the magnetic circuit, (II) the capacitive isolation device and (III) the full bridge resonant converter.

A. The Magnetic Circuit

The magnetics section of the experimental setup was achieved with the help of a series of torroidal air-core inductor structures. The air-core inductors were built with copper traces on a F4 PCB (copper weight 3oz). The behavior and the inductance value of the air-core inductors were obtained with the help of analytical formulae and confirmed with the help of COMSOL Multiphysics 3.4. The layout of a stand-alone inductor and the results of a simulation for a 20 turn inductor are presented in figure 10. The mutual

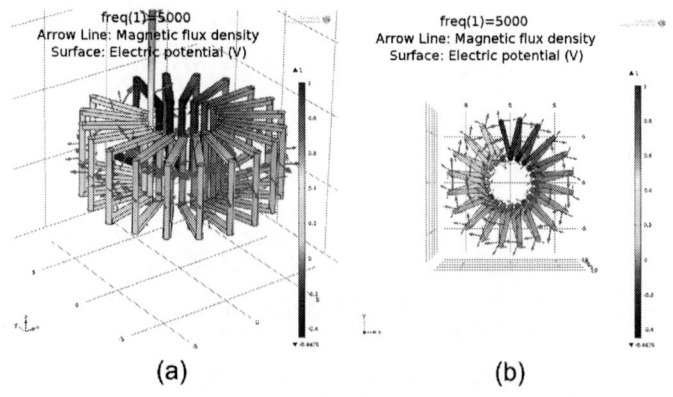

Fig. 10. COMSOL 4.3 simulation results for 20 turn inductor with magnetic field lines: (a) perspective and (b) top view

Fig. 8. Efficiencies of the three models as a function of output load: the two-winding model ($\eta^{(1)}$), the T-model ($\eta^{(2)}$) and the MBT model ($\eta^{(3)}$)

IV. EXPERIMENTAL RESULTS

The experimental results presented provide verification for the simulation results introduced in the previous section. Both the classic transformer model and MBT model were

branch network was designed for the following inductance values: $L_x \approx 300nH$, $L_y \approx 50nH$ and $M_{xy} \approx 85nH$. The inductance values were verified with the help of an

978-1-4799-2706-7/14 $31.00 © 2014 IEEE

LCR meter by measuring the inductance seen between the various terminals. The inductance measurements obtained for the series inductor (L_x) and parallel inductor (L_y) are within 10% of the required values. On the same note, the measured equivalent resistance of the traces was approximately 400mΩ, which is larger than the value obtained with the COMSOL simulations.

B. Capacitive Isolation Device Unit

The capacitive isolation was realized with the help of two Multi-Layer Ceramic Isolation Devices, which were introduced in a previous paper [14]. The purpose of the MLCID is to provide galvanic isolation at high frequencies through capacitive coupling while achieving high transfer power efficiency. The two MLCIDs used are comprised of 5 multi-layer ceramics stacks and were manufactured by TRS Technologies Inc. 150μm Barium Strontium Titanate (BST) and 500μm Alumina layers were used for the dielectric layers. The dielectric layers were sintered individually and bonded together with epoxy resin. Cr/Au (500A/2000A) layers were used for the electrode plates. The devices have a rated voltage of 300V and a rated current of 2A, with a power rating of 600VA and a maximum power density of 920W/cm^3 (see Table III for details).

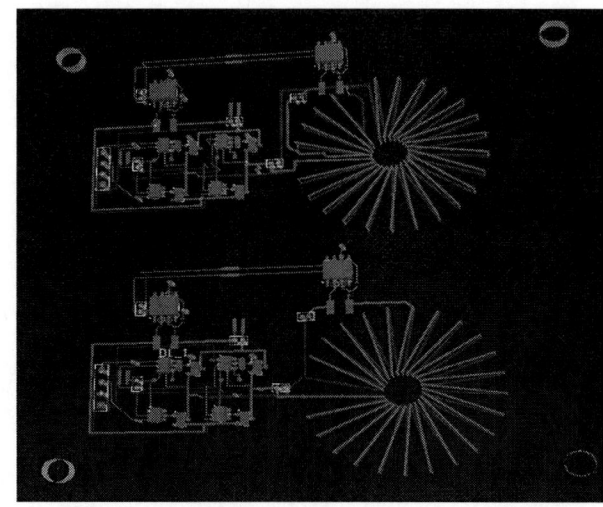

Fig. 11.   Overall layout of converter with mutual branch model (top) and converter with regular transformer (bottom)

TABLE III.   MLCID PARAMETERS

Footprint (mm)	10x10x6.5	Leads Radius	2mm
No. of plates	20	Electrode (Cr/Au)	500A/2000A
BT - ϵ_{high}	1300	Alum - ϵ_{low}	9
Loss Tan (BT)	0.68%	Loss Tan (Alum)	0.15%
Thickness (BT)	150μm	Thickness (Alum)	500μm
Rated Voltage	300V	Rated Current	2A
Power Rating	600VA	Power Density	920W/cm^3

C. Resonant Converter

The purpose of the full bridge converter is to create a 1MHz AC voltage waveform at the load from a DC input. The legs of the full bridge resonant converter were controlled by two half-bridge gate drivers. The half-bridge gate drivers were controlled with phase shifted gating signals generated at 1MHz with the help of an Altera DE2 board. The PCB layout of both converters (the one with the regular transformer and the one with the Mutual Branch Topology) is shown in Fig. 11.

D. Efficiency Results

The experimental results for the efficiency of the two transformer topologies (classic and MBT) are shown in Fig. 12. The experiment was performed for each implementation, for three different loads, with converter operation at 1MHz, and an input voltage of 30V. The efficiencies were obtained by measuring the voltage drop across the output load, computing the output power, and dividing it by the total input power.

Note that while the efficiencies are relatively low for both topologies (mainly due to the increased resistance

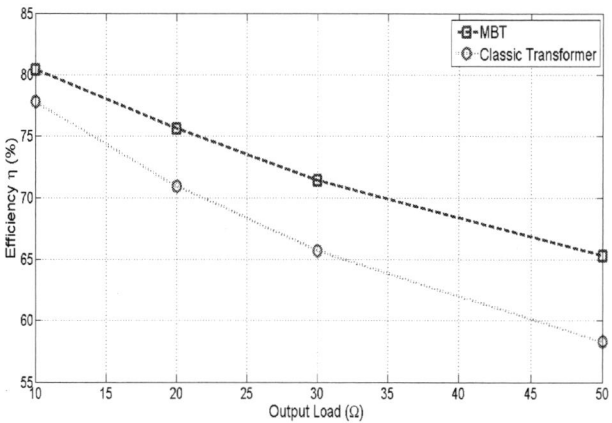

Fig. 12.   Efficiency measurements for the classic transformer and MBT

associated with the copper traces and converter losses), the MBT model has better overall efficiency, especially for lighter loads.

V.   CONCLUSION

This paper has presented an alternative approach to the design and use of high frequency transformer. A comparison was made between the classical two winding transformer and the Mutual Branch Model. The simulation results presented show that the new model is able to provide improved efficiency (as large as 2-3% for high load and 10% for low load) over the conventional two-winding transformer. These simulation results were confirmed by experimental tests in the form of two resonant converters - one using the MBM, and one using the classic transformer. The experimentally performed comparison was done with identical resonant converters and with the two resonant networks having a similar footprint. The main advantage of the Mutual Branch

Model however, remains the lower required input current which implicitly results in lower design constraints on the transformer and on the size of the windings required.

REFERENCES

[1] D. J. Perreault et al., "Opportunities and challenges in very high frequency power conversion," in 24th Annu. IEEE APEC, Washington DC, 2009, pp. 1-14.

[2] P. Pilawa-Podgurski, A. D. Sagneri, J. M. Rivas, D. I. Anderson and D.J. Perreault, "Very-high-frequency resonant boost converters," IEEE Trans. Power Electron., vol. 24, no. 6, pp. 1654-1665, 2009.

[3] O. A. Hassan, C. Klumpner and G. Asher, "Design considerations for core material selection and operating modes for a high frequency transformer used in an isolated DC/DC converter," Proc. 14th EPE, Birmingham, 2011, pp. 1-11.

[4] J. Lu and F. Dawson, "Analysis of Eddy current distribution in high frequency coaxial transformer with Faraday shield," IEEE Trans. Magn., vol. 42, no. 10, pp. 3186-3188, 2006.

[5] W. G. Odendaal and J. A. Ferreira, "A thermal model for high-frequency magnetic components," IEEE Trans. Ind. Appl., vol. 35, no. 4, pp. 924-931, 1999.

[6] M. Kline, I. Izyumin, B. Boser and S. Sanders, "A transformerless galvanically isolated switched capacitor LED driver," in 2012 27th Annu. IEEE APEC, Orlando, FL, pp. 2357 - 2360.

[7] M. Kline, I. Izyumin, B. Boser, and S. Sanders, "Capacitive power transfer for contactless charging," in 2011 26th Annu. IEEE APEC, Forth Worth, TX, pp. 1398-1404.

[8] Y. Han, G. Cheung, A. Li, C. R. Sullivan and D. J. Perrault, "Evaluation of magnetic materials for very high frequency power applications," IEEE Trans. Power Electron., vol. 27, no. 1, pp. 425-435, 2008.

[9] S. Yang, S. Abe, and M. Shoyama, "Design consideration of flat transformer in LLC resonant converter for low core loss," IPEC, pp. 343-348, Sapporo, 2010.

[10] J. Zhang, J. Wang and X. Wu, "A capacitor-isolated LED driver with inherent current balance capability," IEEE Trans. Ind. Electron., vol. 59, no. 4, pp. 1708-1716, 2011.

[11] Y. Shirakata et. al., "High Permeability and Low Loss Ni-Fe Composite Material for High-Frequency Applications," IEEE Trans. Mag., vol. 44, no. 9, pp 2100-2106, Sept. 2008.

[12] A. Urata et al., "Low core loss of non-Si quaternary $Fe_{83.3}B_8P_8Cu_{0.7}$ nanocrystalline alloy with high B_s of 1.7T," J. Appl. Phys., vol. 111, 07A335, 2012.

[13] J. Qiu, H. Syed, C. Sullivan, "Complex Permeability Measurements of Radial-Anisotropy Thin-Film Magnetic Toroidal Cores," ECCE 2013, pp. 1660-1667, Denver, Colorado, Sept. 2013.

[14] A. Z. Amanci, H. E. Ruda, F. P. Dawson, "New Capacitive Coupling Device for High Frequency Applications," ECCE 2013, pp. 3726-3732, Denver, Colorado, Sept. 2013.

978-1-4799-2706-7/14 $31.00 © 2014 IEEE

The 2014 International Power Electronics Conference

Concept and Experimental Evaluation of a Novel DC – 100 MHz Wireless Oscilloscope

Yanick Lobsiger*[†], Gabriel Ortiz*[†], Dominik Bortis*[†], and Johann W. Kolar*

*ETH Zurich
Power Electronic Systems Laboratory
Physikstrasse 3, 8092 Zurich, Switzerland
lobsiger@lem.ee.ethz.ch

[†]Enertronics GmbH
c/o ETH Zurich, Power Electronic Systems Lab.
Physikstrasse 3, 8092 Zurich, Switzerland
info@enertronics.ch

Abstract—A wireless oscilloscope featuring 100 MHz analog bandwidth and ultra-high isolated measurement channels is presented in this paper. In order to visualize the benefits of this potential-free measurement system, the paper starts with a revision of commonly utilized isolated voltage and current measurement concepts. With a clear understanding of the benefits given by a high-bandwidth, high-isolated and highly common mode immune measurement system laid down, the details about the wireless oscilloscope and its potential-free input channels are presented. Experimental measurements comparing the state of the art isolated voltage and current measurement concepts with the proposed wireless oscilloscope regarding common mode rejection ratio and signal-to-noise ratio are presented, showing the excellent performance of the introduced measurement system.

I. INTRODUCTION

During the experimental testing of power electronic circuits, accurate measurement of the converter's voltages and currents is required in order to ensure the correct operation of the system. Moreover, very often these currents and voltages are referred to individual floating potentials, thus the measurement principle must provide a minimum isolation level in order to be safely coupled to the system.

As an example, Fig. 2 (a) shows a state of the art solution for an electric drive system. Here, a cascaded, i.e. series, arrangement of N cells allows the connection of the drive system directly to Medium Voltage (MV), whereby the isolation and voltage adaptation is performed by a high-power DC-DC converter. The outputs of these DC-DC converters are connected in parallel in order to supply the inverter driving the electrical machine.

A detailed internal representation of one converter cell is shown in Fig. 2 (b). This cell consists of an input H-bridge responsible for the rectification of the input voltage while performing power factor correction. At the input of this rectifier, typically a 50 Hz input current with superimposed high-frequency ripple is found along with the cells' switched voltage waveform. It should be noted that, due to the series connection at the input, these voltages and currents are not referred to ground but rather to floating and fast changing potentials. Moreover, within the converter's cell, the DC-DC converter is characterized by an operation at medium-frequency level, whereby fast dynamic behavior of the transformer voltages and currents are expected. Additionally, often a switching performance assessment, involving simultaneous measurement of its emitter current, collector-emitter and gate voltages is desired in order to ensure the safe operation of the semiconductor device. In this case, fast transients are to be expected during the switching process, thus a high-bandwidth, isolated and highly common mode immune measurement concept is required.

This paper provides a revision of the state of the art isolated voltage and current measurement concepts in Section II, whereby the main limitations of these concepts will be highlighted. With the

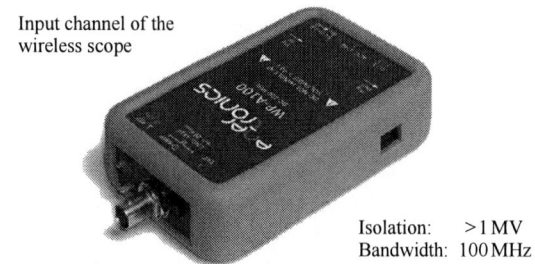

Input channel of the wireless scope

Isolation: >1 MV
Bandwidth: 100 MHz

Fig. 1: Enertronics' wireless oscilloscope channel, featuring 100 MHz analog bandwidth, 400 MS/s sampling rate, no intrinsic limit on isolation and high common mode immunity.

desired features of the new measurement concept outlined, the details of the proposed wireless oscilloscope depicted in Fig. 1 will be shown in Section III. The performance of the proposed oscilloscope in comparison to state of the art voltage and current measurement concepts under typical testing conditions (floating potentials, fast transients and low voltage signals, among others) is presented. The paper finalizes with a summary and outlook of further improvements and extensions of the wireless scope in Section V.

II. STATE OF THE ART MEASUREMENT CONCEPTS

In order to identify the main limitations of today's isolated measurement principles for voltage and current in power electronic systems, a review of these concepts will be presented in the following sections.

A. Isolated Voltage Measurement

The main utilized isolated measurement concepts are *1) Differential Probes*, and *2) Optic-based Concepts*, whereby the details about their operation are presented in the following.

1) Differential Probes: Differential measurement of voltages is the most common method for safely measuring high voltages which are not referred to the scope's ground potential. These differential probes typically feature an high-ohmic input voltage divider followed by an instrumentation amplifier, as shown in Fig. 3 (a). By appropriately selecting the resistors in this circuit, the output voltage v_M of the probe can be described by

$$v_M = v_{IN} \cdot A_{DM} + v_{CM} \cdot A_{CM}, \qquad (1)$$

where v_{IN} is the input voltage to be measured, v_{CM} is the common mode voltage referred to the scope's ground potential, A_{DM} is the probe's differential mode gain and A_{CM} is the probe's common mode

978-1-4799-2706-7/14 $31.00 © 2014 IEEE

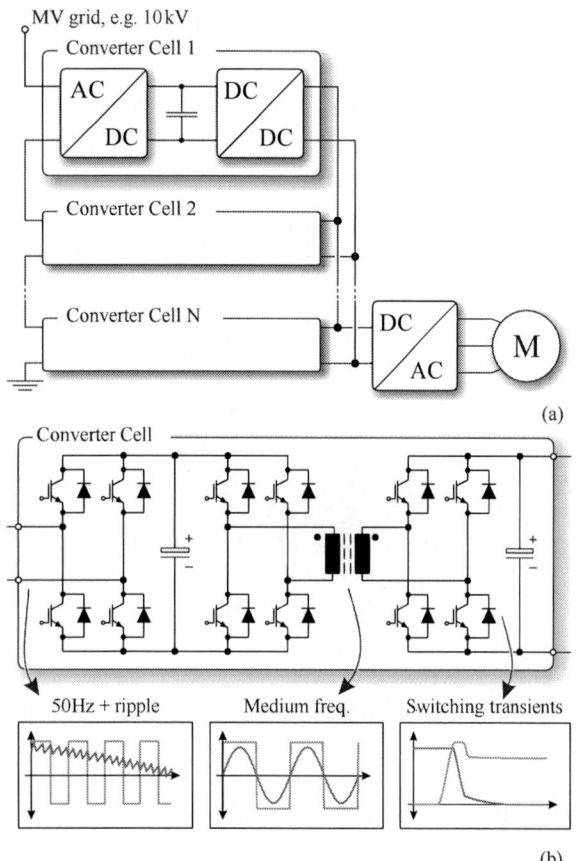

Fig. 2: Typical power electronic system to be measured by isolated voltage and current probes. In (a), a cascaded connection of cells supplying ultimately a machine is presented. These cells, whose detailed view is shown in (b), are characterized by a floating potential given by the aforementioned series connection at the input side.

attenuation. Ideally this common mode attenuation A_{CM} is close to zero, achieving the ideal relation between input and output voltage

$$v_M \approx v_{IN} \cdot A_{DM}. \tag{2}$$

In practice however, due to component tolerances and differences in the AC-characteristics of the probe's inputs, the value of A_{CM} is not zero and deteriorates with increasing frequency, thus a comparably low Common Mode Rejection Ratio (CMRR) is achieved. This introduces errors in the measurement due to a strong presence of common mode voltage components in the output signal v_M. In addition, due to the high input voltage divider, the differential probe suffers from a low Signal-to-Noise Ratio (SNR).

2) Optical-Based Concepts: In order to overcome the limitations concerning SNR, CMRR and input voltage range, isolated measurement systems can be used. In the following, different concepts providing a galvanic isolation between a floating reference voltage and earth ground are presented.

The simplest way to decouple the measurement system from any ground potential is to use a battery powered measurement system, which consequently features an unlimited voltage isolation [1–3]. In addition, the coupling capacitance between the measurement system connected to the floating circuitry and ground is small, thus, the

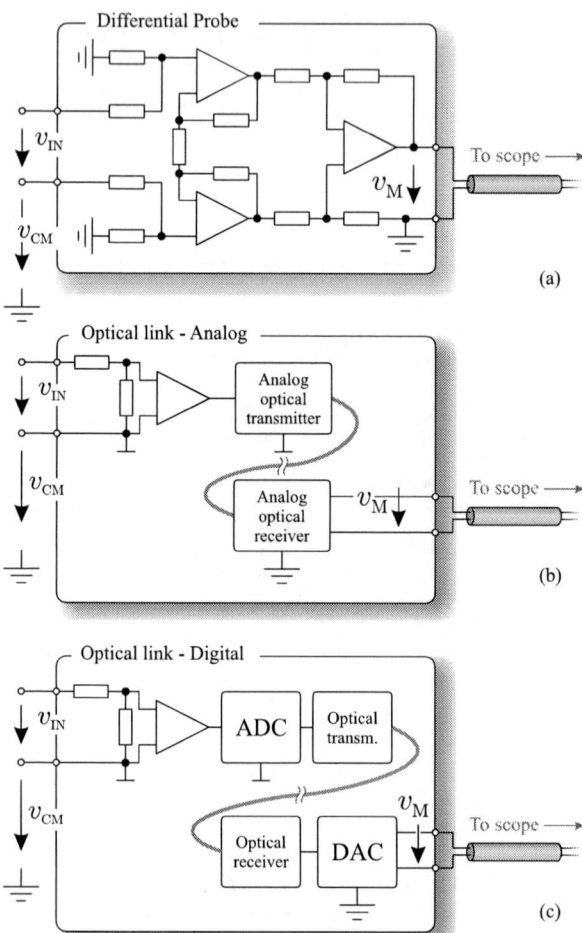

Fig. 3: Isolated voltage measurement concepts: (a) Differential probe based on an input instrumentation amplifier; (b) Analog optical link, the analog signal is modulated and transmitted through an fiber-glass optic link; (c) Digital optical link, the analog signal is converter to digital at the probe side and transmitted via fibre optic to the receiver side where its reconverted to analog.

capacitive current caused by common mode voltage transients and the influence of the measurement system on the current or voltage waveforms can be minimized.

Unlimited voltage isolation of the signal path can be achieved e.g. by means of an analog optical link where the light intensity is modulated as a function of the input signal's amplitude as shown in Fig. 3 (b) and presented in [1]. The major disadvantage of an analog optical transmission systems is the ageing of the optical transmitter and receiver units, which have to be re-calibrated periodically.

If the optical signal transmission is performed digitally as shown by the scheme presented in Fig. 3 (c), these ageing effects are negligible. Then, for the optical link, which is used as a real-time signal isolator of a digital channel, a data transfer rate of several Gigabit per second is needed in order to achieve a reasonable analog bandwidth of around 100 MHz. Even if such fast optical links are commercially available, these transmitters are not suitable for battery powered systems due to their high power consumption - typically in the range of one Watt - which leads either to a reduced battery runtime or a larger measurement system with higher battery storage capacity.

The 2014 International Power Electronics Conference

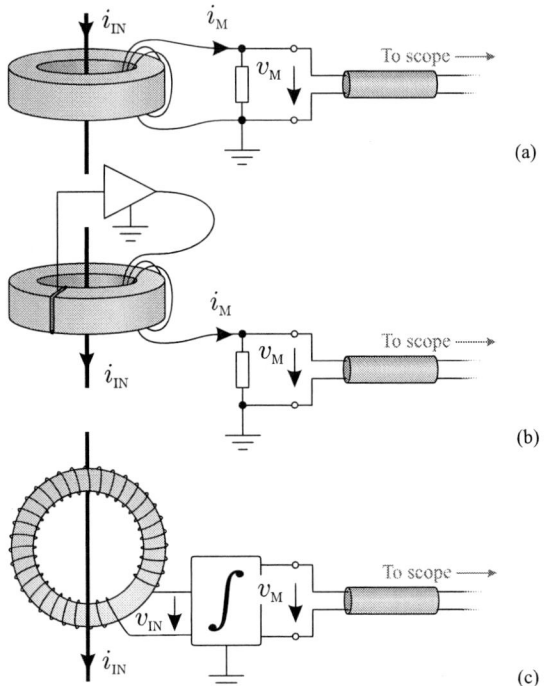

Fig. 4: Isolated current measurement principles: (a) Current transformer with burden resistor; (b) Compensated current transformer for measurement of DC and AC signals; (c) Rogowski coil and its respective integration block.

Furthermore, depending on the measurement system's housing, a higher power consumption could also result in possible overheating of the measurement system. On the other hand, however, if the housing of the measurement system is enlarged, the coupling capacitance of the measurement system to ground is also increased, thus, degrading the measurement quality.

B. Isolated Current Measurements

In case of isolated current measurement principles, the main utilized concepts in power electronics are *1) Current transformer, 2) Compensated Current Transformer* and *3) Rogowski Coil*, all of which are presented in Fig. 4 and will be described in the following.

1) Current Transformer: A current transformer consists of a magnetic core with a secondary winding and a burden resistor. The current to be measured i_{IN} flows through the core's window enabling a magnetic coupling of the input current i_{IN} and the output current i_M. The burden resistor R_B is then responsible for translating this measured current i_M into a voltage signal v_M to be captured by the scope. This way, the relation between input current i_{IN} and the measured voltage v_M is:

$$v_M = R_B \frac{i_{IN}}{N_{CT}}, \qquad (3)$$

where N_{CT} is the number of secondary turns if the current transformer.

However, the expression in (3) is only valid as long as a good coupling between primary and secondary can be ensured, which is not the case at low frequencies [4]. As a consequence, current transformers feature a high-pass characteristic and are therefore not suitable for measurement of DC and/or low frequency signals.

Furthermore, the upper corner frequency of the current transformer, defining the bandwidth of the sensor, is related to the leakage inductance of the transformer, which is typically kept as low as possible by minimizing the size of the sensor. Reducing the size of the sensor, however, deteriorates its isolation performance given the comparatively small distances available for introduction of isolation layers. The result is a clear trade-off between the bandwidth and isolation level of the current sensor.

2) Compensated Current Transformer: As mentioned earlier, a current transformer is not able to measure currents in the low frequency range. In order to overcome this fundamental limitation, a magnetic flux sensing device can be inserted in the air-gap of the core as depicted in Fig. 4 (b). The signal from this sensor is utilized in a closed loop arrangement in order to compensate the low frequency components of the measured current and therefore extending the operation range of the sensor to the low frequency (including DC) range [5], thus enabling measurement of current components from DC to the high MHz range.

The key component in this compensated current sensor arrangement is the flux sensor which is most commonly a semi-conductive Hall-element [6], or a flux-gate-based sensor [7]. The implementation of these flux sensors involves advanced manufacturing techniques and compensation circuits within the sensor. In addition, since the current transformer is still responsible for the measurement of the high-frequency spectra of the current, the same trade-off between isolation and bandwidth is found as in an uncompensated current transformer.

3) Rogowski Coil: A very common current measurement concept in power electronics is the Rogowski coil. This sensor is based on inductive coupling of the input current i_{IN} and an air coil surrounding this conductor as shown in Fig. 4 (c). This inductive coupling results in an coil output voltage which is proportional to the derivative of the input current i_{IN}. It is therefore necessary to integrate this signal in order to obtain a voltage signal v_M proportional to the input current i_{IN},

$$v_M = \int v_{IN} dt = M \int \frac{di_{IN}}{dt} dt, \qquad (4)$$

where M represents the coupling inductance between the input conductor and the Rogowski coil. As can be seen, an integration is required in order to reconstruct the input current's waveform. This integration process represents the main challenge in the implementation of Rogowski-coil-based current probes, as problems associated with drifting due to DC biases are very often encountered. This drifting problems are counteracted with circuits featuring a high-pass characteristic in the low frequency range of the integration circuit, thus no DC signals can be measured [8].

In addition, as well as with the current transformers, the dimensions of the coil itself are directly related to the bandwidth of the sensor, resulting in a tight coupling between isolation and bandwidth.

C. Summary of State of the Art Isolated Measurement Concepts

Isolated voltage and current sensor manufacturers offer solutions with one of the previously described principles featuring different bandwidths and isolation levels. These solutions are conveniently summarized in a bandwidth vs. isolation map in Fig. 5. In case of the voltage sensors, manufacturers [9–13] were selected for the differential probe measurements while [14, 15] were considered for the optically isolated measurement concept. In case of current measurement principles, [9, 10] represent the compensated current transformer concepts while [16] and [13] were selected for the

978-1-4799-2706-7/14 $31.00 © 2014 IEEE

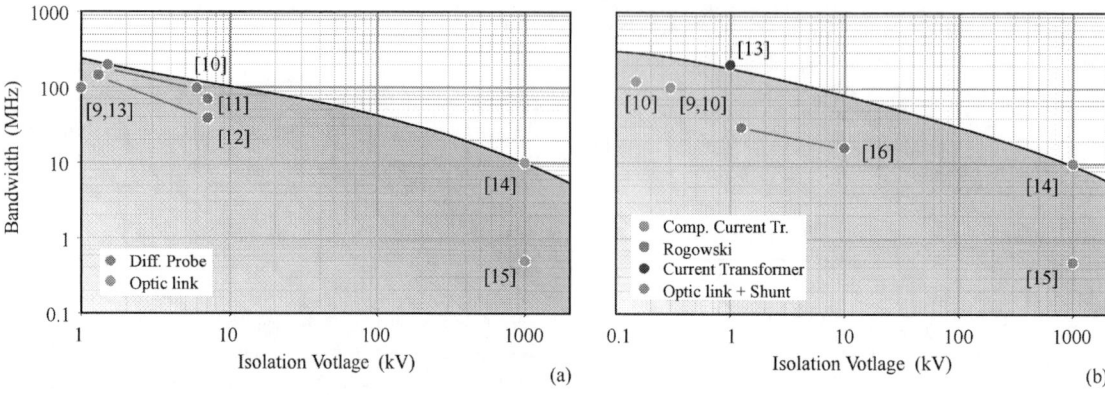

Fig. 5: Summary of state-of-the-art isolated voltage and current measurement concepts displayed in a bandwidth vs. isolation voltage for (a) voltage sensors and (b) current sensors.

Rogowski and current transformer principles respectively. In addition, a optically isolated measurement as the ones described in the voltage measurement principles are considered since a non-isolated current transducer, e.g. a shunt resistor, can be attached at its input, thus enabling isolated current measurement.

From Fig. 5 the aforementioned trade-off between bandwidth and isolation in voltage and current sensing principles can be confirmed by manufacturer-provided data. In the voltage map (cf. Fig. 5 (a)) the highest bandwidth sensor is fabricated by [10] with a maximum specified isolation of 1.5 kV whereas the highest isolation solution is represented by [14] with a bandwidth of 10 MHz. In case of the current map shown in Fig. 5 (b) the highest bandwidth is achieved by the current transformer from [13] with a maximum isolation voltage of 1 kV and the maximum isolation voltage is provided by the coupled non-isolated current measurement with the optic-based solutions, as mentioned earlier.

The bandwidth and isolation limitations represented in Fig. 5 can be overcome by a new concept in oscilloscope data acquisition: If the isolation is provided by each channel of the oscilloscope, the voltage / current probe can be optimized for maximum bandwidth and minimum isolation, thus reaching an unprecedented performance in isolated measurement systems. This is main principle of the wireless voltage probe proposed in this paper, whereby its implementation details and the means to achieve a high bandwidth / sampling rate are detailed in the next section.

III. WIRELESS OSCILLOSCOPE

In contrast to the state of the art isolated voltage and current probes detailed in the last section, a combined high voltage and high bandwidth isolation of an oscilloscope's input channel can be achieved, if the isolation is not provided to the real-time signal / data stream but only to that data, which is recorded / stored by the scope. In this way, the bandwidth requirement on the isolated channel is significantly lower and can be achieved by a digital optical or even a wireless data transmission. This concept is followed by the proposed wireless oscilloscope, which is described in the following.

A schematic overview of the wireless oscilloscope is depicted in Fig. 6. It consists of multiple isolated input channels and a graphical user interface, which exchange the configuration parameters and the measurement data wireless. The battery powered operation of the channels enables a autonomous and fully isolated operation of the system.

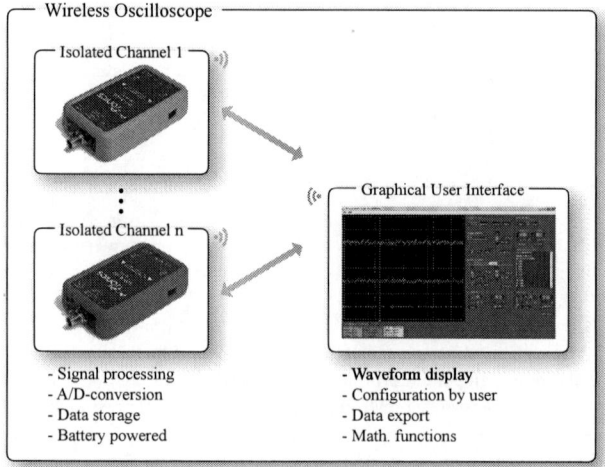

Fig. 6: Schematic overview of the wireless oscilloscope consisting of multiple isolated input channels and a graphical user interface (GUI) running on a computer. The wireless communication between the channels and the GUI is based on the Bluetooth technology.

A. Hardware and Software Implementation

In order to demonstrate the feasibility and finally also to proof the performance of the wireless scope, a hardware prototype has been built and a GUI prototype has been programmed. In the following, these prototypes will be explained in more detail.

1) Isolated Input Channel: The hardware part of the wireless oscilloscope, i.e. an isolated measurement channel, has been implemented in a compact enclosure with silicone cover as shown in Fig. 7. It features a BNC connector for the measurement signal input, as it is the case for a conventional oscilloscope, enabling the use of any kind of external sensor with BNC output. An external passive probe can be connected and calibrated by means of the calibration signal output. If multiple channels are used simultaneously, the internal synchronization platform performs a high accuracy wireless synchronization of typ. ±10 ns in between all probes. This synchronization error can ca. be halved, if the channels are connected among each other using the optical trigger in- and outputs. The internal Li-Ion batteries are charged by means of an external DC power supply. An overview of the internal setup of an input channel is depicted in Fig. 8 and the

The 2014 International Power Electronics Conference

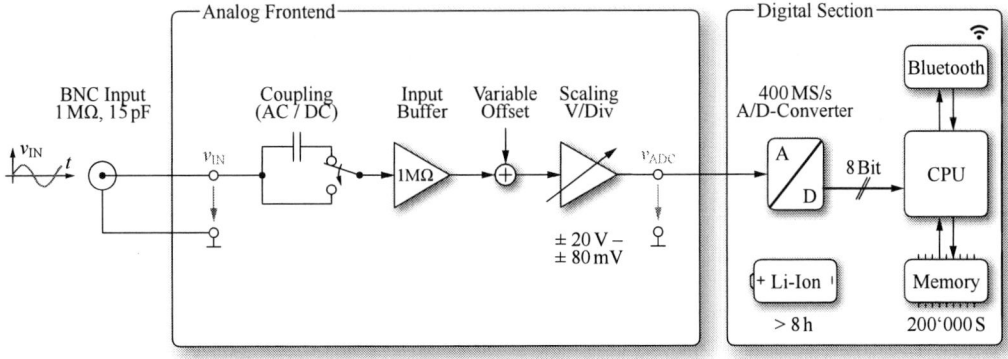

Fig. 8: Schematic overview of an input channel's internal setup.

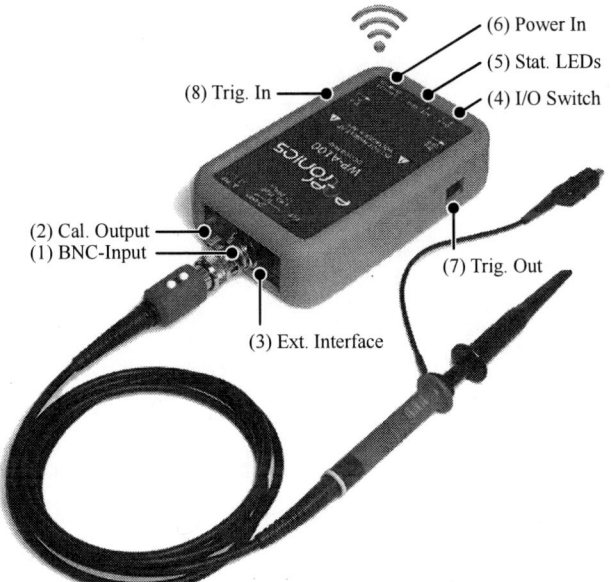

Fig. 7: Prototype of the wireless oscilloscope with physical dimensions of 141 mm × 81 mm × 32 mm. Front side: (1) BNC connector for the input signal, (2) calibration signal output and (3) interface for external sensors; back side: (4) I/O power switch, (5) status and charge indication LEDs and (6) connector for the battery charger; left / right side: (7) optical trigger out- and (8) inputs.

scope's corresponding specifications are listed in Table I.

2) Graphical User Interface (GUI): The counterpart to the isolated channels is the GUI which is similar to the mechanical knobs and the screen of a conventional oscilloscope, where the waveforms of all channels are displayed and where the user can make the configuration for all channels of the wireless oscilloscope. A screen shot of the implemented GUI is depicted in Fig. 9. Contrary to a conventional oscilloscope, where typically only four analog channels can be used, the GUI theoretically enables to connect an unlimited amount of fully isolated input channels. On the left hand side, the waveforms and the corresponding settings of all channels are displayed. On the right hand side, the user can configure the trigger, the individual vertical and horizontal settings of all channels, the cursors and the zoom-function. The GUI also allows to export and import measurement data and settings.

TABLE I: Specification overview of the wireless oscilloscope prototype.

Analog Bandwidth	DC − 100 MHz
Sampling Rate (max.)	400 MS/s
Memory Depth (max.)	200 kS
Resolution	8 Bit
Input Voltage Range	±80 mV . . . ± 20 V
	±800 mV . . . ± 200 V (10:1 passive probe)
	. . .
Input Impedance	1 MΩ, 15 pF
Battery	Li-Ion, Rechargeable
Autonomy (typ.)	8 h
Communication	Bluetooth
Physical Dimensions	141 mm × 81 mm × 32 mm
Weight	350 g

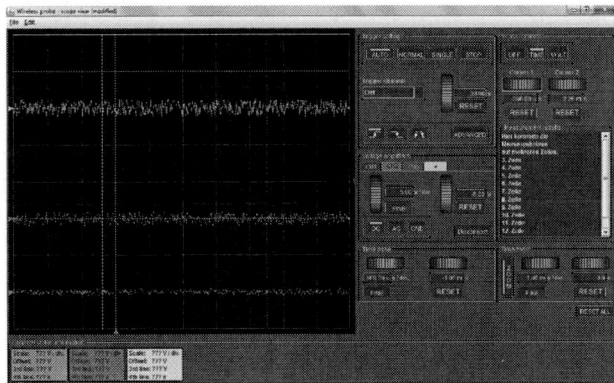

Fig. 9: Graphical User Interface prototype of the wireless oscilloscope implemented in Java. User settings: trigger mode, channel, level and slope; vertical settings like Volts/div, offset, signal coupling; horizontal settings such as Time/div, trigger position; horizontal zoom function; cursors. Outputs: display of waveforms and settings for each channel; cursors; split screen view in zoom-mode; export / import of measurement data and settings.

By means of the implemented wireless oscilloscope prototype, an evaluation of this system and a comparison to commercial isolated measurement systems will be made in the next section.

IV. Experimental Verification

In the following, the performance of the implemented wireless oscilloscope is analysed and compared to commercial isolated measurement systems.

978-1-4799-2706-7/14 $31.00 © 2014 IEEE

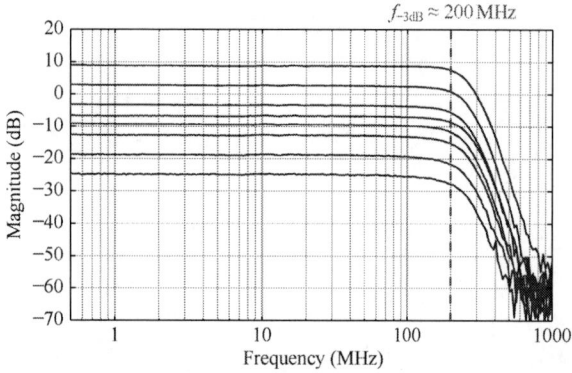

Fig. 10: Measured small-signal frequency characteristic of the wireless scope's analog frontend (from input voltage v_{IN} to the voltage at the A/D-converter v_{ADC}, cf. Fig. 8, for different input attenuations [17].

A. Analog Bandwidth

One of the key performance indicators of a measurement system is the analog bandwidth f_{-3dB}, which basically specifies at which upper frequency the measurement signal is attenuated from nominal gain by $-3\,dB$. The analog bandwidth of the wireless scope is determined by the channels analog frontend, cf. Fig. 8, i.e. by the transfer function from the input voltage v_{IN} to the voltage at the A/D-converter v_{ADC}.

Fig. 10 shows the measurements of this transfer function for the different input attenuations carried out by a network analyser (HP 4396A + HP 85046A). It can be summarized, that the frequency characteristic of the analog frontend is flat, i.e. the deviations are below 1 dB up to 100 MHz, and the analog bandwidth is about 200 MHz for all different input attenuations.

In addition to the analog frontend also the A/D-conversion influences the maximum overall bandwidth of the system, whereby the sample rate limits the signal rise time. If the typical rule of thumb relation between rise time t_{rise} and bandwidth f_{-3dB},

$$t_{rise} \approx \frac{0.35}{f_{-3dB}}, \qquad (5)$$

is considered, reasonably a maximum bandwidth of ca. 140 MHz can be obtained with the sampling rate of 400 MS/s. With some safety margin, an analog bandwidth of DC – 100 MHz is thus specified for the wireless oscilloscope.

B. Common Mode Rejection

In order to achieve accurate measurements on floating potential, the isolated voltage and/or current measurement system must provide a very high Common Mode Rejection Ratio,

$$CMRR_{dB} = 20 \log_{10}\left(\frac{A_{DM}}{A_{CM}}\right) = 20 \log_{10}\left(\frac{v_{CM}}{v_M}\right), \qquad (6)$$

which can also be calculated out of the measurement system's output voltage v_M at shorted probe leads on floating common mode voltage v_{CM}.

In the following, the CM rejection of the wireless scope and three different state of the art differential probes have been measured at $f = 200\,kHz$ in order to evaluate and compare these systems. Fig. 11 shows the measurement setup, which was used for that purpose, and the measured waveforms thereby. The corresponding CMRR values have been calculated according to (6) and are listed in Table II.

It can be stated, that the wireless scope provides an approximately 10- to 100-times higher common mode rejection compared to the

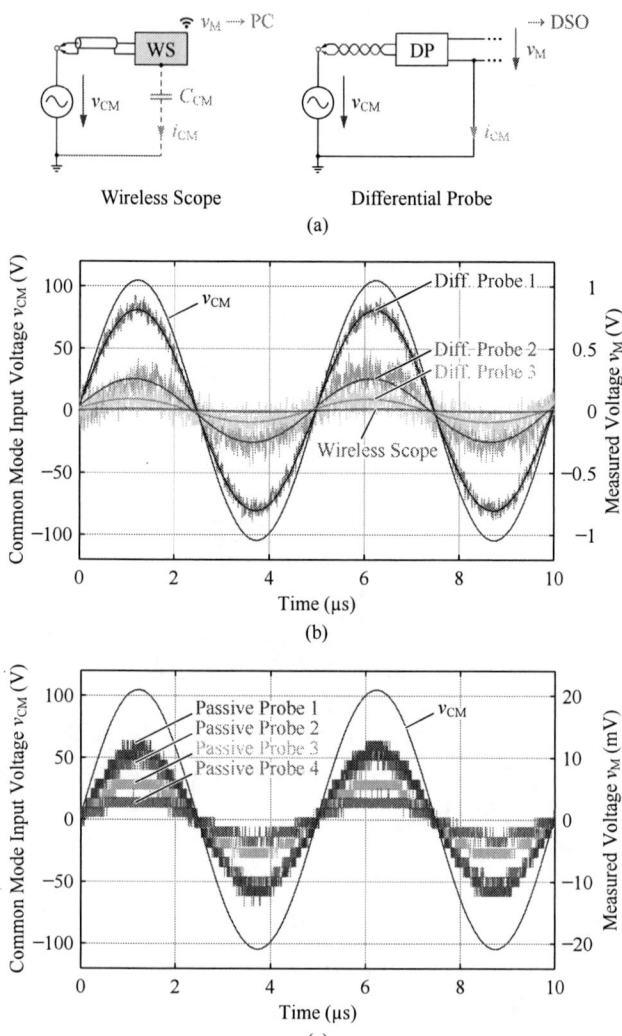

Fig. 11: Measurement of the common mode rejection of the wireless scope prototype and three commercial differential probes by means of shorted probe leads connected to a $f = 200\,kHz$ floating reference voltage v_{CM}. (a) Test setup; Measured voltages v_M by (b) the differential probes (solid: average over 1000 measurements) and the wireless scope and (c) the wireless scope with different 1:10 passive probes.

Measurement System	CMRR
Differential Probe 1, 25 MHz	42 dB
Differential Probe 2, 100 MHz	54 dB
Differential Probe 3, 100 MHz	61 dB
Wireless Scope + 1:10 Passive Probe 1	79 dB
Wireless Scope + 1:10 Passive Probe 2	80 dB
Wireless Scope + 1:10 Passive Probe 3	85 dB
Wireless Scope + 1:10 Passive Probe 4	90 dB
Wireless Scope + 1:1 Passive Probe	100 dB

TABLE II: Common mode rejection ratios at 200 kHz.

The 2014 International Power Electronics Conference

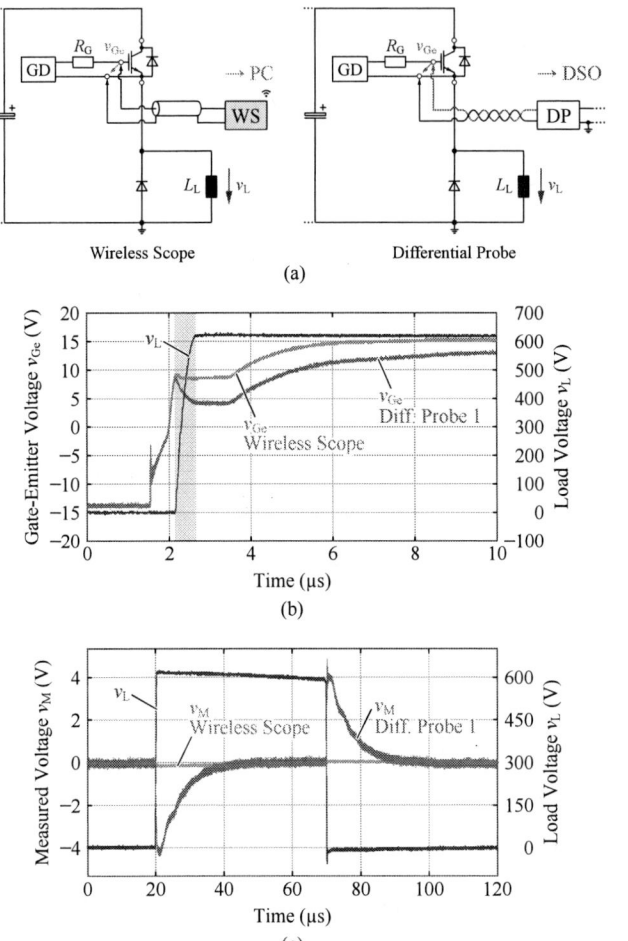

Fig. 12: Measurement of an IGBT's gate voltage v_{Ge} floating on the reference potential v_L by the wireless scope and a commercial differential probe. (a) Test setup; (b) Measured Gate-emitter voltages v_{Ge} and load voltage v_L; (c) Measured voltages v_M for shorted probe leads connected to the load voltage.

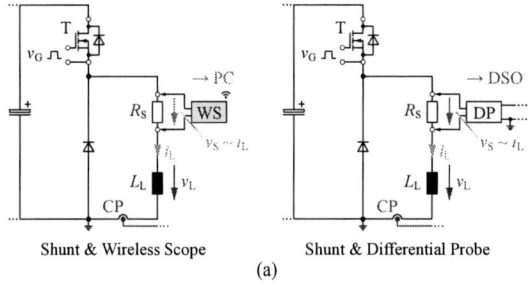

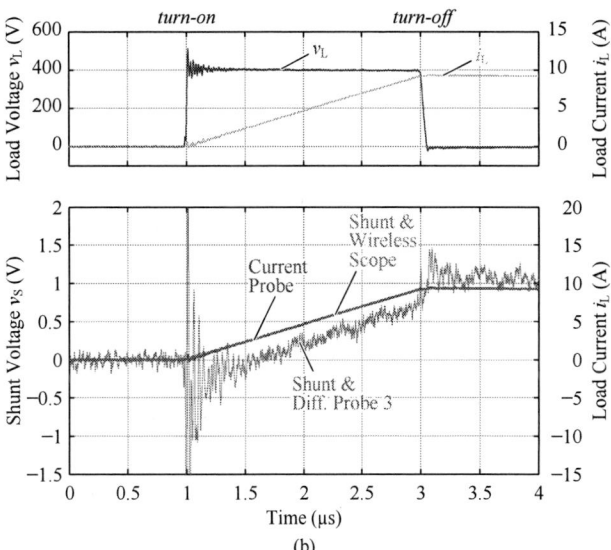

Fig. 13: Load current i_L measurement by means of a floating coaxial shunt $R_S \approx 0.1\,\Omega$ and the wireless scope or a differential probe. (a) Measurement setup; (b) Measured load voltage v_L and load currents i_L of both systems. As a reference, i_L was measured with a 50 MHz current compensated clamp-on probe on the non-floating side of the load.

best differential probe tested. The effective CMRR of the wireless scope finally depends on the utilized passive probe, whose parasitic capacitances, which depend on the mechanical setup, have an impact on the CM current i_{CM} and thus the CMRR.

In the following, the benefits of the wireless scope's superior CM rejection will be illustrated.

C. Voltage Measurement on Floating Potential

A very typical application, which demands an isolated voltage measurement system, is the gate voltage measurement of a floating power semiconductor. In order to test and compare the wireless scope to a differential probe, the Gate-emitter voltage of a floating high-side IGBT was measured by both systems. Fig. 12 (a, b) shows the related test setup and the corresponding measurements of the floating emitter / load voltage v_L as well as the IGBT's Gate-emitter voltage v_{Ge}.

Both systems measure the gate voltage appropriately, as long as the load voltage stays constant. During the CM voltage step of v_L, due to the high CMRR, the wireless scope measures the Miller plateau of v_{Ge} flat as expected. In contrast, a significant error appears at

the measurement of the differential probe, i.e. the gate voltage drops unexpectedly by ca. 4 V. In addition, this error is not only apparent during the CM transients but decays with slow dynamics.

In order to examine this CM error separately, in Fig. 12 (c) the measured voltages v_M for shorted probe leads (0 V differential input voltage) connected to the floating load voltage are shown. The CM error of the wireless scope is considerably small compared to the differential probe. There, the error is injected during the CM step and it only decays exponentially with the time constant $\tau_{CM,DP1} \approx 7\,\mu s$. Another advantage of the wireless scope is the significantly lower noise level compared to the differential probe.

D. Current Measurement on Floating Potential

In addition to voltages also currents need to be measured with respect to floating potentials. Since an input channel of the wireless scope is inherently isolated, it's basically possible to utilize e.g. a non-isolated coaxial shunt on floating potential as an isolated current sensor. The performance of such an arrangement, cf. Fig. 13 (a), was experimentally investigated and compared to combining the shunt with a differential probe in Fig. 13 (b).

The load current waveform of the wireless scope is identical to the reference measurement by the current compensated clamp-on probe. In contrast, the waveform of the differential probe shows a strong

The 2014 International Power Electronics Conference

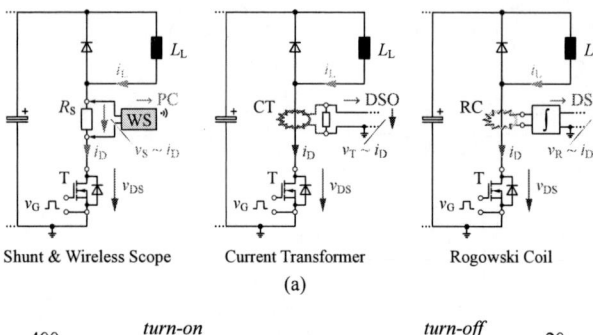

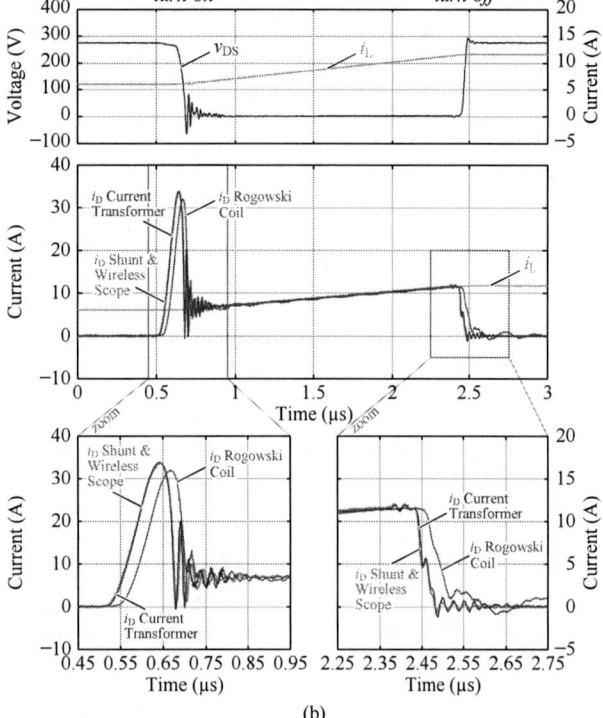

Fig. 14: Low-side Drain current i_D measurement at hard switching by means of the wireless scope and a $0.1\,\Omega$ coaxial shunt, a $100\,\mathrm{MHz}$ passive current transformer and an active $30\,\mathrm{MHz}$ Rogowski coil. (a) Test setup; Measured load voltage v_L and Drain currents i_D of all systems.

ringing and a deviation due to the transient CM voltage step, and also exhibits a comparably large noise level. As a result, in contrast to the differential probe, the wireless scope finally enables to employ low-ohmic coaxial shunt resistors as an isolated current sensor with low CM sensitivity and low noise.

In a next step, the bandwidth of this measurement concept is further investigated by a measurement of a floating MOSFET Drain current exhibiting high frequency components at hard switching, cf. Fig. 14. In addition, this measurement is compared to other isolated current measurement concepts, such as a passive current transformer and an active Rogowski coil.

The $30\,\mathrm{MHz}$ bandwidth of the Rogowski Coil is not high enough to measure the high frequency ringing after the diode reverse recovery snap-off. In addition, the signal of the Rogowski coil is delayed and a ringing is excited after the CM voltage step. Between the measurements of the current transformer and the wireless scope with shunt basically no difference can be visualized. The main benefits of

the wireless scope with shunt over the current transformer is the high voltage isolation and the possibility of measuring DC currents.

V. SUMMARY AND OUTLOOK

The measurement of current and voltage by means of a wireless oscilloscope with high-isolated input channels was identified to have clear advantages with respect to state of the art isolated voltage and current measurement concepts. The implementation details of the key components in the wireless oscilloscope, the isolated measurement channels, and the respective user interface of the system were shown.

The experimental evaluation of the wireless scope confirmed the extraordinary performance of this measurement concept. The isolated channel offers an analog bandwidth of $100\,\mathrm{MHz}$ at basically unlimited voltage isolation and exhibits a 10- to 100-times higher common mode rejection and a significantly higher signal-to-noise ratio than state of the art differential probes. In addition, the isolated input channels enable employing non-isolated / high-bandwidth sensors, e.g. shunt resistors, on floating potential, thus opening new possibilities in measurement and testing of electric circuitry.

REFERENCES

[1] A. Nara, "A measurement of the gate-emitter voltage waveform of IGBT in a motor driver circuit," in *Proc. of the 10th IEEE Instrumentation and Measurement Technology Conf. (IMTC)*, Hamamatsu, Japan, May 1994, pp. 623–626.

[2] W. H. Siew, Y. Wang, and M. Faheem, "Digital wireless electromagnetic interference (EMI) data acquisition system," in *Proc. of the Int. Symp. on Electromagnetic Compatibility (EMC)*, vol. 2, Aug. 2005, pp. 342–345.

[3] Y. Lobsiger, D. Bortis, H. Ertl, and J. W. Kolar, "100 MS/s 10-25 MHz wireless voltage probe," in *Proc. of the Power Conversion and Intelligent Motion Conf. (PCIM Europe)*, Nuremberg, Germany, May 2011, pp. 627–633.

[4] F. Costa, E. Labouree, F. Forest, and C. Gautier, "Wide bandwidth, large AC current probe for power electronics and EMI measurements," *IEEE Trans. Ind. Electron.*, vol. 44, no. 4, pp. 502–511, Aug. 1997.

[5] S. Ziegler, R. C. Woodward, H. H.-C. Iu, and L. J. Borle, "Current Sensing Techniques: A Review," *IEEE Sensors Journal*, vol. 9, no. 4, pp. 354–376, Apr. 2009.

[6] *Isolated current and voltage transducers, 3rd ed.*, LEM Components, Plan-les-Ouates, Switzerland, 2004.

[7] P. Ripka, "Review of Fluxgate Sensors," *Sensors and Actuators A: Physical*, vol. 33, no. 4, pp. 129–141, 1992.

[8] W. F. Ray and C. R. Hewson, "High performance Rogowski current transducers," in *Proc. of the Ind. Appl. Conf.*, vol. 5, Rome, Italy, Oct. 8-12, 2000, pp. 3083–3090.

[9] Lecroy, 2014. [Online]. Available: teledynelecroy.com

[10] Tektronix, 2014. [Online]. Available: www.tek.com

[11] Agilent, 2014. [Online]. Available: www.home.agilent.com

[12] Yokogawa, 2014. [Online]. Available: www.yokogawa.com

[13] Rhode & Schwarz, 2014. [Online]. Available: www.rohde-schwarz.com

[14] Inventronik, 2014. [Online]. Available: www.inventronik.de

[15] OMICRON Lab, 2014. [Online]. Available: www.omicron-lab.com

[16] Power Electronic Measurement, 2014. [Online]. Available: www.pemuk.com

[17] O. Knecht, "Entwicklung eines Tastkopfs zur potentialfreien Spannungsmessung (in German)," Master's thesis, Power Electronic Systems Laboratory, ETH Zurich, Mar. 2013.

978-1-4799-2706-7/14 $31.00 © 2014 IEEE

The 2014 International Power Electronics Conference

Introduction and Effectiveness of STATCOM to the independent Power System of JR East

Masataro Omi, Ryo Kotegawa
Tokyo Electrical Construction and
System Integration Office
East Japan Railway Company (JR East)
Tokyo, Japan

Takeshi Masui
Transportation Systems Department
Mitsubishi Electric Corporation
Kobe, Japan

Masato Ando
Electrical & Signal Network System Department
East Japan Railway Company (JR East)
Tokyo, Japan

Yasuhisa Horita
Power Electronics Department
Toshiba Mitsubishi-Electric Industrial Systems Corporation
Kobe, Japan

Abstract—**This paper describes the overview and effectiveness of STATCOM introduced to the independent power system of JR East. In the independent power system, synchronous phase modifier (Rotary Condenser:RC) was introduced in Musashisakai AC substation. Because of its aging and increase of voltage fluctuation due to increase of loads, RC is replaced to the STATCOM which has higher performance and higher response speed. The various sorts of deliberation, analysis and tests and kind have been done, and the restoration of voltage fluctuation and the improvement of stability of power system are confirmed.**

Keywords— STATCOM, synchronous phase modifier, independent power system of JR East

I. INTRODUCTION

The independent power system of JR East is mainly composed of the Shinanogawa hydraulic power station and Kawasaki thermal power station as electric sources, and backbone transmission lines of 154kV. The system mainly supplies electricity to the railways and stations in the Tokyo area. Fig. 1 shows the schematic figure of independent power system and main railways of JR East. In this figure, Musashisakai AC substation is important because it is the node of the electricity of hydraulic and thermal power stations, and it supplies electricity to the large loads areas like Tokyo and Shinjuku. In this

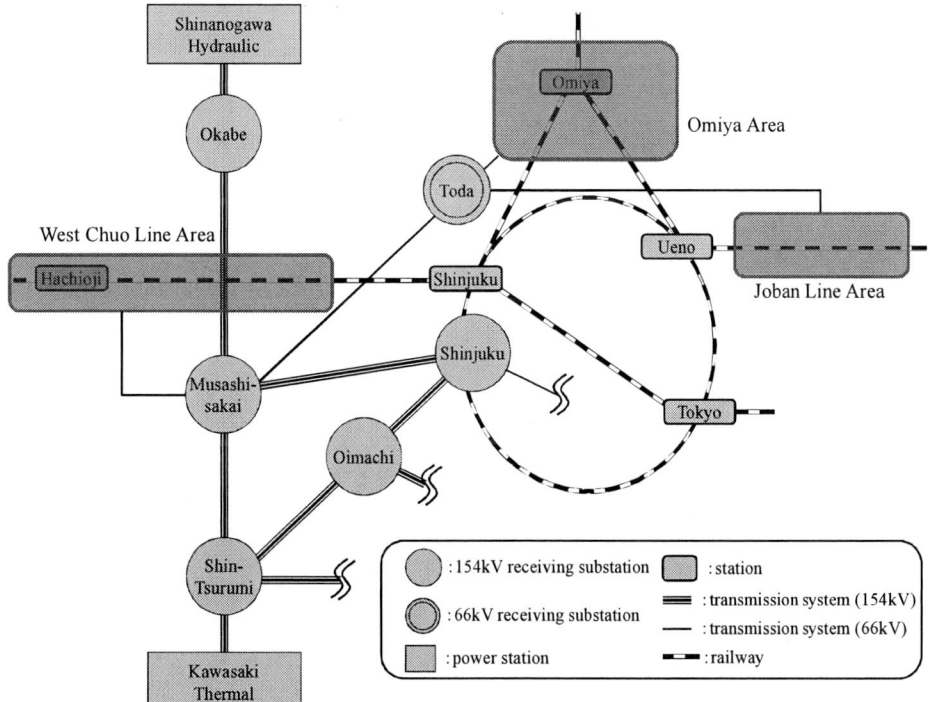

Fig. 1 schematic diagram of independent power system and main railways of JR East

substation, the synchronous phase modifier (Rotary Condenser:RC) has been introduced for the purpose of voltage stability in 1942. But because of increase of operating trains and expansion of supplying electricity to the station buildings, the voltage fluctuation of power system has been increased. So, the high performed phase modifier is demanded. And as RC gets older, the maintenance work is increased and the vibration transmission to the neighbor area is of a problem.

So, the replacement of RC to the STATCOM was done. At this replacement, the various sorts of deliberations, analysis and tests of construction, capacity, pulse number of PWM, effect of railway harmonic current and method of test in the real system and others were done. As a result, the replacement was accomplished in Mar. 2013. And it is confirmed that STATCOM delivers superior performance compared with the RC, so the advancement of voltage stability in the independent power system is achieved. The feature of independent power system of JR East and brief overview and effect of STATCOM is described as follows.

II. INDEPENDENT POWER SYSTEM OF JR EAST

The independent power system of JR East supplies electricity mainly to DC railway system in Tokyo metropolitan region. The feature of the power system is the ratio of railway load is very high (about 70%). Because the railway load changes intermittently (the intervals of trains are less than two minutes in commuting hours) and massiveness of load of trains in Tokyo metropolitan region(more than ten million consumers take railways of JR East), the voltage fluctuation of independent power system of JR East may be high compared with other network of electric power company. So, the voltage stability and reactive power control are very important in this system. And the feature of DC railway system is that it includes harmonics current from rectifier in a characteristic manner. Table 1 shows the typical ratio of harmonic current of 6-pulse and 12-pulse rectifier [1]. In JR East, 5th and 7th order harmonics current is suppressed by installing 12 pulses or equivalent 12 pulses rectifier. But because of the massiveness of DC railway load, the independent power system includes more harmonics compared with other network of electric power companies.

Table 1. Ratio of harmonic current of railway rectifier

	6 pulses rectifier	12 pulses rectifier
Negative phase 5th	17.5%	2.0%
Positive phase 7th	11.0%	1.5%
Negative phase 11th	4.5%	4.5%
Positive phase 13th	3.0%	3.0%
Negative phase 17th	1.5%	0.2%
Positive phase 19th	1.25%	0.15%
Negative phase 23th	0.75%	0.75%
Positive phase 25th	0.75%	0.75%

III. THE HOW AND WHY OF REPLACEMENT

The RC has made a contribution to the voltage stability of independent system of JR East. But by following reason and deliberation, we replaced the RC to the STATCOM.

A. Change of business environment

Recently JR East has increased the number of operating trains and developed stations and station building responding to the increase of population in Tokyo metropolitan area. For example, the number of operating trains of JR East in a day increase 2,000 compared with 25 years ago. So the total load in independent power system has been grown. As a result, the voltage stability becomes more important and faster response phase modifier was demanded.

B. Contemporary issues of the RC

It has been 70 years since installation of the RC. So superannuation of the RC has been progressed. As a result, maintenance expense, such as difficulty of component procurement, obsolescence of overhead traveling cranes for hoisting body of the RC, has been increased.   And as the housing land development around Musashisakai substation has been done, the problems of sound noise and vibration of the RC have become apparent.

C. Replacement of the RC to the STATCOM

Because of the background mentioned above, the replacement of the RC was decided. After several system studies, such as portrait style RC, STATCOM, we selected STATCOM as a new system by following reason.

- high speed response by using inverters
- no vibration
- decrease the sound noise by small size sound barrier compared with other systems
- less space compared with other systems

Next, the measurements of loads of power system for the purpose of deciding the capacity of STATCOM were done. Basically, long term voltage fluctuation should be compensated by fixed output phase modifier, and short term voltage fluctuation should be compensated by STATCOM. First, the measurement of the RC output was done. The short term reactive power fluctuation was about 10~20MVA during commuting hours, and the fluctuations of all day was about 50MVA. As a result, there was decided that the capacity of the STATCOM which compensates short term fluctuation is +/-25MVA and it is installed at the tertiary winding of main transformers (MTr). In parallel, shunt reactor of 30MVA was installed to compensate long term fluctuation. The 3-level inverters were adopted to the STATCOM because they are compact and proven in railway substation of JR-East[2]. And the system is doubled to avoid lack of phase modifier in the system even when one STATCOM is under maintenance. The AVQR cubicle is installed as the panoptic control of two STATCOMs. Fig. 2 shows the picture of RC, fig. 3 shows the picture of STATCOM.

The 2014 International Power Electronics Conference

Fig. 2. Picture of the RC

Fig. 3. Picture of the STATCOM

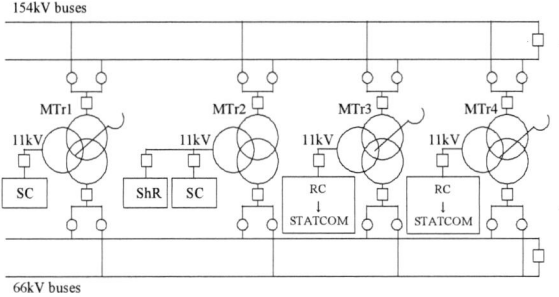

Fig. 4. One-line diagram before and after replacement of Musashisakai substation

And fig. 4 shows the one-line wiring diagram of Musashisakai substation before and after the replacement, Fig 5 shows the diagram of the STATCOM. Next, the deliberation of control of inverters has done. Because 3-level inverter system is connected to the power system containing the characteristic harmonic current of DC susbstation rectifiers, the special considerations are carried out to control the balance of voltage of two capacitors. So, the system analysis has been done and the carrier frequency, phase and controlling constants were decided. Table 2 shows the specifications of the STATCOM.

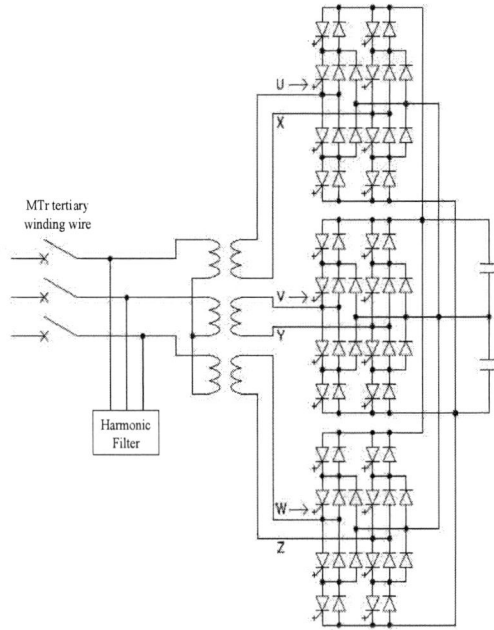

Fig. 5. Architecture of 3-level inverter STATCOM

Table 2. Specs of STATCOM (per a unit)

Rating capacity	+/-25 MVA
Rating voltage	10.75kV
Architectonics of inverters	Single phase 3 level inverters × 3 phase
Appling element	6kV-6kA GCT thyristor
Number of PWM pulses	5 pulses

IV. DELIBERATION OF TEST METHOD

Next, the deliberation of test method of the STATCOM was done. As shown in the fig. 4, STATCOM must work cooperatively with shunt capacitor (SC), shunt reactor (ShR) and on-load tap changer (LTC) in the main transformers. In addition, outputs of STATCOM #1 and #2 must be equal and mutual interference must be avoided.

In consideration of conditions, it is decided to conduct the following tests during night time when the load of substation is light.

● Test of operating only one STATCOM
  ➢ First, the voltage of 66kV bus is measured when CBs for SC and ShR are opened. The measured voltage is set to the control target voltage and the output of STATCOM is set to nearly zero.
  ➢ Measuring the output of STATCOM and confirm the stability of voltage of 66kV bus when the overmuch reactive power is supplied by closing the CB of ShR.
  ➢ Measuring the output of STATCOM and confirm the stability of voltage of 66kV bus when the overmuch reactive power is supplied by closing the CBs of three SCs
  ➢ The output limit of STATCOM is set to

978-1-4799-2706-7/14 $31.00 © 2014 IEEE 1319

2Mvar, and LTC is operated by voltage rise by closing the CB of SC. Confirming that the output of STATCOM is not vibrate or oscillate and keep the secondarily voltage constantly.

- Test of operating two STATCOMs
  - ➢ Confirming the same test cases of one STATCOM
  - ➢ Confirming that the outputs of two STATCOMs are equal and not go wrong when two STATCOMs are started by a time lag.
  - ➢ Confirming that the output of STATCOM is equal to the sum of outputs of STATCOMs when stopping one of STATCOM in operating two STATCOMs.

The circuit of 154kV and 66kV are divided to the operating system and test system as the 154kV and 66kV buses are double circuit systems. For example, Fig. 6 shows the system configuration of substation at the test of the ShR switching. As shown in this figure, the system voltage fluctuation by the ShR switching influences to the STATCOM through 154kV buses. And the protection of trouble of 154kV test system is achieved by opening the tie CB of 154kV buses at the fault of test system. On the other hand, the tie CB of 66kV buses is opened not to influence the 66kV operating system by the voltage fluctuation of 66kV test system.

V. TEST RESULT AND EFFECTIVENESS IN THE ACTUAL OPERATION

The measurement result at the case of only one STATCOM operation is shown in Fig. 7. As shown in Fig. 7 (a) and (b), it is confirmed that STATCOM detects the change of system voltage quickly and supplying the reverse phase reactive power supplied by the ShR and the SC. And it was confirmed that the response times are within 0.5 second and STATCOM controls system voltage within 0.2 kV (0.3%) of control reference voltage.

The result at the time of a LTC examination is shown in Fig. 7 (c). The purpose of this examination is a check of the operation of the STATCOM when the LTC is operated by voltage fluctuation which is exceeding the

capacity of the STATCOM and dead zone of the LTC. The rise of system voltage was imitated by the SC switching. In order to test the transient characteristics of the system voltage at the LTC switching, the STATCOM reactive control is temporary limited in 2Mvar so that the LTC controller operates by the SC switching. From Fig. 7 (c), it is stable even when the tap changes.

However, output restrictions of 2Mvar were applied only by the inverter side, but by the mismatching between the AVQR cubicle which are generalization of control, since a time lag arose about 5 seconds in LTC tap operation of the traction transformer No. 4, and the output fluctuation of SVG, it is actual operation status and carried out the additional check of carrying out

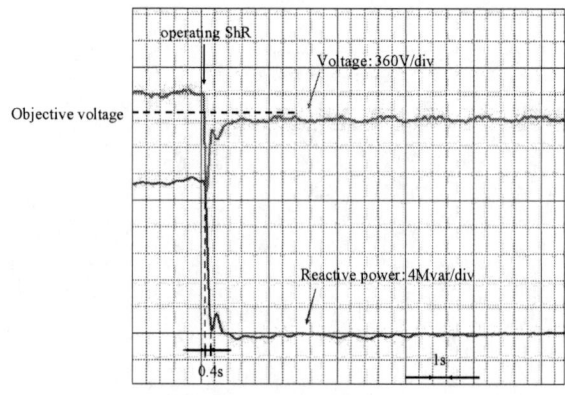

(a) Test result of ShR switching

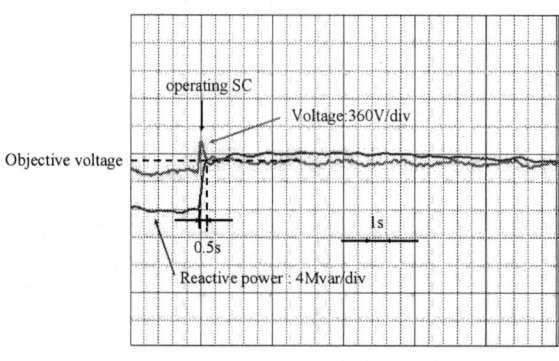

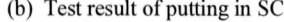

(b) Test result of putting in SC

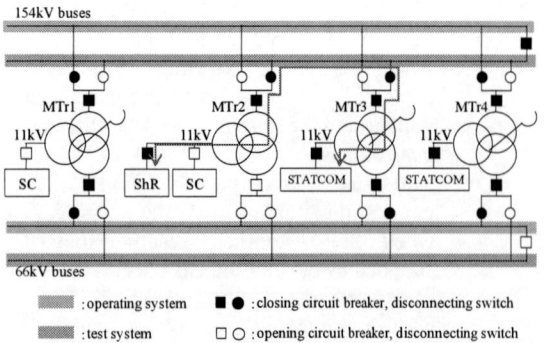

Fig. 6. The system configuration of the substation at the ShR switching test

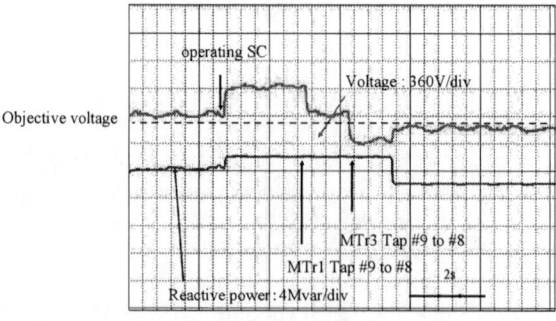

(c) Test result of LTC operation

Fig. 7. Measurement result of STATCOM#1

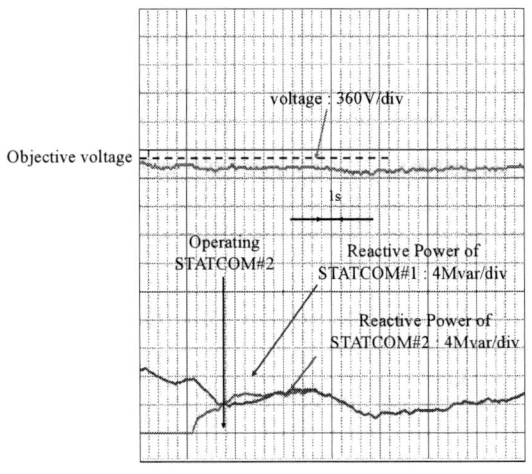

(a) STATCOM#2 delayed operating

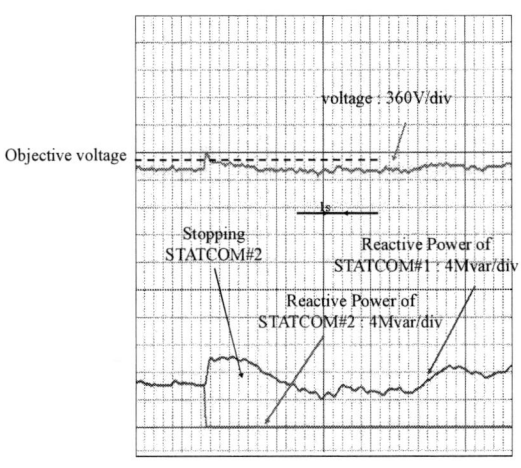

(b) Stop the STATCOM#2 only
Fig 8. Measurement result of parallel operation

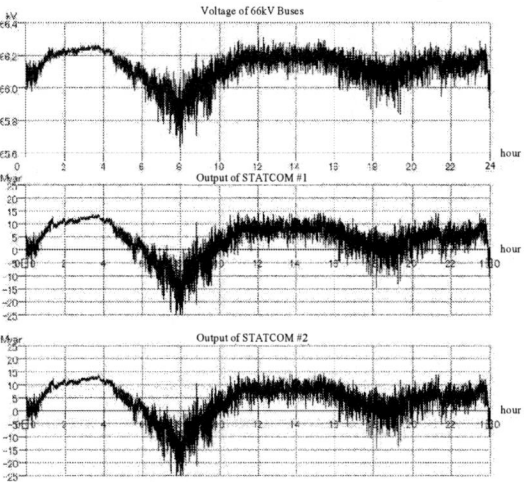

Fig. 9. 66kV bus voltage of substation and output of STATCOMs in one day

neighboring caused by the RC, was solved by adopting the STATCOM. And the sound noise is solved by setting the sound barrier around the SVG transformers.

VI. CONCLUSION

This paper described the overview of STATCOM implemented to the independent power system of JR East. Through the various sorts of analysis and tests, the improvement of power system stability by the STATCOM is achieved and the problem of vibration and sound noise is solved.

REFERENCES

[1] Agency for Natural Resources and Energy of Japan, *Guideline of counter measures taken by users against higher harmonics received at high voltages or extremely high voltages.*, 1994

[2] Tetsuo Uzuka Shouji Ikedo, Keiji Ueda, Yoshifumi Mochinaga, Sadao Funahashi and Koiti Ide "Voltage Fluctuation Compensator for Shinkansen", Trans. Of IEE JAapan, Vol. 125-B, No.9, pp. 885-892, 2005(*in Japanese*)

simultaneous operation and being satisfactory later.

The test results at the time of two-sets operation are shown in Fig. 8. When the STATCOM#2 starts operation, the two values of the STATCOMs reactive power become in agreement within 1.5s. The system voltage fluctuation is slight even when the STATCOM#2 stops operation.

By succession of these tests mentioned above, the STATCOMs started operation by two-set parallel in March, 2013. The 66kV buses voltage after STATCOMs were operated and day-long output reactive power of the STATCOMs are shown in Fig. 9. It turns out that voltage is in less than about 0.4 kV from the reference voltage of 66kV, and STATCOM has contributed to stabilize system voltage. It was confirmed that the STATCOM performs the increase or decrease of the output properly, thereby suppressing the voltage fluctuation range at all times. So, the selection of the STATCOM capacity was also confirmed.

In addition, vibration transmission problem to

The Analysis of Time-Varying Resonances in the Power Supply Line of High Speed Trains

Xi CHU, Fei LIN, Zhongping YANG
School of Electrical Engineering
Beijing Jiaotong University
Beijing, China

Abstract— The harmonic currents generated by the line-side converter of high speed trains may cause the time-varying resonances when the harmonic frequency is close to the resonance frequency of the power supply line. It may destroy the facilities in the power supply range and affect the operation of trains. To avoid the affection, the characteristic of resonance impedance is calculated and simulated with a chain network topology in AT-fed power supply system. To take the coupled relation between the system and trains, a fundamental power source and harmonic voltage source traction drive system combining model is proposed instead of the harmonic current source model. Take the CRH380A and Beijing-Shanghai high speed railway as illustration. In this paper, the regularity of resonance frequencies and the characteristic of resonance voltage are expounded and analyzed in details.

Keywords— *AT fed power supply line, CRH380A, resonance, impedance characteristic*

I. INTRODUCTION

In recent years, with the development of the high speed railway, more and more high speed trains run on rails in China. Meanwhile, the infection of the harmonic currents generated by the line-side converter of the high speed train is more serious than before. The time-varying resonance phenomenon is one of the problems. It is defined that when the harmonic frequency is close to the resonance frequency of the power supply line, the harmonic voltage of the train side will be too high to operate normally. This has happened several times in China. In April 2007, when the CRH_2 ran in Jixian South power supply range, the power supply line voltage fluctuated abnormally from 17kV to 90kV. The high voltage resulted in the protective system tripping. In January 2011, when CRH380 series experimented in a section of Beijing-Shanghai line, the high train side voltage caused the lightning arrester damaging and line voltage losing[1]. The resonance frequencies are widely distributed[4]. Therefore, it's necessary to study on the resonance phenomenon.

The research on the resonance between the power supply line and trains mainly involves two parts. One is to analyze the harmonic characteristic of the power supply line[2]. However, this part is mainly about the modeling method of the power supply line with little analysis about the resonance frequency characteristic of lines. The other is to analyze that of the high speed

train[3][5]. Generally, a model with only one traction wire and one rail is used to analyze the resonance phenomenon. It's not suitable to the complicated power supply system. In this paper, a power supply range model with AT-fed system is built by Matlab/Simulink based on Beijing-Shanghai line. And a high speed train traction drive system simulation model in combination of fundamental power source and harmonic voltage source is built based on the CRH380A. The resonance theory is mentioned. The regularity of resonance frequencies and the characteristic of resonance voltages are expounded and analyzed.

II. THEORY OF RESONANCE

Being equivalent to a RLC circuit, the power supply range can experience resonance at several specific frequencies. When the harmonic current frequencies generated by trains are close to the resonance frequencies, the harmonic voltage will be too high because of the high impedance value in these frequencies. The simple model is shown in Fig.1.

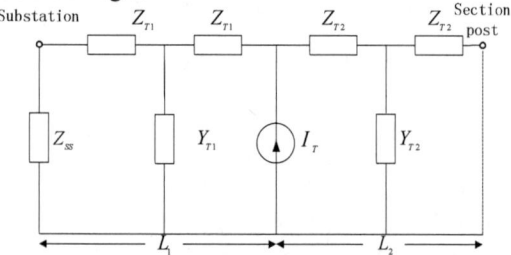

Fig.1 Power supply line simple model

Z_{SS} is the equivalent impedance of substation. Z_{T1} and Y_{T1} are the equivalent impedance and admittance of the contact wire between the substation and the train. Z_{T2} and Y_{T2} are those between the train and the end of the range. I_T represents the harmonic current generated by the train. L_1 is the distance between the train and the substation and L_2 is that of the rest part. The train side equivalent impedance is

$$Z_q = \frac{Z_c \cosh(\gamma L_2)(Z_{ss}\cosh(\gamma L_1) + Z_c \sinh(\gamma L_1))}{Z_{ss}\sinh(\gamma L) + Z_c \cosh(\gamma L)} \quad (1)$$

where $\gamma = \sqrt{Z_0 Y_0}$, $Z_c = \sqrt{\frac{Z_0}{Y_0}}$, $Z_0 = R_0 + j\omega L_0$, $Y_0 = j\omega C_0$.

With exact parameters, the train side impedance is shown as Fig.2 on a location. In this case, it is revealed that there're 4 resonance points with maximum impedance values. The nearby frequencies impedance values are much higher than others'. When the train harmonic current frequencies are closed to the resonance frequencies, they may cause the harmonic voltage with thousands or ten thousands kilovolt.

This simple model is not useful to some complicated power supply systems. In addition, the train is not strictly a current source but a power source. In this paper, a parallel multi-conductor transmission line model and a power source model of the train traction drive system are built for studying the resonance more accurately.

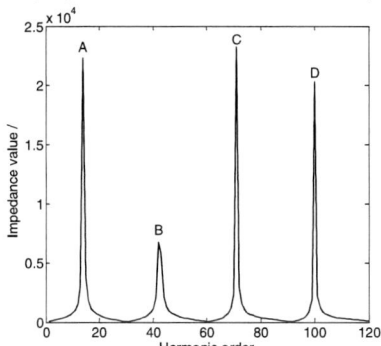

Fig.2 Power supply line simple model

III. HARMONIC MODEL

The high-speed railway system in China is based on 55kV/27.5kV AC auto-transformer power feeding(AT-fed) system including scott transformers, contact wires and rails. The system is shown as Fig.3. The power supply line can be modeled with the chain network topology(Fig.4).

According to the number of parallel conductors (m), the impedance matrix (Z) and admittance matrix (Y), the voltages of the conductors in every section are calculated with the node voltage matrix equation(Equ.2).

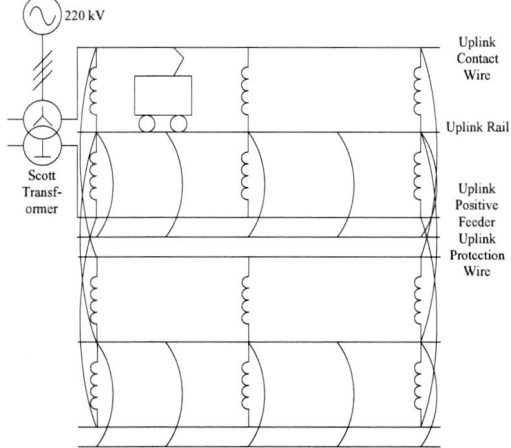

Fig.3 The structure of the power supply system

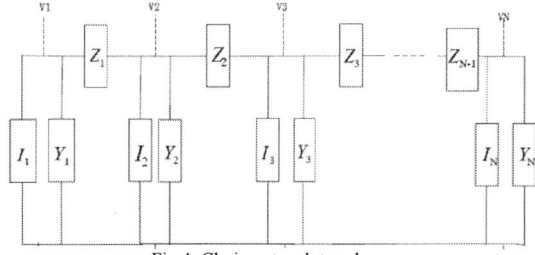

Fig.4 Chain network topology

$$\begin{bmatrix} Y_1+Z_1^{-1} & -Z_1^{-1} \\ -Z_1^{-1} & Z_1^{-1}+Y_2+Z_2^{-1} & -Z_2^{-1} \\ & -Z_2^{-1} & Z_2^{-1}+Y_3+Z_3^{-1} & -Z_3^{-1} \\ & & \ddots & \ddots \\ & & & Z_{N-2}^{-1}+Y_{N-1}+Z_{N-1}^{-1} & -Z_{N-1}^{-1} \\ & & & -Z_{N-1}^{-1} & Z_{N-1}^{-1}+Y_N \end{bmatrix} \begin{bmatrix} U_1 \\ U_2 \\ U_3 \\ \vdots \\ U_{N-1} \\ U_N \end{bmatrix} = \begin{bmatrix} I_1 \\ I_2 \\ I_3 \\ \vdots \\ I_{N-1} \\ I_N \end{bmatrix} \quad (2)$$

A. Harmonic Model of the Line

The structure of the power supply line with AT-fed system is shown in Fig.5. According to the parameters and locations of wires, the impedance matrix and admittance matrix can be calculated per unit length.

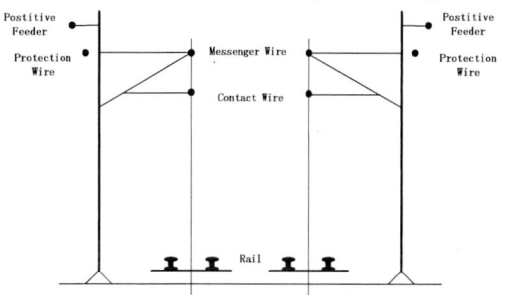

Fig.5 The structure of the traction wires

The 12 wires can be merged based on some principles to simplify the model and reduce the amount of calculation. The power supply line is equaled to the connection of many π-type equivalent circuits(Fig.6).

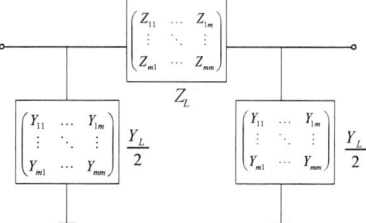

Fig.6 The π-type equivalent circuit

Every circuit stands for one section of wires. The parameters (Z_l, Y_l) are calculated by the T parameters approximation algorithm. The concrete algorithm mentioned above is described in Ref.[2] minutely.

B. Models of Other Parts of the Power Supply System

The equivalent circuit of the scott transformer is shown as Fig.7.The nodal admittance equation is

$$\begin{bmatrix} \dot{I}_a \\ \dot{I}_d \end{bmatrix} = \frac{1}{Z_\alpha} \begin{bmatrix} 1 & -1 \\ -1 & 1 \end{bmatrix} \begin{bmatrix} \dot{U}_a \\ \dot{U}_d \end{bmatrix} + \frac{1}{\sqrt{3}kZ_\alpha} \begin{bmatrix} -2 & 1 & 1 \\ 2 & -1 & -1 \end{bmatrix} \begin{bmatrix} \dot{U}_A \\ \dot{U}_B \\ \dot{U}_C \end{bmatrix} \quad (3)$$

where k is the transformer ratio and Z_a is the equivalent impedance of the AD and ad winding.

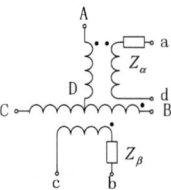

Fig.7 The equivalent circuit of scott transformer

The auto-transformer parallels between the contact wires, the rails and the positive feeders, which is shown as Fig.8. The nodal admittance matrix in the form of three-lines ignoring the excitation branch is

$$Y_{AT} = \frac{1}{4(\sqrt{n}R + jnX)} \begin{bmatrix} 1 & -2 & 1 \\ -2 & 4 & -2 \\ 1 & -2 & 1 \end{bmatrix} \quad (4)$$

where n is the harmonic order, R is the fundamental leakage resistance and X is the fundamental leakage reactance.

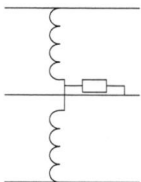

Fig.8 The equivalent circuit of auto-transformer

C. Characteristic of Harmonic Currents Generated by the High Speed Train

Take the structure of CRH-380A traction drive system as an example to analyze the characteristic of harmonic currents. The system consists of one traction transformer and two three-level PWM(Pulse Width Modulation) rectifiers. The rectifiers are controlled with carrier-shifting technology.

In ideal condition, the input current expression of three-level PWM rectifier is

$$i_N(t) = \frac{\sqrt{M^2 U_d^2 - 2U_N^2}}{\omega_m L_N}\cos(\omega_m t + \beta) +$$
$$\sum_{m=2,4\ldots}^{\infty}\sum_{n=\pm1,\pm3\ldots}^{\infty} \frac{2U_d}{m\pi L_N(m\omega_c + n\omega_m)}J_n(m\pi M)\sin\frac{n\pi}{2}$$
$$\sin(m\alpha + n\beta + m\omega_c t + n\omega_m t)$$
$$(5)$$

where U_N is the output voltage virtual value of the transformer winding, U_d is the DC-side voltage value, L_N is the transformer leakage inductance value, M is the modulation depth, ω_m is the modulation wave angle frequency, ω_c is the carrier wave angle frequency, α and β is the angles of modulation wave and carrier wave. $J_n(m\pi M/2)$ is the Bessel function, where n is the order and $m\pi M/2$ is the independent variable[3].

Controlling with the dual PWM theory, the current expression of the transformer primary side winding is

$$i_N(t) = \frac{2\sqrt{M^2 U_d^2 - 2U_N^2}}{k\omega_m L_N}\cos(\omega_m t + \beta) +$$
$$\sum_{m=4,8\ldots}^{\infty}\sum_{n=\pm1,\pm3\ldots}^{\infty} \frac{4U_d}{km\pi L_N(m\omega_c + n\omega_m)}J_n(m\pi M)\sin\frac{n\pi}{2} \quad (6)$$
$$\sin(m\alpha_1 + n\beta + m\omega_c t + n\omega_m t)$$

where α_1 and α_2 are the angles of the rectifier A and rectifier B. And $\alpha_2 = \alpha_1 + \pi/2$[3].

With the equations above, the following conclusions can be drawn:

1) The harmonic frequencies are mainly close to the even multiples of the switching frequency. The higher the frequency band, the lower the harmonic content value;

2) The harmonic content is decided by the DC voltage value, the line-side voltage valid value, the line-side voltage fundamental frequency and so on;

3) The harmonic currents in the frequencies close to 2, 6, 10 times the switching frequency are reduced by dual PWM control, but the harmonic currents in the frequencies close to 4, 8, 12 times the switching frequency are doubled.

D. The Equivalent Fundamental Wave Power Source Model of the High Speed Train

It's more reasonable to treat the train as a power source. The power source model reflects the coupling between lines and trains better. The equivalent circuit of the fundamental power source model is shown in Fig.9.

Take one of the rectifiers as an example, the equation is

$$\dot{U}_1 = \dot{I}_1 Z_1 + k\dot{I}_2 Z_2 + k\frac{P}{\dot{I}_2} \quad (7)$$

$$(1 + Z_1 Y_m)\dot{I}_1 = \frac{2\dot{I}_2}{k} + \dot{U}_1 Y_m \quad (8)$$

where Z_1 is the impedance of the transformer primary winding, Z_2 is the impedance of the transformer secondary winding 1, k is the transformer ratio, U_1 is the voltage of the transformer primary winding. I_1 is the current of the transformer primary winding, I_2 is the current of the transformer secondary winding 1. P is the real power of the train.

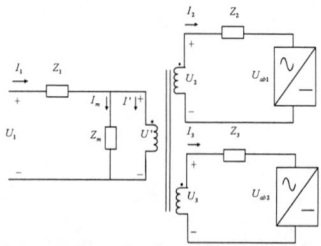

Fig.9 The equivalent circuit of the power source model

E. The Harmonic Model

The equivalent circuit of the harmonic model is shown in Fig.10. The harmonic voltage expression is

$$\dot{U}_{ab1h} = \frac{2U_d}{m\pi}J_n(m\pi M)\angle(m\alpha + n\beta) \quad m = 2,4\ldots \quad n = \pm1,\pm3\ldots(9)$$

where U_d is the DC voltage, $(n\omega_m+m\omega_c)$ is the harmonic order, ω_m is the modulation wave angle frequency, ω_c is the carrier wave angle frequency.

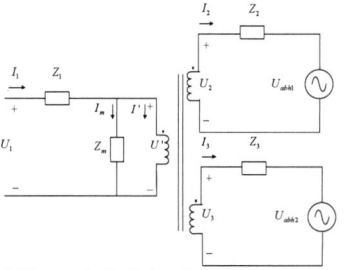

Fig.10 The equivalent circuit of the harmonic model

F. Calculation Flow Process

The calculation flow process of the harmonic analysis model is shown as Fig.11. The process divides in two parts. One is to calculate the fundamental wave value. The other is to calculate the harmonic current and voltage value. It considers the coupling between lines and trains. The train side voltage is calculated based on the load power. And the current is calculated based on the line impedance.

IV. CASE STUDY

A. The Example

The power supply range simulation model is built with AT-fed system. The conventional diagram of the range is shown as Fig.12. The line parameters are based on Beijing-Shanghai line. The model of the high speed train traction drive system is built based on the CRH380A.

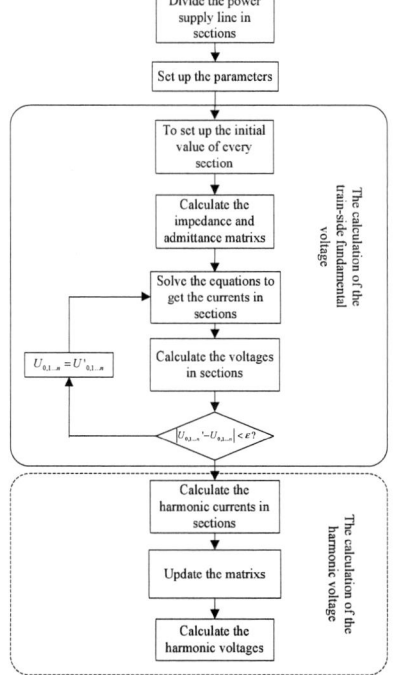

Fig.11 The calculation flow process

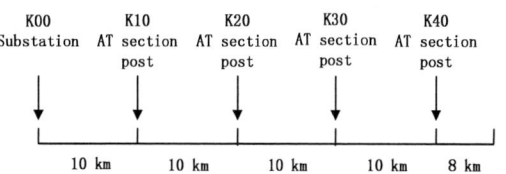

Fig.12 The power supply range

B. Results

a. The impedance characteristic of the power supply line

The impedance of the power supply line is shown in Fig.13. The impedance in different frequencies of the whole line is shown in Fig.14. It's demonstrated that there're several resonance points along the line. The resonance frequencies are widely distributed from low to high frequency range. The impedance in resonance frequencies varies with locations. The maximum and the location in which the impedance appears the maximum are both diverse for one resonance frequency. And the maximums are also different in different resonance frequencies.

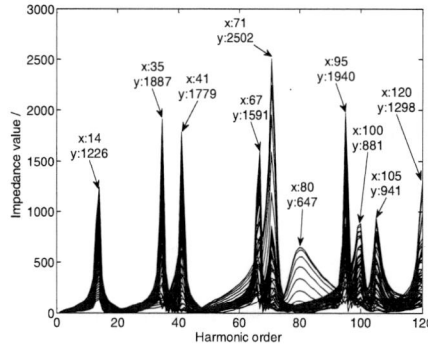

Fig.14 The impedance in different frequencies

b. The affection of the train

The primary winding impedance of the train traction transformer makes an influence on the characteristic of the power supply line. Fig.15 shows the train side impedance comparison result in 5 locations in the power supply range. It's illustrated that the impedance of the traction transformer primary winding makes every resonance frequency a little higher. Because the train parallels between the connect wire and the rail. The impedance of the traction transformer primary winding makes the train side equivalent inductance a little smaller. That's why the resonance frequency gets a little higher. It corresponds with the theoretical analysis. Except for the a little higher resonance frequency, the impedance of the train makes little influence on the rest impedance characteristic of the power supply line without trains. Fig.16 shows the train-side voltage values in 5 locations. The harmonic voltage values in some characteristic frequencies closed to the resonance frequency are shown in Fig.17. Actually, the fundamental voltages are only a little different in different locations with a certain load power. Therefore, the train side voltage peak values vary with locations because of the harmonic voltage. The harmonic values are different in different frequencies and locations. It is illustrated in Fig.17 that the 95th harmonic value is much higher than others. At K24, the 95th harmonic frequency is close to one of the resonance ones. The impedance in 95th harmonic frequency is higher obviously.

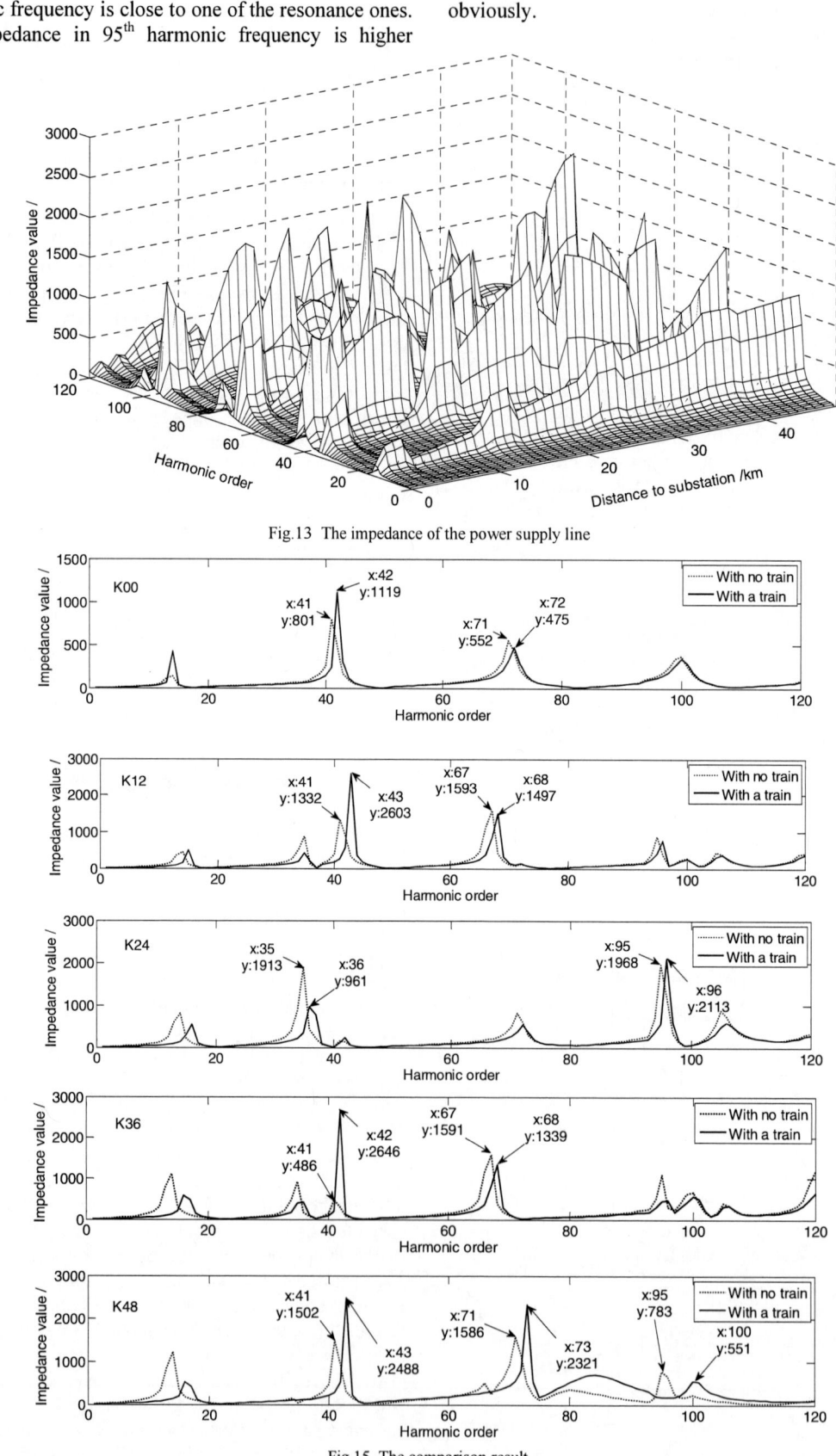

Fig.13 The impedance of the power supply line

Fig.15 The comparison result

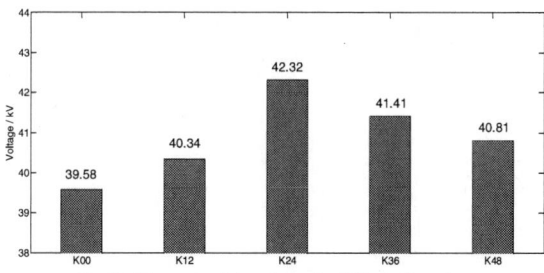

Fig.16 The train-side voltage in different locations

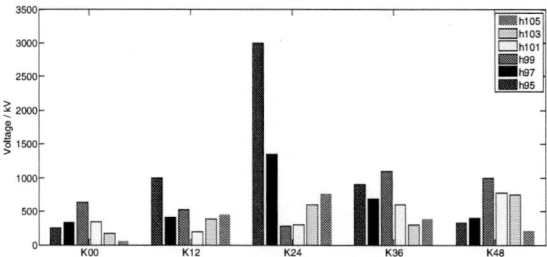

Fig.17 The harmonic voltage values

V. CONCLUSION

In this paper, the resonance phenomenon is simulated and discussed especially the one in high characteristic harmonic frequency. The impedance characteristic of the power supply line is analyzed. In addition, the affection of the train input impedance is mentioned. Conclusions are drawn as follows:

1) There're several resonance frequencies in a certain power supply range. They're mainly decided by the parameters of the wires. The impedance values and maximums in the resonance frequencies vary in locations.

2) The traction transformer impedance of the train makes the resonance frequency a little higher.

3) The train side harmonic voltages vary from locations and frequencies. They are decided by the impedance of the line and the harmonic currents injected to the line by the train.

REFERENCES

[1] J. Q. Liu, Q. L. Yang and T. Q. Zheng, "Harmonic Analysis of Traction Networks Based on the CRH380 Series EMUs Accident," *Transportation Electrification Conference and Expo (ITEC)*, 2012.

[2] M. L. Wu, C. Roberts and S. Hillmansen, "Modeling of AC Feeding Systems of Electric Railways Based on a Uniform Multi-Conductor Chain Circuit Topology," . *Railway Traction Systems (RTS)*, 2010.

[3] B. R. Lin, "Analysis and Implementation of a Three-level PWM Rectifier/Inverter," *IEEE Trans. on Aerospace and Electronic Systems,* vol. 36, no. 3, pp.948-956, August 2002.

[4] J. Holtz and J. O. Krah, "Suppression of Time-Varying Resonances in the Power Supply Line of AC Locomotives by Inverter Control," *IEEE Transactions on Industrial Electronics,* vol. 39, no. 3, pp.223-229, June 1992.

[5] A. M. S. Mendes, R. F. Rocha and A. J. Marques Cardoso, "Analysis of a Railway Power System Based on Four Quadrant Converters Operating Under Faulty Conditions," *International Electric Machines & Drives Conference(IEMDC)*, 2011.

Fuzzy Feed-forward Charge/Discharge Control of Stationary Energy Storage Systems for DC Electric Railways

Takuya KIKUCHI, Hironori TAGA and Ryo TAKAGI

Department of Electrical Engineering
Kogakuin University
1-24-2 Nishi-shinjuku, Shinjuku-ku, Tokyo 163-8677, JAPAN
e-mail: takagi@cc.kogakuin.ac.jp

Abstract— **The feed-forward charge/discharge control scheme of energy storage systems (ESS's) for DC electric railways takes advantage of the fact that the track and speed profiles are already available for all runs of trains before they take place. Using the track and speed profile information, the capacity of the ESS's can be best utilised to obtain maximal effects. This idea can be used for both onboard and stationary ESS's. However, the effect of introducing feed-forward charge/discharge control to stationary ESS's is not as significant as the onboard ESS's, and it is expected to be difficult to prepare and maintain the feed-forward data required to implement such a scheme. In this paper, the authors propose the fuzzy feed-forward control for stationary ESS's, which is expected to make such preparation and maintenance of feed-forward data easier while improving the overall performance of the charge/discharge control.**

Keywords—DC electric railways, energy storage systems, fuzzy control, feed-forward control

I. INTRODUCTION

The energy storage systems (ESS's) are expected to mitigate the problems of the DC railway power feeding systems, especially the occasional loss of line receptivity for the regenerating trains and the large voltage drops. However, despite active developments in the storage technologies such as secondary batteries, the ESS's are still very expensive and the available storage capacity is not always sufficient. This necessitates the development of "cleverer" charge/discharge control scheme that realizes better effects with the introduction of smaller capacity ESS.

As one of such "clever" schemes, the authors' research group has proposed the feed-forward charge/discharge control scheme for the onboard ESS's [1]. The research group has extended this idea to the control of stationary ESS's as well [2]; however, this has not been as successful as the control of onboard ESS's. This is mainly because the conditions of many trains in the feeding network should be considered in controlling a stationary ESS, while an onboard ESS may be controlled by considering only the condition of the train on which the ESS itself is installed.

In this paper, the authors propose the fuzzy feed-forward charge/discharge control for stationary ESS's, which is expected to mitigate these issues. The mathematical details of the control scheme is shown, which is the modification of the non-fuzzy feed-forward control scheme as proposed in [2].

II. FEED-FORWARD CONTROL OF THE STATIONARY ESS

A. SOC Curve: the feed-forward data

The feed-forward charge/discharge control scheme for ESS's in DC electric railways, as proposed in [1] and [2], takes advantage of the fact that the track and speed profiles are already precisely available for all runs of trains before they actually take place. This means that the charge/discharge power of the ESS is dependent on the positions of trains.

In this control scheme, the "SOC (State Of Charge) curve" data is prepared in advance as the feed-forward data. An SOC curve determines "the desirable SOC of an ESS" as the function of train position.

In the case of onboard ESS, the application of this idea is straightforward because only the condition of the train on which the ESS itself is installed should be taken into account in controlling the ESS. Disregarding the losses during charging or discharging, the feed-forward charge power of the ESS, $P_C(x)$, can be instantly calculated using the following equation when the SOC curve is given:-

$$P_C(x) = C \times \frac{dSOC^*(x)}{dx} \times \frac{dx}{dt} \quad \dots\dots\dots\dots\dots\dots(1)$$

where C, x, t and $SOC^*(x)$ are the energy capacity of the ESS, the train position, time and the SOC curve function, respectively.

However, in the case of stationary ESS, more complicated scheme is necessary because generally two or more trains in the feeding system will interact with a stationary ESS.

B. Using SOC curves for stationary ESS's considering multiple trains in the feeding network

If there is a railway line with only one stationary ESS and only one train, one SOC curve can describe the desirable SOC value when the train in question is at any position. If the railway line has more than one ESS and only one train, the desirable SOC values of the ESS's can be described by preparing as many SOC curves as the ESS's. Let us denote an SOC curve for ESS #i when the train is at position x as $SOC_i^*(x)$.

Assume there is a railway line with many stationary ESS's and many trains. For each train, the SOC curves for all ESS's can be prepared based on the assumption that the train is running on its own in the feeding network, i.e. no other trains exist. Let us denote the SOC curve of train #j for ESS #i when the train #j is at position x based on this assumption as $SOC_i^{j*}(x)$. Using this notation, the reference SOC of ESS #i at time t can be calculated using the following equation:-

$$SOC_i^*(t) = T_i^* + \sum_j \{SOC_i^{j*}(x_j(t)) - T_i^*\} \quad \text{...... (2)}$$

where T_i^*, $x_j(t)$ are predefined medium SOC value for ESS #i and position of train #j at time t, respectively. Using this $SOC_i^*(t)$ and disregarding the losses during charging or discharging, the reference charge power $P_{Ci}^*(t)$ of the ESS #i can be derived as follows:-

$$P_{Ci}^*(t) = C_i \times \frac{dSOC_i^*(t)}{dt} \quad \text{.................................. (3)}$$

where C_i is the energy capacity of the ESS.

C. Preparation of SOC curves

SOC curves of each train based on the assumption that the train is running on its own in the feeding network are generated using the process as described hereunder:-

1. For each train, perform single-train simulation to obtain its power profile. The power profile is the electric power at pantograph of a train as the function of position of the train. In single-train simulation, generally the feeding network is not considered, i.e. the pantograph voltage is assumed to be constant. Let us denote the power profile of train #j as $P_{Pj}(x)$.
2. For each train, calculate the aggregated ESS output (discharge) power profile using $P_{Pj}(x)$ under the assumption that the train is running on its own. The aggregated ESS output power profile is the sum of power that is to be output from all ESS's in the feeding network under which the train is running, and is given as the function of train position. Let us denote the aggregated ESS output power profile of train #j as $P_{Ej}(x)$.
3. Define the share of power of each ESS as the function of train position. One example of such profiles are given in Figure 1. In this example, three ESS's are in the network. When the train #j is at position x_a in the figure, i.e. the same position as the location of ESS #2, the charge power of ESS's #1, #2 and #3 are 0, $P_{Ej}(x_a)$ and 0, respectively. If the train moves to position x_b, then

these will change to 0, $P_{Ej}(x_b) / 2$ and $P_{Ej}(x_b) / 2$, respectively. More generally, the reference discharge power of ESS #i for train #j when the train is on its own can be calculated as:-

$$P_{Di}^{j*}(x_j) = s_i(x_j)P_{Ej}(x_j) \quad \text{...........................(4)}$$

where $P_{Di}^{j*}(x_j)$, x_j and $s_i(x_j)$ are the discharge power of ESS #i for train #j at position x_j, position of train #j and the share of power of ESS #i at train position x_j, respectively.

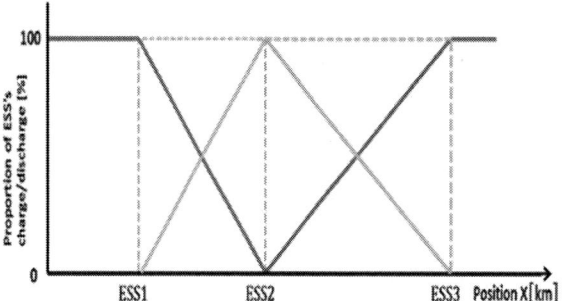

Figure 1. the example of the proportion of ESS's charge/discharge

4. Calculate the energy profile of each ESS by integrating $P_{Di}^{j*}(x_j(t))$:-

$$E_i^*(t) = \int_{\tau=0}^{t} \{-P_{Di}^{j*}(x_j(\tau))\}d\tau \quad \text{..................(5)}$$

where $E_i^*(t)$ is the energy profile of ESS #i. Here, assume that $E_i^*(0) = 0$; let us denote the peak value of $E_i^*(t)$ as E_{iM}^*.

5. Calculate the reference SOC value of each ESS as the function of time by using the following equation:-

$$SOC_i^{j*}(t) = \frac{E_i^*(t) - E_{iM}^* / 2 + T_i^* C_i}{C_i} \quad \text{.......(6)}$$

where $SOC_i^{j*}(t)$ is the reference SOC value of ESS #i for train #j at time t.

6. The reference SOC curve of each ESS can be calculated using the following equation:-

$$SOC_i^{j*}(x) = SOC_i^{j*}(t_j(x)) \quad \text{......................(7)}$$

where $t_j(x)$ is the time on the standard running profile of train #j when the train is at position x.

D. Modification of the voltage-based charge/discharge control scheme

It is generally expected that, when the charge powers of all ESS's in the network are controlled to the values defined by equation (3), there is no possibility that the voltage becomes either too high or too low. However, the conditions of single-train simulation is different from the actual operating condition of the feeding network, and it is the result of the single-train simulation that is used to design the SOC curve. Therefore, the situations can happen when the voltage becomes either too high or too low even when the ESS's are controlled exactly as defined by the SOC curves.

Voltage-based control is the normally used control scheme for many stationary ESS's already in operation. Under this control scheme, the ESS charges power when the voltage exceeds a certain threshold value, and it discharges power when the voltage becomes lower than another threshold value. The control scheme proposed in [2] is implemented by modifying this proven voltage-based control.

Figure 2 shows the characteristics of the charge/discharge current as the function of voltage in the voltage-based control scheme. In the feed-forward control scheme as proposed in [2], the charge/discharge power as defined by Figure 2 and the power defined by the feed-forward control are added together.

Generally, the floating charge/discharge control is combined with the voltage-based scheme; when the voltage is between Vc and Vd in Figure 2 and the SOC of the ESS is apart from a predefined medium value, the floating charge/discharge is done so that the SOC moves closer to this medium value. In the control scheme proposed in [2], this floating charge/discharge will not be activated.

Instead of the floating charge/discharge control, the charge power defined by equation (3) is used.

Also, there may be a situation in which the SOC deviates from the reference value even when the charge/discharge power is controlled exactly to the value defined by equation (3), mainly because of the error in the model. In such situation, the reference follower control is activated to suppress the difference. The charge power of this reference follower can be determined, for example, as in Figure 3, which is the simple P-control.

The final charge power of ESS #i in the control scheme of [2], $P_{Ci}(t)$, can be calculated by the following equation:-

$$P_{Ci}(t) = P_{Vi}(t) + P_{Ci}^{*}(t) + P_{RFi}(t) \quad \text{.....................} (8)$$

where $P_{Vi}(t)$, $P_{Ci}^{*}(t)$ and $P_{RFi}(t)$ are the power defined by the voltage-based control as exemplified in Figure 2, the feed-forward charge power as defined in equation (3), and the reference follower power as exemplified in Figure 3, respectively.

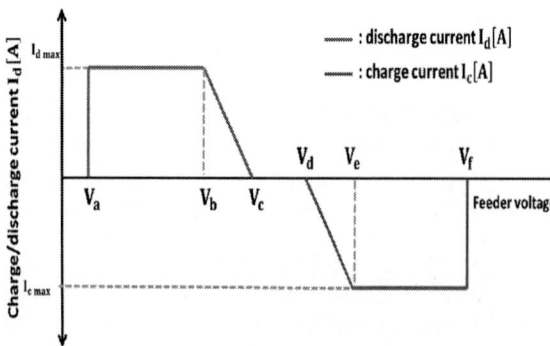

Figure 2. The charge/discharge control depended on feeder voltage.

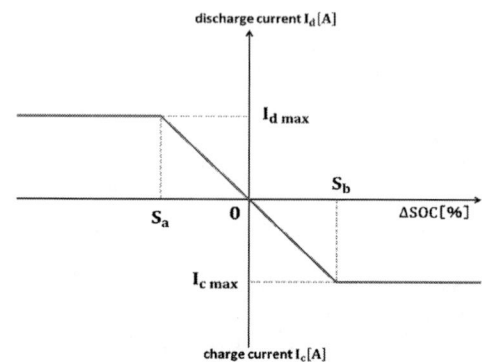

Figure 3. the reference follower power

E. Problems

The feed-forward charge/discharge control scheme for ESS's in DC electric railways, as proposed in [2], has following problems:-

a) It is expected that the preparation and maintenance of many SOC curves are difficult. If there are n ESS's and m classes of trains in the railway line, then $n \times m$ SOC curves must be created and maintained, even when trains of the same class can share the same SOC curves.

b) The effect of using the linear sum of data from the SOC curves designed when only one train exists in the network to obtain the desirable SOC value when multiple trains exist, as defined in equation (2), is not clear. It is expected that the "desirable" SOC value obtained in such a way is not optimal even if the original SOC curves are optimally designed for the single train.

III. FUZZY FEED-FORWARD CHARGE/DISCHARGE CONTROL OF STATIONARY ESS FOR DC ELECTRIC RAILWAYS

A. Fuzzy control

Fuzzy control is the application of fuzzy set theory to the control systems.

A fuzzy set is different from normal sets, in which any elements are allowed to have the conditions of either "belonging" or "not belonging" to a certain set and nothing else. In fuzzy sets, elements are allowed to be "in between these two conditions" of "belonging" and "not belonging", obscuring the boundary of the set. This makes the fuzzy set usable for expressing fuzzy concepts, such as "being big" or "being small", without explicitly specifying the threshold values.

To treat fuzzy sets mathematically, each element in a set is given a "grade", which is a real number between 0 and 1. If the grade is near 0, it means it is closer to "not belonging"; if the grade is near 1, it means it is closer to "belonging". Figure 4 shows an example membership function of a fuzzy set of real numbers, in which the grade of a real number x is expressed as μ(x).

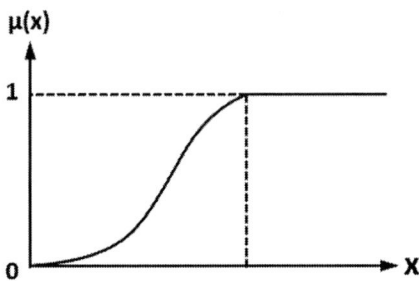

Figure 4.Exmple of membership function

In fuzzy control, the collection of if-then rules is generally used to define the characteristics of the controller.

B. Preparation of the fuzzy feed-forward data

In the control scheme proposd in this paper, the SOC curve function explained in Chapter II is fuzzified. Fuzzification of the SOC curve function makes it difficult to adopt the idea of differentiating the SOC curve function to get the charge/discharge power, as has been done in equation (3); therefore, the charge/discharge power is also given as the pre-calculated feed-forward data.

First, set "P_i" is defined for ESS #i, the elements of which are the fuzzy values representing the charge/discharge power of the ESS. Also, set "S_i" is defined for ESS #i, the elements of which are the fuzzy values representing the SOC values of the ESS. For example:-

$$P_i \equiv \{P_{N2}, P_{N1}, P_{ZR}, P_{P1}, P_{P2}\} \quad\quad\quad (9)$$

$$S_i \equiv \{S_{N2}, S_{N1}, S_{ZR}, S_{P1}, S_{P2}\} \quad\quad\quad (10)$$

In (9), P_{ZR}, P_{Nn} and P_{Pn} are zero, charge and discharge powers, respectively, with the letter n showing the intensity of the charge/discharge. Similarly, in (10), S_{ZR}, S_{Nn} and S_{Pn} are the medium SOC value (e.g. 50%), values lower than the medium value, and values higher than the medium value, respectively, with letter n showing the extent of deviation from the medium value. Each fuzzy value represents a certain value, such as 0 for P_{ZR}, 1000 kW for P_{P1}, etc. This value is called the representative value, and is used in de-fuzzification.

Also, the addition operation of two fuzzy values that belong to the same set is defined. An example of such operation is shown in Table 1.

Table 1. An example addition operation table.

	P2b	P1b	ZRb	N1b	N2b
P2a	P2	P2	P2	P1	ZR
P1a	P2	P2	P1	ZR	N1
ZRa	P2	P1	ZR	N1	N2
N1a	P1	ZR	N1	N2	N2
N2a	ZR	N1	N2	N2	N2

In the control scheme proposed in this paper, the following data are to be pre-calculated for all ESS's and trains in the feeding network, and used in the charge/discharge control:-

(1) SOC curve of ESS #i for train #j, $SOC_{Fi}{}^{j*}(x_j)$, which is the function of train position x_j that returns a fuzzy subset of S_i; and

(2) charge power curve of ESS #i for train #j, $P_{CFi}{}^{j*}(x_j)$, which is the function of train position x_j that returns a fuzzy subset of P_i.

This may seem more complicated than the scheme proposed in [2], especially because the number of "curves" to be prepared and maintained is doubled in the scheme proposed in this paper. However, because the curves are now functions returning the fuzzy set of fuzzy values, the information required to define such "curves" is significantly reduced, and so are the required efforts to maintain such data.

C. Fuzzification of the feed-forward control scheme

The fuzzified SOC curves and power curves can be used in the proposed control scheme, which is the slightly modified version of the scheme proposed in [2], which is detailed in Chapter II.

Firstly, the principles that apply to the calculation of voltage-based charge/discharge power as exemplified in Figure 2 will not change in the proposed scheme.

Secondly, the feed-forward charge/discharge power as defined in equation (3) is fuzzified. In the proposed scheme, $P_{CFi}{}^{j*}(x_j)$ are used to calculate the power.

If a fuzzy subset of P_i, P_{CFi}, is given, the resulting power of ESS #i, $P_{Ci}(P_{CFi})$, can be calculated as:-

$$P_{Ci}(P_{CFi}) = \sum_{x \in P_i} \mu_{CFi}(x) R_{Pi}(x) \quad\quad\quad (11)$$

where $\mu_{CFi}(x)$ and $R_{Pi}(x)$ are the membership function of x in subset P_{CFi} and the representative value of x in set P_i, respectively.

If there is only one train #j in the feeding network, then $P_{CFi}{}^{j*}(x_j)$ can be used as P_{CFi} to calculate equation (11). If two or more trains exist, then the summation of $P_{CFi}{}^{j*}(x_j)$ of all trains will be used as P_{CFi} to calculate equation (11). The addition of two fuzzy subsets A and B of P_i can be defined by the following equation:-

$$\mu_{A+B}(z) = \sum_{(x,y)\in\{(a,b)|a\in P_i, b\in P_i, a+b=z\}} \mu_A(x)\mu_B(y) \quad\quad\quad (12)$$

where $\mu_A(x)$ is the membership function of x in subset A. The addition operation $x + y$ is defined as a table, as exemplified in Table 1.

Finally, the activity of the reference follower remains unchanged; however, the reference SOC value must be calculated (de-fuzzified) from the fuzzified reference SOC value. The de-fuzzified reference SOC value can be calculated in the similar way as equation (11).

If a fuzzy subset of S_i, S_{Fi}, is given, the resulting reference SOC value of ESS #i, $SOC_{Fi}{}^*(S_{Fi})$, can be calculated as:-

$$SOC_{Fi}^{\;*}(S_{Fi}) = \sum_{x \in S_i} \mu_{Fi}(x) R_{Si}(x) \quad\text{.................... (13)}$$

where $\mu_{Fi}(x)$ and $R_{Si}(x)$ are the membership function of x in subset S_{Fi} and the representative value of x in set Si, respectively.

If there is only one train #j in the feeding network, then $SOC_{Fi}{}^{j*}(x_j)$ can be used as S_{Fi} to calculate equation (13). If two or more trains exist, then the summation of $SOC_{Fi}{}^{j*}(x_j)$ of all trains will be used as S_{Fi} to calculate equation (12). The addition of two fuzzy subsets A and B of S_i can be defined in exactly the same way as in equation (11).

Once the voltage-based power, the (fuzzified) feed-forward power and the (fuzzified) reference follower power are calculated, the same equation (8) can be used to get the final charge/discharge power.

D. Application example

Figure 5 is an example speed profile of a train. Here, the train accelerates from position 0 to position A, coasts from position A to position B, and applies brake from position B to position C where the train comes to a halt.

Figure 6 is the example fuzzy charge power curve of this train. It shows that the stationary ESS discharges power from position 0 to position A to assist feeding substations, does nothing from position A to position B because the train is coasting and consume no power, and charges power from position B to position C because the train applies regenerative brake.

Figure 7 is the example fuzzy SOC curve of this train. It starts from medium SOC value, which quickly drops to lower value during the acceleration of the train from position 0 to position A. Then the SOC is kept constant from position A to position B. Finally the value comes back to the original medium value after the train decelerates between positions B and C.

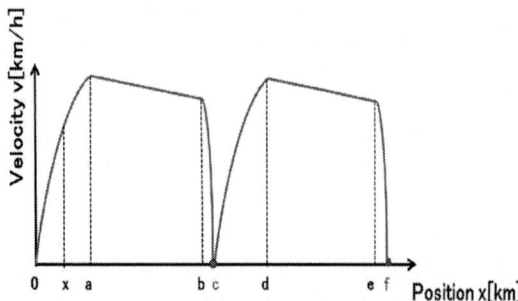

Figure 5.Exmple of the train's speed profile

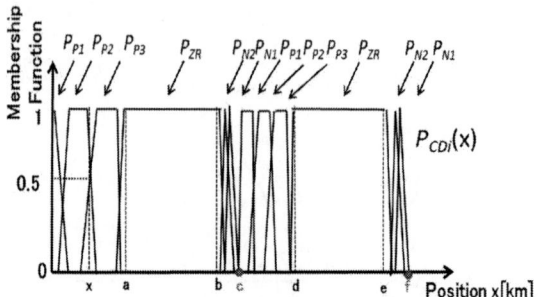

Figure 6.Fuzzified charge/discharge electric power function $P_{CFi}{}^{j*}(x_j)$

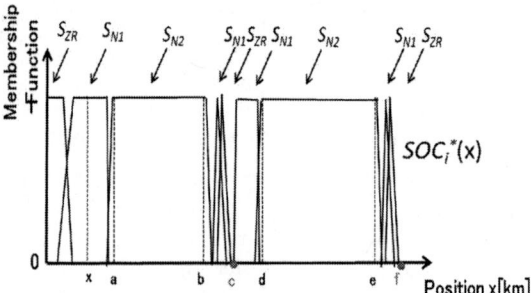

Figure 7.Fuzzified SOC curve function $SOC_{Fi}{}^{j*}(x_j)$

IV. CONCLUSIONS

As one of the "cleverer" charge/discharge control schemes for the stationary ESS for DC electric railways, the fuzzy feed forward charge/discharge control is proposed in this paper, with its in-depth description.

It is ready for implementation in a simulation program of the power feeding network of DC electric railways, such as RTSS [3][4][5][6] maintained within the authors' research group, and the authors expect that test simulations can be conducted in very near future.

It must be noted that the selection of the data, such as the sets of fuzzy values, the representative values for all elements in these sets and addition operation tables, as well as the fuzzy feed-forward data such as the SOC curves and the power curves, will have significant impact on the effectiveness of this control scheme. The test simulations will give some insight into the level of difficulty in selecting the data so that the scheme is effective. It may require some sort of numerical optimization which is recently drawing keen interest, and/or the approach that combines fuzzy logic and other techniques such as neural networks as has been widely tried in the 1990s.

REFERENCES

[1] Takagi, R. And Amano, T.: "Evaluating On-board Energy Storage Systems Using Multi-train Simulator RTSS", *4th Int'l Conf. on Railway Traction Systems (RTS 2010)*, Birmingham, UK (2010).

[2] Terashima, T., *et al.*: "Feed-Forward Charge/Discharge Control of Wayside Energy Storage Systems for DC Electric Railways", *11th Int'l Symposium on Advanced Technology (ISAT-Special)*, Hachioji, Japan (2012).

[3] Takagi, R.: "The Development of the Integrated Intelligent System of Railway Power Feeding and Train Dispatching Subsystems for DC Electric Railways" (in Japanese), The University of Tokyo (PhD Thesis) (1995)

[4] Takagi, R. and Amano, T.: "Evaluating On-board Energy Storage System Using Multi-train Simulator RTSS", *The 4th International Conference on Railway Traction Systems (IET RTS 2010)* (2010)

[5] Shiokawa, K. and Takagi, R.: "Numerical optimisation of the charge/discharge characteristics of wayside energy storage systems by the embedded simulation technique using the railway power network simulator RTSS", *13th International Conference on Design and Operation of Railway Engineering (COMPRAIL 2012)*, Lyndhurst, UK, pp.513-520 (2012)

[6] Takagi, R.: "Preliminary Evaluation of the Energy-saving Effects of the Introduction of Superconducting Cables in the Power Feeding Network for DC Electric Railways Using the Multi-train Power Network Simulator", *IET Electr. Syst. Transp.*, **2**, 3, pp. 103-109 (2012).

Train Group Control for Energy-Saving DC-Electric Railway Operation

Shoichiro WATANABE and Takafumi KOSEKI

Electrical Engineering and Information Systems
The University of Tokyo
Bunkyo-ku, Tokyo, Japan
shoichiro@koseki.t.u-tokyo.ac.jp

Abstract— Effective use of regenerative brake is important to manage energy-saving. This paper presents a running method for energy-saving on DC-electric railways considering restriction of keeping running time. First, two methods are compared for an energy-saving operation. One is applying a low notch-off speed with a strong brake. The other one is applying a fully regenerative brake with an ordinary notch-off speed. Second, we propose a manual train operation assistance method based on a power-limiting brake for an energy-saving train operation. Drivers are assisted to realize the power-limiting brake notch by an on-board system consisting of computers and interface devices. In order to consider the power limitation and resolve assistance operation problems focusing on braking delay time, this method has been checked by experiments on a revenue service line. Third, we propose a method to visualize power consumption on a train diagram to reduce peak power consumption on train rescheduling at operation irregularities. Fourth, we use a train diagram rescheduled by considering passenger flow and visualize the consumed and regenerated powers on the train diagram. Finally, we propose electric circuit models of railway systems to analyze power flow.

Keywords— *Regenerative brake, On-track test and Energy-Saving*

I. INTRODUCTION

Recently we are facing many environment problems and researchers are expected to solve these problems. Railway engineers have studied methods to reduce CO2 emission as follows. The project of ERRAC (European Rail Research Advisory Council) proposed general frame work for reducing CO2 emission [1]. Regenerative braking is a key technology to reduce energy consumption. Furthermore, storage devices are installed [2] and super capacitors are installed in substations to use regenerated energy effectively [3]. However, these methods have drawbacks, such as cost, place and expensive additional electric devices.

We focus on a train operation and propose an energy-saving manual operation method by using an on-board drive assistance system. The system is developed for drivers to assist energy-saving notch operation considering effective use of regenerative brake. Results of on-track tests on revenue service line will be reported to show whether and how this assistance method can increase regenerated energy.

The consideration of energy-saving, peak-off power and rescheduling in train operation still remains an open research issue. This paper shows a trial to visualize power for a train-group control and electric circuit models of railway systems to analyze their electric power flow.

II. ANALYSIS OF ENERGY-SAVING OPERATION FOR A TRAIN

The following two methods, which keep identical traveling time, are compared.

1) Method I (marked as I in Fig. 1): This mode uses hybrid brake control consisting of electric and mechanical brakes, as marked as I in Fig. 1. It can reduce a notch-off speed and keep running time because strong brake is made by the hybrid brake. In this braking control, air supplement control is used and regenerative brake is used priority.

2) Method II (marked as II in Fig. 1): This mode uses only regenerative brake, below the blue curve in Fig. 2. This operation increase regenerated energy but it requests higher notch-off speed because the brake of the method II is weaker than the brake of method I in high speed operation.

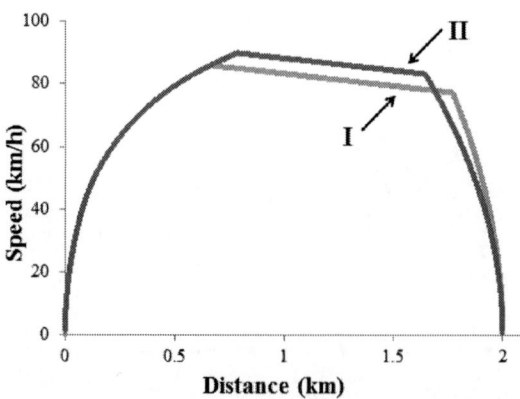

Fig. 1. Train run curves in the analysis for different energy-saving operation.

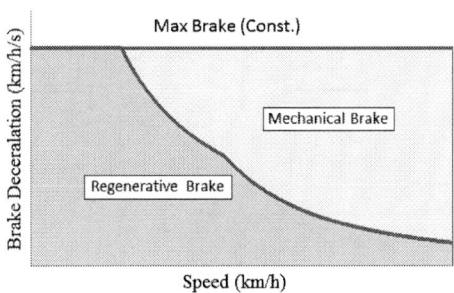

Fig. 2. The relationship of mechanical and regenerative brakes in cooperative brake.

III. NUMERICAL RESULTS AND ANALYSIS OF ENERGY-SAVING OPERATION

The main idea of this simulation is to clear the priority of using regenerated brake. Fig. 3 shows numerical results of relationship between running time and energy consumption of a train. The method I corresponds to cooperation brake and mechanical brake and the method II corresponds to regenerative brake.

The energy consumption increases when a train runs faster because coasting distance and time have to be reduced.

The method II is better than the method I in so many cases.

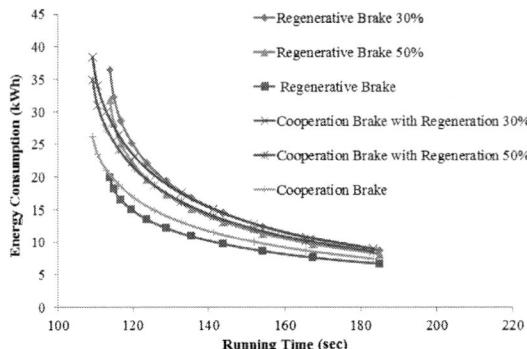

Fig. 3. Relation between running time and energy consumption at full number of passengers.

IV. POWER-LIMITING BRAKE FOR EFFECTIVE USAGE OF REGENERATIVE POWER

The operation method II is useful for effective usage of regenerated energy. In addition, the power-limiting reduces the possibility of regeneration squeezing. Fig. 4. shows the relationship between regenerated power and initial braking speed. Here the regenerative braking power is limited to 70% of the maximal capability. This paper proposes a "best effort method" for energy-saving train operation, which does not use information on other train conditions since train drivers cannot obtain information on actual status of other trains in present railway systems. This idea cannot completely avoid regeneration cancellation, however, it works effectively for energy-saving operation in many practical cases.

The "regeneration ratio" defined in Eq. (1) is

calculated and it will be used for comparing different cases in the following part of this paper.

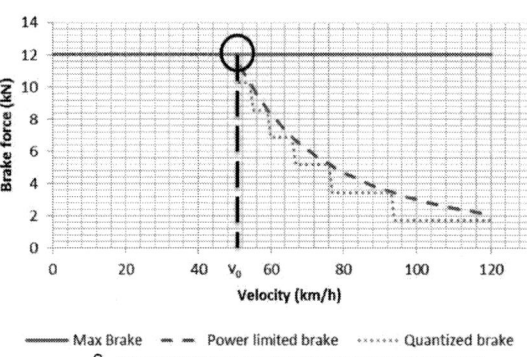

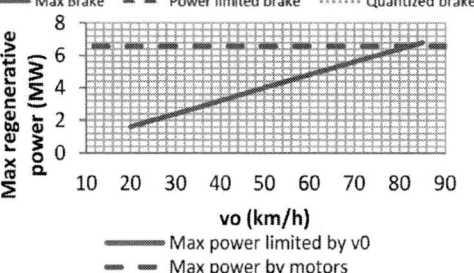

Fig. 4. Relationships between regeneration power and initial braking speed.

$$\text{Regeneration ratio} = \frac{\text{Regenerative energy of motor cars}}{\text{Brake energy of a train set}} \times 100 \ [\%] \quad (1)$$

V. DRIVE ASSIST SYSTEMS FOR EFFECTIVE USING OF POWER-LIMITING BRAKE

A. Proposal of This Assistance Systems

On-track tests have been executed to verify the effectiveness of energy-saving operation on the revenue service line. It is difficult for human drivers to take the braking action according to the regenerative performance curve in Fig. 2. Our project team member [4] have therefore developed on-board operation assistance system and the assistance system has been applied to the on-track tests to realize the best use of regenerative brakes.

B. Configuration of the On-Board System

The assistance system consists of on-board computers, GPS and interface devices shown in Fig. 5. For appropriate and timely assistance, the following data are measured; train speed, position, running time and pantograph voltage.

C. Driver Assistance Methodology

The first purpose of the system is to determine the power-limiting brake running curve which fulfills safety, scheduled running time and planned stopping position by using the measured information. The "braking dictionaries", one of which is shown in Fig. 6, were prepared before the on-track tests, for real-time assistances. It has two-dimensional indices of speed and time, depending on various prospective braking patterns.

The second purpose is to assist drivers. Human braking actions are assisted by interface devices consisting of monitor and speaker. The interface timely informs necessary notch numbers to a driver. In these experiments, the following significant knowledge has been obtained. Female voices and fan shape visualization are effective since driver room is noisy. And notch assistance shall be limited to three steps since drivers usually operate in three notch steps.

D. The Idea for Reducing Operation Delay Time

Proceeding research [4] showed that operational delay, which consists of human driver's delay in response to the assist command and mechanical delay between driver's action and consequent activation of braking force, caused unintentional shortage of braking distance. Consequently, the drivers could not use the assistance effectively. To resolve this problem, assistance based on prognosis of train motion has been introduced as shown in Fig. 7. Firstly, the on-board system measures actual speed and position. Secondly, the on-board computers explore the predictive index tags in a couple of seconds. Finally, the system compares actual speed and scheduled speed and indicates assistance commands.

Fig. 5. The situation of operation assistance

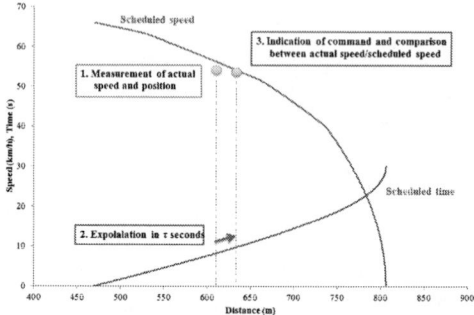

Fig. 7. The method of reduction for delay time.

VI. RESULTS OF ON-TRACK TESTS IN A SIMPLE RUNNING PROFILE

A simple running profile means that train performance pattern is designed with simple brake pattern since there is not speed limitation in braking section.

A. Verification of Advantages of the Proposed Operation

Fig. 8 shows unassisted running curve, assisted running curve, and assist-command curve. Fig. 8 shows that the assistance method considering the delay time has the advantage of low speed deceleration in comparison with conventional operations.

B. Advantages of the Proposed Operation in Regenerated Energy

Figs. 9 show unassisted and assisted running results of pantograph voltage and braking powers in the simple running profile. The pantograph voltages in Figs. 9 indicate that power-limiting brake has a good influence since the voltage of assisted running is lower than unassisted one. It is useful for reducing the possibility of regeneration squeezing.

Figs. 9 also reveal that this method makes a good use of regenerative brake in comparison with the rate of motor car power in train-set power. The motor power at 1200m in Fig. 9 (a) shows that regenerated power is limited to much loss amount of than the braking command. According to Figs. 9 and Table I, the proposed assistance method can increase ratio of regeneration by keeping the regular running time.

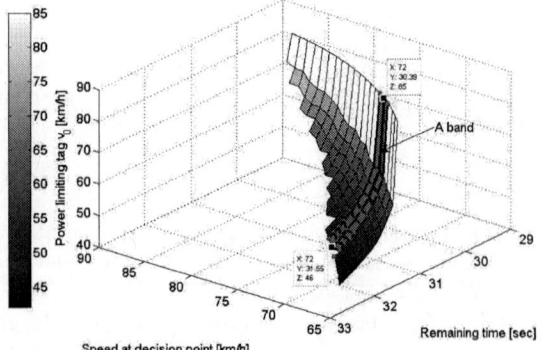

Fig. 6. 2D-table for searching the power limiting tag.

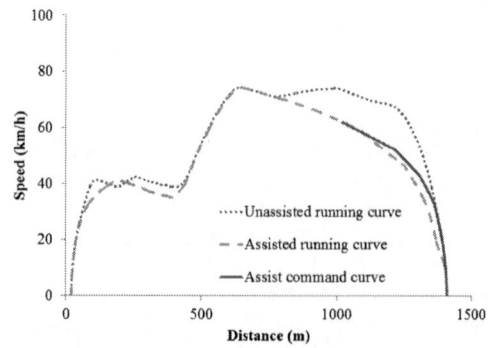

Fig. 8. Assistance pattern running curves under the simple running profile.

The 2014 International Power Electronics Conference

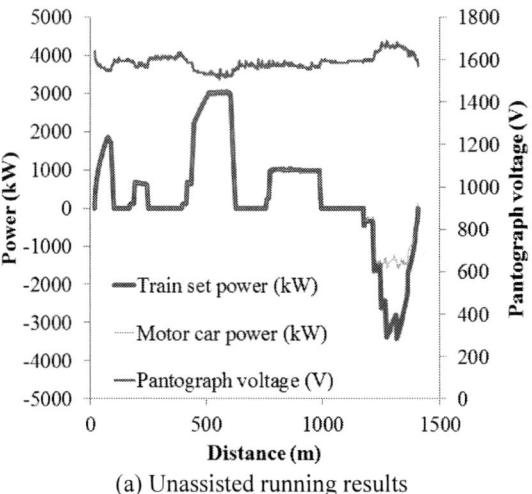

(a) Unassisted running results

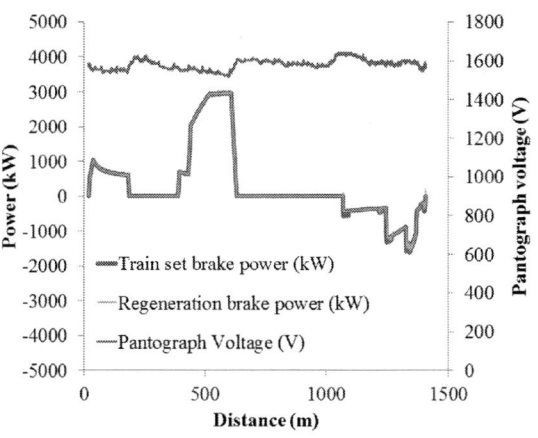

(b) Assisted running results

Fig. 9. Results of pantograph voltage and brake powers in the simple running profile.

TABLE I
RUNNING INFORMATION ON THE SIMPLE RUNNING PROFILE

Assistance pattern	No assistance	Assist considering delay time
Running time	106 sec	119 sec
Acceleration energy	16.54 kWh	13.45 kWh
Braking energy	-9.48 kWh	-6.38 kWh
Regenerative energy	-5.60 kWh	-6.17 kWh
Total energy consumption	10.94 kWh	7.28 kWh
Percentage of regeneration	59.05 %	96.56 %

Regular running time is 120 sec.

VII. RESULTS OF ON-TRACK TESTS IN DIFFICULT BRAKING RUNNING PROFILES

The difficult profile means that train performance pattern is designed without simple brake pattern since there is speed limitation based on shape curve or some

rail environments. For this reason, drivers operate brake notch two times.

A. Additional Assistance Pattern for a Difficult Running Profile

Fig. 10 shows assisted running curve and notch operation under the difficult running profile. This assisted braking section contains a speed limitation. For this reason, on-board systems assists coasting and brake notch two times. In addition, 2 STEP notch is considered by means of questionnaire data obtained from drivers.

Fig. 11 shows manual notch operation by a driver. Fig.11 also indicates that the driver did not use 5 notches since this train has plenty of time for running. According to Table II, this assistance method can keep regular running time.

Fig. 12 reveals unassisted, assisted running curves and assist-command curve under the difficult running profile. This experimental result shows that this method could keep speed limitation.

In this sense, this assistant method was successful.

B. Drawbacks of the Proposed Operation in Regenerated Energy Operation on Pantograph Voltage and Brake Power

Figs. 13 show unassisted and assisted running results of pantograph voltages and brake powers in the difficult running profile. The pantograph voltages in Figs. 13 indicate that power-limiting brake did not have a good influence.

Figs. 13 also reveal that this method made a worse use of regenerative brake in comparison with the rate of motor car power in train-set power. According to Figs. 13 and Table II, the assistance method can keep regular running time but it cannot increase percentage of regeneration.

C. Reasons for a Regeneration Cancellation

The experimental results are understood in the following ways.

1) Coasting Section (iii) in Fig. 13

There is an apparent difference of the pantograph voltages between Fig. 13 (a) and (b). Whereas the voltage in Fig. 13 (a) is standard voltage of 1500V. The voltage of Fig. 13 (b) is higher than the standard voltage. It may mean that other trains were regenerating in the same electric section in the case of Fig. 13 (b).

2) Braking Section (iv) in Fig. 13

There is a significant difference of pantograph voltages in Fig. 13 (a) and (b). A comparison of Figs. 13 (a) and (b) shows that power-limiting brake is effective since the maximal power of the train set has been reduced of 10.3%. The power of the motor car in Fig. 13 (a) is about -3000kW and keeps this power in (iv) section. However, the power of the motor car in Fig. 13 (b) is reduced from -2000kW to -1000kW and the train-set power decreases at the same time in section (iv) since pantograph voltage of Fig. 13 (b) is higher than that of Fig. 13 (a).

978-1-4799-2706-7/14 $31.00 © 2014 IEEE   1337

VIII. SUMMARY OF THE ON-TRACK TESTS

Based on our experiences, we would like to emphasize that the "best effort method" for energy-saving train operation is successful, but we must consider a train-group control for more effective energy-saving. For this reason, power flow among multiple trains is discussed in the following parts.

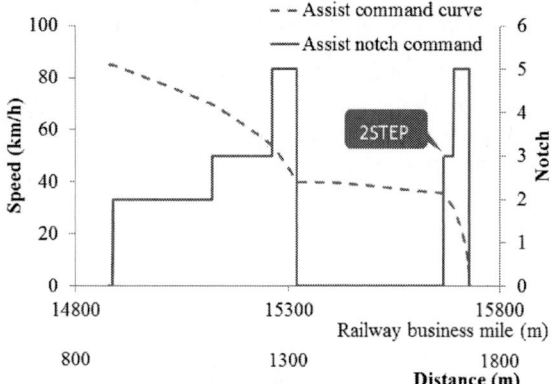

Fig. 10. The assistance running curve and notch operation.

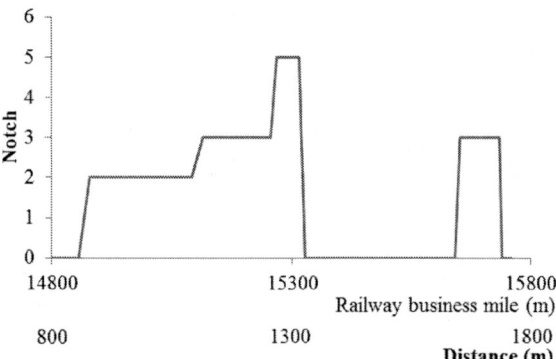

Fig. 11. Notch operation by driver.

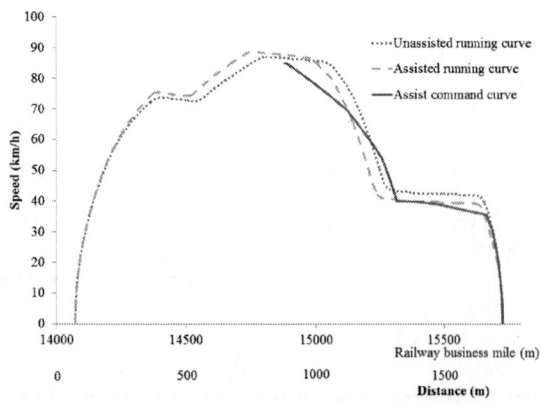

Fig. 12. Unassisted, assisted running curves and assist command curve under the difficult running profile.

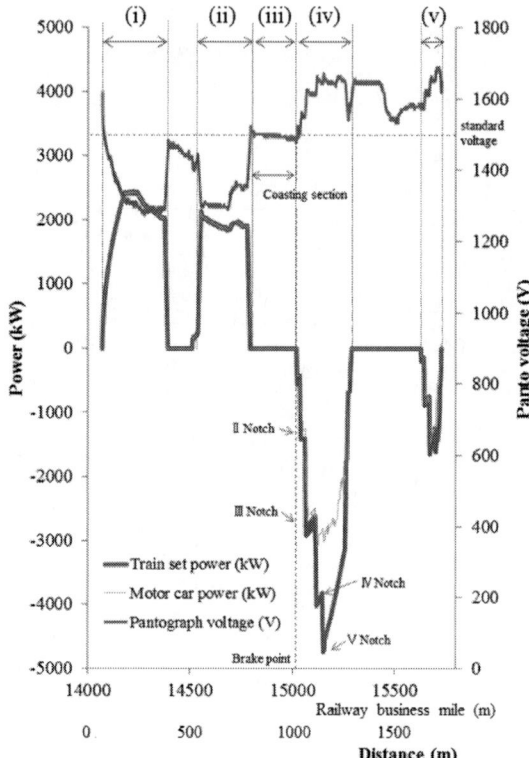

(a) Operation natural for the driver

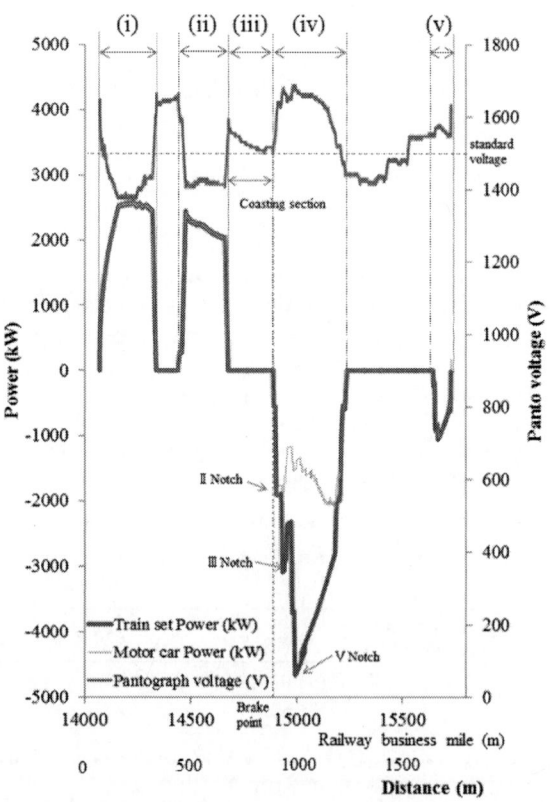

(b) Operation with the 2-STEP assistance

Fig. 13. Regenerative and train-set power and voltage on pantograph at different operation modes.

TABLE II
RUNNING DATA ON THE DIFFICULT RUNNING PROFILE

Assistance pattern	Without assistance	With assist (1 step)	With assist (2 step)
Running time	114 sec	118 sec	115.6
Acceleration energy	18.02 kWh	17.01 kWh	18.45 kWh
Braking energy	-14.64 kWh	-13.84 kWh	15.03 kWh
Regenerative energy	-10.50 kWh	-9.31 kWh	-9.10 kWh
Percentage of regeneration	71.70 %	67.28 %	60.57 %

Regular running time is 124 sec.

IX. TRAIN-GROUP CONTROL AND POWER VISUALIZATION

A. Motivations of Considering Train Group Control

The previous section has described that the "best effort method" for energy-saving train operation is basically successful. However, other train conditions have an influence on the regeneration. It is, therefore, necessary to consider train-group control to reduce total energy consumption in railway systems.

In addition, power flow in electric railway circuit should be considered since each train consumes regenerative power and sends regenerated power to other trains.

B. Approaches and Conditions

Acquisition of the information on system peak power is significant in both cases of scheduled and rescheduled train operation. We propose to visualize power on a running curve as shown in Fig. 14 in order to analyze the power on train diagrams. Red curve means power consumption, and blue curve means power regeneration. This power is calculated using Eq. (1). Where W_p is power of train set, F_m is tractive force, v is speed and η is efficiency. This running curve was calculated under the constant voltage condition of DC 1500V.

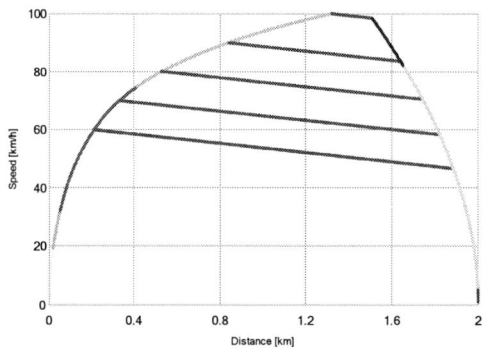

Fig. 14. The running curve with power visualization

$$W_p = F_m v \frac{1}{\eta} \quad (2)$$

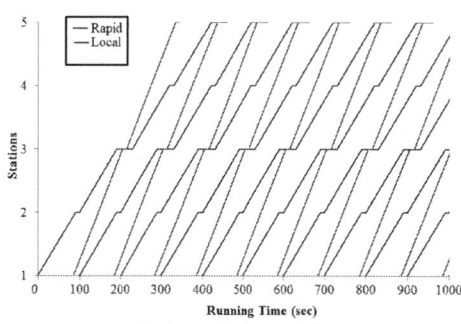

(a) Initial regular train diagram

(b) Rescheduled train diagram considered passenger flow

Fig. 15. Train diagrams consisting 5 stations

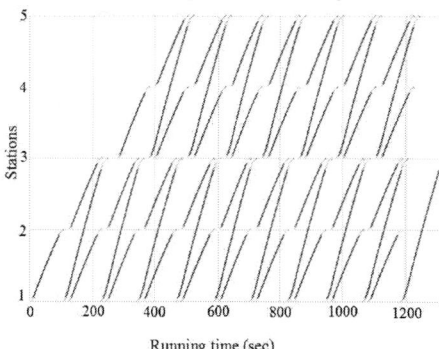

(a) Initial regular train diagram

(b) Rescheduled train diagram considered passenger flow

Fig. 16. The diagram with power visualization

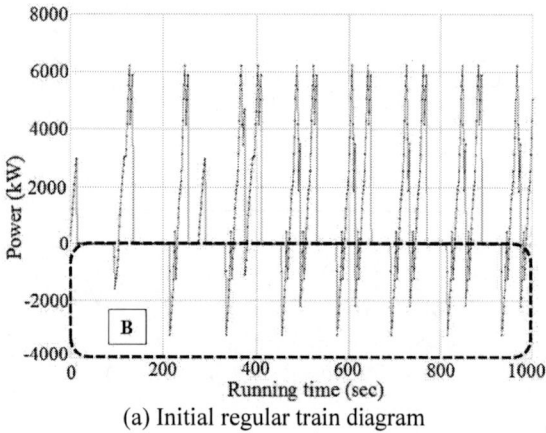

(a) Initial regular train diagram

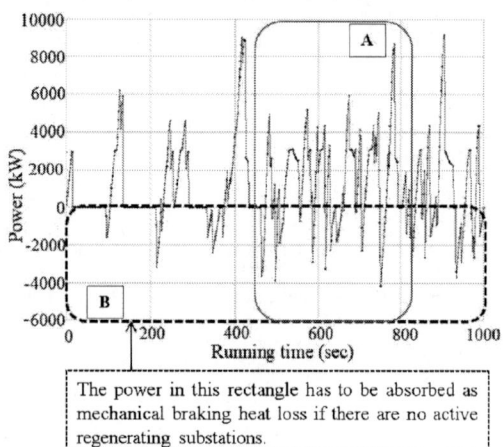

The power in this rectangle has to be absorbed as mechanical braking heat loss if there are no active regenerating substations.

(b) Rescheduled train diagram considered passenger flow

Fig. 17. The peak power on rescheduled train diagram

Figs. 15 show scheduled and rescheduled diagrams [5]. This rescheduled diagram was optimized based on all passengers' travel time by using mixed integer programming. The power is computed for the condition of 1km distance between stations as shown in Figs. 16 to visualize the power on the train-diagrams. Figs. 17 shows time-dependent summation of the train powers in each diagrams of Figs. 16.

C. Analysis of Peak Power

A comparison of Figs. 17 (a) and (b) shows that peak power is larger by 50% in rescheduled diagram since multiple trains are powered simultaneously in Fig. 16 (b). Regenerating of multiple trains are also over rapped in Fig. 16 (b).

Area A in Fig. 17 (b) shows that the power fluctuates more in rescheduled diagram since number of trains per time is higher in rescheduled diagram than in scheduled diagram in the area A in Fig. 16 (b).

The area B in Figs. 17 (a) and (b) shows possibility of improving energy-saving furthermore. The power in this area B has to be absorbed by mechanical brakes as heat loss if there are no active regenerating substations. For this reason, it is possible to use this regenerative power if train group control is considered.

X. ELECTRIC CIRCUIT MODELS FOR POWER AND ENERGY-SAVING IN TRAIN GROUP CONTROL

A. Problems of the Discussion in the Previous Section

The discussions above have suggested the importance of the power flow analysis. However, this was no explicit calculation of power-feeding electric circuits.

B. Modeling of Trains and Electrification Systems

To resolve the problem, the electric circuit models in Fig. 18 and Fig. 19 shall be studied. Fig. 18 shows the model of electric circuit in DC-electric railway systems. A wayside substation consists of constant voltage source, internal resistance and diode. The diode limits the direction of power flow just from the substation to a train.

Trains are modeled as a nonlinear current source, on which the pantograph voltage affects, where the relationships among tractive force, motor voltage, and current is given in Fig. 19.

XI. CONCLUSIONS AND FUTURE WORK

In this paper, a method for energy-saving on DC-electric railways considering restriction of keeping running time has been proposed. The following problems have been solved in this research.

The method of using regenerative brake fully is better. On-track tests give support to the good influence to use power-limiting brake with the assisted operation considering braking delay time. However, we conclude that train group control should be furthermore studied if we want more effective use of regeneration.

Peak power and fluctuation of power are bigger in rescheduled diagram in irregular cases of train group. In addition, mechanical braking heat loss is analyzed in scheduled and rescheduled diagram. It is possible to use this heat loss power if train group control is considered.

It remains a challenge for future research to complete a calculation model for analyzing the power flow, which explicitly includes electric circuit model. The running curves of train group should be designed by using the model and also need to be discussed in more detail. Such theoretical studies shall be a base of energy-saving ATO (Automatic Train Operation) in future.

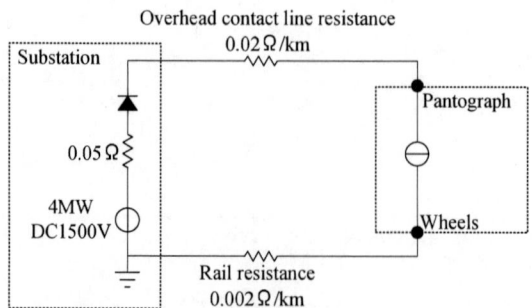

Fig. 18 The electric circuit model of railway systems

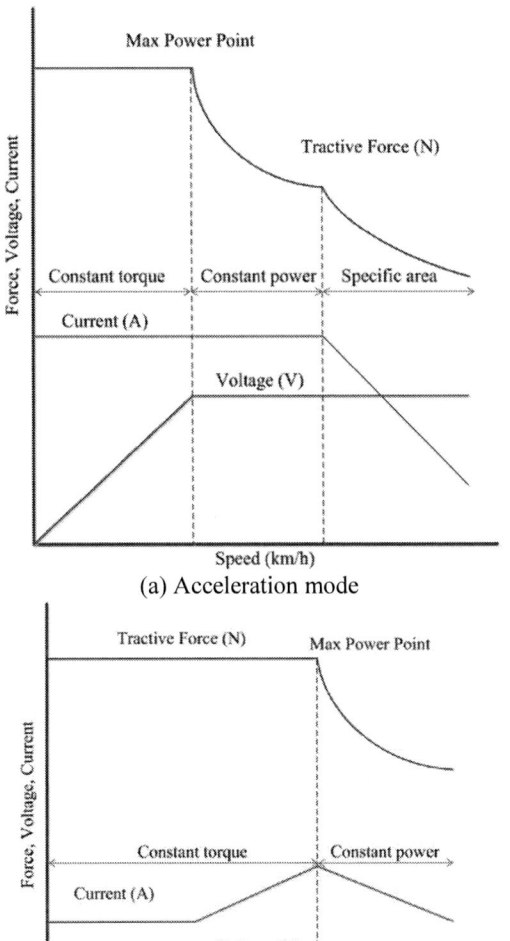

(a) Acceleration mode

(b) Regeneration mode

Fig. 19 Relationships between tractive force, motor voltage and current.

ACKNOWLEDGMENT

The author would like to thank Chiba-University, NTSEL (National Traffic Safety and Environment Laboratory), and Shin-Keisei Electric Railway Co.,Ltd. for their advice.

This work was supported by Japan Railway Construction, Transport and Technology Agency (JRTT) under grant adopted subject (2010-04) and their support is gratefully acknowledged.

REFERENCES

List only one reference per reference number according to the following samples:

[1] Christophe Cheron et al "ERRAC-European railway energy roadmap: towards 2030", WCRR2011
[2] East Japan Railway Company Press, 2012 , http://www.jreast.co.jp/press/2012/20120504.pdf *(in Japanese)*
[3] H. Lee, J. Song, H. Lee and C. Lee "Capacity Optimization of The Supercapacitor Energy Storages on Dc Railway System Using A Railway Powerflow Algorithm" International Journal of Innovative Computing, Information and Control, Vol. 7, 2011.
[4] Z. Yang, S. Watanabe, T. Koseki, T. Mizuma, and Y. Hamazaki, "Theoretical and Experimental Studies on Braking Assistance for Energy-saving Train Operation with Effective Usage of Regenerative Brake," IEEJ TER-13-006, 2013. *(in Japanese)*
[5] T. Mori, T. Koseki "Rescheduling of Railway Operation in Direction-Working Quadruple Track Line Based on Mixed-Integer Programming" IEEJ TER-13-008, 2013. *(in Japanese)*

Transformer-less Unified Power Flow Controller Using the Cascade Multilevel Inverter

Fang Zheng Peng, *Fellow, IEEE*, Shao Zhang, Shuitao Yang, Deepak Gunasekaran, *Student Member, IEEE*, and Ujjwal Karki

Department of Electrical and Computer Engineering
Michigan State University
East Lansing, MI 48824, USA
fzpeng@egr.msu.edu, samzhang@msu.edu, yangsh@msu.edu, gunaseka@msu.edu, karkiujj@msu.edu

Abstract— **The conventional unified power flow controller (UPFC) that consists of two back-to-back inverters requires bulky and often complicated zigzag transformers for isolation and reaching high voltage. This paper proposes a completely transformer-less UPFC based on an innovative configuration of two cascade multilevel inverters (CMIs). The unique configuration and control of the two CMIs as a power flow controller make it possible to independently control active and reactive power flows over a line. The new UPFC offers several advantages over the traditional technology, such as transformer-less, light weight, high efficiency, high reliability, low cost, and fast dynamic response. The transformer-less UPFC is thereby very suited for fast and distributed power flow control, such as wind and solar power transmission. A simulation model is built to demonstrate the operating principle of the proposed transformer-less UPFC.**

Keywords— *AC Transmission, FACTS, Unified Power Flow Controller (UPFC), Cascade Multilevel Inverter.*

I. INTRODUCTION

The conventional ac grid network offers significant benefits in system stability and reliability. However, the network cannot effectively handle the congestion of key transmission lines due to its limited controllability. For instance, some lines are overloaded, but others are lightly-loaded. Furthermore, undesirable loop flows are often observed. Not only does this lead to an inefficient operation of energy markets, but also the first line to reach its capacity limits the capacity of the entire system, even though a majority of the lines in the system are significantly below their limits [1].

On the other hand, the increase of wind and solar energy has been recognized as a global awareness due to the twin problems of global warming and exhaustion of fossil fuels [2, 3]. However, the intermittent power from wind and solar resource has a negative effect on the mesh grid network in terms of voltage control, frequency regulation, transient stability and oscillation damping. Hence, there is a serious technical challenge between the weakened controllability of mesh ac grid and the distributed, unscheduled, and intermittent wind and solar grid-injected power [4, 5].

To address the aforementioned challenge, there are two solutions: 1) to construct new transmission lines (including upgrade the existing lines and building high

voltage direct current (HVDC) transmission lines and 2) to actively control power flow over the lines based on the concept of flexible ac transmission systems (FACTS). The former solution is an expensive and long drawn-out process due to difficulties in permitting, siting and obtaining rights of way [1]. Most importantly, it is difficult with new lines to enhance the controllability of mesh network and ensure the utilization efficiency of the overall grid. The latter one, i.e., power flow control, routes more power onto light loaded paths, and thereby increase the capacity and average utilization of a mesh system.

A number of power flow control solutions are available for practical applications, such as the series reactor / capacitor, the phase shifting transformer [6], the static synchronous compensator (STATCOM) [7], the static series synchronous compensator (SSSC) [8], and the unified power flow controller (UPFC) [9-11]. The UPFC as the ideal FACTS device is literally the combination of a STATCOM and a SSSC that share a common DC-link as shown in Fig. 1 [9, 12]. It provides complete flexibility and controllability to control both active and reactive power flows over a line. It is the most versatile and powerful FACTS device [13] that can do 1) voltage regulation, 2) line impedance compensation, 3) phase shifting, 4) dynamic damping of power oscillation, and 5) simultaneous control of voltage, impedance, phase, and damping.

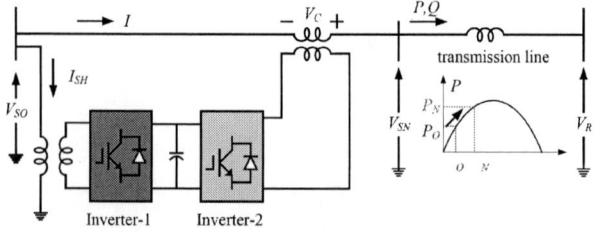

Fig. 1 The conventional unified power flow controller.

The conventional UPFC has been put into several practical applications [14], which requires an interface transformer to isolate each inverter from the transmission line. In addition to isolation, traditional high-voltage high-power inverters that often are 48-pulse for desired high-quality waveforms have to use bulky and complicated zigzag transformers to reach their required VA ratings

978-1-4799-2706-7/14 $31.00 © 2014 IEEE

(from several MVA to hundreds of MVA) [15]. The zigzag transformers are: 1) very expensive (30-40% of total system cost); 2) lossy (50% of the total power losses); 3) bulky (40% of system real estate area and 90% of the system weight); and 4) prone to failure [16]. Moreover, the zigzag transformer-based UPFCs have poor dynamic response due to the six-step operation of each inverter bridge and large time constant of the transformers. Transformer saturation would even worse the control performance [17]. Therefore, the conventional UPFC is not well suited for ultra-fast acting power flow control that is required in transmission of intermittent wind and solar power.

Recently the matrix converter was proposed to replace the back-to-back converter in the conventional UPFC [18], but it still requires the transformer, which inevitably cause the same aforementioned problems, such as bulky, lossy, high cost, and slow in response. Hence, it is highly desirable to eliminate UPFC's zigzag and isolation transformers all together for widespread use of UPFC that is urgently needed for maximizing/optimizing energy transmission over the existing grids, increasing the grid transfer capability, enabling increased penetration of renewable power, and reducing transmission congestion.

The cascade multilevel inverter (CMI) is the only practical inverter technology to reach high-voltage levels without the use of transformers [16]. The CMI-based STATCOMs (up to ±200 Mvar) have been installed in Europe and Asia [19-21]. However, the CMI could not be directly used in the conventional UPFC, because the conventional UPFC requires two inverters connected back-to-back to deal with active power exchange. To address this problem, a UPFC with two face-to-face connected CMIs was developed in [22] to eliminate the zigzag transformers that are needed in the conventional multi-pulse inverter-based UPFC. However, it still required an isolation transformer.

This paper proposes a transformer-less UPFC based on a novel configuration of two CMIs to eliminate transformers completely, which results in significant advantages over the traditional UPFC such as light weight, high efficiency, high reliability, low cost, and a fast dynamic response. The new transformer-less UPFC is also very suited for fast and distributed power flow control of wind and solar power transmission. The theoretical analysis and detailed numerical simulation is carried out to demonstrate the operating principle of the proposed transformer-less UPFC.

II. NOVEL CONFIGURATION OF TRANSFORMER-LESS UPFC

The proposed transformer-less UPFC configuration is shown in Fig. 2 (a), which consists of two CMIs: series CMI and shunt CMI. The series CMI is directly connected in series with the transmission line and the shunt CMI is directly connected in parallel to the line but after the series CMI. Each CMI, as shown in Figs. 2 (b) and (c), is composed of a number of cascade H-bridge or half-bridge

modules. Here are the unique features of the new configuration:

1) Unlike the conventional back-to-back dc link coupling, there is no transformer needed in the CMI based UPFC, hence the new UPFC can achieve low cost, light weight, small size, high efficiency, high reliability, and fast dynamic response;

2) The shunt inverter is connected after the series inverter, which is distinctively different from the traditional UPFC. Each CMI has its own dc capacitor to support dc voltage;

3) There is no active power exchange between the two CMIs and all dc capacitors are floating;

4) The new UPFC uses modular CMIs and their inherent redundancy provides greater flexibility to system design and higher reliability.

III. OPERATING PRINCIPLE OF TRANSFORMER-LESS UPFC

Fig. 3 shows a phasor-diagram explanation of the transformer-less UPFC, where $\vec{V}_{S0}$ and $\vec{V}_R$ are the sending- and receiving-end voltage, respectively. The series CMI is controlled to generate a desired voltage $\vec{V}_C$ for obtaining the new sending-end voltage, $\vec{V}_S$, which in turn, controls active and reactive power flows over the transmission line.

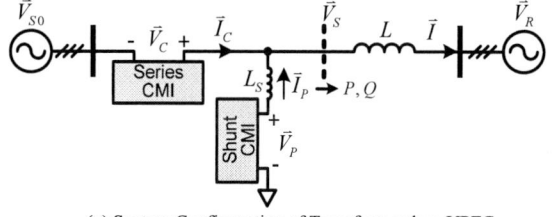

(a) System Configuration of Transformer-less UPFC

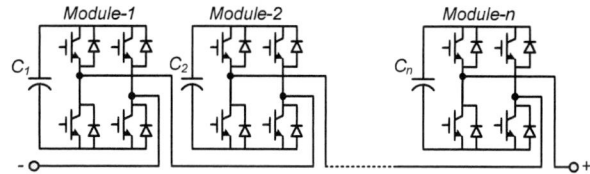

(b) Cascade Multilevel Inverter (CMI) based on H-Bridges

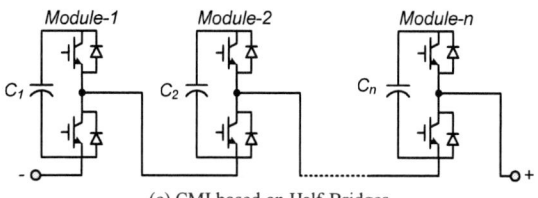

(c) CMI based on Half-Bridges

Fig. 2 The configuration of the proposed transformer-less UFPC.

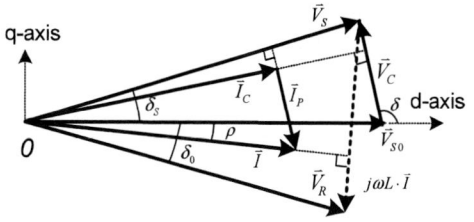

Fig.3 Phasor diagram to explain the transformer-less UPFC.

978-1-4799-2706-7/14 $31.00 © 2014 IEEE   1343

Meanwhile, the shunt CMI injects a current, $\vec{I}_P$ to the new sending-end bus, $\vec{V}_S$ to make zero active power into either CMI, i.e., to make the series CMI current, $\vec{I}_C$ and the shunt CMI current, $\vec{I}_P$ be perpendicular to their voltages, $\vec{V}_C$ and $\vec{V}_S$, respectively. As a result, no active power exchange to either CMI can be achieved and thereby makes the CMI possible for the utilization in the proposed transformer-less UPFC. The detailed operating principle of the transformer-less UPFC can be formulated as follows.

As illustrated in Fig. 3, the active power flowing into the series CMI, P_{se} can be expressed by,

$$P_{se} = \vec{V}_C \cdot \vec{I}_C \tag{1}$$

Meanwhile, the active power flow into the shunt CMI, P_{sh} is,

$$
\begin{aligned}
P_{sh} &= \vec{V}_S \cdot \vec{I}_P = (\vec{V}_{S0} + \vec{V}_C) \cdot (\vec{I} - \vec{I}_C) \\
&= V_{S0} I \cos\rho + V_{S0} I_C \cos(\delta - 90^0) + V_C I \cos(\rho + \delta)
\end{aligned} \tag{2}
$$

where, $\vec{V}_C$ is the voltage generated from the series CMI of the UPFC. $\vec{V}_{S0}$ is the original sending-end voltage phasor, and $\vec{V}_S$ is the new sending-end voltage phasor after the UPFC. As aforementioned, there would be no active power allowed to flow to both the series and shunt CMIs. Hence, the following equations have to be met,

$$
\begin{cases}
P_{se} = 0 \\
P_{sh} = 0
\end{cases} \tag{3}
$$

Substituting P_{se} and P_{sh} in (3) into (1) and (2), then the series CMI current $\vec{I}_C$ can be derived as,

$$\vec{I}_C = \frac{V_{S0} I \cos\rho + V_C I \cos(\delta - \rho)}{V_{S0} \sin\delta} \angle(\delta - 90^0) \tag{4}$$

Since the current, $\vec{I}$ over the transmission line can be calculated according to a commanded power flow, the shunt CMI current, $\vec{I}_P$ can be consequentially calculated as,

$$\vec{I}_P = \vec{I} - \vec{I}_C \tag{5}$$

In summary, the command for the series CMI is the desired voltage phasor $\vec{V}_C$ for controlling power flow over the line according to a power flow command and the command for the shunt CMI is $\vec{I}_P$, which can be calculated from (4) and (5). As a result, the new transformer-less UPFC has the same functionality as the conventional UPFC.

A further theoretical explanation is given as follows: With referring to Fig. 2 and its phasor diagram Fig. 3, the active power, P and reactive power, Q transmitted over the line with the UPFC can be expressed as,

$$
\begin{aligned}
P - jQ &= \vec{V}_R \cdot \left(\frac{\vec{V}_{S0} + \vec{V}_C - \vec{V}_R}{jX}\right)^* \\
&= \frac{V_{S0} V_R}{X} \sin\delta_0 + \frac{V_C V_R}{X} \sin(\delta_0 + \delta) \\
&\quad + j\left(\frac{V_{S0} V_R \cos\delta_0 - V_R^2}{X} + \frac{V_C V_R}{X} \cos(\delta_0 + \delta)\right)
\end{aligned} \tag{6}
$$

where, δ_0 is the original phase angle difference between the original sending end voltage, $\vec{V}_{S0}$ and the receiving end voltage, $\vec{V}_R$. The original active and reactive powers, P_0 and Q_0 over the uncompensated system (without the UPFC) are,

$$
\begin{cases}
P_0 = \dfrac{V_{S0} V_R}{X} \sin\delta_0 \\[2mm]
Q_0 = \dfrac{V_{S0} V_R \cos\delta_0 - V_R^2}{X}
\end{cases} \tag{7}
$$

The net differences between the original (without the UPFC) powers expressed in equation (7) and the new (with the UPFC) powers in equation (6) are the controllable active and reactive powers, P_C and Q_C by the UPFC and can be expressed as,

$$
\begin{cases}
P_C = \dfrac{V_C V_R}{X} \sin(\delta_0 + \delta) \\[2mm]
Q_C = -\dfrac{V_C V_R}{X} \cos(\delta_0 + \delta)
\end{cases} \tag{8}
$$

Because both amplitude V_C and phase angle δ of the UPFC injected voltage phasor $\vec{V}_C$ can be any values as commanded, the new UPFC provides a full controllable range of $(-V_C V_R/X)$ to $(+V_C V_R/X)$ for both active and reactive powers, P_C and Q_C, which are advantageously independent of the original sending end voltage and phase angle δ_0. Therefore, the theory proves that the new transformer-less UPFC has the same functionality as the conventional UPFC.

IV. OPERATING RANGE OF TRANSFORMER-LESS UPFC AND REQUIRED RATINGS FOR SHUNT AND SERIES CMIs

As shown in (8), the UPFC can regulate both the active and reactive power over the line. The voltage phasor $\vec{V}_C$ injected from the series CMI can be derived from (8) as,

$$\vec{V}_C = \frac{X}{V_R} \sqrt{P_C^2 + Q_C^2} \angle(-\delta_0 - \arctan(\frac{P_C}{Q_C})) \tag{9}$$

From (9), it is obvious that the voltage phasor, $\vec{V}_C$ is determined by the controllable active and reactive powers, P_C and Q_C as well as the line impedance, X and the original phase angle difference, δ_0 in the uncompensated system. It should be noted that the amplitude of the series voltage is advantageously small because it is determined by the line impedance, X and the net power changes, P_C and Q_C. For example, a 0.2 pu impedance line requires only 0.2 pu series voltage even for a 1-pu net power change. The command for the series CMI is the required voltage phasor, $\vec{V}_C$ to increase/decrease power flow by P_C and Q_C according to a power flow command. The command for the shunt CMI is its current, $\vec{I}_P$, to guarantee no active power flow into both the series and shunt CMIs of the UPFC. $\vec{I}_P$ in (5) can be rewritten as follows in the form of the d- and q-axis according to the phasor diagram in Fig. 3,

$$I_P \cos(\delta_S - 90^0) = I \cos\rho - I_C \cos(\delta - 90^0) \tag{10}$$

and

$$I_P \sin(\delta_S - 90^0) = I \sin \rho - I_C \sin(\delta - 90^0) \qquad (11)$$

where, $\delta_S = \arctan(V_C \sin\delta/(V_{S0}+V_C \cos\delta))$ is the angle of the new sending-end voltage phasor $\vec{V}_S$. I_P, I and I_C are the magnitude of $\vec{I}_P$, $\vec{I}$ and $\vec{I}_C$, respectively. From (10), I_P can be obtained as,

$$I_P = (I \cos \rho - I_C \sin \delta) \cdot \sqrt{1 + (\frac{V_{S0}+V_C \cos\delta}{V_C \sin\delta})^2} \qquad (12)$$

The detailed derivation of (12) is provided in Appendix A. The main purpose of a UPFC is to control active power, while maintaining reactive power flow minimal over long distance lines. This is because reactive power compensation should be done locally and close to the load. Fig. 4(a) shows required shunt current, I_P in pu in order to increase the active power flow to 1 pu while keeping the reactive power unchanged, that is to make power flow from its original power point (P_0, Q_0) to (1 pu, Q_0). In other words, the controllable active and reactive powers are set at $P_C = (1-P_0)$ and $Q_C = 0$. It can be found from Fig. 4(a) that the shunt CMI's required current is very low for most area (even when the original power flow is 1 pu negative) of the map, however, can be very large when the original active power is already close to 1 pu. This looks counter-intuitive and can be explained as follows: when the original active power, P_0 is already close to 1 pu, only a very small series voltage, V_C and/or a small angle δ is needed to boost power to 1 pu, which in turn result in a very small ($V_C \sin\delta$) and large I_P according to (12). In order to avoid a large shunt current, I_P, it is necessary to make

$$V_C \sin\delta \neq 0 \qquad (13)$$

The controllable reactive power, Q_C can be used to i) minimize the required shunt current, I_P and/or ii) minimize the required series voltage, V_C for a given active power transmission command.

Figs.4 (b) and (c) show how the controllable reactive power, Q_C affects the required shunt current I_P. In Fig. 4 (b), the Q_C is set as 0.05 pu instead of 0, and the shunt CMI current is significantly reduced in the areas close to $P_0 = 1.0$ pu, compared with that in Fig. 4(a). Furthermore, Fig.4(c) shows the shunt current map when the Q_C is set as 0.1 pu. The shunt CMI current can be limited far below 0.5 pu in the entire range of active power flow control.

It is worthwhile to mention that an increase of the Q_C may slightly increase the series voltage, V_C as shown in (9). To examine this, Fig. 5 shows the relationship between the injected voltage of the series CMI and the original power when the controllable reactive power, Q_C is set to 0.1 pu. The increase of the Q_C has limited impact on the series voltage, V_C. This result is consistent with the theoretical analysis in (9).

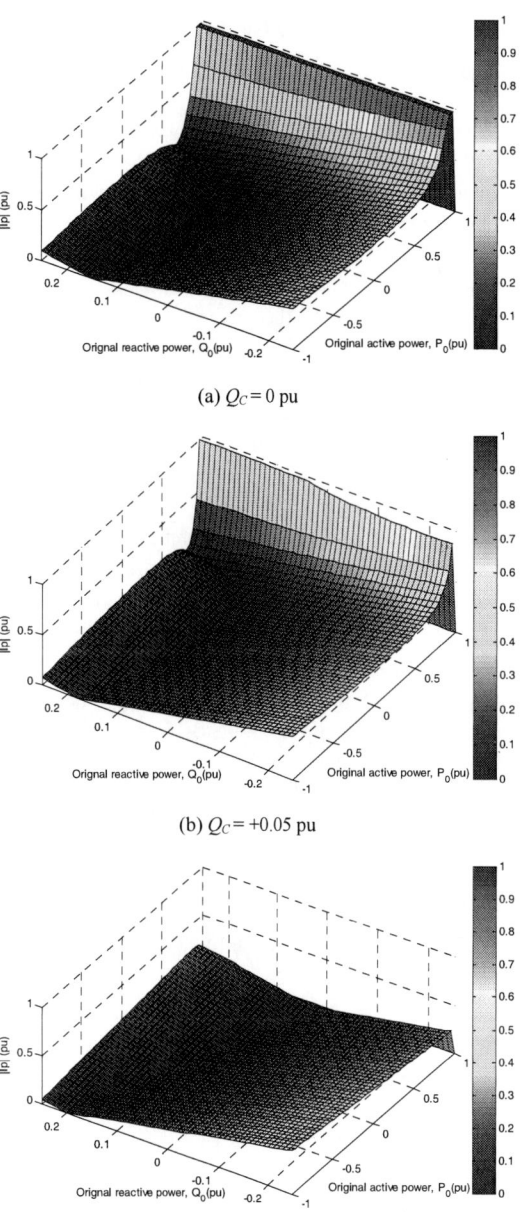

(a) $Q_C = 0$ pu

(b) $Q_C = +0.05$ pu

(c) $Q_C = +0.1$ pu

Fig. 4 The relationship between the shunt CMI current and the original power over the line under different controllable reactive power.

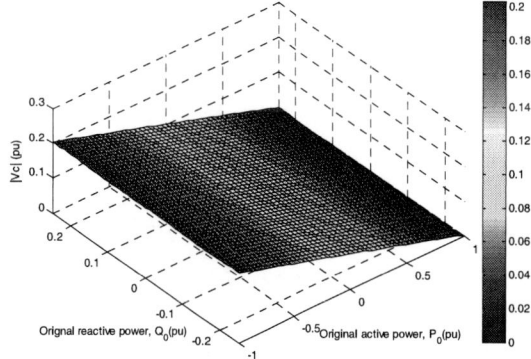

Fig. 5 The relationship between the series CMI voltage and the original power over the line when $Q_C = 0.1$ pu.

V. DC-LINK VOLTAGE CONTROL OF THE H-BRIDGES IN THE TRANSFORMER-LESS UPFC

This section addresses the DC voltage control principle for each H-bridge in the series and shunt CMIs. The overall DC voltage control diagram is shown in Fig. 6 (a). It consists of an average voltage control and an individual voltage control shown in Fig. 6 (b) and (c), respectively.

The average voltage control is to generate active power commends, ΔP_{se} and ΔP_{sh} for compensating respective overall power loss of the series and shunt CMIs. The average DC voltages, $V_{C\_average}$ and $V_{P\_average}$ of the series and shunt CMI are calculated from the measured DC voltage of each H-bridge in the series and shunt CMI, respectively. It is then regulated to track with its reference via a PI regulator in Fig. 6 (b). As discussed in Section III, the voltage commend of the series CMIs, $\vec{V}_C^*$ is determined by the desired active and reactive power flow on the transmission line. Hence, with the generated active power commends, ΔP_{se} and ΔP_{sh}, the current commend of the shunt CMIs, $\vec{I}_P^*$ can be then calculated from (1) and (2), which is illustrated in Fig. 6 (b).

On the other hand, the individual voltage control is for fine tuning control to provide respective active power into each H-bridge to address power loss difference among H-bridges. In practical implementation, the power loss of each H-bridge module may slightly different from each other because of parameter and control deviations. As shown in Fig.6 (c), each H-bridge DC voltage, $V_{individual}$ is fine tuned and controlled independently by adjusting its respective active power flow into each H-bridge module.

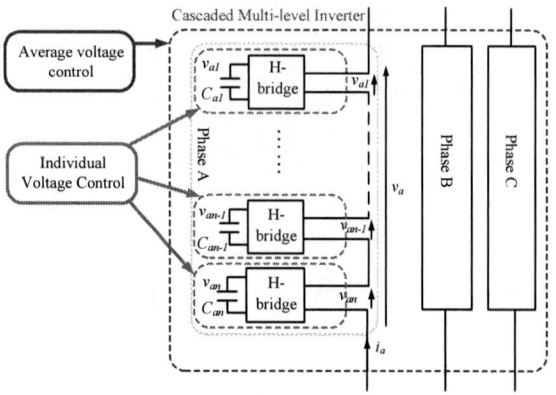

(a) Overall DC voltage control diagram

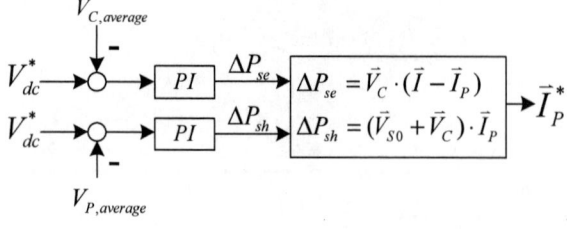

(b) Average voltage control

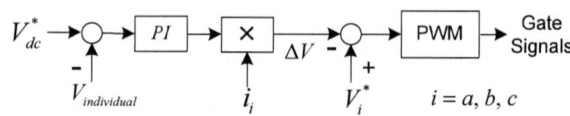

(c) Individual voltage control

Fig.6 DC voltage control principle

VI. SCALABILITY AND MODULARITY OF THE TRANSFORMER-LESS UPFC

For traditional transformer-based multipulse inverters, special transformers and inverters have to be designed and built for every application, because every line/bus has different line impedance, different voltage level, different current rating, and different operation conditions such as phase angle differences. Typical voltage levels (in kV) are 15, 35, 69, 115, 138, 161, 230, 345, 500, and 750. Typical per-circuit current ratings (in A) are from 100 to 800. Typical line impedances range from 5% to 25%. In addition, different lines may have different desired active and reactive power flows. Therefore, for every application/installation, new zigzag transformers have to be designed and built.

The great advantage of CMI is its modularity and scalability. The transformer-less UPFC can fully utilize this CMI advantage to reach any high-voltage level by stacking more or less H-bridge modules. Only a handful of basic H-bridge modules, are needed to cover all voltage and current ratings. In addition to the advantage of no transformers, standardized H-bridge modules can reduce cost even further. Another advantage about CMI is its ease to implement (n+1) redundancy for better reliability. Any faulty module can be bypassed to ensure the whole system's normal continuous operation.

The numbers of H-bridges for the series CMI, n_{se} and for the shunt CMI, n_{sh} are determined by the following equation,

$$\begin{cases} n_{se} = \left\lceil \dfrac{\sqrt{2} \cdot V_{C\max}}{V_{dc}} \right\rceil + 1 \\[2mm] n_{sh} = \left\lceil \dfrac{\sqrt{2} \cdot V_{P\max}}{V_{dc}} \right\rceil + 1 \end{cases} \qquad (14)$$

where, '$\lceil \ \rceil$' is the symbol of ceiling function. V_{Cmax} and V_{Pmax} are voltage ratings for the series and shunt CMIs, respectively. V_{dc} is the dc voltage rating of each H-bridge.

VII. SIMULATION VERIFICATION

A 2-MVA and 13.8-kV simulation model was built to demonstrate the operating performance of the proposed transformer-less UPFC. Each H-bridge DC voltage, V_{dc} is selected as 600 V for the purpose of practical implementation. Hence, from (14), the numbers of H-bridges for the series and shunt CMIs are calculated as 11 and 22, respectively. Table I shows the detailed system parameters of the transformer-less UPFC.

978-1-4799-2706-7/14 $31.00 © 2014 IEEE

Table I
SIMULATION PARAMETERS OF THE TRANSFORMER-LESS UPFC SYSTEM

Parameters	Value
Rated power	2 MVA
V_{S0} (Phase-phase rms)	13.8 kV ∠0, 60 Hz
V_R (Phase-phase rms)	13.68 kV ∠-2.89°, 60 Hz
Transmission resistance R	0.01 pu
Transmission inductance L	0.1 pu
Transmission inductance L_S	0.025 pu
Series CMI per phase	11 H-bridge modules
Shunt CMI per phase	22 H-bridge modules
Series CMI H-bridge capacitance	2300 µF
Shunt CMI H-bridge capacitance	1300 µF
V_{DC} of each H bridge	600 V
Switching frequency	600 Hz

With the parameters listed in Table I, the original system with no UPFC is transmitting active and reactive power from the bus V_{S0} at 0.5 pu and 0.1 pu, respectively, which can also be calculated from (7). The new UPFC is utilized to improve the power transmission of the system. Given that the active and reactive power are demanded to be 1 pu and 0.1 pu, respectively, the voltage of the series CMI, $\vec{V}_C$ can be calculated from (8) as $0.051 \angle 87.11°$. Moreover, the shunt CMI current, $\vec{I}_P$ can be easily obtained from (4) and (5). The following numerical simulation was carried out to verify the aforementioned theoretical analysis.

Fig. 7 shows the simulation waveforms of the voltage and current on the new sending-end V_S as well as the power flow waveforms. From Fig. 7, the original transmission system was delivering active power at 0.5 pu and the reactive power at 0.1 pu till 0.03 seconds. At the time of 0.03 seconds, an active power command of 1 pu was given to the UPFC. The UPFC generated a series voltage $\vec{V}_C$ as calculated by (8) and altered the new sending-end voltage $\vec{V}_S$, which in turn, controlled the active and reactive power flow. As a result, the active power followed the command to 1 pu, while keeping the reactive power unchanged. The dynamic response of the power flow control was within 2 ms, which is much faster than that obtained in the conventional UPFC. It is obvious that the independent control of active and reactive power is achieved. Consequently, the three-phase currents on the receiving-end also changed from the 0.5 pu to 1 pu.

Fig. 8 shows the simulation waveforms of the voltage and current in the series CMI under the aforementioned operating scenario. The series CMI injected the desired voltage to implement the power flow control at the time of 0.03 seconds. It can be found that the voltage magnitude of the series CMI was only 0.05 pu in this simulation condition. The current magnitude was close to 1 pu since the power reference is up to 1 pu after the time 0.03 seconds. The active power flow into the series CMI was zero as shown in Fig. 8.

On the other hand, the voltage and current waveform of the shunt CMI are presented in Fig. 9. The voltage

magnitude of the shunt CMI was close to 1 pu, the same as the new sending-end voltage V_S. The shunt CMI current injected a small current to keep zero active power flow to both series and shunt CMIs. It can be seen from Fig. 9 that the current magnitude was only 0.07 pu. The simulation waveform of the active power flow in the shunt CMI was zero, which is also shown in Fig. 9. Hence, the simulation verified the feasibility of the proposed UPFC circuit.

As shown in Figs. 8 and 9, the voltages were 90° phase shifted from the currents for both series and shunt CMIs. As a consequence, there was no active power flow into the H-bridges, therefore, the dc capacitor of each H-bridge, maintained constant over one cycle. Figs. 10 and 11 show the 11 and 22 dc-capacitor voltage waveforms of the series and shunt CMIs, respectively. All the dc voltages were well controlled and maintained within ±5% of their dc value.

The simulation results agree with the analytical results obtained in the previous section and clearly show that the new transformer-less UPFC is feasible and advantageous over the conventional UPFC in terms of weight, size and cost.

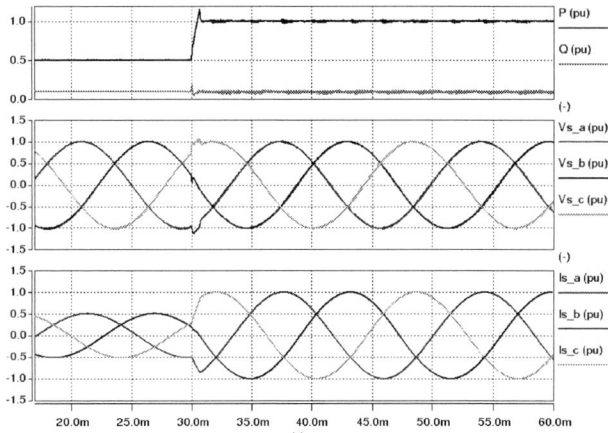

Fig. 7 Simulation waveforms (in pu) of power, voltage and current on the new sending-end V_S

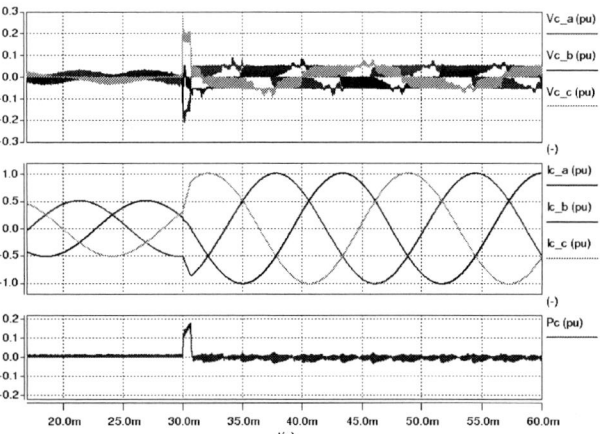

Fig. 8 Simulation waveforms (in pu) of the voltage, current and active power of the series CMI.

The 2014 International Power Electronics Conference

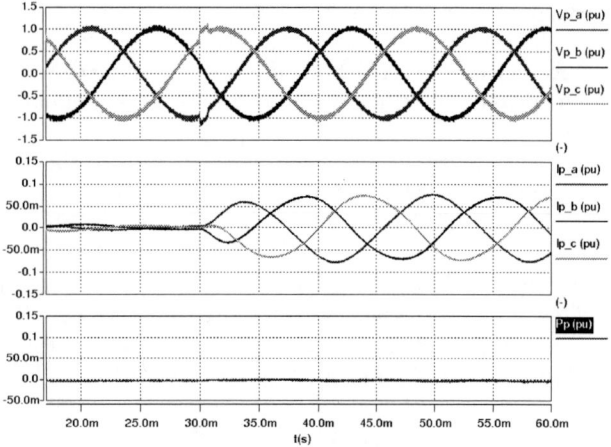

Fig. 9 Simulation waveforms (in pu) of the voltage, current and active power of the shunt CMI.

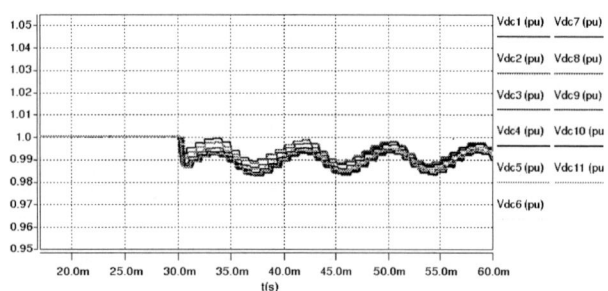

Fig. 10 DC voltage waveforms (in pu) of all 11 H-bridges of the series CMI.

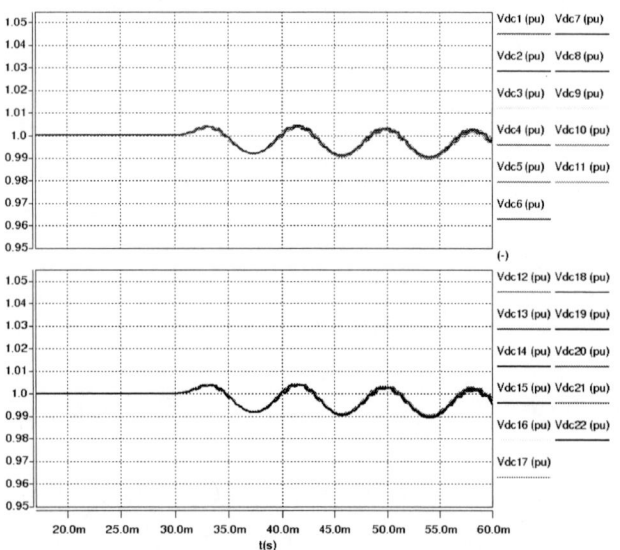

Fig. 11 DC voltage waveforms (in pu) of all 22 H-bridges of the shunt CMI.

VIII. CONCLUSION

In this paper, a new transformer-less UPFC based on a novel configuration and control of two CMIs was proposed. It has been demonstrated that the new UPFC can achieve the same controllability as the traditional UPFC. However, the traditional UPFC consisting of two

back-to-back inverters requires isolation and zigzag transformers. The new UPFC consisting of two CMIs offers several advantages over the traditional UPFC: such as completely transformer-less and highly modular structure, light weight, high efficiency, high reliability, low cost, and fast dynamic response. The new transformer-less UPFC is therefore very suited for fast and distributed power flow control of wind and solar power transmission. The operating principle and performance of the transformer-less UPFC have been fully analyzed by detailed theory and simulation model. A 2-MVA/13.8-kV prototype is being built and will be demonstrated by 2014.

APPENDIX A DERIVATION OF THE EQUATION (12)

From (10), the I_P can be rewritten as,

$$I_P = \frac{(I\cos\rho - I_C\sin\delta)}{\sin\delta_S} \cdot \sqrt{1 + (\frac{V_{S0} + V_C\cos\delta}{V_C\sin\delta})^2} \quad \text{(A.1)}$$

Substituting $\delta_S = \arctan(V_C\sin\delta/(V_{S0}+V_C\cos\delta))$ into (A.1), I_P is then simplified as,

$$I_P = \frac{(I\cos\rho - I_C\sin\delta)}{\sin(\arctan(V_C\sin\delta/(V_{S0}+V_C\cos\delta)))} \quad \text{(A.2)}$$

Based on the following triangle function formula,

$$\sin(\arctan\theta) = \frac{\theta}{\sqrt{1+\theta^2}} = \frac{1}{\sqrt{1+\theta^{-2}}} \quad \text{(A.3)}$$

(A.2) can be further simplified to the form of (12).

ACKNOWLEDGMENT

The authors gratefully acknowledge the funding support of Advanced Research Project Agency - Energy (ARPA-E), Department of Energy (DoE), United States to design, prototype and test the transformer-less UPFC prototype.

REFERENCES

[1] D. Divan and H. Johal, "A Smarter Grid for Improving System Reliability and Asset Utilization," in *Power Electronics and Motion Control Conference, 2006. IPEMC 2006. CES/IEEE 5th International*, 2006, pp. 1-7.

[2] J. M. Carrasco, L. G. Franquelo, J. T. Bialasiewicz, E. Galvan, R. C. P. Guisado, Ma A. M. Prats, J. I. Leon, and N. Moreno-Alfonso, "Power-Electronic Systems for the Grid Integration of Renewable Energy Sources: A Survey," *Industrial Electronics, IEEE Transactions on,* vol. 53, no. 4, pp. 1002-1016, 2006.

[3] V. Hamidi, Li Furong, and Yao Liangzhong, "Value of Wind Power at Different Locations in the Grid," *Power Delivery, IEEE Transactions on,* vol. 26, no. 2, pp. 526-537, 2011.

[4] A. Ipakchi and F. Albuyeh, "Grid of the future," *Power and Energy Magazine, IEEE,* vol. 7, no. 2, pp. 52-62, 2009.

[5] M. Glinkowski, J. Hou, and G. Rackliffe, "Advances in Wind Energy Technologies in the Context of Smart Grid," *Proceedings of the IEEE,* vol. 99, no. 6, pp. 1083-1097, 2011.

[6] B. K. Patel, H. S. Smith, T. S. Hewes, and W. J. Marsh, "Application of Phase Shifting Transformers for Daniel-McKnight 500kV Interconnection," *Power Delivery, IEEE Transactions on,* vol. 1, no. 3, pp. 167-173, 1986.

[7] C. Hochgraf and R. H. Lasseter, "A transformer-less static synchronous compensator employing a multi-level inverter," *Power Delivery, IEEE Transactions on,* vol. 12, no. 2, pp. 881-887, 1997.

978-1-4799-2706-7/14 $31.00 © 2014 IEEE

The 2014 International Power Electronics Conference

[8] L. Gyugyi, C. D. Schauder, and K. K. Sen, "Static synchronous series compensator: a solid-state approach to the series compensation of transmission lines," *Power Delivery, IEEE Transactions on,* vol. 12, no. 1, pp. 406-417, 1997.

[9] L. Gyugyi, "Unified power-flow control concept for flexible AC transmission systems," *Generation, Transmission and Distribution, IEE Proceedings C,* vol. 139, no. 4, pp. 323-331, 1992.

[10] L. Gyugyi, C. D. Schauder, S. L. Williams, T. R. Rietman, D. R. Torgerson, and A. Edris, "The unified power flow controller: a new approach to power transmission control," *Power Delivery, IEEE Transactions on,* vol. 10, no. 2, pp. 1085-1097, 1995.

[11] A. Rajabi-Ghahnavieh, M. Fotuhi-Firuzabad, M. Shahidehpour, and R. Feuillet, "UPFC for Enhancing Power System Reliability," *Power Delivery, IEEE Transactions on,* vol. 25, no. 4, pp. 2881-2890, 2010.

[12] M. Noroozian, L. Angquist, M. Ghandhari, and G. Andersson, "Use of UPFC for optimal power flow control," *Power Delivery, IEEE Transactions on,* vol. 12, no. 4, pp. 1629-1634, 1997.

[13] D. Povh, "Use of HVDC and FACTS," *Proceedings of the IEEE,* vol. 88, no. 2, pp. 235-245, 2000.

[14] C. D. Schauder, L. Gyugyi, M. R. Lund, D. M. Hamai, T. R. Rietman, D. R. Torgerson, and A. Edris, "Operation of the unified power flow controller (UPFC) under practical constraints," *Power Delivery, IEEE Transactions on,* vol. 13, no. 2, pp. 630-639, 1998.

[15] C. Schauder, E. Stacey, M. Lund, L. Gyugyi, L. Kovalsky, A. Keri, A. Mehraban, and A. Edris, "AEP UPFC project: installation, commissioning and operation of the ±160 MVA STATCOM (phase I)," *Power Delivery, IEEE Transactions on,* vol. 13, no. 4, pp. 1530-1535, 1998.

[16] K. Sano and M. Takasaki, "A Transformerless D-STATCOM Based on a Multivoltage Cascade Converter Requiring No DC Sources," *Power Electronics, IEEE Transactions on,* vol. 27, no. 6, pp. 2783-2795, 2012.

[17] B. A. Renz, A. Keri, A. S. Mehraban, C. Schauder, E. Stacey, L. Kovalsky, L. Gyugyi, and A. Edris, "AEP unified power flow controller performance," *Power Delivery, IEEE Transactions on,* vol. 14, no. 4, pp. 1374-1381, 1999.

[18] J. Monteiro, J. F. Silva, S. F. Pinto, and J. Palma, "Matrix Converter-Based Unified Power-Flow Controllers: Advanced Direct Power Control Method," *Power Delivery, IEEE Transactions on,* vol. 26, no. 1, pp. 420-430, 2011.

[19] H. Akagi, S. Inoue, and T. Yoshii, "Control and Performance of a Transformerless Cascade PWM STATCOM With Star Configuration," *Industry Applications, IEEE Transactions on,* vol. 43, no. 4, pp. 1041-1049, 2007.

[20] B. Gultekin, C. O. Gercek, T. Atalik, M. Deniz, N. Bicer, M. Ermis, K. N. Kose, C. Ermis, E. Koc, I. Cadirci, A. Acik, Y. Akkaya, H. Toygar, and S. Bideci, "Design and Implementation of a 154-kV 50-Mvar Transmission STATCOM Based on 21-Level Cascaded Multilevel Converter," *Industry Applications, IEEE Transactions on,* vol. 48, no. 3, pp. 1030-1045, 2012.

[21] B. Gultekin and M. Ermis, "Cascaded Multilevel Converter-Based Transmission STATCOM: System Design Methodology and Development of a 12 kV 12 MVAr Power Stage," *Power Electronics, IEEE Transactions on,* vol. 28, no. 11, pp. 4930-4950, 2013.

[22] Wang Jin and F. Z. Peng, "Unified power flow controller using the cascade multilevel inverter," *Power Electronics, IEEE Transactions on,* vol. 19, no. 4, pp. 1077-1084, 2004.

978-1-4799-2706-7/14 $31.00 © 2014 IEEE

A New Power Flow Controller Using Six Multilevel Cascaded Converters for Distribution Systems

Ryoji Tsuruta, Tatsuya Hosaka, and Hideaki Fujita
Department of Electrical and Electronic Engineering
Tokyo Institute of Technology
Tokyo, Japan
hf@ieee.org

Abstract—**This paper proposes a new power flow controller consisting of six multilevel cascaded converters for stabilizing and regulating the line voltages in distribution networks. Each cascaded converter is composed of series connection of H-bridge converters, and connected between one phase in the sending end and the other phase in receiving end. While each cascaded converter control only reactive power through itself, the active and reactive power flows are controllable. Using multilevel cascaded converter topology, this power flow controller is expected to be applied to more high-voltage and high-power applications and to have fault ride through capabilities. Experimental results are shown to verify the operating principle of the proposed power flow controller and demonstrates the quick and stable power flow control performance.**

Keywords—*grid connected converters, modular multilevel cascade converters, pulse width modulation.*

I. INTRODUCTION

Recently, it is strongly promoted by the Japanese government to install distributed power generation systems using renewable energy resources such as wind turbines and photovoltaic after the Fukushima accident. According to the increase of such distributed power generation systems, voltage difference between distribution feeders is expected to be serious. One of the solution to this problem is to install "power flow controller" at the ends of the power feeders.

Various devices have been already proposed, which have the capability of controlling the power flows among the power systems, such as static synchronous series capacitors (SSSCs) [1], unified power flow controllers (UPFCs) [2], back-to-back converters (BTBs) [3], and so on. The BTBs seems to be the most flexible in active and reactive power flow control as well as in reactive power compensation. However, two sets of full-rating power converters are required and connected in back-to-back. The converter ratings of an SSSC and a UPFC are relatively small, compared with a BTB. The SSSC can control active power flow by injecting an amount of reactive power into the transmission line. However, the SSSC has no capability of injecting active power, and thus, there is some limitation in controlling the active and reactive power flows. On the other hand, the UPFC consists of series and shunt converters. The series device has the capability of injecting both active and reactive power into the transmission line, because the shunt device can take the required active power from the sending

or receiving end. The UPFC has no limitation in active and reactive power flow control unlike the SSSC.

Recent regulations requests fault ride through (FRT) capability to maintain continuous operation of power systems. However, the series device in the UPFC may fall into overvoltage under a fault condition in the transmission line, because its voltage rating is a fraction of the transmission line voltage. Cascade converter topology is attractive to increase the voltage rating and to sustain the voltage applied across the series device in the fault condition. However, a multiple of dc capacitors should be isolated, and it is difficult to provide active power to each isolated dc-capacitor from the shunt device.

This paper proposes a new power flow controller consisting of six multilevel cascaded converters for stabilizing and regulating the line voltages in distribution networks. Each cascaded converter is composed of series connection of H-bridge converters, and connected between one phase in the sending end and the other phase in receiving end [4], [5], [6]. While each cascaded converter control only reactive power through itself, the active and reactive power flows are controllable. As a result, the average value of the dc-capacitor voltage of each H-bridge converter can be controlled at a constant level. This paper discusses control method of active and reactive power flows as well as the capacitor voltage control in detail. Experimental results using a 10-kW downscaled laboratory model are shown to verify the operating principle of the proposed power flow controller and demonstrates the quick and stable power flow control performance.

II. POWER FLOW CONTROLLER USING SIX CASCADED MULTILEVEL CONVERTERS

Fig. 1 shows the system configuration of the new power flow controller proposed in this paper. The power flow controller consists of six cascaded converters which are connected between the sending and receiving ends. Hereafter, the three-phase sending-end voltages are defined by using capital-letter subscripts as v_U, v_V and v_W, and these in receiving ends are as v_u, v_v and v_w using lower cases. Each cascaded converter is connected between one phase in the sending end and the other phase in receiving end. For example, the cascaded converter v_{uV} is connected between the u-phase terminal in the receiving end and the V-phase terminal in the sending end. Thus, the voltage across the converter is denoted as v_{uV}, and its current

The 2014 International Power Electronics Conference

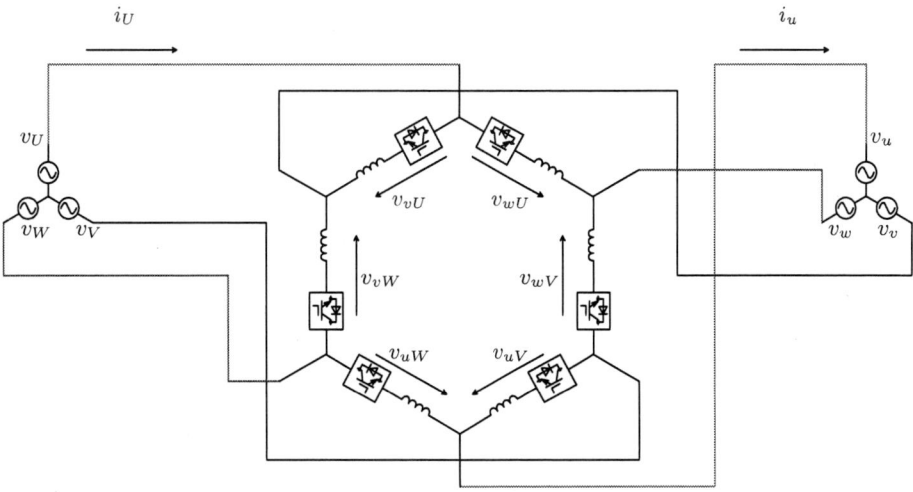

Fig. 1. System configuration of the proposed power flow controller using six cascaded converters.

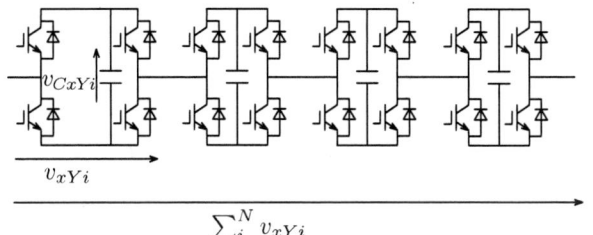

Fig. 2. Composition of a cascaded H-bridge converter.

is referred to as i_{uV}. Note that no converter is connected between the u-phase terminal in the receiving end and the U-phase terminal in the sending end. Therefore, the circuit configuration of the proposed power flow controller can be considered as three sets of two cascaded converters connected a common sending end terminal.

Fig. 2 shows the circuit configuration of each cascaded converter for the proposed power flow converter. Each cascaded converter consists of multiple H-bridge converters, which are connected in series. A dc-capacitor is only attached on the dc side of each H-bridge converter, and neither additional voltage regulating circuit nor dc power supply is connected. Therefore, the dc-capacitor voltages have to be controlled by the voltage and current control in the six cascaded converters.

Here, the ac terminal voltage of an H-bridge converter is defined as v_{xYi}, where x is the phase in receiving end, Y is the phase in sending end, i is the order of the series connected H-bridge converter. The synthesized voltage in a N-series cascade converter connected between Y-phase and x-phase terminals is the sum of the H-bridge converter voltages

from v_{xY1} to v_{xYN}, given by

$$v_{xY} = \sum_{i=1}^{N} v_{xYi}. \tag{1}$$

The dc-capacitor voltage in an H-bridge converter is also defined as v_{CxYi}. Then, the sum of the dc-capacitor voltages is defined as a total dc-capacitor voltage v_{CxY}, which is given by

$$v_{CxY} = \sum_{i=1}^{N} v_{CxYi}. \tag{2}$$

The current flowing through the cascade converter connected between between Y-phase and x-phase terminals is also defined as converter current i_{xY}. Applying Kirchhoff's current law (KCL), the three-phase current at the sending-end are represented by

$$\begin{cases} i_U = i_{vU} + i_{wU} \\ i_V = i_{wV} + i_{uV} \\ i_W = i_{uW} + i_{vW} \end{cases} \tag{3}$$

And the three-phase receiving-end current are given by

$$\begin{cases} i_u = i_{uV} + i_{uW} \\ i_v = i_{vW} + i_{vU} \\ i_w = i_{wU} + i_{wV} \end{cases} \tag{4}$$

From the above equation, cascaded converter currents i_{xY} have six degrees of freedom. Assuming a three-phase three-wire system, the neutral points of the sending and receiving ends are not connected, and no zero-sequence current can flow. Then, the sending- and receiving-end current have the following two restriction.

$$i_U + i_V + i_W = 0 \tag{5}$$

978-1-4799-2706-7/14 $31.00 © 2014 IEEE

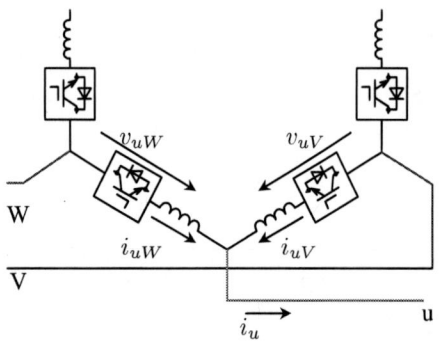

Fig. 3. Two cascaded converters connected with the u phase on the receiving end.

$$i_u + i_v + i_w = 0 \qquad (6)$$

Therefore, they have four degrees of freedom. The proposed power flow controller can control four degrees of freedom, which are the active and reactive power of the sending- and receiving- ends.

III. Operating Principle

Fig. 3 is a part of the proposed power flow controller where two of six cascaded converters connected to the receiving-end u-phase v_{uV} and v_{uW} are shown. The other terminals of the two converters v_{uV} and v_{uW} are connected to the sending-end V-phase and W-phase, respectively. The u-phase receiving-end current is the sum of the current through the two converters as $i_u = i_{uV} + i_{uW}$, where the converter currents are i_{uV} and i_{uW}.

For the sake of simplicity, the sending- and receiving-end voltages are assumed to be three-phase balanced sinusoidal wave shape as

$$v_U = v_u = \sqrt{\frac{2}{3}} V \sin \omega t$$

$$v_V = v_v = \sqrt{\frac{2}{3}} V \sin \left(\omega t - \frac{2\pi}{3}\right) \qquad (7)$$

$$v_W = v_w = \sqrt{\frac{2}{3}} V \sin \left(\omega t + \frac{2\pi}{3}\right),$$

where V is the line-to-line rms voltage. Considering the phase angles, the applied voltage across the two cascaded converter are given by

$$v_{uV} = v_u - v_V = \sqrt{2} V \sin \left(\omega t + \frac{\pi}{6}\right) \qquad (8)$$

$$v_{uW} = v_u - v_W = \sqrt{2} V \sin \left(\omega t - \frac{\pi}{6}\right). \qquad (9)$$

The synthesized receiving-end u-phase current i_u is the sum of the current through two cascaded converters i_{uW} and i_{uV}. Assuming that both converters can make their current

sinusoidal, the converter currents are represented by

$$\begin{aligned} i_{uV} = \quad & \sqrt{2} I_{PuV} \sin \left(\omega t + \frac{\pi}{6}\right) \\ & +\sqrt{2} I_{QuV} \cos \left(\omega t + \frac{\pi}{6}\right) \qquad (10) \\ i_{uW} = \quad & \sqrt{2} I_{PuW} \sin \left(\omega t - \frac{\pi}{6}\right) \\ & +\sqrt{2} I_{QuW} \cos \left(\omega t - \frac{\pi}{6}\right), \qquad (11) \end{aligned}$$

where I_{PuV} and I_{PuW} is the rms current components in phase with their voltages, and I_{QuV} and I_{QuW} are current components leading by 90° from their voltages.

The instantaneous active power flowing out from the converters are given by

$$\begin{aligned} p_{uV} = \quad & V I_{PuV} \left(1 - \cos \left(2\omega t + \frac{\pi}{3}\right)\right) \\ & +V I_{QuV} \cos \left(2\omega t + \frac{\pi}{3}\right) \qquad (12) \\ p_{uW} = \quad & V I_{PuW} \cos \left(1 - \cos \left(2\omega t - \frac{\pi}{3}\right)\right) \\ & +V I_{QuW} \cos \left(2\omega t - \frac{\pi}{3}\right). \qquad (13) \end{aligned}$$

From above equations, the average active power of p_{uV} and p_{uW} are given by

$$P_{uV} = \frac{1}{T} \int_0^T p_{uV} dt = V I_{PuV} \qquad (14)$$

$$P_{uW} = \frac{1}{T} \int_0^T p_{uW} dt = V I_{PuW}, \qquad (15)$$

where the T is the period of the grid voltage cycle. The average active power is flowing into the corresponding H-bridge converters, and charges or discharges the dc capacitors. Thus, it is possible to control the dc-capacitor voltage by adjusting I_{PuV} and I_{PuW}. On the other hand, I_{QuV} and I_{QuW} can be used to control the active and reactive power flows.

Fig. 4 shows the voltage and current phasor diagram of the two cascaded converters. In Fig. 4, the converter currents is assumed $I_{PuV} = I_{PuW} = 0$ and the converter current I_{QuV} and I_{QuW} only flow. Fig. 4 (a) is a phasor diagram of the applied voltage across the two cascaded converters. The converter voltage phasors are given by

$$\vec{V}_{uV} = \vec{V}_u - \vec{V}_V, \qquad (16)$$

$$\vec{V}_{uW} = \vec{V}_u - \vec{V}_W. \qquad (17)$$

Fig. 4 (b) is a phasor diagram of the converter currents where the converter controls a reactive power flow. It is assumed that both converter current $\vec{I}_{QuV}$ and $\vec{I}_{QuW}$ have the same amplitude and leads the converter voltage by 90°. The synthesized current $\vec{I}_u$ is the vector sum of $\vec{I}_{uV}$ and $\vec{I}_{uW}$, and $\vec{I}_u$ is leading by 90° from the grid phase voltage $\vec{V}_u$. Thus, the reactive power flow can be controlled by adjusting the $\vec{I}_{QuV}$ and $\vec{I}_{QuW}$.

Fig. 4 (c) is a phasor diagram of the converter currents where the converter controls an active power flow. In this case, both converter current $\vec{I}_{QuV}$ and $\vec{I}_{QuW}$ also have the same

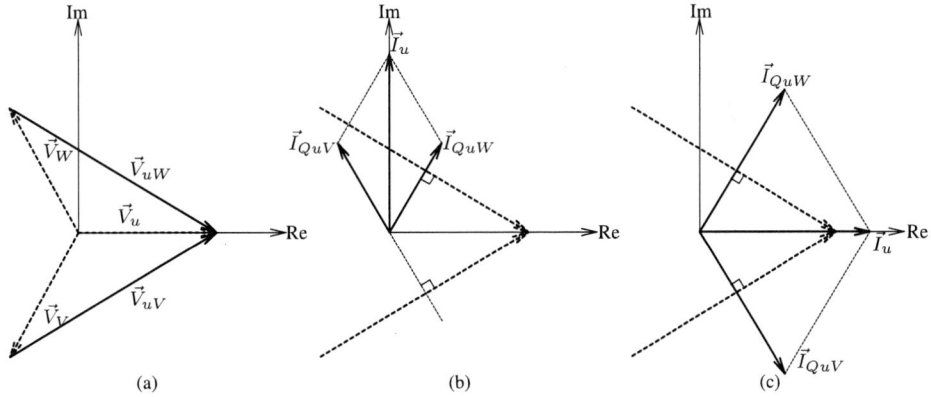

Fig. 4. Phasor diagrams of the proposed power flow controller. (a) terminal voltages, (b) $I_{QuV} > 0$ and $I_{QuW} > 0$, (c) $I_{QuV} < 0$ and $I_{QuW} > 0$.

amplitude. The converter current $\vec{I}_{QuW}$ still leads the converter voltage by 90°, while $\vec{I}_{QuW}$ lags by 90° from its converter voltage $\vec{V}_{uW}$. The synthesized current $\vec{I}_u$ is in phase with the grid phase voltage $\vec{V}_u$. An amount of active power is provided to the receiving end in the case of Fig. 4 (c). Therefore, the proposed system can control the active power flow. although each converter only controls its reactive power.

The synthesized current $\vec{I}_u$ can be controlled at any phase angle and amplitude by adjusting only the amplitude of $\vec{I}_{QuV}$ and $\vec{I}_{QuW}$. Then the converter does not need to control the converter current $\vec{I}_{PuV}$ and $\vec{I}_{PuW}$. Therefore, $\vec{I}_{PuV}$ and $\vec{I}_{PuW}$ can be used only for the dc-capacitor voltage control, and then, the dc-capacitor voltage control is independent from the power flow control.

The complex power flowing into the sending-end u-phase is calculated as

$$
\begin{aligned}
P_u + jQ_u &= \overline{\vec{V}}_u \vec{I}_u = \overline{\vec{V}}_u \vec{I}_{uV} + \overline{\vec{V}}_u \vec{I}_{uW} \\
&= -\frac{V}{2\sqrt{3}}I_{QuV} + j\frac{V}{2}I_{QuV} \\
&\quad + \frac{V}{2\sqrt{3}}I_{QuW} + j\frac{V}{2}I_{QuW},
\end{aligned} \tag{18}
$$

where $\overline{\vec{V}}_u$ is the complex conjugation of $\vec{V}_u$. Assuming the current amplitudes of six cascaded converters as $I_{QuV} = I_{QvW} = I_{QwU} = I_{Q1}$ and $I_{QuW} = I_{QvU} = I_{QwV} = I_{Q2}$, the three-phase complex power flow from the sending end to receiving end is represented by

$$
P = \frac{\sqrt{3}}{2}V(-I_{Q1} + I_{Q2}) \tag{19}
$$

$$
Q = \frac{3}{2}V(I_{Q1} + I_{Q2}). \tag{20}
$$

The above equation implies that the active and reactive power flow can be controlled by adjusting the amplitudes of the orthogonal current through each cascaded converter.

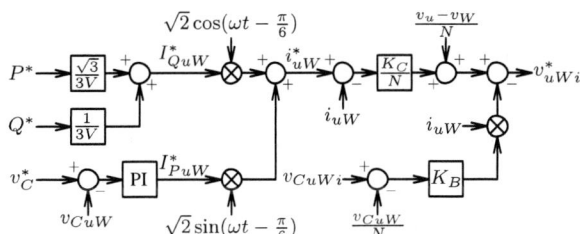

Fig. 5. Control block diagram for one of the H-bridge converters in cascaded converter v_{uW}.

IV. CONTROL METHOD

Fig. 5 shows a control block diagram for one cascaded converter. Fig. 5 calculates the voltage references as v^*_{CuW1} to v^*_{CuW4} of four H-bridge converters that constitute the cascaded converter connected between the u-phase terminal in the receiving end and the W-phase terminal in the sending end. The control block consists of four major parts: current reference calculation, dc-capacitor voltage control, converter current control, and dc-capacitor voltage balancing.

The current reference calculation calculates the reference for the reactive power component in the the converter current, I^*_{QuW} from the active and reactive power references P^* and Q^*. From (19) and (20), the reactive power component the converter current reference for the converter connected between u and W terminals given by

$$
I^*_{QuW} = I^*_{Q2} = \frac{\sqrt{3}P^* + Q^*}{3V}.. \tag{21}
$$

The dc-capacitor voltage control calculates the reference for the active power component in the the converter current, I^*_{PuW} to regulate the total dc-capacitor voltage v_{CuW}. The total dc-capacitor voltage v_{CuW} is compared with its reference v_{CuW}, and proportional and integral gains are applied to manipulate I_{PuW}, as follows:

$$
\begin{aligned}
I^*_{PuW} &= K_{Cp}(v^*_{CuW} - v_{CuW}) \\
&\quad + K_{Ci}\int(v^*_{CuW} - v_{CuW})dt, \tag{22}
\end{aligned}
$$

where K_{Cp} is a proportional gain and K_{Ci} is an integral gain for the dc-capacitor voltage feedback control.

The converter current reference is the sum of the references for the active and reactive components, given by

$$
\begin{aligned}
i_{uW}^* &= i_{PuW}^* + i_{QuW}^* \\
&= \sqrt{2}I_{PuW}^* \sin\left(\omega t - \frac{\pi}{6}\right) \\
&\quad + \sqrt{2}I_{QuW}^* \cos\left(\omega t - \frac{\pi}{6}\right).
\end{aligned} \tag{23}
$$

The converter current control part detects the corresponding converter current i_{uW}, compares it with its reference i_{uW}^*, and manipulate the converter voltage references v_{uW}^* by means of feedback control. The converter voltage reference is represented by

$$
v_{uW}^* = K_C(i_{uW}^* - i_{uW}) + v_u - vW, \tag{24}
$$

where K_C is the current feedback gain, v_u and v_W are the grid phase voltage at the u-phase and W-phase terminals. The first term on the right hand side performs the current feedback control, and the second and third terms acts as a feed forward compensation against the grid voltages v_u and v_W. Note that the feedback gain and the feed forward values are divided by the series number N in Fig.5 to obtain the reference value for one of the N-series H-bridge converters.

The dc-capacitor voltage balancing part is attached to eliminate the voltage difference between the dc capacitors connected to the N-series H-bridge converters. This part detects the corresponding dc-capacitor voltage v_{CuWi}, and compares it with the average capacitor voltage which can be calculated from the total capacitor voltage v_{CuW}. The output of the dc-capacitor voltage balancing part is represented by

$$
r_{uW} = K_B\left(v_{CuWi} - \frac{v_{CuW}}{N}\right), \tag{25}
$$

where K_B is a feedback control gain for dc-capacitor voltage balancing. The voltage reference for one of the N-series H-bridge converters is represented by

$$
v_{uWi}^* = \frac{v_{uW}^*}{N} - r_{uW}i_{uW}. \tag{26}
$$

The second term superposes a additional voltage $r_{uW}i_{uW}$ on the voltage reference v_{uWi}^*. The output of the dc-capacitor voltage balancing part r_{uW} is positive when the corresponding dc-capacitor voltage v_{CuWi} is lower than the average dc-capacitor voltage v_{CuW}/N. Then, the second term acts as a equivalent resistor against the converter current i_{uW}, and thus, an amount of power flows into the H-bridge v_{uWi} to charge the dc-capacitor up. On the other hand, the second term behaves like a equivalent negative resistor when $v_{CuWi} > v_{CuW}/N$, the corresponding dc-capacitor is discharged by the converter current i_{uW}. Therefore, the second term has the capability to balance the dc-capacitor voltages. Note that the sum of the equivalent resistor r_{uW} are always

$$
\sum_{i=1}^{N} r_{uW} = \sum_{i=1}^{N} K_B\left(v_{CuWi} - \frac{v_{CuW}}{N}\right) = 0. \tag{27}
$$

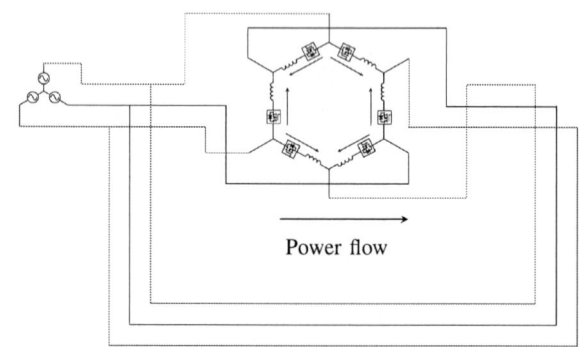

Fig. 6.   Experimental circuit configuration.

This means that the dc-capacitor voltage balancing has no interference on the power flow control and the dc-capacitor voltage control.

V.   Experimental System

Fig. 6 shows circuit configuration of a 10-kVA downscaled laboratory model of power flow controller using six multilevel cascaded converters. The parameter of the laboratory model is shown in Table I. Each cascaded converter consists of four H-bridge converters connected in series, and thus, the whole system is composed of 24 H-bridge converters using 96 power MOSFETs. A phase-shifted triangular carrier signals are used to generate the gate signals in the PWM control for H-bridge converters. The initial angle of each carrier signal is shifted by $\pi/4 = 45°$ between H-bridge converters in a cascaded converter. Then, the switching ripple frequency is increased to $8f_{sw}$. In addition, the initial angle of the carrier signals in adjacent two cascaded converters are also shifted by $\pi/16 = 11.25°$. As a result, the current ripples induced by the two cascaded converters are canceled each other at the sending- and receiving-end terminals.

In the following experiments, the sending and the receiving ends of the power flow controller were connected to a common three-phases power source with a line-to-line rms voltage of $V = 400$ V and a frequency of 50 Hz, as Fig. 6. The reference value for the dc-capacitor voltage was set to $v_C^* = 150$ V, and the triangle carrier frequency for the PWM controller was $f_{sw} = 1$ kHz.

VI.   Experimental Results

Figs. 7-11 show experimental results obtained by the 400-V 10-kVA scale-down laboratory model. Fig. 7 and 8 show experimental waveforms in case of active power flow reference of $P^* = 10$ kW and reactive power reference of $Q^* = 0$. The u-phase receiving-end current i_u had a sinusoidal waveform in phase with the phase voltage v_u, and thus, active power flowed from the sending end to the receiving end. The measured current THD was 2.9% under this operating condition. The converter voltages v_{uV} and v_{uW} were nine-level PWM waveforms due to the phase-shift carrier was used for PWM signal

TABLE I.   CIRCUIT PARAMETERS FOR THE SCALED-DOWN
LABORATORY MODEL

rated power		10 kVA
ac voltage	V	400 V
rated power	I	14.4 A
dc-capacitor voltage	V_C	148 V
dc-capacitor	C	3000 μF
unit capacitance constant	H	78.8 ms (148 V)
AC inductor	L	4 mH (10.5%)
switching frequency	f_{sw}	1 kHz
dead time	T_d	4 μs
current feedback gain	K	20 V/A
dc voltage gain	K_{Cp}	0.01 A/V
dc voltage gain	K_{Ci}	0.01 A/V
voltage balancing gain	K_B	0.01 A^{-1}

on a three-phase, 50-Hz, 400-V, 10 kVA base

generation. As a result, the harmonic component and current ripples in i_{uV} and i_{uW} were well eliminated. Moreover, the phase angles of i_{uV} and i_{uW} were orthogonal to their converter voltages v_{uV} and v_{uW}, respectively. The cascaded converter current i_{uV} and i_{uW} had the same rms value of 14 A which was also equal to the rms current of i_u. Therefore, each cascaded converter produced no active power, and the mean values of the dc-capacitor voltages v_{CuV1} and v_{CuW1} were kept at their reference of $v_C^* = 150$ V. The dc-capacitor voltages v_{CuV1} and v_{CuW1} fluctuated at a frequency twice higher than the line frequency. The ripple voltage in v_{CuV1} and v_{CuW1} were an acceptable level which was less than 10% of the mean value. In Fig. 7 i_{uV} is delayed from v_{uV} by 90°, and i_{uW} leads to v_{uW} by 90° as shown in Fig.4. However, the converter voltage v_{uV} is slightly larger than v_{uW} because of the voltage drop across the ac inductor.

Figs. 9 and 10 show experimental waveforms when the reactive power reference was set to $Q^* = -10$ kvar. The u-phase receiving-end current i_u led by 90° from the u-phase voltage v_u, and thus, the reactive power flow was induced by the power flow controller. The current THD of i_u was as low as 2.6%. The converter voltages v_{uV} and v_{uW} were also nine-level waveforms, and the converter currents i_{uV} and i_{uW} had smaller rms values than those in Fig. 7. The ripple of the dc-capacitor voltages v_{CuV1} and v_{CuW1} also reduced compared with Fig. 8.

Fig. 11 shows experimental waveforms in a transient state, where the active power reference was changed from $P^* = -10$ kW to $+10$ kW for a short duration of 20 ms. Then, the reactive power reference was kept at $Q^* = 0$ kvar. The instantaneous active power followed its reference and changed to $+10$ kW. The dc-capacitor voltages v_{CuV1} and v_{CuW1} were well regulated at their reference value, even under the transient state. Since each cascaded converter controls only the reactive power, the dc-capacitor voltage has almost no voltage error even when the active power flow is suddenly changed.

VII.   CONCLUSION

This paper has discussed control performance of a power flow controller consisting of six multilevel cascaded converters. The operating principle of the power flow controller has

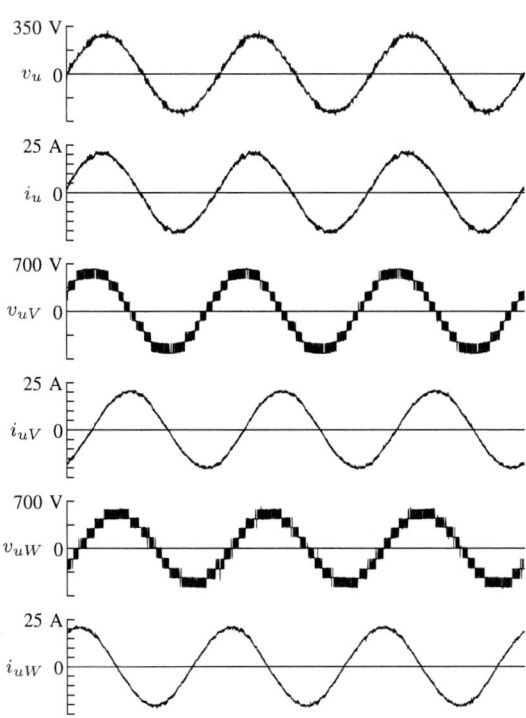

Fig. 7.   Experimental voltage and current waveforms of the cascaded converters when active power reference was set to $P^* = 10$ W.

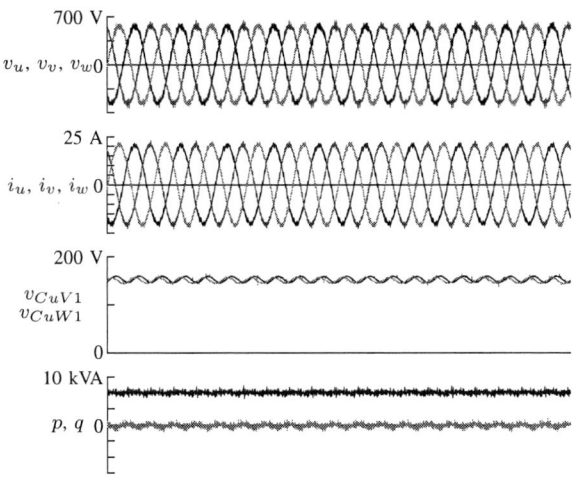

Fig. 8.   Experimental waveforms of the dc-capacitor voltages and power flow when active power reference was set to $P^* = 10$ W.

theoretically been discussed and verified by using a 10-kVA, 400-V scale-down laboratory model. The proposed power flow controller has the capability of controlling active and reactive power flows by controlling only reactive power in the cascaded converter. Then the dc-capacitor voltage of each cascaded converter can be controlled at a constant level, independently

The 2014 International Power Electronics Conference

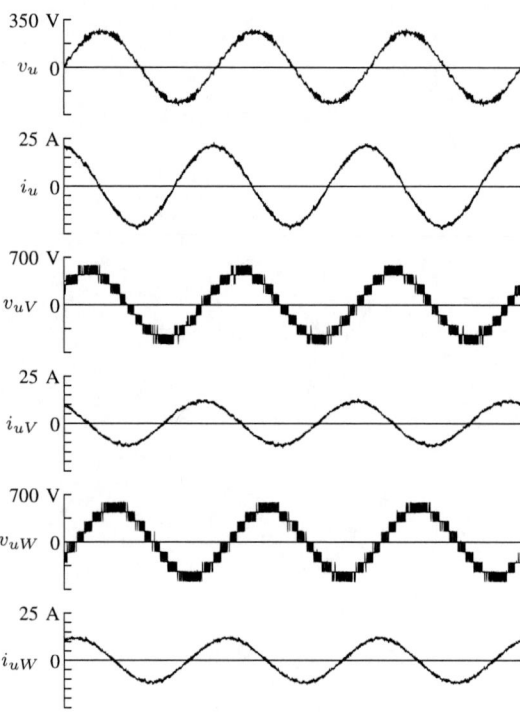

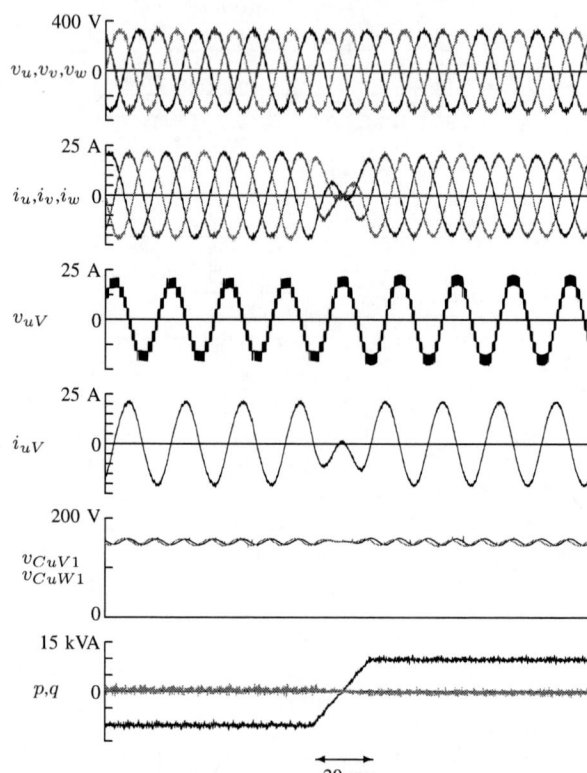

Fig. 9. Experimental voltage and current waveforms of the cascaded converters when active power reference was set to $Q^* = -10$ kvar.

Fig. 11. Transient response for a ramp change in the active power reference.

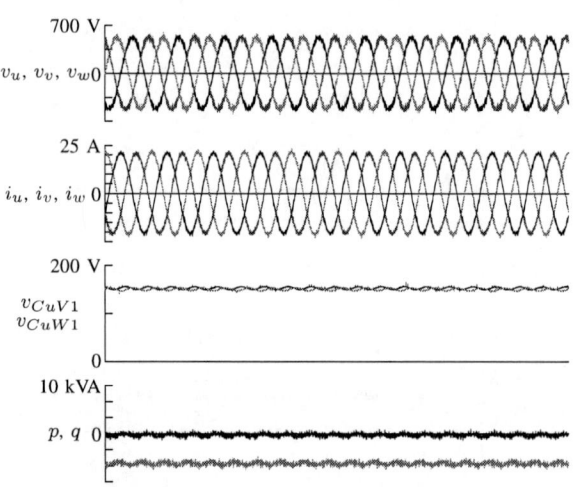

Fig. 10. Experimental waveforms of the dc-capacitor voltages and power flow when active power reference was set to $Q^* = -10$ kvar.

REFERENCES

[1] L. Gyugyi, "Dynamic compensation of ac transmission lines by solid-state synchronous voltage sources," *IEEE Trans. on Power Del.,* vol. 9, no. 2, pp. 904-911, 1994.

[2] L. Gyugyi, C.D. Schauder, S.L. Williams, T.R. Rietman, D.R. Torgerson, A. Edris, "The unified power flow controller: a new approach to power transmission control," *IEEE Trans. on Power Del.,* vol. 10, no. 2, pp. 1085-1097, 1995.

[3] M. Hagiwara, H. Fujita, H. Akagi, "Performance of a self-commutated BTB HVDC link system under a single-line-to-ground fault condition," *IEEE Trans. on Power Electron.,* vol. 18, no. 1, pp. 278-285, 2003.

[4] T. Hosaka, K. Akiba, H. Fujita, "A unified power-flow controller consisting of six cascaded H-bridge converters," *in Proc. of Japan Industry Applications Society Conference,* vol.1, no.35, pp.233–236, 2011 (in Japanese).

[5] L. Baruschka, A. Mertens, "A new 3-phase direct modular multilevel converter," *in Proc. of European Conference on Power Electronics and Applications.* pp. 1-10, 2011.

[6] L. Baruschka, A. Mertens, "A new 3-phase AC/AC modular multilevel converter with six branches in hexagonal configuration," *in Proc. of IEEE Energy Conversion Congress and Exposition,* pp. 4005-4012, 2011.

of the active and reactive power flow control. Experimental results have demonstrated a quick dynamic response of active power flow change from -10 kW to $+10$ kW within 20 ms without any unbalance in the dc-capacitor voltages.

978-1-4799-2706-7/14 $31.00 © 2014 IEEE

The 2014 International Power Electronics Conference

A Proposal of Modular Multilevel Converter Applying Three Winding Transformer

Shunsuke Tamada[*],   Yosuke Nakazawa[*],   Shoichi Irokawa[**]

[*]Power and Industrial Systems R&D Center
Toshiba Corporation
1, Toshiba-Cho, Fuchu-Shi, Tokyo, Japan
shunsuke.tamada@toshiba.co.jp

[**]Transmission & Distribution System Division
Toshiba Corporation
72-34, Horikawa-Cho, Saiwai-Ku, Kawasaki, Japan
shoichi.irokawa@toshiba.co.jp

Abstract— **This paper proposes a new type of Modular Multilevel Converter (MMC) applying a three winding transformer. In general, MMC requires a buffer reactor in each arm, which increases number of components and converter footprint. The proposed MMC with three winding transformer does not require the buffer reactors. We describe mathematical properties of the proposed MMC. We also confirmed it operated as same as typical MMC (with reactor topology) by experimentation using a prototype 10kVA converters.**

Keywords— *Modular Multilevel Converter, High Voltage Direct Current transmission, Flexible AC Transmission Systems, Voltage Source Converter, Transformer*

I. INTRODUCTION

Recently, Modular Multilevel Converter (MMC) has been studied [1]-[8]. MMC consists of series connected half bridge converters, so it enables voltage level and power ratings adjustable. Moreover, MMC can generate approximated sine wave output at AC terminal with reduced switching frequency of individual power devices and does not require AC filters if the number of levels exceeds adequate value. As MMC has these advantages, researches are advancing to high voltage application such as high voltage direct current transmission (HVDC) [9], [10], and it has been already operated in actual transmission system [11], [12].

Grid connected converters such as SVC (TCR) apply air core reactors [13], which are used in MMC also. It has very simple configuration so achieving lower cost and removing risk of magnetic saturation. However, the air core reactor requires wide clearance space for avoiding the effect magnetic field to adjacent equipment or building. In this paper, we propose a new type MMC which does not require the buffer reactors in principle. Most of grid connected converters such as HVDC and FACTS require a transformer for voltage and current scalability and need of grounding to neutral point in three phase. We propose to apply a three winding transformer instead of normal two winding transformer. It can eliminate the buffer reactor, resulting in reducing converter footprint and number of components. In this paper, equations for voltage and current of MMC with three winding transformer are derived analytically.

Furthermore, two sets of prototype 10kVA MMC were manufactured and experimental results of MMC in grid connection (active power, reactive power operation) are presented.

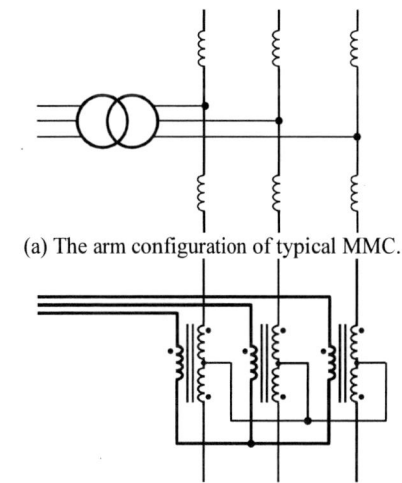

(a) The arm configuration of typical MMC.

(b) The arm configuration of proposed MMC.

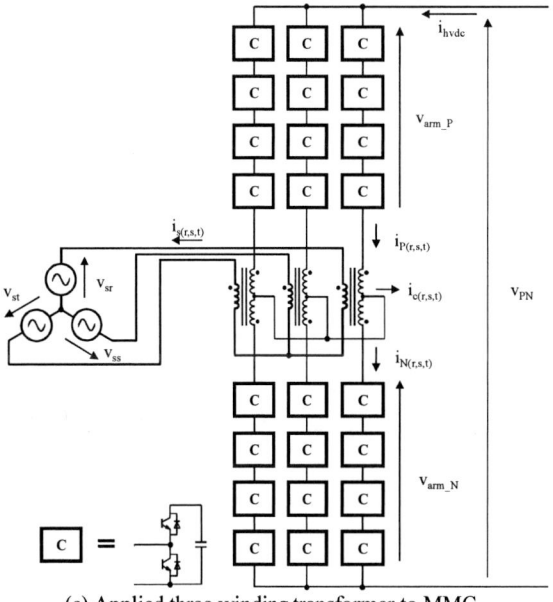

(c) Applied three winding transformer to MMC

Fig.1. Circuit configuration of an MMC with three winding transformer

978-1-4799-2706-7/14 $31.00 © 2014 IEEE   1357

II. MODULAR MULTILEVEL CONVERTER WITH THREE WINDING TRANSFORMER

A. Proposed converter topology

A circuit configuration of proposed topology is shown in figure 1. This MMC consists of cascaded half bridge converters and a three winding transformer. The primary winding of the transformer is connected to AC grid. Secondary and tertiary winding are connected to converter upper arm and lower arm in each legs. These winding are wrapped around common iron core. The secondary winding and the tertiary winding are wound in inverse direction, and negative side of each winding are connected each other. A connected point of secondary and tertiary winding is center of each leg, which becomes neutral point of converter side.

In MMC, a current in each arm contains two components, one is the common component and the other is the distinct component; between upper arm and lower arm [3]-[5]. The common current component carries through both, secondary and tertiary winding. A distinct current component flows through only each winding. Figure 2 shows magnetomotive force at secondary windings and tertiary winding in two cases. In common current component case, the direction of the magnetomotive force at the secondary winding and the tertiary winding are inverse direction and these are cancelled. Thereby magnetic flux by common current component is not generated in common iron core. In differential current component case, the magnetomotive force each windings is same direction, so these are mutually intensified. The common current component is DC, the distinct current component is AC. In the proposed transformer, a DC flows at secondary winding and tertiary winding, but can possible to apply a normal three winding transformer without bigger iron core. The magnetomotive force by the distinct current component is cancelled by the current at primary winding, like an ordinary transformer.

Generally, a transformer has leakage inductance. Leakage inductance between the secondary and the tertiary winding behaves like arm reactor. Thus, the proposed topology does not require the arm reactor. Details are described in the next chapter. Moreover, the neutral point of converter side is possible to connect to the ground which becomes neutral point of the DC link, so this topology can constitute symmetrical monopole in applying to HVDC transmission system. When it is not connected to the ground, it can apply to asymmetrical monopole or bi-pole HVDC transmission system.

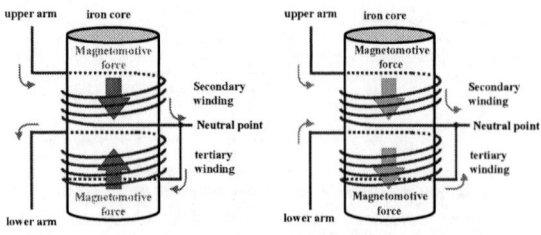

Fig.2. Magnetomotive force at secondary and tertiary winding of three winding transformer

B. Description of three winding transformer

In this chapter, we mathematically describe the proposed three winding transformer. Fig.3 shows flux flow of the proposed transformer. The transformer model has common iron core which is wrapped by three windings. The iron core becomes common magnetic path, so magnetic flux by each winding interlinks between individual windings. Not only its flux but also leakage flux penetrates individual winding. Therefore number of flux interlinkage of each winding φ_s, φ_P, φ_N is given by equation (1).

$$
\begin{bmatrix} \varphi_s \\ \varphi_P \\ \varphi_N \end{bmatrix} = \begin{bmatrix} N_s & 0 & 0 \\ 0 & N_P & 0 \\ 0 & 0 & N_N \end{bmatrix} \begin{bmatrix} \phi_{ls} \\ \phi_{lP} \\ \phi_{lN} \end{bmatrix}
$$
$$
+ \begin{bmatrix} N_s & N_s & N_s \\ N_P & N_P & N_P \\ N_N & N_N & N_N \end{bmatrix} \begin{bmatrix} \phi_s \\ \phi_P \\ \phi_N \end{bmatrix} \quad (1)
$$

where $N_{s,P,N}$ are number of turn in each winding, $\Phi_{ls, lP, lN}$ are leakage flux, $\Phi_{s, P, N}$ are main flux (passing through iron core) by each windings.

When we define R to be magnetic resistance of magnetic circuit through iron core, the main flux of each winding is expressed by (2) using the winding current i_P, i_N and the line current i_s.

$$
\begin{bmatrix} \varphi_s \\ \varphi_P \\ \varphi_N \end{bmatrix} = \begin{bmatrix} N_s & 0 & 0 \\ 0 & N_P & 0 \\ 0 & 0 & N_N \end{bmatrix} \begin{bmatrix} \phi_{ls} \\ \phi_{lP} \\ \phi_{lN} \end{bmatrix}
$$
$$
+ \frac{1}{R} \begin{bmatrix} -N_s^2 i_s + N_s N_P i_P - N_s N_N i_N \\ -N_P N_s i_s + N_P^2 i_P - N_P N_N i_N \\ -N_N N_s i_s + N_N N_P i_P - N_N^2 i_N \end{bmatrix} \quad (2)
$$

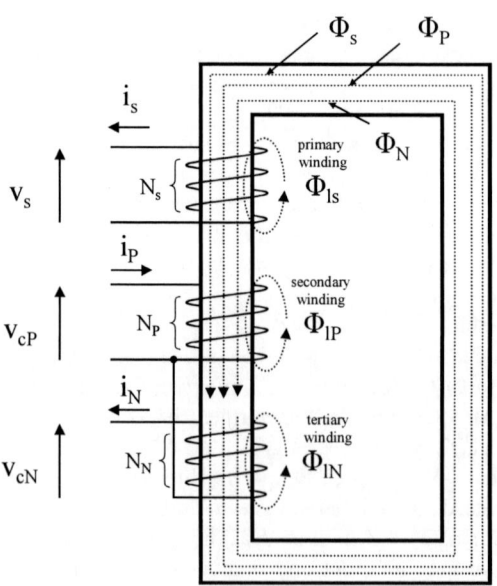

Fig.3. Flux flow of three winding transformer

978-1-4799-2706-7/14 $31.00 © 2014 IEEE

Self and mutual inductance of each winding depend on number of windings and magnetic resistance. These inductances are given by equation (3).

$$
\begin{cases}
L_s = \dfrac{N_s^2}{R} \\[2mm]
L_P = \dfrac{N_P^2}{R} \\[2mm]
L_N = \dfrac{N_N^2}{R}
\end{cases}
,
\begin{cases}
M_{sP} = \dfrac{N_s N_P}{R} \\[2mm]
M_{sN} = \dfrac{N_s N_N}{R} \\[2mm]
M_{PN} = \dfrac{N_P N_N}{R}
\end{cases}
\tag{3}
$$

Leakage flux depends on the winding current and leakage inductance (equation (4)). Accordingly, equation (2) will enable to convert equation (5) using equation (3) and (4).

$$
\begin{bmatrix} \phi_{ls} \\ \phi_{lP} \\ \phi_{lN} \end{bmatrix}
= \begin{bmatrix} -\dfrac{l_s}{N_s} & \dfrac{l_P}{N_P} & -\dfrac{l_N}{N_N} \end{bmatrix}
\begin{bmatrix} i_s \\ i_P \\ i_N \end{bmatrix}
\tag{4}
$$

where $l_{s,P,N}$ are leakage inductance in each windings.

$$
\begin{bmatrix} \varphi_s \\ \varphi_P \\ \varphi_N \end{bmatrix}
= \begin{bmatrix}
-l_s - L_s & M_{sP} & -M_{sN} \\
-M_{sP} & l_P + L_P & -M_{PN} \\
-M_{sN} & M_{PN} & -l_N - L_N
\end{bmatrix}
\begin{bmatrix} i_s \\ i_P \\ i_N \end{bmatrix}
\tag{5}
$$

Relation between the winding current and voltage is obtained to derive from equation (5).

$$
\begin{bmatrix} v_s \\ v_P \\ v_N \end{bmatrix}
= \frac{d}{dt}
\begin{bmatrix}
-l_s - L_s & M_{sP} & -M_{sN} \\
-M_{sP} & l_P + L_P & -M_{PN} \\
-M_{sN} & M_{PN} & -l_N - L_N
\end{bmatrix}
\begin{bmatrix} i_s \\ i_P \\ i_N \end{bmatrix}
\tag{6}
$$

Introducing the transformation ratio n_{sP} and n_{sN} (equation (7)), each mutual-inductance are changed excitation inductance. Hence, the equation (6) is converted to equation (8).

$$
n_{sP} = \frac{N_s}{N_P}, n_{sN} = \frac{N_s}{N_N}, n_{PN} = \frac{N_P}{N_N}
\tag{7}
$$

$$
\begin{bmatrix} v_s \\ v_P \\ v_N \end{bmatrix}
= \frac{d}{dt}
\begin{bmatrix}
-l_s - L_s & \dfrac{1}{n_{sP}}L_s & -\dfrac{1}{n_{sN}}L_s \\[2mm]
-n_{sP}L_P & l_P + L_P & -\dfrac{1}{n_{PN}}L_P \\[2mm]
-n_{sN}L_N & n_{PN}L_N & -l_N - L_N
\end{bmatrix}
\begin{bmatrix} i_s \\ i_P \\ i_N \end{bmatrix}
\tag{8}
$$

This equation provides the equivalent circuit model which is given in figure 4. This circuit model consists of excitation inductance and two ideal transformers and leakage inductance of each winding.

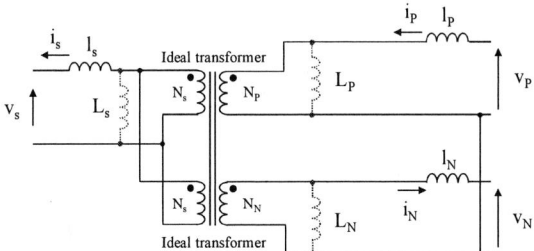

Fig.4. Equivalent circuit of three winding transformer

C. Appling proposed transformer to MMC

In this chapter, we describe the MMC model with the equivalent model of proposed transformer in the case of applying the voltage and current equation of the transformer to MMC model (figure 1). In the arm current of MMC, a common current component and a distinct current component are contain between upper arm current and lower arm current. The common current component and the distinct current component are defined i_{cir}, i_{def} as equation (9), therefore upper arm current i_P and lower arm current i_N are defined as equation (10).

$$
\begin{cases}
i_{cir} = \dfrac{1}{2}(i_P + i_N) \\[2mm]
i_{def} = i_P - i_N
\end{cases}
\tag{9}
$$

$$
\begin{cases}
i_P = i_{cir} + \dfrac{1}{2}i_{def} \\[2mm]
i_N = i_{cir} - \dfrac{1}{2}i_{def}
\end{cases}
\tag{10}
$$

Using the common current component i_{cir} and the distinct current component i_{def}, the voltage and current equation of a three winding transformer is given by equation (11).

$$
\begin{bmatrix} v_s \\ v_P \\ v_N \end{bmatrix}
= \frac{d}{dt}
\begin{bmatrix}
-l_s - L_s & \dfrac{1}{n_{sP}}L_s & -\dfrac{1}{n_{sN}}L_s \\[2mm]
-n_{sP}L_P & l_P + L_P & -\dfrac{1}{n_{PN}}L_P \\[2mm]
-n_{sN}L_N & n_{PN}L_N & -l_N - L_N
\end{bmatrix}
\begin{bmatrix} i_s \\ i_{cir} + i_{def} \\ i_{cir} - i_{def} \end{bmatrix}
\tag{11}
$$

Now, a transformation ratio of primary to secondary n_{sP} and primary to tertiary n_{sN} are made equal to n. Furthermore leakage inductance of the secondary winding l_P and the tertiary winding l_N are made equal to set l, and the voltage and current equation of the transformer is flowed to equation (12).

$$
\begin{bmatrix} v_s \\ v_P \\ v_N \end{bmatrix}
= \frac{d}{dt}
\begin{bmatrix}
-(l + L_s)i_s + \dfrac{1}{n}L_s i_{def} \\[2mm]
(i_{def} - n i_s)\dfrac{L_s}{n^2} + l(i_{cir} + \dfrac{i_{def}}{2}) \\[2mm]
(i_{def} - n i_s)\dfrac{L_s}{n^2} - l(i_{cir} - \dfrac{i_{def}}{2})
\end{bmatrix}
\tag{12}
$$

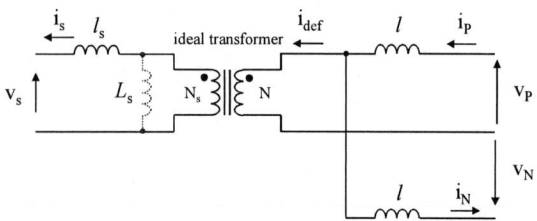

Fig.5. Equivalent circuit of three winding transformer

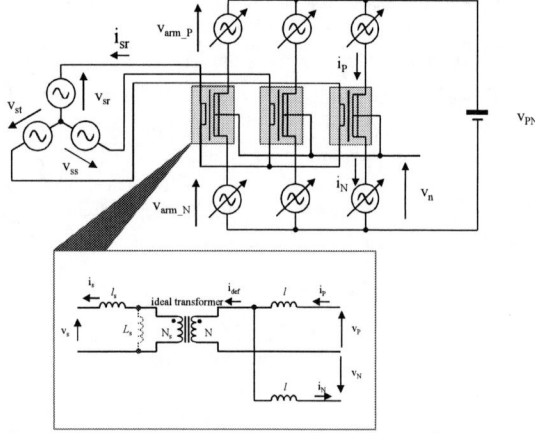

Fig.6. Modular Multilevel Converter with three winding transformer using variable voltage source.

An equivalent circuit configuration of the three winding transformer is shown in figure 5. This equivalent circuit model consists of the primary side leakage inductance l_s, the excitation inductance Ls, ideal transformer and parallel connected two leakage inductances at secondary side of ideal transformer. Figure 6 shows circuit diagram applying three winding transformer model and MMC using variable voltage source. In the converter model, each converter arm is connected to each leakage inductance of secondary side at ideal transformer. The leakage inductance connected to each converter arm behaves like buffer reactor of ordinary MMC. An MMC consists of cascaded converter and transformer. The switching frequency of cascaded converter is relatively higher than output voltage; therefore the cascaded converter can be changed to variable voltage source, to simplify.

D. AC side current dynamics of the proposed MMC

In this chapter, an AC side current dynamics of the proposed MMC is described. Defined three phase AC line voltage is $v_{s(r,s,t)}$, AC line current is $i_{s(r,s,t)}$, and the voltage and current equation of primary side of proposed transformer is given by equation (13).

$$
\begin{bmatrix} v_{sr} \\ v_{ss} \\ v_{st} \end{bmatrix} = -l_s \frac{d}{dt} \begin{bmatrix} i_{sr} \\ i_{ss} \\ i_{st} \end{bmatrix} + n \begin{bmatrix} v_{sec\_r} \\ v_{sec\_s} \\ v_{sec\_t} \end{bmatrix} \tag{13}
$$

where $v_{sec(r,s,t)}$ is secondary voltage of ideal transformer. The secondary voltage of ideal transformer is given by

equation (14), and the secondary voltage is followed from equation (14) to equation (15).

$$
v_{sec} = \begin{cases} v_{PN} - v_{arm\_P} - v_n - l\dfrac{d}{dt}(i_{cir} + \dfrac{i_{def}}{2}) \\ \\ v_{arm\_N} - v_n + l\dfrac{d}{dt}(i_{cir} - \dfrac{i_{def}}{2}) \end{cases} \tag{14}
$$

where v_{PN} is DC link voltage of MMC, v_n is neutral point voltage of connecting point of each transformer, between DC link negative side, v_{arm_P}, v_{arm_N} are total output voltage of consisting converter cells in each arm.

$$
v_{sec} = \frac{1}{2}\left(v_{PN} + v_{arm\_N} - v_{arm\_P} - 2v_n - l\frac{d}{dt}i_{def} \right) \tag{15}
$$

Now, the neutral point voltage of each transformer v_n is considered same voltage level to DC link neutral point, and the secondary voltage is followed to equation (16).

$$
v_{sec} = \frac{1}{2}\left(v_{arm\_N} - v_{arm\_P} - l\frac{d}{dt}i_{def} \right) \tag{16}
$$

Using equation (16), the voltage and current equation of the transformer is followed from equation (13) to equation (17).

$$
\begin{bmatrix} v_{sr} \\ v_{ss} \\ v_{st} \end{bmatrix} = -l_s \frac{d}{dt}\begin{bmatrix} i_{sr} \\ i_{ss} \\ i_{st} \end{bmatrix} + \frac{n}{2}\begin{bmatrix} v_{arm\_N\_r} - v_{arm\_P\_r} \\ v_{arm\_N\_s} - v_{arm\_P\_s} \\ v_{arm\_N\_t} - v_{arm\_P\_t} \end{bmatrix}
$$
$$
- \frac{n}{2}l\frac{d}{dt}\begin{bmatrix} i_{def\_r} \\ i_{def\_s} \\ i_{def\_t} \end{bmatrix} \tag{17}
$$

Generally an excitation inductance of a transformer is huge figure, so a current of a primary side of the transformer and secondary side become the approximately equal.

$$
\begin{bmatrix} v_{sr} \\ v_{ss} \\ v_{st} \end{bmatrix} = -(l_s + \frac{1}{2}n^2 l)\frac{d}{dt}\begin{bmatrix} i_{sr} \\ i_{ss} \\ i_{st} \end{bmatrix} + \frac{n}{2}\begin{bmatrix} v_{arm\_N\_r} - v_{arm\_P\_r} \\ v_{arm\_N\_s} - v_{arm\_P\_s} \\ v_{arm\_N\_t} - v_{arm\_P\_t} \end{bmatrix}
$$
(18)

The subtraction of lower arm voltage v_{arm_N} from upper arm voltage v_{arm_P} is defined as following equation, and the AC line voltage is converted into dq coordinates using transformation matrix C.

$$
\begin{bmatrix} v_{cr} \\ v_{cs} \\ v_{ct} \end{bmatrix} = \frac{n}{2}\begin{bmatrix} v_{arm\_N\_r} - v_{arm\_P\_r} \\ v_{arm\_N\_s} - v_{arm\_P\_s} \\ v_{arm\_N\_t} - v_{arm\_P\_t} \end{bmatrix} \tag{19}
$$

$$[C]\begin{bmatrix} v_{sd} \\ v_{sq} \end{bmatrix} = -(l_s + \frac{1}{2}n^2 l)\frac{d}{dt}\left\{ [C]\begin{bmatrix} i_{sd} \\ i_{sq} \end{bmatrix} \right\} + [C]\begin{bmatrix} v_{cd} \\ v_{cq} \end{bmatrix}$$

$$= -(l_s + \frac{1}{2}n^2 l)\left(\left\{ \frac{d}{dt}[C]\right\}\begin{bmatrix} i_{sd} \\ i_{sq} \end{bmatrix} + [C]\left\{ \frac{d}{dt}\begin{bmatrix} i_{sd} \\ i_{sq} \end{bmatrix}\right\} \right)$$

$$+ [C]\begin{bmatrix} v_{cd} \\ v_{cq} \end{bmatrix} \tag{20}$$

where

$$\begin{bmatrix} v_{sr} \\ v_{ss} \\ v_{st} \end{bmatrix} = [C]\begin{bmatrix} v_{sd} \\ v_{sq} \end{bmatrix}, \quad \begin{bmatrix} i_{sr} \\ i_{ss} \\ i_{st} \end{bmatrix} = [C]\begin{bmatrix} i_{sd} \\ i_{sq} \end{bmatrix}, \quad \begin{bmatrix} v_{cr} \\ v_{cs} \\ v_{ct} \end{bmatrix} = [C]\begin{bmatrix} v_{cd} \\ v_{cq} \end{bmatrix} \tag{21}$$

$$[C] = \begin{bmatrix} \cos\omega t & \sin\omega t \\ \cos(\omega t - \frac{2}{3}\pi) & \cos(\omega t - \frac{2}{3}\pi) \\ \cos(\omega t - \frac{4}{3}\pi) & \cos(\omega t - \frac{4}{3}\pi) \end{bmatrix} \tag{22}$$

Finally, the AC line voltages of dq coordinates are obtained by invers matrix of transformation matrix C.

$$\begin{bmatrix} v_{sd} \\ v_{sq} \end{bmatrix} = -(l_s + \frac{1}{2}n^2 l)[C]^{-1}\left\{ \frac{d}{dt}[C]\right\}\begin{bmatrix} i_{sd} \\ i_{sq} \end{bmatrix}$$

$$-(l_s + \frac{1}{2}n^2 l)[C]^{-1}[C]\left\{ \frac{d}{dt}\begin{bmatrix} i_{sd} \\ i_{sq} \end{bmatrix}\right\} + [C]^{-1}[C]\begin{bmatrix} v_{cd} \\ v_{cq} \end{bmatrix}$$

$$= \begin{bmatrix} \frac{d}{dt}(l_s + \frac{1}{2}n^2 l) & -\omega(l_s + +\frac{1}{2}n^2 l) \\ \omega(l_s + +\frac{1}{2}n^2 l) & \frac{d}{dt}(l_s + \frac{1}{2}n^2 l) \end{bmatrix}\begin{bmatrix} i_{sd} \\ i_{sq} \end{bmatrix} + \begin{bmatrix} v_{cd} \\ v_{cq} \end{bmatrix} \tag{23}$$

where

$$[C]^{-1} = \frac{2}{3}\begin{bmatrix} \cos\omega t & \cos(\omega t - \frac{2}{3}\pi) & \cos(\omega t - \frac{4}{3}\pi) \\ \sin\omega t & \cos(\omega t - \frac{2}{3}\pi) & \cos(\omega t - \frac{4}{3}\pi) \end{bmatrix} \tag{24}$$

$$[C]^{-1}\frac{d}{dt}[C] = \omega\begin{bmatrix} 0 & -1 \\ 1 & 0 \end{bmatrix} \tag{25}$$

AC line current dynamics between it to voltage of converter arm is shown in figure 7 about MMC with equivalent model of proposed transformer as equation (23). The dynamics of AC line current between subtraction voltages of lower arm from upper arm is same as general MMC converter. The added leakage inductance of primary winding and parallel connected leakage inductance of secondary and tertiary winding becomes interconnection reactor to AC line. The AC current dynamics shown in figure 7 is very similar to general six arm grid connected inverter. Hence, the proposed MMC with three winding transformer can operate as same as typical MMC.

E. DC side current dynamics of the proposed MMC

In this chapter, we describe a DC link side current dynamics of the proposed MMC. The voltage and current equation of DC link side is given by equation (26).

$$v_{PN} = v_{arm\_P} + l\frac{d}{dt}i_P + v_{arm\_N} + l\frac{d}{dt}i_N \tag{26}$$

The arm current i_P and i_N are consist of common current component i_{cir} and distinct current component i_{def}, as mentioned before chapter. Using these current, the equation of current and voltage is expressed by equation (27).

$$v_{PN} = v_{arm\_P} + l\frac{d}{dt}(i_{cir} + \frac{i_{def}}{2}) + v_{arm\_N} + l\frac{d}{dt}(i_{cir} - \frac{i_{def}}{2})$$

$$= v_{arm\_P} + v_{arm\_N} + 2l\frac{d}{dt}i_{cir} \tag{27}$$

Finally, the common current component is obtained.

$$i_{cir} = \frac{1}{2l}\int(v_{PN} - v_{arm\_P} - v_{arm\_N})dt \tag{28}$$

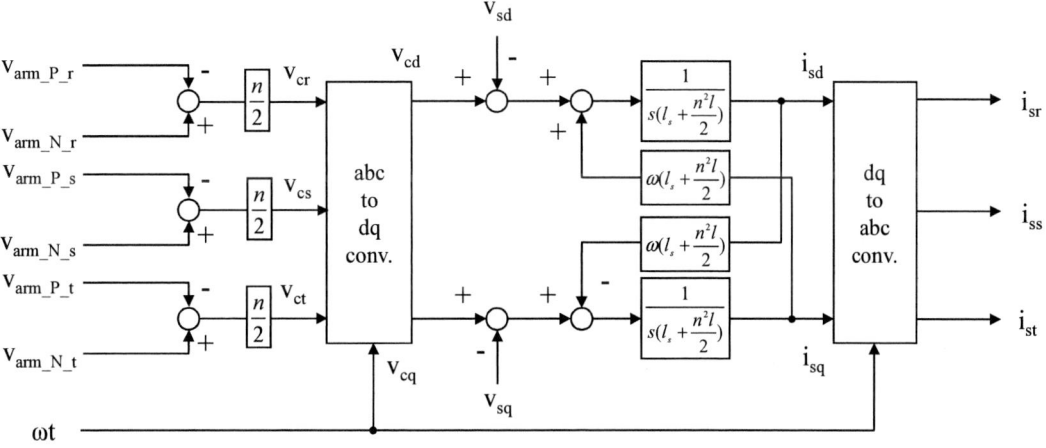

Fig.7. AC line current dynamics between voltage of converter arm

The common current component carries DC link side through upper arm and lower arm of converter. It becomes DC current of proposed MMC. The leakage inductance of secondary l_P and tertiary l_N behave buffer reactor. Therefore, the proposed MMC with three winding transformer operates as same as typical MMC with buffer reactors.

III. EXPERIMENTAL RESULTS

The parameters of the proposed prototype MMC are shown in Table 1. The prototype converters are connected to the same AC grid, and they are connected through DC lines. The transformer used in the experimentation is a normal transformer; its iron core has saturation magnetic density which is 1.85[T] at 116% (233V, 50Hz) of rated voltage level. In rated voltage (200V 50Hz), the magnetic density of iron core is 1.59[T].

Figure 10 shows experimental results in grid connected operation. In all results, good sinusoidal waveforms at ac line current are obtained and the capacitor voltages of each cell are balanced each other. When converter and inverter operation (figure 10 (a), (b)), the converter arm current has dc component. However, the three winding transformer does not occur magnetic saturation in both power flow direction. In also reactive power operation (figure 10 (c), (d)) is same.

Figure 11 shows experimental results in reversing power flow operation by back to back converter connection. The behavior of converter is stable even in during power flow direction changing.

Figure 12 shows experimental results in fault ride through operation at AC line fault. A magnetic saturation does not appear even when AC line voltage is suddenly changing. As these results, the proposed MMC with three winding transformer operates as same as typical MMC with buffer reactors. Furthermore, it does not adversely

Fig.8. The three winding transformer
(single phase)

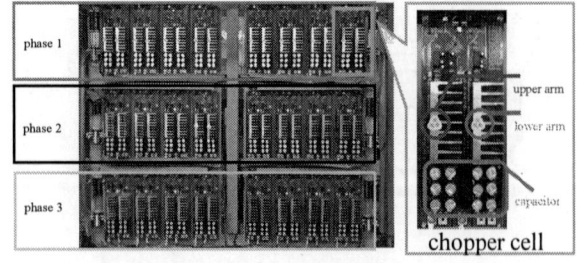

Fig.9. The prototype converter

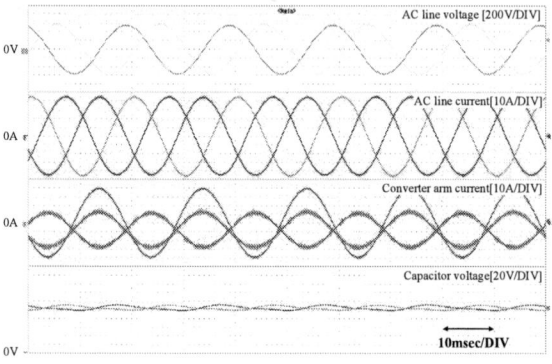

(a) when the direction of power flow is AC to DC

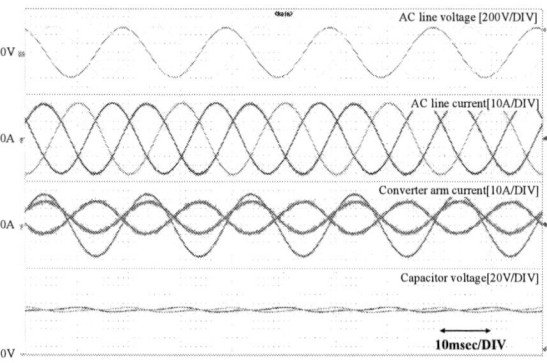

(b) when the direction of power flow is DC to AC

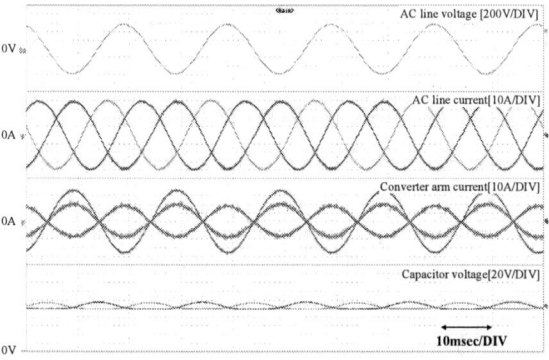

(c) in reactive power lagging operration

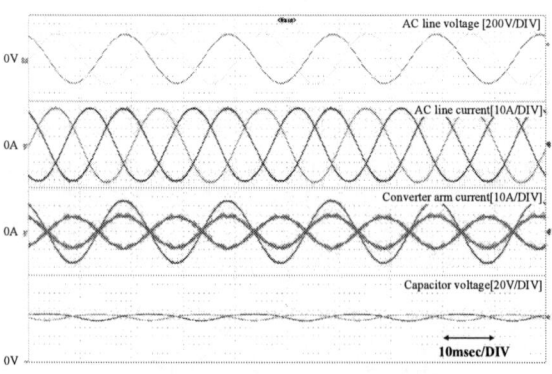

(d) in reacitve power leading operation
Fig. 10. Experimental results of MMC prototype in grid connected operation

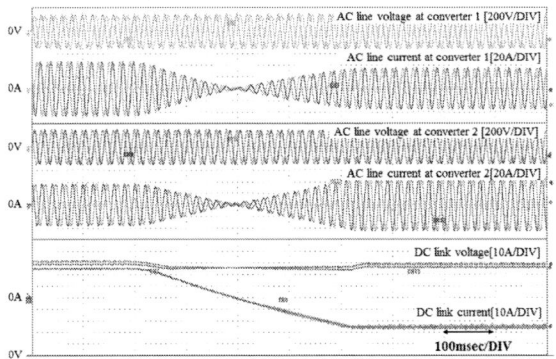

(a)power flow reversal (rectifier is from converter 1 to converter 2)

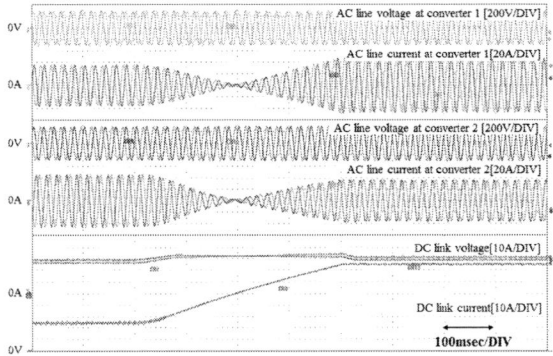

(b)power flow reversal (rectifier is from converter 2 to converter 1)
Fig. 11. Experimental results of MMC prototype in reversing power flow operation.

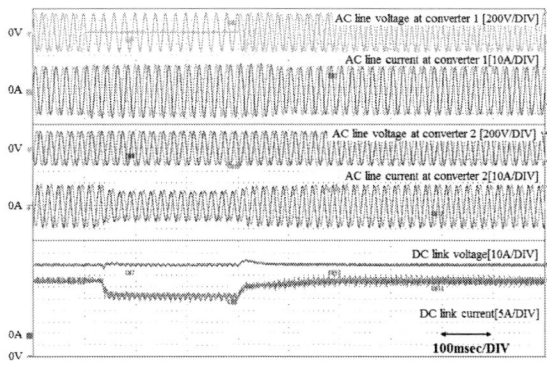

(a) fault ride through operation when 1 line to ground fault

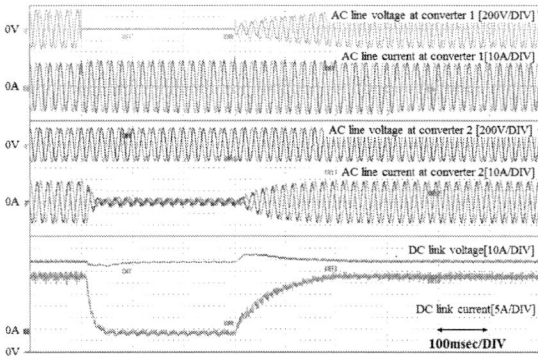

(b) fault ride through operation when 3 lines to ground fault
Fig. 12. Experimental results of MMC prototype in fault ride through operation.

affect the DC bias magnetism at the transformer to merge buffer reactor into a grid connected transformer.

TABLE I
PARAMETERS OF THE MMC PROTOTYPE

Meaning	Value
AC voltage	200V, 50Hz
DC voltage	400V
Power rating	10kVA
Cell capacitor voltage	100V
Number of cell in each arm	4
Switching frequency (in individual devices)	1kHz
Leakage inductance (primary winding)	1.95mH (15.3%Z)
Leakage inductance (secondary, tertiary winding)	1.3mH (8.37%Z)
Capacitance in each cells	3960μF
Transformation ratio	115V(P)/127V(S,T)

IV. CONCLUSIONS

This paper proposes a new type of Modular Multilevel Converter applying a three winding transformer. The proposed topology can merge buffer reactors into a grid connected transformer. In addition, this paper described mathematical property of the proposed MMC with transformer and experimentation by 10kVA prototype shows the converter can operate as same as MMC with buffer reactors. Furthermore, we described that it does not adversely affect the DC bias magnetism at the transformer

to merge buffer reactor into a grid connected transformer and the proposed MMC with three winding transformer does not require a special transformer.

REFERENCES

[1] Lesnicar. A, Marquardt. R, "A new modular voltage source inverter topology", *EPE 2003*, 105, 2003.

[2] Lesnicar. A, Marquardt. R, "An innovative modular multilevel converter topology suitable for a wide power range," *Power Tech Conference Proceedings, 2003 IEEE Bologna*, vol. 3, 2003.

[3] Hagiwara. M, Akagi. H, "Control and Experiment of Pulsewidth-Modulated Modular Multilevel Converters," *IEEE Trans. on Power Electronics*, vol. 24, no. 7, pp. 1737-1746, 2009.

[4] Antonopoulous. A, Angquist. Lennart, Nee. H.-P, "On dynamics and voltage control of the Modular Multilevel Converter", *EPE2009*, 2009.

[5] Angquist, L, Antonopoulos. A, Siemaszko. D, Ilves. K, Vasiladiotis. M, Nee. H.-P., "Inner control of Modular Multilevel Converters – An approach using open-loop estimation of stored energy", *International Power Electoronics Conference 2010*, pp.1579-1585, 2010.

[6] Ilves. K, Antonopoulos. A, Norrga. Staffan, Nee. H.-P., "Steady-State Analysis of Interaction Between Harmonic Components of Arm and Line Quantities of Modular Multilevel Converters", *IEEE transactions on Power Electronics*, vol. 27, no. 1, pp.57-68, 2013

[7] Rohner. S, Bernet. S, Hiller. M, Sommer. R, "Modulation, Losses, and Semiconductor Requirements of Modular Multilevel Converters," *IEEE Trans. on Industrial Electronics*, vol. 57, no. 8, pp. 2633-2642, 2009.

[8] Munch. P, Gorges. D, Izak. M, Liu. Steven, "Integrated Current Control, Energy Control and Energy Balancing of Modular Multilevel Converters," *IECON 2010s*, pp. 150-155, 2010.

[9] Allebrod. S, Hamerski. R, Marquardt. R, "New transformerless, scalable Modular Multilevel Converters for HVDC-transmission,"

978-1-4799-2706-7/14 $31.00 © 2014 IEEE

Power Electronics Specialists Conference 2008, pp. 174-179, 2008.

[10] Saeedifard. M, Iravani. R, "Dynamic Performance of a Modular Multilevel Back-to-Back HVDC system," *IEEE Trans. on Power Delivery*, vol. 25, no. 4, pp. 2903-2912, 2010.

[11] T. Westerweller, K. Friedrich, U. Armonies, A. Orini, D. Parquet and S. When, "Trans bay cable—World\'s first HVDC system using multilevel voltage-sourced converter," *Cigre2010*, B4_101_2010.

[12] M. Pereira, A. Zenkner, M. Claus, "Characteristics and benefits of modular multilevel converters for FACTS," *Cigre2010*, B4_104_2010.

[13] Arai. J, Murao. T, Karube. T, Takagi. K, Ibrahim. M, "Design and Operation of SVC for Voltage Support at Mussafah Substation in Abu Dhabi," *PEDS 2005*, vol. 2, pp. 1356-1360, 2005.

Back-to-Back System for Five-level Converter with Common Flying Capacitors

Isamu Hasegawa, Shota Urushibata, Takeshi Kondo
Kuniaki Hirao, Takashi Kodama,

Research & Development Group
Meidensha Corporation
Numazu, Japan
hasegawa-i@mb.meidensha.co.jp

Hui Zhang
R & D Division
Meiden Singapore
Singapore

Abstract— This study proposes a novel 5-level converter topology in which common flying capacitors (FCs) are shared by three phases. This circuit can output voltage patterns prohibited in the circuits of the previous study. Furthermore, a back-to-back (BTB) configuration can be realized by only increasing the number of the phase units of the 5-level converter. This BTB system employs common FCs for all input and output. This study shows the topology of a 5-level converter and its operational principle as well as the expansion method for realizing the BTB system. A simulation is conducted to verify the control method and operation properly.

Keywords— Flying Capacitor, multilevel converter, ac motor drive, active rectifier.

I. INTRODUCTION

Transformer-less multilevel converters have attracted attention because they are highly efficient and support high-voltages operation [1][2][3][5][6].

Various circuits have been studied, and papers have been published on this subject, such as the modular multilevel cascaded (MMC) [2][5][6] and the active neutral point clamped (ANPC). MMCC is composed of series-connected single phase cells. Each cell has IGBTs and one capacitor. The five-level ANPC [1] is provided with a flying capacitor (FC) in each phase unit.

A previous study has proposed a topology with common FCs shared by three phases [4]. Typically, three capacitors are required to mount an independent FC on each phase; this circuit only requires two.

However, this converter suffers from two problems. First, it has theoretically prohibited combinations of phase voltages. Second, the capacitor voltage control suffers from some limitations. As this circuit needs to avoid these patterns, control-algorithm becomes complicate. In addition, the harmonics of the output current increase because of the combination of the voltages that the circuit cannot output.

This study proposes a new topology which overcomes these problems. Though it still has common FCs shared by all phases, it can output the combinations of phase voltages which are prohibited in the previous topology. Consequently, the control algorithm becomes relatively simple. Furthermore, the harmonics of the output current decrease because the combination of the theoretically

prohibited phase voltage patterns no longer exist. In addition, although only three phase configuration is considered in [4], common FCs have more merits when the number of phase increases more than three. Therefore, this study shows that the phase number of the topology can be expanded beyond three by applying the proposed control. Further, we clarify that the topology can constitute an AC/AC back-to-back (BTB) system by extending the phase number up to six. When the converter is expanded to an AC/AC BTB converter, the utility is improved because only two capacitors are required instead of six.

This study describes the operational principle of the proposed converter and shows that the combination of the prohibited phase voltage patterns do not exist. Further, it shows that the proposed control can increase the phase number and can constitute BTB system. Finally, this paper presents the simulation results and demonstrates that the proposed BTB system possesses operational characteristics that are as good as a three-phase inverter with the same capacitance.

II. THEORY OF TOPOLOGY AND CONTROL METHODS

A. Previous Common FCs Topology

This section shows that the circuit in [4] has theoretically prohibited phase voltage patterns. Fig. 1 shows the circuit. Voltage of C_{DC1} and C_{DC2} are set to 2E and voltage of C_1 and C_2 are set to E. The neutral point of DC Link NP is defined as reference. Table I shows that the comparison of the topology characteristics. ANPC requires three FCs, but circuit in [4] requires only two FCs. However, the three-phase converter contains some prohibited phase voltages because it has common IGBTs, namely, S_{ca}, S_{cb}, S_{cc}, and S_{cd}, and two FCs. For example, IGBT S_{ca} and IGBT S_{cb} should be turned ON simultaneously to output the pattern $V_u = 2E$, $V_v =$ zero, $V_w = -2E$ (Fig. 1). In this case, a short-circuit current flows between the capacitor C_{DC1} and C_1 because both ends of these capacitors are connected by S_{ca} and S_{cb}. Similarly, when IGBT S_{cc} and S_{cd} turn ON simultaneously, a short-circuit current flows in C_{DC2} and C_2. Therefore, the three-phase voltage must not contain combinations such as 2E and zero or -2E and zero. When

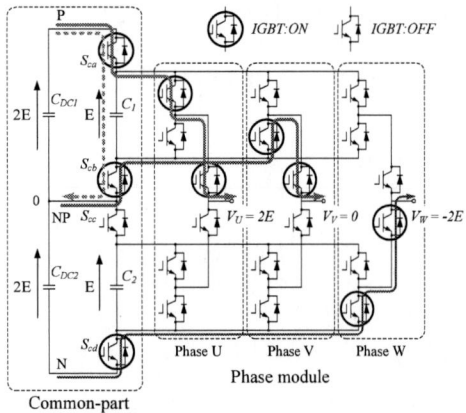

Fig. 1. Topology with Common capacitor and example of prohibited pattern proposed in [4]

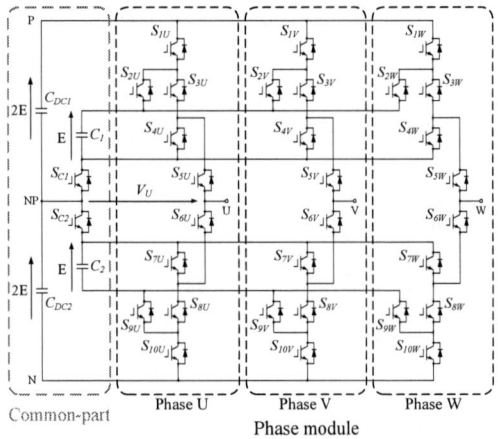

Fig. 2. Proposed topology with common flying capacitor

TABLE I
COMPARISON OF 5-LEVEL TOPOLOGY

	Number of FCs	Prohibit patterns	Limitation of charge/discharge
Fig.1 [4]	2	Yes	Yes
Fig2	2	No	Yes
ANPC [1]	3	No	No

this output voltage reference is given, it outputs other alternative patterns by time sharing [4]. Therefore, the method in [4] further increases the harmonic current and the number of switching compared with a circuit that can directly output these patterns. In addition, the control algorithm becomes complicated.

B. New topology with common FCs

Fig. 2 shows the proposed configuration, where C_{DC1} and C_{DC2} represent the power supplies and C_1 and C_2 represent the common FCs. Voltage of C_{DC1} and C_{DC2} are set to 2E and voltage of C_1 and C_2 are set to E. The neutral point of DC Link is defined as reference. Common IGBTs S_{ca} and S_{cd} are rearranged in S_{2u}, S_{2v}, S_{2w}, S_{9u}, S_{9v} and S_{9w} as shown in Fig. 2. However,

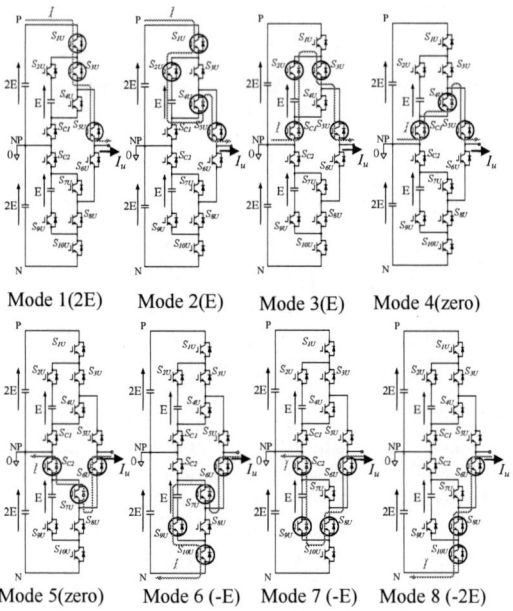

Fig. 3. Switching patterns (one phase)

TABLE II
SWITCHING PATTERNS OF FIG. 2

Switching patterns	Output phase voltage
Mode 1	2E
Mode 2	E (charge $I_u > 0$)
Mode 3	E (discharge $I_u > 0$)
Mode 4	zero
Mode 5	zero
Mode 6	-E (charge $I_u < 0$)
Mode 7	-E (discharge $I_u < 0$)
Mode 8	-2E

common IGBTs S_{c1} and S_{c2} are not changed. Although the number of IGBTs increases in this manner, this converter can directly output even with the prohibited voltage patterns.

Next, we explain about how to operate the proposed circuit. The circuit shown in Fig. 2 contains many IGBTs, which result in many pattern combinations. Therefore, we focus on one phase and assume that all capacitors voltages are constant, and we show the 5 levels of the output phase voltage. We also present the charge and discharge loop of FC using a specific phase from among three and show that control is possible. The output voltage patterns shown in Fig. 3 indicates that the short-circuit of DC capacitors and FCs can be prevented by controlling S_{2u}, S_{2v}, S_{2w}, S_{9u}, S_{9v} and S_{9w} in each phase.

Fig. 3 shows the patterns for a 5-level phase voltage output in one phase. Arrows in Fig.3 illustrates direction of phase current I_u. Mode 1 with 2E, Modes 2 and 3 with E, Modes 4 and 5 with zero, Modes 6 and 7 with −E, and Mode 8 with -2E can be output with a 5-level voltage. Table II shows that if $I_u > 0$, C_1 is charged in

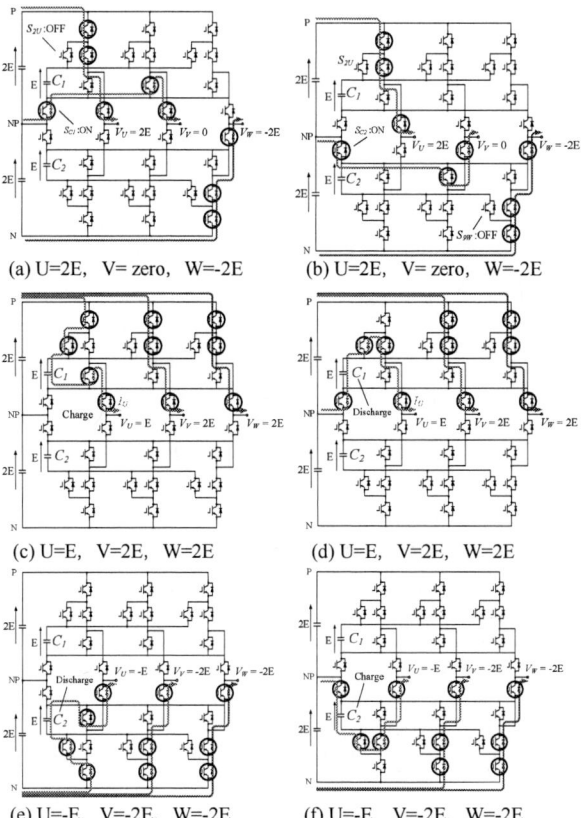

(a) U=2E, V= zero, W=-2E (b) U=2E, V= zero, W=-2E

(c) U=E, V=2E, W=2E (d) U=E, V=2E, W=2E

(e) U=-E, V=-2E, W=-2E (f) U=-E, V=-2E, W=-2E

Fig. 4. Switching patterns (three phases)

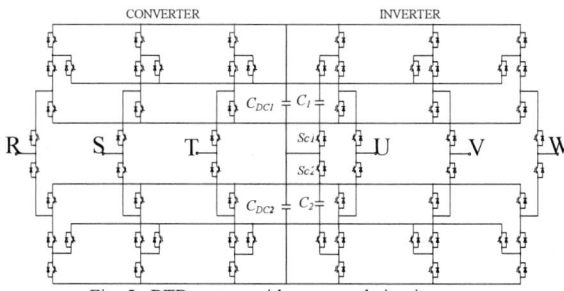

Fig. 5. BTB system with proposed circuit

TABLE III
COMPARISON OF BTB CONFIGURATIONS

	Proposed Topology	ANPC-5L [1]
Number of DC capacitors	2	2
Number of FCs	2	6

Mode 2 and discharged in Mode 3. If $I_u < 0$, C_2 is charged in Mode 6 and discharged in Mode 7. By selecting the charging and discharging of C_1 and C_2 during E and −E, the voltages of C_1 and C_2 can be controlled. Fig. 4 shows the patterns of the combined output voltage in each phase in the three-phase circuit. In Figs. 4(a) and 4(b), the short-circuit of FCs and DC capacitors does not occur even when the converter outputs the voltage combination of 2E, zero and -2E because IGBTs S_{2u} and S_{9w} are turned off, unlike Fig. 1.

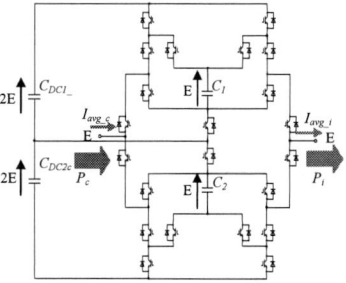

Fig. 6. Relation between power flow and current flow of C_1

Figs. 4(c) and (d) and 4(e) and (f) respectively show that the C_1 and C_2 voltages can be controlled while a 5-level voltage is outputted by combining the switching patterns of the charge and discharge loop of each pair to choose the charge and discharge modes of C_1 and C_2.

A BTB configuration can be realized by only increasing the number of phases to six as shown in Fig. 5. Table III shows comparison of number of FCs required BTB configurations. Specifically, an AC/AC converter configuration with six phases can be realized using three phases each as the converter and the inverter. Therefore, a BTB configuration composed ANPC requires six FCs. Consequently proposed BTB configuration is 4 FCs less than BTB composed ANPC.

However, the topology shown in Fig. 2 suffers from the limitation of charge and discharge control when the phase number is three and more. This problem is described in section II-E.

C. Definition of current flowing through FCs

This section defines current of FCs when number of phases increase to three and more. Current flowing through the C_1 is a sum of phase current, which outputs the phase voltage of E. Similarly, current flowing through the C_2 is a sum of phase currents, which outputs the phase voltage of -E. Therefore, it is difficult to judge either charge or discharge by current direction because FC's current is totalized by several phases current. Furthermore, there is possibility that the judgment is not accurate near zero current because of the current ripple and the noises. Accordingly, this section proposes the method of charge and discharge control without using instantaneous current direction.

The authors of [3] have taken into consideration about current of 5-level converter's capacitors and have focused on one phase considering the average current during E and −E as flowing through capacitors current. Considering the case of Fig. 2, which is to focus on average current during E and –E, we can consider as well as [3]. Fig. 6 shows power flow and average current during outputting phase voltage E with single phase BTB configuration. The inverter outputs voltage as follows

$$v = M_i \sin\theta \qquad (1)$$

where M_i is the modulation index of the inverter and range of θ is defined from 0 to 2π. If output current is the sinusoidal wave, output power of the inverter is constant. Therefore average current I_{avg} is also constant during E.

978-1-4799-2706-7/14 $31.00 © 2014 IEEE

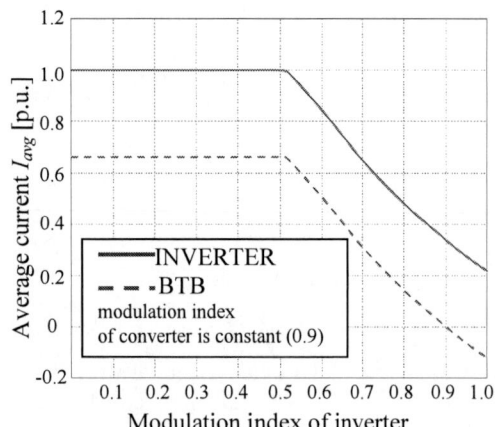

Fig. 7. FCs average current

TABLE IV
RELATION BETWEEN I_{avg} AND FC CHARGE / DISCHARGE

$M_c > M_i$		$M_c < M_i$	
$I_{avg} > 0$		$I_{avg} < 0$	
Charge	Discharge	Discharge	Charge
Mode 2	Mode 3	Mode 2	Mode 3
Mode 6	Mode 7	Mode 6	Mode 7

Now, if the converter is not operated in Fig. 6 and I_{avg} is defined as same direction with I_{avg_i} in Fig. 6 and the carrier frequency of the inverter is very high, per-unit average current I_{avg} is defined as follows

$$
\begin{aligned}
I_{avg} &= I_{avg\_i} \\
&= \frac{1}{\pi M_i} \int_0^\pi (1 - abs(1 - abs(2M_i \sin\theta))) \sin\theta \, d\theta
\end{aligned}
\tag{2}
$$

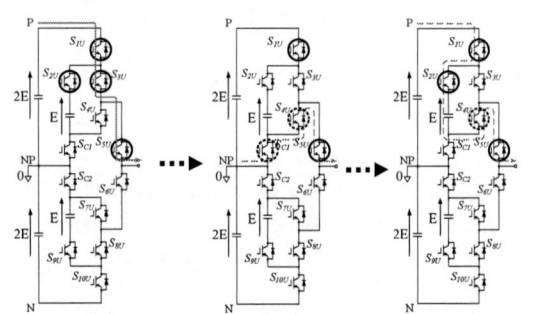

(a) Mode1(2E) (b) Commutation (0) (c) Mode2 (E)

Fig. 8. Level skip caused by commutation.

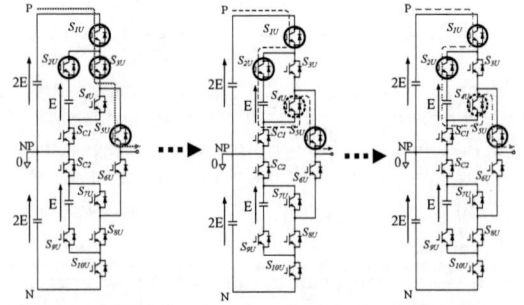

(a) Mode1' (2E) (b) Commutation (E) (c) Mode2 (E)

Fig. 9. Effect of new mode.

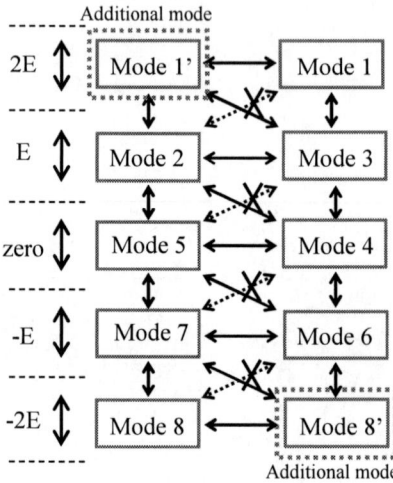

Fig.10. Switching states patterns

Fig. 7 is derived from (2) and shows average current I_{avg} per modulation index M_i. As shown in Fig. 7, polarity of I_{avg} is same even if M_i is changed. Therefore, if duration of the Mode 2 is longer than the Mode 3 during E, C_1 is charged. When the duration of the Mode 3 is longer than the Mode 2 during E, C_1 is discharged. C_2 is as well as C_1. Consequently, to control the rate of duration in Mode 2, Mode 3, FCs voltage can be controlled without using instantaneous current directions.

However, when converter is operated in Fig. 6, because power-polarity of the converter and the inverter is reversed, average current I_{avg} is reversed depending on the modulation index of the inverter and the converter. When the input power of converter equals to the output power of inverter, if converter modulation index defines as M_c, average current of FCs is

$$
\begin{aligned}
I_{avg} &= I_{avg\_i} - I_{avg\_c} \\
&= \int_0^\pi \left\{ \begin{array}{l} \dfrac{1}{\pi M_i}(1 - abs(1 - abs(2M_i \sin\theta))) \\[2mm] -\dfrac{1}{\pi M_c}(1 - abs(1 - abs(2M_c \sin\theta))) \end{array} \right\} \sin\theta \, d\theta
\end{aligned}
\tag{3}
$$

Dashed line of Fig. 7 is the average current of C_1 in the BTB configuration in Fig. 6. M_c is set to 0.9 as a condition in Fig. 7. As shown in Fig 7, direction of I_{avg} becomes negative as a boundary M_c=0.9. Therefore relation of charge/discharge is reversed during $M_i > M_c$. Table IV shows the relation between the modulation index and charge/discharge.

As above mentioned, charge and discharge need to take into account modulation index of the converter and the inverter in the case of BTB configuration. In the following section, we discuss about the charging and discharging of the FC using the average current as defined in this section.

D. Commutation during dead-time

This section considers the commutation during dead-time. If the topology shown in Fig. 2 is composed of ideal switches, no problem arises in the switching patterns in Fig. 3. However, IGBTs have a delay at turn ON and turn

TABLE V
PROHIBIT PATTERNS PROPOSED CIRCUIT

State of FCs voltage ($I_{avg} > 0$)	C_1 charge command	C_2 charge command
$v_{c1} > E$	0	0
$v_{c2} > E$		
$v_{c1} > E$	0	1
$v_{c2} < E$		
$v_{c1} < E$	1	0
$v_{c2} > E$		
$v_{c1} < E$	1 (Prohibition)	1 (Prohibition)
$v_{c1} < E$		

TABLE VI
RELATION BETWEEN SWITCHING PATTERNS AND
THE FC CHARGE COMMANDS

Charge command		2E	E	zero	-E	-2E
C_1	C_2					
0	0	Mode 1	Mode 3	Mode 4 or Mode 5	Mode 7	Mode 8
0	1	Mode 1	Mode 3	Mode 4	Mode 6	Mode 8'
1	0	Mode 1'	Mode 2	Mode 5	Mode 7	Mode 8
1	1					

OFF. Therefore, the IGBTs must add a dead-time at turn ON. The commutation by the free-wheeling diode (FWD) occurs during the dead-time. This phenomenon causes a level skip of the phase voltage. Fig. 8 shows an example of the level skip. The IGBTs marked with circles and FWDs marked with dashed circles are conducting. Fig 8(a) shows the commutation when the switching pattern of the U-phase changes from Mode 1 to Mode 2. The arrows indicate the current flow. As shown in Fig.8(b), if S_{3u} is switched OFF, a two-level skip from 2E to zero during the commutation. This phenomenon must be avoided because it causes voltage error and insulation breakdown in the load.

To avoid this problem, additional new switching modes are introduced for the commutation. In the case of Fig. 8, no two-level skip occurs if the phase voltage level during the commutation is 2E or E. In Fig. 8(b), if the S_{2u} is switched ON during the commutation, the phase voltage level is E. This new pattern for commutation defines Mode 1'. In addition, turning ON S_{9u} is also necessary in case of the transition from Mode 8 to Mode 6. Thus, this new patterns define Mode 8'. Fig. 9 shows the case when Mode 1' is applied. Fig. 9(a) shows that turning ON the IGBTs such as S_{2u} beforehand prevents the problem of changing the phase voltage from 2E to zero during the commutation.

Fig. 10 shows the detailed state patterns, which take into account the commutation shown in Fig. 8
The arrows in Fig. 10 indicate the state flow, and the bidirectional arrows indicate that the state basically can

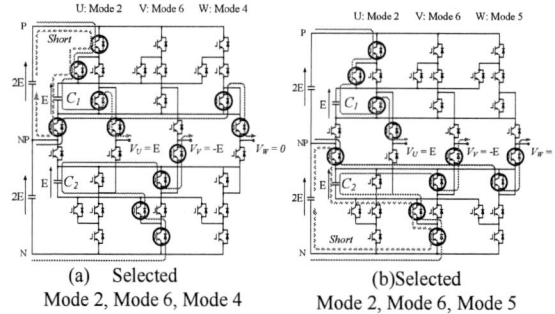

(a) Selected Mode 2, Mode 6, Mode 4 (b) Selected Mode 2, Mode 6, Mode 5

Fig. 11. Example of limitation patterns of Fig.2

be changed to both directions. The state can change its voltage level one by one, such as from 2E to E or E to zero. However, some transitions are prohibited because they cause a phase voltage level skip. The patterns are as follows: from Mode 1 to Mode 2, from Mode 3 to Mode 5, from Mode 4 to Mode 7 and from Mode 6 to Mode 8.

The above method is solving the level-skip problem caused by the dead-time. The next section describes how to realize the state flow shown in Fig. 10

E. Method for Charging / Discharging Limitation Patterns

As explained in section II-B, the proposed topology can control the C_1 and C_2 voltages and can output prohibited phase voltage patterns. Furthermore, FCs C_1, C_2 voltage control also becomes a simple using average current as explained section II-C. However, if the number of phases increases three and more, limitation occurs in the FC charge/discharge control because it has common IGBTs S_{c1} and S_{c2}. This section considers this problem and shows that the proposed circuit can increase the number of phases beyond three.

First, examples of the limited patterns are shown in Fig. 11. When two and more phases select Mode 2 and Mode 6, if one of the phases needs to output zero at the same time the circuit must turn ON common switches S_{c1} or S_{c2}. if Mode 2 and Mode 6 are simultaneously selected, C_{DC1}, C_{DC2} and FCs C_1, C_2 are shorted because of the turning ON of S_{c1} or S_{c2}. Therefore, the number of phases is three and more, and the proposed circuit cannot select combinations of Mode 2, Mode 6, Mode 4, and Mode 2, Mode 6, Mode 5 at the same time. These combinations are avoided by the following rule.

· Prohibition in the simultaneous selection Mode 2 and Mode 6.

This rule does not affect output voltage because the converter can output -E and E by Mode 3 and Mode 7 while it can output zero only by Mode 4 or Mode 5. Therefore, the circuit that applies this condition can operate without any bad influence on the output phase voltage. Then, we describe the method of controlling the voltages of C_1 and C_2 following the condition mentioned above. Considering the condition, if any phase output voltage level of E with Mode 2, the other phases must output a phase voltage level -E with Mode 7. In contrast, for any phase output voltage level of -E with Mode 6, the other phases must output phase voltage level E with Mode 3. Table V and VI list a specific example on how to select the patterns by taking into account the above condition and the switching patterns discussed in section II-*D*. Switching pattern is determined by phase voltage level and C_1 and C_2 charge command. As shown in Table V, when $v_{c1} <E$, the C_1 charge command is one and C_1 is charging. When $v_{c1} >E$, the C_1 charge command becomes zero, and C_1 is discharging. The C_2 charge command is similar to that of the C_1 charge command. The results of C_1 and C_2 charge command are interlocked. Therefore, the interlock prevents the C_1 and $C2$ charge commands to simultaneously become one and prevents the simultaneous selection of Mode 2 and Mode 6.

Next, we describe how to selects the switching patterns. As shown in Table VI, if the phase voltage reference is 2E, the switching pattern is Mode 1 or Mode 1'. When the C_1 charge command is one, Mode 1' is selected. If the C_1 charge command is zero, Mode 1 is selected. The selection of Mode 2, Mode 3, Mode 4, Mode 5, Mode 6, Mode 7, Mode 8, and Mode 8' is performed as well. As a result, depending on the selection status of Mode 2/Mode 3 and Mode 6/Mode 7, Mode 4/Mode 5 can be selected, making it possible for the zero phase voltage level to be output without any problem. Furthermore, this control method can deal with increasing number of phases beyond three because the C_1 and $C2$ charge commands can be independently controlled in all phases.

III. SIMULATION RESULT

Table VII shows simulation conditions. The simulation is performed by using MATLAB/Simulink.

Fig. 12 shows the simulation configuration that the converter is used as an independent three phase DC/AC converter controlled by V/f control. Fig. 13 shows validity of the FCs C_1, C_2 voltage control method. Voltage of FCs was initialized with a value of 0, and operation is started at 0.05s. The FCs voltage reaches reference voltage 2750 V within 30 ms after start-up, as shown in Fig. 13. Consequently, the FCs voltage can control at any value. Fig. 14 illustrates waveforms when modulation index and frequency are ramped up. Fig 13 shows that the proposed circuit can control the voltage of

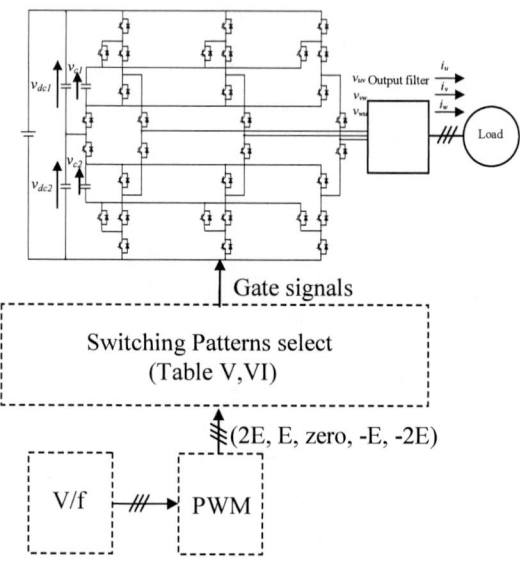

Fig. 12. Simulation inverter-system

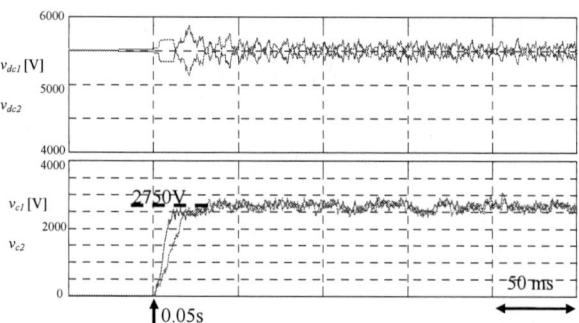

Fig.13. Simulation result of FCs voltage control

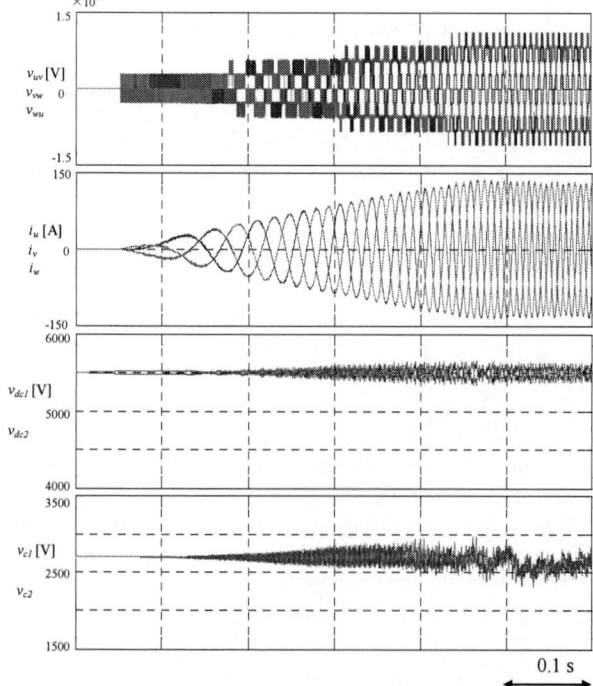

Fig.14. Simulation result of three phase inverter

TABLE VII
SIMULATION CONDITIONS

Supply line to line voltage	Vs	6.6 kVrms
Rated Power	P	1MVA
Supply line frequency	fs	50 Hz
DC-capacitor	C_{DC1}, C_{DC2}	300 μF
Common-capacitor	C_1, C_2	300 μF
Carrier frequency	fc	1 kHz
Inverter modulation index	M_i	0-0.95
Inverter output frequency	f	0-50 Hz

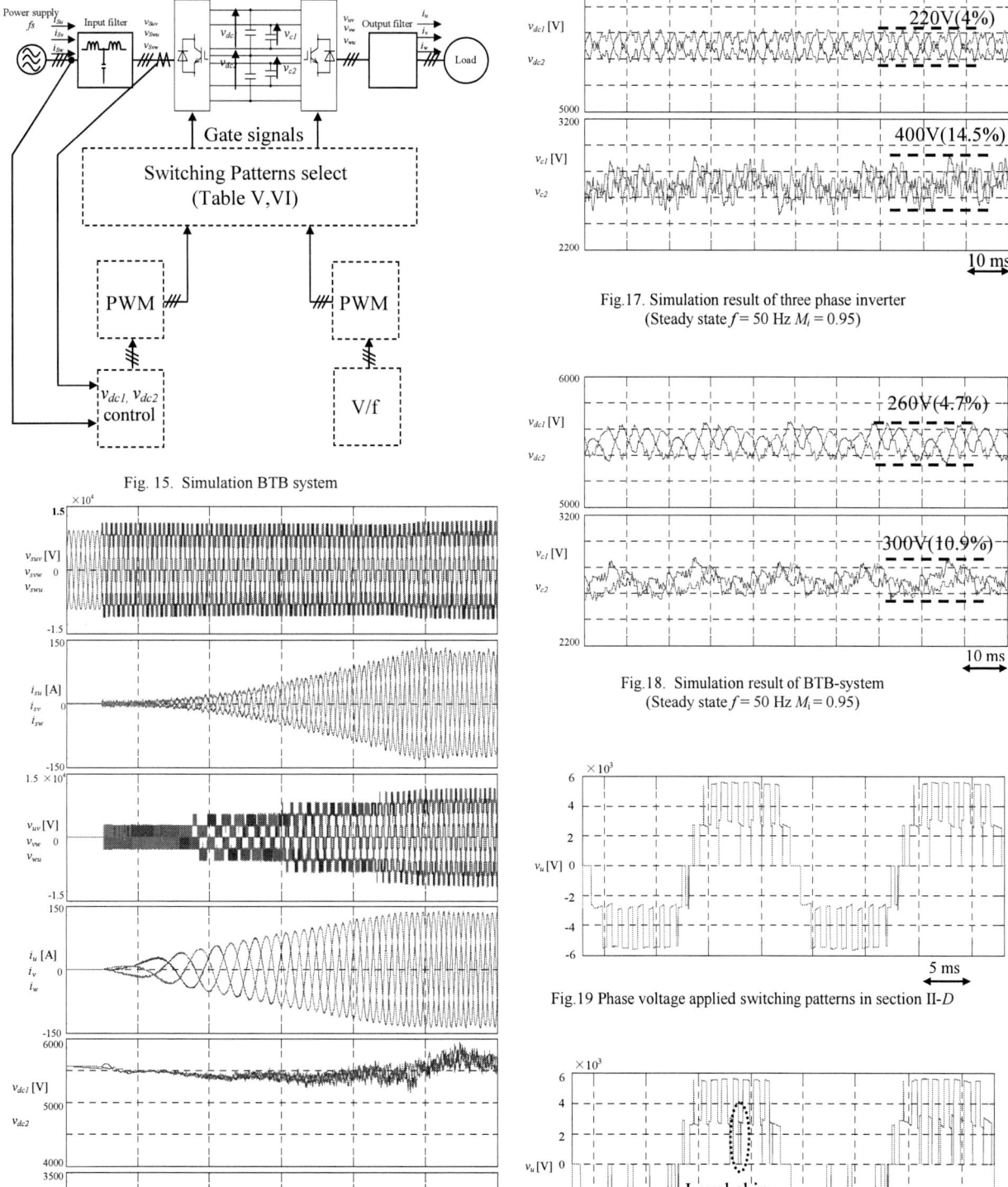

Fig. 15. Simulation BTB system

Fig.17. Simulation result of three phase inverter
(Steady state f = 50 Hz M_i = 0.95)

Fig.18. Simulation result of BTB-system
(Steady state f = 50 Hz M_i = 0.95)

Fig.19 Phase voltage applied switching patterns in section II-D

Level skip
0 → 2E

Fig.20. Phase voltage not applied switching patterns in section II-D

Fig.16. Simulation result of BTB-system

C_1 and C_2 to a constant value while changing the frequency and modulation index.

Fig. 15 shows the simulation configuration the BTB configuration shown in Fig. 5 is used as an AC/AC converter. In this configuration, three phase modules work as converter and the other three works as inverters. The converters connect the system to the AC source and regulate v_{dc1}, v_{dc2}. The inverters are operated by V/f control. Fig. 16 illustrates the waveforms when the modulation index and the fundamental frequency are ramped up. These results show that the converter is properly operated by V/f control while regulating the capacitor voltages at a constant value, even in this dual three-phase BTB configuration.

Fig. 16 and 17 show waveforms of v_{dc1}, v_{dc2}, v_{c1} and v_{c2} in a steady state. Percentage in Fig. 16 and 17 mean the rate of voltage fluctuation to the DC voltage component. Fig.16 is waveforms of three phases inverter and Fig.17 is waveforms of BTB configuration. The comparison between Fig. 16 and Fig. 17 shows that there is no significant difference in the capacitor voltage ripple between the three-phase inverter and the dual three-phase BTB. Therefore, the same capacitance of C_1 and C_2 can be used for the both configurations.

The effectiveness of the method to avoid the level-skip proposed in section II-D is also verified by the simulation study. Fig. 19 and Fig. 20 show the phase voltages obtained from simulations in which the method is applied and not applied, respectively. The level-skip occurs only in the simulation without the method.

This shows that the proposed method prevents the level-skip during dead-time.

IV. CONCLUSION

This study proposes a topology with common FCs shared by all phases. The proposed topology doesn't have prohibited patterns which is problem in [4]. In addition, the control method which can increase the number of phases more than three is also proposed.

The proposed topology requires only two capacitors even when the number of phases is increased to six, while conventional circuits need six capacitors. Moreover, The dual three-phase phase BTB configuration can be operated with the same capacitance of FCs as the three-phase inverter configuration. Therefore, the BTB configuration with the proposed circuit can be fabricated in small size and is lightweight compared to ANPC.

The future tasks of this research are the pattern selection method to reduce switching loss and the control method to reduce voltage fluctuation in the FCs.

REFERENCES

[1] P. Barbosa, P. Steimer, J.Sternke, M. Winkelnkemper, and N.Celanovic , "Active neutral-point-clamped multilevel converters," *inProc. EPE*, p. 10 Sep., 2005

[2] Akagi, H., "Classification, terminology, and application of the modular multilevel cascade converter (MMCC)," *Power Electronics Conference (IPEC), 2010 International* , pp.508-515, June 2010

[3] Zhiguo Pan; Fang Zheng Peng, "A Sinusoidal PWM Method With Voltage Balancing Capability for Diode-Clamped Five-Level Converters," *Industry Applications, IEEE Transactions on* , vol.45, no.3, pp.1028-1034, May-june 2009

[4] Hui Zhang, Wei Yan, Kazuya Ogura, Shota Urushibata, "A Multilevel Converter Topology with Common Flying Capacitors," *ECC*, p. 1274. Sep 2013

[5] Kawamura, W.; Hagiwara, M.; Akagi, H., "A broad range of frequency control for the modular multilevel cascade converter based on triple-star bridge-cells (MMCC-TSBC)," *ECCE*, pp.4014-4021, Sep 2013

[6] Yoshii, T.; Inoue, S.; Akagi, H., "Control and Performance of a Medium-Voltage Transformerless Cascade PWM STATCOM with Star-Configuration," *Industry Applications Conference, 2006. 41st IAS Annual Meeting. Conference Record of the 2006 IEEE* , vol.4, pp.1716,1723, Oct. 2006

The 2014 International Power Electronics Conference

Harmonic Modeling of a Vehicle Traction Circuit Towards the DC Bus

Saeid Haghbin, Andreas Karvonen and Torbjörn Thiringer

Division of Electric Power Engineering, Chalmers University of Technology, Gothenburg, Sweden

E-mails: saeid.haghbin@chalmers.se, andreas.karvonen@chalmers.se, torbjorn.thiringer@chalmers.se

Abstract—Different converters such as the traction inverter and DC/DC converter are connected to the dc bus of an electric or hybrid electric vehicle. Harmonic models of these devices towards the dc bus are needed to investigate different phenomena like the dc bus transients and ripples. A high-frequency harmonic model of a traction circuit, a drive system based on a three-phase inverter connected to an ac motor, is presented and explained. The model is extracted from an analytical approach developed for a three-phase inverter with a sinusoidal PWM controller. In addition to the analytical formulation, simulations and experimental results of a plug-in vehicle are provided to verify the spectrum of the dc bus waveforms.

Index Terms—DC bus quality, frequency spectrum, harmonic model, three-phase inverter, vehicle traction.

I. INTRODUCTION

The concept of more electric vehicle contrives more investigations at different system and component levels within the electrical system of a modern vehicle. Increasing the share of electrical components like the traction circuit or auxiliary systems brings more issues to be considered like component sizing, dc bus system quality, and harmonics. Normally, the vehicle powertrain is the heaviest load in the system; it can be around 50 kW for a typical passenger car. Hence, the impact of traction circuit on the vehicle dc bus is considerable which needs attention.

Several devices are connected to the system dc bus [1] like DC/DC converters, the battery, the battery charger [2], and the traction inverter within the vehicle electrical system. High frequency switching is usually an inherent feature of a controlled modern power electronic device. High frequency components in the voltages and currents are inevitable in the different parts of the system which is considered as an adverse impact. For example, the harmonic currents at the dc bus can result in a considerable amount of voltage ripple that can affect sensitive communication devices. Another disadvantage is the possible battery life reduction due to harmonics in the current. Hence, it is desirable to have a high frequency model in which the complexity and accuracy of the model can be adjusted by considering the phenomena of interest [3]–[6]. In this paper, a high frequency model of traction system towards the dc bus is utilized to investigate the dc bus harmonics.

Fig. 1 shows a simple diagram of traction circuit of a vehicle which is based on a three-phase inverter and an ac motor. In this scheme, the battery is connected to the inverter through a cable that can be of a significant length. Here, it is supposed that the battery is directly connected to the inverter and there

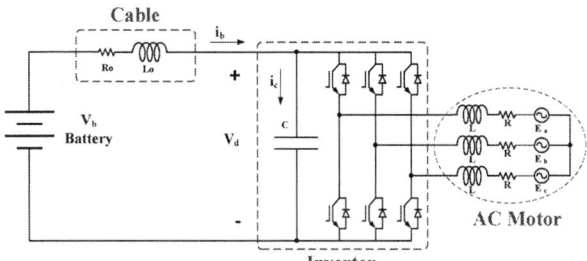

Fig. 1: Vehicle traction system based on a three-phase inverter.

is no DC/DC converter in the path. The battery and the cable are modeled as an ideal voltage source in series with an RL circuit. For a cable with the length of 5-10 m, this model is accurate enough [7] to investigate the dc bus harmonics in a frequency range of dc up to 100 kHz.

The traction system from the dc terminal is usually modeled as a current source with a constant power [8] in which there is no information of the frequency components of the current and voltage. A high frequency model of a three-phase inverter with an ideal sinusoidal currents at the ac side is developed in [9] which is described at the following. The developed harmonic model is applied to a drive system with the rotor field-oriented control (FOC) and sinusoidal pulse width modulation (SPWM).

The main aim of this work is to provide a high-frequency equivalent circuit that can be used to investigate the dc bus voltage quality, for instance its frequency spectrum, for a FOC-based drive system. Simulation results are provided to investigate the dc bus voltage harmonics during steady state operation of the traction system. The battery current is also considered in the presented model. Some practical measurements of a commercial plug-in vehicle are presented.

After this introduction, the FOC scheme of the traction system is shortly presented. The high frequency model of the inverter system towards the dc bus is explained in section III. Simulation and practical measurements are presented in section IV and finally some conclusions are provided in the last section.

II. FIELD-ORIENTED CONTROL OF A PM MOTOR

FOC in the rotor reference frame (dq) is one of the widely used control schemes for traction applications which is shortly

978-1-4799-2706-7/14 $31.00 © 2014 IEEE

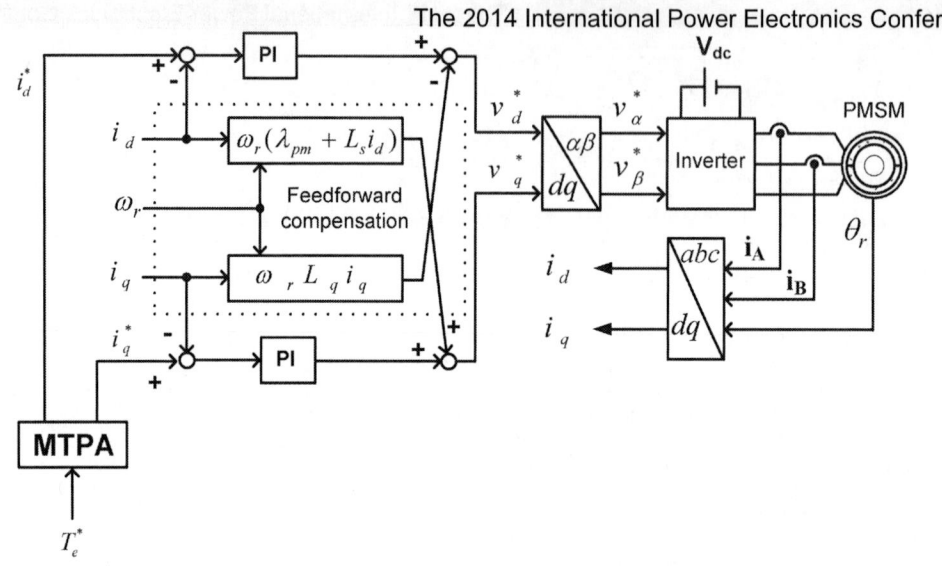

Fig. 2: Field-oriented control of a three-phase ac motor for the vehicle traction.

presented here. The following equations describe the mathematical model of a PM motor in the rotor reference frame:

$$v_d = r_s i_d + L_d \frac{d}{dt} i_d - \omega_r L_q i_q \qquad (1)$$

$$v_q = r_s i_q + L_q \frac{d}{dt} i_q + \omega_r (\lambda_{PM} + L_d i_d) \qquad (2)$$

where v_d, v_q, i_d and i_q are the voltages and currents in d and q axes respectively. Moreover, r_s, L_d, L_q, λ_{PM}, and ω_r are the stator resistance, the dq stator inductances, the permanent magnet flux linkage and the rotor electrical speed, respectively. The developed electromechanical torque can be written as

$$T_e = \frac{P}{2} \frac{3}{2} \{ \lambda_{PM} i_q + (L_d - L_q) i_d i_q \} \qquad (3)$$

where T_e is the torque and P is the number of poles.

A. Structure of the FOC for the IPM

Fig. 2 shows a schematic diagram of the drive structure [10], [11]. The system is a torque controller which is typical in traction applications. As can be seen in the figure, the reference torque is mapped to reference dq currents which form inputs to the high-bandwidth current controllers. A maximum torque per ampere (MTPA) strategy is typically used for this mapping [12] which can be a simple mapping table or a more accurate model [13].

Two independent PI controllers regulate the dq currents. The controller outputs are dq voltages that are compensated by feedforward terms to enhance the system response. The generated voltages are reference signals for the inverter which are transformed to the abc domain by using the Park and Clarke inverse transformation. Then the inverter produces the voltages by using the SPWM scheme. In steady state,

the motor currents constitute symmetrical three-phase waveform with a fundamental frequency and some high frequency components. Consequently, the analytical model explained in the next section is developed for a system with three-phase currents.

B. The DC Bus Model

For the harmonic analysis, rather than an ideal voltage source, a frequency dependent model of the dc source is needed. The battery has an internal impedance that can be a complicated function of the frequency, temperature, current level and state of charge [14]. In this paper a simple RL circuit model is utilized to connect the battery to the inverter. This equivalent circuit includes the cable and battery as shown in Fig. 1. The equations describing the equivalent circuits are

$$i_c = C \frac{dv_d}{dt} \qquad (4)$$

$$V_b = R_o i_b + L_o \frac{di_b}{dt} + V_d \qquad (5)$$

where L_o, R_o, C, V_b and V_d are the cable and source inductance, the cable and source resistance, the inverter capacitance, the battery open circuit voltage and the inverter voltage at the point of connection, respectively.

III. HARMONIC MODEL OF A THREE-PHASE INVERTER TOWARDS THE DC BUS

The dc bus capacitor filters the highly distorted inverter dc side current to provide a smoother current waveform from/to the dc source, a battery for instance. This capacitor is a bulky device that should be kept to a minimum size. However, the quality of the dc waveforms depends on its capacitance. A harmonic analysis in time-frequency domain is presented in

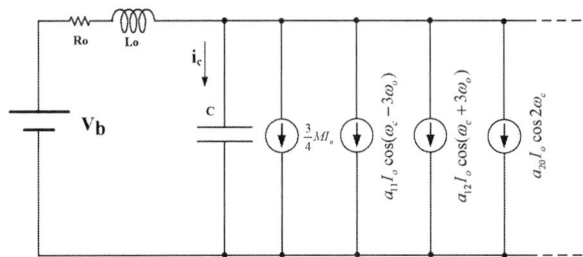

Fig. 3: Proposed high-frequency model of the inverter and load for frequency analysis.

[9] which provides a high-frequency equivalent circuit of a three-phase SPWM inverter towards the dc terminals.

The inverter dc side current, i_{inv}, can be written as [15]

$$i_{inv} = S_A \, i_A + S_B \, i_B + S_C \, i_C \qquad (6)$$

where i_A, i_B and i_C are inverter currents at the ac side and S_A, S_B, and S_C are switches status, either 1 or 0.

Assume that the inverter is connected to a symmetrical three-phase load with pure sinusoidal waveform with a power factor of $cos\phi$. The inverter line currents can be described as

$$i_A = I_o \, cos(\omega_o t + \phi) \qquad (7)$$

$$i_B = I_o \, cos(\omega_o t - \frac{2\pi}{3} + \phi) \qquad (8)$$

$$i_C = I_o \, cos(\omega_o t + \frac{2\pi}{3} + \phi) \qquad (9)$$

where I_o is the magnitude of the current, and ω_o is the angular frequency.

For an inverter with a SPWM control strategy the inverter dc side current, i_{inv}, is [9]

$$
\begin{aligned}
i_b = {} & \frac{3}{4} M I_o cos\phi + I_o \frac{3}{\pi} \{ \\
\ldots + {} & \sqrt{ J_2^2(\frac{\pi}{2}M) + J_4^2(\frac{\pi}{2}M) - 2J_2(\frac{\pi}{2}M)J_4(\frac{\pi}{2}M) \, cos(2\phi)} \\
& cos(\omega_c t - 3\omega_o t + \phi_{11}) \\
+ {} & \sqrt{ J_2^2(\frac{\pi}{2}M) + J_4^2(\frac{\pi}{2}M) - 2J_2(\frac{\pi}{2}M)J_4(\frac{\pi}{2}M) \, cos(2\phi)} \\
& cos(\omega_c t + 3\omega_o t + \phi_{12}) + \ldots \\
\ldots - {} & \frac{2}{2} J_1(2\frac{\pi}{2}M) \, cos(\phi) \, cos(2\omega_c t) + \ldots \\
& \}.
\end{aligned}
\qquad (10)
$$

where $J_\alpha(x)$ is the Bessel function of the first kind, and M is the modulation index.

A simple model of the inverter is to replace the unit by a constant current source. This model can not be used to explain the high frequency ripple on the dc bus. Based on the developed equations of the inverter system at the dc side, as is presented in (10), a more accurate equivalent circuit can be used. Fig. 3 shows the proposed equivalent circuit where coefficients a_{11}, a_{12} and a_{20} are defined by (10).

The dc source and cable are modeled as an ideal voltage source and an RL impedance; a more accurate model can be utilized depend on the type of application. Moreover, the dc

component of the load and high frequency components due to PWM switching are modeled as a summation of current sources. The number of high frequency current sources can be increased if a more accurate model is needed.

The above model includes important frequency information of the system. For example, the resonance frequency of the system due to the interaction of the dc cable that has an inductance value of L_o with the filter capacitor C is

$$f_r = \frac{1}{2\pi} \frac{1}{\sqrt{L_o C}} \qquad (11)$$

where f_r is the resonance frequency. Moreover, the main harmonic components in the system are around multiples of carrier frequencies, i.e. $\omega_c t \pm 3\omega_o t$ and $2\omega_c t$. As is shown in [9] and can be seen later on in the simulations, the higher-order harmonics can be neglected.

IV. SIMULATION RESULTS AND PRACTICAL VERIFICATION

A. Simulation Results

For a commercial plug in vehicle the traction system is simulated with a dc bus voltage of 343 V. To evaluate the drive system performance, software simulations have been performed for a 20 kW system in Matlab/Simulink; ideal switches are assumed for the inverter. Moreover, the dc bus capacitor is ideal (no ESL or ESR) and an RL network is assumed for the battery internal impedance and the cable connecting to the inverter. The parameters of the electric motor used for the simulation are presented in Table I.

TABLE I: Electric motor parameters.

	Value	Unit
Rated power	20	kW
Rated speed	3000	r/min
No of poles	10	-
Permanent magnet flux (λ_{PM})	0.0497	Wb
Stator resistance	0.0117	Ohm
d-axis inductance (L_d)	236	μH
q-axis inductance (L_q)	395	μH

The PWM frequency is 10 kHz and the parameters of the dq controllers are selected according to [11]. For the motor, the values of L_d, L_q and λ_{PM} are considered constant during the simulations. For the harmonic analysis, with a frequency range of $0 - 100$ kHz, it is not expected to have a perfect match between the simulations and measurement because of lack of accurate models and parameters. However, one can study the nature of different phenomena. For example, it is very easy to see how the cable inductance affects the dc bus voltage ripple.

Fig. 4 and Fig. 5 show the motor torque and speed for the simulation case. The speed is almost constant but there are slight ripples on the torque which is typical in such applications. The reference values of the dq currents are set to $i_d^\star = -105$ A and $i_q^\star = 162$ A. The developed torque is around 81 Nm and the motor speed is 2262 rpm; here a load proportional to the speed is assumed.

The 2014 International Power Electronics Conference

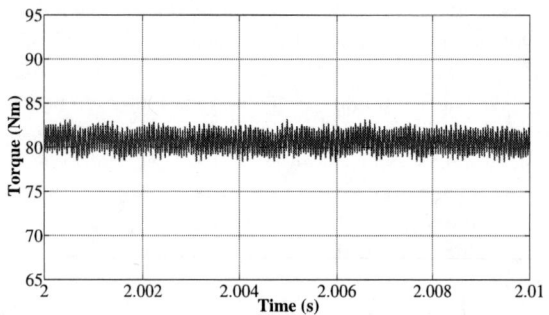

Fig. 4: Simulation: electromagnetic torque of the motor.

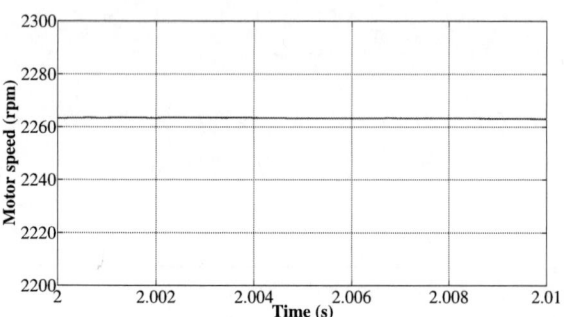

Fig. 5: Simulation: speed of the motor.

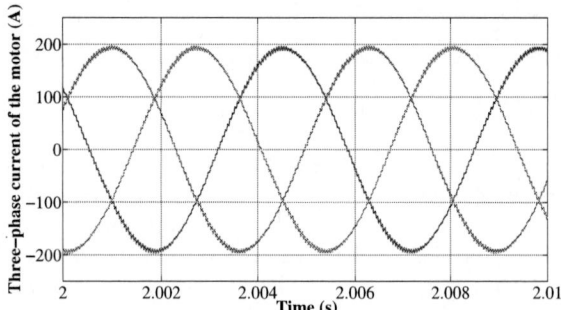

Fig. 6: Simulation: the motor three-phase currents at $2262\ rpm$ and $81\ Nm$.

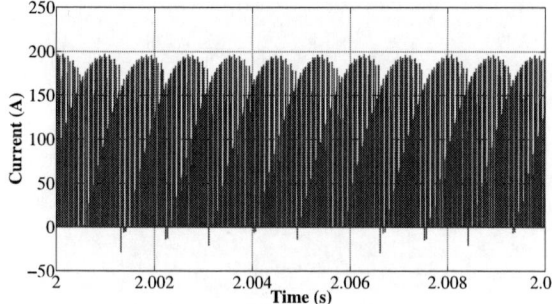

Fig. 7: Simulation: inverter dc side current at $2262\ rpm$ and $81\ Nm$.

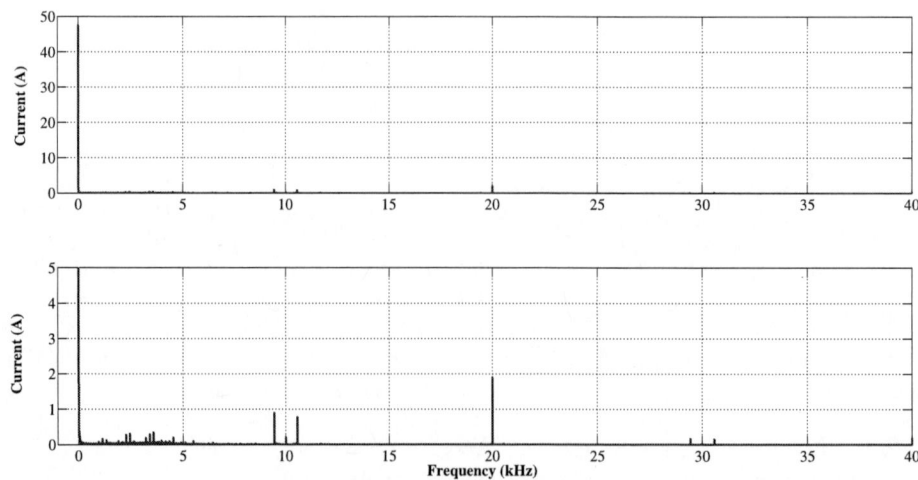

Fig. 8: Simulation: frequency spectrum of the battery current at $2262\ rpm$ and $81\ Nm$: full spectrum (top) and a zoomed plot (bottom).

978-1-4799-2706-7/14 $31.00 © 2014 IEEE   1376

The 2014 International Power Electronics Conference

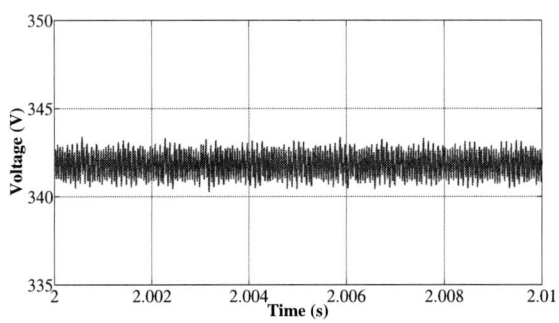

Fig. 9: Simulation: dc bus voltage over the main capacitor at 2262 rpm and 81 Nm.

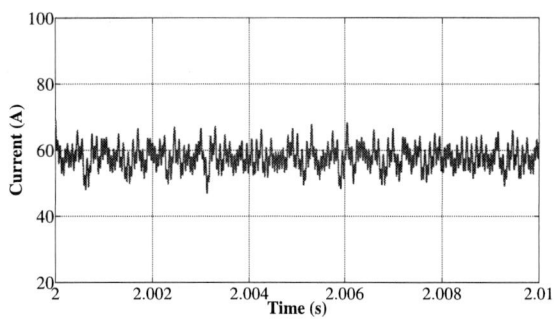

Fig. 10: Simulation: battery current at 2262 rpm and 81 Nm.

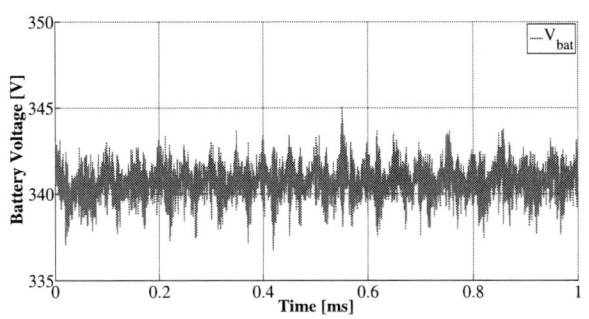

Fig. 11: Measurement: Battery voltage at 2262 rpm and 81 Nm.

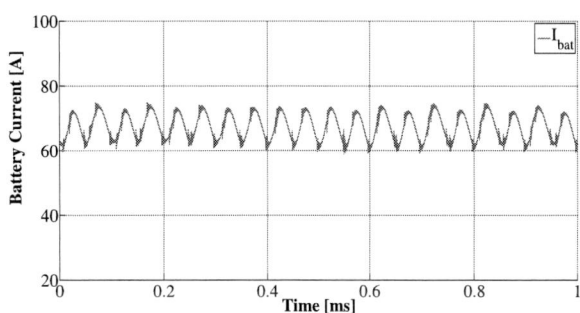

Fig. 12: Measurement: Battery current at 2262 rpm and 81 Nm.

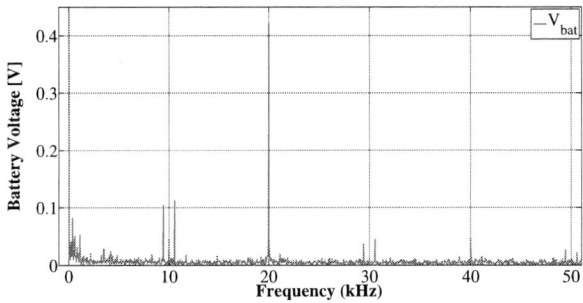

Fig. 13: Measurement: frequency spectrum of the battery voltage at 2262 rpm and 81 Nm.

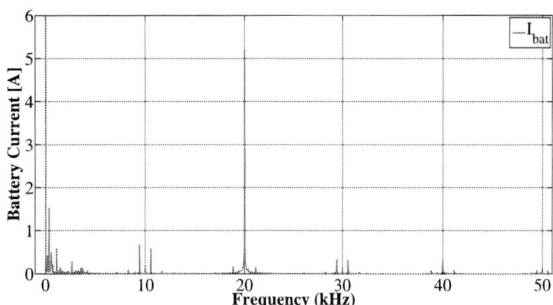

Fig. 14: Measurement: frequency spectrum of the battery current at 2262 rpm and 81 Nm.

The motor three-phase line currents are shown in Fig. 6 and the inverter dc side current is shown in Fig. 7. For an ideal system, the dc bus capacitor absorbs all of the high frequency components (non-dc) of the current, but for a realistic system there are some high frequency terms. The dc bus voltage is shown in Fig. 9. The value of the voltage is 343 V, and there is some ripple due to the switching in the system.

The battery current is depicted in Fig. 10; Fig. 8 shows the frequency spectrum of the battery current. The spectrum of the battery current shows that in addition to the dc component, there is a component around 3.6 kHz due to resonance between the cable and dc bus capacitor and some components

around multiple of the carrier frequencies, i.e. 10 kHz, 20 kHz, and so on. The magnitude of the battery current harmonics for non-dc components is not high.

B. Practical Verification

To investigate different parts of the electrical system of a plug-in vehicle, extensive measurements have been performed, especially on the dc side of the traction circuit where a few results are presented here. Fig. 11 and 12 show the battery voltage and current at a speed of 2262 rpm and a torque level of 81 Nm. Moreover, the FFT of the battery voltage and current are show in Fig. 13 and 14. The plot is zoomed to magnify the harmonic components. So, the dc components

978-1-4799-2706-7/14 $31.00 © 2014 IEEE

TABLE II: Measurement results for a single operation point in steady state.

	Value	Unit
Speed	2262	rpm
Torque	81	Nm
Peak phase current	190	A
Battery voltage	343	V
Battery current	60	A
i_d	-105	A
i_q	162	A
v_d	-75	V
v_q	34	V

is not fully shown in the figure.

As can bee seen from these figures, the practical results are very similar to the simulation values, as is expected from the theoretical formula. The major harmonics in the dc side are around the resonance frequency and multiples of carrier frequency due to PWM switching, i.e. 10 kHz and 20 kHz. Table II summarizes the measured values in steady state.

V. CONCLUSION

To investigate the quality of the dc bus of an electric or hybrid electric vehicle, the high-frequency model of the devices towards the dc terminals are needed. A harmonic model of the traction circuit based on a three-phase inverter and an ac motor towards the dc system is presented. The model is developed for a three-phase inverter with SPWM and a three-phase line current. For a vehicle traction system based on a FOC with a SPWM control strategy, the simulation results are provided and the results show an accurate agreement with the predicted frequency spectrum in the dc bus current and voltage. Measurements of a real vehicle are provided to verify the harmonic pattern in the dc bus, as is expected by using the equivalent circuit and simulation results. The main harmonics are around the resonance frequency of the dc bus capacitor and cable, and multiple of carrier frequency due to the PWM switching that is 10 kHz and 20 kHz.

REFERENCES

[1] M. Anwar, S. Gleason, and T. Grewe, "Design considerations for high-voltage dc bus architecture and wire mechanization for hybrid and electric vehicle applications," in *Energy Conversion Congress and Exposition (ECCE), 2010 IEEE*, 2010, pp. 877–884.

[2] S. Haghbin, S. Lundmark, M. Alakula, and O. Carlson, "Grid-connected integrated battery chargers in vehicle applications: Review and new solution," *Industrial Electronics, IEEE Transactions on*, vol. 60, no. 2, pp. 459–473, 2013.

[3] A. Karvonen and T. Thiringer, "Simulating the emi characteristics of flyback dc/dc converters," in *Telecommunications Energy Conference (INTELEC), 2011 IEEE 33rd International*, 2011, pp. 1–7.

[4] M. Liukkonen, M. Hinkkanen, J. Kyyra, and S. Ovaska, "Modeling of multiport dc busses in power-electronic systems," in *Industrial Technology (ICIT), 2013 IEEE International Conference on*, 2013, pp. 740–745.

[5] S. Sudhoff, S. Glover, P. Lamm, D. H. Schmucker, and D. Delisle, "Admittance space stability analysis of power electronic systems," *Aerospace and Electronic Systems, IEEE Transactions on*, vol. 36, no. 3, pp. 965–973, 2000.

[6] J. Liu, X. Feng, F. Lee, and D. Borojevich, "Stability margin monitoring for dc distributed power systems via perturbation approaches," *Power Electronics, IEEE Transactions on*, vol. 18, no. 6, pp. 1254–1261, 2003.

[7] Z. Peroutka, "Motor insulation breakdowns due to operation of frequency converters," in *Power Tech Conference Proceedings, 2003 IEEE Bologna*, vol. 2, 2003, pp. 8 pp. Vol.2–.

[8] A. Khaligh, A. Rahimi, and A. Emadi, "Negative impedance stabilizing pulse adjustment control technique for dc/dc converters operating in discontinuous conduction mode and driving constant power loads," *Vehicular Technology, IEEE Transactions on*, vol. 56, no. 4, pp. 2005–2016, 2007.

[9] S. Haghbin and T. Thirniger, *High-Frequency Modeling of a Three-Phase Inverter with Sinusoidal Pulse Width Modulation Control*. Chalmers University of Technology, Internal report, 2014.

[10] S. Haghbin, *An Isolated Integrated Charger for Electric or Plug-in Hybrid Vehicles*. Licentiate Thesis, Chalmers University of Technology, 2011.

[11] D. Vindel, S. Haghbin, A. Rabiei, O. Carlson, and R. Ghorbani, "Field-oriented control of a pmsm drive system using the dspace controller," in *Electric Vehicle Conference (IEVC), 2012 IEEE International*, 2012, pp. 1–5.

[12] M. Haque and M. Rahman, "Control trajectories for interior permanent magnet synchronous motor drives," in *Electric Machines Drives Conference, 2007. IEMDC '07. IEEE International*, vol. 1, may 2007, pp. 306 –311.

[13] S. Haghbin, *Integrated Motor Drives and Battery Chargers for Electric or Plug-in Hybrid Electric Vehicles*. PhD Thesis, Chalmers University of Technology, 2013.

[14] F. Savoye, P. Venet, M. Millet, and J. Groot, "Impact of periodic current pulses on li-ion battery performance," *Industrial Electronics, IEEE Transactions on*, vol. 59, no. 9, pp. 3481–3488, Sept 2012.

[15] P. Dahono, A. Purwadi, and T. Kataoka, "A new approach for harmonic analysis of three-phase current-type pwm converters," in *Industry Applications Conference, 1997. Thirty-Second IAS Annual Meeting, IAS '97., Conference Record of the 1997 IEEE*, vol. 2, 1997, pp. 1487–1495 vol.2.

978-1-4799-2706-7/14 $31.00 © 2014 IEEE

AC/DC Converter Based on Instantaneous Power Balance Control for Reducing DC-Link Capacitance

Akira Tokumasu, Hiroshi Taki, Kazuhiro Shirakawa
CORPORATE R&D DIV.2
DENSO CORPORATION
Aichi, Japan
akira_tokumasu@denso.co.jp

Keiji Wada
Dept. of Electrical and Electronic Engineering
Tokyo Metropolitan University
Tokyo, Japan
kj-wada@tmu.ac.jp

Abstract—In AC/DC converters for on-board chargers, the DC-link capacitor on the output side in a power factor correction (PFC) converter uses aluminum electrolytic capacitors in order to obtain a large capacitance with low voltage ripple and small size. Therefore, it is necessary to reduce the capacitance when using film capacitors without increasing the size of the converter in order to realize a capacitor with a long lifetime. This paper describes a control strategy for the AC/DC converter with a reduced DC-link capacitance. This strategy is based on the instantaneous power balance of a PFC converter and a DC/DC converter and controls the AC-side input current without being affected by the large ripple voltage of twice the utility line frequency owing to the reduction in the capacitance. The simulation and experimental results are presented to validate the proposed strategy, and it is confirmed that it is possible to reduce the DC-link capacitance to one-fifth when compared with the conventional AC/DC converter without additional circuits.

Keywords—AC/DC converter, instantaneous power balance, on-board charger, power factor correction

I. INTRODUCTION

In the automotive industry, vehicles with a combustion engine have been converted into electric vehicles (EVs) or plug-in hybrid vehicles (PHVs) with a large-capacity battery [1]. As for the battery charger for EVs/PHVs, a DC quick charger and an on-board charger are used on the market. The on-board charger charges the battery from a utility line at home, as shown in Fig. 1. The charger requires both small volume and a lifetime of more than 50,000 h because it is installed in the limited space of the EVs and used to charge for 8 h per day.

Fig. 2 shows the circuit configuration of the on-board charger. The charger consists of a power factor correction (PFC) converter and an isolated DC/DC converter. DC-link capacitors for suppressing the output voltage ripple of the PFC converter is connected on the DC-link which is the output side of the PFC converter. Aluminum electrolytic capacitors are used for the capacitors in order to obtain a large capacitance with low cost and small size [2]~[4]. However, it is difficult to achieve the required lifetime of more than 50,000 h because the lifetime of

these capacitors is set to less than 8,000 h in the higher-temperature environment of the EVs as compared with the industrial equipment. Therefore, it is necessary to use film capacitors for the DC-link capacitors to achieve these requirements. Generally, the film capacitors have a long lifetime; however, their size is larger than that of the aluminum electrolytic capacitors for the same capacitance and the voltage rating. In order to achieve small size when using film capacitors, it is necessary to reduce the capacitance as small as one-fifth compared with the conventional PFC converter.

During charging of the battery, the input current of the AC/DC converter is controlled to be a sinusoidal waveform in phase with input voltage by the PFC converter for a high power factor in order to satisfy regulations (IEC61000-3-2). Therefore, the frequency of the ripple in the input power waveform is twice the line frequency. On the other hand, the output power is controlled to be a constant power for charging the battery. Therefore, an energy buffer absorbing the power ripple between the input and output is required. In general, the function of the buffer is achieved by charging/discharging the DC-link capacitors at twice the line frequency. Therefore, the DC-link voltage also contains the ripple of twice the line frequency. In the case of decreasing the DC-link capacitance, twice the line frequency ripple of the DC-link voltage is increased. In this case, an increase in twice the line frequency ripple voltage affects the input current distortion, because the DC-link voltage is used for controlling the input current [5].

In order to avoid the influence of the ripple voltage, a control strategy has been proposed for compensating the voltage ripples [6]~[8]. However, the error in manufacturing and deterioration and the approximate calculation cause a reduction in the calculation accuracy, and this distorts the input current. A compensation method for the ripple voltage using additional circuits has been proposed [9]~[12], and it is possible to suppress the voltage ripple to almost zero. However, additional circuits increases the total loss and size of the converters for an on-board charger of more than 3 kW for residential application.

The purpose of this paper is to reduce the DC-link

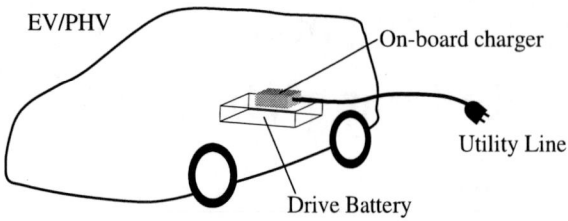

Fig. 1. Charging system of the on-board charger

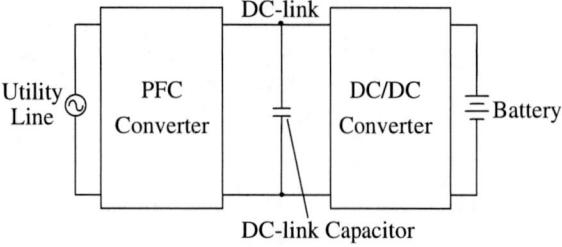

Fig. 2. Configuration of the on-board charger

capacitance without any additional circuit. This paper describes a control strategy for the AC/DC converter by coordinating both a PFC converter and a DC/DC converter. The instantaneous power balance concept of both the PFC converter and the DC/DC converter is used as the control strategy. In order to avoid the influence of the voltage ripple on the input current distortion, this control uses the output power of the DC/DC converter without the ripple component instead of the DC-link voltage. The simulation and experimental results are presented to validate the proposed strategy, and it is confirmed that it is possible to reduce the DC-link capacitance to one-fifth when compared with the conventional AC/DC converter.

II. CONVENTIONAL CONTROL FOR REDUCING THE CAPACITANCE

A. Circuit Configuration of the On-Board Charger

Fig. 3 shows the circuit configuration of the on-board charger that consists of a PFC converter and an isolated DC/DC converter. The intermediate stage is the DC-link part and the DC-link capacitor is arranged this part. Table. I shows the specifications of the circuit.

B. Design of the DC-Link Capacitance

Fig. 4 shows the electric power balance of the input side, the output side and the DC-link part of AC/DC

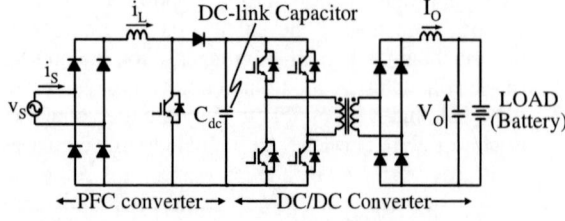

Fig. 3. Circuit configuration of the on-board charger

TABLE I. SPECIFICATION OF ON-BOARD CHARGER

Input voltage	85∼265 V
Max input voltage peak	375 V
Utility line frequency	50/60 Hz
Max input current	15 A
DC output voltage	300 V
DC output power	3.3 kW

converter in the steady state. When the input current i_s is controlled to be a sinusoidal waveform in phase with the input voltage v_s for a unity power factor, and the instantaneous input power p_{in} is expressed as follows:

$$
\begin{aligned}
v_s &= V_s \sin \omega t \\
i_s &= I_s \sin \omega t \\
p_{in} &= V_s I_s \sin^2 \omega t
\end{aligned}
\tag{1}
$$

where V_s means the peak value of the input voltage, I_s is the peak value of the input current, and ω is the utility line angular frequency. As can be observed from (1), the frequency of the ripple in the input power is twice the line frequency. The output power of the on-board charger is controlled to be a constant value for charging the battery, and the output power p_{out} is expressed as follow:

$$
p_{out} = V_o I_o = \frac{1}{2} V_s I_s
\tag{2}
$$

where V_o is the output DC voltage and I_o is the output DC current. In order to absorb the power ripple as twice the line frequency, the instantaneous charge/discharge power p_c of the DC-link capacitors is required as in (3), and the energy power ripple of the capacitor W_c is obtained by (4) from (3), because W_c equal to the charging electric charge of the capacitors in half cycle of the power ripple.

$$
\begin{aligned}
p_c &= p_{in} - p_{out} = -p_{out} \cos 2\omega t \\
W_c &= \int_0^{\frac{\pi}{2}} p_c dt = \frac{p_{out}}{\omega}
\end{aligned}
\tag{3}
\tag{4}
$$

As can be observed from (3), the DC-link capacitors charge/discharge the power at twice the line frequency. Therefore, a voltage that contains ripple of twice the line frequency appears at the DC-link part. Moreover, the W_c is obtained from the relationship between the electric power and the voltage of the capacitor as

$$
W_c = \frac{1}{2} C_{dc} V_{dcmax}^2 - \frac{1}{2} C_{dc} V_{dcmin}^2
\tag{5}
$$

where V_{dcmax} is the allowed maximum voltage, and V_{dcmin} is the allowed minimum voltage of the DC-link part in the steady state and they are determined by the voltage rating of the capacitors and power devices. The DC-link capacitance C_{dc} is calculated by (6) from (4) and

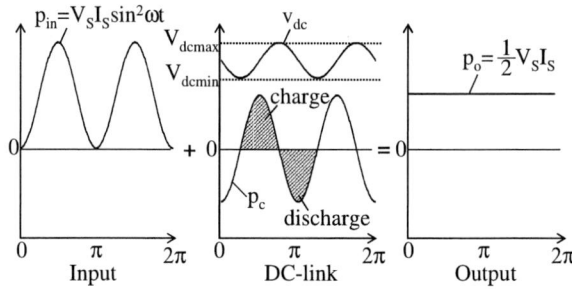

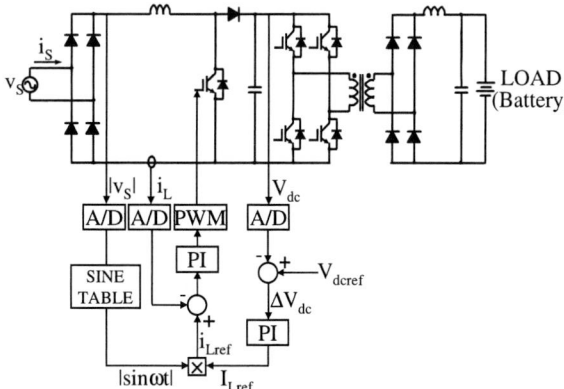

Fig. 4. The electric power of the input side, the output side and the DC-link part

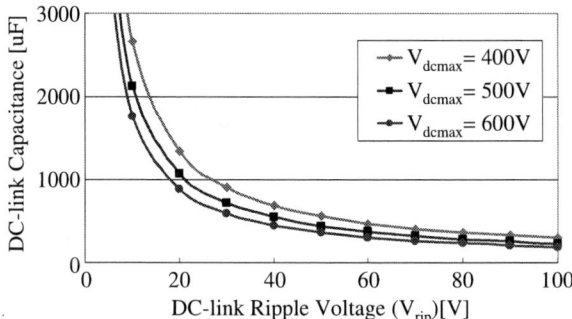

Fig. 5. The relationship between the DC-link voltage and the DC-link capacitance

Fig. 6. Diagram of the conventional control

wave signal $| \sin(\omega t) |$, which is in phase with the input voltage. Therefore, the input current reference i_{Lref} is expressed as follows:

$$i_{Lref} \quad = \quad I_{Lref} \, | \sin(\omega t) | \qquad (8)$$

The power factor of the PFC converter can be controlled to 1.0 because the modulation factor of the PWM control is controlled such that the input current i_L follows i_{Lref} by the PI controller.

As discussed in section II-B, it is necessary to increase twice the line frequency ripple voltage of the DC-link part to reduce the DC-link capacitance. In the case of increasing the ripple voltage, twice the line frequency ripple appears at the error voltage signal Δv_{dc} which is shown Fig. 6. Therefore, the input current amplitude reference I_{Lref} contains twice the line frequency ripple component and the input current reference i_{Lref} is a distorted sinusoidal waveform including a harmonic as shown Fig. 7. The input current is a distortion waveform because the input current i_L is controlled by the distorted current reference. In order to avoid distortion from the DC-link ripple voltage, the response frequency of the voltage control needs to be set sufficiently lower than twice the line frequency. In the case of connecting the utility line, the response frequency is set to less than 10 Hz.

However, the variation of the DC-link voltage at the time of the load change is wider when using the small-capacitance DC-link capacitors and the slow control response. As a result, the voltage rating of the power devices should be increased. Therefore, it is difficult to reduce the capacitance in the conventional control owing to the issues of the input current distortion and the voltage variation during the load change.

(5), and the DC-link ripple voltage V_{rip} is expressed in (7).

$$C_{dc} \quad \geq \quad \frac{2P_{out}}{\omega(V_{dcmax}^2 - V_{dcmin}^2)} \qquad (6)$$

$$V_{rip} \quad = \quad V_{dcmax} - V_{dcmin} \qquad (7)$$

According to (6), DC-link capacitance is determined from the relationship between the output power, utility line angular frequency, and the allowed maximum and minimum voltage of the DC-link part. For reducing the DC-link capacitance, it is necessary to increase V_{dcmax} or to increase the difference between V_{dcmax} and V_{dcmin} by increasing the ripple voltage of the DC-link part. Fig. 5 shows the relationship between the DC-link capacitance C_{dc} and the ripple voltage V_{ripple}. The ripple voltage greatly affects the capacitance as shown in Fig. 5

C. Effect of the DC-Link Ripple Voltage in the Conventional Control

Fig. 6 shows a conventional control diagram of the PFC converter in the AC/DC converter [13]·[14]. The control of the PFC converter has the input current control as the inner loop and the DC-link voltage control as the outer loop. In the voltage control, the amplitude of the input current reference I_{Lref} is produced by a proportional-integral (PI) controller from the error between the DC-link voltage V_{dc} and the DC-link reference voltage V_{dcref}. In the current control, the input current reference i_{Lref} is generated by multiplying I_{Lref} and the absolute sine

III. PROPOSED CONTROL STRATEGY BASED ON THE INSTANTANEOUS POWER BALANCE

A. Principle of the Control

When the input current power factor is set to 1.0 by the PFC converter and the output-voltage and -current are

978-1-4799-2706-7/14 $31.00 © 2014 IEEE

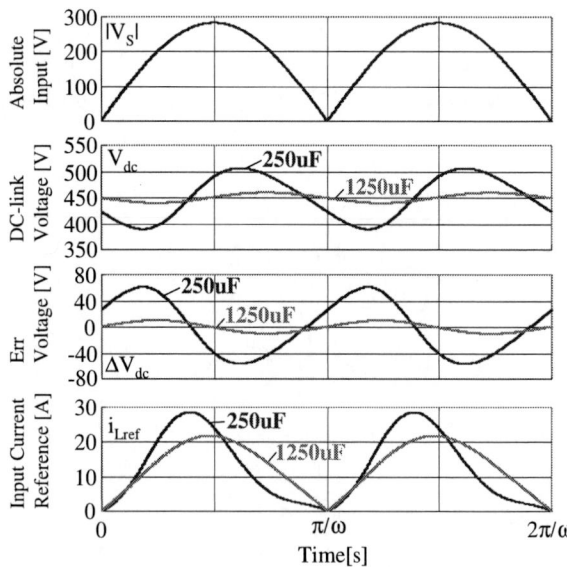

Fig. 7. Distortion of the input current reference by the ripple voltage of the DC-link part ($v_s = 200\,\text{V}/50\,\text{Hz}$, $V_{dcref} = 450\,\text{V}$, $p_{out} = 3\,\text{kW}$)

controlled to be a constant value by the DC/DC converter in the steady state, the instantaneous power balance on both the input and the output sides is expressed as follows from (1)(2):

$$
\begin{aligned}
p_{in} &= p_c + p_{out} \\
&= p_{out}(1 - \cos 2\omega t) \qquad (9)
\end{aligned}
$$

When the input power is controlled to satisfy (9), the input current can be controlled unity power factor.

B. Control Strategy of Instantaneous Power Balance

The propose control realizes a high power factor by controlling the duty ratio to satisfy the equation (9) of the input power. Fig. 8 shows the control diagram of proposed control strategy. The instantaneous input power reference p_{inref} is expressed as

$$
p_{inref} = p_{out}(1 - \cos 2\omega t) \qquad (10)
$$

The input current is controlled to a high power factor because the modulation factor of the PWM control is controlled such that the input power p_{in} follows p_{inref} by the PI controller.

Even though the DC-link capacitance is decreased and the DC-link ripple voltage is increased, the output power of the converter can be controlled to a constant value by the DC/DC converter. Therefore, the input power reference $p_{inref}(= p_{out}(1 - \cos 2\omega t)$ contains few distortion component regardless of the ripple voltage as shown Fig. 9. Thus, the proposed strategy based on the instantaneous power balance can also control the input current to a high power factor in the case of reducing the DC-link capacitance.

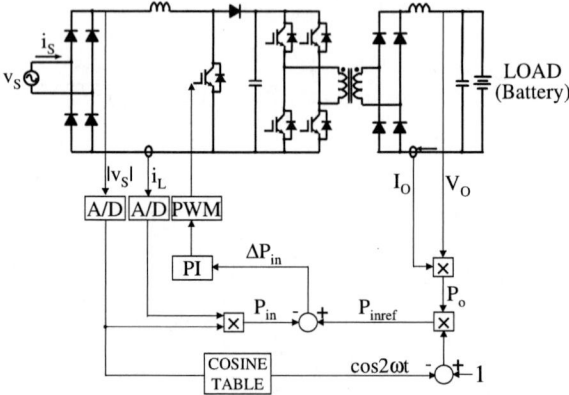

Fig. 8. Diagram of the proposed control

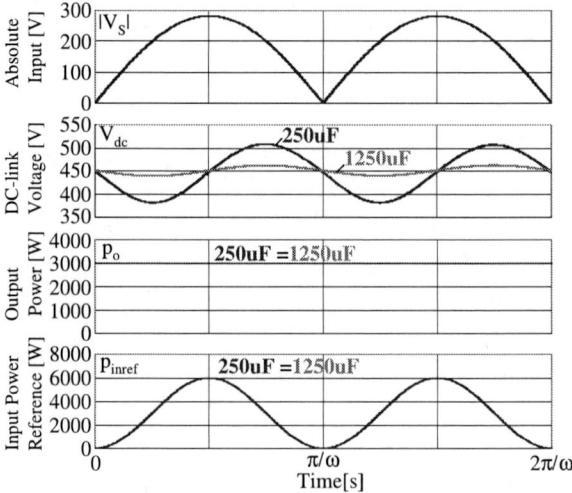

Fig. 9. No effect on input power reference by the ripple voltage of the DC-link ($v_s = 200\,\text{V}/50\,\text{Hz}$, $V_{dcref} = 450\,\text{V}$, $p_{out} = 3\,\text{kW}$)

Moreover, the instantaneous input power reference is changed instantaneously to match the output power at the time of load change because the reference is generated from the output power calculated the output-voltage and -current. Therefore, the input power in the proposed method is controlled to the an appropriate value at faster response, and the buffering energy of the DC-link capacitors is small at the time of load change. Thus, the proposed method can reduce the voltage variation at load change even when using a smaller DC-link capacitance when compared with the conventional control method.

IV. EXPERIMENTAL RESULTS

A. Design Method of the DC-Link Capacitances

As discussed in section II-B, the DC-link capacitance is determined by the output power, the utility line angular frequency and the maximum- and minimum-DC-link voltage. The allowed maximum voltage V_{dcmax} is limited by the voltage rating of power devices and DC-capacitors. The allowed minimum voltage V_{dcmin} is limited by the peak value of the input voltage because the PFC converter operates under the condition of higher DC-link

TABLE II.   THE SIMULATION PARAMETERS

Input voltage	200 V
Utility line frequency	50 Hz
Output voltage	300 V
Output power	3.3 kW
PFC boost inductance	500 uH
DC-link capacitance	220 uF

voltage than the rectified input voltage. In this experiment, V_{dcmax} is determined to be 500 V when using switching devices of the 600 V class, and V_{dcmin} is determined to be 400 V because of the margin from the maximum peak of the input voltage, as summarized in Table I. For these parameters, a DC-link capacitance of 220 uF is required from (6). Therefore, small-capacitance capacitors such as film capacitors can be used instead of the aluminum electrolytic capacitors. Under these conditions, the DC-link ripple voltage is from 400 V to 500 V, and its average value is 452 V.

B. Simulation Results

The operation of the proposed control is demonstrated by the simulation. Table. II shows the simulation parameters. Fig. 10 shows the simulated waveforms with the DC-link capacitors of 1000 uF or 220 uF by the conventional control, respectively. Fig. 11 shows the simulated waveform with that of 220 uF by the proposed control. As can be observed from Fig. 10(b), the input current waveform with 220 uF is the distortion waveform, and its power factor is 0.87 for the conventional control. In contrast, as can be observed from Fig. 11, the input current can be controlled to be a sinusoidal waveform without distortion in the condition of large DC-link Voltage ripple, and its power factor is 0.99 for the proposed control. Fig. 12 shows the effects on the third-harmonic current when changing the DC-link capacitance. In the case of decreasing the DC-link capacitance, the third-harmonic current is increased, and it is close to the regulated limits of the harmonic current (IEC61000-3-2) for the conventional control. In contrast, the third-harmonic current is almost unchanged, and it is much lower than the regulated limits for the proposed control.

Fig. 13 shows the simulated transient responses at the load change by the conventional control and the proposed control. The output load is changed from 3 kW to 1.5 kW. As can be observed from Fig. 13, the input current is controlled at a faster response by the proposed control in comparison with the conventional control. Therefore, the surge voltage of the DC-link is greter than 100 V in the conventional control, whereas it is approximately 10 V or less in the proposed control.

C. Experimental Results

In order to verify the proposed control strategy, a prototype converter of 1 kW has been tested. The experimental parameters are as follows. The input AC voltage is set to 100 V, and the output power is 1 kW. Moreover, the ripple of the DC-link voltage is set to 100 V as same

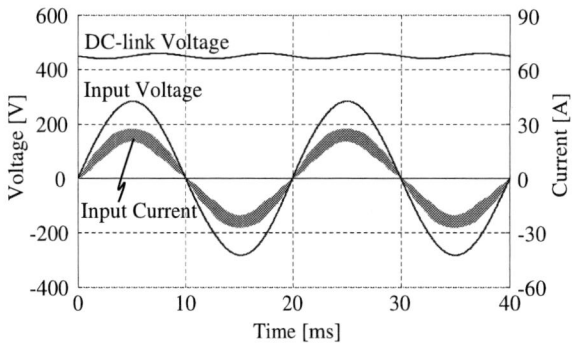

(a) With the DC-link capacitors of 1000 uF

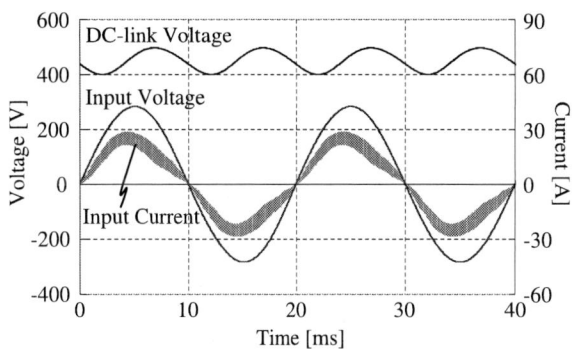

(b) With the DC-link capacitors of 220 uF

Fig. 10.   Simulation waveforms by the conventional control

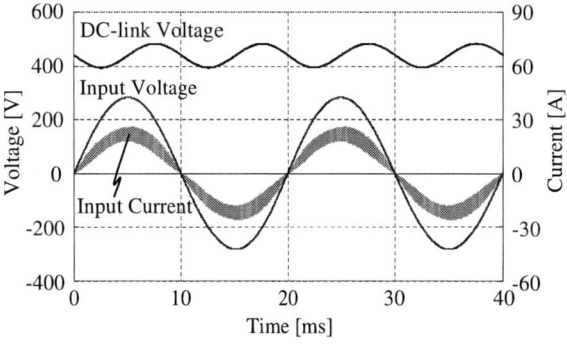

Fig. 11.   Simulation waveforms by the proposed control with the DC-link capacitors of 220 uF

as the simulation, and the DC-link voltage ripple is set to between 270 V and 370 V. The capacitance of the DC-link capacitors is set to 120 uF, and the other parameters are the same as those in simulation conditions.

Fig. 14 shows the experimental waveforms of the current and voltage by the proposed control. The input current is a sinusoidal waveform without distortion, and its power factor can be controlled 0.99. Fig. 15 shows the spectrum of the input current harmonic. As a result, the input current harmonic is much lower than the regulated limits of the harmonic current (IEC61000-3-2), and the

978-1-4799-2706-7/14 $31.00 © 2014 IEEE   1383

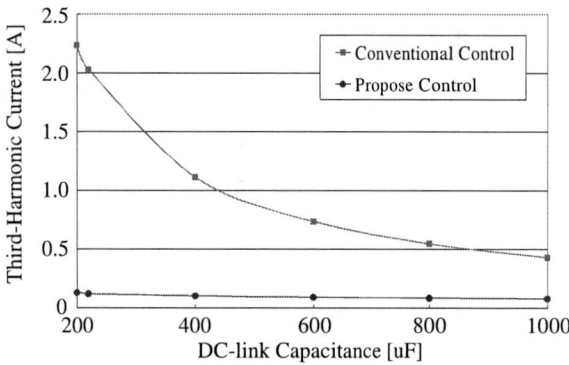

Fig. 12. Third-harmonic current by the conventional control and the proposed control

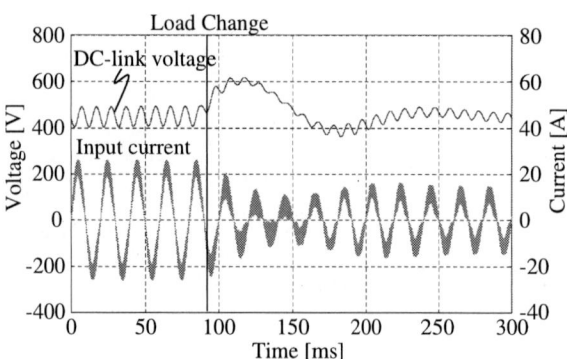

(a) conventional control

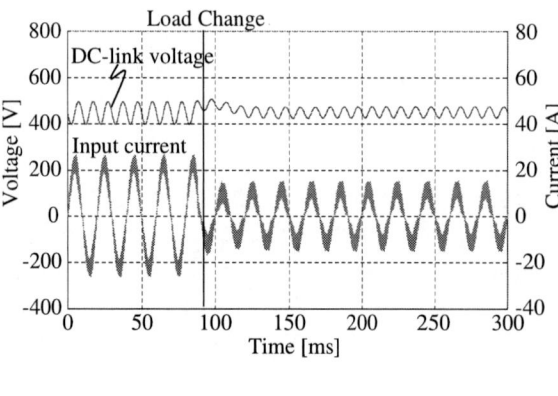

(b) propose control

Fig. 13. Simulation waveforms in the transient response

total harmonic distortion (THD) of the input current can be controlled to 3.1%.

V. CONCLUSION

This paper described an AC/DC control strategy for an on-board charger based on the instantaneous power balance for reducing DC-link capacitance. The proposed control strategy uses the output power of the DC/DC converter instead of the DC-link voltage in order to avoid

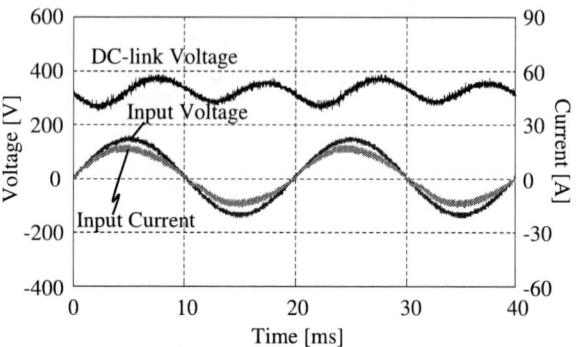

Fig. 14. Experimental waveforms by the proposed control

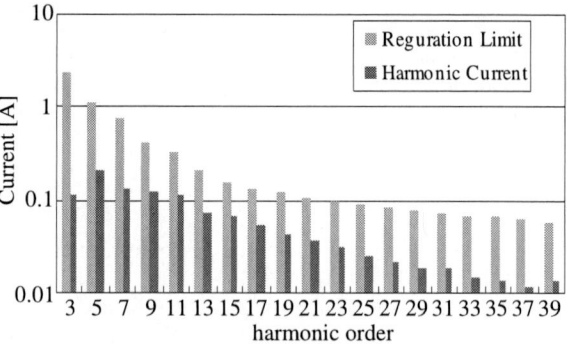

Fig. 15. The harmonic spectrum of the input current in Fig14

the influence to the input current distortion by twice the line frequency ripple of the DC-link voltage. Therefore, it is possible to achieve low input current distortion in the case of a large DC-link voltage ripple owing to reduction in the capacitance. The validity of the proposed strategy is confirmed by the simulation and experimental results. The third-harmonic current is improved to one-eighth as compared with the conventional control when using a DC-link capacitor of only 220 uF, which is approximately one-fifth of DC-link capacitance when compared with conventional on-board chargers. Thus, it is possible to use film capacitors of small capacitance instead of aluminum electrolytic capacitors and to realize an on-board charger with a long lifetime.

In this work, the reduction of the DC-link capacitance is limited four-fifth by the voltage rating of the power devices and the allowed ripple current of the capacitors. However, the proposed control also operates with one-tenth of the capacitance by using ideal devices and increasing the DC-link voltage ripple larger in the simulations. In the future, as improvements in the capacitors and power devices progress, the proposed control will be effective in reducing the DC-link capacitance.

REFERENCES

[1] K. Yamamoto, "The Background of Electric Vehcle Spread", *International Electric Vehicle Technology Conference (EVTeC)*, 2011.

[2] D. Gautam, F. Musavi, M. Edington, W. Eberle, W. Dunford, "Design of on-board charger for plug-in hybrid electric vehicle", *in*

Proc. International Conference on Machines and Drives (PEMD), pp. 1-6, 2010.

[3] M. Grenier, M.G. Hosseini Aghdam, T.Thiringer, "An Automotive Onboard 3.3-kW Battery Charger for PHEV Application", *IEEE Trans. on Vehicular Technology*, vol. 61, no. 11 pp. 1-6, 2012.

[4] H.J. Chae, W.Y. Kim, S.Y.Yun, Y.S. Jeong, J.Y. Lee, H.T.Moon , "3.3 kW on board charger for electric vehicle", *in Proc. International Conference on Power Electronics and ECCE Asia (ICPE & ECCE)*, pp. 2717-2719, 2011.

[5] J. Sebastian, D.G. Lamar, M.M. Hernando, A. Rodriguez-Alonso, A. Fernandez, "Steady-State Analysis and Modeling of Power Factor Correctors With Appreciable Voltage Ripple in the Output-Voltage Feedback Loop to Achieve Fast Transient Response", *IEEE Trans. on Power Electronics*, vol. 24, no. 11, pp. 2555-2566, 2009.

[6] T. Takeshita, Y. Toyoda, N. Matsui, "Harmonic Suppression and DC Voltage Control of Single-phase PFC Converter", *in Proc. Power Electronics Specialists Conference (PESC)*, vol. 2, no. 1, pp. 571-576, 2000.

[7] D.G. Lamar, A. Fernandez, M. Arias, M. Rodriguez, J. Sebastian, M.M. Hernando, "A Unity Power Factor Correction Preregulator With Fast Dynamic Response Based on a Low-Cost Microcontroller", *in Proc. Applied Power Electronics Conference (APEC)*, pp. 186-192, 2007.

[8] A. Pandey, B. Singh, D.P. Kothari, "A simple fast voltage controller for single-phase PFC converters", *in Proc. Conference of the Industrial Electronics Society (IECON)*, vol.2, pp. 1235-1237, 2002.

[9] Y. Ohnuma, J. Itoh, "A Novel Single-phase Buck PFC AC-DC Converter using an Active Buffer", *in Proc. Energy Conversion Congress and Exposition (ECCE)*, pp. 4223-4229, 2012.

[10] H. Wang, H. Chung, "Study of a new technique to reduce the dc-link capacitor in a power electronic system by using a series voltage compensator", *in Proc. Energy Conversion Congress and Exposition (ECCE)*, pp. 4051-4057, 2011.

[11] S. Li, B. Ozpineci, L. M. Tolbert, "Evaluation of a current source active power filter to reduce the DC bus capacitor in a hybrid electric vehicle traction drive", *in Proc. Energy Conversion Congress and Exposition (ECCE)*, pp. 1185-1190, 2009.

[12] S. Li, B. Ozpineci, L. M. Tolbert, "A Novel Concept to Reduce the DC-Link Capacitor in PFC Front-End Power Conversion Systems", *in Proc. Applied Power Electronics Conference and Exposition (APEC)*, pp. 1192-1197, 2012.

[13] J. Chen, A. Prodic, R.W Erickson, D. Maksimovic, "Predictive digital current programmed control", *IEEE Trans. on Power Electronics*, vol. 18, no. 1, pp. 411-419, 2003.

[14] M. Fu, Chen Qing, "A DSP based controller for power factor correction (PFC) in a rectifier circuit", *in Proc. Applied Power Electronics Conference and Exposition*, vol. 1, pp. 144-149, 2001.

The 2014 International Power Electronics Conference

Modular Converter Architecture for Medium Voltage Ultra Fast EV Charging Stations: Dual Half-Bridge-based Isolation Stage

Michail Vasiladiotis[*], Behrooz Bahrani[†], Niklaus Burger[*], and Alfred Rufer[*]
[*]Laboratory of Industrial Electronics (LEI)
École Polytechnique Fédérale de Lausanne (EPFL), Station 11, 1015 Lausanne, Switzerland
michail.vasiladiotis@epfl.ch, niklaus.burger@epfl.ch, alfred.rufer@epfl.ch
[†]School of Electrical and Computer Engineering
Georgia Institute of Technology, Atlanta, GA 30332-0250, USA
behrooz.bahrani@ece.gatech.edu

Abstract—The concept of ultra fast charging of Electric Vehicles (EVs) is expected to contribute significantly to their widespread utilization in long-distance travels. Towards this direction, emerging power electronic topologies will be the key components of the developed charging infrastructures. Lately, a novel converter architecture has been proposed for such a purpose. The latter is based on a Cascaded H-Bridge (CHB) converter with integrated battery energy storage systems, which play the role of a buffer in such a power demanding application. This paper is focusing on the isolated DC/DC conversion stage, which is needed for the achievement of the high charging currents. The dual half-bridge (DHB) converter is chosen, which fulfills several requirements for the studied application. The soft switching operation of the converter under wide input-output voltage variations is discussed. Moreover, a current controller is designed, capable of dealing with the topology-specific challenges. Finally, the development of a down-scaled laboratory prototype is described and experimental results are provided.

Index Terms—DC ultra fast charging, cascaded H-bridge (CHB) converter, battery energy storage system (BESS), isolated DC/DC converter, dual half-bridge (DHB), current control, convex optimization.

I. INTRODUCTION

Over the recent years, a high interest has been given on the development of electric vehicles and their competitive launching into the market over conventional ones based on combustion engines. In addition, the high expectations regarding development of batteries that can withstand very high charging rates [1,2] has pushed industry and academia towards the exploration of their ultra fast charging concept, with the main focus laid on the infrastructure and grid impact [3,4].

In [5], a novel converter architecture has been proposed for the implementation of medium voltage ultra fast EV charging stations, which is shown in Fig. 1. The costly and bulky low frequency transformer is avoided through the utilization of a Cascaded H-Bridge multilevel converter. The latter also offers the possibility to significantly reduce the filtering components on the grid side, since the injected currents exhibit very low harmonic content. On the level of each submodule, battery elements are placed in a split manner. The interface between

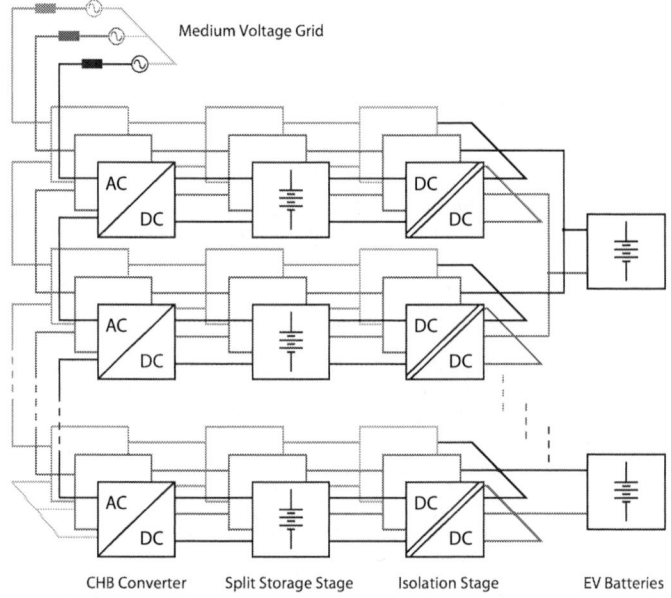

Fig. 1. Proposed detailed converter architecture for ultra fast charging of electric vehicles, based on Cascaded H-Bridge Converter with integrated energy storage and isolated DC/DC converters.

the submodule capacitors and the batteries can be performed through a passive (high-order filter) or an active (power electronics) way, in order to prevent low frequency current ripples from flowing into the latter [5]. The split storage stage plays the role of a power buffer, aiming at the reduction of the instantaneous active power which is extracted from the grid.

Finally and in order to achieve the high output currents as well as meet the galvanic isolation standards, medium frequency transformer-based DC/DC converters are connected in parallel. Since a high number of submodules are needed to block the medium voltage on the grid side, different configurations can be chosen on the parallel-connection level in order to achieve a multi-port output, capable of charging different vehicles simultaneously. This leads to a charging

978-1-4799-2706-7/14 $31.00 © 2014 IEEE

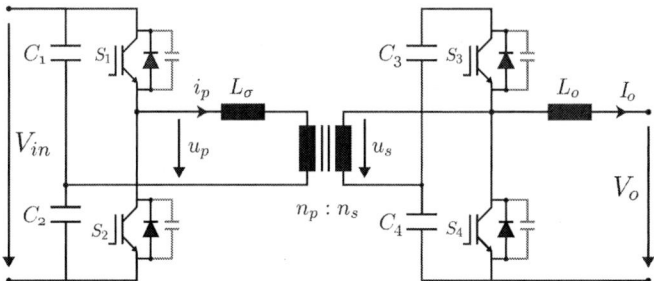

Fig. 2. The dual half-bridge (DHB) topology with current-fed output bridge.

station implementation utilizing a single converter structure.

The control of the Cascaded H-Bridge converter with integrated BESS has been investigated in [6]. It has been shown that an individual power control can be achieved between the submodules of a phase as well as across the three phases, without disturbing the grid symmetry. This is essential in such an application, where each output port will demand a different amount of power to be handled.

This paper deals with the implementation of the isolated DC/DC converter stage, in terms of suitable topology choice and control, and is organized as follows: Section II analyzes the dual half-bridge (DHB) topology, which is employed among a big number of available solutions since it offers several attractive features for such an application. Section III describes the design of an output current control system, which is based on convex optimization, for two cases: high and low output resistance. The parallel connection of several modules is also discussed. Section IV describes the development of a down-scaled laboratory prototype and presents experimental results, leading to Section V, which concludes the work.

II. DUAL HALF-BRIDGE CONVERTER ANALYSIS

The basic requirements that should be fulfilled by the power interface between the CHB converter and the EV batteries are the following: (a) very low charging current ripple, (b) current and voltage control capability, as well as (c) galvanic isolation. Soft switching over a wide operating range is also favorable, since it implies the reduction of passive components through an increase in the switching frequency without affecting significantly the converter efficiency.

In [5], the straightforward solution of a full-bridge phase-shift controlled PWM converter was discussed. However, several drawbacks of such a topology were highlighted, such as unequal thermal stress of the two primary converter legs, as well as incapability of bidirectional power flow without additional external circuitry.

The dual half-bridge (DHB) converter presented in [7,8] exhibits significant advantages and features all aforementioned requirements. In addition, it offers the ability of transferring power from the EV battery to the grid, i.e., V2G operation. This topology is illustrated in Fig. 2. By phase-shifting the transformer secondary voltage in regard to the primary, a specific amount of active power can be transferred between the two transformer sides.

A. ZVS Conditions for Input-Output Voltage Variations

The converter offers natural zero voltage switching (ZVS) transitions for the turn-on of all devices. In order to achieve the same during the turn-off instants as well, snubber capacitors can be added to the circuit. Works such as [9,10] have shown that the duty cycle can be also used as a degree of freedom for extending the ZVS region as well as minimizing the transformer rms current value.

The ZVS conditions can be ensured for any loading conditions when the ratio of $V_o'/(DV_{in})$ is almost unity, where V_o' is the primary-referred output voltage and D the duty cycle of the two half-bridges. In the studied system, both V_{in} and V_o are experiencing significant variations. At the input, V_{in} is buffering a low-order harmonic oscillation of double the grid frequency, which is the result of the intrinsic single-phase submodule nature. At the output, V_o is also varying according to the battery State of Charge (SoC) as well as the EV battery type. In practical cases, the duty cycle change is limited within a specific range, since it implies a different charge level of the four split capacitors C_{1-4}.

The authors of [11] are modeling the second-order harmonic frequency of V_{in} and maintain ZVS in a wide operating range, but the investigated case regards fixed duty cycle and V_o. In the UFCEV system, a duty cycle control is preferable due to the varying V_o and therefore an extension of the analysis is needed. For the extraction of the soft switching conditions, the knowledge of the transformer and output inductor currents i_p and I_o at the switching instants is needed. Such equations can be found in literature for boost converter mode [10,11], but are also derived and given here following the buck-mode conventions of Fig. 2 and for the sake of completeness.

Defining an allowable ripple factor of the input voltage V_{in} as k_u, the four capacitor voltages V_{1-4} can be expressed as

$$V_1 = (1-D)k_u V_{in}, \ V_2 = Dk_u V_{in},$$
$$V_3 = \frac{1-D}{D}V_o, \ V_4 = V_o \tag{1}$$

Referring to the primary, the transformer currents at the four consecutive switching instants are given by

$$i_1 = K\left\{k_u\pi D\left(D-1\right) + \frac{V_o'}{V_{in}}\pi\left(1-D-\phi/\pi\right)\right\}$$
$$i_2 = K\left\{k_u\left(D-1\right)\left(\pi D-\phi\right) + \frac{V_o'}{V_{in}}\pi\left(1-D\right)\right\}$$
$$i_3 = K\left\{k_u\pi D\left(1-D\right) + \frac{V_o'}{V_{in}}\left(D-1\right)\left(\pi-\phi/D\right)\right\}$$
$$i_4 = K\left\{k_u\pi D\left(1-D-\phi/\pi\right) + \frac{V_o'}{V_{in}}\pi\left(1-D\right)\right\} \tag{2}$$

where $K = V_{in}/(\omega_s L_\sigma)$, $V_o' = nV_o$, $n = n_p/n_s$, and ω_s denotes the angular switching frequency.

For $\phi < \min\{2D\pi, 2(1-D)\pi\}$, the average output current is given by

$$\overline{I_o} = \frac{P_o}{V_o} = \frac{k_u V_{in} n\phi}{4\pi\omega_s L_\sigma}\left[4\pi\left(1-D\right) - \frac{\phi}{D}\right] \tag{3}$$

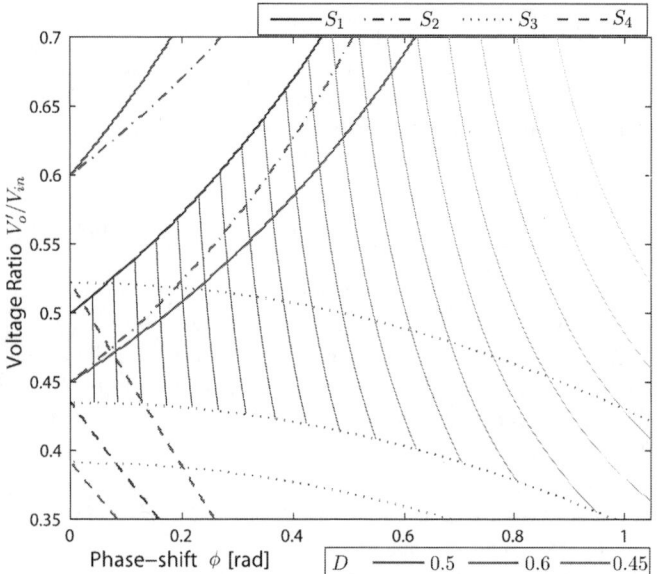

Fig. 3. ZVS regions of the DHB converter in regards with the output/input voltage ratio. The effect of duty cycle change is visible. The contour lines are representing constant power in the case of $D = 0.5$. Output inductance value is set to $L_o = 10\ \mu H$.

By defining the peak-to-peak inductor current ripple as

$$\Delta I_o = \frac{2\pi\left(1 - D\right)V_o}{\omega_s L_o} \qquad (4)$$

the following soft switching conditions are derived for the respective switches turn-on:

$$S_1 : i_1 < 0,\ S_3 : i_2 > I_o^{'min},\ S_2 : i_3 > 0,\ S_4 : I_o^{'max} > i_4 \qquad (5)$$

where $I_o^{'min} = \left(\overline{I_o} - \Delta I_o/2\right)/n$, $I_o^{'max} = \left(\overline{I_o} + \Delta I_o/2\right)/n$.

Since both V_{in} and V_o are varying, it is beneficial to present a graph of the phase-shift ϕ in function of the voltage ratio V_o'/V_{in}. The investigated system parameters are given in Table I and regard a down-scaled laboratory prototype, which aims at testing of the basic control and modulation functions of the proposed ultra fast charger. Fig. 3 shows that an acceptable change in the duty cycle can cover practically all load conditions, providing ZVS for all switches. The latter can be achieved by detecting the converter operating point in real-time and choosing the respective duty cycle value.

TABLE I
DOWN-SCALED PROTOTYPE SYSTEM PARAMETERS

Quantity	Value	Comment
V_{in}	50±10% V	Input voltage (submodule)
V_o	19-31 V	Output voltage (EV battery)
n_p/n_s	1	Transformer ratio
C_{1-4}	2.7 mF	Capacitance
L_σ	1.5 μH	Transformer leakage inductance
L_o	15 μH	Filter inductance
R_o	0.1 Ω	Filter/battery resistance
f_s	50 kHz	Switching frequency
ϕ_n	$\pi/6$ rad	Nominal phase-shift angle

As stated in [10], the output current ripple provides the ZVS conditions for the secondary switches in light load operation. Therefore, there is a trade-off in the inductance value choice between switching and conduction losses. Certain loads, such as EV batteries, are not to be fed with high ripples. In the studied UFCEV system this can be achieved by interleaving the converter outputs, while maintaining the ZVS conditions for each module. Otherwise, additional passive filtering elements are needed to achieve load current ripple suppression [11].

III. CURRENT CONTROL OF A DUAL HALF-BRIDGE

The ultra fast battery charging is typically limited by its internal electrochemical processes and can be therefore executed up to a point of about 80% of its State of Charge (SoC). During this period, a constant current has to be provided to the battery. This implies the need for an accurate control method derivation at the output of the parallel-connected dual half-bridges.

This paper proposes a discrete-time control method, whose design is based on convex optimization. The basic idea is to minimize the absolute error between the open-loop system transfer function and a desired one through the use of nonparametric system models [13,14]. The optimization problem is also subject to appropriate constraints, in order to ensure the closed-loop system stability and desired dynamic performance. Throughout this section, the design process for the studied system is described and analyzed.

A. Nonparametric Model Derivation

The dual half-bridge converter exhibits non-linearities and dynamics of high order, due to the existence of several passive elements. The required nonparametric model of the SISO system, considering the phase-shift ϕ as input and the current I_o as output, can be derived from a duty-cycle dependent analytical average system model, such as by modifying the one given by [11,12]. Since the output current switching-related ripple is high, a low-pass filter is required on a control algorithm level. Thus, the output is defined finally as I_o^f, in order to account for the latter as well. By linearizing it around several operating points, a family of desired models can be derived. The analytical model of the studied system is given in the Appendix.

Alternatively, the nonparametric model of the system can be found by using an identification process. In order to do so, the system is excited by means of a specific signal, e.g., pseudo random binary sequence (PRBS) and the outputs are observed [15]. The system frequency response for a i-th operating point is then identified as

$$G_i(j\omega) = \frac{\mathscr{F}(I_{o,i}^f)}{\mathscr{F}(\phi_i)} \qquad (6)$$

where $\mathscr{F}$ represents the Fourier transform. The same procedure can be repeated for m operating points leading to a family of nonparametric models

$$\mathscr{G} = \{G_i(j\omega); i = 1, ..., m; \omega \in \mathbb{R}\} \qquad (7)$$

The 2014 International Power Electronics Conference

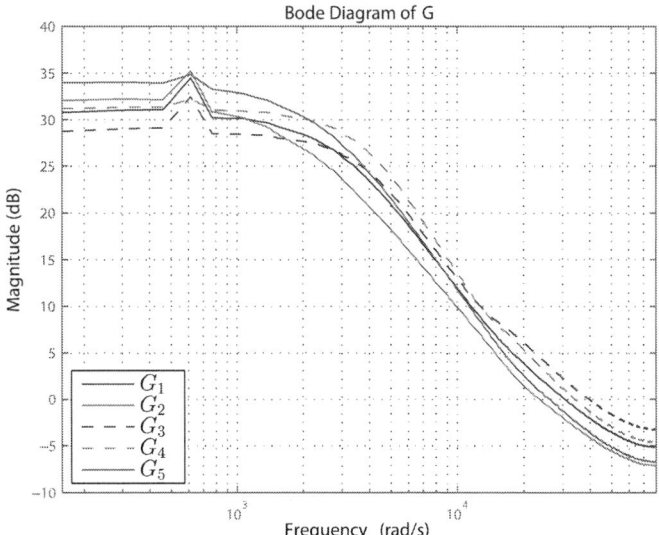

Fig. 4.   Identified nonparametric system models for $R_o = 0.1\ \Omega$.

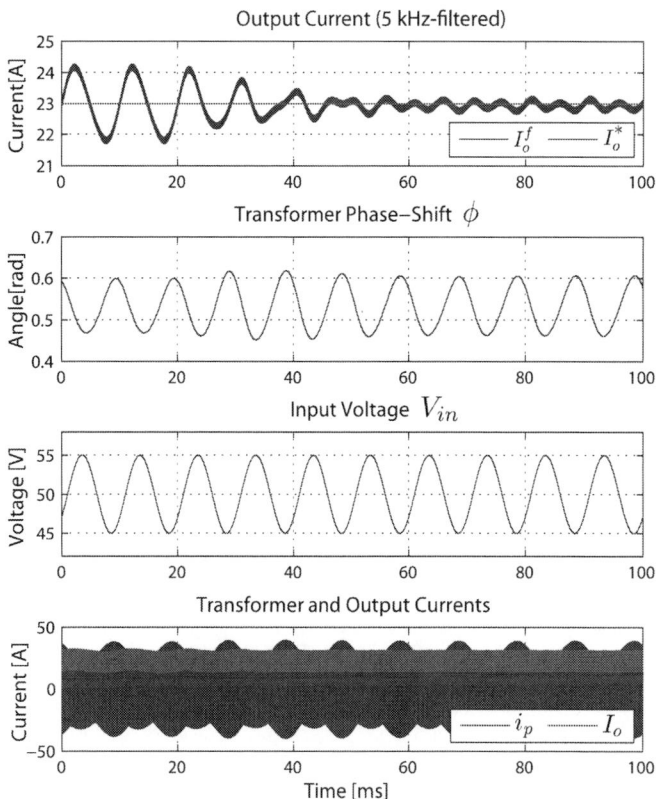

Fig. 5.   Simulation results of the resonant controller activation.

The aforementioned concept has been followed for this work. The down-scaled system, whose parameters are given in Table I, has been identified by means of simulations for five operating points: 1) $D = 0.5$, $V_o = 25$, $\phi = \pi/6$ (G_1), 2) $D = 0.4$, $V_o = 19$, $\phi = \pi/6$ (G_2), 3) $D = 0.6$, $V_o = 31$, $\phi = \pi/6$ (G_3), 4) $D = 0.6$, $V_o = 28$, $\phi = \pi/16$ (G_4), 5) $D = 0.45$, $V_o = 22$, $\phi = \pi/16$ (G_5).

Fig. 4 shows the bode diagrams for the nonparametric models of the studied system. The behavior of the latter does not change significantly in regard to operating conditions. A resonance exists at double the grid frequency, which is a result of the single-phase power pulsation. This sets an additional requirement for the control system design, which regards the decoupling of the second-order input current harmonic (submodule) from the output (EV battery).

B. Controller Class

In order to form the open-loop transfer function of the system to be controlled, the class of the controller needs to be determined. From the curves in Fig. 4 it can be judged that a PI-controller is sufficient for controlling the system with a very good dynamic performance. However, the 100-Hz component also has to be compensated for. The latter can be achieved by adding a resonant term to the controller, tuned at this desired frequency. The overall controller transfer function in z-domain is obtained as [17]

$$K(z, \rho) = \frac{\rho_1 + \rho_2 z^{-1}}{1 - z^{-1}} + \rho_3 \frac{b_1 z^{-1} + b_2 z^{-2}}{1 + a_1 z^{-1} + a_2 z^{-2}} \quad (8)$$

The second term represents a discrete-time resonant controller, where ω_h is the desired frequency and $\zeta = 3/\omega_h$ the damping ratio. The coefficients a_{1-2} and b_{1-2} are determined by

$$b_1 = 1 - \alpha\left(\beta + \frac{\zeta\omega_h}{\omega_b}\eta\right),\ b_2 = \alpha^2 + \alpha\left(\frac{\zeta\omega_h}{\omega_b}\eta - \beta\right)$$
$$\alpha_1 = -2\alpha\beta,\ \alpha_2 = \alpha^2 \quad (9)$$

where $\omega_b = \omega_h\sqrt{1 - \zeta^2}$ for $\zeta < 1$, $\alpha = e^{-\zeta\omega_h T_s}$, $\beta = \cos(\omega_b T_s)$, $\eta = \sin(\omega_b T_s)$ and T_s denotes the controller sampling time, which is set to 100 μs.

The open-loop system transfer function of the i-th operating point therefore becomes $L_i(j\omega, \rho) = K(j\omega, \rho)G_i(j\omega)$, leading to a respective family of open-loop functions

$$\mathscr{L} = \{L_i(j\omega); i = 1, ..., m; \omega \in \mathbb{R}\} \quad (10)$$

C. Optimization-based Loop Shaping

The loop shaping of (10) is performed through an optimization procedure. In order to achieve this, the following cost function should be minimized [13,14],

$$\min_\rho \sum_{i=1}^m \|L_i(\rho) - L_D\|^2 \quad (11)$$

which is the square second norm of the errors between the open-loop transfer function of the system $L_i(j\omega, \rho)$ with a desired one L_D for all systems $i = 1,...,m$. The latter is chosen according to the system requirements, i.e., as the addition of an integral and a resonant terms:

$$L_D(s) = \frac{\omega_{c1}}{s} + \frac{\omega_{c2}}{s^2 + 2\zeta\omega_h s + \omega_h^2} \quad (12)$$

For this paper the values have been set to $\omega_{c1} = 1e3$ and $\omega_{c2} = 40e3$ rad/s, respectively.

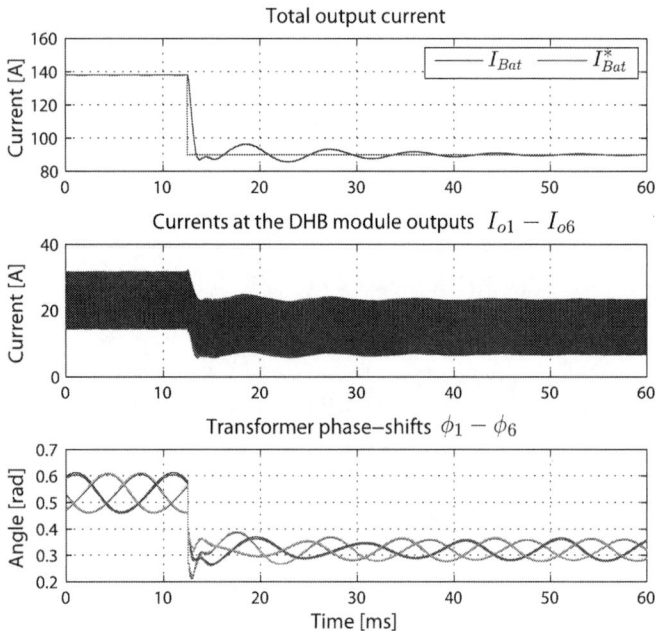

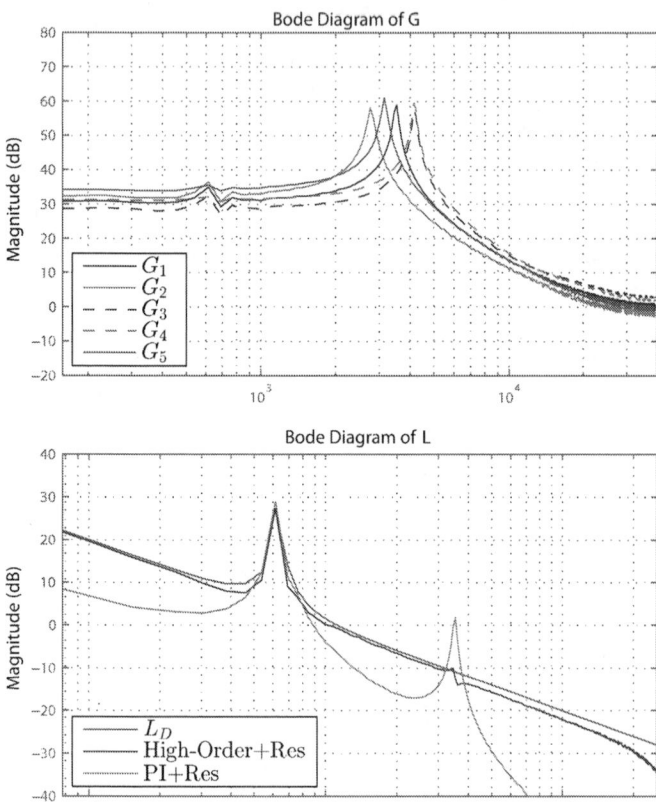

Fig. 6. Simulation results of the six parallel-connected closed-loop controlled DHB modules during a current reference step change.

Fig. 7. Identified nonparametric system models for the case of low output resistance $R_o = 1$ mΩ (upper figure), and the respective results of the loop shaping optimization procedure (lower figure).

The optimization problem is subject to several constraints, which aim at ensuring the stability as well as dynamic performance of the controller [13,14]. The resulting problem is a semi-infinite problem (SIP) including infinite number of constraints and finite number of states. In order to simplify it, only a finite number of frequency points are taken, e.g. in the interval $[0\ \omega_{max}]$. This transforms the problem to a semi-definite problem (SDP), which can be solved utilizing standard respective solvers.

By choosing N linearly-spaced frequencies within the range of $[0\ \omega_{max}] \in \mathbb{R}$, an approximation of the quadratic objective function can be carried out as [13,14]

$$\sum_{i=1}^{m} \| L_i(\rho) - L_D \|^2 \approx \sum_{i=1}^{m} \sum_{k=1}^{N} \| L_i(j\omega_k, \rho) - L_D(j\omega_k) \|_F \tag{13}$$

where $\|.\|_F$ denotes the Frobenius norm. Therefore, the following optimization problem is finally deduced:

$$\min_{\rho} \sum_{i=1}^{m} \sum_{k=1}^{N} \| L_i(j\omega_k, \rho) - L_D(j\omega_k) \|_F \tag{14}$$

subject to the respectively modified discrete linear constraints.

In the present study, the aforementioned technique has been applied only for the case of the identified model G_1 of the rated system. The results of the numerical procedure are: $\rho_1 = 0.0132$, $\rho_2 = -0.0106$ and $\rho_3 = 0.0217$.

For more information on the formulation of constraints as well as several applications of this control method on power electronics systems, the reader is referred to works such as [13,14,16-18].

D. Control Design Validation and DHB Parallel Connection

The effectiveness of the power decoupling for the input and output currents is demonstrated by the simulation results of Fig. 5. Initially, the system is controlled only by means of the PI controller. The output current features a significant 100-Hz component, which should not be absorbed by the EV battery. At $t = 20$ ms, the resonant term is activated and almost eliminates the output current oscillation by changing the amplitude and phase of the respective harmonic in the transformer phase-shift ϕ.

The Cascaded H-Bridge active front end stage gives a significant degree of freedom for choosing the number of submodules feeding each charging port. Except from the resonant controllers, which compensate for the second harmonic locally on the level of each DHB, a parallel connection between different phases is at hand which permits a further reduction. An additional advantage of interconnecting isolated modules from the three-phases comes from the fact that the symmetry of the three-phase currents can be maintained during the EV ultra fast charging without the need for common-mode voltage injection [6].

A simulation model consisting of six parallel-connected dual half-bridge modules has been also built and tested. The connection takes place as two submodules/phase, as for the

The 2014 International Power Electronics Conference

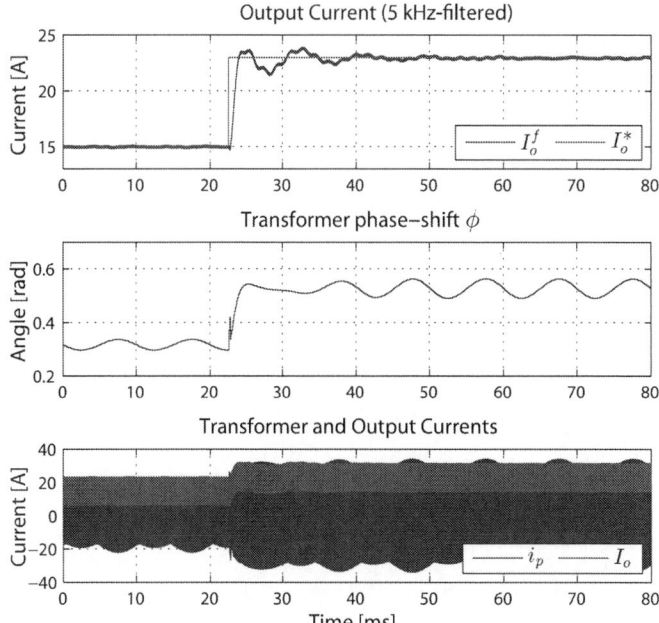

Fig. 8. Simulation results of the low output resistance closed-loop controlled system during a current reference step change.

Fig. 9. The experimental setup.

upper charging port illustrated in Fig.1. By interleaving the outputs, a number of six converters with a duty cycle of $D = 0.5$ lead to a perfect switching ripple compensation. The simulation results for a reference step change are illustrated in Fig. 6. The EV battery current reference I_{Bat}^* is equally split between the individually current-controlled DHBs. The controller presents a good dynamic performance. The worst-case scenario has been simulated in terms of input voltage ripple, i.e., that even after the transient the grid power is kept constant and the excess is charging the stationary submodule batteries. The second harmonic resonance is slightly excited during the transient but it is compensated by the resonant term. Higher resonant control gains might damp the resonance more rapidly, but are not preferred since the latter could lead to implementation issues and the control speed is not the main issue in such an application.

E. Current Control in Low Output Resistance Case

So far, the system has been assumed to have a resistance of $R_o = 0.1\ \Omega$ at its output, considering both the battery and inner inductance resistances. In a high current application, however, the output resistance is expected to be minimized both for efficiency purposes and because of the parallel-connection of EV battery strings which achieves higher capacities. In this paragraph, the analysis is performed once again for an operating scenario of $R_o = 100$ mΩ. The identification procedure is repeated for the same system operating points G_{1-6}. From Fig. 7 it can be observed that a significant undesired resonance exists due to the high-order of the system and the interaction of its passive elements. This is further complicated by the fact that the resonance is depending on the duty cycle value and

is characterized by the following relation:

$$\omega_{res} = D\sqrt{\frac{2}{L_o C}} \qquad (15)$$

This poses difficulties in the control system design. A conventional PI controller can only compensate for the dominant plant pole and is not capable of adequately attenuating the resonance. The solution to the latter comes through the design of a high-order controller. The transfer function of the 3rd-order controller augmented with a resonant term is given in z-domain by

$$K(z, \rho) = \frac{\rho_1 + \rho_2 z^{-1} + \rho_3 z^{-2} + \rho_4 z^{-3}}{1 - z^{-1}}$$
$$+ \rho_5 \frac{b_1 z^{-1} + b_2 z^{-2}}{1 + a_1 z^{-1} + a_2 z^{-2}} \qquad (16)$$

The results of the optimization-based loop shaping of the rated system G_1 are also shown in Fig. 7. It is clear that the PI controller is not capable of following the reference curve L_D and attenuating the resonance, whereas the 3rd-order controller can provide the desired transfer function. The results of the optimization procedure give: $\rho_1 = 0.0092$, $\rho_2 = -0.0033$, $\rho_3 = -0.0162$, $\rho_4 = 0.0131$, and $\rho_5 = 0.0187$.

A current reference step change has been simulated and the results are depicted in Fig. 8. The control system provides a good dynamic performance. The undesired system resonance is visible during the transient as a superimposed component on the second-order harmonic, which is also slightly excited without having a significant impact and is quickly compensated by the resonant term.

The robust controller design for all identified operating-point dependent models G_{1-6} of Fig. 7 will be the task of future work.

IV. EXPERIMENTAL TESTS

This section describes the development of the down-scaled laboratory prototype and presents several preliminary experimental tests. The system parameters are the ones given

The 2014 International Power Electronics Conference

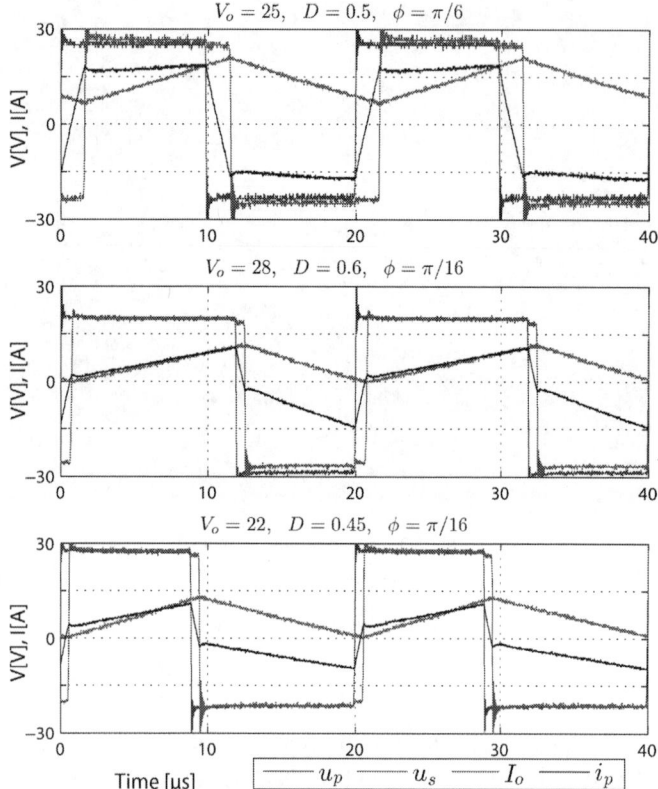

Fig. 10. Experimental results for one DHB module in several operating points of different output voltage V_o, duty cycle D and phase-shift ϕ.

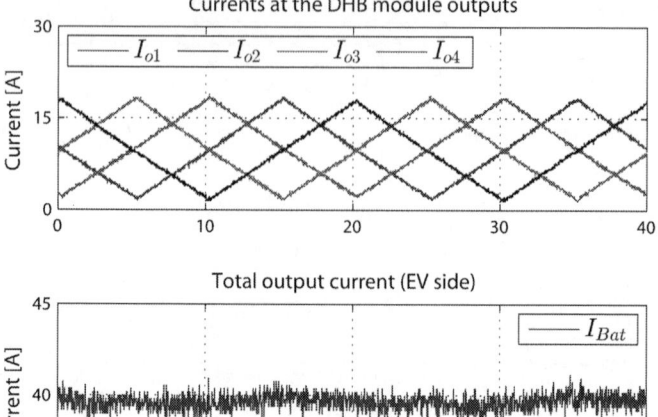

Fig. 11. Experimental results for the parallel connection of the four dual half-bridge modules.

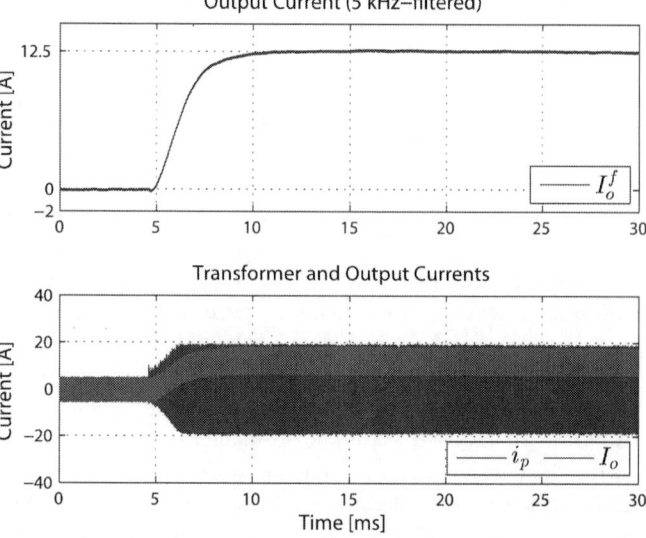

Fig. 12. Experimental results for an output current reference step change from 0 to 12.5 A.

in Table I. A number of four dual half-bridges have been built. The transformers are constructed using ETD49 cores by EPCOS and the windings consist of Litz wire. The achieved leakage inductances $L_{\sigma1-4}$ vary between 1.5 and 1.9 μH. One PCB per module has been designed and the whole system is integrated in a standard rack. A backplane is used, carrying all gate signals, measurements and power supplies. The voltage and current outputs of each module are sensed for control as well as security purposes. The converters are controlled by a modular customized system featuring a master DSP-based board and an FPGA unit. The design of the latter resembles the system described in [19]. The implemented laboratory setup is illustrated in Fig. 9.

A. Experimental Results

Several experimental tests have been carried out, without considering the second-order harmonic of the input current. The EV battery is emulated by a variable voltage source. Since the latter cannot absorb any power, a variable resistance is placed in parallel.

Fig. 10 shows the basic circuit waveforms for one module under different operating point conditions. More specifically, the two phase-shifted transformer voltages u_p and u_s are depicted along with the transformer (i_p) and output (I_o) currents. The input voltage V_{in} is provided by a fixed voltage source and is set to 50 V.

In a second step, all four modules are tested in a series-input/output-parallel configuration, i.e., a single voltage source of 200 V feeds their series-connected inputs. The common output is set to $V_o = 25.6$ V, which refers to the battery nominal voltage. The results of this test are shown in Fig. 11. Since the output inductances are low the individual ripples are high, something which allows ZVS conditions at light loads as mentioned before. However and due to the interleaving of four channels with 0.5 duty cycle each, the total battery current I_{Bat} is completely ripple-free without the use of additional passive elements.

Finally, Fig. 12 validates the closed-loop control design for one DHB module through an output current reference step change from 0 to 12.5 A.

978-1-4799-2706-7/14 $31.00 © 2014 IEEE

V. CONCLUSION

This paper has focused on the parallel-connected isolated DC/DC conversion stage of a recently proposed ultra fast EV charging station architecture. The dual half-bridge topology with current-fed output is an attractive choice due to the advantages that it exhibits for the specific application. A current controller has been designed based on convex optimization. The latter is capable of attenuating the inherent second harmonic of the single-phase nature as well as the undesired converter resonance due to the interaction of its passive elements at low output resistance values. Simulation and experimental results from a down-scaled laboratory prototype verify the discussed and proposed concepts.

APPENDIX
DHB AVERAGE LINEARIZED MODEL

The duty cycle-dependent linearized average state space model of the dual half-bridge converter is derived here with the buck-mode conventions of Fig.2. The model of [11,12] is augmented with the filtered output current I_o^f and linearized using small variations. By defining the state vector as $x = \begin{bmatrix} \Delta I_o & \Delta V_{12} & \Delta V_{34} & \Delta I_o^f \end{bmatrix}^T$, where V_{12} and V_{34} refer to the sum of voltages across the capacitors C_{1-2} and C_{3-4} respectively, and the input vector as $u = \begin{bmatrix} \Delta V_o & \Delta \phi & \Delta V_{in} \end{bmatrix}^T$, the linearized dynamics of the converter are characterized by

$$\dot{x} = Ax + Bu$$
$$y = Cx \tag{17}$$

where

$$A = \begin{pmatrix} -\frac{R_o}{L_o} & 0 & \frac{D}{L_o} & 0 \\ 0 & -\frac{2}{CR_{in}} & \frac{\phi[\phi+4\pi D(D-1)]}{H} & 0 \\ -\frac{2D}{C} & -\frac{\phi[\phi+4\pi D(D-1)]}{H} & 0 & 0 \\ \frac{1}{T_f} & 0 & 0 & -\frac{1}{T_f} \end{pmatrix}$$

$$B = \begin{pmatrix} -\frac{1}{L_o} & 0 & 0 \\ 0 & \frac{DV_o[2\phi+4\pi D(D-1)]}{H} & \frac{2}{CR_{in}} \\ 0 & \frac{-V_{in}[2\phi+4\pi D(D-1)]}{H} & 0 \\ 0 & 0 & 0 \end{pmatrix}$$

$$C = \begin{pmatrix} 0 & 0 & 0 & 1 \end{pmatrix}$$

with $H = 2\pi\omega_s CL_\sigma$, R_{in} representing the modeled inner resistance of the input source and T_f being the time constant of the first-order current measurement filter.

The aforementioned state-space model can be numerically converted into a transfer function from $\Delta\phi$ to I_o^f and used for the control design. Drawing the Bode plots with such a model yields similar results with the identification procedures utilized in this paper.

ACKNOWLEDGMENT

The present study has been supported financially by EOS Holding, Switzerland, in the framework of the Ultra Fast Charging of Electric Vehicles (UFCEV) project.

REFERENCES

[1] B. Kang and G. Ceder, "Battery materials for ultrafast charging and discharging," *Nature*, vol. 458, pp. 190-193, 2009.

[2] H. Zhang, X. Yu, and P. V. Braun, "Three-dimensional bicontinuous ultrafast-charge and -discharge bulk battery electrodes," *Nature Nanotechnology*, vol. 6, pp. 277-281, 2011.

[3] D. Aggeler, F. Canales, H. Zelaya - De La Parra, N. Butcher, and O. Apeldoorn, "Ultra-fast DC-charge infrastructures for EV-mobility and future smart grids," *Proc. of the Innovative Smart Grid Technologies Conference Europe (ISGT)*, pp. 1-8, Oct. 11-13, 2010.

[4] K. Yunus, H. Zelaya - De La Parra, and M. Reza, "Distribution grid impact of Plug-In Electric Vehicles charging at fast charging stations using stochastic charging model," *Proc. of the 14th European Conference on Power Electronics and Applications (EPE)*, pp. 1-11, Aug. 30-Sept. 1, 2011.

[5] M. Vasiladiotis, A. Rufer, and A. Béguin, "Modular converter architecture for medium voltage ultra fast EV charging stations: Global system considerations," *Proc. of the IEEE International Electric Vehicle Conference (IEVC)*, pp. 1-7, Mar. 4-8, 2012.

[6] M. Vasiladiotis and A. Rufer, "Balancing control actions for cascaded H-bridge converters with integrated battery energy storage," *Proc. of the 15th European Conference on Power Electronics and Applications (EPE-ECCE Europe)*, pp. 1-10, Sep. 2-6, 2013.

[7] H. Li, F. Z. Peng, and J. S. Lawler, "A Natural ZVS Medium-Power Bidirectional DC-DC Converter With Minimum Number of Devices," *IEEE Trans. on Industry Applications*, vol. 39, no. 2, pp. 525-535, Mar./Apr. 2003.

[8] F. Z. Peng, H. Li, G.-J. Su, and J. S. Lawler, "A New ZVS Bidirectional DC-DC Converter for Fuel Cell and Battery Application," *IEEE Trans. on Power Electronics*, vol. 19, no. 1, pp. 54-65, Jan. 2004.

[9] Z. Wang and H. Li, "Optimized operating mode of current-fed dual half bridges dc-dc converters for energy storage applications," *Proc. of the IEEE Energy Conversion Congress and Exposition (ECCE)*, pp. 731-737, Sept. 20-24, 2009.

[10] H. Daneshpajooh, A. Bakhshai, and P. Jain, "Optimizing dual half bridge converter for full range soft switching and high efficiency," *Proc. of the IEEE Energy Conversion Congress and Exposition (ECCE)*, pp. 1296-1301, Sept. 17-22, 2011.

[11] X. Liu, H. Li, and Z. Wang, "A Fuel Cell Power Conditioning System With Low-Frequency Ripple-Free Input Current Using a Control-Oriented Power Pulsation Decoupling Strategy," *IEEE Trans. on Power Electronics*, vol. 29, no. 1, pp. 159-169, Jan. 2014.

[12] H. Li and F. Z. Peng, "Modeling of a new ZVS Bi-directional DC-DC Converter," *IEEE Trans. on Aerospace and Electronic Systems*, vol. 40, no. 1, pp. 272-283, Jan. 2004.

[13] A. Karimi and G. Galdos, "Fixed-order H∞ controller design for nonparametric models by convex optimization," *Automatica*, vol. 46, no. 8, pp. 1388-1394, 2010.

[14] G. Galdos, A. Karimi, and R. Longchamp, "H∞ controller design for spectral MIMO models by convex optimization," *Journal of Process Control*, vol. 20, no. 10, pp. 1175-1182, 2010.

[15] I.D. Landau, R. Lozano, M. M'Saad, and A. Karimi, "Adaptive Control: Algorithms, Analysis and Applications," Second Edition, Springer, 2011.

[16] B. Bahrani and A. Rufer, "Optimization-Based Voltage Support in Traction Networks Using Active Line-Side Converters," *IEEE Trans. on Power Electronics*, vol. 28, no. 2, pp. 673-685, Feb. 2013.

[17] B. Bahrani, M. Saeedifard, A. Karimi, and A. Rufer, "A Multivariable Design Methodology for Voltage Control of a Single-DG-Unit Microgrid," *IEEE Trans. on Industrial Informatics*, vol. 9, no. 2, pp. 589-599, May 2013.

[18] B. Bahrani, M. Vasiladiotis, and A. Rufer, "High-Order Vector Control of Grid-Connected Voltage-Source Converters With LCL-Filters," *IEEE Trans. on Industrial Electronics*, vol. 61, no. 6, pp. 2767-2775, June 2014.

[19] N. Cherix, S. Delalay, P. Barrade, and A. Rufer, "Fail-safe modular control platform for power electronic applications in R&D environments," *Proc. of the 15th European Conference on Power Electronics and Applications (EPE)*, pp. 1-10, Sept. 2-6, 2013.

The 2014 International Power Electronics Conference

New Interleaved Current-Fed Resonant Converter with Significantly Reduced High Current Output Filter for EV and HEV application

Dongok Moon, Junsung Park, Sewan Choi, *IEEE Senior Member*
Department of Electrical and Information Engineering
Seoul National University of Science and **Tech**nology
Seoul, Korea
Email: schoi@seoultech.ac.kr

Abstract— This paper proposes a new interleaved current-fed resonant converter with significantly reduced high current output filter which is suitable for EV and HEV applications. The proposed interleaved converter has theoretically zero output filter capacitance, low input current ripple, ZCS turn on and off for all switches and diodes and zero di/dt at turn off of diodes when operated at load independent points. A two-stage power conversion technique is applied for wide input and output voltage range operation of EV and HEV. A 2kW prototype of the proposed converter has been built and tested to verify the validity of the proposed operation.

Keywords— *Interleaved, Current fed resonant converter, Soft-switched, Electric vehicle.*

I. INTRODUCTION

Recently, eco-friendly cars such as Electric Vehicles(EV), Hybrid Electric Vehicles(HEV) or Plug In Hybrid Electric Vehicles(PHEV) are attracting increasing attention as a solution of environmental pollution, global warming and exhaustion of fossil fuels. Configurations of several types of EV and HEV power train systems are described in [1]. The low voltage dc/dc converter(LDC) provides power to 12V loads such as the head lamps, wiper blade motor, electronic power steering (EPS), radio system etc. by charging a 12V auxiliary battery from a high voltage battery(200~400V)[2]. This application requires an efficient(more than 90%), compact and light weight dc-dc converter. Also, due to safety and high step down conversion ratio galvanic isolation is generally required.

The phase shift full-bridge PWM converter is widely used in the dc/dc conversion stage because of its small RMS current and inherent ZVS[3]. Disadvantages of the phase shift full-bridge PWM converter is that turn off current of switches is large and turn off loss associated with the diode reverse recovery is considerable. In order to reduce the turn off losses of the switches and diodes resonant converters with ZVS or ZCS capabilities such as SRC and LLC can be considered as candidates for the low voltage dc/dc converter[4]-[5]. In general, the resonant converter requires output capacitor for suppression of output ripple current while the phase shift full-bridge PWM converter requires output inductor for

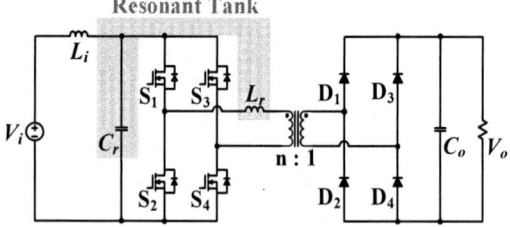

Fig.1 Proposed current-fed resonant converter

suppression of output ripple voltage. The volume of the output filter inductor or capacitor is considerable, especially, in low output voltage and high output current application[6]-[7]. In order to reduce the volume of the output filter interleaved techniques can be applied to the resonant or PWM converters[8]. However, volume reduction of the output filter resulting from interleaving of the conventional resonant or PWM converters may not be satisfactory in the low voltage high current application such as LDC.

This paper proposes a new interleaved current-fed resonant converter with significantly reduced high current output filter which is suitable for EV and HEV applications. The proposed interleaved resonant converter has the following features when operated at $f_s = 0.5f_r$: 1) theoretically zero output filter capacitance 2) low input current ripple 3) ZCS turn on and off for all switches and diodes 4) zero *di/dt* at turn off of diodes, resulting in negligible turn off losses associated with the diode reverse recovery. A two-stage power conversion technique is applied for wide input and output voltage range operation of EV and HEV. A 2kW prototype of the proposed converter has been built and tested to verify the validity of the proposed operation.

I. PROPOSED CURRENT-FED RESONANT CONVERTER

Fig.1 shows the circuit diagram of the proposed current-fed resonant converter. The proposed converter consists of an input filter inductor, four switches, a resonant tank, a transformer, a diode rectifier and an output filter capacitor. The output voltage of the proposed converter is regulated by fixed duty and variable switching frequency.

978-1-4799-2706-7/14 $31.00 © 2014 IEEE

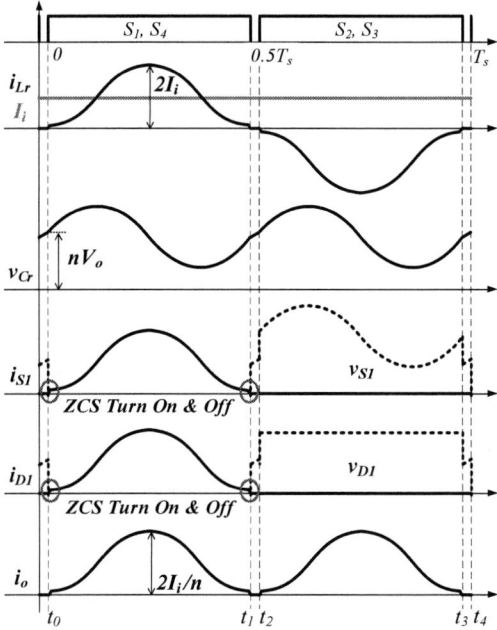

Fig.2 Key waveforms of the proposed converter at $f_s=0.5f_r$

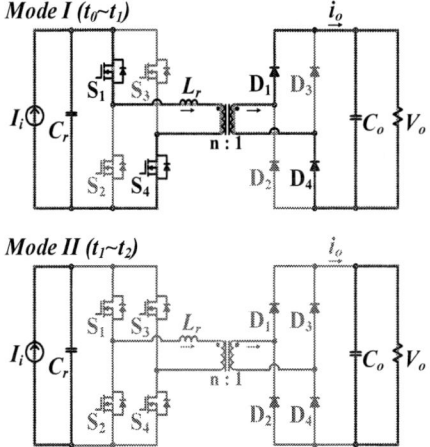

Fig. 3 Operating states of the proposed converter at $f_s=0.5f_r$

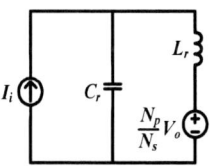

Fig. 4 Equivalent resonant circuit of *Mode I*

A. Operating Principle

The key waveforms and operating states of the proposed converter at switching frequency $f_s = 0.5f_r$ are shown in Figs. 2 and 3, respectively.

Mode I [$t_0 \sim t_1$]: This mode begins with L_r-C_r resonance when switches S_1 and S_4 are turned on at t_0. The equivalent circuit of this mode is shown in Fig. 4.

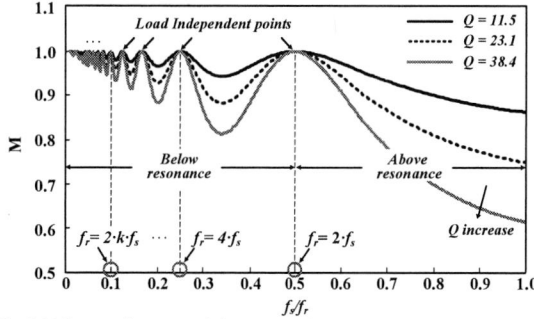

Fig.5 Voltage gain curve of the proposed converter

The resonant voltage and current are determined, respectively, as follow:

$$i_{Lr}(t) = I_i\,(1-\cos\omega_r t) \tag{1}$$

$$v_{Cr}(t) = ZI_i\sin\omega_r t + nV_o \tag{2}$$

where $Z^2 = L_r/C_r$ is characteristic impedance and $\omega_r^2 = 1/L_r\,C_r$ is angular resonant frequency.

Input current I_i is determined, as follow:

$$I_i = \frac{V_o}{n\,R_L} \tag{3}$$

where R_L is load resistance and this mode ends when switches S_1 and S_4 are turned off at t_1.

Mode II [$t_1 \sim t_2$]: During this mode the power is not transferred to the load and the output filter capacitor supplies the load. The voltage of the resonant capacitor increases linearly. The other half of a cycle is repeated in the same fashion. Note that all switch and diode are turned on and off under ZCS condition.

B. Voltage Gain Expression

Assuming that *Mode II*, the dead-time, is neglected, the average voltage across the resonant capacitor that is equal to the input voltage can be expressed as:

$$V_i = V_{Cr.av} = \frac{2}{T_s}\int_0^{\frac{T_s}{2}}(ZI_i\sin\omega_r t + \frac{N_p}{N_s}V_o)dt \tag{4}$$

From eqn. (4) the voltage gain of the proposed converter is determined by:

$$M = \frac{V_o}{V_i} = \frac{(\pi N_p\,\omega_r)(1-\cos\frac{\omega_s}{\omega_r}\pi)N_s}{(\pi N_p\,\omega_r)(1-\cos\frac{\omega_s}{\omega_r}\pi)N_p + Q\omega_s} \tag{5}$$

where Q is the quality factor which determines the slope of voltage gain curve and determined, as follow:

$$Q = \frac{\omega_r L_r}{R_L} \tag{6}$$

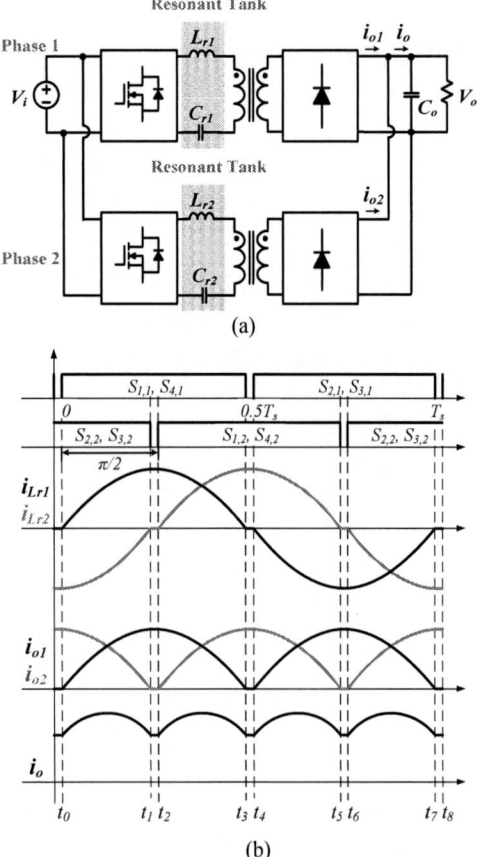

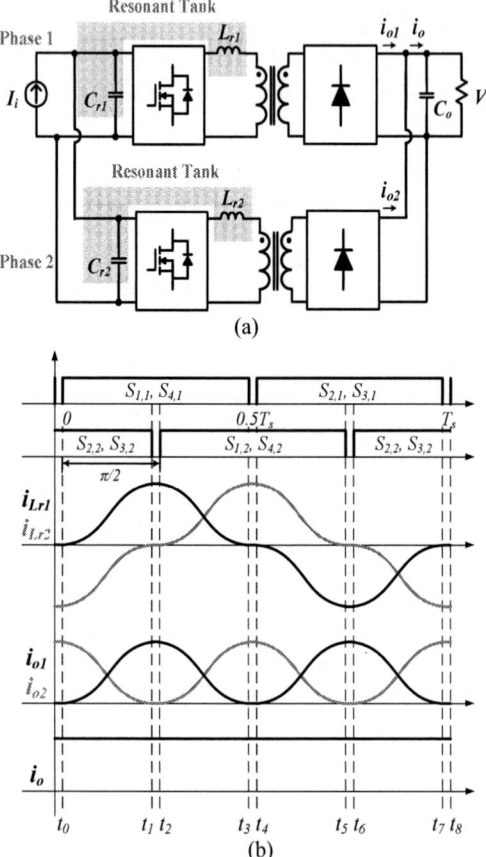

Fig.6 The conventional two-phase interleaved series resonant converter (a) circuit diagram (b) key waveforms at $f_s = f_r$

Fig.7 The proposed two-phase interleaved current-fed resonant converter (a) circuit diagram (b) key waveforms at $f_s = 0.5 f_r$

The higher Q is, the higher slope of the gain curve is. The voltage gain curves of the proposed converter are as shown in Fig. 5. Note that the load independent points of the proposed converter are multiple in the below resonance region and are determined, respectively, as follow:

$$f_r = 2k f_s \qquad (7)$$

where k is the natural number. It should also be noted that the voltage gain of the proposed converter converges to 1 while that of the SRC converges to 0 as the switching frequency decreases in the below resonance region. In the above resonance region, the slopes of the voltage gain curves according to Q of the proposed converter have similar values with those of the SRC.

C. Proposed Interleaved Resonant Converter

The circuit diagram and key waveforms of the conventional two-phase interleaved SRC are shown in Fig. 6. The two SRC are operated at $f_s = f_r$ and phase shifted by $\pi/2$. The output current of each phase is the secondary winding current rectified by the diode bridge and can be expressed as, using Fourier series,

$$i_{o,1} = \frac{I_o}{2} + \sum_{m=1}^{\infty} \frac{I_o}{(1+2m)(1-2m)} \cos(2m\omega t) \qquad (8)$$

$$i_{o,2} = \frac{I_o}{2} + \sum_{m=1}^{\infty} \frac{I_o}{(1+2m)(1-2m)} \cos(2m(\omega t - \frac{\pi}{2})) \qquad (9)$$

The output current of the two-phase interleaved SRC is obtained by,

$$
\begin{aligned}
i_o &= i_{o,1} + i_{o,2} \\
&= I_o + \sum_{m=1}^{\infty} \frac{2 I_o}{(1+4m)(1-4m)} \cos(4m\omega t)
\end{aligned} \qquad (10)
$$

It is seen from eqn. (10) that there exist multiples of fourth harmonic component in the output current. In general, the output current of the conventional n-phase interleaved SRC has ripple contents of multiples of $2n$-th harmonic component as shown in eqn. (11).

$$
\begin{aligned}
i_o &= i_{o,1} + i_{o,2} + i_{o,3} + \cdots + i_{o,N} \\
&= I_o + \sum_{n=1}^{N} \sum_{m=1}^{\infty} \frac{2 I_o}{n(1+2m)(1-2m)} \cos(2a\omega t - \frac{2(m-1)\pi}{N})
\end{aligned} \qquad (11)
$$

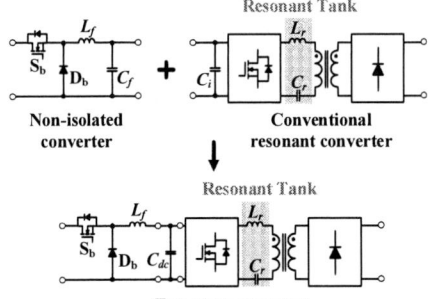

Non-isolated converter + **Conventional resonant converter**

↓

Two-stage converter

Fig. 6 Concept of the conventional two-stage converter

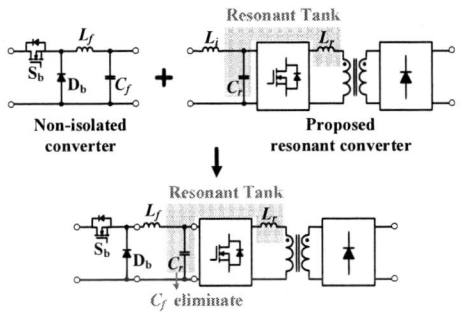

Non-isolated converter + **Proposed resonant converter**

↓

C_f eliminate
Proposed two-stage converter

Fig. 7 Concept of the proposed two-stage converter

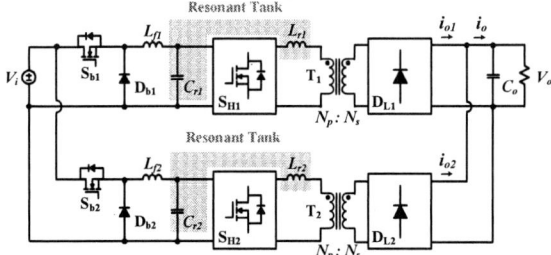

Fig. 8 The PIPO interleaved converter

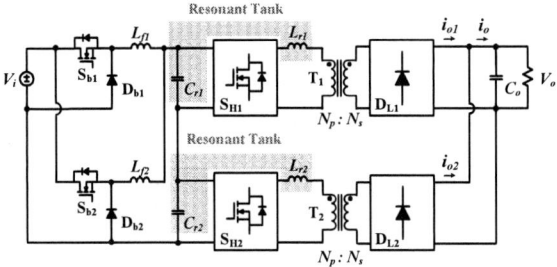

Fig. 9 The SIPO interleaved converter

It means that the output ripple current of the conventional interleaved SRC cannot be completely eliminated.

Figs. 7 (a) and (b) show the circuit diagram and key waveforms of the proposed two-phase interleaved converter. Note that the two current-fed resonant converters are operated at $f_s = 0.5 f_r$ and phase shifted by $\pi/2$. The output current of each phase is the secondary winding current rectified by the diode bridge and can be obtained by,

$$i_{o1} = \frac{1}{2} I_o (1 - \cos \omega t) \tag{12}$$

$$i_{o2} = \frac{1}{2} I_o (1 - \cos(\omega t - \pi)) \tag{13}$$

The output current of the proposed interleaved converter is obtained by,

$$i_o = i_{o1} + i_{o2}$$
$$= \frac{1}{2} I_o (1 - \cos \omega t) + \frac{1}{2} I_o (1 + \cos \omega t) = I_o \tag{14}$$

It should be noted that ac ripple components of the output current are completely eliminated and the output current is ripple-free, meaning that required output capacitance C_o is theoretically zero. The proposed interleaving technique is very effective especially in the low voltage and high current applications where the output filter significantly affects the efficiency and size of the whole system. The concept of interleaving of the proposed current-fed resonant converter can be extended to *n*-phase system. The output current of the *n*-phase interleaved current-fed resonant converter can be obtained by,

$$i_o = i_{o1} + i_{o2} + \cdots + i_{oN}$$
$$= \sum_{m=1}^{N} \frac{I_o}{N} (1 - \cos(\omega t - \frac{2(m-1)\pi}{N})) = I_o \tag{15}$$

A two-stage power conversion technique is employed for wide input and output voltage range operation of EV and HEV.

Figs. 6 and 7 show the concept of the conventional and proposed two-stage converter using a buck converter as a non-isolated converter stage, respectively. It is seen from Fig. 6 that the output capacitor C_f of the non-isolated converter and the input capacitor C_i of the conventional resonant converter are combined into one capacitor C_{dc} in the two-stage converter that is comparatively large since the voltage ripple of the capacitor should be within some specified limit. In the meanwhile, it is seen from Fig. 7 that the output capacitor C_f of the non-isolated converter is eliminated, and the filter inductor L_f of the non-isolated converter and the input inductor L_i of the proposed resonant converter are combined into one inductor L_f in the two-stage converter. Note that the capacitor C_r in the proposed two-stage converter does not need to be large since it is used only as a resonant capacitor.

The two types of the proposed interleaved two-stage converter are shown in Fig. 8 and 9: parallel input and parallel output(PIPO) converter and series input and parallel output(SIPO) converter. The PIPO interleaved converter is better suited to relatively low input voltage application while SIPO interleaved converter is better suited to relatively high input voltage application. The PIPO converter is more vulnerable to current unbalance caused by resonant component tolerances, parasitic component of each converter. Instead, the current unbalance can be alleviated by controlling each of the

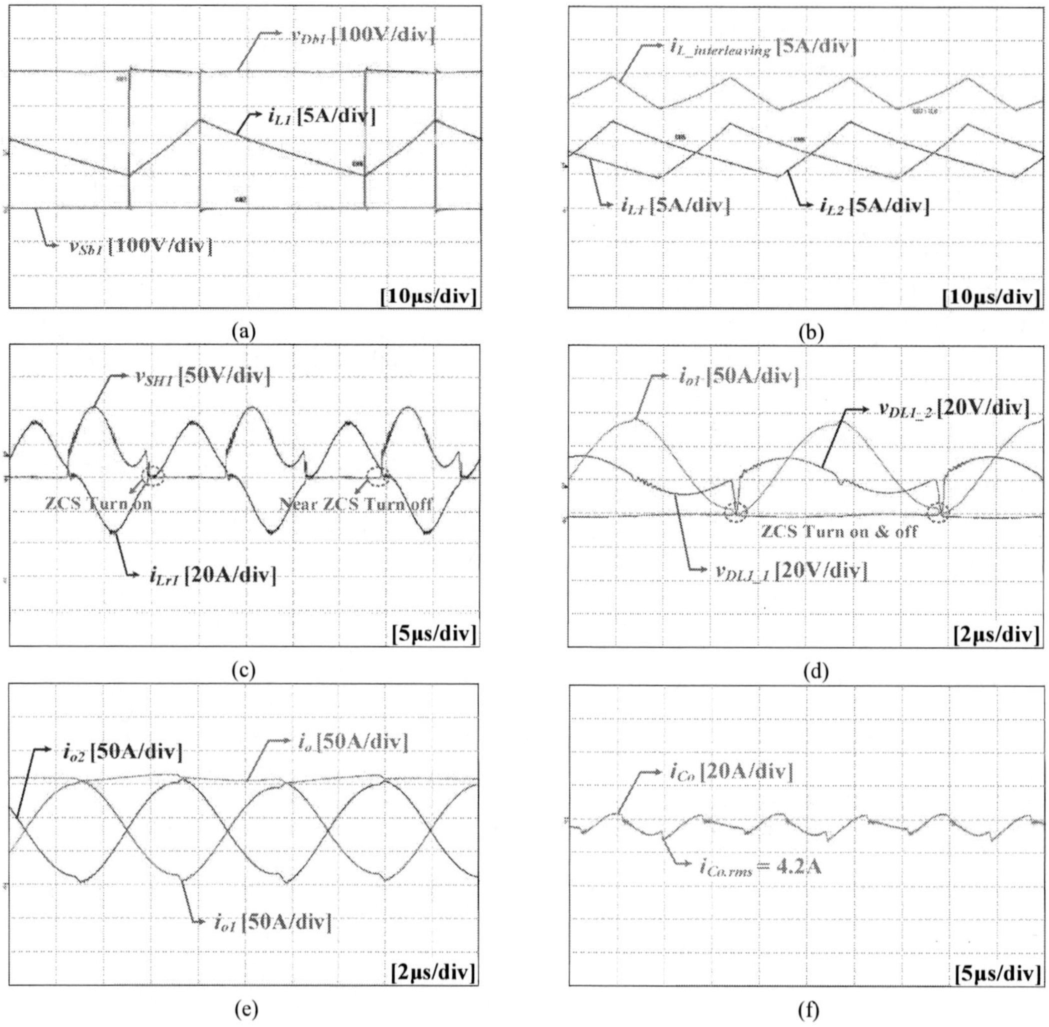

Fig. 10 Experimental waveforms at full load (a) switch voltage V_{Sb1} and diode voltage V_{Db1} and inductor current i_{L1} (b) inductor current i_{L1}, i_{L2} and interleaving current $i_{L\ interleaving}$ (c) switch voltage V_{S1} and resonant current i_{Lr1} (d) diodes voltage $V_{D1.2}$ and rectified current i_{o1} (e) rectified current i_{o1}, i_{o2} and interleaved output current i_o (f) ripple current of output capacitor

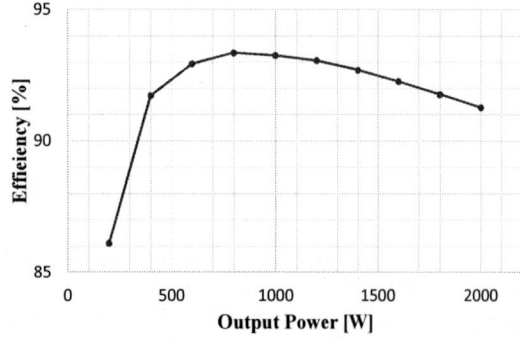

Fig. 11 Measured efficiency as a function of the output power at $V_i = 400\text{V}$

non-isolated stage of the PIPO converter. On the other hand, the SIPO converter is immune to current unbalance caused by resonant component tolerances due

to inherent charging balance of two resonant capacitors[9].

III. EXPERIMENTAL RESULT

A 2-kW prototype of the proposed interleaved converter shown in Fig. 14 was built under the following specification :

$P_o = 2$ kW, $V_i = 400$ V, $V_o = 12$ V, $f_s = 20$ kHz, $f_r = 60$ kHz, $N_p : N_s = 5 : 1$, $L_{f1,f2} = 560$ μH, $C_f = 330$ μF, $L_{r1,r2} = 3.74$ μH, $C_{r1,r2} = 0.47$ uH, $D_dT_s = 100$ ns

Fig. 10 shows key experimental waveforms of the proposed converter at full load. Figs. 10(a) and (b) show non-isolated stage switch voltage and inductor current. Fig. 10(c) and (d) show isolated stage switch voltage and current. It is seen that all switches are turned on and off under ZCS condition. Fig. 10(e) and (f) show interleaved output current and ripple current of the output capacitor. It can

978-1-4799-2706-7/14 $31.00 © 2014 IEEE

Fig. 12 Photograph of the proposed interleaved resonant converter prototype

be seen that output ripple current is significantly reduced to $4.2A_{rms}$. Fig. 11 is the measured efficiency using Yokogawa WT3000. The maximum efficiency is 93.4% at 0.8kW and full load efficiency is 91.3%. Fig. 12 is the prototype of proposed interleaved converter.

IV. CONCLUSIONS

This paper proposes a new interleaved current-fed resonant converter with significantly reduced high current output filter which is suitable for EV and HEV applications. The proposed interleaved converter has theoretically zero output filter capacitance, low input current ripple, ZCS turn on and off for all switches and diodes and zero di/dt at turn off of diodes when operated at $f_s = 0.5f_r$. A two-stage power conversion technique is applied for wide input and output voltage range operation of EV and HEV. A 2kW prototype of the proposed converter has been built and tested to verify the validity of the proposed operation. The maximum efficiency is 93.4% at 0.8kW and full load efficiency is 91.3%. The proposed converter could be a possible option for low voltage high current application such as LDC.

REFERENCES

[1] A. Emadi, L. Joo, and K. Rajashekara, "Power electronics and motor drives in electric, hybrid electric, and plug-in hybrid electric vehicles," *IEEE Trans. Ind. Electron.*, vol. 55, no. 6, pp. 2237–2245, 2008.

[2] Z. Amjadi and S. S. Williamson, "Power-electronics-based solutions for plug-in hybrid electric vehicle energy storage and management systems," *IEEE Trans. Ind. Electron.*, vol. 57, no. 2, pp. 608–616, 2010.

[3] D. Hamza, M. Pahlevaninezhad, and P. Jain, "Implementation of a novel digital active EMI technique in a DSP-based DC–DC digital controller used in electric vehicle (EV) battery charger," *IEEE Trans. Power Electron.*, vol. 28, no. 7, pp. 3126–3137, 2013.

[4] I. Lee and G. Moon, "Analysis and design of a three-level LLC series resonant converter for high-and wide-input-voltage applications," *IEEE trans. Power Electron.*, vol. 27, no. 6, pp. 2966-2979, 2012.

[5] J. Biela, U. Badstuebner, and J. W. Kolar, "Design of a 5-kW, 1-U, 10-kW/dm3 resonant DC–DC converter for telecom applications," *IEEE trans. Power Electron.*, vol. 24, no. 7, pp. 1701–1710, 2009.

[6] S. Cho, I. Lee, J. Kim, and G. Moon, "A new standby structure based on a forward converter integrated with a phase-shift full bridge converter for server power supplies," *IEEE Trans. Power Electron.*,vol. 28, no. 1, pp. 336–346, 2013.

[7] U. Badstuebner, J. Biela, D. Christen, and J. W. Kolar, "Optimization of a 5-kw telecom phase-shift dc–dc converter with magnetically integrated current doubler," *IEEE Trans. Ind. Electron.*, vol. 58, no. 10, pp. 4736–4745, 2011.

[8] K. Yi and G. Moon, "Novel two-phase interleaved LLC series-resonant converter using a phase of the resonant capacitor," IEEE Trans. Ind. Electron., vol. 56, no. 5, pp. 1815-1819, 2009.

[9] B. C. Kim, K. B. Park, C. E. Kim, and G. W. Moon, "Load sharing characteristic of two-phase interleaved LLC resonant converter with parallel and series input structure," in Proc. IEEE ECCE, 2009, pp. 750–753.

15 Phase Induction Motor Drive With 1:3:5 Speed Ratios Using Pole Phase Modulation

Umesh B S, Sivakumar K
Department of Electrical Engineering
Indian Institute of Technology Hyderabad
Andhra Pradesh, India
Emails : umi_umeshbs@yahoo.com, ksiva@iith.ac.in

Abstract—pole phase modulation (PPM) is one of the effective technique to control speed and torque of multiphase machines. This paper presents the winding details and finite element analysis (FEA) of 15 phase squirrel cage induction motor (IM) drive with 1:3:5 speed ratios using PPM. The proposed 15 phase IM with 60 stator slots and full pitched winding is able to run in five different pole-phase combinations with best utilization of copper. The 15 phase IM fed by a 15 leg inverter is operated in 15 phase 4 pole, 5 phase 12 pole, 3 phase 20 pole, 5 phase 4 pole and 3 phase 4 pole combinations. In addition Generalization of PPM is revisited in this paper to add an extra constraint on the pole ratio by showing an exception for the already proposed generalization of PPM. The operation of the proposed drive for five different pole-phase combinations with three possible speeds is verified by FEA simulation using Ansys Maxwell 2D (FEA package for electromagnetic design) and Simplorer for 2kw IM drive prototype with 10Nm load torque.

Keywords—FEA of multiphase machines, 1:3:5 Speed ratios, 15 Phase indction motor drive, Pole phase modulation.

I. INTRODUCTION

Developments in the application areas such as ship propulsion, traction and more electric aircraft accelerated the research on multiphase machines from the beginning of this century. Improved efficiency due to reduction in magnitude of slot harmonics, increased reliability because of fault tolerant capability, reduction in torque pulsations and the capability to handle high power density with reduced power rating of power electronic switches makes the multi-phase machines best suitable for such applications [1-4].

Conventional scalar control (v/f) used for three phase IM drives was extended to multi-phase IM drives in earlier days[5]. Later pole changing using mechanical switches to reconnect the phase windings in series or parallel and star or delta, to change the phase belt are shown in [6-7] to obtain 1:2 and 1:3 speed ratios. The problem of de-energizing the phase windings before reconnection to achieve new pole number (phase belt) limited the discontinuous pole changing techniques.

Pole amplitude modulation (PAM) and PPM are the two efficient continuous pole changing techniques without the need of mechanical switches. PAM with special winding arrangements changes the phase belt there by pole number of multiphase machine. With the clock diagram of special winding arrangements showing phase reversal of currents in some of the phase windings to obtain 1:2 and 1:3 speed ratios are presented in [8-10]. Poor copper utilization and need of different speed ratios with normal operation of induction motor were the issues in PAM. In [11] a 36 slots, 9 full pitched phase windings for 4 pole IM drive is proposed with 1:3 speed ratio using PPM.

Recently for a 9 phase machine with 72 slots, an effective torque control technique during pole change transition is given in [12] with a continuous pole changing advantage. In [13] along with generalization of PPM a 9 phase 36 slot machine is controlled using PPM for 1:3 speed ratio with short pitched conventional and toroidal winding is proposed

In this paper winding details and the FEA simulation of 15 phase IM drive with 1:3:5 speed ratios is discussed with five different pole-phase combinations. Such a wide speed range is required in electric vehicles and special industrial drives. In addition Generalization of PPM is revisited to add an extra constraint on the pole ratio by showing an exception for the already proposed generalization of PPM [13-14].

Rest of the paper is organized as fallows.pole ratio number is generalized with an extra constraint on it for effective PPM to avoid unequal pole width is discussed in next section. Winding details and five different pole phase combinations are discussed in section III. FEA simulation results are discussed in section IV and the paper is concluded in section V.

II. GNERALIZATION OF PPM REVISITED

The basic idea in PPM is to adjust the phase belt by varying both pole and phase number in certain ratio to obtain required speed ratios continuously. The advantage of PPM is physical reconnection of the windings is completly avoided, instead only the phase of the excitation is altered to achive a continious pole change. This basic concept is generalized in [13-14] as fallows. If m_1 and p_1 are the number of phases and number of pole pairs for synchronous speed of a multiphase machine, m_2 and p_2 are another set of phase and pole pair number to obtain second speed using the same multiphase machine. Then number of slots Q with phase belt q is given by equation (1).

$$Q = 2p_1q_1m_1 = 2p_2q_2m_2 \qquad (1)$$

TABLE I.   SHOWS PPM OF PROPOSED IM WITH EACH SLOT ANGLE (IN ELECTRICAL DEGREE) AND EACH PHASE CURRENT DIRECTION WITH POLE PITCH=15 SLOTS, FOR HALF OF THE STATOR PERIPHERY (30 SLOTS)

Slot Number	1	2	3	4	5	6	7	8	9	10	11	12	13	14	15	16	17	18	19	20	21	22	23	24	25	26	27	28	29	30
Up Conductor	↑	↓	↑	↓	↑	↓	↑	↓	↑	↓	↑	↓	↑	↓	↑	↓	↑	↓	↑	↓	↑	↓	↑	↓	↑	↓	↑	↓	↑	↓
Down Conductor	↑	↓	↑	↓	↑	↓	↑	↓	↑	↓	↑	↓	↑	↓	↑	↓	↑	↓	↑	↓	↑	↓	↑	↓	↑	↓	↑	↓	↑	↓
15 Phase 4 Pole Operation																														
Slot Angle*	0	12	24	36	48	60	72	84	96	108	120	132	144	156	168	180	192	204	216	228	240	252	264	276	288	300	312	325	336	348
Phase Current	a	-i	b	-j	c	-k	d	-l	e	-m	f	-n	g	-o	h	-a	i	-b	j	-c	k	-d	l	-e	m	-f	n	-g	o	-h
Poles								N															S							
5 Phase 12 Pole Operation																														
Slot Angle*	0	36	72	108	144	180	216	252	288	324	360	36	72	108	144	180	216	252	288	324	360	36	72	108	144	180	216	252	288	324
Phase Current	p1	-p4	p2	-p5	p3	-p1	p4	-p2	p5	-p3	p1	-p4	p2	-p5	p3	-p1	p4	-p2	p5	-p3	p1	-p4	p2	-p5	p3	-p1	p4	-p2	p5	-p3
Poles			N					S					N					S					N					S		
3 Phase 20 Pole Operation																														
Slot Angle*	0	60	120	180	240	300	360	60	120	180	240	300	360	60	120	180	240	300	360	60	120	180	240	300	360	60	120	180	240	300
Phase Current	A	-C	B	-A	C	-B	A	-C	B	-A	C	-B	A	-C	B	-A	C	-B	A	-C	B	-A	C	-B	A	-C	B	-A	C	-B
Poles		N			S			N			S			N			S			N			S			N			S	
5 Phase 4 Pole Operation																														
Slot Angle*	0	180	0	36	216	36	72	252	72	108	288	108	144	324	144	180	0	180	216	36	216	252	72	252	288	108	288	324	144	324
Phase Current	p1	-p1	p1	-p4	p4	-p4	p2	-p2	p2	-p5	p5	-p5	p3	-p3	p3	-p1	p1	-p1	p4	-p4	p4	-p2	p2	-p2	p5	-p5	p5	-p3	p3	-p3
Poles								N															S							
3 Phase 4 Pole Operation																														
Slot Angle*	0	180	0	180	0	60	240	60	240	60	120	300	120	300	120	180	0	180	0	180	240	60	240	60	240	300	120	300	120	300
Phase Current	A	-A	A	-A	A	-C	C	-C	C	-C	B	-B	B	-B	B	-A	A	-A	A	-A	C	-C	C	-C	C	-B	B	-B	B	-B
Poles								N															S							

Note: * indicates angle in electrical degree

And the pole ratio is given as

$$k = \frac{p_2}{p_1} = \frac{q_1 m_1}{q_2 m_2} \tag{2}$$

The general value of pole ratio k for $p_1 < p_2$ is given as integer ≥ 1. But in the case of $Q = 48$, $m_1 = 12$, $p_1 = 2$, $m_2 = 3$ and $p_2 = 8$ satisfies the generalization with $q = 1$ and $k = 4$ which is ≥ 1. Practically this combination for $1 : 4$ speed ratio and full pitched windings results in poles with unequal width as shown in Fig.1. For $k = 2$ only phase number changes but number of pole pair remains the same.

The above case is a clear exception of the proposed generalization with respect to pole ratio k. To avoid unequal pole width k should be positive odd integer > 1 i.e. $k = 2n + 1$ where $n = 1, 2, 3..$ with proper selection of slot number as given in [12-13].

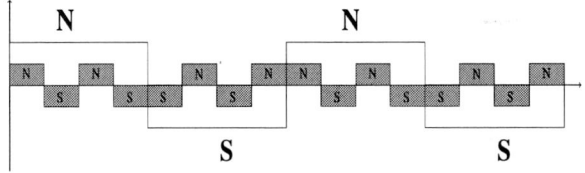

Fig. 1.   Shows the unequal pole width for PPM with $k = 4$.

III.   DESIGN DETAILS OF THE PROPOSED 15 PHASE IM PROTOTYPE

The proposed 2kw 15 phase IM consists of 60 stator slots and 71 rotor slots with a full pitched double layer winding to give 4 poles. The stator and rotor slots are carefully selected to avoid cogging torque and synchronous cups in the torque characteristics of the machine. The air gap length is 0.35mm which is selected according to main dimensions to ensure the best power factor for the proposed IM. The main dimensions of the IM i.e. air gap diameter D=148mm and length of the core L=250mm [15].

One of the phase winding connections out of 15 phases is shown in Fig.2. Pitch of the phase winding is 15 slots which is full pitch for 4 pole 60 slots machine. Two terminals for each phase winding results in total of 30 terminals, out of which 15 terminals on one side of each phase winding are shorted so that machine can be supplied by 15 leg inverter. Five different pole-phase combinations and corresponding speed ratios are possible using the proposed 15 phase 4 pole IM drive.

For pole ratio $k = 3$ the same IM is operated as 5 phase 12 pole machine resulting synchronus speed of 500 rpm. Similarly for pole ratio $k = 5$ the same IM is operated as 3 phase 20 pole machine resulting sunchronus speed of 300 rpm, thus obtaining wide range of speed control. 15 phase 4 pole combination is used for the steady state operation of long duty where as the 5 phase 12 pole and 3 phase 20 pole combinations are used for improving the transient performance of

978-1-4799-2706-7/14 $31.00 © 2014 IEEE

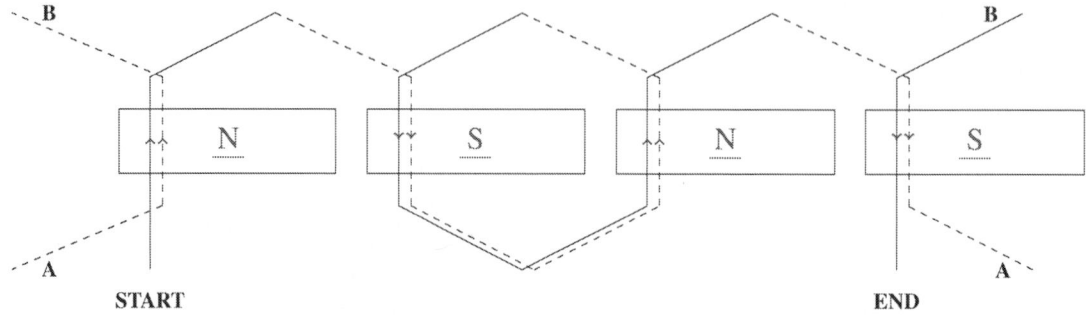

Fig. 2.   Shows one of the phase winding arrangement with pitch equal to 15 slots.

the drive. For example applications like traction and propulsion the drive intially requires large torque which is provided using 5 phase 12 pole or 3 phase 20 pole operation. Once the drive attains steady run torque requirement comes down and speed of ship can be raised by operating in 15 phase 4 pole mode keeping output power of the motor nearly constant. In this paper PPM is demostrated using FEA simulation for a constant torque of 10 Nm and power input to the motor is reduced in case of 5 phase 12 pole and 3 phase 20 pole to maintain the torque constant at 10Nm. The phase belt q for all the three pole-phase combination is $q = 1$. The other two pole phase combinations possible are 5 phase 4 pole and 3 phase 4 pole with phase belt $q = 3$ and 5 respectively. These two combinations don't have any advantageous with respect to speed and torque performances, but require less number of inverter legs compared to first three combinations i.e. 5 and 3 legs respectively [13].

Table I. shows the PPM of proposed drive clearly, with details of each slot angle (in electrical degrees) and the current direction in each conductor of a full pitched double layer winding. In the Table I. the alphabets a-o represents the 15 phase currents with phase difference of 24 electrical degrees, p1-p5 represents the 5 phase currents with phase difference of 72 electrical degrees and A, B, C represents three phase currents with phase difference of 120 electrical degrees. Here the best utilization of copper is justified, as at any given time the top and bottom conductors of the same slot have same phase current. Where as in a short pitched winding and special winding for PAM the top and bottom conductors of the same slot carries two different phase currents for achieving some pole amplitudes leading to underutilization of copper.

IV.   VALIDATION OF PPM FOR PROPOSED DRIVE WITH FEA SIMULATION AND RESULTS

A. FEA simulation details.

Finite element model of the proposed IM is developed using Ansys Maxwell 2D. An automatic meshing technique meshes the model with triangular elements. For better results the meshing near air gap is refined by reducing the length of the triangular elements near airgap as shown in Fig.3. Proposed FEA model is co-simulated

with 15 leg inverter in Ansys Simplorer environment for coupled field analysis of the proposed IM. The phase angle of the inverter output voltage is controlled using sine triangle PWM to achieve required pole phase modulation. Co-simulation confirms all the five different pole-phase combinations are well valid and the results are presented in the next part of this section.

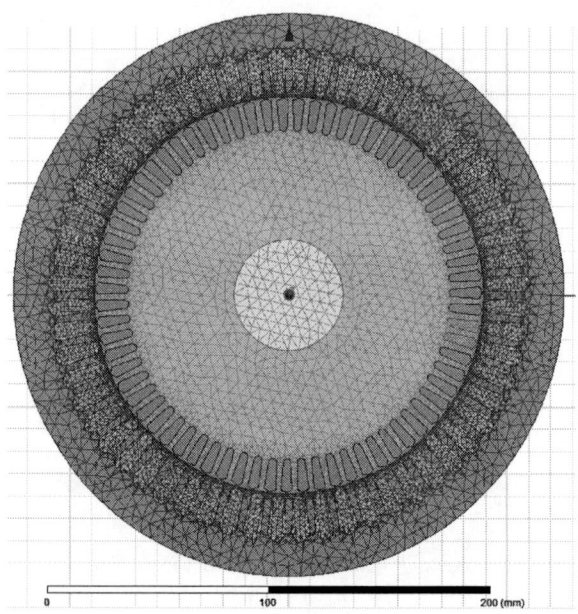

Fig. 3.   Shows triangular meshing refined intensely near air gap.

B. Results and discussion

FEA simulation results presented in this section validates PPM of the proposed 15 phase IM for all the five pole-phase combinations. Fig.4-6. shows the flux line distribution for the first three pole-phase combinations. The legend gives the magnitude of flux lines per meter and the flux lines distribution for the last two pole phase combinations is same as 15 phase 4 pole case as shown in Fig.4.

Fig.7 shows the torque developed for the first three pole phase combinations. The torque curve for 5 phase 12 pole operation and 3 phase 20 pole operation

The 2014 International Power Electronics Conference

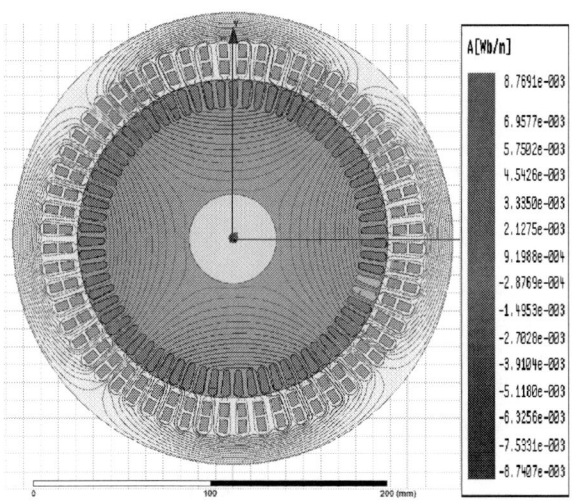

Fig. 4. Shows the flux line distribution for 15 phase 4 pole operation.

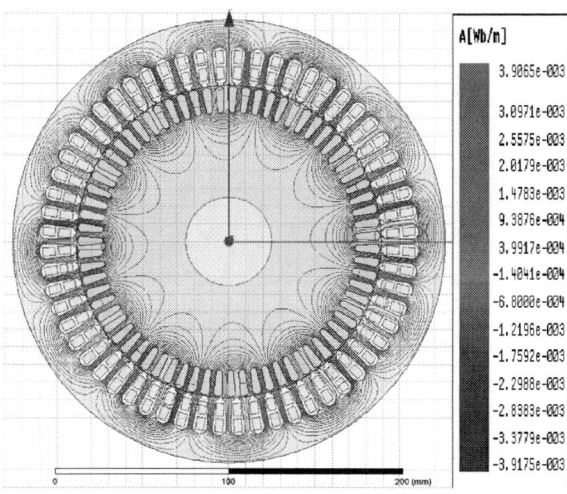

Fig. 5. Shows the flux line distribution for 5 phase 12 pole operation.

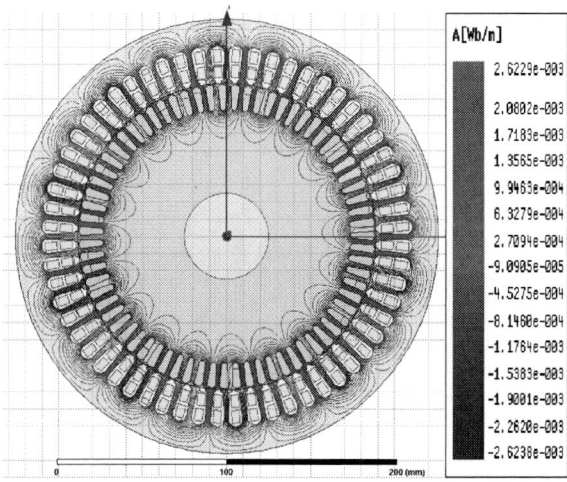

Fig. 6. Shows the flux line distribution for 3 phase 20 pole operation.

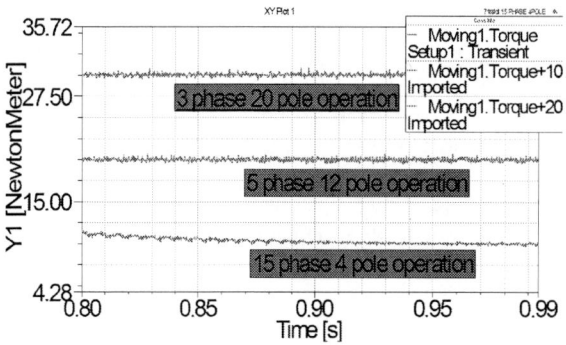

Fig. 7. Shows motor induced torque for first three pole-phase operations (FEA Simulation).

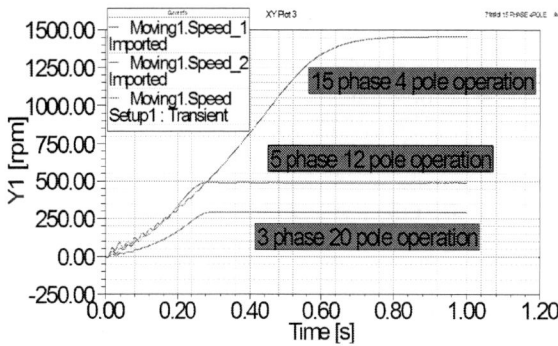

Fig. 8. Shows rotor speed for first three pole-phase operations (FEA Simulation).

are shown with offset of 10 and 20 Nm just to give comparision of torque ripple otherwise load torque for all the combinations is 10Nm. From Fig.7 one important advantage of multiphase is observed that torque ripple in 15 phase is very less compared to the other two combinations [1] and [4]. Fig.8 shows the speed curves for the same first three pole phase combinations.

Table II. compares the performance of the three pole phase combinations with respect to average torque, efficiency, torque ripple, and total losses. All these perfomance indicies are calculated by user defined solution tollkit of Ansys maxwell 2D with user editable python scripting. Efficiency in case of 5 phase 12 pole and 3 phase 20 pole appears to be less in comparision with 15 phase 4 pole case. This is beacause machine

TABLE II. PERFORMANCE INDICES FOR THE FIRST THREE POLE PHASE COMBINATIONS

PERFORMANCE INDICES	15 PHASE 4 POLE	5 PHASE 12 POLE	3 PHASE 20 POLE
INPUT POWER[kw]	1.6301	0.5662	0.4157
OUTPUT POWER[kw]	1.5321	0.4702	0.3069
TORQUE[Nm]	10.0786	9.9925	10.0182
SPEED[rpm]	1451.62	489.75	292.54
POWER FACTOR	0.9860	0.7012	0.4328
PHASE CURRENT[A]	2.3140	1.9045	3.0204
TOTAL LOSS[W]	98	96	108.6
EFFICIENCY[%]	93.98	83.04	73.88
TORQUE RIPPLE[Nm]	0.7187	1.1879	1.38557

is operated less than $1/4th$ of the rated load in case of 12 and 20 pole to meet the given load torque. Full load or near full load operation better efficiency is observed in case of 12 and 20 pole operation. Torque ripple is calculated as peak to peak variation of the average torque over the average value 10 Nm. Torque ripple in case of 15 phase 4 pole operation is least due to reduction in slot hormonics as the number of phases increases.

V. CONCLUSION

The proposed 15 phase IM is operated with Wide speed range in 1:3:5 ratios by applying PPM effectively. All the five pole-phase combinations are validated using the FEA simulation. The torque pulsation is least in 15 phase 4 pole case and well within limits for all other pole phase combinations. Constraint that the pole ratio must be a positive odd integer greater than one brings more clarity in the generalization of PPM. The performance indices table shows the best copper utilization and effectiveness of PPM. Hardware implementation of the proposed drive with a effective control scheme is the future objective of the author.

REFERENCES

[1] E.Levi, R. Bojoi, F. Profumo, H.A. Toliyat and S. WilliamsonG, "Multiphase induction motor drives - a technology status review," IET Electr. Power Appl.,2007, 1, (4), pp. 489-516.

[2] S. Williamson, and A.C. Smith,"Pulsating torque and losses in multiphase induction machines," IEEE Trans. Ind. Appl., 2003, 39, (4), pp. 986-993.

[3] S. Smith, "Developments in power electronics machines and drives," IEE Power Eng. J., 2002, 16, pp. 13-17.

[4] K. gopakumar, S. sathiakumar, S.K Biswas and J. Vithayathil "modified current source inverter fed induction motor drive with reduced torque pulsations," IEE proceedings, vol. 131, Pt. B, No. 4, july 1984

[5] K.N. Pavithran, R. Parimelalagan, and M.R. Krishnamurthy "Studies on inverter-fed five-phase induction motor drive," IEEE Trans. Power Electron., 1988, 3, (2), pp. 224-235.

[6] G. H. Rawcliffe, R.F. Burbidge, "A 2:1 pole-changing induction motor of improved performance," VOL. 104. PART A. No. 18. december 1957.

[7] M.G Say, Alternating Current Machines, Halsted Press,February 1984

[8] Mohamed Osama and Thomas A. Lipo, "Modeling and Analysis of a Wide-Speed-Range Induction Motor Drive Based on Electronic Pole Changing," IEEE Transactions on Industry Applications, vol. 33, no. 5, pp. 1177-1184, September/October 1997.

[9] K.C. Rajaraman, "design of phase-change two-speed windings for induction motors, using pole-amplitude modulation techniques," IEE proceedings, vol. 131, pt. b, no. 5, september 1984.

[10] G. H. Rawcliffe and W. Fong, "Clock diagrams and pole-amplitude modulation," PROC. IEE, Vol. 118, No. 5, may 1971.

[11] G.H. Rawcliffe and B.V. Jayawant, "Development of a New 3:1 Pole-Changing Motor," Proc. IEEE, vol. 103, no. a, pp. 306, 1956.

[12] John W. Kelly and Elias G. Strangas "Torque Control during Pole-Changing Transition of a 3:1 Pole Induction Machine," Proceeding of International Conference on Electrical Machines and Systems 2007, Oct. 8 11, Seoul, Korea.

[13] Dongsen Sun, Baoming Ge, Daqiang Bi "Winding Design for Pole-Phase Modulation of Induction Machines," 978-1-4244-5287-3/10/2010 IEEE

[14] Amrit Gautam and Joesph O. Ojo "Variable Speed Multiphase Induction Machine using Pole Phase Modulation Principle," 978-1-4673-2421-2/12/2012 IEEE

[15] A.K Sawhney, "A course in electrical machine design" Dhanpat Rai and Sons, India, 1984

Mathematical Model of
Novel Wound-Field Synchronous Motor
Self-Excited by Space Harmonics

Masahiro Aoyama/Shizuoka University, Suzuki Motor
Corporation
Department of Environment and Energy System,
Graduate School of Science and Technology
3-5-1 Johoku, Naka-Ku, Hamamatsu,
Shizuoka 432-8561, Japan
aoyamam@hhq.suzuki.co.jp

Toshihiko Noguchi/Shizuoka University
Department of Electrical and Electric Engineering,
Graduate School of Engineering
3-5-1 Johoku, Naka-Ku, Hamamatsu,
Shizuoka 432-8561, Japan
ttnogut@ipc.shizuoka.ac.jp

Abstract—**This paper describes mathematical modeling of a novel wound-field synchronous motor self-excited by space harmonics. The torque equation on the *dq*-reference frame is investigated to clarify the self-excitation mechanism of the proposed motor. It has been analytically explicated how the motor parameters give influences on the self-excited electromagnet torque. Furthermore, the current phase vs. torque characteristics are compared between the FEM based simulation result and the mathematical calculation result. Both of the results showed good agreement, which proves feasibility of the mathematical modeling discussed in the paper.**

Keywords— mathematical model; synchronous motor; self-excitation; space harmonics; concentrated winding.

I. INTRODUCTION

Internal combustion engine based automotive vehicles can be more efficient by introducing electric components such as motors and generators to their drive trains. The advancement of the drive train performance contributes to mitigate the global warming through reduction of CO2 emission and fuel consumption. Even a simple improvement such as an idling-stop system, which is an entry level system of the vehicle electrification, has a remarkable impact on the total efficiency of the vehicles. As the degree of electrification is higher, more advantages can be expected whereas the system becomes more complicated. An electric machine is one of the key components in hybrid vehicles (HEVs) and electric vehicles (EVs) from the viewpoint of dynamic and fuel consumption performances. Traction motors for the HEVs require a wide adjustable speed drive range, high maximum torque, and high power density without sacrificing its power conversion efficiency. Particularly, an IPM (Interior Permanent Magnet) motor is often applied to the HEVs owing to its highly improved efficiency and specific power per physical volume. Permanent magnets used for the IPM motor are relatively expensive because Nd-Fe-B magnets are generally employed to realize high energy density and to improve

fuel efficiency in the low-load operation for street use. Moreover, more expensive rare-earth metals such as Dy and Tb must be added to the Nd-Fe-B magnet to restrain demagnetization caused by the high temperate.

Therefore, varieties of rare-earth-free motors, particularly wound-field synchronous motor which replaces magnets with electromagnets, are focused due to remarkable rise of the Nd-Fe-B magnet market price [1][2]. For example, a separate excitation wound-field synchronous motor is proposed in [2]. This motor is capable to utilize the armature reaction torque by the wound-field torque, and the field magnetization control allows high efficiency operation. An external chopper circuit is, however, indispensable for the wound-field winding. Furthermore, it is rather difficult to transfer the field magnetization power from the primary to the secondary of the motor, and an extra copper loss in the wound-field winding is also a serious problem. Thus, a brushless-excitation technique proposed in [3] is reevaluated by the authors to solve the problems regarding the separate excitation wound-field motors. This classic brushless-excitation based synchronous motor has a stator with distributed windings (3 phase-4 poles) and a salient pole rotor (4 poles) with a single winding connected via a half-bridge rectifier. The harmonic components of an armature magnetomotive force are generated by 2-pole direct current excitation windings, and link to the rotor winding for the field magnetization. Another classic brushless-excitation technique, which has rotor windings with a diode rectifier, is proposed in [4]. This self-excitation technique utilizes the inverted magnetic field generated by auxiliary armature capacitor winding for the rotor magnetization, but the armature copper loss increases because the winding space factor detrimentally decreases.

This paper tries to solve the above problems of the classical wound-field synchronous motor, and proposes a novel configuration and operation mechanism of the self-excitation, focusing on the space harmonics power. Particularly, a mathematical model of the proposed motor

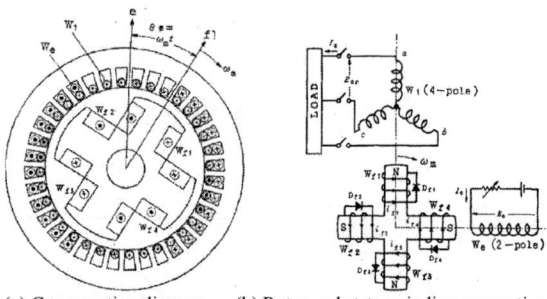

(a) Cross section diagram.   (b) Rotor and stator winding connection.
Fig. 1. Classic self-excitation synchronous motor.

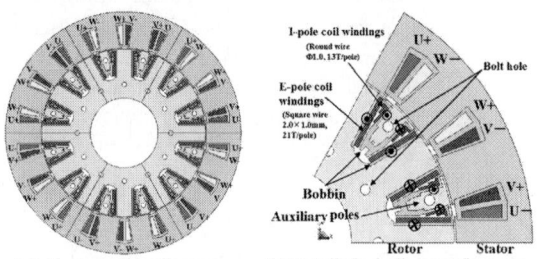

(a) Cross section diagram.   (b) Detailed winding configuration.
Fig. 2. Proposed self-excitation synchronous motor.

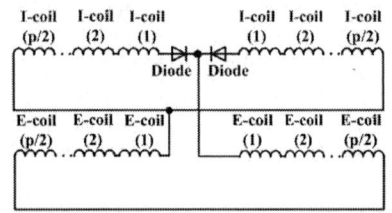

Fig. 3. Rotor winding connection of proposed motor.

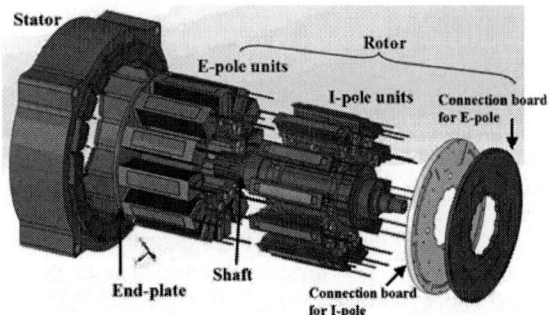

Fig. 4. Mechanical configuration of proposed motor.

is discussed in the paper for the purpose of designing a high-efficiency drive.

II. OUTLINE OF PROPOSED MOTOR

Figure 1 shows a classic blushless-excitation synchronous motor presented in [3]. The motor has a direct current winding connected to an external chopper circuit in the stator slots in addition to the three-phase armature windings for excitation. The direct current winding is indispensable to acquire the second harmonic component for the excitation because the three-phase distributed windings generate only the rotating magnetic field. The proposed motor has, however, concentrated armature windings not to use any special excitation

TABLE I.   SPECIFICATIONS OF MOTOR.

Number of poles	12
Number of slots	18
Stator outer diameter	200 mm
Rotor diameter	138.6 mm
Axial length of core	54 mm
Air gap length	0.7 mm
Maximum current	273 A_{pk} (45 s)
Stator winding resistance	32.1 mΩ / phase
Number of coil-turn	48
Winding connection	6 parallel
I-pole winding resistance	37.0 mΩ / pole
E-pole winding resistance	28.2 mΩ / pole
Thickness of iron core steel plate	0.35 mm

windings in the stator. Conventional common motors dissipate space harmonics power caused by the concentrated stator and the salient pole configuration, whereas the proposed motor positively utilizes the space harmonics power for the excitation. Figure 2 shows the proposed self-excitation motor where the wound-field windings are added to the rotor salient poles and the induction coils are placed in spaces between the rotor salient poles, i.e., rotor slots. Each of the induction poles (I-poles) is a special pole exclusively used to generate the magnetizing power from the third space harmonics. On the other hand, each excitation pole (E-pole) is a salient pole of the rotor magnetized by the I-poles, which uses the retrieved third space harmonics power. Every I-pole and E-pole is connected in series via a diode rectifying circuit as shown in Fig. 3, where p indicates a pole number. The I-pole is an auxiliary pole that induces a voltage proportional to the derivative of the third space harmonic flux, and is designed to be magnetically independent of the main magnetic path to prevent reduction of the saliency. Every I-pole is mechanically held from an axial direction using support ring boards as shown in Fig. 4. Specifications of the proposed motor shown in Fig. 2 are listed in Table I.

III. THEORETICAL ANALYSIS OF PROPOSED MOTOR

A. Principle of Third Space Harmonics Generation

The proposed motor can obtain the field magnetization power from the third space harmonics owing to the slot combination between the rotor pole counts and the stator slot counts. Figure 5 shows the magnetic flux density, the flux lines and the d-axis inductance variation of the 2pole-6slot motor. The d-axis inductance consists of a constant part and a sixth slot harmonics part periodically changes with the rotation. On the other hand, in the case of a fractional slot winding motor such as a 2pole-3slot one, the d-axis inductance variation is caused in a slightly different manner. As illustrated in Fig. 6, one of the rotor salient pole and the other are not always symmetric. If the

The 2014 International Power Electronics Conference

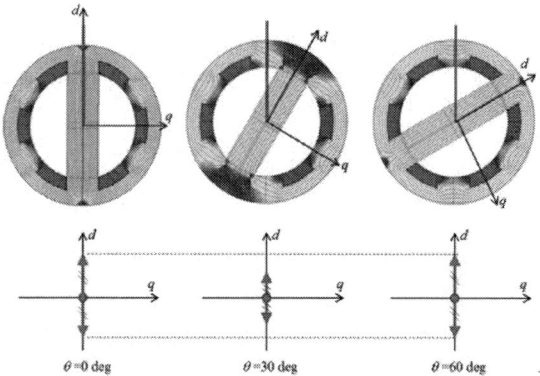

Fig. 5. *d*-axis inductance variation of 2pole-6slot motor.

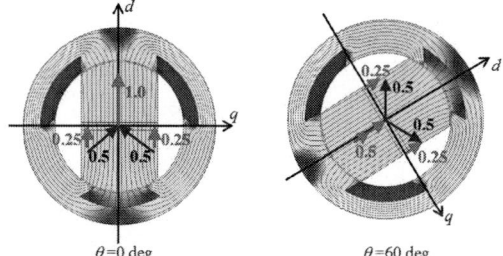

Fig. 6. *d*-axis inductance variation depending on rotor position.

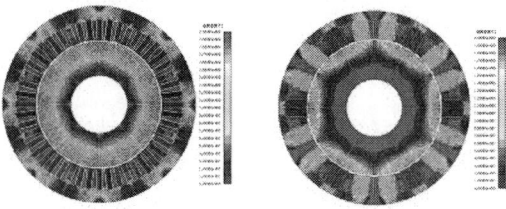

(a) Distributed windings.   (b) Concentrated windings.
Fig. 7. Magnetic flux density and flux lines for solid rotor.

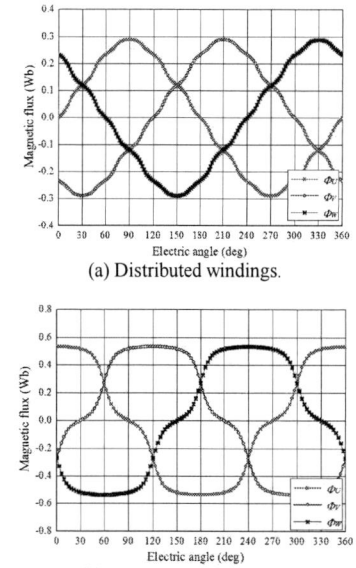

(a) Distributed windings.

(b) Concentrated windings.
Fig. 8. Magnetic flux waveform linking to armature winding.

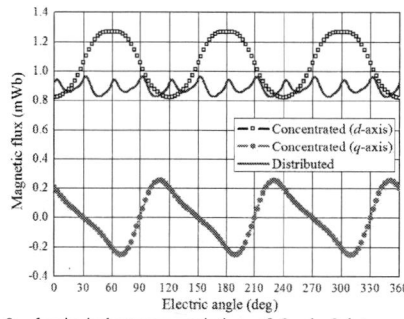

Fig. 9. *d*-axis inductance variation of 2pole-3slot concentrated winding and 8pole-48slot distributed winding configurations.

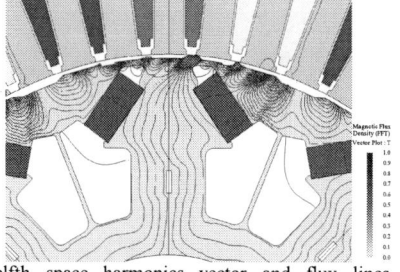

(a) Twelfth space harmonics vector and flux lines of distributed winding configuration.

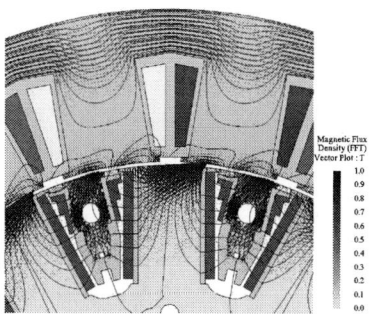

(b) Third space harmonics vector and flux lines of concentrated winding configuration.
Fig. 10. Comparison of space harmonics vector and flux lines.

rotor position is θ=0 deg, the positive direction of the *d*-axis coincides with one of the stator teeth, but the opposite salient pole of the *d*-axis faces with a slot. Therefore, the opposite pole forms a closed magnetic circuit with interference between the *d-q* axes. On the other hand, if the rotor position is θ=30 deg, the positive direction of the *d*-axis is oriented to the slot, where the magnetic reluctance is relatively high. As described above, the 2pole-3slot motor has a *d*-axis inductance composed with a constant part and a periodical third space harmonics part according to the rotation. This periodical variation of the inductance is particularly caused by the doubly salient configuration of the motor. Figure 7 shows magnetic flux density distributions of the distributed winding and the concentrated winding configurations when a bulk solid rotor is used. Figure 8 shows magnetic flux waveforms linking to the armature windings for both winding configurations. As can be seen in the figures, the armature magnetic fluxes of the distributed winding configuration are sinusoidal waveforms, but the fluxes of the concentrated one are close to trapezoidal waveforms, which mainly include the fifth and seventh space harmonics. Figure 9 shows *d*-axis inductance variation with respect to the electrical angle of a 2pole-3slot concentrated winding and an 8pole-48slot distributed winding configurations. As shown in the figure, the periodical variation of the *d*-axis inductance caused by the

978-1-4799-2706-7/14 $31.00 © 2014 IEEE   1407

slot harmonics is observed on the constant DC components, and the third space harmonic component of the concentrated winding configuration is much larger than the ripples of the distributed one. Figure 10 shows the main space harmonics vector and flux lines of the two winding configurations. It is found that the third space harmonic flux, which is caused by the 2 to 3 slot combination, links deeply into the rotor although the twelfth space harmonic flux generated by the distributed winding configuration links around the rotor surface. In addition, the third space harmonic flux goes mainly through the rotor salient pole and the slot. This is the reason why the I-pole must be placed on the q-axis, which is the most efficient way to retrieve the third space harmonic power.

B. Mathematical Model on dq-Reference Frame

The operation principle of the proposed motor can be explicated by voltage equations on the synchronous rotating reference frame. As shown in Fig. 9, the d-axis inductance L_d and the q-axis inductance L_q can be given by

$$L_d(\omega t) = L_{d0} + L_{da}\cos 3\omega t \text{ , and} \tag{1}$$

$$L_q(\omega t) = L_{q0} + L_{qa}\sin\left(-3\left(\omega t - \frac{\pi}{6}\right)\right) = L_{q0} + L_{qa}\sin 3\omega t \text{ ,} \tag{2}$$

where L_{d0} and L_{q0} are constant parts, and L_{da} and L_{qa} are amplitudes of the periodical variations. ω is an electrical synchronous angular velocity. The mathematical model of the proposed motor can be expressed as the following voltage equation:

$$
\begin{bmatrix} v_{sd} \\ v_{sq} \end{bmatrix} = \begin{bmatrix} R_s & 0 \\ 0 & R_s \end{bmatrix}\begin{bmatrix} i_{sd} \\ i_{sq} \end{bmatrix} + \begin{bmatrix} \mathrm{p} & -\omega \\ \omega & \mathrm{p} \end{bmatrix}\begin{bmatrix} \psi_{sd} \\ \psi_{sq} \end{bmatrix}
$$

$$
= \begin{bmatrix} R_s & 0 \\ 0 & R_s \end{bmatrix}\begin{bmatrix} i_{sd} \\ i_{sq} \end{bmatrix} + \begin{bmatrix} \mathrm{p} & -\omega \\ \omega & \mathrm{p} \end{bmatrix}\begin{bmatrix} L_d & 0 & M_d & 0 \\ 0 & L_q & 0 & M_q \end{bmatrix}\begin{bmatrix} i_{sd} \\ i_{sq} \\ i_{rd} \\ i_{rq} \end{bmatrix}, \tag{3}
$$

where v_{sd}, v_{sq}, i_{sd} and i_{sq} are armature voltages and currents, i_{rd} and i_{rq} are a d-axis and a q-axis rotor winding currents, R_s is an armature winding resistance, M_d and M_q are a d-axis and a q-axis mutual inductances, p denotes a differential operator, respectively. The self inductances on the dq-reference frame vary periodically with respect to the time as expressed by Eqs. (1) and (2). Hence, Eq. (3) can be rewritten as follows:

$$
\begin{bmatrix} v_{sd} \\ v_{sq} \end{bmatrix} = \begin{bmatrix} R_s & 0 \\ 0 & R_s \end{bmatrix}\begin{bmatrix} i_{sd} \\ i_{sq} \end{bmatrix} + \begin{bmatrix} L_d & 0 & M_d & 0 \\ 0 & L_q & 0 & M_q \end{bmatrix}\mathrm{p}\begin{bmatrix} i_{sd} \\ i_{sq} \\ i_{rd} \\ i_{rq} \end{bmatrix}
$$

$$
+ \begin{bmatrix} \mathrm{p}L_d & 0 & \mathrm{p}M_d & 0 \\ 0 & \mathrm{p}L_q & 0 & \mathrm{p}M_q \end{bmatrix}\begin{bmatrix} i_{sd} \\ i_{sq} \\ i_{rd} \\ i_{rq} \end{bmatrix} \tag{4}
$$

$$
+ \omega\begin{bmatrix} 0 & -L_q & 0 & -M_q \\ L_d & 0 & M_d & 0 \end{bmatrix}\begin{bmatrix} i_{sd} \\ i_{sq} \\ i_{rd} \\ i_{rq} \end{bmatrix} .
$$

In the above voltage equation, the mutual inductance M_d and M_q can be simply expressed as follows by using the number of turns of the d-axis, q-axis rotor windings and the stator windings N_{rd}, N_{rq} and N_S:

$$M_d = \frac{N_{rd}}{N_S}K_{Ld}L_d = \frac{N_{rd}}{N_S}K_{Ld}\left(L_{d0} + L_{da}\cos 3\omega t\right), \text{ and} \tag{5}$$

$$M_q = \frac{N_{rd}}{N_S}K_{Lq}L_q = \frac{N_{rq}}{N_S}K_{Ld}\left(L_{q0} + L_{qa}\sin 3\omega t\right), \tag{6}$$

where K_{Ld} and K_{Lq} are leakage magnetic flux coefficients of the d-axis and q-axis. Substituting the time derivatives of L_d, L_q, M_d and M_q into the third term of Eq. (4), the mathematical model of the motor is expressed as

$$
\begin{bmatrix} v_{sd} \\ v_{sq} \end{bmatrix} = \begin{bmatrix} R_s & 0 \\ 0 & R_s \end{bmatrix}\begin{bmatrix} i_{sd} \\ i_{sq} \end{bmatrix} + \begin{bmatrix} L_d & 0 & \frac{N_{rd}}{N_S}K_{Ld}L_d & 0 \\ 0 & L_q & 0 & \frac{N_{rq}}{N_S}K_{Lq}L_q \end{bmatrix}\mathrm{p}\begin{bmatrix} i_{sd} \\ i_{sq} \\ i_{rd} \\ i_{rq} \end{bmatrix}
$$

$$
+ \omega\begin{bmatrix} -3L_{da}\left(\dfrac{L_q - L_{q0}}{L_{qa}}\right) & -L_q & * \\ L_d & 3L_{qa}\left(\dfrac{L_d - L_{d0}}{L_{da}}\right) & \\ * & -\dfrac{N_{rd}}{N_S}K_{Ld}L_{da}\left(\dfrac{L_q - L_{q0}}{L_{qa}}\right) & -\dfrac{N_{rq}}{N_S}K_{Lq}L_q \\ & \dfrac{N_{rd}}{N_S}K_{Ld}L_d & \dfrac{N_{rq}}{N_S}K_{Lq}L_{qa}\left(\dfrac{L_d - L_{d0}}{L_{da}}\right) \end{bmatrix}\begin{bmatrix} i_{sd} \\ i_{sq} \\ i_{rd} \\ i_{rq} \end{bmatrix}. \tag{7}
$$

The first term of the above equation is an armature winding resistance voltage drop, the second term is a transformer electromotive force, and the third term is an electromotive force.

C. Field Current

The rotor current i_{rd} and i_{rq} are expressed by using i_{sd} and i_{sq} because the field current i_{rd} and the induced current i_{rq} are caused by the flux linkage from the stator. The induced voltage of the I-pole windings is applied to the E-pole windings for the self excitation through the full-bridge rectifier. Thus, the voltage applied to the E-pole windings is an absolute value of the I-pole winding voltage, and is expressed as Eq. (12), using Fourier series:

$$v_{rd} = |v_{rq}| = a_0 + \sum_{n=1}^{\infty}\left(a_n\cos n\left(\frac{2\pi}{\frac{T}{2}}\right)t + b_n\sin n\left(\frac{2\pi}{\frac{T}{2}}\right)t\right), \tag{8}$$

$$= a_0 + \sum_{n=1}^{\infty}\left(a_n\cos 6n\omega t + b_n\sin 6n\omega t\right)$$

where a_0, a_n and b_n are Fourier series coefficients. In order to derive the I-pole winding voltage v_{rq}, the following magnetic fluxes on the dq-reference frame are considered:

$$\phi_d = L_d i_{sd} = \left(L_{d0} + L_{da}\cos 3\omega t\right)i_{sd} \text{ , and} \tag{9}$$

$$\phi_q = L_q i_{sq} = \left(L_{q0} + L_{qa}\sin 3\omega t\right)i_{sq} . \tag{10}$$

By using the above equations, v_{rq} can be calculated as

$$v_{rq} = -N_{rq}\mathrm{p}\left(K_{d-axis}\phi_d + \phi_q\right)$$

$$= -3\omega N_{rq}\left(-K_{d-axis}L_{da}i_{sd}\sin 3\omega t + L_{qa}i_{sq}\cos 3\omega t\right), \tag{11}$$

where K_{d-axis} is a coefficient expressing interference

between the two axes. This coefficient must be determined by the degree of the magnetic interference from the *d*-axis space harmonic flux onto the I-pole windings. By substituting the above expression into Eq. (8), the voltage applied to the E-pole windings v_{rd} is obtained as follows:

$$v_{rd} = \frac{6\omega}{\pi} \begin{pmatrix} N_{rq} K_{d-axis} L_{da} i_{sd} \left(\frac{1}{2} + \sum_{n=1}^{\infty} \frac{1}{1-4n^2} \cos 6n\omega t \right) \\ + N_{rq} L_{qa} i_{sq} \left(\frac{1}{2} + \sum_{n=1}^{\infty} \frac{1}{1-4n^2} \cos 6n\omega t \right) \end{pmatrix}. \quad (12)$$

The DC component of v_{rd} extracted from Eq. (12) can be given by

$$v_{rd(DC)} = \frac{3\omega N_{rq}}{\pi} \left(K_{d-axis} L_{da} i_{sd} + L_{qa} i_{sq} \right). \quad (13)$$

Since the DC voltage applied to the E-pole windings is a source of the field current i_{rd}, solving the transient response of the rotor winding gives the field current as expressed by the following equation:

$$i_{rd}(t) = \frac{3\omega N_{rq}}{\pi (R_{rd} + R_{rq})} \left(K_{d-axis} L_{da} i_d + L_{qa} i_q \right) \left(1 - e^{\frac{-(R_{rd}+R_{rq})}{L_{rd}}t} \right). \quad (14)$$

D. Torque

The motor model on the *dq*-reference frame can be further rewritten as follows, taking the above discussion into account:

$$\begin{bmatrix} v_d \\ v_q \end{bmatrix} = \begin{bmatrix} R_s & 0 \\ 0 & R_s \end{bmatrix} \begin{bmatrix} i_d \\ i_q \end{bmatrix}$$
$$+ \begin{bmatrix} L_d + 3\omega K_{d-axis} L_{da} K_{Ld} L_d N_{rd} & 3\omega L_{qa} K_E \\ 3\omega K_{d-axis} L_{da} K_{Lq} L_q \frac{N_{rq}}{\pi} K_E & L_q + \frac{3\omega}{\pi} L_{qa} K_E \end{bmatrix} p \begin{bmatrix} i_d \\ i_q \end{bmatrix}$$
$$+ \omega \begin{bmatrix} -3 \left(L_{da} L_{qq} + \omega K_E K_{d-axis} L_{da} \left(N_{rd} K_{Ld} L_{da} L_{qq} + \frac{N_{rq}}{\pi} K_{Lq} L_q \right) \right) \\ L_d + 3\omega K_E K_{d-axis} L_{da} \left(N_{rd} K_{Ld} L_d + \frac{N_{rq}}{\pi} K_{Lq} L_{qa} L_{dd} \right) \end{bmatrix} *$$
$$* \begin{bmatrix} -\left(L_q + 3 K_E K_{d-axis} L_{da} L_{qa} \left(N_{rd} K_{Ld} L_{da} L_{qq} + \frac{N_{rq}}{\pi} K_{Lq} L_q \right) \right) \\ 3 \left(L_q L_{dd} + \omega K_E K_{d-axis} L_{da} L_{qa} \left(N_{rd} K_{Ld} L_d + \frac{N_{rq}}{\pi} K_{Lq} L_{qa} L_{dd} \right) \right) \end{bmatrix} \begin{bmatrix} i_d \\ i_q \end{bmatrix}$$
$$(15)$$

where the coefficients K_E, L_{dd} and L_{qq} are given by Eqs. (20) and (21):

$$K_{E1} = \frac{N_{rq}}{\pi N_S (R_{rd} + R_{rq})}, \quad (16)$$

$$L_{dd} = \frac{L_d - L_{d0}}{L_{da}}, \text{ and } L_{qq} = \frac{L_q - L_{q0}}{L_{qa}}. \quad (17)$$

The output torque of the proposed motor is obtained by the vector product between the armature current and the magnetic flux, which is described in a part of the third term of Eq. (7):

$$T = P_p \begin{bmatrix} i_{sd} & i_{sq} \end{bmatrix} \begin{bmatrix} -3L_{da} L_{qq} & -L_q \\ L_d & 3L_{qa} L_{dd} \end{bmatrix} *$$
$$* \begin{bmatrix} -\frac{N_{rd}}{N_S} K_{Ld} L_{da} L_{qq} & -\frac{N_{rq}}{N_S} K_{Lq} L_q \\ \frac{N_{rd}}{N_S} K_{Ld} L_d & \frac{N_{rq}}{N_S} K_{Lq} L_{qa} L_{dd} \end{bmatrix} \begin{bmatrix} i_{sd} \\ i_{sq} \\ i_{rd} \\ i_{rq} \end{bmatrix}$$
$$= P_p (L_d - L_q) i_{sd} i_{sq} + P_p \left(3 \left(-L_{da} L_{qq} i_{sd}^2 + L_{qa} L_{dd} i_{sq}^2 \right) \right.$$
$$+ \frac{N_{rd}}{N_S} K_{Ld} \left(-L_{da} L_{qq} i_{sd} + L_d i_{sq} \right) i_{rd}$$
$$\left. + \frac{N_{rq}}{N_S} K_{Lq} \left(-L_q i_{sd} + L_{qa} L_{dd} i_{sq} \right) i_{rq} \right)$$
$$(18)$$

where P_p is a pole-pair number. As expressed in the above expression, the output torque is composed of the two terms, i.e., reluctance torque and electromagnet torque. Therefore, output torque on the *dq*-reference frame at the steady-state can be obtained as:

$$T = P_p (L_d - L_q) i_{sd} i_{sq}$$
$$+ P_p \begin{pmatrix} 3 \left(-L_{da} L_{qq} i_{sd}^2 + L_{qa} L_{dd} i_{sq}^2 \right) \\ + 3\omega K_E \left(K_{d-axis} L_{da} i_d + L_{qa} i_q \right) \left(\frac{N_{rd} K_{Ld} \left(-L_{da} L_{qa} i_d + L_d i_q \right)}{\pi} \right) \\ + \frac{N_{rq}}{\pi} K_{Lq} \left(-L_q i_d + L_q L_{dd} i_q \right) \end{pmatrix}.$$
$$(19)$$

Since the field current generating the electromagnet torque is proportional to ω as expressed by Eq. (14), the proposed motor cannot deliver the sufficient electromagnet torque in the low-speed range. In addition, because the electromagnet torque includes the leakage magnetic flux coefficients as shown in Eqs. (5) and (6) as well as the winding turn ratio between the stator and the rotor, their parameter values have significant impact on the torque generation.

IV. VARIFICATION OF OPERATION CHARACTERISTICS

A. Current Phase vs. Torque Characteristics

Figure 11 shows current phase vs. torque (average torque in the steady state) characteristics for 1000 r/min of the proposed motor calculated by FEM based computer simulations. The characteristics are calculated under the condition of use of a sinusoidal current source. Separation of the reluctance torque and the electromagnet torque is performed through the following steps:

1) calculation of the reluctance torque at the current phase of 45 deg without connecting the rotor windings,
2) determination of the current phase vs. reluctance torque characteristic by fitting a $\sin 2\beta$ trigonometric function of which amplitude is a torque value at 45 deg calculated in the previous step,
3) calculation of the current phase vs. total torque characteristic with the rotor windings connected, and
4) subtraction of the reluctance torque from the total torque to obtain the electromagnet torque separately.

As shown in Fig. 11, the electromagnet torque, i.e., an additional torque generated by the self-excitation using the space harmonics, is enlarged as the armature current becomes higher. Figure 12 shows the characteristics

TABLE II. INDUCTANCES USED FOR CALCULATION OF MATHEMATICAL MODEL.

Inductance	L_{d0}			L_{q0}		
Armature current	100 A$_{pk}$	200 A$_{pk}$	273 A$_{pk}$	100 A$_{pk}$	200 A$_{pk}$	273 A$_{pk}$
500 r/min	-	-	7.5 mH	-	-	3.8 mH
1000 r/min	9.0 mH	8.5 mH	7.2 mH	3.6 mH	3.6 mH	3.6 mH
2000 r/min	-	-	6.7 mH	-	-	3.1 mH
3000 r/min	-	-	6.3 mH	-	-	2.7 mH

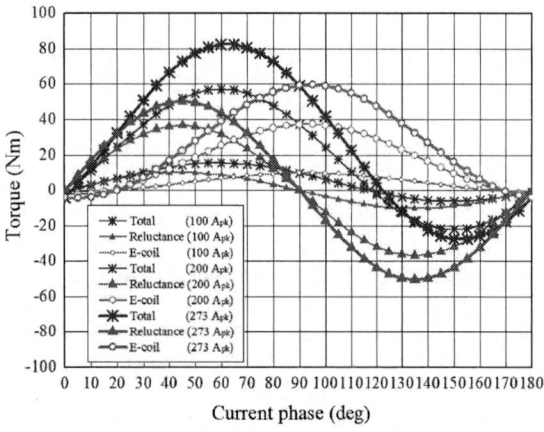

Fig. 11. Current phase vs. torque characteristics at 1000 r/min calculated by FEM model.

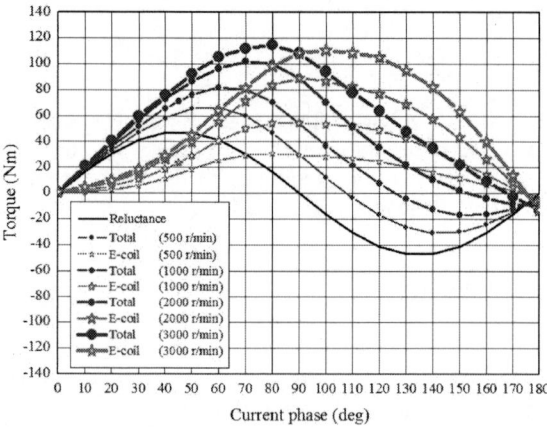

Fig. 13. Current phase vs. torque characteristics for 273 A$_{pk}$ calculated by FEM model.

Fig. 12. Current phase vs. torque characteristics at 1000 r/min calculated by mathematical model.

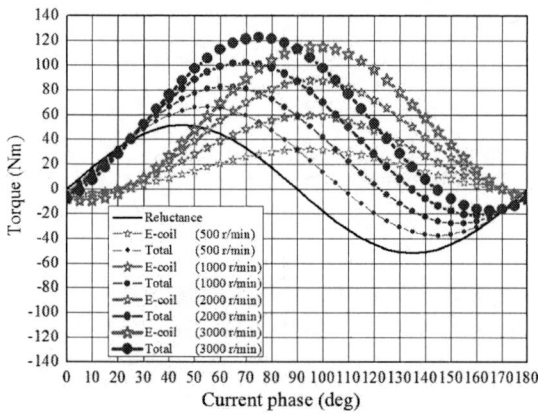

Fig. 14. Current phase vs. torque characteristics for 273 Apk calculated by mathematical model.

calculated by the mathematical model given by Eq. (19). The parameters used in Eq. (19) are listed in TABLES I and II, where the inductance variations caused by the magnetic saturation and the operating speed change are taken into account. The leakage magnetic flux coefficients used in the mathematical model are indicated in TABLE III. The periodical variations of the d-axis and the q-axis inductances are L_{da}=6.2×10^{-3} mH and L_{qa}=8.8×10^{-3} mH, respectively.

As can be seen in Figs. 11 and 12, both of the results have overall similar characteristics with indicating the same tendency. However, some errors of the

electromagnet torque can be found in the region where the armature current phase is not advanced. The error must be caused by having ignored an interference that links from the d-axis space harmonics to the E-pole windings and by having ignored higher order space harmonics more than 2ω (Approximated an ideal magnetomotive force distribution.)

B. Adjustable Speed Drive Characteristics

Figures 13 and 14 show adjustable speed drive characteristics calculated by the FEM simulation and the mathematical model, respectively. Both of the figures indicate very similar characteristics and the same trends.

TABLE III. LEAKAGE INDUCTANCE COEFFICIENTS.

	K_{Ld}	K_{Lq}	$K_{d\text{-}axis}$
500 r/min	0.5	0.5	0.2
1000 r/min	0.5	0.5	0.2
2000 r/min	0.4	0.5	0.2
3000 r/min	0.35	0.5	0.2

As discussed in Eq. (19), the higher electromagnet torque is delivered as the synchronous rotation speed increases because the torque generated by the proposed motor is proportional to the speed. The MTPA (Maximum Torque Per Ampere) control angle advances with the increase of the speed. On the other hand, the electromagnet torque generated by the space harmonics is limited by the magnetic saturation of the rotor teeth or the stator teeth.

V. CONCLUSION

This paper has proposed a novel rare-earth-free motor that can utilize space harmonics power for field magnetization instead of permanent magnets. The operation principle of the proposed motor has been explicated by voltage equations and torque equation on the synchronous rotating reference frame. The comparison of the current phase vs. torque characteristics has been conducted between analyzed results by the FEM model and calculated results by the mathematical model, which verifies feasibility of the mathematical model. The future work is to develop a prototype machine and to examine the various operation characteristics of the machine through experimental tests. Figures 15 to 17 illustrate three-dimensional component exploded views of the prototype. The E-pole windings are mounted on the rotor iron core as shown in Fig. 15 to improve the slot space factor. Both of the E-pole and the I-pole windings are connected with connection boards on the rotor ends as shown in Fig. 17, and the diodes are fixed on the connection boards by using resin mold.

REFERENCES

[1] M. Azuma, M. Hazeyama, M. Morita, Y. Kuroda, and M. Inoue, "Driving Characteristics of a Claw Pole Motor Using Field Excitation for Hybrid Electric Vehicles," *IEEJ Technical Meeting on Vehicle Technology*, pp. 37-40, 2011 (in Japanese).

[2] Y. Kuwahara, T. Kosaka, N. Matsui, Y. Komada, and H, Kajiura, "Drive Performance Evaluation of Wound Field Flux Switching Motor for HV Drives," *IEEJ Technical Meeting on Vehicle Technology*, VT-13-023, pp. 49-54, 2013 (in Japanese).

[3] S. Nonaka and K. Kesamaru, "Blushless Three-Phase Synchronous Generator without Exciter," *IEEJ Trans. on Industry Applications*, vol. 105, no. 10, pp. 851-858, 1985 (in Japanese).

[4] S. Nonaka, "The Brushless Self-Excited Type Single-Phase Synchronous Generator," *IEEJ Trans.*, vol.82, No.883, pp. 627-634, 1962 (in Japanese).

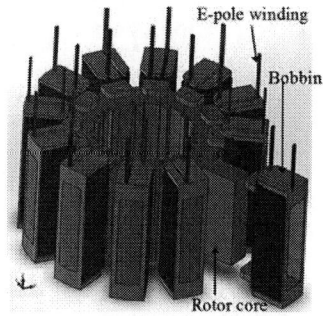

Fig. 15. Mechanical configuration of E-poles.

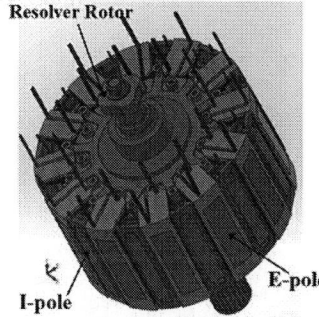

Fig. 16. Rotor assembly without rotor winding connection boards.

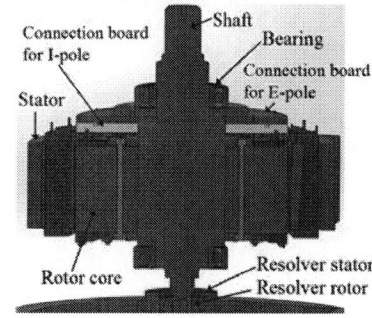

Fig. 17. Vertical cross section diagram of proposed motor.

The 2014 International Power Electronics Conference

Dual purpose no voltage winding design for the bearingless ac homopolar and consequent pole motors

Eric Severson*, Robert Nilssen[†], Tore Undeland[†], and Ned Mohan*
*Department of Electrical and Computer Engineering
University of Minnesota, Minneapolis, Minnesota 55455
Email: sever212@umn.edu
[†]Department of Electric Power Engineering
Norwegian University of Science and Technology, Trondheim, Norway
Email: Robert.Nilssen@ntnu.no

Abstract—**Winding layouts for the bearingless ac homopolar and consequent pole motors that allow the same coils to be used for both radial force and torque production are investigated. This concept enables a "pure" bearingless motor where the same iron and copper are used for both magnetic bearing operation and torque production. For three phase windings, it is shown that these schemes limit 6-pole machines to integral slot windings, but that 8-pole machines support various fractional slot windings. Conclusions are similarly drawn for two phase windings. A winding design procedure is proposed and 3D finite element analysis (FEA) results are presented for an example design.**

Keywords—*Bearingless machines, bridge configured winding, self-bearing motor, single motor winding*

I. INTRODUCTION

By definition, bearingless electric machines are machines which are able to use the same iron to function as both a motor and a magnetic bearing. In conventional bearingless machines, this requires two sets of stator windings: an armature winding for torque production and a suspension winding for magnetic bearing operation. Typically, the suspension winding and the torque winding will share the same stator slots. The slot space allocated to the suspension winding must be large enough to achieve an acceptable current density for a worst-case (highest-possible) suspension current. For machines that are vertically oriented and experience no net radial forces, such as those used in flywheel energy storage, this worse-case suspension current may only occur on rare occasions. An example of such an occasion is during start-up, when the rotor is highly eccentric. For these machines, the normal suspension current is small compared to the worst-case scenario and therefore during typical operation the current density in the suspension winding is very low. This wastes valuable slot space that could otherwise be used for the torque winding, degrading the machine's motor performance.

Recently, several papers have proposed winding methods which use the same coils to produce both radial forces and torque [1], [2], [3], [4], [5]. Such methods

allow for more optimal machine performance as the entire slot space can be used to carry torque-producing current when no magnetic bearing forces are required. In [2], [3], [4], [5] these methods have been explored on bearingless versions of classical machines. In [1], a new type of machine is proposed which inherently has this feature. No investigation has been carried out on applying such techniques to the bearingless ac homopolar or consequent pole motors. This paper investigates applying the two techniques above that result in no motional-EMF for the suspension winding [2], [3], [5], so-called "dual purpose no voltage" windings, to the bearingless ac homopolar motor. The same results are true for the bearingless consequent pole motor, a study of which is omitted because its required drive system is a simpler version of the drive system required for the bearingless ac homopolar motor.

In this paper, the application of "dual purpose no voltage" windings to the bearingless ac homopolar motor is explained, design rules for these windings are developed, a design approach is proposed, and FEA of an example design is presented.

II. BEARINGLESS AC HOMOPOLAR MOTOR

The ac homopolar motor, also known as the homopolar inductor alternator and the synchronous homopolar motor, has a robust rotor structure, uses stator-side current excitation, and has its poles in parallel with the magnetizing flux, which means that no additional ampere-turns of field winding current are required when increasing the number of poles. These features have attracted recent interest in implementing the ac homopolar motor as a high-speed machine, a super-conducting machine, and a high-frequency generator [6], [7], [8]. As a bearingless machine, work has focused on considering this machine in applications where magnetic bearings are typically required, such as in high-speed machinery [9], [10]. The authors of this work are interested in using the machine in a flywheel energy storage system. This is because, in addition to eliminating the need for radial magnetic bearings, the bearingless ac homopolar motor's current-based excitation can easily be adjusted to allow for lower iron losses.

978-1-4799-2706-7/14 $31.00 © 2014 IEEE

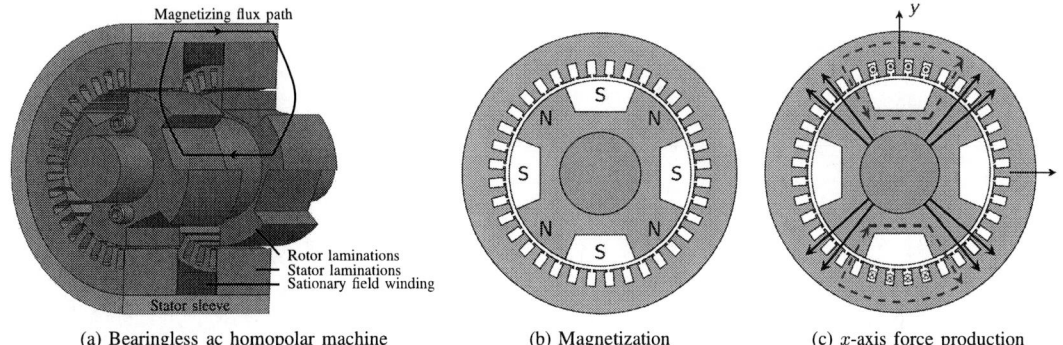

(a) Bearingless ac homopolar machine   (b) Magnetization   (c) x-axis force production

Fig. 1. The bearingless ac homopolar machine: (a) 3-D cross-section, (b) axial view of the magnetization created by the field winding, (c) depiction of flux density created by an x-phase suspension winding for radial force production.

For bearingless versions of classical synchronous motors to create radial force, a $p \pm 1$ pole-pair flux density must be created in the airgap to disrupt the otherwise symmetric p pole-pair flux density. To produce a constant direction force, this $p \pm 1$ pole-pair flux must rotate with the p pole-pair flux. This requires an ac suspension current, high-bandwidth angular position sensors, and imposes stringent control requirements [9]. On the other hand, the bearingless ac homopolar and consequent pole motors with at least 8-poles are able to produce a constant direction force with dc suspension current, regardless of the rotor's angular position. This substantially reduces the cost and complexity of the bearingless motor drive.

A diagram of the bearingless ac homopolar motor explaining torque and radial force production is depicted in Fig. 1. The airgap flux density is created by MMFs from three different windings: a dc field winding which is fixed to the stator and wraps around the rotor circumferentially–$\mathcal{F}_{\mathrm{f}}$, a $2p$-pole synchronous torque winding located in the stator slots and spanning the full axial length of the machine–$\mathcal{F}_{\mathrm{T}}$, and a 2-pole suspension winding which shares slot space with the stator winding but only spans a single rotor/stator section–$\mathcal{F}_{\mathrm{S},\psi}$. The radial airgap flux density can be calculated by imposing each of these MMF's upon an effective airgap length:

$$g_u(\alpha, \theta) = \frac{1}{h_1 + h_2 f(p[\alpha - \theta]))} \quad (1)$$

$$g_l(\alpha, \theta) = \frac{1}{h_1 + h_2 f(p[\alpha - \theta] + \pi))} \quad (2)$$

$$B_\psi(\alpha, \theta) = \mu_0 \frac{\mathcal{F}_{\mathrm{total},\psi}}{g_\psi(\alpha, \theta)} \quad (3)$$

Here, g_u is the effective upper airgap length at the stator location α when the p-pole pair rotor is at the rotational angle θ; f is a function expressed as a summation of the harmonic terms in the airgap; the subscript ψ is used to indicate whether a quantity applies to the upper airgap ($\psi = u$) or the lower airgap ($\psi = l$).

In the simplest case of idealized sinusoidal windings and a sinusoidal effective airgap length:

$$f(\phi) = \cos \phi$$
$$\mathcal{F}_{\mathrm{T}} = \hat{\mathcal{F}}_{\mathrm{T}} \cos(p[\theta - \alpha] + \phi_{\mathrm{T}})$$
$$\mathcal{F}_{\mathrm{S},\psi} = \hat{\mathcal{F}}_{\mathrm{S},\psi} \cos(-\alpha + \phi_{\mathrm{S},\psi})$$

The airgap flux density resulting from each MMF can be expressed as a sum of harmonic terms as:

$$B_u(\alpha, \theta) = B_{\mathrm{F}}(\alpha, \theta) + B_{\mathrm{T}}(\alpha, \theta) + B_{\mathrm{S}}(\alpha, \theta) \quad (4)$$

$$B_{\mathrm{F}}(\alpha, \theta) = \mu_0 h_1 \mathcal{F}_{\mathrm{f}} + \mu_0 h_2 \mathcal{F}_{\mathrm{f}} \cos(p[\alpha - \theta])$$

$$\begin{aligned} B_{\mathrm{T}}(\alpha, \theta) = {}& \mu_0 \frac{h_2}{2} \hat{\mathcal{F}}_{\mathrm{T}} \cos \phi_{\mathrm{T}} \\ & + \mu_0 h_1 \hat{\mathcal{F}}_{\mathrm{T}} \cos(p[\theta - \alpha] + \phi_{\mathrm{T}}) \\ & + \mu_0 \frac{h_2}{2} \hat{\mathcal{F}}_{\mathrm{T}} \cos(2p[\alpha - \theta] - \phi_{\mathrm{T}}) \end{aligned}$$

$$\begin{aligned} B_{\mathrm{S}}(\alpha, \theta) = {}& \mu_0 h_1 \hat{\mathcal{F}}_{\mathrm{S}} \cos(\alpha - \phi_{\mathrm{S}}) \\ & + \frac{\mu_0 h_2}{2} \hat{\mathcal{F}}_{\mathrm{S}} \cos([p-1]\alpha - p\theta + \phi_{\mathrm{S}}) \\ & + \frac{\mu_0 h_2}{2} \hat{\mathcal{F}}_{\mathrm{S}} \cos([p+1]\alpha - p\theta + \phi_{\mathrm{S}}) \end{aligned}$$

The radial forces created by (4) can be calculated using the Maxwell Stress Tensor (5). It can be shown that the only flux density terms which contribute to radial force generation differ in harmonic index by one. Provided that $p \geq 4$, these terms' dependence on the rotor's angular position θ will cancel each other out, proving that the bearingless ac homopolar motor is able to create forces independent of its rotor's angular position. Note that the only torque winding term which contributes to the radial forces is proportional to the current angle: $\cos \phi_{\mathrm{T}}$. If no direct-axis current is present, $\phi_{\mathrm{T}} = 90°$ and the magnetic suspension operation is independent of the motor operation. These results break down for $p < 4$. However, it was shown in [10] that careful winding design can still allow for stable levitation when $p = 3$. Practical suspension and torque windings contain harmonics other than the fundamental which introduces some radial force dependence on the rotor's angular position. More information regarding the theory of operation can be found in [11], [12].

$$F_{x,\psi}(\theta) = \int_0^{2\pi} \frac{1}{2\mu_0} B_\psi(\alpha, \theta)^2 \, lr \cos \alpha \, \mathrm{d}\alpha$$

$$F_{y,\psi}(\theta) = \int_0^{2\pi} \frac{1}{2\mu_0} B_\psi(\alpha, \theta)^2 \, lr \sin \alpha \, \mathrm{d}\alpha \quad (5)$$

III. DUAL PURPOSE NO VOLTAGE BEARINGLESS WINDINGS

Combining the suspension and torque windings into a single set of coils means having a winding with multiple

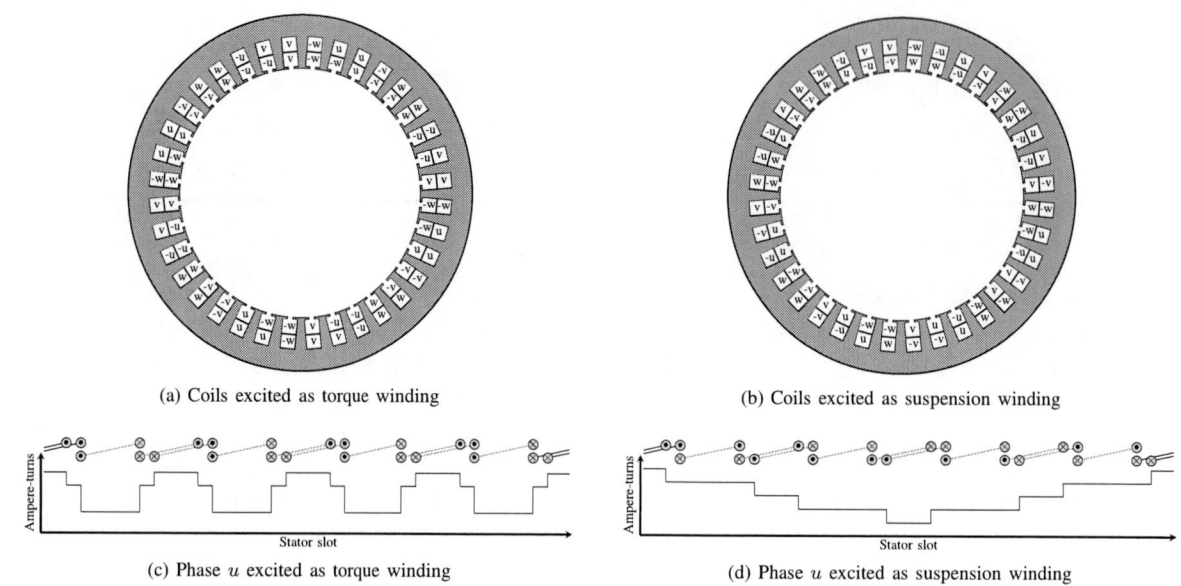

(a) Coils excited as torque winding

(b) Coils excited as suspension winding

(c) Phase u excited as torque winding

(d) Phase u excited as suspension winding

Fig. 2. Depiction of a 36 slot stator winding with coils that when excited from one set of terminals produce an 8-pole MMF and when excited from a separate set of terminals produce a 2-pole MMF.

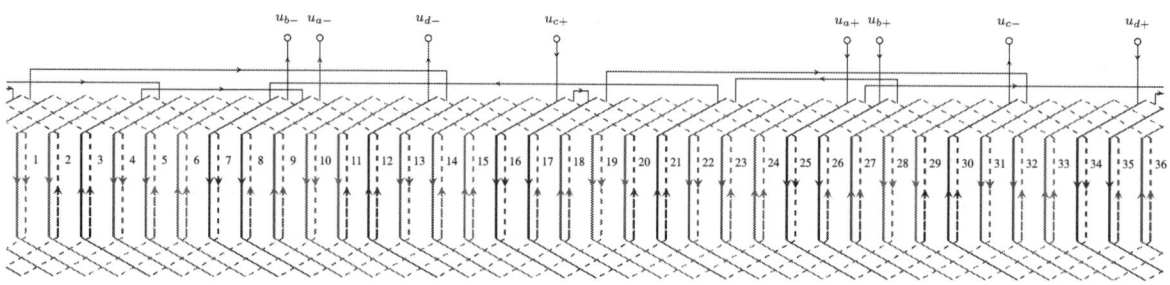

Fig. 3. Winding Diagram – Phase u connections shown.

sets of terminal connections, where flowing current into one set of terminal connections creates a 2-pole MMF (for the creation of suspension forces) and flowing current into another set of terminal connections creates an MMF of the same number of poles as the rotor (for the creation of torque). Furthermore, the phrase "no voltage" means that the set of terminals for the 2-pole winding must experience no motional-EMF from the rotor's rotation.

To understand how a single winding can have these features, an example design of an 8-pole, 36 slot stator is presented here and analyzed in more detail later. Designing a 3-phase, double-layer, fractional-slot winding in the usual way with a phasor diagram and $60°$ phase belts [13], see Fig. 5a, the winding depicted in Fig. 2a is produced. Now, keeping the coils in their same locations, a two-pole winding is constructed by swapping half of the coils' current direction, as shown in Fig. 2b. A single phase of the 8-pole terminals sees the series connection of the phasors shown in Fig. 5b. A two-pole suspension phase sees the series connection of the phasors shown in Fig. 5c, which cancel out at the synchronous frequency but add constructively at the suspension winding's frequency of $p = 1$ pole-pair. A winding schematic that enables these

connections is depicted in Fig. 3 and the interconnection with the two different styles of bearingless motor drives is shown in Fig. 4.

IV. COMPARISON OF THE TWO TOPOLOGIES

Unlike traditional bearingless motors, the bearingless ac homopolar motor has two separately wound stator segments enabling it to provide stable levitation in four axes, and doubling the number of inverters required for suspension. For the bridge style configuration [2], [3], the stator segments can be connected in parallel or in series but for the winding configuration proposed in [4], these segments must be connected in parallel. The bridge winding topology has the advantage of only having to carry the suspension current in the suspension inverter, whereas the parallel winding structure must also carry one-half of the torque winding current associated with each stator segment. This can be expensive in terms of the drive efficiency and the requirements for sizing the suspension winding drive switches as well as exposing some reliability concerns for levitation in the event of a failure in the torque winding inverter.

In terms of the amount of hardware needed to imple-

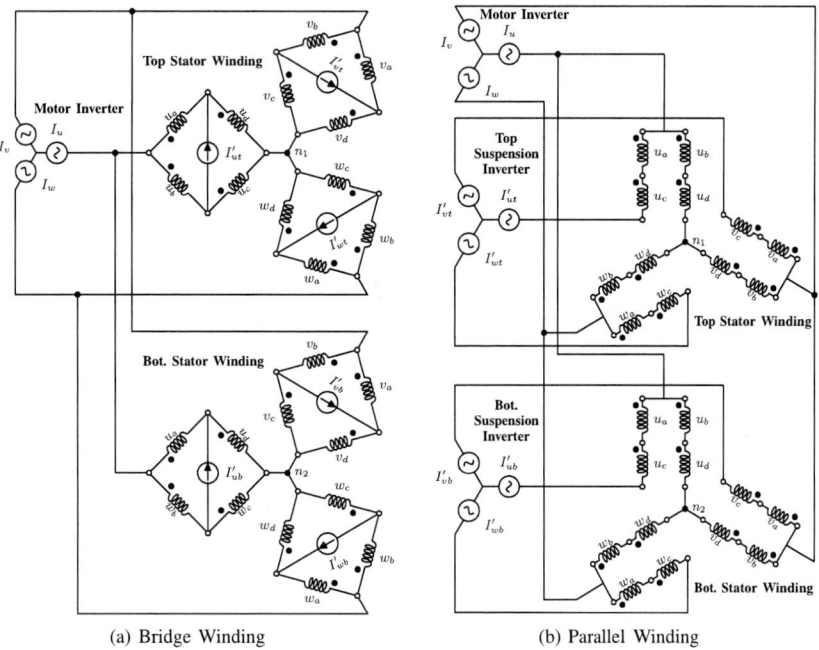

| | (a) Bridge Winding | (b) Parallel Winding |

Fig. 4. Winding connections for the styles of dual-purpose windings considered. The coil names correspond to the labels in Fig. 3 where the dotted terminals match the subscript '+'

ment each inverter, the bridge winding is at a significant disadvantage, as each of the suspension current sources in Fig. 4a must be implemented as an isolated single phase inverter. Furthermore, it will be shown in the following section that the bridge winding is more restrictive on the types of windings that can be used.

V. WINDING DESIGN REQUIREMENTS

Only certain combinations of the number stator coils z_c, stator slots Q, phases m, and rotor pole-pairs p can be used to realize a dual purpose no voltage winding. Restrictions on these combinations are now investigated.

A. Number of coils per phase winding

The bridge winding configuration requires that each phase winding have a multiple of four coils per phase: $z_c/m = 4k$; the parallel winding configuration requires that each phase winding have a multiple of two coils per phase $z_c/m = 2k$.

B. Symmetry conditions

A symmetry condition for fractional slot windings requires that the number of phases, m, be co-prime with the denominator of q (6), the number of slots per pole per phase [13]. Imposing this requirement along with the required number of coils per phase, restricts the types of winding that can be used.

$$q = \frac{Q}{2mp} = \frac{z}{n} \qquad (6)$$

In a single layer winding, the number of coils is $z_c = Q/2$ and in a double layer winding $z_c = Q$. Recalling that k is

TABLE I.   SLOTS PER POLE PER PHASE

Number of layers	Parallel Connection	Bridge Connection
Single	$q = \frac{2k}{p}$	$q = \frac{4k}{p}$
Double	$q = \frac{k}{p}$	$q = \frac{2k}{p}$

TABLE II.   POSSIBLE DENOMINATORS OF q WHEN $p = 4$

Number of layers	Parallel Connection	Bridge Connection
Single	$n = 1$ or 2	$n = 1$
Double	$n = 1, 2,$ or 4	$n = 1$ or 2

an integer, q is restricted to the values shown in Table I. The bearingless ac homopolar motor is most commonly used as a $p = 4$ and, under special circumstances, as a $p = 3$ pole-pair machine for reasons previously described.

For a six pole machine ($p = 3$), it can be seen from Table I that the denominator of q will be always be either 3 (fractional-slot) or 1 (integral slot). This means that a fractional-slot $m = 3$-phase winding cannot be used as it would violate the symmetry condition; however, 6-pole, two-phase fractional slot and three-phase integral slot windings are permissible.

The situation is a bit more complex for an 8-pole machine ($p = 4$) and is summarized in Table II. A single-layer bridge configuration will always yield $n = 1$, which is an integral-slot design, and in all fractional slot designs ($n \neq 1$), n is even. This means that a reduced winding (such as an $m = 2$ phase winding) must be an integral slot design. However, in this case, three-phase fractional slot windings are permissible.

TABLE III. POSSIBLE DUAL PURPOSE NO VOLTAGE WINDINGS

Q/m	q	$p = 4$								q	$p = 3$							
		$m = 3$				$m = 2$					$m = 3$				$m = 2$			
		Q	t	DL P/B	SL P/B	Q	t	DL P/B	SL P/B		Q	t	DL P/B	SL P/B	Q	t	DL P/B	SL P/B
2	$\frac{1}{4}$	6	2	P	-	4	4	-	-	$\frac{1}{3}$	6	3	-	-	4	1	P	-
4	$\frac{1}{2}$	12	4	P, B	P	8	4	-	-	$\frac{2}{3}$	12	3	-	-	8	1	P, B	P
6	$\frac{3}{4}$	18	2	P	-	12	4	-	-	1	18	3	P	-	12	3	P	-
8	1	24	4	P, B	P, B	16	4	P, B	P, B	$\frac{4}{3}$	24	3	-	-	16	1	P, B	P, B
10	$\frac{5}{4}$	30	2	P	-	20	4	-	-	$\frac{5}{3}$	30	3	-	-	20	1	P	-
12	$\frac{3}{2}$	36	4	P, B	P	24	4	-	-	2	36	3	P, B	P	24	3	P, B	P
14	$\frac{7}{4}$	42	2	P	-	28	4	-	-	$\frac{7}{3}$	42	3	-	-	28	1	P	-
16	2	48	4	P, B	P, B	32	4	P, B	P, B	$\frac{8}{3}$	48	3	-	-	32	1	P, B	P, B

C. Restrictions necessary for no suspension voltage

The two-pole suspension phase terminals utilize the same coils as the $2p$-pole torque phase terminals, but have half of the coils connected in the opposite direction in a way that cancels out the motional-EMF of the $2p$-pole rotor. The conditions necessary to facilitate this can be visualized in the winding star of phasors for the synchronous frequency as requiring either:

a) that each phase contain pairs of phasors that are 180 electrical degrees apart from one another, or

b) that each phase contain pairs of phasors that are at the same location.

If this condition is satisfied by criteria a), then careful consideration of end-turn connections must be made to insure that each phasor in a $180°$ pair is part of a separate coil. While this is always true in a double-layer lap winding with $60°$ phase belts, it may not be true for other winding layout approaches or single-layer windings.

VI. WINDING DESIGN

The design of dual purpose no voltage windings is now considered in detail. It is assumed that the machine should first and foremost be a good motor and that the magnetic bearing performance is a secondary concern. With this in mind, the design strategy is to first construct the desired torque winding (which supports a dual purpose winding) and then choose which half of the coils the direction of current flow should be reversed for the suspension terminals. Emphasis is placed on $p = 4$ pole-pole, $m = 3$ phase machines as this is the most common configuration for the bearingless ac homopolar motor.

A. Torque winding

The first step in the winding design is to layout a torque winding which supports a dual purpose winding scheme. Table III provides a list of two and three phase windings for $Q/m \leq 16$ when $p = 3$ and 4 that meet the previously described winding requirements. The table indicates whether a double layer parallel or bridge connection is possible in the "DL P/B" column and whether a single layer parallel of bridge connection is possible in the "SL P/B" column.

The chosen winding should be layed-out in a manner that locates phasors which are displaced by multiples of $180°$ in the same phase. For double-layer windings, this is done when the phasor star method is used with $60°$ phase belts.

B. Selection of suspension winding coils

The second step of the winding design is to determine which coils should be flipped in direction for series connection to the suspension winding terminals. The primary goal is to cancel out the motional-EMF. This requires that each phasor be matched to one which is $180°$ out of phase at the synchronous frequency. By having selected a winding from Table III and layed it out in a manner that guarantees that phasors displaced by multiples of $180°$ are in the same phase and aligned in the same direction (i.e. $+u$ phasor at $0°$ and a $-u$ phasor at $180°$), it is guaranteed to be possible to cancel the motional-EMF. There will typically be several combinations of phasors which meet this requirement. The best combination is determined by optimizing the suspension winding factors to either maximize the amount of force produced per ampere of suspension current (maximizing the fundamental winding factor) or reducing the amount of ripple in the suspension force as a function of rotor position (minimizing the harmonics in the suspension winding).

C. Design when $p = 4$, $m = 3$

When redrawing the phasor star at the frequency of the suspension winding ($p = 1$), the angular distance between each phasor will scale to a factor of $1/p$ of the angular distance in the synchronous frame. For a $p = 4$ winding, t is either 2 or 4. When $t = 2$, two aligned phasors will be 720 electrical degrees apart, meaning that when they are redrawn in the suspension winding frequency, they will be $180°$ degrees apart. Flipping one of these phasors to cancel at the $p = 4$ frequency will cause them to constructively add at the suspension winding's $p = 1$ frequency. In the case of $t = 4$, there will be four layers to the phasor star, meaning that each phasor will be $360°$ apart. When redrawn at the suspension winding frequency, this one phasor will split into four phasors $90°$ apart that when half of which are flipped will create two orthogonal phasors $90°$ apart. For either value of t, this means:

TABLE IV. SUSPENSION WINDING FACTOR MAGNITUDE

	4 slot coil pitch		5 slot coil pitch	
h	winding i	winding ii	winding i	winding ii
1	0.204	0.225	0.252	0.278
3	0.000	0.289	0.000	0.322
5	0.204	0.305	0.170	0.253
7	0.204	0.371	0.027	0.050
9	0.000	0.000	0.500	0.373
11	0.204	0.126	0.316	0.196
13	0.204	0.167	0.119	0.097
15	0.000	0.289	0.000	0.086

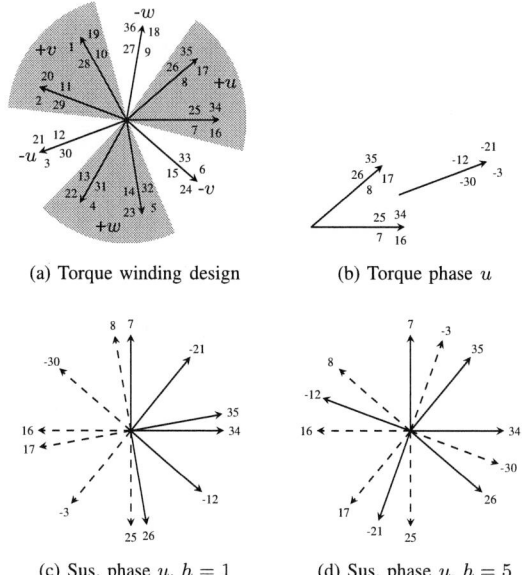

(a) Torque winding design   (b) Torque phase u

(c) Sus. phase u, $h = 1$ (d) Sus. phase u, $h = 5$

Fig. 5. Phasor diagrams for the example $Q = 36$, $p = 4$, $m = 3$ design; (a) torque winding layout using $60°$ phase belts; (b) phase u at the synchronous frequency; (c) phase u at the suspension winding frequency ($p = 1$), dashed phasors correspond to coils whose direction is reversed when connected as a suspension winding; (d) phase u at the 5th harmonic of the suspension frequency, dashed phasors correspond to coils whose direction is reversed when connected as a suspension winding

1) when the phasor star is viewed in the suspension winding frequency, a $180°$ band of phasors are able to be flipped which maximize the fundamental suspension winding factor ($h = 1$)

2) there will be no even harmonics present in the suspension winding, that is: $k_{wh} = 0$ when h is even.

When the torque winding is a fractional slot winding, the choice of the $180°$ band's location in the suspension winding domain can be used to adjust the suspension winding's fundamental winding factor.

D. Example design

As an illustration, consider the double layer example design presented earlier: $Q = 36$, $p = 4$, $m = 3$, $q = 3/2$. This is a common torque winding design for 8-pole, 3-phase motors. The star of phasors used for the torque winding layout is shown in Fig. 5a and a phase of this is redrawn for the $p = 1$ frequency of the suspension winding in Fig. 5c. There are clearly two choices for where the $180°$ band could be located:

(i) starting at the phasor for slot 8 and wrapping around counter-clockwise to the slot 25 phasor,

(ii) starting at the phasor for slot 7 and wrapping around counter-clockwise to the slot 3 phasor.

The resulting winding factors for the suspension winding are summarized in Table IV in the columns corresponding to a coil pitch of 4 slots. The harmonics h are multiples of the $p = 1$ frequency. The harmonic $h = 4$ corresponds to the fundamental harmonic $v = 1$ of the $p = 4$ frequency of the rotor. The winding factors of the suspension winding have low values, meaning that current is used inefficiently. This is a result of the design choice of placing emphasis on the quality of the torque winding and can be understood through the pitching factor: the coils are pitched based on the torque winding design, from the suspension winding perspective this is an extremely short pitch. Notice that while option (ii) has a higher fundamental winding factor than option (i), it also has dramatically higher harmonic winding factors, which will introduce force ripple in the suspension operation.

To understand the importance of winding factor on force ripple, refer back to the discussion in Section II regarding radial force calculation. The airgap flux density will contain the harmonics of each MMF as well as plus or minus the harmonics contained in the effective airgap length function $f(\phi)$ of (1). The suspension winding even harmonics have already been shown to be zero and constraining the suspension phase currents to sum to zero causes the triplen harmonics disappear from the suspension MMF. The torque winding MMF triplen harmonics will also be zero since the torque currents sum to zero. This particular torque winding has even harmonics and no sub-harmonics. The interaction between the first 11 suspension winding harmonics and the magnetizing and torque winding harmonics is summarized in Table V for the simple case where the effective airgap only contains the fundamental p component described in Section II. The interacting harmonic index of the magnetizing or

TABLE V. INTERACTION OF FLUX DENSITY HARMONICS

h	$h - 4$	$h + 4$
1 ($v = 0$)	-3 ($v = 1$)	5 ($v = 1$)
~~2~~	-	-
~~3~~	-	-
~~4~~	-	-
5 ($v = 1$)	1 ($v = 0$)	9 ($v = 2$)
~~6~~	-	-
7 ($v = 2$)	3 ($v = 1$)	11 ($v = 3$)
~~8~~	-	-
~~9~~	-	-
~~10~~	-	-
11 ($v = 3$)	7 ($v = 2$)	15 ($v = 4$)

TABLE VI. FORCE RIPPLE DEPENDENCE ON COIL PITCH

y	Sine rotor			Square rotor		
	$\langle F_x \rangle$	$\frac{\Delta F_x}{\langle F_x \rangle}$	$\frac{\Delta F_y}{\langle F_x \rangle}$	$\langle F_x \rangle$	$\frac{\Delta F_x}{\langle F_x \rangle}$	$\frac{\Delta F_y}{\langle F_x \rangle}$
4 slots	125.9 N	11.1%	11.0%	136.3 N	12.0%	12.2%
5 slots	155.0 N	7.6%	7.3%	167.9 N	8.9%	7.9%

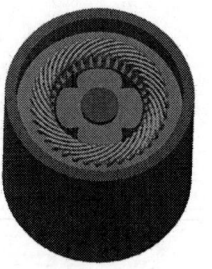

(a) Sine rotor model   (b) Square rotor model

Fig. 6. Models used for 3D FEA, $y = 4$ slots.

torque flux density is written as "$(v = ..)$". Harmonics are crossed out which do not exist in this configuration. The first row of the table is the only row that corresponds to a radial force which does not vary with rotor position– the existence of any other row is undesirable.

From this discussion, it can be concluded that the most important winding factor to minimize for force ripple is $h = 5$ since it is the only harmonic to interact with both main components of the magnetizing flux density, B_f from (4), as well as the torque flux density, B_T. This harmonic is particularly difficult to deal with because it is very near in frequency to the $p = 4$ pole-pair rotor, meaning that the pitching factor at $h = 5$ is relatively large. For a fixed coil pitch, the only way to minimize $h = 5$ is by the winding distribution factor – that is, by selecting the phasors to flip in Fig. 5d in a way that minimizes this harmonic.

One way to improve the winding factors of the suspension winding is to increase the pitch of the torque winding. However, this must be done with caution, as it may lead to an inferior torque winding in terms of winding factor, conduction losses, and end-effects. In this particular example with $q = 3/2$, a coil pitch would ideally be $y = mq = 4.5$ slots. In a typical fractional slot machine this is rounded down 4 slots to obtain shorter coils. However, if this were rounded up to 5 slots, the same magnitude winding factors would be achieved for the torque winding, while the suspension winding factors would be improved considerably, as shown in Table IV.

VII. FE ANALYSIS OF THE EXAMPLE DESIGN

FEA has been conducted for both the example winding design shown in Fig. 2 and the modified coil pitch described in the previous section. Two different rotor structures have been considered: one with an inverted sinusoidal airgap length and one with a square airgap length, where $f(\phi)$ of (1) is $\cos\phi$ and $SQ(\phi)$ respectively. Note that for this winding design, either style of drive can be used, as shown in Fig. 4, and will result in identical coil currents. All simulations have been solved on static, non-linear models using Infolytica's MagNet software. The models corresponding to a coil-pitch of $y = 4$ slots are shown in Fig. 6. Laminated M-18 electric steel is used for the rotor and stator laminations and a solid rotor steel is used for the rotor shaft and stator

sleeve. All models have a minimum airgap length of 2mm and a maximum airgap length of 17.5mm.

Force and torque waveforms for each combination of rotor structure and winding are shown in Fig. 7a and b. In both plots, the same terminal currents are used, resulting in simultaneous torque and x-axis force. Notice that for the same winding, the larger harmonic content of the square rotor structure always results in greater average force, torque, force ripple, and torque ripple. As predicted in the previous section, for the same terminal currents, the winding design with a coil pitch of $y = 5$ slots produces a larger average radial force with the same torque waveform.

The quality of a suspension winding of a bearingless motor is often analyzed by looking at the force ripple along each radial axis (x and y) when a force is desired along only one of the axes. The ripple value is typically normalized by the non-zero average force produced along the desired axis. This situation is depicted in Fig. 7c for the square rotor, where the terminal currents are such that an average force is produced along the x axis and it can be seen that a zero-average force is also produced along the y axis. A numerical comparison of the two winding designs is shown in Table VI. Here, $\Delta F_{x/y}$ is defined as the difference between the maximum and average force produced for that axis. In all cases considered, the same terminal current is used and no torque is created. As expected, the winding design with a coil pitch of $y = 5$ slots outperforms the shorter pitched design in terms of both the amount of average radial force produced and the amount of normalized force ripple.

Finally, to validate that the suspension phase terminals experience no motional-EMF, the flux linkage seen by a single phase of the torque and the suspension terminals is shown in Fig. 7d. In this plot, the suspension and torque currents are all zero while the field current is at its rated value. The torque phase terminal is clearly seeing a varying flux linkage, while the suspension phase terminal is not.

VIII. CONCLUSION

The design of dual purpose no voltage windings for the bearingless ac homopolar and consequent pole motors has been investigated. Design rules and a design approach for these windings has been developed and an example design has been explored using 3D FEA.

Dual purpose no voltage windings allow the same coils to be used for producing radial forces (magnetic bearing operation) and producing torque (motor operation), and thereby alleviate a major design trade-off between designing a machine with a large radial force

The 2014 International Power Electronics Conference

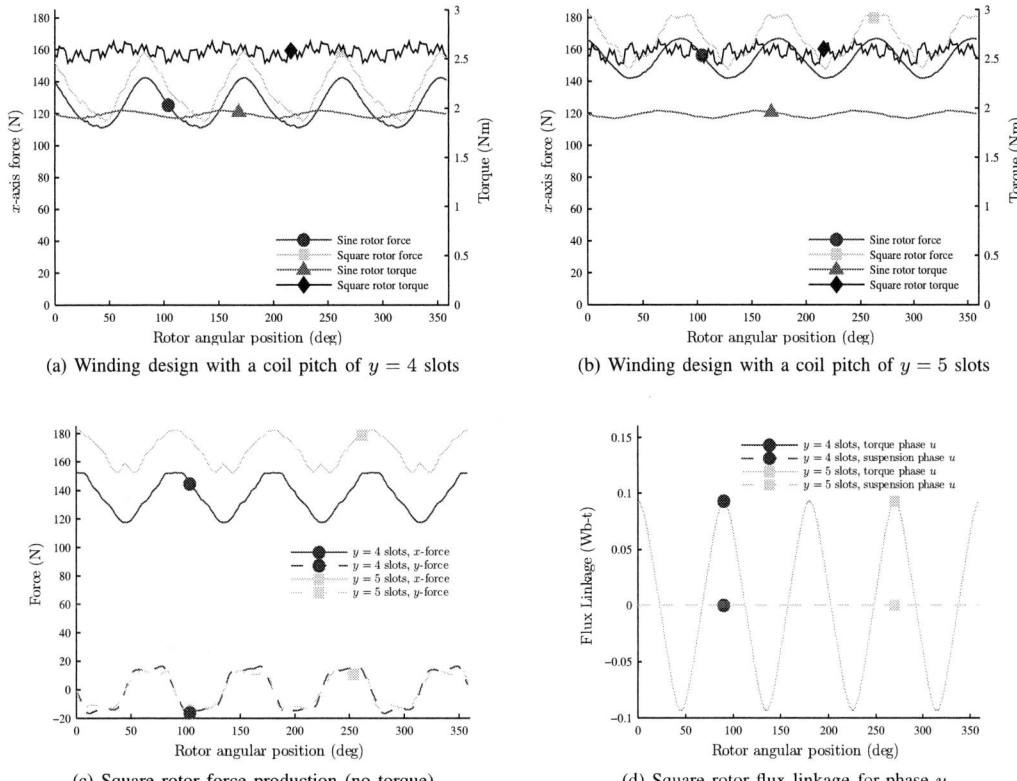

(a) Winding design with a coil pitch of $y = 4$ slots

(b) Winding design with a coil pitch of $y = 5$ slots

(c) Square rotor force production (no torque)

(d) Square rotor flux linkage for phase u

Fig. 7. 3D FEA results comparing the force and torque ripple and coil flux linkage of two variations of the example winding design for the two different rotor shapes shown in Fig. 6. Note that in (d), the each variation of the example winding design produces the same results.

density or a large torque density. This trade-off has been a major obstacle in using bearingless motor technology and for this reason the authors feel that dual purpose windings will enable bearingless motors to be used in a much wider range of applications.

ACKNOWLEDGEMENT

This research was supported by the Office of Naval Research (ONR) and the National Science Foundation Graduate Research Fellowship Program under Grant No. 00006595.

REFERENCES

[1] P. Kascak, R. Jansen, T. Dever, A. Nagorny, and K. Loparo, "Levitation performance of two opposed permanent magnet pole-pair separated conical bearingless motors," in *Energy Conversion Congress and Exposition (ECCE), IEEE*, 2011, pp. 1649–1656.

[2] S. Khoo, R. Fittro, and S. Garvey, "An ac self-bearing rotating machine with a single set of windings," in *Power Electronics, Machines and Drives. International Conference on.* IET, 2002, pp. 292–297.

[3] W. K. S. Khoo, K. Kalita, and S. Garvey, "Practical implementation of the bridge configured winding for producing controllable transverse forces in electrical machines," *Magnetics, IEEE Transactions on*, vol. 47, no. 6, pp. 1712–1718, 2011.

[4] A. Chiba, K. Sotome, Y. Iiyama, and M. Azizur Rahman, "A novel middle-point-current-injection-type bearingless pm synchronous motor for vibration suppression," *Industry Applications, IEEE Transactions on*, vol. 47, no. 4, pp. 1700–1706, 2011.

[5] R. Oishi, S. Horima, H. Sugimoto, and A. Chiba, "A novel parallel motor winding structure for bearingless motors," *Magnetics, IEEE Transactions on*, vol. 49, no. 5, pp. 2287–2290, 2013.

[6] P. Tsao, M. Senesky, and S. Sanders, "An integrated flywheel energy storage system with homopolar inductor motor/generator and high-frequency drive," *Industry Applications, IEEE Transactions on*, vol. 39, no. 6, pp. 1710 – 1725, nov.-dec. 2003.

[7] K. Sivasubramaniam, T. Zhang, M. Lokhandwalla, E. Laskaris, J. Bray, B. Gerstler, M. Shah, and J. Alexander, "Development of a high speed hts generator for airborne applications," *Applied Superconductivity, IEEE Transactions on*, vol. 19, no. 3, pp. 1656 –1661, june 2009.

[8] Z. ao Ren, K. Yu, Z. Lou, and C. Ye, "Investigation of a novel pulse ccps utilizing inertial energy storage of homopolar inductor alternator," *Plasma Science, IEEE Transactions on*, vol. 39, no. 1, pp. 310 –315, jan. 2011.

[9] A. Chiba, T. Fukao, O. Ichikawa, M. Oshima, M. Takemoto, and D. Dorrell, *Magnetic Bearings and Bearingless Drives.* Newnes, 2005.

[10] J. Asama, R. Natsume, H. Fukuhara, T. Oiwa, and A. Chiba, "Optimal suspension winding configuration in a homo-polar bearingless motor," *Magnetics, IEEE Transactions on*, vol. 48, no. 11, pp. 2973 –2976, nov. 2012.

[11] E. Severson, R. Nilssen, T. Undeland, and N. Mohan, "Analysis of the bearingless ac homopolar motor," in *International Conference on Electrical Machines (ICEM), 2012.*, Sep. 2012, pp. 568 – 574.

[12] O. Ichikawa, A. Chiba, and T. Fukao, "Inherently decoupled magnetic suspension in homopolar-type bearingless motors," *Industry Applications, IEEE Transactions on*, vol. 37, no. 6, pp. 1668 – 1674, nov/dec 2001.

[13] J. Pyrhonen, T. Jokinen, and V. Hrabovcová, *Design of rotating electrical machines.* Wiley. com, 2009.

978-1-4799-2706-7/14 $31.00 © 2014 IEEE

Harvesting Energy from Ship Rolling Using an Eccentric Disk Revolving in a Hula-hoop Motion

Yu-Jen Wang

Department of Mechanical Engineering and Institute of
Mechatronics Engineering
National Taiper University of Technology
Taipei, Taiwan
yjwang@ntut.edu.tw

Yu-Ti Hao

Department of Systems and Naval Mechatronic
Engineering
National Cheng Kung University
Tainan, Taiwan
P16014222@mail.ncku.edu.tw

Abstract—**This research developed an eccentric disk and novel circular Halbach array to harvest wave energy from the rolling motion of a ship. The designed eccentric disk revolves in a motion that mimics a hula-hoop at higher angular speeds than eccentric disks oscillate normally having. The produced kinetic energy was converted into electrical energy through electromagnetic induction. To enhance electromagnetic efficiency, this study presents a novel circular Halbach array. The wave energy harvesting device, composed of the eccentric disk and the circular Halbach array, was mounted in a ship's cabin. The simulated power output of the eccentric disk was approximately 2.34 W. The proposed energy harvester can serve as a back-up power source during emergencies.**
Keywords— Energy harvester, Hula-hoop motion and Ship rolling.

I. INTRODUCTION

Ocean wave energy harvesting technology has been in development for a long time. There are many types of wave energy harvester [1]–[3], such as wave energy converters and oscillating body systems. Wave energy converters transform vertical motion to rotational motion, but the technology is structurally complicated, the body is susceptible to seawater corrosion, and the converters are difficult to maintain. Toh et al. [4] applied a technique for harvesting energy on an unmanned surface vehicle (USV) in the open water. Storing the energy in a battery enabled the device to achieve an average useful power output of 0.3 mW. Sharkh et al. [5] proposed a device for harvesting energy from the vertical motion of small ships and yachts, comprising a sprung mass coupled to an electrical generator that involves using a ball screw. In considering the motion of a typical small ship with amplitude of 1 m and a frequency of 0.5 Hz, the device is estimated to produce 75 W. Rastegar et al. [6] presented piezoelectric-based energy harvesters to mount on platforms that vibrate at frequencies lower than those that can usually be harvested efficiently by using piezoelectric elements. Okada et al. [7] introduced a pendulum system to allow a piezoelectric element to vibrate efficiently. Yang [8] used the concept of a flat linear generator to develop a wave energy harvester based on the waves caused by the motion of a ship; according to the

experimental results, the harvester could yield up to 0.124 mW.

This paper proposes an eccentric disk with suitable parameters that revolves in a motion that mimics a hula-hoop motion. Magnets arranged using a circular Halbach array are mounted in the rotor of the eccentric disk to convert kinetic energy into electric energy. Power generation was simulated using the derived physical models.

II. DYNAMIC ANALYSIS OF THE ECCENTRIC DISK

Ship motions are defined according to the six degrees of freedom: heaving, swaying, surging, pitching, rolling, and yawing, as shown in Fig. 1. Chiang [9] analyzed the kinematic energy of a 700-ton fishing ship and discovered that rolling and pitching were the main sources of kinetic energy. Using rolling as an example, the frequency ranges from 0.1–0.3 Hz and its angle is $\pm15°$. The proposed application targets a cargo ship with a metacenter of 11 m and other characteristics listed in Table I. The mounting of the proposed eccentric disk within the cabin of a cargo ship is shown in Fig. 2.

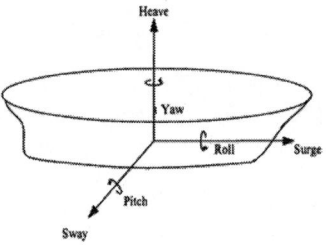

Fig. 1. Definition of ship motion.

TABLE I
Characteristics of 2200 tons cargo ship

The length of ship	185.5 m
The width of ship	30.2 m
The depth of ship	16.6 m
Designed draft	11 m
Ship displaces	42184 mt
Height of center of gravity	13.82 m
Longitudinal center of gravity	-3.79 m
transverse metacentric height	0.32 m
Longitudinal metacentric height	255.07 m

The 2014 International Power Electronics Conference

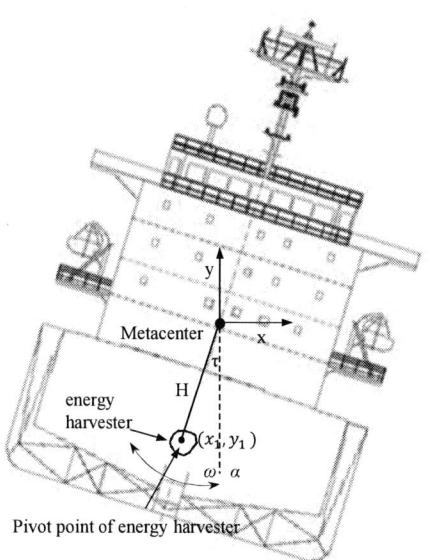

Fig. 2. Schema of an eccentric disk mounted in a cargo ship at ship rolling motion.

To harvest kinetic energy from the rolling motion of a ship, an eccentric disk integrated with magnets was mounted in a ship cabin. The metacenter was the rolling center of the ship. The distance between the metacenter and the pivot point of the eccentric disk was H and the rolling angle of the ship with respect to the fixed coordinates (x, y) was τ. The angular velocity and acceleration of the rolling motion of the ship were ω and α, respectively. A schematic diagram of the eccentric disk is shown in Fig. 3. The eccentric disk can have an arbitrary configuration. A local coordinate $(\bar{x}, \bar{y})$ was located at the pivot of the eccentric disk; the location of the pivot is

$$\begin{cases} X_1 = H \sin \theta \\ Y_1 = H \cos \theta \end{cases} \tag{1}$$

with respect to the fixed coordinate (x, y). The position of an infinitesimal element dm on the eccentric disk was defined according to the distance r and an angle ϕ, as shown in Fig. 3. Variable θ is the swing angle of the eccentric disk. This study further defined an angle φ such that

$$\phi + \theta + \varphi - \tau = \frac{\pi}{2} \tag{2}$$

For an arbitrary swing angle θ, the location of dm with respect to $(\bar{x}, \bar{y})$ is

$$\begin{cases} x_2 = H \sin \tau + L \sin \varphi \\ y_2 = H \cos \tau + L \cos \varphi \end{cases} \tag{3}$$

By differentiating (3) with respect to time t, the velocity of dm can be obtained as follows:

$$\begin{cases} \dot{x}_2 = H \dot{\tau} \cos \tau - L \dot{\varphi} \cos \varphi \\ \dot{y}_2 = -H \dot{\tau} \sin \tau - L \dot{\varphi} \sin \varphi \end{cases} \tag{4}$$

The kinetic energy of the eccentric disk is expressed in (5):

$$T = \frac{1}{2} \int (\dot{x}_2{}^2 + \dot{y}_2{}^2) dm$$
$$= \frac{1}{2} \int [H^2 \dot{\tau}^2 + r^2 \dot{\varphi}^2 + 2 H r \dot{\tau} \dot{\varphi} \cos(\tau - \varphi)] \rho d\bar{V} \tag{5}$$

where ρ is the density of the eccentric disk and $d\bar{V}$ is an infinitesimal volume. The potential energy is

$$V = g \int [H \cos \tau + r \cos \varphi] \rho d\bar{V} \tag{6}$$

where g is the gravitational acceleration. Using the definition of the Lagrangian function and the Euler–Lagrange formulation, the equation of motion is expressed as

$$\int \rho r^2 \ddot{\varphi} \, d\bar{V} + H \ddot{\tau} \int \rho r \cos(\tau - \varphi) \, d\bar{V} - H \dot{\tau}^2 \int \rho r \sin(\tau - \varphi) \, d\bar{V} - g \int \rho r \sin(\varphi) \, d\bar{V} = 0 \tag{7}$$

Substituting (2) into (7) yields

$$(\int \rho r^2 d\bar{V}) \ddot{\theta} + H \dot{\tau}^2 (\int \rho r \sin \phi \, d\bar{V}) \sin \theta = (\int \rho r^2 d\bar{V}) \ddot{\tau} + H \ddot{\tau} (\int \rho r \sin \phi d\bar{V}) \cos \theta + g(\int \rho r \sin \phi \, d\bar{V}) \sin(\theta - \tau) \tag{8}$$

where $\int \rho r^2 d\bar{V} = I$ is the mass moment of inertia and $\int \rho r \sin \phi \, d\bar{V}$ is the eccentric moment. This study defined the characteristic length, L^*, as

$$L^* = \frac{\int \rho r^2 d\bar{V}}{\int \rho r \sin \phi d\bar{V}} \tag{9}$$

and added the damping constant C^* to rewrite (8) as

$$\ddot{\theta} + C^* \dot{\theta} + \frac{H \dot{\tau}^2}{L^*} \sin \theta = \ddot{\tau} + \frac{H \ddot{\tau}}{L^*} \cos \theta + \frac{g}{L^*} \sin(\theta - \tau) \tag{10}$$

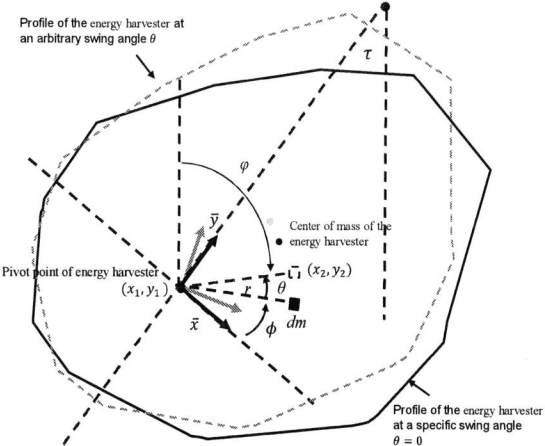

Fig. 3. Diagram of an eccentric disk (energy harvester).

III. HULA-HOOP MOTION

The hula-hoop motion results from the interactive forces between a rotating ring and human motion. This study anticipated that an eccentric disk revolving according to this hula-hoop motion could achieve larger angular speeds than could a disk revolving according to small-amplitude oscillation. According to Faraday's law, electromotive force is proportional to the relative (angular)

978-1-4799-2706-7/14 $31.00 © 2014 IEEE

speeds between the magnets and coils. In other words, an eccentric disk revolving according to this hula-hoop motion is more efficient at kinematic energy transformation than a disk revolving according to small-amplitude oscillation. To observe this heightened efficiency, two set parameters were used: P_1: $L^* = 11.00$ m and P_2: $L^* = 4.18$ m. Both parameters were modeled based on a ship rolling frequency of 0.3 Hz, a rolling angle of $\pm 10°$, and electrical damping of 0.02 N-s/kg/m. Fig. 4 (a) demonstrates that the eccentric disk with set parameters P_1 spins continuously in the same direction, and also indicates that the slope of the angular variation is related to the rolling frequency of the ship. Fig. 4 (b) shows the oscillation of the eccentric disk with set parameters P_2 under the same ship rolling conditions. Fig. 5 shows the transient response of the angular velocity of P_1 and P_2. At a steady state, the average power over a time interval $(t0, t0 + kt)$ is

$$P_{ave} = \frac{1}{kt} \int_{t_0}^{t_0+kt} I \, C^* \, \dot\theta^2 \, dt \qquad (11)$$

Based on the electrical damping of 0.02 N-s/kg/m, the steady-state average power generation over the initial moment (P_{ave}/I) of P_1 and P_2 were 78.7 and 18.3 mW/kg/m^2, respectively. The eccentric disk revolving in a hula-hoop motion clearly exhibits superior performance.

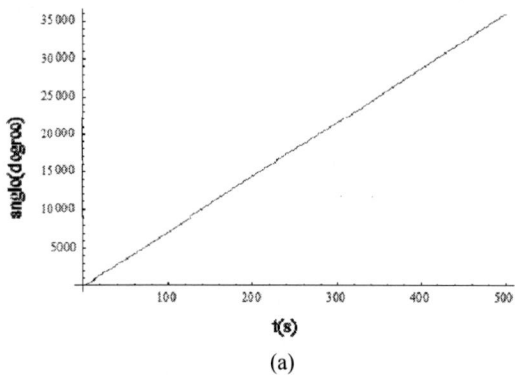

(a)

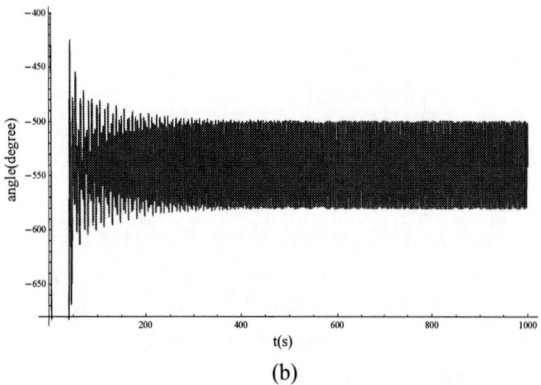

(b)

Fig. 4. Transient response of the rotation angle of the eccentric disk. (a) P_1 produces hula-hoop motion. (b) P_2 produces oscillation motion only.

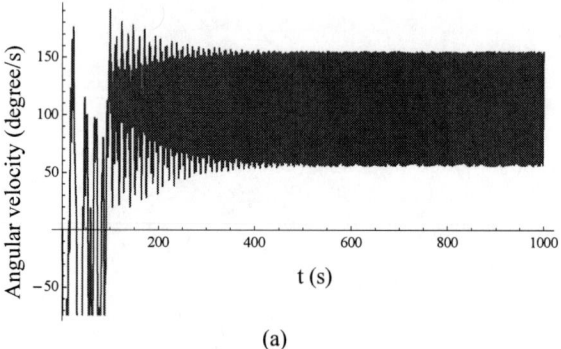

(a)

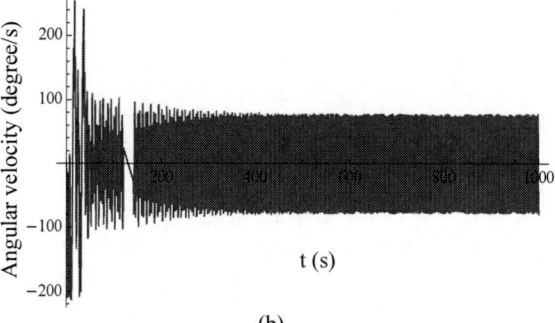

(b)

Fig. 5. Transient response of the angular velocity of the eccentric disk. (a) P_1 produces hula-hoop motion. (b) P_2 produces oscillation motion only.

IV. CIRCULAR HALBACH ARRAY MAGNETIC CIRCUIT

The design of the eccentric disk energy harvester prototype, consisting of a stator and a rotor, is shown in Fig. 6. The stator comprises coils, a bearing, and a base. The rotor is an eccentric disk based on (10). The magnets mounted within the rotor are 16 individual sector-shaped permanent Nd-Fe-B magnets arranged in a circular Halbach array. This circular Halbach array augments the magnetic strength of the side of the array where the coils are located, for more efficient power generation.

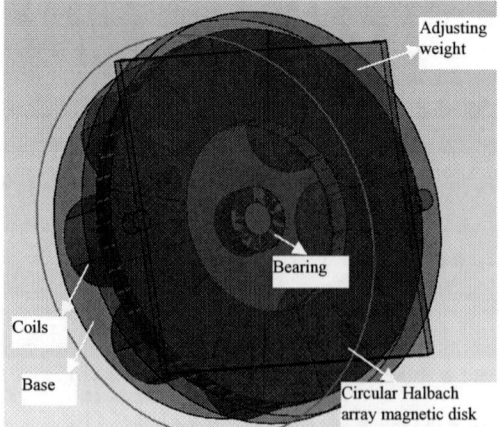

Fig. 6. Prototype of energy harvester with magnets

The average magnetic flux density distributions on the z-axis for the two magnetic circuit designs are shown in Fig. 7. The strength of the z component of the average magnetic flux density for the circular Halbach array disk are 1.25 times greater than those of the normal multipolar magnetic disk. This is beneficial, and the circular Halbach array enhances the EMF. The two magnetic circuits were simulated using identical materials and volumes with COMSOL 4.3.

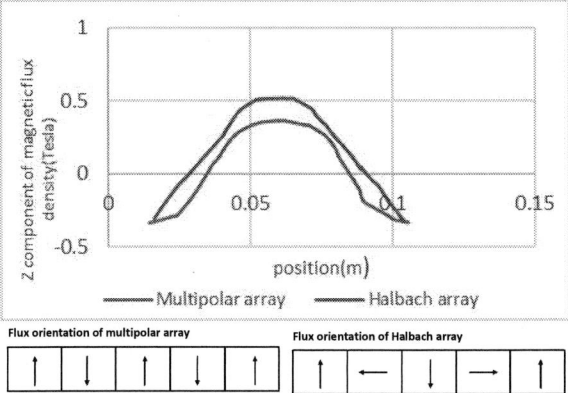

Fig. 7. Simulation results of magnetic flux density in Z-direction

SIMULATION OF HULA-HOOP MOTION MAP

Based on (10), the motion of the eccentric disk is determined by parameters L^*, C^*, the rolling frequency, and the rolling amplitude of the ship. Figs. 8 and 9 show the stability of the hula-hoop motion results obtained using direct numerical integrating (10) given a damping constant of 0.02 N-s/kg/m. The blue points denote the occurrence of hula-hoop motions. As Fig. 7 shows, the rolling angle of the ship was fixed at 8°, and as Fig. 8 shows, the rolling frequency of the ship was fixed at 0.2 Hz. These are the suitable design parameters of an eccentric disk revolving according to hula-hoop motion. The hula-hoop motion region is wider when the rolling frequency and amplitude are greater.

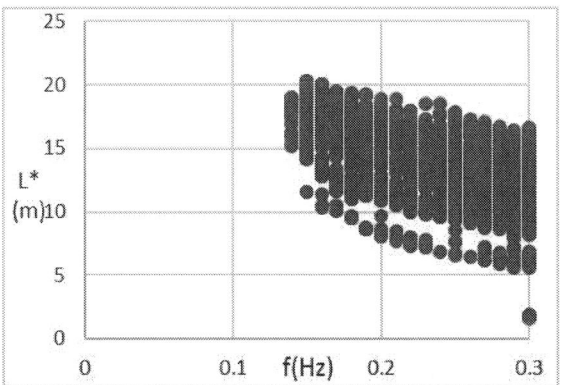

Fig. 8. Ship rolling frequency versus characteristic length for hula-hoop motion occurred.

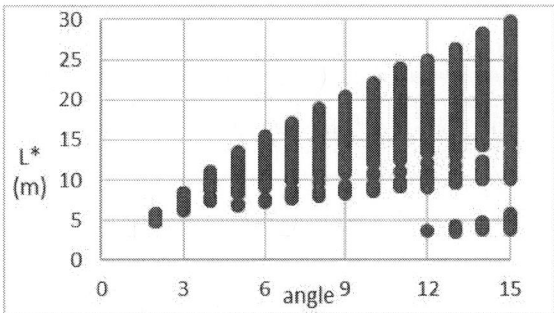

Fig. 9. Ship rolling angle versus characteristic length for hula-hoop motion occurred.

V. CONCLUSION

This paper proposes a novel energy harvester to convert some of the energy generated by the rolling of a ship into electric energy. Based on the proposed models, an eccentric disk with $L^* = 15.22$ m and $I = 65$ kg × m² produces 2.34 W in a regular wave at a frequency of 0.2 Hz and an amplitude of 7°. The proposed novel circular Halbach array design and hula-hoop motion mechanism offer enhanced power generation than the multipolar array design and oscillation motion. This energy harvester can serve as a back-up power source during emergencies; for example, to power a satellite phone during a rescue call.

ACKNOWLEDGMENT

The authors appreciate the support of the National Science Council of the R.O.C., under grants NSC 102-2221-E-027-115 and 102-2622-E-027-016-CC3.

REFERENCES

[1] Antonio F. de O. Falcao, "Wave energy utilization: A review of the technologies," *IDMEC*, Institute Superior Te´cnico, Technical University of Lisbon, 1049-001 Lisbon, Portugal.

[2] Symonds, D., Davis, E., and Ertekin, R.C.,"Low-Power Autonomous Wave Energy Capture Device for Remote Sensing and Communications Applications," *IEEE*, 2010.

[3] Ummaneni, R. B., Nilssen, R., and Eirik, B. J., "Convert Low Frequency Energy from Wave Power Plant to High Frequency Energy in Linear Electrical Generator with Gas Springs," *IEEE*, 2008.

[4] Toh, T. T., Mitcheson, P. D., Dussud, L., Wright, S. W., and Holmes, A. S., "Electronic Resonant Frequency Tuning of a Marine Energy Harvester" Department of Electrical & Electronic Engineering, Imperial College London, UK.

[5] Sharkh, S. M., Hendijanizadeh, M., Moshrefi-Torbati, M., and Russel, M., "An Inertial Coupled Marine Power Generator for Small Ships," *Synopsis to be submitted on the International Conference on Clean Electrical Power*, Ischia, June, 2011.

[6] Rastegar, J., Pereira, C., and Nguyen, H.L., "Piezoelectric-Based Power Sources for Harvesting Energy from Platforms with Low Frequency Vibration," *Proceedings of SPIE - The International Society for Optical Engineering*, Vol. 6171 , 2006.

[7] Naoki Okada, Hironori Fujimoto, Shoichiro Yabe, and Motohiko Murai, "Experiments on floating wave-power generation using piezoelectric elements and pendulums in the water tank," *IEEE*, 2011.

[8] Shiang Chi Yang, The Research On The Technique of Harvesting Wave Energy Through the Ship Motion, *National Cheng Kung University Department of Systems and Naval Mechatronic Engineering.*, Master thesis, 2012.

[9] Jian Huei Chiang, The Study On The Virtual Dynamic Simulation For The Ship Moving in Irregular Wave, *National Cheng Kung University Department of Systems and Naval Mechatronic Engineering.*, Master thesis, 2004.

The 2014 International Power Electronics Conference

Load-Independent Current Output of Inductive Power Transfer Converters with Optimized Efficiency

Wei Zhang, Siu-Chung Wong and Chi K. Tse
Department of Electronic and Information Engineering
The Hong Kong Polytechnic University, Kowloon, Hong Kong

Qianhong Chen
Nanjing University of Aeronautics and Astronautics
Nanjing 210016, China

Abstract—Inductive power transfer (IPT) systems are resonant converters whose output characteristics are mostly load dependent. Conditions for ensuring a load-independent voltage-transfer ratio of an IPT system have been studied. However, load-independent current output can be more desirable for applications such as battery charging and LED driving. This paper studies the characteristics of IPT systems operating at different frequencies as transconductance converters. Operating frequencies for load-independent transconductance are explored, looking for optimal efficiency. The frequencies interested are found to facilitate the design of a current-output IPT converter with efficient power conversion. The analysis is supported by experimental results.

Index Terms—Inductive power transfer, wireless power transmission, compensation topology, efficiency, load-independent transconductance.

I. INTRODUCTION

Wireless power transmission (WPT) technologies are normally classified according to short-range, mid-range, and long-range applications, which are mainly distinguished by the operating principle and the frequency band. The short-range power transmission technology has a typical transmission distance from a few millimeters to hundreds of millimeters. It normally works by means of inductive magnetic coupling at a relatively low frequency (20kHz to 1MHz) [1]. In the recent decade, this short distance wireless power transmission or IPT technology has emerged as an attractive and user-friendly solution to constructing charging platforms for portable electronic products [2] and electric vehicles [3], [4].

Major research work of IPT focuses on voltage to voltage conversion, to select the most appropriate compensation techniques, isolated to specific application requirements [4]–[7]. However, in some applications, a load-independent current output is more desirable. For instance, a constant current output is preferred to drive an LED or to charge battery packs of electric vehicles. An extra stage of current regulator which incurs extra power loss is needed if the IPT converter is an output voltage power source.

Therefore, voltage to current conversion or transconductance IPT systems will be studied in this paper. General calculation and simulation are conducted for the IPT converter using the secondary series- and parallel-compensations, looking for

This work is supported by Hong Kong RGC GRF project PolyU5258/13E.

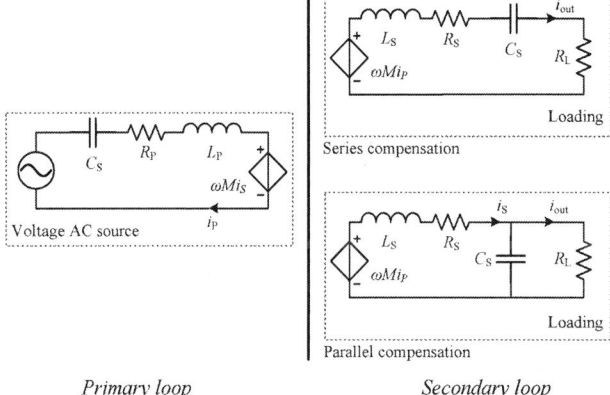

Fig. 1. Series- and parallel-compensation topologies. In this figure, the primary inductance L_P is compensated with a series connected capacitor C_P, while secondary inductance L_S is compensated with either a series or parallel connected capacitor C_S. M is the mutual inductance of the loosely coupled transformer. R_P and R_S are resistances of primary and secondary coils, and R_L is the equivalent loading resistance.

their operating frequencies to achieve maximum efficiency and load-independent current transfer ratio. The maximum efficiencies of the two compensation topologies are compared under the condition of load-independent current output.

II. CIRCUIT MODEL AND EFFICIENCY ANALYSIS

The coils of IPT systems are operating at a frequency well below their self-resonant frequencies [8], and therefore, additional compensation capacitors are needed to form the resonant tanks in both primary and secondary sides of the loosely coupled transformer. The tuned capacitors of the IPT system can be series or parallel connected with the transformer coils. Fig. 1 shows the circuit model for the analysis of steady-state transfer functions. The primary loop of the circuit is driven by a modulated AC voltage source which is an equivalent voltage readily generated from a pulse-width-modulated DC voltage source using either a simple full- or half-bridge switching circuit. For a parallel resonant compensated primary, an equivalent current source is needed. Due to the difficulty of energy storage in the form of a simple current source, extra components will be needed to transform it on demand from a voltage source, creating extra loss. Therefore, series resonant primary compensation will be used in this paper [5], [6]. The secondary loop is compensated with either series connected or parallel connected capacitor to form resonant tanks as shown on the right-hand-side of Fig. 1.

TABLE I
DEFINITION OF VARIABLES.

Coupling coefficient	$k = \frac{M}{\sqrt{L_P L_S}}$
Primary winding quality factor	$Q_P = \frac{\omega L_P}{R_P}$
Secondary winding quality factor	$Q_S = \frac{\omega L_S}{R_R}$
Circuit quality factor of series compensation	$Q_{L-S} = \sqrt{\frac{L_S}{C_S}}\frac{1}{R_L}$
Circuit quality factor of parallel compensation	$Q_{L-P} = \sqrt{\frac{C_S}{L_S}}R_L$
Circuit quality factor of series compensation including R_L	$Q_O = \frac{Q_S Q_{L-S}}{Q_S+Q_{L-S}}$
Resonant frequency of primary loop	$\omega_P = \frac{1}{\sqrt{L_P C_P}}$
Resonant frequency of secondary loop	$\omega_S = \frac{1}{\sqrt{L_S C_S}}$

A thorough efficiency analysis of both secondary series and parallel compensations has been carried out [5], [6]. The system efficiency derived is given as

$$\eta = \eta_P \eta_S = \frac{\Re(Z_r)}{R_P + \Re(Z_r)}\frac{\Re(Z_S) - R_S}{\Re(Z_S)}, \quad (1)$$

where η_P and η_S are the efficiencies of primary and secondary loops. The parameters and variables are summarized in Tables I and II. The tailing subscript $-S$ or $-P$ of Q_L indicates that it is of secondary series or parallel compensation respectively.

For the secondary series compensation, there exists a local maximum of η at $\omega = \omega_M$ which is load-dependent.

$$\omega_M = \frac{\omega_S}{\sqrt{1 - \frac{1}{2Q_O^2}}}. \quad (2)$$

Variable operating frequency is needed for achieving this theoretical maximum efficiency. Fig. 2 shows two types of power transfer efficiency and the corresponding ω_M versus Q_O [5], where

$$Q_{Os} = \sqrt{\frac{1}{2\lambda}\left(1 + \sqrt{1 - 3\lambda}\right)}, \text{ and } \lambda = k^2\frac{Q_P}{Q_S} \quad (3)$$

The two types of efficiency curves are identified as follows:

TABLE II
PARAMETERS NEEDED FOR EFFICIENCY AND G_i CALCULATION.

	Series topology	Parallel topology
Z_S	$j\omega L_S + \frac{1}{j\omega C_S} + R_S + R_L$	$j\omega L_S + R_S + \frac{1}{j\omega C_S}\|R_L$
Z_P	$j\omega L_P + \frac{1}{j\omega C_P} + R_P$	
Z_r	$\frac{\omega^2 M^2}{j\omega L_S + \frac{1}{j\omega C_S} + R_S + R_L}$	$\frac{\omega^2 M^2}{j\omega L_S + R_S + \frac{1}{j\omega C_S}\|R_L}$
i_{out}	$\frac{\omega M}{Z_S}i_p$	$\omega M\frac{\frac{1}{j\omega C_S}\|R_L}{Z_S R_L}i_p$
i_p	$\frac{v_{in}}{Z_P + \frac{\omega^2 M^2}{Z_S}}$	

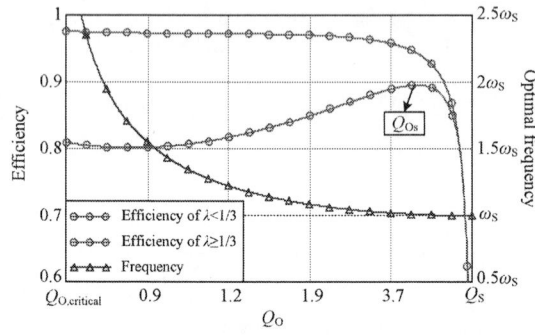

Fig. 2. Simulated efficiency and optimal frequency (ω_M) versus Q_O.

① For $\lambda < \frac{1}{3}$, the optimal efficiency peaks at Q_{Os}. The value of ω_M is just slightly higher than ω_S in a reasonable range of loading (close to Q_{Os}) to achieve high efficiency. Therefore, ω_S, which is load-independent, is an ideal constant operating frequency for the best efficiency operation of the converter.

② For $\lambda \geq \frac{1}{3}$, the optimal efficiency increases monotonically with decreasing Q_O by increasing ω_M to ∞. However, the operating frequency is affected by the self-resonant frequency of the coils, switching losses and voltage stresses. In addition, it can be seen from Fig. 2, when $\lambda \geq \frac{1}{3}$, the efficiency improvement with increasing frequency is insignificant. Therefore, ω_S can be a constant operating frequency to achieve high efficiency.

As a result, operating at constant ω_S, η maximizes at

$$\eta_{max} = \frac{\sigma}{\left(1 + \sqrt{1 + \sigma}\right)^2}, \text{ where } \sigma = k^2 Q_P Q_S. \quad (4)$$

η_{max} is only decided by the value of σ, which is the product of k^2 and the two winding quality factors. Therefore, for an IPT system whose coupling is usually higher than 0.1, the requirement for high Q_P and Q_S is not demanding in order to achieve a high efficiency.

For the secondary parallel compensation, the efficiency maximizes at ω_N, which is load independent.

$$\omega_N = \frac{\omega_S}{(1 + k^2)^{\frac{1}{4}}}. \quad (5)$$

Fig. 3 gives a comparison of maximum achievable efficiencies of secondary series and parallel compensations versus the loading condition Q_L. In the figure, $Q_P = Q_S = 100$. It shows that the two compensation topologies can achieve nearly identical theoretical maximum efficiency except when $\lambda \geq \frac{1}{3}$ (i.e. $k > 0.577$ if $Q_P = Q_S$). For instance, if $k = 0.8$, the theoretical maximum efficiency of the converter with secondary series compensation operating at a much higher operating frequency can be slightly better than that of the secondary parallel compensation converter.

III. TRANSCONDUCTANCE

The output current i_{out} through R_L of the series- and parallel-compensated secondary are calculated with parameters

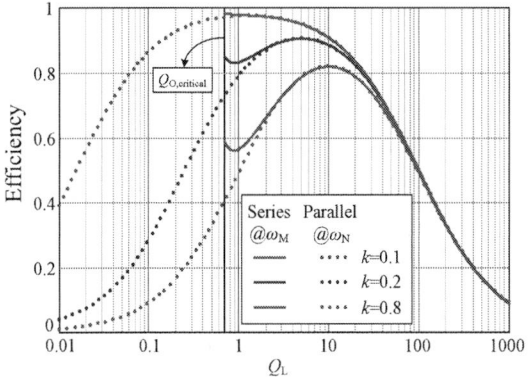

Fig. 3. Efficiency comparison of series and parallel compensations.

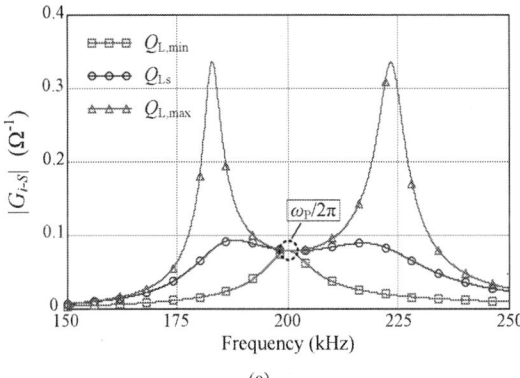

(a)

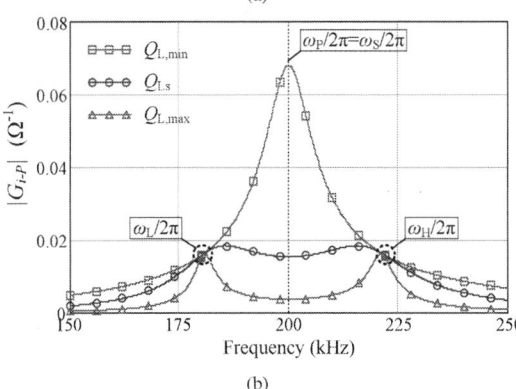

(b)

Fig. 4. Transconductances of (a) series and (b) parallel compensations.

given in Table II. The transconductances G_{i-S} and G_{i-P} of secondary series- and parallel-compensations are expressed in (6) and (7).

$$G_{i-S} = \frac{i_{\text{out}}}{v_{\text{in}}} = \frac{\omega M}{Z_P Z_{S-S} + \omega^2 M^2}, \tag{6}$$

$$G_{i-P} = \frac{i_{\text{out}}}{v_{\text{in}}} = \frac{\omega M}{Z_P Z_{S-P} + \omega^2 M^2} \cdot \frac{1}{1 + j\omega C_S R_L}, \tag{7}$$

where Z_P and Z_S are the impedance of the primary and secondary networks. In this section, R_P and R_S are neglected for their small values and insignificant impact on the transcon-

ductance [6]. From (6), G_{i-S} can be R_L-independent when Z_P equals zero, i.e. the operating frequency ω equals ω_P. This frequency can also be found by solving for ω from the equation $\frac{\partial G_{i-S}}{\partial R_L} = 0$.

The R_L-independent transconductance is a very desirable feature for power converters. It guarantees constant output current while the loading changes. If R_P is neglected, $Z_P = 0$ requires $\omega = \omega_P$. Hence,

$$|G_{i-S}(\omega_P)| = \frac{1}{\omega_P M} = \frac{1}{k\omega_P \sqrt{L_P L_S}}, \tag{8}$$

which is load-independent.

Fig. 4 (a) depicts the SPICE simulation results of transconductance at various loading conditions of the secondary series compensated circuit with zero R_P and R_S. In the simulation, k is set as 0.2 and C_P is selected to resonate with L_P at $\omega_P/(2\pi) = 200$ kHz and $\omega_P = \omega_S$. The $Q_{L\min}$, Q_{Ls} and $Q_{L\max}$ represent different loading conditions of 1.19, 5.00 and 21.05 respectively, which are calculated by the method introduced in [5].

Similarly, for the secondary-parallel-compensated converter, the transconductance in (7) is simplified as

$$G_{i-P} = \frac{1}{\omega M + \frac{j\omega L_S Z_P}{\omega M} + \delta R_L}, \text{ where} \tag{9}$$

$$\delta = j\omega^2 M C_S + \frac{Z_P(1 - \omega^2 L_S C_S)}{\omega M}. \tag{10}$$

From (9), G_{i-P} is R_L-independent when $\delta = 0$. Solving for the roots of (10), the frequencies at which G_{i-P} is R_L-independent can be obtained as

$$\omega_L = \sqrt{\frac{\omega_P^2 + \omega_S^2 - \Delta}{2(1 - k^2)}} \text{ and } \omega_H = \sqrt{\frac{\omega_P^2 + \omega_S^2 + \Delta}{2(1 - k^2)}} \tag{11}$$

$$\Delta = \sqrt{(\omega_P^2 + \omega_S^2)^2 - 4(1 - k^2)\omega_P^2 \omega_S^2} \tag{12}$$

The load-independent transconductances of the converter operating at frequencies ω_L and ω_H are usually not equal to each other except when

$$\omega_P = \omega_S \sqrt{1 - k^2}. \tag{13}$$

Under this condition, we have

$$|G_{i-P}(\omega_L)| = |G_{i-P}(\omega_H)| = \frac{1}{\omega_P \sqrt{L_P L_S}}. \tag{14}$$

The simulated $|G_{i-P}|$ curves are shown in Fig. 4 (b) by using the parameters of series compensation except the condition (13).

IV. COMPARISON OF SERIES AND PARALLEL COMPENSATIONS

The foregoing analyses on the frequencies operating for load-independent transconductance and maximal efficiency are derived independently on Sections II and III, and summarized in Table III.

For a secondary series compensated transconductance converter, the ω_P and ω_S that realize load-independent $|G_i|$ and

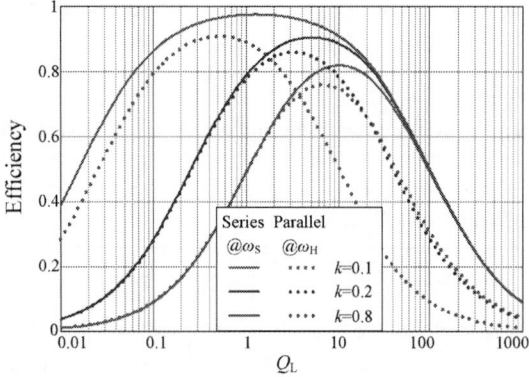

Fig. 5. Efficiency comparison of secondary series compensation converter working at ω_S and ω_H versus loading quality factor.

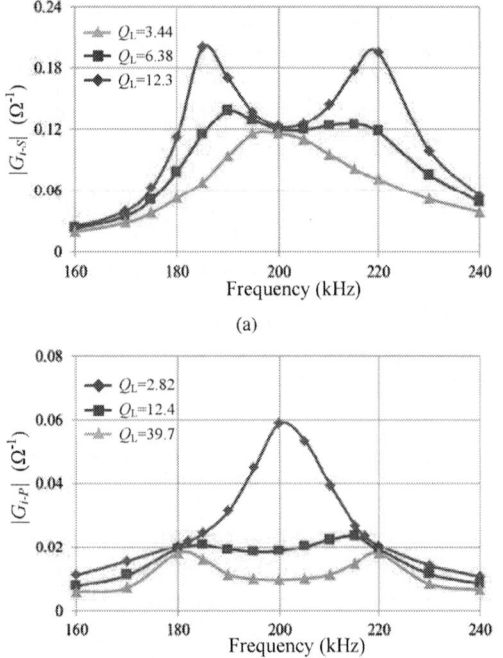

Fig. 6. Transconductances of (a) series compensation and (b) parallel compensation.

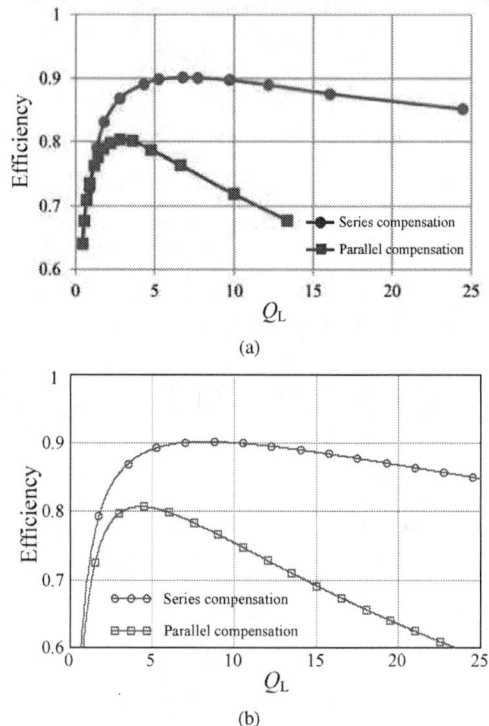

Fig. 7. Efficiencies obtain from (a) measurement and (b) calculation.

TABLE III
FREQUENCIES TO ACHIEVE LOAD-INDEPENDENT $|G_i|$ AND MAXIMUM EFFICIENCY.

Compensation	Load-independent $|G_i|$	Max. efficiency
Series	ω_P	ω_S
Parallel	ω_L or ω_H	ω_N

Therefore, by adjusting the values of C_P and C_S, a secondary series compensated current output IPT converter can have load-independent output and maximum efficiency by operating at $\omega_S = \omega_P$.

V. EXPERIMENTAL EVALUATION

Full-bridge prototype circuits are built to verify the results from the analyses. The prototypes are constructed with a loosely coupled transformer whose parameters are shown in Table IV.

Fig. 6 gives the experimental transconductances of the secondary series and parallel converters versus operating frequency with three output loading conditions.

The efficiencies versus Q_L (adjusted using R_L) at a constant output power of 24 W (adjusted using V_{in}) are measured for the two prototype converters. A constant power operation can maintain a smaller converter temperature variation and thus a smaller variation of converter component parameters, leading to a more consistent efficiency measurement. The calculated and measured efficiencies of the secondary series compensated converter operating at ω_S, as well as secondary parallel compensated converter operating at ω_H, are compared

maximum efficiency are independent. They can be adjusted individually by changing values of compensation capacitors in each side of resonant tank according to their definitions shown in Table I. For a secondary parallel compensated transconductance converter, there is no single frequency that achieves load-independent $|G_i|$ and maximum efficiency as the frequencies never align.

If ω_P is selected equal to ω_S as in [4], [7], the efficiency comparison of operation at ω_S of series compensation and that at ω_H of parallel compensation is depicted in Fig. 5. It is obvious that series compensation has higher efficiency at full load.

TABLE IV
COMPONENTS AND PARAMETERS USED IN THE EXPERIMENTAL
CONVERTERS.

Transformer and compensation	Value
k	0.18
L_P	32.77 μH @200kHz
L_S	31.77 μH @200kHz
C_P	19.32 nF / 630 V
C_S	19.93 nF / 630 V

as shown in Fig. 7. All the efficiencies are measured with soft switching conditions. From the comparison of efficiency curves shown in Fig. 7, and some other measurements with stronger and weaker coupling coefficients, the series compensated IPT current output converter always performs with better efficiency.

VI. CONCLUSION

The operating frequencies of an inductive power transfer output current converter are studied with primary series compensation and secondary series or parallel compensation. Operating frequencies for maximum efficiency and load-independent transconductance are identified and compared with the two secondary compensation techniques. Operating at ω_S of the secondary series compensation and operating at ω_N of the secondary parallel compensation have a similar maximum efficiency. However, for the secondary series compensation, the load-independent transconductance operating frequency ω_P and maximum efficiency operating frequency ω_S can be adjusted independently. Thus, the two desired features can be achieved simultaneously. While for the secondary parallel compensation, the two frequencies never align. The efficiency will be lower than the maximum when it has load-independent current output. Two experimental prototypes are built and the experimental results confirm the analysis.

REFERENCES

[1] G. A. Covic and J. T. Boys, "Inductive power transfer," *Proceedings of the IEEE,* vol. 101, no. 6, pp. 1276–1289, Jun. 2013.

[2] B. Choi, J. Nho, H. Cha, T. Ahn and S. Choi, "Design and implementation of low-profile contactless battery charger using planar printed circuit board windings as energy transfer device," *IEEE Trans. Industrial Electronics,* vol. 51, no. 1, pp. 140–146, Feb 2004.

[3] G. A. Covic, J. T. Boys, M. L. G. Kissin and H. G. Lu, "A three-phase inductive power transfer system for roadway-powered vehicles," *IEEE Trans. Industrial electronics,* vol. 54, no. 6, pp. 3370–3377, Dec. 2007.

[4] C. S. Wang, O. H. Stielau and G. A. Covic, "Design considerations for a contactless electric vehicle battery charger," *IEEE Trans. Industrial electronics,* vol. 52, no. 5, pp. 1308–1314, Oct. 2005.

[5] W. Zhang, S. C. Wong, C. K. Chen, "Design for efficiency optimization and voltage controllability of series-series compensated inductive power transfer systems," *IEEE Trans. Power Electronics,* vol. 29, no. 1, pp. 191–200, Jan. 2014.

[6] W. Zhang, S. C. Wong, C. K. Tse and Q. Chen, "Analysis and comparison of secondary series and parallel compensated inductive power transfer systems operating for optimal efficiency and load-independent voltage transfer ratio," *IEEE Trans. Power Electronics,* to appear.

[7] C. S. Wang, G. A. Covic and O. H. Stielau, "Power transfer capability and bifurcation phenomena of loosely coupled inductive power transfer systems," *IEEE Trans. Industrial Electronics,* vol. 51, no. 1, pp. 148–156, Feb. 2004.

[8] J. Garnica, R. A. Chinga and J. Lin, "Wireless power transmission: from far field to near field," *Proceedings of the IEEE,* vol. 101, no. 6, pp. 1321–1331, Jun. 2013.

Voltage Control of Inductive Contactless Power Transfer System with Coaxial Coreless Transformer for DC Power Distribution

Yushi Miiura, Satoshi Ojika, Tomofumi Ise

Division of Electrical, Electronic and Information Engineering, Graduate School of Engineering
Osaka University
Suita, Osaka, Japan

Abstract— Application of an inductive contactless power transfer system to electric plug and outlet dc power distribution has been proposed. The proposed system consists of two solenoids which function as a coaxial coreless transformer, a half bridge inverter on a primary side and a diode rectifier on a secondary side. This system can be arc-less, small-size and voltage-convertible. However, the rise of output voltage under light load condition was observed in the experiments. In this paper, to suppress the rise of the output voltage, employment of a cascade half bridge inverter and an output voltage estimator is proposed. Voltage control of the proposed system was verified through experiments using a 1 kW class prototype. Additionally, a high efficiency of 93.6% was achieved.

Keywords— *contactless power transfer system, coaxial coreless transformer, dc power distribution system, plug and outlet*

I. INTRODUCTION

Low voltage dc power distribution system [1] for a house or an apartment is one of promising systems for future smart houses. It can reduce power losses in converters because the number of power conversions is reduced [2]. However, it has still had several problems for practical use, and especially for household use, an issue of "electric plug and outlet" are inevitable. If conventional electric plug and outlet with metallic contactors for ac power distribution system are employed in the dc system, they cause problems such as voltage conversion, risk of electric shock, arc discharge and galvanic isolation.

In order to solve these problems, electric plugs and outlets with mechanical interlock switches were proposed for data center applications. Additionally, for low power application, low voltage contactor plug and outlet with protecting plastic cover was also proposed. However, they still need improvement for downsizing and require another electrical circuits for voltage conversion.

From these points of view, we have proposed application of an inductive Contactless Power Transfer (CPT) system to the household dc power distribution system as electric plug and outlet with a voltage conversion function [3]. This concept is illustrated in Fig.

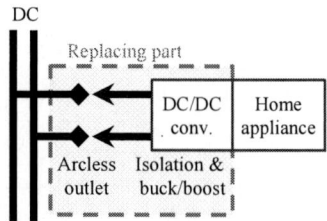

(a) A conventional plug and outlet system with a dc/dc converter.

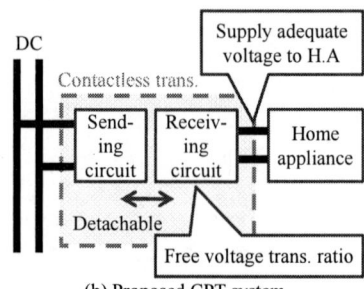

(b) Proposed CPT system.
Fig. 1 Concept of the proposed CPT system.

1. This system is composed of a separable transformer as a plug and outlet system, a half bridge inverter on a primary side and a diode rectifier on a secondary side. The separable transformer consists of two coaxial solenoids which function as primary and secondary windings of a coreless high frequency transformer, and the secondary solenoid is inserted into the primary one like the conventional plug and outlet shown in Fig. 2. Since the system has no contactor between plug and outlet, it is an arc-free system. Moreover, it conducts voltage conversion by choosing the adequate turn ratio between the primary and secondary windings for home appliances which require various voltage levels. Galvanic isolation of home appliances from a power line is also realized. Thus, this CPT system will be safer and smaller comparing with the conventional system.

In the previous research [3], we investigated some types of coreless transformers for the inductive CPT system to improve the system efficiency, and the highest efficiency of 93.2% was obtained through experiments

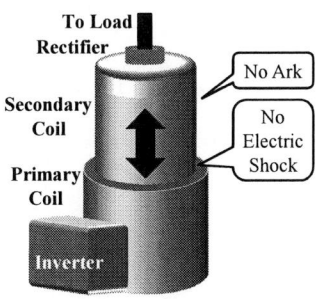

Fig. 2 Proposed CPT plug and outlet.

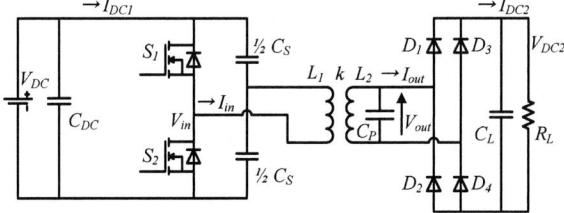

Fig. 3 Circuit configuration of the CPT system with a HB inverter.

TABLE I
MAJOR PARAMETERS OF THE CPT SYSTEM WITH HB INVERTER

Source voltage V_{DC}	340 V
Main switch S_1, S_2	MOSFET TK15A60U (600 V, 15 A)
Rectifier $D_1 - D_4$	FRD STTH8L06D (600 V, 8 A)
Switching frequency f_s	600 kHz
DC capacitor C_{DC}	10 μF
Series capacitor C_S	37 nF
Parallel capacitor C_P	4.7 nF
Smoothing capacitor C_L	1 μF
Transformer turn ratio a	1:1

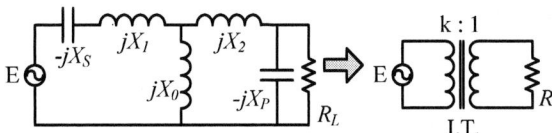

Fig. 4 Equivalent circuit of the transformer with series-parallel (S/P) compensation.

using a 1 kW prototype. However, we found the problem of the rise of output voltage under a light load condition and the output voltage increased more than 130% of a rated value.

In this paper, first we investigate the cause of the voltage rise under light load operation, and secondly, the control method of the output voltage to suppress voltage rise by employing a cascade half bridge inverter is proposed. The proposed voltage control is demonstrated through experiments using the 1 kW class prototype.

II. INDUCTIVE CONTACTLESS POWER TRANSFER SYSTEM FOR DC POWER DISTRIBUTION

In this study, we aim at replacing of the conventional plug and outlet with the proposed CPT system, and therefore the rated power of the system is set at around 1 kW. Differently from other wireless power transfer system for mobile objects using an electric field coupling, an inductive coupling, magnetic resonance [4-6], the plug of a load is fixed in the outlet of the CPT system when it is in use. Therefore, the transfer distance between the primary and secondary windings can be assumed as almost zero. From these viewpoints of relatively high power and zero transfer distance, the inductive coupling is adopted for the CPT system.

A. Circuit configuration

The circuit configuration of the investigated inductive CPT system is shown in Fig. 3 and major parameters are summarized in Table I. The inverter employs a half bridge (HB) topology due to its simplicity and generates ac voltage with a high frequency of 600 kHz on the primary side of the transformer from a dc voltage of 340 V, which is supplied from the power lines of a dc microgrid [1].

The separable transformer of the inductive CPT system, which consists of two coaxial solenoids, can be treated as a normal transformer, however it has relatively large leakage inductance. To compensate the effect of the leakage inductance of the transformer, a series-parallel (S/P) compensation is employed, namely, two compensation capacitors are connected in series on the primary side and in parallel on the secondary side, respectively. An equivalent circuit in the case that the turn ratio of the transformer is unity is shown in Fig. 4. The variables X_1, X_2 and X_0 are leakage reactance of the

primary and secondary sides, and magnetizing reactance of the transformer. Applying the compensation capacitors of X_S and X_P which have the following relationships,

$$X_S = X_1 + \frac{X_0 X_2}{X_0 + X_2}, \qquad (1)$$

$$X_P = X_0 + X_2, \qquad (2)$$

the equivalent impedance Z is expressed as

$$\dot{Z} = \frac{X_0^{\,2} R_L}{(X_0 + X_2)^2} = k^2 R_L \qquad (3)$$

where k is a coupling coefficient of the transformer, and as a result, the transformer with the compensation capacitors can be treated as an ideal transformer independent of the load resistance of R_L.

B. Coaxial Coreless Transformer

A separable transformer is composed of two coaxial solenoids and one solenoid is inserted into the other like a conventional electric plug and outlet. A coreless type is adopted due to cost reduction. We employ a single-layer solenoid with a copper single strand to achieve a high quality factor Q, which is related to a high efficiency. The appearance of the transformer, whose secondary solenoid is inside the primary one, is shown in Fig. 5 and its parameters are summarized in Table II. Outer and

978-1-4799-2706-7/14 $31.00 © 2014 IEEE

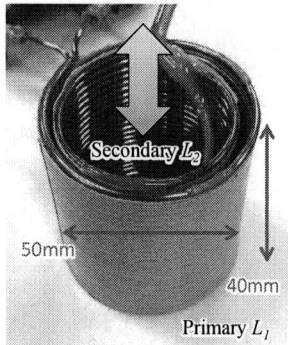

Fig. 5 Appearance of the coaxial coreless transformer.

TABLE II

MAJOR PARAMETERS OF THE COAXIAL CORELESS TRANSFORMER

Wire Type	Dia. 2mm UEW
Layers per winding	1
Target frequency	600 kHz
Outer diameter	50 mm
Inner diameter	43 mm
Length	40 mm
Outer coil turn number	22
Inner coil turn number	22
Outer coil inductance L_1	13.6 μH
Inner coil inductance L_2	11.7 μH
Coupling coefficient k	0.80

inner diameters are 50 mm and 43 mm, respectively. The quality factor Q of the transformer was around 60 at the operation frequency of 600 Hz, which was measured under the condition the secondary solenoid was fully inserted in the primary one.

III. VOLTAGE CONTROL

A. Output Voltage Rise Problem

Numerical simulation of the system was done and waveforms of the transformer under the condition of $R_L = 40\ \Omega$ are shown in Fig.6.

A 1 kW prototype was made and experiments were conducted. Fig.7 shows measured efficiency η and output voltage V_{out} versus output power P_{out}. The efficiency is defined as dc output power P_{out} divided by dc input power P_{in}. In the operation region where P_{out} was over 200 W, the efficiency more than 90% was achieved, and the highest efficiency was 93.2% at the output power of 439 W.

The prototype was operated with open loop control, and the rise of output voltage was observed under light load condition. The output voltage reached up to around 260 V while the rated voltage was 200 V. Such high voltage could damage the circuit devices and it required suppression control for the output voltage.

B. Causes of Output Voltage Rise

Frequency characteristics from input voltage to output voltage of the equivalent circuit in Fig. 4 are shown in

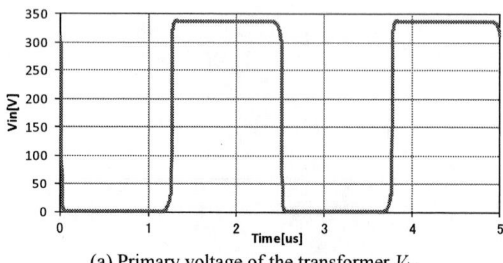

(a) Primary voltage of the transformer V_{in}.

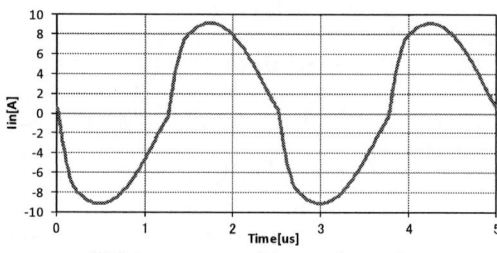

(b) Primary current of the transformer I_{in}.

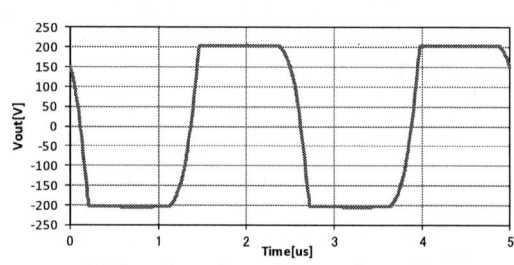

(c) Secondary voltage of the transformer V_{out}.

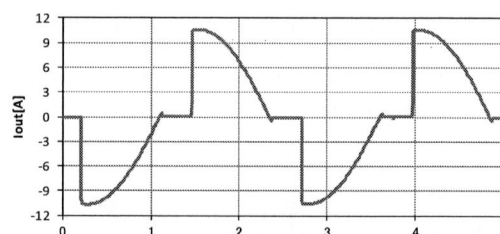

(d) Secondary current of the transformer I_{out}.

Fig. 6 Voltage and current waveforms of the transformer (HB inverter, $R_L = 40\ \Omega$, simulation).

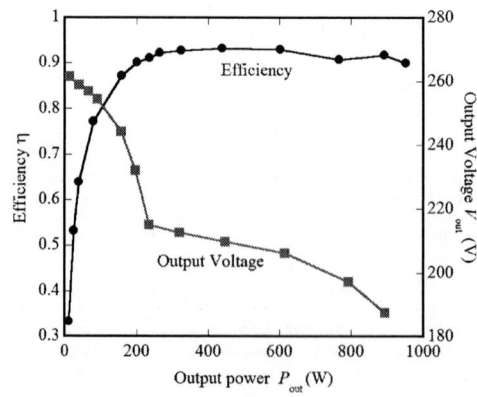

Fig. 7 Measured efficiency and output voltage.

978-1-4799-2706-7/14 $31.00 © 2014 IEEE

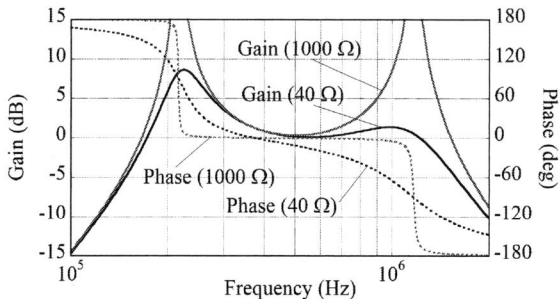

Fig. 8 Transfer characteristic of the CPT system.

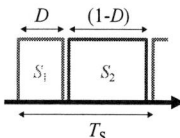

Fig. 9 Switching pattern of duty ratio control of the HB inverter.

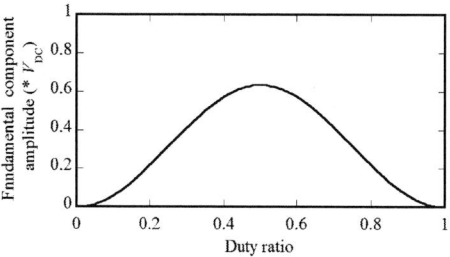

Fig. 10 Amplitude of the fundamental component versus duty ratio of the HB inverter.

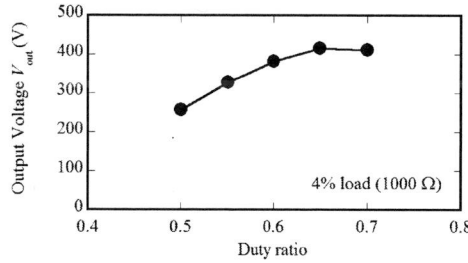

Fig. 11 Output voltage of the CPT system with the HB inverter using duty ratio control ($R_L = 1000\ \Omega$ (4% load), simulation).

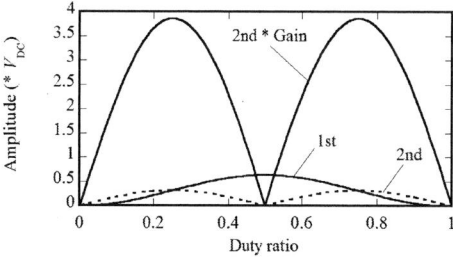

Fig. 12 Effect of 2nd harmonic ($R_L = 1000\ \Omega$ (4% load)).

Fig. 8. At the operation frequency of 600 kHz, gain and phase are around unity and zero deg, respectively, under both light and heavy load conditions. However, the primary side inverter generates almost rectangular voltage waveform on the primary side of the transformer as shown in Fig. 6. Therefore, the waveform contains a large amount of high frequency harmonics, and they would cause the output voltage rise because some harmonics have higher gains than the fundamental component.

C. Duty Ratio Control for Half Bridge Inverter

In order to control output voltage, we first investigated application of duty ratio control to the HB inverter. Using switching pattern during a sample period T_s shown in Fig. 9, the fundamental component of the voltage waveform on the primary side of the transformer V_1 is calculated as

$$V_1 = \frac{V_{DC}}{\pi}\left(1 - \cos 2\pi D\right) \qquad (4)$$

where V_{DC} and D are dc voltage and duty ratio, respectively, and their relationship is shown in Fig. 10. Theoretically, V_1 is controlled with duty ratio D.

To investigate the effect of the voltage control with the duty ratio control, numerical simulation was done using software PSIM. The load resistance was set to 1000 Ω, which was 4% load. Fig. 11 shows simulation results of output voltage versus the duty ratio. Differently from the characteristics in Fig. 10, the output voltage rose as the duty ratio increased.

One of causes of the voltage rise was the 2nd harmonic contained in the inverter output voltage, which was generated except on the condition that the duty ratio was 0.5. The transfer characteristics in Fig. 8 has a higher gain especially at the frequency of the 2nd harmonic than the fundamental frequency. The 2nd harmonic contained the inverter output voltage V_2 can be expressed as

$$V_2 = \frac{\sqrt{2 - 2\cos 4\pi D}}{2\pi} V_{DC} \ . \qquad (5)$$

Under the condition of the simulation, the gain at the 2nd harmonic frequency was around 12. The effect of the 2nd harmonics is shown in Fig.12. The product of the 2nd harmonic amplitude and the gain is much larger than the fundamental component amplitude. These results suggest it is difficult to suppress the output voltage rise

by the HB inverter with duty ratio control.

D. Cascade half Bridge Inverter

In order to suppress the output voltage rise, it is necessary that the voltage waveform of the inverter contains less 2nd harmonic.

One of solutions is application of phase-shift pulse width modulation (PWM) because its waveform does not contain the 2nd harmonic ideally. The phase-shift PWM is frequently applied to a full bridge (FB) inverter because two legs of the inverter should be controlled independently. However, the FB inverter outputs twice the larger voltage than the HB inverter. For applications as power supplies for home appliances, the CPT system mainly requires step-down voltage conversion. Thus, the FB inverter is not suitable.

From these viewpoints, we propose application of a cascade half bridge (CHB) inverter shown in Fig. 13. The CHB inverter can output voltage with the same magnitude as the HB inverter, and because of its cascade structure, switching devices in the CHB inverter have half voltage stresses than the HB inverter. Therefore low-cost and high-performance switching devices can be applied. Moreover, the CHB inverter can employ the phase-shift PWM and therefore its output voltage contains no 2nd harmonic theoretically.

According to applied pulse pattern shown in Fig. 14 (a), the fundamental component in the inverter output voltage waveform can be calculated as

$$V_1 = 2 \frac{V_{DC}}{\pi} \cos \frac{\alpha}{2} \qquad (6)$$

where α is the amount of phase shift (1 p.u. = 2 π), and is shown in Fig. 15. Thus the output voltage can be controlled by phase-shift PWM.

However, the pulse pattern in Fig. 14 (a) has the problem that it causes voltage unbalance between dc capacitors C_{DC1} and C_{DC2}. Figs. 14 (a) and 16 (a) explain the cause of this voltage unbalance. During the period

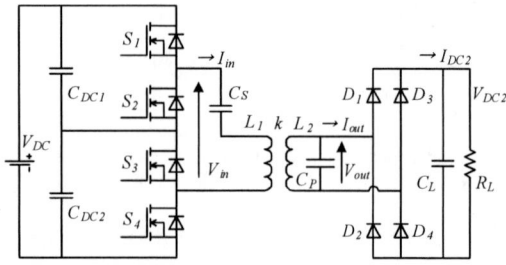

Fig. 13 Circuit configuration of the CPT system with a CHB inverter.

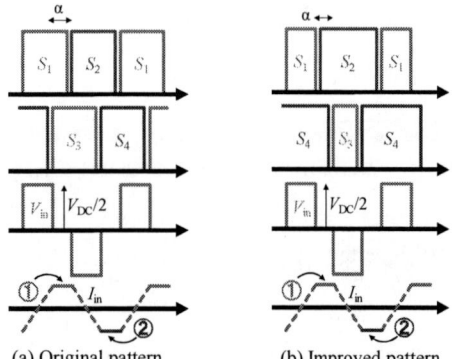

(a) Original pattern. (b) Improved pattern.
Fig. 14 Switching patterns of phase shift PWM of the CHB inverter.

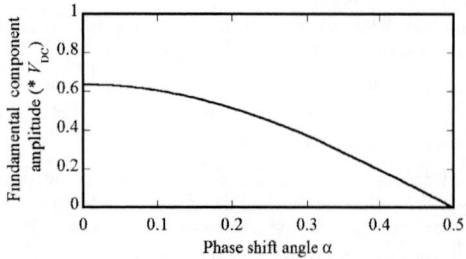

Fig. 15 Amplitude of the fundamental component versus duty ratio with phase-shift PWM.

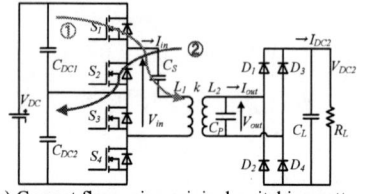

(a) Current flow using original switching pattern.

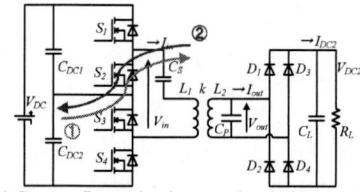

(b) Current flow using improved switching pattern.
Fig. 16 Current flow of the CHB inverter with phase-shift PWM.

when switches S_1 and S_3 are turned on, current flows through C_{DC1}, however during another period when switches S_2 and S_4 are turned on, current flows through C_{DC2}, as a result, electric charge moves from C_{DC1} to C_{DC2} and this brings about voltage unbalance. To solve this problem, we improved pulse pattern generation to make current flow through only one side capacitor as shown in Figs. 14 (b) and 16 (b). In practice, the dc capacitors where current flows are switched alternately with a frequency of 25 kHz to reduce their stresses.

Numerical simulation under the condition of 4% load ($R_L = 1000 \ \Omega$) with open loop control was conducted and waveforms of the CHB inverter are shown in Fig. 17. Output voltage variation versus phase shift is shown in Fig. 18 and it was confirmed that the CHB inverter with phase-shift PWM can suppress the output voltage rise by the increase of the phase shift angle

E. Estimator for Secondary Voltage

In order to control the output voltage, it is necessary to estimate the secondary voltage of the transformer. However, the secondary circuit has no signal communication to the CHB inverter on the primary side.

If we make an observer in the controller of the CHB inverter, a high speed signal processor to monitor and calculate voltage and current which have a high frequency of 600 Hz is required. From the viewpoint of cost reduction, it is undesirable to employ such a high speed signal processor.

To estimate the secondary voltage, we made the estimator using tertiary windings of the transformer, namely, we added a one-turn coil as the tertiary windings at the bottom of the solenoids and used an induced voltage across its terminals. The induced voltage is fed to a rectifier composed of Shottky barrier diodes with a low pass filter to be converted to a dc signal. The circuit configuration, the appearance, the installation position of the estimator are shown in Figs. 19 (a), (b) and (c), respectively. This estimator can be made with low cost and a low-speed controller can be used due to use of the dc signal.

Comparison of the estimator output and the measured

The 2014 International Power Electronics Conference

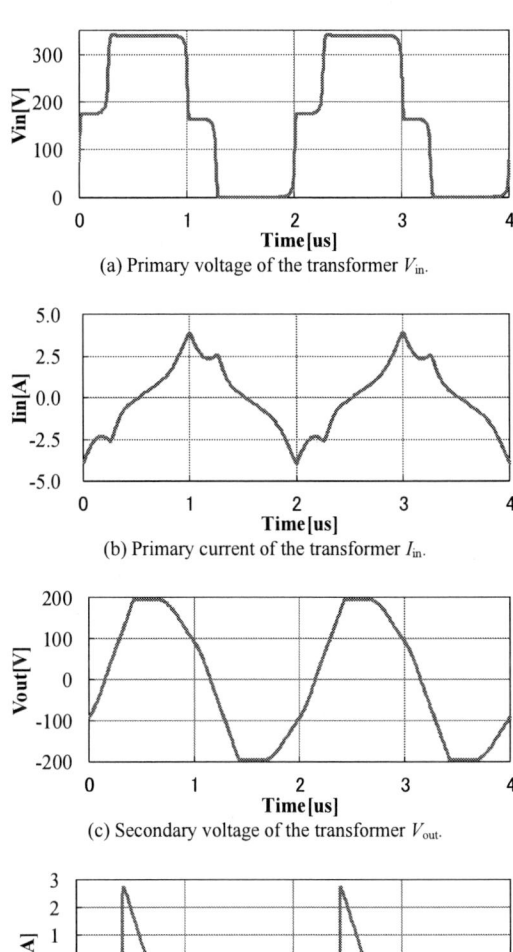

(a) Primary voltage of the transformer V_{in}.

(b) Primary current of the transformer I_{in}.

(c) Secondary voltage of the transformer V_{out}.

(d) Secondary current of the transformer I_{out}.
Fig. 17 Voltage and current waveforms of the transformer (CHB inverter, $R_L = 40\ \Omega$, simulation).

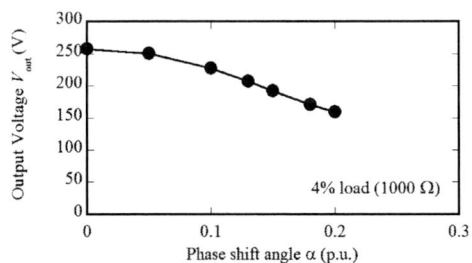

Fig. 18 Output voltage of the CPT system with the HB inverter using duty ratio control ($R_L = 1000\ \Omega$ (4% load), simulation).

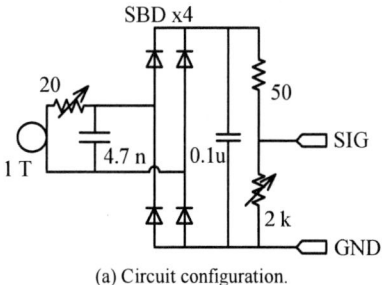

(a) Circuit configuration.

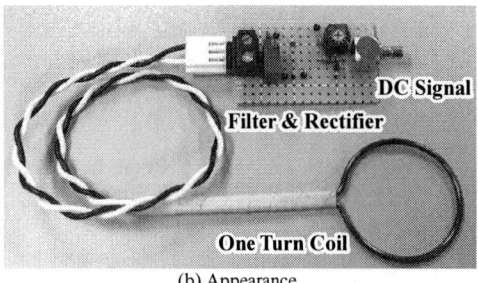

(b) Appearance.

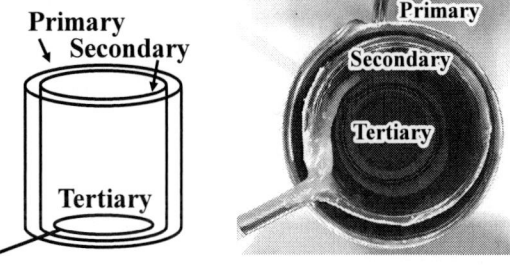

(c) Installation position.
Fig. 19 Circuit configuration, appearance and installation position of the estimator.

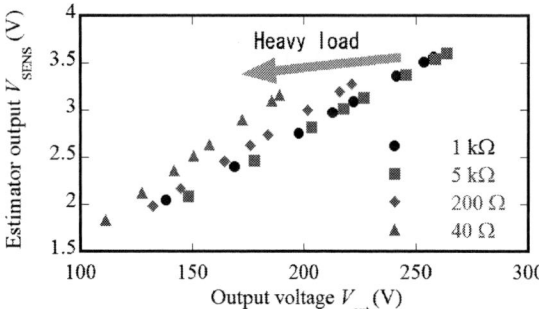

Fig. 20 Comparison of the estimator output and the measured actual output voltage of the transformer.

F. Feedback Control of Output Voltage

Using the estimator of the output voltage mentioned in the previous subsection, we made a feedback controller in a DSP system shown in Fig. 21. The estimator output V_{SENS} is compensated using measured primary side dc current and estimated voltage is fed back to a PI voltage controller. The valuable of V_{REF} is the reference for V_{SENS} and the dc output voltage is controlled by setting an

actual output voltage of the transformer is shown in Fig. 20. The estimator output had linear relationship with the actual output voltage. However, as the load increased, the slope of the line changed due to the effect of leakage inductance of the transformer. Thus, we required a compensator for the estimator output in the controller of the inductive CPT system.

978-1-4799-2706-7/14 $31.00 © 2014 IEEE

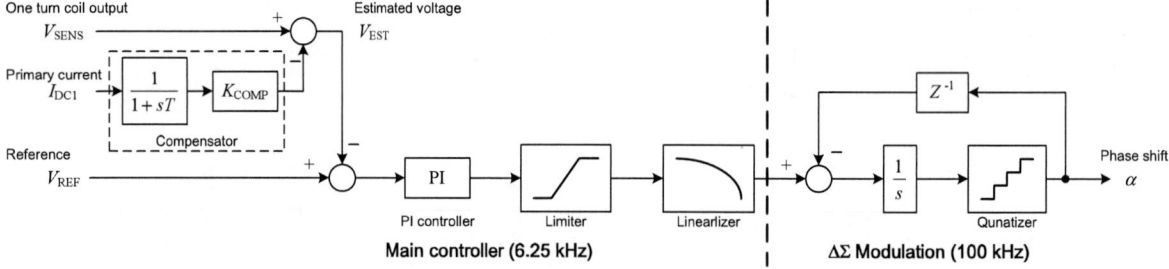

Fig. 21 Control block diagram for the CPT system.

adequate value for V_{REF}. The reference of the phase shift α is obtained through a limiter and a linearizer finally. A $\Delta\Sigma$ modulation was applied because it was easy to be implemented to the DSP system.

IV. EXPERIMENTS

We made a 1 kW class prototype with the CHB inverter and the output voltage estimator. The appearance of the prototype is shown in Fig. 22 and major parameters of the prototype are summarized in Table III.

A. Steady State Characteristics

The experiment with open loop control under the condition of 4% load ($R_L = 1000\ \Omega$) was first conducted and results are shown in Fig. 23. The experiments have a similar result to the numerical simulation results in Fig.

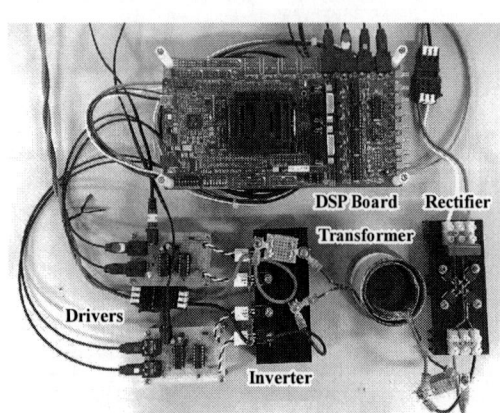

Fig. 22 Appearance of the 1 kW prototype.

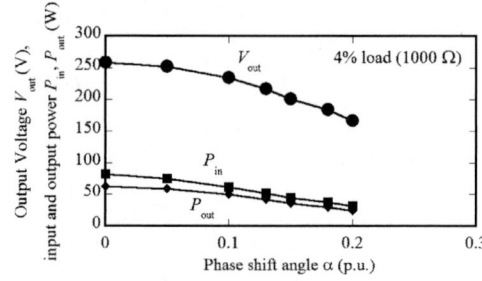

Fig. 23 Output voltage of the CPT system with the CHB inverter using phase shift PWM ($R_L = 1000\ \Omega$ (4% load), experiment).

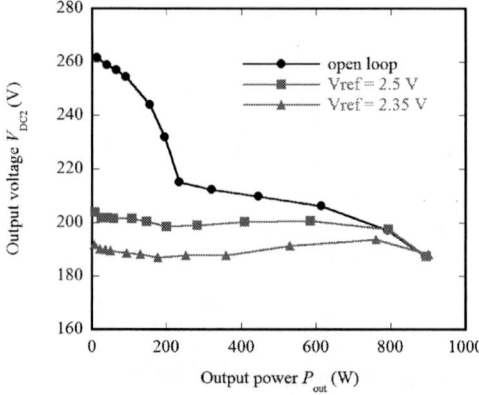

Fig. 24 Output voltage versus output power of the CPT system with the CHB inverter (experiment).

18 and the voltage control by phase-shift PWM was verified.

Secondly, we conducted steady state experiments and results are shown in Fig. 24. The measured dc output voltage V_{DC2} and its variation were 198 V±5% and 191 V±5% in the cases of $V_{REF} = 2.5$ (V) and 2.35 (V), respectively. These results confirmed that the dc output voltage was well controlled and its rise was sufficiently suppressed by the feedback control using the estimator.

B. Transient Characteristics

Since various loads are connected to the CPT system, transient characteristics are also important. Thus we conducted transient experiments and the load changed from 2.4% ($R_L =1667\ \Omega$) to 100% ($R_L = 40\ \Omega$) with a slew rate of 1 A/ms. Results in the case of $V_{REF} = 2.5$ V

TABLE III
MAJOR PARAMETERS OF THE CPT SYSTEM USING THE CHB INVERTER

Source voltage V_{DC}	340 V
Main switch S_1, S_2	MOSFET ST Micro STF17NF25 (250 V, 17 A)
Rectifier $D_1 - D_4$	FRD STTH8L06D (600 V, 8 A)
Switching frequency f_s	600 kHz
DC capacitor C_{DC}	10 μF
Series capacitor C_S	37 nF
Parallel capacitor C_P	4.7 nF
Smoothing capacitor C_L	1 μF
Transformer turn ratio a	1:1

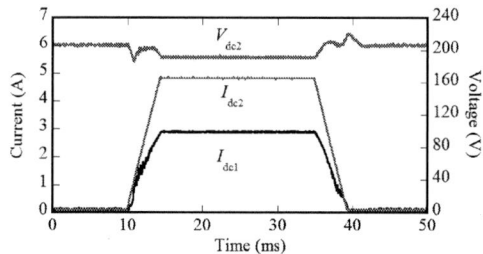

Fig. 25 Transient response of the CPT system with the CHB inverter (load change: 2.4% - 100%, V_{REF} = 2.5 V, experiment).

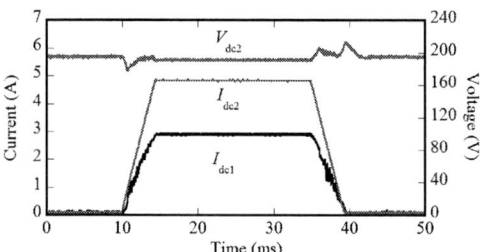

Fig. 26 Transient response of the CPT system with the CHB inverter (load change: 2.4% - 100%, V_{REF} = 2.35 V, experiment).

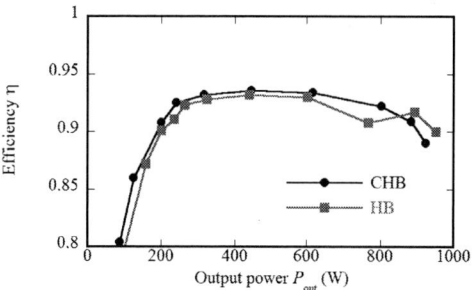

Fig. 27 Efficiency comparison of the CPT system using CHB and HB inverters.

TABLE IV
SWITCHING DEVICES OF HB AND CHB INVERTERS

	Half bridge	Cascade half bridge
MOSFET	Toshiba	ST Micro
	TK15A60U	STF17NF25
Rated Voltage	600 V	250 V
Rated Current	15 A	17 A
$R_{DS(ON)}$	0.24 Ω	0.14 Ω

are shown in Fig. 25. There were less oscillation and settling time was sufficiently short, i.e. within 5 ms. Voltage variation was also small, 36 $V_{peak-to-peak}$. Fig. 26 shows results in the case of V_{REF} = 2.35 V. The settling time was also within 5 ms, and voltage variation was 37 $V_{peak-to-peak}$. Through the experiments using the proposed control, sufficient response and stability were demonstrated.

C. Efficiency

We conducted measurement of the efficiency of this prototype and results are shown together with the efficiency of the HB inverter in Fig. 27. Parameters of employed MOSFETs are summarized in Table IV. The prototype with the CHB inverter had a similar efficiency curve to the one with the HB inverter, however the maximum efficiency of 93.6% was achieved at the output power of 444 W, which was slightly higher than the HB inverter. This improvement was achieved by employment of lower-voltage switching devices in the CHB inverter compared with the HB inverter.

V. CONCLUSIONS

In this paper, we proposed the application of an inductive contactless power transfer system with a coaxial coreless transformer to dc power distribution. In order to solve the problem of output voltage rise, we clarified that it was mainly caused by the harmonics of input voltage and proposed the employment of a cascade half bridge inverter to control the output voltage. In addition, the estimator, which was composed of a one-turn coil and a simple diode rectifier, was proposed to estimate the output voltage on the secondary side of the transformer. A feedback controller with the estimator for the contactless power transfer system was implemented to a 1 kW prototype. Suppression of the dc output voltage rise and achievement of quick response of the CPT system were demonstrated through experiments of the prototype. In addition, the highest efficiency of 93.6% was achieved and it was higher than the system with the half bridge inverter.

From these results, we can conclude that the proposed CPT system is promising for the safe and small-size electric plug and outlet of the dc power distribution.

REFERENCES

[1] H. Kakigano, Y. Miura and T. Ise "Low-Voltage Bipolar-Type DC Microgrid for Super High Quality Distribution" IEEE Transactions on Power Electronics Vol. 25, No. 12, pp. 3066-3075 (2010).

[2] H. Kakigano, M. Nomura, and T. Ise "Loss Evaluation of DC Distribution for Residential Houses Compared with AC System" 2010 International Power Electronics Conference (IPEC), pp.480-486 (2010).

[3] S. Ojika, Y. Miura and T. Ise, "Inductive Contactless Power Transfer System with Coaxial Coreless Transformer for DC Power Distribution," Proc. of the Energy Conversion Congress and Exposition Asia (ECCE-Asia), pp. (2013).

[4] R. Yoshida and others, "The front line of wireless energy transfer technology" published by NTS inc, ISBN 978-4-86043-351-2 (in Japanese).

[5] A. J. Moradewicz, and M. P. Kazmierkowski, "Contactless Energy Transfer System with FPGA-Controlled Resonant Converter" IEEE Transactions on Industrial Electronics, Volume 57, Issue 9, pp. 3181-3190, September 2010.

[6] A Kurs et al., "Wireless Power Transfer via Strongly Coupled Magnetic Resonances" Science, Vol. 317, no. 5834, pp. 83-86 (2007).

The 2014 International Power Electronics Conference

Contactless High Power Transformer Technologies for Railway Vehicles

Keiichiro Kondo, Kohei Yamamoto, Satochi Kitazawa

Graduated School of Chiba University, Faculty of Engineering
Chiba, Japan
1-33 Yayoi, Inage-ku, Chiba-shi, Chiba prefecture
kkondo@faculty.chiba-u.jp

Abstract— **The contactless power transformer systems (CPTs) have recently attracted much attention as it is possible to supply electrical energy to mobile vehicles without cables. 300kW class CPT systems are required for railway vehicles power feeding applications. In this power range, there are some technical problems such as higher reactive power due to longer gap and power control at the misalignment of ground coils and on-board coils. To cope with these problems, authors have been working to establish a reactive power conscious coil designing method and less loss type power conversion system with controlling active power in the secondary side. In this paper, some of these results are presented**

Keywords— *Contactless power transformer, reactive power, PWM rectifier, raiway traction application*

I. INTRODUCTION

Contactless power transfer (CPT) systems have much advantage such as not continuously but frequently and easily power feeding. These features promote the electrification of mobile vehicles, not only internal combustion engine powered automobiles but also diesel motive units of railway vehicles. Power range of CPT systems for automobiles is 3kW to 50kW and some of them are almost in the commercial stage[1]-[3]. On the other hand, 300kW class CPT systems are required for railway vehicles power feeding applications. In this power range, there are some technical issues such as higher reactive power due to longer gap and power control even at the misalignment of ground coils and on-board coils. To cope with these problems, authors have been working to establish a reactive power conscious coil designing method and less loss type power conversion system with controlling active power in the secondary side. In this paper, some of these results are presented.

II. REACTIVE POWER CONSCIOUS COIL DESIGN[3]

A. System Configuration

Fig. 1 shows a contactless power transfer system. In this study, the primary side inverter outputs square wave. Therefore, the primary side capacitor is connected in series. And the secondary side is the constitution to charge secondary battery. Therefore, two circuit

topologies with regard to compensation capacitors, the Series-Parallel (SP) topology as shown in Fig. 2, and the Series-Series (SS) topology as shown in Fig.3 are considered .

The value with an apostrophe in Fig. 2 and Fig. 3 represents the value referred to the primary side. In an actual application, the resistance R_L in Fig.2 is replaced with a choke input rectifier circuit, and the resistance R_L and the secondary capacitor C_{s2} in Fig.3 can be replaced with a PWM rectifier, for example using instantaneous current control method. In this chapter, the characteristics of the reactive power are analyzed using these equivalent circuits.

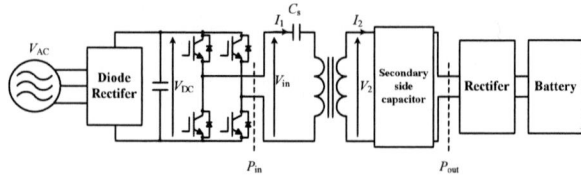

Fig. 1. A battery charging system by contactless power transformer.

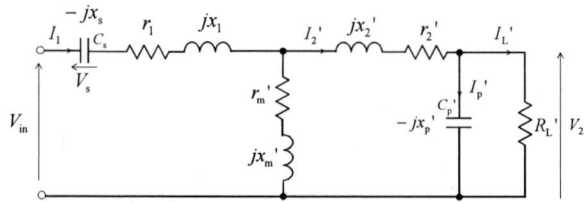

Fig. 2. An equivalent circuit of transformer with series and parallel capacitors.

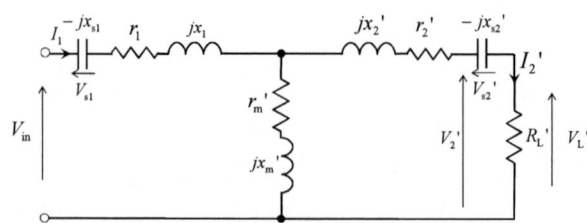

Fig. 3. An equivalent circuit of transformer with series capacitors in both sides.

978-1-4799-2706-7/14 $31.00 © 2014 IEEE   1438

B. Series-Parallel (SP) topology

Because the winding resistances (r_1 and r_2') and the ferrite-core loss r_m' are much lower than the mutual reactance x_m' and leakage reactance (x_1 and x_2'), these are ignored in Fig.2.

And the values of the resonance capacitances are given by [1] [2]

$$\frac{1}{\omega C_p{}'} = x_p{}' = x_m{}' + x_2{}' \quad\text{.......................................(1)}$$

$$\frac{1}{\omega C_s} = x_s = x_1 + \frac{x_2{}'x_m{}'}{x_2{}' + x_m{}'} \quad\text{..........................(2)}$$

Relationship between the primary voltage V_{in} and the secondary voltage V_2' is given by

$$V_2{}' = \frac{x_2{}' + x_m{}'}{x_m{}'} V_{in} \quad\text{.................................(3)}$$

Equation (3) indicates that the secondary voltage also becomes constant by using a constant voltage source.

In the equivalent circuit, x_m', x_1 and x_2' are given by (4). L_m' is the mutual inductance between the transformer windings, l_1 and l_2' denote the leakage inductances of the primary and secondary windings and f is the frequency. L_m', l_1 and l_2' are given by (5). k is the coupling coefficient, L_1 and L_2' denote the self-inductances of the primary and secondary windings.

$$x_m{}' = 2\pi f L_m{}', x_1 = 2\pi f l_1, x_2{}' = 2\pi f l_2{}' \quad\text{...........(4)}$$

$$L_m{}' = k\sqrt{L_1 L_2{}'}, l_1 = L_1 - L_m{}', l_2{}' = L_2{}' - L_m{}' \quad\text{(5)}$$

By defining the coil reactance X as $X=2\pi f L_1$, and the self-inductance ratio of the primary side and the secondary side as $a=L_2/L_1$, (3) can be expressed by the actual value as (6) instead of the value referred to the primary side.

$$V_2 = \frac{a}{k} V_{in} \quad\text{...(6)}$$

The voltage across the primary side capacitor V_s, the primary current I_1, and the current through the secondary side capacitor I_p, can be expressed respectively as (7), (8) and (9). P is the transmission power.

$$V_s = -j(1-k^2)X\frac{P}{V_{in}} \quad\text{...............................(7)}$$

$$I_1 = \frac{P}{V_{in}} \quad\text{...(8)}$$

$$I_p = j\frac{V_{in}}{kaX} \quad\text{.......................................(9)}$$

By defining the product of the root-mean-square value of the voltage and current of the resonant capacitor as the reactive power in this paper, the reactive power on the primary side Q_{1SP} and the reactive power on the secondary side Q_{2SP} can be expressed as (10) and (11).

$$Q_{1SP} = (1-k^2)X\frac{P^2}{V_{in}^2} \quad\text{...............................(10)}$$

$$Q_{2SP} = \frac{V_{in}^2}{k^2 X} \quad\text{.......................................(11)}$$

In Fig. 2, the primary side capacitor mostly compensates for the voltage drop by the leakage reactance, and the secondary side capacitor mostly compensates for the lagging current by the mutual reactance. So the voltage across the primary side capacitor increases due to increasing the leakage reactance, and the current through the secondary side capacitor increases due to decreasing the mutual reactance. In order to decrease the reactive power, it is necessary to increase the value of k. However there is a limit to improve the magnetic coupling due to the restrictions of the gap and the dimension of the coils. So the value of the coil reactance should be focused on. It determines the distribution of the compensating reactive power between the primary side and the secondary side. Therefore, it is possible to determine both of the reactive power on the primary side and the secondary side as realistic values by changing the coil reactance X.

As mentioned above, it is important to consider the distribution of the reactive power between the primary side and the secondary side when high-power contactless power transformers are designed.

C. Series-Series (SS) topology

For the SS topology, the value of the resonance capacitors are given by [1]

$$\frac{1}{\omega C_{s2}{}'} = x_{s2}{}' = x_m{}' + x_2{}' \quad\text{....................................(12)}$$

$$\frac{1}{\omega C_{s1}} = x_{s1} = x_m{}' + x_1 \quad\text{.......................................(13)}$$

The voltage across the primary side capacitor V_{s1}, is given by (14). The current through the primary side is equal to that in the SP topology and given by (8).

$$V_{s1} = -jX\frac{P}{V_{in}} \quad\text{.....................................(14)}$$

The voltage across the secondary side capacitor V_{s2}, is equivalent to V_2 in the SP topology, and the current through the secondary side I_2, is equivalent to I_p in the SP topology. Therefore by using a constant voltage source, the secondary current becomes constant. The reactive power on the primary side Q_{1SS} and the reactive power on the secondary side Q_{2SS} can be expressed as (15) and (16).

$$Q_{1SS} = X\frac{P^2}{V_{in}^2} \quad\text{.......................................(15)}$$

$$Q_{2SS} = \frac{V_{in}^2}{k^2 X} \quad\text{.......................................(16)}$$

The reactive power on the primary side is proportional to the coil reactance. On the other hand, the reactive power on the secondary side is inversely proportional to the coil reactance. These characteristics are similar to the SP topology. Therefore, it is also possible to design both of the reactive power on the primary side and the secondary side to realistic values by optimizing the coil reactance X in the SS topology too.

The reactive power is independent of ratio of primary self-inductance and secondary self-inductance a. In this paper, the value of a for the SS topology is induced, which equalizes the voltage across the secondary coil V_2 and the current through the secondary coil I_2 in both topologies. The relationship between the self-inductance ratio in the SP topology a_{SP} and that in the SS topology a_{SS} is expressed by (18)

$$a_{SS} = \frac{V_{in}^2}{\sqrt{k^4 X^2 P^2 + V_{in}^4}} a_{SP} \quad \text{..............................(18)}$$

From this expression, it is clarified that a_{SS} is lower than a_{SP}, therefore, the self-inductance of the secondary side in the SS topology is lower than that in the SP topology.

D. Designe of 300kW Coil

In this study, the frequency is determined for minimizing the total reactive power of the primary side and the secondary side. The reactive power of the primary side and the secondary side in the SP topology is represented as (10) and (11) respectively. From these equations, the total reactive power is lowest when the reactive power of the primary side is equal to that of the secondary side. And the coil reactance X for the condition is represented as (19).

$$X = \frac{V_{in}^2}{k\sqrt{1-k^2}P} \quad \text{...(19)}$$

From this condition, X that minimizes the total reactive power is set up as 15Ω. Therefore, f is determined as 2.6kHz because L_1 is 0.908mH. This frequency is the value considering switching loss on the converter at 300kW. The coils for the SS topology are designed to equalize the reactive power of the secondary side in both topologies and equalize the voltage across the secondary coil V_2 and the current through the secondary coil I_2 in both topologies. Therefore, the coil reactance X, the self-inductance of the primary side L_1 and the frequency f are equal in both topologies.

As show in (1) to (4), reactive power in the primary side for SS topology Q_{1SS} is higher than ones for SP topology Q_{1SP}, because linkage coefficient between ground coils and on-board coils k is lower than 1.0. From Table I, it is clear that the density of magnetic flux in the SS topology is higher than that of the SP topology.

When designing 300kW rating coils for both topologies under the condition of Fig.4 and with radius less than 400mm, the results are obtained as shown in Table I, Table II, Table III, Fig.5 and Fig.6.

Series capacitor voltage in primary side for SP topology is around 50% of ones for SS topology.

Mass of the coils is designed as shown in Table IV when the maximum flux density of ferrite core is limited by 0.3T. As mentioned above, the reactive power is important index to design high power transfer coils.

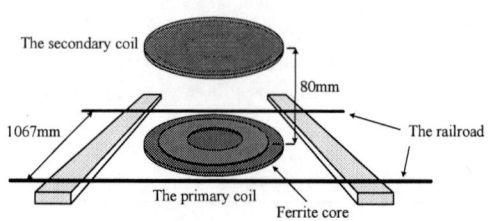

Fig. 4. Dimensions of transformer for charging on board battery.

TABLE I
ANALYTICAL RESULTS

	SP	SS
V_{in}[V]	1500	
I_1[A]	201	201
V_s , V_{s1} [V]	1497	2985
I_2[A]	284	297
V_2[V]	1502	1597
P_{in}[kW]	301	301
P_{out}[kW]	300	300
Q_1[kVar]	301	600
Q_2[kVar]	303	367
B_1[T]	0.30	0.47
B_2[T]	0.30	0.41

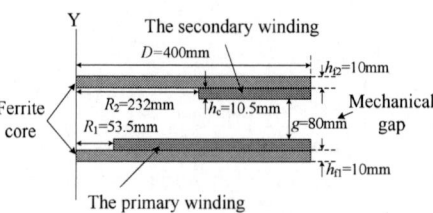

Fig. 5. Analytical model of the SP topology.

TABLE II
ANALYZED VALUE OF INDUCTANCE, COUPLING COEFFICIENT
AND RESISTANCE IN THE SP TOPOLOGY

L_1[mH]	L_2[mH]	k	r_1[mΩ]	r_2[mΩ]
0.908	0.456	0.705	11.7	7.94

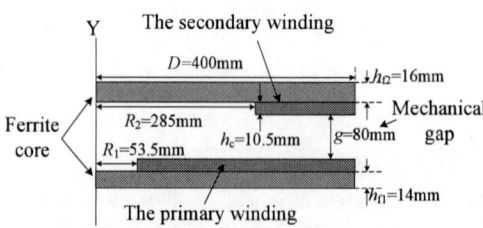

Fig. 6. Analytical model of the SS topology.

TABLE III
ANALYZED VALUE OF INDUCTANCE, COUPLING COEFFICIENT
AND RESISTANCE IN THE SS TOPOLOGY

L_1[mH]	L_2[mH]	k	r_1[mΩ]	r_2[mΩ]
0.908	0.255	0.641	11.7	5.91

TABLE IV
MASS OF THE COILS OF THE SP TOPOLOGY AND THE SS TOPOLOGY

	The SP topology		The SS topology	
	The primary side	The secondary side	The primary side	The secondary side
Mass of the windings [kg]	28.3	19.1	28.3	14.3
Mass of the ferrite core[kg]	24.1	24.1	38.6	33.8
The total mass[kg]	52.5	43.3	66.9	48.0

III. CONTROL METHOD FOR POWER CONVERTER[2]

A. Configurations of Controller

Figure. 7 shows a configuration of the contactless power transfer system. In this figure, power feeding system to energy storage device on board is assumed. A traction inverter can be utilized as a PWM rectifier in the secondary side to control both the active power and the reactive power of the contactless power transfer system. The output voltage of the PWM rectifier in the secondary side is determined to achieve the equivalent circuit as shown in Fig.8 where v_s stands for equivalent input voltage in the primary side including the resonant capacitor and i_s stands for the input current, by measuring the instantaneous current in the secondary side. By this means, the lag reactive power due to the long gap of the contactless power transfer system is supplied by both the input capacitor and the PWM rectifier in the secondary side. In this case, the voltage phasor v_c is expressed by eq. (20) with the detected secondary current phasor i_{con}. The equivalent capacitances of the resonant capacitors C_s and C_c must be determined as eq. (21) and eq. (22), where l_1 is the primary leakage inductance, l_2 is the secondary leakage inductance and l_m is the excitation inductance. Equivalent resistance R_c is determined as expressed in eq.(23) according to i_{conrms} of the rms current of i_{con} and the required power reference P^* in the DC side of the PWM rectifier such as the energy storage devices charging power. ω_0 is the operating angular frequency.

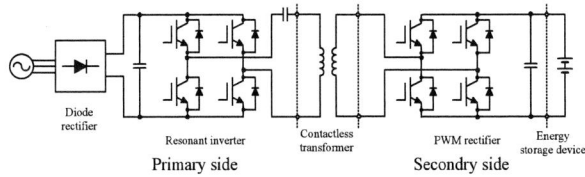

Fig. 7. Configuration of a contactless power transfer system.

$$v_i = (R_c - j\frac{1}{\omega_0 C_c})i_{con} \quad \text{............................} (20)$$

$$C_s = \frac{1}{\omega_0^2(l_m + l_1)} \quad \text{............................} (21)$$

$$C_c = \frac{1}{\omega_0^2(l_m + l_2)} \quad \text{............................} (22)$$

$$R_c = \frac{P^*}{i_{conrms}^2} \quad \text{............................} (23)$$

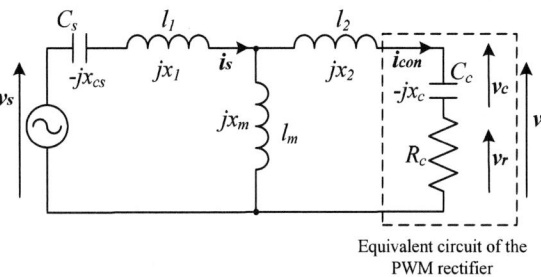

Fig. 8. Equivalent circuit of contactless power transformer system.

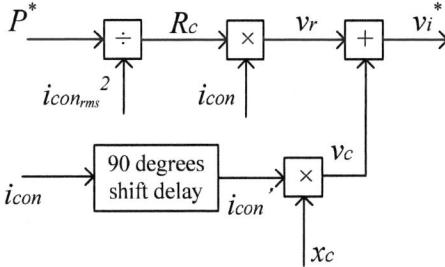

Fig. 9. Block diagram of power control system for PWM rectifier.

Block diagram of power control system for PWM rectifier is shown in Fig.9. To determine the lead phase output voltage component v_c, lead phase current i_{con}' is utilized. The phase lead current i_{con}' can be obtained by shifting 90 degree of the detected secondary current i_{con}

B. Single phase Single-pulse PWM control

The output voltage of the PWM rectifier is modulated by 1-pulse PWM control to reduce the switching loss of the PWM rectifier on board. The block diagram of the single-pulse PWM control is shown in Fig. 10.

The PWM rms value rectifier output voltage V_i is calculated according to the DC side voltage V_d. V_i is restricted by the range in Eq. (24).

$$V_i \le \frac{2\sqrt{2}}{\pi}V_d \quad \text{............................} (24)$$

The modulation index α is compared with the unit cosine wave of the unit amplitude in synchronization with the phase angle θ of the reference voltage value v_i^*, as shown in Fig.11.

978-1-4799-2706-7/14 $31.00 © 2014 IEEE   1441

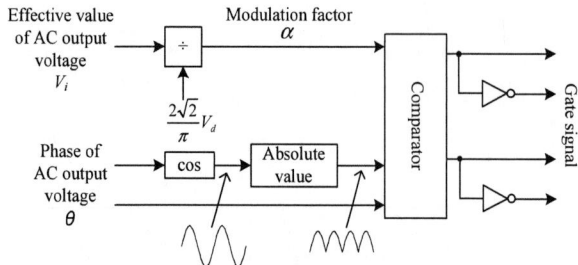

Fig. 10. Block diagram of one-pulse control

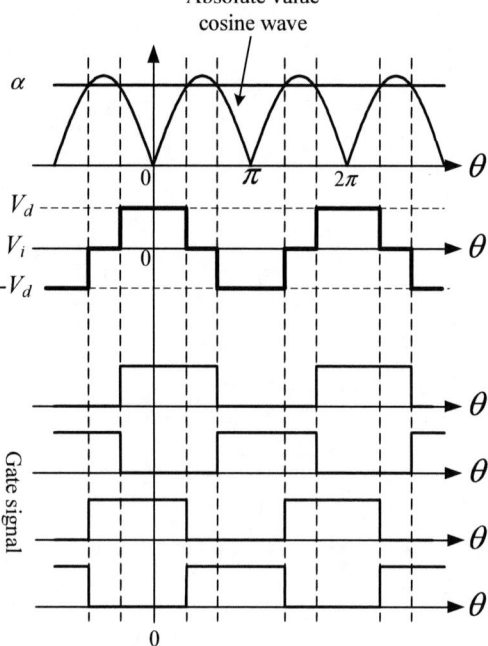

Fig. 11. The AC-side output voltage and each gate signal at the 1-pulse control

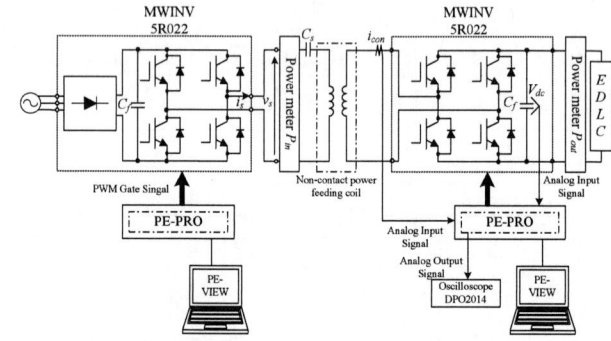

Fig. 12. Experimental system setup.

TABLE. V SPECIFICATIONS OF THE EXPERIMENTAL SYSTEM

Supply RMS voltage V_s [V]	30
Frequency f [Hz]	1546
Initial EDLC voltage V_d [V]	100
Series resonant capacitor C_s [μF]	6.6
Winding resistance $r_1 = r_2$ [mΩ]	454.6
Self-inductance $L_1 = L_2$ [mH]	1.59
Factor of coupling k	0.754
Smoothing capacitor C_f [mF]	13.2
Control operation cycle zh [ms]	0.0269
Power reference value P^* [W]	70

Table VI. Parameters of the experimental coils.

Coil number of turn for coils	80 turn (20 turn × 4 turn)
Wire	0.9φ
Core	JFE ferrite EERX-100K B_s = 0.39 T (100℃) μ_i = 2500 ± 25 % (23℃)

C. Verification of Single phase Single-pulse PWM control

The proposed control system with the PWM rectifier is verified by the 70W scale experimental systems. Figure 12 and Table. III show the experimental system setup and its specifications. The transformer in this experimental test is shown in Fig. 13 and Fig. 14. Table. IV shows specifications of the experimental transformer. B_s and μ_i stand for the saturation magnetic flux density and initial permeability, respectively. The E-shaped ferrite core is used as the core. The coils in primary side and secondary side are identical specifications.

Figure. 15 shows experimental results with the experimental system . The waveforms in Fig.15. stand for input voltage v_s, input current i_s, output voltage of PWM rectifier v_{con}, and secondary current i_{con}, respectively. Since v_s and are v_{con} are requtangular wave forms, i_s and i_{con} are distorted. The power factor value by v_s and i_s are 0.977.

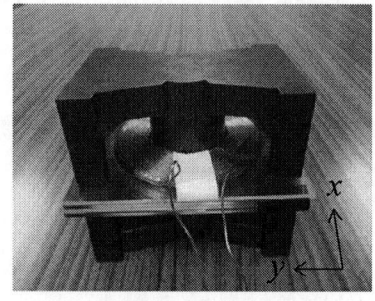

Fig. 13. Appearance of the experimental transformer.

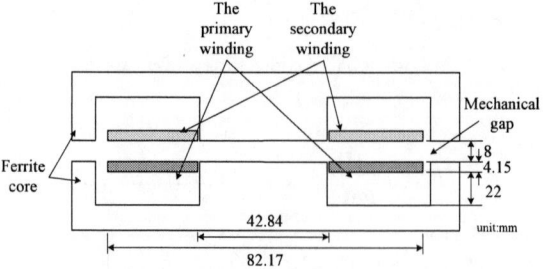

Fig.14. Dimensions of the transformer.

978-1-4799-2706-7/14 $31.00 © 2014 IEEE

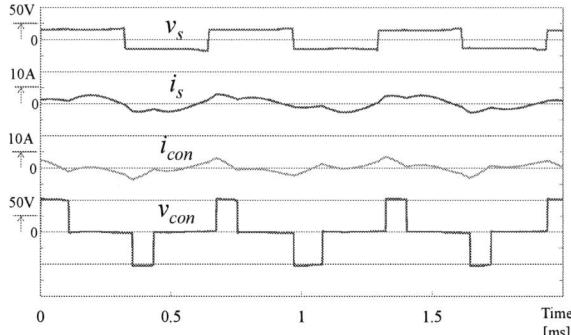

Fig. 15.Experimental waveforms by the proposed control method without the misalignment of coils.

As results, almost unity input power factor is achieved by the proposed controlled method. Furthermore, the output power is measured as 68 W which is very close to the power reference value P^*=70W. According to these results, even if the harmonics are present in the secondary current i_{con}, suitable voltage is output in the AC side of PWM rectifier.

D. Power Transfer Characteristics at Coils Misalignment and the Loss of Each PWM mode.

Only misalignment of the longitudinal direction along rails should be considered in the case of applying this system into the railway vehicle power feeding systems. As show in Fig. 14, the characteristics of transferred power P are examined when the coils misalignment is given. In addition, the loss of the PWM rectifier are also shown in fig.14 with both asynchronous PWM mode control and single-pulse PWM mode control A loss of the PWM rectifier P_{rec_loss} is calculated by Eq.(25).

$$P_{rec\_loss} = P_{in} - (r_1 \cdot I_s^2 + r_2 \cdot I_{con}^2 + P_{out}) \ldots\ldots\ldots\ldots\ldots(25)$$

By calculating equivalent resistance R_c in eq. (23) and controlling the output voltage component for the active power V_r of PWM rectifier, as shown in Fig.11 the active power can be transmitted to the load constantly even when the misalignment of the receiving coil occurs. However, transmitted active power to the load is reduced with 35 mm and more of misalignment of coils in Fig.11. This is caused by the remarkable decrease of the coupling coefficient between the coils. These results show that it is possible to supply the power to the load exactly as the reference value even by single-pulse PWM mode.

Moreover, in Fig.16., the loss at single-pulse PWM mode is lower than that of asynchronous PWM mode since the switching loss decreased. Thus, the increment in the volume of the converter can be prevented in a high power contactless power transfer system.

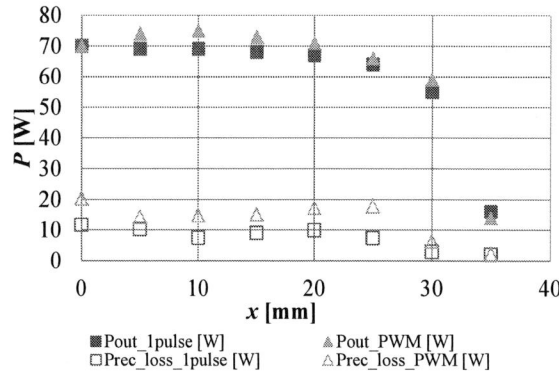

Fig. 16. Transferred power and loss of PWM rectifier at each misalignment of coils.

IV. CONCLUSIONS

Designing method for 300kW class coils and a control method of transmitting power by the secondary side power converter phase control are presented in this paper. The reactive power is important factor to design coils in this power range of high power applications. SP topologies are available from the view point of reducing the reactive power. However, SS topologies are feasible to be combined with voltage source type rectifiers. A control method for voltage source type rectifiers with detection of the secondary current to correct reference active power is presented. The control method is not only for effective to control active power but to feed the increased reactive power even when coil misalignment. In addition, the control method is effective even in the case of applying the single pulse mode to the PWM rectifier. The application of the single pulse mode can reduce the switching loss of the rectifier and may reduce its volume. This feature is appropriate for on-board installation. These technologies may promote the high power application such as power feeding to railway vehicles.

REFERENCES

[1] Chewi-sen Wang, Oskar H. Stielau, and Grant A. Covic, "Design Consideration for a Contactless Electric Vehicle Battery Charger," *IEEE Trans.Ind.Electron.*, vol.52, no.5, pp.1308-1314, Oct.2005

[2] T. Fujita, Y. Kaneko, and S.Abe, "Contactless Power Transfer Systems using Series and Parallel Resonant Capacitors," *IEEJ Trans. IA*, Vol.127, No.2, pp174-180, 2007 (in Japanese)

[3] Y. Kaneko, N. Ehara, T. Iwata, S. Abe, T. Yasuda, and K. Ida, "Comparison of Transformer Winding Methods for Contactless Power Transfer Systems of Electric Vehicle," *IEEJ Trans. IA*, Vol.130, No.6, pp.734-741, 2010 (in Japanese)

[4] Kohei Yamamoto, Satoshi Kitazawa*, Takayuki Kashiwagi, Keiichiro Kondo*, "A Study on the Circuit Configuration with Capacitors to Compensate the Power Factor for High-Power Contactless Power Transformer". Proceedings of ICEMS2013(CD-ROM), 2013.11

[5] Satoshi Kitazawa*, Keiichiro Kondo*, Takayuki Kashiwagi., "Study on a Control Method of the Power Converter for Constant Power Transmission by Single-pulse PWM mode" Proceedings for EPE 2013 (CD-ROM), 2013.09

Two-Switch Voltage Equalizer Based on Half-Bridge Converter with Multi-Stacked Current Doublers for Series-Connected Batteries

Masatoshi Uno and Akio Kukita
Aerospace Research and Development Directorate, Japan Aerospace Exploration Agency
2-1-1 Sengen, Tsukuba city, Ibaraki prefecture, Japan
uno.masatoshi@jaxa.jp

Abstract — A two-switch voltage equalizer using a half-bridge converter with multi-stacked current doublers (MSCDs) is proposed for series-connected batteries in this paper, for which only two switches are required, regardless of the number of batteries connected in series. The proposed equalizer is also capable of relatively large equalization currents without increasing ripple current thanks to the interleaved operation of the MSCDs. Fundamental operational analysis for discontinuous conduction mode, in which currents in the equalizer can be limited to within desired levels without feedback control, is performed. The result of an experimental equalization test performed for four lithium-ion batteries connected in series demonstrated the ability of the proposed equalizer to effectively eliminate voltage imbalance.

Keywords— Current doubler, discontinuous conduction mode (DCM), lithium-ion batery, supercapacitor, voltage equalizer, voltage imbalance

I. INTRODUCTION

In a battery string comprising series-connected energy storage cells/modules (hereafter, simply called as batteries unless otherwise noted), including lithium-ion batteries and supercapacitors (SCs), the voltage mismatch among them gradually grows due to non-uniform individual battery characteristics in terms of capacity, self-discharge rate, and internal impedance. In general, the higher the voltage, the sooner the battery tends to deteriorate. Therefore, each battery in a voltage-mismatched battery string ages non-uniformly, resulting in accelerated ageing for the whole battery string. Furthermore, since all batteries in the string are charged (discharged) in series, some batteries with higher (lower) voltages might be over-charged (or over-discharged) during the charging (discharging) process, potentially triggering hazardous consequences of fire or, at worst, an explosion. Voltage equalization is thus indispensable for battery strings to ensure years of safe operation and optimally exploit battery performance.

Temperature gradient in the battery string is another major cause of voltage mismatch as the self-discharge rate is considerably temperature-dependent. The temperature gradient issue is prone to be very serious, particularly in large-scale energy storage systems, because of the relatively large geometry; the larger the

system, the greater its temperature gradient will be. Cell temperatures in a battery pack comprising several cells connected in series can be easily evened due to its small geometry. Conversely, achieving uniform temperature distribution within a large energy storage system is almost unfeasible, even with sophisticated thermal design. Accordingly, voltage equalization among batteries in a string is more demanding and important than among cells in a battery pack.

Various kinds of voltage equalization techniques have been proposed, demonstrated, and implemented to mitigate or even eliminate such voltage mismatch issues. Typical approaches include equalizers based on individual bidirectional converters, such as buck-boost converters [1]–[4] and switched capacitor converters [5]–[9]. Since the numbers of converters and switches necessary in these equalizers are proportional to the number of batteries connected in series, these equalizers are prone to be complex as the number of series connection increases. Voltage equalizers using a forward or flyback converter with a multi-winding transformer are also among the most popular approaches [10]–[13]. Although the switch count can be reduced to one or two, these equalizers suffer from the design challenge of the multi-winding transformer; parameters for multiple secondary windings must be strictly matched, otherwise voltage mismatch would be exacerbated [14], [15].

Single-switch voltage equalizers using multi-stacked buck-boost converters have been proposed [16], [17], as shown in Fig. 1(a). Although the single-switch configuration without the need for a multi-winding transformer offers simple circuitry and ease of design, a relatively large ripple current flows through the lower batteries when supplying equalization currents for batteries, as shown in Fig. 1(b); the current flowing through upper batteries is superimposed on lower ones, resulting in a large ripple current for the latter. In addition, this ripple current tends to increase with equalization current as well as the number of batteries connected in series. This means a smoothing capacitor with relatively large capacitance needs to be connected to each battery in parallel to decouple the large ripple current, otherwise the application of this equalizer is

978-1-4799-2706-7/14 $31.00 © 2014 IEEE

The 2014 International Power Electronics Conference

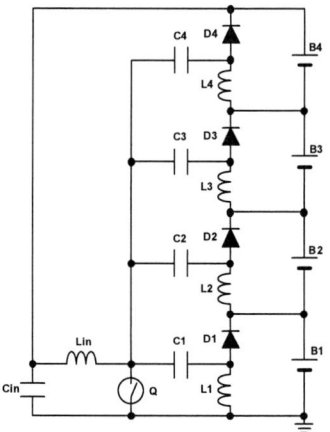

(a) Equalizer using multi-stacked buck-boost converters.

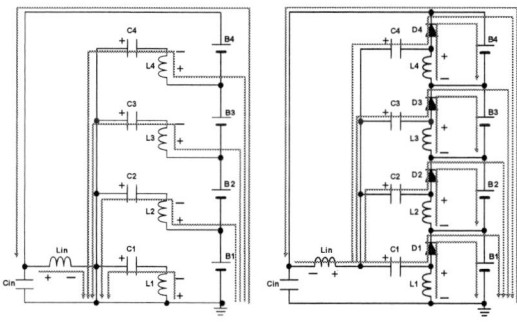

(b) Operation modes.

Fig. 1. Conventional single-switch equalizer using multi-stacked buck-boost converters [6].

likely to be restricted to small-scale energy storage systems comprising few batteries requiring small equalization currents. Given the tendency for considerable voltage imbalance due to temperature gradient in larger systems as aforementioned, an equalizer capable of relatively large equalization current with low ripple is considered desirable for large systems.

A two-switch voltage equalizer based on a half-bridge converter (HBC) with multi-stacked current doublers (MSCDs) is proposed in this paper. The switch count is doubled compared with the conventional equalizer shown in Fig. 1(a), yet remains two. In addition, the proposed equalizer is capable of a large equalization current without increasing the ripple current. Both the derivation procedure and representative topology for four batteries

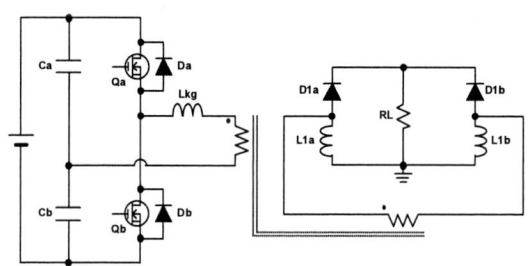

Fig. 2. Conventional current doubler (common-cathode) driven by half-bridge converter.

connected in series are presented in Section II, while operational analyses under voltage-balanced and -imbalanced conditions are performed in Section III. The experimental results of equalizations performed for series-connected SCs and lithium-ion batteries are presented in Section IV.

II. VOLTAGE EQUALIZER USING MULTI-STACKED CURRENT DOUBLERS

A. Circuit Description

A conventional HBC with a common-cathode current doubler, in which cathode pins of two diodes are tied in common, is shown in Fig. 2. The transformer secondary winding is connected to the junctions of diode-inductor pairs and the proposed equalizer can be derived by stacking the current doubler circuits.

The derived voltage equalizer is shown in Fig. 3, although smoothing capacitors connected in parallel with batteries B_1–B_4 are not illustrated for the sake of simplicity. For current doublers to be multi-stacked, the junctions of diode-inductor pairs, L_{1a}-D_{1a}–L_{4a}-D_{4a} and L_{1b}-D_{1b}–L_{4b}-D_{4b}, are tied to the transformer secondary winding through coupling capacitors, C_{1a}–C_{4a} and C_{1b}–C_{4b} that only allow the ac components to flow through them. This means all the inductor-diode pairs are virtually driven by the same square wave voltage produced across the transformer secondary winding, although at different dc voltage levels. The voltage equalization mechanism of the MSCD is qualitatively explained in the next subsection.

Although the MSCDs shown in Fig. 3 are based on common-cathode current doublers, the common-anode configuration is also feasible, similar to conventional current doublers [18].

B. Voltage Equalization Mechanism

As all the current doublers are ac-coupled by coupling capacitors, they can be equivalently separated and grounded as shown in Fig. 4, in which the transformer secondary winding is equivalently illustrated as a square wave generator. Obviously, all the current doublers and

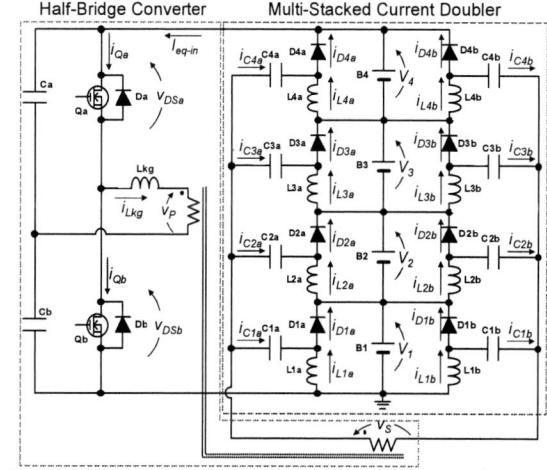

Fig. 3. Proposed two-switch voltage equalizer using a half-bridge converter with multi-stacked current doublers.

978-1-4799-2706-7/14 $31.00 © 2014 IEEE

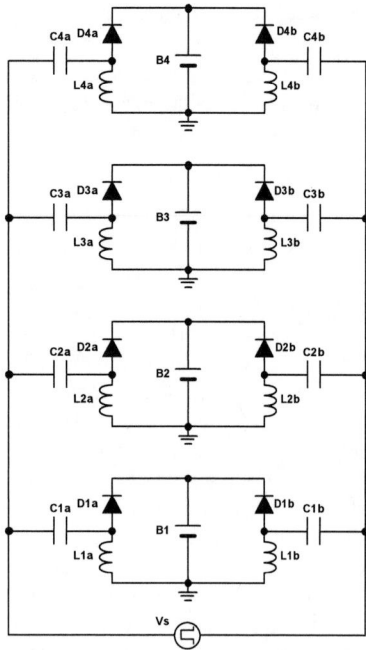

Fig. 4. Equivalent circuit of multi-stacked current doublers.

batteries are paralleled, which means all the current doublers operate identically provided the battery voltages are uniform. Where battery voltages are imbalanced, current is preferentially supplied from the secondary winding to the least charged battery having the lowest voltage, eventually equalizing all the battery voltages.

C. Major Benefits

The number of switches required in the proposed equalizer is only two, regardless of the number of batteries connected in series, which dramatically simplifies the circuitry compared with conventional equalizers using individual bidirectional converters [1]–[9]. In addition to circuit simplicity, the design hurdle can also be significantly lowered thanks to the lack of a multi-winding transformer. Furthermore, the proposed equalizer is capable of providing a relatively large equalization current without being prone to the ripple current issues mentioned in Section I because the MSCDs dramatically reduce ripple current when compared with conventional equalizer shown in Fig. 1. The proposed voltage equalizer is considered more suitable for equalization among batteries rather than among cells, because battery voltages are more prone to be imbalanced and require larger equalization currents, as mentioned in Section I.

III. OPERATIONAL PRINCIPLE

Equalizers must be operated so that no excessive equalization current flows toward batteries. Although measuring and controlling the equalization current for each battery could be possible, numerous current sensors in proportion to the battery count would be necessary, which would impair the simplicity of the proposed equalizer. To limit equalization currents to within desired levels without using current sensors, operation in

discontinuous conduction mode (DCM) is considered in this paper.

A. Operation Waveforms and Current Flows

The theoretical key operation waveforms and current flow directions when the voltage of B_1, V_1, is the lowest are shown in Figs. 5 and 6, respectively. Since Modes 1–4 and 5–8 are symmetrical, current flows in Modes 5–8 are not illustrated for the sake of clarity. In the proposed equalizer, equalization currents preferentially flow toward the least charged battery having the lowest voltage, as mentioned in the previous section. An equalization current supplied to B_i is equal to the sum of diode current of $I_{Dia} + I_{Dib}$ and also equals the sum of the inductor average currents of $I_{Lia} + I_{Lib}$ because the average coupling capacitor currents in the MSCDs must be zero:

$$I_{Dia} = I_{Lia}, \quad I_{Dib} = I_{Lib}. \tag{1}$$

Since no equalization current basically flows toward batteries with higher voltages, the currents of diodes connected to them are zero, meaning these diodes do not conduct while in operation. Equation (1) implies that the average currents of inductors connected to batteries with higher voltages are also zero, though ripple currents still remain as shown in Fig. 5.

In the first mode, 1, currents of L_{1b}–L_{4b}, i_{L1b}–i_{L4b}, linearly increase as the upper switch Q_a is turned on. i_{L1a} and i_{L1b} flow toward B_1 through D_{2a}, whereas other diodes are off. The primary winding current, i_{Lkg}, also linearly increases. As Q_a is turned off, i_{Lkg} is commutated from Q_a to the anti-parallel diode D_b, and Mode 2 begins. D_{2b} starts conducting, and this mode lasts until i_{Lkg} decreases to zero. In Mode 3, no current flows in the HBC, while both D_{1a} and D_{1b} are still conducting. As i_{D1a} reaches zero, the equalizer begins to operate in Mode 4, in which applied voltages of L_{1a}–L_{4a} are zero, and hence, these inductor currents i_{L1a}–i_{L4a} are essentially constant in this mode. Meanwhile, i_{L1b}–i_{L4b} are still linearly

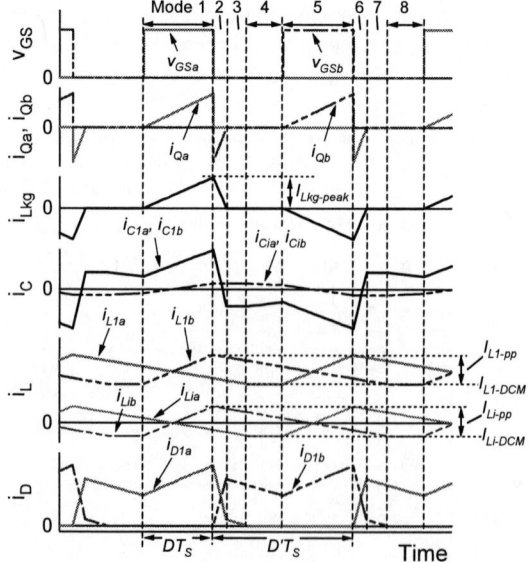

Fig. 5. Key operation waveforms under voltage-imbalanced condition (V_1 is the lowest).

decreasing. As the lower switch Q_b is turned on, the next mode, Mode 5, begins. Operations in Modes 5−8 can be understood as these modes are basically symmetrical to Modes 1−4.

As can be seen in Fig. 6, currents from the equalizer

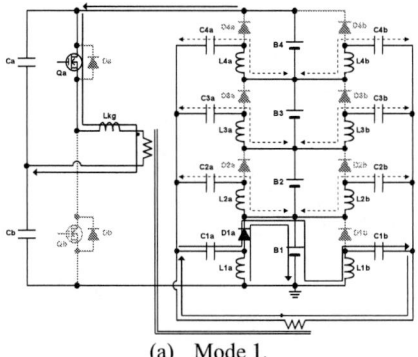

(a) Mode 1.

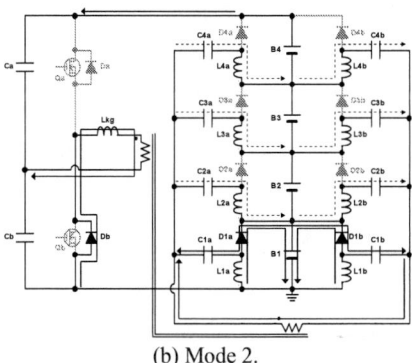

(b) Mode 2.

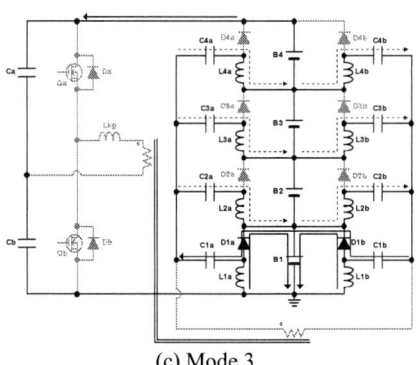

(c) Mode 3.

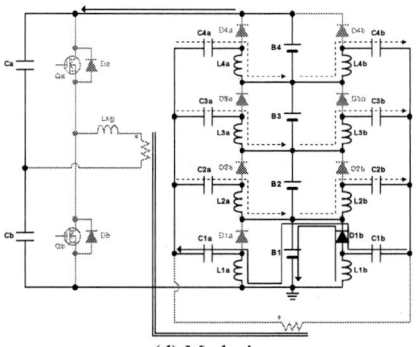

(d) Mode 4.

Fig. 6. Operation modes when B_1 is the least charged battery having the lowest voltage.

flow toward only B_1, the least charged battery with the lowest voltage. In addition, currents from the two inductors (i.e. i_{L1a} and i_{L1b}) are 180° out of phase and supplied to B_1 in interleaved mode by the MSCDs, doubling the substantial switching frequency and reducing the ripple current. The proposed equalizer is thus considered suitable to provide a relatively large equalization current, and the required capacitance of smoothing capacitors, omitted from Fig 6 for the sake of simplicity, can be significantly reduced compared with the conventional equalizer shown in Fig. 1.

B. DCM Operation Criterion

Since average voltages of inductors and transformer windings are zero under a steady-state condition, the sum of voltages of the pair capacitors are also zero (e.g. $V_{C1a} + V_{C1b} = 0$). Based on the volt-second balance on L_{1a} and L_{1b}, the duty cycle of diodes, D', can be yielded as

$$\begin{cases} \left\{ \dfrac{V_{in}}{2N} - (V_1 + V_D) \right\} D = (V_1 + V_D)D' \\ \therefore D' = \dfrac{V_{in} - 2N(V_1 + V_D)}{2N(V_1 + V_D)} D \end{cases}, \quad (2)$$

where D is the duty cycle of each switch, V_{in} is the input voltage or total voltage of V_1–V_4, N is the transformer turns ratio, and V_D is the forward voltage drop of diodes. For the equalizer to operate in DCM, $D' < (1 - D)$ must be ensured, whereupon the critical duty cycle $D_{critical}$ can be obtained as

$$D_{critical} < \frac{2N(V_1 + V_D)}{V_{in}}. \quad (3)$$

C. Operation under Voltage-Imbalanced Condition

Assuming all the inductors of L_{1a}–L_{4a} and L_{1b}–L_{4b} have inductance equivalent to L_i, peak-to-peak currents of i_{L1a} (i_{L1b}) and i_{Lia} (i_{Lib}), I_{L1-pp} and I_{Li-pp}, as designated in Fig. 5, are given by

$$I_{Li-pp} = I_{L1-pp} = \left\{ \frac{V_{in}}{2N} - (V_1 + V_D) \right\} \frac{DT_S}{L_i}. \quad (4)$$

where T_S is the switching period. The average currents of L_{ia} (L_{ib}) and L_{1a} (L_{1b}) are expressed as

$$I_{Li} = \frac{I_{Li-pp}(D + D')}{2} + I_{Li-DCM} = 0, \quad (5)$$

$$I_{L1} = \frac{I_{L1-pp}(D + D')}{2} + I_{L1-DCM}, \quad (6)$$

where I_{Li-DCM} and I_{L1-DCM} are the currents during Mode 4 or 8. From the current flow paths shown in Fig. 6(d),

$$I_{L1-DCM} + 3I_{Li-DCM} = 0. \quad (7)$$

From (4)–(7),

$$\begin{aligned} I_{L1} &= \frac{D + D'}{2} \sum_{i=1}^{4} I_{Li-pp} \\ &= \left\{ \frac{V_{in}}{2N} - (V_1 + V_D) \right\} \frac{D(D + D')T_S}{2} \sum_{i=1}^{4} \frac{1}{L_i}. \end{aligned} \quad (8)$$

By assuming all inductors have inductance equivalent to L_i, (8) can be rewritten as

$$I_{L1} = \left\{ \frac{V_{in}}{2N} - (V_1 + V_D) \right\} \frac{2D(D + D')T_S}{L_i}. \quad (9)$$

The equalization current supplied to B_i is equal to the sum of diode currents of $I_{Dia} + I_{Dib}$ as well as $I_{Lia} + I_{Lib}$, as indicated by (1). Provided that the variation ranges of V_{in} and V_1 are known, (9) indicates that the equalization current supplied to B_1 can be limited to within a desired current level, even at a fixed D.

The peak current of the transformer primary winding, $I_{Lkg-peak}$, is expressed as

$$I_{Lkg-peak} = \frac{3\left(I_{Li-pp} + I_{Li-DCM}\right) + \left(I_{L2-pp} + I_{L2-DCM}\right)}{N}$$
$$= \frac{4I_{Li-pp}}{N} = \left\{\frac{V_{in}}{2N} - \left(V_1 + V_D\right)\right\}\frac{4DT_S}{NL_i} \quad (10)$$

Assuming Mode 2 is negligibly shorter than Mode 1, the average input voltage for the equalizer, I_{in-ave}, can be approximated as

$$I_{in-ave} \approx \frac{I_{Lkg-peak}D}{2} = \left\{\frac{V_{in}}{2N} - \left(V_1 + V_D\right)\right\}\frac{2D^2T_S}{NL_i}. \quad (11)$$

Similar to (9), I_{in-ave} can be limited to within a desired level, even at a fixed D, provided that the variation ranges of V_{in} and V_1 are known.

D. Operation under Voltage-Balanced Condition

In this subsection, currents under a voltage-balanced condition are yielded and compared with those under the voltage-imbalanced condition obtained in the previous subsection. Under a voltage-balanced condition, (5) can be rewritten as

$$I_{Li} = \frac{I_{Li-pp}\left(D + D'\right)}{2} + I_{Li-DCM}, \quad (12)$$

From (4), (7), and (12),

$$I_{L1} = \frac{D + D'}{2}\sum_{i=1}^{4} I_{Li-pp} - \sum_{i=2}^{4} I_{Li}$$
$$= \left\{\frac{V_{in}}{2N} - \left(V_1 + V_D\right)\right\}\frac{D\left(D + D'\right)T_S}{2}\sum_{i=1}^{4}\frac{1}{L_i} - \left(I_{L2} + I_{L3} + I_{L4}\right)$$
$$\qquad (13)$$

This equation implies that average inductor currents are interdependent. If all inductor currents are equal to I_{L1}–I_{L4} = I_{Li} and inductances are uniform, (13) can be simplified to

$$I_{Li} = \left\{\frac{V_{in}}{2N} - \left(V_i + V_D\right)\right\}\frac{D\left(D + D'\right)T_S}{2L_i}. \quad (14)$$

Comparing (9) and (14), I_{Li} under the voltage-balanced condition is a quarter of that under the voltage-imbalanced condition.

As (10) implies, $I_{Lkg-peak}$ is independent on whether voltages are balanced, meaning the average input current I_{in-ave} under a voltage-balanced condition is basically identical to that expressed as (11).

E. Impact of Component Tolerance

The MSCD consists of passive components only, meaning component tolerance would adversely affect equalization performance. As indicated by (3), a battery voltage (V_1 in the case of (3)) depends on V_D, meaning the mismatch in V_D would eventually lead to non-uniform battery voltages. However, if a battery voltage exceeds V_D to a sufficient extent, the impact of this mismatch can

be minimized. In general, where batteries comprise several cells connected in series, with voltages adequately higher than V_D, the mismatch in V_D does not have a significant impact.

The mismatch in coupling capacitors also has no significant impact on equalization performance because their voltages under a steady-state condition can be considered constant, regardless of capacitance mismatch, provided their capacitances are deemed sufficient.

Inductance mismatch would result in non-uniform inductor currents, as indicated by (8) and (13); average inductor currents are dependent inductances. The inductor current mismatch tends to be significant under a voltage-balanced condition because they are interdependent. Under the voltage-imbalanced condition, conversely, the impact of the mismatch is mitigated, since the inductor current is inversely proportional to the sum of all inductances (see (8)). However, regardless of any mismatch in inductances, all battery voltages can be eventually equalized as (2) indicates; a battery voltage is independent on inductances.

IV. EXPERIMENTAL RESULTS

A. Prototype and Experimental Conditions

An 80-W prototype for four batteries connected in series was built, as shown in Fig. 7, with component values listed in Table I. The prototype was operated at a switching frequency of 200 kHz with $D = 0.35$ to ensure DCM operation.

Power conversion efficiencies and key operation waveforms were measured emulating voltage-balanced and -imbalanced (V_1 is the lowest) conditions in the experimental setup shown in Fig. 8. The input and output of the equalizer were separated, and an external power supply V_{ext} was used to power the equalizer. All batteries

Table I. Component values.

Component	Value
C_{1a}–C_{4a}, C_{1b}–C_{4b}	Ceramic Capacitor, 47 μF
C_{out1}–C_{out4}	Ceramic Capacitor, 44 μF
D_{1a}–D_{4a}, D_{1b}–D_{4b}	Schottky Diode, 12CWQ04FN, V_D = 0.48 V
L_{1a}–L_{4a}, L_{1b}–L_{4b}	33 μH
Transformer	N_1:N_2 = 12:15, L_{kg} = 0.3 μH, L_{mg} = 505 μH
Q_a, Q_b	N-Ch MOSFET, FDS86240, R_{on} = 35.3 mΩ
D_a, D_b	Schottky Diode, D3FJ10, V_D = 0.74 V
C_a, C_b	Ceramic Capacitor, 20 μF

Fig. 7. A photograph of 80-W prototype for four batteries connected in series.

The 2014 International Power Electronics Conference

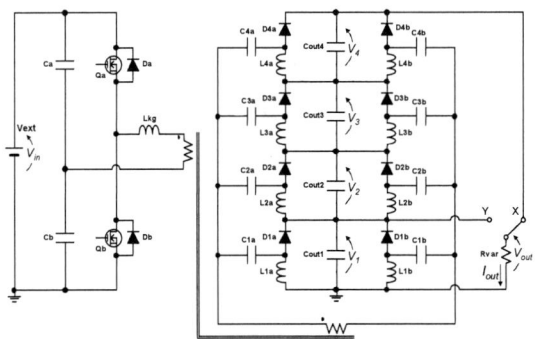

Fig. 8. Experimental setup for efficiency measurement.

were removed, and smoothing capacitors C_{out1}–C_{out4} alone sustained the voltages of V_1–V_4. A variable resistor R_{var} was connected to the MSCD via selectable intermediate taps to emulate voltage-balanced and -imbalanced conditions. When tap X is selected, the current flows under the voltage-balanced condition can be emulated because all the current doublers uniformly supply currents to R_{var}. Meanwhile, when tap Y is

selected, the current is drawn from only the current doubler placed in the lowest position, emulating the current flows under the voltage-imbalanced condition shown in Fig. 6.

B. Fundamental Perfromance

The measured key waveforms at V_{in} = 70 V and V_1 = 15 V under voltage-balanced and -imbalanced conditions are shown in Figs. 9(a) and (b), respectively. Under the voltage-balanced condition shown in Fig. 9(a), i_{L1a}–i_{L4a} and i_{L1b}–i_{L4b} were slightly imbalanced probably due to minor inductance mismatch, as implied by (13). Under the voltage-imbalanced condition shown in Fig. 9(b), conversely, the mismatches between i_{L1a}–i_{L4a} and i_{L1b}–i_{L4b} were very minor because the impact of the mismatch was mitigated under voltage-imbalanced condition, as discussed in Section III-E. Average currents of only i_{L1a} and i_{L1b} exceeded zero, whereas those of others were zero, emulating that B_1 only receives the equalization current. Meanwhile, the measured waveforms related to the HBC under voltage-balanced and -imbalanced conditions were almost similar, verifying that the operation of the HBC is unaffected by whether voltages are balanced or not, as mentioned in Section III-D.

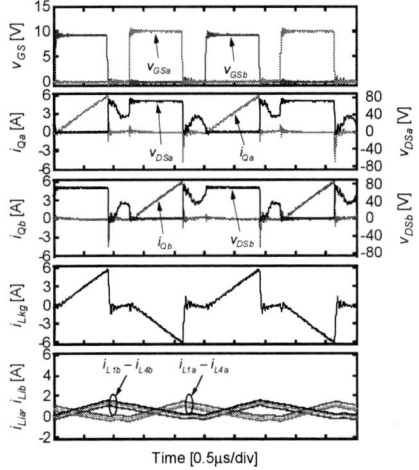

(a)   Under voltage-balanced condition.

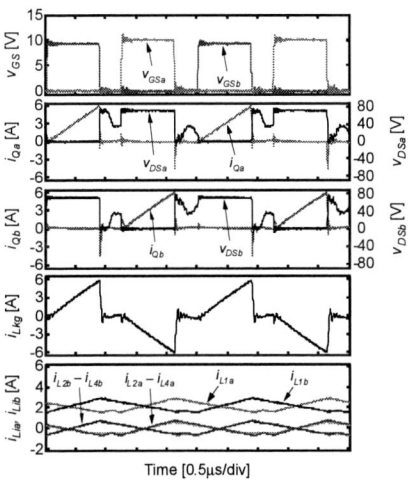

(b) Under voltage-imbalanced condition (V_1 is the lowest).
Fig. 9. Measured key waveforms under (a) voltage-balanced and (b) voltage-imbalanced conditions at V_{in} = 70 V and V_1 = 15 V.

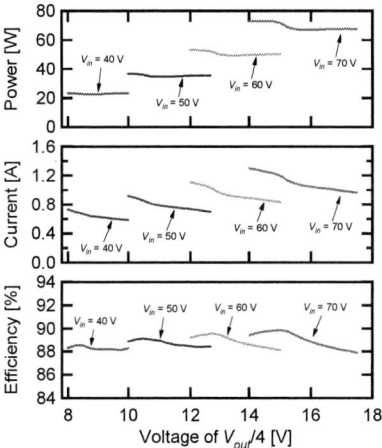

(a)   Under voltage-balanced condition.

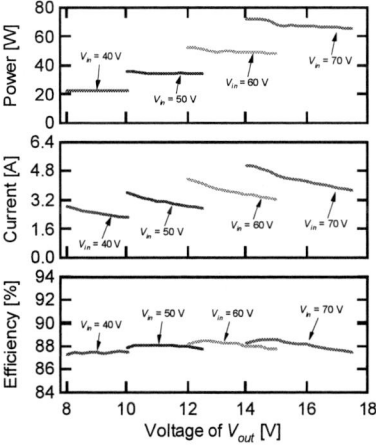

(b) Under voltage-imbalanced condition (V_1 is the lowest).
Fig. 10. Measured efficiencies and output characteristics under (a) voltage-balanced and (b) voltage-imbalanced conditions.

978-1-4799-2706-7/14 $31.00 © 2014 IEEE   1449

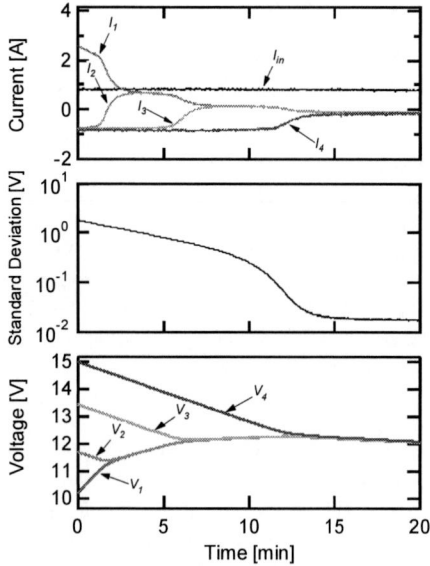

Fig. 11. Experimental equalization profiles of four supercapacitor modules connected in series.

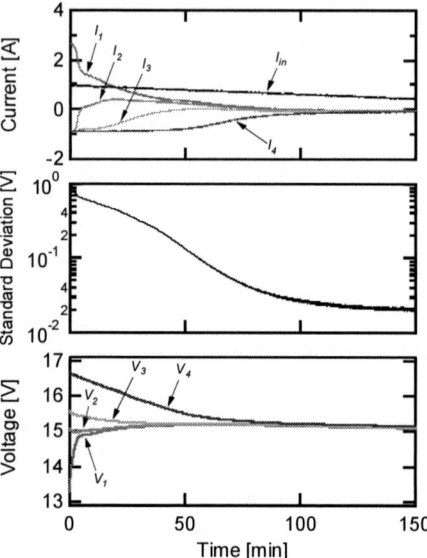

Fig. 12. Experimental equalization profiles of four lithium polymer batteries connected in series.

The measured power conversion efficiencies and output characteristics under voltage-balanced and -imbalance conditions are shown in Figs. 10(a) and (b), respectively. Output currents as well as powers were almost independent of whether the voltages were balanced. Meanwhile, measured efficiencies under the voltage-imbalanced condition were somewhat inferior to those under the voltage-balanced condition. The lower efficiencies under the voltage-imbalanced condition can be attributed to the increased Joule loss due to the current concentration. The currents in the MSCD under the voltage-balanced condition uniformly flow through respective current doublers, whereas those under the voltage-imbalanced condition mainly flowed in the lowest currents doubler, increasing the Joule losses.

C. Equalization for Series-Connected Supercapacitors and Lithium-Ion Batteries

Series-connected SC modules each with capacitance of 220 F at a rated charge voltage of 15 V, were used for the experimental equalization. The initial voltages of SCs were intentionally imbalanced between approximately 10 and 15 V, and the equalization was performed from an initially-voltage-imbalanced condition. The currents of SC modules (I_1–I_4) and the input current of the equalizer I_{in} were also measured.

The resultant equalization profiles are shown in Fig. 11. Since only the least charged module receives an equalization current from the proposed voltage equalizer, only B_1 received the positive (or charging) current of I_1 while V_1 increased at the beginning of the experiment. Meanwhile, other modules supplied the input current of I_{in} for the equalizer's input, hence I_2–I_4 were negative (or discharging) and V_2–V_4 decreased. As V_1 overtook V_2, I_2 gradually changed from negative to positive because both B_1 and B_2 were the least charged modules at this moment. B_3 and B_4, conversely, continued to provide I_{in} for the equalizer, and V_3 and V$_4$ still declined. The energies of

modules with higher initial voltages were therefore redistributed to those with lower initial voltages via the equalizer. The voltage mismatch was gradually eliminated, and all the module voltages ultimately became uniform. The standard deviation of module voltages at the end of the experiment was as low as 20 mV, demonstrating the performance of the proposed voltage equalizer.

The module voltages kept decreasing, even after they had been sufficiently equalized, because the voltage equalizer was still operating. In other words, the voltage equalizer needlessly circulated the energy of modules, some of which was lost in the course of the needless energy redistribution. Although the proposed voltage equalizer can operate without feedback control, voltage sensing would be required to judge the need for equalization and disable the equalizer when not needed.

Similar experimental equalization was also performed for lithium polymer batteries each with capacity of 2500 mAh at a rated charge voltage of 16.8 V from an initially-voltage-imbalanced condition, as shown in Fig. 12. Each battery consisted of four cells connected in series. Although the resultant profiles were somewhat elusive due to the nonlinear characteristics of lithium-ion batteries, the voltage imbalance was gradually eliminated and all the battery voltages were equalized in the same manner as for SC modules.

V. CONCLUSIONS

A two-switch voltage equalizer using HBC with MSCDs was proposed in this paper. Since the proposed equalizer can comprise two switches without the need for a multi-winding transformer, the circuit can be simplified while eliminating the design difficulty associated with the multi-winding transformer. In addition, the proposed equalizer is capable of relatively large equalization currents without increasing ripple current thanks to the interleaved operation of the MSCDs.

To eliminate the feedback control loop and thus further simplify the circuitry, the proposed voltage equalizer was designed to operate in DCM, whereby currents can be automatically limited to within desired levels without feedback control. Fundamental operational analysis in DCM was performed to theoretically express currents in the equalizer.

An 80-W prototype was built for four batteries connected in series. Experimental equalization tests were performed for series-connected SC modules and lithium polymer batteries from initially-voltage-imbalanced conditions. The voltage imbalance was gradually eliminated by the equalizer, and all the voltages eventually became uniform, demonstrating the equalization performance of the proposed equalizer.

ACKNOWLEDGMENTS

This work was supported in part by the Ministry of Education, Culture, Sports, Science, and Technology through Grant-in-Aid for Young Scientists (B) 25820118.

REFERENCES

[1] K. Nishijima, H. Sakamoto, and K. Harada, "A PWM controlled simple and high performance battery balancing system," in Proc. *IEEE Power Electron. Spec. Conf.*, Jun. 2000, pp. 517–520.

[2] Y. S. Lee and M. W. Cheng, "Intelligent control battery equalization for series connected lithium-ion battery strings," *IEEE Trans. Ind. Electron.*, Vol. 52, No. 5, Oct. 2005, pp. 1297–1307.

[3] P. A. Cassani and S. S. Williamson, "Feasibility analysis of a novel cell equalizer topology for plug-in hybrid electric vehicle energy-storage systems," *IEEE Trans. Veh. Technol.*, Vol. 58, No. 8, Oct. 2009, pp. 3938–3946.

[4] C. Pascual and P. T. Krein, "Switched capacitor system for automatic series battery equalization," in Proc. *IEEE Appl. Power Electron. Conf. Expo.*, Feb. 1997, pp. 848–854.

[5] J. W. Kimball, B. T. Kuhn, and P. T. Krein, "Increased performance of battery packs by active equalization," in Proc. *IEEE Veh. Power Propulsion Conf.*, Sep. 2007, pp. 323–327.

[6] A. Baughman and M. Ferdowsi, "Double-tiered switched-capacitor battery charge equalization technique," *IEEE Trans. Ind. Appl.*, Vol. 55, No. 6, Jun. 2008, pp. 2277–2285.

[7] M. Uno and H. Toyota, "Equalization technique utilizing series-parallel connected supercapacitors for energy storage system," in Proc. *IEEE Int. Conf. Sustainable Energy Technology*, Nov. 2008, pp. 999–1003.

[8] M. Uno and K. Tanaka, "Influence of high-frequency charge-discharge cycling induced by cell voltage equalizers on the life performance of lithium-ion cells," *IEEE Trans. Veh. Technol.*, Vol. 60, No. 4, May 2011, pp. 1505–1515.

[9] Y. Yuanmao, K. W. E. Cheng, and Y. P. B. Yeung, "Zero-current switching switched-capacitor zero-voltage-gap automatic equalization system for series battery string," *IEEE Trans. Power Electron.*, Vol. 27, No. 7, Jul. 2012, pp. 3234–3242.

[10] N. H. Kutkut, D. M. Divan, and D. W. Novotny, "Charge equalization for series connected battery strings," *IEEE Trans. Ind. Appl.*, Vol. 31, No. 3, May/Jun. 1995, pp. 562–568.

[11] N. H. Kutkut, H. L. N. Wiegman, D. M. Divan, and D. W. Novotny, "Charge equalization for an electric vehicle battery system," *IEEE Trans. Aerosp. Electron. Syst.*, Vol. 34, No. 1, Jan. 1998, pp. 235–246.

[12] N. H. Kutkut, H. L. N. Wiegman, D. M. Divan, and D. W. Novotny, "Design considerations for charge equalization of an electric vehicle battery system," *IEEE Trans. Ind. Appl.*, Vol. 35, No. 1, Jan. 1999, pp. 28–35.

[13] A. Xu, S. Xie, and X. Liu, "Dynamic voltage equalization for series-connected ultracapacitors in EV/HEV applications," *IEEE Trans. Veh. Technol.*, Vol. 58, No. 8, Oct. 2009, pp. 3981–3987.

[14] J. Cao, N. Schofield, and A. Emadi, "Battery balancing methods: a comprehensive review," in Proc. *IEEE Veh. Power Propulsion Conf.*, Sep. 2008, pp. 1–6.

[15] K. Z. Guo, Z. C. Bo, L. R. Gui, and C. S. Kang, "Comparison and evaluation of charge equalization technique for series connected batteries," in Proc. *IEEE Power Electron. Spec. Conf.*, Jun. 2006, pp. 1–6.

[16] M. Uno and K. Tanaka, "Single-switch cell voltage equalizer using multistacked buck–boost converters operating in discontinuous conduction mode for series-connected energy storage cells," *IEEE Trans. Veh. Technol.*, Vol. 60, No. 8, Oct. 2011, pp. 3635–3645.

[17] M. Uno and K. Tanaka, "Single-switch constant-power equalization charger based on multi-stacked buck-boost converters for series-connected supercapacitors in satellite power systems," in Proc. *IEEE Power Electron. Drive and Systems*, Dec. 2011, pp. 1158–1165.

[18] L. Huber and M. M. Jovanović, "Forward-flyback converter with current-doubler rectifier: analysis, design, and evaluation results," *IEEE Trans. Power Electron.*, Vol. 14, No. 1, Jan. 1999, pp. 184–192.

Optimal Energy Storage System Planning for Microgrids with Contract Capacity Constraint

Shu-Hung Liao[1], Jen-Hao Teng[1], Yung-Ching Huang[2], Dong-Jing Lee[2]

[1]Department of Electrical Engineering,
National Sun Yat-Sen University,
Kaohsiung, Taiwan

[2]Green Energy and Environment Research Laboratories,
Industrial Technology Research Institute
Hsinchu, Taiwan

Abstract- **Due to the inherent variability of Renewable Energy Generation Systems (REGSs), REGSs are commonly uncontrollable. Although microgrids can effectively monitor REGSs, it cannot directly manage and control the intermittent output characteristic of REGSs. Recently, many experts recommend that Energy Storage Systems (ESSs) should be integrated into REGSs to mitigate the instability and further reduce the peak load and contract capacity of microgrids. Although the benefits of ESSs for REGS integrations are significant, their benefits greatly depend on the locations and capacities of ESSs installed. Therefore, this paper proposes an optimal capacity planning for ESS to reduce contract capacity of microgrids. The optimal charging/discharging scheduling of ESSs can also be obtained by the proposed method. An actual microgrid with field measurements is used to test the proposed optimal ESS capacity planning method. Test results demonstrate the validity of the proposed method.**

Keywords: **Renewable Energy Generation System, Microgrid, Energy Storage System, Contract Capacity, Optimal Capacity Planning.**

I. INTRODUCTION

Due to the inherent variability of wind and solar energies, the integration of sizable Renewable Energy Generation Systems (REGSs) is not without challenges [1-2]. To mitigate the impacts of the interconnection of REGSs, the concepts of microgrid have been widely proposed [3-5]. A microgrid is a group of interconnected REGSs and loads within clearly defined electrical boundaries that acts as a single controllable entity with respect to the power grid. Microgrids can connect and disconnect from the power grid to enable it to operate in either grid-connected or islanded mode. During a power grid disturbance, a microgrid can separate and isolate itself from the power grid effortlessly. After the power grid returns to normal, the microgrid can automatically resynchronize and reconnects itself to the grid. Managing a microgrid is like managing a very small vertically integrated electric utility without a transmission system

but with generation and distribution systems. Therefore, the power flow, steady-state and transient conditions, reserve margin with respect to contract capacity, load shedding, demand response, etc. need to be properly designed and realized. A SCADA must be installed to acquire the microgrid measurements all the times and to divide the microgrid into either normal or emergency condition. One of the most key features of a microgrid is its ability to reduce the peak load and contract capacity with the integration of REGSs. Although the concept of microgrid can effectively monitor REGSs, it cannot directly manage and control the intermittent REGSs; especially, for sizable REGSs. Therefore, one of the major challenges in microgrid operation to achieve peak load and contract capacity reduction remains in matching the intermittent energy production with the dynamic power consumption.

A potential remedy that has been considered is to deploy backup diesel generators alongside REGSs, but this basically contradicts the purpose of reducing CO_2 emissions. Instead, a more efficient and robust solution is the utilization of Energy Storage Systems (ESSs), which is making significant progress in helping microgrids to mitigate the instability and further reduce the peak load and contract capacity of microgrids [6-8]. Although the benefits of ESSs for REGS integration are significant, their benefits greatly depend on the locations and capacities of ESSs installed [6-12]. Therefore, this paper proposes an optimal capacity planning for ESS to reduce contract capacity of microgrids. The minimum ESS capacity planning considering the contract capacity of microgrids and the charging/discharging scheduling of ESSs is formulated in this paper. A Particle Swarm Optimization (PSO) based method is then used to solve the proposed optimal problem. The optimal charging/discharging scheduling of ESSs can also be attained by the proposed method. An actual microgrid composed of one wind generation system and two photovoltaic generation systems is used to test the proposed optimal ESS capacity planning method. Test results demonstrate the validity of the proposed method.

II. OPTIMAL ESS PLANNING FOR MICROGRIDS

The paper proposes an optimal ESS capacity planning for microgrids considering the contract capacity constraint. The load profile of a microgrid can be used to calculate the contract capacity. The contract capacity is the maximum value of the average demand per 15 minutes calculated from load profile. Contract capacity charge is used to recover the investment cost of power utility dedicating capacity to meet customers declared capacity requirement. After contract capacity was determined, the pricing rate and power supply voltage are also specified. If the demand exceeds the contract capacity, an extra fine will be charged. Due to the inherent variability of wind and solar energies, the contract capacity of microgrid with sizable REGSs cannot be determined straightforwardly.

The weather data including wind velocity and solar irradiance, the REGS data including types, numbers and locations of wind turbines and photovoltaic modules, the load profile of microgrid and the ESS data including the maximum charging/discharging C rate and State of Charge (SOC), etc. are used as the input data and constraints. The optimal ESS planning for microgrids can then be formulated. The objective function and main constraint are expressed in (1). Other constraints for ESS, REGSs, power quality and so on will be discussed later.

$$\min\ ESS_{kWh} \tag{1a}$$

s.t.

$$P_{MG,i}^{15} \le P_{Contract}\quad i=1\cdots 96 \tag{1b}$$

where ESS_{kWh} is the installed capacity of ESS; $P_{MG,i}^{15}$ is the average demand for the time period i; $P_{Contract}$ is the contract capacity determined by a microgrid. Since the contract capacity is calculated in a 15-minute basis, there are 96 time periods in each day.

The $P_{MG,i}^{15}$ can be calculated by subtracting the average power generated by wind generation systems ($P_{WF,i}^{15}$), photovoltaic generation systems ($P_{PV,i}^{15}$) and

ESSs ($P_{ESS,i}^{15}$) from the total loads in microgrid ($P_{MG,i}^{L,15}$) for time period i. Therefore, $P_{MG,i}^{15}$ can be calculated by

$$P_{MG,i}^{15} = P_{MG,i}^{L,15} - P_{PV,i}^{15} - P_{WF,i}^{15} - P_{ESS,i}^{15} \tag{2}$$

The average power per 15 minutes generated by wind generation systems can be calculated by (3).

$$P_{WF,i}^{15} = \sum_{k=1}^{N_{WF}} P_{WF,i}^{15,k}\left(W_{V,i}^{15}\right) \tag{3a}$$

$$P_{WF,i}^{15,k}\left(W_{V,i}^{15}\right) = \mathit{Eff}_{WG}^{k} * N_{WG}^{k} *$$

$$\begin{cases} 0 & 0 \le W_{V,i}^{15} \le v_{ci}^{l} \\[2mm] P_{WF,rated}^{l} * \dfrac{\left(W_{V,i}^{15} - v_{ci}^{l}\right)}{\left(v_{r}^{l} - v_{ci}^{l}\right)} & v_{ci}^{l} \le W_{V,i}^{15} \le v_{r}^{l} \\[3mm] P_{WF,rated}^{l} & v_{r}^{l} \le W_{V,i}^{15} \le v_{co}^{l} \\[2mm] 0 & v_{co}^{l} \le W_{V,i}^{15} \end{cases} \tag{3b}$$

where $P_{WF,i}^{15,k}$ is the average power for time period i generated by wind generation system k; N_{WF} is the number of wind generation system; N_{WG}^{k} is the number of wind turbine for wind generation system k; $W_{V,i}^{15}$ is the average wind velocity for time period i; Eff_{WG}^{k} is the efficiency of power inverter for wind generation system k; $P_{WF,rated}^{l}$ is the rated power for wind turbine l; v_{ci}^{l}, v_{r}^{l} and v_{co}^{l} are the cut-in, rated and cut-out wind velocities for wind turbine l, respectively.

The average power generated by photovoltaic generation systems can be calculated by (4) [13].

$$P_{PV,i}^{15} = \sum_{k=1}^{N_{PV}} P_{PV,i}^{15,k}\left(S_{PV,i}^{15}\right) \tag{4a}$$

$$T_{cy}^{k} = T_{A}^{k} + S_{PV,i}^{15}\left(\frac{N_{OT}^{k}-20}{0.8}\right)$$

$$I_{y}^{k} = S_{PV,i}^{15} *[I_{sc}^{k} + K_{i}^{k}(T_{cy}^{k}-25)]$$

$$V_{y}^{k} = V_{oc}^{k} - K_{v}^{k} * T_{cy}^{k} \tag{4b}$$

$$FF^{k} = \frac{V_{MPPT}^{k} * I_{MPPT}^{k}}{V_{oc}^{k} * I_{sc}^{k}}$$

$$P_{PV,i}^{15,k}\left(S_{PV,i}^{15}\right) = \mathit{Eff}_{PV}^{k} * N_{p}^{k} * FF^{k} * V_{y}^{k} * I_{y}^{k}$$

where $P_{PV,i}^{15,k}$ is the is the average power for time period i generated by the photovoltaic generation system k; N_{PV} is the number of photovoltaic generation system; $S_{PV,i}^{15}$ is the average solar irradiance per 15 minutes for time period i; T_{cy}^k and T_A^k are the photovoltaic module temperature and ambient temperature for photovoltaic generation system k, respectively; N_{OT}^k, I_{sc}^k and V_{oc}^k are the nominal operating temperature, the short-circuit current and the open-circuit voltage of photovoltaic module installed in photovoltaic generation system k, respectively. K_v^k and K_i^k are the voltage temperature coefficient and current temperature coefficient, respectively; FF^k is the fill factor; N_p^k is the number of photovoltaic modules used in photovoltaic generation system k; I_{MPPT}^k and V_{MPPT}^k are the current and voltage at maximum power point; Eff_{PV}^k is the efficiency of power inverter for photovoltaic generation system k.

The average charging/discharging power per 15 minutes ($P_{ESS,i}^{15}$) of ESS depends on the real-time power consumption of microgrid. The power consumption of microgrid is calculated by subtracting $P_{WF,i}^{15}$ and $P_{PV,i}^{15}$ from $P_{MG,i}^{L,15}$. If the power consumption of microgrid exceeds the contract capacity, then the ESS must supply power to the microgrid (discharging power). On the other hand, if the power consumption of microgrid is less than the contract capacity, then the ESS can absorb power from the microgrid (charging power). Consequently, the time periods for the charging/discharging power of ESS can be scheduled by

$$
\begin{aligned}
&if((P_{contract} - (P_{MG,i}^{L,15} - P_{WF,i}^{15} - P_{PV,i}^{15})) > 0) \\
&then \ \Phi_{ch} = \Phi_{ch} + \{i\} \\
&else \ \Phi_{disch} = \Phi_{disch} + \{i\}
\end{aligned} \tag{5}
$$

where Φ_{ch} and Φ_{disch} are the sets of discharging time periods and probable charging time periods, respectively.

$P_{ESS,i}^{15}$ can therefore be calculated by

$$
P_{ESS,i}^{15} \geq -1*(P_{contract} - (P_{MG,i}^{L,15} - P_{PV,i}^{15} - P_{WF,i}^{15})) \tag{6}
$$

The power converting efficiency of bidirectional

converter/inverter used in ESS should also be taken into account to calculate the actual charging/discharging power of ESS. In the ESS discharging, the power supplied from the battery packs of ESS can be expressed as

$$
Pr_{ESS,i}^{15} = P_{ESS,i}^{15} / Eff_{ESS,disch} \quad i \in \Phi_{disch} \tag{7}
$$

On the contrary, the power absorbed by battery packs of ESS in the ESS charging can be expressed as

$$
Pr_{ESS,i}^{15} = P_{ESS,i}^{15} * Eff_{ESS,ch} \quad i \in \Phi_{ch} \tag{8}
$$

where $Eff_{ESS,disch}$ and $Eff_{ESS,ch}$ are the power converting efficiencies of bidirectional converter/inverter in ESS discharging and charging, respectively. The relations between $P_{ESS,i}^{15}$, $Pr_{ESS,i}^{15}$, $Eff_{ESS,disch}$ and $Eff_{ESS,ch}$ are illustrated in Fig. 1.

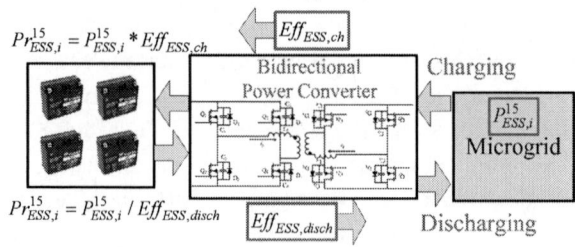

Fig. 1: Relations between $P_{ESS,i}^{15}$ and $Pr_{ESS,i}^{15}$

The constraints for ESS charging/discharging shall at least include the maximum charging/discharging C rate and SOC of ESS. $Pr_{ESS,i}^{15}$ cannot exceed the maximum charging/discharging capacity of ESS; that is

$$
\begin{aligned}
Pr_{ESS,i}^{15} &\leq ESS_{kWH} * CRate_{disch}^{max} \quad i \in \Phi_{disch} \\
Pr_{ESS,i}^{15} &\leq ESS_{kWH} * CRate_{ch}^{max} \quad i \in \Phi_{ch}
\end{aligned} \tag{9}
$$

where $CRate_{disch}^{max}$ and $CRate_{ch}^{max}$ are the maximum discharging and charging C rate of ESS.

The summation of $Pr_{ESS,i}^{15}$ each day cannot exceed the allowable SOC of ESS. That is

$$
\sum_{i \in \Phi_{disch}} Pr_{ESS,i}^{15} \leq ESS_{kWh}(SOC^{max} - SOC^{min}) \tag{10}
$$

where SOC^{max} and SOC^{min} are the allowable maximum and minimum SOC of ESS, respectively.

Besides, in order to make sure that there is adequate charging capacity, the summation of discharging capacity cannot exceed the summation of charging capacity, i.e.

$$\sum_{i \in \Phi_{ch}} Pr_{ESS,i}^{15} \geq \sum_{i \in \Phi_{disch}} Pr_{ESS,i}^{15} \qquad (11)$$

The power quality constraints should at least include the bus voltage and line flow constraints and can be expressed as

$$V^{min} \leq V_i^k \leq V^{max} \quad k=1 \cdots N_B \qquad (12a)$$

$$I^{min} \leq I_i^k \leq I^{max} \quad k=1 \cdots N_L \qquad (12b)$$

where V_i^k is the voltage of bus k at time period i; N_B is the bus number in the microgrid; V^{max} and V^{min} are the allowable maximum and minimum bus voltages, respectively. I_i^k is the current flow of line k at time period i; N_L is the line number in the microgrid; I^{max} and I^{min} are the allowable maximum and minimum line currents, respectively.

This paper uses PSO to find the solution of optimal ESS planning for microgrids with contract capacity constraint formulated in (1)-(12). Due to limited space, the solution procedures of PSO for the proposed ESS planning are not shown in the extended summary, but will be represented in the final manuscript.

III. PRELIMINARY TEST RESULTS

An actual microgrid in National Penghu University of Science and Technology is tested in this paper. The one-line diagram of the microgrid is illustrated in Fig. 2. From Fig. 2, it can be seen that one wind generation system and two photovoltaic generation systems are installed in the microgrid. The wind generation system is composed of four 400W wind turbines, six 1kW wind turbines and nine 2kW wind turbines. The installed capacities for these two photovoltaic generation systems are 50.16kW and 48.3kW. Indoor first farming pond and marine science technology building are the main loads in the microgrid. Historical loads and weather data including solar irradiance and wind velocity in a 15-minute basis are used in the following ESS planning.

The maximum and minimum loads of the microgrid are about 300k and 100kW, respectively. The peak load is happened in 2013/05/28; therefore, the load profile and the corresponding weather data on 2013/05/28 are used for ESS planning first. Fig. 3 illustrates the load profiles of indoor first farming pond and marine science technology building. Figs. 4 and 5 show the solar irradiance and photovoltaic module temperature and the wind velocity, respectively.

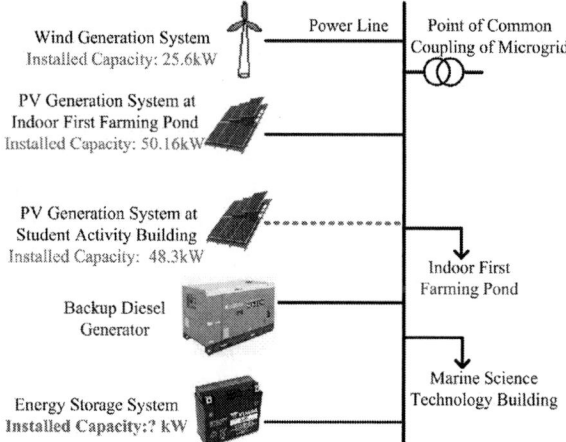

Fig. 2: Microgrid in National Penghu University of Science and Technology

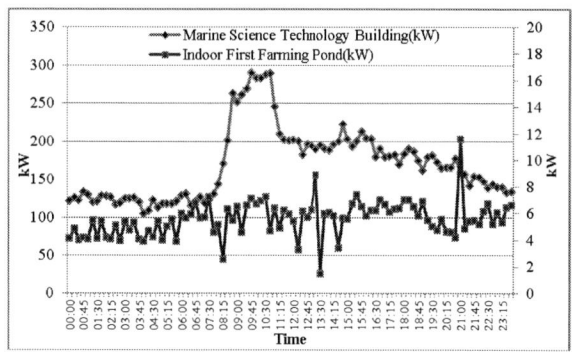

Fig. 3: Main Loads in the Microgrid

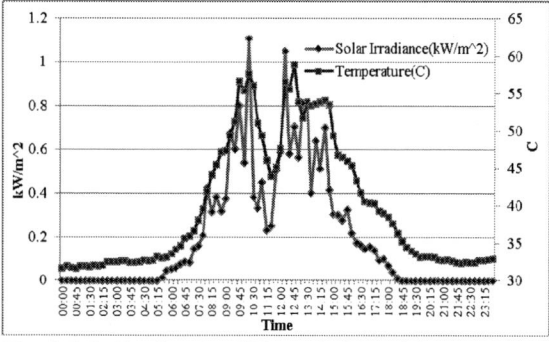

Fig. 4: Solar Irradiance and Photovoltaic Module Temperature

978-1-4799-2706-7/14 $31.00 © 2014 IEEE

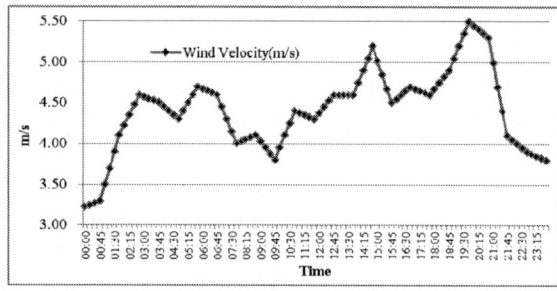

Fig. 5: Wind Velocity

The original contract capacity of the microgrid in National Penghu University of Science and Technology is 300kW. The microgrid needs to reduce its contract capacity to 220kW. In the first test case, only the wind generation system and photovoltaic generation system with installed capacity 50.16kW are considered. Fig, 6 illustrates the total load and total REGS generation in the microgrid. The total installed capacity of REGSs is about 25% of the original microgrid contract capacity. The maximum power generation of REGSs in 2013/05/28 is about 16.7% of the original microgrid contract capacity. The optimal ESS capacity solved by the proposed method is about 94.8kWh. Fig. 7 shows the discharging scheduling of ESS in a 15-minute basis. Fig. 8 illustrates the total demand of the microgrid. From Figs. 6-8, it can be clearly observed that the contract capacity has been successfully reduced to 220kW. Due to limited space, only preliminary test results are shown here, the full test results especially considering the uncertainties of solar irradiance and wind velocity, etc. will be represented in the final manuscript.

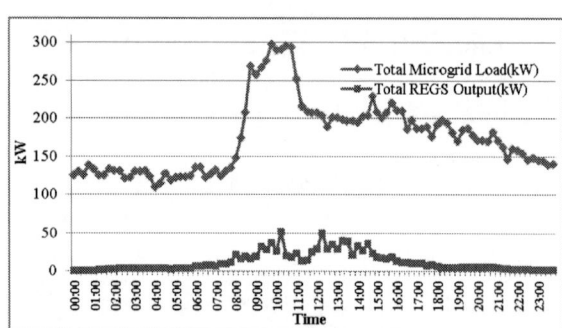

Fig. 6: Total Microgrid Load and REGS Output

ACKNOWLEDGMENT

The authors gratefully acknowledge the financial supports by Bureau of Energy, Ministry of Economic Affairs, R.O.C. under the Project "Grid-Scaled Energy Storage System and Interconnection Technology Development " and by the National Science Council of

Taiwan under Contracts NSC 103-2622-E-110 -003 -.

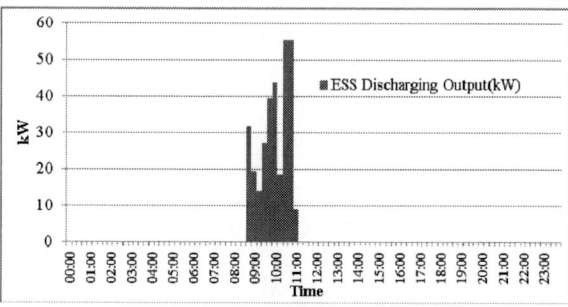

Fig. 7: ESS Discharging Scheduling

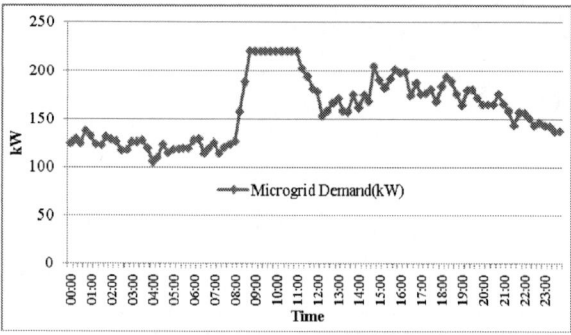

Fig. 8: Microgrid Demand with REGSs and ESS

REFERENCES

[1] "Engineering Guide for Integration of Distributed Generation and Storage into Power Distribution Systems," EPRI Technical Report TR-100419, December 2000.

[2] A.Woyte, V. Van Thong, R. Belmans, and J. Nijs, "Voltage Fluctuations on Distribution Level Introduced by Photovoltaic Systems," IEEE Trans. on Energy Conversion., vol. 21, no. 1, pp. 202–209, Mar. 2006.

[3] Microgrids in New England Technical Challenges & Opportunities, http://energy.pace.edu/sites/default/files/ Panel_2_PACE Microgrids in New England Technical Challenges & Opportunities 201100912.pdf.

[4] CERTS Microgrid Concept, http://certs.lbl.gov/certs-der-micro.html.

[5] Micro-grid technologies in smart community projects by NEDO, http://e2rg.com/microgrid-2012/Sendai_Morozumi. pdf.

[6] "Electric Energy Storage Systems," Cigre Working Group C5.15, 2011.

[7] W. Steeley, "Functional Requirements for Electric Energy Storage Applications on the Power System Grid," EPRI Technical Report 1022544, May 2011.

[8] D. Rastler, "Electricity Energy Storage Technology Options: A White Paper Primer on Applications, Costs, and Benefits," EPRI Technical Report 1020676, Dec. 2010.

[9] Tsung-Ying Lee ; Nanming Chen "Determination of Optimal Contract Capacities and Optimal Sizes of Battery Energy Storage Systems for Time-of-Use Rates Industrial

Customers," IEEE Transactions on Energy Conversion, vol. 10 , no. 3, 1995, pp. 562 – 568.

[10] Zheng, Y. ; Dong, Z.Y. ; Luo, F.J. ; Meng, K. ; Qiu, J. ; Wong, K.P. "Optimal Allocation of Energy Storage System for Risk Mitigation of DISCOs With High Renewable Penetrations," IEEE Transactions on Power Systems, 2013, pp. 1 - 9 (IEEE Early Access Articles).

[11] Jen-Hao Teng, Shang-Wen Luan, Dong-Jing Lee, Yong-Qing Huang, "Optimal Charging/Discharging Scheduling of Battery Storage Systems for Distribution Systems Interconnected with Sizeable PV Generation Systems," IEEE Transactions on Power Systems, vol. 28, No. 2, May 2013, pp. 1425-1433.

[12] Madaeni, S.H. ; Sioshansi, R. ; Denholm, P. "Estimating the Capacity Value of Concentrating Solar Power Plants With Thermal Energy Storage: A Case Study of the Southwestern United States," IEEE Transactions on Power Systems, vol. 28, no. 2, 2013, pp. 1205 – 1215.

[13] Atwa, Y.M.; El-Saadany, E.F.; Salama, M.M.A.; Seethapathy, R.; "Optimal Renewable Resources Mix for Distribution System Energy Loss Minimization," IEEE Trans. on Power Systems, vol. 25, no. 1, 2010, pp. 360 – 370

Optimal Zero Sequence Injection in Multilevel Cascaded H-Bridge Converter under Unbalanced Photovoltaic Power Generation

Yifan Yu, Georgios Konstantinou, Branislav Hredzak and Vassilios G. Agelidis

Australian Energy Research Institute and School of Electrical Engineering and Telecommunications,
The University of New South Wales, Sydney, NSW, 2052, Australia

Abstract—The multilevel cascaded H-bridge converter is a favorable candidate to directly connect large-scale photovoltaic power plants to the medium voltage electricity network. Nevertheless, due to the lack of a common dc link, the unequal power generation among the three phases, primarily caused by non-uniform solar irradiance, affects its operation. The existing power balancing control methods, however, fail to achieve optimal results due to their limitations. This paper proposes an optimal zero sequence injection method to maximize the converter power balancing capability. The feasibility and superiority of the new method are confirmed by simulation results.

Index Terms—ac-dc power converters, cascaded H-bridge converter, multilevel converter, photovoltaics.

I. INTRODUCTION

Large-scale photovoltaic (PV) farms are an attractive way to further increase the renewable energy penetration level in the near future [1], [2]. Compared to small-scale residential applications, the large-scale solution minimizes its potential impact on the grid by establishing solar farms that are more closer to conventional power plants. It thus reduces the overall cost by avoiding the expensive and lengthy infrastructure improvement of the existing distribution network needed to accept more reverse power flow from roof PV modules [3]–[5].

The majority of contemporary converters for PV farms are two-level Voltage Source Converters (VSC) operating with high frequency pulse width modulation. The resultant heat dissipation on the semiconductor switches makes it practically challenging and uneconomic for the power rating of single converter to reach 1MW [6], [7]. For large-scale PV power plants (10 to 100MW), dozens of parallel converters have to be installed. A single converter, able to handle the power produced by the entire farm, is thus a more tempting solution due to the enhanced efficiency and reduced cost [1].

The Cascaded H-Bridge (CHB) converter (Fig. 1) offers an attractive alternative option [8]–[13]. Multiple output voltage levels can be provided, enabling the generation of a stepped waveform with lower harmonic distortion. Furthermore, the presence of multiple separate dc links facilitates independent Maximum Power Point Tracking (MPPT), which further improves the overall efficiency. However, the CHB converter, initially designed for equal power generation in each H-bridge, may suffer when each bridge is required to process different amounts of power [9]–[13]. This unbalance can be because

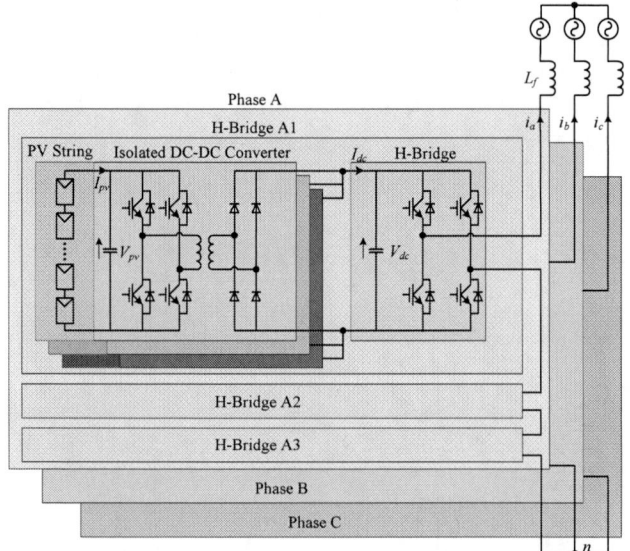

Fig. 1. Three-phase, seven-level cascaded H-bridge converter.

the PV modules connected to each bridge are subject to different solar irradiance, temperature, and/or module parameter deviation.

The injection of a zero sequence to the converter output voltages is necessary to facilitate inter-phase power rebalancing. One way is through the Fundamental Frequency Zero Sequence Injection (FFZSI). It can be derived either by the instantaneous power theory [12] or phasor diagrams [13]. With the fundamental frequency zero sequence injection, the resultant converter output voltages are still sinusoidal yet asymmetrical. Since the sinusoidal waveform is well below its peak for most of the time, it depletes the dc-side voltage in such an inefficient way that the converter easily reaches saturation.

Improved methods, like the Double 1/6 Third Harmonic Injection (DTHI) [13], effectively extend the power balanceable space [13] by making better use of dc voltages. The DTHI, as its name implies, is the combination of two parts. The first part is made up of the three-phase symmetrical voltages optimized with conventional 1/6 third harmonic injection, while the other part consists of the fundamental frequency

978-1-4799-2706-7/14 $31.00 © 2014 IEEE

zero sequence injection required for power balancing control also optimized with its own 1/6 third harmonic. However, the direct superposition of individual optimizations generally does not generate overall optima for nonlinear systems.

This paper investigates this problem from a brand-new perspective. The symmetrical component of three-phase voltages cannot be modified, since it is directly related to the converter output line-to-line voltages and hence the grid currents. The zero sequence injection, on the other hand, can be manipulated arbitrarily, as long as its fundamental frequency component satisfies the power balancing control requirement. Based on this idea, the paper proposes an Optimal Zero Sequence Injection (OZSI), which forces either the upper or lower envelope of resultant voltages (sum of symmetrical and zero sequence components) to be flat. In this way, all voltage overhead provided by the converter can be fully exploited to achieve maximum fundamental frequency component of zero sequence. That is to say, with given dc-side voltages, the power balancing control can be facilitated for severely unbalanced cases which existing methods fail to cope with.

The rest of the paper is organized as follows. Section II is devoted to the review of the power unbalance problem and traditional methods. Section III proposes OZSI as well as its numerical considerations. The power balanceable space of the presented method is displayed and compared with existing methods for a selected scenario in Section IV. Section V provides computer simulation results based on MATLAB/PLECS to verify the effectiveness of the proposed methods, before the conclusions are finally drawn in Section VI.

II. POWER UNBALANCE AND BALANCING CONTROL

A. Problem and Definitions

The CHB is traditionally designed assuming that equal amount of power is generated by each bridge. However, this assumption does not hold for PV plants. Even if the number of PV modules connected to each bridge is the same and they are of the same model type, the instantaneous power of each bridge is unlikely to be equal. This can be caused by partial shading, unequal ambient temperature and non-uniform degradation. As a consequence, the three-phase grid currents become asymmetrical if the control method is left unmodified.

The power balancing control can be realized by zero sequence injection [11]–[13]. It is, nevertheless, achieved at the expense of increased converter output voltage. If the peak voltage exceeds the sum of dc-side voltages, three-phase balanced currents cannot be guaranteed anymore.

A measure of the unbalanced power generation in each phase is given by power generation ratios λ_a, λ_b, and λ_c:

$$\lambda_{a,b,c} = \frac{P_{a,b,c}}{P_{peak}/3}, \tag{1}$$

which compares the actual power of each phase to its peak power $P_{peak}/3$.

Since the ratios can only vary between zero and unity, all the possible operating points $(\lambda_a, \lambda_b, \lambda_c)$ can be represented by a unity cube. All the points that can be rebalanced by

the control strategy form a sub-space of the unity cube. This sub-space is defined as the Power Balanceable Space (PBS) [13]. The volume of PBS, defined as the Power Balanceable Factor (PBF), indicates the power balancing capability of the converter.

B. Fundamental Frequency Zero Sequence Injection (FFZSI)

FFZSI has been proved to be effective in rebalancing grid currents [12], [13]. It can be derived based on either instantaneous power theory [12] or traditional phasor diagram [13] as:

$$v^0 = \sqrt{2}V^0\cos(\omega t + \theta), \tag{2}$$

where:

$$V^0 = \frac{\sqrt{6}\Delta}{3(\lambda_a + \lambda_b + \lambda_c)}V_g, \tag{3}$$

$$\theta = \begin{cases} \sin^{-1}\left(\dfrac{\sqrt{6}\,(\lambda_c - \lambda_b)}{2\Delta}\right) & \text{Case I} \\[2mm] \dfrac{2\pi}{3} + \sin^{-1}\left(\dfrac{\sqrt{6}\,(\lambda_b - \lambda_a)}{2\Delta}\right) & \text{Case II} \\[2mm] \dfrac{4\pi}{3} + \sin^{-1}\left(\dfrac{\sqrt{6}\,(\lambda_a - \lambda_c)}{2\Delta}\right) & \text{Case III} \end{cases}, \tag{4}$$

$$\Delta = \sqrt{(\lambda_a - \lambda_b)^2 + (\lambda_b - \lambda_c)^2 + (\lambda_a - \lambda_c)^2}. \tag{5}$$

The three-phase grid voltages are assumed to be ideal, and V_g denotes their line-to-line rms value. The case determination in (4) has been discussed in detail in [13].

As seen from (2), the injected zero sequence is also sinusoidal. Consequently, the resultant phase voltages (adding the zero sequence to the positive sequence) should still be sinusoidal yet asymmetrical. Since the sinusoidal waveform is well below its peak for most of the time, it depletes the dc-side voltage in such an inefficient way that the converter reaches saturation under slight power unbalance.

The PBS of an example converter with parameters listed in Table I is plotted in Fig. 2(a). All the points that can be rebalanced by FFZSI are included in the cone-like space.

C. Double 1/6 Third Harmonic Injection (DTHI)

DTHI [13] improves the utilisation of dc-side voltages. In this method, the conventional 1/6 third harmonic is injected to the positive sequence of converter output voltages. Meanwhile, the FFZSI (2) is also modified with its own 1/6 third harmonic. These two third harmonic injections reduce the peak value of the resultant phase voltages without affecting the injected fundamental frequency zero sequence. Hence, for same converter parameters, the PBS of DTHI, as plotted in Fig. 2(b), is significantly larger than that of FFZSI.

978-1-4799-2706-7/14 $31.00 © 2014 IEEE

The 2014 International Power Electronics Conference

TABLE I
PARAMETERS OF PV MODULES AND THE THREE-PHASE, SEVEN-LEVEL CHB CONVERTER

PV Parameters	Values
PV Module Type	BP 365
Output Voltage at Maximum Power Point	17.65V
Output Current at Maximum Power Point	3.68A
Maximum Power	65W
Cascaded PV Modules per String	55
Paralleled Strings per H-Bridge	311
Converter Parameters	**Values**
Three-phase Peak Power, P_{peak}	10MW
Grid Voltage, V_g	6600V
Grid Frequency, f	50Hz
Capacitor Voltage, V_{dc}	2200V
IGBT Voltage Rating	3300V
IGBT Current Rating	1500A
Filtering Inductor, L_f	5mH
Phase Shift PWM Carrier Frequency, f_s	600Hz

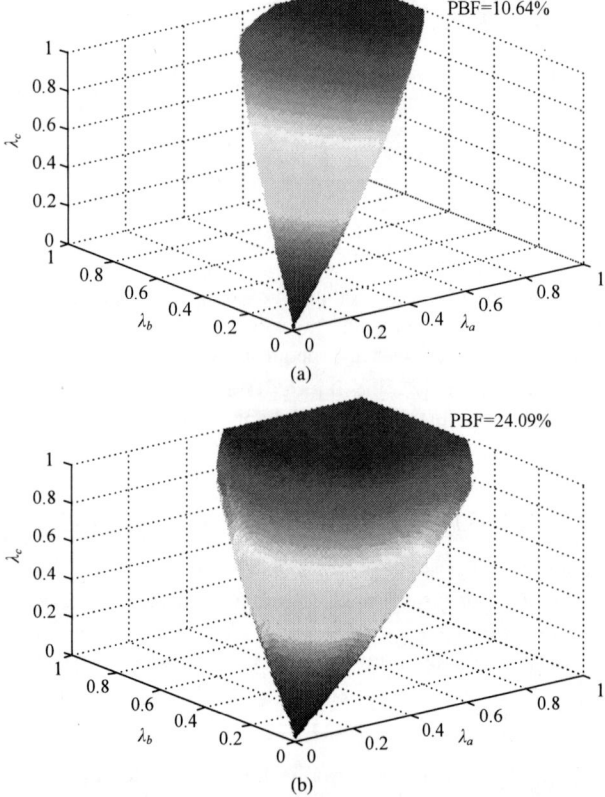

Fig. 2. Power Balanceable Space of (a) FFZSI, (b) DTHI

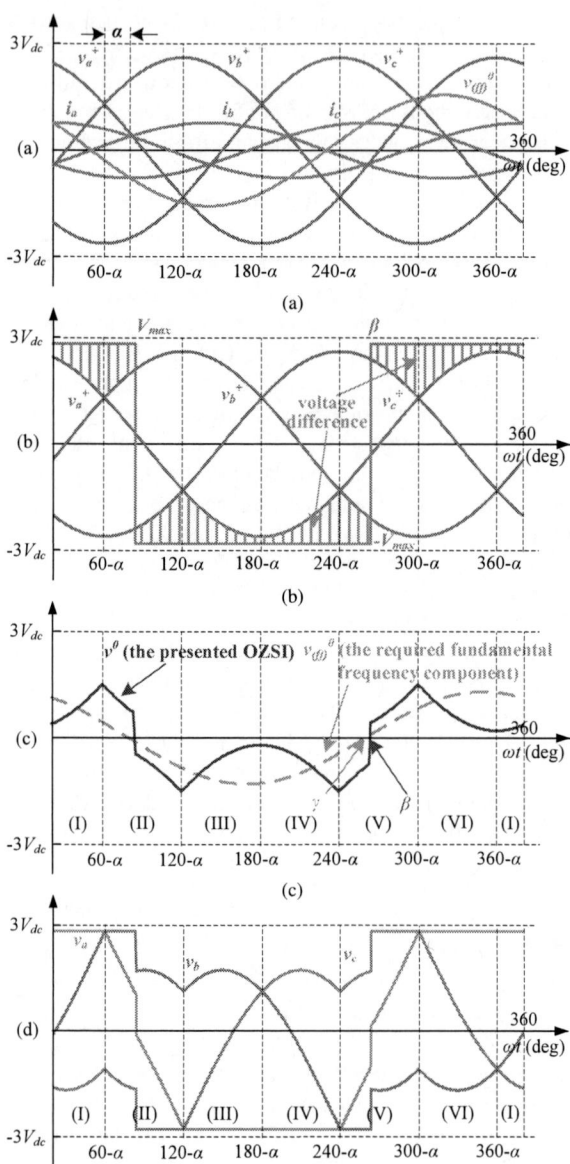

Fig. 3. (a) the positive sequence of the converter output voltages (v_a^+, v_b^+, v_c^+), grid currents (i_a, i_b, i_c), phase shift between converter output voltages and grid voltages (α), and the required fundamental frequency component ($v_{(ff)}^0$), (b) the basic idea of OZSI and the peak resultant voltage V_{max}; (c) the proposed OZSI (v^0) and its fundamental frequency component ($v_{(ff)}^0$); (d) three-phase resultant voltages (v_a, v_b, v_c) under OZSI.

III. OPTIMAL ZERO SEQUENCE INJECTION (OZSI)

If the grid voltages are ideal, only the fundamental frequency component of the injected zero sequence is related to the power balancing control [13]. Therefore, various forms of zero sequence injections (including FFZSI and DTHI) should

have identical fundamental frequency component, although their actual waveforms may differ. Among them, the injection resulting in the lowest peak value of the converter output voltages is the optimal zero sequence injection.

A. Concept

A fundamental frequency zero sequence, as mentioned earlier, has to be injected to the symmetrical component of the converter output voltages, in order to achieve balanced grid currents as shown in Fig. 3(a). The problem of finding the

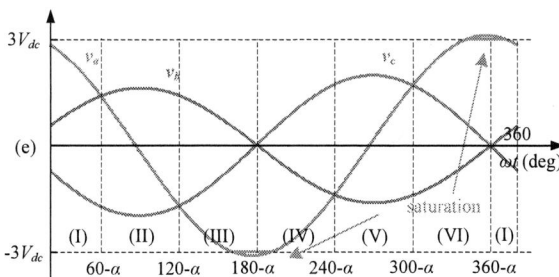

Fig. 4. Three-phase resultant voltages (v_a, v_b, v_c) under FFZSI.

optimal zero sequence injection can then be interpreted as how to generate the largest fundamental frequency zero sequence with a given peak resultant voltage. If the peak resultant voltage is fixed at V_{max}, the maximum voltage difference with regard to the external envelope of the positive sequence is illustrated as the shaded area in Fig. 3(b). If this voltage difference is used as the injected zero sequence (see Fig. 3(c)), the integral over half period is maximized, which, as a result, permits the largest fundamental frequency.

The resultant three-phase voltages are plotted in Fig. 3(d), where the peak voltage rotates among three phases. Unlike the resultant phase voltages under FFZSI where the sinusoidal waveform is well below its peak for most of the time (see Fig. 4), the presented method allows the resultant peak voltage to stay at V_{max}, thus fulfilling the requirement to make full use of the permissible voltage overhead provided.

It is important that both the positive and negative tops cover half a cycle, because only in this way the integral of the presented zero sequence over one period is zero, thus eliminating its dc component.

Please also be aware that the presented method is different to the discontinuous pulse width modulation techniques in which one phase is clamped to the sum of dc-side voltages $3V_{dc}$. In this presented method, the peak voltage V_{max} is smaller than the dc-side voltage limit $3V_{dc}$ and all the devices switch per carrier cycle. The only exception is the critical condition when the limit of this method is reached, and one phase is clamped and does not switch. Furthermore, in discontinuous modulation each phase is clamped to the dc-side voltage limit $3V_{dc}$ for equal amount of time, whereas, in the presented method, the clamping time of each phase is dependent on its power generation ratio. That is to say, the phase with more generated power is more frequently clamped to the peak voltage (V_{max}) than the other two phases.

B. Derivation

As illustrated in Figs. 3(b) and 3(c), the proposed OZSI can be determined by two variables: the peak voltage V_{max} and the negative-to-positive transition angle β. In order to simplify the analysis, the entire period is equally divided into 6 sectors as in Fig. 3(c), according to the positive sequence of three-phase converter output voltages.

For example, if the required fundamental frequency zero

sequence aligns with the Phase A current, β should be in Sector V. In this paper, only this sector is analysed in detail, since similar methods can be applied to other sectors.

The OZSI, as shown in Fig. 3(c), can thus be expressed in the following segregate manner as:

$$v^0 = \begin{cases} V_{max} - v_a^+ & 0 \leq \omega t < \pi/3 - \alpha \\ V_{max} - v_b^+ & \pi/3 - \alpha \leq \omega t < \beta - \pi \\ -V_{max} - v_c^+ & \beta - \pi \leq \omega t < 2\pi/3 - \alpha \\ -V_{max} - v_a^+ & 2\pi/3 - \alpha \leq \omega t < 4\pi/3 - \alpha \;, \\ -V_{max} - v_b^+ & 4\pi/3 - \alpha \leq \omega t < \beta \\ V_{max} - v_c^+ & \beta \leq \omega t < 5\pi/3 - \alpha \\ V_{max} - v_a^+ & 5\pi/3 - \alpha \leq \omega t < 2\pi \end{cases} \quad (6)$$

where α denotes the phase shift between converter output voltages and grid voltages (Fig. 3(a)):

$$\alpha = \arctan \frac{\sqrt{3}\omega L_f I_g}{V_g}. \quad (7)$$

where I_g stands for the rms value of the balanced three-phase grid currents.

Only the fundamental frequency component of v^0 is taken into consideration here, since it is directly related to the balancing control. The dc component should not exist as expected; the harmonic components, on the other hand, are uncontrollable. The fundamental frequency component can be calculated by Fourier analysis as:

$$v_{(ff)}^0 = \sqrt{a_1^2 + b_1^2} \cos\left(\omega t - \arctan \frac{b_1}{a_1}\right), \quad (8)$$

where

$$a_1 = \frac{1}{2\pi}\left(-8V_{max}\sin(\beta + \alpha) - \sqrt{2}\pi V^+ \right. \\ \left. + \sqrt{6}V^+ \cos 2(\beta + \alpha) - \sqrt{6}V^+\right), \quad (9)$$

$$b_1 = \frac{1}{2\pi}\left(8V_{max}\cos(\beta + \alpha) + \sqrt{6}V^+ \sin 2\beta \right. \\ \left. -2\sqrt{6}V^+(\beta + \alpha) + 3\sqrt{6}\pi V^+\right), \quad (10)$$

where V^+ denotes the rms value of the positive sequence of converter output voltages.

The fundamental frequency component (8)-(10) should be identical to the required (2), in order to facilitate power balancing control, hence:

$$\sqrt{a_1^2 + b_1^2} = \sqrt{2}V^0, \quad (11)$$

$$-\arctan \frac{b_1}{a_1} = \theta. \quad (12)$$

By combining (9)-(12), the peak voltage V_{max} and the negative-to-positive transition angle β can be calculated.

C. Numerical Implementation

In order to obtain the required peak voltage V_{max} and negative-to-positive transition angle β, a transcendental equation set derived by (9)-(12) has to be solved. Since it is generally tedious and challenging to analytically solve the transcendental equation set, this section devotes to the derivation of its numerical solution.

The direct iterative equation for the negative-to-positive transition angle β can be written as:

$$\beta_{k+1} = \varphi(\beta_k) \qquad k = 0, 1, 2...$$

$$\varphi(\beta_k) = -\frac{\sqrt{3}V^0}{3V^+}\cos(\beta_k + \alpha)\cot(\beta_k + \alpha)$$
$$- \frac{\sqrt{3}\pi}{6}\cot(\beta_k + \alpha) + \frac{1}{2}\sin 2(\beta_k + \alpha).$$
$$+ \frac{1}{2}(\cos 2(\beta_k + \alpha) - 1)\cot(\beta_k + \alpha)$$
$$+ \frac{\sqrt{3}V^0}{3V^+}\pi\sin(\theta + \alpha) + \frac{3\pi}{2} - \alpha \tag{13}$$

According to (2), the negative-to-positive zero crossing point of the fundamental frequency zero sequence, as shown in Fig. 3(c), can be expressed as:

$$\gamma = \frac{3\pi}{2} - \theta. \tag{14}$$

Observed from Fig. 3(c), γ should be quite close to the negative-to-positive transition angle β; therefore, it is used as the initial value for iteration. Furthermore, γ can be used to determine the sector location of β, since they are always in the same sector.

One selected example is analysed to investigate the iteration convergence. The converter parameters under study are listed in Table I. In this example, we assume that both Phase B and C are generating 68.94% of its peak power, whereas Phase A is producing full power:

$$\lambda_a = 100\% \quad \lambda_b = 68.94\% \quad \lambda_c = 68.94\%. \tag{15}$$

Since the ultimate control aim is to achieve balanced three-phase grid currents, the rms value of grid currents can be calculated as if the converter were under balanced power generation:

$$I_g = \frac{(\lambda_a + \lambda_b + \lambda_c)P_{peak}}{3\sqrt{3}V_g} = 693.6A. \tag{16}$$

The rms value of the positive sequence of converter output voltages V^+ can then be calculated, again assuming total power were equally shared among three phases:

$$V^+ = \sqrt{\left(\frac{V_g}{\sqrt{3}}\right)^2 + (\omega L_f I_g)^2} = 3963.2\text{V}, \tag{17}$$

TABLE II
COMPARISON OF DIRECT ITERATION AND THE STEFFENSEN'S METHOD

β	Direct Iteration	the Steffensen's Method
$\beta_0(\gamma)$	270.0000° ($|\varepsilon| = 0.854\%$)	270.0000° ($|\varepsilon| = 0.854\%$)
β_1	268.8997° ($|\varepsilon| = 1.258\%$)	272.4590° ($|\varepsilon| = 0.049\%$)
β_2	267.3071° ($|\varepsilon| = 1.843\%$)	272.3257° ($|\varepsilon| = 0.000\%$)
β_3	265.0303° ($|\varepsilon| = 2.679\%$)	272.3252° ($|\varepsilon| = 0.000\%$)
β_4	261.8258° ($|\varepsilon| = 3.855\%$)	272.3252° ($|\varepsilon| = 0.000\%$)

and the phase shift α can also be obtained as:

$$\alpha = \arctan\frac{\sqrt{3}\omega L_f I_g}{V_g} = 16.0°. \tag{18}$$

The required fundamental frequency component of the injected zero sequence is derived based on (3)-(5) as:

$$v_{(ff)}^0 = \sqrt{2}V^0\cos(\omega t + \theta) = \sqrt{2} \times 995.1\cos\omega t. \tag{19}$$

Note that the fundamental frequency zero sequence aligns with the positive sequence component of Phase A converter output voltage, since equal amounts of power are generated by Phase B and C. Then the negative-to-positive transition angle β should be within Sector V, since the negative-to-positive zero crossing point γ is calculated by (14) as:

$$\gamma = 270°. \tag{20}$$

The direct iteration based on (13), however, fails to converge as shown in Table II. Hence, an advanced iteration method named the Steffensen's Method [15] is applied to force the iteration to converge. It features quadratic convergence as Newton's method does, but without using derivatives, which enables it to be programmed for a generic function. In this method, the direct iteration function φ has to be evaluated twice for a single iteration:

$$\beta_{k+1} = \psi(\beta_k) \qquad k = 0, 1, 2...$$
$$\psi(\beta_k) = \beta_k - \frac{(\varphi(\beta_k) - \beta_k)^2}{\varphi(\varphi(\beta_k)) - 2\varphi(\beta_k) + \beta_k}, \tag{21}$$

which requires more calculation effort. However, the example shown in Table II indicates that the Steffensen's Method reaches a negligible error in only two iterations. This rapid convergence makes it practical for low-cost digital signal processor implementation.

Finally, the peak voltage V_{max} can be finally derived as:

$$V_{max} = \frac{1}{8\sin(\beta + \alpha)}\left(\sqrt{6}V^+(\cos 2(\beta + \alpha) - 1)\right.$$
$$\left. -2\sqrt{2}\pi V^0\cos(\theta - \alpha) - \sqrt{2}\pi V^+\right) \tag{22}$$

The 2014 International Power Electronics Conference

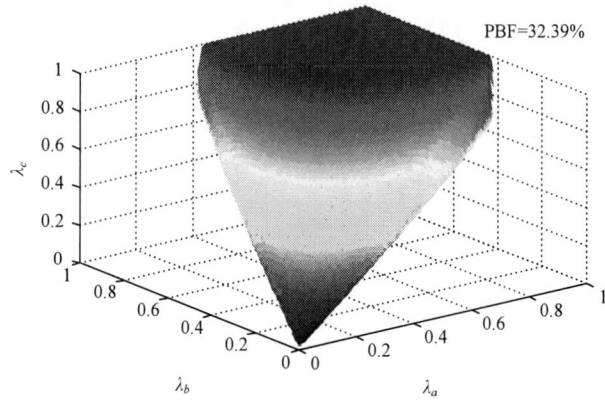

Fig. 5. Power balanceable space of OZSI.

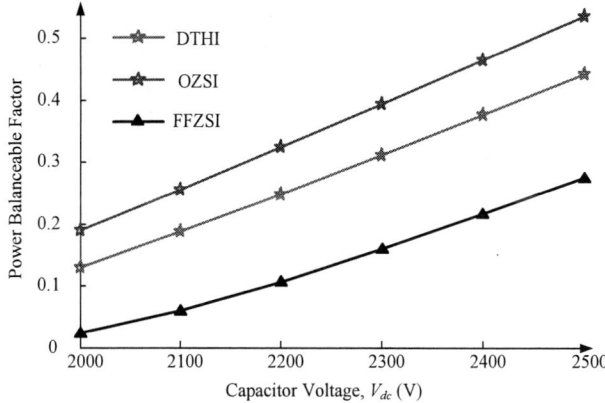

Fig. 6. Power balanceable factor comparison.

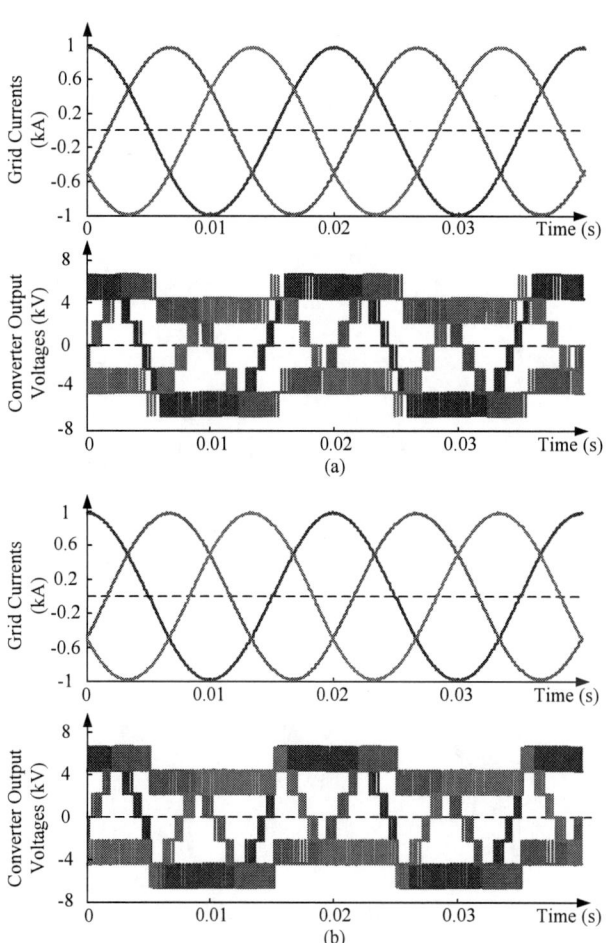

Fig. 7. Power balancing control with (a) DTHI, (b) OZSI.

IV. POWER BALANCEABLE SPACE COMPARISON

As mentioned in Section II, the converter power balancing capability is assessed by PBS and PBF. In this section, the PBSs and PBFs of the existing and newly-presented methods are compared for the same unbalanced case, with converter parameters depicted in Table I. The PBS of the proposed OZSI are displayed in Fig. 5, which features a larger space than conventional methods like DTHI.

To conduct a more thorough comparison, the PBFs of DTHI and OZSI are also demonstrated in Fig. 6. The PBFs are calculated under different capacitor voltages, since the converter is able to provide greater voltage overhead with higher dc-side voltage. Results show that, for the illustrated, OZSI presented in this paper increases the PBF on average by 30%, compared to DTHI. The superiority of the proposed method in improving the converter power balancing control capability is thus confirmed.

V. SIMULATION VERIFICATION

Two selected examples are presented in this section to demonstrate the feasibility and superiority of the proposed method, respectively. Simulations are conducted with PV strings and converter parameters as shown in Table I.

TABLE III
GRID CURRENT DISTORTION (CASE I)

THD	DTHI	OZSI
i_a	0.73%	1.07%
i_b	0.82%	0.85%
i_c	0.77%	0.83%

A. Case I ($\lambda_a = 100\%$, $\lambda_b = 68.94\%$, $\lambda_c = 68.94\%$)

In this case, the solar irradiance on the PV modules of Phase B falls to 700W/m², while the other two phases are subject to uniform irradiance of 1 kW/m². The PV modules connected to Phase A, B, and C then generate 100%, 68.94%, and 68.94% of the peak power, respectively.

For this selected case, both DTHI and OZSI can achieve power rebalance and symmetrical three-phase grid currents as depicted in Fig. 7. The feasibility of the proposed method is thus confirmed. Nevertheless, further investigation of the grid currents distortion shows that OZSI results in slightly higher THD (Table III). This phenomenon is expected, since

978-1-4799-2706-7/14 $31.00 © 2014 IEEE

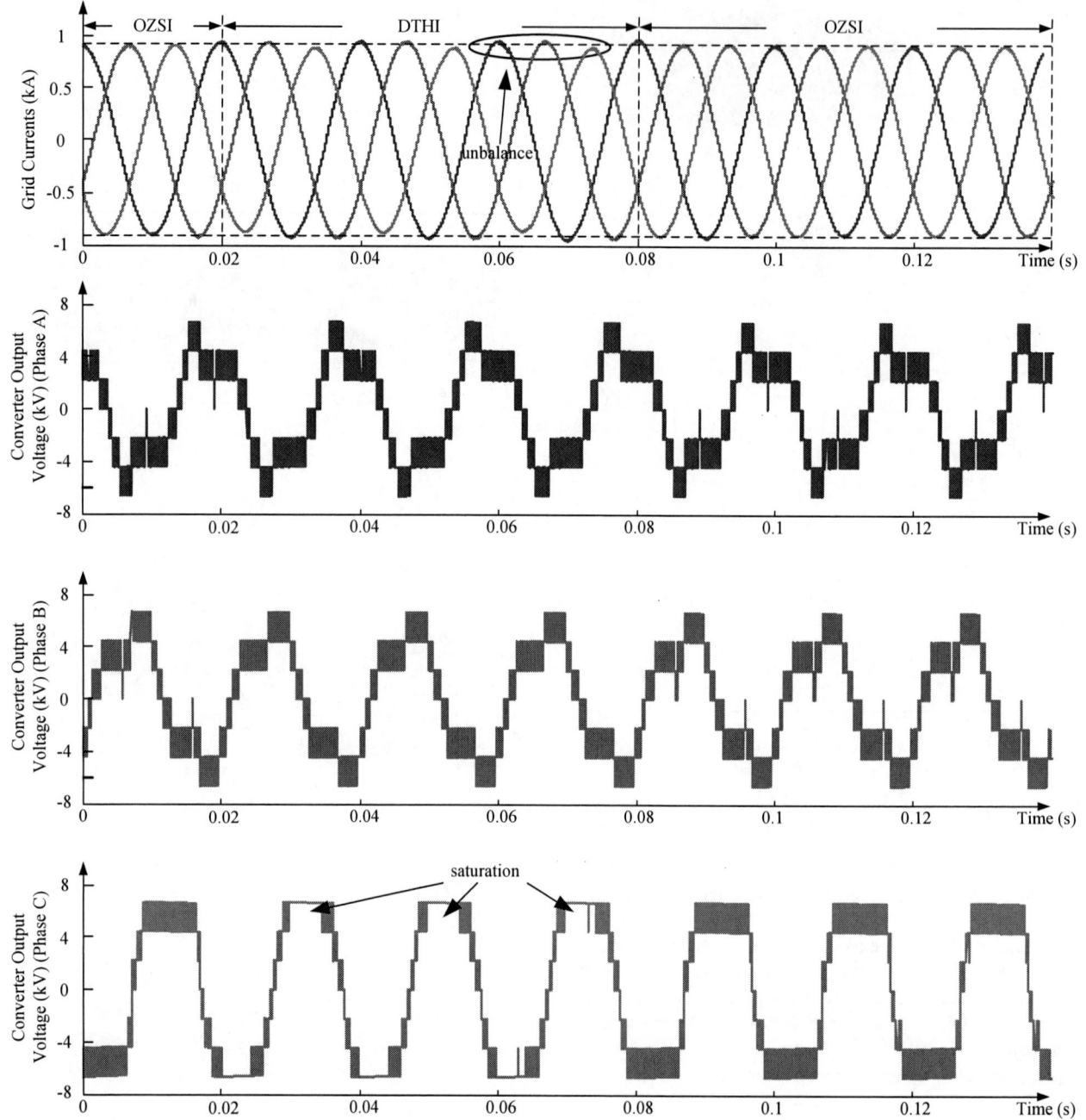

Fig. 8. Power balancing control with DTHI and OZSI (Case II)

harmonic components of OZSI are unregulated.

B. Case II ($\lambda_a = 48.34\%$, $\lambda_b = 68.94\%$, $\lambda_c = 100\%$)

In the second case, the PV modules connected to Phase A, B and C are subject to the solar irradiance of 400 W/m², 700 W/m² and 1000 W/m², respectively. As a result, Phase A, B and C are generating 48.34%, 68.94% and 100% of the peak power, respectively.

Fig. 8 shows the steady state performances of both DTHI and OZSI. At 0.02s, the power balancing control method is switched from OZSI to DTHI for three cycles. After the steady state is reached, DTHI is shown not to be able to provide necessary balanced grid currents, as a result of converter saturation. The grid currents become symmetrical again after the control method is switched back to OZSI at 0.08s. Therefore, the superior power balancing control performance of the presented OZSI is validated.

VI. CONCLUSIONS

The existing methods of extending the power balanceable space of multilevel cascaded H-bridge are derived based on conventional third harmonic injection for balanced three-phase voltages, thus featuring suboptimal performance. This paper proposes an optimal zero sequence injection method. By modifying the zero sequence waveform while keeping the fundamental frequency component unchanged, the largest power balanceable space is achieved. The advanced iteration method presented features fast iteration convergence and is suitable for digital implementation. Results show that the proposed method provides larger power balanceable factor, compared with the existing method. Its superior performance is further verified by computer simulation.

REFERENCES

[1] K. Komoto, C. Breyer, E. Cunow, K. Megherbi, D. Faiman, and P. Vleuten, Energy from the Desert: Practical Proposals for Very Large Scale Photovoltaic Systems, London: Taylor & Francis Group, 2013, pp. 13-15.

[2] T. Kerekes, E. Koutroulis, D. Sera, R. Teodorescu, and M. Katsanevakis, "An Optimization Method for Designing Large PV Plants," *IEEE Journal of Photovoltaics*, vol. 3, no. 2, pp. 814-822, Apr. 2013.

[3] S. Xu, A. Q. Huang, S. Lukic, and M. E. Baran, "On Integration of Solid-State Transformer With Zonal DC Microgrid," *IEEE Trans. Smart Grid*, vol. 3, no. 2, pp. 975-985, Jun. 2012.

[4] F. Blaabjerg, Z. Chen, S. B. Kjaer, "Power electronics as efficient interface in dispersed power generation systems," *IEEE Trans. Power Electron.*, vol. 19, no. 5, pp. 1184-1194, Sep. 2004.

[5] S. Steffel, and A. Dinkel, "Absorbing the rays: advanced inverters help integrate PV into electric utility distribution systems," *IEEE Power and Energy Mag.*, vol. 11, no. 2, pp. 45-54, Mar. 2013.

[6] ABB solar inverters for photovoltaic systems: helping you get more energy out of every day. http://www05.abb.com/global/scot/scot232.nsf/veritydisplay/9d9792bd1a61149bc1257b94003865e4/$file/16728_Solar_Brochure_3AUA0000057127_RevE_lowres.pdf.

[7] SMA Next Innovations 2013. http://files.sma.de/dl/17333/NEXTINNO-KEN132211W.pdf.

[8] W. Zhao, H. Choi, G. Konstantinou, M. Ciobotaru, V. Agelidis, "Cascaded H-bridge multilevel converter for large-scale PV grid-integration with isolated dc-dc stage," in *Proc. IEEE PEDG 2012*, pp. 849-856.

[9] J. Chavarria, D. Biel, F. Guinjoan, C. Meza, and J. Negroni, "Energy-balance control of PV cascaded multilevel grid-connected converters under level-shifted and phase-shifted PWMs," *IEEE Trans. Ind. Electron.*, vol. 60, no. 1, pp. 98-111, Jan. 2013.

[10] E. Villanueva, P. Correa, J. Rodriguez, and M. Pacas, "Control of a single-phase cascaded H-bridge multilevel inverter for grid-connected photovoltaic systems," *IEEE Trans. Ind. Electron.*, vol. 56, no. 11, pp. 4399-4406, Nov. 2009.

[11] S. Rivera, B. Wu, S. Kouro, W. Hong, and D. Zhang, "Cascaded H-bridge multilevel converter topology and three-phase balance control for large scale photovoltaic systems," in *Proc. IEEE PEDG 2012*, pp. 690-697.

[12] C. Townsend, T. Summers, and R. Betz, "Control and modulation scheme for a cascaded H-bridge multi-level converter in large scale photovoltaic systems," in *Proc. IEEE ECCE 2012*, pp. 3707-3714.

[13] Y. Yu, G. Konstantinou, B. Hredzak, and V. Agelidis, "On extending the energy balancing limit of multilevel cascaded H-bridge converters for large-scale photovoltaic farms," in *Proc. of the Australasian Universities Power Engineering Conference*, Hobart, Australia, Sep. 29-Oct. 3, 2013.

[14] D. Holmes and T. A. Lipo, Pulse Width Modulation for Power Converters: Principles and Practice, 3rd ed., New York, NY: John Wiley & Sons, 2003.

[15] G. Dahlquist and Å. Björck, Numerical Methods, Prentice-Hall, Englewood Cliffs, NJ, pp. 230-232, 1974.

Simple Method for Measuring Output Impedance of a Three-Phase Inverter in dq-Domain

Juha Jokipii, Tuomas Messo, and Teuvo Suntio
Department of Electrical Engineering
Tampere University of Technology
Tampere, Finland
juha.jokipii@tut.fi, tuomas.messo@tut.fi, teuvo.suntio@tut.fi

Abstract—**This paper introduces a simple method to measure the output impedance of a three-phase grid-connected inverter in dq-domain. The impedance measurements are most often used for model verification purposes but they can be also utilized to study impedance-based interactions between three-phase converters and the utility grid. Therefore, the methods to model and measure three-phase impedances can provide to the inverter manufacturers valuable information about their products. Implementation of the proposed method requires a three-phase voltage source, a digital signal processor, and a frequency response analyzer which are equipment typically available in most power electronic laboratories. The algorithms used in the method are explained, and the performance of the method is verified by means of frequency response measurements from a small-scale laboratory prototype.**

Keywords—*three-phase grid-connected inverter, three-phase impedance measurement, photovoltaic inverter.*

I. INTRODUCTION

Three-phase grid-connected inverters provide an efficient way to utilize renewable energy resources such as wind and solar. Thus their role as a significant element of future power system is unquestionable [1]. However, it has been observed that large-scale installations of grid-connected inverters may increase the harmonic distortion in grid currents or even lead to instability [2], [3]. As clearly stated in [4], the interactions with the inverter and the grid can be predicted by analyzing their impedance characteristics. According to [4], the output impedance of a grid connected inverter should be as high as possible to maintain harmonic free grid currents and to avoid impedance based interactions in all grid conditions. Thus, the methods to model and measure three-phase impedances are valuable for the inverter manufacturers to validate the performance of their products and to guarantee high quality and uninterrupted power production.

Three-phase impedance can be modeled in phase, sequence or dq-domain. Although the phase and sequence domain modeling and measurements provide clear physical insight into the operation of the grid-connected inverter, they may be difficult to implement in practice, since the large fundamental component easily affects the measurement accuracy at frequencies near the fundamental frequency. When the impedance is modeled in dq-domain the fundamental component is transformed into dc, and can be easily filtered out. Another advantage of dq-domain modeling is that the grid-connected inverters are often modeled and controlled using dq-domain based controllers, and thus the shape of the impedance can be already predicted.

Several different methods to measure three-phase impedances have been introduced. The fundamental issue in these measurements is the injection of an excitation, i.e., a small-signal disturbance, to the grid voltages or currents. The authors in [5] and [6] have proposed a method which is based on the current injection technique. Drawback of the method, when applied to grid-connected inverter measurement, is that a very high current may be required to generate high enough injection in inverter ac side voltages and currents, since the grid-connected inverter has typically very high impedance compared to the grid. The current injection technique also has a limited bandwidth due to the use of an injection transformer. Another method to measure the inverter output impedance is to inject the excitation directly to the grid voltages. The excitation can be excited using a grid simulator [7] or certain type of chopper circuit [8]. The grid-simulator-based approach has been used in [9], where the modeling of the inverter output impedance is carried out in the sequence domain. Similar method has been used in [10] where the impedance of a passive circuit was identified using an existing power electronic converter. Another option is to identify the impedance based on a load transient. Such a method is discussed in [11], where the impedance of an output-voltage-controlled inverter was measured by making a load step and utilizing system identification techniques. Unfortunately, the technique cannot be directly applied to grid connected inverters since the operation mode is different.

In this paper, a simple method to measure the inverter output impedances in dq-domain is proposed, which can be implemented using equipment typically available in power electronic laboratories. Other advantages of the proposed method are that its performance can be easily verified and the same hardware can be used as a grid simulator. The work is an extension to the work presented in [12], which proposes a measurement setup to characterize the VSI-type photovoltaic inverter small-signal behavior. In this paper, the measurement setup is modified to accurately capture the small-signal behavior of the three-phase photovoltaic inverter output impedance.

The rest of the paper is organized as follows: Sec-

978-1-4799-2706-7/14 $31.00 © 2014 IEEE

tion II introduces the proposed impedance measurement method by describing the main components of the measurement setup and the operation of its control system. The procedure to design the control system is also discussed. Frequency response measurements from a small-scale prototype utilizing the proposed method are presented in Section III. Finally, conclusions are drawn in Section IV.

II. PROPOSED THREE-PHASE IMPEDANCE MEASUREMENT TECHNIQUE

Three-phase grid-connected inverters are often analyzed in the dq-domain using space-vectors. Naturally, it is convenient to model the output impedance also in dq-domain. The small-signal behavior of the inverter output terminals can be represented by admittance matrix as in (1), where superscript s stands for synchronous reference frame, i.e. dq-frame. The impedance is presented as an admittance, since in small-signal sense the inverter acts as a current source and grid as an infinite bus (voltage source).

$$\hat{\mathbf{i}}_o^s = \mathbf{Y}_o^s \hat{\mathbf{u}}_o^s \tag{1}$$

Equation (1) may be written also as

$$\begin{bmatrix} \hat{i}_{o\text{-}d} \\ \hat{i}_{o\text{-}q} \end{bmatrix} = \begin{bmatrix} Y_{o\text{-}d} & Y_{o\text{-}qd} \\ Y_{o\text{-}dq} & Y_{o\text{-}q} \end{bmatrix} \begin{bmatrix} \hat{u}_{o\text{-}d} \\ \hat{u}_{o\text{-}q} \end{bmatrix}, \tag{2}$$

from which the corresponding impedance terms can be computed by taking the reciprocal of each term in (2), e.g. $Z_{o\text{-}d} = 1/Y_{o\text{-}d}$. The positive current direction in (1) and (2) is defined to be flowing towards the converter according to Fig. 1.

A. Components of the Measurement Setup

A reliable method to generate the injection to d and q-components of the grid voltage is required to measure the admittances from a physical device. To accurately measure aforementioned admittances, the injection should be present only in one component at a time, i.e. when injection is created to voltage d-component there should be no small-signal injection in voltage q-component and vice versa. Since the output impedance of the inverter depends on the processed power [13], it is also required that the operating point of the inverter under test can be set to an arbitrary value.

The proposed impedance measurement setup is as shown in Fig. 1. The main parts of the system are a three-phase voltage source, a digital signal processor (DSP) and a frequency response analyzer (FRA). The inverter under test is connected to the three-phase voltage source and its ac side currents and voltages are measured using a separate measurement card. The DSP transforms the measured ac side quantities to the rotating dq-frame by utilizing Park's transformation. After the transformation, the dq-domain voltages and current are fed to the frequency response analyzer through two PWM outputs of the DSP. It is worth noting that same delay is present in current and voltage measurement and, thus, it does not affect the results.

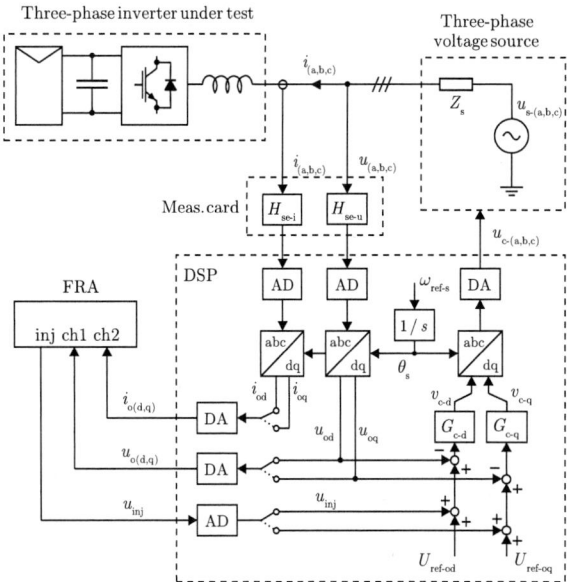

Fig. 1. Overview of the measurement setup.

In addition to the coordinate transformations, the DSP controls the three-phase voltage source to generate ideal grid voltages with an amplitude of $U_{\text{ref-od}}$ and constant angular frequency of $\omega_{\text{ref-s}}$. The reference of the voltage q-component $U_{\text{ref-oq}}$ is set to zero to align the measurement system reference frame correctly in respect to the ideal grid voltages. It should be noted, that the three-phase voltage source should operate in four-quadrants because it has to sink all the power supplied by the inverter. However, as will be shown later, the proposed method gives good results also if resistors are connected between the inverter and three-phase voltage source to dissipate the active power.

The frequency response analyzer generates a small signal voltage injection which is measured using DSP and added to the reference of the voltage d and q-component depending on which impedance is to be measured. A total of four measurements are required to determine the impedances in (2). The method is power invariant and can be scaled up to higher power levels.

B. Controller Design Procedure

According to Fig. 1, two voltage control loops are implemented using the DSP. Their purpose is to regulate the d and q-components of the grid voltages and to decrease the cross-couplings. The voltage controllers are designed based on the measured open-loop transfer functions between the d and q-components of the control voltages $v_{c\text{-}d}$ and $v_{c\text{-}q}$, and measured voltage d and q-components $u_{o\text{-}d}$ and $u_{o\text{-}q}$. The measurement setup used to obtain open-loop transfer functions is illustrated in Fig. 2 and the measurement results in Fig. 3. The measured frequency responses for d and q-component are similar, thus a controller with same parameters can be utilized.

Using the same measurement setup depicted in Fig.

2, the cross-coupling transfer functions, i.e. the transfer function from d-component of the control voltage to measured q-component of the grid voltage and the transfer function from q-component of the control voltage to measured d-component of the grid voltage, was measured. The results are presented in Fig. 4. According to Figs. 3 and 4, the injection to the desired channel is approximately 25 dB higher than the injection to the opposite channel.

During the open-loop measurements, the inverter under test was turned on and operating at the desired operating point since presence of the inverter and its operating mode will affect the transfer functions. Otherwise the voltage control may not perform as expected when the impedances of the inverter under test is measured.

A PID-type controller is designed to regulate the d and q-component of the grid voltages to their reference values. The compensated loop gains of the voltage control loops are illustrated in Fig. 5. The gain and phase margin are approximately 8 dB and 90°, respectively.

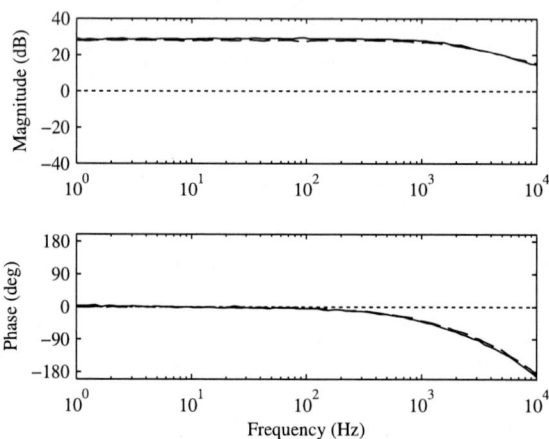

Fig. 3. Measured open-loop frequency responses from d-component of the control voltage to d-component of the measured grid voltage (solid) and from q-component of the control voltage to q-component of the measured grid voltage (dashed).

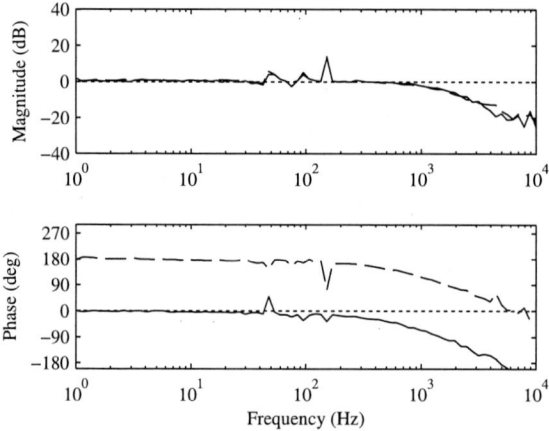

Fig. 4. Measured open-loop cross-coupling frequency responses from d-component of the control voltage to q-component of the measured grid voltage (solid) and from q-component of the control voltage to d-component of the measured grid voltage (dashed).

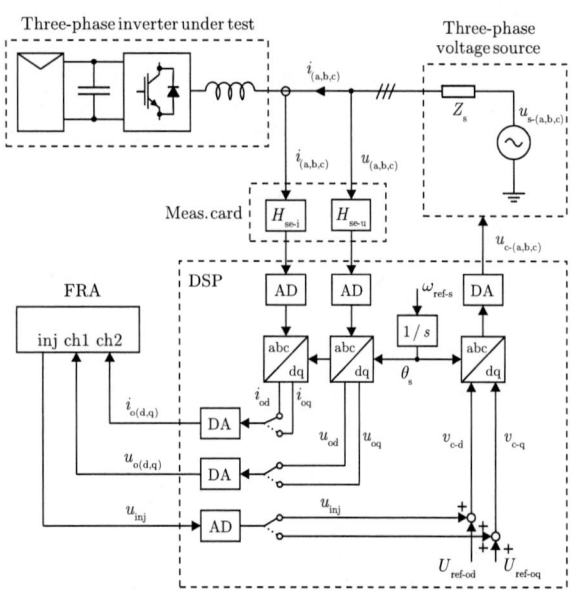

Fig. 2. Measurement setup to obtain the open-loop transfer functions.

III. EXPERIMENTAL EVIDENCE

A small-scale prototype utilizing the proposed impedance measurement method has been constructed. Components of the measurement setup are: a programmable three-phase voltage source Elgar model SW5210A, a digital signal processor eZdsp TMS320F28335 and a frequency response analyzer Venable Model 3120. Contrary to Fig. 1, a resistive load was connected between the inverter under test and the three-phase voltage source to dissipate the generated power due to hardware limitations. A custom measurement board was build to scale the voltages and currents to be suitable for DSP. In addition, a low pass filter with a cut-off frequency set to half of the switching frequency of the inverter under test, was added

in the measurement board. The control system of the measurement setup was designed as discussed in Section II.

The accuracy of the measurement setup was assessed by measuring the closed-loop frequency responses from the d and q-component of the voltage references to the actual d and q-components of the grid voltages. The measurement results shown in Fig. 6 clearly indicate that the three-phase voltage source accurately follows the reference voltages at relatively wide bandwidth. The effect of cross-coupling effect, i.e. the transfer functions from voltage reference d-component to the grid voltage q-component and vice versa are shown in Fig. 7. Compared to Fig. 4, where the measurement setup is operating at open-loop, the cross-coupling effects are now significantly smaller. The control system effectively rejects the injection to the opposite channel.

The 2014 International Power Electronics Conference

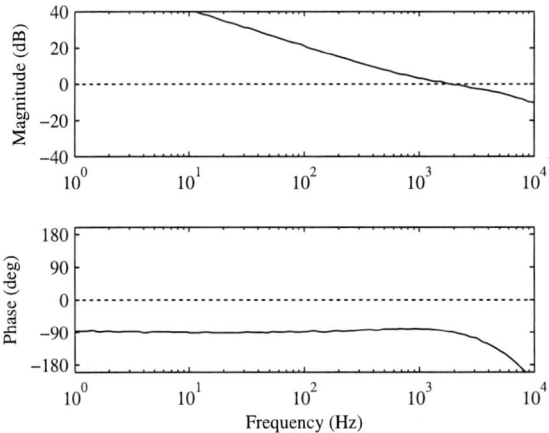

Fig. 5. Compensated loop gains of the voltage control loops.

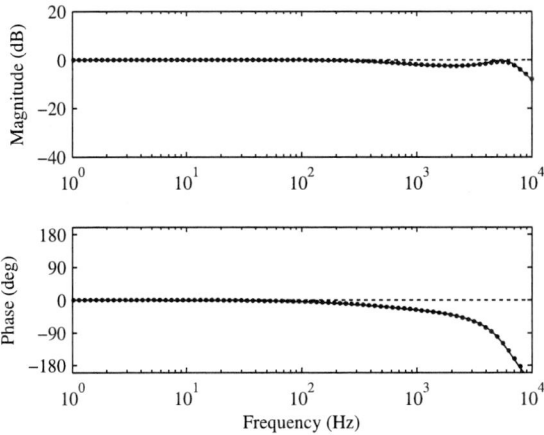

Fig. 6. Measured frequency responses from d-component of the voltage reference to d-component of the measured grid voltage (solid) and from q-component of the voltage reference to q-component of the measured grid voltage (dotted).

To further analyze the performance of the measurement setup, the impedance extraction technique have been tested with a small-scale prototype of a VSI-based photovoltaic inverter. The main parameters of the inverter are collected in Table I [13]. A conventional dq-frame-based control system was implemented, where the d-component current controller is cascaded with the dc voltage control loop and q-component current reference is set to zero. The reference value for the dc voltage was set to 30 V. A standard SRF-PLL was designed to synchronize the inverter control system. The inverter under test was supplied by an Agilent Technologies E4360A solar array simulator and the maximum power point of the solar array was set to 30 V.

TABLE I. PARAMETERS OF THE PHOTOVOLTAIC INVERTER

L	C	f_{sw}	U_{in}	I_{in}	U_o
73 μH	860 μF	100 kHz	30 V	1 A	9 V_{peak} / 50 Hz

Measured and simulated d-component of the output

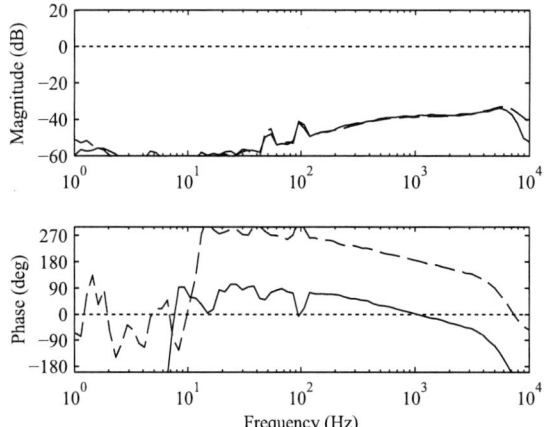

Fig. 7. Measured frequency responses from q-component of the voltage reference to d-component of the measured grid voltage (solid) and from d-component of the voltage reference to q-component of the measured grid voltage (dashed).

impedances are depicted in Fig. 8. The simulated frequency response is extracted from a model implementd in MATLAB/Simulink using the SimPowerSystems Toolbox including all the switching actions and control loops. Correspondingly, the measured and simulated frequency responses of a q-component of the output impedances are depicted in Fig. 9.

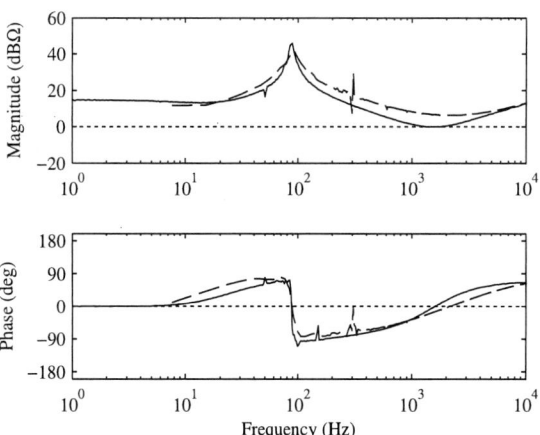

Fig. 8. Measured (solid) and simulated (dashed) d-component of the output impedance $Z_{o\text{-}d} = 1/Y_{o\text{-}d}$.

Figs. 8 and 9 confirm that the proposed method can be used to measure the output impedance of a grid-connected inverter accurately. The differences between the simulated and measured frequency responses are most likely caused by the differences between the component values, i.e. parasitic resistances, between the simulation model and the physical device. As can be seen from the results, the proposed technique can measure dq-domain impedance from 1 Hz to 10 kHz. The measurements also verifies that the q-component of the output impedance resembles negative resistance at low frequencies. This phenomenon,

978-1-4799-2706-7/14 $31.00 © 2014 IEEE

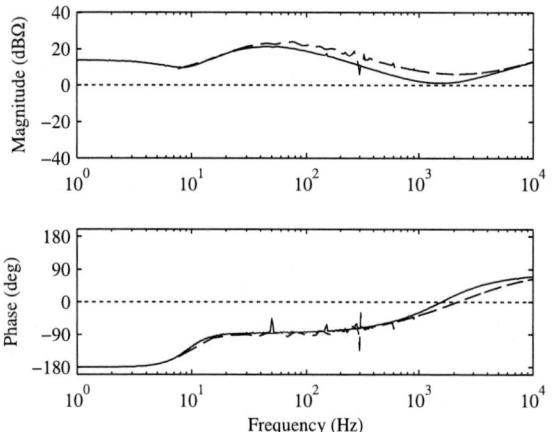

Fig. 9. Measured (solid) and simulated (dashed) q-component of the output impedance $Z_{o\text{-}q} = 1/Y_{o\text{-}q}$.

which is caused by the PLL [13], will remain hidden if the impedance measurement technique cannot measure the low-frequency behavior of the output impedance.

The proposed impedance measurement technique was not able to measure the cross-coupling terms Z_{dq} and Z_{qd} in the output impedance of a grid-connected inverter. The reason is that the levels of the impedance are too low to be sensed with the used frequency response analyzer. The low cross-coupling impedance is a result of a high-bandwidth current control loops implemented in dq-frame that can be considered to be decoupled [14].

IV. CONCLUSION

Impedance measurements provide inverter manufacturers a valuable tool for the design verification process. A simple and accurate method to measure the output impedance of a three-phase inverter in dq-domain is presented. The advantage of the presented method is that the hardware is often already available in most laboratories and the setup can be also used as a grid simulator. The validity of the method was verified using a small-scale prototype by means of frequency response measurements. In addition, the output impedance of a three-phase photovoltaic inverter was presented and results compared to the frequency responses extracted from a simulation model constructed in MATLAB/Simulink. Compared to the other impedance measurement techniques, the proposed method can capture also the low-frequency behavior of the output impedance of a grid-connected inverter. Furthermore, the accuracy of the method is not affected by the components at the fundamental frequency. The method presented in the paper is power invariant and it can be scaled up to higher power levels.

REFERENCES

[1] B. K. Bose, "Global energy scenario and impact of power electronics in 21st century," *IEEE Trans. Ind. Electron.*, vol. 60, no. 7, pp. 2638–2651, Jul. 2013.

[2] J. H. R. Enslin and P. J. M. Heskes, "Harmonic interaction between a large number of distributed power inverters and the distribution network," *IEEE Trans. Power Electron.*, vol. 19, no. 6, pp. 1586–1593, Nov. 2004.

[3] C. Wan, M. Huang, C. K. Tse, and X. Ruan, "Stability of interacting grid-connected power converters," *Journal of Modern Power Systems and Clean Energy*, vol. 1, no. 3, pp. 249–257, Dec. 2013.

[4] J. Sun, "Impedance-based stability criterion for grid-connected inverters," *IEEE Trans. Power Electron.*, vol. 26, no. 11, pp. 3075–3078, Nov. 2011.

[5] J. Huang, K. A. Corzine, and M. Belkhayat, "Small-signal impedance measurement of power-electronics-based AC power systems using line-to-line current injection," *IEEE Trans. Power Electron.*, vol. 24, no. 2, pp. 445–455, Feb. 2009.

[6] G. Francis, R. Burgos, D. Boroyevich, F. Wang, and K. Karimi, "An algorithm and implementation system for measuring impedance in the D-Q domain," in *IEEE Energy Conversion Congress and Exposition (ECCE)*, 2011, pp. 3221–3228.

[7] X. Chen and J. Sun, "Characterization of inverter-grid interactions using a hardware-in-the-loop system test-bed," in *8th International Conference on Power Electronics (ECCE Asia)*, May 2011, pp. 2180–2187.

[8] P. Xiao, G. Venayagamoorthy, and K. Corzine, "A novel impedance measurement technique for power electronic systems," in *Power Electronics Specialists Conference (PESC)*, 2007, pp. 955–960.

[9] M. Cespedes and J. Sun, "Impedance modeling and analysis of grid-connected voltage-source converters," *IEEE Trans. Power Electron.*, vol. 29, no. 3, pp. 1254–1261, Mar. 2014.

[10] D. Martin, I. Nam, J. Siegers, and E. Santi, "Wide bandwidth three-phase impedance identification using existing power electronics inverter," in *Twenty-Eighth Annual IEEE Applied Power Electronics Conference and Exposition (APEC)*, 2013, pp. 334–341.

[11] V. Valdivia, A. Lazaro, A. Barrado, P. Zumel, C. Fernandez, and M. Sanz, "Impedance identification procedure of three-phase balanced voltage source inverters based on transient response measurements," *IEEE Trans. Power Electron. Lett.*, vol. 26, no. 12, pp. 3810–3816, Dec. 2011.

[12] J. Puukko, L. Nousiainen, and T. Suntio, "Three-phase photovoltaic inverter small-signal modelling and model verification," in *Power Electronics, Machines and Drives (PEMD 2012), 6th IET International Conference on*, no. 1, 2012, pp. 1–6.

[13] T. Messo, J. Jokipii, and T. Suntio, "Modeling the grid synchronization induced negative-resistor-like behavior in the output impedance of a three-phase photovoltaic inverter," in *The IEEE 4th International Symposium on Power Electronics for Distributed Generation Systems*, 2013, pp. 1–7.

[14] J. Puukko, T. Messo, L. Nousiainen, J. Huusari, and T. Suntio, "Negative output impedance in three-phase grid connected renewable energy source inverters based on reduced-order model," *IET Conference on Renewable Power Generation (RPG 2011)*, no. 1, pp. 73–73, 2011.

The 2014 International Power Electronics Conference

Analysis and Design of Power Management Scheme for an On-board Solar Energy Storage System

W. Jiang[1], F.Y. Yu[1], Z.Y. Lin[2], G.F. Wu[1], H. Chen[1], S Hashimoto[3]

[1]Department of Electrical Engineering
Yangzhou University, China
Email: jiangwei@yzu.edu.cn

[1]Department of Electrical, Electronic & Power Engineering
Aston University, UK
Email: z.lin@ieee.org

[3]Department of Electronic Engineering
Gunma University, Japan
Email: hashimotos@gunma-u.ac.jp

Abstract—**This paper investigates the power management issues in a mobile solar energy storage system. A multi-converter based energy storage system is proposed, in which solar power is the primary source while the grid or the diesel generator is selected as the secondary source. The existence of the secondary source facilitates the battery state of charge detection by providing a constant battery charging current. Converter modeling, multi-converter control system design, digital implementation and experimental verification are introduced and discussed in details. The prototype experiment indicates that the converter system can provide a constant charging current during solar converter maximum power tracking operation, especially during large solar power output variation, which proves the feasibility of the proposed design.**

I. INTRODUCTION

Photovoltaic is one of the most environmentally friendly power generation technologies. To utilize PV power effectively, two solutions are usually applied. The first solution is to provide ac power conditioning to the solar panel, converting variable dc power to ac power; the load can be either local ac load or the grid [1]-[4] . The other solution is to integrate solar power into the dc power networks for dc applications [5]-[7] , such as sensors, fans and actuators. As the maximum solar power output is affected by the irradiation level and atmosphere conditions, corresponding methods have to be adopted in different applications. Maximum power point tracking (MPPT) algorithms [8][9] are the software solutions to secure efficient usage of solar power, beyond the algorithmic method additional energy storages will be added to balance the load and source power [10][11].

Due to electrification of modern vehicle, the electrical power demand on board is increasing. Solar power as one clear and portable option is gaining more and more attentions in vehicular applications, such as caravan and yacht. With reasonable space and cost, on-board solar power can supply

the balance-of-plant in the form of dc power without leash the power of the drive train.

This paper propose an on-board solar energy storage system for vehicular applications, which is powered by hybrid sources [12]. Multi-converter based structure is proposed, in which solar power is the primary source while the grid or the diesel generator as the secondary source. The existence of the secondary source facilitates the battery state of charge (SOC) detection by providing a constant battery charging current. Converter modeling, multi-converter control system design, digital implementation and experimental verification are introduced and discussed in details. The prototype experiment indicates that the converter system can provide a constant charging current during solar converter maximum power tracking operation, especially during large solar power output variation.

II. ON BOARD SOLAR ENERGY STORAGE SYSTEM

The on-board solar energy storage system for vehicular application is proposed in Figure 1. The battery pack is the main energy buffer in the system. There are two input sources, one is the on-board solar panel; the other one is the multiplexed grid and diesel generator. The solar power is conditioned by a dc-dc converter A and is transferred to the battery pack. The voltage from the grid or diesel generator is converted into a low dc voltage, which is feeding the down-stream battery charger B as well as the load bus. A source discriminator is implemented with two diodes, which naturally commutate based on the availability of the grid/diesel power (external source); when the external power is available, the battery channel is commutated to OFF state, the external source will supply the load bus; if the external source is disconnected, the solar powered battery pack will kick in and supply the on-board load. The local loads are connected to the load bus either directly or by the point-of-load converters.

978-1-4799-2706-7/14 $31.00 © 2014 IEEE

As indicated in the figure, solar converter A and grid/diesel power converter B will supply the load current concurrently or intermittently; therefore, load sharing issue under different operating conditions are of special interest in this research.

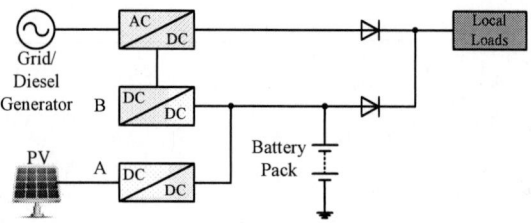

Fig. 1. solar energy storage system

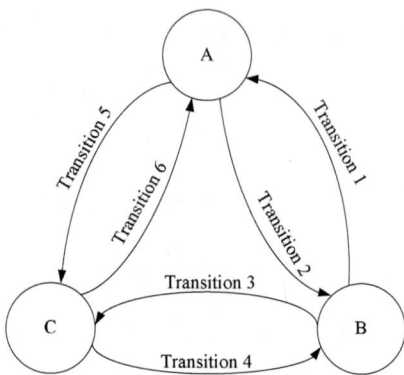

Fig. 2. solar energy storage system

III. Mode of Operation

As multiple converters are involved in energy harvesting and storage tasks, the tri-modal operation is proposed for the solar energy storage system based on the availability and state-of-charge (SOC) of the battery. In mode A, only solar panel is charging the battery, the on-board loads are supplied by the battery, converter A is working under input power control and tracking the maximum power of the solar panel. In mode B, both grid/diesel generator and solar panel are charging the battery, the on-board loads are supplied by the grid/diesel generator converter. In mode C, only grid/diesel generator is charging the battery, the on-board loads are supplied by the grid/diesel generator converter. In order to estimate the SOC more prescisely, terminal voltage look-up method is applied, in which the charging current is kept constant by the grid converter B.

Detailed state flow description are illustrated in Figure 2 and Table I. Due to heavy electrical load on-board, the SOC of the battery is supposed to be not high all the time. Based on the three operational modes, 6 transitional conditions are proposed. During system start-up or acceleration, there is no external plug-in power and diesel generator need to be disconnected to ensure full powertrain power; therefore, battery shall supply the load bus alone with possible assistance of the solar power. When both sources are available, converter A and B will charge the battery. When solar power is not available, the power from diesel or grid will provide power to the battery.

TABLE I.   STATE FLOW CONDITIONS

Transition	Condition
1	system start or acceleration
2	low SOC and both sources available and cruising
3	low SOC and solar unavailable
4	low SOC and both sources available and cruising
5	low SOC and solar unavailable
6	system start or acceleration

IV. Control System Design

The control loop is shown in Figure. 3. Since the battery voltage is kept constant over relatively long period of time,

modulation of duty cycle solar converter A can move the operating point of the solar panel, in which way the maximum power point can be obtained using different tracking method. In this research, perturbation and observation (P&O) algorithm [13] is chosen. Panel terminal voltage and currents are sampled and fed into the MPPT block, duty cycle is updated at the end of each power sweep cycle ($10ms$). The output current of solar converter A varies with solar irradiation, therefore, the grid/diesel converter B is used to keep the output current constant such that accurate battery SOC can be estimated by reading battery voltage under specific grid current. i_{L1} is the output current of solar converter A, while i_{L2} is the output current from grid/diesel converter B. i_L, the charging current into the battery, is contributed by both converters. In order to keep the charging current constant, the current i_L is sensed and controlled by digital signal controller, where $G_{idx}(s)$ ($x = 1, 2$) is the duty cycle to output current transfer function, $H_i(s)$ is the current transducer gain, $C_{i2}(s)$ is the current controller.

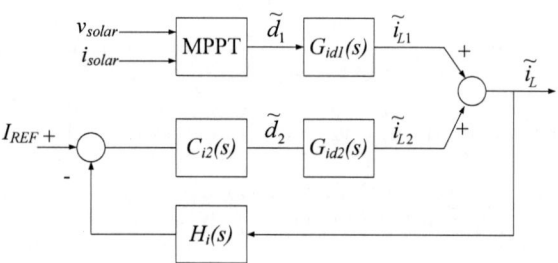

Fig. 3. control loop design

Buck topology is selected for both converters as the $12V$ is one of the common bus voltage for vehicular application. The parameters of the buck converter are listed in Table II. The switching frequency is selected as $50kHz$.

Using state-space averaging method [14], the small signal model of the control plant can be obtained as,

$$G_{id} = \frac{\tilde{i}_L(s)}{\tilde{d}(s)} = \frac{V_S(R_{eq}Cs + 1)}{R_{eq}LCs^2 + Ls + R_{eq}} \quad (1)$$

978-1-4799-2706-7/14 $31.00 © 2014 IEEE   1472

TABLE II.   SYSTEM PARAMETERS

component	Value
C_1	$60 \mu F$
L_1	$40 \mu H$
C_2	$60 \mu F$
L_2	$40 \mu H$

where R_{eq} is the internal impedance of the on-board battery pack.

A PI controller is used to provide load current regulation; the compensated open loop transfer function is shown in the bode plot Figure 4 with $k_p = 0.54$, $k_i = 4360$; as can be seen that the cross-over frequency is $2kHz$ with phase margin of $64°$.

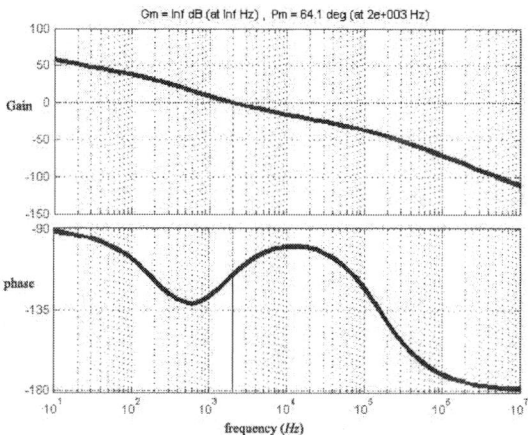

Fig. 4.   open-loop transfer function bode plot

V.   EXPERIMENT RESULTS

Hardware prototype is built and dsPIC33F DSC is used to implement the power management scheme. A $100W$ solar panel with $V_{MPP} = 28V$ is used as the solar energy input, a $30V 10A$ linear power supply is used as the AC/DC converter in Figure 1; the load is a $12V 200AH$ lead-acid battery. The solar panel arrangement and converter test bed is shown in Figure 5.

The MPPT of the solar panel is tested under two typical conditions. The open circuit voltage method[15] is used to set the initial duty cycle while P&O algorithm kicks in after the start-up transient. As shown in the Figure 6(a), with the initial duty cycle setup, solar converter A can directly goes to the vicinity of the maximum power point (MPP), which only takes $7ms$. Figure 6(b) shows MPPT operation under large solar power loss; the voltage sag before T_0 is the fully shaded condition, in which solar panel does not give any output power; the converter A reaches MPP within $1s$ after the solar irradiation recovery.

With converter A and B connected in parallel, the current control capability of converter B is tested. The test is carried on

as follows: the solar converter A is under inductor current (I_{L1}) control, whose current reference varies according to 9th order PRBS. Therefore, under programmed output current reference, converter A is emulating the case when solar irradiation changes randomly. The test results are shown in Figure 7; the output current (charging current) is set to $15A$, as shown in the waveform, converter B output current (I_{L2}) always compensates such that load current I_L is kept at $15A$.

In steady state operation, solar converter A is tracking the MPP of solar panel while converter B compensates the current loss in the total battery charging current. Figure 7 shows this test scenario, as can be found in the waveform, the summation of I_{L1} and I_{L2} keeps constant (curve I_L).

Figure 9 shows the testing scenario that solar panel is undergoing large solar irradiation variation. As indicated in the waveform, the power loss period is the period that the solar panel is fully covered by a blanket; converter B instantaneously balances the load power by providing all $15A$ of load current; it be also found that this transient period is smooth due to effective control action.

(a)

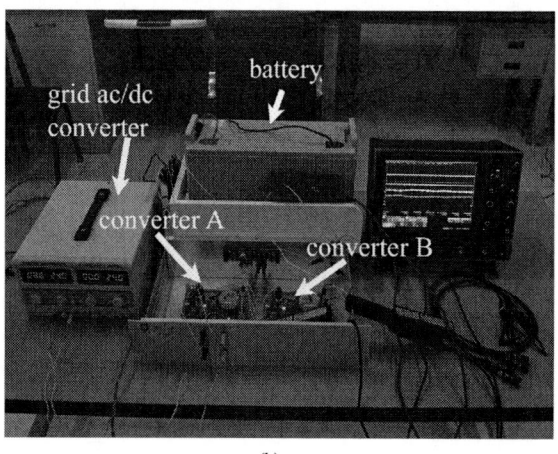

(b)

Fig. 5.   experimental setup (a) mobile solar panel, (b) converter test bed

The 2014 International Power Electronics Conference

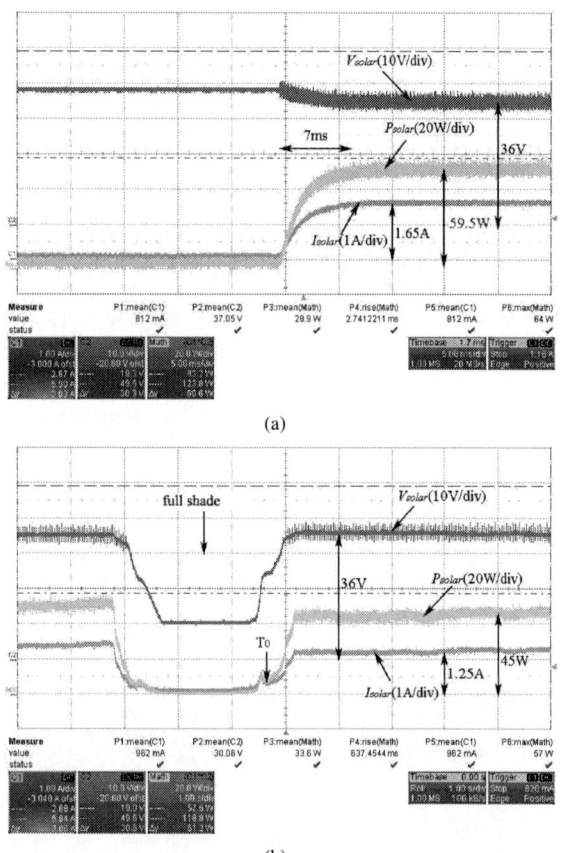

(a)

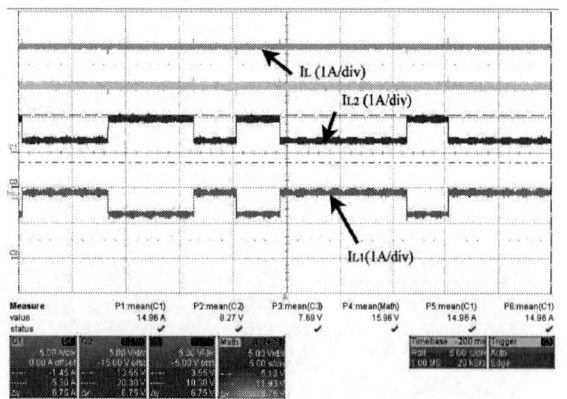

(b)

Fig. 6.   maximum power point tracking test (a) start-up test (5ms/div), (b) abrupt irradiation change test (1s/div)

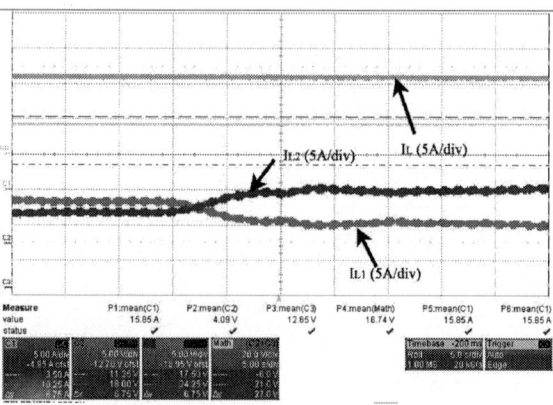

Fig. 8.   steady state test with paralleled operation (5s/div)

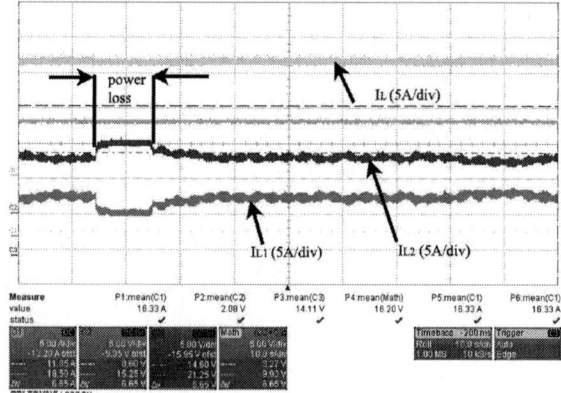

Fig. 9.   Test under large solar irradiation loss (10s/div)

VI.   CONCLUSION

In this paper, an on-board solar energy storage system is proposed for vehicular applications. This system utilize solar panel as one of the input sources, grid or diesel generator as the second source based on the availability. The system structure is proposed and power management scheme is discussed in detail. In order to fulfill constant battery charging requirements, paralleled converter is used to compensate the current variation by taking power from the grid or diesel generator. The hardware is built and the control algorithm is implemented in a DSC. The experiment indicates that the control system can keep the charging current constant during steady state operation as well as large input solar power variations.

ACKNOWLEDGMENT

This work is sponsored by National Natural Science Foundation of China (grant number 51207135), Jiangsu Natural Science Foundation (grant number BK2012266), YZU-Yangzhou City Joint Fund (grant number 2012038-10).

REFERENCES

[1] Q. Li and P. Wolfs, "A review of the single phase photovoltaic module integrated converter topologies with three different dc link configurations," *IEEE Trans. Power Electron.*, vol. 23, no. 3, pp. 1320–1333, Mar. 2008.

Fig. 7.   load sharing test with programmed current (I_{L1}) reference (5s/div)

[2] B. J. Pierquet and D. J. Perreault, "A single-phase photovoltaic inverter topology with a series-connected energy buffer," *IEEE Trans. Power Electron.*, vol. 28, no. 10, pp. 4603–4611, Oct. 2013.

[3] Y. Zhou, L. Liu, and H. Li, "A high-performance photovoltaic module-integrated converter (mic) based on cascaded quasi-z-source inverters (qzsi) using egan fets," *IEEE Trans. Power Electron.*, vol. 28, no. 6, pp. 2727–2738, Jun. 2013.

[4] L. Chen, Amirahmadi, Q. Zhang, N. Kutkut, and I. Batarseh, "Design and implementation of three-phase two-stage grid-connected module integrated converter," *IEEE Trans. Power Electron.*, vol. PP, no. 99.

[5] A. I. Bratcu, I. Munteanu, S. Bacha, D. Picault, and B. Raison, "Cascaded dc-dc converter photovoltaic systems power optimization issues," *IEEE Trans. Ind. Electron.*, vol. 58, no. 2, pp. 403–411, Feb. 2011.

[6] B. Liu, S. Duan, and T. Cai, "Photovoltaic dc-building-module-based bipv system-concept and design considerations," *IEEE Trans. Power Electron.*, vol. 26, no. 5, pp. 1418–1429, May 2011.

[7] Z. Liang, R. Guo, J. Li, and A. Q. Huang, "A high-efficiency pv module-integrated dc-dc converter for pv energy harvest in freedm systems," *IEEE Trans. Power Electron.*, vol. 26, no. 3, pp. 897–909, Mar. 2011.

[8] T. Esram and P. L. Chapman, "Comparison of photovoltaic array maximum power point tracking techniques," *IEEE Trans. Energy Convers.*, vol. 22, no. 2, pp. 439–449, Jun. 2007.

[9] W. Jiang and B. Fahimi, "Maximum Solar Power Transfer in Multiport Power Electronic Interface," 25th *Annual IEEE Applied Power Electronics Conference and Exposition*, pp. 68–73, Feb. 2010.

[10] S. M. Lukic, J. Cao, R. C. Bansal, F. Rodriguez, and A. Emadi, "Energy storage systems for automotive applications," *IEEE Trans. Ind. Electron.*, vol. 55, no. 6, pp. 2258–2267, Jun. 2008.

[11] W. Jiang and B. Fahimi, "Multiport power electronic interface: Concept, modeling and desig," *IEEE Trans. Power Electron.*, vol. 26, no. 7, pp. 1890–1900, Jul. 2011.

[12] ——, "Active current sharing and source management in fuel cell-battery hybrid power system," *IEEE Trans. Ind. Electron.*, vol. 57, no. 2, pp. 752–761, Feb. 2010.

[13] N. Femia, G. Petrone, G. Spagnuolo, and M. Vitelli, "A technique for improving p&o mppt performances of double stage grid-connected photovoltaic systems," *IEEE Trans. Ind. Electron.*, vol. 56, no. 11, p. 4473C4482, Nov. 2009.

[14] A. Emadi, "Modeling of power electronic loads in ac distribution systems using the generalized state-space averaging method," *IEEE Trans. Ind. Electron.*, vol. 51, no. 5, pp. 992–1000, Oct. 2004.

[15] E. I. Ortiz-Rivera and F. Z. Peng, "Analytical model for a photovoltaic module using the electrical characteristics provided by the manufacturer data sheet," 36th *IEEE Power Electronics Specialists Conference*, pp. 2087–2091, Jun. 2005.

978-1-4799-2706-7/14 $31.00 © 2014 IEEE

LVRT Control Strategy of CSC-DPMSG-WGS under Unbalanced Grid Faults

Meiqin Mao,Yong Ding,SHiting Weng
Research Center for Photovoltaic System Engineering
Hefei University of Technology
Hefei 230009, China
mmqmail@163.com

Liuchen Chang
Department of Electrical and Computer Engineering
University of New Brunswick
Fredericton, New Brunswick, Canada E3B 5A3
lchang@unb.ca

Abstract—A control strategy for direct-drive PMSG wind generation system based on three phase current source converter (CSC-DPMSG-WGS) is presented in this paper. By the proposed method, once unbalanced grid voltage fault happens, the generator-side converter limits the power which flows to DC-link rapidly, meanwhile the grid-side converter controls the positive and negative sequence currents respectively with a dc current closed loop to eliminate the power ripples of twice times fundamental frequency caused by unbalanced grid fault to maintain dc current stable. By coordinated control of generator-and grid-side converters, the stability of the CSC-DPMSG-WGS may be improved so that low voltage ride-through capability for CSC-DPMSG-WGS may be realized. Simulation results are provided to verify the proposed control strategy.

Keywords—current source converter(CSC), coordinated power control, positive and negative sequence decomposition, unbalanced grid fault,low voltage ride-through(LVRT)

I. INTRODUCTION

With the high penetration of wind generation system (WGS) into the grid, it is inevitable to develop grid-friendly power electronic interfaces for WGS. Low voltage ride-through (LVRT) requirements have been given more attention by academia as well as industry in the recent decades for that it could keep the wind turbines grid-tied operation during the period of grid voltage dips and supply reactive power to the grid if necessary[1-2]. Up to now, many countries have proposed variable LVRT requirements. Predictably, LVRT capability will be the essential function of the power electronic controllers for WGS.

Direct-drive permanent magnet synchronous generator (DPMSG) wind generation systems are put into application widely at present. And now LVRT technology research on DPMSG wind systems has been mainly focused on the wind generation system based on voltage source converter (VSC-WGS). The various LVRT strategies proposed in current literatures can be divided into two categories. One is to introduce auxiliary hardware circuit such as a crowbar circuit or energy storage system to consume or store the excess energy at the DC-link[1-4]. The other one is to coordinate the

power in the generator-side and grid-side converters to achieve LVRT [5-11]. The former strategies may bring about several problems such as system installation, heat dissipation design, increasing system cost and control system design difficulty. Then some other methods are proposed. For example, linear feedback control of power or the voltage is adopted for balanced grid faults [5][6]. And for the unbalanced faults, reference [7] uses generator output power information in grid-side converter control to stable the DC-link voltage. In addition, balanced voltage method using single-phase voltage delay 60° theory[8], dual current-loop control strategy by introducing the notch filter to the conventional phase-locked loop control[9][10]and reformed instantaneous symmetrical components technology [11]are proposed respectively to achieve LVRT in VSC-WGS.

But all the methods above are proposed for the direct-drive VSC-WGS, thus they could not be applied in DPMSG wind generation system based on current source converter (CSC-DPMSG-WGS) directly. While current source converter (CSC) topologies are especially suitable in large-scale Direct-drive WGS since they possess favorable features, such as simple structure, grid friendly waveforms, controllable power factor, and reliable grid short-circuit protection. CSC-DPMSG-WGS is rarely studied in current literatures, most of which focus on converter topologies, modulation methods, control strategies in CSC-DPMSG-WGSs and issues on the grid integration[12].Reference [13] proposes a control strategy which can improve the dynamic performance of DC-link while reference [14] proposes a novel active/reactive control method for CSC-DPMSG-WGS. Reference [15] provides a unified DC-link current control for LVRT of CSC-DPMSG-WGS by coordinating the two-side converters power with monitoring the grid voltage in real time. However, this method can't be applied to the situation when unbalanced faults occur in the grid.

Power ripples of twice times fundamental frequency will emerge at DC-link if unbalanced faults occur because of the existence of negative sequence currents, leading to fluctuation of dc current and affecting the

978-1-4799-2706-7/14 $31.00 © 2014 IEEE

converters control. Moreover, since the power between two sides converter is not balanced any more, excess energy makes the dc current rise rapidly, which will bring about the damage of DC-link inductance as well as the power devices or even the destruction of entire WGS. Therefore, in order to achieve the LVRT requirement for WGS, it is essential to stable the DC-link current in the system. This paper develops a novel control strategy for a CSC-DPMSG-WGS to realize the LVRT in both balanced and unbalanced grid faults by coordinated power control of the generator-and grid-side converters without external hardware circuits or energy storage system. Based on the proposed strategy, once unbalanced grid voltage fault happens, the generator-side converter limits the power which flows to DC-link rapidly, meanwhile the grid-side converter controls the positive and negative sequence currents respectively with a dc current closed loop to eliminate the power ripples of twice times fundamental frequency caused by unbalanced grid fault to maintain dc current stable and achieve the LVRT control object for WGS. Simulation results are provided to verify the proposed control strategy.

II. MECHANISM OF DC-LINK CURRENT FLUCTUATION

The structure of direct-drive PMSG wind generation system based on three phase current source back to back converters studied in this paper is shown in Fig.1.

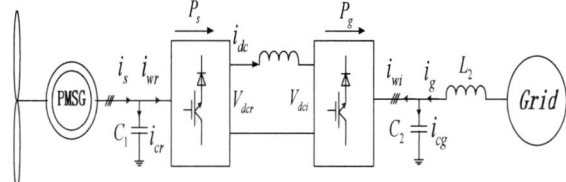

Fig.1. The structure of direct-drive PMSG wind generation system based on three phase current source back to back converter

In Fig.1, i_s is the generator stator current, while i_{wr} and i_{wi} are the generator-side and grid-side converter current respectively; i_{cr} and i_{cg} are the generator-side capacitor current and grid-side capacitor current respectively; i_g is the grid current and i_{dc} is the DC-link current. The power flowing through the DC-link in Fig.1 is:

$$P_s = i_{dc} L \frac{di_{dc}}{dt} = P_S - P_g \qquad (1)$$

Where P_s and P_g are the power flows to DC-link from generator-side and the power flows to grid-side from DC-link respectively, L is the value of DC-link inductance.

When the CSC-DPMSG-WGS operates normally, P_s equals to P_g, hence, the power between generator-and grid side is balanced .While unbalanced grid voltage fault occurs, P_s will not change immediately because of the decoupling effect of the DC-link on grid-side

converter and generator side. But the output power of grid-side converter will drop instantaneously. The unbalanced voltages and currents of the grid can be decomposed into their positive and negative sequence components respectively and are expressed in d-q synchronous frame as follows:

$$\vec{U} = \vec{U}_{dq}^P e^{j\omega t} + \vec{U}_{dq}^N e^{-j\omega t}, \vec{I} = \vec{I}_{dq}^P e^{j\omega t} + \vec{I}_{dq}^N e^{-j\omega t} \qquad (2)$$

$\vec{U}_{dq}^P, \vec{U}_{dq}^N$ and $\vec{I}_{dq}^P, \vec{I}_{dq}^P$ are positive and negative sequence complex vectors of voltage and current respectively.

Then, the active and reactive power of grid-side converter will change to [11]:

$$P_g = P_0 + P_{c2} \cos(2\omega t) + P_{s2} \sin(2\omega t) \qquad (3)$$

$$Q_g = Q_0 + Q_{c2} \cos(2\omega t) + Q_{s2} \sin(2\omega t) \qquad (4)$$

Where,

$$\begin{cases} P_0 = \frac{3}{2}\left(U_d^P I_d^P + U_q^P I_q^P + U_d^N I_d^N + U_q^N I_q^N\right) \\ P_{c2} = \frac{3}{2}\left(U_d^P I_d^N + U_q^P I_q^N + U_d^N I_d^P + U_q^N I_q^P\right) \\ P_{s2} = \frac{3}{2}\left(U_q^N I_d^P - U_d^N I_q^P - U_q^P I_d^N + U_d^P I_q^N\right) \end{cases} \begin{cases} Q_0 = \frac{3}{2}\left(U_q^P I_d^P - U_d^P I_q^P + U_q^N I_d^N - U_d^N I_q^N\right) \\ Q_{c2} = \frac{3}{2}\left(U_q^P I_d^N - U_d^P I_q^N + U_q^N I_d^P - U_d^N I_q^P\right) \\ Q_{s2} = \frac{3}{2}\left(U_d^P I_d^N + U_q^P I_q^N - U_d^N I_d^P + U_q^N I_q^P\right) \end{cases} \qquad (5)$$

$U_d^P, U_q^P, U_d^N, U_q^P$; $I_d^P, I_q^P, I_d^N, I_q^N$ are positive and negative sequence d-q components of the grid voltage and the grid current, respectively. Substitute P_g in (1) with equation (3), then

$$P_L = P_S - P_g = (P_S - P_0) - [P_{c2}\cos(2\omega t) + P_{s2}\sin(2\omega t)] \qquad (6)$$

It is shown in (6) that power ripples of twice times fundamental frequency emerges at DC-link for the reason that an unbalanced fault occurs. If the power ripples are eliminated and the generate-side power P_g can be controlled in keeping with the grid-side power P_s by a proper control method, DC-link current will be stable according to the equation (6). The proposed control strategy in this paper is just based on such a control mechanism.

III. PROPOSED CONTROL STRATEGY

According to the previous analysis, in order to achieve LVRT control of CSC-DPMSG-WGS under unbalanced grid faults, a novel LVRT control strategy applied to CSC-DPMSG-WGS is presented in this paper. The proposed strategy enables to eliminate the power ripples of twice times fundamental frequency by controlling the positive and negative sequence currents respectively, meanwhile, the both-side power could be controlled coordinately and dynamically based on different depth of grid voltage. By this way, the LVDT for CSC-DPMSG-WGS would be achieved.

A. Power control of generator-side converter

According to the characteristics of the wind turbine, the wind generator may be controlled to produce a maximum output power P_{opt} when the wind speed is lower than rated wind speed and P_{opt} can be expressed as:

$$P_{opt} = k\omega_t^3 - \Delta P \qquad (7)$$

Where ω_t represents the mechanical angular velocity of the wind turbine, ΔP is the sum of mechanical loss P_0, stator copper loss P_{cus} and stator iron loss P_{fes}.

When the grid is normal, the generator power command is set as $P^* = P_{opt}$. The generator stator voltage u_s and current i_s can be sampled and the generator power P_s can be calculated by $P_s = u_{sd}i_{sd} + u_{sq}i_{sq}$ in d-q synchronous frame. The calculated generator power is used as the grid-side converter power command. Once unbalanced fault happens, the grid-side power will decrease, then the real-time gird-side active power P_0 is calculated based on the formula (5) and the decreased rate of output power in the generator K_p can be obtained by:

$$K_P = P_0 / P_s \qquad (8)$$

then the generator-side power P^* is set to:

$$P^* = K_P P_{opt} \qquad (9)$$

the balance between the generator-side power and grid-side power can be achieved again and the overcurrent at the DC-link can be avoided.

B. Grid-side converter control strategy

The control goal of grid-side converter is to stabilize the DC current as well as to control the power factor. A dual PI control strategy including a dc current closed loop is proposed in this paper to control the positive and negative sequence currents respectively to eliminate the power ripples of twice times fundamental frequency effectively.

Based on formula (5), the relationship between power command and current command can be arranged in matrix form as:

$$\begin{pmatrix} P_0^* \\ Q_0^* \\ P_{s2}^* \\ P_{c2}^* \end{pmatrix} = \frac{3}{2} \begin{pmatrix} U_d^P & U_q^P & U_d^N & U_q^N \\ U_q^P & -U_d^P & U_q^N & -U_d^N \\ U_q^N & -U_d^N & -U_q^P & U_d^P \\ U_d^P & U_q^N & U_d^P & U_q^P \end{pmatrix} \begin{pmatrix} I_d^{P*} \\ I_q^{P*} \\ I_d^{N*} \\ I_q^{N*} \end{pmatrix} \qquad (10)$$

We set $Q_0^* = P_{s2}^* = P_{c2}^* = 0$, then, the power ripples of twice times fundamental frequency will be suppressed and unit power factor will be achieved. The command value of current fed to the grid would be obtained as:

$$\begin{pmatrix} I_d^{P*} \\ I_q^{P*} \\ I_d^{N*} \\ I_q^{N*} \end{pmatrix} = \begin{pmatrix} U_d^P & U_q^P & U_d^N & U_q^N \\ U_q^P & -U_d^P & U_q^N & -U_d^N \\ U_q^N & -U_d^N & -U_q^P & U_d^P \\ U_q^P & U_q^N & U_d^P & U_q^P \end{pmatrix}^{-1} \begin{pmatrix} \frac{2}{3}P_0 \\ 0 \\ 0 \\ 0 \end{pmatrix} = \frac{2P_0}{3D} \begin{pmatrix} U_d^P \\ U_q^P \\ -U_d^N \\ -U_q^N \end{pmatrix} \qquad (11)$$

Where $D = \left(U_m^P\right)^2 - \left(U_m^N\right)^2$. U_m^P and U_m^N are the amplitude of the positive and negative sequence voltages respectively.

The proposed LVRT control strategy based on(11) is shown in Fig.2.In which, PI controllers can be utilized to control the currents since they are dc values in the positive and negative sequence coordinates. A DC-link current closed loop is introduced and its PI output is added to the grid current positive sequence component command I_{dg}^{P*}, this may adjust the current rapidly under the voltage dips and shorten the system response time.In this approach, by the coordinated control of generator-side and grid-side converters merely, the proposed control strategy can achieve the LVRT requirement well under the unbalanced grid faults.

IV. RESULTS AND DISCUSSION

To verify the correctness and effectiveness of the proposed control strategy of LVRT for the CSC-DPMSG-WGS, this section establishes a 10KW CSC-DPMSG-WGS model in Matlab/Simulink. Performance of DPMSG-WGS with the proposed algorithm under balanced and unbalanced grid faults are investigated by simulation in three cases as follow according to the parameters listed in Table I followed.

TABLE I
SYSTEM SIMULATION PARAMETER

Symbol	Meaning	Value	Symbol	Meaning	Value
U_N	RATED VOLTAGE	220V	C_2	GIRD-SIDE FILTER CAPACITOR	$28\mu F$
P	PAIR OF ROLES	8			
L_d	D-AXIS SYNCHRONOUS INDUCTANCES	$10mH$	L_2	GRID-SIDE FILTER INDUCTANCES	$5mH$
L_q	Q-AXIS SYNCHRONOUS INDUCTANCES	$10mH$	R_f	DAMPING RESISTOR	0.1Ω
ψ_f	GENERATOR ROTOR MAGNETIC FLUX	$2.15Wb$	P_g	RATED GENERATOR-SIDE POWER	6 KW
L_{dc}	DC-LINK INDUCTANCE	$60mH$	I_{dc}^*	GIVEN VALUE OF DC-LINK	25 A
C_1	GENERATOR-SIDE FILTER CAPACITOR	$20\mu F$	v_{wind}	WIND SPEED	10M/S

The 2014 International Power Electronics Conference

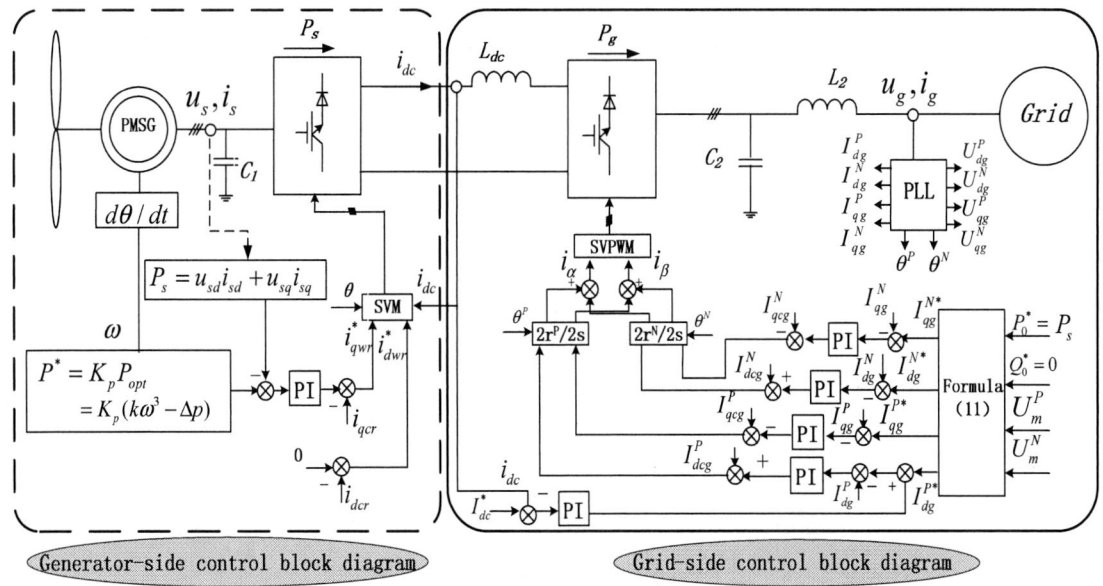

Fig.2. System control block diagram

A. Case 1:Grid voltage balanced fault

In this case, three-phase voltage of the grid drops from 100% to 20% symmetrically at the time 0.15s and recovers from the time 0.3s gradually. Simulation results are presented in Fig.3-5.

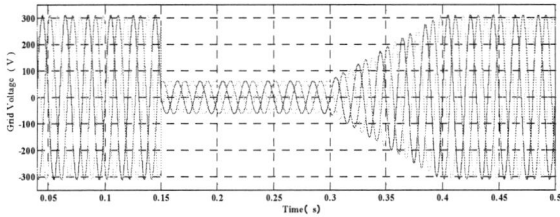

Fig.3. Grid voltage waveform

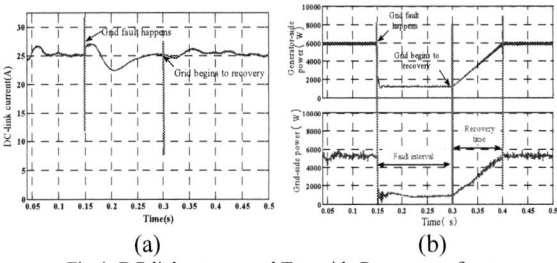

(a)   (b)

Fig.4. DC-link current and Two-side Power waveforms

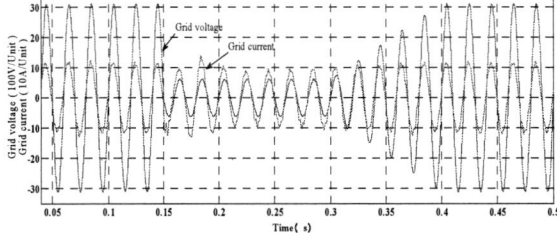

Fig.5. Grid voltage and current waveforms

It can be found in Fig.4(a) that DC-link current reaches a steady value before the fault happens and it will fluctuate during the period of fault. Due to the regulation

of the DC-link current closed-loop, the fluctuation ratio is limited within 9.4% and the current may recover to the given value 25A in a short time. The generator-side and grid-side power are both stable no matter whether the grid is normal or faulty, as shown in Fig.4(b). During the fault time, the power is smaller than the initial power given value. This is caused by the deeply balanced dip which leads to a decrease of grid-side output power, according to the proposed control strategy, when the grid-side power decreases, the output power of the generator-side should be limited to keep the balance of the whole system. Fig.4(b) also shows that the decreased times of power is about 8 which is consistent with the voltage drop depth. The simulation results confirm the feasibility of the strategy for the situations where the grid voltage balanced dips happen.

B. Case 2:Grid voltage unbalanced fault

Case 2.1:Single-phase-to-ground fault

A-phase voltage dips to zero at the time 0.15s because of phase-to-ground fault in this case, and it recovers gradually from the time 0.3s. B and C-phase voltages keep normal. The corresponding simulation results including grid voltages, DC-link current, generator-and grid-side power are shown in Fig.6 to Fig.8.

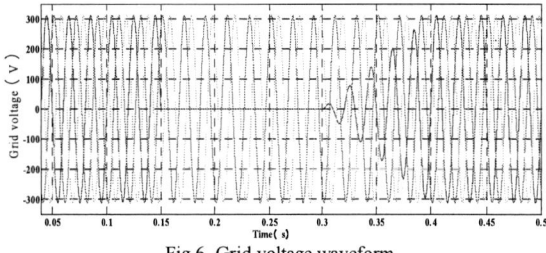

Fig.6. Grid voltage waveform

978-1-4799-2706-7/14 $31.00 © 2014 IEEE   1479

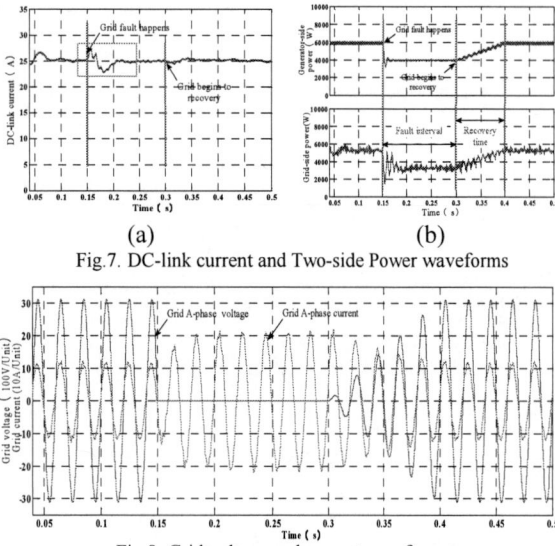

(a)   (b)

Fig.7. DC-link current and Two-side Power waveforms

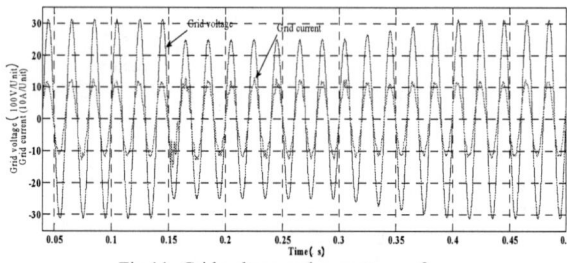

Fig.11. Grid voltage and current waveforms

Fig.8. Grid voltage and current waveforms

Fig.6(a) shows that DC-link current fluctuates slightly at the fault happening time, but it can recover to the given value 25A quickly(almost after 0.05s).Generator-and grid side power waveforms are shown in Fig.6(b), it can be found that two-side power is balanced and stable before single-to-ground fault happens and after that two-side power reduces almost simultaneously. Grid-side power reduction is caused by the grid fault, meanwhile, the generator-side power reduction dues to the proposed control strategy for balancing the power. When the grid recovers to normal, the generator-side power returns to the initial power given value again. Simulation results indict that the control strategy is suitable for the situation that single-to-phase fault happens.

Case 2.2:Three-phase voltage unbalanced dip

The simulation case sets A, B, C-phase voltage drops to 80%, 90% and 20% from 100% respectively at time 0.15s .The fault lasts for 0.15s, and then, the voltage recovers gradually. The corresponding simulation results are shown in Fig.9-11.

Fig.9. Grid voltage waveform

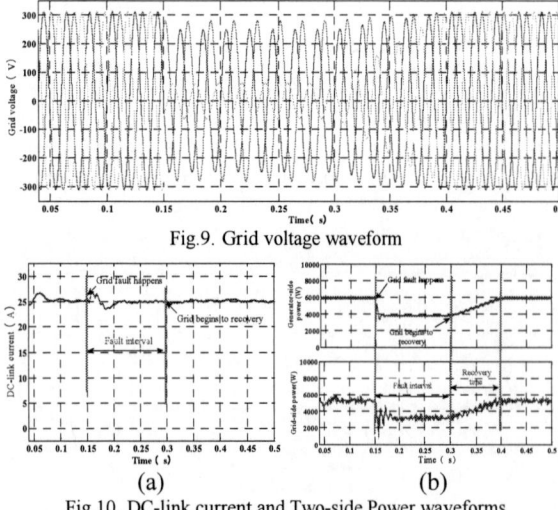

(a)   (b)

Fig.10. DC-link current and Two-side Power waveforms

Fig. 10(a) indicates that when the grid fault happens at time 0.15s, the DC-link current has a large fluctuation, the fluctuation ratio is controlled in 7.2% and it will become stable after 0.05s; when the grid voltage gradually recovers after 0.15s, the DC-link current fluctuates slightly again and the fluctuation ratio is close to zero; Fig.10(b) shows that before the voltage dips, generator-side and grid-side power is almost balanced and stable. Once grid fault occurs at 0.15s, the grid side power begins to decrease, then control strategy makes the grid-side output power decrease rapidly in order to keep the system stable. So during 0.15s to 0.3s the power must be smaller than the given value. However, both of them are still stable and no fluctuation appears;When the grid recovers to normal, the generator-side power returns to the initial power given value again. Grid voltage and grid current waveforms are shown in Fig.11,the simulation results demonstrate that CSC-DPMSG-WGS can achieve LVRT.

V. CONCLUSIONS

Based on the analysis of DC-link current and power fluctuation of CSC-DPMSG-WGS under unbalanced grid faults, a LVRT control strategy without adding external circuits or energy storage systems is put forward in this paper. When the grid fault happens, power of the generator-side converter is limited rapidly and grid-side converter is controlled to stabilize the DC-link current. Particularly, if unbalanced grid fault occurs, the proposed control method can eliminate power ripples of twice times fundamental frequency by adopting a dual PI control method with a dc current closed loop proposed in this paper in detail. A 10kw CSC-DPMSG-WGS model in Matlab/Simulink is presented, and the simulation results under different grid faults including balanced and unbalanced conditions verify the correctness and effectiveness of the proposed control strategy.

ACKNOWLEDGMENT

The authors would like to acknowledge the support partially from Grant International Cooperation Project of MOST (2007DFA71340) and NSFC project (51077033).

REFERENCES

[1] ZHANG Xing, ZHANG Long-yun, YANG Shu-ying, YU Yong, CAO Ren-xian,"Low voltage ride-through technologies in wind turbine generation," Proceedings of The CSU-EPSA,vol.20,no.4, pp.1-8,2008.

[2] J. M. Carrasco, L. G. Franquelo, et al ," Power electronic systems for the grid integration of renewable energy sources: A survey,"

IEEE Trans. Ind. Electron, vol. 53, no. 4, pp. 1002–1016, Jun.2006.

[3] Hu Shuju,LI Jianlin,XU Honghua. "Analysis on the Low-voltage-ride-through Capability of Direct-drive Permanent Magnetic Generator Wind Turbines," Automation of Electric Power Systems, vol.31,no.17, pp.73-77, 2007.

[4] HOU Shi-ying, FANG Yong, ZENG Jian-xing. "Application of supercapacitors to improve wind power system's low Voltage ride through capability," Electric Machines and Control, vol.14, no.5,pp.26-31,2010.

[5] Ki-Hong Kim, Yoon-Cheul Jeung, et al, "LVRT Scheme of PMSG Wind Power Systems Based on Feedback Linearization," IEEE Trans. Power Electronics, vol. 27, no. 5, pp. 2376–2384, May. 2012.

[6] Alan Mullane, Gordon Lightbody, and R. Yacamini,"Wind-Turbine Fault Ride-Through Enhancement," IEEE Trans. Power Systems, vol. 20, no. 4, pp. 1926–1936, Nov. 2005.

[7] Jun Yao,Yong Liao,Kai ZHuan, "A Low Voltage Ride-through Control Strategy of Permanent Magnet Direct-driven Wind Turbine Under Grid Faults," Automation of Electric Power Systems,vol.33,no.12,pp.91-96,2009.

[8] ZHANG Xiao-ying,CHENG Zhi-zhuang et al, "Voltage stability control for direct driven wind turbine with permanent magnet synchronous generator in grid asymmetric faults," Power System Protection and Control,2013.

[9] BAI Shu-hong, "Current Control for the Grid-connected Inverter of Directly-driven Wind Turbine with PMSG During the Asymmetrical Faults," Power Electronics,vol.45,no.8,pp.57-59,2011.

[10] CHENG Hang,et.al, "Research on dual current-loop control strategy for grid-connected inverter of directly-driven wind turbine with permanent magnet synchronous generator under unbalanced network voltage conditions," Power System Protection and Control,vol.40,no.7,pp.66-72,2012.

[11] LI Ge,SONG Xin-fu,CHANG Xi-qiang, "Improved control theory for low voltage ride-through of permanent magnet synchronous generator," Power System Protection and Control,vol.39,no.12,pp.74-83,2011.

[12] Dai Jing-ya, "Current source converters for megawatt wind energy conversion systems,"Ryerson University,2010.

[13] Dai J, Xu D, Wu B, et al, "Dynamic performance analysis and improvements of a current source converter based PMSM wind energy system," Power Electronics Specialists Conference, 2008. PESC 2008. IEEE. IEEE, 2008: 99-105.

[14] Lang Y, Wu B, Zargari N,"A novel reactive power control scheme for CSC based PMSG wind energy system," Industry Applications Society Annual Meeting, 2008. IAS'08. IEEE. IEEE, 2008: 1-6.

[15] Dai J, Xu D, Wu B, et al. "Unified DC-link current control for low-voltage ride-through in current-source-converter-based wind energy conversion systems," Power Electronics, IEEE Transactions on, vol.26,no.1,pp.288-297,2011.

A New Current Control Droop Strategy for VSI-Based Islanded Microgrids

B. Shoeiby,   R. Davoodnezhad,   D. G. Holmes,   B. P. McGrath

School of Electrical and Computer Engineering
RMIT University
Melbourne, Australia
babak.shoeiby@ieee.org   reza.davoodnezhad@ieee.org   grahame.holmes@rmit.edu.au   brendan.mcgrath@rmit.edu.au

Abstract— Microgrids are distributed generation structures that link groups of Distributed Generators (DGs) to either connect to the utility grid, or to operate as a stand-alone (islanded) system. However, the injected power from each DG must be carefully controlled to ensure reliable power distribution, balanced modular operation, and regulated voltages across the microgrid. It is commonly accepted that DG units cannot operate in current-control mode when islanded, because there is no stiff grid voltage available. Consequently they are usually operated in voltage control mode, with individual droop controllers regulating the inverter voltage magnitude and frequency to share active and reactive power. This paper presents a new approach to operate islanded microgrid DG units in current control mode, by combining linear current regulators with predictive voltage control. The result achieves a rapid load response while still maintaining well regulated voltage levels. A current magnitude droop controller is also proposed to maintain equal power sharing between the current controlled DG units.

Keywords— *Current Control, Distributed Generator, Microgrid, Power Sharing.*

I. INTRODUCTION

Renewable distributed energy resources (DERs) have been receiving considerable interest from electrical energy suppliers, due to energy market trends and regulations [1][2]. A microgrid [3] is a small-scale low-voltage distribution network which consists of distributed generators (DGs), loads, and energy storage systems. Microgrids have the ability to operate either in grid-connected or in stand-alone (island) mode. In grid-connected mode, the microgrid exchanges power with the grid, and can compensate the local grid voltage levels. However, in island mode there is no exchange of power between the microgrid and the grid, and the DGs are required to balance their generated active and reactive power with the microgrid loads and consumer demands. In addition, the DGs must maintain their operating frequency and voltages within an acceptable range defined by the grid regulations [4][5]. In order to meet these requirements, it is necessary to control the power flows in the microgrid.

Various control schemes have been proposed to balance the power between the DGs and the loads of a microgrid [6]. Most grid-connected applications control the current of the DG voltage-source inverter (VSI) [7]-[10], since the presence of a stiff grid voltage allows

straightforward control of power by directly commanding the VSI output current. However, in an islanded microgrid, since there is no dominant grid to maintain the voltage level at the inverter connection point, the voltage levels are determined by the output current of the DG units and the microgrid impedances. Consequently these voltages vary as the load impedances change, unless the DG current references are adjusted to maintain the voltage levels. This has led to the belief that the DG units cannot maintain voltage levels if they operate in current control mode [11][12], and so DGs in island mode are usually operated in voltage control mode so that their output voltages can be directly regulated [6]. Typically, a droop controller is also incorporated into each DG unit to adjust their voltage magnitude and frequency to share active and reactive power [13]-[17].

Even though voltage controlled VSIs facilitate the control of microgrid voltages, they have two major drawbacks. Firstly, due to the nature of voltage control, the VSI output current cannot be directly commanded. This makes the VSI output current somewhat unpredictable, since it depends on external grid elements such as distribution line impedances, other DG sources and the connected load. Alternatively, voltage regulators can be implemented as dual loop controllers, with an outer voltage control loop and an inner current control loop. However, the outer voltage control loop must then be tuned to be slower than the inner current loop for stability [12]. This sacrifices the high bandwidth capability of the inner current loop, resulting in a slower dynamic response.

This paper presents a new control scheme based on current controlled VSIs to solve these shortcomings. The strategy employs a predictive voltage controller that continuously adjusts the reference amplitude of the linear current controller to maintain a required DG output voltage level. This allows regulated voltage levels to be achieved across the microgrid without needing a slow voltage feedback loop. The faster dynamics of current regulation also achieve a faster load change response compared to voltage controlled VSIs, and allow the DER injected power to be more rapidly and more accurately matched to the required load.

Finally, a new current based droop controller is proposed to still achieve power sharing between multiple DG units, even when they are operated in current regulation mode.

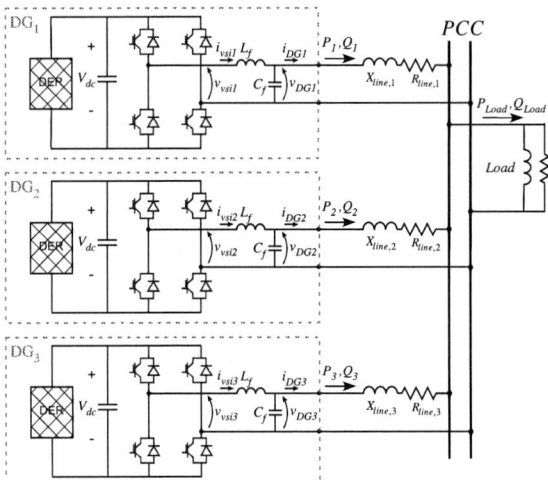

Fig. 1: Islanded microgrid with three VSI-based DG units converting the DER DC voltage to AC voltage.

I. VOLTAGE CONTROLLED MICROGRID

This section reviews the voltage regulators and voltage based droop control that are conventionally used to share power between the DG units of an islanded microgrid. Essentially each DG unit acts as a voltage source and its voltage magnitude and frequency is adjusted according to a droop function in order to achieve the desired power sharing functionality.

A. A Single Voltage Regulated VSI

Fig. 1 shows an islanded microgrid with three VSI-based DG units that convert the DER DC source voltages to AC voltages. Each DER is connected to the AC network through a VSI and an output LC filter which passes through the fundamental AC component of the VSI switched voltage, while reducing the output current ripple to a desired level. In such DG units, the output AC voltage amplitude and frequency is directly controlled using a linear voltage regulator arranged either as a single-loop or a dual-loop feedback controller [18].

1) Single-loop voltage regulator

In the single-loop voltage regulator shown in Fig. 2, the measured DG output voltage is compared against the reference voltage command with a desired magnitude and frequency. This tracking error is then passed on to a linear compensator (Proportional (P), Proportional+Integral (PI), or Proportional+Resonant (PR)) to generate a modulation command (M_a) for the PWM modulator which turns the VSI switches off and on. Generally a PR compensator is used for better tracking accuracy at the fundamental frequency.

2) Dual-loop voltage regulator

To improve the stability of a single-loop voltage controller, a dual-loop regulator based on an inner current control loop and an outer voltage control loop is often used, as shown in Fig. 3. The voltage error is passed through a linear compensator to generate the reference current for comparison against the measured VSI current. The current error is then passed through another linear compensator block to create the modulation command for the PWM modulator. Usually the outer loop uses a PI or PR compensator, and the inner loop uses a P or PI compensator.

B. Microgrid with Voltage Droop Controlled DGs

In an islanded microgrid, it is preferred that each DG unit generates the same amount of active and reactive power. This prevents overloading of the DGs and results in an even distribution of the transient power flows among the DG units. It also reduces any circulating current flowing between DG units, since this current can exceed the DG ratings if not properly controlled.

Fig. 4 shows the block diagram of a typical droop controller to achieve power sharing. At each computational cycle, the output active and reactive powers are calculated and then passed onto the droop function block to generate a reference voltage magnitude and frequency. This voltage reference is then commanded to the DG voltage regulator, usually implemented as a dual-loop voltage controller.

For a low voltage microgrid with mainly resistive distribution lines, the Active Power-Voltage and Reactive Power-Frequency droop controller relationships can be expressed as [16]:

$$V_{ref} = V_{nom} - m \times (P - P_{nom}) \tag{1}$$

$$f_{ref} = f_{nom} - n \times (Q - Q_{nom}) \tag{2}$$

The final commanded voltage reference for each DG unit is then an AC quantity defined by

$$v_{ref}(t) = V_{ref}\sin(\theta_{ref}(t)) \tag{3}$$

where V_{ref} and f_{ref} are determined by (1) and (2), based on the amount of power being supplied by the DG source, and the subscript *nom* denotes the nominal value of each parameter. Finally, $\theta_{ref}(t)$ is then calculated using:

$$\theta_{ref}(t) = 2\pi \int f_{ref}(t)dt \tag{4}$$

Droop control with dual-loop voltage regulators are commonly used in islanded microgrid applications. However, they lack the ability to command a direct

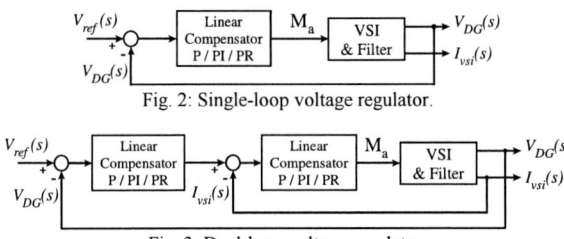

Fig. 2: Single-loop voltage regulator.

Fig. 3: Dual-loop voltage regulator.

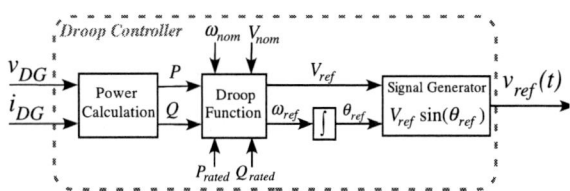

Fig. 4: Voltage droop controller block diagram; $v_{ref}(t)$ is the output of the droop controller, which is commanded to the DG voltage controller.

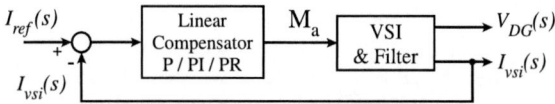

Fig. 5: Conventional linear current controller.

current reference. This makes the control of the DG output power less precise, particularly when the DER available power is limited and variable. In addition, for stability purposes, the outer voltage loop needs to be constrained to act much slower than the inner current loop, which limits the controller's dynamic performance.

II. Proposed Current Controlled Microgrid

In the islanded microgrid shown in Fig. 1, the DG units are in charge of matching their generated power with the microgrid load demands. They must also ensure well regulated voltage levels and frequency across the microgrid. This section develops a new current control structure to regulate both the DG output current and voltage for a stand-alone microgrid with VSI-based DG units. After this development, a current magnitude droop controller is proposed to manage power sharing of multiple DG units in island mode.

A. Current-Controller with Predictive Voltage Regulator

To benefit from the advantages of the conventional current regulation structure shown in Fig. 5, all DGs operate in current control mode. In this case, individual instantaneous current references are used as each DG unit's setpoint.

The block diagram of the proposed controller implemented on DG1 connected to the PCC load is shown in Fig. 6. A predictive voltage regulator calculates the DG output voltage magnitude $v_{DG,pk}$ to adjust the PR regulator current reference magnitude accordingly.

Firstly, the phase angle of the DG unit output voltage, $\theta_{v_{DG}}(t)$, is determined using the second-order generalised integrator (SOGI) phase-locked loop (PLL) [19] shown in Fig. 7. The first part of the PLL measures the DG output voltage $v_{DG}(t)$, and processes it through the SOGI system to create the fundamental in-phase and 90-degree out-of-phase components $v_{alpha}(t)$ and $v_{beta}(t)$ of this voltage. These components are then transformed into d-q synchronous frame components

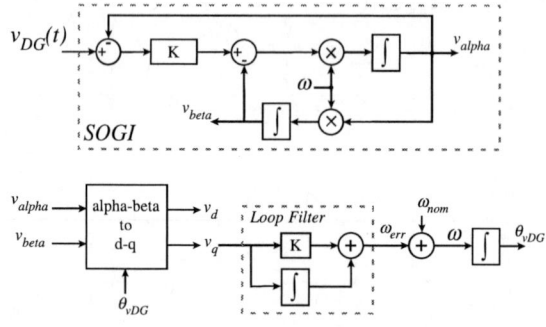

Fig. 7: Structure of the SOGI PLL.

using the matrix transformation of

$$\begin{bmatrix} v_d(t) \\ v_q(t) \end{bmatrix} = \begin{bmatrix} \cos\theta_{v_{DG}} & \sin\theta_{v_{DG}} \\ -\sin\theta_{v_{DG}} & \cos\theta_{v_{DG}} \end{bmatrix} \begin{bmatrix} v_\alpha(t) \\ v_\beta(t) \end{bmatrix} \quad (6)$$

The quadrature component $v_q(t)$ is then passed through a loop filter to calculate a frequency error $\omega_{err}(t)$, and thus determine the frequency $\omega(t)$ of the DG output voltage. This frequency $\omega(t)$ is then integrated to derive the phase angle of the DG output voltage $\theta_{v_{DG}}(t)$, which drives the d-q synchronous frame transformation.

Secondly, the peak magnitude of the DG output voltage is calculated using the instantaneous value of the measured DG output voltage $v_{DG}(t)$ and the calculated phase angle $\theta_{v_{DG}}(t)$, as:

$$v_{DG,pk} = \frac{v_{DG}(t)}{\sin\left(\theta_{v_{DG}}(t)\right)} \quad (5)$$

The ratio of the actual DG output voltage to the desired nominal voltage level is then calculated as follows:

$$V_{ratio}(t) = v_{DG,pk}(t)/V_{nom,pk} \quad (6)$$

This ratio is then used on a sampled basis (denoted by the time index n) as a multiplier to increase/decrease the commanded peak reference current for the inverter, viz.:

$$i_{ref,v,pk}[n] = \frac{i_{ref,v,pk}[n-1]}{V_{ratio}[n]} \quad (7)$$

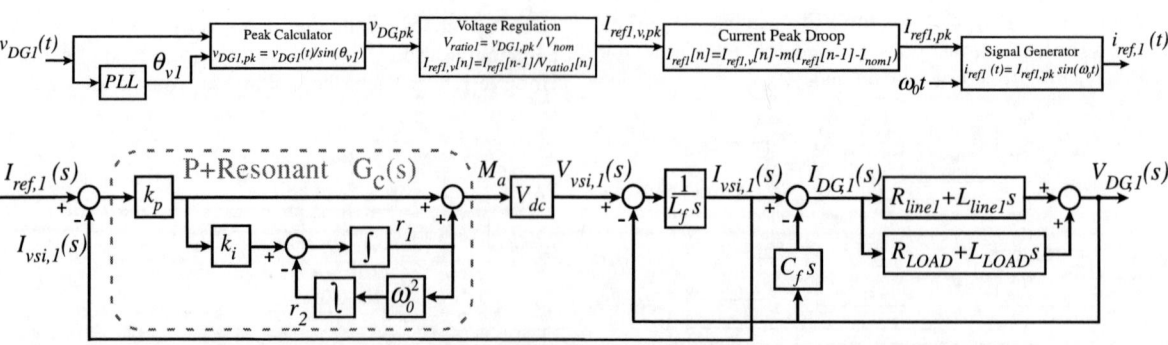

Fig. 6: Proposed current control structure implemented on DG unit #1 connected to the PCC load.

978-1-4799-2706-7/14 $31.00 © 2014 IEEE   1484

where $i_{ref,v,pk}$ is the peak current reference commanded by the voltage regulator for each DG. Both (6) and (7) are independent of the microgrid impedances such as the PCC load and line impedances.

Finally, the reference current and the measured current are compared and passed onto a PR linear current regulator to control the VSI output current. The PR regulator in the continuous domain is defined as [7]:

$$G_c(s) = k_p \left[1 + \frac{k_i s}{s^2 + \omega_0{}^2} \right] \qquad (8)$$

where $\omega_0 = 2\pi f_0$ and f_0 is the nominal microgrid fundamental frequency. The gains of the PR current controller, k_p and k_i, are calculated as described in [7] to achieve the best possible transient response, using

$$k_p = \frac{\omega_{c(max)} L}{V_{DC}} \qquad k_i = \omega_{c(max)}/10 \qquad (9)$$

where $\omega_{c(max)} = \dfrac{\pi/2 - \phi_m}{T_d}$, ϕ_m is the desired controller phase margin and T_d is the controller sampling and transport delay.

B. Proposed Power Sharing Algorithm

To achieve equal power sharing between current controlled DG units, a modified droop control is implemented for each DG unit. This droop controller is derived from conventional droop control in (1), and is modified to suit the proposed current controlled DG units by drooping the commanded current magnitude rather than voltage magnitude as the output apparent power increases.

For each DG unit, the amount of apparent power generated is calculated as

$$S_{DG1} = v_{DG1,pk} \times i_{DG1,pk} \qquad (9)$$

As the predictive voltage regulator always maintains the DG output voltage level at the nominal value the DG apparent power is directly dependent on the DG output current magnitude:

$$S_{DG1} \propto i_{DG1,pk} \qquad (10)$$

Equation (10) shows that in order to achieve equal power sharing, the current magnitudes of all DG units must be equal. To achieve this, a droop term can be added to each DG unit's current reference magnitude as follows:

$$i_{ref1,pk} = i_{ref1,v,pk} - m \times (i_{DG1,pk} - I_{nom1,pk}) \qquad (11)$$

where $i_{ref1,v,pk}$ and $i_{ref1,pk}$ are the current reference peaks calculated by the voltage regulator and the droop function respectively, $i_{DG1,pk}$ is the DG output current peak, $I_{nom1,pk}$ is the nominal DG output current peak, and m is the droop gain. In order to avoid calculation of the DG output current peak value, the commanded current peak can be used since PR regulators result in

close to zero steady state error, ($i_{ref1,pk} \approx i_{DG1,pk}$). Hence (11) can be rewritten as the following current magnitude droop function:

$$i_{ref1,pk} = i_{ref1,v,pk} - m \times (i_{ref1,pk,prev} - I_{nom1,pk}) \qquad (12)$$

where $i_{ref1,pk,prev}$ denotes the reference current magnitude at the previous computational cycle.

Substituting (7) into (12) yields the final reference current magnitude for each DG unit's current regulator:

$$
\begin{aligned}
i_{ref1,pk}[n] = \; & i_{ref1,pk}[n-1] / V_{ratio}[n] \\
& - m \times (i_{ref1,pk}[n-1] - I_{nom1,pk})
\end{aligned} \qquad (13)
$$

where n is the discrete time update point in the DSP controller.

Equation (13) identifies how the reference current peak magnitude for the current regulator is calculated by scaling the commanded peak value by the voltage error, and drooping to share power between the DG units. The droop equation pushes each DG's current reference magnitude to its nominal current magnitude value, while the voltage regulator holds it around a value to maintain nominal DG output voltage. As a result, all DG units push their current magnitude towards their nominal value which results in equal per unit current for all DGs. The droop gain m is tuned to achieve the desired accuracy of power sharing. Adjusting the current amplitude in this way takes into account the effect of apparent power required by resistive and inductive loads, to share both active and reactive power at the same time, as will be presented in the next section. Additionally, this droop algorithm does not use the values of active or reactive power, which avoids the delays and dynamics of power calculation methods.

III. SIMULATION RESULTS

To verify the performance of the proposed control strategy, two different cases are separately investigated using PSIM simulation software, with system parameters as given in Table I. First, the operation of a stand-alone current-controlled DG unit with the proposed control strategy is presented to demonstrate the current control and voltage regulation performance. Next, three DG units forming an islanded microgrid are investigated to verify the proposed power sharing algorithm.

TABLE I. SIMULATION STUDY PARAMETERS

	Single DG Unit	Islanded Microgrid
V_{nom} / $I_{nom\,1,2,3}$	320 Vpk / 10 Apk	320 Vpk / 10 Apk
$V_{dc\,1,2,3}$	400 Vdc	400 Vdc
$f_{0,nom}$ / $f_{sw\,1,2,3}$	50 Hz / 10 kHz	50 Hz / 10 kHz
$L_{f\,1,2,3}$ / $C_{f\,1,2,3}$	4 mH / 5 uF	4 mH / 5 uF
$k_{p\,1,2,3}$ / $k_{i\,1,2,3}$	0.093 / 931	0.093 / 931
$Z_{line\,1}$	0.6 + j0.08 Ω	0.6 + j0.08 Ω
$Z_{line\,2}$	--	1.2 + j0.16 Ω
$Z_{line\,3}$	--	0.9 + j0.12 Ω
$m_{1,2,3}$	0.02	0 / 0.01 / 0.02

A. Stand-alone Single DG Unit

Here only one DG unit of Fig. 1 (DG1) is connected to the PCC through the line impedance Z_{line1} and supplies power to the variable PCC load. The current control algorithm of Fig. 6 governs the DG unit operation. Fig. 8 illustrates the performance of the stand-alone DG unit before and after a PCC load step change of (a) 500W to 1000W, and (b) 500W to 550W + 500VAr at t = 0.065s. The DG unit initially supplies power to a resistive PCC load of 102.4 Ω while regulating the voltage at the nominal value of 320Vpk. In case 8(a), at the instant after the PCC load changes from 102.4 Ω to 51.2 Ω the PCC voltage reduces in magnitude. The controller detects this change in voltage at the DG output and increases the output current magnitude to make up for the voltage drop. This results in fast regulation of the PCC voltage. In case 8(b) a mainly inductive load is added to the PCC, and again the voltage regulator detects the change in the DG voltage magnitude and adjusts the current reference magnitude accordingly. Note that adding an inductive load resulted in an instantaneous phase jump at the PCC voltage and DG output voltage, and the performance of the voltage controller will be limited by the dynamics of the PLL as it catches up with the new phase angle. After this transient, the controller rapidly regulates the load voltage back to the nominal value.

B. Islanded Microgrid

To verify the operation of the proposed power sharing mechanism, three DG units are connected to the PCC through distribution lines (high R/X ratio and different lengths of 2 km, 4 km, and 3 km) similar to the islanded microgrid arrangement in Fig. 1. All DG units use the same controller structure of Fig. 6. The current reference phase angles of the VSIs are synchronised at the zero crossings using a low speed communication that sends out a signal to all DGs at the start of the phase cycle.

Fig. 9 to 11 show the simulation results of the microgrid performance for three different values of current magnitude droop gain (*m = 0.00*, *m = 0.01*, *m = 0.02*). In each case, the microgrid initially has a PCC load of 1kW. Other loads of 1kW and 340W+870VAr are then added to the PCC at t = 0.205s and t = 1.005s, respectively. During all operating conditions the PCC voltage is regulated at the nominal value. However, the power sharing performance is dependent on the chosen gain of the current magnitude droop function. When *m = 0*, the current magnitude droop is essentially deactivated and the microgrid has a poor power sharing performance, as shown in Fig. 9. As the value of *m* increases, the power sharing performance improves as can bee seen in Fig. 10, although the higher sharing gain does result in a slight voltage deviation at the DGs' output terminals. With m = 0.02, the system continues to achieve balanced active and reactive power sharing as shown in Fig. 11, while still maintaining a quite well regulated PCC voltage level as the loading conditions vary.

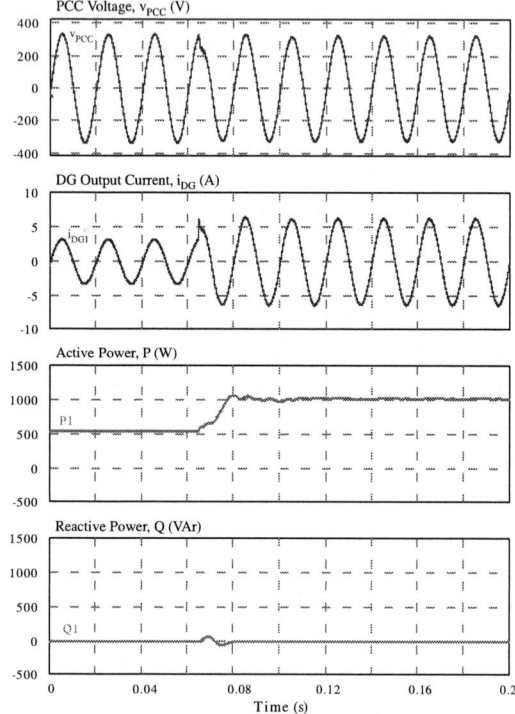

(a): Transient load change from 500W to 1000W.

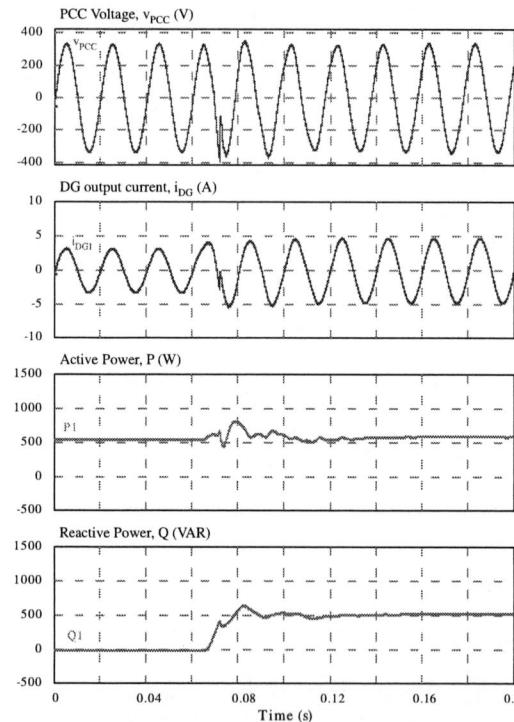

(b): Transient load change from 500W to 550W+500VAr.

Fig. 8: Simulated response of a stand-alone DG unit with load changes.

IV. EXPERIMENTAL RESULTS

The proposed controller has been implemented experimentally on an islanded microgrid with two DG units in order to verify its operation. The experimental setup parameters are given in Table II.

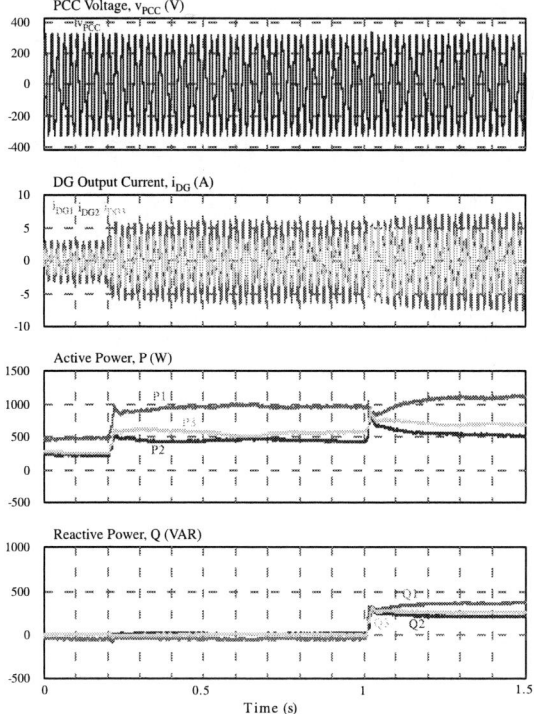

Fig. 9: Simulated power sharing response for an islanded microgrid with droop gain m = 0.00 (Deactivated droop).

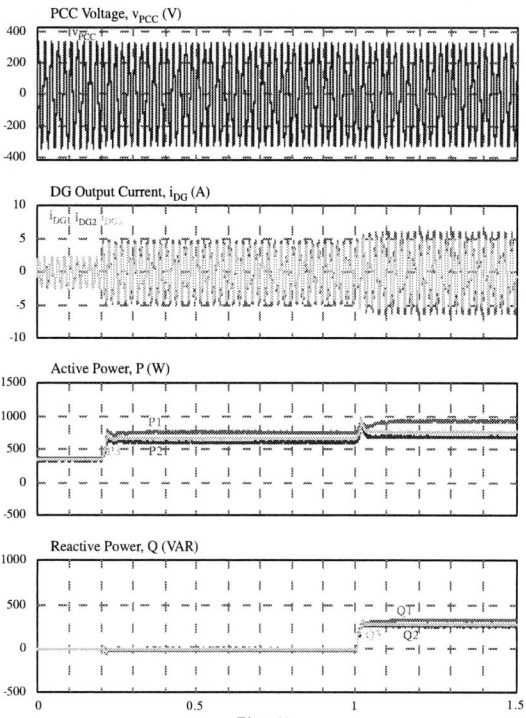

Fig. 10: Simulated power sharing response for an islanded microgrid with droop gain m = 0.01.

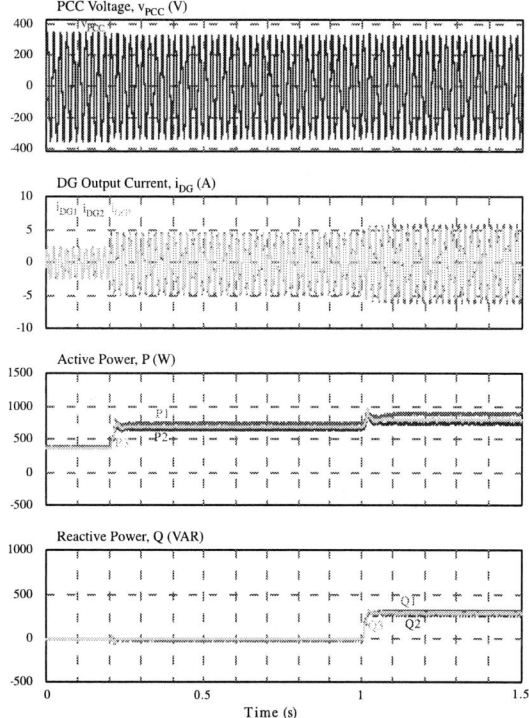

Fig. 11: Simulated power sharing response for an islanded microgrid with droop gain m = 0.02.

A. Stand-alone Single DG Unit

A single stand-alone DG unit is considered feeding a PCC load (same as Section III.A.). The DG unit consists of a H-Bridge VSI fed by a DC voltage source as the DER, and an LC filter on the output of the VSI. The proposed controller in Fig. 6 is implemented on the DG unit and two different PCC load step change scenarios are investigated. Fig. 12 shows the results for a transient load step change of (a) 890W to 1780W, and (b) 890W to 1110W+970VAr. It can be seen that after each transition, the controller adjusts the DG current in order to maintain the nominal voltage level and frequency.

B. Islanded Microgrid

Two DG units are connected to a common PCC load through resistive line impedances, forming an islanded microgrid. The phase angle of the two DG current references are synchronised using a low speed communication strategy. The following investigations demonstrate the performance of the microgrid (Note that these investigations have been carried out at lower voltage levels because of limitations of the available DC voltage sources for each DG unit).

TABLE II. EXPERIMENTAL SETUP PARAMETERS

	Single DG Unit	Islanded Microgrid
V_{nom} / $I_{nom\ 1,2}$	320 Vpk / 10 Apk	270 Vpk / 10 Apk
$V_{dc\ 1,2}$	370 Vdc	370 Vdc
$f_{0,nom}$ / $f_{sw\ 1,2}$	50 Hz / 10 kHz	50 Hz / 10 kHz
$L_{f\ 1,2}$ / $C_{f\ 1,2}$	4 mH / 5 uF	4 mH / 5 uF
$k_{p\ 1,2}$ / $k_{i\ 1,2}$	0.1 / 931	0.1 / 931
$Z_{line\ 1}$	0.5 Ω	0.5 Ω
$Z_{line\ 2}$	--	1 Ω
$m_{1,2}$	0.02	0 / 0.01 / 0.02

The 2014 International Power Electronics Conference

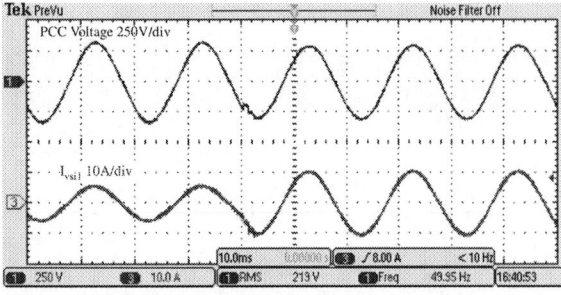

(a) Transient load change from 890W to 1780W.

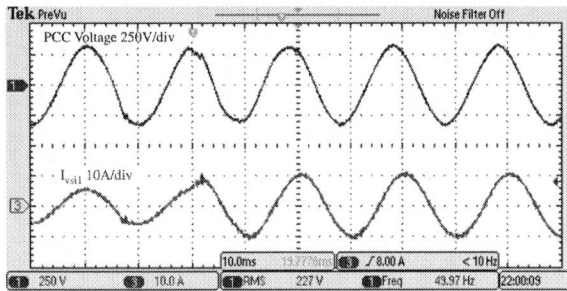

(b) Transient load change from 890W to 1110W+970VAr.
Fig. 12: Stand-alone operation of the DG controller.

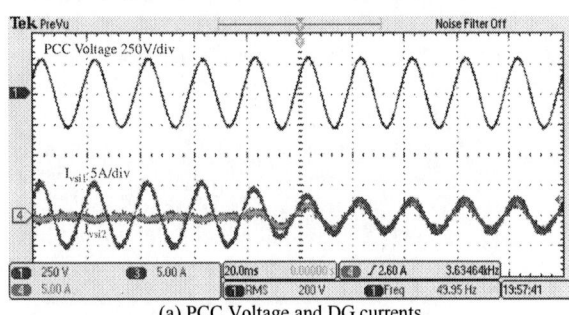

(a) PCC Voltage and DG currents.

(b) DG power traces.
Fig. 13: Activating current magnitude droop in the islanded microgrid.

1) Effect of Current Magnitude Droop

For the responses shown in Fig. 13, the two DG units initially feed the PCC load with their droop controllers (12) inactive. Since there is no control over power sharing, DG1 provides most of the power required by the PCC load, while DG2 provide a small amount of power. By activating the current magnitude droop control on the two DG units, the DG output currents become almost equal in magnitude which results in equal power sharing. Fig. 13 also shows that there is no disturbance in the PCC voltage at the time of activating the droop control.

2) Effect of Droop Gain on Power Sharing

This experiment investigates the effect of the droop

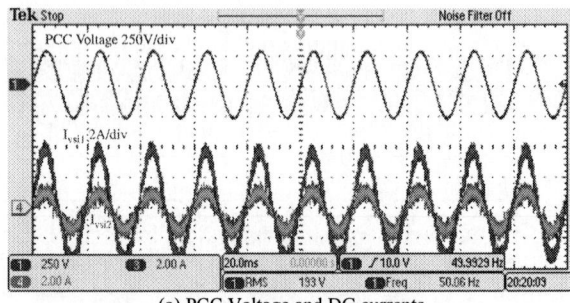

(a) PCC Voltage and DG currents.

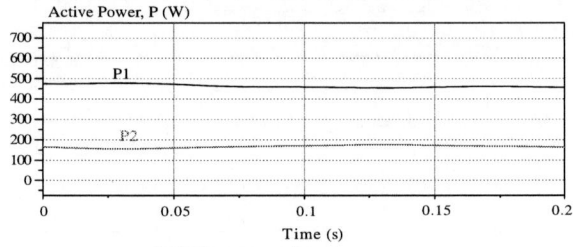

(b) DG source real power responses.
Fig. 14: Experimental power sharing response with m = 0.002.

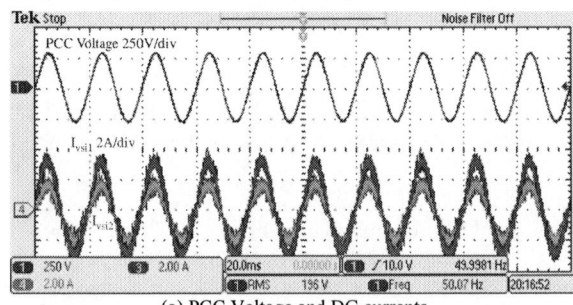

(a) PCC Voltage and DG currents.

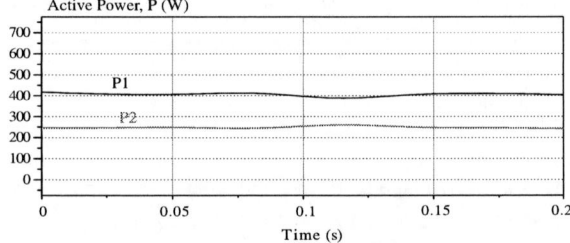

(b) DG source real power responses.
Fig. 15 : Experimental power sharing response with m = 0.01.

gain m on the accuracy of power sharing. Figs. 14, 15 and 16 show the impact on changing the droop gain from $m = 0.002$ to $m = 0.01$, and then $m = 0.02$ respectively. It can be seen that by increasing the droop gain, the VSI currents become closer in magnitude resulting in more accurate sharing of the real power. This increase in the droop gain however, results in a small deviation of the PCC voltage from the nominal value.

3) Transient Response

In this test a PCC load step change is performed to explore the dynamic response of the islanded microgrid. Fig. 17 shows the microgrid response to a transient load change from 1260W to 1900W. After the transition in load, the DG units increase their output current to match the PCC load demand while still sharing power. It can also be seen that the PCC voltage continues to be well maintained with minimal disturbance.

978-1-4799-2706-7/14 $31.00 © 2014 IEEE   1488

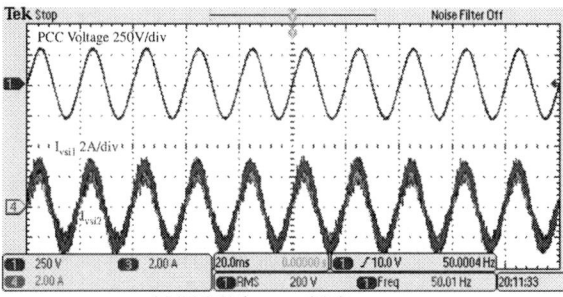

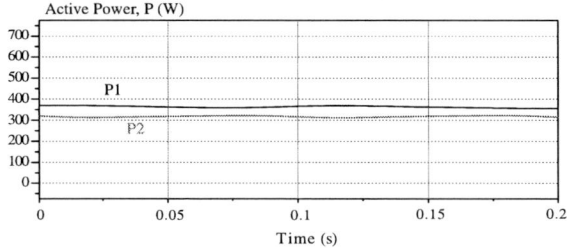

(a) PCC Voltage and DG currents.

(b) DG source real power responses.

Fig. 16 : Experimental power sharing response with m = 0.02.

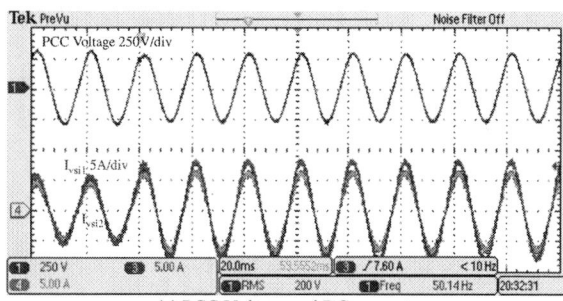

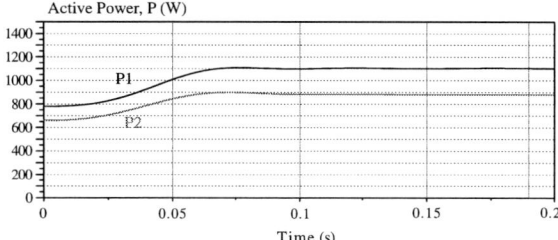

(a) PCC Voltage and DG currents.

(b) DG source real power responses.

Fig. 17: Islanded microgrid with a load change from 1260W to 1900W.

V. CONCLUSION

Current controlled VSIs offer advantages such as fast response and direct control of output current, compared to voltage controlled VSIs. This paper presented an islanded microgrid based on current controlled VSIs. A current control method is proposed to ensure direct control of the VSI output current as well as regulating the output load voltage at the nominal value. A proposed current magnitude droop controller is incorporated into the current controller to balance the output power and ensure equal power sharing between the DG units in the islanded microgrid. Simulation and experimental results are presented to verify the performance of the proposed control strategy.

REFERENCES

[1] T. Ackermann, G. Andersson, and L. Söder, "Distributed generation: a definition," Electric Power Systems Research, vol. 57, pp. 195-204, 2001.

[2] K. A. Nigim and L. Wei-Jen, "Micro Grid Integration Opportunities and Challenges," in Power Engineering Society General Meeting, 2007. IEEE, 2007, pp. 1-6.

[3] R. H. Lasseter and P. Paigi, "Microgrid: a conceptual solution," in Power Electronics Specialists Conference, 2004. PESC 04. 2004 IEEE 35th Annual, 2004, pp. 4285-4290 Vol.6.

[4] P. Piagi and R. H. Lasseter, "Autonomous control of microgrids," in Power Engineering Society General Meeting, 2006. IEEE, 2006, p. 8 pp.

[5] F. Katiraei, R. Iravani, N. Hatziargyriou, and A. Dimeas, "Microgrids management," Power and Energy Magazine, IEEE, vol. 6, pp. 54-65, 2008.

[6] Guerrero, J.M.; Lijun Hang; Uceda, J., "Control of Distributed Uninterruptible Power Supply Systems," Industrial Electronics, IEEE Transactions on , vol.55, no.8, pp.2845,2859, Aug. 2008.

[7] Holmes, D.G.; Lipo, T.A.; McGrath, B.P.; Kong, W.Y., "Optimized Design of Stationary Frame Three Phase AC Current Regulators," Power Electronics, IEEE Transactions on , vol.24, no.11, pp.2417,2426, Nov. 2009.

[8] C. Lascu, L. Asiminoaei, L. Boldea, and F. Blaabjerg, "Frequency response analysis of current controllers for selective harmonic compensation in active power filters," IEEE Trans. Ind. Electron., vol. 56, no. 2, pp. 337–347, Feb. 2009.

[9] R. I. Bojoi, L. R. Limongi, D. Roiu, and A. Tenconi, "Enhanced power quality control strategy for single-phase inverters in distributed generation systems," IEEE Trans. Power Electron., vol. 26, no. 3, pp. 798–806, Mar. 2011.

[10] A. Timbus, M. Liserre, R. Teodorescu, P. Rodriguez, and F. Blaabjerg, "Evaluation of current controllers for distributed power generation systems," IEEE Trans. Power Electron., vol. 24, no. 3, pp. 654–664, Mar. 2009.

[11] Erickson, M.J.; Jahns, T.M.; Lasseter, R.H., "Comparison of PV inverter controller configurations for CERTS microgrid applications," Energy Conversion Congress and Exposition (ECCE), 2011 IEEE , vol., no., pp.659,666, 17-22 Sept. 2011.

[12] J. M. Guerrero, J. C. Vasquez, J. Matas, L. G. de Vicuna, and M. Castilla, "Hierarchical control of droop-controlled ac and dc microgrids—A general approach toward standardization," IEEE Trans. Ind. Electron., vol. 58, no. 1, pp. 158–172, Jan. 2011.

[13] J. M. Guerrero, L. G. de Vicuna, J. Matas, M. Castilla, and J. Miret, "A wireless controller to enhance dynamic performance of parallel inverters in distributed generation systems," Power Electronics, IEEE Transactions on, vol. 19, pp. 1205-1213, 2004.

[14] M. C. Chandorkar, D. M. Divan, and R. Adapa, "Control of parallel connected inverters in standalone AC supply systems," Industry Applications, IEEE Transactions on, vol. 29, pp. 136-143, 1993.

[15] Vandoorn, T.L.; Meersman, B.; Degroote, L.; Renders, B.; Vandevelde, L., "A Control Strategy for Islanded Microgrids With DC-Link Voltage Control," Power Delivery, IEEE Transactions on , vol.26, no.2, pp.703,713, April 2011.

[16] Jinwei He; Yun Wei Li, "Analysis, Design, and Implementation of Virtual Impedance for Power Electronics Interfaced Distributed Generation," Industry Applications, IEEE Transactions on , vol.47, no.6, pp.2525,2538, Nov.-Dec. 2011

[17] Engler, A.; Soultanis, N., "Droop control in LV-grids," Future Power Systems, 2005 International Conference on , vol., no., pp.6 pp.,6, 18-18 Nov. 2005.

[18] Yun Wei Li, "Control and Resonance Damping of Voltage-Source and Current-Source Converters With LC Filters," Industrial Electronics, IEEE Transactions on , vol.56, no.5, pp.1511,1521, May 2009.

[19] Golestan, S.; Monfared, M.; Freijedo, F.D.; Guerrero, J.M., "Dynamics Assessment of Advanced Single-Phase PLL Structures," Industrial Electronics, IEEE Transactions on , vol.60, no.6, pp.2167,2177, June 2013.

Power Exchange Using PFC for Micro Grid

Tomoyasu Sakai, Takashi Takeda, Kazuto Yukita,
Yasuyuki Goto, Katsuhiro Ichiyanagi
Engineering department. Aichi Institute of Technology
Aichi Institute of Technology, AIT
1247 Yachigusa, Yakusa-cho, Toyota City, 470-0392, Japan

Hiroshi Morita
R&D department. Kinden Corporation
Kinden Corporation, Kinden
3-1-1 Saganakadai, Kidugawa City, 619-0223, Japan

Abstract— In the present paper, we examined the power exchange in small-scale systems among multiple power flow controllers. The introduction of renewable-energy power generation systems produces new energy business. For example, on-site power generations, and green electric power supply services, such as wind power generation by local companies and local governments, have been proposed in certain areas. Electric energy (electric power) is being traded as a general amount. Then, as electricity trading becomes more active, electric power exchange is expected to increase. We examined the power exchange in multiple small-scale systems using a power flow controller.

Keywords— *Micro-grid, Power flow controller, DC, Power- exchange, Inverter*

I. INTRODUCTION

Recently, policies such as global warming prevention and feed-in tariffs for renewable energy are being implemented. Power generation methods that use renewable energy have attracted significant attention. In Japan, the target power for photovoltaic systems to be introduced by 2020 is 28 million kW, and an increase of the photovoltaic system to the power system is expected. However, the output of photovoltaic systems will be significantly affected by climate change. Therefore, a stable power supply using such power electronics and power storage devices have been discussed.

On the other hand, the introduction of a renewable-energy power generation system, new energy business, for example, on-site power generation, and green electric power supply services, such as wind power generation by local companies and local governments, have been proposed in certain areas.

Electric energy (electric power) is being traded as a general amount. Then, as the electricity trading becomes more active, electric power exchange is expected to increase. As such, electric power current control technologies, such as the loop power flow controller (LPC), digital grid[1], the unified power flow controller (UPFC), and the HVDC light are investigated.

Then, a renewable-energy power generation system was introduced between systems, and the electric power flexibility control was examined. The primary goals of the present study:

1. Improved operation rate of regional renewable energy power generation systems.
2. Reduce capacity and increased use of accumulation of electricity equipment for electric power quality maintenance introduced into system.
3. Improved supply reliability of renewable energy power generation systems.
4. Advantage of scale for power generation equipment and electricity storage equipment.

Here, power exchange in an AC system or DC system is considered using the electric power flexibility method. There is an advantage in focusing on the DC system. The phase and the frequency aren't existed in DC system. Therefore it can be more easily synchronized than with AC. In addition, DC system is with a high affinity for the photovoltaic system and the storage battery system, and can be expected the energy saving. In the present paper, the power exchange among multiple small-scale systems was examined using a power flow controller (PFC). Specifically, a PFC system is first developed, and this system reports on the effect of power exchange among multiple systems into which the renewable energy power generation system is introduced.

II. POWER FLOW CONTROLLER

Figure 1 shows a block diagram of the PFC, which is composed of two DC/DC converters. The control method is divided at the master and slave, using two DC/DC converters, and performs bidirectional power exchange control. Figure 2 shows the control block of the master and slave with the PFC. Vin_ref is a DC voltage command value between DC/DC converters, and Vin is the DC voltage between DC/DC converters. Moreover, Vdc_ref is the DC system reference voltage of the interconnection system, and Vdc is the voltage of the DC systems of the interconnection systems. First, as shown in Fig. 2(a), the control input of the master derives the voltage difference Vin between DC/DC converters.

$$\Delta V_{in\_m} = V_{in\_ref\_m} - V_{in\_m} \qquad (1)$$

The deviation is then input, and PI control is performed.

978-1-4799-2706-7/14 $31.00 © 2014 IEEE

Next, using the output Iin_vin_ref_m, the difference in the DC current flowing through Idc_m between the DC/DC converters is taken, and the control input of the voltage m_chop_ref(master) is calculated.

$$diin\_m = I_{\text{in\_vin\_ref\_m}} - I_{\text{dc\_m}} \qquad (2)$$

Next, as shown in Fig. 2(b), the control input of the slave calculates the deviation ΔVdc_s of the voltage of the DC systems and the DC system reference voltage of the interconnection systems.

$$\Delta V_{dc\_s} = V_{dc\_ref\_s} - V_{dc\_s} \qquad (3)$$

The deviation ΔVdc_s is input, and PI control is performed. As in the master control, using the output Idc_vdc_ref_s, the difference between the current Idc of the DC system of the interconnection systems is taken, and the control input of the power s_chop_ref(slave) is calculated.

systems are interconnected in a T shape. Fig. 3(b) shows the case in which the three systems are interconnected in a T shape. Fig. 3(b) shows the case in which the three systems are interconnected in a delta shape. The master side and the slave side of the PFC and the control value in each system receive commands from devices that perform central control commands.

In addition, for each system, a microgrid is assumed that has an electric storage system and distributed power. Figure 4 shows the microgrid model for each system. The PFC is assumed to connect the DC system, as shown in Fig. 4.

The control unit monitors the output of the solar power generation and load demand, as well as the capacity of the battery. Then, control unit switches the slave and master in the PFC by State of system.

However, when power exchange with other systems is not possible, the outputs of the power generation facilities within each system are assumed to be stopped or suppressed.

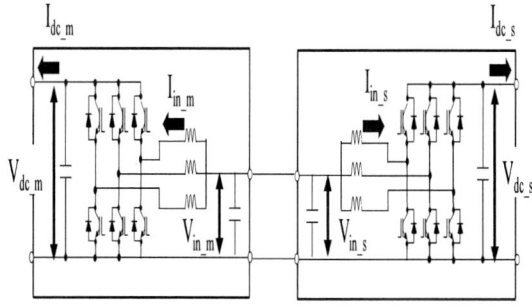

Fig. 1. Configuration of the PFC

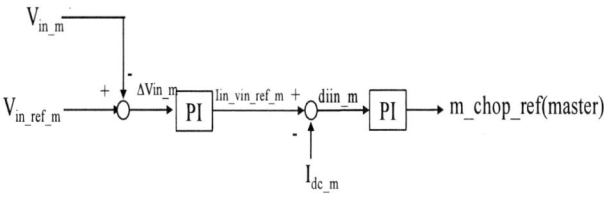

(a) Master side DC/DC converter control block

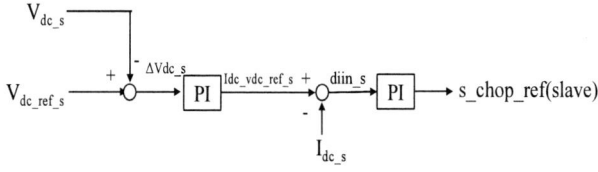

(b) Slave side DC/DC converter control block

Fig. 2. Control block

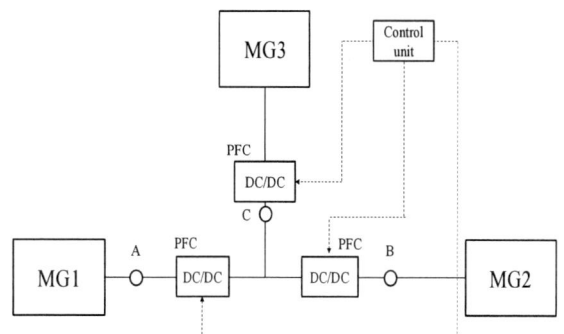

(a) T-shaped connection

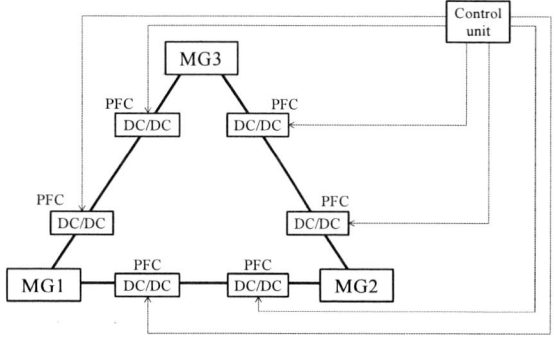

(b) Delta-shaped connection

Fig. 3. System model

III. MODEL SYSTEM

The effectiveness of the PFC used herein was examined using the system model shown in Fig. 3, which shows a model that is grid-connected to a DC system at three systems. Fig. 3(a) shows the case in which the three

IV. EXPERIMENTAL METHODOLOGY

A. Power exchange of the T-shaped connection

Figure 5 shows the time sequence for conducting the power exchange. Figure 6 shows the power exchange direction. As shown in Fig. 5, pattern a in the drawing is reduced the power from 3 kW to 1.5 kW after 60 seconds. Thereafter, the power is reduced to 0 kW to 120 seconds. Pattern b in the drawing is increased to 1.5 kW after 60 seconds. Thereafter, the power is increased from 1.5 kW to 3 kW after 120 seconds.

Fig. 6, A, B, and C indicate the measurement points. Each system is name specified as MG1, MG2, and MG3. The DC/DC converter control block is connected to MG3, which is controlled by master operation, and MG1 and MG2 are controlled by slave operation.

We examined the power exchange in multiple systems. The power exchange control in six patterns is shown below as power exchange that is assumed to the T-shaped model. To power exchange, combine with the time sequence shown in Fig. 5.

Then, the PFC performed the exchange of power according to the time sequence. Furthermore, in order to consider a different system capacity, the maximum DC voltage, Vref, was set to 340 V and 350 V in MG2 and MG1, respectively. The power exchange patterns are classified as follows:

(a) Pattern 1

MG1 performs the power exchange according to the time sequence shown in Fig. 5(a). In addition, MG3 performs the power exchange according to the time sequence shown in Fig. 5(b). Thus, MG2 performs a power exchange to MG1 and MG3.

(b) Pattern 2

MG2 performs the power exchange according to the time sequence shown in Fig. 5(a). In addition, MG3 performs the power exchange according to the time sequence shown in Fig. 5(b). Thus, MG2 and MG3 perform a power exchange to MG1.

(c) Pattern 3

MG1 performs the power exchange according to the time sequence shown in Fig. 5(a). In addition, MG2 performs the power exchange according to the time sequence shown in Fig. 5(b). Thus, MG3 and MG1 perform a power exchange to MG2.

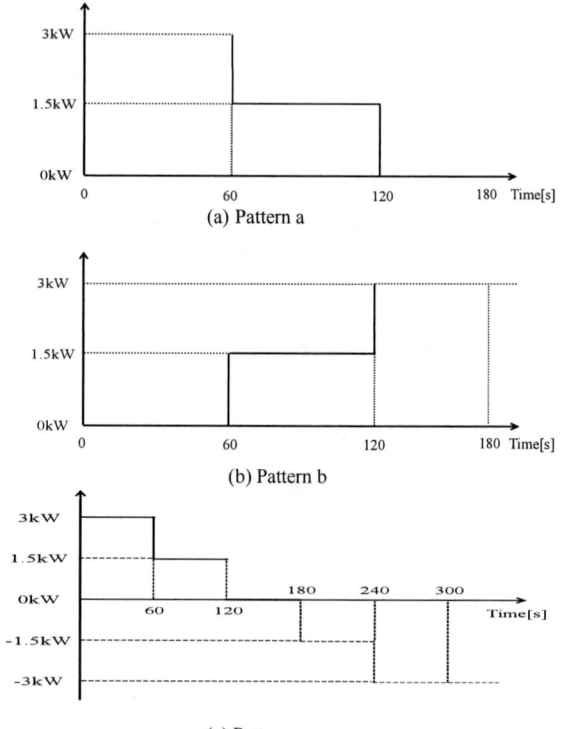

(a) Pattern a

(b) Pattern b

(c) Pattern c

Fig. 5. Time sequence

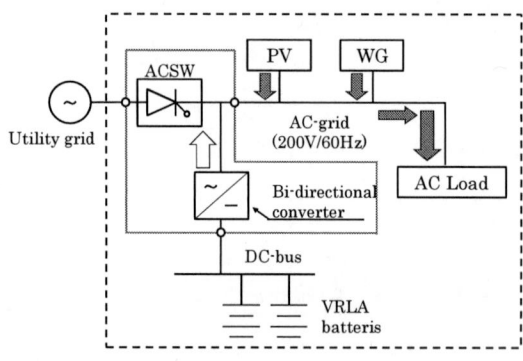

Fig. 4. Model of the microgrid

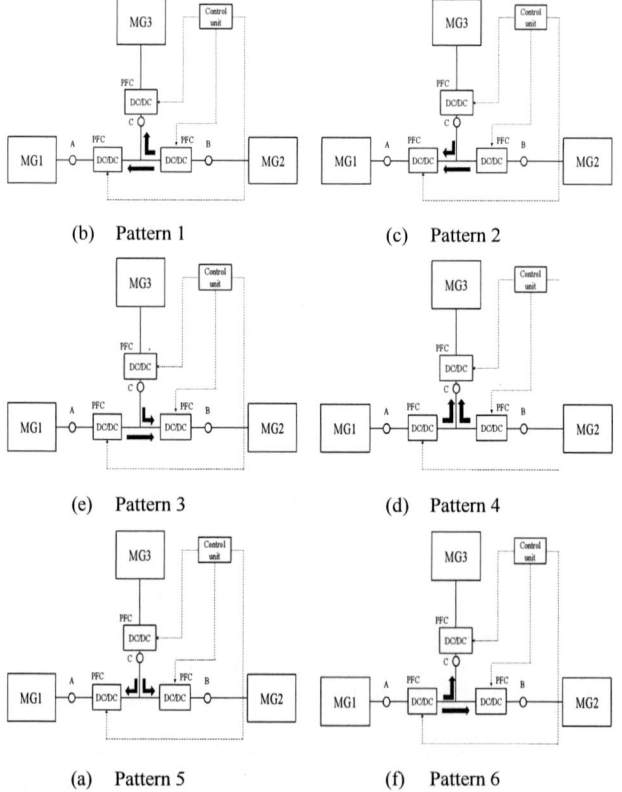

(b) Pattern 1   (c) Pattern 2

(e) Pattern 3   (d) Pattern 4

(a) Pattern 5   (f) Pattern 6

Fig. 6. Power exchange patterns

(d) Pattern 4

MG1 performs the power exchange according to the time sequence shown in Fig. 5(a). In addition, MG2 performs the power exchange according to the time sequence shown in Fig. 5(b). Thus, MG1 and MG2 perform a power exchange to MG3.

(e) Pattern 5

MG1 performs the power exchange according to the time sequence shown in Fig. 5(a). In addition, MG2 performs the power exchange according to the time sequence shown in Fig. 5(b). Thus, MG3 performs a power exchange to MG1 and MG2.

(f) Pattern 6

MG2 performs the power exchange according to the time sequence shown in Fig. 5(a). In addition, MG3 performs the power exchange according to the time sequence shown in Fig. 5(b). Thus, MG1 performs a power exchange to MG2 and MG3.

By performing the power exchange shown in these six patterns, we examined the power exchange using the PFC.

B. Power exchange of the delta-shaped connection

Figure 7 shows System model for delta-sharped. Each system is name specified as MG1, MG2, and MG3. If mutual two systems can exchange power, delta shaped is possible to assume construction. In the present study, we have mutual power exchange between the two systems of MG1 and MG2. The measurement points are indicated by A and B. Power exchange control is performed as indicated by pattern c in Fig. 5. If positive in Fig. 5 the power exchange is performed from MG2 to MG1. As shown in Fig. 5, the power is reduced from 3 kW to 1.5 kW after 60 seconds and is then reduced 1.5 kW to 0 kW after 120 seconds. The power is increased from 1.5 kW to 0 kW after 180 seconds. Then, 3 kW of power is supplied from 240 seconds to 300 seconds.

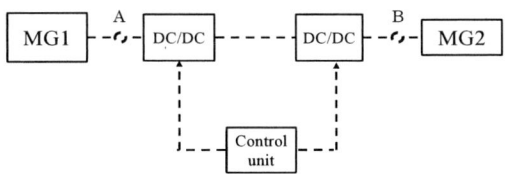

Fig. 7. System model for delta-sharped

V. CONCLUSION

A. Power exchange of the T-shaped connection

Figures 8 through 13 show the power characteristics and voltage characteristics. A positive value in the power characteristics indicates a power exchange. Since the power exchange in the DC/DC converter is always 3 kW, the DC voltage is maintained at 310 V. Moreover, since, in the case of PFC, the power exchange in the DC/DC converter is always 3 kW, the DC voltage is maintained at 310 V.

Figure 8 shows the voltage and power characteristics for pattern 1, in which 3 kW of power is provided from point B. Figure 9 shows the voltage and power characteristics for pattern 2, in which 3 kW of power is

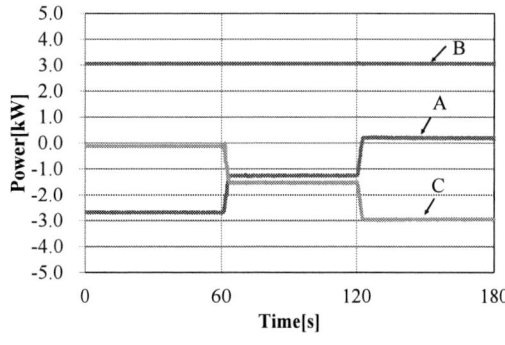

(a) Power characteristics

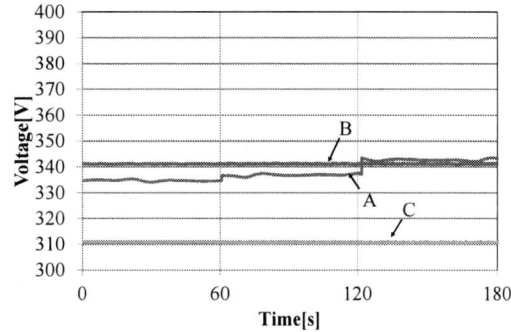

(b) Voltage characteristics
Fig. 8. Control characteristic for pattern 1

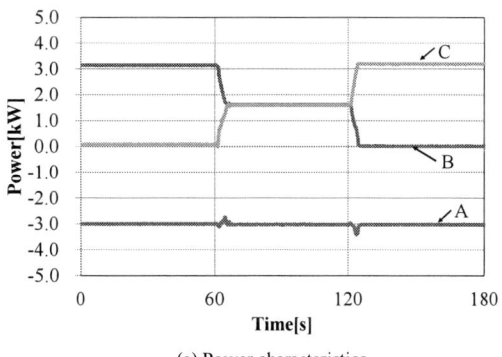

(a) Power characteristics

(b) Voltage characteristics
Fig. 9. Control characteristic for pattern 2

The 2014 International Power Electronics Conference

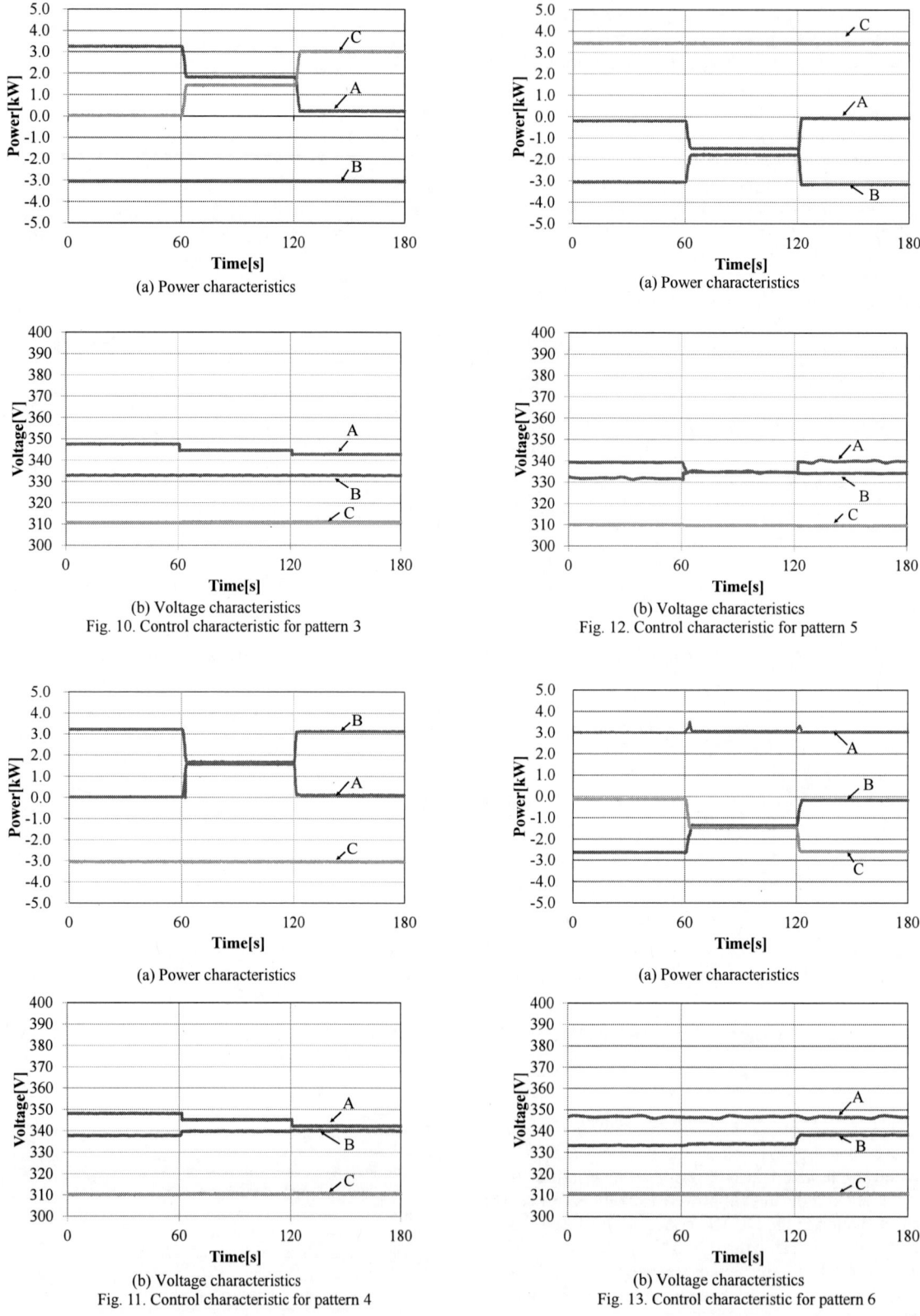

(a) Power characteristics

(b) Voltage characteristics
Fig. 10. Control characteristic for pattern 3

(a) Power characteristics

(b) Voltage characteristics
Fig. 11. Control characteristic for pattern 4

(a) Power characteristics

(b) Voltage characteristics
Fig. 12. Control characteristic for pattern 5

(a) Power characteristics

(b) Voltage characteristics
Fig. 13. Control characteristic for pattern 6

978-1-4799-2706-7/14 $31.00 © 2014 IEEE

supplied, enabling the variation in voltage at point A to be suppressed. PFC of point B is operating in a slave. If the voltage of point B is more than 340V, PFC was performed power exchange. Figure 10 shows the voltage and power characteristics for pattern 3. The voltage and power characteristics are stable, even after power exchange. Figure 11 shows the voltage and power characteristics for pattern 4. A power of 3 kW is supplied to C point. The power and voltage characteristics at points A and B vary until 60 seconds. However, power exchange is performed. Figure 12 shows the voltage and power characteristics for pattern 5. The voltage and power characteristics are stable, even after power exchange. Figure 13 shows the voltage and power characteristics for pattern 6. As shown in Fig. 13, since the response of the power exchange at point B is delayed, the power exchange at point A rises to maximum 500W for a short time.

B. Power exchange of the delta-shaped connection

Figure 14 shows the power and voltage characteristics for the power exchange between two microgrids. The power exchange when the value is positive at points A and B. As the power characteristic of this figure shows, power exchange occurs from MG2 to MG1 up until 120 seconds and from MG1 to MG2 from 180 to 300 seconds.

The voltage characteristic stabilizes because it controls the voltage constancy in the DC bus of the microgrid at point A. PFC of point B performed power exchange when DC system voltage of MG2 exceeds 340V. After 180 seconds, PFC of point B performed power exchange when DC system voltage of MG2 drops below 332V.

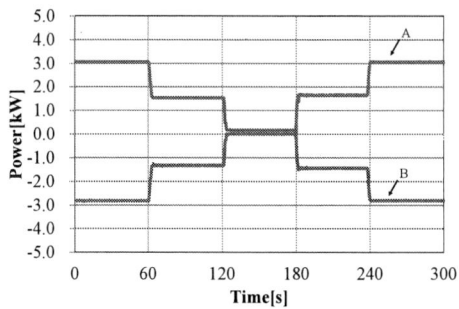

(a) Power characteristics

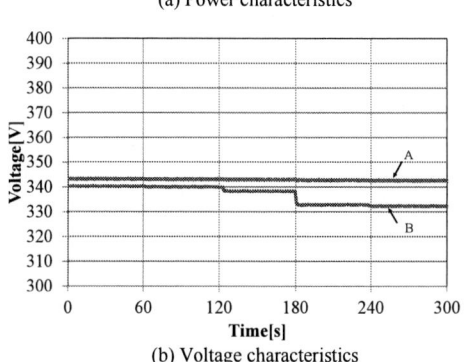

(b) Voltage characteristics
Fig. 14. Control characteristic

VI. SUMMARY

In the present study, we examined power exchange in small-scale systems among multiple systems using PFC. First, we developed the PFC equipment. Next, we examined the power exchange using PFC under the assumption that the distributed power is introduced at three systems. Exchange power control was performed for the case in which the three systems formed a T-shaped interconnection. Moreover, we examined the power exchange in the delta-shaped interconnection. Although the T-shaped connection exhibited a different response, power exchange in multiple small-scale systems was demonstrated. The delta-sharped connection indicated that successful power exchange was achieved between two systems. Therefore, delta shapes can perform the power exchange.

ACKNOWLEDGEMENTS

The present study was supported in part by Heisei 25 fiscal of Grants-in-Aid for Scientific Research (c)

REFERENCES

[1] Rikiya Abe, and Hisao Taoka: "Electric Network Innovation by Digital Grid-Digital Grid Concept-", IEEJ Transactions on Power and Energy, Vol.133, No.2 p.137-140 (2013)

[2] Tomohito Ushirokawa, Keiichi Hirose, KazutoYukita, Katsuhiro Ichiyanagi, Yosiaki Okui: "Control of electric power flow between two micro grids", IEICE Conference, Vol.111 No.400(EE2011 31−51) Page.41-45 (2012.01)

[3] JONES P, RICHARDSON B: "Transmission and distribution Finding the missing LINK ", IEE Power Engineering, Vol.18 No.6 Page.28-31(2004.12)

[4] Ikuo Kurihara:""Power System Technologies to Realize Smart Grid", IEEJ Transactions on Power and Energy,Vol.133 No.4 pp.298-301(2013)

[5] CHAUDHURI Nilanjan Ray, CHAUDHURI Balarko: "Adaptive Droop Control for Effective Power Sharing in Multi-Terminal DC (MTDC) Grids" IEEE Trans Power System, Vol.28 No.1 Page.21-29(2013.02)

[6] Kakigano Hiroaki, Takuya Hashimoto, Yohei Matsumura, Takashi Kurotani, Wataru Iwamoto, Yushi Miura, Toshifumi Ise, Toshinari Momose, Hideki Hayakawa: "Fundamental Characteristics of Laboratory Scale Model DC Microgrid to Exchange Electric Power from Distributed Generations installed in Residential Houses " IEEJ Transactions on Power and Energy, Vol.128, No.9, Page.1099-1110 (2008.09.01)

Gap in pagination due to withheld paper.

Pages 1496-1500

Determination of Rotor Temperature for an Interior Permanent Magnet Synchronous Machine Using a Precise Flux Observer

Andreas Specht, Oliver Wallscheid, Joachim Böcker
Power Electronics and Electrical Drives
University of Paderborn, 33095 Paderborn, Germany
Email: specht@lea.upb.de, wallscheid@lea.upb.de, boecker@lea.upb.de

Abstract—In this paper an enhanced method to determine the rotor temperature of permanent magnet synchronous machines during dynamic operations is presented. The approach is based on a fundamental wave flux observer and is therefore independent of coolant conditions to a large extend. This contribution continues the work presented earlier, where the temperature observer was introduced for steady state operation in exemplary operating points. Hence, an extension to dynamic operation in a wide speed and torque range is proposed. Consequently, the focus now lies on a discrete-time machine modelling as well as on the parameter identification. Measurement results prove the satisfying performance of the observer.

I. Introduction

Traction drives employed in electrical and hybrid motorcars typically rely on interior permanent magnet synchronous machines (IPMSM) using neodymium-iron-boron (NdFeB) as permanent magnet material. This class of machines has two major advantages over the induction machine (IM), which can be considered to be the second most commonly used machine in this type of application. These advantages are the higher power and torque density as well as the better efficiency at least at lower and medium speeds. On the other hand, the drawbacks are, besides safety issues concerning the EMF, mostly the temperature sensitivity of the permanent magnet material. The permanent magnet temperature is of great interest, if the machine should be both maximal utilized and safely operated. As the temperature has direct impact on the remanent flux density of the magnet material monitoring its temperature also helps to increase the accuracy of the estimated torque [1].

Direct measurement of the permanent magnet temperature is in most cases not economically reasonable, as information has to be passed over from rotor to stator preferably via some kind of wireless technology. Alternatively, there are mainly three different ways to acquire the rotor temperature:

- *Thermal modelling:* Thermal modelling is based on the strategy to simplify the complex thermal behaviour inside a machine resulting in Lumped Parameter Thermal Networks (LPTN) with drastically reduced model orders. There are mainly two approaches: In [2] and [3] appropriate model structures representing the dominant heat-flow paths in the machine are defined in advance and

parameters are identified based on measurement data, whereas in [4] and [5] the reduced model structure is derived analytically by calculating thermal parameters based on construction data.

- *Signal injection:* Depending on the geometry of the machine, the rotor temperature may have an impact on the degree of magnetic saturation in the machine. In this case, the rotor temperature can be derived by evaluating step responses of the current towards a voltage injection signal [6] [7].

- *Flux modelling:* The magnetization of the permanent magnet material drops reversibly with rising temperature. This effect is typically modelled by an linear material specific temperature coefficient $k_{Br} \approx -0.1\,\%\mathrm{K}^{-1}$. Using a precise flux model the level of demagnetization can be estimated and used to determine the electrically effective temperature of the permanent magnets [8].

On the one hand, LPTN approaches offer the benefit of being able to estimate not only the rotor temperature, but also the temperature in several critical motor locations (e.g. at the end windings). On the other hand, the LPTN method is time consuming in terms of parameter identification and is based on assumptions regarding the thermal environment of the motor. Furthermore, signal injection methods may not be used in every situation, since they cause a current and consequently a torque ripple. This will lead to additional losses and potentially increased drive train oscillations. Moreover, the operation at higher speeds can be problematic, if the voltage injection signal leads to non-feasible duty cycles. Hence, this contribution focuses on the last approach using a flux observer and continues the work reported in [8]. This fundamental wave method is less sensitive towards varying cooling conditions or DC-link voltages. In addition to [8], the rotor temperature observer is enhanced to be applicable in a wide speed and torque range during dynamic operation. A fundamental prerequisite for guaranteeing accurate temperature observation is a precise machine model, as k_{br} is fairly small and the permanent magnet flux has to be estimated with a precision of 0.5 % in order to obtain a temperature error below 5 K. In this context the performance respectively accuracy of a discrete-time machine model is strongly depending on two aspects:

- *Physical modelling:* The physics of the machine

have to be modelled precisely. Here, saturation and iron losses will be mainly in focus of modelling.

- *Numerical modelling:* Traction drives are mostly designed in a way that the number of control samples during an electrical turn at maximum speed is approximately 10. Here, using the commonly known models as stated in [9] and others will lead to unacceptable numerical errors, which will be shown in the following.

In consequence, a precise discrete-time model of the IPMSM is derived and an appropriate parametrization is presented. This model will then be extended with a feedback mechanism to allow estimation of changes in magnetization and rotor temperature respectively. The necessity of a appropriate time discretization is demonstrated by simulation. The paper concludes with a discussion of validation results obtained at a test bench for automotive traction drives.

II. MACHINE MODEL

In the continuous-time domain, the machine model in rotating coordinates is commonly described by the following set of equations:

$$u_{dq} = R_s i_{dq} + \frac{d}{dt}\psi_{dq} + \omega J \psi_{dq} \tag{1a}$$

$$i_{dq} = \mathbf{f}_{dq}^{-1}\left(\psi_{dq}\right) \tag{1b}$$

$$T = \frac{3}{2}p(\psi_{dq} \times i_{dq}) \tag{1c}$$

The quantities are given by u as the stator voltage, i as the stator current, ω as electrical angular frequency, ψ as the linked flux and T as the torque of the machine. Bold symbols denote vectorial quantities. Furthermore, the parameters are given by R_s as the stator resistance and the usually nonlinear relationship $\mathbf{f}_{dq}^{-1}$ between stator current and linked flux. In this function, the dependency of the linked flux on the rotor temperature as well as effects like saturation and speed dependent iron losses are modelled. Implementing $\mathbf{f}_{dq}^{-1}$ as a lookup table yields good results with limited computational effort. The permutation matrix

$$J = \begin{pmatrix} 0 & -1 \\ 1 & 0 \end{pmatrix}$$

can be considered as the matrix notation of the complex unit j and the index dq indicates the notation in rotor fixed coordinates. As stated in [10] and [11], the direct discretization of (1) using simple numerical integration schemes, like first-order explicit Euler, will lead to discretization errors which may even result in numerical instability of the model especially at higher speeds. The proposed alternative provides a symplectic discrete-time model offering a high numeric accuracy even at high electric frequencies compared to the inverter switching and controller sampling frequency: Starting point is the machine model in stator fixed coordinates:

$$u_{\alpha\beta} = R_s i_{\alpha\beta} + \frac{d}{dt}\psi_{\alpha\beta} \tag{2a}$$

$$\psi_{\alpha\beta} = \mathbf{f}_{\alpha\beta}\left(i_{\alpha\beta}, \varphi\right) \tag{2b}$$

$$T = \frac{3}{2}p\left(\psi_{\alpha\beta} \times i_{\alpha\beta}\right) \tag{2c}$$

Here, the index $\alpha\beta$ indicates the notation in stator fixed coordinates. Due to the saliency of the machine the relationship given in (2b) is additionally dependent on the electrical rotor position φ. The stator voltage equation is discretized by using a simple difference quotient with the sampling time T_a:

$$\frac{dx}{dt} \approx \frac{x_{k+1} - x_k}{T_a} \tag{3}$$

Applying (3) to (2a) leads to

$$u_{\alpha\beta,k} = R_s i_{\alpha\beta,k} + \frac{\psi_{\alpha\beta,k+1} - \psi_{\alpha\beta,k}}{T_a} \tag{4}$$

(2b) and (2c) remain unaffected as they are of purely algebraic nature. Rotating (4) by the rotor position using the rotary matrix

$$\mathcal{R}(\varphi) = \begin{pmatrix} \cos(\varphi) & -\sin(\varphi) \\ \sin(\varphi) & \cos(\varphi) \end{pmatrix} \tag{5}$$

leads to the linked flux difference equation in rotor fixed coordinates:

$$\psi_{dq,k} = \mathcal{R}^{-1}(\Delta\varphi_k)\left[\psi_{dq,k-1} + T_a u_{dq,k-1}\right.$$
$$\left. - T_a \mathcal{R}(\tfrac{\Delta\varphi}{2}_k) R_s i_{dq,k-1}\right] \tag{6a}$$

$$i_{dq,k} = \mathbf{f}_{dq}^{-1}\left(\psi_{dq,k}\right) \tag{6b}$$

$$\Delta\varphi_{k+1} = \varphi_{k+1} - \varphi_k \approx \omega_k T_a \tag{6c}$$

$$\mathcal{R}(\Delta\varphi_{k+1}) = \mathcal{R}(\varphi_{k+1})\,\mathcal{R}^{-1}(\varphi_k) \tag{6d}$$

It has to be noted, that assigning the current in the time interval $[kT_a, (k+1)T_a]$ to the sampling instant kT_a will lead to an error, as the best representative would be at $(k + 0,5)T_a$. Assuming that all quantities are of sinusoidal shape (fundamental wave model) this error can be compensated by a rotation of $\Delta\varphi/2$. This correction has already been applied in (6a). Thus, (6) forms an accurate discrete-time model of the machine capable of being calculated on a micro-controller-based real-time platform. The dependency of the linked flux on the rotor temperature ϑ_r can be modelled as follows:

$$\psi_{dq}(i_{dq}, \vartheta_r) = \psi_{dq}(i_{dq}, \vartheta_{r,0})$$
$$+ \underbrace{\begin{pmatrix} 1 & 0 \\ 0 & 0 \end{pmatrix} k_{Br}(\vartheta_r - \vartheta_{r,0})\psi_{p,0}}_{\Delta\psi_{dq}} \tag{7}$$

Here, the index 0 indicates the model parameters taken at an arbitrary reference temperature. The variable $\psi_p = \psi_d\big|_{i_d=i_q=0}$ is the permanent flux of the machine. From (7) it follows that the drop of the linked flux in direction of the direct axis is proportional to the material specific temperature coefficient of the permanent magnet material. For NdFeB magnets and the typical temperature range in automotive applications the coefficient k_{Br} is assumed to be constant and given as: $k_{Br} \approx -0.1\ \%\mathrm{K}^{-1}$.

III. PARAMETRIZATION OF THE MODEL

Formally speaking, the model consists of three parameters, where only two have to be identified. While R_s is quite simple to identify, the nonlinear function $\mathbf{f}_{dq}^{-1}\left(\psi_{dq}\right)$ takes quite some effort to acquire. The temperature coefficient k_{Br} can be taken from data sheet or experimental measurements at no-load operation.

A. Stator Resistance

Main issue when modelling the stator resistance is its dependency on the stator temperature ϑ_s. The approach of modelling is strongly associated to the specific machine and the cooling method. In the simplest case, a modelling according to the well known linear relationship

$$R_s(\vartheta_s) = R_s(\vartheta_{s,0})\left(1 + \alpha_{\mathrm{cu}}(\vartheta_s - \vartheta_{s,0})\right) \quad (8)$$

using a temperature sensor in the stator windings is sufficient. In other cases, a simple thermal model may have to be set up in order to obtain the electrically effective stator temperature. Depending on the motor design, skin and proximity effects might need to be considered as well. In this instance, (8) has to be extended by the electrical stator frequency.

B. Flux-Current-Relation

The function describing the relation between ψ_{dq} and i_{dq} is again strongly dependent on the type of machine to be identified. As machines for traction drives are optimized with respect to maximum power density, strong saturation is an elemental characteristic of these machines. Hence, the inductance especially in the quadrature axis may drop below 20 % of its nominal value. An analytical approach to model this effect suffers either from a low accuracy or a high online-calculation effort. Practical work experience has shown that measuring the machine's behaviour at different stationary operating points and merging the results in look-up tables leads to good results. In order to obtain $\mathbf{f}_{dq}(i_{dq})$ an evaluation of (6a) under steady-state conditions is performed. In this case, time horizons can be neglected and (6a) reduces to

$$\psi_{dq} = \mathcal{R}^{-1}(\omega T_a)\left[\psi_{dq} + T_a\underbrace{\left(u_{dq} - \mathcal{R}\left(\frac{\omega T_a}{2}\right)R_s i_{dq}\right)}_{u_i}\right] \quad (9)$$

From (9), the calculation rule for the steady state value of the flux can be derived:

$$\psi_d = \frac{u_{i,d}\left(\cos(\omega T_a) - 1\right) + u_{i,q}\sin(\omega T_a)}{2\left(\cos(\omega T_a) - 1\right)}T_a \quad (10a)$$

$$\psi_q = \frac{u_{i,q}\left(\cos(\omega T_a) - 1\right) - u_{i,d}\sin(\omega T_a)}{2\left(\cos(\omega T_a) - 1\right)}T_a \quad (10b)$$

In Fig. 1 and 2, exemplary measurement results from a typical automotive traction drive are depicted. To acquire the correct stator voltage, high-bandwidth-sensors being applicable of measuring pulsating inverter voltages or a precise inverter model for calculating u_{dq} from the duty cycles can be used [12]. Furthermore, it has to be mentioned, that $\mathbf{f}_{dq}^{-1}$ is speed-dependent due to iron losses and frequency-dependent sensor characteristics. As can be seen in Fig. 3 and Fig. 4, the deviation of $\mathbf{f}_{dq}^{-1}$ at different speeds can amount to several percents of the permanent magnet flux. Thus, the flux-current measurement has to be repeated for different motor speeds. As a consequence, the LUTs have to be extended by the frequency dimension:

$$\psi_{dq} = \mathbf{f}_{dq}\left(i_{dq}, \omega\right) \quad (11)$$

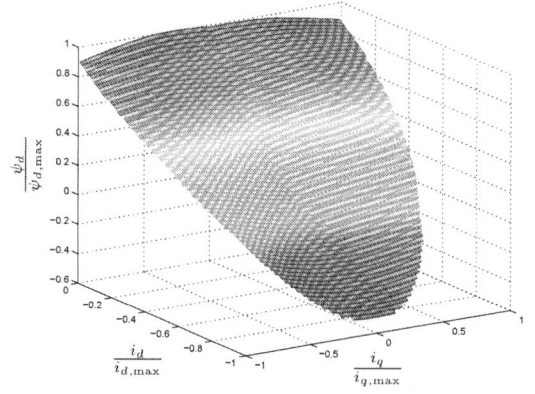

Fig. 1. Relation between direct axis flux and the currents in rotor fixed coordinates

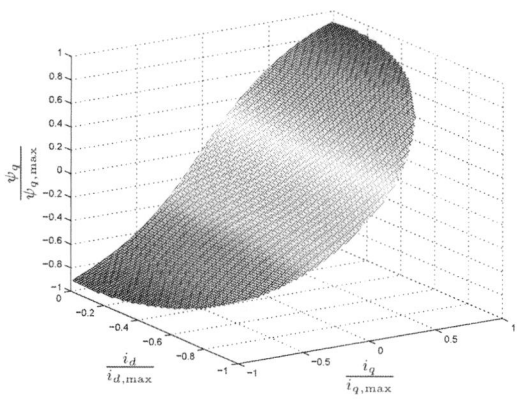

Fig. 2. Relation between quadrature axis flux and the currents in rotor fixed coordinates

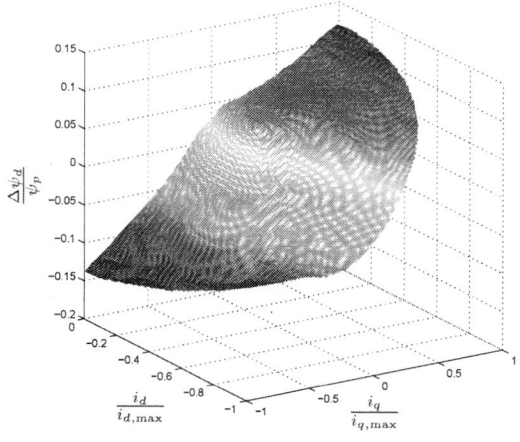

Fig. 3. Deviation of $\mathbf{f}_d^{-1}$ (1500 min^{-1}) towards $\mathbf{f}_d^{-1}$ (465 min^{-1})

The 2014 International Power Electronics Conference

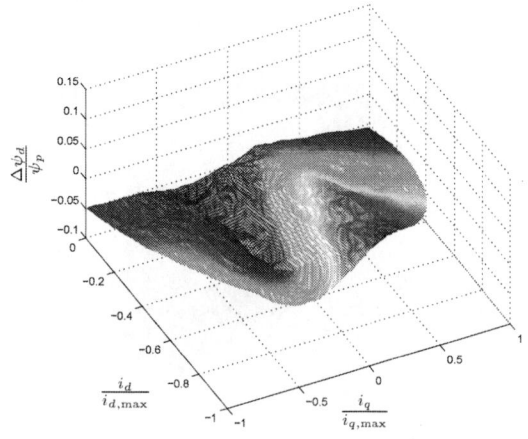

Fig. 4. Deviation of $\mathbf{f}_q^{-1}\left(1500\ \mathrm{min}^{-1}\right)$ towards $\mathbf{f}_q^{-1}\left(465\ \mathrm{min}^{-1}\right)$

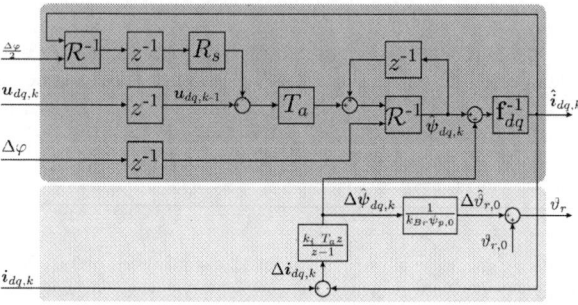

Fig. 5. Structure of the proposed observer: Blue part of the model corresponds to (6); yellow part added to create observer structure

IV. DESIGN OF THE OBSERVER

The rotor temperature observer uses the model according to (6) and adds a feedback based on the the error between observed and measured current:

$$\Delta\hat{\boldsymbol{\psi}}_{dq,k} = \begin{pmatrix} 1 & 0 \\ 0 & 0 \end{pmatrix} k_i T_a (\boldsymbol{i}_{dq,k} - \hat{\boldsymbol{i}}_{dq,k}) + \Delta\hat{\boldsymbol{\psi}}_{dq,k-1} \quad (12)$$

Here, $\hat{\boldsymbol{\psi}}$ and $\hat{\boldsymbol{i}}$ denote the observed quantities. Therefore, the observed currents are calculated as:

$$\hat{\boldsymbol{i}}_{dq,k} = \mathbf{f}_{dq}^{-1}\left(\hat{\boldsymbol{\psi}}_{dq,k} + \Delta\hat{\boldsymbol{\psi}}_{dq,k}, \omega\right) \quad (13)$$

From (7), the absolute rotor temperature can be calculated via

$$\hat{\vartheta}_{r,k} = \frac{\Delta\hat{\psi}_{d,k}}{k_{Br}\psi_{p,0}} + \vartheta_{r,0} \quad (14)$$

The complete structure of the proposed observer is depicted in Fig. 5. Next, the observer dynamics have to be investigated in order to derive an appropriate design rule to choose the feedback factor k_i. The discrete-time transfer function

$$G_{\mathrm{RTO}}(z) = \frac{\Delta\hat{\psi}_d}{\Delta\psi_d} = \frac{m_3 z^3 + m_2 z^2 + m_1 z + m_0}{n_3 z^3 + n_2 z_2 + n_1 z + n_0} \quad (15)$$

is valid if a linear relationship

$$\boldsymbol{\psi}_{dq} = \mathbf{f}_{dq}\left(\boldsymbol{i}_{dq}\right) = \begin{pmatrix} L_d & 0 \\ 0 & L_q \end{pmatrix} \begin{pmatrix} i_d \\ i_q \end{pmatrix} + \begin{pmatrix} \psi_p \\ 0 \end{pmatrix} \quad (16)$$

is postulated, where the inductance L_d and L_q are assumed to be constant at a given operating point. Additionally, the assumption of a constant speed ω has to be fulfilled. The coefficients for (15) are given as:

$$m_3 = 0$$
$$m_2 = -L_q T_a k_i$$
$$m_1 = 2 L_q T_a k_i \cos(\omega T_a) - R_s T_a^2 k_i \cos\left(\frac{\omega T_a}{2}\right)$$
$$m_0 = R_s T_a^2 k_i \cos\left(\frac{\omega T_a}{2}\right) - L_q T_a k_i$$
$$n_3 = L_d L_q$$
$$n_2 = L_d R_s T_a \cos\left(\frac{\omega T_a}{2}\right) - L_q T_a k_i$$
$$\quad - 2 L_d L_q \cos(\omega T_a) - L_d L_q + L_q R_s T_a \cos\left(\frac{\omega T_a}{2}\right)$$
$$n_1 = R_s^2 T_a^2 + L_d L_q + 2 L_d L_q \cos(\omega T_a)$$
$$\quad - R_s T_a^2 k_i \cos\left(\frac{\omega T_a}{2}\right) + 2 L_q T_a k_i \cos(\omega T_a)$$
$$\quad - 2 L_d R_s T_a \cos\left(\frac{\omega T_a}{2}\right) - 2 L_q R_s T_a \cos\left(\frac{\omega T_a}{2}\right)$$
$$n_0 = R_s T_a^2 k_i \cos\left(\frac{\omega T_a}{2}\right) - L_d L_q - L_q T_a k_i - R_s^2 T_a^2$$
$$\quad + L_d R_s T_a \cos\left(\frac{\omega T_a}{2}\right) + L_q R_s T_a \cos\left(\frac{\omega T_a}{2}\right)$$
$$\quad (17)$$

As can be seen from (15) and (17), the behaviour is exclusively dependent on the machine parameters and the speed. The operating point in terms of phase current and voltage is theoretically irrelevant. In practice, though, these quantities will play a certain role mainly due to inaccurate measurement and discretization errors. An exemplary bode diagram for constant parameters is given in Fig. 6. Moreover, the transfer function in (15) can be approximated by a 1st-order low-pass

$$G_{\mathrm{RTO}}(z) \approx \frac{1 - e^{\alpha_{\mathrm{RTO}}}}{z - e^{\alpha_{\mathrm{RTO}}}} \quad (18a)$$

$$\alpha_{\mathrm{RTO}} = -\frac{T_A}{\tau_{\mathrm{RTO}}}, \quad \tau_{\mathrm{RTO}} = \frac{L_d}{k_i} \quad (18b)$$

as deviations occure at a roughly 10 times higher frequency than $\omega_{\mathrm{RTO}} = 1/\tau_{\mathrm{RTO}}$ and high dynamics in $\Delta\psi_{dq}$ are not expected, since the rotor thermal dynamics is rather slow. The dynamics of the rotor temperature observer then depends on the feedback gain k_i and the inductance L_d. By identifying the maximum occurring dynamics of the rotor temperature for a given application,

978-1-4799-2706-7/14 $31.00 © 2014 IEEE

The 2014 International Power Electronics Conference

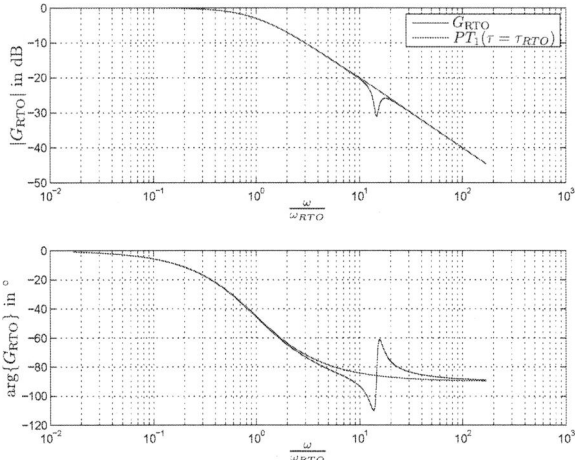

Fig. 6. Bode plot of the transfer function G_{RTO} compared to a 1st-order low-pass with the time constant τ_{RTO}

k_i can be chosen to cover these dynamics while low-pass filtering the observed temperature signal and thus saving an additional filter as well.

V. COMPARISON OF CONVENTIONAL AND SYMPLECTIC MODEL BASED OBSERVER

As mentioned before, the dicretization scheme of the machine model, on which the observer is based on, is of great importance. To demonstrate this circumstance, simulations are carried out using an observer based on a machine model according to (6) and a second observer based on a machine model according to (1a). The discrete-time machine model derived from (1a) is given by:

$$
\begin{aligned}
\boldsymbol{\psi}_{dq,k} =& \boldsymbol{\psi}_{dq,k-1} \\
&+ T_a \left(\boldsymbol{u}_{k-1} - R_s \boldsymbol{i}_{k-1} - \omega \boldsymbol{J} \boldsymbol{\psi}_{k-1} \right)
\end{aligned}
\tag{19}
$$

The simulations have been carried out with two different sampling times T_a. In a first step, an unrealistic small sampling time $T_a = 1\,\mu s$ was chosen (see Fig. 7, top plot). Here, both observers show a good performance with deviations from the nominal machine temperature of 50 °C below ± 1 K. The second simulation (Fig. 7, center plot) has been carried out with a typical sampling time of $T_a = 100\,\mu s$. In this case, the symplectic model-based observer maintains its good performance, while the second observer fails due to discretization errors.

VI. MEASUREMENT RESULTS

The proposed method was implemented on a test bench setup for a typical IPMSM used in automotive traction drives. The machine parameters are given in Tab. I. For validation the IPMSM rotor was equipped

TABLE I. PARAMETER OF THE TESTED MACHINE

Type	IPMSM
Voltage rated	177 V
Current rated (max.)	110(282) A
Power rated (max.)	19, 6(50, 6) kW
Torque rated (max.)	110(283) Nm
Speed rated (max.)	1700(6000) min^{-1}
Number of pole pairs	8
Cooling	Water-Jacket

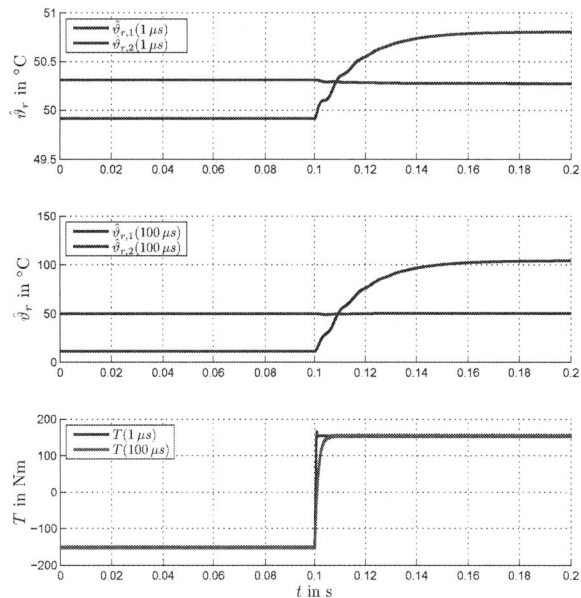

Fig. 7. Comparison of different discretization schemes at constant speed of 1000 min^{-1}. Observer 1 (blue line) is based on symplectic model (6); observer 2 (green line) is based on standard model (19).

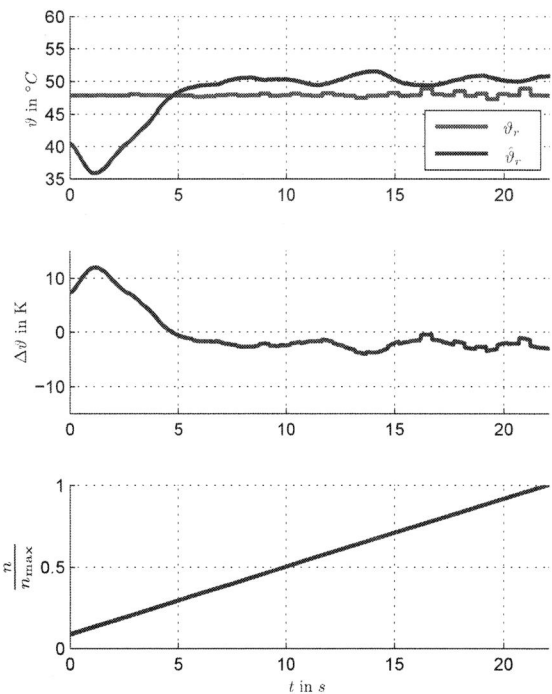

Fig. 8. Measurement results for speed ramp at torque $T = 0.4\,T_{\mathrm{max}}$

with several thermocouples in direct contact to the permanent magnets. In Fig. 8, the rotor temperature is observed for a constant torque applying a speed ramp to the machine. The speed ramp is starting at 5 % of its maximum value. Due to the small voltage magnitude at low speeds the sensitivity of the observer even towards minor voltage acquisition errors is high resulting in an

978-1-4799-2706-7/14 $31.00 © 2014 IEEE

increased error in the observed rotor temperature. As the risk of further rotor temperature increases is low and as the rotor temperature dynamics are diminished at low speeds [3] this observer behaviour is acceptable for the usage in motorcar applications. For the depicted figure, the observed temperature lies in a band of ±10 K around the measured temperature and the highest error occurs only at low speeds while it drops clearly below ±5 K for medium and higher speeds. Additionally, in Fig. 9 a torque ramp given to the machine at a constant speed is shown. Here, it can be seen that the observer is nearly invariant to torque and currents variations, respectively. The observer error is below ±5 K being equivalent to a relative error in the observed flux of 0.5 %.

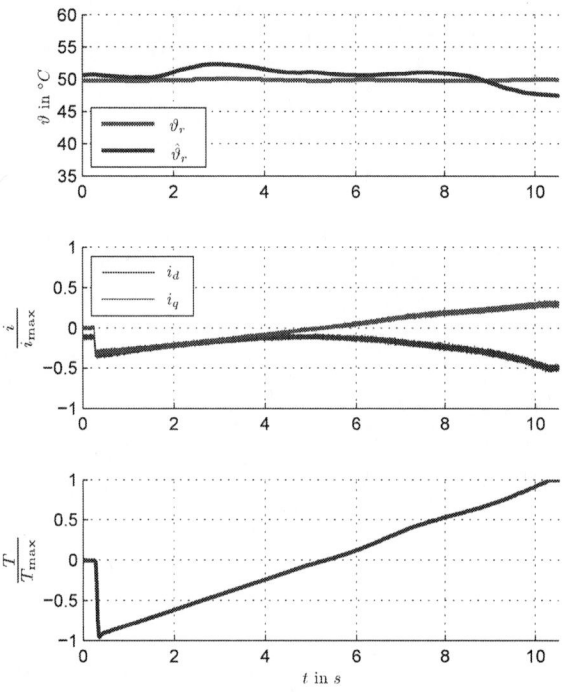

Fig. 9. Measurement results for torque ramp at speed $n = 0.3 \, n_{\max}$

Furthermore, in Fig. 10 measurement results based on the Federal Test Procedure (FTP-75) drive cycle are shown. A simple longitudinal mechanical model was used to calculate the required torque at the rotor shaft for an electric mid-size motorcar. Also, the minimum machine speed was set to $1000 \, \text{min}^{-1}$ since this value is assumed to be the lowest possible one with acceptable observation performance. For speeds below this the induced voltage is too small in contrast to the achievable voltage measurement respectively inverter model accuracy. In future applications k_i can be set to zero to hold the actual observed temperature until the speed increases above the allowed minimum. However, in typical motorcar drive cycles the machine speed can be assumed to return back into the feasible range at regular intervals. Due to the low thermal dynamics in traction applications, it is sufficient to activate the observer occasionally. Additionally, the rotor temperature does not rise in the low speed range by tendency, since the rotor iron losses and the thermal air gap conductance increase with the speed [3] [5] [13]. As

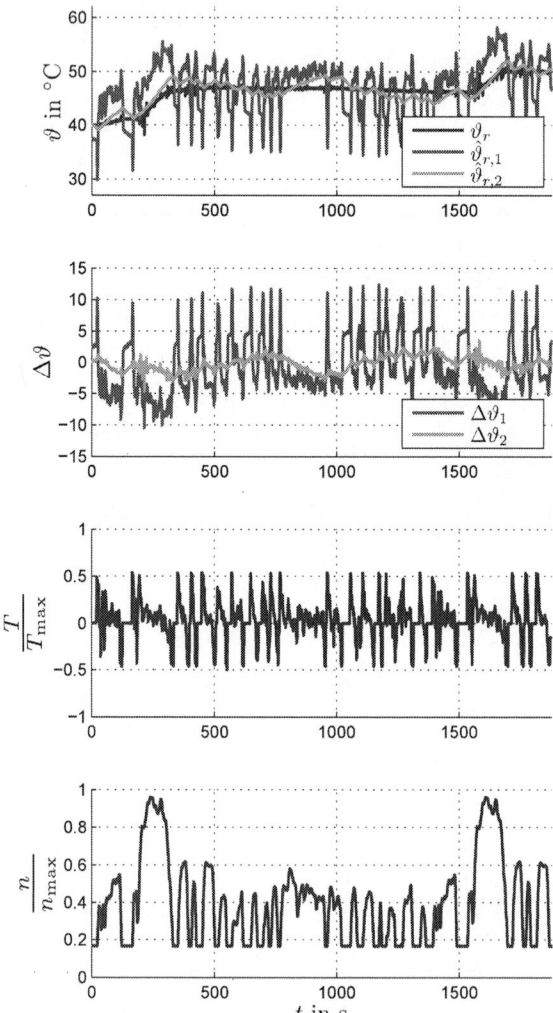

Fig. 10. Measurement results based on a FTP-75 drive cycle derivate

far as temperature-dependent irreversible demagnetization is concerned, this operation region should be considered as not very critical.

Moreover, in Fig. 10 two observer designs are compared: For $\hat{\vartheta}_{r,1}$ the observer gain k_i was set to a rather high value leading to a quasi-stationary observer operation. Thus, the observer dynamics can be assumed to be a couple of times higher than the maximum occurring rotor thermal dynamics. In this case, speed and torque changes have instant impact on the observation accuracy. The maximum absolute observer error is in the range of 10 K. Alternatively, k_i was calculated based on the maximum rotor thermal dynamics for $\hat{\vartheta}_{r,2}$. This design exhibits an inherent low-pass filter effect on the observed temperature leading to a significantly reduced absolute error in the range of 2.5 K. However, the actual improvement by an appropriate choice of k_i is strongly depending on the considered drive cycle.

VII. CONCLUSION

In this contribution, a method to determine the rotor temperature by using a precise flux observer was proposed. As an essential prerequisite a symplectic discrete-time electric motor model was introduced. Simulation results showed a significantly better discretization accuracy in comparison to standard first-order explicit Euler approaches. The investigated method can be easily implemented on commonly used real-time platforms and the effort for parameter identification is acceptable. Measurement results proved satisfying observer performance in a wide speed and torque range. The observer accuracy is strongly correlated to the machine speed and for very low speeds, the method cannot be applied. This is not very critical, as the rotor thermal dynamics is quite slow and do not rise significantly within this speed range. At medium and high speeds an absolute observer error below 5 K was achieved. Thus, the rotor temperature observer seems to be applicable to a broad range of traction applications.

REFERENCES

[1] T. Sebastian, "Temperature Effects on Torque Production and Efficiency of PM Motors Using NdFeB Magnets," *IEEE Transactions on Industry Applications*, vol. 31, no. 2, pp. 353–357, 1995.

[2] C. Kral, A. Haumer, and S. B. Lee, "A Practical Thermal Model for the Estimation of Permanent Magnet and Stator Winding Temperatures," *IEEE Transactions on Power Electronics*, vol. 29, no.1, pp. 455–464, 2014.

[3] T. Huber, W. Peters, and J. Böcker, "Monitoring Critical Temperatures in Permanent Magnet Synchronous Motors Using Low-Order Thermal Models," in *International Power Electronics Conference (IPEC)*, 2014.

[4] G. Dajaku and D. Gerling, "An Improved Lumped Parameter Thermal Model for Electrical Machines," in *17th International Conference on Electrical Machines (ICEM)*, 2006.

[5] G. D. Demetriades, H. Z. D. L. Parra, E. Andersson, and H. Olsson, "A Real-Time Thermal Model of a Permanent-Magnet Synchronous Motor," *IEEE Transactions on Power Electronics*, vol. 25, no. 2, pp. 463–474, 2010.

[6] M. Ganchev, C. Kral, H. Oberguggenberger, and T. Wolbank, "Sensorless Rotor Temperature Estimation of Permanent Magnet Synchronous Motor," in *Annual Conference on IEEE Industrial Electronics Society (IECON)*, 2011.

[7] D. D. Reigosa, F. Briz, P. Garca, J. M. Guerrero, and M. W. Degner, "Magnet Temperature Estimation in Surface PM Machines Using High-Frequency Signal Injection," *IEEE Transactions on Industry Applications*, vol. 45, no. 4, pp. 1468–1475, 2010.

[8] A. Specht and J. Böcker, "Observer for the Rotor Temperature of IPMSM," in *International Power Electronics and Motion Control Conference (EPE/PEMC)*, 2010.

[9] A. Piippo, M. Hinkkanen, and J. Luomi, "Adaptation of Motor Parameters in Sensorless PMSM Drives," *Industry Applications, IEEE Transactions on*, vol. 45, no. 1, pp. 203–212, 2009.

[10] Böcker, "Discrete-Time Model of an Induction Motor," *European Transactions on Electrical Power*, vol. 2, pp. 65–71, 1991.

[11] A. Specht, S. Ober-Blöbaum, O. Wallscheid, C. Romaus, and J. Böcker, "Discrete-Time Model of an IPMSM Based on Variational Integrators," in *IEEE International Electric Machines Drives Conference (IEMDC)*, 2013.

[12] M. Seilmeier, C. Wolz, and B. Piepenbreier, "Modelling and Model Based Compensation of Non-Ideal Characteristics of Two-Level Voltage Source Inverters for Drive Control Application," in *Electric Drives Production Conference (EDPC)*, 2011.

[13] B. Stumberger, A. Hamler, and B. Hribernik, "Analysis of Iron Loss in Interior Permanent Magnet Synchronous Motor Over a Wide-Speed Range of Constant Output Power Operation," *IEEE Transactions on Magnetics*, vol. 36, no. 4, pp. 1846–1849, 2000.

Monitoring Critical Temperatures in Permanent Magnet Synchronous Motors Using Low-Order Thermal Models

Tobias Huber, Wilhelm Peters, Joachim Böcker
Power Electronics and Electrical Drives
University of Paderborn
33098 Paderborn, Germany
huber@lea.upb.de

Abstract—**Monitoring critical temperatures in electric motors is crucial for preventing shortened motor life spans due to excessive thermal stress. With regard to interior permanent magnet synchronous motors (IPMSM), critical temperatures typically occur in the magnets and in the stator end winding. As directly measuring temperatures, especially on the rotating part, is costly, sensitive and thus not applicable with respect to automotive applications, model-based approaches are preferred. In this paper, two low-order thermal models for accurate temperature estimations in the permanent magnets, the winding and the end winding are introduced and compared. The model parameters are estimated solely based on experimental data via a multistep identification approach for linear parameter -varying systems. The model performances are validated by extensive experimental results based on a high-speed PMSM typically used as traction motor in subcompact electric cars.**

Keywords— thermal model, linear parameter-varying, parameter identification, permanent magnet synchronous motor

I. INTRODUCTION

Due to strict weight and size requirements concerning the employed electric motors in automotive traction drives, highly utilized IPMSM are preferred. In order to operate these motors close to their thermal limits while avoiding accelerated thermal ageing or damage, critical motor temperatures need to be permanently monitored. While excessive magnet temperatures involve the risk of irreversible demagnetization when operating the motor in flux weakening mode, exceeding the thermal limits of the winding can destroy the insulation varnish. With regard to distributed winding schemes, the temperature in the winding heads (end-winding) is of particular importance, as it typically exceeds the one in the stator slots significantly. While measuring the PM temperature on the rotor via some kind of contactless technology (e.g. telemetry systems) is costly, sensitive and therefore not feasible in series production, winding and end-winding temperatures can be obtained by means of temperature sensors. However, since these sensors are usually difficult to access, they cannot simply be replaced in case of sensor failures. Therefore, model-based approaches are preferred.

Two ways for estimating PM and winding temperatures are based on electrical machine models [1] or signal injection methods [2, 3]. Both ways, temperatures are determined indirectly by utilizing the thermal properties of electric model parameters, such as the permanent magnetic flux or the winding resistance. A major drawback of this approach originates from the low temperature sensitivity of the commonly employed PM material (NdFeB) as well as the small contributions from the resistive voltage drop to the total terminal voltage in high-efficiency, high-power machines. Accurate temperature estimations therefore always require a great deal of modeling efforts, i.e. machine and inverter parameters need to be precisely determined. Although signal injection methods are generally capable of providing more reliable temperature estimations, they have an invasive character and disturb the motor performance to a certain degree. Moreover, as temperatures are estimated based on the electrically effective (mean) values of the parameters, critical hot spots such as in the end-winding cannot directly be monitored.

An alternative to the previously mentioned electrical approaches are lumped parameter thermal networks (LPTN). In LPTN models, heat transfer processes in the motor are abstracted from a thermal point of view and can be described via equivalent circuit diagrams, as used for describing electric circuits. A major difficulty lies in the proper calibration of those models. Even if the motor geometry and the thermal properties of the used materials are well known, analytical derivation of lumped thermal resistances and capacitances remains a difficult task and requires an experienced designer in order to obtain acceptable results [4]. Nevertheless, some model parameters still need to be tuned based on experimental data due to parameter uncertainties. In literature, several LPTN for IPMSM can be found. Most of them consist of a minimum of 8 nodes and are mainly calibrated based on analytical formulations using geometry and material data of the motor [5, 6]. However, as shown in [7, 8] accurate temperature estimations in the PM and winding can also be achieved with lower model orders, when matching the model to experimental data.

978-1-4799-2706-7/14 $31.00 © 2014 IEEE

In this paper, two low-order thermal models are introduced that guarantee accurate temperature estimations in the PM, winding and end-winding of an IPMSM, typically used as traction motor in subcompact electric cars. Compared to [8], both models are suitable for estimating the critical end-winding temperature as well, without depending on any temperature measurements in the difficult to access stator core. Except for measurements at the inlet of the cooling circuit and in the test cabin, the models rely only on commonly available quantities in the motor ECU, such as motor speed and electric currents. As both models have linear parameter-varying (LPV) structures, the parameterization is performed using a multistep identification approach solely relying on experimental data and rather simple physically-motivated boundary conditions. The models are validated and compared against various experimental results within the entire operating range of the motor. The proposed thermal models in combination with the parameter identification method represent a practical and innovative approach for model-based monitoring of critical motor temperatures, which is applicable to similar IPMSM designs, too.

II. PROPOSED THERMAL MODELS

In the following subsections, two thermal modeling approaches are introduced. Both approaches are based on LPTN networks whose corresponding equivalent circuit diagrams are depicted in Fig. 1 and Fig. 2. The major analogies to the well-known electrical circuit design are summarized in [7]. Both LPTN structures are based on rather simple physical assumptions concerning the dominant heat-flow mechanisms in the motor and, provided that they are appropriately parameterized, guarantee accurate temperature estimations within the relevant operating range of the motor.

A. Thermal Model with Three Nodes

The first presented thermal model (s. Fig. 2) consists of three nodes representing the temperatures that shall be estimated. The winding node T_W is connected to the end-winding node T_{EW} through the resistance R_{W-EW} and to the permanent magnet node T_{PM} through the resistance R_{W-PM}. It can be assumed that R_{W-PM} exhibits a significant speed dependence over the air gap due to convection effects [7, 8]. Moreover, T_W is connected to the inlet temperature of the cooling circuit T_C through the resistance R_{W-C}. In this paper, the resistance R_{W-C} is assumed constant, as the coolant is kept at a nearly constant volume flow rate. The PM node T_{PM} is connected to the environment temperature in the test cabin T_E via R_{PM-E}. In order to consider convection effects between PM and the environment temperature, R_{PM-E} is also assumed speed-dependent. Although the latter heat path constitutes a strong abstraction from the actual physical-thermal conditions, simply neglecting it would lead to a model structure that is not capable of describing thermal equilibrium states in which the win-

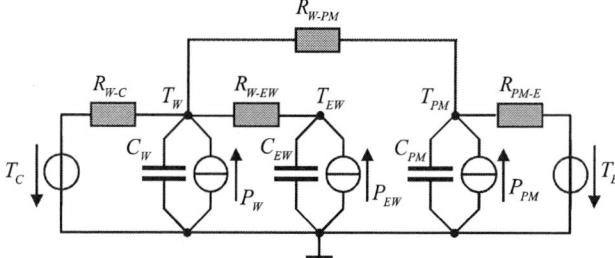

Fig. 1. Proposed LPTN network with three nodes

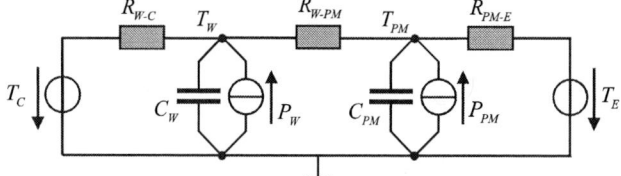

Fig. 2. Proposed LPTN network with two nodes

ding temperature exceeds the PM temperature. Those thermal states are, however, not uncommon and usually occur in operating points with dominant stator losses (e.g. at low speeds and high torque). The thermal model relies on a compact measurement-based loss model, which is calculated based on commonly available quantities in the motor ECU, such as motor speed and electric currents. For the derivation of the loss model, at first a look-up table (LUT) of the total motor losses P_{total} is obtained via measurements. To suppress thermal effects in the LUT, the measurements were performed at constant winding and PM temperatures in an automated way with the help of a state machine. Assuming, that the temperature dependence of the total losses can mainly be allocated to the copper losses in the winding and end-winding, the total losses are split up into a temperature-dependent copper loss portion P_{Cu} and a temperature-independent residual loss portion P_{resid} according to

$$P_{resid}(I,n) = P_{total}(I,n) - P_{Cu}(I,T_W,T_{EW}) \quad (1)$$

In (1), I represents the RMS value of the phase current and n represents the motor speed. The copper losses P_{Cu} are allocated to the winding and end-winding according to their relative effective conductor lengths $\frac{l_W}{l}$ and $\frac{l_{EW}}{l}$ by means of

$$P_{Cu} = P_{Cu,W} + P_{Cu,EW} \quad (2)$$

where

$$
\begin{aligned}
P_{Cu,W} &= 3R_{ref}I^2(1+\alpha_{Cu}\Delta T_W)\frac{l_W}{l} \\
P_{Cu,EW} &= 3R_{ref}I^2(1+\alpha_{Cu}\Delta T_{EW})\frac{l_{EW}}{l}, \\
l &= l_W + l_{EW}, \\
\Delta T_i &= T_i - T_{ref}
\end{aligned} \quad (3)
$$

In (3), R_{ref} represents the phase resistance at a reference temperature of $T_{ref} = 20°$ C and α_{Cu} denotes the linear copper temperature coefficient. The node losses, which are injected into the LPTN are calculated as follows:

$$
\begin{aligned}
P_{EW} &= \delta_1 P_{Cu,EW}(T_{EW}), \\
P_W &= \delta_2 P_{Cu,W}(T_W) + \delta_3 P_{resid}, \\
P_{PM} &= \delta_4 P_{resid}.
\end{aligned} \quad (4)
$$

The loss coefficients δ_i serve the purpose to distribute the dissipated losses in the motor on the specified thermal nodes and are estimated in the parameter identification process. In this context, model inadequacies can also be compensated to some extent [7]. The idea is to provide additional degrees of freedom to the model by assuming certain loss coefficients to vary depending on the currently applied current and speed (I, n). Modeling errors whose thermal impact on the temperature nodes also varies with I and n can then be mapped onto those loss coefficients during the parameter identification procedure implicitly compensating for errors in the temperature estimation that would have occurred otherwise. To what extent modeling errors can be compensated mainly depends on which loss coefficients are assumed to be variable. Here, the coefficients δ_1 and δ_4 were chosen in an empirical manner.

B. Extended Thermal Model with Two Nodes

The second thermal model proposed in this contribution consists of two interdependent models: An LPTN structure with two nodes for estimating the winding and PM temperature and an extension structure for calculating the end-winding temperature. The LPTN structure (s. Fig. 2) is quite similar to the one in Fig. 1, except that the heat path between winding and end-winding is now neglected and the entire copper losses are allocated to the winding node. The node losses, which are injected into the LPTN are calculated as follows:

$$P_W = \gamma_1 \left(P_{Cu,W}(T_W) + P_{Cu,EW}(T_{EW}) \right) + \gamma_2 P_{resid},$$
$$P_{PM} = \gamma_3 P_{resid}. \qquad (5)$$

To compensate for model inadequacies, the loss coefficients γ_1 and γ_3 are assumed to vary with I and n. In order to estimate the critical end-winding temperature, the model is extended via an empirical approach. The idea is to calculate the end-winding temperature directly from the estimated winding temperature via an appropriate conversion look-up table. The estimation accuracy of the end-winding temperature is hereby essentially depending on how well the following assumptions are fulfilled in reality:

1) Strong thermal coupling between winding and end-winding temperature.
2) Winding and end-winding temperature have similar thermal time constants.

Assuming a constant inlet coolant and environment temperature of $T_C = T_E = 20°$ C, a look-up table K_{20} can be obtained from measured thermal equilibriums ($\hat{T}$) in various operating points, according to

$$K_{20}(I, n) = \frac{\hat{T}_{EW}}{\hat{T}_W} \qquad (6)$$

Strictly speaking, as K_{20} is valid only in thermal equilibrium states and only for a constant inlet coolant and environment temperature of $20°$ C, accurate end-winding temperature estimations are limited to these operating conditions, as well. However, with the help of

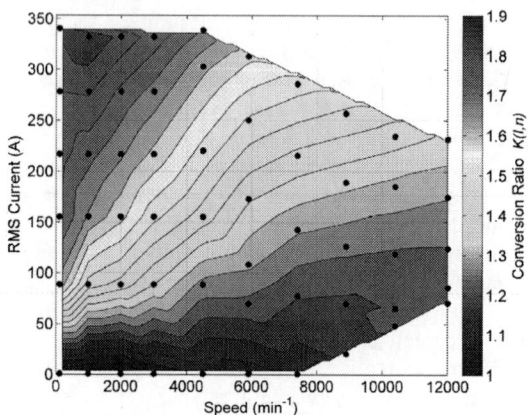

Fig. 3. Interpolated look-up table K for converting winding into end-winding temperature, dots represent measured grid points

some simple measures, the validity of the end-winding temperature estimation can be extended to thermal transients and varying inlet and environment temperatures. To better account for varying inlet and environment temperatures, the calculation of K is modified by means of

$$K(I, n) = \frac{\hat{T}_{EW} - T_{amb}}{\hat{T}_W - T_{amb}}, \qquad (7)$$

whereas

$$T_{amb} = T_C \sigma + (1 - \sigma) T_E, \quad \sigma \in [0,1] \qquad (8)$$

In (8), T_{amb} denotes the ambient temperature, which represents a weighted mean of T_C and T_E. The weighing factor σ is tuned in an empirical manner based on an appropriate validation profile (s. Fig. 6). Although its quantitative physical interpretability is limited, σ roughly indicates the relative degree of thermal coupling between T_C and $T_{(E)W}$ on the one hand and T_E and $T_{(E)W}$ on the other hand. In this work, σ was determined 0.9 indicating a strong coupling between $T_{(E)W}$ and T_C and a relatively weak coupling between $T_{(E)W}$ and T_E. The corresponding linearly interpolated look-up table K is depicted in Fig. 3. The dynamic influence of the coolant and environment temperature on the end-winding temperature can be modeled by filtering T_C and T_E with first-order low-pass filters. The filter time constants τ_C and τ_E are derived directly from the time constants of the winding (τ_W) and from the PM (τ_{PM}) by applying *Thévenin's theorem* to the LPTN in Fig 2:

$$\tau_C = \tau_W = C_W \frac{R_{W-C} R_{W-PM}}{R_{W-C} + R_{W-PM}}, \qquad (9)$$

$$\tau_E = \tau_W + \underbrace{C_{PM} \frac{R_{PM-E} R_{W-PM}}{R_{PM-E} + R_{W-PM}}}_{\tau_{PM}}, \quad \tau_{PM} \gg \tau_W. \quad (10)$$

A further low-pass filter should to be applied to the output of K to improve the model performance during thermal transients. That way, abruptly changing end-winding temperatures due to quickly varying operating points (I, n) can be avoided. Similarly to the weighing factor σ, the filter time constant τ_V can be chosen in an empirical manner within certain limits (here: $\tau_V \approx \frac{\tau_W}{3}$). The entire extension scheme for calculating the end-winding temperature is illustrated in Fig. 4.

As far as the complexity of the LPTN parameterization (s. section III) is concerned, the extended 2-node model approach is somewhat less demanding compared to the more coherent 3-node model approach introduced in the previous section. This is due to the fact, that an explicit axial discretization for describing heat flow between winding and end-winding is neglected in Fig. 2 and is instead substituted by an empirically-motivated model extension for directly calculating the end-winding from the winding temperature. However, this advantage comes at the price of a slightly lower accuracy in the end-winding temperature estimation (s. section VI).

III. PARAMETER IDENTIFICATION

For the parameter identification, a compact mathematical expression describing the heat exchange processes in the proposed LPTN structures in Fig. 1 and Fig. 2 is desired. For this purpose, the following state-space representation can be chosen:

$$\dot{\underline{T}} = \underline{A}(n)\underline{T} + \underline{B}(n,I)\underline{u}(\underline{T}) \qquad (11)$$

While the state vector $\underline{T}$ comprises the temperature nodes, the input vector $\underline{u}$ includes the different loss components as well as the inlet and environment temperature. For the 3-node model, state- and input vector break down to the components

$$\underline{T} = [T_{EW}, T_W, T_{PM}]^T,$$
$$\underline{u} = \left[P_{Cu,EW}(T_{EW}), P_{Cu,W}(T_W), P_{resid}, T_C, T_E \right]^T \quad (12)$$

for the 2-node model to

$$\underline{T} = [T_W, T_{PM}]^T,$$
$$\underline{u} = \left[P_{Cu,W}(T_W, T_{EW}), P_{resid}, T_C, T_E \right]^T. \quad (13)$$

The system- and input matrices $\underline{A}$ and $\underline{B}$ contain the thermal resistances and capacitances and with respect to $\underline{B}$ also the loss coefficients δ_i and λ_i, respectively. As R_{W-PM} and R_{PM-E} vary with n and the loss coefficients even vary with n and I, both models have a linear, but parameter-varying (LPV) character [9].

The applied parameter identification approach is based on the idea to approximate each thermal LPV model with a set of linear time-invariant (LTI) models, where each describes the plant behavior around a particular operating point (I, n). As all LTI models share the same structure, they can be subsumed under one model structure with the varying parameters being stored in look-up tables. The local LTI models can be identified based on a sequence of experiments during which, respectively, the plant and model parameters are held constant. In this context, the inclusion and adaption of constraints throughout the various identification cycles is crucial for obtaining a consistent parameter set and thus a good model performance. As constraints, the thermal state equations in (11), as well as roughly plausible parameter ranges

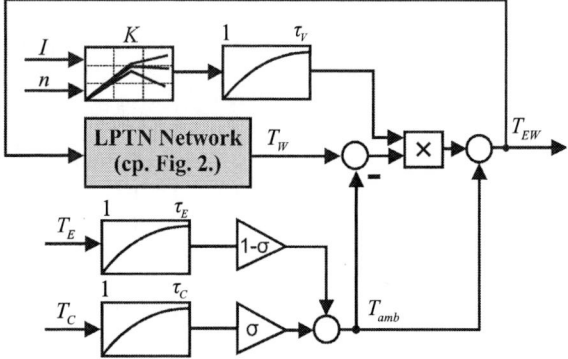

Fig. 4. Extension of 2-node LPTN network for estimation of end-winding temperature

were chosen. The state equations comprise important information concerning the fundamental heat flow mechanisms in the motor (e.g. heat flow results only from temperature differences between the nodes). Neglecting physically-motivated constraints during the identification procedure poses the risk of arbitrary parameter estimations and thus physically meaningless thermal models. Those models might show acceptable performances within the domain of the training data. However, when excited with new data (generalization) the accuracy of the temperature estimation is likely to decrease significantly.

The actual parameter identification is performed with the help of the *Prediction Error Method* (PEM) based on an iterative optimization approach (MATLAB System Identification Toolbox): A cost function $J(\underline{\theta})$, subject to the given constraints $g(\underline{\theta})$ and $h(\underline{\theta})$, is iteratively minimized by varying the thermal parameters in the parameter set $\underline{\theta}$:

$$\min_{\underline{\theta}} J(\underline{\theta}), \quad s.t. \ g(\underline{\theta}) \leq 0, h(\underline{\theta}) = 0 \qquad (14)$$

As cost function, the determinant of the error covariance matrix is chosen:

$$J(\underline{\theta}) = \det\left(\sum_{k=1}^N \left(\underline{e}(k, \underline{\theta}) - \underline{\mu}_e \right) \left(\underline{e}(k, \underline{\theta}) - \underline{\mu}_e \right)^T \right) \ (15)$$

In (15), $\underline{e}(k, \underline{\theta})$ denotes the temperature estimation errors at time step k for a given parameter set $\underline{\theta}$, while $\underline{\mu}_e$ represents their corresponding mean values over the considered identification profile of length N.

The measurement data used for the parameterization as well as the validation of the thermal models was acquired from a 60 kW reference IPMSM with distributed winding scheme (typically employed as traction motor in subcompact electric cars) which was equipped with a variety of temperature sensors in stator and rotor for the purpose of model estimation.

IV. STABILITY ANALYSIS

The following stability analysis is limited to electromechanical steady-state operation, where n and I and thus the model parameters can be assumed as constant. It is noted, however, that for LPV systems in general, the results from such analyses may not directly be adopted for arbitrary parameter variations (dynamic operation). In this case, a more general stability analysis based on *Lyapunov* functions becomes necessary [9].

During the parameter identification, it could be observed that given bounded thermal inputs, the model outputs remain bounded. This fact is consistent with the RC-network analogy and the notion for dissipative thermal systems. It has to be mentioned, though, that during the parameter identification it was convenient to calculate the temperature-dependent copper losses in $\underline{u}$, based on *measured* winding and end-winding temperatures. As it is the goal of this work to employ the identified models as stand-alone simulation models, copper losses need to be calculated based on *estimated* temperatures, however. In this case, $\underline{u}$ is a function of the states T_{EW} and T_W and is thus not exogenous. As far as stability analyses of the simulation models are concerned, it is therefore favorable to eliminate the couplings between $\underline{u}$ and $\underline{T}$ by shifting the non-exogenous model input to the system matrix according to

$$\dot{\underline{T}} = \underbrace{\left(\underline{A}(n) + \underline{B}(n,I)\underline{U}_\Delta\right)}_{\hat{\underline{A}}} \underline{T} + \underline{B}(n,I)\underline{u}^* \quad (16)$$

The (exogenous) input vector $\underline{u}^*$ and the matrix $\underline{U}_\Delta$ can be expressed by means of

$$\underline{u}^* = \left[P_{Cu,0}\frac{l_{EW}}{l}, P_{Cu,0}\frac{l_W}{l}, P_{resid}, T_C, T_E\right]^T, \quad (17)$$

$$\underline{U}_\Delta = \begin{bmatrix} P_{Cu,ref}\alpha_{Cu}\frac{l_{EW}}{l} & 0 & 0 & 0 & 0 \\ 0 & P_{Cu,ref}\alpha_{Cu}\frac{l_W}{l} & 0 & 0 & 0 \\ 0 & 0 & 0 & 0 & 0 \end{bmatrix}^T$$

for the 3-node model and

$$\underline{u}^* = \left[P_{Cu,0} + P_{Cu,ref}\alpha_{Cu}T_{amb}\frac{l_{EW}}{l}(1-K), P_{resid}, T_C, T_E\right]^T$$

$$\underline{U}_\Delta = \begin{bmatrix} P_{Cu,ref}\alpha_{Cu}\left(\frac{l_W}{l} + \frac{l_{EW}}{l}K\right) & 0 & 0 & 0 \\ 0 & & 0 & 0 & 0 \end{bmatrix}^T \quad (18)$$

for the 2-node model, where

$$\begin{aligned} P_{Cu,ref} &= 3R_{ref}I^2, \\ P_{Cu,0} &= P_{Cu,ref}\left(1 - \alpha_{Cu}T_{ref}\right). \end{aligned} \quad (19)$$

In (16), $\hat{\underline{A}}$ denotes the new (extended) system matrix. The eigenvalues of $\hat{\underline{A}}$ can be calculated by evaluating the characteristic equation

$$\det\left(s\underline{I} - \hat{\underline{A}}\right) = 0. \quad (20)$$

Although the eigenvalues of both models vary with I and n, it can be shown that they have solely negative real parts within the relevant operating region of the motor, i.e. the identified models are exponentially stable within the environment of the investigated electromechanical operating points. Instabilities in terms of thermal runaways only occur at currents which, depending on the speed, exceed the maximum allowed continuous motor current by factors of 2 and higher. For operating points, which lie outside of the identified operating region of the motor, the models are linearly extrapolated in terms of

their identified parameter look-up tables. The eigenvalues s_i of the identified thermal models can be expressed in terms of time constants τ_i (of the *modal* transformed system) according to

$$\tau_i(n,I) = -\frac{1}{s_i(n,I)} \quad (21)$$

In the 3-node model, the relevant modal time constants lie within a range of 51 s to 2450 s, in the 2-node model within a range of 95 s to 2250 s. The (locally) exponentially stable behavior of both models reflects the fact, that model parameters are identified consistently and under various physical constraints. That way, it can be ensured that the identified thermal models not only provide good approximations of the system's input/output behavior with respect to the applied training data, but moreover adopt major system-inherent characteristics. On the one hand, this is favorable as far as model performance and stability issues are concerned. On the other hand, model analyses become more meaningful from a physical point of view and help to get a better understanding of the behavior of the thermal reference system [7].

V. MODEL DISCRETIZATION

Before validating the parameterized model on an automotive test bench, the continuous-time thermal state-space models need to be discretized. Supposing, the input vector $\underline{u}^*$ between sampling instants $t = kT_s$ is kept constant, then (16) can be discretized with the transition matrix

$$\underline{\Phi} := e^{\hat{\underline{A}}T_s} \quad (22)$$

and the discrete-time input matrix

$$\underline{H} := \int_0^{T_s} e^{\hat{\underline{A}}T_s} d\tau_s \underline{B} \quad (23)$$

according to

$$\underline{T}((k+1)T_s) = \underline{\Phi}\,\underline{T}(kT_s) + \underline{H}\underline{u}^*(kT_s) \quad (24)$$

As the system matrix $\hat{\underline{A}}$ varies with the motor speed and current, $\underline{\Phi}$ and $\underline{H}$ have to be re-evaluated every time the speed or current changes. Since the evaluation of the transition matrix requires extensive numerical computations, the exact discretization approach above is replaced by the explicit Euler method (RK1). In this approach, matrices $\underline{\Phi}$ and $\underline{H}$ are replaced by their first-order approximations

$$\begin{aligned} \tilde{\underline{\Phi}} &= \underline{I} + T_s\hat{\underline{A}}, \\ \tilde{\underline{H}} &= T_s\underline{B} \end{aligned} \quad (25)$$

where $\underline{I}$ denotes the identity matrix. The resulting RK1 model can then be expressed as follows:

$$\underline{T}((k+1)T_s) = \hat{\underline{A}}\underline{T}(kT_s) + \underline{B}\underline{u}^*(kT_s) + \underline{T}(kT_s). \quad (26)$$

While the RK1 method is computationally less demanding than the exact discretization approach, its numerical stability is crucial. In [10], it was shown that a

The 2014 International Power Electronics Conference

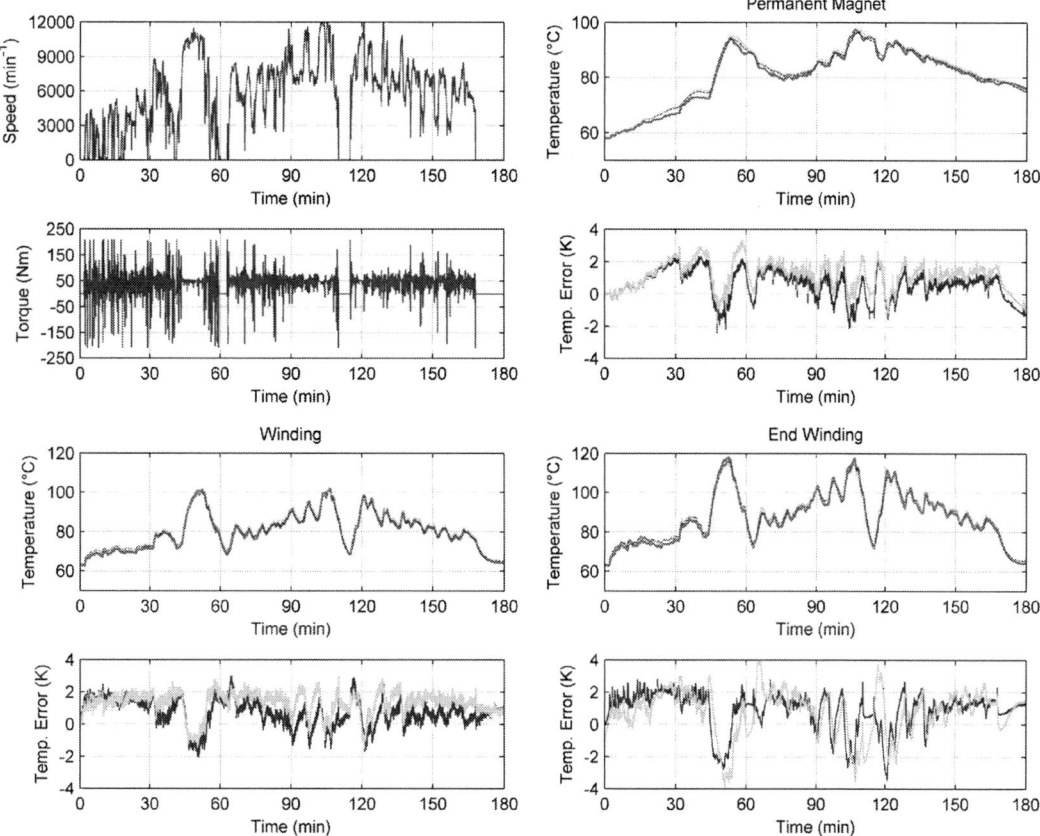

Fig. 5. Validation over real-world driving cycle with constant coolant temperature of 65° C and varying environment temperature in the test cabin [20° C-35° C]; Temperatures: Measured (red), estimated by 3-node model (blue), estimated by extended 2-node model (green)

stable RK1 discretization can only be guaranteed, if all eigenvalues s_i of the continuous-time system are located inside a circle with centre at $-1/T_s$ and radius $1/T_s$:

$$\left| s_i + \frac{1}{T_s} \right| < \frac{1}{T_s}. \qquad (27)$$

With regard to (21) and (27) the upper limit for the sampling period T_s can be derived:

$$T_s < T_{s,\max} = \min_i 2\tau_i. \qquad (28)$$

As for the identified thermal models, the smallest (modal) time constants were determined at 51 s (3-node model) and 95 s (2-node model), allowing maximum sampling periods of $T_{s,\max} \approx 100$ s and $T_{s,\max} \approx 190$ s, respectively. While discretizing the models with $T_{s,\max}$ is acceptable from a stability point of view, it will lead to unacceptable discretization errors in the temperature estimations. In order to keep discretization errors within a desirable ± 1 K band, the sampling times should therefore be reduced to $T_s < 3$ s, according to simulation results. The updates of the model matrices in (26) are based on averaged values of n and I. Both quickly changing quantities are sampled with 10 kHz on the employed dSPACE rapid prototyping environment.

VI. VALIDATION

The validity of the parameterized models within the relevant operating region was investigated on a test bench for automotive electric traction drives. The applied thermal excitation profiles (torque, speed) as well as coolant and environment temperatures hereby essentially differ from the ones used during the parameter identification. To avoid discretization errors, the sampling time of both thermal models was set to $T_s = 1$ s. The model performances are depicted in Fig. 5, Fig. 6 and Fig. 7 where the estimated temperature trajectories are plotted over the measured ones. In Fig. 5, the model performance against real-world driving data with permanent changes in speed and torque is validated. The torque and speed profiles were derived from GPS and acceleration data recorded during a mixed driving cycle (city, highway, German Autobahn) by applying it to an appropriate vehicle model describing the relevant longitudinal force balance equations. Those driving cycles are important for validating the model under realistic driving conditions. However, due to quickly changing electromechanical operating points, those profiles are rather unsuitable for validating a model's complete dynamic thermal behavior. This circumstance is addressed in Fig. 6 and Fig. 7, where time intervals between operating point changes were drastically

The 2014 International Power Electronics Conference

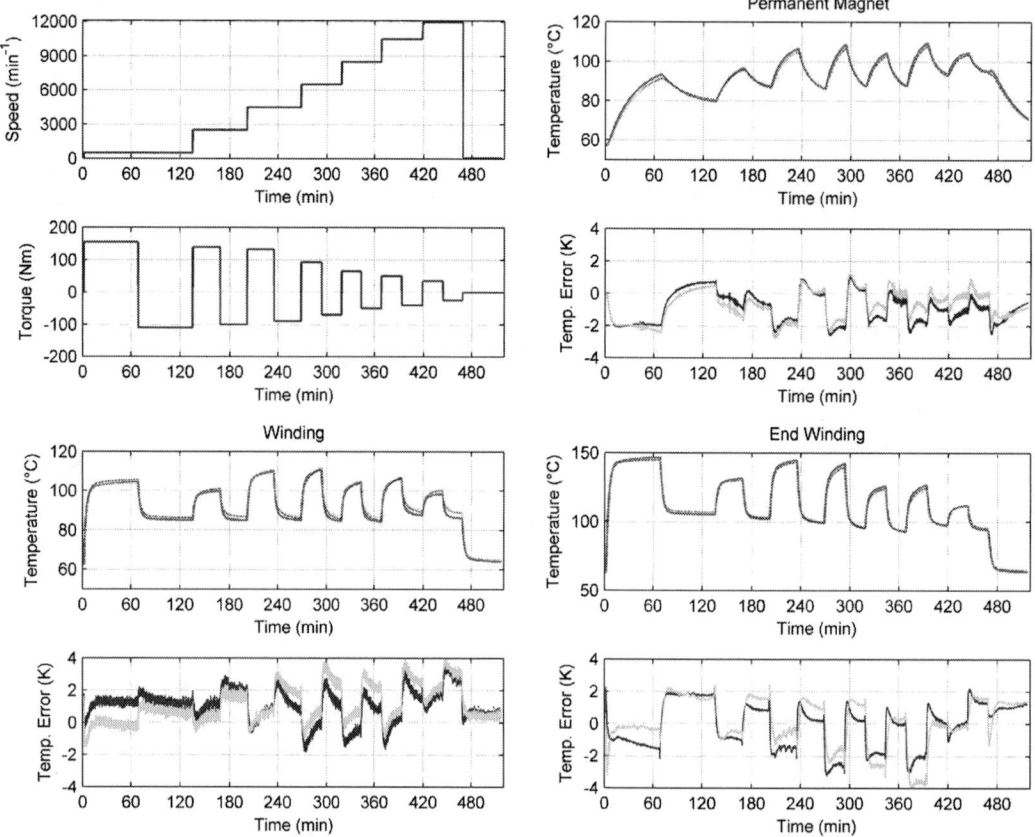

Fig. 6. Validation over staircase-like driving cycle with pronounced thermal transitions, constant coolant (65° C) and varying environment [20° C-30° C] temperatures; Temperatures: Measured (red), estimated by 3-node model (blue), estimated by extended 2-node model (green)

increased demonstrating the model's accuracy in describing pronounced transitions between thermal equilibriums. As the coolant temperature has significant impact on the temperatures in the motor, its influence is investigated in all validation profiles: In Fig. 5 and Fig. 6, its steady-state influence is investigated by applying a constant coolant temperature ($T_C = 65°$ C), which significantly exceeds the one during the parameter identification ($T_C = 20°$ C). In Fig. 7, the dynamic influence of varying coolant temperatures on the model performance is depicted.

While the estimated winding and PM temperatures of both models permanently lie within a ±3 K band around their measured reference values, the accuracy of the temperature estimation in the end-winding varies depending on which modeling approach is chosen. With respect to the 3-node model, the temperature estimation in the end-winding is similarly accurate as in the winding and PM, regardless of the investigated validation profile. As for the extended 2-node model, it becomes obvious that the end-winding estimation accuracy is slightly lower than the one in the winding and PM. This is mainly due to the following reasons:

1) The setup of the extended 2-node model is based on the assumption that winding and end-winding temperatures have similar time constants. As can be seen from the corresponding estimation errors

during pronounced transitions in Fig. 7, this assumption is only valid in a first approximation for the investigated motor. In fact, the actual end-winding time-constant seems to be slightly larger than the one of the winding. This observation is consistent with [7], where the time-constants of the identified 3-node model where explicitly calculated by applying *Thévenin's theorem* to the RC-network in Fig. 1.

2) As the end-winding temperature is calculated directly from the winding temperature through the look-up table K, it exhibits a high sensitivity towards errors in the estimation of the winding temperature. Depending on the current operating point, errors in the estimation of the winding temperature can be amplified by factors of up to 1.9 (cp. Fig. 3) and thus negatively affect the accuracy of the end-winding temperature estimation. This can be seen in Fig. 5, where the maximum errors in the end-winding temperature estimation are almost twice as high (±4 K) as the ones in the winding (±2 K).

VII. CONCLUSION

Two practical and computationally efficient approaches for monitoring critical temperatures in

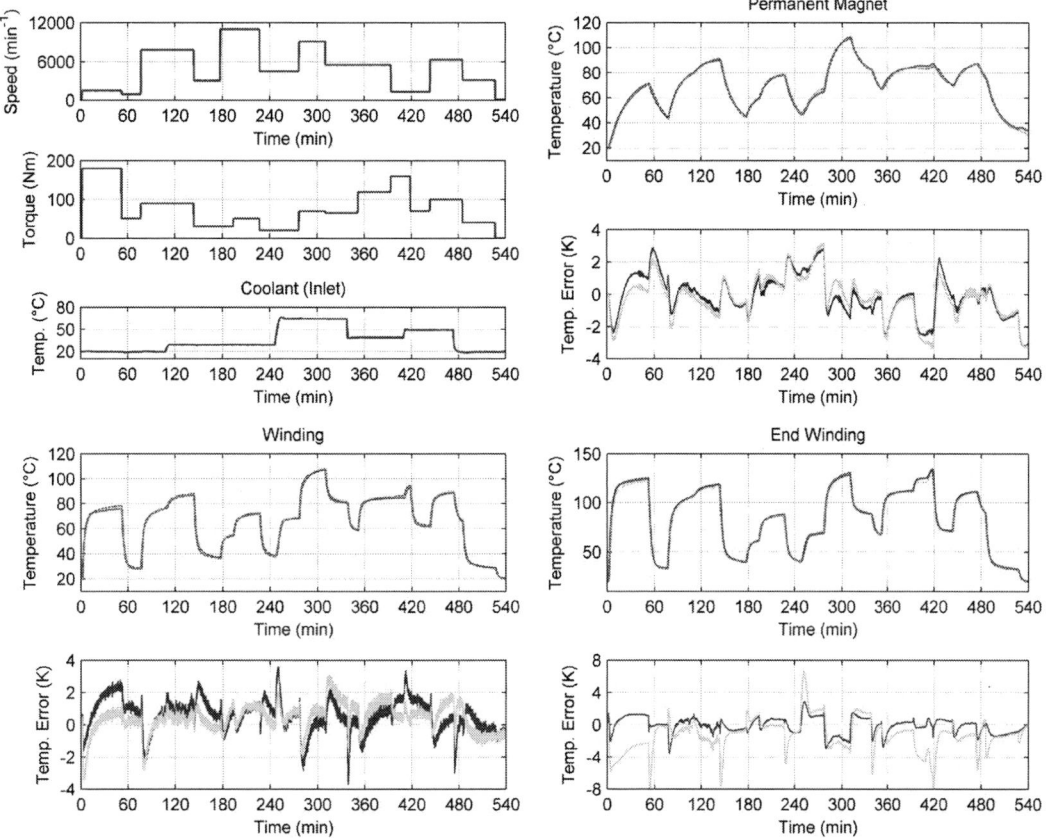

Fig. 7. Validation over driving cycle with pronounced thermal transitions and varying coolant [20° C-65° C] and environment [20° C-35° C] temperatures; Temperatures: Measured (red), estimated by 3-node model (blue), estimated by extended 2-node model (green)

IPMSM were introduced. The proposed low-order thermal models were parameterized solely based on experimental data and rely on loss models that can be easily obtained via measurements. As complicated analytical derivations of thermal parameters based on geometry and material information of the motor were not required, the proposed model structures along with the parameter identification method can easily be applied to similar IPMSM designs, too. Except for the necessary temperature measurements at the inlet of the cooling circuit and in the test cabin, the models rely only on commonly available quantities in the motor ECU, such as motor speed and electric currents. Extensive validations have shown, that both models are suitable for estimating PM, winding and end-winding temperatures reliably and accurately within the relevant operating region of the motor. With respect to the winding and PM temperature, both models provide high estimation accuracies with errors continuously lying in a ±3 K band. As for the end-winding temperature, the lower complexity of the parameterization task with respect to the extended 2-node model comes at the price of a slightly lower estimation accuracy, especially during pronounced thermal transitions. The 3-node model, on the other hand, is capable of maintaining the high accuracy of the winding and PM temperature for the end-winding temperature, as well.

REFERENCES

[1] A. Specht and J. Böcker, "Observer for the Rotor Temperature of IPMSM," *in Proc. 14th Int. Power Electronics and Motion Control Conf. (EPE/PEMC)*, pp. T4-12 - T4-15, 2010.

[2] M. Ganchev, C. Kral and T. Wolbank, "Sensorless rotor temperature estimation of permanent magnet synchronous motor under load conditions," *IECON*, 2012.

[3] S. D. Wilson, P. Stewart, and B. P. Taylor, "Methods of Resistance Estimation in Permanent Magnet Synchronous Motors for Real-Time Thermal Management," *IEEE Transactions on Energy Conversion*, vol. 25, no. 3, pp. 698-707, 2012.

[4] A. Boglietti, A. Cavagnino, D. Staton, M. Shanel, M. Müller and C. Mejuto, "Evolution and Modern Approaches for Thermal Analysis of Electrical Machines," *IEEE Transactions on Power Electronics*, vol. 56, no. 3, pp. 871-882, 2009.

[5] G. Dajaku and D. Gerling, "An Improved Lumped Parameter Thermal Model for Electrical Machines," *IEEE 17th International Conference on Electrical Machines (ICEM)*, pp. 1-6, 2010.

[6] G. D. Demetriades, H. Z. de la Parra, E. Andersson and H. Olsson, "A Real-Time Thermal Model of a Permanent-Magnet Synchronous Motor," *IEEE Transactions on Power Electronics*, vol. 25, no. 2, pp. 463-474, 2010.

[7] T. Huber, W. Peters and J. Böcker, "A Low-Order Thermal Model for Monitoring Critical Temperatures in Permanent Magnet Synchronous Motors," *Power Electronics, Machines And Drives Conf. (PEMD)*, 2014

[8] C. Kral, A. Haumer and S. B. Lee, "A Practical Thermal Model for the Estimation of Permanent Magnet and Stator Winding Temperatures," *IEEE Transactions on Power Electronics*, vol. 29, no. 1, pp. 455-464, 2014.

[9] F. Bruzelius, "Linear Parameter-Varying Systems - An Approach to Gain Scheduling," *Ph.D. Thesis*, Chalmers University of Technololgy, Göteborg, Sweden, 2004

[10] J. Böcker, "Discrete-Time Model of an Induction Motor," *European Transactions on Electrical Power*, pp. 65-71, 1991

Robust Current Control Insensitive to Gain Deviation and Offset of Inverter DC-link Current Sensor for SPMSM

Kei Matsuura, Itaru Ando
Akita National College of Technology
1-1 Bunkyo, Iijima, Akita,
Akita 011-0923, JAPAN
Email: i-ando@ipc.akita-nct.ac.jp

Kiyoshi Ohishi, and Masataka Matsuhashi
Nagaoka University of Technology
1603-1 Kamitomioka, Nagaoka,
Niigata 940-2188, JAPAN
Email: ohishi@vos.nagaokaut.ac.jp

Abstract—An AC servomotor is operated by current sensors. Owing to cost implications, it is often operated by an inverter system based on single DC-link current sensor. However, a current sensor sometimes has offset and gain variation caused by secular and thermal variation. When surface permanent magnet synchronous motor (SPMSM) drive system has the offset and the gain variation of current sensor, SPMSM has the current ripples and torque ripples. This paper proposes a new robust current control system insensitive to deviation and offset of current sensor for SPMSM. The proposed system tunes the current measurement gain deviation and realizes robust current control system. The proposed method suppresses q axis current errors by tuning the current measurement gains. Simulation results and experimental results confirm that the proposed method well suppresses the current sensor offset and the current sensor gain variation. Hence, the robust current control for SPMSM is always completed.

I. Introduction

Surface permanent magnet synchronous motors (SPM motors) have high efficiency, high power density, and less structural distortion. The SPM motors are widely used for high-speed position control, as actuators of electric power steering, and for other applications. A servo system of PMSMs must always maintain fine torque and speed responses, especially in industrial applications.

Generally, an AC servo motor system consists of a feedback control system that has a current control loop in a minor loop. Phase current values detected by the current sensors are used for the current control system. Three current sensors are used to measure three-phase currents, one for each phase. However, in most applications, two current sensors (for example, the u-phase and w-phase) are sufficient to measure the three-phase current. In these methods in which two or three current sensors are used to detect the phase current, there is a problem that the output characteristics of each current sensor are not exactly matched. For example, the current sensor has a characteristic variation such as "offset" and "gain variation." These characteristic variations in the current sensor have a detrimental influence on the current control. The gain offset and unbalanced gain variation cause current and torque ripple

in a motor drive system. If two current sensors each have an offset or gain deviation, the motor torque oscillates at once or twice the motor frequency[1]. Additionally, balanced gain variation causes a q-axis current error.

We have already proposed correction methods for high-performance AC servo drives using a current simulator[2][3]. These methods have relied on current measurement offsets to estimate the parameters of an electric motor[4][5].

To overcome these problems, only one DC-link current sensor is used[6][7]. The DC-link current is measured, and then a three-phase current is reconstructed by using a three-phase current reconstruction algorithm. In this method, when the current sensor has an offset, the reconstructed three-phase current is not affected by that offset. However, if the current sensor exhibits gain variation, the reconstructed three-phase current is affected, and the q-axis current has an DC error component. Thus, it is necessary to identify and compensate in real time the gain variation of the sensor.

In this paper, three-phase current are reconstructed by the single DC-link current sensor and analog reconstruction circuit [8]. By comparison with the detected current from the sensor and the estimated current from the current simulator, the gain variation is compensated. The efficacy of the proposed method has been verified through simulations and experiments.

II. Three Phase Current Reconstruction Method from DC-Link Current

When the inductive load is connected to the three-phase voltage source inverter, the inverter output current does not change rapidly. However, the inverter input current I_{dc} changes rapidly.

The paths of the current flow are shown in Fig. 1(a) when the output voltage vector is 110. The DC-link current is represented as equation (1).

$$ I_{dc} = i_u + i_v \mid_{110} \tag{1} $$

where I_{dc} is DC-link current, i_u is u phase current, i_v is v phase current. Similarly, the paths of the current flow are

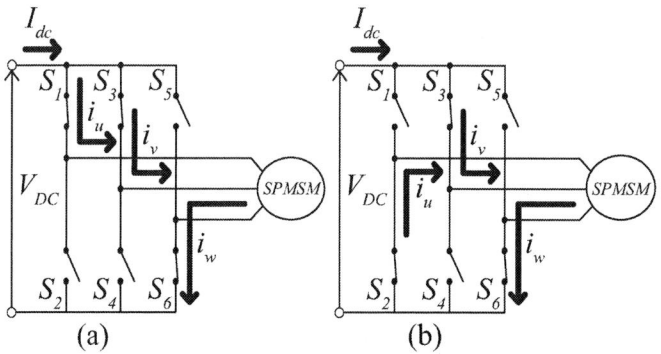

Fig. 1. Current paths in inverter by switching.

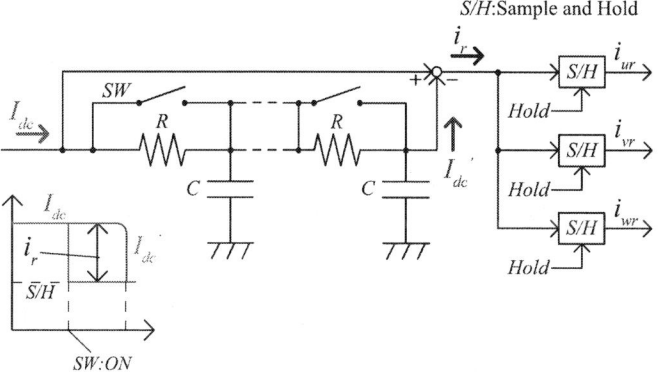

Fig. 2. Current reconstruction block.

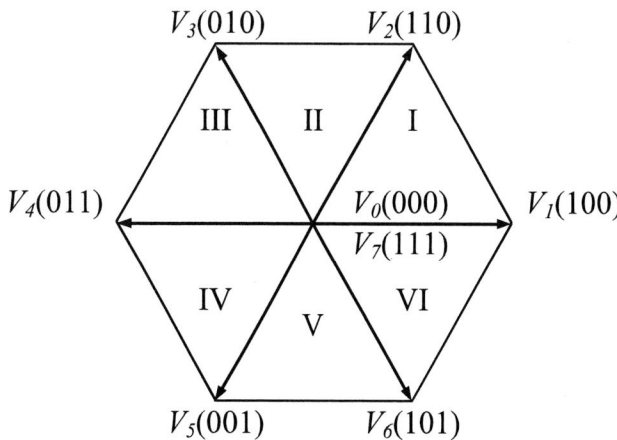

Fig. 3. Space vector of three phase inverter.

shown in Fig. 1(b) when the output voltage vector is 010. Additionally, the DC-link current is represented as equation (2).

$$I_{dc} = i_v |_{010} \qquad (2)$$

In this case, when the output voltage vector is changed from 110 to 010, the variation of the DC-link current is just i_u. Therefore, the u phase current is obtained by only detecting the DC-link current I_{dc}. For the other v or w phase, when only the one-pair switch is changed, each phase current is also obtained.

However, two or three switches are performed simultaneously, and appropriately, each phase current is not obtained. Consequently, in this case, the completion process is performed.

A block diagram of the reconstruction circuit using the DC-link current is shown in Fig. 2. By using multiple RC series circuits, the S curve I'_{dc} is obtained. Then, by comparing stepwise I_{dc} to S curve I'_{dc}, the variation of the DC-link current i_r is obtained before and after switching.

In this method, there is no need to sample and hold both DC-link currents before and after switching. Six sample and hold circuits, rather than eleven circuits, are needed for this system to obtain the DC-link current variation for each switch. It is possible to achieve three-phase current reconstruction with

a simpler circuit than what was used in the other method. Moreover, it is more cost-effective to adopt this technique from the viewpoint of cost and simplicity.

The proposed method has the ability to reconstruct the three-phase current from the DC-link current. By using the SVPWM technique to drive the inverter, only one-pair switches are turned-on or turned-off. Thus, the opportunity to obtain the variation of the DC-link current is increased before and after switching. For example, the output voltage is in a sector I shown in Fig. 3, output voltage vector is changed as follows, from equation (3) to (4).

$$V(000) \rightarrow V(100) \rightarrow V(110) \rightarrow V(111) \qquad (3)$$
$$V(111) \rightarrow V(110) \rightarrow V(100) \rightarrow V(000) \qquad (4)$$

When the output voltage vector is changed from $V(000)$ to $V(100)$, we can detect the u phase current i_u. After the output voltage vector is changed from $V(100)$ to $V(110)$, we can detect the v phase current i_v. The opposite is equally true.

In this method, the current control system is not affected according to the offset and unbalanced gain. Because three-phase currents are reconstructed by the only single DC-link current sensor, other current sensors are not needed.

Additionally, it is possible to reconstruct the three-phase current under conditions of a three-phase unbalanced volatage and unbalanced load. A simulation result of three-phase current reconstruction using the DC-link current is shown in Fig. 4. In this figure, i_u, i_v, and i_w are the actual three phase currents, and i_{ur}, i_{vr}, and i_{wr} are reconstructed three phase currents. It is confirmed that the three phase currents are reconstructed by the DC-link current. The reconstructed three-phase current correspond to the actual phase current.

An experimental result of three-phase current reconstruction by the DC-link current is shown in Fig. 5. It is confirmed that the three-phase currents are well reconstructed by dc-link current. The reconstructed three phase current approximately corresponds to the actual three-phase current.

The 2014 International Power Electronics Conference

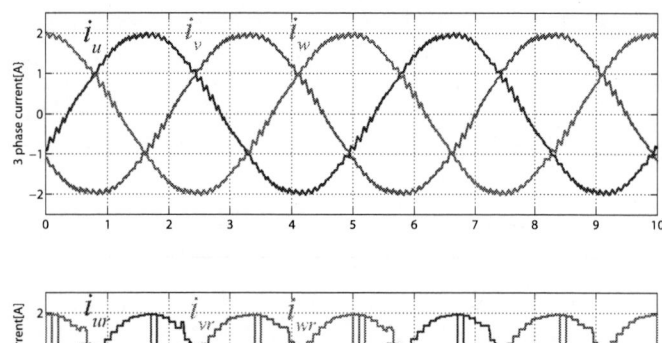

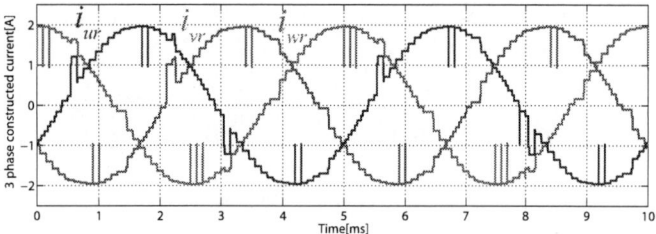

Fig. 4.   Simulation results of current reconstruction.

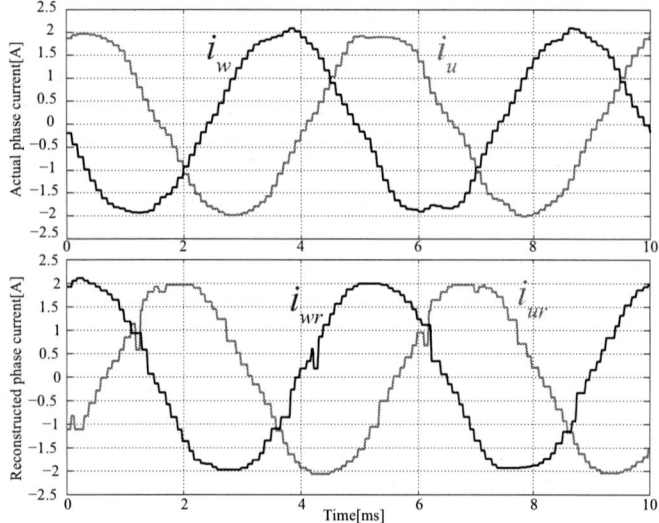

Fig. 5.   Experimental results of current reconstruction.

$$i_d = \frac{1}{R_a + sL_a}\left(v_d + \omega_{re}L_a i_q\right) \quad (5)$$

$$i_q = \frac{1}{R_a + sL_a}\left(v_q - \omega_{re}L_a i_d - \omega_{re}\Phi_{fa}\right) \quad (6)$$

where R_a is the stator winding resistance, L_a is the stator winding self-inductance, v_d is d-axis volatage, v_q is v-axis voltage, ω_{re} is motor speed, Φ_{fa} is the rotor magnetic flux, i_d is d-axis current and i_q is q-axis current. The current simulator consists of nominal motor parameters of R_n, L_n, and Φ_{fa}. By using a zero-order hold, the discrete-state equations of the PMSM are given by equations (7) and (8), which are the equations of the proposed current simulator[4][5].

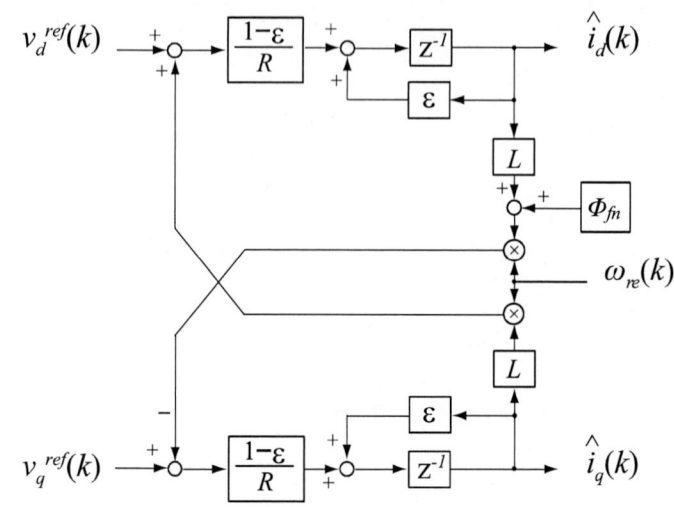

Fig. 6.   Block diagram of current simulator.

III. CURRENT SIMULATOR

The proposed method uses a current simulator to correct the components of the current measurement errors. Motor currents i_d and i_q are obtained from the state equations of the PMSM on the dq-axis. The circuit equations of the SPM motor are represented as equations (5) and (6).

$$\hat{i}_d[k] = \frac{1-\epsilon}{R_n}\{v_d[k-1] + \omega_{re}[k-1]L_n\hat{i}_q[k-1]\}$$
$$+\epsilon\hat{i}_d[k-1] \quad (7)$$

$$\hat{i}_q[k] = \frac{1-\epsilon}{R_n}\{v_q[k-1] - \omega_{re}[k-1]L_n\hat{i}_d[k-1]$$
$$-\omega_{re}[k-1]\Phi_{fn}\} + \epsilon\hat{i}_q[k-1] \quad (8)$$

where,

$$\epsilon = e^{-\frac{R_n}{L_n}T_c}$$

Voltage references v_d^{ref} and v_q^{ref} are inputted to v_d and v_q, and ω_{re} is inputted to the motor electrical speed. Then, the current simulator outputs the estimated currents $\hat{i}_d[k]$ and $\hat{i}_q[k]$ at the present sampling time k. From (7) and (8), the estimated currents under the steady state are expressed as equation (9) and (10).

$$\hat{i}_d = \frac{Rv_d^{ref} + L\omega_{re}v_q^{ref} - L\Phi_{fn}\omega_{re}^2}{R^2 + \omega_{re}^2 L^2} \quad (9)$$

$$\hat{i}_q = \frac{Rv_q^{ref} - L\omega_{re}v_d^{ref} - R\omega_{re}\Phi_{fn}}{R^2 + \omega_{re}^2 L^2} \quad (10)$$

The block diagram of the current simulator is shown in Fig. 6.

IV. TUNING METHOD OF CURRENT SENSOR GAIN VARIATION

Generally, two- or three-phase currents are measured in most applications. The current sensor offset causes a torque

978-1-4799-2706-7/14 $31.00 © 2014 IEEE   1518

ripple at the electrical angular frequency, and the unbalanced gain variation causes a torque ripple at double the electrical angular frequency. Additionally, the balanced gain variation causes a q-axis current error. When the output voltage vector is 110, the DC-link current is represented as equation (11).

$$I_{dc} = i_u + i_v + \Delta I_{dc}\,|_{110} \qquad (11)$$

where ΔI_{dc} is offset of current sensor. When the output voltage vector is 010, the DC-link current is represented as (12).

$$I_{dc} = i_v + \Delta I_{dc}\,|_{010} \qquad (12)$$

In the method of two-phase current detection, d-q axis current ripple of twice the electrical angular frequency caused by unbalanced gains are a problem. However, in the proposed method, torque ripple does not occur because only one sensor is used for current detection.

When the output voltage vector is 110 and gain variation occurs, the DC-link current is represented as equation (13).

$$I_{dc} = K_{dc}(i_u + i_v)\,|_{110} \qquad (13)$$

When the output voltage vector is 010 and gain variation occurs, the DC-link current is represented as (14).

$$I_{dc} = K_{dc}i_v\,|_{010} \qquad (14)$$

In this case, the reconstructed currents of each phase are expressed as equations (15), (16) and (17).

$$i_{ur} = K_{dc}i_u \qquad (15)$$
$$i_{vr} = K_{dc}i_v \qquad (16)$$
$$i_{wr} = K_{dc}i_w \qquad (17)$$

where, i_{ur} is reconstructed u phase current, i_{vr} is reconstructed v phase current, i_{wr} is reconstructed w phase current and K_{dc} is gain variation of current sensor. Each reconstructed current includes an influence of gain variation K_{dc}. The current sensor gain variation has a detrimental influence on the control system. Therefore, current sensor gain variation is compensated by using the current simulator.

In this paper, detected currents i_d^{sense} and i_q^{sense} are expressed as equations (18) and (19).

$$\mathbf{i_d^{sense}} = \mathbf{i_d} + \mathbf{\Delta i_d^{gain}} \qquad (18)$$
$$\mathbf{i_q^{sense}} = \mathbf{i_q} + \mathbf{\Delta i_q^{gain}} \qquad (19)$$

where, i_d is actual d-axis current, Δi_d^{gain} is an error of d-axis current by gain variation, i_q is actual d-axis current, Δi_q^{gain} is an error of q-axis current by gain variation. Here, we assume that i_d is feedback-controled and $i_d = 0$ is satified; therefore, equations (18) and (19) are expressed as equation (20).

$$\mathbf{\Delta i_{dq}^{gain}} = \begin{bmatrix} 0 \\ \frac{\sqrt{3}}{2}I\Sigma\epsilon \end{bmatrix} \qquad (20)$$

where, I is an effective value of motor current.

In equation (20), there is no influence on the d-axis current from the sensor gain variation.

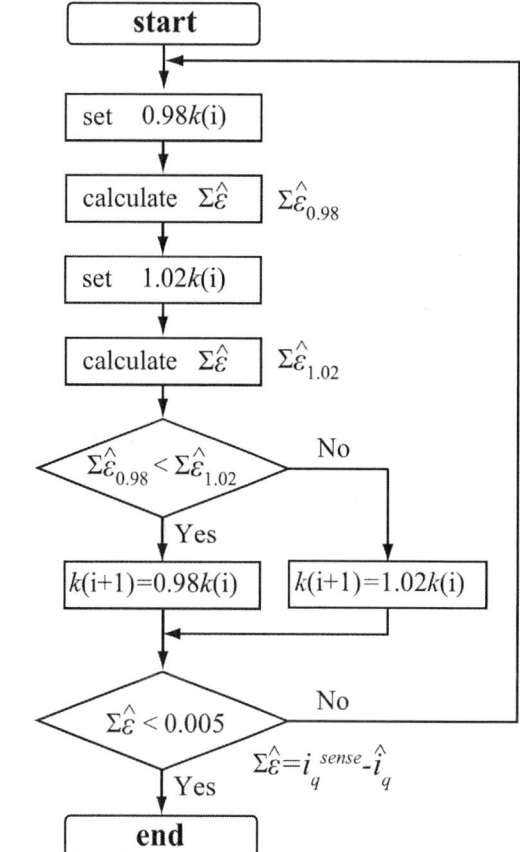

Fig. 7.   Flowchart of current sensor gain tuning.

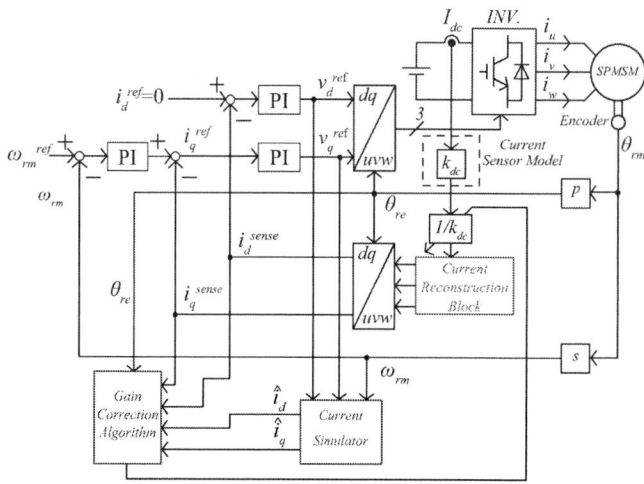

Fig. 8.   Speed control system of the SPM motor by DC-link current detection.

However, it is found that the q-axis current has a current error component. Hence, according to the indices of q-axis current error components, the current sensor gain variation is tuned. Here, in equation (21), the cut-off frequency of the LPF is set to the electrical angular frequency.

$$\Sigma\hat{\epsilon} = [i_q^e]_{LPF} \qquad (21)$$

TABLE I
MOTOR PARAMETERS AND SIMULATION CONDITION

Parameter	Symbol	Value	[Unit]
Number of pole pairs	p	4	[polepairs]
Armature inductance	L_a	9.00	[mH]
Armature Resistance	R_a	2.47	[Ω]
Rated current	I_R	1.4	[A_{RMS}]
Rated speed	N	3000	[min^{-1}]
Sampling period	T_c	100	[$\mu\,sec$]
Current sensor resolution	-	2.44(12bit)	[mA/bit]

where,

$$i_q^e = i_q^{sense} - \hat{i}_q \tag{22}$$

A flowchart of the current sensor gain tuning is shown in Fig. 7. The speed control system of the SPM motor according to the DC-link current detection is shown in Fig. 8. In the proposed method, the gain variation of the DC-link current sensor is tuned.

When the current sensor gain is changed, the tuning performed on the current detection error is reduced. The current sensor gain is tuned in the same positive and negative percentage in every control sequence. The end of the tuning is determined by comparing the threshold value and the value of expression (21). In this paper, the change in the gain is $\pm2\%$. The tuning is finished when the expression is less than 0.005.

V. SIMULATED AND EXPERIMENTAL RESULTS OF SENSOR GAIN TUNING

For the online compensation of the current sensor gain variation, a numerical simulation was carried out. In the simulation, the parameters of the 200-W class SPM motor shown in Table I were used.

In this paper, the simulation was performed under the condition that there were no electrical parameter variations. Additionally, the gain variation of the DC-link current sensor K_{dc} was set to +10%. A three-phase alternating current was reconstructed from the DC-link current by an analog reconstruction circuit. Reconstructed currents were used for current control.

The simulation results before performing the current sensor gain tuning are shown in Fig. 9. It was confirmed that the DC error component attributed to the gain variation of the current sensor affected the q-axis current. There was approximately 10 % error in the q-axis current. Then, simulation waveforms after performing the current sensor gain tuning were shown in Fig. 10. It was found that the DC error component was removed from the detected current of the sensor. Additionally, it was confirmed that the detected q-axis current corresponded to the estimated q-axis current from the current simulator.

In the same situation, an experiment was carried out to verify the efficacy of the proposed method. The experimental results, before the current sensor gain was tuned, are shown in Fig. 11. It was confirmed that the DC error component attributed to the gain variation of the current sensor affected

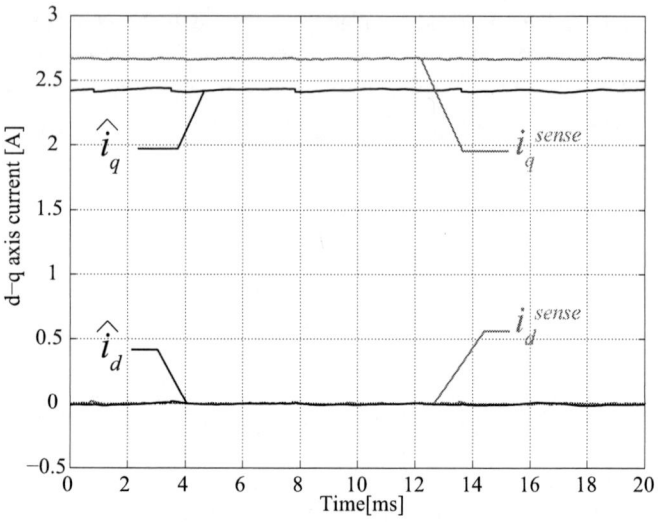

Fig. 9. Simulation result of estimated and detected d-q axis current(before gain tuning).

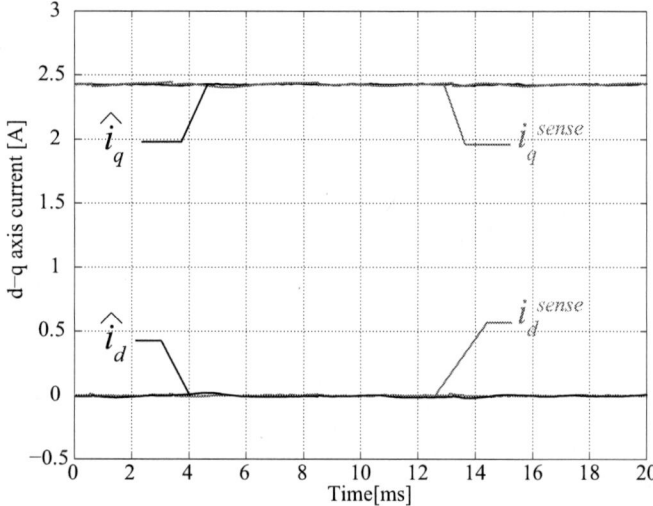

Fig. 10. Simulation result of estimated and detected d-q axis current(after gain tuning).

the q-axis current. The q-axis current also had an error of approximately 10 %. The experimental results after performing the current sensor gain tuning are shown in Fig. 12. It was found that the DC error component was removed from the detected current from the sensor. Additionally, it was confirmed that the detected q-axis current corresponded to the q-axis current estimated by the current simulator.

VI. CONCLUSION

In this paper, a three-phase current was first reconstructed from the DC-link current and analog cicuit. If two current sensors each have an offset or gain deviation, the motor current oscillates at once or twice the motor frequency. However, using the reconstruction technique from the DC-link current, both current ripples do not occur because only one sensor is used for current detection. Additionally, there is no effect on the

The 2014 International Power Electronics Conference

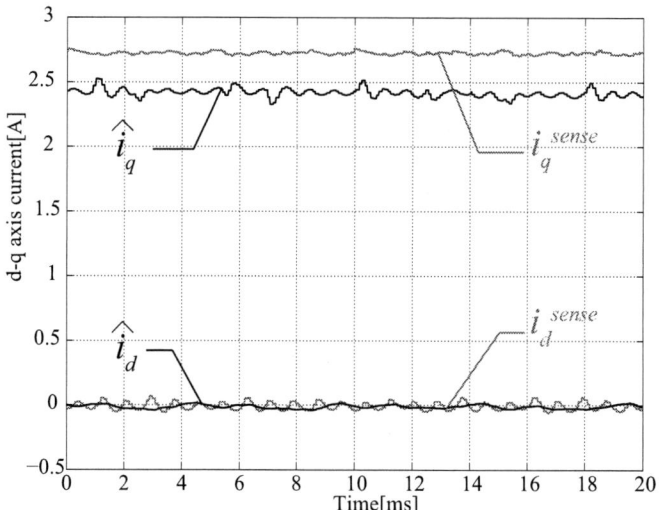

Fig. 11. Experimental result of estimated and detected d-q axis current(before gain tuning).

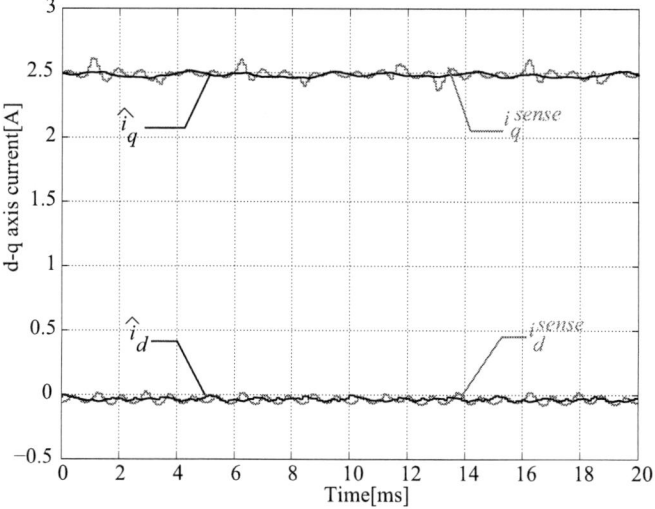

Fig. 12. Experimental result of estimated and detected d-q axis current(after gain tuning).

current sensor offset. However, the q-axis current has a DC error component from the current gain variation. This study applied the online tuning method to compensate for the current measurement gain deviations. It is confirmed that, under rated speed and rated torque, current sensor gain variation is compensated by the gain tuning algorithm. The three phase currents are easily and cost-effectively reconstructed by using the proposed method. The simulated and experimental results confirm that the proposed method suppresses the current offset caused by current measurement errors attributed to the sensor gain variation.

REFERENCES

[1] A.Shimada, K.Kawai, T.Zanma, S.Doki, and M.Ishida: "Sensorless Suppression Control for Frame Vibration of PMSM", T.IEE Japan, Vol.128-D, No.11, pp.1246–1253(2008)[In Japanese]

[2] K. Ohishi and Y. Nakamura, "High Performance Current Sensorless Speed Servo System of PM Motor Based on Current Estimation," Proc. of 2001 IEEE IAS Annual Meeting, pp.1–6, Vol.1(2001)

[3] K. Ohishi and K. Yoshida,"Current Sensor-less Speed Servo System of PM Motor Based on Self-Tuning Current Simulator," Proc. of 2003 IEEE International Electric Machines and Drives Conference, pp.1895–1900, Vol.1(2003)

[4] T. Senko, M. Sazawa, Y. Uenaka and K. Ohishi, "Fine Estimation & Self-tuning of both Current Sensor Offset and Electrical Parameter Variation for PM motor," Proc. of the 35th Annual Conference of the IEEE Industrial Electronics Society IECON'09, pp.1152–1157(2009)

[5] Y. Uenaka, M. Sazawa, K. Ohishi,"Fine Self-tuning Method of both Current Sensor Offset and Electrical Parameter Variations for PM motor," Proc. of the 36th Annual Conference of the IEEE Industrial Electronics Society IECON'10, pp.835–840(2010)

[6] Y.Murai, ett al. : "Three-phase Current-Waveform-Detection on PWM Inverters from DC Link Current-Steps", IPEC YOKOHAMA '95(1995)

[7] S.Aoyagi, et al : "New Detection Method of DC Bus Current Suitable for High Frequency PWM Carrier", Industrial Application Conference of IEEj, No.1-93 pp501-502(2006) (in Japanese)

[8] M.Matsuhashi and K.Ohishi, et al : "A Study on Tuning Method of Current Gain Deviation for SPMSM with Reconstruction of Three Phase AC Current Using DC Link Current Detection," Industrial Application Conference of IEEj, No.3-64 ppIII -319-322(2013) (in Japanese)

Auto-tuning Method of Inductances for Permanent Magnet Synchronous Motors

Naofumi Nomura
Fuji Electric Co.,Ltd.
Hino-City, Tokyo, Japan
nomura-naofumi@fujielectric.co.jp

Shinichi Higuchi
Fuji Electric Co.,Ltd.
Suzuka-City, Mie, Japan
higuchi-s@fujielectric.co.jp

Abstract— This paper presents a novel auto-tuning method of inductances for permanent magnet synchronous motor (PMSM). In the proposed method, inductances are measured injecting high frequency current and voltage to the motor at standstill. On the basis of adaptive control, the current control executing inductance tuning is stable and highly accurate up to the high frequency area, though motor parameters are unknown before parameter tuning is executed. Experimental results of the proposed method are also obtained. According to the proposed method, current control has high accuracy, and inductances measured by auto-tuning are highly accurate.

Keywords— *Adaptive control, auto-tuning, inductance, permanent magnet synchronous motor.*

I. INTRODUCTION

The permanent magnet synchronous motor (PMSM) is a suitable alternative to the induction motor to save energy and reduce the size in various applications such as air conditioners, industrial drives, etc. For lower cost and higher reliability of the PMSM-based drive, some techniques to eliminate the position-sensor (sensorless drive) have been developed [1]-[5]. Moreover, to save energy, to reduce the capacity, and to control torque of the PMSM drive system, some current control methods have been reported [6]-[8]. To implement the sensorless drive and current control methods, motor parameters are necessary, because these control methods are based on the electrical model of the motor. In addition, to achieve stability and high response for the current control, it is necessary to design optimally the current regulator.

Auto-tuning of motor parameters is effective to set up the drive system easily and precisely for a shorter time. Auto-tuning method of motor parameters for PMSM is proposed [9]-[11]. For example, it is shown that pulse voltage is injected to the motor, and inductances are measured from changes in current [9]. In another way, it is shown that a step current change is supplied to the motor, and motor parameters are measured by using least-squares method from the response of the current and the voltage [10][11].

However, there are problems in these inductance tuning methods.

First, for the tuning method of injecting the pulse voltage, the amplitude or injecting time of the pulse voltage should be well adjusted so that the optimal current may flow to measure inductance accurately. To determine the pulse voltage, the rough estimated value of inductance is necessary. Therefore, applying this tuning method to the PMSM whose value of inductance is quite unknown has the disadvantage of a decrease in accuracy of tuning and possible excessive current flow.

On the other hand, for the tuning method to control the current with a current regulator, it is necessary to design the current regulator so that the current control system may be stable and have a quick response. To achieve this, the value of inductance is necessary before inductance tuning is executed. Therefore, this tuning method has problems in the current control system during inductance tuning.

When we replace only an inverter in existing drive system, or we procure an inverter and a motor from another manufacturer, the motor parameters are often unknown. Especially for this case, the proposed auto-tuning methods have some problems already described.

This paper presents a novel auto-tuning method of inductances for the PMSM, which is possible to apply, though motor parameters are unknown. In the proposed method, inductances are measured injecting high frequency current and voltage to the motor at standstill. On the basis of adaptive control, the current control is stable and highly accurate up to the high frequency area, though motor parameters are unknown before parameter tuning is executed. As a result, the proposed auto-tuning method has high accuracy, because it has the ability to control the motor current to the best point for high accurate tuning.

Experimental results of the proposed method are also obtained. According to the proposed method, parameter estimate and current control are stable, current control is highly accurate, and inductances measured by auto-tuning have high accuracy.

II. AUTO-TUNING METHOD OF INDUCTANCES

A. Outline of Auto-tuning

At standstill, the electrical model of the PMSM in the d–q reference frame is given by the following equation.

$$\begin{bmatrix} v_d \\ v_q \end{bmatrix} = \begin{bmatrix} R_a + pL_d & 0 \\ 0 & R_a + pL_q \end{bmatrix} \begin{bmatrix} i_d \\ i_q \end{bmatrix} \qquad (1)$$

where,

i_d, i_q d-axis and q-axis current
v_d, v_q d-axis and q-axis voltage
R_a Stator resistance
L_d, L_q d-axis and q-axis inductance
p   Differential operator

According to the equation (1), d-axis and q-axis are independent of each other, and they are LR series circuits. So, it is possible to measure L_d by injecting high frequency voltage into d-axis, and detecting d-axis high frequency current. Similarly, it is possible to measure L_q by injecting high frequency voltage into q-axis, and detecting q-axis high frequency current.

B. Current Control in d-axis Inductance Tuning

Figure 1 shows the block diagram of the proposed auto-tuning of d-axis inductance for PMSM.

For inductance tuning and current control, the rotor position is necessary. For position-sensorless drive, before starting inductance tuning, it is executed to detect initial rotor position θ_{r0}, and to set rotor position θ_r. The detecting method of initial rotor position is shown in [5].

d-axis current reference i_d^* is obtained from the result of adding up DC component I_{d0}^* and high frequency component i_{dh}^*. To hold a rotor, DC component I_{d0}^* is set to positive constant value.

"High frequency current reference calculator"

generates d-axis high frequency current reference i_{dh}^*, as shown in the following equation.

$$i_{dh}^* = I_{dh}^* \sin \theta_h \qquad (2)$$

where,

$$\theta_h = \int \omega_h dt \qquad (3)$$

ω_h Angular frequency of high frequency current

d-axis current control is executed by feedback control by "d-axis current regulator" and feedforward control by "Feed forward compensator". d-axis voltage reference v_d^* is obtained from the result of adding up output of d-axis current regulator v_{dACR} and output of feed forward compensator v_{dhFF}. v_{dhFF} is calculated by the following equation.

$$v_{dhFF} = R_{aest} I_{dh}^* \sin \theta_h + X_{dhest} I_{dh}^* \cos \theta_h \qquad (4)$$

where,

R_{aest} estimated value of stator resistance
X_{dhest} estimated value of d-axis high frequency reactance

Estimated value R_{aest} and X_{dhest} reach to the true value by "parameter estimator". Detail of the "parameter estimator" will be described in the next section.

As a result, the output of feedforward compensator v_{dhFF} is operated accurately, and d-axis high frequency current reaches to d-axis high frequency current reference i_{dh}^*.

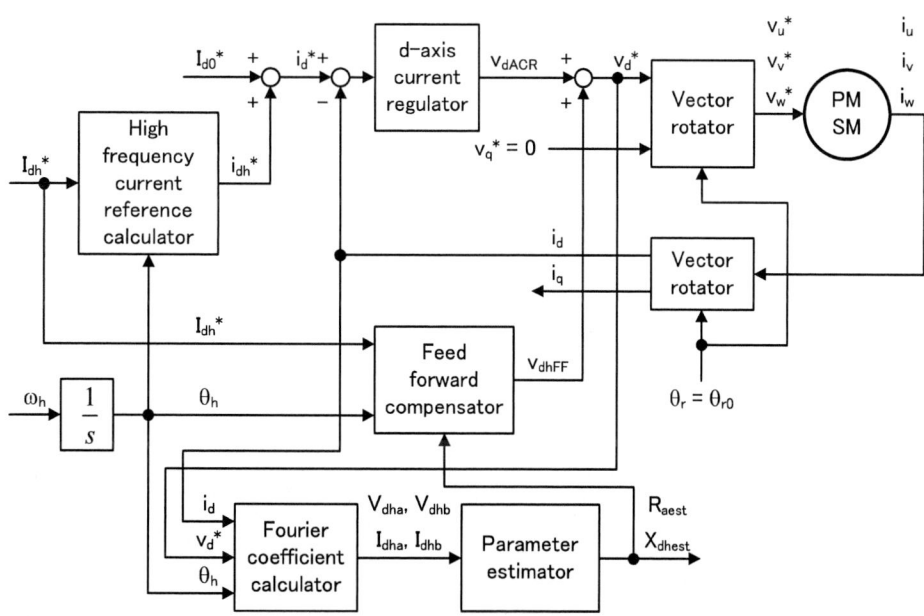

Fig. 1. Block diagram of the proposed auto-tuning of d-axis inductance for PMSM.

Since the feedforward control is operated accurately, "d-axis current regulator" doesn't need high response. So, it is easy to design "d-axis current regulator", even when the inductance is unknown. Therefore, this current control method is effective when motor parameters are unknown.

On the other hand, q-axis voltage reference v_q^* is controlled to zero.

C. Auto-tuning Method of d-axis Inductance

In the case of injecting high frequency voltage to d-axis, high frequency components of d-axis voltage v_{dh} and high frequency components of d-axis current i_{dh} are shown in the following equations.

$$v_{dh} = V_{dhb} \sin \theta_h + V_{dha} \cos \theta_h \tag{5}$$

$$i_{dh} = I_{dhb} \sin \theta_h + I_{dha} \cos \theta_h \tag{6}$$

For high frequency components, d-axis equation with complex impedance is shown in the following equation.

$$\dot{V}_{dh} = \left(R_a + j X_{dh} \right) \dot{I}_{dh} \tag{7}$$

where,

$$X_{dh} = \omega_h L_d \tag{8}$$

$$\dot{V}_{dh} = V_{dhb} + j V_{dha} \tag{9}$$

$$\dot{I}_{dh} = I_{dhb} + j I_{dha} \tag{10}$$

The current in equation (7) is expressed as the following equation.

$$\dot{I}_{dh} = \left(G_{dh} + j B_{dh} \right) \dot{V}_{dh} \tag{11}$$

where,

$$G_{dh} = \frac{R_a}{R_a^2 + X_{dh}^2} \tag{12}$$

$$B_{dh} = \frac{-X_{dh}}{R_a^2 + X_{dh}^2} \tag{13}$$

G_{dh} d-axis high frequency conductance
B_{dh} d-axis high frequency susceptance

Here, the current with the same phase as the voltage is defined as I_{dhpow}, and the current with the 90 degrees delayed phase following the voltage is defined as I_{dhvar}. Figure 2 shows the definition of I_{dhpow} and I_{dhvar} by the vector diagram of the d-axis voltage and the d-axis current on the auto-tuning of d-axis inductance.

From equation (11), I_{dhpow} and I_{dhvar} are shown in the following equation.

$$I_{dhpow} = G_{dh} V_{dhc} \tag{14}$$

$$I_{dh\,var} = \left(- B_{dh} \right) V_{dhc} \tag{15}$$

where,

$$V_{dhc} = \sqrt{V_{dha}^2 + V_{dhb}^2} \tag{16}$$

Equation (14) and (15) are linear models, where the input signal is V_{dhc} and the output signals are I_{dhpow} and I_{dhvar}. Therefore, it is possible to estimate parameters G_{dh} and B_{dh} by the adaptive control theory, and these parameters reach to the true value.

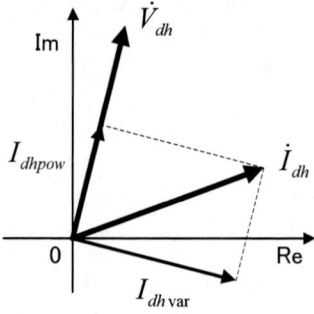

Fig. 2. Definition of I_{dhpow} and I_{dhvar}.

In general, it is impossible to detect I_{dhpow}, I_{dhvar} directly. Instead, it is possible to calculate I_{dhpow}, I_{dhvar} from the following equations.

$$I_{dhpow} = \frac{V_{dhb} I_{dhb} + V_{dha} I_{dha}}{V_{dhc}} \tag{17}$$

$$I_{dh\,var} = \frac{-V_{dhb} I_{dha} + V_{dha} I_{dhb}}{V_{dhc}} \tag{18}$$

As shown in equation (5) and (6), V_{dha}, V_{dhb}, I_{dha}, and I_{dhb} are Fourier coefficients of d-axis voltage and current, and it is possible to calculate them from d-axis voltage and current easily.

Figure 3 shows the block diagram of the G_{dh} and B_{dh} estimator.

Here, the estimated value of I_{dhpow} is defined as $I_{dhpowest}$, and the estimated value of G_{dh} is defined as G_{dhest}. "G_{dh} Estimator" adjusts G_{dhest} so that the deflection between $I_{dhpowest}$ and $I_{dhpowest}$ may become zero based on the adaptive control theory. Similarly, "B_{dh} Estimator" adjusts B_{dhest} so that the deflection between $I_{dhvarest}$ and $I_{dhvarest}$ may become zero.

$I_{dhpowest}$ is calculated from the product of V_{dhc} and G_{dhest}. Deflection between I_{dhpow} and $I_{dhpowest}$ is proportional to the deflection between G_{dh} and G_{dhest}. According to the adaptive control theory, G_{dhest} is calculated from integration of the deflection between I_{dhpow} and $I_{dhpowest}$, as shown in the following equation, and reaches to the true value [12].

$$G_{dhest} = \int \left\{ -\Gamma V_{dhcN} I_{dhpowerrN} \right\} dt \qquad (19)$$

where,

$$V_{dhcN} = \frac{V_{dhc}}{N} \qquad (20)$$

$$I_{dhpowerrN} = \frac{I_{dhpowest} - I_{dhpow}}{N} \qquad (21)$$

$$N = \sqrt{\rho + V_{dhc}^{2}} \qquad (22)$$

Γ gain
ρ constant(positive)

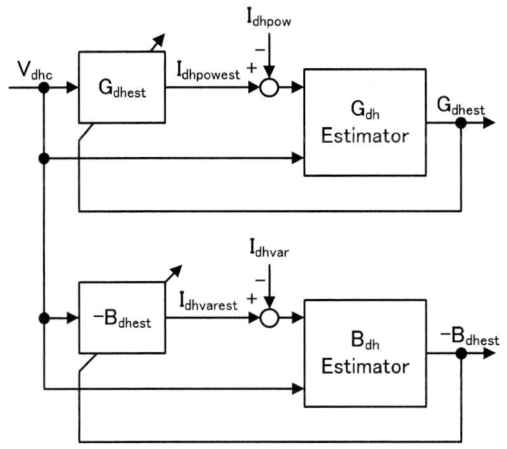

Fig. 3. Block diagram of the G_{dh}, B_{dh} estimator.

From equation (14),(20),(21), and (22), $I_{dhpowest}$ is calculated from the product of V_{dhc} and G_{dhest}, and the following equation can be derived.

$$V_{dhcN} I_{dhpowerrN} = \frac{V_{dhc}^{2}\left(G_{dhest} - G_{dh}\right)}{\rho + V_{dhc}^{2}} \qquad (23)$$

When ρ is enough smaller than V_{dhc}, equation (23) can be approximated to the following equation.

$$V_{dhcN} I_{dhpowerrN} \cong G_{dhest} - G_{dh} \qquad (24)$$

As shown in equation (24), the product of V_{dhcN} and $I_{dhpowestN}$ is approximated to the deflection between G_{dhest} and G_{dh}, and it is independent of voltage and current. As a result, transforming equation (19) by using equation (24), G_{dhest} is expressed as the following equation.

$$G_{dhest}(s) = \frac{1}{\dfrac{s}{\Gamma}+1} G_{dh}(s) \qquad (25)$$

where,
s: Laplace operator

As shown in equation (25), G_{dhest} becomes the first order delay of G_{dh}, and the time constant of G_{dhest} can be designed by gain Γ. Therefore, it is easy to design gain Γ.

Similarly, the estimated value of I_{dhvar} is defined as $I_{dhvarest}$, and the estimated value of B_{dh} is defined as B_{dhest}. $I_{dhvarest}$ is calculated from the product of V_{dhc} and B_{dhest}, and G_{dhest} is calculated from integration of the deflection between I_{dhvar} and $I_{dhvarest}$, as shown in the following equation.

$$-B_{dhest} = \int \left\{ -\Gamma V_{dhcN} I_{dh\,var\,errN} \right\} dt \qquad (26)$$

where,

$$I_{dh\,var\,errN} = \frac{I_{dh\,var\,est} - I_{dh\,var}}{N} \qquad (27)$$

The estimated value of stator resistance R_{aest} and the estimated value of d-axis high frequency reactance X_{dhest} are calculated by the following equations.

$$R_{aest} = \frac{G_{dhest}}{G_{dhest}^{2} + B_{dhest}^{2}} \qquad (28)$$

$$X_{dhest} = \frac{-B_{dhest}}{G_{dhest}^{2} + B_{dhest}^{2}} \qquad (29)$$

The estimated value of d-axis inductance L_{dest} is calculated by the following equation.

$$L_{dest} = \frac{X_{dhest}}{\omega_{h}} \qquad (30)$$

D. Current Control in q-axis Inductance Tuning and Auto-tuning Method of q-axis Inductance

Figure 4 shows the block diagram of the proposed auto-tuning method of q-axis inductance for PMSM.

Auto-tuning of q-axis inductance is achieved by injecting high frequency current and voltage to the motor at standstill. It is similar to the case of d-axis inductance tuning.

The initial rotor position θ_{r0} is set to the rotor position θ_{r}. d-axis current reference i_{d}^{*} is set to DC component I_{d0}^{*} to hold the rotor. q-axis current reference i_{q}^{*} is set to q-axis high frequency current reference i_{qh}^{*}. "High frequency current reference calculator" generates i_{qh}^{*}, as shown in the following equation.

$$i_{qh}^{*} = I_{qh}^{*} \sin \theta_{h} \qquad (31)$$

where, high frequency angle θ_{h} is calculated from equation (2).

d-axis voltage reference v_{d}^{*} is controlled to be equal to the output of d-axis current regulator v_{dACR}. q-axis voltage reference v_{q}^{*} is obtained from the result of adding up the output of q-axis current regulator v_{qACR} and the output of feedforward compensator v_{qhFF}. v_{qhFF} is calculated by the following equation.

$$v_{qhFF} = R_{aest} I_{qh}^* \sin\theta_h + X_{qhest} I_{qh}^* \cos\theta_h \quad (32)$$

where,

R_{aest} estimated value of stator resistance
X_{qhest} estimated value of q-axis high frequency

"Fourier coefficient calculator" calculates Fourier coefficients of q-axis voltage and current V_{qha}, V_{qhb}, I_{qha}, I_{qhb} from q-axis voltage v_q^*, q-axis current i_q, and high frequency angle θ_h. "Parameter estimator" calculates

estimated value R_{aest} and X_{qhest} from these Fourier coefficients of q-axis voltage and current. The method of the parameter estimation is similar to the case of d axis inductance tuning. It can be achieved by replacing the value of d axis with the value of q axis in equation (5) - (30) and Figure 3.

By these control processes, the estimated value R_{aest} and X_{qhest} reach to the true value, and q-axis current reaches to q-axis high frequency current reference i_{qh}^* though motor parameters are unknown.

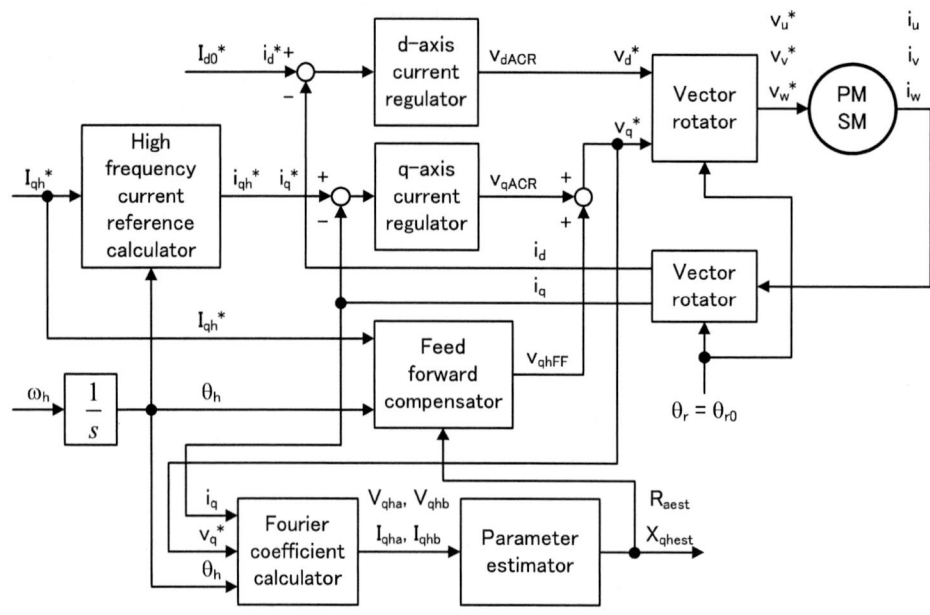

Fig. 4. Block diagram of the proposed auto-tuning of q-axis inductance for PMSM.

III. EXPERIMENTAL RESULTS

Table I shows specifications of the tested PMSM. Table II shows specifications of the inverter, which controls voltage of the tested PMSM and executes inductances tuning. Table III shows conditions of inductances tuning.

TABLE I
SPECIFICATIONS OF THE TESTED PMSM

Rated power	5.5kW
Base speed	1500r/min
Phase	3phase
Pole	6pole
Base frequency	75Hz
Rated current	20A
Rated voltage	185V

TABLE II
SPECIFICATIONS OF THE INVERTER

Input power-supply voltage	AC200V
PWM carrier frequency	4kHz
Control cycle	250μs

TABLE III
CONDITIONS OF INDUCTANCES TUNING

Angular frequency of high frequency current	ω_h	100 Hz
Amplitude of d-axis high frequency current reference	I_{dh}^*	0.38 pu(0-peak)
d-axis DC current reference	I_{d0}^*	0.2 pu
Amplitude of q-axis high frequency current reference	I_{qh}^*	0.53 pu(0-peak)
Gain	Γ	25
Constant	ρ	0.000625 pu

Figure 5 shows the currents and estimated parameters on auto-tuning of d-axis inductance. From the results of figure 5, it is clear that parameter estimate and current control are stable. Parameters and currents become under the steady state in 0.3 sec from the start of injecting high frequency current.

Figure 6 shows d-axis current on auto-tuning of d-axis inductance, expanding time-axis. From the results of figure 6, amplitude of high frequency component is 0.38[pu](0-peak) and frequency is 100[Hz], they reach to the references. Therefore, it is clear that current control is highly accurate though motor parameters are unknown.

Figure 7 shows currents and estimated parameters on

auto-tuning of q-axis inductance. From the results of figure 7 it is clear that parameter estimate and current control are stable, similar to auto-tuning of d-axis inductance.

Figure 8 shows q-axis current on auto-tuning of q-axis inductance, expanding time-axis. From the results of figure 8, the amplitude of high frequency component is 0.53[pu](0-peak) and frequency is 100[Hz], they reach to the references. Therefore, it is clear that current control is highly accurate though motor parameters are unknown, similar to auto-tuning of d-axis inductance.

Table IV shows values of inductances measured by auto-tuning. From the results of table IV, it is clear that the difference of measured inductances is small. Therefore, proposed auto-tuning has high accuracy.

TABLE IV
VALUES OF INDUCTANCES MEASURED BY AUTO-TUNING

Measure number	d-axis [mH]	q-axis [mH]
1	4.10	9.92
2	4.10	9.92
3	4.10	9.92
4	4.08	9.94
5	4.10	9.94

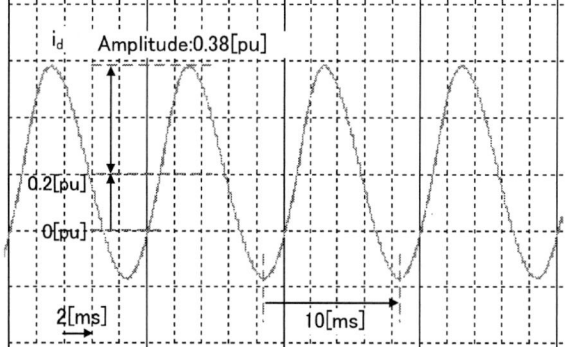

Fig. 6. d-axis current on auto-tuning of d-axis inductance (axis expansion of time).

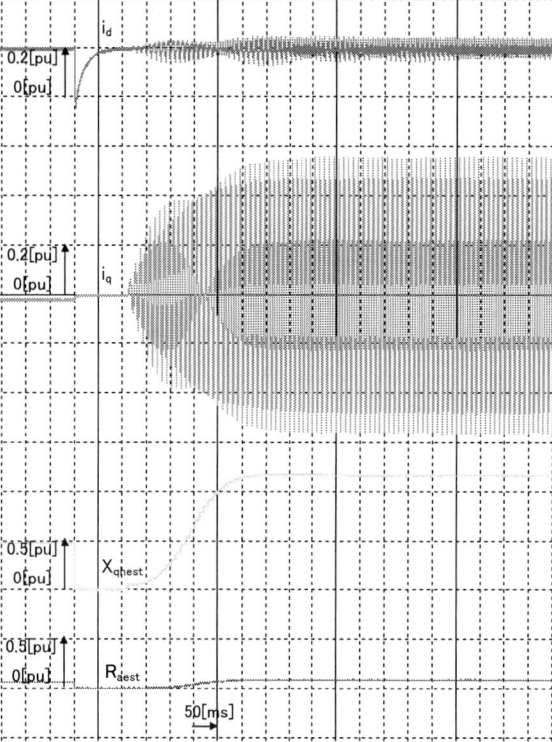

Fig. 7. Currents and estimated parameters on auto-tuning of q-axis inductance.

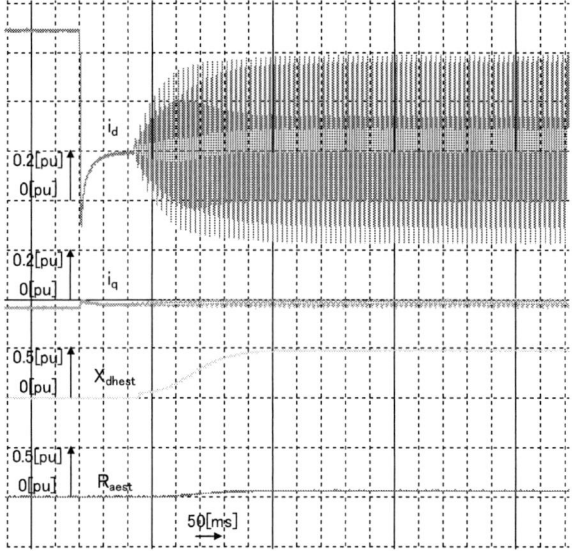

Fig. 5. Currents and estimated parameters on auto-tuning of d-axis inductance.

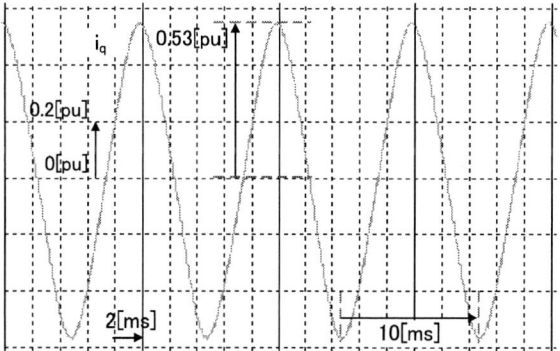

Fig. 8. q-axis current on auto-tuning of q-axis inductance (axis expansion of time).

IV. CONCLUSIONS

A novel auto-tuning method of inductances for the PMSM, which is possible to apply when motor parameters are unknown, has been presented and its validity has been confirmed experimentally. According to the proposed method, parameter estimate and current control are stable, current control is highly accurate up to the high frequency of 100[Hz]. And the difference of measured inductances is small, therefore proposed auto-tuning has high accuracy.

REFERENCES

[1] H. Watanabe, S. Miyazaki, T. Fujii, "A Sensorless Detection Strategy of Rotor Position and Speed on Permanent Magnet Synchronous Motor", IEEJ Trans. IA, Vol.110, No.11, 1990, pp.1193-1200 (in Japanese)

[2] N. Matsui, M. Shigyo, "Brushless DC motor control without position and speed sensors", IEEE Trans. on Industry Applications, Vol.28 , No.1, Jan./Feb., 1992, pp.120-127

[3] M. J. Corley, R. D. Lorenz, "Rotor position and velocity estimation for a salient-pole permanent magnet synchronous machine at standstill and high speeds", IEEE Trans. on Industry Applications, Vol.34 , No.4, July/Aug., 1998, pp.784-789

[4] R. Mizutani, T. Takeshita, N. Matsui, "Current model-based sensorless drives of salient-pole PMSM at low speed and standstill ", IEEE Trans. on Industry Applications, Vol.34 , No.4, July/Aug., 1998, pp.841-846

[5] T. Aihara, A. Toba, T. Yanase, A. Mashimo, K. Endo, "Sensorless Torque Control of Salient-Pole Synchronous Motor at Zero-Speed Operation", IEEE Trans. on Power Electronics, Vol.14, No.1, Jan., 1999, pp.202-208

[6] T. M. Jahns, G. B. Kliman, T. W. Neumann ,"Interior Permanent-Magnet Synchronous Motors for Adjustable-Speed Drives", IIEEE Trans. on Industry Applications, Vol. IA-22, No.4, July/Aug. 1986, pp.738-747

[7] T. M. Jahns, "Flux-weakening Regime Operation of an Interior Permanent Magnet Synchronous Motor Drive", IEEE Trans. on Industry Applications, Vol. IA-23, No.4, July/Aug., 1987, pp.681-689

[8] N. Nomura, A. Toba, T. Yamasaki, S. Ozaki, H. Ohsawa "Position-Sensorless Drive of the Interior Permanent Magnet Synchronous Motor for Wide Speed Range", EPE2001-Graz, 2001

[9] T. Takeshita, A. Usui, A. Sumiya, N. Matsui "Parameter Measurement of Sensorless Permanent Magnet Synchronous Motor", IEEJ Trans. IA., Vol.119, No.10, 1999, p.1184-1191 (in Japanese)

[10] S. Morimoto, M. Sanada, Y. Takeda, "Mechanical Sensorless Drives of IPMSM With Online Parameter Identification", IEEE Trans. on Industry Applications, Vol.42 , No.5, Sep./Oct., 2006, pp.1241 – 1248

[11] S. Morimoto, A. Shimmei, M. Sanada, Y. Yamada, "Position and Speed Sensorless Control System of Permanent Magnet Synchronous Motor with Parameter Identification", IEEJ Trans. IA, Vol.126, No.6, 2006, pp.748-755 (in Japanese)

[12] T. Suzuki, "Adaptive Control", 2001, CORONA PUBLISHING CO., LTD. (in Japanese)

An Impedance-Based Stability Analysis Method for Paralleled Voltage Source Converters

Xiongfei Wang, Frede Blaabjerg, and Poh Chiang Loh
Department of Energy Technology, Aalborg University, Aalborg, Denmark
xwa@et.aau.dk, fbl@et.aau.dk, pcl@et.aau.dk

Abstract— This paper analyses the stability of paralleled voltage source converters in AC distributed power systems. An impedance-based stability analysis method is presented based on the Nyquist criterion for multiloop system. Instead of deriving the impedance ratio as usual, the system stability is assessed based on a series of Nyquist diagrams drawn for the terminal impedance of each converter. Thus, the effect of the right half-plane zeros of terminal impedances in the derivation of impedance ratio for paralleled source-source converters is avoided. The interaction between the terminal impedance of converter and the passive network can also be predicted by the Nyquist diagrams. This method is applied to evaluate the current and voltage controller interactions of converters in both grid-connected and islanded operations. Simulations and experimental results verify the effectiveness of theoretical analysis.

Keywords— *Impedance-based analysis, Nyquist criterion, paralleled voltage source converters, stability*

I. INTRODUCTION

Voltage source converters have commonly been found in renewable energy generation systems, energy-efficient drives, and high-performance electronics equipment. The interactions of the paralleled or cascaded voltage source converters are challenging the stability and power quality in AC distributed power systems [1]. The constant power operation of these converters may destabilize the system with low-frequency oscillations [2]. The inner current or voltage control loops of converters may also interact with each other, and with the resonance conditions brought by the output *LCL* or *LC* filters and parasitic capacitances of power cables, resulting in resonances in a wide frequency range [3]. There is, consequently, an increasing research concern over the interaction of interconnected converters.

The impedance-based analytical approach has widely been used for the stability analysis of power-electronics-based power systems [4]-[8]. The minor-loop gain, which is defined as the terminal impedance ratio of the source and load converters, is proved to be effective to analyze the interactions of interconnected converters [4]. Several stability criteria have been derived based on the minor-loop gain, including the Gain Margin and Phase Margin (GMPM) criterion [5], the opposing argument criterion

[6], the Energy Source Analysis Consortium (ESAC) criterion [7], and the maximum peak criterion [8]. The different forbidden regions are thus defined to derive the impedance specification of the load converter for a given source converter impedance. All of the impedance-based stability criteria assume that the minor-loop gain has no Right Half-Plane (RHP) poles [8]. This is justified in the source-load converter systems, since each converter is designed with a stable terminal behavior. However, in the multiple paralleled source-source converter systems, such as wind farms, photovoltaic power plants, and paralleled uninterruptible power supplies, this prerequisite imposes the constraint on the derivation of impedance ratio. The presence of RHP zeros in the converter impedance may induce the RHP poles in the minor-loop gain [9]-[10].

To mitigate the influence of RHP zeros, two stability criteria have been recently reported, i.e. the passivity-based stability criterion [9], and the impedance sum type criterion [10]. The passivity-based stability criterion is derived from the frequency-domain passivity theory [11], which has been used earlier for current controller design of voltage source converters [12]. Generally, the passivity of the converter impedance is defined that the impedance has no RHP poles and has a positive real part. The system is stable if all the converter impedances are passive. Thus, the derivation of impedance ratio is avoided. This method allows the robust design of controllers for converters, yet is still a sufficient stability condition, since the negative real part of impedance does not indicate the instability of system. In contrast, the impedance sum type criterion is directly based on the characteristic equation of the minor feedback loop, which is the sum of converter impedances [10]. The encirclement of the origin in the complex plane indicates the instability of system. This criterion provides a sufficient and necessary stability condition and works well in the paralleled source-source converter systems. However, by this means, it is difficult to characterize the contribution of each converter to the stability in a system with multiple paralleled converters.

To reveal how the paralleled source-source converters interact with each other and with the passive components, this paper presents an impedance-based stability analysis method by means of the Nyquist criterion for multiloop systems. Instead of deriving the impedance ratio, a series of Nyquist diagrams drawn for the converter impedances and passive components are adopted to predict the system stability. Thus, the effect of RHP zeros in the converter

---

This work was supported by European Research Council (ERC) under the European Union's Seventh Framework Program (FP/2007-2013)/ERC Grant Agreement n. [321149-Harmony].

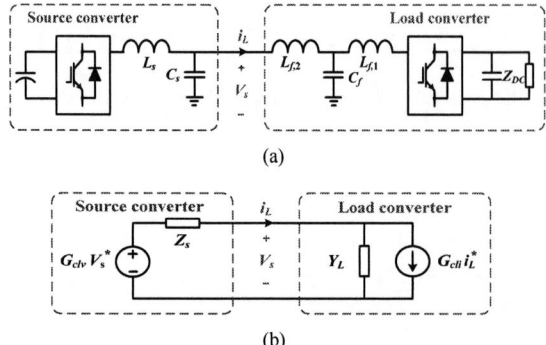

(a)

(b)

Fig. 1. A cascaded source-load converter system and equivalent circuit. (a) Basic configuration. (b) Equivalent circuit.

impedances can be avoided. This approach is applied to evaluate the current and voltage controller interactions of converters in both grid-connected and islanded operations. Simulations and experiments verify the effectiveness of theoretical analysis.

II. IMPEDANCE-BASED STABILITY CRITERION

A. Source-Load Converter System

Fig. 1 shows a typical cascaded source-load converter system and the equivalent circuit to illustrate the basic principle of the impedance-based stability criterion. The closed-loop response of the source converter voltage and the load converter current can be given by

$$i_L(s) = \underbrace{\frac{1}{1+Z_sY_L}}_{Minor\ loop}\Big[G_{cli}i_L^* + Y_L G_{clv}V_s^*\Big] \tag{1}$$

$$V_s(s) = \underbrace{\frac{1}{1+Z_sY_L}}_{Minor\ loop}\Big[G_{clv}V_s^* - Z_s G_{cli}i_L^*\Big] \tag{2}$$

where G_{cli} and Y_L denote the current reference-to-output transfer function and closed-loop input admittance of the load converter, respectively. G_{clv} and Z_s are the voltage reference-to-output transfer function and the closed-loop output impedance of the source converter, respectively. If the converters are designed with stable terminal behavior, i.e. G_{cli} and G_{clv} have no RHP poles, the overall system stability will be merely dependent on the minor feedback loop composed by the impedance product, Z_sY_L, which is also termed as the minor-loop gain.

In this scenario, due to the stable terminal behaviors of converters, the minor-loop gain has no RHP poles and the encirclement of the point $(-1, j0)$ indicates the instability of the system.

B. Source-Source Converter System

Fig. 2 illustrates a paralleled source-source converter system operating in grid-connected and islanded modes. Similarly, the impedance-based model of this system can be derived based on the terminal behaviors of converters, which is shown in Fig. 3. The converters are represented

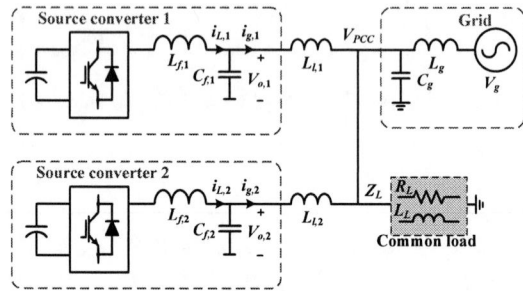

Fig. 2. A paralleled source-source converters system operating in grid-connected and islanded modes.

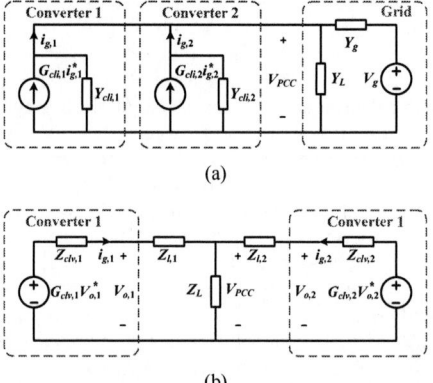

(a)

(b)

Fig. 3. Impedance-based equivalent circuit of paralleled source-source converter system in (a) grid-connected mode and (b) islanded mode.

by the Norton equivalent circuits in the grid-connected mode and the Thevenin equivalent circuits in the islanded mode. Thus, the closed-loop responses of the converters output currents and voltages in grid-connected mode and islanded mode, as well as the Point of Common Coupling (PCC) voltage can be derived in the following

$$i_{g,k} = \frac{G_{cli,k}i_{g,k}^*}{1+Y_{cli,k}/Y_{toi,k}} - \frac{Y_{cli,k}/Y_{toi,k}}{1+Y_{cli,k}/Y_{toi,k}}\Big(G_{cli,j}i_{g,j}^* + Y_g V_g\Big) \tag{3}$$

$$Y_{toi,k} = Y_{cli,j} + Y_g \quad \begin{cases} k,j \in \{1,2\} \\ k \neq j \end{cases} \tag{4}$$

$$V_{PCC} = \frac{1}{1+Y_{cli,k}/Y_{toi,k}} \cdot \frac{G_{cli,k}i_{g,k}^* + G_{cli,j}i_{g,j}^* + Y_g V_g}{Y_{toi,k}} \tag{5}$$

$$V_{o,k} = \frac{G_{clv,k}V_{o,k}^*}{1+Z_{clv,k}/Z_{tov,k}} - \frac{Z_{clv,k}/Z_{tov,k}}{1+Z_{clv,k}/Z_{tov,k}}G_{clv,j}V_{o,j}^* \tag{6}$$

$$Z_{tov,k} = Z_{clv,j} + Z_{l,j} + Z_{l,k} \tag{7}$$

$$i_{g,k} = \frac{1}{1+Z_{clv,k}/Z_{tov,k}} \cdot \frac{G_{clv,k}V_{o,k}^* - G_{clv,j}V_{o,j}^*}{Z_{tov,k}} \tag{8}$$

where $G_{cli,k}$ and $G_{cli,j}$ are the current reference-to-output transfer functions. $Y_{cli,k}$ and $Y_{cli,j}$ denote the closed-loop output admittances of converters in grid-connected mode. $G_{clv,k}$ and $G_{clv,j}$ are the voltage reference-to-output transfer functions, and $Z_{clv,k}$ and $Z_{clv,j}$ are the closed-loop output impedances of converters in the islanded operation. The effect of load, Y_L and Z_L, is disregarded. $Z_{tov,k}$ and $Y_{toi,k}$ are the equivalent system impedance and admittance of the k-th converter seen from the PCC, respectively.

Following (1) and (2), the minor-loop gain for the k-th converter in the grid-connected and islanded modes can be represented by the impedance ratios of $Y_{cli,k}/Y_{toi,k}$ and $Z_{clv,k}/Z_{tov,k}$, respectively. However, unlike the source-load converter systems, the equivalent system impedance may have RHP zeros due to the effect of passive components and the j-th converter. Consequently, the minor-loop gain will have RHP poles, and the system may be stable even if the Nyquist diagram encircles the point $(-1, j0)$. Hence, the stability criteria derived for the cascaded source-load converter system may not be applicable for the paralleled source-source converter system.

C. Presence of RHP Zeros

Fig. 4 shows the control block diagrams of converters in the grid-connected and islanded operations. Tables I and II list the parameters of electrical constants and the controllers which are used in this study.

Since the current and voltage controller interactions of converters are concerned in this work, the power control and grid synchronization loops are neglected. The single-loop grid current control is adopted in the grid-connected mode for the inherent active damping of LCL resonance [13], and the double-loop voltage control scheme is used in the islanded mode. The Proportional Resonant (PR) controller is used for control the grid current and output voltage with zero steady-state error.

Fig. 5 gives a comparison on the pole-zero maps of the system equivalent impedance and the terminal impedance of converter. It is clear that the RHP zeros present in the system impedance and admittance, yet no RHP zeros can be observed in the converter impedance and admittance. This implies that even if the terminal impedances has no RHP zeros, the interaction of the converters impedances and passive components may bring the RHP zeros into the system impedances. It is noted that the presence of RHP zeros implies that the impedance has the negative real part. However, the negative real part of impedance does not necessarily indicate the presence of RHP zeros.

TABLE I
ELECTRICAL PARAMETERS

Symbol	Meaning	Value
V_g	Line-line grid voltage	400 V
f_1	Grid frequency	50 Hz
L_g	Grid inductance	1.5 mH
C_g	Grid capacitance	2 μF
$L_{f,k}$	Filter inductor	1.8 mH
$C_{f,k}$	Filter capacitor	10 μF
$L_{l,k}$	Line inductance	0.9 mH
f_{sw}	Switching frequency	10 kHz
$V_{dc,k}$	DC-link voltage	750 V
R_L	Load resistance	80 Ω
L_L	Load inductance	166 mH

TABLE II
CONTROLLER PARAMETERS

Symbol	Meaning	Value
$K_{pg,k}$	Proportional gain of PR grid current controller	8
$K_{ig,k}$	Integral gain of PR grid current controller	500
$K_{pv,k}$	Proportional gain of PR voltage controller	0.005
$K_{iv,k}$	Integral gain of PR voltage controller	200
$K_{pc,k}$	Proportional converter current controller	8
T_s	Sampling period	100 μs

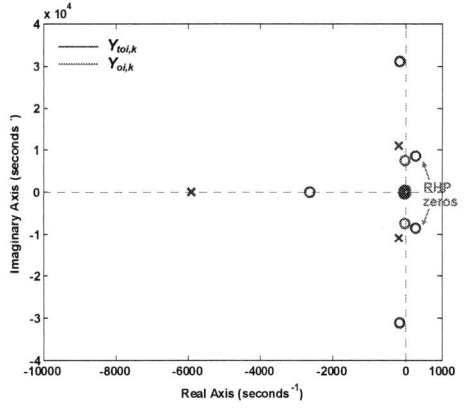

(a)

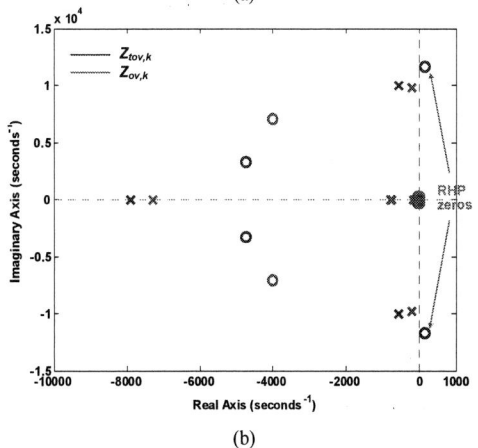

(b)

Fig. 5. Pole-zero maps of the (a) system equivalent admittance $Y_{toi,k}$ (zoom on origin) and (b) system impedance $Z_{tov,k}$ (zoom on origin).

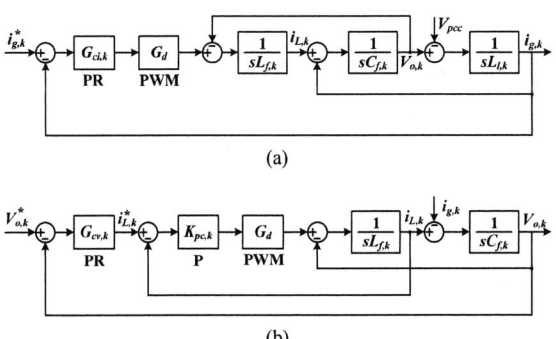

(a)

(b)

Fig. 4. Block diagrams of (a) the current control loop in grid-connected mode, and (b) the voltage control loop in the islanded mode.

III. PROPOSED ANALYSIS METHOD

This section reviews the Nyquist stability criterion for multiloop systems, and then presents an impedance-based stability analysis method to address the influence of RHP zeros in the conventional impedance ratio type criteria.

A. Nyquist Criterion for Multiloop Systems

The Nyquist criterion was generalized to the multiloop systems by Bode [14], which may be stated as follows:

"A linear multiloop system is stable if and only if the total numbers of clockwise and counterclockwise encirclements of the point $(-1, j0)$ are equal to each other in the series of Nyquist diagrams drawn for the individual loops obtained by beginning with all loops open and closing the loops successively in any order to their normal configuration [15]."

From Fig. 3, it can be seen that the impedance-based equivalent model of the interconnected converter system is basically a multiloop system. Thus, instead of deriving the overall open-loop gain of the minor feedback loop, the system stability can also be predicted by the series of Nyquist diagrams of the individual loops according to the Nyquist criterion for multiloop systems. Consequently, the effect of the RHP zeros in the system impedance can be avoided.

B. Stability Analysis of Source-Source Converter System

Fig. 6 illustrates a block diagram representation of the impedance-based equivalent circuit in Fig. 3. The minor feedback loop for the k-th converter is decomposed into two local loops by converter impedances. The stability of the minor feedback loop is thus assessed by successively closing the two loops and analyzing the Nyquist diagrams drawn for them. The system is stable if the total numbers of clockwise and counterclockwise encirclements of the point $(-1, j0)$ are equal to each other in these Nyquist diagrams. Moreover, by this means, how each converter contributes to the system stability can be revealed by the Nyquist diagrams of local loops.

Fig. 7 shows the Nyquist diagrams of two loops in the grid-connected operation. First, the loop that is composed by the grid impedance and the j-th converter admittance, $T_{ik,1}$ is evaluated, and then the loop including the k-th converter admittance, $T_{ik,2}$ is assessed.

$$T_{ik,1}(s) = \frac{Y_{cli,j}}{Y_g} \quad \Rightarrow \quad T_{ik,2}(s) = \frac{Y_{cli,k}}{Y_g + Y_{cli,j}} \qquad (9)$$

It is seen that only the Nyquist diagram of $T_{ik,1}$ encircles the point $(-1, j0)$ once in the clockwise direction, which indicates that the system is unstable. Further, the Nyquist diagram of $T_{ik,1}$ also implies that the interaction between the j-th converter and grid impedance leads to instability when the k-th converter is disconnected. Thus, to attain a stable system, the k-th converter impedance should make the minor feedback loop encircle the point $(-1, j0)$ once in counterclockwise direction. As a consequence, the design specification for the converter admittance can be derived from the Nyquist diagrams of local loops.

Fig. 8 shows the Nyquist diagrams of two loops in the

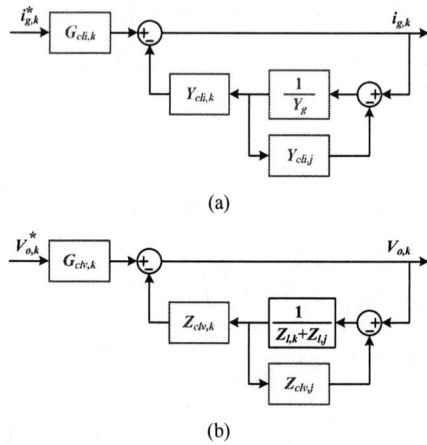

(a)

(b)

Fig. 6. Block diagram of the impedance-based equivalent system model. (a) Grid-connected mode. (b) Islanded mode.

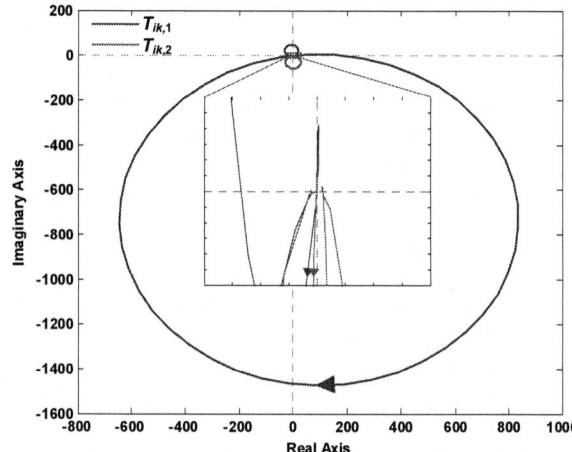

Fig. 7. Nyquist diagrams of two loops in the grid-connected operation.

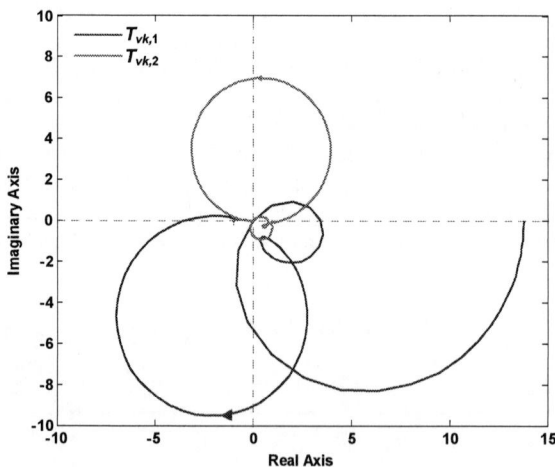

Fig. 8. Nyquist diagrams of two loops in the islanded operation.

islanded operation, in which the loop gains are given by

$$T_{vk,1}(s) = \frac{Z_{clv,j}}{Z_{l,k} + Z_{l,j}} \quad \Rightarrow \quad T_{ik,2}(s) = \frac{Z_{cli,k}}{Z_{l,k} + Z_{l,j} + Z_{clv,j}} \qquad (10)$$

Similarly to Fig. 7, the local feedback loop including the j-th converter impedance and line impedances is first assessed, and then the minor feedback loop with the k-th converter impedance is analyzed with Nyquist diagrams. It is seen that only the Nyquist diagram of $T_{vk,1}$ encircles the point $(-1, j0)$, which implies that the overall system is unstable. Further, the Nyquist diagram of $T_{vk,1}$ shows that the interaction of the j-th converter and line impedances cause instability when the k-th converter is disconnected. Therefore, to stabilize the islanded operation with the k-th converter, the terminal impedance $Z_{clv,k}$ should be shaped so that the Nyquist diagrams of two loops have the same numbers of clockwise and counterclockwise encirclement of the point $(-1, j0)$

Fig. 9 shows the control diagram for the k-th converter to shape the output impedance $Z_{clv,k}$. A feedback of the output voltage of the k-th converter is employed in the filter current control loop. In contrast, the control scheme

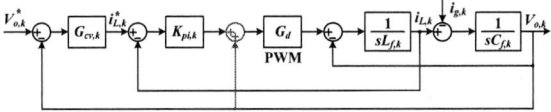

Fig. 9. Block diagram of the k-th converter in the islanded operation with the output voltage feedback in the filter current control loop.

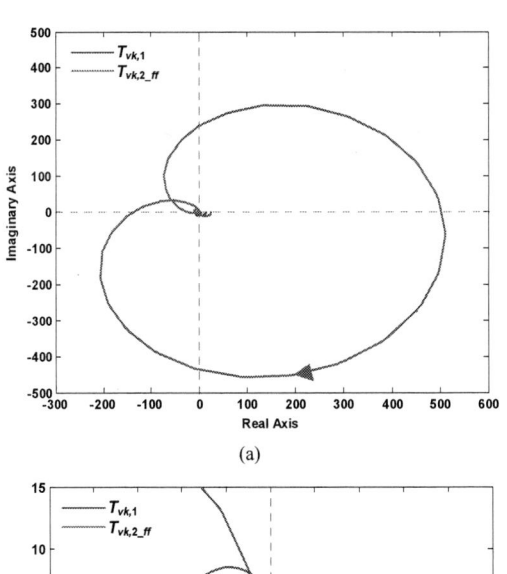

(a)

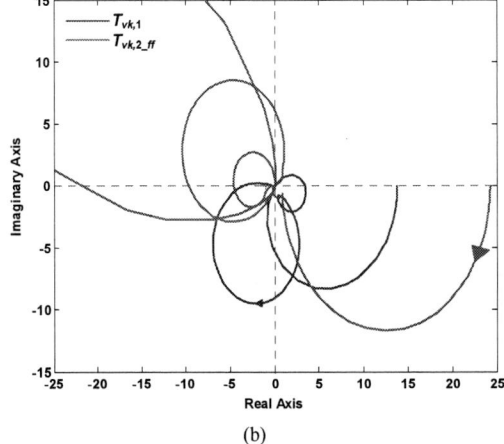

(b)

Fig. 10. Nyquist diagrams of two loops in the islanded operation with the modified control diagram for the k-th converter. (a) Full view. (b) Zoomed out around $(-1, j0)$.

for the j-th converter keeps the same as Fig. 4 (b). Fig. 10 shows the Nyquist diagrams of two loops. The loop $T_{vk,1}$ is the same as Fig. 8, while the loop $T_{vk,2}$ has encircles the point $(-1, j0)$ once in the clockwise and twice in the counterclockwise directions. As a consequence, the total numbers of clockwise and counterclockwise encirclement of the point $(-1, j0)$ are equal to each other. The islanded operation of the system is stable.

It is worth noting that this stability analysis approach can also be generalized to the N-paralleled source-source converter systems. The minor feedback loop for a given source converter can be divided into the N local feedback loops, which include N-1 loops to model the effect of the other $N-1$ paralleled source converters, and the minor feedback loop. Thus, how the source converters interact with each other and with the passive components can be successively assessed by the series of Nyquist diagrams drawn for the N loops.

IV. SIMULATIONS AND EXPERIMENTAL RESULTS

To validate the theoretical analyses, the time-domain simulations using PLECS Blockset and MATLAB, and the experimental tests based on two Danfoss frequency converters are carried out. The converters are powered by the constant DC voltage sources. The control algorithms in experiments are implemented in the DS1006 dSPACE system, in which the DS2004 high-speed Analog/Digital board is used for the sampling and the DS5101 waveform generation board is used for the Pulse Width Modulation (PWM) pulses generation.

A. Grid-Connected Operation

Fig. 11 shows the simulated grid currents of converters and the PCC voltage in the grid-connected operation. The unstable oscillations can be observed, which confirms the stability analysis in Fig. 7. However, due to the presence of RHP zeros in the equivalent system impedance, if only the admittance ratio $Y_{cli,k}/Y_{toi,k}$ is evaluated by the Nyquist diagram, then the instability cannot predicted.

Fig. 12 shows the measured grid currents and the PCC voltage waveforms in the grid-connected operation. It can

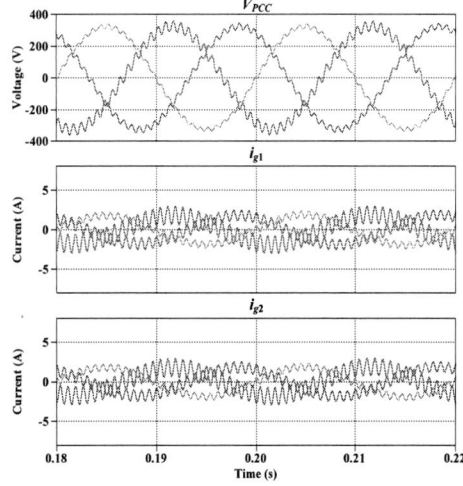

Fig. 11. Simulated grid currents of converters and the PCC voltage in grid-connected operation.

The 2014 International Power Electronics Conference

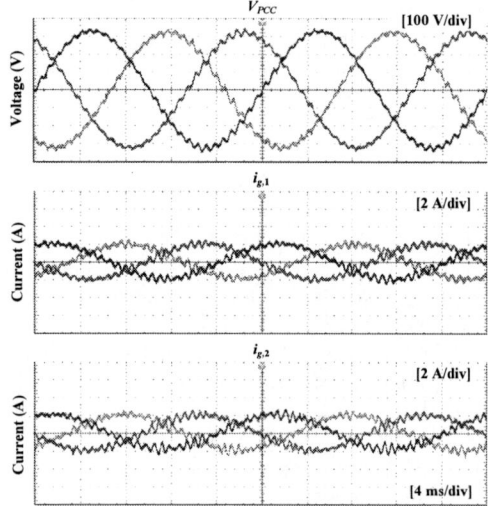

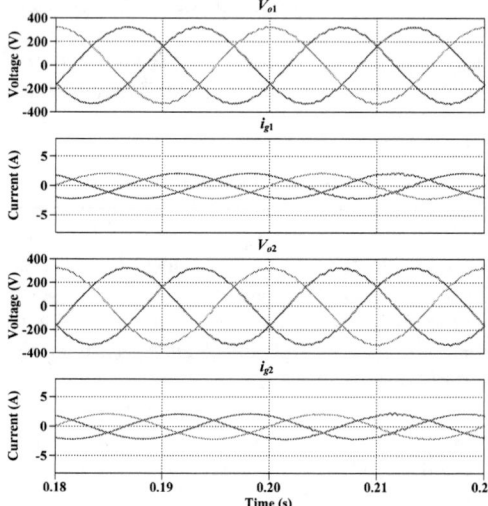

Fig. 12. Measure grid currents of converters and the PCC voltage in the grid-connected operation.

Fig. 14. Simulated converters output voltages and currents based on the control scheme in Fig. 4 (b).

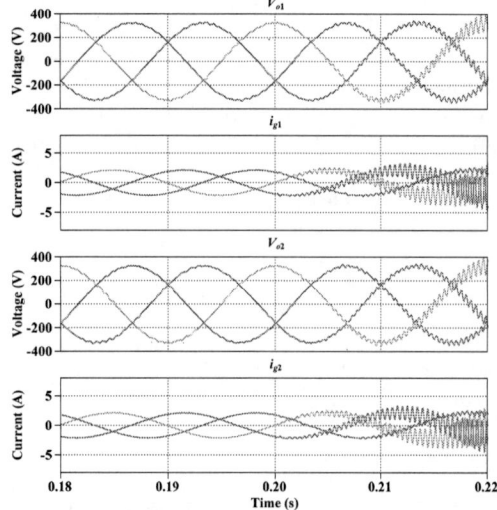

Fig. 13. Simulated converters output voltages and currents based on the control scheme in Fig. 4 (b).

be seen that the experimental tests matches well with the simulation results, which also again verify the theoretical analysis with the Nyquist criterion for multiloop systems.

B. Islanded Operation

Two simulation case studies are carried out to evaluate the system stability in the islanded operation, in order to validate the theoretical analyses shown in Figs. 8 and 10.

Fig. 13 shows the simulated converters output voltages and currents based on the control scheme shown in Fig. 4 (b). The converters are connected in parallel at the time instant of 0.2 s. It is clear that both of the converters are stable when operating standalone, and becomes unstable when they are connected together. This implies that the stable terminal behaviors of converters are designed. The interactions of converters with each other and with the line inductances result in the system instability, which verifies the stability analysis in Fig. 8. Similarly, if only

the Nyquist diagram of the impedance ratio $Z_{clv,k}/Z_{tov,k}$ is assessed following the conventional stability criteria, the opposite conclusion will be drawn.

Fig. 14 shows the simulated waveforms when one of the converters adopts the control scheme shown in Fig. 9. A stable system operation can be observed even when the two converters are connected in parallel. This agrees with the analysis in Fig. 10, and confirms the Nyquist criterion for multiloop systems when there are multiple clockwise and counterclockwise encirclements of the point $(-1, j0)$.

V. CONCLUSIONS

This paper has discussed the stability analysis for the paralleled source-source converter systems. The effect of the RHP zeros in deriving the minor-loop gains of source converters has been analyzed. It has been shown that the equivalent system impedance may have the RHP zeros, due to the interaction of converter impedance and passive components. To reveal how each converter interacts with each other and with passive components, an impedance-based stability analysis approach has been proposed with the Nyquist stability criterion for multiloop systems. The effect of RHP zeros is avoided in this method. Simulation and experimental case studies validate the effectiveness of the theoretical analyses.

REFERENCES

[1] J. Sun, "Small-signal methods for AC distributed power systems–a review," *IEEE Trans. Power Electron.*, vol. 24, pp. 2545-2554, Nov., 2009.

[2] T. Messo, J. Jokipii, J. Puukko, and T. Suntio, "Determining the value of DC-link capacitance to ensure stable operation of a three-phase photovoltaic inverter," *IEEE Trans. Power Electron.*, vol. 29, no. 2, pp. 665-673, Feb. 2014.

[3] X. Wang. F. Blaabjerg, M. Liserre, Z. Chen, J. He, and Y. W. Li, "An active damper for stabilizing power-electronics-based AC systems," *IEEE Trans. Power Electron.*, vol. 29, no. 7, pp. 3318-3329, Jul. 2014.

[4] R. D. Middlebrook, "Input filter design considerations in design and applications of switching regulators," in *Proc. IEEE IAS* 1976, pp. 366-382.

[5] C. M. Wildrick, F. C. Lee, B. H. Cho, and B. Choi, "A method of defining the load impedance specification for a stable distributed power system," *IEEE Trans. Power Electron.*, vol.10, no. 3, pp. 280-285, May 1995.

[6] X. Feng, J. Liu, F. C. Lee, "Impedance specifications for stable DC distributed power systems," *IEEE Trans. Power Electron.*, vol.17, no. 2, pp. 157-162, Mar. 2002.

[7] S. D. Sudhoff, S. F. Glover, P. T. Lamm, D. H. Schmucker, and D. E. Delisle, "Admittance space stability analysis of power electronic systems," *IEEE Trans. Aero. & Electron.*, vol. 36, no. 3, pp. 965-973, Jul. 2000.

[8] S. Vesti, T. Suntio, J. A. Oliver, R. Prieto, and J. A. Cobos, "Impedance-based stability and transient-performance assessment applying maximum peak criteria," *IEEE Trans. Power Electron.*, vol.28, no. 5, pp. 2099-2104, May 2013.

[9] A. Riccobono and E. Santi, "A novel passivity-based stability criterion (PBSC) for switching converter DC distribution systems," in *Proc. IEEE APEC* 2012, pp. 2560-2567.

[10] F. Liu, J. Liu, H. Zhang, and D. Xue, "Stability issues of Z+Z type cascade system in hybrid energy storage system (HESS)," *IEEE Trans. Power Electron.*, in press, 2014.

[11] J. C. Willems, "Dissipative dynamical systems Part I: general theory," *Arch. Ration. Mech. Anal.*, vol. 45, pp. 321-351, 1972.

[12] L. Harnefors, L. Zhang, and M. Bongiorno, "Frequency-domain passivity-based current controller design," *IET Power Electron.*, vol. 1, no. 4, pp. 455-465, Dec. 2008.

[13] J. Yin, S. Duan, and B. Liu, "Stability analysis of grid-connected inverter with LCL filter adopting a digital single-loop controller with inherent damping characteristic" *IEEE Trans. Ind. Inform.*, vol. 9, no. 2, pp. 1104-1112, May 2013.

[14] H. W. Bode, *Network analysis and feedback amplifier design*, Van Nostrand, New York, 1945.

[15] B. J. Lurie and P. J. Enright, *Classical feedback control with MATLAB® and Simulink®*, Boca Raton, FL, CRC Press, Taylor & Francis Group, 2011.

Dynamic Characteristics and Stability Comparisons between Virtual Synchronous Generator and Droop Control in Inverter-Based Distributed Generators

Jia Liu, Yushi Miura, and Toshifumi Ise
Graduate School of Engineering
Osaka University
Suita, Japan

Abstract— Recent years, to replace current source control methods, voltage source control methods such as virtual synchronous generator (VSG) and droop control have been proposed for grid-connected and/or stand-alone inverters. Both methods imitate the behaviors of synchronous generator. With the presence of swing equation, VSG has a virtual inertia, while droop control does not, and the rest of the two control methods are nearly the same. In this paper, the difference brought by the swing equation is studied. Small signal models for both control methods are built to compare the dynamic response during a small loading transition, and state space models are built to analyze the stability of both control methods. The results are verified by simulation in PSCAD/EMTDC.

Keywords—Distributed generator, droop control, virtual synchronous generator.

I. INTRODUCTION

With the development of distributed generation technologies, more and more converter-interfaced distributed generators (DGs) have been integrated into the grid. Meanwhile, various types of control methods for DGs have been proposed. The most commonly used control method, current source inverter (CSI) control, has several disadvantages. For example, CSI controlled DGs are not able to work in stand-alone mode, and due to the lack of inertia, this kind of DGs cannot take a large proportion in the grid. To solve these problems, several voltage source inverter (VSI) control methods have been proposed. As an example, droop control is a well-developed VSI control method [1]-[9], which focuses on the problems of stand-alone operation and power sharing by imitating the stable state characteristics of synchronous generators (SGs). However, no effort is done for transient state in droop control. As a result, DGs equipped with droop control are still lack of inertia, which is prone to cause a frequency instability problem. Therefore, a new concept of VSI control method, called virtual synchronous generator (VSG) [10]-[17], or virtual synchronous machine (VISMA) [18]-[20], or synchronverter [21], has been proposed. VSG imitates

not only the droop characteristics of SGs, but also the swing equation, which makes DGs have a virtual inertia and similar dynamic characteristics as SGs.

In this paper, the comparison of VSG control and droop control is performed. Both dynamic characteristics and stability issues are studied for each method. Dynamic response comparison, in both stand-alone mode and grid-connected mode, are performed through small signal model analysis, followed by stability comparison through state space models. Simulations in PSCAD/EMTDC are executed to verify the theoretical results.

II. PRINCIPLE OF VSG AND DROOP CONTROL

Fig. 1 shows the basic control block of VSG and droop control. It can be observed that for the reactive power control it is the same. Therefore, in this paper, only the difference of active power control is studied.

The swing equation in the module "VSG control" of VSG can be written as

$$P_{in} - P_{out} = J\omega_m \frac{d\omega_m}{dt} + D^* P_{base} \frac{\omega_m - \omega_g}{\omega_0}, \quad (1)$$

with J being the virtual inertia and D^* being the virtual damping factor (in per unit).

In the module "Governor",

$$P_{in} = P_0 - k_p{}^* P_{base} \frac{\omega_m - \omega_0}{\omega_0}, \quad (2)$$

where P_0, ω_0 are the set points of droop and $k_p{}^*$ is the droop coefficient (in per unit).

Let $k_p = \frac{k_p{}^* P_{base}}{\omega_0}$ and $D = \frac{D^* P_{base}}{\omega_0}$, and eliminate P_{in} from (1) and (2), thus

$$P_0 - k_p(\omega_m - \omega_0) - P_{out} = J\omega_m \frac{d\omega_m}{dt} + D(\omega_m - \omega_g). (3)$$

(3) reveals the relation between parameters both from swing equation and from governor. Therefore, dynamic characteristics of VSG can be deduced from (3).

The equation in "P Droop" of droop control is

$$\omega_m = -\frac{1}{k_p}(P_{out} - P_0) + \omega_0. \quad (4)$$

978-1-4799-2706-7/14 $31.00 © 2014 IEEE

Let $J = 0$, $D = 0$, then (3) is equivalent to (4). That is to say, droop control can be considered as a particular case of VSG control, where both inertia and damping factor are equal to zero.

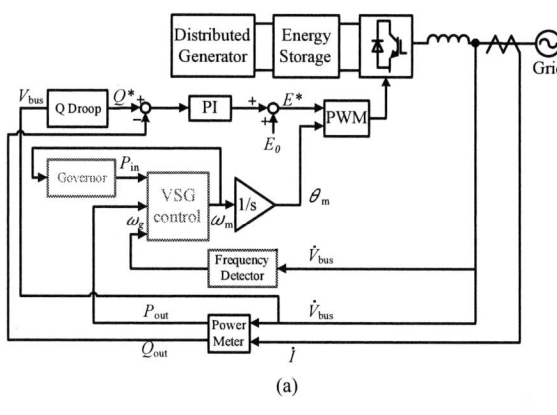

(a)

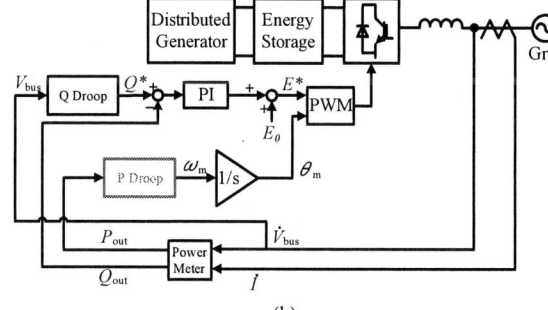

(b)

Fig. 1. Control block of (a) VSG; (b) droop control.

III. COMPARISON OF DYNAMIC CHARACTERISTICS

In this chapter, firstly, small signal models of VSG and droop control are built for both stand-alone mode and grid-connected mode. Then based on these models, the time response of frequency during a loading transition is calculated. The system with slower change of frequency is considered more stable, because high rate of change of frequency is prone to exceed the df/dt threshold of relays during a large load transition and result in unnecessary tripping and load shedding. Moreover, during a fault, systems with slower change of frequency produce smaller maximum frequency excursion within the same fault clearing time.

To simplify the model in order to focus on the active control performance, the reactive power control of both control methods is inactivated in our studies and thus $E^* = E_0$ in Fig. 1. Furthermore, all line losses and A/D conversion delays are neglected.

A. Stand-alone Mode

First of all, a stand-alone network depicted in Fig. 2 is considered. From power flow equation, it can be deduced that

$$\frac{X}{EV_{bus}}\Delta P_{out} = cos\delta_0 \Delta\delta. \tag{5}$$

Knowing that $\Delta\delta = \frac{1}{s}\left(\Delta\omega_m - \Delta\omega_g\right)$ and $\Delta P_{out} = \Delta P_{load}$, let $K = \frac{EV_{bus}}{X}cos\delta_0$, and then (5) can be arranged to

$$\frac{1}{s}\left(\Delta\omega_m - \Delta\omega_g\right) \approx \frac{1}{K}\Delta P_{load}. \tag{6}$$

The small signal model of (3) can be written as

$$-k_p\Delta\omega_m - \Delta P_{out} = J\omega_0 s\Delta\omega_m + D(\Delta\omega_m - \Delta\omega_g). \tag{7}$$

Eliminate $\Delta\omega_g$ from (6) and (7), then

$$\frac{\Delta\omega_m}{\Delta P_{load}} = -\frac{1+\frac{D}{K}s}{k_p + J\omega_0 s}. \tag{8}$$

Let $J = 0$, $D = 0$, the transfer function for droop control can be obtained as

$$\frac{\Delta\omega_m}{\Delta P_{load}} = -\frac{1}{k_p}. \tag{9}$$

Let $s = 0$ in (8), the result is equivalent to (9). This implies that steady state gain of (8) is determined by droop coefficient k_p and is independent of swing equation.

Based on (8) and (9), it is possible to calculate the response of frequency during a small loading transition (Fig. 3). Parameters used for calculation are listed in Table I.

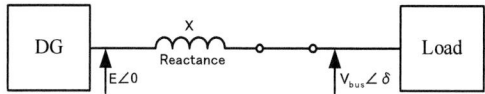

Fig. 2. Stand-alone mode model.

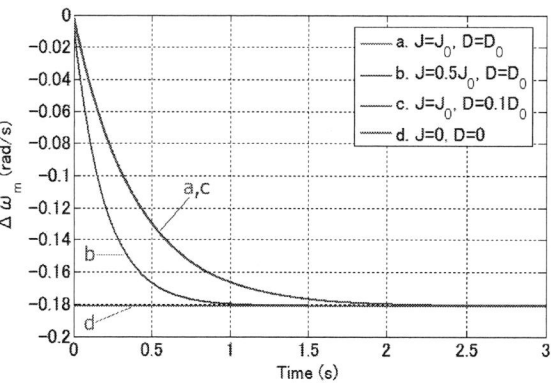

Fig. 3. Time response of DG frequency during a loading transition in stand-alone mode with various parameters.

TABLE I
STAND-ALONE MODE PARAMETERS

Parameter	Value	Parameter	Value
E	6.6kV	J_0	56.3kg·m^2
P_{base}	1MW	$D_0{}^*$	17pu
P_0	1pu	ΔP_{load}	0.0096MW
ω_0	376.99rad/s	V_{bus}	6.585kV
$k_p{}^*$	20pu	δ	0
X	0.1449pu		

By comparing Line (a) (VSG) and Line (d) (droop control) in Fig. 3, it can be observed that the frequency of

VSG changes more slowly. As it is discussed previously, this implies that during the transient state, VSG has better dynamic characteristics than droop control. Moreover, larger value of J results in slower frequency change, and D has barely any influence on dynamic performance in this case. Besides, steady state of all cases seems to be the same, which verifies previous conclusion that steady state only depends on k_p.

B. Grid-connected Mode

For the grid-connected operation, the grid is modeled as a large SG, as depicted in Fig. 4. It is preferred to study the frequency change of SG rather than that of DG because the former is the dominant one.

From the conservation of energy,

$$\Delta P_{out\_dg} + \Delta P_{out\_sg} = \Delta P_{load}, \tag{10}$$

where subscript "dg" indicates parameters of DG and subscript "sg" indicate parameters of SG.

Let $\frac{\Delta P_{out\_dg}}{\Delta \omega_g} = A$ and $\frac{\Delta P_{out\_sg}}{\Delta \omega_g} = B$, (10) can be rewritten as

$$\Delta P_{out\_sg}\left(1 + \frac{A}{B}\right) = \Delta P_{load}. \tag{11}$$

Applying (6) for SG, it can be obtained that

$$\frac{1}{s}\left(\Delta \omega_{m\_sg} - \Delta \omega_g\right) \approx \frac{1}{K_{sg}} \Delta P_{out\_sg}. \tag{12}$$

Eliminate ΔP_{out_sg} from (11) and (12), then

$$\frac{\Delta \omega_{m\_sg}}{\Delta P_{load}} = \frac{K_{sg} + Bs}{K_{sg}(A + B)}. \tag{13}$$

That is to say, if A and B are known, then the required transfer function can be obtained.

Eliminate $\Delta \omega_m$ from (6) and (7), then

$$\frac{\Delta P_{out}}{\Delta \omega_g} = -\frac{k_p + J\omega_0 s}{1 + \frac{k_p + D}{K}s + \frac{J\omega_0}{K}s^2}. \tag{14}$$

(14) can be used for both DG and SG to calculate A and B respectly, as shown in (15) and (16).

$$A = \frac{\Delta P_{out\_dg}}{\Delta \omega_g} = -\frac{k_{p\_dg} + J_{dg}\omega_0 s}{1 + \frac{k_{p\_dg} + D_{dg}}{K_{dg}}s + \frac{J_{dg}\omega_0}{K_{dg}}s^2} \tag{15}$$

$$B = \frac{\Delta P_{out\_sg}}{\Delta \omega_g} = -\frac{k_{p\_sg} + J_{sg}\omega_0 s}{1 + \frac{k_{p\_sg} + D_{sg}}{K_{sg}}s + \frac{J_{sg}\omega_0}{K_{sg}}s^2} \tag{16}$$

Let $J = 0$, $D = 0$ in (15), the transfer function for droop control can be obtained as

$$A = \frac{\Delta P_{out\_dg}}{\Delta \omega_g} = -\frac{k_{p\_dg}}{1 + \frac{k_{p\_dg}}{K_{dg}}s} \tag{17}$$

However, for the SG, K_{sg} and D_{sg} are not constants and both depend on load angle δ_{sg}. They can be calculated by (18) and (19) as following [22]:

$$K_{sg} = \frac{E'_q V_{bus}}{X'_d + X_{sg}}\cos\delta_{sg} - \frac{V_{bus}^2(X_q - X'_d)}{(X_q + X_{sg})(X'_d + X_{sg})}\cos(2\delta_{sg}) \tag{18}$$

$$D_{sg} = V_{bus}^2\left[\frac{T''_d X'_d (X'_d - X''_d)}{X''_d (X'_d + X_{sg})^2}\sin^2\delta_{sg} + \frac{T''_q X'_q (X'_q - X''_q)}{X''_q (X'_q + X_{sg})^2}\cos^2\delta_{sg}\right]$$

$$\tag{19}$$

Knowing that for the calculation of damping factor D_{sg}, it is supposed that mechanical losses can be neglected and the rotor flux linkage is fixed. The latter is not true if automatic voltage regulator (AVR) is applied. However, (19) can be still used to obtain an approximate value.

In order to verify the results through simulation lately, all parameters need to be measurable. Actually, all parameters except load angle δ_{sg}, transient internal emf E'_q, and load voltage V_{bus} are fixed SG parameters which can be found in data sheets. Besides, V_{bus} can be measured directly, while δ_{sg} and E'_q can be calculated through (20) and (21) with the measurement of SG output current I_{sg} and SG power angle φ_{sg}.

$$\delta_{sg} = tan^{-1}\frac{(X_q + X_{sg})I_{sg}\cos\varphi_{sg}}{V_{bus} + (X_q + X_{sg})I_{sg}\sin\varphi_{sg}} \tag{20}$$

$$E'_q = V_{bus}\cos\delta_{sg} + I_{sg}\sin(\delta_{sg} + \varphi_{sg})(X'_d + X_{sg}) \tag{21}$$

With (13)-(21), it is now possible to calculate the frequency change of SG during a small loading transition. The results are shown in Fig. 5 and parameters of SG are listed in Table II.

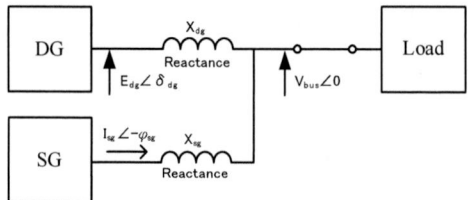

Fig. 4. Grid-connected mode model.

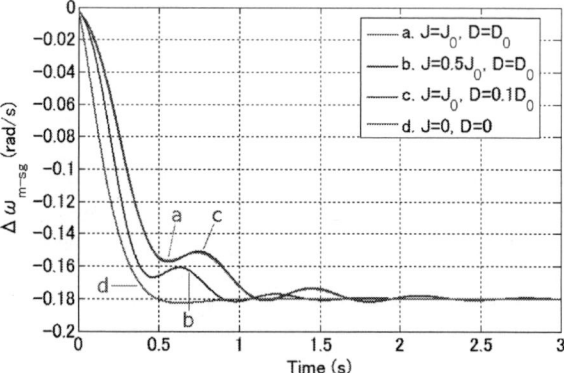

Fig. 5. Time response of SG frequency during a loading transition in grid-connected mode with various parameters.

TABLE II
SG PARAMETERS

Parameter	Value	Parameter	Value
V_{base_sg}	6.6kV	X''_d	0.280pu
P_{base_sg}	1MW	$X_q = X'_q$	0.770pu
P_0	1pu	X''_q	0.375pu
ω_0	376.99rad/s	T''_d	0.0348s
J_{sg}	56.3kg·m^2	T''_q	0.0346s
k_{p_sg}*	20pu	ΔP_{load}	0.0191MW
X_{sg}	0.302pu	V_{bus}	6.56kV
X_d	1.90pu	I_{sg}	43.2A
X'_d	0.314pu	φ_{sg}	-0.1027rad

Fig. 5 indicates that the system with VSG still has a better performance than droop control since the declination of Line (a) is slower than Line (d). Same as stand-alone mode, a larger J leads to better frequency stability. Although it is not very obvious, it can be observed that the system with a larger D (Line (a)) has smaller oscillation than that with a smaller D (Line (c)). Same as stand-alone mode, the changes of parameters of swing equation have no influence on steady state.

C. Effect of Governor Delay

As it is well known, there is always a large time delay in the governor of SG. In previous researches of VSG, this delay has also been imitated when (2) is applied [16], [17]. Meanwhile, for the droop control, a time delay in (4) also exists in the form of a low pass filter to filter the noises in the detected P_{out}. Supposing both delays to be first order delays with a time constant of T_d, k_p^* in (2) has to be replaced by $\frac{k_p^*}{1+T_d s}$ for VSG control, while k_p in (4) has to be replaced by $k_p(1+T_d s)$ for droop control in the small signal model. This difference is caused by the inverse of input and output of the two control methods. The input of VSG is active power and output is frequency, while for droop control they are inversed.

If this effect is taken into account, for VSG, (8) has to be rewritten as

$$\frac{\Delta\omega_m}{\Delta P_{load}} = -\frac{1+\left(T_d+\frac{D}{K}\right)s+\frac{DT_d}{K}s^2}{k_p+J\omega_0 s+J\omega_0 T_d s^2}, \tag{22}$$

and for droop control, (9) becomes

$$\frac{\Delta\omega_m}{\Delta P_{load}} = -\frac{1}{k_p+T_d k_p s}. \tag{23}$$

Meanwhile, for VSG and SG, (14) has to be rewritten as

$$\frac{\Delta P_{out}}{\Delta\omega_g} = -\frac{k_p+J\omega_0 s+T_d J\omega_0 s^2}{1+\left(T_d+\frac{k_p+D}{K}\right)s+\frac{T_d D+J\omega_0}{K}s^2+\frac{T_d J\omega_0}{K}s^3}, \tag{24}$$

and for droop control, (17) becomes

$$\frac{\Delta P_{out\_dg}}{\Delta\omega_g} = -\frac{k_{p\_dg}+k_{p\_dg}T_d s}{1+\frac{k_{p\_dg}}{K_{dg}}s+\frac{k_{p\_dg}T_d}{K_{dg}}s^2}. \tag{25}$$

Based on (22)-(25), time response of frequency can be recalculated, and compared with cases without consideration of delay, as shown in Fig. 6. The time constant T_d is supposed to be 0.1s.

It can be observed that this delay makes the frequency of VSG change faster, while makes that of droop control change more slowly in both stand-alone mode and grid-connected mode. In other words, this delay has a positive effect on droop control while a negative effect on VSG.

It is interesting to compare (8) with (23), and (15) with (25): supposing $T_d = \frac{J\omega_0}{k_p}$, the only difference between (8) and (23) or between (15) and (25) is that equations of VSG have D whereas those of droop control do not. That is to say, with a first order delay $T_d = \frac{J\omega_0}{k_p}$ in the droop

controller, droop control can imitate the performance of VSG with a moment of inertia equal to J. However, droop control cannot replace VSG completely because damping factor D cannot be realized with this method. This inference can be verified by Fig. 7, in which the comparison in Fig. 6(b) is redone with time constant of the first order delay of droop control $T_d = \frac{J\omega_0}{k_p}$. It can be observed that Line (b2) almost overlaps Line (a), although with slightly larger oscillation. This result implies that the moment of inertia of both methods is almost the same. However, droop control has no damping factor to alleviate the oscillation.

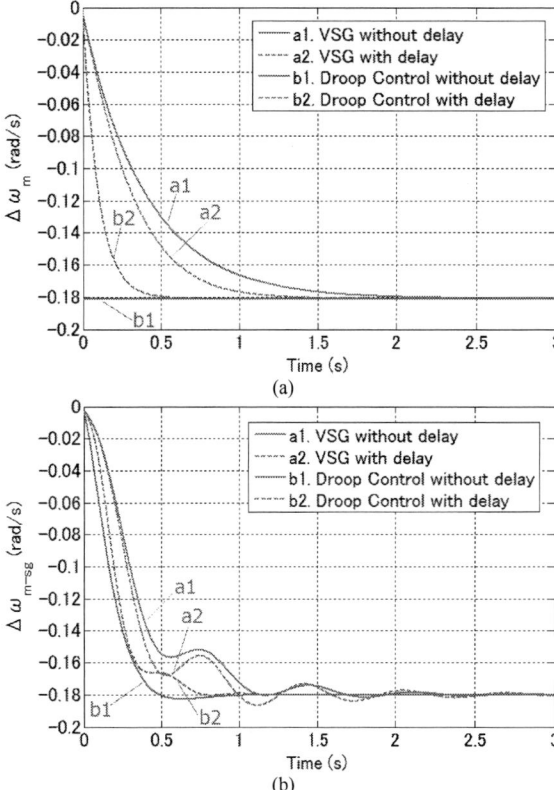

Fig. 6. Effect of delay on (a) time response of DG frequency during a loading transition in stand-alone mode; (b) time response of SG frequency during a loading transition in grid-connected mode.

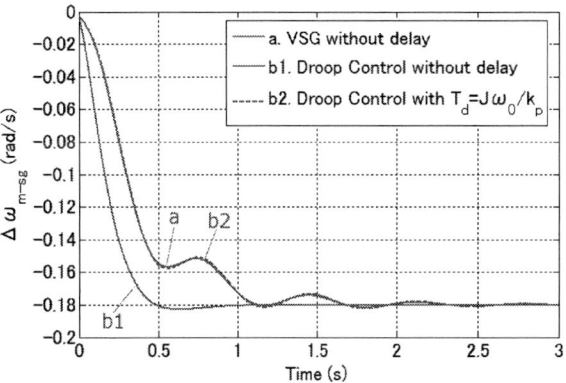

Fig. 7. Time response of SG frequency during a loading transition in grid-connected mode, to verify the possibility to use droop control to imitate VSG control.

IV. COMPARISON OF STABILITY

In order to compare the stability of VSG and that of droop control, state space model is built for each control method in the form of

$$\begin{cases} \dot{x} = Ax + Bu \\ y = Cx + Du \end{cases} \tag{26}$$

In grid-connected mode, for VSG control, vectors and matrix in (26) can be expressed as

$$x = [\Delta\omega_{m\_sg} \quad \Delta\omega_{md\_sg} \quad \Delta P_{out\_sg} \quad \Delta\omega_{m\_dg} \quad \Delta\omega_{md\_dg} \quad \Delta P_{load}]^T \tag{27}$$

$$u = [\Delta\omega_g] \tag{28}$$

$$y = [\Delta\omega_{m\_sg}] \tag{29}$$

$$A = \begin{bmatrix} -\dfrac{D_{sg}}{J_{sg}\omega_0} & -\dfrac{k_{p\_sg}}{J_{sg}\omega_0} & -\dfrac{1}{J_{sg}\omega_0} & 0 & 0 & 0 \\ \dfrac{1}{T_{d\_sg}} & -\dfrac{1}{T_{d\_sg}} & 0 & 0 & 0 & 0 \\ K_{sg} & 0 & 0 & 0 & 0 & 0 \\ 0 & 0 & \dfrac{1}{J_{dg}\omega_0} & -\dfrac{D_{dg}}{J_{dg}\omega_0} & -\dfrac{k_{p\_dg}}{J_{dg}\omega_0} & -\dfrac{1}{J_{dg}\omega_0} \\ 0 & 0 & 0 & \dfrac{1}{T_{d\_dg}} & -\dfrac{1}{T_{d\_dg}} & 0 \\ K_{sg} & 0 & 0 & K_{dg} & 0 & 0 \end{bmatrix} \tag{30}$$

$$B = \begin{bmatrix} \dfrac{D_{sg}}{J_{sg}\omega_0} & 0 & -K_{sg} & \dfrac{D_{dg}}{J_{dg}\omega_0} & 0 & -(K_{sg}+K_{sg}) \end{bmatrix}^T \tag{31}$$

$$C = [1 \quad 0 \quad 0 \quad 0 \quad 0 \quad 0] \tag{32}$$

$$D = [0], \tag{33}$$

and for droop control,

$$x = [\Delta\omega_{m_{sg}} \quad \Delta\omega_{md_{sg}} \quad \Delta P_{out_{sg}} \quad \Delta\omega_{m_{dg}} \quad \Delta P_{load}]^T \tag{34}$$

$$u = [\Delta\omega_g] \tag{35}$$

$$y = [\Delta\omega_{m\_sg}] \tag{36}$$

$$A = \begin{bmatrix} -\dfrac{D_{sg}}{J_{sg}\omega_0} & -\dfrac{k_{p\_sg}}{J_{sg}\omega_0} & -\dfrac{1}{J_{sg}\omega_0} & 0 & 0 \\ \dfrac{1}{T_{d\_sg}} & -\dfrac{1}{T_{d\_sg}} & 0 & 0 & 0 \\ K_{sg} & 0 & 0 & 0 & 0 \\ 0 & 0 & \dfrac{1}{k_{p\_dg}T_{d\_dg}} & -\dfrac{1}{T_{d\_dg}} & -\dfrac{1}{k_{p\_dg}T_{d\_dg}} \\ K_{sg} & 0 & 0 & K_{dg} & 0 \end{bmatrix} \tag{37}$$

$$B = \begin{bmatrix} \dfrac{D_{sg}}{J_{sg}\omega_0} & 0 & -K_{sg} & 0 & -(K_{sg}+K_{sg}) \end{bmatrix}^T \tag{38}$$

$$C = [1 \quad 0 \quad 0 \quad 0 \quad 0] \tag{39}$$

$$D = [0]. \tag{40}$$

With these models, it is possible to study the influence of parameter changes on eigenvalues, as depicted in Fig. 8. It can be observed that when J of VSG increases and T_d of both two control methods increases, eigenvalues converge on the origin, which indicate a stable state. When D of VSG decreases, eigenvalues trend to cross the imaginary axis. However, they are still on the left half plan even when $D = 0$ (the red crosses). Therefore, it can

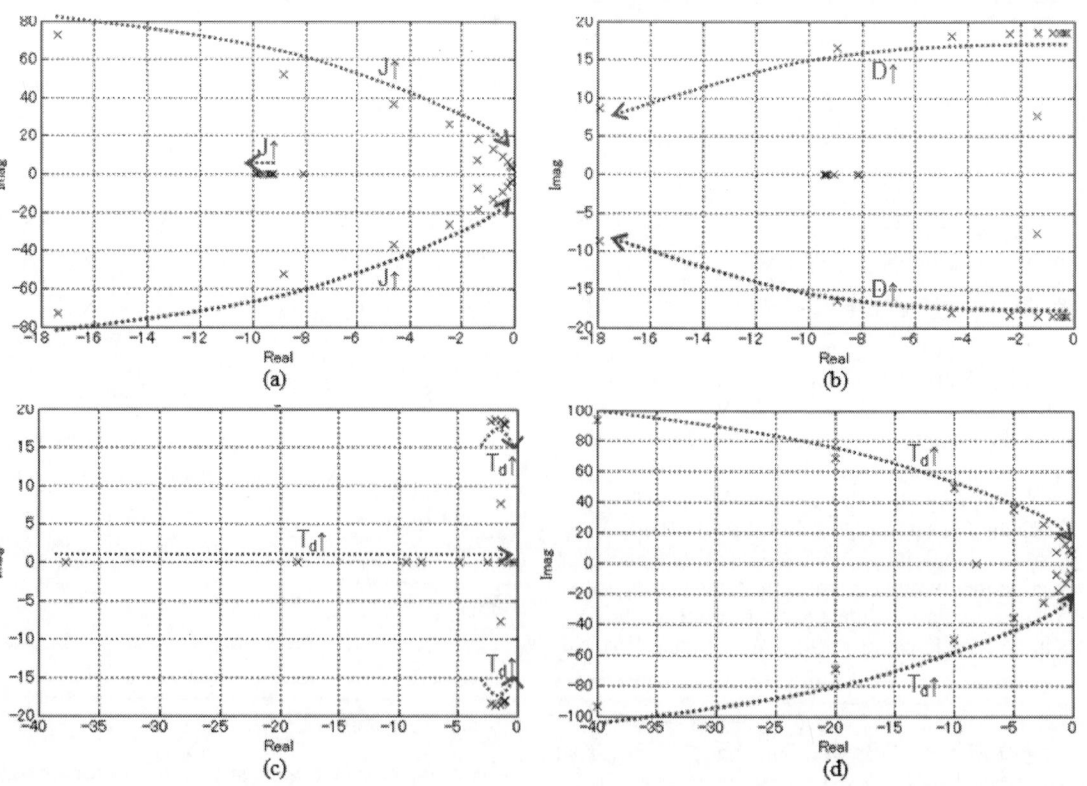

Fig. 8. Eigenvalues for grid-connected mode when (a) J of VSG varies; (b) D of VSG varies; (c) T_d of VSG varies; (d) T_d of droop control varies.

The 2014 International Power Electronics Conference

be concluded that both VSG and droop control are stable systems no matter the parameter changes.

V. SIMULATION RESULTS

To verify the theoretical results, simulations are performed in PSCAD/EMTDC for all cases discussed in Chapter. III, and the results are shown in Fig. 9 and Fig. 10. A three phase three wire system is considered and switch model is used for DGs. All parameters are the same as those in theoretical analyses. It is shown that in all cases, the simulation results (real line) almost overlap the theoretical results (dotted line), which suggests that small signal models are all correct. Furthermore, in all cases, simulation results are slightly delayed compared to theoretical results, probably because in simulation, the input disturbance ΔP_{load}, which is realized by adding more loads, is not an ideal step function due to the existence of line inductance. And it can also be observed that in grid-connected mode (Fig. 9(b), (d) and Fig. 10), the oscillation amplitudes of simulation results are slightly larger than those of theoretical results. This is because in theoretical analyses, damping factor calculated through (19) is slightly smaller than the true value, because the change of flux linkage caused by AVR is neglected, as it is mentioned above.

Another simulation in grid-connected mode is performed to compare the frequency stability during a fault clearing process, and the result is shown in Fig. 11. In this case, the system depicted in Fig. 4 is running with 10kW load when a fault suddenly occurs near the load and the relay between the load and the bus is tripped at $t = 20s$. The fault is cleared within 0.2s and the relay is reconnected at $t = 20.2s$. The result shows that supposing $T_d = 0.1s$, VSG has smaller maximum frequency excursion than droop control owing to a larger inertia and therefore lower rate of change of frequency, although its frequency is more vibratory after fault clearing. The performance of VSG can be further improved if the delay in governor is deleted. As for droop control, if its time constant of low pass filter increases to $T_d = \frac{J\omega_0}{k_p}$, inertia of droop control can reach to the same level as VSG and similar performance can be expected.

To verify the stability of both control methods, a series of simulations are performed, using extreme parameter values for both average model, in which the inverter is modeled as a controlled voltage source, and switch model, in which the output voltage of inverter is generated by PWM. The results are shown in Table III. It is indicated that stability has been proven for all cases in average model. However, unstable issues happen for VSG with very small J, and droop control with very small T_d, when switch model is used. For VSG with very small J, the problem is that when the swing equation (1) is realized by Runge-Kutta method to calculate output frequency, J occurs in the denominator; therefore, small J amplifies the input ripples and leads to system instability. For droop control, large ripples exist in calculated active power P_{out} (see Fig. 1b), and the existence of T_d performs

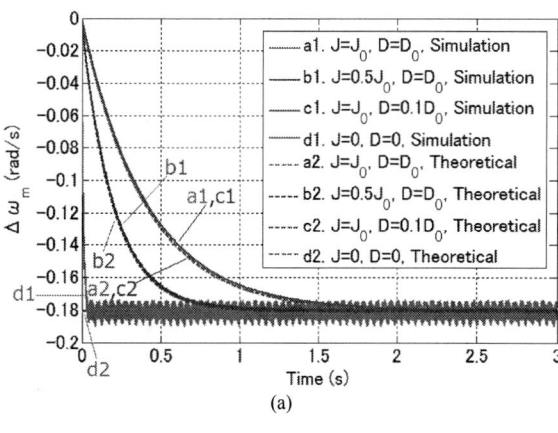

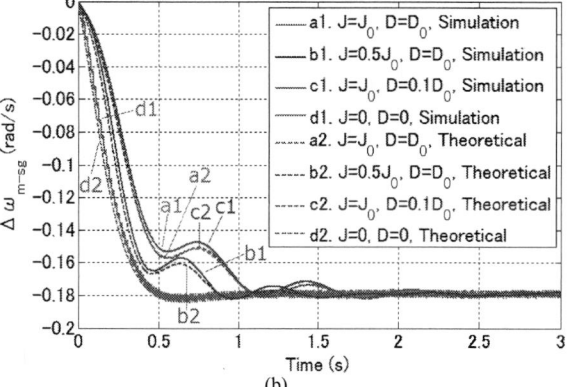

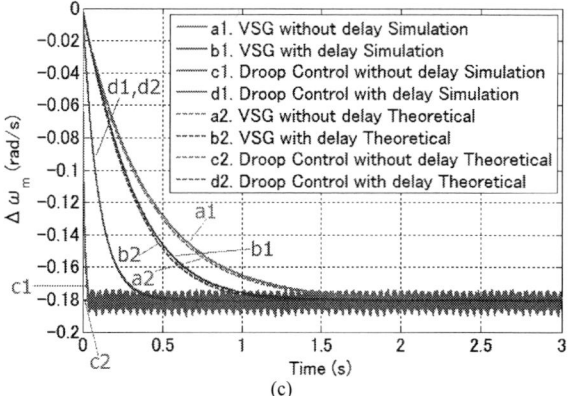

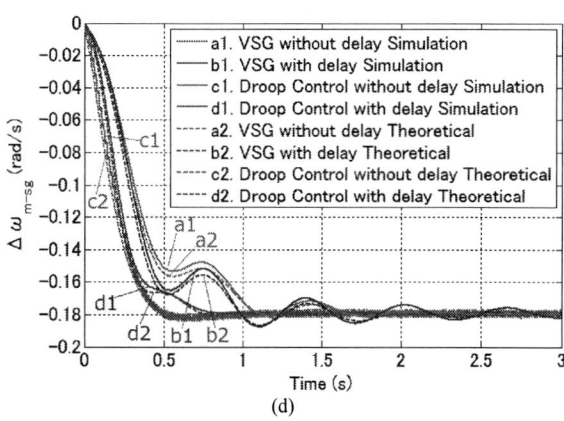

Fig. 9. Simulation results to verify (a) Fig. 3; (b) Fig. 5; (c) Fig. 6(a); (d) Fig. 6(b).

as a low pass filter. A very small T_d allows large ripples to enter into the droop controller to make the system unstable. Knowing that our stability study is based on s domain, these discrete problems are not considered, and the results of average model verify the state space model. However, attention should be paid because it is possible that these problems occur in practice when PWM inverters are used for DGs.

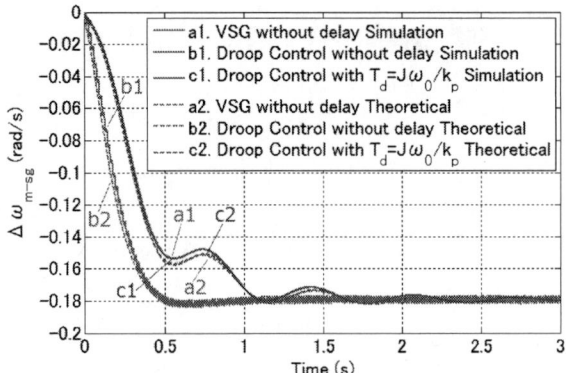

Fig. 10. Simulation results to verify Fig. 7.

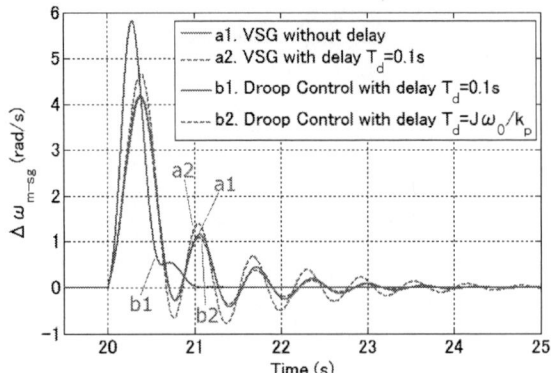

Fig. 11. Simulation results during a fault clearing process.

TABLE III
STABILITY TEST RESULTS

	Case	Switch model	Average model
VSG	$J = 1 \times 10^3 J_0$	stable	stable
	$J = 1 \times 10^{-3} J_0$	unstable	stable
	$D = 1 \times 10^3 D_0$	stable	stable
	$D = 0$	stable	stable
	$T_d = 1 \times 10^3 T_{d0}$	stable	stable
	$T_d = 0$	stable	stable
Droop Control	$T_d = 1 \times 10^3 T_{d0}$	stable	stable
	$T_d = 0$	unstable	stable

VI. CONCLUSION

In this paper, small signal models for VSG control and droop control are built, in both stand-alone mode and grid-connected mode, to compare the dynamic response of the two control methods. Since in VSG control, swing equation is applied to simulate the inertia and damping effect of SG, VSG has smaller rate of change of frequency during a loading transition, which helps to reduce the risk of unnecessary tripping of df/dt relay and to limit the maximum frequency excursion. Furthermore, it is demonstrated that a larger virtual moment of inertia J results in smaller df/dt, and a larger damping factor D contributes to alleviate the oscillation. The integration of governor delay effect into VSG increases df/dt and therefore weakens the advantage of VSG. Contrarily, large delay in droop control decreases df/dt, and is even able to simulate a virtual inertia. However, damping factor cannot be realized with droop control.

Stability analyses are also done through state space model, and both VSG and droop control shows good stability in the aspect of active power control.

Both dynamic characteristics studies and stability studies are verified by simulations results.

REFERENCES

[1] M. C. Chandorkar, D. M. Divan, and R. Adapa, "Control of parallel connected inverters in standalone ac supply systems," *IEEE Trans. Ind. Appl.*, vol. 29, no. 1, pp. 136-143, Jan./Feb. 1993.

[2] J. M. Guerrero, J. C. Vasquez, et al. "Hierarchical control of droop-controlled AC and DC microgrids - A general approach toward standardization," *IEEE Transactions on Industrial Electronics*, vol. 58, no. 1, pp. 158-172, 2011.

[3] J. A. Peças Lopes, C. L. Moreira, and A. G. Madureira, "Defining control strategies for microgrids in islanded operation," *IEEE Trans. Power Syst.*, vol. 21, no. 2, pp. 916-924, May 2006.

[4] K. Debrabandere, B. Bolsens, J. Van den Keybus, A.Woyte, J. Driesen, and R. Belmans, "A voltage and frequency droop control method for parallel inverters," *IEEE Trans. Power Electron.*, vol. 22, no. 4, pp. 1107-1115, Jul. 2007.

[5] R. H. Lasseter, J. H. Eto, B. Schenkman, J. Stevens, H. Vollkommer, D. Klapp, E. Linton, H. Hurtado, and J. Roy, "CERTS microgrid laboratory test bed," *IEEE Trans. Power Del.*, vol. 26, no. 1, pp. 325-332, Jan. 2011.

[6] W. Yao, M. Chen, J. M. Guerrero, and Z.-M. Qian, "Design and analysis of the droop control method for parallel inverters considering the impact of the complex impedance on the power sharing," *IEEE Trans. Ind. Electron.*, vol. 58, no. 2, pp. 576-588, Feb. 2011.

[7] J. M. Guerrero, L. García de Vicuña, J. Matas, M. Castilla, and J. Miret, "Output impedance design of parallel-connected UPS inverters with wireless load-sharing control," *IEEE Trans. Ind. Electron.*, vol. 52, no. 4, pp. 1126-1135, Aug. 2005.

[8] J. M. Guerrero, L. Hang, and J. Uceda, "Control of distributed uninterruptible power supply systems," *IEEE Trans. Ind. Electron.*, vol. 55, no. 8, pp. 2845-2859, Aug. 2008.

[9] Yun Wei Li, Ching-Nan Kao, "An Accurate Power Control Strategy for Power-Electronics-Interfaced Distributed Generation Units Operating in a Low-Voltage Multibus Microgrid," *IEEE Transactions on Power Electronics*, vol. 24, no. 12, pp. 2977-2988, Dec. 2009.

[10] Driesen J and Visscher K. "Virtual Synchronous Generators," *2008 IEEE Power and Energy Society General Meeting - Conversion and Delivery of Electrical Energy in the 21st Century*, pp.1-3, 2008.

[11] M. P. N. Van Wesenbeeck, S. W. H. De Haan, P. Varela, K. Visscher, "Grid tied converter with virtual kinetic storage," *PowerTech, 2009 IEEE Bucharest*, pp. 1-7, 2009.

[12] Torres, M. ; Lopes, L.A.C. "Virtual synchronous generator control in autonomous wind-diesel power systems," *2009 IEEE Electrical Power & Energy Conference (EPEC)*, pp. 1-6, 2009.

[13] V. Karapanos, S. de Haan, and K. Zwetsloot, "Real Time Simulation of a Power System with VSG Hardware in the Loop," *IECON 2011 - 37th Annual Conference on IEEE Industrial Electronics Society*, pp. 3748-3754, 2011.

[14] Yang Xiang-zhen, Su Jian-hui, Ding Ming, Li Jin-wei, and Du Yan, "Control Strategy for Virtual Synchronous Generator in Microgrid", *4th International Conference on Electric Utility*

Deregulation and Restructuring and Power Technologies (DRPT), pp. 1633-1637, 2011.

[15] Y. Hirase, K. Abe, K. Sugimoto, Y. Shido, "A grid connected inverter with virtual synchronous generator model of algebraic type," *IEEJ Trans. on Power and Energy*, vol. 132, No. 4, pp. 341-349, 2012.

[16] K. Sakimoto, Y. Miura and T. Ise, "Stabilization of a power system with a distributed generators by a virtual synchronous generator function," *8th IEEE International Conference on Power Electronics - ECCE Asia*, pp. 1498 - 1505, 2011.

[17] T. Shintai, Y. Miura and T. Ise, "Reactive power control for load sharing with virtual synchronous generator control," *2012 7th International Power Electronics and Motion Control Conference (IPEMC)*, pp. 846-853, 2012.

[18] G. Pepermans, J. Driesen, D. Haeseldonckx, R. Belmans, W. D'Haeseleer, "Distributed generation: definition, benefits and issues," Energy Policy, Vol. 33, Issue 6, pp. 787-798, April, 2005.

[19] Y. Chen; R. Hesse, D. Turschner, H.-P. Beck, "Improving the grid power quality using virtual synchronous machines," *2011 International Conference on Power Engineering, Energy and Electrical Drives (POWERENG)*, pp. 1-6, May 2011.

[20] Y. Chen, R. Hesse, D. Turschner, and H.-P. Beck, "Comparison of methods for implementing virtual synchronous machine on inverters," *Int. Conf. on Renewable Energies and power quality-ICREPQ'12*, Spain, 2012.

[21] Qing-Chang Zhong, G. Weiss, "Synchronverters: Inverters That Mimic Synchronous Generators," *IEEE Transactions on Industrial Electronics*, vol. 58, no. 4, pp. 1259-1267, April 2011.

[22] J. Machowski, J. Bialek, and J. R. Bumby, *Power System Dynamics and Stability*. John Wiley & Sons, NY, USA, 1997.

Embedded Limitations and Protections for Droop-based Control Schemes with Cascaded Loops in the Synchronous Reference Frame

Salvatore D'Arco[*]

Giuseppe Guidi[*]

Jon Are Suul[*,†]

[*]SINTEF Energy Research
7465 Trondheim, Norway
salvatore.darco@sintef.no
Giuseppe.Guidi@sintef.no

[†]Department of Electric Power Engineering
Norwegian University of Science and Technology
7491 Trondheim, Norway
jon.are.suul@ntnu.no

Abstract— Droop based control schemes for power electronic converters have been object of extensive research in the last decade and have arguably become the most popular solution for converter control in microgrid applications. The protection from over-currents and over-voltages are critical features of control schemes for converters required to operate in both stand-alone and parallel-connected modes, but these aspects are marginally treated in literature on microgrids. This paper analyzes possible schemes that can be applied for protection and active limitations of voltages and currents in droop based controls for microgrids based on cascaded voltage and current control loops. Implementation aspects are discussed in detail and performances are compared by numerical simulations.

Keywords— *Droop Control, Cascaded Voltage and Current Controllers, Protections and Limitations of Cascaded Control Loops, Voltage Source Converters*

I. INTRODUCTION

During the last couple of decades, there have been extensive research efforts within the topic of droop based control schemes for power electronic converters. In this process, a wide range of control structures based on active and reactive power droops have been proposed, and the basic principles for such control strategies have arguably become the dominant solution for Uninterruptable Power Supplies (UPS) and microgrid applications [1]- [5]. The droop-based control strategies have recently been complemented by strategies for controlling power electronic converters to emulate the behavior of traditional synchronous machines, i.e. to behave as Virtual Synchronous Machines (VSMs) [6]-[9]. However, the VSM approach has been demonstrated to have a clear equivalence to droop based control [8].

The droop-based control methods, and some of the proposed VSM implementations, result in a control structure where voltage amplitude and frequency, with a corresponding instantaneous phase angle, are given as

reference for converter control. These reference signals can be used directly for the PWM operation of the converter or for a single voltage control loop [4], [7], [5]. However, such control approaches have no explicit way of limiting the converter currents in case of overload or disturbances and faults in the AC system. To ensure simple and explicit implementation of converter current limitations, control systems based on cascaded voltage and current controllers can instead be used [9], [10]-[12]. For three-phase converters, such multilayered control structures are commonly based on nested PI control loops, with decoupling and feed-forward terms, implemented in the Synchronous Reference Frame (SRF) [9], [10].

Protective functions are critical features of any power electronic control scheme, especially if the converter is required to operate both in stand-alone and parallel-connected modes. In such conditions the limitations and protections embedded in the control system must protect the converter from over-currents and over-voltages, and must ensure that unintended over-modulation is avoided. This should be achieved without disrupting the main operation of the converter, even in case of severe transients or temporary faults in the AC system. Additionally, satisfactory continuous operation of the converter also requires that it returns smoothly to normal operation after the disturbance. Since current or voltage limitation during a disturbance will require saturation of at least one of the control loops, seamless transition between operation with activated protections and normal operation requires that controller windup is avoided.

Although cascaded voltage and current control loops allows simple and explicit implementation of limiting functions, the anti-windup strategies required for seamless operation of these control loops during and after a disturbance are marginally treated in literature related to droop-controlled converters. Indeed, strategies for imposing limitations on the current references and for limiting the voltage output of the current controllers to avoid over-modulation and corresponding windup of the current controllers are well known from electric drives and current controlled converters in grid connected applications [13]-[19]. However, such strategies are rarely studied in detail for control systems based on

---

The work of SINTEF Energy Research in this paper was supported by the Blue Sky instrument of SINTEF Energy Research through a Strategic Institute Programme (SIP) funded by the national Basic Funding Scheme of Norway

978-1-4799-2706-7/14 $31.00 © 2014 IEEE

cascaded voltage and current controllers in the synchronous reference frame with voltage references provided by outer loop droop controllers. In such control systems, there are for instance additional challenges compared to the available experience from control of electric drives regarding the priorities between the d- and q-axis components of voltage and currents and regarding strategies for preventing loss of synchronism when the controllers are operating in a saturated condition. Thus, further studies are required to address the challenges of droop-based control schemes with cascaded voltage and current control loops and by that releasing their full potential for applications in stand-alone and grid connected systems.

This paper will first discuss some of the conventional methods used to deal with saturation of multi-axis PI-controllers and nested loops. Then, some possible dedicated schemes for active, coordinated limitations of controllers in droop based control schemes are proposed, which allows for reliable ride-through of grid disturbances. Implementation aspects will be discussed in detail, highlighting advantages and possible drawbacks.

II. METHODS FOR CONTROLLED SATURATION AND ANTI-WINDUP OF PI CONTROLLERS

Linear control theory is normally applied to calculate the proportional and integral gains of PI controllers for achieving the required dynamic performance while ensuring system stability. However, it is well known that their large-signal response is determined by the strategy used to limit the internal state of each regulator when reaching the specified limits of the controller output (saturation). Available strategies for improving the overshoot, settling time, large signal stability etc. in case of temporary saturation of the control loops include anti-windup schemes for single-input single-output (SISO) regulators as well as reference shaping for multiple-input multiple-output (MIMO) and nested structures.

A. Anti-windup schemes for SISO-systems

When the output of a PI regulator in a SISO system reaches its saturation limit, linearity is lost and some action must be undertaken to avoid unwanted increase of the state of the integral action beyond the specified output limitations (windup phenomenon). Different anti-windup methods have therefore been proposed and are routinely applied to such regulators [20], [21]. Some of the most widespread techniques are reported here for reference, with the aim of establishing a framework for extending the same concepts to the more complex control structures for cascaded control of power converters.

1) Integrator clamping (conditional integration)

This is arguably the simplest anti-windup solution, consisting in freezing the integrator action in Fig. 1 when the output of the regulator is limited [20]:

$$\dot{\lambda} = \begin{cases} e & if \ y' = y \\ 0 & otherwise \end{cases} \tag{1}$$

This is generally sufficient to avoid major problems related to integrator windup, but true bumpless response

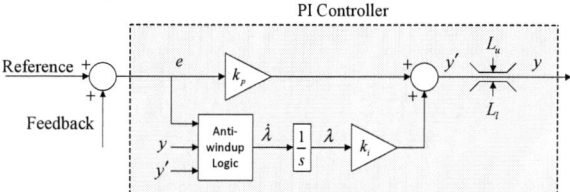

Fig. 1. Basic PI regulator with anti-windup

cannot be achieved without introducing more complex strategies for enabling integration and correcting the initial value of the integral action [21]. Such strategies can be easy to implement for SISO systems, but are undesirable in case of multiple regulators (MIMO and nested loops) as their interaction may create stability problems due to limit cycles and deadlock conditions.

2) Backtracking

One alternative way to avoid windup without recurring to variable-structure regulators is to "backtrack" the limiting state of the regulator. Information about the exceeded limitation is fed back to the input of the integrator according to (2), automatically limiting its increase [21].

$$\dot{\lambda} = e - k_b \cdot \left(y' - y \right) \tag{2}$$

The backtracking gain k_b can be selected to achieve different trade-offs between fast response of the integral inhibition (leading to bumpless transitions) and continuity of the integral state (correct value of the integral state when exiting limitation). To avoid this trade-off, the integrator input can also be switched between the error e signal and the backtracking signal given by $k_b \cdot (y-y')$.

3) Dynamic calculation of integral limits

Another common technique that is easy and efficient to implement on modern digital controllers consists in calculating the limits of the integral action at each sampling period in order to ensure that the sum of proportional plus integral action never exceeds the output limits of the regulator [22]:

$$\begin{cases} L_{\lambda,u}(k) = L_u - \dfrac{k_p}{k_i} \cdot e(k) \\ L_{\lambda,l}(k) = L_l - \dfrac{k_p}{k_i} \cdot e(k) \end{cases} \tag{3}$$

It is noted that this technique can be seen as a special case of backtracking, where the gain k_b tends to infinity. As a result, the large signal response will be perfectly bumpless, but the integral state can be discontinuous. This is undesirable if the system is expected to work consistently around the limit condition.

B. Anti-windup applied to three-phase power electronics converter controllers

In control of three-phase power electronics converters, currents and voltages are usually decomposed in orthogonal components either in fixed or synchronously rotating reference frame. Voltage and current controllers will therefore have two orthogonal inputs and two corresponding outputs. The vector amplitude of the

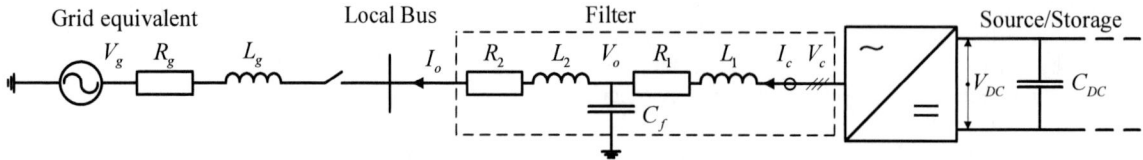

Fig. 3. Schematic of the analyzed configuration

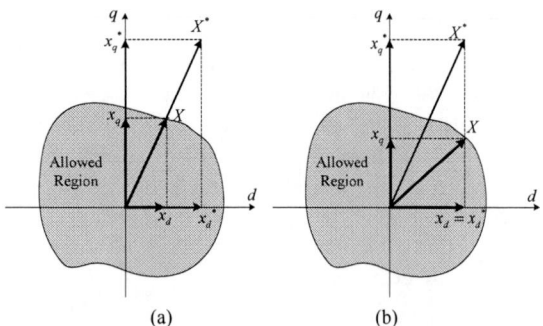

(a)   (b)

Fig. 2. Limitation methods for vector quantities

controller output must be limited within a two-dimensional space whose shape results from the physical constraints on the original phase quantities. Normally, the voltages or currents must be limited so that none of the three phase quantities exceed a certain limit, resulting in a circular operating region mathematically described by:

$$\sqrt{x_d\left(t\right)^2 + x_q\left(t\right)^2} \leq \left|\mathbf{x}\left(t\right)\right|_{\lim} \tag{4}$$

where $\mathbf{x}$ is denoting a generic three-phase quantity and the subscripts $_d$ and $_q$ indicate the orthogonal components used by the MIMO regulator. It can be noted that although the limitation of the PWM is the well-known voltage hexagon, the allowed two-dimensional operating region for the PWM voltage reference will be restricted to a circle when over-modulation must be avoided [23].

It is important to notice that there is a degree of freedom in deciding how to limit each component in order to satisfy (4). Several methods have been proposed in the literature to exploit this degree of freedom in limitation of two-axis vector quantities. The simplest strategy is to keep the direction of the unlimited vector by proportionally limiting the two components according to Fig. 2 (a). Another method is to prioritize one of the components and calculate the limit on the other accordingly as indicated in Fig. 2 (b). This method is often applied in field-weakening control of motor drives where the flux-generating current component is prioritized over the torque-generating one [13]. More general methods can also be devised to optimize the non-

linear transient response of the converter under limited conditions, as for instance in [18]. In the context of converters for microgrid, this degree of freedom can be used to decide the share of active and reactive power that the unit must inject into the grid during contingencies.

C. Anti-windup methods for nested loops

As will be shown in later sections, it is imperative that each loop in a cascaded control structure is aware of whether inner loops are running into limitation. The underlying principle is that the output (and internal states) of outer loops controller must be prevented from increasing when the loop that is taking such output as reference is running into saturation, as the situation would result in a kind of inter-loop windup. In order for the system to operate safely in the event of disturbances, feedback from inner loops to outer loops must therefore be implemented. The output of outer loops should then be modified when inner loops gets saturated. Such approaches are routinely implemented for instance in motor drives designed to operate in field weakening [13], [24]. However, applications to cascaded AC voltage and current control loops with outer loop droop controllers are less established and need further investigation.

III. LIMITATIONS AND ANTI-WINDUP PROTECTION FOR DROOP BASED CONTROL SYSTEMS

A typical configuration of a converter for distributed generation or energy storage required to operate both in grid-connected or local islanded modes is shown in Fig. 3. The system studied in this case includes only one converter with a LCL filter as the interface to the local bus, and a simplified Thévenin equivalent for the grid.

A. Control System Overview

To allow for parallel operation and power sharing among multiple converters operating in stand-alone mode, the control structure in such applications is usually based on droop control of active and reactive power. A typical structure of such a control system with cascaded voltage and current controllers is shown in Fig. 4. In this scheme,

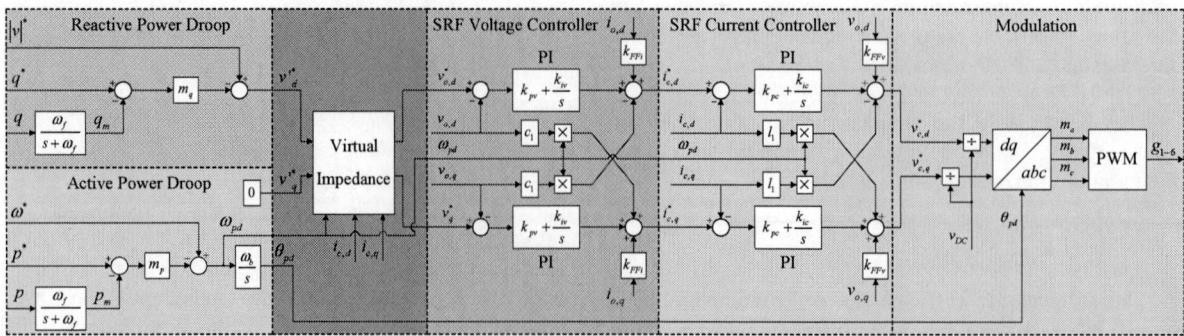

Fig. 4. Droop-based control system with Virtual Impedance and Cascaded Voltage and Current Control Loops

978-1-4799-2706-7/14 $31.00 © 2014 IEEE   1546

The 2014 International Power Electronics Conference

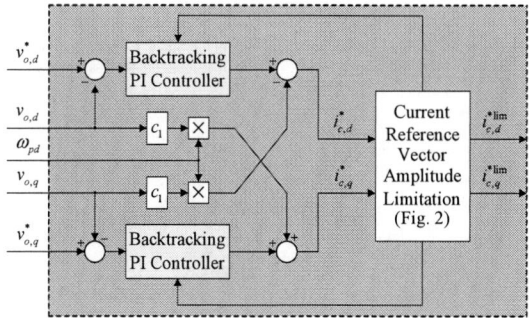

Fig. 5. SRF Voltage Controller with Conventional Limitations

the active power droop controller is providing frequency and phase angle references to the rest of the control system, while the reactive power droop controller is providing a voltage amplitude reference. To shape the characteristics of the control performance and ensure decoupling of the active and reactive power control, the voltage reference is modified by a Virtual Impedance (VI) [3], according to:

$$v_{o,d}^* = \hat{v}_d^{\prime *} - r_{VI} \cdot i_{o,d} + \omega_{pd} \cdot l_{VI} \cdot i_{o,q}$$
$$v_{o,q}^* = 0 - r_{VI} \cdot i_{o,q} - \omega_{pd} \cdot l_{VI} \cdot i_{o,d} \qquad (5)$$

The voltage references from the VI are used as inputs to a PI-based SRF voltage controller controlling the voltages at the filter capacitors of the LCL filter. This voltage control loop is providing the current references to a conventional inner loop SRF current controller, which eventually provides the voltage references for the PWM operation of the converter. The presence of the current control loop ensures that the actual current in the converter semiconductors never exceed the specified current limits.

The structure in Fig. 4 is shown in its general form commonly found in publications discussing design and dynamic operation of microgrid controllers, without any indications of the limitations and anti-windup mechanisms needed to ensure safe operation and recovery from grid disturbances or over-loads. A first assumption is that the inner current controller has properly limited output and internal anti-windup mechanism, so that current control is properly achieved regardless of the status of outer loops, as long as the ratio between AC voltage and DC voltage of the converter does not exceed the controllability limit. The latter may happen during severe AC voltage swell or in case of faults within the generation system causing the DC-side voltage to drop. However, the paper will not deal with those cases as equivalent conditions are well treated for electric drive applications. The further focus will instead be on the less treated issue of interaction between the saturation of the current references, the anti-windup protection of the outer loop voltage controller and the active and reactive droop functions.

B. Voltage Controller with Conventional Current Reference Limitation

A typical implementation of the SRF voltage controller with limitation of the current reference vector amplitude is shown in Fig. 5. To avoid wind-up of the PI

voltage controllers during saturation of the current reference and to ensure bump-less transfer between output-limited and normal operation, the PI controllers must be limited according to the current amplitude limitation. This is achieved by backtracking the two PIs on the individual d-q-axes to follow the saturated output current reference, as indicated by the feedback from the current vector amplitude limitation to the individual PI controllers shown in the figure.

C. Strategies for avoiding loss of synchronization and saturation of outer loops

Cross-axis backtracking as shown in Fig. 5 is effective in avoiding windup of the MIMO voltage controller, but cannot avoid drift of the outer loops providing the references for the voltage controllers. In fact, when the voltage controller is limited due to saturation of the current references, the converter will not be able to follow the power reference, and the droop control will act to increase the frequency and phase angle of the control system in an attempt to increase the power transfer. However, since the converter currents are already saturated, the power balance cannot be reached, resulting in uncontrolled increase of the phase of the voltages across the virtual impedance, as it will be shown in the following sections.

Here, two different methods are proposed to deal with this problem.

1) Switching to PLL-based synchronization and current control during saturation of voltage controller

One possible approach to ensure safe operation of the converter during grid faults, over-load conditions or other disturbances to the system is to switch to conventional current control with grid synchronization based on a Phase Locked Loop (PLL) [1] when saturation of the voltage controller is detected. The current references can then be fixed to suitable values while the synchronization of the system will be maintained by the PLL and not by the active power droop controller. Normal droop operation is restored when the condition that caused the control mode switching (either a grid fault or an overload) disappears. Such strategies can be applied for converters connected to a large scale power system as discussed in [25], but have clear limitations for stand-alone operation where a stable voltage for grid synchronization by a PLL cannot always be ensured. Moreover, reliable operation of a PLL requires that sufficient voltage remains across the filter capacitor at all times during the disturbance.

In the strategy applied in the following, the logic that decides when to switch between droop-based and PLL-based control mode is implemented by a state machine, whose logical outputs are the enable signals en_{vc} and en_{pd} for the SRF voltage controller and the active power droop control respectively. The PLL-based control mode is activated (en_{vc} is set to zero) as soon as the current references from the voltage controller reach the current limitation; in that instant, the output of the voltage regulator is overridden with external references for the d-q currents, as schematically shown in Fig. 6. In principle,

978-1-4799-2706-7/14 $31.00 © 2014 IEEE   1547

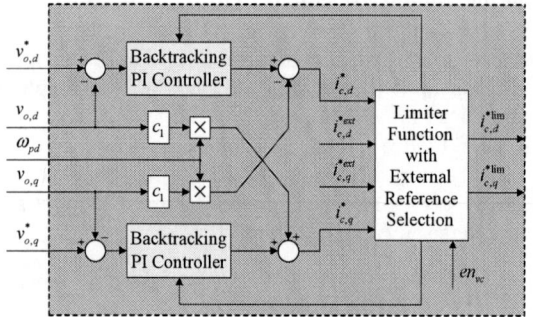

Fig. 6. SRF Voltage Controller with Backtracking and input for external current references

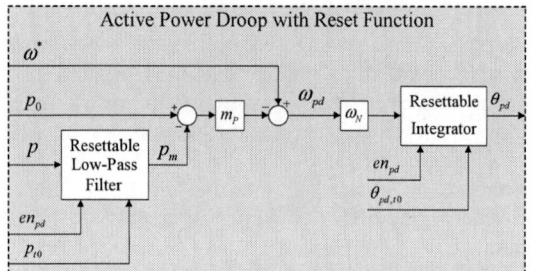

Fig. 7. Active Power Droop with external reset of Integrator and Low-pass filter

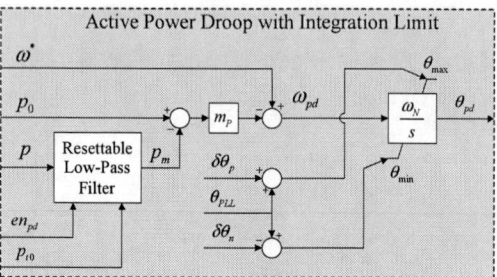

Fig. 8. Active Power Droop with Low-Pass filter reset and integration limit of phase angle for dq transformation

phase difference across the Virtual Impedance before the fault is therefore applied. To allow for some significant margin to the saturated conditions and ensuring that the power flow when entering into the droop-based control mode again avoids overshoot that can cause the system to re-enter into saturation, a phase difference of 0.65 times the initial displacement is used to initialize the integrator in the following. Thus, the phase angle $\theta_{pd,t0}$ for practical initialization of the resettable integrator is expressed by:

$$\theta_{pd,t0} \approx \theta_{PLL,t0} + 0.65 \cdot \left(\theta_{pd,0} - \theta_{PLL,0} \right) \qquad (7)$$

where $\theta_{pd,0}$ and $\theta_{PLL,0}$ are the phase angle from the droop-based power control and the PLL phase angle respectively before the saturation occurred.

2) Limitation of phase angle displacement between power-droop and PLL

The approach described above depends on a state-machine and implies switching between different reference frames and different control structures. As it can be challenging to define unique conditions that ensure proper operation of the state-machine under all kinds of potential disturbances and changes of operating conditions, a simpler and more robust approach is proposed in the following.

The method is based on the consideration that the active power flow resulting from the droop-based control is directly dependent on the phase angle across the Virtual Impedance. Thus, limiting the phase angle integration of the active droop controller to a region around the voltage phase angle provided by a PLL will effectively limit the power flow through the Virtual Impedance. This will ensure that the system never loses synchronism, even during saturation of the current controller references. The method is illustrated in Fig. 8, where the allowed positive and negative phase angle displacement of the VI around the phase of the measured grid voltage is denoted by $\delta\theta_p$ and $\delta\theta_n$ respectively. By this approach, there will never be a discontinuity of the phase angle when going into or out of saturation of the current reference. Depending on the selected filter characteristics, it can however be advantageous to reset the low-pass filter on the measured power to the intended steady-state condition, to limit the transient in the power feedback when going out of saturation.

The limits of the phase angle displacement across the VI can be determined from a similar expression as (6) if a specific share of active and reactive current or power should be achieved for a certain voltage amplitude $|v_0|$ at the filter capacitors. In the following, the limits are

the external current references can be specified according to desired operational characteristics during the disturbance; in this particular implementation, the d-axis current reference is set to the current limit and the q-axis current reference to zero, ensuring close to the maximum active power transfer.

Switching back to normal droop-based control mode takes place when the actual power delivered to the grid is higher than the external reference p^*, meaning that the disturbance is over. A small switching delay is applied in practice, to avoid chattering between the two control modes. When re-enabling the droop control mode (en_{pd} is set active), both the power measurement filter and the phase angle integrator of the active power droop must be re-initialized to achieve smooth mode transition, as schematically shown in Fig. 7. The power measurement filter is simply initialized to the external power reference p^*, while the phase angle of the integrator is initialized to the PLL angle plus the expected voltage phase displacement across the VI. Considering a purely inductive VI, this phase angle can be expressed in steady-state from the active and reactive power flow by (6).

$$\theta_{VI} = \arctan \left(\frac{\omega_{pd} \cdot l_{VI} \, {}^{p}\!/\!{}_{|v_o|}}{|v_o| + \omega_{pd} \cdot l_{VI} \, {}^{q}\!/\!{}_{|v_o|}} \right) \qquad (6)$$

However, since the voltages will not be in steady state immediately after a disturbance, and since the response of the droop-based reactive power control must also be taken into account, a precise initialization of the phase angle from analytical considerations based on reference signals and measured variables after the fault can be unnecessarily complicated. In the following, a simple strategy for initializing the phase angle based on the

TABLE I PARAMETERS OF INVESTIGATED SYSTEM CONFIGURATION

Parameter	Value	Parameter	Value
Rated voltage $V_{S,LL,RMS}$	400 V	Filter inductance l_f	0.08 pu
Rated power S_b	48.5 kVA	Filter resistance r_{lf}	0.003 pu
Rated angular frequency ω_b	$2\pi*50$ rad/s	Filter capacitance c_f	0.074 pu
Virtual Inductance l_{VI}	0.5 pu	Grid inductance l_g	0.20 pu
Virtual Resistance r_{VI}	0.0 pu	Grid resistance r_g	0.01 pu
Frequency reference ω^*	1.0 pu	Grid voltage v_g	1.0 pu
Frequency droop gain m_p	0.02 pu	Voltage reference v^*	1.0 pu
Reactive power reference q^*	0.0 pu	Power filters ω_f	1000 rad/s
Reactive power droop gain k_q	$5*10^{-7}$ pu	Current limit	1.0 pu

however set to be 0.6 rad and equal in both directions to achieve simple and robust implementation that does not depend on any other dynamic feedback signals than the phase angle from the PLL. However, since the method uses the PLL angle to dynamically update the limits of the phase integrator in the active droop controller during the disturbance, there must be sufficient voltage across the filter capacitor for the PLL to keep tracking the voltage at all times.

IV. SIMULATION RESULTS

In order to assess the effectiveness of the methods proposed in the previous section, the system in Fig. 3 with the control structure in Fig. 4 has been simulated in Matlab-Simulink with the main parameters reported in Table I. For the simulations, a sudden drop in grid voltage is imposed as follows:

$$V_g = \begin{cases} 1.0\,p.u. & t < 6s \\ 0.5\,p.u. & 6s \le t < 6.3s \\ 1.0\,p.u. & t \ge 6.3\,s \end{cases} \tag{8}$$

External active and reactive power commands are constant and set to:

$$p^* = 0.8\,p.u. \tag{9}$$
$$q^* = 0.0\,p.u.$$

During to the voltage drop, possibly emulating a fault in the grid, the current required to achieve the commanded power flow will exceed the system capabilities, driving the voltage controller into saturation.

A. Reference Case

Fig. 9 shows the response of the system equipped with the voltage controller of Fig. 5, including internal cross-axis anti-windup but without any coordination between the voltage control loop and the outer droop control. It can be seen that the system goes indeed into current limitation as soon as the grid voltage drops; total current is correctly controlled to its limit throughout the whole low voltage operation, even though the share of active and reactive components is changing because the phase angle of the droop controller is continuously increasing when the power reference cannot be fulfilled.

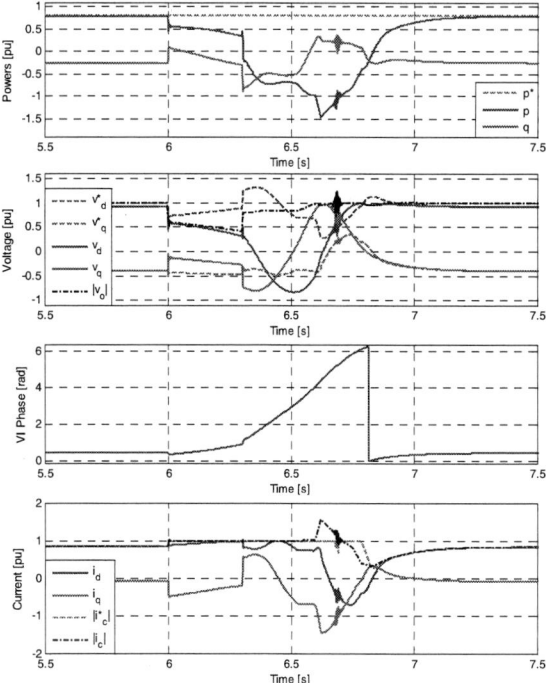

Fig. 9. Simulation results for conventional anti-windup for the case of a sudden voltage drop in the grid

The real problem arises when the grid voltage is restored and the system should in principle be able to resume normal operation. Due to the uncontrolled drift of the voltage phase angle across the virtual impedance that took place during the current limitation, the resulting d-q axes used by the controllers bear no well-defined relationship with active and reactive components of the current, resulting in loss of controllability. Although the current is limited to its maximum value during most of the recovery, the behavior of the system somehow reminds the operation of a synchronous generator during pole-slipping; increasing phase angle and erratic current and power flows during the time needed for the voltage phase angle to undergo a complete revolution and allow for re-synchronization. During this period, the system even loses controllability of the currents for a short time, resulting in current amplitudes exceeding the specified current limitation. When the phase angle resulting from the active power droop is again coming close to the phase angle of the measured voltage at the filter capacitors, the system is able to regain controllability of the voltage and recovers according to its normal dynamics. However, a real implementation of such a system would be tripped long time before it would be able to resynchronize.

B. Switching of current reference

Fig. 10 shows the response of the system equipped with the voltage controller of Fig. 6, including the algorithm for switching between normal droop-control mode and PLL-mode with external current reference. The method in Fig. 7 is used to achieve smooth transition back to normal droop-control mode.

In order to correctly interpret the results, it is emphasized that the switching of control mode from

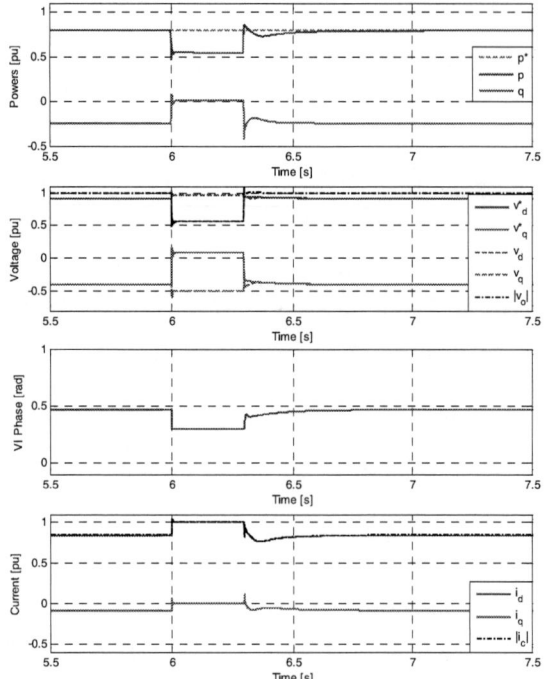

Fig. 10. Simulation results when switching to PLL-based current control with externally specified current reference when saturation is detected for the case of a sudden voltage drop in the grid

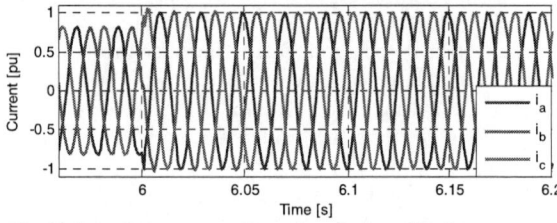

Fig. 11 Converter current when a drop in the grid voltage occurs

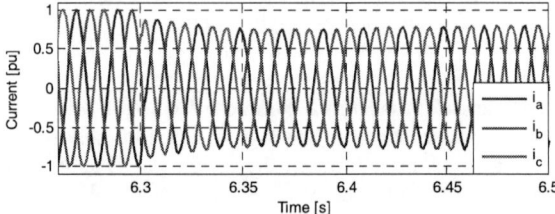

Fig. 12 Converter current when grid voltage is restored

droop-based control to PLL-based grid synchronization inherently implies a change of orientation of the SRF. Thus, there is a discontinuity in how the d-q quantities should be interpreted when the system is operating in the two different modes.

From the simulation results, it is verified that the current references are switched to the saturated values as soon as the voltage controller output reaches the limitation of the current vector amplitude, creating a step-change in the input to the inner current controller. During the whole duration of the voltage drop, rated active current is injected into the grid and the droop controllers are disabled.

As soon as the grid voltage is restored, the active power increases beyond its set point, triggering the control system to switch back to the normal droop control

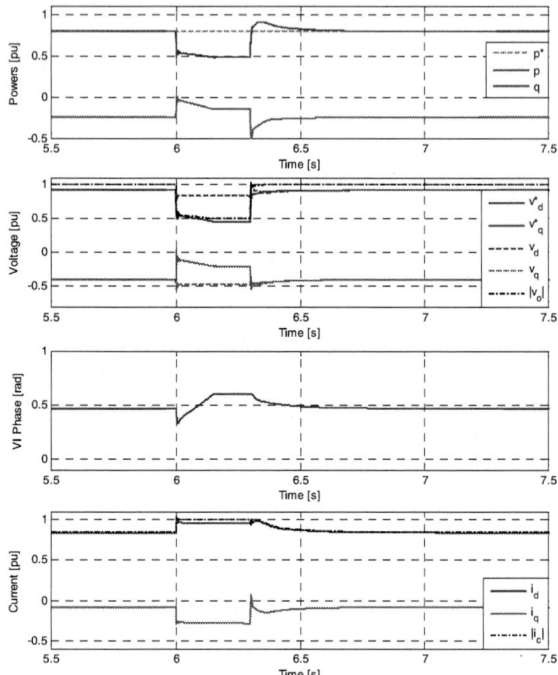

Fig. 13. Simulation results when limiting the maximum phase angle across the virtual impedance when saturation is detected for the case of a sudden voltage drop in the grid

mode. When the transition occurs, the VI phase integrator is properly initialized, resulting in a smooth initialization of the control loops. As ensured by the initialization defined by (7), the system is however starting with a slightly reduced power flow compared to the steady-state operation, and therefore returns to the original operating point through the normal dynamics of the active power droop control.

Fig. 11 shows a magnified view of the currents of the converter immediately following the step-wise drop in grid voltage. In spite of the severe disturbance, the currents are kept under strict control, thus ensuring safe grid fault ride-through without tripping. The same kind of benign behavior is observed when the grid voltage is suddenly restored, as shown in Fig. 12.

C. Limitation of Virtual Impedance Phase Angle

A similar kind of safe and smooth grid fault ride-through as demonstrated in the previous section can be achieved by the strategy described in section III.C.2) based on limitation of the phase angle integrator. As seen in Fig. 13, the system safely limits the current during the voltage drop. The drift of the voltage phase angle across the VI is smoothly limited by the strategy from Fig. 8. When the grid voltage is restored, the voltage phase angle across the VI necessary to achieve the required active and reactive power flow calculated by the droop functions naturally becomes lower than the limit and normal operation is smoothly restored.

Unlike the previous case, no discontinuity is introduced in the control signals and no change in the control structure is necessary, making the system inherently more robust. In this case, the power, however, shows a minor overshoot after the fault since the phase angle across the VI is returning from its maximum value due to the response of the active power control. The currents are nonetheless always limited to safe values.

V. CONCLUSION

In practical implementations of grid-connected converters, safe and predictable operation of the system in case of grid faults, overloads and other disturbances must be ensured. One of the problems that must be considered is the interaction between nested control loops when saturation of the individual controllers occurs. This aspect is, however, only marginally treated for droop-controlled converters where the main research focus has been on transient performances and stability during normal operations. This paper shows how controllers with properly designed local anti-windup algorithms may still give rise to unstable or undesirable system dynamics if the interaction between the individual loops is not considered. In particular, it is demonstrated how the saturation of the voltage controller normally used in droop-based grid connected converters is likely to cause stability problems in case of grid faults resulting in severe voltage drops at the point of coupling.

Two different methods are proposed to ensure coordinated saturation of the current references for the converter and the outer loop droop controllers. The first method uses dynamic switching between droop-based and PLL-based control modes, while the second achieves stable grid fault ride-through by avoiding drift of the integrator present in the active power droop, without any reconfiguration of the control structure. Both methods are shown to be effective in keeping the converter currents under control during the grid fault and in achieving smooth return to normal operation as soon as the fault is cleared. The advantage of the first method is that the currents during fault conditions can be independently selected according to any desirable characteristics, while the main advantage of the second method is the absence of an explicit sequencer for handling contingencies, thus avoiding possible latch-up conditions and chattering between different control modes.

REFERENCES

[1] T. C. Green, M. Prodanović, "Control of inverter-based micro-grids," in *Electric Power System Research*, Vol. 77, No. 9, July 2007, pp. 1204-1213

[2] A. Mohd, E. Ortjohann, D. Morton, O. Omari, "Review of control techniques for inverters parallel operation," in *Electric Power System Research*, Vol. 80, No. 12, December 2010, pp. 1477-1487

[3] J. Rocabert, A. Luna, F. Blaabjerg, P. Rodríguez, "Control of Power Converters in AC Microgrids," in *IEEE Trans. on Power Electronics*, Vol. 27, No. 11, November 2012, pp. 4734-4749

[4] M. C. Chandorkar, D. M. Divan, R. Adapa, "Control of Parallel Connected Inverters in Standalone ac Supply Systems," in *IEEE Transactions on Industry Applications*, Vol. 29, No. 1, January/February 1993, pp. 136-143

[5] J. M. Guerrero, L. G. de Vicuña, J. Miret, J. Matas, J. Cruz, "Output Impedance Performance for Parallel Operation of UPS Inverters Using Wireless and Average Current-Sharing Controllers," in *Proceedings of the 2004 IEEE 35th Annual Power Electronics Specialists Conference*, PESC 2004, Aachen, Germany, 20-25 June 2004, Vol. 4, pp. 2482-2488

[6] Y. Chen, R. Hesse, D. Turschner, H.-P. Beck, "Dynamic Properties of the Virtual Synchronous Machine (VSIMA)" in *Proceedings of the International Conference on Renewable Energies and Power Quality*, ICREPQ'11, Las Palmas, Spain, 13-15 April 2011, 5 pp.

[7] Q.-C. Zhong, G. Weiss, "Synchronverters: Inverters That Mimic Synchronous Generators," *IEEE Transactions on Industrial Electronics*, vol. 58, no. 4, April 2011, pp. 1259-1267

[8] S. D'Arco, J. A. Suul, "Virtual Synchronous Machines – Classification of Implementations and Analysis of Equivalence to Droop Controllers for Microgrids," in *Proceedings of IEEE PowerTech Grenoble 2013*, Grenoble, France, 16-20 June 2013, 7 pp.

[9] S. D'Arco, J. A. Suul, O. B. Fosso, "Control System Tuning and Stability Analysis of Virtual Synchronous Machines," in *Proceedings of the IEEE Energy Conversion Congress & Exposition*, ECCE 2013, Denver, Colorado, USA, 15-19 September 2013, pp. 2664-2671

[10] N. Pogaku, M. Prodanović, T. C. Green , "Modeling, Analysis and Testing of Autonomous Operation of an Inverter-Based Microgrid," *IEEE Transactions on Power Electronics*, Vol. 22, No. 2, March 2007, pp. 613-625

[11] R. Majumder, "Some Aspects of Stability in Microgrids," in *IEEE Transactions on Power Systems*, Vol. 28, No. 3, August 2013, pp. 3243-3252

[12] N. Bottrell, M. Prodanovic, T. C. Green, "Dynamic Stability of a Microgrid with an Active Load," in *IEEE Transactions on Power Electronics*, Vol. 28, No. 11, November 2013, pp. 5107-5119

[13] S.-K. Lim, S.-K. Sul, "Voltage Control Strategy for Maximum Torque Operation of an Induction Machine in the Field-Weakening Region," in *IEEE Transactions on Industrial Electronics*, Vol. 44, No. 4, August 1997, pp. 512-518

[14] H.-B. Shin, "New Antiwindup PI Controller for Variable-Speed Motor Drives," in *IEEE Transactions on Industrial Electronics*, Vol. 45, No. 3, June 1998, pp. 445-450

[15] X.-L. Li, J.-G. Park, H.-B. Shin, "Comparison and Evaluation of Anti-Windup PI Controllers," in *Journal of Power Electronics*, Vol. 11, No. 1, January 2011, pp. 45-50

[16] J. Espina, A. Arias, J. Balcells, C. Ortega, "Speed Anti-Windup PI strategies review for Field Oriented Control of Permanent Magnet Synchronous Motors," in *Proceedings of the 6th International Conference–Workshop –Compatability and Power Electronics*, CPE 2009, Badajoz, Spain, 20-22 May 2009, pp. 279-285

[17] E. J. Bueno, F. Espinosa, F. J. Rodríguez, S. Ureña, S. Cobreces, "Current Control of Voltage Source Converters connected to the grid through an LCL filter," in *Proceedings of the IEEE 35th Annual Power Electronics Specialists Conference*, PESC 2004, Aachen, Germany, 20-25 June 2004, pp. 68-73

[18] R. Bachana, P.S. Sensarma, "Improvement of Current Dynamics During Controller Saturation in a STATCOM," in *Proceedings of the IEEE International Conference on Industrial* Technology, ICIT 2006, Mumbai, India, 15-17 December 2006, pp. 773-778

[19] F. D. Bianchi, O. Gomis-Bellmunt, "Optimal control of voltage source converters for fault operation," in *Proceedings of the 2011 14th European Conference on Power Electronics and Applications*, EPE 2011, Birmingham, UK, 30 August - 1 Sept. 2011, 10 pp.

[20] R. Hanus, M. Kinnaert, and J. L. Henrotte, "Conditioning technique, a general anti-windup and bumpless transfer method," *Automatica*, Vol. 23, No. 6, pp. 729–739, November 1987

[21] K. Åström, and T. Hägglund, PID Controllers, "*Theory, Design, and Tuning,*" Instrument Society of America, Research Triangle Park, NC, USA, January 1995

[22] S. Buso, P. Mattavelli, "*Digital Control in Power Electronics,*" Morgan & Claypool publishers, 2006.

[23] R. Ottersten, J. Svensson, "Vector Current Controlled Voltage Source Converter – Deadbeat Control and Saturation Strategies," *IEEE Transactions on Power Electronics*, Vol. 17, No. 2, March 2002, pp. 279-285

[24] B.J. Seibel, T.M. Rowan, R.J. Kerkman, "Field oriented control of an induction machine with DC link and load disturbance rejection," in *Proceedings of the 1996 IEEE 11th Annual Applied Power Electronics Conference and Exposition*, APEC'96, San Jose, California, USA, 3-7 March 1996, Vol. 1, pp. 387-393

[25] L. Zhang, "*Modeling and Control of VSC-HVDC Links Connected to Weak AC Systems,*" PhD Dissertation, Royal Institute of Technology, KTH, Stockholm, Sweden, 2010

Virtual Synchronous Generator Control with Double Decoupled Synchronous Reference Frame for Single-Phase Inverter

Yuko Hirase, Osamu Noro, Eiji Yoshimura,
and Hidehiko Nakagawa
Kawasaki Technology Co., Ltd.
1-1, Kawasaki-cho, Akashi 673-8666, Japan

Kenichi Sakimoto and Yuji Shindo
Kawasaki Heavy Industries, Ltd.
1-1, Kawasaki-cho, Akashi 673-8666, Japan

Abstract— **We have proposed the Virtual Synchronous Generator control (VSG control) and have tested it using the demonstration equipment [1]. By using the VSG control, three-phase inverters of current control type are able to run both in grid-connecting operation and in grid-disconnecting operation. Furthermore, in order to control a single-phase inverter like a three-phase inverter using the VSG control, we have applied the technique called "Double Decoupled Synchronous Reference Frame" (DDSRF). In this paper, we will show you the simulation results and experimental results of the single-phase inverter using the VSG control with DDSRF.**

Keywords—Double Decoupled Synchronous Reference Frame, Grid disconnecting Operation, Single-phase Inverter, Virtual Synchronous Generator.

I. INTRODUCTION

Installation of distributed power sources are increasing. These power sources are inverter types, such as photovoltaic generation, a fuel cell, a storage battery, and so on. In recent years, it is expected that they will be operated like conventional synchronous generators. Specifically, the inverter is expected to change the operation state without a power failure between grid-connecting operation and grid-disconnecting operation, and to be performed load sharing with other generators with a droop characteristic. The Virtual Synchronous Generator control is proposed as a control scheme of distributed power sources which enable such applications. **The inverter with this control have similar properties to a synchronous generator**, and the various realization techniques are proposed [1]-[4].

We have proposed the Virtual Synchronous Generator control for three-phase grid-connecting inverter of current control type and it has already been proved [1]. This control enables the inverter to run in grid-disconnecting operation as well as in grid-connecting operation. Moreover, the inverter also becomes to be able to run in parallel with private power generators in the absence of the grid. Because each operation above is performed by using the same program, it is possible to change the operation state without blackout. That is true even if system voltage is unbalanced or an unbalanced load is connected.

On the other hand, distributed power source using renewable energy such as solar power has spread, and the need for single-phase inverter for small businesses is accelerating. In order to apply the Virtual Synchronous Generator control to a single-phase inverter like a three-phase inverter, without if possible changing itself, we considered applying the technique of Double Decoupled Synchronous Reference Frame.

In this paper, we will show you simulation results and experimental results of the single-phase inverter using the Virtual Synchronous Generator control with Double Decoupled Synchronous Reference Frame.

Hereafter in this paper, Virtual Synchronous Generator will be referred to as VSG, and Double Decoupled Synchronous Reference Frame will be as referred to as DDSRF.

II. VSG CONTROL

First, we will explain the VSG control of grid-connecting inverter in three-phase. The control was invented by Kawasaki Heavy Industries and was proved by its subsidiary. Next, we will explain the problems when the control is applied to single-phase.

A. VSG Controlled Three-Phase Grid-Connecting Inverter

Fig. 1 shows the demonstration facility of the VSG controlled three-phase grid-connecting inverter of 20 kW. AC 200 V has connected to the commercial grid and DC 144 V has connected to the photovoltaic generation equipment and the Kawasaki nickel-MH battery GIGACELL. In addition, lineup (10 kW, 20 kW, 50 kW and 250 kW) is developed and there is also a type having DC-DC converters.

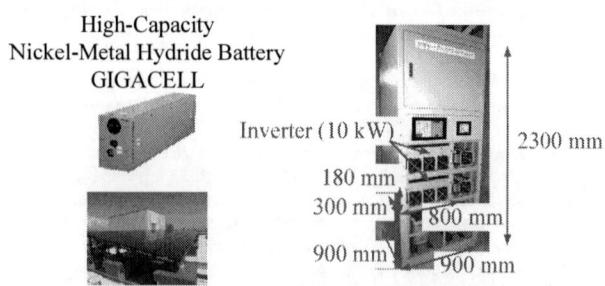

Fig. 1. Demonstration facility.

Fig. 2 shows the hardware which constitutes a three-phase grid-connecting inverter system. It consists of a transformer, an AC filter, an inverter, an electrolytic capacitor, a DC-DC converter, a DC reactor, and a battery. It is possible like the demonstration facility of Fig. 1. to connect arbitrary distributed power sources like photovoltaic generation equipment to the DC side, and DC-DC converters may not be used but may connect a DC source directly to maintain DC capacitor voltage V_{PN}.

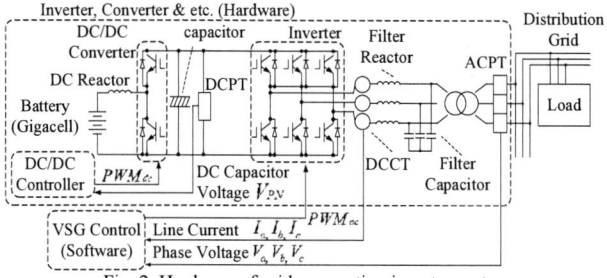

Fig. 2. Hardware of grid-connecting inverter system.

Fig. 3 shows the block diagram of the VSG control in inverter control. The VSG control is the software which consists of a generator controller, a virtual generator model, voltage and current measurement part, and electric power calculation part.

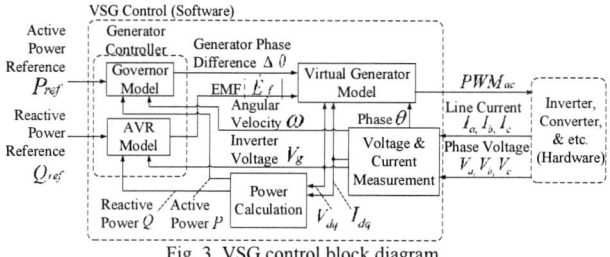

Fig. 3. VSG control block diagram.

The VSG control part of Fig. 3 consists of the governor model and the AVR model and each composition is shown in Fig. 4.

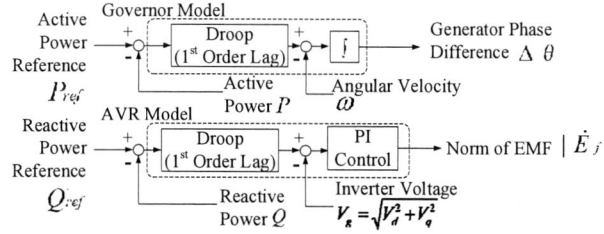

Fig. 4. Governor model and AVR model.

The active power reference P_{ref} inputted into the governor model part means the torque of the generator, which establishes frequency so that it may synchronize with the electrical system. And the reactive power reference Q_{ref}

inputted into the AVR model part means a magnetic field, which adjusts voltage. It is assumed that the motor used as the source of the torque and the magnetization part which generates the magnetic field are constituted in the supervisory control.

Each of the governor model and the AVR model has a 1st order lag function for realizing the droop characteristic. Generator phase difference $\Delta\theta$ calculated in the governor model and the norm of the induced electromotive force vector $\dot{E}_f$ calculated in the AVR model are inputted into the virtual generator model.

Fig. 5 shows the detail of the virtual generator model in Fig. 3. The inverter current reference I_{dq_ref} outputted to the current feedback loop is calculated from the induced electromotive force vector $\dot{E}_f$, the inverter terminal voltage vector $\dot{V}_g$, and the virtual and the actual impedance $r + jx$. In the current feedback loop, the PWM reference PWM_{ac} outputted to the inverter is calculated.

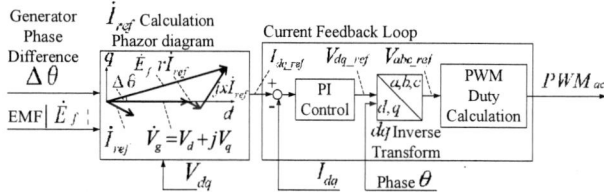

Fig. 5. Virtual generator model.

The left side of Fig. 5 is the phazor diagram of those vectors, and a current reference I_{dq_ref} is calculated as shown in the equations (1) and (2). Here, inverter current means an armature current of a generator.

$$\begin{cases} \Delta V_d = |\dot{E}_f|\cos\Delta\theta - V_d \\ \Delta V_q = |\dot{E}_f|\sin\Delta\theta - V_q \end{cases} \tag{1}$$

$$\begin{cases} I_{ref\_d} = \dfrac{1}{r^2 + x^2}\left(r\Delta V_d + x\Delta V_q\right) \\ I_{ref\_q} = \dfrac{1}{r^2 + x^2}\left(r\Delta V_q - x\Delta V_d\right) \end{cases} \tag{2}$$

B. Problems in a Single-Phased Three-Phase

Fig. 6 shows the detail of the voltage and current measurement part of Fig. 3. In the three-phase system, the line-to-line inverter voltages V_{ab}, V_{bc}, V_{ca} are measured. If the phase angle of V_a is φ, $\sin\varphi, \cos\varphi$ are derived from the equation (3).

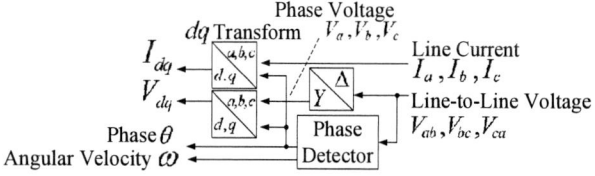

Fig. 6. Voltage and current measurement part of three-phase system.

978-1-4799-2706-7/14 $31.00 © 2014 IEEE

$$\begin{cases} V_a = \dfrac{(V_{ab} - V_{ca})}{3}, \ V_b = \dfrac{(V_{bc} - V_{ab})}{3}, \ V_c = \dfrac{(V_{ca} - V_{bc})}{3} & (3) \\ \sin\varphi = V_{bc}, \ \cos\varphi = V_a \div \sqrt{3} \end{cases}$$

θ is taken as the estimated phase angle outputted from PLL (Phase Locked Loop). If phase detection error $\varphi - \theta$ is close to 0, equation (4) is satisfied.

$$\varphi - \theta \cong \sin(\varphi - \theta) = \sin\varphi\cos\theta - \cos\varphi\sin\theta \qquad (4)$$

Consequently, in the three-phase system, as shown in Fig. 7, $\sin(\varphi - \theta)$ is inputted into PLL as the phase detection error $\varphi - \theta$.

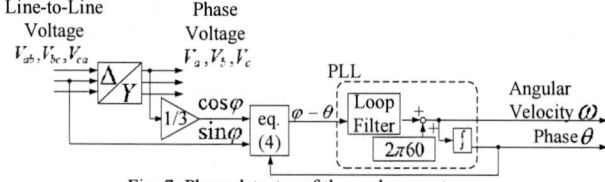

Fig. 7. Phase detector of three-phase system.

If you try to apply the VSG control of the three-phase inverter to a single-phase inverter without changing itself as much as possible, as shown in Fig. 6 and Fig. 7, some new logic for obtaining $V_{dq}, I_{dq}, \omega, \theta$ is required. In order to deal with the problem, we have applied the theory called DDSRF.

III. CHARACTERISTIC OF DDSRF

A. DDSRF for Three-Phase

In order not to affect inverter output current, unbalanced components of the system voltage could be separated from the fundamental components by DDSRF [5]. Fig. 8 shows the block diagram of DDSRF and the phase detector of the three-phase system.

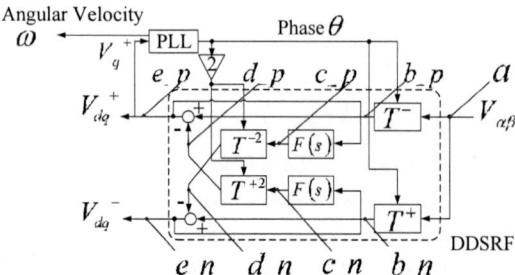

Fig. 8. DDSRF and phase detector of three-phase system.

$V_{\alpha\beta}$ are obtained by the Clarke transformation of the three-phase grid voltage. The Clarke transformation C^T which is the conversion from V_{abc} to $V_{\alpha\beta}$ is given by equation (5).

$$C^T = \sqrt{\dfrac{2}{3}} \begin{bmatrix} 1 & -\dfrac{1}{2} & -\dfrac{1}{2} \\ 0 & \dfrac{\sqrt{3}}{2} & -\dfrac{\sqrt{3}}{2} \end{bmatrix} \qquad (5)$$

The output V_{dq}^+ are the positive sequence components which are remaining after removing negative sequence

components from $V_{\alpha\beta}$. On the contrary, V_{dq}^- are negative sequence components which are remaining after removing positive sequence components from $V_{\alpha\beta}$.

$T^{\pm}$, $T^{\pm 2}$ are rotating matrices given by equations (6) and (7). They convert the signals in stationary frame ($\alpha\beta$ coordinate system) into the ones in reference frame (dq coordinate system). T^- obviously means conventional dq transformation such as used in Fig. 6 and T^+ means dq inverse transformation as used in Fig. 5.

$$T^{\pm} = \begin{bmatrix} \cos(\pm\theta) & -\sin(\pm\theta) \\ \sin(\pm\theta) & \cos(\pm\theta) \end{bmatrix} \qquad (6)$$

$$T^{\pm 2} = \begin{bmatrix} \cos(\pm 2\theta) & -\sin(\pm 2\theta) \\ \sin(\pm 2\theta) & \cos(\pm 2\theta) \end{bmatrix} \qquad (7)$$

$F(s)$ is the LPF which is used in the positive or the negative sequence in order to remove the negative or the positive sequence components, the frequency of which are twice the fundamental frequency. It is a commonly used LPF. If the fundamental frequency is 60Hz, approximately ±40 Hz of cut-off frequencies are suitable.

Fig. 9 shows strength for the frequency of each signal, which is obtained by Fourier transforming the signals named in the blue characters in Fig. 8. $V_{\alpha\beta}$ (a) are assumed to have both of the positive sequence components (red stick) and the negative sequence components (blue stick). The negative sequence components are generated when the three-phase voltage is unbalanced.

When the phase detection error is ideally 0, the output of the positive sequence V_q^+ is close to 0. Hence, it is inputted to PLL. PLL is the same as the one shown in Fig. 7.

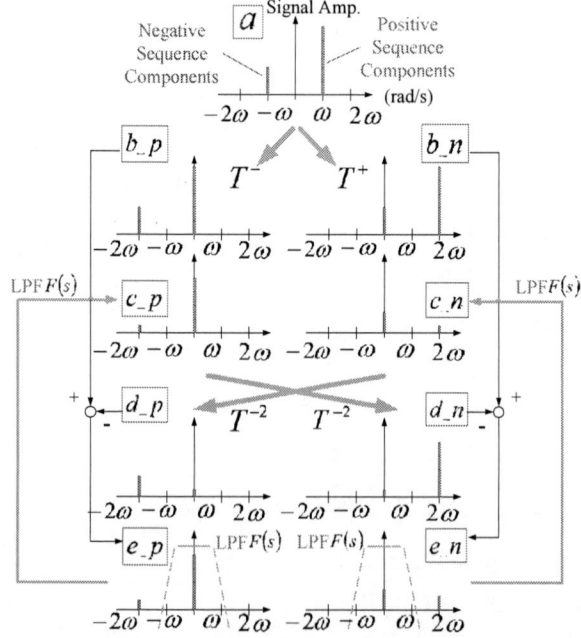

Fig. 9. Image of signal strength for frequency.

B. DDSRF for Single-Phase

If a single-phase system is considered to be a special unbalanced state of a three-phase system, DDSRF is also applicable to a single-phase system.

In order to understand the logic of DDSRF, the calculation results of the three-phase signals given by equations (8), (9), and (10) are shown in Fig. 10, Fig. 11, and Fig. 12, respectively. They have no unit. Here, θ is the phase angle at the constant frequency of 60Hz.

$$V_\alpha = 1.0\cos\theta, \quad V_\beta = 1.0\sin\theta \tag{8}$$

$$V_\alpha = 1.0\cos\theta, \quad V_\beta = 0.5\sin\theta \tag{9}$$

$$V_\alpha = 1.0\cos\theta, \quad V_\beta = 0.0\sin\theta \tag{10}$$

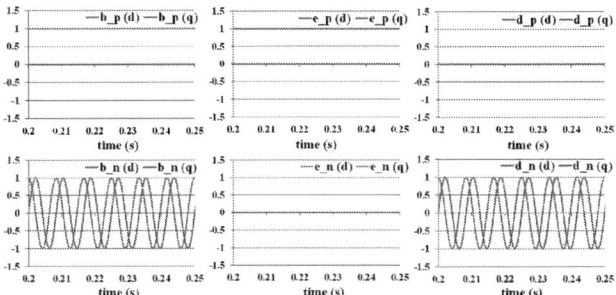

Fig. 10. Time signal of each part in the case of balanced three-phase (Equation (8) V_α=1.0cosθ, V_β=1.0sinθ).

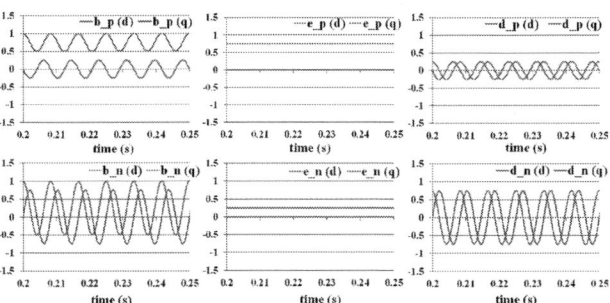

Fig. 11. Time signal of each part in the case of unbalanced three-phase (Equation (9) V_α=1.0cosθ, V_β=0.5sinθ).

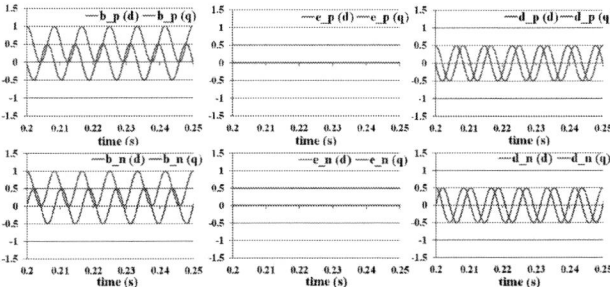

Fig. 12. Time signal of each part in the case of unbalanced three-phase (Equation (10) V_α=1.0cosθ, V_β=0.0sinθ).

As shown in Fig. 10, if the three-phase signal is ideally balanced, there is no difference between b_p and e_p. b_p is a DC signal which is obtained by the conventional dq transformation, and consequently there is no problem even if you use it in the control part.

As shown in Fig. 11 or Fig. 12, if the three-phase signal is unbalanced, b_p is not a DC signal and it oscillates at twice the fundamental frequency in proportion to the strength of the unbalanced components. Therefore, in substitution for b_p, you may use e_p which is a DC signals obtained by removing the vibration.

Equation (10) represents that **the single-phase system is thought as one of the special unbalanced state of three-phase systems.** And as shown in Fig. 12, the strength of DC signal e_p and e_n is equal. It means that **each of the positive and the negative sequence components of a single-phase signal have fundamental frequency ingredients and that the strength of these signals is equal.** Therefore, in the single-phase system, it is sufficient to use only the positive sequence components in the control part and to output what added the negative sequence components to them.

C. Applying DDSRF to VSG Controlled Single-Phase Inverter

The voltage and current measurement part of Fig. 3 and the dq inverse transformation that is in the current feedback loop of the virtual generator model in Fig. 5 should be changed according to a single-phase system.

Fig. 13 is the voltage and current measurement part for the single-phase system. In $DDSRFv$, to which a single-phase signal V_a is inputted, the operation using DDSRF is performed to the voltage. As in the same way, a Single-phase signal I_a is inputted to $DDSRFi$ and the operation is performed to the current. The structure of $DDSRFi$ is the same as that of DDSRF in Fig. 8. On the other hand, the filtered voltage lets the operation of the VSG controlled system is stable. For that reason, the output signals $V_{dq}{}^+$, $V_{dq}{}^-$ of the $DDSRFv$ are not e_p, e_n but c_p, c_n.

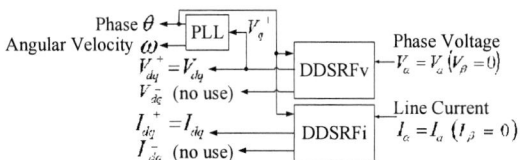

Fig. 13. Voltage and current measurement part for single-phase.

Fig. 14 shows the current feedback loop for the single-phase system. Inputted signals to the rotation matrix T^+ are V_{dq_ref} and each of them is the sum of positive and negative sequence results.

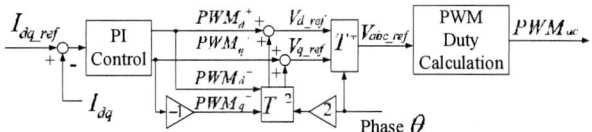

Fig. 14. Current feedback loop for single-phase.

IV. SIMULATION RESULTS OF SINGLE-PHASE INVERTER WITH VSG AND DDSRF

In this chapter, we will show you the simulation results of the single-phase inverter system with the VSG control and DDSRF, which theory is described in detail in previous chapters.

The simulator used is EMTPWorks Ver. 2.5.0 of Powersys.

A. Simulation of a Single System

Fig. 15 is the schematic diagram of the single-phase system simulation. It is assumed that the construction of Fig. 2 and Fig. 3 have changed for a single-phase system.

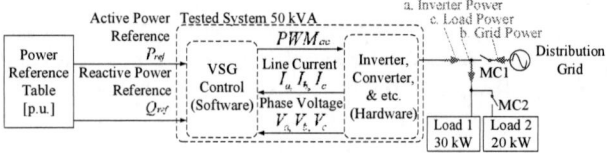

Fig. 15. Schematic diagram of simulation of single system.

The rated capacity of the inverter is 50 kVA=1 p.u., droop is 5 %, and the PWM frequency of the inverter is 8 kHz. In addition, the PWM frequency of the converter is 4 kHz. MC1 is closed from the beginning and an accident at the commercial grid is simulated by opening MC1. Resistive load of 30 kW is connected from the beginning, load of 20 kW is added later when MC2 is closed, and finally the total load electric power is set to 50 kW. The fundamental frequency of the commercial grid is 60Hz.

Those are summarized in TABLE I : the time to start and stop the inverter, the time to change the inverter output power references, the time when the simulated accident occurs, and the time to change the load capacitance. Total simulation time is 16 s and the simulation resolution is 12.5 us per one step.

TABLE I
STATE TRANSITION TIMETABLE OF SIMULATION

time [s]	inverter	Power Ref.	Accident (MC1)	Load (MC2)
0	stop	P_{ref}=0.0 p.u. Q_{ref}=0.0 p.u.	Grid (MC1:On)	30 kW (MC2:Off)
3	servo on	P_{ref}=0.0 p.u. Q_{ref}=0.0 p.u.	Grid (MC1:On)	30 kW (MC2:Off)
5	servo on	P_{ref}=0.2 p.u. Q_{ref}=0.0 p.u.	Grid (MC1:On)	30 kW (MC2:Off)
7	servo on	P_{ref}=0.6 p.u. Q_{ref}=0.0 p.u.	Grid (MC1:On)	30 kW (MC2:Off)
10	servo on	P_{ref}=0.6 p.u. Q_{ref}=0.0 p.u.	Island (MC1:Off)	30 kW (MC2:Off)
12	servo on	P_{ref}=0.6 p.u. Q_{ref}=0.0 p.u.	Island (MC1:Off)	50 kW (MC2:On)

Fig. 16 shows the measurement results of inverter output electric power, grid electric power, and load electric power, which are indicated as *a* to *c* in Fig. 15. Here, a direction feeding the load is defined as plus. And Fig. 17 expresses the system frequency during the time.

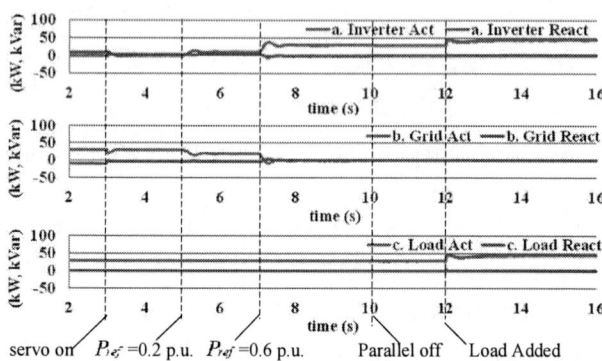

Fig. 16. Power of each part of single system simulation.

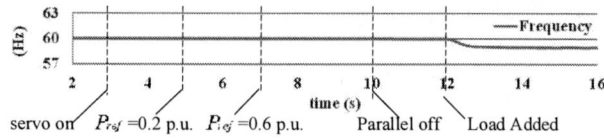

Fig. 17. Frequency of single system simulation.

Fig. 18 shows two charts of the load distribution in a single system. Inclination of the blue line expresses the characteristic of the commercial grid, and it is almost horizontal with a droop close to zero. Inclination of the red line expresses the characteristic of the inverter with droop 5 %. The length of the green line expresses the load capability.

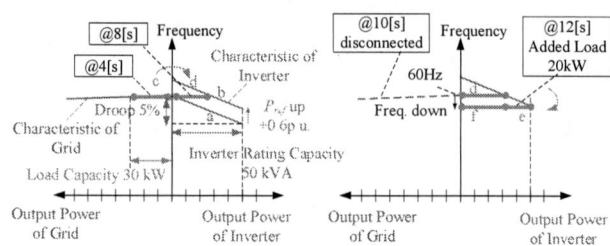

Fig. 18. Image of load distribution in single system.

Immediately after servo on, all the load electric power is fed from the commercial power supply (blue) because the output electric power reference value of the inverter is 0. As shown in Fig. 18, at this time, the inverter characteristic line (red) is at the position of *a*, and the load electric power line (green) is in the position of *c*. If the output electric power reference value of the inverter is raised, the inverter characteristic line (red) moves from *a* to *b*, and a load electric power line (green) moves from *c* to *d* according to it. This means that **the power supply source of the load electric power is shifted to the inverter from the commercial power supply.**

As shown in the right chart in Fig. 18, even if MC1 is opened and the electric power from the commercial power supply is lost, frequency does not decrease because all the load electric power (green line in the position of *d*) is already supplied from the inverter. If MC2 is closed and load electric power (green line in the position of *e*) is added, the total load electric power (green lines in the position of *e* and *f*) is supplied from the inverter. At this time, as shown in Fig. 17, **frequency decreases according to the droop characteristic of the inverter.**

As shown in Fig. 16, Fig. 17, and Fig. 18, **the load electric power is not greatly disturbed and is fed continuously, at any moment of the power supply shift, the parallel off, and the addition of load.**

Fig. 19 shows the results of measuring the AC voltage and the AC current before and after the simulation time 10 s when MC1 is opened (the moment of parallel off). **Neither the AC voltage nor the AC current is also confused greatly at that time.**

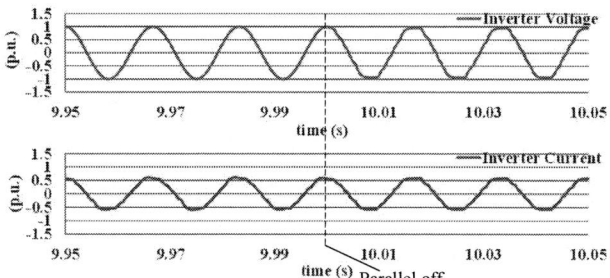

Fig. 19. Voltage and current at parallel off in single system simulation.

B. Simulation of a Multiple System

Fig. 20 is a schematic diagram that two test systems (two inverters) are connecting to the commercial grid. Inverter rating capacity of the system 1 is 50 kVA=1 p.u., and inverter rating capacity of the system 2 is 30 kVA=1 p.u. All other characteristics such as droop constants are the same.

Resistive load of 50 kW is connected from the beginning, load of 30 kW is added later when MC2 is closed, and finally load electric power is set to 80 kW. The unit of output electric power references is p.u. and the value of TABLE I is inputted into each inverter. Other test conditions are the same as the ones for single test system.

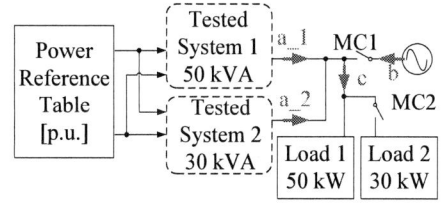

Fig. 20. Schematic diagram of simulation of multiple systems.

In the same manner as Fig. 16, Fig. 21 shows the measurement results of inverter output electric power, grid electric power, and load electric power, which are indicated in Fig. 20. And as Fig. 17, Fig. 22 expresses the system frequency during the time.

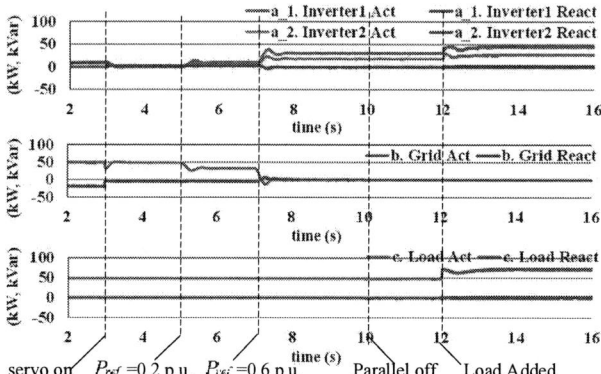

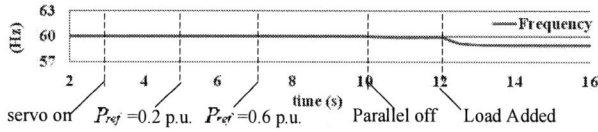

Fig. 21. Power of each part of multiple system simulation.

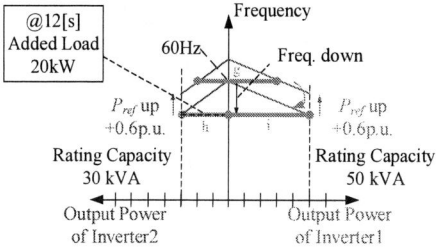

Fig. 22. Frequency of multiple system simulation.

Fig. 23 is an image diagram of load distribution in the multi-system. Inclination of the red lines express the characteristic of the inverter 1 and that of purple lines express the characteristic of the inverter 2. Both have the characteristics of droop 5%.

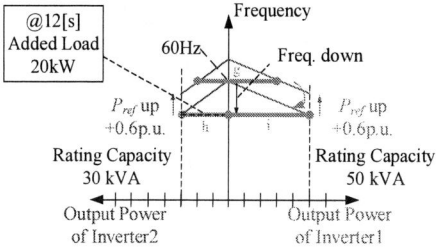

Fig. 23. Image of load distribution in multiple systems.

Concept of load power distribution in the parallel running with the commercial power supply before MC1 is opened is the same as that of the single system simulation (Fig. 18).

After MC1 is opened, **two inverters run in parallel without the commercial power supply.** Immediately after the moment of parallel off, load electric power line (green) is at the position of *g* because the output electric power reference value of each inverter has been raised. If MC2 is closed and load electric power (green line in the position of *h*) is added, the total load electric power (green lines in the position of *h* and *i*) is supplied from two inverters. Then, frequency decreases.

Regardless of the amount of load electric power, **the inverter 1 and the inverter 2 automatically supply the load electric power at the rate of their rating capacity ratios (30 kW : 50 kW).**

In addition to this frequency adjustment function by active power, the VSG controlled inverter also has voltage adjustment function by reactive power.

978-1-4799-2706-7/14 $31.00 © 2014 IEEE   1557

V. EXPERIMENTAL RESULTS OF SINGLE-PHASE INVERTER WITH VSG AND DDSRF

We will show you the experimental results using the testing equipment of 50 kVA. Although multiple systems are verified in the simulation, only a single system is verified in the proving test since we have only one single-phase transformer. Moreover, since the amount of the current of a laboratory is restricted, the load electric power and inverter output electric power are adjusted so that the amount of electric power from the commercial power supply should not become too much large.

A. Experimental Results

Fig. 24 shows the VSG controlled grid-connecting inverter system for the proving test. Its rating capacity is 50 kVA. The photo on the right side shows the state of the system with the lower front door at the bottom of the left side photo opened.

Various switches for safety are accommodated in the upper row. Three power blocks are located in the middle section, which are an inverter to the commercial grid, a converter to the storage battery, and a converter to the photovoltaic generation, respectively.

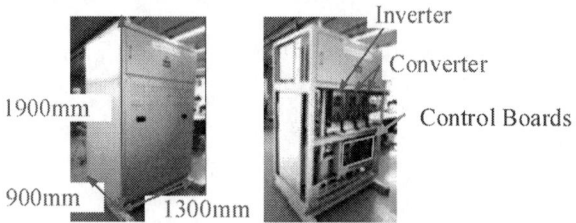

Fig. 24. Grid-connecting inverter (50 kVA) for verification.

In the lower front area, there are control boards disposed vertically, which are communication boards, CPU control boards, and interface boards. They are stacked in three layers and the detail is shown in Fig. 25.

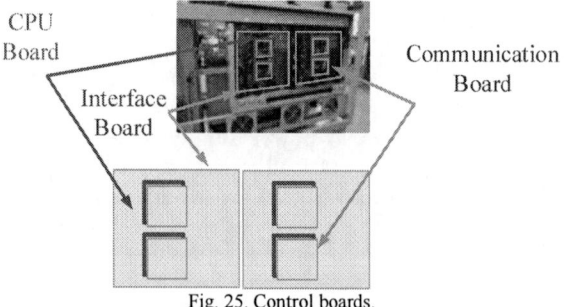

Fig. 25. Control boards.

Fig. 26 is a schematic diagram of the testing equipment. The red arrow shows the measuring point of the inverter output electric power. Here, the direction of feeding the load is defined as plus.

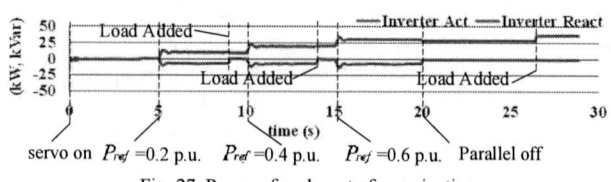

Fig. 26. Schematic diagram of demonstration facility.

TABLE II represents the timetable similar to the simulation. Like a simulation, MC1 is opened manually and an accident is simulated. Resistive load of 10 kW is connected from the beginning. After that, two loads of 10 kW are added when MC2 and MC3 are closed and loads of 7.75 kW are added when MC4 is closed. Finally, the total load electric power is set to 37.75 kW. The fundamental frequency of the commercial grid is 60Hz.

TABLE II
STATE TRANSITION TIMETABLE OF EXAMINATION

time [s]	inverter	Power Ref.	Accident (MC1)	Load
0	servo on	P_{ref}=0.0 p.u. Q_{ref}=0.0 p.u.	Grid (MC1:On)	10 kW (MC2:Off, MC3:Off, MC4:Off)
5	servo on	P_{ref}=0.2 p.u. Q_{ref}=0.0 p.u.	Grid (MC1:On)	10 kW (MC2:Off, MC3:Off, MC4:Off)
9	servo on	P_{ref}=0.2 p.u. Q_{ref}=0.0 p.u.	Grid (MC1:On)	20 kW (MC2:On, MC3:Off, MC4:Off)
10	servo on	P_{ref}=0.4 p.u. Q_{ref}=0.0 p.u.	Grid (MC1:On)	20 kW (MC2:On, MC3:Off, MC4:Off)
14	servo on	P_{ref}=0.4 p.u. Q_{ref}=0.0 p.u.	Grid (MC1:On)	30 kW (MC2:On, MC3:On, MC4:Off)
15	servo on	P_{ref}=0.6 p.u. Q_{ref}=0.0 p.u.	Grid (MC1:On)	30 kW (MC2:On, MC3:On, MC4:Off)
20	servo on	P_{ref}=0.6 p.u. Q_{ref}=0.0 p.u.	Island (MC1:Off)	30 kW (MC2:On, MC3:On, MC4:Off)
26	servo on	P_{ref}=0.6 p.u. Q_{ref}=0.0 p.u.	Island (MC1:Off)	37.75 kW (MC2:On, MC3:On, MC4:On)

In the same manner as the simulation, Fig. 27 expresses the measurement result of inverter output electric power, Fig. 28 express the effective voltage and Fig. 29 express the frequency during the time. These figures express almost the same result as the simulation.

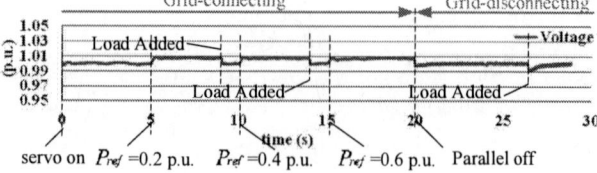

Fig. 27. Power of each part of examination.

Fig. 28. Effective voltage of examination.

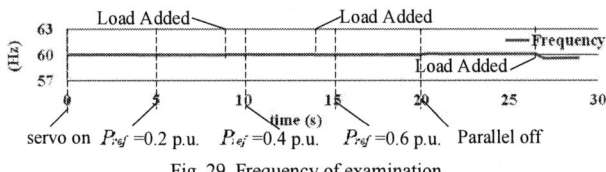

Fig. 29. Frequency of examination.

In the grid-connecting operation, the electric power is outputted from the inverter corresponding to the output power references. Even if an accident occurs at the commercial grid, the electric power supply to the load is continued without any power failure. In the grid-disconnecting operation, frequency decreases according to the droop characteristic of the inverter.

Fig. 28 shows that the voltage of the commercial grid is easy to deviate. It drops when loads are added. On the contrary, when the power supply source of the loads is shifted to the inverter, since the grid voltage becomes higher than the reference value 202 V=1 p.u., the lagging reactive power seen from the grid becomes to be outputted from the inverter. This is clear from Fig. 27. As stated in IV.B, it means that the voltage adjustment function by reactive power works.

Fig. 30 shows the results of measuring the AC voltage and the AC current at the time before and after the MC1 is opened (when the simulated accident occurs). As in the simulation, neither the AC voltage nor the AC current is also confused greatly at that time.

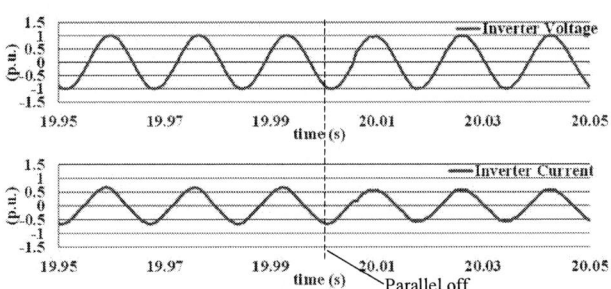

Fig. 30. Voltage and current at parallel off in examination.

B. Challenging Problem

In the system to propose, angular velocity and a phase angle are calculated from the system voltage. Therefore, unbalanced voltage or voltage with harmonics can become an instability factor of a whole system. Unlike there-phase signals, there is no fear of the single-phase signal itself becoming unbalanced. But the estimate of the harmonics should be required.

As shown in Fig. 8 and Fig. 9, DDSRF is originally a method of separating negative sequence components as unbalanced components from the fundamental components. For that reason, we assume that, if DDSRF is used, harmonics ingredients other than fundamental components and unbalanced components will be amplified, and it will become an opposite effect.

To remove the signal of the frequency band of each low order harmonics with comparatively large intensity (3rd, 5th, and 7th order), it will be good to prepare a notch filter. The notch filter should be used for inputted signal of *DDSRFv* or *DDSRFi*.

For example, Fig. 31 is the block diagram of the notch filter of the 3rd order harmonic suppression. Like Fig. 8, $F(s)$ is the commonly used LPF and $T^{\pm 3}$ are the rotating matrices which rotation angles are $\pm 3\theta$. Further investigation is required for that.

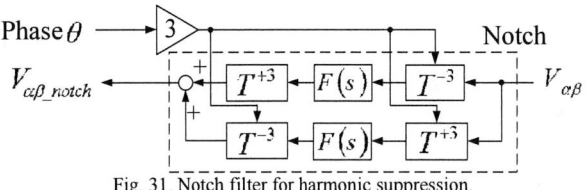

Fig. 31. Notch filter for harmonic suppression.

VI. CONCLUSION

In order to apply, without if possible changing itself, the VSG control into a single-phase grid-connecting inverter like a three-phase grid-connecting inverter, we have considered applying the technique of DDSRF.

When three-phase system voltage is unbalanced, DDSRF have been introduced as a method for separating the unbalanced components. Therefore, by thinking of single-phase system as special unbalanced three-phase system, we have been able to apply DDSRF to single-phase.

In further work, we would like to understand in detail about the stability of the VSG controlled grid-connecting single-phase inverter using DDSRF. And using the testing equipment, we also want to confirm that it can run in parallel with various power sources, such as multiple single-phase inverters with different rating capacities, three-phase inverters, diesel generators and so on.

REFERENCES

[1] Y. Hirase, K. Abe, K. Sugimoto, Y. Shindo, "A grid-connected inverter with virtual synchronous generator model of algebraic type," *Electrical Engineering in Japan*, Vol. 184, Issue 4, pp10–21, Wiley, 2013.

[2] K. Sakimoto, Y. Miura, and T. Ise, "Stabilization of a Power System including Inverter Type Distributed Generators by the Virtual Synchronous Generator," *IEEJ Transaction on Power and Energy*, Vol. 132 No.4, pp.341-349, 2012.

[3] K. Sakimoto, Y. Miura, and T. Ise, "Characteristics of Parallel Operation of Inverter Type Distributed Generators Operated by a Virtual Synchronous Generator," *IEEJ Transaction on Power and Energy*, Vol. 133 No.2, pp.186-194, 2013.

[4] K. Sakimoto, K. Sugimoto, Y. Shindo, "Low Voltage Ride Through Capability of a Grid Connected Inverter based on the Virtual Synchronous Generator," *Proc. of IEEE Intrnl. Conf. of Power Electronics and Drive Systems*, pp.1066-1071, 2013.

[5] D. Siemaszko, A. C. Rufer, "Power Compensation Approach and Double Frame Control for Grid Connected Converters," *Proc. of IEEE Intrnl. Conf. of Power Electronics and Drive Systems*, pp.1263-1268, 2013.

Contactless DC Connector based on GaN LLC Converter for Next Generation Data Centers

Yusuke Hayashi, Hajime Toyoda and Toshifumi Ise
Div. of Electrical Electronic and Information Engineering
Graduate School of Engineering, Osaka University
2–1 Yamada–Oka, Suita, Osaka, Japan
y_hayashi@eei.eng.osaka-u.ac.jp

Akira Matsumoto
Research and Development Headquarters
NTT Facilities, Inc.
2–13–1 Kita–Otsuka, Toshima–ku, Tokyo, Japan

Abstract— **Contactless DC connector has been proposed for the next generation 380 V DC distribution system in data centers. A LLC resonant DC–DC converter topology with Gallium Nitride (GaN) power transistors has been applied to realize the short–distance inductively–coupled connector. A prototype of a 1.2 kW 384 V–192 V connector has been fabricated under 500 kHz operation. The conversion efficiency of 95% has been confirmed, and the approach for realizing higher efficiency has been also shown. The contactless DC connector integrates the functioning of an isolated DC–DC converter into a connector for space–saving. The DC current can be cut off without arc because of the inductive coupling. The proposed connector contributes to realizing highly efficient, space–saving and reliable future DC distribution system.**

Keywords— Contactless Power Supply, DC–DC Converter, Gallium Nitride, DC Distribution.

I. INTRODUCTION

The amount of network traffic in the data centers and the telecommunications buildings has recently been rapidly increasing due to the widespread use of information and communication technology (ICT) equipment [1]. Energy saving in these buildings and data centers will contribute to solving some of our global environmental problems. Since 2008, the NTT (Nippon Telegraph and Telephone) Group has been developing 380 V DC distribution system that goes beyond the conventional 48 V DC distribution system to realize highly efficient and space–saving (high power density) power supply [2] [3].

Now, the advanced DC distribution system has been proposed to realize future low carbonated society [4] [5]. The conversion efficiency has to be improved from 80% in the developed 380 V DC distribution system to 94%. High power density power converters are attractive, and power converters with more than 10 W/cm^3 have been reported [6]–[8].

As the power density of the power converter increases, the volume of peripherals such as connectors will not be negligible. The power density of the connector for several kW output power in 380 V DC distribution system is now at a level of 10 W/cm^3. To realize higher power density 380 V DC distribution system, comprehensive design taking the influence of peripherals into consideration is indispensable.

In this paper, a contactless DC connector is proposed to realize high power density 380 V DC distribution system. The contactless DC connector integrates functions of both an isolated DC–DC converter and a conventional connector in a single instrument to shrink total volume. To integrate functions into a single instrument, short–distance inductively–coupled wireless power transfer technology is applied. An isolated DC–DC converter topology with a gapped transformer is available for short–distance wireless power transfer to achieve high efficiency because the connector in 380 V DC distribution system is not the mobile device.

The DC distribution system using proposed DC connectors is introduced in chapter II. Then, details of the circuit configuration and characteristics of the connector are described in chapter III. The feasibility of the proposed connector is verified experimentally in chapter IV. In chapter V, design consideration for the connector is conducted to achieve higher power density.

II. NEXT GENERATION 380 V DC DISTRIBUTION SYSTEM WITH INDUCTIVELY–COUPLED CONNECTOR

Figure 1 shows the power density roadmap of isolated DC–DC converters and connectors for DC distribution system. In research and development level, power converters with 10 W/cm^3 has been developed for 380 V DC distribution system [9]. The power density of a socket in a connector is now at a level of several tens of

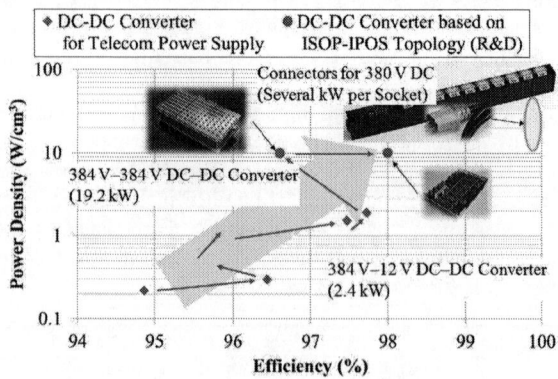

Fig. 1. Power density of isolated DC–DC converters and connectors for DC distribution system.

978-1-4799-2706-7/14 $31.00 © 2014 IEEE   1560

W/cm^3 because the volume of a socket is 100 cm^3 and its rated power is several kW.

Figure 2 shows the schematic diagram of the conventional 380 V DC distribution system for data centers. In Fig. 2, connectors with metal contacts and isolated DC–DC converters are installed into low voltage (e.g. 380 V, 48 V and 12 V) DC distribution line behind the rectifier. The power density of power converters with the rated voltage of DC 380 V is now at a level of 10 W/cm^3 in R&D, and the converters with more than 50 W/cm^3 is now commercially available for DC 48 V and DC 12 V on–board power supply [10]. The influence of peripherals will become obvious in near future.

Figure 3 shows the schematic diagram of the proposed DC distribution system with inductively–coupled contactless DC connectors. Contactless connectors are only installed in the 380 V DC distribution line in front of ICT equipment. A primary and a secondary circuits of an isolated DC–DC converter are utilized as a pair of a socket and a plug respectively. The primary and the secondary circuits are connected via the inductive coupling in a short–distance. The proposed concept for contactless DC connector has following features.

- The contactless connector integrates the functioning of an isolated DC–DC converter and a pair of plug and socket into single instrument. Higher power density DC distribution system will be developed, because the number of installed instruments can be reduced.

- The output voltage can be selected arbitrarily by changing the number of turns at the secondary winding of the transformer. The IPOS (input parallel and output series) and ISOP (input series and output parallel) topology can be also applied to control the output voltage.

- The DC current can be cut off without arc because

of the inductive coupling.

- High efficiency can be accomplished when the primary and the secondary circuits are connected in a short–distance, because they behave as an isolated DC–DC converter with a gapped transformer. The short–distance wireless power transfer is available for this application.

The commercially available isolated DC–DC converter based on LLC circuit topology has achieved the efficiency of 98% and the power density of several tens of W/cm^3 [10]. The IPOS and ISOP topologies are applicable for the LLC resonant converter [11]. To realize the aforementioned inductively–coupled connector, LLC converter topology is applied here.

III. LLC RESONANT CONVERTER TOPOLOGY FOR INDUCTIVELY–COUPLED CONNECTOR

The circuit configuration of the LLC converter is shown in Fig. 4. The availability of LLC converter has been reported and this topology has been applied for wireless power transfer [12] [13]. The zero voltage switching (ZVS) of transistors and the zero current switching (ZCS) of diodes achieve high efficiency. In the case that turn ratio of the transformer n is 1, the voltage gain M and the phase difference between the input voltage of the LLC resonant tank and the input current of the tank θ are calculated by following equations (1) (2).

$$M = |\frac{V_2}{V_1}| = |\frac{s^2 N_1}{s^3 D_1 + s^2 D_2 + s D_3 + D_4}| \cdots\cdots\cdots (1)$$

$$D_1 = \{L_1 L_2 + (L_1 + L_2)L_m\}C, \qquad D_2 = R(L_1 + L_m)C$$
$$D_3 = (L_2 + L_m), \qquad D_4 = R, \qquad N_1 = RL_m C$$

$$\theta = \tan^{-1}\left(\frac{N_2' D_1' - N_1' D_2'}{N_1' D_1' + N_2' D_2'}\right) \cdots\cdots\cdots\cdots\cdots (2)$$

$$D_1' = -\omega^2(L_2 + L_m)C, \qquad D_2' = \omega RC$$
$$N_1' = R - \omega^2 R(L_1 + L_m)C$$

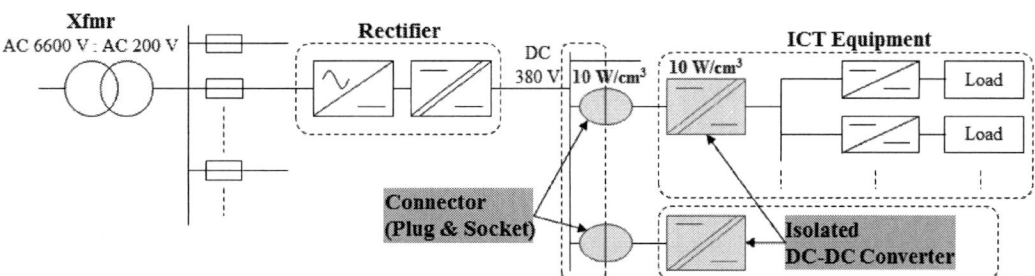

Fig. 2. Configuration of conventional 380 V DC distribution system.

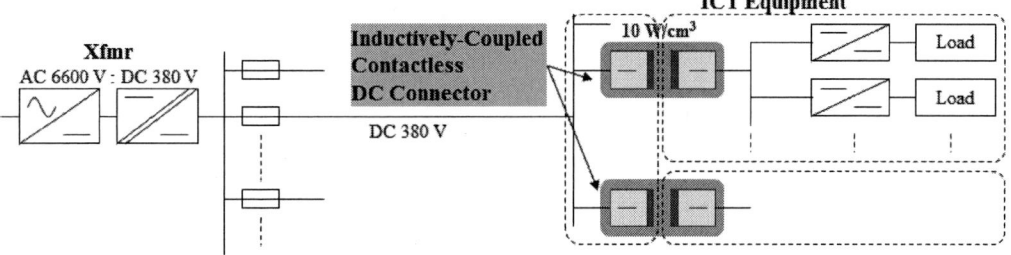

Fig. 3. Configuration of 380 V DC distribution system using contactless DC connector.

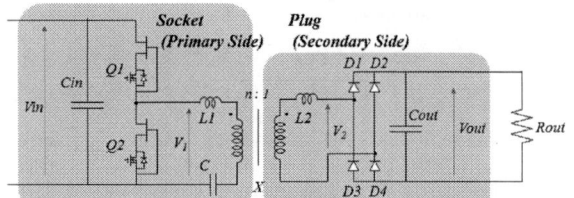

Fig. 4. Circuit configuration of contactless DC connector based on LLC converter topology.

$$N_2' = -\omega^3\{L_1 L_2 + (L_1 + L_2)L_m\}C$$

Parameters V_1 and V_2 in Eq. (1) are respectively input and output voltages of the LLC resonant tank in Fig. 4. Primary and secondary leakage inductances of the transformer are L_1 and L_2 respectively. The magnetizing inductance of the transformer is L_m. The resonant capacitance is C and R means the load resistance in the fundamental frequency analysis. In the case the coupling coefficient of the transformer k is high, the influence of L_2 on the voltage gain is negligible.

Parameters for realizing the connector based on the LLC converter topology are shown in Table 1. Figures 5 and 6 show the calculation results of the voltage gain M and the phase difference θ for the LLC resonant tank. The voltage gain was drawn when the switching frequency f_{SW} varied from 100 kHz to 10 MHz in Fig. 5. The phase difference θ was also drawn for the varied switching frequency in Fig. 6. In Fig. 6, the positive value means the impedance has the inductive characteristics because the phase angle of the LLC input current has the delay against that of the LLC input voltage. The voltage gain and the phase angle were drawn when the load factor changed from 5% to 200%. Here, the load factor 100% means the rated output power of 1,200 W as shown in Table 1.

The frequency f_r which the voltage gain is independent of the load resistance is set at 550 kHz and the switching frequency f_{SW} is 500 kHz. The magnetizing inductance L_m is determined to suppress the magnetizing current i_{Lm} as shown in Eq. (3). The magnetizing current becomes smaller as the magnetizing inductance is getting larger, because the LLC input voltage L_1 is constant value and the time dt is fixed by the frequency f_r. Leakage inductances L_1 and L_2 are calculated by using L_m and the coupling coefficient of the transformer k as shown in Eq. (4). The resonant capacitor C is designed as the resonant frequency caused by L_1, L_2 and C corresponds to the

TABLE I
PARAMETERS FOR CONTACTLESS DC CONNECTOR BASED ON LLC CONVERTER TOPOLOGY.

Symbol	Meaning	Value
P_{out}	Output power	1,200 W
V_{in}, V_{out}	Input / output voltage	384 V / 192 V
f_r	Resonant frequency	550 kHz
f_{sw}	Switching frequency	500 kHz
k	Coupling coefficient	0.98
L_m	Magnetizing inductance	19.6 μH
L_1, L_2	Resonant inductance	3.4 μH, 0.4 μH
C	Resonant capacitance	22.1 nF
n	Turn ratio	1

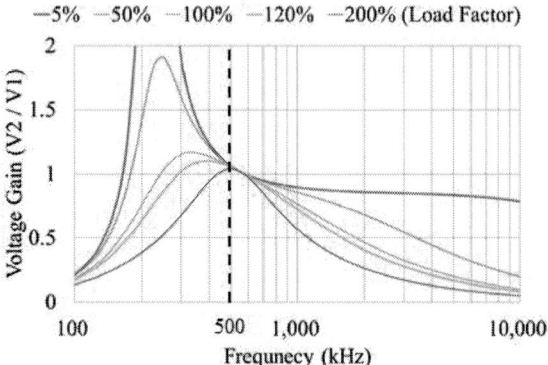

Fig. 5. Voltage gain characteristics of LLC resonant tank in contactless DC connector.

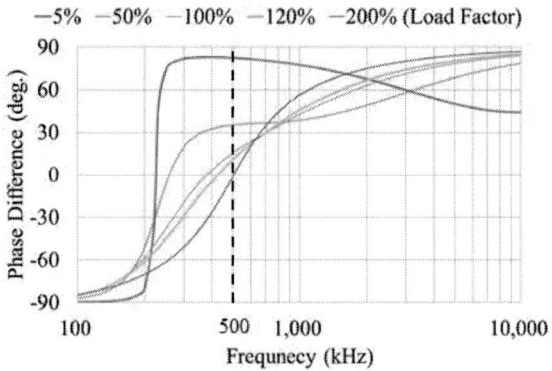

Fig. 6. Phase characteristics of LLC resonant tank (positive: inductive, negative: capacitive).

designated frequency of 550 kHz. Equation (5) is utilized for calculating the capacitor C. In Fig. 6, these parameters make the impedance characteristics inductive (i.e. positive phase angle) at the switching frequency of 500 kHz under the load factor from 5 % to 120 %. This means ZVS can be accomplished at the rated power of 1,200 W.

$$V_1 = L_m \frac{di_{L_m}}{dt} \dots\dots\dots\dots\dots\dots\dots (3)$$

$$L_1 = L_2 = \frac{1-k}{k} L_m \dots\dots\dots\dots\dots (4)$$

$$2\pi f_r = \sqrt{\frac{L_2 + L_m}{\{L_1 L_2 + (L_1 + L_2)L_m\}C}} \dots\dots (5)$$

IV. EXPERIMENTAL VERIFICATION FOR CONTACTLESS DC CONNECTOR

The prototype of the inductively–coupled DC connector was fabricated. Figure 7 shows the experimental apparatus. The output power is 1.2 kW, the input and the output voltages are 384 V and 192 V respectively. Circuit parameters are based on Table I and components shown in Table II were employed to develop the prototype.

A. High–Speed and Ultra–Low Loss Power Devices and Generated Switching Loss

GaN–HEMT (600 V, 16 A) from Transphorm was

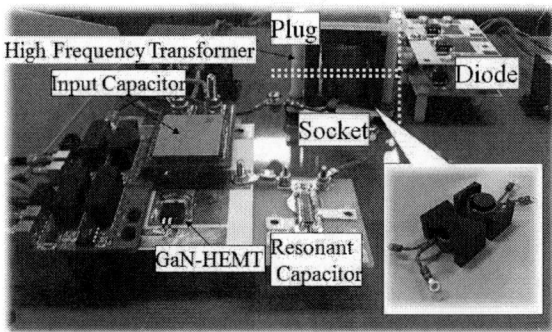

Fig. 7. Experimental apparatus of inductively–coupled contactless DC connector.

TABLE II
HARDWARE FOR EXPERIMENT.

Symbol	Meaning	Specification
Q_1, Q_2	GaN–HEMT	600 V, 16 A (Transphorm)
D_1–D_4	SiC–SBD	600 V, 12A (SiCED)
X	Transformer	MC2 (JFE Ferrite) PQ50 with gap length 200 μm
N_P, N_S	Turn No.	$N_P = 6$ (Primary side) $N_S = 6$ (Secondary side)
C_{in}, C_{out}	MLCC	630 V, 22 μF (Murata)
C	MLCC	1000V, 3.3μF * 7(Murata)

utilized in this experiment. The switching loss energy is smaller than conventional Si power devices and the turn–off energy is independent of the drain current [14] [15]. In LLC converter topology, the turn–on energy can be eliminated by ZVS and the turn–off energy has to be minimized taking the turn–off time into account. The constant turn–off energy for the drain current makes the LLC converter design simple.

Figure 8 shows the measurement result of the switching loss energy for GaN–HEMT, SiC–MOSFET, SiC–JFET and Si–SJ (Super Junction) MOSFET. The double–pulse test was applied to measure switching loss energies under the inductive load condition. The input drain to source voltage is DC 384 V. Here, SiC–SBD is connected for each transistor as a free–wheeling diode. The low–voltage Si–SBD is connected in series for Si–SJ MOSFET taking the behavior of its body–diode into account.

The turn–off switching loss energies for SiC–MOSFET, SiC–JFET and Si–SJ MOSFET increase as the drain current becomes larger. Characteristics which the turn–off energy is independent of the drain current can be seen in the turn–off switching loss energy of GaN–HEMT. The turn–off switching loss P_{OFF} generated from GaN–HEMTs in the connector is simply calculated as $P_{OFF} = f_{SW} \times E_{OFF}$ by using constant value E_{OFF} which means the turn–off switching loss energy shown in Fig. 8.

Moreover, the switching loss energy of the GaN–HEMT is smallest in the aforementioned transistors. Highly efficient connector can be developed by using GaN–HEMT.

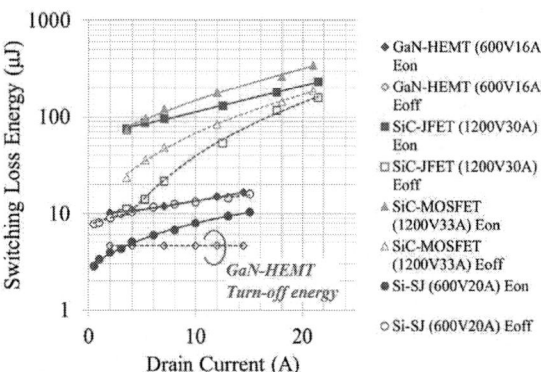

Fig. 8. Measurement results of switching energy losses for GaN–HEMT, SiC–MOSFET, SiC–JFET and Si–SJ MOSFET.

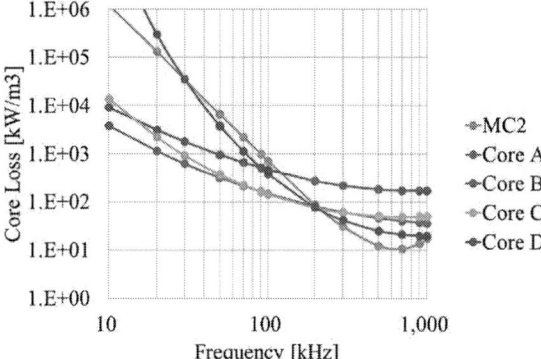

Fig. 9. Core losses generated from magnetic core materials.
($\Delta B \times f_{SW} = 15,000$)

B. Magnetic Core and Winding Characteristics of High Frequency Transformer

Remarkable loss reduction of semiconductor power devices has been achieved by novel power devices such as SiC and GaN. The ratio of power loss generated from magnetic components such as transformers and inductors becomes increasing relatively. To realize highly efficient DC connector, the low loss magnetic core material is indispensable. The core material MC2 from JFE Ferrite was employed in this experiment.

Figure 9 shows the core losses calculated for five types of magnetic core materials. These core materials are commercially available and the core losses were calculated by using published datasheets under the condition that the product of the magnetic flux density ΔB and switching frequency f_{SW} is constant (e.g. 15,000) as the following equation.

$$\Delta B \times f_{SW} = 15,000 \cdots\cdots\cdots (6)$$

As shown in Fig. 9, the core loss generated from the core material of MC2 has the minimum value in the case that the switching frequency is from 200 kHz to 1,000 kHz. High efficiency will be achieved by using MC2 material in the case of the switching frequency of 500 kHz.

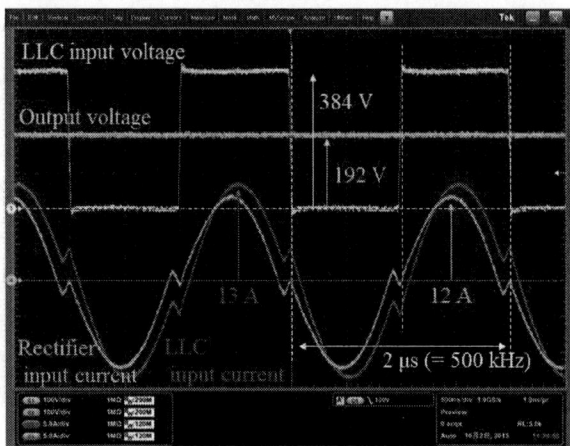

Fig. 10. Steady state waveforms of contactless DC connector under 1,200 W and 500 kHz operation.

C. Steady State Operation and Loss Analysis

Figure 10 shows the steady state waveform under the output power of 1,200 W and the switching frequency of 500 kHz. LLC input voltage means the drain to source voltage of the low–side GaN–HEMT. LLC input current means the current through the leakage inductance L_1. Rectifier input current is through the leakage inductance L_2. The output voltage of 192 V can be seen and the LLC input voltage is pure square wave which changes from 0 V to 384 V periodically. In the steady state operation, the LLC input voltage changes from 384 V to 0 V when the LLC input current is positive. This means the ZVS is achieved for the transistor Q_2. ZCS for the diode D can be also confirmed because the rectifier input current is nearly equal to 0 A when the LLC input voltage changes.

Figure 11 shows the measurement result of conversion efficiency and the estimated efficiency for the contactless DC connector. In this experiment, the maximum efficiency of 95.1% was confirmed at 930 W output power. The conversion efficiency of 94.9% was also shown at the rated output power of 1,200 W. These efficiencies were measured by using the power meter WT3000 from YOKOGAWA.

The conversion efficiency was estimated by calculating power losses generated from the transistor Q, the transformer X, the free–wheeling diode D and the

resonant capacitance C as the following equation (7).

$$
\begin{aligned}
P_T &= P_Q + P_X + P_C + P_D \\
&= (P_{COND} + P_{OFF}) + (P_{Cu} + P_{CORE}) + P_C + P_D \\
&= (R_{ON} \cdot I^2 + P_{OFF}) + \{(R_{WP} + R_{WS}) \cdot I^2 + P_{CORE}\} \\
&\quad + R_C \cdot I^2 + V_D \cdot I \cdots (7)
\end{aligned}
$$

The total power loss is P_T. The power loss generated from transistors, the transformer, the capacitor and the diode are P_Q, P_X, P_C and P_D, respectively. The power loss from the transistor P_Q consists of the conduction loss P_{COND} which depends on the on–resistance of the transistor R_{ON} and the turn–off switching loss P_{OFF}. The loss from the transformer P_X consists of the core loss P_{CORE} and the winding copper loss P_{Cu} which depends on primary and secondary winding resistances of the transformer R_{WP}, R_{WS}. The capacitance loss P_C depends on the equivalent series resistance (ESR) R_C. The diode loss P_D depends on the on–drop voltage V_D. The current I in Eq. (7) means the input current of the LLC resonant tank. The winding resistances R_{WP}, R_{WS} and the ESR R_C were measured by using impedance analyzer PSM1735 from N4L to estimate the power losses exactly.

Figures 12 shows the measurement result of the winding resistance $(R_{WP} + R_{WS})$ and the magnetizing inductance L_m of the transformer. The switching frequency was varied from 1 kHz to 2 MHz and the

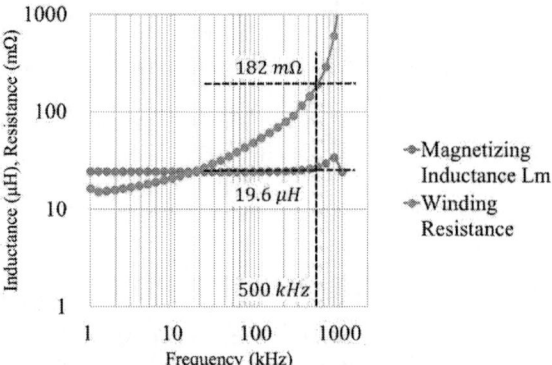

Fig. 12. Measurement result of magnetizing inductance and winding resistance of transformer.

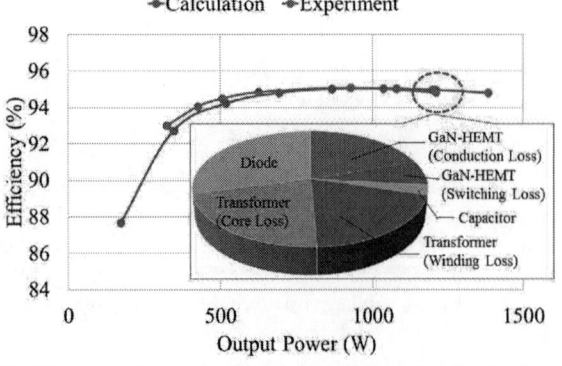

Fig. 11. Measured and estimated conversion efficiencies of contactless DC connector.

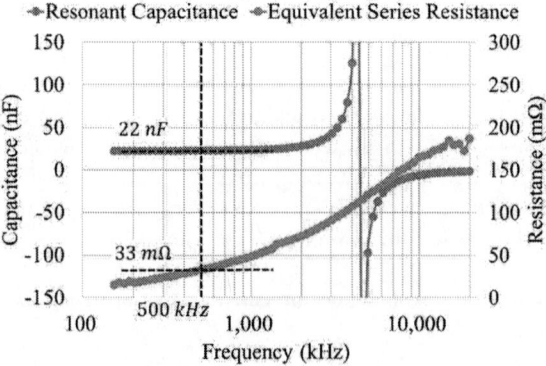

Fig. 13. Measurement result of capacitance and equivalent series resistance of resonant capacitor.

winding resistance increased because of the skin effect and the proximity effect. The magnetizing inductance of 19.6 μH was confirmed at the switching frequency of 500 kHz. Measured resistance 182 mΩ at 500 kHz was utilized to calculate winding loss of the transformer.

Figure 13 shows the measurement result of the equivalent series resistance (ESR) and the capacitance of the resonant capacitor. The frequency was varied from 100 kHz to 20 MHz. MLCC (Multi–Layer Ceramic Capacitor) from Murata Manufacturing was employed to reduce the ESR. The capacitance of 22 nF was confirmed at 500 kHz. Measured resistance 33 mΩ at 500 kHz was utilized to calculate the power loss from the resonant capacitor.

The core loss of the transformer P_{CORE} was calculated using the published datasheet in the case of the operating magnetic flux density of 97.6 mT with the core shape of PQ50. The magnetic flux density of 97.6 mT is calculated by the following equation (8) taking the real circuit operation condition into account. Here, the magnetic flux density is ΔB. The turn number at the primary side of the transformer is N_P and the cross section area of the magnetic core is S_e. The transformer voltage and the voltage induced time are V_X and T_{ON}, respectively.

$$\Delta B = \frac{1}{N_P \cdot S_e} \int_0^{T_{ON}} V_X \, dt$$

$$= \frac{384V/2}{6 \cdot 328mm^2} \cdot \frac{1}{500kHz \cdot 2} = 97.6\text{mT} \quad \cdots (8)$$

In Fig. 11, the experimental results from 300 W to 1,200 W have good agreement with the estimation result. In this estimation, the power loss generated from the transformer was dominant. The transformer design is critical to improve the performance of the proposed contactless connector.

V. IMPROVEMENT FOR HIGHER POWER DENSITY CONTACTLESS DC CONNECTOR

The 1.2 kW prototype of the connector achieved the efficiency of 95.1% under 500 kHz operation. Higher efficiency is recommended because the connectors will be installed in series into highly efficient future 380 V DC distribution system. The power density of 10 W/cm³ has to be achieved because the power density of the conventional connector is now at a level of 10 W/cm³. Here, the design consideration for the connector is conducted to improve the power density.

The design methodology for high power density converters has been proposed [16]–[18]. The trade–off between the power density and the efficiency is essential to evaluate the barrier of power converter performance. The tradeoff curve is drawn simply here [18]. The conversion efficiency η is calculated using the output power P_{OUT} and Eq. (7). The power density D_P is estimated as shown in Eq. (10).

$$\eta = \frac{P_{OUT}}{P_{OUT} + P_T} \times 100 \quad \cdots (9)$$

$$D_P = \frac{P_{OUT}}{Vo_T} = \frac{P_{OUT}}{Vo_X + Vo_C + Vo_{HS(Q,X,C,D)}} \cdots (10)$$

$$Vo_{HS(Q,X,C,D)} = \frac{P_{(Q,X,C,D)}}{k_{HS}} \cdots (11)$$

The total volume of the connector is V_{OT}. The total volume V_{OT} consists of volumes of the transformer V_{OX}, the resonant capacitor V_{OC} and the heat sink V_{OHS}. In this paper, the heat sink volume V_{OHS} is simply estimated by using heat dissipation efficiency k_{HS} [19] and power losses generated from the transistor Q, the transformer X, the capacitor C and the diode D.

Table III shows variable parameters for this design. The magnetic flux density and the core shape of the transformer were varied. Parameters related to the transformer were mainly varied because the power loss generated from the transformer was dominant in the experiment. To change the core shape, only the core width is controlled here. The core depth and the core height are proportion to the core width based on the PQ core shape. Heat dissipation efficiency k_{HS} was 0.45 W/cm³ taking the cooling condition of forced air cooling in data center into account.

Figure 14 shows the calculation result of the power density and the conversion efficiency for the 1.2 kW, 384 V–192 V contactless DC connector under the condition that the operating magnetic flux density varied from 50 mT to 150 mT, the core width varied from 20.0 mm to 50.0 mm and the switching frequency was fixed at 500 kHz. In this design, smaller core volume and higher magnetic flux density contribute to realizing higher power density contact DC connector. The power density of 11.5 W/cm³ and the conversion efficiency of 96.6%

TABLE III
PARAMETERS FOR DESIGN CONSIDERATION.

Parameter	Min.	Max.	Step
Magnetic Flux Density	50 mT	150 mT	10 mT
Transformer Core Shape	PQ20	PQ50	–
Core Width	20.0 mm	50.0 mm	1.0 mm
Core Depth	14.0 mm	32.0 mm	–
Core Height	20.0 mm	50.0 mm	–
Heat Dissipation Efficiency		0.45 W/cm³	

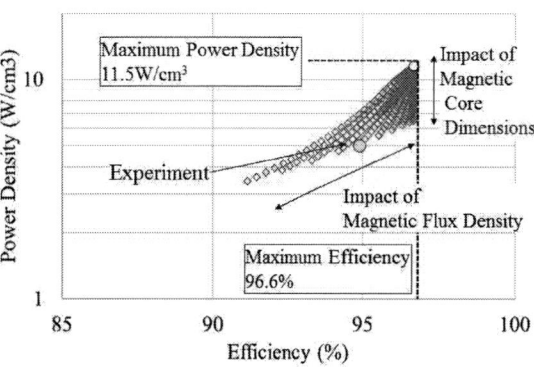

Fig. 14. Power density and efficiency tradeoff for 1.2 kW 384V–192V connector under 500kHz operation.

are maximum values under 500 kHz operation.

The experimental result shown in Fig. 11 was also plotted at the coordinate of the power density of 5.0 W/cm³ and the efficiency of 94.9% at 1,200 W output power. The power density of the experiment was calculated as the following equation by using Eq. (10), taking the heat sink volume based on the heat dissipation efficiency into account. Here, the volume of the resonant capacitor V_{OC} is not considered because its volume is negligible in total converter volume.

$$D_P = \frac{P_{OUT}}{Vo_X + Vo_{HS(Q,X,C,D)}}$$

$$= \frac{1,200 \text{ W}}{100 \text{ cm}^3 + \frac{63.2 \text{ W}}{0.45 \text{ W/cm}^3}}$$

$$= 5.0 \text{ W/cm}^3 \cdots\cdots\cdots (12)$$

In Fig. 14, the power density of over 10 W/cm³ can be achieved by improving the transformer design. The power density will be improved from 5.0 W/cm³ to 11.5 W/cm³ at 500 kHz operation by using smaller magnetic core and operating in higher magnetic flux density. The estimated result means that the contactless DC connector has the potential to be developed in the same volume as the conventional DC connector.

VI. CONCLUSIONS

The contactless DC connector has been proposed to realize the future highly efficient, space–saving and reliable DC distribution system in data centers. The LLC resonant circuit topology was applied for realizing the short–distance and the inductively–coupled contactless DC connector. The prototype with 1,200 W 384 V–192 V was fabricated using GaN–HEMT and the conversion efficiency of 95% under 500 kHz operation was confirmed in the experiment. The design consideration for realizing higher power density connector was conducted and the potential to achieve 10 W/cm³ was shown by improving the transformer design.

REFERENCES

[1] R.R. Schmidt, C. Belady, A. Classen, T. Davidson, M.K. Herrlin, S. Novotny and R. Perry, "Evolution of Data Center Environmental Guidelines", ASHRAE Transactions, Vol. 110, Part1, pp. 559–566, 2004.

[2] T. Babasaki, T. Tanaka, Y. Nozaki, T. Tanaka, T. Aoki and F. Kurokawa, "Developing of Higher Voltage Direct–Current Power–feeding Prototype System" , Proceedings of the International Telecommunications Energy Conference (INTELEC), Incheon, Korea, 2009.

[3] A. Matsumoto, A. Fukui, T. Takeda and M. Yamasaki, "Development of 400–Vdc Output Rectifier for 400-Vdc Power Distribution System in Telecom Sites and Data Centers", Proceedings of the International Telecommunications Energy Conference (INTELEC), Incheon, Korea, 2010.

[4] Y. Sugiyama, "Green ICT toward Low Carbon Society-Green R&D Activities in NTT", Proceedings of 4th International Workshop on Green Communications, Kyoto, Japan, 2011.

[5] T. Ninomiya, Y. Ishizuka, R. Shibahara and S. Abe. "Energy–Saving Technology Using Next–Generation Power Electronics", Proceedings of the IEE–Japan Industry Applications Society Conference, Chiba, Japan, 2012, pp. I15–I20 (in Japanese).

[6] B. Eckardt, A. Hofmann, S. Zeltner, M. Maerz, "Automotive powertrain DC/DC converter with 25 kW/dm³ by using SiC diodes", in Proceedings of 4th International Conference on Integrated Power Systems (CIPS), Naples, Italy, 2006.

[7] J. Sun, M. Xu, Y. Ying, F.C. Lee, "High power density, high efficiency system two-stage power architecture for laptop computers", in Proceedings of 37th IEEE Power Electronics Specialists Conference (PESC), Jeju, Korea, 2006, pp. 231-237.

[8] U. Badstuebner, J. Biela, J.W. Kolar, An optimized, "99% efficient, 5 kW, phase-shift PWM DC-DC converter for data centers and telecom applications", in Proceedings of the International Power Electronics Conference (IPEC), Sapporo, Japan, 2010, pp. 626-634.

[9] Yusuke Hayashi, "High Power Density Rectifier for Highly Efficient Future DC Distribution System", Journal of Electrical Engineering Research (EER), Electrical Engineering Publishing Company Vol. 1, Iss. 3, July 2013.

[10] http://www.vicorpower.com/

[11] Y. Hayashi, and M. Mino, "An Approach to a Higher-Power-Density Power Supply for a 380-V DC Distribution System", Electrical Engineering in Japan, Vol. 186, No. 3, 2014.

[12] T. S. Chan C. Chern–Lin, "LLC resonant converter for wireless energy transmission system with PLL control", Proceedings of IEEE International Conference on Sustainable Energy Technologies (ICSET), Singapore, 2008.

[13] Y. Yang, "A Contactless Charger with LLC Tank and Microcontroller", Proceedings of s the 35th Annual Conference of the IEEE Industrial Electronics Society (IECON), Porto, Portugal, 2009.

[14] Z. Liu, X. Huang, F. C. Lee and Q. Li, "Simulation Model Development and Verification for High Voltage GaN HEMT in Cascode Structure", Proceedings of the IEEE Energy Conversion Congress and Exposition (ECCE), Denver, U.S., 2013.

[15] W. Zhang, Z. Xu, Z. Zhang, F. Wang, L. M. Tolbert, B. J. Blalock, "Evaluation of 600 V Cascode GaN HEMT in Device Characterization and All-GaN-Based LLC Resonant Converter", Proceedings of the IEEE Energy Conversion Congress and Exposition (ECCE), Denver, U.S., 2013.

[16] J. W. Kolar, J. Biela and J. Minibőck, "Exploring the Pareto Front of Multi–Objective Single-Phase PFC Rectifier Design Optimization -99.2% Efficiency vs. 7kW/din3 Power Density", Proceedings of the IEEE 6th International Power Electronics and Motion Control Conference (IPEMC), Wuhan, China, 2009.

[17] H. Ohashi, "Research activities of the power electronics research center with special focus on wide band gap materials", Proceedings of 4th International Conference on Integrated Power Systems (CIPS2006), Naples, Italy, 2006.

[18] Yusuke Hayashi, "Multi–converter approach to higher power density DC–DC converter for 380 V DC distribution system", Journal of Energy and Power Engineering (JEPE), Volume 7, Number 7, pp. 1371–1343, July 2013.

[19] M. Tsukuda, I. Omura, T. Domon, W. Saito and T. Ogura, "Demonstration of High Output Power Density (30 W/cc) Converter using 600 V SiC-SBD and Low Impedance Gate Driver", Proceedings of the 2005 International Power Electronics Conference (IPEC), Niigata, Japan, 2005.

The 2014 International Power Electronics Conference

Analysis of Mis-interruption of Semiconductor Breaker in DC Power Feeding System

Kensuke Murai

Energy and Environment Systems Laboratories
Nippon Telegraph and Telephone Corporation
3-9-11, Midori-cho, Musashino-shi, Tokyo, Japan
murai.kensuke@lab.ntt.co.jp

Koki Asakimori

Energy and Environment Systems Laboratories
Nippon Telegraph and Telephone Corporation
3-9-11, Midori-cho, Musashino-shi, Tokyo, Japan
asakimori.koki@lab.ntt.co.jp

Yasuyuki Kanai

Energy and Environment Systems Laboratories
Nippon Telegraph and Telephone Corporation
3-9-11, Midori-cho, Musashino-shi, Tokyo, Japan
kanai.yasuyuki@lab.ntt.co.jp

Tadatoshi Babasaki

Energy and Environment Systems Laboratories
Nippon Telegraph and Telephone Corporation
3-9-11, Midori-cho, Musashino-shi, Tokyo, Japan
babasaki.tadatoshi@lab.ntt.co.jp

Abstract— **380 VDC power feeding systems are used in telecommunication buildings and datacenters to feed power to loads such as routers and servers.** *Molded case circuit breakers (MCCBs) and fuses are used to protect loads and cables from overcurrent.* **Semiconductor breakers can also protect loads and cables. Because they are faster than MCCBs and fuses and can be controlled remotely, we can improve 380 VDC power feeding systems by using semiconductor breakers. However, semiconductors are sensitive to current. Therefore, mis-interruption of a semiconductor breaker occurs under specific conditions during short circuit or ground fault. A factor in mis-interruption is the X and Y capacitors, which are installed in the loads. We analyzed current flow in the power feeding system during short circuit or ground fault and argue that installing capacitors in a semiconductor breaker is effective in reducing the risk of mis-interruption when short circuit occurs. We also argue ground fault may occur with a pulse current of more than 30 A under certain conditions with ground fault.**

Keywords— *DC power feeding system, semiconcudtor breaker, mis-interruption, telecommunication building, datacentre*

I. Introduction

To operate loads with high reliability and high efficiency, direct current (DC) power feeding systems, which directly connect backup batteries, are used in telecommunication buildings and datacenters. Such systems are suitable for photovoltaic and fuel cells, which are eco-friendly power sources, because they generate electric power with DC.

To adapt to the increase in load power consumption, 380 VDC power feeding systems that use 380 V of feeding voltage, which is larger than the conventional feeding voltage of -48 VDC, have been discussed, standardized, and installed [1] [2] [3] [4].

Overcurrent due to overload and short circuit in DC power feeding systems does not only damage cables but affects the entire system. Therefore, circuit breakers, which interrupt the current in case of overcurrent, are installed in DC power feeding systems. When short circuit and interruption occur, the feeding voltage is changed due to the change in the current. Therefore, circuit breakers are important to maintain system stability and safety.

Molded case circuit breakers (MCCBs) and fuses are passive circuit breakers. these breakers interrupt current in more than 0.1 ms after short circuit occurs. More than 1-kA current flows when short circuit occurs; therefore, semiconductor breakers interrupt a current faster after overcurrent occurs than passive circuit breakers. Only 100-A current flows when short circuit occurs. Therefore, it is easier to stabilize the system with semiconductor breakers. Such breakers have been discussed regarding short and open circuits [5] [6].

Because semiconductor breakers break due to large current, controlling interruption must be done quickly. There are conditions in which large current flows unless there is a short circuit. If semiconductor breakers do not interrupt only short circuit but also other circuits, loads that should operate stop. To avoid mis-interruption, we clarify the factors of mis-interruption and current flow during short circuit and ground fault.

II. System Configuration

The mis-interruption from semiconductor breakers is caused by large current through the semiconductor breaker. The magnitude of the current depends on the system configuration. The 380 VDC power feeding system is discussed in ITU-T as L.1201 "architecture of power feeding system of up to 400 VDC" [7]. Figure 1 shows a typical 380 VDC system consisting of a DC power feeding system and loads. To analyze current

978-1-4799-2706-7/14 $31.00 © 2014 IEEE   1567

caused by faults, we discuss the power feeding system, loads, and semiconductor breaker according to certain standards.

A. Configuration of DC power feeding systems

DC power feeding systems consist of a rectifier, power distribution unit (PDU), and battery.

rectifier

The rectifier converts AC commercial power to DC power. The ITU-T Recommendation L.1200 specifies that referent test voltages are 380 and 300 V. Higher voltage can feed higher power with the same current. Therefore, we analyzed the factors of mis-interruption in which output voltage of a rectifier is 380 V.

PDU

The PDU distributes DC power to multiple branches, which are connected to loads. It is possible to configure PDUs in a multi-level way. The main PDU provides power to several secondary PDUs, and secondary PDUs provide power to small PDUs in a load rack. The PDUs are equipped with circuit breakers, which protect cables running from them to the loads from heat damage when overcurrent occurs. If a multi-level PDU configuration is used, protection coordination should be taken into consideration for ensure high reliability.

battery

The battery is connected to the output of the rectifier. The battery is charged by the rectifier and backs up the power when AC outage occurs. When the rectifier outputs DC power, the battery does not discharge. Therefore, we do not discuss this since the battery does not affect mis-interruption.

B. Earthing arrangement

The earthing arrangement of 380 VDC power feeding systems in telecommunication buildings and datacenters is standardized in EN 301 605 [8]. The standard specifies two types of earthing arrangements; IT system and TN-S system, as shown in Fig. 2. In an IT system, positive and negative lines are connected to a main earthing terminal through high resistance (typically > 50 kΩ), as shown in Fig. 2 (a). In a TN system, the negative line is connected to the main earthing terminal directly. To protect humans from electrical shock, this standard recommends the IT system. Therefore, we analyzed the mis-interruption

factors in which the earthing system is an IT system.

C. Configuration of loads

Recommendation L.1200 specifies load electrical specifications, which include normal operating range and operation under abnormal conditions such as voltage variations, voltage dips, voltage interruptions, and surges. It does not specify noise emission and noise immunity, but CISPR does [9][10].

Figure 3 shows an example configuration of a power supply unit (PSU) of loads to meet the L.1200 requirement. The diode and bulk capacitor enable a converter to operate when the power feeding system cannot feed power due to faults such as short circuit. The X and Y capacitors are set to reduce noise emission and increase noise immunity. The X capacitor is approximately 1 μF, and the Y capacitor is approximately 1000 pF [11].

Since high resistances are installed between power lines and the earthing terminal, as shown in Fig. 2(a), the positive and negative sides of the Y capacitor voltage are

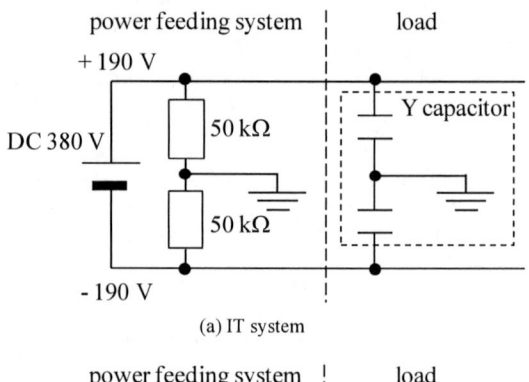

(a) IT system

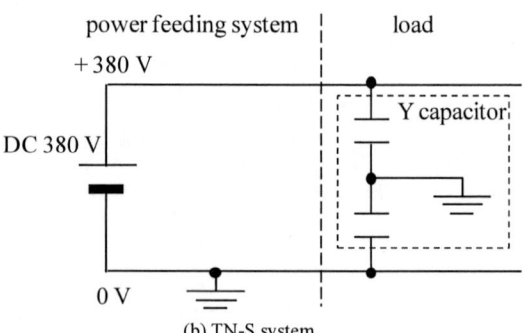

(b) TN-S system

Fig. 2 Earthing arrangement of 380-VDC power feeding system

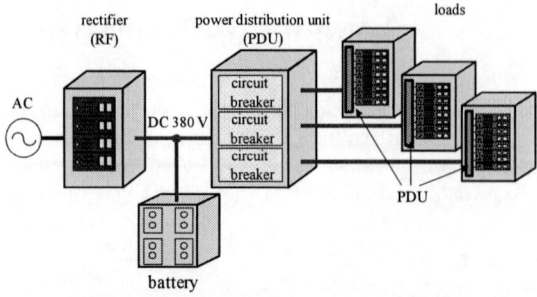

Fig. 1 DC power feeding system and loads

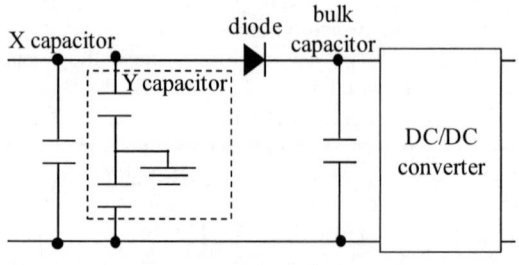

Fig. 3 Example configuration of PSU

equalized during normal operation. Therefore, both sides of the Y capacitor are charged to 190 V under nominal operation conditions.

The rated power specification of a PSU is described in L.1200. However, the typical power of a PSU is between 500 and 2000 W.

(iv) Configuration of semiconductor breaker

Figure 4 shows example configurations of a semiconductor breaker, which consists of a semiconductor switch, control unit, current sensor, and auxiliary circuit. The semiconductor switch includes a body diode. The current sensor detects the current through the semiconductor switch. If the detected current becomes over the threshold current, the control unit determines that an overcurrent has occurred and turns the semiconductor switch off.

When the current through the system decreases due to the turning off of the semiconductor, overvoltage occurs in the semiconductor switch due to cable inductance. To control overvoltage, slowly reducing the current is required. An auxiliary circuit, which bypasses the current, is effective in controlling overvoltage.

Circuit breakers have to control voltage fluctuations in loads under the operating conditions of loads specified in L.1200. The capacitor between the positive and negative lines is effective in controlling voltage change. The semiconductor breaker shown in Figure 4(b) installs a capacitor between the positive and negative lines. Therefore, this semiconductor breaker is more effective in controlling voltage fluctuations than that shown in Figure 4(a). We analyzed the mis-interruption factors in which the semiconductor breaker is configured as shown

in Figure 4(b). Though the semiconductor breaker installs the semiconductor switch in a positive line, it is possible to install the semiconductor switch in a negative line or in both positive and negative lines.

III. Analysis of Mis-interrupt Factors

We next analyzed mis-interruption due to the X and Y capacitors. Current through the high resistances of an IT system is less than 10 mA. The current is much smaller than in a typical feeding current; therefore, it is not taken into account. Figure 5 shows the system state under normal operating conditions. Many loads are connected to a power feeding system. We analyzed the current under conditions such as short circuit and ground fault. Note that we used a two-load system to simplify the analysis. We call the X and Y capacitors in an accidental branch C_{X0} and C_{Y0}, respectively, and those in a normal operating branch C_{X1} and C_{Y1}, respectively.

A. Short circuit

(i) Current not passing through earthing

Figure 6 (a) shows current not passing through earthing when short circuit occurs. When a short circuit occurs, current flows from the rectifier and capacitors toward the short circuit point. Capacitors discharge, and the voltage of the capacitors drop. The current sensor of the accidental branch senses the current that flows through the feeding circuit, and if the control unit detects an overcurrent, the control unit controls the gate voltage and turns off the semiconductor switch.

After the switch turns off, the current path is separated into the upstream and downstream sides of the breaker. At the downstream side of the semiconductor switch, the current flows in a closed loop that consists of a diode, cable, and short-circuit point. At the upstream

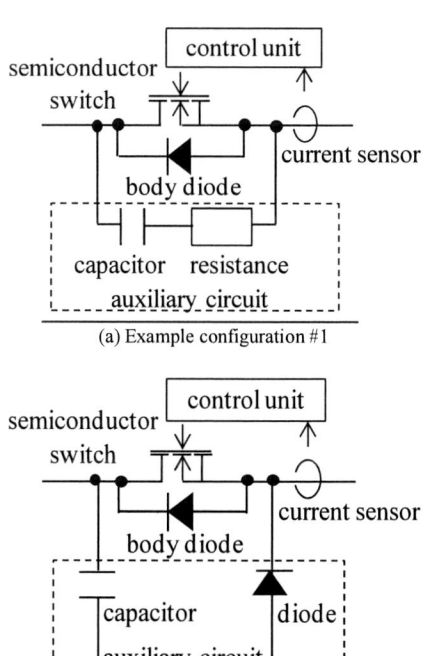

(a) Example configuration #1

(b) Example configuration #2

Fig. 4 Example configuration of semiconductor breaker

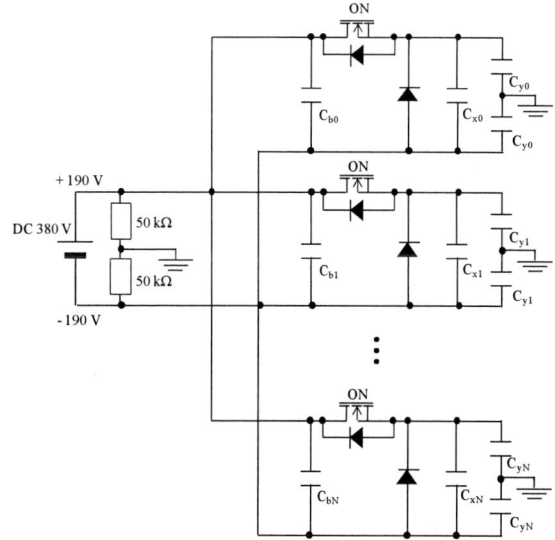

Fig. 5 System circuit figure

978-1-4799-2706-7/14 $31.00 © 2014 IEEE   1569

side, the current from the rectifier mainly flows through C_{b0}, C_{b1}, and C_{X1}. The current through C_{X1} passes through the semiconductor switch in a nominal operating branch. This current may cause mis-interruption from the semiconductor breaker in a nominal operating branch.

(ii) Current passing through earthing

Figure 7(a) shows current passing through earthing

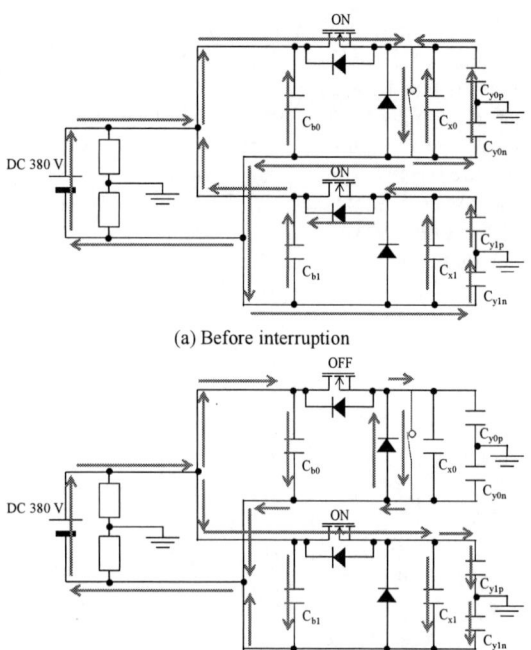

(a) Before interruption

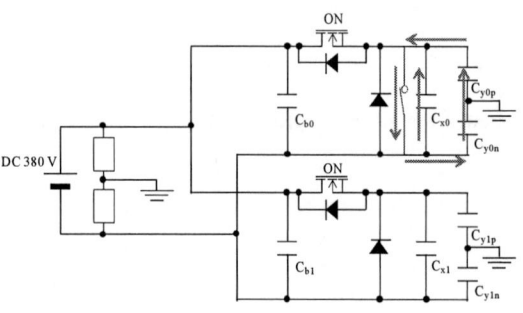

(b) After interruption

Fig. 6 Current not passing through earthing with short circuit

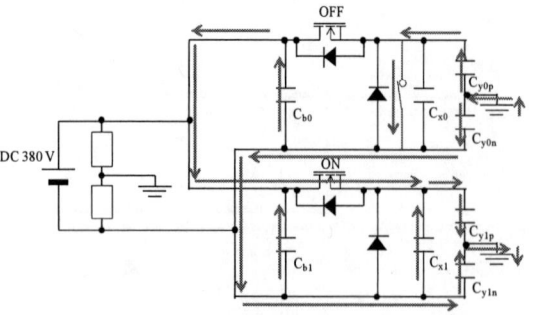

(a) Before interruption

(b) After interruption

Fig. 7 Current passing through earthing with short circuit

when short circuit occurs. Since the inductance between the short-circuit point and load input of an accidental branch is small, C_{X0} and C_{Y0} discharge, and the voltage of the capacitors become approximately 0 V.

Figure 7 (b) shows the current after interruption. The electric potential of the positive and negative lines of C_{Y0} becomes that of the negative line of the power feeding system after interruption. Therefore, C_{Y0} charges until these lines become negative line potential. The current from C_{Y1} flows in the negative line of an accidental branch.

Half of the current from C_{Y0} charges the positive side of C_{Y1}, and the other half discharges the negative side of C_{Y1}. The C_{Y0} charges until the electrical potential of both the positive and negative sides of C_{Y1} become electrical potential of the negative line of the power feeding system. The current through the semiconductor breaker in a normal operating branch does not exceed the current to fully charge C_{Y0} and C_{Y1}. If there are many loads, the total capacitance of the Y capacitors in normal operating branches is large. Since the current through C_{Y0} is separated for Y capacitors in normal operating branches, the current through the semiconductor breakers in a normal operating branch becomes smaller. Therefore, the condition in which the current through a semiconductor breaker becomes large is that there are only two branches and C_{Y0} and C_{Y1} are large. Therefore, large Y capacitance and few branches may result in mis-interruption.

If semiconductor breakers are installed in both positive and negative lines, the current of the negative line cannot flow; therefore, mis-interruption from the semiconductor breaker in a normal operating branch does not occur due to current passing through earthing with short circuit.

B. Positive-line ground fault

When positive line ground fault occurs, the current flows to set the electric potential of the positive line to 0 V and that of the negative line to -380 V. Therefore, the positive side of the Y capacitors is discharged and the negative side is charged. Figures 8 (a) and (b) show the current from C_{Y0} and C_{Y1}, respectively.

The current through the positive side of C_{Y0} directly flows into the ground fault point, and that through the negative side of C_{Y0} flows into the ground fault point through C_{X0}.

The charge and discharge currents of C_{Y1} flow into the ground fault point in an accidental branch. The current from C_{Y1} flows to the positive line through line-to-line capacitors, such as the X capacitors, in both accidental and normal operating branches. If there are other normal operating branches, the current from the Y capacitors in the branches also flows into the ground fault point. Therefore, the current that flows in an accidental branch is the total current to charge and discharge the Y capacitors in normal operating branches.

In the semiconductor switch in an accidental branch, part of the current from the Y capacitors in normal operating branches flows. Therefore, the current through the semiconductor switch becomes large in which many loads are connected to the power feeding system. A

978-1-4799-2706-7/14 $31.00 © 2014 IEEE   1570

smaller C_{X0} also creates a larger current in the semiconductor switch. Therefore, the main mis-interruption factors of the semiconductor breaker in an accidental branch are the Y capacitors in normal operating branches and the capacitance of C_{X0}.

In a normal operating branch, mis-interruption may not occur since current flows through the body diode in the semiconductor switch.

C. Negative-line ground fault

If negative line ground fault occurs, current flows to set the electric potential of the positive line to +380 V and that of the negative line to 0 V. Therefore, the positive side of the Y capacitors is charged and the negative side is discharged. Figures 9 (a) and (b) show the current from C_{Y0} and C_{Y1}, respectively.

The current through the negative side of C_{Y0} directly flows into the ground fault point, and that through the positive side flows into the ground fault point through C_{X0}.

The charge and discharge currents of C_{Y1} flow from the ground fault point in an accidental branch and flows from the negative line through the line-to-line capacitors, such as X capacitors, in both accidental and normal operating branches. If there are other normal operating branches, the current from the Y capacitors in those branches also flow from the ground fault point. Therefore, in an accidental branch, all current to charge and discharge C_{Y1} flows.

In the semiconductor switch in an accidental branch, part of the current to the Y capacitors in normal operating branches flows through the body diode. The magnitude of the current depends on the total Y capacitance of normal operating branches. Large capacitance creates a larger

current through the body diode. Since the current flows through the body diode, mis-interruption does not occur. However, the current, which is larger than the maximum current of the body diode, may flow through the body diode and break the semiconductor switch.

In a normal operating branch, the current flows to charge and discharge C_{Y1}. The magnitude of the current mainly depends on C_{Y1} in each branch. The magnitude of the current in an accidental branch with positive-line fault depends on the total Y capacitance of normal operating branches, but that in a normal operating branch with negative-line fault depends on only the Y capacitance of each normal operating branch. Therefore, the risk of mis-interruption of a semiconductor breaker in a normal operation branch with negative-line fault is less than that in an accidental branch with positive-line fault.

IV. Simulation Results

We now discuss some simulation results. Table I lists the basic simulation conditions.

TABLE I
BASIC SIMULATION CONDITIONS

Parameter	Value
Cable (rectifier to PDU)	length 10 m cross-section area: 200 mm^2
Cable (PDU to ICT equipment)	length 3 m cross-section area: 14 mm^2
Capacitance in circuit breaker: C_b	2.0 μF
Capacitance of X capacitor: C_X	1.0 μF
Capacitance of Y capacitor: C_Y	1000 pF
Interruption Current	100 A
Number of loads (Number of branches)	2 (1 accidental branch and 1 normal operating branch)

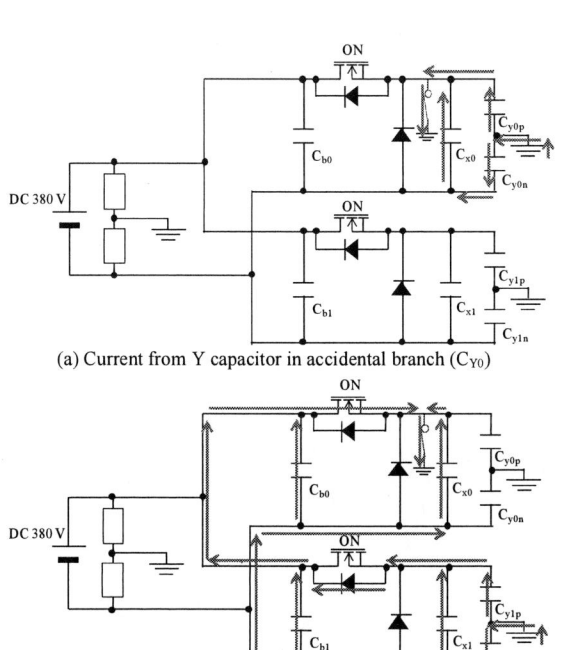

(a) Current from Y capacitor in accidental branch (C_{Y0})

(b) Current from Y capacitor in normal operating branch (C_{Y1})
Fig. 8 Current flow with positive-line ground fault

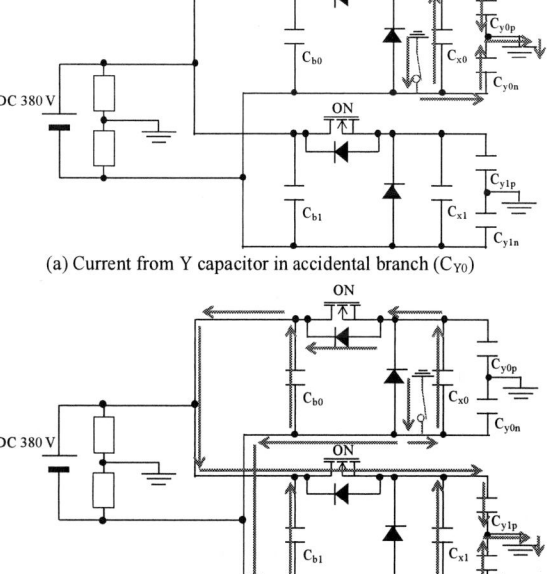

(a) Current from Y capacitor in accidental branch (C_{Y0})

(b) Current from Y capacitor in normal operating branch (C_{Y1})
Fig. 9 Current flow with negative-line ground fault

To clarify the effect of parameters on current through the semiconductor switch in accidental branch and normal operating branches, we conducted a simulation in which certain parameters were changed from the basic simulation conditions.

A. Short circuit

Figure 10(a) shows current through a semiconductor switch in an accidental branch during short circuit in which the breaker capacitance C_b was 2.0 and 0.1 μF. After short circuit occurred, the breaker, X, and Y

capacitors discharged and the current through the semiconductor switch in the accidental branch increased. When the current reached 100 A, the semiconductor switch turned off. Then, the current was interrupted. Figure 10(b) shows current through a semiconductor switch in a normal operating branch. The current flowed from C_{X1} toward the short circuit point while the semiconductor breaker was on state. Therefore, the current was negative. After the semiconductor switch turned off, the current flowed toward C_{X1} to charge the capacitor; therefore, the current became positive. If the

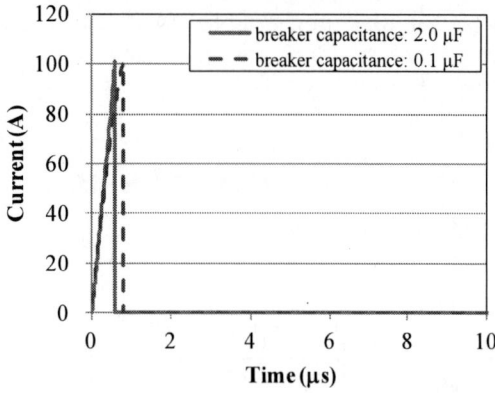

(a) Current through semiconductor switch in accidental branch

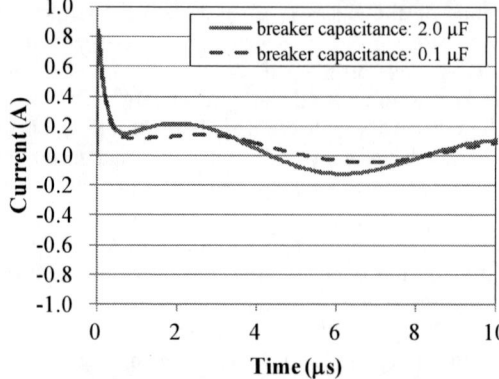

(a) Current through semiconductor switch in accidental branch

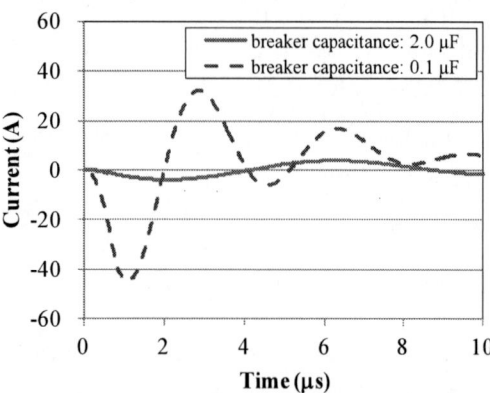

(b) Current through semiconductor switch in normal operating branch

Fig. 10 Current with short circuit

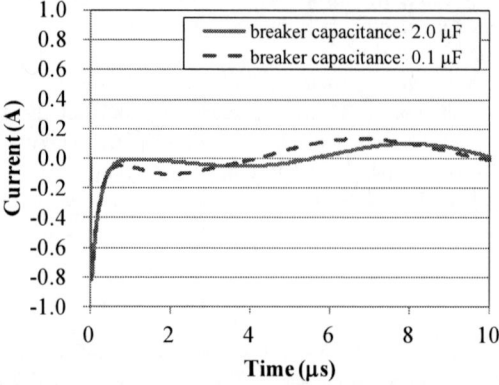

(b) Current through semiconductor switch in normal operating branch

Fig. 12 Current with positive-line ground fault

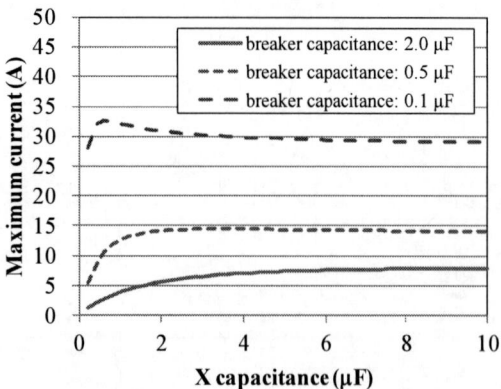

Fig. 11 Relation between C_X and maximum current through semiconductor switch in normal operating branch

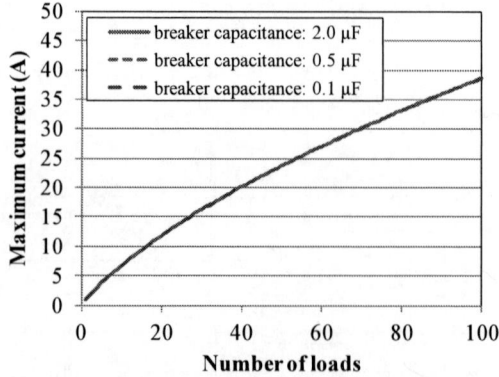

Fig. 13 Relation between number of loads and maximum current through semiconductor switch in accidental branch with positive-line ground fault

threshold current of the semiconductor breaker was less than the current to charge C_{X1}, the semiconductor breaker interrupted the normal operating branch. The maximum current reached more than 5 and 30 A with 2.0 and 0.1 µF of C_b, respectively. This indicates that large capacitance is effective in reducing current in a normal operating branch and prevents the semiconductor breaker from mis-interruption in which short circuit occurs.

Figure 11 shows the relation between X capacitance C_X and maximum current through the semiconductor switch in a normal operating branch. If the X capacitors

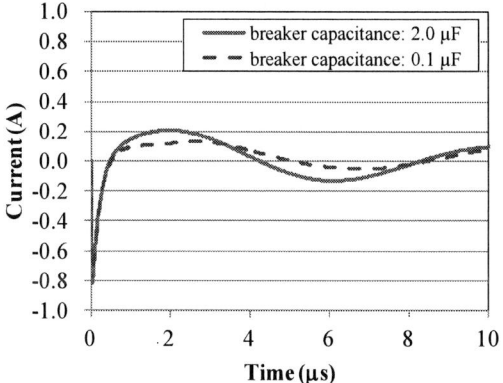

(a) Current through semiconductor switch in accidental branch

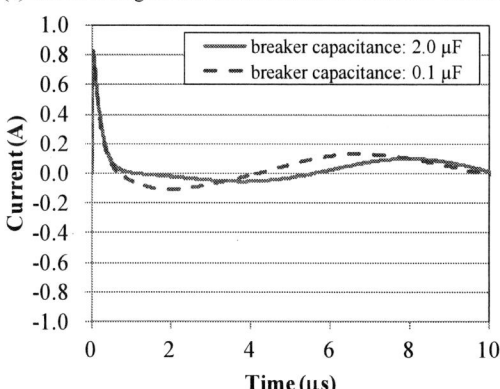

(b) Current through semiconductor switch in normal operating branch

Fig. 14 Current with negative-line ground fault

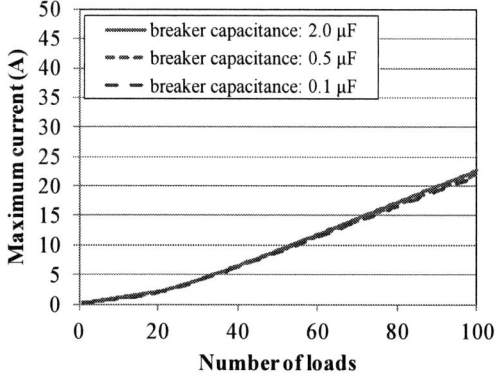

Fig. 15 Relation between number of loads and maximum current through semiconductor switch in accidental branch with negative-line ground fault

are more than 3 µF, the capacitance has little effect on the maximum current. This indicates that the X capacitor is not an important factor in mis-interruption.

B. Positive-line ground fault

Figures 12(a) and (b) shows current through a semiconductor switch in an accidental branch and normal operating branch during positive-line ground fault. After positive-line ground fault occurred, the current increased in both accidental and normal operating branches. However, the direction of current flow was different between accidental and normal operating branches. This is because the current through the semiconductor beaker in the accidental and normal operating branches was from C_{Y1}. By comparing the 2.0 and 0.1 µF of C_b, the maximum current in the accidental branch was the same. This indicates that C_b is not effective in reducing current in an accidental branch.

The positive side of C_{Y1} discharges when positive-line ground fault occurs. After discharge, the current oscillates due to cable inductance, and positive current flows. However, the current is less than the maximum pulse current just before fault occurs. The maximum current is less than 0.2 A. This current does not cause mis-interruption from the semiconductor breaker in a normal operating branch

Figure 13 shows the relation between the number of loads and maximum current through a semiconductor switch in an accidental branch with positive-line ground fault. A larger number of loads increases the maximum current. This is caused by the Y capacitance connected to the power line. If more loads are connected to the power feeding system, the Y capacitance connected to the power line is larger. Therefore, many loads cause large current in an accidental operating branch. If there are one hundred loads, the current reaches 38 A. This may cause mis-interruption from the semiconductor breaker in an accidental branch unless other methods, such as setting the large threshold pulse current of a semiconductor breaker, are not conducted.

C. Negative-line ground fault

Figures 14(a) and (b) shows current through a semiconductor switch in an accidental branch and a normal operation branch during negative-line ground fault. The waveform of the current was almost the same with that of the current during positive-line ground fault. The maximum current in the accidental branch was caused by oscillation. Therefore, the maximum current with negative-line fault was smaller than that with positive-line ground fault. However, as shown in Fig. 15, many loads also caused large current in the accidental branch. Therefore, the number of loads should be taken into consideration to prevent mis-interruption from the semiconductor breaker in an accidental branch.

V. Conclusions

Since semiconductor breakers used for 380 VDC power feeding systems are very sensitive, mis-interruption may occur due to the current to charge and discharge the X and Y capacitors when short circuit or ground fault occur. To develop highly reliable semiconductor breakers, we analyzed the factors of mis-interruption.

When short circuit occurs, current through the semiconductor switch in a normal operating branch is mainly caused by the X capacitor. We clarified that installing a capacitor in the semiconductor breaker is effective in preventing mis-interruption.

When positive-line and negative-line faults occur, the current through a semiconductor switch in an accidental branch is mainly determined by the total capacitance of the Y capacitors in normal operating branches. The maximum current during positive-line ground fault is larger than that during negative-line ground fault. If more than a hundred loads are connected to a power feeding system, the current can reach approximately 40 A. Capacitors installed between positive and negative lines, such as X and breaker capacitors, are not effective in reducing the maximum current during ground fault. Other methods, such as setting the large threshold pulse current of a semiconductor breaker, are required to prevent mis-interruption.

References

[1] D. Marquet, J. Acheen, J. F. Turc, M. Szpek and J. Brunarie, "Pre roll-out field test of 400 VDC power supply: The new alliance of Edison and Tesla towards energy efficiency," IEEE 33rd INTELEC, 2011.

[2] T. Tanaka, K. Hirose, D. Marquet, B. Sonnenberg and M. Szpek, "Analysis of the Wiring Design for the 380 Vdc Power Distribution System in Telecommunications Site," IEEE 35th INTELEC, 2013.

[3] ITU-T Recommendation L.1200, "Direct current power feeding interface up to 400 V at the input to telecommunication and ICT equipment", ITU-T, 2012

[4] ETSI EN 300 132-3-1, "Power supply interface at the input to telecommunications and datacom (ICT) equipment; Part 3: Operated by rectified current source, alternating current source or direct current source up to 400 V; Sub-part 1: Direct current source up to 400", ETSI, 2012

[5] K. Murai, T. Tanaka. T. Babasaki, and Y. Nozaki, "Improvement of PDCs and PDUs with Semiconductor Breakers," IEEE 34th INTELEC, 2012.

[6] R. Mehl and C. Strobl, "Protection of High Voltage DC Next Generation Smart Datacom Power Distribution with Solid State Hybrid Circuit Breakers and ARC Fault Detection Modules (AFDM)," IEEE 34th INTELEC, 2012.

[7] ITU-T SG5 WP3 Q19 http://www.itu.int/en/ITU-T/studygroups/2013-2016/05/Pages/default.aspx

[8] ETSI EN 301 605, "Earthing and Bonding of 400 VDC data and telecom (ICT) equipment," ETSI, 2013.

[9] CISPR 22, "Information technology equipment - Radio disturbance characteristics - Limits and methods of measurement," IEC, 2008.

[10] CISPR 24, "Information technology equipment - Immunity characteristics - Limits and methods of measurement," IEC, 2010.

[11] H. Hoshi, T. Tanaka, M. Noritake, T. Ushirokawa, K. Hirose and M. Mino, "Consideration of inrush current on DC distribution system," IEEE 34th INTELEC, 2012.

A Reliable Electronic Choke with No Need of Gain Adjustment for Wire Communication System

Akihiko Katsuki
Next Generation Switching Power Circuit Course
Nagasaki University
1-14, Bunkyo-machi, Nagasaki, 852-8521 Japan
e-mail: katsuki@nagasaki-u.ac.jp

Tatsuya Nakamura, Tatsuya Mizuki
Dept. of Computer Science and Electronics
Kyushu Institute of Technology
680-4, Kawazu, Iizuka-shi, Fukuoka, 820-8502 Japan
e-mail: katsuki-lab@cse.kyutech.ac.jp

Kohei Shibahara, Tomohiko Abe, Tomohiko Ikeda
Dept. of Computer Science and Electronics
Kyushu Institute of Technology
680-4, Kawazu, Iizuka-shi, Fukuoka, 820-8502 Japan
e-mail: katsuki-lab@cse.kyutech.ac.jp

Shigetaka Maeyama
Technical Center, TDK Corporation
2-15-7, Higashi-Ohwada, Ichikawa-shi, Chiba, 272-8558
Japan
e-mail: maeyama2@jp.tdk.com

Abstract— **Metallic communication lines in wire communication systems are usually used as power lines. Ac communication signal and dc supply power are separated at input part of a powered terminal. Conventional well-known separators utilize the combination of inductors and capacitors. When the signal frequency is low, volume of these components increases. A conventional electronic choke circuit using transistors has the disadvantage of high power dissipation. Furthermore, the direction of the dc current is limited to unidirectional. Our original communication system consists of powering terminals and powered terminals. The direction of the dc current in the powering terminals is opposite to that in the powered terminals. Therefore, it is preferable to have a bidirectional electronic choke. With respect to our low-loss electronic choke with small inductors and an amplifier, wide band frequency characteristics can be obtained. In this paper, a new electronic choke with no need of gain adjustment in amplifier is proposed.**

Keywords— *Electronic choke, Power supply, Transmitter, Wire communication system*

I. INTRODUCTION

Most wire communication systems utilize communication lines as power lines [1]. In the conventional systems, a main power unit supplies all the power consumed in input/output terminal devices [2]. Usually, a choke coil is placed between the main power supply and the communication lines. Choke coil blocks an inflow of communication signal to the main power supply. However, when the lowest frequency of signal is low or the dc output current of power supply is high, the size and the weight of the choke coil increase.

For the purpose of removing the coil, the electronic choke circuit was proposed. This circuit makes use of the feature that output impedance of a common-base bipolar transistor becomes high at the active state [3]. However, the power loss in this circuit increases with dc current.

Originally, the electronic choke circuit is suitable for constant-current power supply in wire telephone system. In addition, the direction of dc current is limited to be unidirectional.

So, we proposed a new electronic choke with small-sized inductors and an amplifier [4]-[6]. This circuit has good features of high efficiency and bidirectional flow of the dc current. We presented a method of enlarging the frequency range suitable for communications by employing special winding structure inductors in electronic choke [7], [8].

To obtain excellent frequency characteristics, the voltage gain of amplifier was desirable to be slightly adjustable. However, this is not preferable from a viewpoint of cost and long-term stability. In this paper, a new electronic choke with no need of gain adjustment is proposed.

II. CONVENTIONAL WIRE COMMUNICATION SYSTEM

Fig. 1 presents an example of conventional system. The impedance looking into the main power supply from the communication lines is especially important with respect to signal. Usually, high-power power supply has the output capacitor of high capacitance and the bypass capacitor for high-frequency use to suppress the ripple or the noise. Therefore, these capacitors have extremely low impedance against the communication signals.

By the way, all terminals are parallel connected to the communication lines. Therefore, when we want to increase the number of terminals, it is important that the impedance looking into the terminals from the communication lines is high against the communication signals.

The impedance looking into the main power supply from the communication lines, that is, the impedance looking into the choke coil or the electronic choke circuit is included in this condition. Electronic choke circuit is

The 2014 International Power Electronics Conference

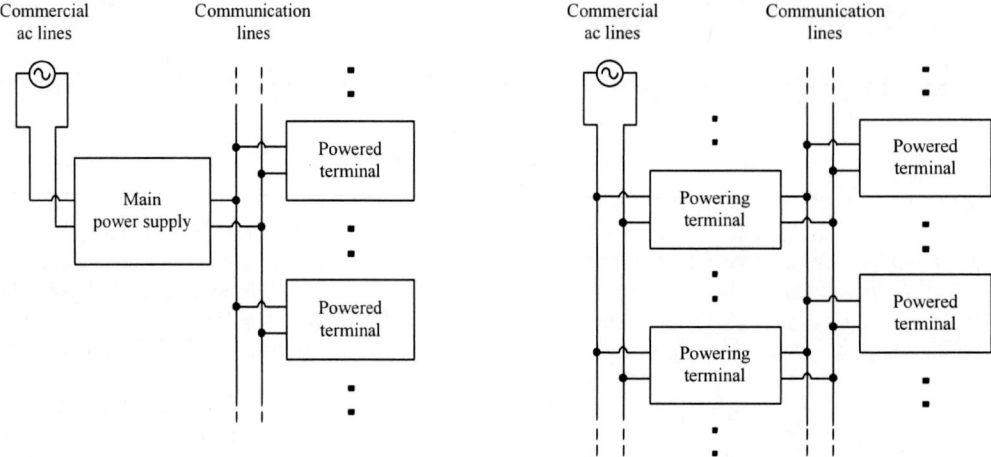

Fig. 1. Conventional wire communication system.

Fig. 2. Proposed wire communication system.

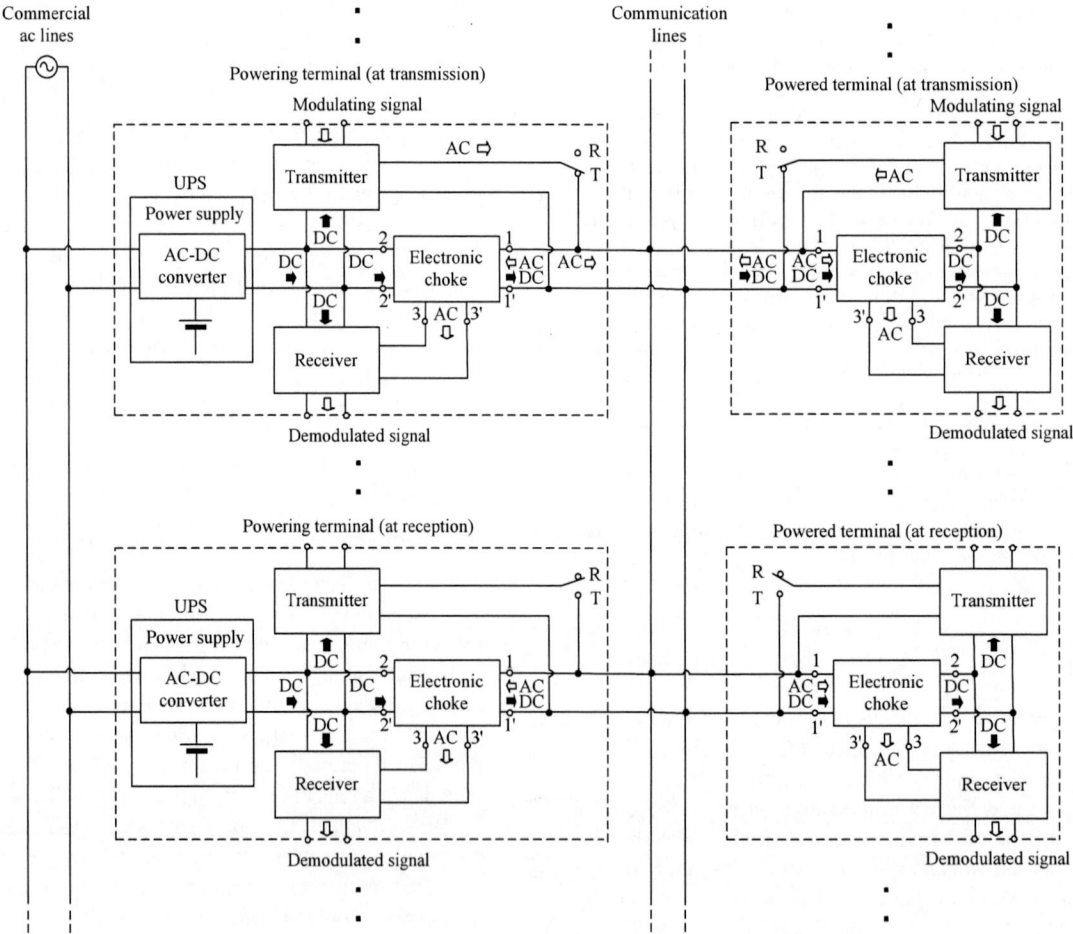

Fig. 3. Operations of the powering terminals and the powered terminals.

978-1-4799-2706-7/14 $31.00 © 2014 IEEE   1576

also useful as the input circuit of the terminals. However, the connection of transistor must be changed, because the direction of dc current is opposite unlike the main power supply.

III. NEW WIRE COMMUNICATION SYSTEM

In the conventional systems, the whole system breaks down, when the main power supply stops operating for trouble or maintenance. Power rating of the main power-supply should be determined after consideration to the number of terminals. Excess of terminals over the estimated number may leads to an exchange of the main power supply to avoid system down. Therefore, reliability and availability of the system are not so high. In addition, these systems have common problems in adaptability to alterations in the system.

To remove these disadvantages, we have proposed a novel wire communication system [9]. Fig. 2 shows the system configuration. In many applications, commercial ac is used as power source of the system. Here, we introduce two types of terminals, that is, the powering terminal and the powered terminal. The former is powered by commercial ac and supplies dc power to the communication lines. The powering terminal has a low-power switched-mode ac-dc converter. These converters construct a parallel-connected power supply system. On the other hand, the latter is powered from the communication lines. The powered terminals are parallel-connected load.

On the power capacity of total system, we assume that the total power rating of power supply system is selected to be more than the total consumed power. In this case, this combination makes a parallel redundant system. The system reliability and availability are much increased in comparison with conventional communication systems. Failure in small number of terminals leads to no system down. Repair or maintenance can be implemented with the system active. When we increase the number of powered terminals, the number of powering terminals should be increased. In this case, there is no need to worry about system down because of shortage in power.

IV. TERMINALS FOR COMMUNICATION

With respect to the new wire communication system, Fig. 3 shows operation of powering terminals and powered terminals. Removing the main power supply makes the dc current through the electronic choke very small.

In powering terminals, the dc power is supplied to the communication lines. Each powering terminal has a power supply of low-power rating. Uninterruptible power supply (UPS) including battery should be placed against ac power failure. About switches, symbols "T" and "R" represent "Transmission" and "Reception," respectively.

Next, in powered terminals, the dc power is received from the communication lines. Arrows with "DC" and "AC" show flow of the dc power-supply current and that of the ac communication signal, respectively.

In the new wire communication system, bi-directional electronic choke is desirable because the same circuit can be applicable for both the powering terminal and the powered terminal. Terminals 1-1' are connected to the communication lines. In the electronic choke, the ac communication signal is input to terminals 1-1' and output from terminals 3-3'. The dc power supply current flows from the terminal 2 to the terminal 1 in the powering terminal. On the other hand, dc current flows from the terminal 1 to the terminal 2 in the powered terminal.

V. NEW ELECTRONIC CHOKE

Fig. 4 shows a basic circuit of new electronic choke. The ac voltage between two input terminals of the operational amplifier with negative feedback is very small. Because the voltage across parallel circuit of the inductor L_1 and the resistor R_2 is almost zero, the input impedance Z_{in} looking into terminals 1-1' becomes very high. Ideally, Z_{in} is determined by the input impedance of the amplifier.

On the dc current, the resistance between the terminal 1 and the terminal 2 is very low, because there exist only winding resistances. Bidirectional flow of the dc current is possible.

VI. EXPERIMENTAL RESULTS

By the use of the experimental circuit shown in Fig. 5, we measured several frequency characteristics, where the dc input voltage V_{in} = 24 V, the dc load current I_o = 10 mA, the ac load resistance R_{AC} = 10 kΩ, the inductance L = 1.3 H and the capacitance C_C = 100 μF. At the electronic choke shown in Fig. 4, the resistance R_1 = 200 kΩ, the inductance L_1 = 10 mH, the capacitances C_1 = 0.1 μF, and C_3 = 1 μF.

In Fig. 6, Fig. 7 and Fig. 8, measured data on the input impedance Z_{in} are depicted. Fig. 6 shows frequency characteristics on influence by the capacitance C_2. Next, frequency characteristics on influence by the resistance R_2 is presented in Fig. 7. Fig. 8 shows frequency characteristics on influence by the inductance L_2.

Finally, Fig. 9 is ac signal reduction characteristics from the terminals 1-1' to the terminals 2-2'. Voltage gain G_v is decibel expression of V_2/V_1, where V_1 and V_2 represent ac input voltage and ac output voltage, respectively. It is desirable that ac signal component is not contained in the voltage across the terminals 2-2'.

From Fig. 6, Fig. 7 and Fig. 8, the conditions of C_2 = 100 μF, R_2 = 10 kΩ and L_2 = 10 mH are recommendable. In this case, input impedance over about 130 kΩ can be obtained from about 10 kHz to about 100 kHz. In this frequency range, it is seen from Fig. 9 that the ac voltage gain from 1-1' to 2-2' is less than -30dB.

VII. CONCLUSION

For existence of non-zero output impedance in the amplifier, its voltage gain does not coincide with the

The 2014 International Power Electronics Conference

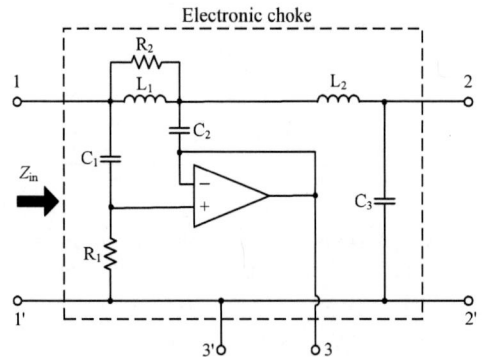

1-1' : AC input and DC input (or output)
2-2' : DC output (or input)
3-3' : AC output

Fig. 4. Basic circuit of the proposed electronic choke.

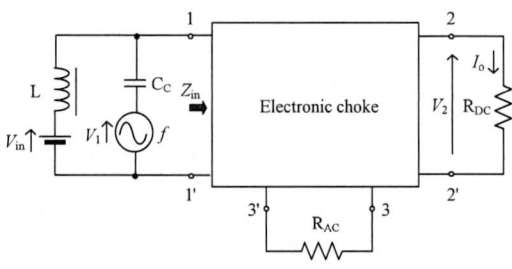

Fig. 5. Experimental circuit.

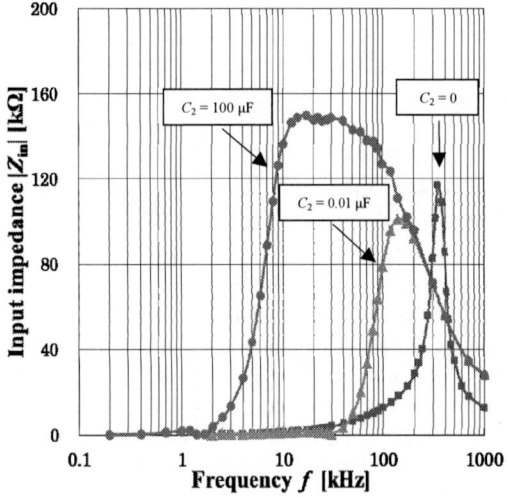

Fig. 6. Frequency characteristics on the input impedance $|Z_{in}|$ with the capacitance C_2 as parameter, where $L_2 = 10$ mH and $R_2 = 10$ kΩ.

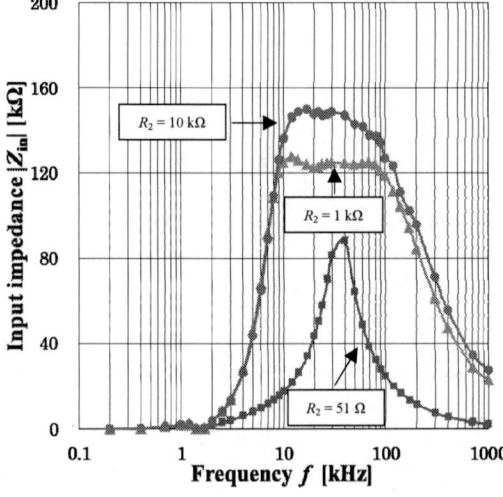

Fig. 7. Frequency characteristics on the input impedance $|Z_{in}|$ with the resistance R_2 as parameter, where $C_2 = 100$ µF and $L_2 = 10$ mH.

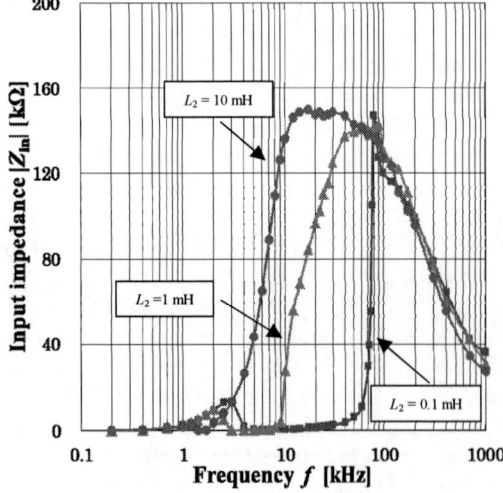

Fig. 8. Frequency characteristics on the input impedance $|Z_{in}|$ with the inductance L_2 as parameter, where $C_2 = 100$ µF and $R_2 = 10$ kΩ.

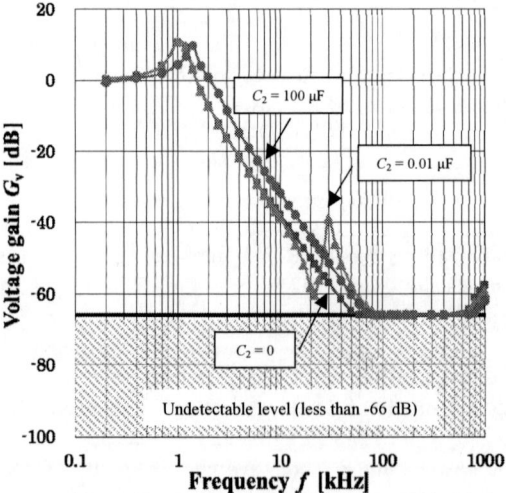

Fig. 9. Frequency characteristics on the ac voltage gain G_v from terminals 1-1' to 2-2' with the capacitance C_2 as parameter, where $L_2 = 10$ mH and $R_2 = 10$ kΩ.

978-1-4799-2706-7/14 $31.00 © 2014 IEEE   1578

unity. This circuit utilizes the feature that the differential-mode input voltage is small in an amplifier with deep negative feedback control.

Newly proposed electronic choke has a good feature of high stability. Because voltage gain adjustment by variable resistor is not required, we can obtain low cost and long-term stability. In comparison with former electronic chokes, low-cost small choke coils for the use of power supplies can be utilized for the inductors L_1 and L_2.

REFERENCES

[1] P. H. Sutterlin, W. R. Bemiss, and G. M. Hey, "Data Communication Network Providing Power and Massage Information," *United States Patent*, 5148144, September 1992.

[2] Y. Nagata, "Simple Type Wire Transmission Device," *Japanese Patent*, 1144477, August 1982. (in Japanese)

[3] S. Nojima, N. Ooba, and K. Yamamoto, "Electronic Choke Circuit," *Japanese Patent*, 2079176, August 1995. (in Japanese)

[4] A. Katsuki, T. Matsumoto, T. Eto, and Y. Hashimoto, "Digital-FM Transmission and Reception in a New Wire-Communication System That Utilizes DC-DC Converter as Transmitter," *Proceedings of INTELEC'03 (The 25th International*

Telecommunications Energy Conference), pp. 615-622, October 2003.

[5] A. Katsuki and T. Matsumoto, "AC Voltage Attenuation Characteristics of a New Electronic Choke Used in the Terminal of a Wire Communication System," *Journal of the Magnetics Society of Japan*, Vol. 28, No. 4, pp. 620-626, April 2004. (in Japanese)

[6] A. Katsuki, T. Matsumoto, S. Watanabe, and M. Fukunaga, "Ripple Reduction Characteristics in the Parallel-Connected DC Power Distribution System Constructed by New Terminals for Wire Communication," *Proceedings of IPEC'05 (The 5th International Power Electronics Conference)*, pp. 2134-2141, April 2005.

[7] A. Katsuki, T. Matsunaga, N. Yamano, and T. Furukawa, "Wide Band Technique in Electronic Choke for Wire Communication System," *Proceedings of INTELEC'09 (The 31st International Telecommunications Energy Conference)*, No. DM1-2, October 2009.

[8] A. Katsuki, N. Yamano, and T. Furukawa, "Frequency Characteristics Analysis of a Wide-Band Electronic Choke for Wire Communication System," *Proceedings of IPEC'10 (The 6th International Power Electronics Conference)*, pp. 1910-1915, June 2010.

[9] A. Katsuki, M. Matsushima, and N. Takimoto, "Wire Communication System Utilizing Ripple-Modulated DC-DC Converter as Signal Transmitter and Power Supply," *Proceedings of INTELEC'99 (The 21st International Telecommunications Energy Conference)*, No. 13-1, June 1999.

Design of New Control Strategies for a Four-Leg Three-Phase Inverter to Eliminate the Neutral Current Under Unbalanced Loads

Zhao-Qin Guo, Sanjib Kumar Panda, I. V. Prasanna
Department of Electrical and Computer Engineering
National University of Singapore
Singapore
guozhaoqin@gmail.com, eleskp@nus.edu.sg, eleivp@nus.edu.sg

Abstract—This paper presents new control strategies for a four-leg three-phase UPS inverter which is used in AC power distribution system in data centers. The IT loads are distributed between the three phase terminals and the neutral terminal of the inverter. Due to the natural unbalance of the three-phase loads in the data center system, there exists a non-zero neutral current if the output voltage of the inverter is controlled to be three-phase balanced. The objective of this work is to eliminate the neutral current such that the thermal losses in the neutral conductor can be reduced and the power usage efficiency (PUE) in date centers can be improved. The elimination of the neutral current is achieved by controlling the inverter output voltages. Three different control methods are designed to dynamically adjust the amplitudes and the mutual phase differences of the three phase voltages, respectively. The effectiveness of the proposed strategies is validated through intensive simulations.

Keywords—Data center system, Neutral current, Unbalanced loads

I. INTRODUCTION

With the increasing reliance on digital data in our society, the number and size of data centers increase rapidly worldwide. The data centers consume large amount of electrical energy and account for a significant fraction of the global carbon footprint [1]-[7]. Therefore, it is in urgent need to put more efforts on enhancing the efficiency of the power supply system in data centers for the purpose of saving energy and promoting a low-carbon society.

Power distribution to IT equipment in a data center or network room can be accomplished using AC or DC power, nevertheless, AC power distribution system is commonly used in data centers [5]. In a typical data center, the energy-consuming equipment include equipment that performs primary IT functions (IT loads) and equipment that ensures continuous operation of IT equipment (non-IT loads). The IT equipment mainly consists of servers, storage devices and network equipment. The supporting equipment mainly consists of cooling systems, power delivery, and other facility infrastructure like lighting.

Two types of AC distribution system are commonly used in data centers [5], as shown in Fig. 1. For type-I

system, which is commonly used in North America, the power goes through a UPS and a transformer-based power distribution unit (PDU) before entering the IT device power supply. There are five principal losses generated in this system: the UPS losses, the primary distribution wiring, the PDU losses, the branch circuit distribution wiring, and the IT power supply. For type II system, which is commonly used outside North America, the UPS is directly connected to the AC powered IT loads, therefore, the PDU transformer and the associated losses can be eliminated.

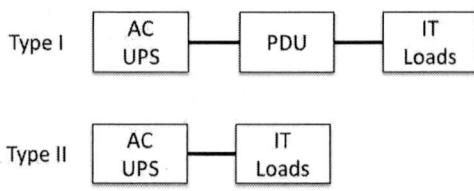

Fig. 1. AC power distribution system in data centers. UPS stands for uninterruptible power supply, PDU stands for power distribution unit.

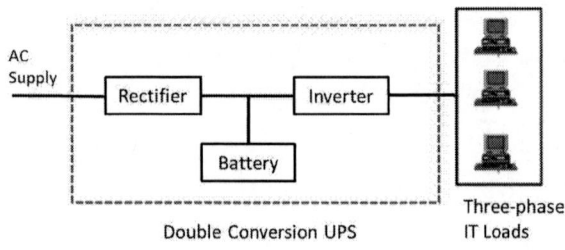

Fig. 2. Configuration of the power supply architecture.

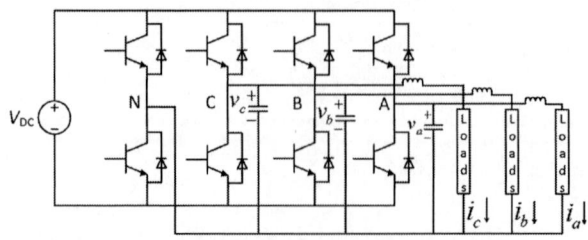

Fig. 3. Four-leg three-phase UPS inverter.

978-1-4799-2706-7/14 $31.00 © 2014 IEEE   1580

In [7], type-II system is adopted to improve the overall efficiency of the power supply system and the configuration of the system architecture is shown in Fig. 2. A double conversion UPS design is adopted. First, a three-phase three-wire AC supply goes through a rectifier, which forms the DC link voltage of the UPS. Next, the DC voltage goes through a four-leg three-phase inverter and generates AC power for loads. The four-leg three-phase inverter is shown in Fig.3. Three of the legs are used as the three phase terminals (A, B and C) and the last leg is used as the neutral terminal (N). The loads are distributed in the three phases, and in each phase, the load is connected between the phase terminal and neutral terminal. The inverter legs are controlled in Sine PWM mode. The output voltages of legs A, B and C can be controlled to follow desired voltages, i.e., reference signals, and the voltage of leg N is controlled to be zero (creating ideal neutral point).

Due to the stochastic variation of the computation loads in different phases, the loads in data centers are asymmetrically distributed in the three phases. A large neutral current would exist if the inverter output voltages are controlled to be three-phase balanced. The existence of the non-zero neutral current would cause undesired thermal loss in the neutral wire and reduce the electrical efficiency of the overall system. To eliminate the neutral current, the inverter output voltages can be controlled to be unbalanced purposely. In [7], a control strategy for the inverter is proposed that the mutual phase angle differences of the inverter output voltages are under control while the amplitudes of the three phase voltages are of the same value and fixed. The mutual phase differences of the three-phase voltage outputs are adjusted dynamically such that zero neutral current can always be achieved in steady-state even if there is a sudden change in a single phase load.

In this work, to eliminate the neutral current, new control strategies are provided for the inverter under different degrees of freedom in control, which can be summarized into three cases: (1) the system only can provide the freedom in changing the mutual phase angle differences among the three phase voltages, i.e., the amplitudes of the inverter output voltages have to be fixed; (2) the system only can provide freedom in changing the amplitudes of the inverter output voltages, i.e., the mutual phase angle differences among the three phase voltages have to be fixed; (3) the system can provide the freedom in changing both the mutual phase differences and amplitudes of the three phase voltages.

The reminder of the paper is organized as the following. In Section II, the problem is formulated. In Section III, control methods are designed for data center system under unbalanced loads. Simulation results are presented in Section IV. Section V summarizes the work.

II. PROBLEM FORMULATION

The three-phase inverter voltages can be expressed as:

$$v_j = V_j \cos(wt + \psi_j), \ (j = a, \ b, \ c). \tag{1}$$

For proper functioning of the data center system, the phase voltages should be maintained in a certain range, denoted as $[V_{\min}, V_{\max}]$, where $V_{\min}$ and $V_{\max}$ are the minimum and maximum peak values of the phase voltages, respectively.

In this work, the loads in the A-B-C phases are considered to be static resistances, i.e., linear loads, which are denoted by R_a, R_b and R_c, respectively. The resulting phase currents are as the following,

$$i_j = \frac{v_j}{R_j} = I_j \cos(wt + \psi_j), \ (j = a, \ b, \ c)$$

The neutral current is as,

$$i_N = i_a + i_b + i_c, \tag{2}$$

that is

$$i_N = I_a \cos(wt + \psi_a) + I_b \cos(wt + \psi_b) + I_c \cos(wt + \psi_c). \tag{3}$$

If the inverter output voltages are three-phase balanced, the neutral current is zero only when the loads in the three phases are identical, i.e., $R_a = R_b = R_c$. However, in real circumstances, the three-phase loads are naturally unbalanced due to the stochastic variation of the computation loads in difference phases. The objective of this work is to maintain zero neutral current in the UPS inverter even when the data center system is under unbalanced three-phase loads.

From the expression of the neutral current as in (3), it can be seen that the value of the neutral current is related with the amplitudes and the phase angles of the phase currents, which are essentially determined by the amplitudes and phase angles of the inventer output voltages, i.e., V_a, V_b, V_c and ψ_a, ψ_b, ψ_c.

III. CONTROL ALGORITHM DESIGN

In this section, we attempt to achieve the zero neutral current by controlling the inverter output voltages in three different ways:

Method I: the amplitudes of the voltages are considered to be fixed, while the mutual phase angle differences of the three-phase voltages are adjusted to eliminate the neutral current;

Method II: the mutual phase angle differences of the three-phase voltages are considered to be fixed, while the amplitudes of the voltages are adjusted to eliminate the neutral current;

Method III: both the amplitudes and the mutual phase angle differences of the three-phase voltages are adjusted to eliminate the neutral current.

For methods I and III, the desired phase angles of the voltages are denoted as ψ_{aN}, ψ_{bN}, ψ_{cN} and the mutual phase angle differences of the three-phase voltages are defined as $\Delta_1 = \psi_{bN} - \psi_{aN}$, $\Delta_2 = \psi_{cN} - \psi_{aN}$. For methods II and III, the desired amplitudes of the voltages are denoted as V_{aN}, V_{bN}, V_{cN}, and the amplitudes of the resulting currents are denoted as I_{aN}, I_{bN} and I_{cN}, respectively.

A. Phase Angle Change Only

By applying "**Method I**", the resulting neutral current is as

$$i_N = I_a \cos(wt + \psi_{aN}) + I_b \cos(wt + \psi_{aN} + \Delta_1)$$
$$+ I_c \cos(wt + \psi_{aN} + \Delta_2). \qquad (4)$$

Through mathematical transformation, the neutral current is represented in the following form

$$i_N = \cos(wt + \psi_{aN})(I_a + I_b \cos\Delta_1 + I_c \cos\Delta_2)$$
$$- \sin(wt + \psi_{aN})(I_b \sin\Delta_1 + I_c \sin\Delta_2). \qquad (5)$$

To make $i_N = 0$, we have

$$I_a + I_b \cos\Delta_1 + I_c \cos\Delta_2 = 0, \qquad (6)$$
$$I_b \sin\Delta_1 + I_c \sin\Delta_2 = 0. \qquad (7)$$

To make the A-B-C phase voltages in positive sequence, we simply let $\sin\Delta_1 < 0$, $\sin\Delta_2 > 0$ and the value of Δ_1 and Δ_2 are specified in the ranges $(-\pi, 0)$ and $(0, \pi)$, respectively. From (6) and (7), the solutions for the desired phase angle differences are obtained as

$$\Delta_1 = \arccos\left(-\frac{I_c^2 - I_a^2 - I_b^2}{2I_aI_b}\right) - \pi, \qquad (8)$$

and

$$\Delta_2 = \arccos\frac{I_b^2 - I_a^2 - I_c^2}{2I_aI_c}. \qquad (9)$$

The solutions for Δ_1 and Δ_2 exist under the following conditions

$$\left|\frac{I_c^2 - I_a^2 - I_b^2}{2I_aI_b}\right| < 1, \quad \left|\frac{I_b^2 - I_a^2 - I_c^2}{2I_aI_c}\right| < 1. \qquad (10)$$

From the above conditions, we have

$$I_a < I_b + I_c, \quad I_b < I_c + I_a, \quad I_c < I_a + I_b. \qquad (11)$$

The constraints as described in (11) can be summarized as

$$I_a + I_b + I_c < 2\max(I_a, I_b, I_c). \qquad (12)$$

Remark 1: Since the values of V_a, V_b, V_c are fixed, the values of I_a, I_b and I_c are inversely proportional to the sizes of R_a, R_b and R_c, respectively. To meet the constraint (12), the sizes of the loads in the three phases should not vary too much. In other words, for circumstances that the sizes of the loads in the three phases varies significantly and the resulting phase currents fail to meet the constraint in (12), the "phase-change-only" method can not perfectly eliminate the neutral current, i.e., the zero neutral current is not achievable.

Without loss of generality, it is assumed that the initial amplitudes of the voltages in the three phases are the same, i.e., $V_a = V_b = V_c$. From (12), a more conservative condition can be obtained as,

$$\max(I_a, I_b, I_c) < 2\min(I_a, I_b, I_c) \qquad (13)$$

that is

$$\frac{V_a}{\min(R_a, R_b, R_c)} < \frac{2V_a}{\max(R_a, R_b, R_c)}. \qquad (14)$$

Subsequently, we have

$$1 \le \frac{\max(R_a, R_b, R_c)}{\min(R_a, R_b, R_c)} < 2. \qquad (15)$$

Perfect elimination of the neutral current is achievable when the loads diversities among the three phases meet the inequality (15).

B. Amplitude Change Only

By applying "**Method II**", the resulting neutral current is as

$$i_N = I_{aN}\cos(wt + \psi_a) + I_{bN}\cos(wt + \psi_b)$$
$$+ I_{cN}\cos(wt + \psi_c). \qquad (16)$$

that is

$$i_N = \cos(wt)(I_{aN}\cos\psi_a + I_{bN}\cos\psi_b + I_{cN}\cos\psi_c)$$
$$- \sin(wt)(I_{aN}\sin\psi_a + I_{bN}\sin\psi_b + I_{cN}\sin\psi_c).$$

To make $i_N = 0$, we have

$$I_{aN}\cos\psi_a + I_{bN}\cos\psi_b + I_{cN}\cos\psi_c = 0, \qquad (17)$$
$$I_{aN}\sin\psi_a + I_{bN}\sin\psi_b + I_{cN}\sin\psi_c = 0. \qquad (18)$$

Subsequently, the desired amplitudes of the three-phase voltages are obtained as

$$V_{aN} = R_aI_{aN} = \frac{V_a}{I_a}I_{aN}, \qquad (19)$$

$$V_{bN} = R_bI_{bN} = \frac{V_b}{I_b}\beta_1 I_{aN} = \beta_1\frac{V_bI_a}{I_bV_a}V_{aN}, \qquad (20)$$

$$V_{cN} = R_cI_{cN} = \frac{V_c}{I_c}\beta_2 I_{aN} = \beta_2\frac{V_cI_a}{I_cV_a}V_{aN}. \qquad (21)$$

with

$$\beta_1 = \frac{\sin(\psi_a - \psi_c)}{\sin(\psi_c - \psi_b)}, \quad \beta_2 = \frac{\sin(\psi_a - \psi_b)}{\sin(\psi_b - \psi_c)}. \qquad (22)$$

and $V_{\min} \le V_{aN}$, V_{bN}, $V_{cN} \le V_{\max}$.

Due to the limitation of the allowable voltage range for the data center system, the following constrains are established for applying the "Method II",

$$\frac{V_{\min}}{V_{\max}} \le \beta_1\frac{V_bI_a}{I_bV_a} = \frac{\sin(\psi_a - \psi_c)}{\sin(\psi_c - \psi_b)} \cdot \frac{R_b}{R_a} \le \frac{V_{\max}}{V_{\min}}. \quad (23)$$

$$\frac{V_{\min}}{V_{\max}} \le \beta_2\frac{V_cI_a}{I_cV_a} = \frac{\sin(\psi_a - \psi_b)}{\sin(\psi_b - \psi_c)} \cdot \frac{R_c}{R_a} \le \frac{V_{\max}}{V_{\min}}. \quad (24)$$

Remark 2: To meet the constraints, the sizes of the loads in the three phases should not vary too much and the mutual phase differences of the voltages should not be too small.

Without loss of generality, it is assumed that the mutual phase differences are the same and the A-B-C phase voltages are in positive sequence, i.e., $\psi_b - \psi_a = -2\pi/3$ and $\psi_c - \psi_a = 2\pi/3$. Accordingly, we have $\beta_1 = 1$ and $\beta_2 = 1$ from (22). It follows that $I_{aN} = I_{bN} = I_{cN}$, and the constraints described in (23)(24) can be summarized into one which is in terms of the allowable load diversities among the three phases, as

$$1 \leq \frac{\max(R_a, \ R_b, \ R_c)}{\min(R_a, \ R_b, \ R_c)} \leq \frac{V_{\max}}{V_{\min}}. \tag{25}$$

C. Both Phase Angle and Amplitude Change

Based on the analysis presented in the previous two subsections, it can be seen that zero neutral current is achievable in spite of the unbalanced loads in the three phases. Nevertheless, the sizes of the loads in difference phases are allowed to vary in limited ranges only. Naturally, we aim to achieve a range which is as large as possible. By applying "**Method III**", both amplitudes and the mutual phase differences of the three phase voltages are considered to be under control. First, the amplitudes of the voltages are regulated to make the amplitudes of the resulting currents meet the constraints as obtained in Section III.A; next, the mutual phase differences are controlled to achieve the zero neutral current.

By applying "**Method III**", the neutral current becomes as

$$i_N = I_{aN}\cos(wt + \psi_{aN}) + I_{bN}\cos(wt + \psi_{bN}) + I_{cN}\cos(wt + \psi_{cN}). \tag{26}$$

The neutral current is represented in the following form

$$i_N = I_{aN}\cos(wt + \psi_{aN}) + I_{bN}\cos(wt + \psi_{aN} + \Delta_1) \\ + I_{cN}\cos(wt + \psi_{aN} + \Delta_2).$$

First, the amplitudes of the voltages can be selected in the range $[V_{\min}, \ V_{\max}]$, denoted as V_{iN} $(i = a, \ b, \ c)$. The amplitudes of the resulting currents are as

$$I_{iN} = \frac{V_{iN}}{V_i}I_i, \quad (i = a, \ b, \ c). \tag{27}$$

As long as the phase currents meet (12), the perfect elimination of neutral current can be achieved by changing the mutual phase differences of the three phase voltages.

Refer to the analysis as in Section III.A, similarly, we can obtain the mutual phase angle differences of the voltages as

$$\Delta_1 = \arccos\left(-\frac{I_{cN}^2 - I_{aN}^2 - I_{bN}^2}{2I_{aN}I_{bN}}\right) - \pi, \tag{28}$$

and

$$\Delta_2 = \arccos\frac{I_{bN}^2 - I_{aN}^2 - I_{cN}^2}{2I_{aN}I_{cN}}. \tag{29}$$

Remark 3: The amplitudes of the voltages are free to assign in the range $[V_{\min}, \ V_{\max}]$, nevertheless, to avoid large phase currents, voltage with a larger amplitude should be applied to the phase which is with a larger size of load.

For the analysis of the allowable load diversity in different phases, extreme cases are considered: (1) even the voltage with the minimum allowable amplitude $V_{\min}$ is applied to the phase which is with the smallest load, the resulting phase current is still with the largest amplitude among the three phases; (2) even the voltage with the maximum allowable amplitude $V_{\max}$ is applied to the phase which is with the smallest load, the resulting phase current is still with the smallest amplitude among the three phases, i.e.,

$$\max(I_a, \ I_b, \ I_c) = \frac{V_{\min}}{\min(R_a, \ R_b, \ R_c)},$$
$$\min(I_a, \ I_b, \ I_c) = \frac{V_{\max}}{\max(R_a, \ R_b, \ R_c)},$$

to meet the conservative condition as in (13), we have

$$\frac{V_{\min}}{\min(R_a, \ R_b, \ R_c)} < \frac{2V_{\max}}{\max(R_a, \ R_b, \ R_c)}, \tag{30}$$

that is

$$1 \leq \frac{\max(R_a, \ R_b, \ R_c)}{\min(R_a, \ R_b, \ R_c)} < \frac{2V_{\max}}{V_{\min}}. \tag{31}$$

Remark 4: From (15)(25)(31), it is evident that by manipulating both the phases and amplitudes of the voltages, a larger load diversity is allowed.

IV. SIMULATION RESULTS

For simulations, the fundamental frequency of the power supply is 50 Hz. For proper functioning of the data center system, the range of the RMS voltage in each phase is 110−240 V, the corresponding range of the peak value of the voltage is 105.27−339.41 V. The simulations are conducted in the following way. Initially, both the output voltages of the inverter and the loads distribution are considered to be three-phase balanced. Thus, the neutral current is zero. Next, the loads are considered to vary in different phases and become unbalanced, which would result in the imbalance of the three-phase current and a non-zero neutral current. The proposed strategies are applied to control the output voltages of the inverter, thus to eliminate the neutral current. The three phase loads are considered to be a three-phase star-connected resistance set.

The initial RMS voltage in each phase is 240 V (peak value: 339.41 V). The phases currents are initially balanced with the peak values as $I_a = 40$ A, $I_b = 40$ A and $I_c = 40$ A. At time $t = 0.02s$, the loads become unbalanced, which results in the unbalanced phase currents with the peak values as $I_a = 40$ A, $I_b = 30$ A and $I_c = 20$ A.

Case 1: At $t = 0.06$ s, "Method I" as introduced in Section III.A is applied to eliminate the neutral current. The new mutual phase differences of the inverter output voltages, i.e., Δ_1 and Δ_2, are computed according to (8) and (9). The simulation results are shown in Fig. 4. It can be seen that, for $t > 0.06$ s, the mutual phase differences of the three phase voltages change and the neutral current becomes zero, which verifies the effectiveness of "Method I".

Case 2: Alternatively, "Method II" as introduced in Section III.B is applied to eliminate the neutral current. The new amplitudes of the voltages are computed according to (19)-(21). The simulation results are shown in

978-1-4799-2706-7/14 $31.00 © 2014 IEEE

The 2014 International Power Electronics Conference

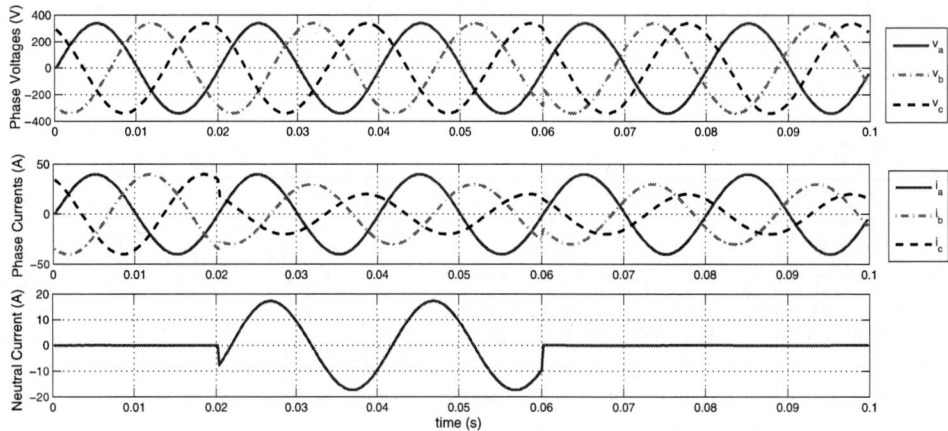

Fig. 4.   Case 1: "Method I" is applied. (a) three-phase inverter line voltages v_a, v_b and v_c; (b) three-phase load currents i_a, i_b and i_c; (c) neutral current i_N.

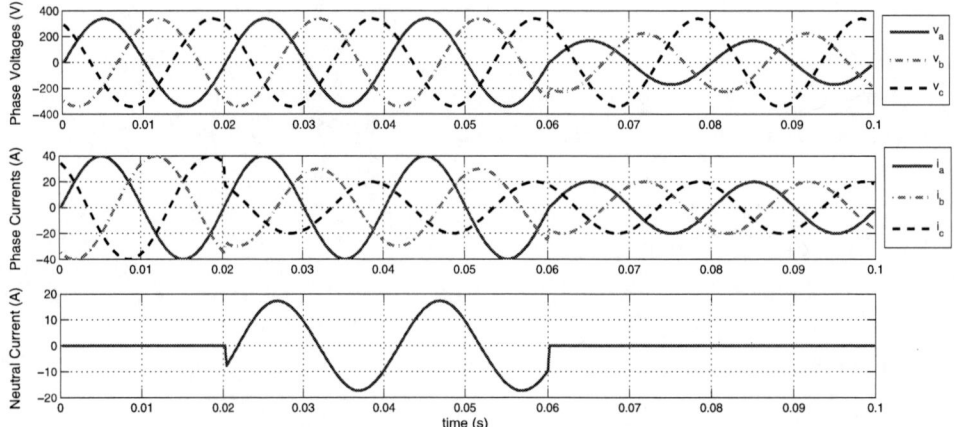

Fig. 5.   Case 2: "Method II" is applied. (a) three-phase inverter line voltages v_a, v_b and v_c; (b) three-phase load currents i_a, i_b and i_c; (c) neutral current i_N.

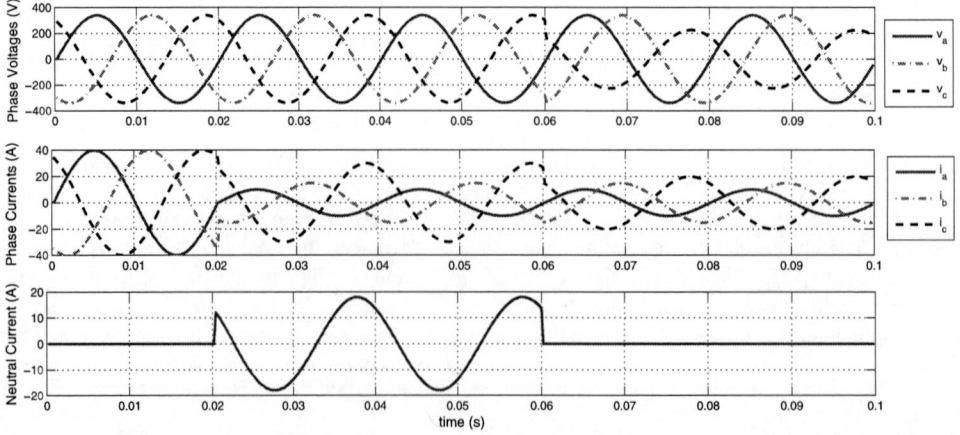

Fig. 6.   Case 3: "Method III" is applied. (a) three-phase inverter line voltages v_a, v_b and v_c; (b) three-phase load currents i_a, i_b and i_c; (c) neutral current i_N.

978-1-4799-2706-7/14 $31.00 © 2014 IEEE   1584

Fig. 5. It can be seen that, for $t > 0.06$ s, the amplitudes of the three phase voltages change and the neutral current becomes zero, which verifies the effectiveness of "Method II".

The system may come to the situation that the phase currents in the three phases vary more significantly than in Case 1, which is essentially due to the large variations of the loads distributed in the three phases. Consider $I_a = 10$ A, $I_b = 15$ A and $I_c = 30$ A, the constraints established for "Method I" and "Method II" as in (12) and (15) are not met while the constraint established for "Method III" as in (25) is still met. Thus, perfect elimination of the neutral current is not achievable by applying "Method I" or "Method II". "Method III" is applied to control both the amplitudes and the mutual phase angle differences of the three phase voltages.

<u>Case 3:</u> We simply keep the same amplitudes of voltages in phases A and B and reduce the amplitude of the voltage in phase C to make $I_{cN} < I_{aN} + I_{bN}$. The desired values for the phase currents are specified as $I_{aN} = I_a = 10$ A, $I_{bN} = I_b = 15$ A and $I_{cN} = 20$ A. Reversely, we can calculate the desired amplitudes of the phase voltages, which are $V_{aN} = 339.41$ V, $V_{bN} = 339.41$ V and $V_{cN} = 226.27$ V. The desired mutual phase differences of the three phase voltages are computed according to (28) and (29). The simulation results are shown in Fig. 6. For $t > 0.06$ s, both the amplitudes and mutual phase differences of the three phase voltages change, and the neutral current becomes zero.

V. CONCLUSION

In this work, strategies are proposed for controlling of a three-phase four-leg inverter to perfectly eliminate the neutral current which exists in a typical data center system with a balanced three-phase power supply but unbalanced loads in the three phases. Based on the information of the initial phase angles and amplitudes of the voltages and currents, the desired phase angles or amplitudes of the voltages are computed reversely. Accordingly, the inverter can be controlled to generate the desired voltages. The conditions for applying the strategies are established in terms of maximum allowable load diversities in different phases. By eliminating the neutral currents, the system power efficiency can be improved. The proposed methods are verified through extensive simulation based studies. Experiment evaluation is under progress at this stage.

REFERENCES

[1] $\cdots$, "Report to congress on server and data center energy efficiency public law 109-431", U.S. Environmental Protection Agency Energy Star Program, 2007.

[2] L. Ganesh, H. Weatherspoon, T. Marian, K. Birman, "Integrated approach to data center power management", IEEE Trans. on Computers, vol. 62, no. 6, pp. 1086-1096, June 2013.

[3] C. Molloy, M. Iqbal, "Improving data-center efficiency for a smarter planet", IBM J. Res. and Dev., vol. 54, no. 4, paper 4, July/Auguest 2010.

[4] M. Ghamkhari, H. Mohsenian-Rad, "Energy and performance management of green data centers: A profit maximization approach", IEEE Trans. on Smart Grid, vol. 4, no. 2, pp. 1017-1025, 2013.

[5] N. Rasmussen, "AC vs. DC power distribution for data centers", Schneider Electric White Paper 63 Revision 6, 2011.

[6] N. Rasmussen, J. Spitaels, "A quantitative comparison of high efficiency AC vs. DC power distribution for data centers," Schneider Electric White Paper 127, 2011.

[7] S. Dasgupta, I. V. Prasanna, S. K. Sahoo, S. K. Panda, "A novel four-leg three-phase inverter control strategy to reduce the Data Center thermal losses: Elimination of neutral current", 38th Annual Conf. on IEEE Industrial Electronics Society (IECON 2012), pp. 3358-3363, 2012.

The 2014 International Power Electronics Conference

Research Trends of Modular Multilevel Cascade Inverter (MMCI-DSCC)-Based Medium-Voltage Motor Drives in a Low-Speed Range

Yuhei Okazaki, *Student Member, IEEE*, Hitoshi Matsui, *Student Member, IEEJ*,
Makoto Hagiwara, *Member, IEEE*, and Hirofumi Akagi, *Fellow, IEEE*
Department of Electrical and Electronic Engineering
Tokyo Institute of Technology, Tokyo, Japan
E-mail: akagi@ee.titech.ac.jp

Abstract—A modular multilevel cascade inverter based on double-star chopper cells (MMCI-DSCC) has been expected as one of the next-generation multilevel PWM inverters for medium-voltage high-power motor drives. This inverter consists of cascaded bidirectional chopper cells with dc capacitors. One of concerns for the motor drive is how to achieve stable operation in a low-speed range. Some kind of solution should be taken to achieve stable operation in a low-speed range, because ac-voltage fluctuations in the dc capacitors are inversely proportional to an inverter frequency. This paper describes state-of-the-art research trends in MMCI-DSCC-based motor drives, especially focusing on the mitigation of the ac-voltage fluctuations. Four kinds of mitigation methods including the proposed ones are summarized, compared, and verified by experiments using a 400-V 15-kW downscaled system. The steady-state waveforms show the validity of the theoretical analysis on the peak current in IGBTs, and start-up waveforms show the stable operation from a standstill to 740 min^{-1} with 60% load torque.

Keywords— AC-voltage fluctuation, medium-voltage motor drives, modular multilevel cascade inverter, mitigation method.

I. INTRODUCTION

Attention has been paid to medium-voltage high-power adjustable-speed motor drives for energy savings without regenerative brakes [1]. While it is possible to reduce an energy consumption of the system by introducing an adjustable-speed motor drive, the implementation rate of the motor drive accounts for only 12% of all the medium-voltage applications, according to [2]. One of the major reasons is that a currently available medium-voltage inverter [1] should be equipped with a multi-windings transformer, which is heavy, bulky, costly, and prone to failure.

A modular multilevel cascade inverter based on double-star chopper cells (MMCI-DSCC) has been expected as one of the next-generation medium-voltage multilevel PWM inverters for medium-voltage high-power motor drives [4]–[15]. This inverter requires no multi-windings transformer. The circuit configuration of the inverter has been proposed in [3], in which the inverter has called as the "Modular Multilevel Converter," or "MMC" for short. Reference [4] has classified the family of modular multilevel cascade converters by its circuit configuration, and named the inverter as the "Modular Multilevel Cascade Inverter based on Double Star Chopper Cells," "MMCI-DSCC," or "DSCC" for short. This paper

follows the naming given in [4], and the inverter is referred to as the "DSCC." The basic control principles of the DSCC have been investigated and verified experimentally in [5]–[8]. The first full scale experimental verification using the 2-MVA, 6.6-kV DSCC and an induction motor has been addressed in [6], where the DSCC is operated at a constant frequency of 50 Hz. Reference [7] has presented a start-up method for an induction motor driven by the DSCC, loaded with a fan- or blower-like loads. It is characterized by starting up a motor with a fixed inverter frequency, e.g., 30 Hz.

Although the control strategy for the DSCC has been widely investigated, the operation of an ac motor driven by the DSCC in a low-speed range remains a challenging issue due to the ac-voltage fluctuation in each dc capacitor. The reason is that the ac-voltage fluctuation, which is not desirable for the inverter, becomes larger as the inverter frequency gets lower because it is inversely proportional to the frequency [7]. Large ac-voltage fluctuation in the dc capacitor should be mitigated to realize stable operation in a low-speed range. Reference [9] has described the operation of a DSCC-based motor drive in a low-speed range, proposing a method to suppress the ac-voltage fluctuation. Then, several papers, following [9] have proposed modified mitigation method of the fluctuation with different applications and control strategies [10]–[15].

This paper presents state-of-the-art research trends of a DSCC-based motor drive for a low-speed operation, especially focusing on a mitigation method of an ac-voltage fluctuation. The investigation developed in this paper points out that the mitigation method has a tradeoff between a reduction in the ac-voltage fluctuation and an increase of a peak current in IGBTs. A reasonable compromise can be achieved by designing the shape of waveforms for a common-mode voltage and ac circulating current used for the mitigation method. This paper theoretically investigates the amount of peak current in the two conventional mitigation methods and two proposed methods. The experimental steady-state waveforms obtained from 400-V, 15-kW system show operating performance of four mitigation methods. The experimental start-up waveforms from a standstill to 740 min^{-1} show that applying a proper mitigation method brings an enhancement of a start-up torque without increasing peak currents.

978-1-4799-2706-7/14 $31.00 © 2014 IEEE

The 2014 International Power Electronics Conference

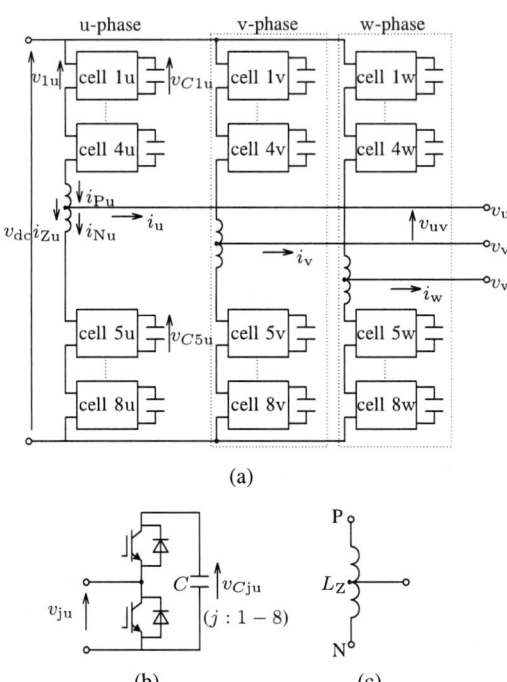

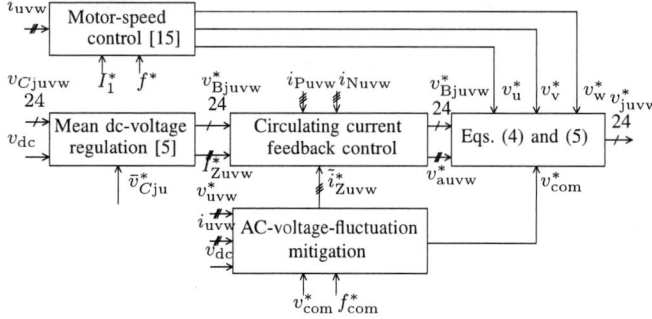

Fig. 2. Overall control block diagram for the DSCC-based motor drive system.

A DSCC has three degrees of freedom: I_Z, $\tilde{i}_Z$, and the common-mode voltage, v_{com}. These freedom are used for the dc-capacitor voltage control.

B. Overall control diagram for the DSCC-based motor drive system

Fig. 2 shows the overall control block diagrams for the DSCC-based motor-drive system used in this paper. The 24 dc-capacitor voltages v_{Cjuvw}, the dc-link voltage v_{dc}, and the six arm currents i_{Puvw} and i_{Nuvw} are detected, and they are input signals for the block diagrams. Note that the three stator currents i_{uvw} are calculated from the detected arm currents. Therefore, no ac current sensors on ac terminals are required.

Several dc-capacitor voltage control has been proposed in [10]–[12]. References [5], [11], and [13] control the circulating currents actively for the dc-capacitor voltage control as follows:

- V_C is regulated by using I_Z.

- $\tilde{v}_{Cju}$ is mitigated by using $\tilde{i}_Z$ and v_{com}.

The regulation of V_C, or the mean dc-voltage regulation can be achieved by using the "arm" balancing control applied to the six arms and the "individual" balancing control applied to the one arm at the same time [5]. The $\tilde{v}_{Cju}$ can be mitigated by the sophisticated control, or the ac-voltage-fluctuation mitigation as discussed in the later section. The mitigation of $\tilde{v}_{Cju}$ is more difficult than regulation of V_C, because the instantaneous power should be controlled using $\tilde{i}_Z$. The circulating-current feedback control yields a command voltage of v_a^*. A motor speed control gives command voltage v_u^*, v_v^*, and v_w^*. Here, any motor speed-control method can be used for DSCC, because both two capacitor voltage control and motor-speed control are independent of each other.

Finally, command u-phase voltages for each chopper cell, v_{ju}^* are given as follows [14]:

$$v_{ju}^* = v_a^* + v_{\text{Bju}}^* - \frac{v_u^* + v_{\text{com}}^*}{4} + \frac{v_{\text{dc}}}{8} \quad (j = 1-4) \quad (4)$$

$$v_{ju}^* = v_a^* + v_{\text{Bju}}^* + \frac{v_u^* + v_{\text{com}}^*}{4} + \frac{v_{\text{dc}}}{8} \quad (j = 5-8). \quad (5)$$

Here, v_{Bju}^* is used to regulate the dc voltage in each arm, v_{com}^* is the command common-mode voltage, and v_{dc} is the dc-link voltage used as feedforward control.

Fig. 1. Circuit configuration of a modular multilevel cascade inverter (MMCI-DSCC). (a) Power circuit. (b) Chopper cell. (c) Center-tapped inductor.

II. CIRCUIT CONFIGURATION AND AC-VOLTAGE FLUCTUATION IN DC CAPACITOR VOLTAGE

A. Circuit configuration

Fig. 1(a) shows the power-circuit configuration for a DSCC. Each leg consists of eight cascaded bidirectional chopper cells shown in Fig. 1(b), and a center-tapped inductor per phase, as shown in Fig. 1(c). The center tap of each inductor is directly connected to each of the stator terminals of a motor, where i_u is the u-phase stator current. The circulating current i_Z is defined as follows [5]:

$$i_{\text{Zu}} \triangleq \frac{1}{2}(i_{\text{Pu}} + i_{\text{Nu}}), \quad (1)$$

where, i_{Pu} and i_{Nu} are the u-phase upper- and lower-arm currents, and are expressed as follows:

$$i_{\text{Pu}} = \frac{i_u}{2} + i_{\text{Zu}} \quad (2)$$

$$i_{\text{Nu}} = -\frac{i_u}{2} + i_{\text{Zu}}. \quad (3)$$

Here, i_{Zu} consists of a dc component I_{Zu} and an ac component $\tilde{i}_Z$. The dc component, I_{Zu} flows from the common dc link to the u-phase leg, while the ac component, $\tilde{i}_{\text{Zu}}$ flows among the three legs. The sum of the three-phase circulating current, $\tilde{i}_{\text{Zu}}$, $\tilde{i}_{\text{Zv}}$, and $\tilde{i}_{\text{Zw}}$ is zero, so that no ac component appears in either the motor current or the dc-link current.

The dc-capacitor voltage v_{Cju} in each chopper cell has a dc component V_C and an ac component, or ac-voltage fluctuation, $\tilde{v}_{Cju}$. These components are determined as a result of a power differences between the power from a dc-link side and the power to a motor side. The dc component, V_C is due to the mean power difference, while the ac-voltage fluctuation, $\tilde{v}_{Cju}$ is due to the instantaneous power difference.

C. AC-Voltage Fluctuation

How to evaluate ac-voltage fluctuation under the control strategy in Fig. 2 has been discussed in [7]. The peak-to-peak ac-voltage fluctuation can be approximated as follows:

$$\tilde{v}_{Cju} \simeq \frac{\sqrt{2}I_1}{4\pi fC}, \qquad (6)$$

where, I_1 is the rms value of a stator current, f is the frequency of a stator current, and C is the capacitance value of each dc capacitor. Note that effects in the ac-voltage fluctuation when i^*_{Zuvw} and v_{com} in Fig. 2 are superimposed, are ignored. According to (6), $\tilde{v}_{Cju}$ is inversely proportional to f, and proportional to I_1. Hence, $\tilde{v}_{Cju}$ increases as f decreases and as I_1 increases. The increase of $\tilde{v}_{Cju}$ introduces the following problems:

- It causes overmodulation to each chopper cell.

- It affects the voltage rating of IGBTs.

Therefore, $\tilde{v}_{Cju}$ should be mitigated as small as possible to achieve stable operation in a low-speed range.

III. LITERATURE REVIEW

How to operate a motor in a low-speed range has been recognized as one of research challenges, because the ac-voltage fluctuation in each dc capacitor voltage is inversely proportional to the motor-current frequency as shown in (6). Therefore, the ac-voltage fluctuation should be mitigated to be an acceptable level. This section summarizes research trends in the mitigation method for an ac-voltage fluctuation. All of the papers referring in this section have confirmed an individual mitigation method by using downscaled experimental system. Here, each chopper cell in a DSCC is considered as an averaging voltage source, therefore no discussions on modulation methods are included.

A. Original Paper

Reference [9] has firstly proposed a control method for a DSCC-driven motor to achieve stable operation in a low-speed range. This method involves the injection of a sinusoidal-wave common-mode voltage and an ac circulating current, i_Z into each leg. This method is called the sinusoidal-wave method in this paper. The ac-voltage fluctuation is attenuated via an instantaneous power resulting from the common-mode voltage and the ac circulating current. The validity of the method has been confirmed by using experimental system with a single-phase $R - L$ load, where the dc-link voltage was 200 V and the rms ac current was 200 mA.

B. Research Trends in DSCC-based motor drives

References [10]–[12] have dealt with a DSCC-based induction motor or synchronous motor drive with different applications.

References [10] focus on an application of a DSCC-driven motor to constant torque applications. The sinusoidal-wave method was employed for mitigating an ac-voltage fluctuation. A downscaled system rated at 11 kW and 380 V has been used to verify steady-state performance, in which the count

of a chopper cell per an arm was set to five. Experimental results have shown that a DSCC-driven motor loaded with a constant rated torque can operate from a standstill to the rated speed. The peak arm current to produce the rated torque at a low speed is about twice that of the rated motor-current amplitude. A reduction for the peak current is also considered by injecting third-order-harmonic voltage into the common-mode voltage to maximize the amplitude of the common-mode voltage superimposed.

Reference [11] and [12] have shown start-up performance of a DSCC-driven induction motor from a standstill with no load torque. The sinusoidal-wave method was employed for mitigating ac-voltage fluctuation. The dc-capacitor voltage controller is basically similar to Fig. 2. Reference [12] consists the dc-capacitor voltage control in the $\alpha - \beta$ coordination for making a controller design easier. A complete decoupling control of the dc-capacitor voltage control from a dc link side and a motor side has been achieved. A downscaled system rated at 17.5 kW and 320 V has been used to verify start-up performance, in which the count of a chopper cell per an arm was set to five. Experimental start-up performance of the induction-motor drive with no load torque, but fast acceleration, e.g., less than 2 sec to reach the rated speed, has shown the stable motor-speed control and dc-capacitor voltage control.

Reference [13] aims to apply the motor drive to the so-called "quadratic-torque" load, the torque of which is proportional to a square of motor mechanical speed. Two kinds of existing mitigation methods for ac-voltage fluctuation were evaluated. One is the square-wave method, which is firstly proposed in [9] and experimentally verified in [14], and the other is the sinusoidal-wave method. A downscaled system rated at 11 kW and 200 V has been used to verify start-up performance, in which the count of a chopper cell per an arm was set to six. Experimental results of the synchronous-motor drive have shown stable start-up performance through all speed range with a quadratic-torque load.

References [14] and [15] aim to apply the motor drive to a quadratic-torque load. Reference [14] has experimentally verified the effectiveness of the square-wave method in terms of mitigation of ac-voltage fluctuation and reduction of arm-current amplitude. Reference [15] has applied minimal stator-current control to the DSCC for a further reduction of the arm-current amplitude. A downscaled system rated at 15 kW and 400 V has been used to verify start-up performance, in which the count of a chopper cell per an arm was set to four. The experimental results of the induction-motor drive have shown stable start-up performance from a standstill to a middle speed of 580 min^{-1} loaded with 60%.

Although the papers successfully operate a motor in a low-speed range with suppressing the ac-voltage fluctuaion by using the mitigation method, an additional ac circulating current introduces not only an additional converter loss, but also increases of the size and volume of the center-tapped inductor. Therefore, the peak value of the ac circulating current in the mitigation methods should be evaluated.

The 2014 International Power Electronics Conference

TABLE I.   COMPARISONS AMONG EXISTING SINUSOIDAL-WAVE METHOD, SQUARE-WAVE METHOD, AND THE PROPOSED METHOD.

	Sinusoidal-wave method	Proposed method	Proposed method + third harmonic current	Square-wave method
Common-mode voltage	sinusoidal wave	square wave	square wave	square wave
AC circulating current	sinusoidal wave	sinusoidal wave	sinusoidal wave + third harmonic current	square wave
Peak ac circulating current	I_{Z0} in (14)	$\frac{\pi}{4}I_{Z0}$	$\frac{9\pi}{38}I_{Z0} \times 0.87$	$\frac{1}{2}I_{Z0}$
Maximum frequency of currents	f_{com}	f_{com}	$3f_{com}$	—
Experimental results	Figs. 4, 9	Fig. 5	Fig. 6	Figs. 8,10
References	[9]–[14]	This paper	This paper	[13][14]

IV. MITIGATION METHODS

This section describes a design guide for the ac-voltage-fluctuation mitigation methods. Firstly, the principles of a mitigation method is summarized. Then, the amount of peak circulating currents of the two conventional mitigation methods and two proposed methods are theoretically evaluated. Finally, comparisons are made for these four mitigation methods.

A. Principles of a mitigation method

The instantaneous power produced by a motor current causes a large ac-voltage fluctuation in a low-speed range. The following assumptions are considered to evaluate an instantaneous power flowing into the dc capacitor:

- v_a^* and v_{Bju}^* are neglected.

- v_u and dc component of i_Z, I_Z are neglected.

- AC-voltage fluctuation produced by v_{com} and $\tilde{i}_Z$ is small.

The first assumption is valid, because the control values for the dc-capacitor voltage control are zero in a steady-state condition. The second assumption is reasonable when a motor operated in a low-speed range. The third assumption is valid when a frequency of v_{com} is set to be much higher than a motor frequency. Finally, the instantaneous power, p can be expressed as follows (see appendix):

$$p = i_{Pu} \times v_{ju} \simeq -\frac{v_{com}}{4}\tilde{i}_{Zu} + \frac{v_{dc}i_u}{16}. \quad (7)$$

The second terms on (7) should be eliminated for a stable operation in a low-speed range. Therefore, the following relationship should be kept:

$$v_{com}\tilde{i}_{Zu} = -\frac{v_{dc}i_u}{4}. \quad (8)$$

The waveforms of v_{com} and $\tilde{i}_{Zu}$ are freely selected as long as (8) is kept.

B. Selection of waveforms for v_{com} and $\tilde{i}_Z$

The selection of waveforms for v_{com} and $\tilde{i}_Z$ are important in terms of a reduction in a peak value of $\tilde{i}_Z$.

The common-mode voltage waveforms are selected as follows in [9] and [14].

$$v_{com} = \sqrt{2}V_{com}\sin\omega_{com}t \quad (9)$$

$$v_{com} = \begin{cases} \sqrt{2}V_{com} & (0 < t < \frac{\pi}{\omega_{com}}) \\ -\sqrt{2}V_{com} & (\frac{\pi}{\omega_{com}}t < \frac{2\pi}{\omega_{com}}), \end{cases} \quad (10)$$

where, ω_{com} is a angular frequency of the common-mode voltage, V_{com} is a rms voltage of v_{com}. Fourier expansion is applied to (10), for the sake of simplicity:

$$v_{com} = \frac{4\sqrt{2}V_{com}}{\pi}\sum_{n=1}^{\infty}\frac{1}{(2n-1)}\sin(2n-1)\omega_{com}t. \quad (11)$$

Once the waveform of v_{com} is determined, $\tilde{i}_Z$ can be chosen freely as long as (8) is hold. The u-phase ac circulating current, $\tilde{i}_{Zu}$ is selected as follows in [9] and [14]:

$$\tilde{i}_{Zu} = I_{Z0}\sin\omega_{com}t \quad (12)$$

$$\tilde{i}_{Zu} = \frac{4}{\pi}I_{Zn}\sum_{n=1}^{\infty}\frac{1}{(2n-1)}\sin(2n-1)\omega_{com}t, \quad (13)$$

where, I_{Z0} and I_{Zn} are the amplitude of $\tilde{i}_{Zu}$.

The sinusoidal-wave method in [9] set v_{com} to (9) and $\tilde{i}_{Zu}$ to (12). The peak ac circulating current superimposed, I_{Z0} is described as follows (see appendix):

$$I_{Z0} = -\left(\frac{V_{dc}I_1}{2V_{com}}\right), \quad (14)$$

where, V_{dc} is a dc-link voltage, I_1 is the rms motor current.

On the other hand, the square-wave method in [14] set v_{com} to (10) and $\tilde{i}_{Zu}$ to (13). The peak value of $\tilde{i}_{Zu}$, I_{Zn} is described as follows:

$$I_{Zn} = \frac{1}{2}I_{Z0}. \quad (15)$$

The selecting v_{com} and $\tilde{i}_{Zu}$ as the square wave can be reduced the peak value of $\tilde{i}_{Zu}$ by the half of (14). However, the circulating current controller should control the all frequency components included in (13) to hold the relationship in (8).

This paper proposes the another mitigation methods. One of the methods set v_{com} to (10) and $\tilde{i}_{Zu}$ to (12). The peak ac circulating current, I_{Z1} is described as follows:

$$I_{Z1} = \frac{\pi}{4}I_{Z0}. \quad (16)$$

The frequency components in $\tilde{i}_{Zu}$ for this mitigation method is the same as the sinusoidal-wave method, but the peak value of $\tilde{i}_{Zu}$ can be reduced.

A further reduction in a peak value of $\tilde{i}_{Zu}$ can be achieved when the third-order-harmonic current is superimposed in (4):

$$\tilde{i}_{Zu} = I_{Z3}\left(\sin\omega_{com}t + \frac{1}{6}\sin 3\omega_{com}t\right). \quad (17)$$

Here, the third-harmonic-current superimposed contributes not only a reduction in a peak current by 87% [14], but also

978-1-4799-2706-7/14 $31.00 © 2014 IEEE

an increase of an average power produced by v_{com} and $\tilde{i}_{\text{Zu}}$, because v_{com} contains the third-order-harmonic voltage. The peak ac circulating current, I_{Z3} can be expressed as follows:

$$I_{\text{Z3}} = \frac{9\pi}{38} I_{\text{Z0}} \times 0.87. \tag{18}$$

The peak ac circulating current can be reduced by 15% compared to the proposed method.

C. Selection of a frequency of the common-mode voltage

Once (8) is satisfied by applying the mitigation methods, remaining ac-voltage fluctuations are all related to the frequency of the common-mode voltage. The freuqnency of v_{com}, f_{com} determines an amplitude of an ac-voltage fluctuation. The higher f_{com} is, the lower amplitude of an ac-voltage fluctuation is. The maximum f_{com} is limited by an one-tenth of a carrier frequency [14].

The following relationship exists in the largest ac-voltage fluctuations of $\tilde{v}_{C\text{ju}}$ after the mitigation methods are applied:

$$\tilde{v}_{C\text{ju}} \propto \frac{1}{f_{\text{com}} - f}, \tag{19}$$

where, f is the motor-current frequency. The ac-voltage fluctuation becomes larger as f increases. Therefore, $f_{\text{com}} - f$ should be controlled to be constant so as to avoid an increase of the ac-voltage fluctuation, which is achieved by increasing f_{com} as f increases.

D. Comparison

Table I shows the comparison among four mitigation methods including two proposed methods in this paper.

The sinusoidal-wave method has the largest in the peak ac circulating current. However, the maximum frequency included in the ac circulating current is the minimum among all mitigation methods.

The proposed method uses the square wave as a common-mode voltage and the sinusoidal wave as the ac circulating current. This method can be considered as the hybrid method of the sinusoidal-wave method and the square-wave method, because the amount of the peak ac circulating current in the proposed method is just between the peak ac circulating current of the sinusoidal-wave method ($= 1$) and square-wave method ($= 0.5$). The theoretical peak ac circulating current in the proposed method is $\pi/4$ ($\simeq 0.79$) times lower than the sinusoidal-wave method. The maximum frequency included in the ac circulating current is the same as the sinusoidal-wave method.

The peak ac circulating current in the proposed method can be decreased by superimposing the third-order-harmonic current into ac circulating current. The idea of injecting the third-order-harmonic current into the ac circulating current has already been proposed in [14], but this method is different in the common-mode voltage waveform. The peak ac circulating current is reduced by 15% compared to the proposed method, and reduced by 64% compared to the sinusoidal-wave method. The maximum frequency included in the ac circulating current is only three times higher than the sinusoidal-wave method.

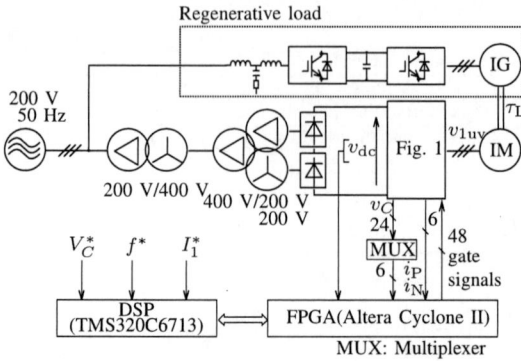

Fig. 3. The 400-V 15-kW downscaled system used in an experiment.

The square-wave method is the lowest in a peak ac circulating current. Although the square-wave method can minimize a peak ac circulating current, all of the frequency components in the ac circulating current should be controlled, which requires a fast circulating current controller. Therefore, when a relationship in (8) is not hold due to the error between a command ac circulating current and an actual ac circulating current, the large ac-voltage fluctuation based on a motor-current frequency appears on a dc-capacitor voltage. This ac-voltage fluctuation can be compensated by superimposing an additional ac circulating current.

V. EXPERIMENTAL RESULTS

A. Experimental Set-Up

Fig. 3 shows the system configuration of the 400-V 15-kW downscaled system. Table II summarizes the circuit parameters used in the experiments. Table III summarizes the specifications of the 380-V 15-kW induction motor tested. Here, a three-phase 12-pulse diode rectifier, consisting of a three-winding transformer with a Δ-Δ-Y connection and two three-phase six-pulse diode rectifiers, is used as the front end. Neither electrolytic capacitor nor film capacitor is connected to the common dc link, because it is not required for a DSCC.

The ac output terminals of the DSCC are directly connected to the induction motor rated at 380 V and 15 kW. The regenerative load in Fig. 3 consists of an induction generator rated at 190 V and 15 kW and two identical PWM converters connected back to back. The field-oriented control is applied to the induction generator, which enables an arbitrary instantaneous torque τ_{L} to be loaded on the induction motor.

The reference dc-capacitor voltage was set to $V_C^* = 140$ V, and the rms value of a common-mode voltage superimposed was set to $V_{\text{com}} = 180$ V. The motor-speed control in [15] was used to control a rms value of a motor current, where the motor-current amplitude is controlled to be the minimal for driving a motor.

B. Steady-State Performance

Figs. 4–8 shows steady-state waveforms when $T_{\text{L}} = 40\%$, $f_{\text{com}} = 50$ Hz and the rms motor current, I_1 was controlled to 13.8 A, which is 43% of the rated rms motor current of 32 A.

TABLE II. CIRCUIT PARAMETERS USED IN AN EXPERIMENT.

Rated active power		15 kW
Rated line-to-line rms voltage	V_S	400 V
Rated dc-link voltage	V_{dc}	560 V
Center-tapped inductor	L_Z	4.0 mH(12%)
DC capacitor of chopper cell	C	3.3 mF
DC-capacitor voltage	V_C	140 V
Unit capacitance constant	H	52 ms
Cell count per leg	N	8
Triangular-wave-carrier frequency	f_C	2 kHz
Equivalent carrier frequency	Nf_C	16 kHz

*The value in () is on a 400-V, 15-kW, and 50-Hz base.

TABLE III. MOTOR PARAMETERS USED IN AN EXPERIMENT.

Rated output power		15 kW
Rated frequency		50 Hz
Rated line-to-line rms voltage	V	380 V
Rated rotating speed	N_{rm}	1460 min^{-1}
Rated stator rms current	I_1	32 A
Rated rms magnetizing current	I_0	18.4 A
Pole-pair number	P	2
Moment of motor inertia	J_M	0.1 kg·m^2
Moment of load inertia	J_L	0.1 kg·m^2

Fig. 4 shows a steady-state waveforms when the sinusoidal-wave method was used. The rms motor current, I_1 was kept constant in its command value of 13.8 A. The peak value of the circulating current i_{Zu} was 37 A, which is defined as the base value (100%) in this paper. The peak-to-peak ac-voltage fluctuation was 35 V, which is 25 % of 140 V base.

Fig. 5 shows steady-state waveforms when the proposed method was used. The peak value of i_{Zu} was 27 A, which is 72% of the peak circulating current in Fig. 4. This peak current was almost $\pi/4$ lower than the peak current in Fig. 4, which agrees well with (12) (= 37 A × $\pi/4$ ≃ 29 A). The peak-to-peak ac-voltage fluctuation was 28 V, which is 20% of 140 V base.

Fig. 6 shows a steady-state waveforms when the proposed method with the third-order-harmonic current was used. The peak value of i_{Zu} was 24 A, which is 65% of the peak circulating current in Fig. 4. This peak current was $9\pi/38 × 0.87$ lower than the peak current in Fig. 4, which agrees well with (18) (= 37 A × $9\pi/38$ × 0.87 ≃ 24 A). The peak-to-peak ac-voltage fluctuation was 25 V, which is 18% of 140 V base.

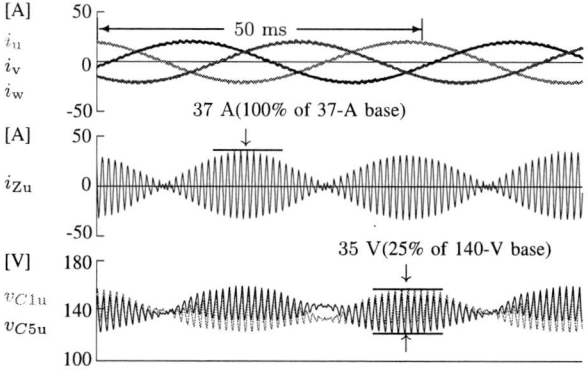

Fig. 4. Experimental steady-state waveforms when $I_1^* =$13.8 A (43%), $f^* = 1$ Hz, and $T_L = 40\%$ with the sinusoidal-wave method.

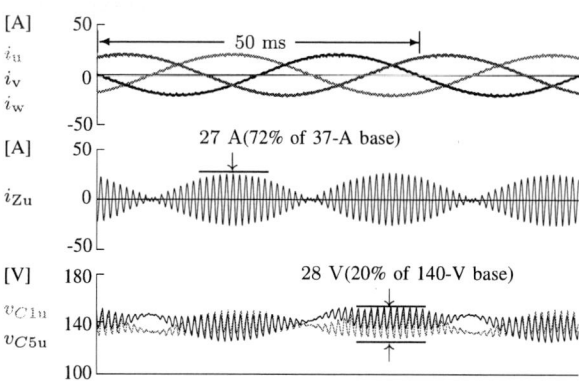

Fig. 5. Experimental steady-state waveforms when $I_1^* =$13.8 A (43%), $f^* = 1$ Hz, and $T_L = 40\%$ with the proposed mitigation method.

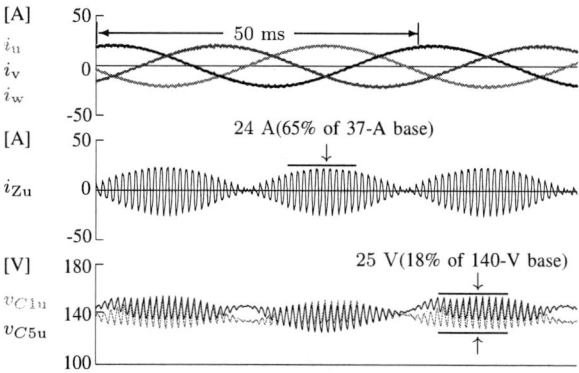

Fig. 6. Experimental steady-state waveforms when $I_1^* =$13.8 A (43%), $f^* = 1$ Hz, and $T_L = 40\%$ with the proposed mitigation method and third-order-harmonic currents.

Fig. 7 shows a steady-state waveforms when the square-wave method was used. The peak value of i_{Zu} was 20 A, which is 54% of the peak circulating current in Fig. 4, which agrees well with (15) (= 37 A × $1/2$ ≃ 19 A). Even though the peak circulating current reduced by almost half value of Fig. 4, the ac-voltage fluctuation becomes larger. The ac-voltage fluctuation contains not only 50 Hz components of the common-mode-voltage frequency, but also 1 Hz components of the motor-current frequency. The motor current waveforms were distorted due to the large ac-voltage fluctuations. The

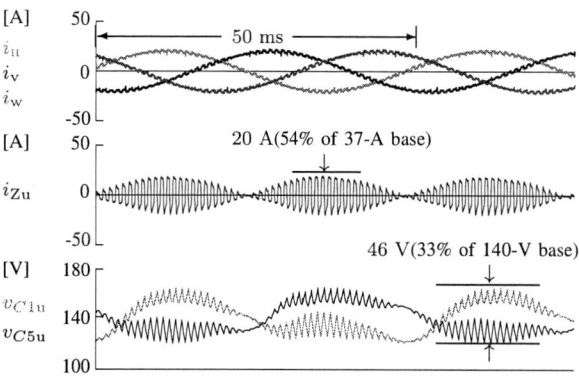

Fig. 7. Experimental steady-state waveforms when $I_1^* =$13.8 A (43%), $f^* = 1$ Hz, and $T_L = 40\%$ with the square-wave method, but without superimposing an additional ac circulating current.

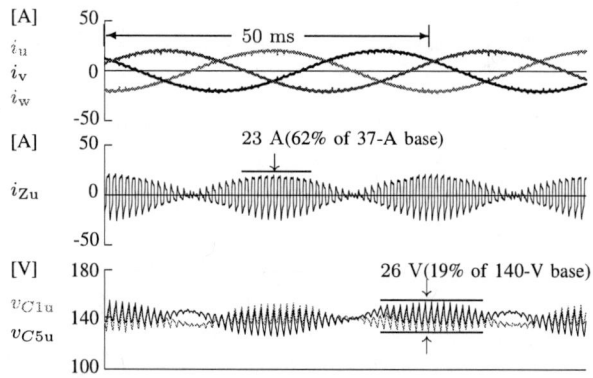

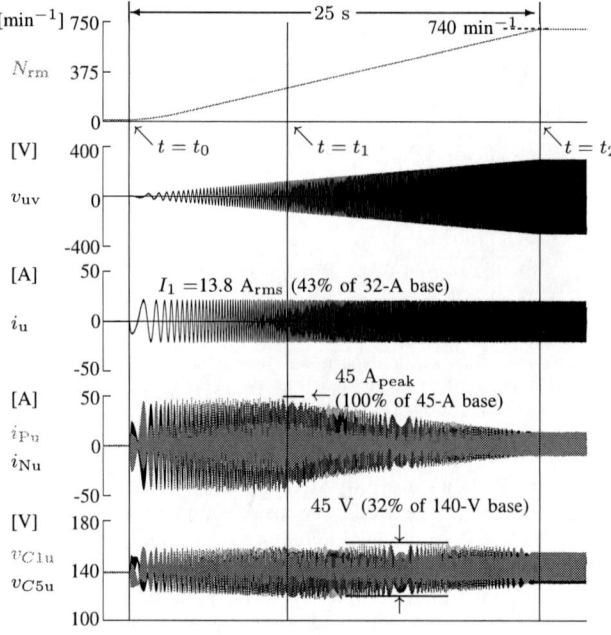

Fig. 8. Experimental steady-state waveforms when $I_1^* = 13.8$ A (43%), $f^* = 1$ Hz, and $T_L = 40\%$ with the square-wave method and an additional ac circulating current.

Fig. 9. Experimental start-up waveforms when the sinusoidal-wave method was used. $I_1^* = 13.8$ A (43%), $T_L = 40\%$, and $f_{com} = 50$ Hz.

peak-to-peak ac-voltage fluctuation was 46 V, which is 33% of 140 V base.

Fig. 8 shows a steady-state waveforms when the square-wave method was used. An additional ac circulating current was superimposed to compensate the ac-voltage fluctuation. The peak value of i_{Zu} was 23 A, which is 62% of the peak circulating current in Fig. 4. Although the peak current of i_{Zu} was only increased 3 A compared to Fig. 7, the peak-to-peak ac-voltage fluctuation was reduced to 26 V, which is 19% of 140 V base, and no 1 Hz components causing from a motor current appeared on the ac-voltage fluctuation.

The experimental results shows that the square-wave method with the additional ac circulating current is the most suitable mitigation method in terms of the peak circulating current.

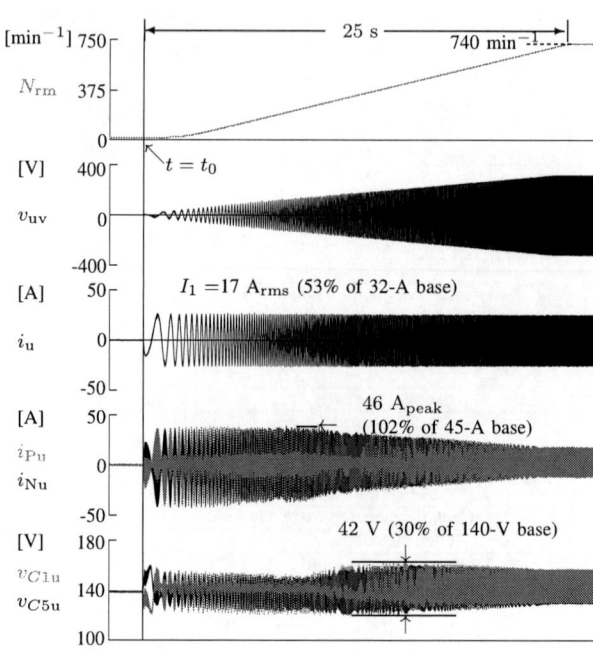

Fig. 10. Experimental start-up waveforms when the square-wave method was used. $I_1^* = 13.8$ A (43%), $T_L = 60\%$, and $f_{com} = 50$ Hz $+ f$.

C. Start-Up Performance

Figs. 9, 10 show start-up performance with different mitigation methods. One of the benefits from the reduction in the peak circulating current is an increase of a start-up torque.

Fig. 9 shows start-up waveforms when a load torque was set to $T_L = 40\%$ and the rms motor current, I_1 was controlled to 13.8 A, which is 43% of the rated rms motor current of 32 A. The motor-current frequency, f increased as a ramp. The common-mode voltage frequency, f_{com} was set to 50 Hz in all speed range. The sinusoidal-wave method was used. Here, the mitigation method switched over as follows [15]:

- $t_0 \leq t \leq t_1$, the rms value of the common-mode voltage V_{com} and the ac circulating currents $\tilde{i}_Z$ were controlled actively to mitigate the ac-voltage fluctuation of each dc-capacitor voltage.

- $t_2 \geq t$, neither V_{com} nor $\tilde{i}_Z$ was superimposed.

During $t_1 \leq t \leq t_2$ Hz, the commands of V_{com} and $\tilde{i}_Z$ decrease linearly in their amplitude. The maximum arm-current amplitude reached to 45 A, which is 100% of the rated stator-current amplitude of 45 A. The peak-to-peak capacitor voltage fluctuation was 51 V, which is 36% of 140 V. The low frequency components appeared on the ac-voltage fluctuations as the motor-current frequency increases, as expected in (19).

Fig. 10 shows start-up waveforms when $T_L = 60\%$ and $I_1^* = 17$ A (53%). The common-mode voltage frequency, f_{com} was increased as a motor-current frequency increases ($f_{com} = 50$ Hz $+ f$). The square-wave method with the additional ac circulating current was used. The maximum arm-current amplitude reached to 45 A (100%). The peak-to-peak ac-voltage fluctuation was 42 V (30%). No low frequency components appeared on the dc-capacitor voltages. The experimental results shows that the start-up torque was enhanced from 40% to 60% with the same arm-current amplitude, compared to Fig. 9.

VI. CONCLUSION

This paper has presented a research trend in the DSCC-based medium-voltage motor drives, especially focusing on the low-speed operation. The tradeoff existed between the mitigation of the ac-voltage fluctuation and the reduction of the peak circulating current has been the main difficulty for operation in a low-speed range. A reasonable compromise can be achieved by designing the shape of waveforms for a common-mode voltage and an ac circulating current used for the mitigation method. The amount of the peak current in the two conventional mitigation methods and the two proposed ones have been theoretically investigated. Experimental results obtained from 400-V, 15-kW system shows the validity of the theoretical analysis on the amount of a peak circulating current, and the practicability of the DSCC-based motor drive for operation in a low-speed range.

REFERENCES

[1] P. W. Hammond, "A new approach to enhance power quality for medium voltage ac drives," *IEEE Trans. Ind. Appl.*, vol. 33, no. 1, pp. 202–208, Jan./Feb. 1997.

[2] T. Geyer and N. Oikonomou, "Model predictive control of industrial drives," *Conf. Rec. IEEE-ECCE Tutorial* 2013.

[3] A. Lesnicar and R. Marquardt, "An innovative modular multilevel converter topology suitable for a wide power range," *IEEE Conf. Rec.*, Bologna PowerTech 2003, CD-ROM.

[4] H. Akagi, "Classification, terminology, and application of the modular multilevel cascade converter (MMCC)," *IEEE Trans. Power Electron.*, vol. 26, no. 11, pp. 3119–3130, Nov. 2011.

[5] M. Hagiwara and H. Akagi, "Control and experiment of pulse-width-modulated modular multilevel converters," *IEEE Trans. Power Electron.*, vol. 24, no. 7, pp. 1737–1746, Jul. 2009.

[6] M. Hiller, D. Krug, R. Sommer, and S. Rohner, "A new highly modular medium voltage converter topology for industrial drive applications," in *Conf. Rec. EPE* 2009, pp. 1–10.

[7] M. Hagiwara, K. Nishimura, and H. Akagi, "A medium-voltage motor drive with a modular multilevel PWM inverter," *IEEE Trans. Power Electron.*, vol. 25, no. 7, pp. 1786–1799, Jul. 2010.

[8] S. Rohner, J. Weber, and S. Bernet, "Continuous Model of Modular Multilevel Converter with Experimental Verification," *Conf. Rec. IEEE-ECCE* 2011, pp. 4021–4028.

[9] A. J. Korn, M. Winkelnkemper, and P. Steimer, "Low output frequency operation of the modular multilevel converter," in *Conf. Rec. IEEE-ECCE* 2010, pp. 3993–3997.

[10] A. Antonopoulos, L. Angquist, S. Norrga, K. Llves, and H. P. Nee, "Modular multilevel converter ac motor drives with constant torque from zero to nominal speed," *IEEE Trans. Ind. Appl.* 2014, early access article.

[11] J. Kolb, F. Kammerer, and M. Braun "Dimensioning and design of a modular multilevel converter for drive applications," in *Conf. Rec. EPE* 2012, CD-ROM.

[12] J. Kolb, F. Kammerer, M. Gommeringer, and M. Braun "Cascaded control system of the modular multilevel converter for feeding variable-speed drives ," in *IEEE Trans. Ind. Appl.* 2014, early access article.

[13] J. J. Jung, H. J. Lee, and S. K. Sul, "Control strategy for improved dynamic performance of variable-speed drives with the modular multilevel converter," in *Conf. Rec. IEEE-ECCE* 2013, pp.1481–1487.

[14] M. Hagiwara, I. Hasegawa, and H. Akagi, "Startup and low-speed operation of an adjustable-speed motor driven by a modular multilevel cascade inverter (MMCI)," *IEEE Trans. Ind. Appl.*, vol. 49, no.4, pp. 1556–1565, Jul./Aug. 2013.

[15] Y. Okazaki, M. Hagiwara, and H. Akagi, "A speed-sensorless start-up method of an induction motor driven by a modular multilevel cascade inverter (MMCI-DSCC)," *IEEE Trans. Ind. Appl.* 2014, early access article.

APPENDIX

A. Derivation of (7)

The instantaneous power, p is described by multiplying (2) by (4):

$$p = \frac{v_{dc}\tilde{i}_{Zu}}{8} - \frac{v_{com}i_u}{8} + \frac{v_{dc}i_u}{16} - \frac{v_{com}\tilde{i}_{Zu}}{4}, \quad (20)$$

where, the v_u and I_Z is neglected. The first and second terms of (20) can be neglected, because the ac-voltage fluctuation from these terms are much smaller than the others. Thus, (7) is derived from (20).

B. Derivation of (14)

The instantaneous power produced by (9) and (12), p_{com} is described as:

$$p_{com} = v_{com} \times \tilde{i}_{Zu} = \sqrt{2}V_{com}I_{Z0}(1 - \cos\omega_{com}t^2). \quad (21)$$

Here, only the average power contributes to hold (8). Therefore, (21) is changed as follows:

$$p_{com} \simeq \sqrt{2}V_{com}I_{Z0}. \quad (22)$$

The peak circulating current superimposed, I_{Z0} is derived by substituting (22) to (8). The same derivations can be applied for the other mitigation methods discussed in this paper.

The 2014 International Power Electronics Conference

An Input Switched Multilevel Inverter for Open-end Winding Induction Motor Drive

B. Zhu[1], Y. Jia[1], U. R. Prasanna[1], K. Rajashekara[1] and H. Kubo[2]

[1]University of Texas at Dallas
Richardson, Texas, USA

[2]Meidensha Corporation
Numazu-shi, Shizuoka, Japan

Abstract - **This paper presents an input switched multilevel inverter topology for controlling a three-phase open-end winding induction motor drive. In this topology, DC link voltage obtained from a single voltage source is divided into '*n*' equal steps, and a single pole multi-throw (SPMT) configuration is used to select any desired voltage level from these possible '*n*' voltage levels. This is followed by H-bridge which can alternate the polarity of voltage obtained at its output and hence generating necessary ac voltage. One salient feature of this multilevel inverter drive system is that the number of switches used is greatly reduced compared with traditional neutral-point-clamped inverters and flying-capacitor inverters based systems. It is possible to obtain two additional voltage levels at the output by adding only two switching devices per phase. And the input switched inverter topology can be easily extended to obtain any desired voltage steps at the machine terminals. A hybrid modulation technique combining symmetrical sinusoidal PWM and traditional SPWM is also proposed. Circuit analysis, modulation technique and software simulation are illustrated with respect to a seven-level input switched inverter drive system.**

Keywords— multilevel inverter, open-end winding induction motor drive, PWM strategies

I. INTRODUCTION

The open-end winding three-phase induction motor is receiving increasing attention for high power applications in order to achieve independent phase current control capability, reduced switching loss, lower THD, and better fault tolerant characteristics in machine windings [6]. Generally, open-end winding induction motor drives are supplied by two three-phase inverters or three H-bridges [4-5]. Compared with conventional two-level inverter drives, multilevel inverter can achieve more voltage levels at machine terminals and results in lower voltage stress on the semiconductor devices [1-3], which eventually results in lower torque ripple on the motor shaft and higher system efficiency.

Various topologies of multilevel inverters for open-end winding motor drives are reported in the literature. Modifications to inverter topologies or variations to modulation strategies in order to eliminate common-mode components and to reduce number of isolated power

supplies are discussed in detail [7-12]. A generalized input switched multilevel inverter consisting of several input switches and an H-bridge inverter is shown in Fig. 1 [13]. It is claimed that this input switched topology archives minimum number of switches for a given number of voltage levels for all kind of multilevel inverters.

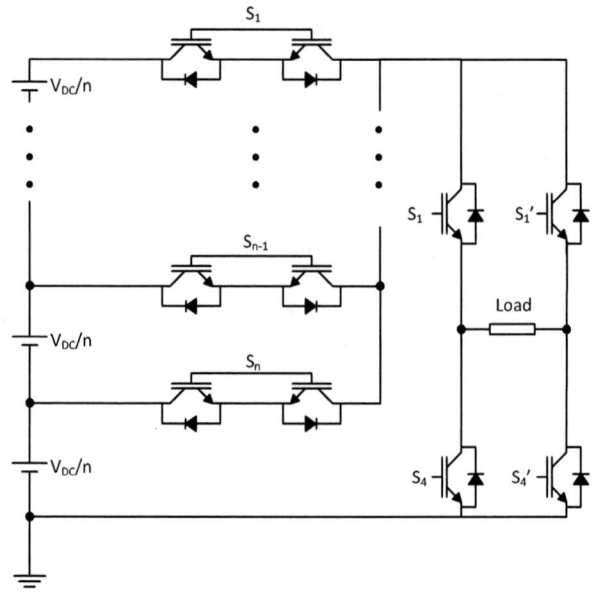

Fig. 1 Power Circuit of a (2*n*+1)-Level Input Switched Inverter [13]

In this paper, an input switched inverter has been used for open-end winding induction motor drive. A detailed illustration for a *n*=3 input switched inverter open-end winding motor drive system is presented including the power circuit topology and implementation of gating signals. The modulation strategy is proposed in such a way that, when operating in low modulation index region, the input switches remain in static ON/OFF states, and the H-bridge is responsible for both PWM voltage regulation and phase commutation. While in a high modulation index region, the input switches are activated to provide PWM voltage regulation, and the H-bridge is used only for phase commutation. It is possible to obtain additional two voltage levels at the output by adding only two switching devices per phase. Switching patterns can be implemented for any

978-1-4799-2706-7/14 $31.00 © 2014 IEEE

level by comparing reference voltage with the multiple level shifted carrier waveforms.

The paper is organized as follows. The configuration of the proposed drive system and its power circuit topology is discussed in Section II. Section III presents the PWM switching strategy of the inverter based on instantaneous modulation index. Section IV illustrates the voltage optimization method according to the switching strategy: three-level, five-level and seven-level working modes of the inverter are discussed. In Section V, the MATLAB/SIMULINK based simulation results of the voltage and current waveforms, as well as speed and torque response of the motor are presented.

II. SYSTEM CONFIGURATION

The block diagram of the open-end winding induction motor drive system using input switched multilevel inverter is shown in Fig. 2. The whole system consists of two parts, power circuit and control logic.

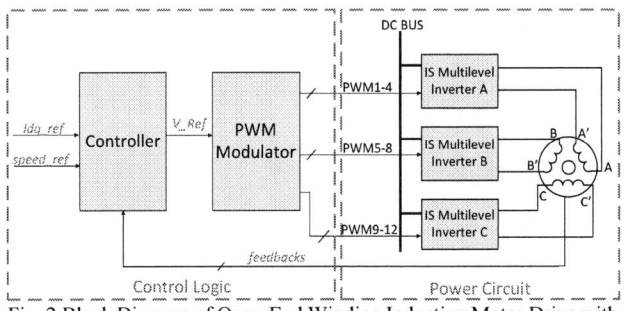

Fig. 2 Block Diagram of Open-End Winding Induction Motor Drive with Input Switched Multilevel Inverter

An open-end winding induction motor fed from a seven level input switched multilevel inverter is shown in Fig. 3. AA′, BB′ and CC′ represent the stator coils of the open-end winding induction motor. Both ends of the winding in each phase are connected to an individual H-bridge. The H-bridge is controlled by 3 bidirectional switches (S_{A1}, S_{A2} and S_{A3}), functioning like a single pole multi-throw switch. Each of the bidirectional switches is connected with a capacitor. With a series connection of the three capacitors, the DC link voltage, which is provided by a single DC source, is split into three equal voltage steps. By gating the bidirectional switches and the H-bridges, a seven-step sinusoidal approximation wave can be generated.

In the control logic, a field oriented control (FOC) is used to regulate the current of the motor. Information of speed and winding currents is fed back to the controller, which consists of three PI regulators to track the speed, d axis current, and q axis current. These controllers generate voltage references for the PWM modulator to generate the PWM signals for switching the inverter devices.

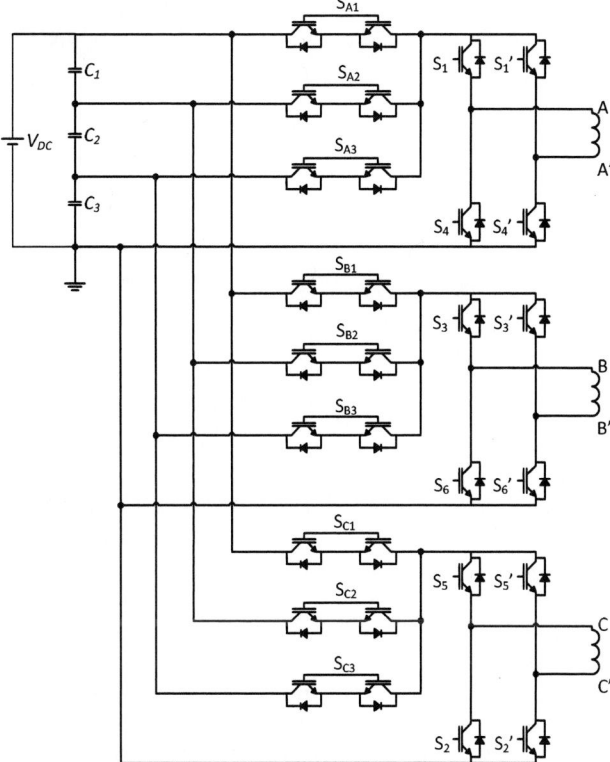

Fig. 3 Three Phase Power Circuit Diagram of the Proposed Motor Drive

III. PWM GENERATION AND SWITCHING STRATEGY

Based on the instantaneous modulation index of the reference wave, the inverter can function from three-level to seven-level configurations. The instantaneous modulation index is defined as:

$$MI = \frac{|V_{ref}(t)|}{V_{DC}} \qquad (1)$$

Where V_{ref} represent the instantaneous value of the reference phase voltage, and V_{DC} is the DC bus voltage. For linear region, MI can range from 0 to 1, and MI is greater than 1 for over modulation.

The PWM signals are generated by comparing the reference wave with three pairs of triangular carriers, which are shown in Fig. 4 as shaded zones in three colors. The minimum and maximum values of each triangular carrier are $\frac{V_{DC}}{3}(i-1)$ and $\frac{V_{DC}}{3}i$, where i varies from 1 to 3 for the seven level topology. Depending on the instantaneous modulation index, the generation of PWM signals can be derived based on the three regions, which are labeled as ①, ② and ③. As there are two parts in the power circuits, i.e. input bidirectional switches and H-bridges, the switching strategy for each part is different.

978-1-4799-2706-7/14 $31.00 © 2014 IEEE

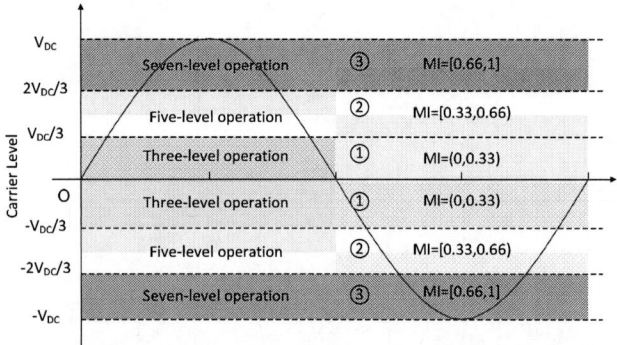

Fig. 4 Relationship between Operation Mode and Modulation Index

A. Switching Strategy for H-bridge

For H-bridge, the gating signals for the switches can be controlled by sinusoidal PWM (SPWM) and symmetrical sinusoidal PWM (SSPWM), the later one is first proposed in [4-5]. The difference between SPWM and SSPWM depends on the number of carrier waves used for modulation; SPWM uses only 1 carrier wave, while SSPWM uses 2. In this paper, SSPWM is chosen for the operation as the switching loss is reduced compared with SPWM [2-3]. In SSPWM, there are $2N$ pulses in a cycle of reference wave, and the whole period of the reference wave is divided into $2N$ equally spaced time intervals T_s. Each pulse is placed at C_j, which is the center point of each switching time interval, where j varies from 1 to N, as shown in Fig. 5.

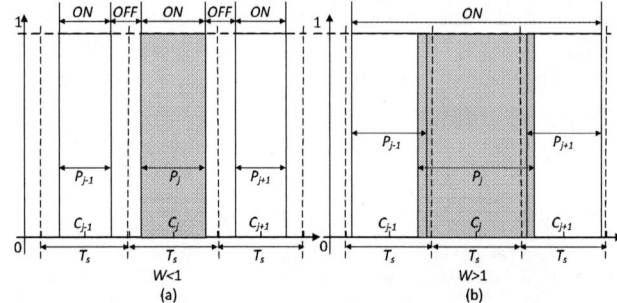

Fig. 5 SSPWM Pulse Placement (a) W<1 (b) W>1

The width of each pulse P_j is obtained according to the value of reference at an angular position equal to C_j. The center point of each pulse C_j can be calculated by:

$$C_j = (2j-1) \times \frac{\pi}{2N}, j = 1, 2, \cdots, N \quad (2)$$

The pulse width of the jth pulse can be obtained by:

$$P_j = T_s W \sin(C(j)), j = 1, 2, \cdots, N \quad (3)$$

And W is defined as width modulation index, which is obtained by the following equation:

$$W = n \cdot MI(C_j) \quad (4)$$

Where $MI(C_j)$ is the instantaneous modulation index at the point C_j. For a seven level inverter configuration, $n = 3$, and $W = 3 \cdot MI(C_j)$. It is clear that when $MI(C_j)$ is less than 1/3, W is less than 1 and P_j is less than T_s, which means the pulse width P_j is still restricted within its own switching interval, as shown in Fig. 5 (a). Hence H-bridge is operating in linear region and switching at carrier frequency. However, if $MI(C_j)$ exceeds 1/3, W is greater than 1 and P_j can be larger than T_s when the sinusoidal wave reaches the peak, which means the jth pulse exceeds its own interval and overlap with its neighboring pulses, as shown in Fig. 5 (b). As a result, the switch is kept conducting during these neighboring intervals. At this moment, H-bridge is working in over-modulation region and hence output is clamped to either positive rail or negative rail.

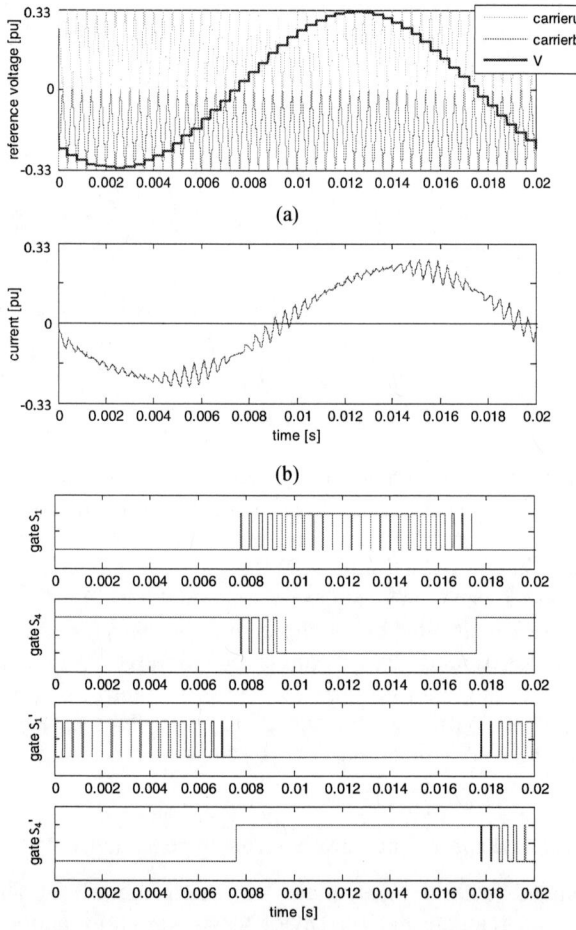

Fig. 6 (a) Reference Wave and Carriers (b) Load Current (c) Gating Signals of the H-bridge

Fig. 6 demonstrates the method to implement SSPWM. For illustration purpose, only two triangular carriers in region ① is shown in Fig. 6 (a), the top one has a maximum value $V_{DC}/3$ and minimum value 0, and the bottom one has a minimum value $-V_{DC}/3$ and maximum value 0. The value of reference wave is zero order hold in

978-1-4799-2706-7/14 $31.00 © 2014 IEEE

each triangular wave period, and it is equal to the pulse width calculated in equation (3). Fig. 6 (b) shows the load current waveform. Based on the sign of the reference wave and the load current, the gating rule for the 4 switches of the H-bridge is listed in Table 1.

TABLE 1
GATING RULE FOR SWITCHES OF H-BRIDGE UNDER SSPWM

	V > 0 I > 0	V > 0 I < 0	V < 0 I > 0	V < 0 I < 0
S_1	PWM	PWM	OFF	OFF
S_4	OFF	$\overline{\text{PWM}}$	ON	ON
S_1'	OFF	OFF	PWM	PWM
S_4'	ON	ON	$\overline{\text{PWM}}$	OFF

Where 'PWM' means the corresponding switch switches at carrier frequency to provide PWM signal, '$\overline{\text{PWM}}$' means the switch switches complementary with the one in the 'PWM' state. 'ON' indicates that the switch is conducting throughout the interval and 'OFF' means that the switch is off during the interval.

B. Switching Strategy for Input Bidirectinal Switches

The switching strategy for input bidirectional switches is based on the maximum instantaneous modulation index MI_{max}. As the maximum instantaneous modulation index varies from 0 to 1, it will fall into three regions, i.e. (0, 0.33), (0.33, 0.66), and (0.66, 1). Depending on the operating region, the switching pattern of the input bidirectional switches varies, which will further result in a multilevel output voltage across the phase winding.

a) $MI_{max} \leq 0.33$

When MI_{max} is less than 0.33, the whole reference wave lies in region ①, at this time, only level 1 carrier is used, and the inverter works in three-level mode. The only function of input side SPMT switch is to maintain the voltage feeding into the H-bridge as $V_{DC}/3$. Hence S_{A1} and S_{A2} are gated off, and S_{A3} is conducting all the time. The PWM signal is provided only by H-bridge.

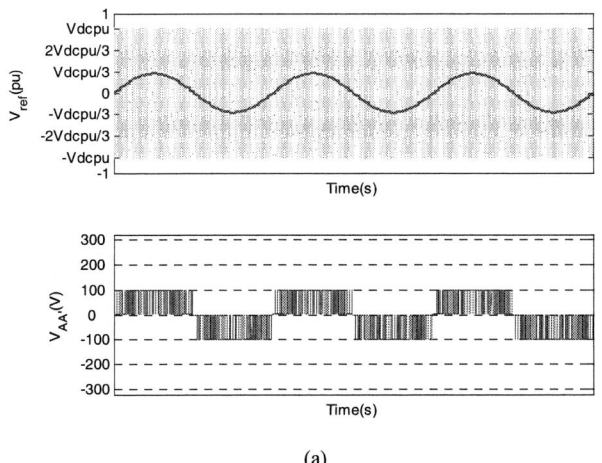

(a)

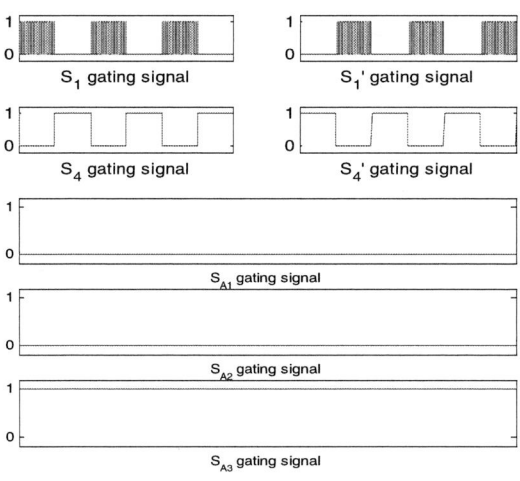

(b)
Fig. 7 Three-level Operation of Input Switched Multilevel Inverter (a) Reference and Output Phase Voltage (b) Gating Signals

Fig. 7 shows the output phase voltage and the gating signals when MI_{max} is less than 0.33, when the inverter works under three-level mode. In this mode, the output phase voltage, which is defined as the voltage across one phase winding, has three voltage levels ($-V_{DC}/3, 0, V_{DC}/3$).

b) $0.33 < MI_{max} \leq 0.66$

When the MI_{max} is between 0.33 and 0.66, the instantaneous modulation index MI varies from 0 to MI_{max}. The reference wave will lie on both region ① and ②, which means both level 1 and level 2 carriers are put into use, level 3 is still left blank. At this time, H-bridge works in both linear region and over modulation region. As long as the instantaneous modulation index MI is in region ①, condition a) still counts. Hence the PWM generation lies on H-bridge during this stage, and the switching manner of the input switches is the same with that when $MI_{max} < 0.33$. Once the reference wave enters region ②, H-bridge will work in over-modulation region, hence S_1 and $S_{4'}$ are conducting when the reference wave is positive, and $S_{1'}$ and S_4 are always gated-on in case of negative reference value. At this moment, the input switches S_{A3} and S_{A2} will provide the PWM signal and this is accomplished by comparing the reference wave with a triangular wave with minimum and maximum value $V_{DC}/3$ and $2V_{DC}/3$. When the reference wave is greater than the triangular carrier, S_{A2} is turned-on. The gating signal to S_{A3} is the complementary of S_{A2} and in this case, S_{A1} is always off. In this mode of operation, SPMT is switched at carrier frequency to switch between $V_{DC}/3$ and $2V_{DC}/3$, whereas devices in the H-bridge function only when $MI < 0.33$. A five level output voltage can be obtained from synthesizing condition a) and b), as shown in Fig. 8. Because S_{A2} is connected with $2V_{DC}/3$ level, the output voltage will have two additional voltage levels, i.e. $2V_{DC}/3$ and $-2V_{DC}/3$.

carrier with its minimum and maximum value $2V_{DC}/3$ and V_{DC}. S_{A1} conducts if the reference is greater than the carrier, and S_{A2} is switched complementary to S_{A1}. Because of the activity of S_{A1}, two voltage levels (V_{DC} and -V_{DC}) are added to the output voltage. By synthesizing condition a), b) and c), a seven-level output voltage can be expected across the phase winding. The output phase voltage and corresponding gating signals are shown in Fig. 9.

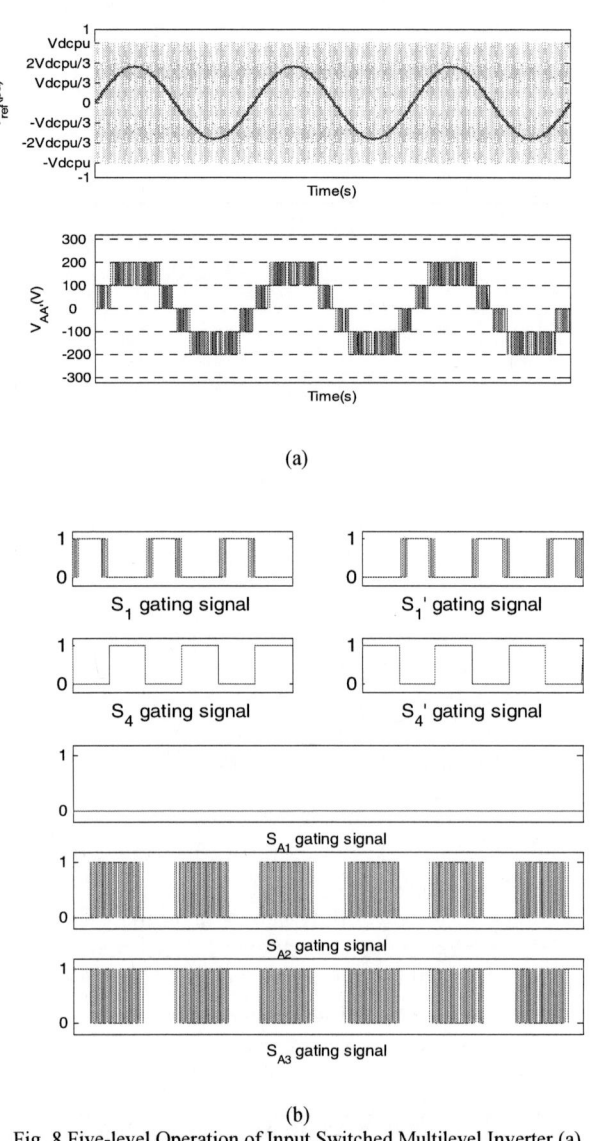

(a)

(b)

Fig. 8 Five-level Operation of Input Switched Multilevel Inverter (a) Reference and Output Phase Voltage (b) Gating Signals

c) *0.66 <MI$_{max}$ ≤ 1*

In this operation mode, MI_{max} is between 0.66 and 1, the instantaneous modulation index MI varies from 0 to MI_{max}. The reference wave will cover all three level carriers. When the reference wave is in region ①, i.e. MI is from 0 to 0.33, condition a) applies. The switches follow the same switching pattern as in condition b) above when the reference wave passes through region ②. When the reference wave enters region ③, the level 3 carrier is being used. At this time, S_1 and $S_{4'}$ conduct when the reference wave is positive, and $S_{1'}$ and S_4 are always ON if the reference wave is negative, that is similar to condition b). As instantaneous MI exceeds 0.66, S_{A3} is 'over modulated' and clamped to off status, S_{A1} takes the role in the modulation process. S_{A1} and S_{A2} are controlled by the PWM signal by comparing the reference wave with a triangular

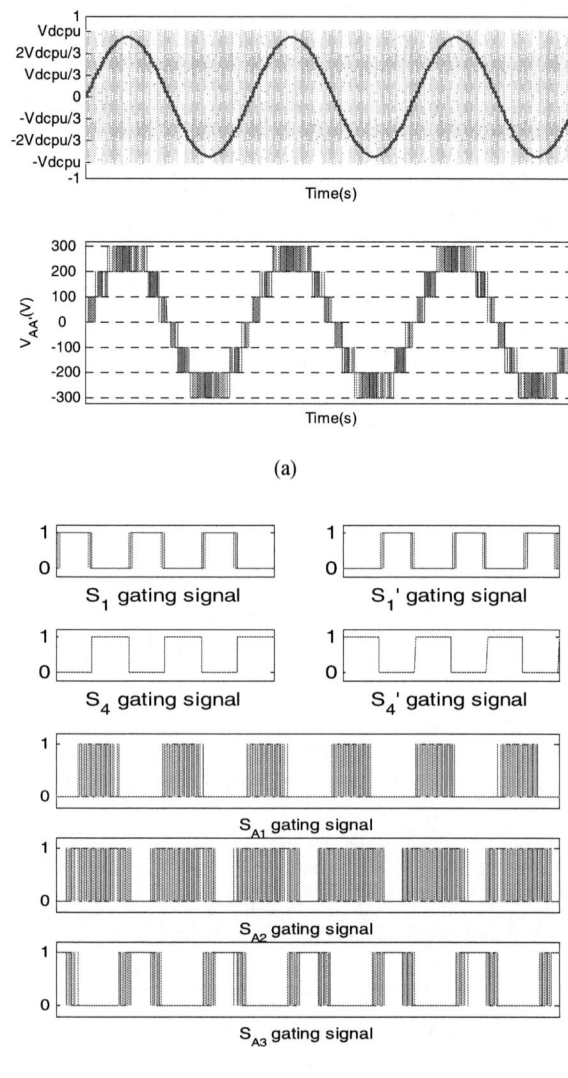

(a)

(b)

Fig. 9 Seven-level Operation of Input Switched Multilevel Inverter (a) Reference and Output Phase Voltage (b) Gating Signals

Table 2 shows the behavior of the H-bridge and input switches in different regions and under different modulation indices, in which "PWM" means the corresponding part will provide PWM signal to the phase winding, "COM" means the H-bridge is responsible for the commutation and remains at a certain state throughout the interval without switching at a high frequency.

978-1-4799-2706-7/14 $31.00 © 2014 IEEE   1598

TABLE 2
BEHAVIOR OF H-BRIDGE AND INPUT SWITCHES UNDER DIFFERENT
CONDITIONS

MI	H-bridge	S_{A1}	S_{A2}	S_{A3}
<0.33	PWM	OFF	OFF	ON
0.33-0.66	COM	OFF	PWM	PWM
>0.66	COM	PWM	PWM	OFF

It should be noted that, as the maximum instantaneous modulation index increases, the main function of H-bridge will shift from providing PWM signal to polarity changing. The PWM generation will mainly lie on the input switches (S_{A1}, S_{A2} and S_{A3}) as MI further increases.

IV. SIMULATION RESULTS

A closed-loop simulation for the proposed drive system for a 5.5 kW open-end winding induction motor and a seven-level input switched inverter has been implemented using MATLAB/SIMULINK. A start-up task is selected to demonstrate both steady-state and dynamic performance of the system. In this simulation, the motor is commanded to start at 0.1second under no load condition. After 1 second, a 0.5 pu step load torque is applied to the motor shaft.

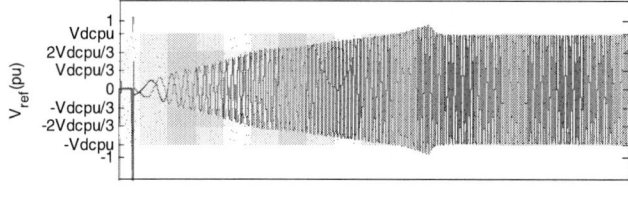

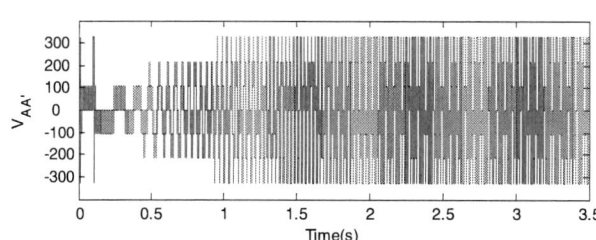

Fig. 10 Three Phase Reference Voltage and Output Phase Voltage of the Open-end Winding Induction Motor Drive

The waveforms of the three phase reference voltage and phase voltage are shown in Fig. 10. During start-up process of the motor, as the output voltage increases, the input switched inverter shifted its mode of operation from three-level, five-level to seven-level. The switching loss of the converter is reduced since less number of switches are activated in three-level and five-level operation.

Fig. 11 and Fig. 12 illustrate the current response, speed and electrical torque characteristics. From Fig. 11, it can be observed that during the whole simulation period, the d axis

current is kept constant, hence the flux is unchanged. For q axis current, during the start-up stage, large q axis current is provided. Once the start-up process is completed, the q axis current drops to 0.5 pu to balance the load torque. The torque and speed waveforms are shown in Fig. 12, in which a step load torque of 0.5 pu is applied to the motor after 1 second. When the acceleration is completed, the electrical torque reduces to 0.5 pu to balance the load.

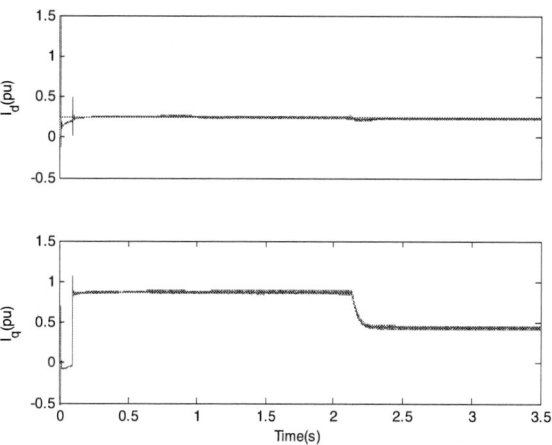

Fig. 11 The d-q Current of the Open-end Winding Induction Motor Drive

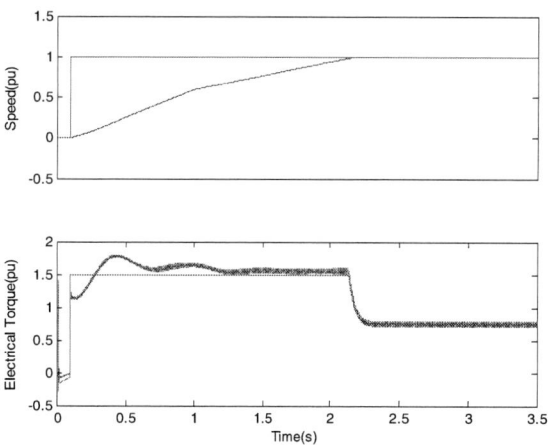

Fig. 12 The Speed and Electro-magnetic Torque Characteristics of the Open-end Winding Induction Motor Drive

V. CONCLUSION

In this paper, a generalized multilevel inverter topology with ($2n+1$) voltage levels is applied on an open-end winding machine. Single DC voltage source is used for generating multilevel output which is being fed to an open-end winding three-phase induction machine. A single pole multi-throw (SPMT) configuration is used to select any desired voltage level from 'n' possible voltage levels obtained from splitting the DC link voltage. Number of voltage levels at the output phase voltage can be incremented by two adding just two switches per phase. In order to illustrate the proposed configuration, a seven-level inverter when $n=3$ has been investigated in detail. Various modes of operation of the input switched multi-level inverter along with the corresponding modulation strategy have been presented. A start-up test of an open-end winding induction motor is launched in MATLAB/SIMULINK to demonstrate the performance of the input switched multilevel inverter based drive system. Simulation results show smooth steady-state and dynamic characteristics of the system in terms of speed reference and load torque variation.

REFERENCES

[1] T.A. Lipo, "New Open Winding Machine Concepts", Course Notes for ECE 511, Department of Electrical and Computer Engineering, University of Wisconsin-Madison, 2013

[2] K.S.Rajashekara and J.Vithayathil, "Microcomputer Based Symme-trical Sinusoidal Pulse Width Modulated Inverter", IEEE IECI Proceedings, San Francisco, USA, PP. 34-38, November 1981.

[3] K.S. Rajashekara and V. Rajagopalan, "Simulation of SSPWM Inverter-Fed Induction Motor", International Symposium on Modeling and Simulation of Electrical Machines, Converters and Power Systems, August 24-25, 1987, Quebec City, Canada.

[4] Jih-Sheng Lai; Fang Zheng Peng, "Multilevel converters-a new breed of power converters," Industry Applications, IEEE Transactions on , vol.32, no.3, pp.509,517, May/Jun 1996

[5] Bum-Seok Suh; Gautam Sinha; Manjrekar, M.D.; Lipo, T.A., "Multilevel Power Conversion - An Overview Of Topologies And Modulation Strategies," Optimization of Electrical and Electronic Equipments, 1998. OPTIM '98. Proceedings of the 6th International Conference on , vol.2, no., pp.AD-11,AD-24, 14-15 May 1998

[6] Rodriguez, J.; Jih-Sheng Lai; Fang Zheng Peng, "Multilevel inverters: a survey of topologies, controls,and applications, "Industrial Electronics, IEEE Transactions on , vol.49, no.4, pp.724,738, Aug 2002

[7] Baiju, M. R.; Mohapatra, K.K.; Kanchan, R. S.; Gopakumar, K., "A dual two-level inverter scheme with common mode voltage elimination for an induction motor drive," Power Electronics, IEEE Transactions on , vol.19, no.3, pp.794,805, May 2004

[8] Mondal, G.; Gopakumar, K.; Tekwani, P. N.; Levi, E., "A five-level inverter scheme with common-mode voltage elimination by cascading conventional two-level and three-level NPC inverters for an induction motor drive," Power Electronics and Applications, 2007 European Conference on , vol., no., pp.1,10, 2-5 Sept. 2007

[9] Mondal, G.; Gopakumar, K.; Tekwani, P. N.; Levi, E., "A Reduced-Switch-Count Five-Level Inverter With Common-Mode Voltage Elimination for an Open-End Winding Induction Motor Drive," Industrial Electronics, IEEE Transactions on , vol.54, no.4, pp.2344,2351, Aug. 2007

[10] Somasekhar, V.T.; Srinivas, S.; Kumar, K.K., "Effect of Zero-Vector Placement in a Dual-Inverter Fed Open-End

Winding Induction-Motor Drive With a Decoupled Space-Vector PWM Strategy," Industrial Electronics, IEEE Transactions on , vol.55, no.6, pp.2497,2505, June 2008

[11] Mondal, G.; Sivakumar, K.; Ramchand, R.; Gopakumar, K.; Levi, E., "A Dual Seven-Level Inverter Supply for an Open-End Winding Induction Motor Drive," Industrial Electronics, IEEE Transactions on , vol.56, no.5, pp.1665,1673, May 2009

[12] Figarado, S.; Sivakumar, K.; Ramchand, R.; Das, A.; Patel, C.; Gopakumar, K., "Five-level inverter scheme for an open-end winding induction machine with less number of switches," Power Electronics, IET , vol.3, no.4, pp.637,647, July 2010

[13] Ebrahimi, J.; Babaei, E.; Gharehpetian, G.B., "A New Multilevel Converter Topology With Reduced Number of Power Electronic Components," Industrial Electronics, IEEE Transactions on , vol.59, no.2, pp.655,667, Feb. 2012

Variable Carrier Frequency Mixed PWM Technique Based on Current Ripple Prediction for Reduced Switching Loss

Hajime Kubo and Yasuhiro Yamamoto
Powertronics Research Department
MEIDENSHA CORPORATION
Numazu, Japan
kubo-ha@mb.meidensha.co.jp

Abstract—Since carrier based pulse width modulation (PWM) has several advantages such as applicability for simple V/f control and equal distribution of stress among all switching devices, it is still widely used in the industrial drives. Switching loss can be reduced by varying the carrier frequency based on current prediction, in such a way that the current ripple always meets the exact requirements. There are two commonly used PWM methods, discontinuous PWM (DPWM) which clamps one of the phases and switches the other two and space vector PWM (SVPWM) which switches all the three during the switching period. Since the locus of the current ripple differs by applying different PWM methods, less switching loss can be obtained by selecting optimal PWM in each control period. A method using combination of DPWM and SVPWM with variable carrier frequency in order to achieve lower switching loss is proposed in this paper. It is shown in the presented simulation results that the proposed method has lower switching frequency compared with those of DPWM and SVPWM under certain modulation index.

Keywords—variable carrier frequency, current ripple prediction, mixed pulse width modulation

I. INTRODUCTION

Voltage source inverter (VSI) controlled by pulse width modulated (PWM) signal has harmonics in the output voltage. The voltage harmonics cause ripple in the load current, which will result in higher copper loss and iron loss in the load machine. Therefore, it is preferable for the machine to increase the switching frequency and reduce the current ripple. However, increase in switching frequency causes higher switching loss in the power semiconductor devices of the VSI reducing efficiency of the system. Since there is trade-off relation between the switching loss and the current ripple, it is required to minimize switching frequency under the limitation of maximum current ripple.

The peak value of the current ripple varies when the carrier frequency is fixed to a constant value. Switching loss can be reduced by varying the carrier frequency based on current ripple prediction in such a way that the current ripple always meets the requirements exactly [1], [2].

Another way to reduce switching frequency while controlling the current ripple within the limitation is using predictive current control (PCC) technique [3]. Every

time the current reaches the boundary, it changes in the direction to retain specific switching state for the maximum duration. The advantage of PCC is fast response in current control. On the other hand, variable carrier frequency PWM has the advantage of the applicability for simple V/f control. In addition, the switching losses are distributed equally among the devices because each device in the six arms has the same switching frequency in carrier based PWM. From the aspect of packaging and thermal design, carrier based PWM is superior due to the full utilization of thermal capacitance in each device.

There are many PWM schemes which have different distribution methods of zero vectors [4], [5]. Several PWM methods which are modified according to the reference voltage (modulation index) and electrical angle have been developed for the purpose of the lowering the torque ripple, reducing the total harmonic distortion and cutting down the switching loss [6]–[10]. Two commonly used PWM methods are discontinuous PWM (DPWM) which clamps one phase and switches the other two phases, and space vector PWM (SVPWM) which switches three phases in such a way that the time span of zero vectors (111 and 000) become equal in a carrier period. Each PWM method has different loci of the current ripple on the two dimensional rotating orthogonal coordinates, I_d-I_q [6]. Therefore, when the carrier frequency is varied to limit the current ripple within the limitation, the switching frequency can be decreased by choosing the PWM method which has optimal locus of the current ripple under certain modulation index and electrical angle. In this paper, combination of DPWM and SVPWM (mixed PWM) is proposed to take best out of the two techniques.

The purpose of this paper is to reduce the switching frequency of devices by alternating between DPWM and SVPWM according to the predicted switching frequency during each switching period. In Section II, the current ripple prediction method for DPWM and SVPWM is introduced. Then the control strategy of carrier frequency based on this current ripple prediction method is proposed. And the implementation of mixed DPWM and SVPWM for reduced switching frequency is illustrated. Simulation results are presented in Section III to verify the effectiveness of the proposed method in reducing the switching frequency. Finally the conclusion is made in

Section IV.

II. METHODOLOGY

The procedures of generating mixed PWM is as below. First, the loci of the current ripple in half period of the carrier wave are predicted from the voltage reference of DPWM and SVPWM respectively. Next, the carrier frequencies of DPWM and SVPWM are decided to fit current ripple within the specified limit. Then the predicted switching frequencies of DPWM and SVPWM are compared, and the PWM scheme with lower switching frequency is chosen for the next control period. Finally, the pulses are generated from the determined PWM which results in lower switching frequency. In this research, the carrier wave is a triangular wave. The reference voltage and carrier frequency are updated at the beginning of maximum and minimum of this triangular carrier wave, which means the control period is half of the carrier period.

The pulses of DPWM and SVPWM are generated by comparing reference voltage to the triangular carrier wave. Zero sequence signals are added to the reference voltage of sinusoidal PWM (SPWM) to obtain the same pulse patterns as DPWM and SVPWM which are implemented using instantaneous SVM [6]. When the reference voltage of the SPWM is V_{ref}, the reference voltage of DPWM $V_{ref,D}$ is obtained by following equations.

$$V_m = [\frac{Vdc}{2}c^T - V_{ref}^T, -\frac{Vdc}{2}c^T - V_{ref}^T]^T \quad (1)$$

$$x = \text{argmin abs}(V_m(x)), \quad (x = 1, 2, \cdots, 6) \quad (2)$$

$$V_{ref,D} = V_{ref} + V_m(x)c. \quad (3)$$

Where, $c = [1, 1, 1]^T$. The reference voltage of SVPWM $V_{ref,SV}$ is obtained as below.

$$V_m = -\frac{\max(V_{ref}) + \min(V_{ref})}{2} \quad (4)$$

$$V_{ref,SV} = V_{ref} + V_m c. \quad (5)$$

Then, the current ripples of the both PWMs are predicted. A Three-phase two level inverter has eight states of switching. Each output voltage vector of these eight states, V_i, $(i = 0, \cdots, 7)$, are shown in the center column of Fig.1. The time span T_i of output voltage V_i is calculated from the comparison between reference voltage and carrier wave as shown in the left column of Fig. 1. The amount of the current change is obtained by multiplying this time span to the time-derivative of the current. The derivative of the current is given by (6).

$$\frac{\mathrm{d}I}{\mathrm{d}t} = \begin{pmatrix} \frac{1}{L_d} & 0 \\ 0 & \frac{1}{L_q} \end{pmatrix} (V - V_{emf}). \quad (6)$$

Where, V is the output voltage of the inverter and V_{emf} is back electromagnetic force (EMF). In calculation, EMF is approximated by reference voltage as $V_{emf} \approx V_{ref}$. Then, the current change vector becomes $T_i\frac{\mathrm{d}I_i}{\mathrm{d}t}$ during the control period while the inverter outputs V_i. The

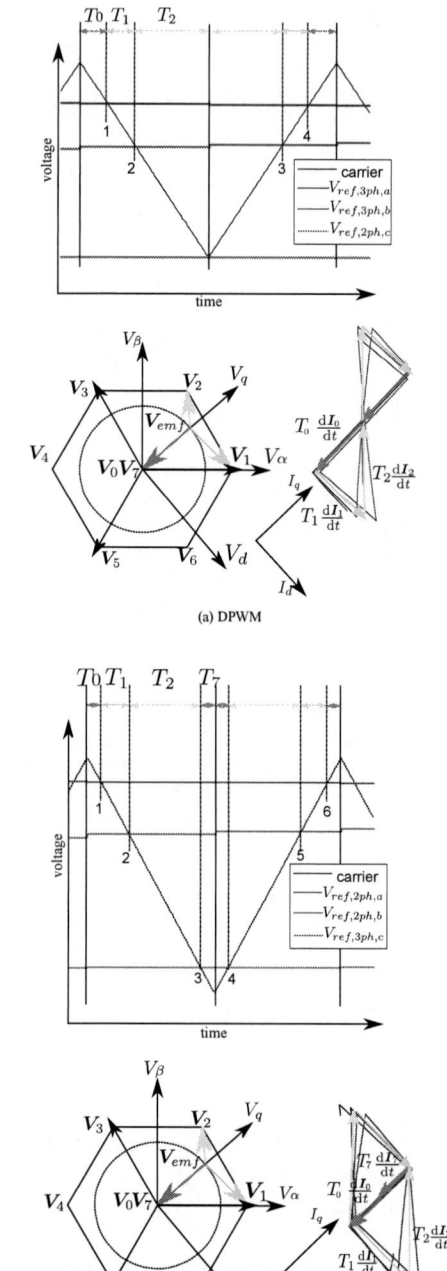

Fig. 1. Reference voltage and triangular carrier wave in one carrier period (left). Voltage vectors on two dimensional fixed orthogonal coordinates α-β (center). Loci of the current ripple on rotating axes d-q (right).

trajectory of the current ripple can be predicted by adding this current change vector sequentially as shown in the right column of Fig. 1.

Based on the predicted magnitude of the current ripple, the carrier frequencies of the two PWM schemes

$$f_{sw,SV} = \frac{6}{T_{c,SV}} = 6f_{c,SV} \qquad (9)$$

Where, $f_{c,D}$, $T_{c,D}$ and $f_{c,SV}$, $T_{c,SV}$ are the updated carrier frequency and period of DPWM and SVPWM, respectively. In mixed PWM, $f_{sw,D}$ and $f_{sw,SV}$ are compared, and the PWM strategy with lower switching frequency is chosen for the next control period.

III. SIMULATION

In the simulation study, the switching frequencies of three variable frequency carrier based PWMs, DPWM, SVPWM and mixed PWM are compared at several modulation indices under the condition of fixed current ripple limit. The definition of modulation index MI is given by (10).

$$MI = \frac{\sqrt{2}V_s}{V_{dc}} \qquad (10)$$

$$V_s = \sqrt{V_\alpha^2 + V_\beta^2}. \qquad (11)$$

Therefore, linear modulation range of PWMs are from 0 to 1. Noted that one switching action means the state change of one leg, and the basic fixed carrier based SPWM has six switching actions in one carrier period. The switching frequency is given by the analogy of the carrier frequency of this SPWM as (12).

$$f_s = \frac{N_{sw,sim}}{6T_{sim}} \qquad (12)$$

T_{sim} is the time span of the simulation and $N_{sw,sim}$ is the total switching counts of the all three legs during T_{sim}.

Simulation is conducted on permanent magnet synchronous machine (PMSM) drive model. The block diagram of the model is shown in Fig.3. The data of steady state for the time span $T_{sim} = 0.5$ is used for the evaluation.

Fig. 4 shows the comparison of the switching frequency among DPWM, SVPWM and mixed PWM at different modulation indices. In low modulation index area, SVPWM has lower switching frequency than DPWM and the same frequency as mixed PWM. This means that no mixture of PWMs occurs in mixed PWM under low modulation index because the switching frequency of SVPWM is always lower than that of DPWM. Figures in the upper row of Fig.5 are (a) reference voltage and chosen PWM, (b) current ripple locus of DPWM and (c) current ripple locus of SVPWM at a low modulation index ($MI = 0.19$). The current ripple locus in one carrier period of DPWM has the same length as that in half carrier period of SVPWM in each of the three directions. In this same time span, DPWM has four switchings while SVPWM has three. Therefore, SVPWM is chosen for this particular interval.

As shown in Fig. 4, mixed PWM has lower switching frequency than DPWM and SVPWM in the region $MI > 0.6$. Therefore the mixture of DPWM and SVPWM occurs in this modulation index range. Figures in the middle and bottom row of Fig.5 are (d)(g) reference voltage and chosen PWM, (e)(h) current ripple locus of DPWM

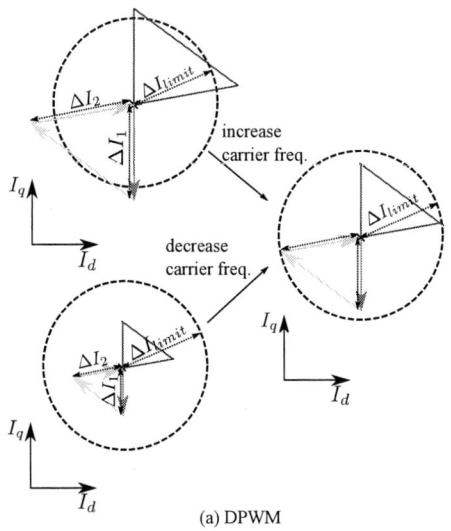

(a) DPWM

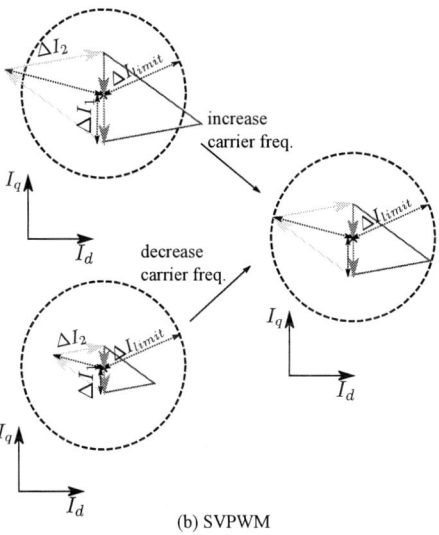

(b) SVPWM

Fig. 2. Locus of current ripple and circle whose radius is the required limitation of the current ripple.

are updated according to equation (7).

$$f_c = f_{c,base} \times \frac{\max(\Delta I_k)}{\Delta I_{limit}} \qquad (7)$$

Were, ΔI_k is the magnitude of current at the kth switching point, $f_{c,base}$ is the initial carrier frequency which is used in current prediction and ΔI_{limit} is the required limit of the current ripple. Since the magnitude of current ripple is proportional to the carrier period, the updated carrier frequency will guarantee the maximum magnitude of the current ripple fitting within the limitation.

Because DPWM has 4 switchings and SVPWM has 6 in one carrier period, the switching frequency of DPWM and SVPWM can be represented as (8) and (9).

$$f_{sw,D} = \frac{4}{T_{c,D}} = 4f_{c,D} \qquad (8)$$

The 2014 International Power Electronics Conference

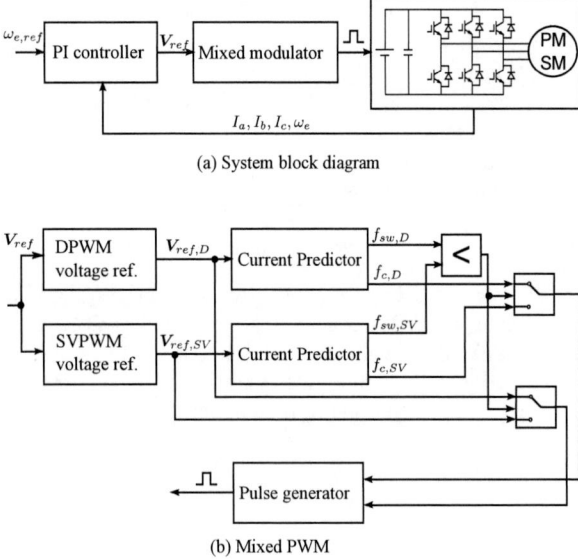

(a) System block diagram

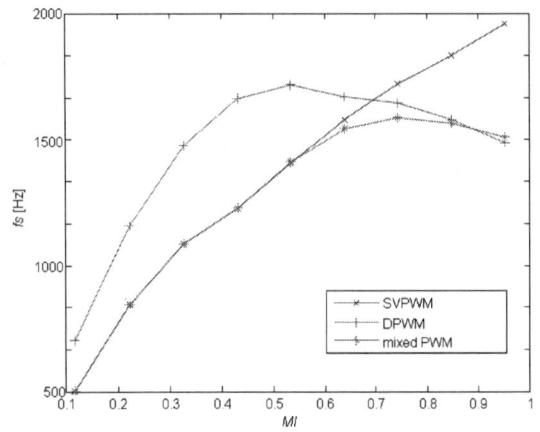

(b) Mixed PWM

Fig. 3. (a) Block diagram of PMSM drive model. (b) Detail of mixed modulator block.

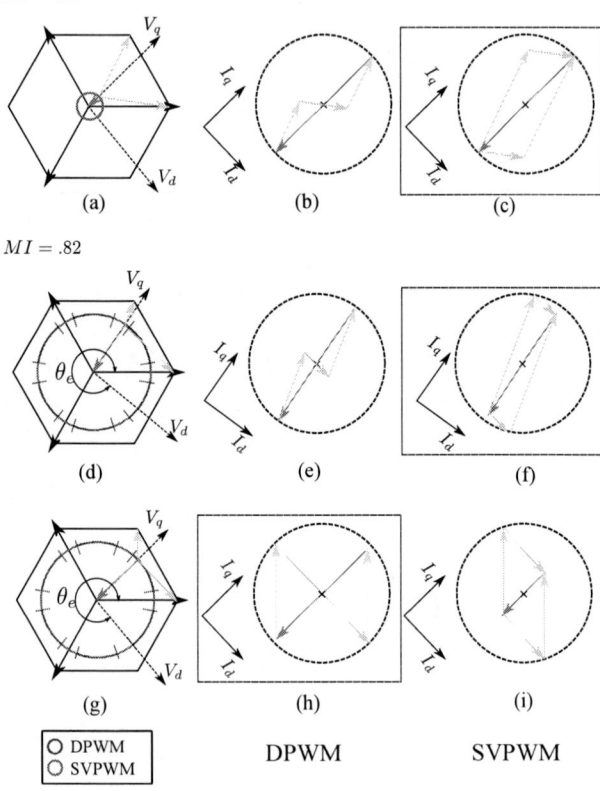

$MI = .19$

(a)   (b)   (c)

$MI = .82$

(d)   (e)   (f)

(g)   (h)   (i)

○ DPWM
○ SVPWM   DPWM   SVPWM

Fig. 5. (a)(d)(g) Amplitude of reference voltage and chosen PWM. (b)(e)(h) Current ripple locus of DPWM. (c)(f)(i) Current ripple locus of SVPWM.

Fig. 4. Comparison of the switching frequency among DPWM, SVPWM and mixed PWM at different modulation indices

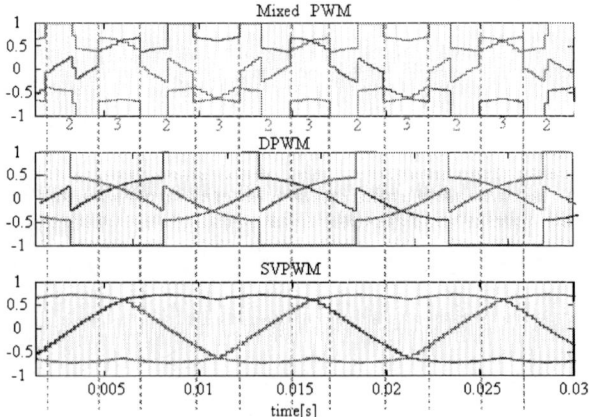

Fig. 6. Reference voltages of mixed PWM, DPWM and SVPWM and their triangular carrier waves.

and (f)(i) current ripple locus of SVPWM at a high modulation index ($MI = 0.82$). As shown in Fig.5(d)(g), DPWM and SVPWM are alternating according to the electrical angle. The middle row of Fig.5 is an example electrical angle in which SVPWM is chosen and the bottom row of Fig. 5 is that in which DPWM is chosen.

Fig. 6 shows the waveforms of reference voltage of mixed PWM, DPWM and SVPWM on their triangular carrier waves. The carrier frequencies of all the three PWM schemes are varying. Since PWM is chosen according to the comparison between $4f_{c,D}$ and $6f_{c,SV}$ in (8) (9), the carrier wave of mixed PWM has higher frequency in the area numbered 2 (DPWM is chosen) than in the area numbered 3 (SVPWM is chosen).

This simulation results shows that switching frequency can be reduced with mixed PWM in range of the modulation index from 0.6 to 0.9.

IV. CONCLUSION

This paper aims at reducing the switching frequency of carrier based PWM. A mixed PWM method is proposed, in which the carrier frequency is varied to fit the current ripple within the required limitation and PWM method is chosen between DPWM and SVPWM at every half cycle of triangular carrier wave. The control of carrier frequency and the determination of PWM in different

978-1-4799-2706-7/14 $31.00 © 2014 IEEE   1604

intervals are executed based on the prediction of the current ripple locus.

In simulation, the switching frequency of DPWM, SVPWM and mixed PWM are evaluated under several modulation indices. All three PWM schemes have variable carrier frequency and fixed current ripple limitation. The simulation results can be concluded as follows:

1) In low modulation index region ($MI < 0.5$), the switching frequency of DPWM is always lower than that of SVPWM. Therefore, the mixture of PWM does not occur in mixed PWM.

2) In high modulation index region ($0.5 < MI < 0.9$), magnitude relation of switching frequency between DPWM and SVPWM changes according to the electrical angle. In this region, the switching frequency of mixed PWM is the lowest among DPWM and SVPWM and mixed PWM because PWM scheme with lowest switching frequency is always chosen based on current ripple prediction.

Result 2) shows the effectiveness of the proposed method in switching frequency reduction.

REFERENCES

[1] X. Mao, R. Ayyanar, and H. Krishnamurthy, "Optimal variable switching frequency scheme for reducing switching loss in single-phase inverters based on time-domain ripple analysis," *Power Electronics, IEEE Transactions on*, vol. 24, no. 4, pp. 991–1001, 2009.

[2] D. Jiang and F. Wang, "Variable switching frequency pwm for three-phase converters based on current ripple prediction," *Power Electronics, IEEE Transactions on*, vol. 28, no. 11, pp. 4951–4961, 2013.

[3] J. Holtz and S. Stadtfeld, "A predictive controller for the stator current vector of ac machines fed from a switched voltage source," in *Conf. Rec. of IPEC-Tokyo*, 1983, pp. 1665–1675.

[4] J. Holtz, "Pulsewidth modulation for electronic power conversion," *Proceedings of the IEEE*, vol. 82, no. 8, pp. 1194–1214, 1994.

[5] A. Hava, R. Kerkman, and T. Lipo, "Simple analytical and graphical methods for carrier-based pwm-vsi drives," *Power Electronics, IEEE Transactions on*, vol. 14, no. 1, pp. 49–61, 1999.

[6] V. Blasko, "Analysis of a hybrid pwm based on modified space-vector and triangle-comparison methods," *Industry Applications, IEEE Transactions on*, vol. 33, no. 3, pp. 756–764, 1997.

[7] G. Narayanan, V. T. Ranganathan, D. Zhao, H. Krishnamurthy, and R. Ayyanar, "Space vector based hybrid pwm techniques for reduced current ripple," *Industrial Electronics, IEEE Transactions on*, vol. 55, no. 4, pp. 1614–1627, 2008.

[8] N. Reddy, T. Reddy, J. Amarnath, and D. SubbaRayudu, "Space vector based minimum switching loss pwm algorithms for vector controlled induction motor drives," in *Power Electronics, Drives and Energy Systems (PEDES) 2010 Power India, 2010 Joint International Conference on*, 2010, pp. 1–6.

[9] D. Zhao, V. S. S. P. K. Hari, G. Narayanan, and R. Ayyanar, "Space-vector-based hybrid pulsewidth modulation techniques for reduced harmonic distortion and switching loss," *Power Electronics, IEEE Transactions on*, vol. 25, no. 3, pp. 760–774, 2010.

[10] M. Reddy and V. Jegathesan, "Open loop v/f control of induction motor based on hybrid pwm with reduced torque ripple," in *Emerging Trends in Electrical and Computer Technology (ICE-TECT), 2011 International Conference on*, 2011, pp. 331–336.

Sliding Mode PWM for Effective Current Control in Switched Reluctance Machine Drives

Iakovos Manolas*, Georgios Papafotiou[†]

ABB AB
*Corporate Research / [†] Drives and Controls
Västerås, Sweden
iakovos.manolas@se.abb.com

Stefanos N. Manias

National Technical University of Athens
Department of Electrical & Computer Engineering
Athens, Greece
manias@central.ntua.gr

Abstract— **In this paper the problem of effective current control in Switched Reluctance Machine (SRM) drives is addressed. A sliding mode pulse width modulator (SMPWM) is introduced to govern the inherently highly nonlinear system dynamics. The proposed SRM drive is analyzed theoretically and subsequently validated through simulation. Satisfactory dynamic performance and precise speed control with reduced torque ripple, uniform power electronic device utilization, and approximately constant switching frequency is obtained. The feasibility of the above control scheme is validated by experimental measurements on a prototype 6/4 SRM drive.**

Keywords— *Current Control, Robust Control, Switched Reluctance Motor Drives, Variable Structure Systems.*

I. INTRODUCTION

The ongoing advances in power electronics and control systems have elevated the Switched Reluctance Machines (SRMs) from obscurity and at present SRM drives offer an interesting alternative for variable speed applications. Given the numerous advantages of the SRMs, extensive research is being carried out to overcome their main disadvantages. The key rests in the precise application of the appropriate phase current waveforms and thus, the development of an effective and robust current control loop is a mandatory prerequisite. Hopefully, this will bring SRM drives closer to expanding their prospects in commercial applications.

Naturally, the SRM is a nonlinear system with characteristics that vary significantly versus time under normal operational conditions. Controlling the current effectively in such a nonlinear system with parameters that depend on rotor angular position, rotational speed, and phase currents is not a simple task. Many researchers in the literature have addressed the problem and have proposed improved current control techniques [1], [2], [11]-[14], [16], [21], [24]-[25]. However, the overwhelming majority of them is a plain extension of the well-known techniques developed for Linear Time – Invariant (LTI) systems and hence, is not appropriate for the control of such nonlinear systems [3], [15], [22]. The convergence of the control algorithm and the controllability of the system are not guaranteed under every operating condition and the performance cannot be held as optimum. On the other hand, intelligent control techniques (e.g. artificial neural networks) present an interesting alternative but lack a solid and cohesive mathematic formulation that will ensure their stability and robustness.

One of the most common SRM current controllers found in the literature is the hysteresis type [11]. It is simple, easy to implement and has guaranteed stability. However, it comes with well-known drawbacks such as variable switching frequency and high ripple currents, making it undesirable for many applications. Moreover, in the modern digital implementations it is necessary to keep a high sampling frequency to control the current effectively, which means need for faster hardware with increased cost. The utilization of the well-known PI current controllers is another option for controlling the current in SRM drives [1], [2], [14]. However, this method has found limited acceptance due to the nonlinear characteristics of the plant. Satisfactory response and stable operation are difficult to achieve in such a time varying system over the full operating range. Many papers present improved methods with PI gain scheduling, back EMF decoupling [24] and hybrid controller structures in order to overcome the aforementioned drawbacks. In most cases, the result is significantly increased complexity and small practical gain. The modulation technique is another issue [1] and in most cases special PWM hardware such as timers and fast comparators are additionally required.

In this paper, a nonlinear and carrierless pulse width modulation technique for current control of SRMs is proposed. The control algorithm is based on sliding mode control theory and is inherently designed for digital implementation. Excellent dynamic response and robustness is obtained, with approximately constant switching frequency, and uniform power electronic device utilization. An experimental 6/4 SRM drive is developed, where the proposed digital control algorithm is implemented and tested.

II. PROPOSED CURRENT CONTROL TECHNIQUE

A. Modeling the 6/4 prototype SRM

In order to acquire an accurate model for the prototype

978-1-4799-2706-7/14 $31.00 © 2014 IEEE

6/4 SRM, an extensive and detailed analysis by means of finite elements was carried out. The FEMM and ANSYS software were used to solve the magnetostatic problem for various rotor positions and current levels. The acquired results for the flux linkage and electromagnetic torque are presented in Figs. 1 and 2, respectively. These characteristics constitute a detailed description (model) of the SRM, which is necessary for simulation purposes.

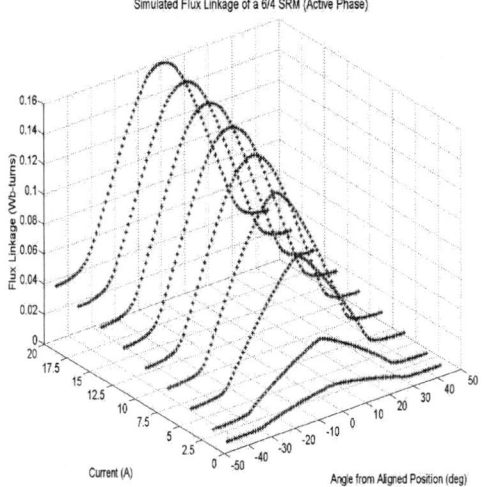

Fig. 1. Finite element analysis results for the flux linkage of the prototype 6/4 SRM.

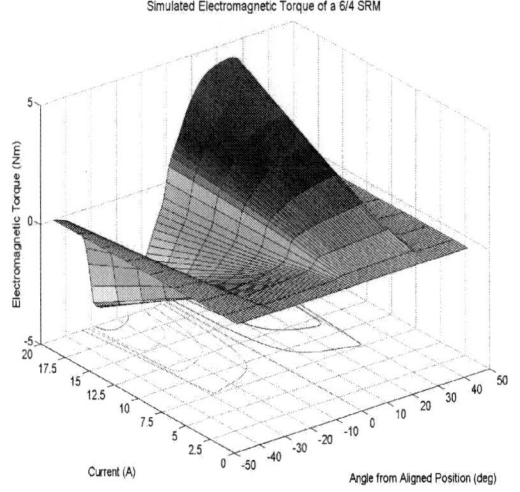

Fig. 2. Finite element analysis results for the produced electromagnetic torque of the prototype 6/4 SRM.

B. Nonlinear Current Control

Switched Reluctance Machines (SRMs) are doubly salient machines, in which torque is produced by the tendency of the movable part to move to a position, where the inductance of the excited winding is maximized [3]. As a result, unipolar phase currents are required to be switched on and off when the rotor is at precise positions, thus a power converter is necessary for their operation. The simplest and more common converter that can meet these requirements is the single

phase asymmetrical inverter. The topology and the four possible switching states are presented in Fig. 3.

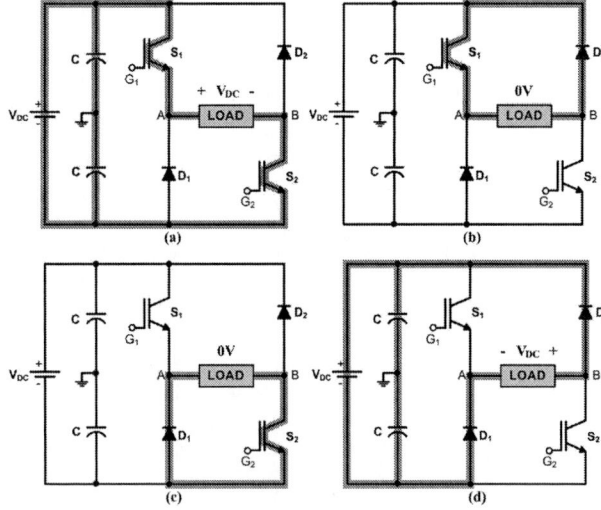

Fig. 3. Switching states of the single phase asymmetrical inverter.

By defining the variables:

$$u_A = \begin{cases} 1, when \quad S_1 = ON \\ -1, when \quad S_1 = OFF \end{cases} \quad u_B = \begin{cases} 1, when \quad S_2 = OFF \\ -1, when \quad S_2 = ON \end{cases} \quad (1)$$

the single phase asymmetrical inverter has the following dynamic model:

$$V_{AB} = \frac{V_{DC}}{2}(u_A - u_B) =$$

$$= \begin{cases} V_{DC} & , S_1 = ON, \ S_2 = ON \ \rightarrow Positive\ Voltage,\ Fig.\ 3(a). \\ 0 & , S_1 = ON, \ S_2 = OFF \ \rightarrow Freewheeling\ 1,\ Fig.\ 3(b). \\ 0 & , S_1 = OFF, \ S_2 = ON \ \rightarrow Freewheeling\ 2,\ Fig.\ 3(c). \\ -V_{DC} & , S_1 = OFF, \ S_2 = OFF \ \rightarrow Negative\ Voltage,\ Fig.\ 3(d). \end{cases} \quad (2)$$

The converter's load is the phase coil of the SRM, which has the following mathematical model:

$$V_m = R_m i_m + \frac{d\Phi_m(\theta, i_m)}{dt} = R_m i_m + L_m(\theta, i_m)\frac{di_m}{dt} + E_m(\theta, i_m, \omega)\ (3)$$

Fig. 4. Equivalent per phase circuit of the SRM as described in detail by (3).

The above equation can lead to the equivalent per phase circuit of the SRM. Evidently, this is a nonlinear model with parameters that are dependent on position, current, and speed. Combining (2) and (3) yields the full per phase mathematical model of the electrical part of the SRM drive, which is formulated in the state space as follows:

$$\frac{di_m}{dt} = \dot{i}_m = -\frac{R_m}{L_m}i_m - \frac{E_m}{L_m} + \frac{V_{DC}}{2L_m}\left(u_A - u_B\right) \quad (4)$$

The proposed current control technique, which is presented in Fig. 5, will have to exhibit a nonlinear pulse width modulation that will provide robust current control over the parameter variations, satisfactory dynamic performance, reduced torque ripple, approximately constant switching frequency, and uniformly distributed losses between the utilized semiconductor switches.

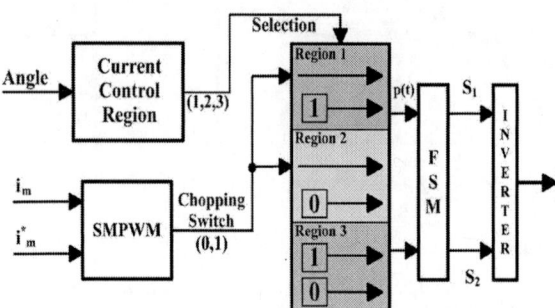

Fig. 5. Block diagram of the proposed current control scheme. FSM stands for Finite State Machine and is analyzed in the following.

By examining SRM's unipolar current control demands, we distinguish three operation regions: The current build-up region, where the phase is energized (Fig. 6-Region 1), the current rapid zero-out region (Fig. 6-Region 2), where the phase is turned off, and finally the phase idle region (Fig. 6- Region 3), where zero voltage is applied. In the first two regions current control can be achieved by soft-switching between positive/negative voltage and freewheeling, while in the third region only zero voltage needs to be applied (freewheeling). At this point it should be noted that there is no essential difference between the two freewheeling modes, excepting the switch utilization. Hence, if the reference current is zero both switches remain turned off and negative voltage is naturally applied through the conducting diodes D_1 and D_2, until the phase coil is fully discharged. Then the applied voltage becomes zero (freewheeling mode) until the phase is energized again. On the other hand, for positive current reference one switch is always on and the other is chopping (Fig. 4). In the latter case equation (4) is formed as follows (regardless of the switch that is always on − S_1 or S_2):

$$\dot{i}_m = -\frac{R_m}{L_m}i_m - \frac{E_m}{L_m} + \frac{V_{DC}}{2L_m}\left(1+u\right) =$$

$$= -\frac{R_m}{L_m}i_m - \frac{E_m}{L_m} + \frac{V_{DC}}{2L_m} + \frac{V_{DC}}{2L_m}u , \ u \in \{-1, +1\} \quad (5)$$

where u is the state variable of the chopping switch. By defining the current error:

$$s_m = i_m^* - i_m \quad (6)$$

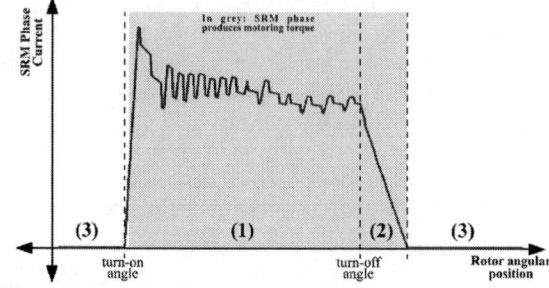

Fig. 6. Current control regions of the SRM.

equation (5) can be formed as follows:

$$\dot{s}_m = -\frac{R_m}{L_m}s_m + \frac{E_m}{L_m} - \frac{V_{DC}}{2L_m} + \frac{R_m}{L_m}i_m^* + \dot{i}_m^* - \frac{V_{DC}}{2L_m}u =$$

$$= -\frac{R_m}{L_m}s_m + \frac{E_m}{L_m} + F - \frac{V_{DC}}{2L_m}u \quad (7)$$

where F stands for the forced response. Considering equation (7) as a forced motion of the state variable s_m and by defining:

$$u = u_x + u_f \quad (8)$$

this motion can be separated into a free and a forced oscillation controlled by u_x and u_f, respectively. The condition for stabilizing both oscillations is:

$$\dot{s}_m s_m \leq 0 \quad (9)$$

If we choose the following variables:

$$u_x = K_x \left|\frac{E_m}{L_m}\right| sgn(s_m) \quad (10)$$

$$u_f = K_f \left|F_{max}\right| sgn(s_m) \quad (11)$$

$$F_{max} = max\left(F\right) \quad (12)$$

$$sgn(x) = \{1, x > 0; \ -1, x < 0\} \quad (13)$$

then the condition (9) for stabilization of the free and the forced oscillation yields:

Free Oscillation:

$$\dot{s}_m = -\frac{R_m}{L_m}s_m + \frac{E_m}{L_m} - \frac{V_{DC}}{2L_m}u_x \xrightarrow{s_m \dot{s}_m \leq 0} K_x \geq \frac{2L_m^{(max)}}{V_{DC}} \quad (14)$$

Forced Oscillation:

$$\dot{s}_m = F - \frac{V_{DC}}{2L_m}u_f \xrightarrow{s_m \dot{s}_m \leq 0} K_f \geq \frac{2L_m^{(max)}}{V_{DC}} \quad (15)$$

Thus, according to Sliding Mode theory the control variable:

$$u = \left(K_x \left|\frac{E_m}{L_m}\right| + K_f \left|F_{max}\right|\right) sgn(s_m) = K_m sgn(s_m) \quad (16)$$

with the conditions (14) and (15) forces the system's motion to slide on $s_m = 0$. The discrete control law u is then approximated by the continuous function:

$$u_m^* = K_m sat\left(\frac{s_m}{\varphi}\right) = \begin{cases} K_m & , when \ s_m \geq \varphi \\ \dfrac{K_m}{\varphi}s_m & , when \ s_m < |\varphi| \\ -K_m & , when \ s_m \leq -\varphi \end{cases} \quad (17)$$

978-1-4799-2706-7/14 $31.00 © 2014 IEEE

C. Sliding Mode PWM

The proposed PWM method for controlling the current given in (5) is the Sliding Mode PWM (SMPWM) technique, depicted in Fig. 7. According to this method the following switching function is selected:

$$\sigma_m = \int_0^t \left(u_m^* - G u_m \right) dt \qquad (18)$$

where

$$u_m = sgn\left(\sigma_m\right) = \begin{cases} +1, \sigma_m > 0 \\ -1, \sigma_m < 0 \end{cases} \qquad (19)$$

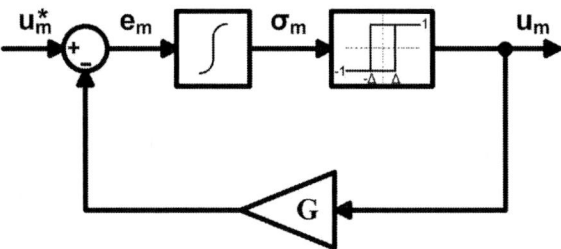

Fig. 7. Sliding Mode PWM block diagram.

Then in the ideal case:

$$\dot{\sigma}_m = u_m^* - G\, sgn\left(\sigma_m\right) \qquad (20)$$

and the demand for stabilization (10) yields the condition:

$$G > \left| u_m^* \right| \qquad (21)$$

Therefore, under the above selections the system's dynamics are governed by the function $\sigma_m = \dot{\sigma}_m = 0$ and the switching frequency is infinite. Actually, in Fig. 7, a more realistic and practical solution is indicated, by utilizing a delta modulator with hysteresis. In this case the sliding motion on $\sigma_m = 0$ is achieved only in average and the switching frequency is given by the formula:

$$f_s = \frac{G}{4\Delta} - \frac{u_m^{*2}}{4\Delta G} \qquad (22)$$

For $\Delta = 0.0001$ and $G=8$, f_s varies between 19678.5 and 20000Hz (i.e. beyond the audible range).
The signal m can then be used for switching purposes:

$$m = \frac{1}{2}\left(1 + u_m\right), \; with\; m \in \{0,1\} \qquad (23)$$

D. Finite State Machine

Finally to achieve uniformly distributed losses between the two switches (S1 and S2) the Finite State Machine (FSM) of Fig. 8 is proposed. The switching between the two states is governed by the function:

$$L = \{1,\; if\; p(t) > p(t-1)\; and\; 0,\; if\; p(t) \le p(t-1)\} \qquad (24)$$

where p(t) is the first input of the FSM (Fig. 5) and p(t-1) is the previous instance of p(t).

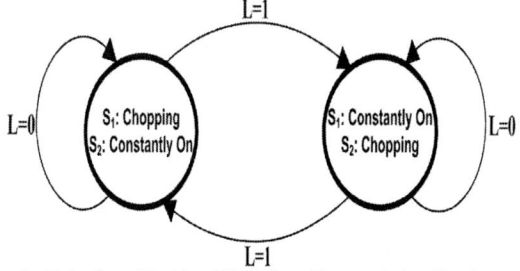

Fig. 8. Finite State Machine (FSM) for uniform switch utilization.

III. APPLICATION ON AN EXPERIMENTAL 6/4 SRM DRIVE

In order to exhibit the validity and verify the feasibility of the proposed current control technique, we choose to present its application on an SRM drive system. The goal is to analyze its performance and robustness under operating conditions and parameter variations that present close resemblance to the ones observed in a typical drive system.

A. Simulation Results

The SRM drive system in question undergoes a detailed computer aided simulation, using the MATLAB/SIMULINK software. The prototype 6/4 SRM is modeled accurately based on the finite element analysis data, depicted previously in Figs. 1 and 2. The drive system is required to follow the demanded speed changes for both directions of rotation, regardless of the load torque variations. In this work the turn-on and turn-off angles are predefined and kept constant for every operation point. Hence, they should not be considered as optimum; however, an optimization is possible [8]. The key simulated waveforms of the SRM drive are depicted in Figs. 9-12.

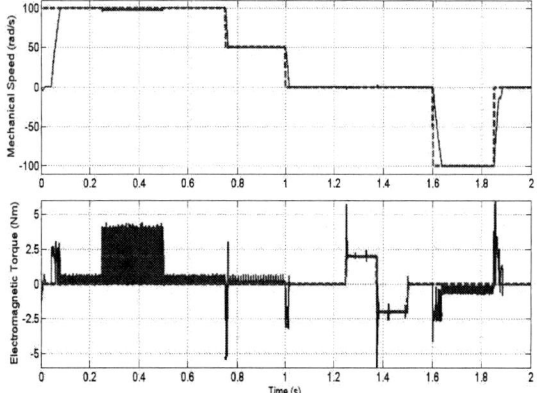

Fig. 9. Simulation results: Mechanical rotational speed and electromagnetic torque of the SRM (the dashed lines represent the speed reference and the load torque, respectively).

From the presented waveforms it may be remarked, that the proposed current control technique exhibits excellent performance under every operational condition. By focusing on the precedent figures, one can notice the nonlinear current profiling produced by the control algorithm, which is a necessity, if we consider the inherently nonlinear model of the SRM. As a result, the

978-1-4799-2706-7/14 $31.00 © 2014 IEEE

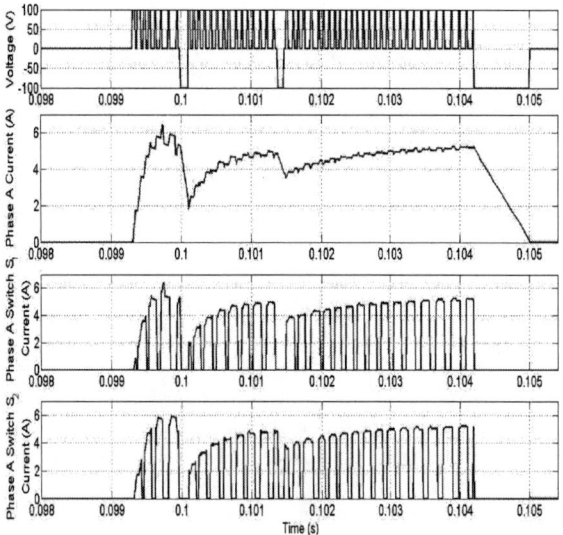

Fig. 10. Simulation results for 100rad/s mechanical rotational speed: Phase A voltage, phase A current, and current through the two IGBT switches (S_1 and S_2, respectively) of phase A (FSM - uniform switch utilization).

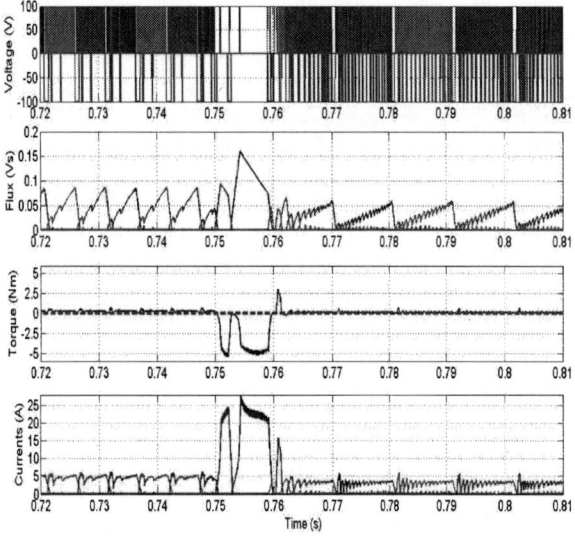

Fig. 11. Simulation results for a speed reference step from 100rad/s to 50rad/s at 0.75s. The SRM's three phase voltages, three phase fluxes, electromagnetic torque (with zero load torque - dashed), and three phase currents are presented in detail.

speed is controlled accurately and the dynamic response to the speed commands is very fast and with no overshoot or oscillations (Figs. 9 and 11). Moreover, load variations within the SRM's rating are successfully compensated, as presented in Fig. 12, and have little effect on the mechanical speed of the SRM. The robustness of the control algorithm as defined theoretically in (11) - (13) is maintained under all operating conditions and despite the SRM drive's parameter variations. Thus, under normal operation the robustness of the proposed technique is guaranteed by defining a proper upper bound for the varying parameters. However, the robustness is sensitive to variations of the supply voltage and therefore can be degraded under faults. Finally, by observing in detail the

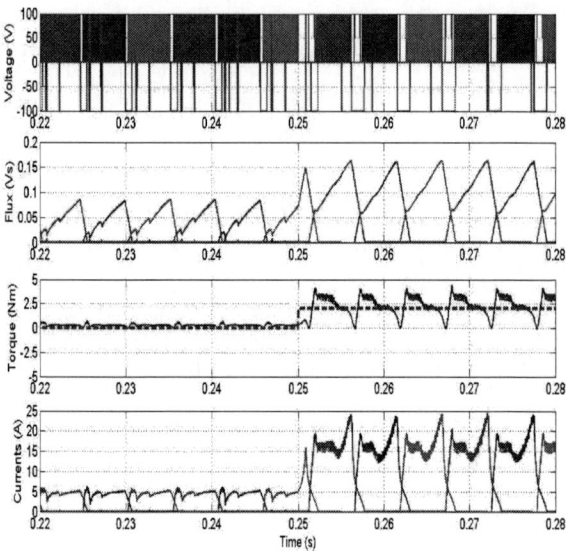

Fig. 12. Simulation results for a load torque step from 0Nm (no load) to 2Nm at 0.25s. The SRM's three phase voltages, three phase fluxes, electromagnetic torque (with the demanded load torque - dashed), and three phase currents are presented in detail.

currents of the two semiconductor switches (IGBTs) of SRM's phase A in Fig. 10, we can confirm the almost constant switching frequency (set in approximately 20kHz in this case) and the uniform utilization of the semiconductor switches, achieved via the use of the FSM. This last merit is very important since it offers uniformly distributed thermal losses and increased reliability.

B. Experimental 6/4 SRM Drive

In order to verify the validity and the feasibility of the proposed control algorithm an experimental 6/4 SRM drive has been developed and is depicted below in Fig. 13.

Fig. 13. Experimental setup of the 6/4 SRM drive.

The 6/4 SRM has been designed and manufactured by our research team and is, therefore, referred to as prototype 6/4 SRM. The basic design parameters and characteristics of the machine are summarized below in Table I.

The drive comprises three single phase asymmetrical

TABLE I
PROTOTYPE 6/4 SRM PROPERTIES

Property	Value
Number of stator poles	6
Stator pole arc	29°
Number of rotor poles	4
Rotor pole arc	30°
Stator inner radius	26.25mm
Air gap	0.25mm
Phase resistance	0.8Ω
Nominal power	0.6kW
Nominal mechanical speed	1000rpm
Nominal voltage	100V

inverters with IGBTs and diodes, the necessary driving circuits, measurement and feedback conditioning circuits. The control algorithm is implemented using a TMS320F2812 Digital Signal Processor on an eZdsp board. The feedbacks required for this control algorithm are the three phase currents and the rotor angular position of the SRM measured through current transducers and an incremental optical encoder, respectively.

The experimental measurements are presented in Figs. 14 and 15. The close resemblance of the measured currents in Fig. 14 to the simulation results verifies the successful modeling of the prototype 6/4 SRM based on detailed Finite Element Analysis data. Moreover, Fig. 15 presents the currents through switches S1 and S2 of phase A while the 6/4 SRM rotates at 100rad/s with no load. By carefully examining these waveforms, one can confirm the almost constant switching frequency (set in approximately 10kHz in this case due to hardware limitations) and the uniform utilization of the semiconductor switches, achieved via the use of the FSM. Therefore, the proposed SMPWM technique presents all the characteristics remarked before: effective and robust current control with approximately constant switching frequency and uniform power electronic device utilization.

IV. CONCLUSIONS

The proposed nonlinear current control technique, when compared to the most commonly used current control techniques for SRM drives (e.g. PI or hysteresis control), exhibits significantly improved performance. This novel digital control algorithm provides a carrierless pulse width modulation, which offers the necessary current profiling. The convergence of the algorithm to the given current reference is mathematically proven and its robustness over the parameter variations is guaranteed by (14) and (15). Moreover, the proposed technique exhibits excellent dynamic performance, approximately constant switching frequency, and uniformly distributed losses between the utilized semiconductor switches. The application in an experimental 6/4 SRM speed control drive verified all the above characteristics.

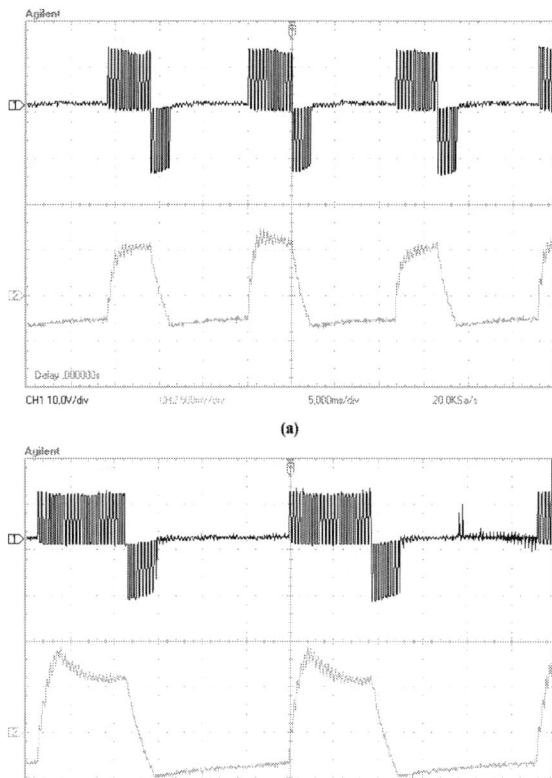

Fig. 14. Experimental results (a) Voltage and current of phase A of the 6/4 SRM while rotating with 100rad/s (no load), (b) Voltage and current of phase A of the 6/4 SRM while rotating with 50rad/s (no load). The voltage probes are set in (x10) and the current probes in 100mV/A.

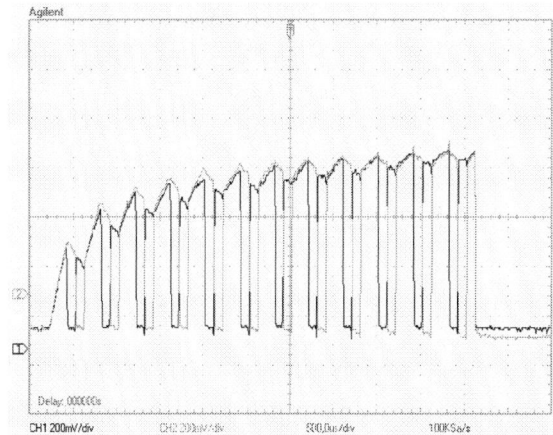

Fig. 15. Experimental results: Currents through switches S_1 and S_2 of phase A while the 6/4 SRM rotates at 100rad/s with no load. These waveforms depict the success of the FSM in achieving uniform switch utilization (current probes set in 100mV/A).

REFERENCES

[1] Blaabjerg F., Kjaer P.C, Rasmussen P.O., Cossar C., "Improved digital current control methods in switched reluctance motor drives", *IEEE Transactions on Power Electronics*, Vol. 14, Issue 3, pp.563-572, May 1999.

[2] Kjaer P.C, Gribble J.J., Miller T.J.E., "High grade control of switched reluctance machines", *IEEE Transactions on Industry Applications*, Vol. 33, Issue 6, pp.1585-1593, Nov.-Dec. 1997.

[3] Miller T.J.E., "Optimal design of switched reluctance motors", *IEEE Transactions on Industrial Electronics*, Vol. 49, Issue 1, pp.15-27, Feb. 2002.

[4] Kaletsanos A., Xepapas F., Xepapas S., Manias S.N., "Nonlinear control technique for three-phase boost AC/DC power converter", *IEEE 34th Annual Power Electronics Specialist Conference (PESC'03)*, Vol. 3, pp.1080-1085, 15-19 June 2003.

[5] Plekhanov S., Shkolnikov I.A., Shtessel Y.B., "High order sigma-delta modulator design via sliding mode control", *Proceedings of the 2003 American Control Conference*, Vol. 1, pp.897-902, 4-6 June 2003.

[6] Utkin V., Gulder J., Shi J., *Sliding Mode Control in Electromechanical Systems*, Ed. Taylor & Francis, 1999.

[7] Cheok A.D., Ertugrul N., "Use of fuzzy logic for modeling, estimation, and prediction in switched reluctance motor drives", *IEEE Transactions on Industrial Electronics*, Vol. 46, Issue 6, pp.1207-1224, Dec. 1999.

[8] Xue X.D., Cheng K.W.E., Ho S.L., Cheung N.C., "Investigation of the effects of the control parameters and outputs on power factor of switched reluctance motor drive systems", *IEEE 33rd Annual Power Electronics Specialist Conference (PESC'02)*, Vol. 3, pp.1469-1474, 23-27 June 2002.

[9] Miller T.J.E., Switched reluctance motors and their control, Magna Physics Pub., 1993.

[10] R. Krishnan, Switched Reluctance Motor Drives, CRC Press, 2004.

[11] Fiedler, J.O., Fuengwarodsakul, N.H., De Doncker, R.W., "Calculation of switching frequency in current hysteresis controlled switched reluctance drives", *IEEE 35th Annual Power Electronics Specialist Conference (PESC'04)*, Vol. 3, pp. 2270 - 2276, 20-25 June 2004.

[12] Lin, Z., Reay, D.S., Williams, B.W., He, X., "High performance current control for switched reluctance motors with on-line modeling", *IEEE 35th Annual Power Electronics Specialist Conference (PESC'04)*, Vol. 2, pp. 1246 - 1251, 20-25 June 2004.

[13] Manolas, I.S., Kaletsanos, A.X., Manias, S.N., "Nonlinear current control technique for high performance switched reluctance machine drives", *IEEE 39th Annual Power Electronics Specialist Conference (PESC'08)*, pp. 1229 - 1234, 15-19 June 2008.

[14] Schulz, S.E., Rahman, K.M., "High performance digital PI current regulator for EV switched reluctance motor drives", *37th Annual IAS Meeting*, Vol. 3, pp. 1617 - 1624, 13-18 Oct. 2002.

[15] Slotine J.J., Li W., *Applied Nonlinear Control*, Ed. Prentice Hall, 1991.

[16] Cheok, A.D., Fukuda, Y., "A new torque and flux control method for switched reluctance motor drives", *IEEE Transactions on Power Electronics*, Vol. 17, Issue 4, pp. 543 - 557, July 2002.

[17] Manolas I.St., Kladas A.G., Manias S.N., "Finite-Element-Based Estimator for High-Performance Switched Reluctance Machine Drives", *IEEE Transactions on Magnetics*, Vol. 45, Issue 3, pp. 1266-1269, March 2009.

[18] S. Paramasivam, S. Vijayan, M. Vasudevan, R. Arumugam, Ramu Krishnan, "Real-time verification of AI based rotor position estimation techniques for a 6/4 pole switched reluctance motor drive", *IEEE Transactions on Magnetics*, vol. 43, no. 7, pp. 3209-3222, July 2007.

[19] Gabriel Gallegos-Lopez, Philip C. Kjaer, T.J.E. Miller, "High-grade position estimation for SRM drives using flux linkage/current correction model", *IEEE Transactions on Industry Applications*, vol. 35, no. 4, pp. 859-869, July/August 1999.

[20] M. Krishnamurthy, B. Fahimi, Chris S. Edrington, "On the measurement of mutual inductance for a switched reluctance machine", *37th Annual IEEE Power Electronics Specialists Conference*, pp. 1-7, June 2006.

[21] Peyrl H., Papafotiou G., Morari M., "Model predictive torque control of a Switched Reluctance Motor", *IEEE International Conference on Industrial Technology ICIT 2009*, pp. 1-6, 10-13 February 2009.

[22] Chen J.H., Chau K.T., Jiang Q., Chan C.C., Jiang S.Z., "Modeling and analysis of chaotic behavior in switched reluctance motor drives", *IEEE 31st Annual Power Electronics Specialist Conference (PESC'00)*, Vol. 3, pp.1551-1556, 18-23 June 2000.

[23] Krishnamurthy M., Fahimi B., Edrington C.S., "Comparison of Various Converter Topologies for Bipolar Switched Reluctance Motor Drives", *IEEE 36th Annual Power Electronics Specialist Conference (PESC'05)*, pp.1858-1864, 16 June 2005.

[24] Lin, Z., Reay, D., Williams, B., He, X., "High-performance current control for switched reluctance motors based on on-line estimated parameters," *Electric Power Applications, IET*, vol.4, no.1, pp.67-74, January 2010.

[25] Xin Cao, Zhiquan Deng, Tian Yao, Jun Cai, Zheng Zhuang, "Analysis and Application of Phase Current in Switched Reluctance Generators", *IEEE Transactions on Applied Superconductivity*, vol.20, no.3, pp.1063-1067, June 2010.

978-1-4799-2706-7/14 $31.00 © 2014 IEEE

Experimental Verification of an EMC Filter Used for PWM Inverter with Wide Band-Gap Devices

Jun-ichi Itoh, Takahiro Araki and Koji Orikawa
Department of Electrical, Electronics and Information Engineering
Nagaoka University of Technology
Nagaoka, Niigata, Japan
itoh@vos.nagaokaut.ac.jp, arakit@stn.nagaokaut.ac.jp, orikawa@vos.nagaokaut.ac.jp

Abstract— This paper discusses a volume of an EMC filter and a cooling system that are used for a PWM inverter with wide band-gap devices. At first, the volume of reactor that is used for EMC filter such as common mode coke coils and differential mode choke coils are estimated by theoretically. Then, the relationship between the carrier frequency of the PWM inverter and the total volume of filter reactors are clarified by simulation. Moreover, the relationship between the carrier frequency and the volume of a cooling system is calculated based on experimental results. As a result, the total volume of the inverter system that contains filter reactors and cooling system will be reduced by 54% at the carrier frequency of 300 kHz by using a two stage filter compared to case of the carrier frequency of 150 kHz by using one stage filter. In addition, an induction motor is driven by a prototype of GaN-FET inverter system with a 300-kHz carrier frequency. As a result, the conduction noise is suppressed below the limit of CISPR. Therefore, the proposed design method for EMC filters is valid in the experiment. Furthermore, the power loss of EMC filter is less than 2% compared with the total loss of the GaN-FET inverter system.

Keywords— EMC filter; High-frequency switching; PWM inverter; Wide Band-Gap Devices

I. INTRODUCTION

Recently, the fast switching and low on-state voltage drop are required for power conversion circuits in order to reduce the power loss. However, silicon based switching devices such as Si-MOSFETs and Si-IGBTs, are difficult to achieve a significant performance improvement because those devices performance is almost reaching the limit that the derived from the physical properties of silicon. On the other hand, the switching devices based on a wide band-gap semiconductor such as gallium nitride (GaN) or silicon carbide (SiC) has been studied in recent years [1-3]. Those wide band-gap devices can perform fast-switching and features low on-voltage drop compared with the normal silicon devices under the high temperature operation.

The one of problems of the wide band-gap device is cost. Therefore, it is expected that the wide band-gap devices are used for general power converter. The power conversion circuit with wide band-gap devices has been studied to achieve high efficiency in a high temperature

operation[4-12]. However, it seems that the performance and the miniaturization effect are not discussed in previous studies.

Authors have been focused on the volume of an EMC filter for PWM inverter. The PWM inverter can control the output voltage and output frequency by using switching devices. However, the noise occurs at the switching because the voltage and current are change rapidly. Recently, this noise becomes lager because the fast switching devices such as wide band-gap devices are applied to the PWM inverter in order to reduce the switching loss. The noise may cause a false operation of surrounding control system. Hence, it is limited by some regulations such as CISPR (Special international committee on radio interference). In order to suppress the noise which is emitted from the PWM inverter, the EMC filter that is constructed by passive components such as inductor and capacitor is added to the input of the inverter system. In addition, the PWM inverter becomes smaller depending on the high performance switching device and development of cooling technique. Thus, the volume of the EMC filter must be considered in order to miniaturize the inverter system.

The volume of the EMC filter is determined by the attenuation rate and frequency of the noise. The noise that is emitted form a PWM inverter is varied by carrier frequency. Therefore, the EMC filter is miniaturized when the switching frequency is higher and higher. In contrast, the switching loss is increased by the high frequency switching. Consequently, the cooling system becomes larger. Therefore, the volume of the cooling system must be considered for minimizing the EMC filter under a high frequency switching operation. Although the EMC filter design method are reported at constant carrier frequency [13-14], it seems that the relation between the power converter volume and the carrier frequency has not been discussed concretely through the experiments in past works.

In this paper, the relationship between the carrier frequency and the total volume of the inverter system is discussed based on simulation and experiments using GaN-FET inverter. At first, the loss and the volume estimation method of the inverter system is mentioned.

Second, the design method of the multi stage EMC filter is clarified. Then, the relationship between the carrier frequency of the PWM inverter and the volume of filter reactor is clarified by simulation, besides the relationship between the carrier frequency and the volume of a cooling system. At last, a prototype of GaN-FET inverter is demonstrated by experiments using an R-L load and an induction motor. Power loss and conducted emission of the system are shown. As a result, it is confirmed that the total volume of GaN-FET inverter system can be reduced to 54% compared to the case of a single stage EMC filter by using a two stage EMC filter.

II. VOLUME ESTIMATION METHOD

A. Power Loss of Switching Devices

Fig. 1 shows the system configuration of the PWM inverter with GaN-FET. The switching loss of each device P_{SW} is calculated by (1) [15].

$$P_{SW} = \frac{V_{DC}I_m}{4\pi V_{DCd}I_{md}}\left(e_{on} + e_{off}\right)f_{carrier} \quad \dots\dots\dots (1)$$

where $f_{carrier}$ is the carrier frequency, V_{DC} is DC link voltage, I_m is the maximum output current, e_{on} and e_{off} are turn on and turn off energy of each switching, and V_{DCd} and I_{md} are voltage and current that described on datasheet for measuring the switching time.

Table1 shows the simulation parameters that are used for design of the EMC filter and cooling system.

The total loss generated in the switching devices P_{loss} which be composed of the conduction loss and the switching loss, is calculated by (2).

$$P_{loss} = 6\left(P_{SW} + P_{CON}\right) = 6\left(P_{SW} + \frac{I_m^2}{2}R_{ON}\right) \quad \dots\dots\dots (2)$$

where R_{ON} is the on-state resistance of the FET.

B. Volume of Cooling Systems

The PWM inverter needs cooling system such as heat sinks and fans because the switching devices are heated by the switching loss and the conduction loss. Generally, cooling system is designed based on a thermal resistance. However, the thermal resistance depends on its volume. Therefore, CSPI (Cooling System Performance Index) is introduced to estimate the volume of cooling system. The CSPI indicates the cooling performance per unit volume of the cooling system which is a reciprocal of the product of the volume and the thermal resistance. It means that a high performance cooling system shows high CSPI. Therefore, the cooling system is miniaturized when CSPI become higher. The volume of the cooling system $vol_{cooling}$ is given by (3) [16].

$$vol_{cooling} = \frac{1}{R_{th} \times CSPI} = \frac{P_{loss}}{\left(T_j - T_a\right) \times CSPI} \quad \dots\dots\dots (3)$$

where R_{th} is the thermal resistance of the cooling system, T_j is the junction temperature of the switching device, and T_a is the ambient temperature.

C. Design method of multi stage EMC filter

Fig. 2 shows the circuit schematics of multi stage EMC

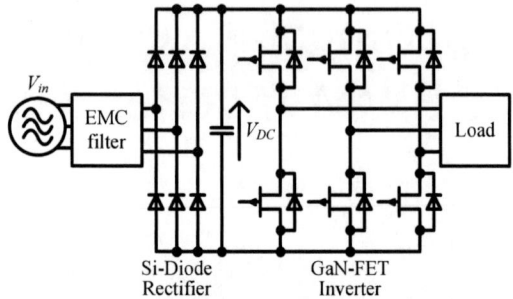

Fig. 1. Circuit configuration of three phase PWM inverter that is constructed by GaN-FET.

Table 1. Simulation parameters that is used for designing the EMC filter and cooling system.

Input voltage V_{in}	200 V
Input frequency f_{in}	50 Hz
DC voltage V_{dc}	282 V
Output voltage V_{out}	173 V
Output current I_{out}	1.9 A
Output frequency f_{out}	50 Hz
Load impedance Z_{load}	53 Ω
Power factor $\cos\phi$	0.99
Modulation ratio α	1
CSPI	3
On-resistance R_{ON}	100 mΩ
Dead time T_d	100 ns
Ambient temperature T_a	20 ℃
Junction temperature T_j	100 ℃
Load factor k	0.1
Lead angle ϕ	10π/180 rad
Leakage current I_{leak}	1 mA

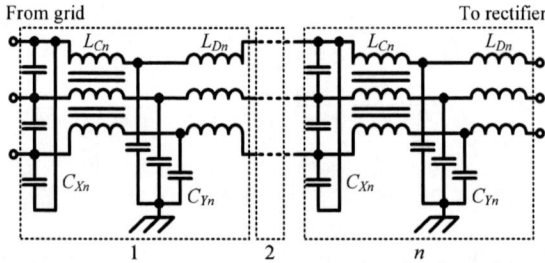

Fig. 2. Circuit schematics of multi stage EMC filter.

filter. In this paper, the multi stage EMC filter is constructed by connecting the single stage EMC filter in series. Each filter that has different number of stages is designed to get a same attenuation. In addition, the capacitance of X capacitors and Y capacitors are divided equally for all stages.

At first, the X capacitor that suppresses the fluctuations of the input voltage is designed by (4) using an allowable lead angle of input current ϕ because it reduces the power factor at the light load.

$$C_{Xn} = \frac{\sqrt{3}kI_{in}\sqrt{\left(1 - \cos\phi^2\right)}}{n\omega V_{in}\cos\phi} \quad \dots\dots\dots\dots (4)$$

where k is the load factor (output power/maximum

978-1-4799-2706-7/14 $31.00 © 2014 IEEE

power), I_{in} is the input current, n is number of filter stage, ω is the input frequency, and V_{in} is the input voltage.

If the lead angle of the input current ϕ is small, (4) is simplified as (5).

$$C_{Xn} = \frac{\sqrt{3}kI_{in}\phi}{n\omega V_{in}} \quad\text{...} (5)$$

Second, the Y capacitor that bypass the common mode current is designed by (6) based on the acceptable leakage current I_{leak}.

$$C_{Yn} = \frac{\sqrt{3}I_{leak}}{n\omega V_{in}} \quad\text{...} (6)$$

Finally, the inductance of differential mode choke coils and common mode choke coils L_n are designed by (7) in order to suppress the conduction noise below the limit of CISPR. Also, the common mode noise and the differential mode noise are separated in order to evaluate the each noise [17].

$$L_n = \frac{1}{\omega_{Att}^2 C_n \left(G_0 - G_f\right)^{\frac{1}{n}}} \quad\text{.................................} (7)$$

where ω_{Att} is the designed frequency of the LC filter, G_0 is the peak value of the conduction noise without the EMC filter, and G_f is the limit of CISPR.

D. Volume of EMC Filter

According to (5) and (6), the capacitance of X capacitor and Y capacitor is not changed by the carrier frequency. Also, the total capacitance is same regardless of the number of EMC filter stages. Therefore, the volume of those capacitors is not considered.

On the other hands, the reactor volume is changed significantly by the parameter of the component. There are several ways to select the core for the reactor. In this paper, the reactors are designed by the Area Product concept [18] using a window area and a cross-sectional area. Therefore, the volume of the reactor vol_L is given by (8).

$$vol_L = K_v \left(\frac{2W}{K_u B_m J}\right)^{\frac{3}{4}} \quad\text{...} (8)$$

where K_v is the constant value depending on the shape of cores, W is the maximum energy of the reactor, K_u is the occupancy of the window, B_m is the maximum flux density of the core, and J is the current density of the wire.

III. VOLUME EVALUATION OF INVERTER

A. Simulation conditions

Fig. 3 shows the conduction noise evaluation system. The capacitors are added between the switching devices and FG, and the output voltage midpoint and FG in order to model the stray capacitances of a general inverter [14]. In this paper, modeled LISN and a spectrum analyzer is used to estimate the conduction noise by the simulation [19].

B. Experiomntal conditions

Table 2 shows the circuit parameters on experiment.

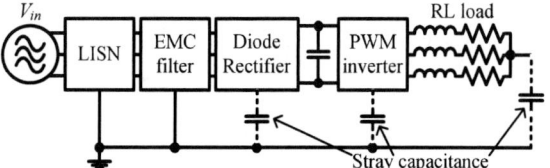

Fig. 3. Conduction noise evaluation system used in simulation.

Table 2. Experimental conditions to measure the power loss of GaN-FET inverter with RL load.

Input voltage V_{in}	100 V
Input frequency f_{in}	50 Hz
DC link voltage V_{DC}	140 V
Modulation ratio α	1
Output frequency f_{out}	20 Hz
Load impedance Z_{load}	51 Ω
Power factor $\cos\phi$	0.99
Ambient temperature T_a	25 ℃
Dead time T_d	100 ns

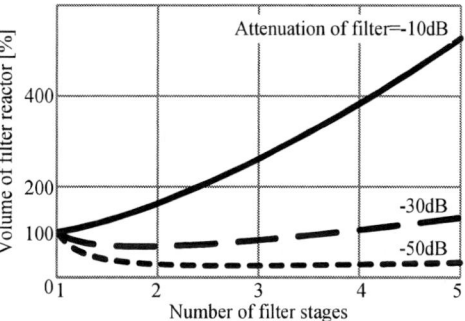

Fig. 4. The relationships between the number of filter stage and volume of filter reactor.

The GaN-FETs (V_{DSmax} = 600 V, I_{Dmax} = 10 A, R_{ON} = 100 mΩ) are used for the PWM inverter to achieve the 300-kHz switching, and the Si-diodes are used as a rectifier part to evaluate the conduction noise generated by the inverter unit. The PWM inverter is controlled by V/f control. The PWM signals are generated by comparing the triangle wave and the output voltage command. The conduction noise generated by control circuit is deducted to evaluate the conduction noise of inverter.

C. Volume of filter reactor

Fig. 4 shows the relationships between the number of filter stage and the volume of differential mode choke coil if the carrier frequency of PWM inverter is 300 kHz. The volume of filter reactor for the case of single EMC filter is applied to the inverter system is used as a standard volume regardless of the attenuation of EMC filter. In addition, the volume of X capacitors and Y capacitors are assumed as same regardless of the number of filter stages because the total capacitance of each filter is same. Owing to this, the EMC filter becomes smallest if the volume of filter reactor is smallest by selecting the suitable number of filter stage.

According to Fig. 4, the filter reactor becomes smallest for the case of two stage filter is applied if the attenuation of filter is 30 dB. On the other hands, the single stage filter is smallest if the attenuation of filter is 10 dB. Moreover, the three stage filter is the smallest if the attenuation of filter is 50 dB. Consequently, the number of filter stage that achieve smallest filter is decided by the required attenuation of the filter.

Fig. 5 shows a relationship between the carrier frequency and the volume of filter reactor. The volume of filter reactor for the case of single EMC filter and 150 kHz-carrier frequency is applied to the inverter system is used as a standard volume.

According to Fig. 5 (a) and Fig.5 (b), the volume of differential mode choke coil and common mode choke coil are reduced by high switching frequency because the attention of LC filter is large at the high frequency region. In addition, the filter reactor becomes smaller regardless of the number of filter stages. As is well known, from 150 kHz to 30 MHz of conduction noise is limited by CISPR. That is, the carrier frequency is match to the lowest frequency of limited band if the carrier frequency is 150 kHz. In other words, the filter reactor becomes largest if the carrier frequency is 150 kHz.

Besides, the choke coils of multi stage filter are miniaturized compared with the single stage filter. If the number of filter stage is increased from one to two, the volume of filter reactor becomes half when the carrier frequency is 150 kHz. Downsizing of EMC filter using multi stage filter can achieve high efficiency because the switching loss does not increase due to without the increasing of switching frequency. On the other hand, the components that construct the EMC filter such as choke coils and capacitors are increased in proportion to the number of filter stages. Because of this, the two stage filter is smaller compared with three stage filter.

The volume of filter reactor is almost same regardless of the number of filter stages when the carrier frequency is 600 kHz. In this case, the filter reactor is not miniaturized by using multi stage filter because filter reactor is already downsized by high switching frequency. Thereby, the multi stage filter is not suitable because number of components is increased when the carrier frequency is 600 kHz.

D. Loss analysis results

Fig. 6 shows the power loss that is measured from the GaN-FET inverter. The maximum efficiency is 94.7% when the carrier frequency is 150 kHz.

The conduction loss is given by y-intercept of approximate formula b because it is not related to carrier frequency. On the other hands, the switching loss of PWM inverter is given by slope of approximate formula. From experimental results, the conduction loss and switching loss on arbitrary output power is calculated by (9) and (10)

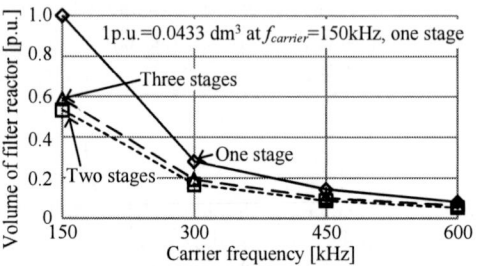

(a) Differential mode choke coil

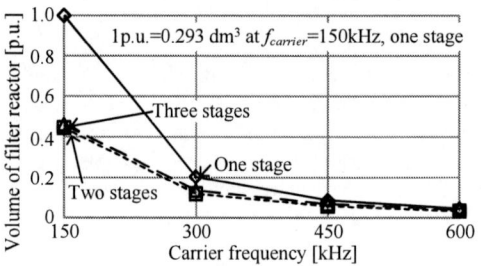

(b) Common mode choke coil
Fig. 5. Volume of filter reactor.

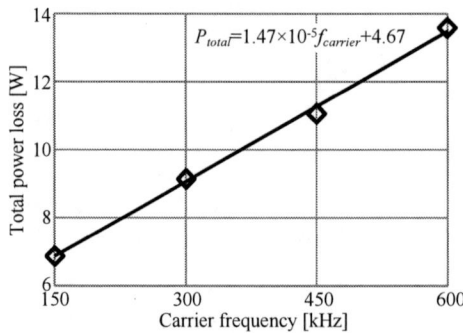

Fig. 6. Measurement result of relationship between carrier frequency and power loss of GaN-FET inverter.

$$P_{CON} = b \times \left(\frac{I_{out\_calc}}{I_{out\_test}} \right)^2 \quad \text{...} (9)$$

$$P_{SW} = a f_{carrier} \frac{V_{DC\_calc}}{V_{DC\_test}} \frac{I_{out\_calc}}{I_{out\_test}} \quad \text{..............................} (10)$$

where I_{out_calc} is the output current on arbitrary output power, and I_{out_test} is the output current when the power loss is measured in experimental.

Fig. 7 shows loss analysis results of GaN-FET inverter based on experimental results. The conduction loss is larger than switching loss when the carrier frequency is 150 kHz. However, the switching loss become larger if the carrier frequency is higher than 300 kHz. In other words, the switching loss has an insignificant effect on total loss in low carrier frequency region. Consequently, the proposed system can downsize the EMC filter without increasing the volume of cooling system in low carrier frequency region. Therefore, the PWM inverter that can achieve low switching loss is suitable for proposed system.

The rate between switching loss and conduction loss of

978-1-4799-2706-7/14 $31.00 © 2014 IEEE

PWM inverter depends on the characteristics of switching device. Thus, the GaN-FET is suitable for proposed system because it can perform fast switching and low switching loss.

Fig. 8 shows the calculation result of power loss if the inverter outputs the rated power. In this case, the conduction loss accounts more portions in a total loss because conduction loss is proportional to the square of the output current.

Fig. 9 shows the relationship between carrier frequency and volume of cooling system. The volume of cooling system is calculated by (3) based on experimental results. The cooling system becomes large at the high frequency region.

E. Power density of GaN-FET inverter

Fig. 10 shows the relationships between the carrier frequency and total volume of inverter system that contains filter reactors and cooling system. The volume of other components is not included because it not depends on carrier frequency. According to Fig.10, the inverter system is miniaturized by high frequency switching because the volume filter reactors are decreased sharply. On the other hand, the cooling system is getting larger when the carrier frequency is higher than 300 kHz. Moreover, the volume of inverter system is reduced by using multi stage filter because the attention of LC filter is significantly increased. As a result, the total volume of GaN-FET inverter system will be reduced by 54% at the carrier frequency of 300 kHz by using a two stage filter compared to case of the carrier frequency of 150 kHz by using one stage filter.

Fig. 11 shows the relationships between the carrier frequency and the power density of the inverter system. It means that a high efficiency and small inverter system shows high power density. In this paper, the power density ρ_{power} is defined as (11) by using the rated power of PWM inverter P_{out} and the total volume of inverter system that contains filter reactors and the cooling system.

$$\rho_{power} = \frac{P_{out}}{vol_{total}} \quad\text{...} (11)$$

The power density is inverse proportion to the total volume of the system. Therefore, the power density becomes the highest at the carrier frequency of 300 kHz.

Similarly, the power density becomes highest if the two stage filter is applied to the inverter system. In this system, the power density is insignificantly changed even if the carrier frequency becomes higher because EMC filter is already downsized. In other words, the downsizing of EMC filter by high frequency switching is most effective if the single stage filter is applied. Generally, the design of multi stage filter is complicated owing to the increasing of components. Therefore, the design of EMC filter becomes easier if the single stage EMC filter is miniaturized by high frequency switching.

Fig. 12 shows Pareto front curves of GaN-FET inverter at the range of carrier frequency from 150 kHz to 600 kHz. The Pareto front curves are used for determine the

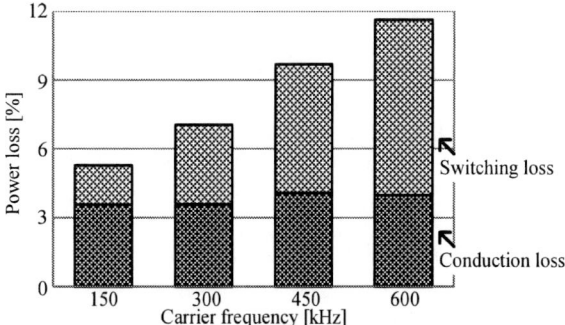

Fig. 7. Breakdown of power loss of GaN-FET inverter based on experimental results.

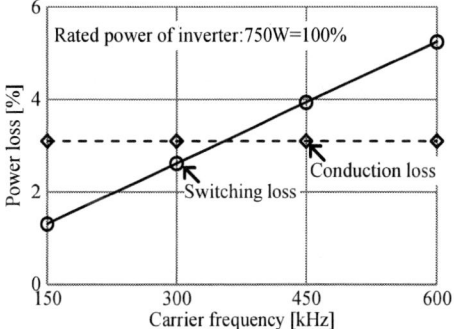

Fig. 8. Calculation result of relationship between carrier frequency and power loss of GaN-FET inverter.

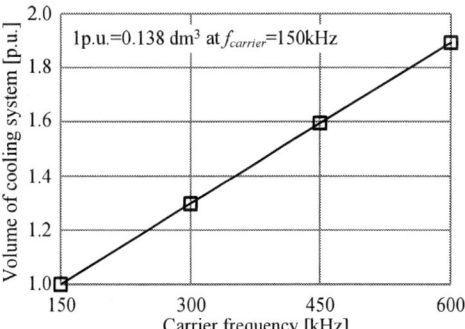

Fig. 9. Relationship between carrier frequency and volume of cooling system.

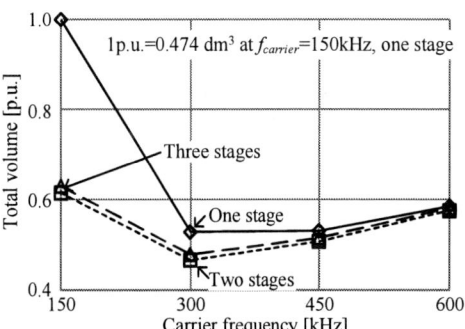

Fig. 10. Relationship between carrier frequency and total volume of GaN-FET inverter system.

optimum point in order to achieve many objectives. The horizontal axis in Fig. 11 indicates the power density, and the vertical axis in Fig. 11 indicates the efficiency of the GaN-FET inverter. Therefore, the inverter system can achieve the high efficiency and the smaller size if the Pareto front curve comes close to top right of the graph.

According to results, the EMC filter should be constructed by two stage filter in order to achieve high power density and high efficiency. In addition, the power density becomes highest at the carrier frequency is 300 kHz, and the efficiency of GaN-FET inverter becomes highest at the carrier frequency is 150 kHz.

F. Measurement results of conduction noise

Fig. 13 shows the prototype of GaN-FET inverter system. Three PCB boards are used to construct the three phase PWM inverter. Two GaN-FETs are mounted on each PCB board and connected in series. This inverter is controlled by Peripheral Interface Controller and logic ICs. The Si-diode rectifier and the GaN-FET inverter are mounted on another heat sink, and both heat sinks are not connected to the earth.

Fig. 14 shows the conduction noise of the PWM inverter with GaN-FET. The red line indicates the limit of CISPR. It is noted that the carrier frequency is 300 kHz. In Fig. 14 (a), the two stage EMC filter is connected to the input of inverter system. Additionally, the resistance and inductance are connected to the output of inverter system as a load. As a result, the conduction noise is suppressed below the limit of CISPR. Therefore, the proposed design method for EMC filters is valid in the experiment.

Fig. 14 (b) shows the conduction noise when the single stage EMC filter is applied to the inverter system. At 9 MHz, the conduction noise is over the limit of CISPR. That is because the attenuation of multi stage filter is larger than that of single stage filter at high frequency region.

On the other hand, the conduction noise at 300 kHz is almost same compared with Fig. 14 (b). It means the attenuation of single stage filter and multi stage filter are same at the designed frequency of the filter. Therefore, the proposed design method of EMC filters is valid regardless of the number of filter stage.

IV. MOTOR DRIVE EXPERIMENT

A. Experimental condition

Fig. 15 shows the system configuration when the induction motor is driven by GaN-FET inverter. The ground terminal of EMC filter and induction motor are connected to the ground terminal of LISN. Similarly, the heat sink of the GaN-FET inverter is connected to the LISN.

Table 3 shows the experimental condition. In this condition, the designed value of differential mode choke coil is smaller than the leakage inductance of common mode choke coil. Thereby, the differential mode choke coil is assumed as leakage inductance of common mode

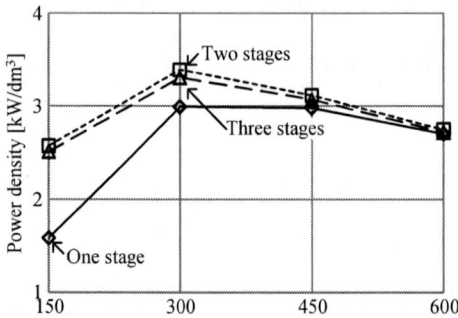

Fig. 11. Relationship between carrier frequency and power density of GaN-FET inverter.

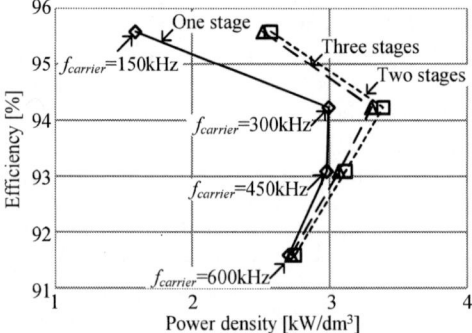

Fig. 12. Pareto front curve of GaN-FET inverter. (Relationship between power density and efficiency of GaN-FET inverter)

Fig. 13. Prototype of GaN-FET inverter.

choke coil.

B. Experimental results

Fig. 16 shows the power loss of the prototype inverter system. It is noted that the power loss of GaN-FET inverter that is proportion to the carrier frequency is defined as switching loss. Also, the power loss that is not proportion to the carrier frequency is defined as conduction loss.

From the experimental results, the efficiency of GaN-FET inverter is low. That is because the power factor is low due to the drive of the induction motor by V/f control at no load. In contrast, the power loss of EMC filter is less than 2% compared with the total loss of the GaN-FET inverter system.

Fig. 17 shows the conduction noise when the induction motor is driven by GaN-FET inverter. From the result, the conduction noise is suppressed below the limit of CISPR. Therefore, the proposed design method for EMC filters is valid in the experiment regardless of the load of inverter system.

In addition, the conduction noise in the low frequency region is increased compared to the Fig. 14 (a) because the DC voltage of the inverter system is high. Hence, the noise generated by the switching becomes larger. Moreover, the conduction noise around 9 MHz is decreased because parasitic capacitance of the load is changed compared to the R-L load.

V. CONCLUSIONS

In this paper, the relationship between the carrier frequency and the total volume of the GaN-FET inverter system is discussed based on simulation and experiments. As a result, the total volume of GaN-FET inverter system that contains an EMC filter and a cooling system will be reduced by 54% at the carrier frequency is 300 kHz by using a two stage filter compared to case of the carrier frequency of 150 kHz by using one stage filter.

In addition, an induction motor is driven by a prototype of GaN-FET inverter system with a 300-kHz carrier frequency. As a result, the conduction noise is suppressed below the limit of CISPR. Therefore, the proposed design method for EMC filters is valid in the experiment. Furthermore, the power loss of EMC filter is less than 2% compared with the total loss of the GaN-FET inverter system.

REFERENCES

[1] T. Funaki, J. C. Balda, J. Junghans, A. S. Kashyap, H. A. Mantooth, F. Barlow, T. Kimoto and T. Hikihara : "Power Conversion With SiC Devices at Extremely High Ambient Temperatures", IEEE Transactions on Power Electronics, Vol.22, No.4, pp.1321-1329 (2007)

[2] F. Xu, T. J. Han, D. Jiang, L. M. Tolbert, F. Wand, J. Nagashima, S. J. Kim and F. Barlow : "Development of a SiC JFET-Based Six-Pack Power Module for a Fully Integrated Inverter", IEEE Transactions on Power Electronics, Vol.23, No.3, pp.1464-1478 (2013)

[3] M. Rodriguez, Y. Zhang and D. Maksimovic : "High-Frequency PWM Buck Converters Using GaN-on-SiC HEMTs", IEEE

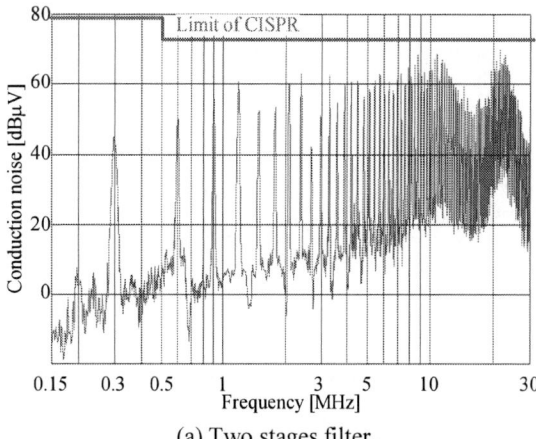

(a) Two stages filter.

(b) One stage filter.

Fig. 14. Conduction noise of the GaN-FET inverter using 300 kHz carrier with designed EMC filter.

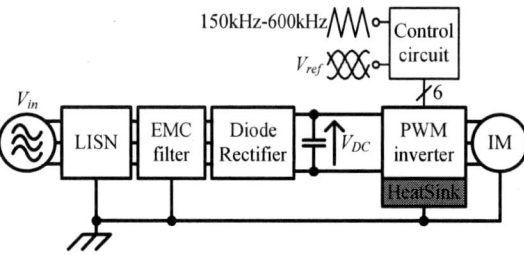

Fig. 15. Experimental configuration when induction motor is driven by GaN-FET inverter.

Table 3. Experimental condition.

Input Voltage V_{in}	200 V
Input frequency f_{in}	50 Hz
DC Voltage V_{DC}	280 V
Output voltage V_{out}	173 V
Output frequency f_{out}	43 Hz
Modulation factor	1
Dead time T_D	100 ns
Common mode reactor L_C	700 μH

Transactions on Power Electronics, Vol.29, No.5, pp.2462-2473 (2014)

[4] T. Friedli, S. D. Round, D. Hassler and J. W. Kolar : "Design and Performance of a 200-kHz All-SiC JFET Current DC-Link Back-to-Back Converter", IEEE Transactions on Industry Applications, Vol.45, No.5, pp.1868-1878 (2009)

[5] J. Rabkowski, D. Peftitsis and H. Nee : "Design Steps Toward a 40-kVA SiC JFET Inverter With Natural-Convection Cooling and an Efficiency Exceeding 99.5%", IEEE Transactions on Industry Applications, Vol.49, No.4, pp.1589-1598 (2013)

[6] Z. Chen, Y. Yao, D. Boroyevich, K. Ngo, P. Mattavelli and K. Rajashekara: "A 1200V, 60A SiC MOSFET Multi-Chip Phase-Leg Module for High-Temperature, High-Frequency Applications" Applied Power Electronics Conference and Exposition, pp.608-615 (2013)

[7] A. Rodriguez, M. Fernandez, A. Vazquez, D. G. Lamar, M. Arias and J. Sebastian: "Optimizing the Efficiency of a DC-DC Boost Converter over 98% by Using Commercial SiC Transistors with Switching Frequencies from 100 kHz to 1 MHz" Applied Power Electronics Conference and Exposition, pp.641-648 (2013)

[8] Y. Hayashi: "Power Density Design of SiC and GaN DC–DC Converters for 380 V DC Distribution System Based on Series-Parallel Circuit Topology" Applied Power Electronics Conference and Exposition, pp.1601-1606 (2013)

[9] H. Nakao, Y. Yonezawa, T. Sugawara, Y. Nakashima, T. Horie, T. Kikkawa, K. Watanabe, K. Shouno, T. Hosoda and Y. Asai: "2.5-kW Power Supply Unit with Semi-Bridgeless PFC Designed for GaN-HEMT" Applied Power Electronics Conference and Exposition, pp.3232-3235 (2013)

[10] A. Rodriguez, M. Fernandez, Marta M. Hernando, Diego G. Lamar, M. Arias and J. Sebastian: "Switching Performance Comparison of the SiC JFET and the SiC JFET/Si MOSFET Cascode Configuration" Energy Conversion Congress and Exposition, pp.472-479 (2013)

[11] S. Hazra, S. Madhusoodhanan, S. Bhattacharya, G. Karimi Moghaddom and K. Hatua: "Design Considerations and Performance Evaluation of 1200 V, 100 A SiC MOSFET Based Converter for High Power Density Application" Energy Conversion Congress and Exposition, pp.4278-4285 (2013)

[12] Xun Gong and J. A. Ferreira, "Comparison and Reduction of Conducted EMI in SiC JFET and Si IGBT-Based Motor Drives" IEEE Transactions of Power Electronics, Vol. 29, No. 4, pp.1757-1767 (2014)

[13] Richard Lee Ozenbaugh, Timothy M. Pullen: "EMI Filter Design, 3rd Edition" CRC Press. (2012)

[14] M. Hartmann, H. Ertl and J. W. Kolar : "EMI Filter Design for a 1 MHz, 10 kW Three-Phase/Level PWM Rectifier", IEEE Transactions on Power Electronics, Vol.26, No.4, pp.1192-1204 (2011)

[15] Y. kashihara, and J. Itoh: "The performance of the multilevel converter topologies for PV inverter", International Conference on Integrated Power Electronics Systems, pp.67-72 (2012)

[16] U. Drofenik, G. Laimer, and J. W. Kolar: "Theoretical Converter Power Density Limits for Forced Convection Cooling" International PCIM Europe Conference, pp.608-619 (2005)

[17] M. L. Heldwein, J. Biela, H. Ertl, T. Nussbaumer and J. W. Kolar : "Novel Three-Phase CM/DM Conducted Emission Separator", IEEE Transactions on Industrial Electronics, Vol.56, No.9, pp.3693-3703 (2009)

[18] Wm T Mclyman: "Transformer and inductor design handbook" Marcel Dekker Inc.(2004)

[19] T. Nussbaumer, M. L. Heldwein and J. W. Kolar : "Differential Mode Input Filter Design for a Three-Phase Buck-Type PWM Rectifier Based on Modeling of the EMC Test Receiver", IEEE Transactions on Industrial Electronics, Vol.53, No.5, pp.1649-1661 (2006)

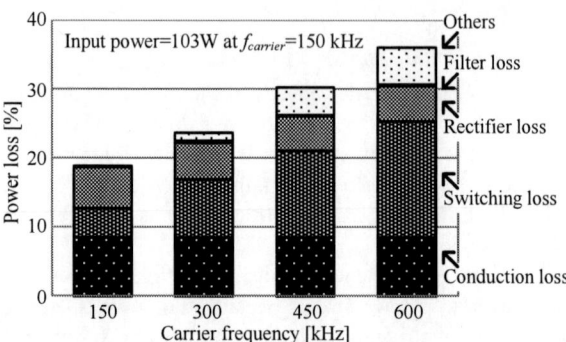

Fig. 16. Power loss of prototype inverter system when drive the induction motor V/f control.

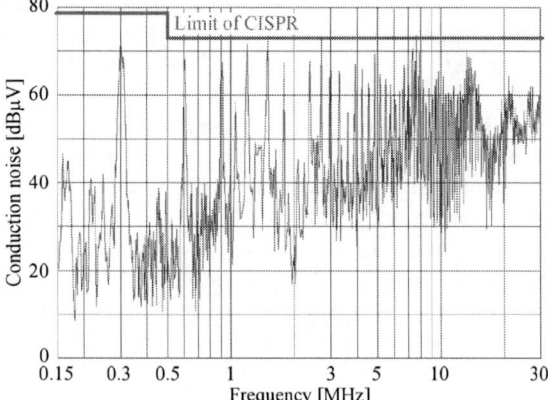

Fig. 17. Conduction noise of the prototype inverter system with induction motor.

Packaging for SiC power device

Tsuyoshi FUNAKI (Osaka university)
Div. Electrical, Electronic and Information Eng.
Graduate School of Eng., Osaka University
2-1, Yamadaoka, Suita, Osaka Japan
funaki@eei.eng.osaka-u.ac.jp

Abstract— The wide band gap nature of SiC semiconductor makes high voltage power device possible to operate at high temperature. The high temperature operation of power device is expected to realize simplification of cooling system and miniaturizing the system size with smaller heat sink and/or less liquid cooling system. However, conventional plastic packaging for Si power device is designed to operate lower than 150 °C, which cannot be used for high temperature operation of SiC power device. Then, this paper develops the ceramic packaging for SiC power device to operate at extremely high temperature. The reliability of the developed ceramic package for long term high temperature exposure test and repetitive heat shock test results are introduced in the paper.

Keywords— SiC power device, high temperature operaiton, ceramic package, reliablity test.

I. INTRODUCTION

The volume and weight ratio of passive components and cooling system in conventional power electronics equipment has been reducing with the development of power device and the improvement of the device performance. The high frequency switching and high temperature operation is necessary to reduce size and cost respectively for inductor, transformer, capacitor and heat sink. However, the usage of conventional Si semiconductor device approaches the limit of physical ceiling in Si semiconductor material. SiC semiconductor has superior characteristics than Si semiconductor in band gap, critical breakdown electric field, thermal conductivity, etc. Then, SiC semiconductor is expected as the candidate material for power switching device, which enables low loss, fast switching and high temperature operation [1, 2]. This paper especially focuses on high temperature operation of SiC power device. There are several researches for high temperature operation and characteristics of SiC power devices [3, 4, 5, 6, 7], but the works related to the high temperature packaging are not many [8, 9]. The conventional power electronics instruments consist of package and peripheral components for Si device with maximum junction temperature of 150 °C. Therefore, they cannot be applied for extremely high temperature operation of SiC power device, e.g. 200 °C or 250 °C. Figure 1 shows the external appearance of SiC Schottky barrier diode (SBD), which

is packaged in a conventional plastic package, which is same as conventional Si power device, and it was operated in high temperature up to 450 °C. The plastic package, which does not have heat resistive capability is carbonized by the high temperature operation and the device lost electrical function stemming from packaging degradation. Therefore, this paper develops heat resistive ceramic package for high temperature operation of SiC power device. The reliability of the developed package to the high temperature condition and heat shock are also evaluated.

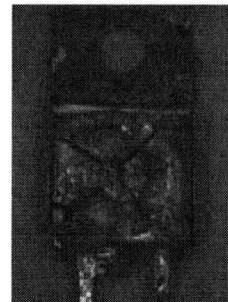

Fig. 1. SiC SBD after high temperature operation.

II. DEVELOPED CERAMIC PACKAGE

Two different types of ceramic packages are designed and developed in this paper cooperatively with Kyocera corp. The size of packages are adjusted to a JEDEC standard TO-254 and TO-3P. The details of the designed package is addressed in the followings.

A. TO-254 package

The structural drawing of the developed TO-254 package is shown in Fig. 2. The leads are placed with spacing of 3.81 mm to have 1 kV withstanding voltage. The rated current of the lead is 20 A. The lead is supported with casing by brazing it. The frame casing to fill encapsulation resin is made from Al_2O_3 ceramic to assure the heat proof. The lead frame is not electrically isolated from die attach layer for preliminary design. Then the metallized thin ceramic plate, which is made from Al_2O_3 with thickness of 250 μm is inserted between

978-1-4799-2706-7/14 $31.00 © 2014 IEEE

lead frame and die attach layer. Then, the lead frame is isolated from the die attach layer, and multiple devices with different voltage level can be directly attached on the same heat sink. The metallic portion of the package is finished with Ni plating of 3 μm, which assures the operation at 350 °C ambient.

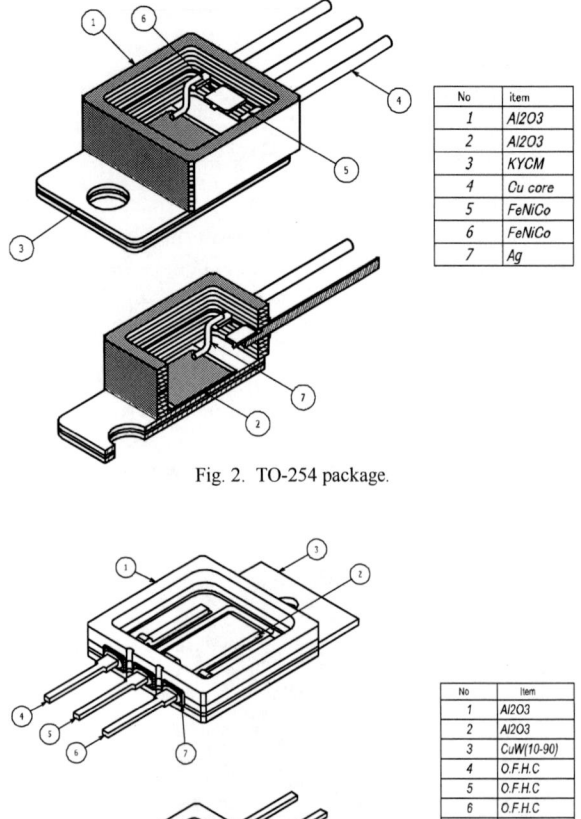

No	item
1	Al2O3
2	Al2O3
3	KYCM
4	Cu core
5	FeNiCo
6	FeNiCo
7	Ag

Fig. 2. TO-254 package.

No	Item
1	Al2O3
2	Al2O3
3	CuW(10-90)
4	O.F.H.C
5	O.F.H.C
6	O.F.H.C
7	FeNiCo
8	CuW(10-90)
9	O.F.H.C

Fig. 3. TO-3P package.

B. TO-3P package

The structural drawing of the developed TO-3P package is shown in Fig. 3. The leads are placed with spacing of 5.45 mm (20 mil) to have 1.5 kV withstanding voltage. The rated current of the lead is 50 A. The lead is also supported with frame casing by brazing it, and the edge of the lead inside the casing is brazed on the bulk oxygen free high conductivity copper (O.F.H.C.) base plate for die attach or wire bonding. The O.F.H.C base plate has the thickness of 0.7 mm. The casing to fill encapsulation resin is also made from Al_2O_3 ceramic for high heat tolerance. The lead frame is made from CuW (10-90) to assure the mechanical strength in high temperature. The thin ceramic plate made from Al_2O_3, whose thickness is 250 μm, is sandwiched with the lead

frame and O.F.H.C base plate, which mitigates the warpage of the package stemming from the stress induced by the mismatch in the coefficient of thermal expansion. The ceramic plate is metalized, and then brazed with the lead frame and base plate. The metallic portion of the package is finished with Ni plating of 3 μm, which assures the operation at 350 °C ambient.

III. RELIABILITY TEST OF THE DEVELOPED PACKAGE

The developed package consist of metal and ceramic. The constituent materials of the package are heat resistive; they are Al_2O_3, Cu-W, O.F.H.C., etc. However, the package has composite body. Therefore, the reliability of the developed package should be assessed to guarantee the operation of SiC power device in harsh environment condition. The two types of high temperature reliability test of the high temperature exposure and heat shock thermal cycle test are performed. The results for the developed TO-3P package are mainly addressed below.

A. The high temperature exposure test

The high temperature exposure test is to evaluate the capability of package usage in high temperature without deterioration of electrical performance. The developed TO-3P package is tested for the exposure of 200 °C in the temperature controlled chamber. The thick aluminum wire (500 um) is bonded for all sample, and heat resistive encapsulation resin (ADEKA, nanotech resin) is potted for some sample before burn in test. The external appearance of the tested sample is shown in Fig. 4. Though, the Ni plate turn color by oxidation for 300 °C in early stage of burn in test, but no observable change of color is fond for 200 °C to the end of test (1000h).

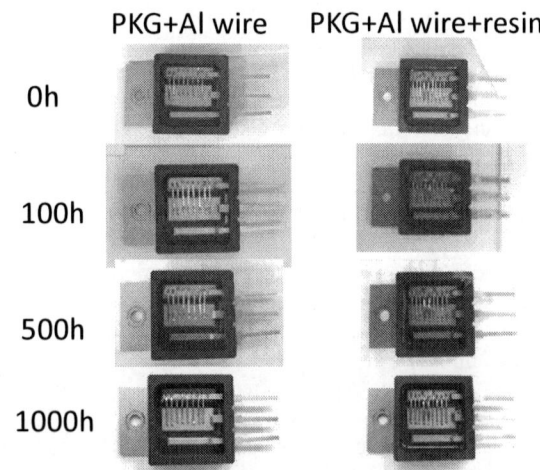

Fig. 4 External appearance of the tested sample.

The strength of wire bonding is evaluated with the peeling strength as shown in Fig. 5. The results for the respective sample are summarized in table I. There is no deterioration in peeling strength of bonding wire is found for high temperature exposure test. Also, no change was observable for potted resin. The deterioration in electrical characteristics of the encapsulated resin is also evaluated with the voltage withstanding capability. The break down voltage between two leads and between lead and base plate before and after burn in test is summarized in Table II. No noticeable break down voltage change is observed.

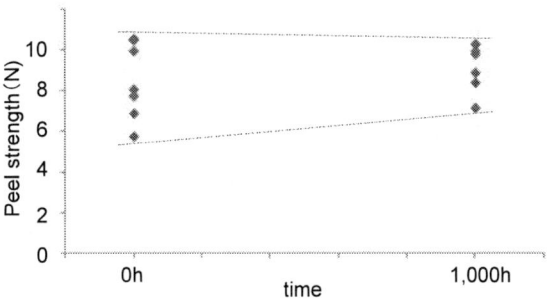

Fig. 5. Peeling strength for high temperature exposure time.

TABLE I
DETERIORATION OF WIRE BONDING FOR HIGH TEMPERATURE EXPOSURE TEST.

No	0h		1000h	
	peel strength (N)	condition	peel strength (N)	condition
1	10.0	2nd bond warpage	7.2	1st bond rift off
2	10.5	bonded neck broken	NA	1st bond rift off
3	10.5	bonded neck broken	8.9	2nd bond rift off
4	8.1	bonded neck broken	9.8	bonded neck broken (Cu side)
5	7.8	wire broken	10	wire broken
6	6.9	1st bond rift off	8.4	bonded neck broken (Cu side)
7	10.3	wire broken	10.3	wire broken
8	5.8	bonded neck broken	10.3	wire broken

TABLE II
BREAKDOWN VOLTAGE FOR HIGH TEMPERATURE EXPOSURE TEST.

Time	Lead - Lead		lead – Base metal	
	with resin	w/o resin	with resin	w/o resin
0h	4.4 kV	4.6 kV	4.2 kV	3.4 kV
1,000h	5.0 kV	4.2 kV	4.0 kV	3.8 kV

B. The heat shock thermal cycle test

The heat shock thermal cycle test is performed to evaluate the capability of package usage for abrupt temperature change, which is caused by self heating of power device loss and/or forced cooling of the heat management system. The applied thermal cycle is made the temperature shift between -25 °C and 300 °C with staying the respective temperature more than 5 min every time. The one portion of observed temperature is shown in Fig. 6, which shows that the rise and fall time of temperature is less than 5 min, and more than 5 min relaxation time in low and high temperature is secured.

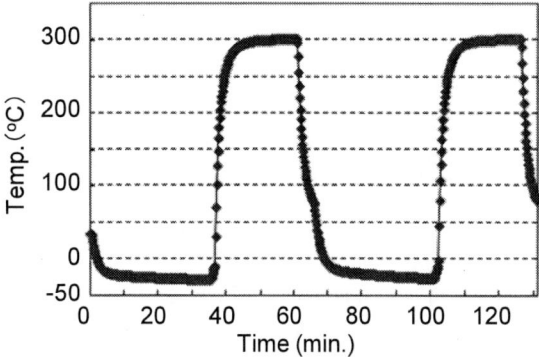

Fig. 6. Temperature change in thermal cycle test.

The external appearance of the tested sample and cross section of the packaging at the metal and ceramic junction by SEM is shown in Fig. 7. The difference in the thermal cycle test from high temperature exposure test is clearly shown in the deterioration of encapsulated resin. The calcification and shrinking of resin is found at 100 cycle for the case of PKG+Al wire+resin as shown in Fig. 7. To make the matter worse, the change of resin color is found at 500 cycle, and cracking is also found. Though, the deterioration in external appearance of resin is severe, but the change in insulating capability of the encapsulated package is small, and only gradual decrement of break down voltage is found as shown in Table III.

Fig. 7. External appearance for thermal cycle test.

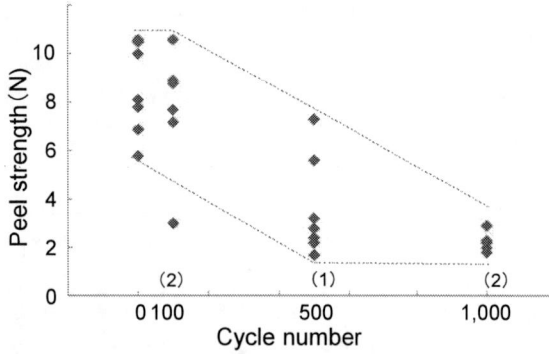

Fig. 8. Peel strength for thermal cycle.

The electron micro scope analysis for the brazed junction between die mounting base metal and insulating ceramic with metallization is also shown in Fig. 7. The brazed joint has complete bond at initial condition (0 cycle). The metallic fatigue and its progression is found with the repetition of heat shock, and the warpage and flake of base metal from ceramic is found for 1000 cycle. Though, there is deterioration in bonding between ceramic and metal, but it does not affect on the insulating capability between leads and between lead and base as shown in Table III.

The strength of wire bonding is also evaluated with the peeling strength as shown in Fig. 8. The peeling strength of wire bonding significantly deteriorates with the applied thermal cycle. Though, the package maintains the electrical property to the thermal cycle test, but the mechanical property of packaging significantly deteriorates.

TABLE III
BREAKDOWN VOLTAGE FOR THERMAL CYCLE TEST.

Cycle	Lead - Lead		lead – Base metal	
	with resin	w/o resin	with resin	w/o resin
0	4.4 kV	4.6 kV	4.2 kV	3.4 kV
100	5.0 kV	3.6 kV	3.8 kV	3.2 kV
500	4.2 kV	3.2 kV	3.8 kV	3.2 kV
1,000	3.8 kV	4.0 kV	3.6 kV	2.8 kV

IV. CONCLUSIONS

This paper developed ceramic package for SiC power device to operate at extremely high temperature. Two types of reliability tests are assessed for the developed package. The high temperature exposure test result indicates that the developed package can be used without any incident for long term 200 °C operation. Also, aluminum wire bonding and encapsulation resin can be used. However, the heat shock thermal cycle test shows some difficulty in junction of base metal and ceramic, aluminum wire bonding, and resin for encapsulation for - 25 – 300 °C, 1000cycle operation.

The future work of device packaging for high temperature operation is summarized as follows.

Conventional soldering for die attach cannot be used for high temperature operaiton, and also it cannot be used for wiring in the circuit. Then, suitable shape of lead frame to avoid soldering is required with electrically stable and low resistance for reliable operation in high temperature. The suitable material and method for adhesive bonding between base metal and ceramic, which mitigates the thermal fatigue by CTE mismatch, is needed. The development of aluminum bonding wire and encapsulation resin is also necessary.

ACKNOWLEDGMENT

This research was supported in part by Environmental Nano-cluster project of the Ministry of Education, Culture, Sports, Sciencesand Technology in Japan, and by the Japan Society for the Promotion of Science (JSPS) through the "Funding Program for World-Leading Innovative R&D on Science and Technology (FIRST Program)," initiated by the Council for Science and Technology Policy (CSTP).

REFERENCES

[1] J.L.Hudgins et al., "An Assessment of Wide Bandgap Semiconductors for Power Devices", IEEE trans. PELS, vol.18, no.3, pp.907-914, 2003.

[2] A.Hefner et al., "Power Diodes Provide Breakthrough Performance for a Wide Range of Applications", IEEE Trans. PELS, vol.16, no.2, pp.273-280, 2001.

[3] B.Ray and R.L.Spyker, "Temperature Design and Testing of a DC-DC Power Converter with Si and SiC Devices", 39th IEEE IAS Annual Meeting, 33p5, Seattle, 3-7 Oct, 2004.

[4] B.Ozpineci et al., "Characterization of SiC Schottky diodes at different temperatures", IEEE trans. PELS Let., Vol.1, No.2, pp.54-57, 2003.

[5] T.Funaki et al., "SiC JFET dc characteristics under extremely high ambient temperatures", IEICE ELEX, Vol. 1, No. 17, pp.523-527, (2004).

[6] T.Funaki et al., "Switching characteristics of SiC JFET and Schottky diode in high-temperature dc-dc power converters", IEICE ELEX, Vol. 2, No. 3, pp.97-102, (2005).

[7] T.Funaki et al., "Power Conversion With SiC Devices at Extremely High Ambient Temperatures", IEEE trans. PELS, Vol. 22, No. 4, pp.1321-1329, (2007).

[8] Y.Jian et al., "High Temperature Embedded SiC Chip Module (ECM) for Power Electronics Applications", IEEE trans. PELS, Vol.22, No.2, pp.392-398, (2007).

[9] M.J.Palmer, "Silicon Carbide Power Modules for High-Temperature Applications", IEEE trans. CPMT, Vol.2, No.2, pp.208-216, (2012).

TABLE IV

DETERIORATION OF WIRE BONDING FOR THERMAL CYCLE TEST.

sample	0 cycle		100 cycle		500 cycle		1000 cycle	
	peel strength (N)	condition	peel strength (N)	condition	peel strength (N)	condition	peel strength (N)	condition
1	10.0	2nd bond warpage	NA	2nd bond rift off	NA	2nd bond rift off	2.0	2nd bond rift off
2	10.5	bonded neck broken	NA	2nd bond rift off	1.7	2nd bond rift off	2.9	2nd bond rift off
3	10.5	bonded neck broken	3.0	2nd bond rift off	3.2	2nd bond rift off	2.2	1st bond rift off
4	8.1	bonded neck broken	7.2	2nd bond rift off	2.8	1st bond rift off	2.3	2nd bond rift off
5	7.8	wire broken	8.8	2nd bond rift off	2.4	1st bond rift off	1.8	1st bond rift off
6	6.9	1st bond rift off	8.9	wire broken	2.2	2nd bond rift off	NA	-
7	10.3	wire broken	10.6	bonded neck borken (Cu side)	5.6	2nd bond rift off	NA	-
8	5.8	bonded neck broken	7.7	1st bond rift off	7.3	2nd bond rift off	NA	-

The 2014 International Power Electronics Conference

Solid State Transformer and MV Grid Tie applications enabled by 15 kV SiC IGBTs and 10 kV SiC MOSFETs based Multilevel Converters

Sachin Madhusoodhanan, Awneesh Tripathi, Dhaval Patel, Krishna Mainali, Arun Kadavelugu, Samir Hazra, Subhashish Bhattacharya
FREEDM Systems Center, Department of Electrical and Computer Engineering, North Carolina State University
Raleigh, NC, USA
sachin, sbhatta4@ncsu.edu

Kamalesh Hatua
Department of Electrical Engineering
Indian Institute of Technology Madras
Chennai, India
kamalesh@ee.iitm.ac.in

Abstract - Recently, medium voltage SiC devices have been developed which can be used for grid tie applications at medium voltage. Two such devices – 15 kV SiC IGBT and 10 kV SiC MOSFET have opened up the possibility of looking into different converter topologies for medium voltage distribution grid interface. These can be used in medium voltage drives, active filter applications or as the active front end converter for Solid State Transformers (SST). Transformer-less Intelligent Power Substation (TIPS) is one such application for these devices. TIPS is proposed as a 3-phase SST interconnecting 13.8 kV distribution grid with 480 V utility grid. The Front End Converter (FEC) of TIPS is made up of 15 kV SiC IGBTs. This paper focuses on the advantages, design considerations and challenges associated with the operation of converters using these devices keeping TIPS as the topology of reference.

Keywords— Active Front End Converter, Medium Voltage Grid Tie Application, Silicon Carbide, Solid State Transformer

I. INTRODUCTION

Traditionally, GTOs and IGCTs are used for medium voltage grid interface applications. These devices have slower switching speed [1, 2]. Silicon (Si) IGBTs have limitation of maximum voltage up to 6.5 kV. Hence, for distribution grid (13.8 kV, 60 Hz) interfacing applications of power electronics converter using Si IGBTs or GTOs, higher order multilevel converters (more than 3-level) were mandatory. Recently, CREE Inc. has developed 15 kV Silicon Carbide (SiC) IGBT [3]. This device has

facilitated the use of power converters at distribution voltage level without employing higher order multilevel converter topologies [3, 4]. Using these devices in three-level Neutral Point Clamped (3L-NPC) structure, a power converter can be interfaced with the 3-phase distribution grid [4]. Furthermore, these devices can be switched at >5 kHz switching frequency [3, 4]. High switching frequency at this voltage level further reduces the size of passive filters, and hence it is possible to build a very high power density converter for medium and high voltage applications [4]. 15 kV SiC IGBTs and 10 kV SiC MOSFETs find applications in medium and high voltage STATCOM, active filters, power converters for renewable energy integration with the distribution grid, FACTs devices and SSTs.

Transformer-less Intelligent Power Substation (TIPS) is a three-phase topology for SST applications [4]. TIPS has many functions such as reactive power compensation at both medium and low voltage grids, power quality improvement at both the grids, integration of ac and dc renewable sources, integration of energy storage devices and bidirectional power flow [4]. TIPS has three stages of power conversion: ac-dc medium voltage Front End Converter (FEC) [5, 6], dc-dc Dual Active Bridge (DAB) [7, 8] and dc-ac low voltage converter [9]. The TIPS topology in detail is shown in Fig. 1. The FEC is a 3L-NPC converter, which is connected to the 13.8 kV distribution grid through L-filter or LCL-filter and has 22 kV output dc bus. In 3L- NPC FEC, each IGBT has to block 11 kV, hence 15 kV SiC IGBTs are suitable for this

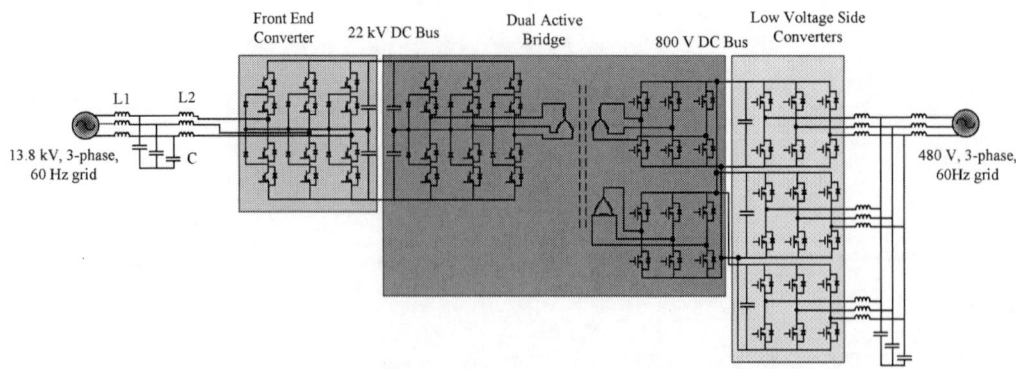

Fig.1. TIPS Complete Schematic

978-1-4799-2706-7/14 $31.00 © 2014 IEEE   1626

The 2014 International Power Electronics Conference

TABLE.I. Rectifier Loss Distribution at 3 kHz, 100 kVA and 11 kV DC Bus

Loss Component		UPF Mode		STATCOM Mode	
		15 kV SiC IGBT	10 kV SiC MOSFET	15 kV SiC IGBT	10 kV SiC MOSFET
Middle Devices	Switching Loss (W)	177	51.96	91.2	24.78
	Conduction Loss(W)	129.6	133.8	108.96	114.9
Top/Bottom Devices	Switching Loss (W)	0	0	91.2	24.78
	Conduction Loss(W)	112.8	112.8	69.42	72
Clamping Diodes (W)		27	27	61.2	61.2
Total Loss (W)		446.4	325.56	421.98	297.66
Efficiency (η%)		99.55	99.67	99.58	99.7

TABLE.II. Rectifier Loss Distribution at 10 kHz, 100 kVA and 11 kV DC Bus

Loss Component		UPF Mode		STATCOM Mode	
		15 kV SiC IGBT	10 kV SiC MOSFET	15 kV SiC IGBT	10 kV SiC MOSFET
Middle Devices	Switching Loss (W)	534	218.28	276.6	108.36
	Conduction Loss(W)	129.6	133.8	108.96	114.9
Top/Bottom Devices	Switching Loss (W)	0	0	276.6	108.36
	Conduction Loss(W)	112.8	112.8	69.42	72
Clamping Diodes (W)		27	27	61.2	61.2
Total Loss (W)		803.4	491.88	792.78	464.82
Efficiency (η%)		99.19	99.51	99.21	99.54

FEC [3]. DAB is the intermediate isolation stage which connects FEC and output converters. DAB is a bidirectional dc-dc converter in which the high voltage side consists of 3L- NPC converter, similar to FEC, and low voltage side consists of two start-delta connected 2-level converters. The high voltage and low voltage side converters are inter-connected using a high frequency transformer that provides galvanic isolation between high voltage and low voltage sides. Transformer primary is star connected and the two secondary sides are star/delta connected [7, 8]. 15 kV SiC IGBTs are used for the high voltage side converter of DAB. 1200 V/100 A SiC MOSFETs are used for low voltage converters of DAB [4, 7, 8]. The third stage of TIPS consists of three low voltage, 2-level converters connected in parallel. This stage is composed of 1200 V/100 A SiC MOSFETs [4, 9].

The control challenges of TIPS FEC using 15 kV SiC IGBTs is presented in [5, 6]. The control structure proposed in [5] includes grid current harmonic compensation, inrush current control during start-up, control under unbalance conditions and dc bus mid-point shift correction. The characteristics of 15 kV SiC IGBT is presented in [3], which discusses the high dv/dt issues associated with IGBT switching and the loss data. Based on these data, the 3-level NPC power circuit is designed. The design and validation of gate driver for the 15 kV SiC IGBT are provided in [10]. The characteristics of 10 kV SiC MOSFET is present in [11]. In this paper, the complete design of TIPS power converters is presented including loss comparisons of 15 kV SiC IGBT and 10 kV SiC MOSFET based converters in various modes of power conversion. Optimal design of high voltage side dc link capacitor and other passive components of FEC and integration challenge of FEC with DAB are also presented. Also it discusses in detail the various design advantages offered by these SiC devices.

II. Rectifier Stage Design And Analysis

15 kV SiC IGBT based 3L-NPC forms the grid tied rectifier or FEC stage of TIPS (Fig.1). The FEC stage handles various functions like Unity Power Factor (UPF) operation at the 13.8 kV grid, reactive power compensation, bidirectional power flow and regulating the high voltage side 22 kV dc bus. The advent of these medium voltage SiC devices has made possible relatively high switching frequency (3 to 10 kHz) operation of the FEC stage. This has resulted in many design benefits which will be discussed in this section. This section also explains in detail the various design considerations and the loss analysis of TIPS FEC.

A. Efficiency Analysis

The rectifier stage of TIPS operates at power factors (p. f) ranging from zero (leading) to one. So the devices in the FEC are mostly hard-switched. 15 kV SiC IGBT and other similar high voltage SiC devices like 10 kV SiC MOSFET results in highly efficient operation even when hard-switched at relatively higher switching frequencies

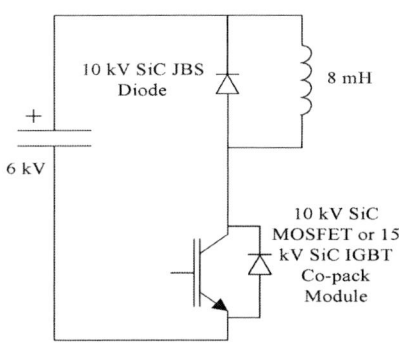

Fig.2. Double Pulse Test Circuit

978-1-4799-2706-7/14 $31.00 © 2014 IEEE   1627

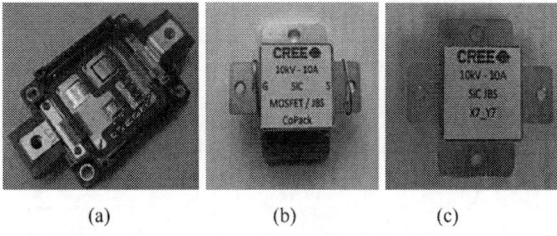

(a)   (b)   (c)

Fig.3.Photos of (a) 15 kV/20 A SiC IGBT/JBS Diode Co-pack Module (b) 10 kV/10 A SiC MOSFET/JBS Diode Co-pack Module and (c) 10 kV/10 A SiC JBS Diode

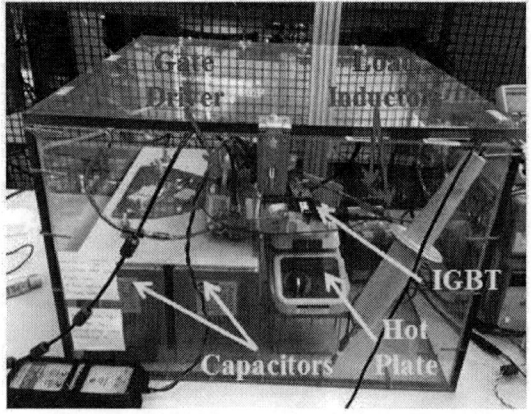

Fig.4. Experimental setup of the Double Pulse Test circuit

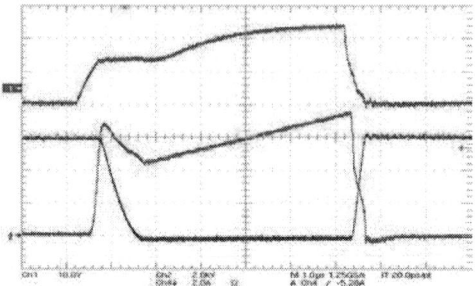

Fig.5. 10 kV SiC MOSFET switching characteristics at 6 kV (Ch1: Gate Voltage; 1 division = 10V, Ch2: Drain-Source Voltage; 1 division = 2 kV, Ch4: Drain Current; 1 division = 2 A)

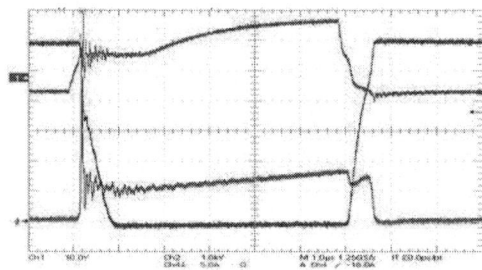

Fig.6. 15 kV SiC IGBT switching characteristics at 6 kV (Ch1: Gate Voltage; 1 division = 10V, Ch2: Drain-Source Voltage; 1 division = 1 kV, Ch4: Drain Current; 1 division = 5 A)

[3, 11]. Tables.I and II give the loss distribution for the rectifier stage at 3 kHz and 10 kHz switching frequencies respectively and 100 kVA in both UPF FEC mode and STATCOM (zero p.f leading) mode. These are calculated using the actual device loss data of these devices measured in lab. The loss comparisons are done for the case with 11 kV on the high voltage side dc bus and input grid at 7.2 kV. This is done to evaluate 10 kV SiC MOSFET within its voltage range. Each device blocks 5.5 kV. So, this is an optimum value for comparing 10 kV SiC MOSFETs with 15 kV SiC IGBTs. It can be seen that the efficiency is more than 99 % in all the cases even when hard-switched. SiC IGBT based FEC has slightly higher loss compared to SiC MOSFET based converter. This is expected to reverse when the power level is scaled up to 1 MVA. The JBS diode switching loss is assumed to be negligible in all the cases. Both SiC MOSFET and SiC IGBT based rectifier shows very high efficiency even when the switching frequency is increased to 10 kHz. High efficiency of these devices demands minimal cooling requirements. The FEC of TIPS is air-cooled using fans. Also this reduces the heat-sink size. Overall the size and weight of TIPS is expected to be one-third of a similar rated line-frequency transformer [4].

The switching loss data for the 10 kV SiC MOSFET and 15 kV SiC IGBT are measured in the lab using a Double Pulse Test (DPT) circuit. Fig. 2 shows the schematic of the standard DPT circuit. In all the cases, the inductance used in the test circuit is 8 mH. On-gate resistance of 100 Ω and off-gate resistance of 10 Ω are considered for these measurements till 6 kV. The actual device switching loss data used for efficiency calculations

is captured using these tests conducted in the laboratory. Figs.3. (a) and (b) show the photos of 15 kV/20 A SiC IGBT and 10 kV/10 A SiC MOSFET co-pack modules respectively from CREE. Fig.3 (c) shows the 10 kV SiC JBS diode module. Fig. 4 shows the experimental set up used for doing the double pulse test. The loss data thus captured is used as the input for the thermal block in Plecs software for calculating the overall system loss. The overall TIPS system is simulated in MATLAB/Plecs.

Figs. 5 and 6 show the switching characteristics of 10 kV SiC MOSFET and 15 kV SiC IGBT measured at 6 kV dc bus voltage. These switching waveforms are also captured at different voltage and current levels for both SiC MOSFET and SiC IGBT. As the load inductor and diode used for the tests are the same with both the devices, the current spike is an indication of dv/dt produced by the corresponding devices. So, ideally the devices have to be compared with different gate resistances so that the dv/dt produced is the same by all of them [3, 10]. Here, a slower gate resistance of 100 Ω is considered for SiC MOSFET

TABLE.III. Vce Drops of All SiC IGBTs Used In TIPS FEC vs Currents

IGBT Position	Vce Drops at Different Currents		
	@5 A	@10 A	@20 A
R-Phase Top IGBT	3.85 V	4.37 V	5.2 V
R-Phase Middle IGBT 1	3.84 V	4.39 V	5.28 V
R-Phase Middle IGBT 2	4.1 V	4.74 V	5.77 V
R-Phase Bottom IGBT	4.65 V	5.88 V	7.04 V
Y-Phase Top IGBT	3.8 V	4.29 V	5.08 V
Y-Phase Middle IGBT 1	4.32 V	5.09 V	6.33 V
Y-Phase Middle IGBT 2	4.46 V	5.31 V	6.66 V
Y-Phase Bottom IGBT	3.92 V	4.97 V	5.39 V
B-Phase Top IGBT	4.61 V	5.5 V	6.9 V
B-Phase Middle IGBT 1	3.81 V	4.35 V	5.24 V
B-Phase Middle IGBT 2	3.81 V	4.28 V	5.04 V
B-Phase Bottom IGBT	4.49 V	5.33 V	6.69 V

978-1-4799-2706-7/14 $31.00 © 2014 IEEE

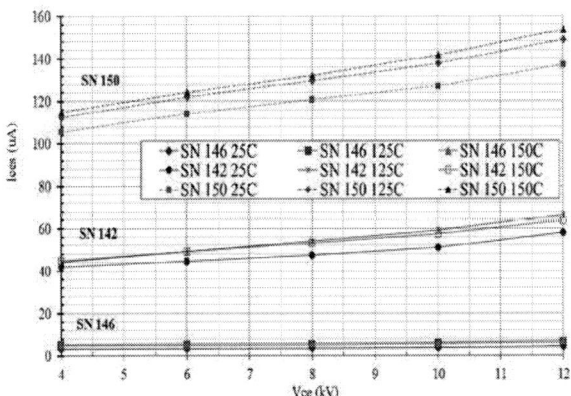

Fig.7. Leakage Current (I_{ces}) Vs Blocking Voltage (V_{ce}) of three 15 kV SiC IGBTs (Serial nos: SN 142, SN 146 and SN 150) at three different temperatures

to have a competitive comparison with the SiC IGBT at relatively lower power levels (< 100 kVA). The buffer layer thickness in the N-IGBT module used for the test is 2 μm [3]. The conduction loss data is calculated from the forward characteristics of the devices as reported in [3, 11]. For the 15 kV SiC IGBT, the forward drop is 6 V at 20 A collector-emitter current with game-emitter voltage at 20 V [3] and for 10 kV SiC MOSFET it is 5.2 V at 10 A [11].

The loss analysis gives a clear picture of the loss distribution across each device in this topology which can be used to come up with an optimum device placement strategy. For example, the top and bottom active devices do not conduct in the 3-level topology in the UPF FEC mode. So if the medium voltage grid tied rectifier is only designed to be operated in the UPF mode, these devices could be placed slightly away from the cooling source compared to the middle devices. But in the STATCOM mode, all the active devices conduct equally and thus, need uniform cooling. These thermal design considerations should be taken care of while designing the rectifier stage of TIPS.

The devices used in the rectifier stage and also in the high voltage side of DAB vary with respect to their losses substantially. Table.III shows the forward voltage drop across each 15 kV SiC IGBT in the rectifier stage of TIPS. Fig. 7 shows the leakage current variation of three of these IGBTs with temperature. From these data it can be concluded that these devices have different losses. Thus, the placement of these devices along a 3-level pole should also be dependent on their loss distribution. As an example, in the UPF mode of operation, it is recommended to use the devices with higher losses as the top and bottom devices since they never get turned on, and only their anti-parallel diodes conduct.

The switching characteristics of 15 kV/20 A SiC IGBT at 11 kV are reported in [3]. Turn-on and turn-off gate resistances used are 200 Ω and 10 Ω respectively at 11 kV. The switching loss data at 11 kV is used for estimating the efficiency of the TIPS FEC operating at 22 kV dc bus, 13.8 kV input grid, 100 kVA and switching at 3 kHz. Table.IV gives the loss distribution of TIPS FEC switching at 3 kHz, both in the UPF mode and STATCOM mode with 22 kV dc bus. The STATCOM mode is found to be slightly more efficient than the UPF FEC mode. Compared to Table.I, the conduction loss is lower in this case as the line current is smaller due to same 100 kVA power but higher grid voltage. The switching loss is higher due to switching from a higher voltage of 11 kV compared to 5.5 kV. It can be seen that the TIPS FEC has very high efficiency (>99%) in both the modes of operation. For TIPS application, the FEC switching frequency is selected to be 3 kHz on the basis of acceptable switching loss for hard switching.

B. AC Side Filter Design

The relatively high switching frequency achievable due to the use of SiC devices has significantly reduced the filter size. Traditionally, for medium voltage, high power applications using GTOs, 0.1-0.15 p.u inductance was required for proper filtering of switching ripples with only L-filter [12]. With higher switching frequencies of 3-10 kHz possible with SiC devices, this can be as low as 0.05 p.u (~ 440 mH for TIPS FEC with 22 kV dc bus, 3 kHz switching ripple and 100 kVA).

The size of filter inductance can be further reduced by using LCL filter. For TIPS FEC, the total filter inductance is designed for 3% drop of the grid voltage (13.8 kV). This is distributed in the ratio of 1:2 for the grid side inductor and converter side inductor. This results in 90 mH for the grid side inductance and 180 mH for the converter side inductance. The converter side inductor has to be designed to withstand the 3 kHz switching ripple. To reduce the losses, the core material has to be either ferrite or nan-crystalline. Fig. 8 shows the photo of a 20 mH, 10 A inductor built using ferrite core. For high voltage insulation, Nomex paper and Kapton tape are used between the windings and the core. Fig.9 shows the plot of inductance and resistance variation of this inductor with frequency from 40 Hz to 10 kHz. It can be seen that even at 5 kHz switching frequency the inductance value is around 20 mH. These inductors can be connected in series on the converter side to get the required inductance for filtering the switching ripple. The grid side inductor will only see fundamental voltage across it. It can also be made of ferrite core, but even iron core inductor can be used to be more economical. Fig. 10

TABLE.IV. TIPS FEC LOSS DISTRIBUTION AT 3 KHZ, 100 KVA AND 22 KV DC BUS

Loss Component	UPF Mode	STATCOM Mode
Total Switching Loss (W)	440	370
Total Conduction Loss (W)	102	79.5
Total Loss (W)	542	449.5
Efficiency (%)	99.46	99.55

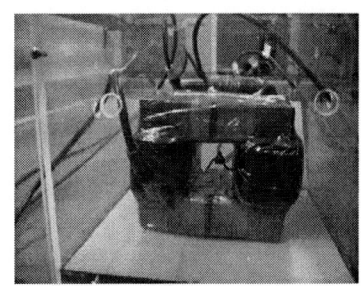

Fig.8. 20 mH Ferrite Core Inductor

The 2014 International Power Electronics Conference

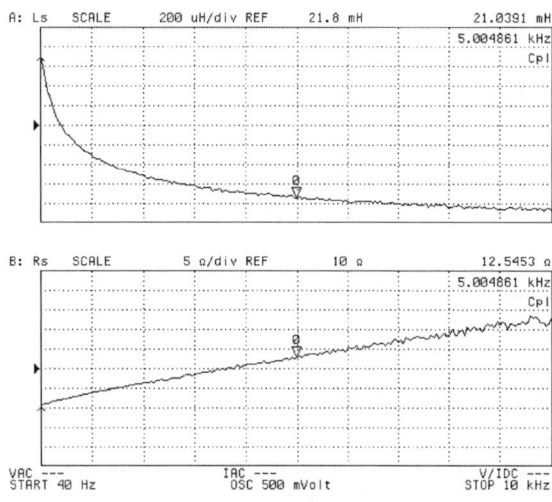

Fig.9. Inductance and Resistance variation of Ferrite Core Inductor with Frequency

Fig.10. 20 mH Iron Core Inductor

shows the 20 mH, 30 A iron core inductor from Hammond Manufacturing which can be connected in series on the grid side to get the required inductance.

The filter capacitor value is then decided based on the desired resonant frequency. The choice of resonant frequency has to be made carefully if the switching frequency is only 3 kHz. It has to be reasonably lower than 3 kHz to eliminate the switching ripple. Also it cannot be near to dominant lower order harmonics like 5th and 7th so as to prevent these harmonics from getting resonated [6]. Thus, an optimum value of 1 kHz is selected. This results in a filter capacitance value of 0.42 µF. Figs. 11 and 12 show the current and voltage waveforms across the filter elements respectively. Table.V gives the design parameters for these filter components.

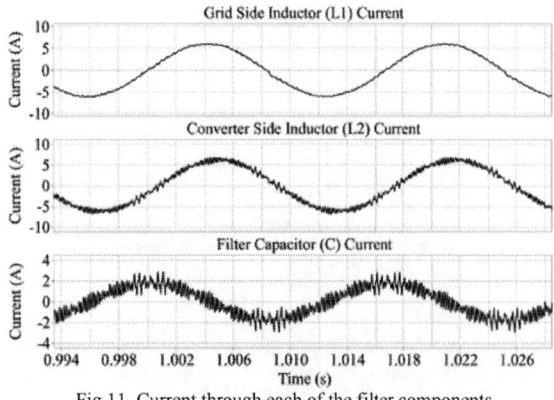

Fig.11. Current through each of the filter components

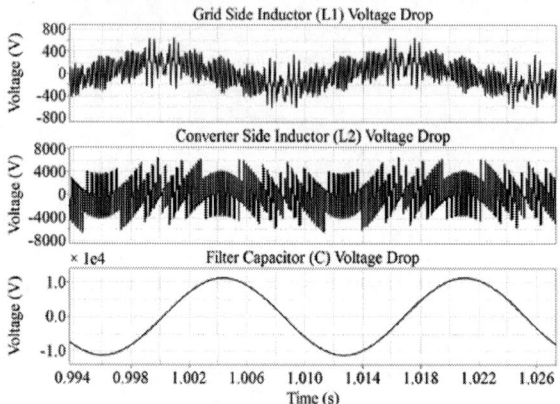

Fig.12. Voltage across each of the filter components

TABLE.V. LCL FILTER DESIGN PARAMETERS

Parameter	Grid Side Inductor	Converter Side Inductor	Filter Capacitor
Value	90 mH	180 mH	0.42 µF
rms Current	4.276 A	4.394 A	1.33 A
Peak current	6.079 A	6.813 A	2.87 A
rms Voltage	217.59 V	2.76 kV	7.95 kV
Peak Voltage	649.88 V	6.52 kV	11.29 kV
THD of Current	2.56 %	9.19 %	32.53 %

C. High Voltage Bus Bar Design

Sandwich bus bar is designed for each pole of the 3L-NPC. This is done to reduce the stray inductance effect. This is necessary to reduce the LC ringing which can occur at higher voltages. These 3-level poles are used for both FEC and DAB stages of TIPS. Fig. 13 shows the photo of bus bar and 3-level pole.

D. Stability of High Voltage DC Link

With these high voltage devices it is possible to have very high voltage dc bus for the front end converter, for example, 22 kV dc bus for the TIPS FEC. As the dc bus voltage increases, it is recommended to use low ESL film capacitors on the dc bus. This is to make sure that the devices see very small ringing due to LC oscillations which can damage these devices. But as the voltage rating goes high, the capacitance value reduces. A total dc bus capacitance of 90 µF is considered for the TIPS FEC. But such a small dc bus capacitance creates instability issues when connected to an active load like DAB at the output of the FEC. With small capacitance it becomes difficult to decouple the two systems and that causes control issues while transferring power in an integrated system like TIPS. For stable operation of any cascaded

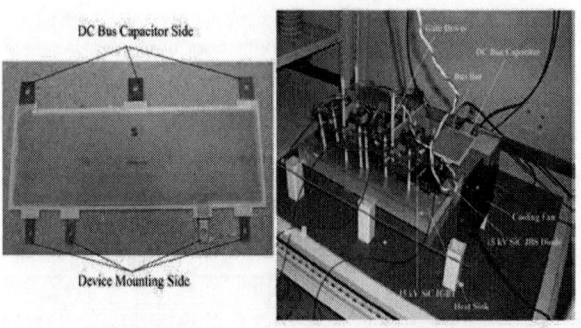

Fig.13. Bus bar and the high voltage 3-level pole of NPC converter

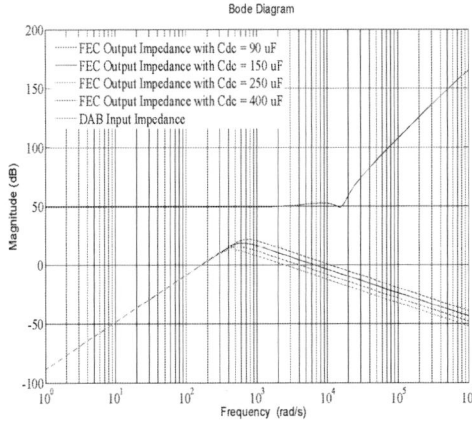

Fig.14. Output impedance variation of FEC with dc bus capacitance along with Input impedance of DAB for phase angle = 41^0

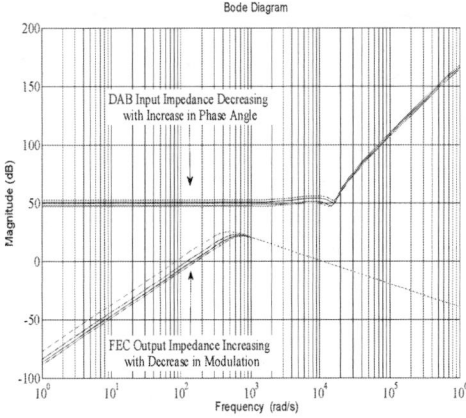

Fig.16. Output impedance variation of FEC with modulation index (0.25 to 0.95) along with Input impedance variation of DAB with phase angle (20^0 to 41^0)

converter system, the closed loop output impedance of the FEC should be lower than the closed loop input impedance of the next stage (DAB for TIPS) for all frequencies [13]. This will ensure that the two systems are totally decoupled and the control systems of the two sections do not interact with each other. The controller design for the FEC and DAB stages of TIPS are discussed in [5-8]. The closed loop output and input impedances can be derived from the control block diagrams. Equations (1) – (2) give the closed loop output impedance of the TIPS FEC using control scheme described in [5, 6].

Here, Z_{out} is the FEC closed loop output impedance, C_{dc} is the dc bus capacitance, K_v is the FEC voltage controller gain, t_v is the FEC voltage controller time constant, m is the modulation index and ω_{ibw} is the inner current control closed loop bandwidth.

$$Z_{out} = \left(\frac{-\omega_{ibw}}{C_{dc}}\right) * \left(\frac{s\left(1+\dfrac{s}{\omega_{ibw}}\right)}{s^3+(\omega_{ibw})*s^2+K'*s+\left(\dfrac{K'}{t_v}\right)}\right) \quad (1)$$

$$K' = \left(\frac{m*K_v*\omega_{ibw}}{C_{dc}}\right) \quad (2)$$

From (1) and (2) it can be seen that the FEC output

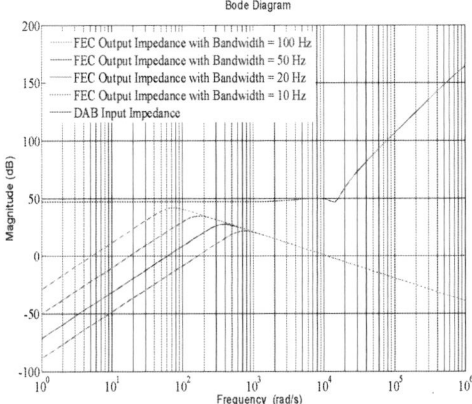

Fig.15. Output impedance variation of FEC with voltage loop bandwidth along with Input impedance of DAB for phase angle = 41^0

impedance is mainly a function of the dc bus capacitance, the dc bus voltage control loop bandwidth through the control parameter K_v and the operating point through the modulation index m. Figs. 14 to 16 show the impedance plots for the TIPS FEC stage and the DAB stage for variation of all these parameters. From Fig.14, it can be seen that as the dc bus capacitance reduces, the closed loop output impedance of FEC increases and might become more than the input impedance of DAB. So an optimum value of 90 μF is considered suitable for TIPS application. From Fig.15, it can be seen that as the voltage control loop bandwidth reduces, the closed loop output impedance of FEC increases and might become more than the input impedance of DAB. The upper bound for voltage loop bandwidth is limited to be around 100 Hz if the switching frequency is as low as 3 kHz. The lower bound is decided by this stability criterion for the high voltage dc link. Fig. 16 gives the variation of these impedances with operating point. DC bus voltage and the power level are the deciding factors. If either the power transferred increases or the dc bus voltage increases, the modulation index reduces and thus, the output impedance of FEC increases. Similarly if the power increases, the phase angle of DAB increases and its output impedance reduces. It is necessary to meet the stability criteria throughout the operating range of TIPS. This gives an indication of the stability of TIPS for various operating points. The high voltage dc link of TIPS is found to be stable throughout its operating range (Fig. 16).

III. DC – DC DAB STAGE

DAB used as the isolation stage for the TIPS topology is a 3-phase DAB. The high voltage side of this DAB is a 3L-NPC based on 15 kV SiC IGBTs [7, 8]. The design considerations for this converter are the same as that for the FEC stage. The low voltage side is formed of two

Fig.17. Single phase high frequency transformer

TABLE.VI. TIPS DAB LOSS DISTRIBUTION AT 10 kHz, 100 kVA AND 22 kV DC BUS TO 800 V DC BUS OPERATION

Loss Component	Value
High Voltage Side Switching Loss (W)	601. 86
High Voltage Side Conduction Loss (W)	84.06
Low Voltage Side Switching Loss (W)	86.34
Low Voltage Side Conduction Loss (W)	511.05
High Frequency Transformer Loss (W)	390
Total Loss (W)	1673.31
Efficiency (%)	98.33

TABLE.VIII. TIPS OVERALL LOSS DISTRIBUTION

Loss Component	FEC (UPF)	DAB	Low Voltage Side (UPF)	Total loss
Switching Loss (W)	440	688.2	297	1425.2
Conduction Loss (W)	102	595.1	738	1435.1
HF Transformer Loss (W)		390		390
Total Loss (W)	542	1673.3	1035	3250.3
Efficiency (%)	99.46	98.33	98.97	96.75

star-delta connected 2-level converters using 1200 V SiC MOSFETs. They are connected in parallel with 30^0 phase shift so as to reduce the circulating current. High voltage and low voltage sides are inter-connected using a high frequency transformer. For TIPS application, this 3-phase high frequency transformer is composed of three single phase transformers (Fig. 17). The main functions of DAB are to control active power flow from FEC to low voltage converters, provide isolation and regulate the 800 V dc bus. DAB is soft switched at 10 kHz [7, 8]. The control of DAB is challenging due to high dv/dt on the input side converters affecting its operation, parasitics in the transformer causing ringing, transformer saturation issues and other factors [8]. At 22 kV dc bus level this effect becomes severe. The control modifications required under such conditions is described in [8]. The design considerations, characterization and specification of this high frequency transformer are given in detail in [14]. Table.VI gives the loss distribution across the DAB stage using actual device loss data of 15 kV SiC IGBT, 1200 V SiC MOSFET and the loss data of the high frequency transformer [3, 9, 14]. Loss is calculated for the rated power of 100 kVA and switching frequency of 10 kHz.

IV. LOW VOLTAGE CONVERTERS

To meet the current requirements, three numbers of low voltage converters are connected in parallel. Each converter is made up of 1200 V, 100 A SiC MOSFETs. The design considerations of these converters are described in detail in [9]. Fig. 18 shows the waveforms captured when two of these converters are connected in parallel at 50 kW, 20 kHz switching frequency and 800 V

dc bus. Table.VII gives the loss distribution for the low voltage side converters using loss data in [9]. Each converter is handling 33 kVA at 17 kHz switching frequency. Three converters are paralleled and tied to the 480 V grid at UPF.

V. TIPS HARDWARE INTEGRATION AND EFFICIENCY ANALYSIS

Each of the previous sections explains in detail the loss distribution across each of the modules of TIPS converter system. Actual device loss data measured through experiments is used for efficiency analysis of TIPS. Table.VIII summarizes these loss results. It can be seen that the overall efficiency of TIPS is projected to be 96.75% at 100 kVA which is very impressive for a solid state converter meant to be operating at medium voltage level. It has to be mentioned that the efficiency can be further improved by decreasing the gate resistance for the 15 kV SiC IGBT from 200 Ω. A trade-off has to be made between the switching loss and the acceptable dv/dt. Also it is possible to parallel more dies in the 1200 V SiC MOSFET module to reduce its conduction loss. Hence, with further optimization TIPS efficiency can be increased to as high as 98 %. Fig. 19 shows the hardware prototypes of the FEC and DAB stages built in the lab. Figs. 20 to 22 shows the experimental results captured at

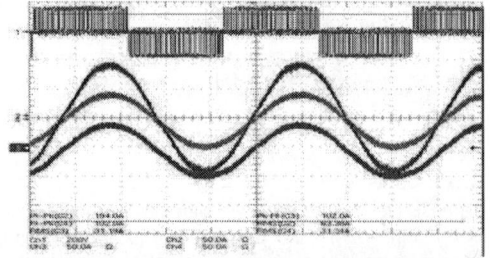

Fig.18. Parallel operation of two 1200 V SiC MOSFET converters supplying total load of 50 kVA (Scale: Load current (Ch2), Converter-1 current (Ch3), Converter-2 current (Ch4) -50A/div, Converter line voltage (Ch1) -1 kV/div

TABLE.VII. TIPS LOW VOLTAGE SIDE LOSS DISTRIBUTION AT 20 kHz, 100 kVA AND 800 V DC BUS

Loss Component	UPF	STATCOM
One Converter Switching Loss (W)	99	105
One Converter Conduction Loss (W)	246	197
One Converter Total Loss (W)	345	302
Total Loss for 3 Converters (W)	1035	906
Efficiency (%)	98.97	99.04

(a)   (b)

Fig.19. (a) TIPS FEC made of 15 kV SiC IGBTs (3 racks showing 3 legs) (b) TIPS DAB showing 3-level NPC made of 15 kV SiC IGBTs (top rack), 1200 V SiC MOSFET based low voltage side converters (middle rack) and the high frequency transformers (bottom rack)

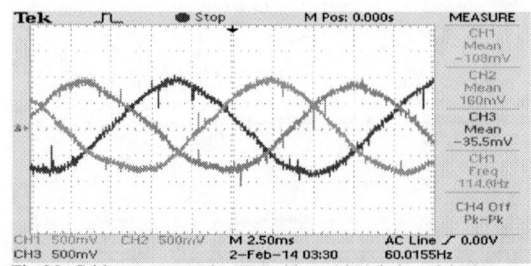

Fig.20. Grid currents on the FEC side at 4 kV dc bus, 2 kW power (Current: 2.5 A/div)

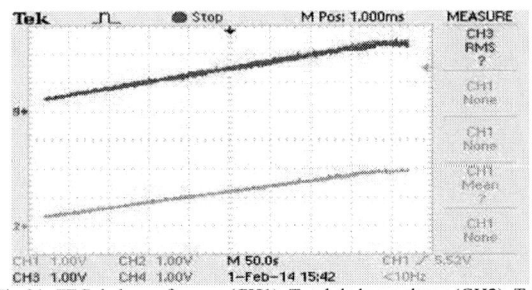

Fig.21. FEC dc bus reference (CH1), Total dc bus voltage (CH2), Top dc bus voltage (CH3) and bottom dc bus voltage while charging from 400 V to 4 kV (CH1, CH2: 2 kV/div, CH3, CH4: 850 V/div)

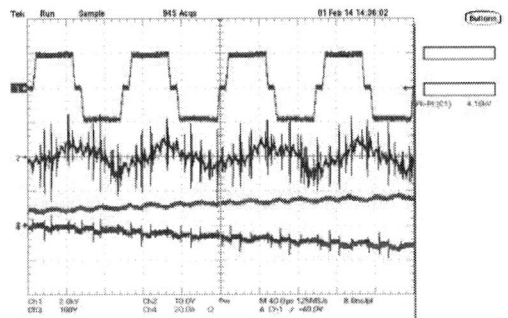

Fig.22. DAB primary side voltage (CH1: 2 kVdiv) and primary side current (CH2: 1 A/div). CH3 and CH4 show the TIPS low voltage converter total line current and voltage zoomed to the 10 kHz scale of DAB

4 kV dc bus and 2.3 kW power transferred. The results are captured with the FEC tied with 480 V grid through a variac for demonstration purpose. The experimental validation of the TIPS topology with all its functions are presented in [15] at lower voltages.

VI. CONCLUSIONS

In this paper, the main focus is to understand the design requirements of grid tied converters built using newly developed 15 kV SiC IGBT and 10 kV SiC MOSFET. These requirements are analyzed using their application in TIPS topology as a reference. Module wise efficiency analysis of the TIPS converter shows that these devices are highly efficient. Efficiency of overall TIPS system is calculated using the actual device loss data measured through experiments and came out to be 96.75 % which is relatively good. It can be increased to as high as 98 % with further optimization. Experimental results are captured for the TIPS system at 4 kV dc bus on the high voltage side.

ACKNOWLEDGEMENTS

This work is supported by US Government through the DOE ARPA-E program under contract no. DE-R0000110. This work made use of FREEDM ERC shared facilities supported by National Science Foundation under award no. EEC-0812121.

REFERENCES

[1] J. Rodriguez, S. Bernet, B. Wu, J. O. Pontt, and S. Kouro, "Multilevel voltage-source-converter topologies for industrial medium-voltage drives", *IEEE Transactions on Ind. Electron.*, vol. 54, no. 6, pp. 2930-2945, Dec 2007.

[2] S. Bernet, R. Teichmann, A. Zuckerberger, and P. K. Steimer, "Comparison of high-power IGBT's and hard-driven GTO's for high-power inverters", *IEEE Transactions on Ind. Appl.*, vol. 35, no. 2, pp. 487-495, Mar./Apr. 1999.

[3] A. Kadavelugu, S. Bhattacharya, S. Ryu, E. V. Brunt, D. Grider, A. Agarwal, and S. Leslie, "Characterization of 15 kV SiC n-IGBT and its application considerations for high power converters", in *proc. 2013 IEEE Energy Conversion Congress and Exposition*, Denver, CO, pp.2528- 2535.

[4] K. Hatua, S. Dutta, A. Tripathi, S. Baek, G. Karimi, and S. Bhattacharya, "Transformer-less intelligent power substation design with 15 kV SiC IGBT for grid interconnection," in *proc. 2011 IEEE Energy Conversion Congress and Exposition*, Phoenix, AZ, pp.4225-4232.

[5] S. Madhusoodhanan, K. Hatua, and S. Bhattacharya, "Control technique for 15 kV SiC IGBT based active front end converter of a 13.8 kV grid tied 100 kVA transformerless intelligent power substation", in *proc. 2013 IEEE Energy Conversion Congress and Exposition*, Denver, CO, pp.4697 - 4704.

[6] S. Madhusoodhanan, K. Hatua, and S. Bhattacharya, "A Unified control scheme for harmonic elimination in the front end converter of a 13.8 kV, 100 kVA transformer-less intelligent power substation grid tied with LCL filter", accepted for publication in the *IEEE Applied Power Electronics Conference and Exposition (APEC) 2014*.

[7] A. K. Tripathi, K. Hatua, H. Mirzaee, and S. Bhattacharya, "A three-phase three winding topology for dual active bridge and its D-Q mode control", in *proc. Applied Power Electronics Conference and Exposition (APEC) 2012*, pp.1368-1372.

[8] A. K. Tripathi, K. Mainali, D. Patel, K. Hatua, S. Bhattacharya, "Closed loop D-Q control of high-voltage high-power three-phase dual active bridge converter in presence of real transformer parasitic parameters", in *proc. 2013 IEEE Energy Conversion Congress and Exposition*, Denver, CO, pp.5488 - 5495.

[9] S. Hazra, S. Madhusoodhanan, G. Karimi-Moghaddam, K. Hatua, and S. Bhattacharya, "Design considerations and performance evaluation of 1200 V, 100 A SiC MOSFET based converter for high power density application", in *proc. 2013 IEEE Energy Conversion Congress and Exposition*, Denver, CO, pp.4278 - 4285.

[10] A. Kadavelugu and S. Bhattcharya, "Design considerations and development of gate driver for 15 kV SiC IGBT", accepted for publication in the *IEEE Applied Power Electronics Conference and Exposition (APEC) 2014*.

[11] J. Wang, T. Zhao, A.Q. Huang, R. Callanan, F. Husna, and A. Agarwal, "Characterization, modeling and application of 10 kV SiC MOSFET", *IEEE Trans. Electron Devices*, vol. 55, no. 8, pp. 1798-1806, Aug. 2008.

[12] B. Wu, High-power converters and AC drives. Hoboken, N.J.: John Wiley & Sons, 2006.

[13] H. Krishnamurthy, R. Ayyanar, "Stability analysis of cascaded converters for bidirectional power flow applications", in *proc. 2008 IEEE Telecommunications Conference*, pp. 1-8, Sept. 2008.

[14] K. Mainali, A. Tripathi, D. Patel, S. Bhattacharya, and T. Challita, "Design, measurement and equivalent circuit synthesis of high power HF transformer for three-phase composite dual active bridge topology", accepted for publication in the *IEEE Applied Power Electronics Conference and Exposition (APEC) 2014*.

[15] S. Madhusoodhanan, A. Tripathi, A. Kadavelugu, S. Hazra, D. Patel, K. Mainali, K. Hatua, Subhashish Bhattacharya, "Experimental validation of the steady state and transient behavior of a transformer-less intelligent power substation", accepted for publication in the *IEEE Applied Power Electronics Conference and Exposition (APEC) 2014*.

AUTHOR INDEX

Abe, Kodai 3153
Abe, Seiya 177, 1179, 2216, 2222, 3652
Abe, Shigeru 1109, 1115
Abe, T. 3007
Abe, Takashi....................................2183, 2189, 3024
Abe, Tomohiko 1575
Abiko, Hiroshi 634
Achara, Pichetjamroen 3687
Adachi, Mitsuo 92
Adhikari, Jeevan 1775
Agelidis, Vassilios G. 1458, 3758, 3764, 3933
Agelidis, Vassilios Georgios 640
Aguglia, Davide 3371
Ahmed, Furqan 480, 790
Ahssanuzzaman, S. M. 3582
Aiso, Kohei 1141
Ajima, Toshiyuki...................................383, 682
Akagi, Hirofumi 750, 1586, 1761, 2290, 2323, 3742
Akagi, Masataka 629
Akahane, Masashi...2302
Akatsu, Kan 1128, 1141, 1234, 2673, 3828
Aketa, M. 2074
Akira 3784
Akiyama, Satoru 2285
Alemi, Payam 1201
Alipoor, Jaber..3298
Aljankawey, A. S. 2156
Allen, Scott..3447
Almer, Stefan...3563
Ama, Naji Rajai Nasri 2413, 2988, 3278
Amanci, Adrian Z. 1303
Amano, Yuki 1824
Amma, Ryosuke 2027
Anazawa, Yoshihisa 3801
Andersen, Michael A. E. 78, 506, 2842, 3352, 3905
Ando, Itaru 1516
Ando, Masato 1317
Anthon, Alexander..78
An-Yeol Ko 796
Aoki, Mutsumi...2400
Aoyagi, Shigehisa 2451
Aoyama, Fumio 2644
Aoyama, Kohei 2266
Aoyama, Masahiro..1405
Aoyama, Tomohiro 3823
Ara, Takahiro...3044
Arai, Haruki 403
Arai, Manabu 3440
Araki, Jun 1728
Araki, Takahiro 1613
Arata, Masanori 1874
Arikawa, S. 3007
Arimatsu, Kenji 415
Arita, Hideaki 2673
Asai, Inami..123

Asakimori, Koki 1567
Asama, Junichi 988
Asano, Katsunori.. 3440
Asano, Yoshinari 1997
Asano, Yuji 3872
Ashikaga, Tadashi 1886
Athab, Hussain S. 3695
Atsushi, Manabe 2745
Awaji, Sosuke.. 3194
Ayano, Hideki 2385
Azuma, M. 1892
Baba, Jumpei 1849
Babasaki, Tadatoshi 1567
Bac Xuan Nguyen 2722
Bafleur, M. 707
Bahman, A. S. 2862
Bahrani, Behrooz 1386
Bak, Claus Leth 3320
Bakran, Mark-M. 2113, 3255
Bang, Deok-Je 2427
Bani Shamseh, Mohammad 2794
Baoquan Liu 1155, 3546
Barater, Davide 433
Barrade, Philippe 1081
Barth, Henry... 2881
Basari, Amat A. 3194
Basu, Kaushik 3061
Bauer, Florian 3898
Bauer, Pavol 1193, 3200
Baumgartner, Thomas 1707
Beczckowski, Szymon 2547
Beczkowski, S. M. 2862
Beczkowski, Szymon 2850
Belanger, Jean 2644
Ben Guo... 3129
Ben, Hongqi 2318
Beres, Remus 3320
Berhouet, S. 707
Berkouk, El Madjid...................................... 560
Bessegato, Luca 1087
Besselmann, Thomas 3563
Bhat, Ashoka K. S. 1721
Bhattacharya, Subhashish 651, 656,
    758, 1626, 2562, 3225, 3286, 3447, 3726
Bianda, Enea 3432
Biela, J. 868
Biela, Jurgen 1788
Bilal, Akin 230
Bin Wu 3482, 3695
Binbin Li 3680
Bizen, Yosio... 2983
Blaabjerg, F. 548, 1912, 2862
Blaabjerg, Frede 216, 857, 1529,
    1634, 1801, 2610, 3320
Blank, Frederic 264

AUTHOR INDEX

Bo Wen.....944
Bocker, J.....2887
Bocker, Joachim.....346, 1501, 1508
Boehm, Andreas.....283
Boillat, David O......1073
Boitier, V......707
Boroyevich, Dushan.....944, 2626, 3850
Bortis, D......1291
Bortis, Dominik.....1309, 2079, 3864
Bosshard, R......2167
Bosshard, Roman.....1904
Boyu Wang.....2893
Braz Cardoso, F......3225
Burger, Niklaus.....1386
Burgos, Rolando.....944, 2626, 3850
Burkart, Ralph M......891, 3460
Buticchi, Giampaolo.....433
Byoungchang Jung.....1185
Byung Moon Han.....937
Byung-Geuk Cho.....2802
Byung-Gyu Yu.....3784
Cai, Zheng-Xiu.....429
Canales, Francisco.....1043, 3432
Cao, Guoen.....2587
Cao, Wei.....567
Cao, Yuan.....647
Cardoso, Braz J......3270
Caris, M. L. A......2954
Carvalho, Eden Luiz.....1276
Casolari, Ronaldo Pedro.....1276
Castellazzi, A......2503
Castellazzi, Alberto.....433, 2920, 3718
Ceballos, Salvador.....3758, 3764
Cha, Honnyong.....110, 480, 790
Chai Feng.....3129
Chang, C.-H......2050
Chang, Chien-Hsuan.....2523, 3333
Chang, Hsiu-Feng.....330
Chang, Kai-Chi.....105
Chang, L......2156
Chang, Yuan-Chih.....330, 1832
Changsheng Hu.....782
Changwoo Kim.....1646
Chao Wang.....2950
Chao-Fu Wang.....2758
Chattopadhyay, Ritwik.....3225
Chen, Ching-Guo.....1734
Chen, H......1471
Chen, Hsin-Chih.....1639
Chen, Hung-Chi.....2580
Chen, Jiann-Jong.....2910
Chen, Jung-Chieh.....677
Chen, Min.....485
Chen, Qianhong.....1425
Chen, Shen-Li.....236

Chen, Wei.....66, 72
Chen, Wenjie.....2950
Chen, Yaow-Ming.....3592
Chen, Ying-Zuo.....351
Chen, Zhe.....3538
Chen-Feng Chuang.....3379
Cheng Deng.....782
Cheng, Chun-An.....2523, 3333
Cheng, Hung-Liang.....2523, 3333
Cheng, Po-Tai.....1261, 1639
Cheng, Shih-Jen.....199, 2593
Cheng, Stone.....3425
Cheng-Chieh Yu.....2910
Cheng-Wei Chen.....3592
Cheol-O Yeon.....1738
Cheon, Jun P......3358
Chia-Chi Chu.....3379
Chiang, Hsin-Wei.....2100
Chiba, Akira.....982, 988, 3513
Chiba, Yoshinori.....634
Chien-Yu Lin.....2758
Chih Wei Chen.....3938
Ching-Hsiang Yang.....1639
Ching-Tasi Pan.....3379
Ching-Wei Wang.....1639
Chiu, Chian-Song.....440
Chiu, Huang-Jen.....172, 199, 2593, 3328
Chiu, Tse-Wei.....440
Cho, Bo-Hyung.....2272, 2575
Choi, Bo H......2232, 3358
Choi, Byungcho.....3638
Choi, Hangseok.....2575
Choi, Seong-Chon.....409
Choi, Sewan.....1394, 2247
Choi, Su Y......1103
Chokchai, Chuenwattanapraniti.....3789
Chou, Tzu-Han.....208, 421
Chow, T. Paul.....2208
Chu, Xi.....1322
Chun, Chang Yoon.....2272
Chung, Tsung-Yuan.....2523
Chung-Chuan Hou.....2821
Chung-Yi Lin.....3185
Chung-Yuen Won.....796, 3532
Chunkag, Viboon.....694
Chu-Shen Chang.....3928
Ciftci, Baris.....3734
Coldevin, Grete H......1861
Colmenares, Juan.....3712
Concari, Carlo.....433
Cortes, Patricio.....3864
Cortizio, Porfirio C......3225
Cosovic, Mirsad.....1148
Daesu Han.....1185
Dahidah, Mohamed S. A......1283

AUTHOR INDEX

Dahono, Pekik Argo 3893
Dai, Wei-Fu 330, 1832
Daikoku, A. 1892
Daikoku, Akihiro 2011, 2673
Dan Chen 3938
Darba, Araz 718
D'Arco, Salvatore 1544
Darus, Rosheila 3758, 3764
Daskalos, Mike 2330
Davoodnezhad, R. 1482
Dawson, Francis P 1303
De Belie, Frederik 718
De Carvalho, Kelly Caroline
Mingorancia 2413, 2618, 2988, 3278
De Doncker, R. W. 3898
De Doncker, Rik W. 736, 2729, 3145
De Haan, Sjoerd 2787
De Mallac, Louis 3371
De Miranda, Rubens Domingos 1276
De Paula, Helder 3225
De S. Brito, Jose A. 3225
De Vega, Angel Ruiz 2547, 2850
De, Ankan 651, 2562, 3286, 3447
De, D. 2503
De, Dipankar 433
Deguchi, Tadayoshi 3440
Dehong Xu 782
Dekka, Apparao 3468
Demetriades, Georgios D. 1220
Deng, Lirong......................................465
Dianguo Xu 3174, 3680
Dianguo, Xu 341
Diduch, C. P. 2156
Dilhac, J-M. 707
Diniz, Rogerio Azevedo 3270
Doki, Shinji 907, 2445, 3079, 3823
Domoto, Kazuhide 3652
Dong Le 1837
Dong-Hee Lee 994, 2693
Dong-Jing Lee 1452
Dongkook Son 2914
Dongouk Kim 925
Dongwook Kim 2914
Dou, Qinyun 3604
Dowaki, Kiyoshi 1207
Do-Yun Kim 796
Drofenik, Uwe.....................................1043
Du Yan 2668
Du, Yimian 1721
Duarte, J. L. 2954
Dujic, Drazen.....................................3476
Durand Estebe, P. 707
Dutta, R. 2679
Endo, Takahisa 2541, 2977
Eni, E. 1912

Enomoto, Toshio 2421
Enomoto, Yuji 1997
Erturk, Feyzullah 3734
Eui-Cheol Nho 2763
Fang Zheng 1342
Fang Zhuo 1155, 3546
Fang, Xiaocun 335
Fassler, Lukas 3864
Fei Lin 807
Fei Meng 2815
Fei Zhang 3857
Fernandes, B. G. 2433
Ferrari, Bruno Augusto 3278
Ferreau, Joachim 3563
Ferreira, J. A. 1935
Ferreira, Jan A. 2787
Figueredo, Ricardo Souza 2413, 2618
Fletcher, J. 2679
Fletcher, John 2926, 2932
Foo, Gilbert 2722
Fosso, Olav B. 1861
Foureaux, Nicole C. 3225
Fournier-Bidoz, Sebastien 3496
Franca, Gleisson J. 3270
Franceschini, Giovanni 433
Franke, Toke 78
Fritz, Dominik 3476
Frohleke, N. 2887
Fujii, Junji 1654
Fujii, Kansuke 1748
Fujii, Toshiyuki 2663
Fujimoto, Hiroshi 1671, 2421
Fujimoto, Takafumi 3857
Fujimoto, Yasutaka 1685, 1968
Fujisaki, Keisuke 289, 2856, 2874
Fujisawa, Hiroyuki 3440
Fujita, Hideaki..........1006, 1160, 1350, 2027, 2042
Fujitsuna, Masami 3079
Fuketa, Hiroshi 2228
Fukuda, Kenji 3440
Fukuhara, Shuhei 289
Fukumoto, Hisao 724, 730, 3067, 3249
Fukuoka, Hiroki 3341
Fukushima, Kentaro 2189
Fulin Zhou 1050
Funabiki, Shigeyuki 2470
Funaki, Tsuyoshi 1621
Funato, Hirohito 1728, 2517
Furukawa, Kimihisa 383
Furukawa, Tatsuya 724, 730, 3067, 3249
Furukawa, Yutaka 2252
Furuta, R. 2120
Gaing, Zwe-Lee 278
Gao Qiang 3050
Gao, Qiang 614

AUTHOR INDEX

Geng, Hua 543
Gerling, Dieter 774
Ghimire, Pramod 2547, 2850
Giaretta, Antonio Ricardo........................1276
Goehler, Lutz 2554
Goh Teck Chiang 1028
Gohara, Hiromichi 671
Goto, Akira 130
Goto, Yasuyuki 1490
Goto, Yuichi 1671
Graus, Johannes 270
Grider, Dave 3726
Gruber, Wolfgang 1691, 1701
Gu, Beom W. 1103
Gueldner, Henry 2554
Guidi, Giuseppe 1544
Gunasekaran, Deepak 1342
Guo, Wei 160, 475
Gurpinar, Emre 433, 3718
Ha, Jung-Ik 3140
Hafner, Jurgen 3667
Haga, Hitoshi 415, 3153
Haghbin, Saeid 1373
Hagiwara, Makoto 1586, 1761, 2323, 3742
Hahn, Ingo 270, 283
Haining Wang 3702
Haitao Yang 782
Hak-Soo Kim 2763
Hakutou, Takuma 2297
Hama, Ryota 2470
Hamasaki, Shin-Ichi 2775, 3674
Hamasaki, Sin-Ichi 3093
Hamazaki, Yasuhiro 2126
Hanada, T. 2074
Hanamoto, Tsuyoshi 538, 1811
Han-Shin Youn 1743
Hao Huang 2967
Hao Yi 1155, 2960, 3546
Hao, Xiang 2950
Hao-Chien Cheng 3379
Hara, Hidenori 1654, 1898
Harada, Shingo 1671
Harada, Shinsuke 3440
Harakawa, Masaya 2638
Hariya, Akinori 3630
Hasegawa, Isamu 1365
Hasegawa, Kohei 3707
Hasegawa, Masaru 183, 907, 2445
Hasegawa, Masataka 2212
Hasegawa, Shinya 294, 299, 2972, 3055, 3159, 3162
Hasegawa, Tomonori 2126
Hashimoto, Kento 1974
Hashimoto, S 1471
Hashimoto, Seiji 3194
Hashimoto, Shizuka 3018

Hassanpoor, Arman 3667
Hatanaka, Ayumu 2285
Hattori, Fumiya 811
Hatua, Kamalesh 758, 1626
Hau-Chen Yen 3928
Hava, Ahmet M. 498, 2034, 3734
Hayase, Masanori 1207
Hayashi, Makoto 1950
Hayashi, Toshihiko 3440
Hayashi, Yusuke 1560
Hayashiya, Hitoshi 1062
Hazeyama, M. 1892
Hazra, Samir 758, 1626, 3447
He, Guofeng 485
Hee-Jun Lee 3532
Hei, Xinhong 647
Hella, Mona 2208
Hermansson, Willy 1220
Hernandez, Juan C 3352
Heung-Geun Kim 790, 2763
Hibino, Shinya 2638
Hidaka, Akira 2216
Hidayat, Nabil M 573, 2529
Higuchi, Shinichi 1522
Higuchi, T. 3007
Higuchi, Tsuyoshi 3024
Hijikata, Hiroki 2673, 3828
Hikita, Masayuki 689
Hinata, Toshifumi 919
Hinkkanen, Marko 2489
Hino, Wataru 3525
Hintz, Andrew 2343
Hira, Yuki 730
Hirahara, Hideaki 3044
Hirai, Junji 1974
Hirakawa, Yuki 3024
Hiraki, Eiji 3292
Hirano, Yosei 1956
Hirao, Kuniaki 191, 1365
Hirase, Yuko 1552
Hirokado, K. 146
Hirose, Toshiro 2252
Hirota, Yukitsugu 1728
Hisada, Yoshihiro 3292
Hisato, Hosoyama 2745
Ho, Kung-Min 3942
Hoene, Eckart 2366
Hoffmann, Stefan 2366
Hofmann, Wilfried 2881
Hojo, Masahide 2152
Hojo, Toshiaki 1276
Hokazono, Hiroaki 2870
Holm, Toni 3432
Holmes, D. G. 1482, 2019, 2372, 3306
Homma, Hiroshi 1880

AUTHOR INDEX

Hong Li 3314
Hong, Ki-Nam 2598
Hong-Hee Lee 1013
Hongqi Ben 3213
Hori, Yoichi 2421
Horiguchi, Takeshi 2290
Horita, Yasuhisa 1317
Hosaka, Tatsuya 1350
Hoshi, Nobukazu 1242
Hosoyamada, Yu 801
Hou, Chih-Hao 1796
Hou, Jiaxin 526
Hou, Lixiang 577
Hredzak, Branislav 1458
Hsieh, Guan-Cyun 526
Hsieh, Hung-I 526, 2380
Hsieh, Min-Fu 278
Hsieh, Yao-Ching 429, 1796
Hsin-Chih Chen 1261
Hsin-Ping Su 2821
Hu, Jia-Sheng 278
Hu, Shang-Hung 2606
Hu, Sheng 555
Hu, Taiyuan 335
Huang, Hsin-Wei 421
Huang, Jia-Wei 3233
Huang, Lang 2950
Huang, Min 2610
Huang, T. D. 2195
Huang, Wen-Nan 1734
Huang, Zhenhui 647
Huang-Jen Chiu 2758, 2810, 3185, 3913
Huber, Jonas E. 766
Huber, Tobias 1508
Hui Liu 1634
Hui Zhang 1365, 3455
Huisman, H. 2954
Hull, Brett 3447
Huu-Nhan Nguyen 1013
Hwang, Seon-Hwan 2427
Hwang, Yuh-Shyan 2910
Hwu, K. I. 204, 2754, 3190, 3392
Hyoyol Yoo 1646
Ichihara, Junichi 2189
Ichiya, Takahiro 370
Ichiyanagi, Katsuhiro 1490
Ide, Kozo 933
Ieda, Jun 2663
Iga, Yuichi 3341
Igarashi, Kazunori 2983
Igarashi, S. 3702
Iida, Mikiya 2977
Iijima, Ryuji 117
Iijima, Yukihia 2095
Ikawa, O. 2569, 3702

Ikeda, Hidehiro 2476
Ikeda, Masahiro 2183
Ikeda, Tomohiko 1575
Ikeda, Y. 2569
Ikeda, Yoshinari 2870
Il-Kuen Won 796
Ilves, Kalle 1087
Imai, Jun 2470
Imakiire, Akihiro 689
Imamura, Yasutaka 863
Imanishi, Takao 2663
Imaoka, Jun 811, 883, 2497
Inamori, Mamiko 3509
In-Dong Kim 2763
Inomata, Kentaro 1654
Inoue, Kaoru 3872
Inoue, Keita 130
Inoue, M. 1892
Inoue, Tatsuki 363
Inoue, Y. 246, 258, 312, 390
Inoue, Yukinori 324, 356, 363, 370, 2183, 3018, 3519
Irokawa, Shoichi 1357
Ise, Tomofumi 1430
Ise, Toshifumi 1536, 1560, 2632, 3298, 3687
Ishibashi, Makoto 724
Ishida, Koichi 2228
Ishida, M. 146
Ishida, Masaaki 3707
Ishida, Masaki 3162
Ishida, Takahito 634
Ishigami, Takashi 1880
Ishigma, Satoru 403
Ishihara, Chio 1984
Ishihara, Yuji 1135
Ishii, Hirotaka 294
Ishikawa, Hiroki 1135, 2183, 2189
Ishikawa, Katsumi 2140, 2285
Ishikawa, Takeo 252, 1697
Ishimaru, Yusuke 92
Ishimori, Hitoshi 3440
Ishitobi, Manabu 811
Ishizuka, Tomotsugu 2644
Ishizuka, Yoichi 2222, 2252, 2737, 3630, 3652
Isida, Takashi 1950
Itako, Kazutaka 3244
Ito, Yasuhide 3823
Ito, Yoichi 403
Itoh, Hideaki 724, 730, 3067, 3249
Itoh, Jum-Ichi 1943
Itoh, Jun-Chi 1253
Itoh, Junichi 130
Itoh, Jun-Ichi 84, 138, 152, 191, 682,
1021, 1028, 1095, 1613, 2277, 3659, 3815
Itoh, Tomomichi 850
Itoh, Youichi 415

AUTHOR INDEX

Itoh, Yuki 883, 2497
Iwaji, Yoshitaka 2451
Iwakami, Tetsuro 817
Iwasaki, Makoto 1665
Iwasaki, Shinya 2663
Iwata, Tetsuki 403
Iyer, Kartik V 3037, 3061
Iyer, Shivkumar 3482
Izumi, Toru 3440
Jacobson, Bjorn 3667
Jae-Bum Lee 1738
Jaeho Choi 2656
Jae-Hun Jung 2763
Jae-Hyun Kim 1738
Jaesig Kim 2656
Jang, Jinhaeng 3638
Jang, Young-Jin 664
Jang-Hwan Kim 925
Jardini, Jose Antonio 1276
Jauch, Felix 1788
Javed, Riffat 624
Jayoon Kang 1185
Jen-Hao Teng 1452
Jenn-Jong Shieh 3190
Jeon, Jin-Yong 166
Jeong, Seog Y 1103
Jeong, Seon-Yeong 2406
Jeongjoong Kim 1185
Jhen-Yu Jian 3928
Jia Liu 1536
Jia, Y. 1594
Jianfeng Li 3718
Jiang, Dawang 647
Jiang, Maoh-Chin 105
Jiang, W. 1471
Jiang, W. Z. 204, 3190
Jiang, Yongjie 458
Jianhui Su 2668
Jiann-Fuh Chen 2714
Jianwen Zhang 3124
Jih-Hua Yu 2910
Jin Miaoxin 3050
Jin, Miaoxin 614
Jin, Xu 341
Jin, Yasuhiro 3207
Jing Bian 3314
Jing-Hsiao Chen 3233
Jing-Yuan Lin 2758
Jinjun Liu 835
Jinno, Masahito 1781, 3333
Jin-Woo Ahn 994, 2693
Jinyong Zhang 3213
Ji-Shiang Lee 3346
Joebges, Philipp 2729
Jokipii, J. 514, 2240

Jokipii, Juha 1466
Jong Kyou Jeong 937
Jonghyung Park 1185
Jonishi, Akihiro 2302
Jou, Sung-Tak 224
Jung, Hochang 1990
Jung, Jae-Jung 1268
Jung, Sang-Yong 1990
Jung, Yong-Chae 166, 409
Junghum Lee 2656
Junjie Feng 835
Juntao Fei 3168
Juyoung Jang 2656
Kabasawa, Yuichiro 2175
Kabiri, R. 3306
Kadavelugu, Arun 758, 1626, 3726
Kai, Masahiko 1054
Kai-Hui Chen 2750, 3346
Kaipia, T. 587
Kajiwara, Kazuhiro 3950
Kakishima, Takeo 3513
Kalogera, Maria 1193, 3200
Kameshiro, Norifumi 2140
Kamikura, Mamoru 2064
Kamnarn, Uthen 694
Kanagawa, Kinji 2983
Kanai, Yasuyuki 1567
Kanamori, Masaki 2541, 2977
Kaneko, Junji 2745
Kaneko, Yasuyoshi 1109, 1115
Kanematsu, Masato 2421
Kanemoto, Daisuke 2737
Kang, Feel-Soon 2260
Kang, Yong 555
Kanno, Hiroshi 2302
Kano, Yoshiaki 2004, 2457
Kanoda, Akihiko 3920
Kanouda, Akihiko 2058
Kantar, Emre 2034
Kanthaphayao, Yutthana 694
Kari, Mat Nasir 573
Karki, Ujjwal 1342
Karvonen, Andreas 1373
Kasai, Makoto 3194, 3194
Kashihara, Yugo 1943
Kasper, Matthias 2079
Katade, Motohumi 130
Katakami, Shuji 3440
Kataoka, Yasuhiro 3801
Katayama, Noboru 1207, 1227
Kato, Hideaki 2972, 3162
Kato, Koji 403, 415
Kato, Shinji 2175
Kato, Takashi 3828
Kato, Taro 2972

AUTHOR INDEX

Kato, Tomohisa 3440
Kato, Toshiji 2183, 2189, 3872
Kato, Yutaka 2644
Katoh, Kaoru....................................2285
Katoh, Shuji....................................850
Katsuki, Akihiko 1575, 3624
Katsura, Seiichiro 1679
Kawachi, Konosuke...............................863
Kawaguchi, Shinichi............................3959
Kawahara, Keiji................................1062
Kawakami, Noriko...............................2095
Kawamura, Atsuo 801, 2266, 2794, 3403
Kawamura, Mitsuhiro 3012
Kawamura, Wataru...............................3742
Kawano, Daisuke................................1671
Kawano, Kenji..................................883
Kawazoe, Yosuke................................2011
Kazuya, Ogura..................................452
Kempen, S......................................2887
Kenji, Matsumoto...............................3218
Kern, Ansgar...................................712
Khan, Ashraf Ali...............................110
Khan, Faisal H.................................2161
Khant, Hlaing Kyi Pyar.........................183
Khomfoi, Surin.................................2392
Kicin, Slavo...................................3432
Kihyun Lee.....................................1646
Kikuchi, Takuya................................1328
Kim, Bong C....................................3358
Kim, Chong-Eun.................................1738, 1743
Kim, Dong-Hun..................................790
Kim, Dong-Rak..................................409
Kim, Hee-Jun...................................2587
Kim, Heung-Geun................................110, 480
Kim, Hyejin....................................2575
Kim, Jae-Hyun..................................1743
Kim, Jang-Mok..................................2406
Kim, Ji H......................................2232
Kim, Ji-Won....................................2427
Kim, Jonghoon..................................619
Kim, Minjae....................................2247
Kim, Seonghye..................................2260
Kim, Su-Han....................................480
Kim, Sungmin...................................1268
Kim, Yong-Jae..................................1990
Kimoto, Tsunenobu..............................3440
Kimura, Hiroshi................................1920
Kimura, Noriyuki...............1299, 1806, 2183, 3341
Kimura, Shota..................................883, 2497
Kinouchi, Shin-Ichi............................750, 2290
Kish, Gregory J................................951
Kitabayashi, Tatsuaki..........................2517
Kitagawa, Wataru...............................2310, 3809
Kitajima, Jun..................................1247
Kitazawa, Satochi..............................1438

Kiyota, Kyohei.................................3513
Kleinecke, John................................2330
Kluge, Andreas.................................2554
Knott, Arnold..................................506
Kobayashi, H...................................2569
Kobayashi, Hiroya..............................2517
Kobayashi, Ryota...............................1115
Kobayashi, Takenori............................1868
Kobayashi, Y...................................2569
Kodama, Takashi................................1365
Kogi, Ryosuke..................................2874
Kogoshi, Sumio.................................1207, 1227
Kohama, Teruhiko...............................522, 2781
Kohno, Yusuke..................................2183
Koiwa, Kazuhiro................................84, 130, 1028
Kolar, J. W....................................1291, 2167
Kolar, Johann W. 766, 821, 891, 899, 975,
    1073, 1309, 1707, 1904, 2079, 2834, 3365,
    3460, 3864
Komada, Satoshi................................1974
Komatsu, Wilson................................1276, 2413, 2988
Komeda, Shohei.................................1160
Komiya, Hiroshi................................2421
Kon, Saytaro...................................3263
Kondo, Keiichiro...............................1438, 2126
Kondo, Seiji...................................415
Kondo, Takeshi.................................1365
Kondou, Masahiko...............................2421
Kono, Y..2120
Kono, Yasuhiko.................................2140
Konoto, Masaaki................................2189
Konstantinou, Georgios.........................1458, 3758, 3764
Korner, Olaf...................................2113
Kosaka, T......................................2438
Kosaka, Takashi................................1984, 1997
Koschik, Stefan................................3145, 3898
Koseki, Takafumi...............................1334, 2126
Kotegawa, Ryo..................................1317
Kotera, Keito..................................3872
Kouki, Matsuse.................................3134
Kouno, Yusuke..................................2189
Kounoto, Masaaki...............................2175
Koyama, Masato.................................750
Krafft, Eberhard...............................2113
Krismer, Florian...............................2834
Kuan-Hsien Chou................................3346
Kubo, H..395, 1594
Kubo, Hajime...................................1601
Kubo, Yuji.....................................3134
Kubota, Hisao..................................919, 1929, 3119, 3134
Kubota, Yutaka.................................2183
Kudo, Takahiro.................................1109
Kuga, Shotaro..................................3955
Kukita, Akio...................................1444, 2351
Kumagai, Shunji................................3194

AUTHOR INDEX

Kumakura, Yoshito....................1715
Kume, Tsuneo....................1898
Kumsuwan, Yuttana....................3417
Kun-Hung Chen....................3592
Kunomura, Ken....................1054
Kuo, Kuan-Yi....................278
Kuperman, A.....................2240
Kurabayashi, Toshiyuki....................1962
Kuribayashi, H.....................2569
Kurihara, Takeshi....................299
Kurihara, Yoshihiro....................1874
Kurita, Nobuyuki....................252, 1697
Kuroda, Y.....................1892
Kurokawa, Fujio....................2108, 3611, 3950
Kusaka, Keisuke....................191
Kusukawa, Jumpei....................2904
Kusunoki, Hironobu....................2330
Kutsuki, Tomohiro....................2064
Kuwahara, Akinobu....................3179
Kuzumaki, Atsuhiko....................1929
Kwasinski, Alexis....................2649
Kwasinski, Andres....................2649
Kwon, Soon-Kurl....................2359
Kyungbae Lim....................2656
Kyungmin Sung....................744
Kyungsub Jung....................1646
Lai, Yen-Shin....................3942
Lamantia, A.....................2503
Lana, A.....................587
Lang, Klaus-Dieter....................2366
Larsson, Tomas....................1220
Laska, Bernd....................2113
Law, Kah Haw....................1283
Le Hoai Nam....................3659
Lee, Chia-Tse....................1639
Lee, Dong-Choon....................1201, 2406
Lee, Eun S.....................1103, 2232, 3358
Lee, Hong-Hee....................2826
Lee, Jae-Bum....................1743
Lee, June-Hee....................493, 532
Lee, June-Seok....................493, 532
Lee, Kyo-Beum....................224, 493, 532
Lee, Min-Hua....................236
Lee, Seong Ryong....................3292
Lee, Shiu-Hui....................1734
Lee, Sung W.....................1103
Lee, Taeck-Kie....................595
Lee, Tzung-Lin....................2606
Lee, Woo-Cheol....................595
Lee, Ya-Ting....................440
Lee, Yuang-Shung....................208, 421
Lehmann, Oliver....................3085
Lehn, Peter W.....................951
Lei, Wanjun....................160, 475
Leibl, Michael....................899

Lelie, Markus....................2729
Leslie, Scott....................3726
Leuenberger, D.....................868
Leuer, Michael....................346
Li Yan....................2899
Li, Ding....................341
Li, Haiqing....................2095
Li, Hong....................2893
Li, Ning....................160, 475
Li, Qian....................2161
Li, Yanxiang....................3002
Lian, K. L.....................2195
Liang Hao....................3174
Liangyi Tang....................3695
Liao, Jhen-Yu....................2580
Lie Guo....................3489
Lin Cheng....................3447
Lin, Chiao-Chien....................3072
Lin, Chia-Yu....................1832
Lin, Chien-Yu....................172
Lin, Chung-Yi....................199, 2593
Lin, Fei....................335, 1322, 2133
Lin, Jing-Yuan....................172
Lin, L.-C.....................2050
Lin, Z. Y.....................1471
Lindberg-Poulsen, Kristian....................2842
Liping Zheng....................1837
Liserre, Marco....................857, 3320
Liu, Baoquan....................577
Liu, Fang....................567
Liu, Fangcheng....................3604
Liu, Fuxin....................458, 2768
Liu, Hanchao....................967
Liu, Jianyu....................614
Liu, Jilong....................66, 72
Liu, Jinjun....................624, 2815, 3604
Liu, Kangzhi....................3568
Liu, Ning....................2156
Liu, Rongqiang....................3099
Liu, Tai-Chun....................105
Liu, Xiankai....................647
Liu, Xiaosheng....................3002
Liu, Yi-Hua....................3233
Liu, Yu-Chen....................199, 2593
Liuchen Chang....................1476, 2668, 3842
Lo, Yu-Kang....................172, 199, 2593
Lobsiger, Yanick....................1309
Loh, Poh Chiang....................216, 1529, 1634, 1801, 2610
Longlong Zhang....................782
Lopez-Arevalo, Saul....................3718
Lovatt, Howard....................2679
Low, K. S.....................446
Lu, Dylan D. C.....................3553
Lu, Kao-Yi....................105
Luo, Guomin....................2145

AUTHOR INDEX

Luthardt, Sven 3029
Ma, K. 548, 2862
Ma, Weigang 647
Madawala, Udaya K. 2722
Madhusoodhanan, Sachin 656, 1626
Maekawa, Sari...............................919, 1929
Maemura, Akihiko.......................................1898
Maeyama, Shigetaka.....................1575, 3624
Maezono, Paulo Koiti....................................1276
Mahdavikhah, Behzad...................................3582
Mainali, Krishna 758, 1626
Makaino, Yuki...914
Makita, Shinji..3823
Mamun, Mostafa 97
Manias, Stefanos N..1606
Mannen, Tomoyuki.......................................2042
Manolas, Iakovos 1606
Mao, Meiqin 2156
Maret, C...3239
Marrero Sosa, Juan Alberto 3476
Martinz, Fernando Ortiz..................2413, 2988, 3278
Maru, Naoki..2285
Marukawa, Yasuhiro......................................1984
Marumori, Hiroki 3055
Maruta, Hidenori 3611
Marz, Andreas..2113
Marzouk, Ahmad Diab 3496
Masaki, Kenji 2663
Mashino, Masahiro 3162
Masic, Semsudin..1148
Maskell, D. L...3598
Masuda, Hiroyuki 92
Masui, Takeshi 1317
Masutomo, Kazufumi 3624
Masuzawa, Hiroshi 1054
Masuzawa, Takashi..2366
Matakas, Lourenco 2413, 2618, 2988, 3278
Matsubara, Masakatsu 1874
Matsuda, Katsuhiro..415
Matsuhashi, Daiki 1886
Matsuhashi, Masataka....................................1516
Matsui, Hitoshi 1586
Matsui, Keiju 183
Matsui, Mikihiko...3489
Matsui, N. 2438
Matsui, Ryota 1128
Matsui, Yoshihiro..2385
Matsui, Yoshinobu 2745
Matsumoto, Akira..1560
Matsumoto, Atsushi.......................................2445
Matsumoto, Kazushi......................................3440
Matsumoto, Satoshi.......................................2216
Matsumoto, Shuhei 1929
Matsumoto, Yasushi 1920
Matsuo, Hirofumi..1781

Matsuo, Keisuke 1886
Matsuo, Yusuke 1671
Matsuoka, Kazumasa.....................................3207
Matsuoka, Yuji 744
Matsushima, Yoshitarou 3801
Matsushita, Makoto 3012
Matsuura, Kei 1516
Matsuura, Ken 3630
Matsuzaki, Ryohei 1978
Mattavelli, Paolo 3850
Mattsson, A. 587
Mauerer, M. 1291
McGrath, B. P..................1482, 2019, 2372, 3306
McLean, Kenneth 3496
Meiqin Mao.........................1476, 2668, 3842
Mekhilef, Saad 560, 3574
Melkebeek, Jan 718
Meng, Fei 624
Meng, Tao 2318
Merahi, Farid 560
Messo, T. 514, 2240
Messo, Tuomas 1466
Mihara, Teruyoshi 1728
Mii, Kenji 2737
Miiura, Yushi 1430
Mikihiko 3784
Mills, Liam 3718
Ming Yang 3174
Mingfei Wu 3553
Mingyan Wang 3129
Mino, Kazuaki 1920
Minoshima, N..2438
Minowa, Masanao 3828
Minsoo Jang 3933
Mira, Maria C..506
Mishima, Tomoakzu 2533
Mishra, Santanu 2707
Mishra, Santanu Kumar 3587
Misu, Daisuke 1874, 3012
Mitterhofer, Hubert 1701
Miura, Yushi 1536, 3298
Miyajima, Hiroki 1054
Miyajima, Takayuki 2421
Miyakawa, Takayuki 2421
Miyama, Yoshihiro 2673
Miyashita, S. 3702
Miyawaki, Satoshi 84
Miyazaki, Hideki 383
Miyazaki, Kensuke 601
Miyazaki, Toshimasa 1956
Miyazaki, Yuji 750
Mizoguchi, Takahiro 1660
Mizukami, Makoto 3440
Mizuki, Tatsuya 1575, 3624
Mizuma, Takeshi 2126

AUTHOR INDEX

Mizuno, Takayuki 1886
Mizusaki, Hiroshi 3093
Moballegh, Shiva 656
Mochikawa, Hiroshi 1929
Mochizuki, Eiji 671, 2870
Mochizuki, K. 2569
Mohamed, Essam Ebaid 3877
Mohamed, Tarek Hassan 3877
Mohan, Ned 1036, 1412, 3037, 3061, 3750
Mohd Arif, Mohd Johari 573
Molinas, Marta 1861
Momose, Fumihiko 671
Moo, Chin-Sien 1796, 3796, 3928
Moon, Dongok 1394
Moon, Gun-Woo 1738, 1743
Moon, Sang-Ho 224
Moorthy, Radha Sree Krishna 2087, 3616
Moraes, Lenin 3225
Mori, Tomohiro 2983
Morikawa, R. 258
Morimoto, Masayuki 3509
Morimoto, S. 246, 258, 312, 390
Morimoto, Shigeo 324, 356, 363, 370, 1997, 3018, 3519
Morimoto, Shinya 1654
Morishita, Shin 130
Morita, Hiroshi 1490
Morita, Kazunori 191, 582
Morita, Kosuke 3624
Morita, M. 1892
Morizane, Toshimitsu 1299, 1806
Morizane, Tosimitsu 3341
Moroi, Takayuki 3134
Morozumi, Akira 671
Mory, David 2554
Motizuki, Shun 2745
Motoi, Naoki 801, 2266
Motomura, Masashi 3611
Mouri, Masayuki 1728
Mrak, Branimir 1701
Mukai, Ryosuke 2775
Mukunoki, Makoto 97, 1950
Munk-Nielsen, Stig 2547, 2850
Murai, Kensuke 1567
Murai, Toshiaki 1122
Murakami, Daichi 1728
Murakami, Kouhei 2385
Murakami, Toshiyuki 1962
Murata, Koji 2108
Murata, Munehiro 1173
Murata, Yuichiro 2064
Musing, Andreas 821
Mustapa, Rijalul Fahmi 2529
Muta, Shoichiro 3067
Nag, Soumya Shubhra 3587

Nagai, Shinichiroh 811
Nagano, Tetsuaki 2638
Nagano, Tsuyoshi 1253
Nagano, Y. 146
Nagashima, Tomohiro 2175
Nagata, Shun 2252
Nagatomo, Yoshinobu 2663
Nagel, Andreas 2113
Nagura, Hirokazu 2451
Nagy, Istvan 2700
Naitoh, Haruo 1135
Nakagawa, Hidehiko 1552
Nakagawa, Yuki 2533
Nakahara, Mizuki 744, 2511
Nakajima, Yoichiro 403
Nakamura, M. 376
Nakamura, Ritaka 92
Nakamura, Sota 2400
Nakamura, T. 2074
Nakamura, Tatsuya 1575
Nakamurame, Fuminori 2632
Nakanishi, Toshiki 1095
Nakano, Y. 2074
Nakao, Hiroshi 2745
Nakao, Noriya 1128, 1141
Nakaoka, Mutsuo 2359, 2533
Nakaoka, Mutuo 3341
Nakashima, Yoshiyasu 2745, 3386
Nakata, Yuki 138
Nakatsu, Kinya 2904
Nakatsugawa, Junnosuke 2451
Nakayama, Koji 3440
Nakayama, Naoyuki 3857
Nakayama, Yasushi 2290
Nakazawa, Yosuke 1357
Nam, Kwang-Hee 664
Narita, Takayoshi 294, 299, 3055, 3159
Nashida, N. 2569
Nayanasiri, D. R. 3598
Nee, Hans-Peter 3712
Neubert, Markus 3145
Nguyen, D. 2679
Nguyen, Quoc Khanh 318
Nguyen, Thanh Hai 2406
Nha, Quang Trong 3913
Nho Van Nguyen 2826
Nian Heng 843
Nicolae, Ileana-Diana 2996
Nicolae, Marian-Stefan 2996
Nicolae, Petre-Marian 2996
Niijima, Koji 1299, 1806
Nilssen, Robert 1412
Nimura, Tomohiro 3079
Ning Liu 2668

AUTHOR INDEX

Ninomiya, Tamotsu 177, 1179, 2216, 2222, 3630, 3652
Nishida, Katsumi 2359
Nishida, Yasuyuki 2189
Nishikata, Shoji 959
Nishimura, T. 3702
Nishimura, Tomohiro 2870
Nishimura, Yoshitaka 671
Nishio, Haruhiko 2302
Nishioka, Tomoya 2152
Nishisu, Koji 2285
Nishiyama, Noriyoshi 2011
Nishizawa, Shinichi 117, 744
Niu, Ruigen 160, 475
Noda, Koji 2541
Noda, Taku 2175
Noguchi, Kenji 2277
Noguchi, S. 2569
Noguchi, Toshihiko 1173, 1405
Noh, Yong-Su 166
Nomura, Naofumi 1522
Nomura, Shinichi 3218
Nonaka, Hirotaka 2737
Norigoe, Isami 117
Noro, Osamu 1552
Norrga, Staffan 1087
Noto, Yasuo 682
Nozaki, Takahiro 1660
Nozawa, Ryosuke 1115
Nussbaumer, Thomas 975, 3365
Nuutinen, P. 587
Oboe, Roberto 1679
O'Byrne, Sean..............................2926, 2932
Oda, Yoshinori 829
Odawara, Shunya....................289, 2856, 2874
Ogasawara, Satoshi..............1728, 2977, 3525
Ogashi, Yoshihiro.................................92
Ogawa, Kazutoshi 2140, 2285
Ogawa, Takashi 2285
Ogura, Kazuya 582, 3455
Ogura, Tsuneo.................................2068
Oh, Min-Seok 166
Ohara, Shinya 850
Ohashi, Hiromichi 117, 744
Ohashi, Shunsuke 3410
Ohchi, Masashi................724, 730, 3067, 3249
Ohishi, Kiyoshi 1247, 1516, 1956, 3153
Ohnishi, Kouhei 1660, 2483
Ohnuma, Takumi 914
Ohnuma, Yoshiya...................................84
Ohse, Naoyuki 3440
Ohtake, Asuka 3857
Oi, Kazunobu 452
Oi, Takeshi...2290
Oishi, K. 376

Oishi, Koji 3012
Oiwa, Takaaki 988
Ojika, Satoshi....................................1430
Oka, T. 376
Oka, Toshiaki 2330
Okamoto, Dai 3440
Okamoto, Masayuki 3292
Okamoto, Shoji 2290
Okamura, Kazuki 3674
Okazaki, Fumihiro 1728
Okazaki, Yuhei 1586
Okitsu, Takashi 1886
Okubo, Toshikazu 2058
Okuma, Jun 1978
Okuma, Yasuhiro 2834
Okumura, Hajime 3440
Okuyama, Yoshihiro 1811
Omata, Shinpei 2944
Omi, Masataro 1317
Omori, Hideki.....................1299, 1806, 3341
Omote, Kenichiro 1950
Omura, Mototsugu 1685
Ong, Andrew 2722
Onishi, Mitsuru 1054
Ooishi, Eiji 183
Ooshima, Masahide 1715
Orikawa, Koji................191, 1613, 2277, 3659
Ortiz, G. 1291
Ortiz, Gabriel....................................1309
Oshima, Ryo..1021
Oshinoya, Yasuo........294, 299, 2972, 3055, 3159, 3162
Oso, Hiroshi 629
Ota, Chiharu 3440
Ota, Satoru 2095
Otsuki, Midori 1054
Ouchi, Takayuki 3920
Ouyang, Shaodi 624
Ouyang, Ziwei 2842
Ozaki, Takayuki 2638
Ozkan, Ziya 498
Pala, Vipindas 2208
Palmour, John 3447
Pan, Miao 582
Panda, S K 1775
Panda, Sanjib Kumar 1580
Pansier, F. 1935
Papafotiou, Georgios 1606
Papastergiou, Konstantinos 1220
Park, Gyeong-Jae 1990
Park, Junsung 1394
Park, Yongsoon 2598
Parker, S. G. 2019, 2372
Partanen, J. 587
Patel, Dhaval 758, 1626
Pedersen, Kristian Bonderup 2547

AUTHOR INDEX

Peftitsis, Dimosthenis.................3712
Peltoniemi, P..........................587
Peng Gao......................2926, 2932
Peng Wang.........................3124
Peng Wen.............................782
Peng, Han...........................2208
Peretti, L...........................3111
Peters, A...........................2887
Peters, Wilhelm.....................1508
Petersen, Lars P....................3352
Petersen, Lars Press................2842
Petrich, Matthias....................318
Pettersson, Sami....................3432
Pham Phu Hieu.......................3913
Pidaparthy, Syam Kumar..............3638
Piepenbreier, Bernhard........1816, 3029
Ping-Heng Wu........................1261
Pires, Igor A..................3225, 3270
Pittet, Serge.......................3371
Pittini, Riccardo...................3905
Po-Chien Chou.......................3425
Poh Chiang Loh.......................857
Po-Jung Tseng..................2810, 3328
Popa, Lucian-Dinut..................2996
Popova, L............................548
Popovic, J..........................1935
Poshtkouhi, Shahab..................2336
Pou, Josep.....................3758, 3764
Prasanna, I. V......................1580
Prasanna, U. R..........230, 395, 1594
Prasanna, Udupi. R..................2343
Prodic, Aleksandar..................3582
Pyrhonen, J..........................548
Qi Zhang............................3489
Qu, Lizhi............................609
Qunzhan Li..........................1050
Rabkowski, Jacek....................3712
Radic, Aleksandar...................3582
Radman, Karlo.......................1691
Rae-Sung Yu..........................925
Rahman, F...........................2679
Rahman, M. A.........................982
Rahman, M. F........................2686
Rajashekara, K................395, 1594
Rajashekara, Kaushik.....230, 2343, 3134
Raju, Siddharth.....................1036
Ramadan, Husam A.....................863
Rambetius, Alexander...........270, 3029
Rannestad, Bjorn....................2547
Rannested, Bjorn....................2850
Rathore, Akshay K...................1775
Rathore, Akshay Kumar..........2087, 3616
Ray, Olive..........................2707
Razik, H............................3239
Reiter, Tomas.......................774

Ren, Kangle..........................465
Riffat, Javid.......................2815
Rikitake, Jungo.....................2216
Rim, Chun T...............1103, 2232, 3358
Rivera, Marco.......................3574
Robbins, William P.............3037, 3061
Rodriguez, Jose.....................3574
Rongfeng Yang.......................3680
Rosekeit, Martin....................2729
Roth-Stielow, Jorg..........264, 318, 3085
Roy, Sudhin..............651, 2562, 3286
Ruan, Xinbo....................458, 2768
Ruda, Harry E.......................1303
Ruderman, Michael...................1665
Rufer, Alfred.................1081, 1386
Ryo, Mina...........................3440
Ryu, Sei-Hyung......................3726
Saarakkala, Seppo E.................2489
Saga, Yasunao.......................1748
Sahoo, Ashish Kumar.................3750
Saikusa, H..........................3007
Saito, Eiichi.......................1679
Saito, Katsuhiko....................2064
Saito, Ryo..........................3397
Saito, Ryoji........................2189
Saito, Takashi.......................671
Saitoh, Ryoh.........................914
Sakaba, Kouichi.....................3159
Sakai, Kazuto........................240
Sakai, Tomoyasu.....................1490
Sakai, Toshifumi....................2451
Sakaino, Sho........................1978
Sakimoto, Kenichi...................1552
Sakurai, Naoki......................2297
Sakurai, Takayasu...................2228
Sampath, Prasad K...................2722
Sanada, M...............246, 258, 312, 390
Sanada, Masayuki.....324, 356, 363, 370, 3018, 3519
Sand, Kjell.........................1861
Sariyildiz, Emre....................2483
Sasaki, Tomotake....................2745
Sasongko, Firman....................1761
Sato, Daisuke.......................3815
Sato, Koji..........................1671
Satria, Andri.......................3893
Sauer, Dirk Uwe.....................2729
Sawada, Tadashi.....................1122
Sayed, Khairy.......................2359
Sayed, Mahmoud A....................3877
Schob, Reto. T......................1691
Schon, Andre........................3255
Schrittwieser, L....................1291
Schupbach, Marcelo..................3447
Schuster, Johannes..................3085
Segaran, D. S.......................2372

AUTHOR INDEX

Segsa, Karl-Heinz 2554
Seilmeier, Markus 1816
Sekiba, Yoichi 2175
Sekisue, Takayuki 2175, 2189
Seo, Gab-Su 2272
Seok-Jin Hong 3532
Seunghoo Song 1646
Seung-Ki Sul 925, 2802
Severson, Eric 1412
Shah, Shahil 843, 967
Shao Zhang 1342
Shaodi Ouyang 2815
Shaofeng Xie 1050
Shaohua Sun 3213
Shaohui Zhong 1155
Shen, Na 582
Shen, Zhiyu 3850
Shenghui Cui 1268
Sheng-Kai Kao 2714
Shi, Hongtao 577
Shi, Rongliang 567
Shibahara, Kohei 1575, 3624
Shibahara, Ryota 2222
Shibanuma, Kenichi 634
Shibata, Yuichiro 3950
Shieh, Hsin-Jang 351
Shieh, Jenn-Jong 204
Shigematsu, Koichi 2183, 2189
Shih, Bing-Jyun 105
Shih, Sheng-Fang 2380
Shih-Jen Cheng 3185
Shimada, Takae..2058
Shimamori, Hiroshi.....................................2745
Shimao, Toshihiro 415
Shimatou, T. 2939
Shimizu, Kyohei 1968
Shimizu, Toshihisa ... 876, 1166, 2944, 2983, 3044, 3771
Shimizu, Toshimasa 1054
Shimode, Daisuke 1122
Shimomura, Junichi 2183
Shimono, Tomoyuki 1685
Shin Shiung Wang 3938
Shin, Hyunhak 110
Shin, Yesl 493
Shinagawa, Syuhei 252
Shinbo, Mitsuo 634
Shindo, Yuji 1552
Shinnaka, Shinji 1824
Shinohara, Atsushi......................................324
Shinozaki, Ikki 1728
Shinozuka, Yasuhiro 2228
Shioda, Masashi 130
Shirakawa, Kazuhiro...........................304, 1379
Shiraki, N. 2120
Shirasawa, Koki 3106

Shishida, Yasuhiro 2644
Shiting Weng 1476
Shixi Hou 3168
Shoeiby, B. 1482
Shoji, Hiroyuki 2058
Shoyama, Masahito 863, 3386
Shu-Hung Liao 1452
Shuitao Yang...1342
Shun-Chung Wang 3233, 3778
Shunke Sui 3680
Shuren Wang 3194
Shu-Wei Kuo 3185
Siemaszko, Daniel 3371
Silva, Marcelo 3864
Silva, Sidelmo M. 3225
Silventoinen, P. 587
Sin, Min-Ho 409
Singh, B. N. 3482
Sintamarean, C. 1912
Sitbon, M. 2240
Sivakumar K 1400
Siwakoti, Yam P. 1801
Skuriat, Robert 2920
Smadi, Issam 1968
Smaka, Senad 1148
Smiththisomboon, Somrat 3885
Sogawa, Yuki 522, 2781
Solomon, Adane Kassa 2920
Sone, Kodai 3525
Song Kejian 640
Song, Z. Q. 2686
Sonoda, Hideki 634
Soo-Cheol Shin 3532
Specht, Andreas 1501
Srirattanawichaikul, Watcharin 3417
Steinert, Daniel 975
Steinke, Gina 1081
Stumpf, Peter 2700
Su, Bonan 614
Su, Hong-Wei 1781
Su, Jianhui 2156
Suetsugu, Tadashi.......................................3955
Sugao, Kazumi 403
Sugimoto, Hiroya..982
Sugiura, Makoto 1135
Suh, Yongsug 1185, 1646
Su-Han Kim 790
Sul, Seung-Ki 1268, 2598
Sumida, Hitoshi 2302
Sun, Jian 843, 967, 2202
Sun, Shaohua 2318
Sun, Wei 609
Sunaga, Keita 3162
Sung, Kyungmin 117, 829
Sunsoon Park 1646

AUTHOR INDEX

Suntio, T................514, 2240
Suntio, Teuvo................1466
Suryadevara, Rohit................2433
Suto, Kenji................3194, 3194
Suul, Jon Are................1544
Suwankawin, Surapong................3885
Suzuki, Genri................1697
Suzuki, Hirokazu................3503
Suzuki, Katsumi................959
Suzuki, Michiaki................883
Suzuki, Nobuyuki................919, 2541
Suzuki, Ryosuke................2972
Suzuki, Shun................1166
Suzuki, Takashi................1062
Suzuki, Toshiki................907
Svensson, Jan R................1220
Tabira, K................2939
Tadano, Yugo................1242
Tadokoro, D................390
Taeck-Kie Lee................3532
Taekyun Kim................3933
Tae-Won Chun................2763
Taga, Hironori................1328
Tajima, G................2438
Takada, Hiromu................1697
Takagi, Ryo................1328
Takahashi, Akiko................2470
Takahashi, Hiroki................152, 1021
Takahashi, Hirotaka................1068
Takahashi, Hisashi................3106
Takahashi, K................2569
Takahashi, Naoya................3920
Takahashi, Nobuhiro................3207
Takahashi, Osamu................2285
Takahashi, Takehiro................3207
Takahashi, Yoshikazu................671, 2870
Takamiya, Makoto................2228
Takao, Kazuto................744, 3440, 3707
Takasaki, Mika................2252
Takashita, Haruomi................3386
Takasu, Shinji................3440
Takatsuka, Yushi................1898
Takayama, Masakazu................3801
Takayanagi, Atsushi................2794
Takeda, Kotaro................1654
Takeda, Masashi................801
Takeda, Takashi................1490
Takei, Manabu................3440
Takemoto, Masatsugu................1000, 3525
Takenaka, Kensuke................3440
Takenami, Fumiaki................2737
Takenoiri, S................2939
Takeshita, Takaharu............123, 601, 2310, 3809, 3877
Takeuchi, Katsutoku................3012
Takeuchi, Shun................3646

Takezaki, Kenichi................3525
Taki, Hiroshi................876, 1379
Takino, Toshiaki................817
Tam Khanh Tu Nguyen................2826
Tamada, Shunsuke................1357
Tamura, Hiroshi................682
Tan, Nadia M. L................750
Tanabe, Ryo................1234
Tanai, Masanobu................829
Tanaka, Daiki................3179
Tanaka, Junya................240
Tanaka, Kiminori................2222
Tanaka, Koutaro................1006
Tanaka, Seiyu................982
Tanaka, Takahide................2302
Tanaka, Toshihiko................3292
Tanaka, Yasunori................3440
Tanaka, Yuichiro................1880
Tanifuji, Hikaru................1115
Taniguchi, Shun................2465
Tao Meng................3213
Tatsuta, Fujio................959
Tauchi, Yuki................3119
Teng, Jen-Hao................677
Teodorescu, R................1912
Tera, Takahiro................304, 876
Terabe, Ryosuke................2638
Terao, Yutaka................2644
Teshima, Masato................1068
Thiringer, Torbjorn................1373
Thogersen, Paul................2547
Thogersen, Paul Bach................2850
Tian, Yanjun................3538
Ting, Pangan................677
Ting, Yeh................2787
Tint Soe Win................3292
Toba, Akio................2011
Toda, Hiroaki................1984
Togashi, Ryo................356
Toi, Takahiro................1109
Tokiwa, Tsuyoshi................2977
Tokuda, Hirokazu................2175
Tokumasu, Akira................1379
Tokuyama, Takeshi................2904
Tominaga, Shinji................2290
Tomioka, Satoshi................3630
Tomita, Mutuwo................907
Tonogi, K................2438
Toru................3784
Tosaka, Shuhei................1207, 1227
Town, Graham E................1801
Toyoda, Hajime................1560
Tran, Q. V................446
Trescases, Olivier................2336, 3496
Trillion Zheng................2893

AUTHOR INDEX

Trintis, Ionut 2547
Tripathi, Awneesh 758, 1626
Trompa, Thomas 2554
Tsai, Jiung-Lin 1781
Tsai, Ming-Hsiao 278
Tsan Chen 2810
Tse, Chi K. 1425
Tseng, K. J. 2145
Tsorng-Juu Liang 2750, 3346
Tsubakidani, Takashi 3674
Tsuboi, Yoshiki 1160
Tsuboi, Yuichi 92
Tsuchida, Kazuo 3397
Tsuda, Junichi 1929
Tsuji, Mineo 2775, 3093, 3674
Tsuji, Satoshi 522, 2781
Tsuji, Toshiaki 1978
Tsukakoshi, Masahiko 92, 2330
Tsuruma, Yoshinori 1054, 2644
Tsuruta, Hironori 629
Tsuruta, Ryoji 1350
Tsuruta, Yukinori 2266, 3403
Tsuyoshi, Hanamoto 2476
Tu, Yunwu 465
Tukiman, Rahayu 2529
Turpin, Santiago 1974
Tuysuz, Arda 1904
Tzou, Ying-Yu 3072
Uddin, Muslem 3574
Ueda, K. 312
Ueda, Tetsuya 403
Ueda, Tetsuzo 2075
Ueda, Yoshinobu 1855
Uemura, Hirofumi 891, 2834
Ukai, Hiroyuki 2400
Umeda, Nobuhiro 2183
Umeno, Masayoshi 183
Umesh B S 1400
Umetani, Kazuhiro 304
Undeland, Tore 1412
Uno, Masatoshi 1444, 2351
Urushibata, Hiroaki 2290
Urushibata, Shota 1365, 3455
Ushiro, Nobumasa 3106
Vaisanen, V. 587
Vajk, Istvan 2700
Van Brunt, Edward 3726
Van Wyk, J. D. 1935
Van-Long Tran 1214
Vasiladiotis, Michail 1386
Vasquez-Arnez, Ricardo Leon 1276
Veerasamy, Balaji 2310
Vieto, Ignacio 843
Viinamaki, J. 2240
Vilathgamuwa, D. M. 2722, 3598

Vogt, T. 2887
Wada, Keiji 744, 1379, 2511, 3646
Wahlstroem, Jonas 3476
Wajima, Kiyoshi 1984
Wallscheid, Oliver 1501
Wang Hui 640
Wang, Bin 2133
Wang, Chao-Fu 172
Wang, Fei 470
Wang, Fusheng 465
Wang, H. 1912
Wang, Hengli 72
Wang, Jun 944
Wang, Lingxiang 465
Wang, Lipeng 458
Wang, Xiaojian 2815
Wang, Xinyu 624, 2815
Wang, Xiongfei 216, 1529, 3320, 3538
Wang, Yanbo 3538
Wang, Yong 470
Wang, Yue 160, 475
Wang, Zhao'An 160, 475
Watanabe, Daisuke 988
Watanabe, Hiroki 84
Watanabe, S. 2939
Watanabe, Shoichiro 1334, 2126
Watashima, T. 2939
Wei Jiang 3194
Wei Liu 1050
Wei Wang 3680
Wei Yan 3455
Wei, Guo 2318
Wei, Sun 341
Weili Dai 3168
Weirong Chen 3695
Wei-Ting Hsu 2910
Wen, Bo 3850
Wen, Chao-Kai 677
Wen, Huiqing 702
Wen-Chien Hsu 2714
Wenjie Chen 2967
Wen-Tai Li 677
Wheeler, Pat. 2920
Won, Chung-Yuen 166, 409
Wong, Siu-Chung 1425
Wonsuk Choi 2914
Woojin Choi 1214
Wu Mingli 640
Wu, Bin 3468
Wu, Chun-Wei 330, 1832
Wu, G. F. 1471
Wu, Gwo-Bin 3796
Wu, T.-F. 2050
Wu, Tsung-Hsi 1796
Wu, Weimin 2610

AUTHOR INDEX

Wu, Weiyang 582
Wu, Wenlong 470
Wu, Wen-Zhe 429
Wunsch, B. 2167
Xia, Huan 2133
Xiangdong Sun 3489
Xiang-Dong Sun 3784
Xiao, D. 2686
Xiao, Fei 66, 72
Xiao, Shuai 543
Xiaojie You 2893
Xiaojie Zhuang 2638
Xiaolong Ma 835
Xiaomei Song 2967
Xie, Ruiliang 2950
Xiong, Li 230
Xiuqin Wei 3955
Xu Cai 1842, 3124
Xu Dianguo 3050
Xu Yang 2967
Xu, David 3099
Xu, Dehong 485
Xu, Dianguo 609, 614, 3002
Xu, Haizhen 567
Xu, Rong 609
Xue, Danhong 3604
Xuling Chen 2768
Yablecki, Jessica 3496
Yachi, Toshiaki 3959
Yakabe, Seichiro 730
Yamada, Hiroaki 538, 1811
Yamada, Kenji 1898
Yamada, Ryuji 1920
Yamada, Takatoshi 2212
Yamada, Tatsuji 3263
Yamagata, Shinichi 829
Yamagishi, Tatsuya 750
Yamaguchi, Shota 3771
Yamaguchi, Takashi 1242
Yamaichi, Katsuya 2517
Yamaji, Masaharu 2302
Yamamoto, Eiji 1654
Yamamoto, Junichi 177, 1179
Yamamoto, Kenji 3106
Yamamoto, Kichiro 689
Yamamoto, Kohei 1438
Yamamoto, Masayoshi 811, 883, 2497
Yamamoto, Shu 3044
Yamamoto, Takashi 3707
Yamamoto, Yasuhiro 1601
Yamamura, N. 146
Yamanaka, Kenji 2152
Yamanaka, Tatsuya 1207, 1227
Yamanoi, Takashi 1062
Yamashita, Nobuyuki 2983

Yamashita, Shigeharu 2745
Yamazaki, Akira 933
Yan Li 3124
Yan Zhang 835
Yanagi, Hiroshige 3630
Yang Chuan 1962
Yang, Cs 199, 2593, 3185
Yang, Daeki 2247
Yang, Geng 543, 582
Yang, Guorun 66
Yang, Hong-Tzer 2100
Yang, Rongfeng 609
Yang, Shih-Sian 2606
Yang, Sihun 863
Yang, Xu 2950
Yang, Zhongping 335, 1322, 2133
Yanhong, Zhang 452
Yano, Yoshihiro 2775
Yanru Zhong 3489
Yaramasu, Venkata 3695
Yashiro, Daisuke 1974
Yashun Li 782
Yasubayashi, Mikio 183
Yasui, Kazuya 2465
Yasumura, Yuji 3568
Yasuno, Takashi 3179
Yau, Y. T. 2754, 3392
Yazdkhasti, Pegah 2156
Yi, Hao 577
Yi-Chun Lin 2714
Yi-Hsun Chiu 3778
Yi-Hua Liu 3778
Yixin Zhu 1155, 3546
Yizhanyi Tang 2977
Yoda, Kazuyuki 1748
Yokoi, Y. 3007
Yokoi, Yuichi 3024
Yokokura, Yuki 1956
Yokoyama, H. 2120
Yokoyama, Natsuki 2285
Yokoyama, Tomoki 3397, 3410
Yonemori, Ryo 689
Yonezawa, Hikaru 3055
Yonezawa, Yoshiyuki 3440
Yonezawa, Yu 2745, 3386
Yong Ding 1476, 3842
Yong, Yu 341
Yong-Cheol Kwon 925
Yongdong Tan 3538
Yongjae Lee 3140
Yoon, Sung Hyun 2272
Yoshida, Morito 3410
Yoshida, S. 2569
Yoshida, T. 2438
Yoshida, Yoshiaki 3503

AUTHOR INDEX

Yoshikawa, Yuichi 2011
Yoshimoto, Kantaro 2421
Yoshimura, Eiji 1552
Yoshino, Teruo 2644, 3834
Yoshino, Yukio 2745
Yoshioka, S. 246
Yoshioka, Takashi..................................1956
Yoshizawa, Daisuke 97, 1950
Young-Do Kim 1738, 1743
Youngjoon Choi 1185
Young-Ryul Kim 796
Yu, Changzhou..567
Yu, F. Y. 1471
Yu, Ling-Chia 208
Yu, Shuai 2318
Yu, Weikai 647
Yu, Yifan 1458
Yu, Yong 567, 609
Yu-Chen Liu 2810, 3328
Yue Chen 2768
Yue, Xiaolong 2960
Yu-Jen Wang 1420
Yu-Kang Lo 2758, 2810, 3185, 3328, 3913
Yuki, Kazuaki 2465
Yukita, Kazuto 1490
Yukutake, Seigo 2140
Yunchang Kwak 2693
Yun-Chu Chiu 3328
Yung-Ching Huang 1452
Yunmei Fang 3168
Yunwei Li 3482
Yura, Masashi 2297
Yu-Shan Cheng 3233, 3778
Yuzurihara, Itsuo 2794
Zaitsu, Toshiyuki 177, 1179
Zanma, Tadanao 3568
Zargari, Navid R. 3468
Zehelein, Matthias..................................3085
Zeliang Shu 1050
Zeljkovic, Sandra 774
Zhang Wei 3050
Zhang Yajing..2899
Zhang, Haodong 3604
Zhang, Huiguo 2202
Zhang, Tao...485
Zhang, Wei 614, 1425
Zhang, Xing.....................................465, 567
Zhang, Xuning..2626
Zhang, Yuzhuo 647
Zhang, Zhe..78
Zhao, Wei..567
Zhao-Qin Guo 1580
Zhe Wang 3129
Zhe Zhang..3905
Zheng Dong 3842

Zheng Li 1842
Zheng, T. Q. 807
Zheng, Trillion Q............................2899, 3314
Zhengzhi Han 3124
Zhenyao Xu 994
Zhongping Yang 807
Zhu, B. 395, 1594
Zhu, Honglin 3002
Zhuo, Fang 577, 2960
Zian Qin 857
Zingerli, Claudius M. 3365
Zitouni, Y. 3239
Zong-Zhen Yang.......................................3778
Zou, Xudong 555
Zwyssig, Christof 1707

CURRAN ASSOCIATES INC.
proceedings
.com

9781479927067